Climate Change 2007

Impacts, Adaptation and Vulnerability

The Intergovernmental Panel on Climate Change (IPCC) was set up jointly by the World Meteorological Organization and the United Nations Environment Programme to provide an authoritative international statement of scientific understanding of climate change. The IPCC's periodic assessments of the causes, impacts and possible response strategies to climate change are the most comprehensive and up-to-date reports available on the subject, and form the standard reference for all concerned with climate change in academia, government and industry worldwide. Through three working groups, many hundreds of international experts assess climate change in this Fourth Assessment Report. The Report consists of three main volumes under the umbrella title *Climate Change 2007,* all available from Cambridge University Press:

Climate Change 2007 – The Physical Science Basis
Contribution of Working Group I to the Fourth Assessment Report of the IPCC
(ISBN 978 0521 88009-1 Hardback; 978 0521 70596-7 Paperback)

Climate Change 2007 – Impacts, Adaptation and Vulnerability
Contribution of Working Group II to the Fourth Assessment Report of the IPCC
(978 0521 88010-7 Hardback; 978 0521 70597-4 Paperback)

Climate Change 2007 – Mitigation of Climate Change
Contribution of Working Group III to the Fourth Assessment Report of the IPCC
(978 0521 88011-4 Hardback; 978 0521 70598-1 Paperback)

Climate Change 2007 – Impacts, Adaptation and Vulnerability provides the most comprehensive and up-to-date scientific assessment of the impacts of climate change, the vulnerability of natural and human environments, and the potential for response through adaptation. The report:

- evaluates evidence that recent observed changes in climate have already affected a variety of physical and biological systems and concludes that these effects can be attributed to global warming;
- makes a detailed assessment of the impacts of future climate change and sea-level rise on ecosystems, water resources, agriculture and food security, human health, coastal and low-lying regions and industry and settlements;
- provides a complete new assessment of the impacts of climate change on major regions of the world (Africa, Asia, Australia/New Zealand, Europe, Latin America, North America, polar regions and small islands);
- considers responses through adaptation;
- explores the synergies and trade-offs between adaptation and mitigation;
- evaluates the key vulnerabilities to climate change, and assesses aggregate damage levels and the role of multiple stresses.

This latest assessment by the IPCC will form the standard scientific reference for all those concerned with the consequences of climate change, including students and researchers in ecology, biology, hydrology, environmental science, economics, social science, natural resource management, public health, food security and natural hazards, and policymakers and managers in governments, industry and other organisations responsible for resources likely to be affected by climate change.

From reviews of the Third Assessment Report – Climate Change 2001:

'This volume makes another significant step forward in the understanding of the likely impacts of climate change on a global scale.'
International Journal of Climatology

'The detail is truly amazing . . . invaluable works of reference . . . no reference or science library should be without a set [of the IPCC volumes]. . . unreservedly recommended to all readers.'
Journal of Meteorology

'This well-edited set of three volumes will surely be the standard reference for nearly all arguments related with global warming and climate change in the next years. It should not be missing in the libraries of atmospheric and climate research institutes and those administrative and political institutions which have to deal with global change and sustainable development.'
Meteorologische Zeitschrift

'The IPCC has conducted what is arguably the largest, most comprehensive and transparent study ever undertaken by mankind . . . The result is a work of substance and authority, which only the foolish would deride.'
Wind Engineering

' . . . the weight of evidence presented, the authority that IPCC commands and the breadth of view can hardly fail to impress and earn respect. Each of the volumes is essentially a remarkable work of reference, containing a plethora of information and copious bibliographies. There can be few natural scientists who will not want to have at least one of these volumes to hand on their bookshelves, at least until further research renders the details outdated by the time of the next survey.'
The Holocene

'The subject is explored in great depth and should prove valuable to policy makers, researchers, analysts, and students.'
American Meteorological Society

From reviews of the Second Assessment Report – Climate Change 1995:

' ... essential reading for anyone interested in global environmental change, either past, present or future. ... These volumes have a deservedly high reputation'
Geological Magazine

'... a tremendous achievement of coordinating the contributons of well over a thousand individuals to produce an authoritative, state-of-the-art review which will be of great value to decision-makers and the scientific community at large ... an indispensable reference.'
International Journal of Climatology

'... a wealth of clear, well-organized information that is all in one place ... there is much to applaud.'
Environment International

Climate Change 2007
Impacts, Adaptation and Vulnerability

Edited by

Martin Parry
Co-Chair,
IPCC Working Group II

Osvaldo Canziani
Co-Chair,
IPCC Working Group II

Jean Palutikof
Head, Technical Support Unit
IPCC Working Group II

Paul van der Linden
Deputy Head, Technical Support Unit
IPCC Working Group II

Clair Hanson
Deputy Head, Technical Support Unit
IPCC Working Group II

**Contribution of Working Group II
to the Fourth Assessment Report of the
Intergovernmental Panel on Climate Change**

Published for the Intergovernmental Panel on Climate Change

CAMBRIDGE UNIVERSITY PRESS
Cambridge, New York, Melbourne, Madrid, Cape Town, Singapore, São Paolo, Delhi

Cambridge University Press
32 Avenue of the Americas, New York, NY 10013-2473, USA

www.cambridge.org
Information on this title: www.cambridge.org/9780521880107

First published 2007

Printed in Canada by Freisens

A catalogue record for this book is available from the British Library

ISBN 978 0521 88010-7 hardback
ISBN 978 0521 70597-4 paperback

Please use the following reference to the whole report:

IPCC, 2007: *Climate Change 2007: Impacts, Adaptation and Vulnerability. Contribution of Working Group II to the Fourth Assessment Report of the Intergovernmental Panel on Climate Change,* M.L. Parry, O.F. Canziani, J.P. Palutikof, P.J. van der Linden and C.E. Hanson, Eds., Cambridge University Press, Cambridge, UK, 976pp.

Cover photo:
© Bjorn Svensson/Science Photo Library

Contents

CD-ROM **Inside back cover:**
This volume: Summary for Policymakers, Technical Summary, Chapters, Appendices, Index
Together with: Supporting material, Chapter supplementary material, Regional and subject database of references, Figures in Powerpoint from SPM and TS

Foreword

The Intergovernmental Panel on Climate Change (IPCC) was established by the World Meteorological Organization and the United Nations Environment Programme in 1988 with the mandate to provide the world community with the most up-to-date and comprehensive scientific, technical and socio-economic information about climate change. The IPCC multivolume assessments have since then played a major role in motivating governments to adopt and implement policies in responding to climate change, including the United Nations Framework Convention on Climate Change and the Kyoto Protocol. The "Climate Change 2007" IPCC Fourth Assessment Report could not be timelier for the world's policy makers to help them respond to the challenge of climate change.

"Climate Change 2007: Impacts, Adaptation and Vulnerability", is the second volume of the IPCC Fourth Assessment Report. After confirming in the first volume on "The Physical Science Basis" that climate change is occurring now, mostly as a result of human activities, this volume illustrates the impacts of global warming already under way and the potential for adaptation to reduce the vulnerability to, and risks of climate change.

Drawing on over 29,000 data series, the current report provides a much broader set of evidence of observed impacts coming from the large number of field studies developed over recent years. The analysis of current and projected impacts is then carried out sector by sector in dedicated chapters. The report pays great attention to regional impacts and adaptation strategies, identifying the most vulnerable areas. A final section provides an overview of the inter-relationship between adaptation and mitigation in the context of sustainable development.

The "Impacts, Adaptation and Vulnerability" report was made possible by the commitment and voluntary labour of a large number of leading scientists. We would like to express our gratitude to all Coordinating Lead Authors, Lead Authors, Contributing Authors, Review Editors and Reviewers. We would also like to thank the staff of the Working Group II Technical Support Unit and the IPCC Secretariat for their dedication in organising the production of another successful IPCC report. Furthermore, we would like to express our thanks to Dr Rajendra K. Pachauri, Chairman of the IPCC, for his patient and constant guidance to the process, and to Drs Osvaldo Canziani and Martin Parry, Co-Chairs of Working Group II, for their skillful leadership.

We also wish to acknowledge and thank those governments and institutions that contributed to the IPCC Trust Fund and supported the participation of their resident scientists in the IPCC process. We would like to mention in particular the Government of the United Kingdom, which funded the Technical Support Unit; the European Commission and the Belgian Government, which hosted the plenary session for the approval of the report; and the Governments of Australia, Austria, Mexico and South Africa, which hosted the drafting sessions to prepare the report.

M. Jarraud
Secretary General
World Meteorological Organisation

A. Steiner
Executive Director
United Nations Environment Director

Preface

This volumes comprises the Working Group II contribution to the IPCC Fourth Assessment (AR4) and contains a Summary for Policymakers, a Technical Summary, the chapters of the Assessment and various annexes. The scope, content and procedures followed are described in the Introduction which follows.

Acknowledgements

This Report is the product of the work of many scientists who acted as Authors, Reviewers or Editors (details are given in the Introduction, Section E). We would like to express our sincere thanks to them for their contribution, and to their institutions for supporting their participation.

We thank the members of the Working Group II Bureau (Edmundo de Alba Alcarez, Abdelkader Allali, Lucka Kajfež-Bogataj, Geoff Love, John Stone and Jean-Pascal van Ypersele), for carrying out their duties with diligence and commitment.

Costs of the Technical Support Unit (TSU) and of Dr Parry were covered by the UK Department for the Environment, Food and Rural Affairs (Defra). The TSU was based in the Met Office Hadley Centre in the UK. We thank David Warrilow (Defra), Dave Griggs and John Mitchell (Met Office) for their support through these agencies.

Four meetings of Authors were held during the preparation of the Report, and the governments of Austria, Australia, Mexico and South Africa, through their Focal Points, kindly agreed to act as hosts. The Approval Session of the Working Group II contribution to the Fourth Assessment was held in Brussels at the generous invitations of the Government of Belgium, through Martine Vanderstraeten, and the European Community, through Lars Mueller. We thank all these governments, institutions and individuals for their hospitality and hard work on behalf of the Working Group II process.

We thank the IPCC Secretary, Renate Christ, and the Secretariat staff Jian Liu, Rudie Bourgeois, Annie Courtin, Joelle Fernandez and Carola Saibante for their efficient and courteous attention to Working Group II needs; and Marc Peeters, WMO Conference Officer, for his work on the organisation of the Brussels Approval Meeting.

Thanks go to ProClim (Forum for Climate and Global Change) and Marilyn Anderson for producing the index to this Report.

Last, but by no means least, we acknowledge the exceptional commitment of the members of the Technical Support Unit throughout the preparation of the Report: Jean Palutikof, Paul van der Linden, Clair Hanson, Norah Pritchard, Chris Sear, Carla Encinas and Kim Mack.

Rajendra Pachauri
Chair IPCC

Martin Parry
Co-Chair IPCC Working Group II

Osvaldo Canziani
Co-Chair IPCC Working Group II

Introduction to the Working Group II Fourth Assessment Report

A. The Intergovernmental Panel on Climate Change

The Intergovernmental Panel on Climate Change (IPCC) was established by the World Meteorological Organization and the United Nations Environment Programme in 1988, in response to the widespread recognition that human-influenced emissions of greenhouse gases have the potential to alter the climate system. Its role is to provide an assessment of the understanding of all aspects of climate change.

At its first session, the IPCC was organised into three Working Groups. The current remits of the three Working Groups are for Working Group I to examine the scientific aspects of the climate system and climate change; Working Group II to address vulnerabilities to, impacts of and adaptations to climate change; and Working Group III to explore the options for mitigation of climate change. The three previous assessment reports were produced in 1990, 1996 and 2001.

B. The Working Group II Fourth Assessment

The decision to produce a Fourth Assessment Report was taken by the 19th Session of the IPCC at Geneva in April 2002. The report was to be more focussed and shorter than before. The Working Group II contribution was to be finalised in mid-2007.

The IPCC Fourth Assessment is intended to be a balanced assessment of current knowledge. Its emphasis is on new knowledge acquired since the IPCC Third Assessment (2001). This required a survey of all published literature, including non-English language and 'grey' literature such as government and NGO reports.

Two meetings were held in 2003 to scope the Fourth Assessment, from which emerged the outline for the Working Group II Assessment submitted to IPCC Plenary 21 in November 2003 for approval and subsequent acceptance.

The Report has twenty chapters which together provide a comprehensive assessment of the climate change literature. These are shown in Table I.1. The opening chapter is on observed changes, and addresses the question of whether observed changes in the natural and managed environment are associated with anthropogenic climate change. Chapter 2 deals with the methods available for impacts analysis, and with the scenarios of future climate change which underpin these analyses. These are followed by the core chapters, which assess the literature on present day and future climate change impacts on systems, sectors and regions, vulnerabilities to these impacts, and strategies for adaptation. Chapters 17 and 18 consider possible responses through adaptation and the synergies with mitigation. The two final chapters look at key vulnerabilities, and the inter-relationships between climate change and sustainability.

Chapters 9 to 16 of the Working Group II Fourth Assessment consider regional climate change impacts. The definitions of these regions are shown in Table I.2.

Table I.1. *The chapters of the Working Group II contribution to the IPCC Fourth Assessment.*

Section A. ASSESSMENT OF OBSERVED CHANGES
1. Assessment of observed changes and responses in natural and managed systems

Section B. ASSESSMENT OF FUTURE IMPACTS AND ADAPTATION: SYSTEMS AND SECTORS
2. New assessment methods and the characterisation of future conditions
3. Freshwater resources and their management
4. Ecosystems, their properties, goods and services
5. Food, fibre and forest products
6. Coastal systems and low-lying areas
7. Industry, settlement and society
8. Human health

Section C. ASSESSMENT OF FUTURE IMPACTS AND ADAPTATION: REGIONS
9. Africa
10. Asia
11. Australia and New Zealand
12. Europe
13. Latin America
14. North America
15. Polar regions (Arctic and Antarctic)
16. Small islands

Section D. ASSESSMENT OF RESPONSES TO IMPACTS
17. Assessment of adaptation practices, options, constraints and capacity
18. Inter-relationships between adaptation and mitigation
19. Assessing key vulnerabilities and the risk from climate change
20. Perspectives on climate change and sustainability

Table I.2. *Countries by region (see Chapters 9 to 16) for the Working Group II Fourth Assessment.*

Africa			
Algeria	Angola	Benin	Botswana
Burkina Faso	Burundi	Cameroon	Central African Republic
Chad	Congo, Republic of	Congo, Democratic Rep. of	Côte d'Ivoire
Djibouti	Egypt	Equatorial Guinea	Eritrea
Ethiopia	Gabon	Ghana	Guinea
Guinea-Bissau	Kenya	Lesotho	Liberia
Libya	Madagascar	Malawi	Mali
Mauritania	Morocco	Mozambique	Namibia
Niger	Nigeria	Reunion	Rwanda
Senegal	Sierra Leone	Somalia	South Africa
Sudan	Swaziland	Tanzania	The Gambia
Togo	Tunisia	Uganda	Zambia
Zimbabwe			
Asia			
Afghanistan	Bahrain	Bangladesh	Bhutan
Brunei Darussalam	Cambodia	China	East Timor
India	Indonesia	Iran, Islamic Republic of	Iraq
Israel	Japan	Jordan	Kazakhstan
Korea, Dem. People's Rep.	Korea, Republic of	Kuwait	Kyrgyz Republic
Laos	Lebanon	Malaysia	Mongolia
Myanmar	Nepal	Oman	Pakistan
Papua New Guinea	Philippines	Qatar	Russia – East of the Urals
Saudi Arabia	Singapore	Sri Lanka	Syria
Tajikistan	Thailand	Turkey	Turkmenistan
United Arab Emirates	Uzbekistan	Vietnam	Yemen
Australia and New Zealand			
Australia	New Zealand		
Europe			
Albania	Andorra	Armenia	Austria
Azerbaijan	Belarus	Belgium	Bosnia and Herzegovina
Bulgaria	Croatia	Czech Republic	Denmark
Estonia	Finland	France	Georgia
Germany	Greece	Hungary	Ireland
Italy	Latvia	Liechtenstein	Lithuania
Luxembourg	Macedonia	Moldova, Republic of	Monaco
Montenegro	Norway	Poland	Portugal
Romania	Russia – West of the Urals	San Marino	Serbia
Slovak Republic	Slovenia	Spain	Sweden
Switzerland	The Netherlands	Ukraine	United Kingdom
Vatican City, State of			
Polar Regions			
Antarctic	North of 60°N (including Greenland and Iceland)		
Latin America			
Argentina	Belize	Bolivia	Brazil
Chile	Colombia	Costa Rica	Ecuador
El Salvador	French Guiana	Guatemala	Guyana
Honduras	Mexico	Nicaragua	Panama
Paraguay	Peru	Suriname	Uruguay
Venezuela			
North America			
Canada	United States of America		
Small islands: non-autonomous small islands are also included in the assessment but are not listed here			
Antigua and Barbuda	Barbados	Cape Verde	Comoros
Cook Islands	Cuba	Cyprus	Dominica
Dominican Republic	Fed. States of Micronesia	Fiji	Grenada
Haiti	Jamaica	Kiribati	Maldives
Malta	Marshall Islands	Mauritius	Nauru
Palau	Saint Kitts and Nevis	Saint Lucia	Saint Vincent & Grenadines
Samoa	São Tomé & Príncipe	Seychelles	Solomon Islands
The Bahamas	Tonga	Trinidad and Tobago	Tuvalu
Vanuatu			

C. Cross-chapter case studies

Early in the writing of the Working Group II contribution to the Fourth Assessment, there emerged themes of environmental importance and widespread interest which are dealt with from different perspectives by several chapters. These themes have been gathered together into 'cross-chapter case studies', which appear in their entirety at the end of the volume and are included in the CD-ROM which accompanies this volume. A 'roadmap' in Table I.3 shows where the cross-chapter case study material appears in the individual chapters.

The four cross-chapter case studies are:

1. The impact of the European 2003 heatwave
2. Impacts of climate change on coral reefs
3. Megadeltas: their vulnerabilities to climate change
4. Indigenous knowledge for adaptation to climate change

D. Regional and subject database of references

This Assessment is based on the review of a very large amount of literature for all parts of the world and for many subjects. For those interested in accessing this literature for a given region or subject, a regional and subject database of references is provided on the CD-ROM which accompanies this volume. The database contains in full all the references in this volume and can be viewed by region and subject.

E. Procedures followed in this Assessment by the authors, reviewers and participating governments

In total, the Working Group II Fourth Assessment involved 48 Coordinating Lead Authors (CLAs), 125 Lead Authors (LAs), and 45 Review Editors (REs), drawn from 70 countries. In addition, there were 183 Contributing Authors and 910 Expert Reviewers.

Each chapter in the Working Group II Fourth Assessment had a writing team of two to four CLAs and six to nine LAs. Led by the CLAs, it was the responsibility of this writing team to produce the drafts and finished version of the chapter. Where necessary, they could recruit Contributing Authors to assist in their task. Three drafts of each chapter were written prior to the production of the final version. Drafts were reviewed in two separate lines of review, by experts and by governments. It was the role of the REs (two to three per chapter) to ensure that the review comments were properly addressed by the authors.

The authors and REs were selected by the Working Group II Bureau from the lists of experts nominated by governments. Due regard was paid to the need to balance the writing team with proper representation from developing and developed countries, and Economies in Transition. In the review by experts, chapters were sent out to experts, including all those nominated by governments but not yet included in the assessment, together with scientists and researchers identified by the Working Group II Co-Chairs and Vice-Chairs from their knowledge of the literature and the global research community.

F. Communication of uncertainty in the Working Group II Fourth Assessment

A set of terms to describe uncertainties in current knowledge is common to all parts of the IPCC Fourth Assessment, based on the *Guidance Notes for Lead Authors of the IPCC Fourth Assessment Report on Addressing Uncertainties*[1], produced by the IPCC in July 2005.

Description of confidence

On the basis of a comprehensive reading of the literature and their expert judgement, authors have assigned a confidence level to the major statements in the Report on the basis of their assessment of current knowledge, as follows:

Terminology	*Degree of confidence in being correct*
Very high confidence	At least 9 out of 10 chance of being correct
High confidence	About 8 out of 10 chance
Medium confidence	About 5 out of 10 chance
Low confidence	About 2 out of 10 chance
Very low confidence	Less than a 1 out of 10 chance

Description of likelihood

Likelihood refers to a probabilistic assessment of some well-defined outcome having occurred or occurring in the future, and may be based on quantitative analysis or an elicitation of expert views. In the Report, when authors evaluate the likelihood of certain outcomes, the associated meanings are:

Terminology	*Likelihood of the occurrence/ outcome*
Virtually certain	>99% probability of occurrence
Very likely	90 to 99% probability
Likely	66 to 90% probability
About as likely as not	33 to 66% probability
Unlikely	10 to 33% probability
Very unlikely	1 to 10% probability
Exceptionally unlikely	<1% probability

[1] http://www.ipcc.ch/activity/uncertaintyguidancenote.pdf

Table I.3. *Cross-chapter Case Studies: location in text.*

The impact of the European 2003 heatwave		
Topic:	*Chapter:*	*Location in chapter:*
Scene-setting and overview		
The European heatwave of 2003	Chapter 12	12.6.1
Impacts on sectors		
Ecological impacts of the European heatwave 2003	Chapter 4	Box 4.1
European heatwave impact on the agricultural sector	Chapter 5	Box 5.1
Industry, settlement and society: impacts of the 2003 heatwave in Europe	Chapter 7	Box 7.1
The European heatwave 2003: health impacts and adaptation	Chapter 8	Box 8.1

Impacts of climate change on coral reefs		
Present-day changes in coral reefs		
Observed changes in coral reefs	Chapter 1	Section 1.3.4.1
Environmental thresholds and observed coral bleaching	Chapter 6	Box 6.1
Future impacts on coral reefs		
Are coral reefs endangered by climate change?	Chapter 4	Box 4.4
Impacts on coral reefs	Chapter 6	Section 6.4.1.5
Climate change and the Great Barrier Reef	Chapter 11	Box 11.3
Impact of coral mortality on reef fisheries	Chapter 5	Box 5.4
Multiple stresses on coral reefs		
Non-climate-change threats to coral reefs of small islands	Chapter 16	Box 16.2

Megadeltas: their vulnerabilities to climate change		
Introduction		
Deltas and megadeltas: hotspots for vulnerability	Chapter 6	Box 6.3
Megadeltas in Asia		
Megadeltas in Asia	Chapter 10	Section 10.6.1, Table 10.10
Climate change and the fisheries of the lower Mekong – an example of multiple stresses on a megadelta fisheries system due to human activity	Chapter 5	Box 5.3
Megadeltas in the Arctic		
Arctic megadeltas	Chapter 15	Section 15.6.2
Case study of Hurricane Katrina		
Hurricane Katrina and coastal ecosystem services in the Mississippi delta	Chapter 6	Box 6.4
Vulnerabilities to extreme weather events in megadeltas in a context of multiple stresses: the case of Hurricane Katrina	Chapter 7	Box 7.4

Indigenous knowledge for adaptation to climate change		
Overview		
Role of local and indigenous knowledge in adaptation and sustainability research	Chapter 20	Box 20.1
Case studies		
Adaptation capacity of the South American highlands´ pre-Colombian communities	Chapter 13	Box 13.2
African indigenous knowledge systems	Chapter 9	Section 9.6.2
Traditional knowledge for adaptation among Arctic peoples	Chapter 15	Section 15.6.1
Adaptation to health impacts of climate change among indigenous populations	Chapter 8	Box 8.6

G. Definitions of key terms

Climate change in IPCC usage refers to any change in climate over time, whether due to natural variability or as a result of human activity. This usage differs from that in the Framework Convention on Climate Change, where *climate change* refers to a change of climate that is attributed directly or indirectly to human activity that alters the composition of the global atmosphere and that is in addition to natural climate variability observed over comparable time periods.

Adaptation is the adjustment in natural or human systems in response to actual or expected climatic stimuli or their effects, which moderates harm or exploits beneficial opportunities.

Vulnerability is the degree to which a system is susceptible to, and unable to cope with, adverse effects of climate change, including climate variability and extremes. *Vulnerability* is a function of the character, magnitude, and rate of climate change and variation to which a system is exposed, the sensitivity and adaptive capacity of that system.

Contribution of Working Group II to the Fourth Assessment Report of the Intergovernmental Panel on Climate Change

Summary for Policymakers

This summary, approved in detail at the Eighth Session of IPCC Working Group II (Brussels, Belgium, 2-5 April 2007), represents the formally agreed statement of the IPCC concerning the sensitivity, adaptive capacity and vulnerability of natural and human systems to climate change, and the potential consequences of climate change.

Drafting Authors:

Neil Adger, Pramod Aggarwal, Shardul Agrawala, Joseph Alcamo, Abdelkader Allali, Oleg Anisimov, Nigel Arnell, Michel Boko, Osvaldo Canziani, Timothy Carter, Gino Casassa, Ulisses Confalonieri, Rex Victor Cruz, Edmundo de Alba Alcaraz, William Easterling, Christopher Field, Andreas Fischlin, Blair Fitzharris, Carlos Gay García, Clair Hanson, Hideo Harasawa, Kevin Hennessy, Saleemul Huq, Roger Jones, Lucka Kajfež Bogataj, David Karoly, Richard Klein, Zbigniew Kundzewicz, Murari Lal, Rodel Lasco, Geoff Love, Xianfu Lu, Graciela Magrín, Luis José Mata, Roger McLean, Bettina Menne, Guy Midgley, Nobuo Mimura, Monirul Qader Mirza, José Moreno, Linda Mortsch, Isabelle Niang-Diop, Robert Nicholls, Béla Nováky, Leonard Nurse, Anthony Nyong, Michael Oppenheimer, Jean Palutikof, Martin Parry, Anand Patwardhan, Patricia Romero Lankao, Cynthia Rosenzweig, Stephen Schneider, Serguei Semenov, Joel Smith, John Stone, Jean-Pascal van Ypersele, David Vaughan, Coleen Vogel, Thomas Wilbanks, Poh Poh Wong, Shaohong Wu, Gary Yohe

This Summary for Policymakers should be cited as:

IPCC, 2007: Summary for Policymakers. In: *Climate Change 2007: Impacts, Adaptation and Vulnerability. Contribution of Working Group II to the Fourth Assessment Report of the Intergovernmental Panel on Climate Change*, M.L. Parry, O.F. Canziani, J.P. Palutikof, P.J. van der Linden and C.E. Hanson, Eds., Cambridge University Press, Cambridge, UK, 7-22.

A. Introduction

This Summary sets out the key policy-relevant findings of the Fourth Assessment of Working Group II of the Intergovernmental Panel on Climate Change (IPCC).

The Assessment is of current scientific understanding of the impacts of climate change on natural, managed and human systems, the capacity of these systems to adapt and their vulnerability.[1] It builds upon past IPCC assessments and incorporates new knowledge gained since the Third Assessment.

Statements in this Summary are based on chapters in the Assessment and principal sources are given at the end of each paragraph.[2]

B. Current knowledge about observed impacts of climate change on the natural and human environment

A full consideration of observed climate change is provided in the Working Group I Fourth Assessment. This part of the Working Group II Summary concerns the relationship between observed climate change and recent observed changes in the natural and human environment.

The statements presented here are based largely on data sets that cover the period since 1970. The number of studies of observed trends in the physical and biological environment and their relationship to regional climate changes has increased greatly since the Third Assessment in 2001. The quality of the data sets has also improved. There is, however, a notable lack of geographical balance in the data and literature on observed changes, with marked scarcity in developing countries.

Recent studies have allowed a broader and more confident assessment of the relationship between observed warming and impacts than was made in the Third Assessment. That Assessment concluded that "there is high confidence[3] that recent regional changes in temperature have had discernible impacts on many physical and biological systems".

From the current Assessment we conclude the following.

Observational evidence from all continents and most oceans shows that many natural systems are being affected by regional climate changes, particularly temperature increases.

With regard to changes in snow, ice and frozen ground (including permafrost),[4] there is high confidence that natural systems are affected. Examples are:
- enlargement and increased numbers of glacial lakes [1.3];
- increasing ground instability in permafrost regions, and rock avalanches in mountain regions [1.3];
- changes in some Arctic and Antarctic ecosystems, including those in sea-ice biomes, and also predators high in the food chain [1.3, 4.4, 15.4].

Based on growing evidence, there is high confidence that the following effects on hydrological systems are occurring:
- increased runoff and earlier spring peak discharge in many glacier- and snow-fed rivers [1.3];
- warming of lakes and rivers in many regions, with effects on thermal structure and water quality [1.3].

There is very high confidence, based on more evidence from a wider range of species, that recent warming is strongly affecting terrestrial biological systems, including such changes as:
- earlier timing of spring events, such as leaf-unfolding, bird migration and egg-laying [1.3];
- poleward and upward shifts in ranges in plant and animal species [1.3, 8.2, 14.2].

Based on satellite observations since the early 1980s, there is high confidence that there has been a trend in many regions towards earlier 'greening'[5] of vegetation in the spring linked to longer thermal growing seasons due to recent warming [1.3, 14.2].

There is high confidence, based on substantial new evidence, that observed changes in marine and freshwater biological systems are associated with rising water temperatures, as well as related changes in ice cover, salinity, oxygen levels and circulation [1.3]. These include:
- shifts in ranges and changes in algal, plankton and fish abundance in high-latitude oceans [1.3];
- increases in algal and zooplankton abundance in high-latitude and high-altitude lakes [1.3];
- range changes and earlier migrations of fish in rivers [1.3].

---

[1] For definitions, see Endbox 1.
[2] Sources to statements are given in square brackets. For example, [3.3] refers to Chapter 3, Section 3. In the sourcing, F = Figure, T = Table, B = Box and ES = Executive Summary.
[3] See Endbox 2.
[4] See Working Group I Fourth Assessment.
[5] Measured by the Normalised Difference Vegetation Index, which is a relative measure of the amount of green vegetation in an area based on satellite images.

The uptake of anthropogenic carbon since 1750 has led to the ocean becoming more acidic, with an average decrease in pH of 0.1 units [IPCC Working Group I Fourth Assessment]. However, the effects of observed ocean acidification on the marine biosphere are as yet undocumented [1.3].

A global assessment of data since 1970 has shown it is likely[6] that anthropogenic warming has had a discernible influence on many physical and biological systems.

Much more evidence has accumulated over the past five years to indicate that changes in many physical and biological systems are linked to anthropogenic warming. There are four sets of evidence which, taken together, support this conclusion:

1. The Working Group I Fourth Assessment concluded that most of the observed increase in the globally averaged temperature since the mid-20th century is very likely due to the observed increase in anthropogenic greenhouse gas concentrations.

2. Of the more than 29,000 observational data series,[7] from 75 studies, that show significant change in many physical and biological systems, more than 89% are consistent with the direction of change expected as a response to warming (Figure SPM.1) [1.4].

3. A global synthesis of studies in this Assessment strongly demonstrates that the spatial agreement between regions of significant warming across the globe and the locations of significant observed changes in many systems consistent with warming is very unlikely to be due solely to natural variability of temperatures or natural variability of the systems (Figure SPM.1) [1.4].

4. Finally, there have been several modelling studies that have linked responses in some physical and biological systems to anthropogenic warming by comparing observed responses in these systems with modelled responses in which the natural forcings (solar activity and volcanoes) and anthropogenic forcings (greenhouse gases and aerosols) are explicitly separated. Models with combined natural and anthropogenic forcings simulate observed responses significantly better than models with natural forcing only [1.4].

Limitations and gaps prevent more complete attribution of the causes of observed system responses to anthropogenic warming. First, the available analyses are limited in the number of systems and locations considered. Second, natural temperature variability is larger at the regional than at the global scale, thus affecting identification of changes due to external forcing. Finally, at the regional scale other factors (such as land-use change, pollution, and invasive species) are influential [1.4].

Nevertheless, the consistency between observed and modelled changes in several studies and the spatial agreement between significant regional warming and consistent impacts at the global scale is sufficient to conclude with high confidence that anthropogenic warming over the last three decades has had a discernible influence on many physical and biological systems [1.4].

Other effects of regional climate changes on natural and human environments are emerging, although many are difficult to discern due to adaptation and non-climatic drivers.

Effects of temperature increases have been documented in the following (medium confidence):

- effects on agricultural and forestry management at Northern Hemisphere higher latitudes, such as earlier spring planting of crops, and alterations in disturbance regimes of forests due to fires and pests [1.3];
- some aspects of human health, such as heat-related mortality in Europe, infectious disease vectors in some areas, and allergenic pollen in Northern Hemisphere high and mid-latitudes [1.3, 8.2, 8.ES];
- some human activities in the Arctic (e.g., hunting and travel over snow and ice) and in lower-elevation alpine areas (such as mountain sports) [1.3].

Recent climate changes and climate variations are beginning to have effects on many other natural and human systems. However, based on the published literature, the impacts have not yet become established trends. Examples include:

- Settlements in mountain regions are at enhanced risk of glacier lake outburst floods caused by melting glaciers. Governmental institutions in some places have begun to respond by building dams and drainage works [1.3].

- In the Sahelian region of Africa, warmer and drier conditions have led to a reduced length of growing season with detrimental effects on crops. In southern Africa, longer dry seasons and more uncertain rainfall are prompting adaptation measures [1.3].

- Sea-level rise and human development are together contributing to losses of coastal wetlands and mangroves and increasing damage from coastal flooding in many areas [1.3].

[6] See Endbox 2.

[7] A subset of about 29,000 data series was selected from about 80,000 data series from 577 studies. These met the following criteria: (1) ending in 1990 or later; (2) spanning a period of at least 20 years; and (3) showing a significant change in either direction, as assessed in individual studies.

Changes in physical and biological systems and surface temperature 1970-2004

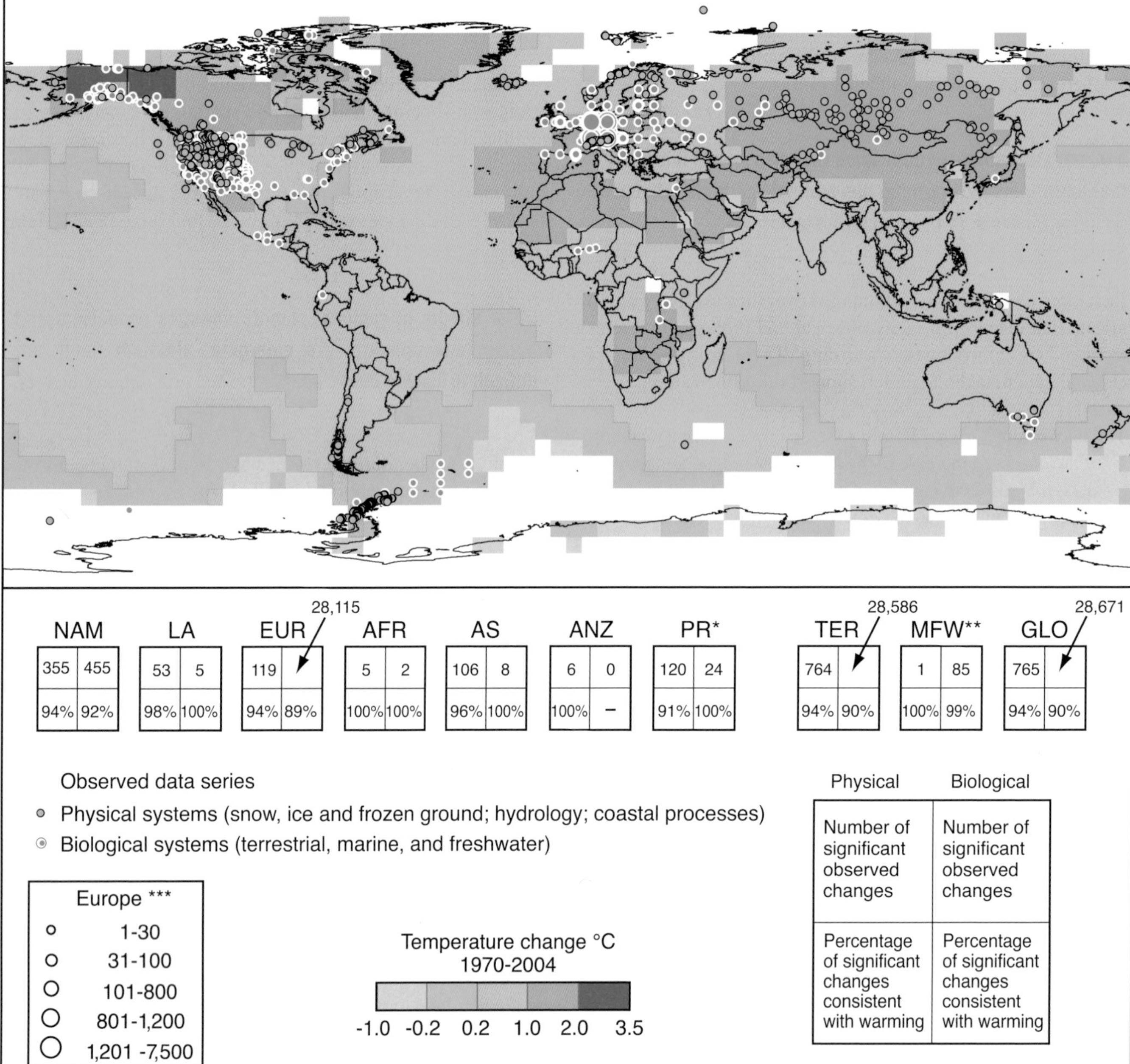

* Polar regions include also observed changes in marine and freshwater biological systems.

** Marine and freshwater includes observed changes at sites and large areas in oceans, small islands and continents. Locations of large-area marine changes are not shown on the map.

*** Circles in Europe represent 1 to 7,500 data series.

Figure SPM.1. *Locations of significant changes in data series of physical systems (snow, ice and frozen ground; hydrology; and coastal processes) and biological systems (terrestrial, marine, and freshwater biological systems), are shown together with surface air temperature changes over the period 1970-2004. A subset of about 29,000 data series was selected from about 80,000 data series from 577 studies. These met the following criteria: (1) ending in 1990 or later; (2) spanning a period of at least 20 years; and (3) showing a significant change in either direction, as assessed in individual studies. These data series are from about 75 studies (of which about 70 are new since the Third Assessment) and contain about 29,000 data series, of which about 28,000 are from European studies. White areas do not contain sufficient observational climate data to estimate a temperature trend. The 2 x 2 boxes show the total number of data series with significant changes (top row) and the percentage of those consistent with warming (bottom row) for (i) continental regions: North America (NAM), Latin America (LA), Europe (EUR), Africa (AFR), Asia (AS), Australia and New Zealand (ANZ), and Polar Regions (PR) and (ii) global-scale: Terrestrial (TER), Marine and Freshwater (MFW), and Global (GLO). The numbers of studies from the seven regional boxes (NAM, ..., PR) do not add up to the global (GLO) totals because numbers from regions except Polar do not include the numbers related to Marine and Freshwater (MFW) systems. Locations of large-area marine changes are not shown on the map. [Working Group II Fourth Assessment F1.8, F1.9; Working Group I Fourth Assessment F3.9b].*

C. Current knowledge about future impacts

The following is a selection of the key findings regarding projected impacts, as well as some findings on vulnerability and adaptation, in each system, sector and region for the range of (unmitigated) climate changes projected by the IPCC over this century[8] judged to be relevant for people and the environment.[9] The impacts frequently reflect projected changes in precipitation and other climate variables in addition to temperature, sea level and concentrations of atmospheric carbon dioxide. The magnitude and timing of impacts will vary with the amount and timing of climate change and, in some cases, the capacity to adapt. These issues are discussed further in later sections of the Summary.

More specific information is now available across a wide range of systems and sectors concerning the nature of future impacts, including for some fields not covered in previous assessments.

Freshwater resources and their management

By mid-century, annual average river runoff and water availability are projected to increase by 10-40% at high latitudes and in some wet tropical areas, and decrease by 10-30% over some dry regions at mid-latitudes and in the dry tropics, some of which are presently water-stressed areas. In some places and in particular seasons, changes differ from these annual figures. ** D[10] [3.4]

Drought-affected areas will likely increase in extent. Heavy precipitation events, which are very likely to increase in frequency, will augment flood risk. ** N [Working Group I Fourth Assessment Table SPM-2, Working Group II Fourth Assessment 3.4]

In the course of the century, water supplies stored in glaciers and snow cover are projected to decline, reducing water availability in regions supplied by meltwater from major mountain ranges, where more than one-sixth of the world population currently lives. ** N [3.4]

Adaptation procedures and risk management practices for the water sector are being developed in some countries and regions that have recognised projected hydrological changes with related uncertainties. *** N [3.6]

Ecosystems

The resilience of many ecosystems is likely to be exceeded this century by an unprecedented combination of climate change, associated disturbances (e.g., flooding, drought, wildfire, insects, ocean acidification), and other global change drivers (e.g., land-use change, pollution, over-exploitation of resources). ** N [4.1 to 4.6]

Over the course of this century, net carbon uptake by terrestrial ecosystems is likely to peak before mid-century and then weaken or even reverse,[11] thus amplifying climate change. ** N [4.ES, F4.2]

Approximately 20-30% of plant and animal species assessed so far are likely to be at increased risk of extinction if increases in global average temperature exceed 1.5-2.5°C. * N [4.4, T4.1]

For increases in global average temperature exceeding 1.5-2.5°C and in concomitant atmospheric carbon dioxide concentrations, there are projected to be major changes in ecosystem structure and function, species' ecological interactions, and species' geographical ranges, with predominantly negative consequences for biodiversity, and ecosystem goods and services e.g., water and food supply. ** N [4.4]

The progressive acidification of oceans due to increasing atmospheric carbon dioxide is expected to have negative impacts on marine shell-forming organisms (e.g., corals) and their dependent species. * N [B4.4, 6.4]

Food, fibre and forest products

Crop productivity is projected to increase slightly at mid- to high latitudes for local mean temperature increases of up to 1-3°C depending on the crop, and then decrease beyond that in some regions. * D [5.4]

At lower latitudes, especially seasonally dry and tropical regions, crop productivity is projected to decrease for even small local temperature increases (1-2°C), which would increase the risk of hunger. * D [5.4]

Globally, the potential for food production is projected to increase with increases in local average temperature over a range of 1-3°C, but above this it is projected to decrease. * D [5.4, 5.6]

---

[8] Temperature changes are expressed as the difference from the period 1980-1999. To express the change relative to the period 1850-1899, add 0.5°C.
[9] Criteria of choice: magnitude and timing of impact, confidence in the assessment, representative coverage of the system, sector and region.
[10] In Section C, the following conventions are used:
Relationship to the Third Assessment:
D *Further development of a conclusion in the Third Assessment*
N *New conclusion, not in the Third Assessment*
Level of confidence in the whole statement:
*** *Very high confidence*
** *High confidence*
* *Medium confidence*
[11] Assuming continued greenhouse gas emissions at or above current rates and other global changes including land-use changes.

Increases in the frequency of droughts and floods are projected to affect local crop production negatively, especially in subsistence sectors at low latitudes. ** D [5.4, 5.ES]

Adaptations such as altered cultivars and planting times allow low- and mid- to high-latitude cereal yields to be maintained at or above baseline yields for modest warming. * N [5.5]

Globally, commercial timber productivity rises modestly with climate change in the short- to medium-term, with large regional variability around the global trend. * D [5.4]

Regional changes in the distribution and production of particular fish species are expected due to continued warming, with adverse effects projected for aquaculture and fisheries. ** D [5.4]

Coastal systems and low-lying areas

Coasts are projected to be exposed to increasing risks, including coastal erosion, due to climate change and sea-level rise. The effect will be exacerbated by increasing human-induced pressures on coastal areas. *** D [6.3, 6.4]

Corals are vulnerable to thermal stress and have low adaptive capacity. Increases in sea surface temperature of about 1-3°C are projected to result in more frequent coral bleaching events and widespread mortality, unless there is thermal adaptation or acclimatisation by corals. *** D [B6.1, 6.4]

Coastal wetlands including salt marshes and mangroves are projected to be negatively affected by sea-level rise especially where they are constrained on their landward side, or starved of sediment. *** D [6.4]

Many millions more people are projected to be flooded every year due to sea-level rise by the 2080s. Those densely-populated and low-lying areas where adaptive capacity is relatively low, and which already face other challenges such as tropical storms or local coastal subsidence, are especially at risk. The numbers affected will be largest in the mega-deltas of Asia and Africa while small islands are especially vulnerable. *** D [6.4]

Adaptation for coasts will be more challenging in developing countries than in developed countries, due to constraints on adaptive capacity. ** D [6.4, 6.5, T6.11]

Industry, settlement and society

Costs and benefits of climate change for industry, settlement and society will vary widely by location and scale. In the aggregate, however, net effects will tend to be more negative the larger the change in climate. ** N [7.4, 7.6]

The most vulnerable industries, settlements and societies are generally those in coastal and river flood plains, those whose economies are closely linked with climate-sensitive resources, and those in areas prone to extreme weather events, especially where rapid urbanisation is occurring. ** D [7.1, 7.3 to 7.5]

Poor communities can be especially vulnerable, in particular those concentrated in high-risk areas. They tend to have more limited adaptive capacities, and are more dependent on climate-sensitive resources such as local water and food supplies. ** N [7.2, 7.4, 5.4]

Where extreme weather events become more intense and/or more frequent, the economic and social costs of those events will increase, and these increases will be substantial in the areas most directly affected. Climate change impacts spread from directly impacted areas and sectors to other areas and sectors through extensive and complex linkages. ** N [7.4, 7.5]

Health

Projected climate change-related exposures are likely to affect the health status of millions of people, particularly those with low adaptive capacity, through:

- increases in malnutrition and consequent disorders, with implications for child growth and development;
- increased deaths, disease and injury due to heatwaves, floods, storms, fires and droughts;
- the increased burden of diarrhoeal disease;
- the increased frequency of cardio-respiratory diseases due to higher concentrations of ground-level ozone related to climate change; and,
- the altered spatial distribution of some infectious disease vectors. ** D [8.4, 8.ES, 8.2]

Climate change is expected to have some mixed effects, such as a decrease or increase in the range and transmission potential of malaria in Africa. ** D [8.4]

Studies in temperate areas[12] have shown that climate change is projected to bring some benefits, such as fewer deaths from cold exposure. Overall it is expected that these benefits will be outweighed by the negative health effects of rising temperatures worldwide, especially in developing countries. ** D [8.4]

The balance of positive and negative health impacts will vary from one location to another, and will alter over time as temperatures continue to rise. Critically important will be factors that directly shape the health of populations such as education, health care, public health initiatives and infrastructure and economic development. *** N [8.3]

[12] Studies mainly in industrialised countries.

More specific information is now available across the regions of the world concerning the nature of future impacts, including for some places not covered in previous assessments.

Africa

By 2020, between 75 million and 250 million people are projected to be exposed to increased water stress due to climate change. If coupled with increased demand, this will adversely affect livelihoods and exacerbate water-related problems. ** D [9.4, 3.4, 8.2, 8.4]

Agricultural production, including access to food, in many African countries and regions is projected to be severely compromised by climate variability and change. The area suitable for agriculture, the length of growing seasons and yield potential, particularly along the margins of semi-arid and arid areas, are expected to decrease. This would further adversely affect food security and exacerbate malnutrition in the continent. In some countries, yields from rain-fed agriculture could be reduced by up to 50% by 2020. ** N [9.2, 9.4, 9.6]

Local food supplies are projected to be negatively affected by decreasing fisheries resources in large lakes due to rising water temperatures, which may be exacerbated by continued over-fishing. ** N [9.4, 5.4, 8.4]

Towards the end of the 21st century, projected sea-level rise will affect low-lying coastal areas with large populations. The cost of adaptation could amount to at least 5-10% of Gross Domestic Product (GDP). Mangroves and coral reefs are projected to be further degraded, with additional consequences for fisheries and tourism. ** D [9.4]

New studies confirm that Africa is one of the most vulnerable continents to climate variability and change because of multiple stresses and low adaptive capacity. Some adaptation to current climate variability is taking place; however, this may be insufficient for future changes in climate. ** N [9.5]

Asia

Glacier melt in the Himalayas is projected to increase flooding, and rock avalanches from destabilised slopes, and to affect water resources within the next two to three decades. This will be followed by decreased river flows as the glaciers recede. * N [10.2, 10.4]

Freshwater availability in Central, South, East and South-East Asia, particularly in large river basins, is projected to decrease due to climate change which, along with population growth and increasing demand arising from higher standards of living, could adversely affect more than a billion people by the 2050s. ** N [10.4]

Coastal areas, especially heavily-populated megadelta regions in South, East and South-East Asia, will be at greatest risk due to increased flooding from the sea and, in some megadeltas, flooding from the rivers. ** D [10.4]

Climate change is projected to impinge on the sustainable development of most developing countries of Asia, as it compounds the pressures on natural resources and the environment associated with rapid urbanisation, industrialisation, and economic development. ** D [10.5]

It is projected that crop yields could increase up to 20% in East and South-East Asia while they could decrease up to 30% in Central and South Asia by the mid-21st century. Taken together, and considering the influence of rapid population growth and urbanisation, the risk of hunger is projected to remain very high in several developing countries. * N [10.4]

Endemic morbidity and mortality due to diarrhoeal disease primarily associated with floods and droughts are expected to rise in East, South and South-East Asia due to projected changes in the hydrological cycle associated with global warming. Increases in coastal water temperature would exacerbate the abundance and/or toxicity of cholera in South Asia. **N [10.4]

Australia and New Zealand

As a result of reduced precipitation and increased evaporation, water security problems are projected to intensify by 2030 in southern and eastern Australia and, in New Zealand, in Northland and some eastern regions. ** D [11.4]

Significant loss of biodiversity is projected to occur by 2020 in some ecologically rich sites including the Great Barrier Reef and Queensland Wet Tropics. Other sites at risk include Kakadu wetlands, south-west Australia, sub-Antarctic islands and the alpine areas of both countries. *** D [11.4]

Ongoing coastal development and population growth in areas such as Cairns and South-east Queensland (Australia) and Northland to Bay of Plenty (New Zealand), are projected to exacerbate risks from sea-level rise and increases in the severity and frequency of storms and coastal flooding by 2050. *** D [11.4, 11.6]

Production from agriculture and forestry by 2030 is projected to decline over much of southern and eastern Australia, and over parts of eastern New Zealand, due to increased drought and fire. However, in New Zealand, initial benefits are projected in western and southern areas and close to major rivers due to a longer growing season, less frost and increased rainfall. ** N [11.4]

The region has substantial adaptive capacity due to well-developed economies and scientific and technical capabilities, but there are considerable constraints to implementation and major challenges from changes in extreme events. Natural systems have limited adaptive capacity. ** N [11.2, 11.5]

Europe

For the first time, wide-ranging impacts of changes in current climate have been documented: retreating glaciers, longer growing seasons, shift of species ranges, and health impacts due to a heatwave of unprecedented magnitude. The observed changes described above are consistent with those projected for future climate change. *** N [12.2, 12.4, 12.6]

Nearly all European regions are anticipated to be negatively affected by some future impacts of climate change, and these will pose challenges to many economic sectors. Climate change is expected to magnify regional differences in Europe's natural resources and assets. Negative impacts will include increased risk of inland flash floods, and more frequent coastal flooding and increased erosion (due to storminess and sea-level rise). The great majority of organisms and ecosystems will have difficulty adapting to climate change. Mountainous areas will face glacier retreat, reduced snow cover and winter tourism, and extensive species losses (in some areas up to 60% under high emission scenarios by 2080). *** D [12.4]

In Southern Europe, climate change is projected to worsen conditions (high temperatures and drought) in a region already vulnerable to climate variability, and to reduce water availability, hydropower potential, summer tourism and, in general, crop productivity. It is also projected to increase health risks due to heat-waves, and the frequency of wildfires. ** D [12.2, 12.4, 12.7]

In Central and Eastern Europe, summer precipitation is projected to decrease, causing higher water stress. Health risks due to heatwaves are projected to increase. Forest productivity is expected to decline and the frequency of peatland fires to increase. ** D [12.4]

In Northern Europe, climate change is initially projected to bring mixed effects, including some benefits such as reduced demand for heating, increased crop yields and increased forest growth. However, as climate change continues, its negative impacts (including more frequent winter floods, endangered ecosystems and increasing ground instability) are likely to outweigh its benefits. ** D [12.4]

Adaptation to climate change is likely to benefit from experience gained in reaction to extreme climate events, specifically by implementing proactive climate change risk management adaptation plans. *** N [12.5]

Latin America

By mid-century, increases in temperature and associated decreases in soil water are projected to lead to gradual replacement of tropical forest by savanna in eastern Amazonia. Semi-arid vegetation will tend to be replaced by arid-land vegetation. There is a risk of significant biodiversity loss through species extinction in many areas of tropical Latin America. ** D [13.4]

In drier areas, climate change is expected to lead to salinisation and desertification of agricultural land. Productivity of some important crops is projected to decrease and livestock productivity to decline, with adverse consequences for food security. In temperate zones soybean yields are projected to increase. ** N [13.4, 13.7]

Sea-level rise is projected to cause increased risk of flooding in low-lying areas. Increases in sea surface temperature due to climate change are projected to have adverse effects on Mesoamerican coral reefs, and cause shifts in the location of south-east Pacific fish stocks. ** N [13.4, 13.7]

Changes in precipitation patterns and the disappearance of glaciers are projected to significantly affect water availability for human consumption, agriculture and energy generation. ** D [13.4]

Some countries have made efforts to adapt, particularly through conservation of key ecosystems, early warning systems, risk management in agriculture, strategies for flood drought and coastal management, and disease surveillance systems. However, the effectiveness of these efforts is outweighed by: lack of basic information, observation and monitoring systems; lack of capacity building and appropriate political, institutional and technological frameworks; low income; and settlements in vulnerable areas, among others. ** D [13.2]

North America

Warming in western mountains is projected to cause decreased snowpack, more winter flooding, and reduced summer flows, exacerbating competition for over-allocated water resources. *** D [14.4, B14.2]

Disturbances from pests, diseases and fire are projected to have increasing impacts on forests, with an extended period of high fire risk and large increases in area burned. *** N [14.4, B14.1]

Moderate climate change in the early decades of the century is projected to increase aggregate yields of rain-fed agriculture by 5-

20%, but with important variability among regions. Major challenges are projected for crops that are near the warm end of their suitable range or which depend on highly utilised water resources. ** D [14.4]

Cities that currently experience heatwaves are expected to be further challenged by an increased number, intensity and duration of heatwaves during the course of the century, with potential for adverse health impacts. Elderly populations are most at risk. *** D [14.4].

Coastal communities and habitats will be increasingly stressed by climate change impacts interacting with development and pollution. Population growth and the rising value of infrastructure in coastal areas increase vulnerability to climate variability and future climate change, with losses projected to increase if the intensity of tropical storms increases. Current adaptation is uneven and readiness for increased exposure is low. *** N [14.2, 14.4]

Polar Regions

In the Polar Regions, the main projected biophysical effects are reductions in thickness and extent of glaciers and ice sheets, and changes in natural ecosystems with detrimental effects on many organisms including migratory birds, mammals and higher predators. In the Arctic, additional impacts include reductions in the extent of sea ice and permafrost, increased coastal erosion, and an increase in the depth of permafrost seasonal thawing. ** D [15.3, 15.4, 15.2]

For human communities in the Arctic, impacts, particularly those resulting from changing snow and ice conditions, are projected to be mixed. Detrimental impacts would include those on infrastructure and traditional indigenous ways of life. ** D [15.4]

Beneficial impacts would include reduced heating costs and more navigable northern sea routes. * D [15.4]

In both polar regions, specific ecosystems and habitats are projected to be vulnerable, as climatic barriers to species invasions are lowered. ** D [15.6, 15.4]

Arctic human communities are already adapting to climate change, but both external and internal stressors challenge their adaptive capacities. Despite the resilience shown historically by Arctic indigenous communities, some traditional ways of life are being threatened and substantial investments are needed to adapt or re-locate physical structures and communities. ** D [15.ES, 15.4, 15.5, 15.7]

Small islands

Small islands, whether located in the tropics or higher latitudes, have characteristics which make them especially vulnerable to the effects of climate change, sea-level rise and extreme events. *** D [16.1, 16.5]

Deterioration in coastal conditions, for example through erosion of beaches and coral bleaching, is expected to affect local resources, e.g., fisheries, and reduce the value of these destinations for tourism. ** D [16.4]

Sea-level rise is expected to exacerbate inundation, storm surge, erosion and other coastal hazards, thus threatening vital infrastructure, settlements and facilities that support the livelihood of island communities. *** D [16.4]

Climate change is projected by mid-century to reduce water resources in many small islands, e.g., in the Caribbean and Pacific, to the point where they become insufficient to meet demand during low-rainfall periods. *** D [16.4]

With higher temperatures, increased invasion by non-native species is expected to occur, particularly on mid- and high-latitude islands. ** N [16.4]

Magnitudes of impact can now be estimated more systematically for a range of possible increases in global average temperature.

Since the IPCC Third Assessment, many additional studies, particularly in regions that previously had been little researched, have enabled a more systematic understanding of how the timing and magnitude of impacts may be affected by changes in climate and sea level associated with differing amounts and rates of change in global average temperature.

Examples of this new information are presented in Figure SPM.2. Entries have been selected which are judged to be relevant for people and the environment and for which there is high confidence in the assessment. All examples of impact are drawn from chapters of the Assessment, where more detailed information is available.

Depending on circumstances, some of these impacts could be associated with 'key vulnerabilities', based on a number of criteria in the literature (magnitude, timing, persistence/reversibility, the potential for adaptation, distributional aspects, likelihood and 'importance' of the impacts). Assessment of potential key vulnerabilities is intended to provide information on rates and levels of climate change to help decision-makers make appropriate responses to the risks of climate change [19.ES, 19.1].

The 'reasons for concern' identified in the Third Assessment remain a viable framework for considering key vulnerabilities. Recent research has updated some of the findings from the Third Assessment [19.3].

Key impacts as a function of increasing global average temperature change

(Impacts will vary by extent of adaptation, rate of temperature change, and socio-economic pathway)

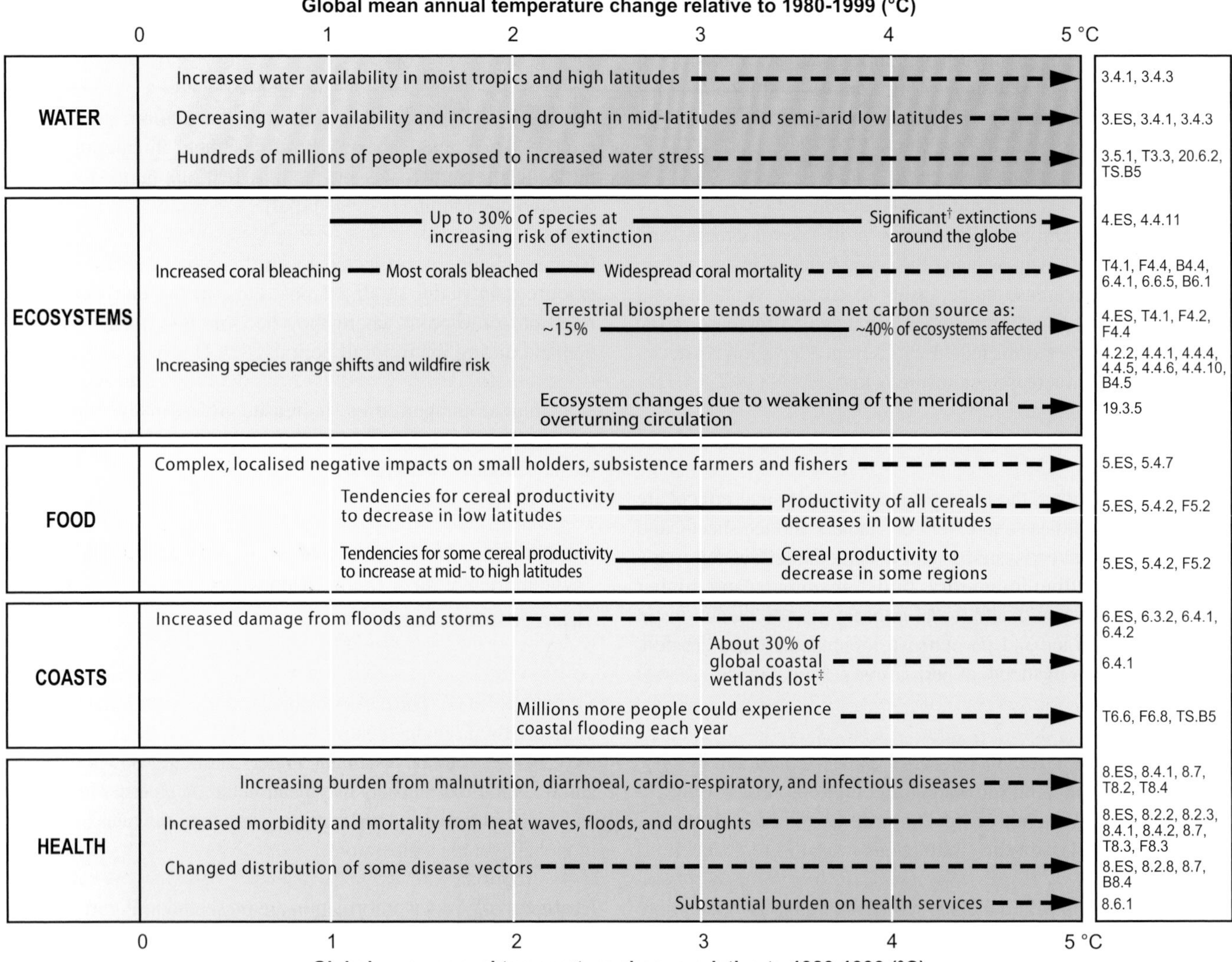

† Significant is defined here as more than 40%.
‡ Based on average rate of sea level rise of 4.2 mm/year from 2000 to 2080.

Figure SPM.2. *Illustrative examples of global impacts projected for climate changes (and sea level and atmospheric carbon dioxide where relevant) associated with different amounts of increase in global average surface temperature in the 21st century [T20.8]. The black lines link impacts, dotted arrows indicate impacts continuing with increasing temperature. Entries are placed so that the left-hand side of the text indicates the approximate onset of a given impact. Quantitative entries for water stress and flooding represent the additional impacts of climate change relative to the conditions projected across the range of Special Report on Emissions Scenarios (SRES) scenarios A1FI, A2, B1 and B2 (see Endbox 3). Adaptation to climate change is not included in these estimations. All entries are from published studies recorded in the chapters of the Assessment. Sources are given in the right-hand column of the Table. Confidence levels for all statements are high.*

Impacts due to altered frequencies and intensities of extreme weather, climate and sea-level events are very likely to change.

Since the IPCC Third Assessment, confidence has increased that some weather events and extremes will become more frequent, more widespread and/or more intense during the 21st century; and more is known about the potential effects of such changes. A selection of these is presented in Table SPM.1.

The direction of trend and likelihood of phenomena are for IPCC SRES projections of climate change.

Some large-scale climate events have the potential to cause very large impacts, especially after the 21st century.

Very large sea-level rises that would result from widespread deglaciation of Greenland and West Antarctic ice sheets imply major changes in coastlines and ecosystems, and inundation of low-lying areas, with greatest effects in river deltas. Relocating populations, economic activity, and infrastructure would be costly and challenging. There is medium confidence that at least partial deglaciation of the Greenland ice sheet, and possibly the West Antarctic ice sheet, would occur over a period of time ranging from centuries to millennia for a global average temperature increase of 1-4°C (relative to 1990-2000), causing a contribution to sea-level rise of 4-6 m or more. The complete melting of the Greenland ice sheet and the West Antarctic ice sheet would lead to a contribution to sea-level rise of up to 7 m and about 5 m, respectively [Working Group I Fourth Assessment 6.4, 10.7; Working Group II Fourth Assessment 19.3].

Based on climate model results, it is very unlikely that the Meridional Overturning Circulation (MOC) in the North Atlantic will undergo a large abrupt transition during the 21st century. Slowing of the MOC during this century is very likely, but temperatures over the Atlantic and Europe are projected to increase nevertheless, due to global warming. Impacts of large-scale and persistent changes in the MOC are likely to include changes to marine ecosystem productivity, fisheries, ocean carbon dioxide uptake, oceanic oxygen concentrations and terrestrial vegetation [Working Group I Fourth Assessment 10.3, 10.7; Working Group II Fourth Assessment 12.6, 19.3].

Impacts of climate change will vary regionally but, aggregated and discounted to the present, they are very likely to impose net annual costs which will increase over time as global temperatures increase.

This Assessment makes it clear that the impacts of future climate change will be mixed across regions. For increases in global mean temperature of less than 1-3°C above 1990 levels, some impacts are projected to produce benefits in some places and some sectors, and produce costs in other places and other sectors. It is, however, projected that some low-latitude and polar regions will experience net costs even for small increases in temperature. It is very likely that all regions will experience either declines in net benefits or increases in net costs for increases in temperature greater than about 2-3°C [9.ES, 9.5, 10.6, T10.9, 15.3, 15.ES]. These observations confirm evidence reported in the Third Assessment that, while developing countries are expected to experience larger percentage losses, global mean losses could be 1-5% GDP for 4°C of warming [F20.3].

Many estimates of aggregate net economic costs of damages from climate change across the globe (i.e., the social cost of carbon (SCC), expressed in terms of future net benefits and costs that are discounted to the present) are now available. Peer-reviewed estimates of the SCC for 2005 have an average value of US$43 per tonne of carbon (i.e., US$12 per tonne of carbon dioxide), but the range around this mean is large. For example, in a survey of 100 estimates, the values ran from US$-10 per tonne of carbon (US$-3 per tonne of carbon dioxide) up to US$350 per tonne of carbon (US$95 per tonne of carbon dioxide) [20.6].

The large ranges of SCC are due in the large part to differences in assumptions regarding climate sensitivity, response lags, the treatment of risk and equity, economic and non-economic impacts, the inclusion of potentially catastrophic losses, and discount rates. It is very likely that globally aggregated figures underestimate the damage costs because they cannot include many non-quantifiable impacts. Taken as a whole, the range of published evidence indicates that the net damage costs of climate change are likely to be significant and to increase over time [T20.3, 20.6, F20.4].

It is virtually certain that aggregate estimates of costs mask significant differences in impacts across sectors, regions, countries and populations. In some locations and among some groups of people with high exposure, high sensitivity and/or low adaptive capacity, net costs will be significantly larger than the global aggregate [20.6, 20.ES, 7.4].

Phenomenon[a] and direction of trend	Likelihood of future trends based on projections for 21st century using SRES scenarios	Examples of major projected impacts by sector			
		Agriculture, forestry and ecosystems [4.4, 5.4]	Water resources [3.4]	Human health [8.2, 8.4]	Industry, settlement and society [7.4]
Over most land areas, warmer and fewer cold days and nights, warmer and more frequent hot days and nights	Virtually certain[b]	Increased yields in colder environments; decreased yields in warmer environments; increased insect outbreaks	Effects on water resources relying on snow melt; effects on some water supplies	Reduced human mortality from decreased cold exposure	Reduced energy demand for heating; increased demand for cooling; declining air quality in cities; reduced disruption to transport due to snow, ice; effects on winter tourism
Warm spells/heat waves. Frequency increases over most land areas	Very likely	Reduced yields in warmer regions due to heat stress; increased danger of wildfire	Increased water demand; water quality problems, e.g., algal blooms	Increased risk of heat-related mortality, especially for the elderly, chronically sick, very young and socially-isolated	Reduction in quality of life for people in warm areas without appropriate housing; impacts on the elderly, very young and poor
Heavy precipitation events. Frequency increases over most areas	Very likely	Damage to crops; soil erosion, inability to cultivate land due to waterlogging of soils	Adverse effects on quality of surface and groundwater; contamination of water supply; water scarcity may be relieved	Increased risk of deaths, injuries and infectious, respiratory and skin diseases	Disruption of settlements, commerce, transport and societies due to flooding; pressures on urban and rural infrastructures; loss of property
Area affected by drought increases	Likely	Land degradation; lower yields/crop damage and failure; increased livestock deaths; increased risk of wildfire	More widespread water stress	Increased risk of food and water shortage; increased risk of malnutrition; increased risk of water- and food-borne diseases	Water shortages for settlements, industry and societies; reduced hydropower generation potentials; potential for population migration
Intense tropical cyclone activity increases	Likely	Damage to crops; windthrow (uprooting) of trees; damage to coral reefs	Power outages causing disruption of public water supply	Increased risk of deaths, injuries, water- and food-borne diseases; post-traumatic stress disorders	Disruption by flood and high winds; withdrawal of risk coverage in vulnerable areas by private insurers, potential for population migrations, loss of property
Increased incidence of extreme high sea level (excludes tsunamis)[c]	Likely[d]	Salinisation of irrigation water, estuaries and freshwater systems	Decreased freshwater availability due to saltwater intrusion	Increased risk of deaths and injuries by drowning in floods; migration-related health effects	Costs of coastal protection versus costs of land-use relocation; potential for movement of populations and infrastructure; also see tropical cyclones above

[a] See Working Group I Fourth Assessment Table 3.7 for further details regarding definitions.
[b] Warming of the most extreme days and nights each year.
[c] Extreme high sea level depends on average sea level and on regional weather systems. It is defined as the highest 1% of hourly values of observed sea level at a station for a given reference period.
[d] In all scenarios, the projected global average sea level at 2100 is higher than in the reference period [Working Group I Fourth Assessment 10.6]. The effect of changes in regional weather systems on sea level extremes has not been assessed.

Table SPM.1. *Examples of possible impacts of climate change due to changes in extreme weather and climate events, based on projections to the mid- to late 21st century. These do not take into account any changes or developments in adaptive capacity. Examples of all entries are to be found in chapters in the full Assessment (see source at top of columns). The first two columns of the table (shaded yellow) are taken directly from the Working Group I Fourth Assessment (Table SPM-2). The likelihood estimates in Column 2 relate to the phenomena listed in Column 1.*

D. Current knowledge about responding to climate change

Some adaptation is occurring now, to observed and projected future climate change, but on a limited basis.

There is growing evidence since the IPCC Third Assessment of human activity to adapt to observed and anticipated climate change. For example, climate change is considered in the design of infrastructure projects such as coastal defence in the Maldives and The Netherlands, and the Confederation Bridge in Canada. Other examples include prevention of glacial lake outburst flooding in Nepal, and policies and strategies such as water management in Australia and government responses to heat-waves in, for example, some European countries [7.6, 8.2, 8.6, 17.ES, 17.2, 16.5, 11.5].

Adaptation will be necessary to address impacts resulting from the warming which is already unavoidable due to past emissions.

Past emissions are estimated to involve some unavoidable warming (about a further 0.6°C by the end of the century relative to 1980-1999) even if atmospheric greenhouse gas concentrations remain at 2000 levels (see Working Group I Fourth Assessment). There are some impacts for which adaptation is the only available and appropriate response. An indication of these impacts can be seen in Figure SPM.2.

A wide array of adaptation options is available, but more extensive adaptation than is currently occurring is required to reduce vulnerability to future climate change. There are barriers, limits and costs, but these are not fully understood.

Impacts are expected to increase with increases in global average temperature, as indicated in Figure SPM.2. Although many early impacts of climate change can be effectively addressed through adaptation, the options for successful adaptation diminish and the associated costs increase with increasing climate change. At present we do not have a clear picture of the limits to adaptation, or the cost, partly because effective adaptation measures are highly dependent on specific, geographical and climate risk factors as well as institutional, political and financial constraints [7.6, 17.2, 17.4].

The array of potential adaptive responses available to human societies is very large, ranging from purely technological (e.g., sea defences), through behavioural (e.g., altered food and recreational choices), to managerial (e.g., altered farm practices) and to policy (e.g., planning regulations). While most technologies and strategies are known and developed in some countries, the assessed literature does not indicate how effective various options[13] are at fully reducing risks, particularly at higher levels of warming and related impacts, and for vulnerable groups. In addition, there are formidable environmental, economic, informational, social, attitudinal and behavioural barriers to the implementation of adaptation. For developing countries, availability of resources and building adaptive capacity are particularly important [see Sections 5 and 6 in Chapters 3-16; also 17.2, 17.4].

Adaptation alone is not expected to cope with all the projected effects of climate change, and especially not over the long term as most impacts increase in magnitude [Figure SPM.2].

Vulnerability to climate change can be exacerbated by the presence of other stresses.

Non-climate stresses can increase vulnerability to climate change by reducing resilience and can also reduce adaptive capacity because of resource deployment to competing needs. For example, current stresses on some coral reefs include marine pollution and chemical runoff from agriculture as well as increases in water temperature and ocean acidification. Vulnerable regions face multiple stresses that affect their exposure and sensitivity as well as their capacity to adapt. These stresses arise from, for example, current climate hazards, poverty and unequal access to resources, food insecurity, trends in economic globalisation, conflict, and incidence of diseases such as HIV/AIDS [7.4, 8.3, 17.3, 20.3]. Adaptation measures are seldom undertaken in response to climate change alone but can be integrated within, for example, water resource management, coastal defence and risk-reduction strategies [17.2, 17.5].

Future vulnerability depends not only on climate change but also on development pathway.

An important advance since the IPCC Third Assessment has been the completion of impacts studies for a range of different development pathways taking into account not only projected climate change but also projected social and economic changes. Most have been based on characterisations of population and income level drawn from the IPCC Special Report on Emission Scenarios (SRES) (see Endbox 3) [2.4].

[13] A table of options is given in the Technical Summary

These studies show that the projected impacts of climate change can vary greatly due to the development pathway assumed. For example, there may be large differences in regional population, income and technological development under alternative scenarios, which are often a strong determinant of the level of vulnerability to climate change [2.4].

To illustrate, in a number of recent studies of global impacts of climate change on food supply, risk of coastal flooding and water scarcity, the projected number of people affected is considerably greater under the A2-type scenario of development (characterised by relatively low per capita income and large population growth) than under other SRES futures [T20.6]. This difference is largely explained, not by differences in changes of climate, but by differences in vulnerability [T6.6].

Sustainable development[14] can reduce vulnerability to climate change, and climate change could impede nations' abilities to achieve sustainable development pathways.

Sustainable development can reduce vulnerability to climate change by enhancing adaptive capacity and increasing resilience. At present, however, few plans for promoting sustainability have explicitly included either adapting to climate change impacts, or promoting adaptive capacity [20.3].

On the other hand, it is very likely that climate change can slow the pace of progress towards sustainable development, either directly through increased exposure to adverse impact or indirectly through erosion of the capacity to adapt. This point is clearly demonstrated in the sections of the sectoral and regional chapters of this report that discuss the implications for sustainable development [See Section 7 in Chapters 3-8, 20.3, 20.7].

The Millennium Development Goals (MDGs) are one measure of progress towards sustainable development. Over the next half-century, climate change could impede achievement of the MDGs [20.7].

Many impacts can be avoided, reduced or delayed by mitigation.

A small number of impact assessments have now been completed for scenarios in which future atmospheric concentrations of greenhouse gases are stabilised. Although these studies do not take full account of uncertainties in projected climate under stabilisation, they nevertheless provide indications of damages avoided or vulnerabilities and risks reduced for different amounts of emissions reduction [2.4, T20.6].

A portfolio of adaptation and mitigation measures can diminish the risks associated with climate change.

Even the most stringent mitigation efforts cannot avoid further impacts of climate change in the next few decades, which makes adaptation essential, particularly in addressing near-term impacts. Unmitigated climate change would, in the long term, be likely to exceed the capacity of natural, managed and human systems to adapt [20.7].

This suggests the value of a portfolio or mix of strategies that includes mitigation, adaptation, technological development (to enhance both adaptation and mitigation) and research (on climate science, impacts, adaptation and mitigation). Such portfolios could combine policies with incentive-based approaches, and actions at all levels from the individual citizen through to national governments and international organisations [18.1, 18.5].

One way of increasing adaptive capacity is by introducing the consideration of climate change impacts in development planning [18.7], for example, by:

- including adaptation measures in land-use planning and infrastructure design [17.2];
- including measures to reduce vulnerability in existing disaster risk reduction strategies [17.2, 20.8].

E. Systematic observing and research

Although the science to provide policymakers with information about climate change impacts and adaptation potential has improved since the Third Assessment, it still leaves many important questions to be answered. The chapters of the Working Group II Fourth Assessment include a number of judgements about priorities for further observation and research, and this advice should be considered seriously (a list of these recommendations is given in the Technical Summary Section TS-6).

[14] The Brundtland Commission definition of sustainable development is used in this Assessment: "development that meets the needs of the present without compromising the ability of future generations to meet their own needs". The same definition was used by the IPCC Working Group II Third Assessment and Third Assessment Synthesis Report.

Endbox 1. Definitions of key terms

Climate change in IPCC usage refers to any change in climate over time, whether due to natural variability or as a result of human activity. This usage differs from that in the Framework Convention on Climate Change, where climate change refers to a change of climate that is attributed directly or indirectly to human activity that alters the composition of the global atmosphere and that is in addition to natural climate variability observed over comparable time periods.

Adaptive capacity is the ability of a system to adjust to climate change (including climate variability and extremes) to moderate potential damages, to take advantage of opportunities, or to cope with the consequences.

Vulnerability is the degree to which a system is susceptible to, and unable to cope with, adverse effects of climate change, including climate variability and extremes. Vulnerability is a function of the character, magnitude, and rate of climate change and variation to which a system is exposed, its sensitivity, and its adaptive capacity.

Endbox 2. Communication of Uncertainty in the Working Group II Fourth Assessment

A set of terms to describe uncertainties in current knowledge is common to all parts of the IPCC Fourth Assessment.

Description of confidence

Authors have assigned a confidence level to the major statements in the Summary for Policymakers on the basis of their assessment of current knowledge, as follows:

Terminology	*Degree of confidence in being correct*
Very high confidence	At least 9 out of 10 chance of being correct
High confidence	About 8 out of 10 chance
Medium confidence	About 5 out of 10 chance
Low confidence	About 2 out of 10 chance
Very low confidence	Less than a 1 out of 10 chance

Description of likelihood

Likelihood refers to a probabilistic assessment of some well-defined outcome having occurred or occurring in the future, and may be based on quantitative analysis or an elicitation of expert views. In the Summary for Policymakers, when authors evaluate the likelihood of certain outcomes, the associated meanings are:

Terminology	*Likelihood of the occurrence/ outcome*
Virtually certain	>99% probability of occurrence
Very likely	90 to 99% probability
Likely	66 to 90% probability
About as likely as not	33 to 66% probability
Unlikely	10 to 33% probability
Very unlikely	1 to 10% probability
Exceptionally unlikely	<1% probability

Endbox 3. The Emissions Scenarios of the IPCC Special Report on Emissions Scenarios (SRES)

A1. The A1 storyline and scenario family describes a future world of very rapid economic growth, global population that peaks in mid-century and declines thereafter, and the rapid introduction of new and more efficient technologies. Major underlying themes are convergence among regions, capacity building and increased cultural and social interactions, with a substantial reduction in regional differences in per capita income. The A1 scenario family develops into three groups that describe alternative directions of technological change in the energy system. The three A1 groups are distinguished by their technological emphasis: fossil intensive (A1FI), non fossil energy sources (A1T), or a balance across all sources (A1B) (where balanced is defined as not relying too heavily on one particular energy source, on the assumption that similar improvement rates apply to all energy supply and end use technologies).

A2. The A2 storyline and scenario family describes a very heterogeneous world. The underlying theme is self reliance and preservation of local identities. Fertility patterns across regions converge very slowly, which results in continuously increasing population. Economic development is primarily regionally oriented and per capita economic growth and technological change more fragmented and slower than other storylines.

B1. The B1 storyline and scenario family describes a convergent world with the same global population, that peaks in mid-century and declines thereafter, as in the A1 storyline, but with rapid change in economic structures toward a service and information economy, with reductions in material intensity and the introduction of clean and resource efficient technologies. The emphasis is on global solutions to economic, social and environmental sustainability, including improved equity, but without additional climate initiatives.

B2. The B2 storyline and scenario family describes a world in which the emphasis is on local solutions to economic, social and environmental sustainability. It is a world with continuously increasing global population, at a rate lower than A2, intermediate levels of economic development, and less rapid and more diverse technological change than in the B1 and A1 storylines. While the scenario is also oriented towards environmental protection and social equity, it focuses on local and regional levels.

An illustrative scenario was chosen for each of the six scenario groups A1B, A1FI, A1T, A2, B1 and B2. All should be considered equally sound.

The SRES scenarios do not include additional climate initiatives, which means that no scenarios are included that explicitly assume implementation of the United Nations Framework Convention on Climate Change or the emissions targets of the Kyoto Protocol.

A report accepted by Working Group II of the Intergovernmental Panel on Climate Change but not approved in detail

'Acceptance' of IPCC Reports at a Session of the Working Group or Panel signifies that the material has not been subject to line-by-line discussion and agreement, but nevertheless presents a comprehensive, objective and balanced view of the subject matter.

Technical Summary

Coordinating Lead Authors:

Martin Parry (UK), Osvaldo Canziani (Argentina), Jean Palutikof (UK)

Lead Authors:

Neil Adger (UK), Pramod Aggarwal (India), Shardul Agrawala (OECD/France), Joseph Alcamo (Germany), Abdelkader Allali (Morocco), Oleg Anisimov (Russia), Nigel Arnell (UK), Michel Boko (Benin), Timothy Carter (Finland), Gino Casassa (Chile), Ulisses Confalonieri (Brazil), Rex Victor Cruz (Philippines), Edmundo de Alba Alcaraz (Mexico), William Easterling (USA), Christopher Field (USA), Andreas Fischlin (Switzerland), Blair Fitzharris (New Zealand), Carlos Gay García (Mexico), Hideo Harasawa (Japan), Kevin Hennessy (Australia), Saleemul Huq (UK), Roger Jones (Australia), Lucka Kajfež Bogataj (Slovenia), David Karoly (USA), Richard Klein (The Netherlands), Zbigniew Kundzewicz (Poland), Murari Lal (India), Rodel Lasco (Philippines), Geoff Love (Australia), Xianfu Lu (China), Graciela Magrín (Argentina), Luis José Mata (Venezuela), Bettina Menne (WHO Regional Office for Europe/Germany), Guy Midgley (South Africa), Nobuo Mimura (Japan), Monirul Qader Mirza (Bangladesh/Canada), José Moreno (Spain), Linda Mortsch (Canada), Isabelle Niang-Diop (Senegal), Robert Nicholls (UK), Béla Nováky (Hungary), Leonard Nurse (Barbados), Anthony Nyong (Nigeria), Michael Oppenheimer (USA), Anand Patwardhan (India), Patricia Romero Lankao (Mexico), Cynthia Rosenzweig (USA), Stephen Schneider (USA), Serguei Semenov (Russia), Joel Smith (USA), John Stone (Canada), Jean-Pascal van Ypersele (Belgium), David Vaughan (UK), Coleen Vogel (South Africa), Thomas Wilbanks (USA), Poh Poh Wong (Singapore), Shaohong Wu (China), Gary Yohe (USA)

Contributing Authors:

Debbie Hemming (UK), Pete Falloon (UK)

Review Editors:

Wolfgang Cramer (Germany), Daniel Murdiyarso (Indonesia)

This Technical Summary should be cited as:

Parry, M.L., O.F. Canziani, J.P. Palutikof and Co-authors 2007: Technical Summary. *Climate Change 2007: Impacts, Adaptation and Vulnerability. Contribution of Working Group II to the Fourth Assessment Report of the Intergovernmental Panel on Climate Change,* M.L. Parry, O.F. Canziani, J.P. Palutikof, P.J. van der Linden and C.E. Hanson, Eds., Cambridge University Press, Cambridge, UK, 23-78.

Table of Contents

Summary of main findings

- Observational evidence from all continents and most oceans shows that many natural systems are being affected by regional climate changes, particularly temperature increases.
- A global assessment of data since 1970 has shown it is likely that anthropogenic warming has had a discernible influence on many physical and biological systems.
- Other effects of regional climate changes on natural and human environments are emerging, although many are difficult to discern due to adaptation and non-climatic drivers.
- More specific information is now available across a wide range of systems and sectors concerning the nature of future impacts, including for some fields not covered in previous assessments.
- More specific information is now available across the regions of the world concerning the nature of future impacts, including for some places not covered in previous assessments.
- Magnitudes of impact can now be estimated more systematically for a range of possible increases in global average temperature.
- Impacts due to altered frequencies and intensities of extreme weather, climate and sea-level events are very likely to change.
- Some large-scale climate events have the potential to cause very large impacts, especially after the 21st century.
- Impacts of climate change will vary regionally but, aggregated and discounted to the present, they are very likely to impose net annual costs which will increase over time as global temperatures increase.
- Some adaptation is occurring now, to observed and projected future climate change, but on a limited basis.
- Adaptation will be necessary to address impacts resulting from the warming which is already unavoidable due to past emissions.
- A wide array of adaptation options is available, but more extensive adaptation than is currently occurring is required to reduce vulnerability to future climate change. There are barriers, limits and costs, but these are not fully understood.
- Vulnerability to climate change can be exacerbated by the presence of other stresses.
- Future vulnerability depends not only on climate change but also on development pathway.
- Sustainable development can reduce vulnerability to climate change, and climate change could impede nations' abilities to achieve sustainable development pathways.
- Many impacts can be avoided, reduced or delayed by mitigation.
- A portfolio of adaptation and mitigation measures can diminish the risks associated with climate change.

TS.1 Scope, approach and method of the Working Group II assessment

The decision to produce a Fourth Assessment Report (AR4) was taken by the 19th Session of the Intergovernmental Panel on Climate Change (IPCC) in April 2002.

The Working Group II Report has twenty chapters. The core chapters (3 – 16) address the future impacts of climate change on sectors and regions, the potential for adaptation and the implications for sustainability. Chapter 1 looks at observed changes and Chapter 2 assesses new methodologies and the characterisation of future conditions. Chapters 17 – 20 assess responses to impacts through adaptation (17), the inter-relationships between adaptation and mitigation (18), key vulnerabilities and risks (19) and, finally, perspectives on climate change and sustainability (20).

The Working Group II Fourth Assessment, in common with all IPCC reports, has been produced through an open and peer-reviewed process. It builds upon past assessments and IPCC Special Reports, and incorporates the results of the past 5 years of climate change impacts, adaptation and vulnerability research. Each chapter presents a balanced assessment of the literature which has appeared since the Third Assessment Report[1] (TAR), including non-English language and, where appropriate, 'grey' literature.[2]

This Assessment aims to describe current knowledge of climate-change impacts, adaptation and vulnerability. Specifically it addresses five questions:

- What is the current knowledge about impacts of climate change which are observable now? (addressed in Section TS.2 of the Technical Summary)
- What new scenarios and research methods have led to improvements in knowledge since the Third Assessment? (addressed in Section TS.3)
- What is the current knowledge about future effects of climate change on different sectors and regions? (addressed in Section TS.4)
- What is the current knowledge about adaptation, the interaction between adaptation and mitigation, key vulnerabilities, and the role of sustainable development in the context of climate change? (addressed in Section TS.5)
- What gaps exist in current knowledge and how best can these be filled? (addressed in Section TS.6).

Each of the twenty chapters of the Working Group II Fourth Assessment had a minimum of two Coordinating Lead Authors, six Lead Authors and two Review Editors. The writing team and review editors were appointed by the IPCC Bureau on the recommendation of the Working Group II Co-Chairs and Vice-Chairs. They were selected from the pool of nominated experts, in consultation with the international community of scientists active in the field, and taking into consideration expertise and experience. In total, the Working Group II Fourth Assessment involved 48 Coordinating Lead Authors, 125 Lead Authors and 45 Review Editors, drawn from 70 countries. In addition there were 183 Contributing Authors and 910 Expert Reviewers.

This Technical Summary is intended to capture the most important scientific aspects of the full Working Group II Assessment. Reducing the information from 800 pages to 50 requires much condensing; consequently every statement in the Summary appears with its source in the Assessment, enabling the reader to pursue more detail. Sourcing information is provided in square brackets in the text (see Box TS.1). Uncertainty information is provided in parentheses (see Box TS.2 for definitions of uncertainty). Key terms are defined in Box TS.3.

TS.2 Current knowledge about observed impacts on natural and managed systems

Observational evidence from all continents and most oceans shows that many natural systems are being affected by regional climate changes, particularly temperature increases (very high confidence). A global assessment of data since 1970 has shown it is likely that anthropogenic warming has had a discernible influence on many physical and biological systems.

The IPCC Working Group II Third Assessment found evidence that recent regional climate changes, particularly temperature increases, have already affected physical and biological systems [1.1.1].[3] The Fourth Assessment has analysed studies since the Third Assessment showing changes in physical, biological and human systems, mainly from 1970 to 2005, in relation to climate drivers, and has found stronger quantitative evidence [1.3, 1.4]. The major focus is on global and regional surface temperature increases [1.2].

Evaluation of evidence on observed changes related to climate change is made difficult because the observed responses of systems and sectors are influenced by many other factors. Non-climatic drivers can influence systems and sectors directly and/or indirectly through their effects on climate variables such as reflected solar radiation and evaporation [1.2.1]. Socio-economic processes, including land-use change (e.g., agriculture

[1] McCarthy, J.J., O.F. Canziani, N.A. Leary, D.J. Dokken and K.S. White, Eds., 2001: *Climate Change 2001: Impacts, Adaptation, and Vulnerability. Contribution of Working Group II to the Third Assessment Report of the Intergovernmental Panel on Climate Change.* Cambridge University Press, Cambridge, UK, 1032 pp.

[2] 'Grey' literature is defined as literature which is not available through traditional commercial publication channels, such as working papers, government reports and theses, which therefore may be difficult to access.

[3] See Box TS.1

Box TS.1. Sourcing in the Technical Summary

For example, source [3.3.2] refers to Chapter 3, Section 3, Sub-section 2. In the sourcing, F = Figure, T = Table, B = Box, ES = Executive Summary.

References to the Working Group I Fourth Assessment are shown as, for example, [WGI AR4 SPM] which refers to the Working Group I Fourth Assessment Summary for Policymakers, [WGI AR4 10.3.2] which refers to Chapter 10 Section 10.3.2, and [WGI AR4 Chapter 10] when the whole chapter is referred to. Where a source refers to both the WGI and WGII Fourth Assessments, these are separated by a semi-colon, for example [WGI AR4 10.2.1; 2.1.4]. References to Working Group III are treated in the same way.

Box TS.2. Communication of uncertainty in the Working Group II Fourth Assessment

A set of terms to describe uncertainties in current knowledge is common to all parts of the IPCC Fourth Assessment, based on the *Guidance Notes for Lead Authors of the IPCC Fourth Assessment Report on Addressing Uncertainties*[4], produced by the IPCC in July 2005.

Description of confidence
On the basis of a comprehensive reading of the literature and their expert judgement, authors have assigned a confidence level to the major statements in the Technical Summary on the basis of their assessment of current knowledge, as follows:

Terminology	*Degree of confidence in being correct*
Very high confidence	At least 9 out of 10 chance of being correct
High confidence	About 8 out of 10 chance
Medium confidence	About 5 out of 10 chance
Low confidence	About 2 out of 10 chance
Very low confidence	Less than a 1 out of 10 chance

Description of likelihood
Likelihood refers to a probabilistic assessment of some well-defined outcome having occurred or occurring in the future, and may be based on quantitative analysis or an elicitation of expert views. In the Technical Summary, when authors evaluate the likelihood of certain outcomes, the associated meanings are:

Terminology	*Likelihood of the occurrence/outcome*
Virtually certain	>99% probability of occurrence
Very likely	90 to 99% probability
Likely	66 to 90% probability
About as likely as not	33 to 66% probability
Unlikely	10 to 33% probability
Very unlikely	1 to 10% probability
Exceptionally unlikely	<1% probability

Box TS.3. Definitions of key terms

Climate change in IPCC usage refers to any change in climate over time, whether due to natural variability or as a result of human activity. This usage differs from that in the Framework Convention on Climate Change, where *climate change* refers to a change of climate that is attributed directly or indirectly to human activity that alters the composition of the global atmosphere and that is in addition to natural climate variability observed over comparable time periods.

Adaptation is the adjustment in natural or human systems in response to actual or expected climatic stimuli or their effects, which moderates harm or exploits beneficial opportunities.

Vulnerability is the degree to which a system is susceptible to, and unable to cope with, adverse effects of climate change, including climate variability and extremes. *Vulnerability* is a function of the character, magnitude and rate of climate change and the variation to which a system is exposed, its sensitivity and its adaptive capacity.

[4] See http://www.ipcc.ch/activity/uncertaintyguidancenote.pdf.

to urban area), land-cover modification (e.g., ecosystem degradation), technological change, pollution, and invasive species constitute some of the important non-climate drivers [1.2.1].

Much more evidence has accumulated over the past 5 years to indicate that the effects described above are linked to the anthropogenic component of warming.[5] There are three sets of evidence which, taken together, support this conclusion (see Box TS.4).

1. There have been several studies that have linked responses in some physical and biological systems to the anthropogenic component of warming by comparing observed trends with modelled trends in which the natural and anthropogenic forcings are explicitly separated [1.4].

2. Observed changes in many physical and biological systems are consistent with a warming world. The majority (>89% of the >29,000 data sets whose locations are displayed in Figure TS.1) of changes in these systems have been in the direction expected as a response to warming [1.4].

3. A global synthesis of studies in this Assessment strongly demonstrates that the spatial agreement between regions of significant regional warming across the globe and the locations of significant observed changes in many systems consistent with warming is very unlikely[6] to be due solely to natural variability of temperatures or natural variability of the systems [1.4].

For physical systems, (i) climate change is affecting natural and human systems in regions of snow, ice and frozen ground, and (ii) there is now evidence of effects on hydrology and water resources, coastal zones and oceans.

The main evidence from regions of snow, ice and frozen ground is found in ground instability in permafrost regions, and rock avalanches; decrease in travel days of vehicles over frozen roads in the Arctic; increase and enlargement of glacial lakes, and destabilisation of moraines damming these lakes, with increased risk of outburst floods; changes in Arctic and Antarctic Peninsula ecosystems, including sea-ice biomes and predators high on the food chain; and limitations on mountain sports in lower-elevation alpine areas (high confidence)[7] [1.3.1]. These changes parallel the abundant evidence that Arctic sea ice, freshwater ice, ice shelves, the Greenland ice sheet, alpine and Antarctic Peninsula glaciers and ice caps, snow cover and permafrost are undergoing enhanced melting in response to global warming (very high confidence) [WGI AR4 Chapter 4].

Recent evidence in hydrology and water resources shows that spring peak discharge is occurring earlier in rivers affected by snow melt, and there is evidence for enhanced glacial melt in the tropical Andes and in the Alps. Lakes and rivers around the world are warming, with effects on thermal structure and water quality (high confidence) [1.3.2].

Sea-level rise and human development are together contributing to losses of coastal wetlands and mangroves and increasing damage from coastal flooding in many areas (medium confidence) [1.3.3.2].

There is more evidence, from a wider range of species and communities in terrestrial ecosystems than reported in the Third Assessment, that recent warming is already strongly affecting natural biological systems. There is substantial new evidence relating changes in marine and freshwater systems to warming. The evidence suggests that both terrestrial and marine biological systems are now being strongly influenced by observed recent warming.

The overwhelming majority of studies of regional climate effects on terrestrial species reveal consistent responses to warming trends, including poleward and elevational range shifts of flora and fauna. Responses of terrestrial species to warming across the Northern Hemisphere are well documented by changes in the timing of growth stages (i.e., phenological changes), especially the earlier onset of spring events, migration, and lengthening of the growing season. Based on satellite observations since the early 1980s, there have been trends in many regions towards earlier 'greening' of vegetation in the spring[8] and increased net primary production linked to longer growing seasons. Changes in abundance of certain species, including limited evidence of a few local disappearances, and changes in community composition over the last few decades have been attributed to climate change (very high confidence) [1.3.5].

Many observed changes in phenology and distribution of marine and freshwater species have been associated with rising water temperatures, as well as other climate-driven changes in ice cover, salinity, oxygen levels and circulation. There have been poleward shifts in ranges and changes in algal, plankton and fish abundance in high-latitude oceans. For example, plankton has moved polewards by 10° latitude (about 1,000 km) over a period of four decades in the North Atlantic. There have also been documented increases in algal and zooplankton abundance in high-latitude and high-altitude lakes, and earlier fish migration and range changes in rivers [1.3]. While there is increasing evidence for climate change impacts on coral reefs, differentiating the impacts of climate-related stresses from other stresses (e.g., over-fishing and pollution) is difficult. The uptake of anthropogenic carbon since 1750 has led to the ocean becoming more acidic, with an average decrease in pH of 0.1 units [WGI AR4 SPM]. However, the effects of observed

[5] Warming over the past 50 years at the continental scale has been attributed to anthropogenic effects [WGI AR4 SPM].
[6] See Box TS-2.
[7] See Box TS-2.
[8] Measured by the Normalised Difference Vegetation Index (NVDI), which is a relative measure of vegetation greenness in satellite images.

Box TS.4. Linking the causes of climate change to observed effects on physical and biological systems

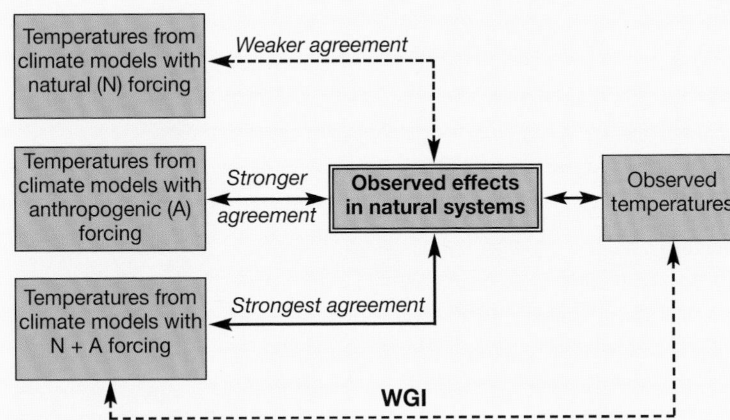

The figure to the left demonstrates the linkages between observed temperatures, observed effects on natural systems, and temperatures from climate model simulations with natural, anthropogenic, and combined natural and anthropogenic forcings. Two ways in which these linkages are utilised in detection and attribution studies of observed effects are described below.

1. Using climate models

The study of causal connection by separation of natural and anthropogenic forcing factors (Set of Evidence 1 on the preceding page) compares observed temporal changes in animals and plants with changes over the same time periods in observed temperatures as well as modelled temperatures using (i) only natural climate forcing; (ii) only anthropogenic climate forcing; and (iii) both forcings combined.

The panel to the right shows the results from a study employing this methodology[9]. The locations for the modelled temperatures were individual grid boxes corresponding to given animal and plant study sites and time periods.

The agreement (in overlap and shape) between the observed (blue bars) and modelled plots is weakest with natural forcings, stronger with anthropogenic forcings, and strongest with combined forcings. Thus, observed changes in animals and plants are likely responding to both natural and anthropogenic climate forcings, providing a direct cause-and-effect linkage [F1.7, 1.4.2.2].

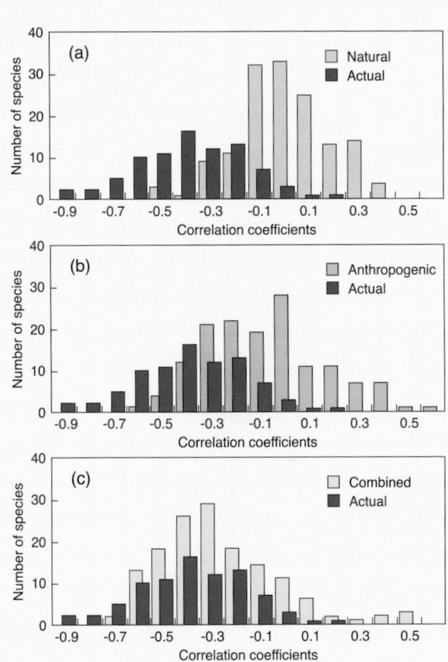

2. Using spatial analysis

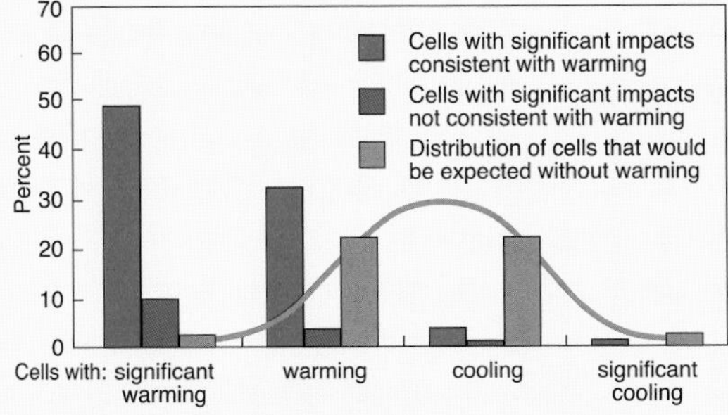

The study of causal connection by spatial analysis (Set of Evidence 3 on the preceding page) follows these stages: (i) it identifies 5° × 5° latitude/longitude cells across the globe which exhibit significant warming, warming, cooling, and significant cooling; (ii) it identifies 5° × 5° cells of significant observed changes in natural systems that are consistent with warming and that are not consistent with warming; and (iii) it statistically determines the degree of spatial agreement between the two sets of cells. In this assessment, the conclusion is that the spatial agreement is significant at the 1% level and is very unlikely to be solely due to natural variability of climate or of the natural systems.

Taken together with evidence of significant anthropogenic warming over the past 50 years averaged over each continent except Antarctica [WGI AR4[10] SPM], this shows a discernible human influence on changes in many natural systems [1.4.2.3].

[9] Plotted are the frequencies of the correlation coefficients (associations) between the timing of changes in traits (e.g., earlier egg-laying) of 145 species and modelled (HadCM3) spring temperatures for the grid-boxes in which each species was examined. (Continues next page after Figure TS.1).

[10] IPCC, 2007: *Climate Change 2007: The Physical Science Basis. Contribution of Working Group I to the Fourth Assessment Report of the Intergovernmental Panel on Climate Change*, S. Solomon, D. Qin, M. Manning, Z. Chen, M. Marquis, K.B. Averyt, M. Tignor and H.L. Miller, Eds., Cambridge University Press, Cambridge, 996 pp.

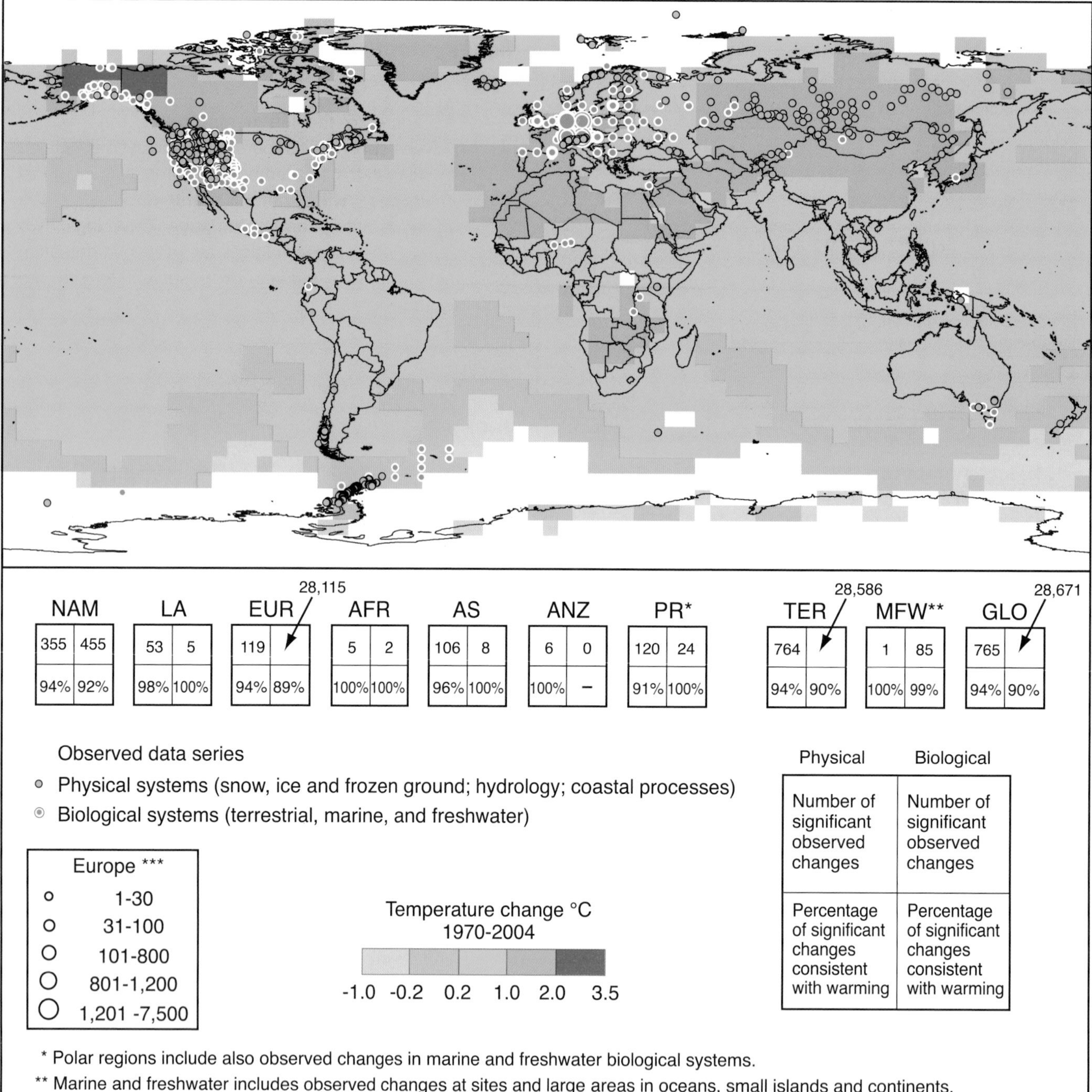

Figure TS.1. *Locations of significant changes in data series of physical systems (snow, ice and frozen ground; hydrology; and coastal processes) and biological systems (terrestrial, marine and freshwater biological systems), are shown together with surface air temperature changes over the period 1970-2004. A subset of about 29,000 data series was selected from about 80,000 data series from 577 studies. These met the following criteria: (i) ending in 1990 or later; (ii) spanning a period of at least 20 years; and (iii) showing a significant change in either direction, as assessed in individual studies. These data series are from about 75 studies (of which about 70 are new since the Third Assessment) and contain about 29,000 data series, of which about 28,000 are from European studies. White areas do not contain sufficient observational climate data to estimate a temperature trend. The 2 × 2 boxes show the total number of data series with significant changes (top row) and the percentage of those consistent with warming (bottom row) for (i) continental regions: North America (NAM), Latin America (LA), Europe (EUR), Africa (AFR), Asia (AS), Australia and New Zealand (ANZ), and Polar Regions (PR); and (ii) global scale: Terrestrial (TER), Marine and Freshwater (MFW), and Global (GLO). The numbers of studies from the seven regional boxes (NAM, ..., PR) do not add up to the global (GLO) totals because numbers from regions except Polar do not include the numbers related to Marine and Freshwater (MFW) systems. Locations of large-area marine changes are not shown on the map. [F1.8, F1.9; Working Group I AR4 F3.9b]*

ocean acidification on the marine biosphere are as yet undocumented [1.3]. Warming of lakes and rivers is affecting abundance and productivity, community composition, phenology and the distribution and migration of freshwater species (high confidence) [1.3.4].

Effects of regional increases in temperature on some managed and human systems are emerging, although these are more difficult to discern than those in natural systems, due to adaptation and non-climatic drivers.

Effects have been detected in agricultural and forestry systems [1.3.6]. Changes in several aspects of the human health system have been related to recent warming [1.3.7]. Adaptation to recent warming is beginning to be systematically documented (medium confidence) [1.3.9].

In comparison with other factors, recent warming has been of limited consequence in the agriculture and forestry sectors. A significant advance in phenology, however, has been observed for agriculture and forestry in large parts of the Northern Hemisphere, with limited responses in crop management such as earlier spring planting in northern higher latitudes. The lengthening of the growing season has contributed to an observed increase in forest productivity in many regions, while warmer and drier conditions are partly responsible for reduced forest productivity and increased forest fires in North America and the Mediterranean Basin. Both agriculture and forestry have shown vulnerability to recent trends in heatwaves, droughts and floods (medium confidence) [1.3.6].

While there have been few studies of observed health effects related to recent warming, an increase in high temperature extremes has been associated with excess mortality in Europe, which has prompted adaptation measures. There is emerging evidence of changes in the distribution of some human disease vectors in parts of Europe and Africa. Earlier onset and increases in the seasonal production of allergenic pollen have occurred in mid- and high latitudes in the Northern Hemisphere (medium confidence) [1.3.7].

Changes in socio-economic activities and modes of human response to climate change, including warming, are just beginning to be systematically documented. In regions of snow, ice and frozen ground, responses by indigenous groups relate to changes in the migration patterns, health, and range of animals and plants on which they depend for their livelihood and cultural identity [1.3.9]. Responses vary by community and are dictated by particular histories, perceptions of change and range, and the viability of options available to groups (medium confidence) [1.3.9].

While there is now significant evidence of observed changes in physical and biological systems in every continent, including Antarctica, as well as from most oceans, the majority of studies come from mid- and high latitudes in the Northern Hemisphere. Documentation of observed changes in tropical regions and the Southern Hemisphere is sparse [1.5].

TS.3 Methods and scenarios

TS.3.1 Developments in methods available to researchers on climate change impacts, adaptation and vulnerability

Since the Third Assessment (TAR), the need for improved decision analysis has motivated an expansion in the number of climate-change impacts, adaptation and vulnerability (CCIAV) approaches and methods in use. While scientific research aims to reduce uncertainty, decision-making aims to manage uncertainty by making the best possible use of the available knowledge [2.2.7, 2.3.4]. This usually involves close collaboration between researchers and stakeholders [2.3.2].

Therefore, although the standard climate scenario-driven approach is used in a large proportion of assessments described in this Report, the use of other approaches is increasing [2.2.1]. They include assessments of current and future adaptations to climate variability and change [2.2.3], adaptive capacity, social vulnerability [2.2.4], multiple stresses and adaptation in the context of sustainable development [2.2.5, 2.2.6].

Risk management can be applied in all of these contexts. It is designed for decision-making under uncertainty; several detailed frameworks have been developed for CCIAV assessments and its use is expanding rapidly. The advantages of risk management include the use of formalised methods to manage uncertainty, stakeholder involvement, use of methods for evaluating policy options without being policy-prescriptive, integration of different disciplinary approaches, and mainstreaming of climate-change concerns into the broader decision-making context [2.2.6].

Stakeholders bring vital input into CCIAV assessments about a range of risks and their management. In particular, how a group or system can cope with current climate risks provides a solid basis for assessments of future risks. An increasing number of

Footnote 9, continued from below Box TS.4. At each location, all of which are in the Northern Hemisphere, the changing trait is compared with modelled temperatures driven by: (a) Natural forcings (pink bars), (b) anthropogenic (i.e., human) forcings (orange bars), and (c) combined natural and anthropogenic forcings (yellow bars). In addition, on each panel the frequencies of the correlation coefficients between the actual temperatures recorded during each study and changes in the traits of 83 species, the only ones of the 145 with reported local-temperature trends, are shown (dark blue bars). On average the number of years species were examined is about 28 with average starting and ending years of 1960 to 1998. Note that the agreement: a) between the natural and actual plots is weaker ($K=60.16$, $p>0.05$) than b) between the anthropogenic and actual ($K=35.15$, $p>0.05$), which in turn is weaker than c) the agreement between combined and actual ($K=3.65$, $p<0.01$). Taken together, these plots show that a measurable portion of the warming regional temperatures to which species are reacting can be attributed to humans, therefore showing joint attribution (see Chapter 1).

assessments involve, or are conducted by, stakeholders. This establishes credibility and helps to confer 'ownership' of the results, which is a prerequisite for effective risk management [2.3.2].

TS.3.2 Characterising the future in the Working Group II IPCC Fourth Assessment

CCIAV assessments usually require information on how conditions such as climate, social and economic development, and other environmental factors are expected to change in the future. This commonly entails the development of scenarios, storylines or other characterisations of the future, often disaggregated to the regional or local scale [2.4.1, 2.4.6].

Scenarios are plausible descriptions, without ascribed likelihoods, of possible future states of the world. Storylines are qualitative, internally consistent narratives of how the future may evolve, which often underpin quantitative projections of future change that, together with the storyline, constitute a scenario [B2.1]. The IPCC Special Report on Emissions Scenarios (SRES), published in 2000, provided scenarios of future greenhouse gas emissions accompanied by storylines of social, economic and technological development that can be used in CCIAV studies (Figure TS.2). Although there can be methodological problems in applying these scenarios (for example, in downscaling projections of population and gross domestic product (GDP) from the four SRES large world regions to national or sub-national scales), they nevertheless provide a coherent global quantification of socio-economic development, greenhouse gas emissions and climate, and represent some of the most comprehensive scenarios presently available to CCIAV researchers. A substantial number of the impact studies assessed in this volume that employed future characterisations made use of the SRES scenarios. For some other studies, especially empirical analyses of adaptation and vulnerability, the scenarios were of limited relevance and were not adopted [2.4.6].

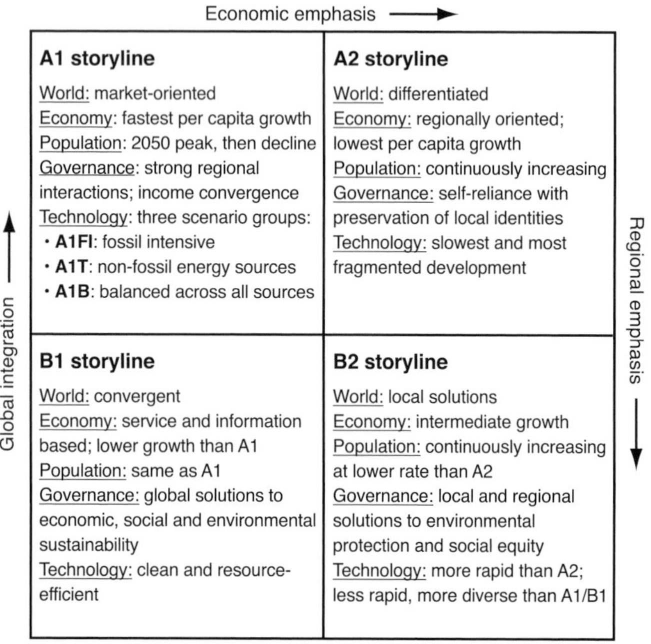

Figure TS.2. *Summary characteristics of the four SRES storylines [F2.5]*

In the future, better integration of climate-related scenarios with those widely adopted by other international bodies (mainstreaming) is desirable, and enhanced information exchange between research and policy communities will greatly improve scenario usage and acceptance. Improved scenarios are required for poorly specified indicators such as future technology and adaptive capacity, and interactions between key drivers of change need to be better specified [2.5].

Characterising future climate

Sensitivity studies
A substantial number of model-based CCIAV studies assessed in this Report employ sensitivity analysis to investigate the behaviour of a system by assuming arbitrary, often regularly spaced, adjustments in important driving variables. Using a range of perturbations allows construction of impact response surfaces, which are increasingly being used in combination with probabilistic representations of future climate to assess risks of impacts [2.4.3, 2.3.1, 2.4.8].

Analogues
Historical extreme weather events, such as floods, heatwaves and droughts, are increasingly being analysed with respect to their impacts and adaptive responses. Such studies can be useful for planning adaptation responses, especially if these events become more frequent and/or severe in the future. Spatial analogues (regions having a present-day climate similar to that expected in a study region in the future) have been adopted as a heuristic device for analysing economic impacts, adaptation needs and risks to biodiversity [2.4.4].

Climate model data
The majority of quantitative CCIAV studies assessed in the AR4 use climate models to generate the underlying scenarios of climate change. Some scenarios are based on pre-SRES emissions scenarios, such as IS92a, or even on equilibrium climate model experiments. However, the greatest proportion is derived from SRES emissions scenarios, principally the A2 scenario (assuming high emissions), for which the majority of early SRES-based climate model experiments were conducted. A few scenario-driven studies explore singular events with widespread consequences, such as an abrupt cessation of the North Atlantic Meridional Overturning Circulation (MOC) [2.4.6.1, 2.4.7].

The CCIAV studies assessed in the Working Group II Fourth Assessment (WGII AR4) are generally based on climate model simulations assessed by Working Group I (WGI) in the TAR. Since the TAR, new simulations have been performed with coupled Atmosphere-Ocean General Circulation Models (AOGCMs) assuming SRES emissions. These are assessed in the WGI AR4, but most were not available for the CCIAV studies assessed for the WGII AR4. Figure TS.3 compares the range of regional temperature and precipitation projections from recent A2-forced AOGCM simulations (assessed by WGI AR4: red bars) with earlier A2-forced simulations assessed in WGI TAR and used for scenario construction in many CCIAV studies assessed for the WGII AR4 (blue bars). The figure supports the WGI AR4 conclusion that the basic pattern of projected warming is little changed from previous assessments (note the positions of the blue and red bars), but confidence in regional projections is now higher for most regions for temperature and in some regions for precipitation (i.e., where red bars are shorter than blue bars) [B2.3].

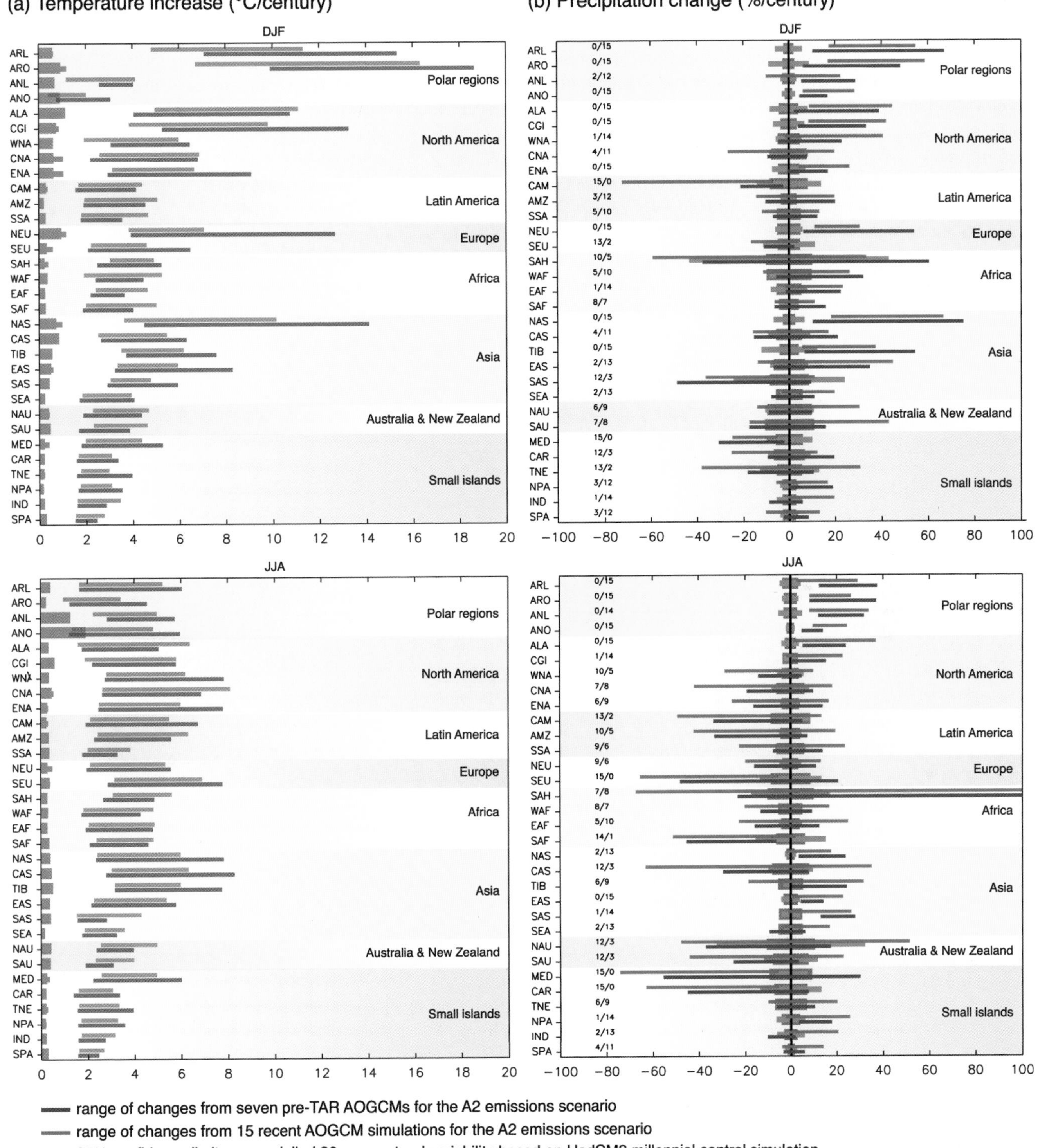

Figure TS.3. *Range of winter and summer temperature and precipitation changes up to the end of the 21st century across recent (fifteen models – red bars) and pre-TAR (seven models – blue bars) AOGCM projections under the SRES A2 emissions scenarios for thirty-two world regions, expressed as rate of change per century. Mauve and green bars show modelled 30-year natural variability. Numbers on precipitation plots show the number of recent A2 runs giving negative/positive precipitation change. DJF: December, January, February; JJA: June, July, August. [F2.6, which includes map of regions]*

Non-climate scenarios

While the CCIAV studies reported in the TAR typically applied one or more climate scenarios, very few applied contemporaneous scenarios of socio-economic, land-use or other environmental changes. Those that did used a range of sources to develop them. In contrast, AR4 studies which include SRES assumptions may now have several estimates, taking into account different storylines. The role of non-climate drivers such as technological change and regional land-use policy is shown in some studies to be more important in determining outcomes than climate change [2.4.6].

Scenarios of CO_2 concentration are required in some studies, as elevated concentrations can affect the acidity of the oceans and the growth and water use of many terrestrial plants. The observed CO_2 concentration in 2005 was about 380 ppm and was projected in the TAR using the Bern-CC model to rise to the following levels by the year 2100 for the SRES marker scenarios – B1: 540 ppm (range 486-681 ppm); A1T: 575 (506-735); B2: 611 (544-769); A1B: 703 (617-918); A2: 836 (735-1,080); A1FI: 958 (824-1,248) ppm. Values similar to these reference levels are commonly adopted in SRES-based impact studies [2.4.6.2]. Moreover, a multi-stressor approach can reveal important regional dependencies between drivers and their impacts (e.g., the

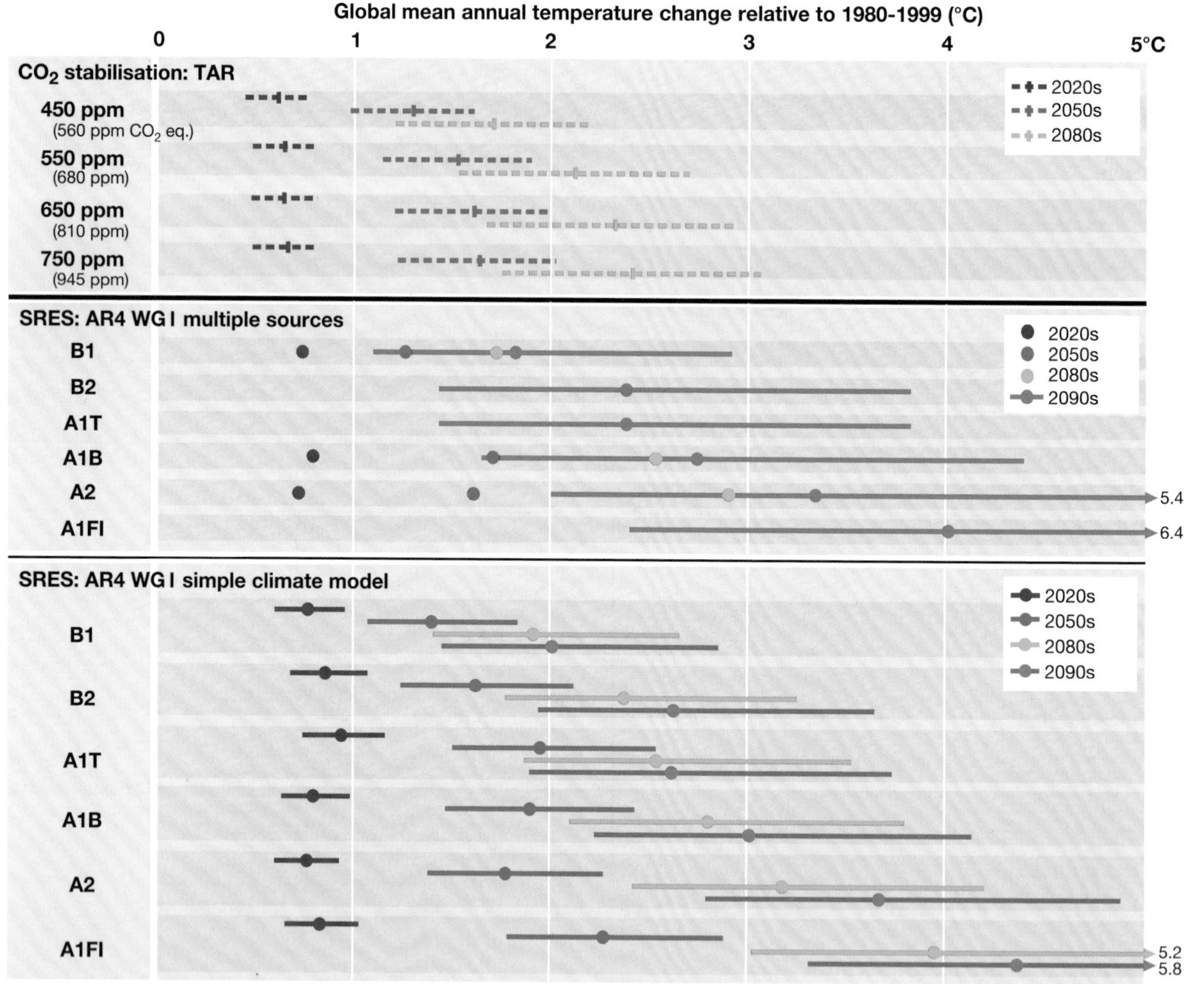

Figure TS.4. *Global temperature changes for selected time periods, relative to 1980-1999, projected for SRES and stabilisation scenarios. To express the temperature change relative to 1850-1899, add 0.5°C. More detail is provided in Chapter 2 [Box 2.8]. Estimates are for the 2020s, 2050s and 2080s, (the time periods used by the IPCC Data Distribution Centre and therefore in many impact studies) and for the 2090s. SRES-based projections are shown using two different approaches.* ***Middle panel:*** *projections from the WGI AR4 SPM based on multiple sources. Best estimates are based on AOGCMs (coloured dots). Uncertainty ranges, available only for the 2090s, are based on models, observational constraints and expert judgement.* ***Lower panel:*** *best estimates and uncertainty ranges based on a simple climate model (SCM), also from WGI AR4 (Chapter 10).* ***Upper panel:*** *best estimates and uncertainty ranges for four CO_2-stabilisation scenarios using an SCM. Results are from the TAR because comparable projections for the 21st century are not available in the AR4. However, estimates of equilibrium warming are reported in the WGI AR4 for CO_2-equivalent stabilisation[11]. Note that equilibrium temperatures would not be reached until decades or centuries after greenhouse gas stabilisation. Uncertainty ranges: middle panel, likely range (> 66% probability); lower panel, range between 19 estimates calculated assuming low carbon-cycle feedbacks (mean - 1 standard deviation) and those assuming high carbon-cycle feedbacks (mean + 1 standard deviation); upper panel, range across seven model tunings for medium carbon-cycle settings.*

[11] Best estimate and likely range of equilibrium warming for seven levels of CO_2-equivalent stabilisation from the WG1 AR4 are: 350 ppm, 1.0°C [0.6–1.4]; 450 ppm, 2.1°C [1.4–3.1]; 550 ppm, 2.9°C [1.9–4.4]; 650 ppm, 3.6°C [2.4–5.5]; 750 ppm, 4.3°C [2.8–6.4]; 1,000 ppm, 5.5°C [3.7–8.3] and 1,200 ppm, 6.3°C [4.2–9.4].

combined effects of extreme weather and air-pollution events on human health). This expansion of scenario scope and application has brought into focus the wide range of potential future impacts and their associated uncertainties [2.2.5, 2.5].

Mitigation/stabilisation scenarios

The SRES storylines assume that no specific climate policies will be implemented to reduce greenhouse gas emissions (i.e. mitigation). Projections of global mean warming during the 21st century for the six SRES scenarios using two different approaches reported by the WGI AR4 (Chapter 10) are depicted in the middle and lower panels of Figure TS-4. Even without assuming explicit climate policies, differences between projections of warming for alternative emissions scenarios by the end of the century can exceed 2°C [B2.8].

CCIAV studies assuming mitigated futures are beginning to assess the benefits (through impacts ameliorated or avoided) of climate policy decisions. Stabilisation scenarios are a type of mitigation scenario describing futures in which emissions reductions are undertaken so that greenhouse gas concentrations, radiative forcing or global average temperature changes do not exceed a prescribed limit. There have been very few studies of the impacts of climate change assuming stabilisation. One reason for this is that relatively few AOGCM stabilisation runs have been completed so far, although the situation is rapidly changing [2.4.6].

Greenhouse gas mitigation is expected to reduce global mean warming relative to baseline emissions, which in turn could avoid some adverse impacts of climate change. To indicate the projected effect of mitigation on temperature during the 21st century, and in the absence of more recent, comparable estimates in the WGI AR4, results from the Third Assessment Report using a simple climate model are reproduced in the upper panel of Figure TS-4. These portray the temperature response for four CO_2-stabilisation scenarios by three dates in the early (2025), mid (2055), and late (2085) 21st century[12] [B2.8].

Large-scale singularities

Very few studies have been conducted on the impacts of large-scale singularities, which are extreme, sometimes irreversible, changes in the Earth system such as an abrupt cessation of the North Atlantic Meridional Overturning Circulation, or rapid global sea-level rise due to Antarctic and/or Greenland ice sheet melting [2.4.7]. Due to incomplete understanding of the underlying mechanisms of these events, or their likelihood, only exploratory studies have been carried out. For example, in terms of exploring the worst-case scenario of abrupt sea-level rise, impact assessments have been conducted for the coastal zone for a 5 m rise, and for a 2.2 m rise by 2100 [2.4.7]. This is the first time these scenarios have been included in any WGII assessment, and the expectation is that many more such studies will become available for assessment in the future.

Probabilistic characterisations

Probabilistic characterisations of future climate and non-climate conditions are increasingly becoming available. A number of studies focused on the climate system have generated probabilistic estimates of climate change, conditional on selected or probabilistic emissions scenarios, the latter being a subject of considerable debate [2.4.8]. Probabilistic futures have been applied in a few CCIAV studies to estimate the risk of exceeding predefined thresholds of impact and the associated timing of such exceedances [2.3.1].

TS.4 Current knowledge about future impacts

This section summarises the main projected impacts in each system and sector (Section TS.4.1) and region (Section TS.4.2) over this century,[13] judged in terms of relevance for people and the environment. It assumes that climate change is not mitigated, and that adaptive capacity has not been enhanced by climate policy. All global temperature changes are expressed relative to 1990 unless otherwise stated.[14] The impacts stem from changes in climate and sea-level changes associated with global temperature change, and frequently reflect projected changes in precipitation and other climate variables in addition to temperature.

TS.4.1 Sectoral impacts, adaptation and vulnerability

A summary of impacts projected for each sector is given in Box TS.5.

Freshwater resources and their management

The impacts of climate change on freshwater systems and their management are mainly due to the observed and projected increases in temperature, evaporation, sea level and precipitation variability (very high confidence).
More than one-sixth of the world's population live in glacier- or snowmelt-fed river basins and will be affected by a decrease in water volume stored in glaciers and snowpack, an increase in the ratio of winter to annual flows, and possibly a reduction in low flows caused by decreased glacier extent or melt-season snow water storage [3.4.1, 3.4.3]. Sea-level rise will extend areas of salinisation of groundwater and estuaries, resulting in a decrease in freshwater availability for humans and ecosystems in coastal areas [3.2, 3.4.2]. Increased precipitation intensity and variability is projected to increase the risk of floods and droughts in many areas [3.3.1]. Up to 20% of the world's population live in river basins that are likely to be affected by increased flood hazard by the 2080s in the course of global warming [3.4.3].

[12] WRE stabilisation profiles were used in the TAR, and a description is given in the TAR Synthesis Report.
[13] Unless otherwise stated.
[14] To express the temperature change relative to pre-industrial (about 1750) levels, add 0.6°C.

The number of people living in severely stressed river basins is projected to increase significantly from 1.4-1.6 billion in 1995 to 4.3-6.9 billion in 2050, for the SRES A2 scenario (medium confidence).
The population at risk of increasing water stress for the full range of SRES scenarios is projected to be: 0.4-1.7 billion, 1.0-2.0 billion and 1.1-3.2 billion, in the 2020s, 2050s and 2080s, respectively [3.5.1]. In the 2050s (A2 scenario), 262-983 million people are likely to move into the water-stressed category [3.5.1]. Water stress is projected to decrease by the 2050s on 20-29% of the global land area (considering two climate models and the SRES scenarios A2 and B2) and to increase on 62-76% of the global land area [3.5.1].

Semi-arid and arid areas are particularly exposed to the impacts of climate change on freshwater (high confidence).
Many of these areas (e.g., Mediterranean Basin, western USA, southern Africa, north-eastern Brazil, southern and eastern Australia) will suffer a decrease in water resources due to climate change (see Figure TS.5) [3.4, 3.7]. Efforts to offset declining surface water availability due to increasing precipitation variability will be hampered by the fact that groundwater recharge is likely to decrease considerably in some already water-stressed regions [3.4.2], where vulnerability is often exacerbated by the rapid increase of population and water demand [3.5.1].

Higher water temperatures, increased precipitation intensity and longer periods of low flows are likely to exacerbate many forms of water pollution, with impacts on ecosystems, human health, and water system reliability and operating costs (high confidence).
These pollutants include sediments, nutrients, dissolved organic carbon, pathogens, pesticides, salt and thermal pollution [3.2, 3.4.4, 3.4.5].

Climate change affects the function and operation of existing water infrastructure as well as water management practices (very high confidence).
Adverse effects of climate on freshwater systems aggravate the impacts of other stresses, such as population growth, changing economic activity, land-use change and urbanisation [3.3.2, 3.5]. Globally, water demand will grow in the coming decades, primarily due to population growth and increased affluence. Regionally, large changes in irrigation water demand as a result of climate change are likely [3.5.1]. Current water management practices are very likely to be inadequate to reduce the negative impacts of climate change on water-supply reliability, flood risk, health, energy and aquatic ecosystems [3.4, 3.5]. Improved incorporation of current climate variability into water-related management is likely to make adaptation to future climate change easier [3.6].

Adaptation procedures and risk management practices for the water sector are being developed in some countries and regions (e.g., Caribbean, Canada, Australia, Netherlands, UK, USA, Germany) that recognise the uncertainty of projected hydrological changes (very high confidence).
Since the IPCC Third Assessment, uncertainties have been evaluated and their interpretation has improved, and new methods (e.g., ensemble-based approaches) are being developed for their characterisation [3.4, 3.5]. Nevertheless, quantitative projections of changes in precipitation, river flows and water levels at the river-basin scale remain uncertain [3.3.1, 3.4].

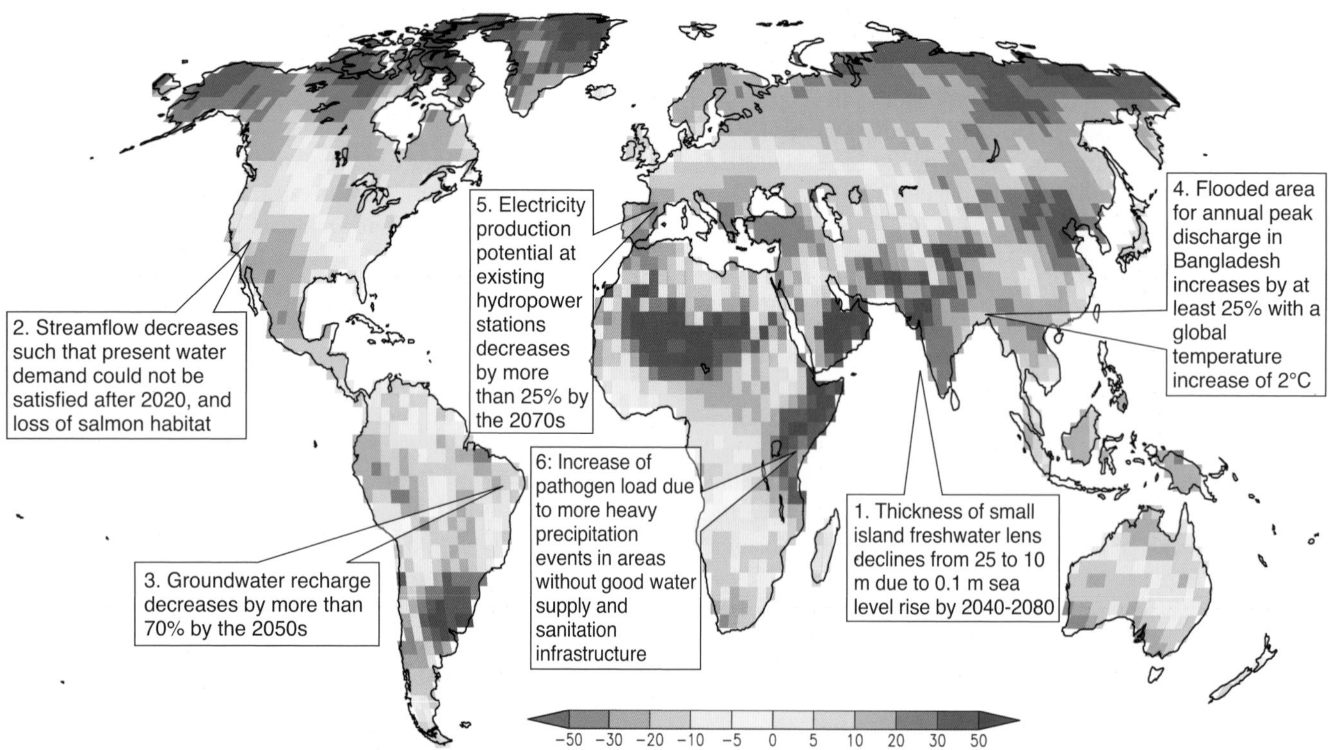

Figure TS.5. *Illustrative map of future climate change impacts on freshwater which are a threat to the sustainable development of the affected regions. Background shows ensemble mean change of annual runoff, in percent, between the present (1981-2000) and 2081-2100 for the SRES A1B emissions scenario; blue denotes increased runoff, red denotes decreased runoff. Underlying map from Nohara et al. (2006) [F3.8].*

The negative impacts of climate change on freshwater systems outweigh its benefits (high confidence).

All IPCC regions show an overall net negative impact of climate change on water resources and freshwater ecosystems. Areas in which runoff is projected to decline are likely to face a reduction in the value of the services provided by water resources. The beneficial impacts of increased annual runoff in other areas is likely to be tempered in some areas by negative effects of increased precipitation variability and seasonal runoff shifts on water supply, water quality and flood risks (see Figure TS.5) [3.4, 3.5].

Ecosystems

Records of the geological past show that ecosystems have some capacity to adapt naturally to climate change [WGI AR4 Chapter 6; 4.2], but this resilience[15] has never been challenged by a large global human population and its multi-faceted demands from and pressures on ecosystems [4.1, 4.2].

The resilience of many ecosystems (their ability to adapt naturally) is likely to be exceeded by 2100 by an unprecedented combination of change in climate, associated disturbances (e.g., flooding, drought, wildfire, insects, ocean acidification), and other global change drivers (e.g., land-use change, pollution, over-exploitation of resources) (high confidence).

Ecosystems are very likely to be exposed to atmospheric CO_2 levels much higher than in the past 650,000 years, and global mean temperatures at least as high as those in the past 740,000 years [WGI AR4 Chapter 6; 4.2, 4.4.10, 4.4.11]. By 2100, ocean pH is very likely to be lower than during the last 20 million years [4.4.9]. Extractive use from and fragmentation of wild habitats are very likely to impair species' adaptation [4.1.2, 4.1.3, 4.2, 4.4.5, 4.4.10]. Exceedance of ecosystem resilience is very likely to be characterised by threshold-type responses, many irreversible on time-scales relevant to human society, such as biodiversity loss through extinction, disruption of species' ecological interactions, and major changes in ecosystem structure and disturbance regimes (especially wildfire and insects) (see Figure TS.6). Key ecosystem properties (e.g., biodiversity) or regulating services (e.g., carbon sequestration) are very likely to be impaired [4.2, 4.4.1, 4.4.2 to 4.4.9, 4.4.10, 4.4.11, F4.4, T4.1].

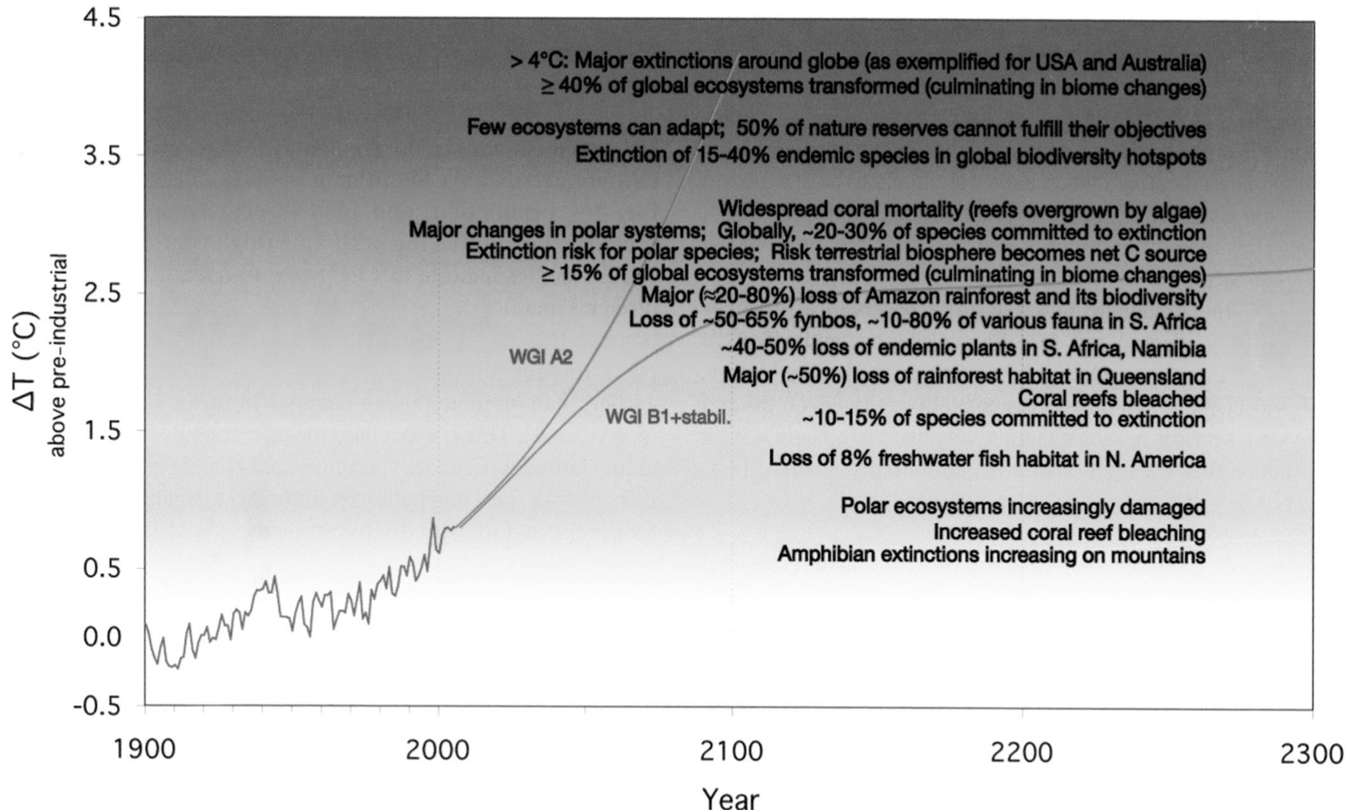

Figure TS.6. *Compendium of projected risks due to critical climate change impacts on ecosystems for different levels of global mean annual temperature rise, ΔT, relative to pre-industrial climate, used as a proxy for climate change. The red curve shows observed temperature anomalies for the period 1900-2005 [WGI AR4 F3.6]. The two grey curves provide examples of the possible future evolution of global average temperature change (ΔT) with time [WGI AR4 F10.4] exemplified by WGI simulated, multi-model mean responses to (i) the A2 radiative forcing scenario (WGI A2) and (ii) an extended B1 scenario (WGI B1+stabil.), where radiative forcing beyond 2100 was kept constant at the 2100 value [WGI AR4 F10.4, 10.7]. White shading indicates neutral, small negative, or positive impacts or risks; yellow indicates negative impacts for some systems or low risks; and red indicates negative impacts or risks that are more widespread and/or greater in magnitude. Illustrated impacts take into account climate change impacts only, and omit effects of land-use change or habitat fragmentation, over-harvesting or pollution (e.g., nitrogen deposition). A few, however, take into account fire regime changes, several account for likely productivity-enhancing effects of rising atmospheric CO_2 and some account for migration effects. [F4.4, T4.1]*

[15] *Resilience* is defined as the ability of a social or ecological system to absorb disturbances while retaining the same basic structure and ways of functioning, the capacity for self-organisation, and the capacity to adapt naturally to stress and change.

The terrestrial biosphere is likely to become a net carbon source by 2100, thus amplifying climate change, given continued greenhouse gas emissions at or above current rates and other unmitigated global changes, such as land-use changes (high confidence).

Several major terrestrial carbon stocks are vulnerable to climate change and/or land-use impacts [F4.1, 4.4.1, F4.2, 4.4.5, 4.4.6, 4.4.10, F4.3]. The terrestrial biosphere currently serves as a variable, but generally increasing, carbon sink (due to CO_2-fertilisation, moderate climate change and other effects) but this is likely to peak before mid-century and then tend towards a net carbon source, thus amplifying climate change [F4.2, 4.4.1, 4.4.10, F4.3, 4.4.11], while ocean buffering capacity begins saturating [WGI AR4, e.g., 7.3.5]. This is likely to occur before 2100, assuming continued greenhouse gas emissions at or above current rates and unmitigated global change drivers including land-use changes, notably tropical deforestation. Methane emissions from tundra are likely to accelerate [4.4.6].

Roughly 20 to 30% (varying among regional biotas from 1% to 80%) of species assessed so far (in an unbiased sample) are likely to be at increasingly high risk of extinction as global mean temperatures exceed 2 to 3°C above pre-industrial levels (medium confidence).

Global losses of biodiversity are of key relevance, being irreversible [4.4.10, 4.4.11, F4.4, T4.1]. Endemic species richness is highest where regional palaeo-climatic changes have been muted, indicating that endemics are likely to be at a greater extinction risk than in the geological past [4.4.5, 4.4.11, F4.4, T4.1]. Ocean acidification is likely to impair aragonite-based shell formation in a wide range of planktonic and shallow benthic marine organisms [4.4.9, B4.4]. Conservation practices are generally ill-prepared for climate change, and effective adaptation responses are likely to be costly to implement [4.4.11, T4.1, 4.6.1]. Although links between biodiversity intactness and ecosystem services remain quantitatively uncertain, there is high confidence that the relationship is qualitatively positive [4.1, 4.4.11, 4.6, 4.8].

Substantial changes in structure and functioning of terrestrial and marine ecosystems are very likely to occur with a global warming of 2 to 3°C above pre-industrial levels and associated increased atmospheric CO_2 (high confidence).

Major biome changes, including emergence of novel biomes, and changes in species' ecological interactions, with predominantly negative consequences for goods and services, are very likely by, and virtually certain beyond, those temperature increases [4.4]. The previously overlooked progressive acidification of oceans due to increasing atmospheric CO_2 is expected to have negative impacts on marine shell-forming organisms (e.g., corals) and their dependent species [B4.4, 6.4].

Food, fibre and forest products

In mid- to high-latitude regions, moderate warming benefits cereal crop and pasture yields, but even slight warming decreases yields in seasonally dry and tropical regions (medium confidence).

Modelling results for a range of sites find that, in temperate regions, moderate to medium increases in local mean temperature (1 to 3°C), along with associated CO_2 increase and rainfall changes, can have small beneficial impacts on crop yields. At lower latitudes, especially the seasonally dry tropics, even moderate temperature increases (1 to 2°C) are likely to have negative yield impacts for major cereals, which would increase the risk of hunger. Further warming has increasingly negative impacts in all regions (medium to low confidence) (see Figure TS.7) [5.4].

Climate change increases the number of people at risk of hunger marginally, with respect to overall large reductions due to socio-economic development (medium confidence).

Compared with 820 million undernourished today, SRES scenarios of socio-economic development, without climate change, project 100-240 million undernourished for the SRES A1, B1 and B2 scenarios (770 million under the A2 scenario) in 2080 (medium confidence). Scenarios with climate change project 100-380 million undernourished for the SRES A1, B1 and B2 scenarios (740-1,300 million under the A2 scenario) in 2080 (low to medium confidence). The ranges here indicate the extent of effects of the exclusion and inclusion of CO_2 effects in the scenarios. Climate change and socio-economics combine to alter the regional distribution of hunger, with large negative effects on sub-Saharan Africa (low to medium confidence) [5.4, T5.6].

Projected changes in the frequency and severity of extreme climate events have significant consequences on food and forestry production, and food insecurity, in addition to impacts of projected mean climate (high confidence).

Recent studies indicate that increased frequency of heat stress, droughts and floods negatively affects crop yields and livestock beyond the impacts of mean climate change, creating the possibility for surprises, with impacts that are larger, and occur earlier, than predicted using changes in mean variables alone [5.4.1, 5.4.2]. This is especially the case for subsistence sectors at low latitudes. Climate variability and change also modify the risks of fires, pest and pathogen outbreaks, negatively affecting food, fibre and forestry (high confidence) [5.4.1 to 5.4.5, 5.ES].

Simulations suggest rising relative benefits of adaptation with low to moderate warming (medium confidence), although adaptation may stress water and environmental resources as warming increases (low confidence).

There are multiple adaptation options that imply different costs, ranging from changing practices in place to changing locations of food, fibre and forest activities [5.5.1]. Adaptation effectiveness varies from only marginally reducing negative impacts to changing a negative impact into a positive one. On average, in cereal-cropping systems, adaptations such as changing varieties and planting times enable avoidance of a 10 to 15% reduction in yield, corresponding to 1 to 2°C local temperature increases. The benefit from adapting tends to increase with the degree of climate change [F5.2]. Changes in policies and institutions are needed to facilitate adaptation. Pressure to cultivate marginal land or to adopt unsustainable cultivation practices may increase land degradation and resource use, and endanger biodiversity of both wild and domestic species

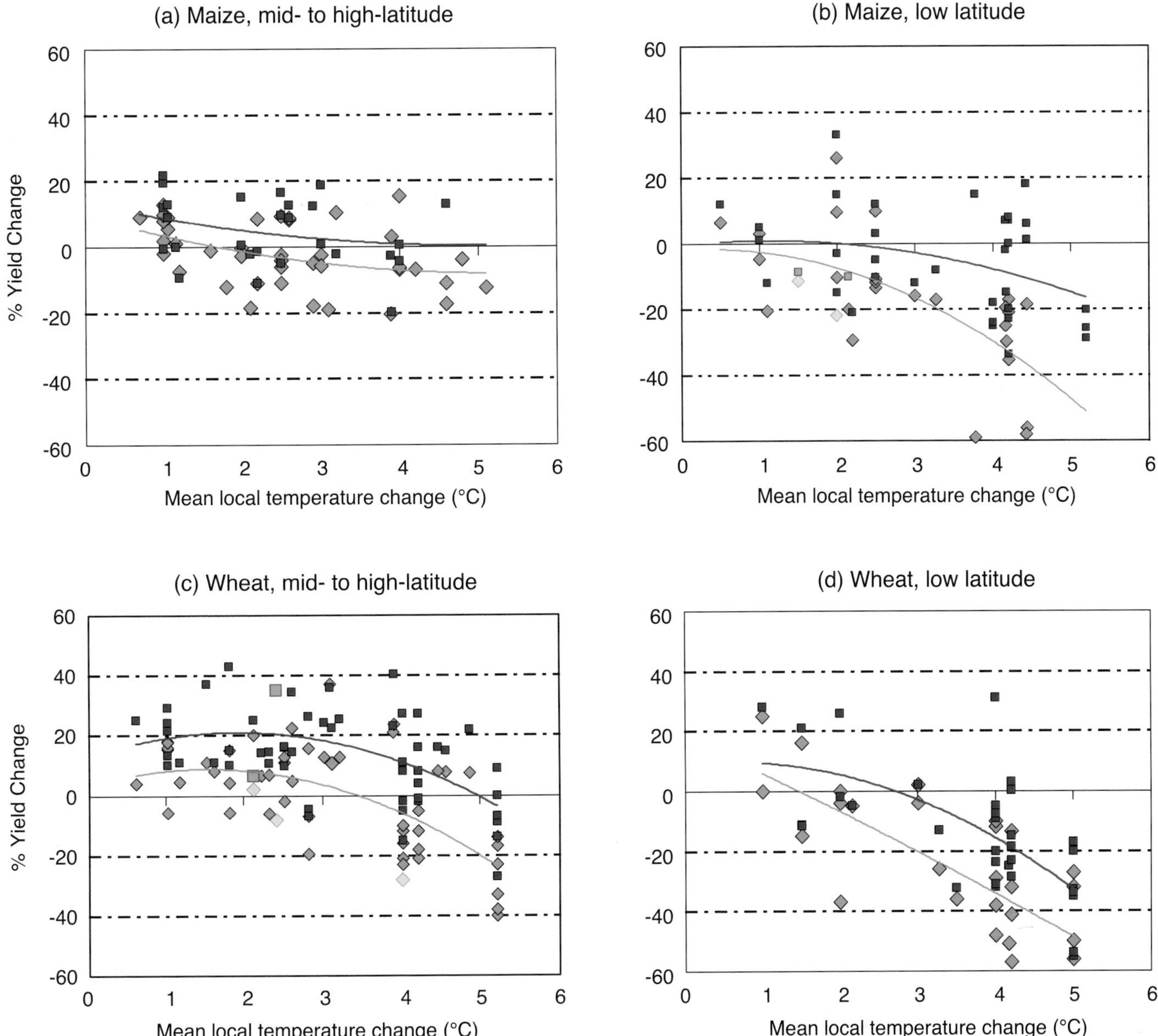

Figure TS.7. *Sensitivity of cereal yield to climate change for maize and wheat. Responses include cases without adaptation (orange dots) and with adaptation (green dots). The studies on which this figure is based span a range of precipitation changes and CO_2 concentrations, and vary in how they represent future changes in climate variability. For instance, lighter-coloured dots in (b) and (c) represent responses of rain-fed crops under climate scenarios with decreased precipitation. [F5.4]*

[5.4.7]. Adaptation measures should be integrated with development strategies and programmes, country programmes and poverty-reduction strategies [5.7].

Smallholder and subsistence farmers, pastoralists and artisanal fisherfolk are likely to suffer complex, localised impacts of climate change (high confidence).

These groups, whose adaptive capacity is constrained, are likely to experience negative effects on yields of tropical crops, combined with a high vulnerability to extreme events. In the longer term, there are likely to be additional negative impacts of other climate-related processes such as snowpack decrease especially in the Indo-Gangetic Plain, sea-level rise, and a spread in the prevalence of human diseases affecting agricultural labour supply (high confidence) [5.4.7].

Globally, forestry production is estimated to change only modestly with climate change in the short and medium term (medium confidence).

The change in global forest product outputs ranges from a modest increase to a slight decrease, although regional and local changes are likely to be large [5.4.5.2]. Production increase is likely to shift from low-latitude regions in the short term, to high-latitude regions in the long term [5.4.5].

Local extinctions of particular fish species are expected at edges of ranges (high confidence).

It is likely that regional changes in the distribution and productivity of particular fish species will continue and local extinctions will occur at the edges of ranges, particularly in freshwater and diadromous species (e.g., salmon, sturgeon). In some cases, ranges and productivity are likely to increase [5.4.6]. Emerging evidence suggests concern that the Meridional Overturning Circulation is slowing down, with potentially serious consequences for fisheries [5.4.6].

Food and forestry trade is projected to increase in response to climate change, with increased food-import dependence of most developing countries (medium to low confidence). While the purchasing power for food is likely to be reinforced in the period to 2050 by declining real prices, it would be adversely affected by higher real prices for food from 2050 to 2080 due to climate change [5.6.1, 5.6.2]. Exports of temperate-zone food products to tropical countries are likely to rise [5.6.2], while the reverse is likely in forestry in the short term [5.4.5].

Experimental research on crop response to elevated CO_2 confirms TAR reviews (medium to high confidence). New results suggest lower responses for forests (medium confidence). Recent reanalyses of free-air carbon dioxide enrichment (FACE) studies indicate that, at 550 ppm CO_2, yields increase under unstressed conditions by 10 to 20% over current concentrations for C3 crops, and by 0 to 10% for C4 crops (medium confidence). Crop model simulations under elevated CO_2 are consistent with these ranges (high confidence) [5.4.1]. Recent FACE results suggest no significant response for mature forest stands and confirm enhanced growth for young tree stands [5.4.1]. Ozone exposure limits CO_2 response in both crops and forests [B5.2].

Coastal systems and low-lying areas

Since the TAR, our understanding of the implications of climate change for coastal systems and low-lying areas (henceforth referred to as 'coasts') has increased substantially, and six important policy-relevant messages emerge.

Coasts are experiencing the adverse consequences of hazards related to climate and sea level (very high confidence). Coasts are highly vulnerable to extreme events, such as storms, which impose substantial costs on coastal societies [6.2.1, 6.2.2, 6.5.2]. Annually, about 120 million people are exposed to tropical cyclone hazards. These killed 250,000 people from 1980 to 2000 [6.5.2]. Throughout the 20th century, the global rise of sea level contributed to increased coastal inundation, erosion and ecosystem losses, but the precise role of sea-level rise is difficult to determine due to considerable regional and local variation due to other factors [6.2.5, 6.4.1]. Late 20th century effects of rising temperature include loss of sea ice, thawing of permafrost and associated coastal retreat at high latitudes, and more frequent coral bleaching and mortality at low latitudes [6.2.5].

Coasts are very likely to be exposed to increasing risks in future decades due to many compounding climate-change factors (very high confidence). Anticipated climate-related changes include: an accelerated rise in sea level of 0.2 to 0.6 m or more by 2100; further rise in sea surface temperatures of 1 to 3°C; more intense tropical and extra-tropical cyclones; generally larger extreme wave and storm surges; altered precipitation/runoff; and ocean acidification [WG1 AR4 Chapter 10; 6.3.2]. These phenomena will vary considerably at regional and local scales, but the impacts are virtually certain to be overwhelmingly negative [6.4, 6.5.3]. Coastal wetland ecosystems, such as salt marshes and mangroves, are very likely threatened where they are sediment-starved or constrained on their landward margin [6.4.1]. The degradation of coastal ecosystems, especially wetlands and coral reefs, has serious implications for the well-being of societies dependent on coastal ecosystems for goods and services [6.4.2, 6.5.3]. Increased flooding and the degradation of freshwater, fisheries and other resources could impact hundreds of millions of people, and socio-economic costs for coasts are virtually certain to escalate as a result of climate change [6.4.2, 6.5.3].

The impact of climate change on coasts is exacerbated by increasing human-induced pressures (very high confidence). Utilisation of the coast increased dramatically during the 20th century and this trend is virtually certain to continue through the 21st century. Under the SRES scenarios, the coastal population could grow from 1.2 billion people (in 1990) to between 1.8 billion and 5.2 billion people by the 2080s, depending on future trends in coastward migration [6.3.1]. Hundreds of millions of people and major assets at risk at the coast are subject to additional stresses by land-use and hydrological changes in catchments, including dams that reduce sediment supply to the coast [6.3]. Three key hotspots of societal vulnerability are: (i) deltas (see Figure TS.8), especially the seven Asian megadeltas with a collective population already exceeding 200 million; (ii) low-lying coastal urban areas, especially those prone to subsidence; and (iii) small islands, especially coral atolls [6.4.3].

Adaptation for the coasts of developing countries is virtually certain to be more challenging than for coasts of developed countries (high confidence). Developing countries already experience the most severe impacts from present coastal hazards [6.5.2]. This is virtually certain to continue under climate change, even allowing for optimum adaptation, with Asia and Africa most exposed [6.4.2, B6.6, F6.4, 6.5.3]. Developing countries have a more limited adaptive capacity due to their development status, with the most vulnerable areas being concentrated in exposed or sensitive settings such as small islands or deltas [6.4.3]. Adaptation in developing countries will be most challenging in these vulnerable 'hotspots' [6.4.3].

Adaptation costs for vulnerable coasts are much less than the costs of inaction (high confidence). Adaptation costs for climate change are virtually certain to be much lower than damage costs without adaptation for most developed coasts, even considering only property losses and human deaths [6.6.2, 6.6.3]. As post-event impacts on coastal businesses, people, housing, public and private social institutions, natural resources and the environment generally go unrecognised in disaster cost accounting, it is virtually certain that the full benefits of adaptation are even larger [6.5.2, 6.6.2]. Without action, the highest sea-level scenarios combined with other climate change (e.g., increased storm intensity) are about as likely as not to make some low-lying islands and other low-lying areas

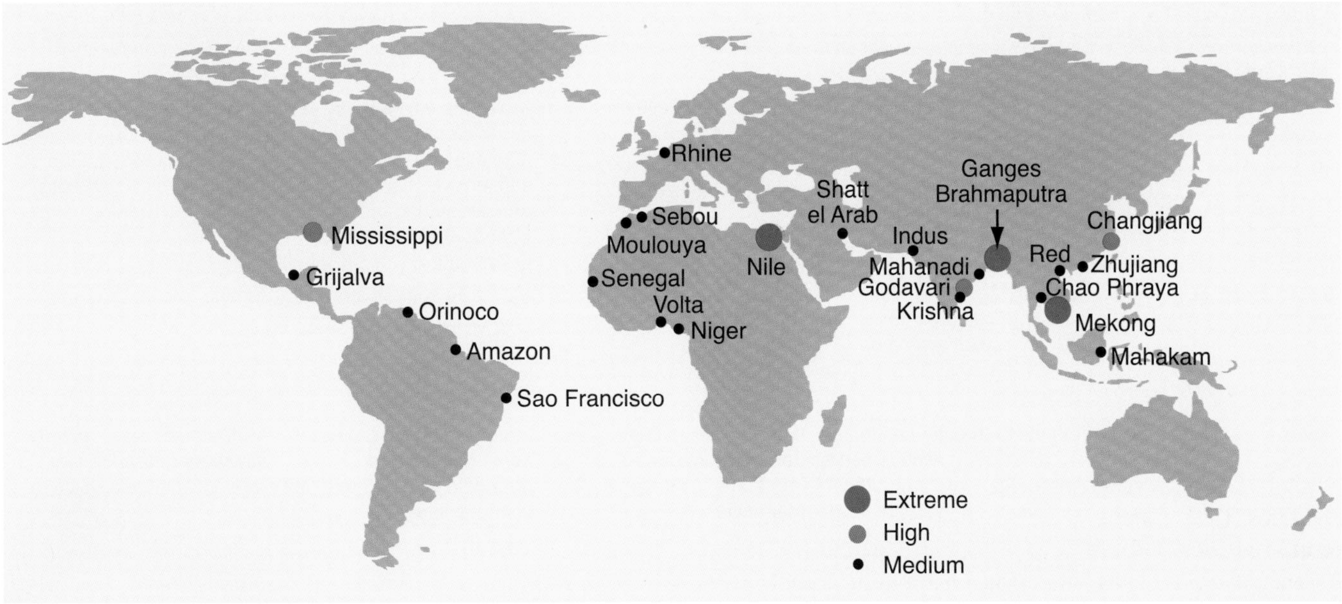

Figure TS.8. *Relative vulnerability of coastal deltas as indicated by estimates of the population potentially displaced by current sea-level trends to 2050 (extreme >1 million; high 1 million to 50,000; medium 50,000 to 5,000) [B6.3]. Climate change would exacerbate these impacts.*

(e.g., in deltas and megadeltas) uninhabitable by 2100 [6.6.3]. Effective adaptation to climate change can be integrated with wider coastal management, reducing implementation costs among other benefits [6.6.1.3].

The unavoidability of sea-level rise, even in the longer term, frequently conflicts with present-day human development patterns and trends (high confidence).
Sea-level rise has substantial inertia and will continue beyond 2100 for many centuries [WG1 AR4 Chapter 10]. Breakdown of the West Antarctic and/or Greenland ice sheets would make this long-term rise significantly larger. For Greenland, the temperature threshold for breakdown is estimated to be about 1.1 to 3.8°C above today's global average temperature. This is likely to happen by 2100 under the A1B scenario [WG1 AR4 Chapter 10]. This questions both the long-term viability of many coastal settlements and infrastructure (e.g., nuclear power stations) across the globe and the current trend of increasing human use of the coastal zone, including a significant coastward migration. This issue presents a challenge for long-term coastal spatial planning. Stabilisation of climate is likely to reduce the risks of ice sheet breakdown, and reduce but not stop sea-level rise due to thermal expansion [B6.6]. Hence, since the IPCC Third Assessment it has become virtually certain that the most appropriate response to sea-level rise for coastal areas is a combination of adaptation to deal with the inevitable rise, and mitigation to limit the long-term rise to a manageable level [6.6.5, 6.7].

Industry, settlement and society

Virtually all of the world's people live in settlements, and many depend on industry, services and infrastructure for jobs, well-being and mobility. For these people, climate change adds a new challenge in assuring sustainable development for societies across the globe. Impacts associated with this challenge will be determined mainly by trends in human systems in future decades as climate conditions exacerbate or ameliorate stresses associated with non-climate systems [7.1.1, 7.4, 7.6, 7.7].

Inherent uncertainties in predicting the path of technological and institutional change and trends in socio-economic development over a period of many decades limit the potential to project future prospects for industry, settlements and society involving *considerable* climate change from prospects involving relatively little climate change. In many cases, therefore, research to date has tended to focus on *vulnerabilities to impacts* rather than on *projections of impacts* of change, saying more about what could happen than about what is expected to happen [7.4].

Key vulnerabilities of industry, settlements and society are most often related to (i) climate phenomena that exceed thresholds for adaptation, related to the rate and magnitude of climate change, particularly extreme weather events and/or abrupt climate change, and (ii) limited access to resources (financial, human, institutional) to cope, rooted in issues of development context (see Table TS.1) [7.4.1, 7.4.3, 7.6, 7.7].

Findings about the context for assessing vulnerabilities are as follows.

Climate change vulnerabilities of industry, settlement and society are mainly to extreme weather events rather than to gradual climate change, although gradual changes can be associated with thresholds beyond which impacts become significant (high confidence).
The significance of gradual climate change, e.g., increases in the mean temperature, lies mainly in variability and volatility, including changes in the intensity and frequency of extreme events [7.2, 7.4].

Climate driven phenomena	Evidence for current impact/vulnerability	Other processes/stresses	Projected future impact/vulnerability	Zones, groups affected
a) Changes in extremes				
Tropical cyclones, storm surge	Flood and wind casualties and damages; economic losses; transport, tourism; infrastructure (e.g., energy, transport); insurance [7.4.2, 7.4.3, B7.2, 7.5].	Land use/population density in flood-prone areas; flood defences; institutional capacities.	Increased vulnerability in storm-prone coastal areas; possible effects on settlements, health, tourism, economic and transportation systems, buildings and infrastructure.	Coastal areas, settlements, and activities; regions and populations with limited capacities and resources; fixed infrastructure; insurance sector.
Extreme rainfall, riverine floods	Erosion/landslides; land flooding; settlements; transportation systems; infrastructure [7.4.2, regional chapters].	Similar to coastal storms plus drainage infrastructure.	Similar to coastal storms plus drainage infrastructure.	Similar to coastal storms.
Heat- or cold-waves	Effects on human health; social stability; requirements for energy, water and other services (e.g., water or food storage); infrastructure (e.g., energy transportation) [7.2, B7.1, 7.4.2.2, 7.4.2.3].	Building design and internal temperature control; social contexts; institutional capacities.	Increased vulnerabilities in some regions and populations; health effects; changes in energy requirements.	Mid-latitude areas; elderly, very young, and/or very poor populations.
Drought	Water availability; livelihoods, energy generation, migration, transportation in water bodies [7.4.2.2, 7.4.2.3, 7.4.2.5].	Water systems; competing water uses; energy demand; water demand constraints.	Water-resource challenges in affected areas; shifts in locations of population and economic activities; additional investments in water supply.	Semi-arid and arid regions; poor areas and populations; areas with human-induced water scarcity.
b) Changes in means				
Temperature	Energy demands and costs; urban air quality; thawing of permafrost soils; tourism and recreation; retail consumption; livelihoods; loss of meltwater [7.4.2.1, 7.4.2.2, 7.4.2.4, 7.4.2.5].	Demographic and economic changes; land-use changes; technological innovations; air pollution; institutional capacities.	Shifts in energy demand; worsening of air quality; impacts on settlements and livelihoods depending on meltwater; threats to settlements/infrastructure from thawing permafrost soils in some regions.	Very diverse, but greater vulnerabilities in places and populations with more limited capacities and resources for adaptation.
Precipitation	Agricultural livelihoods; saline intrusion; water infrastructures; tourism; energy supplies [7.4.2.1, 7.4.2.2, 7.4.2.3].	Competition from other regions/sectors; water resource allocation.	Depending on the region, vulnerabilities in some areas to effects of precipitation increases (e.g., flooding, but could be positive) and in some areas to decreases (see drought above).	Poor regions and populations.
Sea-level rise	Coastal land uses: flood risk, waterlogging; water infrastructures [7.4.2.3, 7.4.2.4].	Trends in coastal development, settlements and land uses.	Long-term increases in vulnerabilities of low-lying coastal areas.	Same as above.

Table TS.1. *Selected examples of current and projected climate-change impacts on industry, settlement and society and their interaction with other processes [for full text, see 7.4.3, T7.4]. Orange shading indicates very significant in some areas and/or sectors; yellow indicates significant; pale brown indicates that significance is less clearly established.*

Aside from major extreme events, climate change is seldom the main factor in considering stresses on sustainability (very high confidence).
The significance of climate change (positive or negative) lies in its interactions with other sources of change and stress, and its impacts should be considered in such a multi-cause context [7.1.3, 7.2, 7.4].

Vulnerabilities to climate change depend considerably on relatively specific geographical and sectoral contexts (very high confidence).
They are not reliably estimated by large-scale (aggregate) modelling and estimation [7.2, 7.4].

Climate change impacts spread from directly impacted areas and sectors to other areas and sectors through extensive and complex linkages (very high confidence).
In many cases, total impacts are poorly estimated by considering only direct impacts [7.4].

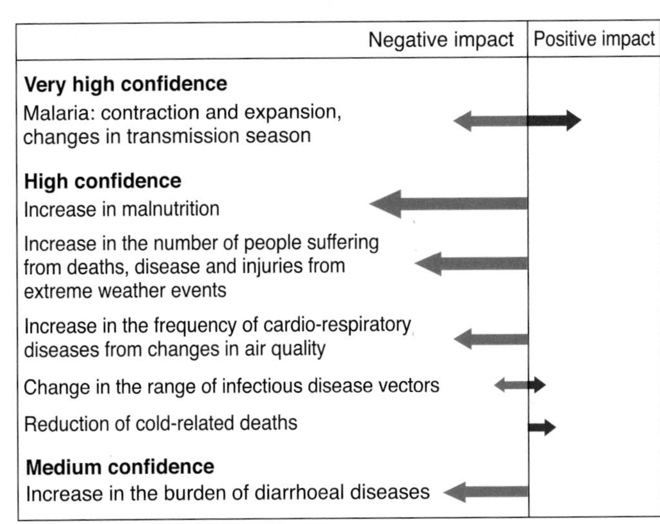

Figure TS.9. *Direction and magnitude of change of selected health impacts of climate change.*

Health

Climate change currently contributes to the global burden of disease and premature deaths (very high confidence).
Human beings are exposed to climate change through changing weather patterns (for example, more intense and frequent extreme events) and indirectly through changes in water, air, food quality and quantity, ecosystems, agriculture and economy. At this early stage the effects are small, but are projected to progressively increase in all countries and regions [8.4.1].

Projected trends in climate-change related exposures of importance to human health will have important consequences (high confidence).
Projected climate-change related exposures are likely to affect the health status of millions of people, particularly those with low adaptive capacity, through:

- increases in malnutrition and consequent disorders, with implications for child growth and development;
- increased deaths, disease and injury due to heatwaves, floods, storms, fires and droughts;
- the increased burden of diarrhoeal disease;
- mixed effects on the range (increases and decreases) and transmission potential of malaria in Africa;
- the increased frequency of cardio-respiratory diseases due to higher concentrations of ground-level ozone related to climate change;
- the altered spatial distribution of some infectious-disease vectors.

This is illustrated in Figure TS.9 [8.2.1, 8.4.1].

Adaptive capacity needs to be improved everywhere (high confidence).
Impacts of recent hurricanes and heatwaves show that even high-income countries are not well prepared to cope with extreme weather events [8.2.1, 8.2.2].

Adverse health impacts will be greatest in low-income countries (high confidence).
Studies in temperate areas (mainly in industrialised countries) have shown that climate change is projected to bring some benefits, such as fewer deaths from cold exposure. Overall it is expected that these benefits will be outweighed by the negative health effects of rising temperatures worldwide, especially in developing countries. The balance of positive and negative health impacts will vary from one location to another, and will alter over time as temperatures continue to rise. Those at greater risk include, in all countries, the urban poor, the elderly and children, traditional societies, subsistence farmers, and coastal populations [8.1.1, 8.4.2, 8.6.1, 8.7].

Current national and international programmes and measures that aim to reduce the burdens of climate-sensitive health determinants and outcomes may need to be revised, reoriented and, in some regions, expanded to address the additional pressures of climate change (medium confidence).
This includes the consideration of climate-change related risks in disease monitoring and surveillance systems, health system planning, and preparedness. Many of the health outcomes are mediated through changes in the environment. Measures implemented in the water, agriculture, food and construction sectors can be designed to benefit human health [8.6, 8.7].

Economic development is an important component of adaptation, but on its own will not insulate the world's population from disease and injury due to climate change (very high confidence).
Critically important will be the manner in which economic growth occurs, the distribution of the benefits of growth, and factors that directly shape the health of populations, such as education, health care, and public health infrastructure [8.3.2].

Box TS.5. The main projected impacts for systems and sectors[16]

Freshwater resources and their management

- Water volumes stored in glaciers and snow cover are very likely to decline, reducing summer and autumn flows in regions where more than one-sixth of the world's population currently live. ** N [3.4.1]
- Runoff and water availability are very likely to increase at higher latitudes and in some wet tropics, including populous areas in East and South-East Asia, and decrease over much of the mid-latitudes and dry tropics, which are presently water-stressed areas. ** D [F3.4]
- Drought-affected areas will probably increase, and extreme precipitation events, which are likely to increase in frequency and intensity, will augment flood risk. Increased frequency and severity of floods and droughts will have implications for sustainable development. ** N [WGI AR4 SPM; 3.4]
- Up to 20% of the world's population live in river basins that are likely to be affected by increased flood hazard by the 2080s in the course of global warming. * N [3.4.3]
- Many semi-arid areas (e.g., Mediterranean Basin, western USA, southern Africa and north-eastern Brazil) will suffer a decrease in water resources due to climate change. *** C [3.4, 3.7]
- The number of people living in severely stressed river basins is projected to increase from 1.4-1.6 billion in 1995 to 4.3-6.9 billion in 2050, for the A2 scenario. ** N [3.5.1]
- Sea-level rise will extend areas of salinisation of groundwater and estuaries, resulting in a decrease in freshwater availability for humans and ecosystems in coastal areas. *** C [3.2, 3.4.2]
- Groundwater recharge will decrease considerably in some already water-stressed regions ** N [3.4.2], where vulnerability is often exacerbated by the rapid increase in population and water demand. *** C [3.5.1]
- Higher water temperatures, increased precipitation intensity and longer periods of low flows exacerbate many forms of water pollution, with impacts on ecosystems, human health, and water system reliability and operating costs. ** N [3.2, 3.4.4, 3.4.5]
- Uncertainties have been evaluated and their interpretation has improved and new methods (e.g., ensemble-based approaches) are being developed for their characterisation *** N [3.4, 3.5]. Nevertheless, quantitative projections of changes in precipitation, river flows and water levels at the river-basin scale remain uncertain. *** D [3.3.1, 3.4]
- Climate change affects the function and operation of existing water infrastructure as well as water management practices *** C [3.6]. Adaptation procedures and risk management practices for the water sector are being developed in some countries and regions that recognise the uncertainty of projected hydrological changes. *** N [3.6]
- The negative impacts of climate change on freshwater systems outweigh the benefits. ** D [3.4, 3.5]
- Areas in which runoff is projected to decline will face a reduction in the value of services provided by water resources *** C [3.4, 3.5]. The beneficial impacts of increased annual runoff in other areas will be tempered by the negative effects of increased precipitation variability and seasonal runoff shifts on water supply, water quality and flood risks. ** N [3.4, 3.5]

Ecosystems

- The following ecosystems are identified as most vulnerable, and are virtually certain to experience the most severe ecological impacts, including species extinctions and major biome changes. On continents: tundra, boreal forest, mountain and Mediterranean-type ecosystems. Along coasts: mangroves and salt marshes. And in oceans: coral reefs and the sea-ice biomes. *** D [4.4, see also Chapters 1, 5, 6, 14, 15; WGI AR4 Chapters 10, 11]
- Initially positive ecological impacts, such as increased net primary productivity (NPP), will occur in ecosystems identified as least vulnerable: savannas and species-poor deserts. However, these positive effects are contingent on sustained CO_2-fertilisation, and only moderate changes in disturbance regimes (e.g., wildfire) and in extreme events (e.g., drought). • D [4.4.1, 4.4.2, B4.2, 4.4.3, 4.4.10, 4.4.11]
- For global mean temperature increases up to 2°C,[17] some net primary productivity increases are projected at high latitudes (contingent to a large degree on effective migration of woody plants), while an NPP decline (ocean and land) is likely at low latitudes. ** D [4.4.1, 4.4.9, 4.4.10]

[16] In the text of Boxes TS.5 and TS.6, the following conventions are used:

Relationship to the TAR		*Confidence in a statement*	
C	*Confirmation*	***	*Very high confidence*
D	*Development*	**	*High confidence*
R	*Revision*	*	*Medium confidence*
N	*New*	•	*Low confidence*

[17] Temperature thresholds/sensitivities in the Ecosystems section (only) are given relative to pre-industrial climate and are a proxy for climate change including precipitation changes. In other sections temperature changes are relative to 1990 as indicated in the first paragraph of Section TS.4.

- Projected carbon sequestration by poleward taiga expansion • D [4.4.5, F4.3] is as likely as not to be offset by albedo changes, wildfire, and forest declines at taiga's equatorial limit ** N/D [4.4.5, F4.3], and methane losses from tundra. * N [4.4.6]
- Tropical forest sequestration, despite recently observed productivity gains, is very likely to depend on land-use change trends *** D [4.2, 4.3, 4.4.10], but by 2100 is likely to be dominated by climate-change impacts, especially in drier regions. ** D [4.4.5, 4.4.10, F4.3]
- Amazon forests, China's taiga, and much of the Siberian and Canadian tundra are very likely to show major changes with global mean temperatures exceeding 3°C ** D [T4.2, 4.4.1, F4.2, 4.4.10, F4.4]. While forest expansions are projected in North America and Eurasia with <2°C warming [4.4.10, F4.4, T4.3], tropical forests are likely to experience severe impacts, including biodiversity losses. * D [4.4.10, 4.4.11, T4.1]
- For global mean temperature increases of about 1.5 to 3°C, the low-productivity zones in sub-tropical oceans are likely to expand by about 5% (Northern) and about 10% (Southern Hemisphere), but the productive polar sea-ice biomes are very likely to contract by about 40% (Northern) and about 20% (Southern Hemisphere). ** N [4.4.9]
- As sea-ice biomes shrink, dependent polar species, including predators such as penguins, seals and polar bears, are very likely to experience habitat degradation and losses. *** D [4.4.6]
- Loss of corals due to bleaching is very likely to occur over the next 50 years *** C [B4.5, 4.4.9], especially for the Great Barrier Reef, where climate change and direct anthropogenic impacts such as pollution and harvesting are expected to cause annual bleaching (around 2030 to 2050) followed by mass mortality. ** D [B4.4, 4.4.9]
- Accelerated release of carbon from vulnerable carbon stocks, especially peatlands, tundra frozen loess ('yedoma'), permafrost soils, and soils of boreal and tropical forests is virtually certain. *** D/N [F4.1, 4.4.1, 4.4.6, 4.4.8, 4.4.10, 4.4.11]
- An intensification and expansion of wildfires is likely globally, as temperatures increase and dry spells become more frequent and more persistent. ** D/N [4.4.2, 4.4.3, 4.4.4, 4.4.5]
- Greater rainfall variability is likely to compromise inland and coastal wetland species through shifts in the timing, duration and depth of water levels. ** D [4.4.8]
- Surface ocean pH is very likely to decrease further, by as much as 0.5 pH units by 2100, with atmospheric CO_2 increases projected under the A1FI scenario. This is very likely to impair shell or exoskeleton formation by marine organisms requiring calcium carbonate (e.g., corals, crabs, squids, marine snails, clams and oysters). ** N [4.4.9, B4.5]

Food, fibre and forest products

- In mid- to high-latitude regions, moderate warming benefits cereal crops and pasture yields, but even slight warming decreases yields in seasonally dry and tropical regions *. Further warming has increasingly negative impacts in all regions [F5.2]. Short-term adaptations may enable avoidance of a 10 to 15% reduction in yield. */• D [F5.2, 5.4]
- Climate change will increase the number of people at risk of hunger marginally, with respect to overall large reductions due to socio-economic development. ** D [5.6.5, T5.6]
- Projected changes in the frequency and severity of extreme climate events, together with increases in risks of fire, pests, and disease outbreak, will have significant consequences on food and forestry production, and food insecurity, in addition to impacts of projected mean climate. ** D [5.4.1 to 5.4.5]
- Smallholder and subsistence farmers, pastoralists and artisanal fisherfolk will suffer complex, localised impacts of climate change. ** N [5.4.7]
- Global food production potential is likely to increase with increases in global average temperature up to about 3°C, but above this it is very likely to decrease. * D [5.6]
- Globally, forestry production is estimated to change only modestly with climate change in the short and medium term. Production increase will shift from low-latitude regions in the short term, to high-latitude regions in the long term. * D [5.4.5]
- Local extinctions of particular fish species are expected at edges of ranges. ** N [5.4.6]
- Food and forestry trade is projected to increase in response to climate change, with increased food-import dependence of most developing countries. */• N [5.6.1, 5.6.2, 5.4.5]
- Experimental research on crop response to elevated CO_2 confirms TAR conclusions * C. New free-air carbon dioxide enrichment (FACE) results suggest a lower response for forests. * D [5.4.1]

Coastal systems and low-lying areas

- Coasts are very likely to be exposed to increasing risks due to climate change and sea-level rise and the effect will be exacerbated by increasing human-induced pressures on coastal areas. *** D [6.3, 6.4]
- It is likely that corals will experience a major decline due to increased bleaching and mortality due to rising sea-water temperatures. Salt marshes and mangroves will be negatively affected by sea-level rise. *** D [6.4]

- All coastal ecosystems are vulnerable to climate change and sea-level rise, especially corals, salt marshes and mangroves. *** D [6.4.1]
- Corals are vulnerable to thermal stress and it is very likely that projected future increases in sea surface temperature (SST) of about 1 to 3°C in the 21st century will result in more frequent bleaching events and widespread mortality, unless there is thermal adaptation or acclimatisation by corals. *** D [B6.1, 6.4.1]
- Coastal wetlands, including salt marshes and mangroves, are sensitive to sea-level rise, with forecast global losses of 33% given a 36 cm rise in sea level from 2000 to 2080. The largest losses are likely to be on the Atlantic and Gulf of Mexico coasts of the Americas, the Mediterranean, the Baltic, and small-island regions. *** D [6.4.1]
- Ocean acidification is an emerging issue with potential for major impacts in coastal areas, but there is little understanding of the details. It is an urgent topic for further research, especially programmes of observation and measurement. ** D [6.2.3, 6.2.5, 6.4.1]
- Coastal flooding in low-lying areas is very likely to become a greater risk than at present due to sea-level rise and more intense coastal storms, unless there is significant adaptation [B6.2, 6.4.2]. Impacts are sensitive to sea-level rise, the socio-economic future, and the degree of adaptation. Without adaptation, more than 100 million people could experience coastal flooding each year by the 2080s due to sea-level rise alone, with the A2 world likely to have the greatest impacts. *** N [F6.2]
- Benefit-cost analysis of responses suggests that it is likely that the potential impacts will be reduced by widespread adaptation. It also suggests that it is likely that impacts and protection costs will fall disproportionately on developing countries. ** C [F6.4, 6.5.3]
- Key human vulnerabilities to climate change and sea-level rise exist where the stresses on natural low-lying coastal systems coincide with low human adaptive capacity and/or high exposure and include: ** D [6.4.2, 6.4.3]
  - deltas, especially Asian megadeltas (e.g., the Ganges-Brahmaputra in Bangladesh and West Bengal);
  - low-lying coastal urban areas, especially areas prone to natural or human-induced subsidence and tropical storm landfall (e.g., New Orleans, Shanghai);
  - small islands, especially low-lying atolls (e.g., the Maldives).
- Regionally, the greatest increase in vulnerability is very likely to be to be in South, South-East and East Asia, and urbanised coastal locations around Africa, and small-island regions. The numbers affected will be largest in the megadeltas of Asia, but small islands face the highest relative increase in risk. ** D [6.4.2]
- Sea-level rise has substantial inertia compared with other climate change factors, and is virtually certain to continue beyond 2100 for many centuries. Stabilisation of climate could reduce, but not stop, sea-level rise. Hence, there is a commitment to adaptation in coastal areas which raises questions about long-term spatial planning and the need to protect versus planned retreat. *** D [B6.6]

Industry, settlement and society

- Benefits and costs of climate change for industry, settlement and society will vary widely by location and scale. Some of the effects in temperate and polar regions will be positive and others elsewhere will be negative. In the aggregate, however, net effects are more likely to be strongly negative under larger or more rapid warming. ** N [7.4, 7.6, 15.3, 15.5]
- Vulnerabilities of industry, infrastructures, settlements and society to climate change are generally greater in certain high-risk locations, particularly coastal and riverine areas, those in areas prone to extreme weather events, and areas whose economies are closely linked with climate-sensitive resources, such as agricultural and forest product industries, water demands and tourism; these vulnerabilities tend to be localised but are often large and growing. For example, rapid urbanisation in most low- and middle-income nations, often in relatively high-risk areas, is placing an increasing proportion of their economies and populations at risk. ** D [7.1, 7.4, 7.5]
- Where extreme weather events become more intense and/or more frequent with climate change, the economic costs of those events will increase, and these increases are likely to be substantial in the areas most directly affected. Experience indicates that costs of major events can range from several percent of annual regional GDP and income in very large regions with very large economies, to more than 25% in smaller areas that are affected by the events. ** N [7.5]
- Some poor communities and households are already under stress from climate variability and climate-related extreme

events; and they can be especially vulnerable to climate change because they tend to be concentrated in relatively high-risk areas, to have limited access to services and other resources for coping, and in some regions to be more dependent on climate-sensitive resources such as local water and food supplies. ** N [7.2, 7.4.5, 7.4.6]

- Growing economic costs from weather-related extreme events are already increasing the need for effective economic and financial risk management. In those regions and locations where risk is rising and private insurance is a major risk management option, pricing signals can provide incentives for adaptation; but protection may also be withdrawn, leaving increased roles for others, including governments. In those regions where private insurance is not widely available, other mechanisms for risk management will be needed. In all situations, poorer groups in the population will need special help in risk management and adaptation. ** D [7.4.2]
- In many areas, climate change is likely to raise social equity concerns and increase pressures on governmental infrastructures and institutional capacities. ** N [7.ES, 7.4.5, 7.6.5]
- Robust and reliable physical infrastructures are especially important to climate-related risk management. Such infrastructures as urban water supply systems are vulnerable, especially in coastal areas, to sea-level rise and reduced regional precipitation; and large population concentrations without infrastructures are more vulnerable to impacts of climate change. ** N [7.4.3 to 7.4.5]

Health

- The projected relative risks attributable to climate change in 2030 show an increase in malnutrition in some Asian countries ** N [8.4.1]. Later in the century, expected trends in warming are projected to decrease the availability of crop yields in seasonally dry and tropical regions [5.4]. This will increase hunger, malnutrition and consequent disorders, including child growth and development, in particular in those regions that are already most vulnerable to food insecurity, notably Africa. ** N [8.4.2]
- By 2030, coastal flooding is projected to result in a large proportional mortality increase; however, this is applied to a low burden of disease so the aggregate impact is small. Overall, a two- to three-fold increase in population at risk of flooding is expected by 2080. ** N [8.4.1]
- Estimates of increases of people at risk of death from heat differ between countries, depending on the place, ageing population, and adaptation measures in place. Overall, significant increases are estimated over this century. ** D [T8.3]
- Mixed projections for malaria are foreseen: globally an estimated additional population at risk between 220 million (A1FI) and 400 million (A2) has been estimated. In Africa, estimates differ from a reduction in transmission in south-east Africa in 2020 and decreases around the Sahel and south-central Africa in 2080, with localised increases in the highlands, to a 16-28% increase in person-months of exposure in 2100 across all scenarios. For the UK, Australia, India and Portugal, some increased risk has been estimated. *** D [T8.2]
- In Canada, a northward expansion of the Lyme-disease vector of approximately 1,000 km is estimated by the 2080s (A2) and a two- to four-fold increase in tick abundance by the 2080s also. In Europe, tick-borne encephalitis is projected to move further north-eastward of its present range but to contract in central and eastern Europe by the 2050s. * N [T8.2]
- By 2030 an increase in the burden of diarrhoeal diseases in low-income regions by approximately 2-5% is estimated ** N [8.4.1]. An annual increase of 5-18% by 2050 was estimated for Aboriginal communities in Australia ** N [T8.2]. An increase in cases of food poisoning has been estimated for the UK for a 1-3°C temperature increase. * N [T8.2]
- In eastern North America under the A2 climate scenario, a 4.5% increase in ozone-related deaths is estimated. A 68% increase in average number of days/summer exceeding the 8-hour regulatory standard is projected to result in a 0.1-0.3% increase in non-accidental mortality and an average 0.3% increase in cardiovascular disease mortality. In the UK, large decreases in days with high particulates and SO_2 and a small decrease in other pollutants have been estimated for 2050 and 2080, but ozone will have increased ** N [T8.4]. The near-term health benefits from reducing air-pollution concentrations (such as for ozone and particulate matter), as a consequence of greenhouse gas reductions, can be substantial. ** D [8.7.1, WGIII AR4]
- By 2085 it is estimated that the risk of dengue from climate change increases to include 3.5 billion people. * N [8.4.1.2]
- Reductions in cold-related deaths due to climate change are projected to be greater than increases in heat-related deaths in the UK. ** D [T8.3]

TS.4.2 Regional impacts, adaptation and vulnerability

A summary of impacts projected for each region is given in Box TS.6.

Africa

Agricultural production in many African countries and regions will likely be severely compromised by climate change and climate variability. This would adversely affect food security and exacerbate malnutrition (very high confidence).
Agricultural yields and dependence on natural resources constitute a large part of local livelihoods in many, but not all, African countries. Agriculture is a major contributor to the current economy of most African countries, averaging 21% and ranging from 10% to 70% of GDP with indications that off-farm income augments the overall contribution of agriculture in some countries [9.2.2, 9.4.4]. Agricultural losses are shown to be possibly severe for several areas (e.g., the Sahel, East Africa and southern Africa) accompanied by changes in length of growing periods impacting mixed rain-fed, arid and semi-arid systems under certain climate projections. In some countries, yields from rain-fed agriculture could be reduced by up to 50% by 2020. At the local level, many people are likely to suffer additional losses to their livelihood when climate change and variability occur together with other stressors (e.g., conflict) [9.2.2, 9.6.1].

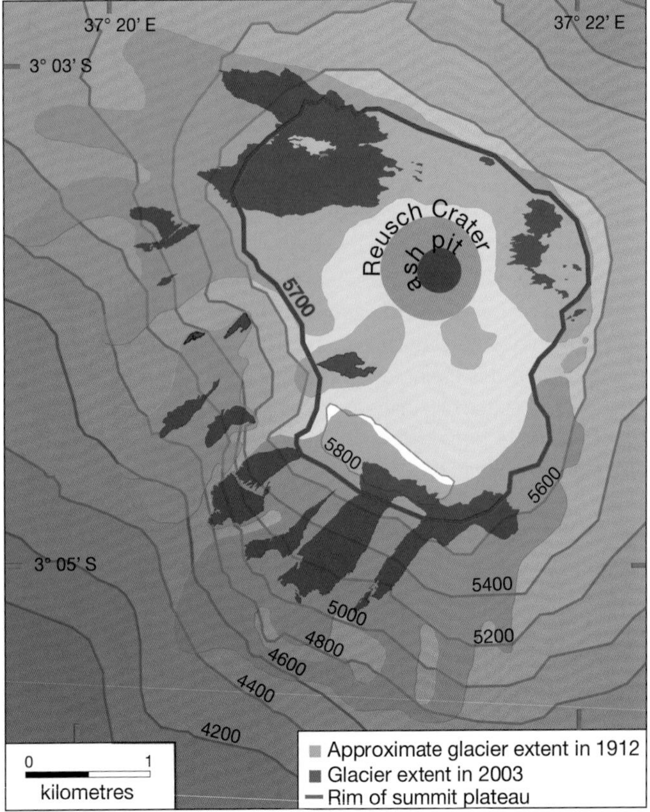

Figure TS.10. *Changes in the Mt. Kilimanjaro ice cap and snow cover over time. Decrease in surface area of Kilimanjaro glaciers from 1912 to 2003. [F9.2]*

Climate change and variability are likely to result in species loss, extinctions and also constrain the 'climate spaces' and ranges of many plants and animals (high confidence).
Changes in a variety of ecosystems are already being detected, particularly in southern African ecosystems, at a faster rate than anticipated as a result of a variety of factors, including the influence of climate, e.g., mountain ecosystems [9.4.5, 4.4.2, 4.4.3, 4.4.8].

In unmanaged environments, multiple, interacting impacts and feedbacks are expected, triggered by changes in climate, but exacerbated by non-climatic factors (high confidence).
Impacts on Kilimanjaro, for example, show that glaciers and snow cover have been retreating as a result of a number of interacting factors (e.g., solar radiation, vegetation changes and human interactions), with a decrease in glacier surface area of approximately 80% between 1912 and 2003 (see Figure TS.10). The loss of 'cloud forests', e.g., through fire, since 1976 has resulted in a 25% annual reduction of water sources derived from fog (equivalent to the annual drinking water supply of 1 million people living around Mt. Kilimanjaro) [9.4.5].

Lack of access to safe water, arising from multiple factors, is a key vulnerability in many parts of Africa. This situation is likely to be further exacerbated by climate change (very high confidence).
By 2020, some assessments project that between 75 and 250 million people are estimated to be exposed to increased water stress due to climate change. If coupled with increased demand, this will adversely affect livelihoods and exacerbate water-related problems. Some assessments, for example, show severe increased water stress and possible increased drought risk for parts of northern and southern Africa and increases in runoff in East Africa. Water access is, however, threatened not only by climate change [9.4.1] but also by complex river-basin management (with several of Africa's major rivers being shared by several countries), and degradation of water resources by abstraction of water and pollution of water sources [9.4.1].

Attributing the contribution of climate change to changes in the risk of malaria remains problematic (high confidence).
Human health, already compromised by a range of factors, could also be further negatively impacted by climate change and climate variability (e.g., in southern Africa and the East African highlands). The debate on climate change attribution and malaria is ongoing and this is an area requiring further research [9.4.3, 8.2.8, 8.4.1].

Africa is one of the most vulnerable continents to climate variability and change because of multiple stresses and low adaptive capacity. The extreme poverty of many Africans, frequent natural disasters such as droughts and floods, and agriculture which is heavily dependent on rainfall, all contribute. Cases of remarkable resilience in the face of multiple stressors have, however, been shown (high confidence).
Africa possesses many examples of coping and adaptation strategies that are used to manage a range of stresses including climate extremes (e.g., droughts and floods). Under possible increases in such stresses, however, these strategies are likely to

be insufficient to adapt to climate variability and change, given the problems of endemic poverty, poor institutional arrangements, poor access to data and information, and growing health burdens [9.2.1, 9.2.2., 9.2.5].

Asia

Observations demonstrate that climate change has affected many sectors in Asia in the past decades (medium confidence).

Evidence of impacts of climate change, variability and extreme events in Asia, as predicted in the Third Assessment, has emerged. The crop yield in most countries of Asia has been observed to be declining, probably partly attributable to rising temperatures. As a likely consequence of warming, the retreat of glaciers and thawing of permafrost in boreal Asia have been unprecedented in recent years. The frequency of occurrence of climate-induced diseases and heat stress in Central, East, South and South-East Asia has increased with rising temperatures and rainfall variability. Observed changes in terrestrial and marine ecosystems have become more pronounced [10.2.3].

Future climate change is expected to affect agriculture through declining production and reductions in arable land area and food supply for fish (medium confidence).

Projected surface warming and shifts in rainfall in most countries of Asia will induce substantial declines in agricultural crop productivity as a consequence of thermal stress and more severe droughts and floods [10.4.1]. The decline in agricultural productivity will be more pronounced in areas already suffering from increasing scarcity of arable land, and will increase the risk of hunger in Asia, particularly in developing countries [10.4.1]. Subsistence farmers are at risk from climate change. Marginal crops such as sorghum and millet could be at the greatest risk, both from a drop in productivity and from a loss of crop genetic diversity [10.4.1]. In response to climate change, it is expected that changes will occur in fish breeding habitats and food supply for fish, and ultimately the abundance of fish populations [10.4.1].

Climate change has the potential to exacerbate water-resource stresses in most regions of Asia (high confidence).

The most serious potential threat arising from climate change in Asia is water scarcity. Freshwater availability in Central, South, East and South-East Asia, particularly in large river basins, is projected to decrease due to climate change which, along with population growth and increasing demand arising from higher standards of living, could adversely affect more than a billion people by the 2050s [10.4.2]. Changes in seasonality of runoff due to rapid melting of glaciers and in some areas an increase in winter precipitation could have significant effects on hydropower generation and on crop and livestock production [10.4.2].

Increases in temperature are expected to result in more rapid recession of Himalayan glaciers and the continuation of permafrost thaw across northern Asia (medium confidence).

If current warming rates are maintained, Himalayan glaciers could decay at very rapid rates (Figure TS.11). Accelerated glacier melt would result in increased flows in some river systems for the next two to three decades, resulting in increased flooding, rock avalanches from destabilised slopes, and disruption of water resources. This would be followed by a decrease in flows as the glaciers recede [10.6.2]. Permafrost degradation can result in ground subsidence, alter drainage characteristics and infrastructure stability, and can result in increased emissions of methane [10.4.4].

Asian marine and coastal ecosystems are expected to be affected by sea-level rise and temperature increases (high confidence).

Projected sea-level rise could result in many additional millions of people being flooded each year [10.4.3.1]. Sea-water intrusion could increase the habitat of brackish-water fisheries but significantly damage the aquaculture industry [10.4.1]. Overall, sea-level rise is expected to exacerbate already declining fish productivity in Asia [10.4.1]. Arctic marine fisheries would be greatly influenced by climate change, with some species, such as cod and herring, benefiting at least for modest temperature increases, and others, such as the northern shrimp, suffering declining productivity [10.4.1].

Climate change is expected to exacerbate threats to biodiversity resulting from land-use/cover change and population pressure in most parts of Asia (high confidence).

Increased risk of extinction for many flora and fauna species in Asia is likely as a result of the synergistic effects of climate change and habitat fragmentation [10.4.4]. Threats to the ecological stability of wetlands, mangroves and coral reefs around Asia would also increase [10.4.3, 10.6.1]. The frequency and extent of forest fires in northern Asia is expected to increase in the future due to climate change and extreme weather events that could likely limit forest expansion [10.4.4].

Figure TS.11. *Projected future changes in the northern Asia permafrost boundary under the SRES A2 scenario for 2100. [F10.5]*

Future climate change is likely to continue to adversely affect human health in Asia (high confidence).
Increases in endemic morbidity and mortality due to diarrhoeal disease primarily associated with floods and droughts are expected in East, South and South-East Asia, due to projected changes in the hydrological cycle associated with global warming [10.4.5]. Increases in coastal water temperature would exacerbate the abundance and/or toxicity of cholera in South Asia [10.4.5]. Natural habitats of vector-borne and water-borne diseases are reported to be expanding [10.4.5].

Multiple stresses in Asia will be further compounded in the future due to climate change (high confidence).
Exploitation of natural resources associated with rapid urbanisation, industrialisation and economic development in most developing countries of Asia has led to increasing air and water pollution, land degradation, and other environmental problems that have placed enormous pressure on urban infrastructure, human well-being, cultural integrity, and socio-economic settings. It is likely that climate change will intensify these environmental pressures and impinge on sustainable development in many developing countries of Asia, particularly in the South and East [10.5.6].

Australia and New Zealand

The region is already experiencing impacts from recent climate change, and adaptation has started in some sectors and regions (high confidence).
Since 1950 there has been a 0.3 to 0.7°C warming in the region, with more heatwaves, fewer frosts, more rain in north-western Australia and south-western New Zealand, less rain in southern and eastern Australia and north-eastern New Zealand, an increase in the intensity of Australian droughts, and a rise in sea level of 70 mm [11.2.1]. Impacts are now evident in water supply and agriculture, changed natural ecosystems, reduced seasonal snow cover and glacier shrinkage [11.2.2, 11.2.3]. Some adaptation has occurred in sectors such as water, agriculture, horticulture and coasts [11.2.5].

The climate of the 21st century is virtually certain to be warmer, with changes in extreme events (medium to high confidence).
Heatwaves and fires are virtually certain to increase in intensity and frequency (high confidence) [11.3]. Floods, landslides, droughts and storm surges are very likely to become more frequent and intense, and snow and frost are likely to become less frequent (high confidence) [11.3.1]. Large areas of mainland Australia and eastern New Zealand are likely to have less soil moisture, although western New Zealand is likely to receive more rain (medium confidence) [11.3].

Without further adaptation, potential impacts of climate change are likely to be substantial (high confidence).

- As a result of reduced precipitation and increased evaporation, water security problems are very likely to intensify by 2030 in southern and eastern Australia and, in New Zealand, in Northland and some eastern regions [11.4.1].
- Significant loss of biodiversity is projected to occur by 2020 in some ecologically rich sites including the Great Barrier Reef and Queensland Wet Tropics. Other sites at risk include Kakadu Wetlands, south-west Australia, sub-Antarctic islands and the alpine areas of both countries [11.4.2].
- Ongoing coastal development and population growth in areas such as Cairns and south-east Queensland (Australia) and Northland to Bay of Plenty (New Zealand) are projected to exacerbate risks from sea-level rise and increases in the severity and frequency of storms and coastal flooding by 2050 [11.4.5, 11.4.7].
- Risks to major infrastructure are likely to markedly increase. By 2030, design criteria for extreme events are very likely to be exceeded more frequently. These risks include the failure of flood protection and urban drainage/sewerage, increased storm and fire damage, and more heatwaves causing more deaths and more black-outs [11.4.1, 11.4.5, 11.4.7, 11.4.10, 11.4.11].
- Production from agriculture and forestry is projected to decline by 2030 over much of southern and eastern Australia, and over parts of eastern New Zealand, due to increased drought and fire. However, in New Zealand, initial benefits to agriculture and forestry are projected in western and southern areas and close to major rivers due to a longer growing season, less frost and increased rainfall [11.4.3, 11.4.4].

Vulnerability is likely to increase in many sectors, but this depends on adaptive capacity.

- Most human systems have considerable adaptive capacity. The region has well-developed economies, extensive scientific and technical capabilities, disaster-mitigation strategies, and biosecurity measures. However, there are likely to be considerable cost and institutional constraints to the implementation of adaptation options (high confidence) [11.5]. Some Indigenous communities have low adaptive capacity (medium confidence) [11.4.8]. Water security and coastal communities are most vulnerable (high confidence) [11.7].
- Natural systems have limited adaptive capacity. Projected rates of climate change are very likely to exceed rates of evolutionary adaptation in many species (high confidence) [11.5]. Habitat loss and fragmentation are very likely to limit species migration in response to shifting climatic zones (high confidence) [11.2.5, 11.5].
- Vulnerability is likely to rise as a consequence of an increase in extreme events. Economic damage from extreme weather is very likely to increase and provide major challenges for adaptation (high confidence) [11.5].
- Vulnerability is likely to be high by 2050 in a few identified hotspots (see Figure TS.12). In Australia, these include the Great Barrier Reef, eastern Queensland, the south-west, Murray-Darling Basin, the Alps and Kakadu; in New Zealand, these include the Bay of Plenty, Northland, eastern regions and the Southern Alps (medium confidence) [11.7].

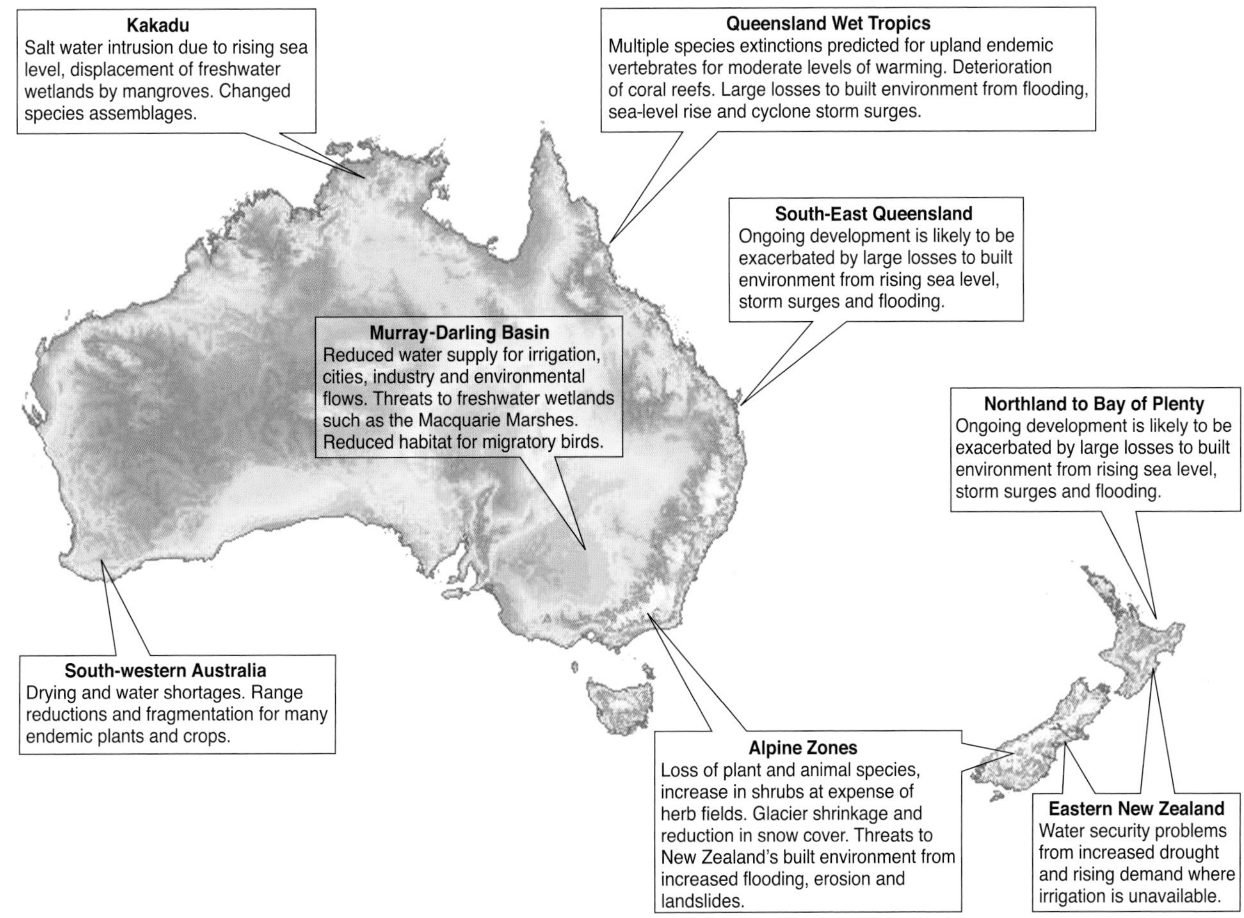

Kakadu
Salt water intrusion due to rising sea level, displacement of freshwater wetlands by mangroves. Changed species assemblages.
Queensland Wet Tropics
Multiple species extinctions predicted for upland endemic vertebrates for moderate levels of warming. Deterioration of coral reefs. Large losses to built environment from flooding, sea-level rise and cyclone storm surges.
South-East Queensland
Ongoing development is likely to be exacerbated by large losses to built environment from rising sea level, storm surges and flooding.
Murray-Darling Basin
Reduced water supply for irrigation, cities, industry and environmental flows. Threats to freshwater wetlands such as the Macquarie Marshes. Reduced habitat for migratory birds.
Northland to Bay of Plenty
Ongoing development is likely to be exacerbated by large losses to built environment from rising sea level, storm surges and flooding.
South-western Australia
Drying and water shortages. Range reductions and fragmentation for many endemic plants and crops.
Alpine Zones
Loss of plant and animal species, increase in shrubs at expense of herb fields. Glacier shrinkage and reduction in snow cover. Threats to New Zealand's built environment from increased flooding, erosion and landslides.
Eastern New Zealand
Water security problems from increased drought and rising demand where irrigation is unavailable.

Figure TS.12. *Key hotspots in Australia and New Zealand, based on the following criteria: large impacts, low adaptive capacity, substantial population, economically important, substantial exposed infrastructure, and subject to other major stresses (e.g., continued rapid population growth, ongoing development, ongoing land degradation, ongoing habitat loss and threats from rising sea level). [11.7]*

Europe

For the first time, wide-ranging impacts of changes in current climate have been documented in Europe (very high confidence).
The warming trend and spatially variable changes in rainfall have affected composition and functioning of the cryosphere (retreat of glaciers and extent of permafrost) as well as natural and managed ecosystems (lengthening of growing season, shift of species and human health due to a heatwave of unprecedented magnitude) [12.2.1]. The European heatwave in 2003 (see Figure TS.13) had major impacts on biophysical systems and society (around 35,000 excess deaths were recorded) [12.6.1]. The observed changes are consistent with projections of impacts due to future climate change [12.4].

Climate-related hazards will mostly increase, although changes will vary geographically (very high confidence).
By the 2020s, increases are likely in winter floods in maritime regions and flash floods throughout Europe [12.4.1]. Coastal flooding related to increasing storminess (particularly in the north-east Atlantic) and sea-level rise are likely to threaten an additional 1.5 million people annually by the 2080s; coastal erosion is projected to increase [12.4.2]. Warmer, drier conditions will lead to more frequent and prolonged droughts (by the 2070s, today's 100-year droughts will return every 50 years or less in southern and south-eastern Europe), as well as a longer fire-season and increased fire risk, particularly in the Mediterranean region [12.3.1, 12.4.4]. A higher frequency of catastrophic fires is also expected on drained peatlands in central and eastern Europe [12.4.5]. The frequency of rockfalls will increase due to destabilisation of mountain walls by rising temperatures and melting of permafrost [12.4.3].

Some impacts may be positive, such as reduced cold-related mortality because of increasing winter temperatures. However, on balance, without adaptive measures, health risks due to more frequent heatwaves, especially in southern, central and eastern Europe, flooding and greater exposure to vector- and food-borne diseases are anticipated to increase [12.4.11].

Climate change is likely to magnify regional differences in Europe's natural resources and assets (very high confidence).
Climate-change scenarios indicate significant warming (A2: 2.5 to 5.5°C; B2: 1 to 4°C), greater in winter in the north and in summer in south and central Europe [12.3.1]. Mean annual precipitation is projected to increase in the north and decrease in the south. Seasonal changes, however, will be more pronounced: summer precipitation is projected to decrease by up to 30 to 45%

over the Mediterranean Basin, and also over eastern and central Europe and, to a lesser degree, over northern Europe even as far north as central Scandinavia [12.3.1]. Recruitment and production of marine fisheries in the North Atlantic are likely to increase [12.4.7]. Crop suitability is likely to change throughout Europe, and crop productivity (all other factors remaining unchanged) is likely to increase in northern Europe, and decrease along the Mediterranean and in south-east Europe [12.4.7]. Forests are projected to expand in the north and retreat in the south [12.4.4]. Forest productivity and total biomass are likely to increase in the north and decrease in central and eastern Europe, while tree mortality is likely to accelerate in the south [12.4.4]. Differences in water availability between regions are anticipated to become more pronounced: annual average runoff increasing in north/north-west, and decreasing in south/south-east Europe (summer low flow is projected to decrease by up to 50% in central Europe and by up to 80% in some rivers in southern Europe) [12.4.1, 12.4.5].

Water stress is likely to increase, as well as the number of people living in river basins under high water stress (high confidence).

Water stress is likely to increase over central and southern Europe. The percentage of area under high water stress is likely to increase from 19% to 35% by the 2070s, and the number of people at risk from 16 to 44 million [12.4.1]. The regions most at risk are southern Europe and some parts of central and eastern Europe [12.4.1]. The hydropower potential of Europe is expected to decline on average by 6%, and by 20 to 50% around the Mediterranean by the 2070s [12.4.8.1].

It is anticipated that Europe's natural systems and biodiversity will be substantially affected by climate change (very high confidence). The great majority of organisms and ecosystems are likely to have difficulty in adapting to climate change (high confidence).

Sea-level rise is likely to cause an inland migration of beaches and loss of up to 20% of coastal wetlands [12.4.2.], reducing the habitat availability for several species that breed or forage in low-lying coastal areas [12.4.6]. Small glaciers will disappear and larger glaciers substantially shrink (projected volume reductions of between 30% and 70% by 2050) during the 21st century [12.4.3]. Many permafrost areas in the Arctic are projected to disappear [12.4.5.]. In the Mediterranean, many ephemeral aquatic ecosystems are projected to disappear, and permanent ones shrink and become ephemeral [12.4.5]. The northward expansion of forests is projected to reduce current tundra areas under some scenarios [12.4.4]. Mountain communities face up to a 60% loss of species under high-emissions scenarios by 2080 [12.4.3]. A large percentage of the European flora (one study found up to 50%) is likely to become vulnerable, endangered or committed to extinction by the end of this century [12.4.6]. Options for adaptation are likely to be limited for many organisms and ecosystems. For example, limited dispersal is very likely to reduce the range of most reptiles and amphibians [12.4.6]. Low-lying, geologically subsiding coasts are likely to be unable to adapt to sea-level rise [12.5.2]. There are no obvious climate adaptation options for either tundra or alpine vegetation [12.5.3]. The adaptive capacity of ecosystems can be enhanced by reducing human stresses [12.5.3, 12.5.5]. New sites for conservation may be needed because climate change is very likely to alter conditions of suitability for many species in current sites (with climate change, to meet conservation goals, the current reserve area in the EU would have to be increased by 41%) [12.5.6].

Nearly all European regions are anticipated to be negatively affected by some future impacts of climate change and these will pose challenges to many economic sectors (very high confidence).

In southern Europe, climate change is projected to worsen conditions (high temperatures and drought) in a region already vulnerable to climate variability. In northern Europe, climate change is initially projected to bring mixed effects, including some benefits, but as climate change continues, its negative effects are likely to outweigh its benefits [12.4].

Agriculture will have to cope with increasing water demand for irrigation in southern Europe due to climate change (e.g., increased water demand of 2 to 4% for maize cultivation and 6 to 10% for potatoes by 2050), and additional restrictions due to increases in crop-related nitrate leaching [12.5.7]. Winter heating demands are expected to decrease and summer cooling demands

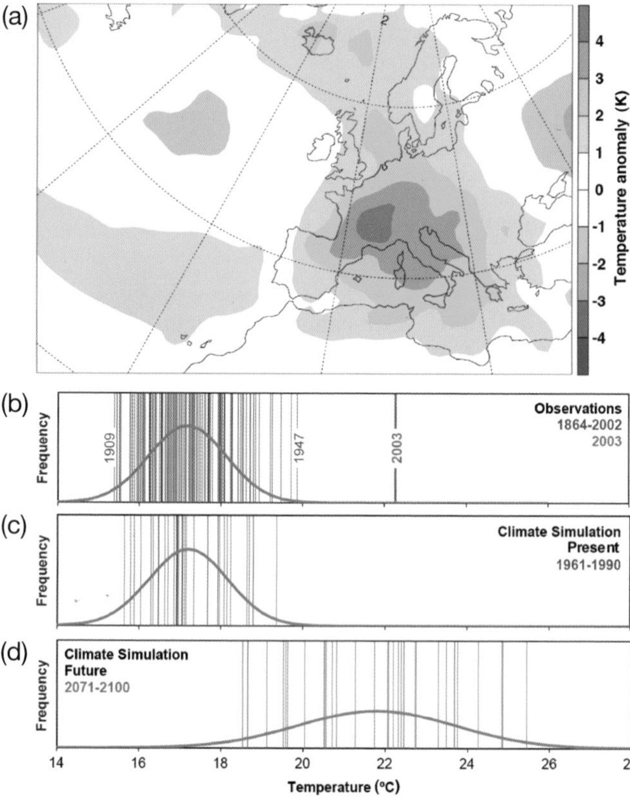

Figure TS.13. *Characteristics of the summer 2003 heatwave: (a) JJA temperature anomaly with respect to 1961-1990; (b-d) June, July, August temperatures for Switzerland; (b) observed during 1864-2003; (c) simulated using a regional climate model for the period 1961-1990; (d) simulated for 2071-2100 under the SRES A2 scenario. The vertical bars in panels (b-d) represent mean summer surface temperature for each year of the time period considered; the fitted Gaussian distribution is indicated in black. Reprinted by permission from Macmillan Publishers Ltd. [Nature] (Schär et al., 2004), copyright 2004, [F12.4].*

to increase due to climate change: around the Mediterranean, 2 to 3 fewer weeks in a year will require heating but an additional 2 to 5 weeks will need cooling by 2050 [12.4.8]. Peak electricity demand is likely to shift in some locations from winter to summer [12.4.8]. Tourism along the Mediterranean is likely to decrease in summer and increase in spring and autumn. Winter tourism in mountain regions is anticipated to face reduced snow cover (the duration of snow cover is expected to decrease by several weeks for each °C of temperature increase in the Alps region) [12.4.9, 12.4.11].

Adaptation to climate change is likely to benefit from experiences gained in reactions to extreme climate events, by specifically implementing proactive climate-change risk management adaptation plans (very high confidence).
Since the TAR, governments have greatly increased the number of actions for coping with extreme climate events. Current thinking about adaptation to extreme climate events has moved away from reactive disaster relief and towards more proactive risk management. A prominent example is the implementation in several countries of early-warning systems for heatwaves (Portugal, Spain, France, UK, Italy, Hungary) [12.6.1]. Other actions have addressed long-term climate change. For example, national action plans have been developed for adapting to climate change [12.5] and more specific plans have been incorporated into European and national policies for agriculture, energy, forestry, transport and other sectors [12.2.3,12.5.2]. Research has also provided new insights into adaptation policies (e.g., studies have shown that crops that become less economically viable under climate change can be profitably replaced by bioenergy crops) [12.5.7].

Although the effectiveness and feasibility of adaptation measures are expected to vary greatly, only a few governments and institutions have systematically and critically examined a portfolio of measures. As an example, some reservoirs used now as a measure for adapting to precipitation fluctuations may become unreliable in regions where long-term precipitation is projected to decrease [12.4.1]. The range of management options to cope with climate change varies largely among forest types, with some types having many more options than others [12.5.5].

Latin America

Climatic variability and extreme events have been severely affecting the Latin America region over recent years (high confidence).
Highly unusual extreme weather events have recently occurred, such as Venezuelan intense rainfall (1999, 2005), flooding in the Argentine Pampas (2000-2002), Amazon drought (2005), hail storms in Bolivia (2002) and the Greater Buenos Aires area (2006), the unprecedented Hurricane Catarina in the South Atlantic (2004), and the record hurricane season of 2005 in the Caribbean Basin [13.2.2]. Historically, climate variability and extremes have had negative impacts on population, increasing mortality and morbidity in affected areas. Recent developments in meteorological forecasting techniques could improve the necessary information for human welfare and security. However, the lack of modern observation equipment and badly-needed upper-air information, the low density of weather stations, the unreliability of their reports, and the lack of monitoring of climate variables hinder the quality of forecasts, with adverse effects on the public, lowering their appreciation of applied meteorological services, as well as their trust in climate records. These shortcomings also affect hydrometeological observing services, with a negative impact on the quality of early warnings and alert advisories (medium confidence) [13.2.5].

During the last few decades, important changes in precipitation and increases in temperature have been observed (high confidence).
Increases in rainfall in south-east Brazil, Paraguay, Uruguay, the Argentine Pampas, and some parts of Bolivia have had impacts on land use and crop yields and have increased flood frequency and intensity. On the other hand, a declining trend in precipitation has been observed in southern Chile, south-west Argentina, southern Peru, and western Central America. Increases in temperature of approximately 1°C in Mesoamerica and South America and of 0.5°C in Brazil have been observed. As a consequence of temperature increases, the trend in glacier retreat reported in the TAR is accelerating (very high confidence). This issue is critical in Bolivia, Peru, Colombia and Ecuador, where water availability has already been compromised either for consumption or hydropower generation [13.2.4]. These problems with supply are expected to increase in the future, becoming chronic if no appropriate adaptation measures are planned and implemented. Over the next decades Andean inter-tropical glaciers are very likely to disappear, affecting water availability and hydropower generation (high confidence) [13.2.4].

Land-use changes have intensified the use of natural resources and exacerbated many of the processes of land degradation (high confidence).
Almost three-quarters of the dryland surface is moderately or severely affected by degradation processes. The combined effects of human action and climate change have brought a decline in natural land cover, which continues to decline at very high rates (high confidence). In particular, rates of deforestation of tropical forests have increased during the last 5 years. There is evidence that biomass-burning aerosols may change regional temperature and precipitation in the southern part of Amazonia (medium confidence). Biomass burning also affects regional air quality, with implications for human health. Land-use and climate changes acting synergistically will increase vegetation fire risk substantially (high confidence) [13.2.3, 13.2.4].

The projected mean warming for Latin America to the end of the 21st century, according to different climate models, ranges from 1 to 4°C for SRES emissions scenario B2 and from 2 to 6°C for scenario A2 (medium confidence).
Most GCM projections indicate rather larger than present (positive and negative) rainfall anomalies for the tropical portions of Latin America and smaller ones for extra-tropical South America. Changes in temperature and precipitation will have especially severe impacts on already vulnerable hotspots,

identified in Figure TS.14. In addition, the frequency of occurrence of weather and climate extremes is likely to increase in the future; as is the frequency and intensity of hurricanes in the Caribbean Basin [13.3.1, 13.3.1].

Under future climate change, there is a risk of significant species extinctions in many areas of tropical Latin America (high confidence).
Gradual replacement of tropical forest by savannas is expected by mid-century in eastern Amazonia and the tropical forests of central and southern Mexico, along with replacement of semi-arid by arid vegetation in parts of north-east Brazil and most of central and northern Mexico, due to increases in temperature and associated decreases in soil water (high confidence) [13.4.1]. By the 2050s, 50% of agricultural lands are very likely to be subjected to desertification and salinisation in some areas (high confidence) [13.4.2]. There is a risk of significant biodiversity loss through species extinction in many areas of tropical Latin America. Seven out of the world's twenty-five most critical places with high endemic species concentrations are in Latin America, and these areas are undergoing habitat loss. Biological reserves and ecological corridors have been either implemented or planned for the maintenance of biodiversity in natural ecosystems, and these can serve as adaptation measures to help protect ecosystems in the face of climate change [13.2.5].

By the 2020s, the net increase in the number of people experiencing water stress due to climate change is likely to be between 7 and 77 million (medium confidence).
For the second half of the 21st century, the potential water availability reduction and the increasing demand from an increasing regional population would increase these figures to between 60 and 150 million [13.4.3].

Generalised reductions in rice yields by the 2020s, as well as increases in soybean yields in temperate zones, are likely when CO_2 effects are considered (medium confidence).
For other crops (wheat, maize), the projected response to climate change is more erratic, depending on the chosen scenario. Assuming low CO_2 fertilisation effects, the number of additional people at risk of hunger under the A2 scenario is likely to reach 5, 26 and 85 million in 2020, 2050 and 2080, respectively (medium confidence). Livestock and dairy productivity is likely to decline in response to increasing temperatures [13.4.2].

The expected increases in sea-level rise, weather and climatic variability and extremes are very likely to affect coastal areas (high confidence).
During the last 10 to 20 years, the rate of sea-level rise increased from 1 to 2-3 mm/year in south-eastern South America [13.2.4]. In the future, sea-level rise is projected to cause an increased risk of flooding in low-lying areas. Adverse impacts would be observed on (i) low-lying areas (e.g., in El Salvador, Guyana, the coast of the province of Buenos Aires), (ii) buildings and tourism (e.g., in Mexico, Uruguay), (iii) coastal morphology (e.g., in Peru), (iv) mangroves (e.g., in Brazil, Ecuador, Colombia, Venezuela), (v) availability of drinking water on the Pacific coast of Costa Rica, Ecuador and the Rio de la Plata estuary [13.4.4].

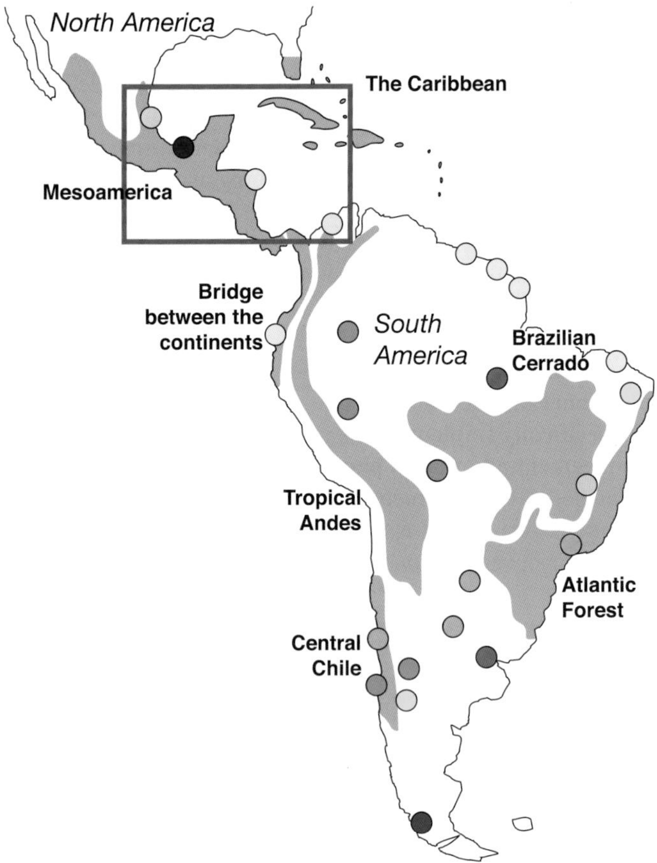

- Coral reefs and mangroves seriously threatened with warmer SST
- Under the worst sea-level rise scenario, mangroves are very likely to disappear from low-lying coastlines
- Amazonia: loss of 43% of 69 tree species by the end of 21st century; savannisation of the eastern part
- Cerrados: Losses of 24% of 138 tree species for a temperature increase of 2°C
- Reduction of suitable lands for coffee
- Increases in aridity and scarcity of water resources
- Sharp increase in extinction of: mammals, birds, butterflies, frogs and reptiles by 2050
- Water availability and hydro-electric generation seriously reduced due to reduction in glaciers
- Ozone depletion and skin cancer
- Severe land degradation and desertification
- Rio de la Plata coasts threatened by increasing storm surges and sea-level rise
- Increased vulnerability to extreme events

Areas in red correspond to sites where biodiversity is currently severely threatened and this trend is very likely to continue in the future

Figure TS.14. *Key hotspots for Latin America, where climate change impacts are expected to be particularly severe. [13.4]*

Future sustainable development plans should include adaptation strategies to enhance the integration of climate change into development policies (high confidence).
Several adaptation measures have been proposed for coastal, agricultural, water and health sectors. However, the effectiveness of these efforts is outweighed by a lack of capacity-building and appropriate political, institutional and technological frameworks, low income, and settlements in vulnerable areas, among others. The present degree of development of observation and monitoring networks necessarily requires improvement, capacity-building, and the strengthening of communication in order to permit the effective operation of environmental observing systems and the reliable dissemination of early warnings. Otherwise, the Latin American countries' sustainable development goals are likely to be seriously compromised, adversely affecting, among other things, their capability to reach the Millennium Development Goals [13.5].

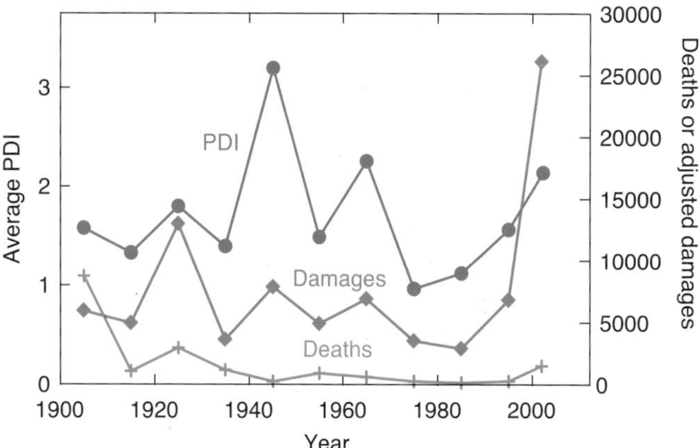

Figure TS.15. *Decadal average (6-year average for 2000-2005) hurricane total dissipated energy (PDI), loss of life, and inflation-adjusted economic damages (in thousands of US$) from hurricanes making landfall in the continental USA since 1900. [F14.1]*

North America

North America has considerable adaptive capacity, which has been deployed effectively at times, but this capacity has not always protected its population from adverse impacts of climate variability and extreme weather events (very high confidence).
Damage and loss of life from Hurricane Katrina in August 2005 illustrate the limitations of existing adaptive capacity to extreme events. Traditions and institutions in North America have encouraged a decentralised response framework where adaptation tends to be reactive, unevenly distributed, and focused on coping with rather than preventing problems. "Mainstreaming" climate change issues into decision making is a key prerequisite for sustainability [14.2.3, 14.2.6, 14.4, 14.5, 14.7].

Emphasis on effective adaptation is critical, because economic damage from extreme weather is likely to continue increasing, with direct and indirect consequences of climate change playing a growing role (very high confidence).
Over the past several decades, economic damage from hurricanes in North America has increased over fourfold (Figure TS.15), due largely to an increase in the value of infrastructure at risk [14.2.6]. Costs to North America include billions of dollars in damaged property and diminished economic productivity, as well as lives disrupted and lost [14.2.6, 14.2.7, 14.2.8]. Hardships from extreme events disproportionately affect those who are socially and economically disadvantaged, especially the poor and indigenous peoples of North America [14.2.6].

Climate change is likely to exacerbate other stresses on infrastructure, and human health and safety in urban centres (very high confidence).
Climate change impacts in urban centres are very likely to be compounded by urban heat islands, air and water pollution, ageing infrastructure, maladapted urban form and building stock, water quality and supply challenges, immigration and population growth, and an ageing population [14.3.2, 14.4.1, 14.4.6].

Coastal communities and habitats are very likely to be increasingly stressed by climate change impacts interacting with development and pollution (very high confidence).
Sea level is rising along much of the coast, and the rate of change is likely to increase in the future, exacerbating the impacts of progressive inundation, storm surge flooding, and shoreline erosion [14.2.3, 14.4.3]. Storm impacts are likely to be more severe, especially along the Gulf and Atlantic coasts [14.4.3]. Salt marshes, other coastal habitats and dependent species are threatened now and increasingly in future decades by sea-level rise, fixed structures blocking landward migration, and changes in vegetation [14.2]. Population growth and rising value of infrastructure in coastal areas increases vulnerability to climate variability and future climate change, with losses projected to increase if the intensity of tropical storms increases. Current adaptation to coastal hazards is uneven and readiness for increased exposure is low [14.2.3, 14.4.3, 14.5].

Warm temperatures and extreme weather already cause adverse human health effects through heat-related mortality, pollution, storm-related fatalities and injuries, and infectious diseases, and are likely, in the absence of effective countermeasures, to increase with climate change (very high confidence).
Depending on progress in health care, infrastructure, technology and access, climate change could increase the risk of heatwave deaths, water-borne diseases and degraded water quality [14.4.1], respiratory illness through exposure to pollen and ozone, and vector-borne infectious diseases (low confidence) [14.2.5, 14.4.5].

Climate change is very likely to constrain North America's already intensively utilised water resources, interacting with other stresses (high confidence).
Diminishing snowpack and increasing evaporation due to rising temperatures are very likely to affect timing and availability of water and intensify competition among uses [B14.2, 14.4.1]. Warming is very likely to place additional stress on groundwater availability, compounding the effects of higher demand from

economic development and population growth (medium confidence) [14.4.1]. In the Great Lakes and some major river systems, lower water levels are likely to exacerbate issues of water quality, navigation, hydropower generation, water diversions, and bi-national co-operation [14.4.1, B14.2].

Disturbances such as wildfire and insect outbreaks are increasing and are likely to intensify in a warmer future with drier soils and longer growing seasons, and to interact with changing land use and development affecting the future of wildland ecosystems (high confidence).
Recent climate trends have increased ecosystem net primary production, and this trend is likely to continue for the next few decades [14.2.2]. However, wildfire and insect outbreaks are increasing, a trend that is likely to intensify in a warmer future [14.4.2, B14.1]. Over the course of the 21st century, the tendency for species and ecosystems to shift northward and to higher elevations is likely to rearrange the map of North American ecosystems. Continuing increases in disturbances are likely to limit carbon storage, facilitate invasives, and amplify the potential for changes in ecosystem services [14.4.2, 14.4.4].

Polar Regions

The environmental impacts of climate change show profound regional differences both within and between the polar regions (very high confidence).
The impacts of climate change in the Arctic over the next hundred years are likely to exceed the changes forecast for many other regions. However, the complexity of responses in biological and human systems, and the fact that they are subject to additive multiple stresses, means that the impacts of climate change on these systems remain difficult to predict. Changes on the Antarctic Peninsula, sub-Antarctic islands and Southern Ocean have also been rapid, and in future dramatic impacts are expected. Evidence of ongoing change over the rest of the Antarctic continent is less conclusive and prediction of the likely impacts is thus difficult. For both polar regions, economic impacts are especially difficult to address due to the lack of available information [15.2.1, 15.3.2, 15.3.3].

There is a growing evidence of the impacts of climate change on ecosystems in both polar regions (high confidence).
There has been a measured change in composition and range of plants and animals on the Antarctic Peninsula and on the sub-Antarctic islands. There is a documented increase in the overall greenness of parts of the Arctic, an increase in biological productivity, a change in species ranges (e.g., shifts from tundra to shrublands), some changes in position of the northern limit of trees, and changes in the range and abundance of some animal species. In both the Arctic and Antarctic, research indicates that such changes in biodiversity and vegetation zone relocation will continue. The poleward migration of existing species and competition from invading species is already occurring, and will continue to alter species composition and abundance in terrestrial and aquatic systems. Associated vulnerabilities are related to loss of biodiversity and the spread of animal-transmitted diseases [15.2.2, 15.4.2].

The continuation of hydrological and cryospheric changes will have significant regional impacts on Arctic freshwater, riparian and near-shore marine systems (high confidence).
The combined discharge of Eurasian rivers draining into the Arctic Ocean shows an increase since the 1930s, largely consistent with increased precipitation, although changes to cryospheric processes (snowmelt and permafrost thaw) are also modifying routing and seasonality of flow [15.3.1, 15.4.1].

The retreat of Arctic sea ice over recent decades has led to improved marine access, changes in coastal ecology/biological production, adverse effects on many ice-dependent marine mammals, and increased coastal wave action (high confidence).
Continued loss of sea ice will produce regional opportunities and problems; reductions in freshwater ice will affect lake and river ecology and biological production, and will require changes in water-based transportation. For many stakeholders, economic benefits may accrue, but some activities and livelihoods may be adversely affected [15.ES, 15.4.7, 15.4.3, 15.4.1].

Around the Antarctic Peninsula, a newly documented decline in krill abundance, together with an increase in salp abundance, has been attributed to a regional reduction in the extent and duration of sea ice (medium confidence).
If there is a further decline in sea ice, a further decline in krill is likely, impacting predators higher up the food chain [15.2.2, 15.6.3].

Warming of areas of the northern polar oceans has had a negative impact on community composition, biomass and distribution of phytoplankton and zooplankton (medium confidence).
The impact of present and future changes on higher predators, fish and fisheries will be regionally specific, with some beneficial and some detrimental effects [15.2.2].

Many Arctic human communities are already adapting to climate change (high confidence).
Indigenous people have exhibited resilience to changes in their local environments for thousands of years. Some indigenous communities are adapting through changes in wildlife management regimes and hunting practices. However, stresses in addition to climate change, together with a migration into small remote communities and increasing involvement in employment economies and sedentary occupations, will challenge adaptive capacity and increase vulnerability. Some traditional ways of life are being threatened and substantial investments are needed to adapt or relocate physical structures and communities [15.4.6, 15.5, 15.7].

A less severe climate in northern regions will produce positive economic benefits for some communities (very high confidence).
The benefits will depend on particular local conditions but will, in places, include reduced heating costs, increased agricultural and forestry opportunities, more navigable northern sea routes and marine access to resources [15.4.2].

The impacts of future climate change in the polar regions will produce feedbacks that will have globally significant consequences over the next hundred years (high confidence).

Current Arctic Conditions

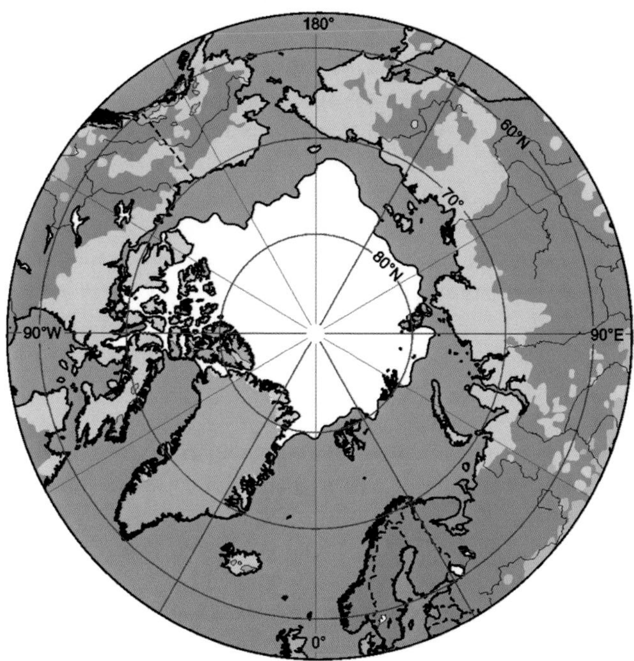

Projected Arctic Conditions

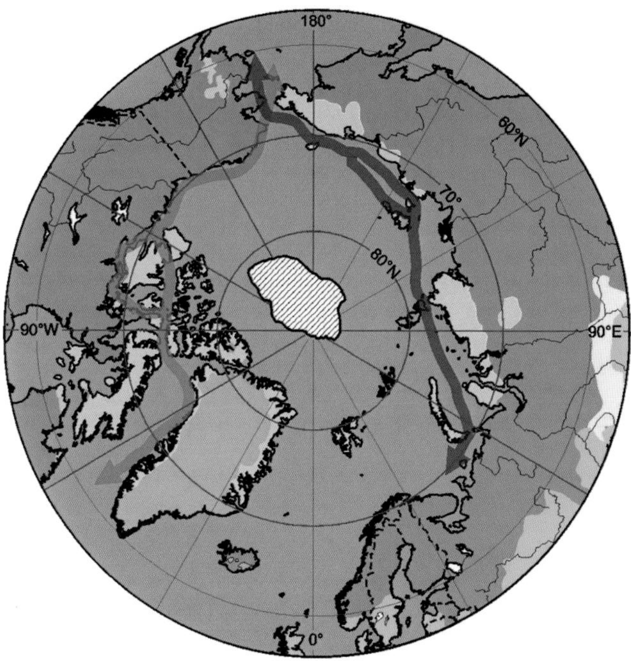

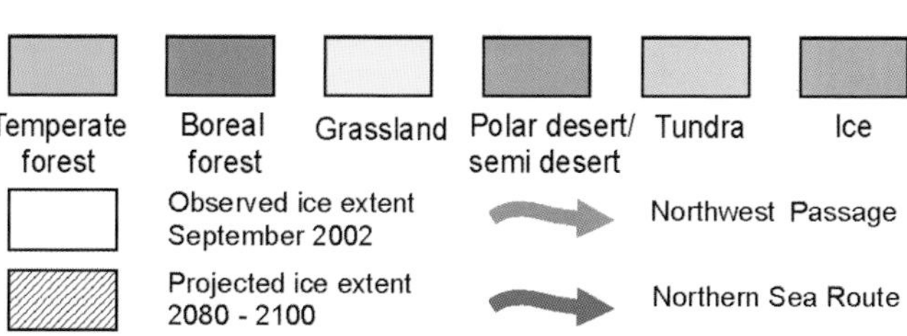

Figure TS.16. *Vegetation of the Arctic and neighbouring regions. Top: present-day, based on floristic surveys. Bottom: modelled for 2090-2100 under the IS92a emissions scenario. [F15.3]*

A continued loss of land-based ice will add to global sea-level rise. A major impact could result from a weakening of the thermohaline circulation due to a net increase in river flow into the Arctic Ocean and the resulting increased flux of freshwater into the North Atlantic. Under CO_2-doubling, total river flow into the Arctic Ocean is likely to increase by up to 20%. Warming will expose more bare ground in the Arctic (Figure TS.16) and on the Antarctic Peninsula, to be colonised by vegetation. Recent models predict a decrease in albedo due to loss of ice and changing vegetation, and that the tundra will be a small sink for carbon, although increased methane emissions from the thawing permafrost could contribute to climate warming [15.4.1, 15.4.2].

Small Islands

Small islands have characteristics which make them especially vulnerable to the effects of climate change, sea-level rise and extreme events (very high confidence).

These include their limited size and proneness to natural hazards and external shocks. They have low adaptive capacity, and adaptation costs are high relative to GDP [16.5].

Sea-level rise is likely to exacerbate inundation, storm surge, erosion and other coastal hazards, thus threatening the vital infrastructure that supports the socio-economic well-being of island communities (very high confidence).

Some studies suggest that sea-level rise could cause coastal land loss and inundation, while others show that some islands are morphologically resilient and are expected to persist [16.4.2]. In the Caribbean and Pacific Islands, more than 50% of the population live within 1.5 km of the shore. Almost without exception, the air and sea ports, major road arteries, communication networks, utilities and other critical infrastructure in the small islands of the Indian and Pacific Oceans and the Caribbean tend to be restricted to coastal locations (Table TS.2). The threat from sea-level rise is likely to be amplified by changes in tropical cyclones [16.4.5, 16.4.7].

There is strong evidence that under most climate-change scenarios, water resources in small islands are likely to be seriously compromised (very high confidence).

Most small islands have a limited water supply. Many small islands in the Caribbean and Pacific are likely to experience increased water stress as a result of climate change [16.4.1]. Predictions under all SRES scenarios for this region show reduced rainfall in summer, so that it is unlikely that demand will be met during low rainfall periods. Increased rainfall in winter will be unlikely to compensate, due to a lack of storage and high runoff during storms [16.4.1].

Climate change is likely to heavily impact coral reefs, fisheries and other marine-based resources (high confidence).

Fisheries make an important contribution to the GDP of many island states. Changes in the occurrence and intensity of El Niño-Southern Oscillation (ENSO) events are likely to have severe impacts on commercial and artisanal fisheries. Increasing sea surface temperature and sea level, increased turbidity, nutrient loading and chemical pollution, damage from tropical cyclones, and decreases in growth rates due to the effects of higher CO_2-

concentrations on ocean chemistry, are very likely to lead to coral bleaching and mortality [16.4.3].

On some islands, especially those at higher latitudes, warming has already led to the replacement of some local species (high confidence).
Mid- and high-latitude islands are virtually certain to be colonised by non-indigenous invasive species, previously limited by unfavourable temperature conditions (see Table TS.2). Increases in extreme events in the short term are virtually certain to affect the adaptation responses of forests on tropical islands, where regeneration is often slow. In view of their small area, forests on many islands can easily be decimated by violent cyclones or storms. On some high-latitude islands it is likely that forest cover will increase [16.4.4, 15.4.2].

It is very likely that subsistence and commercial agriculture on small islands will be adversely affected by climate change (high confidence).
Sea-level rise, inundation, sea-water intrusion into freshwater lenses, soil salinisation and a decline in water supply will very likely adversely impact coastal agriculture. Away from the coast, changes in extremes (e.g., flooding and drought) are likely to have a negative effect on agricultural production. Appropriate adaptation measures may help to reduce these impacts. In some high-latitude islands, new opportunities may arise for increased agricultural production [16.4.3, 15.4.2].

New studies confirm previous findings that the effects of climate change on tourism are likely to be direct and indirect, and largely negative (high confidence).
Tourism is the major contributor to GDP and employment in many small islands. Sea-level rise and increased sea-water temperature are likely to contribute to accelerated beach erosion, degradation of coral reefs and bleaching (Table TS.2). In addition, loss of cultural heritage from inundation and flooding will reduce the amenity value for coastal users. Whereas a warmer climate could reduce the number of people visiting small islands in low latitudes, it could have the reverse effect in mid- and high-latitude islands. However, water shortages and increased incidence of vector-borne diseases are also likely to deter tourists [16.4.6].

There is growing concern that global climate change is likely to impact human health, mostly in adverse ways (medium confidence).
Many small islands lie in tropical or sub-tropical zones with weather conducive to the transmission of diseases such as malaria, dengue, filariasis, schistosomiasis, and food- and water-borne diseases. Outbreaks of climate-sensitive diseases can be costly in terms of lives and economic impact. Increasing temperatures and decreasing water availability due to climate change are likely to increase the burdens of diarrhoeal and other infectious diseases in some small-island states [16.4.5].

Latitude	Region and system at risk	Impacts and vulnerability
High	Iceland and isolated Arctic islands of Svalbard and the Faroe Islands: Marine ecosystem and plant species	• The imbalance of species loss and replacement leads to an initial loss in diversity. Northward expansion of dwarf-shrub and tree-dominated vegetation into areas rich in rare endemic species results in their loss. • Large reduction in, or even a complete collapse of, the Icelandic capelin stock leads to considerable negative impacts on most commercial fish stocks, whales and seabirds.
	High-latitude islands (Faroe Islands): Plant species	• Scenario I (temperature increase 2°C): species most affected by warming are restricted to the uppermost parts of mountains. For other species, the effect will mainly be upward migration. • Scenario II (temperature decrease 2°C): species affected by cooling are those at lower altitudes.
Mid	Sub-Antarctic Marion Islands: Ecosystem	• Changes will directly affect the indigenous biota. An even greater threat is that a warmer climate will increase the ease with which the islands can be invaded by alien species.
	Five islands in the Mediterranean Sea: Ecosystems	• Climate change impacts are negligible in many simulated marine ecosystems. • Invasion into island ecosystems becomes an increasing problem. In the longer term, ecosystems will be dominated by exotic plants irrespective of disturbance rates.
	Mediterranean: Migratory birds (pied flycatchers: *Ficedula hypoleuca*)	• Reduction in nestling and fledgling survival rates of pied flycatchers in two of the southernmost European breeding populations.
	Pacific and Mediterranean: Sim weed (*Chromolaena odorata*)	• Pacific Islands at risk of invasion by sim weed. • Mediterranean semi-arid and temperate climates predicted to be unsuitable for invasion.
Low	Pacific small islands: Coastal erosion, water resources and human settlements	• Accelerated coastal erosion, saline intrusion into freshwater lenses and increased flooding from the sea cause large effects on human settlements. • Lower rainfall coupled with accelerated sea-level rise compounds the threat on water resources; a 10% reduction in average rainfall by 2050 is likely to correspond to a 20% reduction in the size of the freshwater lens on Tarawa Atoll, Kiribati.
	American Samoa, fifteen other Pacific, Islands: Mangroves	• 50% loss of mangrove area in American Samoa; 12% reduction in mangrove area in fifteen other Pacific Islands.
	Caribbean (Bonaire, Netherlands Antilles): Beach erosion and sea-turtle nesting habitats	• On average, up to 38% (±24% standard deviation) of the total current beach could be lost with a 0.5 m rise in sea level, with lower narrower beaches being the most vulnerable, reducing turtle nesting habitat by one-third.
	Caribbean (Bonaire, Barbados): Tourism	• The beach-based tourism industry in Barbados and the marine-diving-based ecotourism industry in Bonaire are both negatively affected by climate change through beach erosion in Barbados and coral bleaching in Bonaire.

Table TS.2. *Range of future impacts and vulnerabilities in small islands [B16.1]. These projections are summarised from studies using a range of scenarios including SRES and Third Assessment Report sea-level rise projections.*

Box TS.6. The main projected impacts for regions

Africa

- The impacts of climate change in Africa are likely to be greatest where they co-occur with a range of other stresses (e.g., unequal access to resources [9.4.1]; enhanced food insecurity [9.6]; poor health management systems [9.2.2, 9.4.3]). These stresses, enhanced by climate variability and change, further enhance the vulnerabilities of many people in Africa. ** D [9.4]
- An increase of 5 to 8% (60 to 90 million ha) of arid and semi-arid land in Africa is projected by the 2080s under a range of climate-change scenarios. ** N [9.4.4]
- Declining agricultural yields are likely due to drought and land degradation, especially in marginal areas. Changes in the length of growing period have been noted under various scenarios. In the A1FI SRES scenario, which has an emphasis on globally-integrated economic growth, areas of major change include the coastal systems of southern and eastern Africa. Under both the A1 and B1 scenarios, mixed rain-fed, semi-arid systems are shown to be heavily affected by changes in climate in the Sahel. Mixed rain-fed and highland perennial systems in the Great Lakes region in East Africa and in other parts of East Africa are also heavily affected. In the B1 SRES scenario, which assumes development within a framework of environmental protection, the impacts are, however, generally less, but marginal areas (e.g., the semi-arid systems) become more marginal, with the impacts on coastal systems becoming moderate. ** D [9.4.4]
- Current stress on water in many areas of Africa is likely to be enhanced by climate variability and change. Increases in runoff in East Africa (possibly floods) and decreases in runoff and likely increased drought risk in other areas (e.g., southern Africa) are projected by the 2050s. Current water stresses are not only linked to climate variations, and issues of water governance and water-basin management must also be considered in any future assessments of water in Africa. ** D [9.4.1]
- Any changes in the primary production of large lakes are likely to have important impacts on local food supplies. For example, Lake Tanganyika currently provides 25 to 40% of animal protein intake for the population of the surrounding countries, and climate change is likely to reduce primary production and possible fish yields by roughly 30% [9.4.5, 3.4.7, 5.4.5]. The interaction of human management decisions, including over-fishing, is likely to further compound fish offtakes from lakes. ** D [9.2.2]
- Ecosystems in Africa are likely to experience major shifts and changes in species range and possible extinctions (e.g., fynbos and succulent Karoo biomes in southern Africa). * D [9.4.5]
- Mangroves and coral reefs are projected to be further degraded, with additional consequences for fisheries and tourism.** D [9.4.5]
- Towards the end of the 21st century, projected sea-level rise will affect low-lying coastal areas with large populations. The cost of adaptation will exceed 5 to 10% of GDP. ** D [B9.2, 9.4.6, 9.5.2]

Asia

- A 1 m rise in sea level would lead to a loss of almost half of the mangrove area in the Mekong River delta (2,500 km^2), while approximately 100,000 ha of cultivated land and aquaculture area would become salt marsh. * N [10.4.3]
- Coastal areas, especially heavily populated megadelta regions in South, East and South-East Asia, will be at greatest risk due to increased flooding from the sea and, in some megadeltas, flooding from the rivers. For a 1 m rise in sea level, 5,000 km^2 of Red River delta, and 15,000 to 20,000 km^2 of Mekong River delta are projected to be flooded, which could affect 4 million and 3.5 to 5 million people, respectively. * N [10.4.3]
- Tibetan Plateau glaciers of under 4 km in length are projected to disappear with a temperature increase of 3°C and no change in precipitation. ** D [10.4.4]
- If current warming rates are maintained, Himalayan glaciers could decay at very rapid rates, shrinking from the present 500,000 km^2 to 100,000 km^2 by the 2030s. ** D [10.6.2]
- Around 30% of Asian coral reefs are expected to be lost in the next 30 years, compared with 18% globally under the IS92a emissions scenario, but this is due to multiple stresses and not to climate change alone. ** D [10.4.3]
- It is estimated that under the full range of SRES scenarios, 120 million to 1.2 billion and 185 to 981 million people will experience increased water stress by the 2020s and the 2050s, respectively. ** D [10.4.2]
- The per capita availability of freshwater in India is expected to drop from around 1,900 m^3 currently to 1,000 m^3 by 2025 in response to the combined effects of population growth and climate change [10.4.2.3]. More intense rain and more frequent flash floods during the monsoon would result in a higher proportion of runoff and a reduction in the proportion reaching the groundwater. ** N [10.4.2]
- It is projected that crop yields could increase up to 20% in East and South-East Asia, while they could decrease up to 30% in Central and South Asia by the mid-21st century. Taken together and considering the influence of rapid population growth and urbanisation, the risk of hunger is projected to remain very high in several developing countries. * N [10.4.1]
- Agricultural irrigation demand in arid and semi-arid regions of East Asia is expected to increase by 10% for an increase in

temperature of 1°C. ** N [10.4.1]
- The frequency and extent of forest fires in northern Asia are expected to increase in the future due to climate change and extreme weather events that would likely limit forest expansion. * N [10.4.4]

Australia and New Zealand

- The most vulnerable sectors are natural ecosystems, water security and coastal communities. ** C [11.7]
- Many ecosystems are likely to be altered by 2020, even under medium-emissions scenarios [11.4.1]. Among the most vulnerable are the Great Barrier Reef, south-western Australia, Kakadu Wetlands, rain forests and alpine areas [11.4.2]. This is virtually certain to exacerbate existing stresses such as invasive species and habitat loss, increase the probability of species extinctions, and cause a reduction in ecosystem services for tourism, fishing, forestry and water supply. * N [11.4.2]
- Ongoing water security problems are very likely to increase by 2030 in southern and eastern Australia and, in New Zealand, in Northland and some eastern regions, e.g., a 0 to 45% decline in runoff in Victoria by 2030 and a 10 to 25% reduction in river flow in Australia's Murray-Darling Basin by 2050. ** D [11.4.1]
- Ongoing coastal development is very likely to exacerbate risk to lives and property from sea-level rise and storms. By 2050, there is very likely to be loss of high-value land, faster road deterioration, degraded beaches, and loss of items of cultural significance. *** C [11.4.5, 11.4.7, 11.4.8]
- Increased fire danger is likely with climate change; for example, in south-east Australia the frequency of very high and extreme fire danger days is likely to rise 4 to 25% by 2020 and 15 to 70% by 2050. ** D [11.3.1]
- Risks to major infrastructure are likely to increase. Design criteria for extreme events are very likely to be exceeded more frequently by 2030. Risks include failure of floodplain levees and urban drainage systems, and flooding of coastal towns near rivers. ** D [11.4.5, 11.4.7]
- Increased temperatures and demographic change are likely to increase peak energy demand in summer and the associated risk of black-outs. ** D [11.4.10]
- Production from agriculture and forestry by 2030 is projected to decline over much of southern and eastern Australia, and over parts of eastern New Zealand, due to increased drought and fire. However, in New Zealand, initial benefits are projected in western and southern areas and close to major rivers due to a longer growing season, less frost and increased rainfall. ** N [11.4]
- In the south and west of New Zealand, growth rates of economically important plantation crops (mainly *Pinus radiata*) are likely to increase with CO_2-fertilisation, warmer winters and wetter conditions. ** D [11.4.4]
- Increased heat-related deaths for people aged over 65 are likely, with an extra 3,200 to 5,200 deaths on average per year by 2050 (allowing for population growth and ageing, but assuming no adaptation). ** D [11.4.11]

Europe

- The probability of an extreme winter precipitation exceeding two standard deviations above normal is expected to increase by up to a factor of five in parts of the UK and northern Europe by the 2080s with a doubling of CO_2. ** D [12.3.1]
- By the 2070s, annual runoff is projected to increase in northern Europe, and decrease by up to 36% in southern Europe, with summer low flows reduced by up to 80% under IS92a. ** D [12.4.1, T12.2]
- The percentage of river-basin area in the severe water stress category (withdrawal/availability higher than 0.4) is expected to increase from 19% today to 34 to 36% by the 2070s. ** D [12.4.1]
- The number of additional people living in water-stressed watersheds in the seventeen western Europe countries is likely to increase from 16 to 44 million based on HadCM3 climate under the A2 and B1 emission scenarios, respectively, by the 2080s. ** D[12.4.1]
- Under A1FI scenarios, by the 2080s an additional 1.6 million people each year are expected to be affected by coastal flooding. ** D [12.4.2]
- By the 2070s, hydropower potential for the whole of Europe is expected to decline by 6%, with strong regional variations from a 20 to 50% decrease in the Mediterranean region to a 15 to 30% increase in northern and eastern Europe. ** D [12.4.8]
- A large percentage of the European flora could become vulnerable, endangered, critically endangered or extinct by the end of the 21st century under a range of SRES scenarios. *** N [12.4.6]
- By 2050, crops are expected to show a northward expansion in area [12.4.7.1]. The greatest increases in climate-related crop yields are expected in northern Europe (e.g., wheat: +2 to +9% by 2020, +8 to +25% by 2050, +10 to +30% by 2080), while the largest reductions are expected in the south (e.g., wheat: +3 to +4% by 2020, –8 to +22% by 2050, –15 to +32% by 2080).*** C [12.4.7]
- Forested area is likely to increase in the north and decrease in the south. A redistribution of tree species is expected, and an elevation of the mountain tree line. Forest-fire risk is virtually certain to greatly increase in southern Europe. ** D [12.4.4]
- Most amphibian (45 to 69%) and reptile (61 to 89%) species are virtually certain to expand their range if dispersal were unlimited. However, if species were unable to disperse, then the range of most species (>97%) would become smaller, especially in the Iberian Peninsula and France. ** N [12.4.6]

- Small Alpine glaciers in different regions will disappear, while larger glaciers will suffer a volume reduction between 30% and 70% by 2050 under a range of emissions scenarios, with concomitant reductions in discharge in spring and summer. *** C [12.4.3]
- Decreased comfort of the Mediterranean region in the summer, and improved comfort in the north and west, could lead to a reduction in Mediterranean summer tourism and an increase in spring and autumn. ** D [12.4.9]
- Rapid shutdown of Meridional Overturning Circulation (MOC), although assigned a low probability, is likely to have widespread severe impacts in Europe, especially in western coastal areas. These include reductions in crop production with associated price increases, increased cold-related deaths, winter transport disruption, population migration to southern Europe and a shift in the economic centre of gravity. * N [12.6.2]

Latin America

- Over the next 15 years, inter-tropical glaciers are very likely to disappear, reducing water availability and hydropower generation in Bolivia, Peru, Colombia and Ecuador. *** C [13.2.4]
- Any future reductions in rainfall in arid and semi-arid regions of Argentina, Chile and Brazil are likely to lead to severe water shortages. ** C [13.4.3]
- By the 2020s between 7 million and 77 million people are likely to suffer from a lack of adequate water supplies, while for the second half of the century the potential water availability reduction and the increasing demand, from an increasing regional population, would increase these figures to between 60 and 150 million. ** D [13.ES, 13.4.3]
- In the future, anthropogenic climate change (including changes in weather extremes) and sea-level rise are very likely to have impacts on ** N [13.4.4]:
    - low-lying areas (e.g., in El Salvador, Guyana, the coast of Buenos Aires Province in Argentina);
    - buildings and tourism (e.g., in Mexico and Uruguay);
    - coastal morphology (e.g., in Peru);
    - mangroves (e.g., in Brazil, Ecuador, Colombia, Venezuela);
    - availability of drinking water in the Pacific coast of Costa Rica and Ecuador.
- Sea surface temperature increases due to climate change are projected to have adverse effects on ** N [13.4.4]:
    - Mesoamerican coral reefs (e.g., Mexico, Belize, Panama);
    - the location of fish stocks in the south-east Pacific (e.g., Peru and Chile).
- Increases of 2°C and decreases in soil water would lead to a replacement of tropical forest by savannas in eastern Amazonia and in the tropical forests of central and southern Mexico, along with replacement of semi-arid by arid vegetation in parts of north-east Brazil and most of central and northern Mexico. ** D [13.4.1]
- In the future, the frequency and intensity of hurricanes in the Caribbean Basin are likely to increase. * D [13.3.1]
- As a result of climate change, rice yields are expected to decline after the year 2020, while increases in temperature and precipitation in south-eastern South America are likely to increase soybean yields if CO_2 effects are considered. * C [13.4.2]
- The number of additional people at risk of hunger under the SRES A2 emissions scenario is likely to attain 5, 26 and 85 million in 2020, 2050 and 2080, respectively, assuming little or no CO_2 effects. * D [13.4.2]
- Cattle productivity is very likely to decline in response to a 4°C increase in temperatures. ** N [13.ES, 13.4.2]
- The Latin American region, concerned with the potential effects of climate variability and change, is trying to implement some adaptation measures such as:
    - the use of climate forecasts in sectors such as fisheries (Peru) and agriculture (Peru, north-eastern Brazil);
    - early-warning systems for flood in the Rio de la Plata Basin based on the 'Centro Operativo de Alerta Hidrológico'.
- The region has also created new institutions to mitigate and prevent impacts from natural hazards, such as the Regional Disaster Information Center for Latin America and the Caribbean, the International Centre for Research on El Niño Phenomenon in Ecuador, and the Permanent Commission of the South Pacific. *** D [13.2.5]

North America

- Population growth, rising property values and continued investment increase coastal vulnerability. Any increase in destructiveness of coastal storms is very likely to lead to dramatic increases in losses from severe weather and storm surge, with the losses exacerbated by sea-level rise. Current adaptation is uneven, and readiness for increased exposure is poor. *** D [14.2.3, 14.4.3]
- Sea-level rise and the associated increase in tidal surge and flooding have the potential to severely affect transportation and infrastructure along the Gulf, Atlantic and northern coasts. A case study of facilities at risk in New York identified surface road and rail lines, bridges, tunnels, marine and airport facilities and transit stations. *** D [14.4.3, 14.4.6, 14.5.1, B14.3]
- Severe heatwaves, characterised by stagnant, warm air masses and consecutive nights with high minimum temperatures, are likely to increase in number, magnitude and duration in cities where they already occur, with potential for adverse health effects. Elderly populations are most at risk. ** D [14.4.5]

- By mid-century, daily average ozone levels are projected to increase by 3.7 ppb across the eastern USA, with the most polluted cities today experiencing the greatest increases. Ozone-related deaths are projected to increase by 4.5% from the 1990s to the 2050s. * D [14.4.5]
- Projected warming in the western mountains by the mid-21st century is very likely to cause large decreases in snowpack, earlier snow melt, more winter rain events, increased peak winter flows and flooding, and reduced summer flows *** D [14.4.1].
- Reduced water supplies coupled with increases in demand are likely to exacerbate competition for over-allocated water resources. *** D [14.2.1, B14.2]
- Climate change in the first several decades of the 21st century is likely to increase forest production, but with high sensitivity to drought, storms, insects and other disturbances. ** D [14.4.2, 14.4.4]
- Moderate climate change in the early decades of the century is projected to increase aggregate yields of rain-fed agriculture by 5 to 20%, but with important variability among regions. Major challenges are projected for crops that are near the warm end of their suitable range or which depend on highly utilised water resources. ** D [14.4]
- By the second half of the 21st century, the greatest impacts on forests are likely to be through changing disturbances from pests, diseases and fire. Warmer summer temperatures are projected to extend the annual window of high fire risk by 10 to 30%, and increase area burned by 74 to 118% in Canada by 2100. *** D [14.4.4, B14.1]
- Present rates of coastal wetland loss are projected to increase with accelerated relative sea-level rise, in part due to structures preventing landward migration. Salt-marsh biodiversity is expected to decrease in north-eastern marshes. ** D [14.4.3]
- Vulnerability to climate change is likely be concentrated in specific groups and regions, including indigenous peoples and others dependent on narrow resource bases, and the poor and elderly in cities. ** D [14.2.6, 14.4.6]
- Continued investment in adaptation in response to historical experience rather than projected future conditions is likely to increase vulnerability of many sectors to climate change [14.5]. Infrastructure development, with its long lead times and investments, would benefit from incorporating climate-change information. *** D [14.5.3, F14.3]

Polar Regions

- By the end of the century, annually averaged Arctic sea-ice extent is projected to show a reduction of 22 to 33%, depending on emissions scenario; and in Antarctica, projections range from a slight increase to a near-complete loss of summer sea ice. ** D [15.3.3]
- Over the next hundred years there will important reductions in thickness and extent of ice from Arctic glaciers and ice caps, and the Greenland ice sheet ***, as a direct response to climate warming; in Antarctica, losses from the Antarctic Peninsula glaciers will continue ***, and observed thinning in part of the West Antarctic ice sheet, which is probably driven by oceanic change, will continue **. These contributions will form a substantial fraction of sea-level rise during this century. *** D [15.3.4, 15.6.3; WGI AR4 Chapters 4, 5]
- Northern Hemisphere permafrost extent is projected to decrease by 20 to 35% by 2050. The depth of seasonal thawing is likely to increase by 15 to 25% in most areas by 2050, and by 50% or more in northernmost locations under the full range of SRES scenarios. ** D [15.3.4]
- In the Arctic, initial permafrost thaw will alter drainage systems, allowing establishment of aquatic communities in areas formerly dominated by terrestrial species ***. Further thawing will increasingly couple surface drainage to the groundwater, further disrupting ecosystems. Coastal erosion will increase. ** D [15.4.1]
- By the end of the century, 10 to 50% of Arctic tundra will be replaced by forest, and around 15 to 25% of polar desert will be replaced by tundra. * D [15.4.2]
- In both polar regions, climate change will lead to decreases in habitat (including sea ice) for migratory birds and mammals [15.2.2, 15.4.1], with major implications for predators such as seals and polar bears ** [15.2, 15.4.3]. Changes in the distribution and abundance of many species can be expected. *** D [15.6.3]
- The climatic barriers that have hitherto protected polar species from competition will be lowered, and the encroachment of alien species into parts of the Arctic and Antarctic are expected. ** D [15.6.3, 15.4.4, 15.4.2]
- Reductions in lake and river ice cover are expected in both polar regions. These will affect lake thermal structures, the quality/quantity of under-ice habitats and, in the Arctic, the timing and severity of ice jamming and related flooding. *** N [15.4.1]
- Projected hydrological changes will influence the productivity and distribution of aquatic species, especially fish. Warming of freshwaters is likely to lead to reductions in fish stock, especially those that prefer colder waters. ** D [15.4.1]
- For Arctic human communities, it is virtually certain that there will be both negative and positive impacts, particularly through changing cryospheric components, on infrastructure and traditional indigenous ways of life. ** D [15.4]
- In Siberia and North America, there may be an increase in agriculture and forestry as the northern limit for these activities shifts by several hundred kilometres by 2050 [15.4.2]. This will benefit some communities and disadvantage others following traditional lifestyles. ** D [15.4.6]

- Large-scale forest fires and outbreaks of tree-killing insects, which are triggered by warm weather, are characteristic of the boreal forest and some forest tundra areas, and are likely to increase. ** N [15.4.2]
- Arctic warming will reduce excess winter mortality, primarily through a reduction in cardiovascular and respiratory deaths and in injuries. *** N [15.4.6]
- Arctic warming will be associated with increased vulnerability to pests and diseases in wildlife, such as tick-borne encephalitis, which can be transmitted to humans. ** N [15.4.6]
- Increases in the frequency and severity of Arctic flooding, erosion, drought and destruction of permafrost, threaten community, public health and industrial infrastructure and water supply. *** N [15.4.6]
- Changes in the frequency, type and timing of precipitation will increase contaminant capture and increase contaminant loading to Arctic freshwater systems. Increased loadings will more than offset the reductions that are expected to accrue from global emissions of contaminants. ** N [15.4.1]
- Arctic human communities are already being required to adapt to climate change. Impacts to food security, personal safety and subsistence activities are being responded to via changes in resource and wildlife management regimes and shifts in personal behaviours (e.g., hunting, travelling). In combination with demographic, socio-economic and lifestyle changes, the resilience of indigenous populations is being severely challenged. *** N [15.4.1, 15.4.2, 15.4.6, 15.6]

Small Islands

- Sea-level rise and increased sea-water temperature are projected to accelerate beach erosion, and cause degradation of natural coastal defences such as mangroves and coral reefs. It is likely that these changes would, in turn, negatively impact the attraction of small islands as premier tourism destinations. According to surveys, it is likely that, in some islands, up to 80% of tourists would be unwilling to return for the same holiday price in the event of coral bleaching and reduced beach area resulting from elevated sea surface temperatures and sea-level rise. ** D [16.4.6]
- Port facilities at Suva, Fiji, and Apia, Samoa, are likely to experience overtopping, damage to wharves and flooding of the hinterland following a 0.5 m rise in sea level combined with waves associated with a 1 in 50-year cyclone. *** D [16.4.7]
- International airports on small islands are mostly sited on or within a few kilometres of the coast, and the main (and often only) road network runs along the coast. Under sea-level rise scenarios, many of them are likely to be at serious risk from inundation, flooding and physical damage associated with coastal inundation and erosion. *** D [16.4.7]
- Coastal erosion on Arctic islands has additional climate sensitivity through the impact of warming on permafrost and massive ground ice, which can lead to accelerated erosion and volume loss, and the potential for higher wave energy. *** D [16.4.2]
- Reduction in average rainfall is very likely to reduce the size of the freshwater lens. A 10% reduction in average rainfall by 2050 is likely to correspond to a 20% reduction in the size of the freshwater lens on Tarawa Atoll, Kiribati. In general, a reduction in physical size resulting from land loss accompanying sea-level rise could reduce the thickness of the freshwater lens on atolls by as much as 29%. *** N [16.4.1]
- Without adaptation, agricultural economic costs from climate change are likely to reach between 2-3% and 17-18% of 2002 GDP by 2050, on high terrain (e.g., Fiji) and low terrain (e.g., Kiribati) islands, respectively, under SRES A2 (1.3°C increase by 2050) and B2 (0.9°C increase by 2050). ** N [16.4.3]
- With climate change, increased numbers of introductions and enhanced colonisation by alien species are likely to occur on mid- and high-latitude islands. These changes are already evident on some islands. For example, in species-poor sub-Antarctic island ecosystems, alien microbes, fungi, plants and animals have been causing a substantial loss of local biodiversity and changes to ecosystem function. ** N [16.4.4]
- Outbreaks of climate-sensitive diseases such as malaria, dengue, filariasis and schistosomiasis can be costly in lives and economic impacts. Increasing temperatures and decreasing water availability due to climate change is likely to increase burdens of diarrhoeal and other infectious diseases in some small-island states. ** D [16.4.5]
- Climate change is expected to have significant impacts on tourism destination selection ** D [16.4.6]. Several small-island countries (e.g., Barbados, Maldives, Seychelles, Tuvalu) have begun to invest in the implementation of adaptation strategies, including desalination, to offset current and projected water shortages. *** D [16.4.1]
- Studies so far conducted on adaptation on islands suggest that adaptation options are likely to be limited and the costs high relative to GDP. Recent work has shown that, in the case of Singapore, coastal protection would be the least-cost strategy to combat sea-level rise under three scenarios, with the cost ranging from US$0.3-5.7 million by 2050 to US$0.9-16.8 million by 2100. ** D [16.5.2]
- Although adaptation choices for small islands may be limited and adaptation costs high, exploratory research indicates that there are some co-benefits which can be generated from pursuing prudent adaptation strategies. For example, the use of waste-to-energy and other renewable energy systems can promote sustainable development, while strengthening resilience to climate change. In fact, many islands have already embarked on initiatives aimed at ensuring that renewables constitute a significant percentage of the energy mix. ** D [16.4.7, 16.6]

TS.4.3 Magnitudes of impact for varying amounts of climate change

Magnitudes of impact can now be estimated more systematically for a range of possible increases in global average temperature.

Since the IPCC Third Assessment, many additional studies, particularly in regions that previously had been little researched, have enabled a more systematic understanding of how the timing and magnitude of impacts is likely to be affected by changes in climate and sea level associated with differing amounts and rates of change in global average temperature.

Examples of this new information are presented in Tables TS.3 and TS.4. Entries have been selected which are judged to be relevant for people and the environment and for which there is at least medium confidence in the assessment. All entries of impact are drawn from chapters of the Assessment, where more detailed information is available. Depending on circumstances, some of these impacts could be associated with 'key vulnerabilities', based on a number of criteria in the literature (magnitude, timing, persistence/reversibility, the potential for adaptation, distributional aspects, likelihood and 'importance' of the impacts). Assessment of potential key vulnerabilities is intended to provide information on rates and levels of climate change to help decision-makers make appropriate responses to the risks of climate change [19.ES, 19.1].

TS.4.4 The impact of altered extremes

Impacts are very likely to increase due to increased frequencies and intensities of extreme weather events.

Since the IPCC Third Assessment, confidence has increased that some weather events and extremes will become more frequent, more widespread or more intense during the 21st century; and more is known about the potential effects of such changes. These are summarised in Table TS.5.

TS.4.5 Especially affected systems, sectors and regions

Some systems, sectors and regions are likely to be especially affected by climate change.

Regarding systems and sectors, these are as follows.

- Some ecosystems especially
  - terrestrial: tundra, boreal forest, mountain, mediterranean-type ecosystems;
  - along coasts: mangroves and salt marshes;
  - in oceans: coral reefs and the sea-ice biomes.

  [4.ES, 4.4, 6.4]
- Low-lying coasts, due to the threat of sea-level rise [6.ES].
- Water resources in mid-latitude and dry low-latitude regions, due to decreases in rainfall and higher rates of evapotranspiration [3.4].
- Agriculture in low-latitude regions, due to reduced water availability [5.4, 5.3].
- Human health, especially in areas with low adaptive capacity [8.3].

Regarding regions, these are as follows.

- The Arctic, because of high rates of projected warming on natural systems [15.3].
- Africa, especially the sub-Saharan region, because of current low adaptive capacity as well as climate change [9.ES, 9.5].
- Small islands, due to high exposure of population and infrastructure to risk of sea-level rise and increased storm surge [16.1, 16.2].
- Asian megadeltas, such as the Ganges-Brahmaputra and the Zhujiang, due to large populations and high exposure to sea-level rise, storm surge and river flooding [T10.9, 10.6].

Within other areas, even those with high incomes, some people can be particularly at risk (such as the poor, young children and the elderly) and also some areas and some activities [7.1, 7.2, 7.4].

TS.4.6 Events with large impacts

Some large-scale climate events have the potential to cause very large impacts, especially after the 21st century.

Very large sea-level rises that would result from widespread deglaciation of Greenland and West Antarctic ice sheets imply major changes in coastlines and ecosystems, and inundation of low-lying areas, with the greatest effects in river deltas. Relocating populations, economic activity and infrastructure would be costly and challenging. There is medium confidence that at least partial deglaciation of the Greenland ice sheet, and possibly the West Antarctic ice sheet, would occur over a period of time ranging from centuries to millennia for a global average temperature increase of 1-4°C (relative to 1990-2000), causing a contribution to sea-level rise of 4-6 m or more. The complete melting of the Greenland ice sheet and the West Antarctic ice sheet would lead to a contribution to sea-level rise of up to 7 m and about 5 m, respectively [WGI AR4 6.4, 10.7; WGII AR4 19.3].

Based on climate model results, it is very unlikely that the Meridional Overturning Circulation (MOC) in the North Atlantic will undergo a large abrupt transition during the 21st century. Slowing of the MOC this century is very likely, but temperatures over the Atlantic and Europe are projected to increase nevertheless, due to global warming. Impacts of large-scale and persistent changes in the MOC are likely to include changes to marine ecosystem productivity, fisheries, ocean CO_2 uptake, oceanic oxygen concentrations and terrestrial vegetation [WGI AR4 10.3, 10.7; WGII AR4 12.6, 19.3].

TS.4.7 Costing the impacts of climate change

Impacts of unmitigated climate change will vary regionally. Aggregated and discounted to the present, they are very likely to impose costs, even though specific estimates are uncertain and should therefore be interpreted very carefully. These costs are very likely to increase over time.

This Assessment (see Tables TS.3 and TS.4) makes it clear that the impacts of future climate change will be mixed across regions. For increases in global mean temperature of less than 1-3°C above 1990 levels, some impacts are projected to produce benefits in some places and some sectors, and produce costs in other places and other sectors. It is, however, projected that some low-latitude and polar regions will experience net costs even for small increases in temperature. It is very likely that all regions will experience either declines in net benefits or increases in net costs for increases in temperature greater than about 2-3°C [9.ES, 9.5, 10.6, T10.9, 15.3, 15.ES]. These observations confirm evidence reported in the Third Assessment that, while developing countries are expected to experience larger percentage losses, global mean losses could be 1-5% of GDP for 4°C of warming [F20.3].

Many estimates of aggregate net economic costs of damages from climate change across the globe (i.e., the social cost of carbon (SCC), expressed in terms of future net benefits and costs that are discounted to the present) are now available. Peer-reviewed estimates of the SCC for 2005 have an average value of US$43 per tonne of carbon (i.e., US$12 per tonne of CO_2) but the range around this mean is large. For example, in a survey of 100 estimates, the values ranged from –US$10 per tonne of carbon (–US$3 per tonne of CO_2) up to US$350 per tonne of carbon (US$95 per tonne of CO_2) [20.6].

The large ranges of SCC are due in large part to differences in assumptions regarding climate sensitivity, response lags, the treatment of risk and equity, economic and non-economic impacts, the inclusion of potentially catastrophic losses, and discount rates. It is very likely that globally aggregated figures underestimate the damage costs because they cannot include many non-quantifiable impacts. Taken as a whole, the range of published evidence indicates that the net damage costs of climate change are likely to be significant and to increase over time [T20.3, 20.6, F20.4].

It is virtually certain that aggregate estimates of costs mask significant differences in impacts across sectors, regions, countries, and populations. In some locations and amongst some groups of people with high exposure, high sensitivity, and/or low adaptive capacity, net costs will be significantly larger than the global aggregate [20.6, 20.ES, 7.4].

TS.5 Current knowledge about responding to climate change

TS.5.1 Adaptation

Some adaptation is occurring now, to observed and projected future climate change, but on a very limited basis.

Societies have a long record of adapting to the impacts of weather and climate through a range of practices that include crop diversification, irrigation, water management, disaster risk management and insurance. But climate change poses novel risks which are often outside the range of experience, such as impacts related to drought, heatwaves, accelerated glacier retreat and hurricane intensity [17.2.1].

There is growing evidence since the TAR that adaptation measures that also consider climate change are being implemented, on a limited basis, in both developed and developing countries. These measures are undertaken by a range of public and private actors through policies, investments in infrastructure and technologies, and behavioural change.

Examples of adaptations to observed changes in climate include:

- partial drainage of the Tsho Rolpa glacial lake (Nepal);
- changes in livelihood strategies in response to permafrost melt by the Inuit in Nunavut (Canada);
- increased use of artificial snow-making by the Alpine ski industry (Europe, Australia and North America);
- coastal defences in the Maldives and the Netherlands;
- water management in Australia;
- government responses to heatwaves in, for example, some European countries.

[7.6, 8.2, 8.6, 17.ES, 16.5, 1.5]

However, all of the adaptations documented were imposed by the climate risk and involve real cost and reduction of welfare in the first instance [17.2.3]. These examples also confirm the observations of attributable climate signals in the impacts of change.

A limited but growing set of adaptation measures also explicitly considers scenarios of future climate change. Examples include consideration of sea-level rise in the design of infrastructure such as the Confederation Bridge in Canada and a coastal highway in Micronesia, as well as in shoreline management policies and flood risk measures, for example in Maine (USA) and the Thames Barrier (UK) [17.2.2].

Adaptation measures are seldom undertaken in response to climate change alone.

Many actions that facilitate adaptation to climate change are undertaken to deal with current extreme events such as heatwaves and cyclones. Often, planned adaptation initiatives are also not undertaken as stand-alone measures, but embedded within broader sectoral initiatives such as water-resource planning, coastal defence, and risk reduction strategies [17.2.2, 17.3.3]. Examples include consideration of climate change in the National Water Plan of Bangladesh, and the design of flood protection and cyclone-resistant infrastructure in Tonga [17.2.2].

Adaptation will be necessary to address impacts resulting from the warming which is already unavoidable due to past emissions.

Past emissions are estimated to involve some unavoidable warming (about a further 0.6°C by the end of the century relative to 1980-1999) even if atmospheric greenhouse gas concentrations remain at 2000 levels (see WGI AR4). There are some impacts for which adaptation is the only available and

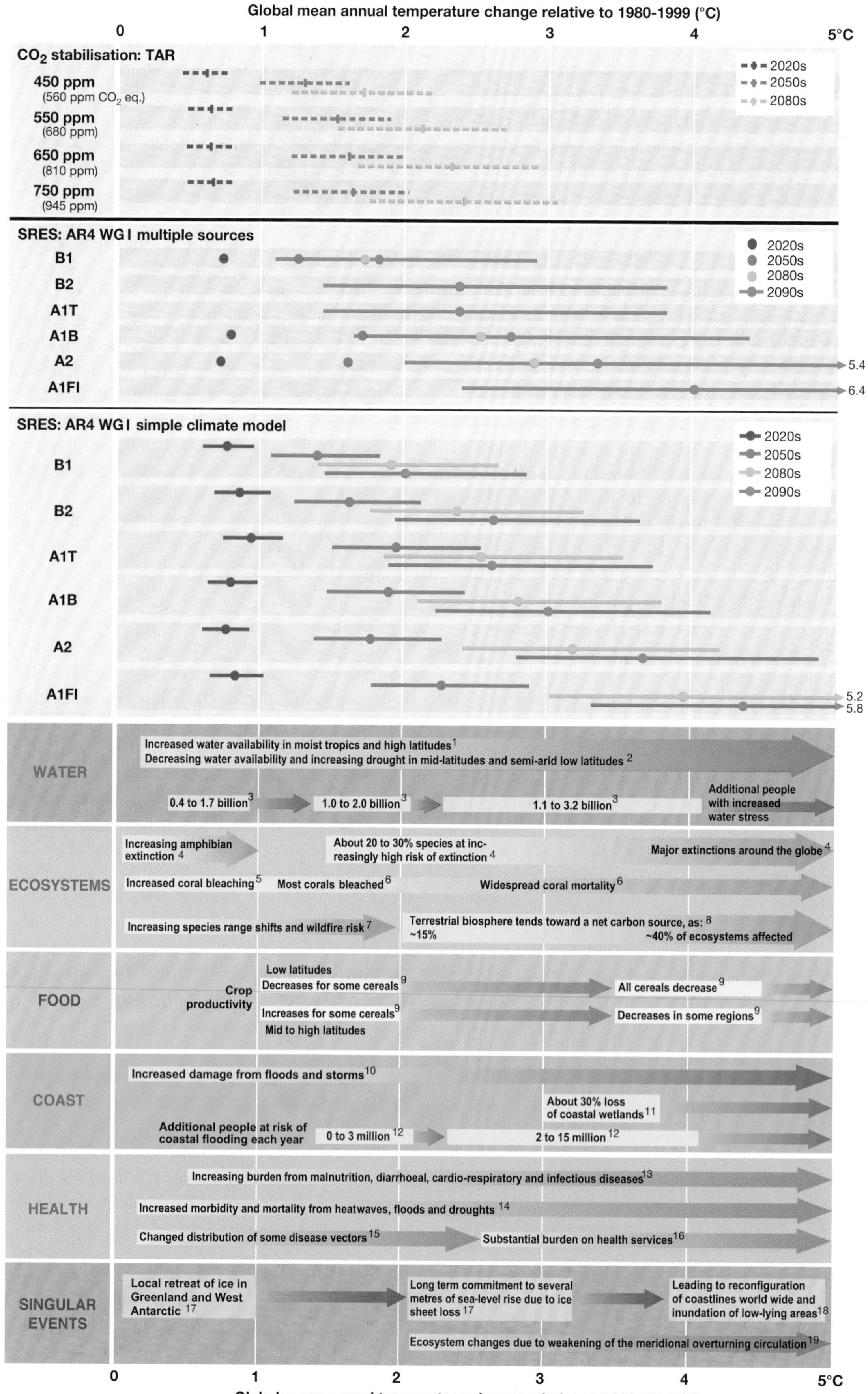

Table TS.3. *Examples of global impacts projected for changes in climate (and sea level and atmospheric CO_2 where relevant) associated with different amounts of increase in global average surface temperature in the 21st century [T20.8]. This is a selection of some estimates currently available. All entries are from published studies in the chapters of the Assessment. (Continues below Table TS.4.)*

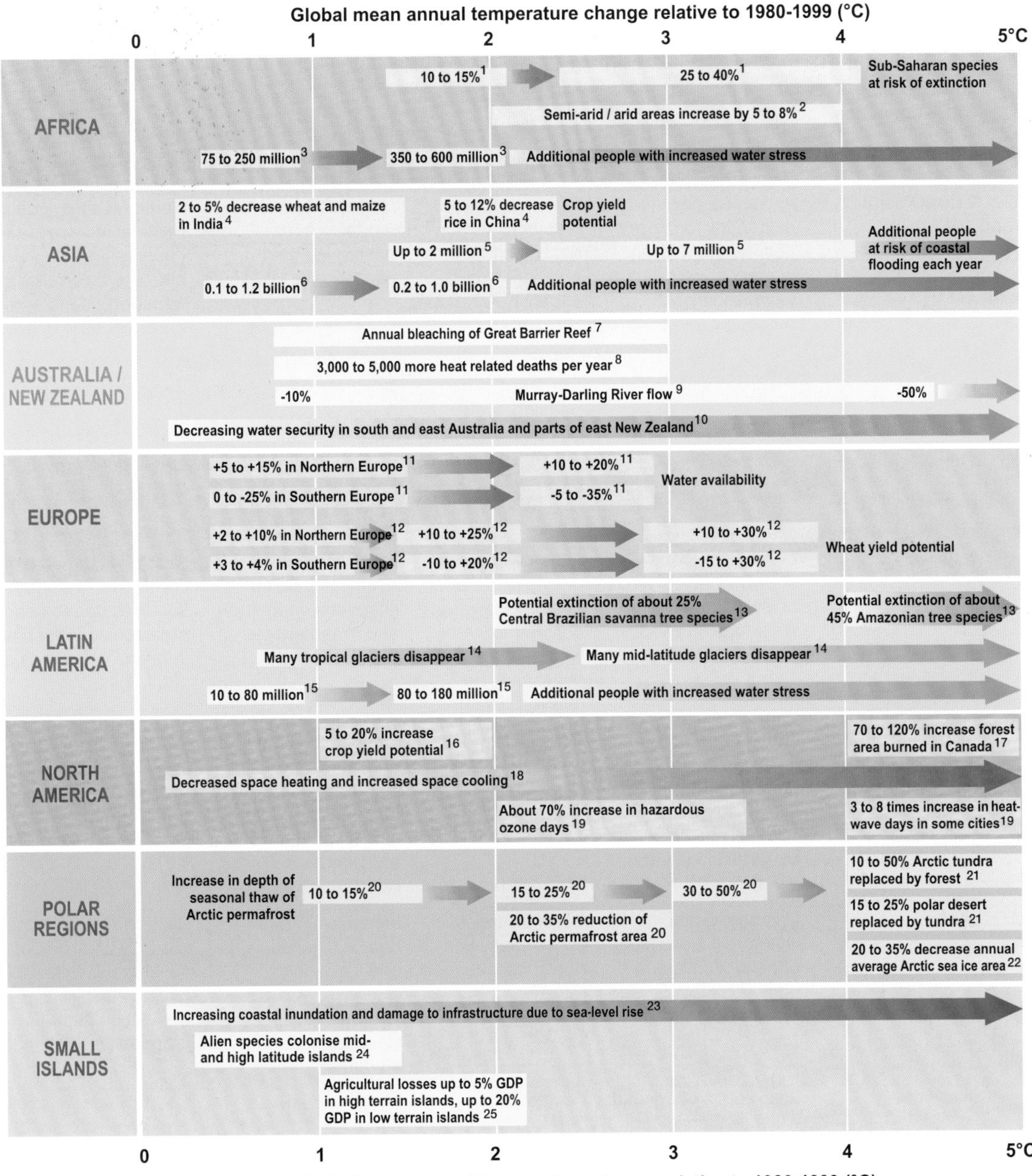

Table TS.4. *Examples of regional impacts [T20.9]. See caption for Table TS.3.*

Table TS.3. (cont.) *Edges of boxes and placing of text indicate the range of temperature change to which the impacts relate. Arrows between boxes indicate increasing levels of impacts between estimations. Other arrows indicate trends in impacts. All entries for water stress and flooding represent the additional impacts of climate change relative to the conditions projected across the range of SRES scenarios A1FI, A2, B1 and B2. Adaptation to climate change is not included in these estimations. For extinctions, 'major' means ~40 to ~70% of assessed species.*

The table also shows global temperature changes for selected time periods, relative to 1980-1999, projected for SRES and stabilisation scenarios. To express the temperature change relative to 1850-1899, add 0.5°C. More detail is provided in Chapter 2 [Box 2.8]. Estimates are for the 2020s, 2050s and 2080s, (the time periods used by the IPCC Data Distribution Centre and therefore in many impact studies) and for the 2090s. SRES-based projections are shown using two different approaches. ***Middle panel:*** *projections from the WGI AR4 SPM based on multiple sources. Best estimates are based on AOGCMs (coloured dots). Uncertainty ranges, available only for the 2090s, are based on models, observational constraints and expert judgement.* ***Lower panel:*** *best estimates and uncertainty ranges based on a simple climate model (SCM), also from WGI AR4 (Chapter 10).* ***Upper panel:*** *best estimates and uncertainty ranges for four CO_2-stabilisation scenarios using an SCM. Results are from the TAR because comparable projections for the 21st century are not available in the AR4. However, estimates of equilibrium warming are reported in the WGI AR4 for CO_2-equivalent stabilisation[18]. Note that equilibrium temperatures would not be reached until decades or centuries after greenhouse gas stabilisation.*

Table TS.3. Sources: 1, 3.4.1; **2,** 3.4.1, 3.4.3; **3,** 3.5.1; **4,** 4.4.11; **5,** 4.4.9, 4.4.11, 6.2.5, 6.4.1; **6,** 4.4.9, 4.4.11, 6.4.1; **7,** 4.2.2, 4.4.1, 4.4.4 to 4.4.6, 4.4.10; **8,** 4.4.1, 4.4.11; **9,** 5.4.2; **10,** 6.3.2, 6.4.1, 6.4.2; **11,** 6.4.1; **12,** 6.4.2; **13,** 8.4, 8.7; **14,** 8.2, 8.4, 8.7; **15,** 8.2, 8.4, 8.7; **16,** 8.6.1; **17,** 19.3.1; **18,** 19.3.1, 19.3.5; **19,** 19.3.5

Table TS.4. Sources: 1, 9.4.5; **2,** 9.4.4; **3,** 9.4.1; **4,** 10.4.1; **5,** 6.4.2; **6,** 10.4.2; **7,** 11.6; **8,** 11.4.12; **9,** 11.4.1, 11.4.12; **10,** 11.4.1, 11.4.12; **11,** 12.4.1; **12,** 12.4.7; **13,** 13.4.1; **14,** 13.2.4; **15,** 13.4.3; **16,** 14.4.4; **17,** 5.4.5, 14.4.4; **18,** 14.4.8; **19,** 14.4.5; **20,** 15.3.4, **21,** 15.4.2; **22,** 15.3.3; **23,** 16.4.7; **24,** 16.4.4; **25,** 16.4.3

[18] Best estimate and likely range of equilibrium warming for seven levels of CO_2-equivalent stabilisation from WGI AR4 are: 350 ppm, 1.0°C [0.6–1.4]; 450 ppm, 2.1°C [1.4–3.1]; 550 ppm, 2.9°C [1.9–4.4]; 650 ppm, 3.6°C [2.4–5.5]; 750 ppm, 4.3°C [2.8–6.4]; 1,000 ppm, 5.5°C [3.7–8.3] and 1,200 ppm, 6.3°C [4.2–9.4].

Phenomenon[a] and direction of trend	Likelihood of future trends based on projections for 21st century using SRES scenarios	Examples of major projected impacts by sector			
		Agriculture, forestry and ecosystems	Water resources	Human health	Industry, settlements and society
Over most land areas, warmer and fewer cold days and nights, warmer and more frequent hot days and nights	Virtually certain[b]	Increased yields in colder environments; decreased yields in warmer environments; increased insect outbreaks [5.8.1, 4.4.5]	Effects on water resources relying on snow melt; effects on some water supply [3.4.1, 3.5.1]	Reduced human mortality from decreased cold exposure [8.4.1, T8.3]	Reduced energy demand for heating; increased demand for cooling; declining air quality in cities; reduced disruption to transport due to snow, ice; effects on winter tourism [7.4.2, 14.4.8, 15.7.1]
Warm spells/ heatwaves. Frequency increases over most land areas	Very likely	Reduced yields in warmer regions due to heat stress; wildfire danger increase [5.8.1, 5.4.5, 4.4.3, 4.4.4]	Increased water demand; water quality problems, e.g., algal blooms [3.4.2, 3.5.1, 3.4.4]	Increased risk of heat-related mortality, especially for the elderly, chronically sick, very young and socially isolated [8.4.2, T8.3, 8.4.1]	Reduction in quality of life for people in warm areas without appropriate housing; impacts on elderly, very young and poor [7.4.2, 8.2.1]
Heavy precipitation events. Frequency increases over most areas	Very likely	Damage to crops; soil erosion, inability to cultivate land due to waterlogging of soils [5.4.2]	Adverse effects on quality of surface and groundwater; contamination of water supply; water stress may be relieved [3.4.4]	Increased risk of deaths, injuries, infectious, respiratory and skin diseases [8.2.2, 11.4.11]	Disruption of settlements, commerce, transport and societies due to flooding; pressures on urban and rural infrastructures; loss of property [T7.4, 7.4.2]
Area affected by drought increases	Likely	Land degradation, lower yields/crop damage and failure; increased livestock deaths; increased risk of wildfire [5.8.1, 5.4, 4.4.4]	More widespread water stress [3.5.1]	Increased risk of food and water shortage; increased risk of malnutrition; increased risk of water- and food-borne diseases [5.4.7, 8.2.3, 8.2.5]	Water shortages for settlements, industry and societies; reduced hydropower generation potentials; potential for population migration [T7.4, 7.4, 7.1.3]
Intense tropical cyclone activity increases	Likely	Damage to crops; windthrow (uprooting) of trees; damage to coral reefs [5.4.5, 16.4.3]	Power outages cause disruption of public water supply [7.4.2]	Increased risk of deaths, injuries, water- and food-borne diseases; post-traumatic stress disorders [8.2.2, 8.4.2, 16.4.5]	Disruption by flood and high winds; withdrawal of risk coverage in vulnerable areas by private insurers, potential for population migrations, loss of property [7.4.1, 7.4.2, 7.1.3]
Increased incidence of extreme high sea level (excludes tsunamis)[c]	Likely[d]	Salinisation of irrigation water, estuaries and freshwater systems [3.4.2, 3.4.4, 10.4.2]	Decreased freshwater availability due to salt-water intrusion [3.4.2, 3.4.4]	Increased risk of deaths and injuries by drowning in floods; migration-related health effects [6.4.2, 8.2.2, 8.4.2]	Costs of coastal protection versus costs of land-use relocation; potential for movement of populations and infrastructure; also see tropical cyclones above [7.4.2]

[a] See WGI AR4 Table 3.7 for further details regarding definitions.
[b] Warming of the most extreme days and nights each year.
[c] Extreme high sea level depends on average sea level and on regional weather systems. It is defined as the highest 1% of hourly values of observed sea level at a station for a given reference period.
[d] In all scenarios, the projected global average sea level at 2100 is higher than in the reference period [WGI AR4 10.6]. The effect of changes in regional weather systems on sea-level extremes has not been assessed.

Table TS.5. *Examples of possible impacts of climate change due to changes in extreme weather and climate events, based on projections to the mid- to late 21st century. These do not take into account any changes or developments in adaptive capacity. Examples of all entries are to be found in chapters in the full Assessment (see sources). The first two columns of this table (shaded yellow) are taken directly from the Working Group I Fourth Assessment (Table SPM.2). The likelihood estimates in column 2 relate to the phenomena listed in column 1. The direction of trend and likelihood of phenomena are for SRES projections of climate change.*

appropriate response. An indication of these impacts can be seen in Tables TS.3 and TS.4.

Many adaptations can be implemented at low cost, but comprehensive estimates of adaptation costs and benefits are currently lacking.

There are a growing number of adaptation cost and benefit-cost estimates at regional and project level for sea-level rise, agriculture, energy demand for heating and cooling, water-resource management, and infrastructure. These studies identify a number of measures that can be implemented at low cost or with high benefit-cost ratios. However, some common adaptations may have social and environmental externalities. Adaptations to heatwaves, for example, have involved increased demand for energy-intensive air-conditioning [17.2.3].

Limited estimates are also available for global adaptation costs related to sea-level rise, and energy expenditures for space heating and cooling. Estimates of global adaptation benefits for the agricultural sector are also available, although such literature does not explicitly consider the costs of adaptation. Comprehensive multi-sectoral estimates of global costs and benefits of adaptation are currently lacking [17.2.3].

Adaptive capacity is uneven across and within societies.

There are individuals and groups within all societies that have insufficient capacity to adapt to climate change. For example, women in subsistence farming communities are disproportionately burdened with the costs of recovery and coping with drought in southern Africa [17.3.2].

The capacity to adapt is dynamic and influenced by economic and natural resources, social networks, entitlements, institutions and governance, human resources, and technology [17.3.3]. For example, research in the Caribbean on hurricane preparedness shows that appropriate legislation is a necessary prior condition to implementing plans for adaptation to future climate change [17.3].

Multiple stresses related to HIV/AIDS, land degradation, trends in economic globalisation, trade barriers and violent conflict affect exposure to climate risks and the capacity to adapt. For example, farming communities in India are exposed to impacts of import competition and lower prices in addition to climate risks; and marine ecosystems over-exploited by globalised fisheries have been shown to be less resilient to climate variability and change (see Box TS.7) [17.3.3].

High adaptive capacity does not necessarily translate into actions that reduce vulnerability. For example, despite a high capacity to adapt to heat stress through relatively inexpensive adaptations, residents in urban areas in some parts of the world, including in European cities, continue to experience high levels of mortality. One example is the 2003 European heatwave-related deaths. Another example is Hurricane Katrina, which hit the Gulf of Mexico Coast and New Orleans in 2005 and caused the deaths of more than 1,000 people, together with very high economic and social costs [17.4.2].

A wide array of adaptation options is available, but more extensive adaptation than is currently occurring is required to reduce vulnerability to future climate change. There are barriers, limits and costs, but these are not fully understood.

The array of potential adaptive responses available to human societies is very large (see Table TS.6), ranging from purely technological (e.g., sea defences), through behavioural (e.g., altered food and recreational choices), to managerial (e.g., altered farm practices) and to policy (e.g., planning regulations). While most technologies and strategies are known and developed in some countries, the assessed literature does not indicate how effective various options are at fully reducing risks, particularly at higher levels of warming and related impacts, and for vulnerable groups.

Although many early impacts of climate change can be effectively addressed through adaptation, the options for successful adaptation diminish and the associated costs increase with increasing climate change. At present we do not have a clear picture of the limits to adaptation, or the cost, partly because effective adaptation measures are highly dependent on specific geographical and climate risk factors as well as institutional, political and financial constraints [7.6, 17.2, 17.4]. There are significant barriers to implementing adaptation. These include both the inability of natural systems to adapt to the rate and magnitude of climate change, as well as formidable environmental, economic, informational, social, attitudinal and behavioural constraints. There are also significant knowledge gaps for adaptation as well as impediments to flows of knowledge and information relevant for adaptation decisions [17.4.1, 17.4.2]. For developing countries, availability of resources and building adaptive capacity are particularly important [see Sections 5 and 6 in Chapters 3 to 16; also 17.2, 17.4]. Some examples and reasons are given below.

a. The large number and expansion of potentially hazardous glacial lakes due to rising temperatures in the Himalayas. These far exceed the capacity of countries in the region to manage such risks.
b. If climate change is faster than is anticipated, many developing countries simply cannot cope with more frequent/intense occurrence of extreme weather events, as this will drain resources budgeted for other purposes.
c. Climate change will occur in the life cycle of many infrastructure projects (coastal dykes, bridges, sea ports, etc.). Strengthening of these infrastructures based on new design criteria may take decades to implement. In many cases, retrofitting would not be possible.
d. Due to physical constraints, adaptation measures cannot be implemented in many estuaries and delta areas.

New planning processes are attempting to overcome these barriers at local, regional and national levels in both developing and developed countries. For example, Least Developed Countries are developing National Adaptation Plans of Action (NAPA) and some developed countries have established national adaptation policy frameworks [17.4.1].

TS.5.2 Interrelationships between adaptation and mitigation

Both adaptation and mitigation can help to reduce the risks of climate change to nature and society.
However, their effects vary over time and place. Mitigation will have global benefits but, owing to the lag times in the climate and biophysical systems, these will hardly be noticeable until around the middle of the 21st century [WGI AR4 SPM]. The benefits of adaptation are largely local to regional in scale but they can be immediate, especially if they also address vulnerabilities to current climate conditions [18.1.1, 18.5.2]. Given these differences between adaptation and mitigation, climate policy is not about making a choice between adapting to and mitigating climate change. If key vulnerabilities to climate change are to be addressed, adaptation is necessary because even the most stringent mitigation efforts cannot avoid further climate change in the next few decades. Mitigation is necessary because reliance on adaptation alone could eventually lead to a magnitude of climate change to which effective adaptation is possible only at very high social, environmental and economic costs [18.4, 18.6].

Many impacts can be avoided, reduced or delayed by mitigation.
A small number of impact assessments have now been completed for scenarios in which future atmospheric concentrations of greenhouse gases are stabilised. Although these studies do not take full account of uncertainties in projected climate under stabilisation – for example, the

	Food, fibre and forestry	Water resources	Human health	Industry, settlement and society
Drying/ Drought	*Crops*: development of new drought-resistant varieties; intercropping; crop residue retention; weed management; irrigation and hydroponic farming; water harvesting *Livestock*: supplementary feeding; change in stocking rate; altered grazing and rotation of pasture *Social*: Improved extension services; debt relief; diversification of income	Leak reduction Water demand management through metering and pricing Soil moisture conservation e.g., through mulching Desalination of sea water Conservation of groundwater through artificial recharge Education for sustainable water use	Grain storage and provision of emergency feeding stations Provision of safe drinking water and sanitation Strengthening of public institutions and health systems Access to international food markets	Improve adaptation capacities, especially for livelihoods Incorporate climate change in development programmes Improved water supply systems and co-ordination between jurisdictions
Increased rainfall/ Flooding	*Crops*: Polders and improved drainage; development and promotion of alternative crops; adjustment of plantation and harvesting schedule; floating agricultural systems *Social*: Improved extension services	Enhanced implementation of protection measures including flood forecasting and warning, regulation through planning legislation and zoning; promotion of insurance; and relocation of vulnerable assets	Structural and non-structural measures. Early-warning systems; disaster preparedness planning; effective post-event emergency relief	Improved flood protection infrastructure "Flood-proof" buildings Change land use in high-risk areas Managed realignment and "Making Space for Water" Flood hazard mapping; flood warnings Empower community institutions
Warming/ Heatwaves	*Crops*: Development of new heat-resistant varieties; altered timing of cropping activities; pest control and surveillance of crops *Livestock*: Housing and shade provision; change to heat-tolerant breeds *Forestry*: Fire management through altered stand layout, landscape planning, dead timber salvaging, clearing undergrowth. Insect control through prescribed burning, non-chemical pest control *Social*: Diversification of income	Water demand management through metering and pricing Education for sustainable water use	International surveillance systems for disease emergence Strengthening of public institutions and health systems National and regional heat warning systems Measures to reduce urban heat island effects through creating green spaces Adjusting clothing and activity levels; increasing fluid intake	Assistance programmes for especially vulnerable groups Improve adaptive capacities Technological change
Wind speed/ Storminess	*Crops*: Development of wind-resistant crops (e.g., vanilla)	Coastal defence design and implementation to protect water supply against contamination	Early-warning systems; disaster preparedness planning; effective post-event emergency relief	Emergency preparedness, including early-warning systems More resilient infrastructure Financial risk management options for both developed and developing regions

Table TS.6. *Examples of current and potential options for adapting to climate change for vulnerable sectors. All entries have been referred to in chapters in the Fourth Assessment. Note that, with respect to ecosystems, generic rather than specific adaptation responses are required. Generic planning strategies would enhance the capacity to adapt naturally. Examples of such strategies are: enhanced wildlife corridors, including wide altitudinal gradients in protected areas. [5.5, 3.5, 6.5, 7.5, T6.5]* sensitivity of climate models to forcing – they nevertheless provide indications of damages avoided or vulnerabilities and risks reduced for different amounts of emissions reduction [2.4, T20.6].

In addition, more quantitative information is now available concerning when, over a range of temperature increases, given amounts of impact may occur. This allows inference of the amounts of global temperature increase that are associated with given impacts. Table TS.3 illustrates the change in global average temperature projected for three periods (2020s, 2050s, 2080s) for several alternative stabilisation pathways and for emissions trends assumed under different SRES scenarios. Reference to Tables TS.3 and TS.4 provides a picture of the impacts which might be avoided for given ranges of temperature change.

A portfolio of adaptation and mitigation measures can diminish the risks associated with climate change.
Even the most stringent mitigation efforts cannot avoid further impacts of climate change in the next few decades, which makes adaptation essential, particularly in addressing near-term impacts. Unmitigated climate change would, in the long term, be likely to exceed the capacity of natural, managed and human systems to adapt [20.7].

This suggests the value of a portfolio or mix of strategies that includes mitigation, adaptation, technological development (to enhance both adaptation and mitigation) and research (on climate science, impacts, adaptation and mitigation). Such portfolios could combine policies with incentive-based approaches and actions at all levels from the individual citizen through to national governments and international organisations [18.1, 18.5].

These actions include technological, institutional and behavioural options, the introduction of economic and policy instruments to encourage the use of these options, and research and development to reduce uncertainty and to enhance the options' effectiveness and efficiency [18.4.1, 18.4.2]. Many different actors are involved in the implementation of these actions, operating on different spatial and institutional scales. Mitigation primarily involves the energy, transportation, industrial, residential, forestry and agriculture sectors, whereas the actors involved in adaptation represent a large variety of sectoral interests, including agriculture, tourism and recreation, human health, water supply, coastal management, urban planning and nature conservation [18.5, 18.6].

Box TS.7. Adaptive capacity to multiple stressors in India

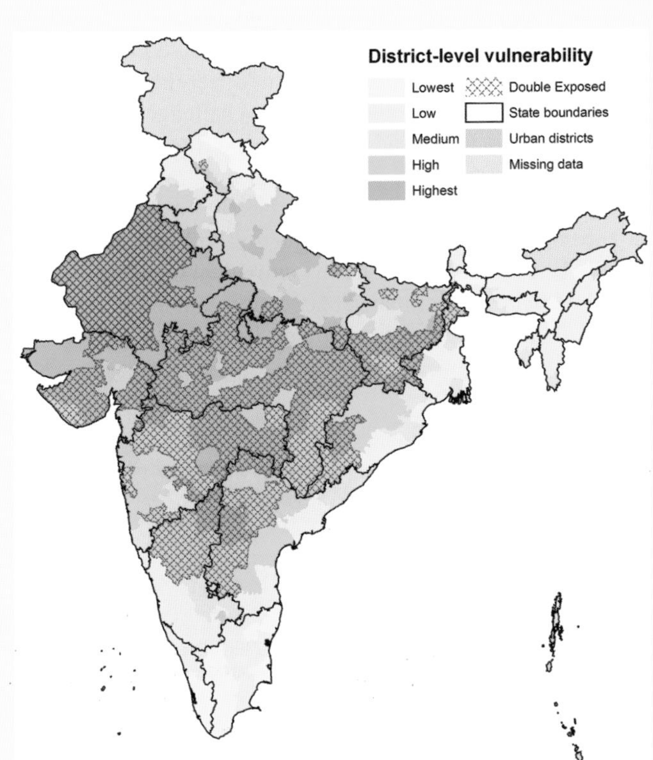

Figure TS.17. *Districts in India that rank highest in terms of (a) vulnerability to climate change and (b) import competition associated with economic globalisation, are considered to be double-exposed (depicted with hatching). From O'Brien et al. (2004) [F17.2].*

The capacity to adapt to climate change is not evenly distributed across or within nations. In India, for example, both climate change and trade liberalisation are changing the context for agricultural production. Some farmers are able to adapt to these changing conditions, including discrete events such as drought and rapid changes in commodity prices, but others are not. Identifying the areas where both processes are likely to have negative outcomes provides a first step in identifying options and constraints in adapting to changing conditions [17.3.2].

Figure TS.17 shows regional vulnerability to climate change, measured as a composite of adaptive capacity and climate sensitivity under exposure to climate change. The superimposed hatching indicates those areas which are doubly exposed through high vulnerability to climate change and high vulnerability to trade liberalisation. The results of this mapping show higher degrees of resilience in districts located along the Indo-Gangetic Plains (except in the state of Bihar), the south and east, and lower resilience in the interior parts of the country, particularly in the states of Bihar, Rajasthan, Madhya Pradesh, Maharashtra, Andhra Pradesh and Karnataka [17.3.2].

One way of increasing adaptive capacity is by introducing the consideration of climate change impacts in development planning [18.7], for example, by:

- including adaptation measures in land-use planning and infrastructure design [17.2];
- including measures to reduce vulnerability in existing disaster risk reduction strategies [17.2, 20.8].

Decisions on adaptation and mitigation are taken at a range of different levels.
These levels include individual households and farmers, private firms and national planning agencies. Effective mitigation requires the participation of the bulk of major greenhouse gas emitters globally, whereas most adaptation takes place at local and national levels. The benefits of mitigation are global, whilst its costs and ancillary benefits arise locally. Both the costs and benefits of adaptation mostly accrue locally [18.1.1, 18.4.2]. Consequently, mitigation is primarily driven by international agreements and the ensuing national public policies, whereas most adaptation is driven by private actions of affected entities and public arrangements of impacted communities [18.1.1, 18.6.1].

Interrelationships between adaptation and mitigation can exist at each level of decision-making.
Adaptation actions can have (often unintended) positive or negative mitigation effects, whilst mitigation actions can have (also often unintended) positive or negative adaptation effects [18.4.2, 18.5.2]. An example of an adaptation action with a negative mitigation effect is the use of air-conditioning (if the required energy is provided by fossil fuels). An example of a mitigation action with a positive adaptation effect could be the afforestation of degraded hill slopes, which would not only sequester carbon but also control soil erosion. Other examples of such synergies between adaptation and mitigation include rural electrification based on renewable energy sources, planting trees in cities to reduce the heat-island effect, and the development of agroforestry systems [18.5.2].

Analysis of the interrelationships between adaptation and mitigation may reveal ways to promote the effective implementation of adaptation and mitigation actions.
Creating synergies between adaptation and mitigation can increase the cost-effectiveness of actions and make them more attractive to potential funders and other decision-makers (see Table TS.7). However, synergies provide no guarantee that resources are used in the most efficient manner when seeking to reduce the risks of climate change. Moreover, essential actions without synergetic effects may be overlooked if the creation of synergies becomes a dominant decision criterion [18.6.1]. Opportunities for synergies exist in some sectors (e.g., agriculture, forestry, buildings and urban infrastructure) but they are rather limited in many other climate-relevant sectors [18.5.2]. A lack of both conceptual and empirical information that explicitly considers both adaptation and mitigation makes it difficult to assess the need for, and potential of synergies in, climate policy [18.7].

Decisions on trade-offs between the immediate localised benefits of adaptation and the longer-term global benefits of mitigation would require information on the actions' costs and benefits over time.
For example, a relevant question would be whether or not investment in adaptation would buy time for mitigation. Global integrated assessment models provide approximate estimates of relative costs and benefits at highly aggregated levels. Intricacies of the interrelationships between adaptation and mitigation become apparent at the more detailed analytical and implementation levels [18.4.2]. These intricacies, including the fact that adaptation and mitigation operate on different spatial, temporal and institutional scales and involve different actors who have different interests and different beliefs, value systems and property rights, present a challenge to the practical implementation of trade-offs beyond the local scale. In particular the notion of an "optimal mix" of adaptation and mitigation is problematic, since it usually assumes that there is a zero-sum budget for adaptation and mitigation and that it would be possible to capture the individual interests of all who will be affected by climate change, now and in the future, into a global aggregate measure of well-being [18.4.2, 18.6.1].

Scale	Adaptation → Mitigation	Mitigation → Adaptation	Parallel decisions affecting adaptation and mitigation	Adaptation and mitigation trade-offs and synergies
Global/policy	Awareness of limits to adaptation motivates mitigation e.g., policy lobbying by ENGOs	CDM trades provide funds for adaptation through surcharges	Allocation of MEA funds or Special Climate Change Fund	Assessment of costs and benefits in adaptation and mitigation in setting targets for stabilisation
Regional/natural strategy/sectoral planning	Watershed planning (e.g., hydroelectricity) and land cover, affect greenhouse gas emissions	Fossil fuel tax increases the cost of adaptation through higher energy prices	National capacity, e.g., self-assessment, supports adaptation and mitigation in policy integration	Testing project sensitivity to mitigation policy, social cost of carbon and climate impacts
Local/biophysical community and individual actions	Increased use of air-conditioning (homes, offices, transport) raises greenhouse gas emissions	Community carbon sequestration affects livelihoods	Local planning authorities implement criteria related to both adaptation and mitigation in land-use planning	Corporate integrated assessment of exposure to mitigation policy and climate impacts

Table TS.7. *Relationships between adaptation and mitigation [F18.2]. ENGO = Environmental Non-Governmental Organisation; CDM = Clean Development Mechanism; MEA = Millennium Ecosystem Assessment.*

People's capacities to adapt and mitigate are driven by similar sets of factors.

These factors represent a generalised response capacity that can be mobilised in the service of either adaptation or mitigation. Response capacity, in turn, is dependent on the societal development pathway. Enhancing society's response capacity through the pursuit of sustainable development pathways is therefore one way of promoting both adaptation and mitigation [18.3]. This would facilitate the effective implementation of both options, as well as their mainstreaming into sectoral planning and development. If climate policy and sustainable development are to be pursued in an integrated way, then it will be important not simply to evaluate specific policy options that might accomplish both goals, but also to explore the determinants of response capacity that underlie those options as they relate to underlying socio-economic and technological development paths [18.3, 18.6.3].

TS.5.3 Key vulnerabilities

Key vulnerabilities are found in many social, economic, biological and geophysical systems.

Vulnerability to climate change is the degree to which geophysical, biological and socio-economic systems are susceptible to, and unable to cope with, adverse impacts of climate change. The term "vulnerability" may therefore refer to the vulnerable system itself (e.g., low-lying islands or coastal cities), the impact to this system (e.g., flooding of coastal cities and agricultural lands or forced migration), or the mechanism causing these impacts (e.g., disintegration of the West Antarctic ice sheet). Based on a number of criteria in the literature (i.e., magnitude, timing, persistence/reversibility, potential for adaptation, distributional aspects, likelihood and 'importance' of the impacts [19.2]), some of these vulnerabilities might be identified as 'key'. Key impacts and resultant key vulnerabilities are found in many social, economic, biological and geophysical systems [19.1.1].

The identification of potential key vulnerabilities is intended to provide guidance to decision-makers for identifying levels and rates of climate change that may be associated with 'dangerous anthropogenic interference' (DAI) with the climate system, in the terminology of the UNFCCC (United Nations Framework Convention on Climate Change) Article 2 [B19.1]. Ultimately, the determination of DAI cannot be based on scientific arguments alone, but involves other judgements informed by the state of scientific knowledge [19.1.1]. Table TS.8 presents an illustrative and selected list of key vulnerabilities.

Key vulnerabilities may be linked to systemic thresholds where non-linear processes cause a system to shift from one major state to another (such as a hypothetical sudden change in the Asian monsoon or disintegration of the West Antarctic ice sheet or positive feedbacks from ecosystems switching from a sink to a source of CO_2). Other key vulnerabilities can be associated with "normative thresholds" defined by stakeholders or decision-makers (e.g., a magnitude of sea-level rise no longer considered acceptable by low-lying coastal dwellers) [19.1.2].

Increasing levels of climate change will result in impacts associated with an increasing number of key vulnerabilities, and some key vulnerabilities have been associated with observed climate change.

Observed climate change to 2006 has been associated with some impacts that can be linked to key vulnerabilities. Among these are increases in human mortality during extreme weather events, and increasing problems associated with permafrost melting, glacier retreat and sea-level rise [19.3.2, 19.3.3, 19.3.4, 19.3.5, 19.3.6].

Global mean temperature changes of up to 2°C above 1990-2000 levels would exacerbate current key vulnerabilities, such as those listed above (high confidence), and cause others, such as reduced food security in many low-latitude nations (medium confidence). At the same time, some systems such as global agricultural productivity at mid- and high-latitudes, could benefit (medium confidence) [19.3.1, 19.3.2, 19.3.3].

Global mean temperature changes of 2 to 4°C above 1990-2000 levels would result in an increasing number of key impacts at all scales (high confidence), such as widespread loss of biodiversity, decreasing global agricultural productivity and commitment to widespread deglaciation of Greenland (high confidence) and West Antarctic (medium confidence) ice sheets [19.3.1, 19.3.4, 19.3.5].

Global mean temperature changes greater than 4°C above 1990-2000 levels would lead to major increases in vulnerability (very high confidence), exceeding the adaptive capacity of many systems (very high confidence) [19.3.1].

Regions already at high risk from observed climate variability and climate change are more likely to be adversely affected in the near future, due to projected changes in climate and increases in the magnitude and/or frequency of already damaging extreme events [19.3.6, 19.4.1].

The "reasons for concern" identified in the Third Assessment remain a viable framework to consider key vulnerabilities. Recent research has updated some of the findings from the Third Assessment.

Unique and threatened systems

There is new and much stronger evidence of the adverse impacts of observed climate change to date on several unique and threatened systems. Confidence has increased that a 1 to 2°C increase in global mean temperature above 1990 levels poses significant risks to many unique and threatened systems, including many biodiversity hotspots [19.3.7].

Extreme events

There is new evidence that observed climate change has likely already increased the risk of certain extreme events such as heatwaves, and it is more likely than not that warming has contributed to intensification of some tropical cyclones, with increasing levels of adverse impacts as temperatures increase [19.3.7].

Key systems or groups at risk	Prime criteria for 'key vulnerability'	Global average temperature change above 1990 0°C	1°C	2°C	3°C	4°C	5°C
Global social systems							
Food supply	Distribution, magnitude		Productivity decreases for some cereals in low latitudes ** Productivity increases for some cereals in mid/high latitudes **	Global production potential increases to around 3°C, decreases above this * **a**	Cereal productivity decreases in some mid/high latitude regions **		
Aggregate market impacts and distribution	Magnitude, distribution	Net benefits in many high latitudes; net costs in many low latitudes * **b**		Benefits decrease, while costs increase. Net global cost * **b**			
Regional system							
Small islands	Irreversibility, magnitude, distribution, low adaptive capacity	Increasing coastal inundation and damage to infrastructure due to sea-level rise **					
Indigenous, poor or isolated communities	Irreversibility, distribution, timing, low adaptive capacity	Some communities already affected ** **c**	Climate change and sea-level rise adds to other stresses **. Communities in low-lying coastal and arid areas are especially threatened ** **d**				
Global biological systems							
Terrestrial ecosystems and biodiversity	Irreversibility, magnitude, low adaptive capacity, persistence, rate of change, confidence	Many ecosystems already affected ***	*c.* 20-30% species at increasingly high risk of extinction *	Terrestrial biosphere tends toward a net carbon source **	Major extinctions around the globe **		
Marine ecosystems and biodiversity	Irreversibility, magnitude, low adaptive capacity, persistence, rate of change, confidence	Increased coral bleaching **	Most corals bleached **	Widespread coral mortality **			
Geophysical systems							
Greenland ice sheet	Magnitude, irreversibility, low adaptive capacity, confidence	Localised deglaciation (already observed due to local warming), extent would increase with temperature *** **e**		Commitment to widespread ** or near-total * deglaciation, 2-7 m sea-level rise[19] over centuries to millennia * **e**		Near-total deglaciation ** **e**	
Meridional Overturning Circulation	Magnitude, persistence, distribution, timing, adaptive capacity, confidence	Variations including regional weakening (already observed but no trend identified) **f**		Considerable weakening **. Commitment to large-scale and persistent change including possible cooling in northern high-latitude areas near Greenland and north-west Europe •, highly dependent on rate of climate change.			
Risks from extreme events							
Tropical cyclone intensity	Magnitude, timing, distribution	Increase in Cat. 4-5 storms */**, with impacts exacerbated by sea-level rise		Further increase in tropical cyclone intensity */**			
Drought	Magnitude, timing	Drought already increasing * **g** Increasing frequency / intensity drought in mid-latitude continental areas ** **h**		Extreme drought increasing from 1% land area to 30% (A2 scenario) * **i** Mid-latitude regions affected by poleward migration of Annular Modes seriously affected ** **j**			

Table TS.8. *Table of selected key vulnerabilities. The key vulnerabilities range from those associated with societal systems, for which the adaptation potential is the greatest, to those associated with biophysical systems, which are likely to have the least adaptive capacity. Adaptation potential for key vulnerabilities resulting from extreme events is associated with the affected systems, most of which are socio-economic. Information is presented where available on how impacts may change at larger increases in global mean temperature (GMT). All increases in GMT are relative to circa 1990. Most impacts are the result of changes in climate, weather and/or sea level, not of temperature alone. In many cases climate change impacts are marginal or synergistic on top of other existing and possibly increasing stresses. Criteria for key vulnerabilities are given in Section TS 5.3. For full details refer to the corresponding text in Chapter 19. Confidence symbol legend: *** very high confidence, ** high confidence, * medium confidence, • low confidence.*
Sources for left hand column are T19.1. Sources for right hand column are T19.1, and are also found in Tables TS.3 and TS.4, with the exception of: ***a****: 5.4.2, 5.6;* ***b****: 20.6, 20.7;* ***c****: 1.3, 11.4.8, 14.2.3, 15.4.5;* ***d****: 3.4, 6.4, 11.4;* ***e****: 19.3.5, T19.1;* ***f****: 19.3.5, 12.6;* ***g****: 1.3.2, 1.3.3, T19.1;* ***h****: WGI 10.3.6.1;* ***i****: WGI AR4 10.3.6.1;* ***j****: WGI AR4 10.3.5.6.*

[19] Range combines results from modelling and analysis of palaeo data.

Distribution of impacts

There is still high confidence that the distribution of climate impacts will be uneven, and that low-latitude, less-developed areas are generally at greatest risk. However, recent work has shown that vulnerability to climate change is also highly variable within individual countries. As a consequence, some population groups in developed countries are also highly vulnerable [19.3.7].

Aggregate impacts

There is some evidence that initial net market benefits from climate change will peak at a lower magnitude and sooner than was assumed in the Third Assessment, and that it is likely there will be higher damages for larger magnitudes of global mean temperature increases than estimated in the Third Assessment. Climate change could adversely affect hundreds of millions of people through increased risk of coastal flooding, reduction in water supplies, increased risk of malnutrition, and increased risk of exposure to climate-dependent diseases [19.3.7].

Large-scale singularities

Since the Third Assessment, the literature offers more specific guidance on possible thresholds for partial or near-complete deglaciation of Greenland and West Antarctic ice sheets. There is medium confidence that at least partial deglaciation of the Greenland ice sheet, and possibly the West Antarctic ice sheet, would occur over a period of time ranging from centuries to millennia for a global average temperature increase of 1-4°C (relative to 1990-2000), causing a contribution to sea-level rise of 4-6 m or more [WGI AR4 6.4, 10.7.4.3, 10.7.4.4; 19.3.5.2].

TS.5.4 Perspectives on climate change and sustainability

Future vulnerability depends not only on climate change but also on development pathway.

An important advance since the Third Assessment has been the completion of impacts studies for a range of different development pathways, taking into account not only projected climate change but also projected social and economic changes. Most have been based on characterisations of population and income levels drawn from the SRES scenarios [2.4].

These studies show that the projected impacts of climate change can vary greatly due to the development pathway assumed. For example, there may be large differences in regional population, income and technological development under alternative scenarios, which are often a strong determinant of the level of vulnerability to climate change [2.4].

To illustrate, Figure TS.18 shows estimates from a recent study of the number of people projected to be at risk of coastal flooding each year under different assumptions of socio-economic development. This indicates that the projected number of people affected is considerably greater under the A2-type scenario of development (characterised by relatively low per capita income and large population growth) than under other SRES futures [T20.6]. This difference is largely explained, not by differences in changes of climate, but by differences in vulnerability [T6.6].

Vulnerability to climate change can be exacerbated by the presence of other stresses.

Non-climate stresses can increase vulnerability to climate change by reducing resilience and can also reduce adaptive capacity because of resource deployment to competing needs. For example, current stresses on some coral reefs include marine pollution and chemical runoff from agriculture as well as increases in water temperature and ocean acidification. Vulnerable regions face multiple stresses that affect their exposure and sensitivity as well as their capacity to adapt. These stresses arise from, for example, current climate hazards, poverty and unequal access to resources, food insecurity, trends in economic globalisation, conflict, and incidence of disease such as HIV/AIDS [7.4, 8.3, 17.3, 20.3].

Climate change itself can produce its own set of multiple stresses in some locations because the physical manifestations of the impacts of climate change are so diverse [9.4.8]. For example, more variable rainfall implies more frequent droughts and more frequent episodes of intense rainfall, whilst sea-level rise may bring coastal flooding to areas already experiencing more frequent wind storm. In such cases, total vulnerability to climate change is greater than the sum of the vulnerabilities to specific impacts considered one at a time in isolation (very high confidence) [20.7.2].

Climate change will very likely impede nations' abilities to achieve sustainable development pathways, as measured, for example, as long-term progress towards the Millennium Development Goals.

Following the lead of the TAR, this Report has adopted the Bruntland Commission definition of sustainable development: "development that meets the needs of the present without compromising the ability of future generations to meet their own needs". Over the next half-century, it is very likely that climate

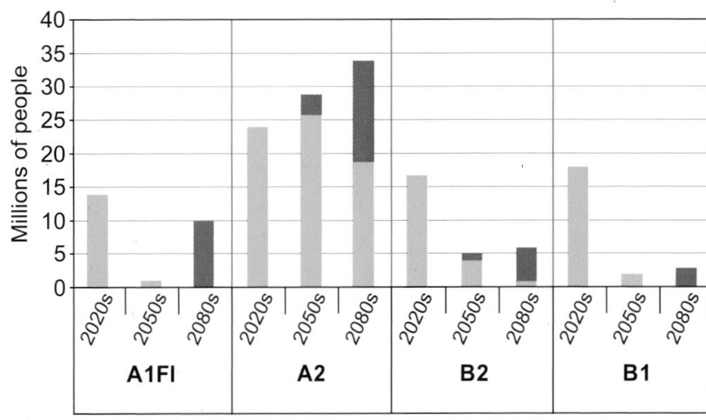

Figure TS.18. *Results from a recent study showing estimated millions of people per annum at risk globally from coastal flooding. Blue bars: numbers at risk without sea-level rise; purple bars: numbers at risk with sea-level rise. [T6.6]*

change will make sustainable development more difficult, particularly as measured by their progress toward achieving Millennium Development Goals for the middle of the century. Climate change will erode nations' capacities to achieve the Goals, calibrated in terms of reducing poverty and otherwise improving equity by 2050, particularly in Africa and parts of Asia (very high confidence) [20.7.1].

Even though there are cases where climate-related extreme events have severely interfered with economic development, it is very unlikely that climate change attributed to anthropogenic sources, per se, will be a significant extra impediment to most nations' reaching their 2015 Millennium Development targets. Many other obstacles with more immediate impacts stand in the way [20.7.1].

Sustainable development can reduce vulnerability to climate change by encouraging adaptation, enhancing adaptive capacity and increasing resilience (very high confidence) [20.3.3]. On the other hand, it is very likely that climate change can slow the pace of progress toward sustainable development either directly through increased exposure to adverse impact or indirectly through erosion of the capacity to adapt. This point is clearly demonstrated in the sections of the sectoral and regional chapters of this Report that discuss implications for sustainable development [see Section 7 in Chapters 3 to 8, 20.3, 20.7]. At present, few plans for promoting sustainability have explicitly included either adapting to climate-change impacts, or promoting adaptive capacity [20.3].

Sustainable development can reduce vulnerability to climate change.

Efforts to cope with the impacts of climate change and attempts to promote sustainable development share common goals and determinants including: access to resources (including information and technology), equity in the distribution of resources, stocks of human and social capital, access to risk-sharing mechanisms and abilities of decision-support mechanisms to cope with uncertainty. Nonetheless, some development activities exacerbate climate-related vulnerabilities (very high confidence).

It is very likely that significant synergies can be exploited in bringing climate change to the development community, and critical development issues to the climate-change community [20.3.3, 20.8.2 and 20.8.3]. Effective communication in assessment, appraisal and action are likely to be important tools both in participatory assessment and governance as well as in identifying productive areas for shared learning initiatives [20.3.3, 20.8.2, 20.8.3]. Despite these synergies, few discussions about promoting sustainability have thus. far explicitly included adapting to climate impacts, reducing hazard risks and/or promoting adaptive capacity [20.4, 20.5, 20.8.3]. Discussions about promoting development and improving environmental quality have seldom explicitly included adapting to climate impacts and/or promoting adaptive capacity [20.8.3]. Most of the scholars and practitioners of development who recognise that climate change is a significant issue at local, national, regional and/or global levels focus their attention almost exclusively on mitigation [20.4, 20.8.3].

Synergies between adaptation and mitigation measures will be effective through the middle of this century, but even a combination of aggressive mitigation and significant investment in adaptive capacity could be overwhelmed by the end of the century along a likely development scenario.

Tables TS.3 and TS.4 track major worldwide impacts for major sectors against temperature increases measured from the 1980 to 1999 period. With very high confidence, no temperature threshold associated with any subjective judgment of what might constitute "dangerous" climate change can be guaranteed to be avoided by anything but the most stringent of mitigation interventions.

As illustrated in Figure TS.19, it is likely that global mitigation efforts designed to cap effective greenhouse gas concentrations at, for example, 550 ppm would benefit developing countries significantly through the middle of this century, regardless of whether the climate sensitivity turns out to be high or low, and especially when combined with enhanced adaptation. Developed countries would also likely see significant benefits from an adaptation-mitigation intervention portfolio, especially for high climate sensitivities and in sectors and regions that are already showing signs of being vulnerable. By 2100, climate change will likely produce significant vulnerabilities across the globe even if aggressive mitigation were implemented in combination with significantly enhanced adaptive capacity [20.7.3].

TS.6 Advances in knowledge and future research needs

TS 6.1 Advances in knowledge

Since the IPCC Third Assessment, the principal advances in knowledge have been as follows.

- Much improved coverage of the impacts of climate change on developing regions, through studies such as the AIACC project (Assessments of Impacts and Adaptations to Climate Change in Multiple Regions and Sectors), although further research is still required, especially in Latin America and Africa [9.ES, 10.ES, 13.ES].
- More studies of adaptation to climate change, with improved understanding of current practice, adaptive capacity, the options, barriers and limits to adaptation [17.ES].
- Much more monitoring of observed effects, and recognition that climate change is having a discernible impact on many natural systems [1.ES, F1.1].
- Some standardisation of the scenarios of future climate change underpinning impact studies, facilitated by centralised data provision through organisations such as the IPCC Data Distribution Centre, thus allowing comparison between sectors and regions [2.2.2].
- Improved understanding of the damages for different levels of global warming, and the link between global warming

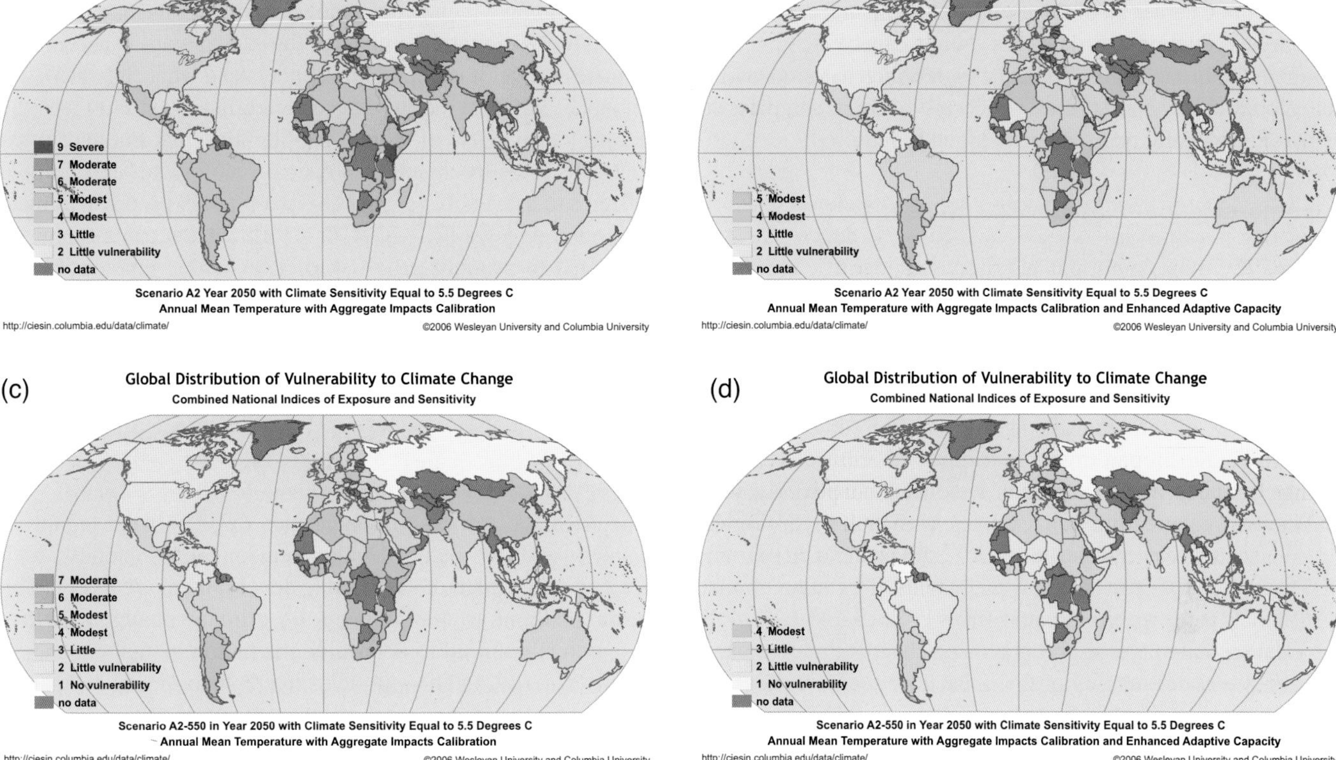

Figure TS.19. *Geographical distribution of vulnerability in 2050 with and without mitigation along an SRES A2 emissions scenario with a climate sensitivity of 5.5°C. Panel (a) portrays vulnerability with a static representation of current adaptive capacity. Panel (b) shows vulnerability with enhanced adaptive capacity worldwide. Panel (c) displays the geographical implications of mitigation designed to cap effective atmospheric concentrations of greenhouse gases at 550 ppm. Panel (d) offers a portrait of the combined complementary effects of mitigation to the same 550 ppmv concentration limit and enhanced adaptive capacity. [F20.6]*

and the probability of stabilising CO_2 at various levels. As a result, we know more about the link between damages and CO_2-stabilisation scenarios [20.7.2, T20.8, T20.9].

However, there has been little advance on:

- impacts under different assumptions about how the world will evolve in future – societies, governance, technology and economic development;
- the costs of climate change, both of the impacts and of response (adaptation and mitigation);
- proximity to thresholds and tipping points;
- impacts resulting from interactions between climate change and other human-induced environmental changes.

TS 6.2 Future research needs

Impacts under different assumptions about future development pathways

Most AR4 studies of future climate change are based on a small number of studies using SRES scenarios, especially the A2 and B2 families [2.3.1]. This has allowed some limited, but incomplete, characterisation of the potential range of futures and their impacts [see Section 4 on key future impacts in all core chapters].

Scenarios are required:

- to describe the future evolution of the world under different and wide-ranging assumptions about how societies, governance, technology, economies will develop in future;
- at the regional and local scales appropriate for impacts analysis;
- which allow adaptation to be incorporated into climate-change impact estimates;
- for abrupt climate change such as the collapse of the North Atlantic Meridional Overturning Circulation, and large sea-level rises due to ice sheet melting [6.8];
- for beyond 2100 (especially for sea-level rise) [6.8, 11.8.1].

Increasingly, climate modellers run model ensembles which allow characterisation of the uncertainty range for each development pathway. Thus, the impacts analyst is faced with very large quantities of data to capture even a small part of the potential range of futures. Tools and techniques to manage these large quantities of data are urgently required [2.3, 2.4].

Damages avoided by different levels of emissions reduction

Very few studies have been carried out to explore the damages avoided, or the impacts postponed, by reducing or stabilising emissions, despite the critical importance of this issue for policy-makers. The few studies which have been performed are reviewed in Chapter 20 of this Report [20.6.2] and show clearly the large reductions in damages which can be achieved by mitigating emissions [T20.4]. Existing research has emphasised the global scale, and studies which are disaggregated to the regional, and even local, scale are urgently required.

Climate-science-related research needs

Two of the most important requirements identified relate to research in climate change science, but have been clearly identified as a hindrance to research in impacts, adaptation and vulnerability.

- The first is that our understanding of the likely future impacts of climate change is hampered by lack of knowledge regarding the nature of future changes, particularly at the regional scale and particularly with respect to precipitation changes and their hydrological consequences on water resources, and changes in extreme events, due in part to the inadequacies of existing climate models at the required spatial scales [T2.5, 3.3.1, 3.4.1, 4.3].
- The second relates to abrupt climate change. Policy-makers require understanding of the impacts of such events as the collapse of the North Atlantic Meridional Overturning Circulation. However, without a better understanding of the likely manifestation of such events at the regional scale, it is not possible to carry out impacts assessments [6.8, 7.6, 8.8, 10.8.3].

Observations, monitoring and attribution

Large-area, long-term field studies are required to evaluate observed impacts of climate change on managed and unmanaged systems and human activities. This will enable improved understanding of where and when impacts become detectable, where the hotspots lie, and why some areas are more vulnerable than others. High-quality observations are essential for full understanding of causes, and for unequivocal attribution of present-day trends to climate change [1.4.3, 4.8].

Timely monitoring of the pace of approaching significant thresholds (such as abrupt climate change thresholds) is required [6.8, 10.8.4].

Multiple stresses, thresholds and vulnerable people and places

It has become clear in the AR4 that the impacts of climate change are most damaging when they occur in the context of multiple stresses arising from the effects, for example, of globalisation, poverty, poor governance and settlement of low-lying coasts. Considerable progress has been made towards understanding which people and which locations may expect to be disproportionately impacted by the negative aspects of climate change. It is important to understand what characteristics enhance vulnerability, what characteristics strengthen the adaptive capacity of some people and places, and what characteristics predispose physical, biological and human systems to irreversible changes as a result of exposure to climate and other stresses [7.1, B7.4, 9.1, 9.ES]. How can systems be managed to minimise the risk of irreversible changes? How close are we to tipping points/thresholds for natural ecosystems such as the Amazon rain forest? What positive feedbacks would emerge if such a tipping point is reached?

Climate change, adaptation and sustainable development

The AR4 recognised that synergies exist between adaptive capacity and sustainable development, and that societies which are pursuing a path of sustainable development are likely to be more resilient to the impacts of climate change. Further research is required to determine the factors which contribute to this synergy, and how policies to enhance adaptive capacity can reinforce sustainable development and vice versa [20.9].

Further understanding of adaptation is likely to require learning-by-doing approaches, where the knowledge base is enhanced through accumulation of practical experience.

The costs of climate change, both the costs of the impacts and of response (adaptation and mitigation)

- Only a small amount of literature on the costs of climate change impacts could be found for assessment [5.6, 6.5.3, 7.5]. Debate still surrounds the topic of how to measure impacts, and which metrics should be used to ensure comparability [2.2.3, 19.3.2.3, 20.9].
- The literature on adaptation costs and benefits is limited and fragmented [17.2.3]. It focuses on sea-level rise and agriculture, with more limited assessments for energy demand, water resources and transport. There is an emphasis on the USA and other OECD countries, with only a few studies for developing countries [17.2.3].

Better understanding of the relative costs of climate change impacts and adaptation allows policy-makers to consider optimal strategies for implementation of adaptation policies, especially the amount and the timing [17.2.3.1].

1

Assessment of observed changes and responses in natural and managed systems

Coordinating Lead Authors:

Cynthia Rosenzweig (USA), Gino Casassa (Chile)

Lead Authors:

David J. Karoly (USA/Australia), Anton Imeson (The Netherlands), Chunzhen Liu (China), Annette Menzel (Germany), Samuel Rawlins (Trinidad and Tobago), Terry L. Root (USA), Bernard Seguin (France), Piotr Tryjanowski (Poland)

Contributing Authors:

Tarekegn Abeku (Ethiopia), Isabelle Côté (Canada), Mark Dyurgerov (USA), Martin Edwards (UK), Kristie L. Ebi (USA), Nicole Estrella (Germany), Donald L. Forbes (Canada), Bernard Francou (France), Andrew Githeko (Kenya), Vivien Gornitz (USA), Wilfried Haeberli (Switzerland), John Hay (New Zealand), Anne Henshaw (USA), Terrence Hughes (Australia), Ana Iglesias (Spain), Georg Kaser (Austria), R. Sari Kovats (UK), Joseph Lam (China), Diana Liverman (UK), Dena P. MacMynowski (USA), Patricia Morellato (Brazil), Jeff T. Price (USA), Robert Muir-Wood (UK), Peter Neofotis (USA), Catherine O'Reilly (USA), Xavier Rodo (Spain), Tim Sparks (UK), Thomas Spencer (UK), David Viner (UK), Marta Vicarelli (Italy), Ellen Wiegandt (Switzerland), Qigang Wu (China), Ma Zhuguo (China)

Review Editors:

Lucka Kajfež-Bogataj (Slovenia), Jan Pretel (Czech Republic), Andrew Watkinson (UK)

This chapter should be cited as:

Rosenzweig, C., G. Casassa, D.J. Karoly, A. Imeson, C. Liu, A. Menzel, S. Rawlins, T.L. Root, B. Seguin, P. Tryjanowski, 2007: Assessment of observed changes and responses in natural and managed systems. *Climate Change 2007: Impacts, Adaptation and Vulnerability. Contribution of Working Group II to the Fourth Assessment Report of the Intergovernmental Panel on Climate Change*, M.L. Parry, O.F. Canziani, J.P. Palutikof, P.J. van der Linden and C.E. Hanson, Eds., Cambridge University Press, Cambridge, UK, 79-131.

Table of Contents

Supplementary material for this chapter is available on the CD-ROM accompanying this report.

Executive summary

Physical and biological systems on all continents and in most oceans are already being affected by recent climate changes, particularly regional temperature increases (very high confidence) [1.3]. Climatic effects on human systems, although more difficult to discern due to adaptation and non-climatic drivers, are emerging (medium confidence) [1.3]. Global-scale assessment of observed changes shows that it is likely that anthropogenic warming over the last three decades has had a discernible influence on many physical and biological systems [1.4].

Attribution of observed regional changes in natural and managed systems to anthropogenic climate change is complicated by the effects of natural climate variability and non-climate drivers (e.g., land-use change) [1.2]. Nevertheless, there have been several joint attribution studies that have linked responses in some physical and biological systems directly to anthropogenic climate change using climate, process and statistical models [1.4.2]. Furthermore, the consistency of observed significant changes in physical and biological systems and observed significant warming across the globe very likely cannot be explained entirely by natural variability or other confounding non-climate factors [1.4.2]. On the basis of this evidence, combined with the likely substantial anthropogenic warming over the past 50 years averaged over each continent except Antarctica (as described in the Working Group I Fourth Assessment Summary for Policymakers), it is likely that there is a discernible influence of anthropogenic warming on many physical and biological systems.

Climate change is strongly affecting many aspects of systems related to snow, ice and frozen ground (including permafrost) [1.3.1]; emerging evidence shows changes in hydrological systems, water resources [1.3.2], coastal zones [1.3.3] and oceans (high confidence) [1.3.4].

Effects due to changes in snow, ice and frozen ground (including permafrost) include ground instability in permafrost regions, a shorter travel season for vehicles over frozen roads in the Arctic, enlargement and increase of glacial lakes in mountain regions and destabilisation of moraines damming these lakes, changes in Arctic and Antarctic Peninsula flora and fauna including the sea-ice biomes and predators higher in the food chain, limitations on mountain sports in lower-elevation alpine areas, and changes in indigenous livelihoods in the Arctic (high confidence). [1.3.1]

The spring peak discharge is occurring earlier in rivers affected by snow melt, and there is evidence for enhanced glacial melt. Lakes and rivers around the world are warming, with effects on thermal structure and water quality (high confidence). [1.3.2]

The effects of sea-level rise, enhanced wave heights, and intensification of storms are found in some coastal regions – including those not modified by humans, e.g., polar areas and barrier beaches – mainly through coastal erosion [1.3.3.1]. Sea-level rise is contributing to losses of coastal wetlands and mangroves, and increased damage from coastal flooding in many areas, although human modification of coasts, such as increased construction in vulnerable zones, plays an important role too (medium confidence). [1.3.3.2]

The uptake of anthropogenic carbon since 1750 has led to the ocean becoming more acidic, with an average decrease in pH of 0.1 units. However, the effects of recent ocean acidification on the marine biosphere are as yet undocumented. [1.3.4]

More evidence from a wider range of species and communities in terrestrial ecosystems and substantial new evidence in marine and freshwater systems show that recent warming is strongly affecting natural biological systems (very high confidence). [1.3.5, 1.3.4]

The overwhelming majority of studies of regional climate effects on terrestrial species reveal consistent responses to warming trends, including poleward and elevational range shifts of flora and fauna. Responses of terrestrial species to warming across the Northern Hemisphere are well documented by changes in the timing of growth stages (i.e., phenological changes), especially the earlier onset of spring events, migration, and lengthening of the growing season. Changes in abundance of certain species, including limited evidence of a few local disappearances, and changes in community composition over the last few decades have been attributed to climate change (very high confidence). [1.3.5]

Many observed changes in phenology and distribution of marine species have been associated with rising water temperatures, as well as other climate-driven changes in salinity, oxygen levels, and circulation. For example, plankton has moved poleward by 10° latitude over a period of four decades in the North Atlantic. While there is increasing evidence for climate change impacts on coral reefs, separating the impacts of climate-related stresses from other stresses (e.g., over-fishing and pollution) is difficult. Warming of lakes and rivers is affecting abundance and productivity, community composition, phenology, distribution and migration of freshwater species (high confidence). [1.3.4]

Although responses to recent climate changes in human systems are difficult to identify due to multiple non-climate driving forces and the presence of adaptation, effects have been detected in forestry and a few agricultural systems [1.3.6]. Changes in several aspects of the human health system have been related to recent warming [1.3.7]. Adaptation to recent warming is beginning to be systematically documented (medium confidence) [1.3.9].

In comparison with other factors, recent warming has been of limited consequence in agriculture and forestry. A significant advance in phenology, however, has been observed for agriculture and forestry in large parts of the Northern Hemisphere, with limited responses in crop management. The lengthening of the growing season has contributed to an observed increase in forest productivity in many regions, while warmer and drier conditions are partly responsible for reduced forest productivity, increased forest fires and pests in North

America and the Mediterranean Basin. Both agriculture and forestry have shown vulnerability to recent trends in heatwaves, droughts and floods (medium confidence). [1.3.6]

While there have been few studies of observed health effects related to recent warming, an increase in high temperature extremes has been associated with excess mortality in Europe, which has prompted adaptation measures. There is emerging evidence of changes in the distribution of some human disease vectors in parts of Europe. Earlier onset and increases in the seasonal production of allergenic pollen have occurred in mid- and high latitudes in the Northern Hemisphere (medium confidence). [1.3.7]

Changes in socio-economic activities and modes of human response to climate change, including warming, are just beginning to be systematically documented. In regions of snow, ice and frozen ground, responses by indigenous groups relate to changes in the migration patterns, health, and range of animals and plants on which they depend for their livelihood and cultural identity. Responses vary by community and are dictated by particular histories, perceptions of change and range, and the viability of options available to groups (medium confidence). [1.3.9]

While there is now significant evidence of observed changes in natural systems in every continent, including Antarctica, as well as from most oceans, the majority of studies come from mid- and high latitudes in the Northern Hemisphere. Documentation of observed changes in tropical regions and the Southern Hemisphere is sparse. [1.5]

1.1 Introduction

The IPCC Working Group II Third Assessment Report (WGII TAR) found evidence that recent regional climate changes, particularly temperature increases, have already affected many physical and biological systems, and also preliminary evidence for effects in human systems (IPCC, 2001a). This chapter focuses on studies since the TAR that analyse significant changes in physical, biological and human systems related to observed regional climate change. The studies are assessed with regard to current functional understanding of responses to climate change and to factors that may confound such relationships, such as land-use change, urbanisation and pollution. The chapter considers larger-scale aggregation of observed changes (across systems and geographical regions) and whether the observed changes may be related to anthropogenic climate forcing. Cases where there is evidence of climate change without evidence of accompanying changes in natural and managed systems are evaluated for insight into time-lag effects, resilience and vulnerability. Managed systems are defined as systems with substantial human inputs, such as agriculture and human health. The chapter assesses whether responses to recent warming are present in a broad range of systems and across varied geographical regions.

1.1.1 Scope and goals of the chapter

The aim of this chapter is to assess studies of observed changes in natural and managed systems related to recent regional climate change, particularly temperature rise in recent decades, and to assess the aggregate changes in regard to potential influence by anthropogenic increase in greenhouse gas concentrations. Temperature rise is selected as the major climate variable because it has a strong and widespread documented signal in recent decades, demonstrates an anthropogenic signal, and has an important influence on many physical and biological processes. Effects of changes in other climate variables related to temperature rise, such as sea-level rise and changes in runoff due to earlier snow melt, are also considered.

The chapter first reviews data sources and methods of detection of observed changes, investigating the roles of climate (including climate extremes and large-scale natural climate variability systems) and non-climate drivers of change (Section 1.2). Evidence of no change, i.e., regions with documented warming trends but with little or no documentation of change in natural and managed systems, is analysed as well.

In Section 1.3, evidence is assessed regarding recent observed changes in natural and managed systems related to regional climate changes: cryosphere (snow, ice and frozen ground – including permafrost), hydrology and water resources, coastal processes and zones, marine and freshwater biological systems, terrestrial biological systems, agriculture and forestry, human health, and disasters and hazards. Evidence regarding other socio-economic effects, including energy use and tourism, is also assessed. The term 'response' is used to denote processes by which natural and managed systems react to the stimuli of changing climate conditions.

In Section 1.4, studies are surveyed that use techniques of larger-scale aggregation (i.e., synthesising studies across systems and regions), including meta-analyses and studies that relate observed changes in natural and managed systems to anthropogenic climate change. From the studies assessed in individual systems in Section 1.3, a subset is selected that fits criteria in regard to length of study and statistically significant changes in a system related to recent changes in temperature or related climate variables, in order to assess the potential influence of anthropogenic climate forcing on observed changes in natural and managed systems.

We consider what observed changes are contributing to the study of adaptation and vulnerability (where there are relevant studies), and address data needs in Section 1.5. There is a notable lack of geographical balance in the data and literature on observed changes in natural and managed systems, with a marked scarcity in many regions. The Supplementary Material[1] (SM) contains additional literature citations and explanatory data relevant to the chapter.

[1] Contained on the CD-ROM which accompanies this volume.

1.1.2 Summary of observed changes in the Third Assessment Report

The Working Group I (WGI) TAR described an increasing body of observations that gave a collective picture of a warming world and other changes in the climate system (IPCC, 2001b). The WGII TAR documented methods of detecting observed changes in natural and managed systems, characterised the processes involved, and summarised the studies across multiple systems (see Sections 2.2, 5.2.1 and 19.1) (IPCC, 2001a). In the TAR, about 60 studies considered about 500 data series in physical or biological systems.

Changes in physical systems:

- Sea ice: Arctic sea-ice extent had declined by about 10 to 15% since the 1950s. No significant trends in Antarctic sea-ice extent were apparent.
- Glaciers and permafrost: mountain glaciers were receding on all continents, and Northern Hemisphere permafrost was thawing.
- Snow cover: extent of snow cover in the Northern Hemisphere had decreased by about 10% since the late 1960s and 1970s.
- Snow melt and runoff: snowmelt and runoff had occurred increasingly earlier in Europe and western North America since the late 1940s.
- Lake and river ice: annual duration of lake- and river-ice cover in Northern Hemisphere mid- and high latitudes had been reduced by about 2 weeks and become more variable.

Changes in biological systems:

- Range: plant and animal ranges had shifted poleward and higher in elevation.
- Abundance: within the ranges of some plants and animals, population sizes had changed, increasing in some areas and declining in others.
- Phenology: timing of many life-cycle events, such as blooming, migration and insect emergence, had shifted earlier in the spring and often later in the autumn.
- Differential change: species changed at different speeds and in different directions, causing a decoupling of species interactions (e.g., predator-prey relationships).

Preliminary evidence for changes in human systems:

- Damages due to droughts and floods: changes in some socio-economic systems had been related to persistent low rainfall in the Sahelian region of Africa and to increased precipitation extremes in North America. Most of the increase in damages is due to increased wealth and exposure. However, part of the increase in losses was attributed to climate change, in particular to more frequent and intense extreme weather events in some regions.

1.2 Methods of detection and attribution of observed changes

In the TAR (Mitchell et al., 2001), *detection* of climate change is the process of demonstrating that an observed change is significantly different (in a statistical sense) from what can be explained by natural variability. The detection of a change, however, does not necessarily imply that its causes are understood. Similarly, *attribution* of climate change to anthropogenic causes involves statistical analysis and the assessment of multiple lines of evidence to demonstrate, within a pre-specified margin of error, that the observed changes are (1) unlikely to be due entirely to natural internal climate variability; (2) consistent with estimated or modelled responses to the given combination of anthropogenic and natural forcing; and (3) not consistent with alternative, physically plausible explanations of recent climate change.

Extending detection and attribution analysis to observed changes in natural and managed systems is more complex. Detection and attribution of observed changes and responses in systems to anthropogenic forcing is usually a two-stage process (IPCC, 2003). First, the observed changes in a system must be demonstrated to be associated with an observed regional climate change within a specified degree of confidence. Second, a measurable portion of the observed regional climate change, or the associated observed change in the system, must be attributed to anthropogenic causes with a similar degree of confidence.

Joint attribution involves both attribution of observed changes to regional climate change and attribution of a measurable proportion of either regional climate change or the associated observed changes in the system to anthropogenic causes, beyond natural variability. This process involves statistically linking climate change simulations from climate models with the observed responses in the natural or managed system. Confidence in joint attribution statements must be lower than the confidence in either of the individual attribution steps alone, due to the combination of two separate statistical assessments.

1.2.1 Climate and non-climate drivers of change

Both climate and non-climate drivers affect systems, making analysis of the role of climate in observed changes challenging. Non-climate drivers such as urbanisation and pollution can influence systems directly and indirectly through their effects on climate variables such as albedo and soil-moisture regimes. Socio-economic processes, including land-use change (e.g., forestry to agriculture; agriculture to urban area) and land-cover modification (e.g., ecosystem degradation or restoration) also affect multiple systems.

1.2.1.1 Climate drivers of change

Climate is a key factor determining different characteristics and distributions of natural and managed systems, including the cryosphere, hydrology and water resources, marine and freshwater biological systems, terrestrial biological systems, agriculture and forestry. For example, temperature is known to strongly influence the distribution and abundance patterns of both plants and animals, due to the physiological constraints of each species (Parmesan and Yohe, 2003; Thomas et al., 2004). Dramatic changes in the distribution of plants and animals during the ice ages illustrate how climate influences the distribution of species. Equivalent effects can be observed in other systems, such as the cryosphere. Hence, changes in temperature due to climate change are expected to be one of the important drivers of change in natural and managed systems.

Many aspects of climate influence various characteristics and distributions of physical and biological systems, including temperature and precipitation, and their variability on all time-scales from days to the seasonal cycle to interannual variations. While changes in many different aspects of climate may at least partially drive changes in the systems, we focus on the role of temperature changes. This is because physical and biological responses to changing temperatures are often better understood than responses to other climate parameters, and the anthropogenic signal is easier to detect for temperature than for other parameters. Precipitation has much larger spatial and temporal variability than temperature, and it is therefore more difficult to identify the impact it has on changes in many systems. Mean temperature (including daily maximum and minimum temperature) and the seasonal cycle in temperature over relatively large spatial areas show the clearest signals of change in the observed climate (IPCC, 2001b).

Large-scale climate variations, such as the Pacific Decadal Oscillation (PDO), El Niño-Southern Oscillation (ENSO) and North Atlantic Oscillation (NAO), are occurring at the same time as the global climate is changing. Consequently, many natural and managed systems are being affected by both climate change and climate variability. Hence, studies of observed changes in regions influenced by an oscillation may be able to attribute these changes to regional climate variations, but decades of data may be needed in order to separate the response to climate oscillations from that due to longer-term climate change.

1.2.1.2 Non-climate drivers of change

Non-climate drivers, such as land use, land degradation, urbanisation and pollution, affect systems directly and indirectly through their effects on climate (Table 1.1). These drivers can operate either independently or in association with one another (Lepers et al., 2004). Complex feedbacks and interactions occur on all scales from local to global.

The socio-economic processes that drive land-use change include population growth, economic development, trade and migration; these processes can be observed and measured at global, regional and local scales (Goklany, 1996). Satellite observations demonstrate that land-use change, including that associated with the current rapid economic development in Asia and Latin America, is proceeding at an unprecedented rate (Rindfuss et al., 2004). Besides influencing albedo and evaporation, land-use changes hamper range-shift responses of species to climate change, leading to an extra loss of biodiversity (Opdam and Wascher, 2004). Additionally, land-use changes have been linked to changes in air quality and pollution that affect the greenhouse process itself (Pielke et al., 2002; Kalnay and Cai, 2003). Land-use and land-cover change can also strongly magnify the effects of extreme climate events, e.g., heat mortality, injuries/fatalities from storms, and ecologically mediated infectious diseases (Patz et al., 2005). Intensification of land use, as well as the extent of land-use change, is also affecting the functioning of ecosystems, and hence emissions of greenhouse gases from soils, such as CO_2 and methane.

There are also a large number of socio-economic factors that can influence, obscure or enhance the observed impacts of climate change and that must be taken into account when

Table 1.1. *Direct and indirect effects of non-climate drivers.*

Non-climate driver	Examples	Direct effects on systems	Indirect effects on climate
Geological processes	Volcanic activity, earthquakes, tsunamis (e.g., Adams et al., 2003)	Lava flow, mudflows (lahars), ash fall, shock waves, coastal erosion, enhanced surface and basal melting of glaciers, rockfall and ice avalanches	Cooling from stratospheric aerosols, change in albedo
Land-use change	Conversion of forest to agriculture (e.g., Lepers et al., 2004)	Declines in wildlife habitat, biodiversity loss, increased soil erosion, nitrification	Change in albedo, lower evapotranspiration, altered water and heat balances (e.g., Bennett and Adams, 2004)
	Urbanisation and transportation (e.g., Kalnay and Cai, 2003)	Ecosystem fragmentation, deterioration of air quality, increased runoff and water pollution (e.g., Turalioglu et al., 2005)	Change in albedo, urban heat island, local precipitation reduction, downwind precipitation increase, lower evaporation (e.g., Weissflog et al., 2004)
	Afforestation (e.g., Rudel et al., 2005)	Restoration or establishment of tree cover (e.g., Gao et al., 2002)	Change in albedo, altered water and energy balances, potential carbon sequestration
Land-cover modification	Ecosystem degradation (desertification)	Reduction in ecosystem services, reduction in biomass, biodiversity loss (e.g., Nyssen et al., 2004)	Changes in microclimate (e.g., Su et al., 2004)
Invasive species	Tamarisk (USA), Alaska lupin (Iceland)	Reduction of biodiversity, salinisation (e.g., Lee et al., 2006)	Change in water balance (e.g., Ladenburger et al., 2006)
Pollution	Tropospheric ozone, toxic waste, oil spills, exhaust, pesticides increased soot emissions (e.g., Pagliosa and Barbosa, 2006)	Reduction in breeding success and biodiversity, species mortality, health impairment, enhanced melting of snow and ice (e.g., Lee et al., 2006)	Direct and indirect aerosol effects on temperature, albedo and precipitation

seeking a climate signal or explaining observations of impacts and even adaptations. For example, the noted effects of sea-level rise and extreme events are much greater when they occur in regions with large populations, inadequate infrastructure, or high property prices (Pielke et al., 2003). The observed impacts of climate change on agriculture are largely determined by the ability of producers to access or afford irrigation, alternate crop varieties, markets, insurance, fertilisers and agricultural extension, or to abandon agriculture for alternate livelihoods (Eakin, 2000). Demography (e.g., the elderly and the very young), poverty (e.g., malnutrition and poor living conditions), preventive technologies (e.g., pest control and immunisation), and healthcare institutions influence the impacts of climate change on humans.

1.2.2 Methods and confidence

Where long data series exist, the detection of trends or changes in system properties that are beyond natural variability has most commonly been made with regression, correlation and time-series analyses. When data exist from two (or more) discontinuous time periods, two-sample tests have frequently been employed. Testing is also done for abrupt changes and discontinuities in a data series. Regression and correlation methods are frequently used in the detection of a relationship of the observed trend with climate variables. Methods also involve studies of process-level understanding of the observed change in relation to a given regional climate change, and the examination of alternative explanations of the observed change, such as land-use change. The analysis sometimes involves comparisons of observations to climate-driven model simulations.

In many biological field studies, species within an area are not fully surveyed, nor is species selection typically based on systematic or random sampling. The selection of species is typically based on a determination of which species might provide information (e.g., on change with warming) in order to answer a particular question. The study areas, however, are often chosen at random from a particular suite of locations defined by the presence of the species being studied. This type of species selection does not provide a well-balanced means for analysing species showing no change. Exceptions are studies that rely on network data, meaning that species information is collected continuously on a large number of species over decades from the same areas; for example, change in spring green-up[2] of a number of plants recorded in phenological botanical gardens across a continent (Menzel and Fabian, 1999). Analysis of change and no-change within network data provides a check on the accuracy of the use of the indicator for global warming and the ability to check for 'false positives', i.e., changes observed where no significant temperature change is measured. The latter can help to elucidate the role of non-climate drivers in the observed changes.

The analysis of evidence of no change is also related to the question of publication or assessment bias. Studies are more likely to be successfully submitted and published when a significant change is found and less likely to be successful when no changes are found, with the result that the 'no change' cases are underrepresented in the published literature. However, in contrast to single-species in single-location studies, multiple species in a single location and single or multiple species in larger-scale studies are less likely to focus only on species showing change. The latter studies often include sub-regions with no-change; for example, no change in the number of frost days in the south-eastern USA (Feng and Hu, 2004), little or no change in spring onset in continental eastern Europe (Ahas et al., 2002; Schleip et al., 2006), or sub-groups of species with no change (Butler, 2003; Strode, 2003).

An accurate percentage of sites exhibiting 'no change' can be assessed reliably by large-scale network studies (see, e.g., Section 1.4.1; Menzel et al., 2006b) for the locations defined by the network. For investigations of a suite of processes or species at numerous locations, the reported ratio of how many species are changing over the total number of species rests on the assumptions that all species in the defined area have been examined and that species showing no change do not have a higher likelihood of being overlooked. Both multi-species network data and studies on groups of species may be used to investigate the resilience of systems and possible time-lag effects. These are important processes in the analysis of evidence of no change.

1.3 Observed changes in natural and managed systems related to regional climate changes

The following sections assess studies that have been published since the TAR of observed changes and their effects related to the cryosphere, hydrology and water resources, coastal processes and zones, freshwater and marine biological systems, terrestrial biological systems, agriculture and forestry, human health, and disasters and hazards related to regional warming. More detailed descriptions of these effects are provided in subsequent chapters of the WGII Fourth Assessment Report (AR4).

In some cases, studies published before the TAR have been included, either because they were not cited in the TAR or because they have been considered to contain relevant information. The sections describe regional climate and non-climate driving forces for the systems, assess the evidence regarding observed changes in key processes, and highlight issues regarding the absence of observed changes and conflicting evidence. An assessment of how the observed changes contribute to understanding of adaptation and vulnerability is found in Sections 1.3.9 and 1.5.

1.3.1 Cryosphere

The cryosphere reacts sensitively to present and past climate changes. The main components of the cryosphere are mountain

[2] Spring green-up is a measure of the transition from winter dormancy to active spring growth.

glaciers and ice caps, floating ice shelves and continental ice sheets, seasonal snow cover on land, frozen ground, sea ice and lake and river ice. In Chapter 4 of WGI, the changes in the cryosphere since the TAR are described in detail, including the description of climate and non-climate forcing factors and mechanisms (Lemke et al., 2007). Chapter 6 of WGI describes glacier changes in the geological past, including Holocene glacier variability (Jansen et al., 2007, Box 6.3). Here we describe the observed effects on the environment and on human activities due to these recent cryospheric changes.

There is abundant evidence that the vast majority of the cryospheric components are undergoing generalised shrinkage in response to warming, with a few cases of growth which have been mainly linked to increased snowfall. The observed recession of glaciers (Box 1.1) during the last century is larger than at any time over at least the last 5,000 years, is outside of the range of normal climate variability, and is probably induced by anthropogenic warming (Jansen et al., 2007). In the Arctic and the Antarctic, ice shelves several thousand years old have started to collapse due to warming (Lemke et al., 2007). In many cases the cryospheric shrinkage shows an increased trend in recent decades, consistent with the enhanced observed warming. Cryospheric changes are described by Lemke et al. (2007), including the contribution of the cryosphere to sea-level rise. Sea-level rise is treated in Section 1.3.3, in the regional chapters of WGII, and in WGI, Chapters 4 and 5 (Bindoff et al., 2007; Lemke et al., 2007).

1.3.1.1 Observed effects due to changes in the cryosphere

Effects of changes in the cryosphere have been documented in relation to virtually all of the cryospheric components, with robust evidence that it is, in general, a response to reduction of snow and ice masses due to enhanced warming.

Mountain glaciers and ice caps, ice sheets and ice shelves

Effects of changes in mountain glaciers and ice caps have been documented in runoff, changing hazard conditions (Haeberli and Burn, 2002) and ocean freshening (Bindoff et al., 2007). There is also emerging evidence of present crustal uplift in response to recent glacier melting in Alaska (Larsen et al., 2005). The enhanced melting of glaciers leads at first to increased river runoff and discharge peaks and an increased melt season (Boon et al., 2003; Hock, 2005; Hock et al., 2005; Juen et al., 2007), while in the longer time-frame (decadal to century scale), glacier wasting should be amplified by positive feedback mechanisms and glacier runoff is expected to decrease (Jansson et al., 2003). Evidence for increased runoff in recent decades due to enhanced glacier melt has already been detected in the tropical Andes and in the Alps. As glaciers disappear, the records preserved in the firn[3] and ice layers are destroyed and disappear due to percolation of melt water and mixing of chemical species and stable isotopes (Table 1.2).

The formation of large lakes is occurring as glaciers retreat from prominent Little Ice Age (LIA) moraines in several steep mountain ranges, including the Himalayas (Yamada, 1998; Mool et al., 2001; Richardson and Reynolds, 2000), the Andes (Ames et al., 1989; Kaser and Osmaston, 2002) and the Alps (Haeberli et al., 2001; Huggel et al., 2004; Kaab et al., 2005) (Table 1.2). Thawing of buried ice also threatens to destabilise the LIA moraines (e.g., Kaser and Osmaston, 2002). These lakes thus have a high potential for glacial lake outburst floods (GLOFs). Governmental institutions in the respective countries have undertaken extensive safety work, and several of the lakes are

Table 1.2. *Selected observed effects due to changes in the cryosphere produced by warming.*

Environmental factor	Observed changes	Time period	Location	Selected references
Glacial lake size	Increase from 0.23 km² to 1.65 km²	1957-1997	Lake Tsho Rolpa, Nepal Himalayas	Agrawala et al., 2005
Glacial lake outburst floods (GLOFs)	Frequency increase from 0.38 events/year in 1950s to 0.54 events/year in 1990s	1934-1998	Himalayas of Nepal, Bhutan and Tibet	Richardson and Reynolds, 2000
Obliteration of firn/ice core record	Percolation, loss of palaeoclimate record	1976-2000	Quelccaya ice cap, Peru	Thompson et al., 2003
Reduction in mountain ice	Loss of ice climbs	1900-2000	Andes, Alps, Africa	Schwörer, 1997; Bowen, 2002
Travel days of vehicles for oil exploration on frozen roads	Decrease from 220 to 130 days	1971-2003	Alaskan tundra	ACIA, 2005
Decreased snow in ski areas at low altitudes	Decrease in number of ski areas from 58 to 17	1975-2002	New Hampshire, north-eastern USA	Hamilton, 2003b
	50% (15%) decrease in snow depth at an elevation of 440 m (2,220 m)	1975-1999	Swiss Alps	Laternser and Schneebeli, 2003
	50% decrease of 1 Dec–30 April snow depth at 1,320 m elevation	1960-2005	Massifs de Chartreuse, Col de Porte, French Pre-Alps,	Francou and Vincent, 2006
	Increase in elevation of starting point of ski lifts from 1,400 to 2,935 m	1950-1987	Central Andes, Chile	Casassa et al., 2003
Increased rockfall after the 2003 summer heatwave	Active layer deepening from 30% to 100% of the depth measured before the heatwave	June-August 2003	Swiss Alps	Noetzli et al., 2003; Gruber et al., 2004; Schär et al., 2004

[3] Firn: ice that is at an intermediate stage between snow and glacial ice.

Box 1.1. Retreat of Chacaltaya and its effects: case study of a small disappearing glacier in Bolivia

The observed general glacier retreat in the warming tropical Andes has increased significantly in recent decades (Francou et al., 2005). Small-sized glaciers are particularly vulnerable in warmer climates, with many of them having already disappeared in several parts of the world during the last century. The Chacaltaya Glacier in Bolivia (16°S) is a typical example of a disappearing small glacier, whose area in 1940 was 0.22 km^2, and which has currently reduced (in 2005) to less than 0.01 km^2 (Figure 1.1) (Ramirez et al., 2001; Francou et al., 2003; Berger et al., 2005), with current estimates showing that it may disappear completely before 2010. In the period 1992 to 2005, the glacier suffered a loss of 90% of its surface area, and 97% of its volume of ice (Berger et al., 2005). Although, in the tropics, glacier mass balance responds sensitively to changes in precipitation and humidity (see Lemke et al., 2007, Section 4.5.3), the fast glacier shrinkage of Chacaltaya is consistent with an ascent of the 0°C isotherm of about 50 m/decade in the tropical Andes since the 1980s (Vuille et al., 2003), resulting in a corresponding rise in the equilibrium line of glaciers in the region (Coudrain et al., 2005).

Ice melt from Chacaltaya Glacier, located in Choqueyapu Basin, provides part of the water resources for the nearby city of La Paz, allowing the release of water stored as ice throughout the long, dry winter season (April-September). Many basins in the tropical Andes have experienced an increase in runoff in recent decades, while precipitation has remained almost constant or has shown a tendency to decrease (Coudrain et al., 2005). This short-term increase in runoff is interpreted as the consequence of glacier retreat, but in the long term there will be a reduction in water supply as the glaciers shrink beyond a critical limit (Jansson et al., 2003).

Chacaltaya Glacier, with a mean altitude of 5,260 m above sea level, was the highest skiing station in the world until a very few years ago. After the accelerated shrinkage of the glacier during the 1990s, enhanced by the warm 1997/98 El Niño, Bolivia lost its only ski area (Figure 1.1), directly affecting the development of snow sports and recreation in this part of the Andes, where glaciers are an important part of the cultural heritage.

Figure 1.1. *Areal extent of Chacaltaya Glacier, Bolivia, from 1940 to 2005. By 2005, the glacier had separated into three distinct small bodies. The position of the ski hut, which did not exist in 1940, is indicated with a red cross. The ski lift, which had a length of about 800 m in 1940 and about 600 m in 1996, was normally installed during the summer months (precipitation season in the tropics) and covered a major portion of the glacier, as indicated with a continuous line. The original location of the ski lift in 1940 is indicated with a segmented line in subsequent epochs. After 2004, skiing was no longer possible. Photo credits: Francou and Vincent (2006) and Jordan (1991).*

now either solidly dammed or drained, but continued vigilance is needed since many tens of potentially dangerous glacial lakes still exist in the Himalayas (Yamada, 1998) and the Andes (Ames, 1998), together with several more in other mountain ranges of the world. The temporary increase in glacier melt can also produce enhanced GLOFs, as has been reported in Chile (Peña and Escobar, 1985), although these have not been linked with any long-term climate trends.

Enhanced colonisation of plants and animals in deglaciated terrain is a direct effect of glacier and snow retreat (e.g., Jones and Henry, 2003). Although changes due to other causes such as introduction by human activities, increased UV radiation, contaminants and habitat loss might be important (e.g., Frenot et al., 2005), 'greening' has been reported in relation to warming in the Arctic and also in the Antarctic Peninsula. Tundra areas in the northern circumpolar high latitudes derived from a 22-year satellite record show greening trends, while forest areas show declines in photosynthetic activity (Bunn and Goetz, 2006). Ice-water microbial habitats have contracted in the Canadian High Arctic (Vincent et al., 2001).

Glacier retreat causes striking changes in the landscape, which has affected living conditions and local tourism in many mountain regions around the world (Watson and Haeberli, 2004; Mölg et al., 2005). Warming produces an enhanced spring-summer melting of glaciers, particularly in areas of ablation, with a corresponding loss of seasonal snow cover that results in an increased exposure of surface crevasses, which can in turn affect, for example, snow runway operations, as has been reported in the Antarctic Peninsula (Rivera et al., 2005). The retreat, enhanced flow and collapse of glaciers, ice streams and ice shelves can lead to increased production of iceberg calving, which can in turn affect sea navigation, although no evidence for this exists as yet.

Snow cover

Spring peak river flows have been occurring 1-2 weeks earlier during the last 65 years in North America and northern Eurasia. There is also evidence for an increase in winter base flow in northern Eurasia and North America. These changes in river runoff are described in detail in Section 1.3.2 and Table 1.3. There is also a measured trend towards less snow at low altitudes, which is affecting skiing areas (Table 1.2).

Frozen ground

Degradation of seasonally frozen ground and permafrost, and an increase in active-layer thickness, should result in an increased importance of surface water (McNamara et al., 1999), with an initial but temporary phase of lake expansion due to melting, followed by their disappearance due to draining within the permafrost, as has been detected in Alaska (Yoshikawa and Hinzman, 2003) and in Siberia (Smith et al., 2005).

Permafrost and frozen ground degradation are resulting in an increased areal extent of wetlands in the Arctic, with an associated 'greening', i.e., plant colonisation (see above). Wetland changes also affect the fauna. Permafrost degradation and wetland increase might produce an increased release of carbon in the form of methane to the atmosphere in the future (e.g., Lawrence and Slater, 2005; Zimov et al., 2006), but this has not been documented.

The observed permafrost warming and degradation, together with an increasing depth of the active layer, should result in mechanical weakening of the ground, and ground subsidence and formation of thermokarst will have a weakening effect on existing infrastructure such as buildings, roads, airfields and pipelines (Couture et al., 2000; Nelson, 2003), but there is no solid evidence for this yet. There is evidence for a decrease in potential travel days of vehicles over frozen roads in Alaska (Table 1.2). Permafrost melting has produced increased coastal erosion in the Arctic (e.g., Beaulieu and Allard, 2003); this is detailed in Section 1.3.3.

Thawing and deepening of the active layer in high-mountain areas can produce slope instability and rock falls (Watson and Haeberli, 2004), which in turn can trigger outburst floods (Casassa and Marangunic, 1993; Carey, 2005), but there is no evidence for trends. A reported case linked to warming is the exceptional rock-fall activity in the Alps during the 2003 summer heatwave (Table 1.2).

Sea ice

Nutritional stresses related to longer ice-free seasons in the Beaufort Sea may be inducing declining survival rates, smaller size, and cannibalism among polar bears (Amstrup et al., 2006; Regehr et al., 2006). Polar bears are entirely dependent on sea ice as a platform to access the marine mammals that provide their nutritional needs (Amstrup, 2003). Reduced sea ice in the Arctic will probably result in increased navigation, partial evidence of which has already been found (Eagles, 2004), and possibly also a rise in offshore oil operations, with positive effects such as enhanced trade, and negative ones such as increased pollution (Chapter 15; ACIA, 2005), but there are no quantitative data to support this.

Increased navigability in the Arctic should also raise issues of water sovereignty versus international access for shipping through the North-west and North-east Passages. Previously uncharted islands and seamounts have been discovered due to a reduction in sea ice cover (Mohr and Forsberg, 2002), which can be relevant for territorial and ocean claims.

Ocean freshening, circulation and ecosystems

There is evidence for freshening in the North Atlantic and in the Ross Sea, which is probably linked to glacier melt (Bindoff et al., 2007). There is no significant evidence of changes in the Meridional Overturning Circulation at high latitudes in the North Atlantic Ocean or in the Southern Ocean, although important changes in interannual to decadal scales have been observed in the North Atlantic (Bindoff et al., 2007). Ocean ecosystem impacts such as a reduction of krill biomass and an increase in salps in Antarctica, decline of marine algae in the Arctic due to their replacement by freshwater species, and impacts on Arctic mammals, are described in Section 1.3.4.2.

Lake and river ice

Seasonal and multi-annual variations in lake and river ice are relevant in terms of freshwater hydrology and for human activities such as winter transportation, bridge and pipeline crossings, but no quantitative evidence of observed effects exists yet. Shortening of the freezing period of lake and river ice by an

Table 1.3. *Observed changes in runoff/streamflow, lake levels and floods/droughts.*

Environmental factor	Observed changes	Time period	Location	Selected references
Runoff/ streamflow	Annual increase of 5%, winter increase of 25 to 90%, increase in winter base flow due to increased melt and thawing permafrost	1935-1999	Arctic Drainage Basin: Ob, Lena, Yenisey, Mackenzie	Lammers et al., 2001; Serreze et al., 2002; Yang et al., 2002
	1 to 2 week earlier peak streamflow due to earlier warming-driven snow melt	1936-2000	Western North America, New England, Canada, northern Eurasia	Cayan et al., 2001; Beltaos, 2002; Stone et al., 2002; Yang et al., 2002; Hodgkins et al., 2003; Ye and Ellison, 2003; Dery and Wood, 2005; McCabe and Clark, 2005; Regonda et al., 2005
Runoff increase in glacial basins in Cordillera Blanca, Peru	23% increase in glacial melt	2001-4 *vs.* 1998-9	Yanamarey Glacier catchment	Mark et al., 2005
	143% increase	1953-1997	Llanganuco catchment	Pouyaud et al., 2005
	169% increase	2000-2004	Artesonraju catchment	Pouyaud et al., 2005
Floods	Increasing catastrophic floods of frequency (0.5 to 1%) due to earlier break-up of river-ice and heavy rain	Last years	Russian Arctic rivers	Smith, 2000; Buzin et al., 2004; Frolov et al., 2005
Droughts	29% decrease in annual maximum daily streamflow due to temperature rise and increased evaporation with no change in precipitation	1847-1996	Southern Canada	Zhang et al., 2001
	Due to dry and unusually warm summers related to warming of western tropical Pacific and Indian Oceans in recent years	1998-2004	Western USA	Andreadis et al., 2005; Pagano and Garen, 2005
Water temperature	0.1 to 1.5°C increase in lakes	40 years	Europe, North America, Asia (100 stations)	Livingstone and Dokulil, 2001; Ozaki et al., 2003; Arhonditsis et al., 2004; Dabrowski et al., 2004; Hari et al., 2006
	0.2 to 0.7°C increase (deep water) in lakes	100 years	East Africa (6 stations)	Hecky et al., 1994; O'Reilly et al., 2003; Lorke et al., 2004; Vollmer et al., 2005
Water chemistry	Decreased nutrients from increased stratification or longer growing period in lakes and rivers	100 years	North America, Europe, Eastern Europe, East Africa (8 stations)	Hambright et al., 1994; Adrian et al., 1995; Straile et al., 2003; Shimaraev and Domysheva, 2004; O'Reilly, 2007
	Increased catchment weathering or internal processing in lakes and rivers.	10-20 years	North America, Europe (88 stations)	Bodaly et al., 1993; Sommaruga-Wograth et al., 1997; Rogora et al., 2003; Vesely et al., 2003; Worrall et al., 2003; Karst-Riddoch et al., 2005

average of 12 days during the last 150 years (Lemke et al., 2007) results in a corresponding reduction in skating activities in the Northern Hemisphere. In Europe there is some evidence for a reduction in ice-jam floods due to reduced freshwater freezing during the last century (Svensson et al., 2006). Enhanced melt conditions could also result in significant ice jamming due to increased break-up events, which can, in turn, result in severe flooding (Prowse and Beltaos, 2002), although there is a lack of scientific evidence that this is already happening.

Changes in lake thermal structure and quality/quantity of under-ice habitation in lakes have been reported, as well as changes in suspended particles and chemical composition (see Section 1.3.2). Earlier ice-out dates can have relevant effects on lake and river ecology, while changes in river-ice dynamics may also have ecological effects (see Section 1.3.4).

1.3.1.2 Summary of cryosphere

There is abundant and significant evidence that most of the cryospheric components in polar regions and in mountains are undergoing generalised shrinkage in response to warming, and that their effects in the environment and in human activities are already detectable. This agrees with the results presented in Chapter 9 of WGI (Hegerl et al., 2007), which concludes that the observed reductions in Arctic sea ice extent, decreasing trend in global snow cover, and widespread retreat and melting of glaciers are inconsistent with simulated internal variability, and consistent with the simulated response to anthropogenic gases. The observed effects of cryosphere reduction include modification of river regimes due to enhanced glacial melt, snowmelt advance and enhanced winter base flow; formation of thermokarst terrain and disappearance of surface lakes in thawing permafrost; decrease in potential travel days of vehicles over frozen roads in the Arctic; enhanced potential for glacier hazards and slope instability due to mechanical weakening driven by ice and permafrost melting; regional ocean freshening; sea-level rise due to glacier and ice sheet shrinkage; biotic colonisation and faunal changes in deglaciated terrain; changes in freshwater and marine ecosystems affected by lake-ice and sea-ice reduction; changes in livelihoods; reduced tourism activities related to skiing, ice climbing and scenic activities in cryospheric areas affected by degradation; and increased ease of ship transportation in the Arctic.

1.3.2 Hydrology and water resources

This section focuses on the relationship of runoff, lake levels, groundwater, floods and droughts, and water quality, with observed climate variability, climate trends, and land-use and land-cover changes reported since the TAR. The time period under consideration is primarily 1975 to 2005, with many studies extending to earlier decades. Observed changes in precipitation and aspects of surface hydrology are described in more detail by Trenberth et al. (2007), Section 3.3.

1.3.2.1 Changes in surface and groundwater systems

Since the TAR there have been many studies related to trends in river flows during the 20th century at scales ranging from catchment to global. Some of these studies have detected significant trends in some indicators of river flow, and some have demonstrated statistically significant links with trends in temperature or precipitation; but no globally homogeneous trend has been reported. Many studies, however, have found no trends, or have been unable to separate the effects of variations in temperature and precipitation from the effects of human interventions in the catchment, such as land-use change and reservoir construction. Variation in river flows from year to year is also very strongly influenced in some regions by large-scale atmospheric circulation patterns associated with ENSO, NAO and other variability systems that operate at within-decadal and multi-decadal time-scales.

At the global scale, there is evidence of a broadly coherent pattern of change in annual runoff, with some regions experiencing an increase at higher latitudes and a decrease in parts of West Africa, southern Europe and southern Latin America (Milly et al., 2005). Labat et al. (2004) claimed a 4% increase in global total runoff per 1°C rise in temperature during the 20th century, with regional variation around this trend, but this has been challenged (Legates et al., 2005) due to the effects of non-climatic drivers on runoff and bias due to the small number of data points. Gedney et al., (2006) gave the first tentative evidence that CO_2 forcing leads to increases in runoff due to the ecophysiological controls of CO_2, although other evidence for such a relationship is difficult to find. The methodology used to search for trends can also influence results, since omitting the effects of cross-correlation between river catchments can lead to an overestimation of the number of catchments showing significant trends (Douglas et al., 2000). Runoff studies that show no trends are listed in the Chapter 1 Supplementary Material (SM).

Runoff in snow basins

There is abundant evidence for an earlier occurrence of spring peak river flows and an increase in winter base flow in basins with important seasonal snow cover in North America and northern Eurasia, in agreement with local and regional climate warming in these areas (Table 1.3). The early spring shift in runoff leads to a shift in peak river runoff away from summer and autumn, which are normally the seasons with the highest water demand, resulting in consequences for water availability (see Chapter 3). See Table SM1.1a for additional changes in runoff/streamflow.

Groundwater

Groundwater in shallow aquifers is part of the hydrological cycle and is affected by climate variability and change through recharge processes (Chen et al., 2002), as well as by human interventions in many locations (Petheram et al., 2001). In the Upper Carbonate Aquifer near Winnipeg, Canada, shallow well hydrographs show no obvious trends, but exhibit variations of 3 to 4 years correlated with changes in annual temperature and precipitation (Ferguson and George, 2003).

Lakes

At present, no globally consistent trend in lake levels has been found. While some lake levels have risen in Mongolia and China (Xinjiang) in response to increased snow and ice melt, other lake levels in China (Qinghai), Australia, Africa (Zimbabwe, Zambia and Malawi), North America (North Dakota) and Europe (central Italy) have declined due to the combined effects of drought, warming and human activities. Within permafrost areas in the Arctic, recent warming has resulted in the temporary formation of lakes due to the onset of melting, which then drain rapidly due to permafrost degradation (e.g., Smith et al., 2005). A similar effect has been reported for a lake formed over an Arctic ice shelf (i.e., an epishelf lake), which disappeared when the ice shelf collapsed (Mueller et al., 2003). Permafrost and epishelf lakes are treated in detail by Le Treut et al. (2007). Observed trends in lake levels are listed in Table SM1.1b.

1.3.2.2 Floods and droughts

Documented trends in floods show no evidence for a globally widespread change. Although Milly et al. (2002) identified an apparent increase in the frequency of 'large' floods (return period >100 years) across much of the globe from the analysis of data from large river basins, subsequent studies have provided less widespread evidence. Kundzewicz et al. (2005) found increases (in 27 cases) and decreases (in 31 cases) and no trend in the remaining 137 cases of the 195 catchments examined worldwide. Table 1.3 shows results of selected changes in runoff/streamflow, lake levels and floods/droughts. Other examples of changes in floods and droughts may be found in Table SM1.2.

Globally, very dry areas (Palmer Drought Severity Index, PDSI ≤ -3.0) have more than doubled since the 1970s due to a combination of ENSO events and surface warming, while very wet areas (PDSI $\geq +3.0$) declined by about 5%, with precipitation as the major contributing factor during the early 1980s and temperature more important thereafter (Dai et al., 2004). The areas of increasing wetness include the Northern Hemisphere high latitudes and equatorial regions. However, the use of PDSI is limited by its lack of effectiveness in tropical regions. Table 1.3 shows the trend in droughts in some regions. Documented trends in severe droughts and heavy rains (Trenberth et al., 2007, Section 3.8.2) show that hydrological conditions are becoming more intense in some regions, consistent with other findings (Huntington, 2006).

1.3.2.3 Changes in physical and chemical aspects of lakes and rivers

Changes in thermal structure and chemistry have been documented in many parts of the world in recent decades.

Thermal structure

Higher water temperatures have been reported in lakes in response to warmer conditions (Table 1.3) (see Table SM1.3 for additional changes in physical water properties). Shorter periods of ice cover and decreases in river- and lake-ice thickness are treated in Section 1.3.1 and Le Treut et al. (2007). Phytoplankton dynamics and primary productivity have also been altered in conjunction with changes in lake physics (see Section 1.3.4.4; Figure 1.2; Table 1.6). Since the 1960s, surface water temperatures have warmed by 0.2 to 2°C in lakes and rivers in Europe, North America and Asia. Along with warming surface waters, deep-water temperatures (which reflect long-term trends) of the large East African lakes (Edward, Albert, Kivu, Victoria, Tanganyika and Malawi) have warmed by 0.2 to 0.7°C since the early 1900s. Increased water temperature and longer ice-free seasons influence the thermal stratification and internal hydrodynamics of lakes. In warmer years, surface water temperatures are higher, evaporative water loss increases, summer stratification occurs earlier in the season, and thermoclines become shallower. In several lakes in Europe and North America, the stratified period has advanced by up to 20 days and lengthened by 2 to 3 weeks, with increased thermal stability.

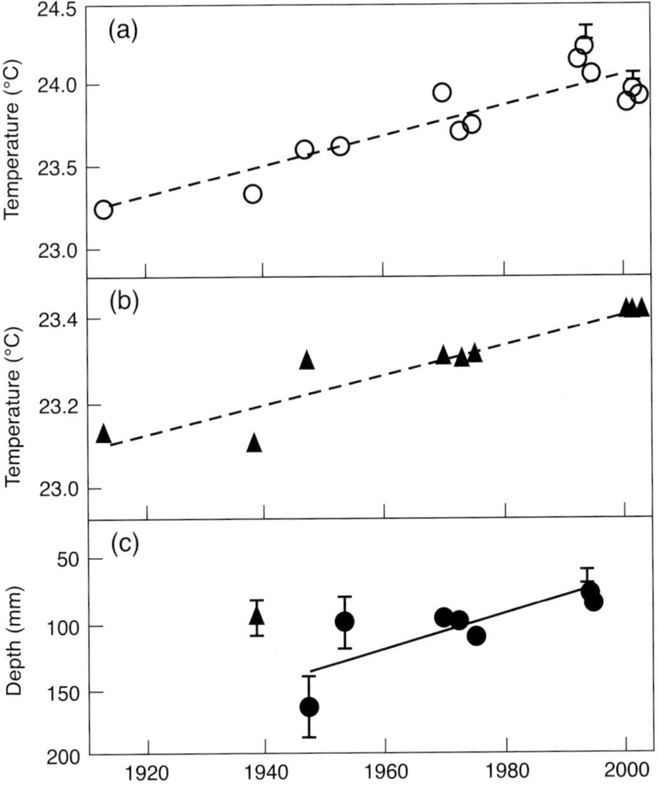

Figure 1.2. *Historical and recent measurements from Lake Tanganyika, East Africa: (a) upper mixed layer (surface water) temperatures; (b) deep-water (600 m) temperatures; (c) depth of the upper mixed layer. Triangles represent data collected by a different method. Error bars represent standard deviations. Reprinted by permission from Macmillan Publishers Ltd. [Nature] (O'Reilly et al., 2003), copyright 2003.*

Chemistry

Increased stratification reduces water movement across the thermocline, inhibiting the upwelling and mixing that provide essential nutrients to the food web. There have been decreases in nutrients in the surface water and corresponding increases in deep-water concentrations of European and East African lakes because of reduced upwelling due to greater thermal stability. Many lakes and rivers have increased concentrations of sulphates, base cations and silica, and greater alkalinity and conductivity related to increased weathering of silicates, calcium and magnesium sulphates, or carbonates, in their catchment. In contrast, when warmer temperatures enhanced vegetative growth and soil development in some high-alpine ecosystems, alkalinity decreased because of increased organic-acid inputs (Karst-Riddoch et al., 2005). Glacial melting increased the input of organochlorines (which had been atmospherically transported to and stored in the glacier) to a sub-alpine lake in Canada (Blais et al., 2001).

Increased temperature also affects in-lake chemical processes (Table 1.3) (also see Table SM1.3 for additional observed changes in chemical water properties). There have been decreases in dissolved inorganic nitrogen from greater phytoplankton productivity (Sommaruga-Wograth et al., 1997; Rogora et al., 2003) and greater in-lake alkalinity generation and increases in pH in soft-water lakes (Psenner and Schmidt, 1992). Decreased solubility from higher temperatures significantly contributed to 11 to 13% of the decrease in aluminium concentration (Vesely et al., 2003), whereas lakes that had warmer water temperatures had increased mercury methylation and higher mercury levels in fish (Bodaly et al., 1993). A decrease in silicon content related to regional warming has been documented in Lake Baikal, Russia. River water-quality data from 27 rivers in Japan also suggest a deterioration in both chemical and biological features due to increases in air temperature.

1.3.2.4 Summary of hydrology and water resources

Changes in river discharge, as well as in droughts and heavy rains in some regions, indicate that hydrological conditions have become more intense. Significant trends in floods and in evaporation and evapotranspiration have not been detected globally. Some local trends in reduced groundwater and lake levels have been reported, but these are likely to be due to human activities rather than climate change. Climate-change signals related to increasing runoff and streamflow have been observed over the last century in many regions, particularly in basins fed by glaciers, permafrost and snow melt. Evidence includes increases in average runoff of Arctic rivers in Eurasia, which has been at least partly correlated with climate warming, and earlier spring snow melt and increase in winter base flow in North America and Eurasia due to enhanced seasonal snow melt associated with climate warming. There are also indications of intensified droughts in drier regions. Lake formation and their subsequent disappearance in permafrost have been reported in the Arctic. Freshwater lakes and rivers are experiencing increased water temperatures and changes in water chemistry. Surface and deep lake waters are warming, with advances and lengthening of periods of thermal stability in some cases associated with physical and chemical changes such as increases in salinity and suspended solids, and a decrease in nutrient content.

1.3.3 Coastal processes and zones

Many coastal regions are already experiencing the effects of relative (local) sea-level rise, from a combination of climate-induced sea-level rise, geological and anthropogenic-induced land subsidence, and other local factors. A major challenge, however, is to separate the different meteorological, oceanographic, geophysical and anthropogenic processes affecting the shoreline in order to identify and isolate the contribution of global warming. An unambiguous attribution of current sea-level rise as a primary driver of shoreline change is difficult to determine at present.

Global sea level has been rising at a rate of about 1.7 to 1.8 mm/yr over the last century, with an increased rate of about 3 mm/yr during the last decade (Church et al., 2004; Holgate and Woodworth, 2004; Church and White, 2006; Bindoff et al., 2007, Section 5.5).

1.3.3.1 Changes in coastal geomorphology

Sea-level rise over the last 100 to 150 years is probably contributing to coastal erosion in many places, such as the East Coast of the USA, where 75% of the shoreline removed from the influence of spits, tidal inlets and engineering structures is eroding (Leatherman et al., 2000; Daniel, 2001; Zhang et al., 2004) (Table 1.4; see Table SM1.4 for observations of changes in storm surges, flood height and areas, and waves). Over the last century, 67% of the eastern coastline of the UK has retreated landward of the low-water mark (Taylor et al., 2004).

In addition to sea-level change, coastal erosion is driven by other natural factors such as wave energy, sediment supply, or local land subsidence (Stive, 2004). In Louisiana, land subsidence has led to high average rates of shoreline retreat (averaging 0.61 m/yr between 1855 and 2002, and increasing to 0.94 m/yr since 1988) (Penland et al., 2005); further erosion occurred after Hurricanes Katrina and Rita in August 2005. These two hurricanes washed away an estimated 562 km^2 of coastal wetlands in Louisiana (USGS, 2006). Climate variability also affects shoreline processes, as documented by shoreline displacement in Estonia associated with increasing severe storms and high surge levels, milder winters, and reduced sea-ice cover (Orviku et al., 2003). Significant sections of glacially rebounding coastlines, which normally would be accreting, are nonetheless eroding, as for example along Hudson Bay, Canada (Beaulieu and Allard, 2003). Reduction in sea-ice cover due to milder winters has also exacerbated coastal erosion, as in the Gulf of St. Lawrence (Bernatchez and Dubois, 2004; Forbes et al., 2004). Degradation and melting of permafrost due to climate warming are also contributing to the rapid retreat of Arctic coastlines in many regions, such as the Beaufort and Laptev Sea coasts (Forbes, 2005).

Anthropogenic activities have intensified beach erosion in many parts of the world, including Fiji, Trinidad and parts of tropical Asia (Mimura and Nunn, 1998; Restrepo et al., 2002; Singh and Fouladi, 2003; Wong, 2003). Much of the observed erosion is associated with shoreline development, clearing of mangroves (Thampanya et al., 2006) and mining of beach sand and coral. Sediment starvation due to the construction of large dams upstream also contributes to coastal erosion (Frihy et al., 1996; Chen et al., 2005b; Georgiou et al., 2005; Penland et al., 2005; Syvitski et al., 2005b; Ericson et al., 2006). Pumping of groundwater and subsurface hydrocarbons also enhances land subsidence, thereby exacerbating coastal erosion (Syvitski et al., 2005a).

1.3.3.2 Changes in coastal wetlands

In the USA, losses in coastal wetlands have been observed in Louisiana (Boesch et al., 1994), the mid-Atlantic region (Kearney et al., 2002), and in parts of New England and New York (Hartig et al., 2002; Hartig and Gornitz, 2004), in spite of recent protective environmental regulations (Kennish, 2001). Many of these marshes have had a long history of anthropogenic modification, including dredging and filling, bulkheading and channelisation, which in turn could have contributed to sediment starvation, eutrophication and ultimately marsh submergence (Donnelly and Bertness, 2001; Bertness et al., 2002). In Europe, losses have been documented in south-east England between 1973 and 1998, although the rate of loss has slowed since 1988 (Cooper et al., 2001); elsewhere there is evidence that not all coastal wetlands are retreating, for example in Normandy, France (Haslett et al., 2003).

Although natural accretion rates of mangroves generally compensate for current rates of sea-level rise, of greater concern at present are the impacts of clearance for agriculture, aquaculture (particularly shrimp), forestry and urbanisation. At least 35% of the world's mangrove forests have been removed in the last two decades but possible sea-level rise effects were not considered (Valiela et al., 2001). In south-eastern Australia, mangrove encroachment inland into salt-marsh environments is probably related to anthropogenic causes and climate variability, rather than sea-level rise (Saintilan and Williams, 1999). Landward replacement of grassy freshwater marshes by more salt-tolerant mangroves in the south-eastern Florida Everglades since the 1940s has been attributed to the combined effects of sea-level rise and water management, resulting in lowered watertables (Ross et al., 2000).

Sea-level rise can have a larger impact on wetland ecosystems when the human land-use pressure in the coastal area is large, e.g., coasts defended by dykes and urbanisation. Wetlands disappear or become smaller when human land use makes inward movement of the ecosystem impossible (Wolters et al., 2005).

1.3.3.3 Changes in storm surges, flood heights and areas, and waves

The vulnerability of the coastal zone to storm surges and waves depends on land subsidence, changes in storminess, and sea-level rise (see Supplementary Material). Along the North American East Coast, although there has been no significant long-term change in storm climatology, storm-surge impacts have increased due to regional sea-level rise (Zhang et al., 2000). The U.S. Gulf Coast is particularly vulnerable to hurricane surges due to low elevation and relative sea-level rise (up to

Table 1.4. *Changes in coastal processes.*

Type of change	Observed changes	Period	Location	References
Shoreline erosion	75% of shoreline, uninfluenced by inlets and structures, is eroding	mid-1800s to 2000	East Coast USA	Zhang et al., 2004
	Shoreline retreat, 0.61 m/yr	1855-2002	Louisiana, USA	Penland et al., 2005
	Shoreline retreat, 0.94 m/yr	1988-2002		
	Beach erosion prevalent due to sea-level rise, mangrove clearance	1960s-1990s	Fiji	Mimura and Nunn, 1998
	Beach erosion due to coral bleaching, mangrove clearance, sand mining, structures	1950s-2000	Tropics: SE Asia, Indian Ocean, Australia, Barbados	Wong, 2003
	19% of studied shoreline is retreating, in spite of land uplift, due to thawing of permafrost	1950-1995	Manitounuk Strait, Canada	Beaulieu and Allard, 2003
	Shoreline erosion, recent acceleration	Pre-1990s to present	Estuary and Gulf of St. Lawrence, Canada	Bernatchez and Dubois, 2004; Forbes et al., 2004
	Increased thermokarst erosion due to climate warming	1970-2000 relative to 1954-1970	Arctic Ocean, Beaufort Sea coasts, Canada	Lantuit and Pollard, 2003
	Beach erosion due to dams across the Nile and reduced river floods due to precipitation changes	Late 20th century	Alexandria, Egypt	Frihy et al., 1996
	Coastal erosion	1843-present	UK coastline	Taylor et al., 2004
Wetland changes	About 1,700 ha of degraded marshes became open water; non-degraded marshes decreased by 1,200 ha	1938-1989	Chesapeake Bay, USA	Kearney et al., 2002
	Decreases in salt marsh area due to regional sea-level rise and human impacts	1920s-1999	Long Island, NY; Connecticut, USA	Hartig et al., 2002; Hartig and Gornitz, 2004
	Salt marshes keep up with sea-level rise with sufficient sediment supply	1880-2000	Normandy, France	Haslett et al., 2003
	Landward migration of cordgrass (*Spartina alterniflora*) due to sea-level rise and excess nitrogen	1995-1999; late 20th century	Rhode Island, USA	Donnelly and Bertness 2001; Bertness et al., 2002
	Decrease from 12,000 to 4,000 ha, from land reclamation, wave-induced erosion and insufficient sediment	1919-2000	Venice, Italy	Day et al., 2005
	Seaward-prograding mudflats replacing sandy beaches, due to increased dredged sediment supply	1897-1999	Queensland coast, Australia	Wolanski et al., 2002
	Wetland losses due to sea-level rise, land reclamation, changes in wind/wave energy, tidal dynamics	1850s-1990s	Greater Thames Estuary, UK	van der Wal and Pye, 2004
	Decreased rates of deltaic wetland progradation due to reduced sediment supply from dam construction	1960s-2003	Yangtze River Delta, Peoples Republic of China	Yang et al., 2005
Coastal vegetation changes	Grassy marshes replaced by mangrove due to sea-level rise, water table changes	1940-1994	South-east Florida, USA	Ross et al., 2000
	Mangrove encroachment into estuarine wetlands due to changing water levels, increased nutrient load, and salt-marsh compaction during drought	1940s-1990s	South-east Australia	Saintilan and Williams,1999; Rogers et al., 2006

1 cm/yr along parts of the Louisiana coast), only part of which is climate-related (Penland et al., 2005). Hurricane Katrina, in August 2005, generated surges over 4 m, with catastrophic consequences (NOAA, 2005). In Venice, Italy, the frequency of surges has averaged around 2 per year since the mid-1960s, compared with only 0.19 surges per year between 1830 and 1930, with land subsidence, which was exacerbated by groundwater pumping between 1930 and 1970 (Carminati et al., 2005), and expanded sea-lagoon interactions (due to channel dredging) playing a greater role than global sea-level rise (Camuffo and Sturaro, 2004). Surges have shown a slight decrease in Brittany, France, in recent decades, largely due to changes in wind patterns (Pirazzoli et al., 2004).

Apparent global increases in extreme high water levels since 1975 are related to mean sea-level rise and to large-scale inter-decadal climate variability (Woodworth and Blackman, 2004). Wave height increases have been documented in the north-east Atlantic Ocean (Woolf et al., 2002), along the US Pacific North-west coast (Allan and Komar, 2006) and in the Maldives (Woodworth and Blackman, 2004), but decreases have been found in some areas of the Mediterranean from 1958 to 2001 (Lionello, 2005; Lionello and Sanna, 2005).

1.3.3.4 Summary of coastal processes and zones

In many coastal regions, particularly in subsiding regions, local sea-level rise exceeds the 20th century global trend of 1.7 to 1.8 mm/yr. Sea-level rise, enhanced wave heights, and increased intensity of storms are affecting some coastal regions distant from human modification, e.g., polar areas and barrier beaches, mainly through coastal erosion. Coastal erosion and losses of wetlands are widespread problems today, under current rates of sea-level rise, although these are largely caused by anthropogenic modification of the shoreline.

1.3.4 Marine and freshwater biological systems

The marine pelagic realm occupies 70% of the planetary surface and plays a fundamental role in modulating the global environment via climate regulation and biogeochemical cycling (Legendre and Rivkin, 2002). Perhaps equally important to global climate change, in terms of modifying the biology of the oceans, is the impact of anthropogenic CO_2 on the pH of the oceans, which will affect the process of calcification for some marine organisms (Feely et al., 2004), but effects of this are as yet undocumented. Other driving forces of change that are operative in marine and freshwater biological systems are over-fishing and pollution from terrestrial runoff (from deforestation, agriculture and urban development) and atmospheric deposition, and human introduction of non-native species.

Observational changes in marine and freshwater environments associated with climate change should be considered against the background of natural variation on a variety of spatial and temporal scales. While many of the biological responses have been associated with rising temperatures, distinguishing the effects of climate change embedded in natural modes of variability such as ENSO and the NAO is challenging.

1.3.4.1 Changes in coral reefs

Concerns about the impacts of climate change on coral reefs centre on the effects of the recent trends in increasing acidity (via increasing CO_2), storm intensity and sea surface temperatures (see Bindoff et al., 2007, Section 5.4.2.3; Trenberth et al., 2007, Sections 3.8.3 and 3.2.2).

Decreasing pH (see Chapter 4, Box 4.4) leads to a decreased aragonite saturation state, one of the main physicochemical determinants of coral calcification (Kleypas et al., 1999). Although laboratory experiments have demonstrated a link between aragonite saturation state and coral growth (Langdon et al., 2000; Ohde and Hossain, 2004), there are currently no data relating altered coral growth *in situ* to increasing acidity.

Storms damage coral directly through wave action and indirectly through light attenuation by suspended sediment and abrasion by sediment and broken corals. Most studies relate to individual storm events, but a meta-analysis of data from 1977 to 2001 showed that coral cover on Caribbean reefs decreased by 17% on average in the year following a hurricane, with no evidence of recovery for at least 8 years post-impact (Gardner et al., 2005). Stronger hurricanes caused more coral loss, but the second of two successive hurricanes caused little additional damage, suggesting a greater future effect from increasing hurricane intensity rather than from increasing frequency (Gardner et al., 2005).

There is now extensive evidence of a link between coral bleaching – a whitening of corals as a result of the expulsion of symbiotic zooxanthellae (see Chapter 6, Box 6.1) – and sea surface temperature anomalies (McWilliams et al., 2005). Bleaching usually occurs when temperatures exceed a 'threshold' of about 0.8-1°C above mean summer maximum levels for at least 4 weeks (Hoegh-Guldberg, 1999). Regional-scale bleaching events have increased in frequency since the 1980s (Hoegh-Guldberg, 1999). In 1998, the largest bleaching event to date is estimated to have killed 16% of the world's corals, primarily in the western Pacific and the Indian Ocean (Wilkinson, 2004). On many reefs, this mortality has led to a loss of structural complexity and shifts in reef fish species composition (Bellwood et al., 2006; Garpe et al., 2006; Graham et al., 2006). Corals that recover from bleaching suffer temporary reductions in growth and reproductive capacity (Mendes and Woodley, 2002), while the recovery of reefs following mortality tends to be dominated by fast-growing and bleaching-resistant coral genera (Arthur et al., 2005).

While there is increasing evidence for climate change impacts on coral reefs, disentangling the impacts of climate-related stresses from other stresses (e.g., over-fishing and pollution; Hughes et al., 2003b) is difficult. In addition, inter-decadal variation in pH (Pelejero et al., 2005), storm activity (Goldenberg et al., 2001) and sea surface temperatures (Mestas-Nunez and Miller, 2006) linked, for example, to the El Niño-Southern Oscillation and Pacific Decadal Oscillation, make it more complicated to discern the effect of anthropogenic climate change from natural modes of variability (Section 1.3.4). An analysis of bleaching in the Caribbean indicates that 70% of the variance in geographic extent of bleaching between 1983 and 2000 could be attributed to variation in ENSO and atmospheric dust (Gill et al., 2006).

1.3.4.2 Changes in marine ecosystems

There is an accumulating body of evidence to suggest that many marine ecosystems, including managed fisheries, are responding to changes in regional climate caused predominately by warming of air and sea surface temperatures (SSTs) and to a lesser extent by modification of precipitation regimes and wind patterns (Table 1.5). The biological manifestations of rising SSTs have included biogeographical, phenological, physiological and species abundance changes. The evidence collected and modelled to date indicates that rising CO_2 has led to chemical changes in the ocean, which in turn have led to the oceans becoming more acidic (Royal Society, 2005). Blended satellite/*in situ* ocean chlorophyll records indicate that global ocean annual primary production has declined by more than 6% since the early 1980s (Gregg et al., 2003), whereas chlorophyll in the North-east Atlantic has increased since the mid-1980s (Raitsos et al., 2005).

In the Pacific and around the British Isles, researchers have found changes to the intertidal communities, where the composition has shifted significantly in response to warmer temperatures (Sagarin et al., 1999; Southward et al., 2005). Similar shifts were also noted in the kelp forest fish communities off the southern Californian coast and in the offshore zooplankton

Table 1.5. *Examples of changes in marine ecosystems and managed fisheries.*

Key changes	Climate link	Location	References
Pelagic productivity/ zooplankton abundance/ plankton assemblages	Biological responses to regional changes in temperature, stratification, upwelling, and other hydro-climatic changes	North Atlantic	Fromentin and Planque, 1996; Reid et al., 1998; Edwards et al., 2002; Beaugrand et al., 2003; Johns et al., 2003; Richardson and Schoeman, 2004
		North Pacific	Roemmich and McGowan, 1995; Walther et al., 2002; Lavaniegos and Ohman, 2003; Chiba and Tadokoro, 2006
		South Atlantic	Verheye et al., 1998
		Southern Ocean	Walther et al., 2002; Atkinson et al., 2004
Pelagic phenology	Earlier seasonal appearance due to increased temperature and trophic mismatch	North Sea	Edwards and Richardson, 2004; Greve, 2004
Pelagic biogeography	Northerly movement of plankton communities due to general warming	Eastern North Atlantic	Beaugrand et al., 2002b
	Southerly movement of boreal plankton in the western North Atlantic due to lower salinities	Western North Atlantic	Johns et al., 2001
Rocky shore/ intertidal communities	Community changes due to regional temperature changes	British Isles	Hawkins et al., 2003; Southward et al., 2005
		North Pacific	Sagarin et al., 1999
Kelp forests/ macroalgae	Effect on communities and spread of warmer-water species due to increased temperatures	North Pacific	Holbrook et al., 1997
		Mediterranean	Walther et al., 2002
Pathogens and invasive species	Geographical range shifts due to increased temperatures	North Atlantic	Harvell et al., 1999; Walther et al., 2002; McCallum et al., 2003
Fish populations and recruitment success	Changes in populations, recruitment success, trophic interactions and migratory patterns related to regional environmental change	British Isles	Attrill and Power, 2002
		North Pacific	McGowan et al., 1998; Chavez et al., 2003
		North Atlantic	Walther et al., 2002; Beaugrand and Reid, 2003; Beaugrand et al., 2003; Brander et al., 2003; Drinkwater et al., 2003
		Barents Sea	Stenseth et al., 2002; Walther et al., 2002
		Mediterranean	Walther et al., 2002
		Bering Sea	Grebmeier et al., 2006
Fish biogeography	Geographical range shifts related to temperature	NE Atlantic	Brander et al., 2003; Beare et al., 2004; Genner et al., 2004; Perry et al., 2005
		NW Atlantic	Rose and O'Driscoll, 2002
		Bering Sea	Grebmeier et al., 2006
Seabirds and marine mammals	Population changes, migratory patterns, trophic interactions and phenology related to regional environmental change, ice habitat loss related to warming	North Atlantic	Walther et al., 2002; Drinkwater et al., 2003; Frederiksen et al., 2004
		North Pacific	McGowan et al., 1998; Hughes, 2000
		Southern Ocean	Barbraud and Weimerskirch, 2001; Walther et al., 2002; Weimerskirch et al., 2003; Forcada et al., 2006; Stirling and Parkinson, 2006
Marine biodiversity	Regional response to general warming	North Atlantic	Beaugrand et al., 2002a

communities (Roemmich and McGowan, 1995; Holbrook et al., 1997; Lavaniegos and Ohman, 2003). These changes are associated with oceanic warming and the resultant geographical movements of species with warmer water affinities. As in the North Atlantic, many long-term biological investigations in the Pacific have established links between changes in the biology and regional climate oscillations such as the ENSO and the Pacific Decadal Oscillation (PDO) (Stenseth et al., 2002). In the case of the Pacific, these biological changes are most strongly associated with El Niño events, which can cause rapid and sometimes dramatic responses to the short-term SST changes (Hughes, 2000). However, recent investigations of planktonic foraminifera from sediment cores encompassing the last 1,400 years has revealed anomalous change in the community structure over the last few decades. The study suggests that ocean warming has already exceeded the range of natural variability (Field et al., 2006). A recent major ecosystem shift in the northern Bering Sea has been attributed to regional climate warming and trends in the Arctic Oscillation (Grebmeier et al., 2006).

The progressive warming in the Southern Ocean has been associated with a decline in krill (Atkinson et al., 2004) and an associated decline in the population size of many seabirds and seals monitored on several breeding sites (Barbraud and Weimerskirch, 2001; Weimerskirch et al., 2003). Some initial observations suggest that changes to the ice habitat via the total thickness of sea ice and its progressively earlier seasonal break-up in the Arctic and Antarctic caused by regional climate warming has had a detrimental impact on marine mammal and seabird populations (Forcada et al., 2005, 2006; Stirling and Parkinson, 2006).

In the North Atlantic, changes in both phytoplankton and zooplankton species and communities have been associated with Northern Hemisphere temperature (NHT) trends and variations in the NAO index. These have included changes in species distributions and abundance, the occurrence of sub-tropical species in temperate waters, changes in overall phytoplankton biomass and seasonal length, changes in the ecosystem functioning and productivity of the North Sea, shifts from cold-adapted to warm-adapted communities, phenological changes, changes in species interactions, and an increase in harmful algal blooms (HABs) (Fromentin and Planque, 1996; Reid et al., 1998; Edwards et al., 2001, 2002, 2006; Reid and Edwards, 2001; Beaugrand et al., 2002a, 2003; Beaugrand and Reid, 2003; Edwards and Richardson, 2004; Richardson and Schoeman, 2004). Over the last decade, numerous other investigations have established links between the NAO and the biology of the North Atlantic, including the benthos, fish, seabirds and whales (Drinkwater et al., 2003) and an increase in the incidence of marine diseases (Harvell et al., 1999). In the Benguela upwelling system in the South Atlantic, long-term trends in the abundance and community structure of coastal zooplankton have been related to large-scale climatic influences (Verheye et al., 1998).

Recent macroscale research has shown that the increase in regional sea temperatures has triggered a major reorganisation in calanoid copepod species composition and biodiversity over the whole North Atlantic Basin (Figure 1.3) (Beaugrand et al., 2002a). During the last 40 years there has been a northerly movement of warmer-water plankton by 10° latitude in the North-East Atlantic and a similar retreat of colder-water plankton to the north. This geographical movement is much more pronounced than any documented terrestrial study, presumably due to advective movements accelerating these processes. In terms of the marine phenological response to climate warming, many plankton taxa have been found to be moving forward in their seasonal cycles (Edwards and Richardson, 2004). In some cases, a shift in seasonal cycles of over six weeks was detected, but more importantly the response to climate warming varied between different functional groups and trophic levels, leading to a mismatch in timing between different trophic levels (Edwards and Richardson, 2004).

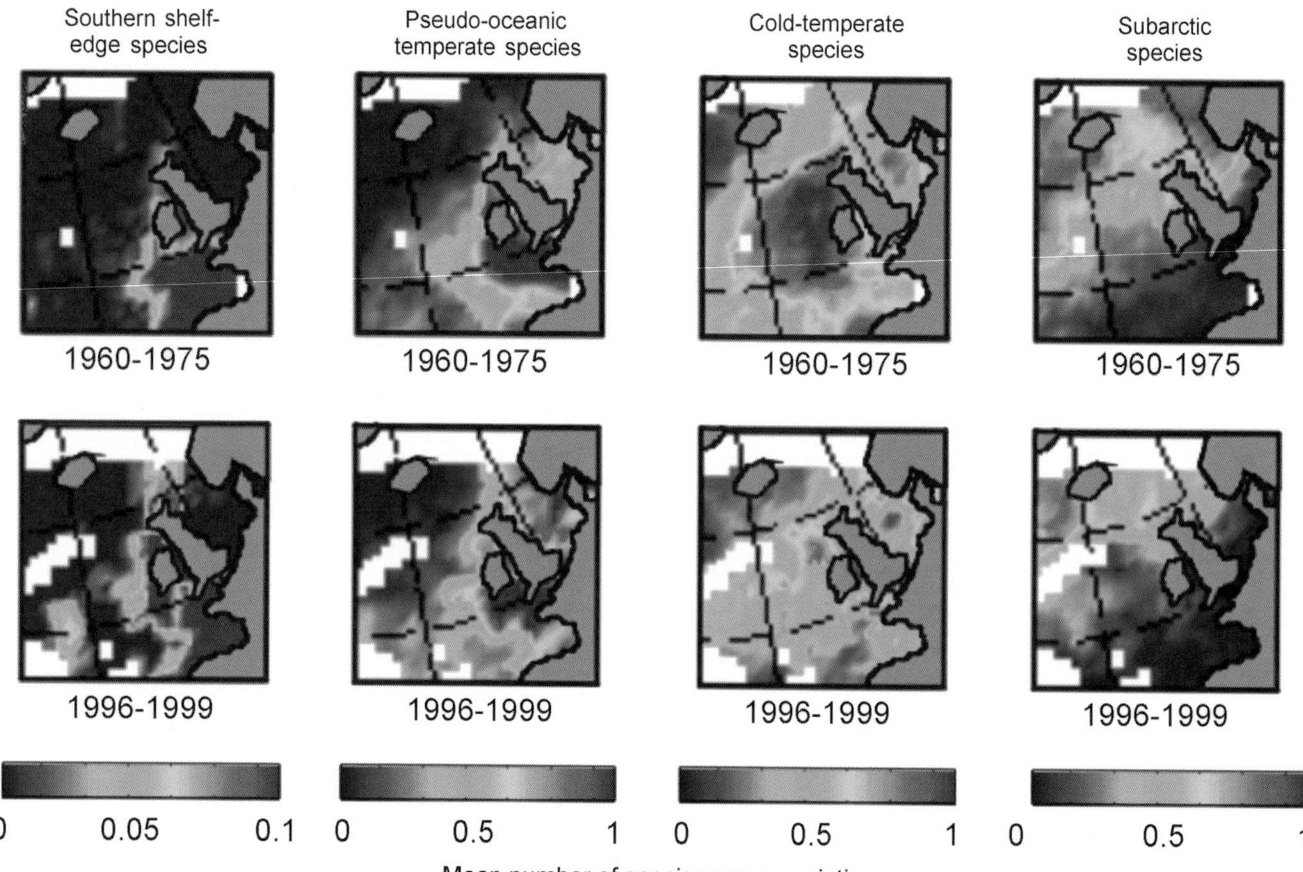

Figure 1.3. *Long-term changes in the mean number of marine zooplankton species per association in the North Atlantic from 1960 to 1975 and from 1996 to 1999. The number of temperate species has increased and the diversity of colder-temperate, sub-Arctic and Arctic species has decreased in the North Atlantic. The scale (0 to 1) indicates the proportion of biogeographical types of species in total assemblages of zooplankton. From Beaugrand et al., 2002b. Reprinted with permission from AAAS.*

1.3.4.3 Changes in marine fisheries

Northerly geographical range extensions or changes in the geographical distribution of fish populations have recently been documented for European Continental shelf seas and along the European Continental shelf edge (Brander et al., 2003; Beare et al., 2004; Genner et al., 2004; Perry et al., 2005). These geographical movements have been related to regional climate warming and are predominantly associated with the northerly geographical movement of fish species (sardines, anchovies, red mullet and bass) with more southern biogeographical affinities. Northerly range extensions of pelagic fish species have also been reported for the Northern Bering Sea region related to regional climate warming (Grebmeier et al., 2006). New records have also been observed over the last decade for some Mediterranean and north-west African species on the south coast of Portugal (Brander et al., 2003). Cooling and freshening of the North-West Atlantic (e.g., in the sub-polar gyre, Labrador Sea and Labrador Current) over the last decade has had an opposite effect, with some groundfish species moving further south (Rose and O'Driscoll, 2002) in the same way as plankton (see 1.3.4.2).

Regional climate warming in the North Sea has affected cod recruitment via changes at the base of the food web (Beaugrand et al., 2003). Key changes in the planktonic assemblage, significantly correlated with the warming of the North Sea over the last few decades, has resulted in a poor food environment for cod larvae, and hence an eventual decline in overall recruitment success. This is an example of how the dual pressures of over-fishing and regional climate warming have combined to negatively affect a commercially important fishery. Recent work on pelagic phenology in the North Sea has shown that plankton communities, including fish larvae, are very sensitive to regional climate warming, with the response varying between trophic levels and functional groups (Edwards and Richardson, 2004). The ability and speed with which fish and planktonic communities adapt to regional climate warming is not yet known.

1.3.4.4 Changes in lakes

Observations indicate that lakes and rivers around the world are warming, with effects on thermal structure and lake chemistry that in turn affect abundance and productivity, community composition, phenology, distribution and migration (see Section 1.3.2.3) (Tables 1.3 and 1.6).

Abundance/productivity

In high-latitude or high-altitude lakes where reduced ice cover has led to a longer growing season and warmer temperatures, many lakes are showing increased algal abundance and productivity over the past century (Schindler et al., 1990; Hambright et al., 1994; Gajewski et al., 1997; Wolfe and Perren, 2001; Battarbee et al., 2002; Korhola et al., 2002; Karst-Riddoch et al., 2005). There have been similar increases in the abundance of zooplankton, correlated with warmer water temperatures and longer growing seasons (Adrian and Deneke, 1996; Straile and Adrian, 2000; Battarbee et al., 2002; Gerten and Adrian, 2002; Carvalho and Kirika, 2003; Winder and Schindler, 2004b; Hampton, 2005; Schindler et al., 2005). For upper trophic levels, rapid increases in water temperature after ice break-up have enhanced fish recruitment in oligotrophic lakes (Nyberg et al., 2001). In contrast to these lakes, some lakes, particularly deep tropical lakes, are experiencing reduced algal abundance and declines in productivity because stronger stratification reduces upwelling of the nutrient-rich deep water (Verburg et al., 2003; O'Reilly, 2007). Primary productivity in Lake Tanganyika may have decreased by up to 20% over the past 200 years (O'Reilly et al., 2003), and for the East African Rift Valley lakes, recent declines in fish abundance have been linked with climatic impacts on lake ecosystems (O'Reilly, 2007).

Community composition

Increases in the length of the ice-free growing season, greater stratification, and changes in relative nutrient availability have generated shifts in community composition. Of potential concern to human health is the increase in relative abundance of cyanobacteria, some of which can be toxic, in some freshwater ecosystems (Carmichael, 2001; Weyhenmeyer, 2001; Briand et al., 2004). Palaeolimnological records have shown widespread changes in phytoplankton species composition since the mid-to-late 1800s due to climate shifts, with increases in chrysophytes and planktonic diatom species and decreases in benthic species (Gajewski et al., 1997; Wolfe and Perren, 2001; Battarbee et al., 2002; Sorvari et al., 2002; Laing and Smol, 2003; Michelutti et al., 2003; Perren et al., 2003; Ruhland et al., 2003; Karst-Riddoch et al., 2005; Smol et al., 2005). These sedimentary records also indicated changes in zooplankton communities (Douglas et al., 1994; Battarbee et al., 2002; Korhola et al., 2002; Brooks and Birks, 2004; Smol et al., 2005). In relatively productive lakes, there was a shift towards more diverse periphytic diatom communities due to increased macrophyte growth (Karst-Riddoch et al., 2005). In lakes where nutrients are becoming limited due to increased stratification, phytoplankton composition shifted to relatively fewer diatoms, potentially reducing food quality for upper trophic levels (Adrian and Deneke, 1996; Verburg et al., 2003; O'Reilly, 2007). Warming has also produced northward shifts in the distribution of aquatic insects and fish in the UK (Hickling et al., 2006).

Phenology

With earlier ice break-up and warmer water temperatures, some species have responded to the earlier commencement of the growing season, often advancing development of spring algal blooms as well as clear-water phases. The spring algal bloom now occurs about 4 weeks earlier in several large lakes (Gerten and Adrian, 2000; Straile and Adrian, 2000; Weyhenmeyer, 2001; Winder and Schindler, 2004b). In many cases where the spring phytoplankton bloom has advanced, zooplankton have not responded similarly, and their populations are declining because their emergence no longer corresponds with high algal abundance (Gerten and Adrian, 2000; Winder and Schindler, 2004a). Zooplankton phenology has also been affected by climate (Gerten and Adrian, 2002; Winder and Schindler, 2004a) and phenological shifts have also been demonstrated for some wild and farmed fish species (Ahas, 1999; Elliott et al., 2000). Because not all organisms respond similarly, differences in the magnitude of phenological responses among species has affected food-web interactions (Winder and Schindler, 2004a).

Table 1.6. *Examples of changes in freshwater ecosystems due to climate warming.*

Environmental factor	Observed changes	Time period considered	Location of lakes/rivers	Total number of lakes/rivers studied	Selected references
Productivity or biomass	Increases associated with longer growing season	100 years	North America, Europe, Eastern Europe	26	Schindler et al., 1990; Adrian and Deneke, 1996; Gajewski et al., 1997; Weyhenmeyer et al., 1999; Straile and Adrian, 2000; Wolfe and Perren, 2001; Battarbee et al., 2002; Gerten and Adrian, 2002; Korhola et al., 2002; Carvalho and Kirika, 2003; Shimaraev and Domysheva, 2004; Winder and Schindler, 2004b; Hampton, 2005; Karst-Riddoch et al., 2005; Schindler et al., 2005
	Decreases due to decreased nutrient availability	100 years	Europe, East Africa	5	Adrian et al., 1995; O'Reilly et al., 2003; Verburg et al., 2003; O'Reilly, 2007
Algal community composition	Shift from benthic to planktonic species	100 to 150 years	North America, Europe	66	Gajewski et al., 1997; Wolfe and Perren, 2001; Battarbee et al., 2002; Sorvari et al., 2002; Laing and Smol, 2003; Michelutti et al., 2003; Perren et al., 2003; Ruhland et al., 2003; Karst-Riddoch et al., 2005; Smol et al., 2005
	Decreased diatom abundance	100 years	East Africa, Europe	3	Adrian and Deneke, 1996; Verburg et al., 2003; O'Reilly, 2007
Phenology	Spring algal bloom up to 4 weeks earlier, earlier clear water phase	45 years	North America, Europe	5	Weyhenmeyer et al., 1999; Gerten and Adrian, 2000; Straile and Adrian, 2000; Gerten and Adrian, 2002; Winder and Schindler, 2004a, 2004b
Fish migration	From 6 days to 6 weeks earlier	20 to 50 years	North America	5	Quinn and Adams, 1996; Huntington et al., 2003; Cooke et al., 2004; Juanes et al., 2004; Lawson et al., 2004

1.3.4.5 Changes in rivers

In rivers, water flow can influence water chemistry, habitat, population dynamics, and water temperature (Schindler et al., 2007). Specific information on the effect of climate change on hydrology can be found in Section 1.3.2. Increasing river temperatures have been associated with increased biological demand and decreased dissolved oxygen, without changes in flow (Ozaki et al., 2003). Riverine dissolved organic carbon concentrations have doubled in some cases because of increased carbon release in the catchment as temperature has risen (Worrall et al., 2003).

Abundance, distribution and migration

Climate-related changes in rivers have affected species abundance, distribution and migration patterns. While warmer water temperatures in many rivers have positively influenced the breeding success of fish (Fruget et al., 2001; Grenouillet et al., 2001; Daufresne et al., 2004), the stressful period associated with higher water temperatures for salmonids has lengthened as water temperatures have increased commensurate with air temperatures in some locations (Bartholow, 2005). In the Rhône River there have been significant changes in species composition as southern, thermophilic fish and invertebrate species have progressively replaced cold-water species (Doledec et al., 1996; Daufresne et al., 2004). Correlated with long-term increases in water temperature, the timing of fish migrations in large rivers in North America has advanced by up to 6 weeks in some years (Quinn and Adams, 1996; Huntington et al., 2003; Cooke et al., 2004; Juanes et al., 2004). Increasing air temperatures have been negatively correlated with smolt production (Lawson et al., 2004), and earlier migrations are associated with greater en-route and pre-spawning mortality (up to 90%) (Cooke et al., 2004). Warming in Alpine rivers caused altitudinal habitat shifts upward for brown trout, and there were increased incidences of temperature-dependent kidney disease at the lower-elevational habitat boundary (Hari et al., 2006).

1.3.4.6 Summary of marine and freshwater biological systems

In marine and freshwater ecosystems, many observed changes in phenology and distribution have been associated with rising water temperatures, as well as changes in salinity, oxygen levels and circulation. While there is increasing evidence for climate change impacts on coral reefs, separating the impacts of climate-related stresses from other stresses (e.g., over-fishing and pollution) is difficult. Globally, freshwater ecosystems are showing changes in organism abundance and productivity, range expansions, and phenological shifts (including earlier fish migrations) that are linked to rising temperatures. Many of these climate-related impacts are now influencing the ways in which marine and freshwater ecosystems function.

1.3.5 Terrestrial biological systems

Plants and animals can reproduce, grow and survive only within specific ranges of climatic and environmental conditions. If conditions change beyond the tolerances of species, then they may respond by:

1. shifting the timing of life-cycle events (e.g., blooming, migrating),
2. shifting range boundaries (e.g., moving poleward) or the

density of individuals within their ranges,

3. changing morphology (e.g., body or egg size), reproduction or genetics,
4. extirpation or extinction.

Additionally, each species has its unique requirements for climatic and environmental conditions. Changes, therefore, can lead to disruption of biotic interaction (e.g., predator/prey) and to changes of species composition as well as ecosystem functioning. Since the TAR, the number of studies finding plants or animals responding to changing climate (associated with varying levels of confidence) has risen substantially, as has the number of reviews (Hughes, 2000; Menzel and Estrella, 2001; Sparks and Menzel, 2002; Walther et al., 2002; Parmesan and Galbraith, 2004; Linderholm, 2006; Parmesan, 2006).

Besides climate affecting species, there are many different types of non-climate driving forces, such as invasive species, natural disturbances (e.g., wildfires), pests, diseases and pollution (e.g., soluble-nitrogen deposition), influencing the changes exhibited by species. Many animal and plant populations have been under pressure from agricultural intensification and land-use change in the past 50 years, causing many species to be in decline. Habitat fragmentation (Hill et al., 1999b; Warren et al., 2001) or simply the absence of suitable areas for colonisation, e.g., at higher elevations, also play an important role (Wilson et al., 2005), especially in species extinction (Williams et al., 2003; Pounds et al., 2006).

1.3.5.1 Changes in phenology

Phenology – the timing of seasonal activities of animals and plants – is perhaps the simplest process in which to track changes in the ecology of species in response to climate change. Observed phenological events include leaf unfolding, flowering, fruit ripening, leaf colouring, leaf fall of plants, bird migration, chorusing of amphibians, and appearance/emergence of butterflies. Numerous new studies since the TAR (reviewed by Menzel and Estrella, 2001; Sparks and Menzel, 2002; Walther et al., 2002; Menzel, 2003; Walther, 2004) and three meta-analyses (Parmesan and Yohe, 2003; Root et al., 2003; Lehikoinen et al., 2004) (see Section 1.4.1) concurrently document a progressively earlier spring by about 2.3 to 5.2 days/decade in the last 30 years in response to recent climate warming.

Although phenological network studies differ with regard to regions, species, events observed and applied methods, their data show a clear temperature-driven extension of the growing season by up to 2 weeks in the second half of the 20th century in mid- and high northern latitudes (see Table 1.7), mainly due to an earlier spring, but partly due also to a later autumn. Remotely-sensed vegetation indices (Myneni et al., 1997; Zhou et al., 2001; Lucht et al., 2002) and analysis of the atmospheric CO_2 signal (Keeling et al., 1996) confirm these findings. A corresponding longer frost-free and climatological growing season is also observed in North America and Europe (see Section 1.3.6.1). This lengthening of the growing season might also account for observed increases in productivity (see Section 1.3.6.2). The signal in autumn is less pronounced and more homogenous. The very few examples of single-station data indicate a much greater lengthening or even a shortening of the growing season (Kozlov and Berlina, 2002; Peñuelas et al., 2002).

Altered timing of spring events has been reported for a broad multitude of species and locations; however, they are primarily from North America, Eurasia and Australia. Network studies where results from all sites/several species are reported, irrespective of their significance (Table 1.8), show that leaf unfolding and flowering in spring and summer have, on average, advanced by 1-3 days per decade in Europe, North America and Japan over the last 30 to 50 years. Earlier flowering implies an earlier start of the pollen season (see Section 1.3.7.4). There are also indications that the onset of fruit ripening in early autumn has advanced in many cases (Jones and Davis, 2000; Peñuelas et al., 2002; Menzel, 2003) (see also Section 1.3.6.1). Spring and summer phenology is sensitive to climate and local weather (Sparks et al., 2000; Lucht et al., 2002; Menzel, 2003). In contrast to autumn phenology (Estrella and Menzel, 2006), their climate signal is fairly well understood: nearly all spring and summer changes in plants, including agricultural crops (Estrella et al., 2007), correlate with spring temperatures in the preceding months. The advancement is estimated as 1 to12 days for every 1°C increase in spring temperature, with average values ranging between 2.5 and 6 days per °C (e.g., Chmielewski and Rotzer, 2001; Menzel, 2003; Donnelly et al., 2004; Menzel et al., 2006b). Alpine species are also partly sensitive to photoperiod (Keller and Korner, 2003) or amount of snowpack (Inouye et al., 2002). Earlier spring events and a longer growing season in Europe are most apparent for time-series ending in the mid-1980s or later (Schaber, 2002; Scheifinger et al., 2002; Dose and Menzel, 2004; Menzel and Dose, 2005), which matches the

Table 1.7. *Changes in length of growing season, based on observations within networks.*

Location	Period	Species/Indicator	Lengthening (days/decade)	References
Germany	1951-2000	4 deciduous trees (LU/LC)	1.1 to 2.3	Menzel et al., 2001; Menzel, 2003
Switzerland	1951-1998	9 spring, 6 autumn phases	2.7*	Defila and Clot, 2001
Europe (Int. Phenological Gardens)	1959-1996 1969-1998	Various spring/autumn phases (LU to LC, LF)	3.5	Menzel and Fabian, 1999; Menzel, 2000; Chmielewski and Rotzer, 2001
Japan	1953-2000	*Gingko biloba* (LU/LF)	2.6	Matsumoto et al., 2003
Northern Hemisphere	1981-1999	Growing season by normalised difference vegetation index (NDVI)	0.7 to 1	Zhou et al., 2001

LU = leaf unfolding; LC = leaf colouring; LF = leaf fall. * indicates mean of significant trends only.

Table 1.8. *Changes in the timing of spring events, based on observations within networks.*

Location	Period	Species/Indicator	Observed changes (days/decade)	References
Western USA	1957-1994	Lilac, honeysuckle (F)	−1.5 (lilac), 3.5 (honeysuckle)	Cayan et al., 2001
North-eastern USA	1965-2001 1959-1993	Lilac (F, LU) Lilac (F)	−3.4 (F) −2.6 (U) −1.7	Wolfe et al., 2005 Schwartz and Reiter, 2000
Washington, DC	1970-1999	100 plant species (F)	−0.8	Abu-Asab et al., 2001
Germany	1951-2000	10 spring phases (F, LU)	−1.6	Menzel et al., 2003
Switzerland	1951-1998	9 spring phases (F, LU)	−2.3 (*)	Defila and Clot, 2001
South-central England	1954-2000	385 species (F)	−4.5 days in 1990s	Fitter and Fitter, 2002
Europe (Int. Phenological Gardens)	1959-1996 1969-1998	Different spring phases (F, LU)	−2.1 −2.7	Menzel and Fabian, 1999; Menzel, 2000; Chmielewski and Rotzer, 2001
21 European countries	1971-2000	F, LU of various plants	−2.5	Menzel et al., 2006b
Japan	1953-2000	*Gingko biloba* (LU)	−0.9	Matsumoto et al., 2003
Northern Eurasia	1982-2004	NDVI	−1.5	Delbart et al., 2006
UK	1976-1998	Butterfly appearance	−2.8 to −3.2	Roy and Sparks, 2000
Europe, N. America	Past 30-60 years	Spring migration of bird species	−1.3 to −4.4	Crick et al., 1997; Crick and Sparks, 1999; Dunn and Winkler, 1999; Inouye et al., 2000; Bairlein and Winkel, 2001; Lehikoinen et al., 2004
N. America (US-MA)	1932-1993	Spring arrival, 52 bird species	+0.8 to −9.6 (*)	Butler, 2003
N. America (US-IL)	1976-2002	Arrival, 8 warbler species	+2.4 to −8.6	Strode, 2003
England (Oxfordshire)	1971-2000	Long-distance migration, 20 species	+0.4 to −6.7	Cotton, 2003
N. America (US-MA)	1970-2002	Spring arrival,16 bird species	−2.6 to −10.0	Ledneva et al., 2004
Sweden (Ottenby)	1971-2002	Spring arrival, 36 bird species	+2.1 to −3.0	Stervander et al., 2005
Europe	1980-2002	Egg-laying, 1 species	−1.7 to −4.6	Both et al., 2004
Australia	1970-1999	11 migratory birds	9 species earlier arrival	Green and Pickering, 2002
Australia	1984-2003	2 spring migratory birds	1 species earlier arrival	Chambers et al., 2005

F = flowering; LU =, leaf-unfolding; – advance; + delay. * indicates mean of significant trends only.

turning points in the respective spring temperature series (Dose and Menzel, 2006).

Records of the return dates of migrant birds have shown changes in recent decades associated with changes in temperature in wintering or breeding grounds or on the migration route (Tryjanowski, 2002; Butler, 2003; Cotton, 2003; Huppop and Huppop, 2003). For example, a 2 to 3 day earlier arrival with a 1°C increase in March temperature is estimated for the swallow in the UK (Sparks and Loxton, 1999) and Ireland (Donnelly et al., 2004). Different measurement methods, such as first observed individual, beginning of sustained migratory period, or median of the migratory period, provide different information about the natural history of different species (Sokolov et al., 1998; Sparks and Braslavska, 2001; Huppop and Huppop, 2003; Tryjanowski et al., 2005).

Egg-laying dates have advanced in many bird species (Hussell, 2003; Dunn, 2004). The confidence in such studies is enhanced when the data cover periods/sites of both local cooling and warming. Flycatchers in Europe (Both et al., 2004) provide such an example, where the trend in egg-laying dates matches trends in local temperatures. Many small mammals have been found to come out of hibernation and to breed earlier in the spring now than they did a few decades ago (Inouye et al., 2000; Franken and Hik, 2004). Larger mammals, such as reindeer, are also showing phenological changes (Post and Forchhammer, 2002), as are butterflies, crickets, aphids and hoverflies (Forister and Shapiro, 2003; Stefanescu et al., 2003; Hickling et al., 2005; Newman, 2005). Increasing regional temperatures are also associated with earlier calling and mating and shorter time to maturity of amphibians (Gibbs and Breisch, 2001; Reading, 2003; Tryjanowski et al., 2003). Despite the bulk of evidence in support of earlier breeding activity as a response to temperature, counter-examples also exist (Blaustein et al., 2001).

Changes in spring and summer activities vary by species and by time of season. Early-season plant species exhibit the strongest reactions (Abu-Asab et al., 2001; Menzel et al., 2001; Fitter and Fitter, 2002; Sparks and Menzel, 2002; Menzel, 2003). Short-distance migrating birds often exhibit a trend towards earlier arrival, while the response of later-arriving long-distance migrants is more complex, with many species showing no change, or even delayed arrival (Butler, 2003; Strode, 2003). Annual plants respond more strongly than congeneric perennials, insect-pollinated more than wind-pollinated plants, and woody less than herbaceous plants (Fitter and Fitter, 2002). Small-scale spatial variability may be due to microclimate, land cover, genetic differentiation, and other non-climate drivers (Menzel et al., 2001; Menzel, 2002). Large-scale geographical variations in the observed changes are found in China with latitude (Chen et al., 2005a), in Switzerland with altitude (Defila and Clot, 2001) and in Europe with magnitude of temperature change (Menzel and Fabian, 1999; Sparks et al., 1999). Spring advance, being more pronounced in maritime western and central Europe than in the continental east (Ahas et al., 2002), is associated with higher spatial variability (Menzel et al., 2006a).

As the North Atlantic Oscillation (NAO) is correlated with temperature (see Trenberth et al., 2007), the NAO has widespread influence on many ecological processes. For example, the speed and pattern (Menzel et al., 2005b), as well as recent trends of spring events in European plants, has also changed consistently with changes seen in the NAO index (Chmielewski and Rotzer, 2001; Scheifinger et al., 2002; Walther et al., 2002; Menzel, 2003). Similarly, earlier arrival and breeding of migratory birds in Europe are often related to warmer local temperatures and higher NAO indices (Hubalek, 2003; Huppop and Huppop, 2003; Sanz, 2003). However, the directions of changes in birds corresponding to NAO can differ across Europe (Hubalek, 2003; Kanuscak et al., 2004). Likewise, the relevance of the NAO index on the phenology of plants differs across Europe, being more pronounced in the western (France, Ireland, UK) and north-western (south Scandinavia) parts of Europe and less distinct in the continental part of Europe (see Figure 1.4a; Menzel et al., 2005b). In conclusion, spring phenological changes in birds and plants and their triggering by spring temperature are often similar, as described in some cross-system studies; however, the NAO influence is weaker than the temperature trigger and is restricted to certain time periods (Walther et al., 2002) (Figure 1.4b).

1.3.5.2 Changes in species distributions and abundances

Many studies of species abundances and distributions corroborate predicted systematic shifts related to changes in climatic regimes, often via species-specific physiological thresholds of temperature and precipitation tolerance. Habitat loss and fragmentation may also influence these shifts. Empirical evidence shows that the natural reaction of species to climate change is hampered by habitat fragmentation and/or loss (Hill et al., 1999b; Warren et al., 2001; Opdam and Wascher, 2004). However, temperature is likely to be the main driver if different species in many different areas, or species throughout broad regions, shift in a co-ordinated and systematic manner. In particular, some butterflies appear to track decadal warming quickly (Parmesan et al., 1999), whereas the sensitivity of tree-line forests to climate warming varies with topography and the tree-line history (e.g., human impacts) (Holtmeier and Broll, 2005). Several different bird species no longer migrate out of Europe in the winter as the temperature continues to warm. Additionally, many species have recently expanded their ranges polewards as these higher-latitude habitats become less marginal (Thomas et al., 2001a). Various studies also found connections between local ecological observations across diverse taxa (birds, mammals, fish) and large-scale climate variations associated with the North Atlantic Oscillation (NAO), El Niño-Southern Oscillation (ENSO), and Pacific Decadal Oscillation (Blenckner and Hillebrand, 2002). For example, the NAO and/or ENSO has been associated with the synchronisation of population dynamics of caribou and musk oxen (Post and Forchhammer, 2002), reindeer calf survival (Weladji and Holand, 2003), fish abundance (Guisande et al., 2004), fish range shifts (Dulčić et al., 2004) and avian demographic dynamics (Sydeman et al., 2001; Jones et al., 2002; Almaraz and Amat, 2004).

Changes in the distribution of species have occurred across a wide range of taxonomic groups and geographical locations

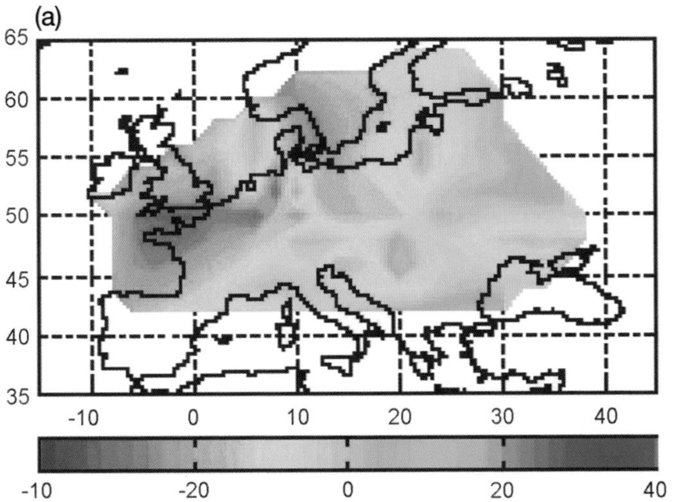

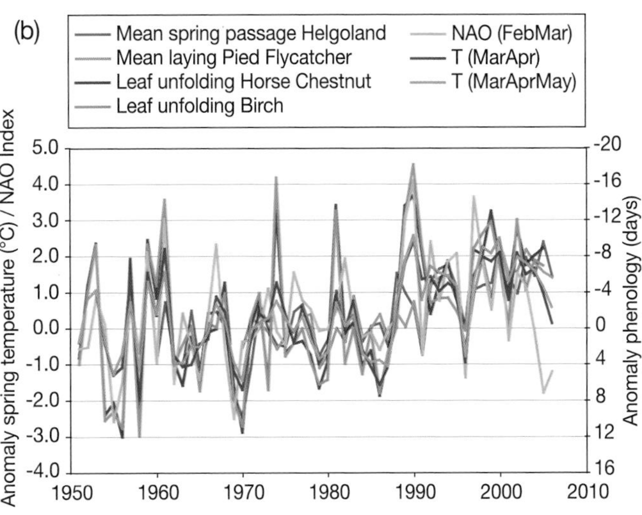

Figure 1.4. *(a) Differences between the mean onset of spring (days) in Europe for the 10 years with the highest (1990, 1882, 1928, 1903, 1993, 1910, 1880, 1997, 1989, 1992) and the lowest (1969, 1936, 1900, 1996, 1960, 1932, 1886, 1924, 1941, 1895) NAO winter and spring index (November to March) drawn from the period 1879 to 1998. After Menzel et al. (2005b). (b) Anomalies of different phenological phases in Germany (mean spring passage of birds at Helgoland, North Sea; mean egg-laying of pied flycatcher in Northern Germany; national mean onset of leaf unfolding of common horse-chestnut (Aesculus hippocastanum) and silver birch (Betula pendula) (negative = earlier)), anomalies of mean spring air temperature T (HadCRUT3v) and North Atlantic Oscillation index (NAO) (http://www.cru.uea.ac.uk/cru/data/). Updated after Walther et al. (2002).*

during the 20th century (Table 1.9). Over the past decades, a poleward extension of various species has been observed, which is probably attributable to increases in temperature (Parmesan and Yohe, 2003). One cause of these expansions is increased survivorship (Crozier, 2004). Many Arctic and tundra communities are affected and have been replaced by trees and dwarf shrubs (Kullman, 2002; ACIA, 2005). In north-western Europe, e.g., in the Netherlands (Tamis et al., 2001) and central Norway (EEA, 2004), thermophilic (warmth-requiring) plant species have become significantly more frequent compared with 30 years ago. In contrast, there has been a small decline in the presence of traditionally cold-tolerant species. These changes in composition are the result of the migration of thermophilic species into these new areas, but are also due to an increased abundance of these species in their current locations.

Altitudinal shifts of plant species have been well documented (Grabherr et al., 2001; Dobbertin et al., 2005; Walther et al., 2005a) (Table 1.9). In several Northern Hemisphere mountain systems, tree lines have markedly shifted to higher elevations during the 20th century, such as in the Urals (Moiseev and Shiyatov, 2003), in Bulgaria (Meshinev et al., 2000), in the Scandes Mountains of Scandinavia (Kullman, 2002) and in Alaska (Sturm et al., 2001). In some places, the position of the tree line has not extended upwards in elevation in the last half-century (Cullen et al., 2001; Masek, 2001; Klasner and Fagre, 2002), which may be due to time-lag effects owing to poor seed production/dispersal, to the presence of 'surrogate habitats' with special microclimates, or to topographical factors (Holtmeier and Broll, 2005). In mountainous regions, climate is a main driver of species composition, but in some areas, grazing,

Table 1.9. *Evidence of significant recent range shifts polewards and to higher elevations.*

Location	Species/Indicator	Observed range shift due to increased temperature (if nothing else stated)	References
California coast, USA	Spittlebug	Northward range shift	Karban and Strauss, 2004
Sweden **Czech Republic**	Tick (*Ixodes ricinus*)	Northward expansion 1982-1996 Expansion to higher altitudes (+300 m)	Lindgren et al., 2000 Daniel et al., 2003
Washington State, USA	Skipper butterfly	Range expansion with increased Tmin	Crozier, 2004
UK	329 species across 16 taxa	Northwards (av. 31-60 km) and upwards (+25 m) in 25 years. Significant northwards and elevational shifts in 12 of 16 taxa. Only 3 species of amphibians and reptiles shifted significantly southwards and to lower elevations	Hickling et al., 2006
UK	Speckled wood (*Pararge aegeria*)	Expanded northern margin, at 0.51-0.93 km/yr, depending on habitat availability	Hill et al., 2001
UK	4 northern butterflies (1970-2004)	2 species retreating 73 and 80 km north, 1 species retreating 149 m uphill	Franco et al., 2006
Central Spain	16 butterfly species	Upward shift of 210 m in the lower elevational limit between 1967-73 and 2004	Wilson et al., 2005
Britain	37 dragonfly and damselfly species	36 out of 37 species shifted northwards (mean 84 km) from 1960-70 to 1985-95	Hickling et al., 2005
Czech Republic	15 of 120 butterfly species	Uphill shifts in last 40 years	Konvicka et al., 2003
Poland	White stork (*Ciconia ciconia*)	Range expansions in elevation, 240 m during last 70 years	Tryjanowski et al., 2005
Australia	3 macropods and 4 feral mammal species	Range expansions to higher altitudes	Green and Pickering, 2002
Australia	Grey-headed flying fox	Contraction of southern boundary poleward by 750 km since 1930s	Tidemann et al., 1999
Senegal, West Africa	126 tree and shrub species (1945-1993)	Up to 600 m/yr latitudinal shift of ecological zones due to decrease in precipitation	Gonzalez, 2001
Russia, Bulgaria, Sweden, Spain, New Zealand, USA	Tree line	Advancement towards higher altitudes	Meshinev et al., 2000; Kullman, 2002; Peñuelas and Boada, 2003; Millar and Herdman, 2004
Canada	Bioclimatic taiga-tundra ecotone indicator	12 km/yr northward shift (NDVI data)	Fillol and Royer, 2003
Alaska	Arctic shrub vegetation	Expansion into previously shrub-free areas	Sturm et al., 2001
European Alps	Alpine summit vegetation	Elevational shift, increased species-richness on mountain tops	Grabherr et al., 2001; Pauli et al., 2001; Walther et al., 2005a
Montana, USA	Arctic-alpine species	Decline at the southern margin of range	Lesica and McCune, 2004
Germany, Scandinavia	English holly (*Ilex aquifolium*)	Poleward shift of northern margin due to increasing winter temperatures	Walther et al., 2005b

logging or firewood collection can be of considerable relevance. In parts of the European Alps, for example, the tree line is influenced by past and present land-use impacts (Theurillat and Guisan, 2001; Carnelli et al., 2004). A climate warming-induced upward migration of alpine plants in the high Alps (Grabherr et al., 2001; Pauli et al., 2001) was observed to have accelerated towards the beginning of the 21st century (Walther et al., 2005a). Species ranges of alpine plants also have extended to higher altitudes in the Norwegian Scandes (Klanderud and Birks, 2003). Species in alpine regions, which are often endemic and of high importance for plant diversity (Vare et al., 2003), are vulnerable to climate warming, most probably because of often restricted climatic ranges, small isolated populations, and the absence of suitable areas at higher elevations in which to migrate (Pauli et al., 2003).

1.3.5.3 Climate-linked extinctions and invasions

Key indicators of a species' risk of extinction (global loss of all individuals) or extirpation (loss of a population in a given location) include the size of its range, the density of individuals within the range, and the abundance of its preferred habitat within its range. Decreases in any of these factors (e.g., declining range size with habitat fragmentation) can lower species population size (Wilson et al., 2004). Each of these factors can be directly affected by rapid global warming, but the causes of extinctions/extirpations are most often multifactorial. For example, a recent extinction of around 75 species of frogs, endemic to the American tropics, was most probably due to a pathogenic fungus (*Batrachochytrium*), outbreaks of which have been greatly enhanced by global warming (Pounds et al., 2006). Other examples of declines in populations and subsequent extinction/extirpation are found in amphibians around the world (Alexander and Eischeid, 2001; Middleton et al., 2001; Ron et al., 2003; Burrowes et al., 2004). Increasing climatic variability, linked to climate change, has been found to have a significant impact on the extinction of the butterfly *Euphydryas editha bayensis* (McLaughlin et al., 2002a, 2002b). Currently about 20% of bird species (about 1,800) are threatened with extinction, while around 5% are already functionally extinct (e.g., small inbred populations) (Sekercioglu et al., 2004). The pika (*Ochotona princeps*), a small mammal found in mountains of the western USA, has been extirpated from many slopes (Beever et al., 2003). New evidence suggests that climate-driven extinctions and range retractions are already widespread, which have been poorly reported due, at least partly, to a failure to survey the distributions of species at sufficiently fine resolution to detect declines and to attribute such declines to climate change (Thomas et al., 2006).

A prominent cause of range contraction or loss of preferred habitat within a species range is invasion by non-native species. Fluctuation in resource availability, which can be driven by climate, has been identified as the key factor controlling invasibility (Davis et al., 2000). The clearest evidence for climate variability triggering an invasion occurs where a suite of species with different histories of introduction spread en-masse during periods of climatic amelioration (Walther, 2000; Walther et al., 2002). Climate change will greatly affect indigenous species on sub-Antarctic islands, primarily due to warmer climates allowing exotic species, such as the house mouse (*Mus musculus*) and springtails (*Collembola* spp.), to become established and proliferate (Smith, 2002). A prominent example is that of exotic thermophilous plants spreading into the native flora of Spain, Ireland and Switzerland (Pilcher and Hall, 2001; Sobrino et al., 2001). Elevated CO_2 might also contribute to the spread of weedy, non-indigenous plants (Hattenschwiler and Korner, 2003).

1.3.5.4 Changes in morphology and reproduction

A change in fecundity is one of the mechanisms altering species distributions (see Section 1.3.5.2). Temperature can affect butterfly egg-laying rate and microhabitat selection; recent warming has been shown to increase egg-laying and thus population size for one species (Davies et al., 2006). The egg sizes of many bird species are changing with increasing regional temperatures, but the direction of change varies by species and location. For example, in Europe, the egg size of pied flycatchers increased with regional warming (Jarvinen, 1994, 1996). In southern Poland, the size of red-backed shrikes' eggs has decreased, probably due to decreasing female body size, which is also associated with increasing temperatures (Tryjanowski et al., 2004). The eggs of European barn swallows are getting larger with increasing temperatures and their breeding season is occurring earlier. Additionally, in the eggs, concentrations of certain maternally supplied nutrients, such as those affecting hatchability, viability and parasite defence, have also increased with warming (Saino et al., 2004). Studies from eastern Poland, Asia, Europe and Japan have found that various birds and mammals exhibit trends toward larger body size, probably due to increasing food availability, with regionally increasing temperatures (Nowakowski, 2002; Yom-Tov, 2003; Kanuscak et al., 2004; Yom-Tov and Yom-Tov, 2004). Reproductive success in polar bears has declined, resulting in a drop in body condition, which in turn is due to melting Arctic Sea ice. Without ice, polar bears cannot hunt seals, their favourite prey (Derocher et al., 2004).

These types of changes are also found in insects and plants. The evolutionary lengthening and strengthening of the wings of some European Orthoptera and butterflies has facilitated their northward range expansion but has decreased reproductive output (Hill et al., 1999a; Thomas et al., 2001a; Hughes et al., 2003a; Simmons and Thomas, 2004). The timing and duration of the pollen season, as well as the amount of pollen produced (Beggs, 2004), have been found to be affected by regional warming (see Section 1.3.7.4).

1.3.5.5 Species community changes and ecosystem processes

In many parts of the world, species composition has changed (Walther et al., 2002), partly due to invasions and distributional changes. The assemblages of species in ecological communities reflect interactions among organisms as well as between organisms and the abiotic environment. Climate change, extreme climatic events or other processes can alter the composition of species in an ecosystem because species differentially track their climate tolerances. As species in a natural community do not respond in synchrony to such external pressures, ecological communities existing today could easily be disaggregated (Root and Schneider, 2002).

Species diversity in various regions is changing due to the number of species shifting, invading or receding (Tamis et al., 2001; EEA, 2004) (see Sections 1.3.5.2 and 1.3.5.3). Average species richness of butterflies per 20 km grid cell in the UK increased between 1970-1982 and 1995-1999, but less rapidly than would have been expected had all species been able to keep up with climate change (Menendez et al., 2006). In non-fragmented Amazon forests, direct effects of CO_2 on photosynthesis, as well as faster forest turnover rates, may have caused a substantial increase in the density of lianas over the last two decades (Phillips et al., 2004). Although many species-community changes are also attributable to landscape fragmentation, habitat modification and other non-climate drivers, many studies show a high correlation between changes in species composition and recent climate change, also via the frequency of weather-based disturbances (Hughes, 2000; Pauli et al., 2001; Parmesan and Yohe, 2003). Examples of altered or stable synchrony in ecosystems via multi-species interactions, e.g., the pedunculate oak–winter moth–tit food chain, are still fairly rare (van Noordwijk et al., 1995; Buse et al., 1999).

1.3.5.6 Species evolutionary processes

Recent evolutionary responses to climate change have been addressed in reviews (Thomas, 2005; Bradshaw and Holzapfel, 2006). Changes have taken place in the plants preferred for egg-laying and feeding of butterflies, e.g., a broadened diet facilitated the colonisation of new habitats during range extension in the UK (Thomas et al., 2001a). The pitcher-plant mosquito in the USA has prolonged development time in late summer by the evolution of changed responses to day length (Bradshaw and Holzapfel, 2001; Bradshaw et al., 2003). The blackcap warbler has recently extended its overwintering range northwards in Europe by evolving a change in migration direction (Berthold et al., 2003). Insects expanding their ranges have undertaken genetically-based changes in dispersal morphology, behaviour and other life-history traits, as 'good colonists' have been at a selective advantage (Hill et al., 1999a; Thomas et al., 2001b; Hughes et al., 2003a; Simmons and Thomas, 2004). Genetic changes in *Drosophila melanogaster* in eastern coastal Australia over 20 years are likely to reflect increasingly warmer and drier conditions (Umina et al., 2005). Evolutionary processes are also demonstrated in the timing of reproduction associated with climate change in North American red squirrels (Berteaux et al., 2004). There is no evidence so far that the temperature response rates of plants have changed over the last century (Menzel et al., 2005a).

1.3.5.7 Summary of terrestrial biological systems

The vast majority of studies of terrestrial biological systems reveal notable impacts of global warming over the last three to five decades, which are consistent across plant and animal taxa: earlier spring and summer phenology and longer growing seasons in mid- and higher latitudes, production range expansions at higher elevations and latitudes, some evidence for population declines at lower elevational or latitudinal limits to species ranges, and vulnerability of species with restricted ranges, leading to local extinctions. Non-climate synergistic factors can significantly limit migration and acclimatisation capacities.

While a variety of methods have been used that provide evidence of biological change over many ecosystems, there remains a notable absence of studies on some ecosystems, particularly those in tropical regions, due to a significant lack of long-term data. Furthermore, not all processes influenced by warming have yet been studied. Nevertheless, in the large majority of studies, the observed trends found in species correspond to predicted changes in response to regional warming in terms of magnitude and direction. Analyses of regional differences in trends reveal that spatio-temporal patterns of both phenological and range changes are consistent with spatio-temporal patterns expected from observed climate change.

1.3.6 Agriculture and forestry

Although agriculture and forestry are known to be highly dependent on climate, little evidence of observed changes related to regional climate changes was noted in the TAR. This is probably due to the strong influence of non-climate factors on agriculture and, to a lesser extent, on forestry, especially management practices and technological changes, as well as market prices and policies related to subsidies (Easterling, 2003). The worldwide trends in increasing productivity (yield per hectare) of most crops over the last 40 years, primarily due to technological improvements in breeding, pest and disease control, fertilisation and mechanisation, also make identifying climate-change signals difficult (Hafner, 2003).

1.3.6.1 Crops and livestock

Changes in crop phenology provide important evidence of responses to recent regional climate change (Table 1.10). Such changes are apparent in perennial crops, such as fruit trees and wine-making varieties of grapes, which are less dependent on yearly management decisions by farmers than annual crops and are also often easier to observe. Phenological changes are often observed in tandem with changes in management practices by farmers. A study in Germany (Menzel et al., 2006c) has revealed that between 1951 and 2004 the advance for agricultural crops (2.1 days/decade) has been significantly less marked than for wild plants or fruit trees (4.4 to 7.1 days/decade). All the reported studies concern Europe, where recent warming has clearly advanced a significant part of the agricultural calendar.

Since the TAR, there has been evidence of recent trends in agro-climatic indices, particularly those with a direct relationship to temperature, such as increases in growing season length and in growing-degree-days during the crop cycle. These increases, associated with earlier last spring frost and delayed autumn frost dates, are clearly apparent in temperate regions of Eurasia (Moonen et al., 2002; Menzel et al., 2003; Genovese et al., 2005; Semenov et al., 2006) and a major part of North America (Robeson, 2002; Feng and Hu, 2004). They are especially detectable in indices applicable to wine-grape cultivation (Box 1.2). In Sahelian countries, increasing temperature in combination with rainfall reduction has led to a reduced length of vegetative period, no longer allowing present varieties to complete their cycle (Ben Mohamed et al., 2002).

However, no detectable change in crop yield directly attributable to climate change has been reported for Europe. For

Table 1.10. *Observed changes in agricultural crop and livestock.*

Agricultural metric	Observed change	Location	Period	References
Phenology	Advance of stem elongation for winter rye (10 days) and emergence for maize (12 days)	Germany	1961-2000	Chmielewski et al., 2004
	Advance in cherry tree flowering (0.9 days/10 years), apple tree flowering (1.1 days/10 years) in response (–5 days/°C) to March/April temperature increase		1951-2000	Menzel, 2003
	Advance in beginning of growing season of fruit trees (2.3 days/10 years), cherry tree blossom (2.0 days/10 years), apple tree blossom (2.2 days/10 years) in agreement with 1.4°C annual air temperature increase		1961-1990	Chmielewski et al., 2004
	Advance of fruit tree flowering of 1-3 weeks for apricot and peach trees, increase in spring frost risks and more frequent occurrence of bud fall or necrosis for sensitive apricot varieties	South of France	1970-2001	Seguin et al., 2004
Management practices, pests and diseases	Advance of seeding dates for maize and sugarbeet (10 days)	Germany	1961-2000	Chmielewski et al., 2004
	Advance of maize sowing dates by 20 days at 4 INRA experimental farms	France	1974-2003	Benoit and Torre, 2004
	Advance of potato sowing date by 5 days, no change for spring cereals	Finland	1965-1999	Hilden et al., 2005
	Partial shift of apple codling moth from 2 to 3 generations	South of France	1984-2003	Sauphanor and Boivin, 2004
Yields	Lower hay yields, in relation to warmer summers	Rothamsted UK	1965-1998	Cannell et al., 1999
	Part of overall yield increase attributed to recent cooling during growing season: 25% maize, 33% soybean	USA county level	1982-1998	Lobell and Asner, 2003
	Decrease of rice yield associated with increase in temperature (0.35°C and 1.13°C for Tmax and Tmin, respectively, during 1979 to 2003)	Philippines	1992-2003	Peng et al., 2004
Livestock	Decrease of measured pasture biomass by 20-30%	Mongolia	1970-2002	Batimaa, 2005
	Decline of NDVI of the third period of 10 days of July by 69% for the whole territory		1982-2002	Erdenetuya, 2004
	Observed increase in animal production related to warming in summer and annual temperature	Tibet	1978-2002	Du et al., 2004

Box 1.2. Wine and recent warming

Wine-grapes are known to be highly sensitive to climatic conditions, especially temperature (e.g., viticulture was thriving in England during the last medieval warm period). They have been used as an indicator of observed changes in agriculture related to warming trends, particularly in Europe and in some areas of North America.

In Alsace, France, the number of days with a mean daily temperature above 10°C (favourable for vine activity) has increased from 170 around 1970 to 210 at the end of the 20th century (Duchêne and Schneider, 2005). An increase associated with a lower year-to-year variability in the last 15 years of the heliothermal index of Huglin (Seguin et al., 2004) has been observed for all the wine-producing areas of France, documenting favourable conditions for wine, in terms of both quality and stability. Similar trends in the average growing-season temperatures (April-October for the Northern Hemisphere) have been observed at the main sites of viticultural production in Europe (Jones, 2005). The same tendencies have also been found in the California, Oregon and Washington vineyards of the USA (Nemani et al., 2001; Jones, 2005).

The consequences of warming are already detectable in wine quality, as shown by Duchêne and Schneider (2005), with a gradual increase in the potential alcohol levels at harvest for Riesling in Alsace of nearly 2% volume in the last 30 years. On a worldwide scale, for 25 of the 30 analysed regions, increasing trends of vintage ratings (average rise of 13.3 points on a 100-point scale for every 1°C warmer during the growing season), with lower vintage-to-vintage variation, has been established (Jones, 2005).

example, the yield trend of winter wheat displays progressive growth from 2.0 t/ha in 1961 to 5.0 t/ha in 2000, with anomalies due to climate variability on the order of 0.2 t/ha (Cantelaube et al., 2004). The same observation is valid for Asia, where the rice production of India has grown over the period 1950-1999 from 20 Mt to over 90 Mt, with only a slight decline during El Niño years when monsoon rainfall is reduced (Selvaraju, 2003). A negative effect of warming for rice production observed by the International Rice Research Institute (IRRI) in the Philippines (yield loss of 15% for 1°C increase of growing-season minimum temperature in the dry season) (Peng et al., 2004) is limited to a local observation for a short time period; a similar effect has been noted on hay yield in the UK (1°C increase in July-August led to a 0.33 t/ha loss) (Cannell et al., 1999). A study at the county level of U.S. maize and soybean yields (Lobell and Asner, 2003) has established a positive effect of cooler and wetter years in the Midwest and hotter and drier years in the North-west plains. In the case of the Sahel region of Africa, warmer and drier conditions have served as a catalyst for a number of other factors that have accelerated a decline in groundnut production (Van Duivenbooden et al., 2002).

For livestock, one study in Tibet reports a significant relationship of improved performance with warming in high mountainous conditions (Du et al., 2004). On the other hand, the pasture biomass in Mongolia has been affected by the warmer and drier climate, as observed at a local station (Batimaa, 2005) or at the regional scale by remote sensing (Erdenetuya, 2004).

1.3.6.2 Forestry

Here we focus on forest productivity and its contributing factors (see Section 1.3.5 for phenological aspects). Rising atmospheric CO_2 concentration, lengthening of the growing season due to warming, nitrogen deposition and changed management have resulted in a steady increase in annual forest CO_2 storage capacity in the past few decades, which has led to a more significant net carbon uptake (Nabuurs et al., 2002). Satellite-derived estimates of global net primary production from satellite data of vegetation indexes indicate a 6% increase from 1982 to 1999, with large increases in tropical ecosystems (Nemani et al., 2003) (Figure 1.5). The study by Zhou et al. (2003), also using satellite data, confirm that the Northern Hemisphere vegetation activity has increased in magnitude by 12% in Eurasia and by 8% in North America from 1981 to 1999. Thus, the overall trend towards longer growing seasons is consistent with an increase in the 'greenness' of vegetation, for broadly continuous regions in Eurasia and in a more fragmented way in North America, reflecting changes in biological activity. Analyses in China attribute increases in net primary productivity, in part, to a country-wide lengthening of the growing season (Fang and Dingbo, 2003). Similarly, other studies find a decrease of 10 days in the frost period in northern China (Schwartz and Chen, 2002) and advances in spring phenology (Zheng et al., 2002).

However, in the humid evergreen tropical forest in Costa Rica, annual growth from 1984 to 2000 was shown to vary inversely with the annual mean of daily minimum temperature, because of increased respiration at night (Clark et al., 2003). For southern Europe, a trend towards a reduction in biomass production has been detected in relation to rainfall decrease (Maselli, 2004), especially after the severe drought of 2003 (Gobron et al., 2005; Lobo and Maisongrande, 2006). A recent

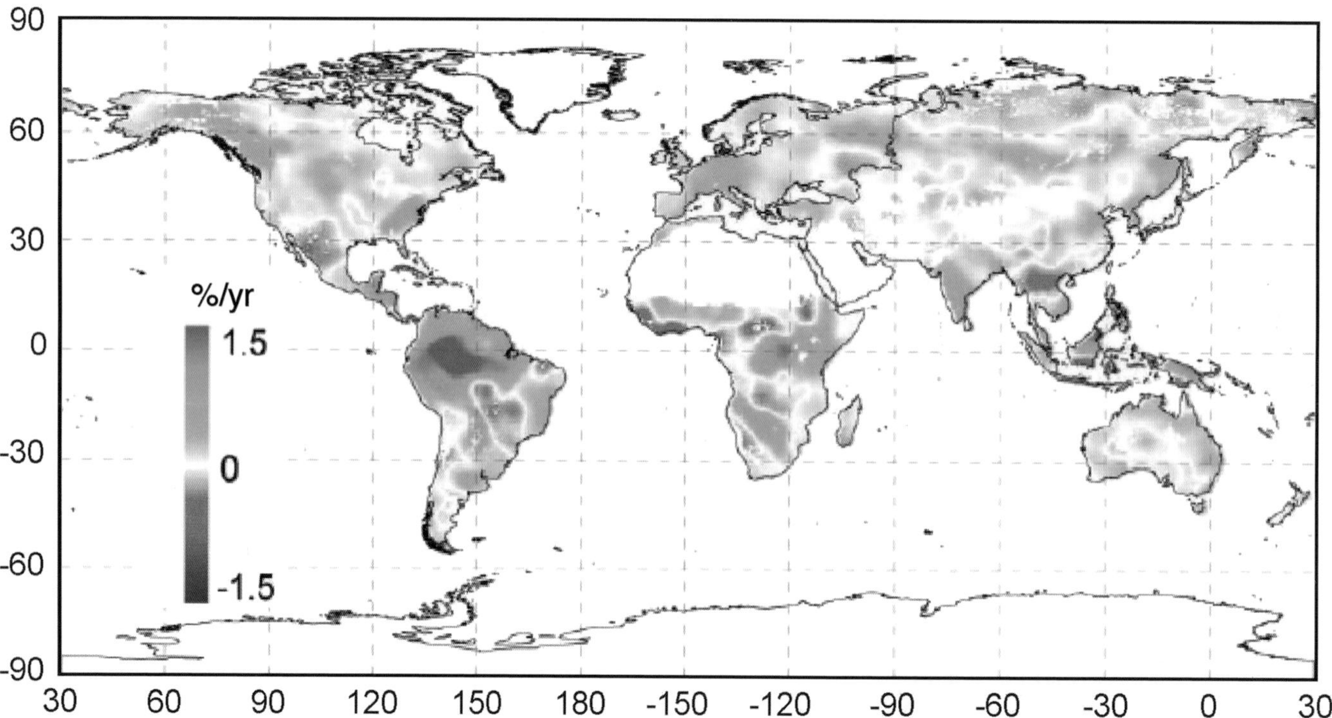

Figure 1.5. *Estimated changes in net primary productivity (NPP) between 1982 and 1999 derived from independent NDVI data sets from the Global Inventory Modeling and Mapping Studies (GIMMS) and Pathfinder Advanced Very High Resolution Radiometer (AVHRR) Land (PAL). An overall increase in NPP is observed, which is consistent with rising atmospheric CO_2 and warming. From Nemani et al., 2003. Reprinted with permission from AAAS.*

study in the mountains of north-east Spain (Jump et al., 2006) shows significantly lower growth of mature beech trees at the lower limit of this species compared with those at higher altitudes. Growth at the lower *Fagus* limit was characterised by a rapid recent decline starting in approximately 1975. By 2003, the growth of mature trees had fallen by 49% when compared with pre-decline levels. Analysis of climate–growth relationships suggests that the observed decline in growth is a result of warming temperatures. For North America, recent observations from satellite imagery (for the period 1982 to 2003) document a decline for a substantial portion of northern forest, possibly related to warmer and longer summers, whereas tundra productivity is continuing to increase (Goetz et al., 2005). They also confirm other results about the effects of droughts (Lotsch et al., 2005), as well those made by ground measurements (D'Arrigo et al., 2004; Wilmking et al., 2004).

Climate warming can also change the disturbance regime of forests by extending the range of some damaging insects, as observed during the last 20 years for bark beetles in the USA (Williams and Liebhold, 2002) or pine processionary moth in Europe (Battisti et al., 2005). The latter has displayed a northward shift of 27 km/decade near Paris, a 70 m/decade upward shift in altitude for southern slopes, and 30 m/decade for northern slopes in Italian mountains.

Trends in disturbance resulting from forest fires are still a subject of controversy. In spite of current management practices that tend to reduce fuel load in forests, climate variability is often the dominant factor affecting large wildfires, given the presence of ignition sources (McKenzie et al., 2004). This is confirmed by an analysis of forest fires in Siberia between 1989 and 1999 (Conard et al., 2002), which detected the significant impacts of two large fires in 1996 and 1998, resulting in 13 million ha burned and 14 to 20% of the annual global carbon emissions from forest fires. The increase in outdoor fires in England and Wales between 1965 and 1998 may be attributable to a trend towards warmer and drier summer conditions (Cannell et al., 1999). Repeated large forest fires during the warm season in recent years in the Mediterranean region and North Africa, as well as in California, have also been linked to drought episodes. One study of forest fires in Canada (Gillett et al., 2004) found that about half of the observed increase in burnt area during the last 40 years, in spite of improved fire-fighting techniques, is in agreement with simulated warming from a general circulation model (GCM). This finding is not fully supported by another study, which found that fire frequency in Canada has recently decreased in response to better fire protection and that the effects of climate change on fire activity are complex (Bergeron et al., 2004). However, it seems to be confirmed by another recent study (Westerling et al., 2006), which established a dramatic and sudden increase in large wildfire activity in the western USA in the mid-1980s closely associated with increased spring and summer temperatures and an earlier spring snow melt.

1.3.6.3 Summary of agriculture and forestry

Trends in individual climate variables or their combination into agro-climatic indicators show that there is an advance in phenology in large parts of North America and Europe, which has been attributed to recent regional warming. In temperate regions, there are clear signals of reduced risk of frost, longer growing season duration, increased biomass, higher quality (for grapevines, a climate-sensitive crop), insect expansion, and increased forest-fire occurrence that are in agreement with regional warming. These effects are hard to detect in aggregate agricultural statistics because of the influence of non-climate factors, particularly where advances in technology confound responses to warming. Although the present effects are of limited economic consequence and appear to lie within the ability of the sectors to adapt, both agriculture and forestry show vulnerability to recent extreme heat and drought events.

1.3.7 Human health

Here we evaluate evidence regarding observed changes in human health, important health exposures, and regional climate change. These observed changes are primarily related to temperature trends and changes in temperature extremes and relate to a range of infectious and non-infectious disease outcomes. These relationships are difficult to separate from the effects of major climate variability systems such as ENSO, which have been shown to be associated with the transmission and occurrence of diseases in certain locations (Kovats et al., 2003; Rodo et al., 2002). Additionally, temperature and rainfall variability can be important determinants of the transmission of vector-borne diseases (Githeko and Ndegwa, 2001).

There is little evidence about the effects of observed climate change on health for two reasons: the lack of long epidemiological or health-related data series, and the importance of non-climate drivers in determining the distribution and intensity of human disease. Studies that have quantified the effect of climate or weather on health outcomes are listed in Table 1.11. There is a wide range of driving forces that can affect and modify the impact of climate change on human health indicators. Consideration of reported trends in a given disease and the attribution to climate change needs to take into account three possible conditions.

1. That the change in disease incidence is real and due to changes in important non-climate determinants which include social factors, such as human population density and behaviour; housing facilities; public health facilities (e.g., water supply and general infrastructure, waste management and vector-control programmes); use of land for food, fuel and fibre supply; and results of adaptation measures (e.g., drug and insecticide use), as well as changed insecticide and drug resistance in pathogens and vector species (Tillman et al., 2001; Githeko and Woodward, 2003; Molyneux, 2003; Sutherst, 2004). Changes in land use and land cover can affect the local climate and ecosystems and should be considered when linking climate and health (Patz et al., 2005).
2. That the change in disease incidence is real and due to changes in climate factors, once all non-climate determinants have been considered and excluded as the main explanation (see, for example, Purse et al., 2006).
3. That the change in disease incidence is not real, but is only apparent due to changed reporting or may be due to changes in other apparent factors such as population growth or movement.

Table 1.11. *Studies of the effects of weather and climate on human health.*

Health effect	Climate effect on health	Other driving forces	Study
Direct impacts of heat or cold	Temperature-related mortality in summers	Declining summer death rates due to air-conditioning adaptation	Diaz et al., 2002; Davis et al., 2003b; Beniston, 2004; Kysely, 2004
Vector-borne diseases	Tick-borne encephalitis (TBE) increases in Sweden with milder climate	Increases in TBE may be due to changes in human and animal behaviour	Randolph, 2001
	High latitudinal spread of ticks – vectors for Lyme disease – with milder winters in Sweden and the Czech Republic		Lindgren et al., 2000; Danielová et al., 2006
Food- and water-borne diseases	Salmonellosis in Australia associated with higher temperatures	*E. coli* and *Cryptosporium* outbreaks could not be attributed to climate change	D'Souza et al., 2004 Charron et al., 2004
Pollen- and dust-related diseases	Increasing pollen abundance and allergenicity have been associated with warming climate	Pollen abundance also influenced by land-use changes	Levetin, 2001; Beggs, 2004

1.3.7.1 Effects of patterns in heat and cold stress

Episodes of extreme heat or cold have been associated with increased mortality (Huynen et al., 2001; Curriero et al., 2002). There is evidence of recent increases in mean surface temperatures and in the number of days with higher temperatures, with the extent of change varying by region (Karl and Trenberth, 2003; Luterbacher et al., 2004; Schär et al., 2004; IPCC, 2007). This increase in heatwave exposures, where heatwaves are defined as temperature extremes of short duration, has been observed in mid-latitudes in Europe and the USA. Individual events have been associated with excess mortality, particularly in the frail elderly, as was dramatically illustrated in the 2003 heatwave in western and central Europe, which was the hottest summer since 1500 (Luterbacher et al., 2004; Chapter 8, Box 8.1).

In general, high-income populations have become less vulnerable to both heat and cold (see Chapter 8, Section 8.2). Studies in Europe and in the USA of mortality over the past 30 to 40 years found evidence of declining death rates due to summer and winter temperatures (Davis et al., 2003a, b; Donaldson et al., 2003). Declines in winter mortality are apparent in many temperate countries primarily due to increased adaptation to cold (Chapter 8, Section 8.2.1.3) (Kunst et al., 1991; Carson et al., 2006). However, the mortality associated with extreme heatwaves has not declined. The 25,000 to 30,000 deaths attributed to the European heatwave is greater than that observed in the last century in Europe (Kosatsky, 2005). Analyses of long-term trends in heatwave-attributable (versus heat-attributable) mortality have not been undertaken.

1.3.7.2 Patterns in vector-borne diseases

Vector-borne diseases are known to be sensitive to temperature and rainfall (as shown by the ENSO effects discussed above). Consideration of these relationships suggests that warmer temperature is likely to have two major kinds of closely related, potentially detectable, outcomes: changes in vectors *per se*, and changes in vector-borne disease outcomes (Kovats et al., 2001). Insect and tick vectors would be expected to respond to changes in climate like other cold-blooded terrestrial species (Table 1.9). There is some evidence that this is occurring in relation to disease vectors, but the evidence for changes in human disease is less clear.

Tick vectors

Changes in the latitudinal distribution and abundance of Lyme disease vectors in relation to milder winters have been well documented in high-latitude regions at the northern limit of the distribution in Sweden (Lindgren et al., 2000; Lindgren and Gustafson, 2001), although the results may have been influenced by changes due to reporting and changes in human behaviour. An increase in TBE in Sweden since the mid-1980s is consistent with a milder climate in this period (Lindgren and Gustafson, 2001), but other explanations cannot be ruled out (Randolph, 2001).

Malaria

Since the TAR, there has been further research on the role of observed climate change on the geographical distribution of malaria and its transmission intensity in African highland areas but the evidence remains unclear. Malaria incidence has increased since the 1970s at some sites in East Africa. Chen et al. (2006) have demonstrated the recent spread of falciparum malaria and its vector *Anopheles arabiensis* in highland areas of Kenya that were malaria-free as recently as 20 years ago. It has yet to be proved whether this is due solely to warming of the environment. A range of studies have demonstrated the importance of temperature variability in malaria transmission in these highland sites (Abeku et al., 2003; Kovats et al., 2001; Zhou et al., 2004) (see Chapter 8, Section 8.2.8.2 for a detailed discussion). While a few studies have shown the effect of long-term upward trends in temperature on malaria at some highland sites (e.g., Tulu, 1996), other studies indicate that an increase in resistance of the malaria parasite to drugs, a decrease in vector-control activities and ecological changes may have been the most likely driving forces behind the resurgence of malaria in recent years. Thus, while climate is a major limiting factor in the spatial and temporal distribution of malaria, many non-climatic factors (drug resistance and HIV prevalence, and secondarily, cross-border movement of people, agricultural activities, emergence of insecticide resistance, and the use of DDT for indoor residual spraying) may alter or override the effects of climate (Craig et al., 2004; Barnes et al., 2005).

There is a shortage of concurrent detailed and long-term historical observations of climate and malaria. Good-quality time-series of malaria records in the East African and the Horn of Africa highlands are too short to address the early effects of climate change. Very few sites have longer data series, and the evidence on the role of climate change is unresolved (Hay et al., 2002a, 2002b; Patz et al., 2002; Shanks et al., 2002), although a recent study has confirmed warming trends at these sites (Pascual et al., 2006).

1.3.7.3 Emerging food- and water-borne diseases

Food- and water-borne diseases (WBD) are major adverse conditions associated with warming and extreme precipitation events. Bacterial infectious diseases are sometimes sensitive to temperature, e.g., salmonellosis (D'Souza et al., 2004), and WBD outbreaks are sometimes caused by extreme rainfall (Casman et al., 2001; Curriero et al., 2001; Rose et al., 2002; Charron et al., 2004; Diergaardt et al., 2004) but, again, no attribution to longer-term trends in climate has been attempted.

1.3.7.4 Pollen- and dust-related diseases

There is good evidence that observed climate change is affecting the timing of the onset of allergenic pollen production. Studies, mostly from Europe, indicate that the pollen season has started earlier (but later at high latitudes) in recent decades, and that such shifts are consistent with observed changes in climate. The results concerning pollen abundance are more variable, as pollen abundance can be more strongly influenced by land-use changes and farming practices (Teranishi et al., 2000; Rasmussen, 2002; Van Vliet et al., 2002; Emberlin et al., 2003; WHO, 2003; Beggs, 2004; Beggs and Bambrick, 2005) (see Section 1.3.5). There is some evidence that temperature changes have increased pollen abundance or allergenicity (Beggs, 2004) (see Chapter 8, Section 8.2.7). Changing agricultural practices, such as the replacement of haymaking in favour of silage production, have also affected the grass-pollen season in Europe.

The impact on health of dust and dust storms has not been well described in the literature. Dust related to African droughts has been transported across the Atlantic to the Caribbean (Prospero and Lamb, 2003), while a dramatic increase in respiratory disease in the Caribbean has been attributed to increases in Sahara dust, which has in turn, been linked to climate change (Gyan et al., 2003).

1.3.7.5 Summary of human health

There is now good evidence of changes in the northward range of some disease vectors, as well as changes in the seasonal pattern of allergenic pollen. There is not yet any clear evidence that climate change is affecting the incidence of human vector-borne diseases, in part due to the complexity of these disease systems. High temperature has been associated with excess mortality during the 2003 heatwave in Europe. Declines in winter mortality are apparent in many temperate countries, primarily due to increased adaptation to cold.

1.3.8 Disasters and hazards

Rapid-onset meteorological hazards with the potential to cause the greatest destruction to property and lives include extreme river floods, intense tropical and extra-tropical cyclone windstorms (along with their associated coastal storm surges), as well as the most severe supercell thunderstorms. Here we assess the evidence for a change in the frequency, geography and/or severity of these high-energy events. By definition, the extreme events under consideration here are rare events, with return periods at a specific location typically in excess of 10 to 20 years, as the built environment is generally sited and designed to withstand the impacts of more frequent extremes. Given that the strong rise in global temperatures only began in the 1970s, it is difficult to demonstrate statistically a change in the occurrence of extreme floods and storms (with return periods of 20 years or more) simply from the recent historical record (Frei and Schar, 2001). In order to identify a change in extreme flood and storm return periods, data has been pooled from independent and uncorrelated locations that share common hazard characteristics so as to search collectively for changes in occurrence. A search for a statistically significant change in occurrence characteristics of relatively high-frequency events (with return periods less than 5 years) can also be used to infer changes at longer return periods.

1.3.8.1 Extreme river floods

The most comprehensive available global study examined worldwide information on annual extreme daily flows from 195 rivers, principally in North America and Europe, and did not find any consistent trends, with the number of rivers showing statistically significant increases in annual extreme flows being approximately balanced by the number showing a decrease (Kundzewicz, 2004) (see Section 1.3.2.2). However, in terms of the most extreme flows, when data were pooled across all the rivers surveyed in Europe, a rising trend was found in the decade of the maximum observed daily flow, with four times as many rivers showing the decade of highest flow in the 1990s rather than in the 1960s.

Again, with a focus only on the most extreme flows, a pooled study examined great floods with return periods estimated as greater than 100 years on very large rivers (with catchments greater than 200,000 km^2) in Asia, North America, Latin America, Europe and Africa (Milly et al., 2002). From the pooled record of all the rivers, the observed trend in the population of 100-year flood events, at a 95% confidence interval averaged across all basins, has been positive since the Mississippi floods in 1993 and can be detected intermittently since 1972. Analysis of available long-term river flow records shows that since 1989 more than half of Scotland's largest rivers (notably those draining from the west) have recorded their highest flows (Werrity et al., 2002). Of sixteen rivers surveyed, with a median record of thirty-nine years, eight had their maximum flow during the period 1989 to 1997, a period of high NAO index (based on the pressure difference between Iceland and the Azores) values consistent with storm tracks bringing high levels of precipitation to the northern UK.

1.3.8.2 North-east Atlantic extra-tropical cyclones

The North-east Atlantic is the region with the deepest observed central pressures of extra-tropical cyclones, and the adjacent margin of north-west Europe has the greatest levels of extra-tropical cyclone historical building damage, forestry windthrow, and storm-surge impacts observed worldwide. Many studies report an increase in the 1980s in the number of deep (and high wind-speed) extra-tropical cyclone storms in this region (see Günther et al., 1998) returning to levels not previously seen since the late 19th century (see Trenberth et al., 2007, Section 3.5.3). Various measures, including increases in the number of deep storms (with central pressures less than 970 hPa) and reductions in the annual pressure minimum of storms crossing the Greenwich Meridian all show evidence for intensification, in particular between 1980 and 1993, when there were a series of major damaging storms. In the North-east Atlantic, wave heights showed significant increases over the period from 1970 to 1995 (Woolf et al., 2002) in parallel with the NAO index, which reached its highest values ever (reflecting deep low pressure over Iceland) in the years of 1989 to 1990. Intense storms returned at the end of the 1990s, when there were three principal damaging storms across western Europe in December 1999. However, since that time, as winter NAO values have continued to fall (through to March 2005), there has been a significant decline in the number of deep and intense storms passing into Europe, to some of the lowest levels seen for more than 30 years. (Other high-latitude regions of extra-tropical cyclone activity also show variations without simple trends: see Trenberth et al., 2007, Section 3.5.3.)

1.3.8.3 Tropical cyclones

While overall numbers of tropical cyclones worldwide have shown little variation over the past 40 years (Pielke et al., 2005), there is evidence for an increase in the average intensity of tropical cyclones in most basins of tropical cyclone formation since 1970 (Webster et al., 2005) as well as in both the number and intensity of storms in the Atlantic (Emanuel, 2005), the basin with the highest volatility in tropical cyclone numbers (see Trenberth et al., 2007, Sections 3.8.3 and 3.8.3.2).

Although the Atlantic record of hurricanes extends back to 1851, information on tracks is only considered comprehensive after 1945 and for intensity assessments it is only complete since the 1970s (Landsea, 2005). From 1995 to 2005, all seasons were above average in the Atlantic, with the exception of the two El Niño years of 1997 and 2002, when activity was suppressed – as in earlier El Niños. The number of intense (CAT 3-5 on the Saffir-Simpson Hurricane Scale) storms in the Atlantic since 1995 was more than twice the level of the 1970 to 1994 period, and 2005 was the most active year ever for Atlantic hurricanes on a range of measures, including number of hurricanes and number of the most intense CAT 5 hurricanes.

In the Atlantic, among the principal reasons for the increases in activity and intensity (Chelliah and Bell, 2004) are trends for increased sea surface temperatures in the tropical North Atlantic. The period since 1995 has had the highest temperatures ever observed in the equatorial Atlantic – "apparently as part of global warming" (see Trenberth et al., 2007, Section 3.8.3.2). The first and only tropical cyclone ever identified in the South Atlantic occurred in March 2004.

While other basins do not show overall increases in activity, observations based on satellite observations of intensity (which start in the 1970s) suggest a shift in the proportion of tropical cyclones that reached the higher intensity (CAT 4 and CAT 5) from close to 20% of the total in the 1970s rising to 35% since the 1990s (Webster et al., 2005). Although challenged by some climatologists based on arguments of observational consistency, as quoted from Trenberth et al., (2007) Section 3.8.3, "the trends found by Emanuel (2005) and Webster et al. (2005) appear to be robust in strong association with higher SSTs". Increases in the population of intense hurricanes in 2005 created record catastrophe losses, principally in the Gulf Coast, USA, and in Florida, when a record four Saffir-Simpson severe (CAT 3-5) hurricanes made landfall, causing more than US$100 billion in damages with almost 2,000 fatalities.

1.3.8.4 Economic and insurance losses

Economic losses attributed to natural disasters have increased from US$75.5 billion in the 1960s to US$659.9 billion in the 1990s (a compound annual growth rate of 8%) (United Nations Development Programme, 2004). Private-sector data on insurance costs also show rising insured losses over a similar period (Munich Re Group, 2005; Swiss Reinsurance Company, 2005). The dominant signal is of significant increase in the values of exposure (Pielke and Hoppe, 2006).

However, as has been widely acknowledged, failing to adjust for time-variant economic factors yields loss amounts that are not directly comparable and a pronounced upward trend through time for purely economic reasons. A previous normalisation of losses, undertaken for U.S. hurricanes by Pielke and Landsea (1998) and U.S. floods (Pielke et al., 2002) included normalising the economic losses for changes in wealth and population so as to express losses in constant dollars. These previous national U.S. assessments, as well as those for normalised Cuban hurricane losses (Pielke et al., 2003), did not show any significant upward trend in losses over time, but this was before the remarkable hurricane losses of 2004 and 2005.

A global catalogue of catastrophe losses was constructed (Muir Wood et al., 2006), normalised to account for changes that have resulted from variations in wealth and the number and value of properties located in the path of the catastrophes, using the method of Landsea et al. (1999). The global survey was considered largely comprehensive from 1970 to 2005 for countries and regions (Australia, Canada, Europe, Japan, South Korea, the USA, Caribbean, Central America, China, India and the Philippines) that had centralised catastrophe loss information and included a broad range of peril types: tropical cyclone, extra-tropical cyclone, thunderstorm, hailstorm, wildfire and flood, and that spanned high- and low-latitude areas.

Once the data were normalised, a small statistically significant trend was found for an increase in annual catastrophe loss since 1970 of 2% per year (see Supplementary Material Figure SM1.1). However, for a number of regions, such as Australia and India, normalised losses show a statistically significant reduction since 1970. The significance of the upward trend is influenced by the losses in the USA and the Caribbean in 2004 and 2005 and is arguably biased by the relative wealth of the USA, particularly relative to India.

1.3.8.5 Summary of disasters and hazards

Global losses reveal rapidly rising costs due to extreme weather-related events since the 1970s. One study has found that while the dominant signal remains that of the significant increases in the values of exposure at risk, once losses are normalised for exposure, there still remains an underlying rising trend. For specific regions and perils, including the most extreme floods on some of the largest rivers, there is evidence for an increase in occurrence.

1.3.9 Socio-economic indicators

The literature on observed changes in socio-economic indicators in response to recent climate change is sparse. Here we summarise some of the few examples related to energy demand and tourism, and some studies on regional adaptations to climate trends. Other relevant indicators include energy supply and markets for natural resources (e.g., timber, fisheries). Indicators of adaptation such as domestic insurance claims, energy demand, and changes in tourism are being defined and tracked for the UK and Europe (Defra, 2003; EEA, 2004).

1.3.9.1 Energy demand

Buildings account for a significant part of total energy use, up to 50% in some developed countries (Lorch, 1990; also see Levine et al., 2007), and the design and energy performance of buildings are related to climate (Steemers, 2003). Work related to climate change and building energy use can be grouped into two major areas – weather data analysis and building energy consumption.

Weather data analysis

A study on 1981 to 1995 weather data (Pretlove and Oreszczyn, 1998) indicated that temperature and solar radiation in the London area (UK) had changed significantly over the period, and climatic data used for energy design calculations could lead to 17% inaccuracies in building energy-use estimates. Based on 1976 to 1995 temperature data from 3 key UK sites, Levermore and Keeble (1998) found that the annual mean dry-bulb temperature had increased by about 1°C over the 19-year period, with milder winters and warmer summers. In sub-tropical Hong Kong SAR, the 40-year period (1961 to 2000) weather data showed an underlying trend of temperature rise, especially during the last 10 years (1991 to 2000) (Lam et al., 2004). The increases occurred largely during the winter months and the impact on peak summer design conditions and cooling requirements, and hence energy use, was considered insignificant. In the 1990s and 2000s, many countries experienced extreme phenomena (notably heatwaves in summer), which induced exceptional peaks of electric power consumption (Tank and Konnen, 2003). These had notable impacts on human mortality (Section 1.3.7) and local socio-economic systems (Easterling et al., 2000; Parmesan et al., 2000; Johnson et al., 2004). Two well-documented cases are the heatwaves in Chicago in 1995 (Karl and Knight, 1997) and in Europe in 2003 (Schär et al., 2004; Trigo et al., 2005).

Building energy consumption

One example related to energy and climate concerns cooling during hot weather. Energy use has been and will continue to be affected by climate change, in part because air-conditioning, which is a major energy use particularly in developed countries, is climate-dependent. However, the extent to which temperature rise has affected energy use for space heating/cooling in buildings is uncertain. There is a concern that energy consumption will increase as air-conditioning is adopted for warmer summers (see Levine et al., 2007). It is likely that certain adaptation strategies (e.g., tighter building energy standards) have been (or would be) taken in response to climate change (e.g., Camilleri et al., 2001; Larsson, 2003; Sanders and Phillipson, 2003; Shimoda, 2003). Adaptation strategies and implementation are strongly motivated by the cost of energy. Besides, in terms of thermal comfort, there is also the question of people adapting to warmer climates (e.g., de Dear and Brager, 1998; Nicol, 2004).

1.3.9.2 Tourism

Climate is a major factor for tourists when choosing a destination (Aguiló et al., 2005) and both tourists and tourism stakeholders are sensitive to fluctuations in the weather and climate (Wall, 1998). Statistical analyses by Maddison (2001), Lise and Tol (2002) and Hamilton (2003a), and a simulation study (Hamilton et al., 2003), have shown the relevance of climatic factors as determinants of tourist demand, next to economic and political conditions, fashion, media attention, and environmental quality. As a result of the complex nature of the interactions that exist between tourism, the climate system, the environment and society, it is difficult to isolate the direct observed impacts of climate change upon tourism activity. There is sparse literature about this relationship at any scale. Responses in skiing have been documented in Switzerland, Austria, the eastern USA and Chile (OECD, 2007; Elsasser and Messerli, 2001; Steininger and Weck-Hannemann, 2002; Beniston, 2003, 2004; Casassa et al., 2003; Hamilton et al., 2005) (see Section 1.3.1.1).

1.3.9.3 Regional adaptation

There are several studies that show societies adapting to climate changes such as drying trends or increasing temperatures (see Chapter 17). For example, responses to recent historical climate variability and change in four locations in southern Africa demonstrated that people were highly aware of changes in the climate, including longer dry seasons and more uncertain rainfall, and were adjusting to change through collective and individual actions that included both short-term coping through switching crops and long-term adaptations such as planting trees, and commercialising and diversifying livelihoods (Thomas and Twyman, 2005; Thomas et al., 2005). One of the most striking conclusions was the importance of local institutions and social capital such as farming associations in initiating and supporting adaptations. The use of climate science in adapting to water management during a long-term drought has been documented in Western Australia (Power et al., 2005).

In Europe, evidence is also accumulating that people are adapting to climate change, either in response to observed changes or in anticipation of predicted change. For example, in the UK, a large number of adaptations have been identified including changes in flood management guidelines (assuming more extremes), hiring of climate change managers, changing nature conservation and disaster plans, climate-proofing buildings, planting different crops and trees, and converting a skiing area to a walking centre in Scotland (West and Gawith, 2005).

Changes in socio-economic activities and modes of human response to climate change, including warming, are just beginning to be systematically documented in the cryosphere

(MacDonald et al., 1997; Krupnik and Jolly, 2002; Huntington and Fox, 2004; Community of Arctic Bay et al., 2005). The impacts associated with these changes are both positive and negative, and are most pronounced in relation to the migration patterns, health and range of animals and plants that indigenous groups depend on for their livelihood and cultural identity. Responses vary by community and are dictated by particular histories, perceptions of change and the viability of options available to groups (Ford and Smit, 2004; Helander and Mustonen, 2004). In Sachs Harbour, Canada, responses include individual adjustments to the timing, location and methods of harvesting animals, as well as adjusting the overall mix of animals harvested to minimise risk (Berkes and Jolly, 2002). Communities that are particularly vulnerable to coastal erosion such as Shishmeref, Alaska, are faced with relocation. Many communities in the North are stepping up monitoring efforts to watch for signs of change so they can respond accordingly in both the long and short term (Fox, 2002). Agent-based simulation models (i.e., models dealing with individual decision making and interactions among individuals) are also being developed to assess adaptation and sustainability in small-scale Arctic communities (Berman et al., 2004). Effective responses will be governed by increased collaboration between indigenous groups, climate scientists and resource managers (Huntington and Fox, 2004).

1.4 Larger-scale aggregation and attribution to anthropogenic climate change

Larger-scale aggregation offers insights into the relationships between the observed changes assessed in Section 1.3 and temperature, by combining results from many studies over multiple systems and larger regions. Aggregation through meta-analysis is described next, followed by joint attribution through climate model studies, and synthesis of the observed changes described in Section 1.3.

1.4.1 Larger-scale aggregation

This section evaluates studies that use techniques that aggregate from individual observations at sites to regional, continental and global scales. Meta-analysis is a statistical method of combining quantitative findings from many studies investigating similar factors for the purpose of finding a general result. The methods used in the various studies, however, need not be similar. The criteria for inclusion of studies in a meta-analysis are determined *a priori*, and rigorously followed to avoid investigator effect.

Several studies have examined the 'fingerprint' of observed warming in recent decades on the phenology and distribution of plants and animal species using meta-analyses (Root and Schneider, 2002; Parmesan and Yohe, 2003; Root et al., 2003). Although the detailed results of these studies are different, because they used different species and different methods, they all conclude that a significant impact of warming is already discernible in animal and plant populations at regional and continental scales in the Northern Hemisphere.

One meta-analysis (Parmesan and Yohe, 2003) of 31 studies of more than 1,700 species showed that recent biological trends matched the expected responses to warming. They estimated northward range shifts of 6.1 km/decade for northern range boundaries of species living in the Northern Hemisphere and advancement of spring events in Northern Hemisphere species by 2.3 days/decade. They also defined a diagnostic fingerprint of temporal and spatial 'sign-switching' responses uniquely predicted by 20th-century observed climate trends. Among long-term, large-scale, multi-species data sets, this diagnostic fingerprint was found for 279 species. They concluded, with 'very high confidence', that climate change is already affecting living systems.

After examining over 2,500 articles on climate change and a wide array of species from around the globe, another study found that 143 studies fitted the criteria for inclusion in their meta-analyses (Root et al., 2003). They focused on only those species showing a significant change and found that about 80% of the species showing change were changing in the direction expected with warming. The types of changes included species expanding their ranges polewards and higher in elevation, and advances in the timing of spring events by about 5 days/decade over the last 30 years. This number is larger than the 2.3 days/decade found by Parmesan and Yohe (2003), because those authors included both changing and not-changing species in their analysis, while Root and co-authors only included changing species. A more recent meta-analysis of bird arrival dates (Lehikoinen et al., 2004) showed strong evidence of earlier arrival. Of 983 data series, 39% were significantly earlier and only 2% significantly later for first arrival dates.

The EU COST725 network analysis project had as its main objective the establishment of a comprehensive European reference data set of phenological observations that could be used for climatological purposes, particularly climate monitoring and the detection of changes (see Box 1.3).

1.4.2 Joint attribution

Joint attribution involves attribution of significant changes in a natural or managed system to regional temperature changes, and attribution of a significant fraction of the regional temperature change to human activities. This has been performed using studies with climate models to assess observed changes in several different physical and biological systems. An assessment of the relationship between significant observed changes from Section 1.3 and significant regional temperature changes is presented in Section 1.4.2.3.

1.4.2.1 Attributing regional temperature change

It is likely that there has been a substantial anthropogenic contribution to surface temperature increases averaged over each continent except Antarctica since the middle of the 20th century (Hegerl et al., 2007, Section 9.4.2). Statistically significant regional warming trends over the last 50 and 30 years are found in many regions of the globe (Spagnoli et al., 2002; Karoly and Wu, 2005; Karoly and Stott, 2006; Knutson et al., 2006; Zhang et

Box 1.3. Phenological responses to climate in Europe: the COST725 project

The COST725 meta-analysis project used a very large phenological network of more than 125,000 observational series of various phases in 542 plant and 19 animal species in 21 European countries, for the period 1971 to 2000. The time-series were systematically (re-)analysed for trends in order to track and quantify phenological responses to changing climate. The advantage of this study is its inclusion of multiple verified nationally reported trends at single sites and/or for selected species, which individually may be biased towards predominant reporting of climate-change-induced impacts. Overall, the phenology of the species (254 national series) was responsive to temperature of the preceding month, with spring/summer phases advancing on average by 2.5 days/°C and leaf colouring/fall being delayed by 1.0 day/°C.

The aggregation of more than 100,000 trends revealed a clear signal across Europe of changing spring phenology with 78% of leaf unfolding and flowering records advancing (31% significantly (sig.)) and only 22% delayed (3% sig.) (Figure 1.6). Fruit ripening was mostly advanced (75% advancing, 25% sig.; 25% delayed, 3% sig.). The signal in farmers' activities was generally smaller (57% advancing, 13% sig.; 43% delayed, 6% sig.). Autumn trends (leaf colouring/fall) were not as strong. Spring and summer exhibited a clear advance by 2.5 days/decade in Europe, mean autumn trends were close to zero, but suggested more of a delay when the average trend per country was examined (1.3 days/decade).

The patterns of observed changes in spring (leafing, flowering and animal phases) were spatially consistent and matched measured national warming across 19 European countries (correlation = –0.69, $P < 0.001$); thus the phenological evidence quantitatively mirrors regional climate warming. The COST725 results assessed the possible lack of evidence at a continental scale as 20%, since about 80% of spring/summer phases were found to be advancing. The findings strongly support previous studies in Europe, confirming them as free from bias towards reporting global climate change impacts (Menzel et al., 2006b).

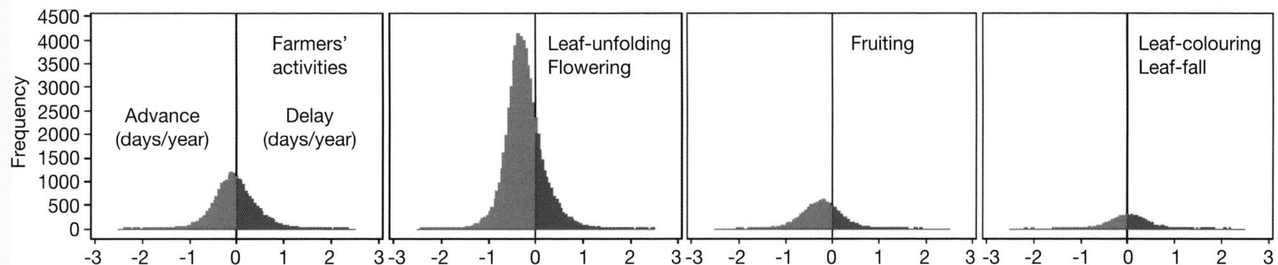

Figure 1.6. *Frequency distributions of trends in phenology (in days/year) over 1971 to 2000 for 542 plant species in 21 European countries. From Menzel et al. (2006b).*

al., 2006; Trenberth et al., 2007, Figure 3.9). These warming trends are consistent with the response to increasing greenhouse gases and sulphate aerosols and likely cannot be explained by natural internal climate variations or the response to changes in natural external forcing (solar irradiance and volcanoes).

Attributing temperature changes on smaller than continental scales and over time-scales of less than 20 years is difficult due to low signal-to-noise ratios at those scales. Attribution of the observed warming to anthropogenic forcing is easier at larger scales because averaging over larger regions reduces the natural variability more, making it easier to distinguish between changes expected from different external forcings, or between external forcing and climate variability.

The influence of anthropogenic forcing has also been detected in various physical systems over the last 50 years, including increases in global oceanic heat content, increases in sea level, shrinking of alpine glaciers, reductions in Arctic sea ice extent, and reductions in spring snow cover (Hegerl et al., 2007).

1.4.2.2 Joint attribution using climate model studies

Several studies have linked the observed responses in some biological and physical systems to regional-scale warming due to anthropogenic climate change using climate models.

One study demonstrated joint attribution by considering changes in wild animals and plants (Root et al., 2005). They found spring phenological data for 145 Northern Hemisphere species from 31 studies. The changes in the timing of these species' spring events (e.g., blooming) are significantly associated with the changes in the actual temperatures recorded as near to the study site as possible and for the same years that the species were observed. If the temperature was warming and the species phenology was getting earlier in the year, then the expected association would be negative, which is what was found for the correlations between the species data and the actual temperatures (Figure 1.7).

Temperature data from the HadCM3 climate model were used to determine whether the changes in the actual temperatures with which the phenological changes in species were associated were due to human or natural causes. Modelled temperature data were derived for each species, over the same years a species was studied and for the grid box within which the study area was located. Three different forcings were used when calculating the modelled values: natural only, anthropogenic only, and combined natural and anthropogenic. Each species' long-term phenological record was correlated with the three differently forced temperatures derived for the location where the species was recorded. The agreement is quite poor between the phenological changes in species and modelled temperatures derived using only natural climatic forcing ($K = 60.16$, $P > 0.05$; Figure 1.7a). A stronger agreement occurs between the same phenological changes in species and temperatures modelled using only anthropogenic forcing ($K = 35.15$, $P > 0.05$; Figure 1.7b). As expected, the strongest agreement is with the modelled temperatures derived using both natural and anthropogenic (combined) forcings ($K = 3.65$, $P < 0.01$; Figure 1.7c). While there is uncertainty in downscaling the model-simulated temperature changes to the areas that would affect the species being examined, these results demonstrate some residual skills, thereby allowing joint attribution to be shown.

Other similar studies have shown that the retreat of two glaciers in Switzerland and Norway cannot be explained by natural variability of climate and the glaciers alone (Reichert et al., 2002), that observed global patterns of changes in streamflow are consistent with the response to anthropogenic climate change (Milly et al., 2005), and that the observed increase in the area of forests burned in Canada over the last four decades is consistent with the response due to anthropogenic climate change (Gillett et al., 2004). Each of these studies has its limitations for joint attribution. For example, the analysis by Reichert used a climate model linked to a local glacier mass balance model through downscaling and showed that the observed glacier retreat over the 20th century could not be explained by natural climate variability. However, they did not show that the observed retreat was consistent with the response to anthropogenic climate change, nor did they eliminate other possible factors, such as changes in dust affecting the albedo of the glacier. Similarly, Gillett and colleagues showed that the observed increases in area of forests burned was consistent with the response to anthropogenic forcing and not consistent with natural climate variability. However, they did not consider changes in forest management as a factor, nor did they consider the climate response to other external forcing factors.

Taken together, these studies show a discernible influence of anthropogenic climate change on specific physical (cryosphere, hydrology) and biological (forestry and terrestrial biology) systems.

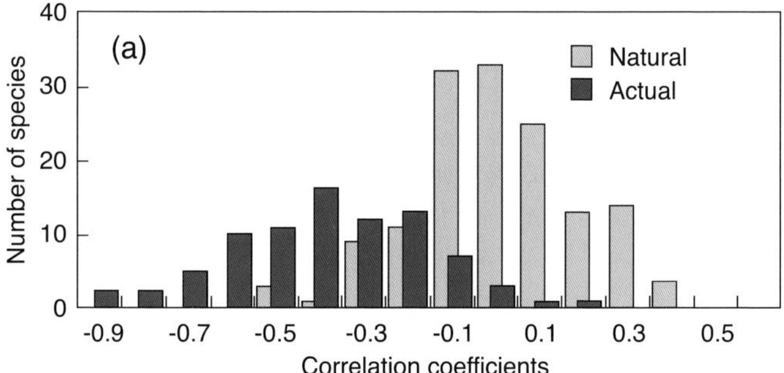

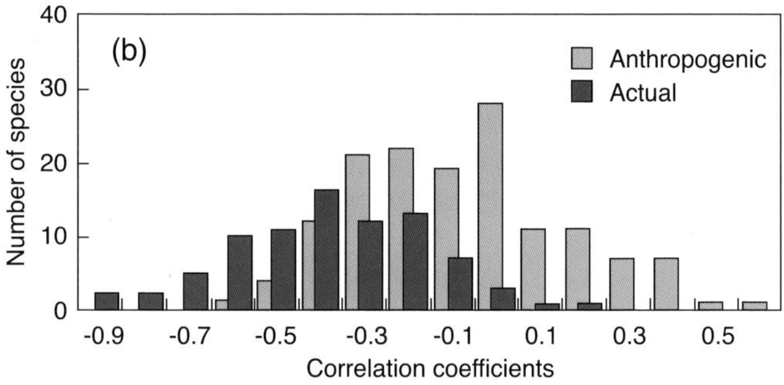

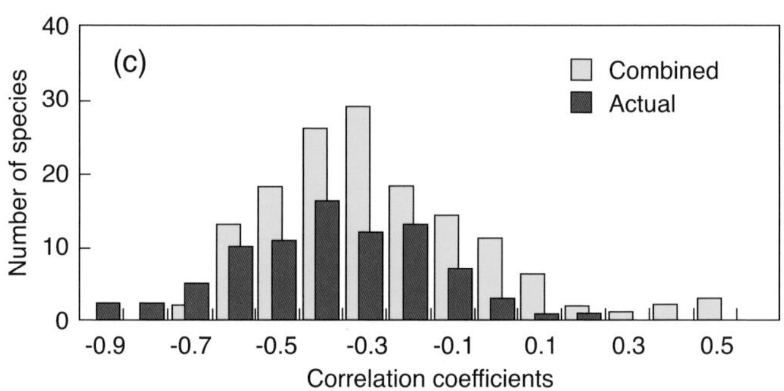

Figure 1.7. *Plotted are the frequencies of the correlation coefficients (associations) between the timing of changes in traits (e.g., earlier egg-laying) of 145 species and modelled (HadCM3) spring temperatures for the grid-boxes in which each species was examined. At each location, all of which are in the Northern Hemisphere, the changing trait is compared with modelled temperatures driven by: (a) natural forcings (purple bars), (b) anthropogenic (i.e., human) forcings (orange bars), and (c) combined natural and anthropogenic forcings (yellow bars). In addition, on each panel the frequencies of the correlation coefficients between the actual temperatures recorded during each study and changes in the traits of 83 species, the only ones of the 145 with reported local-temperature trends, are shown (dark blue bars). On average the number of years that species were examined is about 28, with average starting and ending years of 1960 and 1998. Note that the agreement: (a) between the natural and actual plots is weaker (K = 60.16) than (b) between the anthropogenic and actual (K = 35.15), which in turn is weaker than (c) the agreement between combined and actual (K = 3.65). Taken together, these plots show that a measurable portion of the warming regional temperatures to which species are reacting can be attributed to humans, therefore showing joint attribution (after Root et al., 2005).*

1.4.2.3 Synthesis of studies

Next, a synthesis of the significant observed changes described in Section 1.3 and the observed regional temperatures over the last three decades was performed. Significant observed changes documented since the TAR were divided into the categories of cryosphere, hydrology, coastal processes, marine and freshwater biological systems, terrestrial biological systems, and agriculture and forestry, as assessed in Section 1.3. Studies were selected that demonstrate a statistically significant trend in change in systems related to temperature or other climate change variable as described by the authors, for the period 1970 to 2004 (study periods may be extended later), with at least 20 years of data. Observations in the studies are characterised as 'change consistent with warming' and 'change not consistent with warming'.

Figure 1.8 shows the warming trends over the period 1970 to 2004 (from the GHCN-ERSST dataset; Smith and Reynolds,

2005) and the geographical locations of significant observed changes. A statistical comparison shows that the agreement between the regions of significant and regional warming across the globe and the locations of significant observed changes in systems consistent with warming is very unlikely to be due to natural variability in temperatures or natural variability in the systems (Table 1.12) (see also Supplementary Material).

For regions where there are both significant warming and observed changes in systems, there is a much greater probability of finding coincident significant warming and observed responses in the expected direction. The statistical agreement between the patterns of observed significant changes in systems and the patterns of observed significant warming across the globe very likely cannot be explained by natural climate variability.

Uncertainties in observed change studies at the regional level relate to potential mismatches between climate and system data in temporal and spatial scales and lack of time-series of sufficient length to determine whether the changes are outside normal ranges of variability. The issue of non-climate driving forces is also important. Land-use change, changes in human management practices, pollution and demography shifts are all, along with climate, drivers of environmental change. More explicit consideration of these factors in observed change studies will strengthen the robustness of the conclusions. However, these factors are very unlikely to explain the coherent responses that have been found across the diverse range of systems and across the broad geographical regions considered (Figure 1.9).

Since systems respond to an integrated climate signal, precise assignment of the proportions of natural and anthropogenic forcings in their responses in a specific grid cell is difficult. The observed continent-averaged warming in all continents except Antarctica over the last 50 years has been attributed to anthropogenic causes (IPCC, 2007, Summary for Policymakers). The prevalence of observed changes in physical and biological systems in expected directions consistent with anthropogenic warming on every continent and in some oceans means that anthropogenic climate change is having a discernible effect on physical and biological systems at the global scale.

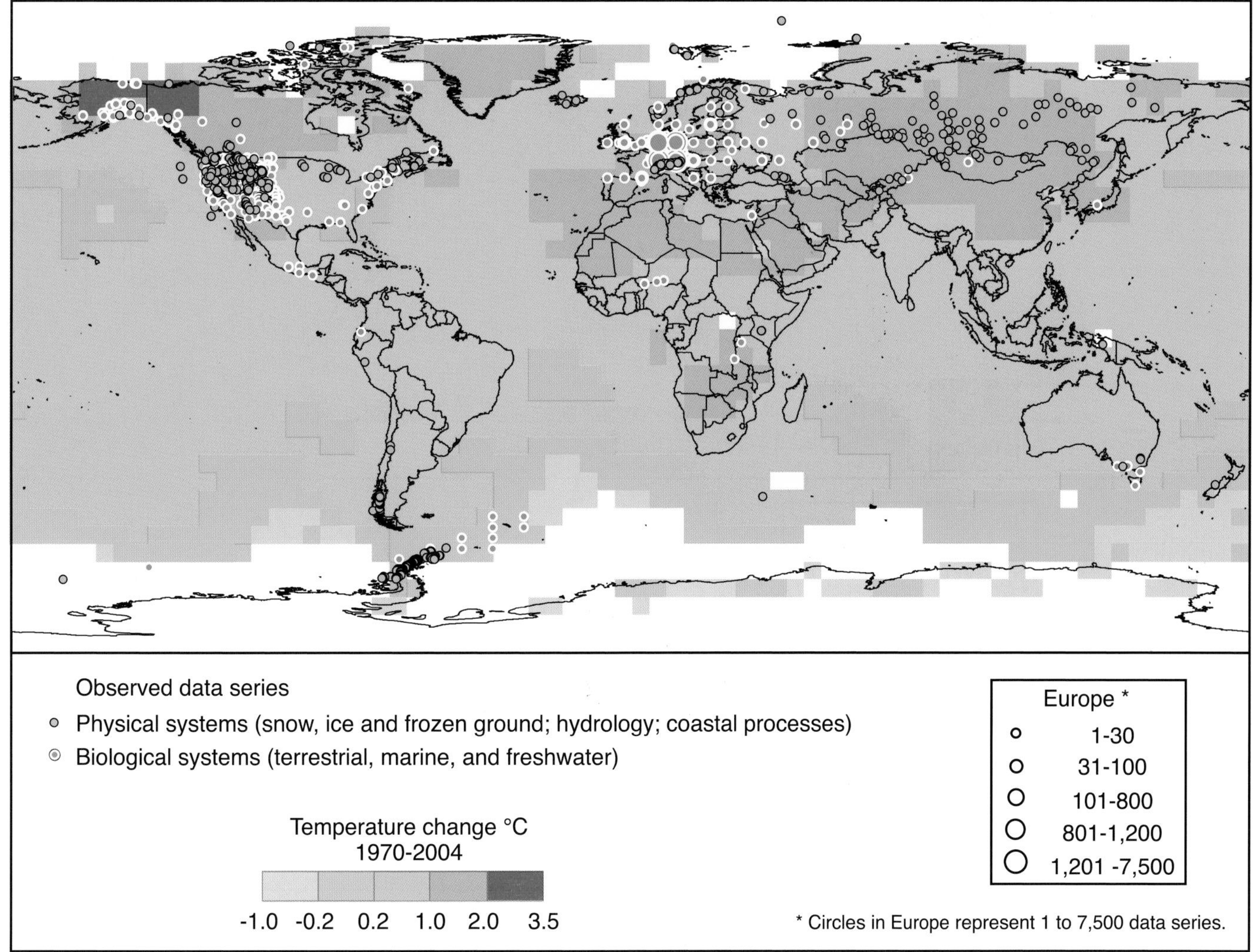

Figure 1.8. *Locations of significant changes in observations of physical systems (snow, ice and frozen ground; hydrology; coastal processes) and biological systems (terrestrial, marine and freshwater biological systems), are shown together with surface air temperature changes over the period 1970 to 2004 (from the GHCN-ERSST datatset). The data series met the following criteria: (1) ending in 1990 or later; (2) spanning a period of at least 20 years; (3) showing a significant change in either direction, as assessed by individual studies. White areas do not contain sufficient observational climate data to estimate a temperature trend.*

Table 1.12. *Global comparison of significant observed changes in physical and biological systems with regional temperature changes. Fraction of 5°×5° cells with significant observed changes in systems (from studies considered in this chapter) and temperature changes (for 1970 to 2004 from the GHCN-ERSST dataset (Smith and Reynolds, 2005)) in different categories (significant warming, warming, cooling, significant cooling). Expected values shown in parentheses are for the null hypotheses:*
(i) significant observed changes in systems are equally likely in each direction,
(ii) temperature trends are due to natural climate variations and are normally distributed,
(iii) there is no relationship between significant changes in systems and co-located warming.

Temperature cells	Cells with significant observed change consistent with warming	Cells with significant observed change not consistent with warming
Significant warming	50% (2.5%)	7% (2.5%)
Warming	34% (22.5%)	6% (22.5%)
Cooling	3% (22.5%)	0% (22.5%)
Significant cooling	0% (2.5%)	0% (2.5%)
Chi-squared value (significance level)		369 (<<1%)

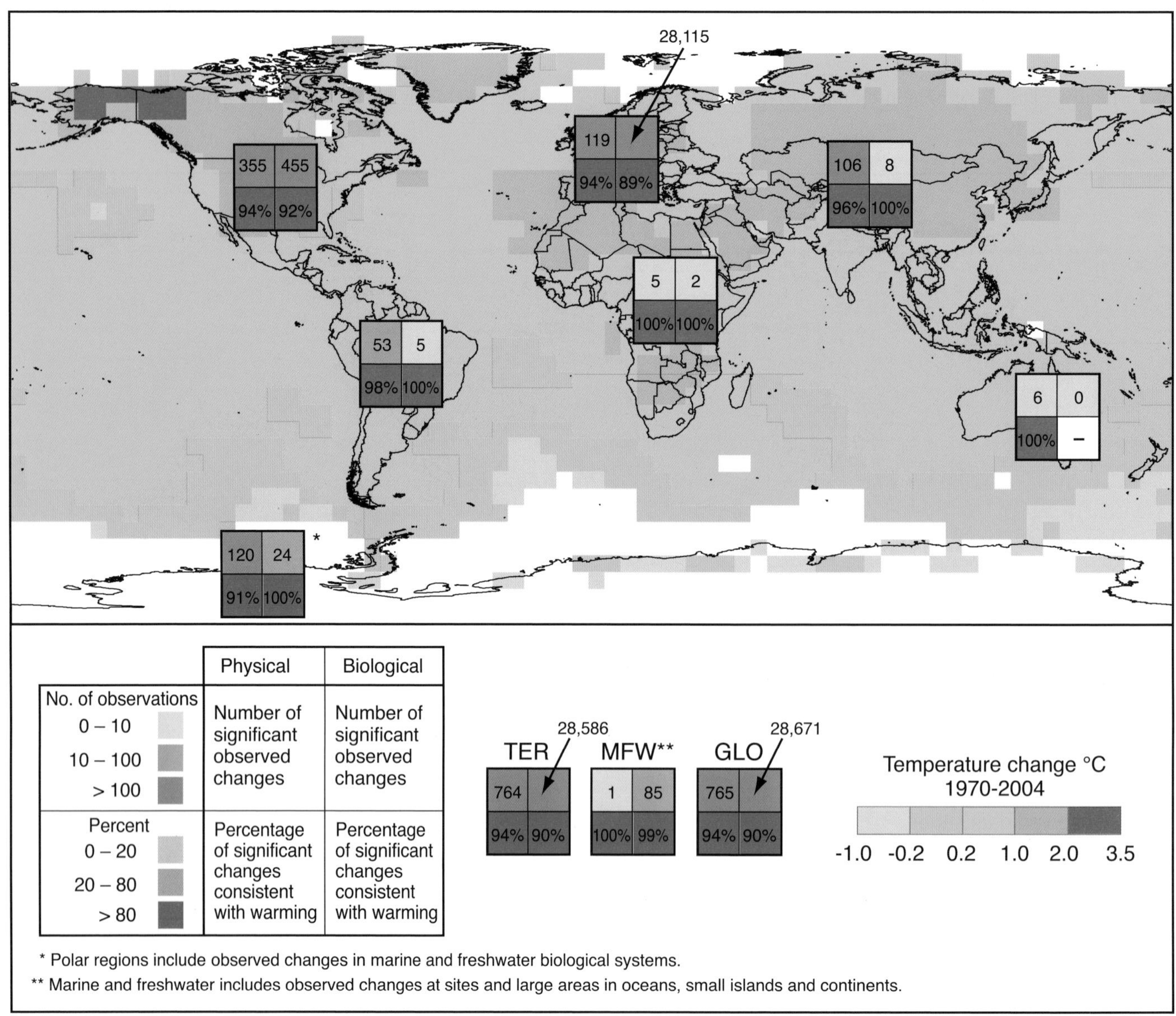

Figure 1.9. *Changes in physical and biological systems and surface temperature. Background shading and the key at the bottom right show changes in gridded surface temperatures over the period 1970 to 2004 (from the GHCN-ERSST dataset). The 2×2 boxes show the total number of data series with significant changes (top row) and the percentage of those consistent with warming (bottom row) for (i) continental regions; North America, Latin America, Europe, Africa, Asia, Australia and New Zealand, and Polar Regions; and (ii) global-scale: Terrestrial (TER), Marine and Freshwater (MFW), and Global (GLO). The numbers of studies from the seven regional boxes do not add up to the global totals because numbers from regions except Polar do not include the numbers related to Marine and Freshwater systems. White areas do not contain sufficient observational climate data to estimate a temperature trend.*

1.5 Learning from observed responses: vulnerability, adaptation and research needs

The great majority of observed changes are consistent with functional understanding and modelled predictions of climate impacts. Examples of expected responses include infrastructure effects of melting in the cryosphere, effects of intensifying droughts and runoff, and effects of rising sea levels. In marine, freshwater and terrestrial biological systems, changes in morphology, physiology, phenology, reproduction, species distribution, community structure, ecosystem processes and species evolutionary processes are, for the most part, in the predicted directions. Agricultural crops have shown similar trends in phenology, and management practices along with the spread of pests and diseases coincide with expected responses to warming. Responses of yields in the few crops with reported changes coincide with model predictions. Temperature-sensitive vectors, e.g., ticks, have spread for some human diseases.

Observed changes are prevalent across diverse physical and biological systems and less prevalent in managed systems and across many, but not all, geographical regions. While there is evidence of observed changes in every continent, including Antarctica, much evidence comes from studies of observed changes in Northern Hemisphere mid- and high latitudes and often from higher altitudes. Significant evidence comes from high-latitude waters in the Northern Hemisphere as well. Evidence is primarily found in places where warming is most pronounced. Documentation of observed changes in tropical regions is still sparse.

The evidence for adaptation and vulnerability to observed climate change is most prevalent in places where warming has been the greatest and in systems that are more sensitive to temperature. Thus, documented changes relating to adaptation in the Arctic and mountain regions include reduced outdoor and tourism activities, and alterations in indigenous livelihoods in the Arctic. Responses to climate change, including warming, vary by community and are beginning to be systematically documented (Section 1.3.9).

In terrestrial biological systems, special conservation measures by resource managers are carried out as an adaptation to the impacts of climate change, focusing on spatial strategies, such as ecological networks, short-term refugia, robust corridors, transnational pathways, or potential future protected areas (Opdam and Wascher, 2004; Thomas, 2005; Gaston et al., 2006). Conservation management for wetlands undergoing erosion has been addressed as well (Wolters et al., 2005).

Documented evidence of adaptation to regional climate trends in the highly managed systems of agriculture and forestry is beginning to emerge, such as shifts of sowing dates of annual crops in Europe (Section 1.3.6). With regard to the assessment of vulnerability, few studies have documented observed effects of warming in subsistence agricultural systems in rural populations in developing countries; there are, however, well-documented studies of adaptive responses and vulnerability to long-term drought in the Sahel.

Vulnerability appears to be high in the case of extreme events or exceptional episodes, even in developed countries, as documented by the agricultural response to, and excess mortality occurring in, the 2003 heatwaves in Europe. The global decline in aggregate deaths and death rates due to extreme weather events during the 20th century suggest that adaptation measures to cope with some of the worst consequences of such events have been successful. However, the 2003 European heatwave and the 2005 hurricane season in the North Atlantic show that, despite possessing considerable adaptive capacity, even developed nations are vulnerable if they do not mobilise adaptation measures in a timely and efficient manner. In human health, air-conditioning has contributed to declines in death rates during the summer in the USA and Europe over the past 30-40 years (Section 1.3.7). Documentation of adaptation and vulnerability in terms of energy and tourism is limited (Section 1.3.9).

There is a notable lack of geographical balance in the data and literature on observed changes in natural and managed systems, with a marked scarcity from developing countries. Regions with climate warming with an accumulation of evidence of observed changes in physical and/or biological systems are Europe, Northern Asia, north-western North America, and the Antarctic Peninsula. Regions with warming where evidence of observed changes is sparse are Africa and Latin America, and evidence is lacking in South-east Asia, the Indian Ocean and regions in the Pacific. Possible reasons for this imbalance are lack of access by IPCC authors, lack of data, research and published studies, lack of knowledge of system sensitivity, differing system responses to climate variables, lag effects in responses, resilience in systems and the presence of adaptation. There is a need to improve the observation networks and to enhance research capability on changes in physical, biological and socio-economic systems, particularly in regions with sparse data. This will contribute to an improved functional understanding of the responses of natural and managed systems to climate change.

References

Abeku, T.A., G.J. van Oortmarssen, G. Borsboom, S.J. de Vlas and J.D.F. Habbema, 2003: Spatial and temporal variations of malaria epidemic risk in Ethiopia: factors involved and implications. *Acta Trop.*, **87**, 331-340.

Abu-Asab, M.S., P.M. Peterson, S.G. Shetler and S.S. Orli, 2001: Earlier plant flowering in spring as a response to global warming in the Washington, DC, area. *Biodivers. Conserv.*, **10**, 597.

ACIA, 2005: *Arctic Climate Impact Assessment*. Cambridge University Press, Cambridge, 1042 pp.

Adams, J.B., M.E. Mann and C.M. Ammann, 2003: Proxy evidence for an El Niño-like response to volcanic forcing. *Nature*, **426**, 274-278.

Adrian, R. and R. Deneke, 1996: Possible impact of mild winters on zooplankton succession in eutrophic lakes of the Atlantic European area. *Freshwater Biol.*, **36**, 757-770.

Adrian, R., R. Deneke, U. Mischke, R. Stellmacher and P. Lederer, 1995: A long-term study of the Heiligensee (1975-1992): evidence for effects of climatic change on the dynamics of eutrophied lake ecosystems. *Arch. Hydrobiol.*, **133**, 315-337.

Agrawala, S., S. Gigli, V. Raksakulthai, A. Hemp, A. Moehner, D. Conway, M. El Raey, A.U. Ahmed, J. Risbey, W. Baethgen and D. Martino, 2005: Climate change and natural resource management: key themes from case studies. *Bridge Over Troubled Waters: Linking Climate Change and Development*, S. Agrawala, Ed., Organisation for Economic Co-operation and Development, Paris, 85-131.

Aguiló, E., J. Alegre and M. Sard, 2005: The persistence of the sun and sand tourism model. *Tourism Manage.*, **26**, 219-231.

Ahas, R., 1999: Long-term phyto-, ornitho- and ichthyophenological time-series analyses in Estonia. *Int. J. Biometeorol.*, **42**, 119-123.

Ahas, R., A. Aasa, A. Menzel, V.G. Fedotova and H. Scheifinger, 2002: Changes in European spring phenology. *Int. J. Climatol.*, **22**, 1727-1738.

Alexander, M.A. and J.K. Eischeid, 2001: Climate variability in regions of amphibian declines. *Conserv. Biol.*, **15**, 930-942.

Allan, J.C. and P.D. Komar, 2006: Climate controls on U.S. West Coast erosion processes. *J. Coastal Res.*, **3**, 511-529.

Almaraz, P. and J. Amat, 2004: Complex structural effects of two hemispheric climatic oscillators on the regional spatio-temporal expansion of a threatened bird. *Ecol. Lett.*, **7**, 547-556.

Ames, A., 1998: A documentation of glacier tongue variations and lake development in the Cordillera Blanca, Peru. *Zeitschrift fur Gletscherkunde and Glazialgeologie*, **34**, 1-26.

Ames, A., S. Dolores, A. Valverde, P. Evangelista, D. Javier, W. Gavnini, J. Zuniga and V. Gomez, 1989: *Glacier Inventory of Peru*. Unidad de Glaciología e Hidrología, Huaraz.

Amstrup, S.C., 2003: Polar bear, *Ursus maritimus*. *Wild Mammals of North America: Biology, Management, and Conservation*, G.A. Feldhamer, B.C. Thompson, J.A. Chapman, Eds., Johns Hopkins University Press, Baltimore, Maryland, 587-610.

Amstrup, S.C., I. Stirling, T.S. Smith, C. Perham and G.W. Thiemann, 2006: Recent observations of intraspecific predation and cannibalism among polar bears in the southern Beaufort Sea. *Polar Biol.*, **29**, 997-1002.

Andreadis, K.M., E.A. Clark, A.W. Wood, A.F. Hamlet and D.P. Lettenmaier, 2005: Twentieth-century drought in the conterminous United States. *J. Hydrometeorol.*, **6**, 985-1001.

Arhonditsis, G.B., M.T. Brett, C.L. DeGasperi and D.E. Schindler, 2004: Effects of climatic variability on the thermal properties of Lake Washington. *Limnol. Oceanogr.*, **49**, 256-270.

Arthur, R., T.J. Done and H. Marsh, 2005: Benthic recovery four years after an El Niño-induced coral mass mortality in the Lakshadweep atolls. *Curr. Sci. India*, **89**, 694-699.

Atkinson, A., V. Siegel, E. Pakhomov and P. Rothery, 2004: Long-term decline in krill stock and increase in salps within the Southern Ocean. *Nature*, **432**, 100-103.

Attrill, M.J. and M. Power, 2002: Climatic influence on a marine fish assemblage. *Nature*, **417**, 275-278.

Bairlein, F. and D.W. Winkel, 2001: Birds and climate change. *Climate of the 21st Century: Changes and Risks*, J.L. Lozán, H. Graßl and P. Hupfer, Eds., Wissenschaftliche Auswertungen, Hamburg, 278.

Barbraud, C. and H. Weimerskirch, 2001: Emperor penguins and climate change. *Nature*, **411**, 183-186.

Barnes, K.I., D.N. Durrheim, F. Little, A. Jackson, U. Mehta, E. Allen, S.S. Dlamini, J. Tsoka, B. Bredenkamp, D.J. Mthembu, N.J. White and B.L. Sharp, 2005: Effect of artemether-lumefantrine policy and improved vector control on malaria burden in KwaZulu–Natal, South Africa. *PLoS Med.*, **2**, 1123-1134.

Bartholow, J.M., 2005: Recent water temperature trends in the Lower Klamath River, California. *N. Am. J. Fish. Manage.*, **25**, 152-162.

Batimaa, P., 2005: The potential impact of climate change and vulnerability and adaptation assessment for the livestock sector of Mongolia. *Assessments of Impacts and Adaptations to Climate Change*, AIACC, Washington, District of Columbia, 20 pp.

Battarbee, R.W., J.A. Grytnes, R. Thompson, P.G. Appleby, J. Catalan, A. Korhola, H.J.B. Birks, E. Heegaard and A. Lami, 2002: Comparing palaeolimnological and instrumental evidence of climate change for remote mountain lakes over the last 200 years. *J. Paleolimnol.*, **28**, 161-179.

Battisti, A., M. Statsmy, A. Schopf, A. Roques, C. Robinet and A. Larsson, 2005: Expansion of geographical range in the pine processionary month caused by increased winter temperature. *Ecol. Appl.*, **15**, 2084-2094.

Beare, D., F. Burns, E. Jones, K. Peach, E. Portilla, T. Greig, E. McKenzie and D. Reid, 2004: An increase in the abundance of anchovies and sardines in the north-western North Sea since 1995. *Glob. Change Biol.*, **10**, 1209-1213.

Beaugrand, G. and P.C. Reid, 2003: Long-term changes in phytoplankton, zooplankton and salmon related to climate. *Glob. Change Biol.*, **9**, 801-817.

Beaugrand, G., F. Ibanez, J.A. Lindley and P.C. Reid, 2002a: Diversity of calanoid copepods in the North Atlantic and adjacent seas: species associations and biogeography. *Mar. Ecol.–Prog. Ser.*, **232**, 179-195.

Beaugrand, G., P.C. Reid, F. Ibanez, J.A. Lindley and M. Edwards, 2002b: Reorganization of North Atlantic marine copepod biodiversity and climate. *Science*, **296**, 1692-1694.

Beaugrand, G., K.M. Brander, J.A. Lindley, S. Souissi and P.C. Reid, 2003: Plankton effect on cod recruitment in the North Sea. *Nature*, **426**, 661-664.

Beaulieu, N. and M. Allard, 2003: The impact of climate change on an emerging coastline affected by discontinuous permafrost: Manitounuk Strait, northern Quebec. *Can. J. Earth Sci.*, **40**, 1393-1404.

Beever, E.A., P.E. Brussard and J. Berger, 2003: Patterns of apparent extirpation among isolated populations of pikas (*Ochotona princeps*) in the Great Basin. *J. Mammal.*, **84**, 37-54.

Beggs, P.J., 2004: Impacts of climate change on aeroallergens: past and future. *Clin. Exp. Allergy*, **34**, 1507-1513.

Beggs, P.J. and H.J. Bambrick, 2005: Is the global rise of asthma an early impact of anthropogenic climate change? *Environ. Health Persp.*, **113**, 915-919.

Bellwood, D.R., A.S. Hoey, J.L. Ackerman and M. Depczynski, 2006: Coral bleaching, reef fish community phase shifts and the resilience of coral reefs. *Glob. Change Biol.*, **12**, 1587-1594.

Beltaos, S., 2002: Effects of climate on mid-winter ice jams. *Hydrol. Process.*, **16**, 789-804.

Ben Mohamed, A., N.V. Duivenbooden and S. Abdoussallam, 2002: Impact of climate change on agricultural production in the Sahel. Part 1. Methodological approach and case study for millet in Niger. *Climatic Change*, **54**, 327-348.

Beniston, M., 2003: Climatic change in mountain regions: a review of possible impacts. *Climatic Change*, **59**, 5-31.

Beniston, M., 2004: The 2003 heat wave in Europe: a shape of things to come? An analysis based on Swiss climatological data and model simulations. *Geophys. Res. Lett.*, **31**, L02202, doi:10.1029/2003GL018857.

Bennett, L.T. and M.A. Adams, 2004: Assessment of ecological effects due to forest harvesting: approaches and statistical issues. *J. Appl. Ecol.*, **41**, 585-598.

Benoit, M. and C.D.L. Torre, 2004: Changement climatique et observation à long terme en unités expérimentales: évolution des pratiques agricoles et des réponses physiologiques des couverts végétaux. Journées MICCES 2004, accessible at: http://www.avignon.inra.fr/les_recherches__1/liste_des_unites/agroclim/mission_changement_climatique_et_effet_de_serre/le_seminaire_de_la_mission_changement_climatique_22_23_janvier_2004/les_posters.

Berger, T., J. Mendoza, B. Francou, F. Rojas, R. Fuertes, M. Flores, L. Noriega, C. Ramallo, E. Ramirez and H. Baldivieso, 2005: Glaciares Zongo – Chacaltaya – Charquini Sur – Bolivia 16°S. Mediciones Glaciológicas, Hidrológicas y Meteorológicas, Año Hidrológico 2004-2005. *Informe Great Ice Bolivia, IRD-IHH-SENMAHI-COBEE*, 171.

Bergeron, Y., M. Flaanigan, S. Gauthier, A. Leduc and P. Lefort, 2004: Past, current and future fire frequency in the Canadian boreal forest: implications for sustainable forest management. *Ambio*, **33**, 356-360.

Berkes, F. and D. Jolly, 2002: Adapting to climate change: social-ecological resilience in a Canadian Western Arctic community. *Conserv. Ecol.*, **5**, 18.

Berman, M., C. Nicolson, G. Kofinas, J. Tetlichi and S. Martin, 2004: Adaptation and sustainability in a small Arctic community: results of an agent-based simulation model. *Arctic*, **57**, 401-414.

Bernatchez, P. and J.-M.M. Dubois, 2004: Bilan des connaissances de la dynamique de l'érosion des côtes du Québec maritime laurentien. *Geogr. Phys. Quatern.*, **58**, 45-71.

Berteaux, D., D. Reale, A.G. McAdam and S. Boutin, 2004: Keeping pace with fast climate change: Can arctic life count on evolution? *Integr. Comp. Biol.*, **44**, 140.

Berthold, P., E. Gwinner and E. Sonnenschein, 2003: *Avian Migration*. Springer-Verlag, Berlin, 565 pp.

Bertness, M.D., P.J. Ewanchuk and B. Reed, 2002: Anthropogenic modification of New England salt marsh landscapes. *P. Natl. Acad. Sci. USA*, **99**, 1395-1398.

Bindoff, N., J. Willebrand, V. Artale, A. Cazenave, J. Gregory, S. Gulev, K. Hanawa, C. Le Quéré, S. Levitus, Y. Nojiri, C.K. Shum, L. Talley and A. Unnikrishnan, 2007: Observations: oceanic climate change and sea level. *Climate Change 2007: The Physical Science Basis. Contribution of Working Group I to the Fourth Assessment Report of the Intergovernmental Panel on Climate Change*, S. Solomon, D. Qin, M. Manning, Z. Chen, M. Marquis, K.B. Averyt, M. Tignor and H.L. Miller, Eds., Cambridge University Press, Cambridge, 385-432.

Blais, J.M., D.W. Schindler, D.C.G. Muir, M. Sharp, D. Donald, M. Lafreniere, E.

Braekevelt and W. M.J. Strachan, 2001: Melting glaciers: a major source of persistent organochlorines to subalpine Bow Lake in Banff National Park, Canada. *Ambio*, **30**, 410-415.

Blaustein, A.R., L.K. Belden, D.H. Olson, D.M. Green, T.L. Root and J.M. Kiesecker, 2001: Amphibian breeding and climate change. *Conserv. Biol.*, **15**, 1804.

Blenckner, T. and H. Hillebrand, 2002: North Atlantic Oscillation signatures in aquatic and terrestrial ecosystems: a meta-analysis. *Glob. Change Biol.*, **8**, 203-212.

Bodaly, R.A., J.W.M. Rudd, R.J.P. Fudge and C.A. Kelly, 1993: Mercury concentrations in fish related to size of remote Canadian shield lakes. *Can. J. Fish. Aquat. Sci.*, **50**, 980-987.

Boesch, D.F., M.N. Josselyn, A.J. Mehta, J.T. Morris, W.K. Nuttle, C.A. Simenstad and D.J.P. Swift, 1994: Scientific assessment of coastal wetland loss, restoration and management in Louisiana. *J. Coast. Res.*, **20**, 1-103.

Boon, S., M. Sharp and P. Nienow, 2003: Impact of an extreme melt event on the runoff and hydrology of a high Arctic glacier. *Hydrol. Process.*, **17**, 1051-1072.

Both, C., A.V. Artemyev, B. Blaauw, R.J. Cowie, A.J. Dekhuijzen, T. Eeva, A. Enemar, L. Gustafsson, E.V. Ivankina, A. Jarvinen, N.B. Metcalfe, N.E.I. Nyholm, J. Potti, P.A. Ravussin, J.J. Sanz, B. Silverin, F.M. Slater, L.V. Sokolov, J. Torok, W. Winkel, J. Wright, H. Zang and M.E. Visser, 2004: Large-scale geographical variation confirms that climate change causes birds to lay earlier. *P. Roy. Soc. Lond. .B Bio.*, **271**, 1657.

Bowen, N., 2002: Canary in a coalmine. *Climbing News*, **208**, 90-97, 138-139.

Bradshaw, W.E. and C.M. Holzapfel, 2001: Genetic shift in photoperiodic response correlated with global warming. *P. Natl. Acad. Sci. USA*, **98**, 14509.

Bradshaw, W.E. and C.M. Holzapfel, 2006: Climate change: evolutionary response to rapid climate change. *Science*, **312**, 1477-1478.

Bradshaw, W.E., M.C. Quebodeaux and C.M. Holzapfel, 2003: Circadian rhythmicity and photoperiodism in the pitcher-plant mosquito: adaptive response to the photic environment or correlated response to the seasonal environment? *Am. Nat.*, **161**, 735.

Brander, K., G. Blom, M.F. Borges, K. Erzini, G. Henderson, B.R. Mackenzie, H. Mendes, J. Ribeiro, A.M.P. Santos and R. Toresen, 2003: Changes in fish distribution in the eastern North Atlantic: are we seeing a coherent response to changing temperature? *ICES Marine Science Symposium*, **219**, 261-270.

Briand, J.F., C. Leboulanger, J.F. Humbert, C. Bernard and P. Dufour, 2004: *Cylindrospermopsis raciborskii* (Cyanobacteria) invasion at mid-latitudes: selection, wide physiological tolerance, or global warming? *J. Phycol.*, **40**, 231-238.

Brooks, S.J. and H.J.B. Birks 2004: The dynamics of Chironomidae (Insecta : Diptera) assemblages in response to environmental change during the past 700 years on Svalbard. *J. Paleolimnol.*, **31**, 483-498.

Bunn, A.G. and S.J. Goetz, 2006: Trends in satellite-observed circumpolar photosynthetic activity from 1982 to 2003: the influence of seasonality, cover type and vegetation density. *Earth Interactions*, **10**, 1-19.

Burrowes, P.A., R.L. Joglar and D.E. Green, 2004: Potential causes for amphibian declines in Puerto Rico. *Herpetologica*, **60**, 141-154.

Buse, A., S.J. Dury, R.J.W. Woodburn, C.M. Perrins and J.E.G. Good, 1999: Effects of elevated temperature on multi-species interactions: the case of pedunculate oak, winter moth and tits. *Funct. Ecol.*, **13**, 74.

Butler, C.J., 2003: The disproportionate effect of global warming on the arrival dates of short-distance migratory birds in North America. *Ibis*, **145**, 484.

Buzin, V.A., A.B. Klaven, Z.D. Kopoliani, V.N. Nikitin and V.I. Teplov, 2004: Results of the studies of the ice jam generation processes and the efficiency of the Lena river hydraulic model at Lensk. *Proceedings of the 17th International Symposium on Ice, St. Petersburg,* Vol. 3.

Camilleri, M., R. Jaques and N. Isaacs, 2001: Impacts of climate change on building performance in New Zealand. *Build. Res. Inf.*, **29**, 440-450.

Camuffo, D. and G. Stararo, 2004: Use of proxy-documentary and instrumental data to assess the risk factors leading to sea flooding in Venice. *Global Planet. Change*, **40**, 93-103.

Cannell, M.G.R., J.P. Palutikof and T.H. Sparks, 1999: *Indicators of Climate Change in the UK*. DETR, London, 87 pp.

Cantelaube, P., J.M. Terres and F.J. Doblas-Reyes, 2004: Climate variability influences on European agriculture: an analysis for winter wheat production. *Climate Res.*, **27**, 135-144.

Carey, M., 2005: Living and dying with glaciers: people's historical vulnerability to avalanches and outburst floods in Peru. *Global Planet. Change*, **47**, 122-134.

Carmichael, W.W., 2001: Health effects of toxin-producing cyanobacteria: "The CyanoHABs". *Hum. Ecol. Risk Assess.*, **7**, 1393-1407.

Carminati, E., C. Doglioni and D. Scroccas, 2005. Long term natural subsidence of Venice: evaluation of its causes and magnitude. *Flooding and Environmental Challenges for Venice and its Lagoon: State of Knowledge 2003*, C. Fletcher and T. Spencer, Eds., Cambridge University Press, Cambridge, 21-28.

Carnelli, A.L., J.P. Theurillat, M. Thinon, G. Vadi and B. Talon, 2004: Past uppermost tree limit in the Central European Alps (Switzerland) based on soil and soil charcoal. *Holocene*, **14**, 393-405.

Carson, C., S. Hajat, B. Armstrong and P. Wilkinson, 2006: Declining vulnerability to temperature-related mortality in London over the 20th century. *Am. J. Epidemiol.*, **164**, 77-84.

Carvalho, L. and A. Kirika, 2003: Changes in shallow lake functioning: response to climate change and nutrient reduction. *Hydrobiologia*, **506**, 789-796.

Casassa, G. and C. Marangunic, 1993: The 1987 Río Colorado rockslide and debris flow, central Andes, Chile. *Bulletin of the Association of Engineering Geologists*, **30**, 321-330.

Casassa, G., A. Rivera, F. Escobar, C. Acuña, J. Carrasco and J. Quintana, 2003: Snow line rise in central Chile in recent decades and its correlation with climate. *Joint Assembly*. Nice, EAE03-A-14395, CR8-1TH2O-007, EGS-AGU-EUG.

Casman, E., B. Fischhoff, M. Small, H. Dowlatabadi, J. Rose and M.G. Morgan, 2001: Climate change and cryptosporidiosis: a qualitative analysis. *Climatic Change*, **50**, 219-249.

Cayan, D.R., S.A. Kammerdiener, M.D. Dettinger, J.M. Caprio and D.H. Peterson, 2001: Changes in the onset of spring in the western United States. *B. Am. Meteorol. Soc.*, **82**, 399-415.

Chambers, L.E., L. Hughes and M.A. Weston, 2005: Climate change and its impact on Australia's avifauna. *Emu*, **105**, 1-20.

Charron, D.F., M.K. Thomas, D. Waltner-Toews, J.J. Aramini, T. Edge, R.A. Kent, A.R. Maarouf and J. Wilson, 2004: Vulnerability of waterborne diseases to climate change in Canada: a review. *J. Toxicol. Env. Heal. A*, **67**, 1667-1677.

Chavez, F.P., J. Ryan, S.E. Lluch-Cota and M. Niquen, 2003: From anchovies to sardines and back: multidecadal change in the Pacific Ocean. *Science*, **299**, 217-221.

Chelliah, M. and G.D. Bell, 2004: Tropical multidecadal and interannual climate variability in the NCEP-NCAR reanalysis. *J. Climate*, **17**, 1777-1803.

Chen, H., X. Chen, B. Hu and R. Yu, 2005a: Spatial and temporal variation of phenological growing season and climate change impacts in temperate eastern China. *Glob. Change Biol.*, **11**, 1118-1130.

Chen, H., A.K. Githeko, G.F. Zhou, J.I. Githure and G.Y. Yan, 2006: New records of *Anopheles arabiensis* breeding on the Mount Kenya highlands indicate indigenous malaria transmission. *Malaria J.*, **5**, doi:10.1186/1475-2875-5-17.

Chen, X., E. Zhang, H. Mu and Y. Zong, 2005b: A preliminary analysis of human impacts on sediment discharges from the Yangtze, China, into the sea. *J. Coastal Res.*, **21**, 515-521.

Chen, Z., S.E. Grasby and K.G. Osadetz, 2002: Predicting average annual groundwater levels from climatic variables: an empirical model. *J. Hydrol.*, **260**, 102-117.

Chiba, S. and K. Tadokoro, 2006: Effects of decadal climate change on zooplankton over the last 50 years in the western subarctic North Pacific. *Glob. Change Biol.*, **12**, 907-920.

Chmielewski, F.M. and T. Rotzer, 2001: Response of tree phenology to climate change across Europe. *Agr. Forest Meteorol.*, **108**, 101-112.

Chmielewski, F.M., A. Muller and E. Bruns, 2004: Climate changes and trends in phenology of fruit trees and field crops in Germany, 1961-2000. *Agr. Forest Meteorol.*, **121**, 69-78.

Church, J.A. and N.J. White, 2006: A 20th century acceleration in global sea-level rise. *Geophys. Res. Lett.*, **33**, L01602, doi:10.1029/2005GL024826.

Church, J.A., N.J. White, R. Coleman, K. Lambeck and J.X. Mitrovica, 2004: Estimates of the regional distribution of sea-level rise over the 1950-2000 period. *J. Climate*, **17**, 2609-2625.

Clark, D.A., S.C. Piper, C.D. Keeling and D.B. Clark, 2003: Tropical rain forest tree growth and atmospheric carbon dynamics linked to interannual temperature variation during 1984-2000. *P. Natl. Acad. Sci. USA*, **100**, 5852-5857.

Community of Arctic Bay, S. Nickels, C. Furgal, J. Akumilik and B.J. Barnes, 2005: *Unikkaaqatigiit: Putting the Human Face on Climate Change – Perspectives from Arctic Bay, Nunavut*. Report of the Workshop held in Arctic Bay, Nunavut, March 3–4, 2004. Joint publication of Inuit Tapiriit Kanatimi, Nasivvik Centre for Inuit Health and Changing Environments at Université Laval and the Ajunnginiq Centre at the National Aboriginal Health Organization, Ottawa, 31 pp, http://www.itk.ca/environment/climate-book/community/ArcticBay.pdf.

Conard, S.G., A.I. Sukhinin, B.J. Stocks, D.R. Cahoon, E.P. Davidenko and G.A. Ivanova, 2002: Determining effects of area burned and fire severity on carbon cycling and emissions in Siberia. *Climatic Change*, **55**, 197-211.

Cooke, S.J., S.G. Hinch, A.P. Farrell, M.F. Lapointe, S.R.M. Jones, J.S. Macdonald, D.A. Patterson, M.C. Healey and G. Van Der Kraak, 2004: Abnormal migration timing and high en route mortality of sockeye salmon in the Fraser River, British Columbia. *Fisheries*, **29**, 22-33.

Cooper, N.J., T. Cooper and F. Burd, 2001: 25 years of salt marsh erosion in Essex: implications from coastal defence and natural conservation. *Journal of Coastal Conservation*, **9**, 31-40.

Cotton, P.A., 2003: Avian migration phenology and global climate change. *P. Natl. Acad. Sci. USA*, **100**, 12219-12222.

Coudrain, A., B. Francou and Z.W. Kundzewicz, 2005: Glacier shrinkage in the Andes and consequences for water resources: editorial. *Hydrolog. Sci. J.*, **50**, 925–932.

Couture, R., S.D. Robinson and M.M. Burgess, 2000: Climate change, permafrost, degradation and infrastructure adaptation: preliminary results from a pilot community study in the Mackenzie Valley. *Geological Survey of Canada Current Research*, **2000**, 1-9.

Craig, M.H., I. Kleinschmidt, D. Le Sueur and B.L. Sharp, 2004: Exploring thirty years of malaria case data in KwaZulu-Natal, South Africa. Part II. The impact of non-climatic factors. *Trop. Med. Int. Health*, **9**, 1258-1266.

Crick, H.Q.P. and T.H. Sparks, 1999: Climate change related to egg-laying trends. *Nature*, **399**, 423-424.

Crick, H.Q.P., C. Dudley, D.E. Glue and D.L. Thomson, 1997: UK birds are laying eggs earlier. *Nature*, **388**, 526.

Crozier, L., 2004: Warmer winters drive butterfly range expansion by increasing survivorship. *Ecology*, **85**, 231-241.

Cullen, L.E., G.H. Stewart, R.P. Duncan and J.G. Palmer, 2001: Disturbance and climate warming influences on New Zealand *Nothofagus* tree-line population dynamics. *J. Ecol.*, **89**, 1061.

Curriero, F.C., J.A. Patz, J.B. Rose and S. Lele, 2001: The association between extreme precipitation and waterborne disease outbreaks in the United States, 1948-1994. *Am. J. Public Health*, **91**, 1194-1199.

Curriero, F., K.S. Heiner, J. Samet, S. Zeger, L. Strug and J.A. Patz, 2002: Temperature and mortality in 11 cities of the Eastern United States. *Am. J. Epidemiol.*, **155**, 80-87.

Dabrowski, M., W. Marszelewski and R. Skowron, 2004: The trends and dependencies between air and water temperatures in lakes in northern Poland in 1961-2000. *Hydrol. Earth Syst. Sc.*, **8**, 79-87.

Dai, A., K.E. Trenberth and T. Qian, 2004: A global data set of Palmer Drought Severity Index for 1870-2002: relationship with soil moisture and effect of surface warming. *J. Hydrometeorol.*, **5**, 1117-1130.

Daniel, H., 2001: Replenishment versus retreat: the cost of maintaining. *Ocean Coast. Manage.*, **44**, 87-104.

Daniel, M., V. Danielová, B. Kříž, A. Jirsa and J. Nožička, 2003: Shift of the tick *Ixodes ricinus* and tick-borne encephalitis to higher altitudes in central Europe. *Eur. J. Clin. Microbiol.*, **22**, 327-328.

Danielová, V., N. Rudenko, M. Daniel, J. Holubova, J. Materna, M. Golovchenko and L. Schwarzova, 2006: Extension of *Ixodes ricinus* ticks and agents of tick-borne diseases to mountain areas in the Czech Republic. *Int. J. Med. Microbiol.*, **296**, 48-53.

D'Arrigo, R.D., R.K. Kaufmann, N. Davi, G.C. Jacoby, C. Laskowski, R.B. Myneni and P. Cherubini, 2004: Thresholds for warming-induced growth decline at elevational tree line in the Yukon Territory, Canada. *Global Biogeochem. Cy.*, **18**, GB3021.

Daufresne, M., M.C. Roger, H. Capra and N. Lamouroux, 2004: Long-term changes within the invertebrate and fish communities of the Upper Rhone River: effects of climatic factors. *Glob. Change Biol.*, **10**, 124-140.

Davies, Z.G., R.J. Wilson, S. Coles and C.D. Thomas, 2006: Changing habitat associations of a thermally constrained species, the silver-spotted skipper butterfly, in response to climate warming. *J. Anim. Ecol.*, **75**, 247-256.

Davis, M.A., J.P. Grime and K. Thompson, 2000: Fluctuating resources in plant communities: a general theory of invasibility. *J. Ecol.*, **88**, 528-534.

Davis, R.E., P.C. Knappenberger, P.J. Michaels and W.M. Novicoff, 2003a: Changing heat-related mortality in the United States. *Environ. Health Persp.*, **111**, 1712-1718.

Davis, R.E., P.C. Knappenberger, W.M. Novicoff and P.J. Michaels, 2003b: Decadal changes in summer mortality in US cities. *Int. J. Biometeorol.*, **47**, 166-175.

Day, J.W., G. Abrami, J. Rybczyk and W. Mitsch, 2005. Venice Lagoon and the Po Delta: system functioning as a basis for sustainable management. *Flooding and Environmental Challenges for Venice and its Lagoon: State of Knowledge*, C.A. Fletcher and T. Spencer, Eds., Cambridge University Press, Cambridge, 445-459.

de Dear, R.J. and G.S. Brager, 1998: Developing an adaptive model of thermal comfort and preference. *ASHRAE Transactions*, **104**, 145-167.

Defila, C. and B. Clot, 2001: Phytophenological trends in Switzerland. *Int. J.Biometeorol.*, **45**, 203-207.

Defra, 2003: *The Impacts of Climate Change: Implications for Defra*. Department for Environment, Food and Rural Affairs, 40 pp, http://www.defra.gov.uk/ENVIRONMENT/climatechange/pubs/impacts/pdf/ccimpacts_defra.pdf.

Delbart, N., T. Le Toan, L. Kergoat and V. Fedotova, 2006: Remote sensing of spring phenology in boreal regions: a free of snow-effect method using NOAA-AVHRR and SPOT-VGT data (1982-2004). *Remote Sens. Environ.*, **101**, 52-62.

Derocher, A.E., N.J. Lunn and I. Stirling, 2004: Polar bears in a warming climate. *Integr. Comp. Biol.*, **44**, 163-176.

Dery, S.J. and E.F. Wood, 2005: Decreasing river discharge in northern Canada. *Geophys. Res. Lett.*, **32**, L10401, doi: 10.1029/2005GL022845.

Diaz, J., A. Jordan, R. Garcia, C. Lopez, J.C. Alberdi, E. Hernandez and A. Otero, 2002: Heat waves in Madrid 1986-1997: effects on the health of the elderly. *Int. Arch. Occ. Env. Hea.*, **75**, 163-170.

Diergaardt, S.M., S.N. Venter, A. Spreeth, J. Theron and V.S. Brozel, 2004: The occurrence of campylobacters in water sources in South Africa. *Water Res.*, **38**, 2589-2595.

Dobbertin, M., N. Hilker, M. Rebetez, N.E. Zimmermann, T. Wohlgemuth and A. Rigling, 2005: The upward shift in altitude of pine mistletoe (*Viscum album* ssp. *austriacum*) in Switzerland: the result of climate warming? *Int. J. Biometeorol.*, **50**, 40-47.

Doledec, S., J. Dessaix and H. Tachet, 1996: Changes within the Upper Rhone River macrobenthic communities after the completion of three hydroelectric schemes: anthropogenic effects or natural change? *Arch. Hydrobiol.*, **136**, 19-40.

Donaldson, G.C., W.R. Keatinge and S. Näyhä, 2003: Changes in summer temperature and heat-related mortality since 1971 in North Carolina, South Finland and Southeast England. *Environ. Res.*, **91**, 1-7.

Donnelly, A., M.B. Jones and J. Sweeney, 2004: A review of indicators of climate change for use in Ireland. *Int. J. Biometeorol.*, **49**, 1-12.

Donnelly, J.P. and M.D. Bertness, 2001: Rapid shoreward encroachment of salt marsh cordgrass in response to accelerated sea-level rise. *P. Natl. Acad. Sci. USA*, **98**, 14218-14223.

Dose, V. and A. Menzel, 2004: Bayesian analysis of climate change impacts in phenology. *Glob. Change Biol.*, **10**, 259-272.

Dose, V. and A. Menzel, 2006: Bayesian correlation between temperature and blossom onset data. *Glob. Change Biol.*, **12**, 1451-1459.

Douglas, E.M., R.M. Vogel and C.N. Kroll, 2000: Trends in floods and low flows in the United States: impacts of spatial correlation. *J. Hydrol.*, **240**, 90-105.

Douglas, M.S.V., J.P. Smol and W. Blake, 1994: Marked post-18th century environmental-change in high-Arctic ecosystems. *Science*, **266**, 416-419.

Drinkwater, K.F., A. Belgrano, A. Borja, A. Conversi, M. Edwards, C.H. Greene, G. Ottersen, A.J. Pershing and H. Walker, 2003: The response of marine ecosystems to North Atlantic climate variability associated with the North Atlantic Oscillation. *The North Atlantic Oscillation: Climate Significance and Environmental Impact*, J. Hurrell, Y. Kushnir, G. Ottersen and M. Visbeck, Eds., Geophysical Monograph 134, American Geophysical Union, Washington, District of Columbia, 211-234.

D'Souza, R.M., N.G. Beeker, G. Hall and K.B.A. Moodie, 2004: Does ambient temperature affect foodborne disease? *Epidemiology*, **15**, 86-92.

Du, M.Y., S. Kawashima, S. Yonemura, X.Z. Zhang and S.B. Chen, 2004: Mutual influence between human activities and climate change in the Tibetan Plateau during recent years. *Global Planet. Change*, **41**, 241-249.

Duchêne, E. and C. Schneider, 2005: Grapevine and climatic changes: a glance at the situation in Alsace. *Agron. Sustain. Dev.*, **25**, 93-99.

Dulčić, J., B. Grbec, L. Lipej, G. Beg-Paklar, N. Supic and A. Smircic, 2004: The effect of the hemispheric climatic oscillations on the Adriatic ichthyofauna. *Fresen. Environ. Bull.*, **13**, 293-298.

Dunn, P., 2004. Breeding dates and reproductive performance. *Adv. Ecol. Res.*, **35**, 69-87.

Dunn, P.O. and D.W. Winkler, 1999: Climate change has affected the breeding date of tree swallows throughout North America. *P. Roy. Soc. Lond. B Bio.*, **266**, 2487.

Eagles, P.F.J., 2004: Trends affecting tourism in protected areas. *Policies, Methods and Tools for Visitor Management: Proceedings of the Second International Conference on Monitoring and Management of Visitor Flows in Recreational and Protected Areas, Rovaniemi, Finland, June 16–20, 2004*, T. Sievänen, J. Erkkonen, J. Jokimäki, J. Saarinen, S. Tuulentie and E. Virtanen, Eds., Working Papers of the Finnish Forest Research Institute 2, 17-25.

Eakin, H., 2000: Smallholder maize production and climate risk: a case study from Mexico. *Climatic Change*, **45**, 19-36.

Easterling, D.R., G.A. Meehl, C. Parmesan, S.A. Changnon, T.R. Karl and L.O. Mearns, 2000: Climate extremes: observations, modeling, and impacts. *Science*, **289**, 2068-2074.

Easterling, W.E., 2003: Observed impact of climate change in agriculture and forestry. *IPCC Workshop on the Detection and Attribution of the Effects of Climate Change*, GISS, New York, 54-55.

Edwards, M. and A.J. Richardson, 2004: Impact of climate change on marine pelagic phenology and trophic mismatch. *Nature*, **430**, 881-884.

Edwards, M., P. Reid and B. Planque, 2001: Long-term and regional variability of phytoplankton biomass in the Northeast Atlantic (1960-1995). *ICES J. Mar. Sci.*, **58**, 39-49.

Edwards, M., G. Beaugrand, P.C. Reid, A.A. Rowden and M.B. Jones, 2002: Ocean climate anomalies and the ecology of the North Sea. *Mar. Ecol.–Prog. Ser.*, **239**, 1-10.

Edwards, M., D.G. Johns, S.C. Leterme, E. Svendsen and A.J. Richardson, 2006: Regional climate change and harmful algal blooms in the northeast Atlantic. *Limnol. Oceanogr.*, **51**, 820-829.

EEA, 2004: *Impacts of Europe's Changing Climate: An Indicator-based Assessment*. European Environment Agency, Copenhagen, Report 2/2004. Office for Official Publications of the European Communities, Luxembourg, 107 pp.

Elliott, J.M., M.A. Hurley and S.C. Maberly, 2000: The emergence period of sea trout fry in a Lake District stream correlates with the North Atlantic Oscillation. *J. Fish Biol.*, **56**, 208-210.

Elsasser, H. and P. Messerli, 2001: The vulnerability of the snow industry in the Swiss Alps. *Mt. Res. Dev.*, **21**, 335-339.

Emanuel, K., 2005: Increasing destructiveness of tropical cyclones over the past 30 years. *Nature*, **434**, 686-688.

Emberlin, J., M. Detandt, R. Gehrig, S. Jaeger, N. Nolard and A. Rantio-Lehtimaki, 2003: Responses in the start of *Betula* (birch) pollen seasons to recent changes in spring temperatures across Europe. *Int. J. Biometeorol.*, **47**, 113-115.

Erdenetuya, M., 2004: Application of remote sensing in climate change study: vulnerability and adaptation assessment for grassland ecosystem and livestock sector in Mongolia project. *AIACC Annual Report*, Washington, DC.

Ericson, J.P., C.J. Vorosmarty, S.L. Dingman, L.G. Ward and M. Meybeck, 2006: Effective sea-level rise and deltas: causes of change and human dimension implications. *Global Planet. Change*, **50**, 63-82.

Estrella, N. and A. Menzel, 2006: Responses of leaf colouring in four deciduous tree species to climate and weather in Germany. *Climate Res.*, **32**, 253-267.

Estrella, N., T.H. Sparks and A. Menzel, 2007: Trends and temperature response in the phenology of crops in Germany. *Glob. Change Biol.*, doi:10.1111/j.1365-2486.2007.01374.x.

Fang, S. and K. Dingbo, 2003: Remote sensing investigation and survey of Qinghai Lake in the past 25 years. *Journal of Lake Sciences*, **15**, 290-296.

Feely, R.A., C.L. Sabine, K. Lee, W. Berelson, J. Kleypas, V.J. Fabry and F.J. Millero, 2004: Impact of anthropogenic CO_2 on the $CaCO_3$ system in the oceans. *Science*, **305**, 362-366.

Feng, S. and Q. Hu, 2004: Changes in agro-meteorological indicators in the contiguous United States: 1951-2000. *Theor. Appl. Climatol.*, **78**, 247-264.

Ferguson, G. and S.S. George, 2003: Historical and estimated ground water levels near Winnipeg, Canada and their sensitivity to climatic variability. *J. Am. Water Resour. As.*, **39**, 1249-1259.

Field, D.B., T.R. Baumgarter, C.D. Charles, V. Ferreira-Bartrina and M.D. Ohman, 2006: Planktonic foraminifera of the California Current reflect 20th-century warming. *Science*, **311**, 63-66.

Fillol, E.J. and A. Royer, 2003: Variability analysis of the transitory climate regime as defined by the NDVI/Ts relationship derived from NOAA-AVHRR over Canada. *IEEE Int.*, **4**, 21-25.

Fitter, A.H. and R.S.R. Fitter, 2002: Rapid changes in flowering time in British plants. *Science*, **296**, 1689.

Forbes, D.L., 2005: Coastal erosion. *Encyclopedia of the Arctic*, M. Nuttall, Ed., Routledge, New York and London, 391-393.

Forbes, D.L., G.S. Parkes, G.K. Manson and L.A. Ketch, 2004: Storms and shoreline retreat in the southern Gulf of St. Lawrence. *Mar. Geol.*, **210**, 169-204.

Forcada, J., P.N. Trathan, K. Reid and E.J. Murphy, 2005: The effects of global climate variability in pup production on Antarctic fur seals. *Ecol. Lett.*, **86**, 2408-2417.

Forcada, J., P.N. Trathan, K. Reid, E.J. Murphy and J.P. Croxall, 2006: Contrasting population changes in sympatric penguin species in association with climate warming. *Glob. Change Biol.*, **12**, 411-423.

Ford, J.D. and B. Smit, 2004: A framework for assessing the vulnerability of communities in the Canadian Arctic to risks associated with climate change. *Arctic*, **57**, 389-400.

Forister, M.L. and A.M. Shapiro, 2003: Climatic trends and advancing spring flight of butterflies in lowland California. *Glob. Change Biol.*, **9**, 1130-1135.

Fox, S., 2002: These things that are really happening: Inuit perspectives on the evidence and impacts of climate change in Nunavut. *The Earth is Faster Now: Indigenous Observations of Arctic Environmental Change*, I. Krupnik and D. Jolly, Eds., Arctic Research Consortium of the US, Fairbanks, Alaska, 12-53.

Franco, A.M.A., J.K. Hill, C. Kitschke, Y.C. Collingham, D.B. Roy, R. Fox, B. Huntley and C.D. Thomas, 2006: Impacts of climate warming and habitat loss on extinctions at species' low-latitude range boundaries. *Glob. Change Biol.*, **12**, 1545-1553.

Francou, B. and C. Vincent, 2006: Les glaciers à l'épreuve du climat. IRD/BELIN, Paris, 274 pp.

Francou, B., M. Vuille, P. Wagnon, J. Mendoza and J.E. Sicart, 2003: Tropical climate change recorded by a glacier of the central Andes during the last decades of the 20th century: Chacaltaya, Bolivia, 16°S. *J. Geophys. Res.*, **108**, 4154.

Francou, B., P. Ribstein, P. Wagnon, E. Ramirez and B. Pouyaud, 2005: Glaciers of the tropical Andes: indicators of the global climate variability. *Global Change and Mountain Regions: A State of Knowledge Overview*, U.M. Huber, H.K.M. Bugmann and M.A. Reasoner, Eds., Advances in Global Change Research, Vol 23, Springer, Berlin, 197-204.

Franken, R.J. and D.S. Hik, 2004: Interannual variation in timing of parturition and growth of collared pikas (*Ochotona collaris*) in the southwest Yukon. *Integr. Comp. Biol.*, **44**, 186.

Frederiksen, M., M.P. Harris, F. Daunt, P. Rothery and S. Wanless, 2004: Scale-dependent climate signals drive breeding phenology of three seabird species. *Glob. Change Biol.*, **10**, 1214-1221.

Frei, C. and C. Schar, 2001: Detection probability of trends in rare events: theory and application to heavy precipitation in the Alpine region. *J. Climate*, **14**, 1568-1584.

Frenot, Y., S.L. Chown, J. Whinam, P.M. Selkirk, P. Convey, M. Skotnicki and D.M. Bergstrom, 2005: Biological invasions in the Antarctic: extent, impacts and implications. *Biol. Rev.*, **80**, 45-72.

Frihy, O., K. Dewidar and M. El Raey, 1996: Evaluation of coastal problems at Alexandria, Egypt. *Ocean Coast. Manage.*, **30**, 281-295.

Frolov, A.V., S.V. Borshch, E.S. Dmitriev, M.V. Bolgov and N.I. Alekseevsky, 2005: Dangerous hydrological events: methods for analysis and forecasting, mitigation of negative results. *Proceedings of the VIth All-Russia Hydrological Congress, Vol.1*, St .Petersburg.

Fromentin, J.M. and B. Planque, 1996: *Calanus* and environment in the eastern North Atlantic. II. Influence of the North Atlantic Oscillation on *Calanus finmarchicus* and *C. hegolandicus*. *Mar. Ecol.–Prog. Ser.*, **134**, 111-118.

Fruget, J.F., M. Centofanti, J. Dessaix, J.M. Olivier, J.C. Druart and P.J. Martinez, 2001: Temporal and spatial dynamics in large rivers: example of a long-term monitoring of the Middle Rhone River. *Ann. Limnol.–Int. J. Lim.*, **37**, 237-251.

Gajewski, K., P.B. Hamilton and R. McNeely, 1997: A high resolution proxy-climate record from an arctic lake with annually-laminated sediments on Devon Island, Nunavut, Canada. *J. Paleolimnol.*, **17**, 215-225.

Gao, Y., G.Y. Qiu, H. Shimizu, K. Tobe, B.P. Sun and J. Wang, 2002: A 10-year study on techniques for vegetation restoration in a desertified salt lake area. *J. Arid Environ.*, **52**, 483-497.

Gardner, T.A., I.M. Côté, J.A. Gill, A. Grant and A.R. Watkinson, 2005: Hurricanes and Caribbean coral reefs: impacts, recovery patterns and role in long-term decline. *Ecology*, **86**, 174-184.

Garpe, K.C., S.A.S. Yahya, U. Lindahl and M.C. Ohman, 2006: Long-term effects of the 1998 coral bleaching event on reef fish assemblages. *Mar. Ecol.–Prog. Ser.*, **315**, 237-247.

Gaston, K.J., K. Charman, S.F. Jackson, P.R. Armsworth, A. Bonn, R.A. Briers, C.S.Q. Callaghan, R. Catchpole, J. Hopkins, W.E. Kunin, J. Latham, P. Opdam, R. Stoneman, D.A. Stroud and R. Tratt, 2006: The ecological effectiveness of protected areas: the United Kingdom. *Biol. Conserv.*, **132**, 76-87.

Gedney, N., P.M. Cox, R.A. Betts, O. Boucher, C. Huntingford and P.A. Stott, 2006: Detection of a direct carbon dioxide effect in continental river runoff records. *Nature*, **439**, 835-838.

Genner, M.J., D.W. Sims, V.J. Wearmouth, E.J. Southall, A.J. Southward, P.A. Henderson and S.J. Hawkins, 2004: Regional climatic warming drives long-term community changes of British marine fish. *P. Roy. Soc. Lond. B Bio.*, **271**, 655-661.

Genovese, G., C. Lazar and F. Micale, 2005: Effects of observed climate fluctuation on wheat flowering as simulated by the European crop growth monitoring system (CGMS). *Proceedings of a Workshop on Adaptation of Crops and Cropping Systems to Climate Change, 7-8 November 2005, Dalum Landbrugsskole, Odense, Denmark*. Nordic Association of Agricultural Scientists, 12 pp.

Georgiou, I.Y., D.M. FitzGerald and G.W. Stone, 2005: The impact of physical processes along the Louisiana coast. *J. Coastal Res.*, **44**, 72-89.

Gerten, D. and R. Adrian, 2000: Climate-driven changes in spring plankton dynamics and the sensitivity of shallow polymictic lakes to the North Atlantic Oscillation. *Limnol. Oceanogr.*, **45**, 1058-1066.

Gerten, D. and R. Adrian, 2002: Species-specific changes in the phenology and peak abundance of freshwater copepods in response to warm summers. *Freshwater Biol.*, **47**, 2163-2173.

Gibbs, J.P. and A.R. Breisch, 2001: Climate warming and calling phenology of frogs near Ithaca, New York, 1900-1999. *Conserv. Biol.*, **15**, 1175.

Gill, J.A., J.P. McWilliams, A.R. Watkinson and I.M. Côté, 2006: Opposing forces of aerosol cooling and El Niño drive coral bleaching on Caribbean reefs. *P. Natl. Acad. Sci. USA*, **103**, 18870-18873.

Gillett, N.P., A.J. Weaver, F.W. Zwiers and M.D. Flannigan, 2004: Detecting the effect of climate change on Canadian forest fires. *Geophys. Res. Lett.*, **31**, L18211, doi:10.1029/2004GL020876.

Githeko, A. and W. Ndegwa, 2001: Predicting malaria epidemics in the Kenya highlands using climate data: a tool for decision makers. *Glob. Change Hum. Health*, **2**, 54-63.

Githeko, A. and A. Woodward, 2003: International consensus on the science of climate and health: the IPCC Third Assessment Report. *Climate Change and Human Health: Risks and Responses*, A.J. McMicheal, D.H. Campbell-Lendrum, C.F. Corvalan, K.L. Ebi, A. Githeko, J.D. Scheraga and A. Woodward, Eds., World Health Organization, Rome, 43-60. www.who.int/globalchange/publications/climatechangechap3.pdf.

Gobron, N., B. Pinty, F. Melin, M. Taberner, M.M. Verstraete, A. Belward, T. Lavergne and J.L. Widlowski, 2005: The state of vegetation in Europe following the 2003 drought. *Int. J. Remote Sens.*, **26**, 2013-2020.

Goetz, S.J., A.G. Bunn, G.J. Fiske and R.A. Houghton, 2005: Satellite-observed photosynthetic trends across boreal North America associated with climate and fire disturbance. *P. Natl. Acad. Sci. USA*, **102**, 13521-13525.

Goklany, I.M., 1996: Factors affecting environmental impacts: the effects of technology on long-term trends in cropland, air pollution and water-related diseases. *Ambio*, **25**, 497-503.

Goldenberg, S.B., C.W. Landsea, A.M. Mestas-Nunez and W.M. Gray, 2001: The recent increase in Atlantic hurricane activity: causes and implications. *Science*, **293**, 474-479.

Gonzalez, P., 2001: Desertification and a shift of forest species in the West African Sahel. *Climate Res.*, **17**, 217-228.

Grabherr, G., M. Gottfried and H. Pauli, 2001: Long-term monitoring of mountain peaks in the Alps. *Biomonitoring: General and Applied Aspects on Regional and Global Scales – Tasks for Vegetation Science*, C.A. Burga and A. Kratochwil, Eds., Kluwer Academic, Dordrecht, 153- 177.

Graham, N.A.J., S.K. Wilson, S. Jennings, N.V.C. Polunin, J.P. Bijoux and J. Robinson, 2006: Dynamic fragility of oceanic coral reef ecosystems. *P. Natl. Acad. Sci. USA*, **103**, 8425-8429.

Grebmeier, J.M., J.E. Overland, S.E. Moore, E.V. Farley, E.C. Carmack, L.W. Cooper, K.E. Frey, J.H. Helle, F.A. McLaughlin and S.L. McNutt, 2006: A major ecosystem shift in the northern Bering Sea. *Science*, **311**, 1461-1464.

Green, K. and C.M. Pickering, 2002: A potential scenario for mammal and bird diversity in the Snowy Mountains of Australia in relation to climate change. *Mountain Biodiversity: A Global Assessment*, C. Korner and E. Spehn, Eds., Parthenon Publishing, London, 241-249.

Gregg, W.W., M.E. Conkright, P. Ginoux, J.E. O'Reilly and N.W. Casey, 2003: Ocean primary production and climate: global decadal changes. *Geophys. Res. Lett.*, **30**, 1809, doi:10.1029/2003GL016889.

Grenouillet, G., B. Hugueny, G.A. Carrel, J.M. Olivier and D. Pont, 2001: Large-scale synchrony and inter-annual variability in roach recruitment in the Rhone River: the relative role of climatic factors and density-dependent processes. *Freshwater Biol.*, **46**, 11-26.

Greve, W., 2004: Aquatic plants and animals. *Phenology: An Integrative Environmental Science*, M. D. Schwartz, Ed., Kluwer, Dordrecht, 385-403.

Gruber, S., M. Hoelzle and W. Haeberli, 2004: Permafrost thaw and destabilization of Alpine rock walls in the hot summer of 2003. *Geophys. Res. Lett.*, **31**, L13504, doi:10.1029/2004GL020051.

Guisande, C., A. Vergara, I. Riveiro and J. Cabanadas, 2004: Climate change and abundance of the Atlantic-Iberian sardine (*Sardina pilchardus*). *Fish. Oceanogr.*, **13**, 91-101.

Günther, H., W. Rosenthal, M. Stawarz, J.C. Carretero, M. Gomez, I. Lozano, O. Serano and M. Reistad, 1998: The wave climate of the Northeast Atlantic over the period 1955-94: the WASA wave hindcast. *The Global Atmosphere and Ocean System*, **6**, 121-163.

Gyan, K., W. Henry, S. Lacaille, A. Laloo, C. Lamsee-Ebanks, S. McKay, R. Antoine and M. Monteil, 2003: African dust clouds are associated with paediatric accident and emergency asthma admissions at the Eric Williams Medical Sciences Complex. *W. Indian Med. J.*, **52**, 46.

Haeberli, W. and C. Burn, 2002: Natural hazards in forests - glacier and permafrost effects as related to climate changes. *Environmental Change and Geomorphic Hazards in Forests*, R.C. Sidle, Ed., IUFRO Research Series, 9, 167-202.

Haeberli, W., A. Kääb, D. Vonder Mühll, and P. Teysseire, 2001: Prevention of outburst floods from periglacial lakes at Grubengletscher, Valais, Swiss Alps. *J. Glaciol.*, **47**, 111-122.

Hafner, S., 2003: Trends in maize, rice and wheat yields for 188 nations over the past 40 years: a prevalence of linear growth. *Agr. Ecosyst. Environ.*, **97**, 275-283.

Hambright, K.D., M. Gophen and S. Serruya, 1994: Influence of long-term climatic changes on the stratification of a subtropical, warm monomictic lake. *Limnol. Oceanogr.*, **39**, 1233-1242.

Hamilton, J.M., 2003a: Climate and the destination choice of German tourists, Working Paper FNU-15 (revised), Centre for Marine and Climate Research, University of Hamburg, 36 pp.

Hamilton, J.M., D.J. Maddison and R.S.J. Tol, 2003: Climate change and international tourism: a simulation study. Working Paper FNU-31, Research Unit Sustainability and Global Change, 41 pp.

Hamilton, J.M., D.J. Maddison and R.S.J. Tol, 2005: Climate change and international tourism: a simulation study. *Global Environ. Chang.*, **15**, 253-266.

Hamilton, L.C., 2003b: Warming winters and New Hampshire's lost ski areas: an integrated case study. *Int. J. Sociol. Social Pol.*, **23**, 52-73.

Hampton, S.E., 2005: Increased niche differentiation between two *Conochilus* species over 33 years of climate change and food web alteration. *Limnol. Oceanogr.*, **50**, 421-426.

Hari, R.E., D.M. Livingstone, R. Siber, P. Burkhardt-Holm and H. Guttinger, 2006: Consequences of climatic change for water temperature and brown trout populations in Alpine rivers and streams. *Glob. Change Biol.*, **12**, 10-26.

Hartig, E.K. and V. Gornitz, 2004: Salt marsh change, 1926-2002 at Marshlands Conservancy, New York. *Proceedings of the Long Island Sound Biennial Research Conference*, 7 pp.

Hartig, E.K., V. Gornitz, A. Kolker, F. Mushacke and D. Fallon, 2002: Anthropogenic and climate-change impacts on salt marshes of Jamaica Bay, New York City. *Wetlands*, **22**, 71-89.

Harvell, C.D., K. Kim, J.M. Burkholder, R.R. Colwell, P.R. Epstein, D.J. Grimes, E.E. Hofmann, E.K. Lipp, A.D.M.E. Osterhaus, R.M. Overstreet, J.W. Porter, G.W. Smith and G.R. Vasta, 1999: Emerging marine diseases: climate links and anthropogenic factors. *Science*, **285**, 1505-1510.

Haslett, S.K., A.B. Cundy, C.F.C. Davies, E.S. Powell and I.W. Croudace, 2003: Salt marsh, sedimentation over the past c. 120 years along the West Cotentin coast of Normandy (France): relationship to sea-level rise and sediment supply. *J. Coastal Res.*, **19**, 609-620.

Hattenschwiler, S. and C. Korner, 2003: Does elevated CO_2 facilitate naturalization of the non-indigenous *Prunus laurocerasus* in Swiss temperate forests? *Funct. Ecol.*, **17**, 778.

Hawkins, S.J., A.J. Southward and M.J. Genner, 2003: Detection of environmental change in a marine ecosystem: evidence from the western English Channel. *Sci. Total Environ.*, **310**, 245-256.

Hay, S.I., J. Cox, D.J. Rogers, S.E. Randolph, D.I. Stern, G.D. Shanks, M.F. Myers and R.W. Snow, 2002a: Climate change and the resurgence of malaria in the East African highlands. *Nature*, **415**, 905-909.

Hay, S.I., D.J. Rogers, S.E. Randolph, D.I. Stern, J. Cox, G.D. Shanks and R.W. Snow, 2002b: Hot topic or hot air? Climate change and malaria resurgence in East African highlands. *Trends Parasitol.*, **18**, 530-534.

Hecky, R.E., F.W.B. Bugenyi, P. Ochumba, J.F. Talling, R. Mugidde, M. Gophen and L. Kaufman, 1994: Deoxygenation of the deep-water of Lake Victoria, East Africa. *Limnol. Oceanogr.*, **39**, 1476-1481.

Hegerl, G.C., F.W. Zwiers, P. Braconnot, N.P. Gillett, Y. Luo, J. Marengo, N. Nicholls, J.E. Penner and P.A. Stott, 2007: Understanding and attributing climate change. *Climate Change 2007: The Physical Science Basis. Contribution of Work-*

ing Group I to the Fourth Assessment Report of the Intergovernmental Panel on Climate Change, S. Solomon, D. Qin, M. Manning, Z. Chen, M. Marquis, K.B. Averyt, M. Tignor and H.L. Miller, Eds., Cambridge University Press, Cambridge, 663-746.

Helander, E. and T. Mustonen, Eds., 2004: *Snowscapes, Dreamscapes: Snowchange Book on Community Voices of Change*. Tampere Polytechnic Publications, Oy, Vaasa, 562 pp.

Hickling, R., D.B. Roy, J.K. Hill and C.D. Thomas, 2005: A northward shift of range margins in British Odonata. *Glob. Change Biol*., **11**, 502-506.

Hickling, R., D.B. Roy, J.K. Hill, R. Fox and C.D. Thomas, 2006: The distributions of a wide range of taxonomic groups are expanding polewards. *Glob. Change Biol*., **12**, 450-455.

Hilden, M., H. Lethtonen, I. Barlund, K. Hakala, T. Kaukoranta and S. Tattari, 2005: The practice and process of adaptation in Finnish agriculture. FINADAPT Working Paper 5, Finnish Environment Institute Mimeographs 335, Helsinki, 28 pp.

Hill, J.K., C.D. Thomas and D.S. Blakeley, 1999a: Evolution of flight morphology in a butterfly that has recently expanded its geographic range. *Oecologia*, **121**, 165-170.

Hill, J.K., C.D. Thomas and B. Huntley, 1999b: Climate and habitat availability determine 20th-century changes in a butterfly's range margin. *P. Roy. Soc. Lond. B Bio*., **266**, 1197.

Hill, J.K., Y.C. Collingham, C.D. Thomas, D.S. Blakeley, R. Fox, D. Moss and B. Huntley, 2001: Impacts of landscape structure on butterfly range expansion. *Ecol. Lett*., **4**, 313.

Hock, R., 2005: Glacier melt: a review on processes and their modelling. *Prog. Phys. Geog*., **29**, 362-391.

Hock, R., P. Jansson and L. Braun, 2005: Modelling the response of mountain glacier discharge to climate warming. *Global Change and Mountain Regions: An Overview of Current Knowledge*, U.M. Huber, H.K.M. Bugmann and M.A. Reasoner, Eds., Advances in Global Change Research Series, Springer Netherlands, Dordrecht, 243-254.

Hodgkins, G.A., R.W. Dudley and T.G. Huntington, 2003: Changes in the timing of high river flows in New England over the 20th century. *J. Hydrol*., **278**, 244-252.

Hoegh-Guldberg, O., 1999: Climate change, coral bleaching and the future of the world's coral reefs. *Mar. Freshwater Res*., **50**, 839-866.

Holbrook, S.J., R.J. Schmitt and J.S. Stephens, 1997: Changes in an assemblage of temperate reef fishes associated with a climate shift. *Ecol. Appl*., **7**, 1299-1310.

Holgate, S.J. and P.L. Woodworth, 2004: Evidence for enhanced coastal sea-level rise during the 1990s. *Geophys. Res. Lett*., **31**, L07305, doi:10.1029/2004GL019626.

Holtmeier, F.K. and G. Broll, 2005: Sensitivity and response of northern hemisphere altitudinal and polar treelines to environmental change at landscape and local scales. *Global Ecol. Biogeogr*., **14**, 395-410.

Hubalek, Z., 2003: Spring migration of birds in relation to North Atlantic Oscillation. *Folia Zool*., **52**, 287-298.

Huggel, C., W. Haeberli, A. Kaab, D. Bieri and S. Richardson, 2004: An assessment procedure for glacial hazards in the Swiss Alps. *Can. Geotech. J*., **41**, 1068-1083.

Hughes, C.L., J.K. Hill and C. Dytham, 2003a: Evolutionary trade-offs between reproduction and dispersal in populations at expanding range boundaries. *P. Roy. Soc. Lond. B Bio*, **270**, S147.

Hughes, L. 2000: Biological consequences of global warming: is the signal already apparent? *Trends Ecol. Evol*., **15**, 56-61.

Hughes, T.P., A.H. Baird, D.R. Bellwood, M. Card, S.R. Connolly, C. Folke R. Grosberg, O. Hoegh-Guldberg, J.B.C. Jackson, J. Kleypas, J.M. Lough, P. Marshall, M. Nystrom, S.R. Palumbi, J.M. Pandolfi, B. Rosen, B. and J. Roughgarden, 2003b: Climate change, human impacts, and the resilience of coral reefs. *Science*, **301**, 929-933.

Huntington, H. and S. Fox, 2004: The changing Arctic: indigenous perspectives. *Arctic Climate Impact Assessment*, ACIA, Cambridge University Press, Cambridge, 62-98.

Huntington, T.G., 2006: Evidence for intensification of the global water cycle: review and synthesis. *J. Hydrol*., **319**, 83-95.

Huntington, T.G., G.A. Hodgkins and R.W. Dudley, 2003: Historical trends in river ice thickness and coherence in hydroclimatological trends in Maine. *Climatic Change*, **61**, 217-236.

Huppop, O. and K. Huppop, 2003: North Atlantic Oscillation and timing of spring migration in birds. *P. Roy. Soc. Lond. B Bio*., **270**, 233-240.

Hussell, D.J.T., 2003: Climate change, spring temperatures and timing of breeding of tree swallows (*Tachycineta bicolor*) in southern Ontario. *Auk*, **120**, 607-618.

Huynen, M., P. Martens, D. Schram, M. Weijenberg and A.E. Kunst, 2001: The impact of cold spells and heatwaves on mortality rates in the Dutch population. *Environ. Health Perspect*., **109**, 463-470.

Inouye, D.W., B. Barr, K.B. Armitage and B.D. Inouye, 2000: Climate change is affecting altitudinal migrants and hibernating species. *P. Natl. Acad. Sci. USA*, **97**, 1630.

Inouye, D.W., M.A. Morales and G.J. Dodge, 2002: Variation in timing and abundance of flowering by *Delphinium barbeyi* Huth (Ranunculaceae): the roles of snowpack, frost and La Niña, in the context of climate change. *Oecologia*, **130**, 543-550.

IPCC, 2001a: *Climate Change 2001: Impacts, Adaptation, and Vulnerability. Contribution of Working Group II to the Third Assessment Report of the Intergovernmental Panel on Climate Change*, J.J. McCarthy, O.F. Canziani, N.A. Leary, D.J. Dokken and K.S. White, Eds., Cambridge University Press, Cambridge, 1032 pp.

IPCC, 2001b: *Climate Change 2001: The Scientific Basis. Contribution of Working Group I to the Third Assessment Report of the Intergovernmental Panel on Climate Change*, J.T. Houghton, Y. Ding, D.J. Griggs, M. Noguer, P.J. van der Linden, X. Dai, K. Maskell and C.A. Johnson, Eds., Cambridge University Press, Cambridge, 881 pp.

IPCC, 2003: *IPCC Workshop Report on the Detection and Attribution of the Effects of Climate Change*, C. Rosenzweig and P.G. Neofotis, Eds., NASA/Goddard Institute for Space Studies, New York, 87 pp.

IPCC, 2007: *Climate Change 2007: The Physical Science Basis. Contribution of Working Group I to the Fourth Assessment Report of the Intergovernmental Panel on Climate Change*, S. Solomon, D. Qin, M. Manning, Z. Chen, M. Marquis, K.B. Averyt, M. Tignor and H.L. Miller, Eds., Cambridge University Press, Cambridge, 996 pp.

Jansen, E., J. Overpeck, K.R. Briffa, J.-C. Duplessy, F. Joos, V. Masson-Delmotte, D.O. Olago, B. Otto-Bliesner, W.R. Peltier, S. Rahmstorf, R. Ramesh, D. Raynaud, D.H. Rind, O. Solomina, R. Villalba and D. Zhang, 2007: Paleoclimate. *Climate Change 2007: The Physical Science Basis. Contribution of Working Group I to the Fourth Assessment Report of the Intergovernmental Panel on Climate Change*, S. Solomon, D. Qin, M. Manning, Z. Chen, M. Marquis, K.B. Averyt, M. Tignor and H.L. Miller, Eds., Cambridge University Press, Cambridge, 433-498.

Jansson, P., R. Hock and T. Schneider, 2003: The concept of glacier storage: a review. *J. Hydrol*., **282**, 116-129.

Jarvinen, A., 1994: Global warming and egg size of birds. *Ecography*, **17**, 108.

Jarvinen, A., 1996: Correlation between egg size and clutch size in the pied flycatcher *Ficedula hypoleuca* in cold and warm summers. *Ibis*, **138**, 620-623.

Johns, D.G., M. Edwards and S.D. Batten, 2001: Arctic boreal plankton species in the Northwest Atlantic. *Can. J. Fish. Aquat. Sci*., **58**, 2121-2124.

Johns, D.G., M. Edwards, A. Richardson and J.I. Spicer, 2003: Increased blooms of a dinoflagellate in the NW Atlantic. *Mar. Ecol.–Prog. Ser*., **265**, 283-287.

Johnson, H., S. Kovats, G. McGregor, J. Stedman, M. Gibbs, H. Walton and L. Cook, 2004: The impact of the 2003 heat wave on mortality and hospital admissions in England. *Epidemiology*, **15**, S126-S126.

Jones, G.A. and G.H.R. Henry, 2003: Primary plant succession on recently deglaciated terrain in the Canadian High Arctic. *J. Biogeogr*., **30**, 277-296.

Jones, G.V., 2005: Climate change in the western United States grape growing regions. *7th International Symposium on Grapevine Physiology and Biotechnology*, **689**, 71-80.

Jones, G.V. and R.E. Davis, 2000: Climate influences on grapevine phenology, grape composition and wine production and quality for Bordeaux, France. *Am. J. Enol. Viticult*., **51**, 249-261.

Jones, I., F. Hunter and G. Roberston, 2002: Annual adult survival of Least Auklets (Aves, Alcidae) varies with large-scale climatic conditions of the North Pacific Ocean. *Oecologia*, **133**, 38-44.

Jordan, E., 1991: Die gletscher der bolivianischen Anden: eine photogrammetrisch-kartographische Bestandsaufnahme der Gletscher Boliviens als Grundlage für klimatische Deutungen und Potential für die wirtschaftliche Nutzung (The glaciers of the Bolivian Andes, a photogrammetric-cartographical inventory of the Bolivian glaciers as a basis for climatic interpretation and potential for economic use). Erdwissenschaftliche Forschung 23, Franz Steiner Verlag, Stuttgart, 401 pp.

Juanes, F., S. Gephard and K. Beland, 2004: Long-term changes in migration timing of adult Atlantic salmon (*Salmo salar*) at the southern edge of the species distribution. *Can. J. Fish. Aquat. Sci*., **61**, 2392-2400.

Juen, I., G. Kaser and C. Georges, 2007: Modelling observed and future runoff from a glacierized tropical catchment (Cordillera Blanca, Perú). *Global Planet. Change*, doi:10.1016/j.gloplacha.2006.11.038.

Jump, A.S., J.M. Hunt and J. Peñuelas, 2006: Rapid climate change-related growth decline at the southern range edge of *Fagus sylvatica*. *Glob. Change Biol.*, **12**, 2163-2174.

Kaab, A., C. Huggel, L. Fischer, S. Guex, F. Paul, I. Roer, N. Salzmann, S. Schlaefli, K. Schmutz, D. Schneider, T. Strozzi and Y. Weidmann, 2005: Remote sensing of glacier- and permafrost-related hazards in high mountains: an overview. *Nat. Hazard. Earth Sys.*, **5**, 527-554.

Kalnay, E. and M. Cai, 2003: Impact of urbanization and land-use change on climate. *Nature*, **423**, 528-531.

Kanuscak, P., M. Hromada, P. Tryjanowski and T. Sparks, 2004: Does climate at different scales influence the phenology and phenotype of the river warbler *Locustella fluviatilis*? *Oecologia*, **141**, 158-163.

Karban, R. and S.Y. Strauss, 2004: Physiological tolerance, climate change and a northward range shift in the spittlebug, *Philaenus spumarius*. *Ecol. Entomol.*, **29**, 251-254.

Karl, T.H. and R.W. Knight, 1997: The 1995 Chicago heat wave: how likely is a recurrence? *B. Am. Meteorol. Soc.*, **78**, 1107-1119.

Karl, T.R. and K.E. Trenberth, 2003: Modern global climate change. *Science*, **302**, 1719-1723.

Karoly, D.J. and Q. Wu, 2005: Detection of regional surface temperature trends. *J. Climate*, **18**, 4337-4343.

Karoly, D.J. and P.A. Stott, 2006: Anthropogenic warming of central England temperature. *Atmos. Sci. Lett.*, **7**, 81-85.

Karst-Riddoch, T.L., M.F.J. Pisaric and J.P. Smol, 2005: Diatom responses to 20th century climate-related environmental changes in high-elevation mountain lakes of the northern Canadian Cordillera. *J. Paleolimnol.*, **33**, 265-282.

Kaser, G. and H. Osmaston, 2002. *Tropical Glaciers*. Cambridge University Press, Cambridge, 227 pp.

Kearney, M., A. Rogers, J. Townshend, E. Rizzo, D. Stutzer, J. Stevenson and K. Sundborg, 2002: Landsat imagery shows decline of coastal marshes in Chesapeake and Delaware Bays. *EOS Transactions*, **83**, 173, doi:10.1029/2002EO000112.

Keeling, C.D., J.F.S. Chin and T.P. Whorf, 1996: Increased activity of northern vegetation inferred from atmospheric CO_2 measurements. *Nature*, **382**, 146-149.

Keller, F. and C. Korner, 2003: The role of photoperiodism in alpine plant development. *Arct. Antarct. Alp. Res.*, **35**, 361-368.

Kennish, M.J., 2001: Coastal salt marsh systems in the US: a review of anthropogenic impacts. *J. Coastal Res.*, **17**, 731-748.

Klanderud, K. and H.J.B. Birks, 2003: Recent increases in species richness and shifts in altitudinal distributions of Norwegian mountain plants. *Holocene*, **13**, 1-6.

Klasner, F.L. and D.B. Fagre, 2002: A half century of change in alpine treeline patterns at Glacier National Park, Montana, USA. *Arct. Antarct. Alp. Res.*, **34**, 49-56.

Kleypas, J.A., R.W. Buddemeier, D. Archer, J.P. Gattuso, C. Langdon and B.N. Opdyke, 1999: Geochemical consequences of increased atmospheric carbon dioxide on coral reefs. *Science*, **284**, 118-120.

Knutson, T.R., T.L. Delworth, K.W. Dixon, I.M. Held, J. Lu, V. Ramaswamy, M.D. Schwarzkopf, G. Stenchikov and R.J. Stouffer, 2006: Assessment of twentieth century regional surface air temperature trends using the GFDL CM2 coupled models. *J. Climate*, **19**, 1624-1651.

Konvicka, M., M. Maradova, J. Benes, Z. Fric and P. Kepka, 2003: Uphill shifts in distribution of butterflies in the Czech Republic: effects of changing climate detected on a regional scale. *Global Ecol. Biogeogr.*, **12**, 403-410.

Korhola, A., S. Sorvari, M. Rautio, P.G. Appleby, J.A. Dearing, Y. Hu, N. Rose, A. Lami and N.G. Cameron, 2002: A multi-proxy analysis of climate impacts on the recent development of subarctic Lake Saanajarvi in Finnish Lapland. *J. Paleolimnol.*, **28**, 59-77.

Kosatsky, T., 2005: The 2003 European heat waves. *Euro Surveillance*, **10**, 148-149.

Kovats, R.S., D.H. Campbell-Lendrum, A.J. McMichael, A. Woodward and J.S. Cox, 2001: Early effects of climate change: do they include changes in vector-borne disease? *Philos. T. Roy. Soc. B*, **356**, 1057-1068.

Kovats, R.S., M.J. Bouma, S. Hajat, E. Worrall and A. Haines, 2003: El Niño and health. *Lancet*, **362**, 1481-1489.

Kozlov, M.V. and N.G. Berlina, 2002: Decline in length of the summer season on the Kola Peninsula, Russia. *Climatic Change*, **54**, 387-398.

Krupnik, I. and D. Jolly, Eds., 2002: *The Earth is Faster Now: Indigenous Observations of Arctic Environmental Change*. Arctic Research Consortium of the United States, Fairbanks, 356 pp.

Kullman, L., 2002: Rapid recent range-margin rise of tree and shrub species in the Swedish Scandes. *J. Ecol.*, **90**, 68-77.

Kundzewicz, Z., 2004: Detection of change in world-wide hydrological time series of maximum annual flow. GRDC Report 32, 36 pp.

Kundzewicz, Z.W., U. Ulbrich, T. Brucher, D. Graczyk, A. Kruger, G.C. Leckebusch, L. Menzel, I. Pinskwar, M. Radziejewski and M. Szwed, 2005: Summer floods in central Europe: climate change track? *Nat. Hazards*, **36**, 165-189.

Kunst, A.E., C.W.N. Looman and J.P. Mackenbach, 1991: The decline in winter excess mortality in the Netherlands. *Int. J. Epidemiol.*, **20**, 971-977.

Kysely, J., 2004: Mortality and displaced mortality during heat waves in the Czech Republic. *Int. J. Biometeorol.*, **49**, 91-97.

Labat, D., Y. Godderis, J.L. Probst and J.L. Guyot, 2004: Evidence for global runoff increases related to climate warming. *Adv. Water Resour.*, **27**, 631-642.

Ladenburger, C.G., A.L. Hild, D.J. Kazmer and L.C. Munn, 2006: Soil salinity patterns in *Tamarix* invasions in the Bighorn Basin, Wyoming, USA. *J. Arid Environ.*, **65**, 111-128.

Laing, T.E. and J.P. Smol, 2003: Late Holocene environmental changes inferred from diatoms in a lake on the western Taimyr Peninsula, northern Russia. *J. Paleolimnol.*, **30**, 231-247.

Lam, J.C., C.L. Tsang and D.H.W. Li, 2004: Long term ambient temperature analysis and energy use implications in Hong Kong. *Energ. Convers. Manage.*, **45**, 315-327.

Lammers, R.B., A.I. Shiklomanov, C.J. Vorosmarty, B.M. Fekete and B.J. Peterson, 2001: Assessment of contemporary Arctic river runoff based on observational discharge records. *J. Geophys. Res.–Atmos.*, **106**(D4), 3321-3334.

Landsea, C.W., 2005: Meteorology: hurricanes and global warming. *Nature*, **438**, E11-E13.

Landsea, C.W., R.A. Pielke Jr., A.M. Mestas-Nuñez and J.A. Knaff, 1999: Atlantic basin hurricanes: indices of climatic change. *Climatic Change*, **42**, 89-129.

Langdon, C., T. Takahashi, C. Sweeney, D. Chipman, J. Goddard, F. Marubini, H. Aceves, H. Barnett and M.J. Atkinson, 2000: Effect of calcium carbonate saturation state on the calcification rate of an experimental coral reef. *Global Biogeochem. Cy.*, **14**, 639-654.

Lantuit, H. and W.H. Pollard, 2003: Remotely sensed evidence of enhanced erosion during the 20th century on Herschel Island, Yukon Territory. *Ber. Polar. Meeresforsch.*, **443**, 54-59.

Larsen, C.F., R.J. Motyka, J.T. Freymueller, K.A. Echelmeyer and E.R. Ivins, 2005: Rapid uplift of southern Alaska caused by recent ice loss. *Geophys. J. Int.*, **158**, 1118-1133.

Larsson, N., 2003: Adapting to climate change in Canada. *Build. Res. Inf.*, **31**, 231-239.

Laternser, M. and M. Schneebeli, 2003: Long-term snow climate trends of the Swiss Alps (1931-99). *Int. J. Climatol.*, **23**, 733-750.

Lavaniegos, B.E. and M.D. Ohman, 2003: Long-term changes in pelagic tunicates of the California Current. *Deep-Sea Res. Pt. II*, **50**, 2473-2498.

Lawrence, D.M. and A.G. Slater, 2005: A projection of severe near-surface permafrost degradation during the 21st century. *Geophys. Res. Lett.*, **32**, L24401, doi:10.1029/2005GL025080.

Lawson, P.W., E.A. Logerwell, N.J. Mantua, R.C. Francis and V.N. Agostini, 2004: Environmental factors influencing freshwater survival and smolt production in Pacific Northwest coho salmon (*Oncorhynchus kisutch*). *Can. J. Fish. Aquat. Sci.*, **61**, 360-373.

Leatherman, S.P., K. Zhang and B.C. Douglas, 2000: Sea-level rise shown to drive coastal erosion. *EOS Transactions*, **81**, 55-57.

Le Treut, H., R. Somerville, U. Cubasch, Y. Ding, C. Mauritzen, A. Mokssit, T. Peterson and M. Prather, 2007: Historical overview of climate change science. *Climate Change 2007: The Physical Science Basis. Contribution of Working Group I to the Fourth Assessment Report of the Intergovernmental Panel on Climate Change*, S. Solomon, D. Qin, M. Manning, Z. Chen, M. Marquis, K.B. Averyt, M. Tignor and H.L. Miller, Eds., Cambridge University Press, Cambridge, 93-128.

Ledneva, A., A.J. Miller-Rushing, R.B. Primack and C. Imbres, 2004: Climate change as reflected in a naturalist's diary, Middleborough, Massachusetts. *Wilson Bull.*, **116**, 224-231.

Lee, C.S., X.D. Li, W.Z. Shi, S.C. Cheung and I. Thornton, 2006: Metal contamination in urban, suburban and country park soils of Hong Kong: a study based on GIS and multivariate statistics. *Sci. Total Environ.*, **356**, 45-61.

Legates, D.R., H.F. Lins and G.J. McCabe, 2005: Comments on "Evidence for global runoff increase related to climate warming" by Labat et al. *Adv. Water Resour.*, **28**, 1310-1315.

Legendre, L. and R.B. Rivkin, 2002: Pelagic food webs: responses to environmental processes and effects on the environment. *Ecol. Res.*, **17**, 143-149.

Lehikoinen, E., T.H. Sparks and M. Zalakevicius, 2004: Arrival and departure dates. *Adv. Ecol. Res.*, **35**, 1-31.

Lemke, P., J. Ren, R. Alley, I. Allison, J. Carrasco, G. Flato, Y. Fujii, G. Kaser, P. Mote, R. Thomas and T. Zhang, 2007: Observations: Changes in Snow, Ice and Frozen Ground. *Climate Change 2007: The Physical Science Basis. Contribution of Working Group I to the Fourth Assessment Report of the Intergovernmental Panel on Climate Change*, S. Solomon, D. Qin, M. Manning, Z. Chen, M. Marquis, K.B. Averyt, M. Tignor and H.L. Miller, Eds., Cambridge University Press, Cambridge, 337-384.

Lepers, E., E.F. Lambin, A.C. Janetos, R. DeFries, F. Achard, N. Ramankutty and R.J. Scholes, 2004: A synthesis of information on rapid land-cover change for the period 1981-2000. *BioScience*, **55**, 115-124.

Lesica, P. and B. McCune, 2004: Decline of arctic-alpine plants at the southern margin of their range following a decade of climatic warming. *J. Veg. Sci.*, **15**, 679-690.

Levermore, G.J. and E. Keeble, 1998: Dry-bulb temperature analyses for climate change at three UK sites in relation to the forthcoming CIBSE Guide to Weather and Solar Data. *Building Services Engineering Research and Technology*, **19**, 175-181.

Levetin, E., 2001: Effects of climate change on airborne pollen. *J. Allergy Clin. Immun.*, **107**, S172-S172.

Levine, M., D. Ürge-Vorsatz, K. Blok, L. Geng, D. Harvey, S. Lang, G. Levermore, A. Mongameli Mehlwana, S. Mirasgedia, A. Novikovam, J. Rilling and H. Yoshino, 2007: Residential and commercial buildings. *Climate Change 2007: Mitigation. Contribution of Working Group III to the Fourth Assessment Report of the Intergovernmental Panel on Climate Change*, B. Metz, O. Davidson, P.Bosch, R. Dave and L. Meyer, Eds., Cambridge University Press, Cambridge, UK.

Linderholm, H.W., 2006: Growing season changes in the last century. *Agr. Forest Meteorol.*, **137**, 1-14.

Lindgren, E. and R. Gustafson, 2001: Tick-borne encephalitis in Sweden and climate change. *Lancet*, **358**, 16-18.

Lindgren, E., L. Talleklint and T. Polfeldt, 2000: Impact of climatic change on the northern latitude limit and population density of the disease-transmitting European tick *Ixodes ricinus*. *Environ. Health Persp.*, **108**, 119-123.

Lionello, P., 2005: Extreme surges in the Gulf of Venice: present and future climate. *Flooding and Environmental Challenges for Venice and its Lagoon: State of Knowledge 2003*, C. Fletcher and T. Spencer, Eds., Cambridge University Press, Cambridge, 59-70.

Lionello, P. and A. Sanna, 2005: Mediterranean wave climate variability and its links with NAO and Indian Monsoon. *Clim. Dynam.*, **25**, 611-623.

Lise, W. and R.S.J. Tol, 2002: Impact of climate on tourist demand. *Climatic Change*, **55**, 429-449.

Livingstone, D.M. and M.T. Dokulil, 2001: Eighty years of spatially coherent Austrian lake surface temperatures and their relationship to regional air temperature and the North Atlantic Oscillation. *Limnol. Oceanogr.*, **46**, 1220-1227.

Lobell, D.B. and G.P. Asner, 2003: Climate and management contributions to recent trends in U.S. agricultural yields. *Science*, **299**, 1032.

Lobo, A. and P. Maisongrande, 2006: Stratified analysis of satellite imagery of SW Europe during summer 2003: the differential response of vegetation classes to increased water deficit. *Hydrol. Earth Syst. Sc.*, **10**, 151-164.

Lorch, R., 1990: Energy and global responsibility. *RIBA J.*, **March**, 50-51.

Lorke, A., K. Tietze, M. Halbwachs and A. Wuest, 2004: Response of Lake Kivu stratification to lava inflow and climate warming. *Limnol. Oceanogr.*, **49**, 778-783.

Lotsch, A.M., A. Friedl, B.T. Anderson and C.J. Tucker, 2005: Response of terrestrial ecosystems to recent Northern Hemispheric drought. *Geophys. Res. Lett.*, **32**, L06705, doi:10.1029/2004GL022043.

Lucht, W., I.C. Prentice, R.B. Myneni, S. Sitch, P. Friedlingstein, W. Cramer, P. Bousquet, W. Buermann and B. Smith, 2002: Climatic control of the high-latitude vegetation greening trend and Pinatubo effect. *Science*, **296**, 1687-1689.

Luterbacher, J., D. Dietrich, E. Xoplaki, M. Grosjean and H. Wanner, 2004: European seasonal and annual temperature variability, trends and extremes since 1500. *Science*, **303**, 1499-1503.

MacDonald, M., L. Arragutainaq and Z. Novalinga, 1997: *Voices from the Bay: Traditional Ecological Knowledge of Inuit and Cree*. Canadian Arctic Resource Committee and Environmental Committee of Municipality of Sanikiluaq, Ottawa, 98 pp.

Maddison, D., 2001: In search of warmer climates? The impact of climate change on flows of British tourists. *Climatic Change*, **49**, 193-208.

Mark, B.G., J.M. McKenzie and J. Gómez, 2005: Hydrochemical evaluation of changing glacier meltwater contribution to stream discharge: Callejon de Huaylas, Peru. *Hydrolog. Sci. J*, **50**, 975-987.

Masek, J.G., 2001: Stability of boreal forest stands during recent climate change: evidence from Landsat satellite imagery. *J. Biogeogr.*, **28**, 967-976.

Maselli, F., 2004: Monitoring forest conditions in a protected Mediterranean coastal area by the analysis of multi-year NDVI data. *Remote Sens. Environ.*, **89**, 423-433.

Matsumoto, K., T. Ohta, M. Irasawa and T. Nakamura, 2003: Climate change and extension of the *Ginkgo biloba* L. growing season in Japan. *Glob. Change Biol.*, **9**, 1634-1642.

McCabe, G. and M.P. Clark, 2005: Trends and variability in snowmelt runoff in the western United States. *J. Hydrometeorol.*, **6**, 476-482.

McCallum, H., D. Harvell and A. Dobson, 2003: Rates of spread of marine pathogens. *Ecol. Lett.*, **6**, 1062-1067.

McGowan, J.A., D.R. Cayan and L.M. Dorman, 1998: Climate-ocean variability and ecosystem response in the northeast Pacific. *Science*, **281**, 210-217.

McKenzie, D., Z. Gedalof, D.L. Peterson and P. Mote, 2004, Climatic change, wildfire and conservation. *Conserv. Biol.*, **18**, 890-902.

McLaughlin, J.F., J.J. Hellmann, C.L. Boggs and P.R. Ehrlich, 2002a: Climate change hastens population extinctions. *P. Natl. Acad. Sci. USA*, **99**, 6070-6074.

McLaughlin, J.F., J.J. Hellmann, C.L. Boggs and P.R. Ehrlich, 2002b: The route to extinction: population dynamics of a threatened butterfly. *Oecologia*, **132**, 538-548.

McNamara, J.P., D.L. Kane and L.D. Hinzman, 1999: An analysis of an arctic channel network using a digital elevation model. *Geomorphology*, **29**, 339-353.

McWilliams, J.P., I.M. Côté, J.A. Gill, W.J. Sutherland and A.R. Watkinson, 2005: Accelerating impacts of temperature-induced coral bleaching in the Caribbean. *Ecology*, **86**, 2055-2060.

Mendes, J.M. and J.D. Woodley, 2002: Effect of the 1995-1996 bleaching event on polyp tissue depth, growth, reproduction and skeletal band formation in *Montastraea annularis*. *Mar. Ecol.–Prog. Ser.*, **235**, 93-102.

Menendez, R., A.G. Megias, J.K. Hill, B. Braschler, S.G. Willis, Y. Collingham, R. Fox, D.B. Roy and C.D. Thomas, 2006: Species richness changes lag behind climate change. *P. Roy. Soc. Lond. B Bio.*, **273**, 1465-1470.

Menzel, A., 2000: Trends in phenological phases in Europe between 1951 and 1996. *Int. J. Biometeorol.*, **44**, 76-81.

Menzel, A., 2002: Phenology: its importance to the global change community. *Climatic Change*, **54**, 379-385.

Menzel, A., 2003: Plant phenological anomalies in Germany and their relation to air temperature and NAO. *Climatic Change*, **57**, 243-263.

Menzel, A. and P. Fabian, 1999: Growing season extended in Europe. *Nature*, **397**, 659-659.

Menzel, A. and N. Estrella, 2001: Plant phenological changes. *"Fingerprints" of Climate Change: Adapted Behaviour and Shifting Species Ranges*, G.-R. Walther, C.A. Burga and P.J. Edwards, Eds., Kluwer Academic/Plenum, New York, 123-137.

Menzel, A. and V. Dose, 2005: Analysis of long-term time series of the beginning of flowering by Bayesian function estimation. *Meteorol. Z.*, **14**, 429-434.

Menzel, A., N. Estrella and P. Fabian, 2001: Spatial and temporal variability of the phenological seasons in Germany from 1951 to 1996. *Glob. Change Biol.*, **7**, 657-666.

Menzel, A., G. Jakobi, R. Ahas, H. Scheifinger and N. Estrella, 2003: Variations of the climatological growing season (1951-2000) in Germany compared with other countries. *Int. J. Climatol.*, **23**, 793-812.

Menzel, A., A. Estrella and A. Testka, 2005a: temperature response rates from long-term phenological records. *Climate Res.*, **30**, 21-28.

Menzel, A., T.H. Sparks, N. Estrella and S. Eckhardt, 2005b: 'SSW to NNE': North Atlantic Oscillation affects the progress of seasons across Europe. *Glob. Change Biol.*, **11**, 909-918.

Menzel, A., T. Sparks, N. Estrella and D.B. Roy, 2006a: Geographic and temporal variability in phenology. *Global Ecol. Biogeogr.*, **15**, 498-504.

Menzel, A., T.H. Sparks, N. Estrella, E. Koch, A. Aasa, R. Ahas, K. Alm-Kübler, P. Bissolli, O. Braslavská, A. Briede, F.M. Chmielewski, Z. Crepinsek, Y. Curnel, Å. Dahl, C. Defila, A. Donnelly, Y. Filella, K. Jatczak, F. Måge, A. Mestre, Ø. Nordli, J. Peñuelas, P. Pirinen, V. Remišová, H. Scheifinger, M. Striz, A. Susnik, A.J.H. van Vliet, F.-E. Wielgolaski, S. Zach and A. Zust, 2006b: European phenological response to climate change matches the warming pattern. *Glob. Change Biol.*, **12**, 1969-1976.

Menzel, A., J. von Vopelius, N. Estrella, C. Schleip and V. Dose, 2006c: Farmers' annual activities are not tracking speed of climate change. *Climate Res.*, **32**, 201-207.

Meshinev, T., I. Apostolova and E. Koleva, 2000: Influence of warming on timberline rising: a case study on *Pinus peuce* Griseb. in Bulgaria. *Phytocoenologia*, **30**, 431-438.

Mestas-Nunez, A.M. and A.J. Miller, 2006: Interdecadal variability and climate change in the eastern tropical Pacific: a review. *Prog. Oceanogr.*, **69**, 267-284.

Michelutti, N., M.S.V. Douglas and J.P. Smol, 2003: Diatom response to recent climatic change in a high arctic lake (Char Lake, Cornwallis Island, Nunavut). *Global Planet. Change*, **38**, 257-271.

Middleton, E.M., J.R. Herman, E.A. Celarier, J.W. Wilkinson, C. Carey and R.J. Rusin, 2001: Evaluating ultraviolet radiation exposure with satellite data at sites of amphibian declines in Central and South America. *Conserv. Biol.*, **15**, 914-929.

Millar, J.S. and E.J. Herdman, 2004: Climate change and the initiation of spring breeding by deer mice in the Kananaskis Valley, 1985-2003. *Can. J. Zool.*, **82**, 1444-1450.

Milly, P.C.D., R.T. Wetherald, K.A. Dunne and T.L. Delworth, 2002: Increasing risk of great floods in a changing climate. *Nature*, **415**, 514-517.

Milly, P.C.D., K.A. Dunne and A.V. Vecchia, 2005: Global pattern of trends in streamflow and water availability in a changing climate. *Nature*, **438**, 347-350.

Mimura, N. and P.D. Nunn, 1998: Trends of beach erosion and shoreline protection in rural Fiji. *J. Coastal Res.*, **14**, 37-46.

Mitchell, J.F.B., D.J. Karoly, M.R. Allen, G. Hegerl, F. Zwiers and J. Marengo, 2001: Detection of climate change and attribution of causes. *Climate Change 2001: The Scientific Basis. Contribution of Working Group I to the Third Assessment Report of the Intergovernmental Panel on Climate Change*, J.T. Houghton, Y. Ding, D.J. Griggs, M. Noguer, P.J. van der Linden, X. Dai, K. Maskell and C.A. Johnson, Eds., Cambridge University Press, Cambridge, 695-738.

Mohr, J.J. and R. Forsberg, 2002: Remote sensing: searching for new islands in sea ice: coastlines concealed in polar seas are now more accessible to cartography. *Nature*, **416**, 35-35.

Moiseev, P.A. and S.G. Shiyatov, Eds., 2003: Vegetation dynamics at the treeline ecotone in the Ural highlands, Russia. *Alpine Biodiversity in Europe: A Europe-wide Assessment of Biological Richness and Change*, L. Nagy, G. Grabherr, C. Körner and D.B.A. Thompson, Eds., Ecological Studies 167, Springer, Berlin, 423-435.

Mölg, T., D.R. Hardy, N. Cullen and G. Kaser, 2005: Tropical glaciers in the context of climate change and society: focus on Kilimanjaro (East Africa). *Contribution to Mountain Glaciers and Society Workshop*. California University Press, Wengen, 28 pp.

Molyneux, D., 2003: Climate change and tropical disease: common themes in changing vector-borne disease scenarios. *T. Roy. Soc. Trop Med. H.*, **97**, 129-32.

Mool, P.K., D. Wangda and S.R. Bajracharya, 2001: *Inventory of Glaciers, Glacial Lakes and Glacial Lake Outburst Floods: Monitoring and Early Warning Systems in the Hindu Kush-Himalayan Region*. ICIMOD, Bhutan, Kathmandu, 227 pp.

Moonen, A.C., L. Ercoli, M. Mariotti and A. Masoni, 2002: Climate change in Italy indicated by agrometeorological indices over 122 years. *Agr. Forest Meteorol.*, **111**, 13-27.

Mueller, D.R., W.F. Vincent and M.O. Jeffries, 2003: Break-up of the largest Arctic ice shelf and associated loss of an epishelf lake. *Geophys. Res. Lett.*, **30**, 2031, doi:10.1029/2003GL017931.

Muir Wood, R., S. Miller and A. Boissonnade, 2006: The search for trends in a global catalogue of normalized weather-related catastrophe losses. *Workshop on Climate Change and Disaster Losses: Understanding and Attributing Trends and Projections*. Hohenkammer, Munich, 188-194.

Munich Re Group, 2005: Annual Review: *Natural Catastrophes 2004*. WKD Offsetdruck GmbH, Munich, 60 pp.

Myneni, R.B., C.D. Keeling, C.J. Tucker, G. Asrar and R.R. Nemani, 1997: Increased plant growth in the northern high latitudes from 1981 to 1991. *Nature*, **386**, 698-702.

Nabuurs, G.J., A. Pussinen, T. Karjalainen, M. Erhard and K. Kramer, 2002: Stemwood volume increment changes in European forests due to climate change: a simulation study with the EFISCEN model. *Glob. Change Biol.*, **8**, 304-316.

Nelson, F.E., 2003: (Un)frozen in time. *Science*, **299**, 1673-1675.

Nemani, R.R., M.A. White, D.R. Cayan, G.V. Jones, S.W. Running, J.C. Coughlan and D.L. Peterson, 2001: Asymmetric warming over coastal California and its impact on the premium wine industry. *Climate Res.*, **19**, 25-34.

Nemani, R.R., C.D. Keeling, H. Hashimoto, W.M. Jolly, S.C. Piper, C.J. Tucker, R.B. Myneni and S.W. Running, 2003: Climate-driven increases in global terrestrial net primary production from 1982 to 1999. *Science*, **300**, 1560-1563.

Newman, J.A., 2005: Climate change and the fate of cereal aphids in Southern Britain. *Glob. Change Biol.*, **11**, 940-944.

Nicol, F., 2004: Adaptive thermal comfort standards in the hot-humid tropics. *Energ. Buildings*, **36**, 628-637.

NOAA, 2005: Preliminary Report: Hurricane Katrina storm tide summary. Center for Operational Oceanographic Products and Services, U.S. Department of Commerce National Ocean Service, Silver Spring, Maryland, 20 pp.

Noetzli, J., M. Hoelzle and W. Haeberli, 2003: Mountain permafrost and recent Alpine rock-fall events: a GIS-based approach to determine critical factors. *8th International Conference on Permafrost, Zürich, Proceedings Vol. 2*, M. Phillips, S. Springman and L. Arenson, Eds., Swets and Zeitlinger, Lisse, 827-832.

Nowakowski, J.J., 2002: Variation of morphometric parameters within the Savi's warbler (*Locustella luscinioides*) population in eastern Poland. *Ring*, **24**, 49.

Nyberg, P., E. Bergstand, E. Degerman and O. Enderlein, 2001: Recruitment of pelagic fish in an unstable climate: studies in Sweden's four largest lakes. *Ambio*, **30**, 559-564.

Nyssen, J., J. Poesen, J. Moeyersons, J. Deckers, M. Haile and A. Lang, 2004: Human impact on the environment in the Ethiopian and Eritrean highlands: a state of the art. *Earth-Sci. Rev.*, **64**, 273-320.

O'Reilly, C.M., 2007: The impact of recent global climate change on the East Africa Rift Valley lakes. *Earth-Sci. Rev.*, submitted.

O'Reilly, C.M., S.R. Alin, P.D. Plisnier, A.S. Cohen and B.A. McKee, 2003: Climate change decreases aquatic ecosystem productivity of Lake Tanganyika, Africa. *Nature*, **424**, 766-768.

OECD, 2007: Climate change in the European Alps. Adapting Winter Tourism and Natural Hazards Management, Shardul Agrawala Ed., OECD Publishing, Paris, 127 pp.

Ohde, S. and M.M.M. Hossain, 2004: Effect of $CaCO_3$ (aragonite) saturation state of seawater on calcification of Porites coral. *Geochem. J.*, **38**, 613-621.

Opdam, P. and D. Wascher, 2004: Climate change meets habitat fragmentation: linking landscape and biogeographical scale levels in research and conservation. *Biol. Conserv.*, **117**, 285-297.

Orviku, K., J. Jaagus, A. Kont, U. Ratas and R. Rivis, 2003: Increasing activity of coastal processes associated with climate change in Estonia. *J. Coastal Res.*, **19**, 364-375.

Ozaki, N., T. Fukushima, H. Harasawa, T. Kojiri, K. Kawashima and M. Ono, 2003: Statistical analyses on the effects of air temperature fluctuations on river water qualities. *Hydrol. Process.*, **17**, 2837-2853.

Pagano, T. and D. Garen, 2005: A recent increase in western US streamflow variability and persistence. *J. Hydrometeorol.*, **6**, 173-179.

Pagliosa, P.R. and F.A.R. Barbosa, 2006: Assessing the environment–benthic fauna coupling in protected and urban areas of southern Brazil. *Biol. Conserv.*, **129**, 408-417.

Parmesan, C., 2006: Ecological and evolutionary responses to recent climate change. *Annu. Rev. Ecol. Evol. S.*, **37**, 637-669.

Parmesan, C. and G. Yohe, 2003: A globally coherent fingerprint of climate change impacts across natural systems. *Nature*, **421**, 37-42.

Parmesan, C. and H. Galbraith, 2004: Observed impacts of global climate change in the US. Report prepared for the Pew Center on Global Climate Change, Washington, District of Columbia, 67 pp.

Parmesan, C., N. Rhyrholm, C. Stefanescu, J.K. Hill, C.D. Thomas, H. Descimon, B. Huntley, L. Kalia, J. Kullberg, T. Tammaru, W.J. Tennent, J.A. Thomas and M. Warren, 1999: Poleward shifts in geographical ranges of butterfly species associated with regional warming. *Nature*, **399**, 579-583.

Parmesan, C., T.L. Root and M.R. Willig, 2000: Impacts of extreme weather and climate on terrestrial biota. *B. Am. Meteorol. Soc.*, **81**, 443-450.

Pascual, M., A. Ahumada, L.F. Chaves, X. Rodo and M.J. Bouma, 2006: Malaria resurgence in the East African highlands: temperature trends revisited. *P. Natl. Acad. Sci. USA*, **103**, 5829-5834.

Patz, J.A., M. Hulme, C. Rosenzweig, T.D. Mitchell, R.A. Goldberg, A.K. Githeko, S. Lele, A.J. McMichael and D. Le Sueur, 2002: Climate change: regional warming and malaria resurgence. *Nature*, **420**, 627-628.

Patz, J.A., D. Campbell-Lendrum, T. Holloway and J.A.N. Foley, 2005: Impact of regional climate change on human health. *Nature*, **438**, 310-317.

Pauli, H., M. Gottfried and G. Grabherr, 2001. High summits of the Alps in a changing climate. *"Fingerprints" of Climate Change: Adapted Behaviour and Shifting Species Ranges*, G.-R. Walther, C.A. Burga and P.J. Edwards, Eds., Kluwer Academic/Plenum, New York, 139-149.

Pauli, H., M. Gottfried, T. Dirnbock, S. Dullinger and G. Grabherr, 2003: Assessing the long-term dynamics of endemic plants at summit habitats. *Alpine Biodiversity in Europe: A Europe-wide Assessment of Biological Richness and Change*, L. Nagy, G. Grabherr, C. Körner and D.B.A. Thompson, Eds., Ecological Studies 167, Springer, Heidelberg, 195-207.

Pelejero, C., E. Calvo, M.T. McCulloch, J.F. Marshall, M.K. Gagan, J.M. Lough and B.N. Opdyke, 2005: Preindustrial to modern interdecadal variability in coral reef pH. *Science*, **309**, 2204-2207.

Peña, H. and F. Escobar, 1985: Análisis de las crecidas del río Paine, XII Región. Publicación Interna E.H. N. 83/7, 78 pp.

Peng, S.B., J.L. Huang, J.E. Sheehy, R.C. Laza, R.M. Visperas, X.H. Zhong, G.S. Centeno, G.S. Khush and K.G. Cassman, 2004: Rice yields decline with higher night temperature from global warming. *P. Natl. Acad. Sci. USA*, **101**, 9971-9975.

Penland, S., P.F. Connor, A. Beall, S. Fearnley and S.J. Williams, 2005: Changes in Louisiana's shoreline: 1855-2002. *J. Coastal Res.*, **44**, S7-S39.

Peñuelas, J. and M. Boada, 2003: A global change-induced biome shift in the Montseny mountains (NE Spain). *Glob. Change Biol.*, **9**, 131-140.

Peñuelas, J., I. Filella and P. Comas, 2002: Changed plant and animal life cycles from 1952 to 2000 in the Mediterranean region. *Glob. Change Biol.*, **8**, 531-544.

Perren, B.B., R.S. Bradley and P. Francus, 2003: Rapid lacustrine response to recent High Arctic warming: a diatom record from Sawtooth Lake, Ellesmere Island, Nunavut. *Arct. Antarct. Alp. Res.*, **35**, 271-278.

Perry, A.L., P.J. Low, J.R. Ellis and J.D. Reynolds, 2005: Climate change and distribution shifts in marine fishes. *Science*, **308**, 1912-1915.

Petheram, C., G. Walker, R. Grayson, T. Thierfelder and L. Zhang, 2001: Towards a framework for predicting impacts of land-use on recharge. *Aust. J. Soil Res.*, **40**, 397-417.

Phillips, O.L., T.R. Baker, L. Arroyo, N. Higuchi, T.J. Killeen, W.F. Laurance, S.L. Lewis, J. Lloyd, Y. Malhi, A. Monteagudo, D.A. Neill, P.N. Vargas, J.N.M. Silva, J. Terborgh, R.V. Martinez, M. Alexiades, S. Almeida, S. Brown, J. Chave, J.A. Comiskey, C.I. Czimczik, A. Di Fiore, T. Erwin, C. Kuebler, S.G. Laurance, H.E.M. Nascimento, J. Olivier, W. Palacios, S. Patino, N.C.A. Pitman, C.A. Quesada, M. Salidas, A.T. Lezama and B. Vinceti, 2004: Pattern and process in Amazon tree turnover, 1976-2001. *Philos. T. Roy. Soc. B*, **359**, 381-407.

Pielke, R.A. and C.W. Landsea, 1998: Normalized hurricane damages in the United States: 1925-95. *Weather Forecast.*, **13**, 621-631.

Pielke, R.A. and P. Hoppe, 2006: Workshop summary report. *Workshop on Climate Change and Disaster Losses: Understanding and Attributing Trends and Projections*. Hohenkammer, Munich, 4-12.

Pielke, R.A., Jr., M.W. Downton and J.Z. Barnard Miller, 2002: *Flood Damage in the United States, 1926-2000: A Reanalysis of National Weather Service Estimates*. National Center for Atmospheric Research (UCAR), Boulder, Colorado, 96 pp.

Pielke, R.A., J. Rubiera, C. Landsea, M.L. Fernandez and R. Klein, 2003: Hurricane vulnerability in Latin America and the Caribbean: normalized damage and loss potentials. *Natural Hazards Review*, **4**, 101-114.

Pielke, R.A., C. Landsea, K. Emanuel, M. Mayfield, J. Laver and R. Pasch, 2005: Hurricanes and global warming. *B. Am. Meteorol. Soc.*, **86**, 1571.

Pilcher, J. and V. Hall, 2001. *Flora Hibernica: The Wild Flowers, Plants and Trees of Ireland*. Collins Press, Cork, 203 pp.

Pirazzoli, P.A., H. Regnauld and L. Lemasson, 2004: Changes in storminess and surges in western France during the last century. *Mar. Geol.*, **210**, S307-S323.

Post, E. and M. Forchhammer, 2002: Synchronization of animal population dynamics by large-scale climate. *Nature*, **420**, 168-171.

Pounds, J.A., M.R. Bustamante, L.A. Coloma, J.A. Consuegra, M.P.L. Fogden, P.N. Foster, E. La Marca, K.L. Masters, A. Merino-Viteri, R. Puschendorf, S.R. Ron, G.A. Sanchez-Azofeifa, C.J. Still and B.E. Young, 2006: Widespread amphibian extinctions from epidemic disease driven by global warming. *Nature*, **439**, 161-167.

Pouyaud, B., M. Zapata, J. Yerren, J. Gomez, G. Rosas, W. Suarez and P. Ribstein, 2005: Avenir des ressources en eau glaciaire de la Cordillère Blanche. *Hydrolog. Sci. J.*, **50**, 999-1022.

Power, S., B. Sadler and N. Nicholls, 2005: The influence of climate science on water management in Western Australia: lessons for climate scientists. *B. Am. Meteorol. Soc.*, **86**, 839-844.

Pretlove, S.E.C. and T. Oreszczyn, 1998: Climate change: impact on the environmental design of buildings. *Building Services Engineering Research and Technology*, **19**, 55-58.

Prospero, J.M. and P.J. Lamb, 2003: African droughts and dust transport to the Caribbean, climate change implications. *Science*, **302**, 1024-27.

Prowse, T.D. and S. Beltaos, 2002: Climatic control of river-ice hydrology: a review. *Hydrol. Process.*, **16**, 805-822.

Psenner, R. and R. Schmidt, 1992: Climate-driven pH control of remote Alpine lakes and effects of acid deposition. *Nature*, **356**, 781-783.

Purse, B.V., P.S. Mellor, D.J. Rogers, A.R. Samuel, P.P. Mertens and M. Baylis, 2006: Climate change and the recent emergence of bluetongue in Europe. *Nat. Rev. Microbiol.*, **4**, 160.

Quinn, T.P. and D.J. Adams, 1996: Environmental changes affecting the migratory timing of American shad and sockeye salmon. *Ecology*, **77**, 1151-1162.

Raitsos, D.E., P.C. Reid, S.J. Lavender, M. Edwards and A.J. Richardson, 2005: Extending the SeaWiFS chlorophyll data set back 50 years in the northeast Atlantic. *Geophys. Res. Lett.*, **32**, L06603.

Ramirez, E., B. Francou, P. Ribstein, M. Descloitres, R. Guerin, J. Mendoza, R. Gallaire, B. Pouyaud and E. Jordan, 2001: Small glaciers disappearing in the tropical Andes: a case-study in Bolivia – Glaciar Chacaltaya (16°S). *J. Glaciol.*, **47**, 187-194.

Randolph, S.E., 2001: The shifting landscape of tick-borne zoonoses: tick-borne encephalitis and Lyme borreliosis in Europe. *Philos. T. Roy. Soc. B*, **356**, 1045-1056.

Rasmussen, A., 2002: The effects of climate change on the birch pollen season in Denmark. *Aerobiologia*, **18**, 253-265.

Reading, C.J., 2003: The effects of variation in climatic temperature (1980-2001) on breeding activity and tadpole stage duration in the common toad, *Bufo bufo*. *Sci. Total Environ.*, **310**, 231-236.

Regehr, E.V., S.C. Amstrup and I. Stirling, 2006: Polar bear population status in the southern Beaufort Sea. U.S. Geological Survey Open-File Report 2000-1337, 20 pp.

Regonda, S.K., B. Rajagopalan, M. Clark and J. Pitlick, 2005: Seasonal cycle shifts in hydroclimatology over the western United States. *J. Climate*, **18**, 372-384.

Reichert, B.K., L. Bengtsson and J. Oerlemans, 2002: Recent glacier retreat exceeds internal variability. *J. Climate*, **15**, 3069-3081.

Reid, P.C. and M. Edwards, 2001: Long-term changes in the pelagos, benthos and fisheries of the North Sea. *Burning Issues of North Sea Ecology*, I. Kröncke, M. Türkay and J. Sündermann, Eds., *Senckenbergiana Maritima*, **31**, 107-115.

Reid, P.C., M. Edwards, H.G. Hunt and A.J. Warner, 1998: Phytoplankton change in the North Atlantic. *Nature*, **391**, 546-546.

Restrepo, J.D., B. Kjerfve, I. Correa and J. Gonzalez, 2002: Morphodynamics of a high discharge tropical delta, San Juan River, Pacific coast of Colombia. *Mar. Geol.*, **192**, 355-381.

Richardson, S.D. and J.M. Reynolds, 2000. An overview of glacial hazards in the Himalayas. *Quatern. Int.*, **65/66**, 31-47.

Richardson, A.J. and D.S. Schoeman, 2004: Climate impact on plankton ecosystems in the Northeast Atlantic. *Science*, **305**, 1609-1612.

Rindfuss, R.R., S.J. Walsh, B.L. Turner, J. Fox and V. Mishra, 2004: Developing a science of land change: challenges and methodological issues. *P. Natl. Acad. Sci. USA*, **101**, 13976-13981.

Rivera, A., G. Casassa, R. Thomas, E. Rignot, R. Zamora, D. Antúnez, C. Acuña and F. Ordenes, 2005: Glacier wastage on Southern Adelaide Island and its impact on snow runway operations. *Ann. Glaciol.*, **41**, 57-62.

Robeson, S.M., 2002: Increasing growing-season length in Illinois during the 20th century. *Climatic Change*, **52**, 219-238.

Rodo, X., M. Pascual, G. Fuchs and A.S.G. Faruque, 2002: ENSO and cholera: a nonstationary link related to climate change? *P. Natl. Acad. Sci. USA*, **99**, 12901-12906.

Roemmich, D. and J. McGowan, 1995: Climatic warming and the decline of zooplankton in the California current. *Science*, **267**, 1324-1326.

Rogers, K., K.M. Wilton and N. Saintilan, 2006: Vegetation change and surface elevation dynamics in estuarine wetlands of southeast Australia. *Estuar. Coast. Shelf S.*, **66**, 559-569.

Rogora, M., R. Mosello and S. Arisci, 2003: The effect of climate warming on the hydrochemistry of Alpine lakes. *Water Air Soil Pollut.*, **148**, 347-361.

Ron, S.R., W.E. Duellman, L.A. Coloma and M.R. Bustamante, 2003: Population decline of the Jambato toad *Atelopus ignescens* (Anura: Bufonidae) in the Andes of Ecuador. *J. Herpetol.*, **37**, 116-126.

Root, T.L. and S.H. Schneider, 2002. Climate change: overview and implications for wildlife. *Wildlife Responses to Climate Change: North American Case Studies*, T.L. Root and S.H. Schneider, Eds., Island Press, Washington, District of Columbia, 1-56.

Root, T.L., J.T. Price, K.R. Hall, S.H. Schneider, C. Rosenzweig and J.A. Pounds, 2003: Fingerprints of global warming on wild animals and plants. *Nature*, **421**, 57-

60.

Root, T.L., D.P. MacMynowski, M.D. Mastrandrea and S.H. Schneider, 2005: Human-modified temperatures induce species changes: Joint attribution. *P. Natl. Acad. Sci. USA*, **102**, 7465-7469.

Rose, G.A. and R.L. O'Driscoll, 2002: Capelin are good for cod: can the northern stock rebuild without them? *ICES J. Mar. Sci*., **59**, 1018-1026.

Rose, J.B., D.E. Huffman and A. Gennaccaro, 2002: Risk and control of waterborne cryptosporidiosis. *FEMS Microbiol. Rev*., **26**, 113-123.

Ross, M.S., J.F. Meeder, J.P. Sah, P.L. Ruiz and G.J. Telesnicki, 2000: The south-east saline everglades revisited: 50 years of coastal vegetation change. *J. Veg. Sci*., **11**, 101-112.

Roy, D.B. and T.H. Sparks, 2000: Phenology of British butterflies and climate change. *Glob. Change Biol*., **6**, 407.

Royal Society, 2005: *Ocean Acidification due to Increasing Atmospheric Carbon Dioxide*. Royal Society, London, 68 pp.

Rudel, T.K., O.T. Coomes, E. Moran, F. Achard, A. Angelsen, J. Xu and E.F. Lambin, 2005: Forest transitions: towards a global understanding of land use change. *Global Environ. Chang*., **15**, 23-31.

Ruhland, K., A. Presnitz and J.P. Smol, 2003: Paleolimnological evidence from diatoms for recent environmental changes in 50 lakes across Canadian arctic treeline. *Arct. Antarct. Alp. Res*., **35**, 110-123.

Sagarin, R.D., J.P. Barry, S.E. Gilman and C.H. Baxter, 1999: Climate-related change in an intertidal community over short and long time scales. *Ecol. Monogr*., **69**, 465-490.

Saino, N., M. Romano, R. Ambrosini, R.P. Ferrari and A.P. Moller, 2004: Timing of reproduction and egg quality covary with temperature in the insectivorous barn swallow, *Hirundo rustica*. *Funct. Ecol*., **18**, 50-57.

Saintilan, N. and R.J. Williams, 1999: Mangrove transgression into saltmarsh environments in south-east Australia. *Global Ecol. Biogeogr*., **8**, 117-124.

Sanders, C.H. and M.C. Phillipson, 2003: UK adaptation strategy and technical measures: the impacts of climate change on buildings. *Build. Res. Inf*., **31**, 210-221.

Sanz, J., 2003: Large-scale effects of climate change on breeding parameters of pied flycatchers in Western Europe. *Ecography*, **26**, 45-50.

Sauphanor, B. and T. Boivin, 2004: Changement climatique et résistance du carpocapse aux insecticides. Journées MICCES 2004, 2 pp.

Schaber, J., 2002: Phenology in Germany in the 20th century: methods, analyses and models, Dissertation, Department of Geoecology, University of Potsdam, 164 pp.

Schär, C., P.L. Vidale, D. Luthi, C. Frei, C. Haberli, M.A. Liniger and C. Appenzeller, 2004: The role of increasing temperature variability in European summer heatwaves. *Nature*, **427**, 332-336.

Scheifinger, H., A. Menzel, E. Koch, C. Peter and R. Ahas, 2002: Atmospheric mechanisms governing the spatial and temporal variability of phenological phases in central Europe. *Int. J. Climatol*., **22**, 1739-1755.

Schindler, D.E., D.E. Rogers, M.D. Scheuerell and C.A. Abrey, 2005: Effects of changing climate on zooplankton and juvenile sockeye salmon growth in southwestern Alaska. *Ecology*, **86**, 198-209.

Schindler, D.E., C. Baldwin, J. Carter, J. Fox, T.B. Francis, S.E. Hampton, G. Holtgrieve, S.P. Johnson, J.W. Moore, W.J. Palen, C. Ruff, J.M. Scheuerell, H. Tallis and M. Winder, 2007: Climate change and responses of freshwater ecosystems. *BioScience*, submitted.

Schindler, D.W., K.G. Beaty, E.J. Fee, D.R. Cruikshank, E.R. Debruyn, D.L. Findlay, G.A. Linsey, J.A. Shearer, M.P. Stainton and M.A. Turner, 1990: Effects of climatic warming on lakes of the central Boreal Forest. *Science*, **250**, 967-970.

Schleip, C., A. Menzel, N. Estrella and V. Dose, 2006: The use of Bayesian analysis to detect recent changes in phenological events throughout the year. *Agr. Forest Meteorol*., **141**, 179-191.

Schwartz, M.D. and B.E. Reiter, 2000: Changes in North American spring. *Int. J. Climatol*., **20**, 929-932.

Schwartz, M.D. and X. Chen, 2002: Examining the onset of spring in China. *Climate Res*., **21**, 157-164.

Schwörer, D.A., 1997: Bergführer und Klimaänderung: eine Untersuchung im Berninagebiet über mögliche Auswirkungen einer Klimaänderung auf den Bergführerberuf (Mountain guides and climate change: an inquiry into possible effects of climatic change on the mountain guide trade in the Bernina region, Switzerland). Diplomarbeit der philosophisch-naturwissenschaftlichen Fakultät der Universität Bern.

Seguin, B., M. Domergue, I.G.D. Cortazar, N. Brisson and D. Ripoche, 2004: Le réchauffement climatique récent: impact sur les arbres fruitiers et la vigne. *Lett. PIGB-PMRC France Changement Global*, **16**, 50-54.

Sekercioglu, C.H., G.C. Daily and P.R. Ehrlich, 2004: Ecosystem consequences of bird declines. *P. Natl. Acad. Sci. USA*, **101**, 18042-18047.

Selvaraju, R., 2003: Impact of El Niño-Southern Oscillation on Indian foodgrain production. *Int. J. Climatol*., **23**, 187-206.

Semenov, S.M., V.V. Yasukevich and E.S. Gel'ver, 2006: *Identification of Climatogenic Changes*. Publishing Center, Meteorology and Hydrology, Moscow, 325 pp.

Serreze, M., D. Bromwich, M. Clark, A. Etringer, T. Zhang and R. Lammers, 2002: Large scale hydroclimatology of the terrestrial Arctic drainage system. *J. Geophys. Res.–Atmos*., **108**, 8160.

Shanks, G.D., S.I. Hay, D.I. Stern, K. Biomndo and R.W. Snow, 2002: Meteorologic influences on *Plasmodium falciparum* malaria in the highland tea estates of Kericho, western Kenya. *Emerg. Infect. Dis*., **8**, 1404-1408.

Shimaraev, M.N. and V.M. Domysheva, 2004: Climate and processes in Lake Baikal ecosystem at present. *The VIth All-Russia Hydrological Congress: Abstracts of Papers* 4, 287-288 (in Russian).

Shimoda, Y., 2003: Adaptation measures for climate change and the urban heat island in Japan's built environment. *Build. Res. Inf*., **31**, 222-230.

Simmons, A.D. and C.D. Thomas, 2004: Changes in dispersal during species' range expansions. *Am. Nat*., **164**, 378.

Singh, B. and A.E. Fouladi, 2003: Coastal erosion in Trinidad in the southern Caribbean: probable causes and solutions. *Coastal Engineering VI: Computer Modelling and Experimental Measurements of Seas and Coastal Regions*, C.A. Brebbia, D. Almorza and F. López-Aguayo, Eds., WIT Press, Southampton, 397-406.

Smith, L.C., 2000: Trends in Russian Arctic river-ice formation and breakup, 1917 to 1994. *Phys. Geogr*., **21**, 46-56.

Smith, L.C., Y. Cheng, G.M. MacDonald and L.D. Hinzman, 2005: Disappearing Arctic lakes. *Science*, **308**, 1429.

Smith, T.M. and R.W. Reynolds, 2005: A global merged land air and sea surface temperature reconstruction based on historical observations (1880-1997). *J. Climate*, **18**, 2021-2036.

Smith, V.R., 2002: Climate change in the sub-Antarctic: an illustration from Marion Island. *Climatic Change*, **52**, 345-357.

Smol, J.P., A.P. Wolfe, H.J.B. Birks, M.S.V. Douglas, V.J. Jones, A. Korhola, R. Pienitz, K. Ruhland, S. Sorvari, D. Antoniades, S.J. Brooks, M.A. Fallu, M. Hughes, B.E. Keatley, T.E. Laing, N. Michelutti, L. Nazarova, M. Nyman, A.M. Paterson, B. Perren, R. Quinlan, M. Rautio, E. Saulnier-Talbot, S. Siitoneni, N. Solovieva and J. Weckstrom, 2005: Climate-driven regime shifts in the biological communities of arctic lakes. *P. Natl. Acad. Sci. USA*, **102**, 4397-4402.

Sobrino, V.E., M.A. González, M. Sanz-Elorza, E. Dana, D. Sánchez-Mata and R. Gavilan, 2001: The expansion of thermophilic plants in the Iberian Peninsula as a sign of climate change. *"Fingerprints" of Climate Change: Adapted Behaviour and Shifting Species Ranges*, G.-R. Walther, C.A. Burga and P.J. Edwards, Eds., Kluwer Academic/Plenum, New York, 163-184.

Sokolov, L.V., M.Y. Markovets, A.P. Shapoval and Y.G. Morozov, 1998: Long-term trends in the timing of spring migration of passerines on the Courish spit of the Baltic Sea. *Avian Ecology and Behaviour*, **1**, 1-21.

Sommaruga-Wograth, S., K.A. Koinig, R. Schmidt, R. Sommaruga, R. Tessadri and R. Psenner, 1997: Temperature effects on the acidity of remote alpine lakes. *Nature*, **387**, 64-67.

Sorvari, S., A. Korhola and R. Thompson, 2002: Lake diatom response to recent Arctic warming in Finnish Lapland. *Glob. Change Biol*., **8**, 171-181.

Southward, A.J., O. Langmead, N.J. Hardman-Mountford, J. Aiken, G.T. Boalch, P.R. Dando, M.J. Genner, I. Joint, M.A. Kendall, N.C. Halliday, R.P. Harris, R. Leaper, N. Mieszkowska, R.D. Pingree, A.J. Richardson, D.W. Sims, T. Smith, A.W. Walne and S.J. Hawkins, 2005: Long-term oceanographic and ecological research in the western English Channel. *Adv. Mar. Biol*., **47**, 1-105.

Spagnoli, B., S. Planton, M. Deque, O. Mestre and J.M. Moisselin, 2002: Detecting climate change at a regional scale: the case of France. *Geophys. Res. Lett*., **29**, doi:10.1029/2001GL014619.

Sparks, T.H. and R.G. Loxton, Eds.,1999: Arrival date of the swallow. *Indicators of Climate Change in the UK*. Centre for Ecology and Hydrology, Huntington, 62-63.

Sparks, T.H. and O. Braslavska, 2001: The effect of temperature, altitude and latitude on the arrival and departure dates of the swallow *Hirundo rustica* in the Slovak Republic. *Int. J. Biometeorol*., **45**, 212-216.

Sparks, T.H. and A. Menzel, 2002: Observed changes in seasons: an overview. *Int. J. Climatol*., **22**, 1715.

Sparks, T.H., H. Heyen, O. Braslavska and E. Lehikoinen, 1999: Are European birds migrating earlier? *BTO News*, **223**, 8.

Sparks, T.H., E.P. Jeffree and C.E. Jeffree, 2000: An examination of the relationship between flowering times and temperature at the national scale using long-term phenological records from the UK. *Int. J. Biometeorol.*, **44**, 82-87.

Steemers, K., 2003: Cities, energy and comfort: a PLEA 2000 review. *Energ. Buildings*, **35**, 1-2.

Stefanescu, C., J. Peñuelas and I. Filella, 2003: Effects of climatic change on the phenology of butterflies in the northwest Mediterranean Basin. *Glob. Change Biol.*, **9**, 1494-1506.

Steininger, K.W. and H. Weck-Hannemann, Eds., 2002: *Global Environmental Change in Alpine Regions: Recognition, Impact, Adaptation and Mitigation*. Edward Elgar, Cheltenham, 296 pp.

Stenseth, N.C., A. Mysterud, G. Ottersen, J.W. Hurrell, K.S. Chan and M. Lima, 2002: Ecological effects of climate fluctuations. *Science*, **297**, 1292-1296.

Stervander, M., K. Lindstrom, N. Jonzen and A. Andersson, 2005: Timing of spring migration in birds: long-term trends, North Atlantic Oscillation and the significance of different migration routes. *J. Avian Biol.*, **36**, 210-221.

Stirling, I. and C.L. Parkinson, 2006: Possible effects of climate warming on selected populations of polar bears (*Ursus maritimus*) in the Canadian Arctic. *Arctic*, **59**, 261-275.

Stive, M.J.F., 2004: How important is global warming for coastal erosion? An editorial comment. *Climatic Change*, **64**, 27-39.

Stone, R.S., E.G., Dutton, J.M. Harris and D. Longenecker, 2002: Earlier spring snowmelt in northern Alaska as an indicator of climate change. *J. Geophys. Res.–Atmos.*, **107**, doi:10.1029/2000JD000286.

Straile, D. and R. Adrian, 2000: The North Atlantic Oscillation and plankton dynamics in two European lakes: two variations on a general theme. *Glob. Change Biol.*, **6**, 663-670.

Straile, D., K. Johns and H. Rossknecht, 2003: Complex effects of winter warming on the physicochemical characteristics of a deep lake. *Limnol. Oceanogr.*, **48**, 1432-1438.

Strode, P.K., 2003: Implications of climate change for North American wood warblers (Parulidae). *Glob. Change Biol.*, **9**, 1137-1144.

Sturm, M., C. Racine and K. Tape, 2001: Climate change: increasing shrub abundance in the Arctic. *Nature*, **411**, 546-547.

Su, Y.Z., H.L. Zhao, T.H. Zhang and X.Y. Zhao, 2004: Soil properties following cultivation and non-grazing of a semi-arid sandy grassland in northern China. *Soil Till. Res.*, **75**, 27-36.

Sutherst, R.W., 2004: Global change and human vulnerability to vector-borne diseases. *Clin. Microbiol. Rev.*, **17**, 136.

Svensson, C., J. Hannaford, Z.W. Kundzewicz and T.J. Marsh, 2006: Trends in river floods: why is there no clear signal in observations? *IAHS/UNESCO Kovacs Colloquium: Frontiers in Flood Research*, **305**, 1-18.

Swiss Reinsurance Company, 2005: Natural catastrophes and man-made disasters in 2004. Swiss Reinsurance Company. Economic Research and Consulting, Zürich, 40 pp.

Sydeman, W.J., M.M. Hester, J.A. Thayer, F. Gress, P. Martin and J. Buffa, 2001: Climate change, reproductive performance and diet composition of marine birds in the southern California Current system, 1969-1997. *Prog. Oceanogr.*, **49**, 309-329.

Syvitski, J.P.M., N. Harvey, E. Wolanski, W.C. Burnett, G.M.E. Perillo, V. Gornitz, 2005a: Dynamics of the coastal zone. *Coastal Fluxes in the Anthropocene*, C.J. Crossland, H.H. Kremer, H.J. Lindeboom, J.I.M. Crossland and M.D.A. Le Tissier, Eds., Springer, Berlin, 39-94.

Syvitski, J.P.M., C.J. Vorosmarty, A.J. Kettner and P. Green, 2005b: Impact of humans on the flux of terrestrial sediment to the global coastal ocean. *Science*, **308**, 376-380.

Tamis, W.L.M., M. Van't Zelfde and R. Van der Meijden, 2001: Changes in vascular plant biodiversity in the Netherlands in the 20th century explained by climatic and other environmental characteristics. *Long-term Effects of Climate Change on Biodiversity and Ecosystem Processes*, H. Van Oene, W.N. Ellis, M.M.P.D. Heijmans, D. Mauquoy, W.L.M. Tamis, F. Berendse, B. Van Geel, R. Van der Meijden and S.A. Ulenberg, Eds., NOP, Bilthoven, 23-51.

Tank, A.M.G.K. and G.P. Konnen, 2003: Trends in indices of daily temperature and precipitation extremes in Europe, 1946-99. *J. Climate*, **16**, 3665-1680.

Taylor, J.A., A.P. Murdock and N.I. Pontee, 2004: A macroscale analysis of coastal steepening around the coast of England and Wales. *Geogr. J.*, **170**, 179-188.

Teranishi, H., Y. Kenda, T. Katoh, M. Kasuya, E. Oura and H. Taira, 2000: Possible role of climate change in the pollen scatter of Japanese cedar *Cryptomeria japonica* in Japan. *Climate Res.*, **14**, 65-70.

Thampanya, U., J.E. Vermaat, S. Sinsakul and N. Papapitukkul, 2006: Coastal erosion and mangrove progradation of southern Thailand. *Estuar. Coast. Shelf Sci.*, **68**, 75-85.

Theurillat, J.P. and A. Guisan, 2001: Potential impact of climate change on vegetation in the European Alps: a review. *Climatic Change*, **50**, 77-109.

Thomas, C.D., 2005: Recent evolutionary effects of climate change. *Climate Change and Biodiversity*, T.E. Lovejoy and L. Hannah, Eds., Yale University Press, New Haven, Connecticut, 75-90.

Thomas, C.D., E.J. Bodsworth, R.J. Wilson, A.D. Simmons, Z.G. Davies, M. Musche and L. Conradt, 2001a: Ecological and evolutionary processes at expanding range margins. *Nature*, **411**, 577-581.

Thomas, C.D., A. Cameron, R.E. Green, M. Bakkenes, L.J. Beaumont, Y.C. Collingham, B.F.N. Erasmus, M.F. de Siqueira, A. Grainger, L. Hannah, L. Hughes, B. Huntley, A.S. van Jaarsveld, G.F. Midgley, L. Miles, M.A. Ortega-Huerta, A.T. Peterson, O.L. Phillips and S.E. Williams, 2004: Extinction risk from climate change. *Nature*, **427**, 145-148.

Thomas, C.D., A.M.A. Franco and J.K. Hill, 2006: Range retractions and extinction in the face of climate warming. *Trends Ecol. Evol.*, **21**, 415-416.

Thomas, D., H. Osbahr, C. Twyman, N. Adger and B. Hewitson, 2005: ADAPTIVE: Adaptations to climate change amongst natural resource-dependant societies in the developing world – across the Southern African climate gradient. Research Technical Report 35, Tyndall Centre for Climate Change, Norwich, 47 pp.

Thomas, D.S.G. and C. Twyman, 2005: Equity and justice in climate change adaptation amongst natural resource-dependant societies. *Global Environ. Chang.*, **15**, 115-124.

Thomas, D.W., J. Blondel, P. Perret, M.M. Lambrechts and J.R. Speakman, 2001b: Energetic and fitness costs of mismatching resource supply and demand in seasonally breeding birds. *Science*, **291**, 2598-2600.

Thompson, L.G., E. Mosley-Thompson, M.E. Davis, P.N. Lin, K. Henderson and T.A. Mashiotta, 2003: Tropical glacier and ice core evidence of climate change on annual to millennial time scales. *Climatic Change*, **59**, 137-155.

Tidemann, C.R., M.J. Vardon, R.A. Loughland and P.J. Brocklehurst, 1999: Dry season camps of flying-foxes (*Pteropus* spp.) in Kakadu World Heritage Area, north Australia. *J. Zool.*, **247**, 155-163.

Tillman, D., J. Fargione, B. Wolff, C. D'Antonio, A. Dobson, R. Howarth, D. Schindler, W. Schlesinger, D. Simberloff and D. Swackhamer, 2001: Forecasting agriculturally driven global environmental change. *Science*, **292**, 281-284.

Trenberth, K.E., P.D. Jones, P.G. Ambenje, R. Bojariu, D.R. Easterling, A.M.G. Klein Tank, D.E. Parker, J.A. Renwick, F. Rahimzadeh, M.M. Rusticucci, B.J. Soden and P.-M. Zhai, 2007: Observations: surface and atmospheric climate change. *Climate Change 2007: The Physical Science Basis. Contribution of Working Group I to the Fourth Assessment Report of the Intergovernmental Panel on Climate Change*, S. Solomon, D. Qin, M. Manning, Z. Chen, M. Marquis, K.B. Averyt, M. Tignor and H.L. Miller, Eds., Cambridge University Press, Cambridge, 235-336.

Trigo, R.M., R. Garcia-Herrera, J. Diaz, I.F. Trigo and M.A. Valente, 2005: How exceptional was the early August 2003 heatwave in France? *Geophys. Res. Lett.*, **32**, L10701, doi:10.1029/2005GL022410.

Tryjanowski, P., 2002: A long-term comparison of laying date and clutch size in the red-backed shrike (*Lanius collurio*) in Silesia, southern Poland. *Acta Zool. Acad. Sci. H.*, **48**, 101-106.

Tryjanowski, P., M. Rybacki and T. Sparks, 2003: Changes in the first spawning dates of common frogs and common toads in western Poland in 1978-2002. *Ann. Zool. Fenn.*, **40**, 459-464.

Tryjanowski, P., T.H. Sparks, J. Ptaszyk and J. Kosicki, 2004: Do white storks *Ciconia ciconia* always profit from an early return to their breeding grounds? *Bird Study*, **51**, 222-227.

Tryjanowski, P., T.H. Sparks and P. Profus, 2005: Uphill shifts in the distribution of the white stork *Ciconia ciconia* in southern Poland: the importance of nest quality. *Diversity and Distributions*, **11**, 219.

Tulu, A.N., 1996: Determinants of malaria transmission in the highlands of Ethiopia: the impact of global warming on morbidity and mortality ascribed to malaria. PhD thesis, University of London, London, 301 pp.

Turalioglu, F.S., A. Nuhoglu and H. Bayraktar, 2005: Impacts of some meteorological parameters on SO_2 and TSP concentrations in Erzurum, Turkey. *Chemosphere*, **59**, 1633-1642.

Umina, P.A., A.R. Weeks, M.R. Kearney, S.W. McKechnie and A.A. Hoffmann, 2005: A rapid shift in a classic clinal pattern in *Drosophila* reflecting climate change. *Science*, **308**, 691-693.

United Nations Development Programme, 2004: *Reducing Disaster Risk: A Challenge for Development*. United Nations Development Programme, New York, 161 pp, http://www.undp.org/bcpr/disred/rdr.htm.

USGS, 2006: Land area changes in coastal Louisiana after the 2005 hurricanes. USGS Open-File Report 2006-1274, 2 pp.

Valiela, I., J.L. Bowen and J.K. York, 2001: Mangrove forests: one of the world's threatened major tropical environments. *BioScience*, **51**, 807-815.

van der Wal, D. and K. Pye, 2004: Patterns, rates and possible causes of saltmarsh erosion in the Greater Thames area (UK). *Geomorphology*, **61**, 373-391.

Van Duivenbooden, N., S. Abdoussalam and A.B. Mohamed, 2002: Impact of climate change on agricultural production in the Sahel. Part 2. Case study for groundnut and cowpea in Niger. *Climatic Change*, **54**, 349-368.

van Noordwijk, A.J., R.H. McCleery and C.M. Perrins, 1995: Selection for the timing of great tit breeding in relation to caterpillar growth and temperature. *J. Anim. Ecol.*, **64**, 451-458.

Van Vliet, A.J.H., A. Overeem, R.S. De Groot, A.F.G. Jacobs and F.T.M. Spieksma, 2002: The influence of temperature and climate change on the timing of pollen release in the Netherlands. *Int. J. Climatol.*, **22**, 1757-1767.

Vare, H., R. Lampinen, C. Humphries and P. Williams, 2003: Taxonomic diversity of vascular plants in the European alpine areas. *Alpine Biodiversity in Europe: A Europe-wide Assessment of Biological Richness and Change*, L. Nagy, G. Grabherr, C. Körner and D.B.A. Thompson, Ecological Studies 167, Springer, Heidelberg, 133-148.

Verburg, P., R.E. Hecky and H. Kling, 2003: Ecological consequences of a century of warming in Lake Tanganyika. *Science*, **301**, 505-507.

Verheye, H.M., A.J. Richardson, L. Hutchings, G. Marska and D. Gianakouras, 1998: Long-term trends in the abundance and community structure of coastal zooplankton in the southern Benguela system, 1951-1996. *S. Afr. J. Mar. Sci.*, **19**, 317-332.

Vesely, J., V. Majer, J. Kopacek and S.A. Norton, 2003: Increasing temperature decreases aluminium concentrations in Central European lakes recovering from acidification. *Limnol. Oceanogr.*, **48**, 2346-2354.

Vincent, W.F., J.A.E. Gibson and M.O. Jeffries, 2001: Ice-shelf collapse, climate change and habitat loss in the Canadian high Arctic. *Polar Rec.*, **37**, 133-142.

Vollmer, M.K., H.A. Bootsma, R.E. Hecky, G. Patterson, J.D. Halfman, J.M. Edmond, D.H. Eccles and R.F. Weiss, 2005: Deep-water warming trend in Lake Malawi, East Africa. *Limnol. Oceanogr.*, **50**, 727-732.

Vuille, M., R.S. Bradley, M. Werner and F. Keimig, 2003: 20th century climate change in the tropical Andes: observations and model results. *Climatic Change*, **59**, 75-99.

Wall, G., 1998: Implications of global climate change for tourism and recreation in wetland areas. *Climatic Change*, **40**, 371-389.

Walther, G.-R., 2000: Climatic forcing on the dispersal of exotic species. *Phytocoenologia*, **30**, 409.

Walther, G.-R., 2004: Plants in a warmer world. *Perspect. Plant Ecol.*, **6**, 169-185.

Walther, G.-R., E. Post, P. Convey, A. Menzel, C. Parmesan, T.J.C. Beebee, J.M. Fromentin, O. Hoegh-Guldberg and F. Bairlein, 2002: Ecological responses to recent climate change. *Nature*, **416**, 389-395.

Walther, G.-R., S. Beissner and C.A. Burga, 2005a: Trends in the upward shift of alpine plants. *J. Veg. Sci.*, **16**, 541-548.

Walther, G.-R., S. Berger and M.T. Sykes, 2005b: An ecological 'footprint' of climate change. *P. Roy. Soc.Lond. B Bio.*, **272**, 1427-1432.

Warren, M.S., J.K. Hill, J.A. Thomas, J. Asher, R. Fox, B. Huntley, D.B. Roy, M.G. Telfer, S. Jeffcoate, P. Harding, G. Jeffcoate, S.G. Willis, J.N. Greatorex-Davies, D. Moss and C.D. Thomas, 2001: Rapid responses of British butterflies to opposing forces of climate and habitat change. *Nature*, **414**, 65-69.

Watson, R.T. and W. Haeberli, 2004: Environmental threats, mitigation strategies and high mountain areas. *Mountain Areas: A Global Resource. Ambio*, **13**, 2-10.

Webster, P.J., G.J. Holland, J.A. Curry and H.R. Chang, 2005: Changes in tropical cyclone number, duration and intensity in a warming environment. *Science*, **309**, 1844-1846.

Weimerskirch, H., P. Inchausti, C. Guinet and C. Barbraud 2003: Trends in bird and seal populations as indicators of a system shift in the Southern Ocean. *Antarct. Sci.*, **15**, 249-256.

Weissflog, L., N. Elansky, E. Putz, G. Krueger, C.A. Lange, L. Lisitzina and A. Pfennigsdorff, 2004: Trichloroacetic acid in the vegetation of polluted and remote areas of both hemispheres. Part II. Salt lakes as novel sources of natural chlorohydrocarbons. *Atmos, Environ,*, **38**, 4197-4204.

Weladji, R. and Ø. Holand, 2003: Global climate change and reindeer: effects of winter weather on the autumn weight and growth of calves. *Oecologia*, **136**, 317-323.

Werrity, A., A. Black, R. Duck, B. Finlinson, N. Thurston, S. Shackley and D. Crichton 2002: *Climate Change: Flooding Occurrences Review*. Scottish Executive Central Research Unit, Edinburgh, 94 pp.

West, C.C. and M.J. Gawith, Eds., 2005: *Measuring Progress: Preparing for Climate Change Through UK Climate Impacts Programme*. UKCIP, Oxford, 72 pp.

Westerling, A.L., H.G. Hidalgo, D.R. Cayan and T.W. Swetnam, 2006: Warming and earlier spring increases Western U.S. forest fire activity. *Science*, **313**, 940-943.

Weyhenmeyer, G.A., 2001: Warmer winters: are planktonic algal populations in Sweden's largest lakes affected? *Ambio*, **30**, 565-571.

Weyhenmeyer, G.A., T. Blenckner and K. Pettersson, 1999: Changes of the plankton spring outburst related to the North Atlantic Oscillation. *Limnol. Oceanogr.*, **44**, 1788-1792.

WHO, 2003: Phenology and human health: allergic disorders. Report on a WHO meeting, Rome. World Health Organization, Rome, 64 pp.

Wilkinson, J.W., 2004. *Status of Coral Reefs of the World*. Australian Institute of Marine Science, Townsville, 580 pp.

Williams, D.W. and A.M. Liebhold, 2002: Climate change and the outbreak range of two North American bark beetles. *Agr. Forest Entomol.*, **4**, 87-99.

Williams, S.E., E.E. Bolitho and S. Fox, 2003: Climate change in Australian tropical rainforests: an impending environmental catastrophe. *P. Roy. Soc. Lond. B Bio.*, **270**, 1887-1892.

Wilmking, M., G.P. Juday, V.A. Barber and H.S.J. Zald, 2004: Recent climate warming forces contrasting growth responses of white spruce at treeline in Alaska through temperature thresholds. *Glob. Change Biol.*, **10**, 1724-1736.

Wilson, R.J., C.D. Thomas, R. Fox, D.B. Roy and W.E. Kunin, 2004: Spatial patterns in species distributions reveal biodiversity change. *Nature*, **432**, 393-396.

Wilson, R.J., D. Gutierrez, J. Gutierrez, D. Martinez, R. Agudo and V.J. Monserrat, 2005: Changes to the elevational limits and extent of species ranges associated with climate change. *Ecol. Lett.*, **8**, 1346-1346.

Winder, M. and D.E. Schindler, 2004a: Climate change uncouples trophic interactions in an aquatic system. *Ecology*, **85**, 3178-3178.

Winder, M. and D.E. Schindler, 2004b: Climatic effects on the phenology of lake processes. *Glob. Change Biol.*, **10**, 1844-1856.

Wolanski, E., S. Spagnol and E.B. Lim, 2002. Fine sediment dynamics in the mangrove-fringed muddy coastal zone. *Muddy Coasts of the World: Processes, Deposits* and *Function*, T. Healy, Y. Wang and J.A. Healy, Eds., Elsevier Science, Amsterdam, 279-292.

Wolfe, A.P. and B.B. Perren, 2001: Chrysophyte microfossils record marked responses to recent environmental changes in high- and mid-arctic lakes. *Can. J. Bot.*, **79**, 747-752.

Wolfe, D.W., M.D. Schwartz, A.N. Lakso, Y. Otsuki, R.M. Pool and N.J. Shaulis, 2005: Climate change and shifts in spring phenology of three horticultural woody perennials in northeastern USA. *Int. J. Biometeorol.*, **49**, 303-309.

Wolters, M., J.P. Bakker, M.D. Bertness, R.L. Jefferies and I. Moller, 2005: Saltmarsh erosion and restoration in south-east England: squeezing the evidence requires realignment. *J. Appl. Ecol.*, **42**, 844-851.

Wong, P.P., 2003: Where have all the beaches gone? Coastal erosion in the tropics. *Singapore. J. Trop. Geo.*, **24**, 111-132.

Woodworth, P.L. and D.L. Blackman, 2004: Evidence for systematic changes in extreme high waters since the mid-1970s. *J. Climate*, **17**, 1190-1197.

Woolf, D.K., P.G. Challenor and P.D. Cotton, 2002: Variability and predictability of the North Atlantic wave climate. *J. Geophys. Res.–Oceans*, **107**, 3145, doi:10.1029/2001JC001124.

Worrall, F., T. Burt and R. Shedden, 2003: Long term records of riverine dissolved organic matter. *Biogeochemistry*, **64**, 165-178.

Yamada, T., 1998: Glacier lake and its outburst flood in the Nepal Himalaya. *Japanese Society of Snow and Ice, Tokyo*, **1**, 96.

Yang, D.Q., D.L. Kane, L.D. Hinzman, X.B. Zhang, T.J. Zhang and H.C. Ye, 2002: Siberian Lena River hydrologic regime and recent change. *J. Geophys. Res.–Atmos.*, **107**(D23), 4694, doi:10.1029/2002JD002542.

Yang, S.L., J. Zhang, J. Zhu, J.P. Smith, S.B. Dai, A. Gao and P. Li, 2005: Impact of dams on Yangtze River sediment supply to the sea and delta intertidal wetland response. *J. Geophys. Res.-Earth*, **110**, F03006, doi:10.1029/2004JF000271.

Ye, H.C. and M. Ellison, 2003: Changes in transitional snowfall season length in northern Eurasia. *Geophys. Res. Lett.*, **30**, 1252, doi:10.1029/2003GL016873.

Yom-Tov, Y., 2003: Body sizes of carnivores commensal with humans have increased over the past 50 years. *Funct. Ecol.*, **17**, 323-327.

Yom-Tov, Y. and S. Yom-Tov, 2004: Climatic change and body size in two species of Japanese rodents. *Biol. J. Linn. Soc.*, **82**, 263-267.

Yoshikawa, K. and L.D. Hinzman, 2003: Shrinking thermokarst ponds and groundwater dynamics in discontinuous permafrost near Council, Alaska. *Permafrost Periglac.*, **14**, 151-160.

Zhang, K.Q., B.C. Douglas and S.P. Leatherman, 2000: Twentieth-century storm activity along the US east coast. *J. Climate*, **13**, 1748-1761.

Zhang, K.Q., B.C. Douglas and S.P. Leatherman, 2004: Global warming and coastal erosion. *Climatic Change*, **64**, 41-58.

Zhang, X.B., K.D. Harvey, W.D. Hogg and T.R. Yuzyk, 2001: Trends in Canadian streamflow. *Water Resour. Res.*, **37**, 987-998.

Zhang, X.B., F.W. Zwiers and P.A. Stott, 2006: Multimodel multisignal climate change detection at regional scale. *J. Climate*, **19**, 4294-4307.

Zheng, J.Y., Q.S. Ge and Z.X. Hao, 2002: Impacts of climate warming on plants phenophases in China for the last 40 years. *Chinese Sci. Bull.*, **47**, 1826-1831.

Zhou, G., N. Minakawa, A. Githeko and G. Yan, 2004: Association between climate variability and malaria epidemics in the East African highlands. *P. Natl. Acad. Sci. USA*, **24**, 2375-2380.

Zhou, L., C.J. Tucker, R.K. Kaufmann, D. Slayback, N.V. Shabanov and R.B. Myneni, 2001: Variations in northern vegetation activity inferred from satellite data of vegetation index during 1981 to 1999. *J. Geophys. Res.–Atmos.*, **106**, 20069-20083.

Zhou, L., R.K. Kaufmann, Y. Tian, R.B. Myneni and C.J. Tucker, 2003: Relation between interannual variations in satellite measures of northern forest greenness and climate between 1982 and 1999. *J. Geophys. Res.*, **108**, 4004, doi:10.1029/2002JD002510.

Zimov, S.A., E.A. Schuur and F.S. Chapin III, 2006: Permafrost and the global carbon budget. *Science*, **312**, 1612-1613.

2

New assessment methods and the characterisation of future conditions

Coordinating Lead Authors:
Timothy R. Carter (Finland), Roger N. Jones (Australia), Xianfu Lu (UNDP/China)

Lead Authors:
Suruchi Bhadwal (India), Cecilia Conde (Mexico), Linda O. Mearns (USA), Brian C. O'Neill (IIASA/USA), Mark D.A. Rounsevell (Belgium), Monika B. Zurek (FAO/Germany)

Contributing Authors:
Jacqueline de Chazal (Belgium), Stéphane Hallegatte (France), Milind Kandlikar (Canada), Malte Meinshausen (USA/Germany), Robert Nicholls (UK), Michael Oppenheimer (USA), Anthony Patt (IIASA/USA), Sarah Raper (UK), Kimmo Ruosteenoja (Finland), Claudia Tebaldi (USA), Detlef van Vuuren (The Netherlands)

Review Editors:
Hans-Martin Füssel (Germany), Geoff Love (Australia), Roger Street (UK)

This chapter should be cited as:
Carter, T.R., R.N. Jones, X. Lu, S. Bhadwal, C. Conde, L.O. Mearns, B.C. O'Neill, M.D.A. Rounsevell and M.B. Zurek, 2007: New Assessment Methods and the Characterisation of Future Conditions. *Climate Change 2007: Impacts, Adaptation and Vulnerability. Contribution of Working Group II to the Fourth Assessment Report of the Intergovernmental Panel on Climate Change,* M.L. Parry, O.F. Canziani, J.P. Palutikof, P.J. van der Linden and C.E. Hanson, Eds., Cambridge University Press, Cambridge, UK, 133-171.

Table of Contents

Executive summary

This chapter describes the significant developments in methods and approaches for climate change impact, adaptation and vulnerability (CCIAV) assessment since the Third Assessment Report (TAR). It also introduces some of the scenarios and approaches to scenario construction that are used to characterise future conditions in the studies reported in this volume.

The growth of different approaches to assessing CCIAV has been driven by the need for improved decision analysis.

The recognition that a changing climate must be adapted to has increased the demand for policy-relevant information. The standard climate scenario-driven approach is used in a large proportion of assessments described in this report, but the use of other approaches is increasing. They include assessments of current and future adaptations to climate, adaptive capacity, social vulnerability, multiple stresses, and adaptation in the context of sustainable development. [2.2.1]

Risk management is a useful framework for decision-making and its use is expanding rapidly.

The advantages of risk-management methods include the use of formalised methods to manage uncertainty, stakeholder involvement, use of methods for evaluating policy options without being policy prescriptive, integration of different disciplinary approaches, and mainstreaming of climate change concerns into the broader decision-making context. [2.2.6]

Stakeholders bring vital inputs into CCIAV assessments about a range of risks and their management.

In particular, how a group or system can cope with current climate risks provides a solid basis for assessments of future risks. An increasing number of assessments involve, or are conducted by, stakeholders. This establishes credibility and helps to confer 'ownership' of the results, which is a prerequisite for effective risk management. [2.3.2]

The impacts of climate change can be strongly modified by non-climate factors.

Many new studies have applied socio-economic, land-use and technology scenarios at a regional scale derived from the global scenarios developed in the IPCC Special Report on Emissions Scenarios (SRES). Large differences in regional population, income and technological development implied under alternative SRES storylines can produce sharp contrasts in exposure to climate change and in adaptive capacity and vulnerability. Therefore, it is best not to rely on a single characterisation of future conditions. [2.4.6.4, 2.4.6.5]

Scenario information is increasingly being developed at a finer geographical resolution for use in CCIAV studies.

A range of downscaling methods have been applied to the SRES storylines, producing new regional scenarios of socio-economic conditions, land use and land cover, atmospheric composition, climate and sea level. Regionalisation methods are increasingly being used to develop high spatial-resolution climate scenarios based on coupled atmosphere-ocean general circulation model (AOGCM) projections. [2.4.6.1 to 2.4.6.5]

Characterisations of the future used in CCIAV studies are evolving to include mitigation scenarios, large-scale singularities, and probabilistic futures.

CCIAV studies assuming mitigated or stabilised futures are beginning to assess the benefits (through impacts ameliorated or avoided) of climate policy decisions. Characterisations of large-scale singularities have been used to assess their potentially severe biophysical and socio-economic consequences. Probabilistic characterisations of future socio-economic and climate conditions are increasingly becoming available, and probabilities of exceeding predefined thresholds of impact have been more widely estimated. [2.4.6.8, 2.4.7, 2.4.8]

2.1 Introduction

Assessments of climate change impacts, adaptation and vulnerability (CCIAV) are undertaken to inform decision-making in an environment of uncertainty. The demand for such assessments has grown significantly since the release of the IPCC Third Assessment Report (TAR), motivating researchers to expand the ranges of approaches and methods in use, and of the characterisations of future conditions (scenarios and allied products) required by those methods. This chapter describes these developments as well as illustrating the main approaches used to characterise future conditions in the studies reported in this volume.

In previous years, IPCC Working Group II[1] has devoted a Special Report and two chapters to assessment methods (IPCC, 1994; Carter et al., 1996; Ahmad et al., 2001). Moreover, the TAR also presented two chapters on the topic of scenarios (Carter et al., 2001; Mearns et al., 2001), which built on earlier descriptions of climate scenario development (IPCC-TGCIA, 1999). These contributions provide detailed descriptions of assessment methods and scenarios, which are not repeated in the current assessment.

In this chapter, an approach is defined as the overall scope and direction of an assessment and can accommodate a variety of different methods. A method is a systematic process of analysis. We identify five approaches to CCIAV in this chapter. Four are conventional research approaches: impact assessment, adaptation assessment, vulnerability assessment, and integrated assessment. The fifth approach, risk management, has emerged as CCIAV studies have begun to be taken up in mainstream policy-making.

Section 2.2 describes developments in the major approaches to CCIAV assessment, followed in Section 2.3 by discussion of a range of new and improved methods that have been applied since the TAR. The critical issue of data needs for assessment is

[1] Hereafter, IPCC Working Groups I, II, and III are referred to as WG I, WG II, and WG III, respectively.

treated at the end of this section. Most CCIAV approaches have a scenario component, so recent advances in methods of characterising future conditions are treated in Section 2.4. Since many recent studies evaluated in this volume use scenarios based on the IPCC Special Report on Emissions Scenarios (SRES; Nakićenović et al., 2000) and derivative studies, boxed examples are presented to illustrate some of these. Finally, in Section 2.5, we summarise the key new findings in the chapter and recommend future research directions required to address major scientific, technical and information deficiencies.

2.2 New developments in approaches

2.2.1 Frameworks for CCIAV assessment

Although the following approaches and methods were all described in the TAR (Ahmad et al., 2001), their range of application in assessments has since been significantly expanded. Factors that distinguish a particular approach include the purpose of an assessment, its focus, the methods available, and how uncertainty is managed. A major aim of CCIAV assessment approaches is to manage, rather than overcome, uncertainty (Schneider and Kuntz-Duriseti, 2002), and each approach has its strengths and weaknesses in that regard. Another important trend has been the move from research-driven agendas to assessments tailored towards decision-making, where decision-makers and stakeholders either participate in or drive the assessment (Wilby et al., 2004a; UNDP, 2005).

The standard approach to assessment has been the climate scenario-driven 'impact approach', developed from the seven-step assessment framework of IPCC (1994).[2] This approach, which dominated the CCIAV literature described in previous IPCC reports, aims to evaluate the likely impacts of climate change under a given scenario and to assess the need for adaptation and/or mitigation to reduce any resulting vulnerability to climate risks. A large number of assessments in this report also follow that structure.

The other approaches discussed are adaptation- and vulnerability-based approaches, integrated assessment, and risk management. All are well represented in conventional environmental research, but they are increasingly being incorporated into mainstream approaches to decision-making, requiring a wider range of methods to fulfil objectives such as (SBI, 2001; COP, 2005):

- assessing current vulnerabilities and experience in adaptation,
- stakeholder involvement in dealing with extreme events,
- capacity-building needs for future vulnerability and adaptation assessments,
- potential adaptation measures,
- prioritisation and costing of adaptation measures,
- interrelationships between vulnerability and adaptation assessments,
- national development priorities and actions to integrate adaptation options into existing or future sustainable development plans.

The adaptation-based approach focuses on risk management by examining the adaptive capacity and adaptation measures required to improve the resilience or robustness of a system exposed to climate change (Smit and Wandel, 2006). In contrast, the vulnerability-based approach focuses on the risks themselves by concentrating on the propensity to be harmed, then seeking to maximise potential benefits and minimise or reverse potential losses (Adger, 2006). However, these approaches are interrelated, especially with regard to adaptive capacity (O'Brien et al., 2006). Integrated approaches include integrated assessment modelling and other procedures for investigating CCIAV across disciplines, sectors and scales, and representing key interactions and feedbacks (e.g., Toth et al., 2003a, b). Risk-management approaches focus directly on decision-making and offer a useful framework for considering the different research approaches and methods described in this chapter as well as confronting, head on, the treatment of uncertainty, which is pervasive in CCIAV assessment. Risk-management and integrated assessment approaches can also be linked directly to mitigation analysis (Nakićenović et al., 2007) and to the joint assessment of adaptation and mitigation (see Chapter 18).

Two common terms used to describe assessment types are 'top-down' and 'bottom-up', which can variously describe the approach to scale, to subject matter (e.g., from stress to impact to response; from physical to socio-economic disciplines) and to policy (e.g., national versus local); sometimes mixing two or more of these (Dessai et al., 2004; see also Table 2.1). The standard impact approach is often described as top-down because it combines scenarios downscaled from global climate models to the local scale (see Section 2.4.6) with a sequence of analytical steps that begin with the climate system and move through biophysical impacts towards socio-economic assessment. Bottom-up approaches are those that commence at the local scale by addressing socio-economic responses to climate, which tend to be location-specific (Dessai and Hulme, 2004). Adaptation assessment and vulnerability assessment are usually categorised as bottom-up approaches. However, assessments have become increasingly complex, often combining elements of top-down and bottom-up approaches (e.g., Dessai et al., 2005a) and decision-making will utilise both (Kates and Wilbanks, 2003; McKenzie Hedger et al., 2006). The United Nations Development Programme's Adaptation Policy Framework (UNDP APF: see UNDP, 2005) has also identified a policy-based approach, which assesses current policy and plans for their effectiveness under climate change within a risk-management framework.

2.2.2 Advances in impact assessment

Application of the standard IPCC impact approach has expanded significantly since the TAR. The importance of providing a socio-economic and technological context for characterising future climate conditions has been emphasised,

[2] The seven steps are: 1. Define problem, 2. Select method, 3. Test method/sensitivity, 4. Select scenarios, 5. Assess biophysical/socio-economic impacts, 6. Assess autonomous adjustments, 7. Evaluate adaptation strategies.

Table 2.1. *Some characteristics of different approaches to CCIAV assessment. Note that vulnerability and adaptation-based approaches are highly complementary.*

	Approach			
	Impact	**Vulnerability**	**Adaptation**	**Integrated**
Scientific objectives	Impacts and risks under future climate	Processes affecting vulnerability to climate change	Processes affecting adaptation and adaptive capacity	Interactions and feedbacks between multiple drivers and impacts
Practical aims	Actions to reduce risks	Actions to reduce vulnerability	Actions to improve adaptation	Global policy options and costs
Research methods	Standard approach to CCIAV Drivers-pressure-state-impact-response (DPSIR) methods Hazard-driven risk assessment	Vulnerability indicators and profiles Past and present climate risks Livelihood analysis Agent-based methods Narrative methods Risk perception including critical thresholds Development/sustainability policy performance Relationship of adaptive capacity to sustainable development		Integrated assessment modelling Cross-sectoral interactions Integration of climate with other drivers Stakeholder discussions Linking models across types and scales Combining assessment approaches/methods
Spatial domains	Top-down Global → Local	Bottom-up Local → Regional (macro-economic approaches are top-down)		Linking scales Commonly global/regional Often grid-based
Scenario types	Exploratory scenarios of climate and other factors (e.g., SRES) Normative scenarios (e.g., stabilisation)	Socio-economic conditions Scenarios or inverse methods	Baseline adaptation Adaptation analogues from history, other locations, other activities	Exploratory scenarios: exogenous and often endogenous (including feedbacks) Normative pathways
Motivation	Research-driven	Research-/stakeholder-driven	Stakeholder-/research-driven	Research-/stakeholder-driven

and scenarios assuming no climate policy to restrict greenhouse gas (GHG) emissions have been contrasted with those assuming GHG stabilisation (e.g., Parry et al., 2001; see also Sections 2.4.6.4 and 2.4.6.8). The use of probabilities in impact assessments, presented as proof-of-concept examples in the TAR (Mearns et al., 2001), is now more firmly established (see examples in Section 2.4.8). Some other notable advances in impact assessment include: a reassessment of bioclimatic niche-based modelling, meta-analyses summarising a range of assessments, and new dynamic methods of analysing economic damages. Nevertheless, the climate-sensitive resources of many regions and sectors, especially in developing countries, have not yet been subject to detailed impact assessments.

Recent observational evidence of climatic warming, along with the availability of digital species distribution maps and greatly extended computer power has emboldened a new generation of bioclimatic niche-based modellers to predict changes in species distribution and prevalence under a warming climate using correlative methods (e.g., Bakkenes et al., 2002; Thomas et al., 2004; see also Chapter 4, Section 4.4.11). However, the application of alternative statistical techniques to the same data sets has also exposed significant variations in model performance that have recently been the subject of intensive debate (Pearson and Dawson, 2003; Thuiller et al., 2004; Luoto et al., 2005; Araújo and Rahbek, 2006) and should promote a more cautious application of these models for projecting future biodiversity.

A global-scale, meta-analysis of a range of studies for different sectors was conducted by Hitz and Smith (2004) to evaluate the aggregate impacts at different levels of global mean temperature. For some sectors and regions, such as agriculture and the coastal zone, sufficient information was available to summarise aggregated sectoral impacts as a function of global warming. For other sectors, such as marine biodiversity and energy, limited information allowed only broad conclusions of low confidence.

Dynamic methods are superseding statistical methods in some economic assessments. Recent studies account, for example, for the role of world markets in influencing climate change impacts on global agriculture (Fischer et al., 2002), the effect on damage from sea-level rise when assuming optimal adaptation measures (Neumann et al., 2000; Nicholls and Tol, 2006), the added costs for adapting to high temperatures due to uncertainties in projected climate (Hallegatte et al., 2007), and increasing long-term costs of natural disasters when explicitly accounting for altered extreme event distributions (Hallegatte et al., 2006). The role of economic dynamics has also been emphasised (Fankhauser and Tol, 2005; Hallegatte, 2005; Hallegatte et al., 2006). Some new studies suggest damage overestimations by previous assessments, while others suggest underestimations, leading to the conclusion that uncertainty is likely to be larger than suggested by the range of previous estimates.

2.2.3 Advances in adaptation assessment

Significant advances in adaptation assessment have occurred, shifting its emphasis from a research-driven activity to one where stakeholders participate in order to improve decision-making. The key advance is the incorporation of adaptation to past and present climate. This has the advantage of anchoring the assessment in what is already known, and can be used to explore adaptation to climate variability and extremes, especially if scenarios of future variability are uncertain or unavailable (Mirza, 2003b; UNDP, 2005). As such, adaptation assessment has accommodated a wide range of methods used in mainstream policy and planning. Chapter 17 of this volume discusses adaptation practices, the processes and determinants of adaptive capacity, and limits to adaptation, highlighting the difficulty of establishing a general methodology for adaptation assessment due to the great diversity of analytical methods employed. These include the following approaches and methods.

- The scenario-based approach (e.g., IPCC, 1994; see also Section 2.2.1), where most impact assessments consider future adaptation as an output.
- Normative policy frameworks, exploring which adaptations are socially and environmentally beneficial, and applying diverse methods, such as vulnerability analysis, scenarios, cost-benefit analysis, multi-criteria analysis and technology risk assessments (UNDP, 2005).
- Indicators, employing models of specific hypothesised components of adaptive capacity (e.g., Moss et al., 2001; Yohe and Tol, 2002; Brooks et al., 2005; Haddad, 2005).
- Economic modelling, anthropological and sociological methods for identifying learning in individuals and organisations (Patt and Gwata, 2002; Tompkins, 2005; Berkhout et al., 2006).
- Scenarios and technology assessments, for exploring what kinds of adaptation are likely in the future (Dessai and Hulme, 2004; Dessai et al., 2005a; Klein et al., 2005).
- Risk assessments combining current risks to climate variability and extremes with projected future changes, utilising cost-benefit analysis to assess adaptation (e.g., ADB, 2005).

Guidance regarding methods and tools to use in prioritising adaptation options include the Compendium of Decision Tools (UNFCCC, 2004), the Handbook on Methods for Climate Change Impact Assessment and Adaptation Strategies (Feenstra et al., 1998), and Costing the Impacts of Climate Change (Metroeconomica, 2004). A range of different methods can also be used with stakeholders (see Section 2.3.2).

The financing of adaptation has received minimal attention. Bouwer and Vellinga (2005) suggest applying more structured decision-making to future disaster management and adaptation to climate change, sharing the risk between private and public sources. Quiggin and Horowitz (2003) argue that the economic costs will be dominated by the costs of adaptation, which depend on the rate of climate change, especially the occurrence of climate extremes, and that many existing analyses overlook these costs (see also Section 2.2.2).

2.2.4 Advances in vulnerability assessment

Since the TAR, the IPCC definition of vulnerability[3] has been challenged, both to account for an expanded remit by including social vulnerability (O'Brien et al., 2004a) and to reconcile it with risk assessment (Downing and Patwardhan, 2005). Different states of vulnerability under climate risks include: vulnerability to current climate, vulnerability to climate change in the absence of adaptation and mitigation measures, and residual vulnerability, where adaptive and mitigative capacities have been exhausted (e.g., Jones et al., 2007). A key vulnerability has the potential for significant adverse affects on both natural and human systems, as outlined in the United Nations Framework Convention on Climate Change (UNFCCC), thus contributing to dangerous anthropogenic interference with the climate system (see Chapter 19). Füssel and Klein (2006) review and summarise these developments.

Vulnerability is highly dependent on context and scale, and care should be taken to clearly describe its derivation and meaning (Downing and Patwardhan, 2005) and to address the uncertainties inherent in vulnerability assessments (Patt et al., 2005). Frameworks should also be able to integrate the social and biophysical dimensions of vulnerability to climate change (Klein and Nicholls, 1999; Polsky et al., 2003; Turner et al., 2003a). Formal methods for vulnerability assessment have also been proposed (Ionescu et al., 2005; Metzger and Schröter, 2006) but are very preliminary.

The methods and frameworks for assessing vulnerability must also address the determinants of adaptive capacity (Turner et al., 2003a; Schröter et al., 2005a; O'Brien and Vogel, 2006; see also Chapter 17, Section 17.3.1) in order to examine the potential responses of a system to climate variability and change. Many studies endeavour to do this in the context of human development, by aiming to understand the underlying causes of vulnerability and to further strengthen adaptive capacities (e.g., World Bank, 2006). In some quantitative approaches, the indicators used are related to adaptive capacity, such as national economic capacity, human resources, and environmental capacities (Moss et al., 2001; see also Section 2.2.3). Other studies include indicators that can provide information related to the conditions, processes and structures that promote or constrain adaptive capacity (Eriksen et al., 2005).

Vulnerability assessment offers a framework for policy measures that focus on social aspects, including poverty reduction, diversification of livelihoods, protection of common property resources and strengthening of collective action (O'Brien et al., 2004b). Such measures enhance the ability to respond to stressors and secure livelihoods under present conditions, which can also reduce vulnerability to future climate change. Community-based interactive approaches for identifying coping potentials provide insights into the underlying causes and structures that shape vulnerability (O'Brien et al., 2004b). Other methods employed in recent regional vulnerability studies include stakeholder elicitation and survey (Eakin et al., 2006; Pulhin et al., 2006), and multi-criteria modelling (Wehbe et al., 2006).

Traditional knowledge of local communities represents an important, yet currently largely under-used resource for CCIAV assessment (Huntington and Fox, 2005). Empirical knowledge from past experience in dealing with climate-related natural

[3] The degree to which a system is susceptible to, or unable to cope with, adverse effects of climate change, including climate variability and extremes. Vulnerability is a function of the character, magnitude, and rate of climate variation to which a system is exposed, its sensitivity, and its adaptive capacity (IPCC, 2001b, Glossary).

disasters such as droughts and floods (Osman-Elasha et al., 2006), health crises (Wandiga et al., 2006), as well as longer-term trends in mean conditions (Huntington and Fox, 2005; McCarthy and Long Martello, 2005), can be particularly helpful in understanding the coping strategies and adaptive capacity of indigenous and other communities relying on oral traditions.

2.2.5 Advances in integrated assessment

Integrated assessment represents complex interactions across spatial and temporal scales, processes and activities. Integrated assessments can involve one or more mathematical models, but may also represent an integrated process of assessment, linking different disciplines and groups of people. Managing uncertainty in integrated assessments can utilise models ranging from simple models linking large-scale processes, through models of intermediate complexity, to the complex, physically explicit representation of Earth systems. This structure is characterised by trade-offs between realism and flexibility, where simple models are more flexible but less detailed, and complex models offer more detail and a greater range of output. No single theory describes and explains dynamic behaviour across scales in socio-economic and ecological systems (Rotmans and Rothman, 2003), nor can a single model represent all the interactions within a single entity, or provide responses to questions in a rapid turn-around time (Schellnhuber et al., 2004). Therefore, integration at different scales and across scales is required in order to comprehensively assess CCIAV. Some specific advances are outlined here; integration to assess climate policy benefits is considered in Section 2.2.6.

Cross-sectoral integration is required for purposes such as national assessments, analysis of economic and trade effects, and joint population and climate studies. National assessments can utilise nationally integrated models (e.g., Izaurralde et al., 2003; Rosenberg et al., 2003; Hurd et al., 2004), or can synthesise a number of disparate studies for policy-makers (e.g., West and Gawith, 2005). Markets and trade can have significant effects on outcomes. For example, a study assessing the global impacts of climate change on forests and forest products showed that trade can affect efforts to stabilise atmospheric carbon dioxide (CO_2) and also affected regional welfare, with adverse effects on those regions with high production costs (Perez-Garcia et al., 2002). New economic assessments of aggregated climate change damages have also been produced for multiple sectors (Tol, 2002a, b; Mendelsohn and Williams, 2004; Nordhaus, 2006). These have highlighted potentially large regional disparities in vulnerability to impacts. Using an integrated assessment general equilibrium model, Kemfert (2002) found that interactions between sectors acted to amplify the global costs of climate change, compared with single-sector analysis.

Integration yields results that cannot be produced in isolation. For example, the Millennium Ecosystem Assessment assessed the impact of a broad range of stresses on ecosystem services, of which climate change was only one (Millennium Ecosystem Assessment, 2005). Linked impact and vulnerability assessments can also benefit from a multiple stressors approach. For instance, the AIR-CLIM Project integrated climate and air pollution impacts in Europe between 1995 and 2100, concluding that that while the physical impacts were weakly coupled, the costs of air pollution and climate change were strongly coupled. The indirect effects of climate policies stimulated cost reductions in air pollution control of more than 50% (Alcamo et al., 2002). Some of the joint effects of extreme weather and air pollution events on human health are described in Chapter 8, Section 8.2.6.

Earth system models of intermediate complexity that link the atmosphere, oceans, cryosphere, land system, and biosphere are being developed to assess impacts (particularly global-scale, singular events that may be considered dangerous) within a risk and vulnerability framework (Rial et al., 2004; see also Section 2.4.7). Global climate models are also moving towards a more complete representation of the Earth system. Recent simulations integrating the atmosphere with the biosphere via a complete carbon cycle show the potential of the Amazon rainforest to suffer dieback (Cox et al., 2004), leading to a positive feedback that decreases the carbon sink and increases atmospheric CO_2 concentrations (Friedlingstein et al., 2006; Denman et al., 2007).

2.2.6 Development of risk-management frameworks

Risk management is defined as the culture, processes and structures directed towards realising potential opportunities whilst managing adverse effects (AS/NZS, 2004). Risk is generally measured as a combination of the probability of an event and its consequences (ISO/IEC, 2002; see also Figure 2.1), with several ways of combining these two factors being possible. There may be more than one event, consequences can range from positive to negative, and risk can be measured qualitatively or quantitatively.

To date, most CCIAV studies have assessed climate change without specific regard to how mitigation policy will influence those impacts. However, the certainty that some climate change will occur (and is already occurring – see Chapter 1) is driving adaptation assessment beyond the limits of what scenario-driven methods can provide. The issues to be addressed include assessing current adaptations to climate variability and extremes before assessing adaptive responses to future climate, assessing the limits of adaptation, linking adaptation to sustainable development, engaging stakeholders, and decision-making under uncertainty. Risk management has been identified as a framework that can deal with all of these issues in a manner that incorporates existing methodologies and that can also accommodate other sources of risk (Jones, 2001; Willows and Connell, 2003; UNDP, 2005) in a process known as mainstreaming.

The two major forms of climate risk management are the mitigation of climate change through the abatement of GHG emissions and GHG sequestration, and adaptation to the consequences of a changing climate (Figure 2.1). Mitigation reduces the rate and magnitude of changing climate hazards; adaptation reduces the consequences of those hazards (Jones, 2004). Mitigation also reduces the upper bounds of the range of potential climate change, while adaptation copes with the lower bounds (Yohe and Toth, 2000). Hence they are complementary

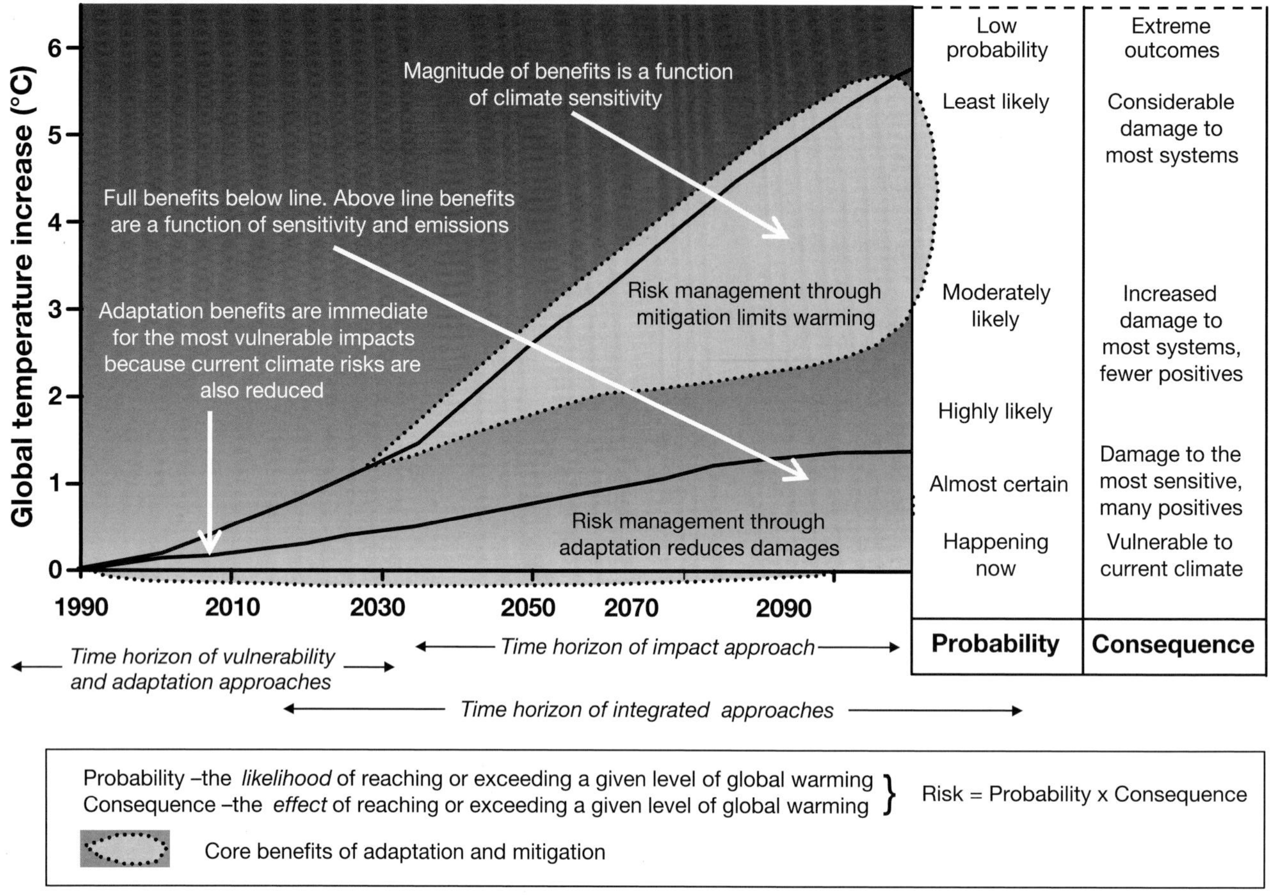

Figure 2.1. *Synthesis of risk-management approaches to global warming. The left side shows the projected range of global warming from the TAR (bold lines) with zones of maximum benefit for adaptation and mitigation depicted schematically. The right side shows likelihood based on threshold exceedance as a function of global warming and the consequences of global warming reaching that particular level based on results from the TAR. Risk is a function of probability and consequence. The primary time horizons of approaches to CCIAV assessment are also shown (modified from Jones, 2004).*

processes, but the benefits will accumulate over different timescales and, in many cases, they can be assessed and implemented separately (Klein et al., 2005). These complementarities and differences are discussed in Section 18.4 of this volume, while integrated assessment methods utilising a risk-management approach are summarised by Nakićenović et al. (2007).

Some of the standard elements within the risk-management process that can be adapted to assess CCIAV are as follows.

- A scoping exercise, where the context of the assessment is established. This identifies the overall approach to be used.
- Risk identification, where what is at risk, who is at risk, the main climate and non-climate stresses contributing to the risk, and levels of acceptable risk are identified. This step also identifies the scenarios required for further assessment.
- Risk analysis, where the consequences and their likelihood are analysed. This is the most developed area, with a range of methods used in mainstream risk assessment and CCIAV assessment being available.
- Risk evaluation, where adaptation and/or mitigation measures are prioritised.
- Risk treatment, where selected adaptation and/or mitigation measures are applied, with follow-up monitoring and review.

Two overarching activities are communication and consultation with stakeholders, and monitoring and review. These activities co-ordinate the management of uncertainty and ensure that clarity and transparency surround the assumptions and concepts being used. Other essential components of risk management include investment in obtaining improved information and building capacity for decision-making (adaptive governance: see Dietz et al., 2003).

Rather than being research-driven, risk management is oriented towards decision-making; e.g., on policy, planning, and management options. Several frameworks have been developed for managing risk, which use a variety of approaches as outlined in Table 2.1. The UNDP Adaptation Policy Framework (UNDP, 2005) describes risk-assessment methods that follow both the standard impact and human development approaches focusing on vulnerability and adaptation (also see Füssel and Klein, 2006). National frameworks constructed to deliver national adaptation strategies include those of the UK (Willows and Connell, 2003) and Australia (Australian Greenhouse Office, 2006). The World Bank is pursuing methods for hazard and risk management that focus on financing adaptation to climate change (van Aalst, 2006) and mainstreaming climate change into natural-hazard risk management (Burton and van Aalst, 2004; Mathur et al., 2004; Bettencourt et al., 2006).

Therefore, risk management is an approach that is being pursued for the management of climate change risks at a range of scales; from the global (mitigation to achieve 'safe' levels of GHG emissions and concentrations, thus avoiding dangerous anthropogenic interference), to the local (adaptation at the scale of impact), to mainstreaming risk with a multitude of other activities.

2.2.7 Managing uncertainties and confidence levels

CCIAV assessments aim to understand and manage as much of the full range of uncertainty, extending from emissions through to vulnerability (Ahmad et al., 2001), as is practicable, in order to improve the decision-making process. At the same time, a primary aim of scientific investigations is to reduce uncertainty through improved knowledge. However, such investigations do not necessarily reduce the uncertainty range as used by CCIAV assessments. A phenomenon or process is usually described qualitatively before it can be quantified with any confidence; some, such as aspects of socio-economic futures, may never be well quantified (Morgan and Henrion, 1990). Often a scientific advance will expand a bounded range of uncertainty as a new process is quantified and incorporated into the chain of consequences contributing to that range. Examples include an expanded range of future global warming due to positive CO_2 feedbacks, from the response of vegetation to climate change (see Section 2.2.5; WG I SPM), and a widened range of future impacts that can be incurred by incorporating development futures in integrated impact assessments, particularly if adaptation is included (see Section 2.4.6.4). In such cases, although uncertainty appears to be expanding, this is largely because the underlying process is becoming better understood.

The variety of different approaches developed and applied since the TAR all have their strengths and weaknesses. The impact assessment approach is particularly susceptible to ballooning uncertainties because of the limits of prediction (e.g., Jones 2001). Probabilistic methods and the use of thresholds are two ways in which these uncertainties are being managed (Jones and Mearns, 2005; see also Section 2.4.8). Another way to manage uncertainties is through participatory approaches, resulting in learning-by-observation and learning-by-doing, a particular strength of vulnerability and adaptation approaches (e.g., Tompkins and Adger, 2005; UNDP, 2005). Stakeholder participation establishes credibility and stakeholders are more likely to 'own' the results, increasing the likelihood of successful adaptation (McKenzie Hedger et al., 2006).

2.3 Development in methods

2.3.1 Thresholds and criteria for risk

The risks of climate change for a given exposure unit can be defined by criteria that link climate impacts to potential outcomes. This allows a risk to be analysed and management options to be evaluated, prioritised, and implemented. Criteria are usually specified using thresholds that denote some limit of tolerable risk. A threshold marks the point where stress on an exposed system or activity, if exceeded, results in a non-linear response in that system or activity. Two types of thresholds are used in assessing change (Kenny et al., 2000; Jones 2001; see also Chapter 19, Section 19.1.2.5):

1. a non-linear change in state, where a system shifts from one identifiable set of conditions to another (systemic threshold);
2. a level of change in condition, measured on a linear scale, regarded as 'unacceptable' and inviting some form of response (impact threshold).

Thresholds used to assess risk are commonly value-laden, or normative. A systemic threshold can often be objectively measured; for example, a range of estimates of global mean warming is reported in Meehl et al. (2007) defining the point at which irreversible melting of the Greenland Ice Sheet would commence. If a policy aim were to avoid its loss, selecting from the given range a critical level of warming that is not to be exceeded would require a value judgement. In the case of an impact threshold, the response is the non-linear aspect; for example, a management threshold (Kenny et al., 2000). Exceeding a management threshold will result in a change of legal, regulatory, economic, or cultural behaviour. Hence, both cases introduce critical thresholds (IPCC, 1994; Parry et al., 1996; Pittock and Jones, 2000), where criticality exceeds, in risk-assessment terms, the level of tolerable risk. Critical thresholds are used to define the coping range (see Section 2.3.3).

Thresholds derived with stakeholders avoid the pitfall of researchers ascribing their own values to an assessment (Kenny et al., 2000; Pittock and Jones, 2000; Conde and Lonsdale, 2005). Stakeholders thus become responsible for the management of the uncertainties associated with that threshold through ownership of the assessment process and its outcomes (Jones, 2001). The probability of threshold exceedance is being used in risk analyses (Jones, 2001, 2004) on local and global scales. For example, probabilities of critical thresholds for coral bleaching and mortality for sites in the Great Barrier Reef as a function of global warming show that catastrophic bleaching will occur biennially with a warming of about 2°C (Jones, 2004). Further examples are given in Section 2.4.8. At a global scale, the risk of exceeding critical thresholds has been estimated within a Bayesian framework, by expressing global warming and sea-level rise as cumulative distribution functions that are much more likely to be exceeded at lower levels than higher levels (Jones, 2004; Mastrandrea and Schneider, 2004; Yohe, 2004). However, although this may be achieved for key global vulnerabilities, there is often no straightforward way to integrate local critical thresholds into a 'mass' damage function of many different metrics across a wide range of potential impacts (Jacoby, 2004).

2.3.2 Stakeholder involvement

Stakeholder involvement is crucial to risk, adaptation, and vulnerability assessments because it is the stakeholders who will be most affected and thus may need to adapt (Burton et al., 2002; Renn, 2004; UNDP, 2005). Stakeholders are characterised as individuals or groups who have anything of value (both monetary and non-monetary) that may be affected by climate change or by the actions taken to manage anticipated climate risks. They might be policy-makers, scientists, communities, and/or managers in the sectors and regions most at risk both now and in the future (Rowe and Frewer, 2000; Conde and Lonsdale, 2005).

Individual and institutional knowledge and expertise comprise the principal resources for adapting to the impacts of climate change. Adaptive capacity is developed if people have time to strengthen networks, knowledge, and resources, and the willingness to find solutions (Cohen, 1997; Cebon et al., 1999; Ivey et al., 2004). Kasperson (2006) argues that the success of stakeholder involvement lies not only in informing interested and affected people, but also in empowering them to act on the enlarged knowledge. Through an ongoing process of negotiation and modification, stakeholders can assess the viability of adaptive measures by integrating scientific information into their own social, economic, cultural, and environmental context (van Asselt and Rotmans, 2002; see also Chapter 18, Section 18.5). However, stakeholder involvement may occur in a context where political differences, inequalities, or conflicts may be raised; researchers must accept that it is not their role to solve those conflicts, unless they want to be part of them (Conde and Lonsdale, 2005). Approaches to stakeholder engagement vary from passive interactions, where the stakeholders only provide information, to a level where the stakeholders themselves initiate and design the process (Figure 2.2).

Current adaptation practices for climate risks are being developed by communities, governments, Non-Governmental Organisations (NGOs), and other organised stakeholders to increase their adaptive capacity (Ford and Smit, 2004; Thomalla et al., 2005; Conde et al., 2006). Indigenous knowledge studies are a valuable source of information for CCIAV assessments, especially where formally collected and recorded data are sparse (Huntington and Fox, 2005). Stakeholders have a part to play in scenario development (Lorenzoni et al., 2000; Bärlund and Carter, 2002) and participatory modelling (e.g., Welp, 2001; van Asselt and Rijkens-Klomp, 2002).

Stakeholders are also central in assessing future needs for developing policies and measures to adapt (Nadarajah and Rankin, 2005). These needs have been recognised in regional and national approaches to assessing climate impacts and adaptation, including the UK Climate Impacts Programme (UKCIP) (West and Gawith, 2005), the US National Assessment (National Assessment Synthesis Team 2000; Parson et al., 2003), the Arctic Climate Impact Assessment (ACIA, 2005), the Finnish National Climate Change Adaptation Strategy (Marttila et al., 2005) and the related FINADAPT research consortium (Kankaanpää et al., 2005), and the Mackenzie Basin Impact Study (Cohen, 1997).

2.3.3 Defining coping ranges

The coping range of climate (Hewitt and Burton, 1971) is described in the TAR as the capacity of systems to accommodate variations in climatic conditions (Smith et al., 2001), and thus serves as a suitable template for understanding the relationship between changing climate hazards and society. The concept of the coping range has since been expanded to incorporate concepts of current and future adaptation, planning and policy horizons, and likelihood (Yohe and Tol, 2002; Willows and Connell, 2003; UNDP, 2005). It can therefore serve as a conceptual model (Morgan et al., 2001) which can be used to integrate analytical techniques with a broader understanding of climate-society relationships (Jones and Mearns, 2005).

The coping range is used to link the understanding of current adaptation to climate with adaptation needs under climate change. It is a useful mental model to use with stakeholders –

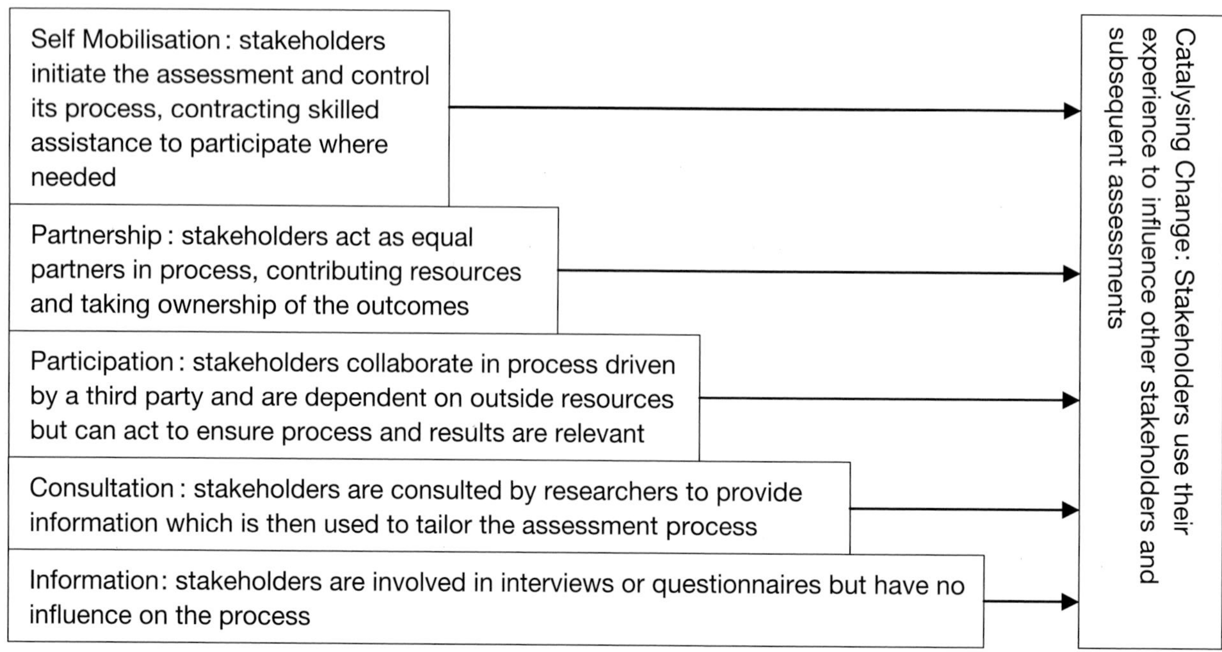

Figure 2.2. *Ladder of stakeholder participation (based on Pretty et al., 1995; Conde and Lonsdale, 2005).*

who often have an intuitive understanding of which risks can be coped with and which cannot – that can subsequently be developed into a quantitative model (Jones and Boer, 2005). It can be depicted as one or more climatic or climate-related variables upon which socio-economic responses are mapped (Figure 2.3). The core of the coping range contains beneficial outcomes. Towards one or both edges of the coping range, outcomes become negative but tolerable. Beyond the coping range, the damages or losses are no longer tolerable and denote a vulnerable state, the limits of tolerance describing a critical threshold (left side of Figure 2.3). A coping range is usually specific to an activity, group, and/or sector, although society-wide coping ranges have been proposed (Yohe and Tol, 2002).

Risk is assessed by calculating how often the coping range is exceeded under given conditions. Climate change may increase the risk of threshold exceedance but adaptation can ameliorate the adverse effects by widening the coping range (right side of Figure 2.3). For example, Jones (2001) constructed critical thresholds for the Macquarie River catchment in Australia for irrigation allocation and environmental flows. The probability of exceeding these thresholds was a function of both natural climate variability and climate change. Yohe and Tol (2002) explored hypothetical upper and lower critical thresholds for the River Nile using current and historical streamflow data. The upper threshold denoted serious flooding, and the lower threshold the minimum flow required to supply water demand. Historical frequency of exceedance served as a baseline from which to measure changing risks using a range of climate scenarios.

2.3.4 Communicating uncertainty and risk

Communicating risk and uncertainty is a vital part of helping people respond to climate change. However, people often rely on intuitive decision-making processes, or heuristics, in solving complicated problems of judgement and decision-making (Tversky and Kahneman, 1974). In many cases, these heuristics are surprisingly successful in leading to successful decisions under information and time constraints (Gigerenzer, 2000; Muramatsu and Hanich, 2005). In other cases, heuristics can lead to predictable inconsistencies or errors of judgement (Slovic et al., 2004). For example, people consistently overestimate the likelihood of low-probability events (Kahneman and Tversky, 1979; Kammen et al., 1994), resulting in choices that may increase their exposure to harm (Thaler and Johnson, 1990). These deficiencies in human judgement in the face of uncertainty are discussed at length in the TAR (Ahmad et al., 2001).

Participatory approaches establish a dialogue between stakeholders and experts, where the experts can explain the uncertainties and the ways they are likely to be misinterpreted, the stakeholders can explain their decision-making criteria, and the two parties can work together to design a risk-management strategy (Fischoff, 1996; Jacobs, 2002; NRC, 2002). Because stakeholders are often the decision-makers themselves (Kelly and Adger, 2000), the communication of impact, adaptation, and vulnerability assessment has become more important (Jacobs, 2002; Dempsey and Fisher, 2005; Füssel and Klein, 2006). Adaptation decisions also depend on changes occurring outside the climate change arena (Turner et al., 2003b).

If the factors that give rise to the uncertainties are described (Willows and Connell, 2003), stakeholders may view that information as more credible because they can make their own judgements about its quality and accuracy (Funtowicz and Ravetz, 1990). People will remember and use uncertainty assessments when they can mentally link the uncertainty and events in the world with which they are familiar; assessments of climate change uncertainty are more memorable, and hence more influential, when they fit into people's pre-existing mental maps of experience of climate variability, or when sufficient detail is provided to help people to form new mental models (Hansen, 2004). This can be aided by the development of visual tools that can communicate impacts, adaptation, and vulnerability to stakeholders while representing uncertainty in an appropriate manner (e.g., Discovery Software, 2003; Aggarwal et al., 2006).

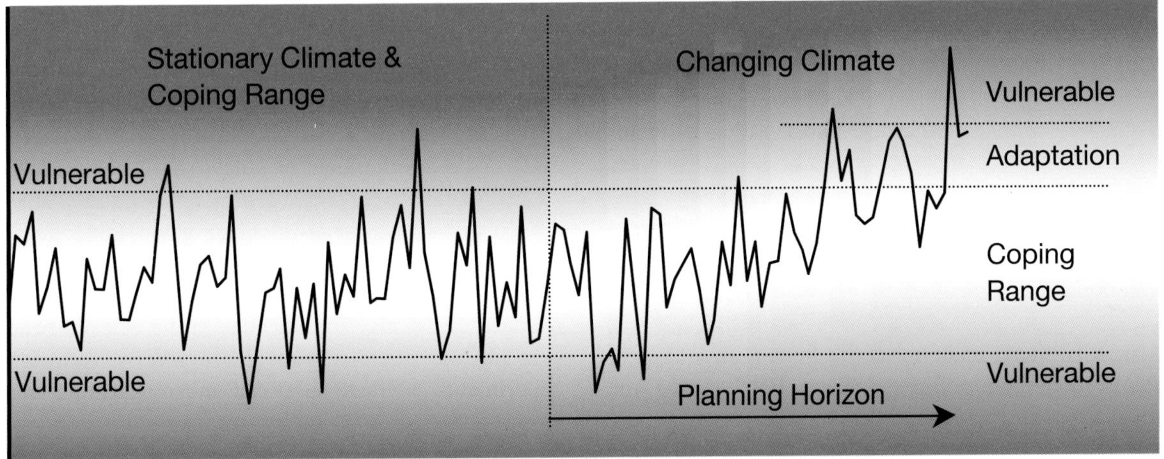

Figure 2.3. *Idealised version of a coping range showing the relationship between climate change and threshold exceedance, and how adaptation can establish a new critical threshold, reducing vulnerability to climate change (modified from Jones and Mearns, 2005).*

2.3.5 Data needs for assessment

Although considerable advances have been made in the development of methods and tools for CCIAV assessment (see previous sections), their application has been constrained by limited availability and access to good-quality data (e.g., Briassoulis, 2001; UNFCCC, 2005; see also Chapter 3, Section 3.8; Chapter 6, Section 6.6; Chapter 7, Section 7.8; Chapter 8, Section, 8.8; Chapter 9, Section 9.5; Chapter 10, Section 10.8; Chapter 12, Section 12.8; Chapter 13, Section 13.5; Chapter 15, Section 15.4; Chapter 16, Section 16.7).

In their initial national communications to the UNFCCC, a large number of non-Annex I countries reported on the lack of appropriate institutions and infrastructure to conduct systematic data collection, and poor co-ordination within and/or between different government departments and agencies (UNFCCC, 2005). Significant gaps exist in the geographical coverage and management of existing global and regional Earth-observing systems and in the efforts to retrieve the available historical data. These are especially acute in developing-country regions such as Africa, where lack of funds for modern equipment and infrastructure, inadequate training of staff, high maintenance costs, and issues related to political instability and conflict are major constraints (IRI, 2006). As a result, in some regions, observation systems have been in decline (e.g., GCOS, 2003; see also Chapter 16, Section 16.7).

Major deficiencies in data provision for socio-economic and human systems indicators have been reported as a key barrier to a better understanding of nature-society dynamics in both developed and developing countries (Wilbanks et al., 2003; but see Nordhaus, 2006). Recognising the importance of data and information for policy decisions and risk management under a changing climate, new programmes and initiatives have been put in place to improve the provision of data across disciplines and scales. Prominent among these, the Global Earth Observation System of Systems (GEOSS) plan (Group on Earth Observations, 2005) was launched in 2006, with a mission to help all 61 involved countries produce and manage Earth observational data. The Centre for International Earth Science Information Network (CIESIN) provides a wide range of environmental and socio-economic data products.[4] In addition, the IPCC Data Distribution Centre (DDC), overseen by the IPCC Task Group on Data and Scenario Support for Impact and Climate Analysis (TGICA), hosts various sets of outputs from coupled Atmosphere-Ocean General Circulation Models (AOGCMs), along with environmental and socio-economic data for CCIAV assessments (Parry, 2002). New sources of data from remote sensing are also becoming available (e.g., Justice et al., 2002), which could fill the gaps where no ground-based data are available but which require resourcing to obtain access. New and updated observational data sets and their deficiencies are also detailed in the WG I report for climate (Trenberth et al., 2007) and sea level (Bindoff et al., 2007).

Efforts are also being made to record human-environment interactions in moderated online databases. For instance, the DesInventar database[5] records climatic disasters of the recent past in Latin America, documenting not only the adverse climatic events themselves, but also the consequences of these events and the parties affected. Information on local coping strategies applied by different communities and sectors is being recorded by the UNFCCC.[6]

Many assessments are now obtaining data through stakeholder elicitation and survey methods. For example, in many traditional societies a large number of social interactions may not be recorded by bureaucratic processes, but knowledge of how societies adapt to climate change, perceive risk, and measure their vulnerability is held by community members (e.g., Cohen, 1997; ACIA, 2005; see also Section 2.3.2). Even in data-rich situations, it is likely that some additional data from stakeholders will be required. However, this also requires adequate resourcing.

2.4 Characterising the future

2.4.1 Why and how do we characterise future conditions?

Evaluations of future climate change impacts, adaptation, and vulnerability require assumptions, whether explicit or implicit, about how future socio-economic and biophysical conditions will develop. The literature on methods of characterising the future has grown in tandem with the literature on CCIAV, but these methods have not been defined consistently across different research communities. Box 2.1 presents a consistent typology of characterisations that expands on the definitions presented in the TAR (Carter et al., 2001), for the purpose of clarifying the use of this terminology in this chapter. Although they may overlap, different types of characterisations of the future can be usefully distinguished in terms of their plausibility and ascription of likelihood, on the one hand, and the comprehensiveness of their representation, on the other (see Box 2.1 for definitions). Since the TAR, comprehensiveness has increased and ascriptions of likelihood have become more common. The following sections make use of the typology in Box 2.1 to address notable advances in methods of characterising the future.

2.4.2 Artificial experiments

The most significant advance in artificial experiments since the TAR is the development of a new set of commitment runs by AOGCMs. These are climate change projections that assume that the radiative forcing at a particular point in time (often the current forcing) is held constant into the future (Meehl et al., 2007). The projections demonstrate the time-lags in the climate response to changes in radiative forcing (due to the delayed penetration of heat into the oceans), and of sea level to warming. Recent experiments estimate a global mean warming commitment

[4] http://www.ciesin.org/index.html
[5] http://www.desinventar.org/desinventar.html
[6] http://maindb.unfccc.int/public/adaptation

Box 2.1. Definitions of future characterisations

Figure 2.4 illustrates the relationships among the categories of future characterisations most commonly used in CCIAV studies. Because definitions vary across different fields, we present a single consistent typology for use in this chapter. Categories are distinguished according to comprehensiveness and plausibility.

Comprehensiveness indicates the degree to which a characterisation of the future captures the various aspects of the socio-economic/biophysical system it aims to represent. Secondarily, it indicates the detail with which any single element is characterised.

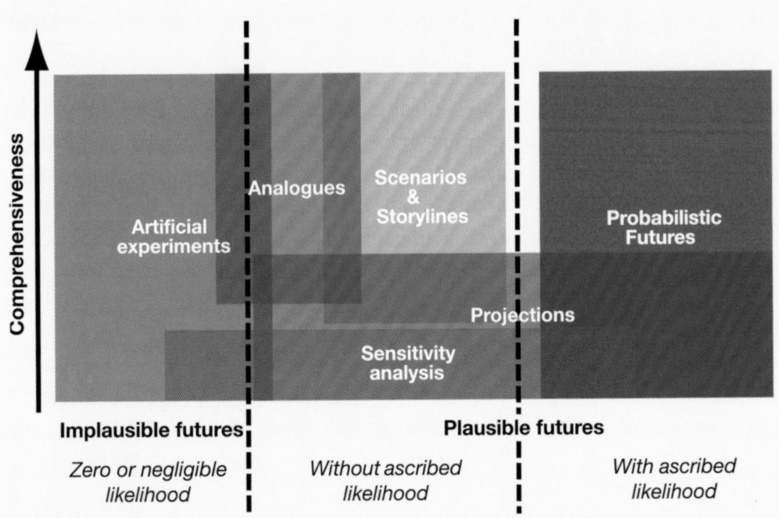

Figure 2.4. *Characterisations of the future.*

Plausibility is a subjective measure of whether a characterisation of the future is possible. Implausible futures are assumed to have zero or negligible likelihood. Plausible futures can be further distinguished by whether a specific likelihood is ascribed or not.

Artificial experiment. A characterisation of the future constructed without regard to plausibility (and hence often implausible) that follows a coherent logic in order to study a process or communicate an insight. Artificial experiments range in comprehensiveness from simple thought experiments to detailed integrated modelling studies.

Sensitivity analysis. Sensitivity analyses employ characterisations that involve arbitrary or graduated adjustments of one or several variables relative to a reference case. These adjustments may be plausible (e.g., changes are of a realistic magnitude) or implausible (e.g., interactions between the adjusted variables are ignored), but the main aim is to explore model sensitivity to inputs, and possibly uncertainty in outputs.

Analogues. Analogues are based on recorded conditions that are considered to adequately represent future conditions in a study region.These records can be of past conditions (temporal analogues) or from another region (spatial analogues). Their selection is guided by information from sources such as AOGCMs; they are used to generate detailed scenarios which could not be realistically obtained by other means. Analogues are plausible in that they reflect a real situation, but may be implausible because no two places or periods of time are identical in all respects.

Scenarios. A scenario is a coherent, internally consistent, and plausible description of a possible future state of the world (IPCC, 1994; Nakićenović et al., 2000; Raskin et al., 2005). Scenarios are not predictions or forecasts (which indicate outcomes considered most likely), but are alternative images without ascribed likelihoods of how the future might unfold. They may be qualitative, quantitative, or both. An overarching logic often relates several components of a scenario, for example a storyline and/or projections of particular elements of a system. Exploratory (or descriptive) scenarios describe the future according to known processes of change, or as extrapolations of past trends (Carter et al., 2001). Normative (or prescriptive) scenarios describe a pre-specified future, either optimistic, pessimistic, or neutral (Alcamo, 2001), and a set of actions that might be required to achieve (or avoid) it. Such scenarios are often developed using an inverse modelling approach, by defining constraints and then diagnosing plausible combinations of the underlying conditions that satisfy those constraints (see Nakićenović et al., 2007).

Storylines. Storylines are qualitative, internally consistent narratives of how the future may evolve. They describe the principal trends in socio-political-economic drivers of change and the relationships between these drivers. Storylines may be stand-alone, but more often underpin quantitative projections of future change that, together with the storyline, constitute a scenario.

Projection. A projection is generally regarded as any description of the future and the pathway leading to it. However, here we define a projection as a model-derived estimate of future conditions related to one element of an integrated system (e.g., an emission, a climate, or an economic growth projection). Projections are generally less comprehensive than scenarios, even if the projected element is influenced by other elements. In addition, projections may be probabilistic, while scenarios do not ascribe likelihoods.

Probabilistic futures. Futures with ascribed likelihoods are probabilistic. The degree to which the future is characterised in probabilistic terms can vary widely. For example, conditional probabilistic futures are subject to specific and stated assumptions about how underlying assumptions are to be represented. Assigned probabilities may also be imprecise or qualitative.

associated with radiative forcing in 2000 of about 0.6°C by 2100 (Meehl et al., 2007). Sea-level rise due to thermal expansion of the oceans responds much more slowly, on a time-scale of millennia; committed sea-level rise is estimated at between 0.3 and 0.8 m above present levels by 2300, assuming concentrations stabilised at A1B levels in 2100 (Meehl et al., 2007). However, these commitment runs are unrealistic because the instantaneous stabilisation of radiative forcing is implausible, implying an unrealistic change in emission rates (see Nakićenović et al., 2007). They are therefore only suitable for setting a lower bound on impacts seen as inevitable (Parry et al., 1998).

2.4.3 Sensitivity analysis

Sensitivity analysis (see Box 2.1) is commonly applied in many model-based CCIAV studies to investigate the behaviour of a system, assuming arbitrary, often regularly spaced, adjustments in important driving variables. It has become a standard technique in assessing sensitivity to climatic variations, enabling the construction of impact response surfaces over multi-variate climate space (e.g., van Minnen et al., 2000; Miller et al., 2003). Response surfaces are increasingly constructed in combination with probabilistic representations of future climate to assess risk of impact (see Section 2.4.8). Sensitivity analysis sampling uncertainties in emissions, natural climate variability, climate change projections, and climate impacts has been used to evaluate the robustness of proposed adaptation measures for water resource management by Dessai (2005). Sensitivity analysis has also been used as a device for studying land-use change, by applying arbitrary adjustments to areas, such as +10% forest, −10% cropland, where these area changes are either spatially explicit (Shackley and Deanwood, 2003) or not (Ott and Uhlenbrook, 2004; van Beek and van Asch, 2004; Vaze et al., 2004).

2.4.4 Analogues

Temporal and spatial analogues are applied in a range of CCIAV studies. The most common of recently reported temporal analogues are historical extreme weather events. These types of event may recur more frequently under anthropogenic climate change, requiring some form of adaptation measure. The suitability of a given climate condition for use as an analogue requires specialist judgement of its utility (i.e., how well it represents the key weather variables affecting vulnerability) and its meteorological plausibility (i.e., how well it replicates anticipated future climate conditions). Examples of extreme events judged likely or very likely by the end of the century (see Table 2.2) that might serve as analogues include the European 2003 heatwave (see Chapter 12, Section 12.6.1) and flooding events related to intense summer precipitation in Bangladesh (Mirza, 2003a) and Norway (Næss et al., 2005). Other extreme events suggested as potential analogues, but about which the likelihood of future changes is poorly known (Christensen et al., 2007a), include El Niño-Southern Oscillation (ENSO)-related events (Glantz, 2001; Heslop-Thomas et al., 2006) and intense precipitation and flooding events in central Europe (Kundzewicz et al., 2005). Note also that the suitability of such analogue events should normally be considered along with information on accompanying changes in mean climate, which may ease or exacerbate vulnerability to extreme events.

Spatial analogues have also been applied in CCIAV analysis. For example, model-simulated climates for 2071 to 2100 have been analysed for selected European cities (Hallegatte et al., 2007). Model grid boxes in Europe showing the closest match between their present-day mean temperatures and seasonal precipitation and those projected for the cities in the future were identified as spatial analogues. These 'displaced' cities were then used as a heuristic device for analysing economic impacts and adaptation needs under a changing climate. A related approach is to seek projected climates (e.g., using climate model simulations) that have no present-day climatic analogues on Earth ('novel' climates) or regions where present-day climates are no longer to be found in the future ('disappearing' climates: see Ohlemüller et al., 2006; Williams et al., 2007). Results from such studies have been linked to risks to ecological systems and biodiversity.

2.4.5 Storylines

Storylines for CCIAV studies (see Box 2.1) are increasingly adopting a multi-sectoral and multi-stressor approach (Holman et al., 2005a, b) over multiple scales (Alcamo et al., 2005; Lebel et al., 2005; Kok et al., 2006a; Westhoek et al., 2006b) and are utilising stakeholder elicitation (Kok et al., 2006b). As they have become more comprehensive, the increased complexity and richness of the information they contain has aided the interpretation of adaptive capacity and vulnerability (Metzger et al., 2006). Storyline development is also subjective, so more comprehensive storylines can have alternative, but equally plausible, interpretations (Rounsevell et al., 2006). The concept of a 'region', for example, may be interpreted within a storyline in different ways – as world regions, nation states, or sub-national administrative units. This may have profound implications for how storylines are characterised at a local scale, limiting their reproducibility and credibility (Abildtrup et al., 2006). The alternative is to link a locally sourced storyline, regarded as credible at that scale, to a global scenario.

Storylines can be an endpoint in their own right (e.g., Rotmans et al., 2000), but often provide the basis for quantitative scenarios. In the storyline and simulation (SAS) approach (Alcamo, 2001), quantification is undertaken with models for which the input parameters are estimated through interpretation of the qualitative storylines. Parameter estimation is often subjective, using expert judgement, although more objective methods, such as pairwise comparison, have been used to improve internal consistency (Abildtrup et al., 2006). Analogues and stakeholder elicitation have also been used to estimate model parameters (e.g., Rotmans et al., 2000; Berger and Bolte, 2004; Kok et al., 2006a). Moreover, participatory approaches are important in reconciling long-term scenarios with the short-term, policy-driven requirements of stakeholders (Velázquez et al., 2001; Shackley and Deanwood, 2003; Lebel et al., 2005).

2.4.6 Scenarios

Advances in scenario development since the TAR address issues of consistency and comparability between global drivers of change, and regional scenarios required for CCIAV assessment (for reviews, see Berkhout et al., 2002; Carter et al., 2004; Parson et al., 2006). Numerous methods of downscaling from global to sub-global scale are emerging, some relying on the narrative storylines underpinning the global scenarios.

At the time of the TAR, most CCIAV studies utilised climate scenarios (many based on the IS92 emissions scenarios), but very few applied contemporaneous scenarios of socio-economic, land-use, or other environmental changes. Those that did used a range of sources to develop them. The IPCC Special Report on Emissions Scenarios (SRES: see Nakićenović et al., 2000) presented the opportunity to construct a range of mutually consistent climate and non-climatic scenarios. Originally developed to provide scenarios of future GHG emissions, the SRES scenarios are also accompanied by storylines of social, economic, and technological development that can be used in CCIAV studies (Box 2.2).

There has been an increasing uptake of the SRES scenarios since the TAR, and a substantial number of the impact studies assessed in this volume that employed future characterisations made use of them.[7] For this reason, these scenarios are highlighted in a series of boxed examples throughout Section 2.4. For some other studies, especially empirical analyses of adaptation and vulnerability, the scenarios were of limited relevance and were not adopted.

While the SRES scenarios were specifically developed to address climate change, several other major global scenario-building exercises have been designed to explore uncertainties and risks related to global environmental change. Recent examples include: the Millennium Ecosystem Assessment scenarios to 2100 (MA: see Alcamo et al., 2005), Global Scenarios Group scenarios to 2050 (GSG: see Raskin et al., 2002), and Global Environment Outlook scenarios to 2032 (GEO-3: see UNEP 2002). These exercises were reviewed and compared by Raskin et al. (2005) and Westhoek et al. (2006a), who observed that many applied similar assumptions to those used in the SRES scenarios, in some cases employing the same models to quantify the main drivers and indicators. All the exercises adopted the storyline and simulation (SAS) approach (introduced in Section 2.4.5). Furthermore, all contain important features that can be useful for CCIAV studies; with some exercises (e.g., MA and GEO-3) going one step further than the original SRES scenarios by not only describing possible emissions under differing socio-economic pathways but also including imaginable outcomes for climate variables and their impact on ecological and social systems. This helps to illustrate risks and possible response strategies to deal with possible impacts.

Five classes of scenarios relevant to CCIAV analysis were distinguished in the TAR: climate, socio-economic, land-use and land-cover, other environmental (mainly atmospheric composition), and sea-level scenarios (Carter et al., 2001). The following sections describe recent progress in each of these classes and in four additional categories: technology scenarios, adaptation scenarios, mitigation scenarios, and scenario integration.

Box 2.2. The SRES global storylines and scenarios

Economic emphasis →

Global integration ↑ | Regional emphasis ↓

A1 storyline	A2 storyline
World: market-oriented Economy: fastest per capita growth Population: 2050 peak, then decline Governance: strong regional interactions; income convergence Technology: three scenario groups: • **A1FI**: fossil intensive • **A1T**: non-fossil energy sources • **A1B**: balanced across all sources	World: differentiated Economy: regionally oriented; lowest per capita growth Population: continuously increasing Governance: self-reliance with preservation of local identities Technology: slowest and most fragmented development
B1 storyline	**B2 storyline**
World: convergent Economy: service and information based; lower growth than A1 Population: same as A1 Governance: global solutions to economic, social and environmental sustainability Technology: clean and resource-efficient	World: local solutions Economy: intermediate growth Population: continuously increasing at lower rate than A2 Governance: local and regional solutions to environmental protection and social equity Technology: more rapid than A2; less rapid, more diverse than A1/B1

← Environmental emphasis

Figure 2.5. *Summary characteristics of the four SRES storylines (based on Nakićenović et al., 2000).*

SRES presented four narrative storylines, labelled A1, A2, B1, and B2, describing the relationships between the forces driving GHG and aerosol emissions and their evolution during the 21st century for large world regions and globally (Figure 2.5). Each storyline represents different demographic, social, economic, technological, and environmental developments that diverge in increasingly irreversible ways and result in different levels of GHG emissions. The storylines assume that no specific climate policies are implemented, and thus form a baseline against which narratives with specific mitigation and adaptation measures can be compared.

The SRES storylines formed the basis for the development of quantitative scenarios using various numerical models that were presented in the TAR. Emissions scenarios were converted to projections of atmospheric GHG and aerosol concentrations, radiative forcing of the climate, effects on regional climate, and climatic effects on global sea level (IPCC, 2001a). However, little regional detail of these projections and no CCIAV studies that made use of them were available for the TAR. Many CCIAV studies have applied SRES-based scenarios since then, and some of these are described in Boxes 2.3 to 2.7 to illustrate different scenario types.

[7] Of 17 chapters surveyed, SRES-based scenarios were used by the majority of impact studies in 5 chapters, and by a large minority in 11 chapters. The most common usage is for climate scenarios, while examples of studies employing SRES-based socio-economic, environmental, or land-use scenarios comprise a small but growing number. The remaining impact studies used either earlier IPCC scenarios (e.g., IS92) or characterisations derived from other sources.

2.4.6.1 Climate scenarios

The most recent climate projection methods and results are extensively discussed in the WG I volume (especially Christensen et al., 2007a; Meehl et al., 2007), and most of these were not available to the CCIAV studies assessed in this volume. Box 2.3 compares recent climate projections from Atmosphere-Ocean General Circulation Models (AOGCMs) with the earlier projections relied on throughout this volume. While AOGCMs are the most common source of regional climate scenarios, other methods and tools are also applied in specific CCIAV studies. Numerous regionalisation techniques[8] have been employed to obtain high-resolution, SRES-based climate scenarios, nearly always using low-resolution General Circulation Model (GCM) outputs as a starting point. Some of these methods are also used to develop scenarios of extreme weather events.

Scenarios from high-resolution models

The development and application of scenarios from high-resolution regional climate models and global atmospheric models (time-slices) since the TAR confirms that improved resolution allows a more realistic representation of the response of climate to fine-scale topographic features (e.g., lakes, mountains, coastlines). Impact models will often produce different results utilising high-resolution scenarios compared with direct GCM outputs (e.g., Arnell et al., 2003; Mearns et al., 2003; Stone et al., 2003; Leung et al., 2004; Wood et al., 2004). However, most regional model experiments still rely on only one driving AOGCM and scenarios are usually available from only one or two regional climate models (RCMs).

More elaborate and extensive modelling designs have facilitated the exploration of multiple uncertainties (across different RCMs, AOGCMs, and emissions scenarios) and how those uncertainties affect impacts. The PRUDENCE project in Europe produced multiple RCM simulations based on the ECHAM/OPYC AOGCM and HadAM3H AGCM simulations for two different emissions scenarios (Christensen et al., 2007b). Uncertainties due to the spatial scale of the scenarios and stemming from the application of different RCMs versus different GCMs (including models not used for regionalisation) were elaborated on in a range of impact studies (e.g., Ekstrom et al., 2007; Fronzek and Carter, 2007; Hingray et al., 2007; Graham et al., 2007; Olesen et al., 2007). For example, Olesen et al. (2007) found that the variation in simulated agricultural impacts was smaller across scenarios from RCMs nested in a single GCM than it was across different GCMs or across the different emissions scenarios.

The construction of higher-resolution scenarios (now often finer than 50 km), has encouraged new types of impact studies. For example, studies examining the combined impacts of increased heat stress and air pollution are now more feasible because the resolution of regional climate models is converging with that of air-quality models (e.g., Hogrefe et al., 2004). Furthermore, scenarios developed from RCMs (e.g., UKMO, 2001) are now being used in many more regions of the world, particularly the developing world (e.g., Arnell et al., 2003; Gao et al., 2003; Anyah and Semazzi, 2004; Government of India, 2004; Rupa Kumar et al., 2006). Results of these regional modelling experiments are reported in Christensen et al. (2007a).

Statistical downscaling (SD)

Much additional work has been produced since the TAR using methods of statistical downscaling (SD) for climate scenario generation (Wilby et al., 2004b; also see Christensen et al., 2007a). Various SD techniques have been used in downscaling directly to (physically-based) impacts and to a greater variety of climate variables than previously (e.g., wind speed), including extremes of variables. For example, Wang et al. (2004) and Caires and Sterl (2005) have developed extreme value models for projecting changes in wave height.

While statistical downscaling has mostly been applied for single locations, Hewitson (2003) developed empirical downscaling for point-scale precipitation at numerous sites and on a 0.1°-resolution grid over Africa. Finally, the wider availability of statistical downscaling tools is being reflected in wider application; for example, the Statistical Downscaling Model (SDSM) tool of Wilby et al. (2002), which has been used to produce scenarios for the River Thames basin (Wilby and Harris, 2006). Statistical downscaling does have some limitations; for example, it cannot take account of small-scale processes with strong time-scale dependencies (e.g., land-cover change). See Christensen et al. (2007a) for a complete discussion of the strengths and weaknesses of both statistical and dynamical downscaling.

Scenarios of extreme weather events

The improved availability of high-resolution scenarios has facilitated new studies of event-driven impacts (e.g., fire risk – Moriondo et al., 2006; low-temperature impacts on boreal forests – Jönsson et al., 2004). Projected changes in extreme weather events have been related to projected changes in local mean climate, in the hope that robust relationships could allow the prediction of extremes on the basis of changes in mean climate alone. PRUDENCE RCM outputs showed non-linear relationships between mean maximum temperature and indices of drought and heatwave (Good et al., 2006), while changes in maximum 1-day and 5-day precipitation amounts were systematically enhanced relative to changes in seasonal mean precipitation across many regions of Europe (Beniston et al., 2007). In a comprehensive review (citing over 200 papers) of the options available for developing scenarios of weather extremes for use in Integrated Assessment Models (IAMs), Goodess et al. (2003) list the advantages and disadvantages of applying direct GCM outputs, direct RCM outputs, and SD techniques. Streams of daily data are the outputs most commonly used from these sources, and these may pose computational difficulties for assessing impacts in IAMs (which

[8] Defined in the TAR as "techniques developed with the goal of enhancing the regional information provided by coupled AOGCMs and providing fine-scale climate information" (Giorgi et al., 2001).

Box 2.3. SRES-based climate scenarios assumed in this report

Not all of the impact studies reported in this assessment employed SRES-based climate scenarios. Earlier scenarios are described in previous IPCC reports (IPCC, 1992, 1996; Greco et al., 1994). The remaining discussion focuses on SRES-based climate projections, which are applied in most CCIAV studies currently undertaken.

In recent years, many simulations of the global climate response to the SRES emission scenarios have been completed with AOGCMs, also providing regional detail on projected climate. Early AOGCM runs (labelled 'pre-TAR') were reported in the TAR (Cubasch et al., 2001) and are available from the IPCC DDC. Many have been adopted in CCIAV studies reported in this volume. A new generation of AOGCMs, some incorporating improved representations of climate system processes and land surface forcing, are now utilising the SRES scenarios in addition to other emissions scenarios of relevance for impacts and policy. The new models and their projections are evaluated in WG I (Christensen et al., 2007a; Meehl et al., 2007; Randall et al., 2007) and compared with the pre-TAR results below. Projections of global mean annual temperature change for SRES and CO_2-stabilisation profiles are presented in Box 2.8.

Pre-TAR AOGCM results held at the DDC were included in a model intercomparison across the four SRES emissions scenarios (B1, B2, A2, and A1FI) of seasonal mean temperature and precipitation change for thirty-two world regions (Ruosteenoja et al., 2003).[9] The inter-model range of changes by the end of the 21st century is summarised in Figure 2.6 for the A2 scenario, expressed as rates of change per century. Recent A2 projections, reported in WG I, are also shown for the same regions for comparison.

Almost all model-simulated temperature changes, but fewer precipitation changes, were statistically significant relative to 95% confidence intervals calculated from 1,000-year unforced coupled AOGCM simulations (Ruosteenoja et al., 2003; see also Figure 2.6). Modelled surface air temperature increases in all regions and seasons, with most land areas warming more rapidly than the global average (Giorgi et al., 2001; Ruosteenoja et al., 2003). Warming is especially pronounced in high northern-latitude regions in the boreal winter and in southern Europe and parts of central and northern Asia in the boreal summer. Warming is less than the global average in southern parts of Asia and South America, Southern Ocean areas (containing many small islands) and the North Atlantic (Figure 2.6a).

For precipitation, both positive and negative changes are projected, but a regional precipitation increase is more common than a decrease. All models simulate higher precipitation at high latitudes in both seasons, in northern mid-latitude regions in boreal winter, and enhanced monsoon precipitation for southern and eastern Asia in boreal summer. Models also agree on precipitation declines in Central America, southern Africa and southern Europe in certain seasons (Giorgi et al., 2001; Ruosteenoja et al., 2003; see also Figure 2.6b).

Comparing TAR projections to recent projections

The WG I report provides an extensive intercomparison of recent regional projections from AOGCMs (Christensen et al., 2007a; Meehl et al., 2007), focusing on those assuming the SRES A1B emissions scenario, for which the greatest number of simulations (21) were available. It also contains numerous maps of projected regional climate change. In summary:

- The basic pattern of projected warming is little changed from previous assessments.
- The projected rate of warming by 2030 is insensitive to the choice of SRES scenarios.
- Averaged across the AOGCMs analysed, the global mean warming by 2090-2099 relative to 1980-1999 is projected to be 1.8, 2.8, and 3.4°C for the B1, A1B, and A2 scenarios, respectively. Local temperature responses in nearly all regions closely follow the ratio of global temperature response.
- Model-average mean local precipitation responses also roughly scale with the global mean temperature response across the emissions scenarios, though not as well as for temperature.
- The inter-model range of seasonal warming for the A2 scenario is smaller than the pre-TAR range at 2100 in most regions, despite the larger number of models (compare the red and blue bars in Figure 2.6a)
- The direction and magnitude of seasonal precipitation changes for the A2 scenario are comparable to the pre-TAR changes in most regions, while inter-model ranges are wider in some regions/seasons and narrower in others (Figure 2.6b).
- Confidence in regional projections is higher than in the TAR for most regions for temperature and for some regions for precipitation.

[9] Scatter diagrams are downloadable at: http://www.ipcc-data.org/sres/scatter_plots/scatterplots_region.html

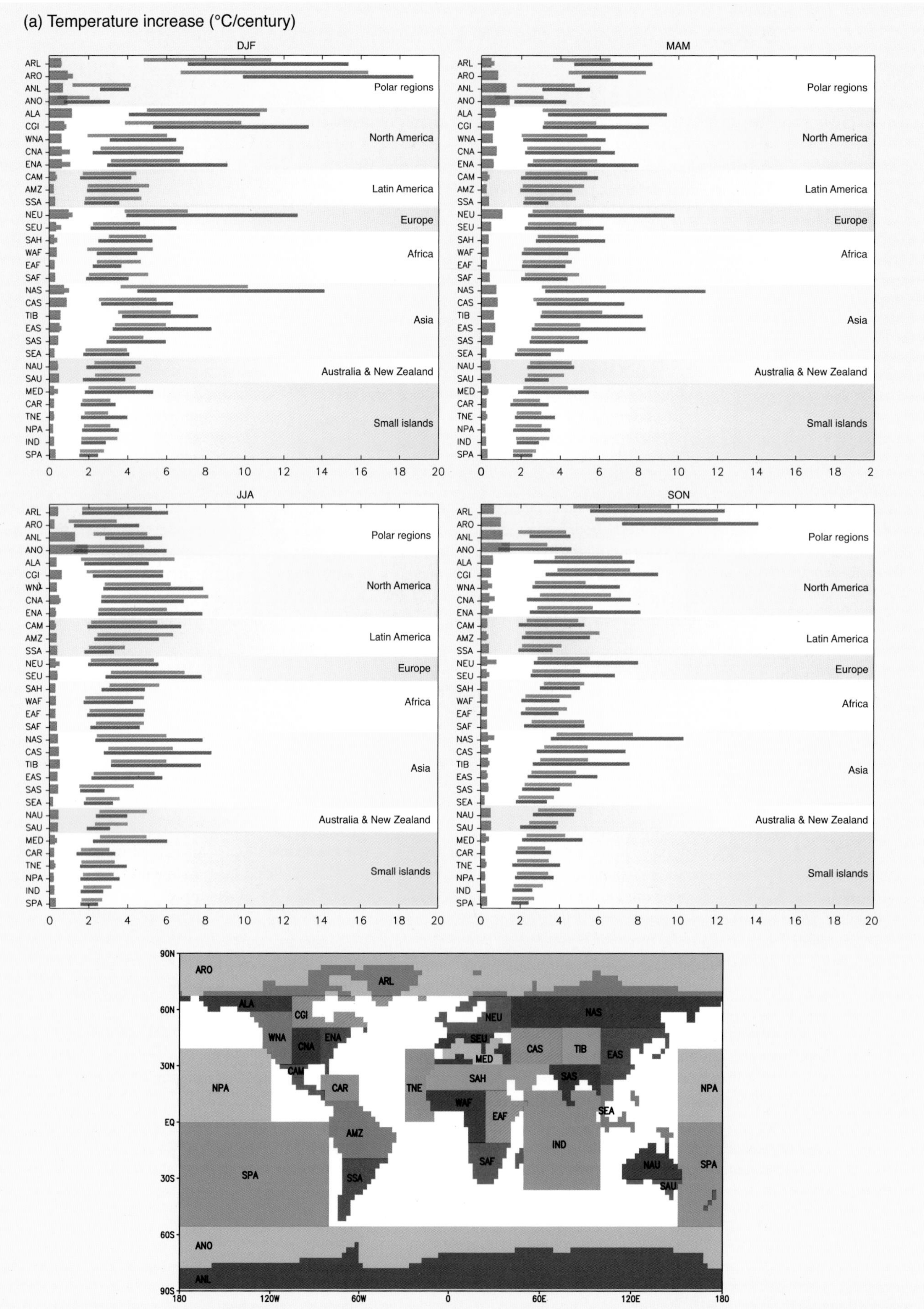
(a) Temperature increase (°C/century)
DJF
MAM
JJA
SON
ARL
ARO
ANL
ANO
ALA
CGI
WNA
CNA
ENA
CAM
AMZ
SSA
NEU
SEU
SAH
WAF
EAF
SAF
NAS
CAS
TIB
EAS
SAS
SEA
NAU
SAU
MED
CAR
TNE
NPA
IND
SPA
Polar regions
North America
Latin America
Europe
Africa
Asia
Australia & New Zealand
Small islands
0 2 4 6 8 10 12 14 16 18 20
90N
60N
30N
EQ
30S
60S
90S
180
120W
60W
0
60E
120E
180

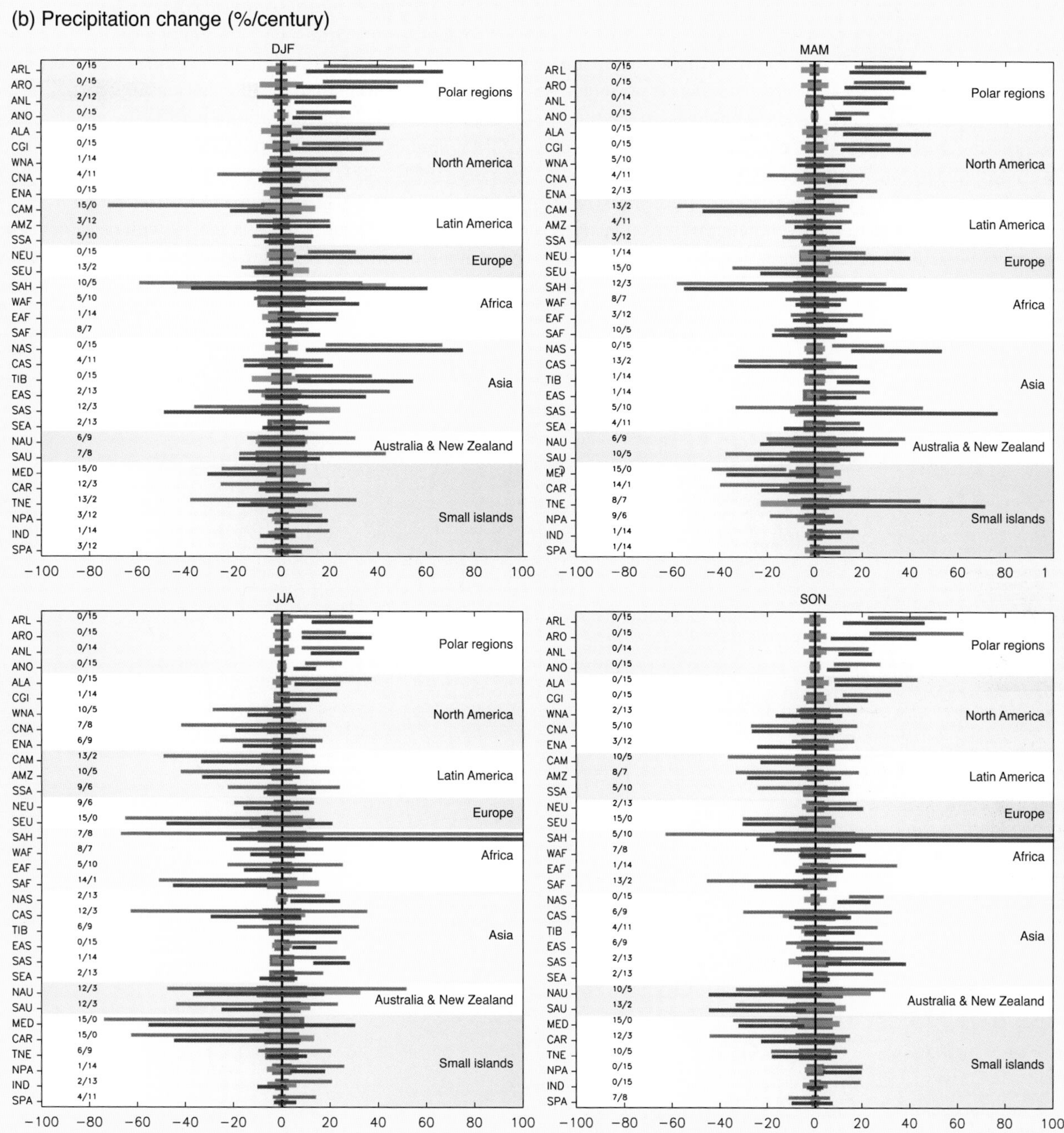

Figure 2.6. *AOGCM projections of seasonal changes in (a) mean temperature (previous page) and (b) precipitation up to the end of the 21st century for 32 world regions. For each region two ranges between minimum and maximum are shown. Red bar: range from 15 recent AOGCM simulations for the A2 emissions scenario (data analysed for Christensen et al., 2007a). Blue bar: range from 7 pre-TAR AOGCMs for the A2 emissions scenario (Ruosteenoja et al., 2003). Seasons: DJF (December–February); MAM (March–May); JJA (June–August); SON (September–November). Regional definitions, plotted on the ECHAM4 model grid (resolution 2.8 × 2.8°), are shown on the inset map (Ruosteenoja et al., 2003). Pre-TAR changes were originally computed for 1961-1990 to 2070-2099 and recent changes for 1979-1998 to 2079-2098, and are converted here to rates per century for comparison; 95% confidence limits on modelled 30-year natural variability are also shown based on millennial AOGCM control simulations with HadCM3 (mauve) and CGCM2 (green) for constant forcing (Ruosteenoja et al., 2003). Numbers on precipitation plots show the number of recent A2 runs giving negative/positive precipitation change. Percentage changes for the SAH region (Sahara) exceed 100% in JJA and SON due to low present-day precipitation.*
Key for (a) and (b):

- range of changes from seven pre-TAR AOGCMs for the A2 emissions scenario
- range of changes from 15 recent AOGCM simulations for the A2 emissions scenario
- 95% confidence limits on modelled 30-year natural variability based on HadCM3 millennial control simulation
- 95% confidence limits on modelled 30-year natural variability based on CGCM2 millennial control simulation

commonly consider only large-scale, period-averaged climate), requiring scenario analysis to be carried out offline. Interpretation of impacts then becomes problematic, requiring a method of relating the large-scale climate change represented in the IAM to the impacts of associated changes in weather extremes modelled offline. Goodess et al. suggest that a more direct, but untested, approach could be to construct conditional damage functions (cdfs), by identifying the statistical relationships between the extreme events themselves (causing damage) and large-scale predictor variables. Box 2.4 offers a global overview of observed and projected changes in extreme weather events.

2.4.6.2 Scenarios of atmospheric composition

Projections of atmospheric composition account for the concurrent effects of air pollution and climate change, which can be important for human health, agriculture and ecosystems. Scenarios of CO_2 concentration ([CO_2]) are needed in some CCIAV studies, as elevated [CO_2] can affect the acidity of the oceans (IPCC, 2007; Chapter 6, Section 6.3.2) and both the growth and water use of many terrestrial plants (Chapter 4, Section 4.4.1; Chapter 5, Section 5.4.1), with possible feedbacks on regional hydrology (Gedney et al., 2006). CO_2 is well mixed in the atmosphere, so concentrations at a single observing site will usually suffice to represent global conditions. Observed [CO_2] in

Box 2.4. SRES-based projections of climate variability and extremes

Possible changes in variability and the frequency/severity of extreme events are critical to undertaking realistic CCIAV assessments. Past trends in extreme weather and climate events, their attribution to human influence, and projected (SRES-forced) changes have been summarised globally by WG I (IPCC, 2007) and are reproduced in Table 2.2.

Table 2.2. *Recent trends, assessment of human influence on the trend, and projections for extreme weather events for which there is an observed late 20th century trend. Source: IPCC, 2007, Table SPM-2.*

Phenomenon and direction of trend	Likelihood[a] that trend occurred in late 20th century (typically post-1960)	Likelihood[a] of a human contribution to observed trend	Likelihood[a] of future trends based on projections for 21st century using SRES scenarios
Warmer and fewer cold days and nights over most land areas	Very likely[b]	Likely[c]	Virtually certain[c]
Warmer and more frequent hot days and nights over most land areas	Very likely[d]	Likely (nights)[c]	Virtually certain[c]
Warm spells/heatwaves. Frequency increases over most land areas	Likely	More likely than not[e]	Very likely
Heavy precipitation events. Frequency (or proportion of total rainfall from heavy falls) increases over most areas	Likely	More likely than not[e]	Very likely
Area affected by droughts increases	Likely in many regions since 1970s	More likely than not	Likely
Intense tropical cyclone activity increases	Likely in some regions since 1970	More likely than not[e]	Likely
Increased incidence of extreme high sea level (excludes tsunamis)[f]	Likely	More likely than not[e,g]	Likely[h]

Notes:

[a] The assessed likelihood, using expert judgement, of an outcome or a result: Virtually certain >99% probability of occurrence, Extremely likely >95%, Very likely >90%, Likely >66%, More likely than not >50%.

[b] Decreased frequency of cold days and nights (coldest 10%).

[c] Warming of the most extreme days and nights each year.

[d] Increased frequency of hot days and nights (hottest 10%).

[e] Magnitude of anthropogenic contributions not assessed. Attribution for these phenomena based on expert judgement rather than formal attribution studies.

[f] Extreme high sea level depends on average sea level and on regional weather systems. It is defined here as the highest 1% of hourly values of observed sea level at a station for a given reference period.

[g] Changes in observed extreme high sea level closely follow the changes in average sea level. It is very likely that anthropogenic activity contributed to a rise in average sea level.

[h] In all scenarios, the projected global average sea level at 2100 is higher than in the reference period. The effect of changes in regional weather systems on sea-level extremes has not been assessed.

2005 was about 379 ppm (Forster et al., 2007) and was projected in the TAR using the Bern-CC model to rise by 2100 to reference, low, and high estimates for the SRES marker scenarios of B1: 540 [486 to 681], A1T: 575 [506 to 735], B2: 611[544 to 769], A1B: 703 [617 to 918], A2: 836 [735 to 1080], and A1FI: 958 [824 to 1248] ppm (Appendix II in IPCC, 2001a). Values similar to these reference levels are commonly adopted in SRES-based impact studies; for example, Arnell et al. (2004) employed levels assumed in HadCM3 AOGCM climate simulations, and Schröter et al. (2005b) used levels generated by the IMAGE-2 integrated assessment model. However, recent simulations with coupled carbon cycle models indicate an enhanced rise in [CO_2] for a given emissions scenario, due to feedbacks from changing climate on the carbon cycle, suggesting that the TAR reference estimates are conservative (Meehl et al., 2007).

Elevated levels of ground-level ozone (O_3) are toxic to many plants (see Chapter 5, Box 5.2) and are strongly implicated in a range of respiratory diseases (Chapter 8, Section 8.2.6). Increased atmospheric concentrations of sulphur dioxide are detrimental to plants, and wet and dry deposition of atmospheric sulphur and nitrogen can lead to soil and surface water acidification, while nitrogen deposition can also serve as a plant fertiliser (Carter et al., 2001; see also Chapter 4, Section 4.4.1; Chapter 5, Section 5.4.3.1). Projections with global atmospheric chemistry models for the high-emissions SRES A2 scenario indicate that global mean tropospheric O_3 concentrations could increase by 20 to 25% between 2015 and 2050, and by 40 to 60% by 2100, primarily as a result of emissions of NO_x, CH_4, CO_2, and compounds from fossil fuel combustion (Meehl et al., 2007). Stricter air pollution standards, already being implemented in many regions, would reduce, and could even reverse, this projected increase (Meehl et al., 2007). Similarly, the range of recent scenarios of global sulphur and NO_x emissions that account for new abatement policies has shifted downwards compared with the SRES emissions scenarios (Smith et al., 2005; Nakićenović et al., 2007).

For the purposes of CCIAV assessment, global projections of pollution are only indicative of local conditions. Levels are highly variable in space and time, with the highest values typically occurring over industrial regions and large cities. Although projections are produced routinely for some regions in order to support air pollution policy using high-resolution atmospheric transport models (e.g., Syri et al., 2004), few models have been run assuming an altered climate, and simulations commonly assume emissions scenarios developed for air pollution policy rather than climate policy (see Alcamo et al., 2002; Nakićenović et al., 2007). Exceptions include regionally explicit global scenarios of nitrogen deposition on a 0.5° latitude × 0.5° longitude grid for studying biodiversity loss in the Millennium Ecosystem Assessment (Alcamo et al., 2005) and simulations based on SRES emissions for sulphur and nitrogen over Europe (Mayerhofer et al., 2002) and Finland (Syri et al., 2004), and for surface ozone in Finland (Laurila et al., 2004).

2.4.6.3 Sea-level scenarios

A principal impact projected under global warming is sea-level rise. Some basic techniques for developing sea-level scenarios were described in the TAR (Carter et al., 2001). Since the TAR, methodological refinements now account more effectively for regional and local factors affecting sea level and, in so doing, produce scenarios that are more relevant for planning purposes. Two main types of scenario are distinguished here: regional sea level and storm surges. A third type, characterising abrupt sea-level rise, is described in Section 2.4.7. Analogue approaches have also been reported (e.g., Arenstam Gibbons and Nicholls, 2006). More details on sea level and sea-level scenarios can be found in Bindoff et al. (2007), Meehl et

Box 2.5. SRES-based sea-level scenarios

At the global level, simple models representing the expansion of sea water and melting/sliding of land-based ice sheets and glaciers were used in the TAR to obtain estimates of globally averaged mean sea-level rise across the SRES scenarios, yielding a range of 0.09 to 0.88 m by 2100 relative to 1990 (Church et al., 2001). This range has been reassessed by WG I, yielding projections relative to 1980-1999 for the six SRES marker scenarios of B1: 0.18 to 0.38 m, A1T: 0.20 to 0.45 m, B2: 0.20 to 0.43 m, A1B: 0.21 to 0.48 m, A2: 0.23 to 0.51 m, and A1FI: 0.26 to 0.59 m (Meehl et al., 2007). Thermal expansion contributes about 60 to 70% to these estimates. Projections are smaller than given in the TAR, due mainly to improved estimates of ocean heat uptake but also to smaller assessed uncertainties in glacier and ice cap changes. However, uncertainties in carbon cycle feedbacks, ice flow processes, and recent observed ice discharge rates are not accounted for due to insufficient understanding (Meehl et al., 2007).

A number of studies have made use of the TAR sea-level scenarios. In a global study of coastal flooding and wetland loss, Nicholls (2004) used global mean sea-level rise estimates for the four SRES storylines by 2025, 2055, and 2085. These were consistent with climate scenarios used in parallel studies (see Section 2.4.6.4). Two subsidence rates were also applied to obtain relative sea level rise in countries already experiencing coastal subsidence. The United Kingdom Climate Impacts Programme adopted the TAR global mean sea-level rise estimates in national scenarios out to the 2080s. Scenarios of high water levels were also developed by combining mean sea-level changes with estimates of future storminess, using a storm surge model (Hulme et al., 2002). SRES-based sea-level scenarios accounting for global mean sea level, local land uplift, and estimates of the water balance of the Baltic Sea were estimated for the Finnish coast up to 2100 by Johansson et al. (2004), along with calculations of uncertainties and extreme high water levels.

al. (2007) and Chapter 6 of this volume. Examples of SRES-based sea-level scenarios are provided in Box 2.5.

Regional sea-level scenarios

Sea level does not change uniformly across the world under a changing climate, due to variation in ocean density and circulation changes. Moreover, long-term, non-climate-related trends, usually associated with vertical land movements, may affect relative sea level. To account for regional variations, Hulme et al. (2002) recommend applying the range of global-mean scenarios ±50% change. Alternative approaches utilise scenario generators. The Dynamic Interactive Vulnerability Assessment (DIVA) model computes relative sea-level rise scenarios using either global-mean or regional patterns of sea-level rise scenarios from CLIMBER-2, a climate model of intermediate complexity (Petoukhov et al., 2000; Ganopolski et al., 2001). CLIMsystems (2005) have developed a software tool that rapidly generates place-based future scenarios of sea-level change during the 21st century, accounting for global, regional, and local factors. Spatial patterns of sea-level rise due to thermal expansion and ocean processes from AOGCM simulations are combined with global-mean sea-level rise projections from simple climate models through the pattern-scaling technique (Santer et al., 1990). Users can specify a value for the local sea-level trends to account for local land movements.

Storm surge scenarios

In many locations, the risk of extreme sea levels is poorly characterised even under present-day climatic conditions, due to sparse tide gauge networks and relatively short records of high measurement frequency. Where such records do exist, detectable trends are highly dependent on local conditions (Woodworth and Blackman, 2004). Box 6.2 in Chapter 6 summarises several recent studies that employ extreme water level scenarios. Two methods were employed to develop these scenarios, one using a combination of stochastic sampling and dynamic modelling, the other using downscaled regional climate projections from global climate models to drive barotropic storm surge models (Lowe and Gregory, 2005).

2.4.6.4 Socio-economic scenarios

Socio-economic changes are key drivers of projected changes in future emissions and climate, and are also key determinants of most climate change impacts, potential adaptations and vulnerability (Malone and La Rovere, 2005). Furthermore, they also influence the policy options available for responding to climate change. CCIAV studies increasingly include scenarios of changing socio-economic conditions, which can substantially alter assessments of the effects of future climate change (Parry, 2004; Goklany, 2005; Hamilton et al., 2005; Schröter et al., 2005b; Alcamo et al., 2006a). Typically these assessments need information at the sub-national level, whereas many scenarios are developed at a broader scale, requiring downscaling of aggregate socio-economic scenario information.

Guidelines for the analysis of current and projected socio-economic conditions are part of the UNDP Adaptation Policy Framework (Malone and La Rovere, 2005). They advocate the use of indicators to characterise socio-economic conditions and prospects. Five categories of indicators are suggested: demographic, economic, natural resource use, governance and policy, and cultural. Most recent studies have focused on the first two of these.

The sensitivity of climate change effects to socio-economic conditions was highlighted by a series of multi-sector impact assessments (Parry et al., 1999, 2001; Parry, 2004; see Table 2.3). Two of these assessments relied on only a single representation of future socio-economic conditions (IS92a), comparing effects of mitigated versus unmitigated climate change (Arnell et al., 2002; Nicholls and Lowe, 2004). The third set considered four alternative SRES-based development pathways (see Box 2.6), finding that these assumptions are often a stronger determinant of impacts than climate change itself (Arnell, 2004; Arnell et al., 2004; Levy et al., 2004; Nicholls, 2004; Parry et al., 2004; van Lieshout et al., 2004). Furthermore, climate impacts can themselves depend on the development pathway, emphasising the limited value of impact assessments of human systems that overlook possible socio-economic changes.

The advantages of being able to link regional socio-economic futures directly to global scenarios and storylines are now being recognised. For example, the SRES scenarios have been used as a basis for developing storylines and quantitative scenarios at national (Carter et al., 2004, 2005; van Vuuren et al., 2007) and sub-national (Berkhout et al., 2002; Shackley and Deanwood, 2003; Solecki and Oliveri, 2004; Heslop-Thomas et al., 2006) scales. In contrast, most regional studies in the AIACC (Assessments of Impacts and Adaptations to Climate Change in Multiple Regions and Sectors) research programme adopted a participatory, sometimes ad hoc, approach to socio-economic scenario development, utilising current trends in key socio-economic indicators and stakeholder consultation (e.g., Heslop-Thomas et al., 2006; Pulhin et al., 2006).

Methods for downscaling quantitative socio-economic information have focused on population and gross domestic product (GDP). The downscaling of population growth has evolved beyond simple initial exercises that made the sometimes unrealistic assumption that rates of population change are uniform over an entire world region (Gaffin et al., 2004). New techniques account for differing demographic conditions and outlooks at the national level (Grübler et al., 2006; van Vuuren et al., 2007). New methods of downscaling to the sub-national level include simple rules for preferential growth in coastal areas (Nicholls, 2004), extrapolation of recent trends at the local area level (Hachadoorian et al., 2007), and algorithms leading to preferential growth in urban areas (Grübler et al., 2006; Reginster and Rounsevell, 2006).

Downscaling methods for GDP are also evolving. The first downscaled SRES GDP assumptions applied regional growth rates uniformly to all countries within the region (Gaffin et al., 2004) without accounting for country-specific differences in initial conditions and growth expectations. New methods assume various degrees of convergence across countries, depending on the scenario; a technique that avoids implausibly high growth for rich countries in developing regions (Grübler et al., 2006; van Vuuren et al., 2007). GDP scenarios have also been downscaled to the sub-national level, either by assuming constant shares of GDP in each grid cell (Gaffin et al., 2004; van Vuuren et al., 2007) or through algorithms that differentiate income across urban and rural areas (Grübler et al., 2006).

Table 2.3. *Key features of scenarios underlying three global-scale, multi-sector assessments: [a] Parry et al. (1999); [b] Arnell et al. (2002); [c] Parry (2004).*

	Impacts of unmitigated emissions [a]	Impacts of stabilisation of CO_2 concentrations [b]	Impacts of SRES emissions scenarios [c]
Emissions scenarios	IS92a (1% per increase in CO_2-equivalent concentrations per year from 1990)	Stabilisation at 750 and 550 ppm	Four SRES emissions scenarios: A1FI, A2, B1, and B2
Climate scenarios (AOGCM-based)	Derived from four ensemble HadCM2 simulations and one HadCM3 simulation forced with IS92a emissions scenarios	Derived from HadCM2 experiments assuming stabilisation at 550 and 750 ppm; comparison with IS92a	Derived from HadCM3 ensemble experiments (number of runs in brackets): A1FI (1), A2 (3), B1 (1), and B2 (2)
Socio-economic scenarios	IS92a-consistent GDP[a] and population projections	IS92a-consistent GDP[a] and population projections	SRES-based socio-economic projections

[a] GDP = Gross Domestic Product.

2.4.6.5 Land-use scenarios

Many CCIAV studies need to account for future changes in land use and land cover. This is especially important for regional studies of agriculture and water resources (Barlage et al., 2002; Klöcking et al., 2003), forestry (Bhadwal and Singh, 2002), and ecosystems (Bennett et al., 2003; Dirnbock et al., 2003; Zebisch et al., 2004; Cumming et al., 2005), but also has a large influence on regional patterns of demography and economic activity (Geurs and van Eck, 2003) and associated problems of environmental degradation (Yang et al., 2003) and pollution (Bathurst et al., 2005). Land-use and land-cover change scenarios have also been used to analyse feedbacks to the climate system (DeFries et al., 2002; Leemans et al., 2002; Maynard and Royer, 2004) and sources and sinks of GHGs (Fearnside, 2000; El-Fadel et al., 2002; Sands and Leimbach, 2003).

The TAR concluded that the use of Integrated Assessment Models (IAMs) was the most appropriate method for developing land-use change scenarios, and they continue to be the only available tool for global-scale studies. Since the TAR, however, a number of new models have emerged that provide fresh insights into regional land-use change. These regional models

Box 2.6. SRES-based socio-economic characterisations

SRES provides socio-economic information in the form of storylines and quantitative assumptions on population, gross domestic product (GDP), and rates of technological progress for four large world regions (OECD-1990, Reforming Economies, Africa + Latin America + Middle East, and Asia). Since the TAR, new information on several of the SRES driving forces has been published (see also the discussion in Nakićenović et al., 2007). For example, the range of global population size projections made by major demographic institutions has reduced by about 1–2 billion since the preparation of SRES (van Vuuren and O'Neill, 2006). Nevertheless, most of the population assumptions used in SRES still lie within the range of current projections, with the exception of some regions of the A2 scenario which now lie somewhat above it (van Vuuren and O'Neill, 2006). Researchers are now producing alternative interpretations of SRES population assumptions or new projections for use in climate change studies (Hilderink, 2004; O'Neill, 2004; Fisher et al., 2006; Grübler et al., 2006).

SRES GDP growth assumptions for the ALM region (Africa, Latin America and Middle East) are generally higher than those of more recent projections, particularly for the A1 and B1 scenarios (van Vuuren and O'Neill, 2006). The SRES GDP assumptions are generally consistent with recent projections for other regions, including fast-growing regions in Asia and, given the small share of the ALM region in global GDP, for the world as a whole.

For international comparison, economic data must be converted into a common unit; the most common choice is US$ based on market exchange rates (MER). Purchasing-power-parity (PPP) estimates, in which a correction is made for differences in price levels among countries, are considered a better alternative for comparing income levels across regions and countries. Most models and economic projections, however, use MER-based estimates, partly due to a lack of consistent PPP-based data sets. It has been suggested that the use of MER-based data results in inflated economic growth projections (Castles and Henderson, 2003). In an ongoing debate, some researchers argue that PPP is indeed a better measure and that its use will, in the context of scenarios of economic convergence, lead to lower economic growth and emissions paths for developing countries. Others argue that consistent use of either PPP- or MER-based data and projections will lead to, at most, only small changes in emissions. This debate is summarised by Nakićenović et al. (2007), who conclude that the impact on emissions of the use of alternative GDP metrics is likely to be small, but indicating alternative positions as well (van Vuuren and Alfsen, 2006). The use of these alternative measures is also likely to affect CCIAV assessments (Tol, 2006), especially where vulnerability and adaptive capacity are related to access to locally traded goods and services.

can generate very different land-use change scenarios from those generated by IAMs (Busch, 2006), often with opposing directions of change. However, the need to define outside influences on land use in regional-scale models, such as global trade, remains a challenge (e.g., Sands and Edmonds, 2005; Alcamo et al., 2006b), so IAMs have an important role to play in characterising the global boundary conditions for regional land-use change assessments (van Meijl et al., 2006).

Regional-scale land-use models often adopt a two-phase (nested scale) approach with an assessment of aggregate quantities of land use for the entire region followed by 'downscaling' procedures to create regional land-use patterns (see Box 2.7 for examples). Aggregate quantities are often based on IAMs or economic models such as General Equilibrium Models (van Meijl et al., 2006) or input-output approaches (Fischer and Sun, 2001). Methods of downscaling vary considerably and include proportional approaches to estimate regional from global scenarios (Arnell et al., 2004), regional-scale economic models (Fischer and Sun, 2001), spatial allocation procedures based on rules (Rounsevell et al., 2006), micro-simulation with cellular automata (de Nijs et al., 2004; Solecki and Oliveri, 2004), linear programming models (Holman et al., 2005a, b), and empirical-statistical techniques (de Koning et al., 1999; Verburg et al., 2002, 2006). In addressing climate change impacts on land use, Agent-Based Models (ABMs: see Alcamo et al., 2006b) aim to provide insight into the decision processes and social interactions that underpin adaptation and vulnerability assessment (Acosta-Michlik and Rounsevell, 2005).

Most land-use scenario assessments are based on gradual changes in socio-economic and climatic conditions, although responses to extreme weather events such as Hurricane Mitch in Central America have also been assessed (Kok and Winograd, 2002). Probabilistic approaches are rare, with the exception being the effects of uncertainty in alternative representations of land-use change for hydrological variables (Eckhardt et al., 2003). Not all land-use scenario exercises have addressed the effects of climate change even though they consider time-frames over which a changing climate would be important. This may reflect a perceived lack of sensitivity to climate variables (e.g., studies on urban land use: see Allen and Lu, 2003; Barredo et al., 2003, 2004; Loukopoulos and Scholz, 2004; Reginster and Rounsevell, 2006), or may be an omission from the analysis (Ahn et al., 2002; Berger and Bolte, 2004).

2.4.6.6 Technology scenarios

The importance of technology has been highlighted specifically for land-use change (Ewert et al., 2005; Rounsevell et al., 2005, 2006; Abildtrup et al., 2006) and for ecosystem service changes, such as agricultural production, water management, or climate regulation (Easterling et al., 2003; Nelson et al., 2005). Technological change is also a principal driver of GHG emissions. Since the TAR, scenarios addressing different technology pathways for climate change mitigation and adaptation have increased in number (see Nakićenović et al., 2007). Technological change can be treated as an exogenous factor to the economic system or be endogenously driven through economic and political incentives. Recent modelling exercises have represented theories on technical and institutional innovation, such as the 'Induced Innovation Theory', in scenario development (Grübler et al., 1999; Grubb et al., 2002), although more work is needed to refine these methods.

For integrated global scenario exercises, the rate and magnitude of technological development is often based on expert judgements and mental models. Storyline assumptions are then used to modify the input parameters of environmental models (e.g., for ecosystems, land use, or climate) prior to conducting model simulations (e.g., Millennium Ecosystem Assessment, 2005; Ewert et al., 2005). Such an approach is useful in demonstrating the relative sensitivity of different systems to technological change, but the role of technology remains a key uncertainty in characterisations of the future, with some arguing that only simple models should be used in constructing scenarios (Casman et al., 1999). In particular, questions such as about the rates of uptake and diffusion of new technologies deserve greater attention, especially as this affects adaptation to climate change (Easterling et al., 2003). However, only a few studies have tackled technology, suggesting an imbalance in the treatment of environmental change drivers within many CCIAV scenario studies, which future work should seek to redress.

2.4.6.7 Adaptation scenarios

Limited attention has been paid to characterising alternative pathways of future adaptation. Narrative information within scenarios can assist in characterising potential adaptive responses to climate change. For instance, the determinants of adaptive capacity and their indicators have been identified for Europe through questionnaire survey (Schröter et al., 2005b). Empirical relationships between these indicators and population and GDP from 1960 to 2000 were also established and applied to downscaled, SRES-based GDP and population projections in order to derive scenarios of adaptive capacity (see Section 2.4.6.4). The SRES storylines have also been interpreted using GDP per capita scenarios to estimate, in one study, the exposure of human populations under climate change to coastal flooding, based on future standards of coastal defences (Nicholls, 2004) and, in a second, access to safe water with respect to the incidence of diarrhoea (Hijioka et al., 2002). The rate of adaptation to climate change was analysed for the agriculture sector using alternative scenarios of innovation uptake (Easterling et al., 2003) by applying different maize yields, representing adaptation scenarios ranging from no adaptation through lagged adaptation rates and responses (following a logistic curve) to perfect (clairvoyant) adaptation (Easterling et al., 2003). This work showed the importance of implied adaptation rates at the farm scale, indicating that clairvoyant approaches to adaptation (most commonly used in CCIAV studies) are likely to overestimate the capacity of individuals to respond to climate change.

One adaptation strategy not considered by Easterling et al. (2003) was land-use change, in the form of autonomous adaptation to climate change driven by the decisions of individual land users (Berry et al., 2006). The land-use change scenarios reported previously can, therefore, be thought of as adaptation scenarios. Future studies, following consultation with key stakeholders, are more likely to include adaptation explicitly

Box 2.7. SRES-based land-use and land-cover characterisations

Future land use was estimated by most of the IAMs used to characterise the SRES storylines, but estimates for any one storyline are model-dependent, and therefore vary widely. For example, under the B2 storyline, the change in the global area of grassland between 1990 and 2050 varies between −49 and +628 million ha (Mha), with the marker scenario giving a change of +167 Mha (Nakićenović et al., 2000). The IAM used to characterise the A2 marker scenario did not include land-cover change, so changes under the A1 scenario were assumed to apply also to A2. Given the differences in socio-economic drivers between A1 and A2 that can affect land-use change, this assumption is not appropriate. Nor do the SRES land-cover scenarios include the effect of climate change on future land cover. This lack of internal consistency will especially affect the representation of agricultural land use, where changes in crop productivity play an important role (Ewert et al., 2005; Audsley et al., 2006). A proportional approach to downscaling the SRES land-cover scenarios has been applied to global ecosystem modelling (Arnell et al., 2004) by assuming uniform rates of change everywhere within an SRES macro-region. In practice, however, land-cover change is likely to be greatest where population and population growth rates are greatest. A mismatch was also found in some of the SRES storylines, and for some regions, between recent trends and projected trends for cropland and forestry (Arnell et al., 2004).

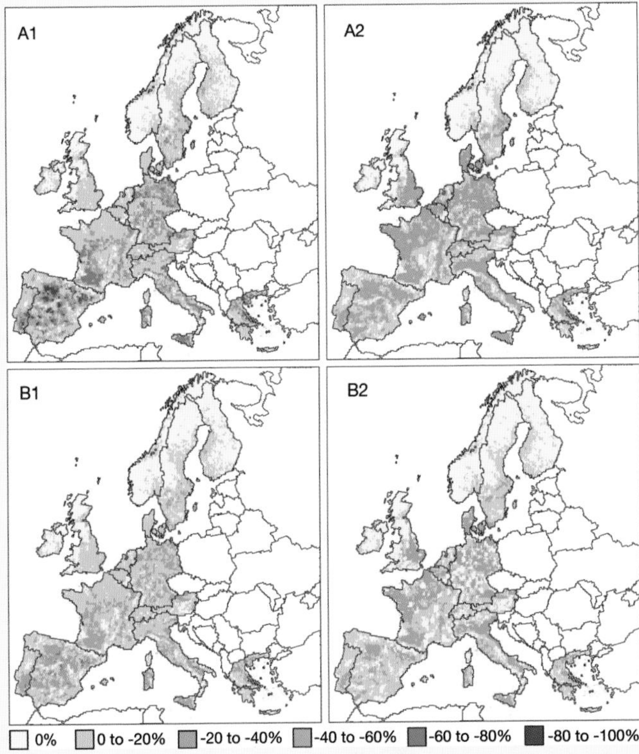

Figure 2.7. *Percentage change in cropland area (for food production) by 2080, compared with the baseline in 2000 for the four SRES storylines (A1FI, A2, B1, B2) with climate calculated by the HadCM3 AOGCM. From Schröter et al., 2005b. Reprinted with permission from AAAS.*

More sophisticated downscaling of the SRES scenarios has been undertaken at the regional scale within Europe (Kankaanpää and Carter, 2004; Ewert et al., 2005; Rounsevell et al., 2005, 2006; Abildtrup et al., 2006; Audsley et al., 2006; van Meijl et al., 2006). These analyses highlighted the potential role of non-climate change drivers in future land-use change. Indeed, climate change was shown in many examples to have a negligible effect on land use compared with socio-economic change (Schröter et al., 2005b). Technology, especially as it affects crop yield development, is an important determinant of future agricultural land use (and much more important than climate change), contributing to declines in agricultural areas of both cropland and grassland by as much as 50% by 2080 under the A1FI and A2 scenarios (Rounsevell et al., 2006). Such declines in land use did not occur within the B2 scenario, which assumes more extensive agricultural management, such as 'organic' production systems, or the widespread substitution of agricultural food and fibre production by bioenergy crops. This highlights the role of policy decisions in moderating future land-use change. However, broad-scale changes often belie large potential differences in the spatial distribution of land-use change that can occur at the sub-regional scale (Schröter et al., 2005b; see also Figure 2.7), and these spatial patterns may have greater effects on CCIAV than the overall changes in land-use quantities (Metzger et al., 2006; Reidsma et al., 2006).

as part of socio-economic scenario development, hence offering the possibility of gauging the effectiveness of adaptation options in comparison to scenarios without adaptation (Holman et al., 2005b).

2.4.6.8 Mitigation/stabilisation scenarios

Mitigation scenarios (also known as climate intervention or climate policy scenarios) are defined in the TAR (Morita et al., 2001), as scenarios that "(1) include explicit policies and/or measures, the primary goal of which is to reduce GHG emissions (e.g., carbon taxes) and/or (2) mention no climate policies and/or measures, but assume temporal changes in GHG emission sources or drivers required to achieve particular climate targets (e.g., GHG emission levels, GHG concentration levels, radiative forcing levels, temperature increase or sea level rise limits)." Stabilisation scenarios are an important subset of inverse mitigation scenarios, describing futures in which emissions reductions are undertaken so that GHG concentrations, radiative forcing, or global average temperature change do not exceed a prescribed limit.

Although a wide variety of mitigation scenarios have been developed, most focus on economic and technological aspects of emissions reductions (see Morita et al., 2001; van Vuuren et al., 2006; Nakićenović et al., 2007). The lack of detailed climate change projections derived from mitigation scenarios has hindered impact assessment. Simple climate models have been used to explore the implications for global mean temperature (see Box 2.8 and Nakićenović et al., 2007), but few AOGCM runs have been undertaken (see Meehl et al., 2007, for recent examples), with few direct applications in regional impact assessments (e.g., Parry et al., 2001). An alternative approach uses simple climate model projections of global warming under stabilisation to scale AOGCM patterns of climate change assuming unmitigated emissions, and then uses the resulting scenarios to assess regional impacts (e.g., Bakkenes et al., 2006).

The scarcity of regional socio-economic, land-use and other detail commensurate with a mitigated future has also hindered impact assessment (see discussion in Arnell et al., 2002). Alternative approaches include using SRES scenarios as surrogates for some stabilisation scenarios (Swart et al., 2002; see Table 2.4), for example to assess impacts on ecosystems (Leemans and Eickhout, 2004) and coastal regions (Nicholls and Lowe, 2004), demonstrating that socio-economic assumptions are a key determinant of vulnerability. Note that WG I reports AOGCM experiments forced by the SRES A1B and B1 emissions pathways up to 2100 followed by stabilisation of concentrations at roughly 715 and 550 ppm CO_2 (equated to 835 and 590 ppm equivalent CO_2, accounting for other GHGs: see Meehl et al., 2007).

A second approach associates impacts with particular levels or rates of climate change and may also determine the emissions and concentration paths that would avoid these outcomes.

Box 2.8. CO_2 stabilisation and global mean temperature response

Global mean annual temperature (GMAT) is the metric most commonly employed by the IPCC and adopted in the international policy arena to summarise future changes in global climate and their likely impacts (see Chapter 19, Box 19.2). Projections of global mean warming during the 21st century for the six SRES illustrative scenarios are presented by WG I (Meehl et al., 2007) and summarised in Figure 2.8. These are baseline scenarios assuming no explicit climate policy (see Box 2.2). A large number of impact studies reported by WG II have been conducted for projection periods centred on the 2020s, 2050s and 2080s[10], but only best estimates of GMAT change for these periods were available for three SRES scenarios based on AOGCMs (coloured dots in the middle panel of Figure 2.8). Best estimates (red dots) and likely ranges (red bars) for all six SRES scenarios are reported only for the period 2090-2099. Ranges are based on a hierarchy of models, observational constraints and expert judgement (Meehl et al., 2007).

A more comprehensive set of projections for these earlier time periods as well as the 2090s is presented in the lower panel of Figure 2.8. These are based on a simple climate model (SCM) and are also reported in WG I (Meehl et al., 2007, Figure 10.26). Although SCM projections for 2090-2099 contributed to the composite information used to construct the likely ranges shown in the middle panel, the projections shown in the middle and lower panels should not be compared directly as they were constructed using different approaches. The SCM projections are included to assist the reader in interpreting how the timing and range of uncertainty in projections of warming can vary according to emissions scenario. They indicate that the rate of warming in the early 21st century is affected little by different emissions scenarios (brown bars in Figure 2.8), but by mid-century the choice of emissions scenario becomes more important for the magnitude of warming (blue bars). By late century, differences between scenarios are large (e.g. red bars in middle panel; orange and red bars in lower panel), and multi-model mean warming for the lowest emissions scenario (B1) is more than 2°C lower than for the highest (A1FI).

GHG mitigation is expected to reduce GMAT change relative to baseline emissions, which in turn could avoid some adverse impacts of climate change. To indicate the projected effect of mitigation on temperature during the 21st century, and in the

[10] 30-year averaging periods for model projections held at the IPCC Data Distribution Centre.

absence of more recent, comparable estimates in the WG I report, results from the Third Assessment Report based on an earlier version of the SCM are reproduced in the upper panel of Figure 2.8 from the Third Assessment Report. These portray the GMAT response for four CO_2-stabilisation scenarios by three dates in the early (2025), mid (2055), and late (2085) 21st century. WG I does report estimates of equilibrium warming for CO_2-equivalent stabilisation (Meehl et al., 2007)[11]. Note that equilibrium temperatures would not be reached until decades or centuries after greenhouse gas stabilisation.

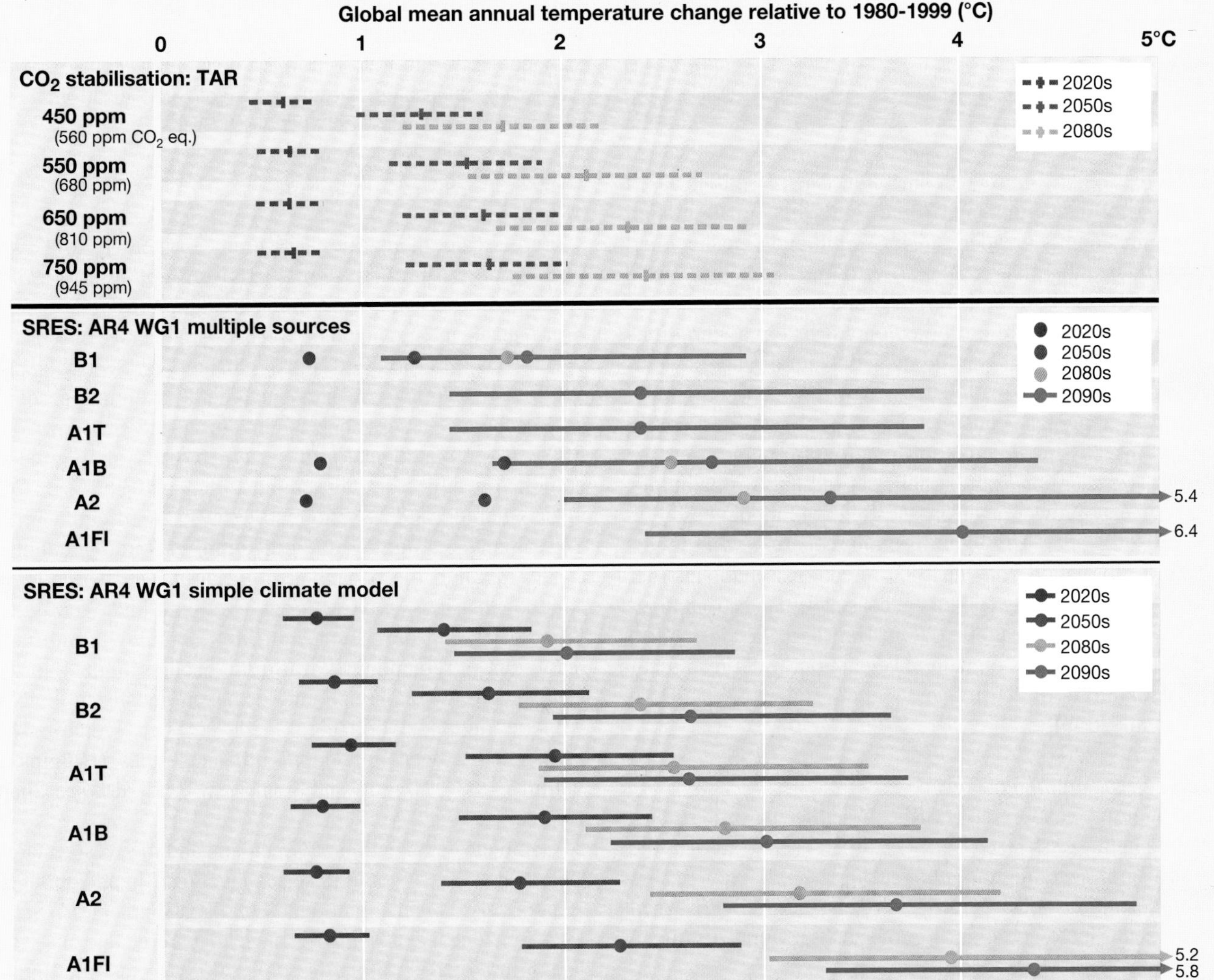

Figure 2.8. *Projected ranges of global mean annual temperature change during the 21st century for CO_2-stabilisation scenarios (upper panel, based on the TAR) and for the six illustrative SRES scenarios (middle and lower panels, based on the WG I Fourth Assessment). Different approaches have been used to obtain the estimates shown in the three panels, which are not therefore directly comparable.* ***Upper panel.*** *Projections for four CO_2-stabilisation profiles using a simple climate model (SCM) tuned to seven AOGCMs (IPCC, 2001c, Figure SPM-6; IPCC, 2001a, Figure 9.17). Broken bars indicate the projected mean (tick mark) and range of warming across the AOGCM tunings by the 2020s (brown), 2050s (blue) and 2080s (orange) relative to 1990. Time periods are based on calculations for 2025, 2055 and 2085. Approximate CO_2-equivalent values – including non-CO_2 greenhouse gases – at the time of CO_2-stabilisation (ppm) are also shown.* ***Middle panel.*** *Best estimates (red dots) and likely range (red bars) of warming by 2090-2099 relative to 1980-1999 for all six illustrative SRES scenarios and best estimates (coloured dots) for SRES B1, A1B and A2 by 2020-2029, 2050-2059 and 2080-2089 (IPCC, 2007, Figure SPM.5).* ***Lower panel.*** *Estimates based on an SCM tuned to 19 AOGCMs for 2025 (representing the 2020s), 2055 (2050s) and 2085 (2080s). Coloured dots represent the mean for the 19 model tunings and medium carbon cycle feedback settings. Coloured bars depict the range between estimates calculated assuming low carbon cycle feedbacks (mean - 1 SD) and those assuming high carbon cycle feedbacks (mean + 1 SD), approximating the range reported by Friedlingstein et al., 2006. Note that the ensemble average of the tuned versions of the SCM gives about 10% greater warming over the 21st century than the mean of the corresponding AOGCMs. (Meehl et al., 2007, Figure 10.26 and Appendix 10.A.1). To express temperature changes relative to 1850-1899, add 0.5°C.*

[11] Best estimate and likely range of equilibrium warming for seven levels of CO_2-equivalent stabilisation: 350 ppm, 1.0°C [0.6–1.4]; 450 ppm, 2.1°C [1.4–3.1]; 550 ppm, 2.9°C [1.9–4.4]; 650 ppm, 3.6°C [2.4–5.5]; 750 ppm, 4.3°C [2.8–6.4]; 1,000 ppm, 5.5°C [3.7–8.3] and 1,200 ppm, 6.3°C [4.2–9.4] (Meehl et al., 2007, Table 10.8).

Table 2.4. *The six SRES illustrative scenarios and the stabilisation scenarios (parts per million CO_2) they most resemble (based on Swart et al., 2002).*

SRES illustrative scenario	Description of emissions	Surrogate stabilisation scenario
A1FI	High end of SRES range	Does not stabilise
A1B	Intermediate case	750 ppm
A1T	Intermediate/low case	650 ppm
A2	High case	Does not stabilise
B1	Low end of SRES range	550 ppm
B2	Intermediate/low case	650 ppm

Climate change and impact outcomes have been identified based on criteria for dangerous interference with the climate system (Mastrandrea and Schneider, 2004; O'Neill and Oppenheimer, 2004; Wigley, 2004; Harvey, 2007) or on meta-analysis of the literature (Hitz and Smith, 2004). A limitation of these types of analyses is that they are not based on consistent assumptions about socio-economic conditions, adaptation and sectoral interactions, and regional climate change.

A third approach constructs a single set of scenario assumptions by drawing on information from a variety of different sources. For example, one set of analyses combines climate change projections from the HadCM2 model based on the S750 and S550 CO_2-stabilisation scenarios with socio-economic information from the IS92a reference scenario in order to assess coastal flooding and loss of coastal wetlands from long-term sea level rise (Nicholls, 2004; Hall et al., 2005) and to estimate global impacts on natural vegetation, water resources, crop yield and food security, and malaria (Parry et al., 2001; Arnell et al., 2002).

2.4.6.9 Scenario integration

The widespread adoption of SRES-based scenarios in studies described in this report (see Boxes 2.2 to 2.7) acknowledges the desirability of seeking consistent scenario application across different studies and regions. For instance, SRES-based downscaled socio-economic projections were used in conjunction with SRES-derived climate scenarios in a set of global impact studies (Arnell et al., 2004; see Section 2.4.6.4). At a regional scale, multiple scenarios for the main global change drivers (socio-economic factors, atmospheric CO_2 concentration, climate factors, land use, and technology), were developed for Europe, based on interpretations of the global IPCC SRES storylines (Schröter et al., 2005b; see Box 2.7).

Nationally, scenarios of socio-economic development (Kaivo-oja et al., 2004), climate (Jylhä et al., 2004), sea level (Johansson et al., 2004), surface ozone exposure (Laurila et al., 2004), and sulphur and nitrogen deposition (Syri et al., 2004) were developed for Finland. Although the SRES driving factors were used as an integrating framework, consistency between scenario types could only be ensured by regional modelling, as simple downscaling from the global scenarios ignored important regional dependencies (e.g., between climate and air pollution and between air pressure and sea level: see Carter et al., 2004). Similar exercises have also been conducted in the east (Lorenzoni et al., 2000) and north-west (Holman et al., 2005b) of England.

Integration across scales was emphasised in the scenarios developed for the Millennium Ecosystem Assessment (MA), carried out between 2001 and 2005 to assess the consequences of ecosystem change for human well-being (Millennium Ecosystem Assessment, 2005). An SAS approach (see Section 2.4.5) was followed in developing scenarios at scales ranging from regional through national, basin, and local (Lebel et al., 2005). Many differed greatly from the set of global MA scenarios that were also constructed (Alcamo et al., 2005). This is due, in part, to different stakeholders being involved in the development of scenarios at each scale, but also reflects an absence of feedbacks from the sub-global to global scales (Lebel et al., 2005).

2.4.7 Large-scale singularities

Large-scale singularities are extreme, sometimes irreversible, changes in the Earth system such as abrupt cessation of the Atlantic Meridional Overturning Circulation (MOC) or melting of ice sheets in Greenland or West Antarctica (see Meehl et al., 2007; Randall et al., 2007; also Chapter 19, Section 19.3.5). With few exceptions, such events are not taken into account in socio-economic assessments of climate change. Shutdown of the MOC is simulated in Earth system models of intermediate complexity subject to large, rapid forcing (Meehl et al., 2007; also Chapter 19, Section 19.3.5.3). Artificial 'hosing' experiments, assuming the injection of large amounts of freshwater into the oceans at high latitudes, also have been conducted using AOGCMs (e.g., Vellinga and Wood, 2002; Wood et al., 2003) to induce an MOC shutdown. Substantial reduction of greenhouse warming occurs in the Northern Hemisphere, with a net cooling occurring mostly in the North Atlantic region (Wood et al., 2003). Such scenarios have subsequently been applied in impact studies (Higgins and Vellinga, 2004; Higgins and Schneider, 2005; also see Chapter 19, Section 19.4.2.5)

Complete deglaciation of Greenland and the West Antarctica Ice Sheet (WAIS) would raise sea level by 7 m and about 5 m, respectively (Meehl et al., 2007; also Chapter 19, Section 19.3.5.2). One recent study assumed an extreme rate of sea level rise, 5 m by 2100 (Nicholls et al., 2005), to test the limits of adaptation and decision-making (Dawson et al., 2005; Tol et al., 2006). A second study employed a scenario of rapid sea level rise of 2.2 m by 2100 by adding an ice sheet contribution to the highest TAR projection for the period, with the increase continuing unabated after 2100 (Arnell et al., 2005). Both studies describe the potential impacts of such a scenario in Europe, based on expert assessments.

2.4.8 Probabilistic futures

Since the TAR, many studies have produced probabilistic representations of future climate change and socio-economic conditions suitable for use in impact assessment. The choices faced in these studies include which components of socio-economic and climate change models to treat probabilistically and how to define the input probability density functions (pdfs) for each component. Integrated approaches derive pdfs of climate change from input pdfs for emissions and for key parameters in models of GHG cycles, radiative forcing, and the climate system. The models then sample repeatedly from the uncertainty distributions for inputs and model parameters, in order to produce a pdf of outcomes, e.g., global temperature and precipitation change. Either simple climate models (e.g., Wigley and Raper, 2001) or climate models of intermediate complexity (Forest et al., 2002) have been applied.

Alternative methods of developing pdfs for emissions are described in Nakićenović et al. (2007), but they all require subjective judgement in the weighting of different future outcomes, which is a matter of considerable debate (Parson et al., 2006). Some argue that this should be done by experts, otherwise decision-makers will inevitably assign probabilities themselves without the benefit of established techniques to control well-known biases in subjective judgements (Schneider, 2001, 2002; Webster et al., 2002, 2003). Others argue that the climate change issue is characterised by 'deep uncertainty' – i.e., system models, parameter values, and interactions are unknown or contested – and therefore the elicited probabilities may not accurately represent the nature of the uncertainties faced (Grübler and Nakićenović, 2001; Lempert et al., 2004).

The most important uncertainties to be represented in pdfs of regional climate change, the scale of greatest relevance for impact assessments, are GHG emissions, climate sensitivity, and inter-model differences in climatic variables at the regional scale. Other important factors include downscaling techniques, and regional forcings such as aerosols and land-cover change (e.g., Dessai, 2005). A rapidly growing literature reporting pdfs of climate sensitivity is providing a significant methodological advance over the long-held IPCC estimate of 1.5°C to 4.5°C for the (non-probabilistic) range of global mean annual temperature change for a doubling of atmospheric CO_2 (see Meehl et al., 2007, for a detailed discussion). For regional change, recent methods of applying different weighting schemes to multi-model ensemble projections of climate are described in Christensen et al. (2007a). Other work has examined the full chain of uncertainties from emissions to regional climate. For example, Dessai et al. (2005b) tested the sensitivity of probabilistic regional climate changes to a range of uncertainty sources including climate sensitivity, GCM simulations, and emissions scenarios. The ENSEMBLES research project is modelling various sources of uncertainty to produce regional probabilities of climate change and its impacts for Europe (Hewitt and Griggs, 2004).

Methods to translate probabilistic climate changes for use in impact assessment (e.g., New and Hulme, 2000; Wilby and Harris, 2006; Fowler et al., 2007) include those assessing probabilities of impact threshold exceedance (e.g., Jones, 2000, 2004; Jones et al., 2007). Wilby and Harris (2006) combined information from various sources of uncertainty (emissions scenarios, GCMs, statistical downscaling, and hydrological model parameters) to estimate probabilities of low flows in the River Thames basin, finding the most important uncertainty to be the differences between the GCMs, a conclusion supported in water resources assessments in Australia (Jones and Page, 2001; Jones et al., 2005). Scholze et al. (2006) quantified risks of changes in key ecosystem processes on a global scale, by grouping scenarios according to ranges of global mean temperature change rather than considering probabilities of individual emissions scenarios. Probabilistic impact studies sampling across emissions, climate sensitivity, and regional climate change uncertainties have been conducted for wheat yield (Howden and Jones, 2004; Luo et al., 2005), coral bleaching (Jones, 2004; Wooldridge et al., 2005), water resources (Jones and Page, 2001; Jones et al., 2005), and freshwater ecology (Preston, 2006).

2.5 Key conclusions and future directions

Climate change impact, adaptation and vulnerability (CCIAV) assessment has now moved far beyond its early status as a speculative, academic endeavour. As reported elsewhere in this volume, climate change is already under way, impacts are being felt, and some adaptation is occurring. This is propelling CCIAV assessment from being an exclusively research-oriented activity towards analytical frameworks that are designed for practical decision-making. These comprise a limited set of approaches (described in Section 2.2), within which a large range of methods can be applied.

The aims of research and decision analysis differ somewhat in their treatment of uncertainty. Research aims to understand and reduce uncertainty, whereas decision analysis seeks to manage uncertainty in order to prioritise and implement actions. Therefore, while improved scientific understanding may have led to a narrowing of the range of uncertainty in some cases (e.g., increased consensus among GCM projections of regional climate change) and a widening in others (e.g., an expanded range of estimates of adaptive capacity and vulnerability obtained after accounting for alternative pathways of socio-economic and technological development), these results are largely a manifestation of advances in methods for treating uncertainty.

Decision makers are increasingly calling upon the research community to provide:

- good-quality information on what impacts are occurring now, their location and the groups or systems most affected,
- reliable estimates of the impacts to be expected under projected climate change,
- early warning of potentially alarming or irreversible impacts,
- estimation of different risks and opportunities associated with a changing climate,
- effective approaches for identifying and evaluating both existing and prospective adaptation measures and strategies,
- credible methods of costing different outcomes and response measures,
- an adequate basis to compare and prioritise alternative response measures, including both adaptation and mitigation.

To meet these demands, future research efforts need to address a set of methodological, technical and information gaps that call for certain actions.

- *Continued development of risk-management techniques.* Methods and tools should be designed both to address specific climate change problems and to introduce them into mainstream policy and planning decision-making.

- *New methods and tools appropriate for regional and local application*. An increasing focus on adaptation to climate change at local scales requires new methods, scenarios, and models to address emerging issues. New approaches are also reconciling scale issues in scenario development; for example by improving methods of interpreting and quantifying regional storylines, and through the nesting of scenarios at different scales.
- *Cross-sectoral assessments*. Limited by data and technical complexity, most CCIAV assessments have so far focused on single sectors. However, impacts of climate change on one sector will have implications, directly and/or indirectly, for others – some adverse and some beneficial. To be more policy-relevant, future analyses need to account for the interactions between different sectors, particularly at national level but also through global trade and financial flows.
- *Collection of empirical knowledge from past experience*. Experience gained in dealing with climate-related natural disasters, documented using both modern methods and traditional knowledge, can assist in understanding the coping strategies and adaptive capacity of vulnerable communities, and in defining critical thresholds of impact to be avoided.
- *Enhanced observation networks and improved access to existing data*. CCIAV studies have increasing requirements for data describing present-day environmental and socio-economic conditions. Some regions, especially in developing countries, have limited access to existing data, and urgent attention is required to arrest the decline of observation networks. Integrated monitoring systems are needed for observing human-environment interactions.
- *Consistent approaches in relation to scenarios in other assessments*. Integration of climate-related scenarios with those widely accepted and used by other international bodies is desirable (i.e., mainstreaming). The exchange of ideas and information between the research and policy communities will greatly improve scenario quality, usage, and acceptance.
- *Improved scenarios for poorly specified indicators*. CCIAV outcomes are highly sensitive to assumptions about factors such as future technology and adaptive capacity that at present are poorly understood. For instance, the theories and processes of technological innovation and its relationship with other indicators such as education, wealth, and governance require closer attention, as do studies of the processes and costs of adaptation.
- *Integrated scenarios*. There are shortcomings in how interactions between key drivers of change are represented in scenarios. Moreover, socio-economic and technological scenarios need to account for the costs and other ancillary effects of both mitigation and adaptation actions, which at present are rarely considered.
- *Provision of improved climate predictions for near-term planning horizons*. Many of the most severe impacts of climate change are manifest through extreme weather and climate events. Resource planners increasingly need reliable information, years to decades ahead, on the risks of adverse weather events at the scales of river catchments and communities.
- *Effective communication of the risks and uncertainties of climate change*. To gain trust and improve decisions, awareness-building and dialogue is necessary between those stakeholders with knowledge to share (including researchers) and with the wider public.

References

Abildtrup, J., E. Audsley, M. Fekete-Farkas, C. Giupponi, M. Gylling, P. Rosato and M.D.A. Rounsevell, 2006: Socio-economic scenario development for the assessment of climate change impacts on agricultural land use: a pairwise comparison approach. *Environ. Sci. Policy*, **9**, 101-115.

ACIA, 2005: *Arctic Climate Impact Assessment*. Cambridge University Press, Cambridge, 1042 pp.

Acosta-Michlik, L. and M.D.A. Rounsevell, 2005: From generic indices to adaptive agents: shifting foci in assessing vulnerability to the combined impacts of climate change and globalization. *IHDP Update: Newsletter of the International Human Dimensions Programme on Global Environmental Change*, **01/2005**, 14-15.

ADB, 2005: *Climate Proofing: A Risk-based Approach to Adaptation*. Pacific Studies Series, Pub. Stock No. 030905, Asian Development Bank, Manila, 191 pp.

Adger, W.N., 2006: Vulnerability. *Global Environ. Chang*., **16**, 268-281.

Aggarwal, P.K., N. Kalra, S. Chander and H. Pathak, 2006: InfoCrop: a dynamic simulation model for the assessment of crop yields, losses due to pests and environmental impact of agro-ecosystems in tropical environments – model description. *Agr. Syst*., **89**, 1-25.

Ahmad, Q.K., R.A. Warrick, T.E. Downing, S. Nishioka, K.S. Parikh, C. Parmesan, S.H. Schneider, F. Toth and G. Yohe, 2001: Methods and tools. *Climate Change 2001: Impacts, Adaptation, and Vulnerability. Contribution of II to the Third Assessment Report of the Intergovernmental Panel on Climate Change*, J.J. McCarthy, O.F. Canziani, N.A. Leary, D.J. Dokken and K.S. White, Eds., Cambridge University Press, Cambridge, 105-143.

Ahn, S.E., A.J. Plantinga and R.J. Alig, 2002: Determinants and projections of land use in the South Central United States. *South. J. Appl. For*., **26**, 78-84.

Alcamo, J., 2001: Scenarios as a tool for international environmental assessments. Environmental Issue Report No 24, European Environmental Agency, 31.

Alcamo, J., P. Mayerhofer, R. Guardans, T. van Harmelen, J. van Minnen, J. Onigkeit, M. Posch and B. de Vries, 2002: An integrated assessment of regional air pollution and climate change in Europe: findings of the AIR-CLIM Project. *Environ. Sci. Policy*, **4**, 257-272.

Alcamo, J., D. van Vuuren, C. Ringler, J. Alder, E. Bennett, D. Lodge, T. Masui, T. Morita, M. Rosegrant, O. Sala, K. Schulze and M. Zurek, 2005: Methodology for developing the MA scenarios. *Ecosystems and Human Well-Being: Scenarios: Findings of the Scenarios Working Group (Millennium Ecosystem Assessment Series)*, S.R. Carpenter, P.L. Pingali, E.M. Bennett and M.B. Zurek, Eds., Island Press, Washington, D.C., 145-172.

Alcamo, J., M. Flörke and M. Märker, 2006a: Changes in Global Water Resources Driven by Socio-economic and Climatic Changes. Research Report, Center for Environmental Systems Research, University of Kassel, Germany, 34 pp.

Alcamo, J., K. Kok, G. Busch, J. Priess, B. Eickhout, M.D.A. Rounsevell, D. Rothman and M. Heistermann, 2006b: Searching for the future of land: scenarios from the local to global scale. *Land Use and Land Cover Change: Local Processes, Global Impacts*, E. Lambin and H. Geist, Eds., Global Change IGBP Series, Springer-Verlag, Berlin, 137-156.

Allen, J. and K. Lu, 2003: Modeling and prediction of future urban growth in the Charleston region of South Carolina: a GIS-based integrated approach. *Conserv. Ecol*., **8**, 2. [Accessed 26.02.07: http://www.consecol.org/vol8/iss2/art2.]

Anyah, R. and F. Semazzi, 2004: Simulation of the sensitivity of Lake Victoria basin climate to lake surface temperatures. *Theor. Appl. Climatol*., **79**, 55-69.

Araújo, M.B. and C. Rahbek, 2006: How does climate change affect biodiversity? *Science*, **313**, 1396-1397.

Arenstam Gibbons, S.J. and R.J. Nicholls, 2006: Island abandonment and sea-level rise: an historical analog from the Chesapeake Bay, USA. *Global Environ. Chang*., **16**, 40-47.

Arnell, N., D. Hudson and R. Jones, 2003: Climate change scenarios from a regional climate model: estimating change in runoff in southern Africa. *J. Geophys.*

Res.–Atmos., **108**(D16), doi:10.1029/2002JD002782.

Arnell, N., E. Tompkins, N. Adger and K. Delaney, 2005: Vulnerability to abrupt climate change in Europe. Technical Report 34, Tyndall Centre for Climate Change Research, Norwich, 63 pp.

Arnell, N.W., 2004: Climate change and global water resources: SRES emissions and socio-economic scenarios. *Global Environ. Chang.*, **14**, 31-52.

Arnell, N.W., M.G.R. Cannell, M. Hulme, R.S. Kovats, J.F.B. Mitchell, R.J. Nicholls, M.L. Parry, M.T.J. Livermore and A. White, 2002: The consequences of CO_2 stabilisation for the impacts of climate change. *Climatic Change*, **53**, 413-446.

Arnell, N.W., M.J.L. Livermore, S. Kovats, P.E. Levy, R. Nicholls, M.L. Parry and S.R. Gaffin, 2004: Climate and socio-economic scenarios for global-scale climate change impacts assessments: characterising the SRES storylines. *Global Environ. Chang.*, **14**, 3-20.

AS/NZS, 2004: *Risk Management*. Australian/New Zealand Standard for Risk Management, AS/NZS 4360:2004. 38 pp.

Audsley, E., K.R. Pearn, C. Simota, G. Cojocaru, E. Koutsidou, M.D.A. Rounsevell, M. Trnka and V. Alexandrov, 2006: What can scenario modelling tell us about future European scale agricultural land use and what not? *Environ. Sci. Policy*, **9**, 148-162.

Australian Greenhouse Office, 2006: *Climate Change Impacts and Risk Management: A Guide for Business and Government*. Prepared for the Australian Greenhouse Office by Broadleaf Capital International and Marsden Jacob Associates, 73 pp.

Bakkenes, M., J. Alkemade, F. Ihle, R. Leemans and J. Latour, 2002: Assessing the effects of forecasted climate change on the diversity and distribution of European higher plants for 2050. *Glob. Change Biol.*, **8**, 390-407.

Bakkenes, M., B. Eickhout and R. Alkemade, 2006: Impacts of different climate stabilisation scenarios on plant species in Europe. *Global Environ. Chang.*, **16**, 19-28.

Barlage, M.J., P.L. Richards, P.J. Sousounis and A.J. Brenner, 2002: Impacts of climate change and land use change on runoff from a Great Lakes watershed. *J. Great Lakes Res.*, **28**, 568-582.

Bärlund, I. and T.R. Carter, 2002: Integrated global change scenarios: surveying user needs in Finland. *Global Environ. Chang.*, **12**, 219-229.

Barredo, J.I., M. Kasanko, N. McCormick and C. Lavalle, 2003: Modelling dynamic spatial processes: simulation of urban future scenarios through cellular automata. *Landscape Urban Plan.*, **64**, 145-160.

Barredo, J.I., L. Demicheli, C. Lavalle, M. Kasanko and N. McCormick, 2004: Modelling future urban scenarios in developing countries: an application case study in Lagos, Nigeria. *Environ. Plann. B*, **31**, 65-84.

Bathurst, J.C., G. Moretti, A. El-Hames, A. Moaven-Hashemi and A. Burton, 2005: Scenario modelling of basin-scale, shallow landslide sediment yield, Valsassina, Italian Southern Alps. *Nat. Hazard. Earth Sys.*, **5**, 189-202.

Beniston, M., D.B. Stephenson, O.B. Christensen, C.A.T. Ferro, C. Frei, S. Goyette, K. Halsnaes, T. Holt, K. Jylhä, B. Koffi, J. Palutikof, R. Schöll, T. Semmler and K. Woth, 2007: Future extreme events in European climate: an exploration of regional climate model projections. *Climatic Change*, **81** (Suppl. 1), 71-95.

Bennett, E.M., S.R. Carpenter, G.D. Peterson, G.S. Cumming, M. Zurek and P. Pingali, 2003: Why global scenarios need ecology. *Front. Ecol. Environ.*, **1**, 322-329.

Berger, P.A. and J.P. Bolte, 2004: Evaluating the impact of policy options on agricultural landscapes: an alternative-futures approach. *Ecol. Appl.*, **14**, 342-354.

Berkhout, F., J. Hertin and A. Jordan, 2002: Socio-economic futures in climate change impact assessment: using scenarios as "learning machines". *Global Environ. Chang.*, **12**, 83-95.

Berkhout, F., J. Hertin and D.M. Gann, 2006: Learning to adapt: organisational adaptation to climate change impacts. *Climatic Change*, **78**, 135-156.

Berry, P.M., M.D.A. Rounsevell, P.A. Harrison and E. Audsley, 2006: Assessing the vulnerability of agricultural land use and species to climate change and the role of policy in facilitating adaptation. *Environ. Sci. Policy*, **9**, 189-204.

Bettencourt, S., R. Croad, P. Freeman, J. Hay, R. Jones, P. King, P. Lal, A. Mearns, G. Miller, I. Pswarayi-Riddihough, A. Simpson, N. Teuatabo, U. Trotz and M. Van Aalst, 2006: *Not If but When: Adapting to Natural Hazards in the Pacific Islands Region: A Policy Note*. The World Bank, East Asia and Pacific Region, Pacific Islands Country Management Unit, Washington, DC, 43 pp.

Bhadwal, S. and R. Singh, 2002: Carbon sequestration estimates for forestry options under different land-use scenarios in India. *Curr. Sci. India*, **83**, 1380-1386.

Bindoff, N., J. Willebrand, V. Artale, A. Cazenave, J. Gregory, S. Gulev, K. Hanawa, C.L. Quéré, S. Levitus, Y. Nojiri, C.K. Shum, L. Talley and A. Unnikrishnan, 2007: Observations: oceanic climate change and sea level. *Climate Change 2007: The Physical Science Basis. Working Group I Contribution to the Intergovernmental Panel on Climate Change Fourth Assessment Report*, S. Solomon, D. Qin, M. Manning, Z. Chen, M. Marquis, K. B. Averyt, M. Tignor and H. L. Miller, Eds., Cambridge University Press, Cambridge, 385-432.

Bouwer, L.M. and P. Vellinga, 2005: Some rationales for risk sharing and financing adaptation. *Water Sci. Technol.*, **51**, 89-95.

Briassoulis, H., 2001: Policy-oriented integrated analysis of land-use change: an analysis of data needs. *Environ. Manage.*, **27**, 1-11.

Brooks, N., W.N. Adger and P.M. Kelly, 2005: The determinants of vulnerability and adaptive capacity at the national level and the implications for adaptation. *Global Environ. Chang.*, **15**, 151-163.

Burton, I. and M. van Aalst, 2004: *Look Before You Leap: A Risk Management Approach for Incorporating Climate Change Adaptation into World Bank Operations*. World Bank, Washington, DC, 47 pp.

Burton, I., S. Huq, B. Lim, O. Pilifosova and E.L. Schipper, 2002: From impacts assessment to adaptation priorities: the shaping of adaptation policy. *Clim. Policy*, **2**, 145-159.

Busch, G., 2006: Future European agricultural landscapes: what can we learn from existing quantitative land use scenario studies? *Agr. Ecosyst. Environ.*, **114**, 121-140.

Caires, S. and A. Sterl, 2005: 100-year return value estimates for ocean wind speed and significant wave height from the ERA-40 data. *J. Climate*, **18**, 1032-1048.

Carter, T.R., M.L. Parry, S. Nishioka, H. Harasawa, R. Christ, P. Epstein, N.S. Jodha, E. Stakhiv and J. Scheraga, 1996: Technical guidelines for assessing climate change impacts and adaptations. *Climate Change 1995: Impacts, Adaptations and Mitigation of Climate Change: Scientific-Technical Analyses. Contribution of Working Group II to the Second Assessment Report of the Intergovernmental Panel on Climate Change*, R.T. Watson, M.C. Zinyowera and R.H. Moss, Eds., Cambridge University Press, Cambridge, 823-833.

Carter, T.R., E.L. La Rovere, R.N. Jones, R. Leemans, L.O. Mearns, N. Nakićenović A.B. Pittock, S.M. Semenov and J. Skea, 2001: Developing and applying scenarios. *Climate Change 2001: Impacts, Adaptation, and Vulnerability. Contribution of Working Group II to the Third Assessment Report of the Intergovernmental Panel on Climate Change*, J.J. McCarthy, O.F. Canziani, N.A. Leary, D.J. Dokken and K.S. White, Eds., Cambridge University Press, Cambridge, 145-190.

Carter, T.R., S. Fronzek and I. Bärlund, 2004: FINSKEN: a framework for developing consistent global change scenarios for Finland in the 21st century. *Boreal Environ. Res.*, **9**, 91-107.

Carter, T.R., K. Jylhä, A. Perrels, S. Fronzek and S. Kankaanpää, Eds., 2005: FINADAPT scenarios for the 21st century: alternative futures for considering adaptation to climate change in Finland. FINADAPT Working Paper 2, Finnish Environment Institute Mimeographs 332, Helsinki, 42 pp. [Accessed 26.02.07: http://www.environment.fi/default.asp?contentid=162966&lan=en]

Casman, E.A., M.G. Morgan and H. Dowlatabadi, 1999: Mixed levels of uncertainty in complex policy models. *Risk Anal.*, **19**, 33-42.

Castles, I. and D. Henderson, 2003: The IPCC emission scenarios: an economic-statistical critique. *Energ. Environ.*, **14**, 159-185.

Cebon, P., U. Dahinden, H.C. Davies, D. Imboden and C.G. Jaeger, 1999: *Views from the Alps: Regional Perspectives on Climate Change*. MIT Press, Boston, Massachusetts, 536 pp.

Christensen, J.H., B. Hewitson, A. Busuioc, A. Chen, X. Gao, I. Held, R. Jones, W.-T. Kwon, R. Laprise, V.M. Rueda, L.O. Mearns, C.G. Menéndez, J. Räisänen, A. Rinke, R.K. Kolli, A. Sarr and P. Whetton, 2007a: Regional climate projections. *Climate Change 2007: The Physical Science Basis. Contribution of Working Group I to the Intergovernmental Panel on Climate Change Fourth Assessment Report*, S. Solomon, D. Qin, M. Manning, Z. Chen, M. Marquis, K. B. Averyt, M. Tignor and H. L. Miller, Eds., Cambridge University Press, Cambridge, 847-940.

Christensen, J.H., T.R. Carter, M. Rummukainen and G. Amanatidis, 2007b: Evaluating the performance and utility of regional climate models: the PRUDENCE project. *Climatic Change*, **81** (Suppl. 1), 1-6.

Church, J.A., J.M. Gregory, P. Huybrechts, M. Kuhn, K. Lambeck, M.T. Nhuan, D. Qin and P.L. Woodworth, 2001: Changes in sea level. *Climate Change 2001: The Scientific Basis. Contribution of Working Group I to the Third Assessment Report of the Intergovernmental Panel on Climate Change*, J.T. Houghton, Y. Ding, D.J. Griggs, M. Noguer, P.J. van der Linden, X. Dai, K. Maskell and C.A. Johnson, Eds., Cambridge University Press, Cambridge, 639-693.

CLIMsystems, 2005: *SimCLIM Sea Level Scenario Generator Overview of Meth-*

ods, R. Warrick, 5 pp. [Accessed 27.02.07: http://www.climsystems.com/site/downloads/?dl=SSLSG_Methods.pdf.]

Cohen, S.J., 1997: Scientist-stakeholder collaboration in integrated assessment of climate change: lessons from a case study of Northwest Canada. *Environ. Model. Assess.*, **2**, 281-293.

Conde, C. and K. Lonsdale, 2005: Engaging stakeholders in the adaptation process. *Adaptation Policy Frameworks for Climate Change: Developing Strategies, Policies and Measures*, B. Lim, E. Spanger-Siegfried, I. Burton, E. Malone and S. Huq, Eds., Cambridge University Press, Cambridge and New York, 47-66.

Conde, C., R. Ferrer and S. Orozco, 2006: Climate change and climate variability impacts on rainfed agricultural activities and possible adaptation measures: a Mexican case study. *Atmosfera*, **19**, 181-194.

COP, 2005: Five-year programme of work of the Subsidiary Body for Scientific and Technological Advice on impacts, vulnerability and adaptation to climate change. Decision -/CP.11, *Proceedings of Conference of the Parties to the United Nations Framework Convention on Climate Change*, Montreal, 5 pp. [Accessed 27.02.07: http://unfccc.int/adaptation/sbsta_agenda_item_adaptation/items/2673.php.]

Cox, P.M., R.A. Betts, M. Collins, P.P. Harris, C. Huntingford and C.D. Jones, 2004: Amazonian forest dieback under climate-carbon cycle projections for the 21st century. *Theor. Appl. Climatol.*, **78**, 137-156.

Cubasch, U., G.A. Meehl, G.J. Boer, R.J. Stouffer, M. Dix, A. Noda, C.A. Senior, S. Raper and K.S. Yap, 2001: Projections of future climate change. *Climate Change 2001: The Scientific Basis. Contribution of Working Group I to the Third Assessment Report of the Intergovernmental Panel on Climate Change*, J.T. Houghton, Y. Ding, D.J. Griggs, M. Noguer, P.J. van der Linden, X. Dai, K. Maskell and C.A. Johnson, Eds., Cambridge University Press, Cambridge, 525-582.

Cumming, G.S., J. Alcamo, O. Sala, R. Swart, E.M. Bennett and M. Zurek, 2005: Are existing global scenarios consistent with ecological feedbacks? *Ecosystems*, **8**, 143-152.

Dawson, R.J., J.W. Hall, P.D. Bates and R.J. Nicholls, 2005: Quantified analysis of the probability of flooding in the Thames Estuary under imaginable worst-case sea level rise scenarios. *Int. J. Water Resour. D.*, **21**, 577-591.

de Koning, G.H.J., P.H. Verburg, A. Veldkamp and L.O. Fresco, 1999: Multi-scale modelling of land use change dynamics in Ecuador. *Agr. Syst.*, **61**, 77-93.

de Nijs, T.C.M., R. de Niet and L. Crommentuijn, 2004: Constructing land-use maps of the Netherlands in 2030. *J. Environ. Manage.*, **72**, 35-42.

DeFries, R.S., L. Bounoua and G.J. Collatz, 2002: Human modification of the landscape and surface climate in the next fifty years. *Glob. Change Biol.*, **8**, 438-458.

Dempsey, R. and A. Fisher, 2005: Consortium for Atlantic Regional Assessment: information tools for community adapation to changes in climate or land use. *Risk Anal.*, **25**, 1495-1509.

Denman, K.L., G. Brasseur, A. Chidthaisong, P. Ciais, P. Cox, R.E. Dickinson, D. Hauglustaine, C. Heinze, E. Holland, D. Jacob, U. Lohmann, S. Ramachandran, P.L. Silva Dias, S.C. Wofsy and X. Zhang, 2007: Couplings between changes in the climate system and biogeochemistry. *Climate Change 2007: The Physical Science Basis. Contribution of Working Group I to the Fourth Assessment Report of the Intergovernmental Panel on Climate Change*, S. Solomon, D. Qin, M. Manning, Z. Chen, M. Marquis, K. B. Averyt, M. Tignor and H. L. Miller, Eds., Cambridge University Press, Cambridge, 499-588.

Dessai, S., 2005: Robust adaptation decisions amid climate change uncertainties. PhD thesis, School of Environmental Sciences, University of East Anglia, Norwich, 281 pp.

Dessai, S. and M. Hulme, 2004: Does climate adaptation policy need probabilities? *Clim. Policy*, **4**, 107-128.

Dessai, S., W.N. Adger, M. Hulme, J.R. Turnpenny, J. Köhler and R. Warren, 2004: Defining and experiencing dangerous climate change. *Climatic Change*, **64**, 11-25.

Dessai, S., X. Lu and J.S. Risbey, 2005a: On the role of climate scenarios for adaptation planning. *Global Environ. Chang.*, **15**, 87-97.

Dessai, S., X. Lu and M. Hulme, 2005b: Limited sensitivity analysis of regional climate change probabilities for the 21st century. *J. Geophys. Res.–Atmos.*, **110**, doi:10.1029/2005JD005919.

Dietz, T., E. Ostrom and P.C. Stern, 2003: The struggle to govern the commons. *Science*, **302**, 1907-1912.

Dirnbock, T., S. Dullinger and G. Grabherr, 2003: A regional impact assessment of climate and land-use change on alpine vegetation. *J. Biogeogr.*, **30**, 401-417.

Discovery Software, 2003: *FloodRanger: Educational Flood Management Game*. [Accessed 27.02.07: http://www.discoverysoftware.co.uk/FloodRanger.htm]

Downing, T.E. and A. Patwardhan, 2005: Assessing vulnerability for climate adaptation. *Adaptation Policy Frameworks for Climate Change: Developing Strategies, Policies and Measures*, B. Lim, E. Spanger-Siegfried, I. Burton, E. Malone and S. Huq, Eds., Cambridge University Press, Cambridge and New York, 67-90.

Eakin, H., M. Webhe, C. Ávila, G.S. Torres and L.A. Bojórquez-Tapia, 2006: A comparison of the social vulnerability of grain farmers in Mexico and Argentina. AIACC Working Paper No. 29, Assessment of Impacts and Adaptation to Climate Change in Multiple Regions and Sectors Program, Washington, DC, 50 pp.

Easterling, W.E., N. Chhetri and X. Niu, 2003: Improving the realism of modeling agronomic adaptation to climate change: simulating technological substitution. *Climatic Change*, **60**, 149-173.

Eckhardt, K., L. Breuer and H.G. Frede, 2003: Parameter uncertainty and the significance of simulated land use change effects. *J. Hydrol.*, **273**, 164-176.

Ekstrom, M., B. Hingray, A. Mezghani and P.D. Jones, 2007: Regional climate model data used within the SWURVE project. 2. Addressing uncertainty in regional climate model data for five European case study areas. *Hydrol. Earth Syst. Sc.*, **11**, 1085-1096.

El-Fadel, M., D. Jamali and D. Khorbotly, 2002: Land use, land use change and forestry related GHG emissions in Lebanon: economic valuation and policy options. *Water Air Soil Poll.*, **137**, 287-303.

Eriksen, S.H., K. Brown and P.M. Kelly, 2005: The dynamics of vulnerability: locating coping strategies in Kenya and Tanzania. *Geogr. J.*, **171**, 287-305.

Ewert, F., M.D.A. Rounsevell, I. Reginster, M. Metzger and R. Leemans, 2005: Future scenarios of European agricultural land use. I. Estimating changes in crop productivity. *Agr. Ecosyst. Environ.*, **107**, 101-116.

Fankhauser, S. and R.S.J. Tol, 2005: On climate change and economic growth. *Resour. Energy Econ.*, **27**, 1-17.

Fearnside, P.M., 2000: Global warming and tropical land-use change: greenhouse gas emissions from biomass burning, decomposition and soils in forest conversion, shifting cultivation and secondary vegetation. *Climatic Change*, **46**, 115-158.

Feenstra, J., I. Burton, J.B. Smith and R.S.J. Tol, Eds., 1998: *Handbook on Methods of Climate Change Impacts Assessment and Adaptation Strategies*. United Nations Environment Programme, Vrije Universiteit Amsterdam, Institute for Environmental Studies, Amsterdam, 464 pp.

Fischer, G. and L.X. Sun, 2001: Model based analysis of future land-use development in China. *Agr. Ecosyst. Environ.*, **85**, 163-176.

Fischer, G., M. Shah and H.V. Velthuizen, 2002: *Climate Change and Agricultural Vulnerability*. International Institute for Applied Systems Analysis, Laxenberg, 152 pp.

Fischoff, B., 1996: Public values in risk research. *Ann. Am. Acad. Polit. SS.*, **45**, 75-84.

Fisher, B.S., G. Jakeman, H.M. Pant, M. Schwoon and R.S.J. Tol, 2006: CHIMP: a simple population model for use in integrated assessment of global environmental change. *Integrated Assess. J.*, **6**, 1-33.

Ford, J. and B. Smit, 2004: A framework for assessing the vulnerability of communities in the Canadian Arctic to risks associated with climate change. *Arctic*, **57**, 389-400.

Forest, C.E., P.H. Stone, A.P. Sokolov, M.R. Allen and M.D. Webster, 2002: Quantifying uncertainties in climate system properties with the use of recent climate observations. *Science*, **295**, 113-117.

Forster, P., V. Ramaswamy, P. Artaxo, T. Berntsen, R.A. Betts, D.W. Fahey, J. Haywood, J. Lean, D.C. Lowe, G. Myhre, J. Nganga, R. Prinn, G. Raga, M. Schulz and R.V. Dorland, 2007: Changes in atmospheric constituents and in radiative forcing. *Climate Change 2007: The Physical Science Basis. Contribution of Working Group I to the Fourth Assessment Report of the Intergovernmental Panel on Climate Change*, S. Solomon, D. Qin, M. Manning, Z. Chen, M. Marquis, K. B. Averyt, M. Tignor and H. L. Miller, Eds., Cambridge University Press, Cambridge, 129-234.

Fowler, H.J., S. Blenkinsop and C. Tebaldi, 2007: Linking climate change modelling to impacts studies: recent advances in downscaling techniques for hydrological modelling. *Int. J. Climatol.*, in press.

Friedlingstein, P., P. Cox, R. Betts, L. Bopp, W. von Bloh, V. Brovkin, P. Cadule, S. Doney, M. Eby, I. Fung, G. Bala, J. John, C. Jones, F. Joos, T. Kato, M. Kawamiya, W. Knorr, K. Lindsay, H.D. Matthews, T. Raddatz, P. Rayner, C. Reick, E. Roeckner, K.-G. Schnitzler, R. Schnur, K. Strassmann, A.J. Weaver, C. Yoshikawa and N. Zeng, 2006: Climate-carbon cycle feedback analysis: results from the C^4MIP model intercomparison. *J. Climate*, **19**, 3337-3353.

Fronzek, S. and T.R. Carter, 2007: Assessing uncertainties in climate change impacts on resource potential for Europe based on projections from RCMs and

GCMs. *Climatic Change*, **81** (Suppl. 1), 357-371.

Funtowicz, S.O. and J.R. Ravetz, 1990: *Uncertainty and Quality in Science for Policy*. Kluwer, Dordrecht, 229 pp.

Füssel, H.-M. and R.J.T. Klein, 2006: Climate change vulnerability assessments: an evolution of conceptual thinking. *Climatic Change*, **75**, 301-329.

Gaffin, S.R., C. Rosenzweig, X. Xing and G. Yetman, 2004: Downscaling and geospatial gridding of socio-economic projections from the IPCC Special Report on Emissions Scenarios (SRES). *Global Environ. Chang*., **14**, 105-123.

Ganopolski, A., V. Petoukhov, S. Rahmstorf, V. Brovkin, M. Claussen, A. Eliseev and C. Kubatzki, 2001: CLIMBER-2: a climate system model of intermediate complexity. Part II. Model sensitivity. *Clim. Dynam*., **17**, 735-751.

Gao, X.J., D.L. Li, Z.C. Zhao and F. Giorgi, 2003: Numerical simulation for influence of greenhouse effects on climatic change of Qinghai-Xizang Plateau along Qinghai-Xizang railway [in Chinese with English abstract]. *Plateau Meteorol*., **22**, 458-463.

GCOS, 2003: The second report on the adequacy of the global observing systems for climate in support of the UNFCCC. Global Climate Observing System, GCOS - 82, WMO/TD No. 1143, World Meteorological Organization, Geneva, 74 pp.

Gedney, N., P.M. Cox, R.A. Betts, O. Boucher, C. Huntingford and P.A. Stott, 2006: Detection of a direct carbon dioxide effect in continental river runoff records. *Nature*, **439**, 835-838.

Geurs, K.T. and J.R.R. van Eck, 2003: Evaluation of accessibility impacts of land-use scenarios: the implications of job competition, land-use and infrastructure developments for the Netherlands. *Environ. Plann. B*., **30**, 69-87.

Gigerenzer, G., 2000: *Adaptive Thinking: Rationality in the Real World*. Oxford University Press, Oxford, 360 pp.

Giorgi, F., B.C. Hewitson, J.H. Christensen, M. Hulme, H. von Storch, P.H. Whetton, R.G. Jones, L.O. Mearns and C.B. Fu, 2001: Regional climate information: evaluation and projections. *Climate Change 2001: The Scientific Basis. Contribution of Working Group I to the Third Assessment Report of the Intergovernmental Panel on Climate Change*, J.T. Houghton, Y. Ding, D.J. Griggs, M. Noguer, P.J. van der Linden, X. Dai, K. Maskell and C.A. Johnson, Eds., Cambridge University Press, Cambridge, 581-638.

Glantz, M.H., 2001: *Once Burned, Twice Shy? Lessons Learned from the 1997-98 El Niño*. United Nations University Press, Tokyo, 294 pp.

Goklany, I., 2005: Is a richer-but-warmer world better than poorer-but-cooler worlds? *25th Annual North American Conference of the US Association for Energy Economics/International Association of Energy Economics, 21-23 September, 2005*.

Good, P., L. Bärring, C. Giannakopoulos, T. Holt and J. Palutikof, 2006: Non-linear regional relationships between climate extremes and annual mean temperatures in model projections for 1961–2099 over Europe. *Climate Res*., **31**, 19-34.

Goodess, C.M., C. Hanson, M. Hulme and T.J. Osborn, 2003: Representing climate and extreme weather events in integrated assessment models: a review of existing methods and options for development. *Integrated Assess. J*., **4**, 145-171.

Government of India, 2004: Vulnerability assessment and adaptation. *India's Initial National Communication to the UNFCCC*, Ministry of Environment and Forests, New Delhi, 57-132.

Graham, L.P., S. Hagemann, S. Jaun and M. Beniston, 2007: On interpreting hydrological change from regional climate models. *Climatic Change*, **81** (Suppl. 1), 97-122.

Greco, S., R.H. Moss, D. Viner and R. Jenne, 1994: *Climate Scenarios and Socio-Economic Projections for IPCC WG II Assessment. Working Document, Intergovernmental Panel on Climate Change*. Working Group II Technical Support Unit, Washington, DC, 67 pp.

Group on Earth Observations, 2005: *Global Earth Observation System of Systems, GEOSS: 10-Year Implementation Plan Reference Document*. GEO 1000R / ESA SP-1284, ESA Publications Division, Noordwijk, 209 pp. [Accessed 27.02.07: http://www.earthobservations.org/docs/10-Year%20Plan%20Reference%20Document%20(GEO%201000R).pdf]

Grubb, M., J. Köhler and D. Anderson, 2002: Induced technical change in energy and environmental modeling: analytic approaches and policy implications. *Annu. Rev. Energ. Env*., **27**, 271-308.

Grübler, A. and N. Nakićenović, 2001: Identifying dangers in an uncertain climate. *Nature*, **412**, 15.

Grübler, A., N. Nakićenović and D.G. Victor, 1999: Modeling technological change: implications for the global environment. *Annu. Rev. Energ. Env*., **24**, 545-569.

Grübler, A., B. O'Neill, K. Riahi, V. Chirkov, A. Goujon, P. Kolp, I. Prommer, S. Scherbov and E. Slentoe, 2006: Regional, national, and spatially explicit scenarios of demographic and economic change based on SRES. *Technol. Forecast. Soc*., doi:10.1016/j.techfore.2006.05.023

Hachadoorian, L., S.R. Gaffin and R. Engleman, 2007: Projecting a gridded population of the world using ratio methods of trend extrapolation. *Human Population: The Demography and Geography of Homo Sapiens and their Implications for Biological Diversity*, R.P. Cincotta, L. Gorenflo and D. Mageean, Eds., Springer-Verlag, Berlin,in press.

Haddad, B.M., 2005: Ranking the adaptive capacity of nations to climate change when socio-political goals are explicit. *Global Environ. Chang*., **15**, 165-176.

Hall, J., T. Reeder, G. Fu, R.J. Nicholls, J. Wicks, J. Lawry, R.J. Dawson and D. Parker, 2005: Tidal flood risk in London under stabilisation scenarios. *Extended Abstract, Symposium on Avoiding Dangerous Climate Change*, Exeter, 1-3 February 2005, 4 pp. [Accessed 27.02.07: http://www.stabilisation2005.com/posters/Hall_Jim.pdf]

Hallegatte, S., 2005: The long timescales of the climate-economy feedback and the climatic cost of growth. *Environ. Model. Assess*., **10**, 277-289.

Hallegatte, S., J.C. Hourcade and P. Dumas, 2006: Why economic dynamics matter in the assessment of climate change damages: illustration extreme events. *Ecol. Econ*., doi:10.1016/j.ecolecon.2006.06.006

Hallegatte, S., J.-C. Hourcade and P. Ambrosi, 2007: Using climate analogues for assessing climate change economic impacts in urban areas. *Climatic Change*, **82**, 47-60.

Hamilton, J.M., D.J. Maddison and R.S.J. Tol, 2005: Climate change and international tourism: a simulation study. *Global Environ. Chang*., **15**, 253-266.

Hansen, J., 2004: Defusing the global warming time bomb. *Sci. Am*., **290**, 68-77.

Harvey, L.D.D., 2007: Dangerous anthropogenic interference, dangerous climate change and harmful climatic change: non-trivial distinctions with significant policy implications. *Climatic Change*, **82**, 1-25.

Heslop-Thomas, C., W. Bailey, D. Amarakoon, A. Chen, S. Rawlins, D. Chadee, R. Crosbourne, A. Owino, K. Polson, C. Rhoden, R. Stennett and M. Taylor, 2006: Vulnerability to dengue fever in Jamaica. AIACC Working Paper No. 27, Assessment of Impacts and Adaptation to Climate Change in Multiple Regions and Sectors Program, Washington, DC, 40 pp.

Hewitson, B., 2003: Developing perturbations for climate change impact assessments. *Eos T. Am. Geophys. Un*., **84**, 337-348.

Hewitt, C.D. and D.J. Griggs, 2004: Ensemble-based predictions of climate changes and their impacts. *Eos*, **85**, 566.

Hewitt, K. and I. Burton, 1971: *The Hazardousness of a Place: A Regional Ecology of Damaging Events*. University of Toronto, Toronto, 154 pp.

Higgins, P.A.T. and M. Vellinga, 2004: Ecosystem responses to abrupt climate change: teleconnections, scale and the hydrological cycle. *Climatic Change*, **64**, 127-142.

Higgins, P.A.T. and S.H. Schneider, 2005: Long-term potential ecosystem responses to greenhouse gas-induced thermohaline circulation collapse. *Glob. Change Biol*., **11**, 699-709.

Hijioka, Y., K. Takahashi, Y. Matsuoka and H. Harasawa, 2002: Impact of global warming on waterborne diseases. *J. Jpn. Soc. Water Environ*., **25**, 647-652.

Hilderink, H., 2004: Population and scenarios: worlds to win? Report 550012001/2004, RIVM, Bilthoven, 74 pp.

Hingray, B., N. Mouhous, A. Mezghani, B. Schaefli and A. Musy, 2007: Accounting for global-mean warming and scaling uncertainties in climate change impact studies: application to a regulated lake system. *Hydrol. Earth Syst. Sc*., **11**, 1207-1226.

Hitz, S. and J. Smith, 2004: Estimating global impacts from climate change. *Global Environ. Chang*., **14**, 201-218.

Hogrefe, C., B. Lynn, K. Civerolo, J.-Y. Ku, J. Rosenthal, C. Rosenzweig, R. Goldberg, S. Gaffin, K. Knowlton and P.L. Kinney, 2004: Simulating changes in regional air pollution over the eastern United States due to changes in global and regional climate and emissions. *J. Geophys. Res*., **109**, doi:10.1029/2004JD004690.

Holman, I.P., M.D.A. Rounsevell, S. Shackley, P.A. Harrison, R.J. Nicholls, P.M. Berry and E. Audsley, 2005a: A regional, multi-sectoral and integrated assessment of the impacts of climate and socio-economic change in the UK. II. Results. *Climatic Change*, **70**, 43-73.

Holman, I.P., M.D.A. Rounsevell, S. Shackley, P.A. Harrison, R.J. Nicholls, P.M. Berry and E. Audsley, 2005b: A regional, multi-sectoral and integrated assessment of the impacts of climate and socio-economic change in the UK. I. Methodology. *Climatic Change*, **70**, 9-41.

Howden, S.M. and R.N. Jones, 2004: Risk assessment of climate change impacts on Australia's wheat industry. *New Directions for a Diverse Planet, Proceedings of the 4th International Crop Science Congress, Brisbane, Australia*, T. Fischer,

Ed. [Accessed 27.02.07: http://www.cropscience.org.au/icsc2004/symposia/6/2/1848_howdensm.htm] [Australia and New Zealand, risk]

Hulme, M., G.J. Jenkins, X. Lu, J.R. Turnpenny, T.D. Mitchell, R.G. Jones, J. Lowe, J.M. Murphy, D. Hassell, P. Boorman, R. McDonald and S. Hill, 2002: *Climate Change Scenarios for the United Kingdom: The UKCIP02 Scientific Report*. Tyndall Centre for Climate Change Research, University of East Anglia, Norwich, 120 pp.

Huntington, H. and S. Fox, 2005: The changing Arctic: indigenous perspectives. *Arctic Climate Impact Assessment*, Cambridge University Press, Cambridge, 61-98.

Hurd, B.H., M. Callaway, J. Smith and P. Kirshen, 2004: Climatic change and US water resources: from modeled watershed impacts to national estimates. *J. Am. Water Resour. As.*, **40**, 129-148.

Ionescu, C., R.J.T. Klein, J. Hinkel, K.S. Kavi Kumar and R. Klein, 2005: Towards a formal framework of vulnerability to climate change. NeWater Working Paper 2, 24 pp. Accessed from http://www.newater.info.

IPCC, 1992: *Climate Change 1992: The Supplementary Report to the IPCC Scientific Assessment*, J.T. Houghton, B.A. Callander and S.K. Varney, Eds., Cambridge University Press, Cambridge, 200 pp.

IPCC, 1994: IPCC Technical guidelines for assessing climate change impacts and adaptations. *IPCC Special Report to the First Session of the Conference of the Parties to the UN Framework Convention on Climate Change, Working Group II, Intergovernmental Panel on Climate Change*, T.R. Carter, M.L. Parry, S. Nishioka and H. Harasawa, Eds., University College London and Center for Global Environmental Research, National Institute for Environmental Studies, Tsukuba, 59 pp.

IPCC, 1996: *Climate Change 1995: The Science of Climate Change. Contribution of Working Group I to the Second Assessment Report of the Intergovernmental Panel on Climate Change*, J.T. Houghton, L.G.M. Filho, B.A. Callander, N. Harris, A. Kattenberg and K. Maskell, Eds., Cambridge University Press, Cambridge, 572 pp.

IPCC, 2001a: *Climate Change 2001: The Scientific Basis. Contribution of Working Group I to the Third Assessment Report of the Intergovernmental Panel on Climate Change*, J.T. Houghton, Y. Ding, D.J. Griggs, M. Noguer, P.J. van der Linden, X. Dai, K. Maskell and C.A. Johnson, Eds., Cambridge University Press, Cambridge, 881 pp.

IPCC, 2001b: *Climate Change 2001: Impacts, Adaptation, and Vulnerability. Contribution of Working Group II to the Third Assessment Report of the Intergovernmental Panel on Climate Change*, J.J. McCarthy, O.F. Canziani, N.A. Leary, D.J. Dokken and K.S. White, Eds., Cambridge University Press, Cambridge, 1032 pp.

IPCC, 2001c: *Climate Change 2001: Synthesis Report. Contribution of Working Groups I, II and III to the Third Assessment Report of the Intergovernmental Panel on Climate Change*, R.T. Watson and the Core Writing Team, Eds., Cambridge University Press, Cambridge, 398 pp.

IPCC, 2007: *Climate Change 2007: The Physical Science Basis. Contribution of Working Group I to the Fourth Assessment Report of the Intergovernmental Panel on Climate Change*, S. Solomon, D. Qin, M. Manning, Z. Chen, M. Marquis, K. B. Averyt, M. Tignor and H. L. Miller, Eds., Cambridge University Press, Cambridge, 996 pp.

IPCC-TGCIA, 1999: *Guidelines on the Use of Scenario Data for Climate Impact and Adaptation Assessment: Version 1*. Prepared by T.R. Carter, M. Hulme and M. Lal, Intergovernmental Panel on Climate Change, Task Group on Scenarios for Climate Impact Assessment, Supporting Material, 69 pp. [Accessed 27.07.02: http://www.ipcc-data.org/guidelines/ggm_no1_v1_12-1999.pdf]

IRI, 2006: *A Gap Analysis for the Implementation of the Global Climate Observing System Programme in Africa*. International Research Institute for Climate and Society, The Earth Institute at Columbia University, Palisades, New York, 47 pp.

ISO/IEC, 2002: Risk management: vocabulary: guidelines for use in standards, PD ISO/IEC Guide 73, International Organization for Standardization/International Electrotechnical Commission, Geneva, 16 pp.

Ivey, J.L., J. Smithers, R.C. de Loë and R.D. Kreutzwiser, 2004: Community capacity for adaptation to climate-induced water shortages: linking institutional complexity and local actors. *Environ. Manage.*, **33**, 36-47.

Izaurralde, R.C., N.J. Rosenberg, R.A. Brown and A.M. Thomson, 2003: Integrated assessment of Hadley Centre (HadCM2) climate-change impacts on agricultural productivity and irrigation water supply in the conterminous United States. Part II. Regional agricultural production in 2030 and 2095. *Agr. Forest Meteorol.*, **117**, 97-122.

Jacobs, K., 2002: *Connecting Science, Policy and Decision-making: A Handbook for Researchers and Science Agencies*. NOAA Office of Global Programs, Washington, DC, 30 pp.

Jacoby, H.D., 2004: Informing climate policy given incommensurable benefits estimates. *Global Environ. Chang.*, **14**, 287-297.

Johansson, M.M., K.K. Kahma, H. Boman and J. Launiainen, 2004: Scenarios for sea level on the Finnish coast. *Boreal Environ. Res.*, **9**, 153-166.

Jones, R.N., 2000: Managing uncertainty in climate change projections: issues for impact assessment. *Climatic Change*, **45**, 403-419.

Jones, R.N., 2001: An environmental risk assessment/management framework for climate change impact assessments. *Nat. Hazards*, **23**, 197-230.

Jones, R.N., 2004: Managing climate change risks. *The Benefits of Climate Policies: Analytical and Framework Issues*, J. Corfee Morlot and S. Agrawala, Eds., OECD, Paris, 251-297.

Jones, R.N. and C.M. Page, 2001: Assessing the risk of climate change on the water resources of the Macquarie River catchment. *Integrating Models for Natural Resources Management Across Disciplines, Issues and Scales (Part 2), Modsim 2001 International Congress on Modelling and Simulation*, F. Ghassemi, P. Whetton, R. Little and M. Littleboy, Eds., Modelling and Simulation Society of Australia and New Zealand, Canberra, 673-678.

Jones, R.N. and R. Boer, 2005: Assessing current climate risks. *Adaptation Policy Frameworks for Climate Change: Developing Strategies, Policies and Measures*, B. Lim, E. Spanger-Siegfried, I. Burton, E. Malone and S. Huq, Eds., Cambridge University Press, Cambridge and New York, 91-118.

Jones, R.N. and L.O. Mearns, 2005: Assessing future climate risks. *Adaptation Policy Frameworks for Climate Change: Developing Strategies, Policies and Measures*, B. Lim, E. Spanger-Siegfried, I. Burton, E. Malone and S. Huq, Eds., Cambridge University Press, Cambridge and New York, 119-143.

Jones, R.N., P. Durack, C. Page and J. Ricketts, 2005: Climate change impacts on the water resources of the Fitzroy River Basin. *Climate Change in Queensland under Enhanced Greenhouse Conditions*, W. Cai et al., Eds., Report 2004–05, CSIRO Marine and Atmospheric Research, Melbourne, 19-58.

Jones, R.N., P. Dettmann, G. Park, M. Rogers and T. White, 2007: The relationship between adaptation and mitigation in managing climate change risks: a regional approach. *Mitig. Adapt. Strat. Glob. Change*, **12**, 685–712.

Jönsson, A.M., M.-L. Linderson, I. Stjernquist, P. Schlyter and L. Bärring, 2004: Climate change and the effect of temperature backlashes causing frost damage in *Picea abies*. *Global Planet. Change*, **44**, 195-207.

Justice, C.O., L. Giglio, S. Korontzi, J. Owens, J.T. Morisette, D. Roy, J. Descloitres, S. Alleaume, F. Petitcolin and Y. Kaufman, 2002: The MODIS fire products. *Remote Sens. Environ.*, **83**, 244-262.

Jylhä, K., H. Tuomenvirta and K. Ruosteenoja, 2004: Climate change projections for Finland during the 21st century. *Boreal Environ. Res.*, **9**, 127-152.

Kahneman, D. and A. Tversky, 1979: Prospect theory: an analysis of decision under risk. *Econometrica*, **47**, 263-291.

Kaivo-oja, J., J. Luukkanen and M. Wilenius, 2004: Defining alternative socio-economic and technological futures up to 2100: SRES scenarios for the case of Finland. *Boreal Environ. Res.*, **9**, 109-125.

Kammen, D., A. Shlyakter and R. Wilson, 1994: What is the risk of the impossible? *J. Franklin I.*, **331**(A), 97-116.

Kankaanpää, S. and T.R. Carter, 2004: Construction of European forest land use scenarios for the 21st century. The Finnish Environment 707, Finnish Environment Institute, Helsinki, 57 pp.

Kankaanpää, S., T.R. Carter and J. Liski, 2005: Stakeholder perceptions of climate change and the need to adapt. FINADAPT Working Paper 14, Finnish Environment Institute, Mimeographs 344, Helsinki, 36 pp. [Accessed 27.02.07: http://www.environment.fi/default.asp?contentid=165486&lan=en]

Kasperson, R.E., 2006: Rerouting the stakeholder express. *Global Environ. Chang.*, **16**, 320-322.

Kates, R.W. and T.J. Wilbanks, 2003: Making the global local: responding to climate change concerns from the ground. *Environment*, **45**, 12-23.

Kelly, P.M. and W.N. Adger, 2000: Theory and practice in assessing vulnerability to climate change and facilitating adaptation. *Climatic Change*, **47**, 325-352.

Kemfert, C., 2002: An integrated assessment model of economy-energy-climate: the model Wiagem. *Integrated Assess.*, **3**, 281-298.

Kenny, G.J., R.A. Warrick, B.D. Campbell, G.C. Sims, M. Camilleri, P.D. Jamieson, N.D. Mitchell, H.G. McPherson and M.J. Salinger, 2000: Investigating climate change impacts and thresholds: an application of the CLIMPACTS integrated assessment model for New Zealand agriculture. *Climatic Change*, **46**, 91-113.

Klein, R. and R.J. Nicholls, 1999: Assessment of coastal vulnerability to climate change. *Ambio*, **28**, 182-187.

Klein, R.J.T., E.L.F. Schipper and S. Dessai, 2005: Integrating mitigation and adaptation into climate and development policy: three research questions. *Environ. Sci. Policy*, **8**, 579-588.

Klöcking, B., B. Strobl, S. Knoblauch, U. Maier, B. Pfutzner and A. Gericke, 2003: Development and allocation of land-use scenarios in agriculture for hydrological impact studies. *Phys. Chem. Earth*, **28**, 1311-1321.

Kok, K. and M. Winograd, 2002: Modeling land-use change for Central America, with special reference to the impact of hurricane Mitch. *Ecol. Model.*, **149**, 53-69.

Kok, K., D.S. Rothman and M. Patel, 2006a: Multi-scale narratives from an IA perspective. Part I. European and Mediterranean scenario development. *Futures*, **38**, 261-284.

Kok, K., M. Patel, D.S. Rothman and G. Quaranta, 2006b: Multi-scale narratives from an IA perspective. Part II. Participatory local scenario development. *Futures*, **38**, 285-311.

Kundzewicz, Z.W., U. Ulbrich, T. Brücher, D. Graczyk, A. Krüger, G.C. Leckebusch, L. Menzel, I. Pińskwar, M. Radziejewski and M. Szwed, 2005: Summer floods in Central Europe: climate change track? *Nat. Hazards*, **36**, 165-189.

Laurila, T., J.-P. Tuovinen, V. Tarvainen and D. Simpson, 2004: Trends and scenarios of ground-level ozone concentrations in Finland. *Boreal Environ. Res.*, **9**, 167-184.

Lebel, L., P. Thongbai, K. Kok, J.B.R. Agard, E. Bennett, R. Biggs, M. Ferreira, C. Filer, Y. Gokhale, W. Mala, C. Rumsey, S.J. Velarde and M. Zurek, 2005: Sub-global scenarios. *Ecosystems and Human Well-Being: Multiscale Assessments, Volume 4, Millennium Ecosystem Assessment*, S.R. Carpenter, P.L. Pingali, E.M. Bennett and M.B. Zurek, Eds., Island Press, Washington, DC, 229-259.

Leemans, R. and B. Eickhout, 2004: Another reason for concern: regional and global impacts on ecosystems for different levels of climate change. *Global Environ. Chang.*, **14**, 219-228.

Leemans, R., B. Eickhout, B. Strengers, L. Bouwman and M. Schaeffer, 2002: The consequences of uncertainties in land use, climate and vegetation responses on the terrestrial carbon. *Sci. China Ser. C*, **45**, 126-142.

Lempert, R., N. Nakićenović, D. Sarewitz and M.E. Schlesinger, 2004: Characterizing climate change uncertainties for decision-makers. *Climatic Change*, **65**, 1-9.

Leung, L.R., Y. Qian, X. Bian, W.M. Washington, J. Han and J.O. Roads, 2004: Mid-century ensemble regional climate change scenarios for the western United States. *Climatic Change*, **62**, 75-113.

Levy, P.E., M.G.R. Cannell and A.D. Friend, 2004: Modelling the impact of future changes in climate, CO_2 concentration and land use on natural ecosystems and the terrestrial carbon sink. *Global Environ. Chang.*, **14**, 21-52.

Lorenzoni, I., A. Jordan, M. Hulme, R.K. Turner and T. O'Riordan, 2000: A co-evolutionary approach to climate change impact assessment. Part I. Integrating socio-economic and climate change scenarios. *Global Environ. Chang.*, **10**, 57-68.

Loukopoulos, P. and R.W. Scholz, 2004: Sustainable future urban mobility: using 'area development negotiations' for scenario assessment and participatory strategic planning. *Environ. Plann. A*, **36**, 2203-2226.

Lowe, J.A. and J.M. Gregory, 2005: The effects of climate change on storm surges around the United Kingdom. *Philos. T. R. Soc. A*, **363**, 1313-1328.

Luo, Q., R.N. Jones, M. Williams, B. Bryan and W.D. Bellotti, 2005: Construction of probabilistic distributions of regional climate change and their application in the risk analysis of wheat production. *Climate Res.*, **29**, 41-52.

Luoto, M., J. Pöyry, R.K. Heikkinen and K. Saarinen, 2005: Uncertainty of bioclimate envelope models based on the geographical distribution of species. *Global Ecol. Biogeogr.*, **14**, 575-584.

Malone, E.L. and E.L. La Rovere, 2005: Assessing current and changing socio-economic conditions. *Adaptation Policy Frameworks for Climate Change: Developing Strategies, Strategies, Policies and Measures*, B. Lim, E. Spanger-Siegfried, I. Burton, E. Malone and S. Huq, Eds., Cambridge University Press, Cambridge and New York, 145-163.

Marttila, V., H. Granholm, J. Laanikari, T. Yrjölä, A. Aalto, P. Heikinheimo, J. Honkatukia, H. Järvinen, J. Liski, R. Merivirta and M. Paunio, 2005: *Finland's National Strategy for Adaptation to Climate Change*. Publication 1a/2005, Ministry of Agriculture and Forestry, Helsinki, 280 pp.

Mastrandrea, M.D. and S.H. Schneider, 2004: Probabilistic integrated assessment of "dangerous" climate change. *Science*, **304**, 571-575.

Mathur, A., I. Burton and M.K. van Aalst, Eds., 2004: *An Adaptation Mosaic: A Sample of the Emerging World Bank Work in Climate Change Adaptation*. The World Bank, Washington, DC, 133 pp.

Mayerhofer, P., B. de Vries, M. den Elzen, D. van Vuuren, J. Onigheit, M. Posch and R. Guardans, 2002: Long-term consistent scenarios of emissions, deposition and climate change in Europe. *Environ. Sci. Policy*, **5**, 273-305.

Maynard, K. and J.F. Royer, 2004: Effects of "realistic" land-cover change on a greenhouse-warmed African climate. *Clim. Dynam.*, **22**, 343-358.

McCarthy, J.J. and M. Long Martello, 2005: Climate change in the context of multiple stressors and resilience. *Arctic Climate Impact Assessment*, Cambridge University Press, Cambridge, 879-922.

McKenzie Hedger, M., R. Connell and P. Bramwell, 2006: Bridging the gap: empowering adaptation decision-making through the UK Climate Impacts Programme. *Clim. Policy*, **6**, 201-215.

Mearns, L.O., M. Hulme, T.R. Carter, R. Leemans, M. Lal and P.H. Whetton, 2001: Climate scenario development. *Climate Change 2001: The Scientific Basis. Contribution of Working Group I to the Third Assessment Report of the Intergovernmental Panel on Climate Change*, J.T. Houghton, Y. Ding, D.J. Griggs, M. Noguer, P.J. van der Linden, X. Dai, K. Maskell and C.A. Johnson, Eds., Cambridge University Press, Cambridge, 739-768.

Mearns, L.O., G. Carbone, E. Tsvetsinskaya, R. Adams, B. McCarl and R. Doherty, 2003: The uncertainty of spatial scale of climate scenarios in integrated assessments: an example from agriculture. *Integrated Assess.*, **4**, 225-235.

Meehl, G.A., T.F. Stocker, W. Collins, P. Friedlingstein, A. Gaye, J. Gregory, A. Kitoh, R. Knutti, J. Murphy, A. Noda, S. Raper, I. Watterson, A. Weaver and Z.-C. Zhao, 2007: Global climate projections. *Climate Change 2007: The Physical Science Basis. Contribution of Working Group I to the Fourth Assessment Report of the Intergovernmental Panel on Climate Change*, S. Solomon, D. Qin, M. Manning, Z. Chen, M. Marquis, K. B. Averyt, M. Tignor and H. L. Miller, Eds., Cambridge University Press, Cambridge, 747-846.

Mendelsohn, R. and L. Williams, 2004: Comparing forecasts of the global impacts of climate change. *Mitig. Adapt. Strat. Glob. Change*, **9**, 315-333.

Metroeconomica, 2004: Costing the Impacts of Climate Change in the UK. UKCIP Technical Report, United Kingdom Climate Impacts Programme Oxford, 90 pp.

Metzger, M. and D. Schröter, 2006: Towards a spatially explicit and quantitative vulnerability assessment of environmental change in Europe. *Reg. Environ. Change*, **6**, 201-206.

Metzger, M.J., M.D.A. Rounsevell, R. Leemans and D. Schröter, 2006: The vulnerability of ecosystem services to land use change. *Agr. Ecosyst. Environ.*, **114**, 69-85.

Millennium Ecosystem Assessment, 2005: *Ecosystems and Human Well-being. Vol. 2: Scenarios: Findings of the Scenarios Working Group, Millennium Ecosystem Assessment*, S.R. Carpenter, P.L. Pingali, E.M. Bennett and M.B. Zurek, Eds., Island Press, Washington, DC, 560 pp.

Miller, N.L., K.E. Bashford and E. Strem, 2003: Potential impacts of climate change on California hydrology. *J. Am. Water Resour. As.*, **39**, 771-784.

Mirza, M.M.Q., 2003a: Three recent extreme floods in Bangladesh, a hydro-meteorological analysis. *Nat. Hazards*, **28**, 35-64.

Mirza, M.M.Q., 2003b: Climate change and extreme weather events: can developing countries adapt? *Clim. Policy*, **3**, 233-248.

Morgan, M.G. and M. Henrion, 1990: *Uncertainty: A Guide to Dealing with Uncertainty in Quantitative Risk and Policy Analysis*. Cambridge University Press, New York, 344 pp.

Morgan, M.G., B. Fischhoff, A. Bostrom and C.J. Atman, 2001: *Risk Communication: A Mental Models Approach*. Cambridge University Press, Cambridge, 366 pp.

Moriondo, M., P. Good, R. Durao, M. Bindi, C. Giannakopoulos and J. Corte-Real, 2006: Potential impact of climate change on forest fire risk in the Mediterranean area. *Climate Res.*, **31**, 85-95.

Morita, T., J. Robinson, A. Adegbulugbe, J. Alcamo, D. Herbert, E.L. La Rovere, N. Nakićenović, H. Pitcher, P. Raskin, K. Riahi, A. Sankovski, V. Sokolov, B. de Vries and D. Zhou, 2001: Greenhouse gas emission mitigation scenarios and implications. *Climate Change 2001: Mitigation. Contribution of Working Group III to the Third Assessment Report of the Intergovernmental Panel on Climate Change*, B. Metz, O. Davidson, R. Swart and J. Pan, Eds., Cambridge University Press, Cambridge, 115-166.

Moss, R.H., A.L. Brenkert and E.L. Malone, 2001: *Vulnerability to Climate Change: A Quantitative Approach*. Pacific Northwest National Laboratory, Richland, Washington.

Muramatsu, R. and Y. Hanich, 2005: Emotions as a mechanism for boundedly rational agents: the fast and frugal way. *J. Econ. Psychol.*, **26**, 201-221.

Nadarajah, C. and J.D. Rankin, 2005: European spatial planning: adapting to climate events. *Weather*, **60**, 190-194.

Næss, L.O., G. Bang, S. Eriksen and J. Vevatne, 2005: Institutional adaptation to climate change: flood responses at the municipal level in Norway. *Global Environ. Chang.*, **15**, 125-138.

Nakićenović, N., J. Alcamo, G. Davis, B. DeVries, J. Fenhann, S. Gaffin, K. Gregory, A. Gruebler, T.Y. Jung, T. Kram, E.L. La Rovere, L. Michaelis, S. Mori, T. Morita, W. Pepper, H. Pitcher, L. Price, K. Riahi, A. Roehrl, H.-H. Rogner, A. Sankovski, M. Schlesinger, P. Shukla, S. Smith, R. Swart, S. VanRooijen, N. Victor and Z. Dadi, 2000: *Special Report on Emissions Scenarios: A Special Report of Working Group III of the Intergovernmental Panel on Climate Change*. Cambridge University Press, Cambridge, 600 pp.

Nakićenović, N., B. Fisher, K. Alfsen, J. Corfee Morlot, F. de la Chesnaye, J.-C. Hourcade, K. Jiang, M. Kainuma, E.L. La Rovere, A. Rana, K. Riahi, R. Richels, D.P. van Vuuren and R. Warren, 2007: Issues related to mitigation in the long-term context. *Climate Change 2007: Mitigation of Climate Change. Contribution of Working Group III to the Fourth Assessment Report of the Intergovernmental Panel on Climate Change*, B. Metz, O. Davidson, P. Bosch, R. Dave and L. Meyer, Eds., Cambridge University Press, Cambridge, UK.

National Assessment Synthesis Team, 2000: *Climate Change Impacts on the United States: The Potential Consequences of Climate Variability and Change, Overview*. Cambridge University Press, Cambridge, 154 pp.

Nelson, G.C., E. Bennett, A. Asefaw Berhe, K.G. Cassman, R. DeFries, T. Dietz, A. Dobson, A. Dobermann, A. Janetos, M. Levy, D. Marco, B. O'Neill, N. Nakićenović, R. Norgaard, G. Petschel-Held, D. Ojima, P. Pingali, R. Watson and M. Zurek, 2005: Drivers of change in ecosystem condition and services. *Ecosystems and Human Well-Being: Scenarios: Findings of the Scenarios Working Group, Millennium Ecosystem Assessment*, S.R. Carpenter, P.L. Pingali, E.M. Bennett and M.B. Zurek, Eds., Island Press, Washington, DC, 173-222.

Neumann, J.E., G. Yohe, R. Nicholls and M. Manion, 2000: Sea-level rise and global climate change: a review of impacts to U.S. coasts. The Pew Center on Global Climate Change, Arlington, Virginia.

New, M. and M. Hulme, 2000: Representing uncertainty in climate change scenarios: a Monte Carlo approach. *Integrated Assess.*, **1**, 203-213.

Nicholls, R.J., 2004: Coastal flooding and wetland loss in the 21st century: changes under the SRES climate and socio-economic scenarios. *Global Environ. Chang.*, **14**, 69-86.

Nicholls, R.J. and J.A. Lowe, 2004: Benefits of mitigation of climate change for coastal areas. *Global Environ. Chang.*, **14**, 229-244.

Nicholls, R.J. and R.S.J. Tol, 2006: Impacts and responses to sea-level rise: A global analysis of the SRES scenarios over the 21st century. *Philos. T. R. Soc.*, **364**, 1073-1095.

Nicholls, R.J., R.S.J. Tol and A.T. Vafeidis, 2005: Global estimates of the impact of a collapse of the West Antarctic Ice Sheet: an application of FUND. Working Paper FNU78, Research Unit on Sustainability and Global Change, Hamburg University, and Centre for Marine and Atmospheric Science, Hamburg, 34 pp.

Nordhaus, W.D., 2006: Geography and macroeconomics: new data and new findings. *P. Natl. Acad. Sci.*, **103**, 3510-3517.

NRC, 2002: *Alerting America: Effective Risk Communication: Summary of a Forum, October 31, 2002*, National Academies Press, Washington, DC, 10 pp.

O'Brien, K. and C.H. Vogel, 2006: Who can eat information? Examining the effectiveness of seasonal climate forecasts and regional climate-risk management strategies. *Climate Res.*, **33**, 111-122.

O'Brien, K., S. Eriksen, A. Schjolden and L. Nygaard, 2004a: What's in a word? Conflicting interpretations of vulnerability in climate change research. Working Paper 2004:04, Centre for International Climate and Environmental Research Oslo, University of Oslo, Oslo, 16 pp.

O'Brien, K., R. Leichenko, U. Kelkar, H. Venema, G. Aandahl, H. Tompkins, A. Javed, S. Bhadwal, S. Barg, L. Nygaard and J. West, 2004b: Mapping vulnerability to multiple stressors: climate change and globalization in India. *Global Environ. Chang.*, **14**, 303-313.

O'Brien, K., S. Eriksen, L. Sygna and L.O. Naess, 2006: Questioning complacency: climate change impacts, vulnerability and adaptation in Norway. *Ambio*, **35**, 50-56.

O'Neill, B.C., 2004: Conditional probabilistic population projections: an application to climate change. *Int. Stat. Rev.*, **72**, 167-184.

O'Neill, B.C. and M. Oppenheimer, 2004: Climate change impacts are sensitive to the concentration stabilization path. *P. Natl. Acad. Sci.*, **101**, 16311-16416.

Ohlemüller, R., E.S. Gritti, M.T. Sykes and C.D. Thomas, 2006: Towards European climate risk surfaces: the extent and distribution of analogous and non-analogous climates 1931-2100. *Global Ecol. Biogeogr.*, **15**, 395-405.

Olesen, J.E., T.R. Carter, C.H. Díaz-Ambrona, S. Fronzek, T. Heidmann, T. Hickler, T. Holt, M.I. Minguez, P. Morales, J. Palutikof, M. Quemada, M. Ruiz-Ramos, G. Rubæk, F. Sau, B. Smith and M. Sykes, 2007: Uncertainties in projected impacts of climate change on European agriculture and terrestrial ecosystems based on scenarios from regional climate models. *Climatic Change*, **81** (Suppl. 1), 123-143.

Osman-Elasha, B., N. Goutbi, E. Spanger-Siegfried, B. Dougherty, A. Hanafi, S. Zakieldeen, A. Sanjak, H.A. Atti and H.M. Elhassan, 2006: Adaptation strategies to increase human resilience against climate variability and change: lessons from the arid regions of Sudan. AIACC Working Paper No. 42, Assessment of Impacts and Adaptation to Climate Change in Multiple Regions and Sectors Program, Washington, DC, 42 pp.

Ott, B. and S. Uhlenbrook, 2004: Quantifying the impact of land-use changes at the event and seasonal time scale using a process-oriented catchment model. *Hydrol. Earth Syst. Sc.*, **8**, 62-78.

Parry, M., N. Arnell, T. McMichael, R. Nicholls, P. Martens, S. Kovats, M. Livermore, C. Rosenzweig, A. Iglesias and G. Fischer, 2001: Millions at risk: defining critical climate change threats and targets. *Global Environ. Chang.*, **11**, 181-183.

Parry, M.L., 2002: Scenarios for climate impact and adaptation assessment. *Global Environ. Chang.*, **12**, 149-153.

Parry, M.L., 2004: Global impacts of climate change under the SRES scenarios. *Global Environ. Chang.*, **14**, 1.

Parry, M.L., T.R. Carter and M. Hulme, 1996: What is dangerous climate change? *Global Environ. Chang.*, **6**, 1-6.

Parry, M.L., N. Arnell, M. Hulme, R.J. Nicholls and M. Livermore, 1998: Adapting to the inevitable. *Nature*, **395**, 741.

Parry, M.L., N.W. Arnell, M. Hulme, P. Martens, R.J. Nicholls and A. White, 1999: The global impact of climate change: a new assessment. *Global Environ. Chang.*, **9**, S1-S2.

Parry, M.L., C. Rosenzweig, A. Iglesias, M. Livermore and G. Fischer, 2004: Effects of climate change on global food production under SRES emissions and socio-economic scenarios. *Global Environ. Chang.*, **14**, 53-67.

Parson, E.A., R.W. Corell, E.J. Barron, V. Burkett, A. Janetos, L. Joyce, T.R. Karl, M.C. MacCracken, J. Melillo, M.G. Morgan, D.S. Schimel and T. Wilbanks, 2003: Understanding climatic impacts, vulnerabilities and adaptation in the United States: building a capacity for assessment. *Climatic Change*, **57**, 9-42.

Parson, E.A., V. Burkett, K. Fisher-Vanden, D. Keith, L. Mearns, H. Pitcher, C. Rosenzweig and M. Webster, 2006: *Global-Change Scenarios: Their Development and Use*, US Climate Change Science Program No. 2.1b (Final draft for public comment), Washington, DC, 339 pp.

Patt, A. and C. Gwata, 2002: Effective seasonal climate forecast applications: examining constraints for subsistence farmers in Zimbabwe. *Global Environ. Chang.*, **12**, 185-195.

Patt, A.G., R. Klein and A. de la Vega-Leinert, 2005: Taking the uncertainties in climate change vulnerability assessment seriously. *C. R. Geosci.*, **337**, 411-424.

Pearson, R. and T. Dawson, 2003: Predicting the impacts of climate change on the distribution of species: are bioclimatic envelope models useful? *Global Ecol. Biogeogr.*, **12**, 361-371.

Perez-Garcia, J., L.A. Joyce, A.D. McGuire and X.M. Xiao, 2002: Impacts of climate change on the global forest sector. *Climatic Change*, **54**, 439-461.

Petoukhov, V., A. Ganopolski, V. Brovkin, M. Claussen, A. Eliseev, C. Kubatzki and S. Rahmstorf, 2000: CLIMBER-2: a climate system model of intermediate complexity. Part I. Model description and performance for present climate. *Clim. Dynam.*, **16**, 1-17.

Pittock, A.B. and R.N. Jones, 2000: Adaptation to what and why? *Environ. Monit. Assess.*, **61**, 9-35.

Polsky, C., D. Schöeter, A. Patt, S. Gaffin, M.L. Martello, R. Neff, A. Pulsipher and H. Selin, 2003: Assessing vulnerabilities to the effects of global change: an eight-step approach. Belford Centre for Science and International Affairs Working Paper, Environment and Natural Resources Program, John F. Kennedy School of Government, Harvard University, Cambridge, Massachusetts, 31 pp.

Preston, B.L., 2006: Risk-based reanalysis of the effects of climate change on U.S. cold-water habitat. *Climatic Change*, **76**, 91-119.

Pretty, J.N., I. Guijt, J. Thompson and I. Scoones, 1995: *Participatory Learning and Action: A Trainer's Guide*. IIED Training Materials Series No. 1. IIED, London, 268 pp.

Pulhin, J., R.J.J. Peras, R.V.O. Cruz, R.D. Lasco, F. Pulhin and M.A. Tapia, 2006: Vulnerability of communities to climate variability and extremes: the Pantabangan-Carranglan watershed in the Philippines. AIACC Working Paper No. 44, Assessment of Impacts and Adaptation to Climate Change in Multiple Regions and Sectors Program, Washington, DC, 56 pp.

Quiggin, J. and J. Horowitz, 2003: Costs of adjustment to climate change. *Aust. J.*

Agr. Resour. Ec., **47**, 429-446.

Randall, D., R.A. Wood, S. Bony, R. Colman, T. Fichefet, J. Fyfe, V. Kattsov, A. Pitman, J. Shukla, J. Srinivasan, R.J. Stouffer, A. Sumi and K. Taylor, 2007: Climate models and their evaluation. *Climate Change 2007: The Physical Science Basis. Contribution of Working Group I to the Fourth Assessment Report of the Intergovernmental Panel on Climate Change*, S. Solomon, D. Qin, M. Manning, Z. Chen, M. Marquis, K. B. Averyt, M. Tignor and H. L. Miller, Eds., Cambridge University Press, Cambridge, 589-662.

Raskin, P., T. Banuri, G. Gallopin, P. Gutman, A. Hammond, R. Kates and R. Swart, 2002: *Great Transition: The Promise and Lure of the Times Ahead*. Global Scenario Group, Stockholm Environment Institute, Boston, Massachusetts, 99 pp.

Raskin, P., F. Monks, T. Ribeiro, D. van Vuuren and M. Zurek, 2005: Global scenarios in historical perspective. *Ecosystems and Human Well-Being: Scenarios: Findings of the Scenarios Working Group, Millennium Ecosystem Assessment*, S.R. Carpenter, P.L. Pingali, E.M. Bennett and M.B. Zurek, Eds., Island Press, Washington, DC, 35-44.

Reginster, I. and M.D.A. Rounsevell, 2006: Future scenarios of urban land use in Europe. *Environ. Plan. B*, **33**, 619-636.

Reidsma, P., T. Tekelenburg, M. van den Berg and R. Alkemade, 2006: Impacts of land-use change on biodiversity: an assessment of agricultural biodiversity in the European Union. *Agr. Ecosyst. Environ.*, **114**, 86-102.

Renn, O., 2004: The challenge of integrating deliberation and expertise: participation and discourse in risk management. *Risk Analysis and Society: An Interdisciplinary Characterization of the Field*, T.L. MacDaniels and M.J. Small, Eds., Cambridge University Press, Cambridge, 289-366.

Rial, J.A., R.A. Pielke, M. Beniston, M. Claussen, J. Canadell, P. Cox, H. Held, N. De Noblet-Ducoudre, R. Prinn, J.F. Reynolds and J.D. Salas, 2004: Nonlinearities, feedbacks and critical thresholds within the Earth's climate system. *Climatic Change*, **65**, 11-38.

Rosenberg, N.J., R.A. Brown, R.C. Izaurralde and A.M. Thomson, 2003: Integrated assessment of Hadley Centre (HadCM2) climate change projections on agricultural productivity and irrigation water supply in the conterminous United States. I. Climate change scenarios and impacts on irrigation water supply simulated with the HUMUS model. *Agr. Forest Meteorol.*, **117**, 73-96.

Rotmans, J. and D. Rothman, Eds., 2003: *Scaling Issues in Integrated Assessment*. Swets and Zeitlinger, Lisse, 374 pp.

Rotmans, J., M.B.A. Van Asselt, C. Anastasi, S.C.H. Greeuw, J. Mellors, S. Peters, D.S. Rothman and N. Rijkens-Klomp, 2000: Visions for a sustainable Europe. *Futures*, **32**, 809-831.

Rounsevell, M.D.A., F. Ewert, I. Reginster, R. Leemans and T.R. Carter, 2005: Future scenarios of European agricultural land use. II. Estimating changes in land use and regional allocation. *Agr. Ecosyst. Environ.*, **107**, 117-135.

Rounsevell, M.D.A., I. Reginster, M.B. Araújo, T.R. Carter, N. Dendoncker, F. Ewert, J.I. House, S. Kankaanpää, R. Leemans, M.J. Metzger, C. Schmit, P. Smith and G. Tuck, 2006: A coherent set of future land use change scenarios for Europe. *Agr. Ecosyst. Environ.*, **114**, 57-68.

Rowe, G. and L. Frewer, 2000: Public participation methods: an evaluative review of the literature. *Sci. Technol. Hum. Val.*, **25**, 3-29.

Ruosteenoja, K., T.R. Carter, K. Jylhä and H. Tuomenvirta, 2003: Future climate in world regions: an intercomparison of model-based projections for the new IPCC emissions scenarios. The Finnish Environment 644, Finnish Environment Institute, Helsinki, 83 pp.

Rupa Kumar, K., A.K. Sahai, K. Krishna Kumar, S.K. Patwardhan, P.K. Mishra, J.V. Revadekar, K. Kamala and G.B. Pant, 2006: High-resolution climate change scenarios for India for the 21st century. *Curr. Sci. India*, **90**, 334-345.

Sands, R.D. and M. Leimbach, 2003: Modeling agriculture and land use in an integrated assessment framework. *Climatic Change*, **56**, 185-210.

Sands, R.D. and J.A. Edmonds, 2005: Climate change impacts for the conterminous USA: an integrated assessment. Part 7. Economic analysis of field crops and land use with climate change. *Climatic Change*, **69**, 127-150.

Santer, B.D., T.M.L. Wigley, M.E. Schlesinger and J.F.B. Mitchell, 1990: Developing climate scenarios from equilibrium GCM results. Report No. 47, Max-Planck-Institut für Meteorologie, Hamburg, 29 pp.

SBI, 2001: National communications from Parties not included in Annex I to the Convention. Report of the Consultative Group of Experts to the Subsidiary Bodies. Subsidiary Body for Implementation, Document FCCC/SBI/2001/15, United Nations Office, Geneva.

Schellnhuber, H.J., R. Warren, A. Hazeltine and L. Naylor, 2004: Integrated assessments of benefits of climate policy. *The Benefits of Climate Change Policies: Analytic and Framework Issues*, J.C. Morlot and S. Agrawala, Eds., Organisation for Economic Cooperation and Development (OECD), Paris, 83-110.

Schneider, S.H., 2001: What is 'dangerous' climate change? *Nature*, **411**, 17-19.

Schneider, S.H., 2002: Can we estimate the likelihood of climatic changes at 2100? *Climatic Change*, **52**, 441-451.

Schneider, S.H. and K. Kuntz-Duriseti, 2002: Uncertainty and climate change policy. *Climate Change Policy: A Survey*, S.H. Schneider, A. Rosencranz and J.O. Niles, Eds., Island Press, Washington, DC, 53-87.

Scholze, M., W. Knorr, N.W. Arnell and I. Prentice, 2006: A climate-change risk analysis for world ecosystems. *P. Natl. Acad. Sci.*, **103**, 13116-13120.

Schröter, D., C. Polsky and A.G. Patt, 2005a: Assessing vulnerabilities to the effects of global change: an eight-step approach. *Mitig. Adapt. Strat. Glob. Change*, **10**, 573-596.

Schröter, D., W. Cramer, R. Leemans, I.C. Prentice, M.B. Araújo, N.W. Arnell, A. Bondeau, H. Bugmann, T.R. Carter, C.A. Garcia, A.C. de la Vega-Leinert, M. Erhard, F. Ewert, M. Glendining, J.I. House, S. Kankaanpää, R.J.T. Klein, S. Lavorel, M. Lindner, M.J. Metzger, J. Meyer, T.D. Mitchell, I. Reginster, M. Rounsevell, S. Sabaté, S. Sitch, B. Smith, J. Smith, P. Smith, M.T. Sykes, K. Thonicke, W. Thuiller, G. Tuck, S. Zaehle and B. Zierl, 2005b: Ecosystem service supply and vulnerability to global change in Europe. *Science*, **310**, 1333-1337

Shackley, S. and R. Deanwood, 2003: Constructing social futures for climate-change impacts and response studies: building qualitative and quantitative scenarios with the participation of stakeholders. *Climate Res.*, **24**, 71-90.

Slovic, P., M.L. Finucane, E. Peters and D.G. MacGregor, 2004: Risk as analysis and risk as feelings: some thoughts about affect, reason, risk and rationality. *Risk Anal.*, **24**, 311-322.

Smit, B. and J. Wandel, 2006: Adaptation, adaptive capacity and vulnerability. *Global Environ. Chang.*, **16**, 282-292.

Smith, J.B., H.-J. Schellnhuber, M.M.Q. Mirza, S. Fankhauser, R. Leemans, L. Erda, L. Ogallo, B. Pittock, R. Richels, C. Rosenzweig, U. Safriel, R.S.J. Tol, J. Weyant and G. Yohe, 2001: Vulnerability to climate change and reasons for concern: a synthesis. *Climate Change 2001: Impacts, Adaptation, and Vulnerability. Contribution of Working Group II to the Third Assessment Report of the Intergovernmental Panel on Climate Change*, J.J. McCarthy, O.F. Canziani, N.A. Leary, D.J. Dokken and K. S. White, Eds., Cambridge University Press, Cambridge, 913-967.

Smith, S.J., H. Pitcher and T.M.L. Wigley, 2005: Future sulfur dioxide emissions. *Climatic Change*, **73**, 267-318.

Solecki, W.D. and C. Oliveri, 2004: Downscaling climate change scenarios in an urban land use change model. *J. Environ. Manage.*, **72**, 105-115.

Stone, M.C., R.H. Hotchkiss and L.O. Mearns, 2003: Water yield responses to high and low spatial resolution climate change scenarios in the Missouri River Basin. *Geophys. Res. Lett.*, **30**, doi:10.1029/2002GL016122.

Swart, R., J. Mitchell, T. Morita and S. Raper, 2002: Stabilisation scenarios for climate impact assessment. *Global Environ. Chang.*, **12**, 155-165.

Syri, S., S. Fronzek, N. Karvosenoja and M. Forsius, 2004: Sulfur and nitrogen oxides emissions in Europe and deposition in Finland during the 21st century. *Boreal Environ. Res.*, **9**, 185-198.

Thaler, R. and E.J. Johnson, 1990: Gambling with the house money and trying to break even: the effects of prior outcomes on risky choice. *Manage. Sci.*, **36**, 643-660.

Thomalla, F., T. Cannon, S. Huq, R.J.T. Klein and C. Schaerer, 2005: Mainstreaming adaptation to climate change in coastal Bangladesh by building civil society alliances. *Proceedings of the Solutions to Coastal Disasters Conference*, L. Wallendorf, L. Ewing, S. Rogers and C. Jones, Eds., Charleston, South Carolina, 8-11 May 2005, American Society of Civil Engineers, Reston, Virginia, 668-684.

Thomas, C.D., A. Cameron, R.E. Green, M. Bakkenes, L.J. Beaumont, Y.C. Collingham, B.F.N. Erasmus, M. Ferreira de Siqueira, A. Grainger, L. Hannah, L. Hughes, B. Huntley, A.S. Van Jaarsveld, G.F. Midgley, L. Miles, M.A. Ortega-Huerta, A.T. Peterson, O.L. Phillips and S.E. Williams, 2004: Extinction risk from climate change. *Nature*, **427**, 145-148.

Thuiller, W., M.B. Araújo, R.G. Pearson, R.J. Whittaker, L. Brotons and S. Lavorel, 2004: Uncertainty in predictions of extinction risk. *Nature*, **430**, 34.

Tol, R.S.J., 2002a: New estimates of the damage costs of climate change. Part I. Benchmark estimates. *Environ. Resour. Econ.*, **21**, 45-73.

Tol, R.S.J., 2002b: New estimates of the damage costs of climate change. Part II. Dynamic estimates. *Environ. Resour. Econ.*, **21**, 135-160.

Tol, R.S.J., 2006: Exchange rates and climate change: an application of FUND. *Climatic Change*, **75**, 59-80.

Tol, R.S.J., M. Bohn, T.E. Downing, M.-L. Guillerminet, E. Hizsnyik, R. Kasperson, K. Lonsdale, C. Mays, R.J. Nicholls, A.A. Olsthoorn, G. Pfeifle, M.

Poumadere, F.L. Toth, A.T. Vafeidis, P.E. van der Werff and I.H. Yetkiner, 2006: Adaptation to five metres of sea level rise. *J. Risk Res.*, **9**, 467-482.

Tompkins, E.L., 2005: Planning for climate change in small islands: insights from national hurricane preparedness in the Cayman Islands. *Global Environ. Chang.*, **15**, 139-149.

Tompkins, E.L. and W.N. Adger, 2005: Defining a response capacity for climate change. *Environ. Sci. Policy*, **8**, 562-571.

Toth, F.L., T. Bruckner, H.-M. Füssel, M. Leimbach and G. Petschel-Held, 2003a: Integrated assessment of long-term climate policies. Part 1. Model presentation. *Climatic Change*, **56**, 37-56.

Toth, F.L., T. Bruckner, H.-M. Füssel, M. Leimbach and G. Petschel-Held, 2003b: Integrated assessment of long-term climate policies. Part 2. Model results and uncertainty analysis. *Climatic Change*, **56**, 57-72.

Trenberth, K.E., P.D. Jones, P.G. Ambenje, R. Bojariu, D.R. Easterling, A.M.G. Klein Tank, D.E. Parker, J.A. Renwick, F. Rahimzadeh, M.M. Rusticucci, B.J. Soden and P.-M. Zhai, 2007: Observations: surface and atmospheric climate change. *Climate Change 2007: The Physical Science Basis. Contribution of Working Group I to the Fourth Assessment Report of the Intergovernmental Panel on Climate Change*, S. Solomon, D. Qin, M. Manning, Z. Chen, M. Marquis, K. B. Averyt, M. Tignor and H. L. Miller, Eds., Cambridge University Press, Cambridge, 235-336.

Turner, B.L., II, R.E. Kasperson, P.A. Matson, J.J. McCarthy, R.W. Corell, L. Christensen, N. Eckley, J.X. Kasperson, A. Luers, M.L. Martello, C. Polsky, A. Pulsipher and A. Schiller, 2003a: A framework for vulnerability analysis in sustainability science. *P. Natl. Acad. Sci.*, **100**, 8074-8079.

Turner, B.L., II, P.A. Matson, J.J. McCarthy, R.W. Corell, L. Christensen, N. Eckley, G.K. Hovelsrud-Broda, J.X. Kasperson, R. Kasperson, A. Luers, M.L. Martello, S. Mathiesen, R. Naylor, C. Polsky, A. Pulsipher, A. Schiller, H. Selin and N. Tyler, 2003b: Illustrating the coupled human-environment system for vulnerability analysis: three case studies. *P. Natl. Acad. Sci.*, **100**, 8080-8085.

Tversky, A. and D. Kahneman, 1974: Judgment under uncertainty: heuristics and biases. *Science*, **211**, 1124-1131.

UKMO, 2001: *The Hadley Centre Regional Climate Modelling System: PRECIS – Providing Regional Climates for Impacts Studies*. UK Meteorological Office, Bracknell, 17 pp.

UNDP, 2005: *Adaptation Policy Frameworks for Climate Change: Developing Strategies, Policies and Measures*. B. Lim, E. Spanger-Siegfried, I. Burton, E. Malone and S. Huq, Eds., Cambridge University Press, Cambridge and New York, 258 pp.

UNEP, 2002: *Global Environment Outlook 3: Past, Present and Future Perspectives*, Earthscan, London, 416 pp.

UNFCCC, 2004: *Compendium of Decision Tools to Evaluate Strategies for Adaptation to Climate Change*. United Nations Framework Convention on Climate Change, Bonn, 49 pp.

UNFCCC, 2005: *Sixth Compilation and Synthesis of Initial National Communications from Parties Not Included in Annex I to the Convention: Executive Summary*. FCCC/SBI/2005/18, United Nations Framework Convention on Climate Change, Bonn, 21 pp. [Accessed 27.02.07: http://unfccc.int/resource/docs/2005/sbi/eng/18.pdf]

van Aalst, M., 2006: *Managing Climate Risk: Integrating Adaptation into World Bank Group Operations*. World Bank Global Environment Facility Program, Washington, DC, 42 pp.

van Asselt, M.B.A. and N. Rijkens-Klomp, 2002: A look in the mirror: reflection on participation in Integrated Assessment from a methodological perspective. *Global Environ. Chang.*, **12**, 167-184.

van Asselt, M.B.A. and J. Rotmans, 2002: Uncertainty in integrated assessment modelling: from positivism to pluralism. *Climatic Change*, **54**, 75-105.

van Beek, L.P.H. and T.W.J. van Asch, 2004: Regional assessment of the effects of land-use change on landslide hazard by means of physically based modelling. *Nat. Hazards*, **31**, 289-304.

van Lieshout, M., R.S. Kovats, M.T.J. Livermore and P. Martens, 2004: Climate change and malaria: analysis of the SRES climate and socio-economic scenarios. *Global Environ. Chang.*, **14**, 87-99.

van Meijl, H., T. van Rheenen, A. Tabeau and B. Eickhout, 2006: The impact of different policy environments on agricultural land use in Europe. *Agr. Ecosyst. Environ.*, **114**, 21-38.

van Minnen, J.G., J. Alcamo and W. Haupt, 2000: Deriving and applying response surface diagrams for evaluating climate change impacts on crop production. *Climatic Change*, **46**, 317-338.

van Vuuren, D.P. and K.H. Alfsen, 2006: PPP versus MER: searching for answers in a multi-dimensional debate. *Climatic Change*, **75**, 47-57.

van Vuuren, D.P. and B.C. O'Neill, 2006: The consistency of IPCC's SRES scenarios to recent literature and recent projections. *Climatic Change*, **75**, 9-46.

van Vuuren, D.P., J. Weyant and F. de la Chesnaye, 2006: Multi-gas scenarios to stabilize radiative forcing. *Energ. Econ.*, **28**, 102-120.

van Vuuren, D.P., P. Lucas and H. Hilderink, 2007: Downscaling drivers of global environmental change scenarios: enabling use of the IPCC SRES scenarios at the national and grid level. *Global Environ. Chang.*, **17**, 114-130.

Vaze, J., P. Barnett, G. Beale, W. Dawes, R. Evans, N.K. Tuteja, B. Murphy, G. Geeves and M. Miller, 2004: Modelling the effects of land-use change on water and salt delivery from a catchment affected by dryland salinity in south-east Australia. *Hydrol. Process.*, **18**, 1613-1637.

Velázquez, A., G. Bocco and A. Torres, 2001: Turning scientific approaches into practical conservation actions: the case of Comunidad Indigena de Nuevo San Juan Parangaricutiro, Mexico. *Environ. Manage.*, **27**, 655-665.

Vellinga, M. and R.A. Wood, 2002: Global climatic impacts of a collapse of the Atlantic thermohaline circulation. *Climatic Change*, **54**, 251-267.

Verburg, P., C.J.E. Schulp, N. Witte and A. Veldkamp, 2006: Downscaling of land use change scenarios to assess the dynamics of European landscapes. *Agr. Ecosyst. Environ.*, **114**, 39-56.

Verburg, P.H., W. Soepboer, A. Veldkamp, R. Limpiada, V. Espaldon and S.S.A. Mastura, 2002: Modeling the spatial dynamics of regional land use: the CLUE-S model. *Environ. Manage.*, **30**, 391-405.

Wandiga, S.O., M. Opondo, D. Olago, A. Githeko, F. Githui, M. Marshall, T. Downs, A. Opere, P.Z. Yanda, R. Kangalawe, R. Kabumbuli, J. Kathuri, E. Apindi, L. Olaka, L. Ogallo, P. Mugambi, R. Sigalla, R. Nanyunja, T. Baguma and P. Achola, 2006: Vulnerability to climate-induced highland malaria in East Africa. AIACC Working Paper No. 25, Assessments of Impacts and Adaptations to Climate Change in Multiple Regions and Sectors Program, Washington, DC, 47 pp.

Wang, X., F. Zwiers and V. Swal, 2004: North Atlantic ocean wave climate change scenarios for the twenty-first century. *J. Climate*, **17**, 2368-2383.

Webster, M.D., M. Babiker, M. Mayer, J.M. Reilly, J. Harnisch, M.C. Sarofim and C. Wang, 2002: Uncertainty in emissions projections for climate models. *Atmos. Environ.*, **36**, 3659-3670.

Webster, M.D., C. Forest, J. Reilly, M. Babiker, D. Kicklighter, M. Mayer, R. Prinn, M.C. Sarofim, A. Sokolov, P. Stone and C. Wang, 2003: Uncertainty analysis of climate change and policy response. *Climatic Change*, **61**, 295-320.

Wehbe, M., H. Eakin, R. Seiler, M. Vinocur, C. Ávila and C. Marutto, 2006: Local perspectives on adaptation to climate change: lessons from Mexico and Argentina. AIACC Working Paper No. 39, Assessment of Impacts and Adaptation to Climate Change in Multiple Regions and Sectors Program, Washington, DC, 37 pp.

Welp, M., 2001: The use of decision support tools in participatory river basin management. *Phys. Chem. Earth Pt B.*, **26**, 535-539.

West, C.C. and M.J. Gawith, Eds., 2005: *Measuring Progress: Preparing for Climate Change through the UK Climate Impacts Programme*. UKCIP, Oxford, 72 pp.

Westhoek, H., B. Eickhout and D. van Vuuren, 2006a: A brief comparison of scenario assumptions of four scenario studies: IPCC-SRES, GEO-3, Millennium Ecosystem Assessment and FAO towards 2030. MNP Report, Netherlands Environmental Assessment Agency (MNP), Bilthoven, 48 pp. [Accessed 27.02.07: http://www.mnp.nl/image/whats_new/]

Westhoek, H.J., M. van den Berg and J.A. Bakkes, 2006b: Scenario development to explore the future of Europe's rural areas. *Agr. Ecosyst. Environ.*, **114**, 7-20.

Wigley, T.M.L., 2004: Choosing a stabilization target for CO_2. *Climatic Change*, **67**, 1-11.

Wigley, T.M.L. and S.C.B. Raper, 2001: Interpretation of high projections for global-mean warming. *Science*, **293**, 451-454.

Wilbanks, T.J., S.M. Kane, P.N. Leiby, R.D. Perlack, C. Settle, J.F. Shogren and J.B. Smith, 2003: Possible responses to global climate change: integrating mitigation and adaptation. *Environment*, **45**, 28-38.

Wilby, R., M. McKenzie Hedger and C. Parker, 2004a: *What We Need to Know and When: Decision-Makers' Perspectives on Climate Change Science*. Report of Workshop, February 2004, Climate Change Unit, Environment Agency, London, 41 pp.

Wilby, R.L. and I. Harris, 2006: A framework for assessing uncertainties in climate change impacts: low-flow scenarios for the River Thames, UK. *Water Resour. Res.*, **42**, W02419, doi:10.1029/2005WR004065.

Wilby, R.L., R. Dawson and E.M. Barrow, 2002: SDSM: a decision support tool for the assessment of regional climate change assessments. *Environ. Modell. Softw.*, **17**, 145-157.

Wilby, R.L., S.P. Charles, E. Zorita, B. Timtal, P. Whetton and L.O. Mearns, 2004b:

Guidelines for Use of Climate Scenarios Developed from Statistical Downscaling Methods. Supporting material of the Intergovernmental Panel on Climate Change, IPCC Task Group on Data and Scenario Support for Impact and Climate Analysis, Cambridge University Press, Cambridge, 27 pp. [Accessed 26.02.07: http://www.ipcc-data.org/guidelines/dgm_no2_v1_09_2004.pdf]

Williams, J.W., S.T. Jackson and J.E. Kutzbach, 2007: Projected distributions of novel and disappearing climates by 2100 AD. *P. Natl. Acad. Sci.*, **104**, 5738-5742

Willows, R. and R. Connell, 2003: Climate adaptation: risk, uncertainty and decision-making. UKCIP Technical Report, UK Climate Impacts Programme, Oxford, 154 pp.

Wood, A.W., L.R. Leung, V. Sridhar and D.P. Letterman, 2004: Hydrologic implications of dynamical and statistical approaches to downscaling climate model outputs. *Climatic Change*, **62**, 189-216.

Wood, R.A., M. Vellinga and R. Thorpe, 2003: Global warming and thermohaline circulation stability. *Philos. T. R. Soc. A*, **361**, 1961-1975.

Woodworth, P.L. and D.L. Blackman, 2004: Evidence for systematic changes in extreme high waters since the mid-1970s. *J. Climate*, **17**, 1190-1197.

Wooldridge, S., T. Done, R. Berkelmans, R. Jones and P. Marshall, 2005: Precursors for resilience in coral communities in a warming climate: a belief network approach. *Mar. Ecol.–Prog. Ser.*, **295**, 157-169.

World Bank, 2006: *World Development Report 2006: Equity and Development*. The World Bank, Washington, DC, and Oxford University Press, Oxford, 336 pp.

Yang, D.W., S. Kanae, T. Oki, T. Koike and K. Musiake, 2003: Global potential soil erosion with reference to land use and climate changes. *Hydrol. Process.*, **17**, 2913-2928.

Yohe, G., 2004: Some thoughts on perspective. *Global Environ. Chang.*, **14**, 283-286.

Yohe, G. and F. Toth, 2000: Adaptation and the guardrail approach to tolerable climate change. *Climatic Change*, **45**, 103-128.

Yohe, G. and R.S.J. Tol, 2002: Indicators for social and economic coping capacity: moving toward a working definition of adaptive capacity. *Global Environ. Chang.*, **12**, 25-40.

Zebisch, M., F. Wechsung and H. Kenneweg, 2004: Landscape response functions for biodiversity: assessing the impact of land-use changes at the county level. *Landscape Urban Plan.*, **67**, 157-172.

3

Freshwater resources and their management

Coordinating Lead Authors:
Zbigniew W. Kundzewicz (Poland), Luis José Mata (Venezuela)

Lead Authors:
Nigel Arnell (UK), Petra Döll (Germany), Pavel Kabat (The Netherlands), Blanca Jiménez (Mexico), Kathleen Miller (USA), Taikan Oki (Japan), Zekai Şen (Turkey), Igor Shiklomanov (Russia)

Contributing Authors:
Jun Asanuma (Japan), Richard Betts (UK), Stewart Cohen (Canada), Christopher Milly (USA), Mark Nearing (USA), Christel Prudhomme (UK), Roger Pulwarty (Trinidad and Tobago), Roland Schulze (South Africa), Renoj Thayyen (India), Nick van de Giesen (The Netherlands), Henk van Schaik (The Netherlands), Tom Wilbanks (USA), Robert Wilby (UK)

Review Editors:
Alfred Becker (Germany), James Bruce (Canada)

This chapter should be cited as:
Kundzewicz, Z.W., L.J. Mata, N.W. Arnell, P. Döll, P. Kabat, B. Jiménez, K.A. Miller, T. Oki, Z. Şen and I.A. Shiklomanov, 2007: Freshwater resources and their management. *Climate Change 2007: Impacts, Adaptation and Vulnerability. Contribution of Working Group II to the Fourth Assessment Report of the Intergovernmental Panel on Climate Change,* M.L. Parry, O.F. Canziani, J.P. Palutikof, P.J. van der Linden and C.E. Hanson, Eds., Cambridge University Press, Cambridge, UK, 173-210.

Table of Contents

Executive summary

The impacts of climate change on freshwater systems and their management are mainly due to the observed and projected increases in temperature, sea level and precipitation variability (very high confidence).

More than one-sixth of the world's population live in glacier- or snowmelt-fed river basins and will be affected by the seasonal shift in streamflow, an increase in the ratio of winter to annual flows, and possibly the reduction in low flows caused by decreased glacier extent or snow water storage (high confidence) [3.4.1, 3.4.3]. Sea-level rise will extend areas of salinisation of groundwater and estuaries, resulting in a decrease in freshwater availability for humans and ecosystems in coastal areas (very high confidence) [3.2, 3.4.2]. Increased precipitation intensity and variability is projected to increase the risks of flooding and drought in many areas (high confidence) [3.3.1].

Semi-arid and arid areas are particularly exposed to the impacts of climate change on freshwater (high confidence).

Many of these areas (e.g., Mediterranean basin, western USA, southern Africa, and north-eastern Brazil) will suffer a decrease in water resources due to climate change (very high confidence) [3.4, 3.7]. Efforts to offset declining surface water availability due to increasing precipitation variability will be hampered by the fact that groundwater recharge will decrease considerably in some already water-stressed regions (high confidence) [3.2, 3.4.2], where vulnerability is often exacerbated by the rapid increase in population and water demand (very high confidence) [3.5.1].

Higher water temperatures, increased precipitation intensity, and longer periods of low flows exacerbate many forms of water pollution, with impacts on ecosystems, human health, water system reliability and operating costs (high confidence).

These pollutants include sediments, nutrients, dissolved organic carbon, pathogens, pesticides, salt, and thermal pollution [3.2, 3.4.4, 3.4.5].

Climate change affects the function and operation of existing water infrastructure as well as water management practices (very high confidence).

Adverse effects of climate on freshwater systems aggravate the impacts of other stresses, such as population growth, changing economic activity, land-use change, and urbanisation (very high confidence) [3.3.2, 3.5]. Globally, water demand will grow in the coming decades, primarily due to population growth and increased affluence; regionally, large changes in irrigation water demand as a result of climate change are likely (high confidence) [3.5.1]. Current water management practices are very likely to be inadequate to reduce the negative impacts of climate change on water supply reliability, flood risk, health, energy, and aquatic ecosystems (very high confidence) [3.4, 3.5]. Improved incorporation of current climate variability into water-related management would make adaptation to future climate change easier (very high confidence) [3.6].

Adaptation procedures and risk management practices for the water sector are being developed in some countries and regions (e.g., Caribbean, Canada, Australia, Netherlands, UK, USA, Germany) that have recognised projected hydrological changes with related uncertainties (very high confidence).

Since the IPCC Third Assessment, uncertainties have been evaluated, their interpretation has improved, and new methods (e.g., ensemble-based approaches) are being developed for their characterisation (very high confidence) [3.4, 3.5]. Nevertheless, quantitative projections of changes in precipitation, river flows, and water levels at the river-basin scale remain uncertain (very high confidence) [3.3.1, 3.4].

The negative impacts of climate change on freshwater systems outweigh its benefits (high confidence).

All IPCC regions (see Chapters 3–16) show an overall net negative impact of climate change on water resources and freshwater ecosystems (high confidence). Areas in which runoff is projected to decline are likely to face a reduction in the value of the services provided by water resources (very high confidence) [3.4, 3.5]. The beneficial impacts of increased annual runoff in other areas will be tempered by the negative effects of increased precipitation variability and seasonal runoff shifts on water supply, water quality, and flood risks (high confidence) [3.4, 3.5].

3.1 Introduction

Water is indispensable for all forms of life. It is needed in almost all human activities. Access to safe freshwater is now regarded as a universal human right (United Nations Committee on Economic, Social and Cultural Rights, 2003), and the Millennium Development Goals include the extended access to safe drinking water and sanitation (UNDP, 2006). Sustainable management of freshwater resources has gained importance at regional (e.g., European Union, 2000) and global scales (United Nations, 2002, 2006; World Water Council, 2006), and 'Integrated Water Resources Management' has become the corresponding scientific paradigm.

Figure 3.1 shows schematically how human activities affect freshwater resources (both quantity and quality) and their management. Anthropogenic climate change is only one of many pressures on freshwater systems. Climate and freshwater systems are interconnected in complex ways. Any change in one

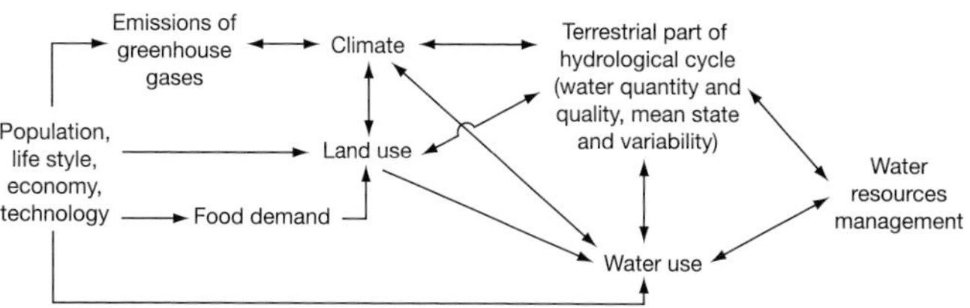

Figure 3.1. *Impact of human activities on freshwater resources and their management, with climate change being only one of multiple pressures (modified after Oki, 2005).*

of these systems induces a change in the other. For example, the draining of large wetlands may cause changes in moisture recycling and a decrease of precipitation in particular months, when local boundary conditions dominate over the large-scale circulation (Kanae et al., 2001). Conversely, climate change affects freshwater quantity and quality with respect to both mean states and variability (e.g., water availability as well as floods and droughts). Water use is impacted by climate change, and also, more importantly, by changes in population, lifestyle, economy, and technology; in particular by food demand, which drives irrigated agriculture, globally the largest water-use sector. Significant changes in water use or the hydrological cycle (affecting water supply and floods) require adaptation in the management of water resources.

In the Working Group II Third Assessment Report (TAR; IPCC, 2001), the state of knowledge of climate change impacts on hydrology and water resources was presented in the light of literature up to the year 2000 (Arnell et al., 2001). These findings are summarised as follows.

- There are apparent trends in streamflow volume, both increases and decreases, in many regions.
- The effect of climate change on streamflow and groundwater recharge varies regionally and between scenarios, largely following projected changes in precipitation.
- Peak streamflow is likely to move from spring to winter in many areas due to early snowmelt, with lower flows in summer and autumn.
- Glacier retreat is likely to continue, and many small glaciers may disappear.
- Generally, water quality is likely to be degraded by higher water temperatures.
- Flood magnitude and frequency are likely to increase in most regions, and volumes of low flows are likely to decrease in many regions.
- Globally, demand for water is increasing as a result of population growth and economic development, but is falling in some countries, due to greater water-use efficiency.
- The impact of climate change on water resources also depends on system characteristics, changing pressures on the system, how the management of the system evolves, and what adaptations to climate change are implemented.
- Unmanaged systems are likely to be most vulnerable to climate change.
- Climate change challenges existing water resource management practices by causing trends not previously experienced and adding new uncertainty.
- Adaptive capacity is distributed very unevenly across the world.

These findings have been confirmed by the current assessment. Some of them are further developed, and new findings have been added. This chapter gives an overview of the future impacts of climate change on freshwater resources and their management, mainly based on research published after the Third Assessment Report. Socio-economic aspects, adaptation issues, implications for sustainable development, as well as uncertainties and research priorities, are also covered. The focus is on terrestrial water in liquid form, due to its importance for freshwater management. Various aspects of climate change impacts on water resources and related vulnerabilities are presented (Section 3.4) as well as the impacts on water-use sectors (Section 3.5). Please refer to Chapter 1 for further information on observed trends, to Chapter 15 (Sections 15.3 and 15.4.1) for freshwater in cold regions and to Chapter 10 of the Working Group I Fourth Assessment Report (Meehl et al., 2007) - Section 10.3.3 for the cryosphere, and Section 10.3.2.3 for impacts on precipitation, evapotranspiration and soil moisture. While the impacts of increased water temperatures on aquatic ecosystems are discussed in this volume in Chapter 4 (Section 4.4.8), findings with respect to the effect of changed flow conditions on aquatic ecosystems are presented here in Section 3.5. The health effects of changes in water quality and quantity are covered in Chapter 8, while regional vulnerabilities related to freshwater are discussed in Chapters 9–16.

3.2 Current sensitivity/vulnerability

With higher temperatures, the water-holding capacity of the atmosphere and evaporation into the atmosphere increase, and this favours increased climate variability, with more intense precipitation and more droughts (Trenberth et al., 2003). The hydrological cycle accelerates (Huntington, 2006). While temperatures are expected to increase everywhere over land and during all seasons of the year, although by different increments, precipitation is expected to increase globally and in many river basins, but to decrease in many others. In addition, as shown in the Working Group I Fourth Assessment Report, Chapter 10, Section 10.3.2.3 (Meehl et al., 2007), precipitation may increase in one season and decrease in another. These climatic changes lead to changes in all components of the global freshwater system.

Climate-related trends of some components during the last decades have already been observed (see Table 3.1). For a number of components, for example groundwater, the lack of data makes it impossible to determine whether their state has changed in the recent past due to climate change. During recent decades, non-climatic drivers (Figure 3.1) have exerted strong pressure on freshwater systems. This has resulted in water pollution, damming of rivers, wetland drainage, reduction in streamflow, and lowering of the groundwater table (mainly due to irrigation). In comparison, climate-related changes have been small, although this is likely to be different in the future as the climate change signal becomes more evident.

Current vulnerabilities to climate are strongly correlated with climate variability, in particular precipitation variability. These vulnerabilities are largest in semi-arid and arid low-income countries, where precipitation and streamflow are concentrated over a few months, and where year-to-year variations are high (Lenton, 2004). In such regions a lack of deep groundwater wells or reservoirs (i.e., storage) leads to a high level of vulnerability to climate variability, and to the climate changes that are likely to further increase climate variability in future. In addition, river basins that are stressed due to non-climatic drivers are likely to be vulnerable to climate change. However, vulnerability to climate change exists everywhere, as water infrastructure (e.g., dikes and pipelines) has been designed for stationary climatic conditions, and water resources management has only just started to take into

Table 3.1. *Climate-related observed trends of various components of the global freshwater system. Reference is given to Chapters 1 and 15 of this volume and to the Working Group I Fourth Assessment Report (WGI AR4) Chapter 3 (Trenberth et al., 2007) and Chapter 4 (Lemke et al., 2007).*

	Observed climate-related trends
Precipitation	Increasing over land north of 30°N over the period 1901–2005. Decreasing over land between 10°S and 30°N after the 1970s (WGI AR4, Chapter 3, Executive summary). Increasing intensity of precipitation (WGI AR4, Chapter 3, Executive summary).
Cryosphere	
Snow cover	Decreasing in most regions, especially in spring (WGI AR4, Chapter 4, Executive summary).
Glaciers	Decreasing almost everywhere (WGI AR4, Chapter 4, Section 4.5).
Permafrost	Thawing between 0.02 m/yr (Alaska) and 0.4 m/yr (Tibetan Plateau) (WGI AR4 Chapter 4 Executive summary; this report, Chapter 15, Section 15.2).
Surface waters	
Streamflow	Increasing in Eurasian Arctic, significant increases or decreases in some river basins (this report, Chapter 1, Section 1.3.2). Earlier spring peak flows and increased winter base flows in Northern America and Eurasia (this report, Chapter 1, Section 1.3.2).
Evapotranspiration	Increased actual evapotranspiration in some areas (WGI AR4, Chapter 3, Section 3.3.3).
Lakes	Warming, significant increases or decreases of some lake levels, and reduction in ice cover (this report, Chapter 1, Section 1.3.2).
Groundwater	No evidence for ubiquitous climate-related trend (this report, Chapter 1, Section 1.3.2).
Floods and droughts	
Floods	No evidence for climate-related trend (this report, Chapter 1, Section 1.3.2), but flood damages are increasing (this section).
Droughts	Intensified droughts in some drier regions since the 1970s (this report, Chapter 1, Section 1.3.2; WGI AR4, Chapter 3, Executive summary).
Water quality	No evidence for climate-related trend (this report, Chapter 1, Section 1.3.2).
Erosion and sediment transport	No evidence for climate-related trend (this section).
Irrigation water demand	No evidence for climate-related trend (this section).

account the uncertainties related to climate change (see Section 3.6). In the following paragraphs, the current sensitivities of components of the global freshwater system are discussed, and example regions, whose vulnerabilities are likely to be exacerbated by climate change, are highlighted (Figure 3.2).

Surface waters and runoff generation

Changes in river flows as well as lake and wetland levels due to climate change depend on changes in the volume, timing and intensity of precipitation (Chiew, 2007), snowmelt and whether precipitation falls as snow or rain. Changes in temperature, radiation, atmospheric humidity, and wind speed affect potential evapotranspiration, and this can offset small increases in precipitation and exaggerate further the effect of decreased precipitation on surface waters. In addition, increased atmospheric CO_2 concentration directly alters plant physiology, thus affecting evapotranspiration. Many experimental (e.g., Triggs et al., 2004) and global modelling studies (e.g., Leipprand and Gerten, 2006; Betts et al., 2007) show reduced evapotranspiration, with only part of this reduction being offset by increased plant growth due to increased CO_2 concentrations. Gedney et al. (2006) attributed an observed 3% rise in global river discharges over the 20th century to CO_2-induced reductions in plant evapotranspiration (by 5%) which were offset by climate change (which by itself would have decreased discharges by 2%). However, this attribution is highly uncertain, among other reasons due to the high uncertainty of observed precipitation time series.

Different catchments respond differently to the same change in climate drivers, depending largely on catchment physiogeographical and hydrogeological characteristics and the amount of lake or groundwater storage in the catchment.

A number of lakes worldwide have decreased in size during the last decades, mainly due to human water use. For some, declining precipitation was also a significant cause; e.g., in the case of Lake Chad, where both decreased precipitation and increased human water use account for the observed decrease in lake area since the 1960s (Coe and Foley, 2001). For the many lakes, rivers and wetlands that have shrunk mainly due to human water use and drainage, with negative impacts on ecosystems, climate change is likely to exacerbate the situation if it results in reduced net water availability (precipitation minus evapotranspiration).

Groundwater

Groundwater systems generally respond more slowly to climate change than surface water systems. Groundwater levels correlate more strongly with precipitation than with temperature, but temperature becomes more important for shallow aquifers and in warm periods.

Floods and droughts

Disaster losses, mostly weather- and water-related, have grown much more rapidly than population or economic growth, suggesting a negative impact of climate change (Mills, 2005). However, there is no clear evidence for a climate-related trend in floods during the last decades (Table 3.1; Kundzewicz et al., 2005; Schiermeier, 2006). However, the observed increase in precipitation intensity (Table 3.1) and other observed climate changes, e.g., an increase in westerly weather patterns during winter over Europe, leading to very rainy low-pressure systems that often trigger floods (Kron and Bertz, 2007), indicate that climate might already have had an impact on floods. Globally,

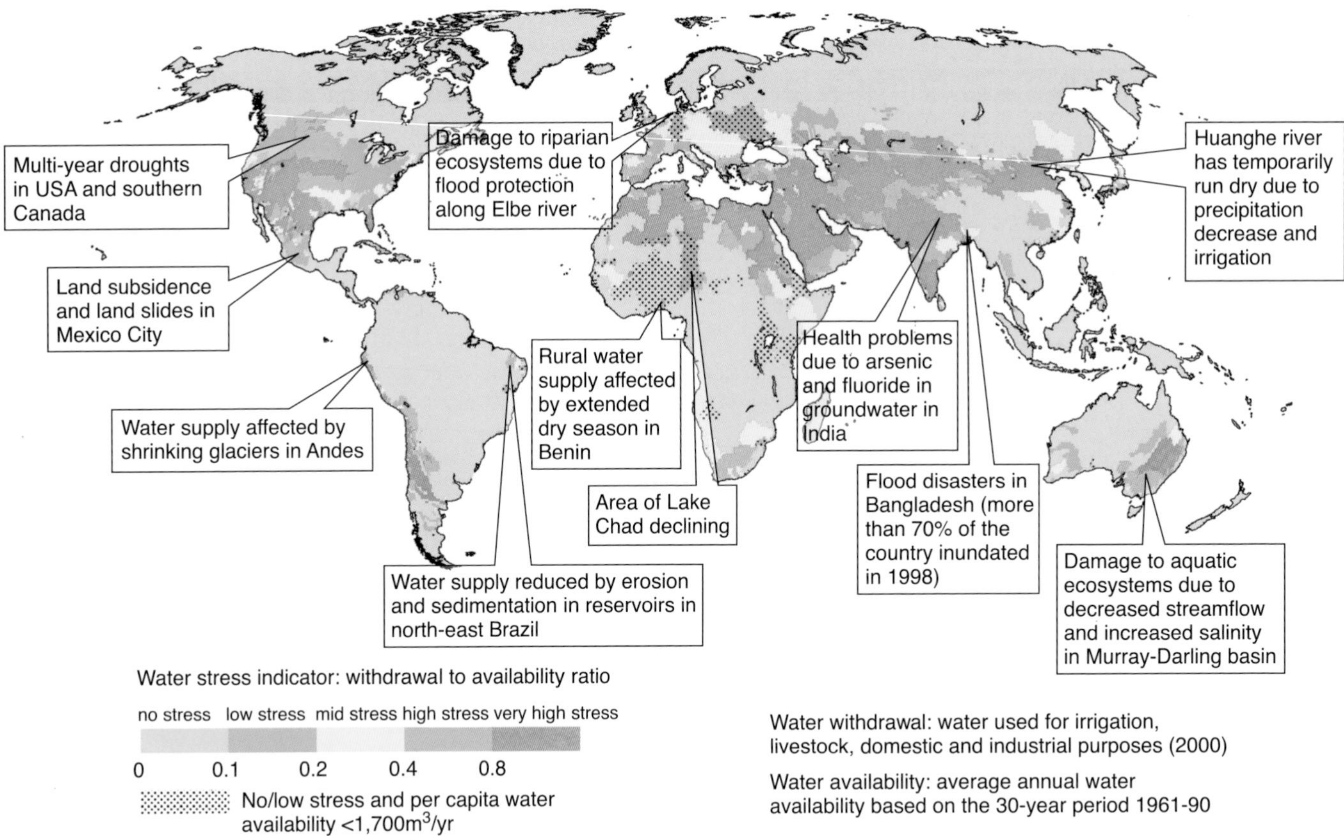

Figure 3.2. *Examples of current vulnerabilities of freshwater resources and their management; in the background, a water stress map based on Alcamo et al. (2003a). See text for relation to climate change.*

the number of great inland flood catastrophes during the last 10 years (between 1996 and 2005) is twice as large, per decade, as between 1950 and 1980, while economic losses have increased by a factor of five (Kron and Bertz, 2007). The dominant drivers of the upward trend in flood damage are socio-economic factors, such as increased population and wealth in vulnerable areas, and land-use change. Floods have been the most reported natural disaster events in Africa, Asia and Europe, and have affected more people across the globe (140 million/yr on average) than all other natural disasters (WDR, 2003, 2004). In Bangladesh, three extreme floods have occurred in the last two decades, and in 1998 about 70% of the country's area was inundated (Mirza, 2003; Clarke and King, 2004). In some river basins, e.g., the Elbe river basin in Germany, increasing flood risk drives the strengthening of flood protection systems by structural means, with detrimental effects on riparian and aquatic ecosystems (Wechsung et al., 2005).

Droughts affect rain-fed agricultural production as well as water supply for domestic, industrial, and agricultural purposes. Some semi-arid and sub-humid regions of the globe, e.g., Australia (see Chapter 11, Section 11.2.1), western USA and southern Canada (see Chapter 14, Section 14.2.1), and the Sahel (Nicholson, 2005), have suffered from more intense and multi-annual droughts, highlighting the vulnerability of these regions to the increased drought occurrence that is expected in the future due to climate change.

Water quality

In lakes and reservoirs, climate change effects are mainly due to water temperature variations, which result directly from climate change or indirectly through an increase in thermal pollution as a result of higher demands for cooling water in the energy sector. This affects oxygen regimes, redox potentials,[1] lake stratification, mixing rates, and biota development, as they all depend on temperature (see Chapter 4). Increasing water temperature affects the self-purification capacity of rivers by reducing the amount of oxygen that can be dissolved and used for biodegradation. A trend has been detected in water temperature in the Fraser River in British Columbia, Canada, for longer river sections reaching a temperature over 20°C, which is considered the threshold beyond which salmon habitats are degraded (Morrison et al., 2002). Furthermore, increases in intense rainfall result in more nutrients, pathogens, and toxins being washed into water bodies. Chang et al. (2001) reported increased nitrogen loads from rivers of up to 50% in the Chesapeake and Delaware Bay regions due to enhanced precipitation.

Numerous diseases linked to climate variations can be transmitted via water, either by drinking it or by consuming crops irrigated with polluted water (Chapter 8, Section 8.2.5). The presence of pathogens in water supplies has been related to extreme rainfall events (Yarze and Chase, 2000; Curriero et al., 2001; Fayer et al., 2002; Cox et al., 2003; Hunter, 2003). In aquifers, a possible relation between virus content and extreme

[1] A change in the redox potential of the environment will mean a change in the reactions taking place in it, moving, for example, from an oxidising (aerobic) to a reducing (anaerobic) system.

rainfall has been identified (Hunter, 2003). In the USA, 20 to 40% of water-borne disease outbreaks can be related to extreme precipitation (Rose et al., 2000). Effects of dry periods on water quality have not been adequately studied (Takahashi et al., 2001), although lower water availability clearly reduces dilution.

At the global scale, health problems due to arsenic and fluoride in groundwater are more important than those due to other chemicals (United Nations, 2006). Affected regions include India, Bangladesh, China, North Africa, Mexico, and Argentina, with more than 100 million people suffering from arsenic poisoning and fluorosis (a disease of the teeth or bones caused by excessive consumption of fluoride) (United Nations, 2003; Clarke and King, 2004; see also Chapter 13, Section 13.2.3).

One-quarter of the global population lives in coastal regions; these are water-scarce (less than 10% of the global renewable water supply) (Small and Nicholls, 2003; Millennium Ecosystem Assessment, 2005b) and are undergoing rapid population growth. Saline intrusion due to excessive water withdrawals from aquifers is expected to be exacerbated by the effect of sea-level rise, leading to even higher salinisation and reduction of freshwater availability (Klein and Nicholls, 1999; Sherif and Singh, 1999; Essink, 2001; Peirson et al., 2001; Beach, 2002; Beuhler, 2003). Salinisation affects estuaries and rivers (Knighton et al., 1992; Mulrennan and Woodroffe, 1998; Burkett et al., 2002; see also Chapter 13). Groundwater salinisation caused by a reduction in groundwater recharge is also observed in inland aquifers, e.g., in Manitoba, Canada (Chen et al., 2004).

Water quality problems and their effects are different in type and magnitude in developed and developing countries, particularly those stemming from microbial and pathogen content (Lipp et al., 2001; Jiménez, 2003). In developed countries, flood-related water-borne diseases are usually contained by well-maintained water and sanitation services (McMichael et al., 2003) but this does not apply in developing countries (Wisner and Adams, 2002). Regretfully, with the exception of cholera and salmonella, studies of the relationship between climate change and micro-organism content in water and wastewater do not focus on pathogens of interest in developing countries, such as specific protozoa or parasitic worms (Yarze and Chase, 2000; Rose et al., 2000; Fayer et al., 2002; Cox et al., 2003; Scott et al., 2004). One-third of urban water supplies in Africa, Latin America and the Caribbean, and more than half in Asia, are operating intermittently during periods of drought (WHO/UNICEF, 2000). This adversely affects water quality in the supply system.

Erosion and sediment transport

Rainfall amounts and intensities are the most important factors controlling climate change impacts on water erosion (Nearing et al., 2005), and they affect many geomorphologic processes, including slope stability, channel change, and sediment transport (Rumsby and Macklin, 1994; Rosso et al., 2006). There is no evidence for a climate-related trend in erosion and sediment transport in the past, as data are poor and climate is not the only driver of erosion and sediment transport. Examples of vulnerable areas can be found in north-eastern Brazil, where the sedimentation of reservoirs is significantly decreasing water storage and thus water supply (De Araujo et al., 2006); increased erosion due to increased precipitation intensities would exacerbate this problem. Human settlements on steep hill slopes, in particular informal settlements in metropolitan areas of developing countries (United Nations, 2006), are vulnerable to increased water erosion and landslides.

Water use, availability and stress

Human water use is dominated by irrigation, which accounts for almost 70% of global water withdrawals and for more than 90% of global consumptive water use, i.e., the water volume that is not available for reuse downstream (Shiklomanov and Rodda, 2003). In most countries of the world, except in a few industrialised nations, water use has increased over the last decades due to demographic and economic growth, changes in lifestyle, and expanded water supply systems. Water use, in particular irrigation water use, generally increases with temperature and decreases with precipitation. There is no evidence for a climate-related trend in water use in the past. This is due to the fact that water use is mainly driven by non-climatic factors and to the poor quality of water-use data in general and time series in particular.

Water availability from surface sources or shallow groundwater wells depends on the seasonality and interannual variability of streamflow, and safe water supply is determined by seasonal low flows. In snow-dominated basins, higher temperatures lead to reduced streamflow and thus decreased water supply in summer (Barnett et al., 2005), for example in South American river basins along the Andes, where glaciers are shrinking (Coudrain et al., 2005). In semi-arid areas, climate change may extend the dry season of no or very low flows, which particularly affects water users unable to rely on reservoirs or deep groundwater wells (Giertz et al., 2006)

Currently, human beings and natural ecosystems in many river basins suffer from a lack of water. In global-scale assessments, basins with water stress are defined either as having a per capita water availability below 1,000 m^3/yr (based on long-term average runoff) or as having a ratio of withdrawals to long-term average annual runoff above 0.4. These basins are located in Africa, the Mediterranean region, the Near East, South Asia, Northern China, Australia, the USA, Mexico, north-eastern Brazil, and the western coast of South America (Figure 3.2). Estimates of the population living in such severely stressed basins range from 1.4 billion to 2.1 billion (Vörösmarty et al., 2000; Alcamo et al., 2003a, b; Oki et al., 2003a; Arnell, 2004b). In water-scarce areas, people and ecosystems are particularly vulnerable to decreasing and more variable precipitation due to climate change. For example, in the Huanghe River basin in China (Yang et al., 2004), the combination of increasing irrigation water consumption facilitated by reservoirs, and decreasing precipitation associated with global El Niño-Southern Oscillation (ENSO) events over the past half century, has resulted in water scarcity (Wang et al., 2006). The irrigation-dominated Murray-Darling Basin in Australia suffers from decreased water inflows to wetlands and high salinity due to irrigation water use, which affects aquatic ecosystems (Goss, 2003; see also Chapter 11, Section 11.7).

Current adaptation

At the Fourth World Water Forum held in Mexico City in 2006, many of the involved groups requested the inclusion of climate change in Integrated Water Resources Management (World Water Council, 2006). In some countries (e.g., Caribbean, Canada, Australia, Netherlands, UK, USA and Germany), adaptation procedures and risk management practices for the water sector have already been developed that take into account climate change impacts on freshwater systems (compare with Section 3.6).

3.3 Assumptions about future trends

In Chapter 2, scenarios of the main drivers of climate change and their impacts are presented. This section describes how the driving forces of freshwater systems are assumed to develop in the future, with a focus on the dominant drivers during the 21st century. Climate-related and non-climatic drivers are distinguished. Assumptions about future trends in non-climatic drivers are necessary in order to assess the vulnerability of freshwater systems to climate change, and to compare the relative importance of climate change impacts and impacts due to changes in non-climatic drivers.

3.3.1 Climatic drivers

Projections for the future

The most dominant climatic drivers for water availability are precipitation, temperature, and evaporative demand (determined by net radiation at ground level, atmospheric humidity, wind speed, and temperature). Temperature is particularly important in snow-dominated basins and in coastal areas (due to the impact of temperature on sea level).

The following summary of future climate change is taken from the Working Group I Fourth Assessment Report (WGI AR4), Chapter 10 (Meehl et al., 2007). The most likely global average surface temperature increase by the 2020s is around 1°C relative to the pre-industrial period, based on all the IPCC Special Report on Emissions Scenarios (SRES; Nakićenović and Swart, 2000) scenarios. By the end of the 21st century, the most likely increases are 3 to 4°C for the A2 emissions scenario and around 2°C for B1 (Figure 10.8). Geographical patterns of projected warming show the greatest temperature increases at high northern latitudes and over land (roughly twice the global average temperature increase) (Chapter 10, Executive summary, see also Figure 10.9). Temperature increases are projected to be stronger in summer than in winter except for Arctic latitudes (Figure 10.9). Evaporative demand is likely to increase almost everywhere (Figures 10.9 and 10.12). Global mean sea-level rise is expected to reach between 14 and 44 cm within this century (Chapter 10, Executive summary). Globally, mean precipitation will increase due to climate change. Current climate models tend to project increasing precipitation at high latitudes and in the tropics (e.g., the south-east monsoon region and over the tropical Pacific) and decreasing precipitation in the sub-tropics (e.g., over much of North Africa and the northern Sahara) (Figure 10.9).

While temperatures are expected to increase during all seasons of the year, although with different increments, precipitation may increase in one season and decrease in another. A robust finding is that precipitation variability will increase in the future (Trenberth et al., 2003). Recent studies of changes in precipitation extremes in Europe (Giorgi et al., 2004; Räisänen et al., 2004) agree that the intensity of daily precipitation events will predominantly increase, also over many areas where means are likely to decrease (Christensen and Christensen, 2003, Kundzewicz et al., 2006). The number of wet days in Europe is projected to decrease (Giorgi et al., 2004), which leads to longer dry periods except in the winters of western and central Europe. An increase in the number of days with intense precipitation has been projected across most of Europe, except for the south (Kundzewicz et al., 2006). Multi-model simulations with nine global climate models for the SRES A1B, A2, and B1 scenarios show precipitation intensity (defined as annual precipitation divided by number of wet days) increasing strongly for A1B and A2, and slightly less strongly for B1, while the annual maximum number of consecutive dry days is expected to increase for A1B and A2 only (WGI AR4, Figure 10.18).

Uncertainties

Uncertainties in climate change projections increase with the length of the time horizon. In the near term (e.g., the 2020s), climate model uncertainties play the most important role; while over longer time horizons, uncertainties due to the selection of emissions scenario become increasingly significant (Jenkins and Lowe, 2003).

General Circulation Models (GCMs) are powerful tools accounting for the complex set of processes which will produce future climate change (Karl and Trenberth, 2003). However, GCM projections are currently subject to significant uncertainties in the modelling process (Mearns et al., 2001; Allen and Ingram, 2002; Forest et al., 2002; Stott and Kettleborough, 2002), so that climate projections are not easy to incorporate into hydrological impact studies (Allen and Ingram, 2002). The Coupled Model Intercomparison Project analysed outputs of eighteen GCMs (Covey et al., 2003). Whereas most GCMs had difficulty producing precipitation simulations consistent with observations, the temperature simulations generally agreed well. Such uncertainties produce biases in the simulation of river flows when using direct GCM outputs representative of the current time horizon (Prudhomme, 2006).

For the same emissions scenario, different GCMs produce different geographical patterns of change, particularly with respect to precipitation, which is the most important driver for freshwater resources. As shown by Meehl et al. (2007), the agreement with respect to projected changes of temperature is much higher than with respect to changes in precipitation (WGI AR4, Chapter 10, Figure 10.9). For precipitation changes by the end of the 21st century, the multi-model ensemble mean exceeds the inter-model standard deviation only at high latitudes. Over several regions, models disagree in the sign of the precipitation change (Murphy et al., 2004). To reduce uncertainties, the use of numerous runs from different GCMs with varying model parameters i.e., multi-ensemble runs (see Murphy et al., 2004), or thousands of runs from a single GCM (as from the climateprediction.net experiment; see Stainforth et al., 2005), is often recommended. This allows the construction of conditional probability scenarios of future changes (e.g., Palmer and

Räisänen, 2002; Murphy et al., 2004). However, such large ensembles are difficult to use in practice when undertaking an impact study on freshwater resources. Thus, ensemble means are often used instead, despite the failure of such scenarios to accurately reproduce the range of simulated regional changes, particularly for sea-level pressure and precipitation (Murphy et al., 2004). An alternative is to consider a few outputs from several GCMs (e.g. Arnell (2004b) at the global scale, and Jasper et al. (2004) at the river basin scale).

Uncertainties in climate change impacts on water resources are mainly due to the uncertainty in precipitation inputs and less due to the uncertainties in greenhouse gas emissions (Döll et al., 2003; Arnell, 2004b), in climate sensitivities (Prudhomme et al., 2003), or in hydrological models themselves (Kaspar, 2003). The comparison of different sources of uncertainty in flood statistics in two UK catchments (Kay et al., 2006a) led to the conclusion that GCM structure is the largest source of uncertainty, next are the emissions scenarios, and finally hydrological modelling. Similar conclusions were drawn by Prudhomme and Davies (2007) regarding mean monthly flows and low flow statistics in Britain.

Incorporation of changing climatic drivers in freshwater impact studies

Most climate change impact studies for freshwater consider only changes in precipitation and temperature, based on changes in the averages of long-term monthly values, e.g., as available from the IPCC Data Distribution Centre (www.ipcc-data.org). In many impact studies, time series of observed climate values are adjusted with the computed change in climate variables to obtain scenarios that are consistent with present-day conditions. These adjustments aim to minimise the error in GCMs under the assumption that the biases in climate modelling are of similar magnitude for current and future time horizons. This is particularly important for precipitation projections, where differences between the observed values and those computed by climate models for the present day are substantial. Model outputs can be biased, and changes in runoff can be underestimated (e.g., Arnell et al. (2003) in Africa and Prudhomme (2006) in Britain). Changes in interannual or daily variability of climate variables are often not taken into account in hydrological impact studies. This leads to an underestimation of future floods, droughts, and irrigation water requirements.

Another problem in the use of GCM outputs is the mismatch of spatial grid scales between GCMs (typically a few hundred kilometres) and hydrological processes. Moreover, the resolution of global models precludes their simulation of realistic circulation patterns that lead to extreme events (Christensen and Christensen, 2003; Jones et al., 2004). To overcome these problems, techniques that downscale GCM outputs to a finer spatial (and temporal) resolution have been developed (Giorgi et al., 2001). These are: dynamical downscaling techniques, based on physical/dynamical links between the climate at large and at smaller scales (e.g., high resolution Regional Climate Models; RCMs) and statistical downscaling methods using empirical relationships between large-scale atmospheric variables and observed daily local weather variables. The main assumption in statistical downscaling is that the statistical relationships identified for the current climate will remain valid under changes in future conditions. Downscaling techniques may allow modellers to incorporate future changes in daily variability (e.g., Diaz-Nieto and Wilby, 2005) and to apply a probabilistic framework to produce information on future river flows for water resource planning (Wilby and Harris, 2006). These approaches help to quantify the relative significance of different sources of uncertainty affecting water resource projections.

3.3.2 Non-climatic drivers

Many non-climatic drivers affect freshwater resources at the global scale (United Nations, 2003). Water resources, both in quantity and quality, are influenced by land-use change, the construction and management of reservoirs, pollutant emissions, and water and wastewater treatment. Water use is driven by changes in population, food consumption, economic policy (including water pricing), technology, lifestyle, and society's views of the value of freshwater ecosystems. Vulnerability of freshwater systems to climate change also depends on water management. It can be expected that the paradigm of Integrated Water Resources Management will be increasingly followed around the world (United Nations, 2002; World Bank, 2003; World Water Council, 2006), which will move water, as a resource and a habitat, into the centre of policy making. This is likely to decrease the vulnerability of freshwater systems to climate change.

Chapter 2 (this volume) provides an overview of the future development of non-climatic drivers, including: population, economic activity, land cover, land use, and sea level, and focuses on the SRES scenarios. In this section, assumptions about key freshwater-specific drivers for the 21st century are discussed: reservoir construction and decommissioning, wastewater reuse, desalination, pollutant emissions, wastewater treatment, irrigation, and other water-use drivers.

In developing countries, new reservoirs will be built in the future, even though their number is likely to be small compared with the existing 45,000 large dams (World Commission on Dams, 2000; Scudder, 2005). In developed countries, the number of dams is very likely to remain stable. Furthermore, the issue of dam decommissioning is being discussed in a few developed countries, and some dams have already been removed in France and the USA (Gleick, 2000; Howard, 2000). Consideration of environmental flow requirements may lead to modified reservoir operations so that the human use of the water resources might be restricted.

Increased future wastewater use and desalination are likely mechanisms for increasing water supply in semi-arid and arid regions (Ragab and Prudhomme, 2002; Abufayed et al., 2003). The cost of desalination has been declining, and desalination has been considered as a water supply option for inland towns (Zhou and Tol, 2005). However, there are unresolved concerns about the environmental impacts of impingement and entrainment of marine organisms, the safe disposal of highly concentrated brines that can also contain other chemicals used in the desalination process, and high energy consumption. These have negative impacts on costs and the carbon footprint, and may hamper the expansion of desalination (Cooley et al., 2006).

Wastewater treatment is an important driver of water quality, and an increase in wastewater treatment in both developed and developing countries could improve water quality in the future. In the EU, for example, more efficient wastewater treatment, as required by the Urban Wastewater Directive and the European Water Framework Directive, should lead to a reduction in point-source nutrient inputs to rivers. However, organic micro-pollutants (e.g., endocrine substances) are expected to occur in increasing concentrations in surface waters and groundwater. This is because the production and consumption of chemicals are likely to increase in the future in both developed and developing countries (Daughton, 2004), and several of these pollutants are not removed by current wastewater treatment technology. In developing countries, increases in point emissions of nutrients, heavy metals, and organic micro-pollutants are expected. With heavier rainfall, non-point pollution could increase in all countries.

Global-scale quantitative scenarios of pollutant emissions tend to focus on nitrogen, and the range of plausible futures is large. The scenarios of the Millennium Ecosystem Assessment expect global nitrogen fertiliser use to reach 110 to 140 Mt by 2050 as compared to 90 Mt in 2000 (Millennium Ecosystem Assessment, 2005a). In three of the four scenarios, total nitrogen load increases at the global scale, while in the fourth, TechnoGarden, scenario (similar to the SRES B1 scenario), there is a reduction of atmospheric nitrogen deposition as compared to today, so that the total nitrogen load to the freshwater system would decrease. Diffuse emissions of nutrients and pesticides from agriculture are likely to continue to be an important water quality issue in developed countries, and are very likely to increase in developing countries, thus critically affecting water quality.

The most important drivers of water use are population and economic development, and also changing societal views on the value of water. The latter refers to such issues as the prioritisation of domestic and industrial water supply over irrigation water supply, and the extent to which water-saving technologies and water pricing are adopted. In all four Millennium Ecosystems Assessment scenarios, per capita domestic water use in 2050 is rather similar in all world regions, around 100 m^3/yr, i.e., the European average in 2000 (Millennium Ecosystem Assessment, 2005b). This assumes a very strong increase in usage in Sub-Saharan Africa (by a factor of five) and smaller increases elsewhere, except for developed countries (OECD), where per capita domestic water use is expected to decline further (Gleick, 2003). In addition to these scenarios, many other plausible scenarios of future domestic and industrial water use exist which can differ strongly (Seckler et al., 1998; Alcamo et al., 2000, 2003b; Vörösmarty et al., 2000).

The future extent of irrigated areas is the dominant driver of future irrigation water use, together with cropping intensity and irrigation water-use efficiency. According to the Food and Agriculture Organization (FAO) agriculture projections, developing countries (with 75% of the global irrigated area) are likely to expand their irrigated area until 2030 by 0.6%/yr, while the cropping intensity of irrigated land will increase from 1.27 to 1.41 crops/yr, and irrigation water-use efficiency will increase slightly (Bruinsma, 2003). These estimates do not take into account climate change. Most of this expansion is projected to occur in already water-stressed areas, such as southern Asia, northern China, the Near East, and North Africa. A much smaller expansion of irrigated areas, however, is assumed in all four scenarios of the Millennium Ecosystem Assessment, with global growth rates of only 0 to 0.18%/yr until 2050. After 2050, the irrigated area is assumed to stabilise or to slightly decline in all scenarios except Global Orchestration (similar to the SRES A1 scenario) (Millennium Ecosystem Assessment, 2005a).

3.4 Key future impacts and vulnerabilities

3.4.1 Surface waters

Since the TAR, over 100 studies of climate change effects on river flows have been published in scientific journals, and many more have been reported in internal reports. However, studies still tend to be heavily focused on Europe, North America, and Australasia. Virtually all studies use a hydrological model driven by scenarios based on climate model simulations, with a number of them using SRES-based scenarios (e.g., Hayhoe et al., 2004; Zierl and Bugmann, 2005; Kay et al., 2006a). A number of global-scale assessments (e.g., Manabe et al., 2004a, b; Milly et al., 2005, Nohara et al., 2006) directly use climate model simulations of river runoff, but the reliability of estimated changes is dependent on the rather poor ability of the climate model to simulate 20th century runoff reliably.

Methodological advances since the TAR have focused on exploring the effects of different ways of downscaling from the climate model scale to the catchment scale (e.g., Wood et al., 2004), the use of regional climate models to create scenarios or drive hydrological models (e.g., Arnell et al., 2003; Shabalova et al., 2003; Andreasson et al., 2004; Meleshko et al., 2004; Payne et al., 2004; Kay et al., 2006b; Fowler et al., 2007; Graham et al., 2007a, b; Prudhomme and Davies, 2007), ways of applying scenarios to observed climate data (Drogue et al., 2004), and the effect of hydrological model uncertainty on estimated impacts of climate change (Arnell, 2005). In general, these studies have shown that different ways of creating scenarios from the same source (a global-scale climate model) can lead to substantial differences in the estimated effect of climate change, but that hydrological model uncertainty may be smaller than errors in the modelling procedure or differences in climate scenarios (Jha et al., 2004; Arnell, 2005; Wilby, 2005; Kay et al., 2006a, b). However, the largest contribution to uncertainty in future river flows comes from the variations between the GCMs used to derive the scenarios.

Figure 3.3 provides an indication of the effects of future climate change on long-term average annual river runoff by the 2050s, across the world, under the A2 emissions scenario and different climate models used in the TAR (Arnell, 2003a). Obviously, even for large river basins, climate change scenarios from different climate models may result in very different projections of future runoff change (e.g., in Australia, South America, and Southern Africa).

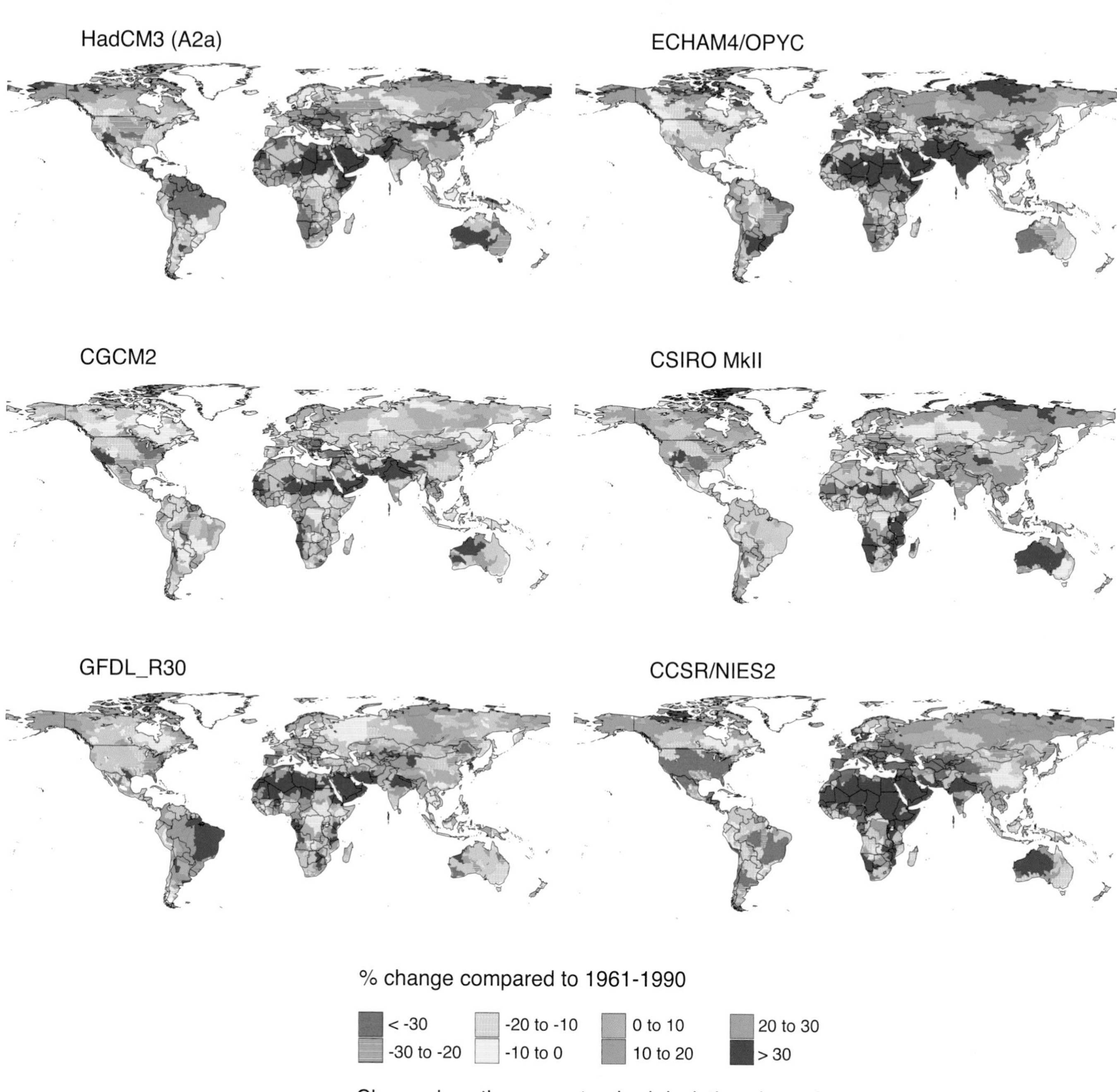

Figure 3.3. *Change in average annual runoff by the 2050s under the SRES A2 emissions scenario and different climate models (Arnell, 2003a).*

Figure 3.4 shows the mean runoff change until 2050 for the SRES A1B scenario from an ensemble of twenty-four climate model runs (from twelve different GCMs) (Milly et al., 2005). Almost all model runs agree at least with respect to the direction of runoff change in the high latitudes of North America and Eurasia, with increases of 10 to 40%. This is in agreement with results from a similar study of Nohara et al. (2006), which showed that the ensemble mean runoff change until the end of the 21st century (from nineteen GCMs) is smaller than the standard deviation everywhere except at northern high latitudes. With higher uncertainty, runoff can be expected to increase in the wet tropics. Prominent regions, with a rather strong agreement between models, of decreasing runoff (by 10 to 30%) include the Mediterranean, southern Africa, and western USA/northern Mexico. In general, between the late 20th century and 2050, the areas of decreased runoff expand (Milly et al., 2005).

A very robust finding of hydrological impact studies is that warming leads to changes in the seasonality of river flows where much winter precipitation currently falls as snow (Barnett et al., 2005). This has been found in projections for the European Alps (Eckhardt and Ulbrich, 2003; Jasper et al., 2004; Zierl and Bugmann, 2005), the Himalayas (Singh, 2003; Singh and Bengtsson, 2004), western North America (Loukas et al.,

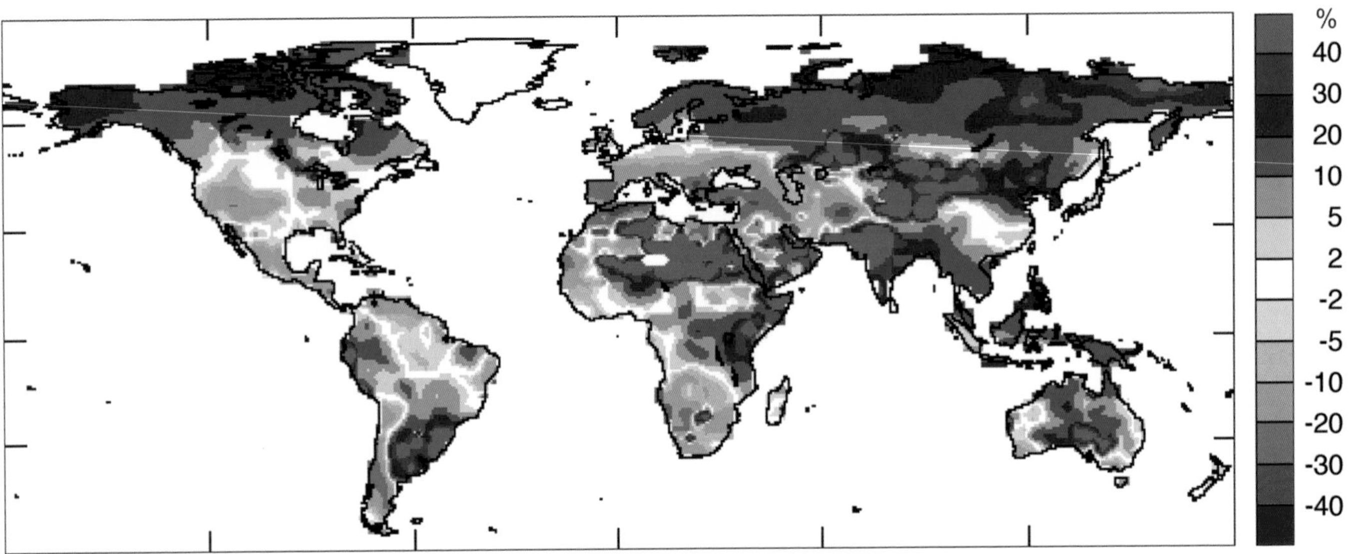

Figure 3.4. *Change in annual runoff by 2041-60 relative to 1900-70, in percent, under the SRES A1B emissions scenario and based on an ensemble of 12 climate models. Reprinted by permission from Macmillan Publishers Ltd. [Nature] (Milly et al., 2005), copyright 2005.*

2002a, b; Christensen et al., 2004; Dettinger et al., 2004; Hayhoe et al., 2004; Knowles and Cayan, 2004; Leung et al., 2004; Payne et al., 2004; Stewart et al., 2004; VanRheenen et al., 2004; Kim, 2005; Maurer and Duffy, 2005), central North America (Stone et al., 2001; Jha et al., 2004), eastern North America (Frei et al., 2002; Chang, 2003; Dibike and Coulibaly, 2005), the entire Russian territory (Shiklomanov and Georgievsky, 2002; Bedritsky et al., 2007), and Scandinavia and Baltic regions (Bergström et al., 2001; Andreasson et al., 2004; Graham, 2004). The effect is greatest at lower elevations (where snowfall is more marginal) (Jasper et al., 2004; Knowles and Cayan, 2004), and in many cases peak flow would occur at least a month earlier. Winter flows increase and summer flows decrease.

Many rivers draining glaciated regions, particularly in the Hindu Kush-Himalaya and the South-American Andes, are sustained by glacier melt during the summer season (Singh and Kumar, 1997; Mark and Seltzer, 2003; Singh, 2003; Barnett et al., 2005). Higher temperatures generate increased glacier melt. Schneeberger et al. (2003) simulated reductions in the mass of a sample of Northern Hemisphere glaciers of up to 60% by 2050. As these glaciers retreat due to global warming (see Chapter 1), river flows are increased in the short term, but the contribution of glacier melt will gradually decrease over the next few decades.

In regions with little or no snowfall, changes in runoff are dependent much more on changes in rainfall than on changes in temperature. A general conclusion from studies in many rain-dominated catchments (Burlando and Rosso, 2002; Evans and Schreider, 2002; Menzel and Burger, 2002; Arnell, 2003b, 2004a; Boorman, 2003a; Booij, 2005) is that flow seasonality increases, with higher flows in the peak flow season and either lower flows during the low flow season or extended dry periods. In most case-studies there is little change in the timing of peak or low flows, although an earlier onset in the East Asian monsoon would bring forward the season of peak flows in China (Bueh et al., 2003).

Changes in lake levels are determined primarily by changes in river inflows and precipitation onto and evaporation from the lake. Impact assessments of the Great Lakes of North America show changes in water levels of between −1.38 m and +0.35 m by the end of the 21st century (Lofgren et al., 2002; Schwartz et al., 2004). Shiklomanov and Vasiliev (2004) suggest that the level of the Caspian Sea will change in the range of 0.5 to 1.0 m. In another study by Elguindi and Giorgi (2006), the levels in the Caspian Sea are estimated to drop by around 9 m by the end of the 21st century, due largely to increases in evaporation. Levels in some lakes represent a changing balance between inputs and outputs and, under one transient scenario, levels in Lake Victoria would initially fall as increases in evaporation offset changes in precipitation, but subsequently rise as the effects of increased precipitation overtake the effects of higher evaporation (Tate et al., 2004).

Increasing winter temperature considerably changes the ice regime of water bodies in northern regions. Studies made at the State Hydrological Institute, Russia, comparing the horizon of 2010 to 2015 with the control period 1950 to 1979, show that ice cover duration on the rivers in Siberia would be shorter by 15 to 27 days and maximum ice cover would be thinner by 20 to 40% (Vuglinsky and Gronskaya, 2005).

Model studies show that land-use changes have a small effect on annual runoff as compared to climate change in the Rhine basin (Pfister et al., 2004), south-east Michigan (Barlage et al., 2002), Pennsylvania (Chang, 2003), and central Ethiopia (Legesse et al., 2003). In other areas, however, such as south-east Australia (Herron et al., 2002) and southern India (Wilk and Hughes, 2002), land-use and climate-change effects may be more similar. In the Australian example, climate change has the potential to exacerbate considerably the reductions in runoff caused by afforestation.

Carbon dioxide enrichment of the atmosphere has two potential competing implications for evapotranspiration, and hence water balance and runoff. First, higher CO_2 concentrations can lead to reduced evaporation, as the stomata, through which evaporation from plants takes place, conduct less water. Second, higher CO_2 concentrations can lead to increased plant growth and thus leaf area, and hence a greater total evapotranspiration from the area. The relative magnitudes of these two effects, however, vary between plant types and also depend on other influences such as the availability of nutrients and the effects of changes in temperature and water availability. Accounting for the effects of CO_2 enrichment on runoff requires the incorporation of a dynamic vegetation model into a hydrological model. A small number of models now do this (Rosenberg et al., 2003; Gerten et al., 2004; Gordon and Famiglietti, 2004; Betts et al., 2007), but are usually at the GCM (and not catchment) scale. Although studies with equilibrium vegetation models suggest that increased leaf area may offset stomatal closure (Betts et al., 1997; Kergoat et al., 2002), studies with dynamic global vegetation models indicate that stomatal responses dominate the effects of leaf area increase. Taking into account CO_2-induced changes in vegetation, global mean runoff under a $2\times CO_2$ climate has been simulated to increase by approximately 5% as a result of reduced evapotranspiration due to CO_2enrichment alone ('physiological forcing') (Betts et al., 2007; Leipprand and Gerten, 2006). This may be compared to (often much larger) changes at the river basin scale (Figures 3.3, 3.4, and 3.7), and global values of runoff change. For example, global mean runoff has been simulated to increase by 5%-17% due to climate change alone in an ensemble of 143 $2\times CO_2$ GCM simulations (Betts et al., 2006).

3.4.2 Groundwater

The demand for groundwater is likely to increase in the future, the main reason being increased water use globally. Another reason may be the need to offset declining surface water availability due to increasing precipitation variability in general and reduced summer low flows in snow-dominated basins (see Section 3.4.3).

Climate change will affect groundwater recharge rates, i.e., the renewable groundwater resource, and groundwater levels. However, even knowledge of current recharge and levels in both developed and developing countries is poor. There has been very little research on the impact of climate change on groundwater, including the question of how climate change will affect the relationship between surface waters and aquifers that are hydraulically connected (Alley, 2001). Under certain circumstances (good hydraulic connection of river and aquifer, low groundwater recharge rates), changes in river level influence groundwater levels much more than changes in groundwater recharge (Allen et al., 2003). As a result of climate change, in many aquifers of the world the spring recharge shifts towards winter, and summer recharge declines. In high latitudes, thawing of permafrost will cause changes in groundwater level and quality. Climate change may lead to vegetation changes which also affect groundwater recharge. Also, with increased frequency and magnitude of floods, groundwater recharge may increase, in particular in semi-arid and arid areas where heavy rainfalls and floods are the major sources of groundwater recharge. Bedrock aquifers in semi-arid regions are replenished by direct infiltration of precipitation into fractures and dissolution channels, and alluvial aquifers are mainly recharged by floods (Al-Sefry et al., 2004). Accordingly, an assessment of climate change impact on groundwater recharge should include the effects of changed precipitation variability and inundation areas (Khiyami et al., 2005).

According to the results of a global hydrological model, groundwater recharge (when averaged globally) increases less than total runoff (Döll and Flörke, 2005). While total runoff (groundwater recharge plus fast surface and sub-surface runoff) was computed to increase by 9% between the reference climate normal 1961 to 1990 and the 2050s (for the ECHAM4 interpretation of the SRES A2 scenario), groundwater recharge increases by only 2%. For the four climate scenarios investigated, computed groundwater recharge decreases dramatically by more than 70% in north-eastern Brazil, south-west Africa and along the southern rim of the Mediterranean Sea (Figure 3.5). In these areas of decreasing total runoff, the percentage decrease of groundwater recharge is higher than that of total runoff, which is due to the model assumption that in semi-arid areas groundwater recharge only occurs if daily precipitation exceeds a certain threshold. However, increased variability of daily precipitation was not taken into account in this study. Regions with groundwater recharge increases of more than 30% by the 2050s include the Sahel, the Near East, northern China, Siberia, and the western USA. Although rising watertables in dry areas are usually beneficial, they might cause problems, e.g., in towns or agricultural areas (soil salinisation, wet soils). A comparison of the four scenarios in Figure 3.5 shows that lower emissions do not lead to significant changes in groundwater recharge, and that in some regions, e.g., Spain and Australia, the differences due to the two climate models are larger than the differences due to the two emissions scenarios.

The few studies of climate impacts on groundwater for various aquifers show very site-specific results. Future decreases of groundwater recharge and groundwater levels were projected for various climate scenarios which predict less summer and more winter precipitation, using a coupled groundwater and soil model for a groundwater basin in Belgium (Brouyere et al., 2004). The impacts of climate change on a chalk aquifer in eastern England appear to be similar. In summer, groundwater recharge and streamflow are projected to decrease by as much as 50%, potentially leading to water quality problems and groundwater withdrawal restrictions (Eckhardt and Ulbrich, 2003). Based on a historical analysis of precipitation, temperature and groundwater levels in a confined chalk aquifer in southern Canada, the correlation of groundwater levels with precipitation was found to be stronger than the correlation with temperature. However, with increasing temperature, the sensitivity of groundwater levels to temperature increases (Chen et al., 2004), particularly where the confining layer is thin. In higher latitudes, the sensitivity of groundwater and runoff to increasing temperature is greater because of increasing biomass and leaf area index (improved growth conditions and increased evapotranspiration). For an unconfined aquifer located in humid north-eastern USA,

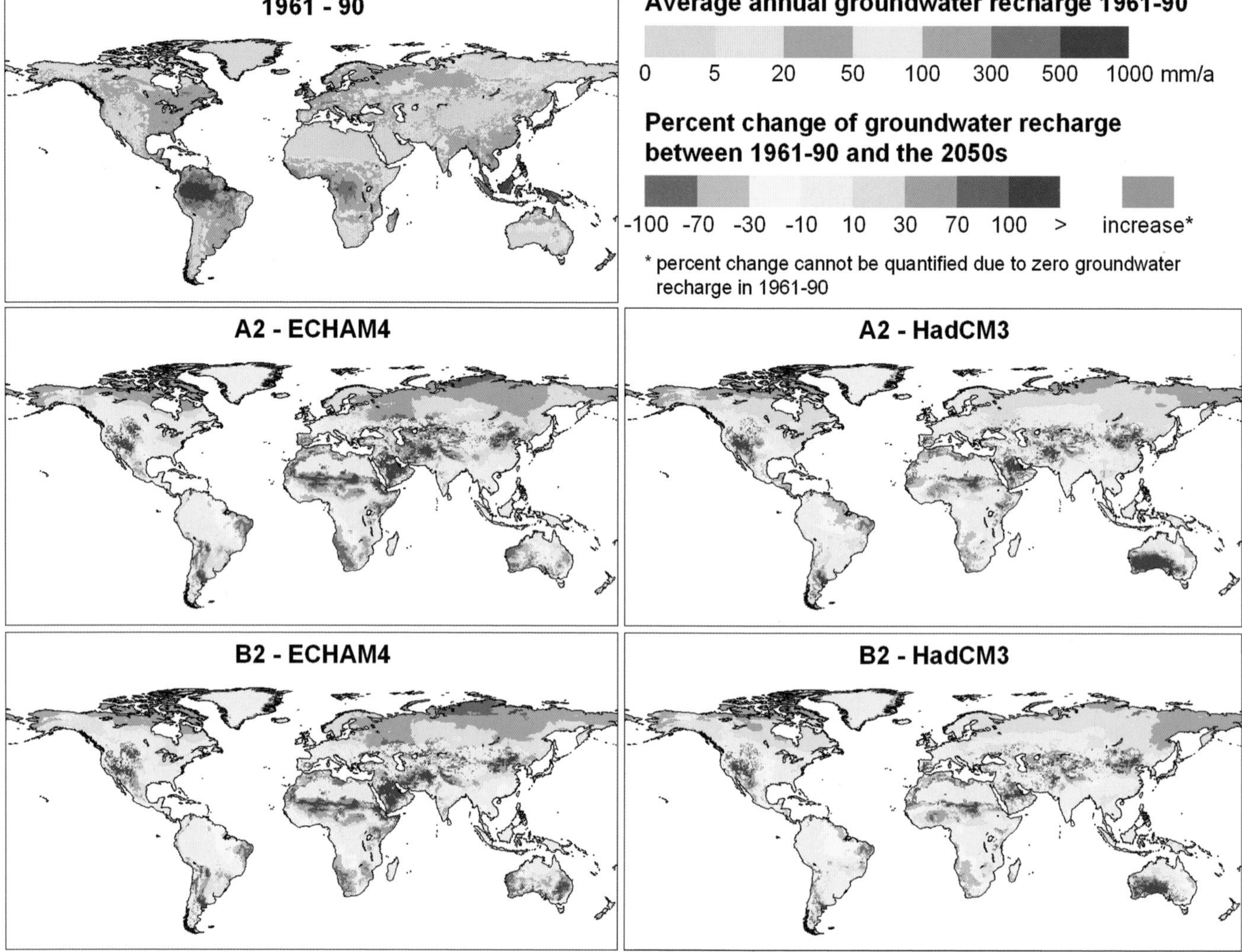

Figure 3.5. *Simulated impact of climate change on long-term average annual diffuse groundwater recharge. Percentage changes of 30 year averages groundwater recharge between present-day (1961 to 1990) and the 2050s (2041 to 2070), as computed by the global hydrological model WGHM, applying four different climate change scenarios (climate scenarios computed by the climate models ECHAM4 and HadCM3), each interpreting the two IPCC greenhouse gas emissions scenarios A2 and B2 (Döll and Flörke, 2005).*

climate change was computed to lead by 2030 and 2100 to a variety of impacts on groundwater recharge and levels, wetlands, water supply potential, and low flows, the sign and magnitude of which strongly depend on the climate model used to compute the groundwater model input (Kirshen, 2002).

Climate change is likely to have a strong impact on saltwater intrusion into aquifers as well as on the salinisation of groundwater due to increased evapotranspiration. Sea level rise leads to intrusion of saline water into the fresh groundwater in coastal aquifers and thus adversely affects groundwater resources. For two small, flat coral islands off the coast of India, the thickness of the freshwater lens was computed to decrease from 25 m to 10 m and from 36 m to 28 m for a sea-level rise of only 0.1 m (Bobba et al., 2000). Any decrease in groundwater recharge will exacerbate the effect of sea-level rise. In inland aquifers, a decrease in groundwater recharge can lead to saltwater intrusion of neighbouring saline aquifers (Chen et al., 2004), and increased evapotranspiration in semi-arid and arid regions may lead to the salinisation of shallow aquifers.

3.4.3 Floods and droughts

A warmer climate, with its increased climate variability, will increase the risk of both floods and droughts (Wetherald and Manabe, 2002; Table SPM2 in IPCC, 2007). As there are a number of climatic and non-climatic drivers influencing flood and drought impacts, the realisation of risks depends on several factors. Floods include river floods, flash floods, urban floods and sewer floods, and can be caused by intense and/or long-lasting precipitation, snowmelt, dam break, or reduced conveyance due to ice jams or landslides. Floods depend on precipitation intensity, volume, timing, antecedent conditions of rivers and their drainage basins (e.g., presence of snow and ice, soil character, wetness, urbanisation, and existence of dikes, dams, or reservoirs). Human encroachment into flood plains and lack of flood response plans increase the damage potential.

The term drought may refer to meteorological drought (precipitation well below average), hydrological drought (low river flows and water levels in rivers, lakes and groundwater), agricultural drought (low soil moisture), and environmental drought (a combination of the above). The socio-economic impacts of droughts may arise from the interaction between natural conditions and human factors, such as changes in land use and land cover, water demand and use. Excessive water withdrawals can exacerbate the impact of drought.

A robust result, consistent across climate model projections, is that higher precipitation extremes in warmer climates are very likely to occur (see Section 3.3.1). Precipitation intensity increases almost everywhere, but particularly at mid- and high latitudes where mean precipitation also increases (Meehl et al., 2005, WGI AR4, Chapter 10, Section 10.3.6.1). This directly affects the risk of flash flooding and urban flooding. Storm drainage systems have to be adapted to accommodate increasing rainfall intensity resulting from climate change (Waters et al., 2003). An increase of droughts over low latitudes and mid-latitude continental interiors in summer is likely (WGI AR4, Summary for Policymakers, Table SPM.2), but sensitive to model land-surface formulation. Projections for the 2090s made by Burke et al. (2006), using the HadCM3 GCM and the SRES A2 scenario, show regions of strong wetting and drying with a net overall global drying trend. For example, the proportion of the land surface in extreme drought, globally, is predicted to increase by the a factor of 10 to 30; from 1-3 % for the present day to 30% by the 2090s. The number of extreme drought events per 100 years and mean drought duration are likely to increase by factors of two and six, respectively, by the 2090s (Burke et al., 2006). A decrease in summer precipitation in southern Europe, accompanied by rising temperatures, which enhance evaporative demand, would inevitably lead to reduced summer soil moisture (Douville et al., 2002) and more frequent and more intense droughts.

As temperatures rise, the likelihood of precipitation falling as rain rather than snow increases, especially in areas with temperatures near to 0°C in autumn and spring (WGI AR4, Summary for Policymakers). Snowmelt is projected to be earlier and less abundant in the melt period, and this may lead to an increased risk of droughts in snowmelt-fed basins in summer and autumn, when demand is highest (Barnett et al., 2005).

With more than one-sixth of the Earth's population relying on melt water from glaciers and seasonal snow packs for their water supply, the consequences of projected changes for future water availability, predicted with high confidence and already diagnosed in some regions, will be adverse and severe. Drought problems are projected for regions which depend heavily on glacial melt water for their main dry-season water supply (Barnett et al., 2005). In the Andes, glacial melt water supports river flow and water supply for tens of millions of people during the long dry season. Many small glaciers, e.g., in Bolivia, Ecuador, and Peru (Coudrain et al., 2005), will disappear within the next few decades, adversely affecting people and ecosystems. Rapid melting of glaciers can lead to flooding of rivers and to the formation of glacial melt-water lakes, which may pose a serious threat of outburst floods (Coudrain et al., 2005). The entire Hindu Kush-Himalaya ice mass has decreased in the last two decades. Hence, water supply in areas fed by glacial melt water from the Hindu Kush and Himalayas, on which hundreds of millions of people in China and India depend, will be negatively affected (Barnett et al., 2005).

Under the IPCC IS92a emissions scenario (IPCC, 1992), which is similar to the SRES A1 scenario, significant changes in flood or drought risk are expected in many parts of Europe (Lehner et al., 2005b). The regions most prone to a rise in flood frequencies are northern and north-eastern Europe, while southern and south-eastern Europe show significant increases in drought frequencies. This is the case for climate change as computed by both the ECHAM4 and HadCM3 GCMs. Both models agree in their estimates that by the 2070s, a 100-year drought of today's magnitude would return, on average, more frequently than every 10 years in parts of Spain and Portugal, western France, the Vistula Basin in Poland, and western Turkey (Figure 3.6). Studies indicate a decrease in peak snowmelt floods by the 2080s in parts of the UK (Kay et al., 2006b) despite an overall increase in rainfall.

Results of a recent study (Reynard et al., 2004) show that estimates of future changes in flood frequency across the UK are now noticeably different than in earlier (pre-TAR) assessments, when increasing frequencies under all scenarios were projected. Depending on which GCM is used, and on the importance of snowmelt contribution and catchment characteristics and location, the impact of climate change on the flood regime (magnitude and frequency) can be both positive or negative, highlighting the uncertainty still remaining in climate change impacts (Reynard et al., 2004).

A sensitivity study by Cunderlik and Simonovic (2005) for a catchment in Ontario, Canada, projected a decrease in snowmelt-induced floods, while an increase in rain-induced floods is anticipated. The variability of annual maximum flow is projected to increase.

Palmer and Räisänen (2002) analysed GCM-modelled differences in winter precipitation between the control run and around the time of CO_2 doubling. A considerable increase in the risk of a very wet winter in Europe and a very wet monsoon season in Asia was found. The probability of total boreal winter precipitation exceeding two standard deviations above normal is projected to increase considerably (even five- to seven-fold) over large areas of Europe, with likely consequences for winter flood hazard.

Milly et al. (2002) demonstrated that, for fifteen out of sixteen large basins worldwide, the control 100-year peak volumes (at the monthly time-scale) are projected to be exceeded more frequently as a result of CO_2 quadrupling. In some areas, what is given as a 100-year flood now (in the control run), is projected to occur much more frequently, even every 2 to 5 years, albeit with a large uncertainty in these projections. Yet, in many temperate regions, the snowmelt contribution to spring floods is likely to decline on average (Zhang et al., 2005). Future changes in the joint probability of extremes have been considered, such as soil moisture and flood risk (Sivapalan et al., 2005), and fluvial flooding and tidal surge (Svensson and Jones, 2005).

Impacts of extremes on human welfare are likely to occur disproportionately in countries with low adaptation capacity (Manabe et al., 2004a). The flooded area in Bangladesh is projected to increase at least by 23-29% with a global temperature rise of 2°C (Mirza, 2003). Up to 20% of the world's population live in river basins that are likely to be affected by increased flood hazard by the 2080s in the course of global warming (Kleinen and Petschel-Held, 2007).

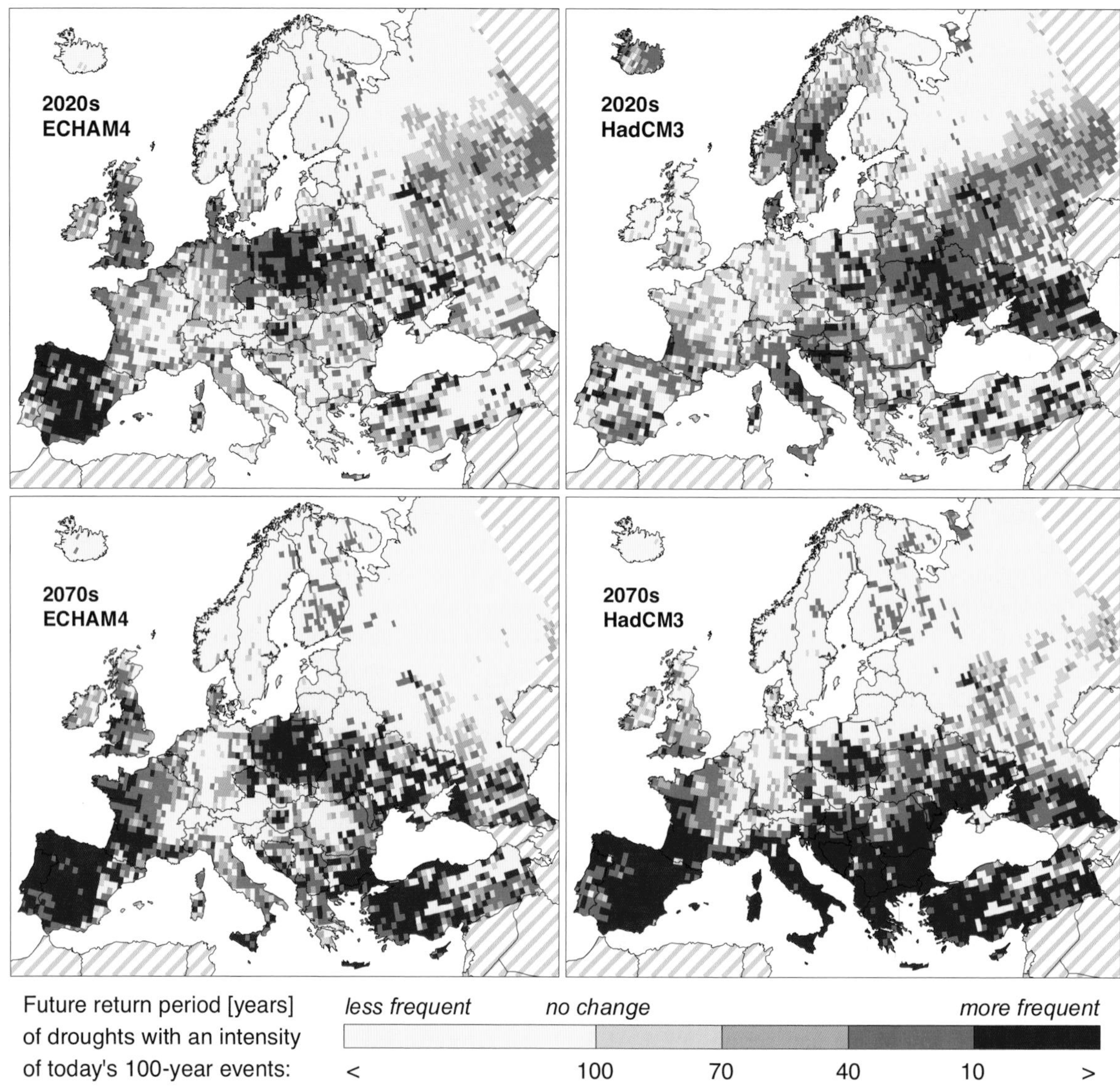

Figure 3.6. *Change in the recurrence of 100-year droughts, based on comparisons between climate and water use in 1961 to 1990 and simulations for the 2020s and 2070s (based on the ECHAM4 and HadCM3 GCMs, the IS92a emissions scenario and a business-as-usual water-use scenario). Values calculated with the model WaterGAP 2.1 (Lehner et al., 2005b).*

3.4.4 Water quality

Higher water temperature and variations in runoff are likely to produce adverse changes in water quality affecting human health, ecosystems, and water use (Patz, 2001; Lehman, 2002; O'Reilly et al., 2003; Hurd et al., 2004). Lowering of the water levels in rivers and lakes will lead to the re-suspension of bottom sediments and liberating compounds, with negative effects on water supplies (Atkinson et al., 1999). More intense rainfall will lead to an increase in suspended solids (turbidity) in lakes and reservoirs due to soil fluvial erosion (Leemans and Kleidon, 2002), and pollutants will be introduced (Mimikou et al., 2000; Neff et al., 2000; Bouraoui et al., 2004).

Higher surface water temperatures will promote algal blooms (Hall et al., 2002; Kumagai et al., 2003) and increase the bacteria and fungi content (Environment Canada, 2001). This may lead to a bad odour and taste in chlorinated drinking water and the occurrence of toxins (Moulton and Cuthbert, 2000; Robarts et al., 2005). Moreover, even with enhanced phosphorus removal in wastewater treatment plants, algal growth may increase with warming over the long term (Wade et al., 2002). Due to the high cost and the intermittent nature of algal blooms, water utilities will be unable to solve this problem with the available technology (Environment Canada, 2001). Increasing nutrients and sediments due to higher runoff, coupled with lower water levels, will negatively affect water quality (Hamilton et al., 2001), possibly rendering a source unusable unless special treatment is introduced (Environment Canada, 2004). Furthermore, higher water temperatures will enhance the transfer of volatile and semi-volatile compounds (e.g., ammonia, mercury, dioxins, pesticides) from surface water bodies to the atmosphere (Schindler, 2001).

In regions where intense rainfall is expected to increase, pollutants (pesticides, organic matter, heavy metals, etc.) will be increasingly washed from soils to water bodies (Fisher, 2000; Boorman, 2003b; Environment Canada, 2004). Higher runoff is expected to mobilise fertilisers and pesticides to water bodies in regions where their application time and low vegetation growth coincide with an increase in runoff (Soil and Water Conservation Society, 2003). Also, acidification in rivers and lakes is expected to increase as a result of acidic atmospheric deposition (Ferrier and Edwards, 2002; Gilvear et al., 2002; Soulsby et al., 2002).

In estuaries and inland reaches with decreasing streamflow, salinity will increase (Bell and Heaney, 2001; Williams, 2001; Beare and Heaney, 2002; Robarts et al., 2005). Pittock (2003) projected the salt concentration in the tributary rivers above irrigation areas in the Murray-Darling Basin in Australia to increase by 13-19% by 2050 and by 21-72% by 2100. Secondary salinisation of water (due to human disturbance of the natural salt cycle) will also threaten a large number of people relying on water bodies already suffering from primary salinisation. In areas where the climate becomes hotter and drier, human activities to counteract the increased aridity (e.g., more irrigation, diversions and impoundments) will exacerbate secondary salinisation (Williams, 2001). Water salinisation is expected to be a major problem in small islands suffering from coastal sea water intrusion, and in semi-arid and arid areas with decreasing runoff (Han et al., 1999; Bobba et al., 2000; Ministry for the Environment, 2001;Williams, 2001; Loáiciga, 2003; Chen et al., 2004; Ragab, 2005). Due to sea-level rise, groundwater salinisation will very likely increase.

Water-borne diseases will rise with increases in extreme rainfall (Hall et al., 2002; Hijioka et al., 2002; D'Souza et al., 2004; see also Chapter 8). In regions suffering from droughts, a greater incidence of diarrhoeal and other water-related diseases will mirror the deterioration in water quality (Patz, 2001; Environment Canada, 2004).

In developing countries, the biological quality of water is poor due to the lack of sanitation and proper potabilisation methods and poor health conditions (Lipp et al., 2001; Jiménez, 2003; Maya et al., 2003; WHO, 2004). Hence, climate change will be an additional stress factor that will be difficult to overcome (Magadza, 2000; Kashyap, 2004; Pachauri, 2004). Regrettably, there are no studies analysing the impact of climate change on biological water quality from the developing countries' perspective, i.e., considering organisms typical for developing countries; the effect of using wastewater to produce food; and Helminthiases diseases, endemic only in developing countries, where low-quality water is used for irrigation (WHO/UNICEF, 2000).

Even in places where water and wastewater treatment plants already exist, the greater presence of a wider variety of micro-organisms will pose a threat because the facilities are not designed to deal with them. As an example, Cryptosporidium outbreaks following intense rainfall events have forced some developed countries to adopt an additional filtration step in drinking-water plants, representing a 20 to 30% increase in operating costs (AWWA, 2006), but this is not universal practice.

Water quality modifications may also be observed in future as a result of:

- more water impoundments for hydropower (Kennish, 2002; Environment Canada, 2004),
- storm water drainage operation and sewage disposal disturbances in coastal areas due to sea-level rise (Haines et al., 2000),
- increasing water withdrawals from low-quality sources,
- greater pollutant loads due to increased infiltration rates to aquifers or higher runoff to surface waters (as result of high precipitation),
- water infrastructure malfunctioning during floods (GEO-LAC, 2003; DFID, 2004),
- overloading the capacity of water and wastewater treatment plants during extreme rainfall (Environment Canada, 2001),
- increased amounts of polluted storm water.

In areas where amounts of surface water and groundwater recharge are projected to decrease, water quality will also decrease due to lower dilution (Environment Canada, 2004). Unfortunately, in some regions the use of such water may be necessary, even if water quality problems already exist (see Section 3.2). For example, in regions where water with arsenic or fluorine is consumed, due to a lack of alternatives, it may still be necessary to consume the water even if the quality worsens.

It is estimated that at least one-tenth of the world's population consumes crops irrigated with wastewater (Smit and Nasr, 1992), mostly in developing countries in Africa, Asia, and Latin America (DFID, 2004). This number will increase with growing populations and wealth, and it will become imperative to use water more efficiently (including reuse). While recognising the convenience of recycling nutrients (Jiménez and Garduño, 2001), it is essential to be aware of the health and environmental risks caused by reusing low-quality water.

In developing countries, vulnerabilities are related to a lack of relevant information, institutional weakness in responding to a changing environment, and the need to mobilise resources. For the world as a whole, vulnerabilities are related to the need to respond proactively to environmental changes under uncertainty. Effluent disposal strategies (under conditions of lower self-purification in warmer water), the design of water and wastewater treatment plants to work efficiently even during extreme climatic conditions, and ways of reusing and recycling water, will need to be reconsidered (Luketina and Bender, 2002; Environment Canada, 2004; Patrinos and Bamzai, 2005).

3.4.5 Erosion and sediment transport

Changes in water balance terms affect many geomorphic processes including erosion, slope stability, channel change, and sediment transport (Rumsby and Macklin, 1994). There are also indirect consequences of geomorphic change for water quality (Dennis et al., 2003). Furthermore, hydromorphology is an influential factor in freshwater habitats.

All studies on soil erosion have suggested that increased rainfall amounts and intensities will lead to greater rates of erosion unless protection measures are taken. Soil erosion rates are expected to change in response to changes in climate for a variety of reasons. The most direct is the change in the erosive power of rainfall. Other reasons include:

- changes in plant canopy caused by shifts in plant biomass production associated with moisture regime;
- changes in litter cover on the ground caused by changes in plant residue decomposition rates driven by temperature, in moisture-dependent soil microbial activity, and in plant biomass production rates;
- changes in soil moisture due to shifting precipitation regimes

and evapotranspiration rates, which changes infiltration and runoff ratios;

- soil erodibility changes due to a decrease in soil organic matter concentrations (which lead to a soil structure that is more susceptible to erosion) and to increased runoff (due to increased soil surface sealing and crusting);
- a shift in winter precipitation from non-erosive snow to erosive rainfall due to increasing winter temperatures;
- melting of permafrost, which induces an erodible soil state from a previously non-erodible one;
- shifts in land use made necessary to accommodate new climatic regimes.

Nearing (2001) used output from two GCMs (HadCM3 and the Canadian Centre for Climate Modelling and Analysis CGCM1) and relationships between monthly precipitation and rainfall erosivity (the power of rain to cause soil erosion) to assess potential changes in rainfall erosivity in the USA. The predicted changes were significant, and in many cases very large, but results between models differed both in magnitude and regional distributions. Zhang et al. (2005) used HadCM3 to assess potential changes in rainfall erosivity in the Huanghe River Basin of China. Increases in rainfall erosivity by as much as 11 to 22% by the year 2050 were projected across the region.

Michael et al. (2005) projected potential increases in erosion of the order of 20 to 60% over the next five decades for two sites in Saxony, Germany. These results are arguably based on significant simplifications with regard to the array of interactions involved in this type of assessment (e.g., biomass production with changing climate). Pruski and Nearing (2002a) simulated erosion for the 21st century at eight locations in the USA using the HadCM3 GCM, and taking into account the primary physical and biological mechanisms affecting erosion. The simulated cropping systems were maize and wheat. The results indicated a complex set of interactions between the several factors that affect the erosion process. Overall, where precipitation increases were projected, estimated erosion increased by 15 to 100%. Where precipitation decreases were projected, the results were more complex due largely to interactions between plant biomass, runoff, and erosion, and either increases or decreases in overall erosion could occur.

A significant potential impact of climate change on soil erosion and sediment generation is associated with the change from snowfall to rainfall. The potential impact may be particularly important in northern climates. Warmer winter temperatures would bring an increasing amount of winter precipitation as rain instead of snow, and erosion by storm runoff would increase. The results described above which use a process-based approach incorporated the effect of a shift from snow to rain due to warming, but the studies did not delineate this specific effect from the general results. Changes in soil surface conditions, such as surface roughness, sealing and crusting, may change with shifts in climate, and hence affect erosion rates.

Zhang and Nearing (2005) evaluated the potential impacts of climate change on soil erosion in central Oklahoma. Monthly projections were used from the HadCM3 GCM, using the SRES A2 and B2 scenarios and GGa1 (a scenario in which greenhouse gases increase by 1%/yr), for the periods 1950 to 1999 and 2070 to 2099. While the HadCM3-projected mean annual precipitation during 2070 to 2099 at El Reno, Oklahoma, decreased by 13.6%, 7.2%, and 6.2% for A2, B2, and GGa1, respectively, the predicted erosion (except for the no-till conservation practice scenario) increased by 18-30% for A2, remained similar for B2, and increased by 67-82% for GGa1. The greater increases in erosion in the GGa1 scenario was attributed to greater variability in monthly precipitation and an increased frequency of large storms in the model simulation. Results indicated that no-till (or conservation tillage) systems can be effective in reducing soil erosion under projected climates.

A more complex, but potentially dominant, factor is the potential for shifts in land use necessary to accommodate a new climatic regime (O'Neal et al., 2005). As farmers adapt cropping systems, the susceptibility of the soil to erosive forces will change. Farmer adaptation may range from shifts in planting, cultivation and harvest dates, to changes in crop type (Southworth et al., 2000; Pfeifer and Habeck, 2002). Modelling results for the upper Midwest U.S. suggest that erosion will increase as a function of future land-use changes, largely because of a general shift away from wheat and maize towards soybean production. For ten out of eleven regions in the study area, predicted runoff increased from +10% to +310%, and soil loss increased from +33% to +274%, in 2040–2059 relative to 1990–1999 (O'Neal et al., 2005). Other land-use scenarios would lead to different results. For example, improved conservation practices can greatly reduce erosion rates (Souchere et al., 2005), while clear-cutting a forest during a 'slash-and-burn' operation has a huge negative impact on susceptibility to runoff and erosion.

Little work has been done on the expected impacts of climate change on sediment loads in rivers and streams. Bouraoui et al. (2004) showed, for southern Finland, that the observed increase in precipitation and temperature was responsible for a decrease in snow cover and increase in winter runoff, which resulted in an increase in modelled suspended sediment loads. Kostaschuk et al. (2002) measured suspended sediment loads associated with tropical cyclones in Fiji, which generated very high (around 5% by volume) concentrations of sediment in the measured flows. The authors hypothesized that an increase in intensity of tropical cyclones brought about by a change in El Niño patterns could increase associated sediment loads in Fiji and across the South Pacific.

In terms of the implications of climate change for soil conservation efforts, a significant realisation from recent scientific efforts is that conservation measures must be targeted at the extreme events more than ever before (Soil and Water Conservation Society, 2003). Intense rainfall events contribute a disproportionate amount of erosion relative to the total rainfall contribution, and this effect will only be exacerbated in the future if the frequency of such storms increases.

3.5 Costs and other socio-economic aspects

Impacts of climate change will entail social and economic costs and benefits, which are difficult to determine. These include the costs of damages and the costs of adaptation (to reduce or avoid damages), as well as benefits that could result from improved water availability in some areas. In addition to uncertainties about the impacts of future climate change on freshwater systems, there are other compounding factors, including demographic, societal, and economic developments, that should be considered when evaluating the costs of climate change. Costs and benefits of climate change may take several forms, including increases or decreases in monetary costs, and human and ecosystem impacts, e.g., displacement of households due to flooding, and loss of aquatic species. So far, very few of these costs have been estimated in monetary terms. Efforts to quantify the economic impacts of climate-related changes in water resources are hampered by a lack of data and by the fact that the estimates are highly sensitive to different estimation methods and to different assumptions regarding how changes in water availability will be allocated across various types of water uses, e.g., between agricultural, urban, or in-stream uses (Changnon, 2005; Schlenker et al., 2005; Young, 2005).

With respect to water supply, it is very likely that the costs of climate change will outweigh the benefits. One reason is that precipitation variability is very likely to increase. The impacts of floods and droughts could be tempered by appropriate infrastructure investments, and by changes in water and land-use management, but all of these responses entail costs (US Global Change Research Program, 2000). Another reason is that water infrastructure, use patterns, and institutions have developed in the context of current conditions (Conway, 2005). Any substantial change in the frequency of floods and droughts or in the quantity and quality or seasonal timing of water availability will require adjustments that may be costly not only in monetary terms, but also in terms of societal impacts, including the need to manage potential conflicts among different interest groups (Miller et al., 1997).

Hydrological changes may have impacts that are positive in some aspects and negative in others. For example, increased annual runoff may produce benefits for a variety of instream and out-of-stream water users by increasing renewable water resources, but may simultaneously generate harm by increasing flood risk. In recent decades, a trend to wetter conditions in parts of southern South America has increased the area inundated by floods, but has also improved crop yields in the Pampa region of Argentina, and has provided new commercial fishing opportunities (Magrin et al., 2005; also see Chapter 13). Increased runoff could also damage areas with a shallow watertable. In such areas, a watertable rise will disturb agricultural use and damage buildings in urban areas. For Russia, for example, the current annual damage caused by shallow watertables is estimated to be US$5-6 billion (Kharkina, 2004) and is likely to increase in the future. In addition, an increase in annual runoff may not lead to a beneficial increase in readily available water resources if the additional runoff is concentrated during the high-flow season.

3.5.1 How will climate change affect the balance of water demand and water availability?

To evaluate how climate change will affect the balance between water demand and water availability, it is necessary to consider the entire suite of socially valued water uses and how the allocation of water across those uses is likely to change. Water is valuable not only for domestic uses, but also for its role in supporting aquatic ecosystems and environmental amenities, including recreational opportunities, and as a factor of production in irrigated agriculture, hydropower production, and other industrial uses (Young, 2005). The social costs or benefits of any change in water availability would depend on how the change affects each of these potentially competing human water demands. Changes in water availability will depend on changes in the volume, variability, and seasonality of runoff, as modified by the operation of existing water control infrastructure and investments in new infrastructure. The institutions that govern water allocation will play a large role in determining the overall social impacts of a change in water availability, as well as the distribution of gains and losses across different sectors of society. Institutional settings differ significantly both within and between countries, often resulting in substantial differences in the efficiency, equity, and flexibility of water use and infrastructure development (Wichelns et al., 2002; Easter and Renwick, 2004; Orr and Colby, 2004; Saleth and Dinar, 2004; Svendsen, 2005).

In addition, quantity of water is not the only important variable. Changes in water quality and temperature can also have substantial impacts on urban, industrial, and agricultural use values, as well as on aquatic ecosystems. For urban water uses, degraded water quality can add substantially to purification costs. Increased precipitation intensity may periodically result in increased turbidity and increased nutrient and pathogen content of surface water sources. The water utility serving New York City has identified heavy precipitation events as one of its major climate-change-related concerns because such events can raise turbidity levels in some of the city's main reservoirs up to 100 times the legal limit for source quality at the utility's intake, requiring substantial additional treatment and monitoring costs (Miller and Yates, 2006).

Water demand

There are many different types of water demand. Some of these compete directly with one another in that the water consumed by one sector is no longer available for other uses. In other cases, a given unit of water may be used and reused several times as it travels through a river basin, for example, providing benefits to instream fisheries, hydropower generators, and domestic users in succession. Sectoral water demands can be expected to change over time in response to changes in population, settlement patterns, wealth, industrial activity, and technology. For example, rapid urbanization can lead to substantial localised growth in water demand, often making it difficult to meet goals for the provision of a safe, affordable, domestic water supply, particularly in arid regions (e.g., Faruqui et al., 2001). In addition, climate change will probably alter the desired uses of water (demands) as well as actual uses (demands in each sector that are actually met). If climate change results in greater water scarcity relative to demand, adaptation may include technical changes that improve water-use efficiency, demand management (e.g., through metering and pricing), and institutional changes that improve the tradability of water rights. It takes time to implement such changes, so they are likely to become more effective as time passes. Because the availability of water for each type of use may be affected by other competing uses of the resource, a complete analysis of the effects of climate change on human water uses should consider cross-sector interactions, including the impacts of changes in water-use efficiency and intentional transfers of the use of water from one sector to another. For example, voluntary water transfers, including short-term water leasing as well as permanent sales of water rights, generally from agricultural to urban or environmental uses, are becoming increasingly common in the western USA. These water-market transactions can be expected to play a role in facilitating adaptation to climate change (Miller et al., 1997; Easter et al., 1998; Brookshire et al., 2004; Colby et al., 2004).

Irrigation water withdrawals account for almost 70% of global water withdrawals and 90% of global consumptive water use (the water fraction that evapotranspires during use) (Shiklomanov and Rodda, 2003). Given the dominant role of irrigated agriculture in global water use, management practices that increase the productivity of irrigation water use (defined as crop output per unit of consumptive water use) can greatly increase the availability of water for other human and environmental uses (Tiwari and Dinar, 2002). Of all sectoral water demands, the irrigation sector will be affected most strongly by climate change, as well as by changes in the effectiveness of irrigation methods. In areas facing water scarcity, changes in irrigation water use will be driven by the combined effects of changes in irrigation water demand, changes in demands for higher value uses (e.g., for urban areas), future management changes, and changes in availability.

Higher temperatures and increased variability of precipitation would, in general, lead to an increased irrigation water demand, even if the total precipitation during the growing season remains the same. As a result of increased atmospheric CO_2 concentrations, water-use efficiency for some types of plants would increase, which would increase the ratio of crop yield to unit of water input (water productivity – 'more crop per drop'). However, in hot regions, such as Egypt, the ratio may even decline as yields decrease due to heat stress (see Chapter 5).

There are no global-scale studies that attempt to quantify the influence of climate-change-related factors on irrigation water use; only the impact of climate change on optimal growing periods and yield-maximising irrigation water use has been modelled, assuming no change in irrigated area and climate variability (Döll, 2002; Döll et al., 2003). Applying the SRES A2 and B2 scenarios as interpreted by two climate models, these authors found that the optimal growing periods could shift in many irrigated areas. Net irrigation requirements of China and India, the countries with the largest irrigated areas worldwide, change by +2% to +15% and by −6% to +5% for the year 2020, respectively, depending on emissions scenario and climate model. Different climate models project different worldwide changes in net irrigation requirements, with estimated increases ranging from 1 to 3% by the 2020s and 2 to 7% by the 2070s. The largest global-scale increases in net irrigation requirements result from a climate scenario based on the B2 emissions scenario.

At the national scale, some integrative studies exist; two modelling studies on adaptation of the agricultural sector to climate change in the USA (i.e., shifts between irrigated and rain-fed production) foresee a decrease in irrigated areas and withdrawals beyond 2030 for various climate scenarios (Reilly et al., 2003; Thomson et al., 2005b). This result is related to a declining yield gap between irrigated and rain-fed agriculture caused by yield reductions of irrigated crops due to higher temperatures, or yield increases of rain-fed crops due to more precipitation. These studies did not take into account the increasing variability of daily precipitation, such that rain-fed yields are probably overestimated. In a study of maize irrigation in Illinois under profit-maximising conditions, it was found that a 25% decrease of annual precipitation had the same effect on irrigation profitability as a 15% decrease combined with a doubling of the standard deviation of daily precipitation (Eheart and Tornil, 1999). This study also showed that profit-maximising irrigation water use responds more strongly to changes in precipitation than does yield-maximising water use, and that a doubling of atmospheric CO_2 has only a small effect.

According to an FAO study in which the climate change impact was not considered (Bruinsma, 2003), an increase in irrigation water withdrawals of 14% is foreseen by 2030 for developing countries. In the four Millennium Ecosystem Assessment scenarios, however, increases at the global scale are much less, as irrigated areas are assumed to increase only between 0% and 6% by 2030 and between 0% and 10% by 2050. The overwhelming water use increases are likely to occur in the domestic and industrial sectors, with increases of water withdrawals by 14-83% by 2050 (Millennium Ecosystem Assessment, 2005a, b). This is based on the idea that the value of water would be much higher for domestic and industrial uses (particularly true under conditions of water stress).

The increase in household water demand (e.g., for garden watering) and industrial water demand due to climate change is likely to be rather small, e.g., less than 5% by the 2050s at selected locations (Mote et al., 1999; Downing et al., 2003). An indirect but small secondary effect on water demand would be the increased electricity demand for cooling of buildings, which would tend to increase water withdrawals for cooling of thermal power plants (see Chapter 7). A statistical analysis of water use in New York City showed that above 25°C, daily per capita water use increases by 11 litres/1°C (roughly 2% of current daily per capita use) (Protopapas et al., 2000).

Water availability for aquatic ecosystems

Of all ecosystems, freshwater ecosystems will have the highest proportion of species threatened with extinction due to climate change (Millennium Ecosystem Assessment, 2005b). In cold or snow-dominated river basins, atmospheric temperature increases do not only affect freshwater ecosystems via the warming of water (see Chapter 4) but also by causing water-flow alterations. In northern Alberta, Canada, for example, a decrease in ice-jam flooding will lead to the loss of aquatic habitat (Beltaos et al., 2006). Where river discharges decrease seasonally, negative impacts on both freshwater ecosystems and coastal marine ecosystems can be expected. Atlantic salmon in north-west England will be affected negatively by climate change because suitable flow depths during spawning time (which now occur all the time) will, under the SRES A2 scenario, only exist for 94% of the time in the 2080s (Walsh and Kilsby, 2007). Such changes will have implications for ecological flow management and compliance with environmental legislation such as the EU Habitats Directive. In the case of decreased discharge in the western USA, by 2050 the Sacramento and Colorado River deltas could experience a dramatic increase in salinity and subsequent ecosystem disruption and, in the Columbia River system, managers will be faced with the choice of either spring and summer releases for salmon runs, or summer and autumn hydroelectric power production. Extinction of some salmon species due to climate change in the Pacific Northwest may take place regardless of water policy (Barnett et al., 2005).

Changed freshwater inflows into the ocean will lead to changes in turbidity, salinity, stratification, and nutrient availability, all of which affect estuarine and coastal ecosystems (Justic et al., 2005). While increased river discharge of the Mississippi would increase the frequency of hypoxia (shortage of oxygen) events in the Gulf of Mexico, increased river discharge into the Hudson Bay would lead to the opposite (Justic et al., 2005). The frequency of bird-breeding events in the Macquarie Marshes in the Murray-Darling Basin in Australia is predicted to decrease with reduced streamflow, as the breeding of colonially nesting water-birds requires a certain minimum annual flow. Climate change and reforestation can contribute to a decrease in river discharge, but before 2070 the largest impact can be expected from a shift in rainfall due to decadal-scale climate variability (Herron et al., 2002).

Water availability for socio-economic activities

Climate change is likely to alter river discharge, resulting in important impacts on water availability for instream and out-of-stream uses. Instream uses include hydropower, navigation, fisheries, and recreation. Hydropower impacts for Europe have been estimated using a macro-scale hydrological model. The results indicate that, by the 2070s, under the IS92a emissions scenario, the electricity production potential of hydropower plants existing at the end of the 20th century will increase, by 15-30% in Scandinavia and northern Russia, where between 19% (Finland) and almost 100% (Norway) of the electricity is produced by hydropower (Lehner et al., 2005a). Decreases by 20-50% or more are computed for Portugal, Spain, Ukraine, Bulgaria, and Turkey, where between 10% (Ukraine, Bulgaria) and 39% of the electricity is produced by hydropower (Lehner et al., 2005a). For the whole of Europe (with a 20% hydropower fraction), hydropower potential shows a decrease of 7-12% by the 2070s. In North America, potential reductions in the outflow of the Great Lakes could result in significant economic losses as a result of reduced hydropower generation at Niagara and on the St. Lawrence River (Lofgren et al., 2002). For a CGCM1 model projection with 2°C global warming, Ontario's Niagara and St. Lawrence hydropower generation would decline by 25-35%, resulting in annual losses of Canadian $240 million to $350 million (2002 prices) (Buttle et al., 2004). With the HadCM2 climate model, however, a small gain in hydropower potential (+ 3%) was computed, worth approximately Canadian $25 million/yr. Another study that examined a range of climate model scenarios found that a 2°C global warming could reduce hydropower-generating capacity on the St. Lawrence River by 1% to 17% (LOSLR, 2006). Increased flood periods in the future will disrupt navigation more often, and low flow conditions that restrict the loading of ships may increase, for the Rhine river, from 19 days under current climate conditions to 26-34 days in the 2050s (Middelkoop et al., 2001).

Out-of-stream uses include irrigation, domestic, municipal, and industrial withdrawals, including cooling water for thermal electricity generation. Water availability for withdrawal is a function of runoff, aquifer conditions, and technical water supply infrastructure (reservoirs, pumping wells, distribution networks, etc.). Safe access to drinking water depends more on the level of technical water supply infrastructure than on the level of runoff. However, the goal of improved safe access to drinking water will be harder to achieve in regions where runoff decreases as a result of climate change. Also, climate change leads to additional costs for the water supply sector, e.g., due to changing water levels affecting water supply infrastructure, which might hamper the extension of water supply services to more people.

Climate-change-induced changes of the seasonal runoff regime and interannual runoff variability can be as important for water availability as changes in the long-term average annual runoff amount if water is not withdrawn from large groundwater bodies or reservoirs (US Global Change Research Program, 2000). People living in snowmelt-fed basins experiencing decreasing snow storage in winter may be negatively affected by decreased river flows in the summer and autumn (Barnett et al., 2005). The Rhine, for example, might suffer from a 5 to 12% reduction in summer low flows by the 2050s, which will negatively affect water supply, in particular for thermal power plants (Middelkoop et al., 2001). Studies for the Elbe River Basin have shown that actual evapotranspiration is projected to increase by 2050 (Krysanova and Wechsung, 2002), while river flow, groundwater recharge, crop yield, and diffuse-source pollution are likely to decrease (Krysanova et al., 2005). Investment and operation costs for additional wells and reservoirs which are required to guarantee reliable water supply under climate change have been estimated for China. This cost is low in basins where the current water stress is low (e.g., Changjiang), and high where it is high (e.g., Huanghe River) (Kirshen et al., 2005a). Furthermore, the impact of climate change on water supply costs will increase in the future, not only because of increasing climate change but also due to increasing demand.

A number of global-scale (Alcamo and Henrichs, 2002; Arnell, 2004b), national-scale (Thomson et al., 2005a), and basin-scale assessments (Barnett et al., 2004) show that semi-arid and arid basins are the most vulnerable basins on the globe with respect to water stress. If precipitation decreases, irrigation water demands, which dominate water use in most semi-arid river basins, would increase, and it may become impossible to satisfy all demands. In the case of the Sacramento-Joaquin River and the Colorado River basins in the western USA, for example, streamflow changes (as computed by basin-scale hydrological models driven by output from a downscaled GCM – the PCM model from the National Center for Atmospheric Research) are so strong that, beyond 2020, not all the present-day water demands (including environmental targets) could be fulfilled even with adapted reservoir management (Barnett et al., 2004). Furthermore, if irrigation use is allowed to increase in response to increased demands, that would amplify the decreases in runoff and streamflow downstream (Eheart and Tornil, 1999). Huffaker (2005) notes that some policies aimed at rewarding improvements in irrigation efficiency allow irrigators to spread a given diversion right to a larger land area. The unintended consequence could be increased consumptive water use that deprives downstream areas of water that would have re-entered the stream as return flow. Such policies could make irrigation no longer feasible in the lower reaches of basins that experience reduced streamflow.

A case study from a semi-arid basin in Canada shows how the balance between water supply and irrigation water demand may be altered due to climate change (see Box 3.1), and how the costs of this alteration can be assessed.

In western China, earlier spring snowmelt and declining glaciers are likely to reduce water availability for irrigated agriculture (see Chapter 10). For an aquifer in Texas, the net income of farmers is projected to decrease by 16-30% by the 2030s and by 30-45% by the 2090s due to decreased irrigation water supply and increased irrigation water demand, but net total welfare due to water use, which is dominated by municipal and industrial use, decreases by less than 2% (Chen et al., 2001). If freshwater supply has to be replaced by desalinated water due to climate change, then the cost of climate change includes the cost of desalination, which is currently around US$1/m^3 for seawater and US$0.6/m^3 for brackish water (Zhou and Tol, 2005), compared to the chlorination cost of freshwater of US$0.02/m^3 and costs between US$0.35 and US$1.9/m^3 for additional supply in a case study in Canada (see Box 3.1). In densely populated coastal areas of Egypt, China, Bangladesh, India, and Southeast Asia (FAO, 2003), desalination costs may be prohibitive.

Most semi-arid river basins in developing countries are more vulnerable to climate change than basins in developed countries, as population, and thus water demand, is expected to grow rapidly in the future and the coping capacity is low (Millennium Ecosystem Assessment, 2005b). Coping capacity is particularly low in rural populations without access to reliable water supply from large reservoirs or deep wells. Inhabitants of rural areas are affected directly by changes in the volume and timing of river discharge and groundwater recharge. Thus, even in semi-arid areas where water resources are not overused, increased climate variability may have a strong negative impact. In humid river basins, people are likely to cope more easily with the impact of climate change on water demand and availability, although they might be less prepared for coping with droughts than people in dry basins (Wilhite, 2001).

Global estimates of the number of people living in areas with high water stress differ significantly among studies (Vörösmarty et al., 2000; Alcamo et al., 2003a, b, 2007; Oki et al., 2003a; Arnell, 2004b). Climate change is only one factor that influences future water stress, while demographic, socio-economic, and technological changes may play a more important role in most time horizons and regions. In the 2050s, differences in the population projections of the four SRES scenarios would have a greater impact on the number of people living in water-stressed river basins (defined as basins with per capita water resources of less than 1,000 m^3/year) than the differences in the emissions scenarios (Arnell, 2004b). The number of people living in severely stressed river basins would increase significantly (Table 3.2). The population at risk of increasing water stress for the full range of SRES scenarios is projected to be: 0.4 to 1.7 billion, 1.0 to 2.0 billion, and 1.1 to 3.2 billion, in the 2020s, 2050s, and 2080s, respectively (Arnell, 2004b). In the 2050s (SRES A2 scenario), 262-983 million people would move into the water-stressed category (Arnell, 2004b). However, using the per capita water availability indicator, climate change would appear to reduce global water stress. This is because increases in runoff are heavily concentrated in the most populous parts of the world, mainly in East and South-East Asia, and mainly occur during high flow seasons (Arnell, 2004b). Therefore, they may not alleviate dry season problems if the extra water is not stored and would not ease water stress in other regions of the world.

If water stress is not only assessed as a function of population and climate change, but also of changing water use, the importance of non-climatic drivers (income, water-use efficiency, water productivity, industrial production) increases (Alcamo et al., 2007). Income growth has a much larger impact than population growth on increasing water use and water stress (expressed as the water withdrawal-to-water resources ratio). Water stress is modelled to decrease by the 2050s on 20 to 29% of the global land area (considering two climate models and the SRES A2 and B2 scenarios) and to increase on 62 to 76% of the global land area. The principal cause of decreasing water stress is the greater availability of water due to increased precipitation, while the principal cause of increasing water stress is growing water withdrawals. Growth of domestic water use as stimulated by income growth was found to be dominant (Alcamo et al., 2007).

The change in the number of people under high water stress after the 2050s greatly depends on emissions scenario: substantial increase is projected for the A2 scenario; the speed of increase will be slower for the A1 and B1 emissions scenarios because of the global increase of renewable freshwater resources and the slight decrease in population (Oki and Kanae, 2006). Nevertheless, changes in seasonal patterns and the increasing probability of extreme events may offset these effects.

Table 3.2. *Impact of population growth and climate change on the number of people (in millions) living in water-stressed river basins (defined as per capita renewable water resources of less than 1,000 m^3/yr) around 2050 (Arnell, 2004b; Alcamo et al., 2007).*

	Estimated millions of people	
	From Arnell, 2004b	From Alcamo et al., 2007
Baseline (1995)	1,368	1,601
2050: A2 emissions scenario	4,351 to 5,747	6,432 to 6,920
2050: B2 emissions scenario	2,766 to 3,958	4,909 to 5,166

Estimates are based on emissions scenarios for several climate model runs. The range is due to the various climate models and model runs that were used to translate emissions scenarios into climate scenarios.

Box 3.1. Costs of climate change in Okanagan, Canada

The Okanagan region in British Columbia, Canada, is a semi-arid watershed of 8,200 km^2 area. The region's water resources will be unable to support an increase in demand due to projected climate change and population growth, so a broad portfolio of adaptive measures will be needed (Cohen and Neale, 2006; Cohen et al., 2006). Irrigation accounts for 78% of the total basin licensed water allocation.

Figure 3.7 illustrates, from a suite of six GCM scenarios, the worst-case and least-impact scenario changes in annual water supply and crop water demand for Trout Creek compared with a drought supply threshold of 30 million m^3/yr (36% of average annual present-day flow) and observed maximum demand of 10 million m^3/yr (Neilsen et al., 2004). For flows below the drought threshold, local water authorities currently restrict water use. High-risk outcomes are defined as years in which water supply is below the drought threshold and water demand above the demand threshold. For all six scenarios, demand is expected to increase and supply is projected to decline. Estimated crop water demand increases most strongly in the HadCM3 A2 emissions scenario in which, by the 2080s, demand exceeds the current observed maximum in every year. For HadCM3 A2, high-risk outcomes occur in 1 out of 6 years in the 2050s, and in 1 out of 3 years in the 2080s. High-risk outcomes occur more often under A2 than under the B2 emissions scenario due to higher crop water demands in the warmer A2 world.

Table 3.3 illustrates the range of costs of adaptive measures currently available in the region, that could either decrease water demand or increase water supply. These costs are expressed by comparison with the least-cost option, irrigation scheduling on large holdings, which is equivalent to US$0.35/m^3 (at 2006 prices) of supplied water. The most expensive options per unit of water saved or stored are metering and lake pumping to higher elevations. However, water treatment requirements will lead to additional costs for new supply options (Hrasko and McNeill, 2006). No single option is expected to be sufficient on its own.

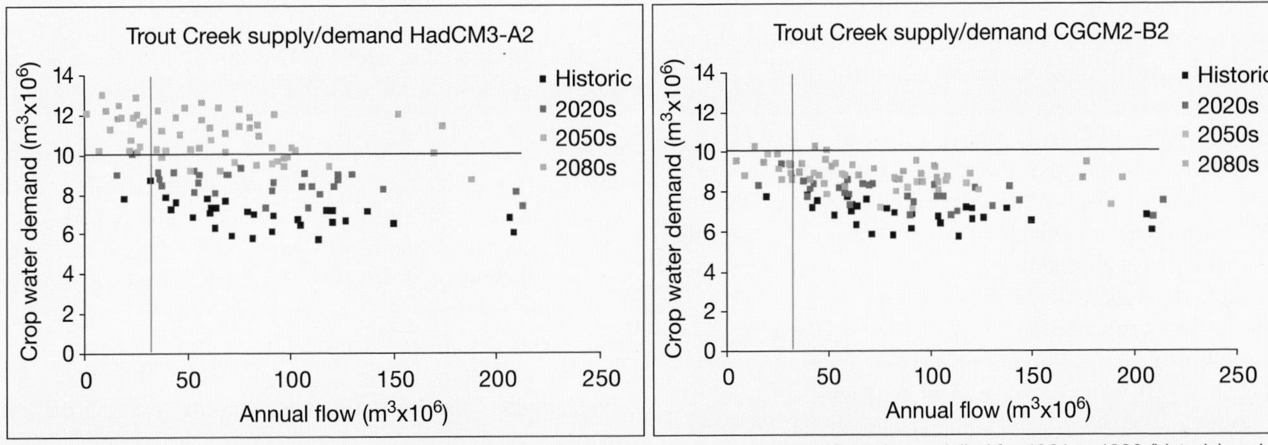

Figure 3.7. *Annual crop water demand and water supply for Trout Creek, Okanagan region, Canada, modelled for 1961 to 1990 (historic) and three 30-year time slices in the future. Each dot represents one year. Drought supply threshold is represented by the vertical line, maximum observed demand is shown as the horizontal line (Neilsen et al., 2004).*

Table 3.3. *Relative costs per unit of water saved or supplied in the Okanagan region, British Columbia (adapted from MacNeil, 2004).*

Adaptation option	Application	Relative unit cost	Water saved or supplied in % of the current supply
Irrigation scheduling	Large holdings to small holdings	1.0 to 1.7	10%
Public education	Large and medium communities	1.7	10%
Storage	Low to high cost	1.2 to 3.0	Limited (most sites already developed)
Lake pumping	Low (no balancing reservoirs) to high cost (with balancing reservoirs)	1.3 to 5.4	0 to 100%
Trickle irrigation	High to medium demand areas	3.0 to 3.3	30%
Leak detection	Average cost	3.1	10 to 15%
Metering	Low to high cost	3.8 to 5.4	20 to 30%

3.5.2 How will climate change affect flood damages?

Future flood damages will depend heavily on settlement patterns, land-use decisions, the quality of flood forecasting, warning and response systems, and the value of structures and other property located in vulnerable areas (Mileti, 1999; Pielke and Downton, 2000; Changnon, 2005), as well as on climatic changes per se (Schiermeier, 2006). Choi and Fisher (2003) estimated the expected change in flood damages for selected USA regions under two climate-change scenarios in which mean annual precipitation increased by 13.5% and 21.5%, respectively, with the standard deviation of annual precipitation either remaining unchanged or increasing proportionally. They used a structural econometric (regression) model based on time series of flood damage, and population, wealth indicator, and annual precipitation as predictors. They found that the mean and standard deviation of flood damage are projected to increase by more than 140% if the mean and standard deviation of annual precipitation increase by 13.5%. The estimates suggest that flood losses are related to exposure because the explanatory power of population and wealth is 82%, while adding precipitation increases the explanatory power to 89%. Another study examined the potential flood damage impacts of changes in extreme precipitation events using the Canadian Climate Centre model and the IS92a emissions scenario for the metropolitan Boston area in the north-eastern USA (Kirshen et al., 2005b). They found that, without adaptation investments, both the number of properties damaged by floods and the overall cost of flood damage would double by 2100 relative to what might be expected with no climate change, and that flood-related transportation delays would become an increasingly significant nuisance over the course of the century. The study concluded that the likely economic magnitude of these damages is sufficiently high to justify large expenditures on adaptation strategies such as universal flood-proofing for all flood plains.

This finding is supported by a scenario study of the damage due to river and coastal flooding in England and Wales in the 2080s (Hall et al., 2005), which combined four emissions scenarios with four scenarios of socio-economic change in an SRES-like framework. In all scenarios, flood damages are predicted to increase unless current flood management policies, practices and infrastructure are changed. For a 2°C temperature increase in a B1-type world, by the 2080s annual damage is estimated to be £5 billion as compared to £1 billion today, while with approximately the same climate change, damage is only £1.5 billion in a B2-type world. In an A1-type world, with a temperature increase of 2°C, the annual damage would amount to £15 billion by the 2050s and £21 billion by the 2080s (Hall et al., 2005; Evans et al., 2004).

The impact of climate change on flood damages can be estimated from modelled changes in the recurrence interval of present-day 20- or 100-year floods, and estimates of the damages of present-day floods as determined from stage-discharge relations (between gauge height (stage) and volume of water per unit of time (discharge)), and detailed property data. With such a methodology, the average annual direct flood damage for three Australian drainage basins was projected to increase by a factor of four to ten under conditions of doubled atmospheric CO_2 concentrations (Schreider et al., 2000).

3.6 Adaptation: practices, options and constraints

3.6.1 The context for adaptation

Adaptation to changing conditions in water availability and demand has always been at the core of water management. Historically, water management has concentrated on meeting the increasing demand for water. Except where land-use change occurs, it has conventionally been assumed that the natural resource base is constant. Traditionally, hydrological design rules have been based on the assumption of stationary hydrology, tantamount to the principle that the past is the key to the future. This assumption is no longer valid. The current procedures for designing water-related infrastructures therefore have to be revised. Otherwise, systems would be over- or under-designed, resulting in either excessive costs or poor performance.

Changing to meet altered conditions and new ways of managing water are autonomous adaptations which are not deliberately designed to adjust with climate change. Drought-related stresses, flood events, water quality problems, and growing water demands are creating the impetus for both infrastructure investment and institutional changes in many parts of the world (e.g., Wilhite, 2000; Faruqui et al., 2001; Giansante et al., 2002; Galaz, 2005). On the other hand, planned adaptations take climate change specifically into account. In doing so, water planners need to recognise that it is not possible to resolve all uncertainties, so it would not be wise to base decisions on only one, or a few, climate model scenarios. Rather, making use of probabilistic assessments of future hydrological changes may allow planners to better evaluate risks and response options (Tebaldi et al., 2004, 2005, 2006; Dettinger, 2005).

Integrated Water Resources Management should be an instrument to explore adaptation measures to climate change, but so far is in its infancy. Successful integrated water management strategies include, among others: capturing society's views, reshaping planning processes, coordinating land and water resources management, recognizing water quantity and quality linkages, conjunctive use of surface water and groundwater, protecting and restoring natural systems, and including consideration of climate change. In addition, integrated strategies explicitly address impediments to the flow of information. A fully integrated approach is not always needed but, rather, the appropriate scale for integration will depend on the extent to which it facilitates effective action in response to specific needs (Moench et al., 2003). In particular, an integrated approach to water management could help to resolve conflicts among competing water users. In several places in the western USA, water managers and various interest groups have been experimenting with methods to promote consensus-based decision making. These efforts include local watershed initiatives and state-led or federally-sponsored efforts to incorporate stakeholder involvement in planning processes (e.g., US Department of the Interior, 2005). Such initiatives can facilitate negotiations among competing interests to achieve mutually satisfactory problem-solving that considers a wide range of factors. In the case of large watersheds, such as the Colorado River Basin, these factors cross several time- and space-scales (Table 3.4).

Lately, some initiatives such as the Dialogue on Water and Climate (DWC) (see Box 3.2) have been launched in order to raise awareness of climate change adaptation in the water sector. The main conclusion out of the DWC initiative is that the dialogue model provides an important mechanism for developing adaptation strategies with stakeholders (Kabat and van Schaik, 2003).

3.6.2 Adaptation options in principle

The TAR drew a distinction between 'supply-side' and 'demand-side' adaptation options, which are applicable to a range of systems. Table 3.5 summarises some adaptation options for water resources, designed to ensure supplies during average and drought conditions.

Each option, whether supply-side or demand-side, has a range of advantages and disadvantages, and the relative benefits of different options depend on local circumstances. In general terms,

Box 3.2. Lessons from the 'Dialogue on Water and Climate'

- The aim of the Dialogue on Water and Climate (DWC) was to raise awareness of climate implications in the water sector. The DWC initiated eighteen stakeholder dialogues, at the levels of a river basin (Lena, Aral Sea, Yellow River, San Pedro, San Juan, Thukela, Murray-Darling, and Nagoya), a nation (Netherlands and Bangladesh), and a region (Central America, Caribbean Islands, Small Valleys, West Africa, Southern Africa, Mediterranean, South Asia, South-east Asia, and Pacific Islands), to prepare for actions that reduce vulnerability to climate change. The Dialogues were located in both developed and developing countries and addressed a wide range of vulnerability issues related to water and climate. Participants included water professionals, community representatives, local and national governments, NGOs, and researchers.
- The results have been substantial and the strong message going out of these Dialogues to governments, donors, and disaster relief agencies is that it is on the ground, in the river basins and in the communities, that adaptation actions have to be taken. The Dialogues in Bangladesh and the Small Valleys in Central America have shown that villagers are well aware that climate extremes are becoming more frequent and more intense. The Dialogues also showed that adaptation actions in Bangladesh, the Netherlands, Nagoya, Murray-Darling, and Small Valleys are under way. In other areas, adaptation actions are in the planning stages (Western Africa, Mekong) and others are still in the initial awareness-raising stages (Southern Africa, Aral Sea, Lena Basin).
- The DWC demonstrated that the Dialogue model provides a promising mechanism for developing adaptation strategies with stakeholders.

Table 3.4. *Cross-scale issues in the integrated water management of the Colorado River Basin (Pulwarty and Melis, 2001).*

Temporal scale	Issue
Indeterminate	Flow necessary to protect endangered species
Long-term	Inter-basin allocation and allocation among basin states
Decadal	Upper basin delivery obligation
Year	Lake Powell fill obligations to achieve equalisation with Lake Mead storage
Seasonal	Peak heating and cooling months
Daily to monthly	Flood control operations
Hourly	Western Area Power Administration's power generation
Spatial Scale	
Global	Climate influences, Grand Canyon National Park
Regional	Prior appropriation (e.g., Upper Colorado River Commission)
State	Different agreements on water marketing within and out of state water district
Municipal and Communities	Watering schedules, treatment, domestic use

Table 3.5. *Some adaptation options for water supply and demand (the list is not exhaustive).*

Supply-side	Demand-side
Prospecting and extraction of groundwater	Improvement of water-use efficiency by recycling water
Increasing storage capacity by building reservoirs and dams	Reduction in water demand for irrigation by changing the cropping calendar, crop mix, irrigation method, and area planted
Desalination of sea water	Reduction in water demand for irrigation by importing agricultural products, i.e., virtual water
Expansion of rain-water storage	Promotion of indigenous practices for sustainable water use
Removal of invasive non-native vegetation from riparian areas	Expanded use of water markets to reallocate water to highly valued uses
Water transfer	Expanded use of economic incentives including metering and pricing to encourage water conservation

however, supply-side options, involving increases in storage capacity or abstraction from water courses, tend to have adverse environmental consequences (which can in many cases be alleviated). Conversely, the practical effectiveness of some demand-side measures is uncertain, because they often depend on the cumulative actions of individuals. There is also a link between measures to adapt water resources and policies to reduce energy use. Some adaptation options, such as desalination or measures which involve pumping large volumes of water, use large amounts of energy and may be inconsistent with mitigation policy. Decreasing water demand in a country by importing virtual water (Allan, 1998; Oki et al., 2003b), in particular in the form of agricultural products, may be an adaptation option only under certain economic and social conditions (e.g., financial means to pay for imports, alternative income possibilities for farmers).

These do not exhaust the range of possibilities. Information, including basic geophysical, hydrometeorological, and environmental data as well as information about social, cultural and economic values and ecosystem needs, is also critically important for effective adaptation. Programmes to collect these data, and use them for effective monitoring and early warning systems, would constitute an important first step for adaptation.

In the western USA, water-market transactions and other negotiated transfers of water from agricultural to urban or environmental uses are increasingly being used to accommodate long-term changes in demand (e.g., due to population growth) as well as short-term needs arising from drought emergencies (Miller, 2000; Loomis et al., 2003; Brookshire et al., 2004; Colby et al., 2004). Water markets have also developed in Chile (Bauer, 2004), Australia (Bjornlund, 2004), and parts of Canada (Horbulyk, 2006), and some types of informal and often unregulated water marketing occur in the Middle East, southern Asia and North Africa (Faruqui et al., 2001). Countries and sub-national jurisdictions differ considerably in the extent to which their laws, administrative procedures, and documentation of water rights facilitate market-based water transfers, while protecting other water users and environmental values (Miller, 2000; Faruqui et al., 2001; Bauer, 2004; Matthews, 2004; Howe, 2005). Where feasible, short-term transfers can provide flexibility and increased security for highly valued water uses such as urban supply, and in some circumstances may prove more beneficial than constructing additional storage reservoirs (Goodman, 2000).

Some major urban water utilities are already incorporating various water-market arrangements in their strategic planning for coping with potential effects of climate change. This is true for the Metropolitan Water District of Southern California (Metropolitan), which supplies wholesale water to urban water utilities in Los Angeles, Orange, San Diego, Riverside, San Bernardino, and Ventura counties. Metropolitan recently concluded a 35-year option contract with Palo Verde Irrigation District. Under the arrangement, the district's landowners have agreed not to irrigate up to 29% of the valley's farm land at Metropolitan's request, thereby creating a water supply of up to 137 Mm^3 for Metropolitan. In exchange, landowners receive a one-time payment per hectare allocated, and additional annual payments for each hectare not irrigated under the programme in that year. The contract also provides funding for community improvement programmes (Miller and Yates, 2006).

Options to counteract an increasing risk of floods can be divided into two categories: either modify the floodwater, for example, via a water conveyance system; or modify the system's susceptibility to flood damage. In recent years, flood management policy in many countries has shifted from protection towards enhancing society's ability to live with floods (Kundzewicz and Takeuchi, 1999). This may include implementing protection measures, but as part of a package including measures such as enhanced flood forecasting and warning, regulations, zoning, insurance, and relocation. Each measure has advantages and disadvantages, and the choice is site-specific: there is no single one-fits-all measure (Kundzewicz et al., 2002).

3.6.3 Adaptation options in practice

Since the TAR, a number of studies have explicitly examined adaptation in real water management systems. Some have sought to identify the need for adaptation in specific catchments or water-management systems, without explicitly considering what adaptation options would be feasible. For example, changes to flow regimes in California would "fundamentally alter California's water rights system" (Hayhoe et al., 2004), the changing seasonal distribution of flows across much of the USA would mean that "additional investment may be required" (Hurd et al., 2004), changing streamflow regimes would "pose significant challenges" to the managers of the Columbia River (Mote et al., 2003), and an increased frequency of flooding in southern Quebec would mean that "important management decisions will have to be taken" (Roy et al., 2001).

A number of studies have explored the physical feasibility and effectiveness of specific adaptation options in specific circumstances. For example, improved seasonal forecasting was shown to offset the effects of climate change on hydropower generation from Folsom Lake, California (Yao and Georgakakos, 2001). In contrast, none of the adaptation options explored in the Columbia River basin in the USA continued to meet all current demands (Payne et al., 2004), and the balance between maintaining power production and maintaining instream flows for fish would have to be renegotiated. Similarly, a study of the Sacramento-San Joaquin basin, California, concluded that "maintaining status quo system performance in the future would not be possible", without changes in demands or expectations (VanRheenen et al., 2004). A review of the implications of climate change for water management in California as a whole (Tanaka et al., 2006) concluded that California's water supply system appears physically capable of adapting to significant changes in climate and population, but that adaptation would be costly, entail significant transfers of water among users, and require some adoption of new technologies. The feasibility of specific adaptation options varies with context: a study of water pricing in the Okanagan catchment in Canada, for example, showed differences in likely success between residential and agricultural areas (Shepherd et al., 2006).

Comprehensive studies into the feasibility of different adaptation options have been conducted in the Netherlands and the Rhine basin (Tol et al., 2003; Middelkoop et al., 2004). It was found that the ability to protect physically against flooding depends on geographical context (Tol et al., 2003). In some cases it is technically feasible to construct flood embankments; in others, high embankments already exist or geotechnical conditions make physical protection difficult. Radical flood management measures, such as the creation of a new flood overflow route for the River Rhine, able to reduce the physical flood risk to the Rhine delta in the Netherlands, would be extremely difficult politically to implement (Tol et al., 2003).

3.6.4 Limits to adaptation and adaptive capacity

Adaptation in the water sector involves measures to alter hydrological characteristics to suit human demands, and measures to alter demands to fit conditions of water availability. It is possible to identify four different types of limits on adaptation to changes in water quantity and quality (Arnell and Delaney, 2006).

- The first is a physical limit: it may not be possible to prevent adverse effects through technical or institutional procedures. For example, it may be impossible to reduce demands for water further without seriously threatening health or livelihoods, it may physically be very difficult to react to the water quality problems associated with higher water temperatures, and in the extreme case it will be impossible to adapt where rivers dry up completely.
- Second, whilst it may be physically feasible to adapt, there may be economic constraints to what is affordable.
- Third, there may be political or social limits to the implementation of adaptation measures. In many countries, for example, it is difficult for water supply agencies to construct new reservoirs, and it may be politically very difficult to adapt to reduced reliability of supplies by reducing standards of service.
- Finally, the capacity of water management agencies and the water management system as a whole may act as a limit on which adaptation measures (if any) can be implemented. The low priority given to water management, lack of coordination between agencies, tensions between national, regional and local scales, ineffective water governance and uncertainty over future climate change impacts constrain the ability of organisations to adapt to changes in water supply and flood risk (Ivey et al., 2004; Naess et al., 2005; Crabbe and Robin, 2006).

These factors together influence the adaptive capacity of water-management systems as well as other determinants such as sensitivities to change, internal characteristics of the system (e.g., education and access to knowledge) and external conditions such as the role of regulation or the market.

3.6.5 Uncertainty and risk: decision-making under uncertainty

Climate change poses a major conceptual challenge to water managers, in addition to the challenges caused by population and land-use change. It is no longer appropriate to assume that past hydrological conditions will continue into the future (the traditional assumption) and, due to climate change uncertainty, managers can no longer have confidence in single projections of the future. It will also be difficult to detect a clear climate-change effect within the next couple of decades, even with an underlying trend (Wilby, 2006). This sub-section covers three issues: developments in the conceptual understanding of sources of uncertainty and how to characterise them; examples of how water managers, in practice, are making climate change decisions under uncertainty; and an assessment of different ways of managing resources under uncertainty.

The vast majority of published water resources impact assessments have used just a small number of scenarios. These have demonstrated that impacts vary among scenarios, although temperature-based impacts, such as changes in the timing and volume of ice-melt-related streamflows, tend to be more robust (Maurer and Duffy, 2005), and the use of a scenario-based approach to water management in the face of climate change is therefore widely recommended (Beuhler, 2003; Simonovic and Li, 2003). There are, however, two problems. First, the large range for different climate-model-based scenarios suggests that adaptive planning should not be based on only a few scenarios (Prudhomme et al., 2003; Nawaz and Adeloye, 2006): there is no guarantee that the range simulated represents the full range. Second, it is difficult to evaluate the credibility of individual scenarios. By making assumptions about the probability distributions of the different drivers of climate change, however, it is possible to construct probability distributions of hydrological outcomes (e.g., Wilby and Harris, 2006), although the resulting probability distributions will be influenced by the assumed initial probability distributions. Jones and Page (2001) constructed probability distributions for water storage, environmental flows and irrigation allocations in the Macquarie River catchment, Australia, showing that the estimated distributions were, in fact, little affected by assumptions about probability distributions of drivers of change.

Water managers in a few countries, including the Netherlands, Australia, the UK, and the USA, have begun to consider the implications of climate change explicitly in flood and water supply management. In the UK, for example, design flood magnitudes can be increased by 20% to reflect the possible effects of climate change (Richardson, 2002). The figure of 20% was based on early impact assessments, and methods are under review following the publication of new scenarios (Hawkes et al., 2003). Measures to cope with the increase of the design discharge for the Rhine in the Netherlands from 15,000 to 16,000 m^3/s must be implemented by 2015, and it is planned to increase the design discharge to 18,000 m^3/s in the longer term, due to climate change (Klijn et al., 2001). Water supply companies in England and Wales used four climate scenarios in their 2004 review of future resource requirements, using a formalised procedure developed by the environmental and economic regulators (Arnell and Delaney, 2006). This procedure basically involved the companies estimating when climate change might impact upon the reliability of supply and, depending on the implementation of different actions, when these impacts would be felt (in most cases estimated effects were too far into the future to cause any changes in practice now, but in some instances the impacts would be soon enough to necessitate undertaking more detailed investigations now).

Dessai et al. (2005) describe an example where water supply managers in Australia were given information on the likelihood of drought conditions continuing, under different assumptions about the magnitude of climate change. They used this information to decide whether to invoke contingency plans to add temporary supplies or to tighten restrictions on water use.

A rather different way of coping with the uncertainty associated with estimates of future climate change is to adopt management measures that are robust to uncertainty (Stakhiv, 1998). Integrated Water Resources Management, for example, is based around the concepts of flexibility and adaptability, using measures which can be easily altered or are robust to changing conditions. These tools, including water conservation, reclamation, conjunctive use of surface and groundwater, and desalination of brackish water, have been advocated as a means of reacting to climate change threats to water supply in California (e.g., Beuhler, 2003). Similarly, resilient strategies for flood management, such as allowing rivers to temporarily flood and reducing exposure to flood damage, are preferable to traditional 'resistance' (protection) strategies in the face of uncertainty (Klijn et al., 2004; Olsen, 2006).

3.7 Conclusions: implications for sustainable development

Most of the seven Millennium Development Goals (MDGs) are related directly or indirectly to water management and climate change, although climate change is not directly addressed in the MDGs. Some major concerns are presented in Table 3.6 (UNDP, 2006).

In many regions of the globe, climate change impacts on freshwater resources may affect sustainable development and put at risk, for example, the reduction of poverty and child mortality. Even with optimal water management, it is very likely that negative impacts on sustainable development cannot be avoided. Figure 3.8 shows some key cases around the world where freshwater-related climate change impacts are a threat to the sustainable development of the affected regions.

'Sustainable' water resources management is generally sought to be achieved by Integrated Water Resources Management. However, the precise interpretation of this term varies considerably. All definitions broadly include the concept of maintaining and enhancing the environment, and in particular the water environment, taking into account competing users, instream ecosystems, and wetlands. Also, wider environmental implications of water management policies, such as implications for land management, or the implications of land management policies for the water environment, are considered. Water and land governance are important components of managing water in order to achieve sustainable water resources for a range of political, socio-economic and administrative systems (GWP, 2002; Eakin and Lemos, 2006).

Energy, equity, health, and water governance are key issues when linking climate change and sustainable development. However, few studies on sustainability have explicitly incorporated the issue of climate change (Kashyap, 2004). Some studies have taken into account the carbon footprint attributable to the water sector. For example, desalination can be regarded as a sustainable water management measure if solar energy is used. Many water management actions and adaptations, particularly those involving pumping or treating water, are very energy-

Table 3.6. *Potential contribution of the water sector to attain the MDGs.*

Goals	Direct relation to water	Indirect relation to water
Goal 1: Eradicate extreme poverty and hunger	Water as a factor in many production activities (e.g., agriculture, animal husbandry, cottage industry) Sustainable production of fish, tree crops and other food brought together in common property resources	Reduced ecosystem degradation improves local-level sustainable development Reduced urban hunger by means of cheaper food from more reliable water supplies
Goal 2: Achieve universal education		Improved school attendance through improved health and reduced water-carrying burdens, especially for girls
Goal 3: Promote gender equity and empower women	Development of gender sensitive water management programmes	Reduce time wasted and health burdens from improved water service leading to more time for income earning and more balanced gender roles
Goal 4: Reduce child mortality	Improved access to drinking water of more adequate quantity and better quality, and improved sanitation reduce the main factors of morbidity and mortality of young children	
Goal 6: Combat HIV/AIDS, malaria and other diseases	Improved access to water and sanitation support HIV/AIDS-affected households and may improve the impact of health care programmes Better water management reduces mosquito habitats and the risk of malaria transmission	
Goal 7: Ensure environmental sustainability	Improved water management reduces water consumption and recycles nutrients and organics Actions to ensure access to improved and, possibly, productive eco-sanitation for poor households Actions to improve water supply and sanitation services for poor communities Actions to reduce wastewater discharge and improve environmental health in slum areas	Develop operation, maintenance, and cost recovery system to ensure sustainability of service delivery

intensive. Their implementation would affect energy-related greenhouse gas emissions, and energy policy could affect their implementation (Mata and Budhooram, 2007). Examples of potential inequities occur where people benefit differently from an adaptation option (such as publicly funded flood protection) or where people are displaced or otherwise adversely impacted in order to implement an adaptation option (e.g., building a new reservoir).

Mitigation measures that reduce greenhouse gas emissions lessen the impacts of climate change on water resources. The number of people exposed to floods or water shortage and potentially affected is scenario-dependent. For example, stabilisation at 550 ppm (resulting in a temperature increase relative to pre-industrial levels of nearly 2°C) only reduces the number of people adversely affected by climate change by 30-50% (Arnell, 2006).

3.8 Key uncertainties and research priorities

There are major uncertainties in quantitative projections of changes in hydrological characteristics for a drainage basin. Precipitation, a principal input signal to water systems, is not reliably simulated in present climate models. However, it is well established that precipitation variability increases due to climate change, and projections of future temperatures, which affect snowmelt, are more consistent, such that useful conclusions are possible for snow-dominated basins.

Uncertainty has two implications. First, adaptation procedures need to be developed which do not rely on precise projections of changes in river discharge, groundwater, etc. Second, based on the studies completed so far, it is difficult to assess in a reliable way the water-related consequences of climate policies and emission pathways. Research on methods of adaptation in the face of these uncertainties is needed. Whereas it is difficult to make concrete projections, it is known that hydrological characteristics will change in the future. Water managers in some countries are already considering explicitly how to incorporate the potential effects of climate change into policies and specific designs.

Research into the water–climate interface is required:

- to improve understanding and estimation, in quantitative terms, of climate change impacts on freshwater resources and their management,
- to fulfil the pragmatic information needs of water managers who are responsible for adaptation.

Among the research issues related to the climate–water interface, developments are needed in the following.

- It is necessary to improve the understanding of sources of uncertainty in order to improve the credibility of projections.
- There is a scale mismatch between the large-scale climatic models and the catchment scale, which needs further resolution. Water is managed at the catchment scale and adaptation is local, while global climate models work on large spatial grids. Increasing the resolution of adequately validated regional climate models and statistical downscaling

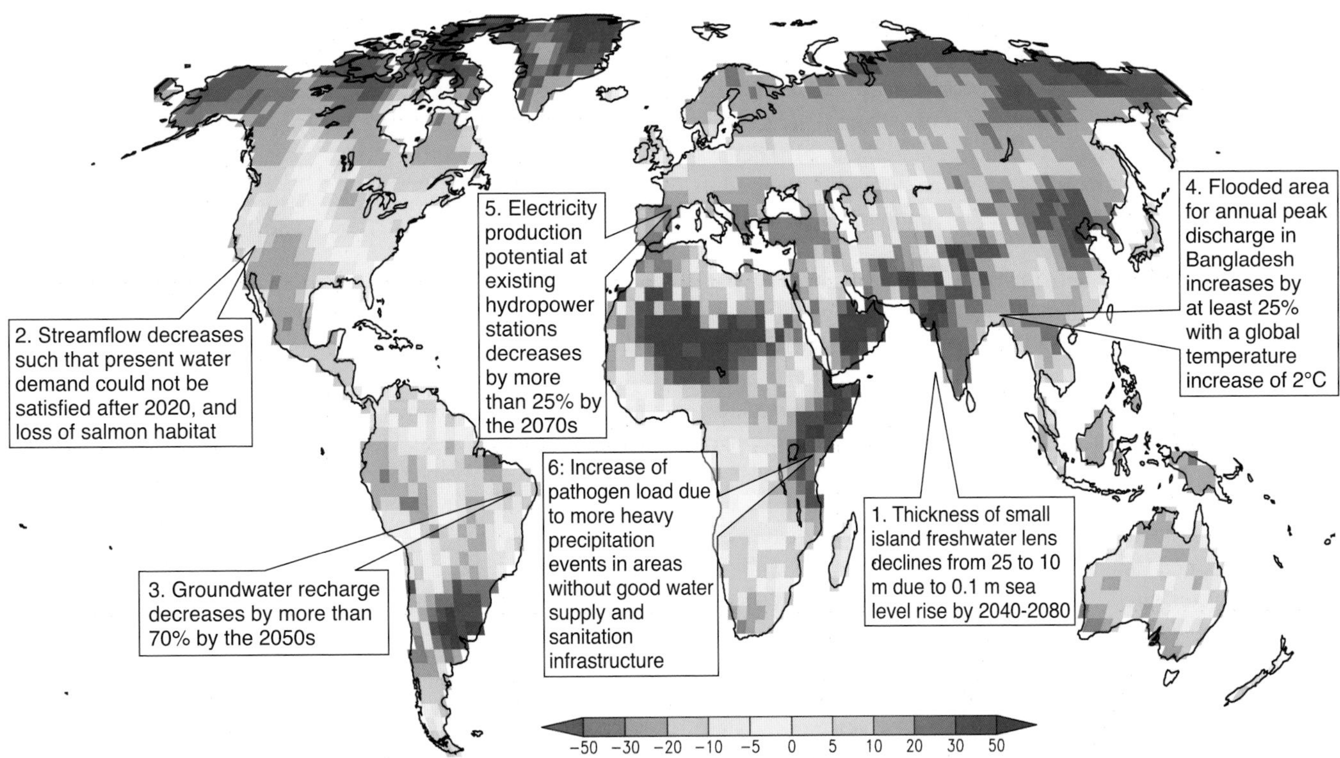

Figure 3.8. *Illustrative map of future climate change impacts on freshwater which are a threat to the sustainable development of the affected regions. 1: Bobba et al. (2000), 2: Barnett et al. (2004), 3: Döll and Flörke (2005), 4: Mirza et al. (2003) 5: Lehner et al. (2005a) 6: Kistemann et al. (2002). Background map: Ensemble mean change of annual runoff, in percent, between present (1981 to 2000) and 2081 to 2100 for the SRES A1B emissions scenario (after Nohara et al., 2006).*

can produce information of more relevance to water management.

- Impacts of changes in climate variability need to be integrated into impact modelling efforts.
- Improvements in coupling climate models with the land-use change, including vegetation change and anthropogenic activity such as irrigation, are necessary.
- Climate change impacts on water quality are poorly understood. There is a strong need for enhancing research in this area, with particular reference to the impacts of extreme events, and covering the needs of both developed and developing countries.
- Relatively few results are available on the economic aspects of climate change impacts and adaptation options related to water resources, which are of great practical importance.
- Research into human-dimension indicators of climate change impacts on freshwater is in its infancy and vigorous expansion is necessary.
- Impacts of climate change on aquatic ecosystems (not only temperatures, but also altered flow regimes, water levels, and ice cover) are not adequately understood.
- Detection and attribution of observed changes in freshwater resources, with particular reference to characteristics of extremes, is a challenging research priority, and methods for attribution of causes of changes in water systems need refinement.
- There are challenges and opportunities posed by the advent of probabilistic climate change scenarios for water resources management.
- Despite its significance, groundwater has received little attention from climate change impact assessments, compared to surface water resources.
- Water resources management clearly impacts on many other policy areas (e.g., energy projections, nature conservation). Hence there is an opportunity to align adaptation measures across different sectors (Holman et al., 2005a, b). There is also a need to identify what additional tools are required to facilitate the appraisal of adaptation options across multiple water-dependent sectors.

Progress in research depends on improvements in data availability, calling for enhancement of monitoring endeavours worldwide, addressing the challenges posed by projected climate change to freshwater resources, and reversing the tendency of shrinking observation networks. Broadening access to available observation data is a prerequisite to improving understanding of the ongoing changes. Relatively short hydrometric records can underplay the full extent of natural variability and confound detection studies, while long-term river flow reconstruction can place recent trends and extremes in a broader context. Data on water use, water quality, and sediment transport are even less readily available.

References

Abufayed, A.A., M.K.A. Elghuel and M. Rashed, 2003: Desalination: available supplemental source of water for the arid states of North Africa. *Desalination*, **152**, 75-81.

Al-Sefry, S.A., Z.Z. Şen, S.A. Al-Ghamdi, W.A. Al-Ashi and W.A. Al-Bardi, 2004: Strategic ground water storage of Wadi Fatimah, Makkah region. Technical Report SGS-TR-2003-2. Saudi Geological Survey, Jeddah, 168 pp.

Alcamo, J. and T. Henrichs, 2002: Critical regions: a model-based estimation of world water resources sensitive to global changes. *Aquat. Sci.*, **64**, 1-11.

Alcamo, J., T. Henrichs and T. Rösch, 2000: World water in 2025: global modeling and scenario analysis for the 21st century. Report A0002. Centre for Environmental Systems Research, University of Kassel, Kassel, 49 pp.

Alcamo, J., P. Döll, T. Henrichs, F. Kaspar, B. Lehner, T. Rösch and S. Siebert, 2003a: Development and testing of the WaterGAP 2 global model of water use and availability. *Hydrol. Sci. J.*, **48**, 317-338.

Alcamo, J., P. Döll, T. Henrichs, F. Kaspar, B. Lehner, T. Rösch and S. Siebert, 2003b: Global estimates of water withdrawals and availability under current and future "business-as-usual" conditions. *Hydrol. Sci. J.*, **48**, 339-348.

Alcamo, J., M. Flörke and M. Märker, 2007: Future long-term changes in global water resources driven by socio-economic and climatic change. *Hydrol. Sci. J.*, **52**, 247-275.

Allan, J.A., 1998: Virtual water: an essential element in stabilizing the political economies of the Middle East. *Transformation of Middle Eastern Natural Environments*, J. Albert, M. Bernhardsson and R. Kenna, Eds., Forestry and Environmental Studies Bulletin No. 103, Yale University, New Haven, Connecticut, 141-149.

Allen, D.M., D.C. Mackie and M. Wei, 2003: Groundwater and climate change: a sensitivity analysis for the Grand Forks aquifer, southern British Columbia, Canada. *Hydrogeol. J.*, **12**, 270-290.

Allen, M.R. and W.J. Ingram, 2002: Constraints on future changes in climate and the hydrologic cycle. *Nature*, **419**, 224-232.

Alley, W.M., 2001: Ground water and climate. *Ground Water*, **39**, 161.

Andreasson, J., S. Bergstörm, B. Carlsson, L.P. Graham and G. Lindström, 2004: Hydrological change: climate change impact simulations for Sweden. *Ambio*, **33**, 228-234.

Arnell, N.W., 2003a: Effects of IPCC SRES emissions scenarios on river runoff: a global perspective. *Hydrol. Earth Syst. Sc.*, **7**, 619-641.

Arnell, N.W., 2003b: Relative effects of multi-decadal climatic variability and changes in the mean and variability of climate due to global warming: future streamflows in Britain. *J. Hydrol.*, **270**, 195-213.

Arnell, N.W., 2004a: Climate-change impacts on river flows in Britain: the UKCIP02 scenarios. *Water Environ. J.*, **18**, 112-117.

Arnell, N.W., 2004b: Climate change and global water resources: SRES scenarios and socio-economic scenarios. *Global Environ. Change*, **14**, 31-52.

Arnell, N.W., 2005: Implications of climate change for freshwater inflows to the Arctic Ocean. *J. Geophys. Res. – Atmos.*, **110**, D07105, doi:10.1029/2004JD005348.

Arnell, N.W., 2006: Climate change and water resources: a global perspective. *Avoiding Dangerous Climate Change*, H.J. Schellnhuber, W. Cramer, N. Nakicenovic, T. Wigley and G. Yohe, Eds., Cambridge University Press, Cambridge, 168-175.

Arnell, N.W. and E.K. Delaney, 2006: Adapting to climate change: public water supply in England and Wales. *Climatic Change*, **78**, 227-255.

Arnell, N.W. C. Liu, R. Compagnucci, L. da Cunha, K. Hanaki, C. Howe, G. Mailu, I Shiklomanov and E. Stakhiv, 2001: Hydrology and water resources. *Climate Change 2001: Impacts, Adaptation and Vulnerability. Contribution of Working Group II to the Third Assessment Report of the Intergovernmental Panel on Climate Change*, J.J. McCarthy, O.F. Canziani, N.A. Leary, D.J. Dokken and K.S. White, Eds., Cambridge University Press, Cambridge, 191-234.

Arnell, N.W., D.A. Hudson and R.G. Jones, 2003: Climate change scenarios from a regional climate model: estimating change in runoff in southern Africa. *J. Geophys. Res. – Atmos.*, **108**(D16), 4519.

Atkinson, J.F., J.V. DePinto and D. Lam, 1999: Water quality. *Potential Climate Change Effects on the Great Lakes Hydrodynamics and Water Quality*, D. Lam and W. Schertzer, Eds., American Society of Civil Engineers, Reston, Virginia.

AWWA [American Water Works Association], 2006: *Optimizing Filtration Operations*. CD-ROM catalogue no. 64275 [available at http://www.awwa.org/bookstore].

Barlage, M.J., P.L. Richards, P.J. Sousounis and A.J. Brenner, 2002: Impacts of climate change and land use change on runoff from a Great Lakes watershed. *J. Great Lakes Res.*, **28**, 568-582.

Barnett, T.P., R. Malone, W. Pennell, D. Stammer, B. Semtner and W. Washington, 2004: The effects of climate change on water resources in the West: introduction and overview. *Climatic Change*, **62**, 1-11.

Barnett, T.P., J.C. Adam and D.P. Lettenmaier, 2005: Potential impacts of a warming climate on water availability in snow-dominated regions. *Nature*, **438**, 303-309.

Bauer, C., 2004: Results of Chilean water markets: empirical research since 1990. *Water Resour. Res.*, **40**, W09S06.

Beach, D., 2002: *Coastal Sprawl: The Effects of Urban Design on Aquatic Ecosystems of the United States*. Pew Oceans Commission, Arlington, Virginia, 40 pp.

Beare, S. and A. Heaney, 2002: Climate change and water resources in the Murray Darling Basin, Australia; impacts and adaptation. ABARE Conference Paper 02.11, 33 pp., Canberra.

Bedritsky, A.I., R.Z. Khamitov, I.A. Shiklomanov and I.S. Zektser, 2007: *Water Resources of Russia and their Use in New Socio-economic Conditions with the Account of Possible Climate Change*. Proceedings of the VIth All-Russia Hydrological Congress. Plenary Reports, St. Petersburg. (in press) (in Russian).

Bell, R. and A. Heaney, 2001: A basin scale model for assessing salinity management options: model documentation, ABARET Technical Working Paper 2000.1, Canberra, 7 pp.

Beltaos, S., T. Prowse, B. Bonsal, R. MacKay, L. Romolo, A. Pietroniro and B. Toth, 2006: Climatic effects on ice-jam flooding of the Peace-Athabasca Delta. *Hydrol. Process.*, **20**, 4031-4050.

Bergström, S., B. Carlsson, M. Gardelin, G. Lindstrom, A. Pettersson and M. Rummukainen, 2001: Climate change impacts on runoff in Sweden: assessments by global climate models, dynamical downscaling and hydrological modelling. *Climate Res.*, **16**, 101-112.

Betts, R.A., P.M. Cox, S.E. Lee and F.I. Woodward, 1997: Contrasting physiological and structural vegetation feedbacks in climate change simulations. *Nature*, **387**, 796-799.

Betts, R.A., O. Boucher, M. Collins, P.M. Cox, P.D. Falloon, N. Gedney, D.L. Hemming, C. Huntingford, C.D. Jones, D.M.H. Sexton and M.J. Webb, 2007: Increase of projected 21st-century river runoff by plant responses to carbon dioxide rise. *Nature*, doi: 10.1038/nature06045.

Beuhler, M., 2003: Potential impacts of global warming on water resources in southern California. *Water Sci. Technol.*, **47**(7-8), 165-168.

Bjornlund, H., 2004: Formal and informal water markets: drivers of sustainable rural communities? *Water Resour. Res.*, **40**, W09S07.

Bobba, A., V. Singh, R. Berndtsson and L. Bengtsson, 2000: Numerical simulation of saltwater intrusion into Laccadive Island aquifers due to climate change. *J. Geol. Soc. India*, **55**, 589-612.

Booij, M.J., 2005: Impact of climate change on river flooding assessed with different spatial model resolutions. *J. Hydrol.*, **303**, 176-198.

Boorman, D.B., 2003a: Climate, hydrochemistry and economics of surface-water systems (CHESS): adding a European dimension to the catchment modelling experience developed under LOIS. *Sci. Total Environ.*, **314**, 411-437.

Boorman, D.B., 2003b: LOIS in-stream water quality modelling. Part 2. Results and scenarios. *Sci. Total Environ.*, **314-316**, 397-409.

Bouraoui, F., B. Grizzetti, K. Granlund, S. Rekolainen and G. Bidoglio, 2004: Impact of climate change on the water cycle and nutrient losses in a Finnish catchment. *Climatic Change*, **66**, 109-126.

Brookshire, D.S., B. Colby, M. Ewers and P.T. Ganderton, 2004: Market prices for water in the semiarid west of the United States. *Water Resour. Res.*, **40**, W09S04.

Brouyere, S., G. Carabin and A. Dassargues, 2004: Climate change impacts on groundwater resources: modelled deficits in a chalky aquifer, Geer basin, Belgium. *Hydrogeol. J.*, **12**, 123-134.

Bruinsma, J., 2003: *World Agriculture: Towards 2015/2030 – An FAO Perspective*. Earthscan, London, 444 pp.

Bueh, C., U. Cubasch and S. Hagemann, 2003: Impacts of global warming on changes in the east Asian monsoon and the related river discharge in a global time-slice experiment. *Climate Res.*, **24**, 47-57.

Burke, E.J., S.J. Brown and N. Christidis, 2006: Modelling the recent evolution of global drought and projections for the 21st century with the Hadley Centre climate model. *J. Hydrometeorol.*, **7**, 1113-1125.

Burkett, V.R., D.B. Zilkoski and D.A. Hart, 2002: Sea-level rise and subsidence: implications for flooding in New Orleans, Louisiana. *US Geological Survey Subsidence Interest Group Conference: Proceedings of the Technical Meeting, Galveston, Texas, 27-29 November 2001*, 63-71.

Burlando, P. and R. Rosso, 2002: Effects of transient climate change on basin hydrology. 2. Impacts on runoff variability in the Arno River, central Italy. *Hydrol. Process.*, **16**, 1177-1199.

Buttle, J., T. Muir and J. Frain, 2004: Economic impacts of climate change on the Canadian Great Lakes hydro-electric power producers: a supply analysis. *Can. Water Resour. J.*, **29**, 89-110.

Chang, H., 2003: Basin hydrologic response to changes in climate and land use: the Conestoga River basin, Pennsylvania. *Phys. Geogr.*, **24**, 222-247.

Chang, H., B. Evans and D. Easterling, 2001: The effects of climate change on streamflow and nutrient loading. *J. Am. Water Resour. As.*, **37**, 973-985.

Changnon, S.A., 2005: Economic impacts of climate conditions in the United States: past, present, and future – an editorial essay. *Climatic Change*, **68**, 1-9.

Chen, C., D. Gillig and B.A. McCarl, 2001: Effects of climatic change on a water-dependent regional economy: a study of the Texas Edwards aquifer. *Climatic Change*, **49**, 397-409.

Chen, Z., S. Grasby and K. Osadetz, 2004: Relation between climate variability and groundwater levels in the upper carbonate aquifer, southern Manitoba, Canada. *J. Hydrol.*, **290**, 43-62.

Chiew, F.H.S., 2007: Estimation of rainfall elasticity of streamflow in Australia. *Hydrol. Sci. J.*, **51**, 613-625.

Choi, O. and A. Fisher, 2003: The impacts of socioeconomic development and climate change on severe weather catastrophe losses: mid-Atlantic region MAR and the US. *Climatic Change*, **58**, 149-170.

Christensen, J.H. and O.B. Christensen, 2003: Severe summertime flooding in Europe. *Nature*, **421**, 805.

Christensen, N.S., A.W. Wood, N. Voisin, D.P. Lettenmaier and R.N. Palmer, 2004: The effects of climate change on the hydrology and water resources of the Colorado River basin. *Climatic Change*, **62**, 337-363.

Clarke, R. and J. King, 2004: *The Atlas of Water*. Earthscan, London, 128 pp.

Coe, M.T. and J.A. Foley, 2001: Human and natural impacts on the water resources of the Lake Chad basin. *J. Geophys. Res. – Atmos.*, **106**(D4), 3349-3356.

Cohen, S. and T. Neale, 2006: Participatory integrated assessment of water management and climate change in the Okanagan Basin, British Columbia. Final Report, Project A846. Natural Resources Canada, Ottawa, Environment Canada and University of British Columbia, Vancouver, 221 pp.

Cohen, S., D. Neilsen, S. Smith, T. Neale, B. Taylor, M. Barton, W. Merritt, Y. Alila, P. Shepherd, R. McNeill, J. Tansey, J. Carmichael and S. Langsdale, 2006: Learning with local help: expanding the dialogue on climate change and water management in the Okanagan Region, British Columbia, Canada. *Climatic Change*, **75**, 331-358.

Colby, B.G., K. Crandall and D.B. Bush, 2004: Water right transactions: market values and price dispersion. *Economics of Water Resources: Institutions, Instruments, and Policies for Managing Scarcity*, K.W. Easter and M.E. Renwick, Eds., Ashgate, Aldershot.

Conway, D., 2005: From headwater tributaries to international river: observing and adapting to climate variability and change in the Nile basin. *Global Environ. Change*, **15**, 99-114.

Cooley, H., P.H. Gleick and G. Wolff, 2006: *Desalination: with a Grain of Salt*. Pacific Institute, Oakland, California, 100 pp.

Coudrain, A., B. Francou and Z.W. Kundzewicz, 2005: Glacier shrinkage in the Andes and consequences for water resources. *Hydrol. Sci. J.*, **50**, 925-932.

Covey, C., K.M. Achuta Rao, U. Cubasch, P. Jones, S.J. Lambert, M.E. Mann, T.J. Phillips and K.E. Taylor, 2003: An overview of results from the coupled model intercomparison project. *Global Planet. Change*, **37**, 103-133.

Cox, P., I. Fisher, G. Kastl, V. Jegatheesan, M. Warnecke, M. Angles, H. Bustamante, T. Chiffings and P.R. Hawkins, 2003: Sydney 1998 – lessons from a drinking water crisis. *J. Am. Water Works Assoc*, **95**, 147-161.

Crabbe, P. and M. Robin, 2006: Institutional adaptation of water resource infrastructures to climate change in eastern Ontario. *Climatic Change*, **78**, 103-133.

Cunderlik, J.M. and S.P. Simonovic, 2005: Hydrological extremes in a southwestern Ontario river basin under future climate conditions. *Hydrol. Sci. J.*, **50**, 631-654.

Curriero, F., J. Patz, J. Rose and S. Lele, 2001: The association between extreme precipitation and waterborne disease outbreaks in the United States, 1948-1994. *Am. J. Public Health*, **91**, 1194-1199.

Daughton, C.G., 2004: Non-regulated water contaminants: emerging research. *Environ. Impact Asses.*, **24**, 711-732.

De Araujo, J.C., A. Güntner and A. Bronstert, 2006: Loss of reservoir volume by sediment deposition and its impact on water availability in semiarid Brazil. *Hydrol. Sci. J.*, **51**, 157-170.

Dennis, I.A., M.G. Macklin, T.J. Coulthard and P.A. Brewer, 2003: The impact of the October–November 2000 floods on contaminant metal dispersal on the River Swale catchment, North Yorkshire. *Hydrol. Process.*, **17**, 1641-1657.

Dessai, S., X. Lu and J.S. Risbey, 2005: On the role of climate scenarios for adaptation planning. *Global Environ. Change*, **15**, 87-97.

Dettinger, M.D., 2005: From climate change spaghetti to climate-change distributions for 21st century California. *San Francisco Estuary and Watershed Science*, **3**, 4.

Dettinger, M.D., D.R. Cayan, M.K. Meyer and A.E. Jeton, 2004: Simulated hydrologic responses to climate variations and change in the Merced, Carson, and American River basins, Sierra Nevada, California, 1900–2099. *Climatic Change*, **62**, 283-317.

DFID [Department for International Development], 2004: *Key Sheet Series on the Impact of Climate Change on Poverty, Focusing on Vulnerability, Health and Pro-poor Growth*. No. 01, 6 pp.

Diaz-Nieto, J. and R. Wilby, 2005: A comparison of statistical downscaling and climate change factor methods: impact on low flows in the River Thames, United Kingdom. *Climatic Change*, **69**, 245-268.

Dibike, Y.B. and P. Coulibaly, 2005: Hydrologic impact of climate change in the Saguenay watershed: comparison of downscaling methods and hydrologic models. *J. Hydrol.*, **307**, 145-163.

Döll, P., 2002: Impact of climate change and variability on irrigation requirements: a global perspective. *Climatic Change*, **54**, 269-293.

Döll, P. and M. Flörke, 2005: Global-scale estimation of diffuse groundwater recharge. Frankfurt Hydrology Paper 03. Institute of Physical Geography, Frankfurt University, 26 pp.

Döll, P., M. Flörke, M. Märker and S. Vassolo, 2003: Einfluss des Klimawandels auf Wasserressourcen und Bewässerungswasserbedarf: eine globale Analyse unter Berücksichtigung neuer Klimaszenarien (Impact of climate change on water resources and irrigation water requirements: a global analysis using new climate change scenarios). *Klima - Wasser - Flussgebietsmanagement: im Lichte der Flut*, H.-B. Kleeberg, Ed., Proceedings of Tag der Hydrologie 2003, Freiburg, Germany, Forum für Hydrologie und Wasserbewirtschaftung, 04.03, 11-14.

Douville, H., F. Chauvin, S. Planton, J.F. Royer, D. Salas-Melia and S. Tyteca, 2002: Sensitivity of the hydrological cycle to increasing amounts of greenhouse gases and aerosols. *Clim. Dynam.*, **20**, 45-68.

Downing, T.E., R.E. Butterfield, B. Edmonds, J.W. Knox, S. Moss, B.S. Piper, E.K. Weatherhead and the CCDeW project team, 2003: Climate change and the demand for water. Research Report, Stockholm Environment Institute, Oxford Office, Oxford.

Drogue, G., L. Pfister, T. Leviander, A. El Idrissi, J.-F. Iffly, P. Matgen, J. Humbert and L. Hoffmann, 2004: Simulating the spatio-temporal variability of streamflow response to climate change scenarios in a mesoscale basin. *J. Hydrol.*, **293**(1-4), 255-269.

D'Souza, R., N. Becker, G. Hall and K. Moodie, 2004: Does ambient temperature affect foodborne disease? *Epidemiology*, **15**, 86-92.

Eakin, H. and M.C. Lemos, 2006: Adaptation and the state: Latin America and the challenge of capacity-building under globalization. *Global Environ. Change*, **16**, 7-18.

Easter, K.W. and M.E. Renwick, Eds., 2004: *Economics of Water Resources: Institutions, Instruments and Policies for Managing Scarcity*. Ashgate, Hampshire, 548 pp.

Easter, K.W., M.W. Rosengrant and A. Dinar, 1998: *Markets for Water: Potential and Performance*. Kluwer Academic, Boston, Massachusetts, 352 pp.

Eckhardt, K. and U. Ulbrich, 2003: Potential impacts of climate change on groundwater recharge and streamflow in a central European low mountain range. *J. Hydrol.*, **284**, 244-252.

Eheart, J.W. and D.W. Tornil, 1999: Low-flow frequency exacerbation by irrigation withdrawals in the agricultural Midwest under various climate change scenarios. *Water Resour. Res.*, **35**, 2237-2246.

Elguindi, N. and F. Giorgi, 2006: Projected changes in the Caspian Sea level for the 21st century based on the latest AOGCM simulations. *Geophys. Res. Lett.*, **33**, L08706.

Environment Canada, 2001: Threats to sources of drinking water and aquatic ecosystems health in Canada. National Water Research Report No.1. National Water Resources Research Institute, Burlington, Ontario, 72 pp.

Environment Canada, 2004: Threats to water availability in Canada. NWRI Scientific Assessment Report No. 3. Prowse and ASCD Science Assessments series No. 1. National Water Research Institute, Burlington, Ontario, 128 pp.

Essink, G., 2001: Improving fresh groundwater supply problems and solutions. *Ocean Coast. Manage.*, **44**, 429-449.

European Union, 2000: EU Water Framework Directive: Directive 2000/60/EC of the European Parliament and of the Council establishing a framework for the Community action in the field of water policy. EU Official Journal (OJ L 327, 22 December 2000).

Evans, E., R. Ashley, J. Hall, E. Penning-Rowsell, A. Saul, P. Sayers, C. Thorne and A. Watkinson, 2004: *Future Flooding: Scientific Summary. Volume 1: Future Risks and Their Drivers*. Foresight, Office of Science and Technology, London [accessed 06.03.07: http://www.foresight.gov.uk/Previous_Projects/Flood_and_Coastal_Defence /Reports_and_Publications/Volume1/Contents.htm]

Evans, J. and S. Schreider, 2002: Hydrological impacts of climate change on inflows to Perth, Australia. *Climatic Change*, **55**, 361-393.

FAO [Food and Agriculture Organization of the United Nations], 2003: *World Agriculture Towards 2015/2030* [accessed 06.03.07: http://www.fao.org/documents/show_cdr.asp?url_file=/docrep/004/y3557e/y 3557e00.htm]

Faruqui, N.I., A.K. Biswas and M.J. Bino, Eds., 2001: *Water Management in Islam*. United Nations University Press, Tokyo, 149 pp.

Fayer, R., J. Trout, E. Lewis, E. Xiao, A. Lal, M. Jenkins and T. Graczyk, 2002: Temporal variability of *Cryptosporidium* in the Chesapeake Bay. *Parasitol. Res.*, **88**, 998-1003.

Ferrier, R. and A. Edwards, 2002: Sustainability of Scottish water quality in the early 21st century. *Sci. Total Environ.*, **294**, 57-71.

Fisher, A., 2000: Preliminary findings from the mid-Atlantic regional assessment. *Climate Res.*, **14**, 261-269.

Forest, C., P. Stone, A. Sokolov, M. Allen and M. Webster, 2002: Quantifying uncertainties in climate system properties with the use of recent climate observations. *Science*, **295**, 113-117.

Fowler, H.J., C.G. Kilsby and J. Stunell, 2007: Modelling the impacts of projected future climate change on water resources in north-west England. *Hydrol. Earth Syst. Sc.*, **11**, 1115-1126.

Frei, A., R.L. Armstrong, M.P. Clark and M.C. Serreze, 2002: Catskill mountain water resources: vulnerability, hydroclimatology and climate-change sensitivity. *Ann. Assoc. Am. Geogr.*, **92**, 203-224.

Galaz, V., 2005: Social-ecological resilience and social conflict: Institutions and strategic adaptation in Swedish water management. *Ambio*, **34**, 567-572.

Gedney, N., P.M. Cox, R.A. Betts, O. Boucher, C. Huntingford and P.A. Stott, 2006: Detection of a direct carbon dioxide effect in continental river runoff records. *Nature*, **439**, 835-838.

GEO-LAC, 2003: *Global Environmental Outlook*. United Nations Environmental Program, 279 pp [accessed 06.03.07: http://www.unep.org/geo/pdfs/GEO__lac2003English.pdf]

Gerten, D., S. Schaphoff, U. Haberlandt, W. Lucht and S. Sitch, 2004: Terrestrial vegetation and water balance: hydrological evaluation of a dynamic global vegetation model. *J. Hydrol.*, **286**, 249-270.

Giansante, C., M. Aguilar, L. Babiano, A. Garrido, A. Gómez, E. Iglesias, W. Lise, L. Moral and B. Pedregal, 2002: Institutional adaptation to changing risk of water scarcity in the Lower Guadalquivir Basin. *Nat. Resour. J.*, **42**, 521-564.

Giertz, S., B. Diekkruger, A. Jaeger and M. Schopp, 2006: An interdisciplinary scenario analysis to assess the water availability and water consumption in the Upper Oum catchment in Benin. *Adv. Geosci.*, **9**, 1-11.

Gilvear, D., K. Heal and A. Stephen, 2002: Hydrology and the ecological quality of Scottish river ecosystems. *Sci. Total Environ.*, **294**, 131-159.

Giorgi, F., B.C. Hewitson, C. Christensen, R. Fu and R.G. Jones, Eds., 2001: Regional climate information: evaluation and projections. *Climate Change 2001: The Scientific Basis. Contribution of Working Group I to the Third*

Assessment Report of the Intergovernmental Panel on Climate Change, J.T. Houghton, Y. Ding, D.J. Griggs, M. Noguer, P.J. van der Linden, X. Dai, K. Maskell and C.A. Johnson, Eds., Cambridge University Press, Cambridge, 583-638.

Giorgi, F., X. Bi and J. Pal, 2004: Mean, interannual variability and trend in a regional climate change experiment over Europe. II. Climate change scenarios 2071–2100. *Clim. Dynam.*, **23**, 839-858.

Gleick, P.H., 2000: The removal of dams: a new dimension to an old debate. *The World's Water 2000–2001*, P.H. Gleick, Ed., Island Press, Washington, District of Columbia, 11 pp.

Gleick, P.H., 2003: Water use. *Annu. Rev. Env. Resour.*, **28**, 275-314.

Goodman, D.J., 2000: More reservoirs or transfers? A computable general equilibrium analysis of projected water shortages in the Arkansas River basin. *J. Agr. Resour. Econ.*, **25**, 698-713.

Gordon, W. and J.S. Famiglietti, 2004: Response of the water balance to climate change in the United States over the 20th and 21st centuries: results from the VEMAP Phase 2 model intercomparisons. *Global Biogeochem. Cy.*, **181**, GB1030.

Goss, K.F., 2003: Environmental flows, river salinity and biodiversity conservation: managing trade-offs in the Murray-Darling basin. *Aust. J. Bot.*, **51**, 619-625.

Graham, L.P., 2004: Climate change effects on river flow to the Baltic Sea. *Ambio*, **33**, 235-241.

Graham, L.P., S. Hagemann, S. Jaun and M. Beniston, 2007a: On interpreting hydrological change from regional climate models. *Climatic Change*, **81**(Suppl. 1), 97-122.

Graham, L.P., J. Andreasson and B. Carlsson, 2007b: Assessing climate change impacts on hydrology from an ensemble of regional climate models, model scales and linking methods: a case study on the Lule River basin. *Climatic Change*, **81**(Suppl. 1), 293-307.

GWP [Global Water Partnership], 2002: *Dialogue on Effective Water Governance*. GWP, 6 pp.

Haines, A., A. McMichael and P. Epstein, 2000: Environment and health. 2. Global climate change and health. *Can. Med. Assoc. J.*, **163**, 729-734.

Hall, G., R. D'Souza and M. Kirk, 2002: Foodborne disease in the new millennium: out of the frying pan and into the fire? *Med. J. Australia*, **177**, 614-618.

Hall, J.W., P.B. Sayers and R.J. Dawson, 2005: National-scale assessment of current and future flood risk in England and Wales. *Nat. Hazards*, **36**, 147-164.

Hamilton, S., N. Crookshank and D. Lam, 2001: Hydrological and hydraulic routing and decision support in the Seymour Watershed. Final Report for the Greater Vancouver Regional District Watershed Management Branch. Joint Report by the Environment Canada, National Research Council and the Canadian Hydraulics Centre, Canada, 96 pp.

Han, M., M. Zhao, D. Li and X. Cao, 1999: Relationship between ancient channel and seawater intrusion in the south coastal plain of the Laizhou Bay. *J. Nat. Disasters*, **8**, 73-80.

Hawkes, P., S. Surendran and D. Richardson, 2003: Use of UKCIP02 climate-change scenarios in flood and coastal defence. *J. Chart. Inst. Water E.*, **17**, 214-219.

Hayhoe, K., D. Cayan, C.B. Field, P.C. Frumhoff, E.P. Maurer, N.L. Miller, S.C. Moser, S.H. Schneider, K.N. Cahill, E.E. Cleland, L. Dale, R. Drapek, R.M. Hanemann, L.S. Kalkstein, J. Lenihan, C.K. Lunch, R.P. Neilson, S.C. Sheridan and J.H. Verville, 2004: Emissions pathways, climate change, and impacts on California. *P. Natl. Acad. Sci. USA*, **101**, 12422-12427.

Herron, N., R. Davis and R. Jones, 2002: The effects of large-scale afforestation and climate change on water allocation in the Macquarie River catchment, NSW, Australia. *J. Environ. Manage.*, **65**, 369-381.

Hijioka, Y., K. Takahashi, Y. Matsuoka and H. Harasawa, 2002: Impact of global warming on waterborne diseases. *J. Jpn. Soc. Water Environ.*, **25**, 647-652.

Holman, I.P., R.J. Nicholls, P.M. Berry, P.A. Harrison, E. Audsley, S. Shackley and M.D.A. Rounsevell, 2005a: A regional, multi-sectoral and integrated assessment of the impacts of climate and socio-economic change in the UK. Part 2. Results. *Climatic Change*, **71**, 43-73.

Holman, I.P., M.D.A. Rounsevell, S. Shackley, P.A. Harrison, R.J. Nicholls, P.M. Berry and E. Audsley, 2005b: A regional, multi-sectoral and integrated assessment of the impacts of climate and socio-economic change in the UK. Part 1. Methodology. *Climatic Change*, **71**, 9-41.

Horbulyk, T.M., 2006: Liquid gold: water markets in Canada. *Eau Canada: The Future of Canada's Water*, K.J. Bakker, Ed., UBC Press, Vancouver, 205-218.

Howard, C.D.D., 2000: *Operation, Monitoring and Decommissioning of Dams: Thematic Review IV.5 Prepared as an Input to the World Commission on Dams, Cape Town*. 127 pp. [Accessed 06.03.07: http://www.dams.org/docs/kbase/thematic/tr45main.pdf]

Howe, C.W., 2005: Property rights, water rights and the changing scene in western water. *Water Institutions: Policies, Performance and Prospects*, C. Gopalakrishnan, C. Tortajada and A.K. Biswas, Eds., Springer, Berlin, 175-185.

Hrasko, B. and R. McNeill, 2006: Costs of adaptation measures: participatory integrated assessment of water management and climate change in the Okanagan Basin, British Columbia. Final Report, Project A846, S. Cohen and T. Neale, Eds., Natural Resources Canada, Ottawa, Environment Canada and University of British Columbia, Vancouver, 43-55.

Huffaker, R., 2005: Finding a modern role for the prior appropriation doctrine in the American West. *Water Institutions: Policies, Performance and Prospects*, C. Gopalakrishnan, C. Tortajada and A.K. Biswas, Eds., Springer, Berlin, 187-200.

Hunter, P., 2003: Climate change and waterborne and vector borne disease. *J. Appl. Microbiol.*, **94**, 37-46.

Huntington, T.G., 2006: Evidence for intensification of the global water cycle: review and synthesis. *J. Hydrol.*, **319**, 83-95.

Hurd, B.H., M. Callaway, J. Smith and P. Kirshen, 2004: Climatic change and US water resources: from modeled watershed impacts to national estimates. *J. Am. Water Resour. As.*, **40**, 129-148.

IPCC, 1992: *Climate Change 1992: The Supplementary Report to the IPCC Scientific Assessment*. J.T. Houghton, B.A. Callander and S.K. Varney, Eds., Cambridge University Press, Cambridge, 200 pp.

IPCC, 2001: *Climate Change 2001: The Scientific Basis. Contribution of Working Group I to the Third Assessment Report of the Intergovernmental Panel on Climate Change,* J.T. Houghton, Y. Ding, D.J. Griggs, M. Noguer, P.J. van der Linden, X. Dai, K. Maskell and C.A. Johnson, Eds., Cambridge University Press, Cambridge, 881 pp.

IPCC, 2007: Summary for Policymakers. *Climate Change 2007: The Physical Science Basis. Contribution of Working Group I to the Fourth Assessment Report of the Intergovernmental Panel on Climate Change,* S. Solomon, D. Qin, M. Manning, Z. Chen, M. Marquis, K.B. Averyt, M. Tignor and H.L. Miller, Eds., Cambridge University Press, Cambridge, 1-18.

Ivey, J.L., J. Smithers, R.C. De Loe and R.D. Kreutzwiser, 2004: Community capacity for adaptation to climate-induced water shortages: linking institutional complexity and local actors. *Environ. Manage.*, **33**, 36-47.

Jasper, K., P. Calanca, D. Gyalistras and J. Fuhrer, 2004: Differential impacts of climate change on the hydrology of two alpine rivers. *Climate Res.*, **26**, 113-125.

Jenkins, G. and J. Lowe, 2003: Handling uncertainties in the UKCIP02 scenarios of climate change. Hadley Centre Technical Note 44, Meteorological Office, Exeter, 15 pp.

Jha, M., Z.T. Pan, E.S. Takle and R. Gu, 2004: Impacts of climate change on streamflow in the Upper Mississippi River Basin: a regional climate model perspective. *J. Geophys. Res. – Atmos.*, **109**(D9), D09105.

Jiménez, B., 2003: Health risks in aquifer recharge with recycle water. *State of the Art Report: Health Risk in Aquifer Recharge Using Reclaimed Water*, R. Aertgeerts and A. Angelakis, Eds., WHO Regional Office for Europe, 54-172.

Jiménez, B. and H. Garduño, 2001: Social, political and scientific dilemmas for massive wastewater reuse in the world. *Navigating Rough Waters: Ethical Issues in the Water Industry*, C. Davis and R. McGin, Eds., American Water Works Association (AWWA), London, 148.

Jones, R.G., M. Noguer, D.C. Hassell, D. Hudson, S.S. Wilson, G.J. Jenkins and J. Mitchell, 2004: *Generating High Resolution Climate Change Scenarios Using PRECIS*. Meteorological Office Hadley Centre, Exeter, 40 pp.

Jones, R.N. and C.M. Page, 2001: Assessing the risk of climate change on the water resources of the Macquarie River catchment. *Integrating Models for Natural Resources Management Across Disciplines: Issues and Scales*, F. Ghassemi, P.H. Whetton, R. Little and M. Littleboy, Eds., Modelling and Simulation Society of Australia and New Zealand, Canberra, 673-678.

Justic, D., N.N. Rabalais and R.E. Turner, 2005: Coupling between climate variability and coastal eutrophication: evidence and outlook for the northern Gulf of Mexico. *J. Sea Res.*, **54**, 25-35.

Kabat, P. and H. van Schaik, 2003: *Climate Changes the Water Rules: How Water*

Managers Can Cope With Today's Climate Variability and Tomorrow's Climate Change. *Dialogue on Water and Climate*. Printfine, Liverpool, 106 pp.

Kanae, S., T. Oki and K. Musiake, 2001: Impact of deforestation on regional precipitation over the Indochina Peninsula. *J. Hydrometeorol*., **2**, 51-70.

Karl, T. and K. Trenberth, 2003: Modern global change. *Science*, **302**, 1719-1722.

Kashyap, A., 2004: Water governance: learning by developing adaptive capacity to incorporate climate variability and change. *Water Sci. Technol*., **19**(7), 141-146.

Kaspar, F., 2003: Entwicklung und Unsicherheitsanalyse eines globalen hydrologischen Modells (Development and uncertainty analysis of a global hydrological model). PhD dissertation, University of Kassel, Germany, 139 pp.

Kay, A., V. Bell and H. Davies, 2006a: *Model Quality and Uncertainty for Climate Change Impact*. Centre for Ecology and Hydrology, Wallingford.

Kay, A., N.S. Reynard and R.N. Jones, 2006b: RCM rainfall for UK flood frequency estimation. II. Climate change results. *J. Hydrol*., **318**, 163-172.

Kennish, M., 2002: Environmental threats and environmental future of estuaries. *Environ. Conserv*., **29**, 78-107.

Kergoat, L., S. Lafont, H. Douville, B. Berthelot, G. Dedieu, S. Planton and J.-F. Royer, 2002: Impact of doubled CO_2 on global-scale leaf area index and evapotranspiration: conflicting stomatal conductance and LAI responses. *J. Geophys. Res. – Atmos*., **107**(D24), 4808.

Kharkina, M.A., 2004: Natural resources in towns. *Energia*, **2**, 44-50.

Khiyami, H.A., Z.Z. Şen, S.G. Al-Harthy, F.A. Al-Ammawi, A.B. Al-Balkhi, M.I. Al-Zahrani and H.M. Al-Hawsawy, 2005: Flood hazard evaluation in Wadi Hali and Wadi Yibah. Technical Report SGS-TR-2004-6, Saudi Geological Survey, Jeddah, 35 pp.

Kim, J., 2005: A projection of the effects of the climate change induced by increased CO_2 on extreme hydrologic events in the western US. *Climatic Change*, **68**, 153-168.

Kirshen, P., M. McCluskey, R. Vogel and K. Strzepek, 2005a: Global analysis of changes in water supply yields and costs under climate change: a case study in China. *Climatic Change*, **68**, 303-330.

Kirshen, P., M. Ruth and W. Anderson, 2005b: Responding to climate change in Metropolitan Boston: the role of adaptation. *New Engl. J. Public Pol*., **20**, 89-104.

Kirshen, P.H., 2002: Potential impacts of global warming on groundwater in eastern Massachusetts. *J. Water Res. Pl. –ASCE*, **128**, 216-226.

Kistemann, T., T. Classen, C. Koch, F. Dagendorf, R. Fischeder, J. Gebel, V. Vacata and M. Exner, 2002: Microbial load of drinking water reservoirs tributaries during extreme rainfall and runoff. *Appl. Environ. Microb*., **68**, 2188-2197.

Klein, R. and R. Nicholls, 1999: Assessment of coastal vulnerability to climate change. *Ambio*, **28**, 182-187.

Kleinen, T. and G. Petschel-Held, 2007: Integrated assessment of changes in flooding probabilities due to climate change. *Climatic Change*, **81**, 283-312.

Klijn, F., J. Dijkman and W. Silva, 2001: Room for the Rhine in the Netherlands. Summary of Research Results. RIZA Report 2001.033.

Klijn, F., M. van Buuren and S.A.M. van Rooij, 2004: Flood-risk management strategies for an uncertain future: living with Rhine river floods in the Netherlands? *Ambio*, **33**, 141-147.

Knighton, A.D., C.D. Woodroffe and K. Mills, 1992: The evolution of tidal creek networks, Mary River, Northern Australia. *Earth Surf. Proc. Land*., **17**, 167-90.

Knowles, N. and D.R. Cayan, 2004: Elevational dependence of projected hydrologic changes in the San Francisco Estuary and watershed. *Climatic Change*, **62**, 319-336.

Kostaschuk, R., J. Terry and R. Raj, 2002: Suspended sediment transport during tropical-cyclone floods in Fiji. *Hydrol. Process*., **17**, 1149-1164.

Kron, W. and G. Berz, 2007: Flood disasters and climate change: trends and options – a (re-)insurer's view. *Global Change: Enough Water for All?* J.L. Lozán, H. Graßl, P. Hupfer, L. Menzel and C.-D. Schönwiese, Eds., Hamburg, 268-273.

Krysanova, V. and F. Wechsung, 2002: Impact of climate change and higher CO_2 on hydrological processes and crop productivity in the state of Brandenburg, Germany. *Climatic Change: Implications for the Hydrological Cycle and for Water Management*, M. Beniston, Ed., Kluwer, Dordrecht, 271-300.

Krysanova, V., F. Hattermann and A. Habeck, 2005: Expected changes in water resources availability and water quality with respect to climate change in the Elbe River basin (Germany). *Nord. Hydrol*., **36**, 321-333.

Kumagai, M., K. Ishikawa and J. Chunmeng, 2003: Dynamics and biogeochemical significance of the physical environment in Lake Biwa. *Lakes Reserv. Res. Manage*., **7**, 345-348.

Kundzewicz, Z.W. and K. Takeuchi, 1999: Flood protection and management: quo vadimus? *Hydrol. Sci. J*., **44**, 417-432.

Kundzewicz, Z.W., S. Budhakooncharoen, A. Bronstert, H. Hoff, D. Lettenmaier, L. Menzel and R. Schulze, 2002: Coping with variability and change: floods and droughts. *Nat. Resour. Forum*, **26**, 263-274.

Kundzewicz, Z.W., U. Ulbrich, T. Brücher, D. Graczyk, A. Krüger, G. Leckebusch, L. Menzel, I. Pińskwar, M. Radziejewski and M. Szwed, 2005: Summer floods in Central Europe: climate change track? *Nat. Hazards*, **36**, 165-189.

Kundzewicz, Z.W., M. Radziejewski and I. Pińskwar, 2006: Precipitation extremes in the changing climate of Europe. *Climate Res*. **31**, 51–58.

Leemans, R. and A. Kleidon, 2002: Regional and global assessment of the dimensions of desertification. *Global Desertification: Do Humans Cause Deserts?* J.F. Reynold and D.S. Smith Eds., Dahlem University Press, Berlin, 215-232.

Legesse, D., C. Vallet-Coulomb and F. Gasse, 2003: Hydrological response of a catchment to climate and land use changes in Tropical Africa: case study South Central Ethiopia. *J. Hydrol*., **275**, 67-85.

Lehman, J., 2002: Mixing patterns and plankton biomass of the St. Lawrence Great Lakes under climate change scenarios. *J. Great Lakes Res*., **28**, 583-596.

Lehner, B., G. Czisch and S. Vassolo, 2005a: The impact of global change on the hydropower potential of Europe: a model-based analysis. *Energ. Policy*, **33**, 839-855.

Lehner, B., P. Döll, J. Alcamo, H. Henrichs and F. Kaspar, 2005b: Estimating the impact of global change on flood and drought risks in Europe: a continental, integrated assessment. *Climatic Change*, **75**, 273-299.

Leipprand, A. and D. Gerten, 2006: Global effects of doubled atmospheric CO_2 content on evapotranspiration, soil moisture and runoff under potential natural vegetation. *Hydrol. Sci. J*., **51**, 171-185.

Lemke, P., J. Ren, R. Alley, I. Allison, J. Carrasco, G. Flato, Y. Fujii, G. Kaser, P. Mote, R. Thomas and T. Zhang, 2007: Observations: changes in snow, ice and frozen ground. *Climate Change 2007: The Physical Science Basis. Contribution of Working Group I to the Fourth Assessment Report of the Intergovernmental Panel on Climate Change*, S. Solomon, D. Qin, M. Manning, Z. Chen, M. Marquis, K.B. Averyt, M. Tignor and H.L. Miller, Eds., Cambridge University Press, Cambridge, 337-384.

Lenton, R., 2004: Water and climate variability: development impacts and coping strategies. *Water Sci. Technol*., **49**(7), 17-24.

Leung, L.R., Y. Qian, X. Bian, W.M. Washington, J. Han and J.O. Roads, 2004: Mid-century ensemble regional climate change scenarios for the western United States. *Climatic Change*, **62**, 75-113.

Lipp, E., R. Kurz, R. Vincent, C. Rodriguez-Palacios, S. Farrah and J. Rose, 2001: The effects of seasonal variability and weather on microbial faecal pollution and enteric pathogens in a subtropical estuary. *Estuaries*, **24**, 226-276.

Loáiciga, H., 2003: Climate change and ground water. *Ann. Assoc. Am. Geogr*., **93**, 30-41.

Lofgren, B., A. Clites, R. Assel, A. Eberhardt and C. Luukkonen, 2002: Evaluation of potential impacts on Great Lakes water resources based on climate scenarios of two GCMs. *J. Great Lakes Res*., **28**, 537-554.

Loomis, J.B., K. Quattlebaum, T.C. Brown and S.J. Alexander, 2003: Expanding institutional arrangements for acquiring water for environmental purposes: transactions evidence for the Western United States. *Water Resour. Dev*., **19**, 21-28.

LOSLR [International Lake Ontario–St. Lawrence River Study Board], 2006: Options for managing Lake Ontario and St. Lawrence River water levels and flows. Final Report to the International Joint Commission, 162 pp.[Accessed 06.03.07: http://www.losl.org/PDF/report-main-e.pdf]

Loukas, A., L. Vasiliades and N.R. Dalezios, 2002a: Climatic impacts on the runoff generation processes in British Columbia, Canada. *Hydrol. Earth Syst. Sc*., **6**, 211-227.

Loukas, A., L. Vasiliades and N.R. Dalezios, 2002b: Potential climate change impacts on flood producing mechanisms in southern British Columbia, Canada using the CGCMA1 simulation results. *J. Hydrol*., **259**, 163-188.

Luketina, D. and M. Bender, 2002: Incorporating long-term trends in water availability in water supply planning. *Water Sci. Technol*., **46**(6-7), 113-120.

MacNeil, R., 2004: Costs of adaptation options. *Expanding the Dialogue on Climate Change and Water Management in the Okanagan Basin*, S. Cohen, D. Neilsen and R. Welbourn, Eds., Final Report, Environment Canada, British Columbia, 161-163.Available from www.ires.ubc.ca, last accessed 01.05.07

Magadza, C., 2000: Climate change impacts and human settlements in Africa: prospects for adaptation. *Environ. Monit. Assess*., **61**, 193-205.

Magrin, G.O., M.I. Travasso and G.R. Rodríguez, 2005: Changes in climate and crops production during the 20th century in Argentina. *Climatic Change*, **72**, 229-249.

Manabe, S., P.C.D. Milly and R. Wetherald, 2004a: Simulated long term changes in river discharge and soil moisture due to global warming. *Hydrol. Sci. J.*, **49**, 625-642.

Manabe, S., R.T. Wetherald, P.C.D. Milly, T.L. Delworth and R.J. Stouffer, 2004b: Century scale change in water availability: CO_2 quadrupling experiment. *Climatic Change*, **64**, 59-76.

Mark, B.G. and G.O. Seltzer, 2003: Tropical glacier meltwater contribution to stream discharge: a case study in the Cordillera Blanca, Peru. *J. Glaciol.*, **49**, 271-281.

Mata, L.J. and J. Budhooram, 2007: Complementarity between mitigation and adaptation: the water sector. *Mitigation and Adaptation Strategies for Global Change*, doi:10.1007/s11027-007-9100-y.

Matthews, O.P., 2004: Fundamental questions about water rights and market reallocation. *Water Resour. Res.*, **40**, W09S08, doi:10.1029/2003WR002836.

Maurer, E.P. and P.B. Duffy, 2005: Uncertainty in projections of streamflow changes due to climate change in California. *Geophys. Res. Lett.*, **32**, L03704.

Maya, C., N. Beltrán, B. Jiménez and P. Bonilla, 2003: Evaluation of the UV disinfection process in bacteria and amphizoic amoebeae inactivation. *Water Sci. Technol.*, **3**(4), 285-291.

McMichael, A., D. Campbell-Lendrum, C. Corvalán, K. Ebi, A. Githeko, J. Scheraga and A. Woodward, Eds., 2003: *Climate Change and Human Health: Risks and Responses*. WHO, Geneva, 322 pp.

Mearns, L., M. Hulme, T. Carter, R. Leemans, M. Lal and P. Whetton, 2001: Climate scenario development. *Climate Change 2001: The Scientific Basis. Contribution of Working Group I to the Third Assessment Report of the Intergovernmental Panel of Climate Change*, J.T. Houghton, Y. Ding, D. Griggs, M. Noguer, P.J. van der Linden, X. Dai, K. Maskell and C.A. Johnson, Eds., Cambridge University Press, Cambridge, 739-768.

Meehl, G.A., J.M. Arblaster and C. Tebaldi, 2005: Understanding future patterns of precipitation intensity in climate model simulations. *Geophys. Res. Lett.*, **32**, L18719, doi:10.1029/2005GL023680.

Meehl, G.A., T.F. Stocker and Co-authors, 2007: Global climate projections. *Climate Change 2007: The Physical Science Basis. Contribution of Working Group I to the Fourth Assessment Report of the Intergovernmental Panel on Climate Change*, S. Solomon, D. Qin, M. Manning, Z. Chen, M. Marquis, K.B. Averyt, M. Tignor and H.L. Miller, Eds., Cambridge University Press, Cambridge, 747-846.

Meleshko, V.P., V.M. Kattsov, V.A. Govorkova, S.P. Malevsky-Malevich, E.D. Nadezhina and P.V. Sporyshev, 2004: Anthropogenic climate change in the XXIst century in North Eurasia. *Meteorologia Hydrologia*, **7**, 5-26 (in Russian).

Menzel, L. and G. Burger, 2002: Climate change scenarios and runoff response in the Mulde catchment (Southern Elbe, Germany). *J. Hydrol.*, **267**, 53-64.

Michael, A., J. Schmidt, W. Enke, T. Deutschlander and G. Malitz, 2005: Impact of expected increase in precipitation intensities on soil loss results of comparative model simulations. *Catena*, **61**, 155-164.

Middelkoop, H., K. Daamen, D. Gellens, W. Grabs, J.C.J. Kwadijk, H. Lang, B.W.A.H. Parmet, B. Schädler, J. Schulla and K. Wilke, 2001: Impact of climate change on hydrological regimes and water resources management in the Rhine basin. *Climatic Change*, **49**, 105-128.

Middelkoop, H., M.B.A. van Asselt, S.A. van 't Klooster, W.P.A. van Deursen, J.C.J. Kwadijk and H. Buiteveld, 2004: Perspectives on flood management in the Rhine and Meuse rivers. *River Res. Appl.*, **20**, 327-342.

Mileti, D.S., 1999: *Disasters by Design: A Reassessment of Natural Hazards in the United States*. Joseph Henry Press, Washington, District of Columbia, 351 pp.

Millennium Ecosystem Assessment, 2005a: *Ecosystems and Human Well-being. Volume 2: Scenarios*. Island Press, Washington, District of Columbia, 515 pp.

Millennium Ecosystem Assessment, 2005b: *Ecosystems and Human Well-being: Synthesis*. Island Press, Washington, District of Columbia, 155 pp.

Miller, K.A., 2000: Managing supply variability: the use of water banks in the western United States. *Drought: A Global Assessment*, D.A. Wilhiten, Ed., Routledge, London, 70-86.

Miller, K.A. and D. Yates, 2006: *Climate Change and Water Resources: A Primer for Municipal Water Providers*. AWWA Research Foundation, Denver, Colorado, 83 pp.

Miller, K.A., S.L. Rhodes and L.J. MacDonnell, 1997: Water allocation in a changing climate: institutions and adaptation. *Climatic Change*, **35**, 157-177.

Mills, E., 2005: Insurance in a climate of change. *Science*, **309**, 1040-1044.

Milly, P.C.D., R.T. Wetherald, K.A. Dunne and T.L. Delworth, 2002: Increasing risk of great floods in a changing climate. *Nature*, **415**, 514-517.

Milly, P.C.D., K.A. Dunne and A.V. Vecchia, 2005: Global pattern of trends in streamflow and water availability in a changing climate. *Nature*, **438**, 347-350.

Mimikou, M., E. Blatas, E. Varanaou and K. Pantazis, 2000: Regional impacts of climate change on water resources quantity and quality indicators. *J. Hydrol.*, **234**, 95-109.

Ministry for the Environment, 2001: *Climate Change Impacts in New Zealand*. Ministry for the Environment, Wellington, 39 pp.

Mirza, M.M.Q., 2003: Three recent extreme floods in Bangladesh: a hydro-meteorological analysis. *Nat. Hazards*, **28**, 35-64.

Mirza, M.M.Q., R.A. Warrick and N.J. Ericksen, 2003: The implications of climate change on floods of the Ganges, Brahmaputra and Meghna Rrivers in Bangladesh. *Climatic Change*, **57**, 287-318.

Moench, M., A. Dixit, S. Janakarajan, M.S. Rathore and S. Mudrakartha, 2003: *The Fluid Mosaic: Water Governance in the Context of Variability, Uncertainty and Change – A Synthesis Paper*. Nepal Water Conservation Foundation, Kathmandu, 71 pp.

Morrison, J., M. Quick and M. Foreman, 2002: Climate change in the Fraser River watershed: flow and temperature projections. *J. Hydrol.*, **263**, 230-244.

Mote, P.W., D.J. Canning, D.L. Fluharty, R.C. Francis, J.F. Franklin, A.F. Hamlet, M. Hershman, M. Holmberg, K.N. Gray-Ideker, W.S. Keeton, D.P. Lettenmaier, L.R. Leung, N.J. Mantua, E.L. Miles, B. Noble, H. Parandvash, D.W. Peterson, A.K. Snover and S.R. Willard, 1999: *Impacts of Climate Variability and Change, Pacific Northwest*. National Atmospheric and Oceanic Administration, Office of Global Programs, and JISAO/SMA Climate Impacts Group, Seattle, Washington, 110 pp.

Mote, P.W., E.A. Parson, A.F. Hamlet, K.N. Ideker, W.S. Keeton, D.P. Lettenmaier, N.J. Mantua, E.L. Miles, D.W. Peterson, D.L. Peterson, R. Slaughter and A.K. Snover, 2003: Preparing for climatic change: the water, salmon, and forests of the Pacific Northwest. *Climatic Change*, **61**, 45-88.

Moulton, R. and D. Cuthbert, 2000: Cumulative impacts/risk assessment of water removal or loss from the Great Lakes–St. Lawrence River system. *Can. Water Resour. J.*, **25**, 181-208.

Mulrennan, M. and C. Woodroffe, 1998: Saltwater intrusions into the coastal plains of the Lower Mary River, Northern Territory, Australia. *J. Environ. Manage.*, **54**, 169-88.

Murphy, J.M., D.M.H. Sexton, D.N. Barnett, G.S. Jones, M.J. Webb, M. Collins and D.A. Stainforth, 2004: Quantification of modelling uncertainties in a large ensemble of climate change simulations. *Nature*, **430**, 768-772.

Naess, L.O., G. Bang, S. Eriksen and J. Vevatne, 2005: Institutional adaptation to climate change: flood responses at the municipal level in Norway. *Global Environ. Change*, **15**, 125-138.

Nakićenović, N. and R. Swart, Eds., 2000: *IPCC Special Report on Emissions Scenarios*. Cambridge University Press, Cambridge, 599 pp.

Nawaz, N.R. and A.J. Adeloye, 2006: Monte Carlo assessment of sampling uncertainty of climate change impacts on water resources yield in Yorkshire, England. *Climatic Change*, **78**, 257-292.

Nearing, M.A., 2001: Potential changes in rainfall erosivity in the United States with climate change during the 21st century. *J. Soil Water Conserv.*, **56**, 229-232.

Nearing, M.A., V. Jetten, C. Baffaut, O. Cerdan, A. Couturier, M. Hernandez, Y. Le Bissonnais, M.H. Nichols, J.P. Nunes, C.S. Renschler, V. Souchère and K. Van Oost, 2005: Modeling response of soil erosion and runoff to changes in precipitation and cover. *Catena*, **61**, 131-154.

Neff, R., H. Chang, C. Knight, R. Najjar, B. Yarnal and H. Walker, 2000: Impact of climate variation and change on Mid-Atlantic Region hydrology and water resources. *Climate Res.*, **14**, 207-218.

Neilsen, D., W. Koch, W. Merritt, G. Frank, S. Smith, Y. Alila, J. Carmichael, T. Neale and R. Welbourn, 2004: Risk assessment and vulnerability: case studies of water supply and demand. *Expanding the Dialogue on Climate Change and Water Management in the Okanagan Basin*, S. Cohen, D. Neilsen and R. Welbourn, Eds., British Columbia, 115-135.

Nicholson, S., 2005: On the question of the "recovery" of the rains in the West African Sahel. *J. Arid Environ.*, **63**, 615-641.

Nohara, D., A. Kitoh, M. Hosaka and T. Oki, 2006: Impact of climate change on river runoff. *J. Hydrometeorol.*, **7**, 1076-1089.

O'Neal, M.R., M.A. Nearing, R.C. Vining, J. Southworth and R.A. Pfeifer, 2005: Climate change impacts on soil erosion in Midwest United States with changes in corn–soybean–wheat management. *Catena*, **61**, 165-184.

O'Reilly, C., S. Alin, P. Plisnier, A. Cohen and B. Mckee, 2003: Climate change decreases aquatic ecosystem productivity of Lake Tanganyika, Africa. *Nature*, **424**, 766-768.

Oki, T., 2005: The hydrologic cycles and global circulation. *Encyclopaedia of Hydrological Sciences*, M.G. Anderson, Ed., John Wiley and Sons, Chichester.

Oki, T. and S. Kanae, 2006: Global hydrological cycles and world water resources. *Science*, **313**, 1068-1072.

Oki, T., Y. Agata, S. Kanae, T. Saruhashi and K. Musiake, 2003a: Global water resources assessment under climatic change in 2050 using TRIP. *Water Resources: Systems Water Availability and Global Change*, S. Franks, G. Blöschl, M. Kumagai, K. Musiake and D. Rosbjerg, Eds., IAHS Publication No. 280, 124-133.

Oki, T., Sato, M., Kawamura, A., Miyaka, M., Kanae, S. and Musiake, K., 2003b: *Virtual Water Trade to Japan and in the World. Value of Water Research Report Series*. IHE, Delft, pp. 221-235.

Olsen, J.R., 2006: Climate change and floodplain management in the United States. *Climatic Change*, **76**, 407-426.

Orr, P. and B. Colby, 2004: Groundwater management institutions to protect riparian habitat. *Water Resour. Res.*, **40**, W12S03, doi:10.1029/2003WR002741.

Pachauri, R., 2004: Climate change and its implications for development: the role of IPCC assessments. *IDS Bull.–I. Dev. Stud.*, **35**, 11.

Palmer, T.N. and J. Räisänen, 2002: Quantifying the risk of extreme seasonal precipitation events in a changing climate. *Nature*, **415**, 512-514.

Patrinos, A. and A. Bamzai, 2005: Policy needs robust climate science. *Nature*, **438**, 285.

Patz, J., 2001: Public health risk assessment linked to climatic and ecological change. *Hum. Ecol. Risk Assess.*, **7**, 1317-1327.

Payne, J.T., A.W. Wood, A.F. Hamlet, R.N. Palmer and D.P. Lettenmaier, 2004: Mitigating the effects of climate change on the water resources of the Columbia River basin. *Climatic Change*, **62**, 233-256.

Peirson, W., R. Nittim, M. Chadwick, K. Bishop and P. Horton, 2001: Assessment of changes to saltwater/freshwater habitat from reductions in flow to the Richmond River estuary, Australia. *Water Sci. Technol.*, **43**(9), 89-97.

Pfeifer, R.A. and M. Habeck, 2002: Farm level economic impacts of climate change. *Effects of Climate Change and Variability on Agricultural Production Systems*, O.C. Doering, J.C. Randolph, J. Southworth and R.A. Pfeifer, Eds., Academic Publishers, Boston, Massachusetts, 159-178.

Pfister, L., J. Kwadijk, A. Musy, A. Bronstert and L. Hoffmann, 2004: Climate change, land use change and runoff prediction in the Rhine-Meuse basins. *River Res. Appl.*, **20**, 229-241.

Pielke, R.A., Jr. and M.W. Downton, 2000: Precipitation and damaging floods: trends in the United States, 1932–97. *J. Climate*, **13**, 3625-3637.

Pittock, B., 2003: *Climate Change: An Australian Guide to the Science and Potential Impacts*. Australian Greenhouse Office, Canberra, 239 pp.

Protopapas, L., S. Katchamart and A. Platonova, 2000: Weather effects on daily water use in New York City. *J. Hydrol. Eng.*, **5**, 332-338.

Prudhomme, C., 2006: GCM and downscaling uncertainty in modelling of current river flow: why is it important for future impacts? Proc. 5th FRIEND World Conf., Havana. *IAHS Publication*, **308**, 375-381.

Prudhomme, C. and H. Davies, 2007: Comparison of different sources of uncertainty in climate change impact studies in Great Britain. *Hydrol. Process. (Special issue on International Workshop "Climatic and Anthropogenic Impacts on Water Resources Variability")*, in press.

Prudhomme, C., D. Jakob and C. Svensson, 2003: Uncertainty and climate change impact on the flood regime of small UK catchments. *J. Hydrol.*, **277**, 1-23.

Pruski, F.F. and Nearing, M.A., 2002a: Climate-induced changes in erosion during the 21st century for eight U.S. locations. *Water Resour. Res.*, **38**, 1298.

Pulwarty, R.S. and T.S. Melis, 2001: Climate extremes and adaptive management on the Colorado River: lessons from the 1997–1998 ENSO event. *J. Environ. Manage.*, **63**, 307-324.

Ragab, R., Ed., 2005: *Advances in Integrated Management of Fresh and Saline Water for Sustainable Crop Production: Modeling and Practical Solutions. Special Issue of Int. J. Agr. Water Manage.* **78**, 1-164.

Ragab, R. and C. Prudhomme, 2002: Climate change and water resources management in arid and semi-arid regions: prospective and challenges for the 21st century. *Biosyst. Eng.*, **81**, 3-34.

Räisänen, J., U. Hansson, A. Ullerstieg, R. Döscher, L.P. Graham, C. Jones, H.E.M. Meier, P. Samuelson and U. Willén, 2004: European climate in the late twenty-first century: regional simulations with two driving global models and two forcing scenarios. *Clim. Dynam.*, **22**, 13-31.

Reilly, J., F. Tubiello, B. McCarl, D. Abler, R. Darwin, K. Fuglie, S. Hollinger, C. Izaurralde, S. Jagtap, J. Jones, L. Mearns, D. Ojima, E. Paul, K. Paustian, S. Riha, N. Rosenberg and C. Rosenzweig, 2003: U.S. agriculture and climate change: new results. *Climatic Change*, **57**, 43-69.

Reynard, N., S. Crooks, R. Wilby and A. Kay, 2004: *Climate Change and Flood Frequency in the UK*, Proceedings of the 39th DEFRA Flood and Coastal Management Conference, York. Defra, London, 11.1.1-11.1.12.

Richardson, D., 2002: Flood risk: the impact of climate change. *P. I. Civil Eng.–Civ. En.*, **150**, 22-24.

Robarts, R., M. Kumagai and C.H. Magadza, 2005: Climate change impacts on lakes: technical report of the session 'Ecosystem Approach to Water Monitoring and Management' organized at the World Water Forum II in Kyoto. *Climatic Change Ecosystem Approach to Water Monitoring and Management*, UNEP Publication, Nairobi.

Rose, J., S. Daeschner, D. Easterling, E. Curriero, L. Lele and J. Patz, 2000: Climate and waterborne outbreaks. *J. Am. Water Works Assoc.*, **92**, 87-97.

Rosenberg, N.J., R.A. Brown, C. Izaurralde and A.M. Thomson, 2003: Integrated assessment of Hadley Centre HadCM2 climate change projections on agricultural productivity and irrigation water supply in the conterminous United States. I. Climate change scenarios and impacts on irrigation water supply simulated with the HUMUS model. *Agr. Forest Meteorol.*, **117**, 73-96.

Rosso, R., M.C. Rulli and G. Vannucchi, 2006: A physically based model for the hydrologic control on shallow landsliding. *Water Resour. Res.*, **42**, W06410.

Roy, L., R. Leconte, F.P. Brisette and C. Marche, 2001: The impact of climate change on seasonal floods of a southern Quebec River Basin. *Hydrol. Process.*, **15**, 3167-3179.

Rumsby, B.T. and M.G. Macklin, 1994: Channel and floodplain response to recent abrupt climate change: the Tyne Basin, Northern England. *Earth Surf. Proc. Land.*, **19**, 499-515.

Saleth, R.M. and A. Dinar, 2004: *Institutional Economics of Water: A Cross-Country Analysis of Institutions and Performance*. Edward Elgar, Cheltenham, 398 pp.

Schiermeier, Q., 2006: Insurers' disaster files suggest climate is culprit. *Nature*, **441**, 674-675.

Schindler, D., 2001: The cumulative effects of climate warming and other human stresses on Canadian freshwaters in the new millennium. *Can. J. Fish. Aquat. Sci.*, **58**, 18-29.

Schlenker, W., W.M. Hanemann and A.C. Fisher, 2005: Will U.S. agriculture really benefit from global warming? Accounting for irrigation in the hedonic approach. *Am. Econ. Rev.*, **95**, 395-406.

Schneeberger, C., H. Blatter, A. Abe-Ouchi and M. Wild, 2003: Modelling changes in the mass balance of glaciers of the northern hemisphere for a transient 2× CO_2 scenario. *J. Hydrol.*, **282**, 145-163.

Schreider, S.Y., D.I. Smith and A.J. Jakeman, 2000: Climate change impacts on urban flooding. *Climatic Change*, **47**, 91-115.

Schwartz, R.C., P.J. Deadman, D.J. Scott and L.D. Mortsch, 2004: Modeling the impacts of water level changes on a Great Lakes community. *J. Am. Water Resour. As.*, **40**, 647-662.

Scott, T., E. Lipp and J. Rose, 2004: The effects of climate change on waterborne disease. *Microbial Waterborne Pathogens*, E. Cloete, J. Rose, L.H. Nel and T. Ford, Eds., IWA Publishing, London.

Scudder, T., 2005: *The Future of Large Dams*. Earthscan, London, 408 pp.

Seckler, D., U. Amarasinghe, D. Molden, R. de Silva and R. Barker, 1998: World water demand and supply, 1990 to 2025: scenarios and issues. Report 19, International Water Management Research Institute, Sri Lanka, 50 pp.

Shabalova, M.V., W.P.A. van Deursen and T.A. Buishand, 2003: Assessing future discharge of the river Rhine using regional climate model integrations and a hydrological model. *Climate Res.*, **23**, 233-246.

Shepherd, P., J. Tansey and H. Dowlatabadi, 2006: Context matters: what shapes adaptation to water stress in the Okanagan? *Climatic Change*, **78**, 31-62.

Sherif, M. and V. Singh, 1999: Effect of climate change on sea water intrusion in coastal aquifers. *Hydrol. Process.*, **13**, 1277-1287.

Shiklomanov, I.A. and V.Y. Georgievsky, 2002: Effect of anthropogenic climate change on hydrological regime and water resources. *Climate Change and its Consequences*. Nauka, St. Petersburg, 152-164 (in Russian).

Shiklomanov, I.A. and J.C. Rodda, Eds., 2003: *World Water Resources at the Beginning of the 21st Century*. Cambridge University Press, Cambridge, 435 pp.

Shiklomanov, I.A. and A.S. Vasiliev, Eds., 2004: *Hydrometeorological Problems of the Caspian Sea Basin*. Hydrometeoizdat, St. Petersburg, 435 pp. (in Russian).

Simonovic, S.P. and L.H. Li, 2003: Methodology for assessment of climate change impacts on large- scale flood protection system. *J. Water Res. Pl.–ASCE*, **129**, 361-371.

Singh, P., 2003: Effect of warmer climate on the depletion of snowcovered area in the Satluj basin in the western Himalayan region. *Hydrol. Sci. J.*, **48**, 413-425.

Singh, P. and N. Kumar, 1997: Impact assessment of climate change on the hydrological response of a snow and glacier melt runoff dominated Himalayan river. *J. Hydrol.*, **193**, 316-350.

Singh, P. and L. Bengtsson, 2004: Hydrological sensitivity of a large Himalayan basin to climate change. *Hydrol. Process.*, **18**, 2363-2385.

Sivapalan, M., G. Bloschl, R. Mertz and D. Gutknecht, 2005: Linking flood-frequency to long-term water balance: incorporating effects of seasonality. *Water Resour. Res.*, **41**, W06012, doi:10.1029/2004WR003439.

Small, C. and R.J. Nicholls, 2003: A global analysis of human settlement in coastal zones. *J. Coastal Res.*, **19**, 584-599.

Smit, J. and J. Nasr, 1992: Urban agriculture for sustainable cities: using wastes and idle land and water bodies as resources. *Environ. Urban.*, **4**, 141-152.

Soil and Water Conservation Society, 2003: Soil erosion and runoff from cropland. Report from the USA, Soil and Water Conservation Society, 63 pp.

Souchere, V., O. Cerdan, N. Dubreuil, Y. Le Bissonnais and C. King, 2005: Modelling the impact of agri-environmental scenarios on runoff in a cultivated catchment Normandy, France. *Catena*, **61**, 229-240.

Soulsby, C., C. Gibbins, A. Wade, R. Smart and R. Helliwell, 2002: Water quality in the Scottish uplands: a hydrological perspective on catchment hydrochemistry. *Sci. Total Environ.*, **294**, 73-94.

Southworth, J., J.C. Randolph, M. Habeck, O.C. Doering, R.A. Pfeifer, D. Gangadhar Rao and J.J. Johnston, 2000: Consequences of future climate change and changing climate variability on maize yields in the mid-western United States. *Agr. Ecosyst. Environ.*, **82**, 139-158.

Stainforth, D.A., T. Aina, C. Christensen, M. Collins, N. Faull, D.J. Frame, J.A. Kettleborough, S. Knight, A. Martin, J.M. Murphy, C. Piani, D. Sexton, L.A. Smith, R.A. Spicer, A.J. Thorpe and M.R. Allen, 2005: Uncertainty in predictions of the climate response to rising levels of greenhouse gases. *Nature*, **433**, 403-406.

Stakhiv, E.Z., 1998: Policy implications of climate change impacts on water resources management. *Water Policy*, **1**, 159-175.

Stewart, I.T., D.R. Cayan and M.D. Dettinger, 2004: Changes in snowmelt runoff timing in western North America under a 'business as usual' climate change scenario. *Climatic Change*, **62**, 217-232.

Stone, M.C., R.H. Hotchkiss, C.M. Hubbard, T.A. Fontaine, L.O. Mearns and J.G. Arnold, 2001: Impacts of climate change on Missouri River Basin water yield. *J. Am. Water Resour. As.*, **37**, 1119-1129.

Stott, P.A. and J.A. Kettleborough, 2002: Origins and estimates of uncertainty in prediction of twenty-first century temperature rise. *Nature*, **416**, 723-726.

Svendsen, M., 2005: *Irrigation and River Basin Management: Options for Governance and Institutions*. CABI Publishing, in association with the International Water Management Institute, Wallingford, 258 pp.

Svensson, C. and D.A. Jones, 2005: Climate change impacts on the dependence between sea surge, precipitation and river flow around Britain. *Proceedings of the 40th Defra Flood and Coastal Management Conference 2005, University of York, 5-7 July 2005*, 6A.3.1-6A.3.10.

Takahashi, K., Y. Matsuoka, Y. Shimada and H. Harasawa, 2001: Assessment of water resource problems under climate change: considering inter-annual variability of climate derived from GCM calculations. *J. Global Environ. Eng.*, **7**, 17-30.

Tanaka, S.K., T. Zhu, J.R. Lund, R.E. Howitt, M.W. Jenkins, M.A. Pulido, M. Tauber, R.S. Ritzema and I.C. Ferreira, 2006: Climate warming and water management adaptation for California. *Climatic Change*, **76**, 361-387.

Tate, E., J. Sutcliffe, P. Conway and F. Farquharson, 2004: Water balance of Lake Victoria: update to 2000 and climate change modelling to 2100. *Hydrol. Sci. J.*, **49**, 563-574.

Tebaldi, C., L.O. Mearns, D. Nychka and R.L. Smith, 2004: Regional probabilities of precipitation change: a Bayesian analysis of multi-model simulations. *Geophys. Res. Lett.*, **32**, L24213, doi:10.1029/2004GL021276.

Tebaldi, C., R.L. Smith, D. Nychka and L.O. Mearns, 2005: Quantifying uncertainty in projections of regional climate change: a Bayesian approach to the analysis of multi-model ensembles. *J. Climate*, **18**, 1524-1540.

Tebaldi, C., K. Hayhoe, J.M. Arblaster and G.A. Meehl, 2006: Going to the extremes: an intercomparison of model-simulated historical and future changes in extreme events. *Climatic Change*, **79**, 185-211.

Thomson, A.M., R.A. Brown, N.J. Rosenberg, R. Srinivasan and R.C. Izaurralde, 2005a: Climate change impacts for the conterminous USA: an integrated assessment. Part 4. Water resources. *Climatic Change*, **69**, 67-88.

Thomson, A.M., N.J. Rosenberg, R.C. Izaurralde and R.A. Brown, 2005b: Climate change impacts for the conterminous USA: an integrated assessment. Part 5. Irrigated agriculture and national grain crop production. *Climatic Change*, **69**, 89-105.

Tiwari, D. and A. Dinar, 2002: Balancing future food demand and water supply: the role of economic incentives in irrigated agriculture. *Q. J. Int. Agr.*, **41**, 77-97.

Tol, R.S.J., N. van der Grijp, A.A. Olsthoorn and P.E. van der Werff, 2003: Adapting to climate: a case study on riverine flood risks in the Netherlands. *Risk Anal.*, **23**, 575-583.

Trenberth, K.E., A.G. Dai, R.M. Rasmussen and D.B. Parsons, 2003: The changing character of precipitation. *B. Am. Meteorol. Soc.*, **84**, 1205-1217.

Trenberth, K.E., P.D. Jones, P.G. Ambenje, R. Bojariu, D.R. Easterling, A.M.G. Klein Tank, D.E. Parker, J.A. Renwick, F. Rahimzadeh, M.M. Rusticucci, B.J. Soden and P.-M. Zhai, 2007: Observations: surface and atmospheric change. *Climate Change 2007: The Physical Science Basis. Contribution of Working Group I to the Fourth Assessment Report of the Intergovernmental Panel on Climate Change*, S. Solomon, D. Qin, M. Manning, Z. Chen, M. Marquis, K.B. Averyt, M. Tignor and H.L. Miller, Eds., Cambridge University Press, Cambridge, 235-336.

Triggs, J.M., B.A. Kimball, P.J. Pinter Jr., G.W. Wall, M.M. Conley, T.J. Brooks, R.L. LaMorte, N.R. Adam, M.J Ottman, A.D. Matthias, S.W. Leavitt and R.S. Cerveny, 2004: Free-air carbon dioxide enrichment effects on energy balance and evapotranspiration of sorghum. *Agr. Forest Meteorol.*, **124**, 63-79.

UNDP [United Nations Development Programme], 2006: *MDG Targets and Indicators* [accessed 06.03.07: http://www.undp.org/mdg/goallist.shtml]

United Nations, 2002: *Johannesburg Plan of Implementation of the World Summit on Sustainable Development*. United Nations, 72 pp.

United Nations, 2003: *World Water Development Report: Water for Life, Water for People*. UNESCO, Paris, and Berghahn Books, Barcelona, Spain, 544 pp.

United Nations, 2006: *World Water Development Report 2: Water, a shared responsibility*. UNESCO, Paris, 601 pp.

United Nations Committee on Economic Social and Cultural Rights, 2003: *General Comment No. 15 (2002). The Right to Water*. E/C.12/2002/11, United Nations Social and Economic Council, 18 pp.

US Department of the Interior, 2005: Water 2025: preventing crisis and conflict in the West. Status Report, 36 pp. [accessed 06.03.07: http://www.doi.gov/water 2025/Water%202025-08-05.pdf]

US Global Change Research Program, 2000: *Water: The Potential Consequences of Climate Variability and Change*. National Water Assessment Group, US Global Change Research Program, US Geological Survey and Pacific Institute, Washington, District ofColumbia, 160 pp.

VanRheenen, N.T., A.W. Wood, R.N. Palmer and D.P. Lettenmaier, 2004: Potential implications of PCM climate change scenarios for Sacramento–San Joaquin River basin hydrology and water resources. *Climatic Change*, **62**, 257-281.

Vörösmarty, C.J., P.J. Green, J Salisbury and R.B. Lammers, 2000: Global water resources: vulnerability from climate change and population growth. *Science*, **289**, 284-288.

Vuglinsky, V. and T. Gronskaya, 2005: *Strategic Forecast up to 2010–2015 on the Effect of Expected Climate Changes on the Economy of Russia*. ROS-HYDROMET, Moscow (in Russian).

Wade, A., P. Whitehead, G. Hornberger and D. Snook, 2002: On modelling the flow controls on macrophyte and epiphyte dynamics in a lowland permeable catchment: the River Kennet, southern England. *Sci. Total Environ.*, **282**, 375-393.

Walsh, C.L. and C.G. Kilsby, 2007: Implications of climate change on flow regime affecting Atlantic salmon. *Hydrol. Earth Syst. Sc.*, **11**, 1127-1143.

Wang, H.J., Z.S. Yang, Y. Saito, J.P. Liu and X.X. Sun, 2006: Interannual and seasonal variation of the Huanghe (Yellow River) water discharge over the past 50 years: connections to impacts from ENSO events and dams. *Global Planet. Change*, **50**, 212-225.

Waters, D., W.E. Watt, J. Marsalek and B.C. Anderson, 2003: Adaptation of a storm drainage system to accommodate increased rainfall resulting from climate change. *J. Environ. Plan. Manage.*, **46**, 755-770.

WDR, 2003: *World Disaster Report: Focus on Ethics in Aid*. International Federation of Red Cross and Red Crescent Societies, Geneva, 240 pp.

WDR, 2004: *World Disaster Report: Focus on Community Resilience*. International Federation of Red Cross and Red Crescent Societies, Geneva, 240 pp.

Wechsung, F., A. Becker and P. Gräfe, Eds., 2005: *Auswirkungen des globalen Wandels auf Wasser, Umwelt und Gesellschaft im Elbegebiet*. Weissensee-Verlag, Berlin, 416 pp.

Wetherald, R.T. and S. Manabe, 2002: Simulation of hydrologic changes associated with global warming. *J. Geophys. Res.*, **107**(D19), 4379, doi:10.1029/2001JD001195.

WHO, 2004: *Guidelines for Drinking Water Quality: Volume 1*, 3rd edn. World Health Organization, Geneva, 540 pp.

WHO/UNICEF, 2000: *Global Water Supply and Sanitation Assessment 2000 Report*. World Health Organization with UNICEF, Geneva, 79 pp.

Wichelns, D., D. Cone and G. Stuhr, 2002: Evaluating the impact of irrigation and drainage policies on agricultural sustainability. *Irrig. Drain. Syst.*, **16**, 1-14.

Wilby, R.L., 2005: Uncertainty in water resource model parameters used for climate change impact assessment. *Hydrol. Process.*, **19**, 3201-3219.

Wilby, R.L., 2006: When and where might climate change be detectable in UK river flows? *Geophys. Res. Lett.*, **33**, L19407.

Wilby, R.L. and I. Harris, 2006: A framework for assessing uncertainties in climate change impacts: low-flow scenarios for the River Thames, UK. *Water Resour. Res.*, **42**, W02419, doi:10.1029/2005WR004065.

Wilhite, D.A., 2000: *Drought: A Global Assessment*. Routledge, London: Vol 1, 396 pp; Vol 2, 304 pp.

Wilhite, D.A., 2001: Moving beyond crisis management. *Forum Appl Res Public Pol*, **16**, 20-28.

Wilk, J. and D.A. Hughes, 2002: Simulating the impacts of land-use and climate change on water resource availability for a large south Indian catchment. *Hydrol. Sci. J.*, **47**, 19-30.

Williams, W., 2001: Salinization: unplumbed salt in a parched landscape. *Water Sci. Technol.*, **43**(4), 85-91.

Wisner, B. and J. Adams, Eds., 2002: *Environmental Health in Emergencies and Disasters*. WHO, Geneva, 272 pp.

Wood, A.W., L.R. Leung, V. Sridhar and D.P. Lettenmaier, 2004: Hydrologic implications of dynamical and statistical approaches to downscaling climate model outputs. *Climatic Change*, **62**, 189-216.

World Bank, 2003: *Water Resources Sector Strategy: Strategic Directions for World Bank Engagement*. The World Bank, Washington, DC.

World Commission on Dams, 2000: *Dams and Development: A New Framework for Decision-Making*. Earthscan, London.

World Water Council, 2006: *Final Report of the 4th World Water Forum*. National Water Commission of Mexico, Mexico City, 262 pp.

Yang, D., C. Li, H. Hu, Z. Lei, S. Yang, T. Kusuda, T. Koike and K. Musiake, 2004: Analysis of water resources variability in the Yellow River of China during the last half century using historical data. *Water Resour. Res.*, **40**, W06502, doi:10.1029/2003WR002763.

Yao, H. and A. Georgakakos, 2001: Assessment of Folsom Lake response to historical and potential future climate scenarios. *J. Hydrol.*, **249**, 176-196.

Yarze, J.C. and M.P. Chase, 2000: E. coli O157:H7 – another waterborne outbreak! *Am. J. Gastroenterol.*, **95**, 1096.

Young, R.A., 2005: *Determining the Economic Value of Water: Concepts and Methods*. Resources for the Future Press, Washington, District of Columbia, 300 pp.

Zhang, G.H., M.A. Nearing and B.Y. Liu, 2005: Potential effects of climate change on rainfall erosivity in the Yellow River basin of China. *T. ASAE*, **48**, 511-517.

Zhang, X.C. and M.A. Nearing, 2005: Impact of climate change on soil erosion, runoff, and wheat productivity in Central Oklahoma. *Catena*, **61**, 185-195.

Zhou, Y. and R.S.J. Tol, 2005: Evaluating the costs of desalination and water transport. *Water Resour. Res.*, **41**, 1-10.

Zierl, B. and H. Bugmann, 2005: Global change impacts on hydrological processes in Alpine catchments. *Water Resour. Res.*, **41**, W02028, doi: 10.11029/2004WR003447.

4

Ecosystems, their properties, goods and services

Coordinating Lead Authors:

Andreas Fischlin (Switzerland), Guy F. Midgley (South Africa)

Lead Authors:

Jeff Price (USA), Rik Leemans (The Netherlands), Brij Gopal (India), Carol Turley (UK), Mark Rounsevell (Belgium), Pauline Dube (Botswana), Juan Tarazona (Peru), Andrei Velichko (Russia)

Contributing Authors:

Julius Atlhopheng (Botswana), Martin Beniston (Switzerland), William J. Bond (South Africa), Keith Brander (ICES/Denmark/UK), Harald Bugmann (Switzerland), Terry V. Callaghan (UK), Jacqueline de Chazal (Belgium), Oagile Dikinya (Australia), Antoine Guisan (Switzerland), Dimitrios Gyalistras (Switzerland), Lesley Hughes (Australia), Barney S. Kgope (South Africa), Christian Körner (Switzerland), Wolfgang Lucht (Germany), Nick J. Lunn (Canada), Ronald P. Neilson (USA), Martin Pêcheux (France), Wilfried Thuiller (France), Rachel Warren (UK)

Review Editors:

Wolfgang Cramer (Germany), Sandra Myrna Diaz (Argentina)

This chapter should be cited as:

Fischlin, A., G.F. Midgley, J.T. Price, R. Leemans, B. Gopal, C. Turley, M.D.A. Rounsevell, O.P. Dube, J. Tarazona, A.A. Velichko, 2007: Ecosystems, their properties, goods, and services. *Climate Change 2007: Impacts, Adaptation and Vulnerability. Contribution of Working Group II to the Fourth Assessment Report of the Intergovernmental Panel on Climate Change*, M.L. Parry, O.F. Canziani, J.P. Palutikof, P.J. van der Linden and C.E. Hanson, Eds., Cambridge University Press, Cambridge, UK, 211-272.

Table of Contents

Executive summary

During the course of this century the resilience of many ecosystems (their ability to adapt naturally) is likely to be exceeded by an unprecedented combination of change in climate, associated disturbances (e.g., flooding, drought, wildfire, insects, ocean acidification) and in other global change drivers (especially land-use change, pollution and over-exploitation of resources), if greenhouse gas emissions and other changes continue at or above current rates (high confidence).

By 2100, ecosystems will be exposed to atmospheric CO_2 levels substantially higher than in the past 650,000 years, and global temperatures at least among the highest of those experienced in the past 740,000 years (very high confidence) [4.2, 4.4.10, 4.4.11; Jansen et al., 2007]. This will alter the structure, reduce biodiversity and perturb functioning of most ecosystems, and compromise the services they currently provide (high confidence) [4.2, 4.4.1, 4.4.2-4.4.9, 4.4.10, 4.4.11, Figure 4.4, Table 4.1]. Present and future land-use change and associated landscape fragmentation are very likely to impede species' migration and thus impair natural adaptation via geographical range shifts (very high confidence) [4.1.2, 4.2.2, 4.4.5, 4.4.10].

Several major carbon stocks in terrestrial ecosystems are vulnerable to current climate change and/or land-use impacts and are at a high degree of risk from projected unmitigated climate and land-use changes (high confidence).

Several terrestrial ecosystems individually sequester as much carbon as is currently in the atmosphere (very high confidence) [4.4.1, 4.4.6, 4.4.8, 4.4.10, 4.4.11]. The terrestrial biosphere is likely to become a net source of carbon during the course of this century (medium confidence), possibly earlier than projected by the IPCC Third Assessment Report (TAR) (low confidence) [4.1, Figure 4.2]. Methane emissions from tundra frozen loess ('yedoma', comprising about 500 Pg C) and permafrost (comprising about 400 Pg C) have accelerated in the past two decades, and are likely to accelerate further (high confidence) [4.4.6]. At current anthropogenic emission rates, the ongoing positive trends in the terrestrial carbon sink will peak before mid-century, then begin diminishing, even without accounting for tropical deforestation trends and biosphere feedback, tending strongly towards a net carbon source before 2100, assuming continued greenhouse gas emissions and land-use change trends at or above current rates (high confidence) [Figure 4.2, 4.4.1, 4.4.10, Figure 4.3, 4.4.11], while the buffering capacity of the oceans will begin to saturate [Denman et al., 2007, e.g., Section 7.3.5.4]. While some impacts may include primary productivity gains with low levels of climate change (less than around 2°C mean global change above pre-industrial levels), synergistic interactions are likely to be detrimental, e.g., increased risk of irreversible extinctions (very high confidence) [4.4.1, Figure 4.2, 4.4.10, Figure 4.3, 4.4.11].

Approximately 20 to 30% of plant and animal species assessed so far (in an unbiased sample) are likely to be at increasingly high risk of extinction as global mean temperatures exceed a warming of 2 to 3°C above pre-industrial levels (medium confidence) [4.4.10, 4.4.11, Figure 4.4, Table 4.1].

Projected impacts on biodiversity are significant and of key relevance, since global losses in biodiversity are irreversible (very high confidence) [4.4.10, 4.4.11, Figure 4.4, Table 4.1]. Endemic species richness is highest where regional palaeoclimatic changes have been muted, providing circumstantial evidence of their vulnerability to projected climate change (medium confidence) [4.2.1]. With global average temperature changes of 2°C above pre-industrial levels, many terrestrial, freshwater and marine species (particularly endemics across the globe) are at a far greater risk of extinction than in the recent geological past (medium confidence) [4.4.5, 4.4.11, Figure 4.4, Table 4.1]. Globally about 20% to 30% of species (global uncertainty range from 10% to 40%, but varying among regional biota from as low as 1% to as high as 80%) will be at increasingly high risk of extinction, possibly by 2100, as global mean temperatures exceed 2 to 3°C above pre-industrial levels [4.2, 4.4.10, 4.4.11, Figure 4.4, Table 4.1]. Current conservation practices are generally poorly prepared to adapt to this level of change, and effective adaptation responses are likely to be costly to implement (high confidence) [4.4.11, Table 4.1, 4.6.1].

Substantial changes in structure and functioning of terrestrial ecosystems are very likely to occur with a global warming of more than 2 to 3°C above pre-industrial levels (high confidence).

Between about 25% (IPCC SRES B1 emissions scenario; 3.2°C warming) and about 40% (SRES A2 scenario; 4.4°C warming) of extant ecosystems will reveal appreciable changes by 2100, with some positive impacts especially in Africa and the Southern Hemisphere arid regions, but extensive forest and woodland decline in mid- to high latitudes and in the tropics, associated particularly with changing disturbance regimes (especially through wildfire and insects) [4.4.2, 4.4.3, 4.4.5, 4.4.10, 4.4.11, Figure 4.3].

Substantial changes in structure and functioning of marine and other aquatic ecosystems are very likely to occur with a mean global warming of more than 2 to 3°C above pre-industrial levels and the associated increased atmospheric CO_2 levels (high confidence).

Climate change (very high confidence) and ocean acidification (medium confidence) will impair a wide range of planktonic and shallow benthic marine organisms that use aragonite to make their shells or skeletons, such as corals and marine snails (pteropods), with significant impacts particularly in the Southern Ocean, where cold-water corals are likely to show large reductions in geographical range this century [4.4.9, Box 4.4]. Substantial loss of sea ice will reduce habitat for dependant species (e.g., polar bears) (very high confidence) [4.4.9, 4.4.6, Box 4.3, 4.4.10, Figure 4.4, Table 4.1, 15.4.3, 15.4.5]. Terrestrial tropical and sub-tropical aquatic systems are at significant risk under at least SRES A2 scenarios; negative impacts across about 25% of Africa by 2100 (especially southern and western Africa) will cause a decline in both water quality and ecosystem goods and services (high confidence) [4.4.8].

Ecosystems and species are very likely to show a wide range of vulnerabilities to climate change, depending on imminence of exposure to ecosystem-specific, critical thresholds (very high confidence).

Most vulnerable ecosystems include coral reefs, the sea-ice biome and other high-latitude ecosystems (e.g., boreal forests), mountain ecosystems and mediterranean-climate ecosystems (high confidence) [Figure 4.4, Table 4.1, 4.4.9, Box 4.4, 4.4.5, 4.4.6, Box 4.3, 4.4.7, 4.4.4, 4.4.10, 4.4.11]. Least vulnerable ecosystems include savannas and species-poor deserts, but this assessment is especially subject to uncertainty relating to the CO_2-fertilisation effect and disturbance regimes such as fire (low confidence) [Box 4.1, 4.4.1, 4.4.2, Box 4.2, 4.4.3, 4.4.10, 4.4.11].

4.1 Introduction

An ecosystem can be practically defined as a dynamic complex of plant, animal and micro-organism communities, and the non-living environment, interacting as a functional unit (Millennium Ecosystem Assessment, Reid et al., 2005). Ecosystems may be usefully identified through having strong interactions between components within their boundaries and weak interactions across boundaries (Reid et al., 2005, part 2). Ecosystems are well recognised as critical in supporting human well-being (Reid et al., 2005), and the importance of their preservation under anthropogenic climate change is explicitly highlighted in Article 2 (The Objective) of the United Nations Framework Convention on Climate Change (UNFCCC).

In this chapter the focus is on the properties, goods and services of non-intensively managed and unmanaged ecosystems and their components (as grouped by widely accepted functional and structural classifications, Figure 4.1), and their potential vulnerability to climate change as based on scenarios mainly from IPCC (see Chapter 2 and IPCC, 2007). Certain ecosystem goods and services are treated in detail in other sectoral chapters (this volume): chapters 3 (water), 5 (food, fibre, fisheries), 6 (coasts) and 8 (health). Key findings from this chapter are further developed in the synthesis chapters 17 to 20 (this volume). Region-specific aspects of ecosystems are discussed in chapters 9 to 16 (this volume). This chapter is based on work published since the Third Assessment Report of the IPCC (TAR) (Gitay et al., 2001). We do not summarise TAR findings here, but refer back to relevant TAR results, where appropriate, to indicate confirmation or revision of major findings.

Projecting the impacts of climate change on ecosystems is complicated by an uneven understanding of the interlinked temporal and spatial scales of ecosystem responses. Processes at large spatial scales, i.e., the biosphere at the global scale, are generally characterised by slow response times on the order of centuries, and even up to millennia (Jansen et al., 2007). However, it is also important to note that some large-scale responses in the palaeorecord (Jansen et al., 2007) and to current climate anomalies such as El Niño events may emerge at much shorter time-scales (Holmgren et al., 2001; Sarmiento and Gruber, 2002; Stenseth et al., 2002; van der Werf et al., 2004). At continental scales, biomes (see Glossary) respond at decadal to millennial time-scales (e.g., Davis, 1989; Prentice et al., 1991; Lischke et al., 2002; Neilson et al., 2005), and groups of organisms forming ecological communities at the regional scale have shorter response times of years to centuries. Responses of populations (i.e., interbreeding individuals of the same species) occur at intermediate temporal scales of months to centuries, and underpin changes in biodiversity. These include changes at the genetic level that may be adaptive, as demonstrated for example for trees (Jump et al., 2006) and corals (Coles and Brown, 2003). Fast physiological response times (i.e., seconds, hours, days, months) of micro-organisms, plants and animals operate at small scales from a leaf or organ to the cellular level; they underlie organism responses to environmental conditions, and are assessed here if they scale up to have a significant impact at broader spatial scales, or where the mechanistic understanding assists in assessing key thresholds in higher level responses.

The spatial distribution of ecosystems at biome scale has traditionally been explained only in terms of climate control (Schimper, 1903), but it is increasingly apparent that disturbance regimes such as fire or insects may strongly influence vegetation structure somewhat independently of climate (e.g., Andrew and

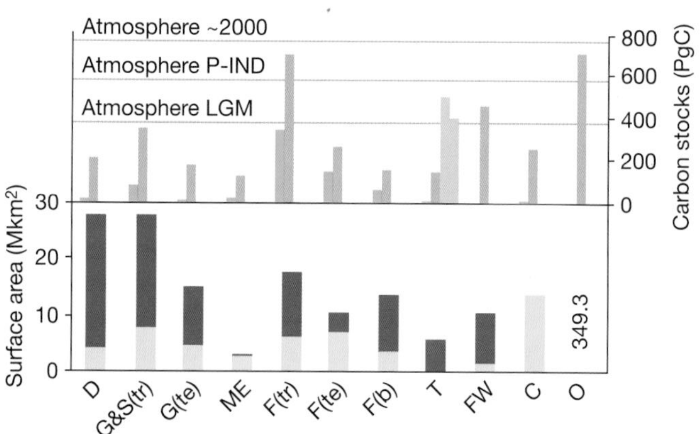

Figure 4.1. *Major ecosystems addressed in this report, with their global areal extent (lower panel, Mkm²), transformed by land use in yellow, untransformed in purple, from Hassan et al. (2005), except for mediterranean-climate ecosystems, where transformation impact is from Myers et al. (2000), and total carbon stores (upper panel, PgC) in plant biomass (green), soil (brown), yedoma/permafrost (light blue). D = deserts, G&S(tr) = tropical grasslands and savannas, G(te) = temperate grasslands, ME = mediterranean ecosystems, F(tr) = tropical forests, F(te) = temperate forests, F(b) = boreal forests, T = tundra, FW = freshwater lakes and wetlands, C = croplands, O = oceans. Data are from Sabine et al. (2004, Table 2.2, p. 23), except for carbon content of yedoma permafrost and permafrost (light blue columns, left and right, respectively, Zimov et al., 2006), ocean organic carbon content (dissolved plus particulate organic; Denman et al., 2007, Section 7.3.4.1), and ocean surface area from Hassan et al. (2005, Summary, Table C2, p. 15, inserted as a number). Figures here update the TAR (Prentice et al., 2001), especially through considering soil C to 3 m depth (Jobbagy and Jackson, 2000), as opposed to 1 m. Approximate carbon content of the atmosphere (PgC) is indicated by the dotted lines for last glacial maximum (LGM), pre-industrial (P-IND) and current (about 2000).*

Hughes, 2005; Bond et al., 2005). Biomes are differentially sensitive to climatic change (e.g., Kirschbaum and Fischlin, 1996; Sala et al., 2000; Gitay et al., 2001), with temperature-limited biomes prone to impacts of warming, and water-limited biomes prone to increasing levels of drought. Some, such as fire-dependent biomes, may be in a meta-stable state that can change rapidly under climate and other environmental changes (Scheffer et al., 2001; Sankaran et al., 2005). Marine biome responses, too, have been shown at decadal scales (Beaugrand et al., 2002), with more rapid regime shifts within decades (Edwards et al., 2002; Richardson and Schoeman, 2004; Edwards et al., 2006). Biomes therefore provide a useful level of ecological organisation at which to summarise climate change impacts, being of large enough extent to conduct a global synthesis, yet having a response time relevant to anthropogenic climate change.

4.1.1 Ecosystem goods and services

Ecosystems provide many goods and services that are of vital importance for the functioning of the biosphere, and provide the basis for the delivery of tangible benefits to human society. Hassan et al. (2005) define these to include supporting, provisioning, regulating and cultural services. In this chapter we divide services into four categories.

i. *Supporting services*, such as primary and secondary production, and biodiversity, a resource that is increasingly recognised to sustain many of the goods and services that humans enjoy from ecosystems. These provide a basis for three higher-level categories of services.
ii. *Provisioning services*, such as products (cf. Gitay et al., 2001), i.e., food (including game, roots, seeds, nuts and other fruit, spices, fodder), fibre (including wood, textiles) and medicinal and cosmetic products (including aromatic plants, pigments; see Chapter 5).
iii. *Regulating services*, which are of paramount importance for human society such as (a) carbon sequestration, (b) climate and water regulation, (c) protection from natural hazards such as floods, avalanches or rock-fall, (d) water and air purification, and (e) disease and pest regulation.
iv. *Cultural services*, which satisfy human spiritual and aesthetic appreciation of ecosystems and their components.

4.1.2 Key issues

Based on new findings for ecosystems since the TAR, we highlight here five overarching key issues pertinent to assessing the vulnerability of ecosystems to anthropogenic climate change, and related adaptation responses.

Firstly, ecosystems are expected to tolerate some level of future climate change and, in some form or another, will continue to persist (e.g., Kirschbaum and Fischlin, 1996; Gitay et al., 2001), as they have done repeatedly with palaeoclimatic changes (Jansen et al., 2007). A primary key issue, however, is whether ecosystem resilience (understood as the disturbance an ecosystem can tolerate before it shifts into a different state, e.g., Scheffer et al., 2001; Cropp and Gabrica, 2002; Folke et al., 2004) inferred from these responses (e.g., Harrison and Prentice, 2003) will be sufficient to tolerate future anthropogenic climate change (e.g., Chapin et al., 2004; Jump and Peñuelas, 2005). The implications of possibly transient increases in productivity for resilience are also very relevant. These may occur in certain terrestrial ecosystems through likely atmospheric CO_2-fertilisation effects and/or modest warming (e.g., Baker et al., 2004; Lewis et al., 2004b; Malhi and Phillips, 2004), and demonstrated consequences of increased radiation due to reduced tropical cloudiness (Nemani et al., 2003). Ecosystem resilience thus seems usefully equivalent to the critical ecosystem property highlighted in Article 2 of the UNFCCC, i.e., an "ability to adapt naturally".

Secondly, ecosystems are increasingly being subjected to other human-induced pressures, such as extractive use of goods, and increasing fragmentation and degradation of natural habitats (e.g., Bush et al., 2004). In the medium term (i.e., decades) especially, climate change will increasingly exacerbate these human-induced pressures, causing a progressive decline in biodiversity (Lovejoy and Hannah, 2005). However, this is likely to be a complex relationship that may also include some region-specific reductions in land-use pressures on ecosystems (e.g., Goklany, 2005; Rounsevell et al., 2006).

A third key issue involves exceeding critical thresholds and triggering non-linear responses in the biosphere that could lead via positive feedback to novel states that are poorly understood. Projected future climate change and other human-induced pressures are virtually certain to be unprecedented (Forster et al., 2007) compared with the past several hundred millennia (e.g., Petit et al., 1999; Augustin et al., 2004; Siegenthaler et al., 2005).

Fourthly, the understanding of time-lags in ecosystem responses is still developing, including, for example, broad-scale biospheric responses or shifting species geographical ranges. Many ecosystems may take several centuries (vegetation) or even possibly millennia (where soil formation is involved) before responses to a changed climate are played out (e.g., Lischke et al., 2002). A better understanding of transient responses and the functioning of ecosystems under continuously changing conditions is needed to narrow uncertainties about critical effects and to develop effective adaptation responses at the time-scale of interest to human society.

A fifth key issue relates to species extinctions, and especially global extinction as distinct from local extinctions, since the former represents irreversible change. This is crucial, especially because of a very likely link between biodiversity and ecosystem functioning in the maintenance of ecosystem services (Duraiappah et al., 2005; Hooper et al., 2005; Diaz et al., 2006; Worm et al., 2006), and thus extinctions critical for ecosystem functioning, be they global or local, are virtually certain to reduce societal options for adaptation responses.

4.2 Current sensitivities

4.2.1 Climatic variability and extremes

The biosphere has been exposed to large variability and extremes of CO_2 and climate throughout geological history (Augustin et al., 2004; Siegenthaler et al., 2005; Jansen et al., 2007), and this provides some insight into the current sensitivities of ecosystems even though it is not possible to match past climate analogues precisely with future warming, due to differences in forcing factors (Overpeck et al., 2006), dominant ecosystems, and species (e.g., Velichko et al., 2002). What can be learned is that, firstly, significant biological changes including species extinctions have accompanied large climate perturbations of the past (e.g., Overpeck et al., 2005). Secondly, endemic biodiversity is concentrated in regions that have experienced lower variability during the Pleistocene (from about 2 million years ago) (Jansson, 2003), during which glacial and inter-glacial conditions have alternated for roughly the past 2 million years. Thirdly, range shifts have been a major species response (Lovejoy and Hannah, 2005), although genetic and physiological responses (Davis and Shaw, 2001) have also occurred, which can be broadly defined as 'natural adaptation' at species level, and by aggregation, at the ecosystem level.

While earlier IPCC reports described several ecosystems to be resilient to warming up to 1°C (e.g., Kirschbaum and Fischlin, 1996), recent studies provide a more differentiated view of ecosystem sensitivity (e.g., Walther et al., 2002) that includes understanding of the role of climatic variability and extremes. Knowledge about climate variability and natural ecosystems has improved with better understanding of the behaviour of decadal-scale climatic oscillations and their impacts, including ENSO (El Niño/Southern Oscillation) and the NAO (North Atlantic Oscillation) (Trenberth et al., 2007, Section 3.6). These low-frequency phenomena indirectly determine vegetation responses, notably through shifts in major controls (temperature, precipitation, snow cover). For example, the European Alps show changes in regional climates that can partly be attributed to NAO variability (Hurrell and van Loon, 1997; Serreze et al., 1997; Wanner et al., 1997; Beniston and Jungo, 2002) such as the lack of snow in the late 1980s and early 1990s (Beniston, 2003). Disruptions of precipitation regimes in the Pacific region and beyond during ENSO events can disrupt vegetation through drought, heat stress, spread of parasites and disease, and more frequent fire (e.g., Diaz and Markgraf, 1992). Similar effects have been reported for NAO (Edwards and Richardson, 2004; Sims et al., 2004; Balzter et al., 2005). Sea surface temperature increases associated with ENSO events have been implicated in reproductive failure in seabirds (Wingfield et al., 1999), reduced survival and reduced size in iguanas (Wikelski and Thom, 2000) and major shifts in island food webs (Stapp et al., 1999).

Many significant impacts of climate change may emerge through shifts in the intensity and the frequency of extreme weather events. Extreme events can cause mass mortality of individuals and contribute significantly to determining which species occur in ecosystems (Parmesan et al., 2000). Drought plays an important role in forest dynamics, driving pulses of tree mortality in the Argentinean Andes (Villalba and Veblen, 1997), North American woodlands (Breshears and Allen, 2002; Breshears et al., 2005), and in the eastern Mediterranean (Körner et al., 2005b). In both the Canadian Rockies (Luckman, 1994) and European Alps (Bugmann and Pfister, 2000) extreme cold through a period of cold summers from 1696 to 1701 caused extensive tree mortality. Heatwaves such as the recent 2003 event in Europe (Beniston, 2004; Schär et al., 2004; Box 4.1) have both short-term and long-term implications for vegetation, particularly if accompanied by drought conditions.

Hurricanes can cause widespread mortality of wild organisms, and their aftermath may cause declines due to the loss of resources required for foraging and breeding (Wiley and Wunderle, 1994). The December 1999 'storm-of-the-century' that affected western and central Europe destroyed trees at a rate of up to ten times the background rate (Anonymous, 2001). Loss of habitat due to hurricanes can also lead to greater conflict with humans. For example, fruit bats (*Pteropus spp.*) declined recently on American Samoa due to a combination of direct mortality events and increased hunting pressure (Craig et al., 1994). Greater storminess and higher return of extreme events will also alter disturbance regimes in coastal ecosystems, leading to changes in diversity and hence ecosystem functioning. Saltmarshes, mangroves and coral reefs are likely to be particularly vulnerable (e.g. Bertness and Ewanchuk, 2002; Hughes et al., 2003).

Assessment of the impacts of climate variability, their trends, and the development of early warning systems has been strongly advanced since the TAR by satellite-based remote sensing efforts. Notable contributions have included insights into phenological shifts in response to warming (e.g., Badeck et al., 2004) and other environmental trends (e.g., Nemani et al., 2003), complex Sahelian vegetation changes (e.g., Prince et al., 1998; Rasmussen et al., 2001; Anyamba and Tucker, 2005; Hein and Ridder, 2006), wildfire impacts (e.g., Isaev et al., 2002; Barbosa et al., 2003; Hicke et al., 2003; Kasischke et al., 2003), coral bleaching events (e.g., Yamano and Tamura, 2004), cryosphere changes (Walsh, 1995; Lemke et al., 2007), ecotone (see Glossary) responses to climate (e.g., Masek, 2001), deforestation (e.g., Asner et al., 2005), and even feedbacks to regional climate (e.g., Durieux et al., 2003), the impacts of extreme climate events (e.g., Gobron et al., 2005; Lobo and Maisongrande, 2006) and monitoring of soil water (Wagner et al., 2003).

4.2.2 Other ecosystem change drivers

Ecosystems are sensitive not only to changes in climate and atmospheric trace gas concentrations but also to other anthropogenic changes such as land use, nitrogen deposition, pollution and invasive species (Vitousek et al., 1997; Mack et al., 2000; Sala et al., 2000; Hansen et al., 2001; Lelieveld et al., 2002; Körner, 2003b; Lambin et al., 2003; Reid et al., 2005). In the recent past, these pressures have significantly increased due to human activity (Gitay et al., 2001). Natural disturbance regimes (e.g., wildfire and insect outbreaks) are also important climate-sensitive drivers of ecosystem change. Projecting the impacts of the synergistic effects of these drivers presents a major challenge, due to the potential for non-linear, rapid, threshold-type responses in ecological systems (Burkett et al., 2005).

Land-use change represents the anthropogenic replacement of one land use type by another, e.g., forest to cultivated land (or the reverse), as well as subtle changes of management practices within a given land use type, e.g., intensification of agricultural practices, both of which are affecting 40% of the terrestrial surface (reviewed by Foley et al., 2005). Land-use change and related habitat loss and fragmentation have long

Box 4.1. Ecological impacts of the European heatwave 2003

Anomalous hot and dry conditions affected Europe between June and mid-August, 2003 (Fink et al., 2004; Luterbacher et al., 2004; Schär et al., 2004). Since similarly warm summers may occur at least every second year by 2080 in a Special Report on Emissions Scenario (SRES; Nakićenović et al, 2000) A2 world, for example (Beniston, 2004; Schär et al., 2004), effects on ecosystems observed in 2003 provide a conservative analogue of future impacts. The major effects of the 2003 heatwave on vegetation and ecosystems appear to have been through heat and drought stress, and wildfires.

Drought stress impacts on vegetation (Gobron et al., 2005; Lobo and Maisongrande, 2006) reduced gross primary production (GPP) in Europe by 30% and respiration to a lesser degree, overall resulting in a net carbon source of 0.5 PgC/yr (Ciais et al., 2005). However, vegetation responses to the heat varied along environmental gradients such as altitude, e.g., by prolonging the growing season at high elevations (Jolly et al., 2005). Some vegetation types, as monitored by remote sensing, were found to recover to a normal state by 2004 (e.g., Gobron et al., 2005), but enhanced crown damage of dominant forest trees in 2004, for example, indicates complex delayed impacts (Fischer, 2005). Freshwater ecosystems experienced prolonged depletion of oxygen in deeper layers of lakes during the heatwave (Jankowski et al., 2006), and there was a significant decline and subsequent poor recovery in species richness of molluscs in the River Saône (Mouthon and Daufresne, 2006). Taken together, this suggests quite variable resilience across ecosystems of different types, with very likely progressive impairment of ecosystem composition and function if such events increase in frequency (e.g., Lloret et al., 2004; Rebetez and Dobbertin, 2004; Jolly et al., 2005; Fuhrer et al., 2006).

High temperatures and greater dry spell durations increase vegetation flammability (e.g., Burgan et al., 1997), and during the 2003 heatwave a record-breaking incidence of spatially extensive wildfires was observed in European countries (Barbosa et al., 2003), with roughly 650,000 ha of forest burned across the continent (De Bono et al., 2004). Fire extent (area burned), although not fire incidence, was exceptional in Europe in 2003, as found for the extraordinary 2000 fire season in the USA (Brown and Hall, 2001), and noted as an increasing trend in the USA since the 1980s (Westerling et al., 2006). In Portugal, area burned was more than twice the previous extreme (1998) and four times the 1980-2004 average (Trigo et al., 2005, 2006). Over 5% of the total forest area of Portugal burned, with an economic impact exceeding €1 billion (De Bono et al., 2004).

Long-term impacts of more frequent similar events are very likely to cause changes in biome type, particularly by promoting highly flammable, shrubby vegetation that burns more frequently than less flammable vegetation types such as forests (Nunes et al., 2005), and as seen in the tendency of burned woodlands to reburn at shorter intervals (Vazquez and Moreno, 2001; Salvador et al., 2005). The conversion of vegetation structure in this way on a large enough scale may even cause accelerated climate change through losses of carbon from biospheric stocks (Cox et al., 2000). Future projections for Europe suggest significant reductions in species richness even under mean climate change conditions (Thuiller et al., 2005b), and an increased frequency of such extremes (as indicated e.g., by Schär et al., 2004) is likely to exacerbate overall biodiversity losses (Thuiller et al., 2005b).

been recognised as important drivers of past and present ecosystem change, particularly of biodiversity (Heywood and Watson, 1995; Fahrig, 2003).

Fire influences community structure by favouring species that tolerate fire or even enhance fire spread, resulting in a relationship between the relative flammability of a species and its relative abundance in a particular community (Bond and Keeley, 2005). As a result, many vegetation types are far from the maximum biomass predicted by regional climate alone (Bond et al., 2005). Geographical shifts in key species or fire may therefore cause fundamental community shifts (Brooks et al., 2004; Schumacher and Bugmann, 2006). Fire-prone vegetation types cover a total of 40% of the world's land surface (Chapin et al., 2002), and are common in tropical and sub-tropical regions (Bond et al., 2005), and the boreal region (Harden et al., 2000) in particular. Intensified wildfire regimes driven at least partly by 20th century climate change (Gillett et al., 2004; Westerling et al., 2006), appear to be changing vegetation structure and composition with shifts from *Picea*- to *Pinus*-dominated communities and 75-95% reductions in tree densities observed in forest-tundra transition in eastern Canada (Lavoie and Sirois, 1998). By contrast, in Quebec, fire frequency appears to have dropped during the 20th century (Bergeron et al., 2001), a trend projected to continue (see Section 4.4.5; Bergeron et al., 2004). Across the entire North American boreal region, however, total burned area from fires increased by a factor of 2.5 between the 1960s and 1990s, while the area burned from human-ignited fires remained constant (Kasischke and Turetsky, 2006). In South-East Asia, by contrast, human activities have significantly altered fire regimes in ways that may be detrimental to the affected ecosystems (Murdiyarso and Lebel, 2007).

Drought facilitated the spread of human-caused fire in tropical regions during the 1997/98 El Niño (Randerson et al., 2005), affecting atmospheric trace gas concentrations such as CO, CH_4 and H_2 (Langenfelds et al., 2002; Novelli et al., 2003; Kasischke et al., 2005), and CO_2 emissions (van der Werf et al., 2004) at hemispheric and global scales. Drought conditions increase Amazon forest flammability (Nepstad et al., 2004). Tropical forest fires are becoming more common (Cochrane, 2003), and have strong negative effects on Amazonian vegetation (Cochrane and Laurance, 2002; Haugaasen et al., 2003), possibly even intensifying rainfall events (Andreae et al., 2004, but see Sections 4.4.1 and 4.4.5 on forest productivity trends).

Significant progress on globally applicable models of fire has been made since the TAR (Thonicke et al., 2001). Modelling suggests increases in wildfire impacts (see Sections 4.4.1 and 4.4.5) during the 21st century under a wide range of scenarios (e.g., Scholze et al., 2006). The implications of the regional and global importance of fire are manifold (Bond et al., 2005). Firstly, fire suppression strategies often have limited impact (Keeley, 2002; Schoennagel et al., 2004; Van Wilgen et al., 2004), and the enhancement of vegetation flammability through more prevalent fire weather (Brown et al., 2004) and the resulting big wildfires threatens human settlements, infrastructure and livelihoods (e.g., Allen Consulting Group, 2005). Secondly, in some ecosystems, including islands, human-caused fires have transformed forests into more flammable shrublands and grasslands (Ogden et al., 1998). Thirdly, the drivers of flammability, such as ecosystem productivity, fuel accumulation and environmental fire risk conditions, are all influenced by climate change (Williams et al., 2001; see Sections 4.4.3, 4.4.4 and 4.4.5).

The spatial impact of insect damage is significant and exceeds that of fire in some ecosystems, but especially in boreal forests (Logan et al., 2003). Spruce bud worm (SBW), for example, defoliated over 20 times the area burned in eastern Ontario between 1941 and 1996 (Fleming et al., 2002). Furthermore, fires tended to occur 3 to 9 years after a SBW outbreak (Fleming et al., 2002), suggesting a greater interaction between these disturbances with further warming. Disturbance by forest tent caterpillar has also increased in western Canada in the past 25 years (Timoney, 2003). In the Mediterranean region, the defoliation of Scots Pine shows a significant association with previous warm winters, implying that future climatic warming may intensify insect damage (Hodar and Zamora, 2004; see Section 4.4.5).

Invasive alien species (IAS) (Chornesky and Randall, 2003) represent a major threat to endemic or native biodiversity in terrestrial and aquatic systems (Sala et al., 2000; Scavia et al., 2002; Occhipinti-Ambrogi and Savini, 2003). Causes of biological invasions are multiple and complex (Dukes and Mooney, 1999), yet some simple models have been developed (Crawley, 1989; Deutschewitz et al., 2003; Chytry et al., 2005; Facon et al., 2006). Alien species invasions also interact with other drivers, sometimes resulting in some unexpected outcomes (Chapuis et al., 2004). Changes in biotic and/or abiotic disturbance regimes are recognised as primary drivers of IAS (Le Maitre et al., 2004), with communities often becoming more susceptible to invasion following extreme events (Smith and Knapp, 1999), such as are projected under future climate change. IAS can also change disturbance regimes through increasing vegetation flammability (Brooks et al., 2004). Overall, ongoing shifts in human-mediated disturbances, insect pests, IAS and fire regimes are very likely to be important in altering regional ecosystem structure, diversity and function (e.g., Timoney, 2003).

4.3 Assumptions about future trends

The work reviewed in this chapter is dependent on assumptions of various types that are important in assessing the level of confidence that can be associated with its results (Moss and Schneider, 2000), but can be challenging to quantify and aggregate. Assumptions and uncertainties associated with climate scenarios (Randall et al., 2007) are not considered here, other than to identify the greenhouse gas emission trends or socio-economic development pathways (e.g., SRES, Nakićenović et al., 2000) assumed in the literature we review (see also Table 4.1, especially scaling methodology and associated uncertainties). Since the TAR, many global or regional scenarios have become available to quantify future impacts (Christensen et al., 2002, 2007; Meehl et al., 2007), and confidence in future climate projections has increased recently (Nakićenović et al., 2000; Randall et al., 2007). However, many assumptions must be made, due to imperfect knowledge, in order to project ecosystem responses to climate scenarios. We provide here a brief outline and guide to the literature of those that are most relevant.

To project impacts of climate change on ecosystems there are basically three approaches: (i) correlative, (ii) mechanistic, and (iii) analogue approaches. For the correlative and mechanistic approaches, studies and insights from the present give rise to the assumption that the same relationships will hold in the future. Three modelling approaches in particular have provided relevant results since the TAR. Firstly, correlative models use knowledge of the spatial distribution of species to derive functions (Guisan and Thuiller, 2005) or algorithms (Pearson et al., 2004) that relate the probability of their occurrence to climatic and other factors (Guisan and Zimmermann, 2000). Criticised for assumptions of equilibrium between species and current climate, an inability to account for species interactions, lack of a physiological mechanism, and inability to account for population processes and migration (see Pearson and Dawson, 2003; Pearson, 2006), these methods have nonetheless proved capable of simulating known species range shifts in the distant (Martinez-Meyer et al., 2004) and recent (Araújo et al., 2005) past, and provide a pragmatic first-cut assessment of risk to species decline and extinction (Thomas et al., 2004a). Secondly, mechanistic models include the modelling of terrestrial ecosystem structure and function. They are based on current understanding of energy, biomass, carbon, nutrient and water relations, and their interacting dynamics with and among species such as primary producers. Such approaches generate projections of future vegetation structure, e.g., as the likely balance of plant functional types (PFTs) after permitting competitive interaction and accounting for wildfire (Woodward and Lomas, 2004b; Lucht et al., 2006; Prentice et al., 2007; but see Betts and Shugart, 2005, for a more complete discussion). Extrapolated to global scale, these are termed Dynamic Global Vegetation Models (DGVMs, see Glossary). An equivalent approach for oceans is lacking (but see Field et al., 1998). Thirdly, Earth system models have begun to incorporate more realistic and dynamic vegetation components, which quantify positive and negative biotic feedbacks by coupling a dynamic biosphere to atmospheric circulations with a focus on the global carbon cycle (Friedlingstein et al., 2003, 2006; Cox et al., 2004, 2006).

Ecosystem- and species-based models are typically applied at scales much finer than are resolved or reliably represented in global climate models. The requisite downscaling techniques of various types (statistical, dynamic) have matured and are increasingly used to provide the necessary spatio-temporal detail (IPCC-TGCIA, 1999; Mearns et al., 2003; Wilby et al., 2004; Christensen et al., 2007). Physically consistent bioclimatic scenarios can now be derived for almost any region, including developing countries (e.g., Jones et al., 2005) and complex, mountainous terrain (e.g., Gyalistras and Fischlin, 1999; Hayhoe et al., 2004). However, major uncertainties relating to downscaling remain in the impact projections presented in this chapter, centring mainly on soil water balance and weather extremes which are key to many ecosystem impacts, yet suffer from low confidence in scenarios for precipitation and climate variability, despite recent improvements (Randall et al., 2007).

Despite the recognised importance of multiple drivers of ecosystem change, they are rarely all included in current climate and ecosystem models used for assessing climate change impacts on ecosystems (Hansen et al., 2001; Levy et al., 2004; Zebisch et al., 2004; Feddema et al., 2005; Holman et al., 2005b; Pielke, 2005). The explicit inclusion of non-climatic drivers and their associated interactions in analyses of future climate change impacts could lead to unexpected outcomes (Hansen et al., 2001; Sala, 2005). Consequently, many impact studies of climate change that ignore land-use and other global change trends (see also Section 4.2.2) may represent inadequate estimates of projected ecosystem responses.

4.4 Key future impacts and vulnerabilities

The scope of this section satisfies that required by the IPCC plenary in relation to future impacts on properties, goods and services of major ecosystems and on biodiversity. However, to assess ecosystem goods and services more completely, issues relating to biogeochemical cycling and other supporting or regulating services are also deemed appropriate for consideration under this heading. Following reviews of impacts on individual ecosystems, impacts that cut across ecosystems (such as large-scale vegetation shifts and migratory species) are elaborated. Finally the overall implications for biodiversity are highlighted in a global synthesis. Within the relevant sub-sections, we describe briefly ecosystem properties, goods and services, we summarise key vulnerabilities as identified by the TAR, and then review what new information is available on impacts, focusing on supporting and regulating services (for provisioning services see Chapters 3, 5 and 6).

4.4.1 Biogeochemical cycles and biotic feedback

The cycling of chemical elements and compounds sustains the function of the biosphere and links ecosystems and climate by regulating chemical concentrations in soil, biota, atmosphere and ocean. Substantial progress has been made since the TAR in understanding the interactive responses of terrestrial ecosystems and the climate system, as determined by plant physiological responses, interactions with the soil, and their scaled-up effects on regional and global biogeochemical cycles (Buchmann, 2002; Cox et al., 2006; Friedlingstein et al., 2006; Gedney et al., 2006). Interactions between ocean and atmosphere and land and oceans are also critical for the future evolution of climate (see Section 4.4.9, but mainly Denman et al., 2007, e.g., Section 7.3.5.4).

Among the most advanced tools to achieve scaling-up of terrestrial systems to the global scale are Dynamic Global Vegetation Models (DGVMs), which simulate time-dependent changes in vegetation distribution and properties, and allow mapping of changes in ecosystem function and services (Schröter et al., 2005; Metzger et al., 2006). Testing at hierarchical levels from leaf to biome and over relevant time-scales has shown encouraging agreement with observations (Lucht et al., 2002; Bachelet et al., 2003; Harrison and Prentice, 2003; Gerten et al., 2004; Joos and Prentice, 2004; Kohler et al., 2005; Peylin et al., 2005), and validation is ongoing (e.g., Woodward and Lomas, 2004b; Prentice et al., 2007). Recently, full coupling between DGVMs and climate models has progressed from earlier work (e.g., Woodward and Lomas, 2001) to explore feedback effects between biosphere and atmospheric processes (Cox et al., 2006; Friedlingstein et al., 2006), that were initially reported as having significant implications for the carbon cycle (Cox et al., 2000).

Key vulnerabilities

Ecosystems are likely to respond to increasing external forcing in a non-linear manner. Most initial ecosystem responses appear to dampen change (Aber et al., 2001), but amplify it if thresholds in magnitude or rate of change are surpassed. Transitions between states may be triggered, or the ecosystem may even 'collapse' i.e., show a rapid transition to a much less productive and/or species-poor assemblage with lower biomass and other impairments such as degrading soils (e.g., Scheffer et al., 2001; Rietkerk et al., 2004; Schröder et al., 2005). Changing fire regimes provide an important example (see Section 4.2.2 for a more complete treatment), as these are of significant concern for the terrestrial carbon balance (Schimel and Baker, 2002; van der Werf et al., 2004; Westerling et al., 2006), especially because they can be self-reinforcing (Bond and Keeley, 2005). However, even less extreme responses of ecosystems are likely to have important ramifications for the biosphere because of their spatial extent.

Based on early versions of DGVMs (equilibrium biogeography models or global biogeochemical models – Neilson et al., 1998), the world's terrestrial ecosystems were projected to continue as a net carbon sink for a number of decades and possibly throughout the 21st century, with an initially 'greening' world due to longer growing seasons, more precipitation and CO_2-fertilisation benefits. Substantial structural changes in biomes were projected towards 2100, with ecosystem shifts towards higher latitudes and altitudes. A reversal of initial carbon sequestration gains was projected during the 21st century, as CO_2-fertilisation benefits approach saturation and temperature effects on respiration and transpiration increase, potentially resulting in net global ecosystem carbon losses relative to today (e.g., Cramer et al., 2001). With feedback from the global carbon cycle to the atmosphere accounted for, dieback of much of the Amazon rainforest due to desiccation was an identified major vulnerability, but with a high degree of uncertainty (Cox et al., 2000). The TAR concluded that the net global terrestrial carbon exchange would be between −6.7 PgC/yr (uptake 1 PgC) and +0.4 PgC/yr, and that anthropogenic CO_2 emissions would remain the dominant determinant of atmospheric CO_2 concentration during the 21st century. Key ecosystem forecasting needs identified in the TAR were for spatially and temporally dynamic models to simulate processes that produce inertia and lags in ecosystem responses. Progress on this issue has now allowed initial assessments of the potential for feedbacks from ecosystems to atmospheric composition and climate change.

Impacts

Observations for global net primary productivity (NPP) from 1982 to 1999 show an increase of 6%, concentrated in the tropics and due virtually certainly to greater solar radiation with reduced cloud cover (Nemani et al., 2003), broadly concurring with the projection in the TAR of an increasing biospheric sink in the initial stages of climate change. Scaled-up effects of direct atmospheric CO_2 enrichment on plant and ecosystem biomass accumulation (CO_2-fertilisation) are largely responsible for the projected continued enhancement of NPP in current global models (Leemans et al., 2002). By contrast, impacts in oceans, especially through acidification, have been largely negative (see Section 4.4.9).

Despite improved experiments, the magnitude of the terrestrial CO_2-fertilisation effect remains uncertain, although improved simulation of major vegetation types (particularly forests and savannas) at the last glacial maximum by incorporating CO_2 effects (Harrison and Prentice, 2003) are encouraging. The three main constraints that have been observed to limit the fertilisation effect are element stoichiometry (nutrient balance), forest tree dynamics, and secondary effects of CO_2 on water relations and biodiversity. Trends in some empirical data suggest caution when estimating future carbon sequestration potentials of the biosphere as a contribution to mitigating climate change, in particular as these benefits may be smaller than the counteracting impacts of land-use change. Persistent grassland responses to elevated CO_2, which range from 0 to 40% biomass gain per season, mainly reflect CO_2-induced water savings induced by scaled-up impacts of reduced stomatal conductance (Morgan et al., 2004; Gerten et al., 2005), and thus rely on current moisture regimes and lack the realistic atmospheric feedback of the future that may negate this benefit. The only replicated test of multiple CO_2 × climate/environment interactions (water, temperature, nutrient supply) yielded no overall CO_2 biomass signal in a grassland system (Shaw et al., 2002), highlighting the significant influence of co-limiting environmental variables.

Similar trends are emerging for forests, although the interpretation is complicated by time-lags in biomass response to the artifactual step-change when initiating CO_2 treatments, requiring longer observation periods before a new steady state (e.g., in terms of leaf area index, fine root dynamics and nutrient cycling) is reached. Three tall forest test systems, loblolly pine plantation (Oren et al., 2001; Schäfer et al., 2003), sweet gum plantation (Norby et al., 2002; Norby and Luo, 2004), and mixed deciduous forest (Körner et al., 2005a) exhibit significant initial biomass stimulation that diminishes with time except for one of the four pairs of test plots (treatment versus control) in the joint Duke pine experiments (Schäfer et al., 2003). A European boreal forest system also showed smaller CO_2 growth stimulation in mature trees under field conditions than expected from results for saplings (Rasmussen et al., 2002). A recent analysis (Norby et al., 2005) suggests that the NPP response of trees to elevated CO_2 is relatively predictable across a broad range of sites, with a stimulation of 23 ± 2% at a median CO_2 of double the pre-industrial level. The logarithmic biotic growth factor derived from this is 0.60 (β-factor, expressing the response as a function of the relative CO_2 increase). Nonetheless, it is uncertain whether test systems with mostly young growing trees provide valid analogies for biomass responses in mature forests with a steady state nutrient cycle and many other factors moderating the response to elevated CO_2 concentrations (e.g., Karnosky, 2003).

It has been suggested that greatest CO_2-fertilisation impacts may be seen in savanna systems post-fire (Bond and Midgley, 2000; Bond et al., 2003), especially where nutrients are less limiting and in systems in which trees require carbon reserves to re-establish after fire (see Section 4.2.2). Scrub oak in Florida shows diminishing CO_2 responses as treatment proceeds (Hungate et al., 2006), even though this is a post-fire regenerating system. For tropical forests, the planet's single largest biomass carbon reservoir, post-industrial atmospheric CO_2 enrichment seems to have enhanced growth dynamics (Phillips et al., 2002; Laurance et al., 2004; Wright et al., 2004). A more dynamic forest might ultimately store less rather than more carbon in future if long-term species compositional changes are realised (Laurance et al., 2004; Malhi et al., 2006), especially given the exceptional CO_2 responsiveness of tropical lianas that may increase tree mortalities and population turnover (Körner, 2004).

Based on experimental data, best estimates of instantaneous CO_2-induced water savings due to reduced stomatal aperture range from 5 to 15% (Wullschleger and Norby, 2001; Cech et al., 2003) for humid conditions, diminishing with drying soils. Desert shrub systems increase production in elevated CO_2 only during exceptional wet periods and not in dry periods (Nowak et al., 2004), contrasting with earlier expectations (Morgan et al., 2004). Evapotranspiration data for temperate zone ecosystems under future CO_2 scenarios suggest that this may be reduced by less than 10% across all weather conditions. Water savings through elevated CO_2 hold limited benefits for trees during drought, because nutrient availability in drying top soil becomes interrupted, and initial water savings are exhausted (Leuzinger et al., 2005). Repeated drought with high temperatures (e.g., Europe in 2003, Box 4.1) may reduce landscape-wide carbon stocks (Ciais et al., 2005). Studies using a land-surface model indicate at least for the past century a hydrological response up to the global scale of increasing runoff (e.g., Gedney et al., 2006) that is consistent with expected stomatal responses to rising CO_2 (e.g., Hetherington and Woodward, 2003; Gedney et al., 2006).

Soil nitrogen availability is key to predicting future carbon sequestration by terrestrial ecosystems (Reich et al., 2006), especially in light of global nitrogen-deposition trends (2-10 fold increase in some industrialised areas – Matson et al., 2002). The future ability of ecosystems overall to sequester additional carbon is very likely to be constrained by levels of nitrogen availability and fixation, and other key nutrients such as phosphorus that may also become increasingly limiting (Hungate et al., 2003). Carbon accumulation and sequestration in critical soil stocks (see Figure 4.1) has been found to be strongly nitrogen-constrained, both because levels well above typical atmospheric inputs are needed to stimulate soil C-sequestration, and because natural N_2-fixation appears to be particularly strongly limited by key nutrients (van Groenigen et al., 2006).

Results from a loblolly pine forest (Lichter et al., 2005) and grassland experiments (Van Kessel et al., 2000) suggest a reduced likelihood for CO_2-fertilisation-driven carbon accumulation in soils, probably because carbon sequestration to humus is more nutrient-demanding (not only nitrogen), than is wood formation, for example (Hungate et al., 2006). Carbon accretion in soil is therefore itself likely to exert negative feedback on plant growth by immobilising soil nutrients (in addition to cation depletion by acidic precipitation), contributing to a faster diminishing of the biospheric sink (see Figure 4.2; Reich et al., 2006) than implemented in model projections (e.g., Scholze et al., 2006; see Figure 4.2).

Accumulation of seasonally transitory soil carbon pools such as in fine roots has been found at elevated ambient CO_2 concentrations, but the general validity of such enhanced C-fluxes and what fraction of these might be sequestered to recalcitrant (see Glossary) soil carbon stocks remains unresolved (Norby et al., 2004). Soil warming may enhance carbon emissions, especially by reducing labile soil organic carbon pools (Davidson and Janssens, 2006). This results in the commonly observed short-term (less than decadal) loss of carbon in warming experiments, followed by the re-establishment of a new equilibrium between inputs and losses of soil carbon (e.g., Eliasson et al., 2005; Knorr et al., 2005). Recent observations indeed show widespread carbon losses from soils (Bellamy et al., 2005; Schulze and Freibauer, 2005) that are consistent with this formulation. However, in regions with thawing permafrost, a decay of historically accumulated soil carbon stocks (yedoma, >10,000 years old, Figure 4.1) due to warming (Zimov et al., 2006) and nutrient deposition (Blodau, 2002; Mack et al., 2004) could release large amounts of carbon to the atmosphere (see also Section 4.4.6). Increased NPP (but see Angert et al., 2005) and vegetation change (see Section 4.4.5 and, e.g., Sturm et al., 2001) may partly counterbalance this carbon release (see Section 4.4.6 and Sitch et al., 2007), thus complicating projections (Blodau, 2002; for a full discussion see Section 4.4.6).

Ecosystem changes associated with land-use and land-cover change (see Section 4.2.2) are complex, involving a number of feedbacks (Lepers et al., 2005; Reid et al., 2005). For example, conversion of natural vegetation to agricultural land drives climate change by altering regional albedo and latent heat flux, causing additional summer warming in key regions in the boreal and Amazon regions, and winter cooling in the Asian boreal zone (Chapin et al., 2005b; Feddema et al., 2005), by releasing CO_2 via losses of biomass and soil carbon (Gitz and Ciais, 2003; Canadell et al., 2004; Levy et al., 2004) and through a 'land-use amplifier effect' (Gitz and Ciais, 2003). In contrast, reforestation, and other land-use or land-management changes such as modifications of agricultural practices, can work to mitigate climate change through carbon sequestration (Lal, 2003, 2004; Jones and Donnelly, 2004; King et al., 2004a; Wang et al., 2004a; de Koning et al., 2005; Nabuurs et al., 2007). This mitigation potential is probably limited to reducing the ultimate atmospheric CO_2 increase by 2100 by between 40 and 70 ppm (House et al., 2002), and by approximately century-long time-lags until mature forests are established (see Sections 4.4.5, 4.4.6 and 4.4.10), and is probably offset by regional warming effects of lower albedo with poleward boreal forest expansion (e.g., Betts, 2000; for a full discussion see Section 4.4.6).

The sequestration and cycling of carbon in terrestrial ecosystems is a key vulnerability, given the above drivers, their generally global extent, their potential irreversibility, and the likely existence of threshold-type impacts. The extent to which the recently discovered methane release from plant foliage (Keppler et al., 2006) can be scaled to biome level is under debate (Houweling et al., 2006), and highlights the currently limited understanding of the methane cycle, and its exclusion from Earth system models (e.g., Betts and Shugart, 2005). Nonetheless, recent work especially with DGVM approaches, has begun to elucidate the likelihood of occurrence of important thresholds, and positive feedback to the atmosphere through diminishing CO_2 sequestration or even net carbon release from ecosystems, thus amplifying climate change (e.g., Friedlingstein et al., 2006; Lucht et al., 2006; Scholze et al., 2006).

Global estimates (IS92a, HACM2-SUL – Cramer et al., 2001) suggest a reduced global sink relative to that expected under CO_2-fertilisation alone, both in 2000 (0.6 ± 3.0 $PgCy^{-1}$) and 2100 (0.3 ± 6.6 $PgCy^{-1}$) as a result of climate change impacts on Net Biome Productivity (NBP) of tropical and Southern Hemisphere ecosystems. According to these models, the rate of NBP increase slows by around 2030 as CO_2-fertilisation itself saturates, and in four of six models shows further, climate-induced, NBP declines, due to increased heterotrophic respiration and declining tropical NPP after 2050. These trends are projected to continue until mid-century, even with stabilised atmospheric CO_2 concentration and instantaneously stabilised climate beyond

2100 (Woodward and Lomas, 2004b; see also next paragraph, Figure 4.2). More recent modelling based on projected deforestation and climate change (for the IS92a emissions scenario and the CGCM1, CSIRO, ECHAM, HadCM3 climate models) in the tropics alone suggests an additional release of 101 to 367 PgC, adding between 29 and 129 ppm to global atmospheric CO_2 by 2100, mainly due to deforestation (Cramer et al., 2004).

Climate scenario uncertainty provides a substantial variance in global terrestrial C balance by 2100, even under a single CO_2 emissions scenario (IS92a, projected to reach 703 ppm atmospheric CO_2 concentration by 2100, excluding vegetation feedback). Using five General Circulation Models (GCMs) to drive DGVMs, global terrestrial C-sequestration is estimated at between −106 and +201 PgC (Schaphoff et al., 2006), though in four out of five, the sink service decreased well before 2060. A risk assessment for terrestrial biomes and biogeochemical cycling shows that a terrestrial carbon source is predicted in almost half of 52 GCM × emissions scenario combinations, and that wildfire frequency increases dramatically even for a warming of <2°C by 2100 (Scholze et al., 2006). Here we show model results for the most recent version of the DGVM Lund-Potsdam-Jena Model (LPJ) (Schaphoff et al., 2006) highlighting changes in biome structure (relative cover of dominant growth forms) and the terrestrial carbon sink under more recent IPCC emissions scenarios SRES A2 and B1 (Nakićenović et al., 2000). This supports projections of diminishing terrestrial C-sequestration as early as 2030 (Figure 4.2) – earlier than suggested in the TAR (Prentice et al., 2001, Figure 3.10) – and substantial shifts in biome structure (Figure 4.3); discussed more fully in Sections 4.4.10 and 4.4.11.

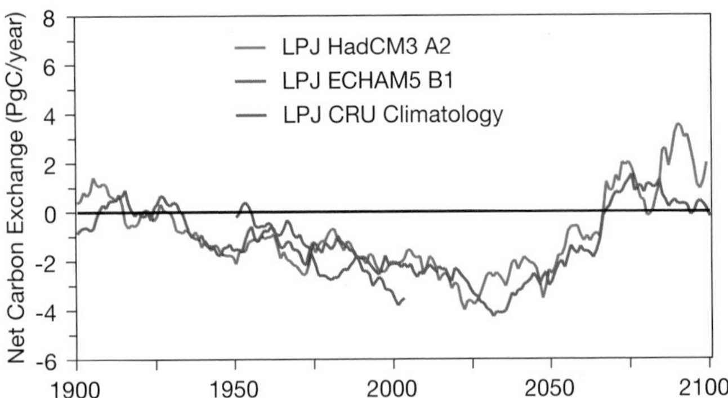

Figure 4.2. *Net carbon exchange of all terrestrial ecosystems as simulated by the DGVM LPJ (Sitch et al., 2003; Gerten et al., 2004 – negative values mean a carbon sink, positive values carbon losses to the atmosphere). Past century data are based on observations and climate model data were normalised to be in accord with these observations for the 1961-1990 data (CRU-PIK). Transient future projections are for the SRES A2 and B1 emissions scenarios (Nakićenović et al., 2000), forcing the climate models HadCM3 and ECHAM5, respectively (cf. Lucht et al., 2006; Schaphoff et al., 2006). In contrast to previous global projections (Prentice et al., 2001 – Figure 3.10), the world's ecosystems sink service saturates earlier (about 2030) and the terrestrial biosphere tends to become a carbon source earlier (about 2070) and more consistently, corroborating other projections of increased forcing from biogenic terrestrial sources (e.g., Cox et al., 2000, 2004; White et al., 2000a; Lucht et al., 2006; Schaphoff et al., 2006; Scholze et al., 2006; see Figure 4.3 for maps on underlying ecosystem changes). Note that these projections assume an effective CO_2-fertilisation (see Section 4.4.1).*

Projections from modelling that dynamically link the physical climate system and vegetation, using Ocean-Atmosphere General Circulation Models (OAGCMs, e.g., Cox et al., 2000), suggest a terrestrial C source that will exacerbate both climate and further vegetation change to at least some degree (e.g., Sarmiento, 2000; Dufresne et al., 2002; Canadell et al., 2004). Impacts include the collapse of the Amazon forest (e.g., White et al., 2000a; Cox et al., 2004), and an overall C source from the tropics that exceeds the boreal C sink (Berthelot et al., 2002), leading to an 18% (Dufresne et al., 2002), 5 to 30% (Friedlingstein et al., 2006), and 40% (Cox et al., 2000) higher atmospheric CO_2 concentration by 2100. Carbon and water cycling, at least, are also affected by shifting biogeographical zones (Gerten et al., 2005) which will be lagged by migration constraints that are not yet incorporated in DGVM approaches (see also Sections 4.4.5 and 4.4.6), leading to a potential overestimation of vegetation C-sequestration potential. This is especially so for boreal regions, due to unrealistically high projections of in-migration rates of trees and shrubs (Neilson et al., 2005).

Changes in air-sea fluxes of dimethyl sulphide (DMS) from −15% to 30% caused by global warming of about 2°C are projected to have a regional radiative and related climatic impact (Bopp et al., 2003, 2004), as DMS is a significant source of cloud condensation nuclei. DMS is produced by coccolithophores, which are sensitive to high sea-water CO_2 (Riebesell et al., 2000). As the largest producers of calcite on the planet (Holligan et al., 1993), reduced calcification by these organisms may also influence the global carbon cycle (Raven et al., 2005) and global albedo (Tyrrell et al., 1999). N_2O of marine origin contributes about 33% of total input to the atmosphere (Enhalt and Prather, 2001). Changes to the concentration and distribution of oxygen in the oceans, either through increased stratification of the surface waters (Sarmiento et al., 1998) or through a decrease in the strength of the thermohaline circulation (IPCC, 2001), will impact the ocean nitrogen cycling, especially the processes of nitrification and denitrification which promote N_2O production.

4.4.2 Deserts

Properties, goods and services

One of the largest terrestrial biomes, deserts cover 27.7 Mkm^2, comprising extra-polar regions with mean annual precipitation <250 mm and an unfavourable precipitation to potential evaporation ratio (Nicholson, 2002; Warner, 2004; Reid et al., 2005). Deserts support on the order of 10 people per km^2, in sparse populations with among the lowest gross domestic product (GDP) of all ecoregions (Reid et al., 2005). Recent estimates suggest that between 10 and 20% of deserts and drylands are degraded due to an imbalance between demand for and supply of ecosystem services (Adeel et al., 2005). Critical provisioning goods and services include wild food sources, forage and rangeland grazing, fuel, building materials, and water for humans and livestock, for irrigation and for sanitation, and genetic resources, especially of arid-adapted species (Adeel et al., 2005; Hassan et al., 2005). Regulating services include air quality, atmosphere composition and climate regulation (Hassan et al., 2005), especially through wind-blown dust and desert albedo influences on regional rainfall, and biogeochemistry of remote terrestrial and marine ecosystems (Warner, 2004).

Key vulnerabilities

The TAR noted several vulnerabilities in drylands (Gitay et al., 2001, p. 239) but chiefly that human overuse and land degradation, exacerbated by an overall lack of infrastructure and investment in resource management, would be very likely to overwhelm climate change impacts, with the exception of impacts of increased dry and wet extremes due to ENSO frequency increase, and negative impacts of projected warming and drying in high biodiversity regions. On the other hand, evidence for region-specific increases in productivity and even community compositional change due to rising atmospheric CO_2 was reported, with associated increased biomass and soil organic matter. Overall impacts of elevated CO_2 were reported as comparable, though usually opposite in sign, to climate change projections. Since the TAR, further work shows that desert biodiversity is likely to be vulnerable to climate change (Reid et al., 2005), with winter-rainfall desert vegetation and plant and animal species especially vulnerable to drier and warmer conditions (Lenihan et al., 2003; Simmons et al., 2004; Musil et al., 2005; Malcolm et al., 2006), and continental deserts vulnerable to desiccation and even soil mobilisation, especially with human land-use pressures (Thomas and Leason, 2005). However, the potentially positive impact of rising atmospheric CO_2 remains a significant uncertainty, especially because it is likely to increase plant productivity, particularly of C_3 plants (Thuiller et al., 2006b) and, together with rainfall change, could even induce wildfires (Bachelet et al., 2001; Hardy, 2003; Duraiappah et al., 2005). The uncertain impact of elevated CO_2 on vegetation productivity and biogeochemical cycling in deserts is an important source of contrasting projections of impacts and vulnerability for different desert regions and vegetation types. Climate change and direct human land-use pressure are likely to have synergistic impacts on desert ecosystems and species that may be offset, at least partly, by vegetation productivity and carbon sequestration gains due to rising atmospheric CO_2. The net effect of these trends is very likely to be region-specific.

Impacts

Deserts are likely to experience more episodic climate events, and interannual variability may increase in future, though there is substantial disagreement between GCM projections and across different regions (Smith et al., 2000; Duraiappah et al., 2005). Continental deserts could experience more severe, persistent droughts (Lioubimtseva and Adams, 2004; Schwinning and Sala, 2004). Vulnerability to desertification will be enhanced due to the indicated increase in the incidence of severe drought globally (Burke et al., 2006). In the Americas, temperate deserts are projected to expand substantially under doubled CO_2 climate scenarios (Lauenroth et al., 2004). However, dry-spell duration and warming trend effects on vegetation productivity may be at least partly offset by rising atmospheric CO_2 effects on plants (Bachelet et al., 2001; Thuiller et al., 2006b), leading to sometimes contrasting projections for deserts that are based on different modelling techniques that either incorporate or ignore CO_2-fertilisation effects.

Elevated CO_2 has been projected to have significant potential impacts on plant growth and productivity in drylands (Lioubimtseva and Adams, 2004). This projection has been confirmed for cool desert shrub species (Hamerlynck et al., 2002), and both desert shrubs and invasive (but not indigenous) grasses in wet years only (Smith et al., 2000). On the whole, evidence for CO_2-fertilisation effects in deserts is conflicting, and species-specific (Lioubimtseva and Adams, 2004; Morgan et al., 2004). In the south-western USA the total area covered by deserts may decline by up to 60% if CO_2-fertilisation effects are realised (Bachelet et al., 2001). Limited direct impacts of atmospheric CO_2 on nitrogen-fixation have been found in soil biological crusts (Billings et al., 2003), but soil microbial activity beneath shrubs has been observed to increase, thus reducing plant-available nitrogen (Billings et al., 2002).

Soil vulnerability to climate change is indicated by shallow desert substrates with high soluble salts and the slow recolonisation of disturbed soil surfaces by different algae components (Evans and Belnap, 1999; Johansen, 2001; Duraiappah et al., 2005). Very low biomass (a drop below a 14% cover threshold) is very likely to make the Kalahari desert dune system in southern Africa susceptible to aeolian erosion (Thomas and Leason, 2005) and, with regional warming of between 2.5 and 3.5°C, most dune fields could be reactivated by 2100 (Thomas and Leason, 2005). Increased dust flux may increase aridity and suppress rainfall outside deserts, with opposite effects under wetting scenarios (Bachelet et al., 2001; Hardy, 2003; Prospero and Lamb, 2003; Lioubimtseva and Adams, 2004), leading to indirect effects on the vulnerability of remote regions to climate change. About one-third of the Sahel was projected to aridify with warming of 1.5 to 2°C by about 2050, with a general equatorward shift of vegetation zones (van den Born et al., 2004; Box 4.2). Alternative climate scenarios show less pronounced changes (van den Born et al., 2004).

Episodic wet periods may increase vulnerability to invasive alien species and subsequent fire outbreaks and this, combined with land overuse, will increase vulnerability to degradation and desertification (Dukes and Mooney, 1999; Dube and Pickup, 2001; Holmgren and Scheffer, 2001; Brooks et al., 2004; Geist and Lambin, 2004; Lioubimtseva and Adams, 2004). Wet spells with elevated humidity and warmer temperatures will increase the prevalence of plant diseases (Harvell et al., 2002).

Desert biodiversity is likely to be vulnerable to climate change (Reid et al., 2005), especially in so-called 'biodiversity hotspots' (Myers et al., 2000). In the Succulent Karoo biome of South Africa, 2,800 plant species face potential extinction as bioclimatically suitable habitat is reduced by 80% with a global warming of 1.5-2.7°C above pre-industrial levels (see Table 4.1). Daytime *in situ* warming experiments suggest high vulnerability of endemic succulent (see Glossary) growth forms of the Succulent Karoo to high-end warming scenarios for 2100 (mean 5.5°C above current ambient temperatures), inducing appreciable mortality in some (but not all) succulent species tested within only a few months (Musil et al., 2005). Desert species that depend on rainfall events to initiate breeding, such as resident birds, and migratory birds whose routes cross deserts,

Box 4.2. Vegetation response to rainfall variability in the Sahel

The Sahel falls roughly between the 100-200 mm/year (northern boundary) and 400-600 mm/year rainfall isohyets (southern boundary), and supports dry savanna vegetation forming transition zones with the Sahara and humid tropical savanna (Nicholson, 2000; Hiernaux and Turner, 2002; Anyamba and Tucker, 2005). These transition zones have historically fluctuated in response to rainfall changes (Hiernaux and Turner, 2002), in the clearest example of multi-decadal variability measured during the past century (Hulme, 2001). Ecosystem responses to past rainfall variability in the Sahel are potentially useful as an analogue of future climate change impacts, in the light of projections that extreme drought-affected terrestrial areas will increase from 1% to about 30% globally by the 2090s (Burke et al., 2006).

During the mid-Holocene, conditions supporting mesic vegetation and abundant wildlife deteriorated rapidly (ECF, 2004; Foley et al., 2003), highlighting the Sahel's sensitivity to forcing effects. The Sahel has shown the largest negative trends in annual rainfall observed globally in the past century, though these reversed somewhat after the late 1970s (Trenberth et al., 2007). Since about 1900, multi-decadal-scale rainfall variability persisted, with drying trends between around 1930-1950 and 1960-1985 (Hulme, 2001; Nicholson, 2001). Conditions apparently improved between 1950 and 1960, with limited evidence suggesting increased human and livestock numbers (Reij et al., 2005). Severe drought prevailed in the early 1980s (Hulme, 2001; Trenberth et al., 2007), and groundwater levels declined, species-specific woody plant mortality increased (mainly of smaller plants), and even dominant perennial C_4 grasses with high water-use efficiency declined. Exposed soil caused increased atmospheric dust loads (Nicholson, 2000, 2001). These events stimulated the concept of desertification and subsequent debates on its causes (Herrmann and Hutchinson, 2005).

The persistence of drought during the latter part of the 20th century prompted suggestions that land-cover change had exerted a positive feedback to reinforce drought conditions, but the modelled vegetation change necessary to induce this effect does not reflect reality (Hulme, 2001). During relatively wet periods (Nicholson et al., 2000; Anyamba and Tucker, 2005; Trenberth et al., 2007) spatially variable regeneration in both the herbaceous and the woody layer have been observed (Gonzalez, 2001; Rasmussen et al., 2001; Hiernaux and Turner, 2002). Remote sensing shows the resilience of Sahelian vegetation to drought, with no directional change in either desert transition zone position or vegetation cover (Nicholson et al., 1998). Sahel green-up between the years 1982 and 1998 (Prince et al., 1998; Hickler et al., 2005) and between 1994 and 2003 (Anyamba and Tucker, 2005) has been noted, but this interpretation has recently been challenged (Hein and Ridder, 2006).

Drivers of Sahel vegetation change remain uncertain (Hutchinson et al., 2005), especially because the correlation between rainfall and Normalised Difference Vegetation Index (NDVI) appear weak, signalling that greening cannot be fully explained by increasing rainfall (Olsson et al., 2005), and greening may not comprise a return to the initial species composition, cover and surface soil conditions (Warren, 2005). Inconclusive interpretations of vegetation dynamics in the Sahel may reflect complex combined effects of human land use and climate variability on arid environments (Rasmussen et al., 2001). It is far from clear how the interactive effect of climate change, land-use activities and rising CO_2 will influence the Sahel in future. Green-up, or desert amelioration (Figure 4.3, vegetation class 4) due to rising CO_2 and enhanced water-use efficiency (as observed by Herrmann et al., 2005) may accrue only in wet years (Morgan et al., 2004).

will be severely affected (Dukes and Mooney, 1999; Hardy, 2003; Box 4.5). The Mountain Wheatear in South Africa was projected to lose 51% of its bioclimatic range by 2050 under an SRES A2 emissions scenario (Simmons et al., 2004). In contrast, desert reptile species could be favoured by warming, depending on rainfall scenario (Currie, 2001).

4.4.3 Grasslands and savannas

Properties, goods and services

Dominated by a spatially and temporally variable mix of grass and tree-growth forms (Sankaran et al., 2005), grasslands and savannas include tropical C_4 grasslands and savannas (C_4 grass-dominated with 10-50% tree cover, about 28 Mkm^2) and temperate C_4 and/or C_3-grass and herb-dominated grasslands (15 million km^2; Bonan, 2002). Generally rich in grazing, browsing and other fauna (especially but not only in Africa), these systems are strongly controlled by fire (Bond et al., 2005) and/or grazing regimes (Scholes and Archer, 1997; Fuhlendorf et al., 2001). Disturbance regimes are often managed (e.g., Sankaran, 2005), although fire regimes depend also on seasonality of ignition events and rainfall-dependent accumulation of flammable material (Brown et al., 2005b). Temperate and tropical systems provide somewhat distinct goods and services. Temperate grasslands contain a substantial soil carbon pool, are important for maintaining soil stability and provide fodder for wild and domestic animals. Tropical savanna systems possess significant wild faunal diversity that supports nature-based tourism revenue (both extractive and non-extractive) and subsistence livelihoods (food, medicinal plants, and construction material), in addition to cultural, regulating and supporting services.

Key vulnerabilities

The structure, productivity and carbon balance of these systems appear more sensitive than indicated in the TAR to variability of, and changes in, major climate change drivers. The direct CO_2-fertilisation impact and warming effect of rising atmospheric CO_2 have contrasting effects on their dominant functional types (trees and C_3 grasses may benefit from rising CO_2 but not from warming; C_4 grasses may benefit from warming, but not from CO_2-fertilisation), with uncertain, non-linear and rapid changes in ecosystem structure and carbon stocks likely. Carbon stocks are very likely to be strongly reduced under more frequent disturbance, especially by fire, and disturbance and drought impacts on cover may exert regional feedback effects. On balance, savannas and grasslands are likely to show reduced carbon sequestration due to enhanced soil respiratory losses through warming, fire regime changes and increased rainfall variability, but possible regional gains in woody cover through direct CO_2-fertilisation, and increased plant carbon stocks, cannot be excluded. Scientific predictive skill is currently limited by very few field-based, multi-factorial experiments, especially in tropical systems. Projected range shifts of mammal species will be limited by fragmented habitats and human pressures, as suggested in the TAR, with declines in species richness likely, especially in protected areas. Because of the important control by disturbance, management options exist to develop adaptive strategies for carbon sequestration and species conservation goals.

Impacts

Ecosystem function and species composition of grasslands and savanna are likely to respond mainly to precipitation change and warming in temperate systems but, in tropical systems, CO_2-fertilisation and emergent responses of herbivory and fire regime will also exert strong control. Very few experimental approaches have assessed ecosystem responses to multi-factorial treatments such as listed above (Norby and Luo, 2004), and experiments on warming, rainfall change or atmospheric CO_2 level are virtually absent in savannas, with many ecosystem studies confined mainly to temperate grasslands (Rustad et al., 2001).

Rainfall change and variability is very likely to affect vegetation in tropical grassland and savanna systems with, for example, a reduction in cover and productivity simulated along an aridity gradient in southern African savanna in response to the observed drying trend of about 8 mm/yr since 1970 (Woodward and Lomas, 2004a). Sahelian woody plants, for example, have shown drought-induced mass mortality and subsequent regeneration during wetter periods (Hiernaux and Turner, 2002). Large-scale changes in savanna vegetation cover may also feed back to regional rainfall patterns. Modelled removal of savannas from global vegetation cover has larger effects on global precipitation than for any other biome (Snyder et al., 2004) and, in four out of five savannas studied globally, modelled savanna-grassland conversion resulted in 10% lower rainfall, suggesting positive feedback between human impacts and changing climate (Hoffmann and Jackson, 2000). At the continental scale, modelled forest-savanna conversion reduced rainfall in tropical African regions, but increased it in central southern Africa (Semazzi and Song, 2001).

Changing amounts and variability of rainfall may also strongly control temperate grassland responses to future climate change (Novick et al., 2004; Zha et al., 2005). A Canadian grassland fixed roughly five times as much carbon in a year with 30% higher rainfall, while a 15% rainfall reduction led to a net carbon loss (Flanagan et al., 2002). Similarly, Mongolian steppe grassland switched from carbon sink to source in response to seasonal water stress, although carbon balance was neutral on an annual basis (Li et al., 2005). Non-linear responses to increasing rainfall variability may be expected, as ecosystem models of mixed C_3/C_4 grasslands show initially positive NPP relationships with increasing rainfall variability, but greater variability ultimately reduces both NPP and ecosystem stability even if the rainfall total is kept constant (Mitchell and Csillag, 2001). Empirical results for C_4 grasslands confirm a similar monotonic (hump-backed) relationship between NPP and rainfall variability (Nippert et al., 2006). Increased rainfall variability was more significant than rainfall amount for tall-grass prairie productivity (Fay et al., 2000, 2002), with a 50% increase in dry-spell duration causing 10% reduction in NPP (Fay et al., 2003) and a 13% reduction in soil respiration (Harper et al., 2005).

The CO_2-fertilisation and warming effect of rising atmospheric CO_2 have generally opposite effects on savanna- and grassland-dominant functional types, with CO_2-fertilisation favouring woody C_3 plants (Ainsworth and Long, 2005), and warming favouring C_4 herbaceous types (Epstein et al., 2002). Simulated heat-wave events increased C_4 dominance in a mixed C_3/C_4 New Zealand grassland within a single growing season, but reduced productivity by over 60% where C_4 plants were absent (White et al., 2000b). Some African savanna trees are sensitive to seasonal high air temperature extremes (Chidumayo, 2001). North American forest vegetation types could spread with up to 4°C warming; but with greater warming, forest cover could be reduced by savanna expansion of up to 50%, partly due to the impacts of fire (Bachelet et al., 2001).

Elevated CO_2 has important effects on production and soil water balance in most grassland types, mediated strongly by reduced stomatal conductance and resulting increases in soil water (Leakey et al., 2006) in many grassland types (Nelson et al., 2004; Niklaus and Körner, 2004; Stock et al., 2005). In short-grass prairie, elevated CO_2 and 2.6°C warming increased production by 26-47% , regardless of grass photosynthetic type (Morgan et al., 2001a). In C_4 tropical grassland, no relative increase in herbaceous C_3 success occurred in double-ambient CO_2 (Stock et al., 2005). Regional climate modelling indicates that CO_2-fertilisation effects on grasslands may scale-up to affect regional climate (Eastman et al., 2001).

Differential effects of rising atmospheric CO_2 on woody relative to herbaceous growth forms are very likely (Bond and Midgley, 2000). Trees and shrubs show higher CO_2 responsiveness than do herbaceous forms (Ainsworth and Long,

2005). Savannas may thus be shifting towards greater tree dominance as atmospheric CO_2 rises, with diminishing grass suppression of faster-growing tree saplings (Bond et al., 2003). Simulations suggest that rising CO_2 may favour C_3 forms at the expense of African C_4 grasses (Thuiller et al., 2006b), even under projected warming. Continuing atmospheric CO_2 rise could increase the resilience of Sahelian systems to drought (Wang and Eltahir, 2002). However, without definitive tests of the CO_2-fertilisation effect on savanna trees, other factors can be invoked to explain widely observed woody plant encroachment in grassland systems (Van Auken, 2000).

Above-ground carbon stocks in savannas are strongly contingent on disturbance regimes. Australian savanna systems are currently a net carbon sink of 1-3 t C/ha/yr, depending on fire frequency and extent (Williams et al., 2004b). Fire exclusion can transform savannas to forests (e.g., Bowman et al., 2001), with an upper (albeit technically unfeasible) global estimate of potential doubling of closed forest cover (Bond et al., 2005). Thus savanna structure and carbon stocks are very likely to be responsive to both individual and interactive effects of the disturbance regime (Bond et al., 2003; Sankaran et al., 2005) and atmospheric CO_2 change (Bond and Midgley, 2000).

There are few factorial experiments on multiple changing factors, but they suggest interactions that are not predictable from single factor experiments – such as the dampening effect of elevated CO_2 on California C_3 grassland responses to increased rainfall, nitrate and air temperature (Shaw et al., 2002). Increasing temperature and rainfall changes are seen to override the potential benefits of rising CO_2 for C_3 relative to C_4 grasses (Winslow et al., 2003), and European C_3 grassland showed minor responses to a 3°C rise in temperature, possibly due to concomitant drying impacts (Gielen et al., 2005). Elevated CO_2 impacts on grassland carbon sequestration also seem to be dependent on management practices (Harmens et al., 2004; Jones and Donnelly, 2004), and are complicated by being species- but not functional-type specific (Niklaus et al., 2001; Hanley et al., 2004).

Soil-mediated responses are important in biogeochemical controls of vegetation response. Long-term CO_2 enrichment of southern African C_4 grassland revealed limited impacts on nitrogen cycling and soil C sequestration (Stock et al., 2005), in contrast to greater C sequestration in short-term studies of grassland ecosystems (e.g., Williams et al., 2004a). Likewise, elevated CO_2 impacts on litter decomposition and soil fauna seem species-specific and relatively minor (Ross et al., 2002; Hungate et al., 2000). Warming of a tall-grass prairie showed increased plant growth that supported enhanced soil fungal success (Zhang et al., 2005). However, complex interactions between plants and fungal symbionts showed potential impacts on soil structure that may predispose them to accelerated erosion (Rillig et al., 2002). Soil respiration shows approximately 20% increase in response to about 2.4°C warming (Norby et al., 2007), although acclimatisation of soil respiration (Luo et al., 2001) and root growth (Edwards et al., 2004) to moderate warming has also been observed. Soil carbon loss from UK soils, many in grasslands, confirm carbon losses of about 2% *per annum* in carbon-rich soils, probably related to regional climate change (Bellamy et al., 2005). In an African savanna system, rainfall after a dry spell generates substantial soil respiration activity and soil respiratory carbon losses (Veenendaal et al., 2004), suggesting strong sensitivity to rainfall variability.

Climate change impact studies for savanna and grassland fauna are few. The proportion of threatened mammal species may increase to between 10 and 40% between 2050 and 2080 (Thuiller et al., 2006a). Changing migration routes especially threaten migratory African ungulates and their predators (Thirgood et al., 2004). Observed population declines in three African savanna ungulates suggest that summer rainfall reductions could result in their local extirpation if regional climate change trends are sustained (Ogutu and Owen-Smith, 2003). For an African arid savanna raptor, population declines have been simulated for drier, more variable rainfall scenarios (Wichmann et al., 2003). A 4 to 98% species range reduction for about 80% of mainly savanna and grassland animal species in South Africa is projected under an IS92a emissions scenario (Erasmus et al., 2002).

4.4.4 Mediterranean ecosystems

Properties, goods and services

Mediterranean-type ecosystems (MTEs) are located in mid-latitudes on all continents (covering about 3.4 Mkm^2), often on nutrient-poor soils and in coastal regions. These biodiverse systems (Cowling et al., 1996) are climatically distinct, with generally wet winters and dry summers (Cowling et al., 2005), and are thus fire-prone (Montenegro et al., 2004). Vegetation structure is mainly shrub-dominated, but woodlands, forests and even grasslands occur in limited regions. Heavily utilised landscapes are dominated by grasses, herbs and annual plant species (Lavorel, 1999). MTEs are valuable for high biodiversity overall (Myers et al., 2000) and thus favour nature-based tourism, but many extractive uses include wildflower harvesting in South Africa and Australia, medicinal herbs and spices, and grazing in the Mediterranean Basin and Chile. Water yield for human consumption and agriculture is critical in South Africa, and these systems provide overall soil-protection services on generally unproductive nutrient-poor soils.

Key vulnerabilities

Mediterranean-type ecosystems were not explicitly reviewed in the TAR, but threats from desertification were projected due to expansion of adjacent semi-arid and arid systems under relatively minor warming and drying scenarios. Warming and drying trends are likely to induce substantial species-range shifts, and imply a need for migration rates that will exceed the capacity of many endemic species. Land use, habitat fragmentation and intense human pressures will further limit natural adaptation responses, and fire-regime shifts may threaten specific species and plant functional types. Vegetation structural change driven by dominant, common or invasive species may also threaten rare species. Overall, a loss of biodiversity and carbon sequestration services may be realised over much of these regions.

Impacts

These systems may be among the most impacted by global change drivers (Sala et al., 2000). Diverse Californian vegetation types may show substantial cover change for temperature increases greater than about 2°C, including desert and grassland expansion at the expense of shrublands, and mixed deciduous forest expansion at the expense of evergreen conifer forest (Hayhoe et al., 2004). The bioclimatic zone of the Cape Fynbos biome could lose 65% of its area under warming of 1.8°C relative to 1961-1990 (2.3°C, pre-industrial), with ultimate species extinction of 23% resulting in the long term (Thomas et al., 2004b). For Europe, only minor biome-level shifts are projected for Mediterranean vegetation types (Parry, 2000), contrasting with between 60 and 80% of current species projected not to persist in the southern European Mediterranean region (global mean temperature increase of 1.8°C – Bakkenes et al., 2002). Inclusion of hypothetical and uncertain CO_2-fertilisation effects in biome-level modelling may partly explain this contrast. Land abandonment trends facilitate ongoing forest recovery (Mouillot et al., 2003) in the Mediterranean Basin, complicating projections. In south-western Australia, substantial vegetation shifts are projected under double CO_2 scenarios (Malcolm et al., 2002b).

Climate change is likely to increase fire frequency and fire extent. Greater fire frequencies are noted in Mediterranean Basin regions (Pausas and Abdel Malak, 2004) with some exceptions (Mouillot et al., 2003). Double CO_2 climate scenarios increase wildfire events by 40-50% in California (Fried et al., 2004), and double fire risk in Cape Fynbos (Midgley et al., 2005), favouring re-sprouting plants in Fynbos (Bond and Midgley, 2003), fire-tolerant shrub dominance in the Mediterranean Basin (Mouillot et al., 2002), and vegetation structural change in California (needle-leaved to broad-leaved trees, trees to grasses) and reducing productivity and carbon sequestration (Lenihan et al., 2003).

Projected rainfall changes are spatially complex (e.g., Sumner et al., 2003; Sanchez et al., 2004; Vicente-Serrano et al., 2004). Rainfall frequency reductions projected for some Mediterranean regions (e.g., Cheddadi et al., 2001) will exacerbate drought conditions, and have now been observed in the eastern Mediterranean (Körner et al., 2005b). Soil water content controls ecosystem water and CO_2 flux in the Mediterranean Basin system (Rambal et al., 2003), and reductions are very likely to reduce ecosystem carbon and water flux (Reichstein et al., 2002). The 2003 European drought had major physiological impacts on Mediterranean vegetation and ecosystems, but most appeared to have recovered from drought by 2004 (Gobron et al., 2005; Box 4.1).

Many MTE species show apparently limited benefits from rising atmospheric CO_2 (Dukes et al., 2005), with constrained increases in above-ground productivity (e.g., Blaschke et al., 2001; Maroco et al., 2002). Yet modelling suggests that under all but extremely dry conditions, CO_2 increases over the past century have already increased NPP and leaf area index (see Glossary) in the Mediterranean Basin, despite warming and drying trends (Osborne et al., 2000). Rising atmospheric CO_2 appears increasingly unlikely to have a major impact in MTEs over the next decades, especially because of consistent projections of reduced rainfall. Elevated CO_2 is projected to facilitate forest expansion and greater carbon storage in California if precipitation increases (Bachelet et al., 2001). In the Mediterranean Basin, CO_2-fertilisation impacts such as increased forest success in the eastern Mediterranean and Turkey and increased shrub cover in northern Africa are simulated if rainfall does not decrease (Cheddadi et al., 2001). There is currently insufficient evidence to project elevated CO_2-induced shifts in ecosystem carbon stocks in MTE, but nutrient-limited systems appear relatively unaffected (de Graaff et al., 2006). Established *Pinus halepensis* (Borghetti et al., 1998) show high drought resistance, but Ponderosa pine forests had reduced productivity and water flux during a 1997 heatwave, and did not recover for the rest of the season, indicating threshold responses to extreme events (Goldstein et al., 2000). Mediterranean Basin pines (Martinez-Vilalta and Pinol, 2002) and other woody species (Peñuelas et al., 2001) showed species-specific drought tolerance under field conditions. Experimental drying differentially reduced productivity of Mediterranean Basin shrub species (Llorens et al., 2003, 2004; Ogaya and Peñuelas, 2004) and tree species (Ogaya and Peñuelas, 2003), but delayed flowering and reduced flower production of Mediterranean Basin shrub species (Llorens and Peñuelas, 2005), suggesting complex changes in species relative success under drying scenarios. Drought may also act indirectly on plants by reducing the availability of soil phosphorus (Sardans and Peñuelas, 2004).

Bioclimatic niche-based modelling studies project reduced endemic species' geographical ranges and species richness in the Cape Floristic region (Midgley et al., 2002, 2003, 2006). Ranges of trees and shrubs may shift unpredictably, and fragment, under IS92a emissions scenarios (Shafer et al., 2001). In southern Europe, species composition change may be high under a range of scenarios (Thuiller et al., 2005b). Range size reductions increase species' extinction risks, with up to 30 to 40% facing increased extinction probabilities beyond 2050 (Thomas et al., 2004a). Species of lowland plains may be at higher risk than montane species both in California (Peterson, 2003) and the Cape Floristic region (Midgley et al., 2003), although in the Mediterranean Basin, montane species show high risk (Thuiller et al., 2005b).

4.4.5 Forests and woodlands

Properties, goods and services

Forests are ecosystems with a dense tree canopy (woodlands have a largely open canopy), covering a total of 41.6 Mkm^2 (about 30% of all land) with 42% in the tropics, 25% in the temperate, and 33% in the boreal zone (Figure 4.1, e.g., Sabine et al., 2004). Forests require relatively favourable environmental conditions and are among the most productive terrestrial ecosystems (Figure 4.1). This makes them attractive both for climate change mitigation (Watson et al., 2000; Nabuurs et al., 2007) and agricultural uses. The latter underlies the currently high deforestation and degradation rates in tropical and sub-tropical regions (Hassan et al., 2005), leading to about one-quarter of anthropogenic CO_2 emissions (e.g., Houghton, 2003a). Nevertheless, forests sequester the largest fraction of terrestrial ecosystem carbon stocks, recently estimated at 1,640 PgC (Sabine et al., 2004; Figure 4.1), equivalent to about 220% of atmospheric carbon. In addition to commercial timber goods (see Chapter 5; Shvidenko et al., 2005, Section 21.5,

p. 600-607) forests provide numerous non-timber forest products, important for subsistence livelihoods (Gitay et al., 2001; Shvidenko et al., 2005). Key ecosystem services include habitat provision for an increasing fraction of biodiversity (in particular where subject to land-use pressures – Hassan et al., 2005; Duraiappah et al., 2005), carbon sequestration, climate regulation, soil and water protection or purification (>75% of globally usable freshwater supplies come from forested catchments – Shvidenko et al., 2005), and recreational, cultural and spiritual benefits (Millennium Ecosystem Assessment, 2005; Reid et al., 2005).

Key vulnerabilities

Forests, especially in the boreal region, have been identified as having a high potential vulnerability to climate change in the long term (Kirschbaum and Fischlin, 1996), but more immediately if disturbance regimes (drought, insects, fire), partly due to climate change, cross critical thresholds (Gitay et al., 2001). Since the TAR, most DGVM models based on A2 emissions scenarios show significant forest dieback towards the end of this century and beyond in tropical, boreal and mountain areas, with a concomitant loss of key services (Figure 4.3). Species-based approaches suggest losses of diversity, in particular in tropical forest diversity hotspots (e.g., north-eastern Amazonia – Miles, 2002) and tropical Africa (McClean et al., 2005), with medium confidence. Mountain forests are increasingly encroached upon from adjacent lowlands, while simultaneously losing high-altitude habitats due to warming (see also Section 4.4.7).

Impacts

Projections for some forests currently limited by their minimum climatic requirements indicate gains from climate change (Figure 4.3, vegetation changes 1 and 2), but many may be impacted detrimentally (Figure 4.3, vegetation change 6), notably for strong warming and its concomitant effects on water availability (Bachelet et al., 2001, 2003; Bergengren et al., 2001; Ostendorf et al., 2001; Smith and Lazo, 2001; Xu and Yan, 2001; Arnell et al., 2002; Enquist, 2002; Iverson and Prasad, 2002; Lauenroth et al., 2004; Levy et al., 2004; Matsui et al., 2004; Izaurralde et al., 2005; Fuhrer et al., 2006; Lucht et al., 2006; Schaphoff et al., 2006; Scholze et al., 2006; cf. Figure 4.3a versus b, vegetation change 6). Productivity gains may result through three mechanisms: (i) CO_2-fertilisation (although the magnitude of this effect remains uncertain in these long-lived systems, see Section 4.4.1); (ii) warming in cold climates, given concomitant precipitation increases to compensate for possibly increasing water vapour pressure deficits; and (iii) precipitation increases under water-limited conditions.

There is growing evidence (see Chapter 5, Section 5.4.1.1) that several factors may moderate direct CO_2 or climate-change effects on net ecosystem productivity in particular, namely nutrient dynamics (e.g., either enrichment or leaching resulting from N deposition), species composition, dynamic age structure effects, pollution and biotic interactions, particularly via soil organisms, (e.g., Karnosky et al., 2003; King et al., 2004b; Heath et al., 2005; Körner et al., 2005a; Section 4.4.1). Climate change impacts on forests will result not only through changes in mean climate, but also through changes in seasonal and diurnal rainfall and temperature patterns (as influenced by the hydrologically relevant surroundings of a forest stand, e.g., Zierl and Bugmann, 2005). Recently observed moderate climatic changes have induced forest productivity gains globally (reviewed in Boisvenue and Running, 2006) and possibly enhanced carbon sequestration, especially in tropical forests (Baker et al., 2004; Lewis et al., 2004a, 2004b; Malhi and Phillips, 2004; Phillips et al., 2004), where these are not reduced by water limitations (e.g., Boisvenue and Running, 2006) or offset by deforestation or novel fire regimes (Nepstad et al., 1999, 2004; Alencar et al., 2006) or by hotter and drier summers at mid- and high latitudes (Angert et al., 2005).

Potential increases in drought conditions have been quantitatively projected for several regions (e.g., Amazon, Europe) during the critical growing phase, due to increasing summer temperatures and precipitation declines (e.g., Cox et al., 2004; Schaphoff et al., 2006; Scholze et al., 2006; Figure 4.3, vegetation change 6). Since all these responses potentially influence forest net ecosystem productivity (NEP), substantive biotic feedbacks may result, either through carbon releases or influences on regional climate, contributing to further major uncertainties (e.g., Betts et al., 2000; Peng and Apps, 2000; Bergengren et al., 2001; Semazzi and Song, 2001; Leemans et al., 2002; Körner, 2003c; Canadell et al., 2004; Cox et al., 2004; Gruber et al., 2004; Heath et al., 2005; Section 4.4.1). Effects of drought on forests include mortality, a potential reduction in resilience (e.g., Lloret et al., 2004; Hogg and Wein, 2005) and can cause major biotic feedbacks (e.g., Ciais et al., 2005; Box 4.1). However, these effects remain incompletely understood and vary from site to site (e.g., Reichstein et al., 2002; Betts et al., 2004). For example, drought impacts can be offset by fertile soils (Hanson and Weltzin, 2000), or if due to a heatwave, drought may even be accompanied by enhanced tree growth at cooler high elevation sites due to a longer growing season and enhanced photosynthetic activity (Jolly et al., 2005; Box 4.1).

Drought conditions further interact with disturbances such as insects (Hanson and Weltzin, 2000; Fleming et al., 2002; Logan et al., 2003; Schlyter et al., 2006; Box 4.1) or fire (Flannigan et al., 2000). Tree-defoliating insects, especially in boreal forests, periodically cause substantial damage (e.g., Gitay et al., 2001, Box 5-10; Logan et al., 2003). Insect pests were found to be at least partly responsible for the decline and ultimate extirpation of stands at the southern margins of the range of their hosts, subjected to warmer and drier conditions (Volney and Fleming, 2000; see also Section 4.2.2). At the poleward ecotone (see Glossary), frosts and general low temperatures appear to limit insect outbreaks (Virtanen et al., 1996; Volney and Fleming, 2000); thus outbreaks currently constrained from northern ranges could become more frequent in the future (Carroll et al., 2004). If climate warms and this ecotone becomes exposed to more droughts, insect outbreaks will become a major factor (Logan et al., 2003; Gan, 2004). With A2 and B2 emissions scenarios downscaled to regional level in northern Europe, projected climate extremes by 2070-2100 will increase the susceptibility of Norway spruce to secondary damage through pests and pathogens, matched by an accelerated life cycle of spruce bark beetle populations (Schlyter et al., 2006).

Climate change is known to alter the likelihood of increased wildfire sizes and frequencies (e.g., Stocks et al., 1998; Podur et al., 2002; Brown et al., 2004; Gillett et al., 2004), while also inducing stress on trees that indirectly exacerbate disturbances (Dale et al., 2000; Fleming et al., 2002; Schlyter et al., 2006). This suggests an increasing likelihood of more prevalent fire disturbances, as has recently been observed (Gillett et al., 2004; van der Werf et al., 2004; Westerling et al., 2006; Section 4.2.2).

Considerable progress has been made since the TAR in understanding fire regimes and related processes (Kasischke and Stocks, 2000; Skinner et al., 2002; Stocks et al., 2002; Hicke et al., 2003; Podur et al., 2003; Gillett et al., 2004) enabling improved projections of future fire regimes (Flannigan et al., 2000; Li et al., 2000; de Groot et al., 2003; Brown et al., 2004; Fried et al., 2004). Some argue (e.g., Harden et al., 2000) that the role of fire regimes in the boreal region has previously been underestimated. About 10% of the 2002/2003 global carbon emission anomaly can be ascribed to Siberian fires by inverse modelling (van der Werf et al., 2004), as supported by remote sensing (Balzter et al., 2005). Climate changes including El Niño events alter fire regimes in fire-prone regions such as Australia (Hughes, 2003; Williams et al., 2004b; Allen Consulting Group, 2005), the Mediterranean region (e.g., Mouillot et al., 2002; see also Section 4.4.4), Indonesia and Alaska (Hess et al., 2001), but also introduce fire into regions where it was previously absent (e.g., Schumacher et al., 2006). Intensified fire regimes are likely to impact boreal forests at least as much as climate change itself (Flannigan et al., 2000), and may accelerate transitions, e.g., between taiga and tundra, through facilitating the invasion of pioneering trees and shrubs into tundra (Landhäusser and Wein, 1993; Johnstone and Chapin, 2006).

Will forest expansions be realised as suggested by DGVMs (Figure 4.3)? Vegetation models project that forest might eventually replace between 11 and 50% of tundra with a doubling of atmospheric CO_2 (White et al., 2000b; Harding et al., 2002; Kaplan et al., 2003; Callaghan et al., 2005; Figure 4.3, vegetation change 1). However, such transitions are likely to be moderated in reality by many processes not yet considered in the models (e.g., Gamache and Payette, 2005; see below). Other studies using a wide range of GCMs and forcing scenarios indicate that forests globally face the risk of major change (non-forested to forested and *vice-versa* within at least 10% of non-cultivated land area) in more than 40% of simulated scenarios if global mean warming remains below 2°C relative to pre-industrial, and in almost 90% of simulated scenarios if global mean warming exceeds 3°C over pre-industrial (Scholze et al., 2006). Those risks have been estimated as especially high for the boreal zone (44% and 88%, respectively) whereas they were estimated as smaller for tropical forests in Latin America (19% and 38%, respectively; see also Figure 4.3).

One key process controlling such shifts is migration (e.g., Higgins and Harte, 2006). Estimates for migration rates of tree species from palaeoecological records are on average 200-300 m/yr, which is a rate significantly below that required in response to anticipated future climate change (≥1 km/yr, Gitay et al., 2001, Box 5-2). However, considerable uncertainties remain:

- although not completely quantified, many species can achieve rapid large-scale migrations (Reid's paradox (see Glossary), e.g., Clark, 1998), but estimates at the low extreme imply a considerable range of lagged responses (Clark et al., 2001; e.g., lag 0-20 years, Tinner and Lotter, 2001; lag several millennia, Johnstone and Chapin, 2003);
- recent genetic analysis (<100 m/yr, McLachlan et al., 2005) indicates that commonly inferred estimates from pollen have overestimated dispersal rates, explaining observed pollen records by multi-front recolonisation from low-density refugees (Pearson, 2006);
- future landscapes will differ substantially from past climate change situations and landscape fragmentation creates major obstacles to migration (e.g., Collingham and Huntley, 2000);
- processes moderating migration such as competition, herbivory and soil formation (land use – Vlassova, 2002; paludification – Crawford et al., 2003; herbivory – Cairns and Moen, 2004; Juday, 2005; pathogens – Moorcroft et al., 2006; Section 4.4.6);
- tree species do not only respond to a changing climate by migration, but also by local adaptation, including genetic adaptation (Davis and Shaw, 2001; Davis et al., 2005).

Modelling studies reconstructing past (e.g., Lischke et al., 2002) or projecting future (Malcolm et al., 2002b; Iverson et al., 2004; Neilson et al., 2005) dispersal all indicate that more realistic migration rates will result in lagged northward shifts of taiga (lag length 150-250 years, Chapin and Starfield, 1997; Skre et al., 2002). While shrubs and the tree line (see Glossary) were found to have advanced polewards in response to recent warming (Sturm et al., 2001; Lloyd, 2005; Tape et al., 2006; Chapter 1), the expected slow encroachment of taiga into tundra is confirmed by satellite data showing no expansion of boreal forest stands (Masek, 2001) indicating century-long time-lags for the forest limit (see Glossary) to move northward (Lloyd, 2005). All these findings suggest considerable uncertainties in how fast forests will shift northwards (e.g., Clark et al., 2003; Higgins et al., 2003; Chapin et al., 2004; Jasinski and Payette, 2005; McGuire et al., 2007) and in the resulting consequences for the climate system (discussed in Section 4.4.6). Lower rates for the majority of species are probably realistic, also because future conditions comprise both unprecedented climate characteristics, including rapid rates of change (Sections 4.2.1 and 4.4.11), and a combination of impediments to local adaptation and migration (with the exception of some generalists).

Compared to the TAR (Gitay et al., 2001), the net global loss due to land-use change in forest cover appears to have slowed further (Stokstad, 2001; FAO, 2001), but in some tropical and sub-tropical regions, notably South-East Asia and similarly the Amazon (e.g., Nepstad et al., 1999), deforestation rates are still high (0.01-2.01%/yr, Lepers et al., 2005; Alcamo et al., 2006), while in some northern regions such as Siberia, degradation rates are increasing largely due to unsustainable logging (Lepers et al., 2005). Though uncertainties in rate estimates are considerable (e.g., FAO, 2001; Houghton, 2003b; Lepers et al., 2005), current trends in pressures (Nelson, 2005) will clearly lead to continued deforestation and degradation in critical areas (historically accumulated loss of 182-199 PgC – Canadell et al., 2004; expected releases in the 21st century of 40-100 PgC – Gruber et al., 2004; Shvidenko et al., 2005) with concomitant implications for biodiversity (Duraiappah et al., 2005) and other supporting services (Hassan et al., 2005). In most industrialised countries, forest areas are expected to increase (e.g., European forests by 2080 up to 6% for the SRES B2 scenario – Karjalainen et al., 2002; Sitch et al., 2005) partly due to intensified agricultural management and climate change.

Although land-use changes may dominate impacts in some areas, climate change generally exacerbates biodiversity risks, especially in biodiversity hotspots and particularly for the first half of the 21st century (montane cloud forests – Foster, 2001; Hawaii – Benning et al., 2002; Costa Rica – Enquist, 2002; Amazonia – Miles, 2002; Australia – Williams et al., 2003). In tropical montane cloud forests, extinctions of amphibian species have been attributed to recent climate change (Pounds et al., 2006; see Section 4.4.7 and Table 4.1, No. 2). In a few exceptions, climate change may increase diversity locally or regionally (Kienast et al., 1998) but in most cases extinction risks are projected to increase.

4.4.6 Tundra and Arctic/Antarctic ecosystems

Properties, goods and services

Tundra denotes vegetation and ecosystems north of the closed boreal forest tree line, covering an area of about 5.6 million km^2, but here we also include ecosystems at circumpolar latitudes, notably the sea-ice biome in both hemispheres (e.g., Arrigo and Thomas, 2004; Section 4.4.9), and sub-Antarctic islands (but see also Chapter 15). Ecosystem services include carbon sequestration, climate regulation, biodiversity and cultural maintenance, fuel, and food and fibre production (Chapin et al., 2005a, p. 721-728). Climate regulation is likely to be dominated by positive feedbacks between climate and albedo changes through diminishing snow cover and, eventually, expanding forests (Chapin et al., 2005b) and net emissions of greenhouse gasses, notably methane. The Arctic significantly contributes to global biodiversity (Chapin et al., 2005a; Usher et al., 2005). Local mixed economies of cash and subsistence depend strongly on the harvest of local resources, food preparation, storage, distribution and consumption. This forms a unique body of cultural knowledge traditionally transmitted from generation to generation (Hassol, 2004a).

Key vulnerabilities

Arctic and sub-Arctic ecosystems (particularly ombrotrophic bog communities, see Glossary) above permafrost were considered likely to be most vulnerable to climatic changes, since impacts may turn Arctic regions from a net carbon sink to a net source (Gitay et al., 2001). Literature since the TAR suggests that changes in albedo and an increased release of methane from carbon stocks (e.g., Christensen et al., 2004), whose magnitudes were previously substantially underestimated, will lead to positive radiative climate forcing throughout the Arctic region (Camill, 2005; Lelieveld, 2006; Walter et al., 2006; Zimov et al., 2006). Adverse impacts, including pollution (see also Chapter 15), were projected for species such as marine birds, seals, polar bears, tundra birds and tundra ungulates (Gitay et al., 2001). Unique endemic biodiversity (e.g., polar bears, Box 4.3) as well as tundra-dependent species such as migratory birds (e.g., waterfowl, Box 4.5, 4.4.8, Table 4.1) have been confirmed to be facing increasing extinction risks, with concomitant threats to the livelihoods and food security for indigenous peoples.

Impacts

Global warming is projected to be most pronounced at high latitudes (Phoenix and Lee, 2004; Meehl et al., 2007; Christensen et al., 2007). Ongoing rapid climatic changes will force tundra polewards at unprecedented rates (Velichko, 2002), causing lagged responses in its slow-growing plant communities (Camill and Clark, 2000; Chapin et al., 2000; Callaghan et al., 2004a, 2004c; Velichko et al., 2004). Movements of some species of habitat-creating plants (edificators) require large spread rates exceeding their migrational capacity (Callaghan et al., 2005). Poleward taiga encroachment into tundra is also likely to lag these changes (see Section 4.4.5 and e.g., Callaghan et al., 2004b). Projections of vegetation changes in the northern Arctic suggest that by about 2080, 17.6% (range 14-23%) replacement of the current polar desert by tundra vegetation will have begun (Callaghan et al., 2005). An eventual replacement of dwarf shrub tundra by shrub tundra is projected for the Canadian Arctic by 2100 (Kaplan et al., 2003). Experimental manipulations of air temperature at eleven locations across the tundra also show that tundra plant communities change substantially through shifts in species dominance, canopy height and diversity (Walker et al., 2006), with cryptogams being particularly vulnerable (Cornelissen et al., 2001; van Wijk et al., 2004). A warming of 1-3°C caused a short-term diversity decrease, but generalisations are unwarranted because of insufficiently long experimentation time (Graglia et al., 2001; Dormann and Woodin, 2002; van Wijk et al., 2004; Walker et al., 2006).

The thermally stable oceanic climate of the sub-Antarctic Marion Island appears to be changing, with a rise in annual mean surface air temperature of 1.2°C between 1969 and 1999. Annual precipitation decreased more or less simultaneously, and the 1990s was the driest in the island's five decades with records (Smith, 2002). These changes may be linked to a shift in phase of the semi-annual oscillation in the Southern Hemisphere after about 1980 (Rouault et al., 2005). Climatic change will directly affect the indigenous biota of sub-Antarctic islands (Smith, 2002; Barnes et al., 2006). Experimental drying of the keystone cushion plant species *Azorella selago* on Marion Island revealed measurable negative impacts after only a few months (Le Roux et al., 2005).

While summer food availability may increase for some vertebrates (Hinzman et al., 2005), formation of ice-crust at critical winter times may reduce abundance of food below snow (Yoccoz and Ims, 1999; Aanes et al., 2002; Inkley et al., 2004). Tundra wetland habitat for migrant birds may dry progressively (Hinzman et al., 2005; Smith et al., 2005). Many species of Arctic-breeding shorebirds and waterfowl are projected to undergo major population declines as tundra habitat shrinks (Box 4.5, Table 4.1). In contrast, northern range expansions of more southern species are expected, e.g., moose and red fox (Callaghan et al., 2005). Some colonisers might ultimately need to be considered 'invasive' species (e.g., North American Mink – Neuvonen, 2004), such as presently-restricted populations of southern shrub species that are likely to spread in a warmer climate (Forbes, 1995) leading to possibly increased carbon sequestration (Sturm et al., 2001; Tape et al., 2006; for a discussion of overall consequences for climate, see end of Section 4.4.6). For arctic species such as the polar bear, increasing risks of extinction are associated with the projected large decrease in the extent of the sea-ice biome and sea-ice cover (Box 4.3).

Significant changes in tundra are of two main types (Velichko et al., 2005), namely in vegetation structure (and related albedo), and in below-ground processes related to a combined increase in temperature, increase in depth of the active layer (see Glossary), and moisture content. These will promote paludification (see Glossary; Crawford et al., 2003), thermokarst processes (see Glossary), and increase the dryness of raised areas. Moisture supply substantially influences the state of permafrost, one of the most important components of the tundra landscape (Anisimov et al., 2002a, 2002b). Increasing active layer instability causes greater mixing and shifting of the soil's mineral matrix, damaging plant roots. Generally this will favour moisture-loving species (e.g., sedges), while the peat-bog vegetation over permafrost could experience drier conditions (Camill, 2005).

Substantial recent upward revisions (Zimov et al., 2006) of carbon stocks (Figure 4.1) in permafrost and yedoma (see Glossary), and measurements of methane releases from north Siberian thaw lakes (Walter et al., 2006), Scandinavian mires (Christensen et al., 2004) and Canadian permafrost (Camill, 2005) now show tundra to be a significantly larger atmospheric methane source than previously recognised. Current estimates of northern wetland methane emissions increase by 10-63% based on northern Siberian estimates alone. This methane source comprises a positive feedback to climate change, as thaw lakes (Walter et al., 2006) and mires (Christensen et al., 2004) are expanding in response to warming. While thermokarst-derived emissions are currently modest relative to anthropogenic sources, a potential stock of about 500 Pg of labile carbon in yedoma permafrost (Figure 4.1) could greatly intensify the positive feedback to high-latitude warming trends that are currently projected (Sazonova et al., 2004; Mack et al., 2004; Lelieveld, 2006; Zimov et al., 2006).

Changes in albedo associated with snow cover loss, and eventual invasion of tundra vegetation by evergreen coniferous trees, is likely to decrease regional albedo significantly and lead to a warming effect greater than the cooling projected from the increased carbon uptake by advancing trees (Section 4.4.5) and shrubs (Betts, 2000; Sturm et al., 2001, 2005; Chapin et al., 2005b; McGuire and Chapin, 2006; McGuire et al., 2007). Remote sensing already shows that tundra has greened over the past 20 years (Sitch et al., 2007). However, the potential for CO_2 sequestration varies from region to region (Callaghan et al., 2005) and model uncertainties are high (Sitch et al., 2007), since

Box 4.3. Polar bears – a species in peril?

There are an estimated 20,000 to 25,000 polar bears (*Ursus maritimus*) worldwide, mostly inhabiting the annual sea ice over the continental shelves and inter-island archipelagos of the circumpolar Arctic, where they may wander for thousands of kilometres per year. They are specialised predators on ice-breeding seals and are therefore dependent on sea ice for survival. Female bears require nourishment after emerging in spring from a 5 to 7 month fast in nursing dens (Ramsay and Stirling, 1988), and are thus very dependent on close proximity between land and sea ice before it breaks up. Continuous access to sea ice allows bears to hunt throughout the year, but in areas where the sea ice melts completely each summer, they are forced to spend several months in tundra fasting on stored fat reserves until freeze-up.

Polar bears face great challenges from the effects of climatic warming (Stirling and Derocher, 1993; Stirling et al., 1999; Derocher et al., 2004), as projected reductions in sea ice will drastically shrink marine habitat for polar bears, ice-inhabiting seals and other animals (Hassol, 2004b). Break-up of the sea ice on the western Hudson Bay, Canada, already occurs about 3 weeks earlier than in the early 1970s, resulting in polar bears in this area coming ashore earlier with reduced fat reserves (a 15% decline in body condition), fasting for longer periods of time and having reduced productivity (Stirling et al., 1999). Preliminary estimates suggest that the Western Hudson Bay population has declined from 1,200 bears in 1987 to fewer than 950 in 2004. Although these changes are specific to one sub-population, similar impacts on other sub-populations of polar bears can be reasonably expected. In 2005, the IUCN Polar Bear Specialist Group concluded that the IUCN Red List classification of the polar bear should be upgraded from *Least Concern* to *Vulnerable* based on the likelihood of an overall decline in the size of the total population of more than 30% within the next 35 to 50 years. The U.S. Fish and Wildlife Service is also considering a petition to list the polar bear as a threatened species based in part on future risks to the species from climate change. If sea ice declines according to some projections (Meehl et al., 2007, Figure 10.13; Figure 4.4, Table 4.1) polar bears will face a high risk of extinction with warming of 2.8°C above pre-industrial (range 2.5-3.0°C, Table 4.1, No. 42). Similar consequences are facing other ice-dependent species, not only in the Arctic but also in the Antarctic (Chapter 1; Barbraud and Weimerskirch, 2001; Croxall et al., 2002).

migration rates (Section 4.4.5), changes in hydrology, fire, insect pest outbreaks and human impacts relevant to the carbon cycle are poorly represented (see also Sections 4.4.1 and 4.4.5).

4.4.7 Mountains

Properties, goods and services

Mountain regions (circa 20-24% of all land, scattered throughout the globe) exhibit many climate types corresponding to widely separated latitudinal belts within short horizontal distances. Consequently, although species richness decreases with elevation, mountain regions support many different ecosystems and have among the highest species richness globally (e.g., Väre et al., 2003; Moser et al., 2005; Spehn and Körner, 2005). Mountain ecosystems have a significant role in biospheric carbon storage and carbon sequestration, particularly in semi-arid and arid areas (e.g., the western U.S., – Schimel et al., 2002; Tibetan plateau – Piao et al., 2006). Mountain ecosystem services such as water purification and climate regulation extend beyond their geographical boundaries and affect all continental mainlands (e.g., Woodwell, 2004). Local key services allow habitability of mountain areas, e.g. through slope stabilisation and protection from natural disasters such as avalanches and rockfall. Mountains increasingly serve as refuges from direct human impacts for many endemic species. They provide many goods for subsistence livelihoods, are home to many indigenous peoples, and are attractive for recreational activities and tourism. Critically, mountains harbour a significant fraction of biospheric carbon (28% of forests are in mountains).

Key vulnerabilities

The TAR identified mountain regions as having experienced above-average warming in the 20th century, a trend likely to continue (Beniston et al., 1997; Liu and Chen, 2000). Related impacts included an earlier and shortened snow-melt period, with rapid water release and downstream floods which, in combination with reduced glacier extent, could cause water shortage during the growing season. The TAR suggested that these impacts may be exacerbated by ecosystem degradation pressures such as land-use changes, over-grazing, trampling, pollution, vegetation destabilisation and soil losses, in particular in highly diverse regions such as the Caucasus and Himalayas (Gitay et al., 2001). While adaptive capacities were generally considered limited, high vulnerability was attributed to the many highly endemic alpine biota (Pauli et al., 2003). Since the TAR, the literature has confirmed a disproportionately high risk of extinction for many endemic species in various mountain ecosystems, such as tropical montane cloud forests or forests in other tropical regions on several continents (Williams et al., 2003; Pounds and Puschendorf, 2004; Andreone et al., 2005; Pounds et al., 2006), and globally where habitat loss due to warming threatens endemic species (Pauli et al., 2003; Thuiller et al., 2005b).

Impacts

Because temperature decreases with altitude by 5-10°C/km, relatively short-distanced upward migration is required for persistence (e.g., MacArthur, 1972; Beniston, 2000; Theurillat and Guisan, 2001). However, this is only possible for the warmer climatic and ecological zones below mountain peaks (Gitay et al., 2001; Peñuelas and Boada, 2003). Mountain ridges, by contrast, represent considerable obstacles to dispersal for many species which tends to constrain movements to slope upward migration (e.g., Foster, 2001; Lischke et al., 2002; Neilson et al., 2005; Pounds et al., 2006). The latter necessarily reduces a species' geographical range (mountain tops are smaller than their bases). This is expected to reduce genetic diversity within species and to increase the risk of stochastic extinction due to ancillary stresses (Peters and Darling, 1985; Gottfried et al., 1999), a hypothesis confirmed by recent genetic analysis showing gene drift effects from past climate changes (e.g., Alsos et al., 2005; Bonin et al., 2006). A reshuffling of species on altitude gradients is to be expected as a consequence of individualistic species responses that are mediated by varying longevities and survival rates. These in turn are the result of a high degree of evolutionary specialisation to harsh mountain climates (e.g., Theurillat et al., 1998; Gottfried et al., 1999; Theurillat and Guisan, 2001; Dullinger et al., 2005; Klanderud, 2005; Klanderud and Totland, 2005; Huelber et al., 2006), and in some cases they include effects induced by invading alien species (e.g., Dukes and Mooney, 1999; Mack et al., 2000). Genetic evidence for *Fagus sylvatica*, e.g., suggests that populations may show some capacity for an *in situ* adaptive response to climate change (Jump et al., 2006). However, ongoing distributional changes (Peñuelas and Boada, 2003) show that this response will not necessarily allow this species to persist throughout its range.

Upper tree lines, which represent the interface between sub-alpine forests and low-stature alpine meadows, have long been thought to be partly controlled by carbon balance (Stevens and Fox, 1991). This hypothesis has been challenged (Hoch and Körner, 2003; Körner, 2003a). Worldwide, cold tree lines appear to be characterised by seasonal mean air temperatures of *circa* 6°C (Körner, 1998; Körner, 2003a; Grace et al., 2002; Körner and Paulsen, 2004; Millar et al., 2004; Lara et al., 2005; Zha et al., 2005). In many mountains, the upper tree line is located below its potential climatic position because of grazing, or disturbances such as wind or fire. In other regions such as the Himalaya, deforestation of past decades has transformed much of the environment and has led to fragmented ecosystems (Becker and Bugmann, 2001). Although temperature control may be a dominant determinant of geographical range, tree species may be unable to migrate and keep pace with changing temperature zones (Shiyatov, 2003; Dullinger et al., 2004; Wilmking et al., 2004).

Where warmer and drier conditions are projected, mountain vegetation is expected to be subject to increased evapotranspiration (Ogaya et al., 2003; Jasper et al., 2004; Rebetez and Dobbertin, 2004; Stampfli and Zeiter, 2004; Jolly et al., 2005; Zierl and Bugmann, 2005; Pederson et al., 2006). This leads to increased drought, which has been projected to induce forest dieback in continental climates, particularly in the interior of mountain ranges (e.g., Fischlin and Gyalistras, 1997; Lischke et al., 1998; Lexer et al., 2000; Bugmann et al., 2005), and mediterranean areas. Even in humid tropical regions, plants and animals have been shown to be sensitive to water stress on mountains (e.g., Borneo – Kitayama, 1996; Costa Rica – Still et al., 1999). There is very high confidence that warming is a driver of amphibian mass extinctions at many highland localities, by creating increasingly favourable conditions for the pathogenic *Batrachochytrium* fungus (Pounds et al., 2006).

The duration and depth of snow cover, often correlated with mean temperature and precipitation (Keller et al., 2005; Monson et al., 2006), is a key factor in many alpine ecosystems (Körner, 2003c; Daimaru and Taoda, 2004). A lack of snow cover exposes plants and animals to frost and influences water supply in spring (Keller et al., 2005). If animal movements are disrupted by changing snow patterns, as has been found in Colorado (Inouye et al., 2000), increased wildlife mortality may result. At higher altitudes, the increased winter precipitation likely to accompany warming leads to greater snowfall, so that earlier arriving altitudinal migrants are confronted with delayed snowmelt (Inouye et al., 2000).

Disturbances such as avalanches, rockfall, fire, wind and herbivore damage interact and are strongly dependent on climate (e.g., Peñuelas and Boada, 2003; Whitlock et al., 2003; Beniston and Stephenson, 2004; Cairns and Moen, 2004; Carroll et al., 2004; Hodar and Zamora, 2004; Kajimoto et al., 2004; Pierce et al., 2004; Schoennagel et al., 2004; Schumacher et al., 2004). These effects may prevent recruitment and thus limit adaptive migration responses of species, and are exacerbated by human land use and other anthropogenic pressures (e.g., Lawton et al., 2001; Dirnböck et al., 2003; Huber et al., 2005).

Ecotonal (see Glossary) sensitivity to climate change, such as upper tree lines in mountains (e.g., Camarero et al., 2000; Walther et al., 2001; Diaz, 2003; Sanz-Elorza et al., 2003), has shown that populations of several mountain-restricted species are likely to decline (e.g., Beever et al., 2003; Florenzano, 2004). The most vulnerable ecotone species are those that are genetically poorly adapted to rapid environmental change, reproduce slowly, disperse poorly, and are isolated or highly specialised, because of their high sensitivity to environmental stresses (McNeely, 1990). Recent findings for Europe, despite a spatially coarse analysis, indicate that mountain species are disproportionately sensitive to climate change (about 60% species loss – Thuiller et al., 2005b). Substantial biodiversity losses are likely if human pressures on mountain biota occur in addition to climate change impacts (Pounds et al., 1999, 2006; Lawton et al., 2001; Pounds, 2001; Halloy and Mark, 2003; Peterson, 2003; Solorzano et al., 2003; Pounds and Puschendorf, 2004).

4.4.8 Freshwater wetlands, lakes and rivers

Properties, goods and services

Inland aquatic ecosystems (covering about 10.3 Mkm^2) vary greatly in characteristics and global distribution. The majority of natural freshwater lakes are located in the higher latitudes, most artificial lakes occur in mid- and lower latitudes, and many saline lakes occur at altitudes up to 5,000 m, especially in the Himalaya and Tibet. The majority of natural wetlands (peatlands) are in the boreal region but most managed wetlands (rice paddies) are in the tropics and sub-tropics (where peatlands also occur). Global estimates of the area under rivers, lakes and wetlands vary greatly depending upon definition (Finlayson et al., 2005). This chapter follows the TAR in considering 'wetlands' as distinct from rivers and lakes. Wetlands encompass a most heterogeneous spectrum of habitats following hydrological and nutrient gradients, and all key processes, including goods and services provided, depend on the catchment level hydrology. Inland waters are subject to many pressures from human activities. Aquatic ecosystems provide a wide range of goods and services (Gitay et al., 2001; Finlayson et al., 2005). Wetlands are often biodiversity 'hotspots' (Reid et al., 2005), as well as functioning as filters for pollutants from both point and non-point sources, and being important for carbon sequestration and emissions (Finlayson et al., 2005). Rivers transport water and nutrients from the land to the oceans and provide crucial buffering capacity during drought spells especially if fed by mountain springs and glaciers (e.g., European summer 2003; Box 4.1; Chapter 12, Section 12.6.1). Closed lakes serve as sediment and carbon sinks (Cohen, 2003), providing crucial repositories of information on past climate changes.

Key vulnerabilities

Gitay et al. (2001) have described some inland aquatic ecosystems (Arctic, sub-Arctic ombrotrophic bog communities on permafrost, depressional wetlands with small catchments, drained or otherwise converted peatlands) as most vulnerable to climate change, and have indicated the limits to adaptations due to the dependence on water availability controlled by outside factors. More recent results show vulnerability varying by geographical region (Van Dam et al., 2002; Stern, 2007). This includes significant negative impacts across 25% of Africa by 2100 (SRES B1 emissions scenario, de Wit and Stankiewicz, 2006) with both water quality and ecosystem goods and services deteriorating. Since it is generally difficult and costly to control hydrological regimes, the interdependence between catchments across national borders often leaves little scope for adaptation.

Impacts

Climate change impacts on inland aquatic ecosystems will range from the direct effects of the rise in temperature and CO_2 concentration to indirect effects through alterations in the hydrology resulting from the changes in the regional or global precipitation regimes and the melting of glaciers and ice cover (e.g., Chapters 1 and 3; Cubasch et al., 2001; Lemke et al., 2007; Meehl et al., 2007).

Studies since the TAR have confirmed and strengthened the earlier conclusions that rising temperature will lower water quality in lakes through a fall in hypolimnetic (see Glossary) oxygen concentrations, release of phosphorus (P) from sediments, increased thermal stability, and altered mixing patterns (McKee et al., 2003; Verburg et al., 2003; Winder and Schindler, 2004; Jankowski et al., 2006). In northern latitudes, ice cover on lakes and rivers will continue to break up earlier and the ice-free periods to increase (Chapter 1; Weyhenmeyer et al., 2004; Duguay et al., 2006). Higher temperatures will negatively affect micro-organisms and benthic invertebrates (Kling et al., 2003) and the distribution of many species of fish (Lake et al., 2000; Poff et al., 2002; Kling et al., 2003);

invertebrates, waterfowl and tropical invasive biota are likely to shift polewards (Moss et al., 2003; Zalakevicius and Svazas, 2005) with some potential extinctions (Jackson and Mandrak, 2002; Chu et al., 2005). Major changes will be likely to occur in the species composition, seasonality and production of planktonic communities (e.g., increases in toxic blue-green algal blooms) and their food web interactions (Gerten and Adrian, 2002; Kling et al., 2003; Winder and Schindler, 2004) with consequent changes in water quality (Weyhenmeyer, 2004). Enhanced UV-B radiation and increased summer precipitation will significantly increase dissolved organic carbon (DOC, see Glossary) concentrations, altering major biogeochemical cycles (Zepp et al., 2003; Phoenix and Lee, 2004; Frey and Smith, 2005). Studies along an altitudinal gradient in Sweden show that NPP can increase by an order of magnitude for a 6°C air temperature increase (Karlsson et al., 2005). However, tropical lakes may respond with a decrease in NPP and a decline in fish yields (e.g., 20% NPP and 30% fish yield reduction in Lake Tanganyika due to warming over the last century – O'Reilly et al., 2003). Higher CO_2 levels will generally increase NPP in many wetlands, although in bogs and paddy fields it may also stimulate methane flux, thereby negating positive effects (Ziska et al., 1998; Schrope et al., 1999; Freeman et al., 2004; Megonigal et al., 2005; Zheng et al., 2006).

Boreal peatlands will be affected most by warming (see also Sections 4.4.5 and 4.4.6) and increased winter precipitation as the species composition of both plant and animal communities will change significantly (Weltzin et al., 2000, 2001, 2003; Berendse et al., 2001; Keller et al., 2004; Sections 4.4.5 and 4.4.6). Numerous arctic lakes will dry out with a 2-3°C temperature rise (Smith et al., 2005; Symon et al., 2005). The seasonal migration patterns and routes of many wetland species will need to change and some may be threatened with extinction (Inkley et al., 2004; Finlayson et al., 2005; Reid et al., 2005; Zalakevicius and Svazas, 2005; Box 4.5).

Small increases in the variability of precipitation regimes will significantly impact wetland plants and animals at different stages of their life cycle (Keddy, 2000). In monsoonal regions, increased variability risks diminishing wetland biodiversity and prolonged dry periods promote terrestrialisation of wetlands as witnessed in Keoladeo National Park, India (Chauhan and Gopal, 2001; Gopal and Chauhan, 2001). In dryland wetlands, changes in precipitation regimes may cause biodiversity loss (Bauder, 2005). Changes in climate and land use will place additional pressures on already-stressed riparian ecosystems along many rivers in the world (Naiman et al., 2005). An increase or decrease in freshwater flows will also affect coastal wetlands (Chapter 6) by altering salinity, sediment inputs and nutrient loadings (Schallenberg et al., 2001; Flöder and Burns, 2004).

4.4.9 Oceans and shallow seas

Properties, goods and services

Oceans cover over 71% of the Earth's surface area from polar to tropical regions to a mean depth of 4,000 m, comprising about 14 billion km^3, are a massive reservoir of inorganic carbon, yet contain only 698-708 Pg organic carbon, 13-23 Pg of which is in living and dead biomass (Figure 4.1; Denman et al., 2007, Section 7.3.4.1). Despite low biomass, phytoplankton carries out almost half of global primary production, and is the basis of the marine food web (Field et al., 1998). Substantial biodiversity exists in both pelagic and benthic realms and along coastlines, in a diverse range of ecosystems from highly productive (e.g., upwelling regions) to those with low productivity (e.g., oceanic gyres). Ocean primary productivity depends on sunlight and nutrients supplied from deep waters (Sarmiento et al., 2004a). Marine ecosystems provide goods and services such as fisheries, provision of energy, recreation and tourism, CO_2 sequestration and climate regulation, decomposition of organic matter and regeneration of nutrients and coastal protection – many of which are critical to the functioning of the Earth system (Chapter 5; Costanza et al., 1997; McLean et al., 2001, Sections 6.3.2, 6.3.4, 6.3.5, 6.4.5 and 6.4.6; Hassan et al., 2005, Table 18.2). Marine biodiversity supports ecosystem function and the services it provides (Worm et al., 2006) with over 1 billion people relying on fish as their main animal protein source, especially in developing nations (Pauly et al., 2005). Coastal zones, particularly low-lying areas, and the highly valuable local and global socioeconomic services they provide (e.g., agricultural land, human settlements and associated infrastructure and industry, aquaculture and fisheries and freshwater supply) are particularly vulnerable to climate change (McLean et al., 2001, Section 6.5; Hassan et al., 2005, Section 19.3.2, Table 19.2).

Key vulnerabilities

Since the TAR, literature has confirmed that salient vulnerable ecosystems are warm-water coral reefs (Box 4.4), cold-water corals, the Southern Ocean and marginal sea-ice ecosystems. Ocean uptake of CO_2, resulting from increasing atmospheric CO_2 concentrations, reduces surface ocean pH and carbonate ion concentrations, an impact that was overlooked in the TAR. This is expected to affect coral reefs, cold water corals, and ecosystems (e.g., the Southern Ocean), where aragonite (used by many organisms to make their shells or skeletons) will decline or become undersaturated. These and other ecosystems where calcareous organisms (e.g., pteropods, see Glossary) play an important role will become vulnerable this century (reviewed by Raven et al., 2005; Haugan et al., 2006; Table 4.1). Synergistic impacts of higher seawater temperatures and declining carbonate make these ecosystems even more vulnerable (e.g., Raven et al., 2005; Turley et al., 2006; Box 4.4). Marginal sea-ice and surrounding ecosystems are vulnerable to warming, particularly in the Northern Hemisphere (Sarmiento et al., 2004b; Christensen et al., 2007).

Impacts

Climate change can impact marine ecosystems through ocean warming (Wang et al., 2004b), by increasing thermal stratification and reducing upwelling (Cox et al., 2000; Sarmiento et al., 2004a), sea level rise (IPCC, 2001), and through increases in wave height and frequency (Monahan et al., 2000; Wang et al., 2004b), loss of sea ice (Sarmiento et al., 2004b; Meehl et al., 2007; Christensen et al., 2007), increased risk of diseases in marine biota (Harvell et al., 2002) and decreases in the pH and carbonate ion concentration of the surface oceans (Caldeira and Wickett, 2003; Feely et al., 2004; Sabine et al., 2004; Raven et al., 2005).

Theoretically, nutrient speciation could be influenced by the lower pH expected this century (Zeebe and Wolf-Gladrow, 2001; Raven et al., 2005). Decreases in both upwelling and formation of deep water and increased stratification of the upper ocean will reduce the input of essential nutrients into the sunlit regions of oceans and reduce productivity (Cox et al., 2000; Loukos et al., 2003; Lehodey et al., 2003; Sarmiento et al., 2004a). In coastal areas and margins, increased thermal stratification may lead to oxygen deficiency, loss of habitats, biodiversity and distribution of species, and impact whole ecosystems (Rabalais et al., 2002). Changes to rainfall and nutrient flux from land may exacerbate these hypoxic events (Rabalais et al., 2002).

Box 4.4. Coral reefs: endangered by climate change?

Reefs are habitat for about a quarter of marine species and are the most diverse among marine ecosystems (Roberts et al., 2002; Buddemeier et al., 2004). They underpin local shore protection, fisheries, tourism (Chapter 6; Hoegh-Guldberg et al., 2000; Cesar et al., 2003; Willig et al., 2003; Hoegh-Guldberg, 2004, 2005) and, though supplying only about 2-5% of the global fisheries harvest, comprise a critical subsistence protein and income source in the developing world (Whittingham et al., 2003; Pauly et al., 2005; Sadovy, 2005).

Corals are affected by warming of surface waters (Chapter 6, Box 6.1; Reynaud et al., 2003; McNeil et al., 2004; McWilliams et al., 2005) leading to bleaching (loss of algal symbionts – Chapter 6, Box 6.1). Many studies incontrovertibly link coral bleaching to warmer sea surface temperature (e.g., McWilliams et al., 2005) and mass bleaching and coral mortality often results beyond key temperature thresholds (Chapter 6, Box 6.1). Annual or bi-annual exceedance of bleaching thresholds is projected at the majority of reefs worldwide by 2030 to 2050 (Hoegh-Guldberg, 1999; Sheppard, 2003; Donner et al., 2005). After bleaching, algae quickly colonise dead corals, possibly inhibiting later coral recruitment (e.g., McClanahan et al., 2001; Szmant, 2001; Gardner et al., 2003; Jompa and McCook, 2003). Modelling predicts a phase switch to algal dominance on the Great Barrier Reef and Caribbean reefs in 2030 to 2050 (Wooldridge et al., 2005).

Coral reefs will also be affected by rising atmospheric CO_2 concentrations (Orr et al., 2005; Raven et al., 2005; Denman et al., 2007, Box 7.3) resulting in declining calcification. Experiments at expected aragonite concentrations demonstrated a reduction in coral calcification (Marubini et al., 2001; Langdon et al., 2003; Hallock, 2005), coral skeleton weakening (Marubini et al., 2003) and strong temperature dependence (Reynaud et al., 2003). Oceanic pH projections decrease at a greater rate and to a lower level than experienced over the past 20 million years (Caldeira and Wickett, 2003; Raven et al., 2005; Turley et al., 2006). Doubling CO_2 will reduce calcification in aragonitic corals by 20%-60% (Kleypas et al., 1999; Kleypas and Langdon, 2002; Reynaud et al., 2003; Raven et al., 2005). By 2070 many reefs could reach critical aragonite saturation states (Feely et al., 2004; Orr et al., 2005), resulting in reduced coral cover and greater erosion of reef frameworks (Kleypas et al., 2001; Guinotte et al., 2003).

Adaptation potential (Hughes et al., 2003) by reef organisms requires further experimental and applied study (Coles and Brown, 2003; Hughes et al., 2003). Natural adaptive shifts to symbionts with +2°C resistance may delay demise of some reefs to roughly 2100 (Sheppard, 2003), rather than mid-century (Hoegh-Guldberg, 2005) although this may vary widely across the globe (Donner et al., 2005). Estimates of warm-water coral cover reduction in the last 20-25 years are 30% or higher (Wilkinson, 2004; Hoegh-Guldberg, 2005) due largely to increasing higher SST frequency (Hoegh-Guldberg, 1999). In some regions, such as the Caribbean, coral losses have been estimated at 80% (Gardner et al., 2003). Coral migration to higher latitudes with more optimal SST is unlikely, due both to latitudinally decreasing aragonite concentrations and projected atmospheric CO_2 increases (Kleypas et al., 2001; Guinotte et al., 2003; Orr et al., 2005; Raven et al., 2005). Coral migration is also limited by lack of available substrate (Chapter 6, Section 6.4.1.5). Elevated SST and decreasing aragonite have a complex synergy (Harvell et al., 2002; Reynaud et al., 2003; McNeil et al., 2004; Kleypas et al., 2005) but could produce major coral reef changes (Guinotte et al., 2003; Hoegh-Guldberg, 2005). Corals could become rare on tropical and sub-tropical reefs by 2050 due to the combined effects of increasing CO_2 and increasing frequency of bleaching events (at 2-3 $\times CO_2$) (Kleypas and Langdon, 2002; Hoegh-Guldberg, 2005; Raven et al., 2005). Other climate change factors (such as sea-level rise, storm impact and aerosols) and non-climate factors (such as over-fishing, invasion of non-native species, pollution, nutrient and sediment load (although this could also be related to climate changes through changes to precipitation and river flow; Chapter 6, Box 6.1; Chapter 11, Box 11.3; Chapter 16)) add multiple impacts on coral reefs (Chapter 16, Box 16.2), increasing their vulnerability and reducing resilience to climate change (Koop et al., 2001; Kleypas and Langdon, 2002; Cole, 2003; Buddemeier et al., 2004; Hallock, 2005).

Projections of ocean biological response to climate warming by 2050 show contraction of the highly productive marginal sea-ice biome by 42% and 17% in Northern and Southern Hemispheres (Sarmiento et al., 2004b; see also Meehl et al., 2007; Christensen et al., 2007). The sea-ice biome accounts for a large proportion of primary production in polar waters and supports a substantial food web. As timing of the spring phytoplankton bloom is linked to the sea-ice edge, loss of sea ice (Walsh and Timlin, 2003) and large reductions of the total primary production in the marginal sea-ice biome in the Northern Hemisphere (Behrenfeld and Falkowski, 1997; Marra et al., 2003) would have strong effects, for example, on the productivity of the Bering Sea (Stabeno et al., 2001). Reductions in winter sea-ice will affect the reproduction, growth and development of fish, krill, and their predators, including seals and seal-dependent polar bears (e.g., Barber and Iacozza, 2004; Box 4.3), leading to further changes in abundance and distribution of marine species (Chapter 15, Section 15.4.3). An expansion by 4.0% (Northern Hemisphere) and 9.4% (Southern), and of the sub-polar gyre biome by 16% (Northern) and 7% (Southern), has been projected for the permanently stratified sub-tropical gyre biome with its low productivity. This effect has now been observed in the North Pacific and Atlantic (McClain et al., 2004; Sarmiento et al., 2004b). A contraction by 11% of the seasonally stratified sub-tropical gyre is also projected in both hemispheres by 2050 due to climate warming. These changes are likely to have significant impacts on marine ecosystem productivity globally, with uncertainties in projections of NPP using six mainly IS92a-based scenarios narrowing to an increase of between 0.7% and 8.1% by mid-century (ΔT_{global} ~1.5-3°C).

Changes to planktonic and benthic community composition and productivity have been observed in the North Sea since 1955 (Clark and Frid, 2001) and since the mid-1980s may have reduced the survival of young cod (Beaugrand et al., 2003). Large shifts in pelagic biodiversity (Beaugrand et al., 2002) and in fish community composition have been seen (Genner et al., 2004; Perry et al., 2005). Changes in seasonality or recurrence of hydrographic events or productive periods could be affected by trophic links to many marine populations, including exploited or cultured populations (Stenseth et al., 2002, 2003; Platt et al., 2003; Llope et al., 2006). Elevated temperatures have increased mortality of winter flounder eggs and larvae (Keller and Klein-Macphee, 2000) and have led to later spawning migrations (Sims et al., 2004). A 2°C rise in sea surface temperature (SST) would result in removal of Antarctic bivalves and limpets from the Southern Ocean (Peck et al., 2004). Tuna populations may spread towards presently temperate regions, based on predicted warming of surface water and increasing primary production at mid- and high latitudes (Loukos et al., 2003).

Marine mammals, birds, cetaceans and pinnipeds (seals, sea lions and walruses), which feed mainly on plankton, fish and squid, are vulnerable to climate change-driven changes in prey distribution, abundance and community composition in response to climatic factors (Learmonth et al., 2006). Changing water temperature also has an effect on the reproduction of cetaceans and pinnipeds, indirectly through prey abundance, either through extending the time between individual breeding attempts, or by reducing breeding condition of the mother (Whitehead, 1997). Current extreme climatic events provide an indication of potential future effects. For example, the warm-water phase of ENSO is associated with large-scale changes in plankton abundance and associated impacts on food webs (Hays et al., 2005), and changes to behaviour (Lusseau et al., 2004), sex ratio (Vergani et al., 2004) and feeding and diet (Piatkowski et al., 2002) of marine mammals.

Melting Arctic ice-sheets will reduce ocean salinities (IPCC, 2001), causing species-specific shifts in the distribution and biomass of major constituents of Arctic food webs, including poleward shifts in communities and the potential loss of some polar species (such as the narwhal, *Monodon monoceros*). Migratory whales (e.g., grey whale, *Eschrichtius robustus*), that spend summer in Arctic feeding grounds, are likely to experience disruptions in their food sources (Learmonth et al., 2006). Nesting biology of sea turtles is strongly affected by temperature, both in timing and in the determination of the sex ratio of hatchlings (Hays et al., 2003), but implications for population size are unknown. A predicted sea-level rise of 0.5 m will eliminate up to 32% of sea-turtle nesting beaches in the Caribbean (Fish et al., 2005).

Surface ocean pH has decreased by 0.1 unit due to absorption of anthropogenic CO_2 emissions (equivalent to a 30% increase in hydrogen ion concentration) and is predicted to decrease by up to a further 0.3-0.4 units by 2100 (Caldeira and Wickett, 2003). This may impact a wide range of organisms and ecosystems (e.g., coral reefs, Box 4.4, reviewed by Raven et al., 2005), including juvenile planktonic, as well as adult, forms of benthic calcifying organisms (e.g., echinoderms, gastropods and shellfish), and will affect their recruitment (reviewed by Turley et al., 2006). Polar and sub-polar surface waters and the Southern Ocean will be aragonite under-saturated by 2100 (Orr et al., 2005) and Arctic waters will be similarly threatened (Haugan et al., 2006). Organisms using aragonite to make their shells (e.g., pteropods) will be at risk and this will threaten ecosystems such as the Southern and Arctic Oceans in which they play a dominant role in the food web and carbon cycling (Orr et al., 2005; Haugan et al., 2006).

Cold-water coral ecosystems exist in almost all the world's oceans and their aerial coverage could equal or exceed that of warm-water coral reefs (Freiwald et al., 2004; Guinotte et al., 2006). They harbour a distinct and rich ecosystem, provide habitats and nursery grounds for a variety of species, including commercial fish and numerous new species previously thought to be extinct (Raven et al., 2005). These geologically ancient, long-lived, slow-growing and fragile reefs will suffer reduced calcification rates and, as the aragonite saturation horizon moves towards the ocean surface, large parts of the oceans will cease to support them by 2100 (Feely et al., 2004; Orr et al., 2005; Raven et al., 2005; Guinotte et al., 2006). Since cold-water corals do not have symbiotic algae but depend on extracting food particles sinking from surface waters or carried by ocean currents, they are also vulnerable to changes to ocean currents, primary productivity and flux of food particles (Guinotte et al., 2006). Warm-water coral reefs are also sensitive to multiple impacts including increased SST and decreasing aragonite concentrations within this century (Box 4.4).

4.4.10 Cross-biome impacts

This section highlights issues that cut across biomes, such as large-scale geographical shifts of vegetation (Figure 4.3) or animal migration patterns (e.g., Box 4.3; Box 4.5), and changes in land use and aquatic systems.

Biome shifts

Boreal forest and Arctic tundra ecosystems are projected generally to show increased growth due to longer and warmer growing seasons (Lucht et al., 2002; Figure 4.3). Woody boreal vegetation is expected to spread into tundra at higher latitudes and higher elevations (Grace et al., 2002; Kaplan et al., 2003; Gerber et al., 2004). At the southern ecotone (see Glossary) with continental grasslands, a contraction of boreal forest is projected due to increased impacts of drought, insects and fires (Bachelet et al., 2001; Scholze et al., 2006), together with a lower rate of sapling survival (Hogg and Schwarz, 1997). Drought stress could partially be counteracted by concurrent CO_2-induced enhanced water-use efficiency (Gerten et al., 2005), small regional increases in precipitation, and an increased depth of permafrost thawing. It is uncertain whether peak summer heat stress on boreal tree species could cause regional transitions to grassland where the continental winter climate remains too cold for temperate forest species to succeed (Gerber et al., 2004; Lucht et al., 2006). In temperate forests, milder winters may reduce winter hardening in trees, increasing their vulnerability to frost (Hänninen et al., 2001; Hänninen, 2006).

Vegetation change in the lower to mid latitudes is uncertain because transitions between tropical desert and woody vegetation types are difficult to forecast. Climate models disagree in pattern and magnitude of projected changes in atmospheric circulation and climate variability, particularly for precipitation (e.g., with respect to the Indian and West African monsoons). For the Sahel and other semi-arid regions, increasing drought is predicted by some models (Held et al., 2005), while increased water-use efficiency is projected to cause more greening (Figure 4.3), though potentially associated with more frequent fires, in others (Bachelet et al., 2003; Woodward and Lomas, 2004b; Ni et al., 2006; Schaphoff et al., 2006). In savannas, woody encroachment is projected to be a consequence of enhanced water-use efficiency and increased precipitation in some regions (Bachelet et al., 2001; Lucht et al., 2006; Ni et al., 2006; Schaphoff et al., 2006; Section 4.4.3; Figure 4.3). The moderate drying, including desert amelioration, as projected in southern Africa, the Sahel region, central Australia, the Arabian Peninsula and parts of central Asia (Figure 4.3) may be due to a positive impact of rising atmospheric CO_2, as noted in eastern Namibia through sensitivity analysis (Thuiller et al., 2006b).

A general increase of deciduous at the expense of evergreen vegetation is predicted at all latitudes, although the forests in both the eastern USA and eastern Asia appear to be sensitive to drought stress and decline under some scenarios (Bachelet et al., 2001; Gerten et al., 2005; Lucht et al., 2006; Scholze et al., 2006). Tropical ecosystems are expected to change, particularly in the Amazon, where a subset of GCMs shows strong to moderate reductions in precipitation with the consequence of transitions of evergreen tropical forest to rain-green forest or grasslands (Cox et al., 2004; Cramer et al., 2004; Woodward and Lomas, 2004b). However, representations of tropical succession remain underdeveloped in current models. The global land biosphere is projected by some models to lose carbon beyond temperature increases of 3°C (Gerber et al., 2004), mainly from temperate and boreal soils, with vegetation carbon declining beyond temperature increases above 5°C (Gerber et al., 2004). Carbon sinks persist mainly in the Arctic and in savanna grasslands (Woodward and Lomas, 2004b; Schaphoff et al., 2006). However, there is large variability between the projections of different vegetation (Cramer et al., 2001) and climate (Schaphoff et al., 2006) models for a given emissions scenario.

Migration patterns

Vagile (see Glossary) animals such as polar bears (sea-ice biome, tundra; Box 4.3) and in particular migratory animals (tundra, wetlands, lakes, tropical forests, savannas, etc.; Box 4.5) respond to impacts both within and across biomes. Many species breed in one area then move to another to spend the non-breeding season (Robinson et al., 2005). Many migratory species may be more vulnerable to climate change than resident species (Price and Root, 2005). As migratory species often move annually in response to seasonal climate changes, their behaviour, including migratory routes, is sensitive to climate. Numerous studies have found that many of these species are arriving earlier (Chapter 1 and e.g., Root et al., 2003). Changes in the timing of biological events are of particular concern because of a potential disconnect between migrants and their food resources if the phenology of each advances at different rates (Inouye et al., 2000; Root et al., 2003; Visser et al., 2004). The potential impact of climate change on migratory birds has been especially well studied (Box 4.5).

Land use

The relative importance of key drivers on ecosystem change varies across regions and biomes (Sala et al., 2000; Sala, 2005). Several global studies suggest that at least until 2050 land-use change will be the dominant driver of terrestrial biodiversity loss in human-dominated regions (Sala et al., 2000; UNEP, 2002; Gaston et al., 2003; Jenkins, 2003; Scharlemann et al., 2004; Sala, 2005). Conversely, climate change is likely to dominate where human interventions are limited, such as in the tundra, boreal, cool conifer forests, deserts and savanna biomes (Sala et al., 2000; Duraiappah et al., 2005). Assessment of impacts on biodiversity differ if other drivers than climate change are taken into account (Thomas et al., 2004a; Sala, 2005; Malcolm et al., 2006). Interactions among these drivers may mitigate or exacerbate the overall effects of climate change (Opdam and Wascher, 2004). The effects of land-use change on species through landscape fragmentation at the regional scale may further exacerbate impacts from climate change (Holman et al., 2005a; Del Barrio et al., 2006; Harrison et al., 2006; Rounsevell et al., 2006).

Global land-use change studies project a significant reduction in native vegetation (mostly forest) in non-industrialised countries and arid regions due to expansion of agricultural or urban land use driven principally by population growth,

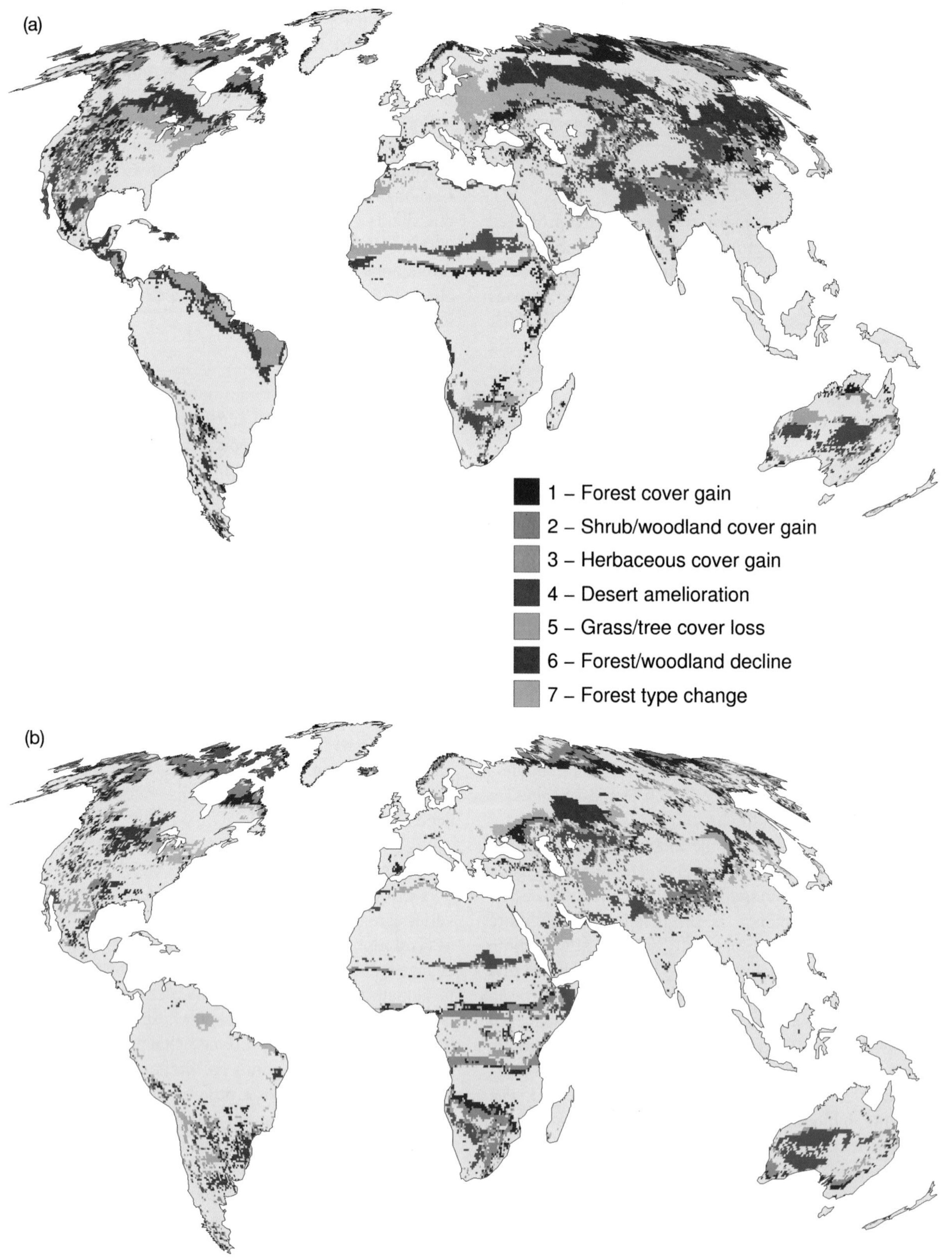

Figure 4.3. *Projected appreciable changes in terrestrial ecosystems by 2100 relative to 2000 as simulated by DGVM LPJ (Sitch et al., 2003; Gerten et al., 2004) for two SRES emissions scenarios (Nakićenović et al., 2000) forcing two climate models: (a) HadCM3 A2, (b) ECHAM5 B1 (Lucht et al., 2006; Schaphoff et al., 2006). Changes are considered appreciable and are only shown if they exceed 20% of the area of a simulated grid cell (see Figure 4.2 for further explanations).*

Box 4.5. Crossing biomes: impacts of climate change on migratory birds

Migratory species can be affected by climate change in their breeding, wintering and/or critical stopover habitats. Models project changes in the future ranges of many species (Peterson et al., 2002; Price and Glick, 2002; Crick, 2004), some suggesting that the ranges of migrants may shift to a greater extent than non-migrants (Price and Root, 2001). In some cases this may lead to a lengthening and in others to a shortening of migration routes. Moreover, changes in wind patterns, especially in relation to seasonal migration timing, could help or hinder migration (Butler et al., 1997). Other expected impacts include continuing changes in phenology, behaviour, population sizes and possibly genetics (reviewed in Crick, 2004; Robinson et al., 2005).

Many migratory species must cross geographical barriers (e.g., the Sahara Desert, oceans) in moving between their wintering and breeding areas. Many species must stop in the Sahel to refuel en route from their breeding to their wintering areas. Degradation of vegetation quality in the Sahel (Box 4.2) could potentially lead to population declines in these species in areas quite remote from the Sahel (Robinson et al., 2005).

More than 80% of the species living within the Arctic Circle winter farther south (Robinson et al., 2005). However, climate-induced habitat change may be greatest in the Arctic (Zöckler and Lysenko, 2000; Symon et al., 2005). For example, the red knot could potentially lose 15%-37% of its tundra breeding habitat by 2100 (HadCM2a1, UKMO). Additionally, at least some populations of this species could also lose critical migratory stopover habitat (Delaware Bay, USA) to sea-level rise (Galbraith et al., 2002).

The breeding areas of many Arctic breeding shorebirds and waterfowl are projected to decline by up to 45% and 50%, respectively (Folkestad et al., 2005) for global temperature increases of 2°C above pre-industrial. A temperature increase of 2.9°C above pre-industrial would cause larger declines of up to 76% for waterfowl and up to 56% for shorebirds. In North America's Prairie Pothole region, models have projected an increase in drought with a 3°C regional temperature increase and varying changes in precipitation, leading to large losses of wetlands and to declines in the populations of waterfowl breeding there (Johnson et al., 2005). Many of these species also winter in coastal areas vulnerable to sea-level rise (Inkley et al., 2004). One review of 300 migrant bird species found that 84% face some threat from climate change, almost half because of changes in water regime (lowered watertables and drought), and this was equal to the summed threats due to all other anthropogenic causes (Robinson et al., 2005).

especially in Africa, South America and in South Asia (Hassan et al., 2005). This reduction in native habitat will result in biodiversity loss (e.g., Duraiappah et al., 2005; Section 4.4.11). Northern-latitude countries and high-altitude regions may become increasingly important for biodiversity and species conservation as the ranges of species distributions move poleward and upward in response to climate change (Berry et al., 2006). Northern-latitude countries and high-altitude regions are also sensitive to the effects of climate change on land use, especially agriculture, which is of particular relevance if those regions are to support adaptation strategies to mitigate the negative effects of future climate and land-use change. Biomes at the highest latitudes that have not already been converted to agriculture are likely to remain relatively unchanged in the future (Duraiappah et al., 2005).

Aquatic systems

Higher CO_2 concentrations lower the nutritional quality of the terrestrial litter (Lindroth et al., 2001; Tuchman et al., 2002, 2003a, 2003b) which in turn will affect the food web relationships of benthic communities in rivers. Greater amounts of DOC (dissolved organic carbon) released in peatlands at higher CO_2 levels are exported to streams and finally reach coastal waters (Freeman et al., 2004).

4.4.11 Global synthesis including impacts on biodiversity

Considerable progress has been made since the TAR in key fields that allow projection of future climate change impacts on species and ecosystems. Two of these key fields, namely climate envelope modelling (also called niche-based, or bioclimatic modelling) and dynamic global vegetation modelling have provided numerous recent results. The synthesis of these results provides a picture of potential impacts and risks that is far from perfect, in some instances apparently contradictory, but overall highlights a wide array of key vulnerabilities (Figures 4.2; 4.4; 4.5, Table 4.1).

Climate envelope modelling has burgeoned recently due to increased availability of species distribution data, together with finer-scale climate data and new statistical methods that have allowed this correlative method to be widely applied (e.g., Guisan and Thuiller, 2005; McClean et al., 2005; Thuiller et al., 2005b). Despite several limitations (Section 4.3 and references cited therein) these models offer the advantage of assessing climate change impacts on biodiversity quantitatively (e.g., Thomas et al., 2004a). Climate envelope models do not simulate dynamic population or migration processes, and results are typically constrained to the regional level, so that the implications for biodiversity at the global level are difficult to infer (Malcolm et al., 2002a).

In modelling ecosystem function and plant functional type response, understanding has deepened since the TAR, though consequential uncertainties remain. The ecophysiological processes affected by climate change and the mechanisms by which climate change may impact biomes, ecosystem components such as soils, fire behaviour and vegetation structure (i.e., biomass distribution and leaf area index) are now explicitly modelled and have been bolstered by experimental results (e.g., Woodward and Lomas, 2004b). One emerging key message is that climate change impacts on the fundamental regulating services may previously have been underestimated (Sections 4.4.1, 4.4.10, Figures 4.2; 4.3; 4.4). Nevertheless, the globally applicable DGVMs are limited inasmuch as the few plant functional types used within the models aggregate numerous species into single entities (Sitch et al., 2003). These are assumed to be entities with very broad environmental tolerances, which are immutable and immune to extinction. Therefore, underlying changes in species richness are not accounted for, and the simultaneous free dispersal of PFTs is assumed (e.g., Neilson et al., 2005; Midgley et al., 2007). The strength of DGVMs is especially in their global application, realistic dynamics and simulation of ecosystem processes including essential elements of the global C-cycle (e.g., Malcolm et al., 2002b). Thus, it is reasonable to equate changes in DGVM-simulated vegetation (e.g., Figure 4.3) to changes in community and population structures in the real world.

What overall picture emerges from the results reviewed here? It appears that moderate levels of atmospheric CO_2 rise and climate change relative to current conditions may be beneficial in some regions (Nemani et al., 2003), depending on latitude, on the CO_2 responsiveness of plant functional types, and on the natural adaptive capacity of indigenous biota (mainly through range shifts that are now being widely observed – see Chapter 1). But as change continues, greater impacts are projected, while

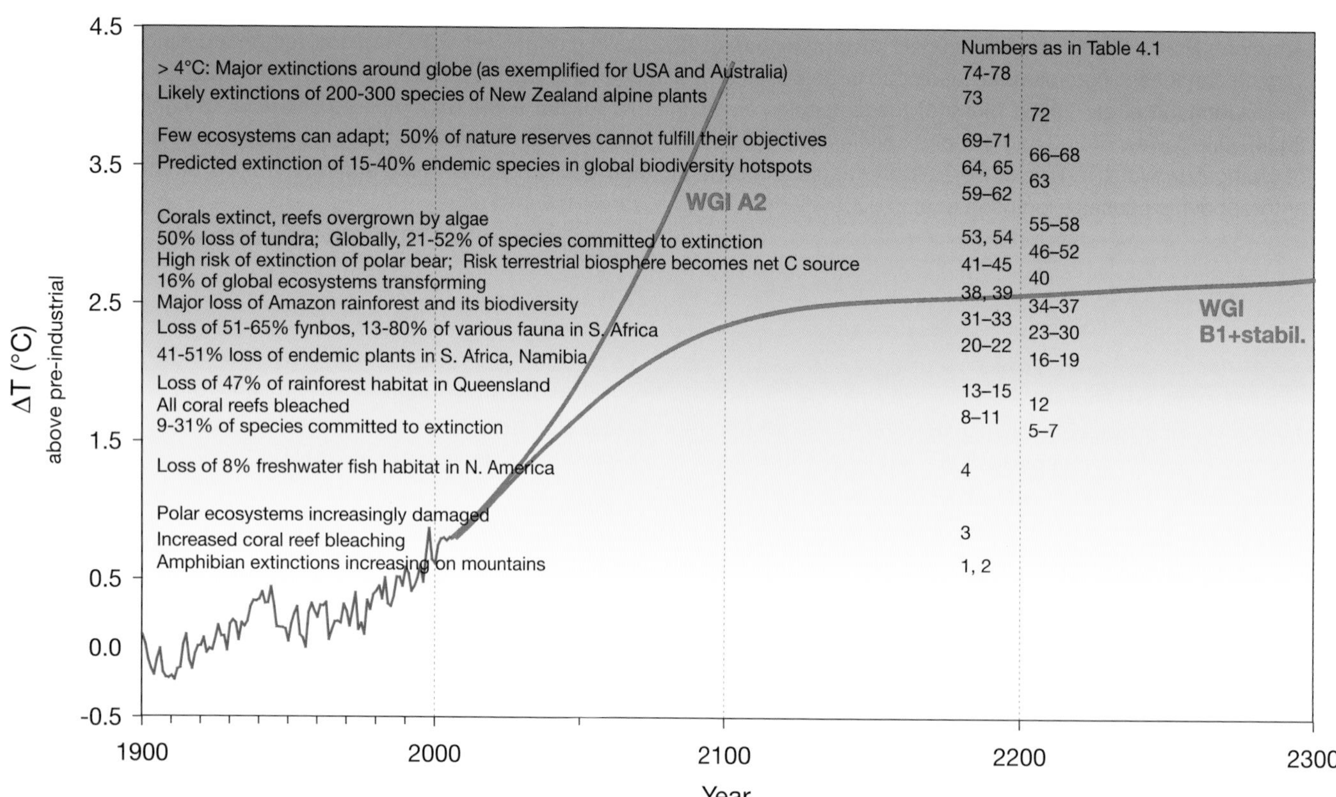

Figure 4.4. *Compendium of projected risks due to critical climate change impacts on ecosystems for different levels of global mean annual temperature rise, ΔT, relative to pre-industrial climate (approach and event numbers as used in Table 4.1 and Appendix 4.1). It is important to note that these impacts do not take account of ancillary stresses on species due to over-harvesting, habitat destruction, landscape fragmentation, alien species invasions, fire regime change, pollution (such as nitrogen deposition), or for plants the potentially beneficial effects of rising atmospheric* CO_2*. The red curve shows observed temperature anomalies for the period 1900-2005 (Brohan et al., 2006, see also Trenberth et al., 2007, Figure 3.6). The two grey curves provide examples of the possible future evolution of temperature against time (Meehl et al., 2007, Figure 10.4), providing examples of higher and lower trajectories for the future evolution of the expected value of ΔT. Shown are the simulated, multi-model mean responses to (i) the A2 emissions scenario and (ii) an extended B1 scenario, where radiative forcing beyond the year 2100 was kept constant to the 2100 value (all data from Meehl et al., 2007, Figure 10.4, see also Meehl et al., 2007, Section 10.7).*

ecosystem and species response may be lagged (Sections 4.4.5, 4.4.6). At key points in time (Figure 4.4), ecosystem services such as carbon sequestration may cease, and even reverse (Figure 4.2). While such 'tipping points' (Kemp, 2005) are impossible to identify without substantial uncertainties, they may lead to irreversible effects such as biodiversity loss or, at the very least, impacts that have a slow recovery (e.g., on soils and corals).

In the two simulations presented in Figure 4.2 (warming of 2.9°C and 5.3°C by 2100 over land relative to the 1961-1990 baseline), the DGVM approach reveals salient changes in a key regulating service of the world's ecosystems: carbon sequestration. Changes in the spatial distributions of ecosystems are given in Figure 4.3 (where it must be stressed that the figure highlights only key vulnerabilities through depicting appreciable vegetation type changes, i.e., PFT change over >20% of the area of any single pixel modelled). In the B1 emissions scenario (Figure 4.3b) about 26% of extant ecosystems reveal appreciable changes by 2100, with some positive impacts especially in Africa and the Southern Hemisphere. However, these positive changes are likely to be due to the assumed CO_2-fertilisation effect (Section 4.4.10, Figure 4.3). By contrast, in mid- to high latitudes on all continents, substantial shifts in forest structure toward more rain-green, summer-green or deciduous rather than evergreen forest, and forest and woodland decline, underlie the overall drop in global terrestrial carbon sequestration potential that occurs post-2030, and approaches a net source by about 2070 (Figure 4.2; 4.3). In the A2 emissions scenario, roughly 37% of extant ecosystems reveal appreciable changes by 2100. Desert amelioration persists in the regions described above, but substantial decline of forest and woodland is seen at northern, tropical and sub-tropical latitudes. In both scenarios the current global sink deteriorates after 2030, and by 2070 (ΔT ~2.5°C over pre-industrial) the terrestrial biosphere becomes an increasing carbon source (Figure 4.2; see also Scholze et al., 2006) with the concomitant risk of positive feedback, developments that amplify climate change. Similar results were obtained by using a wide range of climate models which indicate that the biosphere becomes consistently within this century a net CO_2 source with a global warming of >3°C relative to pre-industrial (Scholze et al., 2006). On the other hand, it must be noted that by about 2100 the modelled biosphere has nevertheless sequestered an additional 205-228 PgC (A2 and B1 emissions scenarios respectively) relative to the year 2000 (Lucht et al., 2006).

Climate envelope modelling suggests that climate change impacts will diminish the areal extent of some ecosystems (e.g., reduction by 2-47% alone due to 1.6°C warming above pre-industrial, Table 4.1, No. 6) and impact many ecosystem properties and services globally. Climate impacts alone will vary regionally and across biomes and will lead to increasing levels of global biodiversity loss, as expressed through area reductions of wild habitats and declines in the abundance of wild species putting those species at risk of extinction (e.g., 3-16% of European plants with 2.2°C warming (Table 4.1, No. 20) or major losses of Amazon rainforest with 2.5°C warming above pre-industrial, Figure 4.4, Table 4.1, No. 36). Globally, biodiversity (represented by species richness and relative abundance) may decrease by 13 to 19% due to a combination of land-use change, climate change and nitrogen deposition under four scenarios by 2050 relative to species present in 1970 (Duraiappah et al., 2005). Looking at projected losses due to land-use change alone (native habitat loss), habitat reduction in tropical forests and woodland, savanna and warm mixed forest accounts for 80% of the species projected to be lost (about 30,000 species – Sala, 2005). The apparent contrast between high impacts shown by projections for species (climate envelope models) relative to PFTs (DGVMs) is likely to be due to a number of reasons – most importantly, real species virtually certainly have narrower climate tolerances than PFTs, a fact more realistically represented by the climate envelope models. DGVM projections reveal some increasing success of broad-range, generalist plant species, while climate envelope model results focus on endemics. Endemics, with their smaller ranges, have been shown to have a greater vulnerability to climate change (Thuiller et al., 2005a), and may furthermore be dependent on keystone species in relationships that are ignored in DGVMs. Therefore, for assessing extinction risks, climate envelope modelling currently appears to offer more realistic results.

As indicated in the TAR, climate changes are being imposed on ecosystems experiencing other substantial and largely detrimental pressures. Roughly 60% of evaluated ecosystems are currently utilised unsustainably and show increasing signs of degradation (Reid et al., 2005; Hassan et al., 2005; Worm et al., 2006). This alone will be likely to cause widespread biodiversity loss (Chapin et al., 2000; Jenkins, 2003; Reid et al., 2005), given that 15,589 species, from every major taxonomic group, are already listed as threatened (Baillie et al., 2006). The likely synergistic impacts of climate change and land-use change on endemic species have been widely confirmed (Hannah et al., 2002a; Hughes, 2003; Leemans and Eickhout, 2004; Thomas et al., 2004a; Lovejoy and Hannah, 2005; Hare, 2006; Malcolm et al., 2006; Warren, 2006), as has over-exploitation of marine systems (Worm et al., 2006; Chapters 5 and 6).

Overall, climate change has been estimated to be a major driver of biodiversity loss in cool conifer forests, savannas, mediterranean-climate systems, tropical forests, in the Arctic tundra, and in coral reefs (Thomas et al., 2004a; Carpenter et al., 2005; Malcolm et al., 2006). In other ecosystems, land-use change may be a stronger driver of biodiversity loss at least in the near term. In an analysis of the SRES scenarios to 2100 (Strengers et al., 2004), deforestation is reported to cease in all scenarios except A2, suggesting that beyond 2050 climate change is very likely to be the major driver for biodiversity loss globally. Due to climate change alone it has been estimated that by 2100 between 1% and 43% of endemic species (average 11.6%) will be committed to extinction (DGVM-based study – Malcolm et al., 2006), whereas following another approach (also using climate envelope modelling-based studies – Thomas et al., 2004a) it has been estimated that on average 15% to 37% of species (combination of most optimistic assumptions 9%, most pessimistic 52%) will be committed to extinction by 2050 (i.e., their range sizes will have begun shrinking and fragmenting in a way that guarantees their accelerated extinction). Climate-change-induced extinction rates in tropical biodiversity hotspots are likely to exceed the predicted extinctions from deforestation during this century (Malcolm et al.,

2006). In the mediterranean-climate region of South Africa, climate change may have at least as significant an impact on endemic Protea species' extinction risk as land-use change does by 2020 (Bomhard et al., 2005). Based on all above findings and our compilation (Figure 4.4, Table 4.1) we estimate that on average 20% to 30% of species assessed are likely to be at increasingly high risk of extinction from climate change impacts possibly within this century as global mean temperatures exceed 2°C to 3°C relative to pre-industrial levels (this chapter). The uncertainties remain large, however, since for about 2°C temperature increase the percentage may be as low as 10% or for about 3°C as high as 40% and, depending on biota, the range is between 1% and 80% (Table 4.1; Thomas et al., 2004a; Malcolm et al., 2006). As global average temperature exceeds 4°C above pre-industrial levels, model projections suggest significant extinctions (40-70% species assessed) around the globe (Table 4.1).

Losses of biodiversity will probably lead to decreases in the provision of ecosystem goods and services with trade-offs between ecosystem services likely to intensify (National Research Council, 1999; Carpenter et al., 2005; Duraiappah et al., 2005). Gains in provisioning services (e.g., food supply, water use) are projected to occur, in part, at the expense of other regulating and supporting services including genetic resources, habitat provision, climate and runoff regulation. Projected changes may also increase the likelihood of ecological surprises that are detrimental for human well-being (Burkett et al., 2005; Duraiappah et al., 2005). Ecological surprises include rapid and abrupt changes in temperature and precipitation, leading to an increase in extreme events such as floods, fires and landslides, increases in eutrophication, invasion by alien species, or rapid and sudden increases in disease (Carpenter et al., 2005). This could also entail sudden shifts of ecosystems to less desired states (Scheffer et al., 2001; Folke et al., 2004; e.g., Chapin et al., 2004) through, for example, the exeedance of critical temperature thresholds, possibly resulting in the irreversible loss of ecosystem services, which were dependent on the previous state (Reid et al., 2005).

Table 4.1. *Projected impacts of climate change on ecosystems and population systems as reported in the literature for different levels of global mean annual temperature rise, ΔT_g, relative to pre-industrial climate – mean and range (event numbers as used in Figure 4.4 and Appendix 4.1). The global temperature change values are used as an indicator of the other associated climate changes that match particular amounts of ΔT_g, e.g., precipitation change and, where considered, change in the concentration of greenhouse gases in the atmosphere. Projections from the literature were harmonised into a common framework by down/upscaling (where necessary) from local to global temperature rise using multiple GCMs, and by using a common global mean temperature reference point for the year 1990 (after Warren, 2006). Whilst some of the literature relates impacts directly to global mean temperature rises or particular GCM scenarios, many studies give only local temperature rises, ΔT_{reg}, and hence require upscaling. The thirteen GCM output data sets used are taken from the IPCC DDC at http://www.ipcc-data.org/.*

No.[i]	ΔT_g above pre-ind[ii]	ΔT_g above pre-ind[iii] (range)	ΔT_{reg} above 1990 (range)	Impacts to unique or widespread ecosystems or population systems	Region	Ref. no.
1	0.6			Increased coral bleaching	Caribbean, Indian Ocean, Great Barrier Reef	2
2	0.6			Amphibian extinctions/extinction risks on mountains due to climate-change-induced disease outbreaks	Costa Rica, Spain, Australia	52, 54
3	<1.0			Marine ecosystems affected by continued reductions in krill possibly impacting Adelie penguin populations; Arctic ecosystems increasingly damaged	Antarctica, Arctic	42, 11, 14
4	1.3	1.1-1.6	1	8% loss freshwater fish habitat, 15% loss in Rocky Mountains, 9% loss of salmon	N. America	13
5	1.6	1.2-2.0	0.7-1.5	9-31% (mean 18%) of species committed to extinction	Globe [iv]	1
6	1.6			Bioclimatic envelopes eventually exceeded, leading to 10% transformation of global ecosystems; loss of 47% wooded tundra, 23% cool conifer forest, 21% scrubland, 15% grassland/steppe, 14% savanna, 13% tundra and 12% temperate deciduous forest. Ecosystems variously lose 2-47% areal extent.	Globe	6
7	1.6	1.1-2.1	1	Suitable climates for 25% of eucalypts exceeded	Australia	12
8	1.7	1-2.3	1°C SST	All coral reefs bleached	Great Barrier Reef, S.E. Asia, Caribbean	2
9	1.7	1.2-2.6		38-45% of the plants in the Cerrado committed to extinction	Brazil	1, 44
10	1.7	1.3-3		2-18% of the mammals, 2-8% of the birds and 1-11% of the butterflies committed to extinction	Mexico	1, 26
11	1.7	1.3-2.4	2	16% freshwater fish habitat loss, 28% loss in Rocky Mountains, 18% loss of salmon	N. America	13
12	<1.9	<1.6-2.4	<1	Range loss begins for golden bowerbird	Australia	4

[i] Same numbers as used in first column in Appendix 4.1.
[ii] The mean temperature change is taken directly from the literature, or is the central estimate of a range given in the literature, or is the mean of upscaling calculations (cf. caption).
[iii] The range of temperature change represents the uncertainty arising from the use of different GCM models to calculate global temperature change.
[iv] 20% of the Earth's land surface covered by study.

No.	ΔT_g above pre-ind	ΔT_g above pre-ind (range)	ΔT_{reg} above 1990 (range)	Impacts to unique or widespread ecosystems or population systems	Region	Ref. no.
13	1.9	1.6-2.4	1	7-14% of reptiles, 8-18% of frogs, 7-10% of birds and 10-15% of mammals committed to extinction as 47% of appropriate habitat in Queensland lost	Australia	1, 7
14	1.9	1.6-2.4	1	Range loss of 40-60% for golden bowerbird	Australia	4
15	1.9	1.0-2.8		Most areas experience 8-20% increase in number ≥7day periods with Forest Fire Weather Index >45: increased fire frequency converts forest and maquis to scrub, leads to more pest outbreaks	Mediterranean	34
16	2.1			41-51% loss in plant endemic species richness	S. Africa, Namibia	39
17	2.1	1.0-3.2	1-2	Alpine systems in Alps can tolerate local temperature rise of 1-2°C, tolerance likely to be negated by land-use change	Europe	8
18	2.1		1.4-2.6	13-23% of butterflies committed to extinction	Australia	1, 30
19	2.1	1.4-2.6		Bioclimatic envelopes of 2-10% plants exceeded, leading to endangerment or extinction; mean species turnover of 48% (spatial range 17-75%); mean species loss of 27% (spatial range 1- 68%)	Europe	22
20	2.2			3-16% of plants committed to extinction	Europe	1
21	2.2	2.1-2.3	1.6-1.8	15-37% (mean 24%) of species committed to extinction	Globe [iv]	1
22	2.2	1.7-3.2		8-12% of 277 medium/large mammals in 141 national parks critically endangered or extinct; 22-25% endangered	Africa	23
23	2.3	1.5-2.7	2°C SST	Loss of Antarctic bivalves and limpets	Southern Ocean	51
24	2.3	2.0-2.5		Fish populations decline, wetland ecosystems dry and disappear	Malawi, African Great Lakes	20
25	2.3	1.5-2.7	2.5-3.0	Extinctions (100% potential range loss) of 10% endemics; 51-65% loss of Fynbos; including 21-40% of Proteaceae committed to extinction; Succulent Karoo area reduced by 80%, threatening 2,800 plant species with extinction; 5 parks lose >40% of plant species	S. Africa	1, 5, 24, 25
26	2.3	2.3-4.0	2.5-3.0	24-59% of mammals, 28-40% of birds, 13-70% of butterflies, 18-80% of other invertebrates, 21-45% of reptiles committed to extinction; 66% of animal species potentially lost from Kruger National Park	S. Africa	1, 27
27	2.3	2.2-4.0		2-20% of mammals, 3-8% of birds and 3-15% of butterflies committed to extinction	Mexico	1, 26
28	2.3	1.6-3.2		48-57% of Cerrado plants committed to extinction	Brazil	1
29	2.3			Changes in ecosystem composition, 32% of plants move from 44% of area with potential extinction of endemics	Europe	16
30	2.3	1.6-3.2	3	24% loss freshwater fish habitat, 40% loss in Rocky Mountains, 27% loss of salmon.	N. America	13
31	2.4			63 of 165 rivers studied lose >10% of their fish species	Globe	19
32	2.4			Bioclimatic range of 25-57% (full dispersal) or 34-76% (no dispersal) of 5,197 plant species exceeded	Sub-Saharan Africa	3
33	>2.5			Sink service of terrestrial biosphere saturates and begins turning into a net carbon source	Globe	55, 56
34	2.5		2°C SST	Extinction of coral reef ecosystems (overgrown by algae)	Indian Ocean	9
35	2.5	1.9-4.3		42% of UK land area with bioclimate unlike any currently found there; in Hampshire, declines in curlew and hawfinch and gain in yellow-necked mouse numbers; loss of montane habitat in Scotland; potential bracken invasion of Snowdonia montane areas		57
36	2.5	2.0-3.0		Major loss of Amazon rainforest with large losses of biodiversity	S. America, Globe	21, 46
37	2.5			20-70% loss (mean 44%) of coastal bird habitat at 4 sites	USA	29
38	2.6	1.6-3.5		Most areas experience 20-34% increase in number ≥7day periods with Forest Fire Weather Index >45: increased fire frequency converts forest and maquis to scrub, causing more pest outbreaks	Mediterranean	34
39	2.6			4-21% of plants committed to extinction	Europe	1
40	2.7			Bioclimatic envelopes exceeded leading to eventual transformation of 16% of global ecosystems: loss of 58% wooded tundra, 31% cool conifer forest, 25% scrubland, 20% grassland/steppe, 21% tundra, 21% temperate deciduous forest, 19% savanna. Ecosystems variously lose 5-66% of their areal extent.	Globe	6

No.	ΔT_g above pre-ind	ΔT_g above pre-ind (range)	ΔT_{reg} above 1990 (range)	Impacts to unique or widespread ecosystems or population systems	Region	Ref. no.
41	2.8	1.2-4.5	1-3	Extensive loss/conversion of habitat in Kakadu wetland due to sea-level rise and saltwater intrusion	Australia	10
42	2.8	2.5-3.0		Multi-model mean 62% (range 40-100%) loss of Arctic summer ice extent, high risk of extinction of polar bears, walrus, seals; Arctic ecosystem stressed	Arctic	11,53
43	2.8	2.3-4.6	2.1-2.5	Cloud-forest regions lose hundreds of metres of elevational extent, potential extinctions ΔT_{reg} 2.1°C for C. America and ΔT_{reg} 2.5°C for Africa	C. America, Tropical Africa, Indonesia	17
44	2.8	2.1-3.1	3	Eventual loss of 9-62% of the mammal species from Great Basin montane areas	USA	32
45	2.8	1.9-3.8	3	38-54% loss of waterfowl habitat in Prairie Pothole region	USA	37, 38
46	2.9		3.2-6.6	50% loss existing tundra offset by only 5% eventual gain; millions of Arctic-nesting shorebird species variously lose up to 5-57% of breeding area; high-Arctic species most at risk; geese species variously lose 5-56% of breeding area	Arctic	14
47	2.9			Latitude of northern forest limits shifts N. by 0.5° latitude in W. Europe, 1.5° in Alaska, 2.5° in Chukotka and 4° in Greenland	Arctic	40
48	2.9	1.6-4.1		Threat of marine ecosystem disruption through loss of aragonitic pteropods	Southern Ocean	49
49	2.9	1.6-4.1		70% reduction in deep-sea cold-water aragonitic corals	Ocean basins	48
50	2.9		2.1-3.9	21-36% of butterflies committed to extinction; >50% range loss for 83% of 24 latitudinally-restricted species	Australia	1,30
51	2.9	2.6-3.3	2.1-2.8	21-52% (mean 35%) of species committed to extinction	Globe [iv]	1
52	2.9			Substantial loss of boreal forest	China	15
53	3.0			66 of 165 rivers studied lose >10% of their fish species	Globe	19
54	3.0	1.9-3.5		20% loss of coastal migratory bird habitat	Delaware, USA	36
55	3.1	2.3-3.7	2°C SST	Extinction of remaining coral reef ecosystems (overgrown by algae)	Globe	2
56	3.1	1.9-4.1	3-4	Alpine systems in Alps degraded	Europe	8
57	3.1	2.5-4.0	2	High risk of extinction of golden bowerbird as habitat reduced by 90%	Australia	4
58	3.1	1.8-4.2	3-4	Risk of extinction of alpine species	Europe	41
59	3.3	2.0-4.5		Reduced growth in warm-water aragonitic corals by 20%-60%; 5% decrease in global phytoplankton productivity	Globe	2, 47, 48
60	3.3	2.3-3.9	2.6-2.9	Substantial loss of alpine zone, and its associated flora and fauna (e.g., alpine sky lily and mountain pygmy possum)	Australia	45
61	3.3	2.8-3.8	2	Risk of extinction of Hawaiian honeycreepers as suitable habitat reduced by 62-89%	Hawaii	18
62	3.3		3.7	4-38% of birds committed to extinction	Europe	1
63	3.4			6-22% loss of coastal wetlands; large loss migratory bird habitat particularly in USA, Baltic and Mediterranean	Globe	35, 36
64	3.5	2.0-5.5		Predicted extinction of 15-40% endemic species in global biodiversity hotspots (case "narrow biome specificity")	Globe	50
65	3.5	2.3-4.1	2.5 – 3.5	Loss of temperate forest wintering habitat of monarch butterfly	Mexico	28
66	3.6	2.6-4.3	3	Bioclimatic limits of 50% of eucalypts exceeded	Australia	12
67	3.6	2.6-3.7		30-40% of 277 mammals in 141 parks critically endangered/extinct; 15-20% endangered	Africa	23
68	3.6	3.0-3.9		Parts of the USA lose 30-57% neotropical migratory bird species richness	USA	43
69	3.7			Few ecosystems can adapt	Globe	6
70	3.7			50% all nature reserves cannot fulfil conservation objectives	Globe	6
71	3.7			Bioclimatic envelopes exceeded leading to eventual transformation of 22% of global ecosystems; loss of 68% wooded tundra, 44% cool conifer forest, 34% scrubland, 28% grassland/steppe, 27% savanna, 38% tundra and 26% temperate deciduous forest. Ecosystems variously lose 7-74% areal extent.	Globe	6
72	3.9			4-24% plants critically endangered/extinct; mean species turnover of 63% (spatial range 22-90%); mean species loss of 42% (spatial range 2.5-86%)	Europe	22

No.	ΔT_g above pre-ind	ΔT_g above pre-ind (range)	ΔT_{reg} above 1990 (range)	Impacts to unique or widespread ecosystems or population systems	Region	Ref. no.
73	4.0	3.0-5.1	3	Likely extinctions of 200-300 species (32-63%) of alpine flora	New Zealand	33
74	>4.0		3.5	38-67% of frogs, 48-80% of mammals, 43-64% of reptiles and 49-72% of birds committed to extinction in Queensland as 85-90% of suitable habitat lost	Australia	1, 7
75	>>4.0		5	Bioclimatic limits of 73% of eucalypts exceeded	Australia	12
76	>>4.0		5	57 endemic frogs/mammal species eventually extinct, 8 endangered	Australia	7
77	>>4.0		7	Eventual total extinction of all endemic species of Queensland rainforest	Australia	7
78	5.2			62-100% loss of bird habitat at 4 major coastal sites	USA	29

Sources by Ref. no.: 1-Thomas et al., 2004a; 2-Hoegh-Guldberg, 1999; 3-McClean et al., 2005; 4-Hilbert et al., 2004; 5-Rutherford et al., 2000; 6-Leemans and Eickhout, 2004; 7-Williams et al., 2003; 8-Theurillat and Guisan, 2001; 9-Sheppard, 2003; 10-Eliot et al., 1999; 11-Symon et al., 2005; 12-Hughes et al., 1996; 13-Preston, 2006; 14-Zöckler and Lysenko, 2000; 15-Ni, 2001; 16-Bakkenes et al., 2002; 17-Still et al., 1999; 18-Benning et al., 2002; 19-Xenopoulos et al., 2005; 20-ECF, 2004; 21-Cox et al., 2004; 22-Thuiller et al., 2005b; 23-Thuiller et al., 2006b; 24-Midgley et al., 2002; 25-Hannah et al., 2002a; 26-Peterson et al., 2002; 27-Erasmus et al., 2002; 28-Villers-Ruiz and Trejo-Vazquez, 1998; 29-Galbraith et al., 2002; 30-Beaumont and Hughes, 2002; 31-Kerr and Packer, 1998; 32-McDonald and Brown, 1992; 33-Halloy and Mark, 2003; 34-Moriondo et al., 2006; 35-Nicholls et al., 1999; 36-Najjar et al., 2000; 37-Sorenson et al., 1998; 38-Johnson et al., 2005; 39-Broennimann et al., 2006; 40-Kaplan et al., 2003; 41-Theurillat et al., 1998; 42-Forcada et al., 2006; 43-Price and Root, 2005; 44-Siqueira and Peterson, 2003; 45-Pickering et al., 2004; 46-Scholze et al., 2006; 47-Raven et al., 2005; 48-Cox et al., 2000; 49-Orr et al., 2005; 50-Malcolm et al., 2006; 51-Peck et al., 2004; 52-Pounds et al., 2006; 53-Arzel et al., 2006; 54-Bosch et al., 2006; 55-Lucht et al., 2006; 56-Schaphoff et al., 2006; 57-Berry et al., 2005.

There is detailed information on the derivation for each entry in Table 4.1 listed in Appendix 4.1.

4.5 Costs and valuation of ecosystem goods and services

There is growing interest in developing techniques for environmental accounting. To that end, definitions of ecosystem goods and services are currently fluid. For example, ecosystem services accrue to society in return for investing in or conserving natural capital (Heal, 2007), or ecosystem services are ultimately the end products of nature, the aspects of nature that people make choices about (Boyd, 2006). Definitions aside, all humans clearly rely on ecosystem services (Reid et al., 2005). While many efforts have been made to use standard economic techniques to estimate the economic value of ecosystem goods and services (Costanza et al., 1997, 2000; Costanza, 2000, 2001; Daily et al., 2000; Giles, 2005; Reid et al., 2005), others argue that such efforts are not only largely futile and flawed (Pearce, 1998; Toman, 1998b; Bockstael et al., 2000; Pagiola et al., 2004), but may actually provide society a disservice (Ludwig, 2000; Kremen, 2005). The estimates from these techniques range from unknown (incomparability cf. Chang, 1997), or invaluable, or infinite (Toman, 1998b) because of lack of human substitutes, to about 38×10^{12} US$ *per annum* (updated to 2000 levels – Costanza et al., 1997; Balmford et al., 2002; Hassan et al., 2005), which is larger but of similar magnitude than the global gross national product (GNP) of 31×10^{12} US$ *per annum* (2000 levels). These monetary estimates are usually targeted at policy-makers to assist assessments of the economic benefits of the natural environment (Farber et al., 2006) in response to cost-benefit paradigms. Some argue (Balmford et al., 2002, 2005; Reid et al., 2005) that unless ecosystem values are recognised in economic terms, ecosystems will continue their decline, placing the planet's ecological health at stake (Millennium Ecosystem Assessment, 2005). Others argue that ecosystems provide goods and services which are invaluable and need to be conserved on more fundamental principles, i.e., the precautionary principle of not jeopardising the conditions for a decent, healthy and secure human existence on this planet (e.g., Costanza et al., 2000; van den Bergh, 2004), or a moral and ethical responsibility to natural systems not to destroy them.

What is sometimes lost in the arguments is that natural capital (including ecosystem goods and services) is part of society's capital assets (Arrow et al., 2004). The question then may be considered as whether one should maximise present value or try to achieve a measure of sustainability. In either case, it is the change in quantities of the capital stock that must be considered (including ecosystem services). One approach in considering valuation of ecosystem services is to calculate how much of one type of capital asset would be needed to compensate for the loss of one unit of another type of capital asset (Arrow et al., 2004). What is not disputed is that factoring in the full value of ecosystem goods and services, whether in monetary or non-monetary terms, distorts measures of economic wealth such that a country may be judged to be growing in wealth according to conventional indicators, while it actually becomes poorer due to the loss of natural resources (Balmford et al., 2002; Millennium Ecosystem Assessment, 2005; Mock, 2005 p. 33-53.). Ignoring such aspects almost guarantees opportunity costs. For instance Balmford et al. (2002) estimated a benefit-cost ratio of at least 100:1 for an effective global conservation programme setting aside 15% of the current Earth's surface if all aspects conventionally ignored are factored in. Additionally, many sectors and industries depend directly or indirectly on ecosystems and their services. The impacts of climate change could hold enormous costs for forests and coastal marine systems, as well as for managed agricultural systems (Epstein and Mills, 2005; Stern, 2007). Multiple industries, such as timber, fisheries, travel, tourism and agriculture, are threatened by disturbances caused by climate change. Impacts on these sectors will influence financial markets, insurance companies and large multinational investors (Mills, 2005).

The United Nations has recognised the need to develop integrated environmental and economic accounting. However, many difficulties remain, especially as ecosystems may be the most difficult of all environmental assets to quantify (Boyd, 2006). There is a growing recognition that national accounting procedures need to be modified to include values for ecosystem goods and services (Heal, 2007). Outside of the techniques mentioned above (often using contingent valuation) others have argued for developing a Green GDP to describe the state of nature and its worth, or an Ecosystem Services Index to account for all of nature's contributions to the welfare of human society (Banzhaf and Boyd, 2005; Boyd, 2006). Ultimately, it may be developing economies that are the most sensitive to the direct impacts of climate change, because they are more dependent on ecosystems and agriculture (Stern, 2007). As such, it is the poor that depend most directly on ecosystem services. Thus the degradation of these systems and their services will ultimately exacerbate poverty, hunger and disease, and obstruct sustainable development (e.g., Millennium Ecosystem Assessment, 2005; Mock, 2005; Mooney et al., 2005; Stern, 2007).

4.6 Acclimation and adaptation: practices, options and constraints

Although climate change is a global issue, local efforts can help maintain and enhance resilience and limit some of the longer-term damages from climate change (e.g., Hughes et al., 2003; Singh, 2003; Opdam and Wascher, 2004). This section discusses adaptation options with respect to natural ecosystems. Adaptation of these ecosystems involves only reactive, autonomous responses to ongoing climate change, including changes in weather variability and extremes. However, ecosystem managers can proactively alter the context in which ecosystems develop. In this way they can improve the resilience, i.e., the coping capacity, of ecosystems (see Glossary). Such ecosystem management involves anticipatory adaptation options. Identifying adaptation responses and adaptation options is a rapidly developing field, so the discussion below is not exhaustive. However, one should realise that beyond certain levels of climate change (Hansen et al., 2003; Table 4.1, Figure 4.4) impacts on ecosystems are severe and largely irreversible.

4.6.1 Adaptation options

As climatic changes occur, natural resource management techniques can be applied to increase the resilience of ecosystems. Increasing resilience is consistent also with the 'ecosystem approach' developed by the Convention on Biological Diversity (CBD) which is a "strategy for management of land, water and living resources that promotes conservation and sustainable use in an equitable way" (Smith and Malthby, 2003). There are many opportunities to increase resilience (Cropp and Gabrica, 2002; Tompkins and Adger, 2003); however, they may only be effective for lower levels of climate change (≤2-3°C, Executive Summary, Figure 4.4, Table 4.1).

Effective responses depend on an understanding of likely regional climatic and ecological changes. Monitoring environmental change, including climate, and associated ecosystem responses is vital to allow for adjustments in management strategies (e.g., Adger et al., 2003; Moldan et al., 2005). Although many adaptation options are available to wildlife managers, uncertainty about the magnitude and timing of climate change and delayed ecosystem responses (e.g., Section 4.4.5) may discourage their application. Nevertheless, 'no regrets' decisions based on the 'precautionary principle' appear preferable. Actions to reduce the impact of other threats, such as habitat fragmentation or destruction, pollution and introduction of alien species, are very likely to enhance resilience to climate change (e.g., Goklany, 1998; Inkley et al., 2004; Opdam and Wascher, 2004). Such proactive approaches would encourage conservation planning that is relevant both today and in the future. Techniques that allow the management of conservation resources in response to climate variability may ultimately prove to be the most beneficial way of preparing for possible abrupt climate change by increasing ecosystem resilience (Bharwani et al., 2005).

A few key options to adapt at least to lower levels of climate change in intensively managed ecosystems (Chapter 5) have been suggested (e.g., Hannah et al., 2002a, 2002b; Hannah and Lovejoy, 2003; Hansen et al., 2003). Expansion of reserve systems can potentially reduce the vulnerability of ecosystems to climate change (McNeely and Schutyser, 2003). Reserve systems may be designed with some consideration of long-term shifts in plant and animal distributions, natural disturbance regimes and the overall integrity of the protected species and ecosystems (e.g., Williams et al., 2005). Ultimately, adaptation possibilities are determined by the conservation priorities of each reserve and by the magnitude and nature of the change in climate. Strategies to cope with climate change are beginning to be considered in conservation (Cowling et al., 1999; Chopra et al., 2005; Scott and Lemieux, 2005), and highlight the importance of planning guided by future climate scenarios.

A primary adaptation strategy to climate change and even current climate variability is to reduce and manage the other stresses on species and ecosystems, such as habitat fragmentation and destruction, over-exploitation, eutrophication, desertification and acidification (Inkley et al., 2004; Duraiappah et al., 2005; Robinson et al., 2005; Worm et al., 2006). Robinson et al. (2005) suggest that this may be the only practical large-scale adaptation policy available for marine ecosystems. In addition to removing other stressors it is necessary to maintain viable, connected and genetically diverse populations (Inkley et al., 2004; Robinson et al., 2005). Small, isolated populations are often more prone to local extirpations than larger, more widespread populations (e.g., Gitay et al., 2002; Davis et al., 2005; Lovejoy and Hannah, 2005). Although connectivity, genetic diversity and population size are important current conservation goals, climate change increases their importance. The reduction and fragmentation of habitats may also be facilitated through increases in agricultural productivity (e.g., Goklany and Trewavas, 2003) reducing pressures on natural ecosystems. However, increasing demand for some types of biofuels may negate this potential benefit (e.g., Busch, 2006).

Reducing stress on ecosystems is difficult, especially in densely populated regions. Recent studies in southern Africa have signalled the need for policy to focus on managing areas outside protected areas (e.g., subsistence rangelands – Von Maltitz et al., 2006). This can, in part, be achieved through the devolution of resource ownership and management to communities, securing community tenure rights and incentives for resource utilisation. This argument is based on the observation that greater species diversity occurs outside protected areas that are more extensive (Scholes et al., 2004). Species migration between protected areas in response to shifting climatic conditions is likely to be impeded, unless assisted by often costly interventions geared towards landscapes with greater ecological connectivity. Strategic national policies could co-ordinate with communal or private land-use systems, especially when many small reserves are involved and would be particularly cost-effective if they address climate change proactively. Finally, migration strategies are very likely to become substantially more effective when they are implemented over larger regions and across national borders (e.g., Hansen et al., 2003).

Controlled burning and other techniques may be useful to reduce fuel load and the potential for catastrophic wildfires. It may also be possible to minimise the effect of severe weather events by, for example, securing water rights to maintain water levels through a drought, or by having infrastructure capable of surviving floods. Maintaining viable and widely dispersed populations of individual species also minimises the probability that localised catastrophic events will cause significant negative effects (e.g., hurricane, typhoon, flood).

Climate change is likely to increase opportunities for invasive alien species because of their adaptability to disturbance (Stachowicz et al., 2002; Lake and Leishman, 2004). Captive breeding for reintroduction and translocation or the use of provenance trials in forestry are expensive and likely to be less successful if climate change is more rapid. Such change could result in large-scale modifications of environmental conditions, including the loss or significant alteration of existing habitat over some or all of a species' range. Captive breeding and translocation should therefore not be perceived as panaceas for the loss of biological diversity that might accompany large changes in the climate. Populations of many species are already perilously small, and further loss of habitat and stress associated with severe climate change may push many taxa to extinction.

A costly adaptation option would be the restoration of habitats currently under serious threat, or creation of new habitats in areas where natural colonisation is unlikely to occur (Anonymous, 2000). In many cases the knowledge of ecosystem interactions and species requirements may be lacking. Engineering habitats to facilitate species movements may call for an entirely new field of study. Engineering interactions to defend coastlines, for example, that change the connectivity of coastal ecosystems, facilitate the spread of non-native species (Bulleri, 2005) as well as warm-temperate species advancing polewards (Helmuth et al., 2006; Mieszkowska et al., 2006).

Ultimately, managers may need to enhance or replace diminished or lost ecosystem services. This could mean manual seed dispersal or reintroducing pollinators. In the case of pest outbreaks, the use of pesticides may be necessary. Enhancing or replacing other services, such as contributions to nutrient cycling, ecosystem stability and ecosystem biodiversity may be much more difficult. The loss or reduced capacity of ecosystem services is likely to be a major source of 'surprises' from climate change.

4.6.2 Assessing the effectiveness and costs of adaptation options

There are few factual studies that have established the effectiveness and costs of adaptation options in ecosystems. Unfortunately, this makes a comprehensive assessment of the avoided damages (i.e., benefits) and costs impossible (see also Section 4.5). But the costs involved in monitoring, increasing the resilience of conservation networks and adaptive management are certainly large. For example, the money spent annually on nature conservation in the Netherlands was recently estimated to be €1 billion (Milieu en Natuurplanbureau, 2005). Of this amount, €285 million was used to manage national parks and reserves and €280 million was used for new reserve network areas and habitat improvement; the main objective being to reduce fragmentation between threatened populations and to respond to other threats. The reserve network planned for the Netherlands (to be established by 2020) will increase the resilience of species, populations and ecosystems to climate change, but at a high cost. Although not defined explicitly in this way, a significant proportion of these costs can be interpreted as climate adaptation costs.

4.6.3 Implications for biodiversity

Many studies and assessments stress the adverse impacts of climate change on biodiversity (e.g., Gitay et al., 2002; Hannah and Lovejoy, 2003; Thomas et al., 2004a; Lovejoy and Hannah, 2005; Schröter et al., 2005; Thuiller et al., 2005b; van Vliet and Leemans, 2006), but comprehensive appraisals of adaptation options to deal with declining biodiversity are rare.

The UN Convention on Biological Diversity (CBD, http://www.biodiv.org) aims to conserve biodiversity, to sustainably use biodiversity and its components and to fairly and equitably share benefits arising from the utilisation of biodiversity. This goes much further than most national biodiversity policies. The CBD explicitly recognises the use of biodiversity, ecosystems and their services and frames this as a developmental issue. As such, it extends beyond UNFCCC's objective of "avoiding dangerous human interference with the climate system at levels where ecosystems cannot adapt naturally". The main tool proposed by the CBD is the ecosystem approach (Smith and Malthby, 2003) based on integrated response options that intentionally and actively address ecosystem services (including biodiversity) and human well-being simultaneously, and involve all stakeholders at different institutional levels. The ecosystem approach resembles sustainable forest management projects (FAO, 2001). In theory, the ecosystem approach helps the conservation and sustainable use of biodiversity, but applications of the approach have had limited success (Brown et al., 2005a). Integrated responses include, however, learning by doing; a proactive approach that should increase the resilience of ecosystems and biodiversity.

4.6.4 Interactions with other policies and policy implications

Formulating integrated policies that cut across multiple UN conventions, such as the UNFCCC, CBD and Convention to Combat Desertification (CCD), could produce win-win situations in addressing climate change, increasing resilience and dealing with other policy issues (Nnadozie, 1998). Strategies aimed at combating desertification, for example, contribute towards increased soil carbon and moisture levels. Mitigation strategies focused on afforestation, including projects under the Clean Development Mechanism (CDM, see Glossary), could help ecosystem adaptation through improved ecological connectivity. The ecosystem approach can fulfil objectives specified by different conventions (Reid et al., 2005) and, in assessing adaptation strategies, such synergies could be identified and promoted.

4.7 Implications for sustainable development

Over the past 50 years, humans have converted and modified natural ecosystems more rapidly and over larger areas than in any comparable period of human history (e.g., Steffen et al., 2004). These changes have been driven by the rapidly growing demands for food, fish, freshwater, timber, fibre and fuel (e.g., Vitousek et al., 1997) and have contributed to substantial net gains in human well-being and economic development, while resulting in a substantial and largely irreversible loss of biodiversity and degradation in ecosystems and their services (Reid et al., 2005).

The consequences of policies to address the vulnerability of ecosystems to climate change at both the national and international level are not yet fully understood. There is growing evidence that significant impacts on the environment may result from perverse or unintended effects of policies from other sectors, which directly or indirectly have adverse consequences on ecosystems and other environmental processes (Chopra et al., 2005). Land re-distribution policies, for example, while designed to increase food self-sufficiency also contribute to reducing carbon sequestration and loss of biodiversity through extensive clear-cutting.

Effective mechanisms to analyse cross-sectoral impacts and to feed new scientific knowledge into policy-making are necessary (Schneider, 2004). There is substantial evidence to suggest that developing and implementing policies and strategies to reduce the vulnerability of ecosystems to climate change is closely linked to the availability of capacity to address current needs (e.g., Chanda, 2001). Thus, prospects for successful adaptation to climate change will remain limited as long as factors (e.g., population growth, poverty and globalisation) that contribute to chronic vulnerability to, for example, drought and floods, are not resolved (Kates, 2000; Reid et al., 2005).

4.7.1 Ecosystems services and sustainable development

Large differences in natural and socio-economic conditions among regions mitigate against simple solutions to the problem of ecosystem degradation and loss of services. Many interactions, lags and feedbacks, including those that operate across a range of spatial, temporal and organisational scales generate complex patterns which are not fully understood. Past actions to slow or reverse the degradation of ecosystems have yielded significant results, but these improvements have generally not kept pace with growing pressures (Reid et al., 2005). However, sound management of ecosystem services provides several cost-effective opportunities for addressing multiple development goals in a synergistic manner (Reid et al., 2005).

Progress achieved in addressing the Millennium Development Goals (MDGs) is unlikely to be sustained if ecosystem services continue to be degraded (Goklany, 2005). The role of ecosystems in sustainable development and in achieving the MDGs involves an array of stakeholders (Jain, 2003; Adeel et al., 2005). Evidence from different parts of the world shows that in most cases it is far from clear who is 'in charge' of the long-term sustainability of an ecosystem, let alone of the situation under future climates. Responding and adapting to the impacts of climate change on ecosystems calls for a clear and structured system of decision making at all levels (Kennett, 2002). Impacts of climate change on ecosystems also show strong interrelationships with ecosystem processes and human activities at various scales over time. Addressing these impacts requires a co-ordinated, integrated, cross-sectoral policy framework with a long-term focus; a strategy that so far has not been easy to implement (Brown, 2003).

4.7.2 Subsistence livelihoods and indigenous peoples

The impacts of climate change on ecosystems and their services will not be distributed equally around the world. Dryland, mountain and mediterranean regions are likely to be more vulnerable than others (Gitay et al., 2001) and ecosystem degradation is largest in these regions (Hassan et al., 2005). Climate change is likely to cause additional inequities, as its impacts are unevenly distributed over space and time and disproportionately affect the poor (Tol, 2001; Stern, 2007). The term 'double exposure' has been used for regions, sectors, ecosystems and social groups that are confronted both by the impacts of climate change and by the consequences of economic globalisation (O'Brien and Leichenko, 2000). Thus special attention needs to be given to indigenous peoples with subsistence livelihoods and groups with limited access to information and few means of adaptation. As a result climate change and sustainable development need to incorporate issues of equity (Kates, 2000; Jain, 2003; Richards, 2003).

4.8 Key uncertainties and research priorities

Key uncertainties listed here are those that limit our ability to project climate change impacts on ecosystems, but only if they have implications at sub-continental and higher spatial scales, are relevant for many species, populations and communities, or significantly weaken a modelling result. In terms of climate uncertainty, it is important to highlight that projections for precipitation carry a significantly higher uncertainty than temperature, yet play a major role for many projections obtained from modelling approaches. In relation to projecting climate change impacts on ecosystems, we find key sources of uncertainty to include:

- inadequate representation of the interactive coupling between ecosystems and the climate system and, furthermore, of the multiple interacting drivers of global change. This prevents a fully integrated assessment of climate change impacts on ecosystem services;
- major biotic feedbacks to the climate system, especially through trace gases from soils in all ecosystems, and methane from labile carbon stocks such as wetlands, peatlands, permafrost and yedoma;
- how aggregation within current DGVMs with respect to the functional role of individual species and the assumption of their instantaneous migration biases impact estimates;
- the net result of changing disturbance regimes (especially through fire, insects and land-use change) on biotic feedbacks to the atmosphere, ecosystem structure, function, biodiversity and ecosystem services;
- the magnitude of the CO_2-fertilisation effect in the terrestrial biosphere and its components over time;
- the limitations of climate envelope models used to project responses of individual species to climate changes, and for deriving estimations of species extinction risks;
- the synergistic role of invasive alien species in both biodiversity and ecosystem functioning;
- the effect of increasing surface ocean CO_2 and declining pH on marine productivity, biodiversity, biogeochemistry and ecosystem functioning;
- the impacts of interactions between climate change and changes in human use and management of ecosystems as well as other drivers of global environmental change.

Guided by the above, the following research needs can be identified as priorities for reducing uncertainties.

- Identify key vulnerabilities in *permafrost–soil–vegetation interactions* at high latitudes, and their potential feedback to the biosphere trace-gas composition. Recent estimates suggest that terrestrial permafrost contains more than 1,000 PgC, which is increasingly emitting CO_2 and more importantly, methane (e.g., Walter et al., 2006; Zimov et al., 2006). The implications of this for abrupt and significant climate forcing are significant (e.g., Schellnhuber, 2002; iLEAPS, 2005; Symon et al., 2005, p. 1015; Lelieveld, 2006; Zimov et al., 2006).
- More robust modelling of interactions between biota and their geophysical environment using several independently developed *DGVMs* and Earth-system models. Validation (Price et al., 2001) beyond model intercomparisons is required, especially also with respect to the methane cycle. The goal should be to narrow uncertainties relating to the vulnerability of the carbon sequestration potential of ecosystems including more realistic estimates of lagged and threshold responses (e.g., Scheffer et al., 2001; iLEAPS, 2005).
- More emphasis on precipitation projections (e.g., Handel and Risbey, 1992) and resulting *water regime* effects. These should emphasise interactions between vegetation and atmosphere, including CO_2-fertilisation effects, in mature forests in the Northern Hemisphere, seasonal tropical forests, and arid or semi-arid grassland and savannas (e.g., Jasienski et al., 1998; Karnosky, 2003).
- Improved understanding of the role of *disturbance regimes*, i.e., frequency and intensity of episodic events (drought, fire, insect outbreaks, diseases, floods and wind-storms) and that of alien species invasions, as they interact with ecosystem responses to climate change itself and pollution (e.g., Osmond et al., 2004; Opdam and Wascher, 2004).
- Development of integrated *large spatial-scale remote sensing with long-term field studies* (May, 1999b; Kräuchi et al., 2000; Morgan et al., 2001b; Osmond et al., 2004; Opdam and Wascher, 2004; Symon et al., 2005, p. 1019) to better address scale mismatches between the climate system and ecosystems (Root and Schneider, 1995).
- Studies on impacts of rising atmospheric CO_2 on *ocean acidification*, and warming on coral reefs and other marine systems (Coles and Brown, 2003; Anonymous, 2004), and widening the range of terrestrial ecosystems for which CO_2-fertilisation responses have been quantified (e.g., Bond et al., 2003).
- Validating species-specific *climate envelope models* by testing model projections against the plethora of range shifts observed in nature (e.g., Walther et al., 2001; Chapter 1).
- Advances in understanding the relationship between *biodiversity* and the *resilience* of ecosystem services at a scale relevant to human well-being, to quote Sir Robert May (1999a): "The relatively rudimentary state of ecological science prevents us from making reliable predictions about how much biological diversity we can lose before natural systems collapse and deprive us of services upon which we depend."
- Improve identification of environmental key factors influencing ecosystem structures that determine functionality and provisioning services of ecosystems together with quantitative information on *economic impacts* (including implications for adaptation costs – Toman, 1998a; Winnett, 1998; Kremen, 2005; Symon et al., 2005, e.g., p. 1019).
- *Integrative vulnerability* studies on adaptive management responses to preserve biodiversity (including conservation and reservation management) and ecosystem services in relation to pressures from land-use change and climate change (Kappelle et al., 1999; Lorenzoni et al., 2005; Stenseth and Hurrell, 2005; Symon et al., 2005).

Appendix 4.1

The table below contains detailed information on models and how the upscaling and downscaling were performed for each entry in Table 4.1 (using the same numbering scheme).

In each case **E** indicates an empirical derivation, **M** indicates a modelling study, a **number** refers to how many GCMs (see Glossary) were used in the original literature (for GCM abbreviations used here see below), other codes indicate whether model projections included respectively, precipitation (**P**), ocean acidification (**pH**), sea ice (**SI**), sea-level rise (**SLR**), sea surface temperature (**SST**) or anthropogenic water use (**W**); dispersal assumptions from the literature (**D** – estimate assumes dispersal; **ND** – estimate assumes no dispersal; **NR** – not relevant since species/ecosystem has nowhere to disperse to in order to escape warming – e.g., habitat is at top of isolated mountain or at southern extremity of austral landmass).

IMAGE, BIOME3, BIOME4, LPJ, MAPSS refer to specific models as used in the study, e.g., **LPJ** denotes the Lund-Potsdam-Jena dynamic global vegetation model (LPJ-DGVM – Sitch et al., 2003; see also Glossary).

Lower case **a-h** refer to how the literature was addressed in terms of up/downscaling (**a** – clearly defined global impact for a specific ΔT against a specific baseline, upscaling not necessary; **b** – clearly defined regional impact at a specific regional ΔT where no GCM used; **c** – clearly defined regional impact as a result of specific GCM scenarios but study only used the regional ΔT; **d** – as c but impacts also the result of regional precipitation changes; **e** – as b but impacts also the result of regional precipitation change; **f** – regional temperature change is off-scale for upscaling with available GCM patterns to 2100, in which case upscaling is, where possible, approximated by using Figures 10.5 and 10.8 from Meehl et al., 2007; **g** – studies which estimate the range of possible outcomes in a given location or region considering a multi-model ensemble linked to a global temperature change. In this case upscaling is not carried out since the GCM uncertainty has already been taken into account in the original literature; **h** – cases where sea surface temperature is the important variable, hence upscaling has been carried out using the maps from Meehl et al. (2007), using Figures 10.5 and 10.8, taking the increases in local annual mean (or where appropriate seasonal, from Figure 10.9) surface air temperature over the sea as equal to the local increases in annual mean or seasonal sea surface temperature. GCM abbreviations used here: **H2** – HadCM2, **H3** – HadCM3, **GF** – GFDL, **EC** – ECHAM4, **CS** – CSIRO, **CG** – CG, **PCM** – NCAR PCM.

The GCM outputs used in this calculation are those used in the Third Assessment Report (IPCC, 2001) and are at 5° resolution: HadCM3 A1FI, A2, B1, B2 where A2 is an ensemble of 3 runs and B2 is an ensemble of 2 runs; ECHAM4 A2 and B2 (not ensemble runs); CSIRO mark 2 A2, B1, B2; NCAR PCM A2 B2; CGCM2 A2 B2 (each an ensemble of 2 runs). Where GCM scenario names only were provided further details were taken from: HadCM2/3 (Mitchell et al., 1995), http://www.ipcc-data.org/ (see also Gyalistras et al., 1994; IPCC-TGCIA, 1999; Gyalistras and Fischlin, 1999; Jones et al., 2005). All used GCMs/AOGCMs have been reviewed here: IPCC (1990), IPCC (1996), Neilson and Drapek (1998), IPCC (2001).

No. [i]	Details on type of study, models, model results and methods used to derive the sensitivities as tabulated in Table 4.1 for each entry
1	M, 4, SST
2	E
3	E, SI
4,11,30	M, 7, ND, c; ref. quotes 13.8% loss in Rocky Mountains for each 1°C rise in JJA temperature, upscaled with CS, PCM, CG
5*	M, D&ND, P, a; 18% matches minimum expected climate change scenarios which Table 3 of ref. (supplementary material) lists as ΔT of 0.9°-1.7°C (mean 1.3°C) above 1961-1990 mean; 8 of the 9 sub-studies used H2, one used H3
6	M, 5, IMAGE, a; authors confirmed temperature baseline is year 2000 which is 0.1°C warmer than 1990
7	M, D, b; upscaled with H3, EC, CS, PCM, CG
8	M, SST, h
9	M, H2, P, ND, d; table 3 of ref. 1 gives global ΔT of 1.35°C above 1961-1990; HHGSDX of H3; downscaled with H3 then upscaled with H3, EC, CS, PCM, CG
10	M, H2, P, D&ND, d; as for No. 9
11	As for No. 4
12,14	M, P, NR, e; upscaled using H3, EC, CS, PCM, CG
13	M, D, b; upscaled using H3, EC, CS, PCM, CG
14	As for No. 12
15	M, P, NR, d; HadRM3PA2 in 2050, figure 13 in ref. shows ΔT matching B2 of H3 of 1.6°C above 1961-1990 mean; downscaled with H3 and upscaled with H3, EC, CS, PCM, CG
16	M, H3, P, D, e; H3 2050 SRES mean
17	E, P, D, b; upscaled using H3, EC, CS, PCM, CG
18	M, 10, P, D, d, g; table 3 of ref. 1 gives global ΔT of 1.35°C above 1961-1990; upscaled with H3, EC, CS, PCM, CG; Uses a local ΔT range across Australia
19	M, H3, P, D&ND, d; ref. gives B1 in 2050 with a ΔT of 1.8°C above the 1961-1990 baseline; downscaled with H3 and then upscaled with H3, EC, CG
20	M, H2, P, D&ND, d; studies used global annual mean ΔT of 1.9°C above 1961-1990 mean
21	M, P, D&ND, a; table 3 of ref. mid-range climate scenarios has a mean ΔT of 1.9°C above 1961-1990
22	M, H2, P, D&ND, d; ref. uses A2 of H3 in 2050 that has a ΔT of 1.9°C above 1961-1990 (Arnell et al., 2004); downscaled with H3 then upscaled with H3, EC, CS, PCM, CG

[i] Same numbers as used in first column in Table 4.1.

No.	Details on type of study, models, model results and methods used to derive the sensitivities as tabulated in Table 4.1 for each entry
23	h; upscaled using maps from WGI, chapter 10
24	E, P, NR, a
25	M, 2, P, NR, d; scenarios on CRU website used with ΔT of 2.0°C above 1961-1990, agrees with Table 3 of ref. 1 which gives ΔT of 2.0°C above 1961-1990 mean; downscaled with H3 then upscaled with H3, EC, CS, PCM, CG
26	M, H2, P, D, d; the 66% is from a suite of 179 representative species, table 3 of ref. 1 lists global ΔT of 3°C above 1961-1990 mean, upscaled with H3, EC, CS, CG
27	M, H2, P, D&ND, d; table 3 of ref. 1 which gives ΔT of 2.0°C above 1961-1990 mean using HHGGAX; downscaled with H3 then upscaled with H3, EC, CS, PCM, CG
28	M, H2, P, ND, d; as for No. 27
29	M, IMAGE, P, D&ND; ref. gives the global temperature change relative to 1990
30	As for No. 4
31	M, H3, W, a; ref. uses B2 of H3 in 2070 that has a ΔT rise of 2.1°C with respect to the 1961-1990 mean
32	M, P, D&ND; ref. uses B1 in H3 in 2080s from (Arnell et al., 2004)
33	M, 2, P, LPJ; upscaled with H3, EC5 (see also Figure 4.2; 4.3)
34	M, SST, h
35	M, P, D, d; UKCIP02 high emissions scenario used as central value; upscaled for Hampshire from UKCIP02 regional maps using H3, EC, CS
36	M, a
37	M, SLR, a; analysis based on transient 50% probability of sea-level rise using the US EPA scenarios for ΔT of 2°C above 1990 baseline
38	M, P, NR, d; see No. 15; HadRM3PA2 in 2050, taken from Figure 13 in ref.
39	M, H2, D&ND, d; ref. uses global ΔT of 2.3°C above 1961-1990 mean
40	As for No. 6
41	M, CS, b; upscaled with H3, EC, CS, PCM, CG
42	M, 15, SI, a; Arzel (Arzel et al., 2006) uses 15 GCMs with A1B for 2080s, ΔT A1B 2080s multi-model mean from Meehl et al., 2007, Figure 10.5 is 2.5°C above 1990; ACIA uses 4 GCMs with B2, multi-model ΔT is 2.2°C over 1961-1990 or 2.0°C above 1990
43	M, GE, P, NR, d; GENESIS GCM with 2.5°C rise for CO_2 doubling from 345 to 690ppm, 345 ppm corresponds quite closely to the 1961-1990 mean; upscaling then gives the range all locations used; variously used H3, EC, CS, CG
44	M, NR, b; upscaled with H3, EC, CS, and CG
45	M, 2, P, d, g; range is due to importance of Δ P, GFDL CO_2 doubling is from 300 ppm which occurred in about 1900, and climate sensitivity in SAR is 3.7; UKMO in 2050 is 1.6°C above 1961-1990 mean, 1.9°C above pre-industrial
46,47	M, H2, BIOME4, P, NR, c; A1 scenario of H2GS has ΔT of 2.6°C relative to 1961-1990 mean
48,49	pH, g; IS92a in 2100 has 788 ppm CO_2 and ΔT of 1.3-3.5°C above 1990 (IPCC, 1996, Figure 6.20)
50	M, 10, P, D, d, g; 2.6°C above 1961-1990 mean.upscaled with H3, EC, CS, CG at lower end, upper end out of range
51	M, P, D&ND, a; Table 3 of ref. maximum climate scenarios have mean ΔT of 2.6°C above 1961-1990 or 2.3°C above 1990; 8 of the 9 sub-studies used H2, one used H3
52	M, BIOME3, P, d, f; H2 2080s with aerosols (HHGSA1) has global ΔT of 2.6°C above 1961-1990 mean
53	M, H3, W, a; ref. uses A2 of H3 in 2070 that has a ΔT of 2.7°C with respect to the 1961-1990 mean and hence 2.5°C with respect to 1990
54	M, 2, SLR, a; IS92a in 2100 has 788 ppm CO_2 and ΔT of 1.3-3.5°C above 1990 (IPCC, 1996, Figure 6.20)
55	M, SST, h
56	E, P, D, e; upscaled with H3, EC, CS
57	M, P, NR, e; upscaled for several sites taken from maps in ref., using H3, EC, CS, CG
58	M, NR
59	pH, a; impact is at CO_2 doubling, T range given by WGI for equilibrium climate sensitivity
60	M, CS, P, d; upscaled with H3, EC, CS, CG
61	M, NR, b; % derived from Table 1 in ref. for all forest areas combined on the 3 islands studied; upscaling considers changes averaged over 3 islands and uses H3, EC, CS, CG
62	M, H3, P, D&ND, d, f; table 3 of ref. lists global ΔT of 3°C above 1961-1990 mean
63	M, H2, SLR, NR, a; H2 2080s without aerosols has global ΔT of 3.4°C above pre-industrial (Hulme et al., 1999)
64	M, 7, BIOME3, MAPSS, P, D&ND, a; uses transient and equilibrium CO_2 doubling scenarios from Neilson & Drapek (1998) table 2; control concentrations were obtained directly from modellers; thus deduced mean global mean ΔT for this study
65	M, 2, P, D, d; study used CO_2 doubling scenarios in equilibrium – CCC ΔT at doubling is 3.5°C relative to 1900 whilst GFDL R30 is 3.3°C relative to 1900; upscaling gives range H3, EC, CG
66	M, D, b; upscaled with H3, EC, CS
67	M, H3, P, D&ND, d; ref. uses A2 in H3 in 2080 that has a ΔT of 3.3°C above 1961-1990 (Arnell et al., 2004)
68	M, CCC, P, D, d; CO_2 equilibrium doubling scenario has ΔT of 3.5°C relative to 1900; downscaled with CGCM and upscaled with H3, EC, CS, CG
69,70,71	M, 5, IMAGE, a; authors confirmed temperature baseline is year 2000 which is 0.1°C warmer than 1990
72	M, H3, P, D&ND, d; ref. lists ΔT of 3.6°C for A1 in 2080 relative to 1961-1990, downscaled with H3 and upscaled with H3, EC, CG
73	M, NR, b; upscaled with H3, EC, CG
74	M, NR, b, f; Meehl et al., 2007, Figures 10.5 and 10.8 suggest global ΔT of 3.5°C relative to 1990
75	M, D, f; Meehl et al., 2007, Figure 10. 5 shows this occurs for ΔT ≥3.5°C above 1990
76	M, NR, b, f; as for No. 75
77	M, NR, b, f
78	M, SLR, a; US EPA scenario of 4.7°C above 1990

References

Aanes, R., B.E. Saether, F.M. Smith, E.J. Cooper, P.A. Wookey and N.A. Oritsland, 2002: The Arctic Oscillation predicts effects of climate change in two trophic levels in a high-arctic ecosystem. *Ecol. Lett.*, **5**, 445-453.

Aber, J., R.P. Neilson, S. McNulty, J.M. Lenihan, D. Bachelet and R.J. Drapek, 2001: Forest processes and global environmental change: predicting the effects of individual and multiple stressors. *BioScience*, **51**, 735-751.

Adeel, Z., U. Safriel, D. Niemeijer, R. White, G. de Kalbermatten, M. Glantz, B. Salem, R. Scholes, M. Niamir-Fuller, S. Ehui and V. Yapi-Gnaore, Eds., 2005: *Ecosystems and Human Well-being: Desertification Synthesis*. Island Press, Washington, District of Columbia, 36 pp.

Adger, W.N., S. Huq, K. Brown, D. Conway and M. Hulme, 2003: Adaptation to climate change in the developing world. *Prog. Dev. Stud.*, **3**, 179-195.

Ainsworth, E.A. and S.P. Long, 2005: What have we learned from 15 years of free-air CO_2 enrichment (FACE)? A meta-analytic review of the responses of photosynthesis, canopy properties and plant production to rising CO_2. *New Phytol.*, **165**, 351-371.

Alcamo, J., K. Kok, G. Busch, J. Priess, B. Eickhout, M. Rounsevell, D. Rothman and M. Heistermann, 2006: Searching for the future of land: scenarios from the local to global scale. *Land-use and Land-cover Change: Local Processes, Global Impacts*, E.F. Lambin and H. Geist, Eds., Springer Verlag, Berlin, 137-156.

Alencar, A., D. Nepstad and M.D.V. Diaz, 2006: Forest understorey fire in the Brazilian Amazon in ENSO and non-ENSO years: area burned and committed carbon emissions. *Earth Interactions*, **10**, 1-17.

Allen Consulting Group, 2005: *Climate Change Risk and Vulnerability: Promoting an Efficient Adaptation Response in Australia*. Australian Greenhouse Office, Department of the Environment and Heritage, Canberra, 159 pp.

Alsos, I.G., T. Engelskjon, L. Gielly, P. Taberlet and C. Brochmann, 2005: Impact of ice ages on circumpolar molecular diversity: insights from an ecological key species. *Mol. Ecol.*, **14**, 2739-2753.

Andreae, M.O., D. Rosenfeld, P. Artaxo, A.A. Costa, G.P. Frank, K.M. Longo and M.A.F. Silva-Dias, 2004: Smoking rain clouds over the Amazon. *Science*, **303**, 1337-1342.

Andreone, F., J.E. Cadle, N. Cox, F. Glaw, R.A. Nussbaum, C.J. Raxworthy, S.N. Stuart, D. Vallan and M. Vences, 2005: Species review of amphibian extinction risks in Madagascar: conclusions from the global amphibian assessment. *Conserv. Biol.*, **19**, 1790-1802.

Andrew, N.R. and L. Hughes, 2005: Herbivore damage along a latitudinal gradient: relative impacts of different feeding guilds. *Oikos*, **108**, 176-182.

Angert, A., S. Biraud, C. Bonfils, C.C. Henning, W. Buermann, J. Pinzon, C.J. Tucker and I. Fung, 2005: Drier summers cancel out the CO_2 uptake enhancement induced by warmer springs. *P. Natl. Acad. Sci. USA*, **102**, 10823-10827.

Anisimov, O.A., N.I. Shiklomanov and F.E. Nelson, 2002a: Variability of seasonal thaw depth in permafrost regions: a stochastic modeling approach. *Ecol. Model.*, **153**, 217-227.

Anisimov, O.A., A.A. Velichko, P.F. Demchenko, A.V. Eliseev, I.I. Mokhov and V.P. Nechaev, 2002b: Effect of climate change on permafrost in the past, present, and future. *Izv. Atmos. Ocean. Phys.*, **38**, S25-S39.

Anonymous, 2000: Potential UK adaptation strategies for climate change. DETR Summary Report 99DPL013, Department of the Environment, Transport and the Regions, Wetherby, West Yorkshire, UK, 19 pp.

Anonymous, 2001: *Lothar: Der Orkan 1999*. Eidg. Forschungsanstalt WSL and Bundesamt für Umwelt, Wald und Landschaft BUWAL, Birmensdorf, Bern, 365 pp.

Anonymous, 2004: Priorities for research on the ocean in a high-CO_2 world. *Symposium on The Ocean in a High-CO_2 World*, Scientific Committee on Oceanic Research (SCOR) and the Intergovernmental Oceanographic Commission (IOC) of UNESCO, UNESCO Headquarters, Paris, 12 pp.

Anyamba, A. and C.J. Tucker, 2005: Analysis of Sahelian vegetation dynamics using NOAA-AVHRR NDVI data from 1981-2003. *J. Arid Environ.*, **63**, 596-614.

Araújo, M.B., R.J. Whittaker, R.J. Ladle and M. Erhard, 2005: Reducing uncertainty in projections of extinction risk from climate change. *Global Ecol. Biogeogr.*, **14**, 529-538.

Arnell, N.W., M.G.R. Cannell, M. Hulme, R.S. Kovats, J.F.B. Mitchell, R.J. Nicholls, M.L. Parry, M.T.J. Livermore and A. White, 2002: The consequences of CO_2 stabilisation for the impacts of climate change. *Climatic Change*, **53**, 413-446.

Arnell, N.W., M.J.L. Livermore, S. Kovats, P.E. Levy, R. Nicholls, M.L. Parry and S.R. Gaffin, 2004: Climate and socio-economic scenarios for global-scale climate change impacts assessments: characterising the SRES storylines. *Global Environ. Chang.*, **14**, 3-20.

Arrigo, K.R. and D.N. Thomas, 2004: Large scale importance of sea ice biology in the Southern Ocean. *Antarct. Sci.*, **16**, 471-486.

Arrow, K., P. Dasgupta, L. Goulder, G. Daily, P. Ehrlich, G. Heal, S. Levin, K.G. Maler, S. Schneider, D. Starett and B. Walker, 2004: Are we consuming too much? *J. Econ. Perspect.*, **18**, 147-172.

Arzel, O., T. Fichefet and H. Goose, 2006: Sea ice evolution over the 20th and 21st centuries as simulated by current AOGCMs. *Ocean Model.*, **12**, 401-415.

Asner, G.P., D.E. Knapp, E.N. Broadbent, P.J.C. Oliveira, M. Keller and J.N. Silva, 2005: Selective logging in the Brazilian Amazon. *Science*, **310**, 480-482.

Augustin, L., C. Barbante, P.R.F. Barnes, J.M. Barnola, M. Bigler, E. Castellano, O. Cattani, J. Chappellaz, D. Dahljensen, B. Delmonte, G. Dreyfus, G. Durand, S. Falourd, H. Fischer, J. Fluckiger, M.E. Hansson, P. Huybrechts, R. Jugie, S.J. Johnsen, J. Jouzel, P. Kaufmann, J. Kipfstuhl, F. Lambert, V.Y. Lipenkov, G.V.C. Littot, A. Longinelli, R. Lorrain, V. Maggi, V. Masson-Delmotte, H. Miller, R. Mulvaney, J. Oerlemans, H. Oerter, G. Orombelli, F. Parrenin, D.A. Peel, J.R. Petit, D. Raynaud, C. Ritz, U. Ruth, J. Schwander, U. Siegenthaler, R. Souchez, B. Stauffer, J.P. Steffensen, B. Stenni, T.F. Stocker, I.E. Tabacco, R. Udisti, R.S.W. van de Wal, M. van den Broeke, J. Weiss, F. Wilhelms, J.G. Winther, E.W. Wolff and M. Zucchelli, 2004: Eight glacial cycles from an Antarctic ice core. *Nature*, **429**, 623-628.

Bachelet, D., R.P. Neilson, J.M. Lenihan and R.J. Drapek, 2001: Climate change effects on vegetation distribution and carbon budget in the United States. *Ecosystems*, **4**, 164-185.

Bachelet, D., R.P. Neilson, T. Hickler, R.J. Drapek, J.M. Lenihan, M.T. Sykes, B. Smith, S. Sitch and K. Thonicke, 2003: Simulating past and future dynamics of natural ecosystems in the United States. *Global Biogeochem. Cy.*, **17**, 1045, doi:10.1029/2001GB001508.

Badeck, F.W., A. Bondeau, K. Bottcher, D. Doktor, W. Lucht, J. Schaber and S. Sitch, 2004: Responses of spring phenology to climate change. *New Phytol.*, **162**, 295-309.

Baillie, J.E.M., C. Hilton-Taylor and S.N. Stuart, Eds., 2006: *2004 IUCN Red List of Threatened Species: A Global Species Assessment*. IUCN: The World Conservation Union, Gland, and Cambridge, 217 pp.

Baker, T.R., O.L. Phillips, Y. Malhi, S. Almeida, L. Arroyo, A. Di Fiore, T. Erwin, N. Higuchi, T.J. Killeen, S.G. Laurance, W.F. Laurance, S.L. Lewis, A. Monteagudo, D.A. Neill, P.N. Vargas, N.C.A. Pitman, J.N.M. Silva and R.V. Martinez, 2004: Increasing biomass in Amazonian forest plots. *Philos. T. Roy. Soc. Lond. B*, **359**, 353-365.

Bakkenes, M., J.R.M. Alkemade, F. Ihle, R. Leemans and J.B. Latour, 2002: Assessing effects of forecasted climate change on the diversity and distribution of European higher plants for 2050. *Global Change Biol.*, **8**, 390-407.

Balmford, A., A. Brunner, P. Cooper, R. Costanza, S. Farber, R.E. Green, M. Jenkins, P. Jefferiss, V. Jessamy, J. Madden, K. Munro, N. Myers, S. Naeem, J. Paavola, M. Rayment, S. Rosendo, J. Roughgarden, K. Trumper and R.K. Turner, 2002: Economic reasons for conserving wild nature. *Science*, **297**, 950-953.

Balmford, A., P. Crane, A. Dobson, R.E. Green and G.M. Mace, 2005: The 2010 challenge: data availability, information needs and extraterrestrial insights. *Philos. T. Roy. Soc. Lond. B*, **360**, 221-228.

Balzter, H., F.F. Gerard, C.T. George, C.S. Rowland, T.E. Jupp, I. McCallum, A. Shvidenko, S. Nilsson, A. Sukhinin, A. Onuchin and C. Schmullis, 2005: Impact of the Arctic Oscillation pattern on interannual forest fire variability in Central Siberia. *Geophys. Res. Lett.*, **32**, L14709, doi:10.1029/2005GL022526.

Banzhaf, S. and J. Boyd, 2005: The architecture and measurement of an ecosystem services index. Discussion paper RFF DP 05-22, Resources for the Future, Washington, District of Columbia, 57 pp.

Barber, D.G. and J. Iacozza, 2004: Historical analysis of sea ice conditions in M'Clintock channel and the Gulf of Boothia, Nunavut: implications for ringed seal and polar bear habitat. *Arctic*, **57**, 1-14.

Barbosa, P., G. Libertà and G. Schmuck, 2003: The European Forest Fires Information System (EFFIS) results on the 2003 fire season in Portugal by the 20th of August. Report 20030820, European Commission, Directorate General Joint Research Centre, Institute for Environment and Sustainability, Land Management Unit, Ispra, 10 pp.

Barbraud, C. and H. Weimerskirch, 2001: Emperor penguins and climate change.

Nature, **411**, 183-186.

Barnes, D.K.A., D.A. Hodgson, P. Convey, C.S. Allen and A. Clarke, 2006: Incursion and excursion of Antarctic biota: past, present and future. *Global Ecol. Biogeogr.*, **15**, 121-142.

Bauder, E.T., 2005: The effects of an unpredictable precipitation regime on vernal pool hydrology. *Freshwater Biol.*, **50**, 2129-2135.

Beaugrand, G., P.C. Reid, F. Ibanez, J.A. Lindley and M. Edwards, 2002: Reorganization of North Atlantic marine copepod biodiversity and climate. *Science*, **296**, 1692-1694.

Beaugrand, G., K.M. Brander, J.A. Lindley, S. Souissi and P.C. Reid, 2003: Plankton effect on cod recruitment in the North Sea. *Nature*, **426**, 661-664.

Beaumont, L.J. and L. Hughes, 2002: Potential changes in the distributions of latitudinally restricted Australian butterfly species in response to climate change. *Global Change Biol.*, **8**, 954-971.

Becker, A. and H. Bugmann, 2001: Global change and mountain regions. IGBP Report, IGBP Secretariat, Royal Swedish Academy of Sciences, Stockholm, 88 pp.

Beever, E.A., P.F. Brussard and J. Berger, 2003: Patterns of apparent extirpation among isolated populations of pikas (*Ochotona princeps*) in the Great Basin. *J. Mammal.*, **84**, 37-54.

Behrenfeld, M.J. and P.G. Falkowski, 1997: A consumer's guide to phytoplankton primary production models. *Limnol. Oceanogr.*, **42**, 1479-1491.

Bellamy, P.H., P.J. Loveland, R.I. Bradley, R.M. Lark and G.J.D. Kirk, 2005: Carbon losses from all soils across England and Wales 1978-2003. *Nature*, **437**, 245.

Beniston, M., 2000: *Environmental Change in Mountains and Uplands*. Arnold, London, 172 pp.

Beniston, M., 2003: Climatic change in mountain regions: a review of possible impacts. *Climatic Change*, **59**, 5-31.

Beniston, M., 2004: The 2003 heat wave in Europe: a shape of things to come? An analysis based on Swiss climatological data and model simulations. *Geophys. Res. Lett.*, **31**, L02202, doi:10.1029/2003GL018857.

Beniston, M. and P. Jungo, 2002: Shifts in the distributions of pressure, temperature and moisture and changes in the typical weather patterns in the Alpine region in response to the behavior of the North Atlantic Oscillation. *Theor. Appl. Climatol.*, **71**, 29-42.

Beniston, M. and D.B. Stephenson, 2004: Extreme climatic events and their evolution under changing climatic conditions. *Global Planet. Change*, **44**, 1-9.

Beniston, M., H.F. Diaz and R.S. Bradley, 1997: Climatic change at high elevation sites: an overview. *Climatic Change*, **36**, 233-251.

Benning, T.L., D. Lapointe, C.T. Atkinson and P.M. Vitousek, 2002: Interactions of climate change with biological invasions and land use in the Hawaiian Islands: modeling the fate of endemic birds using a geographic information system. *P. Natl. Acad. Sci. USA*, **99**, 14246-14249.

Berendse, F., N. van Breemen, H. Kanrydin, A. Buttler, M. Heijmans, M.R. Hoosbeek, J. Lee, E. Mitchell, T. Saarinen, H. Vassander and B. Wallen, 2001: Raised atmospheric CO_2 levels and increased N deposition cause shifts in plant species composition and production in *Sphagnum* bogs. *Global Change Biol.*, **7**, 591-598.

Bergengren, J.C., S.L. Thompson, D. Pollard and R.M. Deconto, 2001: Modeling global climate-vegetation interactions in a doubled CO_2 world. *Climatic Change*, **50**, 31-75.

Bergeron, Y., S. Gauthier, V. Kafka, P. Lefort and D. Lesieur, 2001: Natural fire frequency for the eastern Canadian boreal forest: consequences for sustainable forestry. *Can. J. Forest Res.*, **31**, 384-391.

Bergeron, Y., M. Flannigan, S. Gauthier, A. Leduc and P. Lefort, 2004: Past, current and future fire frequency in the Canadian boreal forest: implications for sustainable forest management. *Ambio*, **33**, 356-360.

Berry, P.M., P.A. Harrison, T.P. Dawson and C.A. Walmsley, 2005: Modelling natural resource responses to climate change (MONARCH): a local approach. UKCIP Technical Report, UK Climate Impacts Programme, Oxford, 24 pp.

Berry, P.M., M.D.A. Rounsevell, P.A. Harrison and E. Audsley, 2006: Assessing the vulnerability of agricultural land use and species to climate change and the role of policy in facilitating adaptation. *Environ. Sci. Policy*, **9**, 189-204.

Berthelot, M., P. Friedlingstein, P. Ciais, P. Monfray, J.L. Dufresne, H. Le Treut and L. Fairhead, 2002: Global response of the terrestrial biosphere to CO_2 and climate change using a coupled climate-carbon cycle model. *Global Biogeochem. Cy.*, **16**, 1084, doi:10.1029/2001GB001827.

Bertness, M.D. and P.J. Ewanchuk, 2002: Latitudinal and climate-driven variation in the strength and nature of biological interactions in New England salt marshes. *Oecologia*, **132**, 392-401.

Betts, R.A., 2000: Offset of the potential carbon sink from boreal forestation by decreases in surface albedo. *Nature*, **408**, 187-190.

Betts, R.A. and H.H. Shugart, 2005: Dynamic ecosystem and Earth system models. *Climate Change and Biodiversity*, T.E. Lovejoy and L. Hannah, Eds., Yale University Press, New Haven, Connecticut, 232-251.

Betts, R.A., P.M. Cox and F.I. Woodward, 2000: Simulated responses of potential vegetation to doubled-CO_2 climate change and feedbacks on near-surface temperature. *Global Ecol. Biogeogr.*, **9**, 171-180.

Betts, R.A., P.M. Cox, M. Collins, P.P. Harris, C. Huntingford and C.D. Jones, 2004: The role of ecosystem-atmosphere interactions in simulated Amazonian precipitation decrease and forest dieback under global climate warming. *Theor. Appl. Climatol.*, **78**, 157-175.

Bharwani, S., M. Bithell, T.E. Downing, M. New, R. Washington and G. Ziervogel, 2005: Multi-agent modelling of climate outlooks and food security on a community garden scheme in Limpopo, South Africa. *Philos. T. Roy. Soc. Lond. B*, **360**, 2183-2194.

Billings, S.A., S.M. Schaeffer, S. Zitzer, T. Charlet, S.D. Smith and R.D. Evans, 2002: Alterations of nitrogen dynamics under elevated carbon dioxide in an intact Mojave Desert ecosystem: evidence from nitrogen-15 natural abundance. *Oecologia*, **131**, 463-467.

Billings, S.A., S.M. Schaeffer and R.D. Evans, 2003: Nitrogen fixation by biological soil crusts and heterotrophic bacteria in an intact Mojave Desert ecosystem with elevated CO_2 and added soil carbon. *Soil Biol. Biochem.*, **35**, 643-649.

Blaschke, L., M. Schulte, A. Raschi, N. Slee, H. Rennenberg and A. Polle, 2001: Photosynthesis, soluble and structural carbon compounds in two Mediterranean oak species (*Quercus pubescens* and *Q. ilex*) after lifetime growth at naturally elevated CO_2 concentrations. *Plant Biol.*, **3**, 288-297.

Blodau, C., 2002: Carbon cycling in peatlands: a review of processes and controls. *Environ. Rev.*, **10**, 111-134.

Bockstael, N.E., A.M. Freeman, R.J. Kopp, P.R. Portney and V.K. Smith, 2000: On measuring economic values for nature. *Environ. Sci. Technol.*, **34**, 1384-1389.

Boisvenue, C. and S.W. Running, 2006: Impacts of climate change on natural forest productivity: evidence since the middle of the 20th century. *Global Change Biol.*, **12**, 862-882.

Bomhard, B., D.M. Richardson, J.S. Donaldson, G.O. Hughes, G.F. Midgley, D.C. Raimondo, A.G. Rebelo, M. Rouget and W. Thuiller, 2005: Potential impacts of future land use and climate change on the Red List status of the Proteaceae in the Cape Floristic Region, South Africa. *Global Change Biol.*, **11**, 1452-1468.

Bonan, G.B., 2002: *Ecological Climatology: Concepts and Applications*. Cambridge University Press, Cambridge, 678 pp.

Bond, W.J. and G.F. Midgley, 2000: A proposed CO_2-controlled mechanism of woody plant invasion in grasslands and savannas. *Global Change Biol.*, **6**, 865-869.

Bond, W.J. and J.J. Midgley, 2003: The evolutionary ecology of sprouting in woody plants. *Int. J. Plant Sci.*, **164**, S103-S114.

Bond, W.J. and J.E. Keeley, 2005: Fire as a global 'herbivore': the ecology and evolution of flammable ecosystems. *Trends Ecol. Evol.*, **20**, 387-394.

Bond, W.J., G.F. Midgley and F.I. Woodward, 2003: The importance of low atmospheric CO_2 and fire in promoting the spread of grasslands and savannas. *Global Change Biol.*, **9**, 973-982.

Bond, W.J., F.I. Woodward and G.F. Midgley, 2005: The global distribution of ecosystems in a world without fire. *New Phytol.*, **165**, 525-537.

Bonin, A., P. Taberlet, C. Miaud and F. Pompanon, 2006: Explorative genome scan to detect candidate loci for adaptation along a gradient of altitude in the common frog (Rana temporaria). *Mol. Biol. Evol.*, **23**, 773-783.

Bopp, L., O. Aumont, S. Belviso and P. Monfray, 2003: Potential impact of climate change on marine dimethyl sulfide emissions. *Tellus B*, **55**, 11-22.

Bopp, L., O. Boucher, O. Aumont, S. Belviso, J.L. Dufresne, M. Pham and P. Monfray, 2004: Will marine dimethylsulfide emissions amplify or alleviate global warming? A model study. *Can. J. Fish. Aquat. Sci.*, **61**, 826-835.

Borghetti, M., S. Cinnirella, F. Magnani and A. Saracino, 1998: Impact of long-term drought on xylem embolism and growth in *Pinus halepensis* Mill. *Trees–Struct. Funct.*, **12**, 187-195.

Bosch, J., L.M. Carrascal, L. Durain, S. Walker and M.C. Fisher, 2006: Climate change and outbreaks of amphibian chytriomycosis in a montane area of central Spain: is there a link? *P. Roy. Soc. Lond. B*, **274**, 253-260.

Bowman, D.M.J.S., A. Walsh and D.J. Milne, 2001: Forest expansion and grassland contraction within a Eucalyptus savanna matrix between 1941 and 1994 at Litchfield National Park in the Australian monsoon tropics. *Global Ecol. Biogeogr.*, **10**, 535-548.

Boyd, J., 2006: The nonmarket benefits of nature: what should be counted in green GDP? *Ecol. Econ.*, **61**, 716-723.

Breshears, D.D. and C.D. Allen, 2002: The importance of rapid, disturbance-induced losses in carbon management and sequestration. *Global Ecol. Biogeogr.*, **11**, 1-5.

Breshears, D.D., N.S. Cobb, P.M. Rich, K.P. Price, C.D. Allen, R.G. Balice, W.H. Romme, J.H. Kastens, M.L. Floyd, J. Belnap, J.J. Anderson, O.B. Myers and C.W. Meyer, 2005: Regional vegetation die-off in response to global-change-type drought. *P. Natl. Acad. Sci. USA*, **102**, 15144-15148.

Broennimann, O., W. Thuiller, G. Hughes, G.F. Midgley, J.M.R. Alkemade and A. Guisan, 2006: Do geographic distribution, niche property and life form explain plants' vulnerability to global change? *Global Change Biol.*, **12**, 1079-1093.

Brohan, P., J.J. Kennedy, I. Haris, S.F.B. Tett and P.D. Jones, 2006: Uncertainty estimates in regional and global observed temperature changes: a new dataset from 1850. *J. Geophys. Res. D*, **111**, D12106, doi:10.1029/2005JD006548.

Brooks, M.L., C.M. D'Antonio, D.M. Richardson, J.B. Grace, J.E. Keeley, J.M. Ditomaso, R.J. Hobbs, M. Pellant and D. Pyke, 2004: Effects of invasive alien plants on fire regimes. *BioScience*, **54**, 677-688.

Brown, K., 2003: Integrating conservation and development: a case of institutional misfit. *Front. Ecol. Environ.*, **1**, 479-487.

Brown, K., J. Mackensen, S. Rosendo, K. Viswanathan, L. Cimarrusti, K. Fernando, C. Morsello, I.M. Siason, S. Singh, I. Susilowati, M. Soccoro Manguiat, G. Bialluch and W.F. Perrin, 2005a: Integrated responses. *Ecosystems and Human Well-being: Volume 3: Policy Responses*, K. Chopra, R. Leemans, P. Kumar and H. Simons, Eds., Island Press, Washington, District of Columbia, 425-464.

Brown, K.J., J.S. Clark, E.C. Grimm, J.J. Donovan, P.G. Mueller, B.C.S. Hansen and I. Stefanova, 2005b: Fire cycles in North American interior grasslands and their relation to prairie drought. *P. Natl. Acad. Sci. USA*, **102**, 8865-8870.

Brown, T.J. and B.L. Hall, 2001: Climate analysis of the 2000 fire season. CE-FA Report 01-02, Program for Climate Ecosystem and Fire Applications, Desert Research Institute, Division of Atmospheric Sciences, Reno, Nevada, 40 pp.

Brown, T.J., B.L. Hall and A.L. Westerling, 2004: The impact of twenty-first century climate change on wildland fire danger in the western United States: an applications perspective. *Climatic Change*, **62**, 365-388.

Buchmann, N., 2002: Plant ecophysiology and forest response to global change. *Tree Physiol.*, **22**, 1177-1184.

Buddemeier, R.W., J.A. Kleypas and R.B. Aronson, 2004: *Coral Reefs and Global Climate Change: Potential Contributions of Climate Change to Stresses on Coral Reef Ecosystems*. Pew Center on Global Climatic Change, Arlington, Virginia, 44 pp.

Bugmann, H. and C. Pfister, 2000: Impacts of interannual climate variability on past and future forest composition. *Reg. Environ. Change*, **1**, 112-125.

Bugmann, H., B. Zierl and S. Schumacher, 2005: Projecting the impacts of climate change on mountain forests and landscapes. *Global Change and Mountain Regions: An Overview of Current Knowledge*, U.M. Huber, H.K.M. Bugmann and M.A. Reasoner, Eds., Springer, Berlin, 477-488.

Bulleri, F., 2005: Role of recruitment in causing differences between intertidal assemblages on seawalls and rocky shores. *Mar. Ecol.–Prog. Ser.*, **287**, 53-64.

Burgan, R.E., P.L. Andrews, L.S. Bradshaw, C.H. Chase, R.A. Hartford and D.J. Latham, 1997: WFAS: wildland fire assessment system. *Fire Management Notes*; **57**, 14-17.

Burke, E.J., S.J. Brown and N. Christidis, 2006: Modelling the recent evolution of global drought and projections for the 21st century with the Hadley Centre climate model. *J. Hydrometeorol.*, **7**, 113-1125.

Burkett, V.R., D.A. Wilcox, R. Stottlemyer, W. Barrow, D. Fagre, J. Baron, J. Price, J.L. Nielsen, C.D. Allen, D.L. Peterson, G. Ruggerone and T. Doyle, 2005: Nonlinear dynamics in ecosystem response to climatic change: case studies and policy implications. *Ecol. Complex.*, **2**, 357-394.

Busch, G., 2006: Future European agricultural landscapes: what can we learn from existing quantitative land use scenario studies? *Agr. Ecosyst. Environ.*, **114**, 121-140.

Bush, M.B., M.R. Silman and D.H. Urrego, 2004: 48,000 years of climate and forest change in a biodiversity hot spot. *Science*, **303**, 827-829.

Butler, R.W., T.D. Williams, N. Warnock and M. Bishop, 1997: Wind assistance: a requirement for shorebird migration. *Auk*, **114**, 456-466.

Cairns, D.M. and J. Moen, 2004: Herbivory influences tree lines. *J. Ecol.*, **92**, 1019-1024.

Caldeira, K. and M.E. Wickett, 2003: Anthropogenic carbon and ocean pH. *Nature*, **425**, 365-365.

Callaghan, T.V., L.O. Bjorn, Y. Chernov, T. Chapin, T.R. Christensen, B. Huntley, R.A. Ims, M. Johansson, D. Jolly, S. Jonasson, N. Matveyeva, N. Panikov, W. Oechel and G. Shaver, 2004a: Effects on the function of arctic ecosystems in the short- and long-term perspectives. *Ambio*, **33**, 448-458.

Callaghan, T.V., L.O. Bjorn, Y. Chernov, T. Chapin, T.R. Christensen, B. Huntley, R.A. Ims, M. Johansson, D. Jolly, S. Jonasson, N. Matveyeva, N. Panikov, W. Oechel, G. Shaver and H. Henttonen, 2004b: Effects on the structure of arctic ecosystems in the short- and long-term perspectives. *Ambio*, **33**, 436-447.

Callaghan, T.V., L.O. Bjorn, Y. Chernov, T. Chapin, T.R. Christensen, B. Huntley, R.A. Ims, M. Johansson, D. Jolly, S. Jonasson, N. Matveyeva, N. Panikov, W. Oechel, G. Shaver, S. Schaphoff and S. Sitch, 2004c: Effects of changes in climate on landscape and regional processes, and feedbacks to the climate system. *Ambio*, **33**, 459-468.

Callaghan, T.V., L.O. Björn, F.S. Chapin III, Y. Chernov, T.R. Christensen, B. Huntley, R. Ims, M. Johansson, D.J. Riedlinger, S. Jonasson, N. Matveyeva, W. Oechel, N. Panikov and G. Shaver, 2005: Arctic tundra and polar desert ecosystems. *Arctic Climate Impact Assessment (ACIA): Scientific Report*, C. Symon, L. Arris and B. Heal, Eds., Cambridge University Press, Cambridge, 243-352.

Camarero, J.J., E. Gutierrez and M.J. Fortin, 2000: Boundary detection in altitudinal treeline ecotones in the Spanish central Pyrenees. *Arct. Antarct. Alp. Res.*, **32**, 117-126.

Camill, P., 2005: Permafrost thaw accelerates in boreal peatlands during late-20th century climate warming. *Climatic Change*, **68**, 135-152.

Camill, P. and J.S. Clark, 2000: Long-term perspectives on lagged ecosystem responses to climate change: permafrost in boreal peatlands and the grassland/woodland boundary. *Ecosystems*, **3**, 534-544.

Canadell, J.G., P. Ciais, P. Cox and M. Heimann, 2004: Quantifying, understanding and managing the carbon cycle in the next decades. *Climatic Change*, **67**, 147-160.

Carpenter, S., P. Pingali, E. Bennett and M. Zurek, Eds., 2005: *Ecosystems and Human Well-being: Volume 2: Scenarios*. Island Press, Washington, District of Columbia, 560 pp.

Carroll, A.L., S.W. Taylor, J. Régnière and L. Safranyik, 2004: Effects of climate change on range expansion by the mountain pine beetle in British Columbia. *Mountain Pine Beetle Symposium: Challenges and Solutions*, T.L. Shore, J.E. Brooks and J.E. Stone, Eds., Natural Resources Canada, Canadian Forest Service, Pacific Forestry Centre, Victoria, British Columbia, 223-232.

Cech, P., S. Pepin and C. Körner, 2003: Elevated CO_2 reduces sap flux in mature deciduous forest trees. *Oecologia*, **137**, 258-268.

Cesar, H., L. Burke and L. Pet-Soede, 2003: *The Economics of Worldwide Coral Reef Degradation*. Cesar Environmental Economics Consulting (CEEC), Arnhem, 23 pp.

Chanda, R., 2001: Safe-guarding the globe or basic well-being? Global environmental change issues and Southern Africa. *Globalization, Democracy and Development in Africa: Challenges and Prospects*, T. Assefa, S.M. Rugumamu and A.G.M. Ahmed, Eds., Organisation of Social Science Research in Eastern and Southern Africa (OSSREA), Addis Ababa, 246-262.

Chang, R., Ed., 1997: *Incommensurability, Incomparability, and Practical Reason*. Harvard University Press, Cambridge, Massachusetts, 303 pp.

Chapin, F.S. and A.M. Starfield, 1997: Time lags and novel ecosystems in response to transient climatic change in arctic Alaska. *Climatic Change*, **35**, 449-461.

Chapin, F.S., E.S. Zavaleta, V.T. Eviner, R.L. Naylor, P.M. Vitousek, H.L. Reynolds, D.U. Hooper, S. Lavorel, O.E. Sala, S.E. Hobbie, M.C. Mack and S. Diaz, 2000: Consequences of changing biodiversity. *Nature*, **405**, 234-242.

Chapin, F.S., P.A. Matson and H.A. Mooney, 2002: *Principles of Terrestrial Ecosystem Ecology*. Springer, New York, 436 pp.

Chapin, F.S., T.V. Callaghan, Y. Bergeron, M. Fukuda, J.F. Johnstone, G. Juday and S.A. Zimov, 2004: Global change and the boreal forest: thresholds, shifting states or gradual change? *Ambio*, **33**, 361-365.

Chapin, F.S., III, M. Berman, T.V. Callaghan, P. Convey, A.S. Crépin, K. Danell, H. Ducklow, B. Forbes, G. Kofinas, A.D. McGuire, M. Nuttall, R. Virginia, O. Young and S.A. Zimov, 2005a: Polar systems. *Ecosystems and Human Well-being: Volume 1: Current State and Trends*, R. Hassan, R. Scholes and N. Ash, Eds., Island Press, Washington, District of Columbia, 717-746.

Chapin, F.S., M. Sturm, M.C. Serreze, J.P. McFadden, J.R. Key, A.H. Lloyd, A.D. McGuire, T.S. Rupp, A.H. Lynch, J.P. Schimel, J. Beringer, W.L. Chapman, H.E. Epstein, E.S. Euskirchen, L.D. Hinzman, G. Jia, C.L. Ping, K.D. Tape, C.D.C. Thompson, D.A. Walker and J.M. Welker, 2005b: Role of land-surface changes in arctic summer warming. *Science*, **310**, 657-660.

Chapuis, J.L., Y. Frenot and M. Lebouvier, 2004: Recovery of native plant communities after eradication of rabbits from the subantarctic Kerguelen Islands, and influence of climate change. *Biol. Conserv.*, **117**, 167-179.

Chauhan, M. and B. Gopal, 2001: Biodiversity and management of Keoladeo National Park (India): a wetland of international importance. *Biodiversity in Wetlands: Assessment, Function and Conservation: Volume 2*, B. Gopal, W.J. Junk and J.A. Davies, Eds., Backhuys Publishers, Leiden, 217-256.

Cheddadi, R., J. Guiot and D. Jolly, 2001: The Mediterranean vegetation: what if the atmospheric CO_2 increased? *Landscape Ecol.*, **16**, 667-675.

Chidumayo, E.N., 2001: Climate and phenology of savanna vegetation in southern Africa. *J. Veg. Sci.*, **12**, 347-354.

Chopra, K., R. Leemans, P. Kumar and H. Simons, Eds., 2005: *Ecosystems and Human Well-being: Volume 3: Policy Responses*. Island Press, Washington, District of Columbia, 621 pp.

Chornesky, E.A. and J.M. Randall, 2003: The threat of invasive alien species to biological diversity: setting a future course. *Ann. Mo. Bot. Gard.*, **90**, 67-76.

Christensen, J., T. Carter and F. Giorgi, 2002: PRUDENCE employs new methods to assess European climate change. *EOS Transactions*, **83**, 147.

Christensen, T.R., T.R. Johansson, H.J. Akerman, M. Mastepanov, N. Malmer, T. Friborg, P. Crill and B.H. Svensson, 2004: Thawing sub-arctic permafrost: effects on vegetation and methane emissions. *Geophys. Res. Lett.*, **31**, L04501, doi:10.1029/2003GL018680.

Christensen, J.H., B. Hewitson, A. Busuioc, A. Chen, X. Gao, I. Held, R. Jones, W.-T. Kwon, R. Laprise, V.M. Rueda, L. Mearns, C.G. Menéndez, J. Räisänen, A. Rinke, R.K. Kolli, A. Sarr and P. Whetton, 2007: Regional climate projections. *Climate Change 2007: The Physical Science Basis. Contribution of Working Group I to the Fourth Assessment Report of the Intergovernmental Panel on Climate Change*, S. Solomon, D. Qin, M. Manning, Z. Chen, M. Marquis, K.B. Averyt, M. Tignor and H.L. Miller, Eds., Cambridge University Press, Cambridge, 847-940.

Chu, C., N.E. Mandrak and C.K. Minns, 2005: Potential impacts of climate change on the distributions of several common and rare freshwater fishes in Canada. *Divers. Distrib.*, **11**, 299-310.

Chytry, M., P. Pysek, L. Tichy, I. Knollová and J. Danihelka, 2005: Invasions by alien plants in the Czech Republic: a quantitative assessment across habitats. *Preslia*, **77**, 339-354.

Ciais, P., M. Reichstein, N. Viovy, A. Granier, J. Ogee, V. Allard, M. Aubinet, N. Buchmann, Chr. Bernhofer, A. Carrara, F. Chevallier, N. De Noblet, A.D. Friend, P. Friedlingstein, T. Grünwald, B. Heinesch, P. Keronen, A. Knohl, G. Krinner, D. Loustau, G. Manca, G. Matteucci, F. Miglietta, J.M. Ourcival, D. Papale, K. Pilegaard, S. Rambal, G. Seufert, J.F. Soussana, M.J. Sanz, E.D. Schulze, T. Vesala and R. Valentini, 2005: Europe-wide reduction in primary productivity caused by the heat and drought in 2003. *Nature*, **437**, 529-533.

Clark, J.S., 1998: Why trees migrate so fast: confronting theory with dispersal biology and the paleorecord. *Am. Nat.*, **152**, 204-224.

Clark, J.S., M. Lewis and L. Horvath, 2001: Invasion by extremes: population spread with variation in dispersal and reproduction. *Am. Nat.*, **157**, 537-554.

Clark, J.S., M. Lewis, J.S. McLachlan and J. Hillerislambers, 2003: Estimating population spread: what can we forecast and how well? *Ecology*, **84**, 1979-1988.

Clark, R.A. and C.L.J. Frid, 2001: Long-term changes in the North Sea ecosystem. *Environ. Rev.*, **9**, 131-187.

Cochrane, M.A., 2003: Fire science for rainforests. *Nature*, **421**, 913-919.

Cochrane, M.A. and W.F. Laurance, 2002: Fire as a large-scale edge effect in Amazonian forests. *J. Trop. Ecol.*, **18**, 311-325.

Cohen, A.S., 2003: *Paleolimnology: The History and Evolution of Lake Systems*. Oxford University Press, Oxford, 528 pp.

Cole, J., 2003: Global change: dishing the dirt on coral reefs. *Nature*, **421**, 705-706.

Coles, S.L. and B.E. Brown, 2003: Coral bleaching: capacity for acclimatization and adaptation. *Adv. Mar. Biol.*, **46**, 183-223.

Collingham, Y.C. and B. Huntley, 2000: Impacts of habitat fragmentation and patch size upon migration rates. *Ecol. Appl.*, **10**, 131-144.

Cornelissen, J.H.C., T.V. Callaghan, J.M. Alatalo, A. Michelsen, E. Graglia, A.E. Hartley, D.S. Hik, S.E. Hobbie, M.C. Press, C.H. Robinson, G.H.R. Henry, G.R. Shaver, G.K. Phoenix, D.G. Jones, S. Jonasson, F.S. Chapin, U. Molau, C. Neill, J.A. Lee, J.M. Melillo, B. Sveinbjornsson and R. Aerts, 2001: Global change and arctic ecosystems: is lichen decline a function of increases in vascular plant biomass? *J. Ecol.*, **89**, 984-994.

Costanza, R., 2000: Social goals and the valuation of ecosystem services. *Ecosystems*, **3**, 4-10.

Costanza, R., 2001: Visions, values, valuation, and the need for an ecological economics. *BioScience*, **51**, 459-468.

Costanza, R., R. D'Arge, R. de Groot, S. Farber, M. Grasso, B. Hannon, K. Limburg, S. Naeem, R.V. O'Neill, J. Paruelo, R.G. Raskin, P. Sutton and M. van den Belt, 1997: The value of the world's ecosystem services and natural capital. *Nature*, **387**, 253-260.

Costanza, R., M. Daly, C. Folke, P. Hawken, C.S. Holling, A.J. McMichael, D. Pimentel and D. Rapport, 2000: Managing our environmental portfolio. *BioScience*, **50**, 149-155.

Cowling, R.M., P.W. Rundel, B.B. Lamont, M.K. Arroyo and M. Arianoutsou, 1996: Plant diversity in Mediterranean-climate regions. *Trends Ecol. Evol.*, **11**, 362-366.

Cowling, R.M., R.L. Pressey, A.T. Lombard, P.G. Desmet and A.G. Ellis, 1999: From representation to persistence: requirements for a sustainable system of conservation areas in the species-rich Mediterranean-climate desert of southern Africa. *Divers. Distrib.*, **5**, 51-71.

Cowling, R.M., F. Ojeda, B. Lamont, P.W. Rundel and R. Lechmere-Oertel, 2005: Rainfall reliability, a neglected factor in explaining convergence and divergence of plant traits in fire-prone Mediterranean-climate ecosystems. *Global Ecol. Biogeogr.*, **14**, 509-519.

Cox, P.M., R.A. Betts, C.D. Jones, S.A. Spall and I.J. Totterdell, 2000: Acceleration of global warming due to carbon-cycle feedbacks in a coupled climate model. *Nature*, **408**, 184-187.

Cox, P.M., R.A. Betts, M. Collins, P.P. Harris, C. Huntingford and C.D. Jones, 2004: Amazonian forest dieback under climate-carbon cycle projections for the 21st century. *Theor. Appl. Climatol.*, **78**, 137-156.

Cox, P.M., C. Huntingford and C.D. Jones, 2006: Conditions for sink-to-source transitions and runaway feedbacks from the land carbon cycle. *Avoiding Dangerous Climate Change*, H.J. Schellnhuber, W. Cramer, N. Nakićenović, T.M.L. Wigley and G. Yohe, Eds., Cambridge University Press, Cambridge, 155-161.

Craig, P., P. Trail and T.E. Morrell, 1994: The decline of fruit bats in American-Samoa due to hurricanes and overhunting. *Biol. Conserv.*, **69**, 261-266.

Cramer, W., A. Bondeau, F.I. Woodward, I.C. Prentice, R.A. Betts, V. Brovkin, P.M. Cox, V. Fisher, J.A. Foley, A.D. Friend, C. Kucharik, M.R. Lomas, N. Ramankutty, S. Sitch, B. Smith, A. White and C. Young-Molling, 2001: Global response of terrestrial ecosystem structure and function to CO_2 and climate change: results from six dynamic global vegetation models. *Global Change Biol.*, **7**, 357-373.

Cramer, W., A. Bondeau, S. Schaphoff, W. Lucht, B. Smith and S. Sitch, 2004: Tropical forests and the global carbon cycle: impacts of atmospheric carbon dioxide, climate change and rate of deforestation. *Philos. T. Roy. Soc. Lond. B*, **359**, 331-343.

Crawford, R.M.M., C.E. Jeffree and W.G. Rees, 2003: Paludification and forest retreat in northern oceanic environments. *Ann. Bot.*, **91**, 213-226.

Crawley, M.J., 1989: Chance and timing in biological invasions. *Biological Invasions: A Global Perspective*, J. Drake, F. Dicastri, R. Groves, F. Kruger, H. Mooney, M. Rejmanek and M. Williamson, Eds., John Wiley, Chichester, 407-423.

Crick, H.Q.P., 2004: The impact of climate change on birds. *Ibis*, **146**, 48-56.

Cropp, R. and A. Gabrica, 2002: Ecosystem adaptation: do ecosystems maximize resilience? *Ecology*, **83**, 2019-2026.

Croxall, J.P., P.N. Trathan and E.J. Murphy, 2002: Environmental change and Antarctic seabird populations. *Science*, **297**, 1510-1514.

Cubasch, U., G.A. Meehl, G.J. Boer, R.J. Stouffer, M. Dix, A. Noda, C.A. Senior, S. Raper and K.S. Yap, 2001: Projections of future climate change. *Climate Change 2001: The Scientific Basis. Contribution of Working Group I to the Third Assessment Report of the Intergovernmental Panel on Climate Change*, J.T. Houghton, Y. Ding, D.J. Griggs, M. Noguer, P.J. van der Linden, X. Dai, K. Maskell and C.A. Johnson, Eds., Cambridge University Press, Cambridge, 525-582.

Currie, D.J., 2001: Projected effects of climate change on patterns of vertebrate and tree species richness in the conterminous United States. *Ecosystems*, **4**, 216-225.

Daily, G.C., T. Söderqvist, S. Aniyar, K. Arrow, P. Dasgupta, P.R. Ehrlich, C. Folke, A. Jansson, B. Jansson, N. Kautsky, S. Levin, J. Lubchenco, K. Mäler, D. Simpson, D. Starrett, D. Tilman and B. Walker, 2000: The value of nature and the nature of value. *Science*, **289**, 395-396.

Daimaru, H. and H. Taoda, 2004: Effects of snow pressure on the distribution of subalpine *Abies mariesii* forests in northern Honshu Island, Japan. *Journal of Agricultural Meteorology*, **60**, 253-261.

Dale, V.H., L.A. Joyce, S. McNulty and R.P. Neilson, 2000: The interplay between climate change, forests, and disturbances. *Sci. Total Environ.*, **262**, 201-204.

Davidson, E.A. and I.A. Janssens, 2006: Temperature sensitivity of soil carbon decomposition and feedbacks to climate change. *Nature*, **440**, 165-173.

Davis, M.B., 1989: Lags in vegetation response to greenhouse warming. *Climatic Change*, **15**, 75-82.

Davis, M.B. and R.G. Shaw, 2001: Range shifts and adaptive responses to Quaternary climate change. *Science*, **292**, 673-679.

Davis, M.B., R.G. Shaw and J.R. Etterson, 2005: Evolutionary responses to changing climate. *Ecology*, **86**, 1704-1714.

De Bono, A., P. Peduzzi, G. Giuliani and S. Kluser, 2004: *Impacts of Summer 2003 Heat Wave in Europe*. Early Warning on Emerging Environmental Threats 2, UNEP: United Nations Environment Programme, Nairobi, 4 pp.

de Graaff, M.A., K.J. van Groenigen, J. Six, B. Hungate and C. van Kessel, 2006: Interactions between plant growth and soil nutrient cycling under elevated CO_2: a meta-analysis. *Global Change Biol.*, **12**, 2077-2091.

de Groot, W.J., P.M. Bothwell, D.H. Carlsson and K.A. Logan, 2003: Simulating the effects of future fire regimes on western Canadian boreal forests. *J. Veg. Sci.*, **14**, 355-364.

de Koning, F., R. Olschewski, E. Veldkamp, P. Benitez, M. Lopez-Ulloa, T. Schlichter and M. de Urquiza, 2005: The ecological and economic potential of carbon sequestration in forests: examples from South America. *Ambio*, **34**, 224-229.

de Wit, M. and J. Stankiewicz, 2006: Changes in surface water supply across Africa with predicted climate change. *Science*, **311**, 1917-1921.

Del Barrio, G., P.A. Harrison, P.M. Berry, N. Butt, M. Sanjuan, R.G. Pearson and T.P. Dawson, 2006: Integrating multiple modelling approaches to predict the potential impacts of climate change on species' distributions in contrasting regions: comparison and implications for policy. *Environ. Sci. Policy,* **9**, 129-147.

Denman, K.L., G. Brasseur, A. Chidthaisong, P. Ciais, P. Cox, R.E. Dickinson, D. Hauglustaine, C. Heinze, E. Holland, D. Jacob, U. Lohmann, S. Ramachandran, P.L. da Silva Dias, S.C. Wofsy and X. Zhang, 2007: Couplings between changes in the climate system and biogeochemistry. *Climate Change 2007: The Physical Science Basis. Contribution of Working Group I to the Fourth Assessment Report of the Intergovernmental Panel on Climate Change*, S. Solomon, D. Qin, M. Manning, Z. Chen, M. Marquis, K.B. Averyt, M. Tignor and H.L. Miller, Eds., Cambridge University Press, Cambridge, 499-587.

Derocher, A.E., N.J. Lunn and I. Stirling, 2004: Polar bears in a warming climate. *Integr. Comp. Biol.*, **44**, 163-176.

Deutschewitz, K., A. Lausch, I. Kuhn and S. Klotz, 2003: Native and alien plant species richness in relation to spatial heterogeneity on a regional scale in Germany. *Global Ecol. Biogeogr.*, **12**, 299-311.

Diaz, H.F., Ed., 2003: *Climate Variability and Change in High Elevation Regions: Past, Present and Future*. Kluwer Academic, Dordrecht, 282 pp.

Diaz, H.F. and V. Markgraf, Eds., 1992: *El Niño Historical and Paleoclimatic Aspects of the Southern Oscillation*. Cambridge University Press, Cambridge, 476 pp.

Diaz, S., J. Fargione, F.S. Chapin and D. Tilman, 2006: Biodiversity loss threatens human well-being. *PLoS Biol.*, **4**, 1300-1305.

Dirnböck, T., S. Dullinger and G. Grabherr, 2003: A regional impact assessment of climate and land-use change on alpine vegetation. *J. Biogeogr.*, **30**, 401-417.

Donner, S.D., W.J. Skirving, C.M. Little, M. Oppenheimer and O. Hoegh-Guldberg, 2005: Global assessment of coral bleaching and required rates of adaptation under climate change. *Global Change Biol.*, **11**, 2251-2265.

Dormann, C.F. and S.J. Woodin, 2002: Climate change in the Arctic: using plant functional types in a meta-analysis of field experiments. *Funct. Ecol.*, **16**, 4-17.

Dube, O.P. and G. Pickup, 2001: Effects of rainfall variability and communal and semi-commercial grazing on land cover in southern African rangelands. *Climate Res.*, **17**, 195-208.

Dufresne, J.L., P. Friedlingstein, M. Berthelot, L. Bopp, P. Ciais, L. Fairhead, H. Le Treut and P. Monfray, 2002: On the magnitude of positive feedback between future climate change and the carbon cycle. *Geophys. Res. Lett.*, **29**, 1405, doi:10.1029/2001GL013777.

Duguay, C.R., T.D. Prowse, B.R. Bonsal, R.D. Brown, M.P. Lacroix and P. Menard, 2006: Recent trends in Canadian lake ice cover. *Hydrol. Process.*, **20**, 781-801.

Dukes, J.S. and H.A. Mooney, 1999: Does global change increase the success of biological invaders? *Trends Ecol. Evol.*, **14**, 135-139.

Dukes, J.S., N.R. Chiariello, E.E. Cleland, L.A. Moore, M.R. Shaw, S. Thayer, T. Tobeck, H.A. Mooney and C.B. Field, 2005: Responses of grassland production to single and multiple global environmental changes. *PLoS Biol.*, **3**, 1829-1837.

Dullinger, S., T. Dirnbock and G. Grabherr, 2004: Modelling climate change-driven treeline shifts: relative effects of temperature increase, dispersal and invasibility. *J. Ecol.*, **92**, 241-252.

Dullinger, S., T. Dirnbock, R. Kock, E. Hochbichler, T. Englisch, N. Sauberer and G. Grabherr, 2005: Interactions among tree-line conifers: differential effects of pine on spruce and larch. *J. Ecol.*, **93**, 948-957.

Duraiappah, A., S. Naeem, T. Agardi, N. Ash, D. Cooper, S. Díaz, D.P. Faith, G. Mace, J.A. McNeilly, H.A. Mooney, A.A. Oteng-Yeboah, H.M. Pereira, S. Polasky, C. Prip, W.V. Reid, C. Samper, P.J. Schei, R. Scholes, F. Schutyser and A. van Jaarsveld, Eds., 2005: *Ecosystems and Human Well-being: Biodiversity Synthesis*. Island Press, Washington, District of Columbia, 100 pp.

Durieux, L., L.A.T. Machado and H. Laurent, 2003: The impact of deforestation on cloud cover over the Amazon arc of deforestation. *Remote Sens. Environ.*, **86**, 132-140.

Eastman, J.L., M.B. Coughenour and R.A. Pielke, 2001: The regional effects of CO_2 and landscape change using a coupled plant and meteorological model. *Global Change Biol.*, **7**, 797-815.

ECF, Ed., 2004: *What is Dangerous Climate Change? Initial Results of a Symposium on Key Vulnerable Regions Climate Change and Article 2 of the UNFCCC*. International Symposium, Beijing, 27-30 October. European Climate Forum, Buenos Aires, 39 pp.

Edwards, E.J., D.G. Benham, L.A. Marland and A.H. Fitter, 2004: Root production is determined by radiation flux in a temperate grassland community. *Global Change Biol.*, **10**, 209-227.

Edwards, M. and A.J. Richardson, 2004: Impact of climate change on marine pelagic phenology and trophic mismatch. *Nature*, **430**, 881-884.

Edwards, M., G. Beaugrand, P.C. Reid, A.A. Rowden and M.B. Jones, 2002: Ocean climate anomalies and the ecology of the North Sea. *Mar. Ecol.–Prog. Ser.*, **239**, 1-10.

Edwards, M., D.G. Johns, S.C. Leterme, E. Svendsen and A.J. Richardson, 2006: Regional climate change and harmful algal blooms in the northeast Atlantic. *Limnol. Oceanogr.*, **51**, 820-829.

Eliasson, P.E., R.E. McMurtrie, D.A. Pepper, M. Stromgren, S. Linder and G.I. Agren, 2005: The response of heterotrophic CO_2 flux to soil warming. *Global Change Biol.*, **11**, 167-181.

Eliot, I., C.M. Finlayson and P. Waterman, 1999: Predicted climate change, sea-level rise and wetland management in the Australian wet-dry tropics. *Wetlands Ecology and Management*, **7**, 63-81.

Enhalt, D. and M. Prather, 2001: Atmospheric chemistry and greenhouse gases. *Climate Change 2001: The Scientific Basis. Contribution of Working Group I to the Third Assessment Report of the Intergovernmental Panel on Climate Change*, J.T. Houghton, Y. Ding, D.J. Griggs, M. Noguer, P.J. van der Linden, X. Dai, K. Maskell and C.A. Johnson, Eds., Cambridge University Press, Cambridge, 239-287.

Enquist, C.A.F., 2002: Predicted regional impacts of climate change on the geographical distribution and diversity of tropical forests in Costa Rica. *J. Biogeogr.*, **29**, 519-534.

Epstein, H.E., R.A. Gill, J.M. Paruelo, W.K. Lauenroth, G.J. Jia and I.C. Burke, 2002: The relative abundance of three plant functional types in temperate grasslands and shrublands of North and South America: effects of projected climate change. *J. Biogeogr.*, **29**, 875-888.

Epstein, P.R. and E. Mills, 2005: *Climate Change Futures: Health, Ecological and Economic Dimensions*. Center for Health and the Global Environment, Harvard Medical School, Boston, Massachusetts, 142 pp.

Erasmus, B.F.N., A.S. Van Jaarsveld, S.L. Chown, M. Kshatriya and K.J. Wessels, 2002: Vulnerability of South African animal taxa to climate change. *Global Change Biol.*, **8**, 679-693.

Evans, R.D. and J. Belnap, 1999: Long-term consequences of disturbance on nitrogen dynamics in an arid ecosystem. *Ecology*, **80**, 150-160.

Facon, B., B.J. Genton, J. Shykoff, P. Jarne, A. Estoup and P. David, 2006: A general eco-evolutionary framework for understanding bioinvasions. *Trends Ecol. Evol.*, **21**, 130-135.

Fahrig, L., 2003: Effects of habitat fragmentation on biodiversity. *Annu. Rev. Ecol. Evol. Syst.*, **34**, 487-515.

FAO, 2001: Global forest resource assessment 2000: main report. FAO Forestry Paper 140, Food and Agriculture Organization of the United Nations, Rome, Italy, 479 pp.

Farber, S., R. Costanza, D.L. Childers, J. Erickson, K. Gross, M. Grove, C.S. Hopkinson, J. Kahn, S. Pincetl, A. Troy, P. Warren and M. Wilson, 2006: Linking ecology and economics for ecosystem management. *BioScience*, **56**, 121-133.

Fay, P.A., J.D. Carlisle, A.K. Knapp, J.M. Blair and S.L. Collins, 2000: Altering rainfall timing and quantity in a mesic grassland ecosystem: design and performance of rainfall manipulation shelters. *Ecosystems*, **3**, 308-319.

Fay, P.A., J.D. Carlisle, B.T. Danner, M.S. Lett, J.K. McCarron, C. Stewart, A.K. Knapp, J.M. Blair and S.L. Collins, 2002: Altered rainfall patterns, gas exchange, and growth in grasses and forbs. *Int. J. Plant Sci.*, **163**, 549-557.

Fay, P.A., J.D. Carlisle, A.K. Knapp, J.M. Blair and S.L. Collins, 2003: Productivity responses to altered rainfall patterns in a C_4-dominated grassland. *Oecologia*, **137**, 245-251.

Feddema, J.J., K.W. Oleson, G.B. Bonan, L.O. Mearns, L.E. Buja, G.A. Meehl and W.M. Washington, 2005: The importance of land-cover change in simulating future climates. *Science*, **310**, 1674-1678.

Feely, R.A., C.L. Sabine, K. Lee, W. Berelson, J. Kleypas, V.J. Fabry and F.J. Millero, 2004: Impact of anthropogenic CO_2 on the $CaCO_3$ system in the oceans. *Science*, **305**, 362-366.

Field, C.B., M.J. Behrenfeld, J.T. Randerson and P. Falkowski, 1998: Primary production of the biosphere: integrating terrestrial and oceanic components. *Science*, **281**, 237-240.

Fink, A.H., T. Brücher, A. Krüger, G.C. Leckebusch, J.G. Pinto and U. Ulbrich, 2004: The 2003 European summer heatwaves and drought: synoptic diagnosis and impacts. *Weather*, **59**, 209-216.

Finlayson, C.M., R. D'Cruz, N. Davidson, J. Alder, S. Cork, R. de Groot, C. Lévêque, G.R. Milton, G. Peterson, D. Pritchard, B.D. Ratner, W.V. Reid, C. Revenga, M. Rivera, F. Schutyser, M. Siebentritt, M. Stuip, R. Tharme, S. Butchart, E. Dieme-Amting, H. Gitay, S. Raaymakers and D. Taylor, Eds., 2005: *Ecosystems and Human Well-being: Wetlands and Water Synthesis*. Island Press, Washington, District of Columbia, 80 pp.

Fischer, R., Ed., 2005: The condition of forests in Europe: 2005 executive report. United Nations Economic Commission for Europe (UN-ECE), Geneva, 36 pp.

Fischlin, A. and D. Gyalistras, 1997: Assessing impacts of climatic change on forests in the Alps. *Global Ecol. Biogeogr. Lett.*, **6**, 19-37.

Fish, M.R., I.M. Cote, J.A. Gill, A.P. Jones, S. Renshoff and A.R. Watkinson, 2005: Predicting the impact of sea-level rise on Caribbean sea turtle nesting habitat. *Conserv. Biol.*, **19**, 482-491.

Flanagan, L.B., L.A. Wever and P.J. Carlson, 2002: Seasonal and interannual variation in carbon dioxide exchange and carbon balance in a northern temperate grassland. *Global Change Biol.*, **8**, 599-615.

Flannigan, M.D., B.J. Stocks and B.M. Wotton, 2000: Climate change and forest fires. *Sci. Total Environ.*, **262**, 221-229.

Fleming, R.A., J.N. Candau and R.S. McAlpine, 2002: Landscape-scale analysis of interactions between insect defoliation and forest fire in Central Canada. *Climatic Change*, **55**, 251-272.

Flöder, S. and C.W. Burns, 2004: Phytoplankton diversity of shallow tidal lakes: influence of periodic salinity changes on diversity and species number of a natural assemblage. *J. Phycol.*, **40**, 54-61.

Florenzano, G.T., 2004: Birds as indicators of recent environmental changes in the Apennines (Foreste Casentinesi National Park, central Italy). *Ital. J. Zool.*, **71**, 317-324.

Foley, J.A., M.T. Coe, M. Scheffer and G.L. Wang, 2003: Regime shifts in the Sahara and Sahel: interactions between ecological and climatic systems in northern Africa. *Ecosystems*, **6**, 524-539.

Foley, J.A., R. Defries, G.P. Asner, C. Barford, G. Bonan, S.R. Carpenter, F.S. Chapin, M.T. Coe, G.C. Daily, H.K. Gibbs, J.H. Helkowski, T. Holloway, E.A. Howard, C.J. Kucharik, C. Monfreda, J.A. Patz, I.C. Prentice, N. Ramankutty and P.K. Snyder, 2005: Global consequences of land use. *Science*, **309**, 570-574.

Folke, C., S. Carpenter, B. Walker, M. Scheffer, T. Elmqvist, L. Gunderson and C.S. Holling, 2004: Regime shifts, resilience, and biodiversity in ecosystem management. *Annu. Rev. Ecol. Evol. Syst.*, **35**, 557-581.

Folkestad, T., M. New, J.O. Kaplan, J.C. Comiso, S. Watt-Cloutier, T. Fenge, P. Crowley and L.D. Rosentrater, 2005: Evidence and implications of dangerous climate change in the Arctic. *Avoiding Dangerous Climate Change*, H.J. Schellnhuber, W. Cramer, N. Nakićenović, T.M.L. Wigley and G. Yohe, Eds., Cambridge University Press, Cambridge, 215-218.

Forbes, B.C., 1995: Effects of surface disturbance on the movement of native and exotic plants under a changing climate. *Global Change and Arctic Terrestrial Ecosystems*, T.V. Callaghan, Ed., European Commission, Directorate-General, Science, Research and Development, Brussels, 209-219.

Forcada, J., P.N. Trathan, K. Reid, E.J. Murphy and J.P. Croxall, 2006: Contrasting population changes in sympatric penguin species in association with climate warming. *Global Change Biol.*, **12**, 411-423.

Forster, P., V. Ramaswamy, P. Araxo, T. Berntsen, R.A. Betts, D.W. Fahey, J. Haywood, J. Lean, D.C. Lowe, G. Myhre, J. Nganga, R. Prinn, G. Raga, M. Schulze and R. Van Dorland, 2007: Changes in atmospheric constituents and in radiative forcing. *Climate Change 2007: The Physical Science Basis. Contribution of Working Group I to the Fourth Assessment Report of the Intergovernmental Panel on Climate Change*, S. Solomon, D. Qin, M. Manning, Z. Chen, M. Marquis, K.B. Averyt, M. Tignor and H.L. Miller, Eds., Cambridge University Press, Cambridge, 130-234.

Foster, P., 2001: The potential negative impacts of global climate change on tropical montane cloud forests. *Earth Sci. Rev.*, **55**, 73-106.

Freeman, C., N. Fenner, N.J. Ostle, H. Kang, D.J. Dowrick, B. Reynolds, M.A. Lock, D. Sleep, S. Hughes and J. Hudson, 2004: Export of dissolved organic carbon from peatlands under elevated carbon dioxide levels. *Nature*, **430**, 195-198.

Freiwald, A., J.H. Fossa, A. Grehan, T. Koslow and J.M. Roberts, 2004: Cold-water coral reefs: out of sight no longer out of mind. UNEP-WCMC No. 22, UNEP World Conservation Monitoring Centre, Cambridge, 86 pp.

Frey, K.E. and L.C. Smith, 2005: Amplified carbon release from vast West Siberian peatlands by 2100. *Geophys. Res. Lett.*, **32**, L09401, doi:10.1029/2004GL02202.

Fried, J.S., M.S. Torn and E. Mills, 2004: The impact of climate change on wildfire severity: a regional forecast for northern California. *Climatic Change*, **64**, 169-191.

Friedlingstein, P., J.L. Dufresne, P.M. Cox and P. Rayner, 2003: How positive is the feedback between climate change and the carbon cycle? *Tellus B*, **55**, 692-700.

Friedlingstein, P., P. Cox, R. Betts, L. Bopp, W. Von Bloh, V. Brovkin, P. Cadule, S. Doney, M. Eby, I. Fung, G. Bala, J. John, C. Jones, F. Joos, T. Kato, M. Kawamiya, W. Knorr, K. Lindsay, H.D. Matthews, T. Raddatz, P. Rayner, C. Reick, E. Roeckner, K.-G. Schnitzler, R. Schnur, K. Strassmann, A.J. Weaver, C. Yoshikawa and N. Zeng, 2006: Climate-carbon cycle feedback analysis: results from the (CMIP)-M-4 model intercomparison. *J. Climate*, **19**, 3337-3353.

Fuhlendorf, S.D., D.D. Briske and F.E. Smeins, 2001: Herbaceous vegetation change in variable rangeland environments: The relative contribution of grazing and climatic variability. *Appl. Veg. Sci*, **4**, 177-188.

Fuhrer, J., M. Beniston, A. Fischlin, C. Frei, S. Goyette, K. Jasper and C. Pfister, 2006: Climate risks and their impact on agriculture and forests in Switzerland. *Climatic Change*, **79**, 79-102.

Galbraith, H., R. Jones, R. Park, J. Clough, S. Herrod-Julius, B. Harrington and G. Page, 2002: Global climate change and sea level rise: potential losses of intertidal habitat for shorebirds. *Waterbirds*, **25**, 173-183.

Gamache, I. and S. Payette, 2005: Latitudinal response of subarctic tree lines to recent climate change in eastern Canada. *J. Biogeogr.*, **32**, 849-862.

Gan, J.B., 2004: Risk and damage of southern pine beetle outbreaks under global climate change. *Forest Ecol. Manag.*, **191**, 61-71.

Gardner, T.A., I.M. Cote, J.A. Gill, A. Grant and A.R. Watkinson, 2003: Long-term region-wide declines in Caribbean corals. *Science*, **301**, 958-960.

Gaston, K.J., T.M. Blackburn and K.K. Goldewijk, 2003: Habitat conversion and global avian biodiversity loss. *P. Roy. Soc. Lond. B*, **270**, 1293-1300.

Gedney, N., P.M. Cox, R.A. Betts, O. Boucher, C. Huntingford and P.A. Stott, 2006: Detection of a direct carbon dioxide effect in continental river runoff records. *Nature*, **439**, 835-838.

Geist, H.J. and E.F. Lambin, 2004: Dynamic causal patterns of desertification. *BioScience*, **54**, 817-829.

Genner, M.J., D.W. Sims, V.J. Wearmouth, E.J. Southall, A.J. Southward, P.A. Henderson and S.J. Hawkins, 2004: Regional climatic warming drives long-term community changes of British marine fish. *P. Roy. Soc. Lond. B*, **271**, 655-661.

Gerber, S., F. Joos and I.C. Prentice, 2004: Sensitivity of a dynamic global vegetation model to climate and atmospheric CO_2. *Global Change Biol.*, **10**, 1223-1239.

Gerten, D. and R. Adrian, 2002: Species-specific changes in the phenology and peak abundance of freshwater copepods in response to warm summers. *Freshwater Biol.*, **47**, 2163-2173.

Gerten, D., S. Schaphoff, U. Haberlandt, W. Lucht and S. Sitch, 2004: Terrestrial vegetation and water balance: hydrological evaluation of a dynamic global vegetation model. *J. Hydrol.*, **286**, 249-270.

Gerten, D., W. Lucht, S. Schaphoff, W. Cramer, T. Hickler and W. Wagner, 2005: Hydrologic resilience of the terrestrial biosphere. *Geophys. Res. Lett.*, **32**, L21408, doi:10.1029/2005GL024247.

Gielen, B., H.J. De Boeck, C.M.H.M. Lemmens, R. Valcke, I. Nijs and R. Ceulemans, 2005: Grassland species will not necessarily benefit from future elevated air temperatures: a chlorophyll fluorescence approach to study autumn physiology. *Physiol. Plant.*, **125**, 52-63.

Giles, J., 2005: Millennium group nails down the financial value of ecosystems. *Nature*, **434**, 547.

Gillett, N.P., A.J. Weaver, F.W. Zwiers and M.D. Flannigan, 2004: Detecting the effect of climate change on Canadian forest fires. *Geophys. Res. Lett.*, **31**, L18211, doi:10.1029/2004GL020876.

Gitay, H., S. Brown, W. Easterling and B. Jallow, 2001: Ecosystems and their goods and services. *Climate Change 2001: Impacts, Adaptation, and Vulnerability. Contribution of Working Group II to the Third Assessment Report of the Intergovernmental Panel on Climate Change*, J.J. McCarthy, O.F. Canziani, N.A. Leary, D.J. Dokken and K.S. White, Eds., Cambridge University Press, Cambridge, 237-342.

Gitay, H., A. Suárez and R.T. Watson, 2002: *Climate Change and Biodiversity*. IPCC Technical Paper, Intergovernmental Panel on Climate Change, Geneva, 77 pp.

Gitz, V. and P. Ciais, 2003: Amplifying effects of land-use change on future atmospheric CO_2 levels. *Global Biogeochem. Cy.*, **17**, 1024-1038, doi:10.1029/2002GB001963.

Gobron, N., B. Pinty, F. Melin, M. Taberner, M.M. Verstraete, A. Belward, T. Lavergne and J.L. Widlowski, 2005: The state of vegetation in Europe following the 2003 drought. *Int. J. Remote Sens.*, **26**, 2013-2020.

Goklany, I.M., 1998: Saving habitat and conserving biodiversity on a crowded planet. *BioScience*, **48**, 941-953.

Goklany, I.M., 2005: A climate policy for the short and medium term: stabilization or adaptation? *Energ. Environ.*, **16**, 667-680.

Goklany, I.M. and A.J. Trewavas, 2003: How technology can reduce our impact on the Earth. *Nature*, **423**, 115-115.

Goldstein, A.H., N.E. Hultman, J.M. Fracheboud, M.R. Bauer, J.A. Panek, M. Xu, Y. Qi, A.B. Guenther and W. Baugh, 2000: Effects of climate variability on the carbon dioxide, water and sensible heat fluxes above a ponderosa pine plantation in the Sierra Nevada (CA). *Agr. Forest Meteorol.*, **101**, 113-129.

Gonzalez, P., 2001: Desertification and a shift of forest species in the West African Sahel. *Climate Res.*, **17**, 217-228.

Gopal, B. and M. Chauhan, 2001: South Asian wetlands and their biodiversity: the role of monsoons. *Biodiversity in Wetlands: Assessment, Function and Conservation: Volume 2*, B. Gopal, W.J. Junk and J.A. Davis, Eds., Backhuys Publishers, Leiden, 257-275.

Gottfried, M., H. Pauli, K. Reiter and G. Grabherr, 1999: A fine-scaled predictive model for changes in species distribution patterns of high mountain plants induced by climate warming. *Divers. Distrib.*, **5**, 241-251.

Grace, J., F. Berninger and L. Nagy, 2002: Impacts of climate change on the tree line. *Ann. Bot.*, **90**, 537-544.

Graglia, E., R. Julkunen-Tiitto, G.R. Shaver, I.K. Schmidt, S. Jonasson and A. Michelsen, 2001: Environmental control and intersite variations of phenolics in *Betula nana* in tundra ecosystems. *New Phytol.*, **151**, 227-236.

Gruber, N., P. Friedlingstein, C.B. Field, R. Valentini, M. Heimann, J.F. Richey, P. Romero, E.D. Schulze and A. Chen, 2004: The vulnerability of the carbon cycle in the 21st century: an assessment of carbon-climate-human interactions. *Global Carbon Cycle: Integrating Humans, Climate, and the Natural World*, C.B. Field and M.R. Raupach, Eds., Island Press, Washington, District of Columbia, 45-76.

Guinotte, J.M., R.W. Buddemeier and J.A. Kleypas, 2003: Future coral reef habitat marginality: temporal and spatial effects of climate change in the Pacific basin. *Coral Reefs*, **22**, 551-558.

Guinotte, J.M., J. Orr, S. Cairns, A. Freiwald, L. Morgan and R. George, 2006: Will human-induced changes in seawater chemistry alter the distribution of deep-sea scleractinian corals? *Front. Ecol. Environ.*, **4**, 141-146.

Guisan, A. and N.E. Zimmermann, 2000: Predictive habitat distribution models in ecology. *Ecol. Model.*, **135**, 147-186.

Guisan, A. and W. Thuiller, 2005: Predicting species distribution: offering more than simple habitat models. *Ecol. Lett.*, **8**, 993-1009.

Gyalistras, D. and A. Fischlin, 1999: Towards a general method to construct regional climatic scenarios for model-based impacts assessments. *Petermann. Geogr. Mitt.*, **143**, 251-264.

Gyalistras, D., H. von Storch, A. Fischlin and M. Beniston, 1994: Linking GCM-simulated climatic changes to ecosystem models: case studies of statistical downscaling in the Alps. *Climate Res.*, **4**, 167-189.

Hallock, P., 2005: Global change and modern coral reefs: new opportunities to understand shallow-water carbonate depositional processes. *Sediment. Geol.*, **175**, 19-33.

Halloy, S.R.P. and A.F. Mark, 2003: Climate-change effects on alpine plant biodiversity: a New Zealand perspective on quantifying the threat. *Arct. Antarct. Alp. Res.*, **35**, 248-254.

Hamerlynck, E.P., T.E. Huxman, T.N. Charlet and S.D. Smith, 2002: Effects of elevated CO_2 (FACE) on the functional ecology of the drought-deciduous Mojave Desert shrub, *Lycium andersonii*. *Environ. Exp. Bot.*, **48**, 93-106.

Handel, M.D. and J.S. Risbey, 1992: Reflections on more than a century of climate change research. *Climatic Change*, **21**, 91-96.

Hanley, M.E., S. Trofimov and G. Taylor, 2004: Species-level effects more important than functional group-level responses to elevated CO_2: evidence from simulated turves. *Funct. Ecol.*, **18**, 304-313.

Hannah, L. and T.E. Lovejoy, 2003: Climate change and biodiversity: synergistic impacts. *Advances in Applied Biodiversity Science*, **4**, 1-123.

Hannah, L., G.F. Midgley, T. Lovejoy, W.J. Bond, M. Bush, J.C. Lovett, D. Scott and F.I. Woodward, 2002a: Conservation of biodiversity in a changing climate. *Conserv. Biol.*, **16**, 264-268.

Hannah, L., G.F. Midgley and D. Millar, 2002b: Climate change-integrated conservation strategies. *Global Ecol. Biogeogr.*, **11**, 485-495.

Hänninen, H., 2006: Climate warming and the risk of frost damage to boreal forest trees: identification of critical ecophysiological traits. *Tree Physiol.*, **26**, 889-898.

Hänninen, H., E. Beuker, Ø.L.I. Johnsen, M. Murray, L. Sheppard and T. Skrøppa, 2001: Impacts of climate change on cold hardiness of conifers. *Conifer Cold Hardiness*, F.J. Bigras, and S.J. Colombo, Eds., Kluwer Academic, Dordrecht, 305-333.

Hansen, A.J., R.P. Neilson, V.H. Dale, C.H. Flather, L.R. Iverson, D.J. Currie, S. Shafer, R. Cook and P. J. Bartlein, 2001: Global change in forests: Responses of species, communities, and biomes. *BioScience*, **51**, 765-779.

Hansen, L.J., J.L. Biringer and J.R. Hoffmann, 2003: *Buying Time: A User's Manual for Building Resistance and Resilience to Climate Change in Natural Systems*. WWF Climate Change Program, Berlin, 246 pp.

Hanson, P.J. and J.F. Weltzin, 2000: Drought disturbance from climate change: response of United States forests. *Sci. Total Environ.*, **262**, 205-220.

Harden, J.W., S.E. Trumbore, B.J. Stocks, A. Hirsch, S.T. Gower, K.P. O'Neill and E.S. Kasischke, 2000: The role of fire in the boreal carbon budget. *Global Change Biol.*, **6**, 174-184.

Harding, R., P. Kuhry, T.R. Christensen, M.T. Sykes, R. Dankers and S. van der Linden, 2002: Climate feedbacks at the tundra-taiga interface. *Ambio Special Report*, **12**, 47-55.

Hardy, J.T., 2003: *Climate Change Causes, Effects and Solutions*. Wiley, Chichester, 247 pp.

Hare, W., 2006: Relationship between increases in global mean temperature and impacts on ecosystems, food production, water and socio-economic systems. *Avoiding Dangerous Climate Change*, H.J. Schellnhuber, W. Cramer, N. Nakićenović, T.M.L. Wigley and G. Yohe, Eds., Cambridge University Press, Cambridge, 77-185.

Harmens, H., P.D. Williams, S.L. Peters, M.T. Bambrick, A. Hopkins and T.W. Ashenden, 2004: Impacts of elevated atmospheric CO_2 and temperature on plant community structure of a temperate grassland are modulated by cutting frequency. *Grass Forage Sci.*, **59**, 144-156.

Harper, C.W., J.M. Blair, P.A. Fay, A.K. Knapp and J.D. Carlisle, 2005: Increased rainfall variability and reduced rainfall amount decreases soil CO_2 flux in a grassland ecosystem. *Global Change Biol.*, **11**, 322-334.

Harrison, P.A., P.M. Berry, N. Butt and M. New, 2006: Modelling climate change impacts on species' distributions at the European scale: implications for conservation policy. *Environ. Sci. Policy*, **9**, 116-128.

Harrison, S.P. and A.I. Prentice, 2003: Climate and CO_2 controls on global vegetation distribution at the last glacial maximum: analysis based on palaeovegetation data, biome modelling and palaeoclimate simulations. *Global Change Biol.*, **9**, 983-1004.

Harvell, C.D., C.E. Mitchell, J.R. Ward, S. Altizer, A.P. Dobson, R.S. Ostfeld and M.D. Samuel, 2002: Climate warming and disease risks for terrestrial and marine biota. *Science*, **296**, 2158-2162.

Hassan, R., R. Scholes and N. Ash, Eds., 2005: *Ecosystems and Human Well-being: Volume 1: Current State and Trends*. Island Press, Washington, District of Columbia, 917 pp.

Hassol, S.J., Ed., 2004a: *Impacts of a Warming Arctic: Arctic Climate Impact Assessment – Highlights*. Cambridge University Press, Cambridge, 20 pp.

Hassol, S.J., Ed., 2004b: *Impacts of a Warming Arctic: Arctic Climate Impact Assessment – Overview Report*. Cambridge University Press, Cambridge, 146 pp.

Haugaasen, T., J. Barlow and C.A. Peres, 2003: Surface wildfires in central Amazonia: short-term impact on forest structure and carbon loss. *Forest Ecol. Manag.*, **179**, 321-331.

Haugan, P.M., C. Turley and H.O. Poertner, 2006: Effects on the marine environment of ocean acidification resulting from elevated levels of CO_2 in the atmosphere. Biodiversity Series 285/2006 DN-utredning 2006-1, OSPAR Commission Convention for the Protection of the Marine Environment of the North-East Atlantic (the 'OSPAR Convention'), London, 33 pp.

Hayhoe, K., D. Cayan, C.B. Field, P.C. Frumhoff, E.P. Maurer, N.L. Miller, S.C. Moser, S.H. Schneider, K.N. Cahill, E.E. Cleland, L. Dale, R. Drapek, R.M. Hanemann, L.S. Kalkstein, J. Lenihan, C.K. Lunch, R.P. Neilson, S.C. Sheridan and J.H. Verville, 2004: Emissions pathways, climate change, and impacts on California. *P. Natl. Acad. Sci. USA*, **101**, 12422-12427.

Hays, G.C., A.C. Broderick, F. Glen and B.J. Godley, 2003: Climate change and sea turtles: a 150-year reconstruction of incubation temperatures at a major marine turtle rookery. *Global Change Biol.*, **9**, 642-646.

Hays, G.C., A.J. Richardson and C. Robinson, 2005: Climate change and marine plankton. *Trends Ecol. Evol.*, **20**, 337-344.

Heal, G., 2007: Environmental accounting for ecosystems. *Ecol. Econ.*, **61**, 693-694.

Heath, J., E. Ayres, M. Possell, R.D. Bardgett, H.I.J. Black, H. Grant, P. Ineson and G. Kerstiens, 2005: Rising atmospheric CO_2 reduces sequestration of root-derived soil carbon. *Science*, **309**, 1711-1713.

Hein, L. and N. Ridder, 2006: Desertification in the Sahel: a reinterpretation. *Global Change Biol.*, **12**, 751-758.

Held, I.M., T.L. Delworth, J. Lu, K.L. Findell and T.R. Knutson, 2005: Simulation of Sahel drought in the 20th and 21st centuries. *P. Natl. Acad. Sci. USA*, **102**, 17891-17896.

Helmuth, B., N. Mieszkowska, P. Moore and S.J. Hawkins, 2006: Living on the edge of two changing worlds: forecasting the responses of rocky intertidal ecosystems to climate change. *Annu. Rev. Ecol. Evol. Syst.*, **37**, 337-404.

Herrmann, S.M. and C.F. Hutchinson, 2005: The changing contexts of the desertification debate. *J. Arid Environ.*, **63**, 538-555.

Herrmann, S.M., A. Anyamba and C.J. Tucker, 2005: Recent trends in vegetation dynamics in the African Sahel and their relationship to climate. *Global Environ. Chang.*, **15**, 394-404.

Hess, J.C., C.A. Scott, G.L. Hufford and M.D. Fleming, 2001: El Niño and its impact on fire weather conditions in Alaska. *Int. J. Wildland Fire*; **10**, 1-13.

Hetherington, A.M. and F.I. Woodward, 2003: The role of stomata in sensing and driving environmental change. *Nature*, **424**, 901-908.

Heywood, V.H. and R.T. Watson, Eds., 1995: *Global Biodiversity Assessment*. Cambridge University Press, Cambridge, 1140 pp.

Hicke, J.A., G.P. Asner, E.S. Kasischke, N.H.F. French, J.T. Randerson, G.J. Collatz, B.J. Stocks, C.J. Tucker, S.O. Los and C.B. Feild, 2003: Postfire response of North American boreal forest net primary productivity analyzed with satellite observations. *Global Change Biol.*, **9**, 1145-1157.

Hickler, T., L. Eklundh, J.W. Seaquist, B. Smith, J. Ardo, L. Olsson, M.T. Sykes and M. Sjostrom, 2005: Precipitation controls Sahel greening trend. *Geophys. Res. Lett.*, **32**, L21415, doi:10.1029/2005GL024370.

Hiernaux, P. and M.D. Turner, 2002: The influence of farmer and pastoral management practices on desertification processes in the Sahel. *Global Desertification. Do Humans Cause Deserts?* J.F. Reynolds and D.M. Stafford-Smith, Eds., Dahlem University Press, Berlin, 135-148.

Higgins, P.A.T. and J. Harte, 2006: Biophysical and biogeochemical responses to climate change depend on dispersal and migration. *BioScience*, **56**, 407-417.

Higgins, S.I., J.S. Clark, R. Nathan, T. Hovestadt, F. Schurr, J.M.V. Fragoso, M.R. Aguiar, E. Ribbens and S. Lavorel, 2003: Forecasting plant migration rates: managing uncertainty for risk assessment. *J. Ecol.*, **91**, 341-347.

Hilbert, D.W., M. Bradford, T. Parker and D.A. Westcott, 2004: Golden bowerbird (*Prionodura newtonia*) habitat in past, present and future climates: predicted extinction of a vertebrate in tropical highlands due to global warming. *Biol. Conserv.*, **116**, 367-377.

Hinzman, L., N. Bettez, W. Bolton, F. Chapin, M. Dyurgerov, C. Fastie, B. Griffith, R. Hollister, A. Hope, H. Huntington, A. Jensen, G. Jia, T. Jorgenson, D. Kane, D. Klein, G. Kofinas, A. Lynch, A. Lloyd, A. McGuire, F. Nelson, W. Oechel, T. Osterkamp, C. Racine, V. Romanovsky, R. Stone, D. Stow, M. Sturm, C. Tweedie, G. Vourlitis, M. Walker, D. Walker, P. Webber, J. Welker, K. Winker and K. Yoshikawa, 2005: Evidence and implications of recent climate change in northern Alaska and other Arctic regions. *Climatic Change*, **72**, 251-298.

Hoch, G. and C. Körner, 2003: The carbon charging of pines at the climatic treeline: a global comparison. *Oecologia*, **135**, 10-21.

Hoch, G. and C. Körner, 2003: The carbon charging of pines at the climatic treeline: a global comparison. *Oecologia*, **135**, 10-21.

Hodar, J.A. and R. Zamora, 2004: Herbivory and climatic warming: a Mediterranean outbreaking caterpillar attacks a relict, boreal pine species. *Biodivers. Conserv.*, **13**, 493-500.

Hoegh-Guldberg, O., 1999: Climate change, coral bleaching and the future of the world's coral reefs. *Mar. Freshwater Res.*, **50**, 839-866.

Hoegh-Guldberg, O., 2004: Coral reefs in a century of rapid environmental change. *Symbiosis*, **37**, 1-31.

Hoegh-Guldberg, O., 2005: Low coral cover in a high-CO_2 world. *J. Geophys. Res. C*, **110**, C09S06, doi:10.1029/2004JC002528.

Hoegh-Guldberg, O., H. Hoegh-Guldberg, D.K. Stout, H. Cesar and A. Timmerman, 2000: *Pacific in Peril: Biological, Economic and Social Impacts of Climate Change on Pacific Coral Reefs*. Greenpeace, Sidney, 72 pp.

Hoffmann, W.A. and R.B. Jackson, 2000: Vegetation–climate feedbacks in the conversion of tropical savanna to grassland. *J. Climate*, **13**, 1593-1602.

Hogg, E.H. and A.G. Schwarz, 1997: Regeneration of planted conifers across climatic moisture gradients on the Canadian prairies: implications for distribution and climate change. *J. Biogeogr.*, **24**, 527-534.

Hogg, E.H. and R.W. Wein, 2005: Impacts of drought on forest growth and regeneration following fire in southwestern Yukon, Canada. *Can. J. Forest Res.*, **35**, 2141-2150.

Holligan, P.M., E. Fernandez, J. Aiken, W.M. Balch, P. Boyd, P.H. Burkill, M. Finch, S.B. Groom, G. Malin, K. Muller, D.A. Purdie, C. Robinson, C.C. Trees, S.M. Turner and P. Vanerwal, 1993: A biogeochemical study of the coccolithophore, *Emiliania huxleyi*, in the North Atlantic. *Global Biogeochem. Cy.*, **7**, 879-900.

Holman, I.P., R.J. Nicholls, P.M. Berry, P.A. Harrison, E. Audsley, S. Shackley and M.D.A. Rounsevell, 2005a: A regional, multi-sectoral and integrated assessment of the impacts of climate and socio-economic change in the UK. Part II. Results. *Climatic Change*, **71**, 43-73.

Holman, I.P., M.D.A. Rounsevell, S. Shackley, P.A. Harrison, R.J. Nicholls, P.M. Berry and E. Audsley, 2005b: A regional, multi-sectoral and integrated assessment of the impacts of climate and socio-economic change in the UK. Part I. Methodology. *Climatic Change*, **71**, 9-41.

Holmgren, M. and M. Scheffer, 2001: El Niño as a window of opportunity for the restoration of degraded arid ecosystems. *Ecosystems*, **4**, 151-159.

Holmgren, M., M. Scheffer, E. Ezcurra, J.R. Gutierrez and G.M.J. Mohren, 2001: El Niño effects on the dynamics of terrestrial ecosystems. *Trends Ecol. Evol.*, **16**, 89-94.

Hooper, D.U., F.S. Chapin, J.J. Ewel, A. Hector, P. Inchausti, S. Lavorel, J.H. Lawton, D.M. Lodge, M. Loreau, S. Naeem, B. Schmid, H. Setala, A.J. Symstad, J. Vandermeer and D.A. Wardle, 2005: Effects of biodiversity on ecosystem functioning: a consensus of current knowledge. *Ecol. Monogr.*, **75**, 3-35.

Houghton, R.A., 2003a: Revised estimates of the annual net flux of carbon to the atmosphere from changes in land use and land management 1850-2000. *Tellus B*, **55**, 378-390.

Houghton, R.A., 2003b: Why are estimates of the terrestrial carbon balance so different? *Global Change Biol.*, **9**, 500-509.

House, J.I., I.C. Prentice and C. Le Quéré, 2002: Maximum impacts of future reforestation or deforestation on atmospheric CO_2. *Global Change Biol.*, **8**, 1047-1052.

Houweling, S., T. Rockmann, I. Aben, F. Keppler, M. Krol, J.F. Meirink, E.J. Dlugokencky and C. Frankenberg, 2006: Atmospheric constraints on global emissions of methane from plants. *Geophys. Res. Lett.*, **33**, L15821, doi:10.1029/2006GL026162.

Huber, U.M., H.K.M. Bugmann and M.A. Reasoner, Eds., 2005: *Global Change and Mountain Regions: An Overview of Current Knowledge*. Springer, Berlin, 650 pp.

Huelber, K., M. Gottfried, H. Pauli, K. Reiter, M. Winkler and G. Grabherr, 2006: Phenological responses of snowbed species to snow removal dates in the Central Alps: implications for climate warming. *Arct. Antarct. Alp. Res.*, **38**, 99-103.

Hughes, L., 2003: Climate change and Australia: trends, projections and impacts. *Austral Ecol.*, **28**, 423-443.

Hughes, L., E.M. Cawsey and M. Westoby, 1996: Climatic range sizes of *Eucalyptus* species in relation to future climate change. *Global Ecol. Biogeogr. Lett.*, **5**, 23-29.

Hughes, T.P., A.H. Baird, D.R. Bellwood, M. Card, S.R. Connolly, C. Folke, R. Grosberg, O. Hoegh-Guldberg, J.B.C. Jackson, J. Kleypas, J.M. Lough, P. Marshall, M. Nystrom, S.R. Palumbi, J.M. Pandolfi, B. Rosen and J. Roughgarden, 2003: Climate change, human impacts, and the resilience of coral reefs. *Science*, **301**, 929-933.

Hulme, M., 2001: Climatic perspectives on Sahelian desiccation: 1973-1998. *Global Environ. Chang.*, **11**, 19-29.

Hulme, M., J. Mitchell, W. Ingram, J. Lowe, T. Johns, M. New and D. Viner, 1999: Climate change scenarios for global impacts studies. *Global Environ. Chang.*, **9**, S3-S19.

Hungate, B.A., C.H. Jaeger III, G. Gamara, F.S. Chapin III and C.B. Field, 2000: Soil microbiota in two annual grasslands: responses to elevated atmospheric CO_2. *Oecologia*, **124**, 589-598.

Hungate, B.A., J.S. Dukes, M.R. Shaw, Y.Q. Luo and C.B. Field, 2003: Nitrogen and climate change. *Science*, **302**, 1512-1513.

Hungate, B.A., D.W. Johnson, P. Dukstra, G. Hymus, P. Stiling, J.P. Megonigal, A.L. Pagel, J.L. Moan, F. Day, J. Li, C.R. Hinkle and B.G. Drake, 2006: Nitrogen cycling during seven years of atmospheric CO_2 enrichment in a scrub oak woodland. *Ecology*, **87**, 26-40.

Hurrell, J.W. and H. van Loon, 1997: Decadal variations in climate associated with the North Atlantic oscillation. *Climatic Change*, **36**, 301-326.

Hutchinson, C.F., S.M. Herrmann, T. Maukonen and J. Weber, 2005: Introduction: the "greening" of the Sahel. *J. Arid Environ.*, **63**, 535-537.

iLEAPS, Ed., 2005: Integrated land ecosystem-atmosphere processes study science plan and implementation strategy. IGBP report No. 54, International Geosphere and Biosphere Program Secretariat, Stockholm, 52 pp.

Inkley, D.B., M.G. Anderson, A.R. Blaustein, V.R. Burkett, B. Felzer, B. Griffith, J. Price and T.L. Root, 2004: *Global Climate Change and Wildlife in North America*. The Wildlife Society, Bethesda, Maryland, 34 pp.

Inouye, D.W., B. Barr, K.B. Armitage and B.D. Inouye, 2000: Climate change is affecting altitudinal migrants and hibernating species. *P. Natl. Acad. Sci. USA*, **97**, 1630-1633.

IPCC, 1990: *IPCC First Assessment Report (1990): Scientific Assessment of Climate Change. Contribution of Working Group I to the Intergovernmental Panel on Climate Change*, J.T. Houghton, G.J. Jenkins and J.J. Ephraums, Eds., Cambridge University Press, Cambridge, 365 pp.

IPCC, 1996: *Climate Change 1995: The Science of Climate Change. Contribution of Working Group I to the Second Assessment Report of the Intergovernmental Panel on Climate Change*, J.T. Houghton, L.G.M. Filho, B.A. Callender, N. Harris, A. Kattenberg and K. Maskell, Eds., Cambridge University Press, Cambridge, 572 pp.

IPCC, 2001: *Climate Change 2001: The Scientific Basis. Contribution of Working Group I to the Third Assessment Report of the Intergovernmental Panel on Climate Change*, J.T. Houghton, Y. Ding, D.J. Griggs, M. Noguer, P.J. van der Linden, X. Dai, K. Maskell and C.A. Johnson, Eds., Cambridge University Press, Cambridge, 881 pp.

IPCC, 2007: *Climate Change 2007: The Physical Science Basis. Contribution of Working Group I to the Fourth Assessment Report of the Intergovernmental Panel on Climate Change*, S. Solomon, D. Qin, M. Manning, Z. Chen, M. Marquis, K.B. Averyt, M. Tignor and H.L. Miller, Eds., Cambridge University Press, Cambridge, 996 pp.

IPCC-TGCIA, 1999: *Guidelines on the Use of Scenario Data for Climate Impact and Adaptation Assessment*. Prepared by T.R. Carter, M. Hulme and M. Lal, Intergovernmental Panel on Climate Change (IPCC): Task Group on Scenarios for Climate Impact Assessment (TGCIA), Norwich, Hamburg, and New York, 69 pp.

Isaev, A.S., G.N. Korovin, S.A. Bartalev, D.V. Ershov, A. Janetos, E.S. Kasischke, H.H. Shugart, N.H.F. French, B.E. Orlick and T.L. Murphy, 2002: Using remote sensing to assess Russian forest fire carbon emissions. *Climatic Change*, **55**, 235-249.

Iverson, L.R. and A.M. Prasad, 2002: Potential redistribution of tree species habitat under five climate change scenarios in the eastern US. *Forest Ecol. Manag.*, **155**, 205-222.

Iverson, L.R., M.W. Schwartz and A.M. Prasad, 2004: How fast and far might tree species migrate in the eastern United States due to climate change? *Global Ecol. Biogeogr.*, **13**, 209-219.

Izaurralde, R.C., A.M. Thomson, N.J. Rosenberg and R.A. Brown, 2005: Climate change impacts for the conterminous USA: an integrated assessment. Part 6. Distribution and productivity of unmanaged ecosystems. *Climatic Change*, **69**, 107-126.

Jackson, D.A. and N.E. Mandrak, 2002: Changing fish biodiversity: predicting the loss of cyprinid biodiversity due to global climate change. *Fisheries in a Changing Climate*. N.A. McGinn, Ed., American Fisheries Society Symposium 32, Phoenix, Arizona. American Fisheries Society, Bethesda, Maryland, 89-98.

Jain, R., 2003: Sustainable development: differing views and policy alternatives. *Clean Technologies and Environmental Policy*, **4**, 197-198.

Jankowski, T., D.M. Livingstone, H. Bührer, R. Forster and P. Niederhaser, 2006: Consequences of the 2003 European heat wave for lake temperature profiles, thermal stability and hypolimnetic oxygen depletion: implications for a warmer world. *Limnol. Oceanogr.*, **51**, 815-819.

Jansen, E., J. Overpeck, K.R. Briffa, J.-C. Duplessy, F. Joos, V. Masson-Delmotte, D.O. Olago, B. Otto-Bliesner, Wm. Richard Pelteir, S. Rahmstorf, R. Ramesh, D. Raynaud, D.H. Rind, O. Solomina, R. Villalba and D. Zhang, 2007: Paleoclimate. *Climate Change 2007: The Physical Science Basis. Contribution of Working Group I to the Fourth Assessment Report of the Intergovernmental Panel on Climate Change*, S. Solomon, D. Qin, M. Manning, Z. Chen, M. Marquis, K.B. Averyt, M. Tignor and H.L. Miller, Eds., Cambridge University Press, Cambridge, 434-496.

Jansson, R., 2003: Global patterns in endemism explained by past climatic change. *P. Roy. Soc. Lond. B Bio*, **270**, 583-590.

Jasienski, M., S.C. Thomas and F.A. Bazzaz, 1998: Blaming the trees: a critique of research on forest responses to high CO_2. *Trends Ecol. Evol.*, **13**, 427.

Jasinski, J.P.P. and S. Payette, 2005: The creation of alternative stable states in the southern boreal forest, Quebec, Canada. *Ecol. Monogr.*, **75**, 561-583.

Jasper, K., P. Calanca, D. Gyalistras and J. Fuhrer, 2004: Differential impacts of climate change on the hydrology of two alpine river basins. *Climate Res.*, **26**, 113-129.

Jenkins, M., 2003: Prospects for biodiversity. *Science*, **302**, 1175-1177.

Jobbagy, E.G. and R.B. Jackson, 2000: The vertical distribution of soil organic carbon and its relation to climate and vegetation. *Ecol. Appl.*, **10**, 423-436.

Johansen, R.J., 2001: Impacts of fire on biological soil crusts. *Biological Soil Crusts: Structure, Function, and Management*, J. Belnap and O.L. Lange, Eds., Springer, Berlin, 386-397.

Johnson, W.C., B.V. Millett, T. Gilmanov, R.A. Voldseth, G.R. Guntenspergen and D.E. Naugle, 2005: Vulnerability of northern prairie wetlands to climate change. *BioScience*, **55**, 863-872.

Johnstone, J.F. and F.S. Chapin, 2003: Non-equilibrium succession dynamics indicate continued northern migration of lodgepole pine. *Global Change Biol.*, **9**, 1401-1409.

Johnstone, J.F. and F.S. Chapin, 2006: Effects of soil burn severity on post-fire tree recruitment in boreal forests. *Ecosystems*, **9**, 14-31.

Jolly, W.M., M. Dobbertin, N.E. Zimmermann and M. Reichstein, 2005: Divergent vegetation growth responses to the 2003 heat wave in the Swiss Alps. *Geophys. Res. Lett.*, **32**, L18409, doi:10.1029/2005GL023252.

Jompa, J. and L.J. McCook, 2003: Coral-algal competition: macroalgae with different properties have different effects on corals. *Mar. Ecol.–Prog. Ser.*, **258**, 87-95.

Jones, M.B. and A. Donnelly, 2004: Carbon sequestration in temperate grassland ecosystems and the influence of management, climate and elevated CO_2. *New Phytol.*, **164**, 423-439.

Jones, P., J. Amador, M. Campos, K. Hayhoe, M. Marín, J. Romero and A. Fischlin, 2005: Generating climate change scenarios at high resolution for impact studies and adaptation: focus on developing countries. *Tropical Forests and Adaptation to Climate Change: In Search of Synergies*, Y. Saloh, C. Robledo and K. Markku, Eds., Island Press, Washington, District of Columbia, 37-55.

Joos, F. and I.C. Prentice, 2004: A paleo-perspective on changes in atmospheric CO_2 and climate. *Global Carbon Cycle: Integrating Humans, Climate, and the Natural World*, Field, C.B. Raupach and M.R. Raupach, Eds., Island Press, Washington, District of Columbia, 165-186.

Juday, G.P., 2005: Forests, land management and agriculture. *Arctic Climate Impact Assessment (ACIA): Scientific Report*, C. Symon, L. Arris and B. Heal, Eds., Cambridge University Press, Cambridge, 781-862.

Jump, A.S. and J. Peñuelas, 2005: Running to stand still: adaptation and the response of plants to rapid climate change. *Ecol. Lett.*, **8**, 1010-1020.

Jump, A.S., J.M. Hunt, J.A. Martinez-Izquierdo and J. Peñuelas, 2006: Natural selection and climate change: temperature-linked spatial and temporal trends in gene frequency in *Fagus sylvatica*. *Mol. Ecol.*, **15**, 3469-3480.

Kajimoto, T., H. Daimaru, T. Okamoto, T. Otani and H. Onodera, 2004: Effects of snow avalanche disturbance on regeneration of subalpine *Abies mariesii* forest, northern Japan. *Arct. Antarct. Alp. Res.*, **36**, 436-445.

Kaplan, J.O., N.H. Bigelow, I.C. Prentice, S.P. Harrison, P.J. Bartlein, T.R. Christensen, W. Cramer, N.V. Matveyeva, A.D. McGuire, D.F. Murray, V.Y. Razzhivin, B. Smith, D.A. Walker, P.M. Anderson, A.A. Andreev, L.B. Brubaker, M.E. Edwards and A.V. Lozhkin, 2003: Climate change and Arctic ecosystems. 2. Modeling, paleodata-model comparisons, and future projections. *J. Geophys. Res. D*, **108**, 8171, doi:10.1029/2002JD002559.

Kappelle, M., M.M.I. Van Vuuren and P. Baas, 1999: Effects of climate change on biodiversity: a review and identification of key research issues. *Biodivers. Conserv.*, **8**, 1383-1397.

Karjalainen, T., A. Pussinen, J. Liski, G.J. Nabuurs, M. Erhard, T. Eggers, M. Sonntag and G.M.J. Mohren, 2002: An approach towards an estimate of the impact of forest management and climate change on the European forest sector carbon budget: Germany as a case study. *Forest Ecol. Manag.*, **162**, 87-103.

Karlsson, J., A. Jonsson and M. Jansson, 2005: Productivity of high-latitude lakes: climate effect inferred from altitude gradient. *Global Change Biol.*, **11**, 710-715.

Karnosky, D.F., 2003: Impacts of elevated atmospheric CO_2 on forest trees and forest ecosystems: knowledge gaps. *Environ. Int.*, **29**, 161-169.

Karnosky, D.F., D.R. Zak, K.S. Pregitzer, C.S. Awmack, J.G. Bockheim, R.E. Dickson, G.R. Hendrey, G.E. Host, J.S. King, B.J. Kopper, E.L. Kruger, M.E. Kubiske, R.L. Lindroth, W.J. Mattson, E.P. McDonald, A. Noormets, E. Oksanen, W.F.J. Parsons, K.E. Percy, G.K. Podila, D.E. Riemenschneider, P. Sharma, R. Thakur, A. Sober, J. Sober, W.S. Jones, S. Anttonen, E. Vapaavuori, B. Mankovska, W. Heilman and J.G. Isebrands, 2003: Tropospheric O_3 moderates responses of temperate hardwood forests to elevated CO_2: a synthesis of molecular to ecosystem results from the Aspen FACE project. *Funct. Ecol.*, **17**, 289-304.

Kasischke, E.S. and B.J. Stocks, Eds., 2000: *Fire, Climate Change and Carbon Cycling in the Boreal Forest*. Springer-Verlag, New York, 461 pp.

Kasischke, E.S. and M.R. Turetsky, 2006: Recent changes in the fire regime across the North American boreal region: spatial and temporal patterns of burning across Canada and Alaska. *Geophys. Res. Lett.*, **33**, L09703, doi:10.1029/2006GL025677.

Kasischke, E.S., J.H. Hewson, B. Stocks, G. van der Werf and J. Randerson, 2003: The use of ATSR active fire counts for estimating relative patterns of biomass burning: a study from the boreal forest region. *Geophys. Res. Lett.*, **30**, 1969, doi:10.1029/2003GL017859.

Kasischke, E.S., E.J. Hyer, P.C. Novelli, L.P. Bruhwiler, N.H.F. French, A.I. Sukhinin, J.H. Hewson and B.J. Stocks, 2005: Influences of boreal fire emissions on Northern Hemisphere atmospheric carbon and carbon monoxide. *Global Biogeochem. Cy.*, **19**, GB1012, doi:10.1029/2004GB002300.

Kates, R.W., 2000: Cautionary tales: adaptation and the global poor. *Climatic Change*, **45**, 5-17.

Keddy, P.A., 2000: *Wetland Ecology: Principles and Conservation*. Cambridge University Press, Cambridge, 614 pp.

Keeley, J.E., 2002: Fire management of California shrubland landscapes. *Environ. Manage.*, **29**, 395-408.

Keller, A.A. and G. Klein-Macphee, 2000: Impact of elevated temperature on the growth, survival, and trophic dynamics of winter flounder larvae: a mesocosm study. *Can. J. Fish. Aquat. Sci.*, **57**, 2382-2392.

Keller, F., S. Goyette and M. Beniston, 2005: Sensitivity analysis of snow cover to climate change scenarios and their impact on plant habitats in alpine terrain. *Climatic Change*, **72**, 299-319.

Keller, J.K., J.R. White, S.D. Bridgham and J. Pastor, 2004: Climate change effects on carbon and nitrogen mineralization in peatlands through changes in soil quality. *Global Change Biol.*, **10**, 1053-1064.

Kemp, M., 2005: Science in culture: inventing an icon. *Nature*, **437**, 1238-1238.

Kennett, S.A., 2002: National policies for biosphere greenhouse gas management: issues and opportunities. *Environ. Manage.*, **30**, 595-608.

Keppler, F., J.T.G. Hamilton, M. Brass and T. Rockmann, 2006: Methane emissions from terrestrial plants under aerobic conditions. *Nature*, **439**, 187-191.

Kerr, J. and L. Packer, 1998: The impact of climate change on mammal diversity in Canada. *Environ. Monit. Assess.*, **49**, 263-270.

Kienast, F., O. Wildi and B. Brzeziecki, 1998: Potential impacts of climate change on species richness in mountain forests: an ecological risk assessment. *Biol. Conserv.*, **83**, 291-305.

King, J.A., R.I. Bradley, R. Harrison and A.D. Carter, 2004a: Carbon sequestration and saving potential associated with changes to the management of agricultural soils in England. *Soil Use Manage.*, **20**, 394-402.

King, J.S., P.J. Hanson, E. Bernhardt, P. Deangelis, R.J. Norby and K.S. Pregitzer, 2004b: A multiyear synthesis of soil respiration responses to elevated atmospheric CO_2 from four forest FACE experiments. *Global Change Biol.*, **10**, 1027-1042.

Kirschbaum, M. and A. Fischlin, 1996: Climate change impacts on forests. *Climate Change 1995: Impacts; Adaptations and Mitigation of Climate Change. Scientific-Technical Analysis. Contribution of Working Group II to the Second Assessment Report of the Intergovernmental Panel of Climate Change*., R. Watson, M.C. Zinyowera and R.H. Moss, Eds., Cambridge University Press, Cambridge, 95-129.

Kitayama, K., 1996: Climate of the summit region of Mount Kinabalu (Borneo) in 1992, an El Niño year. *Mt. Res. Dev.*, **16**, 65-75.

Klanderud, K., 2005: Climate change effects on species interactions in an alpine plant community. *J. Ecol.*, **93**, 127-137.

Klanderud, K. and O. Totland, 2005: Simulated climate change altered dominance hierarchies and diversity of an alpine biodiversity hotspot. *Ecology*, **86**, 2047-2054.

Kleypas, J.A. and C. Langdon, 2002: Overview of CO_2-induced changes in seawater chemistry. *World Coral Reefs in the New Millenium: Bridging Research and Management for Sustainable Development*, M.K. Moosa, S. Soemodihardjo, A. Soegiarto, K. Romimohtarto, A. Nontji and S. Suharsono, Eds., Proceedings of the 9th International Coral Reef Symposium, 2, Bali, Indonesia. Ministry of Environment, Indonesian Institute of Sciences, International Society for Reef Studies, 1085-1089.

Kleypas, J.A., R.W. Buddemeier, D. Archer, J.P. Gattuso, C. Langdon and B.N. Opdyke, 1999: Geochemical consequences of increased atmospheric carbon dioxide on coral reefs. *Science*, **284**, 118-120.

Kleypas, J.A., R.W. Buddemeier and J.P. Gattuso, 2001: The future of coral reefs in an age of global change. *Int. J. Earth Sci.*, **90**, 426-437.

Kleypas, J.A., R.W. Buddemeier, C.M. Eakin, J.P. Gattuso, J. Guinotte, O. Hoegh-Guldberg, R. Iglesias-Prieto, P.L. Jokiel, C. Langdon, W. Skirving and A.E. Strong, 2005: Comment on "Coral reef calcification and climate change: the effect of ocean warming". *Geophys. Res. Lett.*, **32**, L08601, doi:10.1029/2004GL022329.

Kling, J., K. Hayhoe, L.B. Johnsoin, J.J. Magnuson, S. Polasky, S.K. Robinson, B.J. Shuter, M.M. Wander, D.J. Wuebbles and D.R. Zak, 2003: *Confronting Climate Change in the Great Lakes Region: Impacts on our Communities and Ecosystems*. Union of Concerned Scientists and the Ecological Society of America, Cambridge, Massachusetts and Washington, District of Columbia, 92 pp.

Knorr, W., I.C. Prentice, J.I. House and E.A. Holland, 2005: Long-term sensitivity of soil carbon turnover to warming. *Nature*, **433**, 298-301.

Kohler, P., F. Joos, S. Gerber and R. Knutti, 2005: Simulated changes in vegetation distribution, land carbon storage, and atmospheric CO_2 in response to a collapse of the North Atlantic thermohaline circulation. *Clim. Dynam.*, **25**, 689-708.

Koop, K., D. Booth, A. Broadbent, J. Brodie, D. Bucher, D. Capone, J. Coll, W. Dennison, M. Erdmann, P. Harrison, O. Hoegh-Guldberg, P. Hutchings, G.B. Jones, A.W.D. Larkum, J. O'Neil, A. Steven, E. Tentori, S. Ward, J. Williamson and D. Yellowlees, 2001: ENCORE: The effect of nutrient enrichment on coral reefs: synthesis of results and conclusions. *Mar. Pollut. Bull.*, **42**, 91-120.

Körner, C., 1998: Worldwide positions of alpine treelines and their causes. *The Impacts of Climate Variability on Forests*, M. Beniston and J.L. Innes, Eds., Springer, Berlin, 221-229.

Körner, C., 2003a: *Alpine Plant Life: Functional Plant Ecology of High Mountain Ecosystems*. Springer, Berlin, 343 pp.

Körner, C., 2003b: Ecological impacts of atmospheric CO_2 enrichment on terrestrial ecosystems. *Philos. T. Roy. Soc. Lond.*, **361**, 2023-2041.

Körner, C., 2003c: Slow in, rapid out: carbon flux studies and Kyoto targets. *Science*, **300**, 1242-1243.

Körner, C., 2004: Through enhanced tree dynamics carbon dioxide enrichment may cause tropical forests to lose carbon. *Philos. T. Roy. Soc. Lond.*, **359**, 493-498.

Körner, C. and J. Paulsen, 2004: A world-wide study of high altitude treeline temperatures. *J. Biogeogr.*, **31**, 713-732.

Körner, C., R. Asshoff, O. Bignucolo, S. Hättenschwiler, S.G. Keel, S. Peláez-Riedl, S. Pepin, R.T.W. Siegwolf and G. Zotz, 2005a: Carbon flux and growth in mature deciduous forest trees exposed to elevated CO_2. *Science*, **309**, 1360-1362.

Körner, C., D. Sarris and D. Christodoulakis, 2005b: Long-term increase in climatic dryness in the East Mediterranean as evidenced for the island of Samos. *Reg. Environ. Change*, **5**, 27-36.

Kräuchi, N., P. Brang and W. Schonenberger, 2000: Forests of mountainous regions: gaps in knowledge and research needs. *Forest Ecol. Manag.*, **132**, 73-82.

Kremen, C., 2005: Managing ecosystem services: what do we need to know about their ecology? *Ecol. Lett.*, **8**, 468-479.

Lake, J.C. and M.R. Leishman, 2004: Invasion success of exotic plants in natural ecosystems: the role of disturbance, plant attributes and freedom from herbivores. *Biol. Conserv.*, **117**, 215-226.

Lake, P.S., M.A. Palmer, P. Biro, J. Cole, A.P. Covich, C. Dahm, J. Gibert, W. Goedkoop, K. Martens and J. Verhoeven, 2000: Global change and the biodiversity of freshwater ecosystems: impacts on linkages between above-sediment and sediment biota. *BioScience*, **50**, 1099-1107.

Lal, R., 2003: Offsetting global CO_2 emissions by restoration of degraded soils and intensification of world agriculture and forestry. *Land Degrad. Dev.*, **14**, 309-322.

Lal, R., 2004: Soil carbon sequestration to mitigate climate change. *Geoderma*, **123**, 1-22.

Lambin, E.F., H.J. Geist and E. Lepers, 2003: Dynamics of land-use and land-cover change in tropical regions. *Annu. Rev. Ecol. Evol. Syst.*, **28**, 205-241.

Landhäusser, S.M. and R.W. Wein, 1993: Postfire vegetation recovery and tree establishment at the Arctic treeline: climate-change vegetation-response hypotheses. *J. Ecol.*, **81**, 665-672.

Langdon, C., W.S. Broecker, D.E. Hammond, E. Glenn, K. Fitzsimmons, S.G. Nelson, T.H. Peng, I. Hajdas and G. Bonani, 2003: Effect of elevated CO_2 on the community metabolism of an experimental coral reef. *Global Biogeochem. Cy.*, **17**, 1011, doi:10.1029/2002GB001941.

Langenfelds, R.L., R.J. Francey, B.C. Pak, L.P. Steele, J. Lloyd, C.M. Trudinger and C.E. Allison, 2002: Interannual growth rate variations of atmospheric CO_2 and its isotope $\delta^{13}C$, H_2, CH_4, and CO between 1992 and 1999 linked to biomass burning. *Global Biogeochem. Cy.*, **16**, 1048, doi:10.1029/2001GB001466.

Lara, A., R. Villalba, A. Wolodarsky-Franke, J.C. Aravena, B.H. Luckman and E. Cuq, 2005: Spatial and temporal variation in *Nothofagus pumilio* growth at tree line along its latitudinal range (35°40'–55° S) in the Chilean Andes. *J. Biogeogr.*, **32**, 879-893.

Lauenroth, W.K., H.E. Epstein, J.M. Paruelo, I.C. Burke, M.R. Aguiar and O.E. Sala, 2004: Potential effects of climate change on the temperate zones of North and South America. *Rev. Chil. Hist. Nat.*, **77**, 439-453.

Laurance, W.F., A.A. Oliveira, S.G. Laurance, R. Condit, H.E.M. Nascimento, A.C. Sanchez-Thorin, T.E. Lovejoy, A. Andrade, S. D'Angelo, J.E. Riberiro and C.W. Dick, 2004: Pervasive alteration of tree communities in undisturbed Amazonian forests. *Nature*, **428**, 171-175.

Lavoie, L. and L. Sirois, 1998: Vegetation changes caused by recent fires in the northern boreal forest of eastern Canada. *J. Veg. Sci.*, **9**, 483-492.

Lavorel, S., 1999: Ecological diversity and resilience of Mediterranean vegetation to disturbance. *Divers. Distrib.*, **5**, 3-13.

Lawton, R.O., U.S. Nair, R.A. Pielke Sr. and R.M. Welch, 2001: Climatic impact of tropical lowland deforestation on nearby montane cloud forests. *Science*, **294**, 584-587.

Le Maitre, D.C., D.M. Richardson and R.A. Chapman, 2004: Alien plant invasions in South Africa: driving forces and the human dimension. *S. Afr. J. Sci.*, **100**, 103-112.

Le Roux, P.C., M.A. McGeoch, M.J. Nyakatya and S.L. Chown, 2005: Effects of a short-term climate change experiment on a sub-Antarctic keystone plant species. *Global Change Biol.*, **11**, 1628-1639.

Leakey, A.D.B., M. Uribelarrea, E.A. Ainsworth, S.L. Naidu, A. Rogers, D.R. Ort and S.P. Long, 2006: Photosynthesis, productivity, and yield of maize are not affected by open-air elevation of CO_2 concentration in the absence of drought. *Plant Physiol.*, **140**, 779-790.

Learmonth, J.A., C.D. MacLeod, M.B. Santos, G.J. Pierce, H.Q.P. Crick and R.A. Robinson, 2006: Potential effects of climate change on marine mammals. *Oceanogr. Mar. Biol.*, **44**, 431-464.

Leemans, R. and B. Eickhout, 2004: Another reason for concern: regional and global impacts on ecosystems for different levels of climate change. *Global Environ. Chang.*, **14**, 219-228.

Leemans, R., B. Eickhout, B. Strengers, L. Bouwman and M. Schaeffer, 2002: The consequences of uncertainties in land use, climate and vegetation responses on the terrestrial carbon. *Sci. China Ser. C*, **45**, 126-141.

Lehodey, P., F. Chai and J. Hampton, 2003: Modelling climate-related variability of tuna populations from a coupled ocean-biogeochemical-populations dynamics model. *Fish. Oceanogr.*, **12**, 483-494.

Lelieveld, J., 2006: Climate change: a nasty surprise in the greenhouse. *Nature*, **443**, 405-406.

Lelieveld, J., H. Berresheim, S. Borrmann, P.J. Crutzen, F.J. Dentener, H. Fischer, J. Feichter, P.J. Flatau, J. Heland, B. Holzinger, R. Korrmann, M.G. Lawrence, Z. Levin, K.M. Markowicz, N. Milhalopoulos, A. Minikin, V. Ramanathan, M. de Reus, G.J. Roelofs, H.A. Scheeren, J. Sciare, H. Schlager, M. Schultz, P. Siegmund, B. Steil, E.G. Stephanou, P. Stier, M. Traub, C. Warneke, J. Williams and H. Ziereis, 2002: Global air pollution crossroads over the Mediterranean. *Science*, **298**, 794-799.

Lemke, P., J. Ren, R. Alley, I. Allison, J. Carrasco, G. Flato, Y. Fujii, G. Kaser, P. Mote, R. Thomas and T. Zhang, 2007: Observations: changes in snow, ice and frozen ground. *Climate Change 2007: The Physical Science Basis. Contribution of Working Group I to the Fourth Assessment Report of the Intergovernmental Panel on Climate Change*, S. Solomon, D. Qin, M. Manning, Z. Chen, M. Marquis, K.B. Averyt, M. Tignor and H.L. Miller, Eds., Cambridge University Press, Cambridge, 335-383.

Lenihan, J.M., R. Drapek, D. Bachelet and R.P. Neilson, 2003: Climate change effects on vegetation distribution, carbon, and fire in California. *Ecol. Appl.*, **13**, 1667.

Lepers, E., E.F. Lambin, A.C. Janetos, R. Defries, F. Achard, N. Ramankutty and R.J. Scholes, 2005: A synthesis of information on rapid land-cover change for the period 1981-2000. *BioScience*, **55**, 115-124.

Leuzinger, S., G. Zotz, R. Asshoff and C. Körner, 2005: Responses of deciduous forest trees to severe drought in Central Europe. *Tree Physiol.*, **25**, 641-650.

Levy, P.E., M.G.R. Cannell and A.D. Friend, 2004: Modelling the impact of future changes in climate, CO_2 concentration and land use on natural ecosystems and the terrestrial carbon sink. *Global Environ. Chang.*, **14**, 21-30.

Lewis, S.L., Y. Malhi and O.L. Phillips, 2004a: Fingerprinting the impacts of global change on tropical forests. *Philos. T. Roy. Soc. Lond. B*, **359**, 437-462.

Lewis, S.L., O.L. Phillips, T.R. Baker, J. Lloyd, Y. Malhi, S. Almeida, N. Higuchi, W.F. Laurance, D.A. Neill, J.N.M. Silva, J. Terborgh, A.T. Lezama, R.V. Martinez, S. Brown, J. Chave, C. Kuebler, P.N. Vargas and B. Vinceti, 2004b: Concerted changes in tropical forest structure and dynamics: evidence from 50 South American long-term plots. *Philos. T. Roy. Soc. Lond. B*, **359**, 421-436.

Lexer, M.J., K. Honninger, H. Scheifinger, C. Matulla, N. Groll and H. Kromp-Kolb, 2000: The sensitivity of central European mountain forests to scenarios of climatic change: methodological frame for a large-scale risk assessment. *Silva Fenn.*, **34**, 113-129.

Li, C., M.D. Flannigan and I.G.W. Corns, 2000: Influence of potential climate change on forest landscape dynamics of west-central Alberta. *Can. J. Forest Res.*, **30**, 1905-1912.

Li, S.G., J. Asanuma, W. Eugster, A. Kotani, J.J. Liu, T. Urano, T. Oikawa, G. Davaa, D. Oyunbaatar and M. Sugita, 2005: Net ecosystem carbon dioxide exchange over grazed steppe in central Mongolia. *Global Change Biol.*, **11**, 1941-1955.

Lichter, J., S.H. Barron, C.E. Bevacqua, A.C. Finzi, K.F. Irving, E.A. Stemmler and W.H. Schlesinger, 2005: Soil carbon sequestration and turnover in a pine forest after six years of atmospheric CO_2 enrichment. *Ecology*, **86**, 1835-1847.

Lindroth, R.L., B.J. Kopper, W.F.J. Parsons, J.G. Bockheim, D.F. Karnosky, G.R. Hendrey, K.S. Pregitzer, J.G. Isebrands and J. Sober, 2001: Consequences of elevated carbon dioxide and ozone for foliar chemical composition and dynamics in trembling aspen (*Populus tremuloides*) and paper birch (*Betula papyrifera*). *Environ. Pollut.*, **115**, 395-404.

Lioubimtseva, E. and J.M. Adams, 2004: Possible implications of increased carbon dioxide levels and climate change for desert ecosystems. *Environ. Manage.*, **33**, S388-S404.

Lischke, H., A. Guisan, A. Fischlin, J. Williams and H. Bugmann, 1998: Vegetation responses to climate change in the Alps: modeling studies. *Views from the Alps: Regional Perspectives on Climate Change*, P. Cebon, U. Dahinden, H.C. Davies, D.M. Imboden and C.C. Jager, Eds., MIT Press, Boston, Massachusetts, 309-350.

Lischke, H., A.F. Lotter and A. Fischlin, 2002: Untangling a Holocene pollen record with forest model simulations and independent climate data. *Ecol. Model.*, **150**, 1-21.

Liu, X.D. and B.D. Chen, 2000: Climatic warming in the Tibetan Plateau during recent decades. *Int. J. Climatol.*, **20**, 1729-1742.

Llope, M., R. Anadon, L. Viesca, M. Quevedo, R. Gonzalez-Quiros and N.C. Stenseth, 2006: Hydrography of the southern Bay of Biscay shelf-break region: integrating the multiscale physical variability over the period 1993-2003. *J. Geophys. Res. C*, **111**, C09021, doi:10.1029/2005JC002963.

Llorens, L. and J. Peñuelas, 2005: Experimental evidence of future drier and warmer conditions affecting flowering of two co-occurring Mediterranean shrubs. *Int. J. Plant Sci.*, **166**, 235-245.

Llorens, L., J. Peñuelas and M. Estiarte, 2003: Ecophysiological responses of two Mediterranean shrubs, *Erica multiflora* and *Globularia alypum*, to experimentally drier and warmer conditions. *Physiol. Plant.*, **119**, 231-243.

Llorens, L., J. Peñuelas, M. Estiarte and P. Bruna, 2004: Contrasting growth changes in two dominant species of a Mediterranean shrubland submitted to experimental drought and warming. *Ann. Bot.*, **94**, 843-853.

Lloret, F., D. Siscart and C. Dalmases, 2004: Canopy recovery after drought dieback in holm-oak Mediterranean forests of Catalonia (NE Spain). *Global Change Biol.*, **10**, 2092-2099.

Lloyd, A.H., 2005: Ecological histories from Alaskan tree lines provide insight into future change. *Ecology*, **86**, 1687-1695.

Lobo, A. and P. Maisongrande, 2006: Stratified analysis of satellite imagery of SW Europe during summer 2003: the differential response of vegetation classes to increased water deficit. *Hydrol. Earth Syst. Sci.*, **10**, 151-164.

Logan, J.A., J. Regniere and J.A. Powell, 2003: Assessing the impacts of global warming on forest pest dynamics. *Front. Ecol. Environ.*, **1**, 130-137.

Lorenzoni, I., N.F. Pidgeon and R.E. O'Connor, 2005: Dangerous climate change: the role for risk research. *Risk Anal.*, **25**, 1387-1398.

Loukos, H., P. Monfray, L. Bopp and P. Lehodey, 2003: Potential changes in skipjack tuna (*Katsuwonus pelamis*) habitat from a global warming scenario: modelling approach and preliminary results. *Fish. Oceanogr.*, **12**, 474-482.

Lovejoy, T.E. and L. Hannah, Eds., 2005: *Climate Change and Biodiversity*. Yale University Press, New Haven, Connecticut, 418 pp.

Lucht, W., I.C. Prentice, R.B. Myneni, S. Sitch, P. Friedlingstein, W. Cramer, P. Bousquet, W. Buermann and B. Smith, 2002: Climatic control of the high-latitude vegetation greening trend and Pinatubo effect. *Science*, **296**, 1687-1689.

Lucht, W., S. Schaphoff, T. Erbrecht, U. Heyder and W. Cramer, 2006: Terrestrial vegetation redistribution and carbon balance under climate change. *Carbon Balance Manage.*, **1**:6 doi:10.1186/1750-0680-1-6.

Luckman, B., 1994: Using multiple high-resolution proxy climate records to reconstruct natural climate variability: an example from the Canadian Rockies. *Mountain Environments in Changing Climates*, M. Beniston, Ed., Routledge, London, 42-59.

Ludwig, D., 2000: Limitations of economic valuation of ecosystems. *Ecosystems*, **3**, 31-35.

Luo, Y.W.S., D. Hui and L.L. Wallace, 2001: Acclimatization of soil respiration to warming in a tall grass prairie. *Nature*, **413**, 622-625.

Lusseau, D., R. Williams, B. Wilson, K. Grellier, T.R. Barton, P.S. Hammond and P.M. Thompson, 2004: Parallel influence of climate on the behaviour of Pacific killer whales and Atlantic bottlenose dolphins. *Ecol. Lett.*, **7**, 1068-1076.

Luterbacher, J., D. Dietrich, E. Xoplaki, M. Grosjean and H. Wanner, 2004: European seasonal and annual temperature variability, trends, and extremes since 1500. *Science*, **303**, 1499-1503.

MacArthur, R.H., 1972: *Geographical Ecology Patterns in the Distribution of Species*. Harper and Row, New York, 269 pp.

Mack, M.C., E.A.G. Schuur, M.S. Bret-Harte, G.R. Shaver and F.S. Chapin, 2004: Ecosystem carbon storage in arctic tundra reduced by long-term nutrient fertilization. *Nature*, **431**, 440-443.

Mack, R.N., D. Simberloff, W.M. Lonsdale, H. Evans, M. Clout and F. Bazzaz, 2000: Biotic invasions: causes, epidemiology, global consequences and control. *Issues Ecol.*, **5**, 2-22.

Malcolm, J.R., C. Liu, L. Miller, T. Allnut and L. Hansen, Eds., 2002a: *Habitats at Risk: Global Warming and Species Loss in Globally Significant Terrestrial Ecosystems*. WWF World Wide Fund for Nature, Gland, 40 pp.

Malcolm, J.R., A. Markham, R.P. Neilson and M. Garaci, 2002b: Estimated migration rates under scenarios of global climate change. *J. Biogeogr.*, **29**, 835-849.

Malcolm, J.R., C.R. Liu, R.P. Neilson, L. Hansen and L. Hannah, 2006: Global warming and extinctions of endemic species from biodiversity hotspots. *Conserv. Biol.*, **20**, 538-548.

Malhi, Y. and O.L. Phillips, 2004: Tropical forests and global atmospheric change: a synthesis. *Philos. T. Roy. Soc. Lond. B*, **359**, 549-555.

Malhi, Y., D. Wood, T.R. Baker, J. Wright, O.L. Phillips, T. Cochrane, P. Meir, J. Chave, S. Almeida, L. Arroyo, N. Higuchi, T.J. Killeen, S.G. Laurance, W.F. Laurance, S.L. Lewis, A. Monteagudo, D.A. Neill, P.N. Vargas, N.C.A. Pitman, C.A. Quesada, R. Salomao, J.N.M Silva, A.T. Lezama, J. Terborgh, R.V. Martinez and B. Vinceti, 2006: The regional variation of aboveground live biomass in old-growth Amazonian forests. *Global Change Biol.*, **12**, 1107-1138.

Maroco, J.P., E. Breia, T. Faria, J.S. Pereira and M.M. Chaves, 2002: Effects of long-term exposure to elevated CO_2 and N fertilization on the development of photosynthetic capacity and biomass accumulation in *Quercus suber* L. *Plant Cell Environ.*, **25**, 105-113.

Marra, J., C. Ho and C.C. Trees, 2003: An alternative algorithm for the calculation of primary productivity from remote sensing data. LDEO Technical Report LDEO-2003-1, Lamont-Doherty Earth Observatory, Columbia University, Palisades, New York, 27 pp.

Martinez-Meyer, E., A.T. Peterson and W.W. Hargrove, 2004: Ecological niches as stable distributional constraints on mammal species, with implications for Pleistocene extinctions and climate change projections for biodiversity. *Global Ecol. Biogeogr.*, **13**, 305-314.

Martinez-Vilalta, J. and J. Pinol, 2002: Drought-induced mortality and hydraulic architecture in pine populations of the NE Iberian Peninsula. *Forest Ecol. Manag.*, **161**, 247-256.

Marubini, F., H. Barnett, C. Langdon and M.J. Atkinson, 2001: Dependence of calcification on light and carbonate ion concentration for the hermatypic coral *Porites compressa*. *Mar. Ecol.–Prog. Ser.*, **220**, 153-162.

Marubini, F., C. Ferrier-Pages and J.P. Cuif, 2003: Suppression of skeletal growth in scleractinian corals by decreasing ambient carbonate-ion concentration: a cross-family comparison. *P. Roy. Soc. Lond. B*, **270**, 179-184.

Masek, J.G., 2001: Stability of boreal forest stands during recent climate change: evidence from Landsat satellite imagery. *J. Biogeogr.*, **28**, 967-976.

Matson, P., K.A. Lohse and S.J. Hall, 2002: The globalization of nitrogen deposition: consequences for terrestrial ecosystems. *Ambio*, **31**, 113-119.

Matsui, T., T. Yagihashi, T. Nakaya, H. Taoda, S. Yoshinaga, H. Daimaru and N. Tanaka, 2004: Probability distributions, vulnerability and sensitivity in *Fagus crenata* forests following predicted climate changes in Japan. *J. Veg. Sci.*, **15**, 605-614.

May, R., 1999a: How the biosphere is organized. *Science*, **286**, 2091-2091.

May, R., 1999b: Unanswered questions in ecology. *Philos. T. Roy. Soc. Lond. B*, **354**, 1951-1959.

McClain, C.R., S.R. Signorini and J.R. Christian, 2004: Subtropical gyre variability observed by ocean-color satellites. *Deep-Sea Res. Pt. II*, **51**, 281-301.

McClanahan, T.R., N.A. Muthiga and S. Mangi, 2001: Coral and algal changes after the 1998 coral bleaching: interaction with reef management and herbivores on Kenyan reefs. *Coral Reefs*, **19**, 380-391.

McClean, C.J., J.C. Lovett, W. Kuper, L. Hannah, J.H. Sommer, W. Barthlott, M. Termansen, G.E. Smith, S. Tokamine and J.R.D Taplin, 2005: African plant diversity and climate change. *Ann. Mo. Bot. Gard.*, **92**, 139-152.

McDonald, K.A. and J.H. Brown, 1992: Using montane mammals to model extinctions due to global change. *Conserv. Biol.*, **6**, 409-415.

McGuire, A.D. and F.S. Chapin, 2006: Climate feedbacks in the Alaskan boreal forest. *Alaska's Changing Boreal Forest*, F.S. Chapin, M.W. Oswood, K. van Cleve, L.A. Viereck and D.L. Verbyla, Eds., Oxford University Press, New York, 309-322.

McGuire, A.D., F.S. Chapin III, C. Wirth, M. Apps, J. Bhatti, T. Callaghan, T.R. Christensen, J.S. Clein, M. Fukuda, T. Maxima, A. Onuchin, A. Shvidenko and E. Vaganov, 2007: Responses of high latitude ecosystems to global change: Potential consequences for the climate system. *Terrestrial Ecosystems in a Changing World*, J.G. Canadell, D.E. Pataki and L.F. Pitelka, Eds., Springer-Verlag, Berlin, 297-310.

McKee, D., D. Atkinson, S.E. Collings, J.W. Eaton, A.B. Gill, I. Harvey, K. Hatton, T. Heyes, D. Wilson and B. Moss, 2003: Response of freshwater microcosm communities to nutrients, fish, and elevated temperature during winter and summer. *Limnol. Oceanogr.*, **48**, 707-722.

McLachlan, J.S., J.S. Clark and P.S. Manos, 2005: Molecular indicators of tree migration capacity under rapid climate change. *Ecology*, **86**, 2088-2098.

McLean, R.F., A. Tsyban, V. Burkett, J.O. Codignotto, D.L. Forbes, N. Mimura, R.J. Beamish and V. Ittekkot, 2001: Coastal zones and marine ecosystems. *Climate Change 2001: Impacts, Adaptation, and Vulnerability. Contribution of Working Group II to the Third Assessment Report of the Intergovernmental Panel on Climate Change*, J.J. McCarthy, O.F. Canziani, N.A. Leary, D.J. Dokken and K.S. White, Eds., Cambridge University Press, Cambridge, 343-379.

McNeely, J.A., 1990: Climate change and biological diversity: policy implications. *Landscape: Ecological Impact of Climatic Change*, M.M. Boer and R.S. de Groot, Eds., IOS Press, Amsterdam, 406-429.

McNeely, J.A. and F. Schutyser, Eds., 2003: *Protected Areas in 2023: Scenarios for an Uncertain Future*. IUCN: The World Conservation Union, Gland, 51 pp.

McNeil, B.I., R.J. Matear and D.J. Barnes, 2004: Coral reef calcification and climate change: the effect of ocean warming. *Geophys. Res. Lett.*, **31**, L22309, doi:10.1029/2004GL021541.

McWilliams, J.P., I.M. Côté, J.A. Gill, W.J. Sutherland and A.R. Watkinson, 2005: Accelerating impacts of temperature-induced coral bleaching in the Caribbean.

Ecology, **86**, 2055-2060.

Mearns, L.O., F. Giorgi, P. Whetton, D. Pabon, M. Hulme and M. Lal, 2003: *Guidelines for Use of Climate Scenarios Developed from Regional Climate Model Experiments*. Intergovernmental Panel on Climate Change (IPCC): Data Distribution Centre (DDC), Norwich, Hamburg and New York, 38 pp.

Meehl, G.A., T.F. Stocker, W. Collins, P. Friedlingstein, A. Gaye, J. Gregory, A. Kitoh, R. Knutti, J. Murphy, A. Noda, S. Raper, I. Watterson, A. Weaver and Z.-C. Zhao, 2007: Global climate projections. *Climate Change 2007: The Physical Science Basis. Contribution of Working Group I to the Fourth Assessment Report of the Intergovernmental Panel on Climate Change*, S. Solomon, D. Qin, M. Manning, Z. Chen, M. Marquis, K.B. Averyt, M. Tignor and H.L. Miller, Eds., Cambridge University Press, Cambridge, 747-845.

Megonigal, J.P., C.D. Vann and A.A. Wolf, 2005: Flooding constraints on tree (*Taxodium distichum*) and herb growth responses to elevated CO_2. *Wetlands*, **25**, 430-438.

Metzger, M.J., M.D.A. Rounsevell, L. Acosta-Michlik, R. Leemans and D. Schröter, 2006: The vulnerability of ecosystem services to land use change. *Agr. Ecosyst. Environ.*, **114**, 69-85.

Midgley, G.F., L. Hannah, D. Millar, M.C. Rutherford and L.W. Powrie, 2002: Assessing the vulnerability of species richness to anthropogenic climate change in a biodiversity hotspot. *Global Ecol. Biogeogr.*, **11**, 445-451.

Midgley, G.F., L. Hannah, D. Millar, W. Thuiller and A. Booth, 2003: Developing regional and species-level assessments of climate change impacts on biodiversity in the Cape Floristic Region. *Biol. Conserv.*, **112**, 87-97.

Midgley, G.F., R.A. Chapman, B. Hewitson, P. Johnston, M. De Wit, G. Ziervogel, P. Mukheibir, L. Van Niekerk, M. Tadross, B.W. Van Wilgen, B. Kgope, P.D. Morant, A. Theron, R.J. Scholes and G.G. Forsyth, 2005: A status quo, vulnerability and adaptation assessment of the physical and socio-economic effects of climate change in the western Cape. Report to the Western Cape Government, Cape Town, South Africa. CSIR Report No. ENV-S-C 2005-073, CSIR Environmentek, Stellenbosch, 170 pp.

Midgley, G.F., G.O. Hughes, W. Thuiller and A.G. Rebelo, 2006: Migration rate limitations on climate change-induced range shifts in Cape Proteaceae. *Divers. Distrib.*, **12**, 555-562.

Midgley, G.F., W. Thuiller and S.I. Higgins, 2007: Plant species migration as a key uncertainty in predicting future impacts of climate change on ecosystems: progress and challenges. *Terrestrial Ecosystems in a Changing World*, J.G. Canadell, D.E. Pataki and L.F. Pitelka, Eds., Springer-Verlag, Berlin, 149-160.

Mieszkowska, N., M.A. Kendall, S.J. Hawkins, R. Leaper, P. Williamson, N.J. Hardman-Mountford and A.J. Southward, 2006: Changes in the range of some common rocky shore species in Britain: a response to climate change? *Hydrobiologia*, **555**, 241-251.

Miles, L.J., 2002: The impact of global climate change on tropical forest biodiversity in Amazonia. PhD thesis, University of Leeds, Leeds, 328pp.

Milieu en Natuurplanbureau, 2005: *Natuurbalans 2005*. Rapportnummer 408763002, SDU Uitgevers – Wageningen Universiteit en Research Centrum, Rijksinstituut voor Integraal Zoetwaterbeheer en Afvalwaterbehandeling, Rijksinstituut voor Kust en Zee, Den Haag, The Netherlands, 197 pp.

Millar, C.I., R.D. Westfall, D.L. Delany, J.C. King and L.J. Graumlich, 2004: Response of subalpine conifers in the Sierra Nevada, California, USA, to 20th-century warming and decadal climate variability. *Arct. Antarct. Alp. Res.*, **36**, 181-200.

Millennium Ecosystem Assessment, 2005: *Millennium Ecosystem Assessment: Living Beyond our Means: Natural Assets and Human Well-being (Statement from the Board)*. Millennium Ecosystem Assessment, India, France, Kenya, UK, USA, Netherlands, Malaysia, 28 pp. Accessed 13.03.07: http://www.millenniumassessment.org/.

Mills, E., 2005: Insurance in a climate of change. *Science*, **309**, 1040-1044.

Mitchell, J.F.B., T.C. Johns, J.M. Gregory and S.F.B. Tett, 1995: Climate response to increasing levels of greenhouse gases and sulphate aerosols. *Nature*, **376**, 501-504.

Mitchell, S.W. and F. Csillag, 2001: Assessing the stability and uncertainty of predicted vegetation growth under climatic variability: northern mixed grass prairie. *Ecol. Model.*, **139**, 101-121.

Mock, G., Ed., 2005: *World Resources 2005: The Wealth of the Poor – Managing Ecosystems to Fight Poverty*. United Nations Development Programme, United Nations Environment Programme, World Bank, World Resources Institute, Washington, District of Columbia, 266 pp.

Moldan, B., S. Percy, J. Riley, T. Hák, J. Rivera and F.L. Toth, 2005: Choosing responses. *Ecosystems and Human Well-being: Volume 3: Policy Responses*, K. Chopra, R. Leemans, P. Kumar and H. Simons, Eds., Island Press, Washington, District of Columbia, 527-548.

Monahan, A.H., J.C. Fyfe and G.M. Flato, 2000: A regime view of Northern Hemisphere atmospheric variability and change under global warming. *Geophys. Res. Lett.*, **27**, 1139-1142.

Monson, R.K., A.A. Turnipseed, J.P. Sparks, P.C. Harley, L.E. Scott-Denton, K. Sparks and T.E. Huxman, 2002: Carbon sequestration in a high-elevation, subalpine forest. *Global Change Biol.*, **8**, 459-478.

Monson, R.K., D.L. Lipson, S.P. Burns, A.A. Turnipseed, A.C. Delany, M.W. Williams and S.K. Schmidt, 2006: Winter forest soil respiration controlled by climate and microbial community composition. *Nature*, **439**, 711-714.

Montenegro, G., R. Ginochio, A. Segura, J.E. Keeley and M. Gomez, 2004: Fire regimes and vegetation responses in two Mediterranean-climate regions. *Rev. Chil. Hist. Nat.*, **77**, 455-464.

Mooney, H., A. Cropper and W. Reid, 2005: Confronting the human dilemma: how can ecosystems provide sustainable services to benefit society? *Nature*, **434**, 561-562.

Moorcroft, P.R., S.W. Pacala and M.A. Lewis, 2006: Potential role of natural enemies during tree range expansions following climate change. *J. Theor. Biol.*, **241**, 601-616.

Morgan, J.A., D.R. Lecain, A.R. Mosier and D.G. Milchunas, 2001a: Elevated CO_2 enhances water relations and productivity and affects gas exchange in C_{-3} and C_{-4} grasses of the Colorado shortgrass steppe. *Global Change Biol.*, **7**, 451-466.

Morgan, M.G., L.F. Pitelka and E. Shevliakova, 2001b: Elicitation of expert judgments of climate change impacts on forest ecosystems. *Climatic Change*, **49**, 279-307.

Morgan, J.A., D.E. Pataki, C. Körner, H. Clark, S.J. Del Grosso, J.M. Grunzweig, A.K. Knapp, A.R. Mosier, P.C.D. Newton, P.A. Niklaus, J.B. Nippert, R.S. Nowak, W.J. Parton, H.W. Polley and M.R. Shaw, 2004: Water relations in grassland and desert ecosystems exposed to elevated atmospheric CO_2. *Oecologia*, **140**, 11-25.

Moriondo, M., P. Good, R. Durao, M. Bindi, C. Giannakopoulos and J. Corte Real, 2006: Potential impact of climate change on fire risk in the Mediterranean area. *Climate Res.*, **31**, 85-95.

Moser, D., S. Dullinger, T. Englisch, H. Niklfeld, C. Plutzar, N. Sauberer, H.G. Zechmeister and G. Grabherr, 2005: Environmental determinants of vascular plant species richness in the Austrian Alps. *J. Biogeogr.*, **32**, 1117-1127.

Moss, B., D. McKee, D. Atkinson, S.E. Collings, J.W. Eaton, A.B. Gill, I. Harvey, K. Hatton, T. Heyes and D. Wilson, 2003: How important is climate? Effects of warming, nutrient addition and fish on phytoplankton in shallow lake microcosms. *J. Appl. Ecol.*, **40**, 782-792.

Moss, R.H. and S.H. Schneider, 2000: Uncertainties in the IPCC TAR: Recommendations to lead authors for more consistent assessment and reporting. *Guidance Papers on the Cross Cutting Issues of the Third Assessment Report of the IPCC*, R. Pachauri, T. Taniguchi and K. Tanaka, Eds., Geneva, 33-51.

Mouillot, F., S. Rambal and R. Joffre, 2002: Simulating climate change impacts on fire frequency and vegetation dynamics in a Mediterranean-type ecosystem. *Global Change Biol.*, **8**, 423-437.

Mouillot, F., J.P. Ratte, R. Joffre, J.M. Moreno and S. Rambal, 2003: Some determinants of the spatio-temporal fire cycle in a mediterranean landscape (Corsica, France). *Landscape Ecol.*, **18**, 665-674.

Mouthon, J. and M. Daufresne, 2006: Effects of the 2003 heatwave and climatic warming on mollusc communities of the Saône: a large lowland river and of its two main tributaries (France). *Global Change Biol.*, **12**, 441-449.

Murdiyarso, D. and L. Lebel, 2007: Southeast Asian fire regimes and land development policy. *Terrestrial Ecosystems in a Changing World*, J.G. Canadell, D.E. Pataki and L.F. Pitelka, Eds., Springer-Verlag, Berlin, 261-271.

Musil, C.F., U. Schmiedel and G.F. Midgley, 2005: Lethal effects of experimental warming approximating a future climate scenario on southern African quartz-field succulents: a pilot study. *New Phytol.*, **165**, 539-547.

Myers, N., R.A. Mittermeier, C.G. Mittermeier, G.A.B. da Fonseca and J. Kent, 2000: Biodiversity hotspots for conservation priorities. *Nature*, **403**, 853-858.

Nabuurs, G.J., O. Masera, K. Andrasko, P. Benitez-Ponce, R. Boer, M. Dutschke, E. Elsiddig, J. Ford-Robertson, P. Frumhoff, T. Karjalainen, O. Krankina, W. Kurz, M. Matsumoto, W. Oyhantcabal, N.H. Ravindranath, M.S. Sanchez and X. Zhang, 2007: Forestry. *Climate Change 2007: Mitigation. Contribution of Working Group III to the Fourth Assessment Report of the Intergovernmental Panel on Climate Change*, B. Metz, O. Davidson, P. Bosch, R. Dave and L. Meyer, Eds., Cambridge University Press, UK.

Naiman, R.J., H. Dêcamps and M.E. McClain, 2005: *Riparia: Ecology, Conservation and Management of Streamside Communities*. Elsevier, Amsterdam, 448 pp.

Najjar, R.G., H.A. Walker, E.J. Barron, R.J. Bord, J.R. Gibson, V.S. Kennedy, C.G. Knight, J.P. Megonigal, R.E. O'Connor, C.D. Polsky, N.P. Psuty, B.A. Richards, L.G. Sorenson, E.M. Steele and R.S. Swanson, 2000: The potential impacts of climate change on the mid-Atlantic coastal region. *Climate Res.*, **14**, 219-233.

Nakićenović , N., J. Alcamo, G. Davis, B. de Vries, J. Fenhann, S. Gaffin, K. Gregory, A. Grübler, T.Y. Jung, T. Kram, E. Lebre la Rovere, L. Michaelis, S. Mori, T. Morita, W. Pepper, H. Pitcher, L. Price, K. Riahi, A. Roehrl, H.H. Rogner, A. Sankovski, M. Schlesinger, P. Shukla, S. Smith, R. Swart, S. van Rooijen, N. Victor and Z. Dadi, Eds., 2000: *Emissions Scenarios: A Special Report of the Intergovernmental Panel on Climate Change*. Cambridge University Press, Cambridge, 599 pp.

National Research Council, 1999: *Perspectives on Biodiversity: Valuing its Role in an Everchanging World*. National Academies Press, Washington, District of Columbia, 168 pp.

Neilson, R.P. and R.J. Drapek, 1998: Potentially complex biosphere responses to transient global warming. *Global Change Biol.*, **4**, 505-521.

Neilson, R.P., I.C. Prentice, B. Smith, T. Kittel and D. Viner, 1998: Simulated changes in vegetation distribution under global warming. *Regional Impacts of Climatic Change: An Assessment of Vulnerability. A Special Report of the Intergovernmental Panel on Climate Change (IPCC) Working Group II*, R.T. Watson, M.C. Zinyowera, R.H. Moss and D.J. Dokken, Eds., Cambridge University Press, Annex C, 439-456.

Neilson, R.P., L.F. Pitelka, A.M. Solomon, R. Nathan, G.F. Midgley, J.M.V. Fragoso, H. Lischke and K. Thompson, 2005: Forecasting regional to global plant migration in response to climate change. *BioScience*, **55**, 749-759.

Nelson, G.C., 2005: Drivers of ecosystem change: summary chapter. *Ecosystems and Human Well-being: Volume 1: Current State and Trends*, R. Hassan, R. Scholes and N. Ash, Eds., Island Press, Washington, District of Columbia, 73-76.

Nelson, J.A., J.A. Morgan, D.R. Lecain, A. Mosier, D.G. Milchunas and B.A. Parton, 2004: Elevated CO_2 increases soil moisture and enhances plant water relations in a long-term field study in semi-arid shortgrass steppe of Colorado. *Plant Soil*, **259**, 169-179.

Nemani, R.R., C.D. Keeling, H. Hashimoto, W.M. Jolly, S.C. Piper, C.J. Tucker, R.B. Myneni and S.W. Running, 2003: Climate-driven increases in global terrestrial net primary production from 1982 to 1999. *Science*, **300**, 1560-1563.

Nepstad, D., P. Lefebvre, U.L. Da Silva, J. Tomasella, P. Schlesinger, L. Solorzano, P. Moutinho, D. Ray and J.G. Benito, 2004: Amazon drought and its implications for forest flammability and tree growth: a basin-wide analysis. *Global Change Biol.*, **10**, 704-717.

Nepstad, D.C., A. Verissimo, A. Alencar, C. Nobre, E. Lima, P. Lefebvre, P. Schlesinger, C. Potter, P. Moutinho, E. Mendoza, M. Cochrane and V. Brooks, 1999: Large-scale impoverishment of Amazonian forests by logging and fire. *Nature*, **398**, 505-508.

Neuvonen, S., 2004: Spatial and temporal variation in biodiversity in the European North. SCANNET: Scandinavian/North European Network of Terrestrial Field Bases Final Report Work Package 6, Kevo Subarctic Research Istitute University of Turku, Turku, 44 pp.

Ni, J., 2001: Carbon storage in terrestrial ecosystems of China: estimates at different spatial resolutions and their responses to climate change. *Climatic Change*, **49**, 339-358.

Ni, J., S.P. Harrison, I.C. Prentice, J.E. Kutzbach and S. Sitch, 2006: Impact of climate variability on present and Holocene vegetation: a model-based study. *Ecol. Model.*, **191**, 469-486.

Nicholls, R.J., F.M.J. Hoozemans and M. Marchand, 1999: Increasing flood risk and wetland losses due to global sea-level rise: regional and global analyses. *Global Environ. Chang.*, **9**, S69-S87.

Nicholson, S., 2000: Land surface processes and Sahel climate. *Rev. Geophys.*, **38**, 117-139.

Nicholson, S.E., 2001: Climatic and environmental change in Africa during the last two centuries. *Climate Res.*, **17**, 123-144.

Nicholson, S.E., 2002: What are the key components of climate as a driver of desertification? *Global Desertification: Do Humans Cause Deserts?* J.F. Reynolds and D.M. Stafford-Smith, Eds., Dahlem University Press, Berlin, 41-57.

Nicholson, S.E., C.J. Tucker and M.B. Ba, 1998: Desertification, drought, and surface vegetation: an example from the West African Sahel. *B. Am. Meteorol. Soc.*, **79**, 815-829.

Nicholson, S.E., B. Some and B. Kone, 2000: An analysis of recent rainfall conditions in West Africa, including the rainy seasons of the 1997 El Niño and the 1998 La Niña years. *J. Climate*, **13**, 2628-2640.

Niklaus, P.A. and C.H. Körner, 2004: Synthesis of a six-year study of calcareous grassland responses to in situ CO_2 enrichment. *Ecol. Monogr.*, **74**, 491-511.

Niklaus, P.A., P.W. Leadley, C.H. Körner and B. Schmid, 2001: A long-term field study on biodiversity × elevated CO_2 interactions in grassland. *Ecol. Monogr.*, **71**, 341-356.

Nippert, J., A. Knapp and J. Briggs, 2006: Intra-annual rainfall variability and grassland productivity: can the past predict the future? *Plant Ecol.*, **184**, 75-87.

Nnadozie, K.C., 1998: Legal and policy approaches at linking the desertification and climate change agenda in Nigeria. Paper presented at the *Workshop on Climate Change, Biodiversity, and Desertification* held during the 12th Session of the Global Biodiversity Forum (GBF12-Dakar/CCD COP2), Meridien President Dakar, Senegal, 15 pp.

Norby, R.J. and Y.Q. Luo, 2004: Evaluating ecosystem responses to rising atmospheric CO_2 and global warming in a multi-factor world. *New Phytol.*, **162**, 281-293.

Norby, R.J., P.J. Hanson, E.G., O'Neill, T.J. Tschaplinski, J.F. Weltzin, R.A. Hansen, W.X. Cheng, S.D. Wullschleger, C.A. Gunderson, N.T. Edwards and D.W. Johnson, 2002: Net primary productivity of a CO_2-enriched deciduous forest and the implications for carbon storage. *Ecol. Appl.*, **12**, 1261-1266.

Norby, R.J., J. Ledford, C.D. Reilly, N.E. Miller and E.G. O'Neill, 2004: Fine-root production dominates response of a deciduous forest to atmospheric CO_2 enrichment. *P. Natl. Acad. Sci. USA*, **101**, 9689-9693.

Norby, R.J., E.H. Delucia, B. Gielen, C. Calfapietra, C.P. Giardina, J.S. King, J. Ledford, H.R. McCarthy, D.J.P. Moore, R. Ceulemans, P. De Angelis, A.C. Finzi, D.F. Karnosky, M.E. Kubiske, M. Lukac, K.S. Pregitzer, G.E. Scarascia-Mugnozza, W.H. Schlesinger and R. Oren, 2005: Forest response to elevated CO_2 is conserved across a broad range of productivity. *P. Natl. Acad. Sci. USA*, **102**, 18052-18056.

Norby, R.J., L.E. Rustad, J.S. Dukes, D.S. Ojima, W.J. Parton, S.J. Del Grosso, R.E. McMurtrie and D.P. Pepper, 2007: Ecosystem responses to warming and interacting global change factors. *Terrestrial Ecosystems in a Changing World*, J.G. Canadell, D.E. Pataki and L.F. Pitelka, Eds., Springer-Verlag, Berlin, 45-58.

Novelli, P.C., K.A. Masarie, P.M. Lang, B.D. Hall, R.C. Myers and J.W. Elkins, 2003: Reanalysis of tropospheric CO trends: effects of the 1997–1998 wildfires. *J. Geophys. Res. D*, **108**, 4464, doi:10.1029/2002JD003031.

Novick, K.A., P.C. Stoy, G.G. Katul, D.S. Ellsworth, M.B.S. Siqueira, J. Juang and R. Oren, 2004: Carbon dioxide and water vapor exchange in a warm temperate grassland. *Oecologia*, **138**, 259-274.

Nowak, R.S., S.F. Zitzer, D. Babcock, V. Smith-Longozo, T.N. Charlet, J.S. Coleman, J.R. Seemann and S.D. Smith, 2004: Elevated atmospheric CO_2 does not conserve soil water in the Mojave Desert. *Ecology*, **85**, 93-99.

Nunes, M.C.S., M.J. Vasconcelos, J.M.C. Pereira, N. Dasgupta and R.J. Alldredge, 2005: Land cover type and fire in Portugal: do fires burn land cover selectively? *Landscape Ecol.*, **20**, 661-673.

O'Brien, K.L. and R.M. Leichenko, 2000: Double exposure: assessing the impacts of climate change within the context of economic globalization. *Global Environ. Chang.*, **10**, 221-232.

O'Reilly, C.M., S.R. Alin, P.D. Plisnier, A.S. Cohen and B.A. McKee, 2003: Climate change decreases aquatic ecosystem productivity of Lake Tanganyika, Africa. *Nature*, **424**, 766-768.

Occhipinti-Ambrogi, A. and D. Savini, 2003: Biological invasions as a component of global change in stressed marine ecosystems. *Mar. Pollut. Bull.*, **46**, 542-551.

Ogaya, R. and J. Peñuelas, 2003: Comparative seasonal gas exchange and chlorophyll fluorescence of two dominant woody species in a holm oak forest. *Flora*, **198**, 132-141.

Ogaya, R. and J. Peñuelas, 2004: Phenological patterns of *Quercus ilex*, *Phillyrea latifolia*, and *Arbutus unedo* growing under a field experimental drought. *Ecoscience*, **11**, 263-270.

Ogaya, R., J. Peñuelas, J. Martinez-Vilalta and M. Mangiron, 2003: Effect of drought on diameter increment of *Quercus ilex*, *Phillyrea latifolia*, and *Arbutus unedo* in a holm oak forest of NE Spain. *Forest Ecol. Manag.*, **180**, 175-184.

Ogden, J., L. Basher and M. McGlone, 1998: Fire, forest regeneration and links with early human habitation: evidence from New Zealand. *Ann. Bot.*, **81**, 687-696.

Ogutu, J.O. and N. Owen-Smith, 2003: ENSO, rainfall and temperature influences on extreme population declines among African savanna ungulates. *Ecol. Lett.*, **6**, 412-419.

Olsson, L., L. Eklundh and J. Ardo, 2005: A recent greening of the Sahel: trends, patterns and potential causes. *J. Arid Environ.*, **63**, 556-566.

Opdam, P. and D. Wascher, 2004: Climate change meets habitat fragmentation: linking landscape and biogeographical scale level in research and conservation. *Biol. Conserv.*, **117**, 285-297.

Oren, R., D.S. Ellsworth, K.H. Johnsen, N. Phillips, B.E. Ewers, C. Maier, K.V.R. Schäfer, H. McCarthy, G. Hendrey, S.G. McNulty and G.G. Katul, 2001: Soil fertility limits carbon sequestration by forest ecosystems in a CO_2-enriched atmosphere. *Nature*, **411**, 469-472.

Orr, J.C., V.J. Fabry, O. Aumont, L. Bopp, S.C. Doney, R.A. Feely, A. Gnanadesikan, N. Gruber, A. Ishida, F. Joos, R.M. Key, K. Lindsay, E. Maier-Reimer, R. Matear, P. Monfray, A. Mouchet, R.G. Najjar, G.K. Plattner, K.B. Rodgers, C.L. Sabine, J.L. Sarmiento, R. Schlitzer, R.D. Slater, I.J. Totterdell, M.F. Weirig, Y. Yamanaka and A. Yool, 2005: Anthropogenic ocean acidification over the twenty-first century and its impact on calcifying organisms. *Nature*, **437**, 681-686.

Osborne, C.P., P.L. Mitchell, J.E. Sheehy and F.I. Woodward, 2000: Modelling the recent historical impacts of atmospheric CO_2 and climate change on Mediterranean vegetation. *Global Change Biol.*, **6**, 445-458.

Osmond, B., G. Ananyev, J. Berry, C. Langdon, Z. Kolber, G. Lin, R. Monson, C. Nichol, U. Rascher, U. Schurr, S. Smith and D. Yakir, 2004: Changing the way we think about global change research: scaling up in experimental ecosystem science. *Global Change Biol.*, **10**, 393-407.

Ostendorf, B., D.W. Hilbert and M.S. Hopkins, 2001: The effect of climate change on tropical rainforest vegetation pattern. *Ecol. Model.*, **145**, 211-224.

Overpeck, J., J. Cole and P. Bartlein, 2005: A "paleoperspective" on climate variability and change. *Climate Change and Biodiversity*, T.E. Lovejoy and L. Hannah, Eds., Yale University Press, New Haven, Connecticut, 91-108.

Overpeck, J.T., B.L. Otto-Bliesner, G.H. Miller, D.R. Muhs, R.B. Alley and J.T. Kiehl, 2006: Paleoclimatic evidence for future ice-sheet instability and rapid sea-level rise. *Science*, **311**, 1747-1750.

Pagiola, S., K. von Ritter and J.T. Bishop, 2004: Assessing the economic value of ecosystem conservation. Environment Department Paper No. 101, The World Bank Environment Department, Washington, District of Columbia, 66 pp.

Parmesan, C., T.L. Root and M.R. Willig, 2000: Impacts of extreme weather and climate on terrestrial biota. *B. Am. Meteorol. Soc.*, **81**, 443-450.

Parry, M.L., Ed., 2000: Assessment of potential effects and adaptations to climate change in Europe: The Europe Acacia Project. Report of concerted action of the environment programme of the Research Directorate General of the Commission of the European Communities, Jackson Environmental Institute, University of East Anglia, Norwich, 320 pp.

Pauli, H., M. Gottfried, T. Dirnböck, S. Dullinger and G. Grabherr, 2003: Assessing the long-term dynamics of endemic plants at summit habitats. *Alpine Biodiversity in Europe*, L. Nagy, G. Grabherr, C. Körner and D.B.A. Thompson, Eds., Springer-Verlag, Berlin, 195-207.

Pauly, D., J. Alder, A. Bakun, S. Heileman, K.H. Kock, P. Mace, W. Perrin, K. Stergiou, U.R. Sumaila, M. Vierros, K. Freire and Y. Sadovy, 2005: Marine fisheries systems. *Ecosystems and Human Well-being: Volume 1: Current State and Trends*, R. Hassan, R. Scholes and N. Ash, Eds., Island Press, Washington, District of Columbia, 477-511.

Pausas, J.G. and D. Abdel Malak, 2004: Spatial and temporal patterns of fire and climate change in the eastern Iberian Peninsula (Mediterranean Basin). *Ecology, Conservation and Management of Mediterranean Climate Ecosystems of the World*, M. Arianoutsou and V.P. Papanastasis, Eds., 10th International Conference on Mediterranean Climate Ecosystems, Rhodes, Greece. Millpress, Rotterdam, 1-6.

Pearce, D., 1998: Auditing the Earth: the value of the world's ecosystem services and natural capital. *Environment*, **40**, 23-28.

Pearson, R.G., 2006: Climate change and the migration capacity of species. *Trends Ecol. Evol.*, **21**, 111-113.

Pearson, R.G. and T.P. Dawson, 2003: Predicting the impacts of climate change on the distribution of species: are bioclimate envelope models useful? *Global Ecol. Biogeogr.*, **12**, 361-371.

Pearson, R.G., T.P. Dawson and C. Liu, 2004: Modelling species distributions in Britain: a hierarchical integration of climate and land-cover data. *Ecography*, **27**, 285-298.

Peck, L.S., K.E. Webb and D.M. Bailey, 2004: Extreme sensitivity of biological function to temperature in Antarctic marine species. *Funct. Ecol.*, **18**, 625-630.

Pederson, G.T., S.T. Gray, D.B. Fagre and L.J. Graumlich, 2006: Long-duration drought variability and impacts on ecosystem services: a case study from Glacier National Park, Montana. *Earth Interactions*, **10**, 1-28.

Peng, C. and M.J. Apps, 2000: Simulating global soil-CO_2 flux and its response to climate change. *J. Environ. Sci.*, **12**, 257-265.

Peñuelas, J. and M. Boada, 2003: A global change-induced biome shift in the Montseny mountains (NE Spain). *Global Change Biol.*, **9**, 131-140.

Peñuelas, J., F. Lloret and R. Montoya, 2001: Severe drought effects on Mediterranean woody flora in Spain. *Forest Sci.*, **47**, 214-218.

Perry, A.L., P.J. Low, J.R. Ellis and J.D. Reynolds, 2005: Climate change and distribution shifts in marine fishes. *Science*, **308**, 1912-1915.

Peters, R.L. and J.D.S. Darling, 1985: The greenhouse effect and nature reserves: global warming would diminish biological diversity by causing extinctions among reserve species. *BioScience*, **35**, 707-717.

Peterson, A.T., 2003: Projected climate change effects on Rocky Mountain and Great Plains birds: generalities of biodiversity consequences. *Global Change Biol.*, **9**, 647-655.

Peterson, A.T., M.A. Ortega-Huerta, J. Bartley, V. Sanchez-Cordero, J. Soberon, R.H. Buddemeier and D.R.B. Stockwell, 2002: Future projections for Mexican faunas under global climate change scenarios. *Nature*, **416**, 626-629.

Petit, J.R., J. Jouzel, D. Raynaud, N.I. Barkov, J.M. Barnola, I. Basile, M. Benders, J. Chappellaz, M. Davis, G. Delaygue, M. Delmotte, V.M. Kotlyakov, M. Legrand, V.Y. Lipenkov, C. Lorius, L. Pépin, C. Ritz, E. Saltzman and M. Stievenard, 1999: Climate and atmospheric history of the past 420,000 years from the Vostok ice core, Antarctica. *Nature*, **399**, 429-436.

Peylin, P., P. Bousquet, C. Le Quere, S. Sitch, P. Friedlingstein, G. McKinley, N. Gruber, P. Rayner and P. Ciais, 2005: Multiple constraints on regional CO_2 flux variations over land and oceans. *Global Biogeochem. Cy.*, **19**, GB1011, doi:10.1029/2003GB002214.

Phillips, O.L., R.V. Martinez, L. Arroyo, T.R. Baker, T. Killeen, S.L. Lewis, Y. Malhi, A.M. Mendoza, D. Neill, P.N. Vargas, M. Alexiades, C. Ceron, A. Di Fiore, T. Erwin, A. Jardim, W. Palacios, M. Saldias and B. Vinceti, 2002: Increasing dominance of large lianas in Amazonian forests. *Nature*, **418**, 770-774.

Phillips, O.L., T.R. Baker, L. Arroyo, N. Higuchi, T.J. Killeen, W.F. Laurance, S.L. Lewis, J. Lloyd, Y. Malhi, A. Monteagudo, D.A. Neill, P.N. Vargas, J.N.M. Silva, J. Terborgh, R.V. Martinez, M. Alexiades, S. Almeida, S. Brown, J. Chave, J.A. Comiskey, C.I. Czimczik, A. Di Fiore, T. Erwin, C. Kuebler, S.G. Laurance, H.E.M. Nascimento, J. Olivier, W. Palacios, S. Patino, N.C.A. Pitman, C.A. Quesada, M. Salidas, A.T. Lezama and B. Vinceti, 2004: Pattern and process in Amazon tree turnover, 1976-2001. *Philos. T. Roy. Soc. Lond. B*, **359**, 381-407.

Phoenix, G.K. and J.A. Lee, 2004: Predicting impacts of Arctic climate change: past lessons and future challenges. *Ecol. Res.*, **19**, 65-74.

Piao, S.L., J.Y. Fang and J.S. He, 2006: Variations in vegetation net primary production in the Qinghai-Xizang Plateau, China, from 1982 to 1999. *Climatic Change*, **74**, 253-267.

Piatkowski, U., D.F. Vergani and Z.B. Stanganelli, 2002: Changes in the cephalopod diet of southern elephant seal females at King George Island, during El Niño-La Niña events. *J. Mar. Biol. Assoc. UK*, **82**, 913-916.

Pickering, C., R. Good and K. Green, 2004: *Potential Effects of Global Warming on the Biota of the Australian Alps*. Australian Greenhouse Office, Australian Government, Canberra, 51 pp.

Pielke, R.A., Sr., 2005: Land use and climate change. *Science*, **310**, 1625-1626.

Pierce, J.L., G.A. Meyer and A.J. Timothy Jull, 2004: Fire-induced erosion and millennial-scale climate change in northern ponderosa pine forests. *Nature*, **432**, 87-90.

Platt, T., C. Fuentes-Yaco and K.T. Frank, 2003: Spring algal bloom and larval fish survival. *Nature*, 423, 398-399.

Podur, J., D.L. Martell and K. Knight, 2002: Statistical quality control analysis of forest fire activity in Canada. *Can. J. Forest Res.*, **32**, 195-205.

Podur, J., D.L. Martell and F. Csillag, 2003: Spatial patterns of lightning-caused forest fires in Ontario, 1976-1998. *Ecol. Model.*, **164**, 1-20.

Poff, L.N., M.M. Brinson and J.W.J. Day, 2002: *Aquatic Ecosystems and Global Climate Change: Potential Impacts on Inland Freshwater and Coastal Wetland Ecosystems in the United States*. Pew Center on Global Climate Change, Arlington, Virginia, 44 pp.

Pounds, J.A., 2001: Climate and amphibian declines. *Nature*, **410**, 639-640.

Pounds, J.A. and R. Puschendorf, 2004: Ecology: clouded futures. *Nature*, **427**, 107-109.

Pounds, J.A., M.P.L. Fogden and J.H. Campbell, 1999: Biological response to climate change on a tropical mountain. *Nature*, **398**, 611-615.

Pounds, J.A., M.R. Bustamante, L.A. Coloma, J.A. Consuegra, M.P.L. Fogden, P.N. Foster, E. La Marca, K.L. Masters, A. Merino-Viteri, R. Puschendorf, S.R.

Ron, G.A. Sanchez-Azofeifa, C.J. Still and B.E. Young, 2006: Widespread amphibian extinctions from epidemic disease driven by global warming. *Nature*, **439**, 161-167.

Prentice, I.C., P.J. Bartlein and T. Webb III, 1991: Vegetation and climate changes in eastern North America since the last glacial maximum. *Ecology*, **72**, 2038-2056.

Prentice, I.C., G.D. Farquhar, M.J.R. Fasham, M.L. Goulden, M. Heimann, V.J. Jaramillo, H.S. Kheshgi, C. Le Quéré, R.J. Scholes and D.W.R. Wallace, 2001: The carbon cycle and atmospheric carbon dioxide. *Climate Change 2001: The Scientific Basis. Contribution of Working Group I to the Third Assessment Report of the Intergovernmental Panel on Climate Change*, J.T. Houghton, Y. Ding, D.J. Griggs, M. Noguer, P.J. van der Linden, X. Dai, K. Maskell and C.A. Johnson, Eds., Cambridge University Press, Cambridge, 183-237.

Prentice, I.C., A. Bondeau, W. Cramer, S.P. Harrison, T. Hickler, W. Lucht, S. Sitch, B. Smith and M.T. Sykes, 2007: Dynamic global vegetation modelling: quantifying terrestrial ecosystem responses to large-scale environmental change. *Terrestrial Ecosystems in a Changing World*, J.G. Canadell, D.E. Pataki and L.F. Pitelka, Eds., Springer-Verlag, Berlin, 175-192.

Preston, B.L., 2006: Risk-based reanalysis of the effects of climate change on U.S. cold-water habitat. *Climatic Change*, **76**, 91-119.

Price, D.T., N.E. Zimmermann, P.J. van der Meer, M.J. Lexer, P. Leadley, I.T.M. Jorritsma, J. Schaber, D.F. Clark, P. Lasch, S. McNulty, J.G. Wu and B. Smith, 2001: Regeneration in gap models: Priority issues for studying forest responses to climate change. *Climatic Change*, **51**, 475-508.

Price, J.T. and T.L. Root, 2001: Climate change and Neotropical migrants. *T. N. Am. Wildl. Nat. Resour.*, **66**, 371-379.

Price, J.T. and P. Glick, 2002: *The Birdwatcher's Guide to Global Warming*. National Wildlife Federation and American Bird Conservancy, Washington, District of Columbia, 36 pp.

Price, J.T. and T.L. Root, 2005: Potential Impacts of climate change on Neotropical migrants: management implications. *Bird Conservation Implementation and Integration in the Americas*, C.J. Ralph and T.D. Rich, Eds., USDA Forest Service, Arcata, California, 1123-1128.

Prince, S.D., E.B. De Colstoun and L.L. Kravitz, 1998: Evidence from rain-use efficiencies does not indicate extensive Sahelian desertification. *Global Change Biol.*, **4**, 359-374.

Prospero, J.M. and P.J. Lamb, 2003: African droughts and dust transport to the Caribbean: climate change implications. *Science*, **302**, 1024-1027.

Rabalais, N.N., R.E. Turner and W.J. Wiseman, 2002: Gulf of Mexico Hypoxia, a.k.a. "The Dead Zone". *Annu. Rev. Ecol. Syst.*, **33**, 235-263.

Rambal, S., J.M. Ourcival, R. Joffre, F. Mouillot, Y. Nouvellon, M. Reichstein and A. Rocheteau, 2003: Drought controls over conductance and assimilation of a Mediterranean evergreen ecosystem: scaling from leaf to canopy. *Global Change Biol.*, **9**, 1813-1824.

Ramsay, M.A. and I. Stirling, 1988: Reproductive biology and ecology of female polar bears (*Ursus maritimus*). *J. Zool.*, **214**, 601-634.

Randall, D., R. Wood, S. Bony, R. Coleman, T. Fichefet, J. Fyfe, V. Kattsov, A. Pitman, J. Shukla, J. Srinivasan, R.J. Stouffer, A. Sumi and K. Taylor, 2007: Climate models and their evaluation. *Climate Change 2007: The Physical Science Basis. Contribution of Working Group I to the Fourth Assessment Report of the Intergovernmental Panel on Climate Change*, S. Solomon, D. Qin, M. Manning, Z. Chen, M. Marquis, K.B. Averyt, M. Tignor and H.L. Miller, Eds., Cambridge University Press, Cambridge, 589-662.

Randerson, J.T., G.R. van der Werf, G.J. Collatz, L. Giglio, C.J. Still, P. Kasibhatla, J.B. Miller, J.W.C. White, R.S. Defries and E.S. Kasischke, 2005: Fire emissions from C_{-3} and C_{-4} vegetation and their influence on interannual variability of atmospheric CO_2 and $\delta^{13}(CO_2)$. *Global Biogeochem. Cy.*, **19**, GB2019, doi:10.1029/2004GB002366.

Rasmussen, K., B. Fog and J.E. Madsen, 2001: Desertification in reverse? Observations from northern Burkina Faso. *Global Environ. Chang.*, **11**, 271-282.

Rasmussen, L., C. Beier and A. Bergstedt, 2002: Experimental manipulations of old pine forest ecosystems to predict the potential tree growth effects of increased CO_2 and temperature in a future climate. *Forest Ecol. Manag.*, **158**, 179-188.

Raven, J., K. Caldeira, H. Elderfield, O. Hoegh-Guldberg, P. Liss, U. Riebesell, J. Shepherd, C. Turley and A. Watson, 2005: Ocean acidification due to increasing atmospheric carbon dioxide. Policy document 12/05, The Royal Society, The Clyvedon Press Ltd, Cardiff, 68 pp.

Rebetez, M. and M. Dobbertin, 2004: Climate change may already threaten Scots pine stands in the Swiss Alps. *Theor. Appl. Climatol.*, **79**, 1-9.

Reich, P.B., S.E. Hobbie, T. Lee, D.S. Ellsworth, J.B. West, D. Tilman, J.M.H. Knops, S. Naeem and J. Trost, 2006: Nitrogen limitation constrains sustainability of ecosystem response to CO_2. *Nature*, **440**, 922-925.

Reichstein, M., J.D. Tenhunen, O. Roupsard, J.M. Ourcival, S. Rambal, F. Miglietta, A. Peressotti, M. Pecchiari, G. Tirone and R. Valentini, 2002: Severe drought effects on ecosystem CO_2 and H_2O fluxes at three Mediterranean evergreen sites: revision of current hypotheses? *Global Change Biol.*, **8**, 999-1017.

Reid, W.V., H.A. Mooney, A. Cropper, D. Capistrano, S.R. Carpenter, K. Chopra, P. Dasgupta, T. Dietz, A.K. Duraiappah, R. Hassan, R. Kasperson, R. Leemans, R.M. May, A.J. McMichael, P. Pingali, C. Samper, R. Scholes, R.T. Watson, A.H. Zakri, Z. Shidong, N.J. Ash, E. Bennett, P. Kumar, M.J. Lee, C. Raudsepp-Hearne, H. Simons, J. Thonell and M.B. Zurek, Eds., 2005: *Ecosystems and Human Well-being: Synthesis*. Island Press, Washington, District of Columbia, 155 pp.

Reij, C., G. Tappan and A. Belemvire, 2005: Changing land management practices and vegetation on the Central Plateau of Burkina Faso (1968-2002). *J. Arid Environ.*, **63**, 642-659.

Reynaud, S., N. Leclercq, S. Romaine-Lioud, C. Ferrier-Pages, J. Jaubert and J.P. Gattuso, 2003: Interacting effects of CO_2 partial pressure and temperature on photosynthesis and calcification in a scleractinian coral. *Global Change Biol.*, **9**, 1660-1668.

Richards, M., 2003: Poverty reduction, equity and climate change: global governance synergies or contradictions? Globalisation and Poverty Programme. Government Report, Overseas Development Institute, London, 14 pp. Accessed 13.03.07: http://www.odi.org.uk.

Richardson, A.J. and D.S. Schoeman, 2004: Climate impact on plankton ecosystems in the Northeast Atlantic. *Science*, **305**, 1609-1612.

Riebesell, U., I. Zondervan, B. Rost, P.D. Tortell, R.E. Zeebe and F.M.M. Morel, 2000: Reduced calcification of marine plankton in response to increased atmospheric CO_2. *Nature*, **407**, 364-367.

Rietkerk, M., S.C. Dekker, P.C. de Ruiter and J. van de Koppel, 2004: Self-organized patchiness and catastrophic shifts in ecosystems. *Science*, **305**, 1926-1929.

Rillig, M.C., S.F. Wright, M.R. Shaw and C.B. Field, 2002: Artificial climate warming positively affects arbuscular mycorrhizae but decreases soil aggregate water stability in an annual grassland. *Oikos*, **97**, 52-58.

Roberts, C.M., C.J. McClean, J.E.N. Veron, J.P. Hawkins, G.R. Allen, D.E. McAllister, C.G. Mittermeier, F.W. Schueler, M. Spalding, F. Wells, C. Vynne and T.B. Werner, 2002: Marine biodiversity hotspots and conservation priorities for tropical reefs. *Science*, **295**, 1280-1284.

Robinson, R.A., J.A. Learmonth, A.M. Hutson, C.D. Macleod, T.H. Sparks, D.I. Leech, G.J. Pierce, M.M. Rehfisch and H.Q.P. Crick, 2005: Climate change and migratory species. BTO Research Report, Department for Environment, Food and Rural Affairs (Defra), London, 414 pp.

Root, T.L. and S.H. Schneider, 1995: Ecology and climate: research strategies and implications. *Science*, **269**, 334-341.

Root, T.L., J.T. Price, K.R. Hall, S.H. Schneider, C. Rosenzweig and J.A. Pounds, 2003: Fingerprints of global warming on wild animals and plants. *Nature*, **421**, 57-60.

Ross, D.J., K.R. Tate, P.C.D. Newton and H. Clark, 2002: Decomposability of C_3 and C_4 grass litter sampled under different concentrations of atmospheric carbon dioxide at a natural CO_2 spring. *Plant Soil*, **240**, 275-286.

Rouault, M., J.L. Mélice, C.J.C. Reason and J.R.E. Lutijeharms, 2005: Climate variability at Marion Island, Southern Ocean, since 1960. *J. Geophys. Res. C*, **110**, C05007, doi:10.1029/2004JC002492.

Rounsevell, M.D.A., P.M. Berry and P.A. Harrison, 2006: Future environmental change impacts on rural land use and biodiversity: a synthesis of the ACCELERATES project. *Environ. Sci. Policy*, **9**, 93-100.

Rustad, L.E., J.L. Campbell, G.M. Marion, R.J. Norby, M.J. Mitchell, A.E. Hartley, J.H.C. Cornelissen and J. Gurevitch, 2001: A meta-analysis of the response of soil respiration, net nitrogen mineralization, and aboveground plant growth to experimental ecosystem warming. *Oecologia*, **126**, 543-562.

Rutherford, M.C., G.F. Midgley, W.J. Bond, L.W. Powrie, R. Roberts and J. Allsopp, Eds., 2000: *Plant Biodiversity: Vulnerability and Adaptation Assessment*. Department of Environmental Affairs and Tourism, Pretoria, 59 pp.

Sabine, C.L., M. Heimann, P. Artaxo, D.C.E. Bakker, C.T.A. Chen, C.B. Field and N. Gruber, 2004: Current status and past trends of the global carbon cycle. *Global Carbon Cycle: Integrating Humans, Climate, and the Natural World*, C.B. Field and M.R. Raupach, Eds., Island Press, Washington, Distict of Columbia, 17-44.

Sadovy, Y., 2005: Trouble on the reef: the imperative for managing vulnerable and valuable fisheries. *Fish Fish.*, **6**, 167-185.

Sala, O.E., 2005: Biodiversity across scenarios. *Ecosystems and Human Well-*

being: Volume 2: Scenarios, S. Carpenter, P. Pingali, E. Bennett and M. Zurek, Eds., Island Press, Washington, District of Columbia, 375-408.

Sala, O.E., I.F.S. Chapin, J.J. Armesto, E. Berlow, J. Bloomfield, R. Dirzo, E. Huber Sanwald, L.F. Huenneke, R.B. Jackson, A. Kinzig, R. Leemans, D.H. Lodge, H.A. Mooney, M. Oesterheld, N. Leroy Poff, M.T. Sykes, B.H. Walker, M. Walker and D.H. Wall, 2000: Global biodiversity scenarios for the year 2100. *Science*, **287**, 1770-1774.

Salvador, R., F. Lloret, X. Pons and J. Pinol, 2005: Does fire occurrence modify the probability of being burned again? A null hypothesis test from Mediterranean ecosystems in NE Spain. *Ecol. Model.*, **188**, 461-469.

Sanchez, E., C. Gallardo, M.A. Gaertner, A. Arribas and M. Castro, 2004: Future climate extreme events in the Mediterranean simulated by a regional climate model: a first approach. *Global Planet. Change*, **44**, 163-180.

Sankaran, M., 2005: Fire, grazing and the dynamics of tall-grass savannas in the Kalakad-Mundanthurai Tiger Reserve, South India. *Conserv. Soc.*, **3**, 4-25.

Sankaran, M., N.P. Hanan, R.J. Scholes, J. Ratnam, D.J. Augustine, B.S. Cade, J. Gignoux, S.I. Higgins, X. le Roux, F. Ludwig, J. Ardo, F. Banyikwa, A. Bronn, G. Bucini, K.K. Caylor, M.B. Coughenour, A. Diouf, W. Ekaya, C.J. Feral, E.C. February, P.G.H. Frost, P. Hiernaux, H. Hrabar, K.L. Metzger, H.H.T. Prins, S. Ringrose, W. Sea, J. Tews, J. Worden and N. Zambatis, 2005: Determinants of woody cover in African savannas. *Nature*, **438**, 846-849.

Sanz-Elorza, M., E.D. Dana, A. Gonzalez and E. Sobrino, 2003: Changes in the high-mountain vegetation of the central Iberian peninsula as a probable sign of global warming. *Ann. Bot.*, **92**, 273-280.

Sardans, J. and J. Peñuelas, 2004: Increasing drought decreases phosphorus availability in an evergreen Mediterranean forest. *Plant Soil*, **267**, 367-377.

Sarmiento, J., 2000: That sinking feeling. *Nature*, **408**, 155-156.

Sarmiento, J.L. and N. Gruber, 2002: Sinks for anthropogenic carbon. *Phys. Today*, **55**, 30-36.

Sarmiento, J.L., T.M.C. Hughes, R.J. Stouffer and S. Manabe, 1998: Simulated response of the ocean carbon cycle to anthropogenic climate warming. *Nature*, **393**, 245-249.

Sarmiento, J.L., N. Gruber, M. Brzezinksi and J. Dunne, 2004a: High latitude controls of thermocline nutrients and low latitude biological productivity. *Nature*, **426**, 56-60.

Sarmiento, J.L., R. Slater, R. Barber, L. Bopp, S.C. Doney, A.C. Hirst, J. Kleypas, R. Matear, U. Mikolajewicz, P. Monfray, V. Soldatov, S.A. Spall and R. Stouffer, 2004b: Response of ocean ecosystems to climate warming. *Global Biogeochem. Cy.*, **18**, GB3003, doi:10.1029/2003GB002134.

Sazonova, T.S., V.E. Romanovsky, J.E. Walsh and D.O. Sergueev, 2004: Permafrost dynamics in the 20th and 21st centuries along the East Siberian transect. *J. Geophys. Res. D*, **109**, D01108, doi:10.1029/2003JD003680.

Scavia, D., J.C. Field, D.F. Boesch, R.W. Buddemeier, V. Burkett, D.R. Cayan, M. Fogarty, M.A. Harwell, R.W. Howarth, C. Mason, D.J. Reed, T.C. Royer, A.H. Sallenger and J.G. Titus, 2002: Climate change impacts on U.S. coastal and marine ecosystems. *Estuaries*, **25**, 149-164.

Schäfer, K.V.R., R. Oren, D.S. Ellsworth, C.T. Lai, J.D. Herrick, A.C. Finzi, D.D. Richter and G.G. Katul, 2003: Exposure to an enriched CO_2 atmosphere alters carbon assimilation and allocation in a pine forest ecosystem. *Global Change Biol.*, **9**, 1378-1400.

Schallenberg, M., C.J. Hall and C.W. Burns, 2001: Climate change alters zooplankton community structure and biodiversity in coastal wetlands. Report of Freshwater Ecology Group, University of Otago, Hamilton, 9 pp.

Schaphoff, S., W. Lucht, D. Gerten, S. Sitch, W. Cramer and I.C. Prentice, 2006: Terrestrial biosphere carbon storage under alternative climate projections. *Climatic Change*, **74**, 97-122.

Schär, C., P.L. Vidale, D. Lüthi, C. Frei, C. Häberli, M.A. Liniger and C. Appenzeller, 2004: The role of increasing temperature variability in European summer heatwaves. *Nature*, **427**, 332-336.

Scharlemann, J.P.W., R.E. Green and A. Balmford, 2004: Land-use trends in Endemic Bird Areas: global expansion of agriculture in areas of high conservation value. *Global Change Biol.*, **10**, 2046-2051.

Scheffer, M., S. Carpenter, J.A. Foley, C. Folke and B. Walker, 2001: Catastrophic shifts in ecosystems. *Nature*, **413**, 591-596.

Schellnhuber, H.J., 2002: Coping with Earth system complexity and irregularity. *Challenges of a Changing Earth*, W. Steffen, D. Jäger and C. Bradshaw, Eds., Springer, Berlin, 151-156.

Schimel, D. and D. Baker, 2002: Carbon cycle: the wildfire factor. *Nature*, **420**, 29-30.

Schimel, D.S., T.G.F. Kittel, S. Running, R. Monson, A. Turnipseed and D. Anderson, 2002: Carbon sequestration studied in Western US Mountains. *EOS Transactions*, **83**, 445-449.

Schimper, A.F.W., 1903: *Plant Geography on a Physiological Basis*. Clarendon Press, Oxford, 839 pp.

Schlyter, P., I. Stjernquist, L. Bärring, A.M. Jönsson and C. Nilsson, 2006: Assessment of the impacts of climate change and weather extremes on boreal forests in northern Europe, focusing on Norway spruce. *Climate Res.*, **31**, 75-84.

Schneider, S.H., 2004: Abrupt non-linear climate change, irreversibility and surprise. *Global Environ. Chang.*, **14**, 245-258.

Schoennagel, T., T.T. Veblen and W.H. Romme, 2004: The interaction of fire, fuels, and climate across rocky mountain forests. *BioScience*, **54**, 661-676.

Scholes, R.J. and S.R. Archer, 1997: Tree–grass interactions in savannas. *Annu. Rev. Ecol. Syst.*, **28**, 517-544.

Scholes, R.J., G. von Maltitz, M. de Wit, G.O. Hughes, G. Midgley and B. Erasmus, 2004: *Helping Biodiversity Adapt to Climate Change*. LUMONET: Science Development Network, Finnish Environment Institute, Helsinki, 2 pp.

Scholze, M., W. Knorr, N.W. Arnell and I.C. Prentice, 2006: A climate change risk analysis for world ecosystems. *P. Natl. Acad. Sci. USA*, **103**, 13116-13120.

Schröder, A., L. Persson and A.M. De Roos, 2005: Direct experimental evidence for alternative stable states: a review. *Oikos*, **110**, 3-19.

Schrope, M.K., J.P. Chanton, L.H. Allen and J.T. Baker, 1999: Effect of CO_2 enrichment and elevated temperature on methane emissions from rice, *Oryza sativa*. *Global Change Biol.*, **5**, 587-599.

Schröter, D., W. Cramer, R. Leemans, I.C. Prentice, M.B. Araújo, N.W. Arnell, A. Bondeau, H. Bugmann, T.R. Carter, C.A. Gracia, A.C. de la Vega-Leinert, M. Erhard, F. Ewert, M. Glendining, J.I. House, S. Kankaanpää, R.J.T. Klein, S. Lavorel, M. Lindner, M.J. Metzger, J. Meyer, T.D. Mitchell, I. Reginster, M. Rounsevell, S. Sabaté, S. Sitch, B. Smith, J. Smith, P. Smith, M.T. Sykes, K. Thonicke, W. Thuiller, G. Tuck, S. Zaehle and B. Zierl, 2005: Ecosystem service supply and vulnerability to global change in Europe. *Science*, **310**, 1333-1337.

Schulze, E.D. and A. Freibauer, 2005: Carbon unlocked from soils. *Nature*, **437**, 205-206.

Schumacher, S. and H. Bugmann, 2006: The relative importance of climatic effects, wildfires and management for future forest landscape dynamics in the Swiss Alps. *Global Change Biol.*, **12**, 1435-1450.

Schumacher, S., H. Bugmann and D.J. Mladenoff, 2004: Improving the formulation of tree growth and succession in a spatially explicit landscape model. *Ecol. Model.*, **180**, 175-194.

Schumacher, S., B. Reineking, J. Sibold and H. Bugmann, 2006: Modelling the impact of climate and vegetation on fire regimes in mountain landscapes. *Landscape Ecol.*, **21**, 539-554.

Schwinning, S. and O.E. Sala, 2004: Hierarchy of responses to resource pulses in arid and semi-arid ecosystems. *Oecologia*, **141**, 211-220.

Scott, D. and C. Lemieux, 2005: Climate change and protected area policy and planning in Canada. *Forest Chron.*, **81**, 696-703.

Semazzi, F.H.M. and Y. Song, 2001: A GCM study of climate change induced by deforestation in Africa. *Climate Res.*, **17**, 169-182.

Serreze, M.C., F. Carse, R.G. Barry and J.C. Rogers, 1997: Icelandic low cyclone activity: climatological features, linkages with the NAO, and relationships with recent changes in the Northern Hemisphere Circulation. *J. Appl. Syst. Anal.*, **10**, 453-464.

Shafer, S.L., P.J. Bartlein and R.S. Thompson, 2001: Potential changes in the distributions of western North America tree and shrub taxa under future climate scenarios. *Ecosystems*, **4**, 200-215.

Shaw, M.R., E.S. Zavaleta, N.R. Chiariello, E.E. Cleland, H.A. Mooney and C.B. Field, 2002: Grassland responses to global environmental changes suppressed by elevated CO_2. *Science*, **298**, 1987-1990.

Sheppard, C.R.C., 2003: Predicted recurrences of mass coral mortality in the Indian Ocean. *Nature*, **425**, 294-297.

Shiyatov, S.G., 2003: Rates of change in the upper treeline ecotone in the polar Ural Mountains. *PAGES News*, **11**, 8-10.

Shvidenko, A., C.V. Barber and R. Persson, 2005: Forest and woodland systems. *Ecosystems and Human Well-being: Volume 1: Current State and Trends*, R. Hassan, R. Scholes and N. Ash, Eds., Island Press, Washington, District of Columbia, 585-621.

Siegenthaler, U., T.F. Stocker, E. Monnin, D. Luthi, J. Schwander, B. Stauffer, D. Raynaud, J.M. Barnola, H. Fischer, V. Masson-Delmotte and J. Jouzel, 2005: Stable carbon cycle-climate relationship during the late Pleistocene. *Science*, **310**, 1313-1317.

Simmons, R.E., P. Barnard, W.R.J. Dean, G.F. Midgley, W. Thuiller and G. Hughes, 2004: Climate change and birds: perspectives and prospects from southern Africa. *Ostrich*, **75**, 295-308.

Sims, D.W., V.J. Wearmouth, M.J. Genner, A.J. Southward and S.J. Hawkins, 2004: Low-temperature-driven early spawning migration of a temperate marine fish. *J. Anim. Ecol.*, **73**, 333-341.

Singh, H.S., 2003: Vulnerability and adaptability of tidal forests in response to climate change in India. *Indian Forester*, **129**, 749-756.

Siqueira, M.F. and A.T. Peterson, 2003: Consequences of global climate change for geographic distributions of cerrado tree species. *Biota Neotropica*, **3**, 1-14.

Sitch, S., B. Smith, I.C. Prentice, A. Arneth, A. Bondeau, W. Cramer, J.O. Kaplan, S. Levis, W. Lucht, M.T. Sykes, K. Thonicke and S. Venevsky, 2003: Evaluation of ecosystem dynamics, plant geography and terrestrial carbon cycling in the LPJ dynamic global vegetation model. *Global Change Biol.*, **9**, 161-185.

Sitch, S., V. Brovkin, W. von Bloh, D. van Vuuren, B. Assessment and A. Ganopolski, 2005: Impacts of future land cover changes on atmospheric CO_2 and climate. *Global Biogeochem. Cy.*, **19**, GB2013, doi:10.1029/2004GB002311.

Sitch, S., A.D. McGuire, J. Kimball, N. Gedney, J. Gamon, R. Emgstrom, A. Wolf, Q. Zhuang and J. Clein, 2007: Assessing the carbon balance of circumpolar Arctic tundra with remote sensing and process-based modeling approaches. *Ecol. Appl.*, **17**, 213-234.

Skinner, W.R., M.D. Flannigan, B.J. Stocks, D.L. Martell, B.M. Wotton, J.B. Todd, J.A. Mason, K.A. Logan and E.M. Bosch, 2002: A 500 hPa synoptic wildland fire climatology for large Canadian forest fires, 1959-1996. *Theor. Appl. Climatol.*, **71**, 157-169.

Skre, O., R. Baxter, R.M.M. Crawford, T.V. Callaghan and A. Fedorkov, 2002: How will the tundra-taiga interface respond to climate change? *Ambio Special Report*, **12**, 37-46.

Smith, J.B. and J.K. Lazo, 2001: A summary of climate change impact assessments from the US Country Studies Program. *Climatic Change*, **50**, 1-29.

Smith, L.C., Y. Sheng, G.M. MacDonald and L.D. Hinzman, 2005: Disappearing arctic lakes. *Science*, **308**, 1429.

Smith, M.D. and A.K. Knapp, 1999: Exotic plant species in a C_{-4}-dominated grassland: invasibility, disturbance and community structure. *Oecologia*, **120**, 605-612.

Smith, R.D. and E. Malthby, 2003: *Using the Ecosystem Approach to Implement the Convention on Biological Diversity: Key Issues and Case Studies*. Island Press, Chicago, 118 pp.

Smith, S.D., T.E. Huxman, S.F. Zitzer, T.N. Charlet, D.C. Housman, J.S. Coleman, L.K. Fenstermaker, J.R. Seemann and R.S. Nowak, 2000: Elevated CO_2 increases productivity and invasive species success in an arid ecosystem. *Nature*, **408**, 79-82.

Smith, V.R., 2002: Climate change in the sub-Antarctic: an illustration from Marion Island. *Climatic Change*, **52**, 345-357.

Snyder, P.K., C. Delire and J.A. Foley, 2004: Evaluating the influence of different vegetation biomes on the global climate. *Clim. Dynam.*, **23**, 279-302.

Solorzano, S., M.A. Castillo-Santiago, D.A. Navarrete-Gutierrez and K. Oyama, 2003: Impacts of the loss of neotropical highland forests on the species distribution: a case study using resplendent quetzal an endangered bird species. *Biol. Conserv.*, **114**, 341-349.

Sorenson, L.G., R. Goldberg, T.L. Root and M.G. Anderson, 1998: Potential effects of global warming on waterfowl populations breeding in the Northern Great Plains. *Climatic Change*, **40**, 343-369.

Spehn, E. and C. Körner, 2005: A global assessment of mountain biodiversity and its function. *Global Change and Mountain Regions: An Overview of Current Knowledge*, U.M. Huber, H.K.M. Bugmann and M.A. Reasoner, Eds., Springer, Berlin, 393-400.

Stabeno, P.J., N.A. Bond, N.B. Kachel, S.A. Salo and J.D. Schumacher, 2001: On the temporal variability of the physical environment over the south-eastern Bering Sea. *Fish. Oceanogr.*, **10**, 81-98.

Stachowicz, J.J., J.R. Terwin, R.B. Whitlatch and R.W. Osman, 2002: Linking climate change and biological invasions: ocean warming facilitates nonindigenous species invasions. *P. Natl. Acad. Sci. USA*, **99**, 15497-15500.

Stampfli, A. and M. Zeiter, 2004: Plant regeneration directs changes in grassland composition after extreme drought: a 13-year study in southern Switzerland. *J. Ecol.*, **92**, 568-576.

Stapp, P., G.A. Polls and F.S. Pirero, 1999: Stable isotopes reveal strong marine and El Niño effects on island food webs. *Nature*, **401**, 467-469.

Steffen, W., A. Sanderson, P.D. Tyson, J. Jäger, P.A. Matson, B. Moore, F. Oldfield, K. Richardson, H.-J. Schellnhuber, B.L. Turner II and R.J. Wasson, 2004: *Global Change and the Earth System: A Planet Under Pressure*. Springer, Berlin, 336 pp.

Stenseth, N.C. and J.W. Hurrell, 2005: Global climate change: building links between the climate and ecosystem impact research communities. *Climate Res.*, **29**, 181-182.

Stenseth, N.C., A. Mysterud, G. Ottersen, J.W. Hurrell, K.S. Chan and M. Lima, 2002: Ecological effects of climate fluctuations. *Science*, **297**, 1292-1296.

Stenseth, N.C., G. Ottersen, J.W. Hurrell, A. Mysterud, M. Lima, K.S. Chan, N.G. Yoccoz and B. Adlandsvik, 2003: Studying climate effects on ecology through the use of climate indices: the North Atlantic Oscillation, El Niño Southern Oscillation and beyond. *P. Roy. Soc. Lond. B*, **270**, 2087-2096.

Stern, N., 2007: *The Economics of Climate Change: The Stern Review*. Cambridge University Press, Cambridge, 692 pp.

Stevens, G.C. and J.F. Fox, 1991: The causes of treeline. *Annu. Rev. Ecol. Syst.*, **22**, 177-191.

Still, C.J., P.N. Foster and S.H. Schneider, 1999: Simulating the effects of climate change on tropical montane cloud forests. *Nature*, **398**, 608-610.

Stirling, I. and A.E. Derocher, 1993: Possible impacts of climatic warming on polar bears. *Arctic*, **46**, 240-245.

Stirling, I., N.J. Lunn and J. Iacozza, 1999: Long-term trends in the population ecology of polar bears in western Hudson Bay in relation to climatic change. *Arctic*, **52**, 294-306.

Stock, W.D., F. Ludwig, C. Morrow, G.F. Midgley, S.J.E. Wand, N. Allsopp and T.L. Bell, 2005: Long-term effects of elevated atmospheric CO_2 on species composition and productivity of a southern African C_{-4} dominated grassland in the vicinity of a CO_2 exhalation. *Plant Ecol.*, **178**, 211-224.

Stocks, B.J., M.A. Fosberg, T.J. Lynham, L. Mearns, B.M. Wotton, Q. Yang, J.Z. Jin, K. Lawrence, G.R. Hartley, J.A. Mason and D.W. McKenney, 1998: Climate change and forest fire potential in Russian and Canadian boreal forests. *Climatic Change*, **38**, 1-13.

Stocks, B.J., J.A. Mason, J.B. Todd, E.M. Bosch, B.M. Wotton, B.D. Amiro, M.D. Flannigan, K.G. Hirsch, K.A. Logan, D.L. Martell and W.R. Skinner, 2002: Large forest fires in Canada, 1959-1997. *J. Geophys. Res. D*, **108**, 8149, doi:10.1029/2001JD000484.

Stokstad, E., 2001: U.N. report suggests slowed forest losses. *Science*, **291**, 2294.

Strengers, B., R. Leemans, B.J. Eickhout, B. de Vries and A.F. Bouwman, 2004: The land-use projections and resulting emissions in the IPCC SRES scenarios as simulated by the IMAGE 2.2 model. *GeoJournal*, **61**, 381-393.

Sturm, M., C. Racine and K. Tape, 2001: Increasing shrub abundance in the Arctic. *Nature*, **411**, 546-547.

Sturm, M., T. Douglas, C. Racine and G.E. Liston, 2005: Changing snow and shrub conditions affect albedo with global implications. *J. Geophys. Res. G*, **110**, G01004, doi:10.1029/2005JG000013.

Sumner, G.N., R. Romero, V. Homar, C. Ramis, S. Alonso and E. Zorita, 2003: An estimate of the effects of climate change on the rainfall of Mediterranean Spain by the late twenty-first century. *Clim. Dynam.*, **20**, 789-805.

Symon, C., L. Arris and B. Heal, Eds., 2005: *Arctic Climate Impact Assessment (ACIA): Scientific Report*. Cambridge University Press, Cambridge, 1042 pp.

Szmant, A.M., 2001: Why are coral reefs world-wide becoming overgrown by algae? 'Algae, algae everywhere, and nowhere a bite to eat!' *Coral Reefs*, **19**, 299-302.

Tape, K., M. Sturm and C. Racine, 2006: The evidence for shrub expansion in Northern Alaska and the Pan-Arctic. *Global Change Biol.*, **12**, 686-702.

Theurillat, J.P. and A. Guisan, 2001: Potential impact of climate change on vegetation in the European Alps: a review. *Climatic Change*, **50**, 77-109.

Theurillat, J.P., F. Felber, P. Geissler, J.M. Gobat, M. Fierz, A. Fischlin, P. Küpfer, A. Schlüssel, C. Velluti, G.F. Zhao and J. Williams, 1998: Sensitivity of plant and soil ecosystems of the Alps to climate change. *Views from the Alps: Regional Perspectives on Climate Change*, P. Cebon, U. Dahinden, H.C. Davies, D.M. Imboden and C.C. Jager, Eds., MIT Press, Boston, Massachusetts, 225-308.

Thirgood, S., A. Mosser, S. Tham, G. Hopcraft, E. Mwangomo, T. Mlengeya, M. Kilewo, J. Fryxell, A.R.E. Sinclair and M. Borner, 2004: Can parks protect migratory ungulates? The case of the Serengeti wildebeest. *Anim. Conserv.*, **7**, 113-120.

Thomas, C.D., A. Cameron, R.E. Green, M. Bakkenes, L.J. Beaumont, Y.C. Collingham, B.F.N. Erasmus, M.F. de Siqueira, A. Grainger, L. Hannah, L. Hughes, B. Huntley, A.S. van Jaarsveld, G.F. Midgley, L. Miles, M.A. Ortega-Huerta, A.T. Peterson, O.L. Phillips and S.E. Williams, 2004a: Extinction risk from climate change. *Nature*, **427**, 145-148.

Thomas, C.D., S.E. Williams, A. Cameron, R.E. Green, M. Bakkenes, L.J. Beaumont, Y.C. Collingham, B.F.N. Erasmus, M.F de Siqueira, A. Grainger, L. Hannah, L. Hughes, B. Huntley, A.S. van Jaarsveld, G.F. Midgley, L. Miles, M.A. Ortega-Huerta, A.T. Peterson and O.L. Philipps, 2004b: Biodiversity conservation: uncertainty in predictions of extinction risk/Effects of changes in climate and land use/Climate change and extinction risk (reply). *Nature*, **430**, 34.

Thomas, D.S.G. and H.C. Leason, 2005: Dunefield activity response to climate variability in the southwest Kalahari. *Geomorphology*, **64**, 117-132.

Thonicke, K., S. Venevsky, S. Sitch and W. Cramer, 2001: The role of fire disturbance for global vegetation dynamics: coupling fire into a Dynamic Global Vegetation Model. *Global Ecol. Biogeogr.*, **10**, 661-677.

Thuiller, W., S. Lavorel and M.B. Araujo, 2005a: Niche properties and geographical extent as predictors of species sensitivity to climate change. *Global Ecol. Biogeogr.*, **14**, 347.

Thuiller, W., S. Lavorel, M.B. Araujo, M.T. Sykes and I.C. Prentice, 2005b: Climate change threats to plant diversity in Europe. *P. Natl. Acad. Sci. USA*, **102**, 8245-8250.

Thuiller, W., O. Broenniman, G. Hughes, J.R.M. Alkemades, G.F. Midgley and F. Corsi, 2006a: Vulnerability of African mammals to anthropogenic climate change under conservative land transformation assumptions. *Global Change Biol.*, **12**, 424-440.

Thuiller, W., G.F. Midgley, G.O. Hughes, B. Bomhard, G. Drew, M.C. Rutherford and F.I. Woodward, 2006b: Endemic species and ecosystem sensitivity to climate change in Namibia. *Global Change Biol.*, **12**, 759-776.

Timoney, K.P., 2003: The changing disturbance regime of the boreal forest of the Canadian Prairie Provinces. *Forest Chron.*, **79**, 502-516.

Tinner, W. and A.F. Lotter, 2001: Central European vegetation response to abrupt climate change at 8.2 ka. *Geology*, **29**, 551-554.

Tol, R.S.J., 2001: Equitable cost-benefit analysis of climate change policies. *Ecol. Econ.*, **36**, 71-85.

Toman, M., 1998a: Research frontiers in the economics of climate change. *Environ. Resour. Econ.*, **11**, 603-621.

Toman, M., 1998b: Why not to calculate the value of the world's ecosystem services and natural capital. *Ecol. Econ.*, **25**, 57-60.

Tompkins, E.L. and W.N. Adger, 2003: Building resilience to climate change through adaptive management of natural resources. Working Paper 27, Tyndall Centre for Climate Change Research, Norwich, 19 pp.

Trenberth, K.E., P.D. Jones, P.G. Ambenje, R. Bojariu, D.R. Easterling, A.M.G. Klein Tank, D.E. Parker, J.A. Renwick, F. Rahimzadeh, M.M. Rusticucci, B.J. Soden and P.-M. Zhai, 2007: Observations: surface and atmospheric climate change. *Climate Change 2007: The Physical Science Basis. Contribution of Working Group I to the Fourth Assessment Report of the Intergovernmental Panel on Climate Change*, S. Solomon, D. Qin, M. Manning, Z. Chen, M. Marquis, K.B. Averyt, M. Tignor and H.L. Miller, Eds., Cambridge University Press, Cambridge, 235-336.

Trigo, R.M., M.G. Pereira, J.M.C. Pereira, B. Mota, T.J. Calado, C.C. da Camara and F.E. Santo, 2005: The exceptional fire season of summer 2003 in Portugal. *Geophys. Res. Abstracts*, **7**, 09690.

Trigo, R.M., J.M.C. Pereira, M.G. Pereira, B. Mota, T.J. Calado, C.C. Dacamara and F.E. Santo, 2006: Atmospheric conditions associated with the exceptional fire season of 2003 in Portugal. *Int. J. Climatol.*, **26**, 1741-1757.

Tuchman, N.C., R.G. Wetzel, S.T. Rier, K.A. Wahtera and J.A. Teeri, 2002: Elevated atmospheric CO_2 lowers leaf litter nutritional quality for stream ecosystem food webs. *Global Change Biol.*, **8**, 163-170.

Tuchman, N.C., K.A. Wahtera, R.G. Wetzel, N.M. Russo, G.M. Kilbane, L.M. Sasso and J.A. Teeri, 2003a: Nutritional quality of leaf detritus altered by elevated atmospheric CO_2: effects on development of mosquito larvae. *Freshwater Biol.*, **48**, 1432-1439.

Tuchman, N.C., K.A. Wahtera, R.G. Wetzel and J.A. Teeri, 2003b: Elevated atmospheric CO_2 alters leaf litter quality for stream ecosystems: an in situ leaf decomposition study. *Hydrobiologia*, **495**, 203-211.

Turley, C., J. Blackford, S. Widdicombe, D. Lowe and P. Nightingale, 2006: Reviewing the impact of increased atmospheric CO_2 on oceanic pH and the marine ecosystem. *Avoiding Dangerous Climate Change*, H.J. Schellnhuber, W. Cramer, N. Nakićenović, T.M.L. Wigley and G. Yohe, Eds., Cambridge University Press, Cambridge, 65-70.

Tyrrell, T., P.M. Holligan and C.D. Mobley, 1999: Optical impacts of oceanic coccolithophore blooms. *J. Geophys. Res. E*, **104**, 3223-3241, doi:10.1029/1998JC900052.

UNEP, 2002: *Global Environmental Outlook 3: Past, Present and Future Perspectives*. A report from the GEO Project, United Nations Environment Programme, Nairobi, 416 pp.

Usher, M.B., T.V. Callaghan, G. Gilchrist, B. Heal, G.P. Juday, H. Loeng, M.A.K. Muir and P. Prestrud, 2005: Principles of conserving the Arctic's biodiversity. *Arctic Climate Impact Assessment (ACIA): Scientific Report*, C. Symon, L. Arris and B. Heal, Eds., Cambridge University Press, Cambridge, 539-596.

Van Auken, O.W., 2000: Shrub invasions of North American semiarid grasslands. *Annu. Rev. Ecol. Syst.*, **31**, 197-215.

Van Dam, R., H. Gitay, M. Finlayson, N.J. Davidson and B. Orlando, 2002: *Climate Change and Wetlands: Impacts, Adaptation and Mitigation*. Ramsar COP8 DOC. 11, Ramsar Bureau, The Ramsar Convention on Wetlands, Gland, 64 pp.

van den Bergh, J.C.J.M., 2004: Optimal climate policy is a utopia: from quantitative to qualitative cost-benefit analysis. *Ecol. Econ.*, **48**, 385-393.

van den Born, G.J., R. Leemans and M. Schaeffer, 2004: Climate change scenarios for dryland West Africa 1990-2050. *The Impact of Climate Change on Drylands*, A.J. Dietz, R. Ruben and A. Verhagen, Eds., Kluwer Academic, Dordrecht, 43-48.

van der Werf, G.R., J.T. Randerson, G.J. Collatz, L. Giglio, P.S. Kasibhatla, A.F. Arellano, S.C. Olsen and E.S. Kasischke, 2004: Continental-scale partitioning of fire emissions during the 1997 to 2001 El Niño/La Niña period. *Science*, **303**, 73-76.

van Groenigen, K.J., J. Six, B.A. Hungate, M.A. de Graaff, N. van Breemen and C. van Kessel, 2006: Element interactions limit soil carbon storage. *P. Natl. Acad. Sci. USA*, **103**, 6571-6574.

Van Kessel, C., J. Nitschelm, W.R. Horwath, D. Harris, F. Walley, A. Luscher and U. Hartwig, 2000: Carbon-13 input and turn-over in a pasture soil exposed to long-term elevated atmospheric CO_2. *Global Change Biol.*, **6**, 123-135.

van Vliet, A. and R. Leemans, 2006: Rapid species responses to changes in climate requires stringent climate protection target. *Avoiding Dangerous Climate Change*, H.J. Schellnhuber, W. Cramer, N. Nakićenović, T.M.L. Wigley and G. Yohe, Eds., Cambridge University Press, Cambridge, 135-141.

van Wijk, M.T., K.E. Clemmensen, G.R. Shaver, M. Williams, T.V. Callaghan, F.S. Chapin, J.H.C. Cornelissen, L. Gough, S.E. Hobbie, S. Jonasson, J.A. Lee, A. Michelsen, M.C. Press, S.J. Richardson and H. Rueth, 2004: Long-term ecosystem level experiments at Toolik Lake, Alaska, and at Abisko, Northern Sweden: generalizations and differences in ecosystem and plant type responses to global change. *Global Change Biol.*, **10**, 105-123.

Van Wilgen, B.W., N. Govender, H.C. Biggs, D. Ntsala and X.N. Funda, 2004: Response of Savanna fire regimes to changing fire-management policies in a large African National Park. *Conserv. Biol.*, **18**, 1533-1540.

Väre, H., R. Lampinen, C. Humphries and P. Williams, 2003: Taxonomic diversity of vascular plants in the European alpine areas. *Alpine Biodiversity in Europe: A Europe-wide Assessment of Biological Richness and Change*, L. Nagy, G. Grabherr, C.K. Körner and D.B.A. Thompson, Eds., Springer Verlag, Berlin, 133-148.

Vazquez, A. and J.M. Moreno, 2001: Spatial distribution of forest fires in Sierra de Gredos (Central Spain). *Forest Ecol. Manag.*, **147**, 55-65.

Veenendaal, E.M., O. Kolle and J. Lloyd, 2004: Seasonal variation in energy fluxes and carbon dioxide exchange for a broad-leaved semi-arid savanna (Mopane woodland) in Southern Africa. *Global Change Biol.*, **10**, 318-328.

Velichko, A.A., Ed., 2002: *Dynamics of Terrestrial Landscape Components and Inland and Marginal Seas of Northern Eurasia During the Last 130,000 Years*. GEOS Press, Moscow, 296 pp.

Velichko, A.A., N. Catto, A.N. Drenova, V.A. Klimanov, K.V. Kremenetski and V.P. Nechaev, 2002: Climate changes in East Europe and Siberia at the late glacial–Holocene transition. *Quatern. Int.*, **91**, 75-99.

Velichko, A.A., O.K. Borisova, E.M. Zelikson and T.D. Morozova, 2004: Changes in vegetation and soils of the East European Plain to be expected in the 21st century due to the anthropogenic changes in climate. *Geogr. Polonica*, **77**, 35-45.

Velichko, A.A., E.Y. Novenko, V.V. Pisareva, E.M. Zelikson, T. Boettger and F.W. Junge, 2005: Vegetation and climate changes during the Eemian interglacial in Central and Eastern Europe: comparative analysis of pollen data. *Boreas*, **34**, 207-219.

Verburg, P., R.E. Hecky and H. Kling, 2003: Ecological consequences of a century of warming in Lake Tanganyika. *Science*, **301**, 505-507.

Vergani, D.F., Z.B. Stanganelli and D. Bilenca, 2004: Effects of El Niño and La Niña events on the sex ratio of southern elephant seals at King George Island. *Mar. Ecol.-Prog. Ser.*, **268**, 293-300.

Vicente-Serrano, S.M., J.C. Gonzalez-Hidalgo, M. de Luis and J. Raventos, 2004: Drought patterns in the Mediterranean area: the Valencia region (eastern Spain). *Climate Res.*, **26**, 5-15.

Villalba, R. and T.T. Veblen, 1997: Regional patterns of tree population age structures in northern Patagonia: climatic and disturbance influences. *J. Ecol.*, **85**, 113-124.

Villers-Ruiz, L. and I. Trejo-Vazquez, 1998: Impacto del cambio climático en los bosques y áreas naturales protegidas de México (Impact of climatic change in forests and natural protected areas of Mexico). *Interciencia*, **23**, 10-19.

Virtanen, T., S. Neuvonen, A. Nikula, M. Varama and P. Niemelä, 1996: Climate change and the risks of Neodiprion sertifer outbreaks on Scots pine. *Silva Fenn.*, **30**, 169-177.

Visser, M.E., C. Both and M.M. Lambrechts, 2004: Global climate change leads to mistimed avian reproduction. *Adv. Ecol. Res.*, **35**, 89-110.

Vitousek, P.M., H.A. Mooney, J. Lubchenco and J.M. Melillo, 1997: Human domination of Earth's ecosystems. *Science*, **277**, 494-499.

Vlassova, T.K., 2002: Human impacts on the tundra–taiga zone dynamics: the case of the Russian Lesotundra. *Ambio Special Report* , **12**, 30-36.

Volney, W.J.A. and R.A. Fleming, 2000: Climate change and impacts of boreal forest insects. *Agr. Ecosyst. Environ.*, **82**, 283-294.

Von Maltitz, G.P., R.J. Scholes, B. Erasmus and A. Letsoalo, 2006: Adapting conservation strategies to accommodate impacts of climate change in southern Africa. Working paper No. 35, Assessments of Impacts and Adaptations to Climate Change (AIACC), Washington, District of Columbia, 53 pp.

Wagner, W., K. Scipal, C. Pathe, D. Gerten, W. Lucht and B. Rudolf, 2003: Evaluation of the agreement between the first global remotely sensed soil moisture data with model and precipitation data. *J. Geophys. Res. D*, **108**, 4611, doi:10.1029/2003JD003663.

Walker, M.D., C.H. Wahren, R.D. Hollister, G.H.R. Henry, L.E. Ahlquist, J.M. Alatalo, M.S. Bret-Harte, M.P. Calef, T.V. Callaghan, A.B. Carroll, H.E. Epstein, I.S. Jonsdottir, J.A. Klein, B. Magnusson, U. Molau, S.F. Oberbauer, S.P. Rewa, C.H. Robinson, G.R. Shaver, K.N. Suding, C.C. Thompson, A. Tolvanen, O. Totland, P.L. Turner, C.E. Tweedie, P.J. Webber and P.A. Wookey, 2006: Plant community responses to experimental warming across the tundra biome. *P. Natl. Acad. Sci. USA*, **103**, 1342-1346.

Walsh, J.E., 1995: Long-term observations for monitoring of the cryosphere. *Climatic Change*, **31**, 369-394.

Walsh, J.E. and M.S. Timlin, 2003: Northern Hemisphere sea ice simulations by global climate models. *Polar Res.*, **22**, 75-82.

Walter, K.M., S.A. Zimov, J.P. Chanton, D. Verbyla and F.S. Chapin, 2006: Methane bubbling from Siberian thaw lakes as a positive feedback to climate warming. *Nature*, **443**, 71-75.

Walther, G.R., C.A. Burga and P.J. Edwards, Eds., 2001: *"Fingerprints" of Climate Change: Adaptive Behavior and Shifting Species Ranges*. Kluwer, New York, 329 pp.

Walther, G.R., E. Post, P. Convey, A. Menzel, C. Parmesan, T.J.C. Beebee, J.M. Fromentin, O. Hoegh-Guldberg and F. Bairlein, 2002: Ecological responses to recent climate change. *Nature*, **416**, 389-395.

Wang, G.L. and E.A.B. Eltahir, 2002: Impact of CO_2 concentration changes on the biosphere-atmosphere system of West Africa. *Global Change Biol.*, **8**, 1169-1182.

Wang, S.Q., J.Y. Liu, G.R. Yu, Y.Y. Pan, Q.M. Chen, K.R. Li and J.Y. Li, 2004a: Effects of land use change on the storage of soil organic carbon: a case study of the Qianyanzhou Forest Experimental Station in China. *Climatic Change*, **67**, 247-255.

Wang, X.L.L., F.W. Zwiers and V.R. Swail, 2004b: North Atlantic Ocean wave climate change scenarios for the twenty-first century. *J. Climate*, **17**, 2368-2383.

Wanner, H., R. Rickli, E. Salvisberg, C. Schmutz and M. Schüepp, 1997: Global climate change and variability and its influence on Alpine climate: concepts and observations. *Theor. Appl. Climatol.*, **58**, 221-243.

Warner, T.T., 2004: *Desert Meteorology*. Cambridge University Press, Cambridge, 595 pp.

Warren, A., 2005: The policy implications of Sahelian change. *J. Arid Environ.*, **63**, 660-670.

Warren, R., 2006: Impacts of global climate change at different annual mean global temperature increases. *Avoiding Dangerous Climate Change*, H.J. Schellnhuber, W. Cramer, N. Nakićenović, T.M.L. Wigley and G. Yohe, Eds., Cambridge University Press, Cambridge, 93-131.

Watson, R.T., I.R. Noble, B. Bolin, N.H. Ravindranath, D.J. Verardo and D.J. Dokken, Eds., 2000: *Land Use, Land-Use Change, and Forestry: A Special Report of the Intergovernmental Panel on Climate Change*. Cambridge University Press, Cambridge, 377 pp.

Weltzin, J.F., J. Pastor, C. Harth, S.D. Bridgham, K. Updegraff and C.T. Chapin, 2000: Response of bog and fen plant communities to warming and water-table manipulations. *Ecology*, **81**, 3464-3478.

Weltzin, J.F., C. Harth, S.D. Bridgham, J. Pastor and M. Vonderharr, 2001: Production and microtopography of bog bryophytes: response to warming and water-table manipulations. *Oecologia*, **128**, 557-565.

Weltzin, J.F., S.D. Bridgham, J. Pastor, J.Q. Chen and C. Harth, 2003: Potential effects of warming and drying on peatland plant community composition. *Global Change Biol.*, **9**, 141-151.

Westerling, A.L., H.G. Hidalgo, D.R. Cayan and T.W. Swetnam, 2006: Warming and earlier spring increase western US forest wildfire activity. *Science*, **313**, 940-943.

Weyhenmeyer, G.A., 2004: Synchrony in relationships between the North Atlantic Oscillation and water chemistry among Sweden's largest lakes. *Limnol. Oceanogr.*, **49**, 1191-1201.

Weyhenmeyer, G.A., M. Meili and D.M. Livingstone, 2004: Nonlinear temperature response of lake ice breakup. *Geophys. Res. Lett.*, **31**, L07203, doi:10.1029/2004GL019530.

White, A., M.G.R. Cannell and A.D. Friend, 2000a: CO_2 stabilization, climate change and the terrestrial carbon sink. *Global Change Biol.*, **6**, 817-833.

White, T.A., B.D. Campbell, P.D. Kemp and C.L. Hunt, 2000b: Sensitivity of three grassland communities to simulated extreme temperature and rainfall events. *Global Change Biol.*, **6**, 671-684.

Whitehead, H., 1997: Sea surface temperature and the abundance of sperm whale calves off the Galapagos Islands: implications for the effects of global warming. *Report of the International Whaling Commission*, **47**, 941-944.

Whitlock, C., S.L. Shafer and J. Marlon, 2003: The role of climate and vegetation change in shaping past and future fire regimes in the northwestern US and the implications for ecosystem management. *Forest Ecol. Manag.*, **178**, 5-21.

Whittingham, E., J. Campbell and P. Townsley, 2003: *Poverty and Reefs*. DFID-IMM-IOC/UNESCO, Exeter, 260 pp.

Wichmann, M.C., F. Jeltsch, W.R.J. Dean, K.A. Moloney and C. Wissel, 2003: Implication of climate change for the persistence of raptors in arid savanna. *Oikos*, **102**, 186-202.

Wikelski, M. and C. Thom, 2000: Marine iguanas shrink to survive El Niño: changes in bone metabolism enable these adult lizards to reversibly alter their length. *Nature*, **403**, 37-38.

Wilby, R.L., S.P. Charles, E. Zorita, B. Timbal, P. Whetton and L.O. Mearns, 2004: *Guidelines for Use of Climate Scenarios Developed from Statistical Downscaling Methods: IPCC Task Group on Data and Scenario Support for Impact and Climate Analysis (TGICA)*. Intergovernmental Panel on Climate Change (IPCC): Data Distribution Centre (DDC), Norwich, Hamburg and New York, 27 pp.

Wiley, J.W. and J.M. Wunderle, Jr., 1994: The effects of hurricanes on birds, with special reference to Caribbean islands. *Bird Conserv. Int.*, **3**, 319-349.

Wilkinson, C., Ed., 2004: *Status of Coral Reefs of the World: 2004. Vols. 1 and 2*. Global Coral Reef Monitoring Network and Australian Institute of Marine Science, Townsville, 303+557 pp.

Williams, A.A.J., D.J. Karoly and N. Tapper, 2001: The sensitivity of Australian fire danger to climate change. *Climatic Change*, **49**, 171-191.

Williams, M.A., C.W. Rice, A. Omay and C. Owensby, 2004a: Carbon and nitrogen pools in a tallgrass prairie soil under elevated carbon dioxide. *Soil Sci. Soc. Am. J.*, **68**, 148-153.

Williams, P., L. Hannah, S. Andelman, G. Midgley, M. Araujo, G. Hughes, L. Manne, E. Martinez-Meyer and R. Pearson, 2005: Planning for climate change: identifying minimum-dispersal corridors for the Cape proteaceae. *Conserv. Biol.*, **19**, 1063-1074.

Williams, R.J., L.B. Hutley, G.D. Cook, J. Russell-Smith, A. Edwards and X.Y. Chen, 2004b: Assessing the carbon sequestration potential of mesic savannas in the Northern Territory, Australia: approaches, uncertainties and potential impacts of fire. *Funct. Plant Ecol.*, **31**, 415-422.

Williams, S.E., E.E. Bolitho and S. Fox, 2003: Climate change in Australian tropical rainforests: an impending environmental catastrophe. *P. Roy. Soc. Lond. B*, **270**, 1887-1892.

Willig, M.R., D.M. Kaufman and R.D. Stevens, 2003: Latitudinal gradients of biodiversity: pattern, process, scale, and synthesis. *Annu. Rev. Ecol. Evol. Syst.*, **34**, 273-309.

Wilmking, M., G.P. Juday, V.A. Barber and H.S.J. Zald, 2004: Recent climate warming forces contrasting growth responses of white spruce at treeline in Alaska through temperature thresholds. *Global Change Biol.*, **10**, 1724-1736.

Winder, M. and D.E. Schindler, 2004: Climatic effects on the phenology of lake processes. *Global Change Biol.*, **10**, 1844-1856.

Wingfield, J.C., G. Ramos-Fernandez, A.N.D. La Mora and H. Drummond, 1999: The effects of an "El Niño" Southern Oscillation event on reproduction in male and female blue-footed boobies, *Sula nebouxii*. *Gen. Comp. Endocrl.*, **114**, 163-172.

Winnett, S.M., 1998: Potential effects of climate change on US forests: a review. *Climate Res.*, **11**, 39-49.

Winslow, J.C., J. Hunt, E. Raymond and S.C. Piper, 2003: The influence of seasonal water availability on global C_3 versus C_4 grassland biomass and its implications for climate change research. *Ecol. Model.*, **163**, 153-173.

Woodward, F.I. and M.R. Lomas, 2001: Integrating fluxes from heterogeneous vegetation. *Global Ecol. Biogeogr.*, **10**, 595-601.

Woodward, F.I. and M.R. Lomas, 2004a: Simulating vegetation processes along the Kalahari transect. *Global Change Biol.*, **10**, 383-392.

Woodward, F.I. and M.R. Lomas, 2004b: Vegetation dynamics: simulating responses to climatic change. *Biol. Rev.*, **79**, 643-670.

Woodwell, G.M., 2004: Mountains: top down. *Ambio Special Report*, **13**, 35-38.

Wooldridge, S., T. Done, R. Berkelmans, R. Jones and P. Marshall, 2005: Precursors for resilience in coral communities in a warming climate: a belief network approach. *Mar. Ecol.–Prog. Ser.*, **295**, 157-169.

Worm, B., E.B. Barbier, N. Beaumont, J.E. Duffy, C. Folke, B.S. Halpern, J.B.C. Jackson, H.K. Lotze, F. Micheli, S.R. Palumbi, E. Sala, K.A. Selkoe, J.J. Stachowicz and R. Watson, 2006: Impacts of biodiversity loss on ocean ecosystem services. *Science*, **314**, 787-790.

Wright, S.J., O. Calderón, A. Hernandéz and S. Paton, 2004: Are lianas increasing in importance in tropical forests? A 17-year record from Panama. *Ecology*, **85**, 484-489.

Wullschleger, S.D. and R.J. Norby, 2001: Sap velocity and canopy transpiration in a sweetgum stand exposed to free-air CO_2 enrichment (FACE). *New Phytol.*, **150**, 489-498.

Xenopoulos, M.A., D.M. Lodge, J. Alcamo, M. Marker, K. Schulze and D.P. Van Vuuren, 2005: Scenarios of freshwater fish extinctions from climate change and water withdrawal. *Global Change Biol.*, **11**, 1557-1564.

Xu, D.Y. and H. Yan, 2001: A study of the impacts of climate change on the geographic distribution of *Pinus koraiensis* in China. *Environ. Int.*, **27**, 201-205.

Yamano, H. and M. Tamura, 2004: Detection limits of coral reef bleaching by satellite remote sensing: simulation and data analysis. *Remote Sens. Environ.*, **90**, 86-103.

Yoccoz, N.G. and R.A. Ims, 1999: Demography of small mammals in cold regions: the importance of environmental variability. *Animal Responses to Global Change in the North*, A. Hofgaard, J.P. Ball, K. Danell and T.V. Callaghan, Eds., Ecological Bulletin 47, Blackwell, Oxford, 137-144.

Zalakevicius, M. and S. Svazas, 2005: Global climate change and its impact on wetlands and waterbird populations. *Acta Zool. Lituanica*, **15**, 215-217.

Zebisch, M., F. Wechsung and H. Kenneweg, 2004: Landscape response functions for biodiversity: assessing the impact of land-use changes at the country level. *Landscape Urban Plan.*, **67**, 157-172.

Zeebe, R.E. and D. Wolf-Gladrow, 2001: *CO_2 in Seawater: Equilibrium, Kinetics, Isotopes*. Elsevier, Amsterdam, 346 pp.

Zepp, R.G., T.V. Callaghan and D.J. Erickson, 2003: Interactive effects of ozone depletion and climate change on biogeochemical cycles. *Photochem. Photobiol. Sci.*, **2**, 51-61.

Zha, Y., J. Gao and Y. Zhang, 2005: Grassland productivity in an alpine environment in response to climate change. *Area*, **37**, 332-340.

Zhang, W., K.M. Parker, Y. Luo, S. Wan, L.L. Wallace and S. Hu, 2005: Soil microbial responses to experimental warming and clipping in a tallgrass prairie. *Global Change Biol.*, **11**, 266-277.

Zheng, X., Z. Zhou, Y. Wang, J. Zhu, Y. Wang, J. Yue, Y. Shi, K. Kobayashi, K. Inubushi, Y. Huang, S.H. Han, Z.J. Xu, B.H. Xie, K. Butterbach-Bahl and L.X. Yang, 2006: Nitrogen-regulated effects of free-air CO_2 enrichment on methane emissions from paddy rice fields. *Global Change Biol.*, **12**, 1717-1732.

Zierl, B. and H. Bugmann, 2005: Global change impacts on hydrological processes in Alpine catchments. *Water Resour. Res.*, **41**, 1-13.

Zimov, S.A., E.A.G. Schuur and F.S. Chapin, 2006: Permafrost and the global carbon budget. *Science*, **312**, 1612-1613.

Ziska, L.H., T.B. Moya, R. Wassmann, O.S. Namuco, R.S. Lantin, J.B. Aduna, E. Abao, K.F. Bronson, H.U. Neue and D. Olszyk, 1998: Long-term growth at elevated carbon dioxide stimulates methane emission in tropical paddy rice. *Global Change Biol.*, **4**, 657-665.

Zöckler, C. and I. Lysenko, 2000: *Waterbirds on the Edge: First Circumpolar Assessment of Climate Change Impact on Arctic Breeding Water Birds*. WCMC Biodiversity Series No. 11, UNEP and World Conservation Monitoring Centre (WCMC), Cambridge, 27 pp.

5

Food, fibre and forest products

Coordinating Lead Authors:
William Easterling (USA), Pramod Aggarwal (India)

Lead Authors:
Punsalmaa Batima (Mongolia), Keith Brander (ICES/Denmark/UK), Lin Erda (China), Mark Howden (Australia), Andrei Kirilenko (Russia), John Morton (UK), Jean-François Soussana (France), Josef Schmidhuber (FAO/Italy), Francesco Tubiello (USA/IIASA/Italy)

Contributing Authors:
John Antle (USA), Walter Baethgen (Uruguay), Chris Barlow (Lao PDR), Netra Chhetri (Nepal), Sophie des Clers (UK), Patricia Craig (USA), Judith Cranage (USA), Wulf Killmann (FAO/Italy), Terry Mader (USA), Susan Mann (USA), Karen O'Brien (Norway), Christopher Pfeiffer (USA), Roger Sedjo (USA)

Review Editors:
John Sweeney (Ireland), Lucka K. Kajfež-Bogataj (Slovenia)

This chapter should be cited as:
Easterling, W.E., P.K. Aggarwal, P. Batima, K.M. Brander, L. Erda, S.M. Howden, A. Kirilenko, J. Morton, J.-F. Soussana, J. Schmidhuber and F.N. Tubiello, 2007: Food, fibre and forest products. *Climate Change 2007: Impacts, Adaptation and Vulnerability. Contribution of Working Group II to the Fourth Assessment Report of the Intergovernmental Panel on Climate Change*, M.L. Parry, O.F. Canziani, J.P. Palutikof, P.J. van der Linden and C.E. Hanson, Eds., Cambridge University Press, Cambridge, UK, 273-313.

Table of Contents

Executive summary

In mid- to high-latitude regions, moderate warming benefits crop and pasture yields, but even slight warming decreases yields in seasonally dry and low-latitude regions (medium confidence).

Modelling results for a range of sites find that, in mid- to high-latitude regions, moderate to medium local increases in temperature (1-3°C), along with associated carbon dioxide (CO_2) increase and rainfall changes, can have small beneficial impacts on crop yields. In low-latitude regions, even moderate temperature increases (1-2°C) are likely to have negative yield impacts for major cereals. Further warming has increasingly negative impacts in all regions (medium to low confidence) [Figure 5.2]. These results, on the whole, project the potential for global food production to increase with increases in local average temperature over a range of 1 to 3°C, but above this range to decrease [5.4, 5.6].

The marginal increase in the number of people at risk of hunger due to climate change must be viewed within the overall large reductions due to socio-economic development (medium confidence).

Compared to 820 million undernourished today, the IPCC Special Report on Emissions Scenarios (SRES) scenarios of socio-economic development without climate change project a reduction to 100-230 million (range is over A1, B1, B2 SRES scenarios) undernourished by 2080 (or 770 million under the A2 SRES scenario) (medium confidence). Scenarios with climate change project 100-380 million (range includes with and without CO_2 effects and A1, B1, B2 SRES scenarios) undernourished by 2080 (740-1,300 million under A2) (low to medium confidence). Climate and socio-economic changes combine to alter the regional distribution of hunger, with large negative effects on sub-Saharan Africa (low to medium confidence) [Table 5.6].

Projected changes in the frequency and severity of extreme climate events have significant consequences for food and forestry production, and food insecurity, in addition to impacts of projected mean climate (high confidence).

Recent studies indicate that climate change scenarios that include increased frequency of heat stress, droughts and flooding events reduce crop yields and livestock productivity beyond the impacts due to changes in mean variables alone, creating the possibility for surprises [5.4.1, 5.4.2]. Climate variability and change also modify the risks of fires, and pest and pathogen outbreaks, with negative consequences for food, fibre and forestry (FFF) (high confidence) [5.4.1 to 5.4.5].

Simulations suggest rising relative benefits of adaptation with low to moderate warming (medium confidence), although adaptation stresses water and environmental resources as warming increases (low confidence).

There are multiple adaptation options that imply different costs, ranging from changing practices in place to changing locations of FFF activities [5.5.1]. Adaptation effectiveness varies from only marginally reducing negative impacts to changing a negative impact into a positive one. On average, in cereal cropping systems worldwide, adaptations such as changing varieties and planting times enable avoidance of a 10-15% reduction in yield corresponding to 1-2°C local temperature increase. The benefit from adapting tends to increase with the degree of climate change up to a point [Figure 5.2]. Adaptive capacity in low latitudes is exceeded at 3°C local temperature increase [Figure 5.2, Section 5.5.1]. Changes in policies and institutions will be needed to facilitate adaptation to climate change. Pressure to cultivate marginal land or to adopt unsustainable cultivation practices as yields drop may increase land degradation and resource use, and endanger biodiversity of both wild and domestic species [5.4.7]. Adaptation measures must be integrated with development strategies and programmes, country programmes and Poverty Reduction Strategies [5.7].

Smallholder and subsistence farmers, pastoralists and artisanal fisherfolk will suffer complex, localised impacts of climate change (high confidence).

These groups, whose adaptive capacity is constrained, will experience the negative effects on yields of low-latitude crops, combined with a high vulnerability to extreme events. In the longer term, there will be additional negative impacts of other climate-related processes such as snow-pack decrease (especially in the Indo-Gangetic Plain), sea level rise, and spread in prevalence of human diseases affecting agricultural labour supply. [5.4.7]

Globally, commercial forestry productivity rises modestly with climate change in the short and medium term, with large regional variability around the global trend (medium confidence).

The change in the output of global forest products ranges from a modest increase to a slight decrease, although regional and local changes will be large [5.4.5.2]. Production increase will shift from low-latitude regions in the short-term, to high-latitude regions in the long-term [5.4.5].

Local extinctions of particular fish species are expected at edges of ranges (high confidence).

Regional changes in the distribution and productivity of particular fish species are expected due to continued warming and local extinctions will occur at the edges of ranges, particularly in freshwater and diadromous species (e.g., salmon, sturgeon). In some cases ranges and productivity will increase [5.4.6]. Emerging evidence suggests that meridional overturning circulation is slowing, with serious potential consequences for fisheries (medium confidence) [5.4.6].

Food and forestry trade is projected to increase in response to climate change, with increased dependence on food imports for most developing countries (medium to low confidence).

While the purchasing power for food is reinforced in the period to 2050 by declining real prices, it would be adversely affected by higher real prices for food from 2050 to 2080. [5.6.1, 5.6.2].

Exports of temperate zone food products to tropical countries will rise [5.6.2], while the reverse may take place in forestry in the short-term. [5.4.5]

Experimental research on crop response to elevated CO_2 confirms Third Assessment Report (TAR) findings (medium to high confidence). New Free-Air Carbon Dioxide Enrichment (FACE) results suggest lower responses for forests (medium confidence).

Recent re-analyses of FACE studies indicate that, at 550 ppm atmospheric CO_2 concentrations, yields increase under unstressed conditions by 10-25% for C_3 crops, and by 0-10% for C_4 crops (medium confidence), consistent with previous TAR estimates (medium confidence). Crop model simulations under elevated CO_2 are consistent with these ranges (high confidence) [5.4.1]. Recent FACE results suggest no significant response for mature forest stands, and confirm enhanced growth for young tree stands [5.4.1.1]. Ozone exposure limits CO_2 response in both crops and forests.

5.1 Introduction: importance, scope and uncertainty, Third Assessment Report summary, and methods

5.1.1 Importance of agriculture, forestry and fisheries

At present, 40% of the Earth's land surface is managed for cropland and pasture (Foley et al., 2005). Natural forests cover another 30% (3.9 billion ha) of the land surface with just 5% of the natural forest area (FAO, 2000) providing 35% of global roundwood. In developing countries, nearly 70% of people live in rural areas where agriculture is the largest supporter of livelihoods. Growth in agricultural incomes in developing countries fuels the demand for non-basic goods and services fundamental to human development. The United Nations Food and Agriculture Organization (FAO) estimates that the livelihoods of roughly 450 million of the world's poorest people are entirely dependent on managed ecosystem services. Fish provide more than 2.6 billion people with at least 20% of their average per capita animal protein intake, but three-quarters of global fisheries are currently fully exploited, overexploited or depleted (FAO, 2004c).

5.1.2 Scope of the chapter and treatment of uncertainty

The scope of this chapter, with a focus on food crops, pastures and livestock, industrial crops and biofuels, forestry (commercial forests), aquaculture and fisheries, and small-holder and subsistence agriculturalists and artisanal fishers, is to:

- examine current climate sensitivities/vulnerabilities;
- consider future trends in climate, global and regional food security, forestry and fisheries production;
- review key future impacts of climate change in food crops, pasture and livestock production, industrial crops and biofuels, forestry, fisheries, and small-holder and subsistence agriculture;
- assess the effectiveness of adaptation in offsetting damages and identify adaptation options, including planned adaptation to climate change;
- examine the social and economic costs of climate change in those sectors; and,
- explore the implications of responding to climate change for sustainable development.

We strive for consistent treatment of uncertainty in this chapter. Traceable accounts of final judgements of uncertainty in the findings and conclusions are, where possible, maintained. These accounts explicitly state sources of uncertainty in the methods used by the studies that comprise the assessment. At the end of the chapter, we summarise those findings and conclusions and provide a final judgement of their uncertainties.

5.1.3 Important findings of the Third Assessment Report

The key findings of the 2001 Third Assessment Report (TAR; IPCC, 2001) with respect to food, fibre, forestry and fisheries are an important benchmark for this chapter. In reduced form, they are:

Food crops

- CO_2 effects increase with temperature, but decrease once optimal temperatures are exceeded for a range of processes, especially plant water use. The CO_2 effect may be relatively greater (compared to that for irrigated crops) for crops under moisture stress.
- Modelling studies suggest crop yield losses with minimal warming in the tropics.
- Mid- to high-latitude crops benefit from a small amount of warming (about +2°C) but plant health declines with additional warming.
- Countries with greater wealth and natural resource endowments adapt more efficiently than those with less.

Forestry

- Free-air CO_2 enrichment (FACE) experiments suggest that trees rapidly become acclimated to increased CO_2 levels.
- The largest impacts of climate change are likely to occur earliest in boreal forests.
- Contrary to the findings of the Second Assessment Report (SAR), climate change will increase global timber supply and enhance existing market trends of rising market share in developing countries.

Aquaculture and fisheries

- Global warming will confound the impact of natural variation on fishing activity and complicate management.
- The sustainability of the fishing industries of many countries will depend on increasing flexibility in bilateral and multilateral fishing agreements, coupled with international stock assessments and management plans.

- Increases in seawater temperature have been associated with increases in diseases and algal blooms in the aquaculture industry.

5.1.4 Methods

Research on the consequences of climate change on agriculture, forestry and fisheries is addressing deepening levels of system complexity that require a new suite of methodologies to cope with the added uncertainty that accompanies the addition of new, often non-linear, process knowledge. The added realism of experiments (e.g., FACE) and the translation of experimental results to process crop-simulation models are adding confidence to model estimates. Integrated physiological and economic models (e.g., Fischer et al., 2005a) allow holistic simulation of climate change effects on agricultural productivity, input and output prices, and risk of hunger in specific regions, although these simulations rely on a small set of component models. The application of meta-analysis to agriculture, forestry and fisheries in order to identify trends and consistent findings across large numbers of studies has revealed important new information since the TAR, especially on the direct effects of atmospheric CO_2 on crop and forest productivity (e.g., Ainsworth and Long, 2005) and fisheries (Allison et al., 2005). The complexity of processes that determine adaptive capacity dictates an increasing regional focus to studies in order best to understand and predict adaptive processes (Kates and Wilbanks, 2003): hence the rise in numbers of regional-scale studies. This increases the need for more robust methods to scale local findings to larger regions, such as the use of multi-level modelling (Easterling and Polsky, 2004). Further complexity is contributed by the growing number of scenarios of future climate and society that drive inputs to the models (Nakićenović and Swart, 2000).

5.2 Current sensitivity, vulnerability and adaptive capacity to climate

5.2.1 Current sensitivity

The inter-annual, monthly and daily distribution of climate variables (e.g., temperature, radiation, precipitation, water vapour pressure in the air and wind speed) affects a number of physical, chemical and biological processes that drive the productivity of agricultural, forestry and fisheries systems. The latitudinal distribution of crop, pasture and forest species is a function of the current climatic and atmospheric conditions, as well as of photoperiod (e.g., Leff et al., 2004). Total seasonal precipitation as well as its pattern of variability (Olesen and Bindi, 2002) are both of major importance for agricultural, pastoral and forestry systems.

Crops exhibit threshold responses to their climatic environment, which affect their growth, development and yield (Porter and Semenov, 2005). Yield-damaging climate thresholds that span periods of just a few days for cereals and fruit trees include absolute temperature levels linked to particular developmental stages that condition the formation of reproductive organs, such as seeds and fruits (Wheeler et al., 2000; Wollenweber et al., 2003). This means that yield damage estimates from coupled crop–climate models need to have a temporal resolution of no more than a few days and to include detailed phenology (Porter and Semenov, 2005). Short-term natural extremes, such as storms and floods, interannual and decadal climate varizations, as well as large-scale circulation changes, such as the El Niño Southern Oscillation (ENSO), all have important effects on crop, pasture and forest production (Tubiello, 2005). For example, El Niño-like conditions increase the probability of farm incomes falling below their long-term median by 75% across most of Australia's cropping regions, with impacts on gross domestic product (GDP) ranging from 0.75 to 1.6% (O'Meagher, 2005). Recently the winter North Atlantic Oscillation (NAO) has been shown to correlate with the following summer's climate, leading to sunnier and drier weather during wheat grain growth and ripening in the UK and, hence, to better wheat grain quality (Atkinson et al., 2005); but these same conditions reduced summer growth of grasslands through increased drought effects (Kettlewell et al., 2006).

The recent heatwave in Europe (see Box 5.1) and drought in Africa (see Table 5.1) illustrate the potentially large effects of local and/or regional climate variability on crops and livestock.

5.2.2 Sensitivity to multiple stresses

Multiple stresses, such as limited availability of water resources (see Chapter 3), loss of biodiversity (see Chapter 4), and air pollution (see Box 5.2), are increasing sensitivity to climate change and reducing resilience in the agricultural sector

Box 5.1. European heatwave impact on the agricultural sector

Europe experienced a particularly extreme climate event during the summer of 2003, with temperatures up to 6°C above long-term means, and precipitation deficits up to 300 mm (see Trenberth et al., 2007). A record drop in crop yield of 36% occurred in Italy for maize grown in the Po valley, where extremely high temperatures prevailed (Ciais et al., 2005). In France, compared to 2002, the maize grain crop was reduced by 30% and fruit harvests declined by 25%. Winter crops (wheat) had nearly achieved maturity by the time of the heatwave and therefore suffered less yield reduction (21% decline in France) than summer crops (e.g., maize, fruit trees and vines) undergoing maximum foliar development (Ciais et al., 2005). Forage production was reduced on average by 30% in France and hay and silage stocks for winter were partly used during the summer (COPA COGECA, 2003b). Wine production in Europe was the lowest in 10 years (COPA COGECA, 2003a). The (uninsured) economic losses for the agriculture sector in the European Union were estimated at €13 billion, with largest losses in France (€4 billion) (Sénat, 2004).

Table 5.1. *Quantified impacts of selected African droughts on livestock, 1981 to 1999.*

Date	Location	Mortality and species	Source
1981-84	Botswana	20% of national herd	FAO, 1984, cited in Toulmin, 1986
1982-84	Niger	62% of national cattle herd	Toulmin, 1986
1983-84	Ethiopia (Borana Plateau)	45-90% of calves, 45% of cows, 22% of mature males	Coppock, 1994
1991	Northern Kenya	28% of cattle 18% of sheep and goats	Surtech, 1993, cited in Barton and Morton, 2001
1991-93	Ethiopia (Borana)	42% of cattle	Desta and Coppock, 2002
1993	Namibia	22% of cattle 41% of goats and sheep	Devereux and Tapscott, 1995
1995-97	Greater Horn of Africa (average of nine pastoral areas)	20% of cattle 20% of sheep and goats	Ndikumana et al., 2000
1995-97	Southern Ethiopia	46% of cattle 41% of sheep and goats	Ndikumana et al., 2000
1998-99	Ethiopia (Borana)	62% of cattle	Shibru, 2001, cited in Desta and Coppock, 2002

Box 5.2. Air pollutants and ultraviolet-B radiation (UV-B)

Ozone has significant adverse effects on crop yields, pasture and forest growth, and species composition (Loya et al., 2003; Ashmore, 2005; Vandermeiren, 2005; Volk et al., 2006). While emissions of ozone precursors, chiefly nitrous oxide (NO_x) compounds, may be decreasing in North America and Europe due to pollution-control measures, they are increasing in other regions of the world, especially Asia. Additionally, as global ozone exposures increase over this century, direct and indirect interactions with climate change and elevated CO_2 will further modify plant dynamics (Booker et al., 2005; Fiscus et al., 2005). Although several studies confirm TAR findings that elevated CO_2 may ameliorate otherwise negative impacts from ozone (Kaakinen et al., 2004), the essence of the matter should be viewed the other way around: increasing ozone concentrations in future decades, with or without CO_2 increases, with or without climate change, will negatively impact plant production, possibly increasing exposure to pest damage (Ollinger et al., 2002; Karnosky, 2003). Current risk-assessment tools do not sufficiently consider these key interactions. Improved modelling approaches that link the effects of ozone, climate change, and nutrient and water availability on individual plants, species interactions and ecosystem function are needed (Ashmore, 2005): some efforts are under way (Felzer et al., 2004). Finally, impacts of UV-B exposure on plants were previously reviewed by the TAR, which showed contrasting results on the interactions of UV-B exposure with elevated CO_2. Recent studies do not narrow the uncertainty: some findings suggest amelioration of negative UV-B effects by elevated CO_2 (Qaderi and Reid, 2005); others show no effect (Zhao et al., 2003).

(FAO, 2003a). Natural land resources are being degraded through soil erosion, salinisation of irrigated areas, dryland degradation from overgrazing, over-extraction of ground water, growing susceptibility to disease and build-up of pest resistance favoured by the spread of monocultures and the use of pesticides, and loss of biodiversity and erosion of the genetic resource base when modern varieties displace traditional ones (FAO, 2003b). Small-holder agriculturalists are especially vulnerable to a range of social and environmental stressors (see Table 5.2). The total effect of these processes on agricultural productivity is not clear. Additionally, multiple stresses, such as forest fires and insect outbreaks, increase overall sensitivity (see Section 5.4.5). In fisheries, overexploitation of stocks (see Section 5.4.6), loss of biodiversity, water pollution and changes in water resources (see Box 5.3) also increase the current sensitivity to climate.

5.2.3 Current vulnerability and adaptive capacity in perspective

Current vulnerability to climate variability, including extreme events, is both hazard- and context-dependent (Brooks et al., 2005). For agriculture, forestry and fisheries systems, vulnerability depends on exposure and sensitivity to climate conditions (as discussed above), and on the capacity to cope with changing conditions. A comparison of conditions on both sides of the USA–Mexico border reveals how social, political, economic and historical factors contribute to differential vulnerability among farmers and ranchers living within the same biophysical regime (Vasquez-Leon et al., 2003). Institutional and economic reforms linked to globalisation processes (e.g., removal of subsidies, increased import competition) reduce the capacity of some farmers to respond to climate variability (O'Brien et al., 2004). Efforts to reduce vulnerability and facilitate adaptation to climate change are influenced both positively and negatively by changes associated with globalisation (Eakin and Lemos, 2006).

Table 5.2. *Multiple stressors of small-holder agriculture.*

Stressors:	Source:
Population increase driving fragmentation of landholding	Various
Environmental degradation stemming variously from population, poverty, ill-defined property rights	Grimble et al., 2002
Regionalised and globalised markets, and regulatory regimes, increasingly concerned with issues of food quality and food safety	Reardon et al., 2003
Market failures interrupt input supply following withdrawal of government intervention	Kherallah et al., 2002
Continued protectionist agricultural policies in developed countries, and continued declines and unpredictability in the world prices of many major agricultural commodities of developing countries	Lipton, 2004, Various
Human immunodeficiency virus (HIV) and/or acquired immunodeficiency syndrome (AIDS) pandemic, particularly in Southern Africa, attacking agriculture through mass deaths of prime-age adults, which diverts labour resources to caring, erodes household assets, disrupts intergenerational transmission of agricultural knowledge, and reduces the capacity of agricultural service providers	Barnett and Whiteside, 2002
For pastoralists, encroachment on grazing lands and a failure to maintain traditional natural resource management	Blench, 2001
State fragility and armed conflict in some regions	Various

Adaptive capacity with respect to current climate is dynamic, and influenced by changes in wealth, human capital, information and technology, material resources and infrastructure, and institutions and entitlements (see Chapter 17) (Yohe and Tol, 2001; Eakin and Lemos, 2006). The production and dissemination of seasonal climate forecasts has improved the ability of many resource managers to anticipate and plan for climate variability, particularly in relation to ENSO, but with some limitations (Harrison, 2005). However, problems related to infectious disease, conflicts and other societal factors may decrease the capacity to respond to variability and change at the local level, thereby increasing current vulnerability. Policies and responses made at national and international levels also influence local adaptations (Salinger et al., 2005). National agricultural policies are often developed on the basis of local risks, needs and capacities, as well as international markets, tariffs, subsidies and trade agreements (Burton and Lim, 2005).

Box 5.3. Climate change and the fisheries of the lower Mekong – an example of multiple stresses on a megadelta fisheries system due to human activity

Fisheries are central to the lives of the people, particularly the rural poor, who live in the lower Mekong countries. Two-thirds of the basin's 60 million people are in some way active in fisheries, which represent about 10% of the GDP of Cambodia and Lao People's Democratic Republic (PDR). There are approximately 1,000 species of fish commonly found in the river, with many more marine vagrants, making it one of the most prolific and diverse faunas in the world (MRC, 2003). Recent estimates of the annual catch from capture fisheries alone exceed 2.5 Mtonnes (Hortle and Bush, 2003), with the delta contributing over 30% of this.

Direct effects of climate will occur due to changing patterns of precipitation, snow melt and rising sea level, which will affect hydrology and water quality. Indirect effects will result from changing vegetation patterns that may alter the food chain and increase soil erosion. It is likely that human impacts on the fisheries (caused by population growth, flood mitigation, increased water abstractions, changes in land use and over-fishing) will be greater than the effects of climate, but the pressures are strongly interrelated.

An analysis of the impact of climate change scenarios on the flow of the Mekong (Hoanh et al., 2004) estimated increased maximum monthly flows of 35 to 41% in the basin and 16 to 19% in the delta (lower value is for years 2010 to 2138 and higher value for years 2070 to 2099, compared with 1961 to 1990 levels). Minimum monthly flows were estimated to decrease by 17 to 24% in the basin and 26 to 29% in the delta. Increased flooding would positively affect fisheries yields, but a reduction in dry season habitat may reduce recruitment of some species. However, planned water-management interventions, primarily dams, are expected to have the opposite effects on hydrology, namely marginally decreasing wet season flows and considerably increasing dry season flows (World Bank, 2004).

Models indicate that even a modest sea level rise of 20 cm would cause contour lines of water levels in the Mekong delta to shift 25 km towards the sea during the flood season and salt water to move further upstream (although confined within canals) during the dry season (Wassmann et al., 2004). Inland movement of salt water would significantly alter the species composition of fisheries, but may not be detrimental for overall fisheries yields.

Sub-Saharan Africa is one example of an area of the world that is currently highly vulnerable to food insecurity (Vogel, 2005). Drought conditions, flooding and pest outbreaks are some of the current stressors on food security that may be influenced by future climate change. Current response options and overall development initiatives related to agriculture, fisheries and forestry may be constrained by health status, lack of information and ineffective institutional structures, with potentially negative consequences for future adaptations to periods of heightened climate stress (see Chapter 9) (Reid and Vogel, 2006).

5.3 Assumptions about future trends in climate, food, forestry and fisheries

Declining global population growth (UN, 2004), rapidly rising urbanisation, shrinking shares of agriculture in the overall formation of incomes and fewer people dependent on agriculture are among the key factors likely to shape the social setting in which climate change is likely to evolve. These factors will determine how climate change affects agriculture, how rural populations can cope with changing climate conditions, and how these will affect food security. Any assessment of climate change impacts on agro-ecological conditions of agriculture must be undertaken against this background of changing socio-economic setting (Bruinsma, 2003).

5.3.1 Climate

Water balance and weather extremes are key to many agricultural and forestry impacts. Decreases in precipitation are predicted by more than 90% of climate model simulations by the end of the 21st century for the northern and southern sub-tropics (IPCC, 2007a). Increases in precipitation extremes are also very likely in the major agricultural production areas in Southern and Eastern Asia, in East Australia and in Northern Europe (Christensen et al., 2007). It should be noted that climate change impact models for food, feed and fibre do not yet include these recent findings on projected patterns of change in precipitation.

The current climate, soil and terrain suitability for a range of rain-fed crops and pasture types has been estimated by Fischer et al. (2002b) (see Figure 5.1a). Globally, some 3.6 billion ha (about 27% of the Earth's land surface) are too dry for rain-fed agriculture. Considering water availability, only about 1.8% of these dry zones are suitable for producing cereal crops under irrigation (Fischer et al., 2002b).

Changes in annual mean runoff are indicative of the mean water availability for vegetation. Projected changes between now and 2100 (see Chapter 3) show some consistent runoff patterns: increases in high latitudes and the wet tropics, and decreases in mid-latitudes and some parts of the dry tropics (Figure 5.1b). Declines in water availability are therefore projected to affect some of the areas currently suitable for rain-fed crops (e.g., in the Mediterranean basin, Central America and sub-tropical regions of Africa and Australia). Extreme increases in precipitation

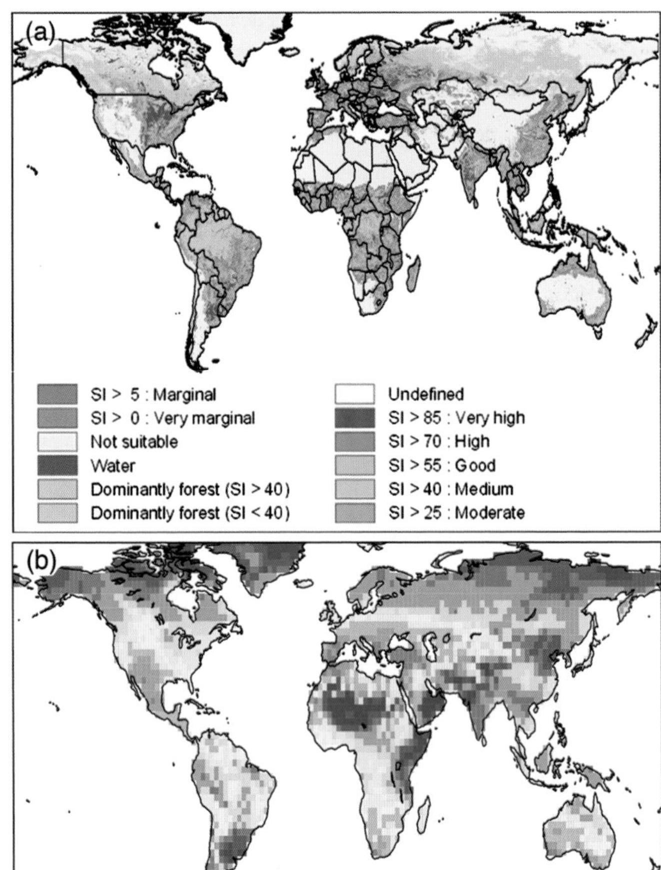

Figure 5.1. *(a) Current suitability for rain-fed crops (excluding forest ecosystems) (after Fischer et al., 2002b). SI = suitability index; (b) Ensemble mean percentage change of annual mean runoff between present (1981 to 2000) and 2100 (Nohara et al., 2006).*

(Christensen et al., 2007) also are very likely in major agricultural production areas (e.g., in Southern and Eastern Asia and in Northern Europe).

5.3.2 Balancing future global supply and demand in agriculture, forestry and fisheries

5.3.2.1 Agriculture

Slower population growth and an increasing proportion of better-fed people who require fewer additional calories are projected to lead to deceleration of global food demand. This slow-down in demand takes the present shift in global food consumption patterns from crop-based to livestock-based diets into account (Schmidhuber and Shetty, 2005). In parallel with the slow-down in demand, FAO (FAO, 2005a) expects growth in world agricultural production to decline from 2.2%/yr during the past 30 years to 1.6%/yr in 2000 to 2015, 1.3%/yr in 2015 to 2030 and 0.8%/yr in 2030 to 2050. This still implies a 55% increase in global crop production by 2030 and an 80% increase to 2050 (compared with 1999 to 2001). To facilitate this growth in output, another 185 million ha of rain-fed crop land (+19%) and another 60 million ha of irrigated land (+30%) will have to be brought into production. Essentially, the entire agricultural land expansion will take place in developing countries with most of it occurring in sub-Saharan Africa and Latin America, which could result in direct trade-offs with ecosystem services (Cassman et al., 2003). In addition to expanded land use, yields are expected to rise. Cereal yields in developing countries are projected to increase from 2.7 tonnes/ha currently to 3.8 tonnes/ha in 2050 (FAO, 2005a).

These improvements in the global supply-demand balance will be accompanied by a decline in the number of undernourished people from more than 800 million at present to about 300 million, or 4% of the population in developing countries, by 2050 (see Table 5.6) (FAO, 2005a). Notwithstanding these overall improvements, important food-security problems remain to be addressed at the local and national levels. Areas in sub-Saharan Africa, Asia and Latin America, with high rates of population growth and natural resource degradation, are likely to continue to have high rates of poverty and food insecurity (Alexandratos, 2005). Cassman et al. (2003) emphasise that climate change will add to the dual challenge of meeting food (cereal) demand while at the same time protecting natural resources and improving environmental quality in these regions.

5.3.2.2 Forestry

A number of long-term studies on supply and demand of forestry products have been conducted in recent years (e.g., Sedjo and Lyon, 1990, 1996; FAO, 1998; Hagler, 1998; Sohngen et al., 1999, 2001). These studies project a shift in harvest from natural forests to plantations. For example, Hagler (1998) suggested the industrial wood harvest produced on plantations will increase from 20% of the total harvest in 2000 to more than 40% in 2030. Other estimates (FAO, 2004a) state that plantations produced about 34% of the total in 2001 and predict this portion may increase to 44% by 2020 (Carle et al., 2002) and 75% by 2050 (Sohngen et al., 2001). There will also be a global shift in the industrial wood supply from temperate to tropical zones and from the Northern to Southern Hemisphere. Trade in forest products will increase to balance the regional imbalances in demand and supply (Hagler, 1998).

Forecasts of industrial wood demand have tended to be consistently higher than actual demand (Sedjo and Lyon, 1990). Actual increases in demand have been relatively small (compare current demand of 1.6 billion m^3 with 1.5 billion m^3 in the early 1980s (FAO, 1982, 1986, 1988, 2005b)). The recent projections of the FAO (1997), Häggblom (2004), Sedjo and Lyon (1996) and Sohngen et al. (2001) forecast similar modest increases in demand to 1.8-1.9 billion m^3 by 2010 to 2015, in contrast to earlier higher predictions of 2.1 billion m^3 by 2015 and 2.7 billion m^3 by 2030 (Hagler, 1998). Similarly, an FAO (2001) study suggests that global fuelwood use has peaked at 1.9 billion m^3 and is stable or declining, but the use of charcoal continues to rise (e.g., Arnold et al., 2003). However, fuelwood use could dramatically increase in the face of rising energy prices, particularly if incentives are created to shift away from fossil fuels and towards biofuels. Many other products and services depend on forest resources; however, there are no satisfactory estimates of the future global demand for these products and services.

Finally, although climate change will impact the availability of forest resources, the anthropogenic impact, particularly land-use change and deforestation in tropical zones, is likely to be extremely important (Zhao et al., 2005). In the Amazon basin, deforestation and increased forest fragmentation may impact water availability, triggering more severe droughts. Droughts combined with deforestation increase fire danger (Laurance and Williamson, 2001): simulations show that during the 2001 ENSO period approximately one-third of Amazon forests became susceptible to fire (Nepstad et al., 2004).

5.3.2.3 Fisheries

Global fish production for food is forecast to increase from now to 2020, but not as rapidly as world demand. Per capita fish consumption and fish prices are expected to rise, with wide variations in commodity type and region. By 2020, wild-capture fisheries are predicted to continue to supply most of the fish produced in sub-Saharan Africa (98%), the USA (84%) and Latin America (84%), but not in India (45%) where aquaculture production will dominate (Delgado et al., 2003). All countries in Asia are likely to produce more fish between 2005 and 2020, but the rate of increase will taper. Trends in capture fisheries (usually zero growth or modest declines) will not unduly endanger overall fish supplies; however, any decline of fisheries is cause for concern given the projected growth in demand (Briones et al., 2004).

5.3.2.4 Subsistence and smallholder agriculture

'Subsistence and smallholder agriculture' is used here to describe rural producers, predominantly in developing countries, who farm using mainly family labour and for whom the farm provides the principal source of income (Cornish, 1998). Pastoralists and people dependent on artisanal fisheries and household aquaculture enterprises (Allison and Ellis, 2001) are also included in this category.

There are few informed estimates of world or regional population in these categories (Lipton, 2004). While not all smallholders, even in developing countries, are poor, 75% of the world's 1.2 billion poor (defined as consuming less than one purchasing power-adjusted dollar per day) live and work in rural areas (IFAD, 2001). They suffer, in varying degrees, problems associated both with subsistence production (isolated and marginal location, small farm size, informal land tenure and low levels of technology), and with uneven and unpredictable exposure to world markets. These systems have been characterised as 'complex, diverse and risk-prone' (Chambers et al., 1989). Risks (Scoones et al., 1996) are also diverse (drought and flood, crop and animal diseases, and market shocks) and may be felt by individual households or entire communities. Smallholder and subsistence farmers and pastoralists often also practice hunting–gathering of wild resources to fulfil energy, clothing and health needs, as well as for direct food requirements. They participate in off-farm and/or non-farm employment (Ellis, 2000).

Subsistence and smallholder livelihood systems currently experience a number of interlocking stressors other than climate change and climate variability (outlined in Section 5.2.2). They also possess certain important resilience factors: efficiencies associated with the use of family labour (Lipton, 2004), livelihood diversity that allows the spreading of risks (Ellis, 2000) and indigenous knowledge that allows exploitation of risky environmental niches and coping with crises (see Cross Chapter Case Study on Indigenous Knowledge). The combinations of stressors and resilience factors give rise to complex positive and negative trends in livelihoods. Rural to urban migration will continue to be important, with urban populations expected to overtake rural populations in less developed regions by 2017 (UNDESA 2004). Within rural areas there will be continued diversification away from agriculture (Bryceson et al., 2000); already non-farm activities account for 30-50% of rural income in developing countries (Davis, 2004). Although Vorley (2002), Hazell (2004) and Lipton (2004) see the possibility, given appropriate policies, of pro-poor growth based on the efficiency and employment generation associated with family farms, it is overall likely that smallholder and subsistence households will decline in numbers, as they are pulled or pushed into other livelihoods, with those that remain suffering increased vulnerability and increased poverty.

5.4 Key future impacts, vulnerabilities and their spatial distribution

5.4.1 Primary effects and interactions

The TAR concluded that climate change and variability will impact food, fibre and forests around the world due to the effects on plant growth and yield of elevated CO_2, higher temperatures, altered precipitation and transpiration regimes, and increased frequency of extreme events, as well as modified weed, pest and pathogen pressure. Many studies since the TAR confirmed and extended previous findings; key issues are described in the following sections.

5.4.1.1 Effects of elevated CO_2 on plant growth and yield

Plant response to elevated CO_2 alone, without climate change, is positive and was reviewed extensively by the TAR. Recent studies confirm that the effects of elevated CO_2 on plant growth and yield will depend on photosynthetic pathway, species, growth stage and management regime, such as water and nitrogen (N) applications (Jablonski et al., 2002; Kimball et al., 2002; Norby et al., 2003; Ainsworth and Long, 2005). On average across several species and under unstressed conditions, recent data analyses find that, compared to current atmospheric CO_2 concentrations, crop yields increase at 550 ppm CO_2 in the range of 10-20% for C_3 crops and 0-10% for C_4 crops (Ainsworth et al., 2004; Gifford, 2004; Long et al., 2004). Increases in above-ground biomass at 550 ppm CO_2 for trees are in the range 0-30%, with the higher values observed in young trees and little to no response observed in mature natural forests (Nowak et al., 2004; Korner et al., 2005; Norby et al., 2005). Observed increase of above-ground production in C_3 pastures is about +10% (Nowak et al., 2004; Ainsworth and Long, 2005). For commercial forestry, slow-growing trees may respond little to elevated CO_2 (e.g., Vanhatalo et al., 2003), and fast-growing trees more strongly, with harvestable wood increases of +15-25% at 550 ppm and high N (Calfapietra et al., 2003; Liberloo et al., 2005; Wittig et al., 2005). Norby et al. (2005) found a mean tree net primary production (NPP) response of 23% in young tree stands; however in mature tree stands Korner et al. (2005) reported no stimulation.

While some studies using re-analyses of recent FACE experimental results have argued that crop response to elevated CO_2 may be lower than previously thought, with consequences for crop modelling and projections of food supply (Long et al., 2005, 2006), others have suggested that these new analyses are, in fact, consistent with previous findings from both FACE and other experimental settings (Tubiello et al., 2007a, 2007b). In addition, simulations of unstressed plant growth and yield response to elevated CO_2 in the main crop-simulation models, including AFRC-Wheat, APSIM, CERES, CROPGRO, CropSyst, LINTULC and SIRIUS, have been shown to be in line with recent experimental data, projecting crop yield increases of about 5-20% at 550 ppm CO_2 (Tubiello et al., 2007b). Within that group, the main crop and pasture models, CENTURY and EPIC, project above-ground biomass production in C_3 species of about 15-20% at 550 ppm CO_2, i.e., at the high end of observed values for crops, and higher than recent observations for pasture. Forest models assume NPP increases at 550 ppm CO_2 in the range 15-30%, consistent with observed responses in young trees, but higher than observed for mature trees stands.

Importantly, plant physiologists and modellers alike recognise that the effects of elevated CO_2 measured in experimental settings and implemented in models may overestimate actual field- and farm-level responses, due to many limiting factors such as pests, weeds, competition for resources, soil, water and air quality, etc., which are neither well understood at large scales, nor well implemented in leading models (Tubiello and Ewert, 2002; Fuhrer, 2003; Karnosky, 2003; Gifford, 2004; Peng et al., 2004; Ziska and George, 2004; Ainsworth and Long, 2005; Tubiello et al., 2007a, 2007b). Assessment studies should therefore include these factors where possible, while analytical capabilities need to be enhanced. It is recommended that yield projections use a range of parameterisations of CO_2 effects to better convey the associated uncertainty range.

5.4.1.2 Interactions of elevated CO_2 with temperature and precipitation

Many recent studies confirm and extend the TAR findings that temperature and precipitation changes in future decades will modify, and often limit, direct CO_2 effects on plants. For instance, high temperature during flowering may lower CO_2 effects by reducing grain number, size and quality (Thomas et al., 2003; Baker, 2004; Caldwell et al., 2005). Increased temperatures may also reduce CO_2 effects indirectly, by increasing water demand. Rain-fed wheat grown at 450 ppm CO_2 demonstrated yield increases with temperature increases of up to 0.8°C, but declines with temperature increases beyond 1.5°C; additional irrigation was needed to counterbalance these negative effects (Xiao et al., 2005). In pastures, elevated CO_2 together with increases in temperature, precipitation and N deposition resulted in increased primary production, with changes in species distribution and litter composition (Shaw et al., 2002; Zavaleta et al., 2003; Aranjuelo et al., 2005; Henry et al., 2005). Future CO_2 levels may favour C_3 plants over C_4 (Ziska, 2003), yet the opposite is expected under associated temperature increases; the net effects remain uncertain.

Importantly, climate impacts on crops may significantly depend on the precipitation scenario considered. In particular, since more than 80% of total agricultural land, and close to 100% of pasture land, is rain-fed, general circulation model (GCM) dependent changes in precipitation will often shape both the direction and magnitude of the overall impacts (Olesen and Bindi, 2002; Tubiello et al., 2002; Reilly et al., 2003). In general, changes in precipitation and, especially, in evaporation-precipitation ratios modify ecosystem function, particularly in marginal areas. Higher water-use efficiency and greater root densities under elevated CO_2 in field and forestry systems may, in some cases, alleviate drought pressures, yet their large-scale implications are not well understood (Schäfer et al., 2002; Wullschleger et al., 2002; Norby et al., 2004; Centritto, 2005).

5.4.1.3 Increased frequency of extreme events

The TAR has already reported on studies that document additional negative impacts of increased climate variability on plant production under climate change, beyond those estimated from changes in mean variables alone. More studies since the TAR have more firmly established such issues (Porter and Semenov, 2005); they are described in detail in Sections 5.4.2 to 5.4.7. Understanding links between increased frequency of extreme climate events and ecosystem disturbance (fires, pest outbreaks, etc.) is particularly important to quantify impacts (Volney and Fleming, 2000; Carroll et al., 2004; Hogg and Bernier, 2005). Although a few models since the TAR have started to incorporate effects of climate variability on plant production, most studies continue to include only effects on changes in mean variables.

5.4.1.4 Impacts on weed and insect pests, diseases and animal health

The importance of weeds and insect pests, and disease interactions with climate change, was reviewed in the TAR. New research confirms and extends these findings, including competition between C_3 and C_4 species (Ziska, 2003; Ziska and George, 2004). In particular, CO_2-temperature interactions are recognised as a key factor in determining plant damage from pests in future decades, though few quantitative analyses exist to date; CO_2-precipitation interactions will be likewise important (Stacey and Fellows, 2002; Chen et al., 2004; Salinari et al., 2006; Zvereva and Kozlov, 2006). Most studies continue to investigate pest damage as a separate function of either CO_2 (Chakraborty and Datta, 2003; Agrell et al., 2004; Chen et al., 2005a, 2005b) or temperature (Bale et al., 2002; Cocu et al., 2005; Salinari et al., 2006). For instance, recent warming trends in the U.S. and Canada have led to earlier spring activity of insects and proliferation of some species, such as the mountain pine beetle (Crozier and Dwyer, 2006; see also Chapter 1). Importantly, increased climate extremes may promote plant disease and pest outbreaks (Alig et al., 2004; Gan, 2004). Finally, new studies, since the TAR, are focusing on the spread of animal diseases and pests from low to mid-latitudes due to warming, a continuance of trends already under way (see Section 5.2). For instance, models project that bluetongue, which mostly affects sheep, and occasionally goat and deer, would spread from the tropics to mid-latitudes (Anon, 2006; van Wuijckhuise et al., 2006). Likewise, White et al. (2003) simulated, under climate change, increased vulnerability of the Australian beef industry to the cattle tick (*Boophilus microplus*). Most assessment studies do not explicitly consider either pest-plant dynamics or impacts on livestock health as a function of CO_2 and climate combined.

5.4.1.5 Vulnerability of carbon pools

Impacts of climate change on managed systems, due to the large land area covered by forestry, pastures and crops, have the potential to affect the global terrestrial carbon sink and to further perturb atmospheric CO_2 concentrations (IPCC, 2001; Betts et al., 2004; Ciais et al., 2005). Furthermore, vulnerability of organic carbon pools to climate change has important repercussions for land sustainability and climate-mitigation actions. The TAR stressed that future changes in carbon stocks and net fluxes would critically depend on land-use planning (set aside policies, afforestation-reforestation, etc.) and management practices (such as N fertilisation, irrigation and tillage), in addition to plant response to elevated CO_2. Recent research confirms that carbon storage in soil organic matter is often increased under elevated CO_2 in the short-term (e.g., Allard et al., 2004); yet the total soil carbon sink may saturate at elevated CO_2 concentrations, especially when nutrient inputs are low (Gill et al., 2002; van Groenigen et al., 2006).

Uncertainty remains with respect to several key issues such as the impacts of increased frequency of extremes on the stability of carbon and soil organic matter pools; for instance, the recent European heatwave of 2003 led to significant soil carbon losses (Ciais et al., 2005). In addition, the effects of air pollution on plant function may indirectly affect carbon storage; recent research showed that tropospheric ozone results in significantly less enhancement of carbon-sequestration rates under elevated CO_2 (Loya et al., 2003), because of the negative effects of ozone on biomass productivity and changes in litter chemistry (Booker et al., 2005; Liu et al., 2005).

Within the limits of current uncertainties, recent modelling studies have investigated future trends in carbon storage over managed land by considering multiple interactions of climate and management variables. Smith et al. (2005) projected small overall carbon increases in managed land in Europe during this century due to climate change. By contrast, also including projected changes in land use resulted in small overall decreases. Felzer et al. (2005) projected increases in carbon storage on croplands globally under climate change up to 2100, but found that ozone damage to crops could significantly offset these gains.

Finally, recent studies show the importance of identifying potential synergies between land-based adaptation and mitigation strategies, linking issues of carbon sequestration, emissions of greenhouse gases, land-use change and long-term sustainability of production systems within coherent climate policy frameworks (e.g., Smith et al., 2005; Rosenzweig and Tubiello, 2007).

5.4.2 Food-crop farming, including tree crops

As noted in Section 5.1.3, the TAR indicated that impacts on food systems at the global scale might be small overall in the first half of the 21st century, but progressively negative after that. Importantly, crop production in (mainly low latitude) developing countries would suffer more, and earlier, than in (mainly mid- to high-latitude) developed countries, due to a combination of adverse agro-climatic, socio-economic and technological conditions (see recent analyses in Alexandratos, 2005).

5.4.2.1 *What is new since the TAR?*

Many studies since the TAR have confirmed key dynamics of previous regional and global projections. These projections indicate potentially large negative impacts in developing regions, but only small changes in developed regions, which causes the globally aggregated impacts on world food production to be small (Fischer et al., 2002b, 2005b; Parry, 2004; Parry et al., 2005). Recent regional assessments have shown the high uncertainty that underlies such findings, and thus the possibility for surprises, by projecting, in some cases, significant negative impacts in key producing regions of developed countries, even before the middle of this century (Olesen and Bindi, 2002; Reilly et al., 2003). Many recent studies have contributed specific new knowledge with respect to several uncertainties and limiting factors at the time of the TAR, often highlighting the possibility for negative surprises, in addition to the impacts of mean climate change alone.

New Knowledge: *Increases in frequency of climate extremes may lower crop yields beyond the impacts of mean climate change.*

More frequent extreme events may lower long-term yields by directly damaging crops at specific developmental stages, such as temperature thresholds during flowering, or by making the timing of field applications more difficult, thus reducing the efficiency of farm inputs (e.g., Antle et al., 2004; Porter and Semenov, 2005). A number of simulation studies performed since the TAR have developed specific aspects of increased climate variability within climate change scenarios. Rosenzweig et al. (2002) computed that, under scenarios of increased heavy precipitation, production losses due to excessive soil moisture would double in the U.S. by 2030 to US$3 billion/yr. Monirul and Mirza (2002) computed an increased risk of crop losses in Bangladesh from increased flood frequency under climate change. In scenarios with higher rainfall intensity, Nearing et al. (2004) projected increased risks of soil erosion, while van Ittersum et al. (2003) simulated higher risk of salinisation in arid and semi-arid regions, due to more water loss below the crop root zone. Howden et al. (2003) focused on the consequences of higher temperatures on the frequency of heat stress during growing seasons, as well on the frequency of frost occurrence during critical growth stages.

New Knowledge: *Impacts of climate change on irrigation water requirements may be large.*

Döll (2002) considered direct impacts of climate change on crop evaporative demand (no CO_2 effects) and computed increases in crop irrigation requirements of +5% to +8% globally by 2070, with larger regional signals (e.g., +15%) in South-East Asia, net of transpiration losses. Fischer et al. (2006) included positive CO_2 effects on crop water-use efficiency and computed increases in global net irrigation requirements of +20% by 2080, with larger impacts in developed versus developing regions, due to both increased evaporative demands and longer growing seasons under climate change. Fischer et al. (2006) and Arnell (2004) also projected increases in water stress (the ratio of irrigation withdrawals to renewable water resources) in the Middle East and South-East Asia. Recent regional studies have also found key climate change and water changes in key irrigated areas, such as North Africa (increased irrigation requirements; Abou-Hadid et al., 2003) and China (decreased requirements; Tao et al., 2003).

New Knowledge: *Stabilisation of CO_2 concentrations reduces damage to crop production in the long term.*

Recent work further investigated the effects of potential stabilisation of atmospheric CO_2 on regional and global crop production. Compared to the relatively small impacts of climate change on crop production by 2100 under business-as-usual scenarios, the impacts were only slightly less under 750 ppm CO_2 stabilization. However, stabilisation at 550 ppm CO_2 significantly reduced production loss (by -70% to –100%) and lowered risk of hunger (–60% to –85%) (Arnell et al., 2002; Tubiello and Fischer, 2006). These same studies suggested that climate mitigation may alter the regional and temporal mix of winners and losers with respect to business-as-usual scenarios, but concluded that specific projections are highly uncertain. In particular, in the first decades of this century and possibly up to 2050, some regions may be worse off with mitigation than without, due to lower CO_2 levels and thus reduced stimulation of crop yields (Tubiello and Fischer, 2006). Finally, a growing body of work has started to analyse potential relations between mitigation and adaptation (see Chapter 18).

TAR Confirmation: *Including effects of trade lowers regional and global impacts.*

Studies by Fischer et al. (2005a), Fischer et al. (2002a), Parry (2004) and Parry et al. (2005) confirm that including trade among world regions in assessment studies tends to reduce the overall projected impacts on agriculture compared to studies that lack an economic component. Yet, despite socio-economic development and trade effects, these and several other regional and global studies indicate that developing regions may be more negatively affected by climate change than other regions (Olesen and Bindi, 2002; Cassman et al., 2003; Reilly et al., 2003; Antle et al., 2004; Mendelsohn et al., 2004). Specific differences among studies depend significantly on factors such as projected population growth and food demand, as well as on trends in production technology and efficiency. In particular, the choice of the SRES scenario has as large an effect on projected global and regional levels of food demand and supply as climate change alone (Parry et al., 2004; Ewert et al., 2005; Fischer et al., 2005a; Tubiello et al., 2007a).

5.4.2.2 *Review of crop impacts versus incremental temperature change*

The increasing number of regional and global simulation studies performed since the TAR make it possible to produce synthesis graphs, showing not only changes in yield for key crops against temperature (a proxy for both time and severity of climate change), but also other important climate and management factors, such as changes in precipitation or adaptation strategies. An important limitation of these syntheses is that they collect single snapshots of future impacts, thereby lacking the temporal and causal dynamics that characterise actual responses in farmers' fields. Yet they are useful to summarise many independent studies.

Figure 5.2 provides an example of such analyses for temperature increases ranging from about 1-2°C, typical of the next several decades, up to the 4-5°C projected for 2080 and beyond. The results of such simulations are generally highly uncertain due to many factors, including large discrepancies in GCM predictions of regional precipitation change, poor representation of impacts of extreme events and the assumed strength of CO_2 fertilisation (5.4.1). Nevertheless, these summaries indicate that in mid- to high-latitude regions, moderate to medium local increases in temperature (1°C to 3°C), across a range of CO_2 concentrations and rainfall changes, can have small beneficial impacts on the main cereal crops. Further warming has increasingly negative impacts (medium to low confidence) (Figure 5.2a, c, e). In low-latitude regions, these simulations indicate that even moderate temperature increases are likely to have negative yield impacts for major cereal crops (Figure 5.2b, d, f). For temperature increases more than 3°C, average impacts are stressful to all crops assessed and to all regions (medium to low confidence) (Figure 5.2). The low and mid-to-high latitude regions encompass the majority of global cereal production area. This suggests that global production potential, defined by Sivakumar and Valentin (1997) as equivalent to crop yield or Net Primary Productivity (NPP), is threatened at +1°C local temperature change and can accommodate no more that +3°C before beginning to decline. The studies summarised in Figure 5.2 also indicate that precipitation changes (and associated changes in precipitation:evaporation ratios), as well as CO_2 concentration, may critically shape crop-yield responses, over and above the temperature signal, in agreement with previous analyses (Section 5.4.1). The effects of adaptation shown in Figure 5.2 are considered in Section 5.5.

5.4.2.3 Research tasks not yet undertaken – ongoing uncertainties

Several uncertainties remain unresolved since the TAR. Better knowledge in several research areas is critical to improve our ability to predict the magnitude, and often even the direction, of future climate change impacts on crops, as well as to better define risk thresholds and the potential for surprises, at local, regional and global scales.

In terms of experimentation, there is still a lack of knowledge of CO_2 and climate responses for many crops other than cereals, including many of importance to the rural poor, such as root crops, millet, brassica, etc., with few exceptions, e.g., peanut (Varaprasad et al., 2003) and coconut (Dash et al., 2002). Importantly, research on the combined effects of elevated CO_2 and climate change on pests, weeds and disease is still insufficient, though research networks have long been put into place and a few studies have been published (Chakraborty and Datta, 2003; Runion, 2003; Salinari et al., 2006). Impacts of climate change alone on pest ranges and activity are also being increasingly analysed (e.g., Bale et al., 2002; Todd et al., 2002; Rafoss and Saethre, 2003; Cocu et al., 2005; Salinari et al., 2006). Finally, the true strength of the effect of elevated CO_2 on crop yields at field to regional scales, its interactions with higher temperatures and modified precipitation regimes, as well as the CO_2 levels beyond which saturation may occur, remain largely unknown.

In terms of modelling, calls by the TAR to enhance crop model inter-comparison studies have remained unheeded; in fact, such activity has been performed with much less frequency after the TAR than before. It is important that uncertainties related to crop-model simulations of key processes, including their spatial-temporal resolution, be better evaluated, as findings of integrated studies will remain dependent upon the particular crop model used. It is still unclear how the implementation of plot-level experimental data on CO_2 responses compares across models; especially when simulations of several key limiting factors, such as soil and water quality, pests, weeds, diseases and the like, remain either unresolved experimentally or untested in models (Tubiello and Ewert, 2002). Finally, the TAR concluded that the economic, trade and technological assumptions used in many of the integrated assessment models to project food security under climate change were poorly tested against observed data. This remains the situation today (see also Section 5.6.5).

5.4.3 Pastures and livestock production

Pastures comprise both grassland and rangeland ecosystems. Grasslands are the dominant vegetation type in areas with low rainfall, such as the steppes of central Asia and the prairies of North America. Grasslands can also be found in areas with higher rainfall, such as north-western and central Europe, New Zealand, parts of North and South America and Australia. Rangelands are found on every continent, typically in regions where temperature and moisture restrictions limit other vegetation types; they include deserts (cold, hot and tundra), scrub, chaparral and savannas.

Pastures and livestock production systems occur under most climates and range from extensive pastoral systems with grazing herbivores, to intensive systems based on forage and grain crops, where animals are mostly kept indoors. The TAR identified that the combination of increases in CO_2 concentration, in conjunction with changes in rainfall and temperature, were likely to have significant impacts on grasslands and rangelands, with production increases in humid temperate grasslands, but decreases in arid and semiarid regions.

5.4.3.1 New findings since TAR

New Knowledge: *Plant community structure is modified by elevated CO_2 and climate change.*

Grasslands consisting of fast-growing, often short lived species, are sensitive to CO_2 and climate change, with the impacts related to the stability and resilience of plant communities (Mitchell and Csillag, 2001). Experiments support the concept of rapid changes in species composition and diversity under climate change. For instance, in a Mediterranean annual grassland after three years of experimental manipulation, plant diversity decreased with elevated CO_2 and nitrogen deposition, increased with elevated precipitation and showed no significant effect from warming (Zavaleta et al., 2003). Diversity responses to both single and combined global change treatments were driven mainly by significant gains and losses of forb[1] species (Zavaleta et al., 2003). Elevated CO_2 influences plant species

[1] Forb: a broad-leaved herb other than grass.

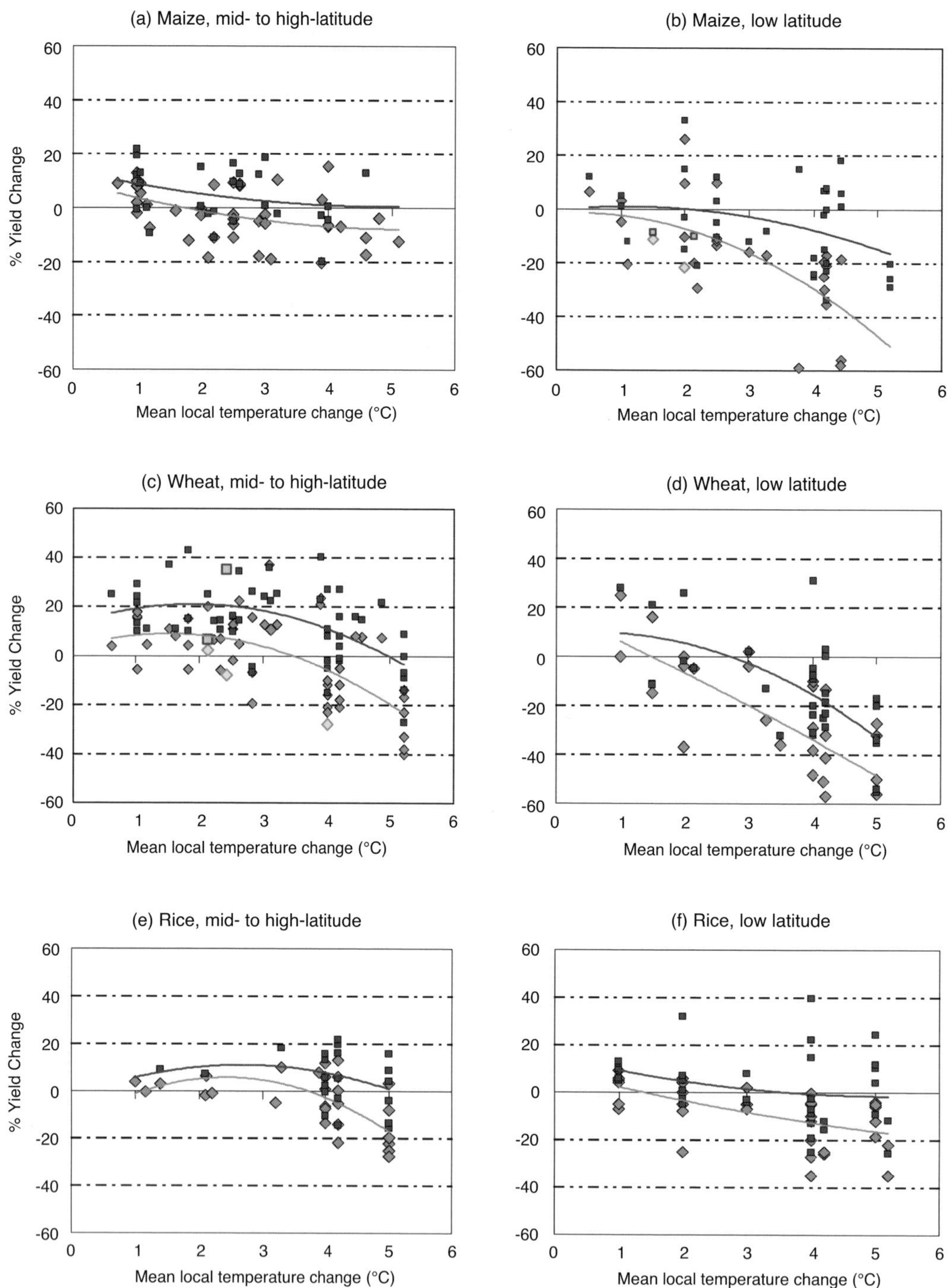

Figure 5.2. *Sensitivity of cereal yield to climate change for maize, wheat and rice, as derived from the results of 69 published studies at multiple simulation sites, against mean local temperature change used as a proxy to indicate magnitude of climate change in each study. Responses include cases without adaptation (red dots) and with adaptation (dark green dots). Adaptations+ represented in these studies include changes in planting, changes in cultivar, and shifts from rain-fed to irrigated conditions. Lines are best-fit polynomials and are used here as a way to summarise results across studies rather than as a predictive tool. The studies span a range of precipitation changes and CO_2 concentrations, and vary in how they represent future changes in climate variability. For instance, lighter-coloured dots in (b) and (c) represent responses of rain-fed crops under climate scenarios with decreased precipitation. Data sources: Bachelet and Gay, 1993; Rosenzweig and Parry, 1994; El-Shaer et al., 1997; Iglesias and Minguez, 1997; Kapetanaki and Rosenzweig, 1997; Matthews et al., 1997; Lal et al., 1998; Moya et al., 1998; Winters et al., 1998; Yates and Strzepek, 1998; Brown and Rosenberg, 1999; Evenson, 1999; Hulme et al., 1999; Parry et al., 1999; Iglesias et al., 2000; Saarikko, 2000; Tubiello et al., 2000; Bachelet et al., 2001; Easterling et al., 2001; Kumar and Parikh, 2001; Aggarwal and Mall, 2002; Alig et al., 2002; Arnell et al., 2002; Chang, 2002; Corobov, 2002; Cuculeanu et al., 2002; Mall and Aggarwal, 2002; Olesen and Bindi, 2002; Parry and Livermore, 2002; Southworth et al., 2002; Tol, 2002; Tubiello and Ewert, 2002; Aggarwal, 2003; Carbone et al., 2003; Chipanshi et al., 2003; Izaurralde et al., 2003; Jones and Thornton, 2003; Luo et al., 2003; Matthews and Wassmann, 2003; Reilly et al., 2003; Rosenberg et al., 2003; Tan and Shibasaki, 2003; Droogers, 2004; Faisal and Parveen, 2004; Adejuwon, 2005; Branco et al., 2005; Butt et al., 2005; Erda et al., 2005; Ewert et al., 2005; Fischer et al., 2005b; Gbetibouo and Hassan, 2005; Gregory et al., 2005; Haque and Burton, 2005; Maracchi et al., 2005; Motha and Baier, 2005; Palmer et al., 2005; Parry et al., 2005; Porter and Semenov, 2005; Sands and Edmonds, 2005; Schröter et al., 2005; Sivakumar et al., 2005; Slingo et al., 2005; Stigter et al., 2005; Thomson et al., 2005a, 2005b; Xiao et al., 2005; Zhang and Liu, 2005; Zhao et al., 2005; Aggarwal et al., 2006.*

composition partly through changes in the pattern of seedling recruitment (Edwards et al., 2001). For sown mixtures, the TAR indicated that elevated CO_2 increased legume development. This finding has been confirmed (Luscher et al., 2005) and extended to temperate semi-natural grasslands using free air CO_2 enrichment (Teyssonneyre et al., 2002; Ross et al., 2004). Other factors such as low phosphorus availability and low herbage use (Teyssonneyre et al., 2002) may, however, prevent this increase in legumes under high CO_2.

How to extrapolate these findings is still unclear. A recent simulation of 1,350 European plant species based on plant species distribution envelopes predicted that half of these species will become classified as 'vulnerable' or 'endangered' by the year 2080 due to rising temperature and changes in precipitation (Thuiller et al., 2005) (see Chapter 4). Nevertheless, such empirical model predictions have low confidence as they do not capture the complex interactions with management factors (e.g., grazing, cutting and fertiliser supply).

New Knowledge: *Changes in forage quality and grazing behaviour are confirmed.*

Animal requirements for crude proteins from pasture range from 7 to 8% of ingested dry matter for animals at maintenance up to 24 % for the highest-producing dairy cows. In conditions of very low N status, possible reductions in crude proteins under elevated CO_2 may put a system into a sub-maintenance level for animal performance (Milchunas et al., 2005). An increase in the legume content of swards may nevertheless compensate for the decline in protein content of the non-fixing plant species (Allard et al., 2003; Picon-Cochard et al., 2004). The decline under elevated CO_2 (Polley et al., 2003) of C_4 grasses, which are a less nutritious food resource than C_3 (Ehleringer et al., 2002), may also compensate for the reduced protein content under elevated CO_2. Yet the opposite is expected under associated temperature increases (see Section 5.4.1.2).

Large areas of upland Britain are already colonised by relatively unpalatable plant species such as bracken, matt grass and tor grass. At elevated CO_2 further changes may be expected in the dominance of these species, which could have detrimental effects on the nutritional value of extensive grasslands to grazing animals (Defra, 2000).

New Knowledge: *Thermal stress reduces productivity, conception rates and is potentially life-threatening to livestock.*

The TAR indicated the negative role of heat stress for productivity. Because ingestion of food and feed is directly related to heat production, any decline in feed intake and/or energy density of the diet will reduce the amount of heat that needs to be dissipated by the animal. Mader and Davis (2004) confirm that the onset of a thermal challenge often results in declines in physical activity with associated declines in eating and grazing (for ruminants and other herbivores) activity. New models of animal energetics and nutrition (Parsons et al., 2001) have shown that high temperatures put a ceiling on dairy milk yield irrespective of feed intake. In the tropics, this ceiling reaches between half and one-third of the potential of the modern (Friesians) cow breeds. The energy deficit of this genotype will exceed that normally associated with the start of lactation, and decrease cow fertility, fitness and longevity (King et al., 2005).

Increases in air temperature and/or humidity have the potential to affect conception rates of domestic animals not adapted to those conditions. This is particularly the case for cattle, in which the primary breeding season occurs in the spring and summer months. Amundson et al. (2005) reported declines in conception rates of cattle (*Bos taurus*) for temperatures above 23.4°C and at high thermal heat index.

Production-response models for growing confined swine and beef cattle, and milk-producing dairy cattle, based on predicted climate outputs from GCM scenarios, have been developed by Frank et al. (2001). Across the entire USA, the percentage decrease in confined swine, beef and dairy milk production for the 2050 scenario averaged 1.2%, 2.0% and 2.2%, respectively, using the CGC (version 1) model and 0.9%, 0.7% and 2.1%, respectively, using the HadCM2 model.

New Knowledge: *Increased climate variability and droughts may lead to livestock loss.*

The impact on animal productivity due to increased variability in weather patterns will likely be far greater than effects associated with the average change in climatic conditions. Lack of prior conditioning to weather events most often results in catastrophic losses in confined cattle feedlots (Hahn et al., 2001), with economic losses from reduced cattle performance exceeding those associated with cattle death losses by several-fold (Mader, 2003).

Many of the world's rangelands are affected by ENSO events. The TAR identified that these events are likely to intensify with climate change, with subsequent changes in vegetation and water availability (Gitay et al., 2001). In dry regions, there are risks that severe vegetation degeneration leads to positive feedbacks between soil degradation and reduced vegetation and rainfall, with corresponding loss of pastoral areas and farmlands (Zheng et al., 2002).

A number of studies in Africa (see Table 5.3) and in Mongolia (Batima, 2003) show a strong relationship between drought and animal death. Projected increased temperature, combined with reduced precipitation in some regions (e.g., Southern Africa) would lead to increased loss of domestic herbivores during extreme events in drought-prone areas. With increased heat stress in the future, water requirements for livestock will increase significantly compared with current conditions, so that overgrazing near watering points is likely to expand (Batima et al., 2005).

5.4.3.2 Impacts of gradual temperature change

A survey of experimental data worldwide suggested that a mild warming generally increases grassland productivity, with the strongest positive responses at high latitudes (Rustad et al., 2001). Productivity and plant species composition in rangelands are highly correlated with precipitation (Knapp and Smith, 2001) and recent findings from IPCC (2007b) (see Figure 5.1) show projected declines in rainfall in some major grassland and rangeland areas (e.g., South America, South and North Africa, western Asia, Australia and southern Europe). Elevated CO_2 can reduce soil water depletion in different native and semi-native temperate and Mediterranean grassland (Morgan et al., 2004). However, increased variability in rainfall may create more severe soil moisture limitation and reduced productivity (Laporte et al., 2002; Fay et al., 2003; Luscher et al., 2005). Other impacts occur directly on livestock through the increase in the thermal heat load (see Section 5.4.3.1).

Table 5.3 summarises the impacts on grasslands for different temperature changes. Warming up to 2°C suggests positive impacts on pasture and livestock productivity in humid temperate regions. By contrast, negative impacts are predicted in arid and semiarid regions. It should be noted that there are very few impact studies for tropical grasslands and rangelands.

5.4.4 Industrial crops and biofuels

Industrial crops include oilseeds, gums and resins, sweeteners, beverages, fibres, and medicinal and aromatic plants. There is practically no literature on the impact of climate change on gums and resins, and medicinal and aromatic plants. Limited new knowledge of climate change impacts on other industrial crops and biofuels has been developed since the TAR. Van Duivenbooden et al. (2002) used statistical models to estimate that rainfall reduction associated with climate change could reduce groundnut production in Niger, a large groundnut producing and exporting country, by 11-25%. Varaprasad et al. (2003) also concluded that groundnut yields would decrease under future warmer climates, particularly in regions where present temperatures are near or above optimum despite increased CO_2.

Impacts of climate change and elevated CO_2 on perennial industrial crops will be greater than on annual crops, as both damages (temperature stresses, pest outbreaks, increased damage from climate extremes) and benefits (extension of latitudinal optimal growing ranges) may accumulate with time (Rajagopal et al., 2002). For example, the cyclones that struck several states of India in 1952, 1955, 1996 and 1998 destroyed so many coconut palms that it will take years before production can be restored to pre-cyclone levels (Dash et al., 2002).

The TAR established large increases in cotton yields due to increases in ambient CO_2 concentration. Reddy et al. (2002), however, demonstrated that such increases in cotton yields were eliminated when changes in temperature and precipitation were also included in the simulations. Future climate change scenarios for the Mississippi Delta estimate a 9% mean loss in fibre yield. Literature still does not exist on the probable impacts of climate change on other fibre crops such as jute and kenaf.

Biofuel crops, increasingly an important source of energy, are being assessed for their critical role in adaptation to climatic change and mitigation of carbon emissions (discussed in IPCC, 2007c). Impacts of climate change on typical liquid biofuel crops such as maize and sorghum, and wood (solid biofuel) are discussed earlier in this chapter. Recent studies indicate that global warming may increase the yield potential of sugar beet, another important biofuel crop, in parts of Europe where drought is not a constraint (Jones et al., 2003; Richter et al., 2006). The annual variability of yields could, however, increase. Studies with other biofuel crops such as switchgrass (*Panicum virgatum L.*), a perennial warm season C_4 crop, have shown yield increases with climate change similar to those of grain crops (Brown et al., 2000). Although there is no information on the impact of climate change on non-food, tropical biofuel crops such as Jatropha and Pongamia, it is likely that their response will be similar to other regional crops.

5.4.5 Key future impacts on forestry

Forests cover almost 4 billion ha or 30% of land; 3.4 billion m^3 of wood were removed in 2004 from this area, 60% as industrial roundwood (FAO, 2005b). Intensively managed forest plantations comprised only 4% of the forest area in 2005, but their area is rapidly increasing (2.5 million ha annually (FAO, 2005b)). In 2000, these forests supplied about 35% of global roundwood;

Table 5.3. *Impacts on grasslands of incremental temperature change. (EXP = experiment; SIM = simulation without explicit reference to a SRES scenario; GMT = global mean temperature.)*

Local temperature change	Sub-sector	Region	Impact trends	Sign of impact	Scenario/Experiment	Source
+0-2°C	Pastures and livestock	Temperate	Alleviation of cold limitation increasing productivity	+	SIM IS92a	Parsons et al., 2001 Riedo et al., 2001
			Increased heat stress for livestock	-	IS92a	Turnpenny et al., 2001
		Semi-arid and Mediterranean	No increase in net primary productivity	0	EXP	Shaw et al., 2002 Dukes et al., 2005
+3°C	Pastures and livestock	Temperate	Neutral to small positive effect (depending on GMT)	0 to +	SIM	Parsons et al., 2001 Riedo et al., 2001
		Temperate	Negative on swine and confined cattle	-	HadCM2 CGCM1	Frank and Dugas, 2001
		Semi-arid and Mediterranean	Productivity decline Reduced ewe weight and pasture growth	-	HadCM3 A2 and B2	Howden et al., 1999 Batima et al., 2005
			More animal heat stress	-		
		Tropical	No effect (no rainfall change assumed)	- to 0	EXP	Newman et al., 2001 Volder et al., 2004
			More animal heat stress	-		

this share is expected to increase to 44% by 2020 (FAO, 2000). This section focuses on commercial forestry, including regional, national and global timber supply and demand, and associated changes in land-use, accessibility for harvesting and overall economic impacts. The ecosystem services of forests are reviewed in Chapter 4, while interactions with climate are discussed in IPCC (2007b). Key regional impacts are further detailed in Chapter 10, Section 10.4.4; Chapter 11, Section 11.4.4; Chapter 12, Section 12.4.4; Chapter 13, Section 13.4.1; and Chapter 14, Section 14.4.4. Finally, bioenergy is discussed in IPCC (2007c).

5.4.5.1 New findings since TAR

***Confirmation of TAR:** Modelling studies predict increased global timber production.*

Simulations with yield models show that climate change can increase global timber production through location changes of forests and higher growth rates, especially when positive effects of elevated CO_2 concentration are taken into consideration (Irland et al., 2001; Sohngen et al., 2001; Alig et al., 2002; Solberg et al., 2003; Sohngen and Sedjo, 2005). For example, Sohngen et al. (2001) and Sohngen and Sedjo (2005) projected a moderate increase of timber yield due to both rising NPP and a poleward shift of the most productive species due to climate change. Changing timber supply will affect the market and could impact supply for other uses, e.g., for biomass energy. Global economic impact assessments predict overall demand for timber production to increase only modestly (see Section 5.3.2.2) with a moderate increase or decrease of wood prices in the future in the order of up to ±20% (Irland et al., 2001; Sohngen et al., 2001; Nabuurs et al., 2002; Perez-Garcia et al., 2002; Solberg et al., 2003; Sohngen and Sedjo, 2005), with benefits of higher production mainly going to consumers. For the U.S., Alig et al. (2002) computed that the net impact of climate change on the forestry sector may be small. Similarly, Shugart et al. (2003) concluded that the U.S. timber markets have low susceptibility to climate change, because of the large stock of existing forests, technological change in the timber industry and the ability to adapt. These and other simulation studies are summarised in Table 5.4.

***New Knowledge:** Increased regional variability; change in non-timber forest products.*

Although models suggest that global timber productivity will likely increase with climate change, regional production will exhibit large variability, similar to that discussed for crops. Mendelsohn (2003), analysing production in California, projected that, at first (2020s), climate change increases harvests by

Table 5.4. *Examples of simulated climate change impacts on forestry.*

Reference; *location*	Scenario and GCM	Production impact	Economic impact
Sohngen et al., 2001; Sohngen and Sedjo, 2005. *Global*	UIUC and Hamburg T-106 for CO_2 topping 550 ppm in 2060	• 2045: production up by 29-38%; reductions in N. America, Russia; increases in S. America and Oceania. • 2145: production up by 30%, increases in N. America, S. America, and Russia.	• 2045: prices reduced, high-latitude loss, low-latitudes gain. • 2145: prices increase up to 80% (no climate change), 50% (with climate change), high-latitude gain, low-latitude loss. Benefits go to consumers.
Solberg et al., 2003. *Europe*	Baseline, 20-40%, increase in forest growth by 2020	• Increased production in W. Europe, • Decreased production in E. Europe.	Price drop with an increase in welfare to producers and consumers. Increased profits of forest industry and forest owners.
Perez-Garcia et al., 2002. *Global*	TEM & CGTM MIT GCM, MIT EPPA emissions	• Harvest increase in the US West (+2 to +11%), New Zealand (+10 to +12%), and S. America (+10 to +13%). • Harvest decrease in Canada.	Demand satisfied; prices drop with an increase in welfare to producers and consumers.
Lee and Lyon, 2004. *Global*	ECHAM-3 (2 × CO_2 in 2060), TSM 2000, BIOME 3, Hamburg model	• 2080s, no climate change: increase of the industrial timber harvest by 65% (normal demand) or 150% (high demand); emerging regions triple their production. • With climate change: increase of the industrial timber harvest by 25% (normal demand) or 56% (high demand), E. Siberia & US South dominate production.	No climate change: • Pulpwood price increases 44% • Solid wood increase 21%. With climate change: • Pulpwood price decrease 25% • Solid wood decrease 34% • Global welfare 4.8% higher than in no climate change scenario.
Nabuurs et al., 2002. *Europe*	HadCM2 under IS92a 1990-2050	18% extra increase in annual stemwood increment by 2030, slowing down on a longer term.	Both decreases or increases in prices are possible.
Schroeter, 2004. *Europe*	IPCC A1FI, A2, B1, B2 up to 2100. Few management scenarios	• Increased forest growth (especially in N. Europe) and stocks, except for A1FI. • 60-80% of stock change is due to management, climate explains 10-30% and the rest is due to land use change.	In the A1FI and A2 scenarios, wood demand exceeds potential felling, particularly in the second half of the 21st century, while in the B1 and B2 scenarios future wood demand can be satisfied.
Alig et al., 2002; Joyce et al., 2001. *USA*	CGCM1+TEM HadCM2+TEM CGCM1+VEMAP HadCM2+VEMAP IS92a	• Increase in timber inventory by 12% (mid-term); 24% (long-term) and small increase in harvest. Major shift in species and an increase in burnt area by 25-50%. • Generally, high elevation and northern forests decline, southern forests expand.	• Reduction in log prices • Producer welfare reduced compared to no climate change scenario • Lower prices; consumers will gain and forest owners will lose

stimulating growth in the standing forest. In the long run, up to 2100, these productivity gains were offset by reductions in productive area for softwoods growth. Climate change will also substantially impact other services, such as seeds, nuts, hunting, resins, plants used in pharmaceutical and botanical medicine, and in the cosmetics industry; these impacts will also be highly diverse and regionalised.

New Knowledge: *CO_2 enrichment effects may be overestimated in models; models need improvement.*

New studies suggest that direct CO_2 effects on tree growth may be revised to lower values than previously assumed in forest growth models. A number of FACE studies in 550 ppm CO_2 showed average NPP increase of 23% in young tree stands (Norby et al., 2005). However, in a 100-year old tree stand, Korner et al. (2005) found little overall stimulation in stem growth over a period of four years. Additionally, the initial increase in growth increments may be limited by competition, disturbance, air pollutants, nutrient limitations and other factors (Karnosky, 2003), and the response is site- and species-specific. By contrast, models often presume larger fertilisation effects: Sohngen et al. (2001) assumed a 35% NPP increase under a 2 × CO_2 scenario. Boisvenue and Running (2006) suggest increasing forest-growth rate due to increasing CO_2 since the middle of the 20th century; however, some of this increase may result from other effects, such as land-use change (Caspersen et al., 2000).

In spite of improvements in forest modelling, model limitations persist. Most of the major forestry models don't include key ecological processes. Development of Dynamic Global Vegetation Models (DGVMs), which are spatially explicit and dynamic, will allow better predictions of climate-induced vegetative changes (Peng, 2000; Bachelet et al., 2001; Cramer et al., 2001; Brovkin, 2002; Moorcroft, 2003; Sitch et al., 2003) by simulating the composition of deciduous and evergreen trees, forest biomass, production, and water and nutrient cycling, as well as fire effects. DGVMs are also able to provide GCMs with feedbacks from changing vegetation, e.g., Cox et al. (2004) found that DGVM feedbacks raise HadCM3LC GCM temperature and decrease precipitation forecasts for Amazonia, leading to eventual loss of rainforests. There are still inconsistencies, however, between the models used by ecologists to estimate the effects of climate change on forest production and composition and those used to predict forest yield. Future development of the models that integrate both the NPP and forestry yield approaches (Nabuurs et al., 2002; Peng et al., 2002) will significantly improve the predictions.

5.4.5.2 Additional factors not included in the models contribute uncertainty

Fire, insects and extreme events are not well modelled. Both forest composition and production are shaped by fire frequency, size, intensity and seasonality. There is evidence of both regional increase and decrease in fire activity (Goldammer and Mutch, 2001; Podur et al., 2002; Bergeron et al., 2004; Girardin et al., 2004; Mouillot and Field, 2005), with some of the changes linked to climate change (Gillett et al., 2004; Westerling et al., 2006). Climate change will interact with fuel type, ignition source and topography in determining future damage risks to the forest industry, especially for paper and pulp operations; fire hazards will also pose health threats (see Chapter 8, Section 8.2) and affect landscape recreational value. There is an uncertainty associated with many studies of climate change and forest fires (Shugart et al., 2003; Lemmen and Warren, 2004); however, current modelling studies suggest that increased temperatures and longer growing seasons will elevate fire risk in connection with increased aridity (Williams et al., 2001; Flannigan et al., 2005; Schlyter et al., 2006). For example, Crozier and Dwyer (2006) indicated the possibility of a 10% increase in the seasonal severity of fire hazard over much of the United States under changed climate, while Flannigan et al. (2005) projected as much as 74-118% increase of the area burned in Canada by the end of the 21st century under a 3 × CO_2 scenario. However, much of this fire increase is expected in inaccessible boreal forest regions, so the effects of climate-induced wildfires on timber production may be more modest.

For many forest types, forest health questions are of great concern, with pest and disease outbreaks as major sources of natural disturbance. The effects vary from defoliation and growth loss to timber damage to massive forest die backs; it is very likely that these natural disturbances will be altered by climate change and will have an impact on forestry (Alig et al., 2004). Warmer temperatures have already enhanced the opportunities for insect spread across the landscape (Carroll et al., 2004; Crozier and Dwyer, 2006). Climate change can shift the current boundaries of insects and pathogens and modify tree physiology and tree defence. Modelling of climate change impacts on insect and pathogen outbreaks remains limited.

The effects of climate extremes on commercial forestry are region-specific and include reduced access to forestland, increased costs for road and facility maintenance, direct damage to trees by wind, snow, frost or ice; indirect damage from higher risks of wildfires and insect outbreaks, effects of wetter winters and early thaws on logging, etc. For example, in January 2005 Hurricane Gudrun, with maximum gusts of 43 m/s, damaged more than 60 million m^3 of timber in Sweden, reducing the country's log trade deficit by 30% (UNECE, 2006). Higher direct and indirect risks could affect timber supplies, market prices and cost of insurance (DeWalle et al., 2003). Globally, model predictions mentioned in the SAR suggested extensive forest die back and composition change; however, some of these effects may be mitigated (Shugart et al., 2003) and changes in forest composition will likely occur gradually (Hanson and Weltzin, 2000).

Interaction between multiple disturbances is very important for understanding climate change impacts on forestry. Wind events can damage trees through branch breaking, crown loss, trunk breakage or complete stand destruction. The damage might increase for faster-growing forests. This damage can be further aggravated by increased damage from insect outbreaks and wildfires (Fleming et al., 2002; Nabuurs et al., 2002). Severe drought increases mortality and is often combined with insect and pathogen damage and wildfires. For example, a positive feedback between deforestation, forest fragmentation, wildfire and increased frequency of droughts appears to exist in the Amazon basin, so that a warmer and drier regional climate may trigger massive deforestation (Laurance and Williamson, 2001; Laurance et al., 2004; Nepstad et al., 2004). Few, if any, models can simulate these effects.

5.4.5.3 *Social and economic impacts*

Climate change impacts on forestry and a shift in production preferences (e.g., towards biofuels) will translate into social and economic impacts through the relocation of forest economic activity. Distributional effects would involve businesses, landowners, workers, consumers, governments and tourism, with some groups and regions benefiting while others experience losses. Net benefits will accrue to regions that experience increased forest production, while regions with declining activity will likely face net losses. If wood prices decline, as most models predict, consumers will experience net benefits, while producers experience net losses. Even though the overall economic benefits are likely to exceed losses, the loss of forest resources may directly affect 90% of the 1.2 billion forest-dependent people who live in extreme poverty (FAO, 2004a). Although forest-based communities in developing countries are likely to have modest impact on global wood production, they may be especially vulnerable because of the limited ability of rural, resource-dependent communities to respond to risk in a proactive manner (Davidson et al., 2003; Lawrence, 2003). Non-timber forest products (NTFP) such as fuel, forest foods or medicinal plants, are equally important for the livelihood of the rural communities. In many rural Sub-Saharan Africa communities, NTFP may supply over 50% of a farmer's cash income and provide the health needs for over 80% of the population (FAO, 2004a). Yet little is known about the possible impacts on NFTP.

5.4.6 Capture fisheries and aquaculture: marine and inland waters

World capture production of fish, crustaceans and molluscs in 2004 was more than twice that of aquaculture (Table 5.5), but since 1997 capture production decreased by 1%, whereas aquaculture increased by 59%. By 2030, capture production and aquaculture are projected to be closer to equality (93 Mt and 83 Mt, respectively) (FAO, 2002). Aquaculture resembles terrestrial animal husbandry more than it does capture fisheries and therefore shares many of the vulnerabilities and adaptations to climate change with that sector. Similarities between aquaculture and terrestrial animal husbandry include ownership, control of inputs, diseases and predators, and use of land and water.

Some aquaculture, particularly of plants and molluscs, depends on naturally occurring nutrients and production, but the rearing of fish and Crustacea usually requires the addition of suitable food, obtained mainly from capture fisheries. Capture fisheries depend on the productivity of the natural ecosystems on which they are based and are therefore vulnerable to changes in primary production and how this production is transferred through the aquatic food chain (climate-induced change in production in natural aquatic ecosystems is dealt with in Chapter 4).

For aquatic systems we still lack the kind of experimental data and models used to predict agricultural crop yields under different climate scenarios; therefore, it is not possible to provide quantitative predictions such as are available for other sectors.

Table 5.5. *World fisheries production in 2004 (source: FAO, Yearbook of Fisheries Statistics http://www.fao.org/fi/statist/statist.asp).*

World production in Mt		Inland	Marine	Total
Capture production	Fish, crustaceans, molluscs, etc.	8.8	85.8	94.6
Aquaculture production	Fish, crustaceans, molluscs, etc.	27.2	18.3	45.5
	Aquatic plants	0.0	13.9	13.9

5.4.6.1 *TAR conclusions remain valid*

The principal conclusions concerning aquaculture and fisheries set out in the TAR (see Section 5.1.3) remain valid and important. The negative impacts of climate change which the TAR identified, particularly on aquaculture and freshwater fisheries, include (i) stress due to increased temperature and oxygen demand and increased acidity (lower pH); (ii) uncertain future water supply; (iii) extreme weather events; (iv) increased frequency of disease and toxic events; (v) sea level rise and conflict of interest with coastal defence needs; and (vi) uncertain future supply of fishmeal and oils from capture fisheries. Positive impacts include increased growth rates and food conversion efficiencies, increased length of growing season, range expansion and use of new areas due to decrease in ice cover.

Information from experimental, observational and modelling studies conducted since the TAR supports these conclusions and provides more detail, especially concerning regional effects.

5.4.6.2 *What is new since the TAR?*

New Knowledge: *Effects of temperature on fish growth.*

One experimental study showed positive effects for rainbow trout (*Oncorhyncus mykiss*) on appetite, growth, protein synthesis and oxygen consumption with a 2°C temperature increase in winter, but negative effects with the same increase in summer. Thus, temperature increases may cause seasonal increases in growth, but also risks to fish populations at the upper end of their thermal tolerance zone. Increasing temperature interacts with other global changes, including declining pH and increasing nitrogen and ammonia, to increase metabolic costs. The consequences of these interactions are speculative and complex (Morgan et al., 2001).

New Knowledge: *Current and future direct effects.*

Direct effects of increasing temperature on marine and freshwater ecosystems are already evident, with rapid poleward shifts in regions, such as the north-east Atlantic, where temperature change has been rapid (see Chapter 1). Further changes in distribution and production are expected due to continuing warming and freshening of the Arctic (ACIA, 2005; Drinkwater, 2005). Local extinctions are occurring at the edges of current ranges, particularly in freshwater and diadromous species[2], e.g., salmon (Friedland et al., 2003) and sturgeon (Reynolds et al., 2005).

New Knowledge: *Current and future effects via the food chain.*

Changes in primary production and transfer through the food chain due to climate will have a key impact on fisheries. Such

[2] Diadromous: migrating between fresh and salt water.

changes may be either positive or negative and the aggregate impact at global level is unknown. Evidence from the Pacific and the Atlantic suggests that nutrient supply to the upper productive layer of the ocean is declining due to reductions in the Meridional Overturning Circulation and upwelling (McPhaden and Zhang, 2002; Curry and Mauritzen, 2005) and changes in the deposition of wind-borne nutrients. This has resulted in reductions in primary production (Gregg et al., 2003), but with considerable regional variability (Lehodey et al., 2003). Further, the decline in pelagic fish catches in Lake Tanganyika since the late 1970s has been ascribed to climate-induced increases in vertical stability of the water column, resulting in reduced availability of nutrients (O'Reilly et al., 2004).

Coupled simulations, using six different models to determine the ocean biological response to climate warming between the beginning of the industrial revolution and 2050 (Sarmiento et al., 2004), showed global increases in primary production of 0.7 to 8.1%, but with large regional differences, which are described in Chapter 4. Palaeological evidence and simulation modelling show North Atlantic plankton biomass declining by 50% over a long time-scale during periods of reduced Meridional Overturning Circulation (Schmittner, 2005). Such studies are speculative, but an essential step in gaining better understanding. The observations and model evidence cited above provide grounds for concern that aquatic production, including fisheries production, will suffer regional and possibly global decline and that this has already begun.

New Knowledge: *Current and future effects of spread of pathogens.*

Climate change has been implicated in mass mortalities of many aquatic species, including plants, fish, corals and mammals, but lack of standard epidemiological data and information on pathogens generally makes it difficult to attribute causes (Harvell et al., 1999) (see Box 5.4). An exception is the northward spread of two protozoan parasites (*Perkinsus marinus* and *Haplosporidium nelsoni*) from the Gulf of Mexico to Delaware Bay and further north, where they have caused mass mortalities of Eastern oysters (*Crassostrea virginica*). Winter temperatures consistently lower than 3°C limit the development of the multinucleated sphere X (MSX) disease caused by *P. marinus* (Hofmann et al., 2001). The poleward spread of this and other pathogens is expected to continue as winter temperatures warm.

New Knowledge: *Economic impacts.*

A recent modelling study predicts that, for the fisheries sector, climate change will have the greatest impact on the economies of central and northern Asian countries, the western Sahel and coastal tropical regions of South America (Allison et al., 2005), as well as some small and medium-sized island states (Aaheim and Sygna, 2000).

Indirect economic impacts of climate change will depend on the extent to which the local economies are able to adapt to new conditions in terms of labour and capital mobility. Change in natural fisheries production is often compounded by decreased harvesting capacity and reduced physical access to markets (Allison et al., 2005).

5.4.6.3 Impacts of decadal variability and extremes

Most of the large global marine-capture fisheries are affected by regional climate variability. Recruitment of the two tropical species of tuna (skipjack and yellowfin) and the sub-tropical albacore (*Thunnus alalunga*) in the Pacific is related to regimes in the major climate indices, ENSO and the Pacific Decadal Oscillation (Lehodey et al., 2003). Large-scale distribution of skipjack tuna in the western equatorial Pacific warm pool can also be predicted from a model that incorporates changes in ENSO (Lehodey, 2001). ENSO events, which are defined by the appearance and persistence of anomalously warm water in the coastal and equatorial ocean off Peru and Ecuador for periods of 6 to 18 months, have adverse effects on Peruvian anchovy production in the eastern Pacific (Jacobson et al., 2001). However, longer term, decadal anomalies appear to have greater long-term consequences for the food-web than the short periods of nutrient depletion during ENSO events (Barber, 2001). Models relating interannual variability, decadal (regional) variability and global climate change must be improved in order to make better use of information on climate change in planning management adaptations.

North Pacific ecosystems are characterised by 'regime shifts' (fairly abrupt changes in both physics and biology persisting for up to a decade). These changes have major consequences for the productivity and species composition of fisheries resources in the region (King, 2005).

Major changes in Atlantic ecosystems can also be related to regional climate indicators, in particular the NAO (Drinkwater et al., 2003; see also Chapter 1 on north-east Atlantic plankton, fish distribution and production). Production of fish stocks, such as cod in European waters, has been adversely affected since the 1960s by the positive trend in the NAO. Recruitment is more sensitive to climate variability when spawning biomass and population structure are reduced (Brander, 2005). In order to reduce sensitivity to climate, stocks may need to be maintained at higher levels.

Climate-related reductions in production cause fish stocks to decline at previously sustainable levels of fishing; therefore the effects of climate must be correctly attributed and taken into account in fisheries management.

Box 5.4. Impact of coral mortality on reef fisheries

Coral reefs and their fisheries are subject to many stresses in addition to climate change (see Chapter 4). So far, events such as the 1998 mass coral bleaching in the Indian Ocean have not provided evidence of negative short-term bio-economic impacts for coastal reef fisheries (Spalding and Jarvis, 2002; Grandcourt and Cesar, 2003). In the longer term, there may be serious consequences for fisheries production that result from loss of coral communities and reduced structural complexity, which result in reduced fish species richness, local extinctions and loss of species within key functional groups of reef fish (Sano, 2004; Graham et al., 2006).

5.4.7 Rural livelihoods: subsistence and smallholder agriculture

The impacts of climate change on subsistence and smallholder agriculture, pastoralism and artisanal fisheries were not discussed explicitly in the TAR, though discussion of these systems is implicit in various sections. A number of case studies of impacts on smallholder livelihood systems in developing countries are beginning to appear, some focussed on recent and current climate variability seen within a climate change context (Thomas et al., 2005a), others using modelling approaches to examine future impacts on key smallholder crops (Abou-Hadid, 2006; Adejuwon, 2006) or ecosystems used by smallholder farmers (Lasco and Boer, 2006). In some cases impacts are discussed within work focussed more on adaptation (Thomas et al., 2005a).

Specific impacts must be examined within the context of whole sets of confounding impacts at regional to local scales (Adger et al., 2003). It is difficult to ascribe levels of confidence to these confounding impacts because livelihood systems are typically complex and involve a number of crop and livestock species, between which there are interactions (for example, intercropping practices (Richards, 1986) or the use of draught-animal power for cultivation (Powell et al., 1998)), and potential substitutions such as alternative crops. Many smallholder livelihoods will also include elements such as use of wild resources, and non-agricultural strategies such as use of remittances. Coping strategies for extreme climatic events such as drought (Davies, 1996; Swearingen and Bencherifa, 2000; Mortimore and Adams, 2001; Ziervogel, 2003) typically involve changes in the relative importance of such elements, and in the interactions between them. Pastoralist coping strategies in northern Kenya and southern Ethiopia are discussed in Box 5.5.

Impacts of climate change upon these systems will include:

- The direct impacts of changes in temperature, CO_2 and precipitation on yields of specific food and cash crops, productivity of livestock and fisheries systems, and animal health, as discussed in Sections 5.4.1 to 5.4.6 above. These will include both impacts of changing means and increased frequency of extreme events, with the latter being more important in the medium-term (to 2025) (Corbera et al., 2006). Positive and negative impacts on different crops may occur in the same farming system. Agrawala et al. (2003) suggest that impacts on maize, the main food crop, will be strongly negative for the Tanzanian smallholder, while impacts on coffee and cotton, significant cash crops, may be positive.
- Other physical impacts of climate change important to smallholders are: (i) decreased water supply from snowcaps for major smallholder irrigation systems, particularly in the Indo-Gangetic plain (Barnett et al., 2005), (ii) the effects of sea level rise on coastal areas, (iii) increased frequency of landfall tropical storms (Adger, 1999) and (iv) other forms of environmental impact still being identified, such as increased forest-fire risk (Agrawala et al., 2003, for the Mount Kilimanjaro ecosystem) and remobilisation of dunes (Thomas et al., 2005b for semi-arid Southern Africa).
- Impacts on human health, like malaria risk (see Chapter 8, Section 8.4.1.2), affect labour available for agriculture and other non-farm rural economic activities, such as tourism (see Chapter 7, Section 7.4.2.2).

For climate change impacts on the three major cereal crops grown by smallholders, we refer to Figure 5.2a-f and discussion in Sections 5.4.2 and 5.5.1. In Section 5.4.1 above we discuss the various negative impacts of increases in climate variability and

Box 5.5. Pastoralist coping strategies in northern Kenya and southern Ethiopia

African pastoralism has evolved in adaptation to harsh environments with very high spatial and temporal variability of rainfall (Ellis, 1995). Several recent studies (Ndikumana et al., 2000; Hendy and Morton, 2001; Oba, 2001; McPeak and Barrett, 2001; Morton, 2006) have focussed on the coping strategies used by pastoralists during recent droughts in northern Kenya and southern Ethiopia, and the longer-term adaptations that underlie them:

- *Mobility* remains the most important pastoralist adaptation to spatial and temporal variations in rainfall, and in drought years many communities make use of fall-back grazing areas unused in 'normal' dry seasons because of distance, land tenure constraints, animal disease problems or conflict. But encroachment on and individuation of communal grazing lands, and the desire to settle to access human services and food aid, have severely limited pastoral mobility.
- Pastoralists engage in *herd accumulation* and most evidence now suggests that this is a rational form of insurance against drought.
- A small proportion of pastoralists now hold some of their wealth in bank accounts, and others use informal savings and credit mechanisms through shopkeepers.
- Pastoralists also use *supplementary feed* for livestock, purchased or lopped from trees, as a coping strategy; they intensify *animal disease management* through indigenous and scientific techniques; they pay for *access to water* from powered boreholes.
- *Livelihood diversification* away from pastoralism in this region predominantly takes the form of shifts into low-income or environmentally unsustainable occupations such as charcoal production, rather than an adaptive strategy to reduce *ex-ante* vulnerability.
- A number of *intra-community mechanisms* distribute both livestock products and the use of live animals to the destitute, but these appear to be breaking down because of the high levels of covariate risk within communities.

frequency of extreme events on yields (see also Porter and Semenov, 2005). Burke et al. (2006) demonstrate the risk of widespread drought in many regions, including Africa. Projected impacts on world regions, some of which are disaggregated into smallholder and subsistence farmers or similar categories, are reviewed in the respective regional chapters. An important study by Jones and Thornton (2003) found that aggregate yields of smallholder rain-fed maize in Africa and Latin America are likely to decrease by almost 10% by 2055, but these results hide enormous regional variability (see also Fischer et al., 2002b) of concern for subsistence agriculture.

With a large body of smallholder and subsistence farming households in the dryland tropics, there is especial concern over temperature-induced declines in crop yields, and increasing frequency and severity of drought. These will lead to the following generalisations (low confidence):

- increased likelihood of crop failure;
- increased diseases and mortality of livestock and/or forced sales of livestock at disadvantageous prices (Morton and de Haan, 2006);
- livelihood impacts including sale of other assets, indebtedness, out-migration and dependency on food relief;
- eventual impacts on human development indicators, such as health and education.

Impacts of climate change will combine with non-climate stressors as listed in Section 5.2.2 above, including the impacts of globalisation (O'Brien and Leichenko, 2000) and HIV and/or AIDS (Gommes et al., 2004; see also Chapter 8).

Modelling studies are needed to understand the interactions between these different forms of climate change impacts and the adaptations they will require. The multi-agent modelling of Bharwani et al. (2005) is one possible approach. Empirical research on how current strategies to cope with extreme events foster or constrain longer-term adaptation is also important (see Davies, 1996). Knowledge of crop responses to climate change also needs to be extended to more crops of interest to smallholders.

Many of the regions characterised by subsistence and smallholder agriculture are storehouses of unexplored biodiversity (Hannah et al., 2002). Pressure to cultivate marginal land or to adopt unsustainable cultivation practices as yields drop, and the break down of food systems more generally (Hannah et al., 2002), may endanger biodiversity of both wild and domestic species. Smallholder and subsistence farming areas are often also environmentally marginal (which does not necessarily conflict with biodiversity) and at risk of land degradation as a result of climate trends, but mediated by farming and livestock-production systems (Dregne, 2000).

5.5 Adaptations: options and capacities

Adaptation is used here to mean both the actions of adjusting practices, processes and capital in response to the actuality or threat of climate change as well as changes in the decision environment, such as social and institutional structures, and altered technical options that can affect the potential or capacity for these actions to be realised (see Chapter 17). Adaptations are divided here into two categories: *autonomous adaptation*, which is the ongoing implementation of existing knowledge and technology in response to the changes in climate experienced, and *planned adaptation*, which is the increase in adaptive capacity by mobilising institutions and policies to establish or strengthen conditions favourable for effective adaptation and investment in new technologies and infrastructure.

The TAR noted agriculture has historically shown high levels of adaptability to climate variations and that while there were many studies of climate change impacts, there were relatively few that had comparisons with and without adaptation. Generally the adaptations assessed were most effective in mid-latitudes and least effective in low-latitude developing regions with poor resource endowments and where ability of farmers to respond and adapt was low. There was limited evaluation of either the costs of adaptation or of the environmental and natural resource consequences of adaptation. Generally, adaptation studies have focussed on situations where climate changes are expected to have net negative consequences: there is a general expectation that if climate improves, then market forces and the general availability of suitable technological options will result in effective change to new, more profitable or resilient systems (e.g., Parson et al., 2003).

5.5.1 Autonomous adaptations

Many of the autonomous adaptation options identified before and since the TAR are largely extensions or intensifications of existing risk-management or production-enhancement activities. For cropping systems there are many potential ways to alter management to deal with projected climatic and atmospheric changes (Aggarwal and Mall, 2002; Alexandrov et al., 2002; Tubiello et al., 2002; Adams et al., 2003; Easterling et al., 2003; Howden et al., 2003; Howden and Jones, 2004; Butt et al., 2005; Travasso et al., 2006; Challinor et al., 2007). These adaptations include:

- altering inputs such as varieties and/or species to those with more appropriate thermal time and vernalisation requirements and/or with increased resistance to heat shock and drought, altering fertiliser rates to maintain grain or fruit quality consistent with the climate and altering amounts and timing of irrigation and other water management practices;
- wider use of technologies to 'harvest' water, conserve soil moisture (e.g., crop residue retention) and to use water more effectively in areas with rainfall decreases;
- water management to prevent waterlogging, erosion and nutrient leaching in areas with rainfall increases;
- altering the timing or location of cropping activities;
- diversifying income by integrating other farming activities such as livestock raising;
- improving the effectiveness of pest, disease and weed management practices through wider use of integrated pest and pathogen management, development and use of varieties and species resistant to pests and diseases, maintaining or improving quarantine capabilities, and sentinel monitoring programs;
- using seasonal climate forecasting to reduce production risk.

If widely adopted, these autonomous adaptations, singly or in combination, have substantial potential to offset negative climate change impacts and take advantage of positive ones. For example, in a modelling study for Modena (Italy), simple, currently practicable adaptations of varieties and planting times to avoid drought and heat stress during the hotter and drier summer months predicted under climate change altered significant negative impacts on sorghum (–48 to –58%) to neutral to marginally positive ones (0 to +12%; Tubiello et al., 2000). We have synthesised results from many crop adaptation studies for wheat, rice and maize (Figure 5.2). The benefits of adaptation vary with crops and across regions and temperature changes; however, on average, they provide approximately a 10% yield benefit when compared with yields when no adaptation is used. Another way to view this is that these adaptations translate to damage avoidance in grain yields of rice, wheat and maize crops caused by a temperature increase of up to 1.5 to 3°C in tropical regions and 4.5 to 5°C in temperate regions. Further warming than these ranges in either region exceeds adaptive capacity. The benefits of autonomous adaptations tend to level off with increasing temperature changes (Howden and Crimp, 2005) while potential negative impacts increase.

While autonomous adaptations such as the above have the potential for considerable damage avoidance from problematic climate changes, there has been little evaluation of how effective and widely adopted these adaptations may actually be, given (i) the complex nature of farm decision-making in which there are many non-climatic issues to manage, (ii) the likely diversity of responses within and between regions in part due to possible differences in climate changes, (iii) the difficulties that might arise if climate changes are non-linear or increase climate extremes, (iv) time-lags in responses and (v) the possible interactions between different adaptation options and economic, institutional and cultural barriers to change. For example, the realisable adaptive capacity of poor subsistence farming and/or herding communities is generally considered to be very low (Leary et al., 2006). These considerations also apply to the livestock, forestry and fisheries.

Adaptations in field-based livestock include matching stocking rates with pasture production, rotating pastures, modifying grazing times, altering forage and animal species/breeds, altering the integration of mixed livestock/crop systems, including the use of adapted forage crops, re-assessing fertiliser applications, ensuring adequate water supplies and using supplementary feeds and concentrates (Daepp et al., 2001; Holden and Brereton, 2002; Adger et al., 2003; Batima et al., 2005). It is important to note, however, that there are often limitations to these adaptations. For example, more heat-tolerant livestock breeds often have lower levels of productivity. Following from the above, in intensive livestock industries, there may be reduced need for winter housing and for feed concentrates in cold climates, but in warmer climates there could be increased need for management and infrastructure to ameliorate heat stress-related reductions in productivity, fertility and increased mortality.

A large number of autonomous adaptation strategies have been suggested for planted forests including changes in management intensity, hardwood/softwood species mix, timber growth and harvesting patterns within and between regions, rotation periods, salvaging dead timber, shifting to species or areas more productive under the new climatic conditions, landscape planning to minimise fire and insect damage, adjusting to altered wood size and quality, and adjusting fire-management systems (Sohngen et al., 2001; Alig et al., 2002; Spittlehouse and Stewart, 2003; Weih, 2004). Adaptation strategies to control insect damage can include prescribed burning to reduce forest vulnerability to increased insect outbreaks, non-chemical insect control (e.g., baculoviruses) and adjusting harvesting schedules, so that those stands most vulnerable to insect defoliation can be harvested preferentially. Under moderate climate changes, these proactive measures may potentially reduce the negative economic consequences of climate change (Shugart et al., 2003). However, as with other primary industry sectors, there is likely to be a gap between the potential adaptations and the realised actions. For example, large areas of forests, especially in developing countries, receive minimal direct human management (FAO, 2000), which limits adaptation opportunities. Even in more intensively managed forests where adaptation activities may be more feasible (Shugart et al., 2003) the long time-lags between planting and harvesting trees will complicate decisions, as adaptation may take place at multiple times during a forestry rotation.

Marine ecosystems are in some respects less geographically constrained than terrestrial systems. The rates at which planktonic ecosystems have shifted their distribution has been very rapid over the past three decades, which can be regarded as natural adaptation to a changing physical environment (see Chapter 1 and Beaugrand et al., 2002). Most fishing communities are dependent on stocks that fluctuate due to interannual and decadal climate variability and consequently have developed considerable coping capacity (King, 2005). With the exception of aquaculture and some freshwater fisheries, the exploitation of natural fish populations, which are common-property resources, precludes the kind of management adaptations to climate change suggested for the crop, livestock and forest sectors. Adaptation options thus centre on altering catch size and effort. Three-quarters of world marine fish stocks are currently exploited at levels close to or above their productive capacity (Bruinsma, 2003). Reductions in the level of fishing are therefore required in many cases to sustain yields and may also benefit fish stocks, which are sensitive to climate variability when their population age-structure and geographic sub-structure is reduced (Brander, 2005). The scope for autonomous adaptation is increasingly restricted as new regulations governing exploitation of fisheries and marine ecosystems come into force. Scenarios of increased levels of displacement and migration are likely to put a strain on communal-level fisheries management and resource access systems, and weaken local institutions and services. Despite their adaptive value for the sustainable use of natural resource systems, migrations can impede economic development (Allison et al., 2005; see Chapter 17, Box 17.8).

5.5.2 Planned adaptations

Autonomous adaptations may not be fully adequate for coping with climate change, thus necessitating deliberate, planned measures. Many options for policy-based adaptation to climate change have been identified for agriculture, forests and fisheries (Howden et al., 2003; Kurukulasuriya and Rosenthal, 2003;

Aggarwal et al., 2004; Antle et al., 2004; Easterling et al., 2004). These can either involve adaptation activities such as developing infrastructure or building the capacity to adapt in the broader user community and institutions, often by changing the decision-making environment under which management-level, autonomous adaptation activities occur (see Chapter 17). Effective planning and capacity building for adaptation to climate change could include:

1. To change their management, enterprise managers need to be convinced that the climate changes are real and are likely to continue (e.g., Parson et al., 2003). This will be assisted by policies that maintain climate monitoring and communicate this information effectively. There could be a case also for targeted support of the surveillance of pests, diseases and other factors directly affected by climate.
2. Managers need to be confident that the projected changes will significantly impact on their enterprise (Burton and Lim, 2005). This could be assisted by policies that support the research, systems analysis, extension capacity, and industry and regional networks that provide this information.
3. There needs to be technical and other options available to respond to the projected changes. Where the existing technical options are inadequate to respond, investment in new technical or management options may be required (e.g., improved crop, forage, livestock, forest and fisheries germplasm, including via biotechnology, see Box 5.6) or old technologies revived in response to the new conditions (Bass, 2005).
4. Where there are major land use changes, industry location changes and migration, there may be a role for governments to support these transitions via direct financial and material support, creating alternative livelihood options. These include reduced dependence on agriculture, supporting community partnerships in developing food and forage banks, enhancing capacity to develop social capital and share information, providing food aid and employment to the more vulnerable and developing contingency plans (e.g., Olesen and Bindi, 2002; Winkels and Adger, 2002; Holling, 2004). Effective planning for and management of such transitions may also result in less habitat loss, less risk of carbon loss (e.g., Goklany, 1998) and also lower environmental costs such as soil degradation, siltation and reduced biodiversity (Stoate et al., 2001).
5. Developing new infrastructure, policies and institutions to support the new management and land use arrangements by addressing climate change in development programs; enhanced investment in irrigation infrastructure and efficient water use technologies; ensuring appropriate transport and storage infrastructure; revising land tenure arrangements, including attention to well-defined property rights (FAO, 2003a); establishment of accessible, efficiently functioning markets for products and inputs (seed, fertiliser, labour, etc.) and for financial services, including insurance (Turvey, 2001).
6. The capacity to make continuing adjustments and improvements in adaptation by understanding what is working, what is not and why, via targeted monitoring of adaptations to climate change and their costs and effects (Perez and Yohe, 2005).

It is important to note that policy-based adaptations to climate change will interact with, depend on or perhaps even be just a subset of policies on natural resource management, human and animal health, governance and political rights, among many others: the 'mainstreaming' of climate change adaptation into policies intended to enhance broad resilience (see Chapter 17).

5.6 Costs and other socio-economic aspects, including food supply and security

5.6.1 Global costs to agriculture

Fischer et al. (2002b) quantify the impact of climate change on global agricultural GDP by 2080 as between -1.5% and +2.6%, with considerable regional variation. Overall, mid- to high-latitudes agriculture stands to benefit, while agriculture in low latitudes will be adversely affected. However, Fischer et al. (2002b) suggest that, taking into account economic adjustment,

Box 5.6. Will biotechnology assist agricultural and forest adaptation?

Breakthroughs in molecular genetic mapping of the plant genome have led to the identification of bio-markers that are closely linked to known resistance genes, such that their isolation is clearly feasible in the future. Two forms of stress resistance especially relevant to climate change are to drought and temperature. A number of studies have demonstrated genetic modifications to major crop species (e.g., maize and soybeans) that increased their water-deficit tolerance (as reviewed by Drennen et al., 1993; Kishor et al., 1995; Pilon-Smits et al., 1995; Cheikh et al., 2000), although this may not extend to the wider range of crop plants. Similarly, there are possibilities for enhanced resistance to pests and diseases, salinity and waterlogging, or for opportunities such as change in flowering times or enhanced responses to elevated CO_2. Yet many research challenges lie ahead. Little is known about how the desired traits achieved by genetic modification perform in real farming and forestry applications. Moreover, alteration of a single physiological process is often compensated or dampened so that little change in plant growth and yield is achieved from the modification of a single physiological process (Sinclair and Purcell, 2005). Although biotechnology is not expected to replace conventional agronomic breeding, Cheikh et al. (2000) and FAO (2004b) argue that it will be a crucial adjunct to conventional breeding (it is likely that both will be needed to meet future environmental challenges, including climate change).

global cereal production by 2080 falls within a 2% boundary of the no-climate change reference production.

Impacts of climate change on world food prices are summarised in Figure 5.3. Overall, the effects of higher global mean temperatures (GMTs) on food prices follow the expected changes in crop and livestock production. Higher output associated with a moderate increase in the GMT likely results in a small decline in real world food (cereals) prices, while GMT changes in the range of 5.5°C or more could lead to a pronounced increase in food prices of, on average, 30%.

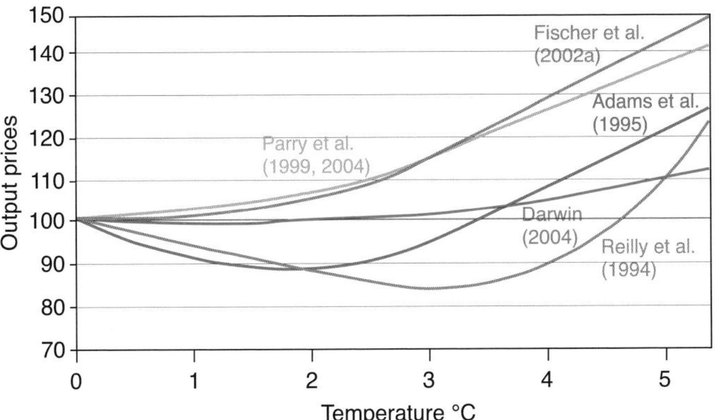

Figure 5.3. *Cereal prices (percent of baseline) versus global mean temperature change for major modelling studies. Prices interpolated from point estimates of temperature effects.*

5.6.2 Global costs to forestry

Alig et al. (2004) suggest that climate variability and climate change may alter the productivity of forests and thereby shift resource management, economic processes of adaptation and forest harvests, both nationally and regionally. Such changes may also alter the supply of products to national and international markets, as well as modify the prices of forest products, impact economic welfare and affect land-use changes. Current studies consider mainly the impact of climate change on forest resources, industry and economy; however, some analyses include feedbacks in the ecological system, including greenhouse gas cycling in forest ecosystems and forest products (e.g., Sohngen and Sedjo, 2005). A number of studies analyse the effects of climate change on the forest industry and economy (e.g., Binkley, 1988; Joyce et al., 1995; Perez-Garcia et al., 1997; Sohngen and Mendelsohn, 1998; Shugart et al., 2003; see Table 5.4 and Section 5.4.5).

If the world develops as the models predict, there will be a general decline of wood raw-material prices due to increased wood production (Perez-Garcia et al., 1997; Sohngen and Mendelsohn, 1998). The same authors conclude that economic welfare effects are relatively small but positive, with net benefits accruing to wood consumers. However, changes in other sectors, such as major shifts in demand and requirements for energy production, will also impact prices in the forest sector. There are no concrete studies on non-wood services from forest resources, but the impacts of climate change on many of these services will likely be spatially specific.

5.6.3 Changes in trade

The principal impact of climate change on agriculture is an increase in production potential in mid- to high-latitudes and a decrease in low latitudes. This shift in production potential is expected to result in higher trade flows of mid- to high-latitude products (e.g., cereals and livestock products) to the low latitudes. Fischer et al. (2002b) estimate that by 2080 cereal imports by developing countries would rise by 10-40%.

5.6.4 Regional costs and associated socio-economic impacts

Fischer et al. (2002b) quantified regional impacts and concluded that globally there will be major gains in potential agricultural land by 2080, particularly in North America (20-50%) and the Russian Federation (40-70%), but losses of up to 9% in sub-Saharan Africa. The regions likely to face the biggest challenges in food security are Africa, particularly sub-Saharan Africa, and Asia, particularly south Asia (FAO, 2006).

Africa

Yields of grains and other crops could decrease substantially across the African continent because of increased frequency of drought, even if potential production increases due to increases in CO_2 concentrations. Some crops (e.g., maize) could be discontinued in some areas. Livestock production would suffer due to deteriorated rangeland quality and changes in area from rangeland to unproductive shrub land and desert.

Asia

According to Murdiyarso (2000), rice production in Asia could decline by 3.8% during the current century. Similarly, a 2°C increase in mean air temperature could decrease rice yield by about 0.75 tonne/ha in India and rain-fed rice yield in China by 5-12% (Lin et al., 2005). Areas suitable for growing wheat could decrease in large portions of south Asia and the southern part of east Asia (Fischer et al., 2002b). For example, without the CO_2 fertilisation effect, a 0.5°C increase in winter temperature would reduce wheat yield by 0.45 ton/ha in India (Kalra et al., 2003) and rain-fed wheat yield by 4-7% in China by 2050. However, wheat production in both countries would increase by between 7% and 25% in 2050 if the CO_2 fertilisation effect is taken into account (Lin et al., 2005).

5.6.5 Food security and vulnerability

All four dimensions of food security, namely food availability (i.e., production and trade), stability of food supplies, access to food, and food utilisation (FAO, 2003a) will likely be affected by climate change. Importantly, food security will depend not only on climate and socio-economic impacts, but also, and critically so, on changes to trade flows, stocks and food-aid policy. Climate change impacts on food production (*food availability*) will be mixed and vary regionally (FAO, 2003b, 2005c). For instance, a reduction in the production potential of tropical developing countries, many of which have poor land and water resources, and are already faced with serious food insecurity, may add to the burden of these countries (e.g., Hitz and Smith, 2004; Fischer et al., 2005a; Parry et al., 2005). Globally, the potential for food production is projected to increase with increases in local average temperature over a range of 1 to 3°C, but above this it is projected to decrease. Changes in the patterns of extreme events, such as increased frequency and intensity of droughts and flooding, will affect the *stability* of, as well as *access* to, food supplies. Food insecurity and loss of livelihood would be further exacerbated by the loss of cultivated land and nursery areas for fisheries through inundation and coastal erosion in low-lying areas (FAO, 2003c).

Climate change may also affect food *utilisation*, notably through additional health consequences (see Chapter 8). For example, populations in water-scarce regions are likely to face decreased water availability, particularly in the sub-tropics, with implications for food processing and consumption; in coastal areas, the risk of flooding of human settlements may increase, from both sea level rise and increased heavy precipitation. This is likely to result in an increase in the number of people exposed to vector-borne (e.g., malaria) and water-borne (e.g., cholera) diseases, thus lowering their capacity to utilise food effectively.

A number of studies have quantified the impacts of climate change on food security at regional and global scales (e.g., Fischer et al., 2002b, 2005b; Parry et al., 2004, 2005; Tubiello and Fischer, 2006). These projections are based on complex modelling frameworks that integrate the outputs of GCMs, agro-ecological zone data and/or dynamic crop models, and socio-economic models. In these systems, impacts of climate change on agronomic production potentials are first computed; then consequences for food supply, demand and consumption at regional to global levels are computed, taking into account different socio-economic futures (typically SRES scenarios). A number of limitations, however, make these model projections highly uncertain. First, these estimates are limited to the impacts of climate change mainly on food availability; they do not cover potential changes in the stability of food supplies, for instance, in the face of changes to climate and/or socio-economic variability. Second, projections are based on a limited number of crop models, and only one economic model (see legend in Table 5.6), the latter lacking sufficient evaluation against observations, and thus in need of further improvements.

Despite these limitations and uncertainties, a number of fairly robust findings for policy use emerge from these studies. *First*, climate change is likely to increase the number of people at risk of hunger compared with reference scenarios with no climate change. However, impacts will depend strongly on projected socio-economic developments (Table 5.6). For instance, Fischer et al. (2002a, 2005b) estimate that climate change will increase the number of undernourished people in 2080 by 5-26%, relative to the no climate change case, or by between 5-10 million (SRES B1) and 120-170 million people (SRES A2). The within-SRES ranges are across several GCM climate projections. Using only one GCM scenario, Parry et al. (2004, 2005) estimated small reductions by 2080, i.e., –5% (–10 [B] to –30 [A2] million people), and slight increases of +13-26% (10 [B2] to 30 [A1] million people).

Second, the magnitude of these climate impacts will be small compared with the impacts of socio-economic development (e.g., Tubiello et al., 2007b). With reference to Table 5.6, these studies suggest that economic growth and slowing population growth projected for the 21st century will, globally, significantly reduce the number of people at risk of hunger in 2080 from current levels. Specifically, compared with FAO estimates of 820 million undernourished in developing countries today, Fischer et al. (2002a, 2005b) and Parry et al. (2004, 2005) estimate reductions by more than 75% by 2080, or by about 560-700 million people, thus projecting a global total of 100-240 million undernourished by 2080 (A1, B1 and B2). By contrast, in A2, the number of the hungry may decrease only slightly in 2080, because of larger population projections compared with other SRES scenarios (Fischer et al., 2002a, 2005b; Parry et al., 2004, 2005; Tubiello and Fischer, 2006). These projections also indicate that, with or without climate change, Millennium Development Goals (MDGs) of halving the proportion of people at risk of hunger by 2015 may not be realised until 2020-2030 (Fischer et al., 2005b; Tubiello, 2005).

Third, sub-Saharan Africa is likely to surpass Asia as the most food-insecure region. However, this is largely independent of climate change and is mostly the result of the projected socio-economic developments for the different developing regions. Studies using various SRES scenarios and model analyses indicate that by 2080 sub-Saharan Africa may account for 40-50% of all undernourished people, compared with about 24% today (Fischer et al., 2002a, 2005b; Parry et al., 2004, 2005); some estimates are as high as 70-75% under the A2 and B2 assumptions of slower economic growth (Fischer et al., 2002a; Parry et al., 2004; Tubiello and Fischer, 2006).

Fourth, there is significant uncertainty concerning the effects of elevated CO_2 on food security. With reference to Table 5.6, under most future scenarios the assumed strength of CO_2 fertilisation would not greatly affect global projections of hunger, particularly when compared with the absolute reductions attributed solely to socio-economic development (Tubiello et al., 2007a,b). For instance, employing one GCM, but assuming no effects of CO_2 on crops, Fischer et al. (2002a, 2005b) and Parry et al. (2004, 2005) projected absolute global numbers of undernourished in 2080 in the range of 120-380 million people across SRES scenarios A1, B1 and B2, as opposed to a range of 100-240 million when account is taken of CO_2 effects. The exception again in these studies is SRES A2, under which scenario the assumption of no CO_2 fertilisation results in a projected range of 950-1,300 million people undernourished in 2080, compared with 740-850 million with climate change and CO_2 effects on crops.

Finally, recent research suggests large positive effects of climate mitigation on the agricultural sector, although benefits, in terms of avoided impacts, may be realised only in the second half of this century due to the inertia of global mean temperature and the easing of positive effects of elevated CO_2 in the mitigated scenarios (Arnell et al., 2002; Tubiello and Fischer, 2006). Even in the presence of robust global long-term benefits, regional and temporal patterns of winners and losers are highly uncertain and critically dependent on GCM projections (Tubiello and Fischer, 2006).

Table 5.6. *The impacts of climate change and socio-economic development paths on the number of people at risk of hunger in developing countries (data from Parry et al., 2004; Tubiello et al., 2007b). The first set of rows in the table depicts reference projections under SRES scenarios and no climate change. The second set (CC) includes climate change impacts, based on Hadley HadCM3 model output, including positive effects of elevated CO_2 on crops. The third (CC, no CO_2) includes climate change, but assumes no effects of elevated CO_2. Projections from 2020 to 2080 are given for two crop-modelling systems: on the left, AEZ (Fischer et al., 2005b); on the right, DSSAT (Parry et al., 2004), each coupled to the same economic and food trade model, BLS (Fischer et al., 2002a, 2005b). The models are calibrated to give 824 million undernourished in 2000, according to FAO data.*

	2020 Millions at risk		2050 Millions at risk		2080 Millions at risk	
Reference	AEZ-BLS	DSSAT-BLS	AEZ-BLS	DSSAT-BLS	AEZ-BLS	DSSAT-BLS
A1	663	663	208	208	108	108
A2	782	782	721	721	768	769
B1	749	749	239	240	91	90
B2	630	630	348	348	233	233
CC	AEZ-BLS	DSSAT-BLS	AEZ-BLS	DSSAT-BLS	AEZ-BLS	DSSAT-BLS
A1	666	687	219	210	136	136
A2	777	805	730	722	885	742
B1	739	771	242	242	99	102
B2	640	660	336	358	244	221
CC, no CO_2	AEZ-BLS	DSSAT-BLS	AEZ-BLS	DSSAT-BLS	AEZ-BLS	DSSAT-BLS
A1	NA	726	NA	308	NA	370
A2	794	845	788	933	950	1320
B1	NA	792	NA	275	NA	125
B2	652	685	356	415	257	384

5.7 Implications for sustainable development

Human societies have, through the centuries, often developed the capacity to adapt to environmental change, and some knowledge about the implications of climate change adaptation for sustainable development can thus be deduced from historical analogues (Diamond, 2004; Easterling et al., 2004).

Unilateral adaptation measures to water shortage related to climate change can lead to competition for water resources and, potentially, to conflict and backlash for development. International and regional approaches are required to develop joint solutions, such as the three-border project Trifinio in Lempa valley between Honduras, Guatemala and El Salvador (Dalby, 2004). Shifts in land productivity may lead to a shift in agriculture and livestock systems in some regions, and to agricultural intensification in others. This results not only in environmental benefits, such as less habitat loss and lower carbon emissions (Goklany, 1998, 2005), but also in environmental costs, such as soil degradation, siltation, reduced biodiversity and others (Stoate et al., 2001).

Adaptive measures in response to habitat and ecosystem shifts, such as expansion of agriculture into previously forested areas, will lead to additional loss and fragmentation of habitats. Currently, deforestation, mainly a result of conversion of forests to agricultural land, continues at a rate of 13 million ha/yr (FAO, 2005b). The degradation of ecosystem services not only poses a barrier to achieving sustainable development in general, but also to meeting specific international development goals, notably the MDGs (Millennium Ecosystem Assessment, 2005). The largest forest losses have occurred in South America and Africa, often in countries marked by high reliance on solid fuels, low levels of access to safe water and sanitation, and the slowest progress towards the MDG targets. Response strategies aimed at minimising such losses will have to focus increasingly on regional and international landscape development (Opdam and Wascher, 2004).

Impacts on trade, economic development and environmental quality, as well as land use, may also be expected from measures to substitute fossil fuels with biofuels, such as the European Biomass Action Plan. It may be necessary to balance competition between the energy and forest products sectors for raw materials, and competition for land for biofuels, food and forestry.

Sustainable economic development and poverty reduction remain top priorities for developing countries (Aggarwal et al., 2004). Climate change could exacerbate climate-sensitive hurdles to sustainable development faced by developing countries (Goklany, 2007). This will require integrated approaches to concurrently advance adaptation, mitigation and sustainable development. Goklany (2007) also offers a portfolio of pro-active strategies and measures, including measures that would simultaneously reduce pressures on biodiversity, hunger and carbon sinks. Moreover, any adaptation measures should be developed as part of, and be closely integrated into, overall and country-specific development programmes and strategies, e.g., into Poverty Reduction Strategy Programmes (Eriksen and Naess, 2003) and pro-poor strategies (Kurukulasuriya and Rosenthal, 2003), and should be understood as a 'shared responsibility' (Ravindranath and Sathaye, 2002).

5.8 Key conclusions and their uncertainties, confidence levels and research gaps

5.8.1 Findings and key conclusions

Projected changes in the frequency and severity of extreme climate events will have more serious consequences for food and forestry production, and food insecurity, than will changes in projected means of temperature and precipitation (high confidence).

Modelling studies suggest that increasing frequency of crop loss due to extreme events, such as droughts and heavy precipitation, may overcome positive effects of moderate temperature increase [5.4.1]. For forests, elevated risks of fires, insect outbreaks, wind damage and other forest-disturbance events are projected, although little is known about their overall effect on timber production [5.4.1].

Climate change increases the number of people at risk of hunger (high confidence). The impact of chosen socio-economic pathways (SRES scenario) on the numbers of people at risk of hunger is significantly greater than the impact of climate change. Climate change will further shift the focus of food insecurity to sub-Saharan Africa.

Climate change alone is estimated to increase the number of undernourished people to between 40 million and 170 million. By contrast, the impacts of socio-economic development paths (SRES) can amount to several hundred million people at risk of hunger [5.6.5]. Moreover, climate change is likely to further shift the regional focus of food insecurity to sub-Saharan Africa. By 2080, about 75% of all people at risk of hunger are estimated to live in this region. The effects of climate mitigation measures are likely to remain relatively small in the early decades; significant benefits of mitigation to the agricultural sector may be realised only in the second half of this century, i.e., once the positive CO_2 effects on crop yields level off and global mean temperature increases become significantly less than in non-mitigated scenarios [5.6.5].

While moderate warming benefits crop and pasture yields in mid- to high-latitude regions, even slight warming decreases yields in seasonally dry and low-latitude regions (medium confidence).

The preponderance of evidence from models suggests that moderate local increases in temperature (to 3°C) can have small beneficial impacts on major rain-fed crops (maize, wheat, rice) and pastures in mid- to high-latitude regions, but even slight warming in seasonally dry and tropical regions reduces yield. Further warming has increasingly negative impacts in all regions [5.4.2 and see Figure 5.2]. These results, on the whole, project the potential for global food production to increase with increases in local average temperature over a range of 1 to 3°C, but above this range to decrease [5.4, 5.6]. Furthermore, modelling studies that include extremes in addition to changes in mean climate show lower crop yields than for changes in means alone, strengthening similar TAR conclusions [5.4.1]. A change in frequency of extreme events is likely to disproportionately impact small-holder farmers and artisan fishers [5.4.7].

Experimental research on crop response to elevated CO_2 confirms Third Assessment Report (TAR) findings (medium to high confidence). New Free-Air Carbon Dioxide Enrichment (FACE) results suggest lower responses for forests (medium confidence). Crop models include CO_2 estimates close to the upper range of new research (high confidence), while forest models may overestimate CO_2 effects (medium confidence).

Recent results from meta-analyses of FACE studies of CO_2 fertilisation confirm conclusions from the TAR that crop yields at CO_2 levels of 550 ppm increase by an average of 15%. Crop model estimates of CO_2 fertilisation are in the range of FACE results [5.4.1.1]. For forests, FACE experiments suggest an average growth increase of 23% for younger tree stands, but little stem-growth enhancement for mature trees. The models often assume higher growth stimulation than FACE, up to 35% [5.4.1.1, 5.4.5].

Globally, commercial timber productivity rises modestly with climate change in the short and medium term, with large regional variability around the global trend (medium confidence).

Overall, global forest products output at 2020 and 2050 changes, ranging from a modest increase to a slight decrease depending on the assumed impact of CO_2 fertilisation and the effect of disturbance processes not well represented in the models (e.g., insect outbreaks), although regional and local changes will be large [5.4.5.2].

Local extinctions of particular fish species are expected at edges of ranges (high confidence).

Regional changes in the distribution and productivity of particular fish species are expected because of continued warming and local extinctions will occur at the edges of ranges, particularly in freshwater and diadromous species (e.g., salmon, sturgeon). In some cases, ranges and productivity will increase [5.4.6]. Emerging evidence suggests concern that the Meridional Overturning Circulation is slowing down, with serious potential consequences for fisheries [5.4.6].

Food and forestry trade is projected to increase in response to climate change, with increased dependence of most developing countries on food imports (medium to low confidence).

While the purchasing power for food is reinforced in the period to 2050 by declining real prices, it would be adversely affected by higher real prices for food from 2050 to 2080 [5.6.1, 5.6.2]. Food security is already challenged in many of the regions expected to suffer more severe yield declines. Agricultural and forestry trade flows are foreseen to rise significantly. Exports of food products from the mid and high latitudes to low latitude countries will rise [5.6.2], while the reverse may take place in forestry [5.4.5].

Simulations suggest rising relative benefits of adaptation with low to moderate warming (medium confidence), although adaptation may stress water and environmental resources as warming increases (low confidence).

There are multiple adaptation options that imply different costs, ranging from changing practices in place to changing locations of food, fibre, forestry and fishery (FFFF) activities [5.5.1]. The potential effectiveness of the adaptations varies from only marginally reducing negative impacts to, in some cases, changing a negative impact into a positive impact. On average in cereal cropping systems adaptations such as changing varieties and planting times enable avoidance of a 10-15% reduction in yield. The benefits of adaptation tend to increase with the degree of climate change up to a point [Figure 5.2]. Pressure to cultivate marginal land or to adopt unsustainable cultivation practices as yields drop may increase land degradation and endanger biodiversity of both wild and domestic species. Climate changes increase irrigation demand in the majority of world regions due to a combination of decreased rainfall and increased evaporation arising from increased temperatures, which, combined with expected reduced water availability, adds another challenge to future water and food security [5.9].

Summary of Impacts and Adaptive Results by Temperature and Time. Major generalisations across the FFFF sectors distilled from the literature are reported either by increments of temperature increase (Table 5.7) or by increments of time (Table 5.8), depending on how the information is originally reported. A global map of regional impacts of FFFF is shown in Figure 5.4.

5.8.2 Research gaps and priorities

Key knowledge gaps that hinder assessments of climate change consequences for FFFF and their accompanying research priorities are listed in Table 5.9.

Table 5.7. *Summary of selected conclusions for food, fibre, forestry, and fisheries, by warming increments.*

Temp. Change	Sub-sector	Region	Finding	Source section
+1 to +2°C	Food crops	Mid- to high-latitudes	- Cold limitation alleviated for all crops - Adaptation of maize and wheat increases yield 10-15%; rice yield no change; regional variation is high	Figure 5.2
	Pastures and livestock	Temperate	- Cold limitation alleviated for pastures; seasonal increased frequency of heat stress for livestock	Table 5.3
	Food crops	Low latitudes	- Wheat and maize yields reduced below baseline levels; rice is unchanged - Adaptation of maize, wheat, rice maintains yields at current levels	Figure 5.2
	Pastures and livestock	Semi-arid	- No increase in NPP; seasonal increased frequency of heat stress for livestock	Table 5.3
	Prices	Global	- Agricultural prices: –10 to –30%	Figure 5.3
+2 to +3°C	Food crops	Global	- 550 ppm CO_2 (approx. equal to +2°C) increases C_3 crop yield by 17%; this increase is offset by temperature increase of 2°C assuming no adaptation and 3°C with adaptation	Figure 5.2
	Prices	Global	- Agricultural prices: –10 to +20%	Figure 5.3
	Food crops	Mid- to high-latitudes	- Adaptation increases all crops above baseline yield	Figure 5.2
	Fisheries	Temperate	- Positive effect on trout in winter, negative in summer	5.4.6.1
	Pastures and livestock	Temperate	- Moderate production loss in swine and confined cattle	Table 5.3
	Fibre	Temperate	- Yields decrease by 9%	5.4.4
	Pastures and livestock	Semi-arid	- Reduction in animal weight and pasture production, and increased heat stress for livestock	Table 5.3
	Food crops	Low latitudes	- Adaptation maintains yields of all crops above baseline; yields drops below baseline for all crops without adaptation	Figure 5.2
+3 to +5°C	Prices and trade	Global	- Reversal of downward trend in wood prices - Agricultural prices: +10 to +40% - Cereal imports of developing countries to increase by 10-40%	5.4.5.1 Figure 5.3 5.6.3
	Forestry	Temperate	- Increase in fire hazard and insect damage	5.4.5.3
		Tropical	- Massive Amazonian deforestation possible	5.4.5
	Food crops	Low latitudes	- Adaptation maintains yields of all crops above baseline; yield drops below baseline for all crops without adaptation	Figure 5.2
	Pastures and livestock	Tropical	- Strong production loss in swine and confined cattle	Table 5.3
	Food crops	Low latitudes	- Maize and wheat yields reduced below baseline regardless of adaptation, but adaptation maintains rice yield at baseline levels	Figure 5.2
	Pastures and livestock	Semi-arid	- Reduction in animal weight and pasture growth; increased animal heat stress and mortality	Table 5.3

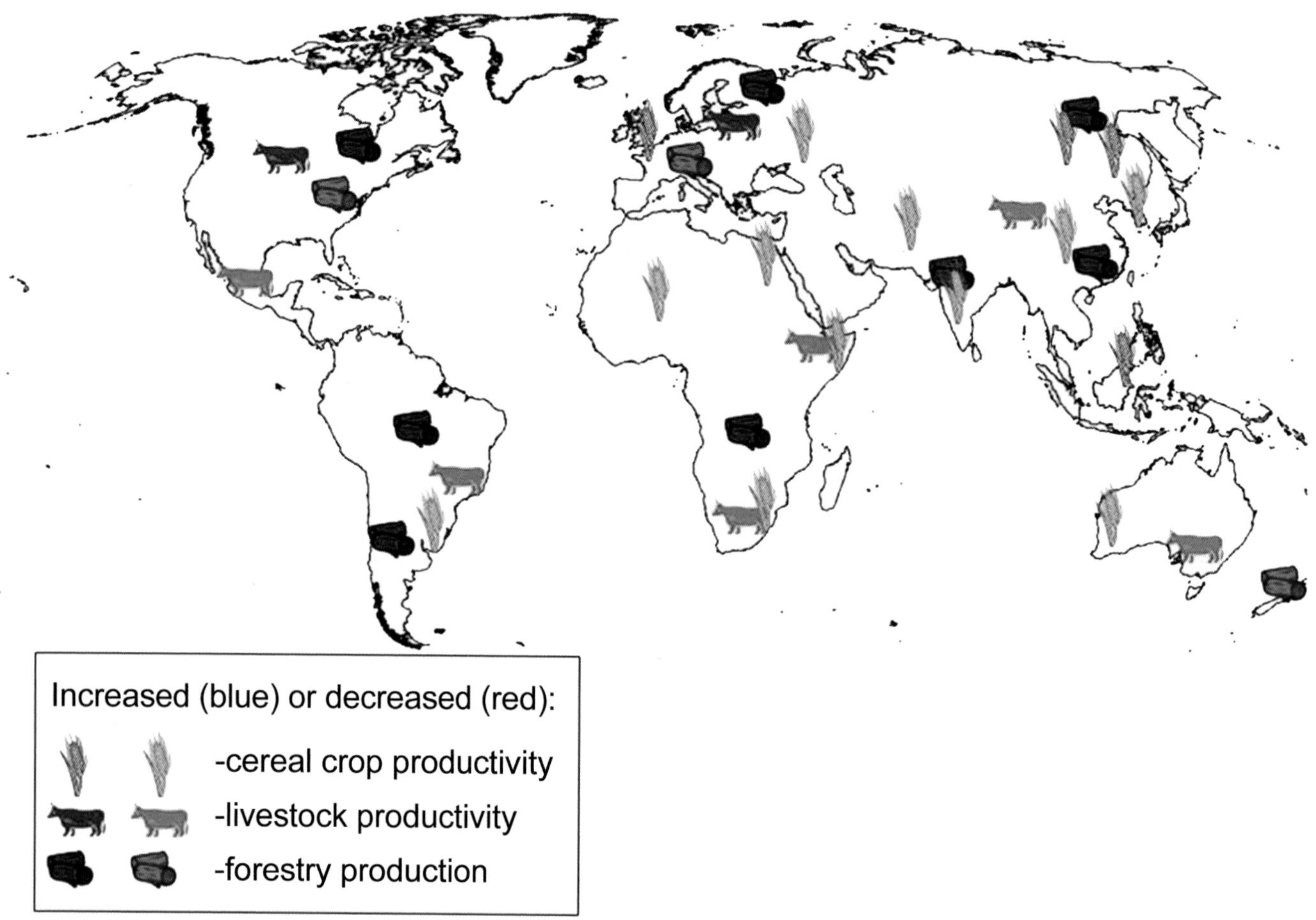

Figure 5.4. *Major impacts of climate change on crop and livestock yields, and forestry production by 2050 based on literature and expert judgement of Chapter 5 Lead Authors. Adaptation is not taken into account.*

Table 5.8. *Summary of selected findings for food, fibre, forestry and fisheries, by time increment.*

Time slice	Sub-sector	Location	Finding	Source
2020	Food crops	USA	- Extreme events, e.g., increased heavy precipitation, cause crop losses to US$3 billion by 2030 with respect to current levels	5.4.2
	Small-holder farming, fishing	Low latitudes, especially east and south Africa	- Decline in maize yields, increased risk of crop failure, high livestock mortality	5.4.7
	Small-holder farming, fishing	Low latitudes, especially south Asia	- Early snow melt causing spring flooding and summer irrigation shortage	5.4.7
	Forestry	Global	- Increased export of timber from temperate to tropical countries - Increase in share of timber production from plantations - Timber production +5 to +15%	5.4.5.2 Table 5.4
2050	Fisheries	Global	- Marine primary production +0.7 to +8.1%, with large regional variation (see Chapter 4)	5.4.6.2
	Food crops	Global	- With adaptation, yields of wheat, rice, maize above baseline levels in mid- to high-latitude regions and at baseline levels in low latitudes.	Figure 5.2
	Forestry	Global	- Timber production +20 to +40%	Table 5.4
2080	Food crops	Global	- Crop irrigation water requirement increases 5-20%, with range due to significant regional variation	5.4.2
	Forestry	Global	- Timber production +20 to +60% with high regional variation	Table 5.4
	Agriculture sector	Global	- Stabilisation at 550 ppm ameliorates 70-100% of agricultural cost caused by unabated climate change	5.4.2

Table 5.9. *Key knowledge gaps and research priorities for food, fibre, forestry, and fisheries (FFFF).*

Knowledge gap	Research priority
There is a lack of knowledge of CO_2 response for many crops other than cereals, including many of importance to the rural poor, such as root crops, millet.	FACE-type experiments needed on expanded range of crops, pastures, forests and locations, especially in developing countries.
Understanding of the combined effects of elevated CO_2 and climate change on pests, weeds and disease is insufficient.	Basic knowledge of pest, disease and weed response to elevated CO_2 and climate change needed.
Much uncertainty of how changes in frequency and severity of extreme climate events with climate change will affect all sectors remains.	Improved prediction of future impacts of climate change requires better representation of climate variability at scales from the short-term (including extreme events) to interannual and decadal in FFFF models.
Calls by the TAR to enhance crop model inter-comparison studies have remained largely unheeded.	Improvements and further evaluation of economic, trade and technological components within integrated assessment models are needed, including new global simulation studies that incorporate new crop, forestry and livestock knowledge in models.
Few experimental or field studies have investigated the impacts of future climate scenarios on aquatic biota.	Future trends in aquatic primary production depend on nutrient supply and on temperature sensitivity of primary production. Both of these could be improved with a relatively small research effort.
In spite of a decade of prioritisation, adaptation research has failed to provide generalised knowledge of the adaptive capacity of FFFF systems across a range of climate and socio-economic futures, and across developed and developing countries (including commercial and small-holder operations).	A more complete range of adaptation strategies must be examined in modelling frameworks in FFFF. Accompanying research that estimates the costs of adaptation is needed. Assessments of how to move from potential adaptation options to adoption taking into account decision-making complexity, diversity at different scales and regions, non-linearities and time-lags in responses and biophysical, economic, institutional and cultural barriers to change are needed. Particular emphasis to developing countries should be given.
The global impacts of climate change on agriculture and food security will depend on the future role of agriculture in the global economy. While most studies available for the Fourth Assessment assume a rapidly declining role of agriculture in the overall generation of income, no consistent and comprehensive assessment was available.	Given the importance of this assumption, more research is needed to assess the future role of agriculture in overall income formation (and dependence of people on agriculture for income generation and food consumption) in essentially all developing countries; such an exercise could also afford an opportunity to review and critique the SRES scenarios.
Relatively moderate impacts of climate change on overall agro-ecological conditions are likely to mask much more severe climatic and economic vulnerability at the local level. Little is known about such vulnerability.	More research is required to identify highly vulnerable micro-environments and associated households and to provide agronomic and economic coping strategies for the affected populations.
The impact of climate change on utilisation of biofuel crops is not well established.	Research on biomass feed stock crops such as switchgrass and short-rotation poplar is needed. Research is needed on the competition for land between bio-energy crops and food crops.

References

Aaheim, A. and L. Sygna, 2000: Economic impacts of climate change on tuna fisheries in Fiji Islands and Kiribati, Cicero Report 4, Cicero, Oslo, 21 pp.

Abou-Hadid, A.F., R. Mougou, A. Mokssit and A. Iglesias, 2003: Assessment of impacts, adaptation, and vulnerability to climate change in North Africa: food production and water resources. AIACC AF90 Semi-Annual Progress Report, 37 pp.

Abou-Hadid, A.F., 2006: Assessment of impacts, adaptation and vulnerability to climate change in North Africa: food production and water resources. Assessments of Impacts and Adaptations to Climate Change, Washington, District of Columbia, 127 pp. [Accessed 19.03.07: http://www.aiaccproject.org/Final%20Reports/Final%20Reports/FinalRept_AIACC_AF90.pdf]

ACIA, 2005: *Arctic Climate Impact Assessment*. Cambridge University Press, Cambridge, 1042 pp.

Adams, R.M., R.A. Fleming, C.C. Chang, B.A. McCarl and C. Rosenzweig, 1995: A reassessment of the economic effects of global climate change in on U.S. agriculture. *Climate Change*, **30**, 147-167.

Adams, R.M., B.A. McCarl and L.O. Mearns, 2003: The effects of spatial scale of climate scenarios on economic assessments: an example from US agriculture. *Climatic Change*, **60**, 131-148.

Adejuwon, J., 2005: Assessing the suitability of the epic crop model for use in the study of impacts of climate variability and climate change in West Africa. *Singapore J. Trop. Geo*., **26**, 44-60.

Adejuwon, J., 2006: Food Security, Climate Variability and Climate Change in Sub Saharan West Africa. Assessments of Impacts and Adaptations to Climate Change, Washington, District of Columbia, 137 pp.[Accessed 19.03.07: http://www.aiaccproject.org/Final%20Reports/Final%20Reports/FinalRept_AIACC_AF23.pdf]

Adger, W.N., 1999: Social vulnerability to climate change and extremes in coastal Vietnam. *World Dev*., **27**, 249-269.

Adger, W.N., S. Huq, K. Brown, D. Conway and M. Hulme, 2003: Adaptation to climate change in the developing world. *Prog. Dev. Stud*., **3**, 179-195.

Aggarwal, P.K., 2003. Impact of climate change on Indian agriculture. *J. Plant Biol*., **30**, 189-198.

Aggarwal, P.K. and P.K. Mall, 2002: Climate change and rice yields in diverse agro-environments of India. II. Effect of uncertainties in scenarios and crop models on impact assessment. *Climatic Change*, **52**, 331-343.

Aggarwal, P.K., P.K. Joshi, J.S. Ingram and R.K. Gupta, 2004: Adapting food systems of the Indo-Gangetic plains to global environmental change: key information needs to improve policy formulation. *Environ. Sci. Policy*, **7**, 487-498.

Aggarwal, P.K., B. Banerjee, M.G. Daryaei, A. Bhatia, A. Bala and S. Rani, 2006: InfoCrop: a dynamic simulation model for the assessment of crop yields, losses due to pests, and environmental impact of agro-ecosystems in tropical environments. II. Performance of the model. *Agr. Syst*., **89**, 47-67.

Agrawala, S., A. Moehner, A. Hemp, M. van Aalst, S. Hitz, J. Smith, H. Meena, S.M. Mwakifwamba, T. Hyera and O.U. Mwaipopo, 2003: *Development and climate change in Tanzania: focus on Mount Kilimanjaro*. Environment Directorate and Development Co-operation Directorate, Organisation for Economic Co-operation and Development, Paris, 72 pp.

Agrell, J., P. Anderson, W. Oleszek, A. Stochmal and C. Agrell, 2004: Combined effects of elevated CO_2 and herbivore damage on alfalfa and cotton. *J. Chem. Ecol*., **30**, 2309-2324.

Ainsworth, E.A. and S.P. Long, 2005: What have we learned from 15 years of free-air CO_2 enrichment (FACE)? A meta-analysis of the responses of photosynthesis, canopy properties and plant production to rising CO_2. *New Phytol.*, **165**, 351-372.

Ainsworth, E.A., A. Rogers, R. Nelson and S.P. Long, 2004: Testing the source–sink hypothesis of down-regulation of photosynthesis in elevated CO_2 in the field with single gene substitutions in Glycine max. *Agr. Forest Meteorol.*, **122**, 85-94.

Alexandratos, N., 2005: Countries with rapid population growth and resources constraints: issues of food, agriculture and development. *Popul. Dev. Rev.*, **31**, 237-258.

Alexandrov, V., J. Eitzinger, V. Cajic and M. Oberforster, 2002: Potential impact of climate change on selected agricultural crops in north-eastern Austria. *Glob. Change Biol.*, **8**, 372-389.

Alig, R.J., D.M. Adams and B.A. McCarl, 2002: Projecting impacts of global climate change on the US forest and agriculture sectors and carbon budgets. *Forest Ecol. Manag.*, **169**, 3-14.

Alig, R.J., D. Adams, L. Joyce and B. Sohngen, 2004: Climate change impacts and adaptation in forestry: responses by trees and market choices, American Agricultural Economics Association, 11 pp. [Accessed 19.03.07: http://www.choicesmagazine.org/2004-3/climate/2004-3-07.htm]

Allard, V., P.C.D. Newton, M. Lieffering, H. Clark, C. Matthew and Y. Gray, 2003: Nitrogen cycling in grazed pastures at elevated CO_2: N returns by ruminants. *Glob. Change Biol.*, **9**, 1731-1742.

Allard, V., P.C.D. Newton, M. Lieffering, J.F. Soussana, P. Grieu and C. Matthew, 2004: Elevated CO_2 effects on decomposition processes in a grazed grassland. *Glob. Change Biol.*, **10**, 1553-1564.

Allison, E.H. and F. Ellis, 2001: The livelihoods approach and management of small-scale fisheries. *Mar. Policy*, **25**, 377-388.

Allison, E.H., W.N. Adger, M.C. Badjeck, K. Brown, D. Conway, N.K. Dulvy, A. Halls, A. Perry and J.D. Reynolds, 2005: Effects of climate change on the sustainability of capture and enhancement fisheries important to the poor: analysis of the vulnerability and adaptability of fisherfolk living in poverty. Project No R4778J, Fisheries Management Science Programme, MRAG for Department for International Development, London, 167 pp. [Accessed 19.03.07: http://www.fmsp.org.uk/Documents/r4778j/R4778J_FTR1.pdf]

Amundson, J.L., T.L. Mader, R.J. Rasby and Q.S. Hu, 2005: Temperature and temperature–humidity index effects on pregnancy rate in beef cattle. *Proc. 17th International Congress on Biometeorology*, Dettscher Wetterdienst, Offenbach, Germany.

Anon, 2006: Bluetongue confirmed in France. News and Reports, *Vet. Rec.*, **159**, 331.

Antle, J.M., S.M. Capalbo, E.T. Elliott and K.H. Paustian, 2004: Adaptation, spatial heterogeneity, and the vulnerability of agricultural systems to climate change and CO_2 fertilization: an integrated assessment approach. *Climate Change*, **64**, 289-315.

Aranjuelo, I., J.J. Irigoyen, P. Perez, R. Martinez-Carrasco and M. Sanchez-Diaz, 2005: The use of temperature gradient tunnels for studying the combined effect of CO_2, temperature and water availability in N_2 fixing alfalfa plants. *Ann. Appl. Biol.*, **146**, 51-60.

Arnell, N.W., 2004: Climate change and global water resources: SRES emissions and socio-economic scenarios. *Global Environ. Change*, **14**, 31-52.

Arnell, N.W., M.G.R. Cannell, M. Hulme, R.S. Kovats, J.F.B. Mitchell, R.J. Nicholls, M.L. Parry, M.T.J. Livermore and A. White, 2002: The consequences of CO_2 stabilisation for the impacts of climate change. *Climatic Change*, **53**, 413-446.

Arnold, M.G., G. Köhlin, R. Persson and G. Shephard, 2003: Fuelwood revisited: what has changed in the last decade? CIFOR Occasional Paper No. 39, Center for International Forestry Research (CIFOR), Indonesia, 35 pp.

Ashmore, M.R., 2005: Assessing the future global impacts of ozone on vegetation. *Plant Cell Environ.*, **29**, 949-964.

Atkinson, M.D., P.S. Kettlewell, P.D. Hollins, D.B. Stephenson and N.V. Hardwick, 2005: Summer climate mediates UK wheat quality response to winter North Atlantic Oscillation. *Agr. Forest Meteorol.*, **130**, 27-37.

Bachelet, D. and C.A. Gay, 1993: The impacts of climate change on rice yield: a comparison of four model performances. *Ecol. Model.*, **65**, 71-93.

Bachelet, D., J.M. Lenihan, C. Daly, R.P. Neilson, D.S. Ojima and W.J. Parton, 2001: MC1: a dynamic vegetation model for estimating the distribution of vegetation and associated carbon, nutrients, and water-technical documentation. Gen. Tech. Rep. PNW-GTR-508. U.S. Department of Agriculture, Forest Service, Pacific Northwest Research Station, Portland, Oregon, 1-95 pp. [Accessed 19.03.07: http://www.fs.fed.us/pnw/pubs/gtr508.pdf]

Baker, J.T., 2004: Yield responses of southern U.S. rice cultivars to CO_2 and temperature. *Agr. For. Meterol.*, **122**, 129-137.

Bale, J.S., G.J. Masters and I.D. Hodkinson, 2002: Herbivory in global climate change research: direct effects of rising temperature on insect herbivores. *Glob. Change Biol.*, **8**, 1-16.

Barber, R., 2001: Upwelling ecosystems. *Encyclopedia of Ocean Sciences*, J.H. Steele, S.A. Thorpe and K.K. Turekian, Eds., Academic Press, London, 3128 pp.

Barnett, A. and A. Whiteside, 2002: *AIDS in the Twenty-First Century; Disease and Globalization*. Palgrave MacMillan, Basingstoke and New York, 432 pp.

Barnett, T.P., J.C. Adam and D.P. Lettenmaier, 2005: Potential impacts of a warming climate on water availability in snow-dominated regions. *Nature*, **438**, 303-309.

Barton, D. and J. Morton, 2001: Livestock marketing and drought mitigation in northern Kenya, *Drought, Planning and Pastoralists: Experiences from Northern Kenya and Elsewhere*, J. Morton, Ed., Natural Resources Institute, Chatham.

Bass, B., 2005: Measuring the adaptation deficit. Discussion on keynote paper: climate change and the adaptation deficit. *Climate Change: Building the Adaptive Capacity*, A. Fenech, D. MacIver, H. Auld, B. Rong, Y.Y. Yin, Environment Canada, Quebec, 34-36.

Batima, P., 2003: Climate change: pasture–livestock. Synthesis report. *Potential Impacts of Climate Change, Vulnerability and Adaptation Assessment for Grassland Ecosystem and Livestock Sector in Mongolia*, ADMON Publishing, Ulaanbaatar, 36-47.

Batima, P., B. Bat, L. Tserendash, S. Bayarbaatar, S. Shiirev-Adya, G. Tuvaansuren, L. Natsagdorj and T. Chuluun, 2005: *Adaptation to Climate Change*, Vol. 90, ADMON Publishing, Ulaanbaatar.

Beaugrand, G., P.C. Reid, F. Ibanez, J.A. Lindley and M. Edwards, 2002: Reorganization of North Atlantic marine copepod biodiversity and climate. *Science*, **296**, 1692-1694.

Bergeron, Y., M. Flannigan, S. Gauthier, A. Leduc and P. Lefort, 2004: Past, current and future fire frequency in the Canadian Boreal Forest: implications for sustainable forest management. *Ambio*, **6**, 356-360.

Betts, R.A., P.M. Cox, M. Collins, P.P. Harris, C. Huntingford and C.D. Jones, 2004: The role of ecosystem–atmosphere interactions in simulated Amazonian precipitation decrease and forest dieback under global climate warming. *Theor. Appl. Climatol.*, **78**, 157-175.

Bharwani, S., M. Bithell, T.E. Downing, M. New, R. Washington and G. Ziervogel, 2005: Multi-agent modelling of climate outlooks and food security on a community garden scheme in Limpopo, South Africa. *Philos. T. Royal Soc. B*, **360**, 2183-2194.

Binkley, C.S., 1988: Economic effects of CO_2 -induced climatic warming on the world's forest sector. *The Impact of Climate Variations on Agriculture: Vol. 1 Assessments in Cool Temperate and Cold Regions*, M. Parry, T. Carter and N. Konijn, Eds., Kluwer Academic Publishers, Dordrecht, 183-218.

Blench, R., 2001: You can't go home again: pastoralism in the new millennium. FAO, Overseas Development Institute, London, 106 pp.

Boisvenue, C. and S.W. Running, 2006: Impacts of climate change on natural forest productivity - evidence since the middle of the 20th century. *Glob. Change Biol.*, **12**, 862-882.

Booker, F.L., S.A. Prior, H.A. Torbert, E.L. Fiscus, W.A. Pursley and S. Hu, 2005: Decomposition of soybean grown under elevated concentrations of CO_2 and O_3. *Glob. Change Biol.*, **11**, 685-698.

Branco, A.D.M., J. Suassuna and S.A. Vainsencher, 2005: Improving access to water resources through rainwater harvesting as a mitigation measure: the case of the Brazilian Semi-Arid Region. *Mitigation and Adaptation Strategies for Global Change*, **10**, 393-409.

Brander, K.M., 2005: Cod recruitment is strongly affected by climate when stock biomass is low. *ICES J. Mar. Sci.*, **62**, 339-343.

Briones, M., M.M. Dey and M. Ahmed, 2004: The future for fish in the food and livelihoods of the poor in Asia. *NAGA, 50 World Fish Center Quarterly*, **27**, 48.

Brooks, N., W.N. Adger and P.M. Kelly, 2005: The determinants of vulnerability and adaptive capacity at the national level and implications for adaptation. *Global Environ. Change*, **15**, 151-163.

Brovkin, V., 2002: Climate-vegetation interaction. *J. Phys. IV*, **12**, 57-72.

Brown, R.A. and N.J. Rosenberg, 1999: Climate change impacts on the potential productivity of corn and winter wheat in their primary United States growing regions. *Climatic Change*, **41**, 73-107.

Brown, R.A., N.J. Rosenberg, C.J. Hays, W.E. Easterling and L.O. Mearns, 2000: Potential production and environmental effects of switchgrass and traditional crops under current and greenhouse-altered climate in the central United States: a simulation study. *Agr. Ecosyst. Environ.*, **78**, 31-47.

Bruinsma, J., 2003: *World Agriculture: Towards 2015/2030: an FAO perspective*. Earthscan, London and FAO, Rome, London, 432 pp.

Bryceson, D.F., C. Kay and J. Mooij, 2000: *Disappearing Peasantries? Rural Labour in Africa, Asia and Latin America*. Intermediate Technology Publications, London, 352 pp.

Burke, E.J., S.J. Brown and N. Christidis, 2006: Modelling the recent evolution of global drought and projections for the 21st century with the Hadley Centre climate model. *J. Hydrometeorol.*, **7**, 1113-1125.

Burton, I. and B. Lim, 2005: Achieving adequate adaptation in agriculture. *Climatic Change*, **70**, 191-200.

Butt, T.A., B.A. McCarl, J. Angerer, P.T. Dyke and J.W. Stuth, 2005: The economic and food security implications of climate change in Mali. *Climatic Change*, **68**, 355-378.

Caldwell, C.R., S.J. Britz and R.M. Mirecki, 2005: Effect of temperature, elevated carbon dioxide, and drought during seed development on the isoflavone content of dwarf soybean [*Glycine max* (L.) Merrill] grown in controlled environments. *J. Agr. Food Chem.*, **53**, 1125-1129.

Calfapietra, C., B. Gielen, A.N.J. Galemma, M. Lukac, P. De Angelis, M.C. Moscatelli, R. Ceulemans and G. Scarascia-Mugnozza, 2003: Free-air CO_2 enrichment (FACE) enhances biomass production in a short-rotation poplar plantation. *Tree Phys.*, **23**, 805-814.

Carbone, G.J., W. Kiechle, C. Locke, L.O. Mearns, L. McDaniel and M.W. Downton, 2003: Response of soybean and sorghum to varying spatial scales of climate change scenarios in the southeastern United States. *Climatic Change*, **60**, 73-98.

Carle, J., P. Vuorinen and A. Del Lungo, 2002: Status and trends in global plantation development. *Forest Prod. J.*, **52**, 1-13.

Carroll, A.L., S.W. Taylor, J. Regniere and L. Safranyik, 2004: Effects of climate change on range expansion by the mountain pine beetle in British Columbia. *Natural Resources Canada, Canadian Forest Service, Pacific Forestry Centre Information Report BC-X-399*, T.L. Shore, J.E. Brooks and J.E. Stone, Eds., Victoria, British Columbia, 223-232.

Caspersen, J.P., S.W. Pacala, J.C. Jenkins, G.C. Hurtt and P.R. Moorcraft, 2000: Contributions of land-use history to carbon accumulation in U.S. forests. *Science*, **290**, 1148-1152.

Cassman, K.G., A. Dobermann, D.T. Walters and H. Yang, 2003: Meeting cereal demand while protecting natural resources and improving environmental quality. *Annu. Rev. Environ. Resour.*, **28**, 315-358.

Centritto, M., 2005: Photosynthetic limitations and carbon partitioning in cherry in response to water deficit and elevated [CO_2]. *Agr. Ecosyst. Environ.*, **106**, 233-242.

Chakraborty, S. and S. Datta, 2003: How will plant pathogens adapt to host plant resistance at elevated CO_2 under a changing climate? *New Phytol.*, **159**, 733-742.

Challinor, A.J., T.R. Wheeler, P.Q. Craufurd, C.A.T. Ferro and D.B. Stephenson, 2007: Adaptation of crops to climate change through genotypic responses to mean and extreme temperatures. *Agric. Ecosys. Environ.*, **119**, 190-204.

Chambers, R., A. Pacey and L.A. Thrupp, 1989: *Farmer First: Farmer Innovation and Agricultural Research*. Intermediate Technology Publications, London, 218 pp.

Chang, C.C., 2002: The potential impact of climate change on Taiwan's agriculture. *Agr. Econ.*, **27**, 51-64.

Cheikh, N., P.W. Miller and G. Kishore, 2000: Role of biotechnology in crop productivity in a changing environment. *Global Change and Crop Productivity*, K.R. Reddy and H.F. Hodges, Eds., CAP International, New York, 425-436.

Chen, F.J., G. Wu and F. Ge, 2004: Impacts of elevated CO_2 on the population abundance and reproductive activity of aphid *Sitobion avenae* Fabricius feeding on spring wheat. *J. Environ. Nutr.*, **128**, 723-730.

Chen, F., G.E. Feng and M.N. Parajulee, 2005a: Impact of elevated CO_2 on tritrophic interaction of *Gossypium hirsutum*, *Aphis gossypii*, and *Leis axyridis*. *Environ. Entomol.*, **34**, 37-46.

Chen, F., G. Wu, F. Ge, M.N. Parajulee and R.B. Shrestha, 2005b: Effects of elevated CO_2 and transgenic Bt cotton on plant chemistry, performance, and feeding of an insect herbivore, the cotton bollworm. *Entomol. Exp. Appl.*, **115**, 341-350.

Chipanshi, A.C., R. Chanda and O. Totolo, 2003: Vulnerability assessment of the maize and sorghum crops to climate change in Botswana. *Climatic Change*, **61**, 339-360.

Christensen, J.H., B. Hewitson, A. Busuioc, A. Chen, X. Gao, I. Held, R. Jones, W.-T. Kwon and Coauthors, 2007: Regional climate projections. *Climate Change 2007: Contribution of Working Group I to the Fourth Assessment Report of the Intergovernmental Panel on Climate Change*, S. Solomon, D. Qin, and M. Manning, Eds., Cambridge University Press, Cambridge, 847-940.

Ciais, P., M. Reichstein, N. Viovy, A. Granier, J. Ogee, V. Allard, M. Aubinet, N. Buchmann and Coauthors, 2005: Europe-wide reduction in primary productivity caused by the heat and drought in 2003. *Nature*, **437**, 529-534.

Cocu, N., R. Harrington, A. Rounsevell, S.P. Worner and M. Hulle, 2005: Geographical location, climate and land use influences on the phenology and numbers of the aphid, *Myzus persicae*, in Europe. *J. Biogeogr.*, **32**, 615-632.

COPA COGECA, 2003a: Committee of Agricultural Organisations in the European Union General Committee for Agricultural Cooperation in the European Union, CDP 03 61 1, Press release, Brussels.

COPA COGECA, 2003b: Assessment of the impact of the heat wave and drought of the summer 2003 on agriculture and forestry. Committee of Agricultural Organisations in the European Union General Committee for Agricultural Cooperation in the European Union, Brussels, 15 pp.

Coppock, D.L., 1994: *The Borana Plateau of Southern Ethiopia: Synthesis of Pastoral Research, Development and Change*, ILCA Systems, Addis Ababa. 393 pp.

Corbera, E., D. Conway, M. Goulden and K. Vincent, 2006: Climate Change in Africa: linking Science and Policy for Adaptation. Workshop Report., The Tyndall Centre and IIED, Norwich and London, 10 pp. [Accessed 20.03.07: http://www.iied.org/CC/documents/ClimateChangeinAfricaWorkshopReport_TyndallIIED_Final.pdf]

Cornish, G.A., 1998: *Modern Irrigation Technologies for Smallholders in Developing Countries*. Intermediate Technology Publications, Wallingford, 96 pp.

Corobov, R., 2002: Estimations of climate change impacts on crop production in the Republic of Moldova. *GeoJournal*, **57**, 195-202.

Cox, P.M., R.A. Betts, M. Collins, P.P. Harris, C. Huntingford and C.D. Jones, 2004: Amazonian forest dieback under climate-carbon cycle projections for the 21st century. *Theor. Appl. Climatol.*, **78**, 137-156.

Cramer, W., A. Bondeau, F.I. Woodward, I.C. Prentice, R.A. Betts, V. Brovkin, P.M. Cox, V. Fisher and Coauthors, 2001: Global response of terrestrial ecosystem structure and function to CO_2 and climate change: results from six dynamic global vegetation models. *Glob. Change Biol.*, **7**, 357-373.

Crozier, L. and G. Dwyer, 2006: Combining population-dynamic and ecophysiological models to predict climate-induced insect range shifts. *Am. Nat.*, **167**, 853-866.

Cuculeanu, V., P. Tuinea and D. Balteanu, 2002: Climate change impacts in Romania: vulnerability and adaptation options. *GeoJournal*, **57**, 203-209.

Curry, R. and C. Mauritzen, 2005: Dilution of the Northern North Atlantic Ocean in recent decades. *Science*, **308**, 1772-1774.

Daepp, M., J. Nosberger and A. Luscher, 2001: Nitrogen fertilization and developmental stage alter the response of *Lolium perenne* to elevated CO_2. *New Phytol.*, **150**, 347-358.

Dalby, S., 2004: Conflict, cooperation and global environment change: advancing the agenda. *International Human Dimensions Programme on Global Environmental Change, Newsletter 'Update'*, **03**, 1-3, Bonn.

Darwin, R., 2004: Effects of greenhouse gas emissions on world agriculture, food consumption, and economic welfare. *Climatic Change*, **66**, 191–238.

Dash, D.K., D.P. Ray and H.H. Khan, 2002: Extent of damage to coconut palms caused by super cyclone in Orissa and pattern of their recovery. *CORD*, **18**, 1-5.

Davidson, D.J., T. Williamson and J.R. Parkins, 2003: Understanding climate change risk and vulnerability in northern forest-based communities. *Can. J. Forest Res.*, **33**, 2252-2261.

Davies, S., 1996: *Adaptable Livelihoods: Coping with Food Insecurity in the Malian Sahel*. Macmillan Press; St. Martin's Press, xxii, 335 pp.

Davis, J.R., 2004: The rural non-farm economy, livelihoods and their diversification: issues and options. NRI Report No. 2753, Natural Resources Institute, Chatham, UK, 39 pp.

Defra, 2000: Impact of Climate Change on Grasslands and Livestock. *Climate Change and Agriculture in the United Kingdom*, Department for Environment, Food and Rural Affairs, London, 43-56.

Delgado, C.L., N. Wada, M.W. Rosegrant, S. Meijer and M. Ahmed, 2003: *Fish to 2020 - Supply and Demand in Changing Global Markets*. Jointly published by the International Food Policy Research Institute (IFPRI) and WorldFish Center, 226 pp.

Desta, S. and D.L. Coppock, 2002: Cattle population dynamics in the Southern Ethiopian Rangelands, 1980-97. *J. Range Manage.*, **55**, 439-451.

Devereux, S. and C. Tapscott, 1995: Coping mechanisms of communal farmers in response to drought. *Coping with Aridity: Drought Impacts and Preparedness in Namibia*, R. Moorsom, J. Franz, and M. Mupotola, Eds., Brandes and Apsel/NEPRU, Frankfurt and Windhoek.

DeWalle, D.R., A.R. Buda and A. Fisher, 2003: Extreme weather and forest management in the mid-Atlantic region of the United States. *North. J. Appl. For.*, **20**, 61-70.

Diamond, J., 2004: *Collapse: How Societies Choose to Fail or Succeed*. Viking, New York, 592 pp.

Döll, P., 2002: Impact of climate change and variability on irrigation requirements: a global perspective. *Climatic Change*, **54**, 269-293.

Dregne, H.E., 2000: Drought and desertification: exploring the linkages. *Drought: A Global Assessment*. Vol. I, D.A. Wilhite, Ed., Routledge, London, 231-240.

Drennen, P.M., M. Smith, D. Goldsworthy and J. van Staten, 1993: The occurrence of trahaolose in the leaves of the desiccation-tolerant angiosperm *Myronthamnus flabellifoliius* Welw. *J. Plant Physiol.*, **142**, 493-496.

Drinkwater, K.F., 2005: The response of Atlantic cod (*Gadus morhua*) to future climate change. *ICES J. Mar. Sci.*, **62**, 1327-1337.

Drinkwater, K.F., A. Belgrano, A. Borja, A. Conversi, M. Edwards, C.H. Greene, G. Ottersen, A.J. Pershing and H. Walker, 2003: The response of marine ecosystems to climate variability associated with the North Atlantic Oscillation. *The North Atlantic Oscillation: Climatic Significance and Environmental Impact. Geophysical Monograph*, J.W. Hurrell, Y. Kushnir, G. Ottersen and M. Visbeck, Eds., American Geophysical Union, Washington, District of Columbia, 211-234.

Droogers, P., 2004: Adaptation to climate change to enhance food security and preserve environmental quality: example for southern Sri Lanka. *Agr. Water Manage.*, **66**, 15-33.

Dukes, J.S., N.R. Chiariello, E.E. Cleland, L.A. Moore, M.R. Shaw, S. Thayer, T. Tobeck, H.A. Mooney and C.B. Field, 2005: Responses of grassland production to single and multiple global environmental changes. *PLOS Biol.*, **3**, 1829-1837.

Eakin, H. and M.C. Lemos, 2006: Adaptation and the state: Latin America and the challenge of capacity-building under globalization. *Global Environ. Change*, **16**, 7-18.

Easterling, W.E. and C. Polsky, 2004: Crossing the complex divide: linking scales for understanding coupled human-environment systems. *Scale and Geographic Inquiry*, R.B. McMaster and E. Sheppard, Eds., Blackwell, Oxford, 66-85.

Easterling, W.E., L.O. Mearns, C.J. Hays and D. Marx, 2001: Comparison of agricultural impacts of climate change calculated from high and low resolution climate change scenarios: Part II. Accounting for adaptation and CO_2 direct effects. *Climatic Change*, **51**, 173-197.

Easterling, W.E., N. Chhetri and X.Z. Niu, 2003: Improving the realism of modeling agronomic adaptation to climate change: simulating technological submission. *Climatic Change*, **60**(1-2), 149-173.

Easterling, W.E., B.H. Hurd and J.B. Smith, 2004: Coping with global climate change: the role of adaptation in the United States, Pew Center on Global Climate Change, Arlington, Virginia, 52 pp. [Accessed 20.03.07: http://www.pewclimate.org/docUploads/Adaptation.pdf]

Edwards, G.R., H. Clark and P.C.D. Newton, 2001: The effects of elevated CO_2 on seed production and seedling recruitment in a sheep-grazed pasture. *Oecologia*, **127**, 383-394.

Ehleringer, J.R., T.E. Cerling and M.D. Dearing, 2002: Atmospheric CO_2 as a global change driver influencing plant-animal interactions. *Integr. Comp. Biol.*, **42**, 424-430.

El-Shaer, H.M., C. Rosenzweig, A. Iglesias, M.H. Eid and D. Hillel, 1997: Impact of climate change on possible scenarios for Egyptian agriculture in the future. *Mitigation and Adaptation Strategies for Global Change*, **1**, 233-250.

Ellis, F., 2000: *Rural Livelihoods and Diversity in Developing Countries*. Oxford University Press, Oxford, 290 pp.

Ellis, J., 1995: Climate variability and complex ecosystem dynamics; implications for pastoral development. *Living with Uncertainty: New Directions in Pastoral Development in Africa*, I. Scoones, Ed., Intermediate Technology Publications, London, 37-46.

Erda, L., X. Wei, J. Hui, X. Yinlong, L. Yue, B. Liping and X. Liyong, 2005: Climate change impacts on crop yield and quality with CO_2 fertilization in China. *Philos. T. Roy. Soc. B*, **360**, 2149-2154.

Eriksen, S. and L.O. Naess, 2003: Pro-poor climate adaptation: Norwegian development cooperation and climate change adaptation: an assessment of issues, strategies and potential entry points, CICERO Report 2003:02. Center for International Climate and Environmental Research, Oslo, p. 8 (table).

Evenson, R.E., 1999: Global and local implications of biotechnology and climate change for future food supplies. *P. Natl. Acad. Sci. USA*, **96**, 5921-5928.

Ewert, F., M.D.A. Rounsevell, I. Reginster, M.J. Metzger and R. Leemans, 2005: Future scenarios of European agricultural land use I. Estimating changes in crop productivity. *Agr. Ecosyst. Environ.*, **107**, 101-116.

Faisal, I. and S. Parveen, 2004: Food security in the face of climate change, population growth, and resource constraints: implications for Bangladesh. *Environ. Manage.*, **34**, 487-498.

FAO, 1982: World forest products: demand and supply 1990 and 2000, FAO Forestry Paper 29, Food and Agriculture Organization of the United Nations, Rome, 366 pp.

FAO, 1984: Report of an animal feed security mission to Botswana. Food and Agriculture Organization of the United Nations, Rome.

FAO, 1986: Forest products: world outlook projections 1985-2000, FAO Forestry Paper 73, Food and Agriculture Organization of the United Nations, Rome, 101 pp.

FAO, 1988: Forest products: world outlook projections - product and country tables 1987-2000, FAO Forestry Paper 84, Food and Agriculture Organization of the United Nations, Rome, 350 pp.

FAO, 1997: *FAO provisional outlook for global forest products consumption, production and trade to 2010*, Food and Agriculture Organization of the United Nations, Rome, 345pp.

FAO, 1998: *Global Fibre Supply Model*. Food and Agriculture Organization of the United Nations, Rome, 72 pp.

FAO, 2000: Global forest resources assessment 2000. FAO Forestry Paper 140, Food and Agriculture Organization of the United Nations, Rome, 511 pp. [Accessed 20.03.07: http://www.fao.org/forestry/site/fra2000report/en/]

FAO, 2001: Forest genomics for conserving adaptive genetic diversity. Paper prepared by Konstantin, V. Krutovskii, and David B. Neale. Forest Genetic Resources Working Papers. Working Paper FGR/3 (July 2001), Forest Resources Development Service, Forest Resources Division, FAO, Rome. [Accessed 20.03.07: http://www.fao.org/DOCREP/003/X6884E/x6884e02.htm #TopOfPage]

FAO, 2002: The state of world fisheries and aquaculture (SOFIA) 2002, FAO Fisheries Department, Food and Agriculture Organization of the United Nations, Rome, 153 pp.

FAO, 2003a: Strengthening coherence in FAO's initiatives to fight hunger (Item 10). Conference Thirty-second Session, 29 November to 10 December, Food and Agriculture Organization of the United Nations, Rome.

FAO, 2003b: Impact of climate change on food security and implications for sustainable food production committee on world food security. Conference Twenty-ninth Session, 12 to 16 May, Food and Agriculture Organization of the United Nations, Rome.

FAO, 2003c: Future climate change and regional fisheries: a collaborative analysis. FAO Fisheries Technical Paper No. 452, Food and Agriculture Organization of the United Nations, Rome, 75 pp.

FAO, 2004a: Trade and Sustainable Forest Management – Impacts and Interactions. Analytic Study of the Global Project GCP/INT/775/JPN. Impact Assessment of Forests Products Trade in the Promotion of Sustainable Forest Management. Food and Agriculture Organization of the United Nations, Rome, 366 pp.

FAO, 2004b: *The state of food and agriculture 2003-04. Agricultural biotechnology: Meeting the needs of the poor*, Food and Agriculture Organization of the United Nations, Rome, 208 pp.

FAO, 2004c: The state of world fisheries and aquaculture (SOFIA) 2004, Fisheries Department, Food and Agriculture Organization of the United Nations, Rome, 153 pp.

FAO, 2005a: World agriculture: towards 2030/2050. Interim report, Global Perspective Studies Unit, Food and Agriculture Organization of the United Nations, Rome, Italy, 71 pp.

FAO, 2005b: Global forest resources assessment 2005. FAO Forestry Paper 147., Food and Agriculture Organization of the United Nations, Rome, 348 pp.

FAO, 2005c: Special event on impact of climate change, pests and diseases on food security and poverty reduction. Background Document. 31st Session of the Committee on World Food Security, 10 pp. [Accessed 20.03.07: ftp://ftp.fao.org/docrep/fao/meeting/009/j5411e.pdf]

FAO, 2006: World agriculture: towards 2030/2050 – Interim report: prospects for food, nutrition, agriculture and major commodity groups, Food and Agriculture Organization of the United Nations, Rome, 78 pp.

Fay, P.A., J.D. Carlisle, A.K. Knapp, J.M. Blair and S.L. Collins, 2003: Productivity responses to altered rainfall patterns in a C-4-dominated grassland. *Oecologia*, **137**, 245-251.

Felzer, B., D. Kicklighter, J. Melillo, C. Wang, Q. Zhuang and R. Prinn, 2004: Effects of ozone on net primary production and carbon sequestration in the conterminous United States using a biogeochemistry model. *Tellus*, **56B**, 230-248.

Felzer, B., J. Reilly, J. Melillo, D. Kicklighter, M. Sarofim, C. Wang, R. Prinn and Q. Zhuang, 2005: Future effects of ozone on carbon sequestration and climate change policy using a global biogeochemical model. *Climatic Change*, **73**, 345-373.

Fischer, G., M. Shah and H. van Velthuizen, 2002a: Climate change and agricultural vulnerability, IIASA Special Report commissioned by the UN for the World Summit on Sustainable Development, Johannesburg 2002. International Institute for Applied Systems Analysis, Laxenburg, Austria, 160 pp.

Fischer, G., H. van Velthuizen, M. Shah and F.O. Nachtergaele, 2002b: Global agro-ecological assessment for agriculture in the 21st century: methodology and results. Research Report RR-02-02. ISBN 3-7045-0141-7., International Institute for Applied Systems Analysis, Laxenburg, Austria, 119 pp and CD-Rom.

Fischer, G., M. Shah, F.N. Tubiello and H. Van Velthuizen, 2005a: Integrated assessment of global crop production. *Philos. T. Roy. Soc, B*, **360**, 2067-2083.

Fischer, G., M. Shah, F.N. Tubiello and H. van Velthuizen, 2005b: Socio-economic and climate change impacts on agriculture: an integrated assessment, 1990-2080. *Philos. T. Roy. Soc. B.*, **360**, 2067-2083.

Fischer, G., F.N. Tubiello, H. van Velthuizen and D. Wiberg, 2006: Climate change impacts on irrigation water requirements: effects of mitigation, 1990-2989. *Technol. Forecast. Soc.*, doi: 10.1016/j.techfore.2006.05.021.

Fiscus, E.L., F.L. Booker and K.O. Burkey, 2005: Crop responses to ozone: uptake, modes of action, carbon assimilation and partitioning. *Plant Cell Environ.*, **28**, 997-1011.

Flannigan, M.D., K.A. Logan, B.D. Amiro, W.R. Skinner and B.J. Stocks, 2005: Future area burned in Canada. *Climatic Change*, **72**, 1-16.

Fleming, R.A., J.N. Candau and R.S. McAlpine, 2002: Landscape-scale analysis of interactions between insect defoliation and forest fire in Central Canada. *Climatic Change*, **55**, 251-272.

Foley, J.A., R. DeFries, G.P. Asner, C. Barford, G. Bonan, S.R. Carpenter, F.S. Chapin, M.T. Coe and Co-authors, 2005: Global consequences of land use. *Science*, **309**, 570-574.

Frank, A.B. and W.A. Dugas, 2001: Carbon dioxide fluxes over a northern, semi-arid, mixed-grass prairie. *Agr. Forest Meteorol.*, **108**, 317-326.

Frank, K.L., T.L. Mader, J.A. Harrington, G.L. Hahn and M.S. Davis, 2001: Climate change effects on livestock production in the Great Plains. *Proc. 6th International Livestock Environment Symposium*, R.R. Stowell, R. Bucklin and R.W. Bottcher, Eds., American Society of Agricultural Engineering, St. Joseph, Michigan, 351-358.

Friedland, K.D., D.G. Reddin, J.R. McMenemy and K.F. Drinkwater, 2003: Multidecadal trends in North American Atlantic salmon (*Salmo salar*) stocks and climate trends relevant to juvenile survival. *Can. J. Fish. Aquat. Sci,*, **60**, 563-583.

Fuhrer, J., 2003: Agroecosystem responses to combination of elevated CO_2, ozone, and global climate change. *Agr. Ecosyst. Environ.*, **97**, 1-20.

Gan, J., 2004: Risk and damage of southern pine beetle outbreaks under global climate change. *Forest Ecol. Manag.*, **191**, 61-71.

Gbetibouo, G.A. and R.M. Hassan, 2005: Measuring the economic impact of climate change on major South African field crops: a Ricardian approach. *Global Planet. Change*, **47**, 143-152.

Gifford, R.M., 2004: The CO_2 fertilising effect – does it occur in the real world? *New Phytol.*, **163**, 221-225.

Gill, R.A., H.W. Polley, H.B. Johnson, L.J. Anderson, H. Maherali and R.B. Jackson, 2002: Nonlinear grassland responses to past and future atmospheric CO_2. *Nature*, **417**(6886), 279-282.

Gillett, N.P., A.J. Weaver, F.W. Zwiers and M.D. Flannigan, 2004: Detecting the effect of climate change on Canadian forest fires. *Geophys. Res. Lett.*, **31**, L12217, doi:10.1029/2004GL020044.

Girardin, M.P., J. Tardif, M.D. Flannigan, B.M. Wotton and Y. Bergeron, 2004: Trends and periodicities in the Canadian Drought Code and their relationships with atmospheric circulation for the southern Canadian boreal forest. *Can. J. For. Res.*, **34**, 103-119.

Gitay, H., S. Brown, W.E. Easterling, B. Jallow, J. Antle, M. Apps, R. Beamish, C. Cerri and Coauthors, 2001: Ecosystems and their services. *Climate Change 2001: Impacts, Adaptation and Vulnerability to Climate Change. Contribution of Working Group II to the Third Assessment Report of the Intergovernmental Panel on Climate Change*, J.J. McCarthy, O.F. Canziani, N.A. Leary, D.J. Dokken and K.S. White, Eds., Cambridge University Press, Cambridge, 236-342.

Goklany, I.M., 1998: Saving habitat and conserving biodiversity on a crowded planet. *BioScience*, **48**, 941-953.

Goklany, I.M., 2005: A climate policy for the short and medium term: stabilization or adaptation? *Energ. Environ.*, **16**, 667-680.

Goklany, I.M., 2007: Integrated strategies to reduce vulnerability and advance adaptation, mitigation, and sustainable development. *Mitigation and Adaptation Strategies for Global Change*,**12**, 755-786.

Goldammer, J.G. and R.W. Mutch, 2001: *Global forest fire assessment 1990-2000*. Food and Agriculture Organization of the United Nations, Rome. [Accessed 22.06.07: http://www.fao.org/docrep/006/AD653E/AD653E00.HTM]

Gommes, R., J. du Guerny and M.H. Glantz, 2004: Climate and HIV/AIDS, a hotspots analysis to contribute to early warning and rapid response systems. UNDP: Bangkok; FAO: Rome; NCAR: Colorado, iv 22 pp. [Accessed 21.03.07: http://www.fao.org/clim/docs/faoclimatehiv.pdf]

Graham, N.A.J., S.K. Wilson, S. Jennings, N.V.C. Polunin, J.P. Bijoux and J. Robinson, 2006: Dynamic fragility of oceanic coral reef ecosystems. *P. Natl. Acad. Sci. USA*, **103**, 8425-8429.

Grandcourt, E.M. and H.S.J. Cesar, 2003: The bio-economic impact of mass coral mortality on the coastal reef fisheries of the Seychelles. *Fish. Res.*, **60**, 539-550.

Gregg, W.W., M.E. Conkright, P. Ginoux, J.E. O'Reilly and N.W. Casey, 2003: Ocean primary production and climate: global decadal changes. *Geophys. Res. Lett.*, **30**, 1809.

Gregory, P.J., J.S.I. Ingram and M. Brklacich, 2005: Climate change and food security. *Philos. T. Roy. Soc. B*, **360**, 2139-2148.

Grimble, R., C. Cardoso and S. Omar-Chowdhury, 2002: *Poor people and the environment: Issues and linkages*. Policy Series No. 16, Natural Resources Institute, Chatham, 49 pp.

Häggblom, R., 2004: Global forest trends. presentation at FINPRO, World Bank: Business Opportunities in Forestry Sector, 7 May 2004, Helsinki, Finland. Jaakko Pöyry Consulting, Vantaa, Finland.

Hagler, R., 1998: The global timber supply/demand balance to 2030: has the equation changed? A Multi-Client Study by Wood Resources International, Reston, VA, 206 pp.

Hahn, L., T. Mader, D. Spiers, J. Gaughan, J. Nienaber, R. Eigenberg, T. Brown-Brandl, Q. Hu and Coauthors, 2001: Heat wave impacts on feedlot cattle: considerations for improved environmental management. *Proc. 6th International Livestock Environment Symposium*, R.R. Stowell, R. Bucklin and R.W. Bottcher, Eds., American Society of Agricultural Engineering, St. Joseph, Michigan, 129-130.

Hannah, L., G.F. Midgley, T. Lovejoy, W.J. Bond, M. Bush, J.C. Lovett, D. Scott and F.I. Woodward, 2002: Conservation of biodiversity in a changing climate. *Conserv. Biol.*, **16**, 264-268.

Hanson, P.J. and J.F. Weltzin, 2000: Drought disturbance from climate change: response of United States forests. *Sci. Total Environ.*, **262**, 205-220.

Haque, C.E. and I. Burton, 2005: Adaptation options strategies for hazards and vulnerability mitigation: an international perspective. *Mitigation and Adaptation Strategies for Global Change*, **10**, 335-353.

Harrison, M., 2005: The development of seasonal and inter-annual climate forecasting. *Climatic Change*, **70**, 210-220.

Harvell, C.D., K. Kim, J.M. Burkholder, R.R. Colwell, P.R. Epstein, D.J. Grimes, E.E. Hofmann, E.K. Lipp and Coauthors, 1999: Emerging marine diseases – climate links and anthropogenic factors. *Science*, **285**, 1505-1510.

Hazell, P., 2004: *Smallholders and pro-poor agricultural growth*. DAC Network on Poverty Reduction, Organisation for Economic Co-operation and Development, Paris, 14 pp. [Accessed 20.03.07: http://www.oecd.org/dataoecd/25/6/36562947.pdf]

Hendy, C. and J. Morton, 2001: Drought-time grazing resources in Northern Kenya. *Pastoralism, Drought and Planning: Lessons from Northern Kenya and*

Elsewhere, Morton, J., Ed., Natural Resources Institute, Chatham, 139-179.

Henry, H.A.L., E.E. Cleland, C.B. Field and P.M. Vitousek, 2005: Interactive effects of elevated CO_2, N deposition and climate change on plant litter quality in a California annual grassland. *Oecologia*, **142**, 465-473.

Hitz, S. and J. Smith, 2004: Estimating global impacts from climate change. *Global Environ. Change*, **14**, 201-218.

Hoanh, C.T., H. Guttman, P. Droogers and J. Aerts, 2004: Will we produce sufficient food under climate change? Mekong Basin (South-east Asia). *Climate Change in Contrasting River Basins: Adaptation Strategies for Water, Food, and Environment*, J.C.J.H. Aerts and P. Droogers, Eds., CABI Publishing, Wallingford, 157-180.

Hofmann, E., S. Ford, E. Powell and J. Klinck, 2001: Modeling studies of the effect of climate variability on MSX disease in eastern oyster (*Crassostrea virginica*) populations. *Hydrobiologia*, **460**, 195-212.

Hogg, E.H. and P.Y. Bernier, 2005: Climate change impacts on drought-prone forests in western Canada. *Forest Chron.*, **81**, 675-682.

Holden, N.M. and A.J. Brereton, 2002: An assessment of the potential impact of climate change on grass yield in Ireland over the next 100 years. *Irish J. Agr. Food Res.*, **41**, 213-226.

Holling, C.S., 2004: From complex regions to complex worlds. *Ecol. Soc.*, 9, 11. [Accessed 20.030.7: http://www.ecologyandsociety.org/vol9/iss1/art11/]

Hortle, K. and S. Bush, 2003: Consumption in the lower Mekong basin as a measure of fish yield. *New Approaches for the Improvement of Inland Capture Fishery Statistics in the Mekong Basin*, T. Clayton, Ed., FAO RAP Publication 2003/01, Bangkok, 76-88.

Howden, M. and R.N. Jones, 2004: Risk assessment of climate change impacts on Australia's wheat industry. *New Directions for a Diverse Planet: Proceedings of the 4th International Crop Science Congress*, T. Fischer, N. Turner, J. Angus, J. McIntyre, L. Robertson, A. Borrell and D. Lloyd, Brisbane, Australia. [Accessed 22.06.07: http://www.cropscience.org.au/icsc2004/symposia/6/2/1848_howdensm.htm]

Howden, S.M. and S. Crimp, 2005: Assessing dangerous climate change impacts on Australia's wheat industry. *MODSIM 2005 International Congress on Modelling and Simulation.*, A. Zerger and R.M. Argent, Eds., Modelling and Simulation Society of Australia and New Zealand, Melbourne, 170-176.

Howden, S.M., W.B. Hall and D. Bruget, 1999: Heat stress and beef cattle in Australian rangelands: recent trends and climate change, *People and Rangelands: Building the Future. Proc. of the VI Intl. Rangeland Congress*, D. Eldringe and D. Freudenberger, Eds., Townsville, Australia, 43-45.

Howden, S.M., A.J. Ash, E.W.R. Barlow, C.S. Booth, R. Cechet, S. Crimp, R.M. Gifford, K. Hennessy and Coauthors, 2003: An overview of the adaptive capacity of the Australian agricultural sector to climate change – options, costs and benefits. Report to the Australian Greenhouse Office, Canberra, Australia, 157 pp.

Hulme, M., E. Barrow, N. Arnell, P. Harrison, T. Johns and T. Downing, 1999: Relative impacts of human-induced climate change and natural climate variability. *Nature*, **397**, 688-691.

IFAD, 2001: *Rural Poverty Report 2001: The Challenge of Ending Rural Poverty*. International Fund for Agricultural Development, Rome, 266 pp.

Iglesias, A. and M.I. Minguez, 1997: Modelling crop–climate interactions in Spain: vulnerability and adaptation of different agricultural systems to climate change. *Mitigation and Adaptation Strategies for Global Change*, **1**, 273-288.

Iglesias, A., C. Rosenzweig and D. Pereira, 2000: Agricultural impacts of climate change in Spain: developing tools for a spatial analysis. *Global Environ. Change*, **10**, 69-80.

IPCC, 2001: *Climate Change 2001: Impacts, Adaptation, and Vulnerability. Contribution of Working Group II to the Third Assessment Report of the Intergovernmental Panel on Climate Change*, J.J. McCarthy, O.F. Canziani, N.A. Leary, D.J. Dokken and K.S. White, Eds., Cambridge University Press, 1032 pp.

IPCC, 2007a: Summary for policy makers. *Climate Change 2007: The Physical Science Basis. Contribution of Working Group I to the Fourth Assessment Report of the Intergovernmental Panel for Climate Change*, S. Solomon, D. Qin, M. Manning, Z. Chen, M. Marquis, K.B. Averyt, M. Tignor and H.L. Miller, Eds., Cambridge University Press, Cambridge, 18 pp.

IPCC, 2007b: *Climate Change 2007: The Physical Science Basis. Contribution of Working Group I to the Fourth Assessment Report of the Intergovernmental Panel on Climate Change*, S. Solomon, D. Qin, M. Manning, Z. Chen, M. Marquis, K.B. Averyt, M. Tignor and H.L. Miller, Eds., Cambridge University Press, Cambridge, 996pp.

IPCC, 2007c: *Climate Change 2007: Mitigation. Contribution of Working Group III to the Fourth Assessment Report of the Intergovernmental Panel on Climate Change*, B. Metz, O. Davidson, P. Bosch, R. Dave and L. Meyer, Eds., Cambridge University Press, Cambridge, UK.

Irland, L.C., D. Adams, R. Alig, C.J. Betz, C.C. Chen, M. Hutchins, B.A. McCarl, K. Skog and B.L. Sohngen, 2001: Assessing socioeconomic impacts of climate change on US forests, wood-product markets, and forest recreation. *BioScience*, **51**, 753-764.

Izaurralde, R.C., N.J. Rosenberg, R.A. Brown and A.M. Thomson, 2003: Integrated assessment of Hadley Center (HadCM2) climate-change impacts on agricultural productivity and irrigation water supply in the conterminous United States, Part II. Regional agricultural production in 2030 and 2095. *Agr. Forest Meteorol.*, **117**, 97-122.

Jablonski, L.M., X. Wang and P.S. Curtis, 2002: Plant reproduction under elevated CO_2 conditions: a meta-analysis of reports on 79 crop and wild species. *New Phytol.*, **156**, 9-26.

Jacobson, L.D., J.A.A. De Oliveira, M. Barange, R. Felix-Uraga, J.R. Hunter, J.Y. Kim, M. ¥iquen, C. Porteiro and Coauthors, 2001: Surplus production, variability, and climate change in the great sardine and anchovy fisheries. *Canadian Journal of Fisheries and Aquatic Sciences*, **58**, 1891.

Jones, P.D., D.H. Lister, K.W. Jaggard and J.D. Pidgeon, 2003: Future climate impact on the productivity of sugar beet *Beta vulgaris* L. in Europe. *Climatic Change*, **58**, 93-108.

Jones, P.G. and P.K. Thornton, 2003: The potential impacts of climate change on maize production in Africa and Latin America in 2055. *Global Environ. Change*, **13**, 51-59.

Joyce, L.A., J.R. Mills, L.S. Heath, A.D. McGuire, R.W. Haynes and R.A. Birdsey, 1995: Forest sector impacts from changes in forest productivity under climate change. *J. Biogeogr.*, **22**, 703-713.

Joyce, L.A., J. Aber, S. McNulty, V. Dale, A. Hansen, L. Irland, R. Neilson and K. Skog, 2001: Potential consequences of climate variability and change for the forests of the United States. *Climate Change Impacts on the United States: The Potential Consequences of Climate Variability and Change*, National Assessment Synthesis Team, Eds., Cambridge University Press, Cambridge, 489-522.

Kaakinen, S., F. Kostiainen, F. Ek, P. Saranpää, M.E. Kubiske, J. Sober, D.F. Karnosky and E. Vapaavuori, 2004: Stem wood properties of *Populus tremuloides*, *Betula papyrifera* and *Acer saccharum* saplings after three years of treatments to elevated carbon dioxide and ozone. *Glob. Change Biol.*, **10**, 1513-1525.

Kalra, N., P.K. Aggarwal, S. Chander, H. Pathak, R. Choudhary, A. Chaudhary, M. Sehgal, U.A. Soni, A. Sharma, M. Jolly, U.K. Singh, O. Ahmed and M.Z. Hussain, 2003: Impacts of climate change on agriculture. *Climate Change and India: Vulnerability Assessment and Adaptation*, P.R. Shukla, S.K. Sharma, N.H. Ravindranath, A. Garg and S. Bhattacharya, Eds., University Press, India, 193-226.

Kapetanaki, G. and C. Rosenzweig, 1997: Impact of climate change on maize yield in central and northern Greece: a simulation study with Ceres-Maize. *Mitigation and Adaptation Strategies for Global Change*, **1**, 251-271.

Karnosky, D.F., 2003: Impact of elevated atmospheric CO_2 on forest trees and forest ecosystems: knowledge gaps. *Environ. Int.*, **29**, 161-169.

Kates, R.W. and T.J. Wilbanks, 2003: Making the global local: responding to climate change concerns from the ground up. *Environment*, **45**, 12-23.

Kettlewell, P.S., J. Easey, D.B. Stephenson and P.R. Poulton, 2006: Soil moisture mediates association between the winter North Atlantic Oscillation and summer growth in the Park Grass Experiment. *P. Roy. Soc. Lond. B Bio.*, **273**, 1149-1154.

Kherallah, M., C. Delgado, E. Gabre-Medhin, N. Minot and M. Johnson, 2002: *Reforming Agricultural Markets in Africa*. Johns Hopkins University Press, Baltimore, Maryland, 224 pp.

Kimball, B.A., K. Kobayashi and M. Bindi, 2002: Responses of agricultural crops to free-air CO_2 enrichment. *Adv Agron.*, **77**, 293-368.

King, J.M., D.J. Parsons, J.R. Turnpenny, J. Nyangaga, P. Bakari and C.M. Wathes, 2005: Ceiling to milk yield on Kenya smallholdings requires rethink of dairy development policy. *British Society of Animal Science Annual Conference*, York.

King, J.R., 2005: Report of the study group on fisheries and ecosystem responses to recent regime shifts. PICES Scientific Report 28, 162 pp.

Kishor, P.B.K., Z. Hong, G. Miao, C. Hu and D. Verma, 1995: Overexpression of Δ1-pyrroline-5-carboxylase synthase increases praline production and confers osmotolerance in transgenic plants. *J. Plant Physiol.*, **108**, 1387-1394.

Knapp, A.K. and M.D. Smith, 2001: Variation among biomes in temporal dynamics of aboveground primary production. *Science*, **291**, 481-484.

Korner, C., R. Asshoff, O. Bignucolo, S. Hattenschwiler, S.G. Keel, S. Pelaez-

Riedl, S. Pepin, R.T.W. Siegwolf and G. Zotz., 2005: Carbon flux and growth in mature deciduous forest trees exposed to elevated CO_2. *Science*, **309**, 1360-1362.

Kumar, K.S.K. and J. Parikh, 2001: Indian agriculture and climate sensitivity. *Global Environ. Change*, **11**, 147-154.

Kurukulasuriya, P. and S Rosenthal, 2003: *Climate Change and Agriculture: A Review of Impacts and Adaptations*. World Bank Climate Change Series, Vol. 91, World Bank Environment Department, Washington, District of Columbia, 96 pp.

Lal, M., K.K. Singh, L.S. Rathore, G. Srinivasan and S.A. Saseendran, 1998: Vulnerability of rice and wheat yields in N.W. India to future changes in climate. *Agr. Forest Meteorol.*, **89**, 101-114.

Laporte, M.F., L.C. Duchesne and S. Wetzel, 2002: Effect of rainfall patterns on soil surface CO_2 efflux, soil moisture, soil temperature and plant growth in a grassland ecosystem of northern Ontario, Canada: implications for climate change. *BMC Ecology*, **2**, 10.

Lasco, R.D. and R. Boer, 2006: An integrated assessment of climate change impacts, adaptations and vulnerability in watershed areas and communities in Southeast Asia. Final report submitted to Assessments of Impacts and Adaptation to Climate Change (AIACC), Project No. AS21, Washington, District of Columbia, 223 pp.

Laurance, W.F. and G.B. Williamson, 2001: Positive feedbacks among forest fragmentation, drought, and climate change in the Amazon. *Conserv. Biol.*, **15**, 1529-1535.

Laurance, W.F., A.K.M. Albernaz, P.M. Fearnside, H.L. Vasconcelos and L.V. Ferreira, 2004: Deforestation in Amazonia. *Science*, **304**, 1109.

Lawrence, A., 2003: No forest without timber? *Int. For. Rev.*, **5**, 87-96.

Leary, N., J. Adejuwon, W. Bailey, V. Barros, M. Caffera, S. Chinvanno, C. Conde, A. De Comarmond amd Co-authors, 2006: For whom the bell tolls: vulnerabilities in a changing climate. A synthesis from the AIACC Project. Working Paper No. 21, Assessments of Impacts and Adaptation to Climate Change (AIACC), 33 pp.

Lee, D.M. and K.S. Lyon, 2004: A dynamic analysis of the global timber market under global warming: an integrated modeling approach. *Southern Econ. J.*, **70**, 467-489.

Leff, B., N. Ramankutty and J.A. Foley, 2004: Geographic distribution of major crops across the world. *Global Biogeochem. Cy.*, **18**, GB1009.

Lehodey, P., 2001: The pelagic ecosystem of the tropical Pacific Ocean: dynamic spatial modelling and biological consequences of ENSO. *Prog. Oceanogr.*, **49**, 439-469.

Lehodey, P., F. Chai and J. Hampton, 2003: Modelling climate-related variability of tuna populations from a coupled ocean biogeochemical-populations dynamics model. *Fish. Oceanogr.*, **12**, 483-494.

Lemmen, D.S. and F.J. Warren, 2004: *Climate Change Impacts and Adaptation: A Canadian Perspective*. Natural Resources Canada, Ottawa, Ontario, 201 pp.

Liberloo, M., S.Y. Dillen, C. Calfapietra, S. Marinari, B.L. Zhi, P. De Angelis and R. Ceulemans, 2005: Elevated CO_2 concentration, fertilization and their interaction: growth stimulation in a short-rotation poplar coppice (EUROFACE). *Tree Physiol.*, **25**, 179-189.

Lin, E., X. Wei, J. Hui, X. Yinlong, L. Yue, B. Liping and X. Liyong, 2005: Climate change impacts on crop yield and quality with CO_2 fertilization in China. *Philos. T. Roy. Soc. B*, **360**, 2149-2154.

Lipton, M., 2004: Crop science, poverty and the family farm in a globalising world. Plenary Paper. *4th International Crop Science Congress*, Brisbane, Australia, 18 pp. [Accessed 20.03.07: http://www.cropscience.org.au/icsc2004/plenary/0/1673_lipton.htm]

Liu, L., J.S. King and C.P. Giardina, 2005: Effects of elevated concentrations of atmospheric CO_2 and tropospheric O_3 on leaf litter production and chemistry in trembling aspen and paper birch communities. *Tree Physiol.*, **25**, 1511-1522.

Long, S.P., E.A. Ainsworth, A. Rogers and D.R. Ort, 2004: Rising atmospheric carbon dioxide: plants FACE the future. *Annu. Rev. Plant Biol.*, **55**, 591-628.

Long, S.P., E.A. Ainsworth, A.D.B. Leakey and P.B. Morgan, 2005: Global food insecurity. Treatment of major food crops with elevated carbon dioxide or ozone under large-scale fully open-air conditions suggests recent models may have overestimated future yields. *Philos. T. Roy. Soc. B*, **360**, 2011-2020.

Long, S.P., E.A. Ainsworth, A.D.B. Leakey, J. Nosberger and D.R. Ort, 2006: Food for thought: lower expected crop yield stimulation with rising CO_2 concentrations. *Science*, **312**, 1918-1921.

Loya, W.M., K.S. Pregitzer, N.J. Karberg, J.S. King and J.P. Giardina, 2003: Reduction of soil carbon formation by tropospheric ozone under increased carbon dioxide levels. *Nature*, **425**, 705-707.

Luo, Q.Y., M.A.J. Williams, W. Bellotti and B. Bryan, 2003: Quantitative and visual assessments of climate change impacts on South Australian wheat production. *Agr. Syst.*, **77**, 172-186.

Luscher, A., J. Fuhrer and P.C.D. Newton, 2005: Global atmospheric change and its effect on managed grassland systems. *Grassland: A Global Resource*, D.A. McGilloway, Ed., Wageningen Academic Publishers, Wageningen, 251-264.

Mader, T.L., 2003: Environmental stress in confined beef cattle. *J. Anim. Sci.*, **81**(electronic suppl. 2), 110-119.

Mader, T.L. and M.S. Davis, 2004: Effect of management strategies on reducing heat stress of feedlot cattle: feed and water intake. *J. Anim. Sci.*, **82**, 3077-3087.

Mall, R.K. and P.K. Aggarwal, 2002: Climate change and rice yields in diverse agro-environments of India. I. Evaluation of impact assessment. *Climatic Change*, **52**, 315-330.

Maracchi, G., O. Sirotenko and M. Bindi, 2005: Impacts of present and future climate variability on agriculture and forestry in the temperate regions: Europe. *Climatic Change*, **70**, 117-135.

Matthews, R. and R. Wassmann, 2003: Modelling the impacts of climate change and methane emission reductions on rice production: a review. *Eur. J. Agron.*, **19**, 573-598.

Matthews, R.B., M.J. Kropff, T. Horie and D. Bachelet, 1997: Simulating the impact of climate change on rice production in Asia and evaluating options for adaptation. *Agr. Syst.*, **54**, 399-425.

McPeak, J.G. and C.B. Barrett, 2001: Differential risk exposure and stochastic poverty traps among East African pastoralists. *Am. J. Agr. Econ.*, **83**, 674-679.

McPhaden, M.J. and D. Zhang, 2002: Slowdown of the meridional overturning circulation in the upper Pacific Ocean. *Nature*, **415**, 603-608.

Mendelsohn, R., 2003: A California model of climate change impacts on timber markets. California Energy Commission, Publication 500-03-058CF, 24 pp. [Accessed 20.03.07: http://www.energy.ca.gov/reports/2003-10-31_500-03-058CF_A12.PDF]

Mendelsohn, R., A. Dinar, A. Basist, P. Kurukulasuriya, M.I. Ajwad, F. Kogan and C. Williams, 2004: Cross-sectional analyses of climate change impacts. World Bank Policy Research Working Paper 3350, Washington, District of Columbia, 97 pp.

Milchunas, D.G., A.R. Mosier, J.A. Morgan, D.R. LeCain, J.Y. King and J.A. Nelson, 2005: Elevated CO_2 and defoliation effects on a shortgrass steppe: forage quality versus quantity for ruminants. *Agr. Ecosyst. Environ.*, **111**, 166-194.

Millennium Ecosystem Assessment, 2005: *Ecosystems and Human Wellbeing: Synthesis*. Island Press, Washington, District of Columbia, 155 pp.

Mitchell, S.W. and F. Csillag, 2001: Assessing the stability and uncertainty of predicted vegetation growth under climatic variability: northern mixed grass prairie. *Ecol. Model.*, **139**, 101-121.

Monirul, M. and Q. Mirza, 2002: Global warming and changes in the probability of occurrence of floods in Bangladesh and implications. *Global Environ. Change*, **12**, 127-138.

Moorcroft, P.R., 2003: Recent advances in ecosystem–atmosphere interactions: an ecological perspective. *P. Roy. Soc. London B Bio.*, **270**, 1215-1227.

Morgan, I., D.G. McDonald and C.M. Wood, 2001: The cost of living for freshwater fish in a warmer, more polluted world. *Glob. Change Biol.*, **7**, 345-355.

Morgan, J.A., D.E. Pataki, C. Korner, H. Clark, S.J. Del Grosso, J.M. Grunzweig, A.K. Knapp and M.R. Shaw, 2004: Water relations in grassland and desert ecosystems exposed to elevated atmospheric CO_2. *Oecologia*, **140**, 11-25.

Mortimore, M.J. and W.M. Adams, 2001: Farmer adaptation, change and 'crisis' in the Sahel. *Global Environ. Change*, **11**, 49-57.

Morton, J., 2006: Pastorilist coping strategies and emergency lifestock market intervention, *Livestock Marketing in Eastern Africa: Research and Policy Challenges*, J.G. McPeak and P.D. Little, Eds., ITDG Publications, Rugby, 227-246.

Morton, J. and C. de Haan, 2006: Community-based drought management for the pastoral livestock sector in sub-Saharan Africa, ALive Initiative Policy Options Paper for ALive Initiative, used as keynote paper for ALive e-conference, 18 pp. [Accessed 20.03.07: http://www.virtualcentre.org/en/ele/ econf_03_alive/download/drought.pdf]

Motha, R.P. and W. Baier, 2005: Impacts of present and future climate change and climate variability on agriculture in the temperate regions: North America. *Climatic Change*, **70**, 137-164.

Mouillot, F. and C.B. Field, 2005: Fire history and the global carbon budget: a 1°× 1 fire history reconstruction for the 20th century. *Glob. Change Biol.*, **11**, 398-420.

Moya, T.B., L.H. Ziska, O.S. Namuco and D. Olszyk, 1998: Growth dynamics and genotypic variation in tropical, field-grown paddy rice (*Oryza sativa* L.) in response to increasing carbon dioxide and temperature. *Glob. Change Biol.*, **4**, 645-656.

MRC, 2003: State of the Basin Report: 2003, Mekong River Commission, Phnom Penh, 300 pp.

Murdiyarso, D., 2000: Adaptation to climatic variability and change: Asian perspectives on agriculture and food security. *Environ. Monit. Assess.*, **61**, 123-131.

Nabuurs, G.J., A. Pussinen, T. Karjalainen, M. Erhard and K. Kramer, 2002: Stemwood volume increment changes in European forests due to climate change – a simulation study with the EFISCEN model. *Glob. Change Biol.*, **8**, 304-316.

Nakićenović, N. and R. Swart, Eds., 2000: *IPCC Special Report on Emissions Scenarios*, Cambridge University Press, Cambridge, 599 pp.

Ndikumana, J., J. Stuth, R. Kamidi, S. Ossiya, R. Marambii and P. Hamlett, 2000: Coping mechanisms and their efficacy in disaster-prone pastoral systems of the Greater Horn of Africa: effects of the 1995-97 drought and the 1997-98 El Niño rains and the responses of pastoralists and livestock. ILRI Project Report. A-AARNET (ASARECA-Animal Agriculture Research Network), Nairobi, Kenya, GL-CRSP LEWS (Global Livestock- Collaboratve Research Support Program Livestock Early Warning System), College Station, Texas, USA, and ILRI (International Livestock Research Institute), Nairobi, Kenya, 124 pp.

Nearing, M.A., F.F. Pruski and M.R. O'Neal, 2004: Expected climate change impacts on soil erosion rates: a review. *J. Soil Water Conserv.*, **59**, 43-50.

Nepstad, D., P. Lefebvre, U.L. Da Silva, J. Tomasella, P. Schlesinger, L. Solorzano, P. Moutinho, D. Ray and J.G. Benito, 2004: Amazon drought and its implications for forest flammability and tree growth: a basin-wide analysis. *Glob. Change Biol.*, **10**, 704-717.

Newman, Y.C., L.E. Sollenberger, K.J. Boote, L.H. Allen and R.C. Littell, 2001: Carbon dioxide and temperature effects on forage dry matter production. *Crop Sci.*, **41**, 399-406.

Nohara, D., A. Kitoh, M. Hosaka and T. Oki, 2006: Impact of climate change on river discharge projected by multimodel ensemble. *J. Hydrometeorol.*, **7**, 1076-1089.

Norby, R.J., J.D. Sholtis, C.A. Gunderson and S.S. Jawdy, 2003: Leaf dynamics of a deciduous forest canopy; no response to elevated CO_2. *Oecologia*, **136**, 574-584.

Norby, R.J., J. Ledford, C.D. Reilly, N.E. Miller and E.G., O'Neill, 2004: Fine-root production dominates response of a deciduous forest to atmospheric CO_2 enrichment. *P. Natl. Acad. Sci. USA*, **101**, 9689-9693.

Norby, R.J., E.H. DeLucia, B. Gielen, C. Calfapietra, C.P. Giardina, J.S. King, J. Ledford, H.R. McCarthy and Co-authors, 2005: Forest response to elevated CO_2 is conserved across a broad range of productivity. *P. Natl. Acad. Sci USA*, **102**, 18052-18056.

Nowak, R.S., D.S. Ellsworth and S.D. Smith, 2004: Tansley review: functional responses of plants to elevated atmospheric CO_2 – Do photosynthetic and productivity data from FACE experiments support early predictions? *New Phytol.*, **162**, 253-280.

O'Brien, K., R. Leichenko, U. Kelkar, H. Venema, G. Aandahl, H. Thompkins, A. Javed, S. Bhadwal and Co-authors, 2004: Mapping vulnerability to multiple stressors: climate change and economic globalization in India. *Global Environ. Change*, **14**, 303-313.

O'Brien, K.L. and R.M. Leichenko, 2000: Double exposure; assessing the impacts of climate change within the context of economic globalization. *Global Environ. Change*, **10**, 221-232.

O'Meagher, B., 2005: Policy for agricultural drought in Australia: an economics perspective. *From Disaster Response to Risk Management: Australia's National Drought Policy*, L.C. Botterill and D. Wilhite, Eds., Springer, Dordrecht, 139-156.

O'Reilly, C.M., S.R. Alin, P.D. Plisnier, A.S. Cohen and B.A. McKee, 2004: Climate change decreases aquatic ecosystem productivity of Lake Tanganyika, Africa. *Nature*, **424**, 766-768.

Oba, G., 2001: The importance of pastoralists' indigenous coping strategies for planning drought management in the arid zone of Kenya. *Nomadic Peoples*, **5**, 89-119.

Olesen, J.E. and M. Bindi, 2002: Consequences of climate change for European agricultural productivity, land use and policy. *Eur. J. Agron.*, **16**, 239-262.

Ollinger, S.V., J.D. Aber, P.B. Reich and R. Freuder, 2002: Interactive effects of nitrogen deposition, tropospheric ozone, elevated CO_2 and land use history on the carbon dynamics of northern hardwood forests. *Glob. Change Biol.*, **8**, 545-562.

Opdam, P. and D. Wascher, 2004: Climate change meets habitat fragmentation: linking landscape and biogeographical scale levels in research and conservation. *Biol. Conserv.*, **117**, 285-297.

Palmer, T.N., F.J. Doblas-Reyes, R. Hagedorn and A. Weisheimer, 2005: Probabilistic prediction of climate using multi-model ensembles: from basics to applications. *Philos. T. Roy. Soc. B*, **360**, 1991-1998.

Parry, M. and M. Livermore, 2002: Climate change, global food supply and risk of hunger. *Global Environ. Change*, **17**, 109-137.

Parry, M., C. Rosenzweig and M. Livermore, 2005: Climate change, global food supply and risk of hunger. *Philos. T. Roy.. Soc. B*, **360**, 2125-2138.

Parry, M.L., 2004: Global impacts of climate change under the SRES scenarios. *Global Environ. Change*, **14**, 1-2.

Parry, M.L., C. Rosenzweig, A. Iglesias, G. Fischer and M. Livermore, 1999: Climate change and world food security: a new assessment. *Global Environ. Change*, **9**, 51-67.

Parry, M.L., C. Rosenzweig, A. Iglesias, M. Livermore and G. Fischer, 2004: Effects of climate change on global food production under SRES emissions and socio-economic scenarios. *Global Environ. Change*, **14**, 53-67.

Parson, E.A., R.W. Corell, E.J. Barron, V. Burkett, A. Janetos, L. Joyce, T.R. Karl, M.C. Maccracken and Co-authors, 2003: Understanding climatic impacts, vulnerabilities, and adaptation in the United States: building a capacity for assessment. *Climatic Change*, **57**, 9-42.

Parsons, D.J., A.C. Armstrong, J.R. Turnpenny, A.M. Matthews, K. Cooper and J.A. Clark, 2001: Integrated models of livestock systems for climate change studies. 1. Grazing systems. *Glob. Change Biol.*, **7**, 93-112.

Peng, C.H., 2000: From static biogeographical model to dynamic global vegetation model: a global perspective on modelling vegetation dynamics. *Ecol. Model.*, **135**, 33-54.

Peng, C.H., J.X. Liu, Q.L. Dang, M.J. Apps and H. Jiang, 2002: TRIPLEX: a generic hybrid model for predicting forest growth and carbon and nitrogen dynamics. *Ecol. Model.*, **153**, 109-130.

Peng, S., J. Huang, J.E. Sheehy, R.C. Laza, R.M. Visperas, X.H. Zhong, G.S. Centeno, G.S. Khush and K.G. Cassman, 2004: Rice yields decline with higher night temperature from global warming. *P. Natl. Acad. Sci. USA of the United States of America*, **101**, 9971-9975.

Perez-Garcia, J., L.A. Joyce, C.S. Binkley and A.D. McGuire, 1997: Economic impacts of climate change on the global forest sector: an integrated ecological/economic assessment. *Crit. Rev. Env. Sci. Tec.*, **27**, 123-138.

Perez-Garcia, J., L.A. Joyce, A.D. McGuire and X. Xiao, 2002: Impacts of climate change on the global forest sector. *Climatic Change*, **54**, 439-461.

Perez, R.T. and G. Yohe, 2005: Continuing the adaptation process. *Adaptation Policy Frameworks for Climate Change*, B. Lim, E. Spanger-Siegfried, I. Burton, E. Malone and S. Huq, Eds., Cambridge University Press, , 205-224.

Picon-Cochard, P., F. Teyssonneyre, J.M. Besle and J.F. Soussana, 2004: Effects of elevated CO_2 and cutting frequency on the productivity and herbage quality of a semi-natural grassland. *Eur. J. Agron.*, **20**, 363-377.

Pilon-Smits, E.A.H., M.J. Ebskamp, M. Ebskamp, M. Paul, M. Jeuken, P. Weisbeek and S. Smeekens, 1995: Improved performance of transgenic fructan-accumulating tobacco under drought stress. *Plant Physiol.*, **107**, 125-130.

Podur, J., D.L. Martell and K. Knight, 2002: Statistical quality control analysis of forest fire activity in Canada. *Can. J. Forest Res.*, **32**, 195-205.

Polley, H.W., H.B. Johnson and J.D. Derner, 2003: Increasing CO_2 from subambient to superambient concentrations alters species composition and increases above-ground biomass in a C_3/C_4 grassland. *New Phytol.*, **160**, 319-327.

Porter, J.R. and M.A. Semenov, 2005: Crop responses to climatic variation. *Philos. T. Royal Soc. B*, **360**, 2021-2035.

Powell, J.M., R.A. Pearson and J.C. Hopkins, 1998: Impacts of livestock on crop production. *Food, Lands and Livelihoods: Setting Research Agendas for Animal Science*, M. Gill, T. Smith, G.E. Pollott, E. Owen and T.L.J. Lawrence, Eds., British Society of Animal Science Occasional Publication No. 21, BSAS, Edinburgh, 53-66.

Qaderi, M.M. and D.M. Reid, 2005: Growth and physiological responses of canola (*Brassica napus*) to UV-B and CO_2 under controlled environment conditions. *Physiol. Plantarum*, **125**, 247-259.

Rafoss, T. and M.G. Saethre, 2003: Spatial and temporal distribution of bioclimatic potential for the Codling moth and the Colorado beetle in Norway: model predictions versus climate and field data from the 1990s. *Agr. Forest Meteorol.*, **5**, 75-85.

Rajagopal, V., K.V. Kasturi Bai and S. Naresh Kumar, 2002: Drought management in plantation crops. *Plantation Crops Research and Development in the New Millennium*, P. Rathinam, H.H. Khan, V.M. Reddy, P.K. Mandal and K.

Suresh, Eds., Coconut Development Board, Kochi, Kerala State, 30-35.

Ravindranath, N.H. and J. Sathaye, 2002: *Climate Change and Developing Countries: Advances in Global Change Research*, Springer, New York, 300 pp.

Reardon, T., C.P. Timmer, C.B. Barrett and J. Berdegue, 2003: The rise of supermarkets in Africa, Asia and Latin America. *Am. J. Agr. Econ*., **85**, 1140-1146.

Reddy, K.R., P.R. Doma, L.O. Mearns, M.Y.L. Boone, H.F. Hodges, A.G. Richardson and V.G. Kakani, 2002: Simulating the impacts of climate change on cotton production in the Mississippi Delta. *Climatic Research*, **22**, 271-281.

Reid, P. and C. Vogel, 2006: Living and responding to multiple stressors in South Africa – Glimpses from KwaZulu-Natal. *Global Environ. Change*, **16**, 195-206.

Reilly, J., N. Hohmann and S. Kanes, 1994: Climate change and agricultural trade. *Global Environ. Change*, **4**, 24-36.

Reilly, J., F.N. Tubiello, B. McCarl, D. Abler, R. Darwin, K. Fuglie, S. Hollinger, C. Izaurralde and Coauthors, 2003: U.S. agriculture and climate change: new results. *Climatic Change*, **57**, 43-69.

Reynolds, J.D., T.J. Webb and L.A. Hawkins, 2005: Life history and ecological correlates of extinction risk in European freshwater fishes. *Can. J. Fish. Aquat. Sci*., **62**, 854-862.

Richards, P., 1986: *Indigenous Agricultural Revolution: Ecology and Food Production in West Africa*. Hutchinson, London, 192 pp.

Richter, G.M., A. Qi, M.A. Semenov and K.W. Jaggard, 2006: Modelling the variability of UK sugar beet yields under climate change and husbandry adaptations. *Soil Use Manage*., **22**, 39-47.

Riedo, M., D. Gyalistras and J. Fuhrer, 2001: Pasture responses to elevated temperature and doubled CO_2 concentration: assessing the spatial pattern across an alpine landscape. *Climate Res*., **17**, 19-31.

Rosenberg, N.J., R.A. Brown, R.C. Izaurralde and A.M. Thomson, 2003: Integrated assessment of Hadley Centre (HadCM2) climate change projections on agricultural productivity and irrigation water supply in the conterminous United States I. Climate change scenarios and impacts on irrigation water supply simulated with the HUMUS model. *Agr. Forest Meteorol*., **117**, 73-96.

Rosenzweig, C. and M.L. Parry, 1994: Potential impact of climate change on world food supply. *Nature*, **367**, 133-138.

Rosenzweig, C. and F.N. Tubiello, 2007: Adaptation and mitigation strategies in agriculture: an analysis of potential synergies. *Mitigation and Adaptation Strategies for Global Change*, **12**, 855-873.

Rosenzweig, C., F.N. Tubiello, R.A. Goldberg, E. Mills and J. Bloomfield, 2002: Increased crop damage in the US from excess precipitation under climate change. *Global Environ. Change*, **12**, 197-202.

Ross, D.J., P.C.D. Newton and K.R. Tate, 2004: Elevated [CO_2] effects on herbage production and soil carbon and nitrogen pools and mineralization in a species-rich, grazed pasture on a seasonally dry sand. *Plant Soil*, **260**, 183-196.

Runion, G.B., 2003: Climate change and plant pathosystem – future disease prevention starts here. *New Phytol*., **159**, 531-533.

Rustad, L.E., J.L. Campbell, G.M. Marion, R.J. Norby, M.J. Mitchell, A.E. Hartley, J.H.C. Cornelissen and J. Gurevitch, 2001: A meta-analysis of the response of soil respiration, net nitrogen mineralization, and aboveground plant growth to experimental ecosystem warming. *Oecologia*, **126**, 543-562.

Saarikko, R.A., 2000: Applying a site based crop model to estimate regional yields under current and changed climates. *Ecol. Model*., **131**, 191-206.

Salinari, F., S. Giosue, F.N. Tubiello, A. Rettori, V. Rossi, F. Spanna, C. Rosenzweig and M.L. Gullino, 2006: Downy mildew epidemics on grapevine under climate change. *Global Change Biol*., **12**, 1-9.

Salinger, M.J., M.V.K. Sivakumar and R. Motha, 2005: Reducing vulnerability of agriculture and forestry to climate variability and change: workshop summary and recommendations. *Climatic Change*, **70**, 341-362.

Sands, R.D. and J.A. Edmonds, 2005: Climate change impacts for the conterminous USA: an integrated assessment Part 7. Economic analysis of field crops and land use with climate change. *Climatic Change*, **69**, 127-150.

Sano, M., 2004: Short-term effects of a mass coral bleaching event on a reef fish assemblage at Iriomote Island, Japan. *Fish. Sci*., **70**, 41-46.

Sarmiento, J.L., R. Slater, R. Barber, L. Bopp, S.C. Doney, A.C. Hirst, J. Kleypas, R. Matear and Co-authors, 2004: Response of ocean ecosystems to climate warming. *Global Biogeochem. Cy*., **18**, GB3003.

Schäfer, K.V.R., R. Oren, C.-T. Lai and G.G. Katul, 2002: Hydrologic balance in an intact temperate forest ecosystem under ambient and elevated atmospheric CO_2 concentration. *Glob. Change Biol*., **8**, 895-911.

Schlyter, P., I. Stjernquist, L. Bärring, A.M. Jönsson and C. Nilsson, 2006: Assessment of the impacts of climate change and weather extremes on boreal forests in northern Europe, focusing on Norway spruce. *Climate Res*., **31**, 75-84.

Schmidhuber, J. and P. Shetty, 2005: The nutrition transition to 2030. Why developing countries are likely to bear the major burden. *Acta Agr. Scand. C-Econ*., **3-4**, 150-166.

Schmittner, A., 2005: Decline of the marine ecosystem caused by a reduction in the Atlantic overturning circulation. *Nature*, **434**, 628-633.

Schroeter, D., 2004: ATEAM, Advanced Terrestrial Ecosystem Analysis and Modelling. Final Report, Potsdam Institute for Climate Impact Research, Postdam, Germany, 139 pp. [Accessed 21.03.07: http://www.pik-potsdam.de/ateam/ateam_final_report_sections_5_to_6.pdf]

Schröter, D., C. Polsky and A.G. Patt, 2005: Assessing vulnerabilities to the effects of global change: an eight step approach. *Mitigation and Adaptation Strategies for Global Change*, **10**, 573-596.

Scoones, I., C. Cibudu, S. Chikura, P. Jeranyama, D. Machaka, W. Machanja, B. Mavedzenge, B. Mombeshora, M. Maxwell, C. Mudziwo, F. Murimbarimba and B. Zirereza, 1996: *Hazards and Opportunities: Farming Livelihoods in Dryland Africa: Lessons from Zimbabwe*. Zed Books in association with IIED, London and New Jersey, xviii+267 pp.

Sedjo, R. and K. Lyon, 1990: *The Long-Term Adequacy of World Timber Supply*. Resources for the Future, Washington, District of Columbia, 230 pp.

Sedjo, R.A. and K.S. Lyon, 1996: Timber supply model 96: a global timber supply model with a pulpwood component. Resources for the Future Discussion Paper 96-15. [Accessed 21.03.07: http://www.rff.org/rff/Documents/RFF-DP-96-15.pdf]

Sénat, 2004: Information report no. 195 – France and the French face the canicule: the lessons of a crisis: appendix to the minutes of the session of February 3, 2004, 59-62. [Accessed 21.03.07: http://www.senat.fr/rap/r03-195/r03-195.html]

Shaw, M.R., E.S. Zavaleta, N.R. Chiariello, E.E. Cleland, H.A. Mooney and C.B. Field, 2002: Grassland responses to global environmental changes suppressed by elevated CO_2. *Science*, **298**, 1987-1990.

Shibru, M., 2001: *Pastoralism and cattle marketing: a case study of the Borana of southern Ethiopia*, Unpublished Masters Thesis, Egerton University.

Shugart, H., R. Sedjo and B. Sohngen, 2003: Forests and global climate change: potential impacts on U.S. forest resources. Pew Center on Global Climate Change, Arlington, Virginia, 64 pp.

Sinclair, T.R. and L.C. Purcell, 2005: Is a physiological perspective relevant in a 'genocentric' age? *J. Exp. Bot*., **56**, 2777-2782.

Sitch, S., B. Smith, I.C. Prentice, A. Arneth, A. Bondeau, W. Cramer, J.O. Kaplan, S. Levis, W. Lucht, M.T. Sykes, K. Thonicke and S. Venevsky, 2003: Evaluation of ecosystem dynamics, plant geography and terrestrial carbon cycling in the LPJ dynamic global vegetation model. *Glob. Change Biol*., **9**, 161-185.

Sivakumar, M.V.K. and C. Valentin, 1997: Agroecological zones and crop production potential. *Phil. Trans. R. Soc. Lond*. B, ***352***, 907-916.

Sivakumar, M.V.K., H.P. Das and O. Brunini, 2005: Impacts of present and future climate variability and change on agriculture and forestry in the arid and semi-arid tropics. *Climatic Change*, **70**, 31-72.

Slingo, J.M., A.J. Challinor, B.J. Hoskins and T.R. Wheeler, 2005: Introduction: food crops in a changing climate. *Philos. T. Roy. Soc. B*, **360**, 1983-1989.

Smith, J., P. Smith, M. Wattenbach, S. Zahele, R. Hiederer, J.A. Jones, L. Montanarella, M.D.A. Rounsevell, I. Reginster and F. Ewert, 2005: Projected changes in mineral soil carbon of European croplands and grasslands, 1990-2080. *Glob. Change Biol*., **11**, 2141-2152.

Sohngen, B. and R. Mendelsohn, 1998: Valuing the market impact of large scale ecological change: the effect of climate change on US timber. *Am. Econ. Rev*., **88**, 689-710.

Sohngen, B. and Sedjo, R., 2005: Impacts of climate change on forest product markets: implications for North American producers. *Forest Chron*., **81**, 669-674.

Sohngen, B., R. Mendelsohn and R.A. Sedjo, 1999: Forest management, conservation, and global timber markets. *Am. J. Agr. Econ*., **81**.

Sohngen, B., R. Mendelsohn and R. Sedjo, 2001: A global model of climate change impacts on timber markets. *J. Agr. Resour. Econ*, **26**, 326-343.

Solberg, B., A. Moiseyev and A.M. Kallio, 2003: Economic impacts of accelerating forest growth in Europe. *Forest Policy Econ*., **5**, 157-171.

Southworth, J., R.A. Pfeifer, M. Habeck, J.C. Randolph, O.C. Doering and D.G. Rao, 2002: Sensitivity of winter wheat yields in the Midwestern United States to future changes in climate, climate variability, and CO_2 fertilization. *Climate Res*., **22**, 73-86.

Spalding, M.D. and G.E. Jarvis, 2002: The impact of the 1998 coral mortality on reef fish communities in the Seychelles. *Mar. Pollut. Bull*., **44**, 309-321.

Spittlehouse, D.L. and R.B. Stewart, 2003: Adaptation to climate change in forest management. *BC J. Ecosys. Manage,*, **4**, 1-11. [Accessed 21.03.07: http://www.forrex.org/publications/jem/ISS21/vol4_no1_art1.pdf]

Stacey, D.A. and M.D.E. Fellows, 2002: Influence of elevated CO_2 on interspecific interactions at higher trophic levels, 2002. *Glob. Change Biol.*, **8**, 668-678.

Stigter, C.J., Z. Dawei, L.O.Z. Onyewotu and M. Xurong, 2005: Using traditional methods and indigenous technologies for coping with climate variability. *Climatic Change*, **70**, 255-271.

Stoate, C., N.D. Boatman, R.J. Borralho, C. Rio Carvalho, G.R. de Snoo and P. Eden, 2001: Ecological impacts of arable intensification in Europe. *J. Environ. Manage.*, **63**, 337-365.

Surtech, 1993: Isiolo meat factory project: feasibility study for Isiolo abattoir. Report prepared for the GTZ-Marsabit Development Programme, PO Box 52514, Nairobi.

Swearingen, W. and A. Bencherifa, 2000: An assessment of the drought hazard in Morocco. *Drought: A Global Assessment*. Vol. I, D.A. Wilhite, Ed., Routledge, London, 279-286.

Tan, G. and R. Shibasaki, 2003: Global estimation of crop productivity and the impacts of global warming by GIS and EPIC integration. *Ecol. Model.*, **168**, 357-370.

Tao, F.L., M. Yokozawa, Y. Hayashi and E. Lin, 2003: Future climate change, the agricultural water cycle, and agricultural production in China. *Agr. Ecosyst. Environ.*, **95**, 203-215.

Teyssonneyre, F., C. Picon-Cochard, R. Falcimagne and J.F. Soussana, 2002: Effects of elevated CO_2 and cutting frequency on plant community structure in a temperate grassland. *Glob. Change Biol.*, **8**, 1034-1046.

Thomas, D., H. Osbahr, C. Twyman, N. Adger and B. Hewitson, 2005a: ADAPTIVE: adaptations to climate change amongst natural resource-dependant societies in the developing world: across the Southern African climate gradient. Technical Report 35, Tyndall Centre, University of East Anglia, Norwich, 47 pp. [Accessed 21.03.07: http://www.tyndall.ac.uk/research/theme3/final_reports/t2_31.pdf]

Thomas, D.S.G., M. Knight and G.F.S. Wiggs, 2005b: Remobilization of southern African desert dune systems by twenty-first century global warming. *Nature*, **435**, 1218-1221.

Thomas, J.M.G., K.J. Boote, L.H. Allen Jr., M. Gallo-Meagher and J.M. Davis, 2003: Elevated temperature and carbon dioxide effects on soybean seed composition and transcript abundance. *Crop Sci.*, **43**, 1548-1557.

Thomson, A.M., R.A. Brown, N.J. Rosenberg, R.C. Izaurralde and V. Benson, 2005a: Climate change impacts for the conterminous USA: an integrated assessment Part 3. Dryland production of grain and forage crops. *Climatic Change*, **69**, 43-65.

Thomson, A.M., N.J. Rosenberg, R.C. Izaurralde and R.A. Brown, 2005b: Climate change impacts for the conterminous USA: an integrated assessment Part 5. Irrigated agriculture and national grain crop production. *Climatic Change*, **69**, 89-105.

Thuiller, W., S. Lavorel, M.B. Araujo, M.T. Sykes and I.C. Prentice, 2005: Climate change threats to plant diversity in Europe. *P. Natl. Acad. Sci. USA*, **102**, 8245-8250.

Todd, M.C., R. Washington, R.A. Cheke and D. Kniveton, 2002: Brown locust outbreaks and climate variability in southern Africa. *J. Appl. Ecol.*, **39**, 31-42.

Tol, R.S., 2002: Estimates of the damage costs of climate change, Part II. Dynamic estimates. *Environ. Resour. Econ.*, **21**, 135-160.

Toulmin, C., 1986: Livestock losses and post-drought rehabilitation in sub-Saharan Africa: policy options and issues. Livestock Policy Unit Working Paper No. 9, International Livestock Centre for Africa, Addis Ababa.

Travasso, M.I., G.O. Magrin, W.E. Baethgen, J.P. Castao, G.R. Rodriguez, R. Rodriguez, J.L. Pires, A. Gimenez, G. Cunha and M. Fernandes, 2006: Adaptation measures for maize and soybean in Southeastern South America. Working Paper No. 28, Assessments of Impacts and Adaptations to Climate Change (AIACC), 38 pp.

Trenberth, K.E., P.D. Jones, P.G. Ambenje, R. Bojariu, D.R. Easterling, A.M.G. Klein Tank, D.E. Parker, J.A. Renwick and Coauthors, 2007: Observations: surface and atmospheric climate change. *Climate Change 2007: The Physical Science Basis. Contribution of Working Group I to the Fourth Assessment Report of the Intergovernmental Panel on Climate Change*, S. Solomon, D. Qin, M. Manning, Z. Chen, M. Marquis, K.B. Averyt, M. Tignor and H.L. Miller, Eds., Cambridge University Press, Cambridge, 235-336.

Tubiello, F.N., 2005: Climate variability and agriculture: perspectives on current and future challenges. *Impact of Climate Change, Variability and Weather Fluctuations on Crops and Their Produce Markets*, B. Knight, Ed., Impact Reports, Cambridge, UK, 45-63.

Tubiello, F.N. and F. Ewert, 2002: Simulating the effects of elevated CO_2 on crops: approaches and applications for climate change. *Eur. J. Agron.*, **18**, 57-74.

Tubiello, F.N. and G. Fischer, 2006: Reducing climate change impacts on agriculture: global and regional effects of mitigation, 2000-2080. *Techol. Forecast. Soc.*, doi: 10.1016/j.techfore.2006.05.027.

Tubiello, F.N., M. Donatelli, C. Rosenzweig and C.O. Stockle, 2000: Effects of climate change and elevated CO_2 on cropping systems: model predictions at two Italian locations. *Eur. J. Agron.*, **13**, 179-189.

Tubiello, F.N., S. Jagtap, C. Rosenzweig, R.A. Goldberg and J.W. Jones, 2002: Effects of climate change on US crop production from the National Assessment. Simulation results using two different GCM scenarios. Part I: Wheat, potato, maize, and citrus. *Climate Res.*, **20**, 259-270.

Tubiello, F.N., J.A. Amthor, W.E. Easterling, G. Fischer, R. Gifford, M. Howden, J. Reilly and C. Rosenzweig, 2007a: Crop response to elevated CO_2 at FACE value. *Science*, in press.

Tubiello, F.N., J.A. Amthor, K. Boote, M. Donatelli, W.E. Easterling, G. Fisher, R. Gifford, M. Howden, J. Reilly and C. Rosenzweig, 2007b: Crop response to elevated CO_2 and world food supply. *Eur. J. Agron.*, **26**, 215-223.

Turnpenny, J.R., D.J. Parsons, A.C. Armstrong, J.A. Clark, K. Cooper and A.M. Matthews, 2001: Integrated models of livestock systems for climate change studies. 2. Intensive systems. *Glob. Change Biol.*, **7**, 163-170.

Turvey, C., 2001: Weather derivatives for specific event risks in agriculture. *Rev. Agr. Econ.*, **23**, 335-351.

UN, 2004: World population to 2300, Document ST/ESA/SER.A/236, United Nations Department of Economic and Social Affairs, Population Division, New York, 254 pp.

UNDESA, 2004: *World Urbanization Prospects: the 2003 Revision*, United Nations Department of Economic and Social Affairs Population Division, New York, 335 pp.

UNECE 2006: Forest products annual market review 2005-2006. Geneva Timber and Forest Study Paper 21. Document ECE/TIM/SP/21. United Nations, New York, Geneva, 163 pp.

van Duivenbooden, N., S. Abdoussalam and A. Ben Mohamed, 2002: Impact of climate change on agricultural production in the Sahel. Part 2. Case study for groundnut and cowpea in Niger. *Climatic Change*, **54**, 349-368.

van Groenigen, K.-J., J. Six, B.A. Hungate, M.-A. de Graaff, N. van Breemen and C. van Kessel, 2006: Element interactions limit soil carbon storage. *P. Natl. Acad. Sci. USA*, **103**, 6571-6574.

van Ittersum, M.K., S.M. Howden and S. Asseng, 2003: Sensitivity of productivity and deep drainage of wheat cropping systems in a Mediterranean environment to changes in CO_2, temperature and precipitation. *Agr. Ecosyst. Environ.*, **97**, 255-273.

van Wuijckhuise, L., D. Dercksen, J. Muskens, J. de Bruyn, M. Scheepers and R. Vrouenraets, 2006: Bluetongue in the Netherlands; description of the first clinical cases and differential diagnosis; Common symptoms just a little different and in too many herds. *Tijdschr. Diergeneesk.*, **131**, 649-654.

Vandermeiren, K., 2005: Impact of rising tropospheric ozone on potato: effects on photosynthesis, growth, productivity and yield quality. *Plant Cell Environ.*, **28**, 982-996.

Vanhatalo, M., J. Back and S. Huttunen, 2003: Differential impacts of long-term (CO_2) and O_3 exposure on growth of northern conifer and deciduous tree species. *Trees-Struct. Funct.*, **17**, 211-220.

Varaprasad, P.V., K.J. Boote, L. Hartwell-Allen and J.M.G. Thomas, 2003: Super-optimal temperatures are detrimental to peanut (*Arachis hypogaea* L.) reproductive processes and yield at both ambient and elevated carbon dioxide. *Glob. Change Biol.*, **9**, 1775-1787.

Vasquez-Leon, M., C.T. West and T.J. Finan, 2003: A comparative assessment of climate vulnerability: agriculture and ranching on both sides of the US–Mexico border. *Global Environ. Change*, **13**, 159-173.

Vogel, C., 2005: "Seven fat years and seven lean years?" Climate change and agriculture in Africa. *IDS Bull-I Dev. Stud.*, **36**, 30-35.

Volder, A., E.J. Edwards, J.R. Evans, B.C. Robertson, M. Schortemeyer and R.M. Gifford, 2004: Does greater night-time, rather than constant, warming alter growth of managed pasture under ambient and elevated atmospheric CO_2? *New Phytol.*, **162**, 397-411.

Volk, M., P. Bungener, F. Contat, M. Montani and J. Fuhrer, 2006: Grassland yield declined by a quarter in 5 years of free-air ozone fumigation. *Glob. Change Biol.*, **12**, 74-83.

Volney, W.J.A. and Fleming, R.A., 2000: Climate change and impacts of boreal forest insects. *Agr. Ecosyst. Environ.*, **82**, 283-294.

Vorley, B., 2002: Sustaining agriculture: policy, governance and the future of family-based farming. A synthesis report of the collaborative research project 'policies that work for sustainable agriculture and regenerating rural livelihoods', International Institute for Environment and Development, London, 189 pp.

Wassmann, R., N.X. Hein, C.T. Hoanh and T.P. Tuong, 2004: Sea level rise affecting the Vietnamese Mekong Delta: water elevation in the flood season and implications for rice production. *Climatic Change*, **66**, 89-107.

Weih, M., 2004: Intensive short rotation forestry in boreal climates: present and future perspectives. *Can. J. For. Res.*, **34**, 1369-1378.

Westerling, A.L., H.G. Hidalgo, D.R. Cayan and T.W. Swetnam, 2006: Warming and earlier spring increase western U.S. forest wildfire activity. *Science*, **313**, 940-943.

Wheeler, T.R., P.Q. Crauford, R.H. Ellis, J.R. Porter and P.V. Vara Prasad, 2000: Temperature variability and the yield of annual crops. *Agr. Ecosyst. Environ.*, **82**, 159-167.

White, N., R.W. Sutherst, N. Hall and P. Whish-Wilson, 2003: The vulnerability of the Australian beef industry to impacts of the cattle tick (*Boophilus microplus*) under climate change. *Climatic Change*, **61**, 157-190.

Williams, A.A.J., D.J. Karoly and N. Tapper, 2001: The sensitivity of Australian fire danger to climate change. *Climatic Change*, **49**, 171-191.

Winkels, A. and W.N. Adger, 2002: Sustainable livelihoods and migration in Vietnam; the importance of social capital as access to resources. *International Symposium on Sustaining Food Security and Managing Natural Resources in Southeast Asia – Challenges for the 21st Century*, Chiang Mai, Thailand, University of Hohenheim, Stuttgart, 15 pp.

Winters, P., R. Murgai, E. Sadoulet, A.D. Janvry and G. Frisvold, 1998: Economic and welfare impacts of climate change on developing countries. *Environ. Resour. Econ.*, **12**, 1-24.

Wittig, V.E., C.J. Bernacchi, X.-G. Zhu, C. Calfapietra, R. Ceulemans, P. Deangelis, B. Gielen, F. Miglietta, P.B. Morgan and S.P. Long, 2005: Gross primary production is stimulated for three Populus species grown under free-air CO_2 enrichment from planting through canopy closure. *Glob. Change Biol.*, **11**, 644-656.

Wollenweber, B., J.R. Porter and J. Schellberg, 2003: Lack of interaction between extreme high-temperature events at vegetative and reproductive growth stages in wheat. *J. Agron. Crop Sci.*, **189**, 142-150.

World Bank, 2004: *Modelled observations on development scenarios in the Lower Mekong Basin*. Mekong Regional Water Resources Assistance Strategy, Prepared for the World Bank with Mekong River Commission cooperation, Washington DC and Vientiane, 142 pp. [Accessed 28.06.07: http://www.mrcmekong.org/free_download/report.htm]

Wullschleger, S.D., T.J. Tschaplinski and R.J. Norby, 2002: Plant water relations at elevated CO_2 – Implications for water-limited environments. *Plant Cell Environ.*, **25**, 319-331.

Xiao, G., W. Liu, Q. Xu, Z. Sun and J. Wang, 2005: Effects of temperature increase and elevated CO_2 concentration, with supplemental irrigation, on the yield of rain-fed spring wheat in a semiarid region of China. *Agr. Water Manage.*, **74**, 243-255.

Yates, D.N. and K.M. Strzepek, 1998: An assessment of integrated climate change impacts on the agricultural economy of Egypt. *Climatic Change*, **38**, 261-287.

Yohe, G. and R.S.J. Tol, 2001: Indicators of social and economic coping capacity: moving toward a working definition of adaptive capacity. *Global Environ. Change*, **12**, 25-40.

Zavaleta, E.S., M.R. Shaw, N.R. Chiariello, B.D. Thomas, E.E. Cleland, C.B. Field and H.A. Mooney, 2003: Grassland responses to three years of elevated temperature, CO_2, precipitation, and N deposition. *Ecol. Monogr.*, **73**, 585-604.

Zhang, X.C. and W.Z. Liu, 2005: Simulating potential response of hydrology, soil erosion, and crop productivity to climate change in Changwu tableland region on the Loess Plateau of China. *Agr. Forest Meteorol.*, **131**, 127-142.

Zhao, D., K.R. Reddy, V.G. Kakani, J.J. Read and J.H. Sullivan, 2003: Growth and physiological responses of cotton (*Gossypium hirsutum* L.) to elevated carbon dioxide and ultraviolet-B radiation under controlled environmental conditions. *Plant Cell Environ.*, **26**, 771-782.

Zhao, Y., C. Wang, S. Wang and L. Tibig, 2005: Impacts of present and future climate variability on agriculture and forestry in the humid and sub-humid tropics. *Climatic Change*, **70**, 73-116.

Zheng, Y.Q., G. Yu, Y.F. Qian, M. Miao, X. Zeng and H. Liu, 2002: Simulations of regional climatic effects of vegetation change in China. *Q. J. Roy. Meteor. Soc.*, **128**, 2089-2114, Part B.

Ziervogel, G., 2003: Targeting seasonal climate forecasts for integration into household level decisions: the case of smallholder farmers in Lesotho. *Geogr. J.*, **170**, 6-21.

Ziska, L.H., 2003: Evaluation of yield loss in field-grown sorghum from a c3 and c4 weed as a function of increasing atmospheric carbon dioxide. *Weed Sci,*, **51**, 914-918.

Ziska, L.H. and K. George, 2004: Rising carbon dioxide and invasive, noxious plants: potential threats and consequences. *World Resource Review*, **16**, 427-447.

Zvereva, E.L and M.V. Kozlov, 2006: Consequences of simultaneous elevation of carbon dioxide and temperature for plant–herbivore interactions: a meta-analysis. *Glob. Change Biol.*, **12**, 27-41.

6

Coastal systems and low-lying areas

Coordinating Lead Authors:

Robert J. Nicholls (UK), Poh Poh Wong (Singapore)

Lead Authors:

Virginia Burkett (USA), Jorge Codignotto (Argentina), John Hay (New Zealand), Roger McLean (Australia), Sachooda Ragoonaden (Mauritius), Colin D. Woodroffe (Australia)

Contributing Authors:

Pamela Abuodha (Kenya), Julie Arblaster (USA/Australia), Barbara Brown (UK), Don Forbes (Canada), Jim Hall (UK), Sari Kovats (UK), Jason Lowe (UK), Kathy McInnes (Australia), Susanne Moser (USA), Susanne Rupp-Armstrong (UK), Yoshiki Saito (Japan), Richard S.J. Tol (Ireland)

Review Editors:

Job Dronkers (The Netherlands), Geoff Love (Australia), Jin-Eong Ong (Malaysia)

This chapter should be cited as:

Nicholls, R.J., P.P. Wong, V.R. Burkett, J.O. Codignotto, J.E. Hay, R.F. McLean, S. Ragoonaden and C.D. Woodroffe, 2007: Coastal systems and low-lying areas. *Climate Change 2007: Impacts, Adaptation and Vulnerability. Contribution of Working Group II to the Fourth Assessment Report of the Intergovernmental Panel on Climate Change*, M.L. Parry, O.F. Canziani, J.P. Palutikof, P.J. van der Linden and C.E. Hanson, Eds., Cambridge University Press, Cambridge, UK, 315-356.

Table of Contents

Executive Summary

Since the IPCC Third Assessment Report (TAR), our understanding of the implications of climate change for coastal systems and low-lying areas (henceforth referred to as 'coasts') has increased substantially and six important policy-relevant messages have emerged.

Coasts are experiencing the adverse consequences of hazards related to climate and sea level (very high confidence).
Coasts are highly vulnerable to extreme events, such as storms, which impose substantial costs on coastal societies [6.2.1, 6.2.2, 6.5.2]. Annually, about 120 million people are exposed to tropical cyclone hazards, which killed 250,000 people from 1980 to 2000 [6.5.2]. Through the 20th century, global rise of sea level contributed to increased coastal inundation, erosion and ecosystem losses, but with considerable local and regional variation due to other factors [6.2.5, 6.4.1]. Late 20th century effects of rising temperature include loss of sea ice, thawing of permafrost and associated coastal retreat, and more frequent coral bleaching and mortality [6.2.5].

Coasts will be exposed to increasing risks, including coastal erosion, over coming decades due to climate change and sea-level rise (very high confidence).
Anticipated climate-related changes include: an accelerated rise in sea level of up to 0.6 m or more by 2100; a further rise in sea surface temperatures by up to 3°C; an intensification of tropical and extra-tropical cyclones; larger extreme waves and storm surges; altered precipitation/run-off; and ocean acidification [6.3.2]. These phenomena will vary considerably at regional and local scales, but the impacts are virtually certain to be overwhelmingly negative [6.4, 6.5.3].

Corals are vulnerable to thermal stress and have low adaptive capacity. Increases in sea surface temperature of about 1 to 3°C are projected to result in more frequent coral bleaching events and widespread mortality, unless there is thermal adaptation or acclimatisation by corals [Box 6.1, 6.4].

Coastal wetland ecosystems, such as saltmarshes and mangroves, are especially threatened where they are sediment-starved or constrained on their landward margin [6.4.1]. Degradation of coastal ecosystems, especially wetlands and coral reefs, has serious implications for the well-being of societies dependent on the coastal ecosystems for goods and services [6.4.2, 6.5.3]. Increased flooding and the degradation of freshwater, fisheries and other resources could impact hundreds of millions of people, and socio-economic costs on coasts will escalate as a result of climate change [6.4.2, 6.5.3].

The impact of climate change on coasts is exacerbated by increasing human-induced pressures (very high confidence).
Utilisation of the coast increased dramatically during the 20th century and this trend is virtually certain to continue through the 21st century. Under the SRES scenarios, the coastal population could grow from 1.2 billion people (in 1990) to 1.8 to 5.2 billion people by the 2080s, depending on assumptions about migration [6.3.1]. Increasing numbers of people and assets at risk at the coast are subject to additional stresses due to land-use and hydrological changes in catchments, including dams that reduce sediment supply to the coast [6.3.2]. Populated deltas (especially Asian megadeltas), low-lying coastal urban areas and atolls are key societal hotspots of coastal vulnerability, occurring where the stresses on natural systems coincide with low human adaptive capacity and high exposure [6.4.3]. Regionally, South, South-east and East Asia, Africa and small islands are most vulnerable [6.4.2]. Climate change therefore reinforces the desirability of managing coasts in an integrated manner [6.6.1.3].

Adaptation for the coasts of developing countries will be more challenging than for coasts of developed countries, due to constraints on adaptive capacity (high confidence).
While physical exposure can significantly influence vulnerability for both human populations and natural systems, a lack of adaptive capacity is often the most important factor that creates a hotspot of human vulnerability. Adaptive capacity is largely dependent upon development status. Developing nations may have the political or societal will to protect or relocate people who live in low-lying coastal zones, but without the necessary financial and other resources/capacities, their vulnerability is much greater than that of a developed nation in an identical coastal setting. Vulnerability will also vary between developing countries, while developed countries are not insulated from the adverse consequences of extreme events [6.4.3, 6.5.2].

Adaptation costs for vulnerable coasts are much less than the costs of inaction (high confidence).
Adaptation costs for climate change are much lower than damage costs without adaptation for most developed coasts, even considering only property losses and human deaths [6.6.2, 6.6.3]. As post-event impacts on coastal businesses, people, housing, public and private social institutions, natural resources, and the environment generally go unrecognised in disaster cost accounting, the full benefits of adaptation are even larger [6.5.2, 6.6.2]. Without adaptation, the high-end sea-level rise scenarios, combined with other climate changes (e.g., increased storm intensity), are as likely as not to render some islands and low-lying areas unviable by 2100, so effective adaptation is urgently required [6.6.3].

The unavoidability of sea-level rise, even in the longer-term, frequently conflicts with present-day human development patterns and trends (high confidence).
Sea-level rise has substantial inertia and will continue beyond 2100 for many centuries. Irreversible breakdown of the West Antarctica and/or Greenland ice sheets, if triggered by rising temperatures, would make this long-term rise significantly larger, ultimately questioning the viability of many coastal settlements across the globe. The issue is reinforced by the increasing human use of the coastal zone. Settlement patterns also have substantial inertia, and this issue presents a challenge for long-term coastal spatial planning. Stabilisation of climate could reduce the risks of ice sheet breakdown, and reduce but

not stop sea-level rise due to thermal expansion [Box 6.6]. Hence, it is now more apparent than it was in the TAR that the most appropriate response to sea-level rise for coastal areas is a combination of *adaptation* to deal with the inevitable rise, and *mitigation* to limit the long-term rise to a manageable level [6.6.5, 6.7].

6.1 Introduction: scope, summary of Third Assessment Report conclusions and key issues

This chapter presents a global perspective on the impacts of climate change and sea-level rise on coastal and adjoining low-lying areas, with an emphasis on post-2000 insights. Here, coastal systems are considered as the interacting low-lying areas and shallow coastal waters, including their human components (Figure 6.1). This includes adjoining coastal lowlands, which have often developed through sedimentation during the Holocene (past 10,000 years), but excludes the continental shelf and ocean margins (for marine ecosystems see Chapter 4). Inland seas are not covered, except as analogues. In addition to local drivers and interactions, coasts are subject to external events that pose a hazard to human activities and may compromise the natural functioning of coastal systems (Figure 6.1). Terrestrial-sourced hazards include river floods and inputs of sediment or pollutants; marine-sourced hazards include storm surges, energetic swell and tsunamis.

In this chapter, we reinforce the findings of the Third Assessment Report (TAR; IPCC, 2001) concerning the potential importance of the full range of climate change drivers on coastal systems and the complexity of their potential effects. The TAR also noted growing interest in adaptation to climate change in coastal areas, a trend which continues to gather momentum, as shown in this assessment. Whereas some coastal countries and communities have the adaptive capacity to minimise the impacts of climate change, others have fewer options and hence are much more vulnerable to climate change. This is compounded as human population growth in many coastal regions is both increasing socio-economic vulnerability

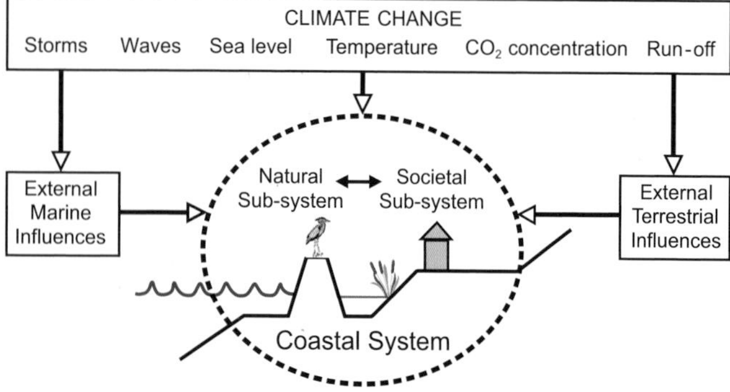

Figure 6.1. *Climate change and the coastal system showing the major climate change factors, including external marine and terrestrial influences.*

and decreasing the resilience of coastal systems. Integrated assessment and management of coastal systems, together with a better understanding of their interaction with socio-economic and cultural development, were presented in the TAR as important components of successful adaptation to climate change.

This chapter builds on and develops these insights in the TAR by considering the emerging knowledge concerning impacts and adaptation to climate change in coastal areas across a wider spectrum of climate change drivers and from local to global scales. Nonetheless, the issue of sea-level rise still dominates the literature on coastal areas and climate change. This chapter includes an assessment of current sensitivity and vulnerability, the key changes that coastal systems may undergo in response to climate and sea-level change, including costs and other socio-economic aspects, the potential for adaptation, and the implications for sustainable development. Given that there are strong interactions both within and between the natural and human sub-systems in the coastal system (Figure 6.1), this chapter takes an integrated perspective of the coastal zone and its management, insofar as the published literature permits.

6.2 Current sensitivity/vulnerability

This section provides key insights into the ways in which coastal systems are presently changing, as context for assessing the impacts of, and early effects attributable to, climate change.

6.2.1 Natural coastal systems

Coasts are dynamic systems, undergoing adjustments of form and process (termed morphodynamics) at different time and space scales in response to geomorphological and oceanographical factors (Cowell et al., 2003a,b). Human activity exerts additional pressures that may dominate over natural processes. Often models of coastal behaviour are based on palaeoenvironmental reconstructions at millennial scales and/or process studies at sub-annual scales (Rodriguez et al., 2001; Storms et al., 2002; Stolper et al., 2005). Adapting to global climate change, however, requires insight into processes at decadal to century scales, at which understanding is least developed (de Groot, 1999; Donnelly et al., 2004).

Coastal landforms, affected by short-term perturbations such as storms, generally return to their pre-disturbance morphology, implying a simple, morphodynamic equilibrium. Many coasts undergo continual adjustment towards a dynamic equilibrium, often adopting different 'states' in response to varying wave energy and sediment supply (Woodroffe, 2003). Coasts respond to altered conditions external to the system, such as storm events, or changes triggered by internal thresholds that cannot be predicted on the basis of external stimuli. This natural variability of coasts can make it difficult to identify the impacts of climate change. For example, most beaches worldwide show evidence of recent erosion but sea-level rise is not necessarily the primary driver. Erosion can result from other factors, such as altered wind patterns (Pirazzoli et al., 2004; Regnauld et al., 2004), offshore bathymetric changes (Cooper and Navas, 2004), or reduced fluvial sediment input (Sections 6.2.4 and 6.4.1.1). A major challenge is determining whether observed changes have resulted from alteration in external factors (such as climate change), exceeding an internal threshold (such as a delta distributary switching to a new location), or short-term disturbance within natural climate variability (such as a storm).

Climate-related ocean-atmosphere oscillations can lead to coastal changes (Viles and Goudie, 2003). One of the most prominent is the El Niño-Southern Oscillation (ENSO) phenomenon, an interaction between pronounced temperature anomalies and sea-level pressure gradients in the equatorial Pacific Ocean, with an average periodicity of 2 to 7 years. Recent research has shown that dominant wind patterns and storminess associated with ENSO may perturb coastal dynamics, influencing (1) beach morphodynamics in eastern Australia (Ranasinghe et al., 2004; Short and Trembanis, 2004), mid-Pacific (Solomon and Forbes, 1999) and Oregon (Allan et al., 2003); (2) cliff retreat in California (Storlazzi and Griggs, 2000); and (3) groundwater levels in mangrove ecosystems in Micronesia (Drexler, 2001) and Australia (Rogers et al., 2005). Coral bleaching and mortality appear related to the frequency and intensity of ENSO events in the Indo-Pacific region, which may alter as a component of climate change (Box 6.1), becoming more widespread because of global warming (Stone et al., 1999). It is likely that coasts also respond to longer term variations; for instance, a relationship with the Pacific Decadal Oscillation (PDO) is indicated by monitoring of a south-east Australian beach for more than 30 years (McLean and Shen, 2006). Correlations between the North Atlantic Oscillation (NAO) and storm frequency imply similar periodic influences on Atlantic coasts (Tsimplis et al., 2005, 2006), and the Indian Ocean Dipole (IOD) may drive similar periodic fluctuations on coasts around the Indian Ocean (Saji et al., 1999).

6.2.2 Increasing human utilisation of the coastal zone

Few of the world's coastlines are now beyond the influence of human pressures, although not all coasts are inhabited (Buddemeier et al., 2002). Utilisation of the coast increased dramatically during the 20th century, a trend that seems certain to continue through the 21st century (Section 6.3.1). Coastal population growth in many of the world's deltas, barrier islands and estuaries has led to widespread conversion of natural coastal landscapes to agriculture, aquaculture, silviculture, as well as industrial and residential uses (Valiela, 2006). It has been estimated that 23% of the world's population lives both within 100 km distance of the coast and <100 m above sea level, and population densities in coastal regions are about three times higher than the global average (Small and Nicholls, 2003) (see also Box 6.6). The attractiveness of the coast has resulted in disproportionately rapid expansion of economic activity, settlements, urban centres and tourist resorts. Migration of people to coastal regions is common in both developed and developing nations. Sixty percent of the world's 39 metropolises with a population of over 5 million are located within 100 km of the coast, including 12 of the world's 16 cities with populations greater than 10 million. Rapid urbanisation has many consequences: for example, enlargement of natural coastal inlets and dredging of waterways for navigation, port facilities, and pipelines exacerbate saltwater intrusion into surface and ground waters. Increasing shoreline retreat and risk of flooding of coastal cities in Thailand (Durongdej, 2001; Saito, 2001), India (Mohanti, 2000), Vietnam (Thanh et al., 2004) and the United States (Scavia et al., 2002) have been attributed to degradation of coastal ecosystems by human activities, illustrating a widespread trend.

The direct impacts of human activities on the coastal zone have been more significant over the past century than impacts that can be directly attributed to observed climate change (Scavia et al., 2002; Lotze et al., 2006). The major direct impacts include drainage of coastal wetlands, deforestation and reclamation, and discharge of sewage, fertilisers and contaminants into coastal waters. Extractive activities include sand mining and hydrocarbon production, harvests of fisheries and other living resources, introductions of invasive species and construction of seawalls and other structures. Engineering structures, such as damming, channelisation and diversions of coastal waterways, harden the coast, change circulation patterns and alter freshwater, sediment and nutrient delivery. Natural systems are often directly or indirectly altered, even by soft engineering solutions, such as beach nourishment and foredune construction (Nordstrom, 2000; Hamm and Stive, 2002). Ecosystem services on the coast are often disrupted by human activities. For example, tropical and subtropical mangrove forests and temperate saltmarshes provide goods and services (they accumulate and transform nutrients, attenuate waves and storms, bind sediments and support rich ecological communities), which are reduced by large-scale ecosystem conversion for agriculture, industrial and urban development, and aquaculture (Section 6.4.2).

6.2.3 External terrestrial and marine influences

External terrestrial influences have led to substantial environmental stresses on coastal and nearshore marine habitats (Sahagian, 2000; Saito, 2001; NRC, 2004; Crossland et al., 2005). As a consequence of activities outside the coastal zone, natural ecosystems (particularly within the catchments draining to the coast) have been fragmented and the downstream flow of water, sediment and nutrients has been disrupted (Nilsson et al., 2005; Section 6.4.1.3). Land-use change, particularly deforestation, and hydrological modifications have had downstream impacts, in addition to localised development on the coast. Erosion in the catchment has increased river sediment load; for example, suspended loads in the Huanghe (Yellow) River have increased 2 to 10 times over the past 2000 years (Jiongxin, 2003). In contrast, damming and channelisation have greatly reduced the supply of sediments to the coast on other rivers through retention of sediment in dams (Syvitski et al., 2005). This effect will likely dominate during the 21st century (Section 6.4.1).

Coasts can be affected by external marine influences (Figure 6.1). Waves generated by storms over the oceans reach the coast as swell; there are also more extreme, but infrequent, high-energy swells generated remotely (Vassie et al., 2004). Tsunamis are still rarer, but can be particularly devastating (Bryant, 2001). Ocean currents modify coastal environments through their influence on heat transfer, with both ecological and geomorphological consequences. Sea ice has physical impacts, and its presence or absence influences whether or not waves reach the coast (Jaagus, 2006). Other external influences include atmospheric inputs, such as dust (Shinn et al., 2000), and invasive species.

6.2.4 Thresholds in the behaviour of coastal systems

Dynamic coastal systems often show complex, non-linear morphological responses to change (Dronkers, 2005). Erosion, transport and deposition of sediment often involve significant time-lags (Brunsden, 2001), and the morphological evolution of sedimentary coasts is the outcome of counteracting transport processes of sediment supply versus removal. A shoreline may adopt an equilibrium, in profile or plan form, where these processes are in balance. However, external factors, such as storms, often induce morphodynamic change away from an equilibrium state. Climate change and sea-level rise affect sediment transport in complex ways and abrupt, non-linear changes may occur as thresholds are crossed (Alley et al., 2003). If sea level rises slowly, the balance between sediment supply and morphological adjustment can be maintained if a saltmarsh accretes, or a lagoon infills, at the same rate. An acceleration in the rate of sea-level rise may mean that morphology cannot keep up, particularly where the supply of sediment is limited, as for example when coastal floodplains are inundated after natural levees or artificial embankments are overtopped. Exceeding the critical sea-level thresholds can initiate an irreversible process of drowning, and other geomorphological and ecological responses follow abrupt changes of inundation and salinity (Williams et al., 1999; Doyle et al., 2003; Burkett et al., 2005). Widespread submergence is expected in the case of the coast of the Wadden Sea if the rate of relative sea-level rise exceeds 10 mm/yr (van Goor et al., 2003). For each coastal system the critical threshold will have a specific value, depending on hydrodynamic and sedimentary characteristics. Abrupt and persistent flooding occurs in coastal Argentina when landward winds (sudestadas) and/or heavy rainfall coincide with storm surges (Canziani and Gimenez, 2002; Codignotto, 2004a), further emphasising non-linearities between several interacting factors. Better understanding of thresholds in, and non-linear behaviour of, coastal systems will enhance the ability of managers and engineers to plan more effective coastal protection strategies, including the placement of coastal buildings, infrastructure and defences.

6.2.5 Observed effects of climate change on coastal systems

Trenberth et al. (2007) and Bindoff et al. (2007) observed a number of important climate change-related effects relevant to coastal zones. Rising CO_2 concentrations have lowered ocean surface pH by 0.1 unit since 1750, although to date no significant impacts on coastal ecosystems have been identified. Recent trend analyses indicate that tropical cyclones have increased in intensity (see Section 6.3.2). Global sea levels rose at 1.7 ± 0.5 mm/yr through the 20th century, while global mean sea surface temperatures have risen about 0.6°C since 1950, with associated atmospheric warming in coastal areas (Bindoff et al., 2007).

Many coasts are experiencing erosion and ecosystem losses (Sections 6.2.1 and 6.4.1), but few studies have unambiguously quantified the relationships between observed coastal land loss and the rate of sea-level rise (Zhang et al., 2004; Gibbons and Nicholls, 2006). Coastal erosion is observed on many shorelines around the world, but it usually remains unclear to what extent these losses are associated with relative sea-level rise due to subsidence, and other human drivers of land loss, and to what extent they result from global warming (Hansom, 2001; Jackson et al., 2002; Burkett et al., 2005; Wolters et al., 2005) (see Chapter 1, Section 1.3.3). Long-term ecological studies of rocky shore communities indicate adjustments apparently coinciding with climatic trends (Hawkins et al., 2003). However, for mid-latitudinal coastal systems it is often difficult to discriminate the extent to which such changes are a part of natural variability; and the clearest evidence of the impact of climate change on coasts over the past few decades comes from high and low latitudes, particularly polar coasts and tropical reefs.

There is evidence for a series of adverse impacts on polar coasts, although warmer conditions in high latitudes can have positive effects, such as longer tourist seasons and improved navigability (see Chapter 15, Section 15.4.3.2). Traditional knowledge also points to widespread coastal change across the North American Arctic from the Northwest Territories, Yukon and Alaska in the west to Nunavut in the east (Fox, 2003). Reduced sea-ice cover means a greater potential for wave generation where the coast is exposed (Johannessen et al., 2002; Forbes, 2005; Kont et al., 2007). Moreover, relative sea-level rise on low-relief, easily eroded, shores leads to rapid retreat, accentuated by melting of permafrost that binds coastal sediments, warmer ground temperatures, enhanced thaw, and subsidence associated with the melting of massive ground ice, as recorded at sites in Arctic Canada (Forbes et al., 2004b; Manson et al., 2006), northern USA (Smith, 2002b; Lestak et al., 2004) and northern Russia (Koreysha et al., 2002; Nikiforov et al., 2003; Ogorodov, 2003). Mid-latitude coasts with seasonal sea ice may also respond to reduced ice cover; ice extent has diminished over recent decades in the Bering and Baltic Seas (ARAG, 1999; Jevrejeva et al., 2004) and possibly in the Gulf of St. Lawrence (Forbes et al., 2002).

Global warming poses a threat to coral reefs, particularly any increase in sea surface temperature (SST). The synergistic effects of various other pressures, particularly human impacts such as over-fishing, appear to be exacerbating the thermal stresses on reef systems and, at least on a local scale, exceeding the thresholds beyond which coral is replaced by other organisms (Buddemeier et al., 2004). These impacts and their likely consequences are considered in Box 6.1, the threat posed by ocean acidification is examined in Chapter 4, Section 4.4.9, the impact of multiple stresses is examined in Box 16.2, and the example of the Great Barrier Reef, where decreases in coral cover could have major negative impacts on tourism, is described in Chapter 11, Section 11.6.

Box 6.1. Environmental thresholds and observed coral bleaching

Coral bleaching, due to the loss of symbiotic algae and/or their pigments, has been observed on many reefs since the early 1980s. It may have previously occurred, but gone unrecorded. Slight paling occurs naturally in response to seasonal increases in sea surface temperature (SST) and solar radiation. Corals bleach white in response to anomalously high SST (~1°C above average seasonal maxima, often combined with high solar radiation). Whereas some corals recover their natural colour when environmental conditions ameliorate, their growth rate and reproductive ability may be significantly reduced for a substantial period. If bleaching is prolonged, or if SST exceeds 2°C above average seasonal maxima, corals die. Branching species appear more susceptible than massive corals (Douglas, 2003).

Major bleaching events were observed in 1982-83, 1987-88 and 1994-95 (Hoegh-Guldberg, 1999). Particularly severe bleaching occurred in 1998 (Figure 6.2), associated with pronounced El Niño events in one of the hottest years on record (Lough, 2000; Bruno et al., 2001). Since 1998 there have been several extensive bleaching events. For example, in 2002 bleaching occurred on much of the Great Barrier Reef (Berkelmans et al., 2004; see Chapter 11, Section 11.6) and elsewhere. Reefs in the eastern Caribbean experienced a massive bleaching event in late 2005, another of the hottest years on record. On many Caribbean reefs, bleaching exceeded that of 1998 in both extent and mortality (Figure 6.2), and reefs are in decline as a result of the synergistic effects of multiple stresses (Gardner et al., 2005; McWilliams et al., 2005; see Box 16.2). There is considerable variability in coral susceptibility and recovery to elevated SST in both time and space, and in the incidence of mortality (Webster et al., 1999; Wilkinson, 2002; Obura, 2005).

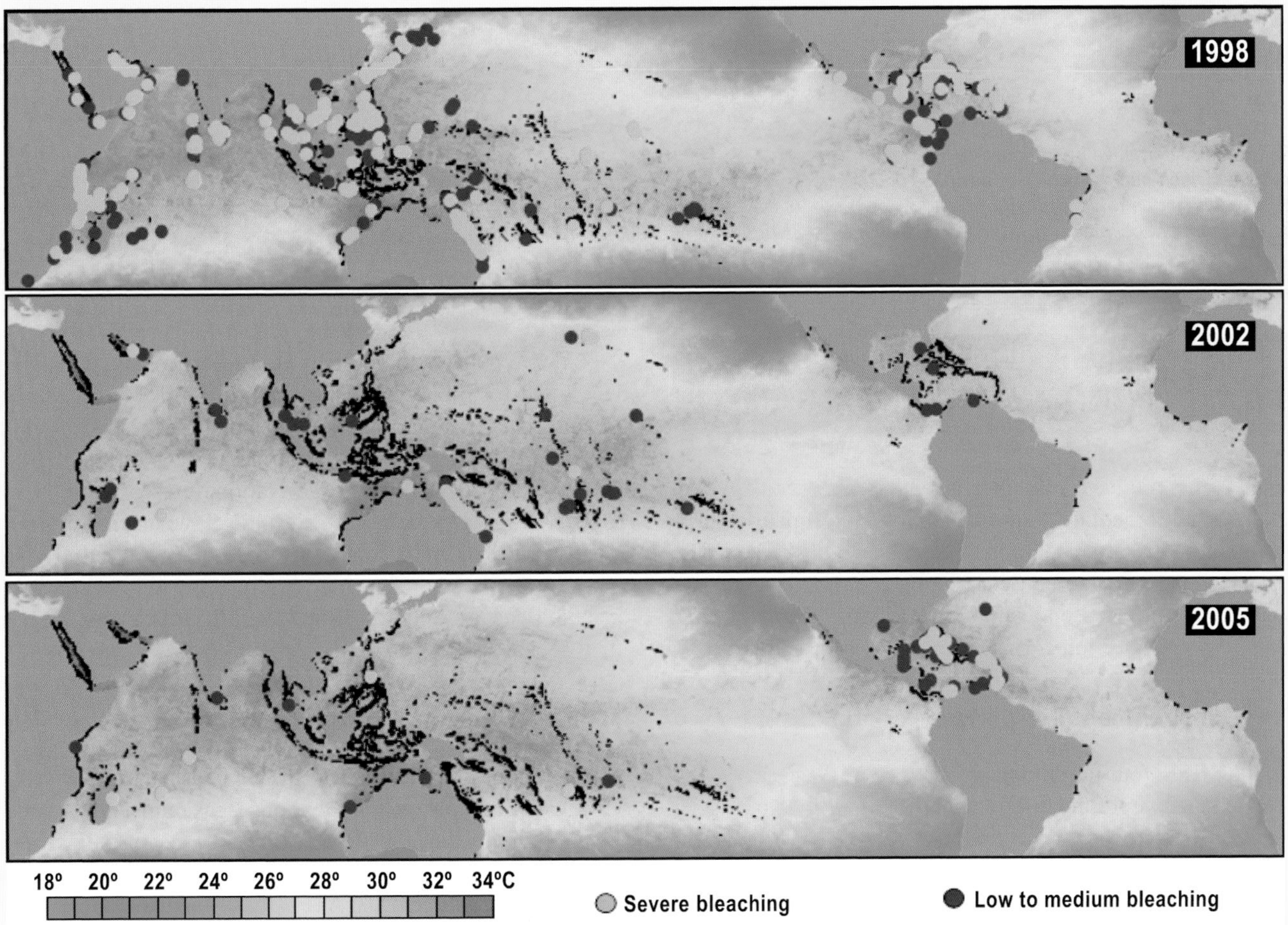

Figure 6.2. *Maximum monthly mean sea surface temperature for 1998, 2002 and 2005, and locations of reported coral bleaching (data source, NOAA Coral Reef Watch (coralreefwatch.noaa.gov) and Reefbase (www.reefbase.org)).*

Global climate model results imply that thermal thresholds will be exceeded more frequently with the consequence that bleaching will recur more often than reefs can sustain (Hoegh-Guldberg, 1999, 2004; Donner et al., 2005), perhaps almost annually on some reefs in the next few decades (Sheppard, 2003; Hoegh-Guldberg, 2005). If the threshold remains unchanged, more frequent bleaching and mortality seems inevitable (see Figure 6.3a), but with local variations due to different susceptibilities to factors such as water depth. Recent preliminary studies lend some support to the adaptive bleaching hypothesis, indicating that the coral host may be able to adapt or acclimatise as a result of expelling one clade[1] of symbiotic algae but recovering with a new one (termed shuffling, see Box 4.4), creating 'new' ecospecies with different temperature tolerances (Coles and Brown, 2003; Buddemeier et al., 2004; Little et al., 2004; Obura, 2005; Rowan, 2004). Adaptation or acclimatisation might result in an increase in the threshold temperature at which bleaching occurs (Figure 6.3b). The extent to which the thermal threshold could increase with warming of more than a couple of degrees remains very uncertain, as are the effects of additional stresses, such as reduced carbonate supersaturation in surface waters (see Box 4.4) and non-climate stresses (see Box 16.2). Corals and other calcifying organisms (e.g., molluscs, foraminifers) remain extremely susceptible to increases in SST. Bleaching events reported in recent years have already impacted many reefs, and their more frequent recurrence is very likely to further reduce both coral cover and diversity on reefs over the next few decades.

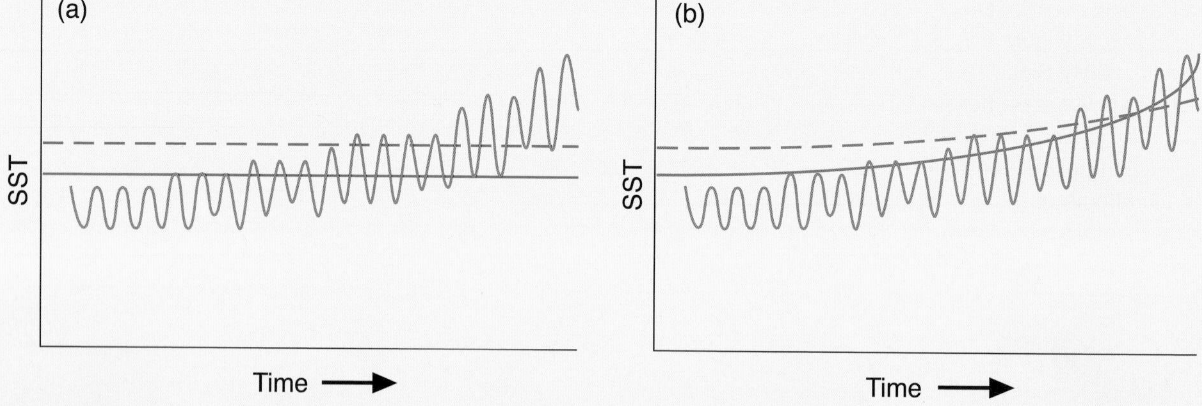

Figure 6.3. *Alternative hypotheses concerning the threshold SST at which coral bleaching occurs; a) invariant threshold for coral bleaching (red line) which occurs when SST exceeds usual seasonal maximum threshold (by ~1°C) and mortality (dashed red line, threshold of 2°C), with local variation due to different species or water depth; b) elevated threshold for bleaching (green line) and mortality (dashed green line) where corals adapt or acclimatise to increased SST (based on Hughes et al., 2003).*

6.3 Assumptions about future trends for coastal systems and low-lying areas

This section builds on Chapter 2 and Section 6.2 to develop relevant environmental, socio-economic, and climate change scenarios for coastal areas through the 21st century. The IPCC Special Report on Emissions Scenarios (SRES; Nakićenović and Swart, 2000) provides one suitable framework (Arnell et al., 2004; Chapter 2, Section 2.4).

National coastal socio-economic scenarios have also been developed for policy analysis, including links to appropriate climate change scenarios. Examples include the UK Foresight Flood and Coastal Defence analysis (Evans et al., 2004a,b; Thorne et al., 2006), and the US National Assessment (NAST, 2000), while model-based methods have been applied to socio-economic futures in the Ebro delta, Spain (Otter, 2000; Otter et al., 2001). However, socio-economic scenarios of coastal areas are underdeveloped relative to climate and sea-level scenarios.

6.3.1 Environmental and socio-economic trends

In the SRES, four families of socio-economic scenarios (A1, A2, B1 and B2) represent different world futures in two distinct dimensions: a focus on economic versus environmental concerns, and global versus regional development patterns. In all four cases, global gross domestic product (GDP) increases substantially and there is economic convergence at differing rates. Global population also increases to 2050 but, in the A1/B1 futures, the population subsequently declines, while in A2/B2 it continues to grow throughout the 21st century (see Chapter 2, Box 2.2). Relevant trends for coastal areas under the SRES scenarios are described in Table 6.1.

6.3.2 Climate and sea-level scenarios

In terms of climate change, the SRES scenarios in Section 6.3.1 translate into six greenhouse-gas emission 'marker' scenarios: one each for the A2, B1 and B2 worlds, and three scenarios for the A1 world – A1T (non-fossil fuel sources), A1B (balanced fuel sources) and A1FI (fossil-intensive fuel sources) (Nakićenović and Swart, 2000). B1 produces the lowest emissions and A1FI produces the highest emissions (see Chapter 2).

Table 6.2 summarises the range of potential drivers of climate change impacts in coastal areas, including the results from Meehl et al. (2007) and Christensen et al. (2007). In most cases

[1] A clade of algae is a group of closely related, but nevertheless different, types.

there will be significant regional variations in the changes, and any impacts will be the result of the interaction between these climate change drivers and other drivers of change, leading to diverse effects and vulnerabilities (Sections 6.2 and 6.4).

Understanding of the relevant climate-change drivers for coastal areas has improved since the TAR. Projected global mean changes under the SRES scenarios are summarised in Table 6.3. As atmospheric CO_2 levels increase, more CO_2 is absorbed by surface waters, decreasing seawater pH and carbonate saturation (Andersson et al., 2003; Royal Society, 2005; Turley et al., 2006). A significant increase in atmospheric CO_2 concentration appears virtually certain (Table 6.3). Sea surface temperatures are also virtually certain to rise significantly (Table 6.3), although less than the global mean temperature rise. The rise will not be spatially uniform, with possible intensification of ENSO and time variability which suggests greater change in extremes with important implications for coral reefs (Box 6.1).

Table 6.1. *Selected global non-climatic environmental and socio-economic trends relevant to coastal areas for the SRES storylines. Regional and local deviations are expected.*

Environmental and socio-economic factors	Non-climatic changes and trends for coastal and low-lying areas (by SRES Future)			
	'A1 World'	'A2 World'	'B1 World'	'B2 World'
Population (2080s) (billions)[a]	1.8 to 2.4	3.2 to 5.2	1.8 to 2.4	2.3 to 3.4
Coastward migration	Most likely	Less likely	More likely	Least likely
Human-induced subsidence[b]	More likely		Less likely	
Terrestrial freshwater/sediment supply (due to catchment management)	Greatest reduction	Large reduction	Smallest reduction	Smaller reduction
Aquaculture growth	Large increase		Smaller increase	
Infrastructure growth	Largest	Large	Smaller	Smallest
Extractive industries	Larger		Smaller	
Adaptation response	More reactive		More proactive	
Hazard risk management	Lower priority		Higher priority	
Habitat conservation	Low priority		High priority	
Tourism growth	Highest	High	High	Lowest

[a] Population living both below 100 m elevation above sea level and within 100 km distance of the coast – uncertainty depends on assumptions about coastward migration (Nicholls, 2004).

[b] Subsidence due to sub-surface fluid withdrawal and drainage of organic soils in susceptible coastal lowlands.

Table 6.2. *Main climate drivers for coastal systems (Figure 6.1), their trends due to climate change, and their main physical and ecosystem effects. (Trend: ↑ increase; ? uncertain; R regional variability).*

Climate driver (trend)	Main physical and ecosystem effects on coastal systems (discussed in Section 6.4.1)
CO_2 concentration (↑)	Increased CO_2 fertilisation; decreased seawater pH (or 'ocean acidification') negatively impacting coral reefs and other pH sensitive organisms.
Sea surface temperature (↑, R)	Increased stratification/changed circulation; reduced incidence of sea ice at higher latitudes; increased coral bleaching and mortality (see Box 6.1); poleward species migration; increased algal blooms
Sea level (↑, R)	Inundation, flood and storm damage (see Box 6.2); erosion; saltwater intrusion; rising water tables/impeded drainage; wetland loss (and change).
Storm intensity (↑, R)	Increased extreme water levels and wave heights; increased episodic erosion, storm damage, risk of flooding and defence failure (see Box 6.2).
Storm frequency (?, R) Storm track (?, R)	Altered surges and storm waves and hence risk of storm damage and flooding (see Box 6.2).
Wave climate (?, R)	Altered wave conditions, including swell; altered patterns of erosion and accretion; re-orientation of beach plan form.
Run-off (R)	Altered flood risk in coastal lowlands; altered water quality/salinity; altered fluvial sediment supply; altered circulation and nutrient supply.

Table 6.3. *Projected global mean climate parameters relevant to coastal areas at the end of the 21st century for the six SRES marker scenarios (from Meehl et al., 2007).*

Climate driver		B1	B2	A1B	A1T	A2	A1FI
Surface ocean pH (baseline today: 8.1)		8.0	7.9	7.9	7.9	7.8	7.7
SST rise (°C) (relative to 1980-1999)		1.5	-	2.2	-	2.6	-
Sea-level rise (relative to 1980-1999)	Best estimate (m)	0.28	0.32	0.35	0.33	0.37	0.43
	Range (m) 5%	0.19	0.21	0.23	0.22	0.25	0.28
	95%	0.37	0.42	0.47	0.44	0.50	0.58

The global mean sea-level rise scenarios (Table 6.3) are based on thermal expansion and ice melt; the best estimate shows an acceleration of up to 2.4 times compared to the 20th century. These projections are smaller than those of Church et al. (2001), reflecting improved understanding, especially of estimates of ocean heat uptake. If recently observed increases in ice discharge rates from the Greenland and Antarctic ice sheets were to increase linearly with global mean temperature change, this would add a 0.05 to 0.11 m rise for the A1FI scenario over the 21st century (Meehl et al., 2007). (Large and long-term sea-level rise beyond 2100 is considered in Box 6.6.)

Importantly, local (or relative) changes in sea level depart from the global mean trend due to regional variations in oceanic level change and geological uplift/subsidence; it is relative sea-level change that drives impacts and is of concern to coastal managers (Nicholls and Klein, 2005; Harvey, 2006a). Meehl et al. (2007) found that regional sea-level change will depart significantly from the global mean trends in Table 6.3: for the A1B scenario the spatial standard deviation by the 2080s is 0.08 m, with a larger rise than average in the Arctic. While there is currently insufficient understanding to develop detailed scenarios, Hulme et al. (2002) suggested that impact analysis should explore additional sea-level rise scenarios of +50% the amount of global mean rise, plus uplift/subsidence, to assess the full range of possible change. Although this approach has been followed in the UK (Pearson et al., 2005; Thorne et al., 2006), its application elsewhere is limited to date.

Furthermore, coasts subsiding due to natural or human-induced causes will experience larger relative rises in sea level (Bird, 2000). In some locations, such as deltas and coastal cities, this effect can be significant (Dixon et al., 2006; Ericson et al., 2006).

Increases of extreme sea levels due to rises in mean sea level and/or changes in storm characteristics (Table 6.2) are of widespread concern (Box 6.2). Meehl et al. (2007) found that models suggest both tropical and extra-tropical storm intensity will increase. This implies additional coastal impacts than attributable to sea-level rise alone, especially for tropical and mid-latitude coastal systems. Increases in tropical cyclone intensity over the past three decades are consistent with the observed changes in SST (Emanuel, 2005; Webster et al., 2005). Changes in other storm characteristics are less certain and the number of tropical and extra-tropical storms might even reduce (Meehl et al., 2007). Similarly, future wave climate is uncertain, although extreme wave heights will likely increase with more intense storms (Meehl et al., 2007). Changes in runoff driven by changes to the hydrological cycle appear likely, but the uncertainties are large. Milly et al. (2005) showed increased discharges to coastal waters in the Arctic, in northern Argentina and southern Brazil, parts of the Indian sub-continent, China and Australia, while reduced discharges to coastal waters are suggested in southern Argentina and Chile, Western and Southern Africa, and in the Mediterranean Basin. The additional effects of catchment management also need to be considered (Table 6.1).

6.4 Key future impacts and vulnerabilities

The following sections characterise the coastal ecosystem impacts that are anticipated to result from the climate change summarised in Figures 6.1 and Table 6.2. The summary of impacts on natural coastal systems and implications for human society (including ecosystem services) leads to the recognition of key vulnerabilities and hotspots.

6.4.1 Natural system responses to climate change drivers

6.4.1.1 Beaches, rocky shorelines and cliffed coasts

Most of the world's sandy shorelines retreated during the past century (Bird, 1985; NRC, 1990; Leatherman, 2001; Eurosion, 2004) and sea-level rise is one underlying cause (see Section 6.2.5 and Chapter 1, Section 1.3.3). One half or more of the Mississippi and Texas shorelines have eroded at average rates of 3.1 to 2.6 m/yr since the 1970s, while 90% of the Louisiana shoreline eroded at a rate of 12.0 m/yr (Morton et al., 2004). In Nigeria, retreat rates up to 30 m/yr are reported (Okude and Ademiluyi, 2006). Coastal squeeze and steepening are also widespread as illustrated along the eastern coast of the United Kingdom where 67% of the coastline experienced a landward retreat of the low-water mark over the past century (Taylor et al., 2004).

An acceleration in sea-level rise will widely exacerbate beach erosion around the globe (Brown and McLachlan, 2002), although the local response will depend on the total sediment budget (Stive et al., 2002; Cowell et al., 2003a,b). The widely cited Bruun (1962) model suggests that shoreline recession is in the range 50 to 200 times the rise in relative sea level. While supported by field data in ideal circumstances (Zhang et al., 2004), wider application of the Bruun model remains controversial (Komar, 1998; Cooper and Pilkey, 2004; Davidson-Arnott, 2005). An indirect, less-frequently examined influence of sea-level rise on the beach sediment budget is due to the infilling of coastal embayments. As sea-level rises, estuaries and lagoons attempt to maintain equilibrium by raising their bed elevation in tandem, and hence potentially act as a major sink of sand which is often derived from the open coast (van Goor et al., 2001; van Goor et al., 2003; Stive, 2004). This process can potentially cause erosion an order of magnitude or more greater than that predicted by the Bruun model (Woodworth et al., 2004), implying the potential for major coastal instability due to sea-level rise in the vicinity of tidal inlets. Several recent studies indicate that beach protection strategies and changes in the behaviour or frequency of storms can be more important than the projected acceleration of sea-level rise in determining future beach erosion rates (Ahrendt, 2001; Leont'yev, 2003). Thus there is not a simple relationship between sea-level rise and horizontal movement of the shoreline, and sediment budget approaches are most useful to assess beach response to climate change (Cowell et al., 2006).

The combined effects of beach erosion and storms can lead to the erosion or inundation of other coastal systems. For example, an increase in wave heights in coastal bays is a secondary effect of sandy barrier island erosion in Louisiana, and increased wave

Box 6.2. Examples of extreme water level simulations for impact studies

Although inundation by increases in mean sea level over the 21st century and beyond will be a problem for unprotected low-lying areas, the most devastating impacts are likely to be associated with changes in extreme sea levels resulting from the passage of storms (e.g., Gornitz et al., 2002), especially as more intense tropical and extra-tropical storms are expected (Meehl et al., 2007). Simulations show that future changes are likely to be spatially variable, and a high level of detail can be modelled (see also Box 11.5 in Christensen et al. (2007).

Figure 6.4. *Increases in the height (m) of the 50-year extreme water level. (a) In the northern Bay of Bengal under the IS92a climate scenario in 2040-2060 (K – Kolkata (Calcutta), C – Chittagong) (adapted from Mitchell et al., 2006). (b) Around the UK for the A2 scenario in the 2080s (L – London; H – Hamburg) (adapted from Lowe and Gregory, 2005).*

Figures 6.4 and 6.5 are based on barotropic surge models driven by climate change projections for two flood-prone regions. In the northern Bay of Bengal, simulated changes in storminess cause changes in extreme water levels. When added to consistent relative sea-level rise scenarios, these result in increases in extreme water levels across the Bay, especially near Kolkata (Figure 6.4a). Around the UK, extreme high sea levels also occur. The largest change near London has important implications for flood defence (Figure 6.4b; Dawson et al., 2005; Lavery and Donovan, 2005). Figure 6.5 shows the change in flooding due to climate change for Cairns (Australia). It is based on a combination of stochastic sampling and dynamic modelling. This assumes a 10% increase in tropical cyclone intensity, implying more flooding than sea-level rise alone would suggest. However, detailed patterns and magnitudes of changes in extreme water levels remain uncertain (e.g., Lowe and Gregory, 2005); better quantification of this uncertainty and further field validation would support wider application of such scenarios.

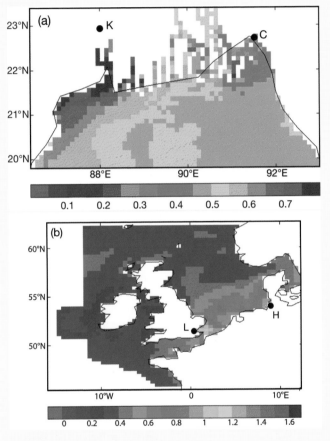

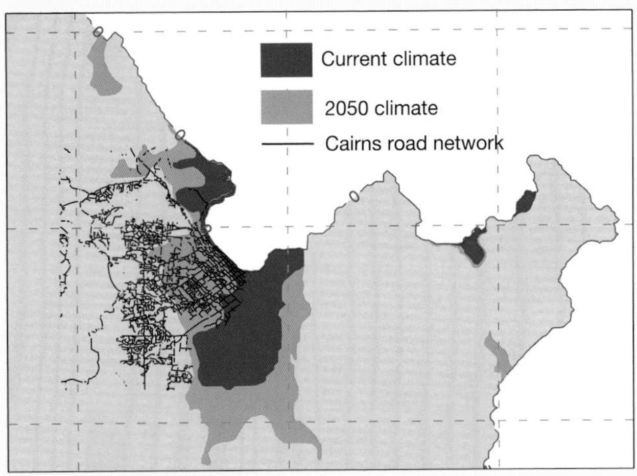

Figure 6.5. *Flooding around Cairns, Australia during the >100 year return-period event under current and 2050 climate conditions based on a 2xCO_2 scenario. The road network is shown in black (based on McInnes et al., 2003).*

heights have enhanced erosion rates of bay shorelines, tidal creeks and adjacent wetlands (Stone and McBride, 1998; Stone et al., 2003). The impacts of accelerated sea-level rise on gravel beaches have received less attention than sandy beaches. These systems are threatened by sea-level rise (Orford et al., 2001, 2003; Chadwick et al., 2005), even under high accretion rates (Codignotto et al., 2001). The persistence of gravel and cobble-boulder beaches will also be influenced by storms, tectonic events and other factors that build and reshape these highly dynamic shorelines (Orford et al., 2001).

Since the TAR, monitoring, modelling and process-oriented research have revealed some important differences in cliff vulnerability and the mechanics by which groundwater, wave climate and other climate factors influence cliff erosion patterns and rates. Hard rock cliffs have a relatively high resistance to erosion, while cliffs formed in softer lithologies are likely to retreat more rapidly in the future due to increased toe erosion resulting from sea-level rise (Cooper and Jay, 2002). Cliff failure and retreat may be amplified in many areas by increased precipitation and higher groundwater levels: examples include UK, Argentina and France (Hosking and McInnes, 2002; Codignotto, 2004b; Pierre and Lahousse, 2006). Relationships between cliff retreat, freeze-thaw cycles and air temperature records have also been described (Hutchinson, 1998). Hence, four physical features of climate change – temperature, precipitation, sea level and wave climate – can affect the stability of soft rock cliffs.

Soft rock cliff retreat is usually episodic with many metres of cliff top retreat occurring locally in a single event, followed by relative quiescence for significant periods (Brunsden, 2001; Eurosion, 2004). Considerable progress has been made in the long-term prediction of cliff-top, shore profile and plan-shape evolution of soft rock coastlines by simulating the relevant physical processes and their interactions (Hall et al., 2002; Trenhaile, 2002, 2004). An application of the SCAPE (Soft Cliff and Platform Erosion) model (Dickson et al., 2005; Walkden and Hall, 2005) to part of Norfolk, UK has indicated that rates of cliff retreat are sensitive to sea-level rise, changes in wave conditions and sediment supply via longshore transport. For soft cliff areas with limited beach development, there appears to be a simple relationship between long-term cliff retreat and the rate of sea-level rise (Walkden and Dickson, 2006), allowing useful predictions for planning purposes.

6.4.1.2 Deltas

Deltaic landforms are naturally shaped by a combination of river, wave and tide processes. River-dominated deltas receiving fluvial sediment input show prominent levees and channels that meander or avulse[2], leaving abandoned channels on the coastal plains. Wave-dominated deltas are characterised by shore-parallel sand ridges, often coalescing into beach-ridge plains. Tide domination is indicated by exponentially tapering channels, with funnel-shaped mouths. Delta plains contain a diverse range of landforms but, at any time, only part of a delta is active, and this is usually river-dominated, whereas the abandoned delta plain receives little river flow and is progressively dominated by marine processes (Woodroffe, 2003).

Human development patterns also influence the differential vulnerability of deltas to the effects of climate change. Sediment starvation due to dams, alterations in tidal flow patterns, navigation and flood control works are common consequences of human activity (Table 6.1). Changes in surface water runoff and sediment loads can greatly affect the ability of a delta to cope with the physical impacts of climatic change. For example, in the subsiding Mississippi River deltaic plain of south-east Louisiana, sediment starvation and increases in the salinity and water levels of coastal marshes due to human development occurred so rapidly that 1565 km^2 of intertidal coastal marshes and adjacent lands were converted to open water between 1978 and 2000 (Barras et al., 2003). By 2050 about 1300 km^2 of additional coastal land loss is projected if current global, regional and local processes continue; the projected acceleration of sea level and increase in tropical storm intensity (Section 6.3.2) would exacerbate these losses (Barras et al., 2003). Much of this land loss is episodic, as demonstrated during the landfall of Hurricane Katrina (Box 6.4).

Deltas have long been recognised as highly sensitive to sea-level rise (Ericson et al., 2006; Woodroffe et al., 2006) (Box 6.3). Rates of relative sea-level rise can greatly exceed the global average in many heavily populated deltaic areas due to subsidence, including the Chao Phraya delta (Saito, 2001), Mississippi River delta (Burkett et al., 2003) and the Changjiang River delta (Liu, 2002; Waltham, 2002), because of human activities. Natural subsidence due to autocompaction of sediment under its own weight is enhanced by sub-surface fluid withdrawals and drainage (Table 6.1). This increases the potential for inundation, especially for the most populated cities on these deltaic plains (i.e., Bangkok, New Orleans and Shanghai). Most of the land area of Bangladesh consists of the deltaic plains of the Ganges, Brahmaputra and Meghna rivers. Accelerated global sea-level rise and higher extreme water levels (Box 6.2) may have acute effects on human populations of Bangladesh (and parts of West Bengal, India) because of the complex relationships between observed trends in SST over the Bay of Bengal and monsoon rains (Singh, 2001), subsidence and human activity that has converted natural coastal defences (mangroves) to aquaculture (Woodroffe et al., 2006).

Whereas present rates of sea-level rise are contributing to the gradual diminution of many of the world's deltas, most recent losses of deltaic wetlands are attributed to human development. An analysis of satellite images of fourteen of the world's major deltas (Danube, Ganges-Brahmaputra, Indus, Mahanadi, Mangoky, McKenzie, Mississippi, Niger, Nile, Shatt el Arab, Volga, Huanghe, Yukon and Zambezi) indicated a total loss of 15,845 km^2 of deltaic wetlands over the past 14 years (Coleman et al., 2005). Every delta showed land loss, but at varying rates, and human development activities accounted for over half of the losses. In Asia, for example, where human activities have led to increased sediment loads of major rivers in the past, the construction of upstream dams is now seriously depleting the supply of sediments to many deltas with increased coastal erosion a widespread consequence (see Chapter 10, Section 10.4.3.2). As an example, large reservoirs constructed on the Huanghe River in China have reduced the annual sediment delivered to its delta from 1.1 billion metric tons to 0.4 billion metric tons (Li et al., 2004). Human influence is likely to continue to increase throughout Asia and globally (Section 6.2.2; Table 6.1).

Sea-level rise poses a particular threat to deltaic environments, especially with the synergistic effects of other climate and human pressures (e.g., Sánchez-Arcilla et al., 2007). These issues are especially noteworthy in many of the largest deltas with an indicative area >10^4 km^2 (henceforth megadeltas) due to their often large populations and important environmental services. The problems of climate change in megadeltas are reflected throughout this report, with a number of chapters considering these issues from complementary perspectives. Box 6.3 considers the vulnerability of delta systems across the globe, and concludes that the large populated Asian megadeltas are especially vulnerable to climate change. Chapter 10, Section 10.6.1 builds on this global

[2] Avulse: when a river changes its course from one channel to another as a result of a flood.

Box 6.3. Deltas and megadeltas: hotspots for vulnerability

Deltas, some of the largest sedimentary deposits in the world, are widely recognised as highly vulnerable to the impacts of climate change, particularly sea-level rise and changes in runoff, as well as being subject to stresses imposed by human modification of catchment and delta plain land use. Most deltas are already undergoing natural subsidence that results in accelerated rates of relative sea-level rise above the global average. Many are impacted by the effects of water extraction and diversion, as well as declining sediment input as a consequence of entrapment in dams. Delta plains, particularly those in Asia (Chapter 10, Section 10.6.1), are densely populated and large numbers of people are often impacted as a result of external terrestrial influences (river floods, sediment starvation) and/or external marine influences (storm surges, erosion) (see Figure 6.1).

Ericson et al. (2006) estimated that nearly 300 million people inhabit a sample of 40 deltas globally, including all the large megadeltas. Average population density is 500 people/km^2 with the largest population in the Ganges-Brahmaputra delta, and the highest density in the Nile delta. Many of these deltas and megadeltas are associated with significant and expanding urban areas. Ericson et al. (2006) used a generalised modelling approach to approximate the effective rate of sea-level rise under present conditions, basing estimates of sediment trapping and flow diversion on a global dam database, and modifying estimates of natural subsidence to incorporate accelerated human-induced subsidence. This analysis showed that much of the population of these 40 deltas is at risk through coastal erosion and land loss, primarily as a result of decreased sediment delivery by the rivers, but also through accentuated rates of sea-level rise. They estimate, using a coarse digital terrain model and global population distribution data, that more than 1 million people will be directly affected by 2050 in three megadeltas: the Ganges-Brahmaputra delta in Bangladesh, the Mekong delta in Vietnam and the Nile delta in Egypt. More than 50,000 people are likely to be directly impacted in each of a further 9 deltas, and more than 5,000 in each of a further 12 deltas (Figure 6.6). This generalised modelling approach indicates that 75% of the population affected live on Asian megadeltas and deltas, and a large proportion of the remainder are on deltas in Africa. These impacts would be exacerbated by accelerated sea-level rise and enhanced human pressures (e.g., Chapter 10, Section 10.6.1). Within the Asian megadeltas, the surface topography is complex as a result of the geomorphological development of the deltas, and the population distribution shows considerable spatial variability, reflecting the intensive land use and the growth of some of the world's largest megacities (Woodroffe et al., 2006). Many people in these and other deltas worldwide are already subject to flooding from both storm surges and seasonal river floods, and therefore it is necessary to develop further methods to assess individual delta vulnerability (e.g., Sánchez-Arcilla et al., 2006).

Figure 6.6. *Relative vulnerability of coastal deltas as shown by the indicative population potentially displaced by current sea-level trends to 2050 (Extreme = >1 million; High = 1 million to 50,000; Medium = 50,000 to 5,000; following Ericson et al., 2006).*

view and examines the Asian megadeltas in more detail. Chapter 5, Box 5.3 considers the threats to fisheries in the lower Mekong and associated delta due to climate change. Hurricane Katrina made landfall on the Mississippi delta in Louisiana, and Box 6.4 and Chapter 7, Box 7.4 consider different aspects of this important event, which gives an indication of the likely impacts if tropical storm intensity continues to increase. Lastly, Section 15.6.2 considers the specific problems of Arctic megadeltas.

6.4.1.3 Estuaries and lagoons

Global mean sea-level rise will generally lead to higher relative coastal water levels and increasing salinity in estuarine systems, thereby tending to displace existing coastal plant and animal communities inland. Estuarine plant and animal communities may persist as sea level rises if migration is not blocked and if the rate of change does not exceed the capacity of natural communities to adapt or migrate. Climate change impacts on one or more 'leverage species', however, can result in sweeping community level changes (Harley et al., 2006).

Some of the greatest potential impacts of climate change on estuaries may result from changes in physical mixing characteristics caused by changes in freshwater runoff (Scavia et al., 2002). A globally intensified hydrologic cycle and regional changes in runoff all portend changes in coastal water quality (Section 6.3.2). Freshwater inflows into estuaries influence water residence time, nutrient delivery, vertical stratification, salinity and control of phytoplankton growth rates. Increased freshwater inflows decrease water residence time and increase vertical stratification, and vice versa (Moore et al., 1997). The effects of altered residence times can have significant effects on phytoplankton populations, which have the potential to increase fourfold per day. Consequently, in estuaries with very short water residence times, phytoplankton are generally flushed from the system as fast as they can grow, reducing the estuary's susceptibility to eutrophication[3] and harmful algal blooms (HABs) (Section 6.4.2.4). Changes in the timing of freshwater delivery to estuaries could lead to a decoupling of the juvenile phases of many estuarine and marine fishery species from the available nursery habitat. In some hypersaline lagoonal systems, such as the Laguna Madre of Mexico and Texas, sea-level rise will increase water depths, leading to increased tidal exchange and hence reduced salinity (cf. Quammen and Onuf, 1993).

Increased water temperature could also affect algal production and the availability of light, oxygen and carbon for other estuarine species (Short and Neckles, 1999). The propensity for HABs is further enhanced by the fertilisation effect of increasing dissolved CO_2 levels. Increased water temperature also affects important microbial processes such as nitrogen fixation and denitrification in estuaries (Lomas et al., 2002). Water temperature regulates oxygen and carbonate solubility, viral pestilence, pH and conductivity, and photosynthesis and respiration rates of estuarine macrophytes[4]. While temperature is important in regulating physiological processes in estuaries (Lomas et al., 2002), predicting the ecological outcome is complicated by the feedbacks and interactions among temperature change and independent physical and biogeochemical processes such as eutrophication (cf. Section 6.2.4).

Decreased seawater pH and carbonate saturation (Mackenzie et al., 2001; Caldeira and Wickett, 2005) has at least two important consequences: the potential for reducing the ability of carbonate flora and fauna to calcify; and the potential for enhanced dissolution of nutrients and carbonate minerals in sediments (Andersson et al., 2003; Royal Society, 2005; Turley et al., 2006). As these potential impacts could be significant, it is important to improve understanding of them.

The landward transgression of natural estuarine shorelines as sea level rises has been summarised by Pethick (2001), who adopted a mass balance approach based on an equilibrium assumption resulting in landward retreat of the entire estuarine system. In this view, sea level rise of 6 mm causes 10 m of retreat of the Blackwater estuary, UK, and only 8 m of retreat for the Humber estuary, UK, due to the steeper gradient of the latter. The Humber estuary will also likely experience a deepening of the main channel, changes in tidal regime and larger waves that will promote further erosion around the margins (Winn et al., 2003). In Venice Lagoon, Italy, the combination of sea-level rise, altered sediment dynamics, and geological land subsidence has lowered the lagoon floor, widened tidal inlets, submerged tidal flats and islands, and caused the shoreline to retreat around the lagoon circumference (Fletcher and Spencer, 2005). In situations where the area of intertidal environments has been reduced by embanking or reclamation, the initial response will be a lowering of remaining tidal flats and infilling of tidal channels. Depending on tidal characteristics, the availability of marine sediment, and the rate of sea-level rise, the remaining tidal flats may either be further drowned, or their relative level in the tidal frame may be maintained, as shown by several tidal basins in the Dutch Wadden Sea (Dronkers, 2005).

A projected increase in the intensity of tropical cyclones and other coastal storms (Section 6.3.2) could alter bottom sediment dynamics, organic matter inputs, phytoplankton and fisheries populations, salinity and oxygen levels, and biogeochemical processes in estuaries (Paerl et al., 2001). The role of powerful storms in structuring estuarine sediments and biodiversity is illustrated in the stratigraphic record of massive, episodic estuary infilling of Bohai Bay, China during the Holocene, with alternating oyster reefs and thick mud deposits (Wang and Fan, 2005).

6.4.1.4 Mangroves, saltmarshes and sea grasses

Coastal vegetated wetlands are sensitive to climate change and long-term sea-level change as their location is intimately linked to sea level. Modelling of all coastal wetlands (but excluding sea grasses) by McFadden et al. (2007a) suggests global losses from 2000 to 2080 of 33% and 44% given a 36 cm and 72 cm rise in sea level, respectively. Regionally, losses would be most severe on the Atlantic and Gulf of Mexico coasts of North and Central America, the Caribbean, the Mediterranean, the Baltic and most small island regions due to

[3] Eutrophication: over-enrichment of a water body with nutrients, resulting in excessive growth of organisms and depletion of oxygen concentration.
[4] Macrophytes: aquatic plants large enough to be visible to the naked eye.

their low tidal range (Nicholls, 2004). However, wetland processes are complex, and Cahoon et al. (2006) developed a broad regional to global geographical model relating wetland accretion, elevation, and shallow subsidence in different plate tectonic, climatic and geomorphic settings for both temperate saltmarshes and tropical mangrove forests. Changes in storm intensity can also affect vegetated coastal wetlands. Cahoon et al. (2003) analysed the elevation responses from a variety of hurricane-influenced coastal settings and found that a storm can simultaneously influence both surface and subsurface soil processes, but with much variability.

Saltmarshes (halophytic grasses, sedges, rushes and succulents) are common features of temperate depositional coastlines. Hydrology and energy regimes are two key factors that influence the coastal zonation of the plant species which typically grade inland from salt, to brackish, to freshwater species. Climate change will likely have its most pronounced effects on brackish and freshwater marshes in the coastal zone through alteration of hydrological regimes (Burkett and Kusler, 2000; Baldwin et al., 2001; Sun et al., 2002), specifically, the nature and variability of hydroperiod and the number and severity of extreme events. Other variables – altered biogeochemistry, altered amounts and pattern of suspended sediments loading, fire, oxidation of organic sediments, and the physical effects of wave energy – may also play important roles in determining regional and local impacts.

Sea-level rise does not necessarily lead to loss of saltmarsh areas, especially where there are significant tides, because these marshes accrete vertically and maintain their elevation relative to sea level where the supply of sediment is sufficient (Hughes, 2004; Cahoon et al., 2006). The threshold at which wetlands drown varies widely depending upon local morphodynamic processes. Saltmarshes of some mesotidal and high tide range estuaries (e.g., Tagus estuary, Portugal) are susceptible to sea-level rise only in a worst-case scenario. Similarly, wetlands with high sediment inputs in the south-east United States would remain stable relative to sea level unless the rate of sea-level rise accelerates to nearly four times its current rate (Morris et al., 2002). Yet, even sediment inputs from frequently recurring hurricanes cannot compensate for subsidence effects combined with predicted accelerations in sea-level rise in rapidly subsiding marshes of the Mississippi River delta (Rybczyk and Cahoon, 2002).

Mangrove forests dominate intertidal subtropical and tropical coastlines between 25°N and 25°S latitude. Mangrove communities are likely to show a blend of positive responses to climate change, such as enhanced growth resulting from higher levels of CO_2 and temperature, as well as negative impacts, such as increased saline intrusion and erosion, largely depending on site-specific factors (Saenger, 2002). The response of coastal forested wetlands to climate change has not received the detailed research and modelling that has been directed towards the saltmarsh coasts of North America (Morris et al., 2002; Reed, 2002; Rybczyk and Cahoon, 2002) and north-west Europe (Allen, 2000, 2003). Nevertheless, it seems highly likely that similar principles are in operation and that the sedimentary response of the shoreline is a function of both the availability of sediment (Walsh and Nittrouer, 2004) and the ability of the organic production by mangroves themselves to fill accommodation space provided by sea-level rise (Simas et al., 2001). Mangroves are able to produce root material that builds up the substrate beneath them (Middleton and McKee, 2001; Jennerjahn and Ittekkot, 2002), but collapse of peat occurs rapidly in the absence of new root growth, as observed after Hurricane Mitch (Cahoon et al., 2003) and after lightning strikes (Sherman et al., 2000). Groundwater levels play an important role in the elevation of mangrove soils by processes affecting soil shrink and swell. Hence, the influence of hydrology should be considered when evaluating the effect of disturbances, sea-level rise and water management decisions on mangrove systems (Whelan et al., 2005). A global assessment of mangrove accretion rates by Saenger (2002) indicates that vertical accretion is variable but commonly approaches 5 mm/yr. However, many mangrove shorelines are subsiding and thus experiencing a more rapid relative sea-level rise (Cahoon et al., 2003).

A landward migration of mangroves into adjacent wetland communities has been recorded in the Florida Everglades during the past 50 years (Ross et al., 2000), apparently responding to sea-level rise over that period. Mangroves have extended landward into saltmarsh over the past five decades throughout south-east Australia, but the influence of sea-level rise in this region is considered minor compared to that of human disturbance (Saintilan and Williams, 1999) and land surface subsidence (Rogers et al., 2005, 2006). Rapid expansion of tidal creeks has been observed in northern Australia (Finlayson and Eliot, 2001; Hughes, 2003). Sea-level rise and salt water intrusion have been identified as a causal factor in the decline of coastal bald cypress (Taxodium disticum) forests in Louisiana (Krauss et al., 2000; Melillo et al., 2000) and die off of cabbage palm (Sabal palmetto) forests in coastal Florida (Williams et al., 1999, 2003).

On balance, coastal wetlands will decline with rising sea levels and other climate and human pressures (reduced sediment inputs, coastal squeeze constraints on landward migration, etc.) will tend to exacerbate these losses. However, the processes shaping these environments are complex and while our understanding has improved significantly over the last 10 years, it remains far from complete. Continued work on the basic science and its application to future prognosis at local, regional and global scales remains a priority (Cahoon et al., 2006; McFadden et al., 2007a).

Sea grasses appear to be declining around many coasts due to human impacts, and this is expected to accelerate if climate change alters environmental conditions in coastal waters (Duarte, 2002). Changes in salinity and temperature and increased sea level, atmospheric CO_2, storm activity and ultraviolet irradiance alter sea grass distribution, productivity and community composition (Short and Neckles, 1999). Increases in the amount of dissolved CO_2 and, for some species, HCO_3 present in aquatic environments, will lead to higher rates of photosynthesis in submerged aquatic vegetation, similar to the effects of CO_2 enrichment on most terrestrial plants, if nutrient availability or other limiting factors do not offset the potential for enhanced productivity. Increases in growth and biomass with elevated CO_2 have been observed for the sea grass *Z. marina* (Zimmerman et al., 1997). Algae growth in lagoons and estuaries may also respond positively to elevated dissolved inorganic carbon (DIC), though marine macroalgae do not appear to be limited by DIC levels (Beer and Koch, 1996). An increase in epiphytic or suspended algae would decrease light available to submerged aquatic vegetation in estuarine and lagoonal systems.

6.4.1.5 Coral reefs

Reef-building corals are under stress on many coastlines (see Chapter 1, Section 1.3.4.1). Reefs have deteriorated as a result of a combination of anthropogenic impacts such as overfishing and pollution from adjacent land masses (Pandolfi et al., 2003; Graham et al., 2006), together with an increased frequency and severity of bleaching associated with climate change (Box 6.1). The relative significance of these stresses varies from site to site. Coral mortality on Caribbean reefs is generally related to recent disease outbreaks, variations in herbivory[5], and hurricanes (Gardner et al., 2003; McWilliams et al., 2005), whereas Pacific reefs have been particularly impacted by episodes of coral bleaching caused by thermal stress anomalies especially during recent El Niño events (Hughes et al., 2003), as well as non-climate stresses.

Mass coral bleaching events are clearly correlated with rises of SST of short duration above summer maxima (Douglas, 2003; Lesser, 2004; McWilliams et al., 2005). Particularly extensive bleaching was recorded across the Indian Ocean region associated with extreme El Niño conditions in 1998 (Box 6.1 and Chapter 11, Section 11.6: Climate change and the Great Barrier Reef case study). Many reefs appear to have experienced similar SST conditions earlier in the 20th century and it is unclear how extensive bleaching was before widespread reporting post-1980 (Barton and Casey, 2005). There is limited ecological and genetic evidence for adaptation of corals to warmer conditions (Boxes 4.4 and 6.1). It is very likely that projected future increases in SST of about 1 to 3°C (Section 6.3.2) will result in more frequent bleaching events and widespread mortality, if there is not thermal adaptation or acclimatisation by corals and their symbionts (Sheppard, 2003; Hoegh-Guldberg, 2004). The ability of coral reef ecosystems to withstand the impacts of climate change will depend on the extent of degradation from other anthropogenic pressures and the frequency of future bleaching events (Donner et al., 2005).

In addition to coral bleaching, there are other threats to reefs associated with climate change (Kleypas and Langdon, 2002). Increased concentrations of CO_2 in seawater will lead to ocean acidification (Section 6.3.2), affecting aragonite saturation state (Meehl et al., 2007) and reducing calcification rates of calcifying organisms such as corals (LeClerq et al., 2002; Guinotte et al., 2003; Chapter 4, Box 4.4). Cores from long-lived massive corals indicate past minor variations in calcification (Lough and Barnes, 2000), but disintegration of degraded reefs following bleaching or reduced calcification may result in increased wave energy across reef flats with potential for shoreline erosion (Sheppard et al., 2005). Relative sea-level rise appears unlikely to threaten reefs in the next few decades; coral reefs have been shown to keep pace with rapid postglacial sea-level rise when not subjected to environmental or anthropogenic stresses (Hallock, 2005). A slight rise in sea level is likely to result in submergence of some Indo-Pacific reef flats and recolonisation by corals, as these intertidal surfaces, presently emerged at low tide, become suitable for coral growth (Buddemeier et al., 2004).

Many reefs are affected by tropical cyclones (hurricanes, typhoons); impacts range from minor breakage of fragile corals to destruction of the majority of corals on a reef and deposition of debris as coarse storm ridges. Such storms represent major perturbations, affecting species composition and abundance, from which reef ecosystems require time to recover. The sequence of ridges deposited on the reef top can provide a record of past storm history (Hayne and Chappell, 2001); for the northern Great Barrier Reef no change in frequency of extremely large cyclones has been detected over the past 5000 years (Nott and Hayne, 2001). An intensification of tropical storms (Section 6.3.2) could have devastating consequences on the reefs themselves, as well as for the inhabitants of many low-lying islands (Sections 6.4.2 and 16.3.1.3). There is limited evidence that global warming may result in an increase of coral range; for example, extension of branching Acropora poleward has been recorded in Florida, despite an almost Caribbean-wide trend for reef deterioration (Precht and Aronson, 2004), but there are several constraints, including low genetic diversity and the limited suitable substrate at the latitudinal limits to reef growth (Riegl, 2003; Ayre and Hughes, 2004; Woodroffe et al., 2005).

The fate of the small reef islands on the rim of atolls is of special concern. Small reef islands in the Indo-Pacific formed over recent millennia during a period when regional sea level fell (Woodroffe and Morrison, 2001; Dickinson, 2004). However, the response of these islands to future sea-level rise remains uncertain, and is addressed in greater detail in Chapter 16, Section 16.4.2. It will be important to identify critical thresholds of change beyond which there may be collapse of ecological and social systems on atolls. There are limited data, little local expertise to assess the dangers, and a low level of economic activity to cover the costs of adaptation for atolls in countries such as the Maldives, Kiribati and Tuvalu (Barnett and Adger, 2003; Chapter 16, Box 16.6).

6.4.2 Consequences for human society

Since the TAR, global and regional studies on the impacts of climate change are increasingly available, but few distinguish the socio-economic implications for the coastal zone (see also Section 6.5). Within these limits, Table 6.4 provides a qualitative overview of climate-related changes on the various socio-economic sectors of the coastal zone discussed in this section.

The socio-economic impacts in Table 6.4 are generally a product of the physical changes outlined in Table 6.2. For instance, extensive low-lying (often deltaic) areas, e.g., the Netherlands, Guyana and Bangladesh (Box 6.3), and oceanic islands are especially threatened by a rising sea level and all its resulting impacts, whereas coral reef systems and polar regions are already affected by rising temperatures (Sections 6.2.5 and 6.4.1). Socio-economic impacts are also influenced by the magnitude and frequency of existing processes and extreme events, e.g., the densely populated coasts of East, South and South-east Asia are already exposed to frequent cyclones, and

[5] Herbivory: the consumption of plants by animals.

Table 6.4. *Summary of climate-related impacts on socio-economic sectors in coastal zones.*

Coastal socio-economic sector	Climate-related impacts (and their climate drivers in Figure 6.1)						
	Temperature rise (air and seawater)	Extreme events (storms, waves)	Floods (sea level, runoff)	Rising water tables (sea level)	Erosion (sea level, storms, waves)	Salt water intrusion (sea level, runoff)	Biological effects (all climate drivers)
Freshwater resources	**X**	**X**	**X**	**X**	–	**X**	x
Agriculture and forestry	**X**	**X**	**X**	**X**	–	**X**	x
Fisheries and aquaculture	**X**	**X**	x	–	x	**X**	**X**
Health	**X**	**X**	**X**	x	–	**X**	**X**
Recreation and tourism	**X**	**X**	x	–	**X**	–	**X**
Biodiversity	**X**	**X**	**X**	**X**	**X**	**X**	**X**
Settlements/ infrastructure	**X**	**X**	**X**	**X**	**X**	**X**	–

X = strong; x= weak; – = negligible or not established.

this will compound the impacts of other climate changes (see Chapter 10). Coastal ecosystems are particularly at risk from climate change (CBD, 2003; Section 6.4.1), with serious implications for the services that they provide to human society (see Section 6.2.2; Box 6.4 and Chapter 4, Section 4.4.9).

Since the TAR, some important observations on the impacts and consequences of climate change on human society at coasts have emerged. First, significant regional differences in climate change and local variability of the coast, including human development patterns, result in variable impacts and adjustments along the coast, with implications for adaptation responses (Section 6.6). Second, human vulnerability to sea-level rise and climate change is strongly influenced by the characteristics of socio-economic development (Section 6.6.3). There are large differences in coastal impacts when comparing the different SRES worlds which cannot be attributed solely to the magnitude of climate change (Nicholls and Lowe, 2006; Nicholls and Tol, 2006). Third, although the future magnitude of sea-level rise will be reduced by mitigation, the long timescales of ocean response (Box 6.6) mean that it is unclear what coastal impacts are avoided and what impacts are simply delayed by the stabilisation of greenhouse gas concentration in the atmosphere (Nicholls and Lowe, 2006). Fourth, vulnerability to the impacts of climate change, including the higher socio-economic burden imposed by present climate-related hazards and disasters, is very likely to be greater on coastal communities of developing countries than in developed countries due to inequalities in adaptive capacity (Defra, 2004; Section 6.5). For example, one quarter of Africa's population is located in resource-rich coastal zones and a high proportion of GDP is exposed to climate-influenced coastal risks (Nyong and Niang-Diop, 2006; Chapter 9). In Guyana, 90% of its population and important economic activities are located within the coastal zone and are threatened by sea-level rise and climate change (Khan, 2001). Low-lying densely populated areas in India, China and Bangladesh (see Chapter 10) and other deltaic areas are highly exposed, as are the economies of small islands (see Chapter 16).

6.4.2.1 Freshwater resources

The direct influences of sea-level rise on freshwater resources come principally from seawater intrusion into surface waters and coastal aquifers, further encroachment of saltwater into estuaries and coastal river systems, more extensive coastal inundation and higher levels of sea flooding, increases in the landward reach of sea waves and storm surges, and new or accelerated coastal erosion (Hay and Mimura, 2005). Although the coast contains a substantial proportion of the world's population, it has a much smaller proportion of the global renewable water supply, and the coastal population is growing faster than elsewhere, exacerbating this imbalance (see Section 6.2.2 and Chapter 3, Section 3.2).

Many coastal aquifers, especially shallow ones, experience saltwater intrusion caused by natural and human-induced factors, and this is exacerbated by sea-level rise (Essink, 2001). The scale of saltwater intrusion is dependent on aquifer dimensions, geological factors, groundwater withdrawals, surface water recharge, submarine groundwater discharges and precipitation. Therefore, coastal areas experiencing increases in precipitation and run-off due to climate change (Section 6.3.2), including floods, may benefit from groundwater recharge, especially on some arid coasts (Khiyami et al., 2005). Salinisation of surface waters in estuaries is also promoted by a rising sea level, e.g., Bay of Bengal (Allison et al., 2003).

Globally, freshwater supply problems due to climate change are most likely in developing countries with a high proportion of coastal lowland, arid and semi-arid coasts, coastal megacities particularly in the Asia-Pacific region, and small island states, reflecting both natural and socio-economic factors that enhance the levels of risks (Alcamo and Henrichs, 2002; Ragab and Prudhomme, 2002). Identifying future coastal areas with stressed freshwater resources is difficult, particularly where there are strong seasonal demands, poor or no metering, and theft of water (Hall, 2003). Overall efficiency of water use is an important consideration, particularly where agriculture is a large consumer, e.g., the Nile delta (see Chapter 9, Box 9.2) and Asian megadeltas.

Based on the SRES emissions scenarios, it is estimated that the increase in water stress would have a significant impact by the 2050s, when the different SRES population scenarios have a clear effect (Arnell, 2004). But, regardless of the scenarios applied, critical regions with a higher sensitivity to water stresses, arising from either increases in water withdrawal or decreases in water available, have been identified in coastal regions that include parts of the western coasts of Latin America and the Algerian coast (Alcamo and Henrichs, 2002).

Box 6.4. Hurricane Katrina and coastal ecosystem services in the Mississippi delta

Whereas an individual hurricane event cannot be attributed to climate change, it can serve to illustrate the consequences for ecosystem services if the intensity and/or frequency of such events were to increase in the future. One result of Hurricane Katrina, which made landfall in coastal Louisiana on 29th August 2005, was the loss of 388 km^2 of coastal wetlands, levees and islands that flank New Orleans in the Mississippi River deltaic plain (Barras, 2006) (Figure 6.7). (Hurricane Rita, which struck in September 2005, had relatively minor effects on this part of the Louisiana coast which are included in this estimate.) The Chandeleur Islands, which lie south-east of the city, were reduced to roughly half of their former extent as a direct result of Hurricane Katrina. Collectively, these natural systems serve as the first line of defence against storm surge in this highly populated region. While some habitat recovery is expected, it is likely to be minimal compared to the scale of the losses. The Chandeleur Islands serve as an important wintering ground for migratory waterfowl and neo-tropical birds; a large population of North American redhead ducks, for example, feed on the rhizomes of sheltered sea grasses leeward of the Chandeleur Islands (Michot, 2000). Historically the region has ranked second only to Alaska in U.S. commercial fisheries production, and this high productivity has been attributed to the extent of coastal marshes and sheltered estuaries of the Mississippi River delta. Over 1800 people lost their lives (Graumann et al., 2005) during Hurricane Katrina and the economic losses totalled more than US$100 billion (NOAA, 2007). Roughly 300,000 homes and over 1,000 historical and cultural sites were destroyed along the Louisiana and Mississippi coasts (the loss of oil production and refinery capacity helped to raise global oil prices in the short term). Post-Katrina, some major changes to the delta's management are being advocated, most notably abandonment of the "bird-foot delta" where artificial levees channel valuable sediments into deep water (EFGC, 2006; NRC, 2006). The aim is to restore large-scale delta building processes and hence sustain the ecosystem services in the long term. Hurricane Katrina is further discussed in Box 7.4 (Chapter 7) and Chapter 14.

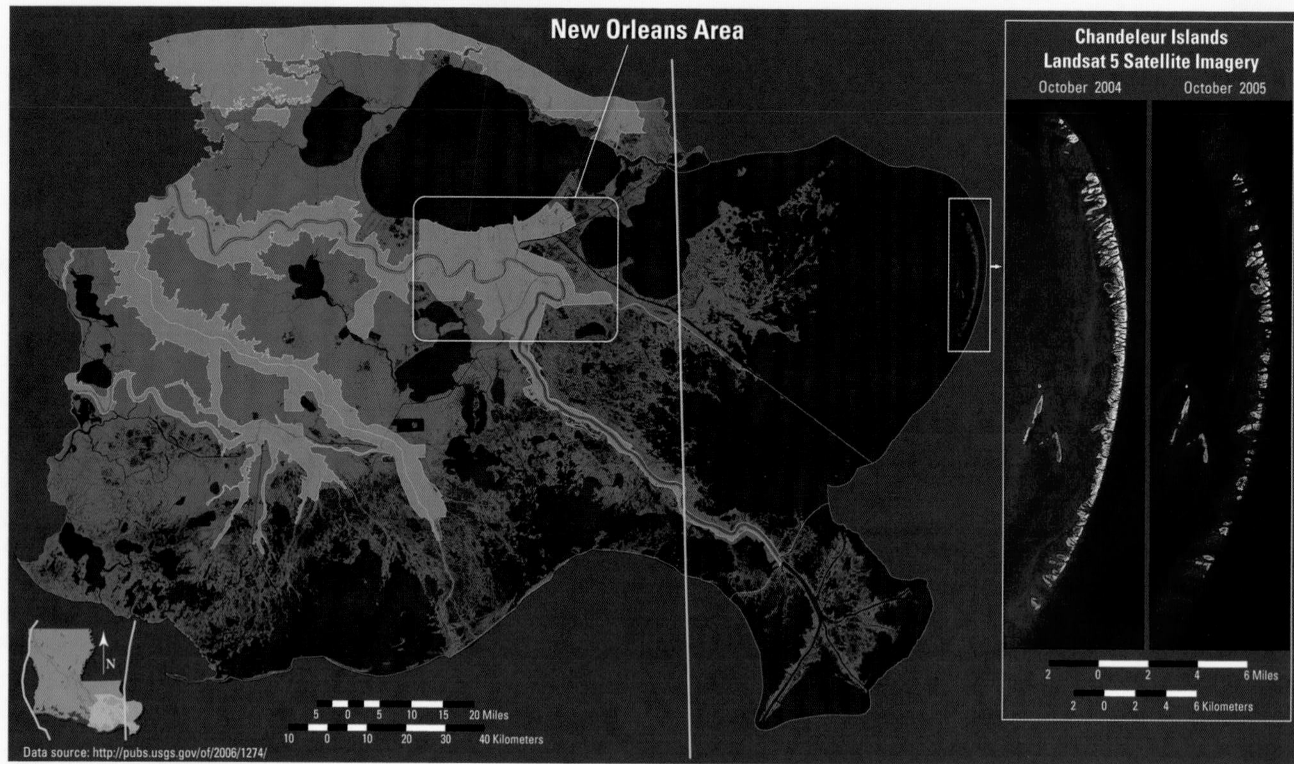

Figure 6.7. *The Mississippi delta, including the Chandeleur Islands. Areas in red were converted to open water during the hurricane. Yellow lines on index map of Louisiana show tracks of Hurricane Katrina on right and Hurricane Rita on left. (Figure source: U.S. Geological Survey, modified from Barras, 2006.)*

6.4.2.2 *Agriculture, forestry and fisheries*

Climate change is expected to have impacts on agriculture and, to a lesser extent, on forestry, although non-climatic factors, such as technological development and management practices can be more significant (Easterling, 2003). Climate variability and change also impacts fisheries in coastal and estuarine waters (Daufresne et al., 2003; Genner et al., 2004), although non-climatic factors, such as overfishing and habitat loss and degradation, are already responsible for reducing fish stocks. Globally an increased agricultural production potential due to climate change and CO_2 fertilisation should in principle add to food security, but the impacts on the coastal areas may differ regionally and locally. For example, in Europe, climate-related increases in crop yields are expected in the north, while the largest reductions are expected in the Mediterranean, the south-west Balkans and southern Russia (Maracchi et al., 2005).

Temperature increases can shorten growing cycles, e.g., those of cotton and mango on the north coast of Peru during the El Niño (see Chapter 13, Section 13.2.2). More frequent extreme climate events during specific crop development stages, together with higher rainfall intensity and longer dry spells, may impact negatively on crop yields (Olesen et al., 2006). Cyclone landfalls causing floods and destruction have negative impacts on coastal areas, e.g., on coconuts in India (see Chapter 5, Section 5.4.4), or on sugar cane and bananas in Queensland (Cyclone Larry in March 2006). Rising sea level has negative impacts on coastal agriculture. Detailed modelling of inundation implies significant changes to the number of rice crops possible in the Mekong delta under 20-40 cm of relative sea-level rise (Wassmann et al., 2004). Rising sea level potentially threatens inundation and soil salinisation of palm oil and coconuts in Benin and Côte d'Ivoire (see Chapter 9, Section 9.4.6) and mangoes, cashew nuts and coconuts in Kenya (Republic of Kenya, 2002).

Coastal forestry is little studied, but forests are easily affected by climatic perturbations, and severe storms can cause extensive losses, e.g., Hurricane Katrina. Plantation forests (mainly *P. radiata*) on the east coast of North Island, New Zealand, are likely to experience growth reductions under projected rainfall decreases (Ministry for the Environment, 2001). Increasing salinity and greater frequency of flooding due to sea-level rise reduces the ability of trees to generate, including mangroves which will also experience other changes (Section 6.4.1.4) (IUCN, 2003).

Future climate change impacts will be greater on coastal than on pelagic species, and for temperate endemics than for tropical species (see Chapter 11, Section 11.4.6). For Europe, regional climate warming has influenced northerly migration of fish species, e.g., sardines and anchovies in the North Sea (Brander et al., 2003a). The biotic communities and productivity of coastal lagoons may experience a variety of changes, depending on the changes in wetland area, freshwater flows and salt intrusion which affect the species. Intensification of ENSO events and increases in SST, wind stress, hypoxia (shortage of oxygen) and the deepening of the thermocline will reduce spawning areas and catches of anchovy off Peru (see Chapter 13, Table 13.7). There is also concern that climate change may affect the abundance and distribution of pathogens and HABs, with implications for aquatic organisms and human health (Section 6.4.2.4). The linkage between temperature changes and HABs is still not robust, and the extent to which coastal eutrophication will be affected by future climate variability will vary with local physical environmental conditions and current eutrophication status (Justic et al., 2005). Ocean acidification is a concern, but impacts are uncertain (Royal Society, 2005). Climate change also has implications for mariculture but again these are not well understood.

6.4.2.3 *Human settlements, infrastructure and migration*

Climate change and sea-level rise affect coastal settlements and infrastructure in several ways (Table 6.4). Sea-level rise raises extreme water levels with possible increases in storm intensity portending additional climate impacts on many coastal areas (Box 6.2), while saltwater intrusion may threaten water supplies. The degradation of natural coastal systems due to climate change, such as wetlands, beaches and barrier islands (Section 6.4.1.1), removes the natural defences of coastal communities against extreme water levels during storms (Box 6.5). Rapid population growth, urban sprawl, growing demand for waterfront properties, and coastal resort development have additional deleterious effects on protective coastal ecosystems.

Much of the coast of many European and East Asian countries have defences against flooding and erosion, e.g., the Netherlands (Jonkman et al., 2005) and Japan (Chapter 10, Section 10.5.3), reflecting a strong tradition of coastal defence. In particular, many coastal cities are heavily dependent upon artificial coastal defences, e.g., Tokyo, Shanghai, Hamburg, Rotterdam and London. These urban systems are vulnerable to low-probability extreme events above defence standards and to systemic failures (domino effects), e.g., the ports, roads and railways along the US Gulf and Atlantic coasts are especially vulnerable to coastal flooding (see Chapter 14, Section 14.2.6). Where these cities are subsiding, there are additional risks of extreme water levels overtopping flood defences, e.g., New Orleans during Hurricane Katrina (Box 6.4). Climate change and sea-level rise will exacerbate flood risk. Hence, many coastal cities require upgraded design criteria for flood embankments and barrages (e.g., the Thames barrier in London, the Delta works in the Netherlands, Shanghai's defences, and planned protection for Venice) (Fletcher and Spencer, 2005) (see Box 6.2 and Section 6.6).

There is now a better understanding of flooding as a natural hazard, and how climate change and other factors are likely to influence coastal flooding in the future (Hunt, 2002). However, the prediction of precise locations for increased flood risk resulting from climate change is difficult, as flood risk dynamics have multiple social, technical and environmental drivers (Few et al., 2004b). The population exposed to flooding by storm surges will increase over the 21st century (Table 6.5). Asia dominates the global exposure with its large coastal population: Bangladesh, China, Japan, Vietnam and Thailand having serious coastal flooding problems (see Section 6.6.2; Chapter 10, Section 10.4.3.1; Mimura, 2001). Africa is also likely to see a substantially increased exposure, with East Africa (e.g., Mozambique) having particular problems due to the combination of tropical storm landfalls and large projected population growth in addition to sea-level rise (Nicholls, 2006).

Table 6.5. *Estimates of the population (in millions) of the coastal flood plain* in 1990 and the 2080s (following Nicholls, 2004). Assumes uniform population growth; net coastward migration could substantially increase these numbers.*

Region	1990 (baseline)	SRES scenarios (and sea-level rise scenario in metres)			
		A1FI (0.34)	A2 (0.28)	B1 (0.22)	B2 (0.25)
Australia	1	1	2	1	1
Europe	25	30	35	29	27
Asia	132	185	376	180	247
North America	12	23	28	22	18
Latin America	9	17	35	16	20
Africa	19	58	86	56	86
Global	197	313	561	304	399

* *Area below the 1 in 1,000 year flood level.*

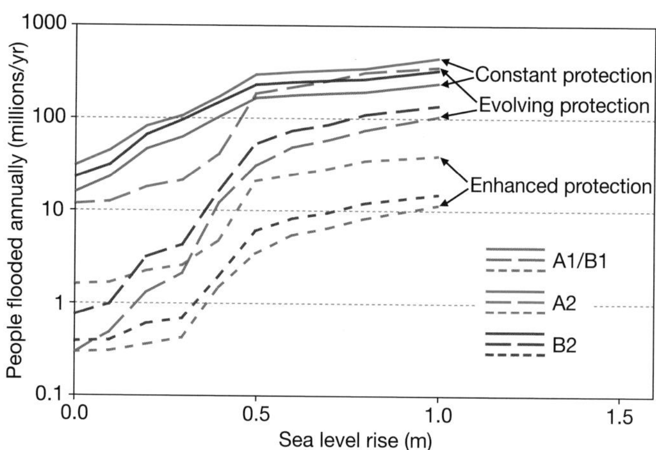

Figure 6.8. *Estimates of people flooded in coastal areas due to sea-level rise, SRES socio-economic scenario and protection response in the 2080s (following Nicholls and Lowe, 2006; Nicholls and Tol, 2006).*

Table 6.6 shows estimates of coastal flooding due to storm surge, taking into account one adaptation assumption. Asia and Africa experience the largest impacts: without sea-level rise, coastal flooding is projected to diminish as a problem under the SRES scenarios while, with sea-level rise, the coastal flood problem is growing by the 2080s, most especially under the A2 scenario. Increased storm intensity would exacerbate these impacts, as would larger rises in sea level, including due to human-induced subsidence (Nicholls, 2004). Figure 6.8 shows the numbers of people flooded in the 2080s as a function of sea-level rise, and variable assumptions on adaptation. Flood impacts vary with sea-level rise scenario, socio-economic situation and adaptation assumptions. Assuming that there will be no defence upgrade has a dramatic impact on the result, with more than 100 million people flooded per year above a 40 cm rise for all SRES scenarios. Upgraded defences reduce the impacts substantially: the greater the upgrade the lower the impacts. This stresses the importance of understanding the effectiveness and timing of adaptation (Section 6.6).

6.4.2.4 Human health

Coastal communities, particularly in low income countries, are vulnerable to a range of health effects due to climate variability and long-term climate change, particularly extreme weather and climate events (such as cyclones, floods and droughts) as summarised in Table 6.7.

The potential impacts of climate change on populations in coastal regions will be determined by the future health status of the population, its capacity to cope with climate hazards and control infectious diseases, and other public health measures. Coastal communities that rely on marine resources for food, in terms of both supply and maintaining food quality (food safety), are vulnerable to climate-related impacts, in both health and economic terms. Marine ecological processes linked to temperature changes also play a role in determining human health risks, such as from cholera, and other enteric pathogens (*Vibrio parahaemolyticus*), HABs, and shellfish and reef fish

Table 6.6. *Estimates of the average annual number of coastal flood victims (in millions) due to sea-level rise (following Nicholls, 2004). Assumes no change in storm intensity and evolving protection**. Range reflects population growth as reported in Table 6.1. Base= baseline without sea-level rise; aSLR = additional impacts due to sea-level rise.*

Region	Case	Timelines, SRES socio-economic (and sea-level rise scenarios in metres)											
		2020s				2050s				2080s			
		A1FI (0.05)	A2 (0.05)	B1 (0.05)	B2 (0.06)	A1FI (0.16)	A2 (0.14)	B1 (0.13)	B2 (0.14)	A1FI (0.34)	A2 (0.28)	B1 (0.22)	B2 (0.25)
Australia	Base	0	0	0	0	0	0	0	0	0	0	0	0
	aSLR	0	0	0	0	0	0	0	0	0	0	0	0
Europe	Base	0	0	0	0	0	0	0	0	0	0	0	0
	aSLR	0	0	0	0	0	0	0	0	2	0	0	0
Asia	Base	9/12	14/20	12/17	9/13	0	15/24	2	1/2	0	11/18	0	0/1
	aSLR	0	0	0	0	0	1/2	0	0	1	4/7	0	0/1
North America	Base	0	0	0	0	0	0	0	0	0	0	0	0
	aSLR	0	0	0	0	0	0	0	0	0	0	0	0
Latin America	Base	0	0	0	0	0	0	0	0	0	0	0	0
	aSLR	0	0	0	0	0	0	0	0	1	0/1	0	0
Africa	Base	1	2/4	1	3/4	0	1/2	0	1/2	0	0/1	0	0
	aSLR	0	0	0	0	0	1	0	0/1	2/5	4/7	1	2/4
Global Total	Base	10/14	17/24	13/18	12/17	0/1	16/26	2	3/4	0	11/19	0	1
	aSLR	0	0	0	0	0	2/3	0	0/1	6/10	9/15	2/3	3/5

** *Protection standards improve as GDP per capita increases, but there is no additional adaptation for sea-level rise.*

Table 6.7. *Health effects of climate change and sea-level rise in coastal areas.*

Exposure/hazard	Health outcome	Sources
(Catastrophic) flooding	Deaths (drowning, other causes), injuries, infectious disease (respiratory, intestinal, skin), mental health disorders, impacts from interruption of health services and population displacement.	Sections 6.4.2, 6.5.2 and 8.2.2; Box 6.4 (Few and Matthies, 2006)
Impairment of food quality and/or food supplies (loss of crop land, decreased fisheries productivity). Climate change effects on HABs.	Food safety: marine bacteria proliferation, shellfish poisoning, ciguatera. Malnutrition and micro-nutrient deficiencies.	Sections 6.4.1.3 6.4.2.2 and 8.2.4
Reduced water quality and/or access to potable water supplies due to salinisation, flooding or drought.	Diarrhoeal diseases (giardia, cholera), and hepatitis, enteric fevers. Water-washed infections.	Sections 6.4.2.1, 7.5 and 8.2.5
Change in transmission intensity or distribution of vector-borne disease. Changes in vector abundance.	Changes in malaria, and other mosquito-borne infections (some *Anopheles* vectors breed in brackish water).	Sections 8.2.8 and 16.4.5
Effects on livelihoods, population movement, and potential "environmental refugees".	Health effects are less well described. Large-scale rapid population movement would have severe health implications.	Section 6.4.2.3 and limited health literature.

poisoning (Pascual et al., 2002; Hunter, 2003; Lipp et al., 2004; Peperzak, 2005; McLaughlin et al., 2006).

Convincing evidence of the impacts of observed climate change on coastal disease patterns is absent (Kovats and Haines, 2005). There is an association between ENSO and cholera risk in Bangladesh (Pascual et al., 2002). Rainfall changes associated with ENSO are known to increase the risk of malaria epidemics in coastal regions of Venezuela and Colombia (Kovats et al., 2003). The projection of health impacts of climate change is still difficult and uncertain (Ebi and Gamble, 2005; Kovats et al., 2005), and socio-economic factors may be more critical than climate. There are also complex relationships between ecosystems and human well-being, and the future coastal ecosystem changes discussed in Section 6.4.1 may affect human health (cf. Butler et al., 2005).

6.4.2.5 Biodiversity

The distribution, production, and many other aspects of species and biodiversity in coastal ecosystems are highly sensitive to variations in weather and climate (Section 6.4.1), affecting the distribution and abundance of the plant and animal species that depend on each coastal system type. Human development patterns also have an important influence on biodiversity among coastal system types. Mangroves, for example, support rich ecological communities of fish and crustaceans, are a source of energy for coastal food chains, and export carbon in the form of plant and animal detritus, stimulating estuarine and nearshore productivity (Jennerjahn and Ittekkot, 2002). Large-scale conversions of coastal mangrove forests to shrimp aquaculture have occurred during the past three decades along the coastlines of Vietnam (Binh et al., 1997), Bangladesh and India (Zweig, 1998), Hong Kong (Tam and Wong, 2002), the Philippines (Spalding et al., 1997), Mexico (Contreras-Espinosa and Warner, 2004), Thailand (Furakawa and Baba, 2001) and Malaysia (Ong, 2001). The additional stressors associated with climate change could lead to further declines in mangroves forests and their biodiversity.

Several recent studies have revealed that climate change is already impacting biodiversity in some coastal systems. Long-term monitoring of the occurrence and distribution of a series of intertidal and shallow water organisms in south-west Britain has shown several patterns of change, particularly in the case of barnacles, which correlate broadly with changes in temperature over the several decades of record (Hawkins et al., 2003; Mieszkowska et al., 2006). It is clear that responses of intertidal and shallow marine organisms to climate change are more complex than simply latitudinal shifts related to temperature increase, with complex biotic interactions superimposed on the abiotic (Harley et al., 2006; Helmuth et al., 2006). Examples include the northward range extension of a marine snail in California (Zacherl et al., 2003) and the reappearance of the blue mussel in Svalbard (Berge et al., 2005).

Patterns of overwintering of migratory birds on the British coast appear to have changed in response to temperature rise (Rehfisch et al., 2004), and it has been suggested that changes in invertebrate distribution might subsequently influence the distribution of ducks and wading birds (Kendall et al., 2004). However, as detailed studies of redshank have shown, the factors controlling distribution are complex and in many cases are influenced by human activities (Norris et al., 2004). Piersma and Lindstrom (2004) review changes in bird distribution but conclude that none can be convincingly attributed to climate change. Loss of birds from some estuaries appears to be the result of coastal squeeze and relative sea-level rise (Hughes, 2004; Knogge et al., 2004). A report by the United Nations Framework Convention on Biodiversity (CBD, 2006) presents guidance for incorporating biodiversity considerations in climate change adaptation strategies, with examples from several coastal regions.

6.4.2.6 Recreation and tourism

Climate change has major potential impacts on coastal tourism, which is strongly dependent on 'sun, sea and sand'. Globally, travel to sunny and warm coastal destinations is the major factor for tourists travelling from Northern Europe to the Mediterranean (16% of world's tourists) and from North America to the Caribbean (1% of world's tourists) (WTO, 2003). By 2020, the total number of international tourists is expected to exceed 1.5 billion (WTO, undated).

Climate change may influence tourism directly via the decision-making process by influencing tourists to choose different destinations; and indirectly as a result of sea-level rise and resulting coastal erosion (Agnew and Viner, 2001). The preferences for climates at tourist destinations also differ among age and income groups (Lise and Tol, 2002), suggesting differential responses. Increased awareness of interactions between ozone depletion and climate change and the subsequent impact on the exposure of human skin to ultraviolet light is another factor influencing tourists' travel choice (Diffey, 2004). In general, air temperature rise is most important to tourism, except where factors such as sea-level rise promote beach degradation and viable adaptation options (e.g., nourishment or recycling) are not available (Bigano et al., 2005). Other likely impacts of climate change on coastal tourism are due to coral reef degradation (Box 6.1; Section 6.4.1.5) (Hoegh-Guldberg et al., 2000). Temperature and rainfall pattern changes may impact water quality in coastal areas and this may lead to more beach closures.

Climate change is likely to affect international tourist flows prior to travel, en route, and at the destination (Becken and Hay, undated). As tourism is still a growth industry, the changes in tourist numbers induced by climate change are likely to be much smaller than those resulting from population and economic growth (Bigano et al., 2005; Hamilton et al., 2005; Table 6.2). Higher temperatures are likely to change summer destination preferences, especially for Europe: summer heatwaves in the Mediterranean may lead to a shift in tourism to spring and autumn (Madisson, 2001) with growth in summer tourism around the Baltic and North Seas (see Chapter 12, Section 12.4.9). Although new climate niches are emerging, the empirical data do not suggest reduced competitiveness of the sun, sea and sand destinations, as they are able to restructure to meet tourists' demands (Aguiló et al., 2005). Within the Caribbean, the rapidly growing cruise industry is not vulnerable to sea-level rise, unlike coastal resorts. On high-risk (e.g., hurricane-prone) coasts, insurance costs for tourism could increase substantially or insurance may no longer be available. This exacerbates the impacts of extreme events or restricts new tourism in high-risk regions (Scott et al., 2005), e.g., four hurricanes in 2004 dealt a heavy toll in infrastructure damage and lost business in Florida's tourism industry (see Chapter 14, Section 14.2.7).

6.4.3 Key vulnerabilities and hotspots

A comprehensive assessment of the potential impacts of climate change must consider at least three components of vulnerability: exposure, sensitivity and adaptive capacity (Section 6.6). Significant regional differences in present climate and expected climate change give rise to different exposure among human populations and natural systems to climate stimuli (IPCC, 2001). The previous sections of this chapter broadly characterise the sensitivity and natural adaptive capacity (or resilience) of several major classes of coastal environments to changes in climate and sea-level rise. Differences in geological, oceanographic and biological processes can also lead to substantially different impacts on a single coastal system at different locations. Some global patterns and hotspots of vulnerability are evident, however, and deltas/estuaries (especially populated megadeltas), coral reefs (especially atolls), and ice-dominated coasts appear most vulnerable to either climate change or associated sea-level rise and changes. Low-lying coastal wetlands, small islands, sand and gravel beaches and soft rock cliffs may also experience significant changes.

An acceleration of sea-level rise would directly increase the vulnerability of all of the above systems, but sea-level rise will not occur uniformly around the world (Section 6.3.2). Variability of storms and waves, as well as sediment supply and the ability to migrate landward, also influence the vulnerability of many of these coastal system types. Hence, there is an important element of local to regional variation among coastal system types that must be considered when conducting site-specific vulnerability assessments.

Our understanding of human adaptive capacity is less developed than our understanding of responses by natural systems, which limits the degree to which we can quantify societal vulnerability in the world's coastal regions. Nonetheless, several key aspects of human vulnerability have emerged. It is also apparent that multiple and concomitant non-climate stresses will exacerbate the impacts of climate change on most natural coastal systems, leading to much larger and detrimental changes in the 21st century than those of the 20th century. Table 6.8 summarises some of the key hotspots of vulnerability that often arise from the combination of natural and societal factors. Note that some examples such as atolls and small islands and deltas/megadeltas recur, stressing their high vulnerability.

While physical exposure is an important aspect of the vulnerability for both human populations and natural systems to both present and future climate variability and change, a lack of adaptive capacity is often the most important factor that creates a hotspot of human vulnerability. Societal vulnerability is largely dependent upon development status (Yohe and Tol, 2002). Developing nations may have the societal will to relocate people who live in low-lying coastal zones but, without the necessary financial resources, their vulnerability is much greater than that of a developed nation in an identical coastal setting. Looking to the scenarios, the A2 SRES world often appears most vulnerable to climate change in coastal areas, again reflecting socio-economic controls in addition to the magnitude of climate change (Nicholls, 2004; Nicholls and Tol, 2006). Hence, development is not only a key consideration in evaluating greenhouse gas emissions and climate change, but is also fundamental in assessing adaptive capacity because greater access to wealth and technology generally increases adaptive capacity, while poverty limits adaptation options (Yohe and Tol, 2002). A lack of risk awareness or institutional capacity can also have an important influence on human vulnerability, as experienced in the United States during Hurricane Katrina.

6.5 Costs and other socio-economic aspects

The costs, benefits and other socio-economic consequences of climate variability and change for coastal and low-lying areas have been determined for many aspects, including heat stress and changes in plant and animal metabolism (see Chapter 4, Section 4.2 and Box 4.4), disease (see Chapter 8, Section 8.5),

Table 6.8. *Key hotspots of societal vulnerability in coastal zones.*

Controlling factors	Examples from this Chapter
Coastal areas where there are substantial barriers to adaptation (economic, institutional, environmental, technical, etc.)	Venice, Asian megadeltas, atolls and small islands, New Orleans
Coastal areas subject to multiple natural and human-induced stresses, such as subsidence or declining natural defences	Mississippi, Nile and Asian megadeltas, the Netherlands, Mediterranean, Maldives
Coastal areas already experiencing adverse effects of temperature rise	Coral reefs, Arctic coasts (USA, Canada, Russia), Antarctic peninsula
Coastal areas with significant flood-plain populations that are exposed to significant storm surge hazards	Bay of Bengal, Gulf of Mexico/Caribbean, Rio de la Plata/Parana delta, North Sea
Coastal areas where freshwater resources are likely to be reduced by climate change	W. Africa, W. Australia, atolls and small islands
Coastal areas with tourist-based economies where major adverse effects on tourism are likely	Caribbean, Mediterranean, Florida, Thailand, Maldives
Highly sensitive coastal systems where the scope for inland migration is limited	Many developed estuarine coasts, low small islands, Bangladesh

water supply (see Chapter 3, Section 3.5), and coastal forests, agriculture and aquaculture (see Chapter 5, Section 5.6). The following section focuses on evaluating the socio-economic consequences of sea-level rise, storm damage and coastal erosion.

6.5.1 Methods and tools for characterising socio-economic consequences

Since the TAR there has been further progress in moving from classical cost-benefit analysis to assessments that integrate monetary, social and natural science criteria. For example, Hughes et al. (2005) report the emergence of a complex systems approach for sustaining and repairing marine ecosystems. This links ecological resilience to governance structures, economics and society. Such developments are in response to the growing recognition of the intricate linkages between physical coastal processes, the diverse coastal ecosystems, and resources at risk from climate change, the many ecological functions they serve and services they provide, and the variety of human amenities and activities that depend on them. Thus a more complete picture of climate change impacts emerges if assessments take into account the locally embedded realities and constraints that affect individual decision makers and community responses to climate change (Moser, 2000, 2005). Increasingly, Integrated Assessment provides an analytical framework, and an interdisciplinary learning and engagement process for experts, decision makers and stakeholders (Turner, 2001). Evaluations of societal and other consequences combine impact-benefit/cost-effectiveness analytical methods with scenario analysis. For example, a recent analysis of managed realignment schemes (Coombes et al., 2004) took into account social, environmental and economic consequences when evaluating direct and indirect benefits.

Direct cost estimates are common across the climate change impact literature as they are relatively simple to conduct and easy to explain. Such estimates are also becoming increasingly elaborate. For example, several studies of sea-level rise considered land and wetland loss, population displacement and coastal protection via dike construction (e.g., Tol, 2007). Socio-economic variables, such as income and population density, are important in estimating wetland value but are often omitted when making such estimations (Brander et al., 2003b). But direct cost estimates ignore such effects as changes in land use and food prices if land is lost. One way to estimate these additional effects is to use a computable general equilibrium (CGE) model to consider markets for all goods and services simultaneously, taking international trade and investment into account (e.g., Bosello et al., 2004). However, the major economic effects of climate change may well be associated with out-of-equilibrium phenomena (Moser, 2006). Also, few CGE models include adequate representations of physical processes and constraints.

Given the recent and anticipated increases in damages from extreme events, the insurance industry and others are making greater use of catastrophe models. These cover event generation (e.g., storm magnitude and frequency), hazard simulation (wind stresses and surge heights), damage modelling (extent of structural damage), and financial modelling (costs) (Muir-Wood et al., 2005). Stochastic modelling is used to generate thousands of simulated events and develop probabilistic approaches to quantifying the risks (Aliff, 2006; Chapter 2).

Methodologically, many challenges remain. Work to date has insufficiently crossed disciplinary boundaries (Visser, 2004). Although valuation techniques are continually being improved, and are now better linked to risk-based decision making, they remain imperfect, and in some instances controversial. This requires a transdisciplinary response from the social and natural sciences.

6.5.2 Socio-economic consequences under current climate conditions

Under current climate conditions, developing countries bear the main human burden of climate-related extreme events (Munich Re Group, 2004; CRED, 2005; UN Secretary General, 2006a). But it is equally evident that developed countries are not insulated from disastrous consequences (Boxes 6.4 and Chapter 7, Box 7.4). The societal costs of coastal disasters are typically quantified in terms of property losses and human deaths. For example, Figure 6.9 shows a significant threshold in real estate damage costs related to flood levels. Post-event impacts on coastal businesses, families and neighbourhoods, public and private social institutions, natural resources, and the environment generally go unrecognised in disaster cost accounting (Heinz Center, 2000; Baxter, 2005). Finding an accurate way to document these unreported or hidden costs is a challenging problem that has received increasing attention in recent years. For example, Heinz Center (2000) showed that family roles and responsibilities after a disastrous coastal storm undergo profound changes associated with household and employment disruption, economic hardship, poor living conditions, and the disruption of pubic services such as education and preventive health care. Indirect costs imposed by health problems (Section 6.4.2.4) result from damaged homes and utilities, extreme temperatures, contaminated food, polluted water, debris- and mud-borne bacteria, and mildew and mould. Within the family, relationships after a disastrous climate-related event can become so stressful that family desertion and divorce may increase. Hence, accounting for the full range of costs is difficult, though essential to the accurate assessment of climate-related coastal hazards.

Tropical cyclones have major economic, social and environmental consequences for coastal areas (Box 6.4). Up to 119 million people are on average exposed every year to tropical cyclone hazard (UNDP, 2004). Worldwide, from 1980 to 2000, a total of more than 250,000 deaths were associated with tropical cyclones, of which 60% occurred in Bangladesh (this is less than the 300,000 killed in Bangladesh in 1970 by a single cyclone). The death toll has been reduced in the past decade due largely to improvements in warnings and preparedness, wider public awareness and a stronger sense of community responsibility (ISDR, 2004). The most-exposed countries have densely populated coastal areas, often comprising deltas and megadeltas (China, India, the Philippines, Japan, Bangladesh) (UNDP, 2004). In Cairns (Australia), cyclone experience and education may have contributed synergistically to a change in risk perceptions and a reduction in the vulnerability of residents to tropical cyclone and storm surge hazards (Anderson-Berry, 2003). In Japan, the annual number of tropical cyclones and typhoons making landfall showed no significant trend from 1950 to 2004, but the number of port-related disasters decreased. This is attributed to increased protection against such disasters. However, annual average restoration expenditures over the period still amount to over US$250 million (Hay and Mimura, 2006).

Between 1980 and 2005, the United States sustained 67 weather-related disasters, each with an overall damage cost of at least US$1 billion. Coastal states in the south-east US experienced the greatest number of such disasters. The total costs including both insured and uninsured losses for the period, adjusted to 2002, were over US$500 billion (NOAA, 2007). There are differing views as to whether climatic factors have contributed to the increasing frequency of major weather-related disasters along the Atlantic and Gulf coasts of the USA (Pielke Jr et al., 2005; Pielke and Landsea, 1998). But the most recent reviews by Trenberth et al. (2007) and Meehl et al. (2007) support the view that storm intensity has increased and this will continue with global warming. Whichever view is correct, the damage costs associated with these events are undisputedly high, and will increase into the future.

Erosion of coasts (Section 6.4.1.1) is a costly problem under present climatic conditions. About 20% of the European Union's coastline suffered serious erosion impacts in 2004, with the area lost or seriously impacted estimated at 15 km^2/yr. In 2001, annual expenditure on coastline protection in Europe was an estimated US$4 billion, up from US$3 billion in 1986 (Eurosion, 2004). The high rates of erosion experienced by beach communities on Delaware's Atlantic coast (USA) are already requiring publicly funded beach nourishment projects in order to sustain the area's attractiveness as a summer resort (Daniel, 2001). Along the east coast of the United States and Canada, sea-level rise over the last century has reduced the return period of extreme water levels, exacerbating the damage to fixed structures from modern storms compared to the same events a century ago (Zhang et al., 2000; Forbes et al., 2004a). These and other studies have raised major questions, including: (i) the feasibility, implications and acceptability of shoreline retreat; (ii) the appropriate type of shoreline protection (e.g., beach nourishment, hard protection or other typically expensive responses) in situations where rates of shoreline retreat are increasing; (iii) doubts as to the longer-term sustainability of such interventions; and (iv) whether insurance provided by the public and private sectors encourages people to build, and rebuild, in vulnerable areas.

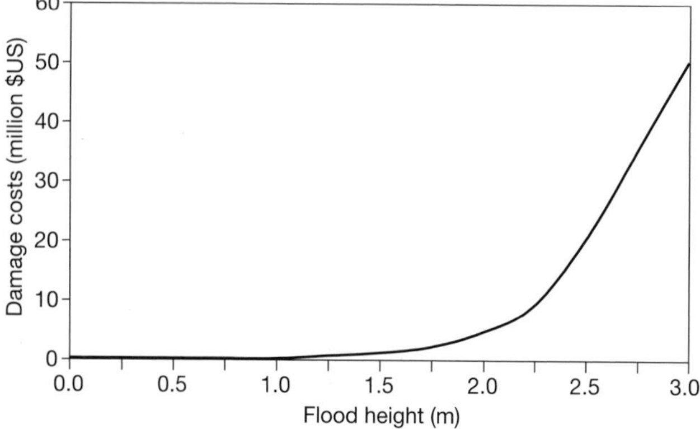

Figure 6.9. *Real estate damage costs related to flood levels for the Rio de la Plata, Argentina (Barros et al., 2006).*

6.5.3 Socio-economic consequences of climate change

Substantial progress has been made in evaluating the socio-economic consequences of climate change, including changes in variability and extremes. In general, the results show that socio-economic costs will likely escalate as a result of climate change, as already shown for the broader impacts (Section 6.4). Most immediately, this will reflect increases in variability and extreme events and only in the longer term will costs (in the widest sense) be dominated by trends in average conditions, such as mean sea-level rise (van Aalst, 2006). The impacts of such changes in climate and sea level are overwhelmingly adverse. But benefits have also been identified, including reduced cold-water mortalities of many valuable fish and shellfish species (see Chapter 15, Section 15.4.3.2), opportunities for increased use of fishing vessels and coastal shipping facilities (see Chapter 15, Section 15.4.3.3), expansion of areas suitable for aquaculture (see Chapter 5, Section 5.4.6.1), reduced hull strengthening and icebreaking costs, and the opening of new ocean routes due to reduced sea ice. Countries with large land areas generally benefit from competitive advantage effects (Bosello et al., 2004).

In the absence of an improvement to protection, coastal flooding could grow tenfold or more by the 2080s, to affect more than 100 million people/yr, due to sea-level rise alone (Figure 6.8). Figure 6.10 shows the consequences and total costs of a rise in sea level for developing and developed countries, and globally. This analysis assumes protection is implemented based on benefit-cost analysis, so the impacts are more consistent with enhanced protection in Figure 6.8, and investment is required for the protection. The consequences of sea-level rise will be far greater for developing countries, and protection costs will be higher, relative to those for developed countries.

Such global assessments are complemented by numerous regional, national and more detailed studies. The number of people in Europe subject to coastal erosion or flood risk in 2020 may exceed 158,000, while half of Europe's coastal wetlands are expected to disappear as a result of sea-level rise (Eurosion, 2004). In Thailand, loss of land due to a sea-level rise of 50 cm and 100 cm could decrease national GDP by 0.36% and 0.69% (US$300 to 600 million) per year, respectively; due to location and other factors, the manufacturing sector in Bangkok could suffer the greatest damage, amounting to about 61% and 38% of the total damage, respectively (Ohno, 2001). The annual cost of protecting Singapore's coast is estimated to be between US$0.3 and 5.7 million by 2050 and between US$0.9 and 16.8 million by 2100 (Ng and Mendelsohn, 2005). In the cities of Alexandria, Rosetta and Port Said on the Nile delta coast of Egypt, a sea-level rise of 50 cm could result in over 2 million people abandoning their homes, the loss of 214,000 jobs and the loss of land valued at over US$35 billion (El-Raey, 1997).

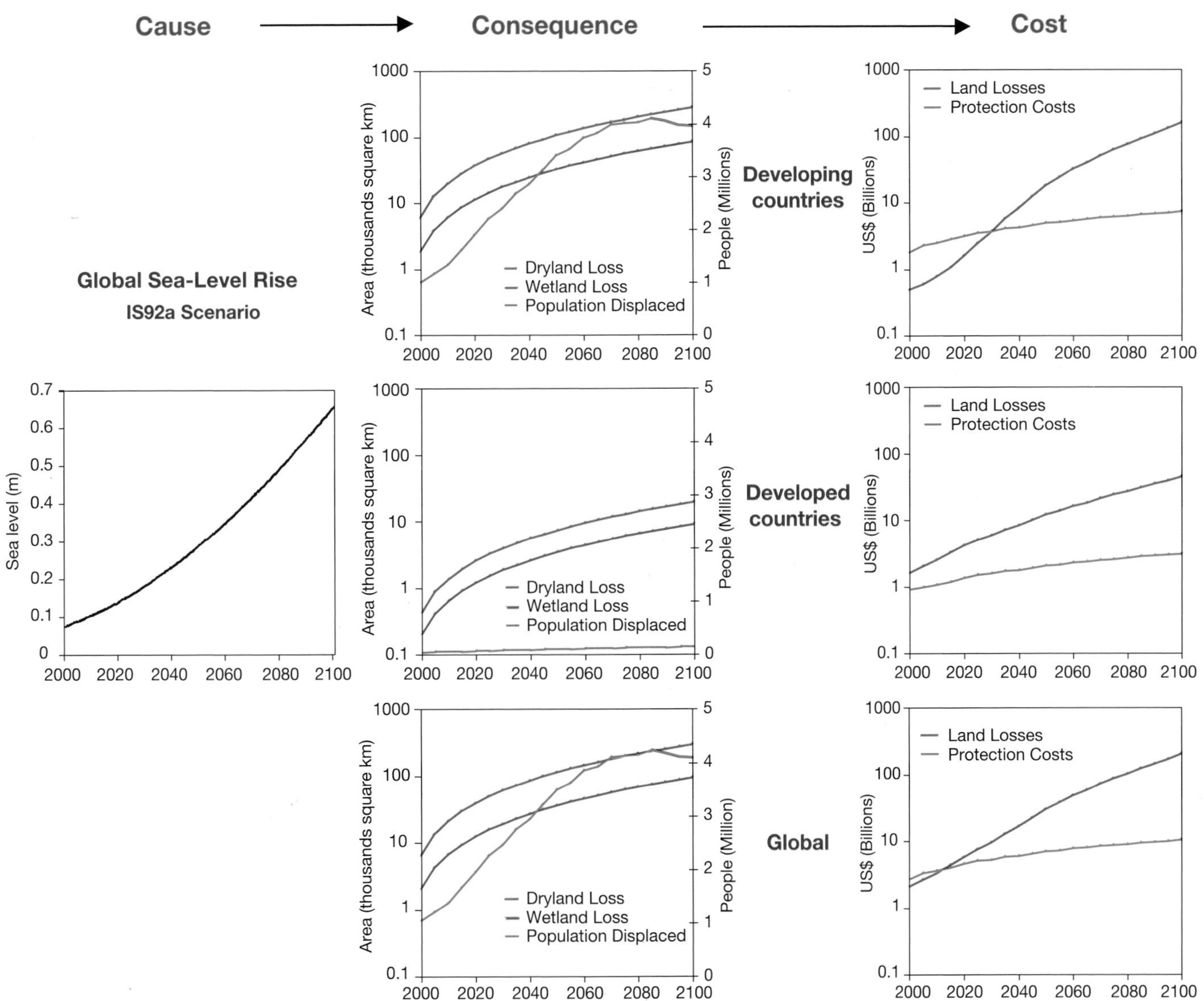

Figure 6.10. *Causes, selected consequences (dryland and wetland loss, people displaced) and the total costs of an assumed sea-level rise, for developing and developed countries, and as a global total (based on Tol, 2007).*

6.6 Adaptation: practices, options and constraints

This section first highlights issues that arise with interventions designed to reduce risks to natural and human coastal systems as a consequence of climate change. As recognised in earlier IPCC assessments (Bijlsma et al., 1996; McLean et al., 2001), a key conclusion is that reactive and standalone efforts to reduce climate-related risks to coastal systems are less effective than responses which are part of integrated coastal zone management (ICZM), including long-term national and community planning (see also Kay and Adler, 2005). Within this context, subsequent sections describe the tools relevant to adaptation in coastal areas, options for adaptation of coastal systems, and current and planned adaptation initiatives. Examples of the costs of, and limits to, coastal adaptation are described, as are the trade-offs. Constraints on, limitations to, and strategies for strengthening adaptive capacity are also described. Finally, the links between coastal adaptation and efforts to mitigate climate change are discussed.

6.6.1 Adaptation to changes in climate and sea level

6.6.1.1 Issues and challenges

Recent extreme events (Box 6.5), whether climate-related or not, have highlighted many of the challenges inherent in adapting to changes in climate and sea level. One constraint on successful management of climate-related risks to coastal systems is the limited ability to characterise in appropriate detail how these systems, and their constituent parts, will respond to climate change drivers and to adaptation initiatives (Sections 6.2.4 and 6.4; Finkl, 2002). Of particular importance is understanding the extent to which natural coastal systems can adapt and therefore continue to provide essential life-supporting services to society. The lack of understanding of the coastal system, including the highly interactive nature and non-linear behaviour (Sections 6.2 and 6.4), means that failure to take an integrated approach to characterising climate-related risks increases the likelihood that the effectiveness of adaptation will be reduced, and perhaps even negated. Despite the growing emphasis on beach nourishment (Hanson et al., 2002), the long-term effectiveness and feasibility of such adaptive measures remains uncertain, especially with the multiple goals explicit within ICZM (Section 6.6.1.2). The question of who pays and who benefits from adaptation is another issue of concern. Public acceptance of the need for adaptation, and of specific measures, also needs to be increased (Neumann et al., 2000). The significant and diverse challenges are summarised in Table 6.9 and discussed further in the identified sections.

6.6.1.2 Integrated coastal zone management (ICZM)

ICZM provides a major opportunity to address the many issues and challenges identified above. Since it offers advantages over purely sectoral approaches, ICZM is widely recognised and promoted as the most appropriate process to deal with climate change, sea-level rise and other current and long-term coastal challenges (Isobe, 2001; Nicholls and Klein, 2005; Harvey, 2006b). Enhancing adaptive capacity is an important part of ICZM. The extent to which climate change and sea-level rise are considered in coastal management plans is one useful measure of commitment to integration and sustainability. Responses to sea-level rise and climate change need to be implemented in the broader context and the wider objectives of coastal planning and management (Kennish, 2002; Moser, 2005). ICZM focuses on integrating and balancing multiple objectives in the planning process (Christie et al., 2005). Generation of equitably distributed social and environmental benefits is a key factor in ICZM process sustainability, but is difficult to achieve. Attention is also paid to legal and institutional frameworks that support integrative planning on local and national scales. Different social groups have contrasting, and often conflicting views on the relative priorities

Box 6.5. Recent extreme events – lessons for coastal adaptation to climate change

Recent extreme events, both climate and non-climate related, that had major consequences for coastal systems, provide important messages for adaptation to climate change. Scientific literature and government reports emanating from hurricane and cyclone impacts (e.g., Cook Islands (Ingram, 2005); Katrina (US Government, 2006); Australia (Williams et al., 2007), flood impacts (e.g., Mumbai (Wisner, 2006)) and the Boxing Day Sumatran tsunami (UNEP, 2005; UNOCHA, 2005) include the following.

- An effective early warning communication and response system can reduce death and destruction;
- Hazard awareness education and personal hazard experience are important contributors to reducing community vulnerability;
- Many factors reduce the ability or willingness of people to flee an impending disaster, including the warning time, access and egress routes, and their perceived need to protect property, pets and possessions;
- Coastal landforms (coral reefs, barrier islands) and wetland ecosystems (mangroves, marshes) provide a natural first line of protection from storm surges and flooding, despite divergent views about the extent to which they reduce destruction;
- Recurrent events reduce the resilience of natural and artificial defences;
- In the aftermath of extreme events, additional trauma occurs in terms of dispossession and mental health;
- Uncoordinated and poorly regulated construction has accentuated vulnerability;
- Effective disaster prevention and response rely on strong governance and institutions, as well as adequate public preparedness.

Table 6.9. *Major impediments to the success of adaptation in the coastal zone.*

Impediment	Example Reference	Section
Lack of dynamic predictions of landform migration	Pethick, 2001	6.6.1.2
Insufficient or inappropriate shoreline protection measures	Finkl, 2002	6.6.1.4
Data exchange and integration hampered by divergent information management systems	Hale et al., 2003	6.6.1.3
Lack of definition of key indicators and thresholds relevant to coastal managers	Rice, 2003	6.6.1.2
Inadequate knowledge of coastal conditions and appropriate management measures	Kay and Adler, 2005	6.6.1.3
Lack of long-term data for key coastal descriptors	Hall, 2002	6.6.1.2
Fragmented and ineffective institutional arrangements, and weak governance	Moser, 2000	6.6.1.3
Societal resistance to change	Tompkins et al., 2005a	6.6.3

to be given to development, the environment and social considerations, as well as short and long-term perspectives (Visser, 2004).

6.6.1.3 Tools for assessing adaptation needs and options

Since the TAR, many more tools have become available to support assessments of the need for adaptation and to identify appropriate interventions (Table 6.10).

6.6.1.4 Adaptation options

Figure 6.11 illustrates the evolution of thinking with respect to planned adaptation practices in the coastal zone. It also provides examples of current adaptation interventions. The capacity of coastal systems to regenerate after disasters, and to continue to produce resources and services for human livelihoods and well-being, is being tested with increasing frequency. This is highlighting the need to consider the resilience of coastal systems at broader scales and for their adaptive capacity to be actively managed and nurtured.

Those involved in managing coastal systems have many practical options for simultaneously reducing risks related to current climate extremes and variability as well as adapting to climate change (Yohe, 2000; Daniel, 2001; Queensland Government, 2001; Townend and Pethick, 2002). This reflects the fact that many disaster and climate change response strategies are the same as those which contribute positively to present-day efforts to implement sustainable development, including enhancement of social equity, sound environmental management and wise resource use (Helmer and Hilhorst, 2006). This will help harmonise coastal planning and climate change adaptation and, in turn, strengthen the anticipatory response capacity of institutions (Few et al., 2004a). The timeframes for development are typically shorter than those for natural changes in the coastal region, though management is starting to address this issue. Examples include restoration and management of the Mississippi River and delta plain (Box 6.4) and management of coastal erosion in Europe (Eurosion, 2004; Defra, 2006; MESSINA, 2006). Identifying and selecting adaptation options can be guided by experience and best practice for reducing the adverse impacts of analogous, though causally unrelated, phenomena such as subsidence (natural and/or human-induced) and tsunami (Olsen et al., 2005). Based on this experience, it is highly advantageous to integrate and mainstream disaster management and adaptation to climate variability and change into wider coastal management, especially given relevant lessons from recent disasters (Box 6.5).

Table 6.10. *Selected tools that support coastal adaptation assessments and interventions.*

Description	Selected examples
Indices of vulnerability to sea-level rise	Thieler and Hammar-Klose, 2000; Kokot et al., 2004
Integrated models and frameworks for knowledge management and adaptation assessment	Warrick et al., 2005; Dinas-Coast Consortium, 2006; Schmidt-Thomé, 2006
Geographic information systems for decision support	Green and King, 2002; Bartlett and Smith, 2005
Scenarios – a tool to facilitate thinking and deciding about the future	DTI, 2002; Ledoux and Turner, 2002
Community vulnerability assessment tool	NOAA Coastal Services Center, 1999; Flak et al., 2002
Flood simulator for flood and coastal defences and other responses	Discovery Software, 2006; Box 6.2
Estimating the socio-economic and environmental effects of disasters	ECLAC, 2003
ICZM process sustainability – a score card	Milne et al., 2003
Monetary economic valuation of the environment	Ledoux et al., 2001; Ohno, 2001
Evaluating and mapping return periods of extreme events	Bernier et al., 2007
Methods and tools to evaluate vulnerability and adaptation	UNFCCC, 2005

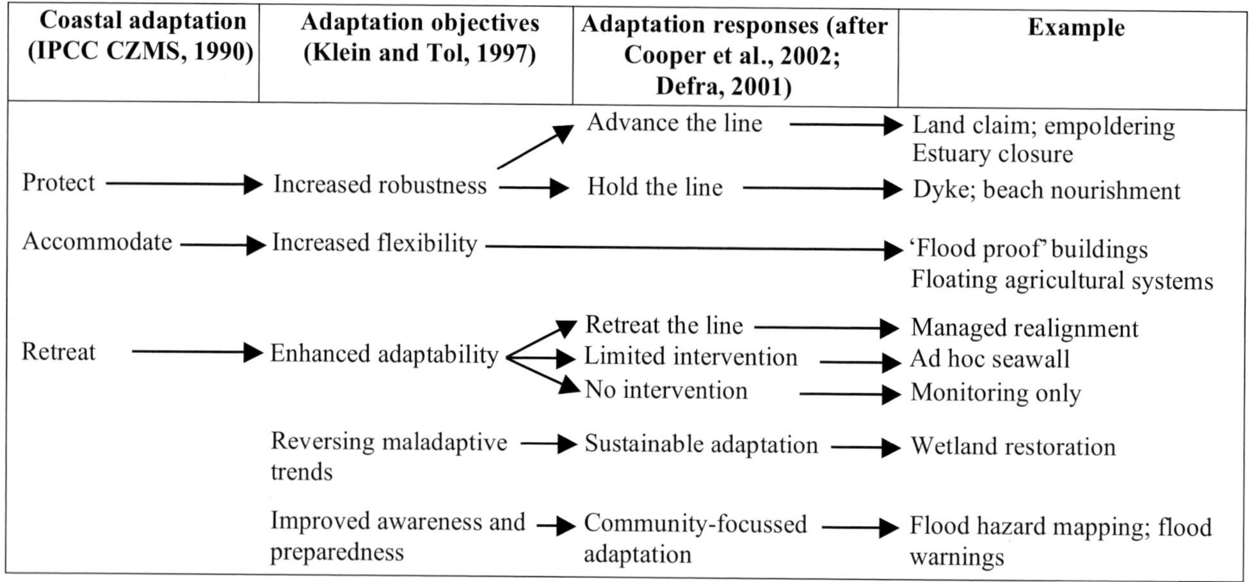

Figure 6.11. *Evolution of planned coastal adaptation practices.*

Klein et al. (2001) describe three trends: (i) growing recognition of the benefits of 'soft' protection and of 'retreat and accommodate' strategies; (ii) an increasing reliance on technologies to develop and manage information; and (iii) an enhanced awareness of the need for coastal adaptation to reflect local natural and socio-economic conditions. The decision as to which adaptation option is chosen is likely to be largely influenced by local socio-economic considerations (Knogge et al., 2004; Persson et al., 2006). It is also important to consider adaptation measures that reduce the direct threats to the survival of coastal ecosystems. These include marine protected areas and 'no take' reserves. Moser (2000) identified several factors that prompted local communities to act against coastal erosion. These included: (i) threats of or actual litigation; (ii) frustration among local officials regarding lack of clarity in local regulations, resulting in confusion as well as exposure to litigation; and (iii) concern over soaring numbers of applications for shoreline-hardening structures, since these are perceived to have negative, often external, environmental impacts. The particular adaptation strategy adopted depends on many factors, including the value of the land or infrastructure under threat, the available financial and economic resources, political and cultural values, the local application of coastal management policies, and the ability to understand and implement adaptation options (Yohe, 2000).

6.6.2 Costs and benefits of adaptation

The body of information on costs of adaptation has increased dramatically since the TAR, covering the range from specific interventions to global aggregations. Most analyses quantify the costs of responses to the more certain and specific effects of sea-level rise. Selected indicative and comparative costs of coastal adaptation measures are presented in Table 6.11. They reveal a wide range in adaptation costs. But in most populated areas such interventions have costs lower than damage costs, even when just considering property losses (Tol, 2002, 2007). Climate change affects the structural stability and performance of coastal defence structures and hence significantly raises the costs of building new structures (Burgess and Townend, 2004) or upgrading existing structures (Townend and Burgess, 2004). Financial cost is not the only criterion on which adaptation should be judged – local conditions and circumstances might result in a more expensive option being favoured, especially where multiple benefits result.

6.6.3 Limits and trade-offs in adaptation

Recent studies suggest that there are limits to the extent to which natural and human coastal systems can adapt even to the more immediate changes in climate variability and extreme events, including in more developed countries (Moser, 2005; Box 6.6). For example, without either adaptation or mitigation, the impacts of sea-level rise and other climate change such as more intense storms (Section 6.3.2) will be substantial, suggesting that some coastal low-lying areas, including atolls, may become unviable by 2100 (Barnett and Adger, 2003; Nicholls, 2004), with widespread impacts in many other areas. This may be reinforced by risk perception and disinvestment from these vulnerable areas. Adaptation could reduce impacts by a factor of 10 to 100 (Hall et al., 2006; Tol, 2007) and, apart from some small island nations, this appears to come at a minor cost compared to the damage avoided (Nicholls and Tol, 2006). However, the analysis is idealised, and while adaptation is likely to be widespread, it remains less clear if coastal societies can fully realise this potential for adaptation (see Box 6.6).

Adaptation for present climate risks is often inadequate and the ability to manage further increases in climate-related risks is frequently lacking. Moreover, increases in coastal development and population will magnify the risks of coastal flooding and other hazards (Section 6.2.2; Pielke Jr et al., 2005). Most measures to compensate and control the salinisation of coastal aquifers are expensive and laborious (Essink, 2001). Frequent floods impose enormous constraints on development. For example, Bangladesh has struggled to put sizeable

Table 6.11. *Selected information on costs and benefits of adaptation.*

Optimal (benefit-cost) coastal protection costs and remaining number of people displaced given a 1 m rise in sea level (Tol, 2002) (see also Figure 6.11).			
Region	**Protection Costs (10^9 US$)**	**Number of People Displaced (10^6)**	
Africa	92	2.74	
OECD Europe	136	0.22	
World	955	8.61	
Construction costs for coastal defence in England and Wales (average total cost in US$/km) (Evans et al., 2004a)			
Earth embankment	970,000	Culverts	3.5 million
Protected embankment	4.7 million	Sea wall	4.7 million
Dunes (excl. replenishment)	93,000	Groynes, breakwater (shingle beach)	9 million
Costs (US$/km) to protect against 1 m in rise in sea level for the USA (Neumann et al., 2000)			
Dike or levee	450,000 – 2.4 million	Sea wall; bulkhead construction	450,000 – 12 million
Capital costs (US$/km) for selected coastal management options in New Zealand (Jenks et al., 2005)			
Sand dune replanting, with community input (maintenance costs minimal)			6,000 – 24,000
Dune restoration, including education programmes (maintenance costs minimal)			15,000 – 35,000
Dune reshaping and replanting (maintenance costs minimal)			50,000 – 300,000
Sea walls and revetments (maintenance costs high – full rebuild every 20 – 40 years)			900,000 – 1.3 million
Direct losses, costs and benefits of adaptation to 65 cm sea-level rise in Pearl Delta, China (Hay and Mimura, 2005)			
Tidal level	Loss (US$ billion)	Cost (US$ billion)	Benefit (US$ billion)
Highest recorded	5.2	0.4	4.8
100 year high water	4.8	0.4	4.4

infrastructure in place to prevent flooding, but with limited success (Ahmad and Ahmed, 2003). Vietnam's transition from state central planning to a more market-oriented economy has had negative impacts on social vulnerability, with a decrease in institutional adaptation to environmental risks associated with flooding and typhoon impacts in the coastal environment (Adger, 2000). In a practical sense adaptation options for coral reefs are limited (Buddemeier, 2001) as is the case for most ecosystems. The continuing observed degradation of many coastal ecosystems (Section 6.2.2), despite the considerable efforts to reverse the trend, suggests that it will also be difficult to alleviate the added stresses resulting from climate change.

Knowledge and skill gaps are important impediments to understanding potential impacts, and thus to developing appropriate adaptation strategies for coastal systems (Crimp et al., 2004). The public often has conflicting views on the issues of sustainability, hard and soft defences, economics, the environment and consultation. Identifying the information needs of local residents, and facilitating access to information, are integral components in the process of public understanding and behavioural change (Myatt et al., 2003; Moser, 2005, 2006; Luers and Moser, 2006).

There are also important trade-offs in adaptation. For instance, while hard protection can greatly reduce the impacts of sea level and climate change on socio-economic systems, this is to the detriment of associated natural ecosystems due to coastal squeeze (Knogge et al., 2004; Rochelle-Newall et al., 2005). Managed retreat is an alternative response, but at what cost to socio-economic systems? General principles that can guide decision making in this regard are only beginning to be developed (Eurosion, 2004; Defra, 2006). Stakeholders will be faced with difficult choices, including questions as to whether traditional uses should be retained, whether invasive alien species or native species increasing in abundance should be controlled, whether planned retreat is an appropriate response to rising relative sea level or whether measures can be taken to reduce erosion. Decisions will need to take into account social and economic as well as ecological concerns (Adam, 2002). Considering these factors, the US Environmental Protection Agency is preparing sea-level rise planning maps that assign all shores along its Atlantic Coast to categories indicating whether shore protection is certain, likely, unlikely, or precluded by existing conservation policies (Titus, 2004). In the Humber estuary (UK) sea-level rise is reducing the standard of protection, and increasing erosion. Adaptation initiatives include creation of new intertidal habitat, which may promote more cost-effective defences and also helps to offset the loss of protected sites, including losses due to coastal squeeze (Winn et al., 2003).

Effective policies for developments that relate to the coast are sensitive to resource use conflicts, resource depletion and to pollution or resource degradation. Absence of an integrated holistic approach to policy-making, and a failure to link the process of policy-making with the substance of policy, results in outcomes that some would consider inferior when viewed within a sustainability framework (Noronha, 2004). Proponents of managed retreat argue that provision of long-term sustainable coastal defences must start with the premise that "coasts need space" (Rochelle-Newall et al., 2005). Some argue that governments must work to increase public awareness, scientific knowledge, and political will to facilitate such a retreat from the "sacrosanct" existing shoreline (Pethick, 2002). Others argue that the highest priority should be the transfer of property rights in lesser developed areas, to allow for changing setbacks in anticipation of an encroaching ocean. This makes inland migration of wetlands and beaches an expectation well before the existing shoreline becomes sacrosanct (Titus, 2001). Property rights and land use often make it difficult to achieve such goals, as shown by the post-Katrina recovery of New Orleans. Economic, social, ecological, legal and political lines of thinking have to be combined in order to achieve meaningful policies for the sustainable development of groundwater reserves and for the protection of subsurface ecosystems (Danielopol et al., 2003). Socio-economic and cultural conditions frequently present barriers to choosing and implementing the most appropriate adaptation to sea-level rise. Many such barriers can often be resolved by way of education at all levels, including local seminars and workshops for relevant stakeholders (Kobayashi, 2004; Tompkins et al., 2005a). Institutional strengthening and other interventions are also of importance (Bettencourt et al., 2005).

6.6.4 Adaptive capacity

Adaptive capacity is the ability of a system to evolve in order to accommodate climate changes or to expand the range of variability with which it can cope (see Chapter 17 for further explanation). The adaptive capacity of coastal communities to cope with the effects of severe climate impacts declines if there is a lack of physical, economic and institutional capacities to reduce climate-related risks and hence the vulnerability of high-risk communities and groups. But even a high adaptive capacity may not translate into effective adaptation if there is no commitment to sustained action (Luers and Moser, 2006).

Current pressures are likely to adversely affect the integrity of coastal ecosystems and thereby their ability to cope with additional pressures, including climate change and sea-level rise. This is a particularly significant factor in areas where there is a high level of development, large coastal populations and high levels of interference with coastal systems. Natural coastal habitats, such as dunes and wetlands, have a buffering capacity which can help reduce the adverse impacts of climate change. Equally, improving shoreline management for non-climate change reasons will also have benefits in terms of responding to sea-level rise and climate change (Nicholls and Klein, 2005). Adopting a static policy approach towards sea-level rise conflicts with sustaining a dynamic coastal system that responds to perturbations via sediment movement and long-term evolution (Crooks, 2004). In the case of coastal megacities, maintaining and enhancing both resilience and adaptive capacity for weather-related hazards are critically important policy and management goals. The dual approach brings benefits in terms of linking analysis of present and future hazardous conditions. It also enhances the capacity for disaster prevention and preparedness, disaster recovery and for adaptation to climate change (Klein et al., 2003).

6.6.4.1 Constraints and limitations

Yohe and Tol (2002) assessed the potential contributions of various adaptation options to improving systems' coping capacities. They suggest focusing attention directly on the underlying determinants of adaptive capacity (see Section 17.3.1). The future status of coastal wetlands appears highly sensitive to societal attitudes to the environment (Table 6.1), and this could be a more important control of their future status than sea-level rise (Nicholls, 2004). This highlights the importance of the socio-economic conditions (e.g., institutional capabilities; informed and engaged public) as a fundamental control of impacts with and without climate change (Tompkins et al., 2005b). Hazard awareness education and personal hazard experience are significant and important contributors to reducing community vulnerability. But despite such experience and education, some unnecessary and avoidable losses associated with tropical cyclone and storm surge hazards are still highly likely to occur (Anderson-Berry, 2003). These losses will differ across socio-economic groups, as has been highlighted recently by Hurricane Katrina. The constraints and limitations on adaptation by coastal systems, both natural and human, highlight the benefits for deeper public discourse on climate risk management, adaptation needs, challenges and allocation and use of resources.

6.6.4.2 Capacity-strengthening strategies

Policies that enhance social and economic equity, reduce poverty, increase consumption efficiencies, decrease the discharge of wastes, improve environmental management, and increase the quality of life of vulnerable and other marginal coastal groups can collectively advance sustainable development, and hence strengthen adaptive capacity and coping mechanisms. Many proposals to strengthen adaptive capacity have been made including: mainstreaming the building of resilience and reduction of vulnerability (Agrawala and van Aalst, 2005; McFadden et al., 2007b); full and open data exchange (Hall, 2002); scenarios as a tool for communities to explore future adaptation policies and practices (Poumadère et al., 2005); public participation, co-ordination among oceans-related agencies (West, 2003); research on responses of ecological and socio-economic systems, including the interactions between ecological, socio-economic and climate systems (Parson et al., 2003); research on linkages between upstream and downstream process to underpin comprehensive coastal management plans (Contreras-Espinosa and Warner, 2004); research to generate useful, usable and actionable information that helps close the science-policy gap (Hay and Mimura, 2006); strengthening institutions and enhancing regional co-operation and co-ordination (Bettencourt et al., 2005); and short-term training for practitioners at all levels of management (Smith, 2002a).

6.6.5 The links between adaptation and mitigation in coastal and low-lying areas

Adaptation (e.g., coastal planning and management) and mitigation (reducing greenhouse gas emissions) are responses to climate change, which can be considered together (King,

2004) (see Chapter 18). The response of sea-level rise to mitigation of greenhouse gas emissions is slower than for other climate factors (Meehl et al., 2007) and mitigation alone will not stop growth in potential impacts (Nicholls and Lowe, 2006). However, mitigation decreases the rate of future rise and the ultimate rise, limiting and slowing the need for adaptation as shown by Hall et al. (2005). Hence Nicholls and Lowe (2006) and Tol (2007) argue that adaptation and mitigation need to be considered together when addressing the consequences of climate change for coastal areas. Collectively these interventions can provide a more robust response to human-induced climate change than consideration of each policy alone.

Adaptation will provide immediate and longer-term reductions in risk in the specific area that is adapting. On the other hand, mitigation reduces future risks in the longer term and at the global scale. Identifying the optimal mix is problematic as it requires consensus on many issues, including definitions, indicators and the significance of thresholds. Importantly, mitigation removes resources from adaptation, and benefits are not immediate, so investment in adaptation may appear preferable, especially in developing countries (Goklany, 2005). The opposite view of the need for urgent mitigation has recently been argued (Stern, 2007). Importantly, the limits to adaptation may mean that the costs of climate change are underestimated (Section 6.6.3), especially in the long term. These findings highlight the need to consider impacts beyond 2100, in order to assess the full implications of different mitigation and adaptation policy mixes (Box 6.6).

6.7 Conclusions: implications for sustainable development

The main conclusions are reported in the Executive Summary and are reviewed here in the context of sustainable development. Coastal ecosystems are dynamic, spatially constrained, and attractive for development. This leads to increasing multiple stresses under current conditions (Section 6.2.2), often resulting in significant degradation and losses, especially to economies highly dependent on coastal resources, such as small islands. Trends in human development along coasts amplify their vulnerability, even if climate does not change. For example, in China 100 million people moved from inland to the coast in the last twenty years (Dang, 2003), providing significant benefits to the national economy, but presenting major challenges for coastal management. This qualitative trend is mirrored in most populated coastal areas and raises the conflict between conservation and development (Green and Penning-Rowsell, 1999). Equally the pattern of development can have tremendous inertia (Klein et al., 2002) and decisions made today may have implications centuries into the future (Box 6.6).

Climate change and sea-level rise increase the challenge of achieving sustainable development in coastal areas, with the most serious impediments in developing countries, in part due to their lower adaptive capacity. It will make achieving the Millennium Development Goals (UN Secretary General, 2006b) more difficult, especially the Goal of Ensuring Environmental Sustainability (reversing loss of environmental resources, and improving lives of slum dwellers, many of whom are coastal). Adapting effectively to climate change and sea-level rise will involve substantial investment, with resources diverted from other productive uses. Even with the large investment possible in developed countries, residual risk remains, as shown by Hurricane Katrina in New Orleans (Box 6.4), requiring a portfolio of responses that addresses human safety across all events (protection, warnings, evacuation, etc.) and also can address multiple goals (e.g., protection of the environment as well as adaptation to climate change) (Evans et al., 2004a; Jonkman et al., 2005). Long-term sea-level rise projections mean that risks will grow for many generations unless there is a substantial and ongoing investment in adaptation (Box 6.6). Hence, sustainability for coastal areas appears to depend upon a combination of adaptation and mitigation (Sections 6.3.2 and 6.6.5).

There will be substantial benefits if plans are developed and implemented in order to address coastal changes due to climate and other factors, such as those processes that also contribute to relative sea-level rise (Rodolfo and Siringan, 2006). This requires increased effort to move from reactive to more proactive responses in coastal management. Strengthening integrated multidisciplinary and participatory approaches will also help improve the prospects for sustaining coastal resources and communities. There is also much to be learnt from experience and retrospective analyses of coastal disasters (McRobie et al., 2005). Technological developments are likely to assist this process, most especially in softer technologies associated with monitoring (Bradbury et al., 2005), predictive modelling and broad-scale assessment (Burgess et al., 2003; Cowell et al., 2003a; Boruff et al., 2005) and assessment of coastal management actions, both present and past (Klein et al., 2001). Traditional practices can be an important component of the coastal management toolkit.

6.8 Key uncertainties, research gaps and priorities

This assessment shows that the level of knowledge is not consistent with the potential severity of the problem of climate change and coastal zones. While knowledge is not adequate in any aspect, uncertainty increases as we move from the natural sub-system to the human sub-system, with the largest uncertainties concerning their interaction (Figure 6.1). An understanding of this interaction is critical to a comprehensive understanding of human vulnerability in coastal and low-lying areas and should include the role of institutional adaptation and public participation (Section 6.4.3). While research is required at all scales, improved understanding at the physiographic unit scale (e.g., coastal cells, deltas or estuaries) would have particular benefits, and support adaptation to climate change and wider coastal management. There also remains a strong focus on sea-level rise, which needs to be broadened to include all the climate drivers in the coastal zone (Table 6.2). Finally, any response to climate change has to address the other non-

Box 6.6. Long-term sea-level rise impacts (beyond 2100)

The timescales of ocean warming are much longer than those of surface air temperature rise. As a result, sea-level rise due to thermal expansion is expected to continue at a significant rate for centuries, even if climate forcing is stabilised (Meehl et al., 2005; Wigley, 2005). Deglaciation of small land-based glaciers, and possibly the Greenland and the West Antarctic ice sheets, may contribute large additional rises, with irreversible melting of Greenland occurring for a sustained global temperature rise of 1.1 to 3.8°C above today's global average temperature: this is likely to happen by 2100 under the A1B scenario, for instance (Meehl et al., 2007). More than 10 m of sea-level rise is possible, albeit over very long time spans (centuries or longer), and this has been termed 'the commitment to sea-level rise'. The potential exposure to these changes, just based on today's socio-economic conditions, is significant both regionally and globally (Table 6.12) and growing (Section 6.3.1). Thus there is a conflict between long-term sea-level rise and present-day human development patterns and migration to the coast (Nicholls et al., 2006).

The rate of sea-level rise is uncertain and a large rise (>0.6 m to 0.7 m/century) remains a low probability/high impact risk (Meehl et al., 2007). Some analyses suggest that protection would be an economically optimum response in most developed locations, even for an arbitrary 2 m/century scenario (Anthoff et al., 2006). However, sea-level rise will accumulate beyond 2100, increasing impact potential (Nicholls and Lowe, 2006). Further, there are several potential constraints to adaptation which are poorly understood (Section 6.4.3; Nicholls and Tol, 2006; Tol et al., 2006). This raises long-term questions about the implications of 'hold the line' versus 'retreat the line' adaptation policies and, more generally, how best to approach coastal spatial planning. While shoreline management is starting to address such issues for the 21st century (Eurosion, 2004; Defra, 2006), the long timescales of sea-level rise suggest that coastal management, including spatial planning, needs to take a long-term view on adaptation to sea-level rise and climate change, especially with long-life infrastructure such as nuclear power stations.

Table 6.12. *Indicative estimates of regional exposure as a function of elevation and baseline (1995) socio-economics. MER – market exchange rates (after Anthoff et al., 2006).*

Region	Exposure by factor and elevation above mean high water								
	Land area (km^2)			Population (millions)			GDP MER (US$ billions)		
	1m	5m	10m	1m	5m	10m	1m	5m	10m
Africa	118	183	271	8	14	22	6	11	19
Asia	875	1548	2342	108	200	294	453	843	1185
Australia	135	198	267	2	3	4	38	51	67
Europe	139	230	331	14	21	30	305	470	635
Latin America	317	509	676	10	17	25	39	71	103
North America	640	1000	1335	4	14	22	103	358	561
Global (Total)	2223	3667	5223	145	268	397	944	1802	2570

climate drivers of coastal change in terms of understanding potential impacts and responses, as they will interact with climate change. As recognised in earlier IPCC assessments and the Millennium Ecosystem and LOICZ Assessments (Agardy et al., 2005; Crossland et al., 2005), these other drivers generally exacerbate the impacts of climate change.

The following research initiatives would substantially reduce these uncertainties and increase the effectiveness and science base of long-term coastal planning and policy development.

Establishing better baselines of actual coastal changes, including local factors and sea-level rise, and the climate and non-climate drivers, through additional observations and expanded monitoring. This would help to better establish the causal links between climate and coastal change which tend to remain inferred rather than observed (Section 6.2.5), and support model development.

- Improving predictive capacity for future coastal change due to climate and other drivers, through field observations, experiments and model development. A particular challenge will be understanding thresholds under multiple drivers of change (Sections 6.2.4; 6.4.1).
- Developing a better understanding of the adaptation of the human systems in the coastal zone. At the simplest this could be an inventory of assets at risk, but much more could be done in terms of deepening our understanding of the qualitative trends suggested in Table 6.1 (see also Section 6.4.2) and issues of adaptive capacity.

- Improving impact and vulnerability assessments within an integrated assessment framework that includes natural-human sub-system interactions. This requires a strong inter-disciplinary approach and the targeting of the most vulnerable areas, such as populated megadeltas and deltas, small islands and coastal cities (Section 6.4.3). Improving systems of coastal planning and zoning and institutions that can enforce regulations for clearer coastal governance is required in many countries.
- Developing methods for identification and prioritisation of coastal adaptation options. The effectiveness and efficiency of adaptation interventions need to be considered, including immediate benefits and the longer term goal of sustainable development (Sections 6.6; 6.7).
- Developing and expanding networks to share knowledge and experience on climate change and coastal management among coastal scientists and practitioners.

These issues need to be explored across the range of spatial scales: from local to global scale assessments and, given the long timescales of sea-level rise, implications beyond the 21st century should not be ignored. Thus this research agenda needs to be taken forward across a broad range of activities from the needs of coastal management and adaptation to global integrated assessments and the benefits of mitigation. While some existing global research efforts are pushing in the direction that is recommended, e.g., the IGBP/IHDP LOICZ Science Plan (Kremer et al., 2004), much more effort is required to achieve these goals, especially those referring to the human, integrated assessment and adaptation goals, and at local to regional scales (Few et al., 2004a).

References

Adam, P., 2002: Saltmarshes in a time of change. *Environ. Conserv.*, **29**, 39-61.

Adger, W.N., 2000: Institutional adaptation to environmental risk under the transition in Vietnam. *Ann. Assoc. Am. Geogr.*, **90**, 738-758.

Agardy, T., J. Alder, P. Dayton, S. Curran, A. Kitchingman, M. Wilson, A. Catenazzi, J. Restrepo and Co-authors, 2005: Coastal systems. *Millennium Ecosystem Assessment: Ecosystems & Human Well-Being, Volume 1: Current State and Trends*, W. Reid, Ed., Island Press, 513-549.

Agnew, M.D., and D. Viner, 2001: Potential impacts of climate change on international tourism. *Tourism and Hospitality Research*, **3**, 37-60.

Agrawala, S., and M. van Aalst, 2005: Bridging the gap between climate change and development. *Bridge over Troubled Waters: Linking Climate Change and Development*, S. Agrawala, Ed., Organisation for Economic Co-operation and Development (OECD). 153pp.

Aguiló, E., J. Alegre and M. Sard, 2005: The persistence of the sun and sand tourism model. *Tourism Manage.*, **26**, 219-231.

Ahmad, Q.K. and A.U. Ahmed, 2003: Regional cooperation in flood management in the Ganges-Brahmaputra-Meghna region: Bangladesh perspective. *Nat. Hazards*, **28**, 181-198.

Ahrendt, K., 2001: Expected effects of climate change on Sylt Island; results from a multidisciplinary German project. *Climate Res.*, **18**, 141-146.

Alcamo, J., and T. Henrichs, 2002: Critical regions: a model-based estimation of world water resources sensitive to global changes. *Aquat. Sci.*, **64**, 352-362.

Aliff, G., 2006: Have hurricanes changed everything? Or is a soft market ahead? *Risk Manage.*, **53**, 12-19.

Allan, J., P. Komar and G. Priest, 2003: Shoreline variability on the high-energy Oregon coast and its usefulness in erosion-hazard assessments. *J. Coastal Res.*, **S38**, 83-105.

Allen, J.R.L., 2000: Morphodynamics of Holocene saltmarshes: a review sketch from the Atlantic and Southern North Sea coasts of Europe. *Quaternary Sci. Rev.*, **19**, 1155-1231.

Allen, J.R.L.,, 2003: An eclectic morphostratigraphic model for the sedimentary response to Holocene sea-level rise in northwest Europe. *Sediment. Geol.*, **161**, 31-54.

Alley, R.B., J. Marotzke, W.D. Nordhaus, J.T. Overpeck, D.M. Peteet, R.A. Pielke, R.T. Pierrehumbert, R.T. Rhines and Co-authors, 2003: Abrupt climate change. *Science*, **299**, 2005-2010.

Allison, M.A., S.R. Khan, S.L. Goodbred Jr. and S.A. Kuehl, 2003: Stratigraphic evolution of the late Holocene Ganges-Brahmaputra lower delta plain. *Sediment. Geol.*, **155**, 317-342.

Anderson-Berry, L.J., 2003: Community vulnerability to tropical cyclones: Cairns, 1996-2000. *Nat. Hazards*, **30**, 209-232.

Andersson, A.J., F.T. Mackenzie and L.M. Ver, 2003: Solution of shallow-water carbonates: an insignificant buffer against rising atmospheric CO_2. *Geology*, **31**, 513-516.

Anthoff, D., R.J. Nicholls, R.S.J. Tol and A.T. Vafeidis, 2006: *Global and Regional Exposure to Large Rises in Sea Level: A Sensitivity Analysis*. Working Paper 96. Tyndall Centre for Climate Change Research, University of East Anglia, Norwich, Norfolk, 31pp.

ARAG, 1999: *The potential consequences of climate variability and change: Alaska. Preparing for a changing climate*. Alaska Regional Assessment Group (ARAG). Center for Global Change and Arctic System Research, University of Alaska, 39 pp.

Arnell, N.W., 2004: Climate change and global water resources: SRES scenarios and socio-economic scenarios. *Glob. Environ. Chang.*, **14**, 31-52.

Arnell, N.W., M.J.L. Livermore, S. Kovats, P.E. Levy, R. Nicholls, M.L. Parry and S.R. Gaffin, 2004: Climate and socio-economic scenarios for global-scale climate change impacts assessments: Characterising the SRES storylines. *Glob. Environ. Chang.*, **14**, 3-20.

Ayre, D.J., and T.P. Hughes, 2004: Climate change, genotypic diversity and gene flow in reef-building corals. *Ecol. Lett.*, **7**, 273-278.

Baldwin, A.H., M.S. Egnotovich and E. Clarke, 2001: Hydrologic change and vegetation of tidal freshwater marshes: Field, greenhouse and seed-bank experiments. *Wetlands*, **21**, 519-531.

Barnett, J., and W.N. Adger, 2003: Climate dangers and atoll countries. *Climatic Change*, **61**, 321-337.

Barras, J. A., cited 2006: Land area change in coastal Louisiana after the 2005 hurricanes - a series of three maps. U.S. Geological Survey Open-File Report 2006-1274. [Accessed 29.03.07: http://pubs.usgs.gov/of/2006/1274/]

Barras, J., S. Beville, D. Britsch, S. Hartley, S. Hawes, J. Johnston, P. Kemp, Q. Kinler and Co-authors, 2003: Historical and projected coastal Louisiana land changes: 1978-2050. Open File Report 03-334. U.S. Geological Survey, 39 pp.

Barros, V., A. Menéndez, C. Natenzon, R.R. Kokot, J.O. Codignotto, M. Re, P. Bronstein, I. Camilloni and Co-authors, 2006: *Vulnerability to floods in the metropolitan region of Buenos Aires under future climate change*. Working Paper 26. Assessments of Impacts and Adaptations to Climate Change (AIACC), 36 pp.

Bartlett, D., and J.L. Smith, Eds., 2005: *GIS for Coastal Zone Management*. CRC Press, 310 pp.

Barton, A.D., and K.S. Casey, 2005: Climatological context for large-scale coral bleaching. *Coral Reefs*, **24**, 536-554.

Baxter, P.J., 2005: The east coast Big Flood, 31 January-1 February 1953: a summary of the human disaster. *Philos. T. Roy. Soc. A*, **363**, 1293-1312.

Becken, S., and J. Hay, undated: *Tourism and Climate Change – Risk and opportunities*. Landcare Research, 292 pp.

Beer, S., and E. Koch, 1996: Photosynthesis of marine macroalgae and seagrass in globally changing CO_2 environments. *Mar. Ecol-Prog. Ser.*, **141**, 199-204.

Berge, J., G. Johnsen, F. Nilsen, B. Gulliksen and D. Slagstad, 2005: Ocean temperature oscillations enable reappearance of blue mussels *Mytilus edulis* in Svalbard after a 1000 year absence. *Mar. Ecol-Prog. Ser.*, **303**, 167-175.

Berkelmans, R., G. De'ath, S. Kininmouth and W.J. Skirving, 2004: A comparison of the 1998 and 2002 coral bleaching events on the Great Barrier Reef: spatial correlation, patterns and predictions. *Coral Reefs*, **23**, 74-83.

Bernier, N.B., K.R. Thompson, J. Ou and H. Ritchie, 2007: Mapping the return periods of extreme sea levels: Allowing for short sea level records, seasonality and climate change. *Global Planet Change*. doi:10.1016/j.gloplacha.2006.11.027.

Bettencourt, S., R. Croad, P. Freeman, J.E. Hay, R. Jones, P. King, P. Lal, A. Mearns and Co-authors, 2005: *Not if but when: Adapting to natural hazards in the Pacific Islands region*. A policy note. Pacific Islands Country Management Unit, East Asia and Pacific Region, World Bank, 46 pp.

Bigano, A., J.M. Hamilton and R.S.J. Tol, 2005: *The Impact of Climate Change on Domestic and International Tourism: A Simulation Study*. Working paper FNU-58, 23 pp.

Bijlsma, L., C.N. Ehler, R.J.T. Klein, S.M. Kulshrestha, R.F. McLean, N. Mimura, R.J. Nicholls, L.A. Nurse and Co-authors, 1996: Coastal zones and small islands. *Climate Change 1995: Impacts, Adaptations and Mitigation of Climate Change: Scientific-Technical Analyses. Contribution of Working Group II to the Second Assessment Report of the Intergovernmental Panel on Climate Change*, R.T. Watson, M.C. Zinyowera and R.H. Moss, Eds., Cambridge University Press, Cambridge, 289-324.

Bindoff, N., J. Willebrand, V. Artale, A. Cazenave, J. Gregory, S. Gulev, K. Hanawa, C. Le Quéré and Co-authors, 2007: Observations: Oceanic climate change and sea level. *Climate Change 2007: The Physical Science Basis. Contribution of Working Group I to the Fourth Assessment Report of the Intergovernmental Panel on Climate Change*, S. Solomon, D. Qin, M. Manning, Z. Chen, M. Marquis, K.B. Averyt, M. Tignor and H.L. Miller, Eds., Cambridge University Press, Cambridge 385-432.

Binh, C.T., M.J. Phillips and H. Demaine, 1997: Integrated shrimp-mangrove farming systems in the Mekong Delta of Vietnam. *Agr. Res.*, **28**, 599-610.

Bird, E.C.F., 1985: *Coastline Changes: A Global Review*. John Wiley and Sons, Chichester, 219 pp.

Bird, E.C.F.,, 2000: *Coastal Geomorphology: An Introduction*. Wiley and Sons, Chichester. 340 pp.

Boruff, B.J., C. Emrich and S.L. Cutter, 2005: Erosion hazard vulnerability of US coastal counties. *J. Coastal Res.*, **21**, 932-942.

Bosello, F., R. Lazzarin and R.S.J. Tol, 2004: *Economy-wide Estimates of the Implications of Climate Change: Sea Level Rise*. Working Paper FNU-38. Research Unit Sustainability and Global Change, Centre for Marina and Climate Research, Hamburg University. 22pp.

Bradbury, A., S. Cope and H. Dalton, 2005: Integration of large-scale regional measurement, data management and analysis programmes, for coastal processes, geomorphology and ecology. *Proceeding of CoastGIS '05*, 21-23 July, Aberdeen, Scotland. 14 pp.

Brander, K., G. Blom, M.F. Borges, K. Erzini, G. Henderson, B.R. MacKenzie, H. Mendes, A.M.P. Santos and P. Toresen, 2003a: Changes in fish distribution in the eastern North Atlantic: are we seeing a coherent response to changing temperature? *ICES Marine Science Symposia*, **219**, 260-273.

Brander, L.M., R.J.G.M. Florax and J.E. Vermaat, 2003b: *The Empirics of Wetland Valuation: A Comprehensive Summary and a Meta-analysis of the Literature*. Report No.W-03/30. Institute for Environmental Studies (IVM), Vrije Universiteit, Amsterdam, 33 pp.

Brown, A.C., and A. McLachlan, 2002: Sandy shore ecosystems and the threats facing them: some predictions for the year 2025. *Environ. Conserv.*, **29**, 62-77.

Bruno, J.F., C.E. Siddon, J.D. Witman, P.L. Colin and M.A. Toscano, 2001: El Niño related coral bleaching in Palau, Western Caroline Islands. *Coral Reefs*, **20**, 127-136.

Brunsden, D., 2001: A critical assessment of the sensitivity concept in geomorphology. *Catena*, **42**, 99-123.

Bruun, P., 1962: Sea-level rise as a cause of shore erosion. *Journal of the Waterways and Harbors Division*, **88**, 117-130.

Bryant, E., 2001: *Tsunami: The Underrated Hazard*. Cambridge University Press, Cambridge, 320 pp.

Buddemeier, R.W., 2001: Is it time to give up? *B. Mar. Sci.*, **69**, 317-326.

Buddemeier, R.W., S.V. Smith, D.P. Swaaney and C.J. Crossland, 2002: *The Role of the Coastal Ocean in the Disturbed and Undisturbed Nutrient and Carbon Cycles*. LOICZ Reports and Studies Series No. 24. 84 pp.

Buddemeier, R.W., J.A. Kleypas and B. Aronson, 2004: *Coral Reefs and Global Climate Change: Potential Contributions of Climate Change to Stresses on Coral Reef Ecosystems*. Report prepared for the Pew Centre of Climate Change, Arlington, Virginia, 56 pp.

Burgess, K., and I. Townend, 2004: The impact of climate change upon coastal defence structures. *39th DEFRA Flood and Coastal Management Conference*, 29 June - 1 July, York UK, Department of Environment, Food and Rural Affairs (Defra), 11.12.11-11.12.12.

Burgess, K., J.D. Orford, K. Dyer, I. Townend and P. Balson, 2003: FUTURECOAST — The integration of knowledge to assess future coastal evolution at a national scale. *Proceedings of the 28th International Conference on Coastal Engineering*, **3**, 3221-3233.

Burkett, V.R., and J. Kusler, 2000: Climate change: potential impacts and interactions in wetlands of the United States. *J. Am. Water Resour. As.*, **36**, 313-320.

Burkett, V.R., D.B. Zilkoski and D.A. Hart, 2003: Sea-level rise and subsidence: Implication for flooding in New Orleans. *U.S. Geological Survey Subsidence Interest Group Conference, proceedings of the technical meeting. Water Resources Division (Open File Report 03-308)*, 27th - 29th November 2001. Galveston, Texas, U.S. Geological Survey, 63-70.

Burkett, V.R., D.A. Wilcox, R. Stottlemeyer, W. Barrow, D. Fagre, J. Baron, J. Price, J. Nielsen and Co-authors, 2005: Nonlinear dynamics in ecosystem response to climate change: Case studies and policy implications. *Ecological Complexity*, **2**, 357-394.

Butler, C.D., C.F. Corvalan and H.S. Koren, 2005: Human health, well-being, and global ecological scenarios. *Ecosystems*, **8**, 153-162.

Cahoon, D.R., P. Hensel, J. Rybczyk, K. McKee, C.E. Proffitt and B. Perez, 2003: Mass tree mortality leads to mangrove peat collapse at Bay Islands, Honduras after Hurricane Mitch. *J. Ecol.*, **91**, 1093-1105.

Cahoon, D.R., P.F. Hensel, T. Spencer, D.J. Reed, K.L. McKee and N. Saintilan, 2006: Coastal wetland vulnerability to relative sea-level rise: wetland elevation trends and process controls. *Wetlands as a Natural Resource, Vol. 1: Wetlands and Natural Resource Management*, J. Verhoeven, D. Whigham, R. Bobbink and B. Beltman, Eds., Springer Ecological Studies series, Chapter 12. 271-292.

Caldeira, K., and M.E. Wickett, 2005: Anthropogenic carbon and ocean pH. *Nature*, **425**, 365.

Canziani, O.F., and J.C. Gimenez, 2002: *Hydrometeorological Aspects of the Impact of Climate Change on the Argentina's Pampas*. EPA Report. United States Environmental Protection Agency.

CBD, 2003: *Interlinkages Between Biological Diversity and Climate Change*. CBD Technical report 10. Secretariat of the Convention on Biological Diversity, 142 pp.

CBD, 2006: *Guidance for Promoting Synergy Among Activities Addressing Biological Diversity, Desertification, Land Degradation and Climate Change*. CBD Technical Series 25. Secretariat of the Convention on Biological Diversity, iv + 43 pp.

Chadwick, A.J., H. Karunarathna, W.R. Gehrels, A.C. Massey, D. O'Brien and D. Dales, 2005: A new analysis of the Slapton barrier beach system, UK. *Proc. Inst. Civ. Eng. Water Maritime Engng*, **158**, 147-161.

Christensen, J.H., B. Hewitson, A. Busuioc, A. Chen, X. Gao, I. Held, R. Jones, W.-T. Kwon and Co-authors, 2007: Regional climate projections. *Climate Change 2007: The Physical Science Basis. Contribution of Working Group I to the Fourth Assessment Report of the Intergovernmental Panel on Climate Change*, S. Solomon, D. Qin, M. Manning, Z. Chen, M. Marquis, K.B. Averyt, M. Tignor and H.L. Miller, Eds., Cambridge University Press, Cambridge, 847-940.

Christie, P., K. Lowry, A.T. White, E.G. Oracion, L. Sievanen, R.S. Pomeroy, R.B. Pollnac, J.M. Patlis and R.L.V. Eisma, 2005: Key findings from a multidisciplinary examination of integrated coastal management process sustainability. *Ocean Coast. Manage.*, **48**, 468-483.

Church, J.A., J.M. Gregory, P. Huybrechts, M. Kuhn, K. Lambeck, M.T. Nhuan, D. Qin and P.L. Woodworth, 2001: Changes in sea level. *Climate Change 2001. The Scientific Basis. Contribution of Working Group I to the Third Assessment Report of the Intergovernmental Panel on Climate Change*, J.T. Houghton, Y. Ding, D.J. Griggs, M. Noguer, P.J. van der Linden and D. Xiaosu, Eds., Cambridge University Press, Cambridge, 639-693.

Codignotto, J.O., 2004a: Sea-level rise and coastal de La Plata River. *Report for the second AIACC "It's raining, it's pouring, ... I''s time to be adapting"; Regional Workshop for Latin America and the Caribbean* 24-27th August 2004, Buenos Aires. 4-18pp.

Codignotto, J.O., 2004b: Diagnóstico del estado actual de las áreas costeras de Argentina. Report for Programa de las Naciones Unidas para el Desarrollo (Argentina) PNUD/ARG 3, 43 pp.

Codignotto, J.O., R.R. Kokot and A.J.A. Monti, 2001: Cambios rápidos en la Costa de Caleta Valdés, Chubut. *Asociación Geológica Argentina*, **56**, 67-72.

Coleman, J.M., O.K. Huh, D.H. Braud, Jr. and H.H. Roberts, 2005: Major World Delta Variability and Wetland Loss. *Gulf Coast Association of Geological Societies (GCAGS) Transactions*, **55**, 102-131

Coles, S.L., and B.E. Brown, 2003: Coral bleaching - capacity for acclimatization and adaptation. *Adv. Mar. Biol.*, **46**, 183-224.

Contreras-Espinosa, F., and B.G. Warner, 2004: Ecosystem characteristics and management considerations for coastal wetlands in Mexico. *Hydrobiologia*, **511**, 233-245.

Coombes, E., D. Burgess, N. Jackson, K. Turner and S. Cornell, 2004: Case study: Climate change and coastal management in practice - A cost-benefit assessment in the Humber, UK. 56 pp. [Accessed 30.05.07: http://www.eloisegroup.org/themes/climatechange/contents.htm]

Cooper, J.A.G., and O.H. Pilkey, 2004: Sea-level rise and shoreline retreat: time to abandon the Bruun Rule. *Global Planet Change*, **43**, 157-171.

Cooper, J.A.G., and F. Navas, 2004: Natural bathymetric change as a control on century-scale shoreline behaviour. *Geology*, **32**, 513-516.

Cooper, N., and H. Jay, 2002: Predictions of large-scale coastal tendency: development and application of a qualitative behaviour-based methodology. *J. Coastal Res.*, **S 36**, 173-181.

Cooper, N., P.C. Barber, M.C. Bray and D.J. Carter, 2002: Shoreline management plans: a national review and an engineering perspective. *Proc. Inst. Civ. Eng. Water Maritime Engng.*, **154**, 221-228.

Cowell, P.J., M.J.F. Stive, A.W. Niedoroda, H.J. De Vriend, D.J.P. Swift, G.M. Kaminsky and M. Capobianco, 2003a: The coastal tract. Part 1: A conceptual approach to aggregated modelling of low-order coastal change. *J. Coastal Res.*, **19**, 812-827.

Cowell, P.J., M.J.F. Stive, A.W. Niedoroda, D.J.P. Swift, H.J. De Vriend, M.C. Buijsman, R.J. Nicholls, P.S. Roy and Co-authors, 2003b: The coastal tract. Part 2: Applications of aggregated modelling of lower-order coastal change. *J. Coastal Res.*, **19**, 828-848.

Cowell, P.J., B.G. Thom, R.A. Jones, C.H. Everts and D. Simanovic, 2006: Management of uncertainty in predicting climate-change impacts on beaches. *J. Coastal Res.*, **22**, 232.

CRED, 2005: *CRED Crunch*. Centre for Research on the Epidemiology of Disasters (CRED), Université catholique de Louvain, Louvain-la-Neuve, 2 pp.

Crimp, S., J. Balston, A.J. Ash, L. Anderson-Berry, T. Done, R. Greiner, D. Hilbert, M. Howden and Co-authors, 2004: *Climate Change in the Cairns and Great Barrier Reef region: Scope and Focus for an Integrated Assessment*. Australian Greenhouse Office, Canberra, 100 pp.

Crooks, S., 2004: The effect of sea-level rise on coastal geomorphology. *Ibis*, **146**, 18-20.

Crossland, C.J., H.H. Kremer, H.J. Lindeboom, J.I. Marshall Crossland and M.D.A. Le Tissier, 2005: *Coastal Fluxes in the Anthropocene*. The Land-Ocean Interactions in the Coastal Zone Project of the International Geosphere-Biosphere Programme Series: Global Change - The IGBP Series, 232 pp.

Dang, N.A., 2003: Internal migration policies in the ESCAP region. *Asia-Pacific Population Journal*, **18**, 27-40.

Daniel, H., 2001: Replenishment versus retreat: the cost of maintaining Delaware's beaches. *Ocean Coast. Manage.*, **44**, 87-104.

Danielopol, D.L., C. Griebler, A. Gunatilaka and J. Notenboom, 2003: Present state and future prospects for groundwater ecosystems. *Environ. Conserv.*, **30**, 104-130.

Daufresne, M., M.C. Roger, H. Capra and N. Lamouroux, 2003: Long-term changes within the invertebrate and fish communities of the Upper Rhône River: effects of climatic factors. *Glob. Change Biol.*, **10**, 124-140.

Davidson-Arnott, R.G.D., 2005: Conceptual model of the effects of sea-level rise on sandy coasts. *J. Coastal Res.*, **21**, 1173-1177.

Dawson, R.J., J.W. Hall, P.D. Bates and R.J. Nicholls, 2005: Quantified analysis of the probability of flooding in the Thames Estuary under imaginable worst case sea-level rise scenarios. *Int. J. Water. Resour. D.*, **21**, 577-591.

de Groot, T.A.M., 1999: Climate shifts and coastal changes in a geological perspective. A contribution to integrated coastal zone management. *Geol. Mijnbouw*, **77**, 351-361.

Defra, 2001: *Shoreline Management Plans: A Guide for Coastal Defence Authorities*. Department for Environment, Food and Rural Affairs (Defra), London, 77pp.

Defra, 2004: *Scientific and Technical Aspects of Climate Change, including Impacts and Adaptation and Associated Costs*. Department for Environment, Food and Rural Affairs (Defra), London, 19 pp.

Defra, 2006: *Shoreline Management Plan Guidance*. Volume 1: Aims and requirements. Department for Environment, Farming and Rural Affairs (Defra), London, 54 pp.

Dickinson, W.R., 2004: Impacts of eustasy and hydro-isostasy on the evolution and landforms of Pacific atolls. *Palaeogeogr., Palaeoclimatol., Palaeoecol.*, **213**, 251-269.

Dickson, M.E., M.J.A. Walkden, J.W. Hall, S.G. Pearson and J. Rees, 2005: Numerical modelling of potential climate change impacts on rates of soft cliff recession, northeast Norfolk, UK. *Proceedings of the International Conference on Coastal Dynamics*, 4-8th April; Barcelona, Spain, American Society of Civil Engineers (ASCE).

Diffey, B., 2004: Climate change, ozone depletion and the impact on ultraviolet exposure of human skin. *Phys. Med. Biol.*, **49**, R1-R11.

Dinas-Coast Consortium, 2006: Dynamic Interactive Vulnerability Assessment (DIVA), Potsdam Institute for Climate Impact Research, Potsdam.

Discovery Software, cited 2006: Software solutions for a dynamic world including STEMgis, FloodRanger, SimCoast and NineNil. [Accessed 10.04.07: www.discoverysoftware.co.uk]

Dixon, T.H., F. Amelung, A. Ferretti, F. Novali, F. Rocca, R. Dokka, G. Sella, S.-W. Kim and Co-authors, 2006: Subsidence and flooding in New Orleans. *Nature*, **441**, 587-588.

Donnelly, J.P., P. Cleary, P. Newby and R. Ettinger, 2004: Coupling instrumental and geological records of sea level change: Evidence from southern New England of an increase in the rate of sea-level rise in the late 19th century. *Geophys. Res. Lett.*, **31**, article no. L05203.

Donner, S.D., W.J. Skirving, C.M. Little, M. Oppenheimer and O. Hoegh-Guldberg, 2005: Global assessment of coral bleaching and required rates of adaptation under climate change. *Glob. Change Biol.*, **11**, 2251-2265.

Douglas, A.E., 2003: Coral bleaching - how and why? *Mar. Pollut. Bull.*, **46**, 385-392.

Doyle, T.W., R.H. Day and J.M. Biagas, 2003: Predicting coastal retreat in the Florida Big Bend region of the Gulf Coast under climate change induced sea-level rise. *Integrated Assessment of the Climate Change Impacts on the Gulf Coast region. Foundation Document*, Z.H. Ning, R.E. Turner, T. Doyle, and K. Abdollahi, Eds., Louisiana State University Press, Baton Rouge, 201-209.

Drexler, J.E., 2001: Effect of the 1997-1998 ENSO-related drought on hydrology and salinity in a Micronesian wetland complex. *Estuaries*, **24**, 343-358.

Dronkers, J., 2005: *Dynamics of Coastal Systems*. Advanced Series on Ocean Engineering. 25. World Scientific Publishing Company, Hackensack, 519 pp.

DTI, 2002: *Foresight Futures 2020. Revised Scenarios and Guidance*. Department of Trade and Industry (DTI), London. 36 pp.

Duarte, C.M., 2002: The future of seagrass meadows. *Environ. Conserv.*, **9**, 192-206.

Durongdej, S., 2001: Land use changes in coastal areas of Thailand. *Proceedings of the APN/SURVAS/LOICZ Joint Conference on Coastal Impacts of Climate Change and Adaptation in the Asia – Pacific Region*, 14-16th November 2000, Kobe, Japan, Asia Pacific Network for Global Change Research, 113-117.

Easterling, W.E., 2003: Observed impacts of climate change in agriculture and forestry. *IPCC Workshop on the Detection and Attribution of the Effects of Climate Change*, June 17-19, New York City, USA.

Ebi, K.L., and J.L. Gamble, 2005: Summary of a workshop on the future of health models and scenarios: strategies for the future. *Environ. Health Persp.*, **113**, 335.

ECLAC, 2003: *Handbook for Estimating the Socio-Economic and Environmental Effects*. LC/MEX/G.5. Economic Commission for Latin America and the Caribbean (ECLAC). 357 pp.

EFGC, 2006: *Envisioning the Future of the Gulf Coast*. Final report and findings. D. Reed, Ed. America's Wetland: Campaign to Save Coastal Louisiana, 11 pp.

El-Raey, M., 1997: Vulnerability assessment of the coastal zone of the Nile delta of Egypt, to the impacts of sea-level rise. *Ocean Coast. Manage.*, **37**, 29-40.

Emanuel, K., 2005: Increasing destructiveness of tropical cyclones over the past 30 years. *Nature*, **436**, 686-688.

Ericson, J.P., C.J. Vorosmarty, S.L. Dingman, L.G. Ward and M. Meybeck, 2006: Effective sea-level rise and deltas: causes of change and human dimension implications. *Global Planet Change*, **50**, 63-82.

Essink, G., 2001: Improving fresh groundwater supply - problems and solutions. *Ocean Coast. Manage.*, **44**, 429-449.

Eurosion, 2004: *Living with Coastal Erosion in Europe: Sediment and Space for Sustainability. Part-1 Major Findings and Policy Recommendations of the EUROSION Project*. Guidelines for implementing local information systems dedicated to coastal erosion management. Service contract B4-3301/2001/329175/MAR/B3 "Coastal erosion – Evaluation of the need for action". Directorate General Environment, European Commission, 54 pp.

Evans, E.P., R.M. Ashley, J. Hall, E. Penning-Rowsell, P. Sayers, C. Thorne and A. Watkinson, 2004a: *Foresight; Future Flooding. Scientific Summary.* Volume II: Managing future risks. Office of Science and Technology, London.

Evans, E.P., R.M. Ashley, J. Hall, E. Penning-Rowsell, A. Saul, P. Sayers, C. Thorne and A. Watkinson, 2004b: *Foresight; Future Flooding. Scientific Summary.* Volume I: Future risks and their drivers. Office of Science and Technology, London.

Few, R., and F. Matthies, 2006: Responses to the health risks from flooding. *Flood Hazards and Health: Responding to present and future risks*, R. Few, and F. Matthies, Eds., Earthscan, London.

Few, R., K. Brown and E.L. Tompkins, 2004a: *Scaling Adaptation: Climate Change Response and Coastal Management in the UK.* Working paper 60. Tyndall Centre for Climate Change Research, University of East Anglia, Norwich, Norfolk, 24 pp.

Few, R., M. Ahern, F. Matthies and S. Kovats, 2004b: *Floods, Health and Climate Change: A Strategic Review.* Working Paper 63. Tyndall Centre for Climate Change Research, University of East Anglia, Norwich, Norfolk, 138 pp.

Finkl, C.W., 2002: Long-term analysis of trends in shore protection based on papers appearing in the Journal of Coastal Research, 1984-2000. *J. Coastal Res.*, **18**, 211-224.

Finlayson, C.M., and I. Eliot, 2001: Ecological assessment and monitoring of coastal wetlands in Australia's wet-dry tropics: a paradigm for elsewhere? *Coast. Manage.*, **29**, 105-115.

Flak, L.K., R.W. Jackson and D.N. Stearn, 2002: Community vulnerability assessment tool methodology. *Natural Hazards Review*, **3**, 163-176.

Fletcher, C.A., and T. Spencer, Eds., 2005: *Flooding and Environmental Challenges for Venice and its Lagoon: State of Knowledge.* Cambridge University Press, Cambridge, 691 pp.

Forbes, D.L., 2005: Coastal erosion. *Encyclopedia of the Arctic*, M. Nutall, Ed., Routledge, 391-393.

Forbes, D.L., G.K. Manson, R. Chagnon, S.M. Solomon, J.J. van der Sanden and T.L. Lynds, 2002: Nearshore ice and climate change in the southern Gulf of St. Lawrence. *Ice in the Environment Proceedings of the 16th International Association of Hydraulic Engineering and Research International Symposium on Ice*, Dunedin, New Zealand, 344-351.

Forbes, D.L., G.S. Parkes, G.K. Manson and L.A. Ketch, 2004a: Storms and shoreline erosion in the southern Gulf of St. Lawrence. *Mar. Geol.*, **210**, 169-204.

Forbes, D.L., M. Craymer, G.K. Manson and S.M. Solomon, 2004b: Defining limits of submergence and potential for rapid coastal change in the Canadian Arctic. *Ber. Polar- Meeresforsch.*, **482**, 196-202.

Fox, S., 2003: *When the Weather is uggianaqtuq: Inuit Observations of Environmental Change.* Cooperative Institute for Research in Environmental Sciences. CD-ROM. University of Colorado, Boulder.

Furakawa, K., and S. Baba, 2001: Effects of sea-level rise on Asian mangrove forests. *Proceedings of the APN/SURVAS/LOICZ Joint Conference on Coastal Impacts of Climate Change and Adaptation in the Asia – Pacific Region*, 14-16th November 2000, Kobe, Japan, Asia Pacific Network for Global Change Research, 219-224.

Gardner, T.A., I.M. Côté, J.A. Gill, A. Grant and A.R. Watkinson, 2003: Long-term region-wide declines in Caribbean corals. *Science*, **301**, 958-960.

Gardner, T.A., I.M. Côté, J.A. Gill, A. Grant and A.R. Watkinson , 2005: Hurricanes and Caribbean coral reefs: impacts, recovery patterns, and role in long-term decline. *Ecology*, **86**, 174-184.

Genner, M.J., D.W. Sims, V.J. Wearmouth, E.J. Southall, A.J. Southward, P.A. Henderson and S.J. Hawkins, 2004: Regional climatic warming drives long-term community changes of British marine fish. *P. Roy. Soc. Lond. B Bio.*, **271**, 655-661.

Gibbons, S.J.A., and R.J. Nicholls, 2006: Island abandonment and sea-level rise: An historical analog from the Chesapeake Bay, USA. *Glob. Environ. Chang.*, **16**, 40-47.

Goklany, I.M., 2005: A climate policy for the short and medium term: Stabilization or adaptation? *Energy Environ.*, **16**, 667-680.

Gornitz, V., S. Couch and E.K. Hartig, 2002: Impacts of sea-level rise in the New York City metropolitan area. *Global Planet Change*, **32**, 61-88.

Graham, N.A.J., S.K. Wilson, S. Jennings, N.V.C. Polunin, J.P. Bijoux and J. Robinson, 2006: Dynamic fragility of oceanic coral reef ecosystems. *Proc. Natl. Acad. Sci. U. S. A.*, **103**, 8425-8429.

Graumann, A.,T. Houston, J. Lawrimore, D. Levinson, N. Lott, S. McCown, S. Stephens and D.Wuertz, 2005: Hurricane Katrina - a climatological perpective. October 2005, updated August 2006. Technical Report 2005-01. 28 p. NOAA National Climate Data Center, available at: http://www.ncdc.noaa.gov /oa/reports/tech-report-200501z.pdf

Green, C., and E. Penning-Rowsell, 1999: Inherent conflicts at the coast. *J. Coast. Conserv.*, **5**, 153-162.

Green, D.R., and S.D. King, 2002: *Coastal and Marine Geo-Information Systems: Applying the Technology to the Environment.* Kluwer, 596 pp.

Guinotte, J.M., R.W. Buddemeier and J.A. Kleypas, 2003: Future coral reef habitat marginality: temporal and spatial effects of climate change in the Pacific basin. *Coral Reefs*, **22**, 551-558.

Hale, S.S., A.H. Miglarese, M.P. Bradley, T.J. Belton, L.D. Cooper, M.T. Frame, C.A. Friel, L.M. Harwell and Co-authors, 2003: Managing troubled data: Coastal data partnerships smooth data integration. *Environ. Monit. Assess.*, **81**, 133-148.

Hall, J.W., R.. Dawson, S. Lavery, R. Nicholls, J. Wicks and D. Parker, 2005: Tidal Flood Risk in London Under Stabilisation Scenarios, *Symposium on Avoiding Dangerous Climate Change*, 1-3 February, Exeter. Poster session paper, extended abstract, 4 pp. [Accessed 12.04.07: *http://www.stabilisation2005.com/posters/Hall_Jim.pdf*]

Hall, J.W., I.C. Meadowcroft, E.M. Lee and P.H.A.J.M. van Gelder, 2002: Stochastic simulation of episodic soft coastal cliff recession. *Coast. Eng.*, **46**, 159-174.

Hall, J.W., P.B. Sayers, M.J.A. Walkden and M. Panzeri, 2006: Impacts of climate change on coastal flood risk in England and Wales: 2030-2100. *Philos. T. Roy. Soc. A*, **364**, 1027-1050.

Hall, M.J., 2003: Global warming and the demand for water. *J. Chart. Inst. Water E.*, **17**, 157-161.

Hall, S.J., 2002: The continental shelf benthic ecosystem: current status, agents for change and future prospects. *Environ. Conserv.*, **29**, 350-374.

Hallock, P., 2005: Global change and modern coral reefs: new opportunities to understand shallow-water carbonate depositional processes. *Sediment. Geol.*, **175**, 19-33.

Hamilton, J.M., D.R. Maddison and R.S.J. Tol, 2005: Climate change and international tourism: a simulation study. *Glob. Environ. Chang.*, **15**, 253-266.

Hamm, L., and M.J.F. Stive, Eds., 2002: Shore nourishment in Europe. *Coast. Eng.*, **47**, 79-263.

Hansom, J.D., 2001: Coastal sensitivity to environmental change: a view from the beach. *Catena*, **42**, 291-305.

Hanson, H., A. Brampton, M. Capobianco, H.H. Dette, L. Hamm, C. Laustrup, A. Lechuga and R. Spanhoff, 2002: Beach nourishment projects, practices, and objectives - a European overview. *Coast. Eng.*, **47**, 81-111.

Harley, C.D.G., A.R. Hughes, K.M. Hultgren, B.G. Miner, C.J.B. Sorte, C.S. Thornber, L.F. Rodriguez, L. Tomanek and S.L. Williams, 2006: The impacts of climate change in coastal marine systems. *Ecol. Lett.*, **9**, 228-241.

Harvey, N., 2006a: Rates and impacts of global sea level change. *New Frontiers in Environmental Research*, M.P. Glazer, Ed., Nova Science Publishers, Hauppage, New York

Harvey, N., Ed., 2006b: *Global Change and Integrated Coastal Management. The Asia-Pacific Region.* Coastal Systems and Continental Margins. 10. Springer, New York. 340 pp.

Hawkins, S.J., A.J. Southward and M.J. Genner, 2003: Detection of environmental change in a marine ecosystem - evidence from the western English Channel. *Sci. Total Environ.*, **310**, 245-256.

Hay, J.E., and N. Mimura, 2005: Sea-level rise: Implications for water resources management. *Mitigation and Adaptation Strategies for Global Change*, **10**, 717-737.

Hay, J.E., and N. Mimura , 2006: Supporting climate change vulnerability and adaptation assessments in the Asia-Pacific Region – An example of sustainability science. *Sustainability Science*, **1**, DOI 10.1007/s11625-11006-10011-11628.

Hayne, M., and J. Chappell, 2001: Cyclone frequency during the last 5000 years at Curacao Island, north Queensland, Australia. *Palaeogeogr., Palaeoclimatol., Palaeoecol.*, **168**, 207-219.

Heinz Center, 2000: *The Hidden Costs of Coastal Hazards: Implications for Risk Assessment and Mitigation.* A multisector collaborative project of the H. John Heinz Center for Science, Economics, and the Environment. Island Press, 220 pp.

Helmer, M., and D. Hilhorst, 2006: Natural disasters and climate change. *Disasters*, **30**, 1-4.

Helmuth, B., N. Mieszkowska, P. Moore and S.J. Hawkins, 2006: Living on the edge of two changing worlds: Forecasting the responses of rocky intertidal ecosystems to climate change. *Annual Review of Ecology Evolution and Systematics*, **37**, 373-404.

Hoegh-Guldberg, O., 1999: Climate change, coral bleaching and the future of the world's coral reefs. *Mar. Freshwater Res.*., **50**, 839-866.

Hoegh-Guldberg, O., 2004: Coral reefs in a century of rapid environmental change.

Symbiosis, **37**, 1-31.

Hoegh-Guldberg, O., 2005: Low coral cover in a high-CO_2 world. *J. Geophys. Res., [Oceans]*, **110**, C09S06.

Hoegh-Guldberg, O., H. Hoegh-Guldberg, H. Cesar and A. Timmerman, 2000: *Pacific in peril: biological, economic and social impacts of climate change on Pacific coral reefs*. Greenpeace, 72 pp.

Hosking, A., and R. McInnes, 2002: Preparing for the impacts of climate change on the Central Southeast of England: A framework for future risk management. *J. Coastal Res.*, **S36**, 381-389.

Hughes, L., 2003: Climatic change and Australia: Trends, projections and impacts. *Austral. Ecol.*, **28**, 423-443.

Hughes, R.J., 2004: Climate change and loss of saltmarshes: Consequences for birds. *Ibis*, **146**, 21-28.

Hughes, T.P., D.R. Bellwood, C. Folke, R.S. Steneck and J. Wilson, 2005: New paradigms for supporting the resilience of marine ecosystems. *Trends Ecol. Evol.*, **20**, 380-386.

Hughes, T.P., A.H. Baird, D.R. Bellwood, M. Card, S.R. Connolly, C. Folke, R. Grosberg, O. Hoegh-Guldberg and Co-authors, 2003: Climate change, human impacts, and the resilience of coral reefs. *Science*, **301**, 929-933.

Hulme, M., G.J. Jenkins, X. Lu, J.R. Turnpenny, T.D. Mitchell, R.G. Jones, J. Lowe, J.M. Murphy and Co-authors, 2002: *Climate Change Scenarios for the United Kingdom: The UKCIP02 Scientific Report*. Tyndall Centre for Climate Change Research, University of East Anglia, Norwich, Norfolk, 120 pp.

Hunt, J.C.R., 2002: Floods in a changing climate: a review. *Philos. T. Roy. Soc. A*, **360**, 1531-1543.

Hunter, P.R., 2003: Climate change and waterborne and vectorborne disease. *J. Appl. Microbiol.*, **94**, 37-46.

Hutchinson, J.N., 1998: A small-scale field check on the Fisher-Lehmann and Bakker-Le Heux cliff degradation models. *Earth Surf. Proc. Land.*, **23**, 913-926.

Ingram, I., 2005: Adaptation behaviour before to the 2004/05 tropical cyclone season in the Cook Islands. Island Climate Update 58. National Climate Center. [Accessed 22.01.07: http://www.niwascience.co.nz/ncc/icu/2005-07/article]

IPCC, 2001: *Climate Change 2001: Impacts, Adaptation and Vulnerability. Contribution of Working Group II to the Third Assessment Report of the Intergovernmental Panel on Climate Change*, J.J. McCarthy, O.F. Canziani, N.A. Leary, D.J. Dokken and K.S. White, Eds., Cambridge University Press, Cambridge, 1032 pp.

IPCC CZMS, 1990: *Strategies for Adaptation to Sea-level rise*. Report of the Coastal Zone Management Subgroup, Response Strategies Working Group of the Intergovernmental Panel on Climate Change. Ministry of Transport, Public Works and Water Management (the Netherlands), x + 122 pp.

ISDR, 2004: *Living With Risk: a Global Review of Disaster Reduction Initiatives*. International Strategy for Disaster Reduction (ISDR).

Isobe, M., 2001: A Theory of Integrated Coastal Zone Management in Japan. Department of Civil Engineering, University of Tokyo, Tokyo, 17 pp.

IUCN, 2003: *Indus Delta, Pakistan: Economic Costs of Reduction in Freshwater Flows*. Case studies in Wetland Valuation 5. The World Conservation Union, Pakistan Country Office, Karachi, 6 pp.

Jaagus, J., 2006: Trends in sea ice conditions in the Baltic Sea near the Estonian coast during the period 1949/1950-2003/2004 and their relationships to large-scale atmospheric circulation. *Boreal Environ. Res.*, **11**, 169-183.

Jackson, N.L., K.F. Nordstrom, I. Eliot and G. Masselink, 2002: 'Low energy' sandy beaches in marine and estuarine environments: a review. *Geomorphology*, **48**, 147-162.

Jenks, G., J. Dah and D. Bergin, 2005: Changing paradigms in coastal protection ideology: The role of dune management. *Presentation to NZ Coastal Society Annual Conference*, 12-14 October, Tutukaka, Northland New Zealand, 27pp.

Jennerjahn, T.C., and V. Ittekkot, 2002: Relevance of mangroves for the production and deposition of organic matter along tropical continental margins. *Naturwissenschaften*, **89**, 23-30.

Jevrejeva, S., V.V. Drabkin, J. Kostjukov, A.A. Lebedev, M. Lepparanta, Y.U. Mironov, N. Schmelzer and M. Sztobryn, 2004: Baltic Sea ice seasons in the twentieth century. *Climate Res.*, **25**, 217-227.

Jiongxin, X., 2003: Sediment flux to the sea as influenced by changing human activities and precipitation: example of the Yellow River, China. *Environ. Manage*, **31**, 328-341.

Johannessen, O.M., L. Bengtsson, M.W. Miles, S.I. Kuzmina, V.A. Semenov, G.V. Aleekseev, A.P. Nagurnyi, V.F. Zakharov and Co-authors, 2002: *Arctic climate change - Observed and modeled temperature and sea ice variability*. Technical Report 218. Nansen Environmental and Remote Sensing Centre, Bergen.

Jonkman, S.N., M.J. Stive and J.K. Vrijling, 2005: New Orleans is a lesson for the Dutch. *J. Coastal Res.*, **21**, xi-xii.

Justic, D., N.N. Rabalais and R.E. Turner, 2005: Coupling between climate variability and coastal eutrophication: evidence and outlook for the northern Gulf of Mexico. *J. Sea Res.*, **54**, 25-35.

Kay, R., and J. Adler, 2005: *Coastal Planning and Management*. 2nd edition. Routeledge. 380 pp.

Kendall, M.A., M.T. Burrows, A.J. Southward and S.J. Hawkins, 2004: Predicting the effects of marine climate change on the invertebrate prey of the birds of rocky shores. *Ibis*, **146**, 40-47.

Kennish, M.J., 2002: Environmental threats and environmental future of estuaries. *Environ. Conserv.*, **29**, 78-107.

Khan, M., 2001: *National Climate Change Adaptation Policy and Implementation Plan for Guyana*. Caribbean: Planning for Adaptation to Global Climate Change, CPACC Component 4. National Ozone Action Unit of Guyana/Hydrometeorological Service, 18 Brickdam, Georgetown, Guyana, 74 pp.

Khiyami, H.A., Z. Şen, S.C. Al-Harthy, F.A. Al-Ammawi, A.B. Al-Balkhi, M.I. Al-Zahrani and H.M. Al-Hawsawy, 2005: *Flood Hazard Evaluation in Wadi Hali and Wadi Yibah*. Technical Report. Saudi Geological Survey, Jeddah, Saudi Arabia, 138 pp.

King, D.A., 2004: Climate change science: adapt, mitigate, or ignore? *Science*, **303**, 176-177.

Klein, R.J.T., and R.S.J. Tol, 1997: *Adaptation to Climate Change: Options and Technologies - An Overview Paper*. Technical Paper FCCC/TP/1997/3. UNFCC Secretariat. 37 pp.

Klein, R.J.T., R.J. Nicholls, S. Ragoonaden, M. Capobianco, J. Aston and E.N. Buckley, 2001: Technological options for adaptation to climate change in coastal zones. *J. Coastal Res.*, **17**, 531-543.

Klein, R.J.T., R.J. Nicholls and F. Thomalla, 2002: The resilience of coastal megacities to weather-related hazards: A review. *Proceedings of The Future of Disaster Risk: Building Safer Cities*, A. Kreimer, M. Arnold and A. Carlin, Eds., World Bank Group, Washington, District of Columbia, 111-137.

Klein, R.J.T., R.J. Nicholls and F. Thomalla , 2003: Resilience to natural hazards: How useful is this concept? *Environmental Hazards*, **5**, 35-45.

Kleypas, J.A., and C. Langdon, 2002: Overview of CO_2-induced changes in seawater chemistry. *Proceedings of the 9th International Coral Reef Symposium*, 23-27th October 2000, Bali, Indonesia, 1085-1089.

Knogge, T., M. Schirmer and B. Schuchardt, 2004: Landscape-scale socio-economics of sea-level rise. *Ibis*, **146**, 11-17.

Kobayashi, H., 2004: Impact evaluation of sea-level rise on Indonesian coastal cities - Micro approach through field survey and macro approach through satellite image analysis. *Journal of Global Environment Engineering*, **10**, 77-91.

Kokot, R.R., J.O. Codignotto and M. Elisondo, 2004: Vulnerabilidad al ascenso del nivel del mar en la costa de la provincia de Río Negro. *Asociación Geológica Argentina Rev*, **59**, 477-487.

Komar, P.D., 1998: *Beach and Nearshore Sedimentation*. Second Edition. Prentice Hall.

Kont, A., J. Jaagus, R. Aunap, U. Ratasa and R. Rivis, 2007: Implications of sea-level rise for Estonia. *J. Coastal Res.*, in press.

Koreysha, M.M., F.M. Rivkin and N.V. Ivanova, 2002: The classification of Russian Arctic coasts for their engineering protection. *Extreme Phenomena in Cryosphere: Basic and Applied Aspects*, Russian Academy of Sciences, 65-66 (in Russian).

Kovats, R.S., and A. Haines, 2005: Global climate change and health: recent findings and future steps. *Can. Med. Assoc. J.*, **172**, 501-502.

Kovats, R.S., M.J. Bouma, S. Hajat, E. Worrall and A. Haines, 2003: El Niño and health. *Lancet*, **362**, 1481-1489.

Kovats, R.S., D. Campbell-Lendrum and F. Matthies, 2005: Climate change and human health: estimating avoidable deaths and disease. *Risk Anal.*, **25**, 1409-1418.

Krauss, K.W., J.L. Chambers, J.A. Allen, D.M. Soileau Jr. and A.S. DeBosier, 2000: Growth and nutrition of baldcypress families planted under varying salinity regimes in Louisiana, USA. *J. Coastal Res.*, **16**, 153-163.

Kremer, H.H., M.D.A. Le Tisser, P.R. Burbridge, L. Talaue-McManus, N.N. Rabalais, J. Parslow, C.J. Crossland and W. Young, 2004: *Land-Ocean Interactions in the Coastal Zone: Science Plan and Implementation Strategy*. IGBP Report 51/IHDP Report 18. IGBP Secretariat, 60 pp.

Lavery, S., and B. Donovan, 2005: Flood risk management in the Thames Estuary:

looking ahead 100 years. *Philos. T. Roy. Soc. A*, **363**, 1455-1474.

Leatherman, S.P., 2001: Social and economic costs of sea-level rise. *Sea-level rise, History and Consequences*, B.C. Douglas, M.S. Kearney, and S.P. Leatherman, Eds., Academic Press, 181-223.

LeClerq, N., J.-P. Gattuso, and J. Jaubert, 2002: Primary production, respiration, and calcification of a coral reef mesocosm under increased CO_2 pressure. *Limnol. Oceanogr.*, **47**, 558-564.

Ledoux, L., and R.K. Turner, 2002: Valuing ocean and coastal resources: a review of practical examples and issues for further action. *Ocean Coast. Manage.*, **45**, 583-616.

Ledoux, L., R.K. Turner, L. Mathieu, and S. Crooks, 2001: Valuing ocean and coastal resources: Practical examples and issues for further action. *Paper presented at the Global Conference on Oceans and Coasts at Rio+10: Assessing Progress, Addressing Continuing and New Challenges*, Paris, UNESCO, 37.

Leont'yev, I.O., 2003: Modelling erosion of sedimentary coasts in the Western Russian Arctic. *Coast. Eng.*, **47**, 413-429.

Lesser, M.P., 2004: Experimental biology of coral reef ecosystems. *J. Exp. Mar. Biol. Ecol.*, **300**, 217-252.

Lestak, L.R., W.F. Manley, and J.A. Maslanik, 2004: Photogrammetric analysis of coastal erosion along the Chukchi coast at Barrow, Alaska. *Ber. Polar- Meeresforsch.*, **482**, 38-40.

Li, C.X., D.D. Fan, B. Deng, and V. Korotaev, 2004: The coasts of China and issues of sea-level rise. *J. Coastal Res.*, **43**, 36-47.

Lipp, E.K., A. Huq, and R.R. Colwell, 2004: Health climate and infectious disease: a global perspective. *Clin. Microbiol. Rev.*, **15**, 757-770.

Lise, W., and R.S.J. Tol, 2002: Impact of climate on tourist demand. *Climatic Change*, **55**, 429-449.

Little, A.F., M.J.H. van Oppen and B. L. Willis, 2004: Flexibility in algal endosymbioses shapes growth in reef corals. *Science*, **304**, 1492-1494.

Liu, Y., 2002: A strategy of ground-water distribution exploitation to mitigate the magnitude of subsidence. *Abstracts from the Denver Annual Meeting*, Denver, The Geological Society of America, Paper no. 116-8. [Accessed 13.04.07: http://gsa.confex.com/gsa/2002AM/finalprogram/abstract_40369.htm]

Lomas, M.W., P.M. Glibert, F. Shiah and E.M. Smith, 2002: Microbial process and temperature in Chesapeake Bay: Current relationships and potential impacts of regional warming. *Glob. Change Biol.*, **8**, 51-70.

Lotze, H.K., H.S. Lenihan, B.J. Bourque, R.H. Bradbury, R.G. Cooke, M.C. Kay, S.M. Kidwell, M.X. Kirby, C.H. Peterson and J.B.C. Jackson, 2006: Depletion, degradation and recovery potential of estuaries and coastal seas. *Science*, **312**, 1806-1809.

Lough, J.M., 2000: 1997-98: unprecedented thermal stress to coral reefs? *Geophys. Res. Lett.*, **27**, 3901-3904.

Lough, J.M., and D.J. Barnes, 2000: Environmental controls on growth of the massive coral porites. *J. Exp. Mar. Biol. Ecol.*, **245**, 225-243.

Lowe, J.A., and J.M. Gregory, 2005: The effects of climate change on storm surges around the United Kingdom. *Philos. T. Roy. Soc. A*, **363**, 1313-1328.

Luers, A.L., and S.C. Moser, 2006: Preparing for the impacts of climate change in California: Opportunities and constraints for adaptation. Report CEC-500-2005-198-SF. California Climate Change Center. 47 pp.

Mackenzie, F.T., A. Lerman and L.M. Ver, 2001: Recent past and future of the global carbon cycle. *Geological Perspectives of Global Climate Change*, L.C. Gerhard, W.E. Harrison and M.M. Hanson, Eds., American Association of Petroleum Geologists (AAPG), 51-82.

Madisson, D., 2001: In search of warmer climates? The impact of climate change on flows of British tourists. *Climatic Change*, **49**, 193-208.

Manson, G.K., S.M. Solomon, D.L. Forbes, D.E. Atkinson and M. Craymer, 2006: Spatial variability of factors influencing coastal change in the western Canadian Arctic. *Geo-Mar. Lett.*, **25**, 138-145.

Maracchi, G., O. Sirotenko and M. Bindi, 2005: Impacts of present and future climate variability on agriculture and forestry in the temperate regions: Europe. *Climatic Change*, **70**, 117-135.

McFadden, L., T. Spencer and R.J. Nicholls, 2007a: Broad-scale modelling of coastal wetlands: What is required? *Hydrobiologia*, **577**, 5-15.

McFadden, L., R.J. Nicholls and E. Penning-Rowsell, Eds., 2007b: *Managing Coastal Vulnerability*, Elsevier, Oxford, 282pp.

McInnes, K.L., K.J.E. Walsh, G.D. Hubbert and T. Beer, 2003: Impact of sea-level rise and storm surges on a coastal community. *Nat. Hazards*, **30**, 187-207.

McLaughlin, J., A. DePaola, C.A. Bopp, K.A. Martinek, N.P. Napolilli, C.G. Allison, S.L. Murray, E.C. Thompson, M.M. Bird and M.D. Middaugh, 2006: Outbreak of *Vibrio parahaemolyticus* gastroenteritis associated with Alaskan oysters. *N. Engl. J. Med.*, **353**, 1463-1470.

McLean, R., A. Tsyban, V. Burkett, J.O. Codignotto, D.L. Forbes, N. Mimura, R.J. Beamish and V. Ittekkot, 2001: Coastal zone and marine ecosystems. *Climate Change 2001: Impacts, Adaptation and Vulnerability. Contribution of Working Group II to the Third Assessment Report of the Intergovernmental Panel on Climate Change*, J.J. McCarthy, O.F. Canziani, N.A. Leary, D.J. Dokken and K.S. White, Eds., Cambridge University Press, Cambridge, 343-380.

McLean, R.F., and J.-S. Shen, 2006: From foreshore to foredune: Foredune development over the last 30 years at Moruya Beach, New South Wales, Australia. *J. Coastal Res.*, **22**, 28-36.

McRobie, A., T. Spencer and H. Gerritsen, 2005: The Big Flood: North Sea storm surge. *Philos. T. Roy. Soc. A*, **363**, 1261-1491.

McWilliams, J.P., I.M. Côté, J.A. Gill, W.J. Sutherland and A.R. Watkinson, 2005: Accelerating impacts of temperature-induced coral bleaching in the Caribbean. *Ecology*, **86**, 2055.

Meehl, G.A., W.M. Washington, W.D. Collins, J.M. Arblaster, A. Hu, L.E. Buja, W.G. Strand and H. Teng, 2005: How much more global warming and sea-level rise? *Science*, **307**, 1769-1772.

Meehl, G.A., T.F. Stocker, W. Collins, P. Friedlingstein, A. Gaye, J. Gregory, A. Kitoh, R. Knutti and Co-authors, 2007: Global climate projections. *Climate Change 2007: The Physical Science Basis. Contribution of Working Group I to the Fourth Assessment Report of the Intergovernmental Panel on Climate Change*, S. Solomon, D. Qin, M. Manning, Z. Chen, M. Marquis, K.B. Averyt, M. Tignor and H.L. Miller, Eds., Cambridge University Press, Cambridge, 747-846.

Melillo, J.M., A.C. Janetos, T.R. Karl, R.C. Corell, E.J. Barron, V. Burkett, T.F. Cecich, K. Jacobs and Co-authors, Eds., 2000: *Climate Change Impacts on the United States: The potential consequences of climate variability and change, Overview*. Cambridge University Press, Cambridge, 154 pp.

MESSINA, cited 2006: Integrating the shoreline into spatial policies. Managing European Shoreline and Sharing Information on Nearshore Areas (MESSINA). 191 pp. [Accessed 13.04.07: http://www.interreg-messina.org/documents/Coastal-Toolkit/Zipped/MESSINA%20-%20Practical%20Guide%20-%20Integrating%20the%20Shoreline.zip]

Michot, T.C., 2000: Comparison of wintering redhead populations in four Gulf of Mexico seagrass beds. *Limnology and Aquatic Birds: Monitoring, Modelling and Management*, F.A. Comin, J.A. Herrera and J. Ramirez, Eds., Universidad Autónoma de Yucatán, Merida, 243-260.

Middleton, B.A., and K.L. McKee, 2001: Degradation of mangrove tissues and implications for peat formation in Belizean island forests. *J. Ecol.*, **89**, 818-828.

Mieszkowska, N., M.A. Kendall, S.J. Hawkins, R. Leaper, P. Williamson, N.J. Hardman-Mountford and A.J. Southward, 2006: Changes in the range of some common rocky shore species in Britain - a response to climate change? *Hydrobiologia*, **555**, 241-251.

Milly, P.C.D., K.A. Dunne and A.V. Vecchia, 2005: Global pattern of trends in streamflow and water availability in a changing climate. *Nature*, **438** 347-350.

Milne, N., P. Christie, R. Oram, R.L. Eisma and A.T. White, 2003: *Integrated Coastal Management Process Sustainability Reference Book*. University of Washington School of Marine Affairs, Silliman University and the Coastal Resource Management Project of the Department of Environment and Natural Resources, Cebu City, 50 pp.

Mimura, N., 2001: Distribution of vulnerability and adaptation in the Asia and Pacific Region. *Global Change and Asia Pacific Coasts. Proceedings of the APN/SURVAS/LOICZ Joint Conference on Coastal Impacts of Climate Change and Adaptation in the Asia – Pacific Region*, 14-16th November 2000, Kobe, Japan, Asia Pacific Network for Global Change Research, 21-25.

Ministry for the Environment, 2001: Climate change impacts on New Zealand. Report prepared by the Ministry for the Environment as part of the New Zealand Climate Change Programme. Ministry for the Environment, 39 pp.

Mitchell, J.F.B., J. Lowe, R.A. Wood and M. Vellinga, 2006: Extreme events due to human-induced climate change. *Philos. T. Roy. Soc. A*, **364**, 2117-2133.

Mohanti, M., 2000: Unprecedented supercyclone in the Orissa Coast of the Bay of Bengal, India. *Cogeoenvironment Newsletter. Commission on Geological Sciences for Environmental Planning of the International Union on Geological Sciences*, **16**, 11-13.

Moore, M.V., M.L. Pace, J.R. Mather, P.S. Murdoch, R.W. Howarth, C.L. Folt, C.Y. Chen, H.F. Hemond, P.A. Flebbe and C.T. Driscoll, 1997: Potential effects of climate change on freshwater ecosystems of the New England/mid-Atlantic region.

Hydrol. Process., **11**, 925-947.

Morris, J.T., P.V. Sundareshwar, C.T. Nietch, B. Kjerfve and D.R. Cahoon, 2002: Responses of coastal wetlands to rising sea level. *Ecology*, **83**, 2869-2877.

Morton, R.A., T.L. Miller and L.J. Moore, 2004: National assessment of shoreline change: Part 1 Historical shoreline changes and associated coastal land loss along the U.S.Gulf Of Mexico. Open File Report 2004-1043. U.S. Geological Survey, 44 pp.

Moser, S., 2000: Community responses to coastal erosion: Implications of potential policy changes to the National Flood Insurance Program. *Evaluation of Erosion Hazards: A Project of The H. John Heinz II Center for Science Economics and the Environment*. Prepared for the Federal Emergency Management Agency. 99 pp.

Moser, S., 2005: Impacts assessments and policy responses to sea-level rise in three U.S. States: An exploration of human dimension uncertainties. *Glob. Environ. Chang.*, **15**, 353-369.

Moser, S., 2006: Climate scenarios and projections: The known, the unknown, and the unknowable as applied to California. *Synthesis Report of a Workshop held at the Aspen Global Change Institute*, Aspen Global Change Institute, Aspen, Colorado, 7 pp.

Muir-Wood, R., M. Drayton, A. Berger, P. Burgess and T. Wright, 2005: Catastrophe loss modelling of storm-surge flood risk in eastern England. *Philos. T. Roy. Soc. A*, **363**, 1407-1422.

Munich Re Group, 2004: *Annual Review Natural Catastrophes in 2003*. Topics Geo. Munich Re Group, Munich, 56 pp.

Myatt, L.B., M.D. Scrimshaw and J.N. Lester, 2003: Public perceptions and attitudes towards a current managed realignment scheme: Brancaster West Marsh, North Norfolk, U.K. *J. Coastal Res.*, **19**, 278-286.

Nakićenović, N., and R. Swart, Eds., 2000: *Emissions Scenarios. Special Report of the Intergovernmental Panel on Climate Change*. Cambridge University Press, Cambridge, 599 pp.

NAST, 2000: *Climate Change Impacts in the United States, Overview*. Report for the U.S. Global Change Research Program. National Assessment Synthesis Team Members (NAST), 154 pp.

Neumann, J.E., G. Yohe, R.J. Nicholls and M. Manion, 2000: *Sea Level Rise and Global Climate Change: A Review of Impacts to U.S. Coasts*. Pew Center on Global Climate Change, Arlington, Virginia43 pp.

Ng, W.S., and R. Mendelsohn, 2005: The impact of sea-level rise on Singapore. *Environ. Dev Econ.*, **10**, 201-215.

Nicholls, R.J., 2004: Coastal flooding and wetland loss in the 21st century: changes under the SRES climate and socio-economic scenarios. *Glob. Environ. Chang.*, **14**, 69-86.

Nicholls, R.J., 2006: Storm surges in coastal areas. *Natural Disaster Hotspots: Case Studies. Disaster Risk Management 6*, M. Arnold, R.S. Chen, U. Deichmann, M. Dilley, A.L. Lerner-Lam, R.E. Pullen and Z. Trohanis, Eds., The World Bank, Washington, District of Columbia, 79-108.

Nicholls, R.J., and R.J.T. Klein, 2005: Climate change and coastal management on Europe's coast. *Managing European Coasts: Past, Present and Future*, J.E. Vermaat, L. Bouwer, K. Turner and W. Salomons, Eds., Springer, Environmental Science Monograph Series. 199-226.

Nicholls, R.J., and J.A. Lowe, 2006: Climate stabilisation and impacts of sea-level rise. *Avoiding Dangerous Climate Change*, H.J. Schellnhuber, W. Cramer, N. Nakićenović, T.M.L. Wigley and G. Yohe, Eds., Cambridge University Press, Cambridge, 195-202.

Nicholls, R.J., and R.S.J. Tol, 2006: Impacts and responses to sea-level rise: a global analysis of the SRES scenarios over the twenty-first century. *Philos. T. Roy. Soc. A*, **364**, 1073-1095.

Nicholls, R.J., S.E. Hanson, J. Lowe, D.A. Vaughan, T. Lenton, A. Ganopolski, R.S.J. Tol and A.T. Vafeidis, 2006: *Metrics for Assessing the Economic Benefits of Climate Change Policies: Sea Level Rise*. Report to the OECD. ENV/EPOC/GSP(2006)3/FINAL. Organisation for Economic Co-operation and Development (OECD). 128 pp.

Nikiforov, S.L., N.N. Dunaev, S.A. Ogorodov and A.B. Artemyev, 2003: Physical geographic characteristics. *The Pechora Sea: Integrated Research*, E.A. Romankevich, A.P. Lisitzin and M.E. Vinogradov, Eds., MOPE, (in Russian).

Nilsson, C., C.A. Reidy, M. Dynesius and C. Revenga, 2005: Fragmentation and flow regulation of the world's large river systems. *Science*, **308**, 405-408.

NOAA, 2007: Billion dollar U.S. weather disasters, 1980-2006. http://www.ncdc.noaa.gov/oa/reports/billionz.html [Accessed 13.04.07]

NOAA Coastal Services Center, 1999: Community Vulnerability Assessment Tool. CD-ROM Available from NOAA Coastal Services Center, Charleston, South Carolina http://www.csc.noaa.gov/products/nchaz/startup.htm. [Accessed 13.04.07]

Nordstrom, K.F., 2000: *Beaches and Dunes of Developed Coasts*. Cambridge University Press, 338 pp.

Noronha, L., 2004: Coastal management policy: observations from an Indian case. *Ocean Coast. Manage.*, **47**, 63-77.

Norris, K., P.W. Atkinson and J.A. Gill, 2004: Climate change and coastal waterbird populations - past declines and future impacts. *Ibis*, **149**, 82-89.

Nott, J., and M. Hayne, 2001: High frequency of 'super-cyclones' along the Great Barrier Reef over the past 5,000 years. *Nature*, **413**, 508-512.

NRC, 1990: *Managing Coastal Erosion*. National Research Council. National Academy Press, Washington, District of Columbia, 204 pp.

NRC , 2004: *River Basins and Coastal Systems Planning within the U.S. Army Corps of Engineers*. National Research Council. National Academy Press, Washington, District of Columbia,167 pp.

NRC , 2006: *Drawing Louisiana's New Map; Addressing Land Loss in Coastal Louisiana*. National Research Council, National Academy Press, Washington, District of Columbia, 204 pp.

Nyong, A., and I. Niang-Diop, 2006: Impacts of climate change in the tropics: the African experience. *Avoiding Dangerous Climate Change*, H.J. Schellnhuber, W. Cramer, N. Nakićenović, T.M.L. Wigley and G. Yohe, Eds., Cambridge University Press, Cambridge, 235-241.

Obura, D.O., 2005: Resilience and climate change: lessons from coral reefs and bleaching in the western Indian Ocean. *Estuar. Coast Shelf Sci.*, **63**, 353-372.

Ogorodov, S.A., 2003: Coastal dynamics in the Pechora Sea under technogenic impact. *Ber. Polar- Meeresforsch*, **443**, 74-80.

Ohno, E., 2001: Economic evaluation of impact of land loss due to sea-level rise in Thailand. *Proceedings of the APN/SURVAS/LOICZ Joint Conference on Coastal Impacts of Climate Change and Adaptation in the Asia – Pacific Region*, 14-16th November 2000, Kobe, Japan, Asia Pacific Network for Global Change Research, 231-235.

Okude, A.S., and I.A. Ademiluyi, 2006: Coastal erosion phenomenon in Nigeria: Causes, control and implications. *World Applied Sciences Journal*, **1**, 44-51.

Olesen, J.E., T.R. Carter, C.H. Díaz-Ambrona, S. Fronzek, T. Heidmann, T. Hickler, T. Holt, M.I. Minguez and Co-authors, 2006: Uncertainties in projected impacts of climate change on European agriculture and terrestrial ecosystems based on scenarios from regional climate models. *Climatic Change*, doi: 10.1007/s10584-006-9216-1.

Olsen, S.B., W. Matuszeski, T.V. Padma and H.J.M. Wickremeratne, 2005: Rebuilding after the tsunami: Getting it right. *Ambio*, **34**, 611-614.

Ong, J.E., 2001: Vulnerability of Malaysia to sea level change. *Proceedings of the APN/SURVAS/LOICZ Joint Conference on Coastal Impacts of Climate Change and Adaptation in the Asia – Pacific Region*, 14-16th November 2000, Kobe, Japan, Asia Pacific Network for Global Change Research, 89-93.

Orford, J.D., S.C. Jennings and D.L. Forbes, 2001: Origin, development, reworking and breakdown of gravel-dominated coastal barriers in Atlantic Canada: future scenarios for the British coast. *Ecology and Geomorphology of Coastal Shingle*, J.R. Packham, R.E. Randall, R.S.K. Barnes and A. Neal, Eds., Westbury Academic and Scientific Publishing, Otley, West Yorkshire.

Orford, J.D., S.C. Jennings and J. Pethick, 2003: Extreme storm effect on gravel-dominated barriers. *Proceedings of the International Conference on Coastal Sediments 2003*, R.A. Davis, Ed., World Scientific Publishing Corporation and East Meets West Productions, Corpus Christi, Texas, CD-ROM.

Otter, H.S., 2000: *Complex Adaptive Land Use Systems: An Interdisciplinary Approach with Agent-based Models*. Unpublished PhD thesis, University of Twente, Enschede, the Netherlands, 245 pp.

Otter, H.S., A. van der Veen and H.J. de Vriend, 2001: Spatial patterns and location behaviour: ABLOOM: An explanation based on agent based modelling. *JASSS*, **4**, 1460-7425.

Paerl, H.W., J.D. Bales, L.W. Ausley, C.P. Buzzelli, L.B. Crowder, L.A. Eby, J.M. Fear, M. Go and Co-authors, 2001: Ecosystem impacts of three sequential hurricanes (Dennis, Floyd, and Irene) on the United States' largest lagoonal estuary, Pamlico Sound, NC. *Proc. Natl. Acad. Sci. U.S.A.*, **98**, 5655-5660.

Pandolfi, J.M., R.H. Bradbury, E. Sala, T.P. Hughes, K.A. Bjorndal, R.G. Cooke, D. McArdle, L. McClenachan and Co-authors, 2003: Global trajectories of the long-term decline of coral reef ecosystems. *Science*, **301**, 955-958.

Parson, E.A., R.W. Corell, E.J. Barron, V. Burkett, A. Janetos, L. Joyce, T.R. Karl, M.C. MacCracken and Co-authors, 2003: Understanding climatic impacts, vulnerabilities, and adaptation in the United States: Building a capacity for assessment. *Climatic Change*, **57**, 9-42.

Pascual, M., M.J. Bouma and A.P. Dobson, 2002: Cholera and climate: revisiting the quantitative evidence. *Microbes Infect*., **4**, 237-246.

Pearson, S., J. Rees, C. Poulton, M. Dickson, M. Walkden, J. Hall, R. Nicholls, M. Mokrech and Co-authors, 2005: *Towards an Integrated Coastal Sediment Dynamics and Shoreline Response Simulator.* Technical report 38. Tyndall Centre for Climate Change Research, University of East Anglia, Norwich, Norfolk, 53 pp.

Peperzak, L., 2005: Future increase in harmful algal blooms in the North Sea due to climate change. *Water Sci. Technol*., **51**, 31-36.

Persson, M., K. Rydell, B. Rankka and E. Uytewaal, cited 2006: Valuing the shoreline: Guideline for socio - economic analyses. Managing European Shoreline and Sharing Information on Nearshore Areas (MESSINA). 72 pp. [Accessed 13.04.07: http://www.interreg-messina.org/documents/CoastalToolkit/Zipped/MESSINA%20-%20Practical%20Guide%20-%20Valuing%20the %20Shoreline.zip]

Pethick, J., 2001: Coastal management and sea-level rise. *Catena*, **42**, 307-322.

Pethick, J 2002: Estuarine and tidal wetland restoration in the United Kingdom: Policy versus practice. *Restor. Ecol*., **10**, 431-437.

Pielke Jr, R.A., C. Landsea, M. Mayfield, J. Laver and R. Pasch, 2005: Hurricanes and global warming. *B. Am. Meteorol. Soc*., **86**, 1571-1575.

Pielke, R.A., and C.W. Landsea, 1998: Normalized hurricane damages in the United States: 1925-95. *Weather Forecast*., **13**, 621-631.

Pierre, G., and P. Lahousse, 2006: The role of groundwater in cliff instability: an example at Cape Blanc-Nez (Pas-de-Calais, France). *Earth Surf. Proc. Land*., **31**, 31-45.

Piersma, T., and A. Lindstrom, 2004: Migrating shorebirds as integrative sentinels of global environmental change. *Ibis*, **146**, 61-69.

Pirazzoli, P.A., H. Regnauld and L. Lemasson, 2004: Changes in storminess and surges in western France during the last century. *Mar. Geol*., **210**, 307-323.

Poumadère, M., C. Mays, G. Pfeifle and A.T. Vafeidis, 2005: *Worst Case Scenario and Stakeholder Group Decision: A 5-6 Meter Sea Level Rise in the Rhone Delta, France*. Working paper FNU-76. Hamburg University and Centre for Marine and Atmospheric Science, Hamburg, 30 pp.

Precht, W.F., and R.B. Aronson, 2004: Climate flickers and range shifts of coral reefs. *Frontiers in Ecology and the Environment*, **2**, 307-314.

Quammen, M.L., and C.P. Onuf, 1993: Laguna Madre - seagrass changes continue decades after salinity reduction. *Estuaries*, **16**, 302-310.

Queensland Government, 2001: *State Coastal Management Plan*. Environmental Protection Agency/Queensland Parks and Wildlife Service, Brisbane, Queensland, 90 pp.

Ragab, R., and C. Prudhomme, 2002: Climate change and water resources management in arid and semi-arid regions: prospective and challenges for the 21st century. *Biosystems Engineering*, **81**, 3-34.

Ranasinghe, R., R. McLoughlin, A.D. Short and G. Symonds, 2004: The Southern Oscillation Index, wave climate, and beach rotation. *Mar. Geol*., **204**, 273-287.

Reed, D.J., 2002: Sea-level rise and coastal marsh sustainability: geological and ecological factors in the Mississippi delta plain. *Geomorphology*, **48**, 233-243.

Regnauld, H., P.A. Pirazzoli, G. Morvan and M. Ruz, 2004: Impact of storms and evolution of the coastline in western France. *Mar. Geol*., **210**, 325-337.

Rehfisch, M.M., G.E. Austin, S.N. Freeman, M.J.S. Armitage and N.H.K. Burton, 2004: The possible impact of climate change on the future distributions and numbers of waders on Britain's non-estuarine coast. *Ibis*, **146**, 70-81.

Republic of Kenya, 2002: *First National Communication of Kenya to the Conference of Parties to the United Nations Framework Convention on Climate Change*. Ministry of Environment and Natural Resources, Nairobi, 155 pp.

Rice, J., 2003: Environmental health indicators. *Ocean Coast. Manage*., **46**, 235-259.

Riegl, B., 2003: Climate change and coral reefs: different effects in two high-latitude areas (Arabian Gulf, South Africa). *Coral Reefs*, **22**, 433-446.

Rochelle-Newall, E., R.J.T. Klein, R.J. Nicholls, K. Barrett, H. Behrendt, T.N.H. Bresser, A. Cieslak, E.F.L.M. de Bruin and Co-authors, 2005: Global change and the European coast – climate change and economic development. *Managing European Coasts: Past, Present and Future*, J.E. Vermaat, L. Ledoux, K. Turner and W. Salomons, Eds., Springer (Environmental Science Monograph Series), New York, New York, 239-254.

Rodolfo, K.S., and F.P. Siringan, 2006: Global sea-level rise is recognised, but flooding from anthropogenic land subsidence is ignored around northern Manila Bay, Philippines. *Disaster Manage*., **30**, 118-139.

Rodriguez, A.B., M.L. Fassell and J.B. Anderson, 2001: Variations in shoreface progradation and ravinement along the Texas coast, Gulf of Mexico. *Sedimentology*, **48**, 837-853.

Rogers, K., N. Saintilan and H. Heinjis, 2005: Mangrove encroachment of salt marsh in Western Port Bay, Victoria: the role of sedimentation, subsidence, and sea-level rise. *Estuaries*, **28**, 551-559.

Rogers, K., K.M. Wilton and N. Saintilan, 2006: Vegetation change and surface elevation dynamics in estuarine wetlands of southeast Australia. *Estuar. Coast Shelf S*., **66**, 559-569.

Ross, M.S., J.F. Meeder, J.P. Sah, P.L. Ruiz and G.J. Telesnicki, 2000: The southeast saline Everglades revisited: 50 years of coastal vegetation change. *J. Veg. Sci*., **11**, 101-112.

Rowan, R., 2004: Coral bleaching - Thermal adaptation in reef coral symbionts. *Nature*, **430**, 742.

Royal Society, 2005: *Ocean Acidification due to Increasing Atmospheric Carbon Dioxide*. The Royal Society, London, 60 pp.

Rybczyk, J.M., and D.R. Cahoon, 2002: Estimating the potential for submergence for two subsiding wetlands in the Mississippi River delta. *Estuaries*, **25**, 985-998.

Saenger, P., 2002: *Mangrove Ecology, Silviculture and Conservation*. Kluwer, 360 pp.

Sahagian, D., 2000: Global physical effects of anthropogenic hydrological alterations: sea level and water redistribution. *Global Planet Change*, **25**, 29-38.

Saintilan, N., and R.J. Williams, 1999: Mangrove transgression into saltmarsh environments in south-east Australia. *Global Ecol. Biogeogr*., **8**, 117-124.

Saito, Y., 2001: Deltas in Southeast and East Asia: their evolution and current problems. *Proceedings of the APN/SURVAS/LOICZ Joint Conference on Coastal Impacts of Climate Change and Adaptation in the Asia – Pacific Region*, 14-16th November 2000, Kobe, Japan, Asia Pacific Network for Global Change Research, 185-191.

Saji, N.H., B.N. Goswami, P.N. Vinayachandran and T. Yamagata, 1999: A dipole mode in the tropical Indian Ocean. *Nature*, **401**, 360-363.

Sánchez-Arcilla, A., J.A. Jiménez and H.I. Valdemoro, 2006: A note on the vulnerability of deltaic coasts. Application to the Ebro delta. *Managing Coastal Vulnerability: An Integrated Approach*, L. McFadden, R.J. Nicholls and E. Penning-Rowsell, Eds., Elsevier Science, Amsterdam, 282 pp.

Sánchez-Arcilla, A., J.A. Jiménez, H.I. Valdemoro and V. Gracia, 2007: Implications of climatic change on Spanish Mediterranean low-lying coasts: The Ebro delta case. *J. Coastal Res*., in press.

Scavia, D., J.C. Field, D.F. Boesch, R. Buddemeier, D.R. Cayan, V. Burkett, M. Fogarty, M. Harwell and Co-authors, 2002: Climate change impacts on U.S. coastal and marine ecosystems. *Estuaries*, **25**, 149-164.

Schmidt-Thomé, P, Ed., 2006: *Sea Level Change Affecting the Spatial Development in the Baltic Sea Region*. Geological Survey of Finland, Special Paper 41, Geological Survey of Finland, Espoo, 154 pp.

Scott, D., G. Wall and G. McBoyle, 2005: Climate change and tourism and recreation in north America: exploring regional risks and opportunities. *Tourism, Recreation and Climate Change*, C.M. Hall and J. Higham, Eds., Channel View, Clevedon, 115-129.

Sheppard, C.R.C., 2003: Predicted recurrences of mass coral mortality in the Indian Ocean. *Nature*, **425**, 294-297.

Sheppard, C.R.C., D.J. Dixon, M. Gourlay, A. Sheppard and R. Payet, 2005: Coral mortality increases wave energy reaching shores protected by reef flats: examples from the Seychelles. *Estuar. Coast Shelf S*., **64**, 223-234.

Sherman, R.E., T.J. Fahey and J.J. Battles, 2000: Small-scale disturbance and regeneration dynamics in a neotropical mangrove forest. *J. Ecol*., **88**, 165-178.

Shinn, E.A., G.W. Smith, J.M. Prospero, P. Betzer, M.L. Hayes, V. Garrison and R.T. Barber, 2000: African dust and the demise of Caribbean coral reefs. *Geophys. Res. Lett*., **27**, 3029-3032.

Short, A.D., and A.C. Trembanis, 2004: Decadal scale patterns in beach oscillation and rotation, Narrabeen Beach, Australia - time series, PCA and wavelet analysis. *J. Coastal Res*., **20**, 523-532.

Short, F.T., and H.A. Neckles, 1999: The effects of global change on seagrasses. *Aquat. Bot*., **63**, 169-196.

Simas, T., J.P. Nunes and J.G. Ferreira, 2001: Effects of global change on coastal salt marshes. *Ecol. Model*., **139**, 1-15.

Singh, O.P., 2001: Cause-effect relationships between sea surface temperature, precipitation and sea level along the Bangladesh cost. *Theor. Appl. Climatol*., **68**, 233-243.

Small, C., and R.J. Nicholls, 2003: A global analysis of human settlement in coastal zones. *J. Coastal Res*., **19**, 584-599.

Smith, H.D., 2002a: The role of the social sciences in capacity building in ocean and coastal management. *Ocean Coast. Manage.*, **45**, 573-582.

Smith, O.P., 2002b: Coastal erosion in Alaska. *Ber. Polar- Meeresforsch.*, **413**, 65-68.

Solomon, S.M., and D.L. Forbes, 1999: Coastal hazards, and associated management issues on South Pacific islands. *Ocean Coast. Manage.*, **42**, 523-554.

Spalding, M., F. Blasco and C. Field, 1997: *World Mangrove Atlas.* The International Society for Mangrove Ecosystems, 178 pp.

Stern, N., 2007: *The Economics of Climate Change: The Stern Review.* Cambridge University Press, Cambridge, 692 pp.

Stive, M.J.F., 2004: How important is global warming for coastal erosion? An editorial comment. *Climatic Change*, **64**, 27-39.

Stive, M.J.F., S.J.C. Aarninkoff, L. Hamm, H. Hanson, M. Larson, K. Wijnberg, R.J. Nicholls and M. Capbianco, 2002: Variability of shore and shoreline evolution. *Coast. Eng.*, **47**, 211-235.

Stolper, D., J.H. List and E.R. Thieler, 2005: Simulating the evolution of coastal morphology and stratigraphy with a new morphological-behaviour model (GEOMBEST). *Mar. Geol.*, **218**, 17-36.

Stone, G.W., and R.A. McBride, 1998: Louisiana barrier islands and their importance in wetland protection: Forecasting shoreline change and subsequent response of wave climate. *J. Coastal Res.*, **14**, 900-916.

Stone, G.W., J.P. Morgan, A. Sheremet and X. Zhang, 2003: *Coastal Land Loss and Wave-surge Predictions During Hurricanes in Coastal Louisiana: Implications for the Oil and Gas Industry.* Coastal Studies Institute, Louisiana State University, Baton Rouge, Louisiana, 67 pp.

Stone, L., A. Huppert, B. Rajagopalan, H. Bhasin and Y. Loya, 1999: Mass coral bleaching: a recent outcome of increased El Niño activity? *Ecol. Lett.*, **2**, 325-330.

Storlazzi, C.D., and G.B. Griggs, 2000: Influence of El Niño-Southern Oscillation (ENSO) events on the evolution of central California's shoreline. *Geol. Soc. Am. Bull.*, **112**, 236-249.

Storms, J.E.A., G.J. Weltje, J.J. van Dijke, C.R. Geel and S.B. Kroonenberg, 2002: Process-response modeling of wave-dominated coastal systems: simulating evolution and stratigraphy on geological timescales. *J. Sediment Res.*, **72**, 226-239.

Sun, G., S.G. McNulty, D.M. Amatya, R.W. Skaggs, L.W. Swift, P. Shepard and H. Riekerk, 2002: A comparison of watershed hydrology of coastal forested wetlands and the mountainous uplands in the Southern US. *J. Hydrol.*, **263**, 92-104.

Syvitski, J.P.M., C.J. Vörösmarty, A.J. Kettner and P. Green, 2005: Impact of humans on the flux of terrestrial sediment to the global coastal ocean. *Science*, **308**, 376-380.

Tam, N.Y.Y., and Y.S. Wong, 2002: Conservation and sustainable exploitation of mangroves in Hong Kong. *Trees*, **16**, 224-229.

Taylor, J.A., A.P. Murdock and N.I. Pontee, 2004: A macroscale analysis of coastal steepening around the coast of England and Wales. *Geogr. J.*, **170**, 179-188.

Thanh, T.D., Y. Saito, D.V. Huy, V.L. Nguyen, T.K.O. Oanh and M. Tateishi, 2004: Regimes of human and climate impacts on coastal changes in Vietnam. *Regional Environmental Change*, **4**, 49-62.

Thieler, E.R., and E.S. Hammar-Klose, 2000: *National Assessment of Coastal Vulnerability to Sea Level Rise: Preliminary Results for the U.S. Pacific Coast.* U.S. Geological Survey Open-File Report 00-178. [Accessed 13.04.07: http://pubs.usgs.gov/of/2000/of00-178/pages/toc.html]

Thorne, C., E. Evans and E. Penning-Rowsell, Eds., 2006: *Future Flooding and Coastal Erosion Risks.* Thomas Telford, London, 350 pp.

Titus, J.G., 2001: Does the U.S. Government realize that the sea is rising? How to restructure Federal programs so that wetland and beaches survive. *Golden Gate University Law Review*, **30**, 717-778.

Titus, J.G., 2004: Maps that depict the business-as-usual response to sea-level rise in the decentralized United States of America. *Global Forum on Sustainable Development: Development and Climate Change. 11-12 November, Paris*, Organisation for Economic Co-operation and Development (OECD). 22 pp.

Tol, R.S.J., 2002: Estimates of the damage costs of climate change. Part II: Dynamic estimates. *Environ. Resour. Econ.*, **21**, 135-160.

Tol, R.S.J., 2007: The double trade-off between adaptation and mitigation for sea-level rise: An application of FUND. *Mitigation and Adaptation Strategies for Global Change*, **12**, 741-753.

Tol, R.S.J., M. Bohn, T.E. Downing, M.L. Guillerminet, E. Hizsnyik, R. Kasperson, K. Lonsdale, C. Mays and Co-authors, 2006: Adaptation to five metres of sea-level rise. *J. Risk Res.*, **9**, 467-482.

Tompkins, E., S. Nicholson-Cole, L. Hurlston, E. Boyd, G. Brooks Hodge, J. Clarke, G. Gray, N. Trotz and L. Varlack, 2005a: *Surviving Climate Change in Small Islands - A guidebook.* Tyndall Centre for Climate Change Research, University of East Anglia, Norwich, Norfolk, 132 pp.

Tompkins, E.L., E. Boyd, S. Nicholson-Cole, K. Weatherhead, N.W. Arnell and W.N. Adger, 2005b: *Linking Adaptation Research and Practice.* Report to DEFRA Climate Change Impacts and Adaptation Cross-Regional Research Programme. Tyndall Centre for Climate Change Research, University of East Anglia, Norwich, Norfolk, 120pp.

Townend, I., and J. Pethick, 2002: Estuarine flooding and managed retreat. *Philos. T. Roy. Soc. A*, **360**, 1477-1495.

Townend, I., and K. Burgess, 2004: Methodology for assessing the impact of climate change upon coastal defence structures. *Proceedings of the 29th International Conference on Coastal Engineering 2004*,19-24 September; Lisbon, Portugal, World Scientific Publishing, 3593-3965.

Trenberth, K.E., P.D. Jones, P.G. Ambenje, R. Bojariu, D.R. Easterling, A.M.G. Klein Tank, D.E. Parker, J.A. Renwick and Co-authors, 2007: Surface and atmospheric climate change. *Climate Change 2007: The Physical Science Basis. Contribution of Working Group I to the Fourth Assessment Report of the Intergovernmental Panel on Climate Change*, S. Solomon, D. Qin, M. Manning, Z. Chen, M. Marquis, K.B. Averyt, M. Tignor and H.L. Miller, Eds., Cambridge University Press, Cambridge, 235-336.

Trenhaile, A.S., 2002: Rock coasts, with particular emphasis on shore platforms. *Geomorphology*, **48**, 7-22.

Trenhaile, A.S., 2004: Modeling the accumulation and dynamics of beaches on shore platforms. *Mar. Geol.*, **206**, 55-72.

Tsimplis, M.N., D.K. Woolf, T.J. Osborn, S. Wakelin, J. Wolf, R.A. Flather, A.G.P. Shaw, P.H. Woodworth, P.G. Challenor, D. Blackman, F. Pert, Z. Yan and S. Jevrejeva, 2005: Towards a vulnerability assessment of the UK and northern European coasts: the role of regional climate variability. *Philos. T. Roy. Soc. A*, **363**, 1329-1358.

Tsimplis, M.N., A.G.P. Shaw, R.A. Flather and D.K. Woolf, 2006: The influence of the North Atlantic Oscillation on the sea level around the northern European coasts reconsidered: the thermosteric effects. *Philos. T. Roy. Soc. A*, **364**, 845-856.

Turley, C., J.C. Blackford, S. Widdicombe, D. Lowe, P.D. Nightingale and A.P. Rees, 2006: Reviewing the impact of increased atmospheric CO_2 on oceanic pH and the marine ecosystem. *Avoiding Dangerous Climate Change*, H.J. Schellnhuber, W. Cramer, N. Nakićenović, T.M.L. Wigley, and G. Yohe, Eds., Cambridge University Press, Cambridge, 65-70.

Turner, R.K., 2001: *Concepts and Methods for Integrated Coastal Management.* Marine Biodiversity and Climate Change. Tyndall Centre for Climate Change Research, University of East Anglia, Norwich, Norfolk, 10 pp.

UN Secretary General, cited 2006a: International co-operation on humanitarian assistance in the field of natural disasters, from relief to development; Strengthening of the coordination of emergency assistance of the United Nations. United Nations General Assembly Sixty-first Session. Sessions A/61. [Accessed 13.04.07: http://www.un.org/ga/61/issues/ha.shtml]

UN Secretary General, cited 2006b: United Nations Millennium Declaration (18th September 2000). United Nations General Assembly Fifty-fifth Session. Agenda item 60(b). 9 pp. [Accessed 13.04.07: http://www.un.org/millennium/declaration/ares552e.pdf]

UNDP, 2004: Reducing disaster risk: A challenge for development, Disaster Reduction Unit, Bureau for Crisis Prevention and Recovery, 161 pp.

UNEP, 2005: *After the Tsunami; Rapid Environmental Assessment.* United Nations Environment Programme (UNEP), 141 pp.

UNFCCC, 2005: *Compendium on Methods and Tools to Evaluate Impacts of, and Vulnerability and Adaptation to, Climate Change.* UNFCC Secretariat, 155 pp.

UNOCHA, 2005: *Regional Workshop on Lessons Learned and Best Practices of Disasters.* United Nations, Economic Commission for Latin America and the Caribbean (ECLAC), 111 pp.

US Government, cited 2006: The Federal response to Hurricane Katrina: Lessons learned. 228 pp. [Accessed 13.04.07: http://www.whitehouse.gov/reports/katrina-lessons-learned.pdf]

Valiela, I., 2006: *Global Coastal Change.* Blackwell, Oxford, 368 pp.

van Aalst, M.K., 2006: The impacts of climate change on the risk of natural disasters. *Disasters*, **30**, 5-18.

van Goor, M.A., M.J. Stive, Z.B. Wang and T.J. Zitman, 2001: Influence of relative sea-level rise on coastal inlets and tidal basins. *Coastal Dynamics 2001. Proceedings of the Fourth Conference on Coastal Dynamics held June 11-15, 2001 in Lund, Sweden*, H. Hanson, Ed., American Society of Civil Engineers (ASCE), Reston, Virginia, 242-251.

van Goor, M.A., T.J. Zitman, Z.B. Wang and M.J. Stive, 2003: Impact of sea-level rise on the morphological equilibrium state of tidal inlets. *Mar. Geol.*, **202**, 211-227.

Vassie, J.M., P.L. Woodworth and M.W. Holt, 2004: An example of North Atlantic deep-ocean swell impacting Ascension and St. Helena Islands in the Central South Atlantic. *J. Atmos. Ocean. Tech.*, **21**, 1095-1103.

Viles, H.A., and A.S. Goudie, 2003: Interannual decadal and multidecadal scale climatic variability and geomorphology. *Earth-Sci. Rev.*, **61**, 105-131.

Visser, L.E., 2004: *Challenging Coasts: Transdisciplinary Excursions into Integrated Coastal Zone Development*. Amsterdam University Press, Amsterdam, 248 pp.

Walkden, M.J., and J.W. Hall, 2005: A predictive mesoscale model of the erosion and profile development of soft rock shores. *Coast. Eng.*, **52**, 535-563.

Walkden, M.J.A., and M.E. Dickson, 2006: *The Response of Soft Rock Shore Profiles to Increased Sea Level Rise*. Working Paper 105. Tyndall Centre for Climate Change Research, University of East Anglia, Norwich, Norfolk, 13 pp.

Walsh, J.P., and C.A. Nittrouer, 2004: Mangrove-bank sedimentation in a mesotidal environment with large sediment supply, Gulf of Papua. *Mar. Geol.*, **208**, 225-248.

Waltham, T., 2002: Sinking cities. *Geology Today*, **18**, 95-100.

Wang, H., and C.F. Fan, 2005: The C14 database (II) on the Circum-Bohai Sea coast. *Quaternary Sciences (in Chinese)*, **25**, 144-155.

Warrick, R., W. Ye, P. Kouwenhoven, J.E. Hay and C. Cheatham, 2005: New developments of the SimCLIM model for simulating adaptation to risks arising from climate variability and change. *MODSIM 2005 International Congress on Modelling and Simulation*, A. Zerger, and R.M. Argent, Eds., Modelling and Simulation Society of Australia and New Zealand, 170-176.

Wassmann, R., X.H. Nguyen, T.H. Chu and P.T. To, 2004: Sea-level rise affecting the Vietnamese Mekong Delta: water elevation in the flood season and implications for rice production. *Climatic Change*, **66**, 89-107.

Webster, P.J., A.M. Moore, J.P. Loschnigg and R.R. Leben, 1999: Coupled ocean-temperature dynamics in the Indian Ocean during 1997-98. *Nature*, **401**, 356-360.

Webster, P.J., G.J. Holland, J.A. Curry and H.-R. Chang, 2005: Changes in tropical cyclone number, duration, and intensity in a warming environment. *Science*, **309**, 1844-1846.

West, M.B., 2003: Improving science applications to coastal management. *Mar. Policy*, **27**, 291-293.

Whelan, K.T., T.J. Smith, D.R. Cahoon, J.C. Lynch and G.H. Anderson, 2005: Groundwater control of mangrove surface elevation: Shrink and swell varies with soil depth. *Estuaries*, **28**, 833-843.

Wigley, T.M.L., 2005: The climate change commitment. *Science*, **307**, 1766-1769.

Wilkinson, C.R., 2002: *Status of Coral Reefs of the World*. Australian Institute of Marine Sciences. 388 pp.

Williams, K., M. MacDonald and L.D.L. Sternberg, 2003: Interactions of storm, drought, and sea-level rise on coastal forest: a case study. *J. Coastal Res.*, **19**, 1116-1121.

Williams, K.L., K.C. Ewel, R.P. Stumpf, F.E. Putz and T.W. Workman, 1999: Sea-level rise and coastal forest retreat on the west coast of Florida. *Ecology*, **80**, 2045-2063.

Williams, M.J., R. Coles, and J.H. Primavera, 2007: A lesson from cyclone Larry: An untold story of the success of good coastal planning. *Estuar. Coast Shelf S.*, **71**, 364-367.

Winn, P.J.S., R.M. Young and A.M.C. Edwards, 2003: Planning for the rising tides: the Humber Estuary Shoreline Management Plan. *Sci. Total Environ.*, **314**, 13-30.

Wisner, B., cited 2006: Monsoon, market, and miscalculation: The political ecology of Mumbai's July 2005 floods. 6 pp. [Accessed 13.04.07: http://www.radixonline.org/resources/wisner-mumbaifloods2005_22-jan2006.doc]

Wolters, M., J.P. Bakker, M.D. Bertness, R.L. Jefferies and I. Möller, 2005: Saltmarsh erosion and restoration in south-east England: squeezing the evidence requires realignment. *J. Appl. Ecol.*, **42**, 844-851.

Woodroffe, C.D., 2003: *Coasts: Form, Process and Evolution*. Cambridge University Press, 623 pp.

Woodroffe, C.D., and R.J. Morrison, 2001: Reef-island accretion and soil development, Makin Island, Kiribati, central Pacific. *Catena*, **44**, 245-261.

Woodroffe, C.D., M. Dickson, B.P. Brooke and D.M. Kennedy, 2005: Episodes of reef growth at Lord Howe Island, the southernmost reef in the southwest Pacific. *Global Planet Change*, **49**, 222-237.

Woodroffe, C.D., R.J. Nicholls, Y. Saito, Z. Chen and S.L. Goodbred, 2006: Landscape variability and the response of Asian megadeltas to environmental change. *Global Change and Integrated Coastal Management: The Asia-Pacific Region*, N. Harvey, Ed., Springer, New York, New York, 277-314.

Woodworth, P.H., J. Gregory and R.J. Nicholls, 2004: Long term sea level changes and their impacts. *The Sea*, vol. 12/13, A. Robinson and K. Brink, Eds., Harvard University Press, Cambridge, Massachusetts.

WTO, 2003: Climate change and tourism. *Proceedings of the 1st International Conference on Climate Change and Tourism*, 9-11th April, Djerba, Tunisia, World Tourism Organization. 55 pp.

WTO, undated: *Long-Term Prospects: Tourism 2020 Vision*. [Accessed 13.04.07: http://www.world-tourism.org/market_research/facts/market_trends.htm]

Yohe, G., 2000: Assessing the role of adaptation in evaluating vulnerability to climate change. *Climatic Change*, **46**, 371-390.

Yohe, G., and R.S.J. Tol, 2002: Indicators for social and economic coping capacity - moving toward a working definition of adaptive capacity. *Glob. Environ. Chang.*, **12**, 25-40.

Zacherl, D., S.D. Gaines and S.I. Lonhart, 2003: The limits to biogeographical distributions: insights from the northward range extension of the marine snail, *Kelletia kelletii* (Forbes, 1852). *J. Biogeogr.*, **30**, 913-924.

Zhang, K.Q., B.C. Douglas and S.P. Leatherman, 2000: Twentieth-century storm activity along the US east coast. *J. Clim.*, **13**, 1748-1761.

Zhang, K.Q., B.C. Douglas and S.P. Leatherman, 2004: Global warming and coastal erosion. *Climatic Change*, **64**, 41-58.

Zimmerman, R.C., D.G. Kohrs, D.L. Steller and R.S. Alberte, 1997: Impacts of CO2 enrichment on productivity and light requirements of eelgrass. *Plant. Physiol.*, **115**, 599-607.

Zweig, R., 1998: *Sustainable Aquaculture: Seizing Opportunities to Meet Global Demand*. Agriculture Technology Notes 22. World Bank, Washington, District of Columbia.

7

Industry, settlement and society

Coordinating Lead Authors:
Tom Wilbanks (USA), Patricia Romero Lankao (Mexico)

Lead Authors:
Manzhu Bao (China), Frans Berkhout (The Netherlands), Sandy Cairncross (UK), Jean-Paul Ceron (France), Manmohan Kapshe (India), Robert Muir-Wood (UK), Ricardo Zapata-Marti (ECLAC/Mexico)

Contributing Authors:
Maureen Agnew (UK), Richard Black (UK), Tom Downing (UK), Stefan Gossling (Sweden), Maria-Carmen Lemos (Brazil), Karen O'Brien (Norway), Christian Pfister (Switzerland), William Solecki (USA), Coleen Vogel (South Africa)

Review Editors:
David Satterthwaite (UK), Y. Dhammika Wanasinghe (Sri Lanka)

This chapter should be cited as:
Wilbanks, T.J., P. Romero Lankao, M. Bao, F. Berkhout, S. Cairncross, J.-P. Ceron, M. Kapshe, R. Muir-Wood and R. Zapata-Marti, 2007: Industry, settlement and society. *Climate Change 2007: Impacts, Adaptation and Vulnerability. Contribution of Working Group II to the Fourth Assessment Report of the Intergovernmental Panel on Climate Change*, M.L. Parry, O.F. Canziani, J.P. Palutikof, P.J. van der Linden and C.E. Hanson, Eds., Cambridge University Press, Cambridge, UK, 357-390.

Table of Contents

Executive summary

Climate-change vulnerabilities of industry, settlement and society are mainly related to extreme weather events rather than to gradual climate change (very high confidence).
The significance of gradual climate change, e.g., increases in the mean temperature, lies mainly in changes in the intensity and frequency of extreme events, although gradual changes can also be associated with thresholds beyond which impacts become significant, such as in the capacities of infrastructures. [7.2, 7.4]

Aside from major extreme events and thresholds, climate change is seldom the main factor in considering stresses on the sustainability of industry, settlements and society (very high confidence).
The significance of climate change (positive or negative) lies in its interactions with other non-climate sources of change and stress, and its impacts should be considered in such a multi-cause context. [7.1.3, 7.2, 7.4]

Vulnerabilities to climate change depend considerably on specific geographic, sectoral and social contexts (very high confidence).
They are not reliably estimated by large-scale (aggregate) modelling and estimation. [7.2, 7.4]

Vulnerabilities of industry, infrastructures, settlements and society to climate change are generally greater in certain high-risk locations, particularly coastal and riverine areas, and areas whose economies are closely linked with climate-sensitive resources, such as agricultural and forest product industries, water demands and tourism; these vulnerabilities tend to be localised but are often large and growing (high confidence).
For example, rapid urbanisation in most low and middle income nations, often in relatively high-risk areas, is placing an increasing proportion of their economies and populations at risk. [7.3, 7.4, 7.5]

Where extreme weather events become more intense and/or more frequent with climate change, the economic and social costs of those events will increase (high confidence).
Experience indicates that costs of major events can range from several percent of annual regional gross domestic product (GDP) and income generation in very large regions with very large economies to more than 25% in smaller areas that are affected by the events. Climate-change impacts spread from directly impacted areas and sectors to other areas and sectors through extensive and complex linkages. [7.4, 7.5]

Poor communities can be especially vulnerable, in particular those concentrated in relatively high-risk areas (high confidence).
They tend to have more limited adaptive capacities, and are more dependent on climate-sensitive resources such as local water and food supplies. [7.2, 7.4, 5.4]

Industry, settlements and society are often capable of considerable adaptation, depending heavily on the competence and capacity of individuals, communities, enterprises and local governments, together with access to financial and other resources (very high confidence).
But that capacity has limits, especially when confronted by climate changes that are relatively extreme or persistent. [7.4.3, 7.6]

Although most adaptations reflect local circumstances, adaptation strategies for industry and settlement and, to a lesser degree, for society, can be supported by linkages with national and global systems that increase potentials and resources for action (very high confidence). [7.6.6]

7.1 Introduction

7.1.1 Key issues

Climate change and sustainable development are linked through their interactions in industries, human settlements and society. Many of the forces shaping carbon *emissions* – such as economic growth, technological transformations, demographic shifts, lifestyles and governance structures – also underlie diverse pathways of development, explaining in part why industrialised countries account for the highest share of carbon emissions. The same drivers are also related to climate-change *impacts*, explaining in part why some regions and sectors, especially from the developing world, are more vulnerable to climate change than others because they lack financial, institutional and infrastructural capacities to cope with the associated stresses (O'Brien and Leichenko, 2003). Settlements and industry are often key focal points for linkages between mitigation and adaptation; for instance, efficient buildings can help in adapting to changing climate by providing protection against warming, while this adaptation may involve increased or decreased energy use and greenhouse gas emissions associated with cooling based on electricity (Hough, 2004); and society is a key to responses based on democratic processes of government.

Industries, settlements and human society are accustomed to variability in environmental conditions, and in many ways they have become resilient to it when it is a part of their normal experience. Environmental changes that are more extreme or persistent than that experience, however, can lead to vulnerabilities, especially if the changes are not foreseen and/or if capacities for adaptation are limited; and the IPCC Third Assessment Report (IPCC, 2001) reported that climate change would increase the magnitude and frequency of weather extremes.

The central issues for industry, settlement and society are whether climate-change impacts are likely to require responses that go beyond normal adaptations to varying conditions, if so, for whom, and under what conditions responses are likely to be sufficient to avoid serious effects on people and the sustainability of their ways of life. Recent experiences such as Hurricane Katrina suggest that these issues are salient for developed as well as developing countries (Figure 7.1).

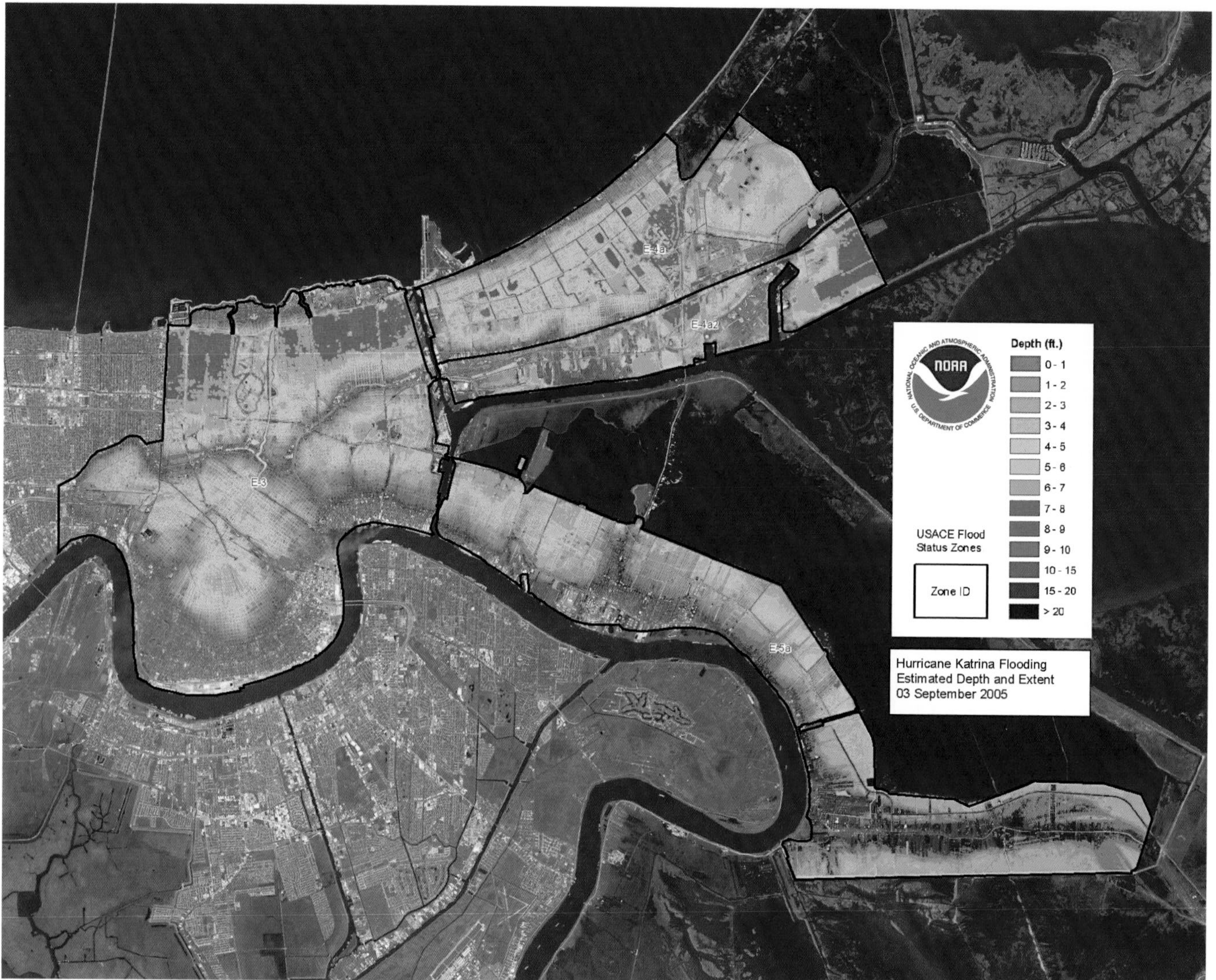

Figure 7.1. *Flood depths in New Orleans, USA, on 3 September, 2005, five days after flooding from Hurricane Katrina, in feet (0.3 m) (Source: www.katrina.noaa.gov/maps/maps.html).*

Scale matters in at least three ways in assessing the impacts of climate change on industry, settlement and society. First, climate change is one of a set of multiple stresses operating at diverse scales in space and through time. Second, both the exposure to climate change and the distribution of climate-sensitive settlements and industrial sectors vary greatly across geographic scale. The primary social and economic conditions that influence adaptive capacity also differ with scale, such as access to financial resources. One could say, for instance, that at a national scale industrialised countries such as the UK and Norway can cope with most kinds of gradual climate change, but focusing on more localised differences can show considerable variability in stresses and capacities to adapt (Environment Canada, 1997; Kates and Wilbanks, 2003; London Climate Change Partnership, 2004; O'Brien et al., 2004; Kirshen et al., 2006). Third, temporal scale is a critical determinant of the capacity of human systems to adapt to climate change; for instance, rapid changes are usually more difficult to absorb without painful costs than gradual change (Section 7.4; Chapter 17).

7.1.2 Scope of the chapter

Guidance for the preparation of the IPCC Fourth Assessment Report requested particular attention by this chapter to five systems of interest: industry, services, utilities/infrastructure, human settlement and social issues. Chapter 5 of the report deals with impacts and adaptation on the food, fibre and forest products sectors, and Chapters 9 to 16 deal with impacts and adaptation in global regions.

Chapter 7's topic of 'industry, settlement and society' is clearly very broad; and many of the components of the chapter, such as industry and services, settlements, financial and social issues, are so heterogeneous that each could be the subject of a separate chapter. Very briefly, however, the chapter will summarise and assess the literature relevant to the impacts of climate change on the structure, functioning, and relationships of all of these components of human systems potentially affected by climate change, positively or negatively.

The chapter (1) identifies current and potential vulnerabilities and positive or negative impacts of climate change on industrial, service and infrastructure sectors, human settlements and human societies; (2) assesses the current knowledge about the costs of possible impacts; and (3) considers possible adaptive responses. In general, it emphasises that climate-change impacts, adaptation potentials, and vulnerabilities are context-specific, related to the characteristics and development pathways of the location or sector involved.

7.1.3 Human systems in context

Human systems include social, economic and institutional structures and processes. Related to industry, settlement and society, these systems are diverse and dynamic, expressed at the individual level through livelihoods. They tend to revolve around such aims of humanity as survival, security, well-being, equity and progress; and in these regards weather and climate are often of secondary importance as sources of benefits or stresses. More important are such issues as access to financial resources, institutional capacities and potentials for conflict (Ocampo and Martin, 2003; Thomas and Twyman, 2005) and such stresses as rapid urbanisation, disease and terrorism. It is in its complex interactions with these kinds of social contexts that climate change can make a difference, easing or aggravating multiple stresses and in some cases potentially pushing a multi-stressed human system across a threshold of sustainability (Wilbanks, 2003b).

In most cases, climate (and thus climate change) affects human systems in three principal ways. First, it provides a context for climate-sensitive human activities ranging from agriculture to tourism. For instance, rivers fed by rainfall enable irrigation and transportation and can enrich or damage landscapes. Second, climate affects the cost of maintaining climate-controlled internal environments for human life and activity; clearly, higher temperatures increase costs of cooling and reduce costs of heating. Third, climate interacts with other types of stresses on human systems, in some cases reducing stresses but in other cases exacerbating them. For example, drought can contribute to rural-urban migration, which, combined with population growth, increases stress on urban infrastructures and socio-economic conditions. In all of these connections, effects can be positive as well as negative; but extreme climate events and other abrupt changes tend to affect human systems more severely than gradual change, because they offer less time for adaptation, although gradual changes may also reach thresholds at which effects are notable.

7.1.4 Conclusions of the IPCC Third Assessment Report

The Third Assessment Report of IPCC Working Group II (TAR) included a chapter on Human Settlements, Energy, and Industry (Scott et al., 2001) and also a separate chapter on Insurance and Other Financial Services (Vellinga et al., 2001). Together, these two chapters in TAR correspond to a part of this one chapter in the Fourth Assessment Report; a substantial part of this chapter is devoted to subject matter not directly addressed in previous IPCC reports (e.g., services, infrastructures and social issues).

The first of the TAR chapters (Chapter 7) was largely devoted to impact issues for human settlements, concluding that settlements are vulnerable to effects of climate change in three major ways: through economic sectors affected by changes in input resource productivity or market demands for goods and services, through impacts on certain physical infrastructures, and through impacts of weather and extreme events on the health of populations. It also concluded that vulnerability tends to be a function mainly of three factors: location (coastal and riverine areas at most risk), economy (those dependent on weather-related sectors at most risk), and size (larger settlements at greater aggregate risk but having more resources for impact prevention and adaptation). The most direct risks are from flooding and landslides due to increases in rainfall intensity and from sea-level rise and storm surges in coastal areas. Although some areas are at particular risk, urban flooding could be a problem in any settlement where drainage infrastructures are inadequate, especially where informal settlement areas lack urban services and adaptive capacities. Rapid urbanisation in relatively high-risk areas is a special concern, because it concentrates people and assets and is generally increasing global and regional vulnerability to climate-change impacts. Other dimensions of vulnerability include general regional vulnerabilities to impacts (e.g., in polar regions), lack of economic diversification and fragile urban infrastructures.

Possible impacts of climate change on financial institutions and risk financing were the focus of a separate chapter (Chapter 8) in the TAR. This chapter concluded that climate change is likely to raise the actuarial uncertainty in catastrophe risk assessment, placing upward pressure on insurance premiums and possibly leading to reductions in risk coverage. It identified a significant rise in the costs of losses from meteorological disasters since the early 1980s which, as has been confirmed by the AR4 (see Chapter 1), appeared to reflect an increase in catastrophe occurrence over and above the rise in values, exposures, and vulnerabilities.

7.2 Current sensitivity/vulnerability

A frequent objective of human societies is to reduce their sensitivity to weather and climate, for example, by controlling the climate in buildings within which people live, shop and work or by controlling the channels and flows of rivers or the configurations of sea coasts. Recent experience with weather variability, however, reminds us that - at least at feasible levels of investment and technological development - human control over climate-related aspects of nature can be limited (see Box 7.4).

In fact, sensitivities of human systems to climate and climate change abound:

1. Environmental quality is a case in point, where weather and climate can affect air and water pollution and, in cases of extreme events, exposures to wastes that are hazardous to health. Consider the interaction between the ambient air

temperature of an urban area and its concentration of ozone, which can have adverse health implications (Hogrefe et al., 2004; Section 7.4.2.4; Chapter 8), or effects of hurricane flooding on exposures to health threats (Marris, 2005).

2. Linkage systems, such as transportation and transmission systems for industry and settlements (e.g., water, food supply, energy, information systems and waste disposal), are important in delivering ecosystem and other services to support human well-being, and can be subject to climate-related extreme events such as floods, landslides, fire and severe storms. Such exposed infrastructures as bridges and electricity transmission networks are especially vulnerable, as in the experience of Hurricane Georges in 1998, which threatened port and oil storage facilities in the Dominican Republic (REC, 2004), or the 2005 experience with Hurricane Katrina (Box 7.4; Section 7.4.2.3).
3. Other physical infrastructures can be affected by weather and climate as well. For example, the rate of deterioration of external shells of building structures is weather-related, depending on the materials used, and buildings are affected by water-logging related to precipitation patterns. Another kind of impact is on demands for physical infrastructures; for instance, demands for water supplies and energy supplies related to temperature.
4. Social systems are also vulnerable, especially to extreme events (e.g., Box 7.1). Storms and floods can damage homes and other shelters and disrupt social networks and means to sustain livelihoods; and risks of such impacts shape structures for emergency preparedness, especially where impacted populations have a strong influence on policy-making. Climate is related to the quality of life in complex ways, including recreational patterns, and changes in temperature and humidity can change health care challenges and requirements (Chapter 8). For instance, it has been estimated that of the 131 million people affected by natural disasters in Asia in 2004, 97% were affected by weather-related disasters. Exposures in highly-populated coastal and riverine areas and small island nations have been especially significant (ADRC et al., 2005). Moreover, some references suggest relationships between weather and climate on the one hand and social stresses on the other, especially in urban areas where the poor lack access to climate-controlled shelters (e.g., the term 'long, hot summers' associated in the 1960s in the United States with summer urban riots; also see Arsenault, 1984 and Box 7.1). In some cases, tolerance for climatic variation is limited, for example in tightly-coupled urban systems where low capacity drinking water systems have limited resilience in the face of drought or population growth, not only in developing countries but also in industrialised countries. Another case is the sensitivity of energy production to heatwaves and drought (Box 7.1; Section 7.4.2.1).
5. Climate can be a factor in an area's comparative advantage for economic production and growth. Climate affects some of an area's assets for economic production and services, from agricultural and fibre products (Chapter 5) to tourist attractions. Climate also affects costs of business operation, e.g., costs of climate control in office, production and storage buildings. Not only can climate affect an area's own economic patterns; it can also affect the competitive position of its markets and competitors, and thus affect prospects for local employment and individual livelihoods. Many workers are 'marginal', whose livelihoods can be especially sensitive to any changes in conditions affecting local economies.
6. Impacts of climate on industry, settlements and society can be either direct or indirect. For instance, temperature

Box 7.1. Impacts of the 2003 heatwave in Europe

The Summer 2003 heatwave in Western Europe affected settlements and economic services in a variety of ways. Economically, this extreme weather event created stress on health, water supplies, food storage and energy systems. In France, electricity became scarce, construction productivity fell, and the cold storage systems of 25-30% of all food-related establishments were found to be inadequate (Létard et al., 2004). The punctuality of the French railways fell to 77%, from 87% twelve months previously, incurring 1 to 3 million (US$1.25 to 3.75 million) in additional compensation payments, an increase of 7-20% compared with the usual annual total. Sales of clothing were 8.9% lower than usual in August, but sales of bottled water increased by 18%, and of ice cream by 14%. The tourist industry in Northern France benefited, but in the South it suffered (Létard et al., 2004).

Impacts of the heatwave were mainly health- and health-service related (see Chapter 8); but they were also associated with settlement and social conditions, from inadequate climate conditioning in buildings to the fact that many of the dead were elderly people, left alone while their families were on vacation. Electricity demand increased with the high heat levels; but electricity production was undermined by the facts that the temperature of rivers rose, reducing the cooling efficiency of thermal power plants (conventional and nuclear) and that flows of rivers were diminished; six power plants were shut down completely (Létard et al., 2004). If the heatwave had continued, as much as 30% of national power production would have been at risk (Létard et al., 2004). The crisis illustrated how infrastructure can be unable to deal with complex, relatively sudden environmental challenges (Lagadec, 2004).

increases can affect air pollutant concentrations in urban areas, which in turn change exposures to respiratory problems in the population, which then impact health care systems (Chapter 8). Tropical storms can affect the livelihoods and economies of coastal communities through effects on coral reefs, mangroves and other coastal ecosystems (Adger et al., 2005a). Tracing out such second, third, and higher-order indirect impacts, especially in advance, is a significant challenge.

7. Impacts are not equally experienced by every portion of an industrial structure or a population. Some industrial sectors and the very young, the very old and the very poor tend to be more vulnerable to climate impacts than the general economy and population (Box 7.1; Section 7.4.2.5). Some of these differences are also regional, more problematic in developing regions and intricately related to development processes (ISDR, 2004).

Current sensitivities to climate change are briefly summarised in Chapter 1 of this Fourth Assessment Report, and in a number of cases they are relevant for the Millennium Development Goals (for a brief discussion of MDGs in the context of possible climate-change impacts on industry, settlement and society see Section 7.6; also Chapter 20).

Tourism is an example of an economic sector where there has been substantial recent research to understand its sensitivity to climate (Besancenot, 1989; Gomez-Martin, 2005); the emphasis on climate change is, however, more recent (Scott et al., 2005a, b). For example, travel decisions are often based on a desire for warm and sunny environments, while winter tourism builds on expectations of snow and snow-covered landscapes (Chapter 14, Section 14.4.7; Chapter 12, Section 12.4.9; Chapter 11, Section 11.4.9). Tourism is thus sensitive to a range of climate variables such as temperature, hours of sunshine, precipitation, humidity, and storm intensity and frequency (Matzarakis and de Frietas, 2001; Matzarakis et al., 2004), along with the consequences that may follow, such as fires, floods, landslides, coastal erosion and disease outbreaks.

7.3 Assumptions about future trends

Defining possible future socio-economic conditions is a key to understanding future vulnerabilities to climatic change and assessing the capacity to adapt in the face of new risks and opportunities. A range of tools, including scenarios and storylines, has been used to develop characterisations of the future (Chapter 2). While specific characterisations have been developed for vulnerability and adaptation studies in certain climate-sensitive sectors (for example, Arnell et al., 2004; Nicholls, 2004), few characterisations have been developed that relate specifically to climate impacts as they could affect industry, settlement and society. Where such characterisations have been done (e.g., NACC, 2000; London Climate Change Partnership, 2004; Raskin et al., 2005), they have common roots in the perspectives embedded in the IPCC Special Report on Emissions Scenarios (SRES; Nakićenović and Swart, 2000; see also Chapter 2, Section 2.4.6). Drivers in the SRES scenarios – population, economic growth, technology and governance – are all highly relevant for the development of industry, settlement and society.

A key future condition, for instance, is human population and its distribution. According to the latest United Nations projections (i.e., post-SRES), even as the rate of population growth continues to decline, the world's total population will rise substantially. The total is expected to reach between 8.7 and 9.3 billion in 2030 (UN, 2004). More than half these people live in urban centres, and practically all live in settlements, many depending on industry, services and infrastructures for jobs, well-being and mobility. Most population growth will take place in cities, largely in urban areas of developing countries, especially from Asia and Africa (Table 7.1). Some mega-cities will grow very substantially, but the major population growth will take place in medium cities of 1 to 5 million people and in small cities of under 500,000 people, which still represent half of the world population (Table 7.1, see also UN-Habitat, 2003).

Table 7.1. *Urban indicators.*

	Percentage urban				Percent of the world's urban population living in the region				Percent of urban population in different size-class of urban centre, 2000				
Year	1950	1975	2000	2030*	1950	1975	2000	2030*	Under 0.5 m	0.5-1 m	1-5 m	5-10 m	10 m +
Northern America	63.9	73.9	79.1	86.7	15.0	11.9	8.8	7.1	37.4	11.0	34.3	5.4	11.9
Latin America and the Caribbean	42.0	61.2	75.4	84.3	9.6	13.0	13.9	12.4	49.8	9.0	21.7	4.9	14.7
Oceania	62.0	71.5	70.5	73.8	1.1	1.0	0.8	0.6	41.9	0	58.1	0	0
Europe	50.5	67.9	71.7	78.3	37.8	29.2	18.4	11.1	67.8	9.8	15.1	5.4	1.9
Asia	16.8	24.0	37.1	54.1	32.0	37.9	47.9	53.7	49.0	10.0	22.6	8.8	9.7
Africa	14.7	25.4	36.2	50.7	4.5	7.0	10.3	15.1	60.2	9.6	22.1	4.6	3.5
WORLD	29.0	37.2	46.8	59.9	100	100	100	100	52.6	9.8	22.4	6.8	8.4

* These are obviously speculative (projections based largely on extrapolating past trends) and, since any nation's or region's level of urbanisation is strongly associated with their per capita income, economic performance between 2000 and 2030 will have a strong influence on the extent to which regional populations continue to urbanise. Source: taken from or derived from statistics in United Nations (2006).

Features of development relevant to adaptation, such as access to resources, location and institutional capacity, are likely to be predominantly urban and to be determined by differences in economic growth and access to assets, which tend to be increasingly unequal (e.g., the income gap between the richest and the poorest 20% of the world population went from a factor of 32 to 78 between 1970 and 2000: UN-Habitat, 2003). It is estimated that one third of the world's urban population (923.9 million) live in "overcrowded and unserviced slums, often situated on marginal and dangerous land" (i.e., steep slopes, food plains, and industrial zones), and that 43% are in developing countries (UN-Habitat, 2003). It is projected that in the next 30 years "the total number of slum dwellers will increase to about 2 billion, if firm and concrete action is not taken" (UN-Habitat, 2003).

Risk-prone settlements such as in coastal areas are expected to experience not only increases in weather-related disasters (CRED, 2005) but also major increases in population, urban area and economic activity, especially in developing countries (Chapter 6). Growing population and wealth in exposed coastal locations could result in increased economic and social damage, both in developing and developed countries (Pielke et al., 2005; Box 7.4).

Global economic growth projections in SRES and SRES-derived scenarios (Chapter 2) vary significantly - more than population projections. Under low-growth scenarios (A2 and B2), world GDP would double by 2020 and increase more than 10-fold by 2100. Under a high-growth scenario (A1), world GDP would nearly triple by 2020 and grow over 25-fold by 2100. Under all these scenarios, more valuable assets and activities are likely to become exposed to climate risks, but it is assumed that the economic potential to respond will also vastly increase. Economic development will be central to adaptive capacity (Toth and Wilbanks, 2004). SRES scenarios also assume convergence of national per capita incomes, which is contrary to historical tendencies for income gaps between the rich and the poor to increase. While the ratio of per capita incomes in developed as compared with developing countries stood at 16.1 in 1990, SRES scenarios assume a narrowing of this ratio to between 8.4 and 6.2 in 2020, and between 3.0 and 1.5 in 2100. Smaller differences in relative incomes are likely to have important consequences for the perception of climate vulnerability and for the pattern of response.

Because it is potentially highly dynamic, the treatment of technology varies greatly between global scenario exercises. For instance, three qualitatively-different technology scenarios were developed for SRES scenario A1 alone (A1FI, A1T and A1B). An even broader universe of technological change scenarios can be developed for global and downscaled national, regional and sectoral scenarios (e.g., Berkhout and Hertin, 2002). In this chapter we make no specific assumptions about the rate and direction of technological change into the future, recognising that very wide ranges of potentials will exist at the local and organisational levels at which climate vulnerability and responses will often be shaped, and also that the knowledge base referenced in the chapter reflects a range of assumptions about future trends. Governance is likewise a topic about which different scenario families make divergent assumptions. The SRES scenarios include both globally-integrated systems of economic and political and sustainability governance, as well as more fragmented, regionalised systems. The Global Scenarios Group set of scenarios include characterisations in which institutions and governance as we know them persist with minor reform; 'barbarisation' scenarios consider futures in which "absolute poverty increases and the gap between rich and poor ...[and] national governments lose relevance and power relative to trans-national corporations and global market forces..." (Gallopin et al., 1997); 'great transitions' scenarios contain storylines in which sustainable development becomes an organising principle in governance. In this chapter we also have made no specific assumptions about the nature of future pattern of governance, while recognising that institutional capacity will be central to adaptive capacity (Section 7.6.5; also see Chapter 2).

7.4 Key future impacts and vulnerabilities

The ability to project how climate change may affect industry, settlement and society is limited by uncertainties about climate change itself at a relatively fine-grained geographical and sectoral scale and also by uncertainties about trends in human systems over the next century *regardless of climate change* (Chapter 2). In some cases, uncertainties about socio-economic factors such as technological and institutional change over many decades undermine the feasibility of comparing future prospects involving *considerable* climate change with prospects involving *relatively little* climate change. Typically, therefore, research often focuses on <u>vulnerabilities</u> to impacts of climate change (defined as the degree to which a system, subsystem or system component is likely to experience harm due to exposure to a perturbation or source of stress (Turner et al., 2003a; also see Clark et al., 2000) rather than on <u>projections of impacts</u> of change on evolving socio-economic systems, especially in the longer run.

Furthermore, climate change will not often be a primary factor in changes for industry, settlement and society. Instead, it will have an impact by modifying other more significant aspects of ongoing socio-economic changes. This may have either an exacerbating or an ameliorating effect in influencing overall vulnerabilities to multi-causal change. It is especially difficult to associate levels of climate-change impacts or their costs with a specified number of degrees of mean global warming or with a particular time horizon such as 2050 or 2080, when so many of the main drivers of impacts and costs are not directly climate-related, even though they may be climate-associated, and when impacts are often highly localised. Some projections have been made for particular sectors or areas and they are cited in appropriate sections below; but in general they should be considered with caution, especially for longer-range futures.

7.4.1 General effects

Certain kinds of effects follow from particular manifestations of climate change, wherever those phenomena occur. For example, increased precipitation in already well-watered areas can increase concerns about drainage and water-logging

(Parkinson and Mark, 2005), while reduced precipitation in areas already subject to water shortages could lead to infrastructure crises. Sea-level rise will affect land uses and physical infrastructures in coastal areas. Changes in conditions can affect requirements for public health services (Chapter 8), water supplies (Chapter 3) and energy services (such as space heating and cooling). Effects can either be cumulative (additive), as in losses of property, or systematic (affecting underlying processes), as in damages to institutions or systems of production (Turner et al., 1990). Even very gradual changes can be associated with thresholds at which the resilience of human systems switches from adequate to inadequate, such as water-supply infrastructures faced with shrinking water availability. Parry et al. (2001), for instance, estimate that many tens of millions of the world's population are at risk of hunger due to climate change, and billions are at risk of water shortages.

Besides gradual changes in climate, human systems are affected by a change in the magnitude, frequency and/or intensity of storms and other extreme weather events, as well as changes in their location. In fact, some assessments suggest that many impact issues are more directly associated with climatic *extremes* than with *averages* (NACC, 2000). Of some concern is the possibility of abrupt climate changes (Chapter 19), which could be associated with locally or regionally catastrophic impacts if they were to occur.

Although localities differ, interactions between climate change and human systems are often substantively different for relatively developed, industrialised countries versus less developed countries and regions. In many cases, it appears that possible negative impacts of climate change pose risks of higher total *monetary* damages in industrialised areas (i.e., currency valuations of property damages) but higher total *human* damages in less-developed areas (i.e., losses of life and dislocations of population) – although such events as Hurricane Katrina show that there are exceptions (Section 7.4.2.5) for developed countries, and monetary damages in developing countries may represent a larger share of their GDP.

Not all implications of possible climate change are negative. For instance, along with possible carbon fertilisation effects and a longer growing season (Chapter 5), many mid- and upper-latitude areas see quality-of-life benefits from winter warming, and some areas welcome changes in precipitation patterns, although such changes could have other social consequences. The greater proportion of the research literature, however, is related to possible adverse impacts. Climate impact concerns include environmental quality (e.g., more ozone, water-logging or salinisation), linkage systems (e.g., threats to water and power supplies), societal infrastructures (e.g., changed energy/water/health requirements, disruptive severe weather events, reductions in resources for other social needs and maintaining sustainable livelihoods, environmental migration (Box 7.2), placing blame for adverse effects, changes in local ecologies that undermine a sense of place), physical infrastructures (e.g., flooding, storm damage, changes in the rate of deterioration of materials, changed requirements for water or energy supply), and economic infrastructures and comparative advantages (e.g., costs and/or risks increased, markets or competitors affected).

Box 7.2. Environmental migration

Migration, usually temporary and often from rural to urban areas, is a common response to calamities such as floods and famines (Mortimore, 1989), and large numbers of displaced people are a likely consequence of extreme events. Their numbers could increase, and so could the likelihood of their migration becoming permanent, if such events increase in frequency. Yet, disaggregating the causes of migration is highly problematic, not least since individual migrants may have multiple motivations and be displaced by multiple factors (Black, 2001). For example, studies of displacement within Bangladesh and to neighbouring India have drawn obvious links to increased flood hazard as a result of climate change. But such migration also needs to be placed in the context of changing economic opportunities in the two countries and in the emerging mega-city of Dhaka, rising aspirations of the rural poor in Bangladesh, and rules on land inheritance and an ongoing process of land alienation in Bangladesh (Abrar and Azad, 2004).

Estimates of the number of people who may become environmental migrants are, at best, guesswork since (a) migrations in areas impacted by climate change are not one-way and permanent, but multi-directional and often temporary or episodic; (b) the reasons for migration are often multiple and complex, and do not relate straightforwardly to climate variability and change; (c) in many cases migration is a longstanding response to *seasonal* variability in environmental conditions, it also represents a strategy to *accumulate* wealth or to seek a route out of poverty, a strategy with benefits for both the receiving and original country or region; (d) there are few reliable censuses or surveys in many key parts of the world on which to base such estimates (e.g., Africa); and (e) there is a lack of agreement on what an environmental migrant is anyway (Unruh et al., 2004; Eakin, 2006).

An argument can also be made that rising ethnic conflicts can be linked to competition over natural resources that are increasingly scarce as a result of climate change, but many other intervening and contributing causes of inter- and intra-group conflict need to be taken into account. For example, major environmentally-influenced conflicts in Africa have more to do with relative abundance of resources, e.g., oil, diamonds, cobalt, and gold, than with scarcity (Fairhead, 2004). This suggests caution in the prediction of such conflicts as a result of climate change.

Economic sectors, settlements and social groups can also be affected by climate change response policies. For instance, certain greenhouse-gas stabilisation strategies can affect economies whose development paths are dependent on abundant local fossil-fuel resources, including economic sectors involved in mining and fuel supply as well as fuel use. In this sense, relationships between climate-change impacts and sustainable development (IPCC Working Group II) are linked with discussions of climate-change mitigation approaches (IPCC Working Group III).

In many cases, the importance of climate-change effects on human systems seems to depend on the geographic (or sectoral) scale of attention (Abler, 2003; Wilbanks, 2003a). At the scale of a large nation or region, at least in most industrialised nations, the economic value of sectors and locations with low levels of vulnerability to climate change greatly exceeds the economic value of sectors and locations with high levels of vulnerability, and the capacity of a complex large economy to absorb climate-related impacts is often considerable. In many cases, therefore, estimates of aggregate damages of climate change (other than major abrupt changes) are often rather small as a percentage of economic production (e.g., Mendelsohn, 2001). On the other hand, at a more detailed scale, from a small region to a small country, many specific localities, sectors and societies can be highly vulnerable, at least to possible low-probability/high-consequence impacts; and potential impacts can amount to very severe damages. It appears that large-regional or national estimates of possible impacts may give a different picture of vulnerabilities than an aggregation of vulnerabilities defined at a small-regional or local scale.

7.4.2 Systems of interest

The specified systems of interest for Chapter 7 are industry, services, utilities/infrastructure, human settlement and social issues.

7.4.2.1 Industry

Industrial sectors are generally thought to be less vulnerable to the impacts of climate change than other sectors, such as agriculture and water services. This is in part because their sensitivity to climatic variability and change is considered to be comparatively lower and, in part, because industry is seen as having a high capacity to adapt in response to changes in climate. The major exceptions are industrial facilities located in climate-sensitive areas (such as coasts and floodplains), industrial sectors dependent on climate-sensitive inputs (such as food processing) and industrial sectors with long-lived capital assets (Ruth et al., 2004).

We define industry as including manufacturing, transport, energy supply and demand, mining, construction and related informal production activities. Other sectors sometimes included in industrial classifications, such as wholesale and retail trade, communications, and real estate and business activities, are included in the categories of services and infrastructure (below). Together, industry and economic services account for more than 95% of GDP in highly-developed economies and between 50 and 80% of GDP in less-developed economies (World Bank, 2006), and they are very often at the heart of the economic base of a location for employment stability and growth.

Industrial activities are, however, vulnerable to direct impacts such as temperature and precipitation changes. For instance, weather-related road accidents translate into annual losses of at least Canadian $1 billion annually in Canada, while more than a quarter of air travel delays in the United States are weather-related (Andrey and Mills, 2003). Buildings are also affected by higher temperatures during hot spells (Livermore, 2005). Moreover, facilities across a range of industrial sectors are often located in areas vulnerable to extreme weather events (including flooding, drought, high winds), as the Hurricane Katrina event clearly demonstrated. Where extreme events threaten linkage infrastructures such as bridges, roads, pipelines or transmission networks, industry can experience substantial economic losses. In other cases, climate change could lead to reductions in the direct vulnerability of industry and infrastructures. For instance, fewer freeze-thaw cycles in temperate regions would lead to less deterioration of road and runway surfaces (Mills and Andrey, 2002). There exist relatively few quantified assessments of these direct impacts, suggesting an important role for new research (Eddowes et al., 2003).

Less direct impacts on industry can also be significant. For instance, sectors dependent on climate-sensitive inputs for their raw materials, such as the food processing and pulp and paper sectors, are likely to experience changes in sources of major inputs. In the longer term, as the impacts of climate change become more pronounced, regional patterns of comparative advantage of industries closely related to climate-sensitive inputs could be affected, influencing regional shifts in production (Easterling et al., 2004). Industrial producers will also be influenced indirectly by regulatory and market changes made in response to climate change. These may influence locational and technology choices, as well affecting costs and demand for goods and services. For instance, increased demand for space cooling may be one result of higher peak summer temperatures (Valor et al., 2001; Giannakopoulos and Psiloglou, 2006). A range of direct (awareness of changing weather-related conditions) and indirect (changing policy, regulation and behaviour) impacts on three different classes of industry is identified in Table 7.2.

In developing countries, besides modern production activities embedded in global supply chains, industry includes a greater proportion of enterprises that are small-scale, traditional and informally organised. Impacts of climate change on these businesses are likely to depend on the determinants identified in the TAR: location in vulnerable areas, dependence on inputs sensitive to climate, and access to resources to support adaptive actions. Many of these activities will be less concerned with climate risks and will have a high capacity to adapt, while others will become more vulnerable to direct and indirect impacts of climate change.

An example of an industrial sector particularly sensitive to climate change is energy (e.g., Hewer, 2006; Chapter 12, Section 12.4.8.1). Climate change is likely to affect both energy use and energy production in many parts of the world. Some of the possible impacts are rather obvious. Where the climate warms due to climate change, less heating will be needed for industrial,

Table 7.2. *Direct and indirect climate change impacts on industry.*

Sector	Direct impacts	Indirect impacts	References
Built Environment: Construction, civil engineering	Energy costs External fabric of buildings Structural integrity Construction process Service infrastructure	Climate-driven standards and regulations Changing consumer awareness and preferences	Consodine, 2000; Graves and Phillipson, 2000; Sanders and Phillipson, 2003; Spence et al., 2004; Brewer, 2005; Kirshen et al., 2006
Infrastructure Industries: Energy, water, telecommunications, transport (see Section 7.4.2.3)	Structural integrity of infrastructures Operations and capacity Control systems	Changing average and peak demand Rising standards of service	Eddowes et al., 2003; UK Water Industry Research, 2004; Fowler et al, 2005
Natural Resource Intensive Industries: Pulp and paper, food processing, etc.	Risks to and higher costs of input resources Changing regional pattern of production	Supply chain shifts and disruption Changing lifestyles influencing demand	Anon, 2004; Broadmeadow et al., 2005

commercial and residential buildings, and cooling demands will increase (Cartalis et al., 2001), with changes varying by region and by season. Net energy demand at a national scale, however, will be influenced by the structure of energy supply. The main source of energy for cooling is electricity, while coal, oil, gas, biomass and electricity are used for space heating. Regions with substantial requirements for both cooling and heating could find that net annual electricity demands increase while demands for other heating energy sources decline (Hadley et al., 2006). Critical factors for the USA are the relative efficiency of space cooling in summer compared to space heating in winter, and the relative distribution of populations within the U.S. in colder northern or warmer southern regions. Seasonal variation in total demand is also important. In some cases, due to infrastructure limitations, peak demand could go beyond the maximum capacity of the transmission system.

Tol (2002a, b) estimated the effects of climate change on the demand for global energy, extrapolating from a simple country-specific (United Kingdom) model that relates the energy used for heating or cooling to degree days, per capita income, and energy efficiency. According to Tol, by 2100 benefits (reduced heating) will be about 0.75% of gross domestic product (GDP) and damages (increased cooling) will be approximately 0.45%, although it is possible that migration from heating-intensive to cooling-intensive regions could affect such comparisons in some areas.

In addition to *demand*-side impacts, energy *production* is also likely to be affected by climate change. Except for impacts of extreme weather events, research evidence is more limited than for energy consumption; but climate change could affect energy production and supply (a) if extreme weather events become more intense, (b) where regions dependent on water supplies for hydropower and/or thermal powerplant cooling face reductions in water supplies, (c) where changed conditions affect facility siting decisions, and (d) where conditions change (positively or negatively) for biomass, windpower or solar energy production.

For instance, the TAR (Chapter 7) concluded that hydropower generation is likely to be impacted because it is sensitive to the amount, timing and geographical pattern of precipitation as well as temperature (rain or snow, timing of melting). Reduced stream flows are expected to jeopardise hydropower production in some areas, whereas greater stream flows, depending on their timing, might be beneficial (Casola et al., 2005; Voisin et al., 2006). According to Breslow and Sailor (2002), climate variability and long term climate change should be considered in siting wind power facilities (also see Hewer, 2006). Extreme weather events could threaten coastal energy infrastructures (e.g., Box 7.4) and electricity transmission and distribution infrastructures. Moreover, soil subsidence caused by the melting of permafrost is a risk to gas and oil pipelines, electrical transmission towers, nuclear-power plants and natural gas processing plants in the Arctic region (Nelson et al., 2001). Structural failures in transportation and industrial infrastructure are becoming more common as a result of permafrost melting in northern Russia, the effects being more serious in the discontinuous permafrost zone (ACIA, 2004).

Policies for reducing greenhouse gas (GHG) emissions are expected to affect the energy sector in many countries. For instance, Kainuma et al. (2004) compared a global reference scenario with six different GHG reduction scenarios. In the reference scenario under which emissions continue to grow, the use of coal increases from 18% in 2000 to 48% in 2100. In aggressive mitigation scenarios, the world's final energy demand drops to nearly one-half of that in the reference scenario in 2100, mainly associated with reducing coal use. Kuik (2003) has found a trade-off between economic efficiency, energy security and carbon dependency for the EU.

7.4.2.2 Services

Services include a wide variety of human needs, activities and systems, related both to meeting consumer needs and to employment in the service activities themselves. This section includes brief discussions of possible climate-change effects on trade, retail and commercial services, tourism and risk financing/insurance as illustrations of the implications of climate change – not implying that these sectors are the only ones that could be affected, negatively or positively.

7.4.2.2.1 Trade

Possible impacts of climate change on inter-regional trade are still rather speculative. Climate change could affect trade by reshaping regional comparative advantage related to (a) general climate-related influences (Figure 7.2), such as on agricultural production, (b) exposure to extreme events combined with a lack of capacity to cope with them, and/or (c) effects of climate-change mitigation policies that might create markets for emission-reduction alternatives. In an era of increased globalisation, small changes in price structures (including transportation costs) could have amplified effects on regional economies and employment. Beyond actual climate-change impacts, a perception of future impacts or regulatory initiatives could also affect investment and trade.

Climate change may also disrupt transport activities that are important to national supplies (and travellers) as well as international trade. For instance, extreme events may temporarily close ports or transport routes and damage infrastructure critical to trade. Increases in the frequency or magnitude of extreme weather events could amplify the costs to transport companies and state authorities from closed roads, train delays and cancellations, and other interruptions of activities (O'Brien et al., 2004). It appears that there could be linkages between climate-change scenarios and international trade scenarios, such as a number of regional and sub-regional free trade agreements, although research on this topic is lacking.

7.4.2.2.2 Retail and commercial services

Retail and other commercial services have often been neglected in climate-change impact studies. Climate change has the potential to affect every link in the supply chain, including the efficiency of the distribution network, the health and comfort of the workforce (Chapter 8), and patterns of consumption. Many of the services can be more difficult to move than industrial facilities, because their locations are focused on where the people are. In addition, climate-change policies could raise industrial and transportation costs, alter world trade patterns, and necessitate changes in infrastructure and design technology. As one example, distribution networks for commercial activities would be affected in a variety of ways by changing winter road conditions (e.g., ACIA, 2004) and negatively affected by an increase in hazardous weather events. Strong winds can unbalance high-sided vehicles on roads and bridges, and may delay the passage of goods by sea. Transportation routes in permafrost zones may be negatively affected by higher temperatures which would shorten the winter-road season (Instanes et al., 2005). Coastal infrastructure and distribution facilities are vulnerable to inundation and flood damage. In contrast, transportation of bulk freight by inland waterways, such as the Rhine, can be disrupted during droughts (Parry, 2000). Further, climate variation creates short-term shifts in patterns of consumption within specific retail markets, such as the clothing and footwear market (Agnew and Palutikof, 1999). However, most impacts entail transfers within the economy (Subak et al., 2000) and are transitory.

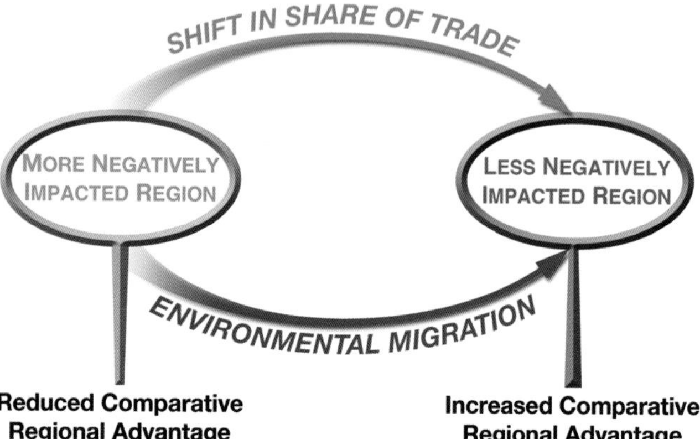

Figure 7.2. *General effects of climate change on international trade: greater net benefits from climate change are likely to show trade benefits, along with environmental in-migration.*

Perishable commodities are one of the most climate-sensitive retail markets (Lin and Chen, 2003). It is possible that climate change will alter the sourcing and processing of agricultural produce; and climate-change policies (e.g., a carbon tax or an emissions offset payment) may further alter the geographical distribution of raw materials and product markets.

7.4.2.2.3 Tourism

A substantial research literature has assessed the consequences of climate change for international tourist flows (e.g., Agnew and Viner, 2001; Hamilton et al., 2005), for the tourist industries of nations (Becken, 2005; Ceron and Dubois, 2005), destinations (Belle and Bramwell, 2005), attractions, such as national parks (Jones and Scott, 2007; Chapter 14, Section 14.4.7), and tourism activities (Perry, 2004; Jones et al., 2006) or sectors of tourism such as ski-tourism (e.g., Elsasser and Burki, 2002; Fukushima et al., 2003; Hamilton et al., 2003).

Likely effects of climate change on tourism vary widely according to location, including both direct and indirect effects. Regarding direct effects, climate change in temperate and high latitude countries seems to mean a poleward shift in conditions favourable to many forms of tourism (Chapter 15). This might, for instance, lead to more domestic tourism in north-west Europe (Chapter 12, Section 12.4.9; Agnew and Viner, 2001; Maddison, 2001) and in the middle latitudes of North America (Chapter 14, Section 14.4.7). If winters turn out to be milder but wet and windy, however, the gains to be expected are less obvious (Ceron, 2000). Areas dependent on the availability of snow are among those most vulnerable to global warming (Chapter 11, Sections 11.4.9; Chapter 12, Section 12.4.9; Chapter 14, Section 14.4.7). In summer, destinations already hot could become uncomfortable (Chapter 12, Section 12.5.9). Tropical destinations might not suffer as much from an increase in temperatures, since tourists might expect warm climates as long as indoor comfort is assured – with implications for greenhouse gas emissions (Gössling and Hall, 2005). For low-lying islands, sea-level rise and increasingly frequent and intense weather extremes might become of great importance in the future (Chapter 16, Section 16.4.2). Extreme climate events, such as tropical storms, could have substantial effects on tourist infrastructure and the economies of small-island states (London, 2004).

Indirect effects include changes in the availability of water and costs of space cooling, but at least as significant could be changes in the landscape of areas of tourist interest, which could be positive or negative (Braun et al.,1999; Uyarra et al., 2005; Chapter 14, Section 14.4.7). Warmer climates open up the possibility of extending exotic environments (such as palm trees in western Europe), which could be considered by some tourists as positive but could lead to a spatial extension and amplification of water- and vector-borne diseases. Droughts and the extension of arid environments (and the effects of extreme weather events) might discourage tourists, although it is not entirely clear what they consider to be unacceptable. In tropical environments, destruction due to extreme weather events (buildings, coral reefs, trees and plants) is a concern, but vegetation and landscape tend to recover relatively quickly with the notable exception of eroded beaches and damaged coral reefs. One indirect factor of considerable importance is energy prices, which affect both the cost of providing comfort in tourist areas and the cost of travelling to them (Becken et al., 2001). This effect can be especially significant for smaller, tourist-oriented countries, often in the developing world; for instance, receipts from international tourism account for 39% of GDP in the Bahamas, but only 2.4% for France (World Tourism Organization, 2003).

The environmental context in which tourism will operate in the future involves considerable uncertainties. The range of possible scenarios is great, and there have been some attempts to link the future of tourist activities to SRES scenarios (Chapter 14, Section 14.4.7; Chapter 11, Section 11.4.9) . In these scenarios, tourist reactions to climate change are assumed to be constant, notwithstanding the fact that these responses are currently not satisfactorily understood.

7.4.2.2.4 Insurance

Insurance is a major service sector with the potential to be directly affected by any increase in damages associated with climate change, such as more intense and/or frequent extreme weather events (see Box 7.3). While a number of lines of insurance have some potential to be affected by catastrophe losses, the principal impacts are expected to be on property lines.

As the actuarial analysis of recent loss experience is typically an inadequate guide to catastrophe risk, since the 1990s probabilistic 'catastrophe' modelling software has become employed by insurers for pricing and managing portfolios of property catastrophe risk (Grossi and Kunreuther, 2005). At the start of 2006, the five-year forward-looking activity rate employed in the most widely used Hurricane Catastrophe Model was increased relative to mean historical rates with an acknowledgement that some contribution to this increase is likely to reflect climate change (Muir Wood et al., 2006).

Within the risk market, reinsurers tend to be more pessimistic about catastrophe risk-costs than the insurers who are ceding the risk, and this perspective has been highlighted by statements from reinsurers going back more than a decade warning of the potential impacts of climate change (Swiss Re, 2004; Munich Re, 2005). However, in 2006, insurers also began to communicate directly with their policyholders regarding the rising costs of claims attributed to climate change (Allianz and World Wildlife Fund, 2006; Crichton, 2006).

The specific insurance risk coverages currently available within a country will have been shaped by the impact of past catastrophes. Because of the high concentration of losses where, over the past 50-60 years, there have been catastrophic floods, private sector flood insurance is generally restricted (or even unavailable), so that in many developed countries governments have put in place alternative state-backed flood insurance schemes (Swiss Re, 1998).

In both developed and developing countries, property insurance coverage will expand with economic growth. If overall risk increases under climate change, the insurance industry can be expected to grow in the volume of premium collected, claims paid and, potentially, income (where insurers overcome consumer and regulatory pressures to restrict increases in insurance rates, and where catastrophe loss cost increases are appropriately anticipated and modelled). However, market dislocations are also likely, as in 2006 when, unable for regulatory reasons to pass on higher technical hurricane risk costs, U.S. insurers declined to cover homeowners and businesses at the highest-risk coastal locations, thereby undermining the real estate market and forcing government intervention in structuring some alternative insurance provision (Freer, 2006).

After a decade of rising losses (from both natural and man-made catastrophes), insurance is generally becoming more restrictive in what is covered. Insurance rates in many areas rose after 2001 so that, while the 2004 year was the worst (up to that time) for U.S. catastrophe losses, it was also the most profitable year ever for U.S. insurers (Dyson, 2005). However, the years 2001 to 2005 were not so profitable for reinsurers, although increases in prices saw significant new capital entering the market in 2002 and 2005, while 2006 appeared a benign year for losses.

Where increased risk costs lead insurers to reduce the availability of insurance, there will be impacts on local and regional economies, including housing and industrial activity,

Box 7.3. The impact of recent hurricane losses

The US$15.5 billion insurance loss of Hurricane Andrew in 1992 (US$45 billion adjusted to 2005 values and exposures) remains an exemplar of the consequences on the insurance industry of a catastrophe more severe than had been anticipated, leading to the insolvency of 12 insurance companies and significant market disruption. However, after major adjustments, including the widespread use of catastrophe models, the private insurance market re-expanded its role, so that in the four hurricanes of 2004 (with a total market loss of around US$29 billion from the U.S., Caribbean and Gulf Energy sectors) only one small U.S. insurance company failed, and there was little impact on reinsurance rates, largely because state-backed insurance and reinsurance mechanisms in Florida absorbed a significant proportion of the loss. However, a far greater proportion of the US$60 billion of insured losses from the 2005 hurricanes in Mexico, the energy sector in the Gulf of Mexico and the USA fell onto the international reinsurance market, leading to at least two situations where medium-sized reinsurers could not remain independently viable. Following more than 250,000 flood claims in 2005 related to Hurricanes Katrina, Rita and Wilma, the U.S. federal National Flood Insurance Program would have gone bankrupt without being given the ability to borrow an additional US$20.8 billion from the U.S. Treasury.

unless government expands its risk protection roles. In particular in developed countries, governments are also likely to be the principal funders of risk mitigation measures (e.g., flood defences) that can help ensure that properties remain insurable. In the developing world, the role of insurers and governments in offering risk protection is generally limited (Mills, 2004).

The use of insurance is far lower in developing and newly-developed countries (Enz, 2000), as insurance reflects wealth protection that typically lags a generation behind wealth generation. As highlighted by events such as 2005 Hurricane Stan in Mexico and Guatemala, individuals bear the majority of the risk and manage it through the solidarity of family and other networks, if at all. However, once development is underway, insurance typically expands faster than the growth in GDP. With this in mind there has been a focus on promoting 'micro-insurance' to reduce people's financial vulnerability when linked with the broader agenda of risk reduction (ProVention Consortium, 2004; Abels and Bullen, 2005), sometimes with the first instalment of the premium paid by the non-governmental organisation (NGO), e.g., in an insurance scheme against cyclones offered in eastern Andhra Pradesh and Orissa.

For the finance sector, climate change-related risks are increasingly considered for specific 'susceptible' sectors such as hydroelectric projects, irrigation and agriculture, and tourism (UNEP, 2002). In high carbon-emitting sectors, such as power generation and petrochemicals, future company valuations could also become affected by threatened litigation around climate-change impacts (Kiernan, 2005). Some specialised investment entities, and in particular hedge funds, take positions around climate related risks, via investments in reinsurance and insurance companies, resource prices such as oil and gas with the potential to be affected by Gulf hurricanes, and through participation in alternative risk transfer products, e.g., insurance-linked securities such as catastrophe bonds and weather derivatives (see Jewson et al., 2005).

7.4.2.3 Utilities/infrastructure

Infrastructures are systems designed to meet relatively general human needs, often through largely or entirely public utility-type institutions. Infrastructures for industry, settlements and society include both 'physical' (such as water, sanitation, energy, transportation and communication systems) and 'institutional' (such as shelter, health care, food supply, security, and fire services and other forms of emergency protection). In many instances, such 'physical' and 'institutional' infrastructures are linked. For example, in New York City adaptations of the physical water supply systems to possible water supply variability are dependent on changes within the institutions that manage them; conversely, institutions such as health care are dependent to some degree on adjustments in physical infrastructures to maintain effective service delivery (Rosenzweig and Solecki, 2001a).

These infrastructures are vulnerable to climate change in different ways and to different degrees, depending on their state of development, their resilience and their adaptability. In general, floods induce more physical damage, while drought and heatwaves tend to have impacts on infrastructure systems that are more indirect.

Often, the institutional infrastructure is less vulnerable as it embodies less fixed investment and is more readily adapted within the time-scale of climate change. Moreover, the effect of climate change on institutional infrastructure can be small or even result in an improvement in its resilience; for example, it could help to trigger an adaptive response (e.g., Bigio, 2003).

There are many points at which impacts on the different infrastructure sectors interact. For instance, failure of flood defences can interrupt power supplies, which in turn puts water and wastewater pumping stations out of action. On the other hand, this means that measures to protect one sector can also help to safeguard the others.

7.4.2.3.1 Water supplies

Climate change, in terms of change in the means or variability, could affect water supply systems in a number of ways. It could affect water *demand*. Increased temperatures and changes in precipitation can contribute to increases in water demand, for drinking, for cooling systems and for garden watering (Kirshen, 2002). If climate change contributes to the failure of small local water sources, such as hand-dug wells, or to inward migration, this may also cause increased demand on regional water supplies. It could also affect water *availability*. Changes in precipitation patterns may lead to reductions in river flows, falling groundwater tables and, in coastal areas, to saline intrusion in rivers and groundwater, and the loss of meltwater will reduce river flows at key times of year in parts of Asia and Latin America (Chapter 3, Section 3.4.3). Furthermore, climate change could *damage the system* itself, including erosion of pipelines by unusually heavy rainfall.

Water supplies have a life of many years and so are designed with spare capacity to respond to future growth in demand. Allowance is also made for anticipated variations in demand with the seasons and with the time of day. From the point of view of the impacts of climate change, therefore, most water supply systems are quite able to cope with the relatively small changes in mean temperature and precipitation which are anticipated for many decades, except at the margin where a change in the mean requires a significant change in the design or technology of the water supply system, e.g., where reduced precipitation makes additional reservoirs necessary (Harman et al., 2005) or leads to saline intrusion into the lower reaches of a river. An example is in southern Africa (Ruosteenoja et al., 2003), where the city of Beira in Mozambique is already extending its 50 km pumping main a further 5 km inland to be certain of freshwater.

More dramatic impacts on water supplies are liable to be felt under extremes of weather that could arise as a result of climate change, particularly drought and flooding. Even where water-resource constraints, rather than system capacity, affect water-supply functioning during droughts, this often results from how the resource is allocated rather than absolute insufficiency. Domestic water consumption, which represents only 2% of global abstraction (Shiklomanov, 2000), is dwarfed by the far greater quantities required for agriculture. Water supply systems, such as those for large coastal cities, are often downstream of other major users and so are the first to suffer when rivers dry up. Under Integrated Water Resource Management, such urban areas would receive priority in allocation, because the value of municipal water use is so much greater than agricultural water use, and therefore they can afford to pay a premium price for the water (Dinar et al., 1997).

In many countries, additional investment is likely to be needed to counter increasing water resource constraints due to climate change. For example, Severn-Trent, one of the nine English water companies, has estimated that its output is likely to fall by 180 Megalitres/day (roughly 9% of the total) by 2030 due to climate change, making a new reservoir necessary to maintain the supply to Birmingham (Environment Agency, 2004). However, such changes will only become a major problem where they are rapid compared to the normal rate of water supply expansion, and where systems have insufficient spare capacity, as in many developing countries.

During the last century, mean precipitation in all four seasons of the year has tended to decrease in all the main arid and semi-arid regions of the world, e.g., northern Chile and the Brazilian North-East, West Africa and Ethiopia, the drier parts of Southern Africa and Western China (Folland et al., 2001). If these trends continue, water resource limitations will become more severe in precisely those parts of the world where they are already most likely to be critical (Rhode, 1999).

Flooding by rivers and tidal surges can do lasting damage to water supplies. Water supply abstraction and treatment works are sited beside rivers, because it is not technically advisable to pump raw water for long distances. They are therefore often the first items of infrastructure to be affected by floods. While sedimentation tanks and filter beds may be solid enough to suffer only marginal damage, electrical switchgear and pump motors require substantial repairs after floods, which cannot normally be accomplished in less than two weeks. In severe riverine floods with high flow velocities, pipelines may also be damaged, requiring more extensive repair work.

7.4.2.3.2 Sanitation and urban drainage

Some of the considerations applying to water supply also apply to sewered sanitation and drainage systems, but in general the effect of climate change on sanitation is likely to be less than on water supply. When water supplies cease to function, sewered sanitation also becomes unusable. Sewer outfalls are usually into rivers or the sea, and so they and any sewage treatment works are exposed to damage during floods (PAHO, 1998). In developing countries, sewage treatment works are usually absent (WHO/Unicef, 2000) or involve stabilisation ponds, which are relatively robust. Sea-level rise will affect the functioning of sea outfalls, but the rise is slow enough for the outfalls to be adapted to the changed conditions at modest expense, by pumping if necessary. Storm drainage systems are also unlikely to suffer serious storm damage, but they will be overloaded more often if heavy storms become more frequent, causing local flooding. The main impact of climate change on on-site sanitation systems such as pit latrines is likely to be through flood damage. However, they are more properly considered as part of the housing stock rather than items of community infrastructure. The main significance of sanitation here is that sanitation infrastructures (or the lack of them) are the main determinant of the contamination of urban flood water with faecal material, presenting a substantial threat of enteric disease (Ahern et al., 2005).

7.4.2.3.3 Transport, power and communications infrastructures

A general increase in temperature and a higher frequency of hot summers are likely to result in an increase in buckled rails and rutted roads, which involve substantial disruption and repair costs (London Climate Change Partnership, 2004). In temperate zones, less salting and gritting will be required, and railway points will freeze less often. Most adaptations to these changes can be made gradually in the course of routine maintenance, for instance by the use of more heat-resistant grades of road metal when resurfacing. Transport infrastructure is more vulnerable to effects of extreme local climatic events than to changes in the mean. For instance, 14% of the annual repair and maintenance budget of the newly-built 760 km Konkan Railway in India is spent repairing damage to track, bridges and cuttings due to extreme weather events such as rain-induced landslides. This amounts to more than Rs. 40 million, or roughly US$1 million annually. In spite of preventive targeting of vulnerable stretches of the line, operations must be suspended for an average of seven days each rainy season because of such damage (Shukla et al., 2005). Parry (2000) provides an assessment of the impact of severe local storms on road transportation, much of which also applies to rail.

Of all the possible impacts on transportation, the greatest in terms of cost is that of flooding. The cost of delays and lost trips would be relatively small compared with damage to the infrastructure and to other property (Kirshen et al., 2006). In the last ten years, there have been four cases when flooding of urban underground rail systems have caused damage worth more than €10 million (US$13 million) and numerous cases of lesser damage (Compton et al., 2002)

Infrastructure for power transmission and communications is subject to much the same considerations. It is vulnerable to high winds and ice storms when in the form of suspended overhead cables and cell phone transmission masts, but is reasonably resilient when buried underground, although burial is significantly more expensive. In developing countries, a common cause of death associated with extreme weather events in urban areas is electrocution by fallen power cables (Few et al., 2004). Such infrastructure can usually be repaired at a fraction of the cost of repairing roads, bridges and railway lines, and in much less time, but its disruption can seriously hinder the emergency response to an extreme event.

7.4.2.4 Human settlement

Climate change is almost certain to affect human settlements, large and small, in a variety of significant ways. Settlements are important because they are where most of the world's population live, often in concentrations that imply vulnerabilities to location-specific events and processes and, like industry and certain other sectors of concern, they are distinctive in the presence of physical capital (buildings, infrastructures) that may be slow to change.

Beyond the general perspectives of TAR (see Section 7.1.4), a growing number of case studies of larger settlements indicate that climate change is likely to increase heat stress in summers while reducing cold-weather stresses in winter. It is likely to change precipitation patterns and water availability, to lead to rising sea levels in coastal locations, and to increase risks of extreme weather events, such as severe storms and flooding, although some kinds of extreme events could decrease, such as blizzards and ice storms (see city references below; Klein et al., 2003; London Climate Change Partnership, 2004; Sherbinin et al., 2006).

Extreme weather events associated with climate change pose particular challenges to human settlements, because assets and populations in both developed and developing countries are increasingly located in coastal areas, slopes, ravines and other risk-prone regions (Freeman and Warner, 2001; Bigio, 2003; UN-Habitat, 2003). The population in the near-coastal zone (i.e., within 100 m elevation and 100 km distance of the coast) has been calculated at between 600 million and 1.2 billion; 10% to 23% of the world's population (Adger et al., 2005b; McGranahan et al., 2006). Globally, coastal populations are expected to increase rapidly, while coastal settlements are at increased risk of climate change-influenced sea-level rise (Chapter 6). Informal settlements within urban areas of developing-country cities are especially vulnerable, as they tend to be built on hazardous sites and to be susceptible to floods, landslides and other climate-related disasters (Cross, 2001; UN-Habitat, 2003).

Several recent assessments have considered vulnerabilities of rapidly growing and/or large urban areas to climate change. Examples include cities in the developed and developing world such as Hamilton City, New Zealand (Jollands et al., 2005), London (London Climate Change Partnership, 2004; Holman et al., 2005), New York (Rosenzweig and Solecki, 2001a, b), Boston (Kirshen et al., 2007), Mumbai, Rio de Janeiro, Shanghai (Sherbinin et al., 2006), Krakow (Twardosz, 1996), Caracas (Sanderson, 2000), Cochin (ORNL/CUSAT, 2003), Greater Santa Fe (Clichevsky, 2003), Mexico City, Sao Paolo, Manila, Tokyo (Wisner, 2003), and Seattle (Office of Seattle Auditor, 2005).

Climate change is likely to interact with and possibly exacerbate ongoing environmental change and environmental pressures in settlements. In areas such as the Gulf Coast of the United States, for example, land subsidence is expected to add to apparent sea-level rise. For New York City, sea-level rise will accelerate the inundation of coastal wetlands, threaten vital infrastructure and water supplies, augment summertime energy demand, and affect public health (Rosenzweig and Solecki, 2001a; Knowlton et al., 2004; Kinney et al., 2006). Significant costs of coastal and riverine flooding are possible in the Boston metropolitan area (Kirshen et al., 2006). Climate change, a city's building conditions, and poor sanitation and waste treatment could coalesce to affect the local quality of life and economic activity of such cities as Mumbai, Rio de Janeiro and Shanghai (Sherbinin et al., 2006). In addition, for cities that play leading roles in regional or global economies, such as New York, effects could be felt at the national and international scales via disruptions of business activities linked to other places (Solecki and Rosenzweig, 2007).

Sea-level rise could raise a wide range of issues in coastal areas. Studies in the New York City metropolitan area have projected that climate-change impacts associated with expectations that sea level will rise, could reduce the return period of the flood associated with the 100-year storm to 19 to 68 years on average, by the 2050s, and to 4 to 60 years by the 2080s (Rosenzweig and Solecki, 2001a), jeopardising low-lying buildings and transportation systems. Similar impacts are expected in the eastern Caribbean, Mumbai, Rio de Janeiro and Shanghai, where coastal infrastructure, population and economic activities could be vulnerable to sea-level rise (Lewsey et al., 2004; Sherbinin et al., 2006). Due to a long coastline and extensive low-lying coastal areas, projected sea-level rise in Estonia and the Baltic Sea region could endanger natural ecosystems, cover beach areas high in recreational value, and cause environmental contamination (Kont et al., 2003).

Another body of evidence suggests that human settlements, coastal and otherwise, are affected by climate change-related shifts in precipitation. Concerns include increased flooding potential from more sizeable rain events (Shepherd et al., 2002). Conversely, as suggested by the TAR, any change in climate that reduces precipitation and impairs underground water resource replenishment would be a very serious concern for some human settlements, particularly in arid and semi-arid areas (Rhode, 1999), in settlements with human-induced water scarcity (Romero Lankao, 2006), and in regions dependent on melted snowpack and glaciers (Chapter 1, Box 1.1; Chapter 12, Section 12.4.3; Chapter 13, Section 13.6.2).

A wider range of health implications of climate change also can affect settlements. For example, besides heat stress and respiratory distress from air quality, changes in temperature, precipitation and/or humidity affect environments for water- and vector-borne diseases and create conditions for disease outbreaks (see Chapters 4 and 8). Projections of climate-change impacts in New York City show significant increases in respiratory-related diseases and hospitalisation (Rosenzweig and Solecki, 2001a).

With growing urbanisation and development of modern industry, air quality and haze have become more salient issues in urban areas. Many cities in the world, especially in developing countries, are experiencing air pollution problems, such as Buenos Aires, London, Chongqing, Lanzhou, Mexico City and São Paulo. How climate change might interact with these problems is not clear as a general rule, although temperature increases would be expected to aggravate ozone pollution in many cities (e.g., Molina and Molina, 2002; Kinney et al., 2006). A study evaluating the effects of changing global climate on regional ozone of 15 cities in the U.S. finds, for instance, that average summertime daily maximum ozone concentrations could increase by 2.7 parts per billion (ppb) for a 5-year span in the 2020s and 4.2 ppb for a 5-year span in the 2050s. As a result, more people (especially the elderly and young) might be forced to restrict outdoor activities (NRDC, 2004).

Another issue is urban heat island (UHI) effects: higher temperatures occur in urban areas than in outlying rural areas because of diurnal cycles of absorption and later re-radiation of solar energy and (to a much lesser extent) heat generation from built/paved physical structures. The causes of UHI are complex, as is the interaction between atmospheric processes at different scales (Oke, 1982). UHI can affect the climatic comfort of the urban population, potentially related to health, labour productivity and leisure activities; there are also economic effects, such as the additional cost of climate control within buildings, and environmental effects, such as the formation of smog in cities and the degradation of green spaces. Even such small coastal towns as Aveiro in Portugal have been shown to create a heat island (Pinho and Orgaz, 2000). Rosenzweig et al. (2005) found that climate change based on downscaled general circulation model (GCM) projections would exacerbate the New York City UHI by increasing baseline temperatures and reducing local wind speeds.

In sum, settlements are vulnerable to impacts that can be exacerbated by direct climate changes (e.g., severe storms and associated coastal and riverine flooding, especially when combined with sea-level rise, snow storms and freezes, and fire). Yet climate change is not the only stress on human settlements, but rather it coalesces with *other* stresses, such as scarcity of water or governance structures that are inadequate even in the absence of climate change (Feng et al., 2006; Sherbinin et al., 2006; Solecki and Rosenzweig, 2007). Such phenomena as unmet resource requirements, congestion, poverty, political and economic inequity, and insecurity can be serious enough in some settlements (UN-Habitat, 2003) that any significant additional stress could be the trigger for serious disruptive events and impacts. Other stresses may include institutional and jurisdictional fragmentation, limited revenue streams for public-sector roles, and inflexible patterns of land use (UNISDR, 2004). These types of stress do not take the same form in every city and community, nor are they equally severe everywhere. Many of the places where people live across the world are under pressure from some combination of continuing growth, pervasive inequity, jurisdictional fragmentation, fiscal strains and aging infrastructure (UN-Habitat, 2003).

7.4.2.5 Social issues

Social system vulnerabilities to impacts of climate variability and change are often related to geographical location. For instance, indigenous societies in polar regions and settlements close to glaciers in Latin America and in Europe are already experiencing threats to their traditional livelihoods (Chapter 12, Section 12.4.3; Chapter 13, Section 13.6.2). Low-lying island nations are also threatened (Chapter 16). Rising temperatures in mountain areas, and in temperate zones needing space-heating during the winter may result in energy cost savings for their populations (Section 7.4.2.1). On the other hand, areas relying on electric fans or air-conditioning may see increased pressures on household budgets as average temperatures rise.

It is increasingly recognised that social impacts associated with climate change will be mainly determined by how the changes interact with economic, social and institutional processes to exacerbate or ameliorate stresses associated with human and ecological systems (Turner et al., 2003b; Adger et al., 2005b; NRC, 2006). As studies undertaken in Latin America, Asia, Africa and the Arctic show, climate change is not the only stress on rural and urban livelihoods. The livelihoods of the Inuit in the Arctic are threatened by multiple stresses (e.g., loss of traditional food sources, growing dependence upon distant fish markets and externally driven values and attitudes). These processes could overtax their adaptive capacity, reduce the role of kinship and family as the centre of social organisation around fishing, and lead to divisions within and between fisher and hunter organisations (Turner et al., 2003b; ACIA, 2004). Rural communities do struggle daily with scarce resources, with insufficient access to commercial markets for their products, and with development policies and other institutional barriers, which frequently limit their ability to cope with extreme climate events (O'Brien et al., 2004; Eakin, 2006). Similarly, in urban settlements, climate change could coalesce with other processes and factors, such as land subsidence due to groundwater withdrawal, the poor condition of many buildings and infrastructures, weak governance structures, and modest income levels, to impact on peoples' livelihoods (Wood and Salway, 2000; Bull-Kamanga et al., 2003; Sherbinin et al., 2006).

The vulnerability of human societies to climate change could vary with economic, social and institutional conditions: particularly socio-economic diversity within urban and rural settlements and their productive sectors, linkage systems and infrastructure (Eakin, 2006; O'Brien et al., 2006). In already-warm areas exposed to further warming, for instance, less-advantaged populations are less likely to have access to air-conditioning in homes and workplaces. Urban neighbourhoods that are well served by health facilities and public utilities, or have additional economic and technical resources, are better equipped to deal with weather extremes than poor and informal settlement areas, and their actions can affect the poor as well (Sherbinin et al., 2006). Relatively-wealthy market-oriented farmers can afford more expensive deep-well pumps. In coastal settlements, large-scale fishing entrepreneurs can afford to relocate or diversify. By contrast, poverty and marginalisation raise serious issues for impacts and responses, including the following:

a. The poor, who make up half of the world's population and earn less than US$2 a day (UN-Habitat, 2003), cannot afford adaptation mechanisms such as air-conditioning, heating or climate-risk insurance (which is unavailable or significantly restricted in most developing countries). The poor depend on water, energy, transportation and other public infrastructures which, when affected by climate-related disasters, are not immediately replaced (Freeman and Warner, 2001). Instead, they base their responses on diversification of their livelihoods or on remittances and other social assets (Klinenberg, 2002; Wolmer and Scoones, 2003; Eakin 2006). In many countries, recent reductions in services and support from central governments have decreased the resources available to provide adequate preparedness and protection (UN-Habitat, 2003; Eakin and Lemos, 2006). This does not necessarily mean that "the poor are lost"; they have other coping mechanisms (see Section 7.6), but climate change might go beyond what traditional coping mechanisms can handle (Wolmer and Scoones, 2003).

b. Especially in developing countries, where more than 90% of the deaths related to natural disasters occur (UNISDR, 2004) and 43% of the urban slums are located (UN-Habitat, 2003), the poor tend to live in informal settlements, with irregular land tenure and self-built substandard houses, lacking adequate water, drainage and other public services and often situated in risk-prone areas (Romero Lankao et al., 2005). Events such as the December 1999 flash floods and landslides in Caracas, killing nearly 30,000, and the 2001 severe flooding in Cape Town, damaging 15,641 informal dwellings, show us that the poor in these countries are the most likely to be killed or harmed by extreme weather-related events (Sherbinin et al., 2006). During 1985 and 1999 the world's wealthiest nations suffered 57.3% of the measured economic losses due to disasters, about 2.5% of their GDP. The world's poorest countries suffered 24.4% of the economic toll of disasters, but this represented 13.4% of their combined GDP (ADRC et al., 2005).

c. Impacts of climate change are likely to be felt most acutely not only by the poor, but also by certain segments of the population, such as the elderly, the very young, the powerless, indigenous people, and recent immigrants, particularly if they are linguistically isolated, i.e., those most dependent on public support. Impacts will also differ according to gender (Cannon, 2002; Klinenberg, 2002; Box 7.4). This happens particularly in developing countries, where gendered cultural expectations, such as women undertaking multiple tasks at home, persist (Wood and Salway, 2000), and the ratios of women affected or killed by climate-related disasters to the total population are already higher than in developed nations (ADRC et al., 2005).

Government/institutional capacities and resources could also be affected by climate change. Examples from Mexico City, Tokyo, Los Angeles and Manila include requirements for public health care, disaster risk reduction, land-use management, social services to the elderly, public transportation, and even public security, where climate-related stresses are associated with uncoordinated planning, legal barriers, staffing shortages and other institutional constrains (Wisner, 2003; UNISDR, 2004). Where budgets of local or regional governments are affected by increased demands, such effects can lead to calls for either increases in revenue bases or reductions in other government expenditures, which implies a vulnerability of governance systems to climate change (Freeman and Warner, 2001). The disruption of social networks and solidarity by extreme weather events and repeated lower impact events can reduce resilience (Thomas and Twyman, 2005). As sources of stress multiply and magnify in consequence of global climate change, the resilience of already overextended economic, political and administrative institutions will tend to decrease, especially in the most impoverished regions. As Hurricane Katrina has shown, it is likely that if things go wrong people will blame "the Government" (Sherbinin et al., 2006). To avoid such outcomes, governance systems are likely to react to perceptions of growing stresses through regulation and strengthening of emergency management systems (Christie and Hanlon, 2001).

7.4.3 Key vulnerabilities

As a general statement about a wide diversity of circumstances, the major climate-change vulnerabilities of industries, settlements and societies are:

1. vulnerabilities to extreme weather and climate events, particularly if abrupt major climate change should occur, along with possible thresholds associated with more gradual changes;
2. vulnerabilities to climate change as one aspect of a larger multi-stress context: relationships between climate change and thresholds of stress in other regards;
3. vulnerabilities of particular geographical areas such as coastal and riverine areas vulnerable to flooding and continental locations where changes have particular impacts on human livelihoods; most vulnerable are likely to be populations in areas where subsistence is at the margin of viability or near boundaries between major ecological zones, such as tundra thawing in polar regions and shifts in ecosystem boundaries along the margins of the Sahel that may undergo significant shifts in climate;
4. vulnerabilities of particular populations with limited resources for coping with and adapting to climate-change impacts;
5. vulnerabilities of particular economic sectors sensitive to climate conditions, such as tourism, risk financing and agro-industry.

All of these concerns can be linked both with direct effects and indirect effects through inter-connections and linkages, both between systems (such as flooding and health) and between locations.

Most key vulnerabilities are related to (a) climate phenomena that exceed thresholds for adaptation, i.e., extreme weather events and/or abrupt climate change, often related to the magnitude and rate of climate change (see Box 7.4), and (b) limited access to resources (financial, technical, human, institutional) to cope, rooted in issues of development context. Most key vulnerabilities are relatively localised, in terms of geographic location, sectoral focus and segments of the population affected, although the literature to support such detailed findings about potential impacts is very limited. Based on the information summarised in the sections above (Table 7.3), key vulnerabilities of industry, settlement and society include the following, each characterised by a level of confidence.

- Interactions between climate change and urbanisation: most notably in developing countries, where urbanisation is often focused in vulnerable areas (e.g., coastal), especially when mega-cities and rapidly growing mid-sized cities approach possible thresholds of sustainability (very high confidence).
- Interactions between climate change and global economic growth: relevant stresses are linked not only to impacts of climate change on such things as resource supply and waste management but also to impacts of climate change response policies, which could affect development paths by requiring higher cost fuel choices (high confidence).
- Increasingly strong and complex global linkages: climate-change effects cascade through expanding series of international trade, migration and communication patterns to produce a variety of indirect effects, some of which may be unanticipated, especially if the globalised economy becomes less resilient and more interdependent (very high confidence).
- Fixed physical infrastructures that are important in meeting human needs: infrastructures susceptible to damage from extreme weather events or sea-level rise and/or infrastructures already close to being inadequate, where an additional source of stress could push the system over a threshold of failure (high confidence).

Table 7.3. *Selected examples of current and projected climate-change impacts on industry, settlement and society and their interaction with other processes.*

Climate Driven Phenomena	Evidence for Current Impact/ Vulnerability	Other Processes/ Stresses	Projected Future Impact/ Vulnerability	Zones, Groups Affected
a) Changes in extremes				
Tropical cyclones, storm surge	Flood and wind casualties and damages; economic losses: transport, tourism, infrastructure (e.g., energy, transport), insurance (7.4.2; 7.4.3; Box 7.3; 7.5)	Land use/ population density in flood-prone areas; flood defences; institutional capacities	Increased vulnerability in storm-prone coastal areas; possible effects on settlements, health, tourism, economic and transportation systems, buildings and infrastructures	Coastal areas, settlements and activities; regions and populations with limited capacities and resources; fixed infrastructures; insurance sector
Extreme rainfall, riverine floods	Erosion/landslides; land flooding; settlements; transportation systems; infrastructure (7.4.2) (see regional Chapters)	As for tropical cyclones and storm surge, plus drainage infrastructure	As for tropical cyclones and storm surge, plus drainage infrastructure	As for tropical cyclones and storm surge, plus flood plains
Heat or cold-waves	Effects on human health; social stability; requirements for energy, water and other services (e.g., water or food storage), infrastructures (e.g., energy transportation) (7.2; Box 7.1; 7.4.2.2; 7.4.2.3)	Building design and internal temperature control; social contexts; institutional capacities	Increased vulnerabilities in some regions and populations; health effects; changes in energy requirements	Mid-latitude areas; elderly, very young, ill and/or very poor populations
Drought	Water availability, livelihoods; energy generation; migration,; transportation in water bodies (7.4.2.2; 7.4.2.3; 7.4.2.5)	Water systems; competing water uses; energy demand; water demand constraints	Water resource challenges in affected areas; shifts in locations of population and economic activities; additional investments in water supply	Semi-arid and arid regions; poor areas and populations; areas with human-induced water scarcity
b) Changes in means				
Temperature	Energy demands and costs; urban air quality; thawing of permafrost soils; tourism and recreation; retail consumption; livelihoods; loss of melt water (7.4.2.1; 7.4.2.2; 7.4.2.4; 7.4.2.5)	Demographic and economic changes; land-use changes; technological innovations; air pollution; institutional capacities	Shifts in energy demand; worsening of air quality; impacts on settlements and livelihoods depending on melt water; threats to settlements/infrastructure from thawing permafrost soils in some regions	Very diverse, but greater vulnerabilities in places and populations with more limited capacities and resources for adaptation
Precipitation	Agricultural livelihoods; saline intrusion; tourism; water infrastructures; energy supplies (7.4.2.1; 7.4.2.2; 7.4.2.3)	Competition from other regions/sectors. Water resource allocation	Depending on the region, vulnerabilities in some areas to effects of precipitation increases (e.g., flooding, but could be positive) and in some areas to decreases (see drought above)	Poor regions and populations
Saline intrusion	Effects on water infrastructures (7.4.2.3)	Trends in groundwater withdrawal	Increased vulnerabilities in coastal areas	Low-lying coastal areas, especially those with limited capacities and resources
Sea-level rise	Coastal land uses; flood risk, water logging; water infrastructures (7.4.2.3; 7.4.2.4)	Trends in coastal development, settlement and land uses	Long-term increases in vulnerabilities of low-lying coastal areas	As for saline intrusion,
c) Abrupt climate change				
	Analyses of potentials	Demographic, economic, and technological changes; institutional developments	Possible significant effects on most places and populations in the world, at least for a limited time	Most zones and groups

Orange shading indicates very significant in some areas and/or sectors; yellow indicates significant; white indicates that significance is less-clearly established.

- Interactions with governmental and social/cultural structures that are already stressed in some places by other kinds of change: examples include population pressure and limited economic resources, where in some cases structures could become no longer viable when climate change is added as a further stress (medium confidence).

In all of these cases, the valuation of vulnerabilities depends considerably on the development context. For instance, vulnerabilities in more developed areas are often focused on physical assets and infrastructures and their economic value and replacement costs, along with linkages to global markets, while vulnerabilities in less developed areas are often focused on human populations and institutions, which need different metrics for valuation. On the other hand, vulnerabilities to physical and economic costs can have a greater proportional impact in developing areas.

Although it would be useful to be able to associate such general vulnerabilities with particular impact criteria, climate-change scenarios, and/or time frames, the current knowledge base does not support such specificity with an adequate level of confidence.

7.5 Costs and other socio-economic issues

Costs or benefits of climate change-related impacts on industry, settlements and society are difficult to estimate. Reasons include the facts that effects to date that are clearly attributable to climate change are limited, most of the relatively small number of estimates of macroeconomic costs of climate change refer to total economies rather than to the more specific subject matter of this chapter, and generalising from scattered cases that are not necessarily representative of the global portfolio of situations is risky. Historical experience is of limited value when the potentially impacted systems are themselves changing (e.g., with global economic restructuring and development, and technological change), and many types of costs – especially to society – are poorly captured by monetary metrics. In many cases, the only current guides to projecting possible costs of climate change are costs associated with recent extreme weather events of types projected to increase in intensity and/or frequency, although this is only one kind of possible impact and cannot be assumed to be representative of aggregate costs and benefits of all aspects of climate change, including more gradual change.

Estimates of aggregate macroeconomic costs of climate change at a global scale (e.g., Smith et al., 2001) are not directly useful for this chapter, other than generally illustrating that because many locations, industrial sectors and settlements are not highly vulnerable, total monetary impacts at that scale might not be large in proportion to the global economy. As Section 7.4 indicates, however, vulnerabilities of or opportunities for particular localities and/or sectors and/or societies could be considerable. A possible example is climate-related contributions to changes already being experienced by societies and settlements in the Arctic, which include destabilised buildings, roads, airports and industrial facilities and other effects of permafrost conditions, requiring substantial rebuilding, maintenance and investments (ACIA, 2004). An impact assessment in the UK projected that annual weather-related damages to land uses and properties could increase by 3 to 9 times by the 2080s (Harman et al., 2005). More generally, as one specific aspect of vulnerabilities to climate change, possible economic costs of sea-level rise have been estimated, since exposures of coastal areas to a specified scenario can be analysed for costs of the change v. costs of protecting against the change; and effects of direct costs in coastal areas can be projected for other parts of a regional or national economy (Nicholls and Tol, 2006; Tol et al., 2006). Generally, these studies conclude that the costs of full protection are greater than the costs of losing land to sea-level rise, although they do not estimate non-monetary costs of social and cultural effects.

Recent climate-related extreme weather events have been associated with cost estimates for countries and economic sectors; and trends in these costs have been examined, especially by the reinsurance industry (e.g., Swiss Re, 2004; Munich Re, 2005; also Chapter 1, Section 1.3.8). According to these estimates, an increase in the intensity and/or frequency of weather-based natural disasters, such as hurricanes, floods or droughts, could be associated with very large costs to targeted regions in terms of economic losses and losses of life and disruptions of livelihoods, depending on such variables as the level of social and economic development, the economic value of property and infrastructure affected, capacities of local institutions to cope with the resulting stresses, and the effective use of risk reduction strategies. Estimates of impacts on a relatively small country's GDP in the year of the event range from 4 to 6% (Mozambique flooding: Cairncross and Alvarinho, 2006) to 3% (El Niño in Central America: www.eclac.cl/mexico/ and follow the link to 'desastres') to 7% (Hurricane Mitch in Honduras: Figure 7.3). Even though these macroeconomic impacts appear relatively minor, countries facing an emergency found it necessary to incur increased public spending and obtain significant support from the international donor community in order to meet the needs of affected populations. This increased fiscal imbalances and current account external deficits in many countries.

For specific regions and locales, of course, the impact on a local economy can be considerably greater (see Box 7.4). Estimates suggest that impacts can exceed GDP and gross capital formation in percentages that vary from less than 10% in larger, more developed and diversified impacted regions to more than 50% in less developed, less diversified, more natural resource-dependent regions (Zapata-Marti, 2004).

It seems likely that if extreme weather events become more intense and/or more frequent with climate change, GDP growth over time could be adversely affected unless investments are made in adaptation and resilience.

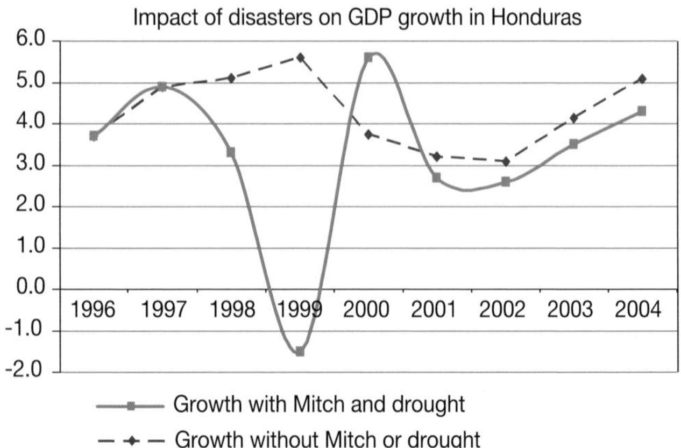

Figure 7.3. *Economic impact of Hurricane Mitch and the 1998 to 1999 drought on Honduras (http://siteresources.worldbank.org/INTDISMGMT/Resources/eclac_LAC&Asia.pdf).*

Box 7.4. Vulnerabilities to extreme weather events in megadeltas in a context of multiple stresses: the case of Hurricane Katrina

It is possible to say with a high level of confidence that sustainable development in some densely populated megadeltas of the world will be challenged by climate change, not only in developing countries but in developed countries also. The experience of the U.S. Gulf Coast with Hurricane Katrina in 2005 is a dramatic example of the impact of a tropical cyclone – of an intensity expected to become more common with climate change – on the demographic, social, and economic processes and stresses of a major city located in a megadelta.

In 2005, the city of New Orleans had a population of about half a million, located on the delta of the Mississippi River along the U.S. Gulf Coast. The city is subject not only to seasonal storms (Emanuel, 2005) but also to land subsidence at an average rate of 6 mm/yr rising to 10-15 mm/year or more (Dixon et al., 2006). Embanking the main river channel has led to a reduction in sedimentation leading to the loss of coastal wetlands that tend to reduce storm surge flood heights, while urban development throughout the 20th century has significantly increased land use and settlement in areas vulnerable to flooding. A number of studies of the protective levee system had indicated growing vulnerabilities to flooding, but actions were not taken to improve protection.

In late August 2005, Hurricane Katrina – which had been a Category 5 storm but weakened to Category 3 before landfall – moved onto the Louisiana and Mississippi coast with a storm surge, supplemented by waves, reaching up to 8.5 m above sea level along the southerly-facing shallow Mississippi Coast (see also Chapter 6, Box 6.4). In New Orleans, the surge reached around 5 m, overtopping and breaching sections of the city's 4.5 m defences, flooding 70 to 80% of New Orleans, with 55% of the city's properties inundated by more than 1.2 m of water and maximum flood depths up to 6 m. In Louisiana 1,101 people died, nearly all related to flooding, concentrated among the poor and elderly.

Across the whole region, there were 1.75 million private insurance claims, costing in excess of US$40 billion (Hartwig, 2006), while total economic costs are projected to be significantly in excess of US$100 billion. Katrina also exhausted the federally-backed National Flood Insurance Program (Hunter, 2006), which had to borrow US$20.8 billion from the Government to fund the Katrina residential flood claims. In New Orleans alone, while flooding of residential structures caused US$8 to 10 billion in losses, US$3 to 6 billion was uninsured. Of the flooded homes, 34,000 to 35,000 carried no flood insurance, including many that were not in a designated flood risk zone (Hartwig, 2006).

Beyond the locations directly affected by the storm, areas that hosted tens of thousands of evacuees had to provide shelter and schooling, while storm damage to the oil refineries and production facilities in the Gulf region raised highway vehicle fuel prices nationwide. Reconstruction costs have driven up the costs of building construction across the southern U.S., and federal government funding for many programmes was reduced because of commitments to provide financial support for hurricane damage recovery. Six months after Katrina, it was estimated that the population of New Orleans was 155,000, with this number projected to rise to 272,000 by September 2008; 56% of its pre-Katrina level (McCarthy et al., 2006).

Research has also considered costs of extreme weather-related events on certain sectors of interest, especially water-supply infrastructures. For instance, if reduced precipitation due to climate change were to result in an interruption of urban water supplies, effects could include disruptions of industrial activity as well as hardships for population, especially the poor, who have the fewest options for alternative supplies. The cost of extending pipelines is considerable, especially if it means that water treatment works also have to be relocated. As a rough working rule, the cost of construction of the abstraction and treatment works and the pumping main for an urban settlement's water supply is about half the cost of the entire system. The cost of flood damage is often even more considerable. For example, the catastrophic flooding of southern Mozambique in 2000 caused damage to water supplies which cost US$13.4 million to repair, or roughly US$50 per person directly affected, of the same order as the cost of providing them with water supplies in the first place (World Bank, 2000). Part of the explanation is that the damaged water supplies also served people whose homes were not directly affected by the flooding; this can be expected to occur in other floods. Nicholls (2004) has estimated that some 10 million people are affected annually by coastal flooding, and that this number is likely to increase until 2020 under all four SRES scenarios, largely because of the increase in the exposed population.

A longer-term concern for industry, settlements and society in developed countries is the prospect of abrupt climate change, which could exceed coping mechanisms in many settlements and societies that would be resilient to gradual climate change

(National Research Council, 2002). In such a case, fixed infrastructures are especially vulnerable, although the research literature is very limited.

Reliable estimates of costs associated with more gradual climate change are scarce, for reasons summarised above, although in some cases cost estimates of adaptation strategies are available (Chapter 17). In general, costs that can be addressed strategically over periods of time have different implications for industry, settlements and society than relatively sudden costs (e.g., Hallegatte et al., 2007). For a combination of gradual changes and extreme events, several recent studies indicate that climate change could reduce the rate of GDP growth over time unless vulnerabilities are addressed (Van Kooten, 2004; Stern, 2007).

The existing literature is, in these ways, useful in considering possible costs of climate change for industry, settlement and society; but it is not sufficient to estimate costs globally or regionally associated with any specific scenario of climate change. What can be said at the present time is that economic costs of extreme weather events at a large national or large regional scale, estimated as a percent of gross product in the year of the event, are unlikely to represent more than several percent of the value of the total economy, except for possible abrupt changes (high confidence), while net aggregate economic costs of extreme event impacts in smaller locations, especially in developing countries, could in the short run exceed 25% of the gross product in that year (high confidence). To the degree that these events increase in intensity and/or frequency, they will represent significant costs due to climate change. For industry, settlements and society, economic valuations of other costs and benefits associated with climate change are generally not yet available.

7.6 Adaptation: practices, options and constraints

7.6.1 General perspectives

Challenges to adapt to variations and changes in environmental conditions have been a part of every phase of human history, and human societies have generally been highly adaptable (Ausubel and Langford, 1997). Adaptations may be anticipatory or reactive, self-induced and decentralised or dependent on centrally-initiated policy changes and social collaboration, gradual and evolutionary or rooted in abrupt changes in settlement patterns or economic activity. Historically, adaptations to climate change have probably been most salient in coastal areas vulnerable to storms and flooding, such as the Netherlands, and in arid areas needing water supplies; but human settlements and activities exist in the most extreme environments on earth, which shows that the capacity to adapt to known conditions, given economic and human resources and access to knowledge, is considerable.

Adaptation strategies vary widely depending on the exposure of a place or sector to dimensions of climate change, its sensitivity to such changes, and its capacities to cope with the changes (Chapter 17). Some of the strategies are multi-sectoral, such as improving climate and weather forecasting at a local scale, emergency preparedness and public education. One example of cross-cutting adaptation is improving information and institutions for emergency preparedness. Systematic disaster preparedness at community level has helped reduce death tolls; for instance, new warning systems and evacuation procedures in Andhra Pradesh, India, reduced deaths from coastal tropical cyclones by 90%, comparing 1979 with 1977 (Winchester, 2000), and poor societies in other parts of the Bay of Bengal area have undertaken practical measures to reduce flood risks due to high levels of awareness and motivation among local communities. However, the effectiveness of such systems in reaching marginal populations, and their responses to such warnings, is uneven; and the timing of decisions to adapt affects the likely benefits.

Other strategies are focused on a sector, such as water, energy, tourism and health (see Chapters 3 and 8). Some are geographically focused, such as coastal area and floodplain adaptations, which can involve such initiatives as changing land uses in highly vulnerable areas and protecting critical areas. Adaptation, in fact, tends very often to be context-specific, within larger market and policy structures (Adger et al., 2005a), although it generally takes place within the larger context of globalisation (Benson and Clay, 2003; Sperling and Szekely, 2005).

There is a considerable literature on adaptations to climate *variation* and on vulnerabilities to *extreme events*, especially in developed countries; but research on potentials and costs of adaptation by industry, settlement and society to *climate change* is still in an early stage (Chapter 17). One challenge is that it is still difficult to project changes in particular places and sectors with much precision, whether by downscaling global climate models or by extrapolating from past experience with climate variation. Uncertainty about the distribution and timing of climate-change impacts at the local level makes judgments about the scale and timing of adaptation actions very difficult. Where there are co-benefits between climate-change adaptation and other economic or social objectives, there will be reasons for early action. In other cases, limits on predictability tend to delay adaptation (Wright and Erickson, 2003). In addition, there is little scientific basis as yet for assessing possible limits of adaptation, especially differences among locations and systems. In particular, the knowledge base about costs of adaptation is less well developed than the knowledge base about possible adaptation benefits. At least in some cases, costs might exceed actual benefits.

7.6.2 Industry

The extent to which potential vulnerabilities of industry are likely to motivate adaptation will depend to a large extent on the flexibility of business and on its capacity to adapt. In general, those industries with longer-lived capital assets (e.g., energy), fixed or weather-dependent resources (mining, food and agriculture), and extended supply chains (e.g., the retail-distribution industry) are likely to be more vulnerable to climate-change impacts. But many of these industries, especially in the industrialised world, are likely to have the technological and economic resources necessary both to recover from the impacts of extreme events (partly by sharing and spreading risk or by moving to safer locations), and to adapt over the longer term to more gradual changes. It is also clear that many other economic and social factors are likely to play a more important role in influencing innovation and change in industry than climate change. For many businesses, climate risk management can be integrated into overall business strategy and operations where it will be regarded as one among many issues that demand attention, to the degree that such adaptation is supported by investors and shareholders.

There is now considerable evidence emerging in Europe, North America and Japan that the construction and transportation sectors are paying attention to climate-change impacts and the need for adaptation (Lisø et al., 2003; Shimoda, 2003; Salagnac, 2004; Chapter 17, Section 17.2.2). As one example, the US$1 billion 12.9 km Confederation Bridge between New Brunswick and Prince Edward Island in Canada, which opened in 1997, was built one metre higher to accommodate anticipated sea-level rise over its 100-year lifespan (McKenzie and Parlee, 2003). A range of technical advice is now available to planners, architects and engineers on climate impacts risk assessment (Willows and Connell, 2003), including specialised advice on options for responding to these risks (Lancaster et al., 2004). A few early estimates of possible costs of adaptation measures are beginning to be available; for instance, O'Connell and Hargreaves (2004) show that measures to reduce wind damage, flood risk and indoor heat would add about 5% to the cost of a typical new house in New Zealand.

Business adaptations will be in response to both direct impacts (involving direct observations of risks and opportunities as a result of changing climatic conditions) and indirect impacts (including changing regulatory pressures and consumer demand) as illustrated in Table 7.2. Adaptations can also take a wide variety of forms. They may include changes in business processes, technologies or business models (Hertin et al., 2003), or changes in the location of activities. Many of these adaptations represent incremental adjustments to current business activities (Berkhout et al., 2006). For instance, techniques already exist for adapting buildings in response to greater risks of ground movement (deeper foundations), higher temperatures (passive and active cooling) and driving rain (building techniques and cladding technologies). Frequently these adaptations are relatively low-cost and represent best practice (ACIA, 2004). For more structural adaptations – such as choice of location for industrial facilities – planning guidance, government policy and risk management by insurers will play major roles.

Awareness, capabilities and access to resources that facilitate adaptation are likely to be much less widely available in less developed contexts, where industrial production often takes place in areas vulnerable to flooding, coastal erosion and land slips. Production is also more likely to be tied to natural resources affected by changing climates. Potentials for adaptation to climate change in informal sectors in developing countries depend largely on the context: e.g., the impacts involved, the sensitivity of the industrial activity to those impacts, and the resources available for coping. Examples of adaptive strategies could include relocating away from risk-prone locations, diversifying production activities, and reducing stresses associated with other operating conditions to add general resiliency. Informal industry employs minimal capital and few fixed assets, so that it usually adapts relatively quickly to gradual changes. But adaptations that are substantial may call for an awareness of threats and responses to them that go beyond historical experience, a willingness to depart from traditional activity patterns, and access to financial resources not normally available to some small producers.

The energy sector can adapt to climate-change vulnerabilities and impacts by anticipating possible impacts and taking steps to increase its resilience, e.g., by diversifying energy supply sources, expanding its linkages with other regions, and investing in technological change to further expand its portfolio of options (Hewer, 2006; Chapter 12, Section 12.5.8). This sector has impressive investment resources and experience with risk management, and it has the potential to be a leader in industrial adaptation initiatives, whether related to reducing risks associated with extreme events or coping with more gradual changes such as in water availability. On the other hand, many energy sector strategies involve high capital costs, and social acceptance of climate-change response alternatives that might imply higher energy prices could be limited. Adaptation prospects are likely to depend considerably on the availability of information about possible climate-change effects to inform decisions about adaptive management.

7.6.3 Services

Concerns about vulnerabilities and impacts for services are likewise concentrated on sectors especially sensitive to climate variation, such as recreation and tourism; and adaptations are also likely to be associated with changes in costs/prices, applications of technology, and attention to risk financing. For instance, wholesale and retail trades are likely to adapt by increasing or reducing space cooling and/or heating, by changing storage and distribution systems to reduce vulnerabilities, and by changing the consumer goods and services offered in particular locations. Some of these adaptations, although by no means all of them, could increase prices of goods and services to consumers.

Where climate change affects comparative advantages for regions in the global economy, trade patterns are likely to adapt largely through market mechanisms as the changes unfold rather than through strategies to reduce risk in anticipation of changes (Figure 7.2). In a general sense, there will be 'winners' and 'losers' as a result, potentially affecting economic growth and employment in both kinds of cases, which suggests the possible value of anticipatory planning and policy discourse. In many cases, building robust ties with the globalising economy could be a useful response to possible climate changes for places and societies built around small-scale social interactions and enterprises, because those ties could open up a wider range of possible alternatives for adaptation.

The short time-scales at which most commercial services operate allow great flexibility for adapting to climate change. Within the retail industry, it is likely that commerce will capitalise on long-term trends in consumer behaviour and lifestyle, relating to climate change through an expansion of markets for cooling equipment, and facilities and goods for outdoor recreation in temperate climates. Large injections of capital may be required to relocate commercial premises from low-lying areas vulnerable to flooding. In addition, technological investment will be required to reduce carbon emissions while maintaining competitive prowess in the global market. The most vulnerable are communities (particularly in developing countries) whose economy is based on the production and distribution of a restricted range of climate-sensitive commodities. For these communities, economic diversification should be a key response to reduce vulnerability.

The tourism sector may in some cases be able to adapt to long-term trends in climate change, such as increasing temperatures, at a cost, for instance by investing in snow-making equipment (see Chapter 14, Section 14.4.7), beach enhancement (see Chapter 6, Sections 6.5.2 and 6.6.1.1), or additional air-conditioning. The sustainability of some adaptation processes may be questionable: air-conditioning because of its energy use, snowmaking for its pressure on water resources or its costs (O'Brien et al., 2004). However, climate change is not likely to be linear, and the frequency and intensity of extreme climatic events, which affect not only the reality of risks, but also the subjective risk-perception of tourists, might become a far greater problem for the tourist industry. There are three categories of adaptation processes: technological, managerial and behavioural. While tourism providers tend to focus on the first two (preserving tourism assets, diversifying supply), tourists might rather change behaviour: they might visit new, suitable locations (for example snow-safe ski resorts at higher altitudes or in other regions) or they might travel during other periods of the year (for example, they might visit a site in spring instead of summer to avoid extreme temperatures). Awareness, adaptive capacities and strategies are likely to vary according to the wealth and the education of different categories of tourists and also among other stakeholders. For example, large tour operators should be able to adapt to changes in tourist destinations, as they are familiar with strategic planning, do not own the infrastructures and can, to some extent, shape demand through marketing.

Perhaps of even greater importance is the role of mobility in future tourism. Increasing prices for fuel and the need to reduce emissions might have substantial effects on transport availability and costs. For instance, the price of air transport, now the means of transport of 42% of all international tourists, is expected to rise in stabilisation scenarios (Gössling and Hall, 2005). This might call for adaptation in terms of leisure lifestyles, such as the substitution of long-distance travel by vacationing at home or nearby (Dubois and Ceron, 2005).

It also seems likely that tourism based on natural environments will see the most substantial changes due to climate change, including changes in economic costs (Gössling and Hall, 2005) and changes in travel flows. Tropical island nations and low-lying coastal areas may be especially vulnerable, as they might be affected by sea-level rise, changes in storm tracks and intensities (Chapter 16; Chapter 4, Section 4.2), changes in perceived climate-related risks, and changes in transport costs, all resulting in concomitant detrimental effects for their often tourism-based economies. In any of these cases, the implications are most notable for areas in which tourism represents a relatively large share of the local or regional economy, and these are areas where adaptation might represent a relatively significant need and a relatively significant cost.

The insurance sector has an important role to play in adaptation (Mills, 2004) as it is in the business of calculating risk costs and has begun to explore how risks can be expected to change into the future (Association of British Insurers, 2002). By communicating risk information to individual stakeholders, as through insurance pricing signals, insurers can help inform appropriate adaptive behaviours, although regulated markets or flat-rated insurance systems obstruct the transmission of the information required to motivate adaptation. Through reductions in premiums charged, insurance can also reward actions taken to reduce risk, such as by fitting hurricane shutters on a building or by the construction of local flood defences.

Where new risks are emerging, or known risks are increasing, new insurance coverages have been designed to help spread losses. Examples include the creation of weather derivatives, crop insurance and expanded property insurance coverage.

Generally, it is recognised that 'ex-ante' (before the fact) funding mechanisms in the form of insurance should be more beneficial for the affected community and the whole country's economy than ex-post (after the fact) mechanisms by means of credit, government subsidies or private donations. Only the ex-ante approach offers the surety of payments as well as the potential to influence the level of risk, through linking insurance prices and conditions with government policy on hazard mitigation, implementation, and supervision of building codes etc., thus reducing a country's financial vulnerability and giving improved prospects for investment and economic growth (Gurenko, 2004). However, in developing countries there are questions about the viability of such approaches, concerning who in a poor country is able to afford an ex-ante premium and how real reductions in risk can be achieved in a society with relatively low risk literacy (Linnerooth-Bayer et al., 2005). Other potential sources of developing country adaptation funding are discussed by Bouwer and Aerts (2006).

Besides incentivising adaptation, the insurance industry itself will need to adapt to stay financially healthy. The main threat is a combination of very high loss events in a short time period (as almost happened in September 2005 with Hurricane Rita heading for the city of Houston after Hurricane Katrina had hit New Orleans). Trends that contribute to increasing the robustness of the sector include better risk management, greater diversification, better risk and capital auditing, greater integration of insurance with other financial services, and improved tools to transfer risks out of the insurance market into the capital markets through catastrophe risk securitisations (European Environment Agency, 2004), which have seen significant increases in value issued since 2004.

The key vulnerability of the current system of risk-bearing concerns the non-availability or withdrawal of private insurance cover, in particular related to flood risk. However, the threat of withdrawal can itself be a spur for adaptation. Following the October-November 2000 floods in England and Wales, the Association of British Insurers negotiated an increased allocation of government expenditure on flood defences and a stakeholder role in decisions around future development in floodplains, by threatening to withdraw flood insurance from locations at greatest risk (Association of British Insurers, 2002). With expectations for rising levels of flood risk in developed countries, political pressures demand that if private insurance is withdrawn, state-backed alternatives should be created leading to increased liabilities for governments. Without such a backstop more significant adaptive measures may be triggered. In the northern Bahaman islands of Abaco and Grand Bahama (hit by three major hurricanes and their associated storm surges between 1999 and 2004), in 2005 flood insurance was withdrawn for some residential developments, ending the ability to raise a bank-loan mortgage. Without a state-backed alternative, houses became abandoned as their value collapsed (Woon and Rose, 2004). Meanwhile, builders have begun to construct new houses in the Bahamian coastal floodplain on concrete stilts, bringing some properties back into the domain of insurability. Similar adaptive outcomes can be expected in other coastal regions affected by increasing flood risk.

7.6.4 Utilities/infrastructure

The most general form of adaptation by infrastructures vulnerable to impacts of climate change is investment in increased resilience, for instance in new sources of water supply for urban areas. Most fields of infrastructure management, including water, sanitation, transportation and energy management, incorporate vulnerabilities to changing trends of supply and demand, and risks of disturbances in their normal planning.

In a situation where climate change, observed or projected, indicates a need for different patterns or priorities in infrastructure planning and investment, common strategies are likely to include increases in reserve margins and other types of backup capacity, attention to system designs that allow adaptation and modification without major redesign and that can handle more extreme conditions for operation. In many cases an issue is tradeoffs between capital costs and operating expenditures.

With regard to infrastructure where adaptation requires long lead times, such as water supply, there is evidence that adaptation to climate change is already taking place. An example would be the planning of British water companies mentioned in Section 7.4.2.3.1 above, undertaken at the behest of the UK Environment Agency (Environment Agency, 2004). Another would be the decision taken in 2004 to install a desalination plant to supplement the dwindling flows available for water supply for the city of Perth, Australia (Chapter 11, Section 11.6).

The infrastructure whose adaptation is especially important for the reduction of key vulnerabilities is that installed for flood protection. For example, London (UK) is protected from major flooding by a combination of tidal defences, including the Thames Barrier, and river defences upstream of the Barrier. The current standard for the tidal defences is about a 2000 to 1 chance of flooding in any year or 0.05% risk of flooding, and this is anticipated to decline to its original design standard of a 1000 to 1 chance, or 0.1% risk of flooding, as sea level rises, by 2030. The defences are being reviewed, in the light of expected climate changes. Preliminary estimates of the cost of providing a 0.1% standard through to the year 2100 show that a major investment in London's flood defence infrastructure of the order of UK£4 billion will be required within the next 40 years (London Climate Change Partnership, 2004). The capacity of storm drainage systems will also need to be increased to prevent local flooding by increasingly intense storms (UK Water Industry Research, 2004).

7.6.5 Human settlement

Adaptation strategies for human settlements, large and small, include assuring effective governance, increasing the resilience of physical and linkage infrastructures, changing settlement locations over a period of time, changing settlement form, reducing heat-island effects, reducing emissions and industry effluents as well as improving waste handling, providing financial mechanisms for increasing resiliency, targeting assistance programmes for especially impacted segments of the population, and adopting sustainable community development practices (Wilbanks et al., 2005). The choice of strategies from among the options depends in part on their relationships with other social and ecological processes (O'Brien and Leichenko, 2000) and the general level of economic development, but recent research indicates that adaptation can make a significant difference; for instance, the New York climate impact assessment projects significant increases in heat-related deaths (Rosenzweig and Solecki, 2001a), based on historical relationships, while the Boston CLIMB assessment (Kirshen et al., 2007) projects that heat-related deaths will decline because of adaptation over the coming century.

The recent case study of London demonstrates that climate change could bring opportunities as well as challenges, depending on socio-economic conditions, institutional settings, and cultural and consumer values (London Climate Change Partnership, 2004). One of the opportunities, especially in growing settlements, is to work towards a more sustainable city and to improve the quality of life for residents (Box 7.5). This can be achieved by making sure that urban planning takes into account the construction density, the distribution and impact of heat emissions, transportation patterns, and green spaces that can reduce not only heat-island effects.

Models have been established to predict the impact of urban thermal property manipulation strategies resulting from albedo and vegetation changes (Akbari et al., 1997) and urban form manipulation (Emmanuel, 2005). The diurnal air temperature inside urban wooded sites and the cooling effect of trees on urban streets and courtyards, and of groves and lawns, has been extensively quantified in Tel-Aviv, Israel (Shashua-Bar and Hoffman, 2002, 2004). For the Los Angeles region, several studies (Taha, 1996; Taha et al., 1997) projected the effects of increasing citywide albedo levels on mitigating the regional heat island (California's South Coast Air Basin, or SoCAB). A doubling of the surface albedo or a doubling of vegetative cover were each projected to reduce air temperature by approximately 2°C. Moreover, the study area was projected to experience a decrease in ozone concentration.

Other adaptive responses by settlements to concerns about climate change tend to focus on institutional development, often including improved structures for co-ordination between individual settlements and other parties, such as enhanced

Box 7.5. Climate-change adaptation and local government

Threats and opportunities presented by climate change are typically focused at a local scale; and it makes sense for local authorities, including mayors, to consider adaptive responses. Climate change can threaten lives, property, environmental quality and future prosperity by increasing the risk of storms, flooding, landslides, heatwaves and drought and by overloading water, drainage and energy supply systems.

Local governments around the world already play a part in climate-change mitigation, but they can also play a role in adaptation (see Chapter 14, Section 14.5.1; Chapter 18, Section 18.7.2), as guarantors of public services and as facilitators, mobilising stakeholders – such as local businesses, developers, utilities, insurers, educational institutions and community organisations – to contribute their technical and even financial resources to a joint initiative, such as the one formed for London (London Climate Change Partnership, 2004).

In many cases, in fact, good governance is a key to climate-change risk management strategies. For example, effective zoning can prevent the encroachment of housing on slopes prone to erosion and landslides; and adequate investment in and maintenance of infrastructure will make the settlement less vulnerable to weather extremes.

regional water supply planning and infrastructure development (Rosenzweig and Solecki, 2001a; Bulkeley and Betsill, 2003). Often, settlements exist in a splintered political landscape that makes coherent collaborative adaptation strategies difficult to contemplate. Policy responses and planning decisions are also hampered by the reactive nature of much policymaking and by the failure to co-ordinate across relevant professional disciplines, related mainly to current obvious problems, when climate change is viewed as a long-term issue with considerable uncertainty.

One approach for improving the understanding of how settlements may respond to climate-change impacts is to consider 'analogues' - circumstances in recent history when those settlements have confronted other environmental management challenges. In Vietnam for example villagers have been forced over the centuries to clean, repair and strengthen their irrigation channels and sea dykes before the start of every annual tropical storm season (UNISDR, 2004). In many cases, settlements have acted under the pressure of immediate crises to seek solutions by going beyond their own borders. Cities such as Mexico City have both drawn upon water from and sent sewage water to hinterlands outside their boundaries to deal with weather-related water scarcity and floods. These actions have imposed externalities on those hinterlands (Romero Lankao, 2006).

7.6.6 Social issues

There has been a recent shift in perceptions of how settlements and society can better adapt to climate related disasters, away from humanitarian and post-disaster actions toward more anticipatory integrative risk reduction measures that include environmental management, structural measures, protection of critical facilities, land-use planning, financial instruments and early warning systems (UNISDR, 2004). These strategies recognise (a) linkages between risks, vulnerability and development, (b) the importance of creating community assets and capacity to face sudden and slow onset disasters, (c) the key role of a democratic implementation of such strategies, and (d) the need to relate those actions to sustainability goals (UNISDR, 2004; Velásquez, 2005). This approach is practised successfully in countries such as the Philippines, Bangladesh, India, Cuba, Vietnam, Malaysia, Switzerland and France (UNISDR, 2004). In Manizales and Medellin, Colombia, and Uganda, for example, the economic damage and death toll due to landslides and floods has diminished noticeably, thanks to actions such as reforestation, improved drainage systems, poverty reduction and decentralisation of risk avoidance planning (Velásquez, 2005). On the other hand, the experience of a disaster is likely to reduce the adaptive capacity of the affected society for a time; adaptive capacity is often reduced during periods of recovery.

The most difficult challenges occur when decision makers lack training and access to information about climate-change implications, risk management and possible responses, when fiscal constrains limit local flexibility, and when infrastructure, technological and institutional capacities for coping with any major challenge are inadequate (UN-Habitat, 2003). However, in the best-case scenarios, policy focusing on adaptation has the potential to create positive synergies between outcomes (better managed natural and social systems) and processes (governance that promotes democratic decision making, participatory management strategies, equity, transparency and accountability), which in turn will result in more resilient systems (UNISDR, 2004; Adger et al., 2005b).

Yet adaptation is not limited to purposeful actions to reduce societies' sensibility to climate change, alter the exposure of the system to it, and increase the resilience of the system to it (Smit et al., 2000). It also includes spontaneous actions which can be implemented at different scales, from individuals to systems, and are not uniform. Individual adaptations may not produce systemic adaptation, and adaptation at a system level may not benefit all individuals (Thomas and Twyman, 2005). Indeed, some adaptations (e.g., warning systems) may not reach poor communities or not fit their information needs (Ferguson, 2003). They may increase the vulnerability of some peoples and places. For example, coastal planning for increased erosion rates includes engineering decisions that potentially impact neighbouring coastal settlements through sediment transport and other physical processes (Adger et al., 2006). As climate change and adaptation becomes a widespread need, there is likely to be competition for resources – investment in one place, sector or risk will reduce the funds available for others, and possibly reduce funding for other social needs (Winchester, 2000).

One challenge to both private (including businesses and NGOs) and public actors is how to build adaptive capacity in the context of current institutional reforms, new trade agreements and changing relationships between the private and public sectors (Lemos and Agrawal, 2003), including roles of environmental organisations. On the one hand, the emergence of new governance structures at the global level (such as the United Nations Framework Convention on Climate Change - UNFCCC) and across the public-private divide (such as public-private partnerships) has provided new tools for policy design and implementation that may build adaptive capacity (Mitchell and Romero Lankao, 2004; Sperling and Szekely, 2005; Eakin and Lemos, 2006). On the other hand, a transfer of authority from the state to lower levels (through decentralisation and privatisation), in some cases related to developments with international regimes and organisations, may have diminished national government capacities to implement adaptation policies (Jessop, 2002; UN-Habitat, 2003). For example, while decentralisation in Latin America, in principle, allows for better decision-making at the local level, it also constrains the state's ability to regulate and distribute critical resources to adaptation (Eakin and Lemos, 2006). Similarly, West African pastoral Peulhs or Fulbes lost access to water and pastures at the hands of settled agricultural people who gained local power in the process of decentralisation (Van Dijk et al., 2004). In contrast, the design of participatory, integrated and decentralised institutions such as in Brazil's recent water reform is likely to build adaptive capacity to climate change in settlements and societies by improving availability and access to technology, involving stakeholders, and encouraging sustainable resource use (Lemos and Oliveira, 2004).

Adaptive capacity is highly uneven across human societies (Adger et al., 2005a, 2006). Among communities that rely on the exploration of natural resources, adaptation practices may benefit some parts of the community more than others. Even within countries with seemingly high capacities to adapt (based on aggregate national indicators for GDP, education levels and technology), there are likely to be some regions and groups that face barriers and constraints to adaptation (O'Brien et al., 2006). For a discussion of strategies for reducing vulnerabilities of the poor to climate change through adaptation, see UNDP et al. (2003).

Among rural communities in Africa and Latin America, one strategy to build adaptive capacity has been to diversify livelihood strategies (Thomas and Twyman, 2005; Eakin, 2006). Rural settlements can cope with a seasonal downturn in rainfall or a mid-season drought by moving livestock, harvesting water, shifting crop mixes and migrating (Scoones et al., 1996); however, without occasional high rainfall periods, and without institutional support, longer-term livelihood sustainability is severely compromised (Eakin, 2006; Eakin and Lemos, 2006). Measures focussed on reducing poverty and increasing access to resources (e.g., the referred landslide management programmes) may enhance the resilience of affected communities or economic sectors.

7.6.7 Key adaptation issues

The central issues for adaptation to climate change by industry, settlements and society are (a) impact types and magnitudes and their associated adaptation requirements, (b) potential contributions by adaptation strategies to reducing stresses and impacts, (c) costs of adaptation strategies relative to benefits, and (d) limits of adaptation in reducing stresses and impacts under realistically conceivable sets of policy and investment conditions (Downing, 2003). Underlying all of these issues, of course, is the larger issue of the adaptive *capacity* of a population, a community, or an organisation: the degree to which it can (or is likely to) act, through individual agency or collective policies, to reduce stresses and increase coping capacities (Chapter 17). In many cases, this capacity differs significantly between developing and developed countries, and it may differ considerably among locations, economic sectors and populations even within the same region (Millennium Ecosystem Assessment, 2005).

Many of the possibly-impacted activities and groups addressed by this chapter are capable of being highly adaptable over time, given information to inform awareness of possible risks and opportunities and financial and human resources for responses (Chapter 17). In some cases, adaptations to possible climate changes can offer opportunities for positive impacts, especially where those actions also address other adaptive management issues (Chapter 20). The knowledge base on disaster response suggests that a number of approaches may be helpful in enhancing and facilitating adaptive behaviour: systems to provide advance warning of changes, especially extreme events; institutional structures that facilitate collective action and provide external linkages; economic systems that offer access to alternatives; increased attention to adaptive structures that are locally appropriate, geographically and/or sectorally; contingency planning and risk financing, which may include strategic stockpiles; incorporating climate-change vulnerability into land-use planning and environmental management for the long term; public awareness/capacity building regarding risks of climate-change impacts; and in some cases physical facility investment, such as flood walls, beach restoration or emergency shelters.

Although the research literatures on adaptation prospects for industry, settlement and society are as yet rather limited, it appears that:

1. Prospects for adaptation depend on the magnitude and rate of climate change: adaptation is more feasible when climate change is moderate and gradual than when it is massive and/or abrupt. However, actual adaptation strategies and measures are often triggered by relatively extreme weather events (high confidence).
2. Climate-change adaptation strategies are inseparable from increasingly strong and complex global linkages. Industrial planning, human settlements and social development are not isolated from changes in other systems or scales. The urban and rural are interconnected, as are developed and developing societies. This issue is becoming more salient as the globalised economy becomes more interdependent. Adaptation decisions for local activities owned or controlled by external systems involve different processes from adaptation decisions for local activities that are under local control (high confidence).
3. Climate change is one of many challenges to human institutions to manage risks. In any society, institutions have developed risk management mechanisms for such purposes,

from family and community self-help to insurance and re-insurance. It is not clear whether, where and to what degree existing risk management structures are adequate for climate change; but these institutions have considerable potential to be foundations for a number of kinds of adaptations (high confidence).

4. Adaptation actions can be effective in achieving their specific goals, but they may have other effects as well. These might be unintended consequences (e.g., increased flood risk downstream), reducing support for mitigation (e.g., higher energy demand with air-conditioning), or reducing resources available to address vulnerabilities elsewhere (e.g., budget constraints affecting other development goals). The benefits of adaptation may be delayed or not realised at all, for example, when design standards are raised to protect against a storm of a certain magnitude that does not occur for another fifty years, if then (medium confidence).

7.7 Conclusions: implications for sustainable development

Sustainable development is largely about people, their well-being, and equity in their relationships with each other, in a context where nature-society imbalances can threaten economic and social stability. Because climate change, its drivers, its impacts and its policy responses will interact with economic production and services, human settlements and human societies, climate change is likely to be a significant factor in the sustainable development of many areas (e.g., Downing, 2002). Simply stated, climate change has the potential to affect many aspects of human development, positively or negatively, depending on the geographic location, the economic sector, and the level of economic and social development already attained (e.g., regarding particular vulnerabilities of the poor, see Dow and Wilbanks, 2003). Because settlements and industry are often focal points for both mitigation and adaptation policy-making and action, these interactions are likely to be at the heart of many kinds of development-oriented responses to concerns about climate change.

In most cases, with the Arctic being a notable exception (ACIA, 2004), these connections between climate change and sustainable development will only *begin* to emerge in the next decade or two (e.g., during the period embraced by the Millennium Development Goals) as a result of significant impacts that can be attributed to climate change. But industry, settlements and societies will be important foci of mitigation actions and adaptations involving land uses and capital investments with relatively long lifetimes. In the meantime, however, actions that address challenges of climate <u>variability</u>, including extreme events, contribute to environmental risk management as well as reducing possible impacts of climate <u>change</u>.

The most serious issues for sustainable development associated with climate-change impacts on the subjects of this chapter are: (a) threats to vulnerable regions and localities from gradual ecological changes leading to impact thresholds and extreme events that could disrupt the sustainability of societies and cultures, with particular attention to coastal areas in current storm tracks and to economies and societies in polar areas, dry land areas and low-lying islands, and (b) threats to fragile social and environmental systems, both from abrupt climate changes and thresholds associated with more gradual climate changes that would exceed the adaptive capacities of affected sectors, locations and societies. Examples include effects on resource supply for urban and industrial growth and waste management (e.g., flooding). As a very general rule, sensitivities of more-developed economies to the implications of climate change are less than in developing economies; but effects of crossing thresholds of sustainability could be especially large in developed economies whose structures are relatively rigid rather than adaptable. In the case of either developed or developing countries, social system inertia may delay adaptive responses when experienced climate change is gradual and moderate.

In general, however, climate change is an issue for sustainable development mainly as *one of many* sources of possible stress (e.g., O'Brien and Leichenko, 2000, 2003; Wilbanks, 2003b). Its significance lies primarily in its interactions with other stresses and stress-related thresholds, such as population growth and redistribution, social and political instability, and poverty and inequity. In the longer run, climate change is likely to affect sustainable development by reshaping the world map of comparative advantage which, in a globalising economy, will support sustainable development in some areas but endanger it in others, especially in areas with limited capacities to adapt. Underlying such questions, of course, are the magnitude and pace of climate change. Most human activities and societies can adapt given information, time and resources, which suggests that actions which moderate the rate of climate change are likely to reduce the negative effects of climate change on sustainable development (Wilbanks, 2003b).

At the same time, development paths may increase or decrease vulnerabilities to climate-change impacts. For instance, development that intensifies land use in areas vulnerable to extreme weather events or sea-level rise adds to risks of climate-change impacts. Another example is development that moves an economy and society toward specialisation in a single economic activity if that activity is climate-sensitive; development that is more diversified is likely to be less risky. In many cases, actions that increase resilience of industry, settlements and society to climate change will also contribute to development with or without climate change by reducing vulnerabilities to climate variation and increasing capacities to cope with other stresses and uncertainties (Wilbanks, 2003b).

Impacts of climate change on development paths also include impacts of climate-change <u>response</u> policies, which can affect a wide range of development-related choices, from energy sources and costs to industrial competitiveness to patterns of tourism. Areas and sectors most heavily dependent on fossil fuels are especially likely to be affected economically, often calling for adaptation strategies that may in some cases require assistance with capacity building, technological development and transition financing.

7.8 Key uncertainties and research priorities

Because research on vulnerabilities and adaptation potentials of human systems has lagged behind research on physical environmental systems, ecological impacts and mitigation, uncertainties dominate the subject matter of this chapter. Key issues include (a) uncertainties about climate-change impacts at a relatively fine-grained geographic and sectoral scale, both harmful and beneficial, which undermine efforts to assess potential benefits from investments in adaptation; (b) improved understanding of indirect second and third order impacts: i.e., the trickle down of primary effects, such as temperature or precipitation change, storm behaviour change and sea-level rise, through interrelationships among human systems; (c) relationships between specific effects in one location and the well-being of other locations, through linkages in inflows/outflows and inter-regional trade and migration flows; (d) uncertainties about potentials, costs and limits of adaptation in keeping stressful impacts within acceptable limits, especially in developing countries and regions (see Parson et al., 2003); and (e) uncertainties about possible trends in societal, economic and technological change with or without climate change. A particular challenge is improving the capacity to provide more quantitative estimates of impacts and adaptation potentials under the sets of assumptions included in SRES and other climate-change scenarios and scenarios of greenhouse gas emissions stabilisation, especially for time horizons of interest to decision makers, such as 2020, 2050 and 2080.

All of these issues are very high priorities for research in both developed and developing countries, with certain differences in emphasis related to the different development contexts. As a broad generalisation, the primary impact issue for developed countries is the possibility of abrupt climate change, which could cause changes too rapid and disruptive even for a relatively developed country to absorb, at least over a period of several decades. High priorities include reducing uncertainty about the potential for adaptation to cope with climate-change impacts in the absence of abrupt climate change, considering possible responses to threats from low-probability/high consequence contingencies, and considering interactions between climate change and other stresses. The primary impact issue for developing countries is the possibility that climate change, combined with other stresses affecting sustainable development, could jeopardise livelihoods and societies in many regions. High priorities include improving the understanding of multiple-stress contexts for sustainable development and improving the understanding of climate-sensitive thresholds for components of sustainable development paths.

Some of these uncertainties call for careful location- and sector-specific research, including better information about the geographic distribution of vulnerabilities of settlements and societies at a relatively localised scale, emphasising especially vulnerable areas, such as coastal areas in lower-income developing countries, and especially vulnerable sectors, such as tourism, and possible financial thresholds regarding the insurability of climate-change impact risks. Others call for attention to cross-sectoral and multi-locational relationships between climate change, adaptation and mitigation (Chapter 18), including both complementarities and trade-offs in policy and investment strategies. Underlying all of these issues for industry, settlement and society are relationships between possible climate-change impact vulnerabilities and adaptation responses and broader processes of sustainable economic and social development, which suggest a need for a much greater emphasis on research that investigates such linkages. In some cases, because of the necessarily speculative nature of research about future contingencies, it is likely to be useful to consider past experiences with climate variability and analogues drawn from other experiences with managing risks and adapting to environmental changes and stresses (e.g., Abler, 2003). In many others, an important step will be to establish mechanisms for monitoring interactions between emerging climate changes and other processes and stresses in order both to learn from the observations and to provide early alerts regarding potential problems or opportunities.

Underlying all of these research needs are often very serious limitations on available data to support valid analysis, especially data on nature-society linkages and data on relatively detailed-scale contexts in both developed and developing countries (e.g., Wilbanks et al., 2003). If information about possible impacts, vulnerabilities and adaptation potentials for industry, settlement and society is to be substantially improved, serious attention is needed towards establishing improved data sources on human-environmental relationships in both developing and developed countries, improving the integration of physical and earth science data from space-based and *in situ* observation systems with socioeconomic data, and improving the ability to associate data systems with high-priority questions.

References

Abels, H. and T. Bullens, 2005: *Microinsurance*, Microinsurance Association of the Netherlands, MIAN.

Abler, R.F., 2003: *Global change in local places: estimating, understanding, and reducing greenhouse gases*, Association of American Geographers Global Change and Local Places Research Group, Eds., Cambridge University Press, Cambridge, 290 pp.

Abrar, C.R. and S. N. Azad, 2004: *Coping with displacement: riverbank erosion in north-west Bangladesh*. University Press Ltd, Dhaka, 132 pp.

ACIA (Arctic Climate Impact Assessment), 2004: *Impacts of a Warming Arctic: Arctic Climate Impact Assessment*, Cambridge University Press, Cambridge, 1042 pp.

Adger, W.N., N.W. Arnell and E.L. Tompkins, 2005a: Successful adaptation to climate change across scales. *Global Environ. Chang.*, **15**, 77-86.

Adger, W.N., T.P. Hughes, C. Folke, S.R. Carpenter and J. Rockström, 2005b: Social-ecological resilience to coastal disasters. *Science*, 309, 1036-1039.

Adger, W.N., J. Paavola, S. Huq and M.J. Mace, Eds., 2006: *Fairness in Adaptation to Climate Change*, MIT Press, Cambridge, 320 pp.

ADRC, Japan, CRED-EMDAT, Université Catholique de Louvain and UNDP, 2005: *Natural Disasters Data Book, 2004*. [Accessed 03.05.07: http://web.adrc.or.jp/publications/databook/databook_2004_eng/]

Agnew, M. and D. Viner, 2001: Potential impact of climate change on international tourism. *Tourism and Hospitality Research*, **3**, 37-60.

Agnew M.D. and J.P. Palutikof, 1999: The impacts of climate on retailing in the UK with particular reference to the anomalously hot summer of 1995. *Int. J. Climatol.*, **19**, 1493-1507.

Ahern, M., R.S. Kovats, P. Wilkinson, R. Few and F. Matthies, 2005: Global health impacts of floods: epidemiologic evidence. *Epidemiol. Rev.*, **27**, 36-46.

Akbari, H., S. Bretz, D.M. Kurn and J. Hanford, 1997: Peak power and cooling energy savings of high-albedo roofs. *Energ. Buildings*, **25**, 117-126.

Allianz and World Wildlife Fund, 2006: Climate change and the financial sector: an agenda for action, 59 pp. [Accessed 03.05.07: http://www.wwf.org.uk/ filelibrary/pdf/allianz_rep_0605.pdf]

Andrey, J. and B.N. Mills, 2003: Climate change and the Canadian transportation system: vulnerabilities and adaptations. *Weather and Transportation in Canada*, J. Andrey and C. Knapper, Eds., Department of Geography publication series 55, University of Waterloo, Waterloo, Ontario,235-279.

Anon, 2004: Government of Canada and Canadian pulp and paper industry agree on blueprint for climate change action. *Forest. Chronic.*, **80**, 9.

Arnell, N.W., M.J.L. Livermore, S. Kovats, P.E. Levy, R. Nicholls, M.L. Parry and S.R. Gaffin, 2004: Climate and socio-economic scenarios for global-scale climate change impacts assessments: characterising the SRES storylines. *Global Environ. Chang.*, **14**, 3-20.

Arsenault, R., 1984: The end of the long hot summer: the air conditioner and southern culture. *J. Southern Hist.*, **50**, 597-628.

Association of British Insurers, 2002: Renewing the partnership: how the insurance industry will work with others to improve protection against floods: a report by the Association of British Insurers, 10 pp.

Ausubel, J. and H.D. Langford, Eds., 1997: *Technological Trajectories and the Human Environment*. National Academy of Sciences, Washington District of Columbia, 214 pp.

Becken, S., 2005: Harmonizing climate change adaptation and mitigation: the case of tourist resorts in Fiji. *Global Environ. Chang.*, **15**, 381-393.

Becken, S., C. Frampton and D. Simmons, 2001: Energy consumption patterns in the accommodation sector: the New Zealand case. *Ecol. Econ.*, **39**, 371-386.

Belle, N. and B. Bramwell, 2005: Climate change and small island tourism: Policymaker and industry perspectives in Barbados. *Journal of Travel Research*, **44**, 32-41.

Benson, C. and E. Clay, 2003: Disasters, vulnerability, and the global economy. *Building Safer Cities: The Future of Disaster Risk*, A. Kreimer, M. Arnold and A. Carlin, Eds., Disaster Risk Management Series No. 3, World Bank, Washington, District of Columbia, 3-32.

Berkhout, F. and J. Hertin, 2002: Foresight futures scenarios: developing and applying a participative strategic planning tool. *Greener Management International*, **37**, 37-52.

Berkhout, F., J. Hertin and D.M. Gann, 2006: Learning to adapt: organisational adaptation to climate change impacts. *Climatic Change*, **78**, 135-156.

Besancenot, J.P., 1989: *Climat et tourisme*. Masson, Paris, 223 pp.

Bigio, A., 2003: Cities and climate change. *Building Safer Cities: The Future of Disaster Risk*, A. Kreimer, M. Arnold and A. Carlin, Eds., World Bank, Washington, District of Columbia, 91-100.

Black, R., 2001: Environmental refugees: myth or reality? New Issues in Refugee Research Working Paper 34, United Nations High Commissioner for Refugees, Geneva, 20 pp.

Bouwer L.M. and J.C.J.H. Aerts, 2006: Financing climate change adaptation. *Disasters*, **30**, 49-63.

Braun, O., M. Lohmann, O. Maksimovic, M. Meyer, A. Merkovic, E. Messerschmidt, A. Reidel and M. Turner, 1999: Potential impact of climate change effects on preferences for tourism destinations: A psychological pilot study. *Climate Res.*, **11**, 2477-2504.

Breslow, P. and D. Sailor, 2002: Vulnerability of wind power resources to climate change in the continental United States. *Renew. Energ.*, **27**, 585–598.

Brewer, T., 2005: U.S. public opinion on climate change issues: implications for consensus building and policymakers. *Clim. Policy*, **4**, 359-376.

Broadmeadow, T., D. Ray and C. Samuel, 2005: Climate change and the future for broadleaved forests in the UK. *Forestry*, **78**, 145-161.

Bulkeley, H., and M.M. Betsill, 2003: *Cities and Climate Change: Urban Sustainability and Global Environmental Governance*. Routledge, New York, 237 pp.

Bull-Kamanga, L., K. Diagne, A. Lavell, E. Leon, F. Lerise, H. MacGregor, A. Maskrey, M. Meshack, M. Pelling, H. Reid, D. Satterthwaite, J. Songsore, K. Westgate and A. Yitambe, 2003: From everyday hazards to disasters: the accumulation of risk in urban areas. *Environ. and Urban.*, **15**, 193-204.

Cairncross, S. and M.J.C. Alvarinho, 2006: The Mozambique floods of 2000: health impact and response. *Flood Hazards and Health: Responding to Present and Future Risks*, R. Few and F. Matthies, Earthscan, London, 111-127.

Cannon, T., 2002: Gender and climate hazards in Bangladesh. *Gender and Development*, **10**, 45-50.

Cartalis, C., A. Synodinou, M. Proedrou, A. Tsangrassoulis and M. Santamouris, 2001: Modifications in energy demand in urban areas as a result of climate changes: an assessment for the southeast Mediterranean region. *Energ. Convers. Manage.*, **42**, 1647-1656.

Casola, J.H., J.E. Kay, A.K. Snover, R.A. Norheim, L.C. Whitely Binder and Climate Impacts Group, 2005: Climate impacts on Washington's hydropower, water supply, forests, fish, and agriculture. Prepared for King County (Washington) by the Climate Impacts Group (Center for Science in the Earth System, Joint Institute for the Study of the Atmosphere and Ocean, University of Washington, Seattle), 44 pp.

Ceron, J.P., 2000: Tourisme et changement climatique. *Impacts Potentiels du Changement Climatique en France au XXIème Siècle*. Premier ministre, Ministère de l'aménagement du territoire et de l'environnement, 1998, deuxième édition 2000, 104-111.

Ceron, J. and G. Dubois, 2005: The potential impacts of climate change on French tourism. *Current Issues in Tourism*, **8**, 125-139.

Christie, F. and J. Hanlon, 2001: *Mozambique and the Great Flood of 2000*. James Currey for the International African Institute, Oxford, 176 pp.

Clark, W.C., and Co-authors, 2000: Assessing Vulnerability to Global Environmental Risks. Discussion Paper 2000-12. *Report of the Workshop on Vulnerability to Global Environmental Change: Challenges for Research, Assessment and Decision Making*. Warrenton, Virginia. Belfer Center for Science and International Affairs (BCSIA), Environment and Natural Resources Program, Kennedy School of Government, Harvard University, 14 pp.

Clichevsky, N., 2003: Urban land markets and disasters: floods in Argentina's Cities. *Building Safer Cities: The Future of Disaster Risk*, A. Kreimer, M. Arnold and A. Carlin, Eds., World Bank, Washington, District of Columbia, 91-100.

Compton, K., T. Ermolieva and J.C. Linnerooth-Bayer, 2002: Integrated flood risk management for urban infrastructure: managing the flood risk to Vienna's heavy rail mass rapid transit system. *Proceedings of the Second Annual International IASA-DPRI Meeting: Integrated disaster risk management: megacity vulnerability and resilience*, Laxenburg, Austria, International Institute for Applied Systems Analysis, 20 pp.

Consodine, T.J., 2000: The impacts of weather variations on energy demand and carbon emissions. *Resour. Energy Econ.*, **22**, 295-314.

CRED (Centre for Research on the Epidemiology of Disasters), 2005: Disaster data: a balanced perspective. [Accessed 03.05.07: www.em-dat.net/documents/ CREDCRUNCH-aug20054.pdf]

Crichton, D., 2006: *Climate change and its effects on small business in the UK*. AXA Insurance UK plc., 41 pp.

Cross, J., 2001: Megacities and small towns: different perspectives on hazard vulnerability. *Environmental Hazards*, **3**, 63-80.

Dinar, A., M.W. Rosegrant and R. Meinzen-Dick, 1997: Water allocation mechanisms: principles and examples. Policy Research Working Paper WPS 1779, World Bank, Washington, District of Columbia, 41 pp.

Dixon, T.H., F. Amelung, A. Ferretti, F. Novali, F. Rocca, R. Dokka, G. Sellall, S.-W. Kim, S. Wdowinski and D. Whitman, 2006: Subsidence and flooding in New Orleans. *Nature*, **441**, 587-588.

Dow, K. and T. Wilbanks, 2003: Poverty and vulnerabilities to climate change. *Briefing Note. Adaptation Research Workshop*, New Delhi, ,5 pp.

Downing, T.E., 2002: Linking sustainable livelihoods and global climate change in vulnerable food systems. *die Erde*, **133**, 363-378.

Downing, T.E., 2003: Lessons from famine early warning systems and food security for understanding adaptation to climate change: toward a vulnerability adaptation science? *Climate Change, Adaptive Capacity and Development*, J.B. Smith, R.J.T. Klein and S. Huq, Eds., Imperial College Press, London, 71-100.

Dubois, G. and J.P. Ceron, 2005: Changes in leisure/tourism mobility patterns facing the stake of global warming: the case of France. *Belgeo*, **1-2**, 103-121.

Dyson, B., 2005: Throwing off the shackles: 2004 insurance performance rankings. *Reactions*, July 2005, 36-41

Eakin, H., 2006: Institutional change, climate risk, and rural vulnerability: cases from Central Mexico. *World Dev.*, **33**, 1923-1938.

Eakin, H. and M.C. Lemos, 2006: Adaptation and expectations of nation-state action: the challenge of capacity-building under globalization. *Global Environ. Chang.*, **16**, 7-18

Easterling, W.E., B.H. Hurd and J.B. Smith, 2004: Coping with global climate change: the role of adaptation in the United States. Pew Center on Global Climate Change, Arlington Virginia, 40 pp.

Eddowes, M.J., D. Waller, P. Taylor, B. Briggs, T. Meade and I. Ferguson, 2003: Railway safety implications of weather, climate and climate change. Report AEAT/RAIR/76148/R03/005 Issue 2, AEA Technology, Warrington, 141 pp.

Elsasser, H. and R. Burki, 2002: Climate change as a threat to tourism in the Alps. *Climate Res.*, **20**, 253-257.

Emanuel, K., 2005: Increasing destructiveness of tropical cyclones over the past 30 years. *Nature*, **434**, 686-688.

Emmanuel, R, 2005: *An Urban Approach to Climate Sensitive Design: Strategies for the Tropics*, Taylor and Francis, Abingdon, Oxfordshire, 172 pp.

Environment Agency, 2004: Maintaining water supply. London, England, 74 pp.

Environment Canada, 1997: The Canada country study: climate impacts and adaptation. Adaptation and Impacts Research Group, Downsview, Ontario.

Enz, R., 2000: The S-curve relation between per-capita income and insurance penetration. *Geneva Pap.*, **25**, 396-406.

European Environment Agency, 2004: Impacts of Europe´s changing climate. European Environment Agency Report No 2/2004, Copenhagen, 100 pp.

Fairhead, J., 2004: Achieving sustainability in Africa. *Targeting Development: Critical Perspectives on the Millennium Goals*, R. Black and H. White, Eds., Routledge, London, 292-306.

Feng, H., L. Yu, and W. Solecki, Eds., 2006: *Urban Dimensions of Environmental Change – Science, Exposure, Policies, and Technologies*. Science Press, Monmouth Junction, New Jersey.

Ferguson J., 2003: From beedees to CDs: snapshots from a journey through India's rural knowledge centres. International Institute for Communication and Development (IICD) Research Brief, No. 4, 8 pp.

Few, R., M. Ahern, F. Matthies and S. Kovats, 2004: Floods, health and climate change; a strategic review. Tyndall Centre for Climate Change Research, Working Paper No. 63, Norwich, 138 pp

Folland, C.K., and Co-authors, 2001: Observed climate variability. *Climate Change 2001: The Scientific Basis. Contribution of Working Group I to the Third Assessment Report of the Intergovernmental Panel on Climate Change*, J.T. Houghton, Y. Ding, D.J. Griggs, M. Noguer, P.J. van der Linden, X. Dai, K. Maskell and C.A. Johnson, Eds., Cambridge University Press, 99-182.

Fowler, H.J., M. Ekstrom, C.G. Kilsby and P.D. Jones, 2005: New estimates of future changes in extreme rainfall across the UK using regional climate model integrations. 1. Assessment of control climate. *J. Hydrol.*, **300**, 212-233.

Freeman, P. and K. Warner, 2001: Vulnerability of infrastructure to climate variability: how does this affect infrastructure lending policies? Disaster Management Facility of The World Bank and the ProVention Consortium, Washington, District of Columbia, 40 pp.

Freer, J., 2006: Insurance woes take toll on building sales. *South Florida Business Journal*. [Accessed 04.05.07: http://southflorida.bizjournals.com/southflorida/stories/2006/09/25/focus1.html?page=1]

Fukushima, T., M. Kureha, N. Ozaki, Y. Fukimori and H. Harasawa, 2003: Influences of air temperature change on leisure industries: case study on ski activities. *Mitigation and Adaptation Strategies for Climate Change*, **7**, 173-189.

Gallopin, G., A. Hammond, P. Raskin and R. Swart, 1997: Branch points: global scenarios and human choice. Resource Paper of the Global Scenario Group, Stockholm Environment Institute (SEI), 55 pp.

Giannakopoulos, C. and B.E. Psiloglou, 2006: Trends in energy load demand in Athens, Greece: weather and non-weather related factors. *Climate Res.*, **31**, 91-108.

Gomez-Martin, B., 2005: Weather, climate and tourism: a geographical perspective. *Ann. Tourism Res.*, **32**, 571-591.

Gössling, S. and C.M. Hall, 2005: An introduction to tourism and global environmental change. *Tourism and Global Environmental Change: Ecological, Social, Economic and Political Interrelationships*, S. Gössling and C.M. Hall, Eds., Routledge, London, 1-34.

Graves, H.M. and M.C. Phillipson, 2000: *Potential Implications of Climate Change in the Built Environment*. FBE Report 2. Building Research Establishment Press, London, 74 pp.

Grossi, P. and H. Kunreuther, Eds., 2005: *Catastrophe Modeling: a New Approach to Managing Risk*, Springer-Verlag, Berlin, 252 pp.

Gurenko, E., 2004: Building effective catastrophe insurance programs at the country level: a risk management perspective. *Catastrophe Risk and Reinsurance: a Country Risk Management Perspective*, Risk Books, London, 3-16.

Hadley, S.W., D.J. Erickson, J.L. Hernandez, C.T. Broniak and T.J. Blasing, 2006: Responses of energy use to climate change: a climate modeling study. *Geophys. Res. Lett.*, **33**, L17703, doi: 10.1029/2006GL026652.

Hallegatte, S., J.-C. Hourcade and P. Dumas, 2007: Why economic dynamics matter in assessing climate change damages: illustration on Extreme Events. *Ecol. Econ.*, **62**, 330-340.

Hamilton, J.M., D.J. Maddison and R.S.J. Tol, 2005: Climate change and international tourism: a simulation study. *Global Environ. Chang.*, **15**, 253-266.

Hamilton, L.C., D.E. Rohall, B.C. Brown, G.F. Hayward and B.D. Keim, 2003: Warming winters and New Hampshire's lost ski areas: an integrated case study. *International Journal of Sociology and Social Policy*, **23**, 52-68.

Harman, J., M. Gawith and M. Calley, 2005: Progress on assessing climate impacts through the UK Climate Impacts Programme. *Weather*, **60**, 258-262.

Hartwig, R., 2006: Hurricane season of 2005: impacts on U.S. P/C insurance markets in 2006 and beyond. Presentation to the Insurance Information Institute, New York, 239 pp. [Accessed 04.05.07: http://www.iii.org/media/presentations/katrina/]

Hertin, J., F. Berkhout, D.M Gann and J. Barlow, 2003: Climate change and the UK house building sector: perceptions, impacts and adaptive capacity. *Build. Res. Inf.*, **31**, 278-290.

Hewer, F., 2006: Climate change and energy management: a scoping study on the impacts of climate change on the UK energy industry. UK Met Office, 18 pp.

Hogrefe, C., J. Biswas, B. Lynn, K. Civerolo, J-Y Ku, J. Rosenthal, C. Rosenweig, R. Goldberg and P.L. Kinney, 2004: Simulating regional-scale ozone climatology over the eastern United States: model evaluation results. *Atmos. Environ.*, **38**, 2627-2638.

Holman, I.P., M. Rounsevell, S. Shackley, P. Harrison, R. Nicholls, P. Berry and E. Audsley, 2005: A regional, multi-sectoral and integrated assessment of the impacts of climate and socio-economic change in the UK, Part I, Methodology, Part II, Results. *Climatic Change*, **71**, 9-41 and 43-73.

Hough, M., 2004: *Cities and Natural Processes: a Basis for Sustainability*. Second Edition. Routledge, London, 292 pp.

Hunter, J.R., 2006: Testimony before the Committee on Banking, Housing and Urban Affairs of the United States Senate regarding proposals to reform the National Flood Insurance Program. Consumer Federation of America, 7 pp.

Instanes, A., O. Anisimov, L. Brigham, D. Goering, L.N. Khrustalev, B. Ladanyi and J.O. Larsen, 2005: Infrastructure: buildings, support systems, and industrial facilities. *Arctic Climate Impact Assessment*, ACIA, Cambridge University Press, Cambridge, 907-944.

IPCC, 2001: *Climate Change 2001: Impacts, Adaptation, and Vulnerability. Contribution of Working Group II to the Third Assessment Report of the Intergovernmental Panel on Climate Change*, J.J. McCarthy, O.F. Canziani, N.A. Leary, D.J. Dokken and K.S. White, Eds., Cambridge University Press, Cambridge, 1032 pp.

ISDR, 2004: *Living With Risk: a Global Review of Disaster Reduction Initiatives*. International Strategy for Disaster Reduction, United Nations, Geneva, 588 pp.

Jessop, B., 2002: Globalization and the national state. *Paradigm Lost: State Theory Reconsidered*, S. Aronowitz and P. Bratsis, Eds., University of Minnesota Press, Minneapolis, Minnesota, 185-220.

Jewson, S., A. Brix and C. Ziehman, 2005: *Weather Derivative Valuation: the Meteorological, Statistical, Financial and Mathematical Foundations*. Cambridge University Press, Cambridge, 390 pp.

Jollands, N., M. Ruth, C. Bernier and N. Golubiewski, 2005: Climate's long-term impacts on New Zealand infrastructure - a Hamilton City case study. *Proc. Ecological Economics in Action*, New Zealand Centre for Ecological Economics, Palmerston North, New Zealand, 30 pp.

Jones, B. and D. Scott, 2007: Implications of climate change to Ontario's provincial parks. *Leisure*, (in press).

Jones, B., D. Scott and H. Abi Khaled, 2006: Implications of climate change for outdoor event planning: a case study of three special events in Canada's National Capital region. *Event Management*, **10**, 63-76

Kainuma, M., Y. Matsuoka, T. Morita, T. Masui and K. Takahashi, 2004: Analysis of global warming stabilization scenarios: the Asian-Pacific Integrated Model. *Energ. Econ.*, **26**, 709-719.

Kates, R. and T. Wilbanks, 2003: Making the global local: responding to climate change concerns from the bottom up. *Environment*, **45**, 12-23.

Kiernan, M., 2005: Climate change, investment risk, and fiduciary responsibility. *The Finance of Climate Change*, K. Tang, Ed., Risk Books, London, 211-226.

Kinney, P.L., J.E. Rosenthal, C. Rosenzweig, C. Hogrefe, W. Solecki, K. Knowlton, C. Small, B. Lynn, K. Civerolo, J.Y. Ku, R. Goldberg, and C. Oliveri. 2006: Assessing the potential public health impacts of changing climate and land use: the New York climate and health project. *Regional Climate Change and Vari-*

ability: Impacts and Responses, M. Ruth, K. Donaghy and P. Kirshen, Eds., Edward Elgar Publishing, Cheltenham, 161-189.

Kirshen, P., M. Ruth and W. Anderson, 2006: Climate's long-term impacts on urban infrastructures and services: the case of Metro Boston. *Regional Climate Change and Variability: Impacts and Responses*, M. Ruth, K. Donaghy and P.H. Kirshen, Eds., Edward Elgar Publishers, Cheltenham, 190-252.

Kirshen, P.H., 2002: Potential impacts of global warming in eastern Massachusetts. *J. Water Res. Pl. and Management*, **128**, 216-226.

Kirshen, P.H., M.R. Ruth and W. Anderson, 2007: Interdependencies of urban climate change impacts and adaptation strategies: a case study of Metropolitan Boston, USA. *Climatic Change*, doi: 10.1007/s10584-007-9252-5.

Klein, R.J., T.J. Nicholls and J. Thomalla, 2003: The resilience of coastal mega cities to weather related hazards. *Building Safer Cities: The Future of Climate Change*, A. Kreimer, M. Arnold and A. Carlin, Eds., World Bank, Washington, District of Columbia, 101-121.

Klinenberg, E., 2002: *Heat Wave: A Social Autopsy of Disaster in Chicago (Illinois)*. The University of Chicago Press, Chicago, Illinois, 238 pp.

Knowlton, K., J.E. Rosenthal, C. Hogrefe, B. Lynn, S. Gaffin, R. Goldberg, C. Rosenzweig, K. Civerolo, J.-Y. Ku and P.L. Kinney, 2004: Assessing ozone-related health impacts under a changing climate. *Environ. Health Persp.*, **112**, 1557-1563.

Kont, A., J. Jaagus and R. Aunap, 2003: Climate change scenarios and the effect of sea-level rise for Estonia. *Global Planet. Change*, **36**, 1-15.

Kuik, O., 2003: Climate change policies, energy security and carbon dependency: trade-offs for the European Union in the longer term, International Environmental Agreements. *Politics, Law and Economics*, **3**, 221-242

Lagadec, P., 2004. Understanding the French 2003 heat wave experience. *Journal of Contingencies and Crisis Management*, **12**, 160-169.

Lancaster, J.W., M. Preene and C.T. Marshall, 2004: Development and flood risk: guidance for the construction industry, Report C624, CIRIA, London, 180 pp.

Lemos, M. and A. Agrawal, 2003: Environmental governance. *Annual Review of Environmental Resources*, **31**, 297-325.

Lemos, M.C. and J.L.F. Oliveira, 2004: Can water reform survive politics? Institutional change and river basin management in Ceará, Northeast Brazil. *World Dev.*, **32**, 2121-2137.

Létard V., H. Flandre and S. Lepeltier, 2004: La France et les Français face à la canicule: les leçons d'une crise. Report No. 195 (2003-2004) to the Sénat, Government of France, 391 pp.

Lewsey, C., G. Cid and E. Kruse, 2004: Assessing climate change impacts on coastal infrastructure in the Eastern Caribbean. *Mar. Policy*, **28**, 393-409.

Lin, C.W.R. and H.Y.S. Chen, 2003: Dynamic allocation of uncertain supply for the perishable commodity supply chain. *Int. J. Prod. Res.*, **41**, 3119-3138.

Linnerooth-Bayer, J., R. Mechler and G. Pflug, 2005: Refocusing disaster aid. *Science*, **309**, 1044-1046.

Lisø, K.R., G. Aandahl, S. Eriksen and K. Alfsen, 2003: Preparing for climate change impacts in Norway's built environment. *Build. Res. Inf.*, **31**, 200-209.

Livermore, M.T.J., 2005: The potential impacts of climate change in Europe: the role of extreme temperature. Phd Thesis, University of East Anglia, 436pp.

London Climate Change Partnership, 2004: *London's Warming: A Climate Change Impacts in London Evaluation Study*, London, 293 pp.

London, J.B., 2004: Implications of climate change on small island developing states: experience in the Caribbean region. *Journal of Environmental Management and Planning*, **47**, 491-501.

Maddison, D., 2001: In search of warmer climates? The impact of climate change on flows of British tourists. *Climatic Change*, **49**, 193-208.

Marris, E., 2005: First tests show flood waters high in bacteria and lead. *News@Nature*, **437**, 301-301.

Matzarakis, A., and C.R. de Freitas, Eds., 2001: *Proceedings of the 1st International Workshop on Climate, Tourism and Recreation*. International Society of Biometeorology, Commission on Climate, Tourism and Recreation, 274 pp.

Matzarakis, A., C.R. de Freitas and D. Scott, Eds., 2004: *Advances in Tourism Climatology*. Berichte des Meteorologischen Institutes der Universität Freiburg N° 12, 260 pp.

McCarthy, K., D.J. Peterson, N. Sastry and M. Pollard, 2006: The repopulation of New Orleans after Hurricane Katrina. Technical Report, Santa Monica, RAND Gulf States Policy Institute, 59 pp.

McGranahan, G., D. Balk and B. Anderson, 2006: Low coastal zones settlements. *Tiempo*, **59**, 23-26.

McKenzie, K. and K. Parlee, 2003: The road ahead – adapting to climate change in Atlantic Canada, *Elements*. [Accessed 08.05.07: http://www.elements.nb.ca/theme/climate03/cciarn/adapting.htm]

Mendelsohn, R., Ed., 2001: *Global Warming and the American Economy: A Regional Assessment of Climate Change Impacts*. Edward Elgar Publisher, Aldershot, Hampshire, 209 pp.

Millennium Ecosystem Assessment, 2005: *Ecosystems and Human Well-being, Volume 4: Multiscale Assessments*, D. Capistrano, C.K. Samper, M.J. Lee and C. Raudsepp-Hearne, Eds., Island Press, Washington, District of Columbia, 412 pp.

Mills, B. and J. Andrey, 2002: Climate change and transportation: potential interactions and impacts. *Summary and Discussion Papers of a Federal Research Partnership Workshop on The Potential Impacts of Climate Change on Transportation*, Washington, District of Columbia, US Department of Transport, 77-88.

Mills E., 2004: Insurance as an adaptation strategy for extreme weather events in developing countries and economies in transition: new opportunities for public-private partnerships. Lawrence Berkeley National Laboratory, Paper LBNL-52220, 134 pp.

Mitchell, R. and P. Romero Lankao, 2004: Institutions, science and technology in a transition to sustainability. *Earth System Analysis for Sustainability*, H.-J. Schellnhuber, P.J. Crutzen, W.C. Clark, M. Claussen and H. Helm, Eds., MIT, Cambridge, Massachusetts, 387-408.

Molina, M.J. and L.T. Molina, 2002: *Air Quality in the Mexico Megacity: An Integrated Assessment*, Kluwer, Dordrecht, Netherlands, 408 pp.

Mortimore, M.J., 1989: *Adapting to Drought: Farmers, Famines, and Desertification in West Africa*. Cambridge University Press, Cambridge, 299 pp.

Muir Wood, R., S. Miller and A. Boissonade, 2006: The Search for Trends in Global Catastrophe Losses. *Final Workshop Report on Climate Change and Disaster Losses: Understanding and Attributing Trends and Projections*, Hohenkammer, Munich, 188-194 [Accessed 08.05.07: http://sciencepolicy.colorado.edu/sparc/research/projects/extreme_events/munich_workshop/workshop_report.html]

Munich Re, 2005: *Weather Catastrophes and Climate Change: is There Still Hope for Us?* Munich, 264 pp.

NACC, 2000: *Climate Change Impacts on the United States: The Potential Consequences of Climate Variability and Change*. U.S. Global Change Research Program, Washington, District of Columbia, 154 pp.

Nakićenović, N., and R. Swart, Eds., 2000: *IPCC Special Report on Emissions Scenarios*,. Cambridge University Press, Cambridge, 599 pp.

National Research Council, 2002: *Abrupt Climate Change: Inevitable Surprises*. Washington, National Academies Press, Washington, District of Columbia, 244 pp.

Nelson, F.E., O.A. Anisimov and N.I. Shiklomanov, 2001: Subsidence risk from thawing permafrost. *Nature*, **410**, 889-890.

Nicholls, R.J., 2004: Coastal flooding and wetland loss in the 21st century: changes under the SRES climate and socio-economic scenarios. *Global Environ. Chang.*, **14**, 69-86.

Nicholls, R. and R. Tol, 2006: Impacts and responses to sea-level rise: a global analysis of the SRES scenarios over the 21st century. *Philos. T. R. Soc. A*, **364**, 1073-1095.

NRC, 2006: *Understanding Multiple Environmental Stresses: Report of a Workshop*. National Academies Press, Washington, District of Columbia, 142 pp.

NRDC, 2004: Heat advisory: how global warming causes more bad air days. National Resources Defense Council, Washington, District of Columbia, 30 pp.

O'Brien, K. and R.M. Leichenko, 2000: Double exposure: assessing the impact of climate change within the context of economic globalization. *Global Environ. Chang.*, **10**, 221-232.

O'Brien, K.L. and R.M. Leichenko, 2003: Winners and losers in the context of global change. *Ann. Assoc. Am. Geogr.*, **93**, 89-103.

O'Brien, K., L. Sygna and J.E. Haugen, 2004: Vulnerable or resilient? a multi-scale assessment of climate impacts and vulnerability in Norway. *Climatic Change*, **62**, 75-113.

O'Brien, K.L., S. Eriksen, L. Sygna and L.O. Næss, 2006: Questioning complacency: climate change impacts, vulnerability and adaptation in Norway. *Ambio*, **35**, 50-56.

Ocampo, J.A. and J. Martin, Eds., 2003: *A Decade of Light and Shadow: Latin America and the Caribbean in the 1990s*. LC/G.2205-P/I, Economic Commission for Latin America and the Caribbean, 355 pp.

O'Connell, M. and R. Hargreaves, 2004: Climate change adaptation: guidance on adapting New Zealand's built environment for the impacts of climate change. Study Report No. 130 (2004) BRANZ Ltd, New Zealand, 51 pp.

Office of Seattle Auditor, 2005: Climate change will impact the Seattle (USA) Department of Transportation. City of Seattle, Washington, 58 pp.

Oke, T.R., 1982: The energetic basis of the urban heat island. *Qu. J. Roy. Meteor. Soc.*, **108**, 1-24.

ORNL/CUSAT 2003: Possible vulnerabilities of Cochin, India, to climate change impacts and response strategies to increase resilience. Oak Ridge National Laboratory and Cochin University of Science and Technology, 44 pp.

PAHO, 1998: *Natural Disaster Mitigation in Drinking Water and Sewerage Systems: Guidelines for Vulnerability Analysis*. Pan-american Health Organization, Washington, District of Columbia, 90 pp.

Parkinson, J. and O. Mark, 2005: *Urban Storm water Management in Developing Countries*. IWA, London, 240 pp.

Parry, M.L., Ed., 2000: Assessment of potential effects and adaptations for climate change in Europe. Jackson Environmental Institute, University of East Anglia, Norwich, 320 pp.

Parry, M., N. Arnell, T. McMichael, R. Nicholls, P. Martens, S. Kovats, M. Livermore, C. Rosenzweig, A. Iglesias and G. Fischer, 2001: Millions at risk: defining critical climate change threats and targets. *Global Environ. Chang.*, **11**, 181-83.

Parson, E.A., and Co-authors, 2003: Understanding climatic impacts, vulnerabilities, and adaptation in the United States: building a capacity for assessment. *Climatic Change*, **57**, 9-42.

Perry, A., 2004: Sports tourism and climate variability. *Advances in Tourism Climatology*, A. Matzarakis, C.R. de Freitas and D. Scott, Eds., Frieburg, Berichte desw Meteorologischen Institutes der Universitat Freiburg, 174-179.

Pielke R.A., C. Landsea, M. Mayfield, J. Lavel and R. Pasch, 2005: Hurricanes and global warming. *B. Am. Meteorol. Soc.*, **86**, 1571-1575.

Pinho, O. and M. Orgaz, 2000: The urban heat island in a small city in coastal Portugal. *International Journal of Biometeorology*, **44**, 198-200.

ProVention Consortium, 2004: Experiences in Micro-Insurance. *Report on ProVention Consortium International Workshop*, Zurich, 11 pp.

Raskin, P., F. Monks, T. Ribiero, D. van Vuuren and M. Zurek, 2005: Global scenarios in historical perspective. *Ecosystems and Human Well-being: Scenarios, Findings of the Scenarios Working Group, Millennium Ecosystem Assessment*, S.R. Carpenter, P.L. Pingali, E.M. Bennett and M.B. Zurek, Eds., Island Press, London, 35-44.

REC (Renewable Energy Centre), 2004: Strategies for reducing the vulnerability of energy systems to natural disasters in the Caribbean. Working Paper Series 2004-1. Barbados Renewable Energy Centre, Bridgetown, Barbados, 16 pp.

Rhode, T.E., 1999: Integrating urban and agriculture water management in southern Morocco. *Arid Lands Newsletter*, **45**.

Romero Lankao, P., 2006: ¿Hacia una gestión sustentable del agua? Alcances y límites de la descentralización hidráulica en la ciudad de México? *La Gestión del Agua Urbana en México*, D. Barkin, Ed., UdeG/UAM Xochimilco, Mexico, 173-196.

Romero-Lankao, P., H. Lopez, A. Rosass, G. Guenter and Z. Correa, Eds., 2005: *Can Cities Reduce Global Warming? Urban Development and Carbon Cycle in Latin America*. IAI/UAM-X/IHDP/GCP, Mexico City, 92 pp.

Rosenzweig, C. and W.D. Solecki, 2001a: *Climate Change and a Global City: The Metropolitan East Coast Regional Assessment*. Columbia Earth Institute, New York, 24 pp.

Rosenzweig, C. and W.D. Solecki, 2001b: Global environmental change and a global city: Lessons for New York. *Environment*, **43**, 8-18.

Rosenzweig, C., W.D. Solecki, L. Parshall, M. Chopping, G. Pope and R. Goldberg, 2005: Characterizing the urban heat island in current and future climates in New Jersey. *Global Environ. Chang.*, **6**, 51-62.

Ruosteenoja, K., T.R. Carter, K. Jylhä and H. Tuomenvirta, 2003: *Future Climate In World Regions: an Intercomparison of Model-Based Projections for the New IPCC Emissions Scenarios*. Finnish Environment Institute, Helsinki, 83 pp.

Ruth, M., B. Davidsdottir and A. Amato, 2004: Climate change policies and capital vintage effects: the case of U.S. pulp and paper, iron and steel, and ethylene. *J. Environ. Manage.*, **70**, 235-252.

Salagnac, J.L., 2004: French perspective on emerging climate change issues. *Build. Res. Inf.*, **32**, 67-70.

Sanders, C.H. and M.C. Phillipson, 2003: UK adaptation strategy and technical measures: the impacts of climate change on buildings. *Build. Res. Inf.*, **31**, 210-221.

Sanderson, D., 2000: Cities, disasters and livelihoods. *Environ. Urban.*, **12**, 93-102.

Scoones, I., and Co-authors, 1996: *Hazards and Opportunities: Farming Livelihoods in Dryland Africa: Lessons from Zimbabwe*. Zed Books, London, 288 pp.

Scott, D., B. Jones and G. McBoyle, 2005a: Climate, tourism and recreation: a bibliography, 1936 to 2005. University of Waterloo, Canada, 38 pp.

Scott, D., G. Wall and G. McBoyle, 2005b: The evolution of the climate change issue in the tourism sector. *Tourism Recreation and Climate Change*, M. Hall and J. Higham, Eds., Channelview Press, London, 44-60.

Scott, M. and Co-authors, 2001: Human settlements, energy, and industry. *Climate Change 2001: Impacts, Adaptation, and Vulnerability. Contribution of Working Group II to the Third Assessment Report of the Intergovernmental Panel on Climate Change*, J.J. McCarthy, O.F. Canziani, N.A. Leary, D.J. Dokken and K.S. White, Eds., Cambridge University Press, Cambridge, 381-416.

Shashua-Bar, L. and M.E. Hoffman, 2002: The Green CTTC model for predicting the air temperature in small urban wooded sites. *Build. Environ.*, **37**, 1279-1288.

Shashua-Bar, L. and M.E. Hoffman, 2004: Quantitative evaluation of passive cooling of the UCL microclimate in hot regions in summer, case study: urban streets and courtyards with trees. *Build. Environ.*, **39**, 1087-1099.

Shepherd, J.M., H. Pierce and A.J. Negri, 2002: Rainfall modification by major urban areas: Observations from spaceborne rain radar on the TRMM satellite. *J. Appl. Meteorol.*, **41**, 689-701.

Sherbinin, A., A. Schiller and A. Pulsiphe, 2006: The vulnerability of global cities to climate hazards. *Environ. Urban.*, **12**, 93-102.

Shiklomanov, I.A., 2000: Appraisal and assessment of world water resources. *Water Int.*, **25**, 11-32.

Shimoda, Y., 2003: Adaptation measures for climate change and the urban heat island in Japan's built environment. *Build. Res. Inf.*, **31**, 222-230.

Shukla, P.R., M. Kapshe and A. Garg, 2005: Development and climate: impacts and adaptation for infrastructure assets in India. *Proc. OECD Global Forum on Sustainable Development: Development and Climate Change*, Organisation for Economic Co-operation and Development, Paris, 38 pp.

Smit, B., I. Burton, R.J.T. Klein and J. Wandel, 2000: An anatomy of adaptation to climate change and variability. *Climatic Change*, **45**, 223-251.

Smith, J.B., and Co-authors, 2001: Vulnerability to climate change and reasons for concern: a synthesis. *Climate Change 2001: Impacts, Adaptation and Vulnerability. Contribution of Working Group II to the Third Assessment Report of the Intergovernmental Panel on Climate Change*, J.J. McCarthy, O.F. Canziani, N.A. Leary, D.J. Dokken and K.S. White, Eds., Cambridge University Press, Cambridge, 913-967.

Solecki, W.D. and C. Rosenzweig, 2007: Climate change and the city: observations from metropolitan New York. *Cities and Environmental Change*, X. Bai, T. Graedel, A. Morishima, Eds., Yale University Press, New York, (in press).

Spence, R., A. Brown, P. Cooper and C.T. Bedford, 2004: Whether to Strengthen? Risk analysis for Strengthening decision-making. *Proc. Henderson Colloquium 2004: Designing for the Consequence of Hazards*, International Association for Bridge and Structural Engineers, Cambridge, Paper 12, 7 pp.

Sperling, F. and F. Szekely, 2005: Disaster risk management in a changing climate. Preprints, *World Conference on Disaster Reduction*, Kobe, Vulnerability and Adaptation Resource Group (VARG), World Bank, Washington, District of Columbia, 45 pp

Stern, N., 2007: *Stern Review on the Economics of Climate Change*, Cambridge University Press, Cambridge, 692 pp.

Subak, S. and Co-authors, 2000: The impact of the anomalous weather of 1995 on the U.K. economy. *Climatic Change*, **44**, 1-26.

Swiss Re, 1998: *Floods: an Insurable Risk*? Swiss Reinsurance Company, Zurich, 51 pp.

Swiss Re, 2004: Tackling climate change. Focus report, Swiss Reinsurance Company, Zurich, 10 pp.

Taha, H., 1996: Modeling the impacts of increased urban vegetation on ozone air quality in the south coast air basin. *Atmos. Environ.*, **30**, 3423-3430.

Taha, H., S. Douglas and J. Haney, 1997: Mesoscale meteorological and air quality impacts of increased urban albedo and vegetation. *Energ. Buildings*, **25**, 169-177.

Thomas, D.S.C. and C. Twyman, 2005: Equity and justice in climate change adaptation among natural-resource dependent societies. *Global Environ. Chang.*, **15**, 115-124.

Tol, R.S.J., 2002a: Estimates of the damage costs of climate change, Part I: Benchmark estimates. *Environ. Resour. Econ.*, **21**, 47-73.

Tol, R.S.J., 2002b: Estimates of the damage costs of climate change, Part II: Dynamic estimates. *Environ. Resour. Econ.*, **21**, 135-160.

Tol, R., and Co-authors, 2006: Adaptation to five metres of sea level rise. *J. Risk Res.*, **9**, 467-482.

Toth, F. and T. Wilbanks, 2004: Considering the technical and socioeconomic assumptions included in the SRES scenario families in AR-4 WGII. IPCC Working Group II Supporting Papers, Fourth Assessment Report, Technical Support Unit, Exeter, 16 pp.

Turner, B., R. Kasperson, W. Meyer, K. Kow, D. Golding, J. Kasperson, R. Mitchell and S. Ratick, 1990: Two types of global environmental change: Definitional and spatial scale issues in their human dimensions. *Global Environ. Chang.*, **1**, 14-22.

Turner, B.L. II, and Co-authors, 2003a: A framework for vulnerability analysis in sustainability science. *P. Natl. Acad. Sci.*, **100**, 8074-8079.

Turner, B.L., and Co-authors, 2003b: Illustrating the coupled human-environment system for vulnerability analysis: three case studies. *P. Natl. Acad. Sci.*, **100**, 8080-8085.

Twardosz, R., 1996: La variabilité des précipitations atmosphériques en Europe Centrale pendant la période 1850-1995. *Publications de l'Association Internationale de Climatologie*, **9**, 520-527.

UK Water Industry Research, 2004: Climate change and the hydraulic design of sewerage systems: summary report. Report 03/CC/10/0, UKWIR, London, 27 pp.

UN, 2004: UN Population Prospects: the 2004 Revision Population Database. [Accessed 09.05.07: http://esa.un.org/unpp]

UNDP, and Co-authors, 2003: *Poverty and Climate Change: Reducing the Vulnerability of the Poor through Adaptation*. African Development Bank, Asian Development Bank, Department for International Development - UK, Directorate-General for Development - European Commission, Federal Ministry for Economic Cooperation and Development - Germany, Ministry of Foreign Affairs - the Netherlands, Organisation for Economic Cooperation and Development, United Nations Development Programme, United Nations Environmental Programme and the World Bank, 43 pp

UNEP, 2002: Climate change and the financial services industry. Report prepared for the UNEP Finance Initiatives Climate Change Working Group by Innovest, UN Environment Programme, 84 pp.

UN-Habitat, 2003: *The Challenge of Slums: Global Report on Human Settlements 2003*. Earthscan Publications, London, 352 pp.

UNISDR, 2004: *Living With Risk: a Global Review of Disaster Risk Reduction Initiatives*. United Nations, 588 pp. [Accessed 09.05.07: http://www.unisdr.org/eng/about_isdr/bd-lwr-2004-eng.htm]

United Nations, 2006: *World Urbanization Prospects: the 2005 Revision*. United Nations Department of Economic and Social Affairs, Population Division, CD-ROM Edition – Data in digital form (POP/DB/WUP/Rev.2005), United Nations, New York.

Unruh, J., M. Krol and N. Kliot, 2004. *Environmental Change and its Implications for Population Migration*. Advances in global change research, Vol. 20, Springer, New York, 313 pp.

Uyarra, M., I. Cote, J. Gill, R. Tinch, D. Viner and A.L Watkinson, 2005: Island-specific preferences of tourists for environmental features: implications of climate change for tourism-dependent states. *Environ. Conserv.*, **32**, 11-19.

Valor, E., V. Meneu and V. Caselles, 2001: Daily temperature and electricity load in Spain. *J. Appl. Meteorol.*, **40**, 1413-1421.

Van Kooten, C., 2004: *Climate Change Economics: Why International Accords Fail*. Edward Elgar, Cheltenham, Gloucestershire, 176 pp.

Van Dijk, H., M. de Bruijn and W. van Beek, 2004: Pathways to mitigate climate variability and climate change. *The Impact of Climate Change on Drylands: With a Focus on West Africa*, A.J. Dietz, R. Ruben and A. Verhagen, Eds., Kluwer Academic Publishers, Dordrecht/Boston/London, 173-206.

Velásquez, L.S., 2005: The Bioplan: decreasing poverty in Manizales, Colombia, through shared environmental management. *Reducing Poverty and Sustaining the Environment*, S. Bass, H. Reid, D. Satterthwaite and P. Steele, Eds., EarthScan Publications, London, 44-72.

Vellinga and Co-authors, 2001: Insurance and other financial services. *Climate Change 2001: Impacts, Adaptation, and Vulnerability. Contribution of Working Group II to the Third Assessment Report of the Intergovernmental Panel on Climate Change*, J.J. McCarthy, O.F. Canziani, N.A. Leary, D.J. Dokken and K.S. White, Eds., Cambridge University Press, Cambridge, 417-450.

Voisin, N., A.F. Hamlet, L.P. Graham, D.W. Pierce, T P. Barnett and D.P. Lettenmaier, 2006: The role of climate forecasts in western U.S. power planning. *J. Appl. Meteorol.*, **45**, 653-673.

WHO/Unicef, 2000: *Global Water Supply and Sanitation Assessment 2000 Report*. World Health Organization with Unicef, Geneva, 79 pp.

Wilbanks, T.J., 2003a: Geographic scaling issues in integrated assessments of climate change. *Scaling Issues in Integrated Assessment*, J. Rotmans and D. Rothman, Eds., Swets and Zeitlinger, Lisse, 5-34.

Wilbanks, T.J., 2003b: Integrating climate change and sustainable development in a place-based context. *Clim. Policy*, **3**, S147-S154.

Wilbanks, T.J., S.M. Kane, P.N. Leiby, R.D. Perlack, C. Settle, J.F. Shogren and J.B. Smith, 2003: Possible responses to global climate change: integrating mitigation and adaptation. *Environment*, **45**, 28-38.

Wilbanks, T.J., P. Leiby, R. Perlack, J.T. Ensminger and S.B. Wright, 2005: Toward an integrated analysis of mitigation and adaptation: some preliminary findings. *Mitigation and Adaptation Strategies for Global Change*.

Willows, R. and R. Connell, Eds., 2003: Climate adaptation: risk, uncertainty and decision-making. UKCIP Technical Report, Oxford, 154 pp.

Winchester, P., 2000: Cyclone mitigation, resource allocation and post-disaster reconstruction in South India: Lessons from two decades of research. *Disasters*, **24**, 18-37.

Wisner, B., 2003: Disaster risk reduction in megacities: making the best of human and social capital. *Building Safer Cities: The Future of Climate Change*, A. Kreimer, M. Arnold and A. Carlin, Eds., World Bank, Washington, District of Columbia, 181-196.

Wolmer, W. and I. Scoones, Eds., 2003: Livelihoods in crisis: new perspectives on governance and rural development in southern Africa. *IDS Bulletin*, **34**.

Wood, G. and S. Salway, 2000: Securing livelihoods in Dhaka slums. *Journal of International Development*, **12**, 669-688.

Woon, G. and D. Rose, 2004: Why the whole island floods now. *Nassau Guardian and Tribune*, November 25, 2004. [Accessed 09.05.07: http://www.unesco.org/csi/smis/siv/Caribbean/bahart3-nassau.htm.]

World Bank, 2000: A preliminary assessment of damage from the flood and cyclone emergency of February-March 2000. Republic of Mozambique, World Bank, 38 pp. [Accessed 09.05.07: http://siteresources.worldbank.org/INTDISMGMT/Resources/WB_flood_damages_Moz.pdf]

World Bank, 2006: *World Development Indicators*. The International Bank for Reconstruction and Development, The World Bank, Washington District of Columbia, 242 pp. [Accessed 09.05.07: http://devdata.worldbank.org/wdi2006/]

World Tourism Organization, 2003: Climate change and tourism. *Proceedings of the 1st International Conference on Climate Change and Tourism*, Djerba, Tunisia, World Tourism Organization, 55 pp.

Wright, E.L. and J.D. Erickson, 2003: Climate variability, economic adaptation and investment timing. *Int. J. Global Env. Issues*, **3**, 357-368.

Zapata-Marti, R., 2004: The 2004 hurricanes in the Caribbean and the tsunami in the Indian Ocean: lessons and policy challenges for development and disaster reduction. Estudios y perspectivas series, **35**, LC/MEX/L.672, 62 pp.

8

Human health

Coordinating Lead Authors:
Ulisses Confalonieri (Brazil), Bettina Menne (WHO Regional Office for Europe/Germany)

Lead Authors:
Rais Akhtar (India), Kristie L. Ebi (USA), Maria Hauengue (Mozambique), R. Sari Kovats (UK), Boris Revich (Russia), Alistair Woodward (New Zealand)

Contributing Authors:
Tarakegn Abeku (Ethiopia), Mozaharul Alam (Bangladesh), Paul Beggs (Australia), Bernard Clot (Switzerland), Chris Furgal (Canada), Simon Hales (New Zealand), Guy Hutton (UK), Sirajul Islam (Bangladesh), Tord Kjellstrom (New Zealand/Sweden), Nancy Lewis (USA), Anil Markandya (UK), Glenn McGregor (New Zealand), Kirk R. Smith (USA), Christina Tirado (Spain), Madeleine Thomson (UK), Tanja Wolf (WHO Regional Office for Europe/Germany)

Review Editors:
Susanna Curto (Argentina), Anthony McMichael (Australia)

This chapter should be cited as:
Confalonieri, U., B. Menne, R. Akhtar, K.L. Ebi, M. Hauengue, R.S. Kovats, B. Revich and A. Woodward, 2007: Human health. *Climate Change 2007: Impacts, Adaptation and Vulnerability. Contribution of Working Group II to the Fourth Assessment Report of the Intergovernmental Panel on Climate Change*, M.L. Parry, O.F. Canziani, J.P. Palutikof, P.J. van der Linden and C.E. Hanson, Eds., Cambridge University Press, Cambridge, UK, 391-431.

Table of Contents

Executive summary

Climate change currently contributes to the global burden of disease and premature deaths (very high confidence).
Human beings are exposed to climate change through changing weather patterns (temperature, precipitation, sea-level rise and more frequent extreme events) and indirectly through changes in water, air and food quality and changes in ecosystems, agriculture, industry and settlements and the economy. At this early stage the effects are small but are projected to progressively increase in all countries and regions. [8.4.1]

Emerging evidence of climate change effects on human health shows that climate change has:

- altered the distribution of some infectious disease vectors (medium confidence) [8.2.8];
- altered the seasonal distribution of some allergenic pollen species (high confidence) [8.2.7];
- increased heatwave-related deaths (medium confidence) [8.2.1].

Projected trends in climate-change-related exposures of importance to human health will:

- increase malnutrition and consequent disorders, including those relating to child growth and development (high confidence) [8.2.3, 8.4.1];
- increase the number of people suffering from death, disease and injury from heatwaves, floods, storms, fires and droughts (high confidence) [8.2.2, 8.4.1];
- continue to change the range of some infectious disease vectors (high confidence) [8.2, 8.4];
- have mixed effects on malaria; in some places the geographical range will contract, elsewhere the geographical range will expand and the transmission season may be changed (very high confidence) [8.4.1.2];
- increase the burden of diarrhoeal diseases (medium confidence) [8.2, 8.4];
- increase cardio-respiratory morbidity and mortality associated with ground-level ozone (high confidence) [8.2.6, 8.4.1.4];
- increase the number of people at risk of dengue (low confidence) [8.2.8, 8.4.1];
- bring some benefits to health, including fewer deaths from cold, although it is expected that these will be outweighed by the negative effects of rising temperatures worldwide, especially in developing countries (high confidence) [8.2.1, 8.4.1].

Adaptive capacity needs to be improved everywhere; impacts of recent hurricanes and heatwaves show that even high-income countries are not well prepared to cope with extreme weather events (high confidence). [8.2.1, 8.2.2]

Adverse health impacts will be greatest in low-income countries. Those at greater risk include, in all countries, the urban poor, the elderly and children, traditional societies, subsistence farmers, and coastal populations (high confidence). [8.1.1, 8.4.2, 8.6.1.3, 8.7]

Economic development is an important component of adaptation, but on its own will not insulate the world's population from disease and injury due to climate change (very high confidence).
Critically important will be the manner in which economic growth occurs, the distribution of the benefits of growth, and factors that directly shape the health of populations, such as education, health care, and public-health infrastructure. [8.3.2]

8.1 Introduction

This chapter describes the observed and projected health impacts of climate change, current and future populations at risk, and the strategies, policies and measures that have been and can be taken to reduce impacts. The chapter reviews the knowledge that has emerged since the Third Assessment Report (TAR) (McMichael et al., 2001). Published research continues to focus on effects in high-income countries, and there remain important gaps in information for the more vulnerable populations in low- and middle-income countries.

8.1.1 State of health in the world

Health includes physical, social and psychological well-being. Population health is a primary goal of sustainable development. Human beings are exposed to climate change through changing weather patterns (for example more intense and frequent extreme events) and indirectly though changes in water, air, food quality and quantity, ecosystems, agriculture, livelihoods and infrastructure (Figure 8.1). These direct and indirect exposures can cause death, disability and suffering. Ill-health increases vulnerability and reduces the capacity of individuals and groups to adapt to climate change. Populations with high rates of disease and debility cope less successfully with stresses of all kinds, including those related to climate change.

In many respects, population health has improved remarkably over the last 50 years. For instance, average life expectancy at birth has increased worldwide since the 1950s (WHO, 2003b, 2004b). However, improvement is not apparent everywhere, and substantial inequalities in health persist within and between countries (Casas-Zamora and Ibrahim, 2004; McMichael et al., 2004; Marmot, 2005; People's Health Movement et al., 2005). In parts of Africa, life expectancy has fallen in the last 20 years, largely as a consequence of HIV/AIDS; in some countries more than 20% of the adult population is infected (UNDP, 2005). Globally, child mortality decreased from 147 to 80 deaths per 1,000 live births from 1970 to 2002 (WHO, 2002b). Reductions were largest in countries in the World Health Organization (WHO) regions of the Eastern Mediterranean, South-East Asia and Latin America. In sixteen countries (fourteen of which are in Africa), current levels of under-five mortality are higher than those observed in 1990 (Anand and Barnighausen, 2004). The Millennium Development Goal (MDG) of reducing under-five mortality rates by two-thirds by 2015 is unlikely to be reached in these countries.

Non-communicable diseases, such as heart disease, diabetes, stroke and cancer, account for nearly half of the global burden of disease (at all ages) and the burden is growing fastest in low- and middle-income countries (Mascie-Taylor and Karim, 2003). Communicable diseases are still a serious threat to public health in many parts of the world (WHO, 2003a) despite immunisation programmes and many other measures that have improved the control of once-common human infections. Almost 2 million deaths a year, mostly in young children, are caused by diarrhoeal diseases and other conditions that are attributable to unsafe water and lack of basic sanitation (Ezzati et al., 2003). Malaria, another common disease whose geographical range may be affected by climate change, causes around 1 million child deaths annually (WHO, 2003b). Worldwide, 840 million people were undernourished in 1998-2000 (FAO, 2002). Progress in overcoming hunger is very uneven. Based on current trends, only Latin America and the Caribbean will achieve the MDG target of halving the proportion of people who are hungry by 2015 (FAO, 2005; UN, 2006a).

8.1.2 Findings from the Third Assessment Report

The main findings of the IPCC TAR (McMichael et al., 2001) were as follows.

- An increase in the frequency or intensity of heatwaves will increase the risk of mortality and morbidity, principally in older age groups and among the urban poor.
- Any regional increases in climate extremes (e.g., storms, floods, cyclones, droughts) associated with climate change would cause deaths and injuries, population displacement, and adverse effects on food production, freshwater availability and quality, and would increase the risks of infectious disease, particularly in low-income countries.
- In some settings, the impacts of climate change may cause social disruption, economic decline, and displacement of populations. The health impacts associated with such socio-economic dislocation and population displacement are substantial.
- Changes in climate, including changes in climate variability, would affect many vector-borne infections. Populations at the margins of the current distribution of diseases might be particularly affected.
- Climate change represents an additional pressure on the world's food supply system and is expected to increase yields at higher latitudes and decrease yields at lower latitudes. This would increase the number of undernourished people in the low-income world, unless there was a major redistribution of food around the world.
- Assuming that current emission levels continue, air quality in many large urban areas will deteriorate. Increases in exposure to ozone and other air pollutants (e.g., particulates) could increase morbidity and mortality.

8.1.3 Key developments since the Third Assessment Report

Overall, research over the last 6 years has provided new evidence to expand the findings of the TAR. Empirical research has further quantified the health effects of heatwaves (see Section 8.2.1). There has been little additional research on the health effects of other extreme weather events. The early effects of climate change on health-relevant exposures have been investigated in the context of changes in air quality and plant and animal phenology (see Chapter 1 and Sections 8.2.7 and 8.2.8). There has been research on a wider range of health issues, including food safety and water-related infections. The contribution made by climate change to the overall burden of disease has been estimated (see Section 8.4.1) (McMichael, 2004). Several countries have conducted health-impact assessments of climate change; either as part of a multi-sectoral study or as a stand-alone project (see Tables 8.1, 8.3 and 8.4). These provide more detailed information on population vulnerability to climate change (see Section 8.4.2). The effect of climate has been studied in the context of other social and environmental determinants of health outcomes (McMichael et al., 2003a; Izmerov et al., 2005). Little advancement has been made in the development of climate–health impact models that project future health effects. Climate change is now an issue of concern for health policy in many countries. Some adaptation measures specific to climate variability have been developed and implemented within and beyond the health sector (see Section 8.6). Many challenges remain for climate- and health-impact and adaptation research. The most important of these is the limited capacity for research and adaptation in low- and middle-income countries.

8.1.4 Methods used and gaps in knowledge

The evidence for the current sensitivity of population health to weather and climate is based on five main types of empirical study:

- health impacts of individual extreme events (e.g., heatwaves, floods, storms, droughts, extreme cold);
- spatial studies where climate is an explanatory variable in the distribution of the disease or the disease vector;
- temporal studies assessing the health effects of interannual climate variability, of short-term (daily, weekly) changes in temperature or rainfall, and of longer-term (decadal) changes in the context of detecting early effects of climate change;
- experimental laboratory and field studies of vector, pathogen, or plant (allergen) biology;
- intervention studies that investigate the effectiveness of public-health measures to protect people from climate hazards.

This assessment of the potential future health impacts of climate change is conducted in the context of:

- limited region-specific projections of changes in exposures of importance to human health;
- the consideration of multiple, interacting and multi-causal health outcomes;
- the difficulty of attributing health outcomes to climate or climate change *per se*;
- the difficulty of generalising health outcomes from one setting to another, when many diseases (such as malaria) have important local transmission dynamics that cannot easily be represented in simple relationships;

- limited inclusion of different developmental scenarios in health projections;
- the difficulty in identifying climate-related thresholds for population health;
- limited understanding of the extent, rate, limiting forces and major drivers of adaptation of human populations to a changing climate.

Table 8.1. *National health impact assessments of climate change published since the TAR.*

Country	Key findings	Adaptation recommendations
Australia (McMichael et al., 2003b)	Increase in heatwave-related deaths; drowning from floods; diarrhoeal disease in indigenous communities; potential change in the geographical range of dengue and malaria; likely increase in environmental refugees from Pacific islands.	Not considered.
Bolivia (Programa Nacional de Cambios Climaticos Componente Salud et al., 2000)	Intensification of malaria and leishmaniasis transmission. Indigenous populations may be most affected by increases in infectious diseases.	Not considered.
Bhutan (National Environment Commission et al., 2006)	Loss of life from frequent flash floods; glacier lake outburst floods; landslides; hunger and malnutrition; spread of vector-borne diseases into higher elevations; loss of water resources; risk of water-borne diseases.	Ensure safe drinking water; regular vector control and vaccination programmes; monitor air and drinking water quality; establishment of emergency medical services.
Canada (Riedel, 2004)	Increase in heatwave-related deaths; increase in air pollution-related diseases; spread of vector- and rodent-borne diseases; increased problems with contamination of both domestic and imported shellfish; increase in allergic disorders; impacts on particular populations in northern Canada.	Monitoring for emerging infectious diseases; emergency management plans; early warning systems; land-use regulations; upgrading water and wastewater treatment facilities; measures for reducing the heat-island effect.
Finland (Hassi and Rytkonen, 2005)	Small increase in heat-related mortality; changes in phenological phases and increased risk of allergic disorders; small reduction in winter mortality.	Awareness-building and training of medical doctors.
Germany (Zebisch et al., 2005)	Observed excess deaths from heatwaves; changing ranges in tick-borne encephalitis; impacts on health care.	Increase information to the population; early warning; emergency planning and cooling of buildings; insurance and reserve funds.
India (Ministry of Environment and Forest and Government of India, 2004)	Increase in communicable diseases. Malaria projected to move to higher latitudes and altitudes in India.	Surveillance systems; vector control measures; public education.
Japan (Koike, 2006)	Increased risk of heat-related emergency visits, Japanese cedar pollen disease patients, food poisoning; and sleep disturbance.	Heat-related emergency visit surveillance.
The Netherlands (Bresser, 2006)	Increase in heat-related mortality, air pollutants; risk of Lyme disease, food poisoning and allergic disorders.	Not considered.
New Zealand (Woodward et al., 2001)	Increases in enteric infections (food poisoning); changes in some allergic conditions; injuries from more intense floods and storms; a small increase in heat-related deaths.	Systems to ensure food quality; information to population and health care providers; flood protection; vector control.
Panama (Autoridad Nacional del Ambiente, 2000)	Increase of vector-borne and other infectious diseases; health problems due to high ozone levels in urban areas; increase in malnutrition.	Not considered.
Portugal (Casimiro and Calheiros, 2002; Calheiros and Casimiro, 2006)	Increase in heat-related deaths and malaria (Tables 8.2, 8.3), food- and water-borne diseases, West Nile fever, Lyme disease and Mediterranean spotted fever; a reduction in leishmaniasis risk in some areas.	Address thermal comfort; education and information as well as early warning for hot periods; and early detection of infectious diseases.
Spain (Moreno, 2005)	Increase in heat-related mortality and air pollutants; potential change of ranges of vector- and rodent-borne diseases.	Awareness-raising; early warning systems for heatwaves; surveillance and monitoring; review of health policies.
Tajikistan (Kaumov and Muchmadeliev, 2002)	Increase in heat-related deaths.	Not considered.
Switzerland (Thommen Dombois and Braun-Fahrlaender, 2004)	Increase of heat-related mortality; changes in zoonoses; increase in cases of tick-borne encephalitis.	Heat information, early warning; greenhouse gas emissions reduction strategies to reduce secondary air pollutants; setting up a working group on climate and health.
United Kingdom (Department of Health and Expert Group on Climate Change and Health in the UK, 2001)	Health impacts of increased flood events; increased risk of heatwave-related mortality; and increased ozone-related exposure.	Awareness-raising.

8.2 Current sensitivity and vulnerability

Systematic reviews of empirical studies provide the best evidence for the relationship between health and weather or climate factors, but such formal reviews are rare. In this section, we assess the current state of knowledge of the associations between weather/climate factors and health outcome(s) for the population(s) concerned, either directly or through multiple pathways, as outlined in Figure 8.1. The figure shows not only the pathways by which health can be affected by climate change, but also shows the concurrent direct-acting and modifying (conditioning) influences of environmental, social and health-system factors.

Published evidence so far indicates that:

- climate change is affecting the seasonality of some allergenic species (see Chapter 1) as well as the seasonal activity and distribution of some disease vectors (see Section 8.2.8);
- climate plays an important role in the seasonal pattern or temporal distribution of malaria, dengue, tick-borne diseases, cholera and some other diarrhoeal diseases (see Sections 8.2.5 and 8.2.8);
- heatwaves and flooding can have severe and long-lasting effects.

8.2.1 Heat and cold health effects

The effects of environmental temperature have been studied in the context of single episodes of sustained extreme temperatures (by definition, heatwaves and cold-waves) and as population responses to the range of ambient temperatures (ecological time-series studies).

8.2.1.1 Heatwaves

Hot days, hot nights and heatwaves have become more frequent (IPCC, 2007a). Heatwaves are associated with marked short-term increases in mortality (Box 8.1). There has been more research on heatwaves and health since the TAR in North America (Basu and Samet, 2002), Europe (Koppe et al., 2004) and East Asia (Qiu et al., 2002; Ando et al., 2004; Choi et al., 2005; Kabuto et al., 2005).

A variable proportion of the deaths occurring during heatwaves are due to short-term mortality displacement (Hajat et al., 2005; Kysely, 2005). Research indicates that this proportion depends on the severity of the heatwave and the health status of the population affected (Hemon and Jougla, 2004; Hajat et al., 2005). The heatwave in 2003 was so severe that short-term mortality displacement contributed very little to the total heatwave mortality (Le Tertre et al., 2006).

Eighteen heatwaves were reported in India between 1980 and 1998, with a heatwave in 1988 affecting ten states and causing 1,300 deaths (De and Mukhopadhyay, 1998; Mohanty and Panda, 2003; De et al., 2004). Heatwaves in Orissa, India, in 1998, 1999 and 2000 caused an estimated 2,000, 91 and 29 deaths, respectively (Mohanty and Panda, 2003) and heatwaves in 2003 in Andhra Pradesh, India, caused more than 3000 deaths (Government of Andhra Pradesh, 2004). Heatwaves in South Asia are associated with high mortality in rural populations, and

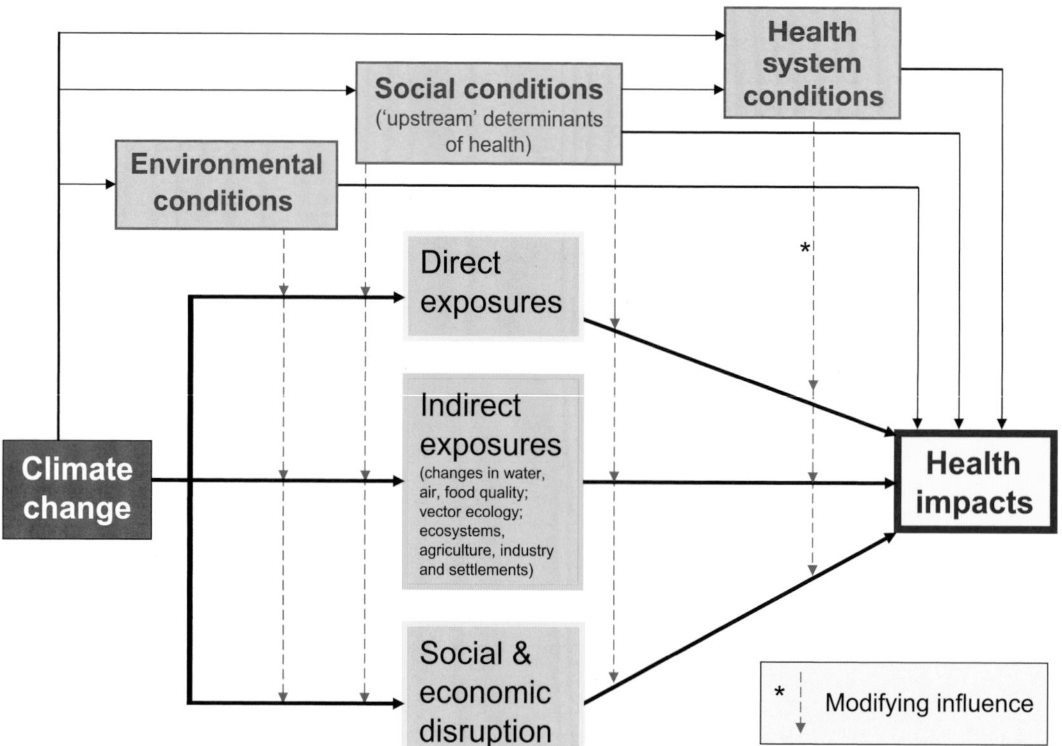

Figure 8.1. *Schematic diagram of pathways by which climate change affects health, and concurrent direct-acting and modifying (conditioning) influences of environmental, social and health-system factors.*

among the elderly and outdoor workers (Chaudhury et al., 2000) (see Section 8.2.9). The mortality figures probably refer to reported deaths from heatstroke and are therefore an underestimate of the total impact of these events.

8.2.1.2 Cold-waves

Cold-waves continue to be a problem in northern latitudes, where very low temperatures can be reached in a few hours and extend over long periods. Accidental cold exposure occurs mainly outdoors, among socially deprived people (alcoholics, the homeless), workers, and the elderly in temperate and cold climates (Ranhoff, 2000). Living in cold environments in polar regions is associated with a range of chronic conditions in the non-indigenous population (Sorogin et al, 1993) as well as with acute risk from frostbite and hypothermia (Hassi et al., 2005). In countries with populations well adapted to cold conditions, cold-

Box 8.1. The European heatwave 2003: impacts and adaptation

In August 2003, a heatwave in France caused more than 14,800 deaths (Figure 8.2). Belgium, the Czech Republic, Germany, Italy, Portugal, Spain, Switzerland, the Netherlands and the UK all reported excess mortality during the heatwave period, with total deaths in the range of 35,000 (Hemon and Jougla, 2004; Martinez-Navarro et al., 2004; Michelozzi et al., 2004; Vandentorren et al., 2004; Conti et al., 2005; Grize et al., 2005; Johnson et al., 2005). In France, around 60% of the heatwave deaths occurred in persons aged 75 and over (Hemon and Jougla, 2004). Other harmful exposures were also caused or exacerbated by the extreme weather, such as outdoor air pollutants (tropospheric ozone and particulate matter) (EEA, 2003), and pollution from forest fires.

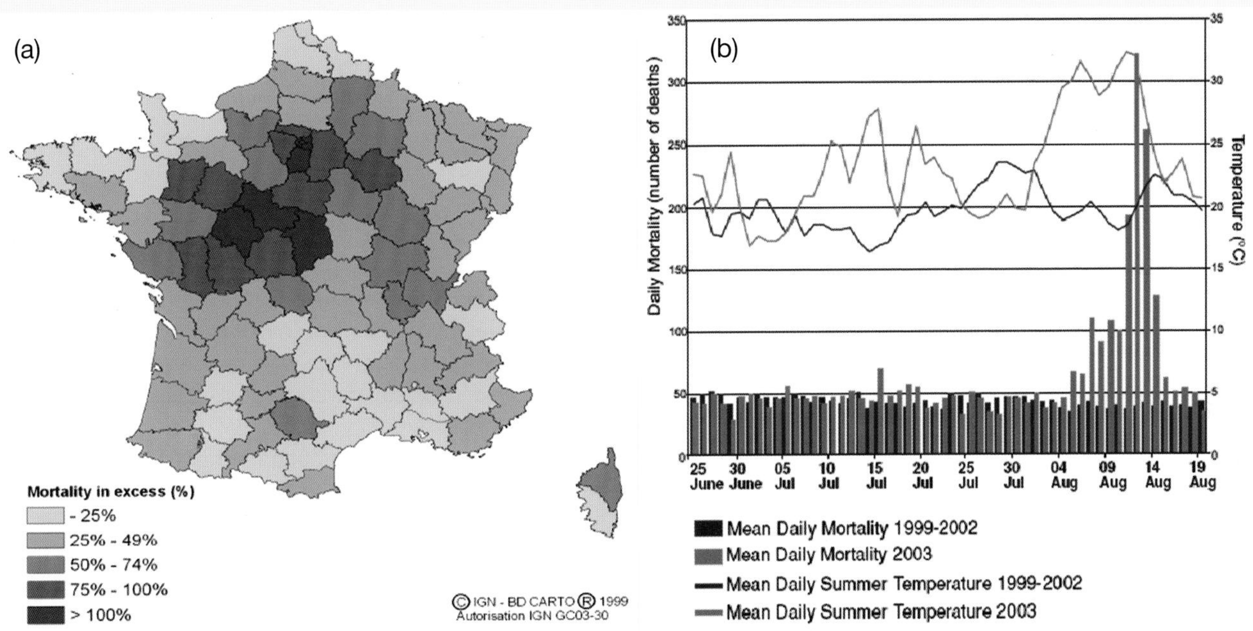

Figure 8.2. *(a) The distribution of excess mortality in France from 1 to 15 August 2003, by region, compared with the previous three years (INVS, 2003); (b) the increase in daily mortality in Paris during the heatwave in early August (Vandentorren and Empereur-Bissonnet, 2005).*

A French parliamentary inquiry concluded that the health impact was 'unforeseen', surveillance for heatwave deaths was inadequate, and the limited public-health response was due to a lack of experts, limited strength of public-health agencies, and poor exchange of information between public organisations (Lagadec, 2004; Sénat, 2004).

In 2004, the French authorities implemented local and national action plans that included heat health-warning systems, health and environmental surveillance, re-evaluation of care of the elderly, and structural improvements to residential institutions (such as adding a cool room) (Laaidi et al., 2004; Michelon et al., 2005). Across Europe, many other governments (local and national) have implemented heat health-prevention plans (Michelozzi et al., 2005; WHO Regional Office for Europe, 2006).

Since the observed higher frequency of heatwaves is likely to have occurred due to human influence on the climate system (Hegerl et al., 2007), the excess deaths of the 2003 heatwave in Europe are likely to be linked to climate change.

waves can still cause substantial increases in mortality if electricity or heating systems fail. Cold-waves also affect health in warmer climates, such as in South-East Asia (EM-DAT, 2006).

8.2.1.3 Estimates of heat and cold effects

Methods for the quantification of heat and cold effects have seen rapid development (Braga et al., 2002; Curriero et al., 2002; Armstrong et al., 2004), including the identification of medical, social, environmental and other factors that modify the temperature–mortality relationship (Basu and Samet, 2002; Koppe et al., 2004). Local factors, such as climate, topography, heat-island magnitude, income, and the proportion of elderly people, are important in determining the underlying temperature–mortality relationship in a population (Curriero et al., 2002; Hajat, 2006). High temperatures contribute to about 0.5 - 2% of annual mortality in older age groups in Europe (Pattenden et al., 2003; Hajat et al., 2006), although large uncertainty remains in quantifying this burden in terms of years of life lost.

The sensitivity of a population to temperature extremes changes over decadal time-scales (Honda et al., 1998). There is some indication that populations in the USA became less sensitive to high temperatures over the period 1964 to 1988 (as measured imprecisely by population- and period-specific thresholds in the mortality response) (Davis et al., 2002, 2003, 2004). Heat-related mortality has declined since the 1970s in South Carolina, USA, and south Finland, but this trend was less clear for the south of England (Donaldson et al., 2003). Cold-related mortality in European populations has also declined since the 1950s (Kunst et al., 1991; Lerchl, 1998; Carson et al., 2006). Cold days, cold nights and frost days have become rarer, but explain only a small part of this reduction in winter mortality; as improved home heating, better general health and improved prevention and treatment of winter infections have played a more significant role (Carson et al., 2006). In general, population sensitivity to cold weather is greater in temperate countries with mild winters, as populations are less well-adapted to cold (Eurowinter Group, 1997; Healy, 2003).

8.2.2 Wind, storms and floods

Floods are low-probability, high-impact events that can overwhelm physical infrastructure, human resilience and social organisation. Floods are the most frequent natural weather disaster (EM-DAT, 2006). Floods result from the interaction of rainfall, surface runoff, evaporation, wind, sea level and local topography. In inland areas, flood regimes vary substantially depending on catchment size, topography and climate. Water management practices, urbanisation, intensified land use and forestry can substantially alter the risks of floods (EEA, 2005). Windstorms are often associated with floods.

Major storm and flood disasters have occurred in the last two decades. In 2003, 130 million people were affected by floods in China (EM-DAT, 2006). In 1999, 30,000 died from storms followed by floods and landslides in Venezuela. In 2000/2001, 1,813 died in floods in Mozambique (IFRC, 2002; Guha-Sapir et al., 2004). Improved structural and non-structural measures, particularly improved warnings, have decreased mortality from floods and storm surges in the last 30 years (EEA, 2005); however, the impact of weather disasters in terms of social and health effects is still considerable and is unequally distributed (see Box 8.2). Flood health impacts range from deaths, injuries, infectious diseases and toxic contamination, to mental health problems (Greenough et al., 2001; Ahern et al., 2005).

In terms of deaths and populations affected, floods and tropical cyclones have the greatest impact in South Asia and Latin America (Guha-Sapir et al., 2004; Schultz et al., 2005). Deaths recorded in disaster databases are from drowning and severe injuries. Deaths from unsafe or unhealthy conditions following the extreme event are also a health consequence, but such information is rarely included in disaster statistics (Combs et al., 1998; Jonkman and Kelman, 2005). Drowning by storm surge is the major killer in coastal storms where there are large numbers of deaths. An assessment of surges in the past 100 years found that major events were confined to a limited number of regions, with many events occurring in the Bay of Bengal, particularly Bangladesh (Nicholls, 2003).

Populations with poor sanitation infrastructure and high burdens of infectious disease often experience increased rates of diarrhoeal diseases after flood events. Increases in cholera (Sur et al., 2000; Gabastou et al., 2002), cryptosporidiosis (Katsumata et al., 1998) and typhoid fever (Vollaard et al., 2004)

Box 8.2. Gender and natural disasters

Men and women are affected differently in all phases of a disaster, from exposure to risk and risk perception; to preparedness behaviour, warning communication and response; physical, psychological, social and economic impacts; emergency response; and ultimately to recovery and reconstruction (Fothergill, 1998). Natural disasters have been shown to result in increased domestic violence against, and post-traumatic stress disorders in, women (Anderson and Manuel, 1994; Garrison et al., 1995; Wilson et al., 1998; Ariyabandu and Wickramasinghe, 2003; Galea et al., 2005). Women make an important contribution to disaster reduction, often informally through participating in disaster management and acting as agents of social change. Their resilience and their networks are critical in household and community recovery (Enarson and Morrow, 1998; Ariyabandu and Wickramasinghe, 2003). After the 1999 Orissa cyclone, most of the relief efforts were targeted at or through women, giving them control over resources. Women received the relief kits, including house-building grants and loans, resulting in improved self-esteem and social status (Briceño, 2002). Similarly, following a disastrous 1992 flood in Pakistan in the Sarghoda district, women were involved in the reconstruction design and were given joint ownership of the homes, promoting their empowerment.

have been reported in low- and middle-income countries. Flood-related increases in diarrhoeal disease have also been reported in India (Mondal et al., 2001), Brazil (Heller et al., 2003) and Bangladesh (Kunii et al., 2002; Schwartz et al., 2006). The floods in Mozambique in 2001 were estimated to have caused over 8,000 additional cases and 447 deaths from diarrhoeal disease in the following months (Cairncross and Alvarinho, 2006).

The risk of infectious disease following flooding in high-income countries is generally low, although increases in respiratory and diarrhoeal diseases have been reported after floods (Miettinen et al., 2001; Reacher et al., 2004; Wade et al., 2004). An important exception was the impact of Hurricanes Katrina and Rita in the USA in 2005, where contamination of water supplies with faecal bacteria led to many cases of diarrhoeal illness and some deaths (CDC, 2005; Manuel, 2006).

Flooding may lead to contamination of waters with dangerous chemicals, heavy metals or other hazardous substances, from storage or from chemicals already in the environment (e.g., pesticides). Chemical contamination following Hurricane Katrina in the USA included oil spills from refineries and storage tanks, pesticides, metals and hazardous waste (Manuel, 2006). Concentrations of most contaminants were within acceptable short-term levels, except for lead and volatile organic compounds (VOCs) in some areas (Pardue et al., 2005). There are also health risks associated with long-term contamination of soil and sediment (Manuel, 2006); however, there is little published evidence demonstrating a causal effect of chemical contamination on the pattern of morbidity and mortality following flooding events (Euripidou and Murray, 2004; Ahern et al., 2005). Increases in population density and accelerating industrial development in areas subject to natural disasters increase the probability of future disasters and the potential for mass human exposure to hazardous materials released during disasters (Young et al., 2004).

There is increasing evidence of the importance of mental disorders as an impact of disasters (Mollica et al., 2004; Ahern et al., 2005). Prolonged impairment resulting from common mental disorders (anxiety and depression) may be considerable. Studies in both low- and high-income countries indicate that the mental-health aspect of flood-related impacts has been insufficiently investigated (Ko et al., 1999; Ohl and Tapsell, 2000; Bokszczanin, 2002; Tapsell et al., 2002; Assanarigkornchai et al., 2004; Norris et al., 2004; North et al., 2004; Ahern et al., 2005; Kohn et al., 2005; Maltais et al., 2005). A systematic review of post-traumatic stress disorder in high-income countries found a small but significant effect following disasters (Galea et al., 2005). There is also evidence of medium- to long-term impacts on behavioural disorders in young children (Durkin et al., 1993; Becht et al., 1998; Bokszczanin, 2000, 2002).

Vulnerability to weather disasters depends on the attributes of the person at risk (including where they live, age, income, education and disability) and on broader social and environmental factors (level of disaster preparedness, health sector responses and environmental degradation) (Blaikie et al., 1994; Menne, 2000; Olmos, 2001; Adger et al., 2005; Few and Matthies, 2006). Poorer communities, particularly slum dwellers, are more likely to live in flood-prone areas. In the USA, lower-income groups were most affected by Hurricane Katrina, and low-income schools had twice the risk of being flooded compared with the reference group (Guidry and Margolis, 2005).

High-density populations in low-lying coastal regions experience a high health burden from weather disasters, such as settlements along the North Sea coast in north-west Europe, the Seychelles, parts of Micronesia, the Gulf Coast of the USA and Mexico, the Nile Delta, the Gulf of Guinea, and the Bay of Bengal (see Chapter 6). Environmentally degraded areas are particularly vulnerable to tropical cyclones and coastal flooding under current climate conditions.

8.2.3 Drought, nutrition and food security

The causal chains through which climate variability and extreme weather influence human nutrition are complex and involve different pathways (regional water scarcity, salinisation of agricultural lands, destruction of crops through flood events, disruption of food logistics through disasters, and increased burden of plant infectious diseases or pests) (see Chapter 5). Both acute and chronic nutritional problems are associated with climate variability and change. The effects of drought on health include deaths, malnutrition (undernutrition, protein-energy malnutrition and/or micronutrient deficiencies), infectious diseases and respiratory diseases (Menne and Bertollini, 2000).

Drought diminishes dietary diversity and reduces overall food consumption, and may therefore lead to micronutrient deficiencies. In Gujarat, India, during a drought in the year 2000, diets were found to be deficient in energy and several vitamins. In this population, serious effects of drought on anthropometric indices may have been prevented by public-health measures (Hari Kumar et al., 2005). A study in southern Africa suggests that HIV/AIDS amplifies the effect of drought on nutrition (Mason et al., 2005). Malnutrition increases the risk both of acquiring and of dying from an infectious disease. A study in Bangladesh found that drought and lack of food were associated with an increased risk of mortality from a diarrhoeal illness (Aziz et al., 1990).

Drought and the consequent loss of livelihoods is also a major trigger for population movements, particularly rural to urban migration. Population displacement can lead to increases in communicable diseases and poor nutritional status resulting from overcrowding, and a lack of safe water, food and shelter (Choudhury and Bhuiya, 1993; Menne and Bertollini, 2000; del Ninno and Lundberg, 2005). Recently, rural to urban migration has been implicated as a driver of HIV transmission (White, 2003; Coffee et al., 2005). Farmers in Australia also appear to be at increased risk of suicide during periods of drought (Nicholls et al., 2005). The range of health impacts associated with a drought event in Brazil are described in Box 8.3.

8.2.3.1 Drought and infectious disease

Countries within the 'Meningitis Belt' in semi-arid sub-Saharan Africa experience the highest endemicity and epidemic frequency of meningococcal meningitis in Africa, although other areas in the Rift Valley, the Great Lakes, and southern Africa are

Box 8.3. Drought in the Amazon

In the dry season of 2005, an intense drought affected the western and central part of the Amazon region, especially Bolivia, Peru and Brazil. In Brazil alone, 280,000 to 300,000 people were affected (see, e.g., Folha, 2006; Socioambiental, 2006). The drought was unusual because it was not caused by an El Niño event, but was linked to a circulation pattern powered by warm seas in the Atlantic – the same phenomenon responsible for the intense Atlantic hurricane season (CPTEC, 2005). There were increased risks to health due to water scarcity, food shortages and smoke from forest fires. Most affected were rural dwellers and riverine traditional subsistence farmers with limited spare resources to mobilise in an emergency. The local and national governments in Brazil provided financial assistance for the provision of safe drinking water, food supplies, medicines and transportation to thousands of people isolated in their communities due to rivers drying up (World Bank, 2005).

also affected. The spatial distribution, intensity and seasonality of meningococcal (epidemic) meningitis appear to be strongly linked to climatic and environmental factors, particularly drought, although the causal mechanism is not clearly understood (Molesworth et al., 2001, 2002a, b, 2003). Climate plays an important part in the interannual variability in transmission, including the timing of the seasonal onset of the disease (Molesworth et al., 2001; Sultan et al., 2005). The geographical distribution of meningitis has expanded in West Africa in recent years, which may be attributable to environmental change driven by both changes in land use and regional climate change (Molesworth et al., 2003).

The transmission of some mosquito-borne diseases is affected by drought events. During droughts, mosquito activity is reduced and, as a consequence, the population of non-immune persons increases. When the drought breaks, there is a much larger proportion of susceptible hosts to become infected, thus potentially increasing transmission (Bouma and Dye, 1997; Woodruff et al., 2002). In other areas, droughts may favour increases in mosquito populations due to reductions in mosquito predators (Chase and Knight, 2003). Other drought-related factors that may result in a short-term increase in the risk for infectious disease outbreaks include stagnation and contamination of drainage canals and small rivers. In the long term, the incidence of mosquito-borne diseases such as malaria decreases because the mosquito vector lacks the necessary humidity and water for breeding. The northern limit of *Plasmodium falciparum* malaria in Africa is the Sahel, where rainfall is an important limiting factor in disease transmission (Ndiaye et al., 2001). Malaria has decreased in association with long-term decreases in annual rainfall in Senegal and Niger (Mouchet et al., 1996; Julvez et al., 1997). Drought events are also associated with dust storms and respiratory health effects (see Section 8.2.6). Droughts are also associated with water scarcity; the risks of water-washed diseases are addressed in Section 8.2.5.

8.2.4 Food safety

Several studies have confirmed and quantified the effects of high temperatures on common forms of food poisoning, such as salmonellosis (D'Souza et al., 2004; Kovats et al., 2004; Fleury et al., 2006). These studies found an approximately linear increase in reported cases with each degree increase in weekly or monthly temperature. Temperature is much less important for the transmission of *Campylobacter* (Kovats et al., 2005; Louis et al., 2005; Tam et al., 2006).

Contact between food and pest species, especially flies, rodents and cockroaches, is also temperature-sensitive. Fly activity is largely driven by temperature rather than by biotic factors (Goulson et al., 2005). In temperate countries, warmer weather and milder winters are likely to increase the abundance of flies and other pest species during the summer months, with the pests appearing earlier in spring.

Harmful algal blooms (HABs) (see Chapter 1, Section 1.3.4.2) produce toxins that can cause human diseases, mainly via consumption of contaminated shellfish. Warmer seas may thus contribute to increased cases of human shellfish and reef-fish poisoning (ciguatera) and poleward expansions of these disease distributions (Kohler and Kohler, 1992; Lehane and Lewis, 2000; Hall et al., 2002; Hunter, 2003; Korenberg, 2004). For example, sea-surface temperatures influence the growth of *Gambierdiscus* spp., which is associated with reports of ciguatera in French Polynesia (Chateau-Degat et al., 2005). No further assessments of the impact of climate change on shellfish poisoning have been carried out since the TAR.

Vibrio parahaemolyticus and *Vibrio vulnificus* are responsible for non-viral infections related to shellfish consumption in the USA, Japan and South-East Asia (Wittmann and Flick, 1995; Tuyet et al., 2002). Abundance is dependent on the salinity and temperature of the coastal water. A large outbreak in 2004 due to the consumption of contaminated oysters (*V. parahaemolyticus*) was linked to atypically high temperatures in Alaskan coastal waters (McLaughlin et al., 2005).

Another example of the implications that climate change can have for food safety is the methylation of mercury and its subsequent uptake by fish and human beings, as observed in the Faroe Islands (Booth and Zeller, 2005; McMichael et al., 2006).

8.2.5 Water and disease

Climate-change-related alterations in rainfall, surface water availability and water quality could affect the burden of water-related diseases (see Chapter 3). Water-related diseases can be classified by route of transmission, thus distinguishing between water-borne (ingested) and water-washed diseases (caused by lack of hygiene). There are four main considerations to take into account when evaluating the relationship between health outcomes and exposure to changes in rainfall, water availability and quality:

- linkages between water availability, household access to improved water, and the health burden due to diarrhoeal diseases;
- the role of extreme rainfall (intense rainfall or drought) in facilitating water-borne outbreaks of diseases through piped water supplies or surface water;
- effects of temperature and runoff on microbiological and chemical contamination of coastal, recreational and surface waters;
- direct effects of temperature on the incidence of diarrhoeal disease.

Access to safe water remains an extremely important global health issue. More than 2 billion people live in the dry regions of the world and suffer disproportionately from malnutrition, infant mortality and diseases related to contaminated or insufficient water (WHO, 2005). A small and unquantified proportion of this burden can be attributed to climate variability or climate extremes. The effect of water scarcity on food availability and malnutrition is discussed in Section 8.2.3, and the effect of rainfall on outbreaks of mosquito-borne and rodent-borne disease is discussed in Section 8.2.8.

Childhood mortality due to diarrhoea in low-income countries, especially in sub-Saharan Africa, remains high despite improvements in care and the use of oral rehydration therapy (Kosek et al., 2003). Children may survive the acute illness but may later die due to persistent diarrhoea or malnutrition. Children in poor rural and urban slum areas are at high risk of diarrhoeal disease mortality and morbidity. Several studies have shown that transmission of enteric pathogens is higher during the rainy season (Nchito et al., 1998; Kang et al., 2001). Drainage and storm water management is important in low-income urban communities, as blocked drains are one of the causes of increased disease transmission (Parkinson and Butler, 2005).

Climate extremes cause both physical and managerial stresses on water supply systems (see Chapters 3 and 7), although well-managed public water supply systems should be able to cope with climate extremes (Nicholls, 2003; Wilby et al., 2005). Reductions in rainfall lead to low river flows, reducing effluent dilution and leading to increased pathogen loading. This could represent an increased challenge to water-treatment plants. During the dry summer of 2003, low flows of rivers in the Netherlands resulted in apparent changes in water quality (Senhorst and Zwolsman, 2005).

Extreme rainfall and runoff events may increase the total microbial load in watercourses and drinking-water reservoirs (Kistemann et al., 2002), although the linkage to cases of human disease is less certain (Schwartz and Levin, 1999; Aramini et al., 2000; Schwartz et al., 2000; Lim et al., 2002). A study in the USA found an association between extreme rainfall events and monthly reports of outbreaks of water-borne disease (Curriero et al., 2001). The seasonal contamination of surface water in early spring in North America and Europe may explain some of the seasonality in sporadic cases of water-borne diseases such as cryptosporidiosis and campylobacteriosis (Clark et al., 2003; Lake et al., 2005). The marked seasonality of cholera outbreaks in the Amazon is associated with low river flow in the dry season (Gerolomo and Penna, 1999), probably due to pathogen concentrations in pools.

Higher temperature was found to be strongly associated with increased episodes of diarrhoeal disease in adults and children in Peru (Checkley et al., 2000; Speelmon et al., 2000; Checkley et al., 2004; Lama et al., 2004). Associations between monthly temperature and diarrhoeal episodes have also been reported in the Pacific islands, Australia and Israel (Singh et al., 2001; McMichael et al., 2003b; Vasilev, 2003).

Although there is evidence that the bimodal seasonal pattern of cholera in Bangladesh is correlated with sea-surface temperatures in the Bay of Bengal and with seasonal plankton abundance (a possible environmental reservoir of the cholera pathogen, *Vibrio cholerae*) (Colwell, 1996; Bouma and Pascual, 2001), winter peaks in disease further inland are not associated with sea-surface temperatures (Bouma and Pascual, 2001). In many countries cholera transmission is primarily associated with poor sanitation. The effect of sea-surface temperatures in cholera transmission has been most studied in the Bay of Bengal (Pascual et al., 2000; Lipp et al., 2002; Rodo et al., 2002; Koelle et al., 2005). In sub-Saharan Africa, cholera outbreaks are often associated with flood events and faecal contamination of the water supplies.

8.2.6 Air quality and disease

Weather at all time scales determines the development, transport, dispersion and deposition of air pollutants, with the passage of fronts, cyclonic and anticyclonic systems and their associated air masses being of particular importance. Air-pollution episodes are often associated with stationary or slowly migrating anticyclonic or high pressure systems, which reduce pollution dispersion and diffusion (Schichtel and Husar, 2001; Rao et al., 2003). Airflow along the flanks of anticyclonic systems can transport ozone precursors, creating the conditions for an ozone event (Lennartson and Schwartz, 1999; Scott and Diab, 2000; Yarnal et al., 2001; Tanner and Law, 2002). Certain weather patterns enhance the development of the urban heat island, the intensity of which may be important for secondary chemical reactions within the urban atmosphere, leading to elevated levels of some pollutants (Morris and Simmonds, 2000; Junk et al., 2003; Jonsson et al., 2004).

8.2.6.1 Ground-level ozone

Ground-level ozone is both naturally occurring and, as the primary constituent of urban smog, is also a secondary pollutant formed through photochemical reactions involving nitrogen oxides and volatile organic compounds in the presence of bright sunshine with high temperatures. In urban areas, transport vehicles are the key sources of nitrogen oxides and volatile organic compounds. Temperature, wind, solar radiation, atmospheric moisture, venting and mixing affect both the emissions of ozone precursors and the production of ozone (Nilsson et al., 2001a, b; Mott et al., 2005). Because ozone formation depends on sunlight, concentrations are typically highest during the summer months, although not all cities have shown seasonality in ozone concentrations (Bates, 2005). Concentrations of ground-level ozone are increasing in most regions (Wu and Chan, 2001; Chen et al., 2004).

Exposure to elevated concentrations of ozone is associated with increased hospital admissions for pneumonia, chronic obstructive pulmonary disease, asthma, allergic rhinitis and other respiratory diseases, and with premature mortality (e.g., Mudway and Kelly, 2000; Gryparis et al., 2004; Bell et al., 2005, 2006; Ito et al., 2005; Levy et al., 2005). Outdoor ozone concentrations, activity patterns and housing characteristics, such as the extent of insulation, are the primary determinants of ozone exposure (Suh et al., 2000; Levy et al., 2005). Although a considerable amount is known about the health effects of ozone in Europe and North America, few studies have been conducted in other regions.

8.2.6.2 Effects of weather on concentrations of other air pollutants

Concentrations of air pollutants in general, and fine particulate matter (PM) in particular, may change in response to climate change because their formation depends, in part, on temperature and humidity. Air-pollution concentrations are the result of interactions between variations in the physical and dynamic properties of the atmosphere on time-scales from hours to days, atmospheric circulation features, wind, topography and energy use (McGregor, 1999; Hartley and Robinson, 2000; Pal Arya, 2000). Some air pollutants demonstrate weather-related seasonal cycles (Alvarez et al., 2000; Kassomenos et al., 2001; Hazenkamp-von Arx et al., 2003; Nagendra and Khare, 2003; Eiguren-Fernandez et al., 2004). Some locations, such as Mexico City and Los Angeles, are predisposed to poor air quality because local weather patterns are conducive to chemical reactions leading to the transformation of emissions, and because the topography restricts the dispersion of pollutants (Rappengluck et al., 2000; Kossmann and Sturman, 2004).

Evidence for the health impacts of PM is stronger than that for ozone. PM is known to affect morbidity and mortality (e.g., Ibald-Mulli et al., 2002; Pope et al., 2002; Kappos et al., 2004; Dominici et al., 2006), so increasing concentrations would have significant negative health impacts.

8.2.6.3 Air pollutants from forest fires

In some regions, changes in temperature and precipitation are projected to increase the frequency and severity of fire events (see Chapter 5). Forest and bush fires cause burns, damage from smoke inhalation and other injuries. Large fires are also accompanied by an increased number of patients seeking emergency services (Hoyt and Gerhart, 2004). Toxic gaseous and particulate air pollutants are released into the atmosphere, which can significantly contribute to acute and chronic illnesses of the respiratory system, particularly in children, including pneumonia, upper respiratory diseases, asthma and chronic obstructive pulmonary diseases (WHO, 2002a; Bowman and Johnston, 2005; Moore et al., 2006). For example, the 1997 Indonesia fires increased hospital admissions and mortality from cardiovascular and respiratory diseases, and negatively affected activities of daily living in South-East Asia (Sastry, 2002; Frankenberg et al., 2005; Mott et al., 2005). Pollutants from forest fires can affect air quality for thousands of kilometres (Sapkota et al., 2005).

8.2.6.4 Long-range transport of air pollutants

Changes in wind patterns and increased desertification may increase the long-range transport of air pollutants. Under certain atmospheric circulation conditions, the transport of pollutants, including aerosols, carbon monoxide, ozone, desert dust, mould spores and pesticides, may occur over large distances and over time-scales typically of 4-6 days, which can lead to adverse health impacts (Gangoiti et al., 2001; Stohl et al., 2001; Buchanan et al., 2002; Chan et al., 2002; Martin et al., 2002; Ryall et al., 2002; Ansmann et al., 2003; He et al., 2003; Helmis et al., 2003; Moore et al., 2003; Shinn et al., 2003; Unsworth et al., 2003; Kato et al., 2004; Liang et al., 2004; Tu et al., 2004). Sources of such pollutants include biomass burning, as well as industrial and mobile sources (Murano et al., 2000; Koe et al., 2001; Jaffe et al., 2003, 2004; Moore et al., 2003).

Windblown dust originating in desert regions of Africa, Mongolia, Central Asia and China can affect air quality and population health in remote areas. When compared with non-dust weather conditions, dust can carry large concentrations of respirable particles, trace elements that can affect human health, fungal spores and bacteria (Claiborn et al., 2000; Fan et al., 2002; Shinn et al., 2003; Cook et al., 2005; Prospero et al., 2005; Xie et al., 2005; Kellogg and Griffin, 2006). However, recent studies have not found statistically significant associations between Asian dust storms and hospital admissions in Canada and Taiwan (Chen and Tang, 2005; Yang et al., 2005a; Bennett et al., 2006). Evidence suggests that local mortality, particularly from cardiovascular and respiratory diseases, is increased in the days following a dust storm (Kwon et al., 2002; Chen et al., 2004).

8.2.7 Aeroallergens and disease

Climate change has caused an earlier onset of the spring pollen season in the Northern Hemisphere (see Chapter 1, Section 1.3.7.4; D'Amato et al., 2002; Weber, 2002; Beggs, 2004). It is reasonable to conclude that allergenic diseases caused by pollen, such as allergic rhinitis, have experienced some concomitant change in seasonality (Emberlin et al., 2002; Burr et al., 2003). There is limited evidence that the length of the pollen season has also increased for some species. Although there are suggestions that the abundance of a few species of air-borne pollens has increased due to climate change, it is unclear whether the allergenic content of these pollen types has changed (pollen content remaining the same or increasing would imply increased exposure) (Huynen and Menne, 2003; Beggs and Bambrick, 2005). Few studies show patterns of increasing exposure for allergenic mould spores or bacteria (Corden et al., 2003; Harrison et al., 2005). Changes in the spatial distribution of natural vegetation, such as the introduction of new aeroallergens into an area, increases sensitisation (Voltolini et al., 2000; Asero, 2002). The introduction of new invasive plant species with highly allergenic pollen, in particular ragweed (*Ambrosia artemisiifolia*), presents important health risks; ragweed is spreading in several parts of the world (Rybnicek and Jaeger, 2001; Huynen and Menne, 2003; Taramarcaz et al., 2005; Cecchi et al., 2006). Several laboratory studies show that increasing CO_2 concentrations and temperatures increase ragweed pollen production and prolong the ragweed pollen season (Wan et al., 2002; Wayne et al., 2002; Singer et al., 2005; Ziska et al., 2005; Rogers et al., 2006a) and increase some plant metabolites that can affect human health (Ziska et al., 2005; Mohan et al., 2006).

8.2.8 Vector-borne, rodent-borne and other infectious diseases

Vector-borne diseases (VBD) are infections transmitted by the bite of infected arthropod species, such as mosquitoes, ticks, triatomine bugs, sandflies and blackflies. VBDs are among the most well-studied of the diseases associated with climate change, due to their widespread occurrence and sensitivity to climatic factors. There is some evidence of climate-change-related shifts in the distribution of tick vectors of disease, of some (non-malarial) mosquito vectors in Europe and North America, and in the phenology of bird reservoirs of pathogens (see Chapter 1 and Box 8.4).

Northern or altitudinal shifts in tick distribution have been observed in Sweden (Lindgren and Talleklint, 2000; Lindgren and Gustafson, 2001) and Canada (Barker and Lindsay, 2000), and altitudinal shifts have been observed in the Czech Republic (Daniel et al., 2004). Geographical changes in tick-borne infections have been observed in Denmark (Skarphedinsson et al., 2005). Climate change alone is unlikely to explain recent increases in the incidence of tick-borne diseases in Europe or North America. There is considerable spatial heterogeneity in the degree of increase of tick-borne encephalitis, for example, within regions of Europe likely to have experienced similar levels of climate change (Patz, 2002; Randolph, 2004; Sumilo et al., 2006). Other explanations cannot be ruled out, e.g., human impacts on the landscape, increasing both the habitat and wildlife hosts of ticks, and changes in human behaviour that may increase human contact with infected ticks (Randolph, 2001).

In north-eastern North America, there is evidence of recent micro-evolutionary (genetic) responses of the mosquito species *Wyeomyia smithii* to increased average land surface temperatures and earlier arrival of spring in the past two decades (Bradshaw and Holzapfel, 2001). Although not a vector of human disease, this species is closely related to important arbovirus vector species that may be undergoing similar evolutionary changes.

Cutaneous leishmaniasis has been reported in dogs (reservoir hosts) further north in Europe, although the possibility of previous under-reporting cannot be excluded (Lindgren and Naucke, 2006). Changes in the geographical distribution of the sandfly vector have been reported in southern Europe (Aransay et al., 2004; Afonso et al., 2005). However, no study has investigated the causes of these changes. The re-emergence of kala-azar (visceral leishmaniasis) in cities of the semi-arid Brazilian north-eastern region in the early 1980s and 1990s was caused by rural–urban migration of subsistence farmers who had lost their crops due to prolonged droughts (Franke et al., 2002; Confalonieri, 2003).

8.2.8.1 Dengue

Dengue is the world's most important vector-borne viral disease. Several studies have reported associations between spatial (Hales et al., 2002), temporal (Hales et al., 1999; Corwin et al., 2001; Gagnon et al., 2001) or spatiotemporal patterns of dengue and climate (Hales et al., 1999; Corwin et al., 2001; Gagnon et al., 2001; Cazelles et al., 2005). However, these reported associations are not entirely consistent, possibly reflecting the complexity of climatic effects on transmission, and/or the presence of competing factors (Cummings, 2004). While high rainfall or high temperature can lead to an increase in transmission, studies have shown that drought can also be a cause if household water storage increases the number of suitable mosquito breeding sites (Pontes et al., 2000; Depradine and Lovell, 2004; Guang et al., 2005).

Climate-based (temperature, rainfall, cloud cover) density maps of the main dengue vector *Stegomyia* (previously called

Box 8.4. Climate change, migratory birds and infectious diseases

Several species of wild birds can act as biological or mechanical carriers of human pathogens as well as of vectors of infectious agents (Olsen et al., 1995; Klich et al., 1996; Gylfe et al., 2000; Friend et al., 2001; Pereira et al., 2001; Broman et al., 2002; Moore et al., 2002; Niskanen et al., 2003; Rappole and Hubalek, 2003; Reed et al., 2003; Fallacara et al., 2004; Hubalek, 2004; Krauss et al., 2004). Many of these birds are migratory species that seasonally fly long distances through different continents (de Graaf and Rappole, 1995; Webster et al., 2002b). Climate change has been implicated in changes in the migratory and reproductive phenology (advancement in breeding and migration dates) of several bird species, their abundance and population dynamics, as well as a northward expansion of their geographical range in Europe (Sillett et al., 2000; Barbraud and Weimerskirch, 2001; Parmesan and Yohe, 2003; Brommer, 2004; Visser et al., 2004; Both and Visser, 2005). Two possible consequences of these phenological changes in birds to the dispersion of pathogens and their vectors are:

1. shifts in the geographical distribution of the vectors and pathogens due to altered distributions or changed migratory patterns of bird populations;
2. changes in the life cycles of bird-associated pathogens due to the mistiming between bird breeding and the breeding of vectors, such as mosquitoes. One example is the transmission of St. Louis encephalitis virus, which depends on meteorological triggers (e.g., precipitation) to bring the pathogen, vector and host (nestlings) cycles into synchrony, allowing an overlap that initiates and facilitates the cycling necessary for virus amplification between mosquitoes and wild birds (Day, 2001).

Aedes) aegypti are a good match with the observed disease distribution (Hopp and Foley, 2003). The model of vector abundance has good agreement with the distribution of reported cases of dengue in Colombia, Haiti, Honduras, Indonesia, Thailand and Vietnam (Hopp and Foley, 2003). Approximately one-third of the world's population lives in regions where the climate is suitable for dengue transmission (Hales et al., 2002; Rogers et al., 2006b).

8.2.8.2 Malaria

The spatial distribution, intensity of transmission, and seasonality of malaria is influenced by climate in sub-Saharan Africa; socio-economic development has had only limited impact on curtailing disease distribution (Hay et al., 2002a; Craig et al., 2004).

Rainfall can be a limiting factor for mosquito populations and there is some evidence of reductions in transmission associated with decadal decreases in rainfall. Interannual malaria variability is climate-related in specific eco-epidemiological zones (Julvez et al., 1992; Ndiaye et al., 2001; Singh and Sharma, 2002; Bouma, 2003; Thomson et al., 2005). A systematic review of studies of the El Niño-Southern Oscillation (ENSO) and malaria concluded that the impact of El Niño on the risk of malaria epidemics is well established in parts of southern Asia and South America (Kovats et al., 2003). Evidence of the predictability of unusually high or low malaria anomalies from both sea-surface temperature (Thomson et al., 2005) and multi-model ensemble seasonal climate forecasts in Botswana (Thomson et al., 2006) supports the practical and routine use of seasonal forecasts for malaria control in southern Africa (DaSilva et al., 2004).

The effects of observed climate change on the geographical distribution of malaria and its transmission intensity in highland regions remains controversial. Analyses of time-series data in some sites in East Africa indicate that malaria incidence has increased in the apparent absence of climate trends (Hay et al., 2002a, b; Shanks et al., 2002). The proposed driving forces behind the malaria resurgence include drug resistance of the malaria parasite and a decrease in vector control activities. However, the validity of this conclusion has been questioned because it may have resulted from inappropriate use of the climatic data (Patz, 2002). Analysis of updated temperature data for these regions has found a significant warming trend since the end of the 1970s, with the magnitude of the change affecting transmission potential (Pascual et al., 2006). In southern Africa, long-term trends for malaria were not significantly associated with climate, although seasonal changes in case numbers were significantly associated with a number of climatic variables (Craig et al., 2004). Drug resistance and HIV infection were associated with long-term malaria trends in the same area (Craig et al., 2004).

A number of further studies have reported associations between interannual variability in temperature and malaria transmission in the African highlands. An analysis of de-trended time-series malaria data in Madagascar indicated that minimum temperature at the start of the transmission season, corresponding to the months when the human–vector contact is greatest, accounts for most of the variability between years (Bouma, 2003). In highland areas of Kenya, malaria admissions have been associated with rainfall and unusually high maximum temperatures 3-4 months previously (Githeko and Ndegwa, 2001). An analysis of malaria morbidity data for the period from the late 1980s until the early 1990s from 50 sites across Ethiopia found that epidemics were associated with high minimum temperatures in the preceding months (Abeku et al., 2003). An analysis of data from seven highland sites in East Africa reported that short-term climate variability played a more important role than long-term trends in initiating malaria epidemics (Zhou et al., 2004, 2005), although the method used to test this hypothesis has been challenged (Hay et al., 2005b).

There is no clear evidence that malaria has been affected by climate change in South America (Benitez et al., 2004) (see Chapter 1) or in continental regions of the Russian Federation (Semenov et al., 2002). The attribution of changes in human diseases to climate change must first take into account the considerable changes in reporting, surveillance, disease control measures, population changes, and other factors such as land-use change (Kovats et al., 2001; Rogers and Randolph, 2006).

Despite the known causal links between climate and malaria transmission dynamics, there is still much uncertainty about the potential impact of climate change on malaria at local and global scales (see also Section 8.4.1) because of the paucity of concurrent detailed historical observations of climate and malaria, the complexity of malaria disease dynamics, and the importance of non-climatic factors, including socio-economic development, immunity and drug resistance, in determining infection and infection outcomes. Given the large populations living in highland areas of East Africa, the limitations of the analyses conducted, and the significant health risks of epidemic malaria, further research is warranted.

8.2.8.3 Other infectious diseases

Recent investigations of plague foci in North America and Asia with respect to the relationships between climatic variables, human disease cases (Enscore et al., 2002) and animal reservoirs (Stapp et al., 2004; Stenseth, 2006) have suggested that temporal variations in plague risk can be estimated by monitoring key climatic variables.

There is good evidence that diseases transmitted by rodents sometimes increase during heavy rainfall and flooding because of altered patterns of human–pathogen–rodent contact. There have been reports of flood-associated outbreaks of leptospirosis (Weil's diseases) from a wide range of countries in Central and South America and South Asia (Ko et al., 1999; Vanasco et al., 2002; Confalonieri, 2003; Ahern et al., 2005). Risk factors for leptospirosis for peri-urban populations in low-income countries include flooding of open sewers and streets during the rainy season (Sarkar et al., 2002).

Cases of hantavirus pulmonary syndrome (HPS) were first reported in Central America (Panama) in 2000, and a suggested cause was the increase in peri-domestic rodents following increased rainfall and flooding in surrounding areas (Bayard et al., 2000), although this requires further investigation. There are climate-related differences in hantavirus dynamics between northern and central Europe (Vapalahti et al., 2003; Pejoch and Kriz, 2006).

The distribution and emergence of other infectious diseases have been affected by weather and climate variability. ENSO-driven bush fires and drought, as well as land-use and land-cover changes, have caused extensive changes in the habitat of some bat species that are the natural reservoirs for the Nipah virus. The bats were driven to farms to find food (fruits), consequently shedding virus and causing an epidemic in Malaysia and neighbouring countries (Chua et al., 2000).

The distribution of schistosomiasis, a water-related parasitic disease with aquatic snails as intermediate hosts, may be affected by climatic factors. In one area of Brazil, the length of the dry season and human population density were the most important factors limiting schistosomiasis distribution and abundance (Bavia et al., 1999). Over a larger area, there was an inverse association between prevalence rates and the length of the dry period (Bavia et al., 2001). Recent studies in China indicate that the increased incidence of schistosomiasis over the past decade may in part reflect the recent warming trend. The critical 'freeze line' limits the survival of the intermediate host (*Oncomelania* water snails) and hence limits the transmission of the parasite *Schistosoma japonicum*. The freeze line has moved northwards, putting an additional 20.7 million people at risk of schistosomiasis (Yang et al., 2005b).

8.2.9 Occupational health

Changes in climate have implications for occupational health and safety. Heat stress due to high temperature and humidity is an occupational hazard that can lead to death or chronic ill-health from the after-effects of heatstroke (Wyndham, 1965; Afanas'eva et al., 1997; Adelakun et al., 1999). Both outdoor and indoor workers are at risk of heatstroke (Leithead and Lind, 1964; Samarasinghe, 2001; Shanks and Papworth, 2001). The occupations most at risk of heatstroke, based on data from the USA, include construction and agriculture/forestry/fishing work (Adelakun et al., 1999; Krake et al., 2003). Acclimatisation in tropical environments does not eliminate the risk, as evidenced by the occurrence of heatstroke in metal workers in Bangladesh (Ahasan et al., 1999) and rickshaw pullers in South Asia (OCHA, 2003). Several of the heatstroke deaths reported in the 2003 and 2006 heatwaves in Paris were associated with occupational exposure (Senat, 2004)

Hot working environments are not just a question of comfort, but a concern for health protection and the ability to perform work tasks. Working in hot environments increases the risk of diminished ability to carry out physical tasks (Kerslake, 1972), diminishes mental task ability (Ramsey, 1995), increases accident risk (Ramsey et al., 1983) and, if prolonged, may lead to heat exhaustion or heatstroke (Hales and Richards, 1987) (see Section 8.5).

8.2.10 Ultraviolet radiation and health

Solar ultraviolet radiation (UVR) exposure causes a range of health impacts. Globally, excessive solar UVR exposure has caused the loss of approximately 1.5 million disability-adjusted life years (DALYs) (0.1% of the total global burden of disease) and 60,000 premature deaths in the year 2000. The greatest burdens result from UVR-induced cortical cataracts, cutaneous malignant melanoma, and sunburn (although the latter estimates are highly uncertain due to the paucity of data) (Prüss-Üstün et al., 2006). UVR exposure may weaken the immune response to certain vaccinations, which would reduce their effectiveness. However, there are also important health benefits: exposure to radiation in the ultraviolet B frequency band is required for the production of vitamin D in the body. Lack of sun exposure may lead to osteomalacia (rickets) and other disorders caused by vitamin D deficiencies.

Climate change will alter human exposure to UVR exposure in several ways, although the balance of effects is difficult to predict and will vary depending on location and present exposure to UVR. Greenhouse-induced cooling of the stratosphere is expected to prolong the effect of ozone-depleting gases, which will increase levels of UVR reaching some parts of the Earth's surface (Beggs, 2005; IPCC/TEAP, 2005). Climate change will alter the distribution of clouds which will, in turn, affect UVR levels at the surface. Higher ambient temperatures will influence clothing choices and time spent outdoors, potentially increasing UVR exposure in some regions and decreasing it in others. If immune function is impaired and vaccine efficacy is reduced, the effects of climate-related shifts in infections may be greater than would occur in the absence of high UVR levels (Zwander, 2002; de Gruijl et al., 2003; Holick, 2004; Gallagher and Lee, 2006; Samanek et al., 2006).

8.3 Assumptions about future trends

The impacts of developmental, climatic and environmental scenarios on population health are important for health-system planning processes. Also, future trends in health are relevant to climate change because the health of populations is an important element of adaptive capacity.

8.3.1 Health in scenarios

The use of scenarios to explore future effects of climate change on population health is at an early stage of development (see Section 8.4.1). Published scenarios describe possible future pathways based on observed trends or explicit storylines, and have been developed for a variety of purposes, including the Millennium Ecosystem Assessment (Millennium Ecosystem Assessment, 2005), the IPCC Special Report on Emissions Scenarios (SRES, Nakićenović and Swart, 2000), GEO3 (UNEP, 2002) and the World Water Report (United Nations World Water Assessment Programme, 2003; Ebi and Gamble, 2005). Examples of the many possible futures that have been described include possible changes in the patterns of infectious diseases, medical technology, and health and social inequalities (Olshansky et al., 1998; IPCC, 2000; Martens and Hilderink, 2001; Martens and Huynen, 2003). Infectious diseases could become more prominent if public-health systems unravel, or if new pathogens arise that are resistant to our current methods of disease control, leading to falling life expectancies and reduced economic productivity (Barrett et al., 1998). An age of expanded medical technology could result from increased economic growth and improvements in technology, which may to some extent offset deteriorations in the physical and social environment, but at the risk of widening current health inequalities (Martens and Hilderink, 2001). Alternatively, an age of sustained health could result from more wide-ranging investment in social and medical services, leading to a reduction in the incidence of disease, benefiting most segments of the population.

Common to these scenarios is a view that major risks to health will remain unless the poorest countries share in the growth and development experienced by richer parts of the world. It is envisaged also that greater mobility and more rapid spread of ideas and technology worldwide will bring a mix of positive and negative effects on health, and that a deliberate focus on sustainability will be required to reduce the impacts of human activity on climate, water and food resources (Goklany, 2002).

8.3.2 Future vulnerability to climate change

The health of populations is an important element of adaptive capacity. Where there is a heavy burden of disease and disability, the effects of climate change are likely to be more severe than otherwise. For example, in Africa and Asia the future course of the HIV/AIDS epidemic will significantly influence how well populations can cope with challenges such as the spread of climate-related infections (vector- or water-borne), food shortages, and increased frequency of storms, floods and droughts (Dixon et al., 2002).

The total number of people at risk, the age structure of the population, and the density of settlement are important variables in any projections of the effects of climate change. Many populations will age appreciably in the next 50 years. This is relevant to climate change because the elderly are more vulnerable than younger age groups to injury resulting from weather extremes such as heatwaves, storms and floods. It is assumed (with a high degree of confidence) that over the course of the 21st century the population will grow substantially in many of the poorest countries of the world, while numbers will remain much the same, or decline, in the high-income countries. The world population will increase from its current 6.4 billion to somewhat below 9 billion by the middle of the century (Lutz et al., 2000), but regional patterns will vary widely. For example, the population density of Europe is projected to fall from 32 to 27 people/km^2, while that of Africa could rise from 26 to 60 people/km^2 (Cohen, 2003). Currently, 70% of all episodes of clinical *Plasmodium falciparum* malaria worldwide occur in Africa, and that fraction will rise substantially in the future (World Bank et al., 2004). Also relevant to considerations of the impacts of climate change is urbanisation, because the effects of higher temperatures and altered patterns of rainfall are strongly modified by the local environment. For instance, during hot weather, temperatures tend to be higher in built-up areas, due to the urban heat-island effect. Almost all the growth in population in the next 50 years is expected to occur in cities (and in particular, cities in poor countries) (Cohen, 2003). These trends in population dominate calculations of the possible consequences of climate change. These are two examples: projections of the numbers of people affected by coastal flooding and the spread of malaria are more sensitive to assumptions about future population trajectories than to the choice of climate-change model (Nicholls, 2004; van Lieshout et al., 2004).

For much of the world's population, the ability to lead a healthy life is limited by the direct and indirect effects of poverty (World Bank et al., 2004). Although the percentage of people living on less than US$1/day has decreased in Asia and Latin America since 1990, in the sub-Saharan region 46% of the population is now living on less than US$1/day and little improvement is expected in the short and medium term. Poverty levels in Europe and Central Asia show few signs of improvement (World Bank, 2004; World Bank et al., 2004). Economic growth in the richest regions has outstripped advances in other parts of the world, meaning that global disparities in income have increased in the last 20 years (UNEP and WCMC, 2002).

In the future, vulnerability to climate will depend not only on the extent of socio-economic change, but also on how evenly the benefits and costs are distributed, and the manner in which change occurs (McKee and Suhrcke, 2005). Economic growth is double-sided. Growth entails social change, and while this change may be wealth-creating, it may also, in the short term at least, cause significant social stress and environmental damage. Rapid urbanisation (leading to plummeting population health) in western Europe in the 19th century, and extensive land clearance (causing widespread ecological damage) in South America and South-East Asia in the 20th century, are two examples of negative consequences of rapid economic growth (Szreter, 2004). Social disorder, conflict, and lack of effective civic institutions will also increase vulnerability to health risks resulting from climate change.

Health services provide a buffer against the hazards of climate variability and change. For instance, access to cheap, effective anti-malarials, insecticide-treated bed nets and indoor spray programmes will be important for future trends in malaria. Emergency medical services have a role (although not a predominant one) in limiting excess mortality due to heatwaves and other extreme climate events.

There are other determinants of vulnerability that relate to particular threats, or particular settings. Heatwaves, for example, are exacerbated by the urban heat-island effect, so that impacts of high temperatures will be modified by the size and design of future cities (Meehl and Tebaldi, 2004). The consequences of changes in food production due to climate change will depend on access to international markets and the conditions of trade. If these conditions exclude or penalise poor countries, then the risks of disease and ill-health due to malnutrition will be much higher than if a more inclusive economic order is achieved. Changes in land-use practices for the production of biofuels in place of grain and other food crops will have benefits for greenhouse gas emissions reductions, but the way in which the fuels are burnt is also important (see Section 8.7.1).

8.4 Key future impacts and vulnerabilities

The impacts of climate change have been projected for a limited range of health determinants and outcomes for which the epidemiologic evidence base is well developed. The studies reviewed in Section 8.4.1 used quantitative and qualitative approaches to project the incidence and geographical range of health outcomes under different climate and socio-economic scenarios. Section 8.4.2 assesses the possible consequences of climate-change-related health impacts on particularly vulnerable populations and regions in the next few decades

Overall, climate change is projected to have some health benefits, including reduced cold-related mortality, reductions in some pollutant-related mortality, and restricted distribution of diseases where temperatures or rainfall exceed upper thresholds for vectors or parasites. However, the balance of impacts will be overwhelmingly negative (see Section 8.7). Most projections suggest modest changes in the burden of climate-sensitive health outcomes over the next few decades, with larger increases beginning mid-century. The balance of positive and negative health impacts will vary from one location to another and will alter over time as temperatures continue to rise.

8.4.1 Projections of climate-change-related health impacts

Projections of climate-change-related health impacts use different approaches to classify the risk of climate-sensitive health determinants and outcomes. For malaria and dengue, results from projections are commonly presented as maps of potential shifts in distribution. Health-impact models are typically based on climatic constraints on the development of the vector and/or parasite, and include limited population projections and non-climate assumptions. However, there are important differences between disease risk (on the basis of climatic and entomological considerations) and experienced morbidity and mortality. Although large portions of Europe and the USA may be at potential risk for malaria based on the distribution of competent disease vectors, locally acquired cases have been virtually eliminated, in part due to vector- and disease-control activities. Projections for other health outcomes often estimate populations-at-risk or person-months at risk.

Economic scenarios cannot be directly related to disease burdens because the relationships between gross domestic product (GDP) and burdens of climate-sensitive diseases are confounded by social, environmental and climate factors (Arnell et al., 2004; van Lieshout et al., 2004; Pitcher et al., 2007). The assumption that increasing per capita income will improve population health ignores the fact that health is determined by factors other than income alone; that good population health in itself is a critical input into economic growth and long-term economic development; and that persistent challenges to development are a reality in many countries, with continuing high burdens from relatively easy-to-control diseases (Goklany, 2002; Pitcher et al., 2007).

8.4.1.1 Global burden of disease study

The World Health Organization conducted a regional and global comparative risk assessment to quantify the amount of premature morbidity and mortality due to a range of risk factors, including climate change, and to estimate the benefit of interventions to remove or reduce these risk factors. In the year 2000, climate change is estimated to have caused the loss of over 150,000 lives and 5,500,000 DALYs (0.3% of deaths and 0.4% of DALYs, respectively) (Campbell-Lendrum et al., 2003; Ezzati et al., 2004; McMichael, 2004). The assessment also addressed how much of the future burden of climate change could be avoided by stabilising greenhouse gas emissions (Campbell-Lendrum et al., 2003). The health outcomes included were chosen based on known sensitivity to climate variation, predicted future importance, and availability of quantitative global models (or the feasibility of constructing them):

- episodes of diarrhoeal disease,
- cases of *Plasmodium falciparum* malaria,
- fatal accidental injuries in coastal floods and inland floods/landslides,
- the non-availability of recommended daily calorie intake (as an indicator for the prevalence of malnutrition).

Limited adjustments for adaptation were included in the estimates.

The projected relative risks attributable to climate change in 2030 vary by health outcome and region, and are largely negative, with most of the projected disease burden being due to increases in diarrhoeal disease and malnutrition, primarily in low-income populations already experiencing a large burden of disease (Campbell-Lendrum et al., 2003; McMichael, 2004). Absolute disease burdens depend on assumptions of population growth, future baseline disease incidence and the extent of adaptation.

The analyses suggest that climate change will bring some health benefits, such as lower cold-related mortality and greater crop yields in temperate zones, but these benefits will be greatly outweighed by increased rates of other diseases, particularly infectious diseases and malnutrition in low-income countries. A proportional increase in cardiovascular disease mortality attributable to climate extremes is projected in tropical regions, and a small benefit in temperate regions. Climate change is projected to increase the burden of diarrhoeal diseases in low-income regions by approximately 2 to 5% in 2020. Countries with an annual GDP per capita of US$6,000 or more are assumed to have no additional risk of diarrhoea. Coastal flooding is projected to result in a large proportional mortality increase under unmitigated emissions; however, this is applied to a low burden of disease, so the aggregate impact is small. The relative risk is projected to increase as much in high- as in low-income countries. Large changes are projected in the risk of *Plasmodium falciparum* malaria in countries at the edge of the current distribution, with relative changes being much smaller in areas that are currently highly endemic for malaria (McMichael et al., 2004; Haines et al., 2006).

8.4.1.2 Malaria, dengue and other infectious diseases

Studies published since the TAR support previous projections that climate change could alter the incidence and geographical range of malaria. The magnitude of the projected effect may be smaller than that reported in the TAR, partly because of advances in categorising risk. There is greater confidence in projected changes in the geographical range of vectors than in changes in disease incidence because of uncertainties about trends in factors other than climate that influence human cases and deaths, including the status of the public-health infrastructure.

Table 8.2 summarises studies that project the impact of climate change on the incidence and geographical range of malaria, dengue fever and other infectious diseases. Models with incomplete parameterisation of biological relationships between temperature, vector and parasite often over-emphasise relative changes in risk, even when the absolute risk is small. Several modelling studies used the SRES climate scenarios, a few applied population scenarios, and none incorporated economic scenarios. Few studies incorporate adequate assumptions about adaptive capacity. The main approaches used are inclusion of current 'control capacity' in the observed climate–health function (Rogers and Randolph, 2000; Hales et al., 2002) and categorisation of the model output by adaptive capacity, thereby separating the effects of climate change from the effects of improvements in public health (van Lieshout et al., 2004).

Malaria is a complex disease to model and all published models have limited parameterisation of some of the key factors that influence the geographical range and intensity of malaria transmission. Given this limitation, models project that, particularly in Africa, climate change will be associated with geographical expansions of the areas suitable for stable *Plasmodium falciparum* malaria in some regions and with contractions in other regions (Tanser et al., 2003; Thomas et al., 2004; van Lieshout et al., 2004; Ebi et al., 2005). Projections also suggest that some regions will experience a longer season of transmission. This may be as important as geographical expansion for the attributable disease burden. Although an increase in months per year of transmission does not directly translate into an increase in malaria burden (Reiter et al., 2004), it would have important implications for vector control.

Few models project the impact of climate change on malaria outside Africa. An assessment in Portugal projected an increase in the number of days per year suitable for malaria transmission; however, the risk of actual transmission would be low or negligible if infected vectors are not present (Casimiro et al., 2006). Some central Asian areas are projected to be at increased risk of malaria, and areas in Central America and around the Amazon are projected to experience reductions in transmission due to decreases in rainfall (van Lieshout et al., 2004). An assessment in India projected shifts in the geographical range and duration of the transmission window for *Plasmodium falciparum* and *P. vivax* malaria (Bhattacharya et al., 2006). An assessment in Australia based on climatic suitability for the main anopheline vectors projected a likely southward expansion of habitat, although the future risk of endemicity would remain low due to the capacity to respond (McMichael et al., 2003a).

Dengue is an important climate-sensitive disease that is largely confined to urban areas. Expansions of vector species that can carry dengue are projected for parts of Australia and New Zealand (Hales et al., 2002; Woodruff, 2005). An empirical model based on vapour pressure projected increases in latitudinal distribution. It was estimated that, in the 2080s, 5-6 billion people would be at risk of dengue as a result of climate change and population increase, compared with 3.5 billion people if the climate remained unchanged (Hales et al., 2002).

The projected impacts of climate change on other vector-borne diseases, including tick-borne encephalitis and Lyme disease, are discussed in the chapters dealing with Europe (Chapter 12) and North America (Chapter 14).

8.4.1.3 Heat- and cold-related mortality

Evidence of the relationship between high ambient temperature and mortality has strengthened since the TAR, with increasing emphasis on the health impacts of heatwaves. Table 8.3 summarises projections of the impact of climate change on heat- and cold-related mortality. There is a lack of information on the effects of thermal stress on mortality outside the industrialised countries.

Reductions in cold-related deaths due to climate change are projected to be greater than increases in heat-related deaths in the UK (Donaldson et al., 2001). However, projections of cold-related deaths, and the potential for decreasing their numbers due to warmer winters, can be overestimated unless they take into account the effects of influenza and season (Armstrong et al., 2004).

Heat-related morbidity and mortality is projected to increase. Heat exposures vary widely, and current studies do not quantify the years of life lost due to high temperatures. Estimates of the burden of heat-related mortality attributable to climate change are reduced, but not eliminated, when assumptions about acclimatisation and adaptation are included in models. On the other hand, increasing numbers of older adults in the population will increase the proportion of the population at risk because a decreased ability to thermo-regulate is a normal part of the aging process. Overall, the health burden could be relatively small for moderate heatwaves in temperate countries, because deaths occur primarily in susceptible persons. Additional research is needed to understand how the balance of heat-related and cold-related mortality could change under different socio-economic scenarios and climate projections.

8.4.1.4 Urban air quality

Background levels of ground-level ozone have risen since pre-industrial times because of increasing emissions of methane, carbon monoxide and nitrogen oxides; this trend is expected to continue over the next 50 years (Fusco and Logan, 2003; Prather et al., 2003). Changes in concentrations of ground-level ozone driven by scenarios of future emissions and/or weather patterns have been projected for Europe and North America (Stevenson et al., 2000; Derwent et al., 2001; Johnson et al., 2001; Taha, 2001; Hogrefe et al., 2004). Future emissions are, of course, uncertain, and depend on assumptions of population growth, economic development, regulatory actions and energy use (Syri et al., 2002; Webster et al., 2002a). Assuming no change in the

Table 8.2. *Projected impacts of climate change on malaria, dengue fever and other infectious diseases.*

Health effect	Metric	Model	Climate scenario, with time slices	Temperature increase and baseline	Population projections and other assumptions	Main results	Reference
Malaria, global and regional	Population at risk in areas where climate conditions are suitable for malaria transmission	Biological model, calibrated from laboratory and field data, for *falciparum* malaria	HadCM3, driven by SRES A1FI, A2, B1, and B2 scenarios. 2020s, 2050s, 2080s		SRES population scenarios; current malaria control status used as an indicator of adaptive capacity	Estimates of the additional population at risk for >1 month transmission range from >220 million (A1FI) to >400 million (A2) when climate and population growth are included. The global estimates are severely reduced if transmission risk for more than 3 consecutive months per year is considered, with a net reduction in the global population at risk under the A2 and B1 scenarios.	van Lieshout et al., 2004
Malaria, Africa	Person-months at risk for stable *falciparum* transmission	MARA/ARMA[a] model of climate suitability for stable *falciparum* transmission	HadCM3, driven by SRES A1FI, A2a, and B1 scenarios. 2020s, 2050s, 2080s	1.1 to 1.3°C in 2020s; 1.9 to 3.0°C in 2050s; 2.6 to 5.3°C in 2080s	Estimates based on 1995 population	By 2100, 16 to 28% increase in person-months of exposure across all scenarios, including a 5 to 7% increase in (mainly altitudinal) distribution, with limited latitudinal expansion. Countries with large areas that are close to the climatic thresholds for transmission show large potential increases across all scenarios.	Tanser et al., 2003
Malaria, Africa	Map of climate suitability for stable *falciparum* transmission [minimum 4 months suitable per year]	MARA/ARMA[a] model of climate suitability for stable *falciparum* transmission	HadCM2 ensemble mean with medium-high emissions. 2020s, 2050s, 2080s		Climate factors only (monthly mean and minimum temperature, and monthly precipitation)	Decreased transmission in 2020s in south-east Africa. By 2050s and 2080s, localised increases in highland and upland areas, and decreases around Sahel and south central Africa.	Thomas et al., 2004
Malaria, Zimbabwe, Africa	Climate suitability for transmission	MARA/ARMA[a] model of climate suitability for stable *falciparum* transmission	16 climate projections from COSMIC. Climate sensitivities of 1.4 and 4.5°C; equivalent CO_2 of 350 and 750 ppm 2100		None	Highlands become more suitable for transmission. The lowlands and regions with low precipitation show varying degrees of change, depending on climate sensitivity, emissions scenario and GCM.	Ebi et al., 2005
Malaria, Britain	Probability of malaria transmission	Statistical multivariate regression, based on historic distributions, land cover, agricultural factors and climate determinants	1 to 2.5°C average temperature increase 2050s	1 to 2.5°C average temperature increase	None. No changes in land cover or agricultural factors.	Increase in risk of local malaria transmission of 8 to 15%; highly unlikely that indigenous malaria will be re-established.	Kuhn et al., 2002
Malaria, Portugal	Percentage days per year with favourable temperature for disease transmission	Transmission risk based on published thresholds	PROMES for 2040s and HadRM2 for 2090s	Average annual temperature increase of 3.3°C in 2040s and 5.8°C in 2090s, compared with 1981-1990 and 2006-2036, respectively	Some assumptions about vector distribution and/or introduction	Significant increase in the number of days suitable for survival of malaria vectors; however, if no infected vectors are present, then the risk is very low for *vivax* and negligible for *falciparum* malaria.	Casimiro and Calheiros, 2002
Malaria, Australia	Geographical area suitable/unsuitable for maintenance of vector	Empirical-statistical model (CLIMEX) based on current distribution, relative abundance, and seasonal phenology of main malaria vector	CSIROMk2 and ECHAM4 driven by SRES B1, A1B, and A1FI emissions scenarios 2020, 2050	0.4 to 2.0°C annual average temperature increase in the 2030s, and 1.0 to 6.0°C in the 2070s, relative to 1990 (CSIRO)	Assumes adaptive capacity; used Australian population projections	'Malaria receptive zone' expands southward to include some regional towns by 2050s. Absolute risk of reintroduction very low.	McMichael et al., 2003b

[a] The Mapping Malaria Risk in Africa/Atlas du Risque de la Malaria en Afrique Project

Table 8.2. *Continued.*

Health effect	Metric	Model	Climate scenario, with time slices	Temperature increase and baseline	Population projections and other assumptions	Main results	Reference
Malaria, India, all states	Climate suitability for *falciparum* and *vivax* malaria transmission	Temperature transmission windows based on observed associations between temperature and malaria cases	HadRM2 driven by IS92a emissions scenario	2 to 4°C increase compared with current climate	None	By 2050s, geographical range projected to shift away from central regions towards south-western and northern states. The duration of the transmission window is likely to widen in northern and western states and shorten in southern states.	Bhattacharya et al., 2006
Dengue, global	Population at risk	Statistical model based on vapour pressure. Baseline number of people at risk is 1.5 billion.	ECHAM4, HadCM2, CCSR/NIES, CGCMA2, and CGCMA1 driven by IS92a emissions scenarios		Population growth based on region-specific projections	By 2085, with both population growth and climate change, global population at risk 5 to 6 billion; with climate change only, global population at risk 3.5 billion.	Hales et al., 2002
Dengue, New Zealand	Map of vector 'hotspots'; dengue currently not present in New Zealand	Threshold model based on rainfall and temperature	DARLAM GCM driven by A2 and B2 emissions scenarios 2050, 2100		None	Potential risk of dengue outbreaks in some regions under the current climate. Climate change projected to increase risk of dengue in more regions.	de Wet et al., 2001
Dengue, Australia	Map of regions climatically suitable for dengue transmission	Empirical model (Hales et al., 2002)	CSIROMk2, ECHAM4, and GFDL driven by high (A2) and low (B2) emissions scenarios and a stabilisation scenario at 450 ppm 2100	1.8 to 2.8°C global average temperature increase compared with 1961-1990	None	Regions climatically suitable increase southwards; size of suitable area varies by scenario. Under the high-emissions scenario, regions as far south as Sydney could become climatically suitable.	Woodruff et al., 2005
Lyme disease, Canada	Geographical range and abundance of Lyme disease vector *Ixodes scapularis*	Statistical model based on observed relationships; tick-abundance model	CGCM2 and HADCM2 driven by SRES A2 and B2 emissions scenarios 2020s, 2050s, 2080s		None	Northward expansion of approximately 200 km by 2020s under both scenarios, and approximately 1000 km by 2080s under A2. Under the A2 scenario, tick abundance increases 30 to 100% by 2020s and 2- to 4-fold by 2080s. Seasonality shifts.	Ogden et al., 2006
Tick-borne encephalitis, Europe	Geographical range	Statistical model based on present-day distribution	HadCM2 driven by low, medium-low, medium-high, and high degrees of change (not further defined) 2020s, 2050s, 2080s	3.45°C increase in mean temperature in 2050s under high scenario, baseline not defined	None	From low to high degrees of climate change, tick-borne encephalitis is pushed further northeast of its present range, only moving westward into southern Scandinavia. Only under the low and medium-low scenarios does tick-borne encephalitis remain in central and eastern Europe by the 2050s.	Randolph and Rogers, 2000
Diarrhoeal disease, global, 14 world regions	Diarrhoea incidence (mortality)	Statistical model, derived from cross-sectional study, including annual average temperature, water supply and sanitation coverage, and GDP per capita	SRES A1B, A2, B1 and B2 emissions scenarios 2025, 2055		SRES population growth	Results vary by region and scenario. Generally, diarrhoeal disease increases with temperature increase.	Hijioka et al., 2002
Diarrhoeal disease, Aboriginal community, central Australia (Alice Springs)	Hospital admissions in children aged under 10	Exposure–response relationship based on published studies	CSIROMk2 and ECHAM4 driven by SRES B1, A1B and A1FI emissions scenarios 2020, 2050	0.4 to 2.0°C annual average temperature increase in the 2030s, and 1.0 to 6.0°C in the 2070s, relative to 1990 (CSIRO)	None	Compared with baseline, no significant increase by 2020 and an annual increase of 5 to 18% by 2050.	McMichael et al., 2003b
Food poisoning, England and Wales	Notified cases of food poisoning (non-specific)	Statistical model, based on observed relationship with temperature	UKCIP scenarios 2020s, 2050s, 2080s	0.57 to 1.38°C in 2020s; 0.89 to 2.44°C in 2050s; 1.13 to 3.47°C in 2080s compared with 1961-1990 baseline	None	For +1, +2 and +3°C temperature increases, absolute increases of approximately 4,000, 9,000, and 14,000 notified cases of food poisoning	Department of Health and Expert Group on Climate Change and Health in the UK, 2001

Table 8.3. *Projected impacts of climate change on heat- and cold-related mortality.*

Area	Health effect	Model	Climate scenario, time slices	Temperature increase and baseline	Population projections and other assumptions	Main results	Reference
UK	Heat- and cold-related mortality	Empirical-statistical model derived from observed mortality	UKCIP scenarios 2020s, 2050s, 2080s	0.57 to 1.38°C in 2020s; 0.89 to 2.44°C in 2050s; 1.13 to 3.47°C in 2080s compared with 1961-1990 baseline	Population held constant at 1996. No acclimatisation assumed.	Annual heat-related deaths increase from 798 in 1990s to 2,793 in 2050s and 3,519 in the 2080s under the medium-high scenario. Annual cold-related deaths decrease from 80,313 in 1990s to 60,021 in 2050s and 51,243 in 2080s under the medium-high scenario.	Donaldson et al., 2001
Germany, Baden-Wuertemberg	Heat- and cold-related mortality	Thermo-physiological model combined with conceptual model for adaptation	ECHAM4-OPYC3 driven by SRES A1B emissions scenario. 2001-2055 compared with 1951-2001		Population growth and aging and short-term adaptation and acclimatisation.	About a 20% increase in heat-related mortality. Increase not likely to be compensated by reductions in cold-related mortality.	Koppe, 2005
Lisbon, Portugal	Heat-related mortality	Empirical-statistical model derived from observed summer mortality	PROMES and HadRM2 2020s, 2050s, 2080s	1.4 to 1.8°C in 2020s; 2.8 to 3.5°C in 2050s; 5.6 to 7.1°C in 2080s, compared with 1968-1998 baseline	SRES population scenarios. Assumes some acclimatisation.	Increase in heat-related mortality from baseline of 5.4 to 6 deaths/100,000 to 5.8 to 15.1 deaths/100,000 by the 2020s, 7.3 to 35.9 deaths/100,000 by the 2050s, 19.5 to 248.4 deaths/100,000 by the 2080s	Dessai, 2003
Four cities in California, USA (Los Angeles, Sacramento, Fresno, Shasta Dam)	Annual number of heatwave days, length of heatwave season, and heat-related mortality	Empirical-statistical model derived from observed summer mortality	PCM and HadCM3 driven by SRES B1 and A1FI emissions scenarios 2030s, 2080s	1.35 to 2.0°C in 2030s; 2.3 to 5.8°C in 2080s compared with 1961-1990 baseline	SRES population scenarios. Assumes some adaptation.	Increase in annual number of days classified as heatwave conditions. By 2080s, in Los Angeles, number of heatwave days increases 4-fold under B1 and 6 to 8-fold under A1FI. Annual number of heat-related deaths in Los Angeles increases from about 165 in the 1990s to 319 to 1,182 under different scenarios.	Hayhoe, 2004
Australian capital cities (Adelaide, Brisbane, Canberra, Darwin, Hobart, Melbourne, Perth, Sydney)	Heat-related mortality in people older than 65 years	Empirical-statistical model, derived from observed daily mortality	CSIROMk2, ECHAM4, and HADCM2 driven by SRES A2 and B2 emissions scenarios and a stabilisation scenario at 450 ppm 2100	0.8 to 5.5°C increase in annual maximum temperature in the capital cities, compared with 1961-1990 baseline	Population growth and population aging. No acclimatisation.	Increase in temperature-attributable death rates from 82/100,000 across all cities under the current climate to 246/100,000 in 2100; death rates decreased with implementation of policies to mitigate GHG.	McMichael et al., 2003b

emissions of ozone precursors, the extent to which climate change affects the frequency of future 'ozone episodes' will depend on the occurrence of the required meteorological conditions (Jones and Davies, 2000; Sousounis et al., 2002; Hogrefe et al., 2004; Laurila et al., 2004; Mickley et al., 2004). Table 8.4 summarises projections of future morbidity and mortality based on current exposure–mortality relationships applied to projected ozone concentrations. An increase in ozone concentrations will affect the ability of regions to achieve air-quality targets. There are no projections for cities in low- or middle-income countries, despite the heavier pollution burdens in these populations.

There are few models of the impact of climate change on other pollutants. These tend to emphasise the role of local abatement strategies in determining the future levels of, primarily, particulate matter, and tend to project the probability of air-quality standards being exceeded instead of absolute concentrations (Jensen et al., 2001; Guttikunda et al., 2003; Hicks, 2003; Slanina and Zhang, 2004); the results vary by region. The severity and duration of summertime regional air pollution episodes (as diagnosed by tracking combustion carbon monoxide and black carbon) are projected to increase in the north-eastern and Midwest USA by 2045-2052 because of climate-change-induced decreases in the frequency of surface cyclones (Mickley et al., 2004). A UK study projected that climate change will result in a large decrease in days with high particulate concentrations due to changes in meteorological conditions (Anderson et al., 2001). Because transboundary

Table 8.4. *Projected impacts of climate change on ozone-related health effects.*

Area	Health effect	Model	Climate scenario, time slices	Temperature increase and baseline	Population projections and other assumptions	Main results	Reference
New York metropolitan region, USA	Ozone-related deaths by county	Concentration response function from published epidemiological literature. Gridded ozone concentrations from CMAQ (Community Multiscale Air Quality model).	GISS driven by SRES A2 emissions scenario downscaled using MM5 2050s	1.6 to 3.2°C in 2050s compared with 1990s	Population and age structure held constant at year 2000. Assumes no change from United States Environmental Protection Agency (USEPA) 1996 national emissions inventory and A2-consistent increases in NO_x and VOCs by 2050s.	A2 climate only: 4.5% increase in ozone-related deaths. Ozone elevated in all counties. A2 climate and precursors: 4.4% increase in ozone-related deaths. (Ozone not elevated in all areas due to NO_x interactions.)	Knowlton et al., 2004
50 cities, eastern USA	Ozone-related hospitalis-ations and deaths	Concentration response function from published epidemiological literature. Gridded ozone concentrations from CMAQ.	GISS driven by SRES A2 emissions scenario downscaled using MM5 2050s	1.6 to 3.2°C in 2050s compared with 1990s	Population and age structure held constant at year 2000. Assumes no change from USEPA 1996 national emissions inventory and A2-consistent increases in NO_x and VOCs by 2050s.	Maximum ozone concentrations increased for all cities, with the largest increases in cities with currently higher concentrations. 68% increase in average number of days/summer exceeding the 8-hour regulatory standard, resulting in 0.11 to 0.27% increase in non-accidental mortality and an average 0.31% increase in cardiovascular disease mortality.	Bell et al., 2007
England and Wales	Exceedance days (ozone, particulates, NO_x)	Statistical, based on meteorological factors for high-pollutant days (temperature, wind speed).	UKCIP scenarios 2020s, 2050s, 2080s	0.57 to 1.38°C in 2020s; 0.89 to 2.44°C in 2050s; 1.13 to 3.47°C in 2080s compared with 1961-1990 baseline	Emissions held constant.	Over all time periods, large decreases in days with high particulates and SO_2, small decrease in other pollutants except ozone, which may increase.	Anderson et al., 2001

transport of pollutants plays a significant role in determining local to regional air quality (Holloway et al., 2003; Bergin et al., 2005), changing patterns of atmospheric circulation at the hemispheric to global level are likely to be just as important as regional patterns for future local air quality (Takemura et al., 2001; Langmann et al., 2003).

8.4.2 Vulnerable populations and regions

Human health vulnerability to climate change was assessed based on a range of scientific evidence, including the current burdens of climate-sensitive health determinants and outcomes, projected climate-change-related exposures, and trends in adaptive capacity. Box 8.5 describes trends in climate-change-related exposures of importance to human health. As highlighted in the following sections, particularly vulnerable populations and regions are more likely to suffer harm, have less ability to respond to stresses imposed by climate variability and change, and have exhibited limited progress in reducing current vulnerabilities. For example, all persons living in a flood plain are at risk during a flood, but those with lowered ability to escape floodwaters and their consequences (such as children and the infirm, or those living in sub-standard housing) are at higher risk.

8.4.2.1 Vulnerable urban populations

Urbanisation and climate change may work synergistically to increase disease burdens. Urban populations are growing faster in low-income than in high-income countries. The urban population increased from 220 million in 1900 to 732 million in 1950, and is estimated to have reached 3.2 billion in 2005 (UN, 2006b). In 2005, 74% of the population in more-developed regions was urban, compared with 43% in less-developed regions. Approximately 4.9 billion people are projected to be urban dwellers in 2030, about 60% of the global population, including 81% of the population in more-developed regions and 56% of the population in less-developed regions.

Urbanisation can positively influence population health; for example, by making it easier to provide safe water and improved sanitation. However, rapid and unplanned urbanisation is often

Box 8.5. Projected trends in climate-change-related exposures of importance to human health

Heatwaves, floods, droughts and other extreme events: IPCC (2007b) concludes, with high confidence, that heatwaves will increase, cold days will decrease over mid- to low-latitudes, and the proportion of heavy precipitation events will increase, with differences in the spatial distribution of the changes (although there will be a few areas with projected decreases in absolute numbers of heavy precipitation events) (Meehl et al., 2007). Water availability will be affected by changes in runoff due to alterations in the rainy and dry seasons.

Air quality: Climate change could affect tropospheric ozone by modifying precursor emissions, chemistry and transport; each could cause positive or negative feedbacks to climate change. Future climate change may cause either an increase or a decrease in background tropospheric ozone, due to the competing effects of higher water vapour and higher stratospheric input; increases in regional ozone pollution are expected, due to higher temperatures and weaker circulation. Future climate change may cause significant air-quality degradation by changing the dispersion rate of pollutants, the chemical environment for ozone and aerosol generation, and the strength of emissions from the biosphere, fires and dust. The sign and magnitude of these effects are highly uncertain and will vary regionally (Denman et al., 2007).

Crop yields: Chapter 5 concluded that crop productivity is projected to increase slightly at mid- to high latitudes for local mean temperature increases of up to 1-3°C depending on the crop, and then decrease beyond that in some regions. At lower latitudes, especially seasonally dry and tropical regions, crop productivity is projected to decrease for even small local temperature increases (1-2°C), which would increase the risk of hunger, with large negative effects on sub-Saharan Africa. Smallholder and subsistence farmers, pastoralists and artisanal fisherfolk will suffer complex, localised impacts of climate change.

associated with adverse health outcomes. Urban slums and squatter settlements are often located in areas subject to landslides, floods and other natural hazards. Lack of water and sanitation in these settlements are not only problems in themselves, but also increase the difficulty of controlling disease reservoirs and vectors, facilitating the emergence and re-emergence of water-borne and other diseases (Obiri-Danso et al., 2001; Akhtar, 2002; Hay et al., 2005a). Combined with declining economies, unplanned urbanisation may affect the burden and control of malaria, with the disease burden increasing among urban dwellers (Keiser et al., 2004). Currently, approximately 200 million people in Africa (24.6% of the total population) live in urban settings where they are at risk of malaria. In India, unplanned urbanisation has contributed to the spread of *Plasmodium vivax* malaria (Akhtar et al., 2002) and dengue (Shah et al., 2004). In addition, noise, overcrowding and other possible features of unplanned urbanisation may increase the prevalence of mental disorders, such as depression, anxiety, chronic stress, schizophrenia and suicide (WHO, 2001). Problems associated with rapid and unplanned urbanisation are expected to increase over the next few decades, especially in low-income countries.

Populations in high-density urban areas with poor housing will be at increased risk with increases in the frequency and intensity of heatwaves, partly due to the interaction between increasing temperatures and urban heat-island effects (Wilby, 2003). Adaptation will require diverse strategies which could include physical modification to the built environment and improved housing and building standards (Koppe et al., 2004).

8.4.2.2 Vulnerable rural populations

Climate change could have a range of adverse effects on some rural populations and regions, including increased food insecurity due to geographical shifts in optimum crop-growing conditions and yield changes in crops, reduced water resources for agriculture and human consumption, flood and storm damage, loss of cropping land through floods, droughts, a rise in sea level, and increased rates of climate-sensitive health outcomes. Water scarcity itself is associated with multiple adverse health outcomes, including diseases associated with water contaminated with faecal and other hazardous substances (including parasites), vector-borne diseases associated with water-storage systems, and malnutrition (see Chapter 3). Water scarcity constitutes a serious constraint to sustainable development particularly in savanna regions: these regions cover approximately 40% of the world land area (Rockstrom, 2003).

8.4.2.3 Food insecurity

Although the International Food Policy Research Institute's International Model for Policy Analysis of Agricultural Commodities and Trade projects that global cereal production could increase by 56% between 1997 and 2050, primarily in temperate regions, and livestock production by 90% (Rosegrant and Cline, 2003), expert assessments of future food security are generally pessimistic over the medium term. There are indications that it will take approximately 35 additional years to reach the World Food Summit 2002 target of reducing world hunger by half by 2015 (Rosegrant and Cline, 2003; UN Millennium Project, 2005). Child malnutrition is projected to persist in regions of low-income countries, although the total global burden is expected to decline without considering the impact of climate change.

Attribution of current and future climate-change-related malnutrition burdens is problematic because the determinants of malnutrition are complex. Due to the very large number of people that may be affected, malnutrition linked to extreme climatic events may be one of the most important consequences of climate change. For example, climate change is projected to increase the percentage of the Malian population at risk of hunger from 34% to between 64% and 72% by the 2050s, although this could be substantially reduced by the effective implementation of a range of adaptive strategies (Butt et al., 2005). Climate-change models project that those likely to be adversely affected are the regions already most vulnerable to food insecurity, notably Africa, which may lose substantial agricultural land. Overall, climate change is projected to increase the number of people at risk of hunger (FAO, 2005).

8.4.2.4 Populations in coastal and low-lying areas

One-quarter of the world's population resides within 100 km distance and 100 m elevation of the coastline, with increases likely over the coming decades (Small and Nicholls, 2003). Climate change could affect coastal areas through an accelerated rise in sea level; a further rise in sea-surface temperatures; an intensification of tropical cyclones; changes in wave and storm surge characteristics; altered precipitation/runoff; and ocean acidification (see Chapter 6). These changes could affect human health through coastal flooding and damaged coastal infrastructure; saltwater intrusion into coastal freshwater resources; damage to coastal ecosystems, coral reefs and coastal fisheries; population displacement; changes in the range and prevalence of climate-sensitive health outcomes; amongst others. Although some Small Island States and other low-lying areas are at particular risk, there are few projections of the health impact of climate variability and change. Climate-sensitive health outcomes of concern in Small Island States include malaria, dengue, diarrhoeal diseases, heat stress, skin diseases, acute respiratory infections and asthma (WHO, 2004a).

A model of a 4°C increase of the summer temperature maximum in the Netherlands in 2100, in combination with water column stratification, projected a doubling of the growth rates of selected species of potentially harmful phytoplankton in the North Sea, increasing the frequency and intensity of algal blooms that can negatively affect human health (Peperzak, 2005). Projections of impacts are complex because of substantial differences in the sensitivity to increasing ocean temperatures of phytoplankton harmful to human health.

The population at risk of flooding by storm surges throughout the 21st century has been projected based on a range of global mean sea-level rise and socio-economic scenarios (Nicholls, 2004). Under the baseline conditions, it was estimated that in 1990 about 200 million people lived beneath the 1-in-1,000-year storm surge height (e.g., people in the hazard zone), and about 10 million people/yr experienced flooding. Across all time slices, population growth increased the number of people living in a hazard zone under the four SRES scenarios (A1FI, A2, B1 and B2). Assuming that defences are upgraded against existing risks as countries become wealthier, but sea level rise is ignored, the number of people affected by flooding decreases by the 2080s under the A1FI, B1 and B2 scenarios. Under the A2 scenario, a two-to-three-fold increase is projected in the number of people flooded per year in the 2080s compared with 1990. Island regions are especially vulnerable, particularly in the A1FI world, especially in South-East Asia, South Asia, the Indian Ocean coast of Africa, the Atlantic coast of Africa and the southern Mediterranean (Nicholls, 2004).

Densely populated regions in low-lying areas are vulnerable to climate change. In Bangladesh, it is projected that 4.8% of people living in unprotected dryland areas could face inundation by a water depth of 30 to 90 cm based on assumptions of a 2°C temperature increase, a 30 cm increase in sea level, an 18% increase in monsoon precipitation, and a 5% increase in monsoon discharge into major rivers (BCAS/RA/Approtech, 1994). This could increase to 57% of people based on assumptions of a 4°C temperature increase, a 100 cm increase in sea level, a 33% increase in monsoon precipitation, and a 10% increase in monsoon discharge into major rivers. Some areas could face higher levels of inundation (90 to 180 cm).

Studies in industrialised countries indicate that densely populated urban areas are at risk from sea-level rise (see Chapter 6). As demonstrated by Hurricane Katrina, areas of New Orleans (USA) and its vicinity are 1.5 to 3 m below sea level (Burkett et al., 2003). Considering the rate of subsidence and using the TAR mid-range estimate of 480 mm sea-level rise by 2100, it is projected that this region could be 2.5 to 4.0 m or more below mean sea level by 2100, and that a storm surge from a Category 3 hurricane (estimated at 3 to 4 m without waves) could be 6 to 7 m above areas that were heavily populated in 2004 (Manuel, 2006).

8.4.2.5 Populations in mountain regions

Changes in climate are affecting many mountain glaciers, with rapid glacier retreat documented in the Himalayas, Greenland, the European Alps, the Andes Cordillera and East Africa (WWF, 2005). Changes in the depth of mountain snowpacks and glaciers, and changes in their seasonal melting, can have significant impacts on the communities from mountains to plains that rely on freshwater runoff. For example, in China, 23% of the population live in the western regions where glacial melt provides the principal dry season water source (Barnett et al., 2005). A long-term reduction in annual glacier snow melt could result in water insecurity in some regions.

Little published information is available on the possible health consequences of climate change in mountain regions. However, it is likely that vector-borne pathogens could take advantage of new habitats at altitudes that were formerly unsuitable, and that diarrhoeal diseases could become more prevalent with changes in freshwater quality and availability (WHO Regional Office for South-East Asia, 2006). More extreme rainfall events are likely to increase the number of floods and landslides. Glacier lake outburst floods are a risk unique to mountain regions; these are associated with high morbidity and mortality and are projected to increase as the rate of glacier melting increases.

8.4.2.6 Populations in polar regions

The approximately 10% of the circumpolar population that is indigenous is particularly vulnerable to climate change (ACIA, 2005). Factors contributing to their vulnerability include their close relationship with the land, location of communities in coastal regions, reliance on the local environment for aspects of their diet and economy, and socio-economic and other factors

(Berner and Furgal, 2005). The interactions of climate change with underlying social, cultural, economic and political trends are projected to have significant impacts on Arctic residents (Curtis et al., 2005).

Increasing winter temperatures in Arctic regions are projected to reduce excess winter mortality, primarily through a reduction in cardiovascular and respiratory deaths. A reduction in cold-related injuries is projected, assuming that cold protection, including human behavioural factors, does not change (Nayha, 2005). Observations in northern Canadian Aboriginal communities suggest that the number of land-based accidents and injuries associated with unpredictable environmental conditions such as thinning and earlier break-up of sea ice are likely to increase (e.g., Furgal et al., 2002a, b). Diseases transmitted by wildlife and insects are projected to have a longer season in some regions such as the north-western North American Arctic, resulting in increased burdens of disease in key animal species (e.g., marine mammals, birds, fish and shellfish) that can be transmitted to humans (Bradley et al., 2005; Parkinson and Butler, 2005). The traditional diet of circumpolar residents is likely to be negatively affected by changes in animal migrations and distribution, and human access to them, partly because of the impacts of increasing temperatures on snow and ice timing and distribution. Further, increasing temperatures may indirectly influence human exposure to environmental contaminants in some foods (e.g., marine mammal fats). Temperature increases in the North Atlantic are projected to increase rates of mercury methylation in fish and marine mammals, thus increasing human exposure via consumption (Booth and Zeller, 2005).

8.5 Costs

Studies focusing on the welfare costs (and benefits) of climate-change impacts aggregate the 'damage' costs of climate change (Tol, 1995, 1996, 2002a, b; Fankhauser and Tol, 1997; Fankhauser et al., 1997) or estimate the costs and benefits of measures to reduce climate change (Nordhaus, 1991; Cline, 1992, 2004; Nordhaus and Boyer, 2000). The global economic value of loss of life due to climate change ranges between around US$6 billion and US$88 billion, in 1990 dollar prices (Tol, 1995, 1996, 2002a, b; Fankhauser and Tol, 1997; Fankhauser et al., 1997). The economic methods for estimating welfare costs (and benefits) have several shortcomings; the studies include only a limited number of health outcomes, generally heat- and cold-related mortality and malaria. Some assessments of the direct costs of health impacts at the national level have been undertaken, but the evidence base for estimating the health effects is relatively weak (IGCI, 2000; Turpie et al., 2002; Woodruff et al., 2005). Where they have been estimated, the welfare costs of health impacts contribute substantially to the total costs of climate change (Cline, 1992; Tol, 2002a). Given the importance of these types of assessments, further research is needed.

Mortality attributable to climate change is projected to be greatest in low-income countries, where economists traditionally assign a lower value to life (van der Pligt et al., 1998; Hammitt and Graham, 1999; Viscusi and Aldy, 2003). Some estimates suggest that replacing national values with a 'global average value' would increase the mortality costs by as much as five times (Fankhauser et al., 1997). Climate change is also likely to have important direct effects on productivity via exposure of workers to heat stress (see Section 8.2.9). Estimates of economic impacts via changes in productivity ignore important health impacts in children and the elderly. Further research is needed to estimate productivity costs.

8.6 Adaptation: practices, options and constraints

Adaptation is needed now in order to reduce current vulnerability to the climate change that has already occurred and additional adaptation is needed in order to address the health risks projected to occur over the coming decades. Current levels of vulnerability are partly a function of the programmes and measures in place to reduce burdens of climate-sensitive health determinants and outcomes, and partly a result of the success of traditional public-health activities, including providing access to safe water and improved sanitation to reduce diarrhoeal diseases, and implementing surveillance programmes to identify and respond to outbreaks of malaria and other infectious diseases. Weak public-health systems and limited access to primary health care contribute to high levels of vulnerability and low adaptive capacity for hundreds of millions of people.

Current national and international programmes and measures that aim to reduce the burdens of climate-sensitive health determinants and outcomes may need to be revised, reoriented and, in some regions, expanded to address the additional pressures of climate change. The degree to which programmes will need to be augmented will depend on factors such as the current burden of climate-sensitive health outcomes, the effectiveness of current interventions, projections of where, when and how the burden could change with changes in climate and climate variability, access to the human and financial resources needed to implement activities, stressors that could increase or decrease resilience to impacts, and the social, economic and political context within which interventions are implemented (Yohe and Ebi, 2005; Ebi et al., 2006a). Some recent programmes and measures implemented to address climate variability and change are highlighted in the examples that follow.

The planning horizon of public-health decision-makers is short relative to the projected impacts of climate change, which will require modification of current risk-management approaches that focus only on short-term risks (Ebi et al., 2006b). A two-tiered approach may be needed, with modifications to incorporate current climate change concerns into ongoing programmes and measures, along with regular evaluations to determine a programme's likely effectiveness to cope with projected climate risks. For example, epidemic malaria is a public-health problem in most areas in Africa, with programmes in place to reduce the morbidity and mortality associated with these epidemics. Some projections suggest that climate change may facilitate the spread of malaria further up some highland areas (see Section 8.4.1.2). Therefore, programmes should not only continue their current focus, but should also consider where and when to implement additional surveillance to identify and prevent epidemics if the *Anopheles* vector changes its range.

How public health and other infrastructure will develop is a key uncertainty (see Section 8.3) that is not determined by GDP per capita alone. Public awareness, effective use of local resources, appropriate governance arrangements and community participation are necessary to mobilise and prepare for climate change (McMichael, 2004). These present particular challenges in low-income countries. Furthermore, the status of and trends in other sectors affect public health, particularly water quantity, quality and sanitation (see Chapter 3), food quality and quantity (see Chapter 5), the urban environment (see Chapter 7), and ecosystems (see Chapter 4). These sectors will also be affected by climate change, creating feedback loops that can increase or decrease population vulnerability, particularly in low-income countries (Figure 8.1).

8.6.1 Approaches at different scales

Pro-active adaptation strategies, policies and measures need to be implemented by regional and national governments, including Ministries of Health, by international organisations such as the World Health Organization, and by individuals. Because the range of possible health impacts of climate change is broad and the local situations diverse, the examples that follow are illustrative and not comprehensive.

8.6.1.1 National- and regional-level responses

Climate-based early warning systems for heatwaves and malaria outbreaks have been implemented at national and local levels to alert the population and relevant authorities that a disease outbreak can be expected based on climatic and environmental forecasts (Abeku et al., 2004; Teklehaimanot et al., 2004; Thomson et al., 2005; Kovats and Ebi, 2006). To be effective in reducing health impacts, such systems must be coupled with a specific intervention plan and have an ongoing evaluation of the system and its components (Woodruff et al., 2005; Kovats and Ebi, 2006).

Seasonal forecasts can be used to increase resilience to climate variability, including to weather disasters. For example, the Pacific ENSO Application Center (PEAC) alerted governments, when a strong El Niño was developing in 1997/1998, that severe droughts could occur, and that some islands were at unusually high risk of tropical cyclones (Hamnett, 1998). The interventions launched, such as public education and awareness campaigns, were effective in reducing the risk of diarrhoeal and vector-borne diseases. For example, despite the water shortage in Pohnpei, fewer children were admitted to hospital with severe diarrhoeal disease than normal because of frequent public-health messages about water safety. However, the interventions did not eliminate all negative health impacts, such as micronutrient deficiencies in pregnant women in Fiji.

Participatory approaches that include governments, researchers and community residents are increasingly being used to build awareness of climate-related health impacts and adaptation options, and to take advantage of local knowledge and perspectives (see Box 8.6).

8.6.1.2 Responses by international organisations and agencies

Improvements in international surveillance systems facilitate national and regional preparedness and reduce future vulnerability to epidemic-prone diseases. At present, surveillance systems in many parts of the world are incomplete and slow to respond to disease outbreaks. It is expected that this will improve through the implementation of the International Health Regulations. Improvements in the responsiveness and accuracy of current surveillance programmes, including addressing spatial and temporal limitations, are needed to account for and anticipate the increased pressures on disease-control programmes that are projected to result from climate change. Earth observations, monitoring and surveillance, such as remote sensing and biosensors, may increase the accuracy and precision of some of these activities (Maynard, 2006).

Donors, international and national aid agencies, emergency relief agencies, and a range of non-governmental organisations play key roles through direct aid, support of research and development, and other approaches developed in conjunction with national Ministries of Health to improve current public-health responses and to more effectively incorporate climate-change-related risks into the design, implementation and evaluation of disease-control policies and measures.

Box 8.6. Cross-cutting case study: indigenous populations and adaptation

A series of workshops organised by the national Inuit organisation in Canada, Inuit Tapiriit Kantami, documented climate-related changes and impacts, and identified and developed potential adaptation measures for local response (Furgal et al., 2002a, b; Nickels et al., 2003). The strong engagement of Inuit community residents will facilitate the successful adoption of the adaptation measures identified, such as using netting and screens on windows and house entrances to prevent bites from mosquitoes and other insects that have become more prevalent.

Another example is a study of the links between malaria and agriculture that included participation and input from a farming community in Mwea division, Kenya (Mutero et al., 2004). The approach facilitated identification of opportunities for long-term malaria control in irrigated rice-growing areas through the integration of agro-ecosystem practices aimed at sustaining livestock systems within a broader strategy for rural development.

Two or more countries can develop international responses jointly when adverse health outcomes and their drivers cross borders. For example, flood prevention guidelines were developed through the United Nations Economic Commission for Europe for countries along the Elbe, Danube, Rhine and other transboundary rivers where floods have intensified due to human alteration of the environment (UN, 2000). The guidelines recognise that co-operation is needed both within and between riparian countries in order to reduce current impacts and increase future resilience.

8.6.1.3 Individual-level responses

The effectiveness of warning systems for extreme events depends on individuals taking appropriate actions, such as responding to heat alerts and flood warnings. Individuals can reduce their personal exposure by adjusting clothing and activity levels in response to high ambient temperatures and by modifying built environments, such as by the use of fans, to reduce the heat load (Davis et al., 2004; Kovats and Koppe, 2005). Weather can partially determine cultural practices that may affect exposure.

8.6.1.4 Adaptation in health systems

Health systems need to plan for and respond to climate change (Menne and Bertollini, 2005). There are effective interventions for many of the most common causes of ill-health, but frequently these interventions do not reach those who could benefit most. One way of promoting adaptation and reducing vulnerability to climate change is to promote the uptake of effective clinical and public-health interventions in high-need cities and regions of the world. For example, health in Africa must be treated as a high priority investment in the international development portfolio (Sachs, 2001). Funding health programmes is a necessary step towards reducing vulnerability but will not be enough on its own (Brewer and Heymann, 2004; Regidor, 2004a, b; de Vogli et al., 2005; Macintyre et al., 2005). Progress depends also on strengthening public institutions; building health systems that work well, treating people fairly and providing universal primary health care; providing adequate education, generating demand for better and more accessible services; and ensuring that there are enough staff to do the required work (Haines and Cassels, 2004). Health-service infrastructure needs to be resilient to extreme events (EEA, 2005). Efforts are needed to train health professionals to understand the threats posed by climate change.

8.6.2 Integration of responses across scales

Adaptation responses to specific health risks will often cut across scales. For example, an integrated response to heatwaves could include, in addition to measures already discussed, consideration of climate change projections in the design and construction of new buildings and in the planning of new urban areas (Kovats and Koppe, 2005). In addition, national energy efficiency programmes and transport policies could include approaches for reducing both urban heat islands and emissions of ozone and other air pollutants.

Interventions designed to increase the adaptive capacity of a community or region could also facilitate the achievement of greenhouse gas mitigation targets. For example, measures to reduce the urban heat-island effect, such as planting trees, roof gardens, growth planned to reduce urban heat islands, and other measures, increase the resilience of communities to heatwaves while reducing energy requirements. Increasing the proportion of energy derived from solar, wind and other renewable resources would reduce emissions of greenhouse gases and other air pollutants from the burning of fossil fuels.

8.6.3 Limits to adaptation

Constraints to adaptation arise when one or more of the prerequisites for public-health prevention have not been met: an awareness that a problem exists; a sense that the problem matters; an understanding of what causes the problem; the capability to influence; and the political will to influence the problem (Last, 1998). Decision-makers will choose which adaptations to implement where, when and how, based on assessments of the balance between competing priorities (Scheraga et al., 2003). For example, different regions may make different assessments of the public-health and environmental-welfare implications of the ecological consequences of draining wetlands to reduce vector-breeding sites. Local laws and social customs may constrain adaptation options. For example, although the application of pesticides for vector control may be an effective adaptation measure, residents may object to spraying, even in communities with regulations to assure appropriate use. Increasing awareness of climate-change-related health impacts and knowledge diffusion of adaptation options are of fundamental importance to better decision-making.

Although specific limits will vary by health outcome and region, fundamental constraints exist in low-income countries where adaptation will partially depend on development pathways in the public-health, water, agriculture, transport, energy and housing sectors. Poverty is the most serious obstacle to effective adaptation. Despite economic growth, low-income countries are likely to remain poor and vulnerable over the medium term, with fewer options than high-income countries for adapting to climate change. Therefore, adaptation strategies should be designed in the context of development, environment, and health policies. Many of the options that can be used to reduce future vulnerability are of value in adapting to current climate and can be used to achieve other environmental and social objectives. However, because resources used for adaptation will be shared across other problems of concern to society, there is the potential for conflicts among stakeholders with differing priorities. Questions also will arise about equity (i.e., a decision that leads to differential health impacts among different demographic groups), efficiency (i.e., targeting those programmes that will yield the greatest improvements in public health), and political feasibility (McMichael et al., 2003a).

8.6.4 Health implications of adaptation strategies, policies and measures

Because adaptation strategies, policies and measures can have inadvertent short- and long-term negative health consequences, potential risks should be evaluated before implementation. For example, a microdam and irrigation programme in Ethiopia developed to increase resilience to famine increased local malaria mortality by 7.3-fold (Ghebreyesus et al., 1999). Increased ambient temperatures due to climate change could further exacerbate the problem. In another example, air-conditioning of private and public spaces is a primary measure used in the USA to reduce heat-related morbidity and mortality (Davis et al., 2003); however, depending on the energy source used to generate electricity, an increased use of air conditioning can increase greenhouse gas emissions, air pollution and the urban heat island.

Measures to combat water scarcity, such as the re-use of wastewater for irrigation, have implications for human health (see Chapter 3). Irrigation is currently an important determinant of the spread of infectious diseases such as malaria and schistosomiasis (Sutherst, 2004). Strict water-quality guidelines for wastewater irrigation are designed to prevent health risks from pathogenic organisms and to guarantee crop quality (Steenvoorden and Endreny, 2004). However, in rural and peri-urban areas of most low-income countries, the use of sewage and wastewater for irrigation, a common practice, is a source of faecal–oral disease transmission. The use of wastewater for irrigation is likely to increase with climate change, and the treatment of wastewater remains unaffordable for low-income populations (Buechler and Scott, 2000).

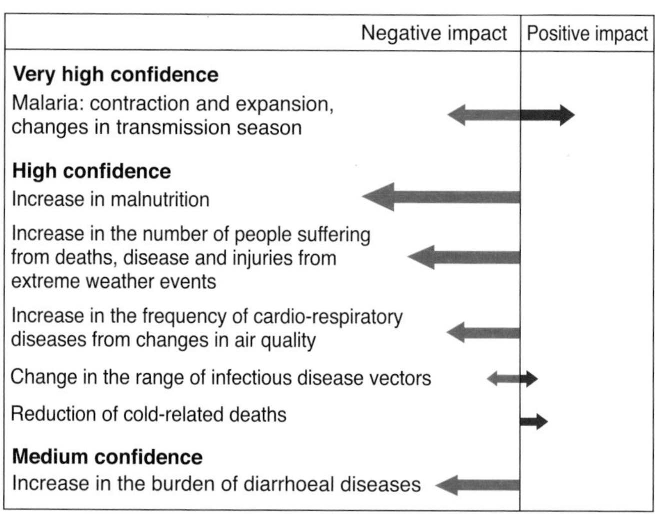

Figure 8.3. *Direction and magnitude of change of selected health impacts of climate change (confidence levels are assigned based on the IPCC guidelines on uncertainty, see http://www.ipcc.ch/activity/uncertaintyguidancenote.pdf).*

8.7 Conclusions: implications for sustainable development

Evidence has grown that climate change already contributes to the global burden of disease and premature deaths. Climate change plays an important role in the spatial and temporal distribution of malaria, dengue, tick-borne diseases, cholera and other diarrhoeal diseases; is affecting the seasonal distribution and concentrations of some allergenic pollen species; and has increased heat-related mortality. The effects are unequally distributed, and are particularly severe in countries with already high disease burdens, such as sub-Saharan Africa and Asia.

The projected health impacts of climate change are predominately negative, with the most severe impacts being seen in low-income countries, where the capacity to adapt is weakest. Vulnerable groups in developed countries will also be affected (Haines et al., 2006). Projected increases in temperature and changes in rainfall patterns can increase malnutrition; disease and injury due to heatwaves, floods, storms, fires and droughts; diarrhoeal illness; and the frequency of cardio-respiratory diseases due to higher concentrations of ground-level ozone. There are expected to be some benefits to health, including fewer deaths due to exposure to the cold and reductions in climate suitability for vector-borne diseases in some regions. Figure 8.3 summarises the relative direction and magnitude of projected health impacts, taking into account the likely numbers of people at risk and potential adaptive capacity.

Health is central to the achievement of the Millennium Development Goals and to sustainable development, both directly (in the case of child mortality, maternal health, HIV/AIDS, malaria and other diseases) and indirectly (ill-health contributes to extreme poverty, hunger and lower educational achievements) (Haines and Cassels, 2004). Rapid and intense climate change is likely to delay progress towards achieving development targets in some regions. Recent events demonstrate that populations and health systems may be unable to cope with increases in the frequency and intensity of extreme events. These events can reduce the resilience of communities, affect vulnerable regions and localities, and overwhelm the coping capacities of most societies.

There is a need to develop and implement adaptation strategies, policies and measures at different levels and scales. Current national and international programmes and measures that aim to reduce the burdens of climate-sensitive health determinants and outcomes may need to be revised, reoriented and, in some regions, expanded to address the additional pressures of climate change. This includes the consideration of climate-change-related risks in disease monitoring and surveillance systems, health system planning, and preparedness. Many of the health outcomes are mediated through changes in the environment. Measures implemented in the water, agriculture, food, and construction sectors should be designed to benefit human health. However, adaptation is not enough.

8.7.1 Health and climate protection: clean energy

There is general agreement that health co-benefits from reduced air pollution as a result of actions to reduce GHG emissions can be substantial and may offset a substantial fraction of mitigation costs (Barker et al., 2001, 2007; Cifuentes et al., 2001; West et al., 2004). In addition, actions to reduce methane emissions will decrease global concentrations of surface ozone. A portfolio of actions, including energy efficiency, renewable energy, and transport measures, is needed in order to achieve these reductions (see IPCC, 2007c).

In many low-income countries, access to electricity is limited. Over half of the world's population still relies on biomass fuels and coal to meet their energy needs (WHO, 2006). These biomass fuels have low combustion efficiency and a significant, but unknown, portion is harvested non-renewably, thus contributing to net carbon emissions. The products of incomplete combustion from small-scale biomass combustion contain a number of health-damaging pollutants, including small particles, carbon monoxide, polyaromatic hydrocarbons and a range of toxic volatile organic compounds (Bruce et al., 2000). Human exposures to these pollutants within homes are large in comparison with outdoor air pollution exposures. Current best estimates, based on published epidemiological studies, are that biomass fuels in households are responsible annually for approximately 0.7 to 2.1 million premature deaths in low-income countries (from a combination of lower-respiratory infections, chronic obstructive pulmonary disease and lung cancer). About two-thirds occur in children under the age of five and most of the rest occur in women (Smith et al., 2004).

Clean development and other mechanisms could require calculation of the co-benefits for health when taking decisions about energy projects, including the development of alternative fuel sources (Smith et al., 2000, 2005). Projects promoting co-benefits in low-income populations show promise to help achieve cost-effective, long-term protection from climate impacts as well as promoting immediate sustainable development goals (Smith et al., 2000).

8.8 Key uncertainties and research priorities

More empirical epidemiological research on the observed health effects of climate change have been published since the TAR, and the few national health impact assessments that have been conducted have provided valuable information on population vulnerability. However, the lack of appropriate longitudinal health data makes attribution of adverse health outcomes to observed climate trends difficult. Further, most studies have focused on middle- and high-income countries. Gaps in information persist on trends in climate, health and environment in low-income countries, where data are limited and other health priorities take precedence for research and policy development. Climate-change-related health impact assessments in low- and middle-income countries will be instrumental in guiding adaptation projects and investments.

Advances have been made in the development of climate–health impact models that project the health effects of climate change under a range of climate and socio-economic scenarios. The models are still limited to a few infectious diseases, thermal extremes and air pollution. Considerable uncertainties surround the projections, including uncertainty about how population health is likely to evolve based on changes in the level of commitment to preventing avoidable ill-health, technological developments, economic growth and other factors; the rate and intensity of future climate change; uncertainty about how the climate–health relationship might change over time; and uncertainty about the extent, rate, limiting forces and major drivers of adaptation (McMichael et al., 2004). Uncertainties include not just whether the key health outcomes described in this chapter will improve, but how fast, where, when, at what cost, and whether all population groups will be able to share in these developments. Significant advances will occur by improving social and economic development, governance and resources. It is apparent that these problems will only be solved over time-frames longer than decades.

Considerable uncertainty will remain about projected climate change at geographical and temporal scales of relevance to decision-makers, increasing the importance of risk management approaches to climate risks. However, no matter what the degree of preparedness is, projections suggest that some future extreme events will be catastrophic because of the unexpected intensity of the event and the underlying vulnerability of the affected population. The European heatwave in 2003 and Hurricane Katrina are examples. The consequences of particularly severe extreme events will be greater in low-income countries. A better understanding is needed of the factors that convey vulnerability and, more importantly, the changes that need to be made in health care, emergency services, land use, urban design and settlement patterns to protect populations against heatwaves, floods, and storms.

Key research priorities include addressing the major challenges for research on climate change and health in the following ways.

- Development of methods to quantify the current impacts of climate and weather on a range of health outcomes, particularly in low- and middle-income countries.
- Development of health-impacts models for projecting climate-change-related impacts under different climate and socio-economic scenarios.
- Investigations on the costs of the projected health impacts of climate change; effectiveness of adaptation; and the limiting forces, major drivers and costs of adaptation.

Low-income countries face additional challenges, including limited capacity to identify key issues, collect and analyse data, and design, implement and monitor adaptation options. There is a need to strengthen institutions and mechanisms that can more systematically promote interactions among researchers, policy-makers and other stakeholders to facilitate the appropriate incorporation of research findings into policy decisions in order to protect population health no matter what the climate brings (Haines et al., 2004).

References

Abeku, T., G. van Oortmarssen, G. Borsboom, S. de Vlas and J. Habbema, 2003: Spatial and temporal variations of malaria epidemic risk in Ethiopia: factors involved and implications. *Acta Trop.*, **87**, 331-340.

Abeku, T.A., S.I. Hay, S. Ochola, P. Langi, B. Beard, S.J. de Vlas and J. Cox, 2004: Malaria epidemic early warning and detection in African highlands. *Trends Parasitol.*, **20**, 400-4005.

ACIA, 2005: *Arctic Climate Impact Assessment*. Cambridge University Press, New York, 1042 pp.

Adelakun, A., E. Schwartz and L. Blais, 1999: Occupational heat exposure. *Appl. Occup. Environ. Hyg.*, **14**, 153-154.

Adger, W., T. Hughes, C. Folke, S. Carpenter and J. Rockstrom, 2005: Social-ecological resilience to coastal disasters. *Science*, **309**, 1036-1039.

Afanas'eva, R.F., N.A. Bessonova, M.A. Babaian, N.V. Lebedeva, T.K. Losik and V.V. Subbotin, 1997: Kobosnovaniiu reglamentatsii termicheskoi nagruzki sredy na rabotaiushchikh v nagrevaiushchem mikroklimate (na primere staleplavil'nogo proizvodstva) [Substantiation of the regulation of environmental heat load for workers exposed to heating microclimate (for example, steel smelting)] (in Russian). *Med. Tr. Prom. Ekol.*, 30-34.

Afonso, M.O., L. Campino, S. Cortes and C. Alves-Pires, 2005: The phlebotomine sandflies of Portugal. XIII. Occurrence of *Phlebotomus sergenti* Parrot, 1917 in the *Arrabida leishmaniasis* focus. *Parasite*, **12**, 69-72.

Ahasan, M.R., G. Mohiuddin, S. Vayrynen, H. Ironkannas and R. Quddus, 1999: Work-related problems in metal handling tasks in Bangladesh: obstacles to the development of safety and health measures. *Ergonomics*, **42**, 385-396.

Ahern, M.J., R.S. Kovats, P. Wilkinson, R. Few and F. Matthies, 2005: Global health impacts of floods: epidemiological evidence. *Epidemiol. Rev.*, **27**, 36-45.

Akhtar, R., 2002: *Urban Health in the Third World*. APH Publications, New Delhi, 454 pp.

Akhtar, R., A. Dutt and V. Wadhwa, 2002: Health planning and the resurgence of malaria in urban India. *Urban Health in the Third World*, R. Akhtar, Ed., AHP Publications, New Delhi, 65-92.

Alvarez, E., F. de Pablo, C. Tomas and L. Rivas, 2000: Spatial and temporal variability of ground-level ozone in Castilla-Leon (Spain). *Int. J. Biometeorol.*, **44**, 44-51.

Anand, S. and T. Barnighausen, 2004: Human resources and health outcomes: cross-country econometric study. *Lancet*, **364**, 1603-1609.

Anderson, H.R., R.G. Derwent and J. Stedman, 2001: Air pollution and climate change. *Health Effects of Climate Change in the UK*. Department of Health, London, 193-217.

Anderson, K. and G. Manuel, 1994: Gender differences in reported stress response to the Loma Prieta earthquake. *Sex Roles*, **30**, 9-10.

Ando, M., S. Yamamato and S. Asanuma, 2004: Global warming and heatstroke. *Japan. J. Biometeorol.*, **41**, 45.

Ansmann, A., J. Bosenberg, A. Chaikovsky, A. Comeron, S. Eckhardt, R. Eixmann, V. Freudenthaler, P. Ginoux, L. Komguem, H. Linne, M.A.L. Marquez, V. Matthias, I. Mattis, V. Mitev, D. Muller, S. Music, S. Nickovic, J. Pelon, L. Sauvage, P. Sobolewsky, M.K. Srivastava, A. Stohl, O. Torres, G. Vaughan, U. Wandinger and M. Wiegner, 2003: Long-range transport of Saharan dust to northern Europe: the 11–16 October 2001 outbreak observed with EARLINET. *J. Geophys. Res. D*, **108**, 4783.

Aramini, J., M. McLean, J. Wilson, B. Allen, W. Sears and J. Holt, 2000: *Drinking Water Quality and Health Care Utilization for Gastrointestinal Illness in Greater Vancouver*. Centre for Infectious Disease Prevention and Control, Foodborne, Waterborne and Zoonotic Infections Division, Health Canada, Ottawa, 79 pp.

Aransay, A.M., J.M. Testa, F. Morillas-Marquez, J. Lucientes and P.D. Ready, 2004: Distribution of sandfly species in relation to canine leishmaniasis from the Ebro Valley to Valencia, northeastern Spain. *Parasitol. Res.*, **94**, 416-420.

Ariyabandu, M. and M. Wickramasinghe, 2003: *Gender Dimensions in Disaster Management: A Guide for South Asia*. ITGD South Asia, Colombo, Sri Lanka, 176 pp.

Armstrong, B., P. Mangtani, A. Fletcher, R.S. Kovats, A.J. McMichael, S. Pattenden and P. Wilkinson, 2004: Effect of influenza vaccination on excess deaths occurring during periods of high circulation of influenza: cohort study in elderly people. *Brit. Med. J.*, **329**, 660-663.

Arnell, N.W., M.T. Livermore, R.S. Kovats, P. Levy, R.J. Nicholls, M.L. Parry and S.R. Gaffin, 2004: Climate and socio-economic scenarios for global-scale climate change impacts assessments: characterising the SRES storylines. *Global Environ. Chang.*, **14**, 3-20.

Asero, R., 2002: Birch and ragweed pollinosis north of Milan: a model to investigate the effects of exposure to "new" airborne allergens. *Allergy*, **57**, 1063-1066.

Assanarigkornchai, S., S.N. Tangboonngam and J.G. Edwards, 2004: The flooding of Hat Yai: predictors of adverse emotional responses to a natural disaster. *Stress Health*, **20**, 81-89.

Autoridad Nacional del Ambiente, 2000: *Primera comunicacion nacional sobre cambio climatico, Panama 2000 [First National Communication on Climate Change Panama 2000]*. ANAM, Panama, 136 pp. http://unfccc.int/resource/docs/natc/pannc1/index.html.

Aziz, K.M.A., B.A. Hoque, S. Huttly, K.M. Minnatullah, Z. Hasan, M.K. Patwary, M.M. Rahaman and S. Cairncross, 1990: Water supply, sanitation and hygiene education: Report of a health impact study in Mirzapur, Bangladesh. Water and Sanitation Report Series, No. 1, World Bank, Washington, District of Columbia, 99 pp.

Barbraud, C. and H. Weimerskirch, 2001: Emperor penguins and climate change. *Nature*, **411**, 183-186.

Barker, I.K. and L.R. Lindsay, 2000: Lyme borreliosis in Ontario: determining the risks. *Can. Med. Assoc. J.*, **162**, 1573-1574.

Barker, T., L. Srivastava and B. Metz, 2001: Sector costs and ancillary benefits of mitigation. *Climate Change 2001: Mitigation. Contribution of Working Group III to the Third Assessment Report of the Intergovernmental Panel on Climate Change*, B. Metz, O. Davidson, R. Swart and J. Pan, Eds., Cambridge University Press, Cambridge, 561-600.

Barker, T., I. Bashmakov, A. Alharthi, M. Amann, L. Cifuentes, J. Drexhage, M. Duan, O. Edenhofer, B. Flannery, M. Grubb, M. Hoogwijk, F.I. Ibitoye, C.J. Jepma, W.A. Pizer and K. Yamaji, 2007: Mitigation from a cross-sectoral perspective. *Climate Change 2007: Mitigation. Contribution of Working Group III to the Fourth Assessment Report of the Intergovernmental Panel on Climate Change*, B. Metz, O. Davidson , P. Bosch, R. Dave and L. Meyer, Eds., Cambridge University Press, Cambridge, UK.

Barnett, T.P., J.C. Adam and D.P. Lettenmaier, 2005: Potential impacts of a warming climate on water availability in snow-dominated regions. *Nature*, **438**, 303-309.

Barrett, R., C. Kuzawa, T. McDade and G. Armelagos, 1998: Emerging and re-emerging infectious diseases: the third epidemiologic transition. *Annu. Rev. Anthropol.*, **27**, 247-271.

Basu, R. and J.M. Samet, 2002: Relation between elevated ambient temperature and mortality: a review of the epidemiologic evidence. *Epidemiol. Rev.*, **24**, 190-202.

Bates, D.V., 2005: Ambient ozone and mortality. *Epidemiology*, **16**, 427-429.

Bavia, M.E., L.F. Hale, J.B. Malone, D.H. Braud and S.M. Shane, 1999: Geographic information systems and the environmental risk of schistosomiasis in Bahia, Brazil. *Am. J. Trop. Med. Hyg.*, **60**, 566-572.

Bavia, M.E., J.B. Malone, L. Hale, A. Dantas, L. Marroni and R. Reis, 2001: Use of thermal and vegetation index data from earth observing satellites to evaluate the risk of schistosomiasis in Bahia, Brazil. *Acta Trop.*, **79**, 79-85.

Bayard, V., E. Ortega, A. Garcia, L. Caceres, Z. Castillo, E. Quiroz, B. Armien, F. Gracia, J. Serrano, G. Guerrero, R. Kant, E. Pinifla, L. Bravo, C. Munoz, I.B. de Mosca, A. Rodriguez, C. Campos, M.A. Diaz, B. Munoz, F. Crespo, I. Villalaz, P. Rios, E. Morales, J.M.T. Sitton, L. Reneau-Vernon, M. Libel, L. Castellanos, L. Ruedas, D. Tinnin and T. Yates, 2000: Hantavirus pulmonary syndrome: Panama, 1999–2000 [Reprinted from *MMWR*, **49**, 205-207, 2000]. *JAMA–J. Am. Med. Assoc.*, **283**, 2232.

BCAS/RA/Approtech, 1994: Vulnerability of Bangladesh to climate change and sea level rise: concepts and tools for calculating risk in integrated coastal zone management. Technical Report, Bangladesh Centre for Advanced Studies (BCAS), Dhaka, 80 pp.

Becht, M.C., M.A.L. van Tilburg, A.J.J.M. Vingerhoets, I. Nyklicek, J. de Vries, C. Kirschbaum, M.H. Antoni and G.L. van Heck, 1998: Watersnood: een verkennend onderzoek naar de gevolgen voor het welbevinden en de gezondheid van volwassenen en kinderen [Flood: a pilot study on the consequences for well-being and health of adults and children]. *Tijdsch. Psychiat.*, **40**, 277-289.

Beggs, P.J., 2004: Impacts of climate change on aeroallergens: past and future. *Clin. Exp. Allergy*, **34**, 1507-1513.

Beggs, P.J., 2005: Admission to hospital for sunburn and drug phototoxic and photoallergic responses, New South Wales, 1993–94 to 2000–01. *N.S.W. Public Health Bull.*, **16**, 147-150.

Beggs, P.J. and H.J. Bambrick, 2005: Is the global rise of asthma an early impact of anthropogenic climate change? *Environ. Health Persp.*, **113**, 915-919.

Bell, M.L., F. Dominici and J.M. Samet, 2005: A meta-analysis of time-series studies of ozone and mortality with comparison to the national morbidity, mortality, and air pollution study. *Epidemiology*, **16**, 436-445.

Bell, M.L., R.D. Peng and F. Dominici, 2006: The exposure-response curve for ozone and risk of mortality and the adequacy of current ozone regulations. *Environ. Health Persp.*, **114**, 532-536.

Bell, M.L., R. Goldberg, C. Hogrefe, P.L. Kinney, K. Knowlton, B. Lynn, J. Rosenthal, C. Rosenzweig and J.A. Patz, 2007: Climate change, ambient ozone, and health in 50 US cities. *Climatic Change*, **82**, 61-76.

Benitez, T.A., A. Rodriquez and M. Sojo, 2004: Descripcion de un brote epidemico de malaria de altura en un areas originalmente sin malaria del Estado Trujillo,

Venezuela. *Bol. Malariol. Salud Amb.*, **XLIV**, 999.

Bennett, C.M., I.G. McKendry, S. Kelly, K. Denike and T. Koch, 2006: Impact of the 1998 Gobi dust event on hospital admissions in the Lower Fraser Valley, British Columbia. *Sci. Total Environ.*, **366**, 918-925.

Bergin, M.S., J.J. West, T.J. Keating and A.G. Russell, 2005: Regional atmospheric pollution and transboundary air quality management. *Annu. Rev. Env. Resour.*, **30**, 1-37.

Berner, J. and C. Furgal, 2005: Human health. *Arctic Climate Impact Assessment*, Arctic Climate Impact Assessment (ACIA), Cambridge University Press, Cambridge, 863-906.

Bhattacharya, S., C. Sharma, R.C. Dhiman and A.P. Mitra, 2006: Climate change and malaria in India. *Curr. Sci.*, **90**, 369-375.

Blaikie, P., T. Cannon, I. Davis and B. Wisner, 1994: *At Risk: Natural Hazards, People's Vulnerability and Disasters*, 2nd ed., Routledge, New York, 320 pp.

Bokszczanin, A., 2000: Psychologiczne konsekwencje powodzi u dzieci i mlodziezy [Psychological consequences of floods in children and youth]. *Psychol. Wychowawcza*, **43**, 172-181.

Bokszczanin, A., 2002: Long-term negative psychological effects of a flood on adolescents. *Polish Psychol. Bull.*, **33**, 55-61.

Booth, S. and D. Zeller, 2005: Mercury, food webs, and marine mammals: implications of diet and climate change for human health. *Environ. Health Persp.*, **113**, 521-526.

Both, C. and M.E. Visser, 2005: The effect of climate change on the correlation between avian life-history traits. *Glob. Change Biol.*, **11**, 1606-1613.

Bouma, M. and C. Dye, 1997: Cycles of malaria associated with El Niño in Venezuela. *J. Am. Med. Assoc.*, **278**, 1772-1774.

Bouma, M.J., 2003: Methodological problems and amendments to demonstrate effects of temperature on the epidemiology of malaria: a new perspective on the highland epidemics in Madagascar, 1972–1989. *T. Roy. Soc. Trop. Med. H.*, **97**, 133-139.

Bouma, M.J. and M. Pascual, 2001: Seasonal and interannual cycles of endemic cholera in Bengal 1891–1940 in relation to climate and geography. *Hydrobiologia*, **460**, 147-156.

Bowman, D.M.J.S. and F.H. Johnston, 2005: Wildfire smoke, fire management, and human health. *EcoHealth*, **2**, 76-80.

Bradley, M., S.J. Kutz, E. Jenkins and T.M. O'Hara, 2005: The potential impact of climate change on infectious diseases of Arctic fauna. *Int. J. Circumpolar Health*, **64**, 468-477.

Bradshaw, W.E. and C.M. Holzapfel, 2001: Genetic shift in photoperiodic response correlated with global warming. *P. Natl. Acad. Sci. USA*, **98**, 14509-14511.

Braga, A., A. Zanobetti and J. Schwartz, 2002: The effect of weather on respiratory and cardiovascular deaths in 12 US cities. *Environ. Health Persp.*, **110**, 859-863.

Bresser, A., 2006: *The Effect of Climate Change in the Netherlands*. Netherlands Environmental Assessment Agency, MNP, Bilthoven, 112 pp.

Brewer, T.F. and S.J. Heymann, 2004: The long journey to health equity. *J. Am. Med. Assoc.*, **292**, 269-271.

Briceño, S., 2002: Gender mainstreaming in disaster reduction. Commission on the Status of Women. Panel presentation. Secretariat of the International Strategy for Disaster Reduction. United Nations, International Strategy for Disaster Reduction, Geneva, 12 pp.

Broman, T., H. Palmgren, S. Bergstrom, M. Sellin, J. Waldenstrom, M.L. Danielsson-Tham and B. Olsen, 2002: Campylobacter jejuni in black-headed gulls (*Larus ridibundus*): prevalence, genotypes, and influence on *C. jejuni* epidemiology. *J. Clin. Microbiol.*, **40**, 4594-4602.

Brommer, J.E., 2004: The range margins of northern birds shift polewards. *Ann. Zool. Fenn.*, **41**, 391-397.

Bruce, N., R. Perez-Padilla and R. Albalak, 2000: Indoor air pollution in developing countries: a major environmental and public health challenge. *B. World Health Organ.*, **78**, 1097-1092.

Buchanan, C.M., I.J. Beverland and M.R. Heal, 2002: The influence of weather-type and long-range transport on airborne particle concentrations in Edinburgh, UK. *Atmos. Environ.*, **36**, 5343-5354.

Buechler, S.J. and C.A. Scott, 2000: *For Us, This is Life: Irrigating under Adverse Conditions*. IWMI Latin American Series No. 20. International Water Management Institution, Bierstalpad.

Burkett, V.R., D.B. Zilkoski and D.A. Hart, 2003: Sea level rise and subsidence: implications for flooding in New Orleans. *U.S. Geological Survey Subsidence Interest Group Conference, Proceedings of the Technical Meeting, Galveston, Texas, November 27–29, 2001*, 63-70.

Burr, M.L., J.C. Emberlin, R. Treu, S. Cheng and N.E. Pearce, 2003: Pollen counts in relation to the prevalence of allergic rhinoconjunctivitis, asthma and atopic eczema in the International Study of Asthma and Allergies in Childhood (ISAAC). *Clin. Exp. Allergy*, **33**, 1675-1680.

Butt, T., B. McCarl, J. Angerer, P. Dyke and J. Stuth, 2005: The economic and flood security implications of climate change in Mali. *Climatic Change*, **68**, 355-378.

Cairncross, S. and M. Alvarinho, 2006: The Mozambique floods of 2000: health impact and response. *Flood Hazards and Health: Responding to Present and Future Risks*, R. Few and F. Matthies, Eds., Earthscan, London, 111-127.

Calheiros, J. and E. Casimiro, 2006: Saude humana [Human health]. *Alteracoes climaticas em Portugal: Cenarios, impactos e medias de adapacao – Projecto SIAM [Climate Change in Portugal: Scenarios, Impacts and Adaptation measures – SIAM Project]*, F. Santos and P. Miranda, Eds., Gravida, Lisbon, 451-462.

Campbell-Lendrum, D., A. Pruss-Ustun and C. Corvalan, 2003: How much disease could climate change cause? *Climate Change and Human Health: Risks and Responses*, A. McMichael, D. Campbell-Lendrum, C. Corvalan, K. Ebi, A. Githeko, J. Scheraga and A. Woodward, Eds., WHO/WMO/UNEP, Geneva, 133-159.

Carson, C., S. Hajat, B. Armstrong and P. Wilkinson, 2006: Declining vulnerability to temperature-related mortality in London over the twentieth century. *Am. J. Epidemiol.*, **164**, 77-84.

Casas-Zamora, J.A. and S.A. Ibrahim, 2004: Confronting health inequity: the global dimension. *Am. J. Public Health*, **94**, 2055.

Casimiro, E. and J. Calheiros, 2002: Human health. *Climate Change in Portugal: Scenarios, Impacts and Adaptation Measures – SIAM Project*, F. Santos, K. Forbes and R. Moita, Eds., Gradiva, Lisbon, 241-300.

Casimiro, E., J. Calheiros, D. Santos and S. Kovats, 2006: National assessment of human health impacts of climate change in Portugal: approach and key findings. *Environ. Health Persp.*, **114**, 1950-1956. doi:10.1289/ehp.8431.

Cazelles, B., M. Chavez, A.J. McMichael and S. Hales, 2005: Nonstationary influence of El Niño on the synchronous dengue epidemics in Thailand. *PLoS Med.*, **2**, e106.

CDC, 2005: Vibrio illnesses after Hurricane Katrina: multiple states, August–September 2005. *MMWR–Morb. Mortal. Wkly. Rep.*, **54**, 928-931.

Cecchi, L., M. Morabito, P. Domeneghetti M., A. Crisci, M. Onorari and S. Orlandini, 2006: Long distance transport of ragweed pollen as a potential cause of allergy in central Italy. *Ann. Allergy. Asthma. Im.*, **96**, 86-91.

Chan, C., L.Y. Chan, K.S. Lam, Y.S. Li, J.M. Harris and S.J. Oltmans, 2002: Effects of Asian air pollution transport and photochemistry on carbon monoxide variability and ozone production in subtropical coastal south China. *J. Geophys. Res. D*, **107**, 4746.

Chase, J.M. and T.M. Knight, 2003: Drought-induced mosquito outbreaks in wetlands. *Ecol. Lett.*, **6**, 1017-1024.

Chateau-Degat, M.-L., M. Chinain, N. Cerf, S. Gingras, B. Hubert and E. Dewailly, 2005: Seawater temperature, *Gambierdiscus* spp. variability and incidence of ciguatera poisoning in French Polynesia. *Harmful Algae*, **4**, 1053–1062.

Chaudhury, S.K., J.M. Gore and K.C.S. Ray, 2000: Impact of heat waves in India. *Curr. Sci.*, **79**, 153-155.

Checkley, W., L.D. Epstein, R.H. Gilman, D. Figueroa, R.I. Cama, J.A. Patz and R.E. Black, 2000: Effects of El Niño and ambient temperature on hospital admissions for diarrhoeal diseases in Peruvian children. *Lancet*, **355**, 442-450.

Checkley, W., R.H. Gilman, R.E. Black, L.D. Epstein, L. Cabrera, C.R. Sterling and L.H. Moulton, 2004: Effect of water and sanitation on childhood health in a poor Peruvian peri-urban community. *Lancet*, **363**, 112-118.

Chen, K., Y. Ho, C. Lai, Y. Tsai and S. Chen, 2004: Trends in concentration of ground-level ozone and meteorological conditions during high ozone episodes in the Kao-Ping Airshed, Taiwan. *J. Air Waste Manage.*, **54**, 36-48.

Chen, Y. and C. Tang, 2005: Effects of Asian dust storm events on daily hospital admissions for cardiovascular disease in Taipei, Taiwan. *J. Toxicol. Env. Heal. A*, **68**, 1457-1464.

Choi, G.Y., J.N. Choi and H.J. Kwon, 2005: The impact of high apparent temperature on the increase of summertime disease-related mortality in Seoul: 1991–2000. *J. Prev. Med. Pub. Health*, **38**, 283-290.

Choudhury, A.Y. and A. Bhuiya, 1993: Effects of biosocial variable on changes in nutritional status of rural Bangladeshi children, pre- and post-monsoon flooding. *J. Biosoc. Sci.*, **25**, 351-357.

Chua, K.B., W.J. Bellini, P.A. Rota, B.H. Harcourt, A. Tamin, S.K. Lam, T.G. Ksiazek, P.E. Rollin, S.R. Zaki, W. Shieh, C.S. Goldsmith, D.J. Gubler, J.T. Roehrig, B. Eaton, A.R. Gould, J. Olson, H. Field, P. Daniels, A.E. Ling, C.J. Peters, L.J.

Anderson and B.W. Mahy, 2000: Nipah virus: a recently emergent deadly paramyxovirus. *Science*, **288**, 1432-1435.

Cifuentes, L., V.H. Borja-Aburto, N. Gouveia, G. Thurston and D.L. Davis, 2001: Climate change. Hidden health benefits of greenhouse gas mitigation. *Science*, **293**, 1257-1259.

Claiborn, C.S., D. Finn, T.V. Larson and J.Q. Koenig, 2000: Windblown dust contributes to high PM2.5 concentrations. *J. Air Waste Manage.*, **50**, 1440-1445.

Clark, C.G., L. Price, R. Ahmed, D.L. Woodward, P.L. Melito, F.G. Rogers, D. Jamieson, B. Ciebin, A. Li and A. Ellis, 2003: Characterization of water borne disease outbreak associated *Campylobacter jejuni*, Walkerton, Ontario. *Emerg. Infect. Dis.*, **9**, 1232-1241.

Cline, W., 1992: *Economics of Global Warming*. Institute for International Economics, Washington, District of Columbia, 424 pp.

Cline, W., 2004: Meeting the challenge of global warming: reply to Manne and Mendelsohn. Copenhagen Consensus Challenge Paper, Opponents Notes Reply, 8 pp. http://www.copenhagenconsensus.com/Files/Filer/CC/Papers/Reply_-_Cline_-_Climate_Change_180504.pdf.

Coffee, M., G. Garnett, M. Mlilo, H. Voeten, S. Chandiwana and S. Gregson, 2005: Patterns of movement and risk of HIV infection in rural Zimbabwe. *J. Infect. Dis.*, **191**, S159-S167.

Cohen, J.C., 2003: Human population: the next half century. *Science*, **302**, 1172-1175.

Colwell, R.R., 1996: Global climate and infectious disease: the cholera paradigm. *Science*, **274**, 2025-2031.

Combs, D.L., L.E. Quenenmoen and R.G. Parrish, 1998: Assessing disaster attributable mortality: development and application of definition and classification matrix. *Int. J. Epidemiol.*, **28**, 1124-1129.

Confalonieri, U., 2003: Climate variability, vulnerability and health in Brazil. *Terra Livre*, **19**, 193-204.

Conti, S., P. Meli, G. Minelli, R. Solimini, V. Toccaceli, M. Vichi, C. Beltrano and L. Perini, 2005: Epidemiologic study of mortality during the Summer 2003 heat wave in Italy. *Environ. Res.*, **98**, 390-399.

Cook, A.G., P. Weinstein and J.A. Centeno, 2005: Health effects of natural dust: role of trace elements and compounds. *Biol. Trace Elem. Res.*, **103**, 1-15.

Corden, J.M., W.M. Millington and J. Mullins, 2003: Long-term trends and regional variation in the aeroallergens in Cardiff and Derby UK: are differences in climate and cereal production having an effect? *Aerobiologia*, **19**, 191.

Corwin, A.L., R.P. Larasati, M.J. Bangs, S. Wuryadi, S. Arjoso, N. Sukri, E. Listyaningsih, S. Hartati, R. Namursa, Z. Anwar, S. Chandra, B. Loho, H. Ahmad, J.R. Campbell and K.R. Porter, 2001: Epidemic dengue transmission in southern Sumatra, Indonesia. *T. Roy. Soc. Trop. Med. H.*, **95**, 257-265.

CPTEC, 2005: *Climanálise : Boletim de Monitorimento e Análise Climática*, Vol. 20, No 9, Setembro. http://www.cptec.inpe.br/products/climanalise/0905/index.html.

Craig, M.H., I. Kleinschmidt, J.B. Nawn, D. Le Sueur and B. Sharp, 2004: Exploring 30 years of malaria case data in KwaZulu-Natal, South Africa. Part I. The impact of climatic factors. *Trop. Med. Int. Health*, **9**, 1247.

Cummings, D.A., 2004: Travelling waves in the occurrence of dengue haemorrhagic fever in Thailand. *Nature*, **427**, 344.

Curriero, F., J.A. Patz, J.B. Rose and S. Lele, 2001: The association between extreme precipitation and waterborne disease outbreaks in the United States, 1948–1994. *Am. J. Publ. Health*, **91**, 1194-1199.

Curriero, F., K.S. Heiner, J. Samet, S. Zeger, L. Strug and J.A. Patz, 2002: Temperature and mortality in 11 cities of the Eastern United States. *Am. J. Epidemiol.*, **155**, 80-87.

Curtis, T., S. Kvernmo and P. Bjerregaard, 2005: Changing living conditions, life style and health. *Int. J. Circumpolar Health*, **64**, 442-450.

D'Amato, G., G. Liccardi, M. D'Amato and M. Cazzola, 2002: Outdoor air pollution, climatic changes and allergic bronchial asthma. *Eur. Respir. J.*, **20**, 763-776.

D'Souza, R., N. Becker, G. Hall and K. Moodie, 2004: Does ambient temperature affect foodborne disease? *Epidemiology*, **15**, 86-92.

Daniel, M., V. Danielova, B. Kriz and I. Kott, 2004: An attempt to elucidate the increased incidence of tick-borne encephalitis and its spread to higher altitudes in the Czech Republic. *Int. J. Med. Microbiol.*, **293**, 55-62.

DaSilva, J., B. Garanganga, V. Teveredzi, S. Marx, S. Mason and S. Connor, 2004: Improving epidemic malaria planning, preparedness and response in Southern Africa. *Malaria J.*, **3**, 37.

Davis, R., P. Knappenberger, W. Novicoff and P. Michaels, 2002: Decadal changes in heat related human mortality in the eastern United States. *Climate Res.*, **22**, 175-184.

Davis, R., P. Knappenberger, P. Michaels and W. Novicoff, 2003: Changing heat-related mortality in the United States. *Environ. Health Persp.*, **111**, 1712 -1718.

Davis, R., P. Knappenberger, P. Michaels and W. Novicoff, 2004: Seasonality of climate-human mortality relationships in US cities and impacts of climate change. *Climate Res.*, **26**, 61-76.

Day, J.F., 2001: Predicting St. Louis encephalitis virus epidemics: lessons from recent, and not so recent, outbreaks. *Annu. Rev. Entomol.*, **46**, 111-138.

de Graaf, R. and J. Rappole, 1995: *Neotropical Migratory Birds: Natural History, Distribution and Population Change*. Cornell University Press, Ithaca, New York, 676 pp.

de Gruijl, F., J. Longstreth, C. Norval, A. Cullen, H. Slaper, M. Kripke, Y. Takizawa and J. van der Leun, 2003: Health effects from stratospheric ozone depletion and interactions with climate change. *Photochem. Photobio. S.*, **2**, 16-28.

De, U.S. and R.K. Mukhopadhyay, 1998: Severe heat wave over the Indian subcontinent in 1998, in perspective of global climate. *Curr. Sci.*, **75**, 1308-1315.

De, U.S., M. Khole and M. Dandekar, 2004: Natural hazards associated with meteorological extreme events. *Nat. Hazards*, **31**, 487-497.

de Vogli, R., R. Mistry, R. Gnesotto and G.A. Cornia, 2005: Has the relation between income inequality and life expectancy disappeared? Evidence from Italy and top industrialised countries. *J. Epidemiol. Commun. H.*, **59**, 158-162.

de Wet, N., W. Ye, S. Hales, R.A. Warrick, A. Woodward and P. Weinstein, 2001: Use of a computer model to identify potential hotspots for dengue fever in New Zealand. *New Zeal. Med. J.*, **11**, 420-422.

del Ninno, C. and M. Lundberg, 2005: Treading water: the long term impact of the 1998 flood on nutrition in Bangladesh. *Econ. Hum. Biol.*, **3**, 67-96.

Denman, K.L., G. Brasseur, A. Chidthaisong, P. Ciais, P.M. Cox, R.E. Dickinson, D. Hauglustaine, C. Heinze, E. Holland, D. Jacob, U. Lohmann, S. Ramachandran, P.L. da Silva Dias, S.C. Wofsy and X. Zhang, 2007: Couplings between changes in the climate system and biogeochemistry. *Climate Change 2007: The Physical Science Basis. Contribution of Working Group I to the Fourth Assessment Report of the Intergovernmental Panel on Climate Change*, S. Solomon, D. Qin, M. Manning, Z. Chen, M. Marquis, K.B. Averyt, M. Tignor and H.L. Miller, Eds., Cambridge University Press, Cambridge, 499-588.

Department of Health and Expert Group on Climate Change and Health in the UK, 2001: *Health Effects of Climate Change in the UK*. Department of Health, London, 238 pp.

Depradine, C.A. and E.H. Lovell, 2004: Climatological variables and the incidence of dengue fever in Barbados. *Int. J. Environ. Heal. R.*, **14**, 429-441.

Derwent, R.G., W.J. Collins, C.E. Johnson and D.S. Stevenson, 2001: Transient behaviour of tropospheric ozone precursors in a global 3-D CTM and their indirect greenhouse effects. *Climatic Change*, **49**, 463-487.

Dessai, S., 2003: Heat stress and mortality in Lisbon. Part II. An assessment of the potential impacts of climate change. *Int. J. Biometeorol.*, **48**, 37-44.

Dixon, S., S. McDonald and J.A. Roberts, 2002: The impact of HIV and AIDS on Africa's economic development. *Brit. Med. J.*, **324**, 232-234.

Dominici, F., R.D. Peng, M.L. Bell, L. Pham, A. McDermott, S.L. Zeger and J.M. Samet, 2006: Fine particulate air pollution and hospital admission for cardiovascular and respiratory diseases. *J. Am. Med. Assoc.*, **295**, 1127-1134.

Donaldson, G.C., R.S. Kovats, W.R. Keatinge and A. McMichael, 2001: Heat-and-cold-related mortality and morbidity and climate change. *Health Effects of Climate Change in the UK*, Department of Health, London, 70-80.

Donaldson, G.C., W.R. Keatinge and S. Nayha, 2003: Changes in summer temperature and heat-related mortality since 1971 in North Carolina, South Finland, and Southeast England. *Environ. Res.*, **91**, 1-7.

Durkin, M.S., N. Khan, L.L. Davidson, S.S. Zaman and Z.A. Stein, 1993: The effects of a natural disaster on child behaviour: evidence for posttraumatic stress. *Am. J. Public Health*, **83**, 1549-1553.

Ebi, K.L. and J.L. Gamble, 2005: Summary of a workshop on the development of health models and scenarios: a strategy for the future. *Environ. Health Persp.*, **113**, 335-338.

Ebi, K.L., J. Hartman, N. Chan, J. McConnell, M. Schlesinger and J. Weyant, 2005: Climate suitability for stable malaria transmission in Zimbabwe under different climate change scenarios. *Climatic Change*, **73**, 375-393.

Ebi, K.L, I. Burton and B. Menne, 2006a: Policy implications for climate change related health risks. *Climate Change Adaptation Strategies and Human Health*, B. Menne and K. Ebi, Eds., Steinkopff, Darmstadt, 297-310.

Ebi, K.L., J. Smith, I. Burton and J.S. Scheraga, 2006b: Some lessons learned from public health on the process of adaptation. *Mitigation and Adaptation Strategies for Global Change*, **11**, 607-620.

EEA, 2003: Air pollution by ozone in Europe in summer 2003: overview of exceedances of EC ozone threshold values during the summer season April–August 2003 and comparisons with previous years. Topic Report No 3/2003, European Economic Association, Copenhagen, 33 pp. http://reports.eea.europa.eu/topic_report_2003_3/en.

EEA, 2005: Climate change and river flooding in Europe. *EEA Briefing*, **1**, 1-4.

Eiguren-Fernandez, A., A. Miguel, J. Froines, S. Thurairatnam and E. Avol, 2004: Seasonal and spatial variation of polycyclic aromatic hydrocarbons in vapor-phase and PM2.5 in Southern California urban and rural communities. *Aerosol Sci. Tech.*, **38**, 447 - 455.

EM-DAT, 2006: *The OFDA/CRED International Disaster Database*. http://www.em-dat.net.

Emberlin, J., M. Detandt, R. Gehrig, S. Jaeger, N. Nolard and A. Rantio-Lehtimaki, 2002: Responses in the start of *Betula* (birch) pollen seasons to recent changes in spring temperatures across Europe. *Int. J. Biometeorol.*, **46**, 159-170.

Enarson, E. and B. Morrow, Eds., 1998: *The Gendered Terrain of Disaster: Through Women's Eyes*. Praeger, Westport, Connecticut and London, 288 pp.

Enscore, R., B. Biggerstaff, T. Brown, R. Fulgham, P. Reynolds, D. Engelthaler, C. Levy, R. Parmenter, J. Montenieri, J. Cheek, R. Grinnell, P. Ettestad and K. Gage, 2002: Modeling relationships between climate and the frequency of human plague cases in the southwestern United States, 1960–1997. *Am. J. Trop. Med. Hyg.*, **66**, 186-196.

Euripidou, E. and V. Murray, 2004: Public health impacts of floods and chemical contamination. *J. Public Health*, **26**, 376-383.

Eurowinter Group, 1997: Cold exposure and winter mortality from ischaemic heart disease, cerebrovascular disease, respiratory disease, and all causes in warm and cold regions of Europe. *Lancet*, **349**, 1341-1346.

Ezzati, M., S.V. Hoorn, A. Rodgers, A.D. Lopez, C.D. Mathers and C.J. Murray, 2003: Estimates of global and regional potential health gains from reducing multiple major risk factors. *Lancet*, **362**, 271-280.

Ezzati, M., A. Lopez, A. Rodgers and C. Murray, Eds., 2004: *Comparative Quantification of Health Risks: Global and Regional Burden of Disease due to Selected Major Risk Factors, Vols. 1 and 2*. World Health Organization, Geneva, 2235 pp.

Fallacara, D.M., C.M. Monahan, T.Y. Morishita, C.A. Bremer and R.F. Wack, 2004: Survey of parasites and bacterial pathogens from free-living waterfowl in zoological settings. *Avian Dis.*, **48**, 759-767.

Fan, G.C., C.N. Chang, Y.S. Wu, S.C. Lu, P.P. Fu, S.C. Chang, C.D. Cheng and W.H. Yuen, 2002: Concentration of atmospheric particulates during a dust storm period in central Taiwan, Taichung. *Sci. Total Environ.*, **287**, 141-145.

Fankhauser, S. and R.S.J. Tol, 1997: The social costs of climate change: the IPCC Second Assessment Report and beyond. *Mitigation and Adaptation Strategies for Global Change*, **1**, 385.

Fankhauser, S., R.S.J. Tol and D. Pearce, 1997: The aggregation of climate change damages: a welfare theoretic approach. *Environ. Resour. Econ.*, **10**, 249-266.

FAO, 2002: *The State of Food Insecurity in the World 2002*. http://www.fao.org/docrep/005/y7352e/y7352e00.HTM.

FAO, 2005: *The State of Food Insecurity around the World: Eradicating Hunger – Key to Achieving the Millennium Development Goals*. FAO, Rome, 40 pp. http://www.fao.org/docrep/008/a0200e/a0200e00.htm.

Few, R. and F. Matthies, 2006: *Flood Hazards and Health: Responding to Present and Future Risks*. Earthscan, London, 240 pp.

Fleury, M., D.F. Charron, J.D. Holt, O.B. Allen and A.R. Maarouf, 2006: A time series analysis of the relationship of ambient temperature and common bacterial enteric infections in two Canadian provinces. *Int. J. Biometeorol.*, **50**, 385-391.

Folha, 2006: Seca nos rios da Amazônia leva animais em extinção à morte. http://www1.folha.uol.com.br/folha/ciencia/ult306u13869.shtml.

Fothergill, A., 1998: The neglect of gender in disaster work: an overview of the literature. *The Gendered Terrain of Disaster: Through Women's Eyes*, E. Enarson and B. Morrow, Eds., Praeger, Westport, Connecticut and London, 9-25.

Franke, C.R., M. Ziller, C. Staubach and M. Latif, 2002: Impact of the El Niño/Southern Oscillation on visceral leishmaniasis, Brazil. *Emerg. Infect. Dis.*, **8**, 914-917.

Frankenberg, E., D. McKee and D. Thomas, 2005: Health consequences of forest fires in Indonesia. *Demography*, **42**, 109-129.

Friend, M., R.G. McLean and F.J. Dein, 2001: Disease emergence in birds: challenges for the twenty-first century. *Auk*, **118**, 290-303.

Furgal, C., D. Martin and P. Gosselin, 2002a: Climate change in Nunavik and Labrador: lessons from Inuit knowledge. *The Earth is Faster Now: Indigenous Observations on Arctic Environmental Change,* I. Krupnik and D. Jolly, Eds., ARCUS, Washington, District of Columbia, 266-300.

Furgal, C.M., D. Martin, P. Gosselin, A. Viau, Nunavik Regional Board of Health and Social Services (NRBHSS) and Labrador Inuit Association (LIA), 2002b: Climate change in Lunavik and Labrador: what we know from science and Inuit ecological knowledge. Final Report prepared for Climate Change Action Fund, WHO/PAHO Collaborating Center on Environmental and Occupational Health Impact Assessment and Surveillance, Centre Hospitalier Universitaire de Quebec (CHUQ), Beauport, Quebec, 141 pp.

Fusco, A.C. and J.A. Logan, 2003: Analysis of 1970-1995 trends in tropospheric ozone at Northern Hemisphere midlatitudes with the GEOS-CHEM model. *J. Geophys. Res. D*, **108**, 4449.

Gabastou, J.M., C. Pesantes, S. Escalante, Y. Narvaez, E. Vela, L. Garcia, D. Zabala and Z.E. Yadon, 2002: Characteristics of the cholera epidemic of 1998 in Ecuador during El Niño (in Spanish). *Rev. Panam. Salud Publ.*, **12**, 157-164.

Gagnon, A.S., A.B.G. Bush and K.E. Smoyer-Tomic, 2001: Dengue epidemics and the El Niño Southern Oscillation. *Climate Res.*, **19**, 35-43.

Galea, S., A. Nandi and D. Vlahov, 2005: The epidemiology of post-traumatic stress disorders after disasters. *Epidemiol. Rev.*, **27**, 78-91.

Gallagher, R.P. and T.K. Lee, 2006: Adverse effects of ultraviolet radiation: a brief review. *Prog. Biophys. Mol. Bio.*, **92**, 119-131.

Gangoiti, G., M.M. Millan, R. Salvador and E. Mantilla, 2001: Long-range transport and re-circulation of pollutants in the western Mediterranean during the project Regional Cycles of Air Pollution in the West-Central Mediterranean Area. *Atmos. Environ.*, **35**, 6267-6276.

Garrison, C.Z., E.S. Bryant, C.L. Addy, P.G. Spurrier, J.R. Freedy and D.G. Kilpatrick, 1995: Posttraumatic stress disorder in adolescents after Hurricane Andrew. *J. Am. Acad. Child Psy.*, **34**, 1193-1201.

Gerolomo, M. and M.F. Penna, 1999: Os primeiros cinco anos da setima pandemia de cólera no Brasil. *Informe Epid. SUS*, **8**, 49-58.

Ghebreyesus, T.A., M. Haile, K.H. Witten, A. Getachew, A.M. Yohannes, M. Yohannes, H.D. Teklehaimanot, S.W. Lindsay and P. Byass, 1999: Incidence of malaria among children living near dams in northern Ethiopia: community based incidence survey. *Brit. Med. J.*, **319**, 663-666.

Githeko, A.K. and W. Ndegwa, 2001: Predicting malaria epidemics in the Kenyan Highlands using climate data: a tool for decision makers. *Global Change Human Health*, **2**, 54-63.

Goklany, I., 2002: The globalization of human well-being. *Policy Anal.*, **447**, 1-20.

Goulson, D., L.C. Derwent, M. Hanley, D. Dunn and S. Abolins, 2005: Predicting calyptrate fly populations from the weather, and the likely consequences of climate change. *J. Appl. Ecol.*, **42**, 784-794.

Government of Andhra Pradesh, 2004: Report of the state level committee on heat wave conditions in Andhra Pradesh State. Revenue (Disaster Management) Department. Hyderabad, India. 67pp.

Greenough, G., M.A. McGeehin, S.M. Bernard, J. Trtanj, J. Riad and D. Engelberg, 2001: The potential impacts of climate variability and change on health impacts of extreme weather events in the United States. *Environ. Health Persp.*, **109**, 191-198.

Grize, L., A. Huss, O. Thommen, C. Schindler and C. Braun-Fahrländer, 2005: Heat wave 2003 and mortality in Switzerland. *Swiss Med. Wkly.*, **135**, 200–205.

Gryparis, A., B. Forsberg, K. Katsouyanni, A. Analitis, G. Touloumi, J. Schwartz, E. Samoli, S. Medina, H.R. Anderson, E.M. Niciu, H.E. Wichmann, B. Kriz, M. Kosnik, J. Skorkovsky, J.M. Vonk and Z. Dortbudak, 2004: Acute effects of ozone on mortality from the "Air Pollution and Health: A European Approach" project. *Am. J. Respir. Crit. Care Med.*, **170**, 1080-1087.

Guang, W., W. Qing and M. Ono, 2005: Investigation on *Aedes aegypti* and *Aedes albopictus* in the north-western part of Hainan Province. *China Trop. Med.*, **5**, 230-233.

Guha-Sapir, P., D. Hargitt and H. Hoyois, 2004: *Thirty Years of Natural Disasters 1974–2003: The Numbers*. UCL, Presses Universitaires de Louvrain, Louvrain-la Neuve, 188 pp.

Guidry, V.T. and L.H. Margolis, 2005: Unequal respiratory health risk: using GIS to explore hurricane related flooding of schools in Eastern North Carolina. *Environ. Res.*, **98**, 383-389.

Guttikunda, S.K., G.R. Carmichael, G. Calori, C. Eck and J.-H. Woo, 2003: The contribution of megacities to regional sulfur pollution in Asia. *Atmos. Environ.*, **37**, 11-22.

Gylfe, A., S. Bergstrom, J. Lunstrom and B. Olsen, 2000: Epidemiology: reactivation of *Borrelia* infection in birds. *Nature*, **403**, 724-725.

Haines, A. and A. Cassels, 2004: Can the Millennium Development Goals be attained? *Brit. Med. J.*, **329**, 394-397.

Haines, A., S. Kuruvilla and M. Borchert, 2004: Bridging the implementation gap between knowledge and action for health. *B. World Health Organ.*, **82**, 724-732.

Haines, A., R.S. Kovats, D. Campbell-Lendrum and C. Corvalan, 2006: Climate change and human health: impacts, vulnerability, and mitigation. *Lancet*, **367**, 2101-2109.

Hajat, S., 2006: Heat- and cold-related deaths in England and Wales: who is at risk? *Occup. Environ. Med.*, **64**, 93-100.

Hajat, S., B. Armstrong, N. Gouveia and P. Wilkinson, 2005: Comparison of mortality displacement of heat-related deaths in Delhi, Sao Paulo and London. *Epidemiology*, **16**, 613-620.

Hajat, S., B. Armstrong, M. Baccini, A. Biggeri, L. Bisanti, A. Russo, A. Paldy, B. Menne and T. Kosatsky, 2006: Impact of high temperatures on mortality: is there an added "heat wave" effect? *Epidemiology*, **17**, 632-638.

Hales, J. and D. Richards, 1987: Heat stress: physical exertion and environment. *Proceedings of the 1st World Conference on Heat Stress, Physical Exertion and Environment, Sydney, Australia, 27 April–1 May 1987*. Excerpta Medica, Amsterdam and New York, 558 pp.

Hales, S., P. Wienstein, Y. Souares and A. Woodward, 1999: El Niño and the dynamics of vectorborne disease transmission. *Environ. Health Persp.*, **107**, 99-102.

Hales, S., N. de Wet, J. Maindonald and A. Woodward, 2002: Potential effect of population and climate changes on global distribution of dengue fever: an empirical model. *Lancet*, **360**, 830-834.

Hall, G.V., R.M. D'Souza and M.D. Kirk, 2002: Foodborne disease in the new millennium: out of the frying pan and into the fire. *Med. J. Australia*, **177**, 614-618.

Hammitt, J.K. and J.D. Graham, 1999: Willingness to pay for health protection: inadequate sensitivity to probability? *J. Risk Uncertainty*, **18**, 33-62.

Hamnett, M.P., 1998: The Pacific ENSO Applications Centre and the 1997-98 El Niño. *Pacific ENSO Update*, **4**.

Hari Kumar, R., K. Venkaiah, N. Arlappa, S. Kumar, G. Brahmam and K. Vijayaraghavan, 2005: Diet and nutritional status of the population in the severely drought affected areas of Gujarat. *J. Hum. Ecol.*, **18**, 319-326.

Harrison, R.M., A.M. Jones, P.D. Biggins, N. Pomeroy, C.S. Cox, S.P. Kidd, J.L. Hobman, N.L. Brown and A. Beswick, 2005: Climate factors influencing bacterial count in background air samples. *Int. J. Biometeorol.*, **49**, 167-178.

Hartley, S. and D.A. Robinson, 2000: A shift in winter season timing in the Northern Plains of the USA as indicated by temporal analysis of heating degree days. *Int. J. Climatol.*, **20**, 365-379.

Hassi, J. and M. Rytkonen, 2005: Climate warming and health adaptation in Finland. FINADAPT Working Paper 7, Finnish Environment Institute Mimeographs 337, Helsinki, 28 pp.

Hassi, J., M. Rytkonen, J. Kotaniemi and H. Rintamaki, 2005: Impacts of cold climate on human heat balance, performance and health in circumpolar areas. *Int. J. Circumpolar Health*, **64**, 459-467.

Hay, S.I., D.J. Rogers, S.E. Randolph, D.I. Stern, J. Cox, G.D. Shanks and R.W. Snow, 2002a: Hot topic or hot air? Climate change and malaria resurgence in East African highlands. *Trends Parasitol.*, **18**, 530-534.

Hay, S.I., J. Cox, D.J. Rogers, S.E. Randolph, D.I. Stern, G.D. Shanks, M.F. Myers and R.W. Snow, 2002b: Climate change and the resurgence of malaria in the East African highlands. *Nature*, **415**, 905-909.

Hay, S.I., C.A. Guerra, A.J. Tatem, P.M. Atkinson and R.W. Snow, 2005a: Urbanization, malaria transmission and disease burden in Africa. *Nat. Rev. Microbiol.*, **3**, 81-90.

Hay, S.I., G.D. Shanks, D.I. Stern, R.W. Snow, S.E. Randolph and D.J. Rogers, 2005b: Climate variability and malaria epidemics in the highlands of East Africa. *Trends Parasitol.*, **21**, 52-53.

Hayhoe, K., 2004: Emissions pathways, climate change, and impacts on California. *P. Natl. Acad. Sci. USA*, **101**, 12422.

Hazenkamp-von Arx, M.E., T. Gotschi Fellmann, L. Oglesby, U. Ackermann-Liebrich, T. Gislason, J. Heinrich, D. Jarvis, C. Luczynska, A.J. Manzanera, L. Modig, D. Norback, A. Pfeifer, A. Poll, M. Ponzio, A. Soon, P. Vermeire and N. Kunzli, 2003: PM2.5 assessment in 21 European study centers of ECRHS II: method and first winter results. *J. Air Waste Manage*, **53**, 617-628.

He, Z., Y.J. Kim, K.O. Ogunjobi and C.S. Hong, 2003: Characteristics of PM2.5 species and long-range transport of air masses at Taean background station, South Korea. *Atmos. Environ.*, **37**, 219-230.

Healy, J.D., 2003: Excess winter mortality in Europe: a cross country analysis identifying key risk factors. *J. Epidemiol. Commun. H.*, **57**, 784-789.

Hegerl, G.C., F.W. Zwiers, P. Braconnot, N.P. Gillett, Y. Luo, J.A. Marengo Orsini, N. Nicholls, J.E. Penner and P.A. Stott, 2007: Understanding and attributing climate change. *Climate Change 2007: The Physical Science Basis. Contribution of Working Group I to the Fourth Assessment Report of the Intergovernmental Panel on Climate Change*, S. Solomon, D. Qin, M. Manning, Z. Chen, M. Marquis, K.B. Averyt, M. Tignor and H.L. Miller, Eds., Cambridge University Press, Cambridge, 663-746.

Heller, L., E. Colosimo and C. Antunes, 2003: Environmental sanitation conditions and health impact: a case-control study. *Rev. Soc. Bras. Med. Tro.*, **36**, 41-50.

Helmis, C.G., N. Moussiopoulos, H.A. Flocas, P. Sahm, V.D. Assimakopoulos, C. Naneris and P. Maheras, 2003: Estimation of transboundary air pollution on the basis of synoptic-scale weather types. *Int. J. Climatol.*, **23**, 405-416.

Hemon, D. and E. Jougla, 2004: La canicule du mois d'aout 2003 en France [The heatwave in France in August 2003]. *Rev. Epidemiol. Santé*, **52**, 3-5.

Hicks, B.B., 2003: Planning for air quality concerns of the future. *Pure Appl. Geophys.*, **160**, 57-74.

Hijioka, Y., K. Takahashi, Y. Matsuoka and H. Harasawa, 2002: Impact of global warming on waterborne diseases. *J. Jpn. Soc. Water Environ.*, **25**, 647-652.

Hogrefe, C., J. Biswas, B. Lynn, K. Civerolo, J.Y. Ku, J. Rosenthal, C. Rosenzweig, R. Goldberg and P.L. Kinney, 2004: Simulating regional-scale ozone climatology over the eastern United States: model evaluation results. *Atmos. Environ.*, **38**, 2627.

Holick, M.F., 2004: Sunlight and vitamin D for bone health and prevention of autoimmune diseases, cancers and cardiovascular disease. *Am. J. Clin. Nutr.*, **80**, 1678S-1688S.

Holloway, T., A. Fiore and M.G. Hastings, 2003: Intercontinental transport of air pollution: will emerging science lead to a new hemispheric treaty? *Environ Sci. Technol.*, **37**, 4535-4542.

Honda, Y., M. Ono, A. Sasaki and I. Uchiyama, 1998: Shift of the short term temperature mortality relationship by a climate factor: some evidence necessary to take account of in estimating the health effect of global warming. *J. Risk Res.*, **1**, 209-220.

Hopp, M.J. and J.A. Foley, 2003: Worldwide fluctuations in dengue fever cases related to climate variability. *Climate Res.*, **25**, 85-94.

Hoyt, K.S. and A.E. Gerhart, 2004: The San Diego County wildfires: perspectives of health care. *Disaster Manage. Response*, **2**, 46-52.

Hubalek, Z., 2004: An annotated checklist of pathogenic microorganisms associated with migratory birds. *J. Wildlife Dis.*, **40**, 639-659.

Hunter, P.R., 2003: Climate change and waterborne and vectorborne disease. *J. Appl. Microbiol.*, **94**, 37-46.

Huynen, M. and B. Menne, 2003: Phenology and human health: allergic disorders. Report of a WHO meeting in Rome, Italy, 16–17 January 2003. Health and Global Environmental Series, EUR/03/5036791. World Health Organization, Copenhagen, 64 pp.

Ibald-Mulli, A., H.E. Wichmann, W. Kreyling and A. Peters, 2002: Epidemiological evidence on health effects of ultrafine particles. *J. Aerosol Med.*, **15**, 189-201.

IFRC, 2002: *World Disaster Report 2002*. International Federation of Red Cross and Red Crescent Societies, Geneva, 240 pp.

IGCI, 2000: *Climate Change Vulnerability and Adaptation Assessment for Fiji. Pacific Islands Climate Change Assistance Programme*. International Global Change Institute, University of Waikato, Hamilton.

INVS, 2003: *Impact sanitaire de la vague de chaleur d'août 2003 en France. Bilan et perpectives [Health Impact of the Heatwave in August 2003 in France]*. Institut de Veille Sanitaire, Saint-Maurice, 120 pp.

IPCC, 2000: *Emissions Scenarios: A Special Report of Working Group III of the Intergovernmental Panel on Climate Change*, N. Nakićenović and R. Swart, Eds., Cambridge University Press, New York, 570 pp.

IPCC, 2007a: *Summary for Policymakers. Climate Change 2007: The Physical Science Basis. Contribution of Working Group I to the Fourth Assessment Report of the Intergovernmental Panel on Climate Change*, S. Solomon, D. Qin, M. Manning, Z. Chen, M. Marquis, K.B. Averyt, M.Tignor and H.L. Miller, Eds., Cambridge University Press, Cambridge, 18 pp.

IPCC, 2007b: *Climate Change 2007: The Physical Science Basis. Contribution of Working Group I to the Fourth Assessment Report of the Intergovernmental Panel on Climate Change*, S. Solomon, D. Qin, M. Manning, Z. Chen, M. Marquis, K.B. Averyt, M. Tignor and H.L. Miller, Eds., Cambridge University Press, Cambridge, 996 pp.

IPCC, 2007c: *Climate Change 2007: Mitigation. Contribution of Working Group III to the Fourth Assessment Report of the Intergovernmental Panel on Climate*

Change, B. Metz, O. Davidson, P. Bosch, R. Dave and L. Meyer, Eds., Cambridge University Press, Cambridge, UK.

IPCC/TEAP, 2005: *Special Report: Safeguarding the Ozone Layer and the Global Climate System: Issues Related to Hydrofluorocarbons and Perfluorocarbons*, B. Metz, L. Kuijpers, S. Solomon, S.O. Andersen, O. Davidson, J. Pons, D. de Jager, T. Kestin, M. Manning and L. Meyer, Eds., Cambridge University Press, New York, 468 pp.

Ito, K., S.F. De Leon and M. Lippmann, 2005: Associations between ozone and daily mortality: analysis and meta-analysis. *Epidemiology*, **16**, 446-457.

Izmerov, N.F., B.A. Revich and E.I. Korenberg, 2005: Climate changes and health of population in Russia in XXI century (in Russian). *Med. Tr. Prom. Ekol.* 1-6.

Jaffe, D., I. McKendry, T. Anderson and H. Price, 2003: Six "new" episodes of trans-Pacific transport of air pollutants. *Atmos. Environ.*, **37**, 91-404.

Jaffe, D., I. Bertschi, L. Jaegle, P. Novelli, J.S. Reid, H. Tanimoto, R. Vingarzan and D.L. Westphal, 2004: Long-range transport of Siberian biomass burning emissions and impact on surface ozone in western North America. *Geophys. Res. Lett.*, **31**, L16106.

Jensen, S., R. Berkowicz, M. Winther, F. Palmgren and Z. Zlatev, 2001: Future air quality in Danish cities due to new emission and fuel quality directives of the European Union. *Int. J. Vehicle Des.*, **27**, 195-208.

Johnson, C.E., D.S. Stevenson, W.J. Collins and R.G. Derwent, 2001: Role of climate feedback on methane and ozone studied with a coupled ocean-atmosphere-chemistry model. *Geophys. Res. Lett.*, **28**, 1723-1726.

Johnson, H., R.S. Kovats, G.R. McGregor, J.R. Stedman, M. Gibbs, H. Walton, L. Cook and E. Black, 2005: The impact of the 2003 heatwave on mortality and hospital admissions in England. *Health Statistics Q.*, **25**, 6-12.

Jones, J.M. and T.D. Davies, 2000: The influence of climate on air and precipitation chemistry over Europe and downscaling applications to future acidic deposition. *Climate Res.*, **14**, 7-24.

Jonkman, S.N. and I. Kelman, 2005: An analysis of the causes and circumstances of flood disaster deaths. *Disasters*, **29**, 75-97.

Jonsson, P., C. Bennet, I. Eliasson and E. Selin Lindgren, 2004: Suspended particulate matter and its relations to the urban climate in Dar es Salaam, Tanzania. *Atmos. Environ.*, **38**, 4175.

Julvez, J., M. Develoux, A. Mounkaila and J. Mouchet, 1992: Diversity of malaria in the Sahelo-Saharan region: a review apropos of the status in Niger, West Africa (in French). *Ann. Soc. Belg. Med. Tr.*, **72**, 163-177.

Julvez, J., J. Mouchet, A. Michault, A. Fouta and M. Hamidine, 1997: The progress of malaria in Sahelian eastern Niger: an ecological disaster zone (in French). *B. Soc. Pathol. Exot.*, **90**, 101-104.

Junk, J., A. Helbig and J. Luers, 2003: Urban climate and air quality in Trier, Germany. *Int. J. Biometeorol.*, **47**, 230-238.

Kabuto, M., Y. Honda and H. Todoriki, 2005: A comparative study of daily maximum and personally exposed temperatures during hot summer days in three Japanese cities (in Japanese). *Nippon Koshu Eisei Zasshi*, **52**, 775-784.

Kang, G., B.S. Ramakrishna, J. Daniel, M. Mathan and V. Mathan, 2001: Epidemiological and laboratory investigations of outbreaks of diarrhoea in rural South India: implications for control of disease. *Epidemiol. Infect.*, **127**, 107-112.

Kappos, A.D., P. Bruckmann, T. Eikmann, N. Englert, U. Heinrich, P. Hoppe, E. Koch, G.H.M. Krause, W.G. Kreyling, K. Rauchfuss, P. Rombout, V. Schulz-Klemp, W.R. Thiel and H.E. Wichmann, 2004: Health effects of particles in ambient air. *Int. J. Hyg. Envir. Heal.*, **207**, 399-407.

Kassomenos, P., A. Gryparis, E. Samoli, K. Katsouyanni, S. Lykoudis and H.A. Flocas, 2001: Atmospheric circulation types and daily mortality in Athens, Greece. *Environ. Health Persp.*, **109**, 591-596.

Kato, S., Y. Kajiia, R. Itokazu, J. Hirokawad, S. Kodae and Y. Kinjof, 2004: Transport of atmospheric carbon monoxide, ozone, and hydrocarbons from Chinese coast to Okinawa island in the Western Pacific during winter. *Atmos. Environ.*, **38**, 2975-2981.

Katsumata, T., D. Hosea, E.B. Wasito, S. Kohno, K. Hara, P. Soeparto and I.G. Ranuh, 1998: Cryptosporidiosis in Indonesia: a hospital-based study and a community-based survey. *Am. J. Trop. Med. Hyg.*, **59**, 628-632.

Kaumov, A. and B. Muchmadeliev, 2002: *Climate Change and its Impacts on Human Health*. Dushanbe, Avesto, 172 pp.

Keiser, J., J. Utzinger, M.C. De Castro, T.A. Smith, M. Tanner and B.H. Singer, 2004: Urbanization in sub-Saharan Africa and implication for malaria control. *Am. J. Trop. Med. Hyg.*, **71**, 118-127.

Kellogg, C.A. and D.W. Griffin, 2006: Aerobiology and the global transport of desert dust. *Trends Ecol. Evol.*, **21**, 638-644.

Kerslake, D., 1972: *The Stress of Hot Environments*. Cambridge University Press, Cambridge, 326 pp.

Kistemann, T., T. Classen, C. Koch, F. Dangendorf, R. Fischeder, J. Gebel, V. Vacata and M. Exner, 2002: Microbial load of drinking water reservoir tributaries during extreme rainfall and runoff. *Appl. Environ. Microbiol.*, **68**, 2188-2197.

Klich, M., M.W. Lankester and K.W. Wu, 1996: Spring migratory birds (Aves) extend the northern occurrence of blacklegged tick (Acari: Ixodidae). *J. Med. Entomol.*, **33**, 581-585.

Knowlton, K., J.E. Rosenthal, C. Hogrefe, B. Lynn, S. Gaffin, R. Goldberg, C. Rosenzweig, K. Civerolo, J.Y. Ku and P.L. Kinney, 2004: Assessing ozone-related health impacts under a changing climate. *Environ. Health Persp.*, **112**, 1557-1563.

Ko, A.I., M. Galvao Reis, C.M. Ribeiro Dourado, W.D. Johnson Jr. and L.W. Riley, 1999: Urban epidemic of severe leptospirosis in Brazil. Salvador Leptospirosis Study Group. *Lancet*, **354**, 820-825.

Koe, L., A.J. Arellano and J. McGregor, 2001: Investigating the haze transport from 1997 biomass burning in Southeast Asia: its impact upon Singapore. *Atmos. Environ.*, **35**, 2723-2734.

Koelle, K., X. Rodo, M. Pascal, M. Yunus and G. Mostafa, 2005: Refractory periods and climate forcing in cholera dynamics. *Nature*, **436**, 696.

Kohler, S. and C. Kohler, 1992: Dead bleached coral provides new surfaces for dinoflagellates implicated in ciguatera fish poisonings. *J. Env. Biol. Fish.*, **35**, 413-416.

Kohn, R., I. Levav, I. Donaire, M. Machuca and R. Tamashiro, 2005: Reacciones psicologicas y psicopatologicas en Honduras despues del huracan Mitch: implicaciones para la planificacion de los servicios [Psychological and psychopathological reactions in Honduras following Hurricane Mitch: implications for service planning] (in Spanish). *Rev. Panam. Salud Publ.*, **18**, 287-295.

Koike, I., 2006: State of the art findings of global warming: contributions of the Japanese researchers and perspective in 2006. Second Report of the Global Warming Initiative, Climate Change Study Group, Ministry of Environment, Tokyo, 165-173.

Koppe, C., 2005: *Gesundheitsrelevante Bewertung von thermischer Belastung unter Berücksichtigung der kurzfristigen Anpassung der Bevölkerung and die lokalen Witterungsverhältnisse [Evaluation of Health Impacts of Thermal Exposure under Consideration of Short-term Adaptation of Populations to Local Weather] (in German)*. Berichte des Deutschen Wetterdienstes 226, Offenbach am Main, 167 pp.

Koppe, C., G. Jendritzky, R.S. Kovats and B. Menne, 2004: *Heat-waves: Impacts and Responses*. Health and Global Environmental Change Series, No. 2. World Health Organization, Copenhagen, 123 pp.

Korenberg, E., 2004: Environmental causes for possible relationship between climate change and changes of natural foci of diseases and their epidemiologic consequences. *Climate Change and Public Health in Russia in the XXI Century: Proceedings of the International Workshop*, Moscow, 54-67.

Kosek, M., C. Bern and R.L. Guerrent, 2003: The global burden of diarrhoeal disease, as estimated from studies published between 1992 and 2000. *B. World Health Organ.*, **81**, 197-204.

Kossmann, M. and A. Sturman, 2004: The surface wind field during winter smog nights in Christchurch and coastal Canterbury, New Zealand. *Int. J. Climatol.*, **24**, 93-108.

Kovats, R.S. and C. Koppe, 2005: Heatwaves past and future impacts on health. *Integration of Public Health with Adaptation to Climate Change: Lessons Learned and New Directions*, K. Ebi, J. Smith and I. Burton, Eds., Taylor and Francis, Lisse, 136-160.

Kovats, R.S., D. Campbell-Lendrum, A. McMichael, A. Woodward and J. Cox, 2001: Early effects of climate change: do they include changes in vector-borne disease? *Philos. T. Roy. Soc. Lond. B*, **356**, 1057-1068.

Kovats, R.S. and K.L. Ebi, 2006: Heatwaves and public health in Europe. *Eur. J. Public Health*, **16**, 592-599. doi:10.1093/eurpub/ckl049.

Kovats, R.S., M.J. Bouma, S. Hajat, E. Worrall and A. Haines, 2003: El Nino and health. *Lancet*, **362**, 1481-1489.

Kovats, R.S., S. Edwards, S. Hajat, B. Armstrong, K.L. Ebi and B. Menne, 2004: The effect of temperature on food poisoning: time series analysis in 10 European countries. *Epidemiol. Infect.*, **132**, 443-453.

Kovats, R.S., S.J. Edwards, D. Charron, J. Cowden, R.M. D'Souza, K.L. Ebi, C. Gauci, P. Gerner-Smidt, S. Hajat, S. Hales, G.H. Pezzi, B. Kriz, K. Kutsar, P. McKeown, K. Mellou, B. Menne, S. O'Brien, W. van Pelt and H. Schmid, 2005: Climate variability and campylobacter infection: an international study. *Int. J.*

Biometeorol., **49**, 207-214.

Krake, A., J. McCullough and B. King, 2003: Health hazards to park rangers from excessive heat at Grand Canyon National Park. *Appl. Occup. Environ. Hyg.*, **18**, 295-317.

Krauss, S., D. Walker, S.P. Pryor, L. Niles, C.H. Li, V.S. Hinshaw and R.G. Webster, 2004: Influenza A viruses of migrating wild aquatic birds in North America. *Vector-Borne Zoonot*, **4**, 177-189.

Kuhn, K., D. Campbell-Lendrum and C.R. Davies, 2002: A continental risk map for malaria mosquito (Diptera: Culicidae) vectors in Europe. *J. Med. Entomol.*, **39**, 621-630.

Kunii, O., S. Nakamura, R. Abdur and S. Wakai, 2002: The impact on health and risk factors of the diarrhoea epidemics in the 1998 Bangladesh floods. *Public Health*, **116**, 68-74.

Kunst, A.E., C.W. Looman and J.P. Mackenbach, 1991: The decline in winter excess mortality in the Netherlands. *Int. J. Epidemiol.*, **20**, 971-977.

Kwon, H.J., S.H. Cho, Y. Chun, F. Lagarde and G. Pershagen, 2002: Effects of the Asian dust events on daily mortality in Seoul, Korea. *Environ. Res.*, **90**, 1-5.

Kysely, J., 2005: Mortality and displaced mortality during heat waves in the Czech Republic. *Int. J. Biometeorol.*, **49**, 91-97.

Laaidi, K., M. Pascal, M. Ledrans, A. Le Tertre, S. Medina, C. Caserio, J.C. Cohen, J. Manach, P. Beaudeau and P. Empereur-Bissonnet, 2004: *Le système français d'alerte canicule et santé (SACS 2004): Un dispositif intégéré au Plan National Canicule [The French Heatwave Warning System and Health: An Integrated National Heatwave Plan]*. Institutde Veille Sanitaire, 35 pp.

Lagadec, P., 2004: Understanding the French 2003 heat wave experience: beyond the heat, a multi-layered challenge. *J. Contingencies Crisis Management*, **12**, 160-169.

Lake, I., G. Bentham, R.S. Kovats and G. Nichols, 2005: Effects of weather and river flow on cryptosporidiosis. *Water Health*, **3**, 469-474.

Lama, J.R., C.R. Seas, R. León-Barúa, E. Gotuzzo and R.B. Sack, 2004: Environmental temperature, cholera, and acute diarrhoea in adults in Lima, Peru. *J. Health Popul. Nutr.*, **22**, 399-403.

Langmann, B., S. Bauer and I. Bey, 2003: The influence of the global photochemical composition of the troposphere on European summer smog. Part 1. Application of a global to mesoscale model chain. *J. Geophys. Res. D*, **108**, 4146.

Last, J.M., 1998: *Public Health and Human Ecology*. Prentice Hall International, London, 464 pp.

Laurila, T., J. Tuovinen, V. Tarvainen and D. Simpson, 2004: Trends and scenarios of ground-level ozone concentrations in Finland. *Boreal Environ. Res.*, **9**, 167-184.

Le Tertre, A., A. Lefranc, D. Eilstein, C. Declercq, S. Medina, M. Blanchard, B. Chardon, P. Fabre, L. Filleul, J.F. Jusot, L. Pascal, H. Prouvost, S. Cassadou and M. Ledrans, 2006: Impact of the 2003 heatwave on all-cause mortality in 9 French cities. *Epidemiology*, **17**, 75-79.

Lehane, L. and R.J. Lewis, 2000: Ciguatera: recent advances but the risk remains. *Int. J. Food Microbiol.*, **61**, 91-125.

Leithead, C. and A. Lind, 1964: *Heat Stress and Heat Disorders*. Cassell, London, 304 pp.

Lennartson, G. and M. Schwartz, 1999: A synoptic climatology of surface-level ozone in Eastern Wisconsin, USA. *Climate Res.*, **13**, 207-220.

Lerchl, A., 1998: Changes in the seasonality of mortality in Germany from 1946 to 1995: the role of temperature. *Int. J. Biometeorol.*, **42**, 84-88.

Levy, J.I., S.M. Chemerynski and J.A. Sarnat, 2005: Ozone exposure and mortality: an empiric bayes metaregression analysis. *Epidemiology*, **16**, 458-468.

Liang, Q., L. Jaegle, D. Jaffe, P. Weiss-Penzias, A. Heckman and J. Snow, 2004: Long-range transport of Asian pollution to the northeast Pacific: seasonal variations and transport pathways of carbon monoxide. *J. Geophys. Res. D*, **109**, D23S07.

Lim, G., J. Aramini, M. Fleury, R. Ibarra and R. Meyers, 2002: *Investigating the Relationship between Drinking Water and Gastro-enteritis in Edmonton, 1993–1998*. Division of Enteric, Food-borne and Waterborne Diseases, Health Canada, Ottawa, 61 pp.

Lindgren, E. and L. Talleklint, 2000: Impact of climatic change on the northern latitude limit and population density of the disease-transmitting European tick *Ixodes ricinus*. *Environ. Health Persp.*, **108**, 119-123.

Lindgren, E. and R. Gustafson, 2001: Tick-borne encephalitis in Sweden and climate change. *Lancet*, **358**, 16-18.

Lindgren, E. and T. Naucke, 2006: Leishmaniasis: influences of climate and climate change epidemiology, ecology and adaptation measures. *Climate Change and Adaptation Strategies for Human Health*, B. Menne and K. Ebi, Eds., Steinkopff, Darmstadt, 131-156.

Lipp, E.K., A. Huq and R.R. Colwell, 2002: Effects of global climate on infectious disease: the cholera model. *Clin. Microbiol. Rev.*, **15**, 757.

Louis, V.R., I.A. Gillespie, S.J. O'Brien, E. Russek-Cohen, A.D. Pearson and R.R. Colwell, 2005: Temperature-driven Campylobacter seasonality in England and Wales. *Appl. Environ. Microbiol.*, **71**, 85-92.

Lutz, W., W. Sanderson and S. Scherbov, 2000: Doubling of world population unlikely. *Nature*, **387**, 803-805.

Macintyre, S., L. McKay and A. Ellaway, 2005: Are rich people or poor people more likely to be ill? Lay perceptions, by social class and neighbourhood, of inequalities in health. *Soc. Sci. Med.*, **60**, 313-317.

Maltais, D., L. Lachance, A. Brassard and M. Dubois, 2005: Social support, coping and psychological health after a flood (in French). *Sci. Soc. Santé*, **23**, 5-38.

Manuel, J., 2006: In Katrina's wake. *Environ. Health Persp.*, **114**, A32-A39.

Marmot, M., 2005: Social determinants of health inequalities. *Lancet*, **365**, 1099-1104.

Martens, P. and H.B. Hilderink, 2001: Human health in transition: towards more disease or sustained health? *Transitions in a Globalising World*, P. Martens and J. Rotmans, Eds., Swets and Zeitlinger, Lisse, 61-84.

Martens, P. and M. Huynen, 2003: A future without health? Health dimension in global scenario studies. *B. World Health Organ.*, **81**, 896-901.

Martin, B., H. Fuelberg, N. Blake, J. Crawford, L. Logan, D. Blake and G. Sachse, 2002: Long-range transport of Asian outflow to the equatorial Pacific. *J. Geophys. Res. D*, **108**, 8322.

Martinez-Navarro, F., F. Simon-Soria and G. Lopez-Abente, 2004: Valoracion del impacto de la ola de calor del verano de 2003 sobre la mortalidad [Evaluation of the impact of the heatwave in the summer of 2003 on mortality]. *Gac. Sanit.*, **18**, 250-258.

Mascie-Taylor, C.G. and E. Karim, 2003: The burden of chronic disease. *Science*, **302**, 1921-1922.

Mason, J.B., A. Bailes, K.E. Mason, O. Yambi, U. Jonsson, C. Hudspeth, P. Hailey, A. Kendle, D. Brunet and P. Martel, 2005: AIDS, drought, and child malnutrition in southern Africa. *Public Health Nutr.*, **8**, 551-563.

Maynard, N.G., 2006: Satellites, settlements and human health. *Remote Sensing of Human Settlements: Manual of Remote Sensing*, M. Ridd and J.D. Hipple, Eds., American Society of Photogrammetry and Remote Sensing, Bethesda, Maryland, 379-399.

McGregor, G.R., 1999: Basic meteorology. *Air Pollution and Health*, S.T. Holgate, J. Samet, H. Koren and R.L. Maynard, Eds., Academic Press, San Diego, California, 21-49.

McKee, M. and M. Suhrcke, 2005: Commentary: health and economic transition. *Int. J. Epidemiol.*, **34**, 1203-1206.

McLaughlin, J.B., A. DePaola, C.A. Bopp, K.A. Martinek, N.P. Napolilli, C.G. Allison, S.L. Murray, E.C. Thompson, M.M. Bird and J.P. Middaugh, 2005: Outbreak of *Vibrio parahaemolyticus* gastroenteritis associated with Alaskan oysters. *New Engl. J. Med.*, **353**, 1463-1470.

McMichael, A., 2004: Climate change. *Comparative Quantification of Health Risks: Global and Regional Burden of Disease due to Selected Major Risk Factors, Vol. 2*, M. Ezzati, A. Lopez, A. Rodgers and C. Murray, Eds., World Health Organization, Geneva, 1543-1649.

McMichael, A., A. Githeko, R. Akhtar, R. Carcavallo, D.J. Gubler, A. Haines, R.S. Kovats, P. Martens, J. Patz, A. Sasaki, K. Ebi, D. Focks, L.S. Kalkstein, E. Lindgren, L.R. Lindsay and R. Sturrock, 2001: Human population health. *Climate Change 2001: Impacts, Adaptation, and Vulnerability. Contribution of Working Group II to the Third Assessment Report of the Intergovernmental Panel on Climate Change*, J.J. McCarthy, O.F. Canziani, N.A. Leary, D.J. Dokken and K.S. White, Eds., Cambridge University Press, Cambridge, 453-485.

McMichael, A.J., D. Campbell-Lendrum, C. Corvalan, K. Ebi, A. Githeko, J. Scheraga and A. Woodward, Eds., 2003a: *Climate Change and Human Health: Risk and Responses*. World Health Organization, Geneva, 333 pp.

McMichael, A.J., R. Woodruff, P. Whetton, K. Hennessy, N. Nicholls, S. Hales, A. Woodward and T. Kjellstrom, 2003b: *Human Health and Climate Change in Oceania: Risk Assessment 2002*. Department of Health and Ageing, Canberra, 128 pp.

McMichael, A.J., M. McKee, V. Shkolnikov and T. Valkonen, 2004: Mortality trends and setbacks: global convergence or divergence? *Lancet*, **363**, 1155-1159.

McMichael, A.J., R.E. Woodruff and S. Hales, 2006: Climate change and human health: present and future risks. *Lancet*, **367**, 859-869.

Meehl, G.A. and C. Tebaldi, 2004: More intense, more frequent and longer lasting heat waves in the 21st century. *Nature*, **305**, 994-997.

Meehl, G.A., T.F. Stocker, W.D. Collins, P. Friedlingstein, A.T. Gaye, J.M. Gregory, A. Kitoh, R. Knutti, J.M. Murphy, A. Noda, S.C.B. Raper, I.G. Watterson, A.J. Weaver and Z.-C. Zhao, 2007: Global climate projections. *Climate Change 2007: The Physical Science Basis. Contribution of Working Group I to the Fourth Assessment Report of the Intergovernmental Panel on Climate Change*, S. Solomon, D. Qin, M. Manning, Z. Chen, M. Marquis, K.B. Averyt, M. Tignor and H.L. Miller, Eds., Cambridge University Press, Cambridge, 747-846.

Menne, B., 2000: Floods and public health consequences, prevention and control measures. UN 2000 (MP.WAT/SEM.2/1999/22).

Menne, B. and R. Bertollini, 2000: The health impacts of desertification and drought. *Down to Earth*, **14**, 4-6.

Menne, B. and R. Bertollini, 2005: Health and climate change: a call for action. *Brit. Med. J.*, **331**, 1283-1284.

Michelon, T., P. Magne and F. Simon-Delavelle, 2005: Lessons from the 2003 heat wave in France and action taken to limit the effects of future heat wave. *Extreme Weather Events and Public Health Responses*, W. Kirch, B. Menne and R. Bertollini, Eds., Springer, Berlin, 131-140.

Michelozzi, P., F. de Donato, G. Accetta, F. Forastiere, M. D'Ovido and L.S. Kalkstein, 2004: Impact of heat waves on mortality: Rome, Italy, June–August 2003. *J. Am. Med. Assoc.*, **291**, 2537-2538.

Michelozzi, P., F. de Donato, L. Bisanti, A. Russo, E. Cadum, M. Demaria, M. D'Ovidio, G. Costa and C. Perucci, 2005: The impact of the summer 2003 heat waves on mortality in four Italian cities. *EuroSurveillance*, **10**, 161-165.

Mickley, L.J., D.J. Jacob, B.D. Field and D. Rind, 2004: Effects of future climate change on regional air pollution episodes in the United States. *Geophys. Res. Lett.*, **30**, L24103.

Miettinen, I.T., O. Zacheus, C.H. von Bonsdorff and T. Vartiainen, 2001: Waterborne epidemics in Finland in 1998-1999. *Water Sci. Technol.*, **43**, 67-71.

Millennium Ecosystem Assessment, 2005: *Ecosystems and Human Well-Being: Scenarios*. Findings of the Scenarios Working Group Millennium Ecosystem Assessment Series, Island Press, Washington, District of Columbia, 515 pp.

Ministry of Environment and Forest and Government of India, 2004: *India's Initial National Communication to the United National Framework Convention on Climate Change*. Government of India, New Delhi, 292 pp.

Mohan, J.E., L.H. Ziska, W.H. Schlesinger, R.B. Thomas, R.C. Sicher, K. George and J.S. Clark, 2006: Biomass and toxicity responses of poison ivy (*Toxicodendron radicans*) to elevated atmospheric CO_2. *P. Natl. Acad. Sci. USA*, **103**, 9086-9089.

Mohanty, P. and U. Panda, 2003: *Heatwave in Orissa: A Study Based on Heat Indices and Synoptic Features – Heatwave Conditions in Orissa*. Regional Research Laboratory, Institute of Mathematics and Applications, Bubaneshwar, 15 pp.

Molesworth, A.M., M.H. Djingary and M.C. Thomson, 2001: Seasonality in meningococcal disease in Niger, West Africa: a preliminary investigation. *GEOMED 1999, Paris*. Elsevier, 92-97.

Molesworth, A.M., L.E. Cuevas, A.P. Morse, J.R. Herman and M.C. Thomson, 2002a: Dust clouds and spread of infection. *Lancet*, **359**, 81-82.

Molesworth, A.M., M.C. Thomson, S.J. Connor, M.P. Cresswell, A.P. Morse, P. Shears, C.A. Hart and L.E. Cuevas, 2002b: Where is the meningitis belt? Defining an area at risk of epidemic meningitis in Africa. *T. Roy. Soc. Trop. Med. H.*, **96**, 242-249.

Molesworth, A.M., L.E. Cuevas, S.J. Connor, A.P. Morse and M.C. Thomson, 2003: Environmental risk and meningitis epidemics in Africa. *Emerg. Infect. Dis.*, **9**, 1287-1293.

Mollica, R.F., B.L. Cardozo, H. Osofsky, B. Raphael, A. Ager and P. Salama, 2004: Mental health in complex emergencies. *Lancet*, **364**, 2058-2067.

Mondal, N., M. Biswas and A. Manna, 2001: Risk factors of diarrhoea among flood victims: a controlled epidemiological study. *Indian J. Public Health*, **45**, 122-127.

Moore, D., R. Copes, R. Fisk, R. Joy, K. Chan and M. Brauer, 2006: Population health effects of air quality changes due to forest fires in British Columbia in 2003: Estimates from physician-visit billing data. *Can. J. Public Health*, **97**, 105-108.

Moore, J.E., D. Gilpin, E. Crothers, A. Canney, A. Kaneko and M. Matsuda, 2002: Occurrence of *Campylobacter* spp. and *Cryptosporidium* spp. in seagulls (*Larus* spp.). *Vector-Borne Zoonot.*, **2**, 111-114.

Moore, K., A. Clarke, V. Kapustin and S. Howell, 2003: Long-range transport of continental plumes over the Pacific Basin: aerosol physiochemistry and optical properties during PEM-Tropics A and B. *J. Geophys. Res. D*, **108**, 8236.

Moreno, J., 2005: A preliminary assessment of the impacts in Spain due to the effects of climate change. ECCE Project Final Report. Universidad de Castilla-La Mancha, Ministry of the Environment, Madrid, 741 pp.

Morris, C.J.G. and I. Simmonds, 2000: Associations between varying magnitudes of the urban heat island and the synoptic climatology in Melbourne, Australia. *Int. J. Climatol.*, **20**, 1931-1954.

Mott, J.A., D.M. Mannino, C.J. Alverson, A. Kiyu, J. Hashim, T. Lee, K. Falter and S.C. Redd, 2005: Cardiorespiratory hospitalizations associated with smoke exposure during the 1997 Southeast Asian forest fires. *Int. J. Hyg. Envir. Heal.*, **208**, 75-85.

Mouchet, J., O. Faye, J. Juivez and S. Manguin, 1996: Drought and malaria retreat in the Sahel, West Africa. *Lancet*, **348**, 1735-1736.

Mudway, I.S. and F.J. Kelly, 2000: Ozone and the lung: a sensitive issue. *Mol. Aspects Med.*, **21**, 1-48.

Murano, K., H. Mukai, S. Hatakeyama, E. Jang and I. Uno, 2000: Trans-boundary air pollution over remote islands in Japan: observed data and estimates from a numerical model. *Atmos. Environ.*, **34**, 5139-5149.

Mutero, C.M., C. Kabutha, V. Kimani, L. Kabuage, G. Gitau, J. Ssennyonga, J. Githure, L. Muthami, A. Kaida, L. Musyoka, E. Kiarie and M. Oganda, 2004: A transdisciplinary perspective on the links between malaria and agroecosystems in Kenya. *Acta Trop.*, **89**, 171-186.

Nagendra, S. and M. Khare, 2003: Diurnal and seasonal variations of carbon monoxide and nitrogen dioxide in Delhi city. *Int. J. Environ. Pollut.*, **19**, 75-96.

Nakićenović, N. and R. Swart, Eds., 2000: *Special Report on Emissions Scenarios: A Special Report of Working Group III of the Intergovernmental Panel on Climate Change*. Cambridge University Press, Cambridge, 599 pp.

National Environment Commission, Royal Government of Bhutan, UNDP and GEF, 2006: *Bhutan National Adaptation Programme of Action*. National Environment Commission, Royal Government of Bhutan, Thimphu, 95 pp.

Nayha, S., 2005: Environmental temperature and mortality. *Int. J. Circumpolar Health*, **64**, 451-458.

Nchito, M., P. Kelly, S. Sianongo, N.P. Luo, R. Feldman, M. Farthing and K.S. Baboo, 1998: Cryptosporidiosis in urban Zambian children: an analysis of risk factors. *Am. J. Trop. Med. Hyg.*, **59**, 435-437.

Ndiaye, O., J.Y. Hesran, J.F. Etard, A. Diallo, F. Simondon, M.N. Ward and V. Robert, 2001: Climate variability and number of deaths at attributable to malaria in Niakhar area, Senegal, from 1984 to 1996 (in French). *Santé*, **11**, 25-33.

Nicholls, N., C. Butler and I. Hanigan, 2005: Inter-annual rainfall variations and suicide in New South Wales, Australia, 1964 to 2001. *Int. J. Biometeorol.*, **50**, 139-143.

Nicholls, R.J., 2003: An expert assessment of storm surge "hotspots". Interim Report to Center for Hazards and Risk Research, Lamont-Doherty Observatory, Columbia University. Flood Hazard Research Centre, University of Middlesex, London, 10 pp.

Nicholls, R.J., 2004: Coastal flooding and wetland loss in the 21st century: changes under the SRES climate and socio-economic scenarios. *Global Environ. Chang.*, **14**, 69-86.

Nickels, S., C. Furgal and J. Castleden, 2003: Putting the human face on climate change through community workshops: Inuit knowledge, partnerships and research. *The Earth is Faster Now: Indigenous Observations of Arctic Environmental Change*, I. Krupnik and D. Jolly, Eds., Arctic Studies Centre, Smithsonian Institution, Washington, District of Columbia, 300-344.

Nilsson, E., J. Paatero and M. Boy, 2001a: Effects of air masses and synoptic weather on aerosol formation in the continental boundary layer. *Tellus B*, **53**, 462-478.

Nilsson, E., U. Rannik, M. Kulmala and G. Buzorius, 2001b: Effects of continental boundary layer evolution, convection, turbulence and entrainment, on aerosol formation. *Tellus B*, **53**, 441-461.

Niskanen, T., J. Waldenstrom, M. Fredriksson-Ahomaa, B. Olsen and H. Korkeala, 2003: virF-positive *Yersinia pseudotuberculosis* and *Yersinia enterocolitica* found in migratory birds in Sweden. *Appl. Environ. Microbiol.*, **69**, 4670-4675.

Nordhaus, W.D., 1991: To slow or not to slow: the economics of the greenhouse effect. *Econ. J.*, **101**, 920-937.

Nordhaus, W.D. and J. Boyer, 2000: *Warming the World: Economic Models of Global Warming*. MIT Press, Cambridge, Massachusetts, 246 pp.

Norris, F.H., A.D. Murphy, C.K. Baker and J.L. Perilla, 2004: Postdisaster PTSD over four waves of a panel study of Mexico's 1999 flood. *J. Trauma. Stress*, **17**, 283-292.

North, C.S., A. Kawasaki, E.L. Spitznagel and B.A. Hong, 2004: The course of

PTSD, major depression, substance abuse, and somatization after a natural disaster. *J. Nerv. Ment. Dis.*, **192**, 823-829.

Obiri-Danso, K., N. Paul and K. Jones, 2001: The effects of UVB and temperature on the survival of natural populations and pure cultures of *Campylobacter jejuni, Camp. coli, Camp. lari* and urease-positive thermophilic campylobacters (UPTC) in surface waters. *J. Appl. Microbiol.*, **90**, 256-267.

OCHA, 2003: India: Heat Wave – Occurred: 20 May 2003–5 June 2003. OCHA Situation Report No.1. http://cidi.org/disaster/03a/ixl131.html.

Ogden, N.H., A. Maarouf, I.K. Barker, M. Bigras-Poulin, L.R. Lindsay, M.G. Morshed, C.J. O'Callaghan, F. Ramay, D. Waltner-Toews and D.F. Charron, 2006: Climate change and the potential for range expansion of the Lyme disease vector *Ixodes scapularis* in Canada. *Int. J. Parasitol.*, **36**, 63-70.

Ohl, C.A. and S. Tapsell, 2000: Flooding and human health. *Brit. Med. J.*, **321**, 1167-1168.

Olmos, S., 2001: Vulnerability and adaptation to climate change: concepts, issues, assessment methods. Climate Change Knowledge Network Foundation Paper, Oslo, 20 pp.

Olsen, B., D.C. Duffy, T.G.T. Jaenson, A. Gylfe, J. Bonnedahl and S. Bergstrom, 1995: Transhemispheric exchange of Lyme-disease spirochetes by seabirds. *J. Clin. Microbiol.*, **33**, 3270-3274.

Olshansky, S.J., B.A. Carnes and C. Cassel, 1998: The future of long life. *Science*, **281**, 1612-3, 1613-1615.

Pal Arya, S., 2000: Air pollution meteorology and dispersion. *Bound.-Lay. Meteorol.*, **94**, 171-172.

Pardue, J., W. Moe, D. McInnis, L. Thibodeaux, K. Valsaraj, E. Maciasz, I. van Heerden, N. Korevec and Q. Yuan, 2005: Chemical and microbiological parameters in New Orleans floodwater following Hurricane Katrina. *Environ. Sci. Technol.*, **39**, 8591-8599.

Parkinson, A.J. and J.C. Butler, 2005: Potential impacts of climate change on infectious diseases in the Arctic. *Int. J. Circumpolar Health*, **64**, 478-486.

Parmesan, C. and G. Yohe, 2003: A globally coherent fingerprint of climate change impacts across natural systems. *Nature*, **421**, 37-42.

Pascual, M., X. Rodo, S.P. Ellner, R. Colwell and M.J. Bouma, 2000: Cholera dynamics and El Niño Southern Oscillation. *Science*, **289**, 1766-1767.

Pascual, M., J.A. Ahumada, L.F. Chaves, X. Rodo and M. Bouma, 2006: Malaria resurgence in the East African highlands: temperature trends revisited. *P. Natl. Acad. Sci. USA*, **103**, 5829-5834.

Pattenden, S., B. Nikiforov and B.G. Armstrong, 2003: Mortality and temperature in Sofia and London. *J. Epidemiol. Commun. H.*, **57**, 628-633.

Patz, J.A., 2002: A human disease indicator for the effects of recent global climate change. *P. Natl. Acad. Sci. USA*, **99**, 12506-12508.

Pejoch, M. and B. Kriz, 2006: Ecology, epidemiology and prevention of Hantavirus in Europe. *Climate Change and Adaptation Strategies for Human Health,* B. Menne and K.L. Ebi, Eds., Steinkopff, Darmstadt, 243-265.

People's Health Movement, Medact, Global Equity Gauge Alliance and Zed Books, 2005: *Global Health Watch 2005–2006: An Alternative World Health Report*. London and New York, 368 pp.

Peperzak, L., 2005: Future increase in harmful algal blooms in the North Sea due to climate change. *Water Sci. Technol.*, **51**, 31.

Pereira, L.E., A. Suzuki, T. Lisieux, M. Coimbra, R.P. de Souza and E.L.B. Chamelet, 2001: *Ilheus arbovirus* in wild birds (*Sporophila caerulescens* and *Molothrus bonariensis*). *Rev. Saude Publ.*, **35**, 119-123.

Pitcher, H., K. Ebi and A. Brenkert, 2007: Population health model for integrated assessment models. *Climatic Change*. doi: 10.1007/s10584-007-9286-8.

Pontes, R.J., J. Freeman, J.W. Oliveira-Lima, J.C. Hodgson and A. Spielman, 2000: Vector densities that potentiate dengue outbreaks in a Brazilian city. *Am. J. Trop. Med. Hyg.*, **62**, 378-383.

Pope, C.A., R.T. Burnett, M.J. Thun, E.E. Calle, D. Krewski, K. Ito and G.D. Thurston, 2002: Lung cancer, cardiopulmonary mortality, and long-term exposure to fine particulate air pollution. *J. Am. Med. Assoc.*, **287**, 1132-1141.

Prather, M., M. Gauss, T. Berntsen, I. Isaksen, J. Sundet, I. Bey, G. Brasseur, F. Dentener, R. Derwent, D. Stevenson, L. Grenfell, D. Hauglustaine, L. Horowitz, D. Jacob, L. Mickley, M. Lawrence, R. von Kuhlmann, J.-F. Muller, G. Pitari, H. Rogers, M. Johnson, J. Pyle, K. Law, M. van Weele and O. Wild, 2003: Fresh air in the 21st century? *Geophys. Res. Lett.*, **30**, 1100.

Programa Nacional de Cambios Climaticos Componente Salud, Viceministerio de Medio Ambiente and Recursos Naturales y Desarrollo Forestal, 2000: *Vulnerabilidad y adaptacion de al salud humana ante los efectos del cambio climatico en Bolivia [Vulnerability and Adaptation to Protect Human Health from Effects of Climate Change in Bolivia]*. Programa Nacional de Cambios Climaticos Componente Salud, Viceministerio de Medio Ambiente, Recursos Naturales y Desarrollo Forestal, 111 pp.

Prospero, J.M., E. Blades, G. Mathison and R. Naidu, 2005: Interhemispheric transport of viable fungi and bacteria from Africa to the Caribbean with soil dust. *Aerobiologia*, **21**, 1-19.

Prüss-Üstün, A., H. Zeeb, C. Mathers and M. Repacholi, Eds., 2006: *Solar Ultraviolet Radiation: Global Burden of Disease from Ultraviolet Radiation*. Environmental Burden of Disease Series, Vol. 13. World Health Organization, Geneva, 285 pp.

Qiu, D., T. Tanihata, H. Aoyama, T. Fujita, Y. Inaba and M. Minowa, 2002: Relationship between a high mortality rate and extreme heat during the summer of 1999 in Hokkaido Prefecture, Japan. *J. Epidemiol.*, **12**, 254-257.

Ramsey, J., 1995: Task performance in heat: a review. *Ergonomics*, **38**, 154-165.

Ramsey, J., C. Burford, M. Beshir and R. Hensen, 1983: Effects of workplace thermal conditions on safe working behavior. *J. Safety Res.*, **14**, 105-114.

Randolph, S.E., 2001: The shifting landscape of tick-borne zoonoses: tick-borne encephalitis and Lyme borreliosis in Europe. *Philos. T. Roy. Soc. Lond. B*, **356**, 1045-1056.

Randolph, S.E., 2004: Evidence that climate change has caused 'emergence' of tick-borne diseases in Europe? *Int. J. Med. Microbiol.*, **293**, S5-S15.

Randolph, S.E. and D.J. Rogers, 2000: Fragile transmission cycles of tick-borne encephalitis virus may be disrupted by predicted climate change. *Philos. T. Roy. Soc. Lond. B*, **267**, 1741-1744.

Ranhoff, A.H., 2000: Accidental hypothermia in the elderly. *Int. J. Circumpolar Health*, **59**, 255-259.

Rao, S., J. Ku, S. Berman, D. Zhang and H. Mao, 2003: Summertime characteristics of the atmospheric boundary layer and relationships to ozone levels over the eastern United States. *Pure Appl. Geophys.*, **160**, 21-55.

Rappengluck, B., P. Oyola, I. Olaeta and P. Fabian, 2000: The evolution of photochemical smog in the Metropolitan Area of Santiago de Chile. *J. Appl. Meteorol.*, **39**, 275-290.

Rappole, J.H. and Z. Hubalek, 2003: Migratory birds and West Nile virus. *J. Appl. Microbiol.*, **94**, 47S-58S.

Reacher, M., K. McKenzie, C. Lane, T. Nichols, I. Kedge, A. Iverson, P. Hepple, T. Walter, C. Laxton and J. Simpson, 2004: Health impacts of flooding in Lewes: a comparison of reported gastrointestinal and other illness and mental health in flooded and non flooded households. *Communicable Disease and Public Health*, **7**, 1-8.

Reed, K.D., J.K. Meece, J.S. Henkel and S.K. Shukla, 2003: Birds, migration and emerging zoonoses: West Nile virus, Lyme disease, influenza A and enteropathogens. *Clin. Med. Res.*, **1**, 5-12.

Regidor, E., 2004a: Measures of health inequalities: part 2. *J. Epidemiol. Commun. H.*, **58**, 900-903.

Regidor, E., 2004b: Measures of health inequalities: part 1. *J. Epidemiol. Commun. H.*, **58**, 858-861.

Reiter, P., C.J. Thomas, P. Atkinson, S.E. Randolph, D.J. Rogers, G.D. Shanks, R.W. Snow and A. Spielman, 2004: Global warming and malaria: a call for accuracy. *Lancet Infect. Dis.*, **4**, 323.

Riedel, D., 2004: Human health and well-being. *Climate Change: Impacts and Adaptation A – Canadian Perspective*, D. Lemmen and F. Warren, Eds., Climate Change Impacts and Adaptation Directorate, Natural Resources Canada, Ottawa, 151-171.

Rockstrom, J., 2003: Water for food and nature in drought-prone tropics: vapour shift in rain-fed agriculture. *Philos. T. Roy. Soc. Lond. B*, **358**, 1997-2009.

Rodo, X., M. Pascual, G. Fuchs and A.S.G. Faruque, 2002: ENSO and cholera: a nonstationary link related to climate change? *P. Natl. Acad. Sci. USA*, **99**, 12901-12906.

Rogers, C., P. Wayne, E. Macklin, M. Muilenberg, C. Wagner, P. Epstein and F. Bazzaz, 2006a: Interaction of the onset of spring and elevated atmospheric CO_2 on ragweed (*Ambrosia artemisiifolia* L.) pollen production. *Environ. Health Persp.*, **114**, 865-869. doi:10.1289/ehp.8549.

Rogers, D.J. and S.E. Randolph, 2000: The global spread of malaria in a future, warmer world. *Science*, **289**, 1763-1765.

Rogers, D.J. and S.E. Randolph, 2006: Climate change and vector-borne diseases. *Adv. Parasitol.*, **62**, 345-381.

Rogers, D.J., A.J. Wilson, S.I. Hay and A.J. Graham, 2006b: The global distribution of yellow fever and dengue. *Adv. Parasitol.*, **62**, 181-220.

Rosegrant, M.W. and S.A. Cline, 2003: Global food security: challenges and policies. *Science*, **302**, 1917-1919.

Ryall, D.B., R.G. Derwent, A.J. Manning, A.L. Redington, J. Corden, W. Millington, P.G. Simmonds, S. O'Doherty, N. Carslaw and G.W. Fuller, 2002: The origin of high particulate concentrations over the United Kingdom, March 2000. *Atmos. Environ.*, **36**, 1363-1378.

Rybnicek, O. and S. Jaeger, 2001: Ambrosia (ragweed) in Europe. *ACI International*, **13**, 60-66.

Sachs, J., 2001: *Macroeconomics and Health: Investing in Health for Economic Development*. Report of the Commission on Macro Economics and Health, World Health Organization, Geneva, 208 pp.

Samanek, A.J., E.J. Croager, P. Giesfor, E. Milne, R. Prince, A.J. McMichael, R.M. Lucas and T. Slevin, 2006: Estimates of beneficial and harmful sun exposure times during the year for major Australian population centres. *Med. J. Australia*, **184**, 338-341.

Samarasinghe, J., 2001: Heat stroke in young adults. *Trop. Doct.*, **31**, 217-219.

Sapkota, A., J.M. Symons, J. Kleissl, L. Wang, M.B. Parlange, J. Ondov, P.N. Breysse, G.B. Diette, P.A. Eggleston and T.J. Buckley, 2005: Impact of the 2002 Canadian forest fires on particulate matter air quality in Baltimore city. *Environ. Sci. Technol.*, **39**, 24-32.

Sarkar, U., S.F. Nascimento, R. Barbosa, R. Martins, H. Nuevo, I. Kalafanos, I. Grunstein, B. Flannery, J. Dias, L.W. Riley, M.G. Reis and A.I. Ko, 2002: Population-based case-control investigation of risk factors for leptospirosis during an urban epidemic. *Am. J. Trop. Med. Hyg.*, **66**, 605-610.

Sastry, N., 2002: Forest fires, air pollution, and mortality in Southeast Asia. *Demography*, **39**, 1-23.

Scheraga, J.S., K.L. Ebi, J. Furlow and A.R. Moreno, 2003: From science to policy: developing responses to climate change. *Climate Change and Human Health: Risks and Responses*, A. McMichael, D. Campbell-Lendrum, C. Corvalan, K.L. Ebi, A.K. Githeko, J.S. Scheraga and A. Woodward, Eds., World Health Organization, Geneva, 237-266.

Schichtel, B. and R. Husar, 2001: Eastern North American transport climatology during high- and low-ozone days. *Atmos. Environ.*, **35**, 1029-1038.

Schultz, J.M., J. Russell and Z. Espine, 2005: Epidemiology of tropical cyclones: the dynamics of disaster, disease and development. *Epidemiol. Rev.*, **27**, 21-35.

Schwartz, B.S., J.B. Harris, A.I. Khan, R.C. Larocque, D.A. Sack, M.A. Malek, A.S. Faruque, F. Qadri, S.B. Calderwood, S.P. Luby and E.T. Ryan, 2006: Diarrheal epidemics in Dhaka, Bangladesh, during three consecutive floods: 1988, 1998, and 2004. *Am. J. Trop. Med. Hyg.*, **74**, 1067-1073.

Schwartz, J. and R. Levin, 1999: Drinking water turbidity and health. *Epidemiology*, **10**, 86-89.

Schwartz, J., R. Levin and R. Goldstein, 2000: Drinking water turbidity and gastrointestinal illness in the elderly of Philadelphia. *J. Epidemiol. Commun. H.*, **54**, 45-51.

Scott, G.M. and R.D. Diab, 2000: Forecasting air pollution potential: a synoptic climatological approach. *J. Air Waste Manage.*, **50**, 1831-1842.

Semenov, S.M., E.S. Gelver and V.V. Yasyukevich, 2002: Temperature conditions for development of two species of malaria pathogens in Russia in 20th century. *Dokl. Akad. Nauk*, **387**, 131-136.

Sénat, 2004: La France et les Français face a la canicule: les leçons d'une crise [France and the French facing the heat wave: lessons from a crisis]. Rapport d'Information No. 195 (2003–2004) de Mme Letard, M.M. Flandre, S. Lepeltier, fait au nom de la mission commune d'information du Senat, depose le 3 Fevrier 2004. http://www.senat.fr/rap/r03-195/r03-1951.pdf.

Senhorst, H.A. and J.J. Zwolsman, 2005: Climate change and effects on water quality: a first impression. *Water Sci. Technol.*, **51**, 53-59.

Shah, I., G.C. Deshpande and P.N. Tardeja, 2004: Outbreak of dengue in Mumbai and predictive markers for dengue shock syndrome. *J. Trop. Pediatrics*, **50**, 301-305.

Shanks, G.D., S.I. Hay, D.I. Stern, K. Biomndo and R.W. Snow, 2002: Meteorologic influences on *Plasmodium falciparum* malaria in the highland tea estates of Kericho, Western Kenya. *Emerg. Infect. Dis.*, **8**, 1404-1408.

Shanks, N. and G. Papworth, 2001: Environmental factors and heatstroke. *Occup. Med.*, **51**, 45-49.

Shinn, E.A., D.W. Griffin and D.B. Seba, 2003: Atmospheric transport of mold spores in clouds of desert dust. *Arch. Environ. Health*, **58**, 498-504.

Sillett, T.S., R.T. Holmes and T.W. Sherry, 2000: Impacts of a global climate cycle on population dynamics of a migratory songbird. *Science*, **288**, 2040-2042.

Singer, B.D., L.H. Ziska, D.A. Frenz, D.E. Gebhard and J.G. Straka, 2005: Increasing Amb a 1 content in common ragweed (*Ambrosia artemisiifolia*) pollen as a function of rising atmospheric CO_2 concentration. *Funct. Plant Biol.*, **32**, 667-670.

Singh, N. and V.P. Sharma, 2002: Patterns of rainfall and malaria in Madhya Pradesh, central India. *Ann. Trop. Med. Parasit.*, **965**, 349-359.

Singh, R., S. Hales, N. de Wet, R. Raj, M. Hearnden and P. Weinstein, 2001: The influence of climate variation and change on diarrhoeal disease in the pacific islands. *Environ. Health Persp.*, **109**, 155-159.

Skarphedinsson, S., P.M. Jensen and K. Kristiansen, 2005: Survey of tick borne infections in Denmark. *Emerg. Infect. Dis.*, **11**, 1055-1061.

Slanina, S. and Y. Zhang, 2004: Aerosols: connection between regional climate change and air quality. IUPAC Technical Report. *Pure Appl. Chem.*, **76**, 1241-1253.

Small, C. and R.J. Nicholls, 2003: A global analysis of human settlement in coastal zones. *J. Coastal Res.*, **19**, 584-599.

Smith, K.R., J. Zhang, R. Uma, V.V.N. Kishore and M.A.K. Khalil, 2000: Greenhouse implications of household fuels: an analysis for India. *Annu. Rev. Energ. Env.*, **25**, 741-763.

Smith, K.R., S. Mehta and M. Maeusezahl-Feuz, 2004: Indoor air pollution from household use of solid fuels. *Comparative Quantification of Health Risks: Global and Regional Burden of Disease Attributable to Selected Major Risk Factors*, M. Ezzati, A.D. Lopez, A. Rodgers and C.J.L. Murray, Eds., World Health Organisation, Geneva, 1435-1494.

Smith, K.R., J. Rogers and S.C. Cowlin, 2005: *Household Fuels and Ill-Health in Developing Countries: What Improvements can be Brought by LP Gas?* World LP Gas Association and Intermediate Technology Development Group, Paris, 59 pp.

Socioambiental, 2006: Seca na Amazônia: alguma coisa está fora da ordem. http://www.socioambiental.org/nsa/detalhe?id=2123.

Sorogin, V.P. and Co-authors, 1993: *Problemy Ohrany Zdoroviya i Socialnye Aspecty Osvoeniya Gazovyh i Neftyanyh Mestorozhdenij v Arcticheskih Regionah [Problems of Public Health and Social Aspects of Exploration of Oil and Natural Gas Deposits in Arctic Regions]*. Nadym.

Sousounis, J., C. Scott and M. Wilson, 2002: Possible climate change impacts on ozone in the Great Lakes region: some implications for respiratory illness. *J. Great Lakes Res.*, **28**, 626-642.

Speelmon, E.C., W. Checkley, R.H. Gilman, J. Patz, M. Calderon and S. Manga, 2000: Cholera incidence and El Niño-related higher ambient temperature. *J. Am. Med. Assoc.*, **283**, 3072-3074.

Stapp, P., M. Antolin and M. Ball, 2004: Patterns of extinction in prairie dog metapopulations: plague outbreaks follow El Niño events. *Front. Ecol. Environ.*, **2**, 235-240.

Steenvoorden, J. and T. Endreny, 2004: *Wastewater Re-use and Groundwater quality*. IAHS Publication 285, Wallingford, Oxfordshire, 112 pp.

Stenseth, N., 2006: Plague dynamics are driven by climate variations. *P. Natl. Acad. Sci. USA*, **1003**, 13110-13115.

Stevenson, D.S., C.E. Johnson, W.J. Collins, R.G. Derwent and J.M. Edwards, 2000: Future estimates of tropospheric ozone radiative forcing and methane turnover: the impact of climate change. *Geophys. Res. Lett.*, **27**, 2073-2076.

Stohl, A., L. Haimberger, M. Scheele and H. Wernli, 2001: An intercomparison of results from three trajectory models. *Meteorol. Appl.*, **8**, 127-135.

Suh, H.H., T. Bahadori, J. Vallarino and J.D. Spengler, 2000: Criteria air pollutants and toxic air pollutants. *Environ. Health Persp.*, **108**, 625-633.

Sultan, B., K. Labadi, J.F. Guegan and S. Janicot, 2005: Climate drives the meningitis epidemics onset in west Africa. *PLoS Med.*, **2**, e6. doi: 10.1371/journal.pmed.0020006

Sumilo, D., A. Bormane, L. Asokliene, I. Lucenko, V. Vasilenko and S. Randolph, 2006: Tick-borne encephalitis in the Baltic States: identifying risk factors in space and time. *Int. J. Med. Microbiol.*, **296**, 76-79.

Sur, D., P. Dutta, G.B. Nair and S.K. Bhattacharya, 2000: Severe cholera outbreak following floods in a northern district of West Bengal. *Indian J. Med. Res.*, **112**, 178-182.

Sutherst, R.W., 2004: Global change and human vulnerability to vector-borne diseases. *Clin. Microbiol. Rev.*, **17**, 136.

Syri, S., N. Karvosenoja, A. Lehtila, T. Laurila, V. Lindfors and J.P. Tuovinen, 2002: Modeling the impacts of the Finnish Climate Strategy on air pollution. *Atmos. Environ.*, **36**, 3059-3069.

Szreter, S., 2004: Industrialization and health. *Brit. Med. Bull.*, **69**, 75-86.

Taha, H., 2001: *Potential Impacts of Climate Change on Tropospheric Ozone in California: A Preliminary Episodic Modeling Assessment of the Los Angeles Basin and the Sacramento Valley*. Lawrence Berkeley National Laboratories,

Berkeley, California, 39 pp.

Takemura, T., T. Nakajima, T. Nozawa and K. Aoki, 2001: Simulation of future aerosol distribution, radioactive forcing, and long-range transport in East Asia. *J. Meteorol. Soc. Jpn.*, **79**, 1139-1155.

Tam, C., L. Rodrigues, S. O'Brien and S. Hajat, 2006: Temperature dependence of reported *Campylobacter* infection in England, 1989–1999. *Epidemiol. Infect.*, **134**, 119-125.

Tanner, P. and P. Law, 2002: Effects of synoptic weather systems upon the air quality in an Asian megacity. *Water Air Soil Pollut*, **136**, 105-124.

Tanser, F.C., B. Sharp and D. Le Sueur, 2003: Potential effect of climate change on malaria transmission in Africa. *Lancet Infect. Dis.*, **362**, 1792-1798.

Tapsell, S., E. Penning-Rowsell, S. Tunstall and T. Wilson, 2002: Vulnerability to flooding: health and social dimensions. *Philos. T. Roy. Soc. Lond. A*, **360**, 1511-1525.

Taramarcaz, P., B. Lambelet, B. Clot, C. Keimer and C. Hauser, 2005: Ragweed (Ambrosia) progression and its health risks: will Switzerland resist this invasion? *Swiss Med. Wkly.*, **135**, 538-548.

Teklehaimanot, H., J. Schwartz, A. Teklehaimanot and A. Lipsitch, 2004: Weather-based prediction of *Plasmodium falciparum* malaria in epidemic-prone regions of Ethiopia. *Malaria J.*, **3**.

Thomas, C.J., G. Davies and C.E. Dunn, 2004: Mixed picture for changes in stable malaria distribution with future climate in Africa. *Trends Parasitol.*, **20**, 216-220.

Thommen Dombois, O. and C. Braun-Fahrlaender, 2004: *Gesundheitliche Auswirkungen der Klimaaenderung mit Relevanz fuer die Schweiz [Health Impacts of Climate Change with Relevance for Switzerland]*. Insititut fuer Sozial- und Preventivmedizin der Universitaet Basel, Bundesamt fuer Gesundheit, Bundesamt fuer Umwelt, Wald und Landschaft, Basel, 85 pp.

Thomson, M.C., S.J. Mason, T. Phindela and S.J. Connor, 2005: Use of rainfall and sea surface temperature monitoring for malaria early warning in Botswana. *Am. J. Trop. Med. Hyg.*, **73**, 214-221.

Thomson, M.C., F.J. Doblas-Reyes, S.J. Mason, R. Hagedorn, S.J. Connor, T. Phindela, A.P. Morse and T.N. Palmer, 2006: Malaria early warnings based on seasonal climate forecasts from multi-model ensembles. *Nature*, **439**, 576-579.

Tol, R.S., 1995: The damage costs of climate change toward more comprehensive calculations. *Environ. Resour. Econ.*, **5**, 353-374.

Tol, R.S., 1996: The damage costs of climate change towards a dynamic representation. *Ecol. Econ.*, **19**, 67-90.

Tol, R.S., 2002a: Estimates of the damage costs of climate change. Part II. Dynamic estimates. *Environ. Resour. Econ.*, **21**, 135-160.

Tol, R.S., 2002b: Estimates of the damage costs of climate change. Part I. Benchmark estimates. *Environ. Resour. Econ.*, **21**, 47-73.

Tu, F., D. Thornton, A. Brandy and G. Carmichael, 2004: Long-range transport of sulphur dioxide in the central Pacific. *J. Geophys. Res. D*, **109**, D15S08.

Turpie, J., H. Winkler, R. Spalding-Fecher and G. Midgley, 2002: *Economic Impacts of Climate Change in South Africa: A Preliminary Analysis of Unmitigated Costs*. Southern Waters Ecological Research and Consulting, Energy and Development Research Centre, University of Cape Town, Cape Town, 64 pp.

Tuyet, D.T., V.D. Thiem, L. Von Seidlein, A. Chowdhury, E. Park, D.G. Canh, B.T. Chien, T. Van Tung, A. Naficy, M.R. Rao, M. Ali, H. Lee, T.H. Sy, M. Nichibuchi, J. Clemens and D.D. Trach, 2002: Clinical, epidemiological, and socioeconomic analysis of an outbreak of *Vibrio parahaemolyticus* in Khanh Hoa Province, Vietnam. *J. Infect. Dis.*, **186**, 1615-1620.

UN, 2000: *Convention on the Protection and Use of Transboundary Watercourses and International Lakes*. 32 pp. http://www.unece.org/env/water/pdf/ watercon.pdf.

UN, 2006a: *The Millennium Development Goals Report 2006*. United Nations Department of Economic and Social Affairs, DESA, New York, 32 pp.

UN, 2006b: Executive summary, fact sheets, data tables. *World Urbanization Prospects: The 2005 Revision*. United Nations, New York, 210 pp.

UN Millennium Project, 2005: *Investing in Development: A Practical Plan to Achieve the Millennium Development Goals*. Earthscan, London, 329 pp.

UNDP, 2005: *World Population Prospects: The 2004 Revision*. III. UNDP, New York, 105 pp.

UNEP, 2002: *Synthesis GEO-3: Global Environmental Outlook 3*. United Nations Environment Programme. Earthscan, London, 20 pp.

UNEP and WCMC, 2002: *Human Development Report 2002: Deepening Democracy in a Fragmented World*. Oxford University Press, New York and Oxford, 276 pp.

United Nations World Water Assessment Programme, 2003: *Water for People: Water for Life*. United Nations World Water Development Report. United Nations Educational Scientific and Cultural Organization (UNESCO), Berghan Books, Barcelona, 529 pp.

Unsworth, J., R. Wauchope, A. Klein, E. Dorn, B. Zeeh, S. Yeh, M. Akerblom, K. Racke and B. Rubin, 2003: Significance of the long range transport of pesticides in the atmosphere. *Pest. Manag. Sci.*, **58**, 314.

van der Pligt, J., E.C.M. van Schie and R. Hoevenagel, 1998: Understanding and valuing environmental issues: the effects of availability and anchoring on judgment. *Z. Exp. Psychol.*, **45**, 286-302.

van Lieshout, M., R.S. Kovats, M.T.J. Livermore and P. Martens, 2004: Climate change and malaria: analysis of the SRES climate and socio-economic scenarios. *Global Environ. Chang.*, **14**, 87-99.

Vanasco, N.B., S. Fusco, J.C. Zanuttini, S. Manattini, M.L. Dalla Fontana, J. Prez, D. Cerrano and M.D. Sequeira, 2002: Human leptospirosis outbreak after an inundation at Reconquista (Santa Fe), 1998 (in Spanish). *Rev. Argent. Microbiol.*, **34**, 124.

Vandentorren, S. and P. Empereur-Bissonnet, 2005: Health impact of the 2003 heatwave in France. *Extreme Weather Events and Public Health Responses*, W. Kirch, B. Menne and R. Bertollini, Eds., Springer, 81-88.

Vandentorren, S., F. Suzan, S. Medina, M. Pascal, A. Maulpoix, J.-C. Cohen and M. Ledrans, 2004: Mortality in 13 French cities during the August 2003 heatwave. *Am. J. Public Health*, **94**, 1518-1520.

Vapalahti, O., J. Mustonen, A. Lundkvist, H. Henttonen, A. Plyusnin and A. Vaheri, 2003: Hantavirus infections in Europe. *Lancet Infect. Dis.*, **3**, 653-661.

Vasilev, V., 2003: Variability of *Shigella flexneri* serotypes during a period in Israel, 2000–2001. *Epidemiol. Infect.*, **132**, 51-56.

Viscusi, W.K. and J.E. Aldy, 2003: The value of a statistical life: a critical review of market estimates throughout the world. *J. Risk Uncertainty*, **27**, 5-76.

Visser, M.E., C. Both and M.M. Lambrechts, 2004: Global climate change leads to mistimed avian reproduction. *Adv. Ecol. Res.*, **35**, 89-110.

Vollaard, A.M., S. Ali, H.A.G.H. van Asten, S. Widjaja, L.G. Visser, C. Surjadi and J.T. van Dissel, 2004: Risk factors for typhoid and paratyphoid fever in Jakarta, Indonesia. *J. Am. Med. Assoc.*, **291**, 2607-2615.

Voltolini, S., P. Minale, C. Troise, D. Bignardi, P. Modena, D. Arobba and A. Negrini, 2000: Trend of herbaceous pollen diffusion and allergic sensitisation in Genoa, Italy. *Aerobiologia*, **16**, 245-249.

Wade, T.J., S.K. Sandhu, D. Levy, S. Lee, M.W. LeChevallier, L. Katz and J.M. Colford, 2004: Did a severe flood in the Midwest cause an increase in the incidence of gastrointestinal symptoms? *Am. J. Epidemiol.*, **159**, 398-405.

Wan, S.Q., T. Yuan, S. Bowdish, L. Wallace, S.D. Russell and Y.Q. Luo, 2002: Response of an allergenic species *Ambrosia psilostachya* (Asteraceae), to experimental warming and clipping: implications for public health. *Am. J. Bot.* **89**, 1843-1846.

Wayne, P., S. Foster, J. Connolly, F. Bazzaz and P. Epstein, 2002: Production of allergenic pollen by ragweed (*Ambrosia artemisiifolia* L.) is increased in CO_2-enriched atmospheres. *Ann. Allergy. Asthma Im.*, **88**, 279-282.

Weber, R.W., 2002: Mother Nature strikes back: global warming, homeostasis, and implications for allergy. *Ann. Allergy. Asthma Im.*, **88**, 251-252.

Webster, M.D., M. Babiker, M. Mayer, J.M. Reilly, J. Harnisch, R. Hyman, M.C. Sarofim and C. Wang, 2002a: Uncertainty in emissions projections for climate models. *Atmos. Environ.*, **36**, 3659-3670.

Webster, M.S., P.P. Marra, S.M. Haig, S. Bensch and R.T. Holmes, 2002b: Links between worlds: unravelling migratory connectivity. *Trends Ecol. Evol.*, **17**, 76-83.

West, J.J., P. Osnaya, I. Laguna, J. Martinez and A. Fernandez, 2004: Co-control of urban air pollutants and greenhouse gases in Mexico City. *Environ. Sci. Technol.*, **38**, 3474-3481.

White, R., 2003: Commentary: What can we make of an association between human immunodeficiency virus prevalence and population mobility? *Int. J. Epidemiol.*, **32**, 753-754.

WHO, 2001: *World Health Report 2001: Mental Health – New Understanding, New Hope*. World Health Organization, Geneva, 178 pp.

WHO, 2002a: *Injury Chart Book: Graphical Overview of the Burden of Injuries*. World Health Organization, Geneva, 81 pp.

WHO, 2002b: *World Health Report 2002: Reducing Risks, Promoting Healthy Life*. World Health Organization, Geneva, 268 pp.

WHO, 2003a: *Global Defence against the Infectious Disease Threat: Progress Report, Communicable Diseases 2002*. World Health Organization, Geneva, 233 pp.

WHO, 2003b: *The World Health Report 2003: Shaping the Future*. World Health

Organization, Geneva, 210 pp.
WHO, 2004a: *Synthesis Workshop on Climate Variability, Climate Change and Health in Small Island States*. WHO, WMO, UNEP, WHO/SDE/OEH/04.02, 95 pp.
WHO, 2004b: Malaria epidemics: forecasting, prevention, early warning and control – from policy to practice. Report of an informal consultation, Leysin, Switzerland, 8–10 December 2003. World Health Organization, Geneva, 52 pp.
WHO, 2005: Ecosystems and human well-being: health synthesis. A report of the Millennium Ecosystem Assessment, World Health Organization, Geneva, 54 pp.
WHO, 2006: *Preventing Disease through Healthy Environments: Towards an Estimate of the Environmental Burden of Disease*. World Health Organization, Geneva, 106 pp.
WHO Regional Office for Europe, 2006: 1st meeting of the project "Improving Public Health Responses to Extreme Weather/Heat-waves". EuroHEAT Report on a WHO Meeting in Rome, Italy, 20–22 June 2005. WHO Regional Office for Europe, Copenhagen, 52 pp.
WHO Regional Office for South-East Asia, 2006: Human health impacts from climate variability and climate change in the Hindu Kush-Himalaya region. Report of an Inter-Regional Workshop, Mukteshwar, India, October 2005. WHO, 49 pp.
Wilby, R., 2003: Past and projected trends in London's urban heat island. *Weather*, **58**, 251-260.
Wilby, R., M. Hedger and H.G. Orr, 2005: Climate change impacts and adaptation: a science agenda for the Environment Agency of England and Wales. *Weather*, **60**, 206-211.
Wilson, J., B. Philips and D. Neal, 1998: Domestic violence after disaster. *The Gendered Terrain of Disaster: Through Women's Eyes,* E. Enearson and B.H. Morrow, Eds., International Hurricane Center, Florida International University, Miami, Florida, 115-122.
Wittmann, R. and G. Flick, 1995: Microbial contamination of shellfish: prevalence, risk to human health and control strategies. *Annu. Rev. Publ. Health*, **16**, 123-140.
Woodruff, R.E., 2005: Epidemic early warning systems: Ross River virus disease in Australia. *Integration of Public Health with Adaptation to Climate Change: Lessons Learned and New Directions*, K. Ebi, J. Smith and I. Burton, Eds., Taylor and Francis, Leiden, 91-113.
Woodruff, R.E., C.S. Guest, M.G. Garner, N. Becker, J. Lindesay, T. Carvan and K. Ebi, 2002: Predicting Ross River virus epidemics from regional weather data. *Epidemiology*, **13**, 384-393.
Woodruff, R.E., S. Hales, C. Butler and A. McMichael, 2005: Climate change and health impacts in Australia: effects of dramatic CO_2 emission reductions. Report for the Australia Conservation Foundation and the Australian Medical Association. Australian National University, Canberra, 45 pp.
Woodward, A., S. Hales and N. de Wet, 2001: *Climate Change: Potential Effects on Human Health in New Zealand*. Ministry for the Environment, Wellington, New Zealand, 27 pp.
World Bank, 2004: *World Development Report 2004: Making Services Work for Poor People*. World Bank, New York, 32 pp.
World Bank, 2005: Drought in the Amazon: scientific and social aspects. Report of a World Bank Seminar, December 12, 2005. Brasília, Brazil, 14 pp.
World Bank, African Development Bank, Asian Development Bank, DFID, Directorate-Generale for Development European Commission, Federal Ministry for Economic Cooperation and Development Germany, Ministry of Foreign Affairs Netherlands, UNDP and UNEP, 2004: *Poverty and Climate Change: Reducing the Vulnerability of the Poor through Adaptation*. World Bank, New York, 43 pp.
Wu, H. and L. Chan, 2001: Surface ozone trends in Hong Kong in 1985–1995. *Environ. Int*., **26**, 213-222.
WWF, 2005: An overview of glaciers, glacier retreat, and subsequent impacts in Nepal, India and China. World Wildlife Fund Nepal Program, 79 pp. http://assets.panda.org/downloads/himalayaglaciersreport2005.pdf.
Wyndham, C., 1965: A survey of causal factors in heat stroke and of their prevention in gold mining industry. *J. S. Afr. I. Min. Metall*., **66**, 125-155.
Xie, S.D., T. Yu, Y.H. Zhang, L.M. Zeng, L. Qi and X.Y. Tang, 2005: Characteristics of PM_{10}, SO_2, NO_x and O_3 in ambient air during the dust storm period in Beijing. *Sci. Total Environ*., **345**, 153-164.
Yang, C.Y., Y.S. Chen, H.F. Chiu and W.B. Goggins, 2005a: Effects of Asian dust storm events on daily stroke admissions in Taipei, Taiwan. *Environ. Res*., **99**, 79-84.
Yang, G.J., P. Vounatsou, X.N. Zhou, M. Tanner and J. Utzinger, 2005b: A potential impact of climate change and water resource development on the transmission of *Schistosoma japonicum* in China. *Parassitologia*, **47**, 127-134.
Yarnal, B., A.C. Comrie, B. Frakes and D.P. Brown, 2001: Developments and prospects in synoptic climatology. *Int. J. Climatol*., **21**, 1923-1950.
Yohe, G. and K. Ebi, 2005: Approaching adaptation: parallels and contrasts between the climate and health communities. *A Public Health Perspective on Adaptation to Climate Change,* K. Ebi and I. Burton, Eds., Taylor and Francis, Leiden, 18-43.
Young, S., L. Balluz and J. Malilay, 2004: Natural and technologic hazardous material releases during and after natural disasters: a review. *Sci. Total Environ*., **322**, 3-20.
Zebisch, M., T. Grothmann, D. Schroeter, C. Hasse, U. Fritsch and W. Cramer, 2005: *Climate Change in Germany. Vulnerability and Adaptation of Climate Sensitive Sectors*. Federal Environmental Agency (Umweltbundesamt), Dessau, 205 pp.
Zhou, G., N. Minakawa, A.K. Githeko and G. Yan, 2004: Association between climate variability and malaria epidemics in the East African highlands. *P. Natl. Acad. Sci. USA*, **101**, 2375.
Zhou, G., N. Minakawa, A.K. Githeko and G. Yan, 2005: Climate variability and malaria epidemics in the highlands of East Africa. *Trends Parasitol*., **21**, 54-56.
Ziska, L.H., S.D. Emche, E.L. Johnson, K. George, D.R. Reed and R.C. Sicher, 2005: Alterations in the production and concentration of selected alkaloids as a function of rising atmospheric carbon dioxide and air temperature: implications for ethno-pharmacology. *Glob. Change Biol*., **11**, 1798-1807.
Zwander, H., 2002: Der Pollenflug im Klagenfurter Becken (Kaernten) 1980–2000. *Carinthia II*, **192**, 197-214.

9

Africa

Coordinating Lead Authors:

Michel Boko (Benin), Isabelle Niang (Senegal), Anthony Nyong (Nigeria), Coleen Vogel (South Africa)

Lead Authors:

Andrew Githeko (Kenya), Mahmoud Medany (Egypt), Balgis Osman-Elasha (Sudan), Ramadjita Tabo (Chad), Pius Yanda (Tanzania)

Contributing Authors:

Francis Adesina (Nigeria), Micheline Agoli-Agbo (Benin), Samar Attaher (Egypt), Lahouari Bounoua (USA), Nick Brooks (UK), Ghislain Dubois (France), Mukiri wa Githendu (Kenya), Karim Hilmi (Morocco), Alison Misselhorn (South Africa), John Morton (UK), Imoh Obioh (Nigeria), Anthony Ogbonna (UK), Hubert N'Djafa Ouaga (Chad), Katharine Vincent (UK), Richard Washington (South Africa), Gina Ziervogel (South Africa)

Review Editors:

Frederick Semmazzi (USA), Mohamed Senouci (Algeria)

This chapter should be cited as:

Boko, M., I. Niang, A. Nyong, C. Vogel, A. Githeko, M. Medany, B. Osman-Elasha, R. Tabo and P. Yanda, 2007: Africa. *Climate Change 2007: Impacts, Adaptation and Vulnerability. Contribution of Working Group II to the Fourth Assessment Report of the Intergovernmental Panel on Climate Change*, M.L. Parry, O.F. Canziani, J.P. Palutikof, P.J. van der Linden and C.E. Hanson, Eds., Cambridge University Press, Cambridge UK, 433-467.

Table of Contents

Executive summary

Africa is one of the most vulnerable continents to climate change and climate variability, a situation aggravated by the interaction of 'multiple stresses', occurring at various levels, and low adaptive capacity (high confidence).
Africa's major economic sectors are vulnerable to current climate sensitivity, with huge economic impacts, and this vulnerability is exacerbated by existing developmental challenges such as endemic poverty, complex governance and institutional dimensions; limited access to capital, including markets, infrastructure and technology; ecosystem degradation; and complex disasters and conflicts. These in turn have contributed to Africa's weak adaptive capacity, increasing the continent's vulnerability to projected climate change. [9.2.2, 9.5, 9.6.1]

African farmers have developed several adaptation options to cope with current climate variability, but such adaptations may not be sufficient for future changes of climate (high confidence).
Human or societal adaptive capacity, identified as being low for Africa in the Third Assessment Report, is now better understood and this understanding is supported by several case studies of both current and future adaptation options. However, such advances in the science of adaptation to climate change and variability, including both contextual and outcome vulnerabilities to climate variability and climate change, show that these adaptations may be insufficient to cope with future changes of climate. [9.2, 9.4, 9.5, 9.6.2, Table 9.2]

Agricultural production and food security (including access to food) in many African countries and regions are likely to be severely compromised by climate change and climate variability (high confidence).
A number of countries in Africa already face semi-arid conditions that make agriculture challenging, and climate change will be likely to reduce the length of growing season as well as force large regions of marginal agriculture out of production. Projected reductions in yield in some countries could be as much as 50% by 2020, and crop net revenues could fall by as much as 90% by 2100, with small-scale farmers being the most affected. This would adversely affect food security in the continent. [9.2.1, 9.4.4, 9.6.1]

Climate change will aggravate the water stress currently faced by some countries, while some countries that currently do not experience water stress will become at risk of water stress (very high confidence).
Climate change and variability are likely to impose additional pressures on water availability, water accessibility and water demand in Africa. Even without climate change, several countries in Africa, particularly in northern Africa, will exceed the limits of their economically usable land-based water resources before 2025. About 25% of Africa's population (about 200 million people) currently experience high water stress. The population at risk of increased water stress in Africa is projected to be between 75-250 million and 350-600 million people by the 2020s and 2050s, respectively. [9.2.1, 9.2.2, 9.4.1]

Changes in a variety of ecosystems are already being detected, particularly in southern African ecosystems, at a faster rate than anticipated (very high confidence).
Climate change, interacting with human drivers such as deforestation and forest fires, are a threat to Africa's forest ecosystems. Changes in grasslands and marine ecosystems are also noticeable. It is estimated that, by the 2080s, the proportion of arid and semi-arid lands in Africa is likely to increase by 5-8%. Climate change impacts on Africa's ecosystems will probably have a negative effect on tourism as, according to one study, between 25 and 40% of mammal species in national parks in sub-Saharan Africa will become endangered. [9.2.2, 9.4.4, 9.4.5]

Climate variability and change could result in low-lying lands being inundated, with resultant impacts on coastal settlements (high confidence).
Climate variability and change, coupled with human-induced changes, may also affect ecosystems e.g., mangroves and coral reefs, with additional consequences for fisheries and tourism. The projection that sea-level rise could increase flooding, particularly on the coasts of eastern Africa, will have implications for health. Sea-level rise will probably increase the high socio-economic and physical vulnerability of coastal cities. The cost of adaptation to sea-level rise could amount to at least 5-10% of gross domestic product. [9.4.3, 9.4.6, 9.5.2]

Human health, already compromised by a range of factors, could be further negatively impacted by climate change and climate variability, e.g., malaria in southern Africa and the East African highlands (high confidence).
It is likely that climate change will alter the ecology of some disease vectors in Africa, and consequently the spatial and temporal transmission of such diseases. Most assessments of health have concentrated on malaria and there are still debates on the attribution of malaria resurgence in some African areas. The need exists to examine the vulnerabilities and impacts of future climate change on other infectious diseases such as dengue fever, meningitis and cholera, among others. [9.2.1.2, 9.4.3 9.5.1]

9.1 Introduction

9.1.1 Summary of knowledge assessed in the Third Assessment Report

The Third Assessment Report (TAR) of the IPCC identified a range of impacts associated with climate change and variability, including decreases in grain yields; changes in runoff and water availability in the Mediterranean and southern countries of Africa; increased stresses resulting from increased droughts and floods; and significant plant and animal species extinctions and associated livelihood impacts. Such factors were shown, moreover, to be aggravated by low adaptive capacity (IPCC, 2001). Many of these conclusions, as shown below, remain valid for this Fourth Assessment Report[1].

9.1.2 New advances and approaches used in the Fourth Assessment Report

Recent scientific efforts, including a focus on both an *impacts-led* approach as well as a *vulnerability-led* approach (see Adger et al., 2004, for a summary), have enabled a more detailed assessment of the interacting roles of climate and a range of other factors driving change in Africa. This approach has been used to frame much of what follows in this chapter and has enabled a greater sensitivity to, and a deeper understanding of, the role of 'multiple stresses' in heightening vulnerability to climate stress. Several of these stresses (outlined in Sections 9.2.1, 9.2.2 and 9.4) are likely to be compounded by climate change and climate variability in the future. Recent additional case studies on adaptation have also been undertaken, providing new insights (see Section 9.5, Table 9.2).

9.2 Current sensitivity/vulnerability

9.2.1 Current sensitivity to climate and weather

The climate of the continent is controlled by complex maritime and terrestrial interactions that produce a variety of climates across a range of regions, e.g., from the humid tropics to the hyper-arid Sahara (see Christensen et al., 2007). Climate exerts a significant control on the day-to-day economic development of Africa, particularly for the agricultural and water-resources sectors, at regional, local and household scales. Since the TAR, observed temperatures have indicated a greater warming trend since the 1960s. Although these trends seem to be consistent over the continent, the changes are not always uniform. For instance, decadal warming rates of 0.29°C in the African tropical forests (Malhi and Wright, 2004) and 0.1 to 0.3°C in South Africa (Kruger and Shongwe, 2004) have been observed. In South Africa and Ethiopia, minimum temperatures have increased slightly faster than maximum or mean temperatures (Conway et al., 2004; Kruger and Shongwe, 2004). Between 1961 and 2000, there was an increase in the number of warm spells over southern and western Africa, and a decrease in the number of extremely cold days (New et al., 2006). In eastern Africa, decreasing trends in temperature from weather stations located close to the coast or to major inland lakes have been observed (King'uyu et al., 2000).

For precipitation, the situation is more complicated. Rainfall exhibits notable spatial and temporal variability (e.g., Hulme et al., 2005). Interannual rainfall variability is large over most of Africa and, for some regions, multi-decadal variability is also substantial. In West Africa (4°-20°N; 20°W-40°E), a decline in annual rainfall has been observed since the end of the 1960s, with a decrease of 20 to 40% noted between the periods 1931-1960 and 1968-1990 (Nicholson et al., 2000; Chappell and Agnew, 2004; Dai et al., 2004). In the tropical rain-forest zone, declines in mean annual precipitation of around 4% in West Africa, 3% in North Congo and 2% in South Congo for the period 1960 to 1998 have been noted (e.g., Malhi and Wright, 2004). A 10% increase in annual rainfall along the Guinean coast during the last 30 years has, however, also been observed (Nicholson et al., 2000). In other regions, such as southern Africa, no long-term trend has been noted. Increased interannual variability has, however, been observed in the post-1970 period, with higher rainfall anomalies and more intense and widespread droughts reported (e.g., Richard et al., 2001; Fauchereau et al., 2003). In different parts of southern Africa (e.g., Angola, Namibia, Mozambique, Malawi, Zambia), a significant increase in heavy rainfall events has also been observed (Usman and Reason, 2004), including evidence for changes in seasonality and weather extremes (Tadross et al., 2005a; New et al., 2006). During recent decades, eastern Africa has been experiencing an intensifying dipole rainfall pattern on the decadal time-scale. The dipole is characterised by increasing rainfall over the northern sector and declining amounts over the southern sector (Schreck and Semazzi, 2004).

Advances in our understanding of the complex mechanisms responsible for rainfall variability have been made (see Reason et al., 2005; Warren et al., 2006; Washington and Preston, 2006; Christensen et al., 2007). Understanding how possible climate-regime changes (e.g., in El Niño-Southern Oscillation (ENSO) events) may influence future climate variability is critical in Africa and requires further research. The drying of the Sahel region since the 1970s has, for example, been linked to a positive trend in equatorial Indian Ocean sea-surface temperature (SST), while ENSO is a significant influence on rainfall at interannual scales (Giannini et al., 2003; Christensen et al., 2007). In the same region, the intensity and localisation of the African Easterly Jet (AEJ) and the Tropical Easterly Jet (TEJ) also influence rainfall variability (Nicholson and Grist, 2003), as well as SSTs in the Gulf of Guinea (Vizy and Cook, 2001), and a relationship has also been identified between the warm Mediterranean Sea and abundant rainfall (Rowell, 2003). The influence of ENSO decadal variations has also been recognised in south-west Africa, influenced in part by the North Atlantic Oscillation (NAO) (Nicholson and Selato, 2000). Changes in the ways these mechanisms influence regional weather patterns have been identified in southern Africa, where severe droughts have been linked to regional atmospheric-oceanic anomalies before the 1970s but to ENSO in more recent decades (Fauchereau et al., 2003).

Several studies also have highlighted the importance of terrestrial vegetation cover and the associated dynamic feedbacks on the physical climate (see Christensen et al., 2007). An increase in vegetation density, for example, has been suggested to result in a year-round cooling of 0.8°C in the

[1] Note that several authors (e.g., Agoumi, 2003; Legesse et al., 2003; Conway, 2005, Thornton et al., 2006) caution against over-interpretation of results owing to the limitations of some of the projections and models used.

tropics, including tropical areas of Africa (Bounoua et al., 2000). Complex feedback mechanisms, mainly due to deforestation/land-cover change and changes in atmospheric dust loadings, also play a role in climate variability, particularly for drought persistence in the Sahel and its surrounding areas (Wang and Eltahir, 2000, 2002; Nicholson, 2001; Semazzi and Song, 2001; Prospero and Lamb, 2003; Zeng, 2003). The complexity of the interactions precludes 'simple interpretations'; for instance, the role of human-induced factors (e.g., migration), together with climate, can contribute to changes in vegetation in the Sahel that feed back into the overall physical system in complex ways (see, e.g., Eklundh and Olsson, 2003; Held et al., 2005; Herrmann et al., 2005; Olsson et al., 2005). Mineral dust is the largest cause of uncertainty in the radiative forcing of the planet and the key role of the Sahara has long been known. Better quantitative estimates of Saharan dust loadings and controls on emissions have now emerged from both satellite and field campaigns (e.g., Washington and Todd, 2005; Washington et al., 2006).

Finally, changes in extreme events, such as droughts and floods, have major implications for numerous Africans and require further attention. Droughts, notwithstanding current limitations in modelling capabilities and understanding of atmospheric system complexity, have attracted much interest over the past 30 years (AMCEN/UNEP, 2002), particularly with reference to impacts on both ecological systems and on society. Droughts have long contributed to human migration, cultural separation, population dislocation and the collapse of prehistoric and early historic societies (Pandey et al., 2003). One-third of the people in Africa live in drought-prone areas and are vulnerable to the impacts of droughts (World Water Forum, 2000). In Africa, for example, several million people regularly suffer impacts from droughts and floods. These impacts are often further exacerbated by health problems, particularly diarrhoea, cholera and malaria (Few et al., 2004). During the mid-1980s the economic losses from droughts totalled several hundred million U.S. dollars (Tarhule and Lamb, 2003). Droughts have mainly affected the Sahel, the Horn of Africa and southern Africa, particularly since the end of the 1960s (see Section 9.6.2; Richard et al., 2001; L'Hôte et al., 2002; Brooks, 2004; Christensen et al., 2007; Trenberth et al., 2007). Floods are also critical and impact on African development. Recurrent floods in some countries are linked, in some cases, with ENSO events. When such events occur, important economic and human losses result (e.g., in Mozambique – see Mirza, 2003; Obasi, 2005). Even countries located in dry areas (Algeria, Tunisia, Egypt, Somalia) have not been flood-free (Kabat et al., 2002).

9.2.1.1 *Sensitivity/vulnerability of the water sector*

The water sector is strongly influenced by, and sensitive to, changes in climate (including periods of prolonged climate variability). Evidence of interannual lake-level fluctuations and lake-level volatility, for example, has been observed since the 1960s, probably owing to periods of intense droughts followed by increases in rainfall and extreme rainfall events in late 1997 (e.g., in Lakes Tanganyika, Victoria and Turkana; see Riebeek, 2006). After the 1997 flood, Lake Victoria rose by about 1.7 m by 1998, Lake Tanganyika by about 2.1 m, and Lake Malawi by about 1.8 m, and very high river-flows were recorded in the Congo River at Kinshasha (Conway et al., 2005). The heavy rains and floods have been possibly attributed to large-scale atmosphere-ocean interactions in the Indian Ocean (Mercier et al., 2002).

Changes in runoff and hydrology linked to climate through complex interactions also include those observed for southern Africa (Schulze et al., 2001; New, 2002), south-central Ethiopia (Legesse et al., 2003), Kenya and Tanzania (Eriksen et al., 2005) and the wider continent (de Wit and Stankiewicz, 2006; Nkomo et al., 2006). Fewer assessments of impacts and vulnerabilities with regard to groundwater and climate interactions are available, and yet these are clearly of great concern for those dependent on groundwater for their water supply.

About 25% of the contemporary African population experiences high water stress. About 69% of the population lives under conditions of relative water abundance (Vörösmarty et al., 2005). However, this relative abundance does not take into account other equally important factors such as access to clean drinking water and sanitation, which effectively reduces the quantity of freshwater available for human use. Despite the considerable improvements in access to freshwater in the 1990s, only about 62% of the African population had access to improved water supplies in 2000 (WHO/UNICEF, 2000; Vörösmarty, 2005). As illustrated in Section 9.2.2, issues that affect access to water, including water governance, also need to be considered in any discussion of vulnerability to water stress in Africa.

9.2.1.2 *Sensitivity/vulnerability of the health sector*

Assessments of health in Africa show that many communities are already impacted by health stresses that are coupled to several causes, including poor nutrition. These assessments repeatedly pinpoint the implications of the poor health status of many Africans for future development (Figure 9.1a-d) (e.g., Sachs and Malaney, 2002; Sachs, 2005). An estimated 700,000 to 2.7 million people die of malaria each year and 75% of those are African children (see http://www.cdc.gov/malaria/; Patz and Olson, 2006). Incidences of malaria, including the recent resurgence in the highlands of East Africa, however, involve a range of multiple causal factors, including poor drug-treatment implementation, drug resistance, land-use change, and various socio-demographic factors including poverty (Githeko and Ndegwa, 2001; Patz et al., 2002; Abeku et al., 2004; Zhou et al., 2004; Patz and Olson, 2006). The economic burden of malaria is estimated as an average annual reduction in economic growth of 1.3% for those African countries with the highest burden (Gallup and Sachs, 2001).

The resurgence of malaria and links to climate and/or other causal 'drivers' of change in the highlands of East Africa has recently attracted much attention and debate (e.g., Hay et al., 2002a; Pascual et al., 2006). There are indications, for example, that in areas that have two rainy seasons – March to June (MAMJ) and September to November (SON) – more rain is falling in SON than previously experienced in the northern sector of East Africa (Schreck and Semazzi, 2004). The SON period is relatively warm, and higher rainfall is likely to increase malaria transmission because of a reduction in larval development duration. The spread of malaria into new areas (for

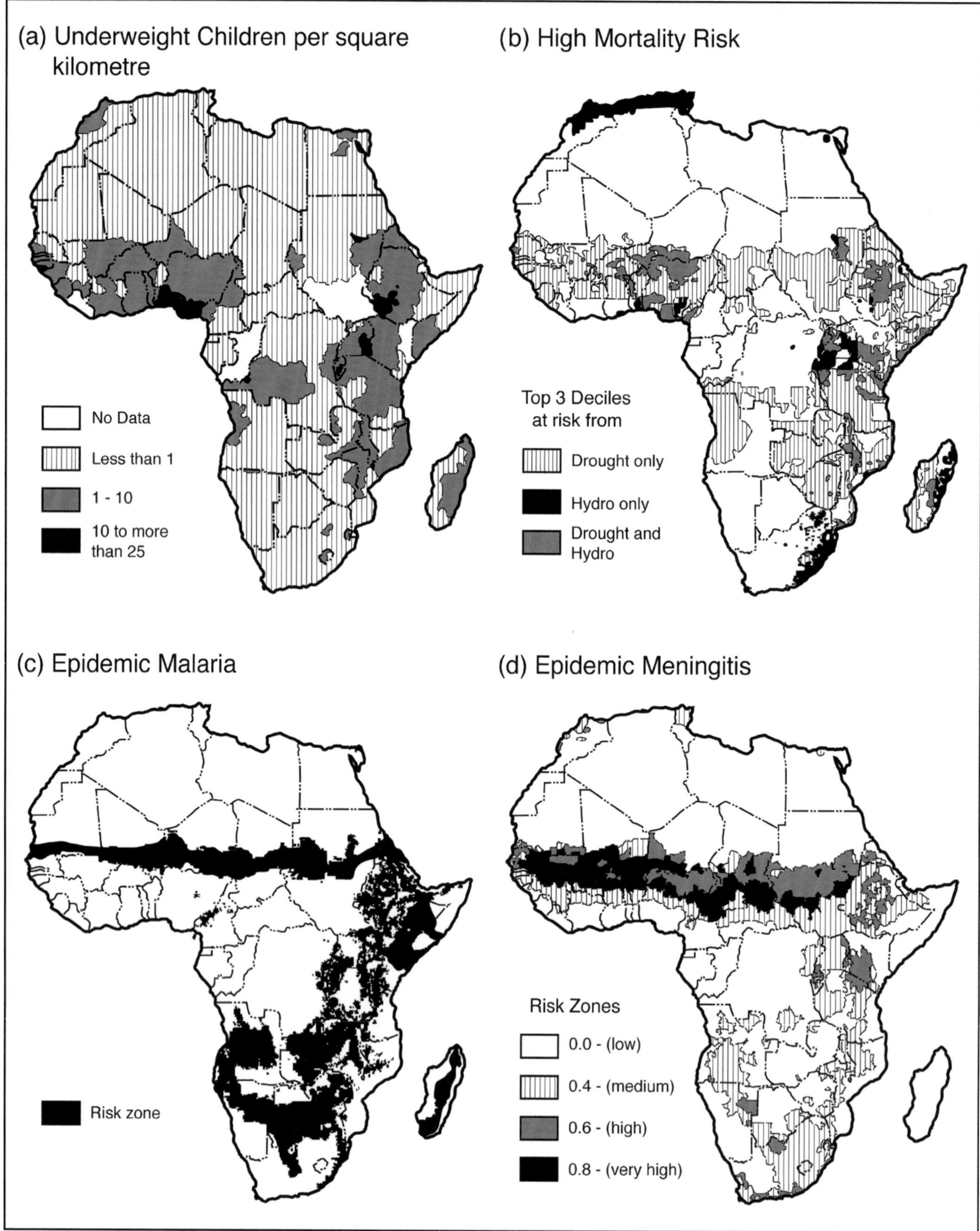

Figure 9.1. *Examples of current 'hotspots' or risk areas for Africa: (a) 'hunger'; (b) 'natural hazard-related disaster risks'; (c) regions prone to malaria derived from historical rainfall and temperature data (1950-1996); and (d) modelled distribution of districts where epidemics of meningococcal meningitis are likely to occur, based on epidemic experience, relative humidity (1961-1990) and land cover (adapted from IRI et al., 2006, p. 5; for further details see also Molesworth et al., 2003; Balk et al., 2005; Dilley et al., 2005; Center for International Earth Science Information Network, 2006; Connor et al., 2006).*

example, observations of malaria vector *Anopheles arabiensis* in the central highlands of Kenya, where no malaria vectors have previously been recorded) has also been documented (Chen et al., 2006). Recent work (e.g., Pascual et al., 2006) provides further new insights into the observed warming trends from the end of the 1970s onwards in four high-altitude sites in East Africa. Such trends may have significant biological implications for malaria vector populations.

New evidence regarding micro-climate change due to land-use changes, such as swamp reclamation for agricultural use and deforestation in the highlands of western Kenya, suggests that suitable conditions for the survival of *Anopheles gambiae* larvae are being created and therefore the risk of malaria is increasing (Munga et al., 2006). The average ambient temperature in the deforested areas of Kakamega in the western Kenyan highlands, for example, was 0.5°C higher than that of the forested area over a 10-month period (Afrane et al., 2005). Mosquito pupation rates and larval-to-pupal development have been observed to be significantly faster in farmland habitats than in swamp and forest habitats (Munga et al., 2006). Floods can also trigger malaria epidemics in arid and semi-arid areas (e.g., Thomson et al., 2006).

Other diseases are also important to consider with respect to climate variability and change, as links between variations in climate and other diseases, such as cholera and meningitis, have also been observed. About 162 million people in Africa live in areas with a risk of meningitis (Molesworth et al., 2003; Figure 9.1d). While factors that predispose populations to meningococcal meningitis are still poorly understood, dryness, very low humidity and dusty conditions are factors that need to be taken into account. A recent study, for example, has demonstrated that wind speeds in the first two weeks of February explained 85% of the variation in the number of meningitis cases (Sultan et al., 2005).

9.2.1.3 Sensitivity/vulnerability of the agricultural sector

The agricultural sector is a critical mainstay of local livelihoods and national GDP in some countries in Africa (Mendelsohn et al., 2000a, b; Devereux and Maxwell, 2001). The contribution of agriculture to GDP varies across countries but assessments suggest an average contribution of 21% (ranging from 10 to 70%) of GDP (Mendelsohn et al., 2000b). This sector is particularly sensitive to climate, including periods of climate variability (e.g., ENSO and extended dry spells; see Usman and Reason, 2004). In many parts of Africa, farmers and pastoralists also have to contend with other extreme natural-resource challenges and constraints such as poor soil fertility, pests, crop diseases, and a lack of access to inputs and improved seeds. These challenges are usually aggravated by periods of prolonged droughts and/or floods and are often particularly severe during El Niño events (Mendelsohn et al., 2000a, b; Biggs et al., 2004; International Institute of Rural Reconstruction, 2004; Vogel, 2005; Stige et al., 2006).

9.2.1.4 Sensitivity/vulnerability of ecosystems

Ecosystems are critical in Africa, contributing significantly to biodiversity and human well-being (Biggs et al., 2004; Muriuki et al., 2005). The rich biodiversity in Africa, which occurs principally outside formally conserved areas, is under threat from climate variability and change and other stresses (see Chapter 4, Section 4.2). Africa's social and economic development is constrained by climate change, habitat loss, over-harvesting of selected species, the spread of alien species, and activities such as hunting and deforestation, which threaten to undermine the integrity of the continent's rich but fragile ecosystems (UNEP/GRID-Arendal, 2002; Thomas et al., 2004).

Approximately half of the sub-humid and semi-arid parts of the southern African region are at moderate to high risk of desertification (e.g., Reich et al., 2001; Biggs et al., 2004). In West Africa, the long-term decline in rainfall from the 1970s to the 1990s caused a 25-35 km southward shift of the Sahelian, Sudanese and Guinean ecological zones in the second half of the 20th century (Gonzalez, 2001). This has resulted in a loss of grassland and acacia, the loss of flora/fauna, and shifting sand-dunes in the Sahel (ECF and Potsdam Institute, 2004).

The 1997/1998 coral bleaching episode observed in the Indian Ocean and Red Sea was coupled to a strong ENSO. In the western Indian Ocean region, a 30% loss of corals resulted in reduced tourism in Mombasa and Zanzibar, and caused financial losses of about US$12-18 million (Payet and Obura, 2004). Coral reefs are also exposed to other local anthropogenic threats, including sedimentation, pollution and over-fishing, particularly when they are close to important human settlements such as towns and tourist resorts (Nelleman and Corcoran, 2006). Recent outbreaks of the 'crown-of-thorns' starfish have occurred in Egypt, Djibouti and western Somalia, along with some local bleaching (Kotb et al., 2004).

Observed changes in ecosystems are not solely attributable to climate. Additional factors, such as fire, invasive species and land-use change, interact and also produce change in several African locations (Muriuki et al., 2005). Sensitive mountain environments (e.g., Mt. Kilimanjaro, Mt. Ruwenzori) demonstrate the complex interlinkages between various atmospheric processes including solar radiation micro-scale processes, glacier-climate interactions, and the role of vegetation changes and climate interactions (Kaser et al., 2004). For example, the drop in atmospheric moisture at the end of the 19th century, and the drying conditions that then occurred, have been used to explain some of the observed glacier retreat on Kilimanjaro (Kaser et al., 2004). Ecosystem change, also induced by complex land-use/climate interactions, including the migration of species and the interaction with fire (e.g., Hemp, 2005), produces a number of feedbacks or 'knock-on' impacts. Changes in the range of plant and animal species, for example, are already occurring because of forest fires on Kilimanjaro, and may place additional pressure on ecosystem services (Agrawala, 2005). The loss of 'cloud forests' through fire since 1976 has resulted in an estimated 25% annual reduction in 'fog water' (the equivalent of the annual drinking water demand of 1 million people living on Kilimanjaro) and is another critical impact in this region (see Chapter 4, Section 4.2; Box 9.1; Agrawala, 2005; Hemp, 2005).

9.2.1.5 Sensitivity/vulnerability of settlements and infrastructure

Impacts on settlements and infrastructure are well recorded for recent extreme climate events (e.g., the 2000 flooding event

Box 9.1. Environmental changes on Mt. Kilimanjaro

There is evidence that climate is modifying natural mountain ecosystems via complex interactions and feedbacks including, for example, solar radiation micro-scale processes on Mt. Kilimanjaro (Mölg and Hardy, 2004; Lemke et al., 2007). Other drivers of change are also modifying environments on the mountain, including fire, vegetation changes and human modifications (Hemp, 2005). During the 20th century, the areal extent of Mt. Kilimanjaro's ice fields decreased by about 80% (Figure 9.2). It has been suggested that if current climatological conditions persist, the remaining ice fields are likely to disappear between 2015 and 2020 (Thompson et al., 2002).

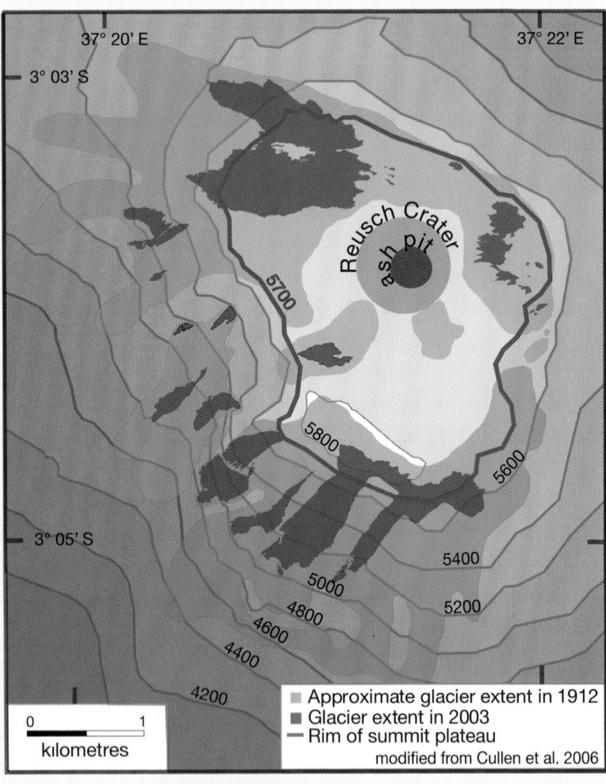

Figure 9.2. *Decrease in surface area of Mt. Kilimanjaro glaciers from 1912 to 2003 (modified from Cullen et al., 2006).*

in Mozambique – Christie and Hanlon, 2001; IFRCRCS, 2002; see also various infrastructural loss estimates from severe storm events in the western Cape, South Africa – http://www.egs.uct.ac.za/dimp/; and southern Africa – Reason and Keibel, 2004). Large numbers of people are currently at risk of floods (see, for example, UNDP, 2004; UNESCO-WWAP, 2006), particularly in coastal areas, where coastal erosion is already destroying infrastructure, housing and tourism facilities (e.g., in the residential region of Akpakpa in Benin (Niasse et al., 2004; see also Chapter 7, Section 7.2.).

9.2.2 Current sensitivity and vulnerability to other stresses

Complex combinations of socio-economic, political, environmental, cultural and structural factors act and interact to affect vulnerability to environmental change, including climate change and variability. Economic development in Africa has been variable (Ferguson, 2006). African economies have recently registered a significant overall increase in activity (growing by more than 5% in 2004 – OECD, 2004/2005; World Bank, 2006a, b). Sub-Saharan Africa, for example, has shown an increase of 1.2%/yr growth in average income since 2000 (UNDP, 2005). Despite this positive progress, boosted in part by increases in oil exports and high oil prices, several African economies, including informal and local-scale economic activities and livelihoods, remain vulnerable to regional conflicts, the vagaries of the weather and climate, volatile commodity prices and the various influences of globalisation (see, e.g., Devereux and Maxwell, 2001; OECD, 2004/2005; Ferguson, 2006). Certain countries in sub-Saharan Africa suffer from deteriorating food security (Figure 9.1a) and declines in overall real wealth, with estimates that the average person in sub-Saharan Africa becomes poorer by a factor of two every 25 years (Arrow et al., 2004; Sachs, 2005). The interaction between economic stagnation and slow progress in education has been compounded by the spread of HIV/AIDS. In 2003, 2.2 million Africans died of the disease and an estimated 12 million children in sub-Saharan Africa lost one or both parents to HIV/AIDS (UNAIDS, 2004; Ferguson, 2006). This has produced a 'freefall' in the Human Development Index ranking, with southern African countries accounting for some of the steepest declines (UNDP, 2005). Indeed, some commentators have noted that sub-Saharan Africa is the only region in the world that has become poorer in this generation (Devereux and Maxwell, 2001; Chen and Ravallion, 2004).

A large amount of literature exists on the various factors that influence vulnerability to the changes taking place in Africa (e.g., to climate stress), and this section outlines some of the key issues (see, for example, Figure 9.1a-d). However, these factors do not operate in isolation, and usually interact in complex and 'messy' ways, frustrating attempts at appropriate interventions to increase resilience to change.

9.2.2.1 Globalisation, trade and market reforms

There are important macro-level processes that serve to heighten vulnerability to climate variability and change across a range of scales in Africa (Sachs et al., 2004; UNDP, 2005; Ferguson, 2006). Issues of particular importance include globalisation, trade and equity (with reference to agriculture, see FAO, 2005; Schwind, 2005) and modernity and social justice (e.g., Ferguson, 2006). Numerous 'structural' factors are 'driving' and 'shaping' poverty and livelihoods (Hulme and Shepherd, 2003) and changing the face of rural Africa (e.g., intensification versus extensification, see Bryceson, 2004; Section 9.6.1). Structural adjustment accompanied by complex market reforms and market liberalisation (e.g., access to credit and subsidy arrangements) has aggravated the vulnerability of many in Africa, particularly those engaged in agriculture (see, e.g., Eriksen, 2004; Kherallah et al., 2004). Fertiliser prices, for example, have risen in response to subsidy removal, resulting in some mixed responses to agricultural reforms (Kherallah et al., 2004; Institute of Development Studies, 2005). Market-related and structural issues can thus serve to reduce people's agricultural productivity and reduce resilience to further agricultural stresses associated with climate change.

9.2.2.2 Governance and institutions

Complex institutional dimensions are often exposed during periods of climate stress. Public service delivery is hampered by poor policy environments in some sectors which provide critical obstacles to economic performance (Tiffen, 2003). Africa is also characterised by institutional and legal frameworks that are, in some cases, insufficient to deal with environmental degradation and disaster risks (Sokona and Denton, 2001; Beg et al., 2002). Various actors, structures and networks are therefore required to reconfigure innovation processes in Africa (e.g., in agriculture) to improve responses to climate variability and change in both rural and urban contexts (Tiffen, 2003; Scoones, 2005; Reid and Vogel, 2006; see also Section 9.5).

9.2.2.3 Access to capital, including markets, infrastructure and technology

Constraints in technological options, limited infrastructure, skills, information and links to markets further heighten vulnerability to climate stresses. In the agricultural sector, for example, many African countries depend on inefficient irrigation systems (UNEP, 2004) which heighten vulnerability to climate variability and change. Africa has been described as the world's great laggard in technological advance in the area of agriculture (Sachs et al., 2004). For instance, most of the developing world experienced a Green Revolution: a surge in crop yields in the 1970s through to the 1990s as a result of scientific breeding that produced high-yielding varieties (HYVs), combined with an increased use of fertilisers and irrigation. Africa's uptake of HYVs was the lowest in the developing world. The low levels of technological innovation and infrastructural development in Africa result in the extraction of natural resources for essential amenities such as clean water, food, transportation, energy and shelter (Sokona and Denton, 2001). Such activities degrade the environment and compound vulnerability to a range of stresses, including climate-related stress. Sub-Saharan African countries also have extremely low per capita densities of rail and road infrastructure (Sachs, 2005). As a result, cross-country transport connections within Africa tend to be extremely poor and are in urgent need of extension in order to reduce intra-regional transport costs and promote cross-border trade (Sachs, 2005). Such situations often exacerbate drought and flood impacts (see, for example, the role of information access in IFRCRCS, 2005) as well as hindering adaptation to climate stresses (see Section 9.5; Chapter 17, Section 17.3.2).

9.2.2.4 Population and environment interactions

Notwithstanding the range of uncertainties related to the accuracy of census data, the African continent is witnessing some of the most rapid population growth, particularly in urban areas (Tiffen, 2003). During the period 1950 to 2005, the urban population in Africa grew by an average annual rate of 4.3% from 33 million to 353 million (ECA, 2005; Yousif, 2005). Complex migration patterns, which are usually undertaken to ensure income via remittances (Schreider and Knerr, 2000) and which often occur in response to stress-induced movements linked to conflict and/or resource constraints, can further trigger a range of environmental and socio-economic changes. Migration is also associated with the spread of HIV/AIDS and other diseases. Several studies have shown that labour migrants tend to have higher HIV infection rates than non-migrants (UNFPA, 2003). Increases in population also exert stresses on natural resources. Agricultural intensification and/or expansion into marginal lands can trigger additional conflicts, cause crop failure, exacerbate environmental degradation (e.g., Olsson et al., 2005) and reduce biodiversity (Fiki and Lee, 2004), and this then, in turn, feeds back, via complex pathways, into the biophysical system. Variations in climate, both short and long term, usually aggravate such interactions. Changes in rain-fed livestock numbers in Africa, a sector often noted for exerting noticeable pressure on the environment, are already strongly coupled with variations in rainfall but are also linked to other socio-economic and cultural factors (see, for example, Little et al., 2001; Turner, 2003; Boone et al., 2004; Desta and Coppock, 2004; Thornton et al., 2004).

9.2.2.5 Water access and management

Water access and water resource management are highly variable across the continent (Ashton, 2002; van Jaarsveld et al., 2005; UNESCO-WWAP, 2006). The 17 countries in West Africa that share 25 transboundary rivers have notably high water interdependency (Niasse, 2005). Eastern and southern African countries are also characterised by water stress brought about by climate variability and wider governance issues (Ashton, 2002; UNESCO-WWAP, 2006). Significant progress has, however, been recorded in some parts of Africa to improve this situation, with urban populations in the southern African region achieving improved water access over recent years (van Jaarsveld et al., 2005). Despite this progress, about 35 million people in the region are still using unimproved water sources; the largest proportion being in Mozambique, followed by Angola, South Africa, Zambia and Malawi (Mutangadura et al., 2005). When water is available it is often of poor quality, thus contributing to a range of health problems including diarrhoea, intestinal worms and trachoma. Much of the suffering from lack of access to safe drinking water and sanitation is borne by the poor, those who live in degraded environments, and overwhelmingly by women and children. The relevance of the problem of water scarcity is evident in North Africa, considering that estimates for the average annual growth of the population are the world's highest: 2.9% for the period 1990-2002. The Water Exploitation Index[2] is high in several countries in the sub-region: >50% for Tunisia, Algeria, Morocco and Sudan, and

[2] Water Exploitation Index: total water abstraction per year as percentage of long-term freshwater resources.

>90% for Egypt and Libya (Gueye et al., 2005). Until recently, these countries have adopted a supply-oriented approach to managing their water resources. However, managing the supply of water cannot in itself ensure that the needs of a country can be met in a sustainable way.

Attributing sensitivity and vulnerability in the water sector solely to variations in climate is problematic. The complex interactions between over-fishing, industrial pollution and sedimentation, for example, are also degrading local water sources such as Lake Victoria (Odada et al., 2004), which impacts on catches. Integrated analyses of climate change in Egypt, moreover, show that population changes, land-use changes and domestic growth strategies may be more important in water management decision-making than a single focus on climate change (Conway, 2005).

9.2.2.6 Health management

In much the same way as the aforementioned sectors, the health sector is affected by the interaction of several 'human dimensions', e.g., inadequate service management, poor infrastructure, the stigma attached to HIV/AIDS, and the 'brain drain'. HIV/AIDS is contributing to vulnerability with regard to a range of stresses (Mano et al., 2003; USAID, 2003; Gommes et al., 2004). Maternal malaria, for example, has been shown to be associated with a twice as high HIV-1 viral concentration (ter Kuile et al., 2004) and infection rates are estimated to be 5.5% and 18.8% in populations with a HIV prevalence of 10% and 40%, respectively. The deadly duo of HIV/AIDS and food insecurity in southern Africa are key drivers of the humanitarian crisis (Gommes et al., 2004; see also Section 9.6). While infectious diseases such as cholera are being eradicated in other parts of the world, they are re-emerging in Africa. A major challenge facing the continent is the relative weakness in disease surveillance and reporting systems, which hampers the detection and control of cholera epidemics, and, as a side effect, makes it difficult to obtain the long-term linked data sets on climate and disease that are necessary for the development of early warning systems (WHO, 2005).

9.2.2.7 Ecosystem degradation

Human 'drivers' are also shaping ecosystem services that impact on human well-being (e.g., Muriuki et al., 2005; van Jaarsveld et al., 2005). Several areas, for example, Zimbabwe, Malawi, eastern Zambia, central Mozambique as well as the Congo Basin rainforests in the Democratic Republic of Congo, underwent deforestation at estimated rates of about 0.4% per year during the 1990s (Biggs et al., 2004). Further threats to Africa's forests are also posed by the high dependency on fuelwood and charcoal, major sources of energy in rural areas, that are estimated to contribute about 80 to 90% of the residential energy needs of low-income households in the majority of sub-Saharan countries (IEA, 2002). Moreoever, fire incidents represent a huge threat to tropical forests in Africa. An estimated 70% of detected forest fires occur in the tropics, with 50% of them being in Africa. More than half of all forested areas were estimated to have burned in Africa in 2000 (Dwyer et al., 2000; Kempeneers et al., 2002). Bush fires are a particular threat to woodlands, causing enormous destruction of both flora and fauna in eastern and southern Africa (for an extensive and detailed review on the role of fire in southern Africa, see SAFARI, 2004). The African continent also suffers from the impacts of desertification. At present, almost half (46%) of Africa's land area is vulnerable to desertification (Granich, 2006).

9.2.2.8 Energy

Access to energy is severely constrained in sub-Saharan Africa, with an estimated 51% of urban populations and only about 8% of rural populations having access to electricity. This is compared with about 99% of urban populations and about 80% of rural populations who have access in northern Africa (IEA, 2002). Other exceptions also include South Africa, Ghana and Mauritius. Extreme poverty and the lack of access to other fuels mean that 80% of the overall African population relies primarily on biomass to meet its residential needs, with this fuel source supplying more than 80% of the energy consumed in sub-Saharan Africa (Hall and Scrase, 2005). In Kenya, Tanzania, Mozambique and Zambia, for example, nearly all rural households use wood for cooking and over 90% of urban households use charcoal (e.g., IEA, 2002, p. 386; van Jaarsveld et al., 2005). Dependence on biomass can promote the removal of vegetation. The absence of efficient and affordable energy services can also result in a number of other impacts including health impacts associated with the carrying of fuelwood, indoor pollution and other hazards (e.g., informal settlement fires - IEA, 2002). Further challenges from urbanisation, rising energy demands and volatile oil prices further compound energy issues in Africa (ESMAP, 2005).

9.2.2.9 Complex disasters and conflicts

The juxtaposition of many of the complex socio-economic factors outlined above and the interplay between biophysical hazards (e.g., climate hazards - tropical cyclones, fire, insect plagues) is convincingly highlighted in the impacts and vulnerabilities to disaster risks and conflicts in several areas of the continent (see, for example, several reports of the International Federation of the Red Cross and Red Crescent Societies (IFRCRCS) of the past few years, available online at http://www.ifrc.org/; and several relevant documents such as those located on http://www.unisdr.org/) (see Figure 9.1b). Many disasters are caused by a combination of a climate stressor (e.g., drought, flood) and other factors such as conflict, disease outbreaks and other 'creeping' factors e.g., economic degradation over time (Benson and Clay, 2004; Reason and Keibel, 2004; Eriksen et al., 2005). The role of these multiple interactions is well illustrated in the case of Malawi and Mozambique. In 2000 in Malawi, agriculture accounted for about 40% of the GDP, a drop of about 4% from 1980. The real annual fluctuations in agricultural, non-agricultural and total GDP for 1980 to 2001 show that losses during droughts (e.g., as occurred in the mid-1990s) were more severe than disaster losses during the floods in 2001 (Benson and Clay, 2004) (for more details on structural causes and drought interactions and impacts, e.g., food security, see Section 9.6.1). Likewise, the floods in Mozambique in 2000 revealed a number of existing vulnerabilities that were heightened by the floods. These included: poverty (an estimated 40% of the population lives on less than US$1 per day and another 40% on less than US$2 per day); the debt problem, which is one of the biggest challenges facing the country; the fact that most of the floodwaters originated in cross-border shared basins; the poor disaster risk-reduction strategies with regard to dam design and management; and the poor communication networks (Christie and Hanlon, 2001; IFRCRCS, 2002; Mirza, 2003).

Conflicts, armed and otherwise, have recently occurred in the Greater Horn of Africa (Somalia, Ethiopia and Sudan) and the Great Lakes region (Burundi, Rwanda and the Democratic Republic of Congo) (Lind and Sturman, 2002; Nkomo et al., 2006). The causes of such conflicts include structural inequalities, resource mismanagement and predatory States. Elsewhere, land distribution and land scarcity have promoted conflict (e.g., Darfur, Sudan; see, for example, Abdalla, 2006), often exacerbated by environmental degradation. Ethnicity is also often a key driving force behind conflict (Lind and Sturman, 2002; Balint-Kurti, 2005; Ron, 2005). Climate change may become a contributing factor to conflicts in the future, particularly those concerning resource scarcity, for example, scarcity of water (Ashton, 2002; Fiki and Lee, 2004),).

It is against this background that an assessment of vulnerability to climate change and variability has to be contextualised. Although the commonly used indicators have limitations in capturing human well-being (Arrow et al., 2004), some aggregated proxies for national-level vulnerability to climate change for the countries in Africa have been developed (e.g., Vincent, 2004; Brooks et al., 2005). These indicators include elements of economy, health and nutrition, education, infrastructure, governance, demography, agriculture, energy and technology. The majority of countries classified as vulnerable in an assessment using such proxies were situated in sub-Saharan Africa (33 of the 50 assessed by Brooks et al., 2005, were sub-Saharan African countries). At the local level, several case studies similarly show that it is the interaction of such 'multiple stresses', including composition of livelihoods, the role of social safety nets and other social protection measures, that affects vulnerability and adaptive capacity in Africa (see Section 9.5).

9.3 Assumptions about future trends

9.3.1 Climate-change scenarios

In this section, the limits of the regions are those defined by Ruosteenoja et al. (2003). Very few regional to sub-regional climate change scenarios using regional climate models or empirical downscaling have been constructed in Africa mainly due to restricted computational facilities and lack of human resources (Hudson and Jones, 2002; Swart et al., 2002) as well as problems of insufficient climate data (Jenkins et al., 2002). Under the medium-high emissions scenario (SRES A1B, see the Special Report on Emissions Scenarios: Nakićenović et al., 2000), used with 20 General Circulation Models (GCMs) for the period 2080-2099, annual mean surface air temperature is expected to increase between 3 and 4°C compared with the 1980-1999 period, with less warming in equatorial and coastal areas (Christensen et al., 2007). Other experiments (e.g., Ruosteenoja et al., 2003) indicate higher levels of warming with the A1FI emissions scenario and for the 2070-2099 period: up to 9°C for North Africa (Mediterranean coast) in June to August, and up to 7°C for southern Africa in September to November. Regional Climate Model (RCM) experiments generally give smaller temperature increases (Kamga et al., 2005). For southern Africa (from the equator to 45°S and from 5° to 55°E, which includes parts of the surrounding oceans), Hudson and Jones (2002), using the HadRM3H RCM with the A2 emissions scenario, found for the 2080s a 3.7°C increase in summer (December to February) mean surface air temperature and a 4°C increase in winter (June to August). As demonstrated by Bounoua et al. (2000), an increase in vegetation density, leading to a cooling of 0.8°C/yr in the tropics, including Africa, could partially compensate for greenhouse warming, but the reverse effect is simulated in the case of land cover conversion, which will probably increase in the next 50 years (DeFries et al., 2002). A stabilisation of the atmospheric CO_2 concentration at 550 ppm (by 2150) or 750 ppm (by 2250) could also delay the expected greenhouse gas-induced warming by 100 and 40 years, respectively, across Africa (Arnell et al., 2002). For the same stabilisation levels in the Sahel (10°-20°N, 20°W-40°E) the expected annual mean air temperature in 2071-2100 (5°C) will be reduced, respectively, by 58% (2.1°C) and 42% (2.9°C) (Mitchell et al., 2000; Christensen et al., 2007).

Precipitation projections are generally less consistent with large inter-model ranges for seasonal mean rainfall responses. These inconsistencies are explained partly by the inability of GCMs to reproduce the mechanisms responsible for precipitation including, for example, the hydrological cycle (Lebel et al., 2000), or to account for orography (Hudson and Jones, 2002). They are also explained partly by model limitations in simulating the different teleconnections and feedback mechanisms which are responsible for rainfall variability in Africa. Other factors that complicate African climatology include dust aerosol concentrations and sea-surface temperature anomalies, which are particularly important in the Sahel region (Hulme et al., 2001; Prospero and Lamb, 2003) and southern Africa (Reason, 2002), deforestation in the equatorial region (Semazzi and Song, 2001; Bounoua et al., 2002), and soil moisture in southern Africa (New et al., 2006). These uncertainties make it difficult to provide any precise estimation of future runoff, especially in arid and semi-arid regions where slight changes in precipitation can result in dramatic changes in the runoff process (Fekete et al., 2004). Nonetheless, estimations of projected future rainfall have been undertaken.

With the SRES A1B emissions scenario and for 2080-2099, mean annual rainfall is very likely to decrease along the Mediterranean coast (by 20%), extending into the northern Sahara and along the west coast to 15°N, but is likely increase in tropical and eastern Africa (around +7%), while austral winter (June to August) rainfall will very probably decrease in much of southern Africa, especially in the extreme west (up to 40%) (Christensen et al., 2007). In southern Africa, the largest changes in rainfall occur during the austral winter, with a 30% decrease under the A2 scenario, even though there is very little rain during this season (Hudson and Jones, 2002). There are, however, differences between the equatorial regions (north of 10°S and east of 20°E), which show an increase in summer (December to February) rainfall, and those located south of 10°S, which show a decrease in rainfall associated with a decrease in the number of rain days and in the average intensity of rainfall. Recent downscaling experiments for South Africa indicate increased summer rainfall over the convective region of the central and eastern plateau and the Drakensberg Mountains (Hewitson and Crane, 2006). Using RCMs, Tadross et al. (2005b), found a decrease in early summer (October to December) rainfall and an increase in late summer (January to March) rainfall over the eastern parts of southern Africa.

For the western Sahel (10 to 18°N, 17.5°W to 20°E), there are still discrepancies between the models: some projecting a significant drying (e.g., Hulme et al., 2001; Jenkins et al., 2005) and others simulating a progressive wetting with an expansion of vegetation into the Sahara (Brovkin, 2002; Maynard et al., 2002; Claussen et al., 2003; Wang et al., 2004; Haarsma et al., 2005; Kamga et al., 2005; Hoerling et al., 2006). Land-use changes and degradation, which are not simulated by some models, could induce drier conditions (Huntingford et al., 2005; Kamga et al., 2005). The behaviour of easterly jets and squall lines is also critical for predicting the impacts of climate change on the sub-region, given the potential links between such phenomena and the development of the rainy season (Jenkins et al., 2002; Nicholson and Grist, 2003).

Finally, there is still limited information available on extreme events (Christensen et al., 2007), despite frequent reporting of such events, including their impacts (see Section 9.2.1). A recent study using four GCMs for the Sahel region (3.75 to 21.25°N, 16.88°W to 35.63°E) showed that the number of extremely dry and wet years will increase during the present century (Huntingford et al., 2005). Modelling of global drought projections for the 21st century, based on the SRES A2 emissions scenario, shows drying for northern Africa that appears consistent with the rainfall scenarios outlined above, and wetting over central Africa (Burke et al., 2006). On a global basis, droughts were also estimated to be slightly more frequent and of much longer duration by the second half of the 21st century relative to the present day. Other experiments indicate that in a warmer world, and by the end of the century (2080-2100), there could also be more frequent and intense tropical storms in the southern Indian Ocean (e.g., McDonald et al., 2005). Tropical cyclones are likely to originate over the Seychelles from October to June due to the southward displacement of the Near Equatorial Trough (Christensen et al., 2007). There could very probably be an increase of between 10 and 20% in cyclone intensity with a 2-4°C SST rise (e.g., Lal, 2001), but this observation is further complicated by the fact that SST does not account for all the changes in tropical storms (McDonald et al., 2005).

9.3.2 Socio-economic scenarios

The SRES scenarios adopt four storylines or 'scenario families' that describe how the world populations, economies and political structures may evolve over the next few decades (Nakićenović et al., 2000). The 'A' scenarios focus on economic growth, the 'B' scenarios on environmental protection, the '1' scenarios assume more globalisation, and the '2' scenarios assume more regionalisation. While some authors have criticised the population and economic details included in the SRES scenarios, the scenarios still provide a useful baseline for studying impacts related to greenhouse gas emissions (Tol et al., 2005). The situation for the already-vulnerable region of sub-Saharan Africa still appears bleak, even in the absence of climate change and variability. For example, 24 countries in sub-Saharan Africa are projected to be unable to meet several of the Millennium Development Goals (MDGs), and not one sub-Saharan country with a significant population is on track to meet the target with respect to child and maternal health (UNDP, 2005). The sub-Saharan share of the global total of those earning below US$1/day is also estimated to rise sharply from 24% today to 41% by 2015 (UNDP, 2005). It is within this context, coupled with the multiple stresses presented in Section 9.2.2, that the following summary of key future impacts and vulnerabilities associated with possible climate change and variability needs to be assessed.

9.4 Expected key future impacts and vulnerabilities, and their spatial variation

Having provided some background on existing sensitivities/vulnerabilities generated by a range of factors, including climate stress, some of the impacts and vulnerabilities that may arise under a changing climate in Africa, using the various scenarios and model projections as guides, are presented for various sectors. Note that several authors (e.g., Agoumi, 2003; Legesse et al., 2003; Conway, 2005, Thornton et al., 2006) warn against the over-interpretation of results, owing to the limitations of some of the projections and models used. For other assessments see also Biggs et al. (2004), Muriuki et al. (2005) and Nkomo et al. (2006).

9.4.1 Water

Climate change and variability have the potential to impose additional pressures on water availability, water accessibility and water demand in Africa. Even in the absence of climate change (see Section 9.2.2), present population trends and patterns of water use indicate that more African countries will exceed the limits of their "economically usable, land-based water resources before 2025" (Ashton, 2002, p. 236). In some assessments, the population at risk of increased water stress in Africa, for the full range of SRES scenarios, is projected to be 75-250 million and 350-600 million people by the 2020s and 2050s, respectively (Arnell, 2004). However, the impact of climate change on water resources across the continent is not uniform. An analysis of six climate models (HadCM3, ECHAM4-OPYC, CSIRO-Mk2, CGCM2, GFDL_r30 and CCSR/NIES2) and the SRES scenarios (Arnell, 2004) shows a likely increase in the number of people who could experience water stress by 2055 in northern and southern Africa (Figure 9.3). In contrast, more people in eastern and western Africa will be likely to experience a reduction rather than an increase in water stress (Arnell, 2006a).

Clearly these estimations are at macro-scales and may mask a range of complex hydrological interactions and local-scale differences (for other assessments on southern Africa, where some of these interacting scalar issues have been addressed, see Schulze et al., 2001). Detailed assessments in northern Africa based on temperature increases of 1-4°C and reductions in precipitation of between 0 and 10% show that the Ouergha watershed in Morocco is likely to undergo changes for the period 2000-2020. A 1°C increase in temperature could change runoff by of the order of 10%, assuming that the precipitation levels remain constant. If such an annual decrease in runoff were to occur in other watersheds, the impacts in such areas could be equivalent to the loss of one large dam per year (Agoumi, 2003). Further interactions between climate and other factors influencing water resources have also been well highlighted for Egypt (Box 9.2).

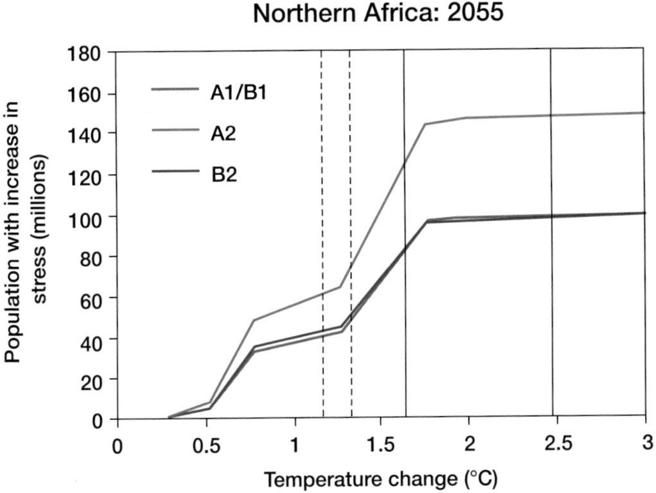

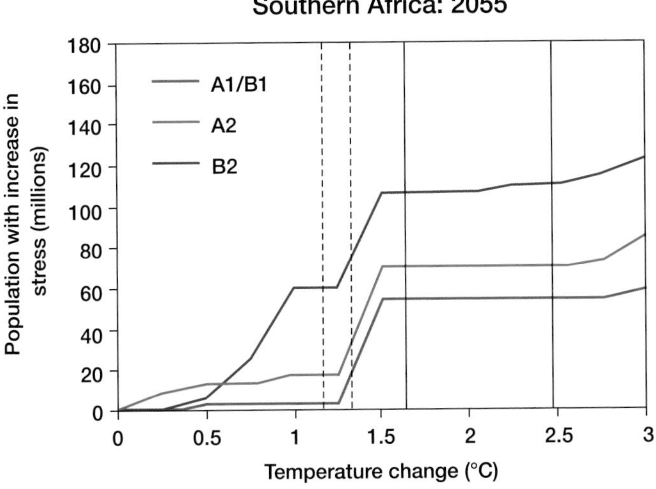

Figure 9.3. *Number of people (millions) with an increase in water stress (Arnell, 2006b). Scenarios are all derived from HadCM3 and the red, green and blue lines relate to different population projections.*

Box 9.2. Climate, water availability and agriculture in Egypt

Egypt is one of the African countries that could be vulnerable to water stress under climate change. The water used in 2000 was estimated at about 70 km³ which is already far in excess of the available resources (Gueye et al., 2005). A major challenge is to close the rapidly increasing gap between the limited water availability and the escalating demand for water from various economic sectors. The rate of water utilisation has already reached its maximum for Egypt, and climate change will exacerbate this vulnerability.

Agriculture consumes about 85% of the annual total water resource and plays a significant role in the Egyptian national economy, contributing about 20% of GDP. More than 70% of the cultivated area depends on low-efficiency surface irrigation systems, which cause high water losses, a decline in land productivity, waterlogging and salinity problems (El-Gindy et al., 2001. Moreover, unsustainable agricultural practices and improper irrigation management affect the quality of the country's water resources. Reductions in irrigation water quality have, in their turn, harmful effects on irrigated soils and crops.

Institutional water bodies in Egypt are working to achieve the following targets by 2017 through the National Improvement Plan (EPIQ, 2002; ICID, 2005):

- improving water sanitation coverage for urban and rural areas;
- wastewater management;
- optimising use of water resources by improving irrigation efficiency and agriculture drainage-water reuse.

However, with climate change, an array of serious threats is apparent.

- Sea-level rise could impact on the Nile Delta and on people living in the delta and other coastal areas (Wahab, 2005).
- Temperature rises will be likely to reduce the productivity of major crops and increase their water requirements, thereby directly decreasing crop water-use efficiency (Abou-Hadid, 2006; Eid et al., 2006).
- There will probably be a general increase in irrigation demand (Attaher et al., 2006).
- There will also be a high degree of uncertainty about the flow of the Nile.
- Based on SRES scenarios, Egypt will be likely to experience an increase in water stress, with a projected decline in precipitation and a projected population of between 115 and 179 million by 2050. This will increase water stress in all sectors.
- Ongoing expansion of irrigated areas will reduce the capacity of Egypt to cope with future fluctuation in flow (Conway, 2005).

Using ten scenarios derived by using five climate models (CSIRO2, HadCM3, CGCM2, ECHAM and PCM) in conjunction with two different emissions scenarios, Strzepek and McCluskey (2006) arrived at the following conclusions regarding impacts of climate change on streamflow in Africa. First, the possible range of Africa-wide climate-change impacts on streamflow increases significantly between 2050 and 2100. The range in 2050 is from a decrease of 15% in streamflow to an increase of 5% above the 1961-1990 baseline. For 2100, the range is from a decrease of 19% to an increase of 14%. Second, for southern Africa, almost all countries except South Africa will probably experience a significant reduction in streamflow. Even for South Africa, the increases under the high emissions scenarios are modest at under 10% (Strzepek and McCluskey, 2006).

Additional assessments of climate change impacts on hydrology, based on six GCMs and a composite ensemble of African precipitation models for the period 2070-2099 derived from 21 fully coupled ocean-atmosphere GCMs, show various drainage impacts across Africa (de Wit and Stankiewicz, 2006). A critical 'unstable' area is identified for some parts, for example, the east-west band from Senegal to Sudan, separating the dry Sahara from wet Central Africa. Parts of southern Africa are projected to experience significant losses of runoff, with some areas being particularly impacted (e.g., parts of South Africa) (New, 2002; de Wit and Stankiewicz, 2006). Other regional assessments report emerging changes in the hydrology of some of the major water systems (e.g., the Okavango River basin) which could be negatively impacted by changes in climate; impacts that could possibly be greater than those associated with human activity (Biggs et al., 2004; Anderssen et al., 2006).

Assessments of impacts on water resources, as already indicated, currently do not fully capture multiple future water uses and water stress and must be approached with caution (see, e.g., Agoumi, 2003; Conway, 2005). Conway (2005) argues that there is no clear indication of how Nile flow will be affected by climate change because of the uncertainty about rainfall patterns in the basin and the influence of complex water management and water governance structures. Clearly, more detailed research on water hydrology, drainage and climate change is required. Future access to water in rural areas, drawn from low-order surface water streams, also needs to be addressed by countries sharing river basins (see de Wit and Stankiewicz, 2006). Climate change should therefore be considered among a range of other water governance issues in any future negotiations to share Nile water (Conway, 2005; Stern, 2007).

9.4.2 Energy

There are remarkably few studies available that examine the impacts of climate change on energy use in Africa (but see a recent regional assessment by Warren et al., 2006). However, even in the absence of climate change, a number of changes are expected in the energy sector. Africa's recent and rapid urban growth (UNEP, 2005) will lead to increases in aggregate commercial energy demand and emissions levels (Davidson et al., 2003), as well as extensive land-use and land-cover changes, especially from largely uncontrolled urban, peri-urban and rural settlements (UNEP/GRID-Arendal, 2002; du Plessis et al., 2003). These changes will alter existing surface microclimates and hydrology and will possibly exacerbate the scope and scale of climate-change impacts.

9.4.3 Health

Vigorous debate among those working in the health sector has improved our understanding of the links between climate variability (including extreme weather events) and infectious diseases (van Lieshout et al., 2004; Epstein and Mills, 2005; McMichael et al., 2006; Pascual et al., 2006; Patz and Olson, 2006). Despite various contentious issues (see Section 9.2.1.2), new assessments of the role of climate change impacts on health have emerged since the TAR. Results from the "Mapping Malaria Risk in Africa" project (MARA/ARMA) show a possible expansion and contraction, depending on location, of climatically suitable areas for malaria by 2020, 2050 and 2080 (Thomas et al., 2004). By 2050 and continuing into 2080, for example, a large part of the western Sahel and much of southern central Africa is shown to be likely to become unsuitable for malaria transmission. Other assessments (e.g., Hartmann et al., 2002), using 16 climate-change scenarios, show that by 2100, changes in temperature and precipitation could alter the geographical distribution of malaria in Zimbabwe, with previously unsuitable areas of dense human population becoming suitable for transmission. Strong southward expansion of the transmission zone will probably continue into South Africa.

Using parasite survey data in conjunction with results from the HadCM3 GCM, projected scenarios estimate a 5-7% potential increase (mainly altitudinal) in malaria distribution, with little increase in the latitudinal extent of the disease by 2100 (Tanser et al., 2003). Previously malaria-free highland areas in Ethiopia, Kenya, Rwanda and Burundi could also experience modest incursions of malaria by the 2050s, with conditions for transmission becoming highly suitable by the 2080s. By this period, areas currently with low rates of malaria transmission in central Somalia and the Angolan highlands could also become highly suitable. Among all scenarios, the highlands of eastern Africa and areas of southern Africa are likely to become more suitable for transmission (Hartmann et al., 2002).

As the rate of malaria transmission increases in the highlands, the likelihood of epidemics may increase due to the lack of protective genetic modifications in the newly-affected populations. Severe malaria-associated disease is more common in areas of low to moderate transmission, such as the highlands of East Africa and other areas of seasonal transmission. An epidemic in Rwanda, for example, led to a four-fold increase in malaria admissions among pregnant women and a five-fold increase in maternal deaths due to malaria (Hammerich et al., 2002). The social and economic costs of malaria are also huge and include considerable costs to individuals and households as well as high costs at community and national levels (Holding and Snow, 2001; Utzinger et al., 2001; Malaney et al., 2004).

Climate variability may also interact with other background stresses and additional vulnerabilities such as immunocompromised populations (HIV/AIDS) and conflict and war (Harrus and Baneth, 2005) in the future, resulting in increased susceptibility and risk of other infectious diseases (e.g., cholera) and malnutrition. The potential for climate change to intensify or alter flood patterns may become a major additional driver of future health risks from flooding (Few et al., 2004). The probability that sea-level rise could increase flooding, particularly on the coasts of eastern Africa (Nicholls, 2004), may also have implications for health (McMichael et al., 2006).

Relatively fewer assessments of possible future changes in animal health arising from climate variability and change have been undertaken. The demographic impacts on trypanosomiasis, for example, can arise through modification of the habitats suitable for the tsetse fly. These modifications can be further exacerbated by climate variability and climate change. Climate change is also expected to affect both pathogen and vector habitat suitability through changes in moisture and temperature (Baylis and Githeko, 2006). Changes in disease distribution, range, prevalence, incidence and seasonality can all be expected. However, there is low certainty about the degree of change. Rift Valley Fever epidemics, evident during the 1997/98 El Niño event in East Africa and associated with flooding, could increase with a higher frequency of El Niño events. Finally, heat stress and drought are likely to have further negative impacts on animal health and production of dairy products, as already observed in the USA (St-Pierre et al., 2003; see also Warren et al., 2006).

9.4.4 Agriculture

Results from various assessments of impacts of climate change on agriculture based on various climate models and SRES emissions scenarios indicate certain agricultural areas that may undergo negative changes. It is estimated that, by 2100, parts of the Sahara are likely to emerge as the most vulnerable, showing likely agricultural losses of between 2 and 7% of GDP. Western and central Africa are also vulnerable, with impacts ranging from 2 to 4%. Northern and southern Africa, however, are expected to have losses of 0.4 to 1.3% (Mendelsohn et al., 2000b).

More recent assessments combining global- and regional-scale analysis, impacts of climate change on growing periods and agricultural systems, and possible livelihood implications, have also been examined (Jones and Thornton, 2003; Huntingford et al., 2005; Thornton et al., 2006). Based on the A1FI scenario, both the HadCM3 and ECHAM4 GCMs agree on areas of change in the coastal systems of southern and eastern Africa (Figure 9.4). Under both the A1 and B1 scenarios, mixed rain-fed semi-arid systems are shown to be affected in the Sahel, as well as mixed rain-fed and highland perennial systems in the Great Lakes region and in other parts of East Africa. In the B1 world, marginal areas (e.g., semi-arid lands) become more marginal, with moderate impacts on coastal systems (Thornton et al., 2006; see Chapter 5, Section 5.4.2). Such changes in the growing period are important, especially when viewed against

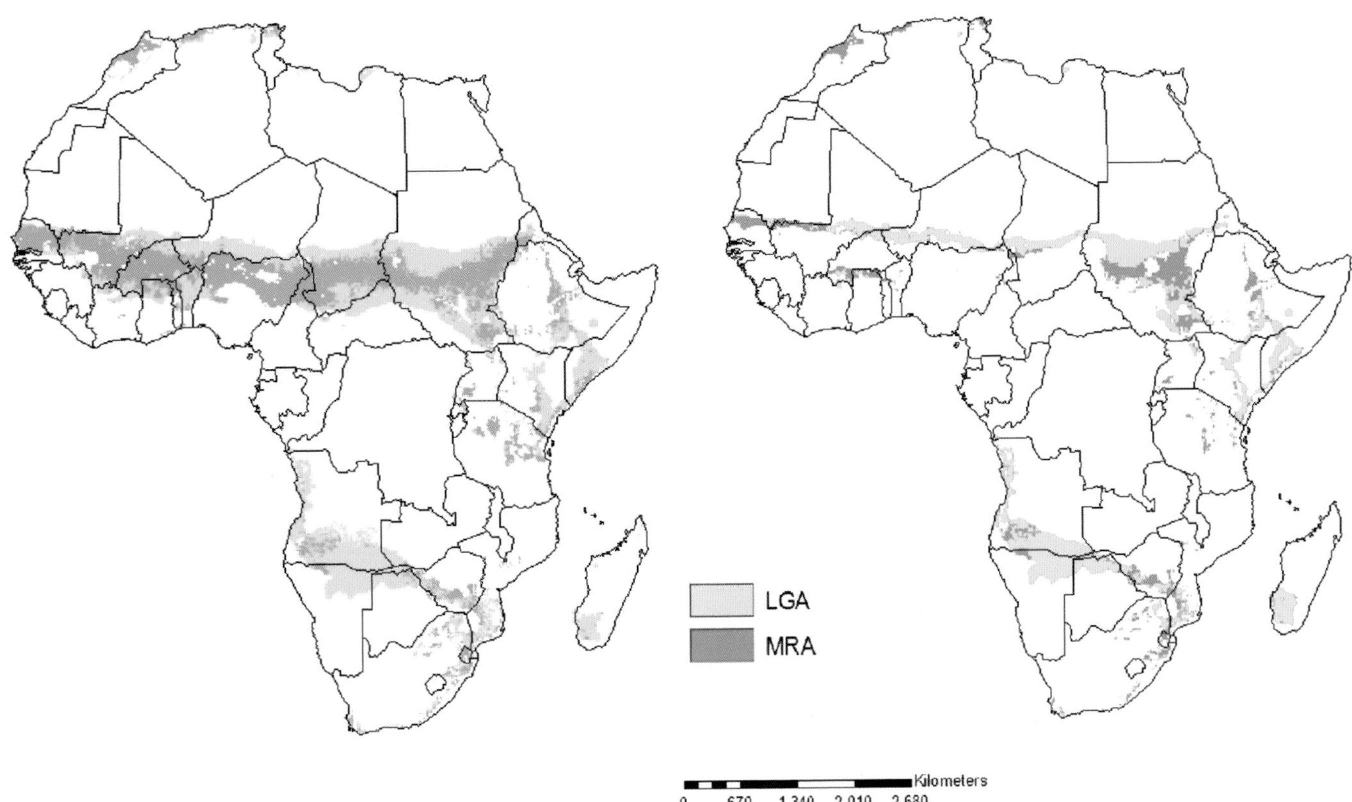

Figure 9.4. *Agricultural areas within the livestock-only systems (LGA) in arid and semi-arid areas, and rain-fed mixed crop/livestock systems (MRA) in semi-arid areas, are projected by the HadCM3 GCM to undergo >20% reduction in length of growing period to 2050, SRES A1 (left) and B1 (right) emissions scenarios, after Thornton et al. (2006).*

possible changes in seasonality of rainfall, onset of rain days and intensity of rainfall, as indicated in Sections 9.2.1 and 9.3.1.

Other recent assessments using the FAO/IIASA Agro-Ecological Zones model (AEZ) in conjunction with IIASA's world food system or Basic Linked System (BSL), as well as climate variables from five different GCMs under four SRES emissions scenarios, show further agricultural impacts such as changes in agricultural potential by the 2080s (Fischer et al., 2005). By the 2080s, a significant decrease in suitable rain-fed land extent and production potential for cereals is estimated under climate change. Furthermore, for the same projections, for the same time horizon the area of arid and semi-arid land in Africa could increase by 5-8% (60-90 million hectares). The study shows that wheat production is likely to disappear from Africa by the 2080s. On a more local scale, assessments have shown a range of impacts. Southern Africa would be likely to experience notable reductions in maize production under possible increased ENSO conditions (Stige et al., 2006).

In other countries, additional risks that could be exacerbated by climate change include greater erosion, deficiencies in yields from rain-fed agriculture of up to 50% during the 2000-2020 period, and reductions in crop growth period (Agoumi, 2003). A recent study on South African agricultural impacts, based on three scenarios, indicates that crop net revenues will be likely to fall by as much as 90% by 2100, with small-scale farmers being the most severely affected. However, there is the possibility that adaptation could reduce these negative effects (Benhin, 2006). In Egypt, for example, climate change could decrease national production of many crops (ranging from –11% for rice to –28% for soybeans) by 2050 compared with their production under current climate conditions (Eid et al., 2006). Other agricultural activities could also be affected by climate change and variability, including changes in the onset of rain days and the variability of dry spells (e.g., Reason et al., 2005; see also Chapter 5).

However, not all changes in climate and climate variability will be negative, as agriculture and the growing seasons in certain areas (for example, parts of the Ethiopian highlands and parts of southern Africa such as Mozambique), may lengthen under climate change, due to a combination of increased temperature and rainfall changes (Thornton et al., 2006). Mild climate scenarios project further benefits across African croplands for irrigated and, especially, dryland farms. However, it is worth noting that, even under these favourable scenarios, populated regions of the Mediterranean coastline, central, western and southern Africa are expected to be adversely affected (Kurukulasuriya and Mendelsohn, 2006a).

Fisheries are another important source of revenue, employment and proteins. They contribute over 6% of Namibia's and Senegal's GDP (Njaya and Howard, 2006). Climate-change impacts on this sector, however, need to be viewed together with other human activities, including impacts that may arise from governance of fresh and marine waters (AMCEN/UNEP, 2002). Fisheries could be affected by different biophysical impacts of climate change, depending on the resources on which they are based (Niang-Diop, 2005; Clark, 2006). With a rise in annual global temperature (e.g. of the order of 1.5 to 2.0°C) fisheries in North West Africa and the East African lakes are shown to be impacted (see ECF and Potsdam Institute, 2004; Warren et al., 2006). In coastal regions that have major lagoons or lake systems, changes in freshwater flows and a greater intrusion of salt water into lagoons will affect the species that are the basis of inland fisheries or aquaculture (République de Côte d'Ivoire, 2000; République du Congo, 2001; Cury and Shannon, 2004). In South Africa, fisheries could be affected by changes in estuaries, coral reefs and upwelling; with those that are dependent on the first two ecosystems being the most vulnerable (Clark, 2006). Recent simulations based on the NCAR GCM under a doubling of carbon dioxide indicate that extreme wind and turbulence could decrease productivity by 50-60%, while turbulence will probably bring about a 10% decline in productivity in the spawning grounds and an increase of 3% in the main feeding grounds (Clark et al., 2003).

The impact of climate change on livestock farming in Africa was examined by Seo and Mendelsohn (2006a, b). They showed that a warming of 2.5°C could increase the income of small livestock farms by 26% (+US$1.4 billion). This increase is projected to come from stock expansion. Further increases in temperature would then lead to a gradual fall in net revenue per animal. A warming of 5°C would probably increase the income of small livestock farms by about 58% (+US$3.2 billion), largely as a result of stock increases. By contrast, a warming of 2.5°C would be likely to decrease the income of large livestock farms by 22% (–US$13 billion) and a warming of 5°C would probably reduce income by as much as 35% (–US$20 billion). This reduction in income for large livestock farms would probably result both from a decline in the number of stock and a reduction in the net revenue per animal. Increased precipitation of 14% would be likely to reduce the income of small livestock farms by 10% (–US$ 0.6 billion), mostly due to a reduction in the number of animals kept. The same reduction in precipitation would be likely to reduce the income of large livestock farms by about 9% (–US$5 billion), due to a reduction both in stock numbers and in net revenue per animal.

The study by Seo and Mendelsohn (2006a) further shows that higher temperatures are beneficial for small farms that keep goats and sheep because it is easy to substitute animals that are heat-tolerant. By contrast, large farms are more dependent on species such as cattle, which are not heat-tolerant. Increased precipitation is likely to be harmful to grazing animals because it implies a shift from grassland to forests and an increase in harmful disease vectors, and also a shift from livestock to crops.

Assessing future trends in agricultural production in Africa, even without climate change, remains exceedingly difficult (e.g., contributions to GDP and impacts on GDP because of climate variability and other factors - see, for example, Mendelsohn et al., 2000b; Tiffen, 2003; Arrow et al., 2004; Desta and Coppock, 2004; Ferguson, 2006). While agriculture is a key source of livelihood in Africa, there is evidence that off-farm incomes are also increasing in some areas - up to 60 to 80% of total incomes in some cases (Bryceson, 2002). Urbanisation and off-farm increases in income also seem to be contributing to reduced farm sizes. Future scenarios and projections may thus need to include such changes, as well as relevant population estimates, allowing for the impact of HIV/AIDS, especially on farm labour productivity (Thornton et al., 2006).

9.4.5 Ecosystems

A range of impacts on terrestrial and aquatic ecosystems has been suggested under climate change (see, for example, Leemans and Eickhout, 2004), some of which are summarised in Table 9.1 (for further details see Chapter 4; Nkomo et al., 2006; Warren et al., 2006).

Mountain ecosystems appear to be undergoing significant observed changes (see Section 9.2.1.4), aspects of which are likely to be linked to complex climate-land interactions and which may continue under climate change (e.g., IPCC, 2007a). By 2020, for example, indications are that the ice cap on Mt. Kilimanjaro could disappear for the first time in 11,000 years (Thompson et al., 2002). Changes induced by climate change are also likely to result in species range shifts, as well as in changes in tree productivity, adding further stress to forest ecosystems (UNEP, 2004). Changes in other ecosystems, such as grasslands, are also likely (for more detail, see assessments by Muriuki et al., 2005; Levy, 2006).

Mangroves and coral reefs, the main coastal ecosystems in Africa, will probably be affected by climate change (see Chapter 4, Box 4.4; Chapter 6, Section 6.4.1, Box 6.1). Endangered species associated with these ecosystems, including manatees and marine turtles, could also be at risk, along with migratory birds (Government of Seychelles, 2000; Republic of Ghana, 2000; République Démocratique du Congo, 2000). Mangroves could also colonise coastal lagoons because of sea-level rise (République du Congo, 2001; Rocha et al., 2005).

The coral bleaching following the 1997/1998 extreme El Niño, as mentioned in Section 9.2.1, is an indication of the potential impact of climate change-induced ocean warming on coral reefs (Lough, 2000; Muhando, 2001; Obura, 2001); disappearance of low-lying corals and losses of biodiversity could also be expected (République de Djibouti, 2001; Payet and Obura, 2004). The proliferation of algae and dinoflagellates during these warming events could increase the number of people affected by toxins (such as ciguatera) due to the consumption of marine food sources (Union des Comores, 2002; see also Chapter 16, Section 16.4.5). In the long term, all these impacts will have negative effects on fisheries and tourism (see also Chapter 5, Box 5.4). In South Africa, changes in estuaries are expected mainly as a result of reductions in river runoff and the inundation of salt marshes following sea-level rise (Clark, 2006).

The species sensitivity of African mammals in 141 national parks in sub-Saharan Africa was assessed using two climate-change scenarios (SRES A2 and B2 emissions scenarios with the HadCM3 GCM, for 2050 and 2080), applying a simple IUCN Red List assessment of potential range loss (Thuiller et al., 2006). Assuming no migration of species, 10-15% of the species were projected to fall within the IUCN Critically Endangered or Extinct categories by 2050, increasing to 25-40% of species by 2080. Assuming unlimited species migration, the results were less extreme, with these proportions dropping to approximately 10-20% by 2080. Spatial patterns of loss and gain showed contrasting latitudinal patterns, with a westward range shift of species around the species-rich equatorial transition zone in central Africa, and an eastward shift in southern Africa; shifts which appear to be related mainly to the latitudinal aridity gradients across these ecological transition zones.

Table 9.1. *Significant ecosystem responses estimated in relation to climate change in Africa. These estimations are based on a variety of scenarios (for further details on models used and impacts see Chapter 4, Section 4.4 and Table 4.1).*

Ecosystem impacts	Area affected	Scenario used and source
About 5,000 African plant species impacted: substantial reductions in areas of suitable climate for 81-97% of the 5,197 African plants examined, 25-42% lose all area by 2085.	Africa	HadCM3 for years 2025, 2055, 2085, plus other models – shifts in climate suitability examined (McClean et al., 2005)
Fynbos and succulent Karoo biomes: losses of between 51 and 61%.	South Africa	Projected losses by 2050, see details of scenarios (Midgley et al., 2002; see Chapter 4, Section 4.4, Table 4.1)
Critically endangered taxa (e.g. Proteaceae): losses increase, and up to 2% of the 227 taxa become extinct.	Low-lying coastal areas	4 land use and 4 climate change scenarios (HadCM2 IS92aGGa) (Bomhard et al., 2005)
Losses of nyala and zebra: Kruger Park study estimates 66% of species lost.	Malawi South Africa (Kruger Park)	(Dixon et al., 2003) Hadley Centre Unified Model, no sulphates (Erasmus et al., 2002; see Chapter 4, Section 4.4.3)
Loss of bird species ranges: (restriction of movements). An estimated 6 species could lose substantial portions of their range.	Southern African bird species (Nama-Karoo area)	Projected losses of over 50% for some species by 2050 using the HadCM3 GCM with an A2 emissions scenario (Simmons et al., 2004; see Chapter 4, Section 4.4.8)
Sand-dune mobilisation: enhanced dune activity.	Southern Kalahari basin – northern South Africa, Angola and Zambia. For details in Sahel, see Section 9.6.2 and Chapter 4, Section 4.3.	Scenarios: HadCM3 GCM, SRES A2, B2 and A1fa, IS92a. By 2099 all dune fields shown to be highly dynamic (Thomas et al., 2005; see Chapter 4, Section 4.4.2)
Lake ecosystems, wetlands	Lake Tanganyika	Carbon isotope data show aquatic losses of about 20% with a 30% decrease in fish yields. It is estimated that climate change may further reduce lake productivity (O'Reilly et al., 2003; see Chapter 4, Section 4.4.8)
Grasslands	Complex impacts on grasslands including the role of fire (southern Africa)	See detailed discussion Chapter 4, Section 4.4.3

9.4.6 Coastal zones

In Africa, highly productive ecosystems (mangroves, estuaries, deltas, coral reefs), which form the basis for important economic activities such as tourism and fisheries, are located in the coastal zone. Forty percent of the population of West Africa live in coastal cities, and it is expected that the 500 km of coastline between Accra and the Niger delta will become a continuous urban megalopolis of more than 50 million inhabitants by 2020 (Hewawasam, 2002). By 2015, three coastal megacities of at least 8 million inhabitants will be located in Africa (Klein et al., 2002; Armah et al., 2005; Gommes et al., 2005). The projected rise in sea level will have significant impacts on these coastal megacities because of the concentration of poor populations in potentially hazardous areas that may be especially vulnerable to such changes (Klein et al., 2002; Nicholls, 2004). Cities such as Lagos and Alexandria will probably be impacted. In very recent assessments of the potential flood risks that may arise by 2080 across a range of SRES scenarios and climate change projections, three of the five regions shown to be at risk of flooding in coastal and deltaic areas of the world are those located in Africa: North Africa, West Africa and southern Africa (see Nicholls and Tol, 2006; for more detailed assessments, see Warren et al., 2006).

Other possible direct impacts of sea-level rise have been examined (Niang-Diop et al., 2005). In Cameroon, for example, indications are that a 15% increase in rainfall by 2100 would be likely to decrease the penetration of salt water in the Wouri estuary (République de Côte d'Ivoire, 2000). Alternatively, with an 11% decrease in rainfall, salt water could extend up to about 70 km upstream. In the Gulf of Guinea, sea-level rise could induce overtopping and even destruction of the low barrier beaches that limit the coastal lagoons, while changes in precipitation could affect the discharges of rivers feeding them. These changes could also affect lagoonal fisheries and aquaculture (République de Côte d'Ivoire, 2000). Indian Ocean islands could also be threatened by potential changes in the location, frequency and intensity of cyclones; while East African coasts could be affected by potential changes in the frequency and intensity of ENSO events and coral bleaching (Klein et al., 2002). Coastal agriculture (e.g., plantations of palm oil and coconuts in Benin and Côte d'Ivoire, shallots in Ghana) could be at risk of inundation and soil salinisation. In Kenya, losses for three crops (mangoes, cashew nuts and coconuts) could cost almost US$500 million for a 1 m sea-level rise (Republic of Kenya, 2002). In Guinea, between 130 and 235 km^2 of rice fields (17% and 30% of the existing rice field area) could be lost as a result of permanent flooding, depending on the inundation level considered (between 5 and 6 m) by 2050 (République de Guinée, 2002). In Eritrea, a 1 m rise in sea level is estimated to cause damage of over US$250 million as a result of the submergence of infrastructure and other economic installations in Massawa, one of the country's two port cities (State of Eritrea, 2001). These results confirm previous studies stressing the great socio-economic and physical vulnerability of settlements located in marginal areas.

9.4.7 Tourism

Climate change could also place tourism at risk, particularly in coastal zones and mountain regions. Important market changes could also result from climate change (World Tourism Organization, 2003) in such environments. The economic benefits of tourism in Africa, which according to 2004 statistics accounts for 3% of worldwide tourism, may change with climate change (World Tourism Organization, 2005). However, very few assessments of projected impacts on tourism and climate change are available, particularly those using scenarios and GCM outputs. Modelling climate changes as well as human behaviour, including personal preferences, choices and other factors, is exceedingly complex. Although scientific evidence is still lacking, it is probable that flood risks and water-pollution-related diseases in low-lying regions (coastal areas), as well as coral reef bleaching as a result of climate change, could impact negatively on tourism (McLeman and Smit, 2004). African places of interest to tourists, including wildlife areas and parks, may also attract fewer tourists under marked climate changes. Climate change could, for example, lead to a poleward shift of centres of tourist activity and a shift from lowland to highland tourism (Hamilton et al., 2005).

9.4.8 Settlements, industry and infrastructure

Climate variability, including extreme events such as storms, floods and sustained droughts, already has marked impacts on settlements and infrastructure (Freeman and Warner, 2001; Mirza, 2003; Niasse et al., 2004; Reason and Keibel, 2004). Indeed, for urban planners, the biggest threats to localised population concentrations posed by climate variability and change are often expected to be from little-characterised and unpredictable rapid-onset disasters such as storm surges, flash floods and tropical cyclones (Freeman, 2003). Negative impacts of climate change could create a new set of refugees, who may migrate into new settlements, seek new livelihoods and place additional demands on infrastructure (Myers, 2002; McLeman and Smit, 2005). A variety of migration patterns could thus emerge, e.g., repetitive migrants (as part of ongoing adaptation to climate change) and short-term shock migrants (responding to a particular climate event). However, few detailed assessments of such impacts using climate as a driving factor have been undertaken for Africa.

In summary, a range of possible impacts of climate change has been discussed in this section (for other summaries, see Epstein and Mills, 2005; Nkomo et al., 2006). The roles of some other stresses that may compound climate-induced changes have also been considered. Clearly, several areas require much more detailed investigation (particularly in the energy, tourism, settlement and infrastructure sectors). Despite the uncertainty of the science and the huge complexity of the range of issues outlined, initial assessments show that several regions in Africa may be affected by different impacts of climate change (Figure 9.5). Such impacts, it is argued here, may further constrain development and the attainment of the MDGs in Africa. Adaptive capacity and adaptation thus emerge as critical areas for consideration on the continent.

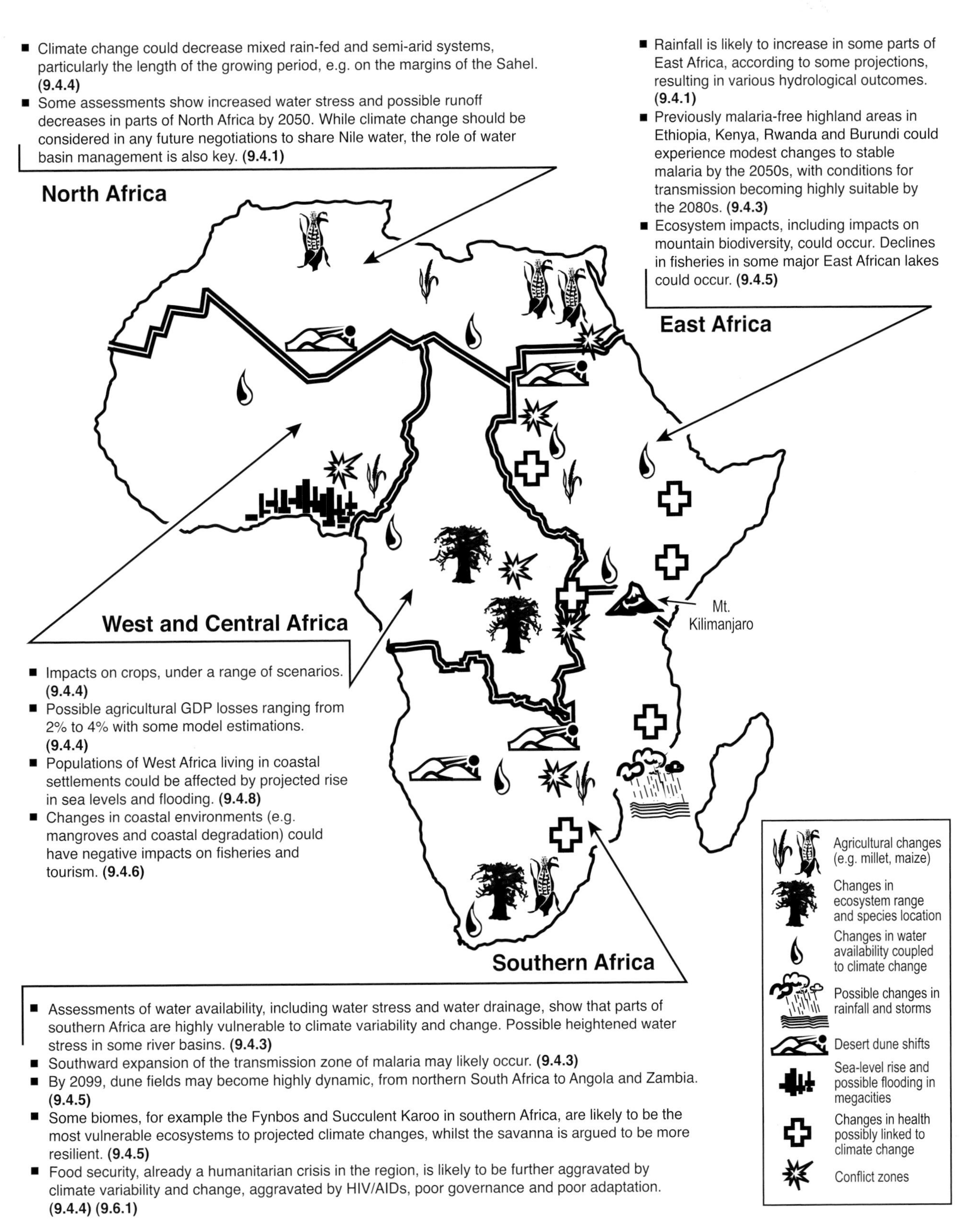

■ Climate change could decrease mixed rain-fed and semi-arid systems, particularly the length of the growing period, e.g. on the margins of the Sahel. (9.4.4)
■ Some assessments show increased water stress and possible runoff decreases in parts of North Africa by 2050. While climate change should be considered in any future negotiations to share Nile water, the role of water basin management is also key. (9.4.1)
North Africa
■ Rainfall is likely to increase in some parts of East Africa, according to some projections, resulting in various hydrological outcomes. (9.4.1)
■ Previously malaria-free highland areas in Ethiopia, Kenya, Rwanda and Burundi could experience modest changes to stable malaria by the 2050s, with conditions for transmission becoming highly suitable by the 2080s. (9.4.3)
■ Ecosystem impacts, including impacts on mountain biodiversity, could occur. Declines in fisheries in some major East African lakes could occur. (9.4.5)
East Africa
Mt. Kilimanjaro
West and Central Africa
■ Impacts on crops, under a range of scenarios. (9.4.4)
■ Possible agricultural GDP losses ranging from 2% to 4% with some model estimations. (9.4.4)
■ Populations of West Africa living in coastal settlements could be affected by projected rise in sea levels and flooding. (9.4.8)
■ Changes in coastal environments (e.g. mangroves and coastal degradation) could have negative impacts on fisheries and tourism. (9.4.6)
Southern Africa
■ Assessments of water availability, including water stress and water drainage, show that parts of southern Africa are highly vulnerable to climate variability and change. Possible heightened water stress in some river basins. (9.4.3)
■ Southward expansion of the transmission zone of malaria may likely occur. (9.4.3)
■ By 2099, dune fields may become highly dynamic, from northern South Africa to Angola and Zambia. (9.4.5)
■ Some biomes, for example the Fynbos and Succulent Karoo in southern Africa, are likely to be the most vulnerable ecosystems to projected climate changes, whilst the savanna is argued to be more resilient. (9.4.5)
■ Food security, already a humanitarian crisis in the region, is likely to be further aggravated by climate variability and change, aggravated by HIV/AIDs, poor governance and poor adaptation. (9.4.4) (9.6.1)
Agricultural changes (e.g. millet, maize)
Changes in ecosystem range and species location
Changes in water availability coupled to climate change
Possible changes in rainfall and storms
Desert dune shifts
Sea-level rise and possible flooding in megacities
Changes in health possibly linked to climate change
Conflict zones

Figure 9.5. *Examples of current and possible future impacts and vulnerabilities associated with climate variability and climate change for Africa (for details see sections highlighted in bold). Note that these are indications of possible change and are based on models that currently have recognised limitations.*

9.5 Adaptation constraints and opportunities

The covariant mix of climate stresses and other factors in Africa means that for many in Africa adaptation is not an option but a necessity (e.g., Thornton et al., 2006). A growing cohort of studies is thus emerging on adaptation to climate variability and change in Africa, examples of which are given below (see also Chapter 18). Owing to constraints of space, not all cases nor all details can be provided here. A range of factors including wealth, technology, education, information, skills, infrastructure, access to resources, and various psychological factors and management capabilities can modify adaptive capacity (e.g., Block and Webb, 2001; Ellis and Mdoe, 2003; Adger and Vincent, 2005; Brooks et al., 2005; Grothmann and Patt, 2005). Adaptation is shown to be successful and sustainable when linked to effective governance systems, civil and political rights and literacy (Brooks et al., 2005).

9.5.1 Adaptation practices

Of the emerging range of livelihood adaptation practices being observed (Table 9.2), diversification of livelihood activities, institutional architecture (including rules and norms of governance), adjustments in farming operations, income-generation projects and selling of labour (e.g., migrating to earn an income – see also Section 9.6.1) and the move towards off- or non-farm livelihood incomes in parts of Africa repeatedly surface as key adaptation options (e.g., Bryceson, 2004; Benhin, 2006; Osman-Elasha et al., 2006). As indicated in Section 9.2.1, reducing risks with regard to possible future events will depend on the building of stronger livelihoods to ensure resilience to future shocks (IFRCRCS, 2002). The role of migration as an adaptive measure, particularly as a response to drought and flood, is also well known. Recent evidence, however, shows that such migration is not only driven by periods of climate stress but is also driven by a range of other possible factors (see, for example, Section 9.2.2.9). Migration is a dominant mode of labour (seasonal migration), providing a critical livelihood source. The role of remittances derived from migration provides a key coping mechanism in drought and non-drought years but is one that can be dramatically affected by periods of climate shock, when adjustments to basic goods such as food prices are impacted by food aid and other interventions (Devereux and Maxwell, 2001).

Institutions and their effective functioning play a critical role in successful adaptation; it is therefore important to understand the design and functioning of such institutions (Table 9.2). The role of institutions at more local scales, both formal and informal institutions, also needs to be better understood (e.g., Reid and Vogel, 2006)

Other opportunities for adaptation that can be created include many linked to technology. The role of seasonal forecasts, and their production, dissemination, uptake and integration in model-based decision-making support systems, has been fairly extensively examined in several African contexts (Table 9.2). Significant constraints, however, include the limited support for climate risk management in agriculture and therefore a limited demand for such seasonal forecast products (e.g., O'Brien and Vogel, 2003).

Enhanced resilience to future periods of drought stress may also be supported by improvements in existing rain-fed farming systems (Rockström, 2003), such as water-harvesting systems to supplement irrigation practices in semi-arid farming systems ('more crop per drop' strategies, see Table 9.2). Improved early warning systems and their application may also reduce vulnerability to future risks associated with climate variability and change. In malaria research, for example, it has been shown that, while epidemics in the highlands have been associated with positive anomalies in temperature and rainfall (Githeko and Ndegwa, 2001; as discussed in Section 9.4.3), those in the semi-arid areas are mainly associated with excessive rainfall (Thomson et al., 2006). Using such climate information it may be possible to give outlooks with lead times of between 2 and 6 months before the onset of an event (Thomson et al., 2006). Such lead times provide opportunities for putting interventions in place and for preventing excessive morbidity and mortality during malaria epidemics.

In Africa, biotechnology research could also yield tremendous benefits if it leads to drought- and pest-resistant rice, drought-tolerant maize and insect-resistant millet, sorghum and cassava, among other crops (ECA, 2002). Wheat grain yield cultivated under current and future climate conditions (for example, increases of 1.5 and 3.6°C) in Egypt highlight a number of adaptation measures, including various technological options that may be required under an irrigated agriculture system (e.g., Abou-Hadid, 2006). A detailed study of current crop selection as an adaptation strategy to climate change in Africa (Kurukulasuriya and Mendelsohn, 2006b) shows that farmers select sorghum and maize-millet in the cooler regions of Africa, maize-beans, maize-groundnut and maize in moderately warm regions, and cowpea, cowpea-sorghum and millet-groundnut in hot regions. The study further shows that farmers choose sorghum and millet-groundnut when conditions are dry, cowpea, cowpea-sorghum, maize-millet and maize when medium-wet, and maize-beans and maize-groundnut when very wet. As the weather becomes warmer, farmers tend to shift towards more heat-tolerant crops. Depending upon whether precipitation increases or decreases, farmers will shift towards water-loving or drought-tolerant crops, respectively.

The design and use of proactive rather than reactive strategies can also enhance adaptation. Proactive, *ex ante*, interventions, such as agricultural capital stock and extension advice in Zimbabwe (Owens et al., 2003), can raise household welfare and heighten resilience during non-drought years. In many cases these interventions can also be coupled with disaster risk-reduction strategies (see several references on http://www.unisdr.org/). Capital and extension services can also increase net crop incomes without crowding-out net private transfers. Other factors that could be investigated to enhance resilience to shocks such as droughts include: national grain reserves, grain future markets, weather insurance, the role of food price subsidies, cash transfers and school feeding schemes (for a detailed discussion, see Devereux, 2003).

Table 9.2. *Some examples of complex adaptations already observed in Africa in response to climate and other stresses (adapted from the initial categorisation of Rockström, 2003).*

Theme	Emerging characteristics of adaptation	Authors
	Social resilience	
Social networks and social capital	• Perceptions of risks by rural communities are important in configuring the problem (e.g., climate risk). Perceptions can shape the variety of adaptive actions taken. • Networks of community groups are also important. • Local savings schemes, many of them based on regular membership fees, are useful financial 'stores' drawn down during times of stress.	Ellis and Bahiigwa, 2003; Quinn et al., 2003; Eriksen et al., 2005; Grothmann and Patt, 2005.
Institutions	• Role and architecture of institutional design and function is critical for understanding and better informing policies/measures for enhanced resilience to climate change. • Interventions linked to governance at various levels (state, region and local levels) either enhance or constrain adaptive capacity.	Batterbury and Warren, 2001; Ellis and Mdoe, 2003; Owuor et al., 2005; Osman-Elasha et al., 2006; Reid and Vogel, 2006.
	Economic resilience	
Equity	• Issues of equity need to be viewed on several scales • ***Local scale***: (within and between communities) - Interventions to enhance community resilience can be hampered by inaccessibility of centres for obtaining assistance (aid/finance) • ***Global scale***: see IPCC, 2007b, re CDMS etc.	Sokona and Denton, 2001; AfDB et al., 2002; Thomas and Twyman, 2005.
Diversification of livelihoods	• Diversification has been shown to be a very strong and necessary economic strategy to increase resilience to stresses. • Agricultural intensification, for example based on increased livestock densities, the use of natural fertilisers, soil and water conservation, can be useful adaptation mechanisms.	Ellis, 2000; Toulmin et al., 2000; Block and Webb, 2001; Mortimore and Adams, 2001; Ellis, 2003; Ellis and Mdoe, 2003; Eriksen and Silva, 2003; Bryceson, 2004; Chigwada, 2005.
Technology	• Seasonal forecasts, their production, dissemination, uptake and integration in model-based decision-making support systems have been examined in several African contexts (see examples given). • Enhanced resilience to future periods of drought stress may also be supported by improvements in present rain-fed farming systems through: - water-harvesting systems; - dam building; - water conservation and agricultural practices; - drip irrigation; - development of drought-resistant and early-maturing crop varieties and alternative crop and hybrid varieties.	Patt, 2001; Phillips et al., 2001; Roncoli et al., 2001; Hay et al., 2002b; Monyo, 2002; Patt and Gwata, 2002; Archer, 2003; Rockström, 2003; Ziervogel and Calder, 2003; Gabre-Madhin and Haggblade, 2004; Malaney et al., 2004; Ziervogel, 2004; Ziervogel and Downing, 2004; Chigwada, 2005; Orindi and Ochieng, 2005; Patt et al., 2005; Matondo et al., 2005; Seck et al., 2005; Van Drunen et al., 2005; Ziervogel et al., 2005; Abou-Hadid, 2006; Osman-Elasha et al., 2006.
Infrastructure	• Improvements in the physical infrastructure may improve adaptive capacity. • Improved communication and road networks for better exchange of knowledge and information. • General deterioration in infrastructure threatens the supply of water during droughts and floods.	Sokona and Denton, 2001.

9.5.2 Adaptation costs, constraints and opportunities

Many of the options outlined above come with a range of costs and constraints, including large transaction costs. However, deriving quantitative estimates of the potential costs of the impacts of climate change (or those associated with climate variability, such as droughts and floods) and costs without adaptation (Yohe and Schlesinger, 2002) is difficult. Limited availability of data and a variety of uncertainties relating to future changes in climate, social and economic conditions, and the responses that will be made to address those changes, frustrate precise cost and economic loss inventories. Despite these problems, some economic loss inventories and estimations have been undertaken (e.g., Mirza, 2003). In some cases (e.g., Egypt and Senegal), assessments have attempted to measure costs that may arise with and without adaptation to climate-change impacts. Large populations are estimated to be at risk of impacts linked to possible climate change. Assessments of the impacts of sea-level rise in coastal countries show that costs of adaptation could amount to at least 5-10% of GDP (Niang-Diop, 2005). However, if no adaptation is undertaken, then the losses due to climate change could be up to 14% GDP (Van Drunen et al., 2005). In South Africa, initial assessments of the costs of adaptation in the Berg River Basin also show that the costs of not adapting to climate change can be much greater than the costs of including flexible and efficient approaches to adapting to climate change into management options (see Stern, 2007).

Despite some successes (see examples in Table 9.2), there is also evidence of an erosion of coping and adaptive strategies as a result of varying land-use changes and socio-political and cultural stresses. Continuous cultivation, for example, at the expense of soil replenishment, can result in real 'agrarian dramas' (e.g., Rockström, 2003). The interaction of both social (e.g., access to food) and biophysical (e.g., drought) stresses thus combine to aggravate critical stress periods (e.g., during and after ENSO events). Traditional coping strategies (see Section 9.6.2) may not be sufficient in this context, either currently or in the future, and may lead to unsustainable responses in the longer term. Erosion of traditional coping responses not only reduces resilience to the next climatic shock but also to the full range of shocks and stresses to which the poor are exposed (DFID, 2004). Limited scientific capacity and other scientific resources are also factors that frustrate adaptation (see, e.g., Washington et al., 2004, 2006).

As shown in several sections in this chapter, the low adaptive capacity of Africa is due in large part to the extreme poverty of many Africans, frequent natural disasters such as droughts and floods, agriculture that is heavily dependent on rainfall, as well as a range of macro- and micro-structural problems (see Section 9.2.2). The full implications of climate change for development are, however, currently not clearly understood. For example, factors heightening vulnerability to climate change and affecting national-level adaptation have been shown to include issues of local and national governance, civil and political rights and literacy (e.g., Brooks et al., 2005). The most vulnerable nations in the assessment undertaken by Brooks et al. (2005) (using mortality from climate-related disasters as an indication of climate outcomes) were those situated in sub-Saharan Africa and those that have recently experienced conflict. At the more local level, the poor often cannot adopt diversification as an adaptive strategy and often have very limited diversification options available to them (e.g., Block and Webb, 2001; Ellis and Mdoe, 2003). Micro-financing and other social safety nets and social welfare grants, as a means to enhance adaptation to current and future shocks and stresses, may be successful in overcoming such constraints if supported by local institutional arrangements on a long-term sustainable basis (Ellis, 2003; Chigwada, 2005).

Africa needs to focus on increasing adaptive capacity to climate variability and climate change over the long term. *Ad hoc* responses (e.g., short-term responses, unco-ordinated processes, isolated projects) are only one type of solution (Sachs, 2005). Other solutions that could be considered include mainstreaming adaptation into national development processes (Huq and Reid, 2004; Dougherty and Osman, 2005). There may be several opportunities to link disaster risk reduction, poverty and development (see, for example, several calls and plans for such action such as the Hyogo Declaration - http://www.unisdr.org/wcdr/intergover/official-doc/L-docs/Hyo-go-declaration-english.pdf). Where communities live with various risks, coupling risk reduction and development activities can provide additional adaptation benefits (e.g., Yamin et al., 2005). Unprecedented efforts by governments, humanitarian and development agencies to collaborate in order to find ways to move away from reliance on short-term emergency responses to food insecurity to longer-term development-oriented strategies that involve closer partnerships with governments, are also increasing (see food insecurity case study below and SARPN - http://www.sarpn.org/ - for several case studies and examples; see also Table 9.2 for other possible adaptation options).

Notwithstanding these efforts and suggestions, the context and the realities of the causes of vulnerability to a range of stresses, not least climate change and variability, must be kept at the forefront, including a deeper and further examination of the causes of poverty (both structural and other) at international, national and local levels (Bryceson, 2004). The causes, impacts and legacies of various strategies - including liberalisation policies, decades of structural adjustment programmes (SAP) and market conditions - cannot be ignored in discussions on poverty alleviation and adaptation to stresses, including climate change. Some of the complex interactions of such drivers and climate are further illustrated in the two case studies below.

9.6 Case studies

9.6.1 Food insecurity: the role of climate variability, change and other stressors

It has long been recognised that climate variability and change have an impact on food production, (e.g., Mendelsohn et al., 2000a, b; Devereux and Maxwell, 2001; Fischer et al., 2002; Kurukulasuriya and Rosenthal, 2003), although the extent and nature of this impact is as yet uncertain. Broadly speaking, food security is less seen in terms of sufficient global and national agricultural food production, and more in terms of livelihoods that are sufficient to provide enough food for individuals and households (Devereux and Maxwell, 2001; Devereux, 2003; Gregory et al., 2005). The key recognition in this shifting focus is that there are multiple factors, at all scales, that impact on an individual or household's ability to access sufficient food: these include household income, human health, government policy, conflict, globalisation, market failures, as well as environmental issues (Devereux and Maxwell, 2001; Marsland, 2004; Misselhorn, 2005).

Building on this recognition, three principal components of food security may be identified:

i. the *availability* of food (through the market and through own production);
ii. adequate purchasing and/or relational power to acquire or *access* food;
iii. the acquisition of sufficient nutrients from the available food, which is influenced by the ability to digest and absorb *nutrients* necessary for human health, access to safe drinking water, environmental hygiene and the nutritional content of the food itself (Swaminathan, 2000; Hugon and Nanterre, 2003).

Climate variability, such as periods of drought and flood as well as longer-term change, may – either directly or indirectly – profoundly impact on all these three components in shaping food security (Ziervogel et al., 2006; Figure 9.6).

The potential impacts of climate change on *food access* in Figure 9.6 may, for example, be better understood in the light of changes in Africa's livelihoods landscape. A trajectory of diversification out of agricultural-based activities – 'deagrarianisation' – has been found in the livelihoods of rural people in many parts of sub-Saharan Africa. Less reliance on food production as a primary source of people's food security runs counter to the assumption that people's food security in Africa derives solely (or even primarily) from their own

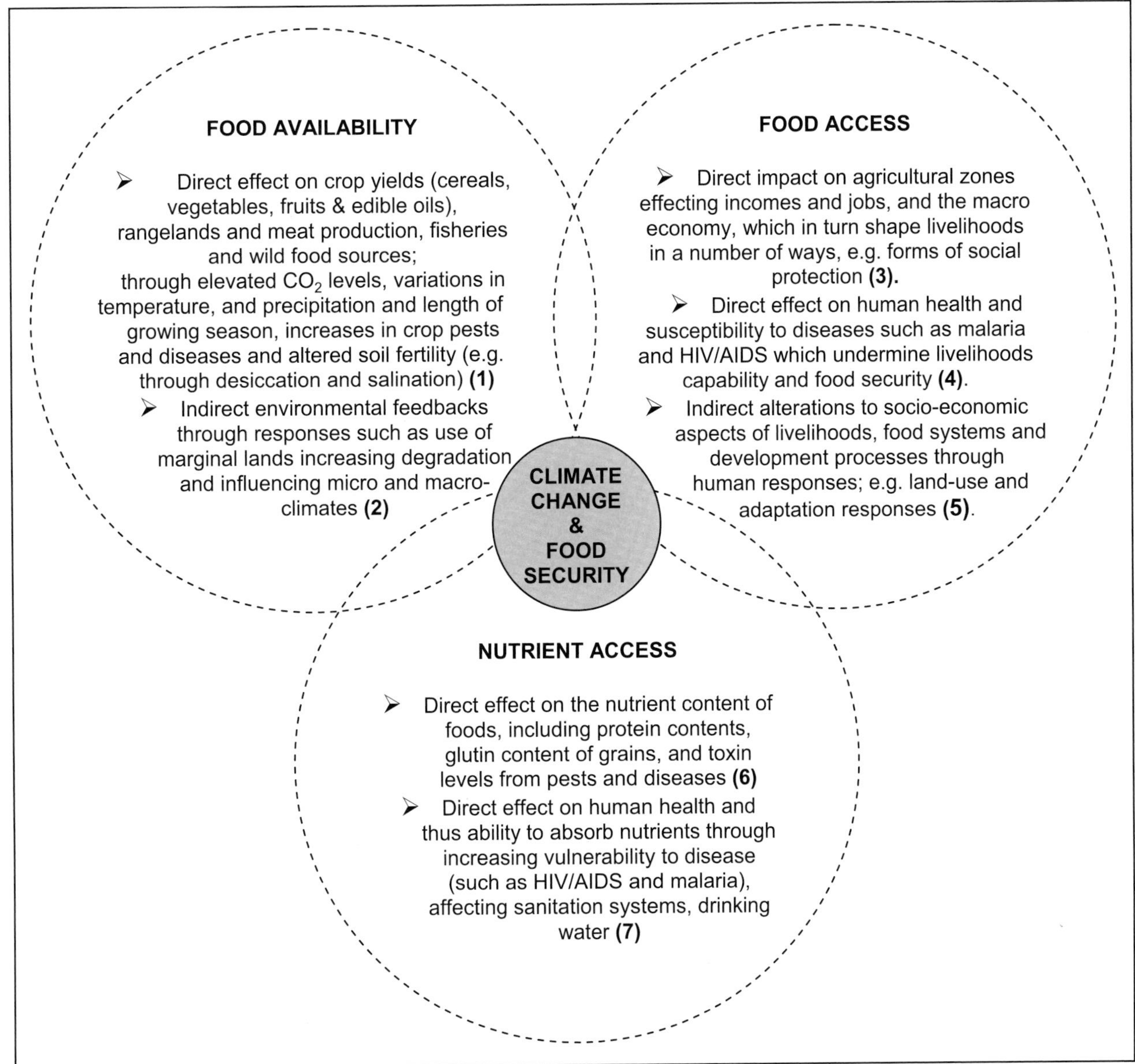

Figure 9.6. *Linkages identified between climate change in Africa and three major components of food security. Adapted from inputs of (1) Swaminathan, 2000; Fischer et al., 2002; Turpie et al., 2002; Rosegrant and Cline, 2003; Slingo et al., 2005. (2) Fischer et al., 2002; Slingo et al., 2005. (3) Turpie et al., 2002; African Union, 2005. (4) Piot and Pinstrup-Anderson, 2002; Turpie et al., 2002; Mano et al., 2003; USAID, 2003; Gommes et al., 2004; van Lieshout et al., 2004. (5) Adger and Vincent, 2005; Brooks et al., 2005; Gregory et al., 2005; Thomas and Twyman, 2005; O'Brien, 2006. (6) Slingo et al., 2005. (7) Swaminathan, 2000; Schulze et al., 2001; Gommes et al., 2004.*

agricultural production (Bryceson, 2000, 2004; Bryceson and Fonseca, 2006). At the same time, however, for the continent as a whole, the agriculture sector, which is highly dependent on precipitation, is estimated to account for approximately 60% of total employment, indicating its crucial role in livelihoods and food security derived through food access through purchase (Slingo et al., 2005).

There are a number of other illustrative impacts that climate variability and change have on livelihoods and food access, many of which also impact on food availability and nutrient access aspects of food security. These include impacts on the tourism sector (e.g., Hamilton et al., 2005), and on market access, which both affect the ability of farmers to obtain agricultural inputs, sell surplus crops, and purchase alternative foods. These impacts affect food security through altering or restraining livelihood strategies, while also affecting the variety of foods available and nutritional intake (Kelly et al., 2003). Market access is influenced not only by broader socio-economic and political factors, but also by distance from markets and the condition of the infrastructure, such as roads, which can be damaged during climate events (e.g., Abdulai and Crolerees, 2001; Ellis, 2003).

The key issues, therefore, in relation to the potential impacts of climate variability and change on food security in Africa encompass not only a narrow understanding of such impacts on food production but also a wider understanding of how such changes and impacts might interact with other environmental, social, economic and political factors that determine the vulnerability of households, communities and countries, as well as their capacity to adapt (Swaminathan, 2000; Adger and Vincent, 2005; Brooks et al., 2005). The impact of climate variability and change on food security therefore cannot be considered independently of the broader issue of human security (O'Brien, 2006). The inclusion of climate variability and change in understanding human vulnerability and adaptation is being increasingly explored at household and community levels, as well as though regional agro-climatological studies in Africa (e.g., Verhagen et al., 2001).

A number of studies have been undertaken that show that resource-poor farmers and communities use a variety of coping and adaptive mechanisms to ensure food security and sustainable livelihoods in the face of climate change and variability (see also Table 9.2). Adaptive capacity and choices, however, are based on a variety of complex causal mechanisms. Crop choices, for example, are not based purely on resistance to drought or disease but on factors such as cultural preferences, palatability, and seed storage capacity (Scoones et al., 2005). Research elsewhere in the world also indicates that elements of social capital (such as associations, networks and levels of trust) are important determinants of social resilience and responses to climate change, but how these develop and are used in mitigating vulnerability remains unclear.

While exploring the local-level dynamics of people's vulnerability to climate change, of which adaptive capacity is a key component, it is important to find ways to embed such findings into wider scales of assessment (e.g., country and regional scales) (Brooks et al., 2005). A number of recent studies are beginning to probe the enormous challenges of developing scenarios of adaptive capacity at multiple scales. From these studies, a complex range of factors, including behavioural economics (Grothmann and Patt, 2005), national aspirations and socio-political goals (Haddad, 2005), governance, civil and political rights and literacy, economic well-being and stability, demographic structure, global interconnectivity, institutional stability and well-being, and natural resource dependence (Adger and Vincent, 2005), are all emerging as powerful determinants of vulnerability and the capacity to adapt to climate change. Such determinants permeate through food 'systems' to impact on food security at various levels. Attainment of the Millennium Development Goals, particularly the first goal of eradicating extreme poverty and hunger, in the face of climate change will therefore require science that specifically considers food insecurity as an integral element of human vulnerability within the context of complex social, economic, political and biophysical systems, and that is able to offer usable findings for decision-makers at all scales.

9.6.2 Indigenous knowledge systems

The term 'indigenous knowledge' is used to describe the knowledge systems developed by a community as opposed to the scientific knowledge that is generally referred to as 'modern' knowledge (Ajibade, 2003). Indigenous knowledge is the basis for local-level decision-making in many rural communities. It has value not only for the culture in which it evolves, but also for scientists and planners striving to improve conditions in rural localities. Incorporating indigenous knowledge into climate-change policies can lead to the development of effective adaptation strategies that are cost-effective, participatory and sustainable (Robinson and Herbert, 2001).

9.6.2.1 Indigenous knowledge in weather forecasting

Local communities and farmers in Africa have developed intricate systems of gathering, predicting, interpreting and decision-making in relation to weather. A study in Nigeria, for example, shows that farmers are able to use knowledge of weather systems such as rainfall, thunderstorms, windstorms, harmattan (a dry dusty wind that blows along the north-west coast of Africa) and sunshine to prepare for future weather (Ajibade and Shokemi, 2003). Indigenous methods of weather forecasting are known to complement farmers' planning activities in Nigeria. A similar study in Burkina Faso showed that farmers' forecasting knowledge encompasses shared and selective experiences. Elderly male farmers formulate hypotheses about seasonal rainfall by observing natural phenomena, while cultural and ritual specialists draw predictions from divination, visions or dreams (Roncoli et al., 2001). The most widely relied-upon indicators are the timing, intensity and duration of cold temperatures during the early part of the dry season (November to January). Other forecasting indicators include the timing of fruiting by certain local trees, the water level in streams and ponds, the nesting behaviour of small quail-like birds, and insect behaviour in rubbish heaps outside compound walls (Roncoli et al., 2001).

9.6.2.2 Indigenous knowledge in mitigation and adaptation

African communities and farmers have always coped with changing environments. They have the knowledge and practices to cope with adverse environments and shocks. The enhancement of indigenous capacity is a key to the empowerment of local communities and their effective participation in the development process (Leautier, 2004). People are better able to adopt new ideas when these can be seen in the context of existing practices. A study in Zimbabwe observed that farmers' willingness to use seasonal climate forecasts increased when the forecasts were presented in conjunction with and compared with the local indigenous climate forecasts (Patt and Gwata, 2002).

Local farmers in several parts of Africa have been known to conserve carbon in soils through the use of zero-tilling practices in cultivation, mulching, and other soil-management techniques (Dea and Scoones, 2003). Natural mulches moderate soil temperatures and extremes, suppress diseases and harmful pests, and conserve soil moisture. The widespread use of indigenous plant materials, such as agrochemicals to combat pests that normally attack food crops, has also been reported among small-scale farmers (Gana, 2003). It is likely that climate change will alter the ecology of disease vectors, and such indigenous practices of pest management would be useful adaptation strategies. Other indigenous strategies that are adopted by local farmers include: controlled bush clearing; using tall grasses such as *Andropogon gayanus* for fixing soil surface nutrients washed away by runoff; erosion-control bunding to reduce significantly the effects of runoff; restoring lands by using green manure; constructing stone dykes; managing low-lying lands and protecting river banks (AGRHYMET, 2004).

Adaptation strategies that are applied by pastoralists in times of drought include the use of emergency fodder, culling of weak livestock for food, and multi-species composition of herds to survive climate extremes. During drought periods, pastoralists and agro-pastoralists change from cattle to sheep and goat husbandry, as the feed requirements of the latter are lower (Seo and Mendelsohn, 2006b). The pastoralists' nomadic mobility reduces the pressure on low-capacity grazing areas through their cyclic movements from the dry northern areas to the wetter southern areas of the Sahel.

African women are particularly known to possess indigenous knowledge which helps to maintain household food security, particularly in times of drought and famine. They often rely on indigenous plants that are more tolerant to droughts and pests, providing a reserve for extended periods of economic hardship (Ramphele, 2004; Eriksen, 2005). In southern Sudan, for example, women are directly responsible for the selection of all sorghum seeds saved for planting each year. They preserve a spread of varieties of seeds that will ensure resistance to the range of conditions that may arise in any given growing season (Easton and Roland, 2000).

9.7 Conclusion: links between climate change and sustainable development

African people and the environment have always battled the vagaries of weather and climate (see Section 9.2.1). These struggles, however, are increasingly waged alongside a range of other stresses, such as HIV/AIDS, conflict and land struggles (see Section 9.2.2). Despite good economic growth in some countries and sectors in Africa (OECD, 2004/2005), large inequalities still persist, and some sources suggest that hopes of reaching the MDGs by 2015 are slipping (UNDP, 2005). While climate change may not have featured directly in the setting of the MDGs, it is clear from the evidence presented here that climate change and variability may be an additional impediment to achieving them (Table 9.3; Thornton et al., 2006).

Although future climate change seems to be marginally important when compared to other development issues (Davidson et al., 2003), it is clear that climate change and variability, and associated increased disaster risks, will seriously hamper future development. On an annual basis, for example, developing countries have already absorbed US$35 billion in direct losses from natural disasters (Mirza, 2003). However, these figures do not include livelihood assets and losses and overall emotional and other stresses that are often more difficult to assess. A challenge, therefore, is to shape and manage development that also builds resilience to shocks, including those related to climate change and variability (Davidson et al., 2003; Adger et al., 2004).

9.8 Key uncertainties, confidence levels, unknowns, research gaps and priorities

While much is being discovered about climate variability and change, the impacts and possible responses to such changes result in significant areas that require more concerted effort and learning.

9.8.1 Uncertainties, confidence levels and unknowns

- While climate models are generally consistent regarding the direction of warming in Africa, projected changes in precipitation are less consistent.
- The role of land-use and land-cover change (i.e., land architecture in various guises) emerges as a key theme. The links between land-use changes, climate stress and possible feedbacks are not yet clearly understood.
- The contribution of climate to food insecurity in Africa is still not fully understood, particularly the role of other multiple stresses that enhance impacts of droughts and floods and possible future climate change. While drought may affect production in some years, climate variability alone does not explain the limits of food production in Africa. Better models and methods to improve understanding of multiple stresses, particularly at a range of scales, e.g., global, regional and local, and including the role of climate change and variability, are therefore required.
- Several areas of debate and contention, some shown here, also exist, with particular reference to health, the water sector and certain ecosystem responses, e.g., in mountain environments. More research on such areas is clearly needed.
- Impacts in the water sector, while addressed by global- and regional-scale model assessments, are still relatively poorly researched, particularly for local assessments and for groundwater impacts. Detailed 'systems' assessments, including hydrological systems assessments, also need to be expanded upon.
- Several of the impacts and vulnerabilities presented here derived from global models do not currently resolve local-level changes and impacts. Developing and improving regional and local-level climate models and scenarios could improve the confidence attached to the various projections.
- Local-scale assessments of various sorts, including adaptation studies, are still focused on understanding current vulnerabilities and adaptation strategies. Few comprehensive, comparable studies are available within regions, particularly those focusing on future options and pathways for adaptation.
- Finally, there is still much uncertainty in assessing the role of climate change in complex systems that are shaped by interacting multiple stressors. Preliminary investigations give some indications of these interactions, but further analysis is required.

Table 9.3. *Potential impacts of climate change on the Millennium Development Goals (after AfDB et al., 2002; Thornton et al., 2006).*

Millennium Development Goals: climate change as a cross-cutting issue	
Potential impacts	***Millennium Development Goal****
Climate Change (CC) may reduce poor people's livelihood assets, for example health, access to water, homes and infrastructure. It may also alter the path and rate of economic growth due to changes in natural systems and resources, infrastructure and labour productivity. A reduction in economic growth directly impacts poverty through reduced income opportunities. In addition to CC, expected impacts on regional food security are likely, particularly in Africa, where food security is expected to worsen (see Sections 9.4.1, 9.4.3, 9.4.4 and 9.4.8).	Eradicate extreme poverty and hunger (**Goal 1**)
Climate change is likely to directly impact children and pregnant women because they are particularly susceptible to vector- and water-borne diseases, e.g., malaria is currently responsible for a quarter of maternal mortality. Other expected impacts include: • increased heat-related mortality and illness associated with heatwaves (which may be balanced by less winter-cold-related deaths in some countries); • increased prevalence of some vector-borne diseases (e.g., malaria, dengue fever), and vulnerability to water, food or person-to-person diseases (e.g. cholera, dysentery) (see Section 9.4.3); • declining quantity and quality of drinking water, which worsens malnutrition, since it is a prerequisite for good health; • reduced natural resource productivity and threatened food security, particularly in sub-Saharan Africa (see Sections 9.4.3, 9.4.3, 9.4.4, 9.6.1).	Health-related goals: • reduce infant mortality (**Goal 4**); • improve maternal health (**Goal 5**); • combat major diseases (**Goal 6**).
Direct impacts: • Climate change may alter the quality and productivity of natural resources and ecosystems, some of which may be irreversibly damaged, and these changes may also decrease biological diversity and compound existing environmental degradation (see Section 9.4.4). • Climate change would alter the ecosystem-human interfaces and interactions that may lead to loss of biodiversity and hence erode the basic support systems for the livelihood of many people in Africa (see Section 9.4, Table 9.1 and Chapter 4).	Ensure environmental sustainability (**Goal 7**)
Indirect impacts: links to climate change include: • Loss of livelihood assets (natural, health, financial and physical capital) may reduce opportunities for full time education in numerous ways. • Natural disasters and drought reduce children's available time (which may be diverted to household tasks), while displacement and migration can reduce access to education opportunities (see Sections 9.2.1 and 9.2.2).	Achieve universal primary education (**Goal 2**)
One of the expected impacts of climate change is that it could exacerbate current gender inequalities, through impacting on the natural resource base, leading to decreasing agricultural productivity. This may place additional burdens on women's health, and reduce time available to participate in decision-making and for practicing income-generation activities. Climate-related disasters have been found to impact female-headed households, particularly where they have fewer assets (see Section 9.7.1, Table 9.2).	Promote gender equality and empower women (**Goal 3**)
Global climate change is a global issue, and responses require global co-operation, especially to help developing countries adapt to the adverse impacts of climate change.	Global partnerships (**Goal 8**)

* The order in which the Millennium Development Goals are listed here places the goals that could be directly impacted first, followed by those that are indirectly impacted.

9.8.2 Research gaps and priorities

As shown at the outset of this chapter, there has been a substantial shift from an *impacts-led* approach to a *vulnerability-led* approach in climate-change science. Despite this shift, much of the climate-change research remains focused on impacts. For Africa, however, as this chapter has attempted to show, a great deal more needs to be done in order to understand and show the interactions between vulnerability and adaptation to climate change and variability and the consequences of climate variability and change both in the short and long term.

9.8.2.1 Climate

Notwithstanding the marked progress made in recent years, particularly with model assessments (e.g., in parts of Africa, see Christensen et al., 2007), the climate of many parts of Africa is still not fully understood. Climate scenarios developed from GCMs are very coarse and do not usually adequately capture important regional variations in Africa's climate. The need exists to further develop regional climate models and sub-regional models at a scale that would be meaningful to decision-makers and to include stakeholders in framing some of the issues that may require more investigation. A further need is an improved understanding of climate variability, including an adequate representation of the climate system and the role of regional oceans and diverse feedback mechanisms.

9.8.2.2 Water

Detailed, regional-scale research on the impact of, and vulnerability to, climate change and variability with reference to water is needed; e.g., for African watersheds and river basins including the complex interactions of water governance in these areas. Water quality and its relation to water-usage patterns are also important issues that need to be incorporated into future projections. Further research on the impacts of climate variability and change on groundwater is also needed.

9.8.2.3 Energy

There is very little detailed information on the impacts and vulnerabilities of the energy sector in Africa specific to climate change and variability, particularly using and applying SRES scenarios and GCMs outputs. There is also a need to identify and assess the barriers (technical, economic and social) to the transfer and adoption of alternative and renewable energy sources, specifically solar energy, as well as the design, implications, impacts and possible benefits of current mitigation options (e.g., Clean Development Mechanisms (CDMs), including carbon sequestration).

9.8.2.4 Ecosystems

There is a great need for a well-established programme of research and technology development in climate prediction, which could assess the risks and impacts of climate change on ecosystems. Assessment of the impacts of climate variability and change on important, sensitive and unique ecosystems in Africa (hotspots), on the rainforests of the Congo Basin, on other areas of mountain biodiversity, as well as inland and on marine fish stocks, still requires further research.

9.8.2.5 Tourism

There is a need to enhance practical research regarding the vulnerability and impacts of climate change on tourism, as tourism is one of the most important and highly promising economic activities in Africa. Large gaps appear to exist in research on the impacts of climate variability and change on tourism and related matters, such as the impacts of climate change on coral reefs and how these impacts might affect ecotourism.

9.8.2.6 Health

Most assessments on health have concentrated on malaria, and there are still debates on the attribution of malaria resurgence in some African areas. The need exists to examine the impacts of future climate change on other health problems, e.g., dengue fever, meningitis, etc, and their associated vulnerabilities. There is also an urgent need to begin a dialogue and research effort on the heightened vulnerabilities associated with HIV/AIDS and periods of climate stress and climate change.

9.8.2.7 Agriculture

More regional and local research is still required on a range of issues, such as the study of the relationship between CO_2-enrichment and future production of agricultural crops in Africa, salt-tolerant plants, and other trees and plants in coastal zones. Very little research has been conducted on the impacts of climate change on livestock, plant pests and diseases. The livestock sector is very important in Africa and is considered very vulnerable to climate variability and change. Research on the links between agriculture, land use, and carbon sequestration and agricultural use in biofuels also needs to be expanded.

9.8.2.8 Adaptation

There is a need to improve our understanding of the role of complex socio-economic, socio-cultural and biophysical systems, including a re-examination of possible myths of environmental change and of the links between climate change, adaptation, and development in Africa. Such investigations arguably underpin much of the emerging discourse on adaptation. There is also a need to assess current and expected future impacts and vulnerabilities, and the future adaptation options and pathways that may arise from the interaction of multiple stressors on the coping capacities of African communities.

9.8.2.9 Vulnerability and risk reduction

While there are some joint activities that involve those trying to enhance risk-reduction activities, there is still little active engagement between communities that are essentially researching similar themes. The need exists, therefore, to enhance efforts on the coupling and drawing together of disaster risk-reduction activities, vulnerability assessments, and climate change and variability assessments. There is also a need to improve and continue to assess the means (including the institutional design and requirements) by which scientific knowledge and advanced technological products (e.g., early warning systems, seasonal forecasts) could be used to enhance the resilience of vulnerable communities in Africa in order to improve their capacity to cope with current and future climate variability and change.

9.8.2.10 Enhancing African capacity

A need exists for African recognised 'hubs' or centres of excellence established by Africans and developed by African scientists. There is the need to also enhance institutional 'absorptive capacity' in the various regions, providing opportunities for young scientists to improve research in the fields of climate-change impacts, vulnerability and adaptation.

9.8.2.11 Knowledge for action

Much of the research on climate has been driven by the atmospheric sciences community, including, more recently, greater interaction with biophysical scientists (e.g., global change programmes including IGBP/WCRP). However, this chapter has shown that there is much to be gained from a more nuanced approach, which includes those working in the sociological and economic sciences (e.g., IHDP and a range of others). Moreover, the growing interest in partnerships, both public and private, as well as the inclusion of large corporations, formal and informal business, and wider civic society requires more inclusive processes and activities. Such activities, however, may not be sufficient, particularly if change is rapid. For this reason, more 'urgent' and 'creative' interactions (e.g., greater interactions between users and producers of science, stakeholder interactions, communication, institutional design, etc.) will be required. Much could also be gained by greater interactions between those from the disaster risk-reduction, development, and climate-science communities.

Finally, despite the shift in focus from 'impacts-led' research to 'vulnerability-led' research, there are still few studies that clearly show the interaction between multiple stresses and adaptation to such stresses in Africa. The role of land-use and land-cover change is one area that could be further explored to enhance such an understanding. Likewise, while there is evidence of researchers grappling with various paradigms of research, e.g., disaster risk-reduction and climate change, there are still few detailed and rich compendia of studies on 'human dimensions' interactions, adaptation and climate change (of both a historical, current, and future-scenarios nature). The need for more detailed local-level analyses of the role of multiple interacting factors, including development activities and climate risk-reduction in the African context, is evident from much of this chapter.

References

Abdalla, A.A., 2006: Environmental degradation and conflict in Darfur: experiences and development options. *Environmental Degradation as a Cause for Conflict in Darfur: Conference Proceedings*, Khartoum, Sudan, December 2004, University for Peace, Addis Ababa, 112 pp. http://www.unsudanig.org/darfur-jam/trackII/data/cluster/development/Environmental%20Degradation%20as%20a%20cause%20of%20conflict%20in%20Darfur%20(.pdf.

Abdulai, A. and A. Crolerees, 2001: Determinants of income diversification amongst rural households in Southern Mali. *Food Policy*, **26**, 437-452.

Abeku, T.A., S.J. De Vlas, G. Borsboom, A. Tadege, Y. Gebreyesus, H. Gebreyohannes, D. Alamirew, A. Seifu and Co-authors, 2004: Effects of meteorological factors on epidemic malaria in Ethiopia: a statistical modelling approach based on theoretical reasoning. *Parasitology*, **128**, 585-593.

Abou-Hadid, A.F., 2006: Assessment of impacts, adaptation and vulnerability to climate change in North Africa: food production and water resources. AIACC Final Report (AF 90), Washington, District of Columbia, 148 pp.

Adger, N. and K. Vincent, 2005: Uncertainty in adaptive capacity. *C. R. Geosci.*, **337**, 399-410.

Adger, N., N. Brooks, M. Kelly, G. Bentham, M. Agnew and S. Eriksen, 2004: New indicators of vulnerability and adaptive capacity. Technical Report 7, Tyndall Centre for Climate Change Research, University of East Anglia, Norwich, 128 pp.

AfDB (African Development Bank), Asian Development Bank, UK DFID, Netherlands DGIS, European Commission DG for Development, German BMZ, OECD, UNDP, UNEP and World Bank, 2002: *Poverty and Climate Change: Reducing Vulnerability of the Poor Through Adaptation*, UNEP, 56 pp. http://povertymap.net/publications/doc/PovertyAndClimateChange_WorldBank.pdf.

Afrane, Y.A., B.W. Lawson, A.K. Githeko and G. Yan, 2005: Effects of microclimatic changes caused by land use and land cover on duration of gonotrophic cycles of *Anopheles gambiae* (Diptera: Culicidae) in western Kenya highlands. *J. Med. Entomol.*, **42**, 974-980.

African Union, 2005: Status of food security and prospects for agricultural development in Africa: 2005. AU Ministerial Conference of Ministers of Agriculture, Addis Ababa, 26 pp.

Agoumi, A., 2003: Vulnerability of North African countries to climatic changes: adaptation and implementation strategies for climatic change. Developing Perspectives on Climate Change: Issues and Analysis from Developing Countries and Countries with Economies in Transition. IISD/Climate Change Knowledge Network, 14 pp. http://www.cckn.net//pdf/north_africa.pdf.

Agrawala, S., Ed., 2005: *Bridge over Troubled Waters: Linking Climate Change and Development*. OECD, Paris, 154 pp.

AGRHYMET, 2004: Rapport synthèse de l'enquête générale sur les itinéraires d'adaptation des populations locales à la variabilité et aux changements climatiques conduite sur les projets pilotes par AGRHYMET et l'UQAM, par Hubert N'Djafa Ouaga, 13 pp.

Ajibade, L.T., 2003: A methodology for the collection and evaluation of farmers' indigenous environmental knowledge in developing countries. *Indilinga: African Journal of Indigenous Knowledge Systems*, **2**, 99-113.

Ajibade, L.T. and O. Shokemi, 2003: Indigenous approaches to weather forecasting in Asa L.G.A., Kwara State, Nigeria. *Indilinga: African Journal of Indigenous Knowledge Systems*, **2**, 37-44.

AMCEN/UNEP, 2002: *Africa Environment Outlook: Past, Present and Future Perspectives*. Earthprint, Stevenage, 410 pp.

Anderssen, L., J. Wilk, M.C. Todd, D.A. Hughes, A. Earle, D. Kniveton, R. Layberry, H.H.G. Savenije, 2006: Impact of climate change and development scenarios on flow patterns in the Okavango River, *J. Hydrol.*, **331**, 43-57.

Archer, E.R.M., 2003: Identifying underserved end-user groups in the provision of climate information. *B. Am. Meteorol. Soc.*, **84**, 1525-1532.

Armah, K.A., G. Wiafe and D.G. Kpelle, 2005: Sea-level rise and coastal biodiversity in West Africa: a case study from Ghana. *Climate Change and Africa*, P.S. Low, Ed., Cambridge University Press, Cambridge, 204-217.

Arnell, N.W., 2004: Climate change and global water resources: SRES emissions and socio-economic scenarios. *Global Environ. Chang.*, **14**, 31-52.

Arnell, N.W., 2006a: Global impacts of abrupt climate change: an initial assessment. Working Paper 99, Tyndall Centre for Climate Change Research, University of East Anglia, Norwich, 37 pp.

Arnell, N.W., 2006b: Climate change and water resources: a global perspective. *Avoiding Dangerous Climate Change*, H.J. Schellnhuber, W. Cramer, N. Nakićenović, T. Wigley and G. Yohe, Eds., Cambridge University Press, Cambridge, 167-175.

Arnell, N.W., M.G.R. Cannell, M. Hulme, R.S. Kovats, J.F.B. Mitchell, R.J. Nicholls, M.L. Parry, M.T.J. Livermore and A. White, 2002: The consequences of CO_2 stabilisation for the impacts of climate change. *Climatic Change*, **53**, 413-446.

Arrow, K., P. Dasgupta, L. Goulder, G. Daily, P. Ehrlich, G. Heal, S. Levin, K.-G. Mäler and Co-authors, 2004: Are we consuming too much? *J. Econ. Perspect.*, **18**, 147-172.

Ashton, P.J., 2002: Avoiding conflicts over Africa's water resources. *Ambio*, **31**, 236-242.

Attaher, S., M.A. Medany, A.A. Abdel Aziz and A. El-Gindy, 2006: Irrigation-water demands under current and future climate conditions in Egypt. *Misr. Journal of Agricultural Engineering*, **23**, 1077-1089.

Balint-Kurti, D., 2005: Tin trade fuels Congo War. *News24*, 07/03/2005. http://www.news24.com/News24/Africa/Features/0,,2-11-37_1672558,00.html.

Balk, D., A. Storeygard, M. Levy, J. Gaskell, M. Sharma and R. Flor, 2005: Child hunger in the developing world: an analysis of environmental and social correlates. *Food Policy*, **30**, 584-611.

Batterbury, S. and A. Warren, 2001: The African Sahel 25 years after the great drought: assessing progress and moving towards new agendas and approaches. *Global Environ. Chang.*, **11**, 1-8.

Baylis, M. and A.K. Githeko, 2006: The effects of climate change on infectious diseases of animals. UK Foresight Project, Infectious Diseases: Preparing for the Future. Office of Science and Innovation, London, 35 pp.

Beg, N., J. Corfee Morlot, O. Davidson, Y. Afrane-Okesse, L. Tyani, F. Denton, Y. Sokona, J.P. Thomas and Co-authors, 2002: Linkages between climate change and sustainable development. *Clim. Policy*, **2**, 129-144.

Benhin, J.K.A., 2006: Climate change and South African agriculture: impacts and adaptation options. CEEPA Discussion Paper No. 21, Special Series on Climate Change and Agriculture in Africa. Centre for Environmental Economics and Policy in Africa, University of Pretoria, Pretoria, 78 pp.

Benson, C. and E.J. Clay, 2004: *Understanding the Economic and Financial Impacts of Natural Disasters*. Disaster Risk Management Series No. 4. World Bank Publications, Washington, District of Columbia, 130 pp.

Biggs, R., E. Bohensky, P.V. Desanker, C. Fabricius, T. Lynam, A.A. Misselhorn, C. Musvoto, M. Mutale and Co-authors, 2004: *Nature Supporting People: The Southern African Millennium Ecosystem Assessment Integrated Report*. Millennium Ecosystem Assessment, Council for Scientific and Industrial Research, Pretoria, 68 pp.

Block, S. and P. Webb, 2001: The dynamics of livelihood diversification in post-famine Ethiopia. *Food Policy*, **26**, 333-350.

Bomhard, B., D.M. Richardson, J.S. Donaldson, G.O. Hughes, G.F. Midgley, D.C. Raimondo, A.G. Rebelo, M. Rouget and W. Thuiller, 2005: Potential impacts of future land use and climate change on the Red List status of the Proteaceae in the Cape Floristic Region, South Africa. *Glob. Change Biol.*, **11**, 1452-1468.

Bounoua, L., G.J. Collatz, S.O. Los, P.J. Sellers, D.A. Dazlich, C.J. Tucker and D.A. Randall, 2000: Sensitivity of climate to changes in NDVI. *J. Climate*, **13**, 2277-2292.

Bounoua, L., R. DeFries, G.J. Collatz, P. Sellers and H. Khan, 2002: Effects of land cover conversion on surface climate. *Climatic Change*, **52**, 29-64.

Brooks, N., 2004: Drought in the African Sahel: long-term perspectives and future prospects. Working Paper 61, Tyndall Centre for Climate Change Research, University of East Anglia, Norwich, 31 pp.

Brooks, N., W.N. Adger and P.M. Kelly, 2005: The determinants of vulnerability and adaptive capacity at the national level and the implications for adaptation. *Global Environ. Chang.*, **15**, 151-163.

Brovkin, V., 2002: Climate-vegetation interaction. *J. Phys. IV*, **12**, 57-72.

Bryceson, D., 2000: Rural Africa at the crossroads: livelihood practices and policies. *Natural Resource Perspectives*, **52**, http://www.odi.org.uk/nrp/52.pdf.

Bryceson, D.F., 2002: The scramble in Africa: reorienting rural livelihoods. *World Dev.*, **30**, 725-739.

Bryceson, D.F., 2004: Agrarian vista or vortex: African rural livelihood policies. *Review of African Political Economy*, **31**, 617-629.

Bryceson, D.F. and J. Fonseca, 2006: Risking death for survival: peasant responses to hunger and HIV/AIDS in Malawi. *World Dev.*, **34**, 1654-1666.

Burke, E.J., S.J. Brown and N. Christidis, 2006: Modelling the recent evolution of global drought and projections for the twenty-first century with the Hadley Centre climate model. *J. Hydrometeorol.*, **7**, 1113-1125.

Center for International Earth Science Information Network, 2006: *Where the Poor Are: An Atlas of Poverty*. Columbia University, New York, 57 pp. http://www.ciesin.org/povmap/downloads/maps/atlas/atlas.pdf.

Chappell, A. and C.T. Agnew, 2004: Modelling climate change in West African Sahel rainfall (1931-90) as an artifact of changing station locations. *Int. J. Climatol.*, **24**, 547-554.

Chen, H., A.K. Githeko, G. Zhou, J.I. Githure and G. Yan, 2006: New records of *Anopheles arabiensis* breeding on the Mount Kenya Highlands indicate indigenous malaria transmission. *Malaria J.*, **5**, 17.

Chen, S. and M. Ravallion, 2004: How have the world's poorest fared since the early 1980s? *The World Bank Research Observer*, **19**, 141-169.

Chigwada, J., 2005: Climate proofing infrastructure and diversifying livelihoods in Zimbabwe. *IDS Bull.*, **36**, S103-S116.

Christensen, J.H., B. Hewitson, A. Busuioc, A. Chen, X. Gao, I. Held, R. Jones, R.K. Koli, W.-T. Kwon, R. Laprise, V.M. Rueda, L. Mearns, C.G. Menéndez, J. Räisänen, A. Rinke, A. Sarr and P. Whetton, 2007: Regional climate projections. *Climate Change 2007: The Physical Science Basis. Contribution of Working Group I to the Fourth Assessment Report of the Intergovernmental Panel on Climate Change*, S. Solomon, D. Qin, M. Manning, Z. Chen, M. Marquis, K.B. Averyt, M. Tignor and H.L. Miller, Eds., Cambridge University Press, Cambridge, 847-940.

Christie, F. and J. Hanlon, 2001: *Mozambique and the Great Flood of 2000*. Indiana University Press, Indiana, 176 pp.

Clark, B.M., 2006: Climate change: a looming challenge for fisheries management in southern Africa. *Mar. Policy*, **30**, 84-95.

Clark, B.M., S. Young and A. Richardson, 2003: Likely effects of global climate change on the purse seine fishery for Cape anchovy *Engraulis capensis* off the west coast of Southern Africa (SE Atlantic). http://swfsc.noaa.gov/uploadedFiles/Education/lasker_events.doc.

Claussen, M., V. Brovkin, A. Ganopolski, C. Kubatzki and V. Petoukhov, 2003: Climate change in northern Africa: the past is not the future. *Climatic Change*, **57**, 99-118.

Connor, S.J., P. Ceccato, T. Dinku, J. Omumbo, E.K. Grover-Kopec and M.C. Thomson, 2006: Using climate information for improved health in Africa: relevance, constraints and opportunities. *Geospatial Health*, **1**, 17-31.

Conway, D., 2005: From headwater tributaries to international river: observing and adapting to climate variability and change in the Nile Basin. *Global Environ. Chang.*, **15**, 99-114.

Conway, D., C. Mould and W. Bewket, 2004: Over one century of rainfall and temperature observations in Addis Ababa, Ethiopia. *Int. J. Climatol.*, **24**, 77-91.

Conway, D., E. Allison, R. Felstead and M. Goulden, 2005: Rainfall variability in East Africa: implications for natural resources management and livelihoods. *Philos. T. Roy. Soc. A*, **363**, 49-54.

Coughenour, J. Hudson, P. Weisberg, C. Vogel and J.E. Ellis, 2004: Ecosystem modeling adds value to South African climate forecasts. *Climatic Change*, **64**, 317-340.

Cullen, N.J., T. Mölg, G. Kaser, K. Hussein, K. Steffen and D.R. Hardy, 2006: Kilimanjaro glaciers: recent areal extent from satellite data and new interpretation of observed 20th century retreat rates. *Geophys. Res. Lett.*, **33**, L16502, doi:10.1029/2006GL027084.

Cury, P. and L. Shannon, 2004: Regime shifts in upwelling ecosystems: observed changes and possible mechanisms in the northern and southern Benguela. *Prog. Oceanogr.*, **60**, 223-243.

Dai, A., P.J. Lamb, K.E. Trenberth, M. Hulme, P.D. Jones and P. Xie, 2004: The recent Sahel drought is real. *Int. J. Climatol.*, **24**, 1323-1331.

Davidson, O., K. Halsnaes, S. Huq, M. Kok, B. Metz, Y. Sokona and J. Verhagen, 2003: The development and climate nexus: the case of sub-Saharan Africa. *Clim. Policy*, **3**, 97-113.

de Wit, M. and J. Stankiewicz, 2006: Changes in water supply across Africa with predicted climate change. *Science*, **311**, 1917-1921.

Dea, D. and I. Scoones, 2003: Networks of knowledge: how farmers and scientists understand soils and their fertility: a case study from Ethiopia. *Oxford Development Studies*, **31**, 461-478.

DeFries, R.S., L. Bounoua and G.J. Collatz, 2002: Human modification of the landscape and surface climate in the next fifty years. *Glob. Change Biol.*, **8**, 438-458.

Desta, S. and L. Coppock, 2004: Pastoralism under pressure: tracking system change in southern Ethiopia. *Hum. Ecol.*, **32**, 465-486.

Devereux, S., 2003: Policy options for increasing the contribution of social protection to food security. Forum for Food Security in Southern Africa, Overseas Development Institute, London, 35 pp.

Devereux, S. and S. Maxwell, Eds., 2001: *Food Security in Sub-Saharan Africa*. ITDG Publishing, Pietermaritzburg, 361 pp.

DFID (Department for International Development), 2004: The impact of climate change on the vulnerability of the poor. Policy Division, Global Environmental Assets, Key sheet 3, 6 pp. http://www.dfid.gov.uk/pubs/files/climatechange/3vulnerability.pdf.

Dilley, M., R.S. Chen, U. Deichmann, A.L. Lerner-Lam, M. Arnold, J. Agwe, P. Buys, O. Kjekstad and Co-authors, 2005: *Natural Disaster Hotspots: A Global Risk Analysis*. Disaster Risk Management Series No. 5. The World Bank, Washington, District of Columbia, 148 pp. http://www.proventionconsortium.org/themes/default/pdfs/Hotspots.pdf.

Dixon, R.K., J. Smith and S. Guill, 2003: Life on the edge: vulnerability and adaptation of African ecosystems to global climate change. *Mitigation and Adaptation Strategies for Global Change*, **8**, 93-113.

Dougherty, W. and B. Osman, 2005: Mainstreaming adaptation into national development plans. *AIACC Second Regional Workshop for Africa and Indian Ocean Islands*, Dakar, Senegal. http://www.aiaccproject.org/meetings/meetings.html.

du Plessis C, D.K. Irurah and R.J. Scholes, 2003: The built environment and climate change in South Africa. *Build. Res. Inf.*, **31**, 240–256.

Dwyer, E.S., S. Pinnock, J.M. Gregoire and J.M.C. Pereira, 2000: Global spatial and temporal distribution of vegetation fire as determined from satellite observations. *Int. J. Remote Sensing*, **21**, 1289-1302.

Easton, P. and M. Roland, 2000: Seeds of life: women and agricultural biodiversity in Africa. IK Notes 23. World Bank, Washington, District of Columbia, 4 pp.

ECA (Economic Commission for Africa), 2002: Harnessing technologies for sustainable development. Economic Commission for Africa Policy Research Report, Addis Ababa, 178 pp.

ECA (Economic Commission for Africa), 2005: Assessing sustainable development in Africa. Africa's Sustainable Development Bulletin. Economic Commission for Africa, Addis Ababa, 59 pp.

ECF (European Climate Forum) and Potsdam Institute, 2004: What is dangerous climate change? *Initial Results of a Symposium on Key Vulnerable Regions, Climate Change and Article 2 of the UNFCCC*, Buenos Aires. Beijing, 39 pp. http://www.european-climate-forum.net/.

2006: Assessing the impacts of climate change on agriculture in Egypt: a ricardian approach. Centre for Environmental Economics and Policy in Africa (CEEPA) Discussion Paper No. 16, Special Series on Climate Change and Agriculture in Africa, University of Pretoria, Pretoria, 1-33.

Eklundh, L. and L. Olsson, 2003: Vegetation trends for the African Sahel 1982–1999. *Geophys. Res. Lett.*, **30**, 1430, doi:10.1029/2002GL016772.

El-Gindy, A., A.A. Abdel Azziz and E.A. El-Sahaar, 2001: Design of irrigation and drainage networks. Faculty of Agriculture lectures, Ain Shams University, 28 pp. (in Arabic).

Ellis, F., 2000: *Rural Livelihoods and Diversity in Developing Countries*. Oxford University Press, Oxford, 296 pp.

Ellis, F., 2003: Human vulnerability and food insecurity: policy implications. Forum for Food Security in Southern Africa, Overseas Development Group, University of East Anglia, Norwich, 47 pp.

Ellis, F. and G. Bahiigwa, 2003: Livelihoods and rural poverty reduction in Uganda. *World Dev.*, **31**, 997-1013.

Ellis, F. and N. Mdoe, 2003: Livelihoods and rural poverty reduction in Tanzania. *World Dev.*, **31**, 1367-1384.

EPIQ (Environmental Policy and Institutional Strengthening Indefinite Quantity, Water Policy Reform Activity, Agricultural Policy Reform Programme and Market-Based Incentives Team), 2002: Economic Instruments for Improved Water Resources Management in Egypt, Prepared for the United States Agency for International Development/Egypt, No. PCE-I-00-96-00002-00, 173 pp.

Epstein, P.R. and E. Mills, Eds., 2005: *Climate Change Futures: Health, Ecological and Economic Dimensions*. Center for Health and the Global Environment, Harvard Medical School, Boston, Massachusetts, 142 pp.

Erasmus, B.F.N., A.S. van Jaarsveld, S.L. Chown, M. Kshatriya and K.J. Wessels, 2002: Vulnerability of South Africa animal taxa to climate change. *Glob. Change Biol.*, **8**, 679-693.

Eriksen, S., 2004: Building adaptive capacity in a 'glocal' world: examples from Norway and Africa. *ESS Bulletin*, **2**, 18-26.

Eriksen, S., 2005: The role of indigenous plants in household adaptation to climate change: the Kenyan experience. *Climate Change and Africa*, P.S Low, Ed., Cambridge University Press, Cambridge, 248-259.

Eriksen, S.H. and J. Silva, 2003: The impact of economic liberalisation on climate vulnerability among farmers in Mozambique. IHDP Open Meeting, 16–18 October 2003, Montreal, 15 pp.

Eriksen, S.H., K. Brown and P.M. Kelly, 2005: The dynamics of vulnerability: locating coping strategies in Kenya and Tanzania. *Geogr. J.*, **171**, 287-305.

ESMAP (Energy Sector Management Assistance Program), 2005: The vulnerability of African countries to oil price shocks: major factors and policy options – the case of oil importing countries. World Bank Report 308/5, Washington, District of Columbia, 76 pp.

FAO, 2004: Locust crisis to hit northwest Africa again: situation deteriorating in the Sahel. FAO News Release, 17 September 2004. http://www.fao.org/newsroom/en/news/2004/50609/.

FAO, 2005: *The State of Food and Agriculture*. Food and Agriculture Organization of the United Nations, Rome, 211 pp.

Fauchereau, N., S. Trzaska, Y. Richard, P. Roucou and P. Camberlin, 2003: Sea-surface temperature co-variability in the southern Atlantic and Indian Oceans and its connections with the atmospheric circulation in the southern hemisphere. *Int. J. Climatol.*, **23**, 663-677.

Fekete, B.M., C.J. Vörösmarty, J.O. Roads and C.J. Willmott, 2004: Uncertainties in precipitation and their impact on runoff estimates. *J. Climate*, **17**, 294-304.

Ferguson, J., 2006: *Global Shadows: Africa in the Neoliberal World Order*. Duke University Press, Durham and London, 257 pp.

Few R., M. Ahern, F. Matthies and S. Kovats, 2004: Floods, health and climate change: a strategic review. Working Paper 63, Tyndall Centre for Climate Change Research, University of East Anglia, Norwich, 138 pp.

Fiki, O.C. and B. Lee, 2004: Conflict generation, conflict management and self-organizing capabilities in drought-prone rural communities in north eastern Nigeria: a case study. *Journal of Social Development in Africa*, **19**, 25-48.

Fischer, G., M. Shah and H. van Velthuizen, 2002: *Climate Change and Agricultural Vulnerability*. International Institute for Applied Systems Analysis, Laxenberg, 152 pp.

Fischer, G., M. Shah, F.N. Tubiello and H. van Velthuizen, 2005: Socio-economic and climate change impacts on agriculture: an integrated assessment, 1990–2080. *Philos. T. Roy. Soc. B*, **360**, 2067-2083.

Freeman, P.K., 2003: Natural hazard risk and privatization. *Building Safer Cities: The Future of Disaster Risk*, A. Kreimer, M. Arnold and A. Carlin, Eds., World Bank Disaster Management Facility, Washington, District of Columbia, 33-44.

Freeman, P. and K. Warner, 2001: Vulnerability of infrastructure to climate variability: how does this affect infrastructure lending policies? Report commissioned by the Disaster Management Facility of the World Bank and the ProVention Consortium, Washington, District of Columbia, 42 pp.

Frenken, K., Ed., 2005: Irrigation in Africa in figures – AQUASTAT Survey – 2005. Food and Agricultural Organisation of the United Nations, Rome, 89 pp.

Gabre-Madhin, E.Z. and S. Haggblade, 2004: Successes in African agriculture: results of an expert survey. *World Dev.*, **32**, 745-766.

Gallup, J.L. and J.D. Sachs, 2001: The economic burden of malaria. *Am. J. Trop. Med. Hyg.*, **64**, 85-96.

Gana, F.S., 2003: The usage of indigenous plant materials among small-scale farmers in Niger state agricultural development project: Nigeria. *Indilinga: African Journal of Indigenous Knowledge Systems*, **2**, 53-60.

Giannini, A., R. Saravanan and P. Chang, 2003: Oceanic forcing of Sahel rainfall on interannual to interdecadal time scales. *Science*, **302**, 1027-1030.

Githeko, A.K. and W. Ndegwa, 2001: Predicting malaria epidemics in Kenyan highlands using climate data: a tool for decision makers. *Global Change and Human Health*, **2**, 54-63.

Gommes, R., J. du Guerny, M.H. Glantz, L.-N. Hsu and J. White, 2004: Climate and HIV/AIDS: a hotspots analysis for early warning rapid response systems. UNDP, NCAR and FAO, Bangkok, 36 pp.

Gommes, R., J. du Guerny, F.O. Nachtergalle and R. Brinkman, 2005: Potential impacts of sea-level rise on populations and agriculture. *Climate Change and Africa*, P.S Low, Ed., Cambridge University Press, Cambridge, 191-203.

Gonzalez, P., 2001: Desertification and a shift of forest species in the West African Sahel. *Climate Res.*, **17**, 217-228.

Government of Seychelles, 2000: Initial National Communication under the United Nations Framework Convention on Climate Change (UNFCCC), Ministry of Environment and Transport, 145 pp. http://unfccc.int/resource/docs/natc/seync1.pdf.

Granich, S., 2006: Deserts and desertification. *Tiempo*, **59**, 8-11.

Gregory, P.J., J.S.I. Ingram and M. Brklacich, 2005: Climate change and food security. *Philos. T. Roy. Soc. B*, **360**, 2139-2148.

Grothmann, T. and A. Patt, 2005: Adaptive capacity and human cognition: the process of individual adaptation to climate change. *Global Environ. Chang.*, **15**, 199-213.

Gueye, L., M. Bzioul and O. Johnson, 2005: Water and sustainable development in the countries of Northern Africa: coping with challenges and scarcity. *Assessing sustainable development in Africa*, Africa's Sustainable Development Bulletin, Economic Commission for Africa, Addis Ababa, 24-28.

Haarsma, R.J., F.M. Selten, S.L. Weber and M. Kliphuis, 2005: Sahel rainfall variability and response to greenhouse warming. *Geophys. Res. Lett.*, **32**, L17702, doi:10.1029/2005GL023232.

Haddad, B.M., 2005: Ranking the adaptive capacity of nations to climate change when socio-political goals are explicit. *Global Environ. Chang.*, **15**, 165-176.

Hall, D.O. and J.I. Scrase, 2005: Biomass energy in sub-Saharan Africa. *Climate Change and Africa*, P.S Low, Ed., Cambridge University Press, Cambridge, 107-112.

Hamilton, J.M., D.J. Maddison and R.S.J. Tol, 2005: Effects of climate change on international tourism. *Climate Res.*, **29**, 245-254.

Hammerich, A., O.M.R. Campbell and D. Chandramohan, 2002: Unstable malaria transmission and maternal mortality: experiences from Rwanda. *Trop. Med. Int. Health*, **7**, 573-576.

Harrus, S. and G. Baneth, 2005: Drivers for the emergence and re-emergence of vector-borne protozoal and bacterial diseases. *Int. J. Parasitol.*, **35**, 1309-1318.

Hartmann, J., K. Ebi, J. McConnell, N. Chan and J.P. Weyant, 2002: Climate suitability: for stable malaria transmission in Zimbabwe under different climate change scenarios. *Global Change and Human Health*, **3**, 42-54.

Hay, S.I., J. Cox, D.J. Rogers, S.E. Randolph, D.I. Stern, G.D. Shanks, M.F. Myers and R.W. Snow, 2002a: Climate change and the resurgence of malaria in the East African Highlands. *Nature*, **415**, 905-909.

Hay, S.I., D.J. Rogers, S.E. Randolph, D.I. Stern, J. Cox, G.D. Shanks and R.W. Snow, 2002b: Hot topic or hot air? Climate change and malaria resurgence in East African highlands. *Trends in Parasitology*, **18**, 530-534.

Held, I.M., T.L. Delworth, J. Lu, K.L. Findell and T.R. Knuston, 2005: Simulation of Sahel drought in the 20th and 21st centuries. *P. Natl. Acad. Sci. USA*, **102**, 17891-17896.

Hemp, A., 2005: Climate change-driven forest fires marginalize the impact of ice cap wasting on Kilimanjaro. *Glob. Change Biol.*, **11**, 1013-1023.

Herrmann, S.M., A. Anyamba and C.J. Tucker, 2005: Recent trends in vegetation dynamics in the African Sahel and their relationship to climate. *Global Environ. Chang.*, **15**, 394-404.

Hewawasam, I., 2002: Managing the marine and coastal environment of sub-Saharan Africa: strategic directions for sustainable development. World Bank, Washington, District of Columbia, 57 pp.

Hewitson, B.C. and R.G. Crane, 2006: Consensus between GCM climate change projections with empirical downscaling: precipitation downscaling over South Africa. *Int. J. Climatol.*, **26**, 1315-1337.

Hoerling, M., J. Hurrell, J. Eischeid and A. Phillips, 2006: Detection and attribution of twentieth-century northern and southern African rainfall change. *J. Climate*, **19**, 3989-4008.

Holding, P.A. and R.W. Snow, 2001: Impact of *Plasmodium falciparum* malaria on performance and learning: review of the evidence. *Am. J. Trop. Med. Hyg.*, **64**, 68-75.

Hudson, D.A. and R.G. Jones, 2002: Regional climate model simulations of present day and future climates of Southern Africa. Technical Note 39, Hadley Centre, Bracknell, 42 pp.

Hugon, P. and P.X. Nanterre, 2003: Food insecurity and famine in southern Africa, an economic debate: lack of availabilities, market failures, inequities of rights, effects of shocks or systemic risks? Preprints, *SARPN Meeting: Food Security in Southern Africa*, 18 March 2003, Pretoria.

Hulme, D. and A. Shepherd, 2003: Conceptualizing chronic poverty. *World Dev.*, **31**, 403-423.

Hulme, M., R.M. Doherty, T. Ngara, M.G. New and D. Lister, 2001: African climate change: 1900–2100. *Climate Res.*, **17**, 145-168.

Hulme, M., R. Doherty, T. Ngara and M. New, 2005: Global warming and African climate change. *Climate Change and Africa*, P.S Low, Ed., Cambridge University Press, Cambridge, 29-40.

Huntingford, C., F.H. Lambert, J.H.C. Gash, C.M. Taylor and A.J. Challinor, 2005: Aspects of climate change prediction relevant to crop productivity. *Philos. T. Roy. Soc. B*, **360**, 1999-2009.

Huq, S. and H. Reid, 2004: Mainstreaming adaptation in development. *IDS Bull.*, **35**, 15-21.

ICID (International Commission on Irrigation and Drainage, New Delhi), 2005: Water Policy Issues of Egypt, Country Policy Support Programme, 36 pp.

IEA (International Energy Association), 2002: *World Energy Outlook 2002*. International Energy Agency, Paris, 533 pp. http://www.iea.org/textbase/nppdf/free/2000/weo2002.pdf.

IFRCRCS (International Federation of Red Cross and Red Crescent Societies), 2002: *World Disasters Report: Focusing on Reducing Risk*. Geneva, 240 pp.

IFRCRCS (International Federation of Red Cross and Red Crescent Societies), 2005: *World Disasters Report: Focus on Information in Disasters*. Geneva, 251 pp.

Institute of Development Studies, 2005: *New Directions for African Agriculture*, I. Scoones, A. deGrassi, S. Devereux and L. Haddad, Eds., *IDS Bull.*, **36**, 160 pp.

International Institute of Rural Reconstruction, 2004: *Drought Cycle Management: A Toolkit for the Drylands of the Greater Horn*. Cordiad and Acacia Consultants, Nairobi, 253 pp.

IPCC, 2001: *Climate Change 2001: Impacts, Adaptation, and Vulnerability. Contribution of Working Group II to the Third Assessment Report of the Intergovernmental Panel on Climate Change*, J.J. McCarthy, O.F. Canziani, N.A. Leary, D.J. Dokken and K.S. White, Eds., Cambridge University Press, Cambridge, 1032 pp.

IPCC, 2007a: *Climate Change 2007: The Physical Science Basis. Contribution of Working Group I to the Fourth Assessment Report of the Intergovernmental Panel on Climate Change*, S. Solomon, D. Qin, M. Manning, Z. Chen, M. Marquis, K.B. Averyt, M. Tignor and H.L. Miller, Eds., Cambridge University Press, Cambridge, 996 pp.

IPCC, 2007b: *Climate Change 2007: Mitigation. Contribution of Working Group III to the Fourth Assessment Report of the Intergovernmental Panel on Climate Change*, B. Metz, O. Davidson, P. Bosch, R. Dave and L. Meyer, Eds., Cambridge University Press, Cambridge, UK.

IRI, GCOS, DfID and ECA, 2006: A gap analysis for the implementation of the Global Climate Observing System programme in Africa, IRI (International Research Institute for Climate and Society) Technical Report, IRI-TR/06/1, 52 pp.

Jenkins, G.S., G. Adamou and S. Fongang, 2002: The challenges of modelling climate variability and change in West Africa. *Climatic Change*, **52**, 263-286.

Jenkins, G.S., A.T. Gaye and B. Sylla, 2005: Late 20th century attribution of drying trends in the Sahel from the Regional Climate Model (RegCM3). *Geophys. Res. Lett.*, **32**, L22705, doi:10.1029/2005GL024225.

Jones, P.G. and P.K. Thornton, 2003: The potential impacts of climate change on maize production in Africa and Latin America in 2055. *Global Environ. Chang.*, **13**, 51-59.

Kabat, P., R.E. Schulze, M.E. Hellmuth and J.A. Veraart, Eds., 2002: Coping with impacts of climate variability and climate change in water management: a scoping paper. DWC-Report No. DWCSSO-01 (2002). Dialogue on Water and Climate, Wageningen, 114 pp.

Kamga, A.F., G.S. Jenkins, A.T. Gaye, A. Garba, A. Sarr and A. Adedoyin, 2005: Evaluating the National Center for Atmospheric Research climate system model over West Africa: present-day and the 21st century A1 scenario. *J. Geophys. Res.–Atmos.*, **110**, D03106, doi:10.1029/2004JD004689.

Kaser, G., D.R. Hardy, T. Mölg, R.S. Bradley and T.M. Hyera, 2004: Modern glacier retreat on Kilimanjaro as evidence of climate change: observations and facts. *Int. J. Climatol.*, **24**, 329-339.

Kelly, V., A.A. Adesina and A. Gordon, 2003: Expanding access to agricultural inputs in Africa: a review of recent market development experience. *Food Policy*, **28**, 379-404.

Kempeneers, P., E. Swinnen and F. Fierens, 2002: GLOBSCAR Final Report, TAP/N7904/FF/FR-001 Version 1.2, VITO, Belgium, 25 pp. http://geofront.vgt.vito.be/geosuccess/documents/Final%20Report.pdf;jsessionid=668127C51913A12DB357A8569F4FE74F.

Kherallah, M., C. Delgado, E. Gabre-Madhin, N. Minot and M. Johnson, 2004: The road half travelled: agricultural market reform in Sub-Saharan Africa. Policy Report, International Food Policy Research Institute, Washington, District of Columbia, 4 pp.

King'uyu, S.M., L.A. Ogallo and E.K. Anyamba, 2000: Recent trends of minimum and maximum surface temperatures over Eastern Africa. *J. Climate*, **13**, 2876-2886.

Klein, R.J.T., R.J. Nicholls and F. Thomalla, 2002: The resilience of coastal megacities to weather-related hazards. *Building Safer Cities: The Future of Disaster Risk*, A. Kreimer, M. Arnold and A. Carlin, Eds., The World Bank Disaster Management Facility, Washington, District of Columbia, 101-120.

Kotb, M., M. Abdulaziz, Z. Al-Agwan, K. Al-Shaikh, H. Al-Yami, A. Banajah, L. DeVantier, M. Eisinger and Co-authors, 2004: Status of coral reefs in the Red Sea and Gulf of Aden in 2004. *Status of Coral Reefs of the World: 2004*, C. Wilkinson, Ed., Volume 1, Australian Institute of Marine Science, Townsville, 137-154.

Kruger, A.C. and S. Shongwe, 2004: Temperature trends in South Africa: 1960–2003. *Int. J. Climatol.*, **24**, 1929-1945.

Kurukulasuriya, P. and S. Rosenthal, 2003: Climate change and agriculture: a review of impacts and adaptations. Climate Change Series Paper 91, World Bank, Washington, District of Columbia, 106 pp.

Kurukulasuriya, P. and R. Mendelsohn, 2006a: A Ricardian analysis of the impact of climate change on African cropland. Centre for Environmental Economics and Policy in Africa (CEEPA) Discussion Paper No. 8. University of Pretoria, Pretoria, 58 pp.

Kurukulasuriya, P. and R. Mendelsohn, 2006b: Crop selection: adapting to climate change in Africa. Centre for Environmental Economics and Policy in Africa (CEEPA) Discussion Paper No. 26. University of Pretoria, Pretoria, 28 pp.

Lal, M., 2001: Tropical cyclones in a warmer world. *Current Sci.*, **80**, 1103-1104.

Leautier, F., 2004: Indigenous capacity enhancement: developing community knowledge. *Indigenous Knowledge: Local Pathways to Global Development*. The World Bank, Washington, District of Columbia, 4-8.

Lebel, T., F. Delclaux, L. Le Barbe and J. Polcher, 2000: From GCM scales to hydrological scales: rainfall variability in West Africa. *Stoch. Env. Res. Risk A.*, **14**, 275-295.

Leemans, R. and B. Eickhout, 2004: Another reason for concern: regional and global impacts on ecosystems for different levels of climate change. *Global Environ. Chang.*, **14**, 219-228.

Legesse, D., C. Vallet-Coulomb and F. Gasse, 2003: Hydrological response of a catchment to climate and land use changes in Tropical Africa: case study South Central Ethiopia. *J. Hydrol.*, **275**, 67-85.

Lemke, P., J. Ren, R.B. Alley, I. Allison, J. Carrasco, G. Flato, Y. Fujii, G. Kaser, P. Mote, R.H. Thomas and T. Zhang, 2007: Observations: Changes in snow, ice and frozen ground. *Climate Change 2007: The Physical Science Basis. Contribution of Working Group I to the Fourth Assessment Report of the Intergovernmental Panel on Climate Change*, S. Solomon, D. Qin, M. Manning, Z. Chen, M. Marquis, K.B. Averyt, M. Tignor and H.L. Miller, Eds., Cambridge University Press, Cambridge, 337-384.

Levy, P., 2006: Regional climate change impacts on global vegetation. *Understanding the Regional Impacts of Climate Change: Research Report prepared for the Stern Review on the Economics of Climate Change*, R. Warren, N. Arnell, R. Nicholls, P. Levy and J. Price, Eds., Working Paper 90, Tyndall Centre for Climate Change Research, University of East Anglia, Norwich, 99-108.

L'Hôte, Y., G. Mahé, B. Some and J.P. Triboulet, 2002: Analysis of a Sahelian annual rainfall index from 1896 to 2000: the drought continues. *Hydrolog. Sci. J.*, **47**, 563-572.

Lind, J. and K. Sturman, Eds., 2002: *Scarcity and Surfeit: The Ecology of Africa's Conflicts*. Institute for Security Studies, Pretoria, 388 pp.

Little, P.D., H. Mahmoud and D.L. Coppock, 2001: When deserts flood: risk management and climate processes among East African pastoralists. *Climate Res.*, **19**, 149-159.

Lough, J.M., 2000: 1997-98: Unprecedented thermal stress to coral reefs? *Geophys. Res. Lett.*, **27**, 3901-3904.

Malaney, P., A. Spielman and J. Sachs, 2004: The malaria gap. *Am. J. Trop. Med. Hyg.*, **71**, 141-146.

Malhi, Y. and J. Wright, 2004: Spatial patterns and recent trends in the climate of tropical rainforest regions. *Philos. T. Roy. Soc. B*, **359**, 311-329.

Mano, R., B. Isaacson and P. Dardel, 2003: Identifying policy determinants of food security response and recovery in the SADC region: the case of the 2002 food emergency. FANRPAN Policy Paper, 31 pp. http://www.reliefweb.int/ library/documents/2003/fews-sad-13may.pdf .

Marsland, N., 2004: Development of food security and vulnerability information systems in Southern Africa: the experience of Save the Children UK. Save the Children, 64 pp.

Matondo, J.I., G. Peter and K.M. Msibi, 2005: Managing water under climate change for peace and prosperity in Swaziland. *Phys. Chem. Earth*, **30**, 943-949.

Maynard, K., J.-F. Royer and F. Chauvin, 2002: Impact of greenhouse warming on the West African summer monsoon. *Clim. Dynam.*, **19**, 499-514.

McClean, C.J., J.C. Lovett, W. Küper, L. Hannah, J.H. Sommer, W. Barthlott, M. Termansen, G.F. Smith and Co-authors, 2005: African plant diversity and climate change. *Ann. Mo. Bot. Gard.*, **92**, 139-152.

McDonald, R.E., D.G. Bleaken, D.R. Cresswell, V.D. Pope and C.A. Senior, 2005: Tropical storms: representation and diagnosis in climate models and the impacts of climate change. *Clim. Dynam.*, **25**, 19-36.

McLeman, R. and B. Smit, 2004: Climate change, migration and security. Canadian Security Intelligence Service, Commentary No. 86. http://www.csis-scrs.gc.ca/en/publications/commentary/com86.asp.

McLeman, R. and B. Smit, 2005: Assessing the security implications of climate change-related migration. Preprint, *Human Security and Climate Change: An International Workshop*, Oslo, 20 pp.

McMichael, A.J., R.E. Woodruff and S. Hales, 2006: Climate change and human health: present and future risks. *Lancet*, **367**, 859-869.

Mendelsohn, R., W. Morrison, M.E. Schlesinger and N.G. Andronova, 2000a: Country-specific market impacts from climate change. *Climatic Change*, **45**, 553-569.

Mendelsohn, R., A. Dinar and A. Dalfelt, 2000b: Climate change impacts on African agriculture. Preliminary analysis prepared for the World Bank, Washington, District of Columbia, 25 pp.

Mercier, F., A. Cazenave and C. Maheu, 2002: Interannual lake level fluctuations (1993–1999) in Africa from Topex/Poseidon: connections with ocean-atmosphere interactions over the Indian Ocean. *Global Planet. Change*, **32**, 141-163.

Midgley, G.F., L. Hannah, D. Millar, M.C. Rutherford and L.W. Powrie, 2002: Assessing the vulnerability of species richness to anthropogenic climate change in a biodiversity hotspot. *Global Ecol. Biogeogr.*, **11**, 445-451.

Mirza, M.M.Q., 2003: Climate change and extreme weather events: can developing countries adapt? *Clim. Policy*, **3**, 233-248.

Misselhorn, A.A., 2005: What drives food insecurity in Southern Africa? A meta-analysis of household economy studies. *Global Environ. Chang.*, **15**, 33-43.

Mitchell, J.F.B., T.C. Johns, W.J. Ingram and J.A. Lowe, 2000: The effect of stabilising the atmospheric carbon dioxide concentrations on global and regional climate change. *Geophys. Res. Lett.*, **27**, 2977-2980.

Molesworth A.M., L.E. Cuevas, S.J. Connor, A.P. Morse and M.C. Thomson, 2003: Environmental risk and meningitis epidemics in Africa. *Emerg. Infect. Dis.*, **9**, 1287-1293.

Mölg, T. and D.R. Hardy, 2004: Ablation and associated energy balance of a horizontal glacier surface on Kilimanjaro. *J. Geophys. Res.*, **109**, D16104, doi:10.1029/2003JD004338.

Monyo, E.S., 2002: Pearl millet cultivars released in the SADC region. ICRISAT, Bulawayo, 40 pp.

Mortimore, M.J. and W.M. Adams, 2001: Farmer adaptation, change and 'crisis' in the Sahel. *Global Environ. Chang.*, **11**, 49-57.

Muhando, C.A., 2001: The 1998 coral bleaching and mortality event in Tanzania: implications for coral reef research and management. *Marine Science Development in Tanzania and Eastern Africa: Proc. 20th Anniversary Conference on Advances in Marine Science in Tanzania*, M.D. Richmond and J. Francis, Eds., Institute of Marine Science/Western Indian Ocean Marine Science Association, Zanzibar, 329-342.

Munga, S., N. Minakawa, G. Zhou, E. Mushinzimana, O.O.J. Barrack, A.K. Githeko and G. Yan, 2006: Association between land cover and habitat productivity of malaria vectors in western Kenyan highlands. *Am. J. Trop. Med. Hyg.*, **74**, 69-75.

Muriuki, G.W., T.J. Njoka, R.S. Reid and D.M. Nyariki, 2005: Tsetse control and land-use change in Lambwe valley, south-western Kenya. *Agr. Ecosyst. Environ.*, **106**, 99-107.

Mutangadura, G., S. Ivens and S.M. Donkor, 2005: Assessing the progress made by southern Africa in implementing the MDG target on drinking water and sanitation. *Assessing sustainable development in Africa*, Africa's Sustainable Development Bulletin, Economic Commission for Africa, Addis Ababa, 19-23.

Myers, N., 2002: Environmental refugees: a growing phenomenon of the 21st century. *Philos. T. Roy. Soc. B*, **357**, 609-613.

Nakićenović, N., J. Alcamo, G. Davis, B. de Vries, J. Fenhann, S. Gaffin, K. Gregory, A. Grübler, T.Y. Jung, T. Kram, E. Lebre la Rovere, L. Michaelis, S. Mori, T. Morita, W. Pepper, H. Pitcher, L. Price, K. Riahi, A. Roehrl, H.H. Rogner, A. Sankovski, M. Schlesinger, P. Shukla, S. Smith, R. Swart, S. van Rooijen, N. Victor and Z. Dadi, Eds., 2000: *Emissions Scenarios: A Special Report of the Intergovernmental Panel on Climate Change (IPCC)*. Cambridge University Press, Cambridge, 509 pp.

Nelleman, C. and E. Corcoran, Eds., 2006: Our precious coasts: marine pollution, climate change and the resilience of coastal ecosystems. United Nations Environment Programme, GRID- Arendal, Norway, 40 pp.

New, M., 2002: Climate change and water resources in the southwestern Cape, South Africa. *S. Afr. J. Sci.*, **98**, 369-373.

New, M., B. Hewitson, D.B. Stephenson, A. Tsiga, A. Kruger, A. Manhique, B. Gomez, C.A.S. Coelho and Co-authors, 2006: Evidence of trends in daily climate extremes over southern and west Africa. *J. Geophys. Res.–Atmos.*, **111**, D14102, doi:10.1029/2005JD006289.

Niang-Diop, I., 2005: Impacts of climate change on the coastal zones of Africa. *Coastal Zones in Sub-Saharan Africa: A Scientific Review of the Priority Issues Influencing Sustainability and Vulnerability in Coastal Communities*, IOC, Ed., IOC Workshop Report No. 186. ICAM Dossier No. 4, 27-33.

Niang-Diop, I., M. Dansokho, A.T. Diaw, S. Faye, A. Guisse, I. Ly, F. Matty, A. Sene and Co-authors, 2005: Senegal. *Climate Change in Developing Countries: An Overview of Study Results from the Netherlands Climate Change Studies Assistance Programme*, M.A. van Drunen, R. Lasage and C. Dorland, Eds., IVM, Amsterdam, 101-109.

Niasse, M., 2005: Climate-induced water conflict risks in West Africa: recognizing and coping with increasing climate impacts on shared watercourses. Preprint, *Human Security and Climate Change: An International Workshop*, Oslo, 15 pp.

Niasse, M., A. Afouda and A. Amani, Eds., 2004: *Reducing West Africa's Vulnerability to Climate Impacts on Water Resources, Wetlands and Desertification: Elements for a Regional Strategy for Preparedness and Adaptation*. International Union for Conservation of Nature and Resources (IUCN), Cambridge, 84 pp.

Nicholls, R.J., 2004: Coastal flooding and wetland loss in the 21st century: changes under the SRES climate and socio-economic scenarios. *Global Environ. Chang.*, **14**, 69-86.

Nicholls, R.J. and R.S.J. Tol, 2006: Impacts and responses to sea-level rise: a global analysis of the SRES scenarios over the twenty-first century. *Philos. T. Roy. Soc. A*, **364**, 1073-1095.

Nicholson, S.E., 2001: Climatic and environmental change in Africa during the last two centuries. *Climate Res.*, **17**, 123-144.

Nicholson, S.E. and J.C. Selato, 2000: The influence of La Niña on African rainfall. *Int. J. Climatol.*, **20**, 1761-1776.

Nicholson, S.E. and J.P. Grist, 2003: The seasonal evolution of the atmospheric circulation over West Africa and Equatorial Africa. *J. Climate*, **16**, 1013-1030.

Nicholson, S.E., B. Some and B. Kone, 2000: An analysis of recent rainfall conditions in West Africa, including the rainy season of the 1997 El Niño and the 1998 La Niña years. *J. Climate*, **13**, 2628-2640.

Njaya, F. and C. Howard, 2006: Climate and African fisheries. *Tiempo*, **59**, 13-15.

Nkomo, J.C., A.O. Nyong and K. Kulindwa, 2006: The impacts of climate change in Africa. Report prepared for the Stern Review on the Economics of Climate Change, 51 pp. http://www.hm-treasury.gov.uk/media/8AD/9E/ Chapter_5_The_Impacts_of_Climate_Change_in_Africa-5.pdf.

Obasi, G.O.P., 2005: The impacts of ENSO in Africa. *Climate Change and Africa*, P.S Low, Ed., Cambridge University Press, Cambridge, 218-230.

O'Brien, K., 2006: Are we missing the point? Global environmental change as an issue of human security. *Global Environ. Chang.*, **16**, 1-3.

O'Brien, K. and C. Vogel, Eds., 2003: *Coping with Climate Variability: The Use of Seasonal Climate Forecasts in Southern Africa*. Ashgate Publishing, Aldershot, 176 pp.

Obura, D.O., 2001: Differential bleaching and mortality of eastern African corals. *Marine Science Development in Tanzania and Eastern Africa: Proc. 20th Anniversary Conference on Advances in Marine Science in Tanzania*, M.D. Richmond and J. Francis, Eds., Institute of Marine Science/Western Indian Ocean Marine Science Association, Zanzibar, 301-317.

Odada, E.O., D.O. Olago, K. Kulindwa, M. Ntiba and S. Wandiga, 2004: Mitigation of environmental problems in Lake Victoria, East Africa: causal chain and policy option analysis. *Ambio*, **33**, 13-23.

OECD, 2004/2005: *African Economic Outlook: What's New for Africa?* Joint Project of the African Development Bank and OECD Development Centre, Paris, 540 pp.

Olsson, L., L. Eklundh and J. Ardö, 2005: A recent greening of the Sahel: trend, patterns and potential causes. *J. Arid Environ.*, **63**, 556-566.

O'Reilly, C.M., S.R. Alin, P.D. Plisnier, A.S. Cohen and B.A. McKee, 2003: Climate change decreases aquatic ecosystem productivity of Lake Tanganyika, Africa. *Nature*, **424**, 766-768.

Orindi, V.A. and A. Ochieng, 2005: Seed Fairs as a drought recovery strategy in Kenya. *IDS Bull.*, **36**, 87-102.

Osman-Elasha, B., N. Goutbi, E. Spanger-Siegfried, W. Dougherty, A. Hanafi, S. Zakieldeen, A. Sanjak, H. Abdel Atti and H.M. Elhassan, 2006: Adaptation strategies to increase human resilience against climate variability and change: lessons from the arid regions of Sudan. Working Paper 42, AIACC, 44 pp.

Owens, T., J. Hoddinot and B. Kinsey, 2003: Ex-ante actions and ex-post public responses to drought shocks: evidence and simulations from Zimbabwe. *World Dev.*, **31**, 1239-1255.

Owuor, B., S. Eriksen and W. Mauta, 2005: Adapting to climate change in a dryland mountain environment in Kenya. *Mt. Res. Dev.*, **25**, 310-315.

Pandey, D.N., A.K. Gupta and D.M. Anderson, 2003: Rainwater harvesting as an adaptation to climate change. *Curr. Sci. India*, **85**, 46-59.

Parry, M.L., C.A. Rosenzweig, A. Iglesias, M. Livermore and G. Fisher, 2004: Effects of climate change on global food production under SRES emissions and socioeconomic scenarios. *Global Environ. Chang.*, **14**, 53-67.

Pascual, M., J.A. Ahumada, L.F. Chaves, X. Rodó and M. Bouma, 2006: Malaria resurgence in the East African highlands: temperature trends revisited. *P. Natl. Acad. Sci. USA*, **103**, 5829-5834.

Patt, A., 2001: Understanding uncertainty: forecasting seasonal climate for farmers in Zimbabwe. *Risk, Decision and Policy*, **6**, 105-119.

Patt, A. and C. Gwata, 2002: Effective seasonal climate forecast applications: examining constraints for subsistence farmers in Zimbabwe. *Global Environ. Chang.*, **12**, 185-195.

Patt, A., P. Suarez and C. Gwata, 2005: Effects of seasonal climate forecasts and participatory workshops among subsistence farmers in Zimbabwe. *P. Natl. Acad. Sci. USA*, **102**, 12623-12628.

Patz, J.A. and S.H. Olson, 2006: Climate change and health: global to local influences on disease risk. *Ann. Trop. Med. Parasit.*, **100**, 535-549.

Patz, J.A., M. Hulme, C. Rosenzweig, T.D. Mitchell, R.A. Goldberg, A.K. Githeko, S. Lele, A.J. McMichael and D. Le Sueur, 2002: Climate change (communications arising): Regional warming and malaria resurgence. *Nature*, **420**, 627-628.

Payet, R. and D. Obura, 2004: The negative impacts of human activities in the Eastern African region: an international waters perspective. *Ambio*, **33**, 24-33.

Phillips, J.G., E. Makaudze and L. Unganai, 2001: Current and potential use of climate forecasts for resource-poor farmers in Zimbabwe. *Impacts of El Niño and Climate Variability on Agriculture*, C. Rosenzweig, K.J. Boote, S. Hollinger, A. Iglesias and J.G. Phillips, Eds., American Society of Agronomy Special Publication 63, Madison, Wisconsin, 97-100.

Piot, P. and P. Pinstrup-Anderson, 2002: AIDS: the new challenge to food security. International Food Policy Research Institute, 2001-2002 Annual Report, 11-17.

Prospero, J.M. and P.J. Lamb, 2003: African droughts and dust transport to the Caribbean: climate change implications. *Science*, **302**, 1024-1027.

Quinn, C.H., M. Huby, H. Kiwasila and J.C. Lovett, 2003: Local perceptions of risk to livelihood in semi-arid Tanzania. *J. Environ. Manage.*, **68**, 111-119.

Ramphele, M., 2004: Women's indigenous knowledge: building bridges between the traditional and the modern. *Indigenous Knowledge: Local Pathways to Development*, The World Bank, Washington, District of Columbia, 13-17.

Reason, C.J.C., 2002: Sensitivity of the southern African circulation to dipole sea-surface temperature patterns in the south Indian Ocean. *Int. J. Climatol.*, **22**, 377-393.

Reason, C.J.C. and A. Keibel, 2004: Tropical cyclone Eline and its unusual penetration and impacts over the southern African mainland. *Weather Forecast.*, **19**, 789-805.

Reason, C.J.C., S. Hachigonta and R.F. Phaladi, 2005: Interannual variability in rainy season characteristics over the Limpopo region of southern Africa. *Int. J. Climatol.*, **25**, 1835-1853.

Reich, P.F., S.T. Numbem, R.A. Almaraz and H. Eswaran, 2001: Land resource stresses and desertification in Africa. *Agro-Science*, **2**, 1-10.

Reid, P. and C. Vogel, 2006: Living and responding to multiple stressors in South Africa: glimpses from KwaZulu Natal. *Global Environ. Chang.*, **16**, 195-206.

Republic of Ghana, 2000: *Initial National Communication to the United Nations Framework Convention on Climate Change*. Ministry of Environment, Science and Technology, Accra, 171 pp.

Republic of Kenya, 2002: *First National Communication of Kenya to the Conference of Parties to the United Nations Framework Convention on Climate Change*. Ministry of Environment and Natural Resources, Nairobi, 155 pp.

République de Côte d'Ivoire, 2000: *Communication Initiale de la Côte d'Ivoire*. Ministère de l'Environnement, de l'Eau et de la Forêt, Abidjan, 97 pp.

République de Djibouti, 2001: *Communication Nationale Initiale de la Republique des Nations Unies sur les Changements Climatiques*. Ministere de l'Habitat, de l'Urbanisme, de l'Environment et de l'Amenagement de Territoire, 91 pp.

République de Guinée, 2002: *Communication Initiale de la Guinée à la Convention Cadre des Nations Unies sur les Changements Climatiques*. Ministère des Mines, de la Géologie et de l'Environnement, Conakry, 78 pp.

République Démocratique du Congo, 2000: *Communication Nationale Initiale de la République Démocratique du Congo (année 1994)*. Ministère des Affaires Foncières, Environnement et Développement Touristique, Kinshasa, 207 pp.

République du Congo, 2001: *Communication Nationale Initiale*. Ministère de l'Industrie Minière et de l'Environnement, Brazzaville, 56 pp.

Richard, Y., N. Fauchereau, I. Poccard, M. Rouault and S. Trzaska, 2001: 20th century droughts in Southern Africa: spatial and temporal variability, teleconnections with oceanic and atmospheric conditions. *Int. J. Climatol.*, **21**, 873-885.

Riebeek, H., 2006: Lake Victoria's falling waters. Earth Observatory Features, NASA. http://earthobservatory.nasa.gov/Study/Victoria/printall.php.

Robinson, J.B. and D. Herbert, 2001: Integrating climate change and sustainable development. *Int. J. Global Environ.*, **1**, 130-149.

Rocha, L.A., D.R. Robertson, C.R. Rocha, J.L. Van Tassell, M.T. Craig and B.W. Bowens, 2005: Recent invasion of the tropical Atlantic by an Indo-Pacific coral reef fish. *Mol. Ecol.*, **14**, 3921-3928.

Rockström, J., 2003: Resilience building and water demand management for drought mitigation. *Phys. Chem. Earth*, **28**, 869-877.

Ron, J., 2005: Paradigm in distress? Primary commodities and civil war. *J. Conflict Resolut.*, **49**, 443-450.

Roncoli, C., K. Ingram and P. Kirshen, 2001: The costs and risks of coping with drought: livelihood impacts and farmers' responses in Burkina Faso. *Climate Res.*, **19**, 119-132.

Rosegrant, M.W. and S.A. Cline, 2003: Global food security: challenges and policies. *Science*, **302**, 1917-1919.

Rowell, D.P., 2003: The impact of Mediterranean SSTs on the Sahelian rainfall season. *J. Climate*, **16**, 849-862.

Ruosteenoja, K., T.R. Carter, K. Jylhä and H. Tuomenvirt, 2003: Future climate in world regions: an intercomparison of model-based projections for the new IPCC emissions scenarios. Finnish Environment Institute No. 644, Helsinki, 83 pp.

Sachs, J.D., 2005: *The End of Poverty: Economic Possibilities for our Time*. Penguin, New York, 416 pp.

Sachs, J. and P. Malaney, 2002: The economic and social burden of malaria. *Nature*, **415**, 680-685.

Sachs, J.D., J.W. McArthur, G. Schmidt-Traub, C. Kruk, M. Bahadur, M. Faye and G. McCord, 2004: Ending Africa's poverty trap. *Brookings Papers on Economic Activity*, **1**, 117-240.

SAFARI, 2004: South African Regional Science Initiative (CSIR, YB, NASA, ZMD and WITS, 2004): SAFARI 2000; reprinted from *J. Geophys. Res., AGU Special Collection*, **108**, 2003. http://www.agu.org/contents/sc/ViewCollection.do?collectionCode=SAF1&journalCode=JD.

Schreck, C.J. and F.H.M. Semazzi, 2004: Variability of the recent climate of eastern Africa. *Int. J. Climatol.*, **24**, 681-701.

Schreider, G. and B. Knerr, 2000: Labor migration as a social security mechanism for smallholder households in sub-Saharan Africa: the case of Cameroon. *Oxford Development Studies*, **28**, 223-236.

Schulze, R., J. Meigh and M. Horan, 2001: Present and potential future vulnerability of eastern and southern Africa's hydrology and water resources. *S. Afr. J. Sci.*, **97**, 150-160.

Schwind, K., 2005: Going local on a global scale: rethinking food trade in the era of climate change, dumping and rural poverty. *BackGrounder*, **11**, 1-4.

Scoones, I., 2005: Governing technology development: challenges for agricultural research in Africa. *IDS Bull.*, **36**, 109-114.

Scoones, I., S. Devereux and L. Haddad, 2005: Introduction: New directions for African agriculture. *IDS Bull.*, **36**, 1-12.

Seck, M., M.N.A. Mamouda and S. Wade, 2005: Case Study 4: Adaptation and mitigation through "produced environments": the case for agriculture intensification in Senegal. *IDS Bull.*, **36**, 71-86.

Semazzi, F.H.M. and Y. Song, 2001: A GCM study of climate change induced by deforestation in Africa. *Climate Res.*, **17**, 169-182.

Seo, S.N. and R. Mendelsohn, 2006a: Climate change impacts on animal husbandry in Africa: a Ricardian analysis. Centre for Environmental Economics and Policy in Africa (CEEPA) Discussion Paper No. 9, University of Pretoria, Pretoria, 42 pp.

Seo, S.N. and R. Mendelsohn, 2006b: Climate change adaptation in Africa: a microeconomic analysis of livestock choice. Centre for Environmental Economics and Policy in Africa (CEEPA) Discussion Paper No.19, University of Pretoria, Pretoria, 37 pp.

Simmons, R.E., P. Barnard, W.R.J. Dean, G.F. Midgley, W. Thuiller and G. Hughes, 2004: Climate change and birds: perspectives and prospects from southern Africa. *Ostrich: Journal of African Ornithology*, **75**, 295-308.

Slingo, J.M., A.J. Challinor, B.J. Hoskins and T.R. Wheeler, 2005: Introduction: food crops in a changing climate. *Philos. T. Roy. Soc. B*, **360**, 1983-1989.

Sokona, Y. and F. Denton, 2001: Climate change impacts: can Africa cope with the challenges? *Clim. Policy*, **1**, 117-123.

St-Pierre, N.R., B. Cobanov and G. Schnitkey, 2003: Economic losses from heat stress by US livestock industries. *J. Dairy Sci.*, **86**, 52-77.

State of Eritrea, 2001: *Eritrea's Initial National Communication under the United Nations Framework Convention on Climate Change (UNFCCC)*. Ministry of Land, Water and Environment, Asmara, 99 pp. http://unfccc.int/resource/docs/natc/erinc1.pdf.

Stern, N., 2007: *The Economics of Climate Change: The Stern Review*. Cambridge University Press, Cambridge, 692 pp.

Stige, L.C., J. Stave, K.S. Chan, L. Ciannelli, N. Pretorelli, M. Glantz, H.R. Herren and N.C. Stenseth, 2006: The effect of climate variation on agro-pastoral production in Africa. *P. Natl. Acad. Sci. USA*, **103**, 3049-3053.

Strzepek, K. and A. McCluskey, 2006: District level hydro-climatic time series and scenario analysis to assess the impacts of climate change on regional water resources and agriculture in Africa. Centre for Environmental Economics and Policy in Africa (CEEPA) Discussion Paper No.13, University of Pretoria, Pretoria, 59 pp.

Sultan, B., K. Labadi, J.F. Guégan and S. Janicot, 2005: Climate drives the meningitis epidemics onset in West Africa. *PLoS Med.*, **2**, 43-49. http://medicine.plosjournals.org/archive/1549-1676/2/1/pdf/10.1371_journal.pmed.0020006-S.pdf.

Swaminathan, M.S., 2000: Climate change and food security. *Climate Change and Development*, L. Gomez-Echeverri, Ed., UNDP Regional Bureau for Latin America and the Caribbean and Yale School of Forestry and Environmental Studies, 103-114. http://environment.yale.edu/doc/786/climate_change_and_development/.

Swart, R., J. Mitchell, T. Morita and S. Raper, 2002: Stabilisation scenarios for climate impacts assessment. *Global Environ. Chang.*, **12**, 155-165.

Tadross, M.A., B.C. Hewitson and M.T. Usman, 2005a: The interannual variability of the onset of the maize growing season over South Africa and Zimbabwe. *J. Climate*, **18**, 3356-3372.

Tadross, M., C. Jack and B. Hewitson, 2005b: On RCM-based projections of change in southern African summer climate. *Geophys. Res. Lett.*, **32**, L23713, doi:10.1029/2005GL024460.

Tanser, F.C., B. Sharp and D. le Sueur, 2003: Potential effect of climate change on malaria transmission in Africa. *Lancet*, **362**, 1792-1798.

Tarhule, A. and P.J. Lamb, 2003: Climate research and seasonal forecasting for West Africans: perceptions, dissemination, and use? *B. Am. Meteorol. Soc.*, **84**, 1741-1759.

ter Kuile, F.O., M.E. Parise, F.H. Verhoeff, V. Udhayakumar, R.D. Newman, A.M. van Eijk, S.J. Rogerson and R.W. Steketee, 2004: The burden of co-infection with human immunodeficiency virus type 1 and malaria in pregnant women in sub-Saharan Africa. *Am. J. Trop. Med. Hyg.*, **71**, 41-54.

Thomas, C.D., A. Cameron, R.E. Green, M. Bakkenes, L.J. Beaumont, Y.C. Collingham, B.F.N. Erasmus, M.F. de Siqueira and Co-authors, 2004: Extinction from climate change. *Nature*, **427**, 145-148.

Thomas, D.S.G. and C. Twyman, 2005: Equity and justice in climate change adaptation amongst natural-resource-dependent societies. *Global Environ. Chang.*, **15**, 115-124.

Thomas, D.S.G., M. Knight and G.F.S. Wiggs, 2005: Remobilization of southern African desert dune systems by twenty-first century global warming. *Nature*, **435**, 1218-1221.

Thompson, L.G., E. Mosley-Thompson, M.E. Davis, K.A. Henderson, H.H. Brecher, V.S. Zagorodnov, T.A. Mashiotta, P.N. Lin and Co-authors, 2002: Kilimanjaro ice core records: evidence of Holocene change in tropical Africa. *Science*, **298**, 589-593.

Thomson, M.C., F.J. Doblas-Reyes, S.J. Mason, R. Hagedorn, S.J. Connor, T. Phindela, A.P. Morse and T.N. Plamer, 2006: Malaria early warnings based on seasonal climate forecasts from multi-model ensembles. *Nature*, **439**, 576-579.

Thornton, P.K., R.H. Fawcett, K.A. Galvin, R.B. Boone, J.W. Hudson and C.H. Vogel, 2004: Evaluating management options that use climate forecasts: modelling livestock production systems in the semi-arid zone of South Africa. *Climate Res.*, **26**, 33-42.

Thornton, P.K., P.G. Jones, T.M. Owiyo, R.L. Kruska, M. Herero, P. Kristjanson, A. Notenbaert, N. Bekele and Co-authors, 2006: Mapping Climate Vulnerability and Poverty in Africa. Report to the Department for International Development, ILRI, Nairobi, 200 pp.

Thuiller, W., O. Broennimann, G. Hughes, J.R.M. Alkemade, G.F. Midgley and F. Corsi, 2006: Vulnerability of African mammals to anthropogenic climate change under conservative land transformation assumptions. *Glob. Change Biol.*, **12**, 424-440.

Tiffen, M., 2003: Transition in sub-Saharan Africa: agriculture, urbanization and income growth. *World Dev.*, **31**, 1343-1366.

Tol, R.S.J., B. O'Neill and D.P. van Vuuren, 2005: A critical assessment of the IPCC SRES scenarios. http://www.uni-hamburg.de/Wiss/FB/15/Sustainability/ensemblessres.pdf.

Toulmin, C., R. Leonard, K. Brock, N. Coulibaly, G. Carswell and D. Dea, 2000: Diversification of livelihoods: evidence from Mali and Ethiopia. Research Report 47, Institute of Development Studies, Brighton, 60 pp.

Trenberth, K.E., P.D. Jones, P. Ambenje, R. Bojariu, D. Easterling, A. Klein Tank, D. Parker, F. Rahimzadeh, J.A. Renwick, M. Rusticucci, B. Soden and P. Zhai, 2007: Observations: Surface and atmospheric climate change. *Climate Change 2007: The Physical Science Basis. Contribution of Working Group I to the Fourth Assessment Report of the Intergovernmental Panel on Climate Change*, S. Solomon, D. Qin, M. Manning, Z. Chen, M. Marquis, K.B. Averyt, M. Tignor and H.L. Miller, Eds., Cambridge University Press, Cambridge, 235-336.

Turner, M., 2003: Environmental science and social causation in the analysis of Sahelian pastoralism. *Political Ecology: An Integrative Approach to Geography and Environment-Development Studies*, K.S. Zimmerer and T.J. Bassett, Eds., Guilford Press, New York, 159-178.

Turpie, J., H. Winkler, R. Spalding-Fecher and G. Midgley, 2002: Economic impacts of climate change in South Africa: a preliminary analysis of unmitigated damage costs. Energy and Development Research Centre, University of Cape Town, Cape Town, 64 pp.

UNAIDS, 2004: 2004 Report on the Global AIDS Epidemic. 4th Global Report, Joint United Nations Programme on HIV/AIDS, Rome, 236 pp. http://www.unaids.org/bangkok2004/report_pdf.html.

UNDP, 2004: *Reducing Disaster Risk: a Challenge for Development*. UNDP Bureau for Crisis Prevention and Recovery, New York, 161 pp. http://www.undp.org/bcpr/disred/documents/publications/rdr/english/rdr_english.pdf.

UNDP, 2005: Human Development Report 2005. United Nations Development Programme, New York, 388 pp. http://hdr.undp.org/reports/global/2005/pdf/HDR05_complete.pdf.

UNEP, 2004: GEO Year Book 2003. The United Nations Environment Programme Global Environmental Outlook Report, 76 pp. http://www.unep.org/geo/yearbook/yb2003/index.htm.

UNEP, 2005: GEO Year BOOK 2004/5. The United Nations Environment Programme Global Environmental Outlook Report, 96 pp. http://www.unep.org/geo/pdfs/GEO%20YEARBOOK%202004%20(ENG).pdf.

UNEP/GRID-Arendal, 2002: Vital Climate Graphics. United Nations Environment Programme, GRID-Arendal, 65 pp. http://grida.no/climate/vital/index.htm.

UNESCO-WWAP, 2006: Water: A Shared Responsibility. United Nations World Water Development Report 2. United Nations Educational, Scientific and Cultural Organisation (UNESCO) and the United Nations World Water Assessment Programme, 601 pp.

UNFPA, 2003: *The Impact of HIV/AIDS: A Population and Development Perspective*. Population and Development Strategies No. 9, United Nations Population Fund, 147 pp.

Union des Comores, 2002: *Communication Nationale Initiale. Convention Cadre des Nations Unies sur les Changements Climatiques*. Ministère du Développement, des Infrastructures, des Postes et Télécommunications et des Transports Internationaux, Moroni, 72 pp.

USAID, 2003: RCSA food security strategic option: synthesis of selected readings. Report prepared by Nathan and Associates for USAID Regional Centre for South Africa, 175 pp.

Usman, M.T. and C.J.C. Reason, 2004: Dry spell frequencies and their variability over southern Africa. *Climate Res.*, **26**, 199-211.

Utzinger, J., Y. Tozan, F. Doumani and B.H. Singer, 2001: The economic payoffs of integrated malaria control in the Zambian Copper Belt between 1930 and 1950. *Trop. Med. Int. Health*, **7**, 657-677.

Van Drunen, M.A., R. Lasage and C. Dorland, Eds., 2005: *Climate Change in Developing Countries: An Overview of Study Results from the Netherlands Climate Change Studies Assistance Programme*. CABI Publishing, Amsterdam, 320 pp.

van Jaarsveld, A.S., R. Biggs, R.J. Scholes, E. Bohensky, B. Reyers, T. Lynam, C. Musvoto and C. Fabricius, 2005: Measuring conditions and trends in ecosystem services at multiple scales: the Southern African Millennium Ecosystem Assessment (SAfMA) experience. *Philos. T. Roy. Soc. B*, **360**, 425-441.

van Lieshout, M., R.S. Kovats, M.T.J. Livermore and P. Martens, 2004: Climate change and malaria: analysis of the SRES climate and socio-economic scenarios. *Global Environ. Chang.*, **14**, 87-99.

Verhagen, A., A.J. Dietz and R. Ruben, 2001: Impact of climate change on water availability, agriculture and food security in semi-arid regions, with special focus on West Africa. Global Change NOP-NRP Report 410200076. Dutch National Institute for Public Health and Environment, Bilthoven, 140 pp.

Vincent, K., 2004: Creating an index of social vulnerability to climate change for Africa. Working Paper 56, Tyndall Centre for Climate Change Research, University of East Anglia, Norwich, 50 pp.

Vizy, E.K. and K.H. Cook, 2001: Mechanisms by which Gulf of Guinea and eastern North Atlantic sea surface temperature anomalies influence African rainfall. *J. Climate*, **14**, 795-821.

Vogel, C., 2005: "Seven fat years and seven lean years?" Climate change and agriculture in Africa. *IDS Bull.*, **36**, 30-35.

Vörösmarty, C.J., E.M. Douglas, P.A. Green and C. Revenga, 2005: Geospatial indicators of emerging water stress: an application to Africa. *Ambio*, **34**, 230-236.

Wahab, H.M., 2005: The impact of geographical information system on environmental development, unpublished MSc Thesis, Faculty of Agriculture, Al-Azhar University, Cairo, 148 pp.

Wang, G. and E.A.B. Eltahir, 2000: Role of vegetation dynamics in enhancing the low-frequency variability of the Sahel rainfall. *Water Resour. Res.*, **36**, 1013-1022.

Wang, G. and E.A.B. Eltahir, 2002: Impact of CO_2 concentration changes on the biosphere-atmosphere system of West Africa. *Glob. Change Biol.*, **8**, 1169-1182.

Wang, G., E.A.B. Eltahir, J.A. Foley, D. Pollard and S. Levis, 2004: Decadal variability of rainfall in the Sahel: results from the coupled GENESIS-IBIS atmosphere-biosphere model. *Clim. Dynam.*, **22**, 625-637.

Warren, R., N. Arnell, R. Nicholls, P. Levy and J. Price, 2006: Understanding the regional impacts of climate change: research report prepared for the Stern Review on the Economics of Climate Change. Tyndall Centre for Climate Change Research, Working Paper 90, University of East Anglia, Norwich, 223 pp.

Washington, R. and M.C. Todd, 2005: Atmospheric controls on mineral dust emission from the Bodélé depression, Chad: the role of the Low Level Jet. *Geophys. Res. Lett.*, **32**, L17701, doi:10.1029/2005GL023597.

Washington, R. and A. Preston, 2006: Extreme wet years over southern Africa: role of Indian Ocean sea surface temperatures. *J. Geophys. Res.– Atmos.*, **111**, D15104, doi:10.1029/2005JD006724.

Washington, R., M. Harrison and D. Conway, 2004: African climate report. Report commissioned by the UK Government to review African climate science, policy and options for action, 45 pp.

Washington, R., M. Harrison, D. Conway, E. Black, A. Challinor, D. Grimes, R. Jones, A. Morse and Co-authors, 2006: African climate change: taking the shorter route. *B. Am. Meteorol. Soc.*, **87**, 1355-1366.

WHO, 2005: *Using Climate to Predict Infectious Disease Epidemics*. World Health Organization, Geneva, 56 pp. http://www.who.int/entity/globalchange/publications/infectdiseases.pdf.

WHO/UNICEF, 2000: Global water supply and sanitation assessment: 2000 report. World Health Organization, Geneva, 87 pp. http://www.who.int/entity/water_sanitation_health/monitoring/jmp2000.pdf.

World Bank, 2006a: *World Development Indicators*. The International Bank for Reconstruction and Development, The World Bank, Washington, District of Columbia, 242 pp. http://devdata.worldbank.org/wdi2006/.

World Bank, 2006b: *World Development Report, 2006: Equity and Development*. Co-publication of The World Bank, Washington, District of Columbia and Oxford University Press, Oxford, 336 pp.

World Tourism Organization, 2003: *Climate Change and Tourism: Proc. 1st International Conference on Climate Change and Tourism*, Djerba, Tunisia. World Tourism Organization, 55 pp. http://www.world-tourism.org/sustainable/climate/final-report.pdf.

World Tourism Organization, 2005: *Tourism Vision 2020*. World Trade Organisation, Washington, District of Columbia. http://www.world-tourism.org/facts/wtb.html.

World Water Forum, 2000: *The Africa Water Vision for 2025: Equitable and Sustainable Use of Water for Socioeconomic Development*. UN Water/Africa, 34 pp.

Yamin, F., A. Rahman and S. Huq, 2005: Vulnerability, adaptation and climate disasters: a conceptual overview. *IDS Bull.*, **36**, 1-14.

Yohe, G. and M. Schlesinger, 2002: The economic geography of the impacts of climate change. *J. Econ. Geogr.*, **2**, 311-341.

Yousif, H.M., 2005: Rapid urbanisation in Africa: impacts on housing and urban poverty. , in ECA, 2005, Assessing sustainable development in Africa. *Assessing sustainable development in Africa*, Africa's Sustainable Development Bulletin, Economic Commission for Africa, Addis Ababa, 55-59.

Zeng, N., 2003: Drought in the Sahel. *Science*, **302**, 999-1000.

Zhou, G., N. Minakawa, A.K. Githeko and G. Yan, 2004: Association between climate variability and malaria epidemics in the East African highlands. *P. Natl. Acad. Sci. USA*, **101**, 2375-2380.

Ziervogel, G., 2004: Targeting seasonal climate forecasts for integration into household level decisions: the case of smallholder farmers in Lesotho. *Geogr. J.*, **170**, 6-21.

Ziervogel, G. and R. Calder, 2003: Climate variability and rural livelihoods: assessing the impact of seasonal climate forecasts in Lesotho. *Area*, **35**, 403-417.

Ziervogel, G. and T.E. Downing, 2004: Stakeholder networks: improving seasonal forecasts. *Climatic Change*, **65**, 73-101.

Ziervogel, G., M. Bithell, R. Washington and T. Downing, 2005: Agent-based social simulation: a method for assessing the impact of seasonal climate forecast applications among smallholder farmers. *Agr. Syst.*, **83**, 1-26.

Ziervogel, G., A.O. Nyong, B. Osman, C. Conde, S. Cortés and T. Downing, 2006: Climate variability and change: implications for household food security. AIACC Working Paper No. 20, 25 pp. http://www.aiaccproject.org/working_papers/Working%20Papers/AIACC_WP_20_Ziervogel.pdf.

10

Asia

Coordinating Lead Authors:

Rex Victor Cruz (Philippines), Hideo Harasawa (Japan), Murari Lal (India), Shaohong Wu (China)

Lead Authors:

Yurij Anokhin (Russia), Batima Punsalmaa (Mongolia), Yasushi Honda (Japan), Mostafa Jafari (Iran), Congxian Li (China), Nguyen Huu Ninh (Vietnam)

Contributing Authors:

Shiv D. Atri (India), Joseph Canadell (Australia), Seita Emori (Japan), Daidu Fan (China), Hui Ju (China), Shuangcheng Li (China), Tushar K. Moulik (India), Faizal Parish (Malaysia), Yoshiki Saito (Japan), Ashok K. Sharma (India), Kiyoshi Takahashi (Japan), Tran Viet Lien (Vietnam), Qiaomin Zhang (China)

Review Editors:

Daniel Murdiyarso (Indonesia), Shuzo Nishioka (Japan)

This chapter should be cited as:

Cruz, R.V., H. Harasawa, M. Lal, S. Wu, Y. Anokhin, B. Punsalmaa, Y. Honda, M. Jafari, C. Li and N. Huu Ninh, 2007: Asia. *Climate Change 2007: Impacts, Adaptation and Vulnerability. Contribution of Working Group II to the Fourth Assessment Report of the Intergovernmental Panel on Climate Change*, M.L. Parry, O.F. Canziani, J.P. Palutikof, P.J. van der Linden and C.E. Hanson, Eds., Cambridge University Press, Cambridge, UK, 469-506.

Table of Contents

Executive summary

New evidences show that climate change has affected many sectors in Asia (medium confidence).
The crop yield in many countries of Asia has declined, partly due to rising temperatures and extreme weather events. The retreat of glaciers and permafrost in Asia in recent years is unprecedented as a consequence of warming. The frequency of occurrence of climate-induced diseases and heat stress in Central, East, South and South-East Asia has increased with rising temperatures and rainfall variability. Observed changes in terrestrial and marine ecosystems have become more pronounced (medium confidence). [10.2.3, 10.2.4]

Future climate change is likely to affect agriculture, risk of hunger and water resource scarcity with enhanced climate variability and more rapid melting of glaciers (medium confidence).
About 2.5 to 10% decrease in crop yield is projected for parts of Asia in 2020s and 5 to 30% decrease in 2050s compared with 1990 levels without CO_2 effects (medium confidence) [10.4.1.1]. Freshwater availability in Central, South, East and South-East Asia, particularly in large river basins such as Changjiang, is likely to decrease due to climate change, along with population growth and rising standard of living that could adversely affect more than a billion people in Asia by the 2050s (high confidence) [10.4.2]. It is estimated that under the full range of Special Report on Emissions Scenarios (SRES) scenarios, 120 million to 1.2 billion will experience increased water stress by the 2020s, and by the 2050s the number will range from 185 to 981 million people (high confidence) [10.4.2.3]. Accelerated glacier melt is likely to cause increase in the number and severity of glacial melt-related floods, slope destabilisation and a decrease in river flows as glaciers recede (medium confidence) [10.2.4.2, 10.4.2.1]. An additional 49 million, 132 million and 266 million people of Asia, projected under A2 scenario without carbon fertilisation, could be at risk of hunger by 2020, 2050 and 2080, respectively (medium confidence) [10.4.1.4].

Marine and coastal ecosystems in Asia are likely to be affected by sea-level rise and temperature increases (high confidence).
Projected sea-level rise is very likely to result in significant losses of coastal ecosystems and a million or so people along the coasts of South and South-East Asia will likely be at risk from flooding (high confidence) [10.4.3.1]. Sea-water intrusion due to sea-level rise and declining river runoff is likely to increase the habitat of brackish water fisheries but coastal inundation is likely to seriously affect the aquaculture industry and infrastructure particularly in heavily-populated megadeltas (high confidence) [10.4.1.3, 10.4.3.2]. Stability of wetlands, mangroves and coral reefs around Asia is likely to be increasingly threatened (high confidence) [10.4.3.2, 10.6.1]. Recent risk analysis of coral reef suggests that between 24% and 30% of the reefs in Asia are likely to be lost during the next 10 years and 30 years, respectively (medium confidence) [10.4.3.2].

Climate change is likely to affect forest expansion and migration, and exacerbate threats to biodiversity resulting from land use/cover change and population pressure in most of Asia (medium confidence).
Increased risk of extinction for many flora and fauna species in Asia is likely as a result of the synergistic effects of climate change and habitat fragmentation (medium confidence) [10.4.4.1]. In North Asia, forest growth and northward shift in the extent of boreal forest is likely (medium confidence) [10.4.4]. The frequency and extent of forest fires in North Asia is likely to increase in the future due to climate change that could likely limit forest expansion (medium confidence) [10.4.4].

Future climate change is likely to continue to adversely affect human health in Asia (high confidence).
Increases in endemic morbidity and mortality due to diarrhoeal disease primarily associated with climate change are expected in South and South-East Asia (high confidence). Increases in coastal water temperature would exacerbate the abundance and/or toxicity of cholera in south Asia (high confidence). Natural habitats of vector-borne and water-borne diseases in north Asia are likely to expand in the future (medium confidence). [10.4.5]

Multiple stresses in Asia will be compounded further due to climate change (high confidence).
It is likely that climate change will impinge on sustainable development of most developing countries of Asia as it compounds the pressures on natural resources and the environment associated with rapid urbanisation, industrialisation and economic development. Mainstreaming sustainable development policies and the inclusion of climate-proofing concepts in national development initiatives are likely to reduce pressure on natural resources and improve management of environmental risks (high confidence) [10.7].

10.1 Summary of knowledge assessed in the Third Assessment Report

10.1.1 Climate change impacts in Asia

Climate change and variability.

Extreme weather events in Asia were reported to provide evidence of increases in the intensity or frequency on regional scales throughout the 20th century. The Third Assessment Report (TAR) predicted that the area-averaged annual mean warming would be about 3°C in the decade of the 2050s and about 5°C in the decade of the 2080s over the land regions of Asia as a result of future increases in atmospheric concentration of greenhouse gases (Lal et al., 2001a). The rise in surface air temperature was projected to be most pronounced over boreal Asia in all seasons.

Climate change impacts.

An enhanced hydrological cycle and an increase in area-averaged annual mean rainfall over Asia were projected. The increase in annual and winter mean precipitation would be highest in boreal Asia; as a consequence, the annual runoff of major Siberian Rivers would increase significantly. A decline in summer precipitation was likely over the central parts of arid and semi-arid Asia leading to expansion of deserts and periodic severe water stress conditions. Increased rainfall intensity, particularly during the summer monsoon, could increase flood-prone areas in temperate and tropical Asia.

10.1.2 Vulnerabilities and adaptive strategies

Vulnerable sectors. Water and agriculture sectors are likely to be most sensitive to climate change-induced impacts in Asia. Agricultural productivity in Asia is likely to suffer severe losses because of high temperature, severe drought, flood conditions, and soil degradation. Forest ecosystems in boreal Asia would suffer from floods and increased volume of runoff associated with melting of permafrost regions. The processes of permafrost degradation resulting from global warming strengthen the vulnerability of all relevant climate-dependent sectors affecting the economy in high-latitude Asia.

Vulnerable regions. Countries in temperate and tropical Asia are likely to have increased exposure to extreme events, including forest die back and increased fire risk, typhoons and tropical storms, floods and landslides, and severe vector-borne diseases. The stresses of climate change are likely to disrupt the ecology of mountain and highland systems in Asia. Glacial melt is also expected to increase under changed climate conditions. Sea-level rise would cause large-scale inundation along the vast Asian coastline and recession of flat sandy beaches. The ecological stability of mangroves and coral reefs around Asia would be put at risk.

Adaptation strategies. Increases in income levels, education and technical skills, and improvements in public food distribution, disaster preparedness and management, and health care systems through sustainable and equitable development could substantially enhance social capital and reduce the vulnerability of developing countries of Asia to climate change. Development and implementation of incremental adaptation strategies and policies to exploit 'no regret' measures and 'win-win' options were to be preferred over other options. Adaptations to deal with sea-level rise, potentially more intense cyclones, and threats to ecosystems and biodiversity were recommended as high priority actions in temperate and tropical Asian countries. It was suggested that the design of an appropriate adaptation programme in any Asian country must be based on comparison of damages avoided with costs of adaptation.

Advances since the TAR. Aside from new knowledge on the current trends in climate variability and change – including the extreme weather events – more information is now available that confirms most of the key findings on impacts, vulnerabilities and adaptations for Asia. This chapter assesses the state of knowledge on impacts, vulnerabilities and adaptations for various regions in Asia.

10.2 Current sensitivity and vulnerability

10.2.1 Asia: regional characteristics

Asia is the most populous continent (Figure 10.1). Its total population in 2002 was reported to be about 3,902 million, of which almost 61% is rural and 38.5% lives within 100 km of the coast (Table 10.1). The coastline of Asia is 283,188 km long (Duedall and Maul, 2005). In this report, Asia is divided into seven sub-regions, namely North Asia, Central Asia, West Asia, Tibetan Plateau, East Asia, South Asia and South-East Asia (for further details on boundaries of these sub-regions see Table 10.5).

North Asia, located in the Boreal climatic zone, is the coldest region of the northern hemisphere in winter (ACIA, 2005). One of the world's largest and oldest lakes, Baikal, located in this region contains as much as 23,000 km^3 of freshwater and holds nearly 20% of the world surface freshwater resources (Izrael and Anokhin, 2000). *Central* and *West Asia* include several countries of predominantly arid and semi-arid region. *Tibetan Plateau* can be divided into the eastern part (forest region), the northern part (open grassland), and the southern and central part (agricultural region). *East Asia* stretches in the east-west direction to about 5,000 km and in the north-south to about 3,000 km including part of China, Japan and Korea. *South Asia* is physiographically diverse and ecologically rich in natural and crop-related biodiversity. The region has five of the 20 megacities of the world (UN-HABITAT, 2004). *South-East Asia* is characterised by tropical rainforest, monsoon climates with high and constant rainfall, heavily-leached soils, and diverse ethnic groups. Table 10.1 lists the key socio-economic and natural resource features of the countries of Asia (WRI, 2003; FAO, 2004a, b, c; World Bank, 2005).

10.2.2 Observed climate trends, variability and extreme events

Past and present climate trends and variability in Asia are generally characterised by increasing surface air temperature which is more pronounced during winter than in summer. Increasing trends have been observed across the seven sub-regions of Asia. The observed increases in some parts of Asia during recent decades ranged between less than 1°C to 3°C per century. Increases in surface temperature are most pronounced in North Asia (Savelieva et al., 2000; Izrael et al., 2002a; Climate Change in Russia, 2003; Gruza and Rankova, 2004).

Interseasonal, interannual and spatial variability in rainfall trend has been observed during the past few decades all across Asia. Decreasing trends in annual mean rainfall are observed in Russia, North-East and North China, coastal belts and arid plains of Pakistan, parts of North-East India, Indonesia, Philippines and some areas in Japan. Annual mean rainfall exhibits increasing trends in Western China, Changjiang Valley and the South-Eastern coast of China, Arabian Peninsula, Bangladesh and along the western coasts of the Philippines. Table 10.2 lists more details on observed characteristics in surface air temperature and rainfall in Asian sub-regions.

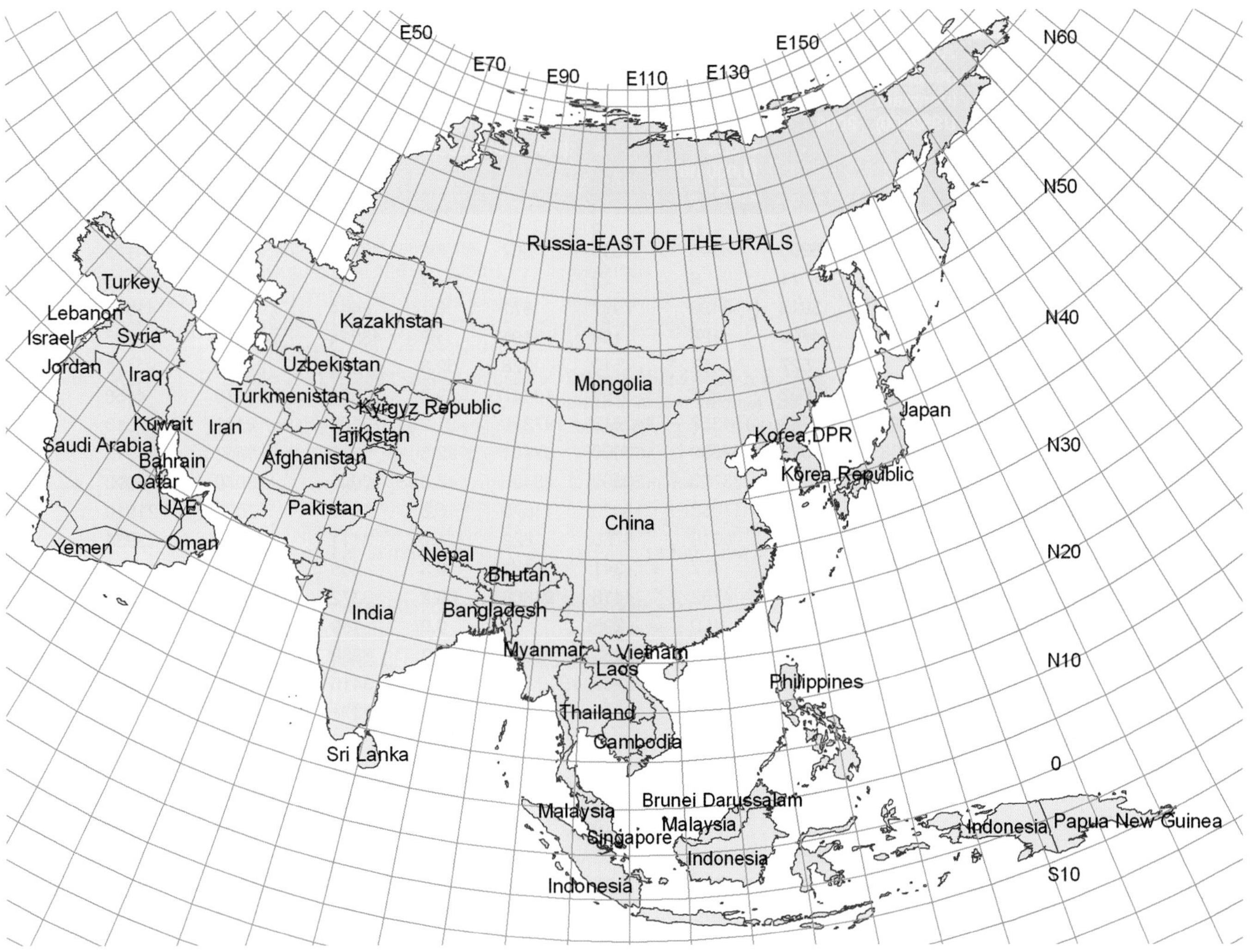

Figure 10.1. *Location of countries covered under Asia included in this chapter.*

10.2.3 Observed changes in extreme climatic events

New evidence on recent trends, particularly on the increasing tendency in the intensity and frequency of extreme weather events in Asia over the last century and into the 21st century, is briefly discussed below and summarised in Table 10.3. In South-East Asia, extreme weather events associated with El-Niño were reported to be more frequent and intense in the past 20 years (Trenberth and Hoar, 1997; Aldhous, 2004).

Significantly longer heatwave duration has been observed in many countries of Asia, as indicated by pronounced warming trends and several cases of severe heatwaves (De and Mukhopadhyay, 1998; Kawahara and Yamazaki, 1999; Zhai et al., 1999; Lal, 2003; Zhai and Pan, 2003; Ryoo et al., 2004; Batima et al., 2005a; Cruz et al., 2006; Tran et al., 2005).

Generally, the frequency of occurrence of more intense rainfall events in many parts of Asia has increased, causing severe floods, landslides, and debris and mud flows, while the number of rainy days and total annual amount of precipitation has decreased (Zhai et al., 1999; Khan et al., 2000; Shrestha et al., 2000; Izrael and Anokhin, 2001; Mirza, 2002; Kajiwara et al., 2003; Lal, 2003; Min et al., 2003; Ruosteenoja et al., 2003; Zhai and Pan, 2003; Gruza and Rankova, 2004; Zhai, 2004). However, there are reports that the frequency of extreme rainfall in some countries has exhibited a decreasing tendency (Manton et al., 2001; Kanai et al., 2004).

Increasing frequency and intensity of droughts in many parts of Asia are attributed largely to a rise in temperature, particularly during the summer and normally drier months, and during ENSO events (Webster et al., 1998; Duong, 2000; PAGASA, 2001; Lal, 2002, 2003; Batima, 2003; Gruza and Rankova, 2004; Natsagdorj et al., 2005).

Recent studies indicate that the frequency and intensity of tropical cyclones originating in the Pacific have increased over the last few decades (Fan and Li, 2005). In contrast, cyclones originating from the Bay of Bengal and Arabian Sea have been noted to decrease since 1970 but the intensity has increased (Lal, 2001). In both cases, the damage caused by intense cyclones has risen significantly in the affected countries, particularly India, China, Philippines, Japan, Vietnam and Cambodia, Iran and Tibetan Plateau (PAGASA, 2001; ABI, 2005; GCOS, 2005a, b).

Table 10.3. *Summary of observed changes in extreme events and severe climate anomalies*

Country/Region	Key trend	Reference
Heatwaves		
Russia	Heatwaves broke past 22-year record in May 2005	Shein, 2006
Mongolia	Heatwave duration has increased by 8 to 18 days in last 40 years; coldwave duration has shortened by 13.3 days	Batima et al., 2005a
China	Increase in frequency of short duration heatwaves in recent decade, increasing warmer days and nights in recent decades	Zhai et al., 1999; Zhai and Pan, 2003
Japan	Increasing incidences of daily maximum temperature >35°C, decrease in extremely low temperature	Kawahara and Yamazaki, 1999; Japan Meteorological Agency, 2005
Korea	Increasing frequency of extreme maximum temperatures with higher values in 1980s and 1990s; decrease in frequency of record low temperatures during 1958 to 2001	Ryoo et al., 2004
India	Frequency of hot days and multiple-day heatwave has increased in past century; increase in deaths due to heat stress in recent years	De and Mukhopadhyay, 1998; Lal, 2003
South-East Asia	Increase in hot days and warm nights and decrease in cold days and nights between 1961 and 1998	Manton et al., 2001; Cruz et al., 2006; Tran et al., 2005
Intense Rains and Floods		
Russia	Increase in heavy rains in western Russia and decrease in Siberia; increase in number of days with more than 10 mm rain; 50 to 70% increase in surface runoff in Siberia	Gruza et al., 1999; Izrael and Anokhin, 2001; Ruosteenoja et al., 2003; Gruza and Rankova, 2004
China	Increasing frequency of extreme rains in western and southern parts including Changjiang river, and decrease in northern regions; more floods in Changjiang river in past decade; more frequent floods in North-East China since 1990s; more intense summer rains in East China; severe flood in 1999; seven-fold increase in frequency of floods since 1950s	Zhai et al., 1999; Ding and Pan, 2002; Zhai and Pan, 2003; Zhai, 2004
Japan	Increasing frequency of extreme rains in past 100 years attributed to frontal systems and typhoons; serious flood in 2004 due to heavy rains brought by 10 typhoons; increase in maximum rainfall during 1961 to 2000 based on records from 120 stations	Kawahara and Yamazaki, 1999; Isobe, 2002; Kajiwara et al., 2003; Kanai et al., 2004
South Asia	Serious and recurrent floods in Bangladesh, Nepal and north-east states of India during 2002, 2003 and 2004; a record 944 mm of rainfall in Mumbai, India on 26 to 27 July 2005 led to loss of over 1,000 lives with loss of more than US$250 million; floods in Surat, Barmer and in Srinagar during summer monsoon season of 2006; 17 May 2003 floods in southern province of Sri Lanka were triggered by 730 mm rain	India Meteorological Department, 2002 to 2006; Dartmouth Flood Observatory, 2003.
South-East Asia	Increased occurrence of extreme rains causing flash floods in Vietnam; landslides and floods in 1990 and 2004 in the Philippines, and floods in Cambodia in 2000	FAO/WFP, 2000; Environment News Service, 2002; FAO, 2004a; Cruz et al., 2006; Tran et al., 2005
Droughts		
Russia	Decreasing rain and increasing temperature by over 1°C have caused droughts; 27 major droughts in 20th century have been reported	Golubev and Dronin, 2003; Izrael and Sirotenko, 2003
Mongolia	Increase in frequency and intensity of droughts in recent years; droughts in 1999 to 2002 affected 70% of grassland and killed 12 million livestock	Batima, 2003; Natsagdorj et al., 2005
China	Increase in area affected by drought has exceeded 6.7 Mha since 2000 in Beijing, Hebei Province, Shanxi Province, Inner Mongolia and North China; increase in dust storm affected area	Chen et al., 2001; Yoshino, 2000, 2002; Zhou, 2003
South Asia	50% of droughts associated with El Niño; consecutive droughts in 1999 and 2000 in Pakistan and N-W India led to sharp decline in watertables; consecutive droughts between 2000 and 2002 caused crop failures, mass starvation and affected ~11 million people in Orissa; droughts in N-E India during summer monsoon of 2006	Webster et al., 1998; Lal, 2003; India Meteorological Department, 2006
South-East Asia	Droughts normally associated with ENSO years in Myanmar, Laos, Philippines, Indonesia and Vietnam; droughts in 1997 to 98 caused massive crop failures and water shortages and forest fires in various parts of Philippines, Laos and Indonesia	Duong, 2000; Kelly and Adger, 2000; Glantz, 2001; PAGASA, 2001
Cyclones/Typhoons		
Philippines	On an average, 20 cyclones cross the Philippines Area of Responsibility with about 8 to 9 landfall each year; with an increase of 4.2 in the frequency of cyclones entering PAR during the period 1990 to 2003	PAGASA, 2001
China	Number and intensity of strong cyclones increased since 1950s; 21 extreme storm surges in 1950 to 2004 of which 14 occurred during 1986 to 2004	Fan and Li, 2005
South Asia	Frequency of monsoon depressions and cyclones formation in Bay of Bengal and Arabian Sea on the decline since 1970 but intensity is increasing causing severe floods in terms of damages to life and property	Lal, 2001, 2003
Japan	Number of tropical storms has two peaks, one in mid 1960s and another in early 1990s, average after 1990 and often lower than historical average	Japan Meteorological Agency, 2005

Table 10.4. *Recent trends in permafrost temperatures measured at different locations (modified from Romanovsky et al., 2002 and Izrael et al., 2006)*

Country	Region	Permafrost temperature change/trends	References
Russia	East Siberia (1.6 to 3.2 m), 1960 to 1992	+0.03°C/year	Romanovsky et al., 2001
	West Siberia (10 m), 1960 to 2005	+0.6°C/year	Izrael et al., 2006
China	Qinghai-Tibet Plateau (1975 to 1989)	+0.2 to +0.3°C	Cheng and Wu, 2007
Kazakhstan	Northern Tian Shan (1973 to 2003)	+0.2° to +0.6°C	Marchenko, 2002
Mongolia	Khentei and Khangai Mountains, Lake Hovsgol (1973 to 2003)	+0.3° to +0.6°C	Sharkhuu, 2003

10.2.4.2 Hydrology and water resources

Rapid thawing of permafrost (Table 10.4) and decrease in depths of frozen soils (4 to 5 m in Tibet according to Wang et al., 2004b) due largely to rising temperature has threatened many cities and human settlements, has caused more frequent landslides and degeneration of some forest ecosystems, and has resulted in increased lake-water levels in the permafrost region of Asia (Osterkamp et al., 2000; Guo et al., 2001; Izrael and Anokhin, 2001; Jorgenson et al., 2001; Izrael et al., 2002b; Fedorov and Konstantinov, 2003; Gavriliev and Efremov, 2003; Melnikov and Revson, 2003; Nelson, 2003; ACIA, 2005).

In drier parts of Asia, melting glaciers account for over 10% of freshwater supplies (Meshcherskaya and Blazhevich, 1990; Fitzharris, 1996; Meier, 1998). Glaciers in Asia are melting faster in recent years than before, as reported in Central Asia, Western Mongolia and North-West China, particularly the Zerafshan glacier, the Abramov glacier and the glaciers on the Tibetan Plateau (see Section 10.6.2) (Pu et al., 2004). As a result of rapid melting of glaciers, glacial runoff and frequency of glacial lake outbursts causing mudflows and avalanches have increased (Bhadra, 2002; WWF, 2005). A recent study in northern Pakistan, however, suggests that glaciers in the Indus Valley region may be expanding, due to increases in winter precipitation over western Himalayas during the past 40 years (Archer and Fowler, 2004).

In parts of China, the rise in temperature and decreases in precipitation (Ma and Fu, 2003; Wang and Zhai, 2003), along with increasing water use have caused water shortages that led to drying up of lakes and rivers (Liu et al., 2006; Wang and Jin, 2006). In India, Pakistan, Nepal and Bangladesh, water shortages have been attributed to rapid urbanisation and industrialisation, population growth and inefficient water use, which are aggravated by changing climate and its adverse impacts on demand, supply and water quality. In arid Central and West Asia, changes in climate and its variability continue to challenge the ability of countries in the arid and semi-arid region to meet the growing demands for water (Abu-Taleb, 2000; UNEP, 2002; Bou-Zeid and El-Fadel, 2002; Ragab and Prudhomme, 2002). Decreasing precipitation and increasing temperature commonly associated with ENSO have been reported to increase water shortage, particularly in parts of Asia where water resources are already under stress from growing water demands and inefficiencies in water use (Manton et al., 2001).

10.2.4.3 Oceans and coastal zones

Global warming and sea-level rise in the coastal zone of Boreal Asia have influenced sea-ice formation and decay, thermo-abrasion process, permafrost and the time of river freeze-up and break-up in recent decades (ACIA, 2005; Leont'yev, 2004). The coastlines in monsoon Asia are cyclone-prone with ~42% of the world's total tropical cyclones occurring in this region (Ali, 1999). The combined extreme climatic and non climatic events caused coastal flooding, resulting in substantial economic losses and fatalities (Yang, 2000; Li et al., 2004a). Wetlands in the major river deltas have been significantly altered in recent years due to large scale sedimentation, land-use conversion, logging and human settlement (Lu, 2003). Coastal erosion in Asia has led to loss of lands at rates dependent on varying regional tectonic activities, sediment supply and sea-level rise (Sin, 2000). Salt water from the Bay of Bengal is reported to have penetrated 100 km or more inland along tributary channels during the dry season (Allison et al., 2003). Severe droughts and unregulated groundwater withdrawal have also resulted in sea-water intrusion in the coastal plains of China (Ding et al., 2004).

Over 34% of the vast and diverse coral reefs of Asia that are of immense ecological and economic importance to this region (Spalding et al., 2001; Burke et al., 2002; Zafar, 2005) particularly in South, South-East and East Asia are reported to have been lost in 1998, largely due to coral bleaching induced by the 1997/98 El Niño event (Wilkinson, 2000; Arceo et al., 2001; Wilkinson, 2002; Ministry of the Environment and Japanese Coral Reef Society, 2004; Yamano and Tamura, 2004). The destructive effects of climate change compound the human-induced damages on the corals in this region. A substantial portion of the vast mangroves in South and South-East Asian regions has also been reportedly lost during the last 50 years of the 20th century, largely attributed to human activities (Zafar, 2005). Evidence of the impacts of climate-related factors on mangroves remain limited to the severe destruction of mangroves due to reduction of freshwater flows and salt-water intrusion in the Indus delta and Bangladesh (IUCN, 2003a).

10.2.4.4 Natural ecosystems

Increasing intensity and spread of forest fires in Asia were observed in the past 20 years, largely attributed to the rise in temperature and decline in precipitation in combination with increasing intensity of land uses (Page et al., 2002; De Grandi et al., 2003; Goldammer et al., 2003; FFARF, 2004; Isaev et al., 2004; Murdiyarso et al., 2004; Shoigu, 2004; Vorobyov, 2004; Achard et al., 2005; Murdiyarso and Adiningsih, 2006). During the last decade, 12,000 to 38,000 wild fires annually hit the boreal forests in North Asia affecting some 0.3 to 3 million hectares (Dumnov et al., 2005; Malevski-Malevich et al., 2005; FNCRF, 2006). Recent studies have also shown a dramatic increase of fires in Siberian peatlands (of which 20 million ha were burnt in 2003) linked to increased human activities combined with changing climate conditions, particularly the increase in temperature. Fires in peatlands of Indonesia during the 1997 to 98 El Niño dry season affected over 2 million ha and emitted an estimated 0.81 to 2.57 PgC to the atmosphere (Page et al., 2002). In the past 10 years about 3 million ha of peatland in South-East Asia have been burnt, releasing between 3 to 5 PgC, and drainage of peat has affected an additional 6 million ha and released a further 1 to 2 PgC. As a consequence of a 17% decline in spring precipitation and a rise in surface temperature by 1.5°C during the last 60 years, the frequency and aerial extent of the forest and steppe fires in Mongolia have significantly increased over a period of 50 years (Erdnethuya, 2003). The 1997/98 ENSO event in Indonesia triggered forest and brush fires in 9.7 million hectares, with serious domestic and trans-boundary pollution consequences. Thousands of hectares of second growth and logged-over forests were also burned in the Philippines during the 1997/98 ENSO events (Glantz, 2001; PAGASA, 2001).

With the gradual reduction in rainfall during the growing season for grass, aridity in Central and West Asia has increased in recent years, reducing growth of grasslands and increasing bareness of the ground surface (Bou-Zeid and El-Fadel, 2002). Increasing bareness has led to increased reflection of solar radiation, such that more soil moisture is evaporated and the ground has become increasingly drier in a feedback process, thus adding to the acceleration of grassland degradation (Zhang et al., 2003).

Wetlands in Asia are being increasingly threatened by warmer climate in recent decades. The precipitation decline and droughts in most delta regions of Pakistan, Bangladesh, India and China have resulted in the drying up of wetlands and severe degradation of ecosystems. The recurrent droughts from 1999 to 2001, as well as the building of an upriver reservoir and improper use of groundwater, have led to drying up of the Momoge Wetland located in the Songnen Plain (Pan et al., 2003).

10.2.4.5 Biodiversity

Biodiversity in Asia is being lost as a result of development activities and land degradation (especially overgrazing and deforestation), pollution, over-fishing, hunting, infrastructure development, species invasion, land-use change, climate change and the overuse of freshwater (UNEP, 2002; Gopal, 2003). Though evidence of climate-related biodiversity loss in Asia remains limited, a large number of plant and animal species are reported to be moving to higher latitudes and altitudes as a consequence of observed climate change in many parts of Asia in recent years (Yoshio and Ishii, 2001; IUCN, 2003a). Changes in the flowering date of Japanese Cherry, a decrease in alpine flora in Hokkaido and other high mountains and the expansion of the distribution of southern broad-leaved evergreen trees have also been reported (Oda and Ishii, 2001; Ichikawa, 2004; Kudo et al., 2004; Wada et al., 2004).

10.2.4.6 Human health

A large number of deaths due to heatwaves – mainly among the poor, elderly and labourers such as rural daily wage earners, agricultural workers and rickshaw pullers – have been reported in the Indian state of Andhra Pradesh, Orissa and elsewhere during the past five years (Lal, 2002). Serious health risks associated with extreme summer temperatures and heatwaves have also been reported in Siberian cities (Zolotov and Caliberny, 2004).

In South Asia, endemic morbidity and mortality due to diarrhoeal disease is linked to poverty and hygiene behaviour compounded by the effect of high temperatures on bacterial proliferation (Checkley et al., 2000). Diarrhoeal diseases and outbreaks of other infectious diseases (e.g., cholera, hepatitis, malaria, dengue fever) have been reported to be influenced by climate-related factors such as severe floods, ENSO-related droughts, sea-surface temperatures and rainfall in association with non-climatic factors such as poverty, lack of access to safe drinking water and poor sewerage system (Durkin et al., 1993; Akhtar and McMichael, 1996; Bouma and van der Kaay, 1996; Colwell, 1996; Bangs and Subianto, 1999; Lobitz et al., 2000; Pascual et al., 2000; Bouma and Pascual, 2001; Glantz, 2001; Pascual et al., 2002; Rodo et al., 2002).

10.3 Assumptions about future trends

10.3.1 Climate

Table 10.5 provides a snapshot of the projections on likely increase in area-averaged seasonal surface air temperature and percent change in area-averaged seasonal precipitation (with respect to the baseline period 1961 to 1990) for the seven sub-regions of Asia. The temperature projections for the 21st century, based on Fourth Assessment Report (AR4) Atmosphere-Ocean General Circulation Models (AOGCMs), and discussed in detail in Working Group I Chapter 11, suggest a significant acceleration of warming over that observed in the 20th century (Ruosteenoja et al., 2003; Christensen et al., 2007). Warming is least rapid, similar to the global mean warming, in South-East Asia, stronger over South Asia and East Asia and greatest in the continental interior of Asia (Central, West and North Asia). In general, projected warming over all sub-regions of Asia is higher during northern hemispheric winter than during summer for all time periods. The most pronounced warming is projected at high latitudes in North Asia. Recent modelling experiments suggest that the warming would be significant in Himalayan Highlands including the Tibetan Plateau and arid regions of Asia (Gao et al., 2003).

The consensus of AR4 models, as discussed in Chapter 2 and in Christensen et al. (2007) and confirmed in several studies using regional models (Lal, 2003; Rupa Kumar et al., 2003; Kwon et al., 2004; Boo et al., 2004; Japan Meteorological Agency, 2005; Kurihara et al., 2005), indicates an increase in annual precipitation in most of Asia during this century; the relative increase being largest and most consistent between models in North and East Asia. The sub-continental mean winter precipitation will very likely increase in northern Asia and the Tibetan Plateau and likely increase in West, Central, South-East and East Asia. Summer precipitation will likely increase in North, South, South-East and East Asia but decrease in West and Central Asia. The projected decrease in mean precipitation in Central Asia will be accompanied by an increase in the frequency of very dry spring, summer and autumn seasons. In South Asia, most of the AR4 models project a decrease of precipitation in December, January and February (DJF) and support earlier findings reported in Lal et al. (2001b).

An increase in occurrence of extreme weather events including heatwave and intense precipitation events is also projected in South Asia, East Asia, and South-East Asia (Emori et al., 2000; Kato et al., 2000; Sato, 2000; Lal, 2003; Rupa Kumar et al., 2003; Hasumi and Emori, 2004; Ichikawa, 2004; May, 2004b; Walsh, 2004; Japan Meteorological Agency, 2005; Kurihara et al., 2005) along with an increase in the interannual variability of daily precipitation in the Asian summer monsoon (Lal et al., 2000; May, 2004a; Giorgi and Bi, 2005). Results of regional climate model experiments for East Asia (Sato, 2000; Emori et al., 2000; Kato et al., 2000; Ichikawa, 2004; Japan Meteorological Agency, 2005; Kurihara et al., 2005) indicate that heatwave conditions over Japan are likely to be enhanced in the future (Figure 10.2). Extreme daily precipitation, including that associated with typhoon, would be further enhanced over Japan due to the increase in atmospheric moisture availability (Hasumi and Emori, 2004). The increases in annual temperature and precipitation over Japan are also projected regionally using regional climate model (Figure 10.3; Japan Meteorological Agency, 2005; Kurihara et al., 2005).

An increase of 10 to 20% in tropical cyclone intensities for a rise in sea-surface temperature of 2 to 4°C relative to the current threshold temperature is likewise projected in East Asia, South-East Asia and South Asia (Knutson and Tuleya, 2004). Amplification in storm-surge heights could result from the occurrence of stronger winds, with increase in sea-surface temperatures and low pressures associated with tropical storms resulting in an enhanced risk of coastal disasters along the coastal regions of East, South and South-East Asian countries. The impacts of an increase in cyclone intensities in any location will be determined by any shift in the cyclone tracks (Kelly and Adger, 2000).

In coastal areas of Asia, the current rate of sea-level rise is reported to be between 1 to 3 mm/yr which is marginally greater than the global average (Dyurgerov and Meier, 2000; Nerem and Mitchum, 2001; Antonov et al., 2002; Arendt et al., 2002; Rignot et al., 2003; Woodworth et al., 2004). A rate of sea-level rise of 3.1 mm/yr has been reported over the past decade compared to 1.7 to 2.4 mm/yr over the 20th century as a whole (Arendt et al., 2002; Rignot et al., 2003), which suggests that the rate of sea-level rise has accelerated relative to the long-term average.

10.3.2 Socio-economics

In the SRES framework narrative storylines were developed which provide broadly qualitative and quantitative descriptions of regional changes on socio-economic development (e.g., population, economic activity), energy services and resource availability (e.g., energy intensities, energy demand, structure of energy use), land use and land cover, greenhouse gases (GHG) and sulphur emissions, and atmospheric composition (Nakićenović and Swart, 2000). In Asia, GHG emissions were quantified reflecting socio-economic development such as energy use, land-use changes, industrial production processes, and so on. The population growth projections for Asia range between 1.54 billion people in 2050 and 4.5 billion people in 2100 (Nakićenović and Swart, 2000). The economic growth is estimated to range between 4.2-fold and 3.6-fold of the current gross domestic product (GDP), respectively.

10.4 Key future impacts and vulnerabilities

Key future climate change impacts and vulnerabilities for Asia are summarised in Figure 10.4. A detailed discussion of these impacts and vulnerabilities are presented in the sections below.

10.4.1 Agriculture and food security

10.4.1.1 Production

Results of recent studies suggest that substantial decreases in cereal production potential in Asia could be likely by the end of this century as a consequence of climate change. However, regional differences in the response of wheat, maize and rice yields to projected climate change could likely be significant (Parry et al., 1999; Rosenzweig et al., 2001). Results of crop yield projection using HadCM2 indicate that crop yields could likely increase up to 20% in East and South-East Asia while it could decrease up to 30% in Central and South Asia even if the direct positive physiological effects of CO_2 are taken into account. As a consequence of the combined influence of fertilisation effect and the accompanying thermal stress and

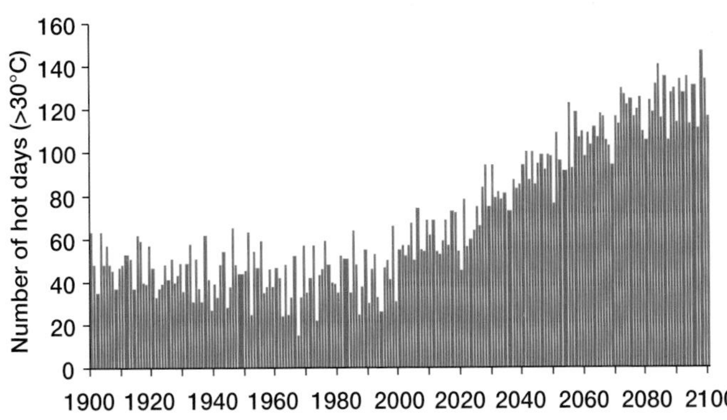

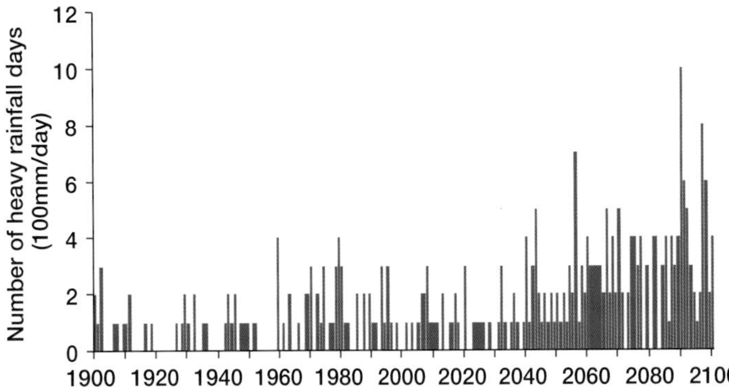

Figure 10.2. *Projected number of hot days (>30°C) and days of heavy rainfall (>100 mm/day) by the high resolution general circulation model (Hasumi and Emori, 2004).*

Table 10.5. *Projected changes in surface air temperature and precipitation for sub-regions of Asia under SRES A1FI (highest future emission trajectory) and B1 (lowest future emission trajectory) pathways for three time slices, namely 2020s, 2050s and 2080s.*

Sub-regions	Season	2010 to 2039				2040 to 2069				2070 to 2099			
		Temperature °C		Precipitation %		Temperature °C		Precipitation %		Temperature °C		Precipitation %	
		A1FI	B1	A1FI	B1	A1FI	B1	A1FI	B1	A1FI	B1	A1FI	B1
North	DJF	2.94	2.69	16	14	6.65	4.25	35	22	10.45	5.99	59	29
Asia	MAM	1.69	2.02	10	10	4.96	3.54	25	19	8.32	4.69	43	25
(50.0N-67.5N;	JJA	1.69	1.88	4	6	4.20	3.13	9	8	6.94	4.00	15	10
40.0E-170.0W)	SON	2.24	2.15	7	7	5.30	3.68	14	11	8.29	4.98	25	15
Central	DJF	1.82	1.52	5	1	3.93	2.60	8	4	6.22	3.44	10	6
Asia	MAM	1.53	1.52	3	-2	3.71	2.58	0	-2	6.24	3.42	-11	-10
(30N-50N;	JJA	1.86	1.89	1	-5	4.42	3.12	-7	-4	7.50	4.10	-13	-7
40E-75E)	SON	1.72	1.54	4	0	3.96	2.74	3	0	6.44	3.72	1	0
West	DJF	1.26	1.06	-3	-4	3.1	2.0	-3	-5	5.1	2.8	-11	-4
Asia	MAM	1.29	1.24	-2	-8	3.2	2.2	-8	-9	5.6	3.0	-25	-11
(12N-42N;	JJA	1.55	1.53	13	5	3.7	2.5	13	20	6.3	2.7	32	13
27E-63E)	SON	1.48	1.35	18	13	3.6	2.2	27	29	5.7	3.2	52	25
Tibetan	DJF	2.05	1.60	14	10	4.44	2.97	21	14	7.62	4.09	31	18
Plateau	MAM	2.00	1.71	7	6	4.42	2.92	15	10	7.35	3.95	19	14
(30N-50N;	JJA	1.74	1.72	4	4	3.74	2.92	6	8	7.20	3.94	9	7
75E-100E)	SON	1.58	1.49	6	6	3.93	2.74	7	5	6.77	3.73	12	7
East	DJF	1.82	1.50	6	5	4.18	2.81	13	10	6.95	3.88	21	15
Asia	MAM	1.61	1.50	2	2	3.81	2.67	9	7	6.41	3.69	15	10
(20N-50N;	JJA	1.35	1.31	2	3	3.18	2.43	8	5	5.48	3.00	14	8
100E-150E)	SON	1.31	1.24	0	1	3.16	2.24	4	2	5.51	3.04	11	4
South	DJF	1.17	1.11	-3	4	3.16	1.97	0	0	5.44	2.93	-16	-6
Asia	MAM	1.18	1.07	7	8	2.97	1.81	26	24	5.22	2.71	31	20
(5N-30N;	JJA	0.54	0.55	5	7	1.71	0.88	13	11	3.14	1.56	26	15
65E-100E)	SON	0.78	0.83	1	3	2.41	1.49	8	6	4.19	2.17	26	10
South-East	DJF	0.86	0.72	-1	1	2.25	1.32	2	4	3.92	2.02	6	4
Asia	MAM	0.92	0.80	0	0	2.32	1.34	3	3	3.83	2.04	12	5
(10S-20N;	JJA	0.83	0.74	-1	0	2.13	1.30	0	1	3.61	1.87	7	1
100E-150E)	SON	0.85	0.75	-2	0	1.32	1.32	-1	1	3.72	1.90	7	2

water scarcity (in some regions) under the projected climate change scenarios, rice production in Asia could decline by 3.8% by the end of the 21st century (Murdiyarso, 2000). In Bangladesh, production of rice and wheat might drop by 8% and 32%, respectively, by the year 2050 (Faisal and Parveen, 2004). For the warming projections under A1FI emission scenarios (see Table 10.5), decreases in crop yields by 2.5 to 10% in 2020s and 5 to 30% in 2050s have been projected in parts of Asia (Parry et al., 2004). Doubled CO_2 climates could decrease rice yields, even in irrigated lowlands, in many prefectures in central and southern Japan by 0 to 40% (Nakagawa et al., 2003) through the occurrence of heat-induced floret sterility (Matsui and Omasa, 2002). The projected warming accompanied by a 30% increase in tropospheric ozone and 20% decline in humidity is expected to decrease the grain and fodder productions by 26% and 9%, respectively, in North Asia (Izrael, 2002).

Crop simulation modelling studies based on future climate change scenarios indicate that substantial loses are likely in rain-fed wheat in South and South-East Asia (Fischer et al., 2002). For example, a 0.5°C rise in winter temperature would reduce wheat yield by 0.45 tonnes per hectare in India (Lal et al., 1998; Kalra et al., 2003). More recent studies suggest a 2 to 5% decrease in yield potential of wheat and maize for a temperature rise of 0.5 to 1.5°C in India (Aggarwal, 2003). Studies also suggest that a 2°C increase in mean air temperature could decrease rain-fed rice yield by 5 to 12% in China (Lin et al., 2004). In South Asia, the drop in yields of non-irrigated wheat and rice will be significant for a temperature increase of beyond 2.5°C incurring a loss in farm-level net revenue of between 9% and 25% (Lal, 2007). The net cereal production in South Asian countries is projected to decline at least between 4 to 10% by the end of this century under the most conservative climate

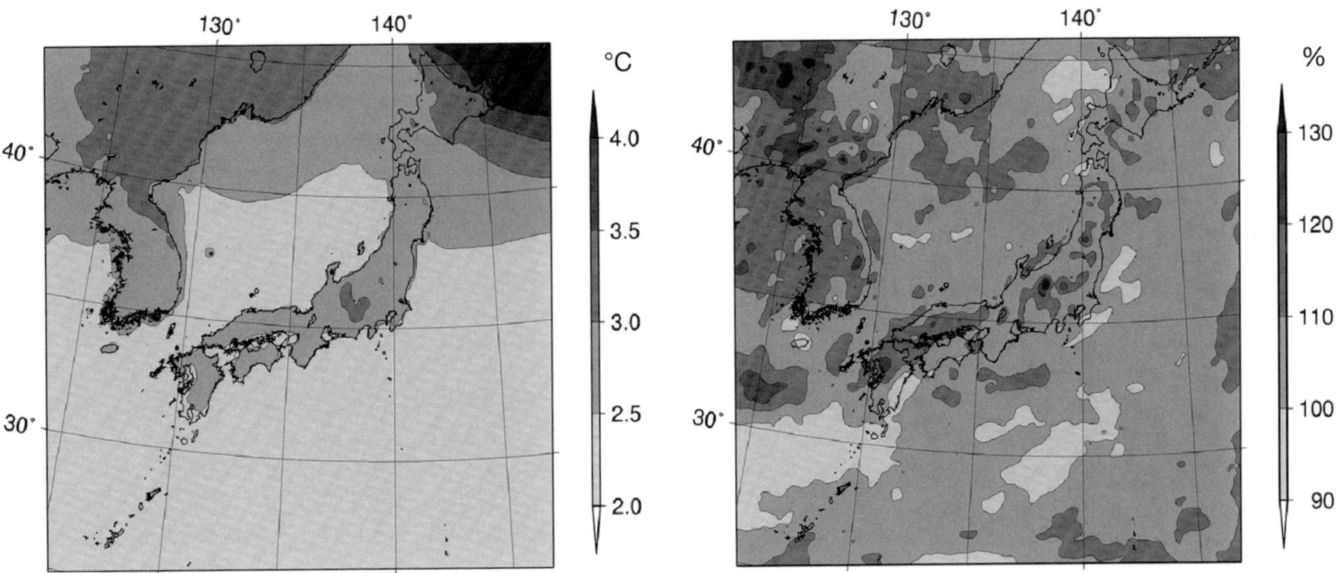

Figure 10.3. *Projected change in annual mean surface air temperature (°C) and rainfall (%) during 2081 to 2100 period compared to 1981 to 2000 period simulated by a high resolution regional climate model (left: annual temperature, right: annual precipitation, Japan Meteorological Agency, 2005; Kurihara et al., 2005).*

change scenario (Lal, 2007). The changes in cereal crop production potential indicate an increasing stress on resources induced by climate change in many developing countries of Asia.

10.4.1.2 Farming system and cropping areas

Climate change can affect not only crop production per unit area but also the area of production. Most of the arable land that is suitable for cultivation in Asia is already in use (IPCC, 2001).

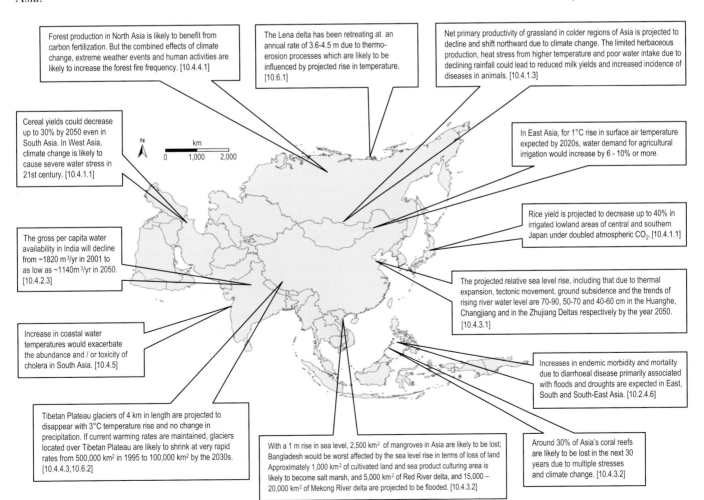

Forest production in North Asia is likely to benefit from carbon fertilization. But the combined effects of climate change, extreme weather events and human activities are likely to increase the forest fire frequency. [10.4.4.1]

The Lena delta has been retreating at an annual rate of 3.6-4.5 m due to thermo-erosion processes which are likely to be influenced by projected rise in temperature. [10.6.1]

Net primary productivity of grassland in colder regions of Asia is projected to decline and shift northward due to climate change. The limited herbaceous production, heat stress from higher temperature and poor water intake due to declining rainfall could lead to reduced milk yields and increased incidence of diseases in animals. [10.4.1.3]

Cereal yields could decrease up to 30% by 2050 even in South Asia. In West Asia, climate change is likely to cause severe water stress in 21st century. [10.4.1.1]

N

km

0 1,000 2,000

In East Asia, for 1°C rise in surface air temperature expected by 2020s, water demand for agricultural irrigation would increase by 6 - 10% or more.

Rice yield is projected to decrease up to 40% in irrigated lowland areas of central and southern Japan under doubled atmospheric CO_2. [10.4.1.1]

The gross per capita water availability in India will decline from ~1820 m³/yr in 2001 to as low as ~1140m³/yr in 2050. [10.4.2.3]

The projected relative sea level rise, including that due to thermal expansion, tectonic movement, ground subsidence and the trends of rising river water level are 70-90, 50-70 and 40-60 cm in the Huanghe, Changjiang and in the Zhujiang Deltas respectively by the year 2050. [10.4.3.1]

Increase in coastal water temperatures would exacerbate the abundance and / or toxicity of cholera in South Asia. [10.4.5]

Increases in endemic morbidity and mortality due to diarrhoeal disease primarily associated with floods and droughts are expected in East, South and South-East Asia. [10.2.4.6]

Tibetan Plateau glaciers of 4 km in length are projected to disappear with 3°C temperature rise and no change in precipitation. If current warming rates are maintained, glaciers located over Tibetan Plateau are likely to shrink at very rapid rates from 500,000 km² in 1995 to 100,000 km² by the 2030s. [10.4.4.3,10.6.2]

With a 1 m rise in sea level, 2,500 km² of mangroves in Asia are likely to be lost; Bangladesh would be worst affected by the sea level rise in terms of loss of land Approximately 1,000 km² of cultivated land and sea product culturing area is likely to become salt marsh, and 5,000 km² of Red River delta, and 15,000 – 20,000 km² of Mekong River delta are projected to be flooded. [10.4.3.2]

Around 30% of Asia's coral reefs are likely to be lost in the next 30 years due to multiple stresses and climate change. [10.4.3.2]

Figure 10.4. *Hotspots of key future climate impacts and vulnerabilities in Asia.*

A northward shift of agricultural zones is likely, such that the dry steppe zone in eastern part of Mongolia would push the forest-steppe to the north resulting in shrinking of the high mountainous and forest-steppe zones and expansion of the steppe and desert steppe (Tserendash et al., 2005). Studies suggest that by the middle of this century in northern China, tri-planting boundary will likely shift by 500 km from Changjiang valley to Huanghe basin, and double planting regions will move towards the existing single planting areas, while single planting areas will shrink by 23% (Wang, 2002). Suitable land and production potentials for cereals could marginally increase in the Russian Federation and in East Asia (Fischer et al., 2002).

More than 28 Mha in South and East Asia require a substantial increase in irrigation for sustained productivity (FAO, 2003). Agricultural irrigation demand in arid and semi-arid regions of Asia is estimated to increase by at least 10% for an increase in temperature of 1°C (Fischer et al., 2002; Liu, 2002). The rain-fed crops in the plains of North and North-East China could face water-related challenges in coming decades, due to increases in water demands and soil-moisture deficit associated with projected decline in precipitation (Tao et al., 2003b).

As land for agriculture becomes limited, the need for more food in South Asia could likely be met by increasing yields per unit of land, water, energy and time, such as through precision farming. Enhanced variability in hydrological characteristics will likely continue to affect grain supplies and food security in many nations of Asia. Intensification of agriculture will be the most likely means to meet the food requirements of Asia, which is likely to be invariably affected by projected climate change.

10.4.1.3 Livestock, fishery, aquaculture

Consumption of animal products such as meat and poultry has increased steadily in comparison to milk and milk products-linked protein diets in the past few decades (FAO, 2003). However, in most regions of Asia (India, China, and Mongolia) pasture availability limits the expansion of livestock numbers. Cool temperate grassland is projected to shift northward with climate change and the net primary productivity will decline (Sukumar et al., 2003; Christensen et al., 2004; Tserendash et al., 2005). The limited herbaceous production, heat stress from higher temperature, and limited water intake due to a decrease in rainfall could cause reduced milk yields in animals and an increased incidence of some diseases.

The Asia-Pacific region is the world's largest producer of fish, from both aquaculture and capture fishery sectors. Recent studies suggest a reduction of primary production in the tropical oceans because of changes in oceanic circulation in a warmer atmosphere. The tuna catch of East Asia and South-East Asia is nearly one-fourth of the world's total. A modelling study showed significant large-scale changes of skipjack tuna habitat in the equatorial Pacific under projected warming scenario (Loukos et al., 2003). Marine fishery in China is facing threats from over fishing, pollution, red tide, and other climatic and environmental pressures. The migration route and migration pattern and, hence, regional catch of principal marine fishery species, such as ribbon fish, small and large yellow croakers, could be greatly affected by global climate change (Su and Tang, 2002; Zhang and Guo, 2004). Increased frequency of El Niño events could likely lead to measurable declines in fish larvae abundance in coastal waters of South and South-East Asia. These phenomena are expected to contribute to a general decline in fishery production in the coastal waters of East, South and South-East Asia. Arctic marine fishery would also be greatly influenced by climate change. Moderate warming is likely to improve the conditions for some economically gainful fisheries, such as cod and herring. Higher temperatures and reduced ice cover could increase productivity of fish-prey and provide more extensive habitats. In contrast, the northern shrimp will likely decrease with rise in sea-surface temperatures (ACIA, 2005).

The impact of climate change on Asian fishery depends on the complicated food chains in the surrounding oceans, which are likely to be disturbed by the climate change. Fisheries at higher elevations are likely to be adversely affected by lower availability of oxygen, due to a rise in surface air temperatures. In the plains, the timing and amount of precipitation could also affect the migration of fish species from the river to the floodplains for spawning, dispersal and growth (FAO, 2003). Future changes in ocean currents, sea level, sea-water temperature, salinity, wind speed and direction, strength of upwelling, the mixing layer thickness and predator response to climate change have the potential to substantially alter fish breeding habitats and food supply for fish and ultimately the abundance of fish populations in Asian waters (IPCC, 2001).

10.4.1.4 Future food supply and demand

Half the world's population is located in Asia. There are serious concerns about the prevalence of malnutrition among poorer and marginal groups, particularly rural children, and about the large number of people below the poverty line in many countries. Large uncertainties in our understanding as to how the regional climate change will impact the food supply and demand in Asia continue to prevail in spite of recent scientific advances. Because of increasing interdependency of global food system, the impact of climate change on future food supply and demand in Asia as a whole as well as in countries located in the region depends on what happens in other countries. For example, India's surplus grain in past few years has been used to provide food aid to drought-affected Cambodia (Fischer et al., 2002). However, increasing urbanisation and population in Asia will likely result in increased food demand and reduced supply due to limited availability of cropland area and yield declines projected in most cases (Murdiyarso, 2000; Wang, 2002; Lin et al., 2004).

Food supply or ability to purchase food directly depends on income and price of the products. The global cereal prices have been projected to increase more than three-fold by the 2080s as a consequence of decline in net productivity due to projected climate change (Parry et al., 2004). Localised increases in food prices could be frequently observed. Subsistence producers growing crops, such as sorghum, millet, etc., could be at the greatest risk, both from a potential drop in productivity as well as from the danger of losing crop genetic diversity that has been preserved over generations. The risk of hunger, thus, is likely to remain very high in several developing countries with an additional 49 million, 132 million and 266 million people of Asia projected under A2 scenario without carbon fertilisation that could be at risk of hunger by 2020, 2050 and 2080, respectively (Parry et al., 2004). In terms of percent increase in risk hunger, it is projected under A2 scenario without CO_2 fertilisation that an increase of 7 to 14% by 2020s, 14 to 40% by 2050s and 14 to 137% by 2080s are likely (Parry et al., 2004).

Some recent studies (PAGASA, 2001; Sukumar et al., 2003; Batima et al., 2005b) confirm TAR findings that grasslands, livestock and water resources in marginal areas of Central Asia and South-East Asia are likely to be vulnerable to climate change. Food insecurity and loss of livelihood are likely to be further exacerbated by the loss of cultivated land and nursery areas for fisheries by inundation and coastal erosion in low-lying areas of the tropical Asia. Management options, such as better stock management and more integrated agro-ecosystems could likely improve land conditions and reduce pressures arising from climate change.

10.4.1.5 Pests and diseases

Some studies (Rosenzweig et al., 2001; FAO, 2004c) agree that higher temperatures and longer growing seasons could result in increased pest populations in temperate regions of Asia. CO_2 enrichment and changes in temperature may also affect ecology, the evolution of weed species over time and the competitiveness of C_3 v. C_4 weed species (Ziska, 2003). Warmer winter temperatures would reduce winter kill, favouring the increase of insect populations. Overall temperature increases may influence crop pathogen interactions by speeding up pathogen growth rates which increases reproductive generations per crop cycle, by decreasing pathogen mortality due to warmer winter temperatures, and by making the crop more vulnerable.

Climate change, as well as changing pest and disease patterns, will likely affect how food production systems perform in the future. This will have a direct influence on food security and poverty levels, particularly in countries with a high dependency on agriculture. In many cases, the impact will likely be felt directly by the rural poor, as they are often closely linked to direct food systems outcomes for their survival and are less able to substitute losses through food purchases. The urban poor are also likely to be affected negatively by an increase in food prices that may result from declining food production.

10.4.2 Hydrology and water resources

10.4.2.1 Water availability and demand

The impacts of climate change on water resources in Asia will be positive in some areas and negative in others. Changes in seasonality and amount of water flows from river systems are likely to occur due to climate change. In some parts of Russia, climate change could significantly alter the variability of river runoff such that extremely low runoff events may occur much more frequently in the crop growing regions of the south west (Peterson et al., 2002). Changes in runoff of river basins could have a significant effect on the power output of hydropower generating countries like Tajikistan, which is the third-highest producer in the world (World Bank, 2002). Likewise, surface water availability from major rivers like the Euphrates and Tigris may also be affected by alteration of riverflows. In Lebanon the annual net usable water resources will likely decrease by 15% in response to a general circulation model (GCM) estimated average rise in temperature of 1.2°C under doubled CO_2 climate, while the flows in rivers are likely to increase in winter and decrease in spring (Bou-Zeid and El-Fadel, 2002) which could negatively affect existing uses of river waters. In North China, irrigation from surface and groundwater sources will meet only 70% of the water requirement for agricultural production, due to the effects of climate change and increasing demand (Liu et al., 2001; Qin, 2002). The maximum monthly flow of the Mekong is estimated to increase by 35 to 41% in the basin and by 16 to 19% in the delta, with lower value estimated for years 2010 to 38 and higher value for years 2070 to 99, compared with 1961 to 90 levels. In contrast, the minimum monthly flows are estimated to decline by 17 to 24% in the basin and 26 to 29% in the delta (see Chapter 5, Box 5.3; Hoanh et al., 2004) suggesting that there could be increased flooding risks during wet season and an increased possibility of water shortage in dry season. Flooding could increase the habitat of brackish water fisheries but could also seriously affect the aquaculture industry and infrastructure, particularly in heavily-populated megadeltas. Decrease in dry season flows may reduce recruitment of some species.

In parts of Central Asia, regional increases in temperature will lead to an increased probability of events such as mudflows and avalanches that could adversely affect human settlements (Iafiazova, 1997). Climate change-related melting of glaciers could seriously affect half a billion people in the Himalaya-Hindu-Kush region and a quarter of a billion people in China who depend on glacial melt for their water supplies (Stern, 2007). As glaciers melt, river runoff will initially increase in winter or spring but eventually will decrease as a result of loss of ice resources. Consequences for downstream agriculture, which relies on this water for irrigation, will be likely unfavourable in most countries of South Asia. The thawing volume and speed of snow cover in spring is projected to accelerate in North-West China and Western Mongolia and the thawing time could advance, which will increase some water sources and may lead to floods in spring, but significant shortages in wintertime water availability for livestock are projected by the end of this century (Batima et al., 2004, 2005b).

10.4.2.2 Water quality

Over-exploitation of groundwater in many countries of Asia has resulted in a drop in its level, leading to ingress of sea water in coastal areas making the sub-surface water saline. India, China and Bangladesh are especially susceptible to increasing salinity of their groundwater as well as surface water resources, especially along the coast, due to increases in sea level as a direct impact of global warming (Han et al., 1999). Rising sea level by 0.4 to 1.0 m can induce salt-water intrusion 1 to 3 km further inland in the Zhujiang estuary (Huang and Xie, 2000). Increasing frequency and intensity of droughts in the catchment area will lead to more serious and frequent salt-water intrusion in the estuary (Xu, 2003; Thanh et al., 2004; Huang et al., 2005) and thus deteriorate surface and groundwater quality.

10.4.2.3 Implications of droughts and floods

Global warming would cause an abrupt rise of water quantity as a result of snow or glacier melting that, in turn, would lead to floods. The floods quite often are caused by rise of river water level due to blockage of channels by drifting ice, as happened in Central Siberia, Lensk, or enormous precipitation from destructive shower cyclones, as it was in the North Asia Pacific coast, Vladivostok (Izrael et al., 2002a). A projected increase in surface air temperature in North-West China will result in a 27% decline in glacier area (equivalent to the ice volume of 16,184 km^3), a 10 to 15% decline in frozen soil area, an increase in flood and debris flow, and more severe water shortages (Qin, 2002). The duration of seasonal snow cover in alpine areas, namely the Tibetan Plateau, Xinjiang and Inner Mongolia of China, will shorten and snow cover will thaw out in advance of the spring season, leading to a decline in volume and resulting in severe spring droughts. Between 20 to 40% reduction of runoff per capita in Ningxia, Xinjiang and Qinghai Province is likely by the end of 21st century (Tao et al., 2005). However, the pressure due to increasing population and socio-economic development on water resources is likely to grow. Higashi et al. (2006) project that future flood risk in Tokyo, Japan between 2050 to 2300 under SRES A1B is likely to be 1.1 to 1.2 times higher than the present condition.

The gross per capita water availability in India will decline from about 1,820 m^3/yr in 2001 to as low as about 1,140 m^3/yr in 2050 (Gupta and Deshpande, 2004). India will reach a state of water stress before 2025 when the availability falls below 1000 m^3 per capita (CWC, 2001). The projected decrease in the winter precipitation over the Indian subcontinent would reduce the total seasonal precipitation during December, January and February implying lesser storage and greater water stress during the lean monsoon period. Intense rain occurring over fewer days, which implies increased frequency of floods during the monsoon, will also result in loss of the rainwater as direct runoff, resulting in reduced groundwater recharging potential.

Expansion of areas under severe water stress will be one of the most pressing environmental problems in South and South-East Asia in the foreseeable future as the number of people living under severe water stress is likely to increase substantially in absolute terms. It is estimated that under the full range of SRES scenarios, 120 million to 1.2 billion, and 185 to 981 million people will experience increased water stress by the 2020s, and the 2050s, respectively (Arnell, 2004). The decline in annual flow of the Red River by 13 to 19% and that of Mekong River by 16 to 24% by the end of 21st century will contribute in increasing water stress (ADB, 1994).

10.4.3 Coastal and low lying areas

10.4.3.1 Coastal erosion and inundation of coastal lowland

Average global sea-level rise over the second half of the 20th century was 1.8 ± 0.3 mm/yr, and sea-level rise of the order of 2 to 3 mm/yr is considered likely during the early 21st century as a consequence of global warming (Woodroffe et al., 2006). However, the sea-level rise in Asia is geographically variable and an additional half a metre of sea-level rise is projected for the Arctic during this century (ACIA, 2005). The rising rates of sea level vary considerably from 1.5 to 4.4 mm/yr along the East Asia coast, due to regional variation in land surface movement (Mimura and Yokoki, 2004). The projected rise of mean high-water level could be greater than that of mean sea level (Chen, 1991; Zhang and Du, 2000). The projected relative sea-level rise (RSLR), including that due to thermal expansion, tectonic movement, ground subsidence and the trend of rising river water level, is 40 to 60 cm, 50 to 70 cm and 70 to 90 cm in the Zhujiang, Changjiang and Huanghe Deltas, respectively by the year 2050 (Li et al., 2004a, b). Choi et al. (2002) has reported that the regional sea-level rise over the north-western Pacific Ocean would be much more significant compared with the global average mainly due to exceptionally large warming near the entrance of the Kuroshio extension. The slope of the land and land surface movement would also affect the relative sea-level rise in the Asian Arctic (ACIA, 2005).

In Asia, erosion is the main process that will occur to land as sea level continues to rise. As a consequence, coast-protection structures built by humans will usually be destroyed by the sea while the shoreline retreats. In some coastal areas of Asia, a 30 cm rise in sea level can result in 45 m of landward erosion. Climate change and sea-level rise will tend to worsen the currently eroding coasts (Huang and Xie, 2000). In Boreal Asia, coastal erosion will be enhanced as rising sea level and declining sea ice allow higher wave and storm surge to hit the shore (ACIA, 2005). The coastal recession can add up to 500 to 600 m in 100 years, with a rate of between 4 to 6 m/yr. The coastal recession by thermal abrasion is expected to accelerate by 1.4 to 1.5 times in the second half of the 21st century as compared to the current rate (Leont'yev, 2004). In monsoonal Asia, decreasing sediment flux is generally a main cause of coastal erosion. Available evidence suggests a tendency of river sediment to further decline that will tend to worsen coastal erosion in Asia (Liu et al., 2001).

Projected sea-level rise could flood the residence of millions of people living in the low lying areas of South, South-East and East Asia such as in Vietnam, Bangladesh, India and China (Wassmann et al., 2004; Stern, 2007). Even under the most conservative scenario, sea level will be about 40 cm higher than today by the end of 21st century and this is projected to increase the annual number of people flooded in coastal populations from 13 million to 94 million. Almost 60% of this increase will occur in South Asia (along coasts from Pakistan, through India, Sri Lanka and Bangladesh to Burma), while about 20% will occur in South-East Asia, specifically from Thailand to Vietnam including Indonesia and the Philippines (Wassmann et al., 2004). The potential impacts of one metre sea-level rise include inundation of 5,763 km^2 and 2,339 km^2 in India and in some big cities of Japan, respectively (TERI, 1996; Mimura and Yokoki, 2004). For one metre sea-level rise with high tide and storm surge, the maximum inundation area is estimated to be 2,643 km^2 or about 1.2% of total area of the Korean Peninsula (Matsen and Jakobsen, 2004). In China, a 30 cm sea-level rise would inundate 81,348 km^2 of coastal lowland (Du and Zhang, 2000).

The coastal lowlands below the elevation of 1,000-year storm surge are widely distributed in Bangladesh, China, Japan, Vietnam and Thailand, where millions of people live (Nicholls, 2004). In Japan, an area of 861 km^2 of coastal lowland is located below high water level mainly in large cities like Tokyo, Osaka and Nagoya. A one metre rise in sea level could put up to 4.1

million people at risk (Mimura and Yokoki, 2004). Using a coarse digital terrain model and global population distribution data, it is estimated that more than 1 million people will be directly affected by sea-level rise in 2050 in each of the Ganges-Brahmaputra-Meghna delta in Bangladesh, the Mekong delta in Vietnam and the Nile delta in Egypt (see Chapter 6, Box 6.3; Ericson et al., 2005). Damages in flooded areas are largely dependent on the coastal protection level. It can be much less in highly protected coasts like in Japan but can be very high such as in coastal areas of South Asia where the protection level is low. A 30 cm rise in sea level will increase coastal flooding areas by five or six times in both the 'with' and 'without protection' scenarios in the Changjiang and Zhujiang deltas. Similarly, the flooding areas in the Huanghe delta for a 100 cm rise in sea level are almost the same under the 'without protection' and 'existing protection' scenarios. These two cases indicate that the current protection level is insufficient to protect the coasts from high sea-level rise (Du and Zhang, 2000; Li et al., 2004a). Further climate warming may lead to an increase in tropical cyclone destructive potential, and with an increasing coastal population substantial increase in hurricane-related losses in the 21st century is likely (Emanuel, 2005).

In summary, all coastal areas in Asia are facing an increasing range of stresses and shocks, the scale of which now poses a threat to the resilience of both human and environmental coastal systems, and are likely to be exacerbated by climate change. The projected future sea-level rise could inundate low lying areas, drown coastal marshes and wetlands, erode beaches, exacerbate flooding and increase the salinity of rivers, bays and aquifers. With higher sea level, coastal regions would also be subject to increased wind and flood damage due to storm surges associated with more intense tropical storms. In addition, warming would also have far reaching implications for marine ecosystems in Asia.

10.4.3.2 Deltas, estuaries, wetland and other coastal ecosystems

Future evolution of the major deltas in monsoonal Asia depends on changes in ocean processes and river sediment flux. Coastal erosion of the major deltas will be caused by sea-level rise, intensifying extreme events (e.g., storm surge) due to climate change and excessive pumping of groundwater for irrigation and reservoir construction upstream. In the Tibetan Plateau and adjoining region, sediment starvation is generally the main cause of shrinking of deltas. Annual mean sediment discharge in the Huanghe delta during the 1990s was only 34% of that observed during the 1950s and 1970s. The Changjiang sediment discharge will also be reduced by 50% on average after construction of the Three-Gorges Dam (Li et al., 2004b). Saltwater intrusion in estuaries due to decreasing river runoff can be pushed 10 to 20 km further inland by the rising sea level (Shen et al., 2003; Yin et al., 2003; Thanh et al., 2004).

Many megacities in Asia are located on deltas formed during sea-level change in the Holocene period (Hara et al., 2005). These Asian megacities with large populations and intensified socio-economic activities are subject to threats of climate change, sea-level rise and extreme climate event. For a 1 m rise in sea level, half a million square hectares of Red River delta and from 15,000 to 20,000 km^2 of Mekong River delta is projected to be flooded. In addition, 2,500 km^2 of mangrove will be completely lost, while approximately 1,000 km^2 of cultivated farm land and sea product culturing area will become salt marshes (Tran et al., 2005).

Rise in water temperatures and eutrophication in the Zhujiang and Changjiang estuaries have led to the formation of the bottom oxygen-deficient horizon and an increase in the frequency and intensity of red tides (Hu et al., 2001). Projected increases in the frequency and intensity of extreme weather events will exert adverse impacts on aquatic ecosystems, and existing habitats will be redistributed, affecting estuarine flora distribution (Short and Neckles, 1999; Simas et al., 2001; Lu, 2003; Paerl et al., 2003).

Recent risk analysis of coral reefs suggests that between 24% and 30% of the reefs in Asia are projected to be lost during the next 2 to 10 years and 10 to 30 years, respectively (14% and 18% for global), unless the stresses are removed and relatively large areas are protected (Table 10.6). In other words, the loss of reefs in Asia may be as high as 88% (59% for global) in the next 30 years under IS92a emission scenario (IPCC, 1992; Sheppard, 2003; Wilkinson, 2004). If conservation measures receive increasing attention, large areas of the reefs could recover from the direct and indirect damage within the next 10 years. However, if abnormally high sea-surface temperatures (SST) continue to cause major bleaching events (see Chapter 6, Section 6.2.5, Box 6.1), and reduce the capacity of reefs to calcify due to CO_2 increase, most human efforts will be futile (Kleypas et al., 1999; Wilkinson, 2002).

A new study suggests that coral reefs, which have been severely affected by abnormally high SST in recent years, contain some coral species and their reef-associated micro-algal symbionts that show far greater tolerance to higher SST than others. Bleaching thresholds may be more realistically visualised as a broad spectrum of responses, rather than a single bleaching threshold for all coral species (Hughes et al., 2003; Baker et al., 2004). This corals' adaptive response to climate change may protect devastated reefs from extinction or significantly prolong the extinction of surviving corals beyond previous assumption.

Net growth rates of coral reef, which can reach up to 8 to 10 mm/year, may exceed the projected rates of future sea-level rise in the South China Sea, so that coral reefs could not be at risk due merely to sea-level rise. Water depth increased by sea-level rise would lead to storminess and destruction of coral reefs (Knowlton, 2001; Wang, 2005).

10.4.4 Natural ecosystems and biodiversity

10.4.4.1 Structure, production and function of forests

Up to 50% of the Asia's total biodiversity is at risk due to climate change. Boreal forests in North Asia would move further north. A projected large increase in taiga is likely to displace tundra, while the northward movement of the tundra will in turn decrease polar deserts (see Chapter 15, Section 15.2.2, Figure 15.3; Callaghan et al., 2005; Juday et al., 2005). Large populations of many other species could also be extirpated as a result of the synergistic effects of climate change and habitat fragmentation (Ishigami et al., 2003, 2005). Projections under doubled-CO_2 climate using two GCMs show that 105 to 1,522 plant species and 5 to 77 vertebrates in China and 133 to 2,835

Table 10.6. *The 2004 status of coral reefs in selected regions of Asia (Wilkinson, 2004).*

Region	Coral reef area (km²)	Destroyed reefs (%)	Reefs recovered since 1998 (%)	Reefs at critical stage (%)	Reefs at threatened stage (%)	Reefs at low or no threat level (%)
Red Sea	17,640	4	2	2	10	84
The Gulfs	3,800	65	2	15	15	5
South Asia	19,210	45	13	10	25	20
S-E Asia	91,700	38	8	28	29	5
E & N Asia	5,400	14	3	23	12	51
Total	137,750	34.4	7.6	21.6	25.0	19.0
Asia	(48.4%)					

Note: Destroyed reefs: 90% of the corals lost and unlikely to recover soon; Reefs at a critical stage: 50% to 90% of corals lost or likely to be destroyed in 10 to 20 years; Reefs at threatened stage: 20 to 50% of corals lost or likely to be destroyed in 20 to 40 years.

plants and 10 to 213 vertebrates in Indo-Burma could become extinct (Malcolm et al., 2006).

As a consequence of climate change, no significant change in spatial patterns of productivity of the forest ecosystems in North-East China is projected (Liu et al., 1998). The areal coverage of broad-leaved Korean pine forests is projected to decrease by 20 to 35% with a significant northward shift (Wu, 2003). About 90% of the suitable habitat for a dominant forest species, beech tree (*Fagus crenata*), in Japan could disappear by the end of this century (Matsui et al., 2004a, b). The impact of elevated atmospheric CO_2 on plant biomass production is influenced by the availability of soil nitrogen and deposition of atmospheric nitrogen (Oren et al., 2001; Hajima et al., 2005; Kitao et al., 2005; Reich et al., 2006). The overall impact of climate change on the forest ecosystems of Pakistan could be negative (Siddiqui et al., 1999).

The observations in the past 20 years show that the increasing intensity and spread of forest fires in North and South-East Asia were largely related to rises in temperature and declines in precipitation in combination with increasing intensity of land uses (see Section 10.2.4.4). Whether this trend will persist in the future or not is difficult to ascertain in view of the limited literature on how the frequency and severity of forest and brush fires will likely respond to expected increase in temperature and precipitation in North and South-East Asia (see Section 10.3.1). The uncertainty lies on whether the expected increase in temperature would be enough to trigger more frequent and severe fires despite the projected increase in precipitation. One study on the impacts of climate change on fires show that for an average temperature increase of 1°C, the duration of wild fire season in North Asia could increase by 30% (Vorobyov, 2004), which could have varying adverse and beneficial impacts on biodiversity, forest structure and composition, outbreaks of pest and diseases, wildlife habitat quality and other key forest ecosystem functions.

10.4.4.2 Grasslands, rangelands and endangered species

The natural grassland coverage and the grass yield in Asia, in general, are projected to decline with a rise in temperature and higher evaporation (Lu and Lu, 2003). Large decreases in the natural capital of grasslands and savannas are likely in South Asia as a consequence of climate change. A rise in surface air temperature and decline in precipitation is estimated to reduce pasture productivity in the Mongolian steppe by about 10 to 30%, except in high mountains and in Gobi where a marginal decrease in pasture productivity is projected by the end of this century (Tserendash et al., 2005). Traditional land-use systems should provide conditions that would promote greater rangeland resilience and provide a better management strategy to cope with climate change in the region to offset the potential decrease of carbon storage and grassland productivity in the Mongolian Steppe under various climate scenarios (Ojima et al., 1998).

The location and areas of natural vegetation zone on the Tibetan Plateau will substantially change under the projected climate scenarios. The areas of temperate grassland and cold-temperate coniferous forest could expand, while temperate desert and ice-edge desert may shrink. The vertical distribution of vegetation zone could move to higher altitude. Climate change may result in a shift of the boundary of the farming-pastoral transition region to the south in North-East China, which can increase the grassland areas and provide favourable conditions for livestock production. However, as the transition area of farming-pastoral region is also the area of potential desertification, if protection measures are not taken in the new transition area, desertification may occur (Li and Zhou, 2001; Qiu et al., 2001). More frequent and prolonged droughts as a consequence of climate change and other anthropogenic factors together will result in the increasing trends of desertification in Asia.

10.4.4.3 Permafrost

The permafrost thawing will continue over vast territories of North Asia under the projected climate change scenarios (Izrael et al., 2002b). The transient climate model simulations (Pavlov and Ananjeva-Malkova, 2005; FNCRF, 2006) show that the perennially frozen rocks and soils (eastern part of the permafrost terrain) and soils (western part of the terrain) may be completely degraded within the present southern regions of North Asia (see Figure 10.5). In northern regions, mean annual temperature of frozen soil and rocks and the depth of seasonal thawing will increase in 2020 by as much as 4°C for the depth of 0.8 m and by at most 2.2°C for the depth of 1.6 m (FNCRF, 2006; Izrael et al., 2006). The change in the rock and soil temperatures will result in a change in the strength characteristics, bearing capacity, and compressibility of the frozen rocks and soils, thaw settlement strains, frozen ground exploitability in the course of excavation and mining, generation of thermokarst, thermal erosion and some other geocryological processes (Climate Change, 2004).

Permafrost degradation will lead to significant ground surface subsidence and pounding (Osterkamp et al., 2000; Jorgenson et al., 2001). Permafrost thawing on well-drained portions of slopes and highlands in Russia and Mongolia will improve the drainage conditions and lead to a decrease in the groundwater content (Hinzman et al., 2003; Batima et al., 2005b). On the Tibetan Plateau, in general, the permafrost zone is expected to decrease in size, move upward and face degradation by the end of this century (Wu et al., 2001). For a rise in surface temperature of 3°C and no change in precipitation, most Tibetan Plateau glaciers shorter than 4 km in length are projected to disappear and the glacier areas in the Changjiang Rivers will likely decrease by more than 60% (Shen et al., 2002).

Figure 10.5. *The projected shift of permafrost boundary in North Asia due to climate change by 2100 (FNCRF, 2006).*

10.4.5 Human health

Climate change poses substantial risks to human health in Asia. Global burden (mortality and morbidity) of climate-change attributable diarrhoea and malnutrition are already the largest in South-East Asian countries including Bangladesh, Bhutan, India, Maldives, Myanmar and Nepal in 2000, and the relative risks for these conditions for 2030 is expected to be also the largest (McMichael et al., 2004), although in some areas, such as southern states in India, there will be a reduction in the transmission season by 2080 (Mitra et al., 2004). An empirical model projected that the population at risk of dengue fever (the estimated risk of dengue transmission is greater than 50%) will be larger in India and China (Hales et al., 2002). Also in India and China, the excess mortality due to heat stress is projected to be very high (Takahashi et al., 2007), although this projection did not take into account possible adaptation and population change. There is already evidence of widespread damage to human health by urban air quality and enhanced climate variability in Asia. Throughout newly industrialised areas in Asia, such as Chongqing, China, and Jakarta, Indonesia, air quality has deteriorated significantly and will likely contribute to widespread heat stress and smog induced cardiovascular and respiratory illnesses in the region (Patz et al., 2000). Also, the number of patients of Japanese cedar pollen disease is likely to increase when the summer temperature rises (Takahashi and Kawashima, 1999; Teranishi et al., 2000).

The negative influence of temperature anomalies on public health has been established in Russia (Izmerov et al., 2004) and in the semi-arid city, Beirut (El-Zein et al., 2004). Exposure to higher temperatures appears to be a significant risk factor for cerebral infarction and cerebral ischemia during the summer months (Honda et al., 1995). Natural habitats of vector-borne diseases are reported to be expanding (Izmerov et al., 2004). Prevalence of malaria and tick-borne encephalitis has also increased over time in Russia (Yasukevich and Semenov, 2004). The distribution of vector-borne infectious diseases such as malaria is influenced by the spread of vectors and the climate dependence of the infectious pathogens. There are reports on the possible effects of pesticide resistance of a certain type of mosquito on the transition of malaria type (Singh et al., 2004). The insect-borne infectious diseases strongly modulated by future climate change include malaria, schistosomiasis, dengue fever and other viral diseases (Kovats et al., 2003). *Oncomelania* is strongly influenced by climate and the infection rate of schistosomiasis is the highest in the temperature range of 24°C to 27°C. Temperature can directly influence the breeding of malaria protozoa and suitable climate conditions can intensify the invasiveness of mosquito (Tong and Ying, 2000). A warmer and more humid climate would be favourable for propagation and invasiveness of infectious insect vector. Serious problems are connected with the impact of air pollution due to Siberian forest fires on human health (Rachmanin et al., 2004).

Warmer sea-surface temperatures along coastlines of South and South-East Asia would support higher phytoplankton blooms. These phytoplankton blooms are excellent habitats for survival and spread of infectious bacterial diseases such as cholera (Pascual et al., 2002). Water-borne diseases including cholera and the suite of diarrhoeal diseases caused by organisms such as *Giardia, Salmonella* and *Cryptosporidium* could also become common with the contamination of drinking water. Precipitation increase and frequent floods, and sea-level rise in the future will degrade the surface water quality owing to more pollution and, hence, lead to more water-borne infectious diseases such as dermatosis, cardiovascular disease and gastrointestinal disease. For preventive actions, assessment of climate change impacts on nutritional situation, drinking water supply, water salinity and ecosystem damage will be necessary. The risk factor of climate-related diseases will depend on improved environmental sanitation, the hygienic practice and medical treatment facilities.

10.4.6 Human dimensions

Study of social vulnerability provides a complementary approach to the study of climate impacts based on model projections and biophysical simulations. Adger et al. (2001)

illustrate the approach through theoretical discussion and case studies based in Vietnam. The following sections detail specific examples of the human dimension of general relevance within Asia.

10.4.6.1 Population growth

As of mid-2000, over 3.6 billion people, roughly three-fifths of the total population of the globe, resided in Asia. Seven of the world's 10 most populous countries - China, India, Indonesia, Russia, Pakistan, Bangladesh and Japan - are located within Asia (ADB, 2002). The majority of the region's population growth is forecast to come from South Asia, which expects to add 570 million people in India, 200 million in Pakistan and 130 million in Bangladesh over the next 50 years (UN-DESA-PD, 2002). Population growth, particularly in countries with already high population densities, is inextricably associated with the increasing pressure on the natural resources and the environment as the demands for goods and services expand. Some of the key impacts of increasing population include those linked with the intensification of use of natural forests including mangroves, agriculture, industrialisation and urbanisation. In Asia, the pressure on land in the 21st century will increase, due to the increasing food grain demand for the growing population, the booming economic development, as well as climate change. This will be exacerbated by the increasing scarcity of arable lands as a result of using vast agricultural lands to support industrialisation and urbanisation in pursuit of economic development (Zeqiang et al., 2001).

In the developing regions, the remaining natural flood plains are disappearing at an accelerating rate, primarily as a result of changes in land use and hydrological cycle, particularly changes in streamflows due to climatic and human-related factors. The future increase of human population will lead to further degradation of riparian areas, intensification of the land and water use, increase in the discharge of pollutants, and further proliferation of species invasions. The most threatened flood plains will be those in South and South-East Asia.

In some parts of South-East Asia, population growth, particularly in the uplands, continues to exert pressure on the remaining forests in the region. Encroachment into forest zones for cultivation, grazing, fuel wood and other purposes has been a major cause of changes in natural forests. In the Philippines, forest degradation has been attributed partly to upland farming (Pulhin et al., 2006).

10.4.6.2 Development activities

Development, to a large extent, is responsible for much of the greenhouse gases emitted into the atmosphere that drives climate change. On the other hand, development greatly contributes in reducing vulnerability to climate change and in enhancing the adaptive capacity of vulnerable sectors.

Demands for biological resources caused by population and increased consumption have grown with increasing economy and fast development all over Asia in recent years. Rates of both total forest loss and forest degradation are higher in Asia than anywhere else in the world. The conversion of forested area to agriculture in Asia during the past two decades occurred at a rate of 30,900 km^2/yr. In many developing countries of Asia, small scale fuel wood collection and industrial logging for exports of timber and conversion of forests into estate crop plantation (i.e., oil palm) and mining are also responsible for deforestation. It is likely that climate change would aggravate the adverse impacts of forest cover loss.

10.4.6.3 Climate extremes and migration

In Asia, migration accounts for 64% of urban growth (Pelling, 2003). Total population, international migration and refugees in Asia and the Pacific region are currently estimated to be 3,307 million, 23 million, and 4.8 million, respectively (UN-HABITAT, 2004). Future climate change is expected to have considerable impacts on natural resource systems, and it is well-established that changes in the natural environment can affect human sustenance and livelihoods. This, in turn, can lead to instability and conflict, often followed by displacement of people and changes in occupancy and migration patterns (Barnett, 2003).

Climate-related disruptions of human populations and consequent migrations can be expected over the coming decades. Such climate-induced movements can have effects in source areas, along migration routes and in the receiving areas, often well beyond national borders. Periods when precipitation shortfalls coincide with adverse economic conditions for farmers (such as low crop prices) would be those most likely to lead to sudden spikes in rural-to-urban migration levels in China and India. Climatic changes in Pakistan and Bangladesh would likely exacerbate present environmental conditions that give rise to land degradation, shortfalls in food production, rural poverty and urban unrest. Circular migration patterns, such as those punctuated by shocks of migrants following extreme weather events, could be expected. Such changes would likely affect not only internal migration patterns, but also migration movements to other western countries.

Food can be produced on currently cultivated land if sustainable management and adequate inputs are applied. Attaining this situation would also require substantial improvements of socio-economic conditions of farmers in most Asian countries to enable access to inputs and technology. Land degradation, if continued unchecked, may further exacerbate land scarcities in some countries of Asia. Concerns for the environment as well as socio-economic considerations may infringe upon the current agricultural resource base and prevent land and water resources from being developed for agriculture (Tao et al., 2003b). The production losses due to climate change may drastically increase the number of undernourished in several developing countries in Asia, severely hindering progress against poverty and food insecurity (Wang et al., 2006).

10.4.6.4 Urban development, infrastructure linkages, industry and energy

The compounding influence of future rises in temperature due to global warming, along with increases in temperature due to local urban heat-island effects, makes cities more vulnerable to higher temperatures than would be expected due to global warming alone (Kalnay and Cai, 2003; Patz et al., 2005). Existing stresses in urban areas include crime, traffic congestion, compromised air and water quality, and disruptions due to development and deterioration of infrastructure. Climate change

Table 10.7. *A summary of projected impacts of global warming on industries and energy sectors identified in Japan.*

Changes in climate parameters	Impacts
1°C temperature increase in June to August	About 5% increase of consumption of summer products
Extension of high temperature period	Increase of consumption of air-conditioners, beer, soft drinks, ice creams
Increase in thunder storms	Damage to information devices and facilities
1°C temperature increase in summer	Increase in electricity demand by about 5 million kW Increase in electricity demand in factories to enhance production
Increase in annual average temperature	Increase of household electricity consumption in southern Japan Decrease in total energy consumption for cooling, warming in northern Japan
Change in amount and pattern of rainfall	Hydroelectric power generation, management and implementation of dams, cooling water management
1°C increase in cooling water temperature	0.2 to 0.4% reduction of generation of electricity in thermal power plants, 1 to 2% reduction in nuclear power plant

is likely to amplify some of these stresses (Honda et al., 2003), although much of the interactions are not yet well understood. For example, it has been suggested that climate change will exacerbate the existing heat-island phenomenon in cities of Japan by absorbing increased solar radiation (Shimoda, 2003). This will lead to further increases in temperatures in urban areas with negative implications for energy and water consumption, human health and discomfort, and local ecosystems. Vulnerabilities of urban communities in megacities of Asia to long-term impacts of projected climate change need to be assessed in terms of energy, communication, transportation, water run-off and water quality, as well as the interrelatedness of these systems, and implications for public health (McMichael et al., 2003).

Nature-based tourism is one of the booming industries in Asia, especially ski resorts, beach resorts and ecotourist destinations which are likely vulnerable to climate change; yet only a few assessment studies are on hand for this review. Fukushima et al. (2002) reported a drop of more than 30% in skiers in almost all ski areas in Japan except in the northern region (Hokkaido) and high altitude regions (centre of the Main Island) in the event of a 3°C increase in air temperature. If the mean June to August temperature rises by 1°C in Japan, consumption of summer products such as air-conditioners, beer, soft drinks, clothing and electricity are projected to increase about 5% (Harasawa and Nishioka, 2003). Table 10.7 lists a summary of projected impacts of global warming on industries and energy sectors identified in Japan.

Limited studies on the impacts of climate change on the energy sector in Asia suggest that this sector will be affected by climate change. In particular, South Asia is expected to account for one-fifth of the world's total energy consumption by the end of 21st century (Parikh and Bhattacharya, 2004). An increase in the energy consumption of industry, residential and transport sectors could be significant as population, urbanisation and industrialisation rise. It is likely that climate change will influence the pattern of change in energy consumption that could have significant effects on CO_2 emission in this region.

10.4.6.5 Financial aspects

The cost of damages from floods, typhoons and other climate-related hazards will likely increase in the future. According to the European insurer Munich Re, the annual cost of climate change-related claims could reach US$300 billion annually by 2050. The Association of British Insurers examined the financial implications of climate change through its effects on extreme storms (hurricanes, typhoons and windstorms) using an insurance catastrophe model (ABI, 2005). Annual insured losses from hurricanes in United States, typhoons in Japan and windstorms in Europe are projected to increase by two-thirds to US$27 billion by the 2080s. The projected increase in insured losses due to the most extreme storms (with current return periods of 100 to 250 years) by the 2080s would be more than twice the reported losses of the 2004 typhoon season, the costliest in terms of damage during the past 100 years. The cost of direct damage in Asia caused by tropical cyclones has increased more than five times in the 1980s as compared with those in the 1970s and about 35 times more in the early 1990s than in 1970s (Yoshino, 1996). Flood-related damages also increased by about three times and eight times respectively in the 1990s, relative to those in the 1980s and 1970s. These trends are likely to persist in the future.

10.4.6.6 Vulnerability of the poor

Social vulnerability is the exposure of groups of people or individuals to stress as a result of the impacts of environmental change including climate change (Adger, 2000). Social vulnerability emphasises the inequitable distribution of damages and risks amongst groups of people (Wu et al., 2002) and is a result of social processes and structures that constrain access to the resources that enable people to cope with impacts (Blaikie et al., 1994). The poor, particularly in urban and urbanising cities of Asia, are highly vulnerable to climate change because of their limited access to profitable livelihood opportunities and limited access to areas that are fit for safe and healthy habitation. Consequently, the poor sector will likely be exposed to more risks from floods and other climate-related hazards in areas they are forced to stay in (Adger, 2003). This also includes the rural poor who live in the lower Mekong countries and are dependent on fisheries as their major livelihood, along with those living in coastal areas who are likely to suffer heavy losses without appropriate protection (see Table 10.10; MRC, 2003). Protection from the social forces that create inequitable exposure to risk will be as important if not more important than structural protection from natural hazards in reducing the vulnerability of the poor (Hewitt, 1997).

10.5 Adaptation: sector-specific practices, options and constraints

10.5.1 Agriculture and food security

Many studies (Parry, 2002; Ge et al., 2002; Droogers, 2004; Lin et al., 2004; Vlek et al., 2004; Wang et al., 2004a; Zalikhanov, 2004; Lal, 2007; Batima et al., 2005c) on the impacts of climate change on agriculture and possible adaptation options have been published since the TAR. More common adaptation measures that have been identified in the above-mentioned studies are summarised in Table 10.8. Generally, these measures are intended to increase adaptive capacity by modifying farming practices, improving crops and livestock through breeding and investing in new technologies and infrastructure. Specific examples include adaptation of grassland management to the actual environmental conditions as well as the practice of reasonable rotational grazing to ensure the sustainability of grassland resources (Li et al., 2002; Wang et al., 2004a; Batima et al., 2005c), improvement of irrigation systems and breeding of new rice varieties to minimise the risk of serious productivity losses caused by climate change (Ge et al., 2002), and information, education and communication programmes to enhance the level of awareness and understanding of the vulnerable groups.

Changes in management philosophy could also enhance adaptive capacity. This is illustrated by integrating fisheries and aquaculture management into coastal zone management to increase the coping ability of small communities in East Asia, South Asia and South-East Asia to sea-level rise (Troadec, 2000).

The ability of local populations to adapt their production systems to cope with climate change will vary across Asia and will be largely influenced by the way government institutions and policies mediate the supply of, and access to, food and related resources. The adaptive capacity of poor subsistence farming/herding communities is commonly low in many developing countries of Asia. One of the important and effective measures to enhance their adaptive capacity is through education and the provision of easy access to climate change-related information.

10.5.2 Hydrology and water resources

In some parts of Asia, conversion of cropland to forest (grassland), restoration and re-establishment of vegetation, improvement of the tree and herb varieties, and selection and cultivation of new drought-resistant varieties are effective measures to prevent water scarcity due to climate change. Water saving schemes for irrigation should be enforced to avert water scarcity in regions already under water stress (Wang, 2003). In North Asia, recycling and reuse of municipal wastewater (Frolov et al., 2004), increasing efficiency of water used for irrigation and other purposes (Alcamo et al., 2004), reduction of hydropower production (Kirpichnikov et al., 2004) and improved use of rivers for navigation (Golitsyn and Yu, 2002) will likely help avert water scarcity.

Table 10.8. *Adaptation measures in agriculture.*

Sectors	Adaptation measures
1°C temperature increase in June to August	Choice of crop and cultivar: • Use of more heat/drought-tolerant crop varieties in areas under water stress • Use of more disease and pest tolerant crop varieties • Use of salt-tolerant crop varieties • Introduce higher yielding, earlier maturing crop varieties in cold regions Farm management: • Altered application of nutrients/fertiliser • Altered application of insecticide/pesticide • Change planting date to effectively use the prolonged growing season and irrigation • Develop adaptive management strategy at farm level
Livestock production	• Breeding livestock for greater tolerance and productivity • Increase stocks of forages for unfavourable time periods • Improve pasture and grazing management including improved grasslands and pastures • Improve management of stocking rates and rotation of pastures • Increase the quantity of forages used to graze animals • Plant native grassland species • Increase plant coverage per hectare • Provide local specific support in supplementary feed and veterinary service
Fishery	• Breeding fish tolerant to high water temperature • Fisheries management capabilities to cope with impacts of climate change must be developed
Development of agricultural bio-technologies	• Development and distribution of more drought, disease, pest and salt-tolerant crop varieties • Develop improved processing and conservation technologies in livestock production • Improve crossbreeds of high productivity animals
Improvement of agricultural infrastructure	• Improve pasture water supply • Improve irrigation systems and their efficiency • Improve use/store of rain and snow water • Improve information exchange system on new technologies at national as well as regional and international level • Improve sea defence and flood management • Improve access of herders, fishers and farmers to timely weather forecasts

There are many adaptation measures that could be applied in various parts of Asia to minimise the impacts of climate change on water resources and use: several of which address the existing inefficiency in the use of water. Modernisation of existing irrigation schemes and demand management aimed at optimising physical and economic efficiency in the use of water resources and recycled water in water stressed countries of Asia could be useful in many agricultural areas in Asia, particularly in arid and semi-arid countries. Public investment policies which are aimed at improving access to available water resources, integrated water management, respect for the environment and promotion of better practices for wise use of water in agriculture, including recycled waste water could potentially enhance adaptive capacity. As an adaptation measure, apart from meeting non-potable water demands, recycled water can be used for recharging groundwater aquifers and augmenting surface water reservoirs. Recycled water can also be used to create or enhance wetlands and riparian habitats. While water recycling is a sustainable approach towards adaptation to climate change and can be cost-effective in the long term, the treatment of wastewater for reuse, such as that being practiced now in Singapore, and the installation of distribution systems, can be initially expensive compared to such water supply alternatives as imported water or groundwater, but are potentially important adaptive options in many countries of Asia. Reduction of water wastage and leakages, which in some cities like Damascus can be substantial, could be practiced to cushion the decrease in water supply due to decline in precipitation and increase in temperature. The use of market-oriented approaches to reduce wasteful water uses could also be effective in reducing effects of climate change on water resources (Ragab and Prudhomme, 2002). In rivers like the Mekong where wet season riverflows are estimated to increase and the dry season flows projected to decrease, planned water management interventions could marginally decrease wet season flows and substantially increase dry season flows (World Bank, 2004).

10.5.3 Coastal and low lying areas

The response to sea-level rise could mean protection, accommodation and retreat. As substantial socio-economic activities and populations are currently highly concentrated in the coastal zones in Asia, protection should remain a key focus area in Asia. Coastal protection constructions in Asia for 5-year to 1,000-year storm-surge elevations need to be considered. Most megacities of Asia located in coastal zones need to ensure that future constructions are done at elevated levels (Nicholls, 2004; Nishioka and Harasawa, 1998; Du and Zhang, 2000). The dike heightening and strengthening has been identified as one of the adaptation measures for coastal protection (Du and Zhang, 2000; Huang and Xie, 2000; Li et al., 2004a, b).

Integrated Coastal Zone Management (ICZM) provides an effective coastal protection strategy to maximise the benefits provided by the coastal zone and to minimise the conflicts and harmful effects of activities on social, cultural and environmental resources to promote sustainable management of coastal zones (World Bank, 2002). The ICZM concept is being embraced as a central organising concept in the management of fisheries, coral reefs, pollution, megacities and individual coastal systems in China, India, Indonesia, Japan, Korea, the Philippines, Sri Lanka, Vietnam and Kuwait. It has been successfully applied for prevention and control of marine pollution in Batangas Bay of the Philippines and Xiamen of China over the past few years (Chua, 1999; Xue et al., 2004). The ICZM concept and principle could potentially promote sustainable coastal area protection and management in other countries of Asia.

10.5.4 Natural ecosystems and biodiversity

The probability of significant adverse impacts of climate change on Asian forests is high in the next few decades (Isaev et al., 2004). Improved technologies for tree plantation development and reforestation could likely enhance adaptation especially in vulnerable areas such as the Siberian forests. Likewise improvement of protection from fires, insects and diseases could reduce vulnerability of most forests in Asia to climate change and variability.

Comprehensive intersectoral programs that combine measures to control deforestation and forest degradation with measures to increase agricultural productivity and sustainability will likely contribute more to reducing vulnerability of forests to climate change, land use change and other stress factors than independent sectoral initiatives. Other likely effective adaptation measures to reduce the impacts of climate change on forest ecosystems in Asia include extending rotation cycles, reducing damage to remaining trees, reducing logging waste, implementing soil conservation practices, and using wood in a more carbon-efficient way such that a large fraction of their carbon is conserved.

10.5.5 Human health

Assessment of the impact of climate change is the first step for exploring adaptation strategy. The disease monitoring system is essential as the basic data source. Specifically, the monitoring of diseases along with related ecological factors is required because the relation between weather factors and vector-borne diseases are complicated and delicate (Kovats et al., 2003). Also, disease monitoring is necessary in assessing the effectiveness and efficiency of the adaptation measures (Wilkinson et al., 2003). For effective adaptation measures, the potential impacts of climate variability and change on human health need to be identified, along with barriers to successful adaptation and the means of overcoming such barriers.

The heat watch and warning system in the USA was evaluated to be effective (Ebi et al., 2004). Also, a similar system was operated in Shanghai, China (Tan et al., 2004). Implementation of this type of heat watch and warning system and other similar monitoring systems in other parts of Asia will likely be helpful in reducing the impacts of climate change on human health.

10.5.6 Human dimensions

Rapid population growth, urbanisation and weak land-use planning and enforcement are some of the reasons why poor people move to fragile and high-risk areas which are more exposed to natural hazards. Moreover, the rapid growth of industries in urban areas has induced rural-urban migration. Rural development together with networking and advocacy, and building alliances among communities is a prerequisite for reducing the migration of people to cities and coastal areas in most developing countries of Asia (Kelly and Adger, 2000). Raising awareness about the dangers of natural disasters, including those due to climate extremes, is also crucial among the governments and people so that mitigation and preparedness measures could be strengthened. Social capital has been paid attention to build adaptive capacity (Allen, 2006). For example, a community-based disaster management programme was introduced to reduce vulnerability and to strengthen people's capacity to cope with hazards by the Asian Disaster Preparedness Centre, Bangkok (Pelling, 2003).

Tourism is one of the most important industries in Asia, which is the third centre of tourism activities following Europe and North America. Sea-level rise, warming sea temperatures and extreme weather events are likely to have impacts on the regions' islands and coasts which attract considerable number of visitors from countries such as Japan and Taiwan (World Tourism Organization, 2003; Hamilton et al., 2005). Relevant adaptation measures in this case include designing and building appropriate infrastructures to protect tourists, installation and maintenance of weather prediction and hazard warning systems, especially during rainy and tropical storm seasons. Conservation of mangroves is considered as effective natural protection against storm surges, coastal erosion and strong wave actions (Mazda et al., 1997, 2006; Vermaat and Thampanya, 2006). To minimise the anticipated impact of global warming on the ski industry, development of new leisure industries more resistant to or suited to a warmer atmosphere, thus avoiding excessive reliance on the ski industry, e.g., grass-skiing, hiking, residential lodging and eco-tourism, could be helpful in compensating for the income reduction due to snow deterioration (Fukushima et al., 2002).

To minimise the risks of heat stress that are most pronounced in large cities due to the urban heat-island effect in summer (Kalnay and Cai, 2003) urban planning should consider: reducing the heat island in summer, the heat load on buildings, cooling load and high night-time temperature, and taking climate change into account in planning new buildings and setting up new regulations on buildings and urban development. Planting trees, building houses with arcades and provision for sufficient ventilation could help in reducing heat load (Shimoda, 2003). The use of reflective surfaces, control of solar radiation by vegetation and blinds, earth tubes, the formation of air paths for natural ventilation, and rooftop planting could reduce the cooling load.

10.5.7 Key constraints and measures to strengthen adaptation

Effective adaptation and adaptive capacity in Asia, particularly in developing countries, will continue to be limited by several ecological, social and economic, technical and political constraints including spatial and temporal uncertainties associated with forecasts of regional climate, low level of awareness among decision makers of the local and regional impacts of El Niño, limited national capacities in climate monitoring and forecasting, and lack of co-ordination in the formulation of responses (Glantz, 2001).

Radical climate change may cause alterations of the physical environment in an area that may limit adaptation possibilities (Nicholls and Tol, 2006). For example, migration is the only option in response to sea-level rise that inundates islands and coastal settlements (see Chapter 17, Section 17.4.2.1). Likewise, impacts of climate change may occur beyond certain thresholds in the ability of some ecosystems to adapt without dramatic changes in their functions and resilience. The inherent sensitivity of some ecosystems, habitats and even species with extremely narrow ranges of biogeographic adaptability will also limit the options and effectiveness of adaptation.

Poverty is identified as the largest barrier to developing the capacity to cope and adapt (Adger et al., 2001). The poor usually have a very low adaptive capacity due to their limited access to information, technology and other capital assets which make them highly vulnerable to climate change. Poverty also constrains the adaptation in other sectors. Poverty, along with infrastructural limitations and other socioeconomic factors, will continue to limit the efforts to conserve biodiversity in South-East Asia (Sodhi et al., 2004). Adaptive capacity in countries where there is a high incidence of poverty will likely remain limited.

Insufficient information and knowledge on the impacts of climate change and responses of natural systems to climate change will likely continue to hinder effective adaptation particularly in Asia. The limited studies on the interconnections between adaptation and mitigation options, costs and benefits of adaptation, and trade-offs between various courses of actions will also likely limit adaptation in Asia. The deficiency in available information and knowledge will continue to make it difficult to enhance public perception of the risks and dangers associated with climate change. In addition, the absence of information on adaptation costs and benefits makes it difficult to undertake the best adaptation option. This limiting factor will be most constraining in developing countries where systems for monitoring and research on climate and responses of natural and human systems to climate are usually lacking. More relevant information such as on the crop yield benefits linked to changes in planting dates for various regions, as reported by Tan and Shibasaki (2003), and on the optimal levels and cost of coastal protection investment in Vietnam, Cambodia and other countries, as reported by Nicholls and Tol (2006), will be needed.

Based on the discussion in Chapter 17, Section 17.4.2.4, it is very likely that in countries of Asia facing serious domestic conflicts, pervasive poverty, hunger, epidemics, terrorism and other pressing and urgent concerns, attention may be drawn away from the dangers of climate change and the need to implement adaptation. The slow change in political and institutional landscape in response to climate change could also be a major limitation to future adaptation. The existing legal and institutional framework in most Asian countries remains inadequate to facilitate implementation of comprehensive and integrated response to climate change in synergy with the pursuit of sectoral development goals.

To address the constraints discussed above and strengthen adaptation in Asia, some of the measures suggested by Stern (2007) could be useful. These include improving access to high-quality information about the impacts of climate change; adaptation and vulnerability assessment by setting in place early warning systems and information distribution systems to enhance disaster preparedness; reducing the vulnerability of livelihoods and infrastructure to climate change; promoting good governance including responsible policy and decision making; empowering communities and other local stakeholders so that they participate actively in vulnerability assessment and implementation of adaptation; and mainstreaming climate change into development planning at all scales, levels and sectors.

10.6 Case studies

10.6.1 Megadeltas in Asia

There are 11 megadeltas with an area greater than 10,000 km^2 (Table 10.10) in the coastal zone of Asia that are continuously being formed by rivers originating from the Tibetan Plateau (Milliman and Meade, 1983; Penland and Kulp, 2005). These megadeltas are vital to Asia because these are home to millions of people, especially the seven megacities that are located in these deltas (Nicholls, 1995; Woodroffe et al., 2006). The megadeltas, particularly the Zhujiang delta, Changjiang delta and Huanghe delta, are also economically important, accounting for a substantial proportion of China's total GDP (Niou, 2002; She, 2004). Ecologically, the Asian megadeltas are critical diverse ecosystems of unique assemblages of plants and animals located in different climatic regions (IUCN, 2003b; ACIA, 2005; Macintosh, 2005; Sanlaville and Prieur, 2005). However, the megadeltas of Asia are vulnerable to climate change and sea-level rise that could increase the frequency and level of inundation of megadeltas due to storm surges and floods from river drainage (Nicholls, 2004; Woodroffe et al., 2006) putting communities, biodiversity and infrastructure at risk of being damaged. This impact could be more pronounced in megacities located in megadeltas where natural ground subsidence is enhanced by human activities, such as in Bangkok in the Chao Phraya delta, Shanghai in the Changjiang delta, Tianjin in the old Huanghe delta (Nguyen et al., 2000; Li et al., 2004a; Jiang, 2005; Li et al., 2005; Woodroffe et al., 2006). Climate change together with human activities could also enhance erosion that has, for example, caused the Lena delta to retreat at a rate of 3.6 to 4.5 m/yr (Leont'yev, 2004) and has affected the progradation and retreat of megadeltas fed by rivers originating from the Tibetan Plateau (Li et al., 2004b; Thanh et al., 2004; Shi et al., 2005; Woodroffe et al., 2006). The adverse impacts of salt-water intrusion on water supply in the Changjiang delta and Zhujiang delta, mangrove forests, agriculture production and freshwater fish catch, resulting in a loss of US\$125x10^6 per annum in the Indus delta could also be aggravated by climate change (IUCN, 2003a, b; Shen et al., 2003; Huang and Zhang, 2004).

Externally, the sediment supplies to many megadeltas have been reduced by the construction of dams and there are plans for many more dams in the 21st century (Chapter 6, Box 6.3; Woodroffe et al., 2006). Reduction of sediment supplies make these systems much more vulnerable to climate change and sea-level rise. When considering all the non-climate pressures, there is very high confidence that the group of populated Asian megadeltas is highly threatened by climate change and responding to this threat will present important challenges (see also Chapter 6, Box 6.3). The sustainability of megadeltas in Asia in a warmer climate will rest heavily on policies and programmes that promote integrated and co-ordinated development of the megadeltas and upstream areas, balanced use and development of megadeltas for conservation and production goals, and comprehensive protection against erosion from river flow anomalies and sea-water actions that combines structural with human and institutional capability building measures (Du and Zhang, 2000; Inam et al., 2003; Li et al., 2004b; Thanh et al., 2004; Saito, 2005; Wolanski, 2007; Woodroffe et al., 2006).

10.6.2 The Himalayan glaciers

Himalayan glaciers cover about three million hectares or 17% of the mountain area as compared to 2.2% in the Swiss Alps. They form the largest body of ice outside the polar caps and are the source of water for the innumerable rivers that flow across the Indo-Gangetic plains. Himalayan glacial snowfields store about 12,000 km^3 of freshwater. About 15,000 Himalayan glaciers form a unique reservoir which supports perennial rivers such as the Indus, Ganga and Brahmaputra which, in turn, are the lifeline of millions of people in South Asian countries (Pakistan, Nepal, Bhutan, India and Bangladesh). The Gangetic basin alone is home to 500 million people, about 10% of the total human population in the region.

Glaciers in the Himalaya are receding faster than in any other part of the world (see Table 10.9) and, if the present rate continues, the likelihood of them disappearing by the year 2035 and perhaps sooner is very high if the Earth keeps warming at the current rate. Its total area will likely shrink from the present 500,000 to 100,000 km^2 by the year 2035 (WWF, 2005).

The receding and thinning of Himalayan glaciers can be attributed primarily to the global warming due to increase in anthropogenic emission of greenhouse gases. The relatively high population density near these glaciers and consequent deforestation and land-use changes have also adversely affected these glaciers. The 30.2 km long Gangotri glacier has been receding alarmingly in recent years (Figure 10.6). Between 1842 and 1935, the glacier was receding at an average of 7.3 m every year; the average rate of recession between 1985 and 2001 is about 23 m per year (Hasnain, 2002). The current trends of glacial melts suggest that the Ganga, Indus, Brahmaputra and other rivers that criss-cross the northern Indian plain could likely become seasonal rivers in the near future as a consequence of climate change and could likely affect the economies in the region. Some other glaciers in Asia – such as glaciers shorter than 4 km length in the Tibetan Plateau – are projected to disappear and the glaciated areas located in the headwaters of the Changjiang River will likely decrease in area by more than 60% (Shen et al., 2002).

Table 10.9. *Record of retreat of some glaciers in the Himalaya.*

Glacier	Period	Retreat of snout (metre)	Average retreat of glacier (metre/year)
Triloknath Glacier (Himachal Pradesh)	1969 to 1995	400	15.4
Pindari Glacier (Uttaranchal)	1845 to 1966	2,840	135.2
Milam Glacier (Uttaranchal)	1909 to 1984	990	13.2
Ponting Glacier (Uttaranchal)	1906 to 1957	262	5.1
Chota Shigri Glacier (Himachal Pradesh)	1986 to 1995	60	6.7
Bara Shigri Glacier (Himachal Pradesh)	1977 to 1995	650	36.1
Gangotri Glacier (Uttaranchal)	1977 to 1990	364	28.0
Gangotri Glacier (Uttaranchal)	1985 to 2001	368	23.0
Zemu Glacier (Sikkim)	1977 to 1984	194	27.7

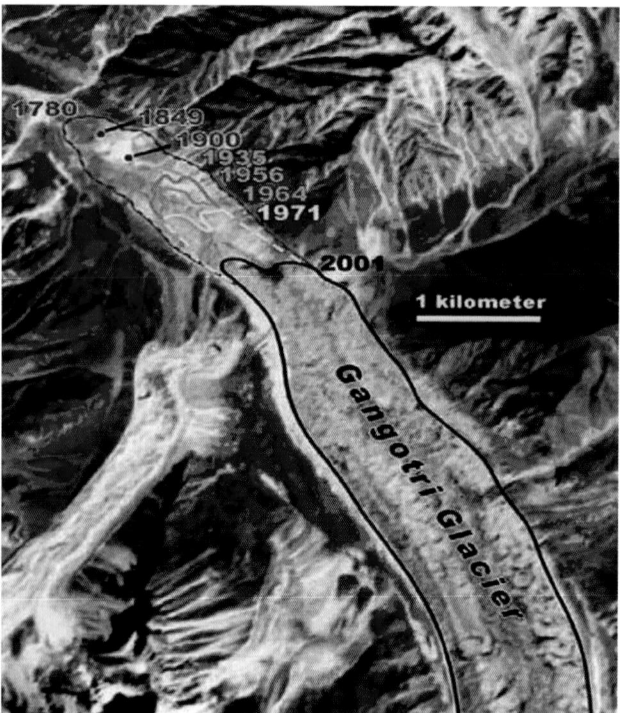

Figure 10.6. *Composite satellite image showing how the Gangotri Glacier terminus has retracted since 1780 (courtesy of NASA EROS Data Center, 9 September 2001).*

10.7 Implications for sustainable development

Chapter 20, Section 20.1 of this volume uses the succinct definition of the Bruntland Commission to describe sustainable development as *"development that meets the needs of the present without compromising the ability of future generations to meet their own needs"*. Sustainable development represents a balance between the goals of environmental protection and human economic development and between the present and future needs. It implies equity in meeting the needs of people and integration of sectoral actions across space and time. This section focuses mainly on how the impacts of projected climate change on poverty eradication, food security, access to water and other key concerns described above will likely impinge on the pursuit of sustainable development in Asia. In most instances, the reference to sustainable development will be confined to a specific country or sub-region, primarily due to the existing difficulty of aggregating responses to climate change and other stressors across the whole of Asia.

10.7.1 Poverty and illiteracy

A significant proportion of the Asian population is living below social and economic poverty thresholds. Asia accounts for more than 65% of all people living in rural areas without access to sanitation, of underweight children, of people living on less than a dollar a day and of TB cases in the world. It accounts for over 60% of all malnourished people, people without access to sanitation in urban areas and people without access to water in rural areas (UN-ESCAP, 2006). Most of the world's poor reside in South Asia and, within South Asia, the majority resides in rural areas (Srinivasan, 2000). Greater inequality could both undermine the efficiency with which future growth could reduce poverty and make it politically more difficult to pursue pro-poor policies (Fritzen, 2002).

Coupled with illiteracy, poverty subverts the ability of the people to pursue the usually long-term sustainable development goals in favour of the immediate goal of meeting their daily subsistence needs. This manifests in the way poverty drives poor communities to abusive use of land and other resources that lead to onsite degradation and usually macroscale environmental deterioration. In the absence of opportunities for engaging in stable and gainful livelihood, poverty stricken communities are left with no option but to utilise even the disaster-prone areas, unproductive lands and ecologically fragile lands that have been set aside for protection purposes such as conservation of biodiversity, soil and water. With climate change, the poor sectors will be most vulnerable and, without appropriate measures, climate change will likely exacerbate the poverty situation and continue to slow down economic growth in developing countries of Asia (Beg et al., 2002).

10.7.2 Economic growth and equitable development

Rapid economic growth characterised by increasing urbanisation and industrialisation in several countries of Asia (i.e.,

China, India and Vietnam) will likely drive the increase in the already high demand for raw materials such as cement, wood, steel and other construction materials in Asia. Consequently, the use of forests, minerals and other natural resources will increase along with the increase in carbon emission. The challenge here is finding the development pathways wherein GHG emission is minimised while attaining high economic growth (Jiang et al., 2000). Equally vital in this regard is the promotion of equity in spreading the benefits that will arise from economic growth so as to uplift the condition of the poor sector to a state of enhanced capacity to adapt to climate change. Another concern related to economic growth is the increase in the value of land to a level where it becomes economically less profitable to farm agricultural land than using the land for industrial and commercial purposes. In the absence of appropriate regulatory intervention, this can undermine the production of adequate food supply and further jeopardise the access of the poor to food support.

Sustaining economic growth in the context of changing climate in many Asian countries will require the pursuit of enhancing preparedness and capabilities in terms of human, infrastructural, financial and institutional dimensions with the aim in view of reducing the impacts of climate change on the economy. For instance, in many developing countries, instituting financial reforms could likely result in a more robust economy that is likely to be less vulnerable to changing climate (Fase and Abma, 2003). In countries with predominantly agrarian economies, climate change, particularly an increase in temperature and reduction in precipitation, could, in the absence of adequate irrigation and related infrastructural interventions, dampen the economic growth by reducing agricultural productivity (Section 10.4.1).

10.7.3 Compliance with and governance of Multilateral Environmental Agreements

Many countries in Asia are signatories to one or more of the Multilateral Environmental Agreements (MEAs) that seek to address common concerns such as biodiversity conservation and sustainable forest management, climate change, international water resources, over-exploitation of regional fisheries, trans-boundary air pollution, and pollution of regional seas. Some of these MEAs include the United Nations Framework Convention on Climate Change (UNFCCC), the Convention on Biological Diversity (CBD), the Convention to Combat Desertification (CCD), the Convention on International Trade of Endangered Fauna and Flora (CITES), the Ramsar Convention to protect Mangroves and Wetlands, the Montreal and Kyoto Protocols to address problems of the breakdown in the Earth's protective ozone layer and global warming, International Tropical Timber Organization (ITTO) that governs the exploitation of tropical forests and conservation of biodiversity, and International Convention for the Prevention of Pollution from Ships for control of pollution of regional seas. The major challenge for Asian countries is how to take advantage of opportunities in designing integrated and synergistic responses in adherence to and compliance with the terms and conditions of MEAs and improve environmental quality without unduly hampering economic development (Beg et al., 2002).

10.7.4 Conservation of natural resources

Natural resources utilisation could intensify in several parts of Asia in response to increasing demands. In South-East Asia, intensification of forest utilisation could likely increase further the already high rate of deforestation that could lead to the loss of much of its original forests and biodiversity by 2100 (Sodhi et al., 2004). To sustain development in this region, measures to minimise deforestation and enhance restoration of degraded forests will be required. The challenge in Asia will be in countries with developing economies where the need to maximise production could lead to increased perturbations of the ecosystems and the environment that could be aggravated by climate change. In the same manner, the use of water will continue to increase as the population and economies of countries grow. This will likely put more stress on water that could be exacerbated by climate change as discussed above. Integrated responses to cope with the impacts of climate change and other stressors on the supply and demand side will likely contribute in the attainment of sustainable development in many countries in the West, South and South-East Asia.

10.8 Key uncertainties, research gaps and priorities

10.8.1 Uncertainties

The base for future climate change studies is designing future social development scenarios by various models and projecting future regional and local changes in climate and its variability, based on those social development scenarios so that most plausible impacts of climate change could be assessed. The emission scenarios of greenhouse gases and aerosols are strongly related to the socio-economics of the countries in the region and could be strongly dependent on development pathways followed by individual nations. Inaccurate description on future scenarios of socio-economic change, environmental change, land-use change and technological advancement and its impacts will lead to incorrect GHG emissions scenarios. Therefore factors affecting design of social development scenarios need to be examined more carefully to identify and properly respond to key uncertainties.

The large natural climate variability in Asia adds a further level of uncertainty in the evaluation of a climate change simulation. Our current understanding of the precise magnitude of climate change due to anthropogenic factors is relatively low, due to imperfect knowledge and/or representation of physical processes, limitations due to the numerical approximation of the model's equations, simplifications and assumptions in the models and/or approaches, internal model variability, and inter-model or inter-method differences in the simulation of climate response to given forcing. Current efforts on climate variability and climate change studies increasingly rely upon diurnal, seasonal, latitudinal and vertical patterns of temperature trends to provide evidence for anthropogenic signatures. Such approaches require increasingly detailed understanding of the

Table 10.10. *Megadeltas of Asia.*

Features	Lena	Huanghe-Huaihe	Changjiang	Zhujiang	Red River	Mekong	Chao Phraya	Irrawaddy	Ganges-Brahmaputra	Indus	Shatt-el-Arab (Arvand Rud)
Area (10^3 km^2)	43.6	36.3	66.9	10	16	62.5	18	20.6	100	29.5	18.5
Water discharge (10^9 m^3/yr)	520	33.3	905	326	120	470	30	430	1330	185	46
Sediment load (10^6 t/yr)	18	849	433	76	130	160	11	260	1969	400	100
Delta growth (km^2/yr)	--	21.0	16.0	11.0	3.6	1.2		10.0	5.5 to 16.0	PD30	
Climate zone	Boreal	Temperate	Sub-tropical	Sub-tropical	Tropical	Tropical	Tropical	Tropical	Tropical	Semi-arid	Arid
Mangroves (10^3 km^2)	None	None	None	None		5.2	2.4	4.2	10	1.6	None
Population (10^6) in 2000	0.000079	24.9 (00)	76 (03)	42.3 (03)	13.3	15.6	11.5	10.6	130	3.0	0.4
Population increase by 2015	None	18	-	176	21	21	44	15	28	45	--
GDP (US$$10^9$)		58.8 (00)	274.4 (03)	240.8 (03)	9.2 (04)	7.8 (04)	--	--	--	--	--
Megacity	None	Tianjin	Shanghai	Guangzhou	--	--	Bangkok	--	Dhaka	Karachi	--
Ground subsidence (m)	None	2.6 to 2.8	2.0 to 2.6	X	XX	--	0.2 to 1.6	--	0.6 to 1.9 mm/a		--
SLR (cm) in 2050	10 to 90 (2100)	70 to 90	50 to 70	40 to 60	--	--	--	--	--	20 to 50	--
Salt-water intrusion (km)	--	--	100	--	30 to 50	60 to 70	--	--	100	80	--
Natural hazards	--	FD	CS, SWI, FD	CS, FD, SWI	CS, FD, SWI	SWI	--	--	CS, FD, SWI	CS, SWI	--
Area inundated by SLR (10^3 km^2). Figure in brackets indicates amount SLR.	--	21.3 (0.3m)	54.5 (0.3m)	5.5 (0.3m)	5 (1m)	20 (1m)	--	--	--	--	--
Coastal protection	No protection	Protected	Protected	Protected	Protected	Protected	Protected	Protected	Protected	Partial Protection	Partial protection

PD: Progradation of coast; CS: Tropical cyclone and storm surge; FD: Flooding; SLR: Sea-level rise; SWI: Salt water intrusion; DG: Delta growth in area; XX: Strong ground subsidence; X: Slight ground subsidence; --: No data available

spatial variability of all forcing mechanisms and their connections to global, hemispheric and regional responses.

Uncertainty in assessment methodologies per se is also one of the main sources of uncertainty. In model-based assessments, results on impacts of climate change, in fact, accumulate errors from the methodologies for establishment of socio-economic scenarios, environmental scenarios, climate scenarios and climate impact assessment (Challinor et al., 2005).

10.8.2 Confidence levels and unknowns

The vulnerability of key sectors to the projected climate change for each of the seven sub-regions of Asia based on currently available scientific literature referred to in this assessment have been assigned a degree of confidence which is listed in Table 10.11. The assigned confidence levels could provide guidance in weighing which of the sectors ought to be

Table 10.11. *Vulnerability of key sectors to the impacts of climate change by sub-regions in Asia.*

Sub-regions	Food and fibre	Biodiversity	Water resource	Coastal ecosystem	Human health	Settlements	Land degradation
North Asia	+1 / H	-2 / M	+1 / M	-1 / M	-1 / M	-1 / M	-1 / M
Central Asia and West Asia	-2 / H	-1 / M	-2 / VH	-1 / L	-2 / M	-1 / M	-2 / H
Tibetan Plateau	+1 / L	-2 / M	-1 / M	Not applicable	No information	No information	-1 / L
East Asia	-2 / VH	-2 / H	-2 / H	-2 / H	-1 / H	-1 / H	-2 / H
South Asia	-2 / H	-2 / H	-2 / H	-2 / H	-2 / M	-1 / M	-2 / H
South-East Asia	-2 / H	-2 / H	-1 / H	-2 / H	-2 / H	-1 / M	-2 / H

Vulnerability: -2 – *Highly vulnerable*
-1 – *Moderately vulnerable*
0 – *Slightly or not vulnerable*
+1 – *Moderately resilient*
+2 – *Most resilient*

Level of confidence: *VH - Very high*
H - High
M - Medium
L - Low
VL - Very low

the priority concerns based on the most likely future outcomes. However, some of the greatest concerns emerge not from the most likely future outcomes but rather from possible 'surprises'. Growing evidence suggests the ocean-atmosphere system that controls the world's climate can lurch from one state to another, such as a shutdown of the 'ocean conveyor belt' in less than a decade. Certain threshold events may become more probable and non-linear changes and surprises should be anticipated, even if they cannot be predicted with a high degree of confidence. Abrupt or unexpected changes pose great challenges to our ability to adapt and can thus increase our vulnerability to significant impacts (Preston et al., 2006).

The spotlight in climate research is shifting from gradual to rapid or abrupt change. There is some risk that a catastrophic collapse of the ice sheet could occur over a couple of centuries if polar water temperatures warm by a few degrees. Scientists suggest that such a risk has a probability of between 1 and 5% (Alley, 2002). Because of this risk, as well as the possibility of a larger than expected melting of the Greenland Ice Sheet, a recent study estimated that there is a 1% chance that global sea level could rise by more than 4 metres in the next two centuries (Hulbe and Payne, 2001).

10.8.3 Research gaps and priorities

A number of fundamental scientific questions relating to the build-up of greenhouse gases in the atmosphere and the behaviour of the climate system need to be critically addressed. These include (a) the future usage of fossil fuels, (b) the future emissions of methane (Slingo et al., 2005; Challinor et al., 2006), (c) the fraction of the future fossil-fuel carbon that will remain in the atmosphere and provide radiative forcing versus exchange with the oceans or net exchange with the land biosphere, (d) details of the regional and local climate change given an overall level of global climate change, (e) the nature and causes of the natural variability of climate and its interactions with forced changes, and (f) the direct and indirect effects of the changing distributions of aerosols.

An effective strategy for advancing the understanding of adverse impacts of climate change in Asia will require strengthening the academic and research institutions to conduct innovative research on the response of human and natural systems to multiple stresses at various levels and scales. Key specific research-related priorities for Asia are:

- basic physiological and ecological studies on the effects of changes in atmospheric conditions;
- enhancing capability to establish and maintain observation facilities and to collect, and compile, climatic, social and biophysical data;
- improvement of information-sharing and data networking on climate change in the region;
- impacts of extreme weather events such as disasters from flood, storm surges, sea-level rise, heatwaves, plant diseases and insect pests;
- identification of social vulnerability to multiple stressors due to climate change and environmental change;
- adaptation researches concerning agro-technology, water resources management, integrated coastal zone management; pathology and diseases monitoring and control;
- sectoral interaction such as between irrigation and water resources, agricultural land use and natural ecosystem, water resources and cropping, water resources and livestock farming, water resources and aquaculture, water resource and hydropower, sea-level rise and land use, sea-water invasion and land degradation;
- mainstreaming science of climate change impacts, adaptation and vulnerability in policy formulation; and
- identification of the critical climate thresholds for various regions and sectors.

References

ABI, 2005: Financial Risks of Climate Change, Summary Report. Association of British Insurers, London, 40 pp. http://www.abi.org.uk/Display/File/Child/552/Financial_Risks_of_Climate_Change.pdf.

Abu-Taleb, M.F., 2000: Impacts of global climate change scenarios on water supply and demand in Jordan. *Water International*, **25**, 457-463.

Achard, F., H.J. Stibig, L. Laestadius, V. Roshchanka, A. Yaroshenko and D. Aksenov, Eds., 2005: Identification of 'hotspot areas' of forest cover changes in boreal Eurasia. Office for Official Publication of the European Communities,

ences of air temperature change on leisure industries: case study on ski activities. *Mitigation and Adaptation Strategies for Global Change*, **7**, 173-189.

Gao, X.J., D.L. Li, Z.C. Zhao and F. Giorgi, 2003: Climate change due to greenhouse effects in Qinghai-Xizang Plateau and along Qianghai-Tibet Railway. *Plateau Meteorology*, **22**, 458-463 (In Chinese with English abstract).

Gavriliev, P.P. and P.V. Efremov, 2003: Effects of cryogenic processes on Yakutian landscapes under climate warming. *Proc. 7th International Conference on Permafrost*, Québec, 277-282.

GCOS, 2005a: GCOS Regional Action Plan for South and South-East Asia. http://www.wmo.int/pages/prog/gcos/documents/GCOS_ESEA_RAP_FINAL-DRAFT_Sept2005.pdf

GCOS, 2005b: GCOS Regional Action Plan for South and South-West Asia. http://www.wmo.int/pages/prog/gcos/documents/GCOS_SSWA_RAP_FINAL-DRAFT_Sept2005.pdf

Ge, D.K., Z.O. Jin, C.L. Shi and L.Z. Gao, 2002: Gradual impacts of climate change on rice production and adaptation strategies in southern China, Jiangsu. *J. Agr. Sci.*, **18**, 1-8.

Giorgi, F. and X. Bi, 2005: Regional changes in surface climate interannual variability for the 21st century from ensembles of global model simulations. *Geophys. Res. Lett.*, **32**, L13701, doi:10.1029/2005GL023002.

Glantz, M.H., Ed., 2001: *Once Burned, Twice Shy? Lessons Learned from the 1997-98 El Niño*. United Nations University, Tokyo, 294 pp.

Goldammer, J.G., A.I. Sukhinin and I. Csiszar, 2003: The current fire situation in the Russian Federation. *International Forest Fire News*, **29**, 89-111.

Golitsyn, G.S. and I.A. Yu, Eds., 2002: *Global Climate Change and Its Impacts on Russia*. Ministry of Industry and Science, Moscow, 465 pp. (in Russian)

Golubev, G. and N. Dronin, 2003: Geography of droughts and food problems in Russia of the twentieth century. Research Monograph of the Centre for Environmental Systems Research, University of Kassel and Department of Geography, Moscow State University, Moscow, 25 pp.

Gopal, B., 2003: Future of wetlands in Asia. *Abstract, 5th Int. Conf. on Environmental Future, Zürich*. http://www.icef.eawag.ch/abstracts/Gopal.pdf.

Gruza, G. and E. Rankova, 2004: Detection of changes in climate state, climate variability and climate extremity. *Proc. World Climate Change Conference, Moscow*, Y. Izrael, G. Gruza, S. Semenov and I. Nazarov, Eds., Institute of Global Climate and Ecology, Moscow, 90-93.

Gruza, G., E. Rankova, V. Razuvaev and O. Bulygina, 1999: Indicators of climate change for the Russian Federation. *Climatic Change*, **42**, 219-242.

Guo, Q.X., J.L. Li, J.X. Liu and Y.M. Zhang, 2001: The scientific significance of the forest vegetation ecotone between Daxing'an and Xiaoxing'an mountains to global climate change study. *J. Forest.*, **29**, 1-4.

Gupta, S.K. and R.D. Deshpande, 2004: Water for India in 2050: first-order assessment of available options. *Curr. Sci. India*, **86**, 1216-1224.

Hajima T., Y. Shimizu, Y. Fujita and K. Omasa, 2005: Estimation of net primary production in Japan under nitrogen-limited scenario using BGGC Model. *Journal of Agricultural Meteorology*, **60**, 1223-1225.

Hales, S, N., de Wet, J. Maindonald and A. Woodward, 2002: Potential effect of population and climate changes on global distribution of dengue fever: an empirical model. *Lancet*, **360**, 830-834.

Hamilton, J.M., D.J. Maddison and R.S.J. Tol, 2005: Climate change and international tourism: a simulation study. *Global Environ. Chang.*, **15**, 253-266.

Han, M., M.H. Zhao, D.G. Li and X.Y. Cao, 1999: Relationship between ancient channel and seawater intrusion in the south coastal plain of the Laizhou Bay. *Journal of Natural Disasters*, **8**,73-80.

Hara, Y., K. Takeuchi and S. Okubo, 2005: Urbanization linked with past agricultural landuse patterns in the urban fringe of a deltaic Asian mega-city: a case study in Bangkok. *Landscape Urban Plan.*, **73**, 16-28.

Harasawa, H. and S. Nishioka, Eds., 2003: Climate Change on Japan, KokonShoin Publications, Tokyo (In Japanese).

Hasnain, S.I., 2002: Himalayan glaciers meltdown: impacts on South Asian Rivers. *FRIEND 2002-Regional Hydrology: Bridging the Gap between Research and Practice*, H.A.J. van Lanen and S. Demuth, Eds., IAHS Publications, Wallingford, No. 274, 417-423.

Hasumi, H. and S. Emori, Eds., 2004: K-1 coupled model (MIROC) description. K-1 Technical Report 1, Centre for Climate System Research, University of Tokyo, Tokyo, 34 pp.

Hewitt, K., 1997: *Regions of Risk: a Geographical Introduction to Disaster*. Longman, 389 pp.

Higashi, H., K. Dairaku and T. Matuura, 2006: Impacts of global warming on heavy precipitation frequency and flood risk. *Journal of Hydroscience and Hydraulic Engineering*, **50**, 205-210.

Hinzman, L.D., D.L. Kane, K. Yoshikawa, A. Carr, W.R. Bolton and M. Fraver, 2003: Hydrological variations among watersheds with varying degrees of permafrost. *Proc. 7th International Conference on Permafrost*, Québec, 407-411.

Ho, C.H., J.Y. Lee, M.H. Ahn and H.S. Lee, 2003: A sudden change in summer rainfall characteristics in Korea during the late 1970s. *Int. J. Climatol.*, **23**, 117-128.

Hoanh, C.T., H. Guttman, P. Droogers and J. Aerts, 2004: Will we produce sufficient food under climate change? Mekong Basin (South-east Asia). *Climate Change in Contrasting River Basins: Adaptation Strategies for Water, Food, and Environment*, J.C.J.H. Aerts and P. Droogers, Eds., CABI Publishing, Wallingford, 157-180.

Honda, Y., M. Ono, I. Uchiyama and A. Sasaki, 1995: Relationship between daily high temperature and mortality in Kyushu, Japan. *Nippon Koshu Eisei Zasshi*, **42**, 260-268.

Honda, Y., H. Nitta and M. Ono, 2003: Low level carbon monoxide and mortality of persons aged 65 or older in Tokyo, Japan, 1976-1990. *J. Health Sci.*, **49**, 454-458.

Hu, D.X., W.Y. Han and S. Zhang, 2001: *Land-Ocean Interaction in Changjiang and Zhujiang Estuaries and Adjacent Sea Areas*. China Ocean Press, Beijing, 218 pp. (in Chinese).

Hu, Z.Z., S. Yang and R. Wu, 2003: Long-term climate variations in China and global warming signals. *J. Geophys. Res.*, **108**, 4614, doi:10.1029/2003JD003651.

Huang, H.J., F. Li, J.Z. Pang, K.T. Le and S.G. Li, 2005: *Land-Ocean Interaction Between Huanghe Delta and Bohai Gulf and Yellow Sea*. China Science Press, Beijing, 313 pp. (in Chinese).

Huang, Z.G. and X.D. Xie, 2000: *Sea Level Changes in Guangdong and Its Impacts and Strategies*. Guangdong Science and Technology Press, Guangzhou, 263 pp.

Huang, Z.G. and W.Q. Zhang, 2004: Impacts of artificial factors on the evolution of geomorphology during recent 30 years in the Zhujiang Delta. *Quaternary Research*, **24**, 394-401.

Hughes, T.P., A.H. Baird, D.R. Bellwood, M. Card, S.R. Connolly, C. Folke, R. Grosberg, O. Hoegh-Guldberg and Co-authors, 2003: Climate change, human impacts, and the resilience of coral reefs. *Science*, **301**, 929-933.

Hulbe, C.L. and A.J. Payne, 2001: Numerical modelling of the West Antarctic Ice Sheet. *Antarctic Research Series*, **77**, 201-220.

Iafiazova, R.K., 1997: Climate change impact on mud flow formation in Trans-Ili Alatay mountains. *Hydrometeorology and Ecology*, **3**, 12-23. (in Russian).

Ichikawa, A., Ed., 2004: *Global Warming – The Research Challenges: A Report of Japan's Global Warming Initiative*. Springer, 160 pp.

Inam, A., T.M. Ali Khan, A.R. Tabrez, S. Amjad, M. Danishb and S.M. Tabrez, 2003: Natural and man-made stresses on the stability of Indus deltaic Eco region. *Extended Abstract, 5th International Conference on Asian Marine Geology*, Bangkok, Thailand (IGCP475/APN).

India Meteorological Department, 2002 to 2006: Southwest Monsoons of 2002, 2003, 2004, 2005 and 2006 - end of season reports. Published by India Meteorological Department, Government of India, Pune.

India Meteorological Department, 2006: Southwest Monsoon 2006 - end of season report. Published by India Meteorological Department, Government of India, Pune.

IPCC, 1992: *Climate Change 1992: The Supplementary Report to the IPCC Scientific Assessment*. J.T. Houghton, B.A. Callander and S.K. Varney, Eds., Cambridge University Press, Cambridge, 200 pp.

IPCC, 2001: *Climate Change 2001: Impacts, Adaptation and Vulnerability. Contribution of Working Group II to the Third Assessment Report of the Intergovernmental Panel on Climate Change*, J.J. McCarthy, O.F. Canziani, N.A. Leary, D.J. Dokken and K.S. White, Eds., Cambridge University Press, Cambridge, 1032 pp.

IRIMO, 2006a: Country Climate Analysis in year 2005, Islamic Republic of Iran Meteorological Organization, Tehran.

IRIMO, 2006b: Country Climate Analysis in spring 2006, Islamic Republic of Iran Meteorological Organization, Tehran.

Isaev, A., V. Stolbovoi, V. Kotlyakov, S. Nilsson and I. McCallum, 2004: Climate change and land resources of Russia. *Proc. World Climate Change Conference, Moscow*, Y. Izrael, G. Gruza, S. Semenov and I. Nazarov, Eds., Institute of Global Climate and Ecology, Moscow, 234-240.

Ishigami, Y., Y. Shimizu and K. Omasa, 2003: Projection of climatic change effects on potential natural vegetation distribution in Japan. *Journal of Agricultural Meteorology*, **59**, 269-276 (in Japanese with an English abstract).

Ishigami, Y., Y. Shimizu and K. Omasa, 2005: Evaluation of the risk to natural vegetation from climate change in Japan. *Journal of Agricultural Meteorology*, **61**, 69-75 (in Japanese with an English abstract).

Isobe, H., 2002: Trends in precipitation over Japan. *Proc. 6th Symposium on Water Resources*, 585-590.

IUCN (The World Conservation Union), 2003a: Indus Delta, Pakistan: economic costs of reduction in freshwater flows. Case Studies in Wetland Valuation No. 5, Pakistan Country Office, Karachi, 6 pp. Accessed 24.01.07: www.waterandnature.org/econ/CaseStudy05Indus.pdf.

IUCN (The World Conservation Union), 2003b: The lower Indus river: balancing development and maintenance of wetland ecosystems and dependent livelihoods. Water and Nature Initiative, 5 pp. Accessed 24.01.07: www.iucn.org/themes/wani/flow/cases/Indus.pdf.

Izmerov, N.F., B.A. Revich and E.I. Korenberg, Eds., 2004: *Proc. Intl. Conf. Climate Change and Public Health in Russia in the 21st Century*. Russian Academy of Medical Sciences - Russian Regional Committee for Cooperation with UNEP Centre for Demography and Human Ecology Russian Regional Environmental Centre Non-Governmental Organisation 'Environmental Defence', Moscow, 461 pp.

Izrael, Y. A., Ed., 2002: Scientific aspects of environmental problems in Russia. *Proc. All-Russia Conference*, Nauka, Moscow, Vol. 1, 622 pp. and Vol. 2, 650 pp. (in Russian).

Izrael, Y.A. and Y.A. Anokhin, 2000: Monitoring and assessment of the environment in the Lake Baikal region. *Aquatic Ecosystem Health and Management*, **3**, 199-202.

Izrael, Y.A. and Y.A. Anokhin, 2001: Climate change impacts on Russia. *Integrated Environmental Monitoring*, Nauka, Moscow, 112-127 (in Russian with an English abstract).

Izrael, Yu. A. and O.D. Sirotenko, 2003, Modeling climate change impact on agriculture of Russia. *Meteorology and Hydrology*, **6**, 5-17.

Izrael, Y.A., and others, Eds. 2002a: *Third National Communication of the Russian Federation*. Inter-agency Commission of the Russian Federation of Climate Change, Moscow, 142 pp. http://unfccc.int/resource/docs/natc/rusnce3.pdf.

Izrael, Y.A., Y.A. Anokhin and A.V. Pavlov, 2002b: Permafrost evolution and the modern climate change. *Meteorology and Hydrology*, **1**, 22-34.

Izrael, Yu., A.V. Pavlov, Yu. A. Anokhin, L.T. Miach and B.G. Sherstiukov, 2006: Statistical evaluation of climate change on permafrost terrain in the Russian Federation. *Meteorology and Hydrology*, **5**, 27-38.

Japan Meteorological Agency, 2005: Global Warming Projection, Vol.6 - with the RCM20 and with the UCM, 58 pp. (in Japanese).

Jiang, H.T., 2005: Problems and discussion in the study of land subsidence in the Suzhou-Wuxi-Changzhou Area. *Quaternary Res.*, **25**, 29-33.

Jiang, K., T. Masui, T. Morita and Y. Matsuoka, 2000: Long-term GHG emission scenarios for Asia -Pacific and the world. *Technol. Forecast. Soc.*, **63**, 207-229.

Jin, Z.Q., C.L. Shi, D.K. Ge and W. Gao, 2001: Characteristic of climate change during wheat growing season and the orientation to develop wheat in the lower valley of the Yangtze River, Jiangsu. *J. Agr. Sci.*, **17**, 193-199.

Jorgenson, M.T., C.H. Racine, J.C. Walters and T.E. Osterkamp, 2001: Permafrost degradation and ecological changes associated with a warming climate in central Alaska. *Climatic Change*, **48**, 551-571.

Juday, G.P., V. Barber, P. Duffy, H. Linderholm, S. Rupp, S. Sparrow, E. Vaganov and J. Yarie, 2005: Forests, Land Management and agriculture. *Arctic Climate Impact Assessment*. Cambridge University Press, Cambridge, 781-862.

Jung, H.S., Y. Choi, J.-H. Oh and G.H. Lim, 2002: Recent trends in temperature and precipitation over South Korea. *Int. J. Climatol.*, **22**, 1327-1337.

Kajiwara, M., T. Oki and J. Matsumoto, 2003: Inter-annual variability of the frequency of severe rainfall in the past 100 years over Japan, *Extended abstract, Bi-annual meeting of the Meteorological Society of Japan* (in Japanese).

Kalnay, E. and M. Cai, 2003: Impact of urbanization and land-use change on climate. *Nature*, **423**, 528-531.

Kalra, N., P.K. Aggarwal, S. Chander, H. Pathak, R. Choudhary, A. Chaudhary, S. Mukesh, H.K. Rai, U.A. Soni, S. Anil, M. Jolly, U.K. Singh, A. Owrs and M.Z. Hussain, 2003: Impacts of climate change on agriculture. *Climate Change and India: Vulnerability Assessment and Adaptation*, P.R. Shukla, S.K. Sharma, N.H. Ravindranath, A. Garg and S. Bhattacharya, Eds., Orient Longman Private, Hyderbad, 193-226.

Kanai, S., T. Oki and A. Kashida, 2004: Changes in hourly precipitation at Tokyo from 1890 to 1999. *J. Meteor. Soc. Japan*, **82**, 241-247.

Kato, H., K. Nishizawa, H. Hirakuchi, S. Kadokura, N. Oshima and F. Giorgi, 2000: Performance of RegCM2.5/NCAR-CSM nested system for the simulation of climate change in East Asia caused by global warming. *J. Meteor. Soc. Japan*, **79**, 99-121.

Kawahara, M. and N. Yamazaki, 1999: Long-term trend of incidences of extreme high or low temperatures in Japan, *Extended Abstract, Bi-annual meeting of the Meteorological Society of Japan* (in Japanese).

Kelly, P.M. and W.N. Adger, 2000: Theory and Practice in Assessing Vulnerability to Climate Change and Facilitating Adaptation. *Climatic Change*, **47**, 325-352.

Khan, T.M.A., O.P. Singh and M.D. Sazedur Rahman, 2000: Recent sea level and sea surface temperature trends along the Bangladesh coast in relation to the frequency of intense cyclones. *Marine Geodesy*, **23**, 103-116.

Kirpichnikov, M., B. Reutov and A. Novikov, 2004: Scientific and technical projections of the Russian energy sector development as a factor of reduction of the negative climate change. *Proc. World Climate Change Conference, Moscow*, Y. Izrael, G. Gruza, S. Semenov and I. Nazarov, Eds., Institute of Global Climate and Ecology, Moscow, . 413-417.

Kitao M., T. Koike, H. Tobita and Y. Maruyama, 2005: Elevated CO_2 and limited nitrogen nutrition can restrict excitation energy dissipation in photosystem II of Japanese white birch (*Betula platyphylla var. japonica*) leaves. *Physiologia Plantarum*, **125**, 64-73.

Kleypas, J.A., R.W. Buddemeier, D. Archer, J.P. Gattuso, C. Langdon and B.N. Opdyke, 1999: Geochemical consequences of increased atmospheric carbon dioxide on coral reefs. *Science*, **284**, 118-120.

Knowlton, N., 2001: The future of coral reefs. *P. Natl. Acad. Sci. USA*, **98**, 5419–5425.

Knutson, T.R. and R.E. Tuleya, 2004: Impacts of CO_2 induced warming on simulated hurricane intensities and precipitation: sensitivity to the choice of climate model and convective parameterization. *J. Climate*, **17**, 3477-3495.

Kovats, R.S., M.J. Bouma, S. Hajat, E. Worrall and A. Haines, 2003: El Niño and health. *Lancet*, **362**, 1481-89.

Kripalani, R.H., S.R. Inamdar and N.A. Sontakke, 1996: Rainfall variability over Bangladesh and Nepal: comparison and connection with features over India. *Int. J. Climatol.*, **16**, 689–703.

Kudo, G., Y. Nishikawa, T. Kasagi and S. Kosuge, 2004: Does seed production of spring ephemerals decrease when spring comes early? *Ecol. Res.*, **19**, 255-259.

Kurihara, K., K. Ishihara, H. Sakai, Y. Fukuyama, H. Satou, I. Takayabu, K. Murazaki, Y. Sato, S. Yukimoto and A. Noda, 2005: Projections of climatic change over Japan due to global warming by high resolution regional climate model in MRI. *SOLA*, **1**, 97-100.

Kwon, W.T., I.C. Shin, H.J. Baek, Y. Choi, K.O. Boo, E.S. Im, J.H. Oh and S.H. Lee, 2004: The development of regional climate change scenario for the National climate change Report (III), (In Korean), METRI Technical Report.

Lal, M., 2001: Tropical cyclones in a warmer world. *Curr. Sci. India*, **80**, 1103-1104.

Lal, M., 2002: Global climate change: India's monsoon and its variability. Final Report under Country Studies Vulnerability and Adaptation Work Assignment with Stratus Consulting's Contract of the U.S. Environmental Protection Agency, 58 pp.

Lal, M., 2003: Global climate change: India's monsoon and its variability. *Journal of Environmental Studies and Policy*, **6**, 1-34.

Lal, M., 2007: Implications of climate change on agricultural productivity and food security in South Asia. *Key vulnerable regions and climate change - Identifying thresholds for impacts and adaptation in relation to Article 2 of the UNFCCC*, Springer, Dordrecht, in press.

Lal, M., G. Srinivasan and U. Cubasch, 1996: Implications of global warming on the diurnal temperature cycle of the Indian subcontinent. *Curr. Sci. India*, **71**, 746-752.

Lal, M., K.K. Singh, L.S. Rathore, G. Srinivasan and S.A. Saseendran, 1998: Vulnerability of rice and wheat yields in NW - India to future changes in climate. *Agri. & Forest Meteorol.*, **89**, 101-114.

Lal, M., G.A. Meehl and J.M. Arblaster, 2000: Simulation of Indian summer monsoon rainfall and its intraseasonal variability. *Reg. Environ. Change*, **1**, 163-179.

Lal, M., H. Harasawa, D. Murdiyarso, W.N. Adger, S. Adhikary, M. Ando, Y. Anokhin, R.V. Cruz and Co-authors, 2001a: Asia. *Climate Change 2001: Impacts, Adaptation, and Vulnerability. Contribution of Working Group II to the*

Third Assessment Report of the Intergovernmental Panel on Climate Change, J.J. McCarthy, O.F. Canziani, N.A. Leary, D.J. Dokken and K.S. White, Eds., Cambridge University Press, Cambridge, 533-590.

Lal, M., T. Nozawa, S. Emori, H. Harasawa, K. Takahashi, M. Kimoto, A. Abe-Ouchi, T. Nakajima, T. Takemura and A. Numaguti, 2001b: Future climate change: implications for Indian summer monsoon and its variability. *Curr. Sci. India*, **81**, 1196-1207.

Leont'yev, I.O., 2004: Coastal profile modelling along the Russian Arctic coast. *Coast. Eng*., **51**, 779–794.

Li, B.L. and C.H. Zhou, 2001: Climatic variation and desertification in West Sandy Land of Northeast China Plain. *Journal of Natural Resources*, **16**, 234-239.

Li, C.X, D.D. Fan, B. Deng and V. Korotaev, 2004a: The coasts of China and issues of sea level rise. *J. Coastal Res*., **43**, 36-47.

Li, C.X., S.Y. Yang, D.D. Fan and J. Zhao, 2004b: The change in Changjiang suspended load and its impact on the delta after completion of Three-Gorges Dam. *Quaternary Sciences*, **24**, 495-500 (in Chinese with an English abstract).

Li, J., J.Y. Zang, Y. Saito, X.W. Xu, Y.J. Wang, E. Matsumato and Z.Y. Zhang, 2005: Several cooling events over the Hong River Delta, Vietnam during the past 5000 years. *Advances in Marine Science*, **23**, 43-53 (in Chinese with an English abstract).

Li, Q.F., F.S. Li and L. Wu, 2002: A primary analysis on climatic change and grassland degradation in Inner Mongolia. *Agricultural Research in the Arid Areas*, **20**, 98-102.

Lin, E.D., Y.L. Xu, H. Ju and W. Xiong, 2004: Possible adaptation decisions from investigating the impacts of future climate change on food and water supply in China. Paper presented at the 2nd AIACC Regional Workshop for Asia and the Pacific, 2-5 November 2004, Manila, Philippines. http://www.aiaccproject.org /meetings/Manila_04/Day2/erda_nov3.doc.

Liu, C.M. and Z.K. Chen, 2001: *Assessment on Water Resources Status and Analysis on Supply and Demand Growth in China*. China Water Conservancy and Hydropower Press, Beijing, 168pp.

Liu, C.Z., 2002: Suggestion on water resources in China corresponding with global climate change. *China Water Resources*, **2**, 36-37.

Liu, S.G., C.X. Li, J. Ding, X.Z. Li and V.V. Ivanov, 2001: The rough balance of progradation and erosion of the Yellow River delta and its geological meaning. *Marine Geology & Quaternary Geology*, **21**, 13-17.

Liu, S.R., Q.S. Guo and B. Wang, 1998: Prediction of net primary productivity of forests in China in response to climate change. *Acta Ecologica Sinica*, **18**, 478-483.

Liu, Y.B. and Y.N. Chen, 2006: Impact of population growth and land-use change on water resources and ecosystems of the arid Tarim River Basin in western China. *Int. J. Sust. Dev. World*, **13**, 295-305.

Lobitz, B., L. Beck, A. Huq, B.L. Wood, G. Fuchs, A.S.G. Faruque and R. Colwell, 2000: Climate and infectious disease: use of remote sensing for detection of Vibrio cholerae by indirect measurement. *P. Natl. Acad. Sci. USA*, **97**, 1438-1443.

Loukos, H, P. Monfray, L. Bopp and P. Lehodey, 2003: Potential changes in skipjack tuna (*Katsuwonus pelamis*) habitat from a global warming scenario: modelling approach and preliminary results. *Fish. Oceanogr.*, **12**, 474-482.

Lu, J.J., 2003: *Estuarine Ecology*. China Ocean Press, Beijing, 318 pp (in Chinese).

Lu, X.R. and X.Y. Lu, 2003: Climate tendency analysis of warming and drying in grassland of Northeast Qingzang Plateau of China. *Grassland of China*, **24**, 8-13.

Ma, Z.G. and C.B. Fu, 2003: Interannual characteristics of the surface hydrological variables over the arid and semi-arid areas of northern China. *Global Planet. Change*, **37**, 189-200.

Macintosh, D., 2005: Asia, eastern, coastal ecology. *Encyclopedia of Coastal Science*, M. Schwartz, Ed., Springer, Dordrecht, 56-67.

Malcolm, J.R., C. Liu, R.P. Neilson, L. Hansen and L. Hannah, 2006: Global warming and extinctions of endemic species from biodiversity hotspots. *Conserv. Bio*., **20**, 538–548.

Malevski-Malevich S.P., E.K. Molketin, E.D. Nadezdina and O.B. Skliarevich, 2005: Assessment of forest fire scales in the Russia under the projected climate change by 2100. *Meteorology and Hydrology*, **3**, 36-44.

Manton, M.J., P.M. Della-Marta, M.R. Haylock, K.J. Hennessy, N. Nicholls, L.E. Chambers, D.A. Collins, G. Daw and Co-authors, 2001: Trends in extreme daily rainfall and temperature in Southeast Asia and the South Pacific; 1961–1998. *Int. J. Climatol*., **21**, 269–284.

Marchenko, S.S., 2002: Results of monitoring of the active layer in the northern Tien Shan mountains. *Journal Earth Cryosphere*, **4**, 25-34. (in Russian).

Matsen, H. and F. Jakobsen, 2004: Cyclone induced storm surge and flood forecasting in the northern Bay of Bengal. *Coast. Eng*., **51**, 277-296.

Matsui, T. and K. Omasa, 2002: Rice (*Oryza sativa L.*) cultivars tolerant to a high temperature at flowering: anther characteristics. *Ann. Bot.-London*, **89**, 683-687.

Matsui, T., T. Yagihashi, T. Nakaya, H. Taoda, S. Yoshinaga, H. Daimaru and N. Tanaka, 2004a: Probability distributions, *Fagus crenata* forests following vulnerability and predicted climate sensitivity in changes in Japan. *J. Veg. Sci*., **15**, 605-614.

Matsui, T., T. Yagihashi, T. Nakaya, N. Tanaka, and H. Taoda, 2004b: Climate controls on distribution of *Fagus crenata* forests in Japan. *J. Veg. Sci*., **15**, 57-66.

May, W., 2004a: Simulation of the variability and extremes of daily rainfall during the Indian summer monsoon for present and future times in a global time-slice experiment. *Clim. Dynam*., **22**, 183-204.

May, W., 2004b: Potential future changes in the Indian summer monsoon due to greenhouse warming: analysis of mechanisms in a global time-slice experiment. *Clim. Dynam*., **22**, 389-414.

Mazda Y., M. Magi, M. Kogo and P.N. Hong, 1997: Mangroves as a costal protection from waves in the Tong King Delta, Vietnam. *Mangroves and Salt Marshes*, **1**, 127-135.

Mazda Y., M. Magi, Y. Ikeda, T. Kurokawa and T. Asano, 2006: Wave reduction in a mangrove forest dominated by *Sonneratia sp. Wetlands Ecology and Management*, **14**, 365-378.

McMichael, A.J., D.H. Campbell-Lendrum, C.F. Corvalan, K.L. Ebi, A.K. Githeko, J.D. Scheraga and A. Woodward, Eds., 2003: *Climate Change and Human Health - Risks and Responses*, World Health Organization, Geneva, 333 pp.

McMichael, A.J., D. Campbell-Lendrum, S. Kovats, S. Edwards, P. Wilkinson, T. Wilson, R. Nicholls and Co-authors, 2004: Global climate change. *Comparative Quantification of Health Risks: Global and Regional Burden of Disease due to Selected Major Risk Factors*, M. Ezzati, A. Lopez, A. Rodgers and C. Murray, Eds., World Health Organization, Geneva, 1543-1649.

Meier, M., 1998: Land ice on Earth: a beginning of a global synthesis, unpublished transcript of the 1998 Walter B. Langbein Memorial Lecture, American Geophysical Union Spring Meeting - 26 May 1998, Boston, MA.

Melnikov, B.V. and A.L. Revson, 2003: Remote sensing of northern regions of West Siberia. *Cryosphere of Earth*, **4**, 37-48 (in Russian).

Meshcherskaya, A.V. and V.G. Blazhevich, 1990: Catalogues of temperature-humidity characteristics with the account being taken, distribution over economic regions of the principal grain-growing zone of the USSR (1891-1983), *Gidrometeoizdat*, **3**, Leningrad.

Milliman, J.D. and R.H. Meade, 1983: World-wide delivery of river sediment to the oceans. *J. Geol*., **90**, 1-21.

Mimura, N. and H. Yokoki, 2004: Sea level changes and vulnerability of the coastal region of East Asia in response to global warming. SCOPE/START Monsoon Asia Rapid Assessment Report.

Min, S.K., W.T. Kwon, E.H. Park and Y. Choi, 2003: Spatial and temporal comparisons of droughts over Korea with East Asia. *Int. J. Climatol*., **23**, 223-233.

Ministry of the Environment and Japanese Coral Reef Society, 2004: *Coral Reefs of Japan*. Ministry of the Environment, 356 pp.

Mirza, M.Q., 2002: Global warming and changes in the probability of occurrence of floods in Bangladesh and implications. *Global Environ. Chang*., **12**, 127-138.

Mirza, M.Q. and A. Dixit, 1997: Climate change and water management in the GBM Basins. *Water Nepal*, **5**, 71-100.

Mitra, A., S. Bhattacharya, R.C. Dhiman, K.K. Kumar and C. Sharma, 2004: Impact of climate change on health: a case study of malaria in India. *Climate Change and India: Vulnerability Assessment and Adaptation*, P.R. Shukla, S.K. Sharma, N.H. Ravindranath, A. Garg and S. Bhattacharya, Eds., Orient Longman Private, Hyderbad, 360-388.

MRC, 2003: *State of the Basin Report: 2003*. Mekong River Commission, Phnom Penh, 300 pp.

Murdiyarso, D., 2000: Adaptation to climatic variability and change: Asian perspectives on agriculture and food security. *Environ. Monit. Assess*., **61**, 123-131.

Murdiyarso, D. and E. Adiningsih, 2006: Climatic anomalies, Indonesian vegetation fires and terrestrial carbon emissions. *Mitigation and Adaptation Strategies for Global Change*, **12**, 101-112.

Murdiyarso, D., L. Lebel, A.N. Gintings, S.M.H. Tampubolon, A. Heil and M. Wasson, 2004: Policy responses to complex environmental problems: insights from a science-policy activity on transboundary haze from vegetation fires in

Southeast Asia. *Agr. Ecosyst. Environ.*, **104**, 47-56.

Nakagawa, H., T. Horie and T. Matsui, 2003: Effects of climate change on rice production and adaptive technologies. *Rice Science: Innovations and Impact for Livelihood*, T.W. Mew, D.S. Brar, S. Peng, D. Dawe and B. Hardy, Eds., International Rice Research Institute, Manila 635-658.

Nakićenović, N., and R. Swart, Eds., 2000: *Special Report on Emissions Scenarios. A Special Report of Working Group III of the Intergovernmental Panel on Climate Change*. Cambridge University Press, Cambridge, 599 pp.

Natsagdorj, L., P. Gomboluudev and P. Batima, 2005: Climate change in Mongolia. *Climate Change and its Projections*, P. Batima and B. Myagmarjav, Eds., Admon publishing, Ulaanbaatar, 39-84.

Nelson, F.E., 2003: Geocryology: enhanced: (un)frozen in time. *Science*, **299**, 1673-1675.

Nerem, R.S. and G.T. Mitchum, 2001: Observations of sea level change from satellite altimetry. *Sea Level Rise: History and Consequences*, B.C. Douglas, M.S. Kearney and S.P. Leatherman, Eds., Academic, San Diego, California, 121-163.

Nguyen V.L., T.K.O. Ta and M. Tateishib, 2000: Late Holocene depositional environments and coastal evolution of the Mekong River Delta, Southern Vietnam. *J. Asian Earth Sci.*, **18**, 427-439.

Nicholls, R.J., 1995: Coastal mega-cities and climate change. *GeoJournal*, **37**, 369-379.

Nicholls, R.J., 2004: Coastal flooding and wetland loss in the 21st century: changes under the SRES climate and socio-economic scenarios. *Global Environ. Chang.*, **14**, 69-86.

Nicholls, R.J. and R.S.J. Tol, 2006: Impacts and responses to sea-level rise: a global analysis of the SRES scenarios over the 21st Century. *Philos. T. R. Soc. A.*, **364**, 1073-1095.

Niou, Q.Y., 2002: *2001-2002 Report on Chinese Metropolitan Development*. Xiyuan Press, Beijing, 354 pp (in Chinese).

Nishioka, S. and H. Harasawa, Eds., 1998: *Global Warming - The Potential Impact on Japan*. Springer-Verlag, Tokyo, 244 pp.

Oda, K. and M. Ishii, 2001: Body color polymorphism in nymphs and ntaculatus (*Orthoptera*: *Tettigoniidae*) adults of a katydid, Conocephalus. *Appl. Entomol. Zool.*, **36**, 345-348.

Ojima, D.S., X. Xiangming, T. Chuluun and X.S. Zhang, 1998: Asian grassland biogeochemistry: factors affecting past and future dynamics of Asian grasslands. *Asian Change in the Context of Global Climate Change*, J.N. Galloway and J.M. Melillo, Eds., *IGBP Book Series*, **3**, Cambridge University Press, Cambridge, 128-144.

Oren, R., D.S. Ellsworth, K.H. Johnsen, N. Phillips, B.E. Ewers, C. Maier, K.V.R. Schafer, H. McCarthy, G. Hendrey, S.G. McNulty and G.G. Katul, 2001: Soil fertility limits carbon sequestration by forest ecosystems in a CO_2-enriched atmosphere. *Nature*, **411**, 469-472.

Osterkamp, T.E., L. Vierek, Y. Shur, M.T. Jorgenson, C. Racine, A. Doyle and R.D. Boone, 2000: Observations of thermokarst and its impact on boreal forests in Alaska, U.S.A. *Arct., Antarct. Alp. Res.*, **32**, 303-315.

Paerl, H.W., J. Dyble, P.H. Moisander, R.T. Noble, M.F. Piehler, J.L. Pinckney, T.F. Steppe, L. Twomey and L.M. Valdes, 2003: Microbial indicators of aquatic ecosystem change: current applications to eutrophication studies. *FEMS Microbiol. Ecol.*, **46**, 233-246.

PAGASA (Philippine Atmospheric, Geophysical and Astronomical Services Administration), 2001: Documentation and analysis of impacts of and responses to extreme climate events. Climatology and Agrometeorology Branch Technical Paper No. 2001-2, Philippine Atmospheric, Geophysical and Astronomical Services Administration, Quezon City, 55 pp.

Page, S.E., F. Siegert, J.O. Rieley, H.D.V. Boehm and A. Jaya, 2002: The amount of carbon released from peat and forest fires in Indonesia during 1997. *Nature*, **420**, 61-65.

Pan, X.L., W. Deng and D.Y. Zhang, 2003: Classification of hydrological landscapes of typical wetlands in Northeast China and their vulnerability to climate change. *Research of Environmental Sciences*, **16**, 14-18.

Parikh, J. and K. Bhattacharya, 2004: South Asian energy and emission perspectives for 21st century. Report of the IIASA-World Energy Council Project on Global Energy Perspectives for the 21st Century, International Institute of Applied Systems Analysis, 33 pp. http://www.iiasa.ac.at/Research/ECS/IEW2003/Abstracts/2003A_bhattacharya.pdf.

Parry, M.L., 2002: Scenarios for climate impacts and adaptation assessment. *Global Environ. Chang.*, **12**, 149-153.

Parry, M.L., C. Rosenzweig, A. Iglesias, G. Fischer and M. Livermore, 1999: Climate change and world food security: A new assessment. *Global Environ. Chang.*, **9**, 51-67.

Parry, M.L., C. Rosenzweig, A. Iglesias, M. Livermore and G. Fischer, 2004: Effects of climate change on global food production under SRES emissions and socio-economic scenarios. *Global Environ. Chang.*, **14**, 53-67.

Pascual, M., X. Rodo, S.P. Ellner, R. Colwell and M.J. Bouma, 2000: Cholera dynamics and El Niño Southern Oscillation. *Science*, **289**, 1766-1767.

Pascual, M., M.J. Bouma and A.P. Dobson, 2002: Cholera and climate: revisiting the quantitative evidence. *Microbes Infect.*, **4**, 237-245.

Patz, J.A., M.A. McGeehin, S.M. Bernard, K.L. Ebi, P.R. Epstein, A. Grambsch, D.J. Gubler, P. Reiter, I. Romieu, J.B. Rose, J.M. Samet and J. Trtanj, 2000: The potential health impacts of climate variability and change for the United States: executive summary of the report of the health sector of the U.S. National Assessment. *Environ. Health Persp.*, **108**, 367-376.

Patz, J.A., D. Cambell-Lendrum, T. Holloway and J.A. Foley, 2005: Impact of regional climate on human health. *Nature*, **438**, 310-317.

Pavlov, A.V. and G.V. Ananjeva-Malkova, 2005: Small-scale mapping of contemporary air and ground temperature changes in northern Russia. *Priorities in the Earth Cryosphere Research*. Pushchino, 62-73.

Pelling, M., 2003: *The Vulnerability of Cities: Natural Disasters and Social Resilience*. Earthscan, London, 212 pp.

Peng, S., J. Huang, J.E. Sheehy, R.E. Laza, R.M. Visperas, X. Zhong, G.S. Centeno, G.S. Khush and K.G. Cassman, 2004: Rice yields decline with higher night temperature from global warming. *P. Natl. Acad. Sci. USA*, **101**, 9971-9975.

Penland, S. and M.A. Kulp, 2005: Deltas. *Encyclopedia of Coastal Science*, M.L. Schwartz, Ed., Springer, Dordrecht, 362-368.

Peterson, B.J., R.M. Holmes, J.W. McClelland, C.J. Vorosmarty, R.B. Lammers, A.I. Shiklomanov, I.A. Shiklomanov and S. Rahmstorf, 2002: Increasing river discharge to the Arctic Ocean. *Science*, **298**, 137-143.

Preston, B.L., R. Suppiah, I. Macadam and J. Bathols, 2006: Climate change in the Asia/Pacific region. Report prepared for the Climate Change and Development Roundtable, Commonwealth Scientific and Industrial Research Organisation, CSIRO Publishing, Collingwood, Victoria, 89 pp.

Pu, J.C., T.D. Yao, N.L. Wang, Z. Su and Y.P. Shen, 2004: Fluctuations of the glaciers on the Qinghai-Tibetan Plateau during the past century. *Journal of Glaciology and Geocryology*, **26**, 517-522.

Pulhin, J.M., U. Chokkalingam, R.T. Acosta, A.P. Carandang, M.Q. Natividad and R.A. Razal, 2006: Historical overview of forest rehabilitation in the Philippines. *One Century of Forest Rehabilitation in the Philippines: Approaches, Outcomes and Lessons*, U. Chokkalingam, A.P. Carandang, J.M. Pulhin, R.D. Lasco and R.J.J. Peras, Eds., CIFOR-CFNR UPLB-FMB DENR, Bogor, 6-41.

Qin, D.H., 2002: Assessment of environment change in Western China, 2nd Volume. *Prediction of Environment Change in Western China*, Science Press, Beijing, 64-161.

Qiu, G.W., Y.X. Hao and S.L. Wang, 2001: The impacts of climate change on the interlock area of farming – pastoral region and its climatic potential productivity in Northern China. *Arid Zone Research*, **18**, 23-28.

Rachmanin, Y.A. and Co-authors, 2004: Criteria for assessment of influence of weather and air pollution on public health. *Proc. Intl. Conf. Climate, Change and Public Health in Russia in the 21st Century*, N.F. Izmerov, B.A. Revich and E.I. Korenberg, Eds., Russian Academy of Medical Sciences - Russian Regional Committee for Cooperation with UNEP Centre for Demography and Human Ecology Russian Regional Environmental Centre Non-Governmental Organisation 'Environmental Defence', Moscow, 171-175.

Ragab, R. and C. Prudhomme, 2002: Climate change and water resources management in arid and semi-arid regions: prospective and challenges for the 21st century. *Biosystems Engineering*, **81**, 3-34.

Rahimzadeh, F., 2006: Study of precipitation variability in Iran, Research Climatology Institute, IRIMO, Tehran.

Reich, P.B., S.E. Hobbie, T. Lee, D.S. Ellsworth, J.B. West, D. Tilman, J.M.H. Knops, S. Naeem and J. Trost, 2006: Nitrogen limitation constrains sustainability of ecosystem response to CO_2. *Nature*, **440**, 922-925.

Rignot, E., A. Rivera and G. Casassa, 2003: Contribution of the Patagonia icefields of South America to sea level rise. *Science*, **302**, 434-437.

Rodo, X., M. Pascual, G. Fuchs and A.S.G. Faruque, 2002: ENSO and cholera: a nonstationary link related to climate change? *P. Natl. Acad. Sci. USA*, **99**, 12901-12906.

Romanovsky, V., M. Burgess, S. Smith, K. Yoshikawa and J. Brown, 2002: Permafrost temperature records: indicators of climate change. *EOS Transactions*, **83**,

589-594. doi:10.1029/2002EO000402.

Romanovsky, V.E., N.I. Shender, T.S. Sazonova, V.T. Balobaev, G.S. Tipenko and V.G. Rusakov, 2001: Permafrost temperatures in Alaska and East Siberia: past, present and future. *Proc. 2nd Russian Conference on Geocryology (Permafrost Science)*, Moscow, 301-314.

Rosenzweig, C., A. Iglesias, X.B. Yang, P.R. Epstein and E. Chivian, 2001: Climate change and extreme weather events: implications for food production, plant diseases and pests. *Global Change and Human Health*, **2**, 90-104.

Ruosteenoja, K., T.R. Carter, K. Jylhä and H. Tuomenvirta, 2003: Future climate in world regions: an intercomparison of model-based projections for the new IPCC emissions scenarios. The Finnish Environment 644, Finnish Environment Institute, Helsinki, 83 pp.

Rupa Kumar, K., K. Kumar, V. Prasanna, K. Kamala, N.R. Desphnade, S.K. Patwardhan and G.B. Pant, 2003: Future climate scenario. *Climate Change and India: Vulnerability Assessment and Adaptation*, P.R. Shukla, S.K. Sharma, N.H. Ravindranath, A. Garg and S. Bhattacharya, Eds., Orient Longman Private, Hyderbad, 69-127.

Ryoo, S.B., W.T. Kwon and J.G. Jhun, 2004: Characteristics of wintertime daily and extreme temperature over South Korea. *Int. J. Climatol.*, **24**, 145-160.

Saito, Y., 2005: Mega-deltas in Asia: characteristics and human influences. *Mega-Deltas of Asia: Geological Evolution and Human Impact*, Z.Y. Chen, Y. Saito, S.L. Goodbred, Jr., Eds., China Ocean Press, Beijing, 1-8.

Sanlaville P. and A. Prieur, 2005: Asia, Middle East, coastal ecology and geomorphology. *Encyclopedia of Coastal Science*, M.L. Schwartz, Ed., Springer, Dordrecht, 71-83.

Sato, Y., 2000: Climate change prediction in Japan. *Tenki*, **47**, 708-716. (in Japanese)

Savelieva, I.P., Semiletov, L.N. Vasilevskaya and S.P. Pugach, 2000: A climate shift in seasonal values of meteorological and hydrological parameters for Northeastern Asia. *Prog. Oceanogr.*, **47**, 279–297.

Sharkhuu, N., 2003: Recent changes in the permafrost of Mongolia. *Proc. 7th International Conference on Permafrost*, Québec, 1029-1034.

She, Z.X., 2004: Human-land interaction and socio-economic development, with special reference to the Changjiang Delta. *Proc. Xiangshan Symposium on Human-land Coupling System of River Delta Regions: Past, Present and Future*, Beijing (in Chinese).

Shein, K.A., Ed., 2006: State of the climate in 2005. *B. Am. Meteorol. Soc.*, **87**, S1-S102.

Shen, X.T., Z.C. Mao and J.R. Zhu, 2003: Saltwater intrusion in the Changjiang Estuary. China Ocean Press, Beijing, 175 pp (in Chinese).

Shen, Y.P., G.X. Wang, Q.B. Wu and S.Y. Liu, 2002: The impact of future climate change on ecology and environments in the Changjiang - Yellow Rivers source region. *Journal of Glaciology and Geocryology*, **24**, 308-313.

Sheppard, C.R.C., 2003: Predicted recurrences of mass coral mortality in the Indian Ocean. *Nature*, **425**, 294 -297.

Shi, L.Q., J.F. Li, M. Ying, W.H. Li, S.L. Chen and G.A. Zhang, 2005: Advances in researches on the modern Huanghe Delta development and evolution. *Advances in Marine Science*, **23**, 96-104.

Shi, Y.F., Y.P. Shen and R.J. Hu, 2002: Preliminary study on signal, impact and foreground of climatic shift from warm-dry to warm-humid in Northwest China. *Journal of Glaciology and Geocryology*, **24**, 219-226.

Shimoda, Y., 2003: Adaptation measures for climate change and the urban heat island in Japan's built environment. *Build. Res. Inf.*, **31**, 222-230.

Shoigu, S., 2004: Global climate changes and emergencies in Russia. *Proc. World Climate Change Conference,* Moscow, Y. Izrael, G. Gruza, S. Semenov and I. Nazarov, Eds., Institute of Global Climate and Ecology, Moscow, 73-85.

Short, F.T. and H.A. Neckles, 1999: The effects of global climate change on seagrasses. *Aquat. Bot.*, **63**, 169-196.

Shrestha, A.B., 2004: Climate change in Nepal and its impact on Himalayan glaciers. *Presented European Climate Forum Symposium on "Key vulnerable regions and climate change: Identifying thresholds for impacts and adaptation in relation to Article 2 of the UNFCCC"*, Beijing.

Shrestha, A.B., C.P. Wake, J.E. Dibb and P.A. Mayewski, 2000: Precipitation fluctuations in the Nepal Himalaya and its vicinity and relationship with some large scale climatological parameters. *Int. J. Climatol.*, **20**, 317-327.

Siddiqui, K.M., I. Mohammad and M. Ayaz, 1999: Forest ecosystem climate change impact assessment and adaptation strategies for Pakistan. *Climate Res.*, **12**, 195–203.

Simas, T., J.P. Nunes and J.G. Ferreira, 2001: Effects of global climate change on coastal salt marshes. *Ecol. Model.*, **139**, 1–15.

Sin, S., 2000: Late Quaternary geology of the Lower Central Plain, Thailand. *J. Asian Earth Sci.*, **18**, 415-426.

Singh, N. and N.A. Sontakke, 2002: On climatic fluctuations and environmental changes of the Indo-Gangetic plains, India. *Climatic Change*, **52**, 287-313.

Singh, N., A.C. Nagpal, A. Saxena and A.P. Singh, 2004: Changing scenario of malaria in central India, the replacement of Plasmodium vivax by Plasmodium falciparum (1986-2000). *Trop. Med. Int. Health*, **9**, 364-371.

Slingo, J.M., A.J. Challinor, B.J. Hoskins and T.R. Wheeler, 2005: Food crops in a changing climate. *Phil. T. Roy. Soc. B.*, **360**, 1983-1989.

Sodhi, N.S., L.P. Koh, B.W. Brook and P.K.L. Ng, 2004: Southeast Asian biodiversity: an impending disaster. *Trends Ecol. Evol.*, **19**, 654-660.

Spalding, M., C. Ravilious and E.P. Green, 2001: *World Atlas of Coral Reefs*. University of California Press, Los Angeles, 424 pp.

Srinivasan, T.N., 2000: Poverty and undernutrition in South Asia. *Food Policy*, **25**, 269-282.

Stern, N., 2007: *Stern Review on the Economics of Climate Change*. Cambridge University Press. Cambridge, 692 pp.

Su, J.L. and Q.S. Tang, 2002: *Study on Marine Ecosystem Dynamics in Coastal Ocean. II Processes of the Bohai Sea Ecosystem Dynamics*. Science Press, Beijing, 445 pp.

Sukumar, R., K.G. Saxena and A. Untawale, 2003: Climate change impacts on natural ecosystem. *Climate Change and India: Vulnerability Assessment and Adaptation*, P.R. Shukla, S.K. Sharma, N.H. Ravindranath, A. Garg and S. Bhattacharya, Eds., Orient Longman Private, Hyderbad, 266-290.

Takahashi, K., Y. Honda and S. Emori, 2007: Estimation of changes in mortality due to heat stress under changed climate. *Risk Res.*, **10**, 339-354.

Takahashi, Y. and S. Kawashima, 1999: A new prediction method for the total pollen counts of *Cryprtomeria japonica* based on variation in annual summertime temperature. *Allergy*, **48**, 1217-1221.

Tan, G. and R. Shibasaki, 2003: Global estimation of crop productivity and the impacts of global warming by GIS and EPIC integration. *Ecol. Model.*, **168**, 357-370.

Tan, J., L.S. Kalkstein, J. Huang, S. Lin, H. Yin and D. Shao, 2004: An operational heat/health warning system in Shanghai. *Int. J. Biometeorol.*, **48**, 157-62.

Tao, F., M. Yokozawa, Y. Hayashi and E. Lin, 2003a: Changes in agricultural water demands and soil moisture in China over the last half-century and their effects on agricultural production. *Agr. Forest Meteorol.*, **118**, 251–261.

Tao, F., M. Yokozawa, Y. Hayashi and E. Lin, 2003b: Future climate change, the agricultural water cycle, and agricultural production in China. *Agriculture, Ecosystems and Environment*, **95**, 203-215.

Tao, F., M. Yokozawa, Z. Zhang, Y. Hayashi, H. Grassl and C. Fu, 2004: Variability in climatology and agricultural production in China in association with the East Asia summer monsoon and El Niño South Oscillation. *Climate Res.*, **28**, 23-30.

Tao, F., M. Yokozawa, Y. Hayashi and E. Lin, 2005: A perspective on water resources in China: interactions between climate change and soil degradation. *Climatic Change*, **68**, 169-197.

Teranishi, L., Y. Kenda, T. Katoh, M. Kasuya, E. Oura and H. Taira, 2000: Possible role of climate change in the pollen scatter of Japanese cedar *Cryptomeria japonica* in Japan. *Climate Res.*, **14**, 65-70.

TERI, 1996: Tata Energy Research Institute, New Delhi, Report No 93/GW/52, submitted to the Ford Foundation.

Thanh, T.D., Y. Saito, D.V. Huy, V.L. Nguyen, T.K.O. Ta and M. Tateish, 2004: Regimes of human and climate impacts on coastal changes in Vietnam. *Reg. Environ. Change*, **4**, 49-62.

Tong, S.L. and L.V. Ying, 2000: Global Climate Change and Epidemic Disease. *Journal of Disease Control*, **4**, 17-19.

Tran, V.L., D.C. Hoang and T.T. Tran, 2005: Building of climate change scenario for Red River catchments for sustainable development and environmental protection. Preprints, *Science Workshop on Hydrometeorological Change in Vietnam and Sustainable Development*, Hanoi, Vietnam, Ministry of Natural Resource and Environment, Hanoi, 70-82.

Trenberth, K.E. and T.J. Hoar, 1997: El Niño and climate change. *Geophys. Res. Lett.*, **24**, 3057-3060.

Troadec, J.P., 2000: Adaptation opportunities to climate variability and change in the exploitation and utilisation of marine living resources. *Environ. Monit. Assess.*, **61**, 101-112.

Tserendash, S., B. Bolortsetseg, P. Batima, G. Sanjid, M. Erdenetuya, T. Ganbaatar and N. Manibazar, 2005: Climate change impacts on pasture. *Climate Change*

Impacts, P. Batima and B. Bayasgalan, Eds., Admon publishing, Ulaanbaatar, 59-115.

UNEP (United Nations Environment Programme), 2002: *Global Environment Outlook 3*, Earthscan, London, 426 pp.

UN-DESA-PD, 2002: World Population Prospects: World Population Prospects-2002, United Nations Department of Economic and Social Affairs - Population Division, UN, New York.

UN-ESCAP (United Nations Economic and Social Commission for Asia and the Pacific), 2006: Achieving MDGs in Asia: a case for more aid? UN-ESCAP, New York, 37 pp.

UN-HABITAT (United Nations Human Settlements Programme), 2004: *The State of the World's Cities 2004/2005 - Globalization and Urban Culture*. Earthscan, London, 198 pp.

Vermaat, J. and U. Thampanya, 2006: Mangroves mitigate tsunami damage: a further response. *Estuar. Coast. Shelf S.*, **69**, 1-3.

Vlek, P.L.G., K.G. Rodriguez and R. Sommer, 2004: Energy use and CO_2 production in tropical agriculture and means and strategies for reduction or mitigation. *Environment Development and Sustainability*, **6**, 213-233.

Vorobyov, Y., 2004: Climate change and disasters in Russia. *Proc. World Climate Change Conference, Moscow*, Y. Izrael, G. Gruza, S. Semenov and I. Nazarov, Eds., Institute of Global Climate and Ecology, Moscow, 293-298.

Wada, N., K. Watanuki, K. Narita, S. Suzuki, G. Kudo and A. Kume, 2004: Climate change and shoot elongation of Alpine Dwarf Pine (*Pinus Pumila*): comparisons among six Japanese mountains. *6th International Symposium on Plant Responses to Air Pollution and Global Changes*, Tsukuba, 215.

Walsh, K.J.E., 2004: Tropical cyclones and climate change: unresolved issues. *Climate Res.*, **27**, 77-84.

Wang, F.T., 2002: Advances in climate warming impacts research in China in recent ten years. *Journal of Applied Meteorological Science*, **13**, 766.

Wang, G.Z., 2005: Global sea-level change and coral reefs of China. *Journal of Palaeogeography*, **7**, 483-492 (in Chinese with an English abstract).

Wang, M., Y. Li and S. Pang, 2004a: Influences of climate change on sustainable development of the hinterland of Qinghai-Tibet Plateau, Chinese Journal of *Population, Resources and Environment*, **14**, 92-95.

Wang, T., 2003: Study on desertification in China, 2 - Contents of desertification research. *Journal of Desert Research*, **23**, 477-482.

Wang, X., F. Chen and Z. Dong, 2006: The relative role of climatic and human factors in desertification in semiarid China. *Global Environ. Chang.*, **16**, 48-57.

Wang, X.C. and P.K. Jin, 2006: Water shortage and needs for wastewater re-use in the north China. *Water Sci. Technol.*, **53**, 35-44.

Wang, Y.B., G.X. Wang and J. Chang, 2004b: Impacts of human activity on permafrost environment of the Tibetan Plateau. *Journal of Glaciology and Geocryology*, **26**, 523-527.

Wang, Z.W. and P.M. Zhai, 2003: Climate change in drought over northern China during 1950–2000. *Acta Geographica Sinica*, **58**, 61–68 (in Chinese).

Wassmann, R., N.X. Hien, C.T. Hoanh and T.P. Tuong, 2004: Sea level rise affecting the Vietnamese Mekong Delta: water elevation in the flood season and implications for rice production. *Climatic Change*, **66**, 89-107.

Webster, P.J., V.O. Magana, T.N. Palmer, J. Shukla, R.A. Tomas, M. Yanagi and T. Yasunari, 1998: Monsoons: processes, predictability and the prospects for prediction. *J. Geophys. Res.*, **103**, 14451-14510.

Wijeratne, M.A., 1996: Vulnerability of Sri Lanka tea production to global climate change. *Water Air Soil Poll.*, **92**, 87-94.

Wilkinson, C., Ed., 2000: *Status of Coral Reefs of the World: 2000*. Australian Institute of Marine Science, Townsville, 363 pp.

Wilkinson, C., Ed., 2002: *Status of Coral Reefs of the World: 2002*. Australian Institute of Marine Science, Townsville, 378 pp.

Wilkinson, C., Ed., 2004: *Status of Coral Reefs of the World: 2004, Volume 1*. Australian Institute of Marine Science, Townsville, 302 pp.

Wilkinson, P., D.H. Campbell-Lendrum and C.L. Bartlett, 2003: Monitoring the health effects of climate change. *Climate Change and Human Health - Risks and Responses*, A.J. McMichael, D.H. Campbell-Lendrum, C.F. Corvalan, K.L. Ebi, A.K. Githeko, J.D. Scheraga, A. Woodward, Eds., World Health Organization, Geneva, 204-219.

Wolanski, E., 2007: Protective functions of coastal forests and trees against natural hazards. Coastal protection in the aftermath of the Indian Ocean tsunami: what role for forests and trees? *Proc. FAO Regional Technical Workshop, Khao Lak, Thailand*, 28-31 August 2006. FAO, Bangkok.

Woodroffe, C.D., R.J. Nicholls, Y. Saito, Z. Chen and S.L. Goodbred, 2006: Landscape variability and the response of Asian megadeltas to environmental change. *Global Change and Integrated Coastal Management: The Asia-Pacific Region*, N. Harvey, Ed., Springer, 277-314.

Woodworth, P.L., J.M. Gregory and R.J. Nicholls, 2004: Long term sea level changes and their impacts. *The Sea*, A. Robinson and K. Brink, Eds., Harvard Univ. Press, Cambridge, Massachusetts, 717-752.

World Bank, 2002: World Development Indicators 2002, CD-ROM, Washington, District of Columbia.

World Bank, 2004: Modelled Observations on Development Scenarios in the Lower Mekong Basin, Prepared for the World Bank with cooperation of the Mekong Regional Water Resources Assistance Strategy, World Bank, Vientiane, 142 pp. http://www.catchment.crc.org.au/cgi-bin/WebObjects/toolkit.woa/1/wa/publicationDetail?wosid=5Ir4xCtBn8NbvG2w3KZWrw&publicationID=1000068.

World Bank, 2005: World Development Indicators, World Bank, Washington, District of Columbia. http://web.worldbank.org/WBSITE/EXTERNAL/DATASTATISTICS/0,,contentMDK:20523710~hlPK:1365919~menuPK:64133159~pagePK:64133150~piPK:64133175~theSitePK:239419,00.html

World Tourism Organization, 2003: Climate change and tourism. *Proc. 1st International Conference on Climate Change and Tourism*, Djerba, Tunisia, World Tourism Organization, 55 pp.

WRI (World Resource Institute), 2003: *World Resources 2002-2004*. World Resources Institute, Washington, District of Columbia, 315 pp.

Wu, Q.B., X. Li and W.J. Li, 2001: The response model of permafrost along the Qinghai – Tibetan Highway under climate change. *Journal of Glaciology and Cryology*, **23**, 1-6.

Wu, S.Y., B. Yarnal and A. Fisher, 2002: Vulnerability of coastal communities to sea-level rise: a case study of Cape May County, New Jersey, USA. *Climate Res.*, **224**, 255-270.

Wu, Z.F., 2003: Assessment of ecoclimatic suitability and climate change impacts on broad leaved Korean pine forest in Northeast China. *Chinese Journal of Applied Ecology*, **14**, 771-775.

WWF (World Wildlife Fund), 2005: An overview of glaciers, glacier retreat, and subsequent impacts in Nepal, India and China. World Wildlife Fund, Nepal Programme, 79 pp.

Xu, C.X., 2003: *China National Offshore and Coastal Wetlands Conservation Action Plan*. China Ocean Press, Beijing, 116 pp (in Chinese).

Xue, X.Z., H.S. Hong and A.T. Charles, 2004: Cumulative environmental impacts and integrated coastal management: the case of Xiamen, China. *J. Environ. Manage.*, **71**, 271-283.

Yamano, H. and M. Tamura, 2004: Detection limits of coral reef bleaching by satellite remote sensing: simulation and data analysis. *Remote Sens. Environ.*, **90**, 86-103.

Yang, G.S., 2000: Historic change and future trends of the storm surge disaster in China's coastal area. *Journal of Natural Disasters*, **9**, 23-30 (in Chinese with an English abstract).

Yao, T.D., X.D. Liu, N.L. Wang and Y.F. Shi, 2000: Amplitude of climatic changes in Qinghai-Tibetan Plateau. *Chinese Sci. Bull.*, **45**, 98-106.

Yasukevich, V.V. and S.M. Semenov, 2004: Simulation of climate-induced changes in natural habitats of malaria in Russia and neighboring countries, *Proc. Intl. Conf. on Climate Change and Public Health in Russia*, N.F. Izmerov, B.A. Revich and E.I. Korenberg, Eds., Russian Academy of Sciences, Moscow, 147-153.

Yin, Y.Y., Q.L. Miao and G.S. Tian, 2003: *Climate Change and Regional Sustainable Development*. Science Press, Beijing and New York, 224 pp.

Yoshino, M., 1996: Change of global environment and insurance. *Saigainokennkyuu*, **26**, 84-100 (in Japanese).

Yoshino, M., 2000: Problems in climatology of dust storm and its relation to human activities in Northwest China. *Journal of Arid Land Studies*, **10**, 171-181.

Yoshino, M., 2002: Kosa (Asian dust) related to Asian Monsoon System. *Korean Journal of Atmospheric Sciences*, **5**, S93-S100.

Yoshio, M. and M. Ishii, 2001: Relationship between cold hardiness and northward invasion in the great mormon butterfly, *Papilio memnon L.* (*Lepidoptera: Papilionidae*) in Japan. *Appl. Entomol. Zool.*, **36**, 329-335.

Zafar, A., 2005: Training and capacity building for managing our mangroves resources- UNU's role to meet regional challenges. Environment and Sustainable Development Programme, Tokyo, Japan, Technical Notes of United Nations University, 5 pp. http://www.inweh.unu.edu/inweh/Training/UNU_Man

groves_course.pdf.
Zalikhanov, M., 2004: Climate change and sustainable development in the Russian Federation. *Proc. World Climate Change Conference, Moscow, September 29-October 2003*, Yu. Izrael, G. Gruza, S. Semenov and I. Nazarov, Eds., Institute of Global Climate and Ecology, Moscow, 466-477.
Zeqiang, F., C. Yunlong, Y. Youxiao and D. Erfu, 2001: Research on the relationship of cultivated land change and food security in China. *Journal of Natural Resources*, **16**, 313-319.
Zhai, P. and X. Pan, 2003: Trends in temperature extremes during 1951-1999 in China. *Geophys. Res. Lett.*, **30**, 1913, doi:10.1029/2003GL018004.
Zhai, P.M., 2004: Climate change and meteorological disasters. *Science and Technology Reviews*, **7**, 11-14.
Zhai, P.M., A. Sun, F. Ren, X. Liu, B. Gao and Q. Zhang, 1999: Changes of climate extremes in China. *Climatic Change*, **42**, 203-218.
Zhang, J.W. and B.L. Du, 2000: The trend of tidal range enlarging along the coast of the Yellow Sea. *Marine Science Bulletin*, **19**, 1-9.
Zhang, Q. and G. Guo, 2004: The spatial and temporal features of drought and flood disasters in the past 50 years and monitoring and warning services in China. *Science and Technology Reviews*, **7**, 21-24.
Zhang, Y., W. Chen and J. Cihlar, 2003: A process-based model for quantifying the impact of climate change on permafrost thermal regimes. *J. Geophys. Res.*, **108**, 4695 doi:10.1029/2002JD003354.
Zhao, L., C.L. Ping, D. Yang, G. Cheng, Y. Ding and S. Liu, 2004: Changes of climate and seasonally frozen ground over the past 30 years in Qinghai–Xizang (Tibetan) Plateau, China. *Global Planet. Change*, **43**, 19-31.
Zhou, Y.H., 2003: Characteristics of weather and climate during drought periods in South China. *Journal of Applied Meteorological Science*, **14**, S118-S125.
Ziska, L.H., 2003: Evaluation of the growth response of six invasive species to past, present and future atmospheric carbon dioxide. *J. Exp. Bot.*, **54**, 395-404.
Zolotov, P.A. and Caliberny, 2004: Human physiological functions and public health Ultra-Continental Climate. *Proc. Intl. Conf. on Climate Change and Public Health in Russia*, N.F. Izmerov, B.A. Revich and E.I. Korenberg, Eds., Russian Academy of Sciences, Moscow, 212-222.

11

Australia and New Zealand

Coordinating Lead Authors:
Kevin Hennessy (Australia), Blair Fitzharris (New Zealand)

Lead Authors:
Bryson C. Bates (Australia), Nick Harvey (Australia), Mark Howden (Australia), Lesley Hughes (Australia), Jim Salinger (New Zealand), Richard Warrick (New Zealand)

Contributing Authors:
Susanne Becken (New Zealand), Lynda Chambers (Australia), Tony Coleman (Australia), Matt Dunn (New Zealand), Donna Green (Australia), Roddy Henderson (New Zealand), Alistair Hobday (Australia), Ove Hoegh-Guldberg (Australia), Gavin Kenny (New Zealand), Darren King (New Zealand), Guy Penny (New Zealand), Rosalie Woodruff (Australia)

Review Editors:
Michael Coughlan (Australia), Henrik Moller (New Zealand)

This chapter should be cited as:
Hennessy, K., B. Fitzharris, B.C. Bates, N. Harvey, S.M. Howden, L. Hughes, J. Salinger and R. Warrick, 2007: Australia and New Zealand. *Climate Change 2007: Impacts, Adaptation and Vulnerability. Contribution of Working Group II to the Fourth Assessment Report of the Intergovernmental Panel on Climate Change,* M.L. Parry, O.F. Canziani, J.P. Palutikof, P.J. van der Linden and C.E. Hanson, Eds., Cambridge University Press, Cambridge, UK, 507-540.

Table of Contents

Executive summary

Literature published since the IPCC Third Assessment Report confirms and extends its main findings (high confidence).
There is more extensive documentation of observed changes to natural systems, major advances in understanding potential future climate changes and impacts, more attention to the role of planned adaptation in reducing vulnerability, and assessments of key risks and benefits [11.1].

Regional climate change has occurred (very high confidence).
Since 1950, there has been 0.4 to 0.7°C warming, with more heatwaves, fewer frosts, more rain in north-west Australia and south-west New Zealand, less rain in southern and eastern Australia and north-eastern New Zealand, an increase in the intensity of Australian droughts, and a rise in sea level of about 70 mm [11.2.1].

Australia and New Zealand are already experiencing impacts from recent climate change (high confidence).
These are now evident in increasing stresses on water supply and agriculture, changed natural ecosystems, reduced seasonal snow cover, and glacier shrinkage [11.2.1, 11.2.3].

Some adaptation has already occurred in response to observed climate change (high confidence).
Examples come from sectors such as water, natural ecosystems, agriculture, horticulture and coasts [11.2.5]. However, ongoing vulnerability to extreme events is demonstrated by substantial economic losses caused by droughts, floods, fire, tropical cyclones and hail [11.2.2].

The climate of the 21st century is virtually certain to be warmer, with changes in extreme events.
Heatwaves and fires are virtually certain to increase in intensity and frequency (high confidence). Floods, landslides, droughts and storm surges are very likely to become more frequent and intense, and snow and frost are very likely to become less frequent (high confidence). Large areas of mainland Australia and eastern New Zealand are likely to have less soil moisture, although western New Zealand is likely to receive more rain (medium confidence) [11.3.1].

Potential impacts of climate change are likely to be substantial without further adaptation.

- As a result of reduced precipitation and increased evaporation, water security problems are projected to intensify by 2030 in southern and eastern Australia and, in New Zealand, in Northland and some eastern regions (high confidence) [11.4.1].
- Ongoing coastal development and population growth, in areas such as Cairns and south-east Queensland (Australia) and Northland to Bay of Plenty (New Zealand), are projected to exacerbate risks from sea-level rise and increases in the severity and frequency of storms and coastal flooding by 2050 (high confidence) [11.4.5, 11.4.7].
- Significant loss of biodiversity is projected to occur by 2020 in some ecologically rich sites, including the Great Barrier Reef and Queensland Wet Tropics. Other sites at risk include Kakadu wetlands, south-west Australia, sub-Antarctic islands and alpine areas of both countries (very high confidence) [11.4.2].
- Risks to major infrastructure are likely to increase. By 2030, design criteria for extreme events are very likely to be exceeded more frequently. Risks include failure of floodplain protection and urban drainage/sewerage, increased storm and fire damage, and more heatwaves, causing more deaths and more blackouts (high confidence) [11.4.1, 11.4.5, 11.4.7, 11.4.10, 11.4.11].
- Production from agriculture and forestry is projected to decline by 2030 over much of southern and eastern Australia, and over parts of eastern New Zealand, due to increased drought and fire. However, in New Zealand, initial benefits to agriculture and forestry are projected in western and southern areas and close to major rivers due to a longer growing season, less frost and increased rainfall (high confidence) [11.4.3, 11.4.4].

Vulnerability is likely to increase in many sectors, but this depends on adaptive capacity.

- *Most human systems have considerable adaptive capacity:* The region has well-developed economies, extensive scientific and technical capabilities, disaster mitigation strategies, and biosecurity measures. However, there are likely to be considerable cost and institutional constraints to the implementation of adaptation options (high confidence) [11.5]. Some Indigenous communities have low adaptive capacity (medium confidence) [11.4.8]. Water security and coastal communities are the most vulnerable sectors (high confidence) [11.7].
- *Natural systems have limited adaptive capacity:* Projected rates of climate change are very likely to exceed rates of evolutionary adaptation in many species (high confidence) [11.5]. Habitat loss and fragmentation are very likely to limit species migration in response to shifting climatic zones (high confidence) [11.2.5, 11.5].
- *Vulnerability is likely to rise due to an increase in extreme events:* Economic damage from extreme weather is very likely to increase and provide major challenges for adaptation (high confidence) [11.5].
- *Vulnerability is likely to be high by 2050 in a few identified hotspots:* In Australia, these include the Great Barrier Reef, eastern Queensland, the South-West, Murray-Darling Basin, the Alps and Kakadu wetlands; in New Zealand, these include the Bay of Plenty, Northland, eastern regions and the Southern Alps (medium confidence) [11.7].

11.1 Introduction

The region is defined here as the lands and territories of Australia and New Zealand. It includes their outlying tropical, mid-latitude and sub-Antarctic islands and the waters of their Exclusive Economic Zones. New Zealand's population was 4.1 million in 2006, growing by 1.6%/yr (Statistics New Zealand, 2006). Australia's population was 20.1 million in 2004, growing by 0.9%/yr (ABS, 2005a). Many of the social, cultural and economic aspects of the two countries are comparable. Both countries are relatively wealthy and have export-based economies largely dependent on natural resources, agriculture, manufacturing, mining and tourism. Many of these are climatically sensitive.

11.1.1 Summary of knowledge from the Third Assessment Report (TAR)

In the IPCC Third Assessment Report (TAR; Pittock and Wratt, 2001), the following impacts were assessed as important for Australia and New Zealand.

- Water resources are likely to become increasingly stressed in some areas of both countries, with rising competition for water supply.
- Warming is likely to threaten the survival of species in some natural ecosystems, notably in alpine regions, south-western Australia, coral reefs and freshwater wetlands.
- Regional reductions in rainfall in south-west and inland Australia and eastern New Zealand are likely to make agricultural activities particularly vulnerable.
- Increasing coastal vulnerability to tropical cyclones, storm surges and sea-level rise.
- Increased frequency of high-intensity rainfall, which is likely to increase flood damage.
- The spread of some disease vectors is very likely, thereby increasing the potential for disease outbreaks, despite existing biosecurity and health services.

The overall conclusions of the TAR were that: (i) climate change is likely to add to existing stresses to the conservation of terrestrial and aquatic biodiversity and to achieving sustainable land use, and (ii) Australia has significant vulnerability to climate change expected over the next 100 years, whereas New Zealand appears more resilient, except in a few eastern areas.

11.1.2 New findings of this Fourth Assessment Report (AR4)

The scientific literature published since 2001 supports the TAR findings. Key differences from the TAR include (i) more extensive documentation of observed changes in natural systems consistent with global warming, (ii) significant advances in understanding potential future impacts on water, natural ecosystems, agriculture, coasts, Indigenous people and health, (iii) more attention to the role of adaptation, and (iv) identification of the most vulnerable sectors and hotspots. Vulnerability is given more attention – it is dependent on the exposure to climate change, the sensitivity of sectors to this exposure, and their capacity to adapt.

11.2 Current sensitivity/vulnerability

11.2.1 Climate variability and 20th-century trends

In this section, climate change is taken to be due to both natural variability and human activities. The relative proportions are unknown unless otherwise stated. The strongest regional driver of climate variability is the El Niño-Southern Oscillation (ENSO). In New Zealand, El Niño brings stronger and cooler south-westerly airflow, with drier conditions in the north-east of the country and wetter conditions in the south-west (Gordon, 1986; Mullan, 1995). The converse occurs during La Niña. In Australia, El Niño tends to bring warmer and drier conditions to eastern and south-western regions, and the converse during La Niña (Power et al., 1998). The positive phase of the Inter-decadal Pacific Oscillation (IPO) strengthens the ENSO-rainfall links in New Zealand and weakens links in Australia (Power et al., 1999; Salinger et al., 2004; Folland et al., 2005).

In New Zealand, mean air temperatures have increased by 1.0°C over the period 1855 to 2004, and by 0.4°C since 1950 (NIWA, 2005). Local sea surface temperatures have risen by 0.7°C since 1871 (Folland et al., 2003). From 1951 to 1996, the number of cold nights and frosts declined by 10-20 days/yr (Salinger and Griffiths, 2001). From 1971 to 2004, tropical cyclones in the south-west Pacific averaged nine/year, with no trend in frequency (Burgess, 2005) or intensity (Diamond, 2006). The frequency and strength of extreme westerly winds have increased significantly in the south. Extreme easterly winds have decreased over land but have increased in the south (Salinger et al., 2005a). Relative sea-level rise has averaged 1.6 ± 0.2 mm/yr since 1900 (Hannah, 2004). Rainfall has increased in the south-west and decreased in the north-east (Salinger and Mullan, 1999) due to changes in circulation linked to the IPO, with extremes showing similar trends (Griffiths, 2007). Pan evaporation has declined significantly at six out of nineteen sites since the 1970s, with no significant change at the other thirteen sites (Roderick and Farquhar, 2005). Snow accumulation in the Southern Alps shows considerable interannual variability but no trend since 1930 (Owens and Fitzharris, 2004).

In Australia, from 1910 to 2004, the average maximum temperature rose 0.6°C and the minimum temperature rose 1.2°C, mostly since 1950 (Nicholls and Collins, 2006). It is very likely that increases in greenhouse gases have significantly contributed to the warming since 1950 (Karoly and Braganza, 2005a, b). From 1957 to 2004, the Australian average shows an increase in hot days (≥35°C) of 0.10 days/yr, an increase in hot nights (≥20°C) of 0.18 nights/yr, a decrease in cold days (≤15°C) of 0.14 days/yr and a decrease in cold nights (≤5°C) of 0.15 nights/yr (Nicholls and Collins, 2006). Due to a shift in climate around 1950, the north-western two-thirds of Australia has seen an increase in summer monsoon rainfall, while southern and eastern Australia have become drier (Smith, 2004b). While the causes of decreased rainfall in the east are unknown, the decrease in the south-west is probably due to a combination of increased greenhouse gas concentrations, natural climate variability and land-use change, whilst the increased rainfall in the north-west may be due to increased aerosols resulting from human activity, especially in Asia (Nicholls, 2006). Droughts have become hotter since about

1973 because temperatures are higher for a given rainfall deficiency (Nicholls, 2004). From 1950 to 2005, extreme daily rainfall has increased in north-western and central Australia and over the western tablelands of New South Wales (NSW), but has decreased in the south-east, south-west and central east coast (Gallant et al., 2007). Trends in the frequency and intensity of most extreme temperature and rainfall events are rising faster than the means (Alexander et al., 2007). South-east Australian snow depths at the start of October have declined 40% in the past 40 years (Nicholls, 2005). Pan evaporation averaged over Australia from 1970 to 2005 showed large interannual variability but no significant trend (Roderick and Farquhar, 2004; Jovanovic et al., 2007; Kirono and Jones, 2007). There is no trend in the frequency of tropical cyclones in the Australian region from 1981 to 2003, but there has been an increase in intense systems (very low central pressure) (Kuleshov, 2003; Hennessy, 2004). Relative sea-level rise around Australia averaged 1.2 mm/yr from 1920 to 2000 (Church et al., 2004).

The offshore islands of Australia and New Zealand have recorded significant warming. The Chatham Islands (44°S, 177°W) have warmed 1°C over the past 100 years (Mullan et al., 2005b). Macquarie Island (55°S, 159°E) has warmed 0.3°C from 1948 to 1998 (Tweedie and Bergstrom, 2000), along with increases in wind speed, precipitation and evapotranspiration, and decreases in air moisture content and sunshine hours since 1950 (Frenot et al., 2005). Campbell Island (53°S, 169°E) has warmed by 0.6°C in summer and 0.4°C in winter since the late 1960s. Heard Island (53°S, 73°E) shows rapid glacial retreat and a reduced area of annual snow cover from 1948 to 2001 (Bergstrom, 2003).

11.2.2 Human systems: sensitivity/vulnerability to climate and weather

Extreme events have severe impacts in both countries (Box 11.1). In Australia, around 87% of economic damage due to natural disasters (storms, floods, cyclones, earthquakes, fires and landslides) is caused by weather-related events (BTE, 2001). From 1967 to 1999, these costs averaged US$719 million/yr, mostly due to floods, severe storms and tropical cyclones. In New Zealand, floods are the most costly natural disasters apart from earthquakes and droughts, and total flood damage costs averaged about US$85 million/yr from 1968 to 1998 (NZIER, 2004).

11.2.3 Natural systems: sensitivity/vulnerability to climate and weather

Some species and natural systems in Australia and New Zealand are already showing evidence of recent climate-associated change (Table 11.1). In many cases, the relative contributions of other factors such as changes in fire regimes and land use are not well understood.

11.2.4 Sensitivity/vulnerability to other stresses

Human and natural systems are sensitive to a variety of stresses independent of those produced by climate change. Growing populations and energy demands have placed stress on energy supply infrastructure. In Australia, energy consumption has increased 2.5%/yr over the past 20 years (PB Associates, 2007). Increases in water demand have placed stress on supply capacity for irrigation, cities, industry and environmental flows. Increased water demand in New Zealand has been due to agricultural intensification (Woods and Howard-Williams, 2004)

Box 11.1. Examples of extreme weather events in Australia and New Zealand*

Droughts: In Australia, the droughts of 1982-1983, 1991-1995 and 2002-2003 cost US$2.3 billion, US$3.8 billion and US$7.6 billion, respectively (Adams et al., 2002; BoM, 2006a). In New Zealand, the 1997-1998 and 1998-1999 droughts had agricultural losses of US$800 million (MAF, 1999).

Sydney hailstorm, 14 April 1999: With the exception of the droughts listed above, this is the most expensive natural disaster in Australian history, costing US$1.7 billion, of which US$1.3 billion was insured (Schuster et al., 2005).

Eastern Australian heatwave, 1 to 22 February 2004: About two-thirds of continental Australia recorded maximum temperatures over 39°C. Temperatures reached 48.5°C in western New South Wales. The Queensland ambulance service recorded a 53% increase in ambulance call-outs (Steffen et al., 2006).

Canberra fire, 19 January 2003: Wildfires caused US$261 million damage (Lavorel and Steffen, 2004; ICA, 2007). About 500 houses were destroyed, four people were killed and hundreds injured. Three of the city's four dams were contaminated for several months by sediment-laden runoff.

South-east Australian storm, 2 February 2005: Strong winds and heavy rain led to insurance claims of almost US$152 million (ICA, 2007). Transport was severely disrupted and beaches were eroded.

Tropical cyclone Larry, 20 March 2006: Significant damage or disruption to houses, businesses, industry, utilities, infrastructure (including road, rail and air transport systems, schools, hospitals and communications), crops and state forests, costing US$263 million. Fortunately, the 1.75 m storm surge occurred at low tide (BoM, 2006b; Queensland Government, 2006).

New Zealand floods: The 10 April 1968 Wahine storm cost US$188 million, the 26 January 1984 Southland floods cost US$80 million, and the February 2004 North Island floods cost US$78 million (Insurance Council of New Zealand, 2005).

* All costs are adjusted to 2002-2006 values.

and has seen the irrigated area of New Zealand increase by around 55% each decade since the 1960s (Lincoln Environmental, 2000). Per capita daily water consumption is 180-300 litres in New Zealand and 270 litres for Australia (Robb and Bright, 2004). In Australia, dryland salinity, alteration of river flows, over-allocation and inefficient use of water resources, land clearing, intensification of agriculture, and fragmentation of ecosystems still represent major stresses (SOE, 2001; Cullen, 2002). From 1985 to 1996, Australian water demand increased by 65% (NLWRA, 2001). Invasive plant and animal species pose significant environmental problems in both countries, particularly for agriculture and forestry (MfE, 2001; SOE, 2001); for example, *Cryptostegia grandiflora* (Kriticos et al., 2003a, b).

Table 11.1. *Examples of observed changes in species and natural systems linked to changing climate in Australia, New Zealand and their sub-Antarctic islands.*

Taxa or system	Observed change	References
Australia		
Rainforest and woodland ecotones	Expansion of rainforest at the expense of eucalypt forest and grassland in Northern Territory, Queensland and New South Wales, linked to changes in rainfall and fire regimes.	Bowman et al., 2001; Hughes, 2003
Sub-alpine vegetation	Encroachment by snow gums into sub-alpine grasslands at higher elevations.	Wearne and Morgan, 2001
Freshwater swamps and floodplains	Saltwater intrusion into freshwater swamps since the 1950s in Northern Territory accelerating since the 1980s, possibly associated with sea level and precipitation changes.	Winn et al., 2006
Coral reefs	Eight mass bleaching events on the Great Barrier Reef since 1979, triggered by unusually high sea surface temperatures; no serious events known prior to 1979 (see Section 11.6). Most widespread events appear to have occurred in 1998 and 2002, affecting up to 50% of reefs within the Great Barrier Reef Marine Park.	Hoegh-Guldberg, 1999; Done et al., 2003; Berkelmans et al., 2004
Birds	Earlier arrival of migratory birds; range shifts and expansions for several species; high sea surface temperatures associated with reduced reproduction in wedge-tailed shearwaters.	Smithers et al., 2003; Chambers, 2005; Chambers et al., 2005; Beaumont et al., 2006
Mammals	Increased penetration of feral mammals into alpine and high sub-alpine areas and prolonged winter presence of macropods.	Green and Pickering, 2002
Insects	Change in genetic constitution of *Drosophila*, equivalent to a 4° latitude shift (about 400 km).	Umina et al., 2005
New Zealand		
Birds	Earlier egg laying in the welcome swallow.	Evans et al., 2003
Southern beech	Seed production increase in *Nothofagus* (1973 to 2002) along elevational gradient related to warming during flower development.	Richardson et al., 2005
Fish	Westward shift of Chilean jack mackerel in the Pacific and subsequent invasion into New Zealand waters in the mid-1980s associated with increasing El Niño frequency.	Taylor, 2002
Glaciers	Ice volume decreased from about 100 km^3 to 53 km^3 over the past century. Loss of at least one-quarter of glacier mass since 1950. Mass balance of Franz Josef glacier decreased 0.02 m/yr from 1894 to 2005.	Chinn, 2001; Clare et al., 2002; Anderson, 2004
Sub-Antarctic Islands		
Birds	Population increases in black-browed albatross and king penguin on Heard Island; population declines on Campbell Island of rockhopper penguins, grey-headed albatross and black-browed albatross related to ocean warming and changed fishing practices.	Waugh et al., 1999; Woehler et al., 2002; Weimerskirch et al., 2003
Vertebrates	Population increases in fur seals on Heard Island and elephant seals on Campbell Island, linked to changes in food supply, warming and oceanic circulation; rats moving into upland herb-fields and breeding more often on Macquarie Island.	Budd, 2000; Weimerskirch et al., 2003; Frenot et al., 2005
Plant communities	Plant colonisation of areas exposed by glacial retreat on Heard Island; decline in area of sphagnum moss since 1992 on Macquarie Island associated with drying trend.	Whinam and Copson, 2006

11.2.5 Current adaptation

Since vulnerability is influenced by adaptation, a summary of current adaptation is given here rather than in Section 11.5 (which looks at future adaptation). Adaptation refers to planned and autonomous (or spontaneous) adjustments in natural or human systems in response to climatic stimuli. Adaptation can reduce harmful effects or exploit opportunities (see Chapter 17). An example of autonomous adaptation is the intensification of grazing in the rangelands of north-west Australia over the last 30 years, as graziers have exploited more reliable and better pasture growth following an increase in monsoon rainfall (Ash et al., 2006). However, there is currently insufficient information to comprehensively quantify this capacity. While planned adaptation usually refers to specific measures or actions, it can also be viewed as a *dynamic process* that evolves over time, involving five major pre-conditions for encouraging implementation (Figure 11.1). This section assesses how well Australia and New Zealand are engaged in the adaptation process.

Provision of knowledge, data and tools.

Since the TAR, the New Zealand Foundation for Research, Science and Technology has created a separate strategic fund for global change research (FRST, 2005). Operational research and development related to climate impacts on specific sectors have also increased over the last 10 years (e.g., agricultural impacts, decision-support systems and extension activities for integration with farmers' knowledge) (Kenny, 2002; MAF, 2006). One of Australia's four National Research Priorities is "an environmentally sustainable Australia", which includes "responding to climate change and variability" (DEST, 2004). The Australian Climate Change Science Programme and the National Climate Change Adaptation Programme are part of this effort (Allen Consulting Group, 2005). All Australian state and territory governments have greenhouse action plans that include development of knowledge, data and tools.

Risk assessments

A wide range of regional and sectoral risk assessments has been undertaken since 2001 (see Section 11.4). Both countries occasionally produce national reports that synthesise these assessments and provide a foundation for adaptation (MfE, 2001; Warrick et al., 2001; Howden et al., 2003a; Pittock, 2003). Regionally relevant guidelines are available for use in risk assessments (Wratt et al., 2004; AGO, 2006).

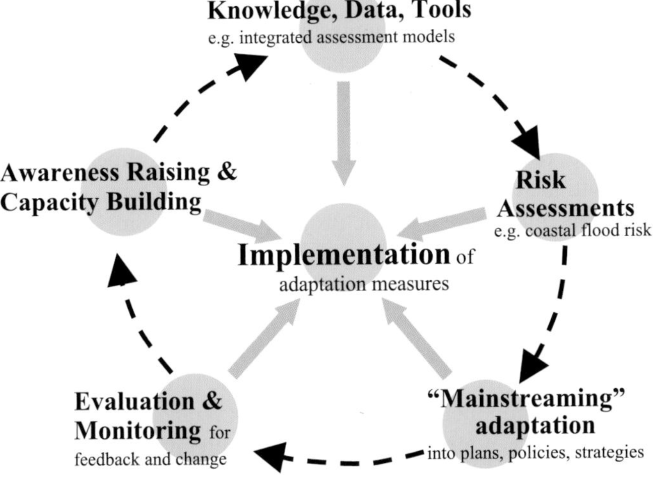

Figure 11.1. *Adaptation as a process (Warrick, 2000, 2006).*

Mainstreaming

Climate change issues are gradually being 'mainstreamed' into policies, plans and strategies for development and management. For example, in New Zealand, the Coastal Policy Statement included consideration of sea-level rise (DoC, 1994), the Resource Management (Energy and Climate Change) Amendment Act 2004 made explicit provisions for the effects of climate change, and the Civil Defence and Emergency Management Act 2002 requires regional and local government authorities (LGAs) to plan for future natural hazards. New Zealand farmers, particularly in the east, implemented a range of adaptation measures in response to droughts in the 1980s and 1990s and as a result of the removal of almost all subsidies. Increasing numbers of farmers are focusing on building long-term resilience with a diversity of options (Kenny, 2005; Salinger et al., 2005b). In Australia, climate change is included in several environmentally focused action plans, including the National Agriculture and Climate Change Action Plan (NRMMC, 2006) and the National Biodiversity and Climate Change Action Plan. A wide range of water adaptation strategies has been implemented or proposed (Table 11.2), including US$1.5 billion for the National Water Fund from 2004 to 2009 and US$1.7 billion for drought relief from 2001 to 2006.

Climate change is listed as a Key Threatening Process under the Commonwealth Environment Protection and Biodiversity Conservation Act 1999. Climate change has been integrated into several state-based and regional strategies, such as the Queensland Coastal Management Plan, the Great Barrier Reef Climate Change Action Plan, the Victorian Sustainable Water Strategy and South Australia's Natural Resources Management Plan. The Wild Country (The Wilderness Society), Gondwana Links (Western Australia) and Nature Links (South Australia) and Alps to Atherton (Victoria, NSW, Queensland) initiatives promote connectivity of landscapes and resilience of natural systems in recognition of the fact that some species will need to migrate as climate zones shift. Guidelines prepared for the coastal and ocean engineering profession for implementing coastal management strategies include consideration of climate change (Engineers Australia, 2004).

Evaluation and monitoring

The New Zealand Climate Committee monitors the present state of knowledge of climate science, climate variability and current and future climate impacts, and makes recommendations about research and monitoring needs, priorities and gaps regarding climate, its impacts and the application of climate information (RSNZ, 2002). In Australia, the Australian Greenhouse Office (AGO) monitors and evaluates performance against objectives in the National Greenhouse Strategy. The AGO and state and territory governments commission research to assess current climate change knowledge, gaps and priorities for research on risk and vulnerability (Allen Consulting Group,

Table 11.2. *Examples of government adaptation strategies to cope with water shortages in Australia.*

Government	Strategy	Investment	Source
Australia	Drought aid payments to rural communities	US$1.7 billion from 2001 to 2006	DAFF, 2006b
Australia	National Water Initiative, supported by the Australian Water Fund	US$1.5 billion from 2004 to 2009	DAFF, 2006a
Australia	Murray-Darling Basin Water Agreement	US$0.4 billion from 2004 to 2009	DPMC, 2004
Victoria	Melbourne's Eastern Treatment Plant to supply recycled water	US$225 million by 2012	Melbourne Water, 2006
Victoria	New pipeline from Bendigo to Ballarat, water recycling, interconnections between dams, reducing channel seepage, conservation measures	US$153 million by 2015	Premier of Victoria, 2006
Victoria	Wimmera Mallee pipeline replacing open irrigation channels	US$376 million by 2010	Vic DSE, 2006
NSW	NSW Water Savings Fund supports projects which save or recycle water in Sydney	US$98 million for Round 3, plus more than US$25 million to 68 other projects	DEUS, 2006
Queensland (Qld)	Qld Water Plan 2005 to 2010 to improve water-use efficiency and quality, recycling, drought preparedness, new water pricing	Includes US$182 million for water infrastructure in south-east Qld, and US$302 million to other infrastructure programmes	Queensland Government, 2005
South Australia	Water Proofing Adelaide project is a blueprint for the management, conservation and development of Adelaide's water resources to 2025	N/A	Government of South Australia, 2005
Western Australia (WA)	State Water Strategy (2003) and State Water Plan (proposed) WA Water Corporation doubled supply from 1996 to 2006	US$500 million spent by WA Water Corporation from 1996 to 2006, plus US$290 million for the Perth desalination plant	Government of Western Australia, 2003, 2006; Water Corporation, 2006

2005). The National Land and Water Resources Audit (NLWRA, 2001) and State of the Environment Report (SOE, 2001) also have climate-change elements.

Awareness raising and capacity building

In New Zealand, efforts are underway for transferring scientific information to LGAs and facilitating exchange of information between LGAs. The New Zealand Climate Change Office has held a number of workshops for LGAs (MfE, 2002, 2004b), supported case studies of 'best practice' adaptation by LGAs, and has commissioned guidance documents for LGAs on integrating climate change adaptation into their functions (MfE, 2004c). The AGO, the Australian Bureau of Meteorology, the Commonwealth Scientific and Industrial Research Organisation (CSIRO) and most Australian state and territory governments have developed products and services for raising awareness about climate change. Government-supported capacity-building programmes, such as the Australian National Landcare Programme, enhance resilience to climate change via mechanisms such as whole-farm planning.

In general, the domestic focus of both countries has, until recently, been on mitigation, while adaptation has had a secondary role in terms of policy effort and government funding for implementation (MfE, 2004b). However, since the TAR, recognition of the necessity for adaptation has grown and concrete steps have been taken to bolster the pre-conditions for adaptation, as discussed above. Initiatives such as the Australia-New Zealand Bilateral Climate Change Partnership (AGO, 2003) explicitly include adaptation. Overall, in comparison to most other countries, New Zealand and Australia have a relatively high and growing level of adaptive capacity, which has the potential to be implemented systematically on a wide scale.

11.3 Assumptions about future trends

11.3.1 Climate

Regional climate change projections are provided in Chapter 11 of the Working Group I Fourth Assessment Report (Christensen et al., 2007). For Australia and New Zealand, these projections are limited to averages over two very broad regions: northern Australia and southern Australia (including New Zealand). More detailed regional projections are required to assess local impacts and are described below. Developed over the past five years, these are similar to those presented in the TAR, and include the full range of emissions scenarios from the IPCC Special Report on Emissions Scenarios (SRES: Nakićenović and Swart, 2000) (see Chapter 2.4.6). Some SRES scenarios have been suggested as surrogates for CO_2 concentration stabilisation scenarios: the SRES B1, B2 and A1B emissions scenarios are similar to the CO_2 stabilisation scenarios for 550 ppm by 2150, 650 ppm by 2200, and 750 ppm by 2250, respectively. Projected changes will be superimposed on continued natural variability including ENSO and the IPO. There is uncertainty about projected changes in ENSO as discussed in Chapter 10 of the Working Group I Fourth Assessment Report (Meehl et al., 2007).

In New Zealand, a warming of 0.1 to 1.4°C is likely by the 2030s and 0.2 to 4.0°C by the 2080s (Table 11.3). The mid-range projection for the 2080s is a 60% increase in the annual mean westerly component of wind speed (Wratt et al., 2004). Consequently, a tendency for increased precipitation is likely except in the eastern North Island and the northern South Island. Due to the projected increased winter precipitation over the

Table 11.3. *Projected changes in New Zealand annual precipitation and mean temperature for the 2030s and 2080s, relative to 1990. The ranges are based on results from forty SRES emission scenarios and six climate models for various locations in each region (Wratt et al., 2004).*

Temperature change (°C)	2030s	2080s
Western North Island	+0.2 to 1.3	+0.3 to 4.0
Eastern North Island	+0.2 to 1.4	+0.5 to 3.8
Northern South Island	+0.1 to 1.4	+0.4 to 3.5
Western South Island	+0.1 to 1.3	+0.2 to 3.5
Eastern South Island	+0.1 to 1.4	+0.4 to 3.4
Rainfall change (%)	**2030s**	**2080s**
Western North Island	-4 to +14	-6 to +26
Eastern North Island	-19 to +7	-32 to +2
Northern South Island	-7 to +3	-7 to +5
Western South Island	-4 to +15	+1 to +40
Eastern South Island	-12 to +13	-21 to +31

Southern Alps, it is less clear whether snow will be reduced (MfE, 2004a), although snowlines are likely to be higher (Fitzharris, 2004). By 2100, there is likely to be a 5 to 20 day decrease in frosts in the lower North Island, 10 to 30 fewer frost days in the South Island, and a 5 to 70 day increase in the number of days with temperatures over 30°C (Mullan et al., 2001). The frequency of heavy rainfall is likely to increase, especially in western areas (MfE, 2004a).

In Australia, within 800 km of the coast, a mean warming of 0.1 to 1.3°C is likely by the year 2020, relative to 1990, 0.3 to 3.4°C by 2050, and 0.4 to 6.7°C by 2080 (Table 11.4). In temperate areas, this translates to 1 to 32 more days/yr over 35°C by 2020 and 3 to 84 more by 2050, with 1 to 16 fewer days/yr below 0°C by 2020 and 2 to 32 fewer by 2050 (Suppiah et al., 2007). A tendency for decreased annual rainfall is likely over most of southern and sub-tropical Australia, with a tendency for increases in Tasmania, central Northern Territory and northern NSW (Table 11.4). The 15-model average shows decreasing rainfall over the whole continent (Suppiah et al., 2007). A decline in runoff in southern and eastern Australia is also likely (see Section 11.4.1).

The area of mainland Australia with at least one day of snow cover per year is likely to shrink by 10 to 40% by 2020 and by 22 to 85% by 2050 (Hennessy et al., 2003). Increases in extreme daily rainfall are likely where average rainfall either increases or decreases slightly. For example, the intensity of the 1-in-20 year daily rainfall event is likely to increase by up to 10% in parts of South Australia by the year 2030 (McInnes et al., 2002), by 5 to 70% by the year 2050 in Victoria (Whetton et al., 2002), by up to 25% in northern Queensland by 2050 (Walsh et al., 2001) and by up to 30% by 2040 in south-east Queensland (Abbs, 2004). In NSW, the intensity of the 1-in-40 year event increases by 5 to 15% by 2070 (Hennessy et al., 2004). The frequency of severe tropical cyclones (Categories 3, 4 and 5) on the east Australian coast increases 22% for the IS92a scenario (IPCC, 1992) from 2000 to 2050, with a 200 km southward shift in the cyclone genesis region, leading to greater exposure in south-east Queensland and north-east NSW (Leslie and Karoly, 2007). For tripled pre-industrial CO_2 conditions, there is a 56% increase in the number of simulated tropical cyclones over north-eastern Australia with peak winds greater than 30 m/s (Walsh et al., 2004). Decreases in hail frequency are simulated for Melbourne and Mt. Gambier (Niall and Walsh, 2005).

Potential evaporation (or evaporative demand) is likely to increase (Jones, 2004a). Projected changes in rainfall and evaporation have been applied to water-balance models, indicating that reduced soil moisture and runoff are very likely over most of Australia and eastern New Zealand (see Section 11.4.1 and Meehl et al., 2007). Up to 20% more droughts (defined as the 1-in-10 year soil moisture deficit from 1974 to 2003) are simulated over most of Australia by 2030 and up to 80% more droughts by 2070 in south-western Australia (Mpelasoka et al., 2007). Projected increases in the Palmer Drought Severity Index for the SRES A2 scenario are indicated over much of eastern Australia between 2000 and 2046 (Burke et al., 2006). In New Zealand, severe droughts (the current 1-in-20 year soil moisture deficit) are likely to occur every 7 to 15 years by the 2030s, and every 5 to 10 years by the 2080s, in the east of both islands, and parts of Bay of Plenty and Northland (Mullan et al., 2005a). The drying of pastures in eastern New Zealand in spring is very likely to be advanced by one month, with an expansion of droughts into both spring and autumn.

An increase in fire danger in Australia is likely to be associated with a reduced interval between fires, increased fire intensity, a decrease in fire extinguishments and faster fire spread (Tapper, 2000; Williams et al., 2001; Cary, 2002). In south-east Australia, the frequency of very high and extreme fire danger days is likely to rise 4-25% by 2020 and 15-70% by 2050 (Hennessy et al., 2006). By the 2080s, 10-50% more days with very high and extreme fire danger are likely in eastern areas of New Zealand, the Bay of Plenty, Wellington and Nelson regions (Pearce et al., 2005), with increases of up to 60% in some western areas. In both Australia and New Zealand, the fire season length is likely to be extended, with the window of opportunity for controlled burning shifting toward winter.

Relative to the year 2000, the global-mean projection of sea-level rise by 2100 is 0.18 to 0.59 m, excluding uncertainties in

Table 11.4. *Projected changes in annual average rainfall and temperature for 2020, 2050 and 2080, relative to 1990, for Australia. The ranges are based on results from forty SRES emission scenarios and fifteen climate models for various locations in each region (Suppiah et al., 2007).*

Temperature change (°C)	2020	2050	2080
0 to 400 km inland of coast	+0.1 to 1.0	+0.3 to 2.7	+0.4 to 5.4
400 to 800 km inland	+0.2 to 1.3	+0.5 to 3.4	+0.8 to 6.7
Central Australia	+0.2 to 1.5	+0.5 to 4.0	+0.8 to 8.0
Rainfall change (%)	**2020**	**2050**	**2080**
Within 400 km of western and southern coasts	-15 to 0	-40 to 0	-80 to 0
Sub-tropics (latitudes 20-28°S) except west coast and inland Queensland	-10 to +5	-27 to +13	-54 to +27
Northern NSW, Tasmania and central Northern Territory (NT)	-5 to +10	-13 to +27	-27 to +54
Central South Australia, southern NSW and north of latitude 20°S, except central NT	-5 to +5	-13 to +13	-27 to +27
Inland Queensland	-10 to +10	-27 to +27	-54 to +54

carbon cycle feedbacks and the possibility of faster ice loss from Greenland and Antarctica (Meehl et al., 2007). These values would apply to Australia and New Zealand, but would be further modified by as much as ±25% due to regional differences in thermal expansion rates, oceanic circulation changes (as derived from atmosphere-ocean general circulation model experiments; Gregory et al., 2001) and by local differences in relative sea-level changes due to vertical land movements. An increase in westerly winds is probable south of latitude 45°S, with a strengthening of the East Australian Current and southern mid-latitude ocean circulation (Cai et al., 2005).

11.3.2 Population, energy and agriculture

The Australian population is projected to grow from 20 million in 2003 to 26.4 million in 2051, then stabilise (ABS, 2003a). This is under medium assumptions, including a fall in the number of children per woman from 1.75 at present to 1.6 from 2011 onward, net immigration of 100,000/yr, and a 10% increase in life expectancy by 2051 (ABS, 2003a). A greater concentration of the population is likely in Sydney, Melbourne, Perth, Brisbane and south-east Queensland. The proportion of people aged 65 and over is likely to increase from 13% in 2003 to 27% in 2051 (ABS, 2003a). Population growth is likely to intensify the urban heat island effect, exacerbating greenhouse-induced warming (Torok et al., 2001). Up to at least 2020, Australian energy consumption is projected to grow 2.1%/yr on average (ABARE, 2004). New energy sources will be needed to meet peak energy demands in Victoria, NSW, Queensland and South Australia between 2007 and 2010 (NEMMCO, 2006). Agriculture is likely to contribute about 3% of national gross domestic product (GDP).

In New Zealand, under medium assumptions, the population is likely to grow from 4.1 million in 2004 to 5.05 million in 2051 (Statistics New Zealand, 2005b). These assumptions include a net immigration of 10,000/yr, a drop in fertility rate from 2.01 in 2004 to 1.85 from 2016 onward and a 10% increase in life expectancy by 2051. The proportion aged 65 and over is likely to grow from 12% in 2004 to 25% in 2051. Total energy demand is likely to grow at an average rate of 2.4%/yr from 2005 to 2025 (Electricity Commission, 2005). Agriculture is likely to continue contributing about 5% of GDP (MFAT, 2006).

11.4 Key future impacts and vulnerabilities

This section discusses potential impacts of climate change, mostly based on climate projections consistent with those described in Section 11.3. It does not take into account adaptation; this is discussed in Section 11.5 and in more detail in Chapter 17. Conclusions are drawn from the available literature. Very little information is available on social and economic impacts. Further details on potential impacts can be found in various synthesis reports (MfE, 2001; Pittock, 2003).

11.4.1 Freshwater resources

11.4.1.1 Water security

The impact of climate change on water security is a significant cross-cutting issue. In Australia, many new risk assessments have been undertaken since the TAR (Table 11.5). The Murray-Darling Basin is Australia's largest river basin, accounting for about 70% of irrigated crops and pastures (MDBC, 2006). Annual streamflow in the Basin is likely to fall 10-25% by 2050 and 16-48% by 2100 (Table 11.5). Little is known about future impacts on groundwater in Australia.

In New Zealand, annual flow from larger rivers with headwaters in the Southern Alps is likely to increase. Proportionately more runoff is very likely from South Island rivers in winter, and less in summer (Woods and Howard-Williams, 2004). This is very likely to provide more water for hydro-electric generation during the winter peak demand period, and reduce dependence on hydro-storage lakes to transfer generation into the next winter. However, industries dependent on irrigation are likely to experience negative effects due to lower water availability in spring and summer, their time of peak demand. Increased drought frequency is very likely in eastern areas, with potential losses in agricultural production. The effects of climate change on flood and drought frequency are virtually certain to be modulated by phases of the ENSO and IPO (McKerchar and Henderson, 2003; see Section 11.2.1). The groundwater aquifer for Auckland City has spare capacity to accommodate recharge under all scenarios

Table 11.5. *Impacts on Australian water security. SRES scenarios are specified where possible.*

Year	Impacts
2030	• Change in annual runoff: -5 to +15% on the north-east coast, ±15% on the east coast, a decline of up to 20% in the south-east, ±10% in Tasmania, a decline of up to 25% in the Gulf of St Vincent (South Australia), and -25 to +10% in the south-west (Chiew and McMahon, 2002). • Decline in annual runoff: 6-8% in most of eastern Australia and 14% in south-west Australia in the period 2021 to 2050 relative to 1961 to 1990 for the A2 scenario (Chiew et al., 2003). • Burrendong dam (NSW): inflows change by +10% to 30% across all SRES scenarios, but the 90% confidence interval is 0% to -15% (Jones and Page, 2001). • Victoria: runoff in 29 catchments declines by 0-45% (Jones and Durack, 2005).
2050	• Murray Darling Basin: for B1, streamflow drops 10-19% and salinity changes -6 to +16%; for A1, streamflow drops 14-25% and salinity changes -8 to +19% (Beare and Heaney, 2002). • Melbourne: a risk assessment using ten climate models (driven by the SRES B1, A1B and A1FI scenarios) indicated that average streamflow is likely to decline 7-35% (Howe et al., 2005); however, planned demand-side and supply-side actions are likely to alleviate water shortages through to 2020 (Howe et al., 2005).
2070	• Burrendong Dam (NSW): inflows change by +5 to -35% across all SRES scenarios, for the 90% confidence interval (Jones and Page, 2001).
2100	• Murray-Darling Basin: for B1, streamflow declines 16 to 30%, salinity changes -16 to +35%, agricultural costs US$0.6 billion; for A1, streamflow declines 24 to 48%, salinity changes -25 to +72%, agricultural costs US$0.9 billion (Beare and Heaney, 2002).

examined (Namjou et al., 2005). Base flows in principal streams and springs are very unlikely to be compromised unless many dry years occur in succession.

11.4.1.2 Flood and waste water management

Little quantitative information is available about potential changes in flood risk in Australia. Sufficient capacity exists within the Melbourne sewerage and drainage systems to accommodate moderate increases (up to 20%) in storm rainfall totals with minimal surcharging (Howe et al., 2005). For the Albert-Logan Rivers system near the Gold Coast in Queensland, each 1% increase in rainfall intensity is likely to produce a 1.4% increase in peak runoff (Abbs et al., 2000). However, increases in runoff and flooding are partially offset by a reduction in average rainfall, which reduces soil wetness prior to storms. A high-resolution atmospheric model of storm events coupled with a non-linear flood event model has been applied to flooding around the Gold Coast caused by tropical cyclone Wanda in 1974. If the same event occurred in 2050 with a 10 to 40 cm rise in mean sea level, the number of dwellings and people affected is likely to increase by 3 to 18% (Abbs et al., 2000).

In New Zealand, rain events are likely to become more intense, leading to greater storm runoff, but with lower river levels between events. This is likely to cause greater erosion of land surfaces, more landslides (Glade, 1998; Dymond et al., 2006), redistribution of river sediments (Griffiths, 1990) and a decrease in the protection afforded by levees. Increased demands for enhancement of flood protection works are likely, as evidenced by the response to large floods in 2004 (MCDEM, 2004; CAE, 2005). Flood risk to Westport has been assessed using a regional atmospheric model, a rainfall-runoff model for the Buller River, projected sea-level rise and a detailed inundation model. Assuming the current levee configuration, the proportion of the town inundated by a 1-in-50 year event is currently 4.3%, but rises to 13 to 30% by 2030, and 30 to 80% by 2080 (Gray et al., 2005). Peak flow increases 4% by 2030 and 40% by 2080. In contrast, a flood risk study for Auckland using 2050 climate scenarios with 1 to 2°C global warming indicated only minor increases in flood levels (Dayananda et al., 2005). Higher flows and flood risk are likely in the Wairau catchment in North Shore City (URS, 2004).

11.4.1.3 Water quality

In Australia, there is a 50% chance by 2020 of the average salinity of the lower Murray River exceeding the 800 EC threshold set for desirable drinking and irrigation water (MDBMC, 1999). There are no integrated assessments of the impacts of climate change on runoff quantity and quality, salt interception and revegetation policies, and water pricing and trading policies. Eutrophication is a major water-quality problem (Davis, 1997; SOE, 2001). Toxic algal blooms are likely to become more frequent and to last longer due to climate change. They can pose a threat to human health, for both recreation and consumptive water use, and can kill fish and livestock (Falconer, 1997). Simple, resource-neutral, adaptive management strategies, such as flushing flows, can substantially reduce their occurrence and duration in nutrient-rich, thermally stratified water bodies (Viney et al., 2003).

In New Zealand, lowland waterways in agricultural catchments are in a relatively poor state and these streams are under pressure from land-use intensification and increasing water abstraction demands (Larned et al., 2004). There is no literature on impacts of climate change on water quality in New Zealand.

11.4.2 Natural ecosystems

The flora and fauna of Australia and New Zealand have a high degree of endemism (80 to 100% in many taxa). Many species are at risk from rapid climate change because they are restricted in geographical and climatic range. Most species are well-adapted to short-term climate variability, but not to longer-term shifts in mean climate and increased frequency or intensity of extreme events. Many reserved areas are small and isolated, particularly in the New Zealand lowlands and in the agricultural areas of Australia. Bioclimatic modelling studies generally project reductions and/or fragmentation of existing climatic ranges. Climate change will also interact with other stresses such as invasive species and habitat fragmentation. The most vulnerable include the Wet Tropics and Kakadu wetlands, alpine areas, tropical and deep-sea coral reefs, south-east Tasman Sea, isolated habitats in the New Zealand lowlands, coastal and freshwater wetlands and south-west Australian heathlands (Table 11.6). There is little research on the impacts of climate change on New Zealand species or natural ecosystems, with the exception of the alpine zone and some forested areas.

Major changes are expected in all vegetation communities. In the Australian rangelands (75% of total continental land area), shifts in rainfall patterns are likely to favour establishment of woody vegetation and encroachment of unpalatable woody shrubs. Interactions between CO_2, water supply, grazing practices and fire regimes are likely to be critical (Gifford and Howden, 2001; Hughes, 2003). In New Zealand, fragmented native forests of drier lowland areas (Northland, Waikato, Manawatu) and in the east (from East Cape to Southland) are likely to be most vulnerable to drying and changes in fire regimes (McGlone, 2001; MfE, 2001). In alpine zones of both countries, reductions in duration and depth of snow cover are likely to alter distributions of communities, for example favouring an expansion of woody vegetation into herbfields (Pickering et al., 2004). More fires are likely in alpine peatlands (Whinam et al., 2003). Alpine vertebrates dependent on snow cover for hibernation are likely to be at risk of extinction (Pickering et al., 2004). In regions such as south-western Australia, many narrow-ranged endemic species will be vulnerable to extinction with relatively small amounts of warming (Hughes, 2003). Saltwater intrusion as a result of sea-level rise, decreases in river flows and increased drought frequency, are very likely to alter species composition of freshwater habitats, with consequent impacts on estuarine and coastal fisheries (Bunn and Arthington, 2002; Hall and Burns, 2002; Herron et al., 2002; Schallenberg et al., 2003). In marine ecosystems, ocean acidification is likely to decrease productivity and diversity of plankton communities around Australia, while warmer oceans are likely to lead to further southward movement of fish and kelp communities (Poloczanska et al., 2007).

Table 11.6. *Examples of projected impacts on species and ecosystems, relative to 1990.*

Year	Potential Impacts	Source
2020	Bleaching and damage to the Great Barrier Reef equivalent to that in 1998 and 2002 in up to 50% of years.	Berkelmans et al., 2004; Crimp et al., 2004
	60% of the Great Barrier Reef regularly bleached.	Jones, 2004b
	Habitat lost for marine invertebrates currently confined to cool waters (>10% of Victoria's total).	O'Hara, 2002; Watters et al., 2003
	63% decrease in golden bowerbird habitat in northern Australia.	Hilbert et al., 2004
	50% decrease in montane tropical rainforest area in northern Australia.	Hilbert et al., 2001
2030	58 to 81% of the Great Barrier Reef bleached every year.	Jones, 2004b
	Hard coral reef communities widely replaced by algal communities.	Wooldridge et al., 2005
	88% of Australian butterfly species' core habitat decreases.	Beaumont and Hughes, 2002
	97% of Wet Tropics endemic vertebrates have reduced core habitat.	Williams et al., 2003
2050	97% of the Great Barrier Reef bleached every year.	Jones, 2004b
	92% of butterfly species' core habitat decreases.	Beaumont and Hughes, 2002
	98% decrease in golden bowerbird habitat in northern Australia.	Hilbert et al., 2004
	80% loss of freshwater wetlands in Kakadu for a 30 cm sea-level rise.	Hare, 2003
2080	Catastrophic mortality of coral species annually.	Jones, 2004b
	95% decrease in distribution of Great Barrier Reef species.	Jones et al., 2004
	65% loss of Great Barrier Reef species in the Cairns region.	Crimp et al., 2004
	46% of Wet Tropics endemic vertebrates lose core habitat.	Williams et al., 2003
	200 to 300 indigenous New Zealand alpine plant species may become extinct.	Halloy and Mark, 2003
	Reduced calcification for 70% of the area where deep sea corals occur, loss of endemic species.	Poloczanska et al., 2007

On the sub-Antarctic Islands, likely impacts include increased mortality of burrowing petrels, increased invasions by disturbance-tolerant alien plants such as *Poa annua*, increased abundance of existing rats, mice and rabbits on islands, and reduced distribution of *Sphagnum* moss (Bergstrom and Selkirk, 1999; Frenot et al., 2005).

11.4.3 Agriculture

11.4.3.1 Cropping

Since the TAR, there has been further assessment of potential impacts of climate and CO_2 changes at local, regional and national scales in both Australia and New Zealand. Overall, these emphasise the vulnerability of cropping and the potential for regional differences. Impacts of climate change on pests, diseases and weeds, and their effects on crops, remain uncertain, since few experimental or modelling studies have been performed (Chakraborty et al., 2002).

In New Zealand, for C_3 crops such as wheat, the CO_2 response is likely to more than compensate for a moderate increase in temperature (Jamieson et al., 2000) (see Section 5.4). The net impact in irrigation areas depends on the availability of water (Miller and Veltman, 2004). For maize (a C_4 crop), reduction in growth duration reduces crop water requirements, providing closer synchronisation of development with seasonal climatic conditions (Sorensen et al., 2000).

In Australia, the potential impacts of climate change on wheat vary regionally, as shown by a study which used the full range of CO_2 and climate change in the IPCC SRES scenarios (Howden and Jones, 2004), in conjunction with a crop model recently validated for its CO_2 response for current wheat varieties (Reyenga et al., 2001; Asseng et al., 2004). South-western Australian regions are likely to have significant yield reductions by 2070 (increased yield very unlikely). In contrast, regions in north-eastern Australia are likely to have moderate increases in yield (unlikely to have substantial yield reductions). Nationally, median crop yields dropped slightly. There is a substantial risk to the industry as maximum potential increases in crop value are limited (to about 10% or US$0.3 billion/yr) but maximum potential losses are large (about 50% or US$1.4 billion/yr) (Figure 11.2). However, adaptation through changing planting dates and varieties is likely to be highly effective: the median benefit is projected to be US$158 million/yr but with a range of US$70 million to over US$350 million/yr (Howden and Jones, 2004) (Figure 11.2).

Climate change is likely to change land use in southern Australia, with cropping becoming non-viable at the dry margins if rainfall is reduced substantially, even though yield increases from elevated CO_2 partly offset this effect (Sinclair et al., 2000; Luo et al., 2003). In contrast, cropping is likely to expand into the wet margins if rainfall declines. In the north of Australia, climate change and CO_2 increases are likely to enable cropping to persist (Howden et al., 2001a). Observed warming trends are already reducing frost risk and increasing yields (Howden et al., 2003b).

Grain quality is also likely to be affected. Firstly, elevated CO_2 reduces grain protein levels (Sinclair et al., 2000). Significant increases in nitrogenous fertiliser application or increased use of pasture legume rotations would be needed to maintain protein levels (Howden et al., 2003c). Secondly, there is increased risk of development of undesirable heat-shock proteins in wheat grain in both northern and southern cropping zones with temperature increases greater than 4°C (Howden et al., 1999d).

Land degradation is likely to be affected by climate change. Elevated atmospheric CO_2 concentrations slightly reduce crop evapotranspiration. This increases the risk of water moving below the root zone of crops (deep drainage), potentially exacerbating three of Australia's most severe land degradation problems across agricultural zones: waterlogging, soil acidification and dryland salinity. In Western Australia, deep drainage is simulated to increase 1 to 10% when CO_2 is raised to 550 ppm, but deep drainage decreases 8 to 29% for a 3°C warming (van Ittersum et al., 2003). Deep drainage is reduced by

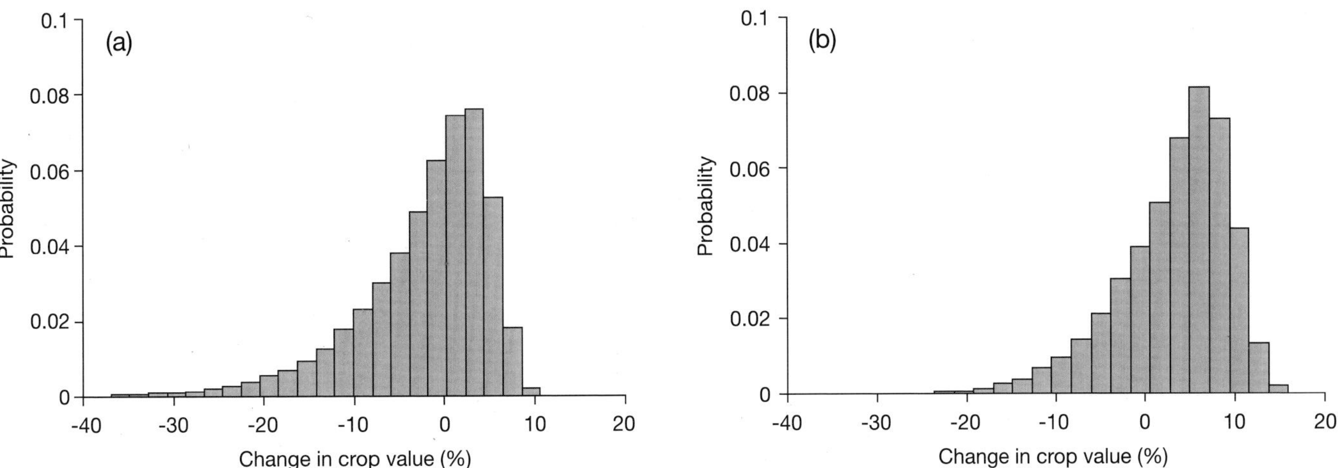

Figure 11.2: *Change in national gross value of wheat from historical baseline values (%) for 2070 as a result of increases in CO_2 and changes in temperature and rainfall: (a) without adaptation and (b) with adaptations of changed planting dates and varieties (Howden and Jones, 2004).*

up to 94% in low precipitation scenarios. However, the changes in deep drainage were not correlated with changes in productivity or gross margin.

11.4.3.2 Horticulture

Australian temperate fruits and nuts are all likely to be negatively affected by warmer conditions because they require winter chill or vernalisation. Crops reliant on irrigation are likely to be threatened where irrigation water availability is reduced. Climate change is likely to make a major horticultural pest, the Queensland fruit fly *Bactrocera tryoni*, a significant threat to southern Australia. Warming scenarios of 0.5, 1.0 and 2.0°C suggest expansion from its endemic range in the north and north-east across most of the non-arid areas of the continent, including the currently quarantined fruit fly-free zone (Sutherst et al., 2000). Apple, orange and pear growers in endemic Queensland fruit fly areas are likely to have cost increases of 42 to 82%, and 24 to 83% in the current fruit fly-free zone (Sutherst et al., 2000).

In New Zealand, warmer summer temperatures for Hayward kiwifruit are likely to increase vegetative growth at the expense of fruit growth and quality (Richardson et al., 2004). Kiwifruit budbreak is likely to occur later, reducing flower numbers and yield in northern zones (Hall et al., 2001). Production of current kiwifruit varieties is likely to become uneconomic in Northland by 2050 because of a lack of winter chilling, and be dependent on dormancy-breaking agents and varieties bred for warmer winter temperatures in the Bay of Plenty (Kenny et al., 2000). In contrast, more areas in the South Island are likely to be suitable (MfE, 2001). Apples, another major crop, are very likely to flower and reach maturity earlier, with increased fruit size, especially after 2050 (Austin et al., 2000). New Zealand is likely to be more susceptible to the establishment of new horticultural pests. For example, under the current climate, only small areas in the north are suitable for the oriental fruit fly, but by the 2080s it is likely to expand to much of the North Island (Stephens et al., 2007).

Viticulture has expanded rapidly in both countries. Earlier ripening and reductions in grape quality and value are likely by 2030, e.g., in Australia, price per tonne drops 4 to 10% in the Yarra Valley and 16 to 52% in the Riverina (Webb et al., 2006). In cooler Australian climates, warming is likely to allow alternative varieties to be grown. With warming and a longer growing season in New Zealand, red wine production is increasingly likely to be practised in the south, with higher yields (Salinger et al., 1990). Higher CO_2 levels increase vine vegetative growth, and subsequent shading is likely to reduce fruitfulness. Distribution of vines is likely to change depending upon suitability compared with high-yield pasture and silviculture, and with future irrigation water availability and cost (Hood et al., 2002).

11.4.3.3 Pastoral and rangeland farming

In western, southern and higher-altitude areas of New Zealand, higher temperatures, a longer growing season, higher CO_2 concentrations and less frost are very likely to increase annual pasture production by 10 to 20% by 2030, although gains may decline thereafter (MfE, 2001). In eastern New Zealand and Northland, pasture productivity is likely to decline by 2030 due to increased drought frequency (see Section 11.3.1). Sub-tropical pastoral species with lower feed quality such as *Paspalum* are likely to spread southwards, reducing productivity (Clark et al., 2001), particularly in the Waikato district. The range and incidence of many pests and diseases are likely to increase. Drought and water security problems are likely to make irrigated agriculture vulnerable, e.g., intensive dairying in Canterbury (Jenkins, 2006).

In Australia, a rise in CO_2 concentration is likely to increase pasture growth, particularly in water-limited environments (Ghannoum et al., 2000; Stokes and Ash, 2006; see also Section 5.4). However, if rainfall is reduced by 10%, this CO_2 benefit is likely to be offset (Howden et al., 1999d; Crimp et al., 2002). A 20% reduction in rainfall is likely to reduce pasture productivity by an average of 15% and liveweight gain in cattle by 12%, substantially increasing variability in stocking rates and reducing farm income (Crimp et al., 2002). Elevated concentrations of CO_2 significantly decrease leaf nitrogen content and increase non-structural carbohydrate, but cause little change in digestibility (Lilley et al., 2001). In farming systems with high nitrogen forage (e.g., temperate pastures), these effects are likely to increase energy availability, nitrogen processing in the rumen and productivity. In contrast, where nitrogen is deficient (e.g., rangelands), higher temperatures are likely to exacerbate existing problems by decreasing non-structural carbohydrate concentrations and digestibility, particularly in tropical C_4 grasses (see Section 5.4.3). Doubled CO_2 concentrations and warming are likely to result in only limited changes in the distributions of native C_3 and C_4 grasses (Howden et al., 1999b).

Climatic changes are likely to increase major land-degradation problems such as erosion and salinisation (see Section 11.4.3.1). They are also likely to increase the potential distribution and abundance of exotic weeds, e.g., *Acacia nilotica* and *Cryptostegia grandiflora* (Kriticos et al., 2003a, b) and native woody species, e.g., *A. aneura* (Moore et al., 2001). This is likely to increase competition with pasture grasses, reducing livestock productivity. However, the same CO_2 and climate changes are likely to provide increased opportunities for woody weed control through increased burning opportunities (Howden et al., 2001b). A warming of 2.5°C is likely to lead to a 15 to 60% reduction in rabbit populations in some areas via the impact on biological control agents, e.g., myxomatosis and rabbit haemorrhagic disease virus (Scanlan et al., 2006).

Heat stress already affects livestock in many Australian regions, reducing production and reproductive performance and enhancing mortality (see Section 5.4.3). Increased thermal stress on animals is very likely (Howden et al., 1999a). In contrast, less cold-stress is likely to reduce lamb mortality in both countries. Impacts of the cattle tick (*Boophilus microplus*) on the Australian beef industry are likely to increase and move southwards (White et al., 2003). If breakdown of quarantine occurs, losses in live-weight gain from tick infestation are projected to increase 30% in 2030 and 120% in 2100 (in the absence of adaptation). The net present value of future tick losses is estimated as 21% of farm cash income in Queensland, the state currently most severely affected.

11.4.4 Forestry

In Australia, the value of wood and wood products in 2001-2002 was US$5 billion/yr. About 164 million ha are classified as forest, with 1% as plantation forests and 7% available for timber production in state-managed, multiple-use native forests (BRS, 2003). New Zealand's indigenous forests cover 6.4 million ha, with 1.7 million ha of planted production exotic forests, the latter providing substantial export income (MAF, 2001). Research since the TAR confirms that climate change is likely to have both positive and negative impacts on forestry in both countries. Productivity of exotic softwood and native hardwood plantations is likely to be increased by CO_2 fertilisation effects, although the amount of increase will be limited by projected increases in temperature, reductions in rainfall and by feedbacks such as nutrient cycling (Howden et al., 1999c; Kirschbaum, 1999a, b).

Where trees are not water-limited, warming expands the growing season in southern Australia, but pest damage is likely to negate some gains (see Section 5.4.5). Reduction in average runoff in some regions (see Section 11.4.1) and increased fire risk (see Section 11.3.1) are very likely to reduce productivity, whilst increased rainfall intensity is likely to exacerbate soil erosion problems and pollution of streams during forestry operations (Howden et al., 1999c). In *Pinus radiata* and *Eucalyptus* plantations, fertile sites are likely to have increased productivity for moderate warming, whereas infertile sites are likely to have decreased production (Howden et al., 1999c).

In New Zealand, the growth rates for plantation forestry (mainly *P. radiata*) are likely to increase in response to elevated CO_2 and wetter conditions in the south and west. Studies of pine seedlings confirm that the growth and wood density of *P. radiata* are enhanced during the first two years of artificial CO_2 fertilisation (Atwell et al., 2003). Tree growth reductions are likely for the east of the North Island due to projected rainfall decreases and increased fire risk (see Section 11.3.1). However, uncertainties remain regarding increased water-use efficiency with elevated CO_2 (MfE, 2001), and whether warmer and drier conditions could increase the frequency of upper mid-crown yellowing and winter fungal diseases (MfE, 2001).

11.4.5 Coasts

Over 80% of the Australian population lives in the coastal zone, with significant recent non-metropolitan population growth (Harvey and Caton, 2003). About 711,000 addresses (from the National Geo-coded Address File) are within 3 km of the coast and less than 6 m above sea level, with more than 60% located in Queensland and NSW (Chen and McAneney, 2006). These are potentially at risk from long-term sea-level rise and large storm surges.

Rises in sea level, together with changes to weather patterns, ocean currents, ocean temperature and storm surges are very likely to create differences in regional exposure (Walsh, 2002; MfE, 2004a; Voice et al., 2006). In New Zealand, there are likely to be more vigorous and regular swells on western coasts (MfE, 2004a). In northern Australia, tropical cyclones are likely to become more intense (see Section 11.3). The area of Cairns at risk of inundation by a 1-in-100 year storm surge is likely to more than double by 2050 (McInnes et al., 2003). Major impacts are very likely for coral reefs, particularly the Great Barrier Reef (see Section 11.6).

Future effects on coastal erosion include climate-induced changes in coastal sediment supply and storminess. In Pegasus Bay (New Zealand), shoreline erosion of up to 50 m is likely between 1980 and 2030 near the Waipara River if southerly waves are reduced by 50%, and up to 80 m near the Waimakariri River if river sand is reduced by 50% (Bell et al., 2001). In New Zealand, emphasis has been placed on providing information, guidelines and tools such as zoning and setbacks to local authorities for risk-based planning and management of coastal hazards affected by climate change and variability (Bell et al., 2001; MfE, 2004a) (see Section 11.6). In Australia, linkages between the IPO, ENSO and changes in coastal geomorphology have been demonstrated for the northern NSW coast (Goodwin, 2005; Goodwin et al., 2006) and between historic beach erosion and ENSO for Narabeen Beach (NSW) (Ranasinghe et al., 2004).

Sea-level rise is virtually certain to cause greater coastal inundation, erosion, loss of wetlands and salt-water intrusion into freshwater sources (MfE, 2004a), with impacts on infrastructure, coastal resources and existing coastal management programmes. Model simulations indicate that the loss of wetlands and mangroves in Spencer Gulf due to sea-level rise is influenced largely by elevation and exposure (Bryan et al., 2001). At Collaroy/Narrabeen beach (NSW), a sea-level rise of 0.2 m by 2050 combined with a 50-year storm event leads to coastal recession exceeding 110 m and causing losses of US$184 million (Hennecke et al., 2004). Investigations for metropolitan coasts reveal increased costs of protection for existing management systems (Bell et al., 2001). Mid-range sea-level rise projections for 2005 to 2025 are likely to increase the cost of sand replenishment on the Adelaide metropolitan coast by at least US$0.94 million/yr (DEH, 2005). Uncertainties in projected impacts can be managed through a risk-based approach involving stochastic simulation (Cowell et al., 2006). Coasts are also likely to be affected by changes in pollution and sediment loads from changes in the intensity and seasonality of river flows, and future impacts of river regulation (Kennish, 2002). In the next 50 to 100 years, 21% of the Tasmanian coast is at risk of erosion and significant recession from predicted sea-level rise (Sharples, 2004).

11.4.6 Fisheries

In Australia, the gross value of fisheries production is US$1.7 billion annually, of which 68% is wild-catch and 32% is aquaculture. In New Zealand, the combined value of fisheries production is US$0.8 billion, of which 80% is from the commercial catch and 20% from the growing aquaculture sector (Seafood Industry Council, 2006), which continues to grow. Little research has been completed on impacts of climate change on freshwater fisheries and aquaculture.

Marine fisheries around the world are threatened by over-exploitation. In Australia, of 74 stocks considered in 2005, 17 were over-fished, 17 were not over-fished, and 40 were of uncertain status (ABARE, 2005). In New Zealand, of 84 stocks of demersal fish where landings were greater than 500 tonnes/yr, 5 were regarded as over-fished, 24 were assessed as not over-fished, and 55 were of uncertain status (Ministry of Fisheries Science Group, 2006). Climate change will be an additional stress (Hobday and Matear, 2005). The key variables expected to drive impacts on marine fisheries are changes in ocean temperature, currents, winds, nutrient supply, acidification and rainfall. Changes in four emergent biological properties are likely as a result of climate change, the first of which is best understood: (i) distribution and abundance of impacted species, (ii) phenology, (iii) community composition, and (iv) community structure and dynamics (including productivity). Few climate-change impact studies have been undertaken, so this assessment mostly relies on extrapolation of observed relationships between climate variability and fisheries. With sea-level rise, increasing marine intrusions are highly likely to affect coastal fisheries and inshore sub-tidal breeding and nursery areas (Schallenberg et al., 2003). Overall, future climate-change impacts are likely to be greater for temperate endemics than for tropical species (Francis, 1994, 1996) and on coastal and demersal fisheries relative to pelagic and deep-sea fisheries (Hobday and Matear, 2005).

Changes in sea surface temperature or currents are likely to affect the distribution of several commercial pelagic (e.g., tuna) fisheries in the region (Lehodey et al., 1997; Lyne, 2000; Sims et al., 2001; Hobday and Matear, 2005). In particular, circulation changes may increase the availability of some species and reduce others, as has been demonstrated in Western Australia for the Leeuwin Current. Different management regimes are likely to be required: fishers will be faced with relocation or face reduced catches *in situ*. Recruitment is likely to be reduced in cool-water species. For example, for New Zealand species such as red cod, recruitment is correlated with cold autumn and winter conditions associated with El Niño events (Beentjes and Renwick, 2001; Annala et al., 2004). In contrast, for snapper, relatively high recruitment and faster growth rate of juveniles and adults are correlated with warmer conditions during La Niña events (Francis, 1994; Maunder and Watters, 2003), with decreases in larval recruitment during El Niño events (Zeldis et al., 2005). A similar pattern of recruitment exists for gemfish (Renwick et al., 1998). Regarding physiological changes, temperature has a major influence on the population genetics of ectotherms, selecting for changes in abundance of temperature-sensitive alleles and genotypes and their adaptive capacity. For New Zealand snapper, differences in allele frequencies at one enzyme marker are found among year classes from warm and cold summers (Smith, 1979). If species cannot adapt to the pace of climate change, then major changes in distribution are likely, particularly for species at the edges of suitable habitats (Richardson and Schoeman, 2004; Hampe and Petit, 2005).

Projected changes in Southern Ocean circulation (see Section 11.3.1) are likely to affect fisheries. Seasonal to interannual variability of westerly winds and strong wind events are associated with recruitment and catch rates in several species (Thresher et al., 1989, 1992; Thresher, 1994). A decline in wind due to a poleward shift in climate systems underlies recent stock declines off south-eastern Australia and western Tasmania, and these are linked to changes in larval growth rates and recruitment of juveniles in two fish species around Tasmania (Koslow and Thresher, 1999; Thresher, 2002). Reductions in upwelling of nutrients and extension of warm water along the east Australian coast are likely to reduce krill and jack mackerel abundance, upon which many other species are reliant, including tuna, seals and seabirds (CSIRO, 2002).

11.4.7 Settlements, industry and societies

Settlements, industry and societies are sensitive to extreme weather events, drought and sea-level rise (see Chapter 7). Many planning decisions for settlements and infrastructure need to account for new climatic conditions and higher sea-levels, but little research has been done on climate change impacts. The planning horizon for refurbishing major infrastructure is 10 to 30 years, while major upgrades or replacements have an expected lifetime of 50 to 100 years (PIA, 2004). Substantial infrastructure is at risk from projected climate change. About US$1,125 billion of Australia's wealth is locked up in homes, commercial buildings, ports and physical assets, which is equivalent to nine times the current national budget or twice the GDP (Coleman et al., 2004). In New Zealand, homes are valued at about US$280 billion, which is equivalent to about triple the national GDP (QVL, 2006). The average life of a house is 80 years and some last for 150 years or more (O'Connell and Hargreaves, 2004).

For infrastructure, design criteria for extreme events are very likely to be exceeded more frequently. Increased damage is likely for buildings (e.g., concrete joints, steel, asphalt, protective cladding, sealants), transport structures (e.g., roads, railways, ports, airports, bridges, tunnels), energy services (see Section 11.4.10), telecommunications (e.g., cables, towers, manholes), and water services (see Section 11.4.1) (PIA, 2004; BRANZ, 2007; Holper et al., 2007). In Victoria, water infrastructure is at significant risk for the B1 scenario by 2030, while power, telecommunications, transport and buildings are all at significant risk for the A1FI emission scenario by 2030 (Holper et al., 2007).

Climate change is very likely to affect property values and investment through disclosure of increased hazards and risk, as well as affecting the price and availability of insurance. In many Australian jurisdictions, flood hazard liability is not mandatory, or is poorly quantified (Yeo, 2003). Governments sometimes provide financial relief to the uninsured from large natural disasters (Box 11.1) and such costs are likely to rise. Insurance costs are very likely to rise in areas with increased risk. Hail damage accounts for 50% of the 20 highest insurance payouts in Australia (ICA, 2007), but there is limited information about potential changes in hail frequency (see Section 11.3).

Despite the economic significance of mining in Australia (5% of GDP and 35% of export earnings; ABS, 2005c), there is little information regarding climate change impacts on mining. However, in northern Australia, projected increases in extreme events, such as floods and cyclones, have the potential to increase erosion, slow down re-vegetation, shift capping materials and expose tailings in the area that includes Ranger and Jabiluka mines. These impacts have not been adequately considered in long-term mine planning (Wasson et al., 1988; Parliament of Australia, 2003). The traditional owners, the Mirrar, are concerned that these impacts may detrimentally affect land between Madjinbardi Billabong and the East Alligator River and the lowlands on the floodplain margins that lie downstream from these mine sites (Kyle, 2006).

There are major implications for amenities, cultural heritage, accessibility, and health of communities. These include costs, injury and trauma due to increased storm intensity and higher extreme temperatures, damage to items and landscapes of cultural significance, degraded beaches due to sea-level rise and larger storm surges, and higher insurance premiums (PIA, 2004). Increased demand for emergency services is likely. By 2100, costs of road maintenance in Australia are estimated to rise 31% for the SRES A2 scenario in a CSIRO climate simulation (Austroads, 2004).

Climate change may contribute to destabilising unregulated population movements in the Asia-Pacific region, providing an additional challenge to national security (Dupont and Pearman, 2006; Preston et al., 2006). Population growth and a one-metre rise in sea-level are likely to affect 200-450 million people in the Asia-Pacific region (Mimura, 2006). An increase in migrations from the Asia-Pacific region to surrounding nations such as New Zealand and Australia is possible (Woodward et al., 2001). Displacement of Torres Strait Islanders to mainland Australia is also likely (Green, 2006b).

11.4.8 Indigenous people

Indigenous people comprise about 15% of the New Zealand population (Statistics New Zealand, 2005a) and 2.4% of the Australian population (including about 30% of the Northern Territory population) (ABS, 2002).

Changes in New Zealand's climate over the next 50 to 100 years are likely to challenge the Māori economy and influence the social and cultural landscapes of Māori people (Packman et al., 2001). Some Māori have significant investment in fishing, agriculture and forestry and the downstream activities of processing and marketing (NZIER, 2003), as well as being important stakeholders in New Zealand's growing tourist industry (McIntosh, 2004) and in the energy sector. Economic performance and opportunities in these primary industries are likely to be influenced by climate-induced changes to production rates, product quality, pest and disease prevalence, drought, fire-risk and biodiversity, which, in turn, will affect the ability to raise development capital in these industries (MAF, 2001; Cottrell et al., 2004). While the majority of Māori live in urban environments, they also occupy remote and rural areas where the economy and social and cultural systems are strongly tied to natural environmental systems (e.g., traditional resource use, tourism), and where vital infrastructure and services are vulnerable to extreme weather events (e.g., flooding, landslides) (Harmsworth and Raynor, 2005). The capacity of the Māori people to plan and respond to threats of climate change to their assets (i.e., buildings, farms, forests, native forest, coastal resources, businesses) varies greatly, and is likely to be limited by access to funds, information and human capital, especially in Northland and on the East Coast, where there are large populations of Māori (TPK, 2001) and increased risks of extreme weather are likely (Mullan et al., 2001). Other pressures include the unclear role of local authorities with regard to rules, regulations and strategies for adaptation; multiple land-ownership and decision-making processes can be complex, often making it difficult to reach consensus and implement costly or non-traditional adaptation measures; and the high spiritual and cultural value placed on traditional lands/resources that can restrict or rule out some adaptation options such as relocation (NZIER, 2003). Many rural Māori also rely on the use of public and private land and coastal areas for hunting and fishing to supplement household food supplies, recreation, and the collection of firewood and cultural resources. The distribution and abundance of culturally important flora and fauna is likely to be adversely influenced by climate change, so the nature of such activities and the values associated with these resources are likely to be adversely affected, including spiritual well-being and cultural affirmation (NIWA, 2006). These challenges compound the sensitivity of the Māori to climate change.

Indigenous communities in remote areas of Australia often have inadequate infrastructure, health services and employment (Braaf, 1999; Ring and Brown, 2002; IGWG, 2004; Arthur and Morphy, 2005). Consequently, many of these communities show features of social and economic disadvantage (Altman, 2000; ABS, 2005b). Existing social disadvantage reduces coping ability and may restrict adaptive capacity (Woodward et al., 1998; Braaf, 1999), affecting these communities' resilience to climate hazards

(Watson and McMichael, 2001; Ellemor, 2005). Many of these communities strongly connect the health of their 'country' to their cultural, mental and physical well-being (Smith, 2004a; Jackson, 2005). Direct biophysical impacts, such as increases in temperature, rainfall extremes or sea-level rise, are likely to have significant indirect impacts on the social and cultural cohesion of these communities. There is recent recognition of the untapped resource of Indigenous knowledge about past climate change (Rose, 1996; Lewis, 2002; Orlove, 2003) which could be used to inform adaptation options. However, the oral tradition of recording this knowledge has, until recently, largely hindered non-Indigenous scientists from using this expertise to inform their science (Webb, 1997; Hill, 2004). Climate-change impacts identified for remote Indigenous communities include increases in the number of days of extreme heat, which may affect disease vectors, reproduction and survival of infectious pathogens, and heat stress (Green, 2006a; McMichael et al., 2006); extreme rainfall events and flooding, causing infrastructure damage (Green and Preston, 2006); salt inundation of freshwater aquifers and changes in mangrove ecology (UNEP-WCMC, 2006); changing fire regimes; sea-level rise and coastal erosion (Bessen Consulting Services, 2005; Green and Preston, 2006). King tides[1] in 2005 and 2006 in the Torres Strait have highlighted the need to revisit short-term coastal protection and long-term relocation plans for up to 2,000 Australians living on the central coral cays and north-west islands (Mulrennan, 1992; Green, 2006b).

11.4.9 Tourism and recreation

Tourism contributes 4.5% of Australian GDP and represents 11.2% of exports (Allen Consulting Group, 2005), and even more in New Zealand (about 5% of GDP and 16% of exports). The main tourism centres are the Gold Coast and tropical north Queensland in Australia, and Queenstown and Rotorua in New Zealand. Most tourism and recreation in Australia and New Zealand rely on resources of the natural environment. In Australia's Wet Tropics, the value of ecosystem goods and services, including tourism, is about US$132 to 148 million/yr (Curtis, 2004).

Few regional studies have assessed potential impacts on tourism, but elsewhere there is evidence that climate change has direct impacts (Agnew and Palutikof, 2001; Maddison, 2001). Some tourist destinations may benefit from drier and warmer conditions, e.g., for beach activities, viewing wildlife and geothermal activity, trekking, camping, climbing, wine tasting and fishing. However, greater risks to tourism are likely from increases in hazards such as flooding, storm surges, heatwaves, cyclones, fires and droughts (World Tourism Organisation, 2003; Scott et al., 2004; Becken, 2005; Hall and Higham, 2005; Becken and Hay, 2007). These adversely affect transport, personal safety, communication, water availability and natural attractions such as coral reefs, beaches, freshwater wetlands, snow, glaciers and forests. Changes in species distribution and ecosystems in National Parks (see Section 11.4.2) are likely to alter their tourism appeal. Tropical Australian destinations are particularly vulnerable to climate impacts (Allen Consulting Group, 2005). Queensland tourism is likely to be negatively affected by more intense tropical cyclones and by degradation of the Great Barrier Reef (see Section 11.6) and beaches (PIA, 2004).

Skiing attracts many tourists to New Zealand and south-eastern Australia. For the full range of SRES scenarios, by 2020 in south-east Australia, there are likely to be 5 to 40 fewer days of snow cover per year, a rise in the snowline of 30 to 165 m, and a reduction in the total snow-covered area of 10 to 40% (Hennessy et al., 2003). By 2050, the duration of snow cover reduces by 15 to 100 days, the maximum snow depth reduces by 10 to 99%, the snowline rises 60 to 570 m and the total area of snow cover shrinks by 20 to 85%. Similarly, in New Zealand, changes in seasonal snow cover are likely to have a significant impact on the ski industry. The snow line is likely to rise by 120 to 270 m based on scenarios for the 2080s (Fitzharris, 2004). Tourist flows from Australia to New Zealand might grow as a result of the relatively poorer snow conditions in Australia. Numerical modelling of the Franz Josef glacier reveals that temperature is the dominant control on glacier length for New Zealand's maritime glaciers (Anderson and Mackintosh, 2006). Noticeable shrinkage and retreat is very likely for even small temperature increases (Anderson, 2004; Anderson et al., 2006), and is likely to reduce visitor flows through tourism-dependent towns such as Fox and Franz Josef.

11.4.10 Energy

Energy consumption is projected to grow due to demographic and socio-economic factors (see Section 11.3.2). However, average and peak energy demands are also linked to climatic conditions. Increases in peak energy demand due to increased air-conditioner use are likely to exceed increases for base load. The risk of line outages and blackouts is likely to increase (PB Associates, 2007). More peak generating capacity is likely to be needed beyond that for underlying economic growth (Howden and Crimp, 2001). For a 2°C warming, *peak* demand increases 4% in Brisbane and 10% in Adelaide, but decreases 1% in Melbourne and Sydney (Howden and Crimp, 2001). About 10% of the existing asset levels may be required to allow for climate-related increases in peak demand by 2030 (PB Associates, 2007). However, *annual total* demand may be less sensitive to warming; a likely reduction in winter heating demand counteracts the increasing summer demand, e.g., New Zealand electricity demand decreases by 3%/°C increase in mean winter temperature (Salinger, 1990).

Climate change is likely to affect energy infrastructure in Australia and New Zealand through impacts of severe weather events on wind power stations, electricity transmission and distribution networks, oil and gas product storage and transport facilities, and off-shore oil and gas production (see Chapter 7). There are also likely to be costs and damages that can be avoided by adaptation and mitigation (see Section 18.4). An assessment of potential risks for Australia (PB Associates, 2007) found (i) increased peak and average temperatures are likely to reduce electricity generation efficiency, transmission line capacity, transformer capacity and the life of switchgear and other components; (ii) if climate changes gradually, both the generation utilities and the equipment manufacturers are likely to have enough time to adjust their standards and specifications; and (iii)

[1] King tide: any high tide well above average height.

vulnerability to the above impacts is low, but there is medium vulnerability to a decline in water supply for large-scale coal, hydro and gas turbine power generation.

In New Zealand, increased westerly wind speed is very likely to enhance wind generation and spill-over precipitation into major South Island hydro catchments, and to increase winter rain in the Waikato catchment (Wratt et al., 2004). Warming is virtually certain to increase snow melt, the ratio of rainfall to snowfall, and river flows in winter and early spring. This is very likely to assist hydroelectric generation at the time of highest energy demand for heating.

11.4.11 Human health

One of the most significant health impacts of climate change is likely to be an increase in heat-related deaths. Assuming no planned adaptation, the number of deaths is likely to rise from 1,115/yr at present in Adelaide, Melbourne, Perth, Sydney and Brisbane to 2,300 to 2,500/yr by 2020, and 4,300 to 6,300/yr by 2050, for all SRES scenarios, including demographic change (McMichael et al., 2003). In Auckland and Christchurch, a total of 14 heat-related deaths occur per year in people aged over 65, but this is likely to rise to 28, 51 and 88 deaths for warmings of 1, 2 and 3°C, respectively (McMichael et al., 2003). Demographic change is likely to amplify these figures. By 2100, the Australian annual death rate in people aged over 65 is estimated to increase from a 1999 baseline of 82 per 100,000 to 131-246 per 100,000, for the SRES B2 and A2 scenarios and the 450 ppm stabilisation scenario (Woodruff et al., 2005). Australian temperate cities are likely to experience higher heat-related deaths than tropical cities, and the winter peak in deaths is likely to be overtaken by heat-related deaths in nearly all cities by 2050 (McMichael et al., 2003). In New Zealand, the winter peak in deaths is likely to decline.

There are likely to be alterations in the geographical range and seasonality of some mosquito-borne infectious diseases. Fewer but heavier rainfall events are likely to affect mosquito breeding and increase the variability in annual rates of Ross River disease, particularly in temperate and semi-arid areas (Woodruff et al., 2002, 2006). The risk of establishment of dengue fever is likely to increase through changes in climate and population sensitivity in both tropical and temperate latitudes (Sutherst, 2004). Dengue is a substantial threat in Australia: the climate of the far north already supports *Aedes aegypti* (the major mosquito vector of the dengue virus); and outbreaks of dengue have occurred with increasing frequency and magnitude in far-northern Australia over the past decade. Projected climate changes in north-eastern Australia, combined with population growth, are likely to increase the average annual number of people living in areas suitable for supporting the dengue vector (an additional 0.1 to 0.3 million exposed in 2020, and 0.6 to 1.4 million in 2050) (McMichael et al., 2003). Malaria is unlikely to become established unless there is a dramatic deterioration in the public health response (McMichael et al., 2003). In New Zealand, parts of the North Island are likely to become suitable for breeding of the major dengue vector, while much of the country becomes receptive to other less-efficient vector species (De Wet et al., 2001; Woodward et al., 2001). The risk of dengue in New Zealand is likely to remain below the threshold for local transmission beyond 2050 (McMichael et al., 2003).

Warmer temperatures and increased rainfall variability are likely to increase the intensity and frequency of food-borne (D'Souza et al., 2004) and water-borne (Hall et al., 2002) diseases in both countries. Indigenous people living in remote communities are likely to be at increased risk due to their particular living conditions and poor access to services. The annual number of diarrhoeal hospital admissions among Aboriginal children living in central Australia is likely to increase 10% by 2050, assuming no change in current health standards (McMichael et al., 2003). The relationship between drought, suicide and severe mental health impacts in rural communities (Nicholls et al., 2006) suggests that parts of Australia are likely to experience increased mental health risks in future. Impacts on aeroallergens and photochemical smog in cities remain uncertain. High concentrations of bushfire smoke play a role in increasing hospital presentations of asthma (Johnston et al., 2002), so projected increases in fire risk may lead to more asthma.

11.4.12 Synthesis

Climate change adds new dimensions to the challenges already facing communities, businesses, governments and individuals. Assessment of the information given in this section leads to the conclusion that climate change is likely to give rise to six key *risks* in specific sectors (Table 11.7): natural systems, water security, coastal communities, agriculture and forestry, major infrastructure and health. Some extreme events can trigger multiple and simultaneous impacts across systems, e.g., heatwaves leading to heat-related deaths, fires, smoke pollution, respiratory illness, blackouts, and buckling of railways. There are also four key *benefits* for particular sectors: (i) in New Zealand, initial benefits to agriculture and forestry are projected in western and southern areas and close to major rivers due to a longer growing season, less frost and increased rainfall; (ii) reduced energy demand is very likely in winter; (iii) tourism is likely to directly benefit from drier and warmer weather in some areas; and (iv) flows in New Zealand's larger mountain-fed rivers are likely to increase, benefiting hydroelectricity generation and irrigation supply. Adaptation can alleviate or delay vulnerability in some sectors, as well as allowing benefits to accrue more rapidly (see Section 11.5).

11.5 Adaptation constraints and opportunities

Planned adaptation can greatly reduce vulnerability (see examples in Chapter 17). Since the TAR, Australia and New Zealand have taken notable steps in building adaptive capacity (see Section 11.2.5) by increasing support for research and knowledge, expanding assessments for decision makers of the risks of climate change, infusing climate change into policies and plans, promoting awareness, and better dealing with climate issues. However, there remain formidable environmental,

Table 11.7. *Six key risks in Australia (Aus) and New Zealand (NZ) (assuming no new adaptation). Underlying climate projections (see Section 11.3) include higher temperatures, sea-level rise, heavier rainfall, greater fire risk, less snow cover, reduced runoff over southern and eastern Australia and in the smaller lowland rivers of eastern New Zealand, more intense tropical cyclones and larger storm surges.*

System	Impacts	Identified hotspots
Natural systems	Damage to coral reefs, coasts, rainforests, wetlands and alpine areas. Increased disturbance, loss of biodiversity including possible extinctions, changed species ranges and interactions, loss of ecosystem services (e.g., for tourism and water). Potentially catastrophic for some systems (e.g., reefs may be dominated by macroalgae by 2050, extinctions of endemic vertebrates in Queensland Wet Tropics). Shrinking glaciers create slope instability.	Aus: Great Barrier Reef, Kakadu, east Queensland, alpine zones, Murray-Darling Basin (MDB) and south-western Aus; NZ: Southern Alps and their National Parks, eastern lowlands
Water security	Reduction in water supply for irrigation, cities, industry and riverine environment in those areas where streamflow declines, e.g., in the Murray-Darling Basin, annual mean flow may drop 10 to 25% by 2050 and 16 to 48% by 2100.	Southern and eastern Aus; Northland and parts of eastern lowlands of NZ
Coastal communities	Greater coastal inundation and erosion, especially in regions exposed to cyclones and storm surges. Coastal development is exacerbating the climate risks.	Tropical and south-east Queensland (Aus) and from Bay of Plenty to Northland in NZ
Agriculture and forestry	Reduced crop, pastoral and rangeland production over much of southern and eastern Australia and parts of eastern New Zealand. Reduced grain and grape quality. A southward shift of pests and disease vectors. Increased fire risk for forests.	Southern and eastern Aus and eastern NZ
Major infrastructure	Design criteria for extreme climatic events, floods and storm surges very likely to be exceeded more frequently. Increased damage likely for buildings, transport structures, telecommunications, energy services and water services.	Large cities, floodplains of major rivers, coastal communities, north-eastern parts of both countries
Health	By 2050, 3,200 to 5200 more heat-related deaths/yr, and 0.6 to 1.4 million more people exposed to dengue fever.	Large cities of both countries

economic, informational, social, attitudinal and political barriers to the implementation of adaptation.

For many natural ecosystems, impacts have limited reversibility. Planned adaptation opportunities for offsetting potentially deleterious impacts are often limited due to fixed habitat regions (e.g., the Wet Tropics and upland rainforests in Australia and the alpine zone in both Australia and New Zealand). One adaptive strategy is to provide corridors to facilitate migration of species under future warming. This will require changes in land tenure in many regions, with significant economic costs, although schemes to promote such connectivity are already under way in some Australian states (see Section 11.2.5). Another strategy is translocation of species. This is a very expensive measure, but it may be considered desirable for some iconic, charismatic or particularly vulnerable species.

For water, planned adaptation opportunities lie in the inclusion of risks due to climate change on both the demand and supply side (Allen Consulting Group, 2005; Table 11.2). In urban catchments, better use of storm and recycled water can augment supply, although existing institutional arrangements and technical systems for water distribution constrain implementation. Moreover, there is community resistance to the use of recycled water for human consumption (e.g., in such cities as Toowoomba in Queensland and Goulburn in NSW). Desalination schemes are being considered in some Australian capital cities. Installation of rainwater tanks is another adaptation response and is now actively pursued through incentive policies and rebates. For rural activities, more flexible arrangements for allocation are required, via the expansion of water markets, where trading can increase water-use efficiency (Beare and Heaney, 2002). Existing attitudes toward water pricing and difficulties with structural adjustment are significant barriers.

For agriculture, there are opportunities for planned adaptation via improvements in crop varieties (Figure 11.2), rotations, farm technology, farm practices and land-use mix. Cropping can be extended to historically wetter regions. Implementation will require new investment and significant managerial changes (Howden et al., 2003a). Farmers in eastern New Zealand are engaging in local discussion of risks posed by future climate change and how to enhance adaptation options (Kenny, 2005). They stress the need for support and education for 'bottom-up' adaptation (Kenny, 2007). Farming of marginal land at the drier fringe is likely to be increasingly challenging, especially in those regions of both countries with prospective declines in rainfall.

In coastal areas, there is solid progress in risk assessments and in fashioning policies and plans at the local and regional level in New Zealand. However, there remain significant challenges to achieving concrete actions that reduce risks. Consistent implementation of adaptation measures (e.g., setback lines, planned retreat, dune management (Dahm et al., 2005), building designs, prohibition of new structures and siting requirements that account for sea-level rise) has been difficult. Differences in political commitment, lack of strong and clear guidelines from government, and legal challenges by property owners are major constraints (MfE, 2003).

Considering all sectors, four broad barriers to adaptation are evident.

1. A lack of methods for integrated assessment of impacts and adaptation that can be applied on an area-wide basis. While sector-specific knowledge and tools have steadily progressed, the vulnerability of water resources, coasts, agriculture and ecosystems of local areas and regions are interconnected and need to be assessed accordingly (see Section 11.8).
2. Lack of well-developed evaluation tools for assessing planned adaptation options, such as benefit-cost analysis, incorporating climate change and adapted for local and regional application.

3. Ongoing scepticism about climate change science, uncertainty in regional climate change projections, and a lack of knowledge about how to promote adaptation. This is despite 87% of Australians being more concerned about climate change impacts than terrorism (Lowy Institute, 2006). Application of risk-based approaches to adaptation (e.g., upgrading urban storm-water infrastructure design; Shaw et al., 2005) demonstrate how developments can be 'climate-proofed' (ADB, 2005). While a risk-based method for planned adaptation has been published for Australia (AGO, 2006), there are few examples of where it has been applied.
4. Weak linkages between the various strata of government, from national to local, regarding adaptation policy, plans and requirements. Stronger guidance and support are required from state (in Australia) and central government (in New Zealand) to underpin efforts to promote adaptation locally. For example, the New Zealand Coastal Policy Statement recommends that regional councils should take account of future sea-level rise. But there is a lack of guidance as to how this should be accomplished and little support for building capacity to undertake the necessary actions. As a consequence, regional and local responses have been limited, variable and inconsistent.

11.6 Case studies

The following case studies (Boxes 11.2 to 11.4) illustrate regions where climate change has already occurred, impacts are evident and planned adaptation is being considered or implemented.

Box 11.2. Adaptation of water supplies in cities

In capital cities such as Perth, Brisbane, Sydney, Melbourne, Adelaide, Canberra and Auckland, concern about population pressures and the impact of climate change is leading water planners to implement a range of adaptation options (Table 11.2). For example, the winter rainfall-dominated region of south-west Western Australia has experienced a substantial decline in May to July rainfall since the mid-20th century. The effects of the decline on natural runoff have been severe, as evidenced by a 50% drop in annual inflows to reservoirs supplying the city of Perth (Figure 11.3). Similar pressures have been imposed on groundwater resources and wetlands. This has been accompanied by a 20% increase in domestic usage in 20 years, and a population growth of 1.7%/yr (IOCI, 2002). Climate simulations indicate that at least some of the observed drying is due to the enhanced greenhouse effect (IOCI, 2002). To ensure water security, a US$350 million programme of investment in water source development was undertaken by the WA Water Corporation (WA Water Corporation, 2004) from 1993 to 2003. In 2004, the continuation of low streamflow led to the decision to construct a seawater desalination plant, which will provide 45 Gl of water each year, at a cost of US$271 million. Energy requirements (24 MW) will be met by 48 wind turbines.

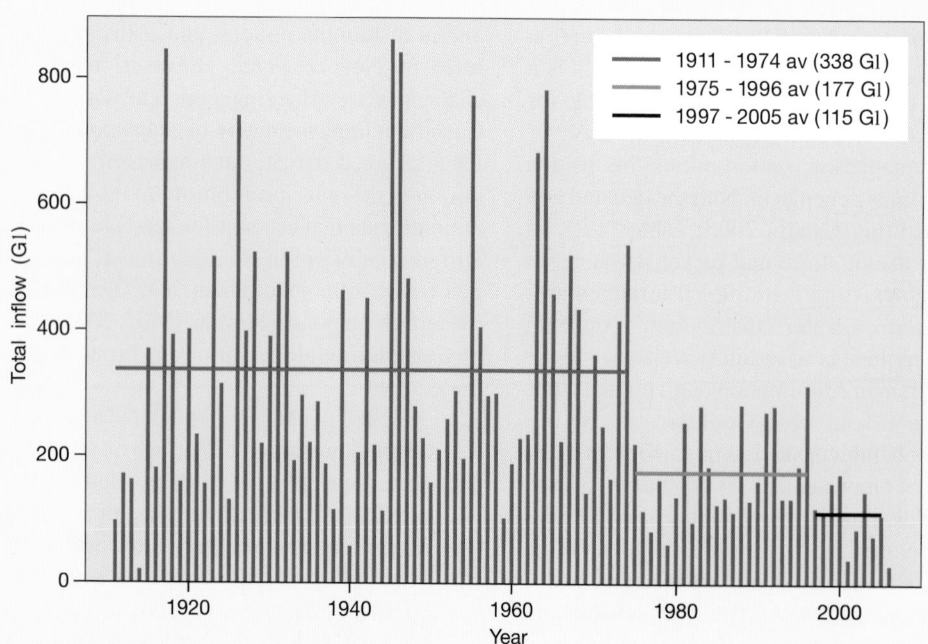

Figure 11.3. *Annual inflow to Perth Water Supply System from 1911 to 2005. Horizontal lines show averages. Source: www.watercorporation.com.au/D/dams_streamflow.cfm. Courtesy of the Water Corporation of Western Australia.*

Box 11.3. Climate change and the Great Barrier Reef

The Great Barrier Reef (GBR) is the world's largest continuous reef system (2,100 km long) and is a critical storehouse of Australian marine biodiversity and a breeding ground for seabirds and other marine vertebrates such as the humpback whale. Tourism associated with the GBR generated over US$4.48 billion in the 12-month period 2004/5 and provided employment for about 63,000 full-time equivalent persons (Access Economics, 2005). The two greatest threats from climate change to the GBR are (i) rising sea temperatures, which are almost certain to increase the frequency and intensity of mass coral bleaching events, and (ii) ocean acidification, which is likely to reduce the calcifying ability of key organisms such as corals. Other factors, such as droughts and more intense storms, are likely to influence reefs through physical damage and extended flood plumes (Puotinen, 2006).

Sea temperatures on the GBR have warmed by about 0.4°C over the past century (Lough, 2000). Temperatures currently typical of the northern tip of the GBR are very likely to extend to its southern end by 2040 to 2050 (SRES scenarios A1, A2) and 2070 to 2090 (SRES scenarios B1, B2) (Done et al., 2003). Temperatures only 1°C above the long-term summer maxima already cause mass coral bleaching (loss of symbiotic algae). Corals may recover but will die under high or prolonged temperatures (2 to 3°C above long-term maxima for at least 4 weeks). The GBR has experienced eight mass bleaching events since 1979 (1980, 1982, 1987, 1992, 1994, 1998, 2002 and 2006); there are no records of events prior to 1979 (Hoegh-Guldberg, 1999). The most widespread and intense events occurred in the summers of 1998 and 2002, with about 42% and 54% of reefs affected, respectively (Done et al., 2003; Berkelmans et al., 2004). Mortality was distributed patchily, with the greatest effects on near-shore reefs, possibly exacerbated by osmotic stress caused by floodwaters in some areas (Berkelmans and Oliver, 1999). The 2002 event was followed by localised outbreaks of coral disease, with incidence of some disease-like syndromes increasing by as much as 500% over the past decade at a few sites (Willis et al., 2004). While the impacts of coral disease on the GBR are currently minor, experiences in other parts of the world suggest that disease has the potential to be a threat to GBR reefs. Effects from thermal stress are likely to be exacerbated under future scenarios by the gradual acidification of the world's oceans, which have absorbed about 30% excess CO_2 released to the atmosphere (Orr et al., 2005; Raven et al., 2005). Calcification declines with decreasing carbonate ion concentrations, becoming zero at carbonate ion concentrations of approximately 200 μmol/kg (Langdon et al., 2000; Langdon, 2002). These occur at atmospheric CO_2 concentrations of approximately 500 ppm. Reduced growth due to acidic conditions is very likely to hinder reef recovery after bleaching events and will reduce the resilience of reefs to other stressors (e.g., sediment, eutrophication).

Even under a moderate warming scenario (A1T, 2°C by 2100), corals on the GBR are very likely to be exposed to regular summer temperatures that exceed the thermal thresholds observed over the past 20 years (Done et al., 2003). Annual bleaching is projected under the A1FI scenario by 2030, and under A1T by 2050 (Done et al., 2003; Wooldridge et al., 2005). Given that the recovery time from a severe bleaching-induced mortality event is at least 10 years (and may exceed 50 years for full recovery), these models suggest that reefs are likely to be dominated by non-coral organisms such as macroalgae by 2050 (Hoegh-Guldberg, 1999; Done et al., 2003). Substantial impacts on biodiversity, fishing and tourism are likely. Maintenance of hard coral cover on the GBR will require corals to increase their upper thermal tolerance limits at the same pace as the change in sea temperatures driven by climate change, i.e. about 0.1-0.5°C/decade (Donner et al., 2005). There is currently little evidence that corals have the capacity for such rapid genetic change; most of the evidence is to the contrary (Hoegh-Guldberg, 1999, 2004). Given that recovery from mortality can be potentially enhanced by reducing local stresses (water quality, fishing pressure), management initiatives such as the Reef Water Quality Protection Plan and the Representative Areas Programme (which expanded totally protected areas on the GBR from 4.6% to over 33%) represent adaptation options to enhance the ability of coral reefs to endure the rising pressure from rapid climate change.

Box 11.4. Climate change adaptation in coastal areas

Australia and New Zealand have very long coastlines with ongoing development and large and rapidly growing populations in the coastal zone. This situation is placing intense pressure on land and water resources and is increasing vulnerability to climatic variations, including storm surges, droughts and floods. A major challenge facing both countries is how to adapt to changes in climate, reduce vulnerability, and yet achieve sustainable development. Two examples illustrate this challenge.

Bay of Plenty, North Island, New Zealand. This bay is characterised by a narrow coastal zone with two of the fastest-growing districts of New Zealand. Combined population growth was 13.4% over the period 1996 to 2001, centred on the cities of Tauranga and Whakatane. By 2050, the population is projected to increase 2 to 3 times. Beachfront locations demand the highest premiums on the property market, but face the highest risks from storm surge flooding and erosion. Substantial efforts have been made to reduce the risks. For the purpose of delineation of hazard zones and design of adaptation measures, the Environment Bay of Plenty regional council explicitly included IPCC projections of sea-level rise in its Regional Coastal Environment Plan. This identified 'areas sensitive to coastal hazards within the next 100 years'. Implementation of such policy and plans by local government authorities has been repeatedly challenged by property developers, commercial interests and individual homeowners with different interpretations of the risks.

Sunshine Coast and Wide Bay-Burnett, Queensland, Australia. Between 2001 and 2021, the Sunshine Coast population is projected to grow from 277,987 to 479,806 (QDLGP, 2003), and the Wide Bay-Burnett population is projected to grow from 236,500 to 333,900 (ABS, 2003b). Sandy beaches and dunes are key biophysical characteristics of this coastline, including Fraser Island which is the largest sand island in the world. These natural features and the human populations they attract are vulnerable to sea-level rise, flooding, storm surges and tropical cyclones. Many estuaries and adjacent lowlands have been intensively developed, some as high-value canal estates. Local government is clearly becoming aware of climate-change risks. This topic is included in the agenda of the Sea-Change Taskforce, made up of coastal councils throughout Australia. At the regional planning level, climate change was recently embedded at a policy level into the strategic planning processes for the Wide Bay-Burnett region.

11.7 Conclusions: implications for sustainable development

An assessment of aggregate vulnerability for key sectors of the region is given in Figure 11.4, as a function of potential global warming. It synthesises relevant information in Sections 11.2 to 11.5 about current sensitivity, coping ranges, potential impacts, adaptive capacity and vulnerability. It follows similar diagrams and concepts published elsewhere (Jones et al., 2007) and emulates the 'Reasons for Concern' diagram (Figure SPM-3) in the TAR Synthesis Report. Since most impact assessments in the available literature do not allow for adaptation, the yellow band in Figure 11.4 is indicative only. In line with Chapter 19, vulnerability is assessed using criteria of: magnitude of impact, timing, persistence and reversibility, likelihood and confidence, potential for planned adaptation, geographical distribution and importance of the vulnerable system. Ecosystems, water security and coastal communities of the region have a narrow coping range. Even if adaptive capacity is realised, vulnerability becomes significant for 1.5 to 2.0°C of global warming. Energy security, health (heat-related deaths), agriculture and tourism have larger coping ranges and adaptive capacity, but they become vulnerable if global warming exceeds 3.0°C. The three key vulnerability factors identified in Article 2 of the United Nations Framework Convention on Climate Change (UNFCCC) – natural ecosystems, sustainable development and food security – are also shown in Figure 11.4.

When these climate change impacts are combined with other non-climate trends (see Section 11.3.2), there are some serious implications for sustainability in both Australia and New Zealand. Climate change is very likely to threaten natural ecosystems, with extinction of some species. There are limited planned adaptation options, but the resilience of many ecosystems can be enhanced by reducing non-climatic stresses such as water pollution, habitat fragmentation and invasive species. In river catchments, where increasing urban and rural water demand has already exceeded sustainable levels of supply, ongoing and proposed adaptation strategies (see Section 11.2.5) are likely to buy some time. Continued rates of coastal development are likely to require tighter planning and regulation if they are to remain sustainable. Climate change is very likely to increase peak energy demand during heatwaves, posing challenges for sustainable energy supply. A substantial public health and community response is likely to be needed in order to avoid an increase of several thousand heat-related deaths per year.

Large shifts in the geographical distribution of agriculture and its services are very likely. Farming of marginal land in drier regions is likely to become unsustainable due to water shortages, new biosecurity hazards, environmental degradation and social disruption. In areas that are likely to become wetter and less frosty, it may be possible to grow new crops or those displaced from other regions. Adaptation has the capacity to capture these benefits; they are unlikely to accrue without investment in the adaptation process. Food security is very likely to remain robust,

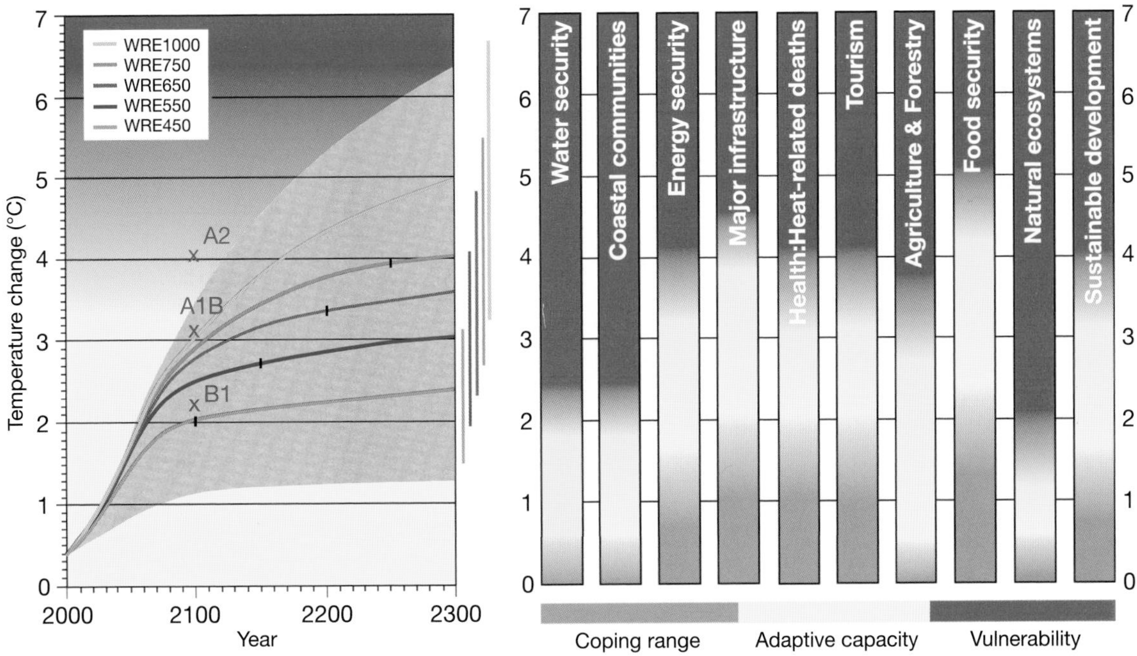

Figure 11.4. *Vulnerability to climate change aggregated for key sectors in the Australia and New Zealand region, allowing for current coping range and adaptive capacity. Right-hand panel is a schematic diagram assessing relative coping range, adaptive capacity and vulnerability. Left-hand panel shows global temperature change taken from the TAR Synthesis Report (Figure SPM-6). The coloured curves in the left panel represent temperature changes associated with stabilisation of CO_2 concentrations at 450 ppm (WRE450), 550 ppm (WRE550), 650 ppm (WRE650), 750 ppm (WRE750) and 1,000 ppm (WRE1000). Year of stabilisation is shown as black dots. It is assumed that emissions of non-CO_2 greenhouse gases follow the SRES A1B scenario until 2100 and are constant thereafter. The shaded area indicates the range of climate sensitivity across the five stabilisation cases. The narrow bars show uncertainty at the year 2300. Crosses indicate warming by 2100 for the SRES B1, A1B and A2 scenarios.*

with both countries able to produce more food than they require for internal consumption, although imports of selected foods may be needed temporarily to cover shortages due to extreme events. Climate changes are also likely to bring benefits in some areas for hydro-generation, winter heating requirements and tourism.

Figure 11.5 assesses key hotspots identified for the region, where vulnerability to climate change is likely to be high. Their selection is based on the following criteria: large impacts, low adaptive capacity, substantial population, economically important, substantial exposed infrastructure and subject to other major stresses (e.g., continued rapid population growth, ongoing development, ongoing land degradation, ongoing habitat loss, threats from rising sea level). Their development at current rates and accustomed supply of ecosystem services are unlikely to be sustainable with ongoing climate change, unless there is considerable planned adaptation.

For Australia and New Zealand, the magnitude of investment in adaptation is overshadowed by that in mitigation. The latter is intended to slow global warming. However, there is unlikely to be any noticeable climate effect from reducing greenhouse gases until at least 2040 (see Chapter 18). In contrast, the benefits of adaptation can be immediate, especially when they also address climate variability. Many adaptation options can be implemented now for Australia and New Zealand at personal, local and regional scales. Enhancing society's response capacity through the pursuit of sustainable development pathways is one way of promoting both adaptation and mitigation (see Chapter 18).

11.8 Key uncertainties and research priorities

Assessment of impacts is hampered because of uncertainty in climate change projections at the local level (e.g., in rainfall, rate of sea-level rise and extreme weather events). Research priorities for these are identified in the IPCC Working Group I Fourth Assessment Report (IPCC, 2007). Other uncertainties stem from an incomplete knowledge of natural and human system dynamics, and limited knowledge of adaptive capacity, constraints and options (Allen Consulting Group, 2005). More needs to be done to assess vulnerability within a risk-assessment framework. Based on the information presented in this Chapter, the main research priorities are assessed to fall into four categories:

11.8.1 Assessing impacts of climate change and vulnerability for critical systems

- *Water:* Impacts and optimum adaptation strategies for projected changes in drought and floods, and implications for water security within an integrated catchment framework. This includes impacts on long-term groundwater levels, water quality, environmental flows and future requirements for hydroelectricity generation, irrigation and urban supply.
- *Natural ecosystems:* Identification of thresholds including rates at which autonomous adaptation is possible; identification of the most vulnerable species (including key

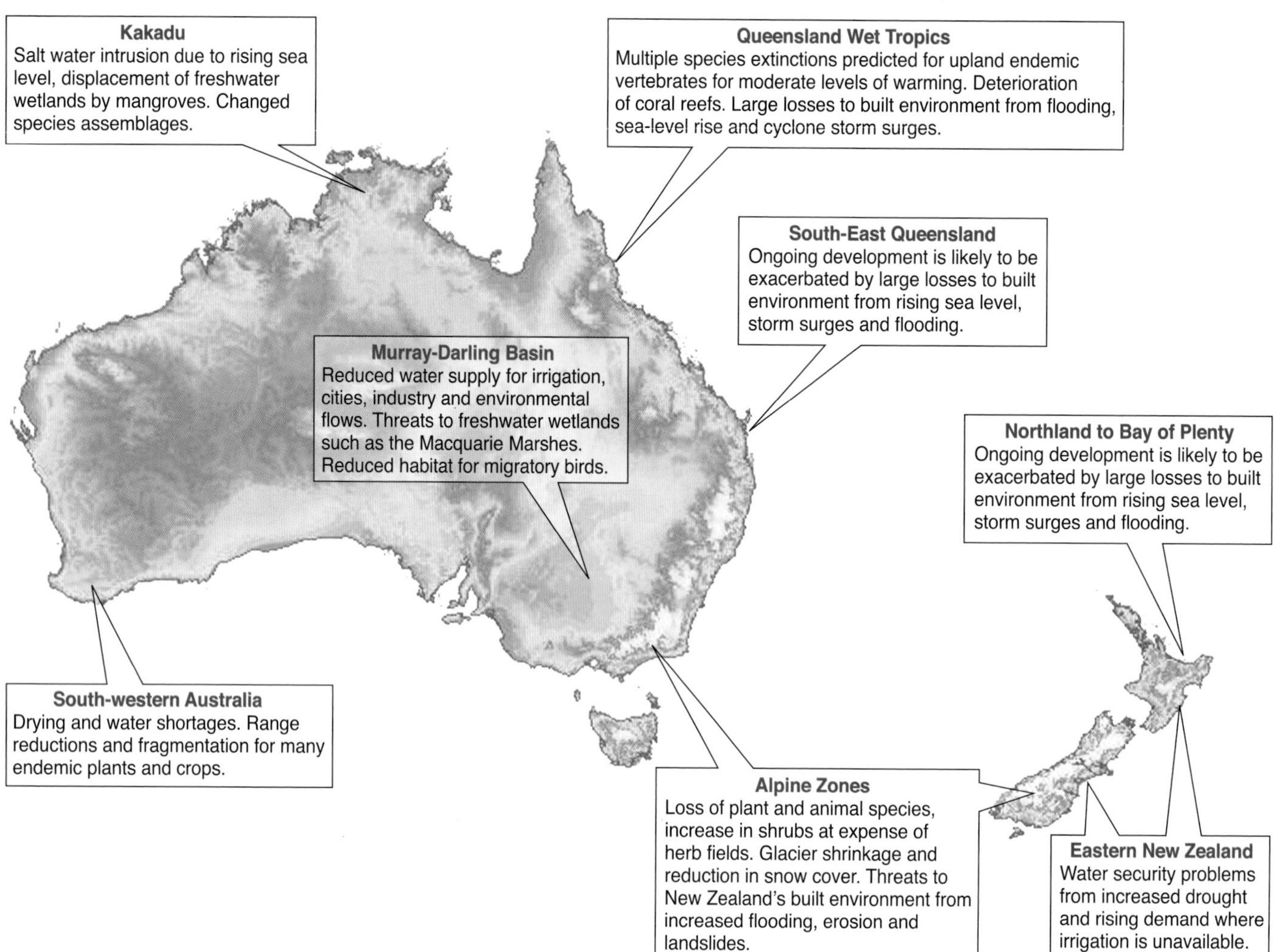

Kakadu
Salt water intrusion due to rising sea level, displacement of freshwater wetlands by mangroves. Changed species assemblages.
Queensland Wet Tropics
Multiple species extinctions predicted for upland endemic vertebrates for moderate levels of warming. Deterioration of coral reefs. Large losses to built environment from flooding, sea-level rise and cyclone storm surges.
South-East Queensland
Ongoing development is likely to be exacerbated by large losses to built environment from rising sea level, storm surges and flooding.
Murray-Darling Basin
Reduced water supply for irrigation, cities, industry and environmental flows. Threats to freshwater wetlands such as the Macquarie Marshes. Reduced habitat for migratory birds.
Northland to Bay of Plenty
Ongoing development is likely to be exacerbated by large losses to built environment from rising sea level, storm surges and flooding.
South-western Australia
Drying and water shortages. Range reductions and fragmentation for many endemic plants and crops.
Alpine Zones
Loss of plant and animal species, increase in shrubs at expense of herb fields. Glacier shrinkage and reduction in snow cover. Threats to New Zealand's built environment from increased flooding, erosion and landslides.
Eastern New Zealand
Water security problems from increased drought and rising demand where irrigation is unavailable.

Figure 11.5. *Key hotspots identified for Australia and New Zealand, assuming a medium emissions scenario for 2050.*

indicator species), long-term monitoring; modelling of potential impacts on key ecosystems; interactions with stresses such as invasive species; improved bioclimatic modelling; and management options to reduce vulnerability.

- *Agriculture:* Impacts and adaptation strategies for a complete range of farming systems, including both costs and benefits for rural livelihoods. Analyses should address changes in the industry supply chain and regional land use, and the threat of new pests and diseases.
- *Oceans and fisheries:* Potential impacts of changes in climate, ENSO and IPO on physical oceanography, marine life and fish stocks in the waters that surround Australia and New Zealand.
- *Settlements, especially coastal communities:* Comprehensive assessments of vulnerability and adaptation options so as to provide improved guidance for planning and hazard management. Investigation of local and regional costs of projected changes in extreme weather events and adaptation planning for scenarios of sea-level rise beyond 2100.
- *Climate extremes and infrastructure:* Risks to building, transport, water, communication, energy and mining infrastructure, and insurance protection from an increase in extreme weather events. A re-evaluation is required of probable maximum precipitation and design floods[2] for dams, bridges, river protection, major urban infrastructure and risks of glacier outburst floods.
- *Tourism:* Improved understanding as to how direct and indirect impacts of climate change affect human behaviour with respect to recreation patterns and holiday destination choice.
- *Climate surprises:* Impacts of abrupt climate change, faster than expected sea-level rise and sudden changes in ocean circulation. Little is known about potential impacts and vulnerability on the region beyond 2100.

11.8.2 Fostering the process of adaptation to climate change

Australia and New Zealand have few integrated regional and sectoral assessments of impacts, adaptation and socio-economic risk. More are desirable, especially when set within the wider context of other multiple stresses. Methods to incorporate adaptation into environmental impact assessments and other regional planning and development schemes need to be

[2] A hypothetical flood representing a specific likelihood of occurrence, e.g., the 100-year or 1% probability flood.

developed. More research is required as to how local communities can shape adaptation (Kenny, 2005) and of adaptation options for Māori and Indigenous Australian communities, especially for those on traditional lands. Priority should be given to reducing the vulnerability of 'hotspot' areas through:

- identification of mechanisms that governments might use to reduce vulnerability,
- better understanding of societal preparedness and of the limitations and barriers to adaptation,
- better definition of costs and benefits of adaptation options, including benefits of impacts avoided, co-benefits, side effects, limits and better modelling,
- analyses of various options for social equity and fairness, the impacts of different discount rates, price incentives, delayed effects and inter-generational equity.

11.8.3 Assessing risks and opportunities of climate change for different scenarios

Impact scenarios underpin policy decisions about adaptation options and emission reduction targets. The following analyses are required for the full range of SRES and CO_2 stabilisation scenarios:

- definition of the probabilities of exceeding critical biophysical and socio-economic thresholds and assessment of consequent vulnerability or new opportunities,
- assessment of net costs and benefits for key economic sectors and for each country,
- better modelling of land-use change as climatic boundaries shift, and assessment of the implications for regional development, social change, food security and sustainability.

11.8.4 Analysing global trade, immigration and security for climate change outcomes

Impacts of climate change and adaptation elsewhere in the world are very likely to change global interactions, and especially trade in commodities. The implications are large for the strongly export-based economies of Australia and New Zealand. Further studies are needed in order to assess the impacts of climate change on the region's competitiveness and export mix. In the Asia-Pacific region, adverse effects on food, disease, water, energy and coastal settlements are likely (Dupont and Pearman, 2006), but implications for immigration and security in Australia and New Zealand are poorly understood.

References

ABARE, 2004: *Securing Australia's Energy Future*. Australian Bureau of Resource Economics, Canberra, 104 pp.

ABARE, 2005: *Australian Fisheries Statistics 2004*. Australian Bureau of Agricultural and Resource Economics, Canberra, 65 pp. http://www.abareconomics.com/publications_html/fisheries/fisheries_04/fs04.pdf.

Abbs, D.J., 2004: A high-resolution modelling study of the effect of climate change on the intensity of extreme rainfall events. *Proc. Staying Afloat: Floodplain Management Authorities of NSW 44th Annual Conference*, Floodplain Management Authorities of NSW, Coffs Harbour, New South Wales, 17-24.

Abbs, D.J., S. Maheepala, K.L. McInnes, G. Mitchell, B. Shipton and G. Trinidad, 2000: Climate change, urban flooding and infrastructure. *Proc. Hydro 2000: 3rd International Hydrology and Water Resources Symposium of the Institution of Engineers, Australia*: Perth, Western Australia, 686-691.

ABS, 2002: *Population Distribution, Indigenous Australians, 2001*. Report 4705.0, Australian Bureau of Statistics, Canberra. http://abs.gov.au/Ausstats/abs@.nsf/lookupMF/14E7A4A075D53A6CCA2569450007E46.

ABS, 2003a: *Population Projections Australia 2004-2101*. Report 3222.0, Australian Bureau of Statistics, Canberra. http://www.abs.gov.au/Ausstats/abs@.nsf/0e5fa1cc95cd093c4a2568110007852b/0cd69ef8568dec8eca2568a900139392!OpenDocument.

ABS, 2003b: *Queensland Government Population Projections*. Report 3218.0, Australian Bureau of Statistics, Canberra. http://www.abs.gov.au/AUSSTATS/abs@.nsf/DetailsPage/3218.02004-05?OpenDocument.

ABS, 2005a: *Australian Social Trends 2005*. Report 4102.0, Australian Bureau of Statistics, Canberra. http://www.abs.gov.au/AUSSTATS/abs@.nsf/Lookup/4102.0I-Note12002?OpenDocument.

ABS, 2005b: *The Health and Welfare of Australia's Aboriginal and Torres Strait Islander Peoples*. Report 4704.0, Australian Bureau of Statistics, Canberra. http://www.abs.gov.au/ausstats/abs@.nsf/mf/4704.0.

ABS, 2005c: *Year Book Australia. Industry Structure and Performance: 100 Years of Change in Australian Industry*. Report 1301.0, Australian Bureau of Statistics, Canberra. http://www.abs.gov.au/Ausstats/abs@.nsf/Previousproducts/1301.0 Feature%20Article212005?opendocument&tabname=Summary&prodno=1301.0&issue=2005&num=&view=.

Access Economics, 2005: *Measuring the Economic and Financial Value of the Great Barrier Reef Marine Park*. Report by Access Economics for Great Barrier Reef Marine Park Authority, June 2005, 61 pp. http://www.accesseconomics.com.au/publicationsreports/showreport.php?id=10&searchfor=Economic%20Consulting&searchby=area.

Adams, P.D., M. Horridge, J.R. Masden and G. Wittwer, 2002: Drought, regions and the Australian economy between 2001-02 and 2004-05. *Australian Bulletin of Labour*, **28**, 233-249.

ADB, 2005: *Climate Proofing: A Risk-based Approach to Adaptation*. Pacific Studies Series. Asian Development Bank, Manila, 191 pp. http://www.adb.org/Documents/Reports/Climate-Proofing/default.asp.

Agnew, M.D. and J.P. Palutikof, 2001: Impacts of climate on the demand for tourism. *Proceedings of the First International Workshop on Climate, Tourism and Recreation*, A. Matzarakis and C.R.d. Freitas, Eds., International Society of Biometeorology, Commission on Climate Tourism and Recreation, Porto Carras, Halkidiki, 1-10.

AGO, 2003: *Australia–New Zealand Bilateral Climate Change Partnership*. http://www.greenhouse.gov.au/international/partnerships/index.html#newzealand.

AGO, 2006: *Vulnerability to Climate Change of Australia's Coastal Zone: Analysis of gaps in methods, data and system thresholds*.. Australian Greenhouse Office, Commonwealth Government, 120 pp. http://www.greenhouse.gov.au/impacts/publications/pubs/coastal-vulnerability.pdf.

Alexander, L., P. Hope, D. Collins, B. Trewin, A. Lynch and N. Nicholls, 2007: Trends in Australia's climate means and extremes: a global context. *Aust. Meteorol. Mag.*, **56**, 1-18.

Allen Consulting Group, 2005: *Climate Change Risk and Vulnerability: Promoting an Efficient Adaptation Response in Australia*. Report to the Australian Greenhouse Office by the Allen Consulting Group, 159 pp. http://www.greenhouse.gov.au/impacts/publications/risk-vulnerability.html.

Altman, J., 2000: The economic status of Indigenous Australians. Discussion Paper #193, Centre for Aboriginal Economic Policy Research, Australian National University, 18 pp. http://eprints.anu.edu.au/archive/00001001/.

Anderson, B., 2004: The response of Ko Roimate o Hine Hukatere Franz Josef Glacier to climate change. PhD thesis, University of Canterbury, Christchurch.

Anderson, B. and A. Mackintosh, 2006: Temperature change is the major driver of late-glacial and Holocene glacier fluctuations in New Zealand. *Geology*, **34**, 121-124.

Anderson, B., W. Lawson, I. Owens and B. Goodsell, 2006: Past and future mass balance of Ka Roimata o Hine Hukatere (Franz Josef Glacier). *Int. J. Glaciol.*, **52**, 597-607.

Annala, J.H., K.J. Sullivan, N.W.M. Smith, M.H. Griffiths, P.R. Todd, P.M. Mace and A.M. Connell, 2004: *Report from the Fishery Assessment Plenary, May 2004: stock assessment and yield estimates*. Unpublished report held in NIWA library, Wellington, 690 pp.

Arthur, B. and F. Morphy, 2005: *Macquarie Atlas of Indigenous Australia: Culture and Society through Space and Time*. Macquarie University, Macquarie, New South Wales, 278 pp.

Ash, A., L. Hunt, C. Petty, R. Cowley, A. Fisher, N. MacDonald and C. Stokes, 2006: Intensification of pastoral lands in northern Australia. *Proceedings of the 14th Biennial Conference of the Australian Rangeland Society*, P. Erkelenz, Ed., Renmark, Australia, 43.

Asseng, S., P.D. Jamieson, B. Kimball, P. Pinter, K. Sayre, J.W. Bowden and S.M. Howden, 2004: Simulated wheat growth affected by rising temperature, increased water deficit and elevated atmospheric CO_2. *Field Crop. Res.*, **85**, 85-102.

Atwell, B.J., M.L. Henery and D. Whitehead, 2003: Sapwood development in *Pinus radiata* trees grown for three years at ambient and elevated carbon dioxide partial pressures. *Tree Physiol.*, **23**, 13-21.

Austin, P.T., A.J. Hall, W.P. Snelgar and M.J. Currie, 2000: Earlier maturity limits increasing apple fruit size under climate change scenarios. *New Zealand Institute of Agricultural Science and the New Zealand Society for Horticultural Science Annual Convention 2000*, 77pp.

Austroads, 2004: *Impact of Climate Change on Road Infrastructure*. AP-R243/04, Austroads, Sydney, 124 pp. http://www.onlinepublications.austroads.com.au /script/Details.asp?DocN=AR0000048_0904.

Beare, S. and A. Heaney, 2002: Climate change and water resources in the Murray Darling Basin, Australia, impacts and adaptation. Conference Paper 02.11, Australian Bureau of Agricultural and Resource Economics, 33 pp. http://www.abarepublications.com/product.asp?prodid=12389.

Beaumont, L. and L. Hughes, 2002: Potential changes in the distributions of latitudinally restricted Australian butterflies in response to climate change. *Glob. Change Biol.*, **8**, 954-971.

Beaumont, L.J., I.A.W. McAllan and L. Hughes, 2006: A matter of timing: changes in the first date of arrival and last date of departure of Australian migratory birds. *Glob. Change Biol.*, **12**, 1339-1354.

Becken, S., 2005: Harmonizing climate change adaptation and mitigation: the case of tourist resorts in Fiji. *Global Environ. Chang.*, **15**, 381-393.

Becken, S. and J. Hay, 2007: *Tourism and Climate Change: Risks and Opportunities*. Climate Change, Economies and Society series, Channel View Publications, Clevedon, 352 pp.

Beentjes, M.P. and J.A. Renwick, 2001: The relationship between red cod, *Pseudophycis bachus*, recruitment and environmental variables in New Zealand. *Environ. Biol. Fish.*, **61**, 315-328.

Bell, R.G., T.M. Hume and D.M. Hicks, 2001: *Planning for Climate Change Effects on Coastal Margins*. National Institute of Water and Atmospheric Research and New Zealand Climate Change Programme, Hamilton. http://www.climate change.govt.nz/resources/reports/effect-coastal-sep01/index.html.

Bergstrom, D., 2003: Impact of climate change on terrestrial Antarctic and sub-Antarctic biodiversity. *Climate Change Impacts on Biodiversity in Australia*, M. Howden, L. Hughes, M. Dunlop, I. Zethoven, D.W. Hilbert and C. Chilcott, Eds., CSIRO Sustainable Ecosystems, Canberra, 55-57.

Bergstrom, D.M. and P.M. Selkirk, 1999: Bryophyte propagule banks in a feldmark on sub-Antarctic Macquarie Island. *Arct. Antarct. Alp. Res.*, **31**, 202-208. http://www.environment.gov.au/biodiversity/publications/greenhouse/index.html.

Berkelmans, R. and J.K. Oliver, 1999: Large-scale bleaching of corals on the Great Barrier Reef. *Coral Reefs*, **18**, 55-60.

Berkelmans, R., G. De'ath, S. Kininmonth and W.J. Skirving, 2004: A comparison of the 1998 and 2002 coral bleaching events of the Great Barrier Reef: spatial correlation, patterns and predictions. *Coral Reefs*, **23**, 74-83.

Bessen Consulting Services, 2005: *Land and Sea Management Strategy for Torres Strait*. Bessen Consulting Services and Torres Strait NRM Reference Group (TSRA), 97 pp. http://www.tsra.gov.au/pdf/Torres%20Strait%20 Land%20& %20Sea%20Mgt%20Strategy%202006.pdf.

BoM, 2006a: *Living with Drought*. Australian Bureau of Meteorology. http://www.bom.gov.au/climate/drought/livedrought.shtml.

BoM, 2006b: *Severe Tropical Cyclone Larry*. Queensland Regional Office, Australian Bureau of Meteorology. http://www.bom.gov.au/weather/qld/cyclone/ tc_larry/.

Bowman, D.M.J.S., A. Walsh and D.J. Milne, 2001: Forest expansion and grassland contraction within a Eucalyptus savanna matrix between 1941 and 1994 at Litchfield National Park in the Australian monsoon tropics. *Global Ecol. Biogeogr.*, **10**, 535-538.

Braaf, R., 1999: Improving impact assessment methods: climate change and the health of indigenous Australians. *Global Environ. Chang.*, **9**, 95-104.

BRANZ, 2007: An assessment of the need to adapt buildings for the unavoidable consequences of climate change. Report EC0893, Building Research Australia and New Zealand. http://www.greenhouse.gov.au. In press.

BRS, 2003: *State of the Forest Report*. Bureau of Rural Sciences, Australian Government Printer, Canberra, 408 pp. http://www.affa.gov.au/stateoftheforests.

Bryan, B., N. Harvey, T. Belperio and B. Bourman, 2001: Distributed process modeling for regional assessment of coastal vulnerability to sea-level rise. *Environ. Model. Assess.*, **6**, 57-65.

BTE, 2001: Economic costs of natural disasters in Australia. Report 103, Bureau of Transport Economics, Canberra. http://www.btre.gov.au/docs/reports/r103/ r103.aspx.

Budd, G.M., 2000: Changes in Heard Island glaciers, king penguins and fur seals since 1947. *Papers and Proceedings of the Royal Society of Tasmania*, **133**, 47-60.

Bunn, S.E. and A.H. Arthington, 2002: Basic principles and ecological consequences of altered flow regimes for aquatic biodiversity. *Environ. Manage.*, **30**, 492-507.

Burgess, S.M., 2005: 2004-05 Tropical cyclone summary. *Island Climate Update*, **57**, 6.

Burke, E.J., S.J. Brown and N. Christidis, 2006: Modelling the recent evolution of global drought and projections for the 21st century with the Hadley Centre climate model. *J. Hydrometeorol.*, **7**, 1113-1125.

CAE, 2005: Managing flood risk: the case for change. Report prepared for the Flood Risk Management Governance Group by the Centre for Advanced Engineering, University of Canterbury. http://www.caenz.com/ info/MFR/MFR.html.

Cai, W., G. Shi, T. Cowan, D. Bi and J. Ribbe, 2005: The response of the Southern annular mode, the East Australian current, and the southern mid-latitude ocean circulation to global warming. *Geophys. Res. Lett.*, **32**, L23706, doi:10.1029/2005GL024701.

Cary, G.J., 2002: Importance of a changing climate for fire regimes in Australia. *Flammable Australia: Fire Regimes and Biodiversity of a Continent*, R.A. Bradstock, J.E. Williams and A.M. Gill, Eds., Cambridge University Press, Melbourne, 26-49.

Chakraborty, S., G. Murray and N. White, 2002: *Impact of Climate Change on Important Plant Diseases in Australia*. RIRDC Publication No. W02/010. http://www.rirdc.gov.au/reports/AFT/02-010sum.html.

Chambers, L.E., 2005: Migration dates at Eyre Bird Observatory: links with climate change? *Climate Res.*, **29**, 157-165.

Chambers, L.E., L. Hughes and M.A. Weston, 2005: Climate change and its impact on Australia's avifauna. *Emu*, **105**, 1-20.

Chen, K. and J. McAneney, 2006: High-resolution estimates of Australia's coastal population with validations of global population, shoreline and elevation datasets. *Geophys. Res. Lett.*, **33**, L16601. doi:10.1029/2006GL026981.

Chiew, F.H.S. and T.A. McMahon, 2002: Modelling the impacts of climate change on Australian stream-flow. *Hydrol. Process.*, **16**, 1235-1245.

Chiew, F.H.S., T.I. Harrold, L. Siriwardenena, R.N. Jones and R. Srikanthan, 2003: Simulation of climate change impact on runoff using rainfall scenarios that consider daily patterns of change from GCMs. *MODSIM 2003: International Congress on Modelling and Simulation: Proceedings*, D.A. Post, Ed., Modelling and Simulation Society of Australia and New Zealand, Canberra, 154-159.

Chinn, T.J., 2001: Distribution of the glacial water resources of New Zealand. *Journal of Hydrolology (NZ)*, **40**, 139–187.

Christensen, J.H., B. Hewitson, A. Busuioc, A. Chen, X. Gao, I. Held, R. Jones, R.K. Kolli, W.-T. Kwon, R. Laprise, V. Magaña Rueda, L. Mearns, C.G. Menéndez, J. Räisänen, A. Rinke, A. Sarr and P. Whetton, 2007: Regional climate projections. *Climate Change 2007:The Physical Science Basis, Contribution of: Working Group I to the Fourth Assessment Report of the Intergovernmental Panel on Climate Change*. S. Solomon, D. Qin, M. Manning, Z. Chen, M. Marquis, K.B. Averyt, M. Tignor and H.L. Miller, Eds., Cambridge University Press, Cambridge, 847-940.

Church, J., J. Hunter, K. McInnes and N.J. White, 2004: Sea-level rise and the frequency of extreme event around the Australian coastline. *Coast to Coast '04: Australia's National Coastal Conference, conference proceedings,* Hobart, Tasmania. 8 pp.

Clare, G.R., B.B. Fitzharris, T.J. Chinn and M.J. Salinger, 2002: Interannual variation in end-of-summer snow-lines of the Southern Alps of New Zealand, in response to changes in Southern Hemisphere atmospheric circulation and sea surface temperature patterns. *Int. J. Climatol.*, **22**, 107-120.

Clark, H., N.D. Mitchell, P.C.D. Newton and B.D. Campbell, 2001: The sensitivity of New Zealand's managed pastures to climate change. *The Effects of Climate Change and Variation in New Zealand: An Assessment using the CLIMPACTS System*, R.A. Warrick, G.J. Kenny and J.J. Harman, Eds., International Global Change Institute (IGCI), University of Waikato, Hamilton, 65-77.

Coleman, T., O. Hoegh-Guldberg, D. Karoly, I. Lowe, T. McMichael, C.D. Mitchell, G.I. Pearman, P. Scaife and J. Reynolds, 2004: *Climate Change: Solutions for Australia*. Australian Climate Group, 35 pp. http://www.wwf.org.au/publications/acg_solutions.pdf.

Cottrell, B., C. Insley, R. Meade and J. West, 2004: Report of the Climate Change Maori Issues Group. New Zealand Climate Change Office, Ministry for the Environment, Wellington. http://www.climatechange.govt.nz/resources/.

Cowell, P.J., B.G. Thom, R.A. Jones, C.H. Evert and D. Simanovic, 2006: Management of uncertainty in predicting climate-change impacts on beaches. *J. Coastal Res.*, **22**, 232-245.

Crimp, S.J., N.R. Flood, J.O. Carter, J.P. Conroy and G.M. McKeon, 2002: *Evaluation of the Potential Impacts of Climate Change On Native Pasture Production – Implications for livestock carrying capacity*: Final Report for the Australian Greenhouse Office, 60pp.

Crimp, S., J. Balston, A. Ash, L. Anderson-Berry, T. Done, R. Greiner, D. Hilbert, M. Howden, R. Jones, C. Stokes, N. Stoeckl, B. Sutherst and P. Whetton, 2004: *Climate Change in the Cairns and Great Barrier Reef Region: Scope and Focus for an Integrated Assessment*. Australian Greenhouse Office, 100 pp. http://www.greenhouse.gov.au/impacts/publications/pubs/gbr.pdf.

CSIRO, 2002: Climate change and Australia's coastal communities. CSIRO Atmospheric Research, Aspendale, Victoria, 8 pp. http://www.cmar.csiro.au/e-print/open/CoastalBroch2002.pdf.

Cullen, P., 2002: Living with water: sustainability in a dry land. *Adelaide Festival of Arts, Getting it Right Symposium, 1-12 March, 2002*. http://freshwater.canberra.edu.au/Publications.nsf/0/3c1c61a1a89d5798ca256f0b001b7a30?OpenDocument.

Curtis, I.A., 2004: Valuing ecosystem goods and services: a new approach using a surrogate market and the combination of a multiple criteria analysis and a Delphi panel to assign weights to the attributes. *Ecol. Econ.*, **50**, 163-194.

DAFF, 2006a: *National Water Initiative*. Department of Agriculture, Forestry and Fisheries, Australia. http://www.pmc.gov.au/water_reform/nwi.cfm.

DAFF, 2006b: *Contours*. Department of Agriculture, Forestry and Fisheries, 24 pp. http://www.daff.gov.au/__data/assets/pdf_file/0020/98201/contours-dec-06.pdf.

Dahm, J., G. Jenks and D. Bergin, 2005: Community-based dune management for the mitigation of coastal hazards and climate change effects: a guide for local authorities. Report for the New Zealand Ministry for the Environment, 36 pp. http://www.envbop.govt.nz/media/pdf/Report_Coastalhazardsandclimate.pdf.

Davis, J.R., Ed., 1997: *Managing Algal Blooms: Outcomes from CSIRO's Multi-Divisional Blue-Green Algae Program*. CSIRO Land and Water, Canberra, 113 pp.

Dayananda, K.G., J. Pathirana, M.D. Davis, M.J. Salinger, A.B. Mullan, P. Kinley and G. Paterson, 2005: Flood risk under existing and future climate scenarios in Auckland City (New Zealand). *World Water and Environmental Resources Congress 2005*, American Society of Civil Engineers, Reston, Virginia, doi: 10.1061/40792(173)477.

De Wet, N., W. Ye, S. Hales, R. Warrick, A. Woodward and P. Weinstein, 2001: Use of a computer model to identify potential hotspots for dengue fever in New Zealand. *New Zeal. Med. J.*, **114**, 420-422.

DEH, 2005: *Adelaide's Living Beaches: A Strategy for 2005-2025*. South Australian Department for Environment and Heritage, Government of South Australia, 220 pp. http://www.environment.sa.gov.au/coasts/pdfs/alb_technical_report.pdf.

DEST, 2004: *National Research Priorities*. http://www.dest.gov.au/sectors/research_sector/policies_issues_reviews/key_issues/national_research_priorities/default.htm.

DEUS, 2006: *NSW Government Water Savings Fund*. Department of Energy, Utilities and Sustainability, 17 pp. http://www.deus.nsw.gov.au/Publications/WaterSavingsFundR3Guide.pdf.

Diamond, H., 2006: Review of recent tropical cyclone research. *Island Climate Update*, **72**, 6.

DoC, 1994: *New Zealand Coastal Policy Statement 1994*. Department of Conservation, Wellington, 30 pp. http://www.doc.govt.nz/upload/documents/conservation/marine-and-coastal/coastal-management/nz-coastal-policy-statement.pdf.

Done, T., P. Whetton, R. Jones, R. Berkelmans, J. Lough, W. Skirving and S. Wooldridge, 2003: *Global climate change and coral bleaching on the Great Barrier Reef*. Final Report to the State of Queensland Greenhouse Taskforce through the Department of Natural Resources and Mining, Townsville, 49pp. http://www.longpaddock.qld.gov.au/ClimateChanges/pub/CoralBleaching.pdf.

Donner, S.D., W.J. Skirving, C.M. Little, M. Oppenheimer and O. Hoegh-Guldberg, 2005: Global assessment of coral bleaching and required rates of adaptation under climate change. *Glob. Change Biol.*, **11**, 1-15.

DPMC, 2004: *Water Reform*. Department of Prime Minister and Cabinet, Australia. http://www.dpmc.gov.au/nwi/index.cfm.

D'Souza, R.M., N.G. Becker, G. Hall and K. Moodie, 2004: Does ambient temperature affect foodborne disease? *Epidemiology*, **15**, 86-92.

Dupont, A. and G. Pearman, 2006: Heating up the planet: climate change and security. Paper 12, Lowy Institute for International Policy. Longueville Media, 143 pp.www.lowyinstitute.org/PublicationGet.asp?i=391.

Dymond, J.R., A.G. Ausseil, J.D. Shepherd and L. Buetttner, 2006: Validation of a region-wide model of landslide susceptibility in the Manawatu-Wanganui region of New Zealand. *Geomorphology*, **74**, 70-79.

Electricity Commission, 2005: Initial statement of opportunities: Electricity Commission, Wellington, 158 pp. http://www.electricitycommission.govt.nz/opdev/transmis/soo/pdfssoo/SOOv2.pdf.

Ellemor, H., 2005: Reconsidering emergency management and Indigenous communities in Australia. *Environmental Hazards*, **6**, 1-7.

Engineers Australia, 2004: *Guidelines for Responding to the Effects of Climate Change in Coastal and Ocean Engineering*. National Committee on Coastal and Ocean Engineering, 64 pp. http://www.engineersaustralia.org.au/.

Evans, K.L., C. Tyler, T.M. Blackburn and R.P. Duncan, 2003: Changes in the breeding biology of the welcome swallow (*Hirundo tahitica*) in New Zealand since colonisation. *Emu*, **103**, 215-220.

Falconer, I.R., 1997: Blue-green algae in lakes and rivers: their harmful effects on human health. *Australian Biologist*, **10**, 107-110.

Fitzharris, B.B., 2004: Possible impact of future climate change on seasonal snow of the Southern Alps of New Zealand. *A Gaian World: Essays in Honour of Peter Holland*, G. Kearsley and B. Fitzharris, Eds., Department of Geography, School of Social Science, University of Otago, Dunedin, 231-241.

Folland, C.K., M.J. Salinger, N. Jiang and N. Rayner, 2003: Trends and variations in South Pacific island and ocean surface temperatures. *J. Climate*, **16**, 2859-2874.

Folland, C.K., B. Dong, R.J. Allan, H. Meinke and B. Bhaskaran, 2005: The Interdecadal Pacific Oscillation and its climatic impacts. *AMS 16th Conference on Climate Variability and Change*. http://ams.confex.com/ams/Annual2005/techprogram/paper_86167.htm.

Francis, M.P., 1994: Growth of juvenile snapper, *Pagrus auratus* (Sparidae). *New Zeal. J. Mar. Fresh.*, **28**, 201-218.

Francis, M.P., 1996: Geographic distribution of marine reef fishes in the New Zealand region. *New Zeal. J. Mar. Fresh.*, **30**, 35-55.

Frenot, Y., S.L. Chown, J. Whinam, P.M. Selkirk, P. Convey, M. Skotnicki and D. Bergstrom, 2005: Biological invasions in the Antarctic: extent, impacts and implications. *Biol. Rev.*, **80**, 45-72.

FRST, 2005: *Target Outcomes and Themes for the Foundation's Investment Portfolios*. Foundation for Research Science and Technology, 39 pp. http://www.frst.govt.nz/research/downloads/ArchivedDocuments/20050500_Investment_Portfolio-Target_Outcomes_ and_Themes_May05.pdf.

Gallant, A., K. Hennessy and J. Risbey, 2007: A re-examination of trends in rainfall indices for six Australian regions. *Aust. Meteorol. Mag.*, accepted.

Ghannoum, O., S. von Caemmerer, L.H. Ziska and J.P. Conroy, 2000: The growth response of C4 plants to rising atmospheric CO_2 partial pressure: a reassessment. *Plant Cell Environ.*, **23**, 931-942.

Gifford, R.M. and S.M. Howden, 2001: Vegetation thickening in an ecological perspective. significance to national greenhouse-gas inventories. *Environ. Sci. Policy*, **2-3**, 59-72.

Glade, T., 1998: Establishing the frequency and magnitude of landslide-triggering rainstorm events in New Zealand. *Environ. Geol.*, **35**, 160-174.

Goodwin, I.D., 2005: A mid-shelf, mean wave direction climatology for south-eastern Australia, and its relationship to the El Niño-Southern Oscillation since 1878 AD. *Int. J. Climatol.*, **25**, 1715-1729.

Goodwin, I.D., M.A. Stables and J.M. Olley, 2006: Wave climate, sand budget and shoreline alignment evolution of the Iluka–Woody Bay sand barrier, northern New South Wales, Australia, since 3000 yr BP. *Mar. Geol.*, **226**, 127-144.

Gordon, N.D., 1986: The Southern Oscillation and New Zealand weather. *Mon. Weather Rev.*, **114**, 371-387.

Government of South Australia, 2005: *Water Proofing Adelaide: A Thirst for Change 2005–2025*. Government of SA, 64 pp. http://www.waterproofingadelaide.sa.gov.au/pdf/wpa_Strategy.pdf.

Government of Western Australia, 2003: *Securing our Water Future: A State Water Strategy for Western Australia*. Government of WA, 64 pp. http://dows.lincdigital.com.au/files/State_Water_Strategy_complete_001.pdf.

Government of Western Australia, 2006: *Draft State Water Plan*. Government of WA, 88 pp. http://dows.lincdigital.com.au/files/Draft%20State%20Water%20Plan.pdf.

Gray, W., R. Ibbitt, R. Turner, M. Duncan and M. Hollis, 2005: A methodology to assess the impacts of climate change on flood risk in New Zealand. NIWA Client Report CHC2005-060, New Zealand Climate Change Office, Ministry for the Environment. NIWA, Christchurch, 36 pp. http://www.mfe.govt.nz/ publications/climate/impact-climate-change-flood-risk-jul05/html/page8.html.

Green, D., 2006a: Climate Change and health: impacts on remote Indigenous communities in northern Australia. CSIRO Research Paper No. 12, Melbourne, 17 pp. http://sharingknowledge.net.au/files/climateimpacts_health_report.pdf.

Green, D., 2006b: How might climate change impact island culture in the in the Torres Strait? CSIRO Research Paper No. 11, Melbourne, 20 pp. www.dar.csiro.au/sharingknowledge/files/climateimpacts_TSIculture_report.pdf

Green, D. and B. Preston, 2006: Climate change impacts on remote Indigenous communities in northern Australia. CSIRO Working Paper, Melbourne. http://sharingknowledge.net.au/regions.html

Green, K. and C.M. Pickering, 2002: A potential scenario for mammal and bird diversity in the Snowy Mountains of Australia in relation to climate change. *Mountain Biodiversity: A Global Assessment*, C. Korner and E.M. Spehn, Eds., Parthenon Press, London, 241-249.

Gregory, J.M., J.A. Church, G.J. Boer, K.W. Dixon, G.M. Flato, D.R. Jackett, J.A. Lowe, S.P. O'Farrell, E. Roeckner, G.L. Russell, R.J. Stouffer and M. Winton, 2001: Comparison of results from several AOGCMs for global and regional sea-level change 1900-2100. *Clim. Dynam.*, **18**, 225-240.

Griffiths, G.A., 1990: Water resources. *Climatic Change: Impacts on New Zealand*. Ministry for the Environment, Wellington, 38-43.

Griffiths, G.M, 2007: Changes in New Zealand daily rainfall extremes 1930-2004. *Weather Climate*, **27**, 47-66.

Hall, A.J., G.J. Kenny, P.T. Austin and H.G. McPherson, 2001: Changes in kiwifruit phenology with climate. *The Effects of Climate Change and Variation in New Zealand: An Assessment using the CLIMPACTS System*, R.A. Warrick, G.J. Kenny, and J.J. Harman, Eds., International Global Change Institute (IGCI), Hamilton, 33-46. http://www.waikato.ac.nz/igci/ climpacts/Linked%20documents/Chapter_3.pdf.

Hall, C.J. and C.W. Burns, 2002: Mortality and growth responses of *Daphnia carinata* to increases in temperature and salinity. *Freshwater Biol.*, **47**, 451-458.

Hall, G.V., R.M. D'Souza and M.D. Kirk, 2002: Food-borne disease in the new millennium: out of the frying pan and into the fire? *Med. J. Australia*, **177**, 614-618.

Hall, M. and J. Higham, 2005: *Tourism, Recreation and Climate Change: Aspects of Tourism*. Channel View Publications, Clevedon, 309 pp.

Halloy, S.R.P. and A.F. Mark, 2003: Climate change effects on alpine plant biodiversity: a New Zealand perspective on quantifying the threat. *Arct. Antarct. Alp. Res.*, **35**, 248-254.

Hampe, A. and R.J. Petit, 2005: Conserving biodiversity under climate change: the rear edge matters. *Ecol. Lett.*, **8**, 461-467.

Hannah, J., 2004: An updated analysis of long-term sea-level change in New Zealand. *Geophys. Res. Lett.*, **31**, L03307. doi:10.1029/2003GL019166.

Hare, W., 2003: *Assessment of Knowledge on Impacts of Climate Change: Contribution to the Specification of Article 2 of the UNFCCC*. WGBU, Berlin, 106 pp. http://www.wbgu.de/wbgu_sn2003_ex01.pdf.

Harmsworth, G.R. and B. Raynor, 2005: Cultural consideration in landslide risk perception. *Landslide Hazard and Risk*, T. Glade, M. Anderson and M. Crozier, Eds., John Wiley and Sons, Chichester, 219-249.

Harvey, N. and B. Caton, 2003: *Coastal Management in Australia*. Oxford University Press, Melbourne, 342 pp.

Hennecke, W., C. Greve, P. Cowell and B. Thom, 2004: GIS-based coastal behaviour modeling and simulation of potential land and property loss: implications of sea-level rise at Colloroy/Narrabeen beach, Sydney (Australia). *Coast. Manage.*, **32**, 449-470.

Hennessy, K., C. Lucas, N. Nicholls, J. Bathols, R. Suppiah and J. Ricketts, 2006: *Climate Change Impacts on Fire-Weather in South-East Australia*. Consultancy Report for the New South Wales Greenhouse Office, Victorian Department of Sustainability and Environment, Tasmanian Department of Primary Industries, Water and Environment, and the Australian Greenhouse Office. CSIRO Atmospheric Research and Australian Government Bureau of Meteorology, 78 pp. http://www.greenhouse.gov.au/impacts/publications/pubs/bushfire-report.pdf.

Hennessy, K.J., 2004: Climate change and Australian storms. *Proceedings of the International Conference on Storms*, Brisbane, 8.

Hennessy, K.J., P.H. Whetton, J. Bathols, M. Hutchinson and J. Sharples, 2003: *The Impact of Climate Change on Snow Conditions in Australia*. Consultancy Report for the Victorian Dept of Sustainability and Environment, NSW National Parks and Wildlife Service, Australian Greenhouse Office and the Australian Ski Areas Association., CSIRO Atmospheric Research, 47 pp. http://www.cmar.csiro.au/e-print/open/hennessy_2003a.pdf.

Hennessy, K.J., K.L. McInnes, D.J. Abbs, R.N. Jones, J.M. Bathols, R. Suppiah, J.R. Ricketts, A.S. Rafter, D. Collins and D. Jones, 2004: *Climate Change in New South Wales. Part 2, Projected Changes in Climate Extremes*. Consultancy Report for the New South Wales Greenhouse Office, CSIRO Atmospheric Research, 79 pp. http://www.cmar.csiro.au/e-print/ open/hennessy_2004c.pdf.

Herron, N., R. Davis and R. Jones, 2002: The effects of large-scale afforestation and climate change on water allocation in the Macquarie river catchment, NSW, Australia. *Journal of Environmental Management Australia*, **65**, 369-381.

Hilbert, D.H., B. Ostendorf and M.S. Hopkins, 2001: Sensitivity of tropical forests to climate change in the humid tropics of North Queensland. *Austral Ecol.*, **26**, 590-603.

Hilbert, D.W., M. Bradford, T. Parker and D.A. Westcott, 2004: Golden bowerbird (*Prionodura newtonia*) habitat in past, present and future climates: predicted extinction of a vertebrate in tropical highlands due to global warming. *Biol. Conserv.*, **116**, 367-377. http://www.cababstractsplus.org/google/abstract.asp?AcNo=20043043667.

Hill, R., 2004: *Yalanji Warranga Kaban: Yalanji People of the Rainforest Fire Management Book*. Little Ramsay Press, Cairns, 110 pp.

Hobday, A. and R. Matear, Eds., 2005: *Review of climate impacts on Australian fisheries and aquaculture: implications for the effects of climate change*. Report to the Australian Greenhouse Office, Canberra.

Hoegh-Guldberg, O., 1999: Climate change, coral bleaching and the future of the world's coral reefs. *Mar. Freshwater Res.*, **50**, 839-866.

Hoegh-Guldberg, O., 2004: Coral reefs in a century of rapid environmental change. *Symbiosis*, **37**, 1-31.

Holper, P.N., S. Lucy, M. Nolan, C. Senese and K. Hennessy, 2007: *Infrastructure and Climate Change Risk Assessment for Victoria*. Consultancy Report to the Victorian Government prepared by CSIRO, Maunsell AECOM and Phillips Fox, 84 pp. http://www.greenhouse.vic.gov.au/greenhouse/wcmn302.nsf/childdocs/-9440F41741A0AF31CA2571A80011CBB6?open.

Hood, A., H. Hossain, V. Sposito, L. Tiller, S. Cook, C. Jayawardana, S. Ryan, A. Skelton, P. Whetton, B. Cechet, K. Hennessy and C. Page, 2002: *Options for Victorian Agriculture in a "New" Climate: A Pilot Study Linking Climate Change Scenario Modelling and Land Suitability Modelling. Vol. 1: Concepts and Analysis. Vol. Two: Modelling Outputs*. Department of Natural Resources and Environment, Victoria, 83pp. http://www.dar.csiro.au/cgi-bin/abstract_srch.pl?_ abstract_available_2256.

Howden, M. and R.N. Jones, 2004: Risk assessment of climate change impacts on Australia's wheat industry. *New Directions for a Diverse Planet: Proceedings of the 4th International Crop Science Congress*, Brisbane. http://www.cropscience.org.au/icsc2004/symposia/6/2/1848_howdensm.htm.

Howden, S.M. and S. Crimp, 2001: Effect of climate change on electricity demand in Australia. *Proceedings of Integrating Models for Natural Resource Management across Disciplines, Issues and Scales (2). MODSIM 2001 International Congress on Modelling and Simulation*, F. Ghassemi, P. Whetton, R. Little and M. Littleboy, Eds., Modelling and Simulation Society of Australia and New Zealand, Canberra, 655-660.

Howden, S.M., W.B. Hall and D. Bruget, 1999a: Heat stress and beef cattle in Australian rangelands: recent trends and climate change. *People and Rangelands: Building the Future. Proceedings of the VI International Rangeland Congress*, D. Eldridge and D. Freudenberger, Eds., Townsville, 43-45.

Howden, S.M., G.M. McKeon, J.O. Carter and A. Beswick, 1999b: Potential global change impacts on C3-C4 grass distribution in eastern Australian rangelands. *People and Rangelands: Building the Future. Proceedings of the VI In-*

ternational Rangeland Congress, D. Eldridge and D. Freudenberger, Eds., Townsville, 41-43.

Howden, S.M., P.J. Reyenga and J.T. Gorman, 1999c: Current evidence of global change and its impacts: Australian forests and other ecosystems. CSIRO Wildlife and Ecology Working Paper 99/01, Canberra, 23 pp.

Howden, S.M., P.J. Reyenga and H. Meinke, 1999d: Global change impacts on Australian wheat cropping. CSIRO Wildlife and Ecology Working Paper 99/04, Report to the Australian Greenhouse Office, Canberra, 121 pp. http://www.cse.csiro.au/publications/1999/globalchange-99-01.pdf.

Howden, S.M., G.M. McKeon, H. Meinke, M. Entel and N. Flood, 2001a: Impacts of climate change and climate variability on the competitiveness of wheat and beef cattle production in Emerald, northeast Australia. *Environ. Int.*, **27**, 155-160.

Howden, S.M., J.L. Moore, G.M. McKeon and J.O. Carter, 2001b: Global change and the mulga woodlands of southwest Queensland: Greenhouse emissions, impacts and adaptation. *Environ. Int.*, **27**, 161-166.

Howden, S.M., A.J. Ash, E.W.R. Barlow, T. Booth, S. Charles, R. Cechet, S. Crimp, R.M. Gifford, K. Hennessy, R.N. Jones, M.U.F. Kirschbaum, G.M. McKeon, H. Meinke, S. Park, R. Sutherst, L. Webb and P.J. Whetton, 2003a: *An Overview of the Adaptive Capacity of the Australian Agricultural Sector to Climate Change: Options, Costs and Benefits*. Report to the Australian Greenhouse Office, Canberra, 157 pp.

Howden, S.M., H. Meinke, B. Power and G.M. McKeon, 2003b: Risk management of wheat in a non-stationary climate: frost in Central Queensland. Integrative modelling of biophysical, social and economic systems for resource management solutions. *Proceedings of the International Congress on Modelling and Simulation*, D.A. Post, Ed., Townsville, 17-22.

Howden, S.M., P.J. Reyenga and H. Meinke, 2003c: Managing the quality of wheat grain under global change. *Integrative Modelling of Biophysical, Social and Economic Systems for Resource Management Solutions: Proceedings of the International Congress on Modelling and Simulation*, D.A. Post, Ed., Townsville, 35-41.

Howe, C., R.N. Jones, S. Maheepala and B. Rhodes, 2005: *Implications of Potential Climate Change for Melbourne's Water Resources*. CSIRO Urban Water, CSIRO Atmospheric Research, and Melbourne Water, Melbourne, 36 pp. http://www.melbournewater.com.au/content/library/news/whats_new/Climate_Change_Study.pdf

Hughes, L., 2003: Climate change and Australia: trends, projections and impacts. *Austral Ecol.*, **28**, 423-443.

ICA, 2007: *Catastrophe Information*. Insurance Council of Australia. http://www.insurancecouncil.com.au/Catastrophe-Information/default.aspx

IGWG, 2004: *Environmental Health Needs of Indigenous Communities in Western Australia: The 2004 Survey and its Findings*. Western Australia Department of Indigenous Affairs, 240 pp. http://www.dia.wa.gov.au/Publications/.

Insurance Council of New Zealand, 2005: *Current Issues: The Cost of Weather Losses: Claims History*. NZ Insurance Council, Wellington. http://www.icnz.org.nz/current/wx.php.

IOCI, 2002: *Climate Variability and Change in Southwest Western Australia*. Indian Ocean Climate Initiative, Perth, 36 pp. http://www.ioci.org.au/publications/pdf/IOCI_CVCSW02.pdf.

IPCC, 1992: *Climate Change 1992: The Supplementary Report to the IPCC Scientific Assessment*. J.T. Houghton, B.A. Callander and S.K. Varney, Eds., Cambridge University Press, Cambridge, 200 pp.

IPCC, 2007: *Climate Change 2007: The Physical Science Basis, Contribution of: Working Group I to the Fourth Assessment Report of the Intergovernmental Panel on Climate Change*. S. Solomon, D. Qin, M. Manning, Z. Chen, M. Marquis, K.B. Averyt, M. Tignor and H.L. Miller, Eds., Cambridge University Press, Cambridge, 18 pp.

Jackson, S., 2005: A burgeoning role for Aboriginal knowledge. ECOS CSIRO Publication, June-July 2005, 11-12.

Jamieson, P.D., J. Berntsen, F. Ewert, B.A. Kimball, J.E. Olesen, P.J. Pinter Jr, J.R. Porter and M.A. Semenov, 2000: Modelling CO_2 effects on wheat with varying nitrogen supplies. *Agr. Ecosyst. Environ.*, **82**, 27-37.

Jenkins, B., 2006: Overview of Environment Canterbury water issues: managing drought in a changing climate. *Royal Society of New Zealand Drought Workshop, 10 April 2006*, Christchurch. http://www.rsnz.org/advisory/nz_climate/workshopApr2006/.

Johnston, F.H., A.M. Kavanagh, D.M. Bowman and R.K. Scott, 2002: Exposure to bushfire smoke and asthma: an ecological study. *Med. J. Australia*, **176**, 535–538.

Jones, R. and P. Durack, 2005: Estimating the impacts of climate change on Victoria's runoff using a hydrological sensitivity model. Consultancy Report for the Victorian Department of Sustainability and Environment, Victoria, 50 pp. http://www.greenhouse.vic.gov.au/CSIRO%20Report%20-%20Runoff.pdf.

Jones, R.J., J. Bowyer, O. Hoegh-Guldberg and L.L. Blackall, 2004: Dynamics of a temperature-related coral disease outbreak. *Mar. Ecol.-Prog. Ser.*, **281**, 63-77.

Jones, R.N., Ed., 2004a: Applying a climate scenario generator to water resource policy issues. *Final Report*. CSIRO Atmospheric Research, Hassall and Associates, NSW Department of Infrastructure, Planning and Natural Resources, RIRDC Project No. CSM-5A, 48 pp.

Jones, R.N., 2004b: Managing climate change risks. *The Benefits of Climate Policies: Analytical and Framework Issues*, J. Corfee Morlot and S. Agrawala, Eds., OECD, 251-297. http://www.oecd.org/document/35/0,2340,en_ 2649_34361_34086819_1_1_1_1,00.html.

Jones, R.N. and C.M. Page, 2001: Assessing the risk of climate change on the water resources of the Macquarie river catchment. *Integrating Models for Natural Resources Management across Disciplines, Issues and Scales. Part 2. MODSIM 2001 International Congress on Modelling and Simulation*, P. Ghassemi, P. Whetton, R. Little and M. Littleboy, Eds., Modelling and Simulation Society of Australia and New Zealand, Canberra, 673-678. http://www.aiaccproject.org/resources/ele_lib_docs/rjones.pdf.

Jones, R.N., P. Dettman, G. Park, M. Rogers and T. White, 2007: The relationship between adaptation and mitigation in managing climate change risks: a regional approach. *Mitigation Adaptation Strategies Global Change*, **12**, 685-712.

Jovanovic, B., D.A. Jones and D. Collins, 2007: A high-quality monthly pan evaporation dataset for Australia. *Climatic Change,* accepted.

Karoly, D.J. and K. Braganza, 2005a: Attribution of recent temperature changes in the Australian region. *J. Climate*, **18**, 457-464.

Karoly, D.J. and K. Braganza, 2005b: A new approach to detection of anthropogenic temperature changes in the Australian region. *Meteorol. Atmos. Phys.*, **89**, 57-67.

Kennish, M.J., 2002: Environmental threats and environmental future of estuaries. *Environ. Conserv.*, **29**, 78-107.

Kenny, G., 2005: *Adapting to Climate Change in Eastern New Zealand: A Farmer Perspective*. Earthwise Consulting, Hastings, 148 pp. http://www.earthlimited.org.

Kenny, G., 2007: *Adapting to Climate Change: A View from the Ground*. Parliamentary Commissioner for the Environment, New Zealand, 23 pp., in press.

Kenny, G.J., 2002: Climate change and land management in Hawke's Bay: a pilot study on adaptation. Ministry for the Environment, Wellington. http://www.maf.govt.nz/mafnet/publications/rmupdate/rm10/rm-update-june-2002-01.htm.

Kenny, G.J., R.A. Warrick, B.D. Campbell, G.C. Sims, M. Camilleri, P.D. Jamieson, N.D. Mitchell, H.G. McPherson and M.J. Salinger, 2000: Investigating climate change impacts and thresholds: an application of the CLIMPACTS integrated assessment model for New Zealand agriculture. *Climatic Change*, **46**, 91-113.

Kirono, D.G.C. and R.N. Jones, 2007: A bivariate test for detecting inhomogeneities in pan evaporation time series. *Aust. Meteorol. Mag.*, accepted.

Kirschbaum, M.U.F., 1999a: Forest growth and species distributions in a changing climate. *Tree Physiol.*, **20**, 309-322.

Kirschbaum, M.U.F., 1999b: Modelling forest growth and carbon storage with increasing CO_2 and temperature. *Tellus B*, **51**, 871-888.

Koslow, A. and R.E. Thresher, 1999: Climate and fisheries on the south-east Australian continental shelf and slope. Final Report, FRDC Project Number 96/111. http://www.cvap.gov.au/newfsKoslow.html.

Kriticos, D.J., R.W. Sutherst, J.R. Brown, S.W. Adkins and G.F. Maywald, 2003a: Climate change and biotic invasions: a case history of a tropical woody vine. *Biol. Invasions*, **5**, 147-165.

Kriticos, D.J., R.W. Sutherst, J.R. Brown, S.W. Adkins and G.F. Maywald, 2003b: Climate change and the potential distribution of an invasive alien plant: *Acacia nilotica* spp. *indica* in Australia. *J. Appl. Ecol.*, **40**, 111-124.

Kuleshov, Y.A., 2003: *Tropical Cyclone Climatology for the Southern Hemisphere. Part 1. Spatial and Temporal Profiles of Tropical Cyclones in the Southern Hemisphere*. National Climate Centre, Australian Bureau of Meteorology, 1-22.

Kyle, G., 2006: Impacts of climate change on the Mirarr estates, Northern Territory. Report #100406, Gundjeihmi Aboriginal Corporation report, 7 pp.

Langdon, C., 2002: Review of experimental evidence for effects of CO_2 on calcification of reef builders. *Proceedings of the 9th International Coral Reef Symposium*, Ministry of Environment, Indonesian Institute of Science, International Society for Reef Studies, Bali 23–27 October, 1091-1098.

Langdon, C., T. Takahashi, F. Marubini, M. Atkinson, C. Sweeney, H. Aceves, H. Barnett, D. Chipman and J. Goddard, 2000: Effect of calcium carbonate saturation state on the calcification rate of an experimental coral reef. *Global Biogeochem. Cy.*, **14**, 639-654.

Larned, S.T., M.R. Scarsbrook, T.H. Snelder, N.J. Norton and B.J.F. Biggs, 2004: Water quality in low-elevation streams and rivers of New Zealand: recent state and trends in contrasting land-cover classes. *New Zeal. J. Mar. Fresh.*, **38**, 347-366.

Lavorel, S. and W. Steffen, 2004: Cascading impacts of land use through time: the Canberra bushfire disaster. *Global Change and the Earth System: A Planet Under Pressure*, W. Steffen, A. Sanderson, P.D. Tyson, J. Jäger, P.A. Matson, B. Moore III, F. Oldfield, K. Richardson, H.J. Schellnhuber, B.L. Turner II and R.J. Wasson, Eds., IGBP Global Change Series. Springer-Verlag, Berlin, 186-188.

Lehodey, P., M. Mertignac, J. Hampton, A. Lewis and J. Picaut, 1997: El Niño Southern Oscillation and tuna in the Western Pacific. *Nature*, **389**, 715-718.

Leslie, L.M. and D.J. Karoly, 2007: Variability of tropical cyclones over the southwest Pacific Ocean using a high-resolution climate model. *Meteorol. Atmos. Phys.*, doi:10.1007/s00703-006-0250-3.

Lewis, D., 2002: *Slower Than the Eye Can See: Environmental Change in Northern Australia's Cattle Lands, A Case Study from the Victoria River District, Northern Territory*. Tropical Savannas CRC, Darwin, Northern Territory. http://savanna.ntu.edu.au/publications/landscape_change.html.

Lilley, J.M., T.P. Bolger, M.B. Peoples and R.M. Gifford, 2001: Nutritive value and the dynamics of *Trifolium subterrraneum* and *Phalaris aquatica* under warmer, high CO_2 conditions. *New Phytol.*, **150**, 385-395.

Lincoln Environmental, 2000: Information on water allocation in New Zealand. Report No 4375/1, Prepared for Ministry for the Environment by Lincoln Ventures, Canterbury, 190 pp. http://www.mfe.govt.nz/publications/water/water-allocation-apr00.pdf.

Lough, J.M., 2000: 1997-98: Unprecedented thermal stress to coral reefs? *Geophys. Res. Lett.*, **27**, 3901-3904.

Lowy Institute, 2006: *Public Opinion on Foreign Policy in the Asia-Pacific*. Chicago Council on Global Affairs in partnership with Asia Society and in association with East Asia Institute and Lowy Institute for International Policy, 91 pp. http://www.lowyinstitute.org/Publication.asp?pid=483.

Luo, Q., M.A.J. Williams, W. Bellotti and B. Bryan, 2003: Quantitative and visual assessments of climate change impacts on South Australian wheat production. *Agr. Syst.*, **77**, 173-186.

Lyne, V.D., 2000: Development, evaluation and application of remotely sensed ocean colour data by Australian fisheries. Final Report for 1994/045, Fisheries Research and Development Corporation, Canberra, 88 pp. http://www.frdc.com.au/.

Maddison, D., 2001: In search of warmer climates? The impact of climate change on flows of British tourists. *Climatic Change*, **49**, 193-208.

MAF, 1999: 1999 Post-election brief: the agriculture, horticultural and forestry sectors. http://www.maf.govt.nz/mafnet/publications/1999-post-election-brief/pstelbrf99-05.htm.

MAF, 2001: A national exotic forest description. Ministry of Agriculture and Forestry, 63 pp. http://www.maf.govt.nz/statistics/primaryindustries /forestry/forest-resources/national-exotic-forest-2001/nefd-2001.pdf.

MAF, 2006: How climate change is likely to affect agriculture and forestry in New Zealand. Ministry of Agriculture and Forestry. http://www.maf.govt.nz/mafnet/rural-nz/sustainable-resource-use/climate/index.htm.

Maunder, M.N. and B.M. Watters, 2003: A general framework for integrating environmental time series into stock assessment models: model description, simulation testing, and example. *Fish. B.–NOAA*, **101**, 89-99.

MCDEM, 2004: Review of the February 2004 flood event. Review Team Report. Director of Civil Defence and Emergency Management, Wellington, 93 pp. http://www.civildefence.govt.nz/memwebsite.nsf/Files/FINAL_review_of_the_February_2004_floods/$file/FINAL_review_of_the_February_2004_floods.pdf.

McGlone, M.S., 2001: Linkages between climate change and biodiversity in New Zealand. Landcare Research Contract Report LC0102/014, Ministry for the Environment, Wellington, 36 pp. http://www.mfe.govt.nz/publications/climate/biodiversity-sep01/biodiversity-sep01.pdf.

McInnes, K.L., R. Suppiah, P.H. Whetton, K.J. Hennessy and R.N. Jones, 2002: Climate change in South Australia. Report on assessment of climate change, impacts and possible adaptation strategies relevant to South Australia, CSIRO Atmospheric Research, Aspendale, 61 pp. http://www.cmar.csiro.au /e-print /open/mcinnes_2003a.pdf.

McInnes, K.L., K.J.E. Walsh, G.D. Hubbert and T. Beer, 2003: Impact of sea-level rise and storm surges on a coastal community. *Nat. Hazards*, **30**, 187-207.

McIntosh, A.J., 2004: Tourists' appreciation of Maori culture in New Zealand. *Tourism Manage.*, **25**, 1-15.

McKerchar, A.I. and R.D. Henderson, 2003: Shifts in flood and low-flow regimes in New Zealand due to inter-decadal climate variations. *Hydrolog. Sci. J.*, **48**, 637-654.

McMichael, A., R. Woodruff, P. Whetton, K. Hennessy, N. Nicholls, S. Hales, A. Woodward and T. Kjellstrom, 2003: *Human Health and Climate Change in Oceania: A Risk Assessment 2002*. Commonwealth Department of Health and Ageing, 126 pp. http://www.health.gov.au/internet/wcms/Publishing.nsf/Content/health-pubhlth-publicat-document-metadata-env_climate.htm.

McMichael, A.J., R.E. Woodruff and S. Hales, 2006: Climate change and human health: present and future risks. *Lancet*, **367**, 859-869.

MDBC, 2006: *Basin Statistics*. Murray-Darling Basin Commission. http://www.mdbc.gov.au/about/basin_statistics.

MDBMC, 1999: *The Salinity Audit of the Murray-Darling Basin*. Murray-Darling Basin Ministerial Council, 48 pp. http://www.mdbc.gov.au/salinity/salinity_audit_1999.

Meehl, G.A., T.F. Stocker, W.D. Collins, P. Friedlingstein, A.T. Gaye, J.M. Gregory, A. Kitoh, R. Knutti, J.M. Murphy, A. Noda, S.C.B. Raper, I.G. Watterson, A.J. Weaver and Z.-C. Zhao, 2007: Global climate projections. *Climate Change 2007: The Physical Science Basis, Contribution of: Working Group I to the Fourth Assessment Report of the Intergovernmental Panel on Climate Change*. S. Solomon, D. Qin, M. Manning, Z. Chen, M. Marquis, K.B. Averyt, M. Tignor and H.L. Miller, Eds., Cambridge University Press, Cambridge, 747-846.

Melbourne Water, 2006: Eastern Treatment plant: treating sewage from Melbourne's south-eastern and eastern suburbs. http://www.melbournewater .com.au/content/sewerage/eastern_treatment_plant/eastern_treatment_plant.asp?bhcp=1.

MFAT, 2006: A joint study investigating the benefits of a closer economic partnership (CEP) agreement between Thailand and New Zealand: overview of the Thailand and New Zealand economies. http://www.mfat.govt.nz/foreign /tnd/ceps/nzthaicep/thainzeconomies.html.

MfE, 2001: Climate change impacts on New Zealand. Report prepared by the Ministry for the Environment as part of the New Zealand Climate Change Programme, Ministry for the Environment, Wellington, 39 pp. http://www.climatechange.govt.nz/resources/reports/impacts-report/impacts-report-jun01.pd.

MfE, 2002: Adapting to the impacts of climate change: what role for local government? *Proceedings of a Workshop Organised by the Ministry for the Environment as Part of the New Zealand Climate Change Programme*, 31 pp. http://www.lgnz.co.nz/library/files/store_001/impacts-of-climate-change.pdf.

MfE, 2003: Climate change case study. Local government adaptation to climate change: Environment Bay of Plenty and coastal hazards: "Issues, barriers and solutions". Prepared for the NZ Climate Change Office (Ministry for the Environment) by Lawrence Cross and Chapman and the International Global Change Institute, in conjunction with Environment Bay of Plenty, Tauranga District Council and Western Bay of Plenty District Council, 13 pp. http://www.mfe.govt.nz/publications/climate/ebop-coastal-hazards-jul03/ebop-coastal-hazards-jul03.pdf.

MfE, 2004a: *Coastal Hazards and Climate Change: A Guidance Manual for Local Government in New Zealand*. Ministry for the Environment, 13 pp. http://www.mfe.govt.nz/publications/climate/ebop-coastal-hazards-jul03/ebop-coastal-hazards-jul03.pdf.

MfE, 2004b: Workshop summary report. *International Workshop on Adaptation Practices and Strategies in Developed Countries, 11-13 October 2004, Wellington*, A. Reisinger and H. Larsen, Eds., New Zealand Climate Change Office, NZCCO of the Ministry for the Environment. http://www.mfe.govt.nz/issues/climate/resources/workshops/preparing-climate-change/index.html.

MfE, 2004c: *Preparing for Climate Change: A Guide for Local Government in New Zealand*. New Zealand Climate Change Office, Ministry for Environment, Wellington, 37 pp.. http://www.mfe.govt.nz/publications/climate/preparing-for-climate-change-jul04/index.html

Miller, M.G. and A. Veltman, 2004: Proposed Canterbury Natural Resources Plan for river and groundwater allocation policies and the implications for irrigation dependent farming in Canterbury. *Proceedings of the New Zealand Grassland Association*, **66**, 11-23.

Mimura, N., 2006: State of the environment in Asia and Pacific coastal zones. *Global Change and Integrated Coastal Management: The Asia Pacific Region*, N. Harvey, Ed., Springer, 339 pp.

Ministry of Fisheries Science Group, 2006: Report from the Fishery Assessment Plenary, May 2006: stock assessment and yield estimates. Report held in NIWA Library, Wellington, 875 pp.

Moore, J.L., S.M. Howden, G.M. McKeon, J.O. Carter and J.C. Scanlan, 2001: The dynamics of grazed woodlands in southwest Queensland, Australia and their effect on greenhouse gas emissions. *Environ. Int.*, **27**, 147-153.

Mpelasoka, F., K.J. Hennessy, R. Jones and J. Bathols, 2007: Comparison of suitable drought indices for climate change impacts assessment over Australia towards resource management. *Int. J. Climatol.*, accepted.

Mullan, A.B., 1995: On the linearity and stability of Southern Oscillation–climate relationships for New Zealand. *Int. J. Climatol.*, **15**, 1356-1386.

Mullan, A.B., M.J. Salinger, C.S. Thompson and A.S. Porteous, 2001: The New Zealand climate: present and future. *The Effects of Climate Change and Variation in New Zealand: An Assessment using the CLIMPACTS System*, R.A. Warrick, G.J. Kenny and J.J. Harman, Eds., International Global Change Institute (IGCI), University of Waikato, Hamilton, 11-31. http://www.waikato.ac.nz/igci/climpacts/Linked%20documents/Chapter_2.pdf.

Mullan, A.B., A. Porteous, D. Wratt and M. Hollis, 2005a: Changes in drought risk with climate change. NIWA Report WLG2005-23, 58 pp. http://www.climatechange.govt.nz/resources/reports/drought-risk-may05/drought-risk-climate-change-may05.pdf.

Mullan, A.B., J. Salinger, D. Ramsay and M. Wild, 2005b: Chatham Islands Climate Change. NIWA Report WLG2005-35, 42 pp. http://www.climatechange.govt.nz/resources/reports/chatham-islands-climate-change-jun05/chatham-islands-climate-change-jun05.pdf.

Mulrennan, M., 1992: Coastal management: challenges and changes in the Torres Strait islands. Australian National University, North Australia Research Unit, Discussion Paper 5, 40 pp.

Nakićenović, N. and R. Swart, Eds., 2000: *Special Report on Emissions Scenarios: A Special Report of Working Group III of the Intergovernmental Panel on Climate Change*. Cambridge University Press, Cambridge, 599 pp.

Namjou, P., G. Strayton, A. Pattle, M.D. Davis, P. Kinley, P. Cowpertwait, M.J. Salinger, A.B. Mullan, G. Paterson and B. Sharman, 2005: The integrated catchment study of Auckland City (New Zealand): long-term groundwater behaviour and assessment. *EWRI Conference*, 13.

NEMMCO, 2006: 2006 Statement of Opportunities. National Electricity Market Management Company. http://www.nemmco.com.au/nemgeneral/040-0042.htm.

Niall, S. and K. Walsh, 2005: The impact of climate change on hailstorms in southeastern Australia. *Int. J. Climatol.*, **25**, 1933-1952.

Nicholls, N., 2004: The changing nature of Australian droughts. *Climatic Change*, **63**, 323-336.

Nicholls, N., 2005: Climate variability, climate change and the Australian snow season. *Aust. Meteorol. Mag.*, **54**, 177-185.

Nicholls, N., 2006: Detecting and attributing Australian climate change: a review. *Aust. Meteorol. Mag.*, **55**, 199-211.

Nicholls, N. and D. Collins, 2006: Observed change in Australia over the past century. *Energy and Environment*, **17**, 1-12.

Nicholls, N., C.D. Butler and I. Hanigan, 2006: Inter-annual rainfall variations and suicide in New South Wales, Australia, 1964-2001. *Int. J. Biometeorol.*, **50**, 139-143.

NIWA, 2005: *Past Climate Variations over New Zealand*. National Institute of Water and Atmospheric Research, Wellington. http://www.niwascience.co.nz/ncc/clivar/pastclimate#y140.

NIWA, 2006: *Proceedings of the Second Maori Climate Forum, 24 May 2006*, Hongoeka Marae, Plimmerton. http://www.niwascience.co.nz/ncc/maori/2006-05/.

NLWRA, 2001: *Australian Water Resources Assessment 2000*. National Land and Water Resources Audit, Land and Water Australia. http://www.nlwra.gov.au/.

NRMMC, 2006: *National Agriculture and Climate Change Action Plan 2006-2009*. National resource management Ministerial Council, 20 pp. http://www.daffa.gov.au/__data/assets/pdf_file/33981/nat_ag_clim_chang_action_plan2006.pdf.

NZIER, 2003: *Maori Economic Development: Te Ohanga Whakaketanga Maori*. New Zealand Institute of Economic Research, Wellington. http://www.nzier.org.nz/.

NZIER, 2004: Economic impacts on New Zealand of climate change-related extreme events: focus on freshwater floods. Report to Climate Change Office, New Zealand Institute of Economic Research, Wellington, 73 pp. http://www.climatechange.govt.nz/resources/reports/economic-impacts-extreme-events-jul04/economic-impacts-extreme-events-jul04.pdf.

O'Connell, M. and R. Hargreaves, 2004: Climate change adaptation. guidance on adapting New Zealand's built environment for the impacts of climate change. BRANZ Study Report No. 130, 44 pp. http://www.branz.co.nz/branzltd/publications/pdfs/SR130.pdf.

O'Hara, T.D., 2002: Endemism, rarity and vulnerability of marine species along a temperate coastline. *Invertebr. Syst.*, **16**, 671-679.

Orlove, B., 2003: *How People Name Seasons*, S. Strauss and B. Orlove, Eds., Berg Publishers, Oxford, 121-141.

Orr, J.C., V.J. Fabry, O. Aumont, L. Bopp, S.C. Doney, R.A. Feely, A. Gnanadesikan, N. Gruber, A. Ishida, F. Joos, R.M. Key, K. Lindsay, E. Maier-Reimer, R.J. Matear, P. Monfray, A. Mouchet, R.G. Najjar, G.-K. Plattner, K.B. Rodgers, C.L. Sabine, J.L. Sarmiento, S. R, R.D. Slater, I.J. Totterdell, M.-F. Weirig, Y. Yamanaka and A. Yool, 2005: Anthropogenic ocean acidification over the twenty-first century and its impact on calcifying organisms. *Nature*, **437**, 681-686.

Owens, I.F. and B.B. Fitzharris, 2004: Seasonal snow and water. *Waters of New Zealand*, M.P. Mosley, C. Pearson and B. Sorrell, Eds., New Zealand Hydrological Society and New Zealand Limnological Society, 5.1-5.12.

Packman, D., D. Ponter and T. Tutua-Nathan, 2001: Maori issues. Climate Change Working Paper, New Zealand Climate Change Office, Ministry for the Environment, Wellington, 18 pp. http://www.climatechange.govt.nz/resources/.

Parliament of Australia, 2003: *Regulating the Ranger, Jabiluka, Beverly and Honeymoon Uranium Mines*. Parliament of Australia, Committee Address: Environment, Communications, Information Technology and the Arts Legislation Committee S1, Canberra, ACT. http://www.aph.gov.au/senate/ committee/ecita_ctte/completed_inquiries/2002-04/uranium/report/index.htm.

PB Associates, 2007: Assessment of the vulnerability of Australia's energy infrastructure to the impacts of climate change. Issues Paper, 158292A-REP-002.doc, 78 pp., in press.

Pearce, G., A.B. Mullan, M.J. Salinger, T.W. Opperman, D. Woods and J.R. Moore, 2005: Impact of climate variability and change on long-term fire danger. Report to the New Zealand Fire Service Commission, 75 pp. http://www.fire.org.nz/research/reports/reports/Report_50.pdf.

PIA, 2004: Sustainable regional and urban communities adapting to climate change. Issues Paper, Planning Institute of Australia Queensland Division, Brisbane, 88 pp. http://www.greenhouse.sa.gov.au/PDFs/issuespaperjune_21.pdf.

Pickering, C., R. Good and K. Green, 2004: Potential effects of global warming on the biota of the Australian Alps. Report to the Australian Greenhouse Office, Canberra, 51 pp. http://www.greenhouse.gov.au/impacts/publications/ alps.html.

Pittock, B., 2003: *Climate Change: An Australian Guide to the Science and Potential of Impacts*. Department for the Environment and Heritage, Australian Greenhouse Office, Canberra, ACT, 250 pp. http://www.greenhouse. gov.au/science/guide/index.html.

Pittock, B. and D. Wratt, 2001: Australia and New Zealand. *Climate Change 2001: Impacts, Adaptation, and Vulnerability. Contribution of Working Group II to the Third Assessment Report of the Intergovernmental Panel on Climate Change*, J.J. McCarthy, O.F. Canziani, N.A. Leary, D.J. Dokken and K.S. White, Eds., Cambridge University Press, Cambridge, 591-639.

Poloczanska, E.S., R.C. Babcock, A. Butler, A.J. Hobday, O. Hoegh-Guldberg, T.J. Kunz, R. Matear, D. Milton, T.A. Okey and A.J. Richardson, 2007: Climate change and Australian marine life. *Oceanography and Marine Biology: An Annual Review, Volume 45*, R.N. Gibson, R.J.A. Atkinson and J.D.M. Gordon, Eds., CRC Press, Cleveland, 544 pp.

Power, S., F. Tseitkin, S. Torok, B. Lavery, R. Dahni and B. McAvaney, 1998: Australian temperature, Australian rainfall and the Southern Oscillation, 1910-1992: coherent variability and recent changes. *Aust. Meteorol. Mag.*, **47**, 85-101.

Power, S., T. Casey, C. Folland, C. A. and V. Mehta, 1999: Inter-decadal modulation of the impact of ENSO on Australia. *Clim. Dynam.*, **15**, 319-324.

Premier of Victoria, 2006: Ballarat's future water supplies secured by major Bracks government action plan. Media release, 17 October 2006. http://www.premier.vic.gov.au/newsroom/news_item.asp?id=978.

Preston, B., R. Suppiah, I. Macadam and J. Bathols, 2006: Climate change in the Asia/Pacific region. Consultancy Report prepared for the Climate Change and

Development Roundtable, CSIRO Marine and Atmospheric Research, Aspendale, 92 pp. http://www.csiro.au/csiro/content/file/pfkd.html.

Puotinen, M.L., 2006: Modelling the risk of cyclone wave damage to coral reefs using GIS: a case study of the Great Barrier Reef, 1969-2003. *Int. J. Geogr. Inf. Sci.*, **21**, 97-120.

QDLGP, 2003: *Queensland's Future Population*. Queensland Department of Local Government and Planning (Medium Series). http://www.lgp.qld.gov.au/?id=1216.

Queensland Government, 2005: *Queensland Water Plan 2005-2010*. Queensland Government, 27 pp. http://www.nrw.qld.gov.au/water/pdf/qld_water_plan_05_10.pdf.

Queensland Government, 2006: *TC Larry Report, 28 March 2006*. http://www.disaster.qld.gov.au/news/view.asp?id=1323.

QVL, 2006: *Residential Property:Price Movement*. Quotable Value Limited. http://www.qv.co.nz/onlinereports/marketstatistics/.

Ranasinghe, R., R. McLoughlin, A.D. Short and G. Symonds, 2004: The Southern Oscillation Index, wave climate, and beach rotation. *Mar. Geol.*, **204**, 273-287.

Raven, J., K. Caldeira, H. Elderfield, O. Hoegh-Guldberg, P. Liss, U. Riebesell, J. Shepherd, C. Turley and A. Watson, 2005: Ocean acidification due to increasing carbon dioxide. Report to the Royal Society, London, 68 pp. http://www.royalsoc.ac.uk/displaypagedoc.asp?id=13539.

Renwick, J.A., R.J. Hurst and J.W. Kidson, 1998: Climatic influences on the recruitment of southern gemfish (*Rexea solandri, Gempylidae*) in New Zealand waters. *Int. J. Climatol.*, **18**, 1655-1667.

Reyenga, P.J., S.M. Howden, H. Meinke and W.B. Hall, 2001: Global change impacts on wheat production along an environmental gradient in South Australia. *Environ. Int.*, **27**, 195-200.

Richardson, A.C., K.B. Marsh, H.L. Boldingh, A.H. Pickering, S.M. Bulley, N.J. Frearson, A.R. Ferguson, S.E. Thornber, K.M. Bolitho and E.A. Macrae, 2004: High growing temperatures reduce fruit carbohydrate and vitamin C in fruit. *Plant Cell Environ.*, **27**, 423-435.

Richardson, A.J. and D.S. Schoeman, 2004: Climate impact on plankton ecosystems in the North-East Atlantic. *Science*, **305**, 1609-1612.

Richardson, S.J., R.B. Allen, D. Whitehead, F.E. Carswell, W.A. Ruscoe and K.H. Platt, 2005: Climate and net carbon availability determine temporal patterns of seed productivity by *Nothofagus*. *Ecology*, **86**, 972-981.

Ring, T. and N. Brown, 2002: Indigenous health: chronically inadequate responses to damning statistics. *Med. J. Australia*, **177**, 629-631.

Robb, C. and J. Bright, 2004: Values and uses of water. *Freshwaters of New Zealand*, J.S. Harding, M.P. Mosley, C.P. Pearson and B.K. Sorrell, Eds., New Zealand Hydrological Society and New Zealand Limnological Society, Caxton Press, Christchurch, 42.1-42.14.

Roderick, M.L. and G.D. Farquhar, 2004: Changes in Australian pan evaporation from 1970 to 2002. *Int. J. Climatol.*, **24**, 1077-1090.

Roderick, M.L. and G.D. Farquhar, 2005: Changes in New Zealand pan evaporation since the 1970s. *Int. J. Climatol.*, **25**, 2031-2039.

Rose, D., 1996: *Nourishing Terrains: Australian Aboriginal Views of Landscape and Wilderness*. Australian Heritage Commission, Canberra, 95 pp. http://www.ahc.gov.au/publications/generalpubs/nourishing/index.html.

RSNZ, 2002: *New Zealand Climate Committee: Terms of Reference*. Royal Society of New Zealand, New Zealand Climate Committee. http://www.rsnz.org/advisory/nz_climate/terms.php/.

Salinger, M.J., 1990: Climate change and the electricity industry. New Zealand Meteorological Service Report, 59 pp.

Salinger, M.J. and A.B. Mullan, 1999: New Zealand climate: temperature and precipitation variations and their links with atmospheric circulation 1930-1994. *Int. J. Climatol.*, **19**, 1049-1071.

Salinger, M.J. and G.M. Griffiths, 2001: Trends in New Zealand daily temperature and rainfall extremes. *Int. J. Climatol.*, **21**, 1437-1452.

Salinger, M.J., W.M. Williams, J.M. Williams and R.J. Martin, 1990: Agricultural resources. *Climate Change: Impacts on New Zealand*. Ministry for the Environment, Wellington, 108-132.

Salinger, M.J., W. Gray, B. Mullan and D. Wratt, 2004: Atmospheric circulation and precipitation. *Freshwaters of New Zealand*, J. Harding, P. Mosley, C. Pearson and B. Sorrell, Eds., Caxton Press, Christchurch, 2.1-2.18.

Salinger, M.J., G.M. Griffiths and A. Gosai, 2005a: Extreme daily pressure differences and winds across New Zealand. *Int. J. Climatol.*, **25**, 1301-1330.

Salinger, M.J., M.V.K. Sivakumar and R. Motha, 2005b: Reducing vulnerability of agriculture and forestry to climate variability and change: workshop summary and recommendations. *Climatic Change*, **70**, 341-362.

Scanlan, J.C., D.M. Berman and W.E. Grant, 2006: Population dynamics of the European rabbit (*Oryctolagus cuniculus*) in north eastern Australia: simulated responses to control. *Ecol. Model.*, **196**, 221-236.

Schallenberg, M., C.J. Hall and C.W. Burns, 2003: Consequences of climate-induced salinity increases on zooplankton abundance and diversity in coastal lakes. *Mar. Ecol.-Prog. Ser.*, **251**, 181-189.

Schuster, S.S., R.J. Blong, R.J. Leigh and K.J. McAneney, 2005: Characteristics of the 14 April 1999 Sydney hailstorm based on ground observations, weather radar, insurance data and emergency calls. *Nat. Hazard. Earth Sys.*, **5**, 613-620.

Scott, D., G. McBoyle and M. Schwartzentruber, 2004: Climate change and the distribution of climatic resources for tourism in North America. *Climate Res.*, **27**, 105-117.

Seafood Industry Council, 2006: *New Zealand's Seafood Business*. http://www.seafood.co.nz/sc-business/.

Sharples, C., 2004: Indicative mapping of Tasmania's coastal vulnerability to climate change and sea-level rise. Explanatory Report, Tasmanian Department of Primary Industries, Water and Environment, Hobart, 126 pp. http://www.dpiw.tas.gov.au/inter.nsf/Publications/PMAS-6B59RQ?open.

Shaw, H., A. Reisinger, H. Larsen and C. Stumbles, 2005: Incorporating climate change into stormwater design: why and how? *Proceedings of the 4th South Pacific Conference on Stormwater and Aquatic Resource Protection*, May 2005, Auckland, 18 pp. http://www.climatechange.govt.nz/resources/local-govt/stormwater-design-mar05/stormwater-design-mar05.pdf.

Sims, D.W., M.J. Genner, A.J. Southward and S.J. Hawkins, 2001: Timing of squid migration reflects North Atlantic climate variability. *P. Roy. Soc. Lond. B*, 2607-2611.

Sinclair, T.R., P.J. Pinter Jr, B.A. Kimball, F.J. Adamsen, R.L. LaMorte, G.W. Wall, D.J. Hunsaker, N. Adam, T.J. Brooks, R.L. Garcia, T. Thompson, S. Leavitt and A. Matthias, 2000: Leaf nitrogen concentration of wheat subjected to elevated [CO_2] and either water or N deficits. *Agr. Ecosyst. Environ.*, **79**, 53-60.

Smith, B.R., 2004a: Some natural resource management issues for Indigenous people in northern Australia. *Arafura Timor Research Facility Forum*, ANU Institute for Environment, The National Australia University. http://www.anu.edu.au/anuie/.

Smith, I.N., 2004b: Trends in Australian rainfall: are they unusual? *Aust. Meteorol. Mag.*, **53**, 163-173.

Smith, P.J., 1979: Esterase gene frequencies and temperature relationships in the New Zealand snapper *Chryophrys auratus*. *Mar. Biol.*, **53/4**, 305-310.

Smithers, B.V., D.R. Peck, A.K. Krockenberger and B.C. Congdon, 2003: Elevated sea-surface temperature, reduced provisioning and reproductive failure of wedge-tailed shearwaters (*Puffinus pacificus*) in the Southern Great Barrier Reef, Australia. *Mar. Freshwater Res.*, **54**, 973-977.

SOE, 2001: *Australia State of the Environment 2001: Independent Report to the Commonwealth Minister for the Environment and Heritage*. Australian State of the Environment Committee, CSIRO Publishing on behalf of the Department of the Environment and Heritage, 129 pp. http:/www.ea.gov.au/soe/2001.

Sorensen, I., P. Stone and B. Rogers, 2000: Effect of time of sowing on yield of a short and a long-season maize hybrid. *Agronomy New Zealand*, **30**, 63-66.

Statistics New Zealand, 2005a: *Demographic Trends*. Statistics New Zealand, New Zealand Government. http://www.stats.govt.nz/products-and-services/reference-reports/demographic-trends.htm.

Statistics New Zealand, 2005b: *National Population Projections 2004-2051*. Statistics New Zealand, New Zealand Government. http://www.stats.govt.nz/datasets/population/population-projections.htm.

Statistics New Zealand, 2006: *QuickStats About New Zealand's Population and Dwellings 2006 Census*. Statistics New Zealand, 11 pp. http://www.stats.govt.nz/NR/rdonlyres/CCA37BF2-2E49-44D4-82AB-35538608DEFD/0/2006censusquickstatsaboutnzspopanddwellings.pdf.

Steffen, W., G. Love and P.H. Whetton, 2006: Approaches to defining dangerous climate change: an Australian perspective. *Avoiding Dangerous Climate Change*, H.J. Schellnhuber, W. Cramer, N. Nakićenović, T. Wigley and G. Yohe, Eds., Cambridge University Press, Cambridge, 392 pp.

Stephens, A.E.A., D.J. Kriticos and A. Leriche, 2007: The current and future distribution of the oriental fruit fly, *Bactrocera dorsalis* (Diptera: Tephritidae). *B. Entomol. Res.*, in press.

Stokes, C. and A. Ash, 2006: Impacts of climate change on marginal tropical production systems. *Agroecosystems in a Changing Climate*, P. Newton, A. Carran and G. Edwards, Eds., Taylor and Francis, Florida, 181-188.

Suppiah, R., K.J. Hennessy, P.H. Whetton, K. McInnes, I. Macadam, J. Bathols and J. Ricketts, 2007: Australian climate change projections derived from simulations performed for the IPCC 4th Assessment Report. *Aust. Meteorol. Mag.*, accepted.

Sutherst, R.W., 2004: Global change and human vulnerability to vector-borne diseases. *Clin. Microbiol. Rev.*, **17**, 136-173.

Sutherst, R.W., B.S. Collyer and T. Yonow, 2000: The vulnerability of Australian horticulture to the Queensland fruit fly, *Bactrocera (Dacus) tryoni*, under climate change. *Aust. J. Agr. Res.*, **51**, 467-480.

Tapper, N., 2000: Atmospheric issues for fire management in Eastern Indonesia and northern Australia. *Proceedings of an International Workshop: Fire and Sustainable Agricultural and Forestry Development in Eastern Indonesia and Northern Australia*, J. Russell-Smith, G. Hill, S. Djoeroemana and B. Myers, Eds., Northern Territory University, Darwin, 20-21. http://www.arts. monash.edu.au/ges /who/pdf/Atmospheric%20Issues%20for%20Fire% 20Management.pdf.

Taylor, P.R., 2002: Stock structure and population biology of the Peruvian jack mackerel, *Trachurus symmetricus murphyi*. Report 2002/21, New Zealand Fisheries Assessment, 79 pp.

Thresher, R.E., 1994: Climatic cycles may help explain fish recruitment in South-East Australia. *Australian Fisheries*, **53**, 20-22.

Thresher, R.E., 2002: Solar correlates of Southern Hemisphere climate variability. *Int. J. Climatol.*, **22**, 901-915.

Thresher, R.E., B.D. Bruce, D.M. Furnali and J.S. Gunn, 1989: Distribution, advection, and growth of larvae of the southern temperate gadoid, *Macruronus novaezelandiae* (Teleostei: Merlucciidae), in Australian coastal waters. *Fish. B.–NOAA*, **87**, 29-48.

Thresher, R.E., P.D. Nichols, J.S. Gunn, B.D. Bruce and D.M. Furlani, 1992: Seagrass detritus as the basis of a coastal planktonic food-chain. *Limnol. Oceanogr.*, **37**, 1754-1758.

Torok, S.J., C.J.G. Morris, C. Skinner and N. Plummer, 2001: Urban heat island features of southeast Australian towns. *Aust. Meteorol. Mag.*, **50**, 1-13.

TPK, 2001: *Te Maori I Nga Rohe: Maori Regional Diversity*. Te Puni Kokiri, Ministry of Maori Development, Wellington, 168 pp. http://www.tpk.govt.nz/ publications/docs/MRD_full.pdf.

Tweedie, C.E. and D.M. Bergstrom, 2000: A climate change scenario for surface air temperature at sub-Antarctic Macquarie Island. *Antarctic Ecosystems: Models for Wider Ecological Understanding*, W. Davison, C. Howard-Williams and P.A. Broady, Eds., New Zealand Natural Sciences, Christchurch, 272-281.

Umina, P.A., A.R. Weeks, M.R. Kearney, S.W. McKechnie and A.A. Hoffmann, 2005: A rapid shift in a classic clinal pattern in *Drosophila* reflecting climate change. *Science*, **308**, 691-693.

UNEP-WCMC, 2006: In the front line: shoreline protection and other ecosystem services from mangroves and coral reefs. United Nations Environment Programme: World Conservation Monitoring Centre, Cambridge, 36 pp. http://www.unep.org/pdf/infrontline_06.pdf.

URS, 2004: Climate change and infrastructure design case study: Wairau Catchment, North Shore City. Prepared for North Shore City Council, August 2004, 26 pp. http://www.climatechange.govt.nz/resources/local-govt/case-study-wairau-catchment-aug04/case-study-wairau-catchment-aug04.pdf.

van Ittersum, M.K., S.M. Howden and S. Asseng, 2003: Sensitivity of productivity and deep drainage of wheat cropping systems in a Mediterranean environment to changes in CO_2, temperature and precipitation. *Agr. Ecosyst. Environ.*, **97**, 255-273.

Vic DSE, 2006: *Wimmera Mallee Pipeline*. Department of Sustainability and Environment, Victoria. http://www.dse.vic.gov.au/DSE/wcmn202.nsf/LinkView /77BBB217C2024716CA256FE2001F7B2460A84DA0F2283EA4CA256FDD 00136E35.

Viney, N.R., B.C. Bates, S.P. Charles, I.T. Webster, M. Bormans and S.K. Aryal, 2003: Impacts of climate variability on riverine algal blooms. *Proceedings of the International Congress on Modelling and Simulation, MODSIM 2003 14-17 July*, Modelling and Simulation Society of Australia and New Zealand, 23-28. http://mssanz.org.au/modsim03/Media/Articles/Vol%201%20Articles/23-28.pdf.

Voice, M., N. Harvey and K. Walsh, Eds., 2006: *Vulnerability to Climate Change of Australia's Coastal Zone: Analysis of Gaps in Methods, Data and System Thresholds*. Report to the Australian Greenhouse Office, Department of the Environment and Heritage, Canberra, 120 pp. http://www.greenhouse.gov.au/impacts/publications/pubs/coastal-vulnerability.pdf

WA Water Corporation, 2004: *Perth Seawater Desalination Project*. http://www.watercorporation.com.au/D/desalination_environment.cfm.

Walsh, K., 2002: Climate change and coastal response. *A Theme Report from the Coast to Coast 2002 National Conference, November, 2002*. Coastal CRC, Gold Coast, Queensland, 37 pp. http://www.coastal.crc.org.au/pdf/coast2coast 2002papers/climate.pdf.

Walsh, K., K. Hennessy, R. Jones, K.L. McInnes, C.M. Page, A.B. Pittock, R. Suppiah and P. Whetton, 2001: *Climate Change in Queensland Under Enhanced Greenhouse Conditions: Third Annual Report, 1999-2000*. CSIRO Consultancy Report for the Queensland Government, Aspendale, 108 pp. http://www.cmar.csiro.au/e-print/open/walsh_2001a.pdf.

Walsh, K.J.E., K.C. Nguyen and J.L. McGregor, 2004: Finer-resolution regional climate model simulations of the impact of climate change on tropical cyclones near Australia. *Clim. Dynam.*, **22**, 47-56.

Warrick, R.A., 2000: Strategies for vulnerability and adaptation assessment in the context of national communications. *Asia-Pacific Journal for Environmental Development*, **7**, 43-51.

Warrick, R.A., 2006: Climate change impacts and adaptation in the Pacific: recent breakthroughs in concept and practice. *Confronting Climate Change: Critical Issues for New Zealand*, R. Chapman, J. Boston and M. Schwass, Eds., Victoria University Press, Wellington, 189-196.

Warrick, R.A., G.J. Kenny and J.J. Harman, Eds., 2001: *The Effects of Climate Change and Variation in New Zealand: An Assessment using the CLIMPACTS System*. International Global Change Institute (IGCI), University of Waikato, Hamilton. http://www.waikato.ac.nz/igci/climpacts/Assessment.htm.

Wasson, R.J., I. White, B. Mackey and M. Fleming, 1988: The Jabiluka project: environmental issues that threaten Kakadu National Park. Submission to UNESCO World Heritage Committee delegation to Australia, 22 pp. http://www.deh.gov.au/ssd/uranium-mining/arr-mines/jabiluka/pubs/appn2of main.pdf.

Water Corporation, 2006: *Planning for New Sources of Water*. West Australian Water Corporation. http://www.watercorporation.com.au/W/water_sources _new.cfm.

Watson, R. and A. McMichael, 2001: Global climate change – the latest assessment: does global warming warrant a health warning? *Global Change and Human Health*, **2**, 64-75.

Watters, G.M., R.J. Olson, R.C. Francis, P.C. Fiedler, J.J. Polovina, S.B. Reilly, K.Y. Aydin, C.H. Boggs, T.E. Essington, C.J. Walters and J.F. Kitchell, 2003: Physical forcing and the dynamics of the pelagic ecosystem in the eastern tropical Pacific: simulations with ENSO-scale and global warming climate drivers. *Can. J. Fish. Aquat. Sci.*, **60**, 1161-1175.

Waugh, S.M., H. Weimerskirch, P.J. Moore and P.M. Sagar, 1999: Population dynamics of black-browed and grey-headed albatrosses *Diomedea melanophrys* and *D. chrysostoma* at Campbell Island, New Zealand, 1942-96. *Ibis*, **141**, 216-225.

Wearne, L.J. and J.W. Morgan, 2001: Recent forest encroachment into sub-alpine grasslands near Mount Hotham, Victoria, Australia. *Arct. Antarct. Alp. Res.*, **33**, 369-377.

Webb, E.K., 1997: *Windows on Meteorology: Australian Perspective*. CSIRO Publishing, Collingwood, 342 pp.

Webb, L., P. Whetton and S. Barlow, 2006: Impact on Australian viticulture from greenhouse-induced climate change. *Bulletin of the Australian Meteorological and Oceanographic Society*, **19**, 15-17.

Weimerskirch, H., P. Inchausti, C. Guinet and C. Barbraud, 2003: Trends in bird and seal populations as indicators of a system shift in the Southern Ocean. *Antarct. Sci.*, **15**, 249-256.

Whetton, P.H., R. Suppiah, K.L. McInnes, K.J. Hennessy and R.N. Jones, 2002: Climate change in Victoria: high-resolution regional assessment of climate change impacts. CSIRO Consultancy Report for the Department of Natural Resources and Environment, Victoria, 44 pp. http://www.greenhouse. vic.gov.au/climatechange.pdf.

Whinam, J. and G. Copson, 2006: Sphagnum moss: an indicator of climate change in the Sub-Antarctic. *Polar Rec.*, **42**, 43-49.

Whinam, J., G.S. Hope, B.R. Clarkson, R.P. Buxton, P.A. Alspach and P. Adam, 2003: Sphagnum in peat-lands of Australasia: their distribution, utilisation and management. *Wetlands Ecology and Management*, **11**, 37-49.

White, N., R.W. Sutherst, N. Hall and P. Whish-Wilson, 2003: The vulnerability of the Australian beef industry to impacts of the cattle tick (*Boophilus microplus*) under climate change. *Climatic Change*, **61**, 157-90.

Williams, A.A., D.J. Karoly and N. Tapper, 2001: The sensitivity of Australian fire danger to climate change. *Climatic Change*, **49**, 171-191.

Williams, S.E., E.E. Bolitho and S. Fox, 2003: Climate change in Australian tropical forests: an impending environmental catastrophe. *P. Roy. Soc. Lond. B*, **270**, 1887-1892.

Willis, B.L., C.A. Page and E.A. Dinsdale, 2004: Coral disease on the Great Barrier Reef. *Coral Health and Disease*, E. Rosenberg and Y. Loya, Eds., Springer, Berlin, 69-104.

Winn, K.O., M.J. Saynor, M.J. Eliot and I. Eliot, 2006: Saltwater intrusion and morphological change at the mouth of the East Alligator River, Northern Territory. *J. Coastal Res.*, **22**, 137-149.

Woehler, E.J., H.J. Auman and M.J. Riddle, 2002: Long-term population increase of black-browed albatrosses *Thalassarche melanophrys* at Heard Island, 1947/1948–2000/2001. *Polar Biol.*, **25**, 921-927.

Woodruff, R.E., C.S. Guest, M.G. Garner, N. Becker, J. Lindesay, T. Carvan and K. Ebi, 2002: Predicting Ross River virus epidemics from regional weather data. *Epidemiology*, **13**, 384-393.

Woodruff, R.E., S. Hales, C. Butler and A.J. McMichael, 2005: *Climate Change Health Impacts in Australia: Effects of Dramatic CO_2 Emission Reductions*. Australian Conservation Foundation and the Australian Medical Association, Canberra, 44 pp. http://www.acfonline.org.au/uploads/res_AMA_ACF_Full_Report.pdf.

Woodruff, R.E., C.S. Guest, M.G. Garner, N. Becker and M. Lindsay, 2006: Early warning of Ross River virus epidemics: combining surveillance data on climate and mosquitoes. *Epidemiology*, **17**, 569-575.

Woods, R.A. and C. Howard-Williams, 2004: Advances in freshwater sciences and management. *Freshwaters of New Zealand*, J.S. Harding, M.P. Mosley, C.P. Pearson and B.K. Sorrell, Eds., New Zealand Hydrological Society and New Zealand Limnological Society, Caxton Press, Christchurch, 764 pp.

Woodward, A., S. Hales and P. Weinstein, 1998: Climate change and human health in the Asia-Pacific region: who will be most vulnerable? *Climate Res.*, **11**, 31-38.

Woodward, A., S. Hales and N. de Wet, 2001: *Climate Change: Potential Effects on Human Health in New Zealand*. Ministry for the Environment, Wellington, 21 pp. http://www.mfe.govt.nz/publications/climate/effect-health-sep01/index. html.

Wooldridge, S., T. Done, R. Berkelmans, R. Jones and P. Marshall, 2005: Precursors for resilience in coral communities in a warming climate: a belief network approach. *Mar. Ecol.-Prog. Ser.*, **295**, 157-169.

World Tourism Organisation, 2003: Climate change and tourism. *Proceedings of the First International Conference on Climate Change and Tourism, Djerba 9-11 April*. World Tourism Organisation, Madrid, 55 pp. http://www.world-tourism.org/sustainable/climate/final-report.pdf.

Wratt, D.S., A.B. Mullan, M.J. Salinger, S. Allan, T. Morgan and G. Kenny, 2004: *Climate Change Effects and Impacts Assessment: A Guidance Manual for Local Government in New Zealand*. New Zealand Climate Change Office, Ministry for the Environment, Wellington, 140 pp. http://www.mfe.govt.nz/publications/climate/effects-impacts-may04/effects-impacts-may04.pdf.

Yeo, S., 2003: Effects of disclosure of flood-liability on residential property values. *Australian Journal of Emergency Management*, **18**, 35-44.

Zeldis, J., J. Oldman, S. Ballara and L. Richards, 2005: Physical fluxes, pelagic ecosystem structure, and larval fish survival in Hauraki Gulf, New Zealand. *Can. J. Fish. Aquat. Sci.*, **62**, 593-610.

12

Europe

Coordinating Lead Authors:

Joseph Alcamo (Germany), José M. Moreno (Spain), Béla Nováky (Hungary)

Lead Authors:

Marco Bindi (Italy), Roman Corobov (Moldova), Robert Devoy (Ireland), Christos Giannakopoulos (Greece), Eric Martin (France), Jørgen E. Olesen (Denmark), Anatoly Shvidenko (Russia)

Contributing Authors:

Miguel Araújo (Portugal), Abigail Bristow (UK), John de Ronde (The Netherlands), Sophie des Clers (UK), Andrew Dlugolecki (UK), Phil Graham (Sweden), Antoine Guisan (Switzerland), Erik Jeppesen (Denmark), Sari Kovats (UK), Petro Lakyda (Ukraine), John Sweeney (Ireland), Jelle van Minnen (The Netherlands)

Review Editors:

Seppo Kellomäki (Finland), Ivan Nijs (Belgium)

This chapter should be cited as:

Alcamo, J., J.M. Moreno, B. Nováky, M. Bindi, R. Corobov, R.J.N. Devoy, C. Giannakopoulos, E. Martin, J.E. Olesen, A. Shvidenko, 2007: Europe. *Climate Change 2007: Impacts, Adaptation and Vulnerability. Contribution of Working Group II to the Fourth Assessment Report of the Intergovernmental Panel on Climate Change,* M.L. Parry, O.F. Canziani, J.P. Palutikof, P.J. van der Linden and C.E. Hanson, Eds., Cambridge University Press, Cambridge, UK, 541-580.

Table of Contents

Executive summary

Many of the results reported here are based on a range of emissions scenarios extending up to the end of the 21st century and assume no specific climate policies to mitigate greenhouse gas emissions.

For the first time, wide ranging impacts of changes in current climate have been documented in Europe (very high confidence).

The warming trend and spatially variable changes in rainfall have affected composition and functioning of both the cryosphere (retreat of glaciers and extent of permafrost) as well as natural and managed ecosystems (lengthening of growing season, shift of species) [12.2.1]. Another example is the European heatwave in 2003 which had major impacts on biophysical systems and society [12.6.1]. The observed changes are consistent with projections of impacts due to climate change [12.4].

Climate-related hazards will mostly increase, although changes will vary geographically (very high confidence).

Winter floods are likely to increase in maritime regions and flash floods are likely to increase throughout Europe [12.4.1]. Coastal flooding related to increasing storminess and sea-level rise is likely to threaten up to 1.6 million additional people annually [12.4.2]. Warmer, drier conditions will lead to more frequent and prolonged droughts, as well as to a longer fire season and increased fire risk, particularly in the Mediterranean region [12.3.1.2, 12.4.4.1]. During dry years, catastrophic fires are expected on drained peatlands in central Europe [12.4.5]. The frequency of rock falls will increase due to destabilisation of mountain walls by rising temperatures and melting of permafrost [12.4.3].Without adaptive measures, risks to health due to more frequent heatwaves, particularly in central and southern Europe, and flooding, and greater exposure to vector- and food-borne diseases are anticipated to increase [12.3.1.2, 12.6.1]. Some impacts may be positive, as in reduced risk of extreme cold events because of increasing winter temperatures. However, on balance, health risks are very likely to increase [12.4.11].

Climate change is likely to magnify regional differences of Europe's natural resources and assets (very high confidence).

Climate scenarios indicate significant warming, greater in winter in the North and in summer in southern and central Europe [12.3.1]. Mean annual precipitation is projected to increase in the North and decrease in the South [12.3.1]. Crop suitability is likely to change throughout Europe, and crop productivity (all other factors remaining unchanged) is likely to increase in northern Europe, and decrease along the Mediterranean and in south-eastern Europe [12.4.7.1]. Forests are projected to expand in the North and retreat in the South [12.4.4.1]. Forest productivity and total biomass is likely to increase in the North and decrease in central Europe, while tree mortality is likely to accelerate in the South [12.4.4.1]. Differences in water availability between regions are anticipated to become sharper (annual average runoff increases in the North and North-west, and to decrease in the South and South-east) [12.4.1].

Water stress will increase, as well as the number of people living in river basins under high water stress (high confidence).

Water stress will increase over central and southern Europe. The percentage area under high water stress is likely to increase from 19% today to 35% by the 2070s, and the additional number of people affected by the 2070s is expected to be between 16 millions and 44 millions [12.4.1]. The most affected regions are southern Europe and some parts of central and eastern Europe, where summer flows may be reduced by up to 80% [12.4.1]. The hydropower potential of Europe is expected to decline on average by 6% but by 20 to 50% around the Mediterranean by the 2070s [12.4.8.1].

It is anticipated that Europe's natural (eco)systems and biodiversity will be substantially affected by climate change (very high confidence). The great majority of organisms and ecosystems are likely to have difficulty in adapting to climate change (high confidence).

Sea-level rise is likely to cause an inland migration of beaches and the loss of up to 20% of coastal wetlands [12.4.2], reducing habitat availability for several species that breed or forage in low-lying coastal areas [12.4.6]. Small glaciers will disappear and larger glaciers substantially shrink during the 21st century [12.4.3]. Many permafrost areas in the Arctic are projected to disappear [12.4.5]. In the Mediterranean, many ephemeral aquatic ecosystems are projected to disappear, and permanent ones to shrink [12.4.5]. Recruitment and production of marine fisheries in the North Atlantic are likely to increase [12.4.7.2]. The northward expansion of forests is projected to reduce current tundra areas under some scenarios [12.4.4]. Mountain plant communities face up to a 60% loss of species under high emissions scenarios [12.4.3]. A large percentage of the European flora is likely to become vulnerable, endangered, or committed to extinction by the end of this century [12.4.6]. Options for adaptation are likely to be limited for many organisms and ecosystems. For example, limited dispersal is very likely to reduce the range of most reptiles and amphibians [12.4.6]. Low-lying, geologically subsiding coasts are likely to be unable to adapt to sea-level rise [12.5.2]. There are no obvious climate adaptation options for either tundra or alpine vegetation [12.5.3]. The adaptive capacity of ecosystems can be enhanced by reducing human stresses [12.5.3, 12.5.5]. New sites for conservation may be needed because climate change is very likely to alter conditions of suitability for many species in current sites [12.5.6].

Climate change is estimated to pose challenges to many European economic sectors and is expected to alter the distribution of economic activity (high confidence).

Agriculture will have to cope with increasing water demand for irrigation in southern Europe, and with additional restrictions due to increases in crop-related nitrate leaching [12.5.7]. Winter heating demands are expected to decrease and summer cooling demands to increase: around the Mediterranean, two to three fewer weeks in a year will require heating but an additional two to five weeks will need cooling by 2050 [12.4.8.1]. Peak electricity demand is likely to shift in some locations from winter to summer [12.4.8.1]. Tourism along the Mediterranean is likely to decrease in summer and increase in spring and autumn. Winter tourism in mountain regions is anticipated to face reduced snow cover [12.4.9].

Adaptation to climate change is likely to benefit from experiences gained in reaction to extreme climate events, by specifically implementing proactive climate change risk management adaptation plans (high confidence).
Since the Third Assessment Report, governments have increased greatly the number of actions for coping with extreme climate events. Current thinking about adaptation to extreme climate events has moved away from reactive disaster relief towards more proactive risk management. A prominent example is the implementation in several countries of early warning systems for heatwaves [12.6.1]. Other actions have addressed long-term climate changes. For example, national action plans have been developed for adapting to climate change [12.5] and more specific plans have been incorporated into European and national policies for agriculture, energy, forestry, transport, and other sectors [12.2.3, 12.5.2]. Research has also provided new insights into adaptation policies (e.g., studies showed that crops that become less economically viable under climate change can be replaced profitably by bioenergy crops) [12.5.7].

Although the effectiveness and feasibility of adaptation measures are expected to vary greatly, only a few governments and institutions have systematically and critically examined a portfolio of measures (very high confidence).
As an example, some reservoirs used now as a measure for adapting to precipitation fluctuations may become unreliable in regions where long-term precipitation is projected to decrease [12.4.1]. In terms of forestry, the range of management options to cope with climate change varies largely among forest types, some having many more options than others [12.5.4].

12.1 Introduction

This chapter reviews existing literature on the anticipated impacts, adaptation and vulnerability of Europe to climate change during the 21st century. The area covered under Europe in this report includes all countries from Iceland in the west to Russia (west of the Urals) and the Caspian Sea in the east, and from the northern shores of the Mediterranean and Black Seas and the Caucasus in the south to the Arctic Ocean in the north. Polar issues, however, are covered in greater detail in Chapter 15.

12.1.1 Summary of knowledge from the Third Assessment Report

Climate trends in the 20th century

During the 20th century, most of Europe experienced increases in average annual surface temperature (average increase over the continent 0.8°C), with stronger warming over most regions in winter than in summer. The 1990s were the warmest in the instrumental record. Precipitation trends in the 20th century showed an increase in northern Europe (10 to 40%) and a decrease in southern Europe (up to 20% in some parts). The latest data reported in this assessment have confirmed these trends.

Climate change scenarios

The most recent climate modelling results available to the Third Assessment Report (TAR) showed an increase in annual temperature in Europe of 0.1 to 0.4°C/decade over the 21st century based on a range of scenarios and models. The models show a widespread increase in precipitation in the north, small decreases in the south, and small or ambiguous changes in central Europe. It is likely that the seasonality of precipitation will change and the frequency of intense precipitation events will increase, especially in winter. The TAR noted a very likely increase in the intensity and frequency of summer heatwaves throughout Europe, and one such major heatwave has occurred since the TAR.

Sensitivities to climate

With regards to its current sensitivities to climate, Europe was found to be most sensitive to the following conditions:

- extreme seasons, in particular exceptionally hot and dry summers and mild winters,
- short-duration events such as windstorms and heavy rains, and
- slow, long-term changes in climate which, among other impacts, will put particular pressure on coastal areas e.g., through sea-level rise.

More information is now available on the geographic variability of Europe's sensitivity to changes in climate.

Variability of impacts in regions and on social groups

Impacts of climate change will vary substantially from region to region, and from sector to sector within regions. More adverse impacts are expected in regions with lower economic development which is often related to lower adaptive capacity. Climate change will have greater or lesser impacts on different social groups (e.g., age classes, income groups, occupations).

Economic effects

The TAR identified many climate change impacts on Europe's economy:

- sea-level rise will affect important coastline industries,
- increasing CO_2 concentrations may increase agricultural yields, although this may be counteracted by decreasing water availability in southern and south-eastern Europe,
- recreation preferences are likely to change (more outdoor activity in the north, less in the south),
- the insurance industry should expect increased climate-related claims, and
- warmer temperatures and higher CO_2 levels may increase the potential timber harvest in northern Europe, while warmer temperatures may increase forest fire risk in southern Europe.

12.2 Current sensitivity/vulnerability

12.2.1 Climate factors and trends

The warming trend throughout Europe is well established (+0.90°C for 1901 to 2005; updated from Jones and Moberg, 2003). However, the recent period shows a trend considerably higher than the mean trend (+0.41°C/decade for the period 1979 to 2005; updated from Jones and Moberg, 2003). For the 1977 to 2000 period, trends are higher in central and north-eastern Europe and in mountainous regions, while lower trends are found in the Mediterranean region (Böhm et al., 2001; Klein Tank, 2004). Temperatures are increasing more in winter than summer (Jones and Moberg, 2003). An increase of daily temperature variability is observed during the period 1977 to 2000 due to an increase in warm extremes, rather than a decrease of cold extremes (Klein Tank et al., 2002; Klein Tank and Können, 2003).

Precipitation trends are more spatially variable. Mean winter precipitation is increasing in most of Atlantic and northern Europe (Klein Tank et al., 2002). In the Mediterranean area, yearly precipitation trends are negative in the east, while they are non-significant in the west (Norrant and Douguédroit, 2006). An increase in mean precipitation per wet day is observed in most parts of the continent, even in some areas which are becoming drier (Frich et al., 2002; Klein Tank et al., 2002; Alexander et al., 2006). Some of the European systems and sectors have shown particular sensitivity to recent trends in temperature and (to a lesser extent) precipitation (Table 12.1).

12.2.2 Non-climate factors and trends

Europe has the highest population density (60 persons/km^2) of any continent. Of the total European population, 73% lives in urban areas (UN, 2004), with 67% in southern Europe and 83% in northern Europe. The 25 countries belonging to the European Union (EU25) have stable economies, high productivity and integrated markets. Economic conditions among the non-EU countries are more varied. European income (as annual gross domestic product (GDP) per capita based on market exchange rate) ranges from US$1,760 in Moldova to US$55,500 in Luxembourg (World Bank, 2005). The EU25 cover 60% of the total European population, but only 17% of the total European land area and 36% of its agricultural area. In 2003, the European Union (EU) with its then 15 countries (EU15), contributed 20% of global GDP and 40% of global exports of goods and services (IMF, 2004). Central and Eastern Europe (CEE) plus European Russia constituted 16% of global GDP.

Since 1990, countries in CEE have undergone dramatic economic and political change towards a market economy and democracy and, for some countries, also integration in the EU. Annual GDP growth rates have exceeded 4% for all CEE countries and Russia, as compared to 2% in the EU (IMF, 2004).

Energy use in Europe constituted *circa* 30% of global energy consumption in 2003 (EEA, 2006a). More than 60% of this consumption occurred in the Organisation for Economic Co-operation and Development (OECD) countries (EEA, 2006a), whereas oil resources in Russia alone are more than four times higher than those of OECD Europe. Combustion of fossil fuels accounts for almost 80% of total energy consumption and 55% of electricity production in EU25 (EEA, 2006a). The large reliance on external fossil fuel resources has led to an increasing focus on renewable energy sources, including bioenergy (EEA, 2006a, b). In 2003, renewable energy contributed 6% and 13% to total energy and gross electricity consumption in EU25, respectively (EEA, 2006a).

The EU25 in 2002 had average greenhouse gas emissions of 11 tonnes CO_2 per capita (EEA, 2004a) and this is projected to increase to 12 tonnes CO_2 per capita in 2030 under baseline conditions (EEA, 2006a). Most European countries have ratified the Kyoto Protocol, and the EU15 countries have a common reduction target between 2008 and 2012 of 8% (Babiker and Eckaus, 2002). From 1990 to 2003 EU25 greenhouse gas emissions, excluding Land Use, Land Use Change and Forestry (LULUCF), decreased by 5.5%, but emissions in the transport sector grew 23% in the EU15 (EEA, 2005).

The hydrological characteristics of Europe are very diverse, as well as its approaches to water use and management. Of the total withdrawals of 30 European countries (EU plus adjacent countries) 32% are for agriculture, 31% for cooling water in power stations, 24% for the domestic sector and 13% for manufacturing (Flörke and Alcamo, 2005). Freshwater abstraction is stable or declining in northern Europe and growing slowly in southern Europe (Flörke and Alcamo, 2005). There are many pressures on water quality and availability including those arising from agriculture, industry, urban areas, households and tourism (Lallana et al., 2001). Recent floods and droughts have placed additional stresses on water supplies and infrastructure (Estrela et al., 2001).

Europe is one of the world's largest and most productive suppliers of food and fibre (in 2004: 21% of global meat production and 20% of global cereal production). About 80% of this production occurred in the EU25. The productivity of European agriculture is generally high, in particular in western Europe: average cereal yields in the EU are more than 60% higher than the global average. During the last decade the EU Common Agricultural Policy (CAP) has been reformed to reduce overproduction, reduce environmental impacts and improve rural development. This is not expected to greatly affect agricultural production in the short term (OECD, 2004). However, agricultural reforms are expected to enhance the current process of structural adjustment leading to larger and fewer farms (Marsh, 2005).

The forested areas of Europe are increasing and annual fellings are considerably below sustainable levels (EEA, 2002). Forest policies have been modified during the past decade to promote multiple forest services at the expense of timber production (Kankaanpää and Carter, 2004). European forests are a sink of atmospheric CO_2 of about 380 Tg C/yr (mid 1990s) (Janssens et al., 2003). However, CO_2 emissions from the agricultural and peat sectors reduce the net carbon uptake in Europe's terrestrial biosphere to between 135 and 205 Tg C/yr, equivalent to 7 to 12% of European anthropogenic emissions in 1995 (Janssens et al., 2003).

Despite policies to protect fish, over-fishing has put many fish stocks in European waters outside sustainable limits (62-

Table 12.1. *Attribution of recent changes in natural and managed ecosystems to recent temperature and precipitation trends. See Chapter 1, Section 1.3 for additional data.*

Region	Observed change	Reference
Coastal and marine systems		
North-east Atlantic, North Sea	Northward movement of plankton and fish	Brander and Blom, 2003; Edwards and Richardson, 2004; Perry et al., 2005
Terrestrial ecosystems		
Europe	Upward shift of the tree line	Kullman, 2002; Camarero and Gutiérrez, 2004; Shiyatov et al., 2005; Walther et al., 2005a
Europe	Phenological changes (earlier onset of spring events and lengthening of the growing season); increasing productivity and carbon sink during 1950 to 1999 of forests (in 30 countries)	Menzel et al., 2006a Nabuurs et al., 2003, Shvidenko and Nilsson, 2003; Boisvenue and Running, 2006
Alps	Invasion of evergreen broad-leaved species in forests; upward shift of *Viscum album*	Walther, 2004; Dobbertin et al., 2005
Scandinavia	Northward range expansion of *Ilex aquifolium*	Walter et al., 2005a
Fennoscandian mountains and sub-Artic	Disappearance of some types of wetlands (palsa mires[1]) in Lapland; increased species richness and frequency at altitudinal margin of plant life	Klanderud and Birks, 2003; Luoto et al., 2004
High mountains	Change in high mountain vegetation types and new occurrence of alpine vegetation on high summits.	Grabherr et al., 2001; Kullman, 2001; Pauli et al., 2001; Klanderud and Birks, 2003; Peñuelas and Boada, 2003; Petriccione, 2003; Sanz Elorza and Dana, 2003; Walther et al., 2005a
Agriculture		
Northern Europe	Increased crop stress during hotter, drier summers; increased risk to crops from hail	Viner et al., 2006
Britain, southern Scandinavia	Increased area of silage maize (more favourable conditions due to warmer summer temperatures)	Olesen and Bindi, 2004
France	Increases in growing season of grapevine; changes in wine quality	Jones and Davis, 2000; Duchene and Schneider, 2005
Germany	Advance in the beginning of growing season for fruit trees	Menzel, 2003; Chmielewski et al., 2004
Cryosphere		
Russia	Decrease in thickness and areal extent of permafrost and damages to infrastructure	Frauenfeld et al., 2004; Mazhitova et al., 2004
Alps	Decrease in seasonal snow cover (at lower elevation)	Laternser and Schneebeli, 2003; Martin and Etchevers, 2005
Europe	Decrease in glacier volume and area (except some glaciers in Norway)	Hoelzle et al., 2003
Health		
North, East	Movement of tick vectors northwards, and possibly to high altitudes	Lindgren and Gustafson, 2001; Randolph, 2002; Beran et al., 2004; Danielova et al., 2004; Izmerov, 2004; Daniel et al., 2005; Materna et al., 2005
Mediterranean, West, South	Northward movement of *Visceral Leishmaniasis* in dogs and humans [low confidence]	Molyneux, 2003; Kuhn et al., 2004; WHO, 2005; Lindgren and Naucke, 2006
Mediterranean, Atlantic, Central	Heatwave mortality	Fischer et al., 2004; Kosatsky, 2005; Nogueira et al., 2005, Pirard et al., 2005
Atlantic, Central, East, North	Earlier onset and extension of season for allergenic pollen	Huynen and Menne, 2003; van Vliet et al., 2003; Beggs, 2004 [Chapter 1.3.7.5]

92% of commercial fish stocks in north-eastern Atlantic, 100% in the western Irish Sea, 75% in the Baltic Sea, and 65-70% in the Mediterranean) (EEA, 2002; Gray and Hatchard, 2003). Aquaculture is increasing its share of the European fish market leading to possible adverse environmental impacts in coastal waters (Read and Fernandes, 2003).

Increasing urbanisation and tourism, as well as intensification of agriculture, have put large pressures on land resources (EEA, 2004a), yet there is increasing political attention given to the sustainable use of land and natural resources. Despite general reductions in the extent of air pollution in Europe over the last decades, significant problems still remain with acidification,

[1] Palsa mire: a type of peatland typified by high mounds with permanently frozen cores and separated by wet depressions; they form where the ground surface is only frozen for part of the year.

terrestrial nitrogen deposition, ozone, particulate matter and heavy metals (WGE, 2004). Environmental protection in the EU has led to several directives such as the Emissions Ceilings Directive and the Water Framework Directive. The EU Species and Habitats Directive and the Wild Birds Directive have been integrated in the Natura 2000 network, which protects nature in over 18% of the EU territory. Awareness of environmental issues is also growing in CEE (TNS Opinion and Social, 2005).

12.2.3 Current adaptation and adaptive capacity

It is apparent that climate variability and change already affects features and functions of Europe's production systems (e.g., agriculture, forestry and fisheries), key economic sectors (e.g., tourism, energy) and its natural environment. Some of these effects are beneficial, but most are estimated to be negative (EEA, 2004b). European institutions have recognised the need to prepare for an intensification of these impacts even if greenhouse gas emissions are substantially reduced (e.g., EU Environmental Council meeting, December 2004).

The sensitivity of Europe to climate change has a distinct north-south gradient, with many studies indicating that southern Europe will be more severely affected than northern Europe (EEA, 2004b). The already hot and semi-arid climate of southern Europe is expected to become warmer and drier, and this will threaten its waterways, agricultural production and timber harvests (e.g., EEA, 2004b). Nevertheless, northern countries are also sensitive to climate change.

The Netherlands is an example of a country highly susceptible to both sea-level rise and river flooding because 55% of its territory is below sea level where 60% of its population lives and 65% of its Gross National Product (GNP) is produced. As in other regions, natural ecosystems in Europe are more vulnerable to climate change than managed systems such as agriculture and fisheries (Hitz and Smith, 2004). Natural ecosystems usually take decades or longer to become established and therefore adapt more slowly to climatic changes than managed systems. The expected rate of climate change in Europe is likely to exceed the current adaptive capacity of various non-cultivated plant species (Hitz and Smith, 2004). Sensitivity to climate variability and change also varies across different ecosystems. The most sensitive natural ecosystems in Europe are located in the Arctic, in mountain regions, in coastal zones (especially the Baltic wetlands) and in various parts of the Mediterranean (WBGU, 2003). Ecosystems in these regions are already affected by an increasing trend in temperature and decreasing precipitation in some areas and may be unable to cope with expected climate change.

The possible consequences of climate change in Europe have stimulated efforts by the EU, national governments, businesses, and Non-Governmental Organisations (NGOs) to develop adaptation strategies. The EU is supporting adaptation research at the pan-European level while Denmark, Finland, Hungary, Portugal, Slovakia, Spain and the UK are setting up national programmes for adapting to climate change. Plans for adaptation to climate change have been included in flood protection plans of the Czech Republic and coastal protection plans of the Netherlands and Norway.

12.3 Assumptions about future trends

12.3.1 Climate projections

12.3.1.1 Mean climate

Results presented here and in the following sections are for the period 2070 to 2099 and are mostly based on the IPCC Special Report on Emissions Scenarios (SRES: Nakićenović and Swart, 2000; see also Section 12.3.2) using the climate normal period (1961 to 1990) as a baseline.

Europe undergoes a warming in all seasons in both the SRES A2 and B2 emissions scenarios (A2: 2.5 to 5.5°C, B2: 1 to 4°C; the range of change is due to different climate modelling results). The warming is greatest over eastern Europe in winter (December to February: DJF) and over western and southern Europe in summer (June to August: JJA) (Giorgi et al., 2004). Results using two regional climate models under the PRUDENCE project (Christensen and Christensen, 2007) showed a larger warming in winter than in summer in northern Europe and the reverse in southern and central Europe. A very large increase in summer temperatures occurs in the south-western parts of Europe, exceeding 6°C in parts of France and the Iberian Peninsula (Kjellström, 2004; Räisänen et al., 2004; Christensen and Christensen, 2006; Good et al., 2006).

Generally for all scenarios, mean annual precipitation increases in northern Europe and decreases further south, whilst the change in seasonal precipitation varies substantially from season to season and across regions in response to changes in large-scale circulation and water vapour loading. Räisänen et al. (2004) identified an increase in winter precipitation in northern and central Europe. Likewise, Giorgi et al. (2004) found that increased Atlantic cyclonic activity in DJF leads to enhanced precipitation (up to 15-30%) over much of western, northern and central Europe. Precipitation during this period decreases over Mediterranean Europe in response to increased anticyclonic circulation. Räisänen et al. (2004) found that summer precipitation decreases substantially (in some areas up to 70% in scenario A2) in southern and central Europe, and to a smaller degree in northern Europe up to central Scandinavia. Giorgi et al. (2004) identified enhanced anticyclonic circulation in JJA over the north-eastern Atlantic, which induces a ridge over western Europe and a trough over eastern Europe. This blocking structure deflects storms northward, causing a substantial and widespread decrease in precipitation (up to 30-45%) over the Mediterranean Basin as well as over western and central Europe. Both the winter and summer changes were found to be statistically significant (very high confidence) over large areas of the regional modelling domain. Relatively small precipitation changes were found for spring and autumn (Kjellström, 2004; Räisänen et al., 2004).

Change in mean wind speed is highly sensitive to the differences in large-scale circulation that can result between different global models (Räisänen et al., 2004). From regional simulations based on ECHAM4 and the A2 scenario, mean annual wind speed increases over northern Europe by about 8% and decreases over Mediterranean Europe (Räisänen et al., 2004; Pryor et al., 2005). The increase for northern Europe is largest in winter and early spring, when the increase in the average north-south pressure gradient is largest. Indeed, the simulation of DJF mean pressure indicates an increase in average westerly flow over northern Europe when the ECHAM4 global model is used, but a slight decrease when the HadAM3H model (Gordon et al., 2000) is used. For France and central Europe, all four of the simulations documented by Räisänen et al. (2004) indicate a slight increase in mean wind speeds in winter and some decrease in spring and autumn. None of the reported simulations show significant change during summer for northern Europe.

12.3.1.2 Extreme events

The yearly maximum temperature is expected to increase much more in southern and central Europe than in northern Europe (Räisänen et al., 2004; Kjellström et al., 2007). Kjellström (2004) shows that, in summer, the warming of large parts of central, southern and eastern Europe may be more closely connected to higher temperatures on warm days than to a general warming. A large increase is also expected for yearly minimum temperature across most of Europe, which at many locations exceeds the average winter warming by a factor of two to three. Much of the warming in winter is connected to higher temperatures on cold days, which indicates a decrease in winter temperature variability. An increase in the lowest winter temperatures, although large, would primarily mean that current cold extremes would decrease. In contrast, a large increase in the highest summer temperatures would expose Europeans to unprecedented high temperatures.

Christensen and Christensen (2003), Giorgi et al. (2004) and Kjellström (2004) all found a substantial increase in the intensity of daily precipitation events. This holds even for areas with a decrease in mean precipitation, such as central Europe and the Mediterranean. Impact over the Mediterranean region during summer is not clear due to the strong convective rainfall component and its great spatial variability (Llasat, 2001). Palmer and Räisänen (2002) estimate that the probability of extreme winter precipitation exceeding two standard deviations above normal would increase by a factor of five over parts of the UK and northern Europe, while Ekström et al. (2005) have found a 10% increase in short duration (1 to 2 days) precipitation events across the UK. Lapin and Hlavcova (2003) found an increase in short duration (1 to 5 days) summer rainfall events in Slovakia of up to 40% for a 3.5°C summer warming.

The combined effects of warmer temperatures and reduced mean summer precipitation would enhance the occurrence of heatwaves and droughts. Schär et al. (2004) conclude that the future European summer climate would experience a pronounced increase in year-to-year variability and thus a higher incidence of heatwaves and droughts. Beniston et al. (2007) estimated that countries in central Europe would experience the same number of hot days as currently occur in southern Europe and that Mediterranean droughts would start earlier in the year and last longer. The regions most affected could be the southern Iberian Peninsula, the Alps, the eastern Adriatic seaboard, and southern Greece. The Mediterranean and even much of eastern Europe may experience an increase in dry periods by the late 21st century (Polemio and Casarano, 2004). According to Good et al. (2006), the longest yearly dry spell could increase by as much as 50%, especially over France and central Europe. However, there is some recent evidence (Lenderink et al., 2007) that these projections for droughts and heatwaves may be slightly over-estimated due to the parameterisation of soil moisture (too small soil storage capacity resulting in soil drying out too easily) in regional climate models.

Regarding extreme winds, Rockel and Woth (2007) and Leckebusch and Ulbrich (2004) found an increase in extreme wind speeds for western and central Europe, although the changes were not statistically significant for all months of the year. Beniston et al. (2007) found that extreme wind speeds increased for the area between 45°N and 55°N, except over and south of the Alps. Woth et al. (2005) and Beniston et al. (2007) conclude that this could generate more North Sea storms leading to increases in storm surges along the North Sea coast, especially in the Netherlands, Germany and Denmark.

12.3.2 Non-climate trends

The European population is expected to decline by about 8% over the period from 2000 to 2030 (UN, 2004). The relative overall stability of the population of Europe is due to population growth in western Europe alone, mainly from immigration (Sardon, 2004). Presently, CEE and Russia have a surplus of deaths over births, with the balance of migration being positive only in Russia. Fertility rates vary considerably across the continent, from 1.10 children per woman in Ukraine to 1.97 in Ireland. There is a general decline in old-age mortality in most European countries (Janssen et al., 2004), although there has been a reduction in life expectancy in Russia during the 1990s. The low birth rate and increase in duration of life lead to an overall older population. The proportion of the population over 65 years of age in the EU15 is expected to increase from 16% in 2000 to 23% in 2030, which will likely affect vulnerability in recreational (see Section 12.4.9) and health aspects (see Section 12.4.11).

The SRES scenarios (see Chapter 2 Section 2.4.6) for socio-economic development have been adapted to European conditions (Parry, 2000; Holman et al., 2005; Abildtrup et al., 2006). Electricity consumption in the EU25 is projected to continue growing twice as fast as the increase in total energy consumption (EEA, 2006a), primarily due to higher comfort levels and larger dwellings increasing demand for space heating and cooling, which will have consequences for electricity demand during summer (see Section 12.4.8.1).

Assumptions about future European land use and the environmental impact of human activities depend greatly on the development and adoption of new technologies. For the SRES scenarios it has been estimated that increases in crop productivity relative to 2000 could range between 25 and 163% depending on the time slice (2020 to 2080) and scenario (Ewert et al., 2005). These increases were found to be smallest for the B2 and highest for the A1FI scenario. Temporally and spatially explicit future scenarios of European land use have been developed for the four core SRES scenarios (Schröter et al., 2005; Rounsevell et al., 2006). These scenarios show large declines in agricultural land area, resulting primarily from the assumptions about technological development and its effect on crop yield (Rounsevell et al., 2005), although climate change may also play a role (see Section 12.5.7). The expansion of urban area is similar between the scenarios, and forested areas also increase in all scenarios (Schröter et al., 2005). The scenarios showed decreases in European cropland for 2080 of 28 to 47% and decreases in grassland of 6 to 58% (Rounsevell et al., 2005). This decline in agricultural area will mean that land resources will be available for other uses such as biofuel production and nature reserves. Over the shorter term (up to 2030) changes in agricultural land area may be small (van Meijl et al., 2006).

12.4 Key future impacts and vulnerabilities

The wide range of climate change impacts and vulnerabilities expected in Europe is summarised in Figure 12.3 and Table 12.4.

12.4.1 Water resources

It is likely that climate change will have a range of impacts on water resources. Projections based on various emissions scenarios and General Circulation Models (GCMs) show that annual runoff increases in Atlantic and northern Europe (Werritty, 2001; Andréasson et al., 2004), and decreases in central, Mediterranean and eastern Europe (Chang et al., 2002; Etchevers et al., 2002; Menzel and Bürger, 2002; Iglesias et al., 2005). Most of the hydrological impact studies reported here are based on global rather than regional climate models. Annual average runoff is projected to increase in northern Europe (north of 47°N) by approximately 5 to 15% up to the 2020s and 9 to 22% up to the 2070s, for the SRES A2 and B2 scenarios and climate scenarios from two different climate models (Alcamo et al., 2007) (Figure 12.1). Meanwhile, in southern Europe (south of 47°N), runoff decreases by 0 to 23% up to the 2020s and by 6 to 36% up to the 2070s (for the same set of assumptions). The projected changes in annual river basin discharge by the 2020s are likely to be affected as much by climate variability as by climate change. Groundwater recharge is likely to be reduced in central and eastern Europe (Eitzinger et al., 2003), with a larger reduction in valleys (Krüger et al., 2002) and lowlands (e.g., in the Hungarian steppes) (Somlyódy, 2002).

Studies show an increase in winter flows and decrease in summer flows in the Rhine (Middelkoop and Kwadijk, 2001), Slovakian rivers (Szolgay et al., 2004), the Volga and central and eastern Europe (Oltchev et al., 2002). It is likely that glacier retreat will initially enhance summer flow in the rivers of the

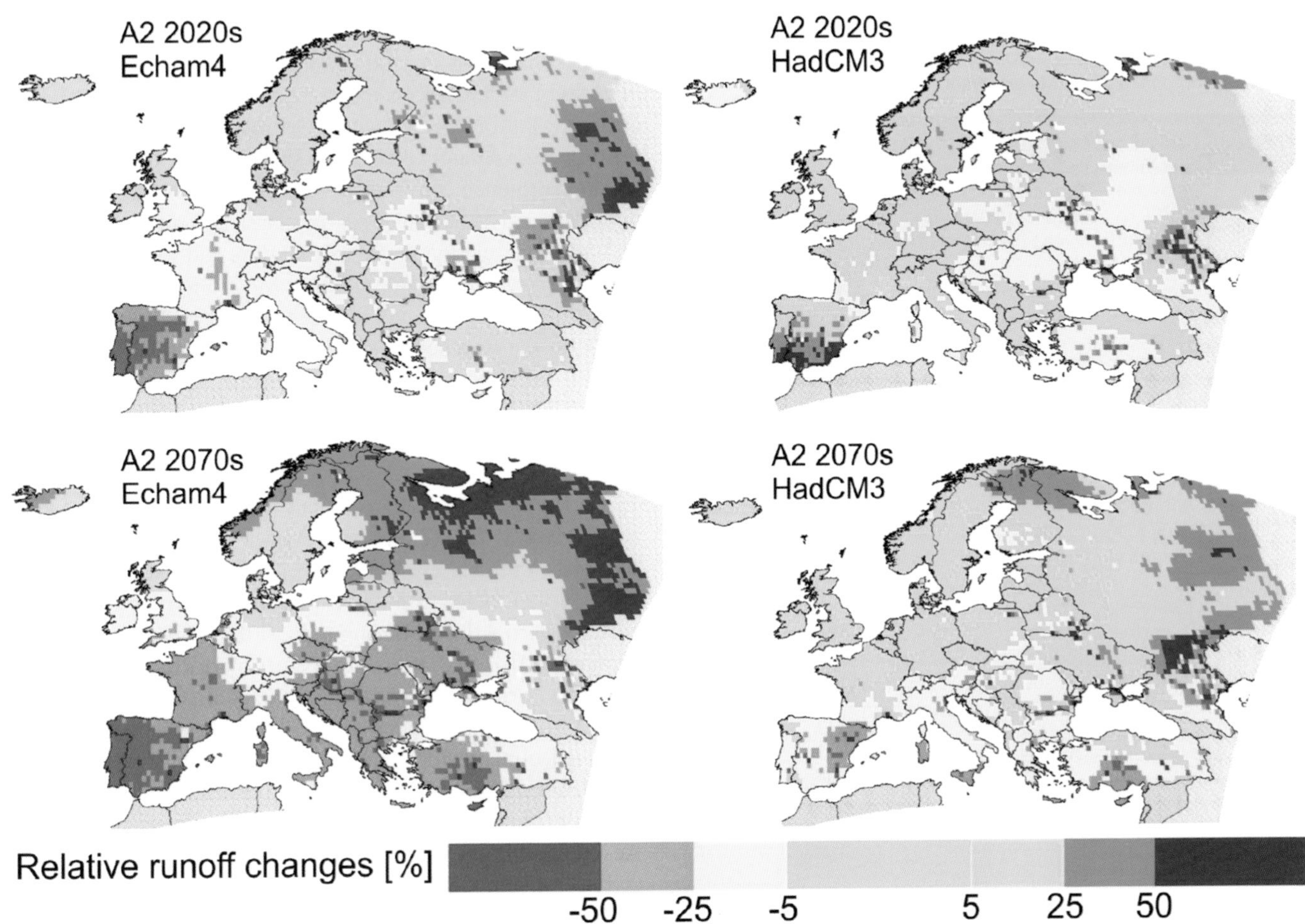

Figure 12.1. *Change in annual river runoff between the 1961-1990 baseline period and two future time slices (2020s and 2070s) for the A2 scenarios (Alcamo et al., 2007).*

Alps; however, as glaciers shrink, summer flow is likely to be significantly reduced (Hock et al., 2005), by up to 50% (Zierl and Bugmann, 2005). Summer low flow may decrease by up to 50% in central Europe (Eckhardt and Ulbrich, 2003), and by up to 80% in some rivers in southern Europe (Santos et al., 2002).

Changes in the water cycle are likely to increase the risk of floods and droughts. Projections under the IPCC IS92a scenario (similar to SRES A1B; IPCC, 1992) and two GCMs (Lehner et al., 2006) indicate that the risk of floods increases in northern, central and eastern Europe, while the risk of drought increases mainly in southern Europe (Table 12.2). Increase in intense short-duration precipitation in most of Europe is likely to lead to increased risk of flash floods (EEA, 2004b). In the Mediterranean, however, historical trends supporting this are not extensive (Ludwig et al., 2003; Benito et al., 2005; Barrera et al., 2006).

Increasing flood risk from climate change could be magnified by increases in impermeable surface due to urbanisation (de Roo et al., 2003) and modified by changes in vegetation cover (Robinson et al., 2003) in small catchments. The effects of land use on floods in large catchments are still being debated. The more frequent occurrence of high floods increases the risk to areas currently protected by dykes. The increasing volume of floods and peak discharge would make it more difficult for reservoirs to store high runoff and prevent floods.

Table 12.2. *Impact of climate change on water availability, drought and flood occurrence in Europe for various time slices and under various scenarios based on the ECHAM4 and HadCM3 models.*

Time slice	Water availability and droughts	Floods
2020s	Increase in annual runoff in northern Europe by up to 15% and decrease in the south by up to 23%.[a] Decrease in summer flow.[b]	Increasing risk of winter flood in northern Europe, of flash flooding across all of Europe. Risk of snowmelt flood shifts from spring to winter.[c]
2050s	Decrease in annual runoff by 20-30% in south-eastern Europe.[d]	
2070s	Increase in annual runoff in the north by up to 30% and decrease by up to 36% in the south.[a] Decrease in summer low flow by up to 80%.[d, b] Decreasing drought risk in northern Europe, increasing drought risk in western and southern Europe. Today's 100-year droughts return every 50 years (or less) in southern and south-eastern Europe (Portugal, all Mediterranean countries, Hungary, Romania, Bulgaria, Moldova, Ukraine, southern Russia).[c]	Today's 100-year floods occur more frequently in northern and north-eastern Europe (Sweden, Finland, northern Russia), in Ireland, in central and eastern Europe (Poland, Alpine rivers), in Atlantic parts of southern Europe (Spain, Portugal), and less frequently in large parts of southern Europe.[c]

[a] Alcamo et al., 2007; [b] Santos et al., 2002; [c] Lehner et al., 2006; [d] Arnell, 2004

Increasing drought risk for western Europe (e.g., Great Britain; Fowler and Kilsby, 2004) is primarily caused by climate change; for southern and eastern Europe increasing risk from climate change would be amplified by an increase in water withdrawals (Lehner et al., 2006). The regions most prone to an increase in drought risk are the Mediterranean (Portugal, Spain) and some parts of central and eastern Europe, where the highest increase in irrigation water demand is projected (Döll, 2002; Donevska and Dodeva, 2004). Irrigation requirements are likely to become substantial in countries (e.g., Ireland) where demand now hardly exists (Holden et al., 2003). It is likely that, due to both climate change and increasing water withdrawals, the river-basin area affected by severe water stress (withdrawal : availability >0.40) will increase and lead to increasing competition for available water resources (Alcamo et al., 2003; Schröter et al., 2005). Under the IS92a scenario, the percentage of river basin area in the severe water stress category increases from 19% today to 34-36% by the 2070s (Lehner et al., 2001). The number of additional people living in water-stressed watersheds in the EU15 plus Switzerland and Norway is likely to increase to between 16 million and 44 million, based on climate projected by the HadCM3 GCM under the A2 and B1 emissions scenarios, respectively (Schröter et al., 2005).

12.4.2 Coastal and marine systems

Climate variability associated with the North Atlantic Oscillation (NAO) determines many physical coastal processes in Europe (Hurrell et al., 2003, 2004), including variations in the seasonality of coastal climates, winter wind speeds and patterns of storminess and coastal flooding in north-west Europe (Lozano et al., 2004; Stone and Orford, 2004; Yan et al., 2004). For Europe's Atlantic coasts and shelf seas, the NAO also has a strong influence on the dynamic sea-surface height and geographic distribution of sea-level rise (Woolf et al., 2003), as well as some relation to coastal flooding and water levels in the Caspian Sea (Lal et al., 2001). Most SRES-based climate scenarios show a continuation of the recent positive phase of the NAO for the first decades of the 21st century with significant impacts on coastal areas (Cubasch et al., 2001; Hurrell et al., 2003).

Wind-driven waves and storms are seen as the primary drivers of short-term coastal processes on many European coasts (Smith et al., 2000). Climate simulations using the IS92a and A2 and B2 SRES scenarios (Meier et al., 2004; Räisänen et al., 2004) reinforce existing trends in storminess. These indicate some further increase in wind speeds and storm intensity in the north-eastern Atlantic during at least the early part of the 21st century (2010 to 2030), with a shift of storm centre maxima closer to European coasts (Knippertz et al., 2000; Leckebusch and Ulbrich, 2004; Lozano et al., 2004). These experiments also show a decline in storminess and wind intensity eastwards into the Mediterranean (Busuioc, 2001; Tomozeiu et al., 2007), but with localised increased storminess in parts of the Adriatic, Aegean and Black Seas (Guedes Soares et al., 2002).

Ensemble modelling of storm surges and tidal levels in shelf seas, particularly for the Baltic and southern North Sea, indicate fewer but more extreme surge events under some SRES emissions scenarios (Hulme et al., 2002; Meier et al., 2004;

Lowe and Gregory, 2005). In addition, wave simulations show higher significant wave heights of >0.4m in the north-eastern Atlantic by the 2080s (Woolf et al., 2002; Tsimplis et al., 2004a; Wolf and Woolf, 2006). Higher wave and storm-surge elevations will be particularly significant because they will cause erosion and flooding in estuaries, deltas and embayments (Flather and Williams, 2000; Lionello et al., 2002; Tsimplis et al., 2004b; Woth et al., 2005; Meier et al., 2007).

Model projections of the IPCC SRES scenarios give a global mean sea-level rise of 0.09 to 0.88 m by 2100, with sea level rising at rates *circa* 2 to 4 times faster than those of the present day (EEA, 2004b; Meehl et al., 2007). In Europe, regional influences may result in sea-level rise being up to 50% higher than these global estimates (Woodworth et al., 2005). The impact of the NAO on winter sea levels provides an additional uncertainty of 0.1 to 0.2 m to these estimates (Hulme et al., 2002; Tsimplis et al., 2004a). Furthermore, the sustained melting of Greenland ice and other ice stores under climate warming, coupled with the impacts of a possible abrupt shut-down of the Atlantic meridional overturning circulation (MOC) after 2100, provide additional uncertainty to sea-level rise for Europe (Gregory et al., 2004; Levermann et al., 2005; Wigley, 2005; Meehl et al., 2007).

Sea-level rise can have a wide variety of impacts on Europe's coastal areas; causing flooding, land loss, the salinisation of groundwater and the destruction of built property and infrastructures (Devoy, 2007; Nicholls and de la Vega-Leinert, 2007). Over large areas of formerly glaciated coastlines the continued decline in isostatic land uplift is bringing many areas within the range of sea-level rise (Smith et al., 2000). For the Baltic and Arctic coasts, sea-level rise projections under some SRES scenarios indicate an increased risk of flooding and coastal erosion after 2050 (Johansson et al., 2004; Meier et al., 2004, 2006; Kont et al., 2007). In areas of coastal subsidence or high tectonic activity, as in the low tidal range Mediterranean and Black Sea regions, climate-related sea-level rise could significantly increase potential damage from storm surges and tsunamis (Gregory et al., 2001). Sea-level rise will also cause an inland migration of Europe's beaches and low-lying, soft sedimentary coasts (Sánchez-Arcilla et al., 2000; Stone and Orford, 2004; Hall et al., 2007). Coastal retreat rates are currently 0.5 to 1.0 m/yr for parts of the Atlantic coast most affected by storms and under sea-level rise these rates are expected to increase (Cooper and Pilkey, 2004; Lozano et al., 2004).

The vulnerability of marine and nearshore waters and of many coasts is very dependent on local factors (Smith et al., 2000; EEA, 2004b; Swift et al., 2007). Low-lying coastlines with high population densities and small tidal ranges will be most vulnerable to sea-level rise (Kundzewicz et al., 2001). Coastal flooding related to sea-level rise could affect large populations (Arnell et al., 2004). Under the SRES A1FI scenario up to an additional 1.6 million people each year in the Mediterranean, northern and western Europe, might experience coastal flooding by 2080 (Nicholls, 2004). Approximately 20% of existing coastal wetlands may disappear by 2080 under SRES scenarios for sea-level rise (Nicholls, 2004; Devoy, 2007). Impacts of climate warming upon coastal and marine ecosystems are also likely to intensify the problems of eutrophication and stress on these biological systems (EEA, 2004b; Robinson et al., 2005; SEPA, 2005; SEEG, 2006).

12.4.3 Mountains and sub-Arctic regions

The duration of snow cover is expected to decrease by several weeks for each °C of temperature increase in the Alps region at middle elevations (Hantel et al., 2000; Wielke et al., 2004; Martin and Etchevers, 2005). An upward shift of the glacier equilibrium line is expected from 60 to 140 m/°C (Maisch, 2000; Vincent, 2002; Oerlemans, 2003). Glaciers will experience a substantial retreat during the 21st century (Haeberli and Burn, 2002). Small glaciers will disappear, while larger glaciers will suffer a volume reduction between 30% and 70% by 2050 (Schneeberger et al., 2003; Paul et al., 2004). During the retreat of glaciers, spring and summer discharge will decrease (Hagg and Braun, 2004). The lower elevation of permafrost is likely to rise by several hundred metres. Rising temperatures and melting permafrost will destabilise mountain walls and increase the frequency of rock falls, threatening mountain valleys (Gruber et al., 2004). In northern Europe, lowland permafrost will eventually disappear (Haeberli and Burns, 2002). Changes in snowpack and glacial extent may also alter the likelihood of snow and ice avalanches, depending on the complex interaction of surface geometry, precipitation and temperature (Martin et al., 2001; Haeberli and Burns, 2002).

It is virtually certain that European mountain flora will undergo major changes due to climate change (Theurillat and Guisan, 2001; Walther, 2004). Change in snow-cover duration and growing season length should have much more pronounced effects than direct effects of temperature changes on metabolism (Grace et al., 2002; Körner, 2003). Overall trends are towards increased growing season, earlier phenology and shifts of species distributions towards higher elevations (Kullman 2002; Körner, 2003; Egli et al., 2004; Sandvik et al., 2004; Walther, 2004). Similar shifts in elevation are also documented for animal species (Hughes, 2000). The treeline is predicted to shift upward by several hundred metres (Badeck et al., 2001). There is evidence that this process has already begun in Scandinavia (Kullman, 2002), the Ural Mountains (Shiyatov et al., 2005), West Carpathians (Mindas et al., 2000) and the Mediterranean (Peñuelas and Boada, 2003; Camarero and Gutiérrez, 2004). These changes, together with the effect of abandonment of traditional alpine pastures, will restrict the alpine zone to higher elevations (Guisan and Theurillat, 2001; Grace et al., 2002; Dirnböck et al., 2003; Dullinger et al., 2004), severely threatening nival flora[2] (Gottfried et al., 2002). The composition and structure of alpine and nival communities are very likely to change (Guisan and Theurillat, 2000; Walther, 2004). Local plant species losses of up to 62% are projected for Mediterranean and Lusitanian mountains by the 2080s under the A1 scenario (Thuiller et al., 2005). Mountain regions may additionally experience a loss of endemism due to invasive species (Viner et al., 2006). Similar extreme impacts are

[2] Nival flora: growing in or under snow.

expected for habitat and animal diversity as well, making mountain ecosystems among the most threatened in Europe (Schröter et al., 2005).

12.4.4 Forests, shrublands and grasslands

12.4.4.1 Forests

Forest ecosystems in Europe are very likely to be strongly influenced by climate change and other global changes (Shaver et al., 2000; Blennow and Sallnäs, 2002; Askeev et al., 2005; Kellomäki and Leinonen, 2005; Maracchi et al., 2005). Forest area is expected to expand in the north (Kljuev, 2001; MNRRF, 2003; Shiyatov et al., 2005), decreasing the current tundra area by 2100 (White et al., 2000), but contract in the south (Metzger et al., 2004). Native conifers are likely to be replaced by deciduous trees in western and central Europe (Maracchi et al., 2005; Koca et al., 2006). The distribution of a number of typical tree species is likely to decrease in the Mediterranean (Schröter et al., 2005). Tree vulnerability will increase as populations/plantations are managed to grow outside their natural range (Ray et al., 2002; Redfern and Hendry, 2002; Fernando and Cortina, 2004).

In northern Europe, climate change will alter phenology (Badeck et al., 2004) and substantially increase net primary productivity (NPP) and biomass of forests (Jarvis and Linder, 2000; Rustad et al., 2001; Strömgren and Linder, 2002; Zheng et al., 2002; Freeman et al., 2005; Kelomäki et al., 2005; Boisvenue and Running, 2006). In the boreal forest, soil CO_2 fluxes to the atmosphere increase with increased temperature and atmospheric CO_2 concentration (Niinisto et al., 2004), although many uncertainties remain (Fang and Moncrieff, 2001; Ågren and Bosatta, 2002; Hyvönen et al., 2005). Climate change may induce a reallocation of carbon to foliage (Magnani et al., 2004; Lapenis et al., 2005) and lead to carbon losses (White et al., 2000; Kostiainen et al., 2006; Schaphoff et al., 2006). Climate change may alter the chemical composition and density of wood while impacts on wood anatomy remain uncertain (Roderick and Berry, 2001; Wilhelmsson et al., 2002; Kostiainen et al., 2006).

In the northern and maritime temperate zones of Europe, and at higher elevations in the Alps, NPP is likely to increase throughout the century. However, by the end of the century (2071 to 2100) in continental central and southern Europe, NPP of conifers is likely to decrease due to water limitations (Lasch et al., 2002; Lexer et al., 2002; Martínez-Vilalta and Piñol, 2002; Freeman et al., 2005; Körner et al., 2005) and higher temperatures (Pretzch and Dursky, 2002). Negative impacts of drought on deciduous forests are also likely (Broadmeadow et al., 2005). Water stress in the south may be partially compensated by increased water-use efficiency (Magnani et al., 2004), elevated CO_2 (Wittig et al., 2005) and increased leaf area index (Kull et al., 2005), although this is currently under debate (Medlyn et al., 2001; Ciais et al., 2004).

Abiotic hazards for forest are likely to increase, although expected impacts are regionally specific and will be substantially dependent on the forest management system used (Kellomäki and Leinonen, 2005). A substantial increase in wind damage is not predicted (Barthod, 2003; Nilsson et al., 2004; Schumacher and Bugmann, 2006). In northern Europe, snow cover will decrease, and soil frost-free periods and winter rainfall increase, leading to increased soil waterlogging and winter floods (Nisbet, 2002; KSLA, 2004). Warming will prevent chilling requirements from being met[3], reduce cold-hardiness during autumn and spring, and increase needle loss (Redfern and Hendry, 2002). Frost damage is expected to be reduced in winter, unchanged in spring and more severe in autumn due to later hardening (Linkosalo et al., 2000; Barklund, 2002; Redfern and Hendry, 2002), although this may vary among regions and species (Jönsson et al., 2004). The risk of frost damage to trees may even increase after possible dehardening and growth onset during mild spells in winter and early spring (Hänninen, 2006). Fire danger, length of the fire season, and fire frequency and severity are very likely to increase in the Mediterranean (Santos et al., 2002; Pausas, 2004; Moreno, 2005; Pereira et al., 2005; Moriondo et al., 2006), and lead to increased dominance of shrubs over trees (Mouillot et al., 2002). Albeit less, fire danger is likely to also increase in central, eastern and northern Europe (Goldammer et al., 2005; Kellomäki et al., 2005; Moriondo et al., 2006). This, however, does not translate directly into increased fire occurrence or changes in vegetation (Thonicke and Cramer, 2006). In the forest-tundra ecotone, increased frequency of fire and other anthropogenic impacts is likely to lead to a long-term (over several hundred years) replacement of forest by low productivity grassy glades or wetlands over large areas (Sapozhnikov, 2003). The range of important forest insect pests may expand northward (Battisti, 2004), but the net impact of climate and atmospheric change is complex (Bale et al., 2002; Zvereva and Kozlov, 2006).

12.4.4.2 Shrublands

The area of European shrublands has increased over recent decades, particularly in the south (Moreira et al., 2001; Mouillot et al., 2003; Alados et al., 2004). Climate change is likely to affect its key ecosystem functions such as carbon storage, nutrient cycling, and species composition (Wessel et al., 2004). The response to warming and drought will depend on the current conditions, with cold, moist sites being more responsive to temperature changes, and warm, dry sites being more responsive to changes in rainfall (Peñuelas et al., 2004). In northern Europe, warming will increase microbial activity (Sowerby et al., 2005), growth and productivity (Peñuelas et al., 2004), hence enabling higher grazing intensities (Wessel et al., 2004). Encroachment by grasses (Werkman and Callaghan, 2002) and elevated nitrogen leaching (Emmet et al., 2004; Gorissen et al., 2004; Schmidt et al., 2004) are also likely. In southern Europe, warming and, particularly, increased drought, are likely to lead to reduced plant growth and primary productivity (Ogaya et al., 2003; Llorens et al., 2004), reduced nutrient turnover and nutrient availability (Sardans and Peñuelas, 2004, 2005), altered plant recruitment (Lloret et al., 2004; Quintana et al., 2004), changed phenology (Llorens and Peñuelas, 2005), and changed species interactions (Maestre and Cortina, 2004; Lloret et al.,

[3] Many plants, and most deciduous fruit trees, need a period of cold temperatures (the chilling requirement) during the winter in order for the flower buds to open in the spring.

2005). Shrubland fires are likely to increase due to their higher propensity to burn (Vázquez and Moreno, 2001; Mouillot et al., 2005; Nunes et al., 2005; Salvador et al., 2005). Furthermore, increased torrentiality (Giorgi et al., 2004) is likely to lead to increased erosion risk (de Luis et al., 2003) due to reduced plant regeneration after frequent fires (Delitti et al., 2005).

12.4.4.3 Grasslands

Permanent pastures occupied 37% of the agricultural area in Europe in 2000 (FAOSTAT, 2005). Grasslands are expected to decrease in area by the end of this century, the magnitude varying depending on the emissions scenario (Rounsevell et al., 2006). Climate change is likely to alter the community structure of grasslands in ways specific to their location and type (Buckland et al., 2001; Lüscher et al., 2004; Morecroft et al., 2004). Management and species richness may increase resilience to change (Duckworth et al., 2000). Fertile, early succession grasslands were found to be more responsive to climate change than more mature and/or less fertile grasslands (Grime et al., 2000). In general, intensively-managed and nutrient-rich grasslands will respond positively to both increased CO_2 concentration and temperature, given that water and nutrient supply is sufficient (Lüscher et al., 2004). Nitrogen-poor and species-rich grasslands may respond to climate change with small changes in productivity in the short-term (Winkler and Herbst, 2004). Overall, productivity of temperate European grassland is expected to increase (Byrne and Jones, 2002; Kammann et al., 2005). Nevertheless, warming alone is likely to have negative effects on productivity and species mixtures (Gielen et al., 2005; de Boeck et al., 2006). In the Mediterranean, changes in precipitation patterns are likely to negatively affect productivity and species composition of grasslands (Valladares et al., 2005).

12.4.5 Wetlands and aquatic ecosystems

Climate change may significantly impact northern peatlands (Vasiliev et al., 2001). The common hypothesis is that elevated temperature will increase productivity of wetlands (Dorrepaal et al., 2004) and intensify peat decomposition, which will accelerate carbon and nitrogen emissions to the atmosphere (Vasiliev et al., 2001; Weltzin et al., 2003). However, there are opposing results, reporting decreasing radiative forcing for drained peatlands in Finland (Minkkinen et al., 2002). Loss of permafrost in the Arctic (ACIA, 2004) will likely cause a reduction of some types of wetlands in the current permafrost zone (Ivanov and Maximov, 2003). During dry years, catastrophic fires are expected on drained peatlands in European Russia (Zeidelman and Shvarov, 2002; Bannikov et al., 2003). Processes of paludification[4] are likely to accelerate in northern regions with increasing precipitation (Lavoie et al., 2005).

Throughout Europe, in lakes and rivers that freeze in the winter, warmer temperatures may result in earlier ice melt and longer growing seasons. A consequence of these changes could be a higher risk of algal blooms and increased growth of toxic cyanobacteria in lakes (Moss et al., 2003; Straile et al., 2003; Briers et al., 2004; Eisenreich, 2005). Higher precipitation and reduced frost may enhance nutrient loss from cultivated fields (Eisenreich, 2005). These factors may result in higher nutrient loadings (Bouraoui et al., 2004; Kaste et al., 2004; Eisenreich, 2005) and concentrations of dissolved organic matter in inland waters (Evans and Monteith, 2001; ACIA, 2004; Worrall et al., 2006). Higher nutrient loadings may intensify the eutrophication of lakes and wetlands (Jeppesen et al., 2003). Streams in catchments with impermeable soils may have increased runoff in winter and deposition of organic matter in summer, which could reduce invertebrate diversity (Pedersen et al., 2004).

Inland waters in southern Europe are likely to have lower volume and increased salinisation (Williams, 2001; Zalidis et al., 2002). Many ephemeral ecosystems may disappear, and permanent ones shrink (Alvarez Cobelas et al., 2005). Although an overall drier climate may decrease the external loading of nutrients to inland waters, the concentration of nutrients may increase because of the lower volume of inland waters (Zalidis et al., 2002). Also an increased frequency of high rainfall events could increase nutrient discharge to some wetlands (Sánchez Carrillo and Alvarez Cobelas, 2001).

Warming will affect the physical properties of inland waters (Eisenreich, 2005; Livingstone et al., 2005). The thermocline of summer-stratified lakes will descend, while the bottom-water temperature and duration of stratification will increase, leading to higher risk of oxygen depletion below the thermocline (Catalán et al., 2002; Straile et al., 2003; Blenckner, 2005). Higher temperatures will also reduce dissolved oxygen saturation levels and increase the risk of oxygen depletion (Sand-Jensen and Pedersen, 2005).

12.4.6 Biodiversity

Climate change is affecting the physiology, phenology and distribution of European plant and animal species (e.g., Thomas et al., 2001; Warren et al., 2001; van Herk et al., 2002; Walther et al., 2002; Parmesan and Yohe, 2003; Root et al., 2003, 2005; Brommer, 2004; Austin and Rehfisch, 2005; Hickling et al., 2005, 2006; Robinson et al., 2005; Learmonth et al., 2006; Menzel et al., 2006a, b). A Europe-wide assessment of the future distribution of 1,350 plant species (nearly 10% of the European flora) under various SRES scenarios indicated that more than half of the modelled species could become vulnerable, endangered, critically endangered or committed to extinction by 2080 if unable to disperse (Thuiller et al., 2005). Under the most severe climate scenario (A1), and assuming that species could adapt through dispersal, 22% of the species considered would become critically endangered, and 2% committed to extinction. Qualitatively-similar results were obtained by Bakkenes et al. (2002). According to these analyses, the range of plants is very likely to expand northward and contract in southern European mountains and in the Mediterranean Basin. Regional studies (e.g., Theurillat and Guisan, 2001; Walther et al., 2005b) are consistent with Europe-wide projections.

[4] Peat bog formation.

An assessment of European fauna indicated that the majority of amphibian (45% to 69%) and reptile (61% to 89%) species could expand their range under various SRES scenarios if dispersal was unlimited (Araújo et al., 2006). However, if unable to disperse, then the range of most species (>97%) would become smaller, especially in the Iberian Peninsula and France. Species in the UK, south-eastern Europe and southern Scandinavia are projected to benefit from a more suitable climate, although dispersal limitations may prevent them from occupying new suitable areas (Figure 12.2). Consistent with these results, another Europe-wide study of 47 species of plants, insects, birds and mammals found that species would generally shift from the south-west to the north-east (Berry et al., 2006; Harrison et al., 2006). Endemic plants and vertebrates in the Mediterranean Basin are also particularly vulnerable to climate change (Malcolm et al., 2006). Habitat fragmentation is also likely to increase because of both climate and land-use changes (del Barrio et al., 2006).

Currently, species richness in inland freshwater systems is highest in central Europe declining towards the south and north because of periodic droughts and salinisation (Declerck et al., 2005). Increased projected runoff and lower risk of drought in the north will benefit the fauna of these systems (Lake, 2000; Daufresne et al., 2003), but increased drought in the south will have the opposite effect (Alvarez Cobelas et al., 2005). Higher temperatures are likely to lead to increased species richness in freshwater ecosystems in northern Europe and decreases in parts of south-western Europe (Gutiérrez Teira, 2003). Invasive species may increase in the north (McKee et al., 2002). Woody plants may encroach upon bogs and fens (Weltzin et al., 2003). Cold-adapted species will be forced further north and upstream; some may eventually disappear from Europe (Daufresne et al., 2003; Eisenreich, 2005).

Sea-level rise is likely to have major impacts on biodiversity. Examples include flooding of haul-out sites used for breeding nurseries and resting by seals (Harwood, 2001). Increased sea temperatures may also trigger large scale disease-related mortality events of dolphins in the Mediterranean and of seals in Europe (Geraci and Lounsbury, 2002). Seals that rely on ice for breeding are also likely to suffer considerable habitat loss (Harwood, 2001). Sea-level rise will reduce habitat availability for bird species that nest or forage in low-lying coastal areas. This is particularly important for the populations of shorebirds that breed in the Arctic and then winter on European coasts (Rehfisch and Crick, 2003). Lowered water tables and increased anthropogenic use and abstraction of water from inland wetlands are likely to cause serious problems for the populations of migratory birds and bats that use these areas while on migration within Europe and between Europe and Africa (Robinson et al., 2005).

12.4.7 Agriculture and fisheries

12.4.7.1 Crops and livestock

The effects of climate change and increased atmospheric CO_2 are expected to lead to overall small increases in European crop productivity. However, technological development (e.g., new

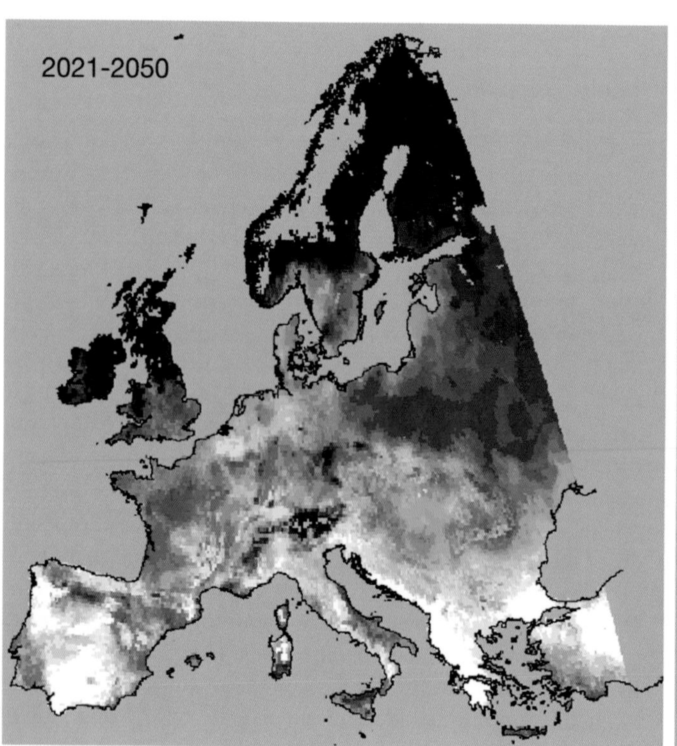

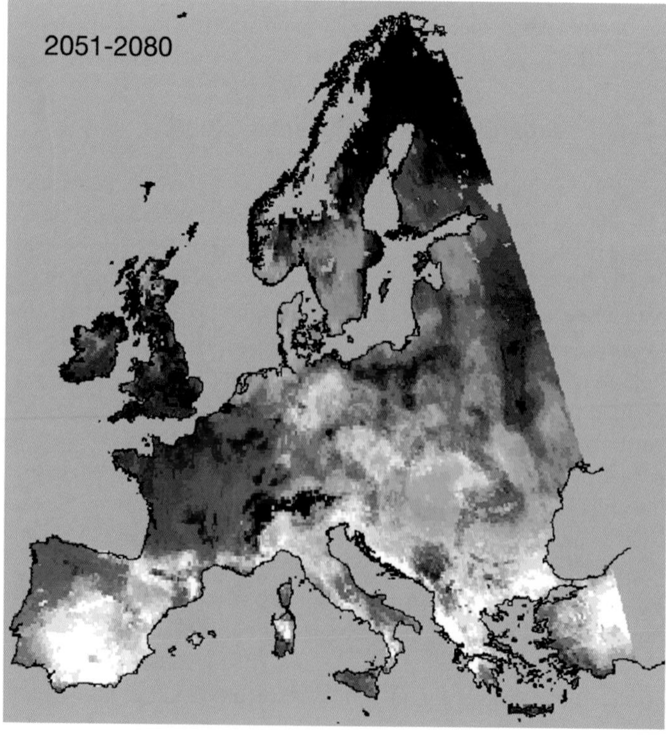

Figure 12.2. *Change in combined amphibian and reptile species richness under climate change (A1FI emissions; HadCM3 GCM), assuming unlimited dispersal. Depicted is the change between current and future species richness projected for two 30-year periods (2021 to 2050 and 2051 to 2080), using artificial neural networks. Increasing intensities of purple indicate a decrease in species richness, whereas increasing intensities of green represent an increase in species richness. Black, white and grey cells indicate areas with stable species richness: black grid cells show low species richness in both periods; white cells show high species richness; grey cells show intermediate species richness (Araújo et al., 2006).*

crop varieties and better cropping practices) might far outweigh the effects of climate change (Ewert et al., 2005). Combined yield increases of wheat by 2050 could range from 37% under the B2 scenario to 101% under the A1 scenario (Ewert et al., 2005). Increasing crop yield and decreasing or stabilising food and fibre demand could lead to a decrease in total agricultural land area in Europe (Rounsevell et al., 2005). Climate-related increases in crop yields are expected mainly in northern Europe, e.g., wheat: +2 to +9% by 2020, +8 to +25% by 2050, +10 to +30% by 2080 (Alexandrov et al., 2002; Ewert et al., 2005; Audsley et al., 2006; Olesen et al., 2007), and sugar beet +14 to +20% until the 2050s in England and Wales (Richter and Semenov, 2005), while the largest reductions of all crops are expected in the Mediterranean, the south-west Balkans and in the south of European Russia (Olesen and Bindi, 2002; Alcamo et al., 2005; Maracchi et al., 2005). In southern Europe, general decreases in yield (e.g., legumes -30 to + 5%; sunflower -12 to +3% and tuber crops -14 to +7% by 2050) and increases in water demand (e.g., for maize +2 to +4% and potato +6 to +10% by 2050) are expected for spring sown crops (Giannokopoulos et al., 2005; Audsley et al., 2006). The impacts on autumn sown crops are more geographically variable; yield is expected to strongly decrease in most southern areas, and increase in northern or cooler areas (e.g., wheat: +3 to +4% by 2020, -8 to +22% by 2050, -15 to +32% by 2080) (Santos et al., 2002; Giannakopoulos et al., 2005; Audsley et al., 2006; Olesen et al., 2007).

Some crops that currently grow mostly in southern Europe (e.g., maize, sunflower and soybeans) will become viable further north or at higher-altitude areas in the south (Audsley et al., 2006). Projections for a range of SRES scenarios show a 30 to 50% increase in the area suitable for grain maize production in Europe by the end of the 21st century, including Ireland, Scotland, southern Sweden and Finland (Hildén et al., 2005; Olesen et al., 2007). By 2050 energy crops (e.g, oilseeds such as rape oilseed and sunflower), starch crops (e.g., potatoes), cereals (e.g., barley) and solid biofuel crops (such as sorghum and Miscanthus) show a northward expansion in potential cropping area, but a reduction in southern Europe (Tuck et al., 2006). The predicted increase in extreme weather events, e.g., spells of high temperature and droughts (Meehl and Tebaldi, 2004; Schär et al., 2004; Beniston et al., 2007), is expected to increase yield variability (Jones et al., 2003) and to reduce average yield (Trnka et al., 2004). In particular, in the European Mediterranean region, increases in the frequency of extreme climate events during specific crop development stages (e.g., heat stress during flowering period, rainy days during sowing time), together with higher rainfall intensity and longer dry spells, are likely to reduce the yield of summer crops (e.g., sunflower). Climate change will modify other processes on agricultural land. Projections made for winter wheat showed that climate change beyond 2070 may lead to a decrease in nitrate leaching from agricultural land over large parts of eastern Europe and some smaller areas in Spain, and an increase in the UK and in other parts of Europe (Olesen et al., 2007).

An increase in the frequency of severe heat stress in Britain is expected to enhance the risk of mortality of pigs and broiler chickens grown in intensive livestock systems (Turnpenny et al., 2001). Increased frequency of droughts along the Atlantic coast (e.g., Ireland) may reduce the productivity of forage crops such that they are no longer sufficient for livestock at current stocking rates without irrigation (Holden and Brereton, 2002, 2003; Holden et al., 2003). Increasing temperatures may also increase the risk of livestock diseases by (i) supporting the dispersal of insects, e.g., *Culicoides imicola*, that are main vectors of several arboviruses, e.g., bluetongue (BT) and African horse sickness (AHS); (ii) enhancing the survival of viruses from one year to the next; (iii) improving conditions for new insect vectors that are now limited by colder temperatures (Wittmann and Baylis, 2000; Mellor and Wittmann, 2002; Colebrook and Wall, 2004; Gould et al., 2006).

12.4.7.2 Marine fisheries and aquaculture

An assessment of the vulnerability of the north-east Atlantic marine ecoregion concluded that climate change is very likely to produce significant impacts on selected marine fish and shellfish (Baker, 2005). Temperature increase has a major effect on fisheries production in the North Atlantic, causing changes in species distribution, increased recruitment and production in northern waters and a marked decrease at the southern edge of current ranges (Clark et al., 2003; Dutil and Brander, 2003; Hiscock et al., 2004; Perry et al., 2005). High fishing pressure is likely to exacerbate the threat to fisheries, e.g., for Northern cod (Brander, 2005). Sea-surface temperature changes as low as 0.9°C over the 45 years to 2002 have affected the North Sea phytoplankton communities, and have led to mismatches between trophic levels (see Glossary) throughout the community and the seasonal cycle (Edwards and Richardson, 2004). Together with fishing pressure, these changes are expected to influence most regional fisheries operating at trophic levels close to changes in zooplankton production (Anadón et al., 2005; Heath, 2005). Long-term climate variability is an important determinant of fisheries production at the regional scale (see Klyashtorin, 2001; Sharp, 2003), with multiple negative and positive effects on ecosystems and livelihoods (Hamilton et al., 2000; Eide and Heen, 2002; Roessig et al., 2004). Our ability to assess biodiversity impacts, ecosystem effects and socio-economic costs of climate change in coastal and marine ecosystems is still limited but is likely to be substantial for some highly dependent communities and enterprises (Gitay et al., 2002; Pinnegar et al., 2002; Robinson and Frid, 2003; Anadón et al, 2005; Boelens, et al., 2005). The overall interactions and cumulative impacts on the marine biota of sea-level rise (coastal squeeze with losses of nursery and spawning habitats), increased storminess, changes in the NAO, changing salinity, acidification of coastal waters, and other stressors such as pollutants, are likely but little known.

Marine and freshwater fish and shellfish aquaculture represented 33% of the total EU fishery production value and 17% of its volume in 2002 (EC, 2004). Warmer sea temperatures have increased growing seasons, growth rates, feed conversion and primary productivity (Beaugrand et al., 2002; Edwards et al., 2006), all of which will benefit shellfish production. Opportunities for new species will arise from expanded geographic distribution and range (Beaugrand and Reid, 2003), but increased temperatures will increase stress and susceptibility to pathogens (Anadón et al., 2005). Ecosystem changes with new invasive or non-native species such as gelatinous zooplankton and medusa, toxic algal blooms, increased fouling and decreased dissolved oxygen events, will increase operation costs. Increased storm-induced damage to equipment and facilities will increase capital costs. Aquaculture has its own local environmental impacts derived from particulate organic wastes and the spread of pathogens to wild populations, which are likely to compound climate-induced ecosystem stress (SECRU, 2002; Boelens et al., 2005).

12.4.8 Energy and transport

12.4.8.1 Energy

Under future climate change, demand for heating decreases and demand for cooling increases relative to 1961 to 1990 levels (Santos et al., 2002; Livermore, 2005; López Zafra et al., 2005; Hanson et al., 2006). In the UK and Russia, a 2°C warming by 2050 is estimated to decrease space heating needs in winter, thus decreasing fossil fuel demand by 5 to 10% and electricity demand by 1 to 3% (Kirkinen et al., 2005). Wintertime heating demand in Hungary and Romania is expected to decrease by 6 to 8% (Vajda et al., 2004) and by 10% in Finland (Venalainen et al., 2004) by the period 2021 to 2050. By 2100, this decrease rises from 20 to 30% in Finland (Kirkinen et al., 2005) to around 40% in the case of Swiss residential buildings (Frank, 2005; Christenson et al., 2006). Around the Mediterranean, two to three fewer weeks a year will require heating but an additional two to three (along the coast) to five weeks (inland areas) will need cooling by 2050 (Giannakopoulos et al., 2005). Cartalis et al. (2001) estimated up to 10% decrease in energy heating requirements and up to 28% increase in cooling requirements in 2030 for the south-east Mediterranean region. Fronzek and Carter (2007) reported a strong increase in cooling requirements for central and southern Europe (reaching 114% for Madrid) associated with an increase in inter-annual variability by 2071 to 2100. Summer space cooling needs for air conditioning will particularly affect electricity demand (Valor et al., 2001; Giannakopoulos and Psiloglou, 2006) with increases of up to 50% in Italy and Spain by the 2080s (Livermore, 2005). Peaks in electricity demand during summer heatwaves are very likely to equal or exceed peaks in demand during cold winter periods in Spain (López Zafra et al., 2005).

The current key renewable energy sources in Europe are hydropower (19.8% of electricity generated) and wind. By the 2070s, hydropower potential for the whole of Europe is expected to decline by 6%, translated into a 20 to 50% decrease around the Mediterranean, a 15 to 30% increase in northern and eastern Europe and a stable hydropower pattern for western and central Europe (Lehner et al., 2005). There will be a small increase in the annual wind energy resource over Atlantic and northern Europe, with more substantial increases during the winter season by 2071 to 2100 (Pryor et al., 2005). Biofuel production is largely determined by the supply of moisture and the length of the growing season (Olesen and Bindi, 2002). By the 22nd century, land area devoted to biofuels may increase by a factor of two to three in all parts of Europe (Metzger et al., 2004). More solar energy will be available in the Mediterranean region (Santos et al., 2002). Climate change could have a negative impact on thermal power production since the availability of cooling water may be reduced at some locations because of climate-related decreases (Arnell et al., 2005) or seasonal shifts in river runoff (Zierl and Bugmann, 2005). The distribution of energy is also vulnerable to climate change. There is a small increase in line resistance with increasing mean temperatures (Santos et al., 2002) coupled with negative effects on line sag and gas pipeline compressor efficiency due to higher maximum temperatures (López Zafra et al., 2005). All these combined effects add to the overall uncertainty of climate change impacts on power grids.

12.4.8.2 Transport

Higher temperatures can damage rail and road surfaces (AEAT, 2003; Wooller 2003; Mayor of London, 2005) and affect passenger comfort. There is likely to be an increased use of air conditioning in private vehicles and where public transport is perceived to be uncomfortable, modal switch may result (London Climate Change Partnership, 2002). The likely increase in extreme weather events may cause flooding, particularly of underground rail systems and roads with inadequate drainage (London Climate Change Partnership, 2002; Defra, 2004a; Mayor of London, 2005). High winds may affect the safety of air, sea and land transport whereas intense rainfall can also impact adversely on road safety although in some areas this may be offset to a degree by fewer snowy days (Keay and Simmonds, 2006). Reduced incidences of frost and snow will also reduce maintenance and treatment costs. Droughts and the associated reduced runoff may affect river navigation on major thoroughfares such as the Rhine (Middelkoop and Kwadijk, 2001) and shrinkage and subsidence may damage infrastructure (Highways Agency, 2005a). Reduced sea ice and thawing ground in the Arctic will increase marine access and navigable periods for the Northern Sea Route; however, thawing of ground permafrost will disrupt access through shorter ice road seasons and cause damage to existing infrastructure (ACIA, 2004).

12.4.9 Tourism and recreation

Tourism is closely linked to climate, in terms of the climate of the source and destination countries of tourists and climate seasonality, i.e., the seasonal contrast that drives demand for summer vacations in Europe (Viner, 2006). Conditions for tourism as described by the Tourism Comfort Index (Amelung and Viner, 2006) are expected to improve in northern and western Europe (Hanson et al., 2006). Hamilton et al. (2005) indicated that an arbitrary climate change scenario of 1°C would lead to a gradual shift of tourist destinations further north and up mountains affecting the preferences of sun and beach lovers from western and northern Europe. Mountainous parts of France, Italy and Spain could become more popular because of their relative coolness (Ceron and Dubois, 2000). Higher summer temperatures may lead to a gradual decrease in summer tourism in the Mediterranean but an increase in spring and perhaps autumn (Amelung and Viner, 2006). Maddison (2001) has shown that Greece and Spain will experience a lengthening and a flattening of their tourism season by 2030. Occupancy rates associated with a longer tourism season in the Mediterranean will spread demand evenly and thus alleviate the pressure on summer water supply and energy demand (Amelung and Viner, 2006).

The ski industry in central Europe is likely to be disrupted by significant reductions in natural snow cover especially at the beginning and end of the ski season (Elsasser and Burki, 2002). Hantel et al. (2000) found at the most sensitive elevation in the Austrian Alps (600 m in winter and 1400 m in spring) and with no snowmaking adaptation considered, a 1°C rise leads to four fewer weeks of skiing days in winter and six fewer weeks in spring. Beniston et al. (2003) calculated that a 2°C warming with no precipitation change would reduce the seasonal snow cover at a Swiss Alpine site by 50 days/yr, and with a 50% increase in precipitation by 30 days.

12.4.10 Property insurance

Insurance systems differ widely between countries (e.g., in many countries flood damage is not insured) and this affects the vulnerability of property to climate change. The value of property at risk also varies between countries. The damage from a wind speed of 200 km/h varies from 0.2% of the value of insured property in Austria, to around 1.2% in Denmark (Munich Re, 2002). While insurers are able in principle to adapt quickly to new risks such as climate change, the uncertainty of future climate impacts has made it difficult for them to respond to this new threat.

The uncertainty of future climate as well as socio-economic factors leads to a wide range of estimates for the costs of future flood damage. For instance, annual river flood damage in the UK is expected to increase by the 2080s between less than twice the current level of damages under the B2 scenario to greater than twenty times more under the A1 scenario (ABI, 2004). Moreover, future insurance costs will rise significantly if current rare events become more common. This is because the costs of infrequent catastrophic events are much higher than more frequent events, e.g., in the UK, the cost of a 1000-year extreme climate event is roughly 2.5 times larger than the cost of a 100-year event (Swiss Re, 2000), and in Germany, insurance claims increase as the cube of maximum wind speed (Klawa and Ulbrich, 2003).

12.4.11 Human health

Countries in Europe currently experience mortality due to heat and cold (Beniston, 2002; Ballester et al., 2003; Crawford et al., 2003; Keatinge and Donaldson, 2004). Heat-related deaths are apparent at relatively moderate temperatures (Huynen et al., 2001; Hajat et al., 2002; Keatinge, 2003; Hassi 2005; Páldy et al., 2005), but severe impacts occur during heatwaves (Kosatsky, 2005; Pirard et al. 2005; Kovats and Jendritzky, 2006; WHO, 2006; see also Section 12.6.1). Over the next century, heatwaves are very likely to become more common and severe (Meehl and Tebaldi, 2004). Heat-related deaths are likely to increase, even after assuming acclimatisation (Casimiro and Calheiros, 2002; Department of Health, 2002). Cold mortality is a problem in mid-latitudes (Keatinge et al., 2000; Nafstad et al., 2001; Mercer, 2003; Hassi, 2005) but is likely to decline with milder winters (Department of Health, 2002; Dessai, 2003). Major determinants of winter mortality include respiratory infections and poor quality housing (Aylin et al., 2001; Wilkinson et al., 2001, 2004; Mitchell et al., 2002; Izmerov et al., 2004; Díaz et al., 2005). Climate change is likely to increase the risk of mortality and injury from wind storms, flash floods and coastal flooding (Kirch et al., 2005). The elderly, disabled, children, women, ethnic minorities and those on low incomes are more vulnerable and need special consideration (Enarson and Fordham, 2001; Tapsell and Tunstall, 2001; Hajat et al., 2003; WHO, 2004, 2005; Penning-Rowsell et al., 2005; Ebi, 2006).

Changes in tick distribution consistent with climate warming have been reported in several European locations, although evidence is not conclusive (Kovats et al., 2001; Lindgren and Gustafson, 2001; Department of Health, 2002; Bröker and Gniel, 2003; Hunter, 2003; Butenco and Larichev, 2004; Korenberg, 2004; Kuhn et al., 2004). The effect of climate variability on tick-borne encephalitis (TBE) or Lyme disease incidence is still unclear (Randolph, 2002; Beran et al., 2004; Izmerov et al., 2004; Daniel et al., 2006; Lindgren and Jaenson, 2006; Rogers and Randolph, 2006). Future changes in tick-host habitats and human-tick contacts may be more important for disease transmission than changes in climate (Randolph, 2004). Visceral leishmaniasis is present in the Mediterranean region and climate change may expand the range of the disease northwards (Department of Health, 2002; Molyneux, 2003; Korenberg, 2004; Kuhn et al., 2004; Lindgren and Naucke, 2006). The re-emergence of endemic malaria in Europe due to climate change is very unlikely (Reiter, 2000, 2001; Semenov et al., 2002; Yasukevich, 2003; Kuhn et al., 2004; Reiter et al., 2004; Sutherst, 2004; van Lieshout et al., 2004). The maintenance of the current malaria situation is projected up to 2025 in Russia (Yasyukevich, 2004). An increased risk of localised outbreaks is possible due to climate change, but only if suitable vectors are present in sufficient numbers (Casimiro and Calheiros, 2002; Department of Health 2002). Increases in malaria outside Europe may affect the risk of imported cases. Diseases associated with rodents are known to be sensitive to climate variability, but no assessments on the impacts of climate change have been published for Europe.

Climate change is also likely to affect water quality and quantity in Europe, and hence the risk of contamination of public and private water supplies (Miettinen et al., 2001; Hunter, 2003; Elpiner, 2004; Kovats and Tirado, 2006). Higher temperatures have implications for food safety, as transmission of salmonellosis is temperature sensitive (Kovats et al, 2004; Opopol and Nicolenco, 2004; van Pelt et al. 2004). Both extreme rainfall and droughts can increase the total microbial loads in freshwater and have implications for disease outbreaks and water quality monitoring (Howe et al., 2002; Kistemann et al., 2002; Opopol et al. 2003; Knight et al., 2004; Schijven and de Roda Husman, 2005).

Important climate change effects on air quality are likely in Europe (Casimiro and Calheiros, 2002; Sanderson et al., 2003; Langner et al., 2005; Stevenson et al., 2006). Climate change may increase summer episodes of photochemical smog due to increased temperatures, and decreased episodes of poor air quality associated with winter stagnation (Hennessy, 2002; Revich and Shaposhnikov, 2004; Stedman, 2004; Kislitsin et al., 2005), but model results are inconsistent. Stratospheric ozone depletion and warmer summers influence human exposure to ultra-violet radiation and therefore increase the risk of skin cancer (Inter-Agency Commission, 2002; van der Leun and de Gruijl, 2002; de Gruijl et al., 2003; Diffey, 2004). Pollen phenology is changing in response to observed climate change, especially in central Europe, and at a wide range of elevations (Emberlin et al., 2002; Bortenschlager and Bortenschlager, 2005). Earlier onset and extension of the allergenic pollen seasons are likely to affect some allergenic diseases (van Vliet et al., 2002; Verlato et al., 2002; Huynen and Menne, 2003; Beggs, 2004; Weiland et al., 2004).

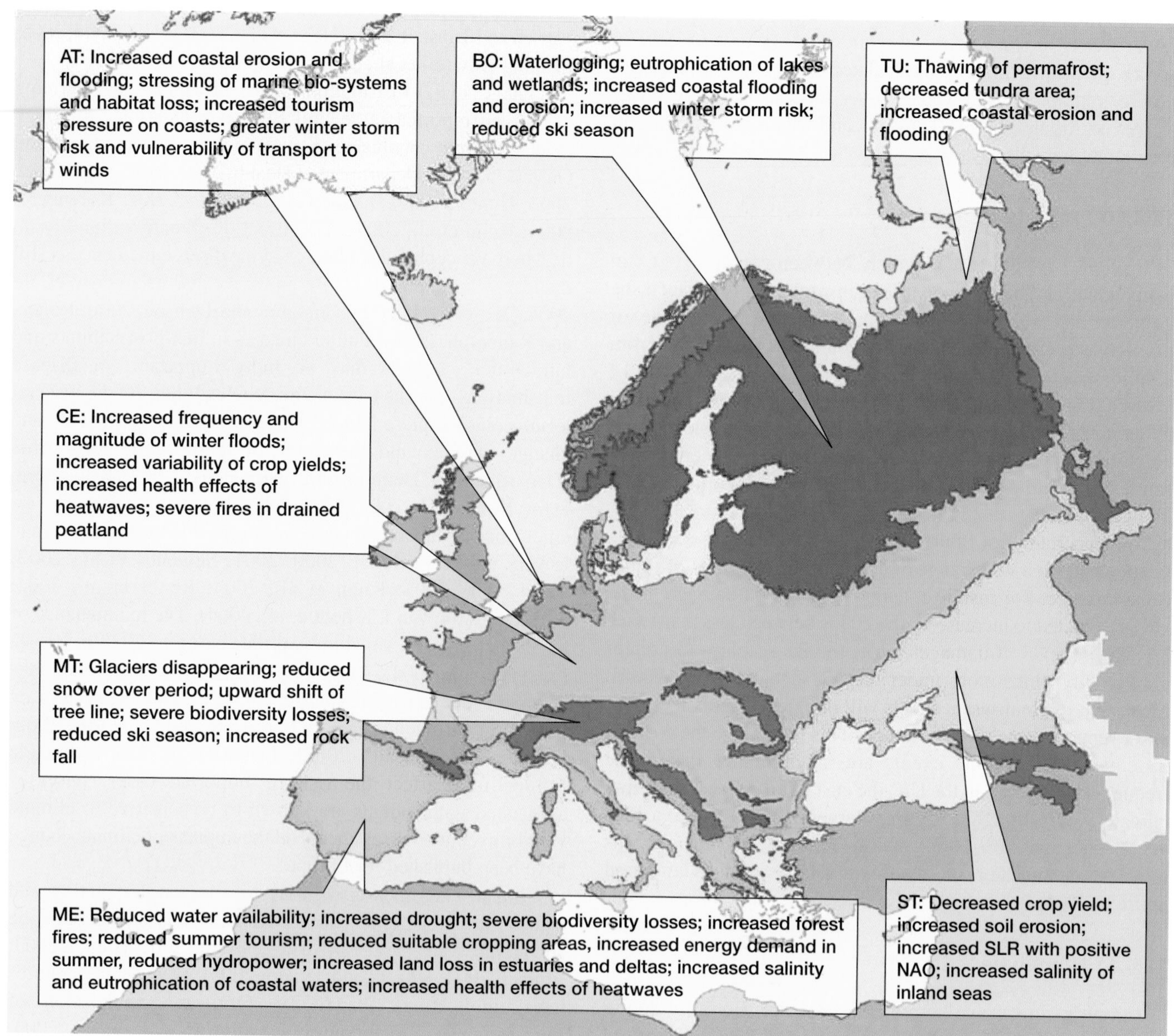

Figure 12.3. *Key vulnerabilities of European systems and sectors to climate change during the 21st century for the main biogeographic regions of Europe (EEA, 2004a): TU: Tundra, pale turquoise. BO: Boreal, dark blue. AT: Atlantic, light blue. CE: Central, green; includes the Pannonian Region. MT: Mountains, purple. ME: Mediterranean, orange; includes the Black Sea region. ST: Steppe, cream. SLR: sea-level rise. NAO: North Atlantic Oscillation. Copyright EEA, Copenhagen. http://www.eea.europa.eu*

12.5 Adaptation: practices, options and constraints

12.5.1 Water resources

Climate change will pose two major water management challenges in Europe: increasing water stress mainly in south-eastern Europe, and increasing risk of floods throughout most of the continent. Adaptation options to cope with these challenges are well-documented (IPCC, 2001). The main structural measures to protect against floods are likely to remain reservoirs and dykes in highland and lowland areas respectively (Hooijer et al., 2004). However, other planned adaptation options are becoming more popular such as expanded floodplain areas (Helms et al., 2002), emergency flood reservoirs (Somlyódy, 2002), preserved areas for flood water (Silander et al., 2006), and flood warning systems, especially for flash floods. Reducing risks may have substantial costs.

To adapt to increasing water stress the most common and planned strategies remain supply-side measures such as impounding rivers to form in-stream reservoirs (Santos et al., 2002; Iglesias et al., 2005). However, new reservoir construction is being increasingly constrained in Europe by environmental regulations (Barreira, 2004) and high investment costs (Schröter et al., 2005). Other supply-side approaches such as wastewater reuse and desalination are being more widely considered but their popularity is reduced by health concerns related to using wastewater (Geres, 2004) and the high energy costs of desalination (Iglesias et al., 2005). Some planned demand-side strategies are also feasible (AEMA, 2002), such as household, industrial and agricultural water conservation, the reduction of leaky municipal and irrigation water systems (Donevska and Dodeva, 2004; Geres, 2004), and water pricing (Iglesias et al., 2005). Irrigation water demand may be reduced by introducing crops more suitable to the changing climate. As is the case for the supply-side approaches, most demand-side approaches are not specific to Europe. An example of a unique European approach to adapting to water stress is that regional and watershed-level strategies to adapt to climate change are being incorporated into plans for integrated water management (Kabat et al., 2002; Cosgrove et al., 2004; Kashyap, 2004) while national strategies are being designed to fit into existing governance structures (Donevska and Dodeva, 2004).

12.5.2 Coastal and marine systems

Strategies for adapting to sea-level rise are well documented (Smith et al., 2000; IPCC, 2001; Vermaat et al., 2005). Although a large part of Europe's coastline is relatively robust to sea-level rise (Stone and Orford, 2004), exceptions are the subsiding, geologically 'soft', low-lying coasts with high populations, as in the southern North Sea and coastal plains/deltas of the Mediterranean, Caspian and Black Seas. Adaptation strategies on low-lying coasts have to address the problem of sediment loss from marshes, beaches and dunes (de Groot and Orford, 2000; Devoy et al., 2000). The degree of coastal erosion that may result from sea-level rise is very uncertain (Cooper and Pilkey, 2004), though feedback processes in coastal systems do provide a means of adaptation to such changes (Devoy, 2007). Modelling changes in coastal sediment flux under climate warming scenarios shows some 'soft' coasts responding with beach retreat rates of >40 m/100 years, contrasting with gains in others by accretion of about 10 m/100 years (Walkden and Hall, 2005; Dickson et al., 2007).

The development of adaptation strategies for coastal systems has been encouraged by an increase in public and scientific awareness of the threat of climate change to coastlines (Nicholls and Klein, 2004). Many countries in north-west Europe have adopted the approach of developing detailed shoreline management plans that link adaptation measures with shoreline defence, accommodation and retreat strategies (Cooper et al., 2002; Defra, 2004b; Hansom et al., 2004). Parts of the Mediterranean and eastern European regions have been slower to follow this pattern and management approaches are more fragmented (Tol et al., 2007).

A key element of adaptation strategies for coastlines is the development of new laws and institutions for managing coastal land (de Groot and Orford, 2000; Devoy, 2007). For example, no EU Directive exists for coastal management, although EU member governments were required to develop and publish coastal policy statements by 2006. The lack of a Directive reflects the complexity of socio-economic issues involved in coastal land use and the difficulty of defining acceptable management strategies for the different residents, users and interest groups involved with the coastal region (Vermaat et al., 2005).

12.5.3 Mountains and sub-Arctic regions

Mountainous and sub-Arctic regions have only a limited number of adaptation options. In northern Europe it will become necessary to factor in the dissipation and eventual disappearance of permafrost in infrastructure planning (Nelson, 2003) and building techniques (Mazhitova et al., 2004). There are few obvious adaptation options for either tundra or alpine vegetation. It may be possible to preserve many alpine species in managed gardens at high elevation since many mountain plants are likely to survive higher temperatures if they are not faced with competition from other plants (Guisan and Theurillat, 2005). However, this option remains very uncertain because the biotic factors determining the distribution of mountain plant species are not well known. Another minimal adaptation option is the reduction of other stresses on high elevation ecosystems, e.g., by lessening the impact of tourism (EEA, 2004b). Specific management strategies have yet to be defined for mountain forests (Price, 2005).

12.5.4 Forests, shrublands and grasslands

Since forests are managed intensively in Europe, there is a wide range of available management options that can be employed to adapt forests to climate change. General strategies for adaptation include changing the species composition of forest stands and planting forests with genetically improved seedlings adapted to a new climate (if the risk of genetically modified species is considered acceptable) (KSLA, 2004). Extending the rotation period of commercially important tree species may increase sequestration and/or the storage of carbon, and can be viewed as an adaptation measure (Kaipainen et al., 2004). Adaptive forest management could substantially decrease the risk of forest destruction by wind and other extreme weather events (Linder, 2000; Olofsson and Blennow 2005; Thurig et al., 2005). Strategies for coniferous forests include the planting of deciduous trees better adapted to the new climate as appropriate, and the introduction of multi-species planting into currently mono-species coniferous plantations (Fernando and Cortina, 2004; Gordienko and Gordienko, 2005).

Adaptation strategies need to be specific to different parts of Europe. The range of alternatives is constrained, among other factors, by the type of forest. Forests that are already moisture limited (Mediterranean forests) or temperature limited (boreal forests) will have greater difficulty in adapting to climate change than other forests, e.g., in central Europe (Gracia et al., 2005). Fire protection will be important in Mediterranean and boreal forests and includes the replacement of highly flammable species, regulation of age-class distributions, and widespread management of accumulated fuel, eventually through prescribed burning (Baeza et al., 2002; Fernandes and Botelho, 2004). Public education, development of advanced systems of forest inventories, and forest health monitoring are important prerequisites of adaptation and mitigation.

Productive grasslands are closely linked to livestock production. Dairy and cattle farming may become less viable because of climate risks to fodder production and therefore grasslands could be converted to cropland or other uses (Holman et al., 2005). Grassland could be adapted to climate change by changing the intensity of cutting and grazing, or by irrigating current dryland pastures (Riedo et al., 2000). Another option is to take advantage of continuing abandonment of cropland in Europe (Rounsevell et al., 2005) to establish new grassland areas.

12.5.5 Wetlands and aquatic ecosystems

Better management practices are needed to compensate for possible climate-related increases in nutrient loading to aquatic ecosystems from cultivated fields in northern Europe (Ragab and Prudhomme, 2002; Viner et al., 2006). These practices include 'optimised' fertiliser use and (re-)establishment of wetland areas and river buffer zones as sinks for nutrients (Olesen et al., 2004). New wetlands could also dampen the effects of increased frequency of flooding. A higher level of treatment of domestic and industrial sewage and reduction in farmland areas can further reduce nutrient loadings to surface waters and also compensate for climate-related increases in these loadings. Practical possibilities for adaptation in northern wetlands are limited and may only be realised as part of integrated landscape management including the minimisation of unregulated anthropogenic pressure, avoiding the physical destruction of surface and applying appropriate technologies for infrastructure development on permafrost (Ivanov and Maximov, 2003). Protection of drained peatlands against fire in European Russia is an important regional problem which requires the restoration of drainage systems and the regulation of water regimes in such territories (Zeidelman and Shvarov, 2002).

In southern Europe, to compensate for increased climate-related risks (lowering of the water table, salinisation, eutrophication, species loss) (Williams, 2001; Zalidas et al., 2002), a lessening of the overall human burden on water resources is needed. This would involve stimulating water saving in agriculture, relocating intensive farming to less environmentally sensitive areas and reducing diffuse pollution, increasing the recycling of water, increasing the efficiency of water allocation among different users, favouring the recharge of aquifers and restoring riparian vegetation, among others (Alvarez Cobelas et al., 2005).

12.5.6 Biodiversity

Climate change threatens the assumption of static species ranges which underpins current conservation policy. The ability of countries to meet the requirements of EU Directives and other international conventions is likely to be compromised by climate change, and a more dynamic strategy for conservation is required for sustaining biodiversity (Araújo et al., 2004; Brooker and Young, 2005; Robinson et al., 2005; Harrison et al., 2006). Conservation strategies relevant to climate change can take at least two forms: *in situ* involving the selection, design and management of conservation areas (protected areas, nature reserves, NATURA 2000 sites, wider countryside), and *ex situ* involving conservation of germplasm in botanical gardens, museums and zoos. A mixed strategy is the translocation of species into new regions or habitats (e.g., Edgar et al., 2005). In Europe, appropriate *in situ* and *ex situ* conservation measures for mitigating climate change impacts have not yet been put in place. Conservation experts have concluded that an expansion of reserve areas will be necessary to conserve species in Europe. For example, Hannah et al. (2007) calculated that European protected areas need to be increased by 18% to meet the EU goal of providing conditions by which 1,200 European plant species can continue thriving in at least 100 km^2 of habitat. To meet this goal under climate change they estimated that the current reserve area must be increased by 41%. They also point out that it would be more cost effective to expand protected areas proactively rather than waiting for climate change impacts to occur and then acting reactively. Dispersal corridors for species are another important adaptation tool (Williams et al., 2005), although large heterogeneous reserves that maximise microclimate variability might sometimes be a suitable alternative. Despite the importance of modifying reserve areas, some migratory species are vulnerable to loss of habitat outside Europe (e.g., Viner et al., 2006). For these migratory species, trans-continental conservation policies need to be put in place.

12.5.7 Agriculture and fisheries

Short-term adaptation of agriculture in southern Europe may include changes in crop species (e.g., replacing winter with spring wheat) (Mínguez et al., 2007), cultivars (higher drought resistance and longer grain-filling) (Richter and Semenov,

2005) or sowing dates (Olesen et al., 2007). Introducing new crops and varieties are also an alternative for northern Europe (Hildén et al., 2005), even if this option may be limited by soil fertility, e.g., in northern Russia. A feasible long-term adaptation measure is to change the allocation of agricultural land according to its changing suitability under climate change. Large-scale abandonment of cropland in Europe estimated under the SRES scenarios (Rounsevell et al., 2006) may provide an opportunity to increase the cultivation of bioenergy crops (Schröter et al., 2005). Moreover, Schröter et al. (2005) and Berry et al. (2006) found that different types of agricultural adaptation (intensification, extensification and abandonment) may be appropriate under different IPCC SRES scenarios and at different locations. It is indisputable that the reform of EU agricultural policies will be an important vehicle for encouraging European agriculture to adapt to climate change (Olesen and Bindi, 2002) and for reducing the vulnerability of the agricultural sector (Metzger et al., 2006).

At the small scale there is evidence that fish and shellfish farming industries are adapting their technology and operations to changing climatic conditions, for example, by expanding offshore and selecting optimal culture sites for shellfish cages (Pérez et al., 2003). However, adaptation is more difficult for smaller coastal-based fishery businesses which do not have the option to sail long distances to new fisheries as compared to larger businesses with long distance fleets. At the larger scale, adaptation options have not yet been considered in important policy institutions such as the European Common Fisheries Policy (CFP) although its production quotas and technical measures provide an ideal platform for such adaptation actions. Another major adaptation option is to factor the long-term potential impacts of climate change into the planning for new Marine Protected Areas (Soto, 2001). Adaptation strategies should eventually be integrated into comprehensive plans for managing coastal areas of Europe. However, these plans are lacking, especially around the Mediterranean, and need to be developed urgently (Coccossis, 2003).

12.5.8 Energy and transport

A wide variety of adaptation measures are available in the energy sector ranging from the redesign of the energy supply system to the modification of human behaviour (Santos et al., 2002). The sensitivity of European energy systems to climate change could be reduced by enhancing the interconnection capacity of electricity grids and by using more decentralised electric generation systems and local micro grids (Arnell et al., 2005). Another type of adaptation would be to reduce the exposure of energy users and producers to impacts of unfavourable climate through the mitigation of greenhouse gas emissions, for example by reducing overall energy use. This can be accomplished through various energy conservation measures such as energy-saving building codes and low-electricity standards for new appliances, increasing energy prices and through training and public education. Over the medium to long term, shifting from fossil fuels to renewable energy use will be an effective adaptive measure (Hanson et al., 2006).

Clearly, one aspect of adaptation may be through measures to mitigate emissions from transport through cleaner technologies and adapting behaviour (National Assessment Synthesis Team, 2001; AEAT, 2003; Highways Agency, 2005a). There is clearly a need for capacity building in the response to incidents, risk assessments, developments in maintenance, renewal practice and design standards for new infrastructure (Highways Agency, 2005b; Mayor of London, 2005). Assessment of the costs and benefits of adapting existing infrastructure or raising standards in the design of new vehicles and infrastructure to improve system resilience and reliability to the range of potential impacts should consider the wider economic and social impacts of disruption to the transport system.

12.5.9 Tourism and recreation

A variety of adaptation measures are available to the tourism industry (WTO, 2003, Hanson et al., 2006). Regarding winter tourism, compensating for reduced snowfall by artificial snowmaking is already common practice for coping with year-to-year snow pack variability. However, this adaptation strategy is likely to be economic only in the short term, or in the case of very high elevation resorts in mountain regions, and may be ecologically undesirable. New leisure industries, such as grass-skiing or hiking could compensate for any income decrease experienced by the ski industry due to snow deterioration (Fukushima et al., 2002). Regarding coastal tourism, the protection of resorts from sea-level rise may be feasible by constructing barriers or by moving tourism infrastructure further back from the coast (Pinnegar et al., 2006). In the Mediterranean region, the likely reduction of tourism during the hotter summer months may be compensated for by promoting changes in the temporal pattern of seaside tourism, for example by encouraging visitors during the cooler months (Amelung and Viner, 2006). The increasing, new climate-related risks to health, availability of water, energy demand and infrastructure are likely to be dealt with through efficient co-operation with local governments. Another adaptive measure for European tourism, in general, is promoting new forms of tourism such as eco-tourism or cultural tourism and placing greater emphasis on man-made rather than natural attractions, which are less sensitive to weather conditions (Hanson et al., 2006). It is also likely that people will adapt autonomously and reactively by changing their recreation and travel behaviour in response to the new climatic conditions (Sievanen et al., 2005).

12.5.10 Property insurance

The insurance industry has several approaches for adapting to the growing climate-related risk to property. These include raising the cost of insurance premiums, restricting or removing coverage, reinsurance and improved loss remediation (Dlugolecki, 2001). Insurers are beginning to use Geographical Information Systems (GIS) to provide information needed to adjust insurance tariffs to climate-related risks (Dlugolecki, 2001; Munich Re, 2004) although the uncertainty of future climate change is an obvious problem in making these adjustments. Insurers are also involved in discussions of measures for climate change mitigation and adaptation, including measures such as more stringent control of floodplain development and remedial measures for damages derived from weather action and extreme events (ABI, 2000; Dlugolecki and Keykhah, 2002).

An obvious adaptation measure against property damage is to improve construction techniques so that buildings and infrastructure are more robust to extreme climate events. However, even if building techniques are immediately improved, the benefits will not be instantaneous because current building stock has a long remaining lifetime. Hence these buildings would not be replaced for many years by more resilient structures unless they are retrofitted. While retrofitting can be an effective adaptation measure it also has drawbacks. Costs are often high, residents are disrupted and poor enforcement of building regulations and construction practices could lead to unsatisfactory results.

12.5.11 Human Health

Risks posed by weather extremes are the most important in terms of requiring society's preparedness (Ebi, 2005; Hassi and Rytkönen, 2005; Menne, 2005; Menne and Ebi, 2006). Primary adaptation measures to heatwaves include the development of health early warning systems and preventive emergency plans (Garssen et al., 2005; Nogueira et al., 2005; Pirard et al., 2005). Many European countries and cities have developed such measures, especially after the summer of 2003 (Koppe et al., 2004; Ministerio de Sanidad y Consumo, 2004; Menne, 2005; see also Chapter 8 Box 8.1). Other measures are aimed at the mitigation of 'heat islands' through urban planning, the adaptation of housing design to local climate and expanding air conditioning, shifts in work patterns and mortality monitoring (Keatinge et al., 2000; Ballester et al., 2003; Johnson et al., 2005; Marttila et al., 2005; Penning-Rowsell et al., 2005).

Principal strategies to lessen the risks of flooding include public flood warning systems, evacuations from lowlands, waterproof assembling of hospital equipment and the establishment of decision hierarchies between hospitals and administrative authorities (Ohl and Tapsell, 2000; Hajat et al., 2003; EEA, 2004b; WHO, 2004; Hedger, 2005; Marttila et al., 2005; Penning-Rowsell et al., 2005).

12.6 Case studies

12.6.1 Heatwave of 2003

A severe heatwave over large parts of Europe in 2003 extended from June to mid-August, raising summer temperatures by 3 to 5 °C in most of southern and central Europe (Figure 12.4). The warm anomalies in June lasted throughout the entire month (increases in monthly mean temperature of up to 6 to 7°C), but July was only slightly warmer than on average (+1 to +3°C), and the highest anomalies were reached between 1st and 13th August (+7°C) (Fink et al., 2004). Maximum temperatures of 35 to 40°C were repeatedly recorded and peak temperatures climbed well above 40°C (André et al., 2004; Beniston and Díaz, 2004).

Average summer (June to August) temperatures were far above the long-term mean by up to five standard deviations (Figure 12.4), implying that this was an extremely unlikely event under current climatic conditions (Schär and Jendritzky, 2004). However, it is consistent with a combined increase in mean temperature and temperature variability (Meehl and Tebaldi, 2004; Pal et al., 2004; Schär et al., 2004) (Figure 12.4). As such, the 2003 heatwave resembles simulations by regional climate models of summer temperatures in the latter part of the 21st century under the A2 scenario (Beniston, 2004). Anthropogenic warming may therefore already have increased the risk of heatwaves such as the one experienced in 2003 (Stott et al., 2004).

The heatwave was accompanied by annual precipitation deficits up to 300 mm. This drought contributed to the estimated 30% reduction in gross primary production of terrestrial ecosystems over Europe (Ciais et al., 2005). This

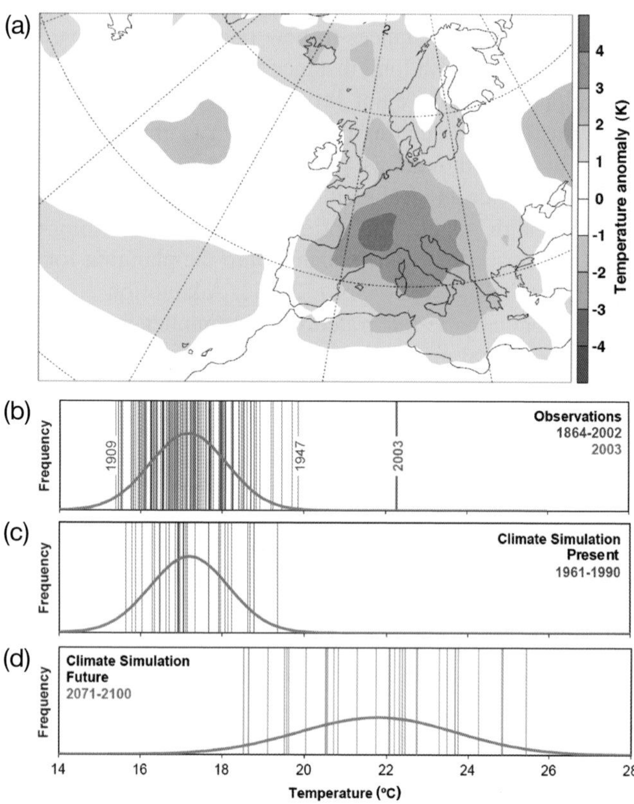

Figure 12.4. *Characteristics of the summer 2003 heatwave (adapted from Schär et al., 2004). (a) JJA temperature anomaly with respect to 1961 to 1990. (b) to (d): JJA temperatures for Switzerland observed during 1864 to 2003 (b), simulated using a regional climate model for the period 1961 to 1990 (c) and simulated for 2071 to 2100 under the A2 scenario using boundary data from the HadAM3H GCM (d). In panels (b) to (d): the black line shows the theoretical frequency distribution of mean summer temperature for the time-period considered, and the vertical blue and red bars show the mean summer temperature for individual years. Reprinted by permission from Macmillan Publishers Ltd. [Nature] (Schär et al., 2004), copyright 2004.*

reduced agricultural production and increased production costs, generating estimated damages of more than €13 billion (Fink et al., 2004; see also Chapter 5 Box 5.1). The hot and dry conditions led to many very large wildfires, in particular in Portugal (390,000 ha: Fink et al., 2004; see also Chapter 4 Box 4.1). Many major rivers (e.g., the Po, Rhine, Loire and Danube) were at record low levels, resulting in disruption of inland navigation, irrigation and power-plant cooling (Beniston and Díaz, 2004; Zebisch et al., 2005; see also Chapter 7 Box 7.1). The extreme glacier melt in the Alps prevented even lower river flows in the Danube and Rhine (Fink et al., 2004).

The excess deaths due to the extreme high temperatures during the period June to August may amount to 35,000 (Kosatsky, 2005), elderly people were among those most affected (WHO, 2003; Kovats and Jendritzky, 2006; see also Chapter 8 Box 8.1). The heatwave in 2003 has led to the development of heat health-watch warning systems in several European countries including France (Pascal et al., 2006), Spain (Simón et al., 2005), Portugal (Nogueira, 2005), Italy (Michelozzi et al., 2005), the UK (NHS, 2006) and Hungary (Kosatsky and Menne, 2005).

12.6.2 Thermohaline circulation changes in the North Atlantic: possible impacts for Europe

Earlier studies of the possible impacts of rapid change in Meridional Overturning Circulation (MOC), also known as the thermohaline circulation (THC), in the North Atlantic are now being updated (Vellinga and Wood 2002, 2006; Alley et al., 2003; Jacob et al., 2005; Rahmstorf and Ziekfeld, 2005; Stouffer et al., 2006; Schlesinger et al., 2007). Model simulations of an abrupt shut-down of the Atlantic MOC indicate that this is unlikely to occur before 2100 and that the impacts on European temperatures of any slowing in circulation before then are likely to be offset by the immediate effects of positive radiative forcings under increasing greenhouse gases (Arnell et al., 2005; Gregory et al., 2005; Vellinga and Wood, 2006; Meehl et al., 2007). Under slowing or full Atlantic MOC shut-down, temperatures on Europe's western margin would be most affected, together with further rises in relative sea level on European coasts (Vellinga and Wood, 2002, 2006; Jacob et al., 2005; Levermann et al., 2005; Wood et al., 2006; Meehl et al., 2007). Although there are no indications of an imminent change in the North Atlantic THC (Dickson et al., 2003; Curry and Mauritzen, 2005) it is recognised that MOC shut-down, should it occur, is likely to have potential socio-economic impacts for Europe and more widely (Table 12.3). Hence, it would be valuable to consider these impacts in developing climate policy (Defra, 2004c; Keller et al., 2004; Arnell et al., 2005; Schneider et al., 2007). Such policies are currently difficult to quantify (Manning et al., 2004; Parry, 2004). Assessment of the likely impacts of an abrupt Atlantic MOC shut-down on different economic and social sectors in Europe has been made using integrated assessment models, e.g., FUND (Tol, 2002, 2006; Link and Tol, 2004). Results suggest that the repercussions for socio-economic factors are likely to be less severe than was previously thought.

12.7 Conclusions: implications for sustainable development

The fraction of total plant growth or the net primary production appropriated by humans (HANPP) is a measure widely used to assess the 'human domination of Earth's ecosystems' (Haberl et al., 2002). Currently, HANPP in western Europe (WE) amounts to 2.86 tonnes carbon/capita/yr, which is 72.2% of its terrestrial net primary production. This exceeds, by far, the global average of 20% (Imhoff et al., 2004). The 'ecological footprint' (EF) is an estimate of the territory required to provide resources consumed by a given population (Wackernagel et al., 2002). In 2001, the EF of central and eastern Europe (CEE) was 3.8 ha/capita, and of WE 5.1 ha/capita (WWF, 2004). These values also far exceed the global average of 2.2 ha/capita (WWF, 2004). WE is one of the largest 'importers' of land, an expression of the net trade balance for agricultural products (van Vuuren and Bouwman, 2005). Globally, by 2050 the total EF is very likely to increase by between 70% (B2 scenario) and 300% (A1B scenario), thus placing an additional burden on a planet which some consider is already at an unsustainable level (Wackernagel et al., 2002; Wilson, 2002). Large changes in demand for land in regions with high population growth and changing consumption habits are expected, which is likely to result in a (need to) decrease WE imports (van Vuuren and Bouwman, 2005). The per capita EF of WE and CEE is projected to converge by the middle of this century, at which time values for WE become slightly lower (B2 scenario) or larger (A1B scenario) than current ones, and those of CEE increase to reach those of WE. In any case, European EF is very likely to remain much higher than the global average (van Vuuren and Bouwman, 2005).

Table 12.3. *Main types of impact for Europe following a rapid shut-down of the Meridional Overturning Circulation relative to the 'pre-industrial' climate (after: Arnell et al., 2005; Levermann et al., 2005; Vellinga and Wood, 2006).*

- Reductions in runoff and water availability in southern Europe; major increase in snowmelt flooding in western Europe.
- Increased sea-level rise on western European and Mediterranean coasts.
- Reductions in crop production with consequent impacts on food prices.
- Changes in temperature affecting ecosystems in western Europe and the Mediterranean (e.g., affecting biodiversity, forest products and food production).
- Disruption to winter travel opportunities and increased icing of northern ports and seas.
- Changes in regional patterns of increases versus decreases in cold- and heat-related deaths and ill-health.
- Movement of populations to southern Europe and a shift in the centre of economic gravity.
- Requirement to refurbish infrastructure towards Scandinavian standards.

Climate change in Europe is likely to have some positive effects (e.g., increased forest area, increased crop yield in northern Europe), or offer new opportunities (e.g., 'surplus land'). However, many changes are very likely to increase vulnerability due to reduced supply of ecosystem services (declining water availability, climate regulation potential or biodiversity), increase of climate-related hazards and disruption in productive sectors, among others (Schröter et al., 2005; Metzger et al., 2006) (Table 12.4). Therefore, additional pressures are very likely to be exerted upon Europe's environment, which is already subject to substantial pressures (EEA, 2003), and social and economic systems. Furthermore, climate change is likely to magnify regional differences in terms of Europe's natural resources and assets since impacts are likely to be unevenly distributed geographically, with the most negative impacts occurring in the south and east (Table 12.4). Adaptive capacity is high, although it varies greatly between countries (higher in the north than in the south and east) due to their different socio-economic systems (Yohe and Tol, 2002). Adaptive capacity is expected to increase in the future, yet, differences among countries will persist (Metzger et al., 2004, 2006). Hence, climate change is likely to create additional imbalances since negative impacts are likely to be largest where adaptive capacity is lowest.

The integration of sustainability goals into other sectoral policy areas is progressing, for instance, through national, regional and local sustainable development strategies and plans. However, these have not yet had a decisive effect on policies (EEA, 2003). Although climate change and sustainable development policies have strong linkages, they have evolved in parallel, at times they even compete with one another. Climate change is very likely to challenge established sustainability goals. Tools, such as integrated modelling approaches (Holman et al., 2005; Berry et al., 2006), integration frameworks (Tschakert and Olsson, 2005) and scenario build-up (Wiek et al., 2006) can help bridge the gap in the limited understanding we have on how climate change will ultimately affect sustainability. Pursuit of sustainable development goals might be a better avenue for achieving climate change policy goals than climate change policies themselves (Robinson et al., 2006).

12.8 Key uncertainties and research priorities

Uncertainties in future climate projections are discussed in great detail in Working Group I Section 10.5 (Meehl et al., 2007). For Europe, a major uncertainty is the future behaviour of the NAO and North Atlantic THC. Also important, but not specific to Europe, are the uncertainties associated with the still insufficient resolution of GCMs (e.g., Etchevers et al., 2002; Bronstert, 2003), and with downscaling techniques and regional climate models (Mearns et al., 2003; Haylock et al., 2006; Déqué et al., 2007). Uncertainties in climate impact assessment also stem from the uncertainties of land-use change and socio-economic development (Rounsevell et al., 2005, 2006) following European policies (e.g., CAP), and European Directives (Water Framework Directive, European Maritime Strategy Directive). Although most impact studies use the SRES scenarios, the procedures for scenario development are the subject of debate (Castle and Henderson, 2003a, b; Grübler et al., 2004; Holtsmark and Alfsen, 2005; van Vuuren and Alfsen, 2006). While current scenarios appear to reflect well the course of events in the recent past (van Vuuren and O'Neill, 2006), further research is needed to better account for the range of possible scenarios (Tol, 2006). This might be important for Europe given the many economies in transition.

Uncertainties in assessing future climate impacts also arise from the limitations of climate impact models including (i) structural uncertainty due to the inability of models to capture all influential factors, e.g., the models used to assess health impacts of climate change usually neglect social factors in the spread of disease (Kuhn et al., 2004; Reiter et al., 2004; Sutherst, 2004), and climate-runoff models often neglect the direct effect of increasing CO_2 concentration on plant transpiration (Gedney et al., 2006), (ii) lack of long-term representative data for model evaluation, e.g., current vector-monitoring systems are often unable to provide the reliable identification of changes (Kovats et al., 2001). Hence, more attention should be given to structural improvement of models and intensifying efforts of long-term monitoring of the environment, and systematic testing of models against observed data in field trials or catchment monitoring programmes (Hildén et al., 2005). Another way to address the uncertainty of deterministic models is to use probabilistic modelling which can produce an ensemble of scenarios, (e.g., Wilby and Harris, 2006; Araújo and New, 2007; ENSEMBLES project, http://ensembles-eu.metoffice.com/).

Until now, most impact studies have been conducted for separate sectors even if, in some cases, several sectors have been included in the same study (e.g., Schröter et al., 2005). Few studies have addressed impacts on various sectors and systems including their possible interactions by integrated modelling approaches (Holman et al., 2005; Berry et al., 2006). Even in these cases, there are various levels (supra-national, national, regional and sub-regional) that need to be jointly considered, since, if adaptation measures are to be implemented, knowledge down to the lowest decision level will be required. The varied geography, climate and human values of Europe pose a great challenge for evaluation of the ultimate impacts of climate change.

Although there are some good examples, such as the ESPACE-project (Nadarajah and Rankin, 2005), national-scale programmes, such as the FINADAPT project, studies of adaptation to climate change and of adaptation costs are at an early stage and need to be carried out urgently. These studies need to match adaptation measures to specific climate change impacts (e.g., targeted to alleviating impacts on particular types of agriculture, water management or on tourism at specific locations). They need to take into account regional differences in adaptive capacity (e.g., wide regional differences exist in Europe in the style and application of coastal management). Adaptation studies need to consider that in some cases both positive and negative impacts may occur as a result of climate change (e.g., the productivity of some crops may increase, while others decrease at the same location, e.g., Alexandrov et al., 2002). Key research priorities for impacts of climate change, adaptation and implications are included in Table 12.5.

Table 12.4. *Summary of the main expected impacts of climate change in Europe during the 21st century, assuming no adaptation.*

Sectors and Systems	Impact	Area				
		North	Atlantic	Central	Mediterr.	East
Water resources	Floods	↓↓	↓↓	↓↓	↓	↓↓↓
	Water availability	↑↑	↑↑	↓	↓↓↓	↓↓
	Water stress	↑↑	↑↑	↓	↓↓↓	↓↓
Coastal and marine systems	Beach, dune: low-lying coast erosional 'coastal squeeze'	↓↓↓	↓↓↓	na	↓↓	↓↓
	SLR- and surge-driven flooding	↓↓↓	↓↓	na	↓↓	↓↓↓
	River sediment supply to estuaries and deltas	↓↓	↓	na	↓↓↓	↓
	Saltwater intrusion to aquifers	↓	↓	na	↓↓	↓
	Northward migration of marine biota	↑	↑↑↑	na	↑	↑
	Rising SSTs, eutrophication and stress on biosystems	↓↓↓	↓↓	na	↓↓	↓
	Development of ICZM	↑↑	↑↑	na	↑↑	↑
	Deepening and larger inshore waters	↑↑	↑	na	↑	↑↑
Mountains, cryosphere	Glacier retreat	↓↓↓	↓	↓↓↓	↓↓↓	↓↓↓
	Duration of snow cover	↓↓↓	↓↓↓	↓↓↓	↓↓↓	↓↓↓
	Permafrost retreat	↓↓↓	↓	↓	na	↓
	Tree line upward shift	↑↑↑	↑↑↑	↑↑↑	↑	↑↑↑
	Nival species losses	↓↓↓	↓↓↓	↓↓↓	↓↓↓	↓↓↓
Forest, shrublands and grasslands	Forest NPP	↑↑↑	↑↑	↑ to ↓	↓	↑ to ↓
	Northward/inland shift of tree species	↑↑↑	↑↑	↑↑	↑ to ↓	↓↓
	Stability of forest ecosystems	↓↓	↓	↓	↓↓↓	↓↓↓
	Shrublands NPP	↑↑↑	↑↑↑	↑	↓↓↓	↓↓
	Natural disturbances (e.g., fire, pests, wind-storm)	↓	↓	↓	↓↓↓	↓↓
	Grasslands NPP	↑↑↑	↑↑	↑ to ↓	↓↓↓	↑
Wetlands and aquatic ecosystems	Drying/transformation of wetlands	↓↓	↓	↓	↓↓↓	↓↓↓
	Species diversity	↑ to ↓	↑	??	↓↓	↓
	Eutrophication	↓	↓↓	↓↓	↓↓↓	↓
	Disturbance of drained peatlands	↓↓↓	↓	↓↓	na	↓↓↓
Biodiversity	Plants	↓↓	↓↓	↓↓↓(Mt)	↓↓↓	↓
	Amphibians	↓↓	↓↓↓	↑↑	↓↓↓(SW) ↑↑(SE)	↑↑↑
	Reptiles	↓↓	↓↓	↑↑	↓↓↓(SW) ↑↑↑(SE)	↑↑↑
	Marine mammals	↓↓↓	??	na	↓↓↓	??
	Low-lying coastal birds	↓↓↓	↓↓↓	na	↓↓↓	??
	Freshwater biodiversity	↑ to ↓	??	??	↓↓↓	??
Agriculture and fisheries	Suitable cropping area	↑↑↑	↑↑	↑	↓↓	↓
	Agricultural land area	↓↓	↓↓	↓↓	↓↓	↓↓
	Summer crops (maize, sunflower)	↑↑↑	↑↑	↑	↓↓↓	↓↓
	Winter crops (winter wheat)	↑↑↑	↑↑	↑ to ↓	↓↓	↑
	Irrigation needs	na	↑ to ↓	↓↓	↓↓↓	↓
	Energy crops	↑↑↑	↑↑	↑	↓↓	↓
	Livestock	↑ to ↓	↓	↓↓	↓↓	↓↓
	Marine fisheries	↑↑	↑	na	↓	na
Energy and transport	Energy supply and distribution	↑	↑↑	↑	↓	↑
	Winter energy demand	↑↑	↑↑	↑	↑↑	↑
	Summer energy demand	↓	↓	↓↓	↓↓↓	↓↓
	Transport	↑	↓	↓	↓	↑
Tourism	Winter (including ski) tourism	↑↑	↓	↓↓↓	↑↑↑	↓↓
	Summer tourism	↑	↑↑	↑	↓↓	↑
Property insurance	Flooding claims	??	↓↓	↓↓	??	??
	Storms claims	↓	↓↓	↓↓	??	??
Human health	Heat-related mortality/morbidity	↓	↓↓	↓↓	↓↓↓	↓↓
	Cold-related mortality/morbidity	↑	↑↑	↑↑	↑	↑↑↑
	Health effects of flooding	↓	↓↓	↓↓	↓↓	↓↓
	Vector-borne diseases	↓	↓	↓	↓↓	↓↓
	Food safety/Water-borne diseases	↓	↓	↓	↓↓	↓↓
	Atopic diseases, due to aeroallergens	↓	↓	↓	↓	↓

Scoring has taken into account: a) geographical extent of impact/number of people exposed; b) intensity and severity of impact. The projected magnitude of impact increases with the number of arrows (one to three). Type of impact: positive (upward, blue); negative (downward, red); a change in the type of impact during the course of the century is marked with 'to' between arrows. na=not applicable; ??=insufficient information; North=boreal and Arctic; Central, Atlantic and Mediterranean as in Figure 12.3, including their mountains; East=steppic Russia, the Caucasus and the Caspian Sea; Mt=Mountains; SW=Southwest; SE=Southeast; SLR=Sea-Level Rise; ICZM=Integrated Coastal Zone Management; SST=Sea-Surface Temperature; NPP=Net Primary Productivity.

Table 12.5. *Key uncertainties and research needs.*

Impact of climate change
• Improved long-term monitoring of climate-sensitive physical (e.g., cryosphere), biological (e.g., ecosystem) and social sectors (e.g., tourism, human health). • Improvement of climate impact models, including better understanding of mechanism of climate impacts, e.g., of heat/cold morbidity, differences between impacts due to short-term climate variability and long-term climate change, and the effects of extreme events, e.g., heatwaves, droughts, on longer-term dynamics of both managed and natural ecosystems. • Simultaneous consideration of climatic and non-climatic factors, e.g., the synergistic effect of climate change and air pollution on buildings, or of climate change and other environmental factors on the epidemiology of vector-borne diseases; the validation and testing of climate impact models through the enhancement of experimental research; increased spatial scales; long-term field studies and the development of integrated impact models. • Enhancement of climate change impact assessment in areas with little or no previous investigation, e.g., groundwater, shallow lakes, flow regimes of mountain rivers, renewable energy sources, travel behaviour, transport infrastructure, tourist demand, major biogeochemical cycles, stability, composition and functioning of forests, natural grasslands and shrublands), nutrient cycling and crop protection in agriculture. • More integrated impact studies, e.g., of sensitive ecosystems including human dimensions. • Better understanding of the socio-economic consequences of climate change for different European regions with different adaptive capacity.
Adaptation measures
• The comprehensive evaluation (i.e., of effectiveness, economy and constraints) of adaptation measures used in past in different regions of Europe to reduce the adverse impacts of climate variability and extreme meteorological events. • Better understanding, identification and prioritisation of adaptation options for coping with the adverse effects of climate change on crop productivity, on the quality of aquatic ecosystems, on coastal management and the capacity of health services. • Evaluation of the feasibility, costs and benefits of potential adaptation options, measures and technologies. • Quantification of bio-climatic limitations of prevalent plant species. • Continuation of studies on the regional differences in adaptive capacity.
Implementation
• Identification of populations at risk and the lag time of climate change impacts. • Approaches for including climate change in management policy and institutions. • Consideration of non-stationary climate in the design of engineering structures. • Identification of the implications of climate change for water, air, health and environmental standards. • Identification of the pragmatic information needs of managers responsible for adaptation.

References

ABI (Association of British Insurers), 2000: *Inland Flooding Risk*. Association of British Insurers, London, 16 pp.

ABI (Association of British Insurers), 2004: *A Changing Climate for Insurance*. Association of British Insurers, London, 24 pp.

Abildtrup, J., E. Audsley, M. Fekete-Farkas, C. Giupponi, M. Gylling, P. Rosato and M.D.A. Rounsevell, 2006: Socio-economic scenario development for the assessment of climate change impacts on agricultural land use. *Environ. Sci. Pol*. **9**, 101-115.

ACIA (Arctic Climate Impact Assessment), 2004: *Impacts of a warming Arctic*. Cambridge University Press, Cambridge, 144 pp.

AEAT, 2003: *Railway Safety Implications of Weather, Climate and Climate Change*. Final Report to the Railway Safety and Standards Board. Retrieved 12.10.2006 from http://www.railwaysafety.org.uk/pdf/ClimateChangeFR.pdf.

AEMA, 2002: *Uso sostenible del agua en Europa. Gestión de la demanda*. Ministerio de Medio Ambiente, Madrid, 94 pp.

Ågren, G.I. and E. Bosatta, 2002: Reconciling differences in predictions of temperature response of soil organic matter. *Soil Bio. Biochem*., **34**, 129-132.

Alados, C.I., Y. Pueyo, O. Barrantes, J. Escós, L. Giner and A.B., Robles, 2004: Variations in landscape patterns and vegetation cover between 1957 and 1994 in a semiarid Mediterranean ecosystem. *Lands. Ecol*., **19**, 543-559.

Alcamo, J., P. Döll, T. Heinrichs, F. Kaspar, B. Lehner, T. Rösch and S. Siebert, 2003: Global estimates of water withdrawals and availability under current and future business-as-usual conditions. *Hydrological Sci. J*., **48**, 339-348.

Alcamo, J., M. Endejan, A.P. Kirilenko, G.N., Golubev and N.M. Dronin, 2005: Climate Change and its Impact on Agricultural Production in Russia. In: *Understanding Land-Use and Land-Cover Change in Global and Regional Context*, E. Milanova, Y. Himiyama, and I. Bicik Eds., Science Publishers, Plymouth, Devon, 35-46.

Alcamo, J., M. Floerke and M. Maerker, 2007: Future long-term changes in global water resources driven by socio-economic and climatic changes. *Hydrological Sciences*, **52**, 247-275.

Alexander, L.V., X. Zhang, T.C. Peterson, J. Caesar, B. Gleason, A.M.G. Klein Tank, M. Haylock, D. Collins, B. Trewin, F. Rahimzadeh, A. Tagipour, K. Rupa Kumar, J. Revadekar, G. Griffiths, L. Vincent, D.B. Stephenson, J. Burn, E. Aguilar, M. Brunet, M. Taylor, M. New, P. Zhai, M. Rusticucci and J.L. Vázquez Aguirre, 2006: Global observed changes in daily climate extremes of temperature and precipitation. *J. Geophys. Res*., **111**, D05109, doi:10.1029/2005JD006290.

Alexandrov, V., J. Eitzinger, V. Cajic and M. Oberforster, 2002: Potential impact of climate change on selected agricultural crops in north-eastern Austria. *Glob. Change Biol*., **8**, 372-389.

Alley, R.B., J. Marotzke, W.D. Nordhaus, J.T. Overpeck, D.M. Peteet, R.A Pielke Jr., R.T. Pierrehumbert, P.B. Rhines, T.F. Stocker, L.D. Talley and J.M. Wallace, 2003: Abrupt climate change. *Science*, **299**, 2005-2010.

Álvarez Cobelas, M., J. Catalán and D. García de Jalón, 2005: Impactos sobre los ecosistemas acuáticos continentales. *Evaluación Preliminar de los Impactos en España por Efecto del Cambio Climático*, J.M. Moreno, Ed., Ministerio de Medio Ambiente, Madrid, 113-146 .

Amelung, B. and D. Viner, 2006: Mediterranean tourism: Exploring the future with the tourism climatic index. *J. Sust. Tour.,* **14**, 349-366.

Anadón, R., C.M. Duarte and C. Fariña, 2005: Impactos sobre los ecosistemas marinos y el sector pesquero. *Evaluación Preliminar de los Impactos en España por Efecto del Cambio Climático*, J.M. Moreno, Ed., Ministerio de Medio Ambiente, Madrid, 147-182.

André, J.-C., M. Déqué, P. Rogel and S. Planton, 2004: The 2003 summer heatwave and its seasonal forecasting. *C. R. Geosci*., **336**, 491-503.

Andréasson, J., S. Bergström, B. Carlsson, L.P. Graham and G. Lindström, 2004: Hydrological change – climate impact simulations for Sweden. *Ambio*, **33**, 228-234.

Araújo, M.B. and M. New, 2007: Ensemble forecasting of species distributions. *Trends Ecol. Evol*., **22**, 42-47.

Araújo, M.B., M. Cabeza, W. Thuiller, L. Hannah and P.H. Williams, 2004: Would climate change drive species out of reserves? An assessment of existing reserve-selection methods. *Glob. Change Biol.,* **10**, 1618-1626.

Araújo, M.B., W. Thuiller and R.G. Pearson, 2006: Climate warming and the decline of amphibians and reptiles in Europe. *J. Biogeogr*., **33**, 1677-1688.

Arnell, N.W., 2004: Climate change and global water resources: SRES emissions and socio-economic scenarios. *Glob. Environ. Change*, **14**, 31-52.

Arnell, N.W., M.J.L. Livermore, S. Kovats, P.E. Levy, R. Nicholls, M.L. Parry and S.R. Gaffin, 2004: Climate and socio-economic scenarios for global-scale climate change impacts and assessments: characterising the SRES storylines. *Glob. Environ. Change*, **14**, 3-20.

Arnell, N., E. Tomkins, N. Adger and K. Delaney, 2005: *Vulnerability to Abrupt Climate Change in Europe*. ESRC/ Tyndall Centre Technical Report N° 20, Tyndall Centre for Climate Change Research, University of East Anglia, Norwich, 63 pp.

Askeev, O.V., D. Tischin, T.H. Sparks and I.V. Askeev, 2005: The effect of climate on the phenology, acorn crop and radial increment of pedunculate oak (*Quercus robur*) in the middle Volga region, Tatarstan, Russia. *Int. J. Biometeorol.*, **49**, 262-266.

Audsley, E., K.R. Pearn, C. Simota, G. Cojocaru, E. Koutsidou, M.D.A. Rounsevell, M. Trnka and V. Alexandrov, 2006: What can scenario modelling tell us about future European scale agricultural land use, and what not? *Environ. Sci. Pol.*, **9**, 148-162.

Austin, G.E. and M.M. Rehfisch, 2005: Shifting nonbreeding distributions of migratory fauna in relation to climatic change. *Glob. Change Biol.*, **11**, 31-38.

Aylin, P., S. Morris, J. Wakefield, A. Grissinho, L. Jarup and P. Elliott, 2001: Temperature, housing, deprivation and their relationship to excess winter mortality in Great Britain, 1986–1996. *Int. J. Epidemiol.*, **30,** 1100–1108.

Babiker, M.H. and R.S. Eckaus, 2002: Rethinking the Kyoto emissions targets. *Climatic Change*, **54**, 399-414.

Badeck, F.-W., H. Lischke, H. Bugmann, T. Hicker, K. Höniger, P. Lasch, M.J. Lexer, F. Mouillot, J. Schaber and B. Smith, 2001. Tree species composition in European pristine forests: Comparison of stand data to model predictions. *Climatic Change,* **51**, 307-347.

Badeck, F.-W., A. Bondeau, K. Bottcher, D. Doktor, W.Lucht, J. Schaber and S. Sitch, 2004: Responses of spring phenology to climate change. *New Phytol.*, **162**, 295-309.

Baeza, M.J., M. de Luís, J. Raventós and A. Escarré, 2002: Factors influencing fire behaviour in shrublands of different stand ages and the implications for using prescribed burning to reduce wildfire risk. *J. Environ. Manage.*, **65**, 199-208.

Baker, T., 2005: *Vulnerability Assessment of the North-East Atlantic Shelf Marine Ecoregion to Climate Change*, Workshop Project Report, WWF, Godalming, Surrey,79 pp.

Bakkenes, M., R.M. Alkemade, F. Ihle, R. Leemans and J.B. Latour, 2002: Assessing effects of forecasted climate change on the diversity and distribution of European higher plants for 2050. *Glob. Change Biol.*, **8**, 390-407.

Bale, J.S., G.J. Masters, I.D. Hodkinson, C. Awmack, T.M. Bezemer, V.K. Brown, J.Butterfield, A. Buse, J.C. Coulson, J. Farrar, J.E.G., Good, R. Harrington, S. Hartley, T.H. Jones, R.L. Lindroth, M.C. Press, I. Symrnioudis, A.D. Watt and J.B. Whittaker, 2002: Herbivory in global climate change research: direct effects of rising temperature on insect herbivores. *Glob. Change Biol.*, **8**, 1-16.

Ballester, F., P. Michelozzi and C. Iñiguez, 2003: Weather, climate, and public health. *J. Epidemiol. Community Health,* **57**, 759–760.

Bannikov, M.V., A.B. Umarova and M.A. Butylkina, 2003: Fires on drained peat soils of Russia: Causes and effects. *Int. Forest Fire News*, **28**, 29-32.

Barklund, P., 2002: Excador I Europa. Skogsstyrelsen, Jönköping, Rapport No1 (in Swedish).

Barreira, A., 2004: *Dams in Europe. The Water Framework Directive and the World Commission on Dam Recommendations: A Legal and Policy Analysis*. Retrieved 05.11.2006 from http://assets.panda.org/downloads/wfddamsineurope.pdf

Barrera, A., M.C. Llasat and M. Barriendos, 2006: Estimation of extreme flash flood evolution in Barcelona County from 1351 to 2005. *Nat. Haz. Earth Syst. Sci.*, **6**, 505-518.

Barthod, C., 2003: *Forests for the Planet: Reflections on the Vast Storms in France in 1999*. Proceedings of the XII World Forestry Congress, September 2003, Quebec, Canada, Volume B, 3-9.

Battisti, A., 2004: Forests and climate change – lessons from insects. *Forest,* **1**, 17-24.

Beaugrand, G. and P.C. Reid, 2003: Long-term changes in phytoplankton, zooplankton and salmon related to climate. *Glob. Change Biol.*, **9,** 801-817.

Beaugrand, G., P.C. Reid, F. Ibáñez, J.A. Lindley and M. Edwards, 2002: Reorganization of north atlantic marine copepod biodiversity and climate. *Science,* **296**, 1692-1694.

Beggs, P.J., 2004: Impacts of climate change on aeroallergens: past and future. *Clin. Exp. Allergy*, **34**, 1507-1513.

Beniston, M., 2002: Climatic change: possible impacts on human health. *Swiss Med. Wkly.*, **132**, 332–337.

Beniston, M., 2004: The 2003 heatwave in Europe: a shape of things to come? an analysis based on swiss climatological data and model simulations. *Geophys. Res. Lett.*, **31**, L02202 doi:10.1029/2003GL018857.

Beniston, M. and H.F. Díaz, 2004: The 2003 heatwave as an example of summers in a greenhouse climate? Observations and climate model simulations for Basel, Switzerland. *Glob. Planet. Change*, **44**, 73-81.

Beniston, M., F. Keller and S. Goyette, 2003: Snow pack in the Swiss Alps under changing climatic conditions: an empirical approach for climate impact studies. *Theor. Appl. Climatol.*, **74**, 19-31.

Beniston, M., D.B. Stephenson, O.B. Christensen, C.A.T. Ferro, C. Frei, S. Goyette, K. Halsnaes, T. Holt, K. Jylhä, B. Koffi, J. Palutikof, R. Schöll, T. Semmler and K. Woth, 2007: Future extreme events in European climate: an exploration of regional climate model projections. *Climatic Change*, **81**, S71-S95.

Benito, G., M. Barriendos, M.C. Llasat, M. Machado and V. Thorndycraft, 2005: Impactos sobre los riesgos naturales de origen climático: riesgo de crecidas fluviales. *Evaluación Preliminar de los Impactos en España por efecto del Cambio Climático,* J.M. Moreno, Ed., Ministerio de Medio Ambiente, Madrid, 527-548.

Beran, J., L. Asokliene and I. Lucenko, 2004: Tickborne encephalitis in Europe: Czech Republic, Lithuania and Latvia. *Euro Surveill. Weekly Release*, **8**.

Berry, P.M., M.D.A. Rounsevell, P.A. Harrison and E. Audsley, 2006: Assessing the vulnerability of agricultural land use and species to climate change and the role of policy in facilitating adaptation. *Environ. Sci. Pol.*, **9**, 189-204.

Blenckner, T., 2005: A conceptual model of climate-related effects on lake ecosystems. *Hydrobiologia*, **533**, 1-14.

Blennow, K. and O. Sallnäs, 2002: Risk perception among non-industrial private forest owners. *Scand. J. Forest Res.*, **17**, 472-479.

Boelens, R., D. Minchin and G. O'Sullivan, 2005: *Climate Change - Implications for Ireland's Marine Environment and Resources*.Marine Institute, Galway, 48 pp.

Böhm, R., I. Auer, M. Brunetti, M. Maugeri, T. Nanni and W. Schöner, 2001: Regional temperature variability in the European Alps: 1760-1998 from homogenized instrumental time series. *Int. J. Climatol.* **21**, 1779-1801.

Boisvenue, C. and S.W. Running, 2006: Impacts of climate change on natural forest productivity-evidence since the middle of the 20th century. *Glob. Change Biol.*, **12**, 862-882.

Bortenschlager, S. and I. Bortenschlager, 2005: Altering airborne pollen concentrations due to the Global Warming. A comparative analysis of airborne pollen records from Innsbruck and Obergurgl (Austria) for the period 1980–2001. *Grana*, **44**, 172-180.

Bouraoui, F., B. Grizzetti, K. Granlund, S. Rekolainen and G. Bidoglio, 2004: Impact of climate change on the water cycle and nutrient losses in a Finnish catchment. *Climatic Change,* **66**, 109-126.

Brander, K.M., 2005: Cod recruitment is strongly affected by climate when stock biomass is low. *ICES J. Mar. Sci.*, **62**, 339-343.

Brander, K. M. and G. Blom, 2003: Changes in fish distribution in the Eastern North Atlantic: Are we seeing a coherent response to changing temperature? *ICES Marine Science Symposia*, **219**, 261-270.

Briers, R.A., J.H.R. Gee and R. Geoghegan, 2004: Effects of North Atlantic oscillation on growth and phenology of stream insects. *Ecography*, **27**, 811-817.

Broadmeadow, M.S.J., D. Ray and C.J.A. Samuel, 2005. Climate change and the future for broadleaved tree species in Britain. *Forestry*, **78**, 145-161.

Bröker, M. and D. Gniel, 2003: New foci of tick-borne encephalitis virus in Europe: consequences for travelers from abroad. *Travel Med. Infect. Dis.*, **1**, 181-184.

Brommer, J.E., 2004: The range margins of northern birds shift polewards. *Ann. Zool. Fenn.*, **41**, 391-397.

Bronstert, A., 2003: Floods and climate change: interactions and impacts. *Risk Anal.*, **3**, 545-557.

Brooker, R. and J. Young, Eds., 2005: *Climate Change and Biodiversity in Europe: a Review of Impacts, Policy, Gaps in Knowledge and Barriers to the Exchange of Information between Scientists and Policy Makers*. NERC Centre for Ecology and Hydrology, Banchory Research Station, Banchory, Aberdeenshire, 31 pp. Retrieved 28.10.2006 from http://www.ceh.ac.uk/sections/ed/documents/Backgroundpaper_final.pdf.

Buckland, S.M., K. Thompson, J.G. Hodgson and J.P. Grime, 2001: Grassland invasions: effects of manipulations of climate and management. *J. Appl. Ecol.*, **38**, 301-309.

Busuioc, A., 2001: Large-scale mechanisms influencing the winter Romanian climate variability. *Detecting and Modelling Regional Climate Change*. M. Bruner and D. López, Eds., Springer-Verlag, Berlin, 333-344.

Butenko, A.M. and V.F. Larichev, 2004: Climate influence on the activity and areas of Crimea hemorrhagic fever in the north part of its virus' natural habitat. In: *Climate Change and Public Health in Russia in the XXI century*. Proceeding of the workshop, April 5-6, 2004, Moscow, Publishing Company "Adamant", Moscow, 134-138, (in Russian).

Byrne, C. and M.B. Jones, 2002: Effects of Elevated CO_2 and Nitrogen Fertilizer on Biomass Productivity, Community Structure and Species Diversity of a Semi-Natural Grassland in Ireland. *Proceeding of the Royal Irish Academy*, 141-150.

Camarero, J. J. and E. Gutiérrez, 2004: Pace and pattern of recent treeline dynamics response of ecotones to climatic variability in the Spanish Pyrenees. *Climatic Change*, **63**, 181-200.

Cartalis, C., A. Synodinou, M. Proedrou, A. Tsangrassoulis and M. Santamouris, 2001: Modifications in energy demand in urban areas as a result of climate changes: an assessment for the southeast Mediterranean region. *Energy Conv. Manag.*, **42**, 1647-1656.

Casimiro, E. and J.M. Calheiros, 2002: Human Health. *Climate Change in Portugal: Scenarios, Impacts, and Adaptation Measures,* F.D. Santos, K. Forbes and R. Moita, Eds., SIAM project, Gradiva Publisher, Lisbon, 245-300.

Castle, I. and D. Henderson, 2003a: The IPCC emission scenarios: an economic-statistical critique. *Ener. Environ.*, **14**, 159-185.

Castle, I. and D. Henderson, 2003b: Economics, emission scenarios and the work of the IPCC. *Ener. Environ.*, **14**, 415-436.

Catalán, J., S. Pla, M. Rieradevall., M. Felip, M. Ventura, T. Buchaca, L. Camarero, A. Brancelj, P.G. Appleby, A. Lami, J.A. Grytnes, A. Agustí-Panareda and R. Thompson, 2002: Lake Redó ecosystem response to an increasing warming in the Pyrenees during the twentieth century. *J. Paleolimnol.*, **28**, 129-145.

Ceron, J.-P. and G. Dubois, 2000: Tourisme et changement climatique. In: *Impacts Potentiels du Changement Climatique en France au XXIème Siècle*. 2d ed. Premier ministre, Ministère de l'aménagement du territoire et de l'environnement, 1998, 104-111.

Chang, H., C.G. Knight, M.P. Staneva and D. Kostov, 2002: Water resource impacts of climate change in southwestern Bulgaria. *GeoJournal*, **57**, 159-168.

Chmielewski, F.-M., A. Müller and E. Bruns, 2004: Climate changes and trends in phenology of fruit trees and field crops in Germany, 1961-2000. *Agric. For. Meteorol.*, **121**, 69-78.

Christensen, J.H. and O.B. Christensen, 2003: Severe summertime flooding in Europe. *Nature,* **421**, 805-806.

Christensen, J.H. and O.B. Christensen, 2007: A summary of the PRUDENCE model projections of changes in European climate during this century. *Climatic Change*, **81**, S7-S30.

Christenson, M., H. Manz and D. Gyalistras, 2006: Climate warming impact on degree-days and building energy demand in Switzerland. *Energy Conv. Manag.*, **47**, 671-686.

Ciais, Ph., I. Janssens, A. Shvidenko, C. Wirth, Y. Malhi, J. Grace, E.-D. Schulze, M. Herman, O. Phillips and H. Dolman, 2004: The potential for rising CO_2 to account for the observed uptake of carbon by tropical, temperate, and boreal forest biomes. *The Carbon Balance of Forest Biomes,* H. Griffiths and P.J. Jarvis, Eds., Garland Science/BIOS Scientific Publishers, Abingdon, Oxfordshire, 109-149.

Ciais, Ph., M. Reichstein, N. Viovy, A. Granier, J. Ogée, V. Allard, M. Aubinet, N. Buchmann, C. Bernhofer, A. Carrara, F. Chevallier, N. de Noblet, A.D. Friend, P. Friedlingstein, T. Grünwald, B. Heinesch, P. Keronen, A. Knohl, G. Krinner, D. Loustau, G. Manca, G. Matteucci, F. Miglietta, J.M. Ourcival, D. Papale, K. Pilegaard, S. Rambal, G. Seufert, J.F. Soussana, M.J. Sanz, E.D. Schulze, T. Vesala and R. Valentini, 2005: Europe-wide reduction in primary productivity caused by the heat and drought in 2003. *Nature,* **437**, 529-533.

Clark, R.A., J.F. Fox, D. Vinerand M. Livermore, 2003: North Sea cod and climate change - modelling the effects of temperature on population dynamics. *Glob. Change Biol.*, **9**, 1660-1690.

Coccossis, H., 2003: *Towards a regional legal framework for Integrated Coastal Area Management in the Mediterranean*. Accessed 10.03.2006 from http://www.globaloceans.org/globalconferences/2003/pdf/PreconferenceProceedingsVolume.pdf

Colebrook, E. and R. Wall, 2004: Ectoparasites of livestock in Europe and the Mediterranean region. *Vet. Parasitol.*, **120**, 251-274.

Cooper, J.A.G. and O.H. Pilkey, 2004: Sea-level rise and shoreline retreat: time to abandon the Bruun Rule. *Glob. Planet. Change*, **43**, 157-171.

Cooper, N.J., P.G. Barber, M.C.Bray and D.J. Carter, 2002: Shoreline management plans: a national review and an engineering perspective. *Proc. Inst. Civil. Eng.-Marit. Eng.*, **154**, 221-228.

Cosgrove, W., R. Connor and J. Kuylenstierna, 2004: Workshop 3 (synthesis): Climate variability, water systems and management options. *Water Sci. Technol*, **7**, 129-132.

Crawford, V.L.S., M. McCann and R.W. Stout, 2003: Changes in seasonal deaths from myocardial infarction. *Q. J. Med.*, **96,** 45-52.

Cubasch, U., G.A. Meehl, G.J. Boer, R.J. Stouffer, M. Dix, A. Noda, C.A. Senior, S. Raper and K.S. Yap, 2001: Projections of future climate change. *Climate Change 2001: The Scientific Basis. Contribution of Working Group I to the Third Assessment Report of the Intergovernmental Panel on Climate Change*, J.T. Houghton, Y. Ding, D.J. Griggs, M. Noguer, P.J. van der Linden, X. Dai, K. Maskell and C.A. Johnson, Eds., Cambridge University Press, Cambridge, 525-582.

Curry, R. and C. Mauritzen, 2005: Dilution of the northern North Atlantic Ocean in recent decades. *Science*, **308**, 1772-1774.

Daniel, M., B. Kriz, V. Danielová, J. Materna, N. Rudenko, J. Holubová, L. Schwarzová and M. Golovchenko, 2005: Occurrence of ticks infected by tickborne encephalitis virus and Borrelia genospecies in mountains of the Czech Republic. Euro Surveill. **10**, E050331.1. Accessed 05.11.2006 at: http://www.eurosurveillance.org/ew/2005/050331.asp#1.

Daniel, M., V. Danielová, B. Křiž and Č. Beneš, 2006: Tick-borne encephalitis. *Climate Change and Adaptation Strategies for Human Health,* B. Menne and K.L. Ebi, Eds., Steinkopff Verlag, Darmstadt, 189-205.

Danielova, V., B. Kriz, M. Daniel, C. Benes, J. Valter and I. Kott, 2004: Effects of climate change on the incidence of tick-borne encephalitis in the Czech Republic in the past two decades. *Epidemiol. Mikrobiol. Imunol.*, **53**, 174-181.

Daufresne, M., M.C. Roger, H. Capra and N. Lamouroux, 2003: Long-term changes within invertebrate and fish communities of Upper Rhone river: effects of climate factors. *Glob. Change Biol.,* **10**, 124-140.

de Boeck, H.J., C.M.H.M. Lemmens, H. Bossuyt, S. Malchair, M. Carnol, R. Merckx, I. Nijs and R. Ceulemans, 2006: How do climate warming and plant species richness affect water use in experimental grasslands? *Plant Soil*, 10.1007/s11104-006-9112-5.

de Groot, Th.A.M. and J.D. Orford, 2000: Implications for coastal zone Management. *Sea Level Change and Coastal Processes: Implications for Europe,* D.E. Smith, S.B. Raper, S. Zerbini and A. Sánchez-Arcilla, Eds., Office for Official Publications of the European Communities, Luxembourg, 214-242.

de Gruijl, F.R., J. Longstreth, M. Norval, A.P. Cullen, H. Slaper, M.L. Kripke, Y. Takizawa and J.C. van der Leun, 2003: Health effects from stratospheric ozone depletion and interactions with climate change**.** *Photochem. Photobiol. Sci.*, **2**, 16-28.

de Luis, M., J.C. González-Hidalgo and J. Raventós, 2003: Effects of fire and torrential rainfall on erosion in a Mediterranean gorse community. *Land Deg. Devel.*, **14**, 203-213.

de Roo, A., G. Schmuck, V. Perdigao and J. Thielen, 2003: The influence of historic land use change and future planned land use scenarios on floods in the Oder catchment. *Phys. Chem. Earth, Parts A/B/C*, **28**, 1291-1300.

Declerck, S., J. Vandekerkhove, L.S. Johansson, K. Muylaert, J.M. Conde-Porcuna, K. van der Gucht, C. Pérez-Martínez, T.L. Lauridsen, K. Schwenk, G. Zwart, W. Rommens, J. López-Ramos, E. Jeppesen, W. Vyverman, L. Brendonck and L. de Meester, 2005: Multi-group biodiversity in shallow lakes along gradients of phosphorus and water plant cover. *Ecology*, **86**, 1905-1915.

Defra, 2004a: *Scientific and Technical Aspects of Climate Change, including Impacts and Adaptation and Associated Costs*. Department for Environment, Food and Rural Affairs, London.. Retrieved 12.10.2006 from http://www.defra.gov.uk /ENVIRONMENT/climatechange/pubs/index.htm.

Defra, 2004b: *Making Space for Water: Developing a New Government Strategy for Flood and Coastal Erosion Risk Management in England: A Consultation Exercise*. Department for Environment, Food and Rural Affairs, London, 154 pp.

Defra, 2004c: *Scoping Uncertainty in the Social Cost of Carbon*. Final Project Report (Draft), Department for Environment, Food and Rural Affairs, London, 41pp.

del Barrio, G., P.A. Harrison, P.M. Berry, N. Butt, M. Sanjuan, R.G. Pearson and T. Dawson, 2006: Integrating multiple modelling approaches to predict the potential impacts of climate change on species' distributions in contrasting regions: comparison and implications for policy. *Environ. Sci. Pol.*, **9**, 129-147.

Delitti, W., A. Ferrán, L. Trabaud and V.R. Vallejo, 2005: Effects of fire recurrence in *Quercus coccifera* L. shrublands of the Valencia Region (Spain): I. plant composition and productivity. *Plant Ecol.*, **177**, 57-70.

Department of Health, 2002: *Health Effects of Climate Change in the UK: Report of the Expert Advisory Group on Climate change and Health*. London, 238 pp.

Déqué, M., D.P. Roswell, D. Lüthi, F. Giorgi, J.H. Christensen, B. Rockel, D. Jacob, E. Kjellström. M. de Castro and B. van den Hurk, 2007: An intercomparison of climate simulations for Europe: assessing uncertainties in model projections. *Climatic Change*, **81**, S53-S70.

Dessai, S., 2003: Heat stress and mortality in Lisbon. Part II: An assessment of the potential impacts of climate change. *Int. J. Biometeorol.*, **48**, 37-44.

Devoy, R.J.N., 2007: Coastal vulnerability and the implications of sea-level rise for Ireland. *J. Coast. Res.*, (in press).

Devoy, R.J.N., T.A. de Groot, S. Jelgersma, I. Marson, R.J. Nicholls, R. Paskoff and H.-P. Plag, 2000: Sea-level change and coastal processes: recommendations.*Sea Level Change and Coastal Processes: Implications for Europe,* D.E. Smith, S.B. Raper, S. Zerbini and A. Sánchez-Arcilla, Eds.,Office for Official Publications of the European Communities, Luxembourg, 1-8.

Díaz, J., R. García, C. López, C. Linares, A. Tobías and L. Prieto, 2005: Mortality impact of extreme winter temperatures. *Int. J. Biometeorol.*, **49**, 179-83.

Dickson, M.E., M.J.A. Walkden, J.W. Hall, S.G. Pearson and J.G. Rees, 2007: Numerical modelling of potential climate-change impacts on rates of soft-cliff recession, north-east Norfolk, UK. *Proc. Coastal Dyn.* (in press).

Dickson, R.R., R. Curry and I. Yashayaev, 2003: Recent changes in the North Atlantic. *Philos. Trans. Roy. Soc. London.*, **361**, 1917-1934.

Diffey, B., 2004: Climate change, ozone depletion and the impact on ultraviolet exposure of human skin. *Phys. Med. Biol.,* **49**, 1-11.

Dirnböck, T., S. Dullinger and G. Grabherr, 2003: A regional impact assessment of climate and land-use change on alpine vegetation. *J. Biogeogr.,* **30**, 401-417.

Dlugolecki, A., Ed., 2001: *Climate Change and Insurance*. Chartered Insurance Institute, London, 110 pp.

Dlugolecki, A. and M. Keykhah, 2002: Climate change and the insurance sector: its role in adaptation and mitigation. *Greener Management International*, **39**, 83-98.

Dobbertin, M., N. Hilker, M. Rebetez, N. E. Zimmermann, T. Wohlgemuth and A. Rigling, 2005: The upward shift in altitude of pine mistletoe (*Viscum album* ssp. *Austriacum*) in Switzerland – the result of climate warming? *Int. J. Biometeorol.*, **50**, 40-47.

Döll, P., 2002: Impact of climate change and variability on irrigation requirements: a global perspective. *Climatic Change,* **54**, 269-293.

Donevska, K. and S. Dodeva, 2004: Adaptation measures for water resources management in case of drought periods. In: *Proceedings, XXIInd Conference of the Danubian Countries on the Hydrological Forecasting and Hydrological Bases of Water Management*. Brno, 30 August-2 September 2004, CD-edition.

Dorrepaal, E.R., R. Aerts, J.H.C. Cornelissen, T.V. Callaghan and R.S.P. van Logtestijn, 2004: Summer warming and increased winter snow cover affect *Sphagnum fuscum* growth, structure and production in a sub-arctic bog. *Glob. Change Biol.*, **10**, 93-104.

Duchene, E. and C. Schneider, 2005: Grapevine and climatic changes: a glance at the situation in Alsace. *Agron. Sust. Dev.*, **25**, 93-99.

Duckworth, J.C., R.G.H. Bunce and A.J.C. Malloch, 2000: Modelling the potential effects of climate change on calcareous grasslands in Atlantic Europe. *J. Biogeogr.,* **27**, 347-358.

Dullinger, S., T. Dirnbock and G. Grabherr, 2004: Modelling climate change-driven treeline shifts: relative effects of temperature increase, dispersal and invasibility. *J. Ecol.*, **92**, 241-252.

Dutil, J.D. and K. Brander, 2003: Comparing productivity of North Atlantic cod (*Gadus morhua*) stocks and limits to growth production. *Fish Oceanogr.*, **12**, 502–512.

Ebi, K.L., 2005: Improving public health responses to extreme weather events. *Extreme weather events and public health responses,* W. Kirch, B. Menne and R. Bertollini, Eds., Springer Verlag, Berlin, 47-56.

Ebi, K.L., 2006: Floods and human health. In: *Climate Change and Adaptation Strategies for Human Health*, B. Menne and K.L. Ebi, Eds., Steinkopff Verlag, Darmstadt, 99-120.

EC, 2004: *Facts and Figures on the CFP*. Luxembourg, European Communities, 40 pp.

Eckhardt, K. and U. Ulbrich, 2003: Potential impacts of climate change on groundwater recharge and streamflow in a central European low mountain range. *J. Hydrol.*, **284**, 244-252.

Edgar, P.W., R.A. Griffiths and J.P. Foster, 2005: Evaluation of translocation as a tool for mitigating development threats to great crested newts (*Triturus cristatus*) in England, 1990–2001. *Biol. Conserv*, **122**, 45-52.

Edwards, M. and A.J. Richardson, 2004: Impact of climate change on marine pelagic phenology and trophic mismatch. *Nature*, **430**, 881-884.

Edwards, M., D.G. Johns, P. Licandro, A.W.G. John and D.P. Stevens, 2006: *Ecological Status Report: Results from the CPR Survey 2004/2005*. SAHFOS Technical Report, **3**, 1-8.

EEA, 2002: *Environmental Signals 2002. Benchmarking the Millenium*. Environmental Assessment Report No. 9, European Environmental Agency, Copenhagen, 147 pp.

EEA, 2003: *Europe's Environment: The Third Assessment*. Environmental Assessment Report No. 10, European Environment Agency, Copenhagen, 344 pp.

EEA, 2004a: *Environmental signals 2004. A European Environment Agency Update on Selected Issues*. European Environmental Agency, Copenhagen, 36 pp.

EEA, 2004b: *Impacts of Europe's Changing Climate: An Indicator-Based Assessment*. EEA Report No 2/2004, European Environment Agency, Copenhagen (or: Luxembourg, Office for Official Publications of the EC), 107 pp.

EEA, 2005: *Annual European Community Greenhouse Gas Inventory 1990-2003 and Inventory Report 2005. Submission to the UNFCCC Secretariat*. EEA Technical report No 4/2005. European Environmental Agency, Copenhagen, 87 pp.

EEA, 2006a: *Energy and Environment in the European Union. Tracking Progress towards Integration*. European Environment Agency, Copenhagen, 52 pp.

EEA, 2006b: *How much Bioenergy can Europe Produce without Harming the Environment?* EEA Report No 7/2006, European Environmental Agency, Copenhagen, 67 pp.

Egli, M., C. Hitz, P. Fitze and A. Mirabella, 2004: Experimental determination of climate change effects on above ground and below-ground organic matter in alpine grasslands by translocation of soil cores. *J. Plant Nutr. Soil Sci.*, **167**, 457-470.

Eide, A. and K. Heen, 2002: Economic impact of global warming. A study of the fishing industry in North Norway. *Fish Res.*, **56**, 261-274.

Eisenreich, S.J., Ed., 2005: *Climate change and the European Water Dimension*. Report to the European Water Directors. European Commission-Joint Research Centre, Ispra, 253 pp.

Eitzinger, J., M. Stastna, Z. Zalud and M. Dubrovsky, 2003: A simulation study of the effect of soil water balance and water stress in winter wheat production under different climate change scenarios. *Agric. Water Manage.*, **61**, 195-217.

Ekström, M., H.J. Fowler, C.G. Kilsby and P.D. Jones, 2005: New estimates of future changes in extreme rainfall across the UK using regional climate model integrations. 2. Future estimates and use in impact studies. *J. of Hydrol.*, **300**, 234-251.

Elpiner, L.I., 2004: Scenarios of human health changes under global hydroclimatic transformations. In: *Climate Change and Public Health in Russia in the XXI Century*. Proceeding of the workshop, April 5-6, 2004, Moscow, Publishing Company "Adamant", 195-199 (in Russian).

Elsasser, H. and R. Burki, 2002: Climate change as a threat to tourism in the Alps. *Clim. Res.*, **20**, 253-257.

Emberlin, J., M. Detandt, R. Gehrig, S. Jaeger, N. Nolard and A. Rantio-Lehtimaki, 2002: Responses in the start of *Betula* (birch) pollen seasons to recent changes in spring temperatures across Europe. *Int. J. of Biometeorol.*, **46**, 159-170.

Emmett, B.A., C. Beier, M. Estiarte, A. Tietema, H.L. Kristensen, D. Williams, J. Peñuelas, I. Schmidt and A. Sowerby, 2004: The response of soil processes to climate change: results from manipulation studies of shrublands across an environmental gradient. *Ecosystems*, **7**, 625-637.

Enarson, E. and M. Fordham, 2001: Lines that divide, ties that bind: race, class and gender in women's flood recovery in the US and UK. *Aust. J. Emerg. Manag.*, **15**, 43–52.

Estrela, T., M. Menéndez, M. Dimas, C. Marcuello, G. Rees, G. Cole, K. Weber, J. Grath, J. Leonard, N.B. Ovesen and J. Fehér, 2001: *Sustainable Water Use in Europe. Part 3: Extreme Hydrological Events: Floods and Droughts*. Environmental issue report No 21, European Environment Agency, Copenhagen, 84 pp.

Etchevers, P., C. Golaz, F. Habets and J. Noilhan, 2002: Impact of a climate change on the Rhone river catchment hydrology. *J. Geophys. Res.*, **107**, 4293, doi:10.1029/2001JD000490.

Evans, C.D. and D.T. Monteith, 2001: Chemical trends at lakes and streams in the UK Acid Waters Monitoring Network 1988-2000: evidence for recent recovery at a national scale. *Hydr. Earth Syst. Sci.,* **5**, 283-297.

Ewert, F., M.D.A. Rounsevell, I. Reginster, M.J. Metzger and R. Leemans, 2005: Future scenarios of European agricultural land use I. Estimating changes in crop productivity. *Agr. Ecosyst. Environ.,* **107**, 101-116.

Fang, C. and J.B. Moncrieff, 2001: The dependence of soil CO_2 efflux on temperature. *Soil Bio. Biochem.*, **33**, 155-165.

FAOSTAT, 2005. http://faostat.fao.org/faostat. Last accessed 01.01.2006.

Fernandes, P. and H. Botelho, 2004: Analysis of the prescribed burning practice in the pine forest of northwestern Portugal. *J. Environ. Manage.*, **70**, 15-26.

Fernando, T.M. and J. Cortina, 2004: Are *Pinus halepensis* plantations useful as a restoration tool in semiarid Mediterranean areas? *For. Ecol. Manage.*, **198**, 303-317.

Fink, A.H., T. Brücher, A. Krüger, G.C. Leckebusch, J.G. Pinto and U. Ulbrich, 2004: The 2003 European summer heatwaves and drought - Synoptic diagnosis and impact. *Weather*, **59**, 209-216.

Fischer, P.H., B. Brunekreef and E. Lebret, 2004: Air pollution related deaths during the 2003 heat wave in the Netherlands. *Atmos. Env.*, **38**, 1083-1085.

Flather, R.A. and J. Williams, 2000: Climate change effects on storm surges: methodologies and results. *Climate Scenarios for Water-Related and Coastal Impacts*, J. Beersma, M. Agnew, D. Viner and M. Hulme, Eds., ECLAT - 2 Workshop Report No. 3, Climate Research Unit, University of East Anglia, Norwich, 66-72.

Flörke, M. and J. Alcamo, 2005: *European Outlook On Water Use*. Prepared for the European Environment Agency, 83 pp.

Fowler, H.J. and C.G. Kilsby, 2004: Future increase in UK water resource drought projected by a regional climate model. *Proceedings of the BHS International Conference on Hydrology: Science & Practice for the 21st Century*. London, 12-16 July, British Hydrological Society, 15-21.

Frank, T., 2005: Climate change impacts on building heating and cooling energy demand in Switzerland. *Energy Build.*, **37**, 1175-1185.

Frauenfeld, O.W., T. Zhang, R.G. Barry and D. Gilichinsky, 2004: Interdecadal changes in seasonal freeze and thaw depths in Russia. *J. Geophys. Res.*, **109**, doi:10.1029/2003JD004245.

Freeman, M., A.S. Morén, M. Strömgren and S. Linder, 2005: Climate change impacts on forests in Europe: biological impact mechanisms. *Management of European Forest under Changing Climatic Conditions*, S. Kellomäki and S. Leinonen, Eds., Research Notes 163, University of Joensuu, Joensuu, 46-115.

Frich, P., L.V. Alexander, P. Della-Marta, B. Gleason, M. Haylock, A.M.G.K. Tankand T. Peterson, 2002: Observed coherent changes in climatic extremes during the second half of the twentieth century. *Clim. Res.*, **19**, 193-212.

Fronzek, S and T.R. Carter, 2007: Assessing uncertainties in climate change impacts on resource potential for Europe based on projections from RCMs and GCMs. *Climatic Change*, **81**, S357-S371.

Fukushima, T., M. Kureha, N. Ozaki, Y. Fujimori and H. Harasawa, 2002: Influences of air temperature change on leisure industries: case study of ski activities. *Mitigation and Adaptation Strategies for Global Change*, **7**, 173–189.

Garssen, J., C. Harmsen and J. de Beer, 2005: The effect of the summer 2003 heatwave on mortality in the Netherlands. *Euro Surveill.*, **10**, 165-168.

Gedney, N., P.M. Cox, R.A. Betts, O. Boucher, C. Huntingford and P. Stott, 2006: Detection of a direct carbon dioxide effect in continental river runoff records. *Nature*, **439,** 835-838.

Geraci, J. R., and V. Lounsbury, 2002: Marine mammal health: holding the balance in an ever changing sea. *Marine mammals: biology and conservation*, P.G.H. Evans and J.A. Raga, Eds., Kluwer Academic/Plenum Publishers, New York, 365-384.

Geres, D., 2004: Analysis of the water demand management. *Proceedings, XXIInd Conference of the Danubian Countries on the Hydrological Forecasting and Hydrological Bases of Water Management*, Brno, 30 August-2 September 2004, CD-edition.

Giannakopoulos, C. and B.E. Psiloglou, 2006: Trends in energy load demand for Athens, Greece: weather and non-weather related factors. *Clim. Res.*, **13**, 97-108.

Giannakopoulos, C., M. Bindi, M. Moriondo, P. LeSager and T. Tin, 2005: *Climate Change Impacts in the Mediterranean Resulting from a 2°C Global Temperature Rise*. WWF report, Gland Switzerland. Accessed 01.10.2006 at http://assets.panda.org/downloads/medreportfinal8july05.pdf.

Gielen, B., H. de Boeck, C.M.H.M. Lemmens, R. Valcke, I. Nijs and R. Ceulemans, 2005: Grassland species will not necessarily benefit from future elevated air temperatures: a chlorophyll fluorescence approach to study autumn physiology. *Physiol. Plant.*, **125**, 52–63.

Giorgi, F., X. Bi and J. Pal, 2004: Mean interannual and trends in a regional climate change experiment over Europe. II: Climate Change scenarios (2071-2100). *Climate Dyn.*, **23**, 839-858.

Gitay, H., A. Suarez and R. Watson, Eds., 2002: *Climate Change and Biodiversity*. IPCC Technical Paper V, 86 pp.

Goldammer, J.G., A. Shukhinin and I. Csiszar, 2005: The current fire situation in the Russian Federation: implications for enhancing international and regional co-operation in the UN framework and the global programs on fire monitoring and assessment. *Int. Forest Fire News*, **32**, 13-42.

Good, P., L. Barring, C. Giannakopoulos, T. Holt and J.P. Palutikof, 2006: Non-linear regional relationships between climate extremes and annual mean temperatures in model projections for 1961-2099 over Europe. *Clim. Res.*, **13**, 19-34.

Gordienko, M.I. and N.M. Gordienko, 2005: *Silvicultural Properties of Tree Plants*. "Visti" publ., Kiyiv, Ukraine, 818 pp. (in Ukrainian).

Gordon, C., C. Cooper, C.A. Senior, H. Banks, J.M. Gregory, T.C. Johns, J.F.B. Mitchell and R.A. Wood, 2000: The simulation of SST, sea ice extents and ocean heat transports in a version of the Hadley Centre coupled model without flux adjustments. *Climate Dyn.*, **16**, 147-168.

Gorissen, A., A. Tietema, N.N. Joosten, M. Estiarte, J. Peñuelas, A. Sowerby, B.A. Emmet and C. Beier, 2004: Climate change affects carbon allocation to the soil in shrublands. *Ecosystems*, **7**, 650-661.

Gottfried, M., H. Pauli, K. Reiter and G. Grabherr, 2002: Potential effects of climate change on alpine and nival plants in the Alps. *Mountain Biodiversity - a Global Assessmnt*, C. Korner and E. Spehn. Eds., *Parthenon* Publishing, London, 213-223.

Gould, E.A., S. Higgs, A. Buckley and T.S. Gritsun, 2006: Potential arbovirus emergence and implications for the United Kingdom. *Emerg. Infect. Dis.*, **12**, 549-555.

Grabherr, G., M. Gottfried and H. Pauli, 2001: Long term monitoring of mountain peaks in the Alps. *Tasks Veg. Sci.*, **35**, 153-177.

Grace, J., F. Berninger and L. Nagy, 2002: Impacts of climate change on the tree line. *Ann. Bot.*, 90, 537-544.

Gracia, C., L. Gil and G. Montero, 2005: Impactos sobre el Sector Forestal. *Evaluación Preliminar de los Impactos en España for Efecto del Cambio Climático*, J.M. Moreno, Ed., Ministerio de Medio Ambiente, Madrid, 399-435.

Gray, T. and J. Hatchard, 2003: The 2002 reform of the Common Fisheries governance - rhetoric or reality? *Marine Policy*, **27**, 545-554.

Gregory, J.M., J.E. Church, G.J. Boer, K.W. Dixon, G.M. Flato, D.R. Jackett, J.A. Lowe, S.P. O'Farrell, E. Roekner, G.L. Russell, R.J. Stouffer and M. Winton, 2001: Comparison of results from several AOGCMs for global and regional sea-level change 1900-2100. *Climate Dyn.*, **18**, 225-240.

Gregory, J.M., P. Huybrechts and S.C.B. Raper, 2004: Threatened loss of the Greenland ice-sheet. *Nature*, **428**, 616.

Gregory, J.M.., K.W. Dixon, R.J. Stouffer, A.J. Weaver, E. Driesschaert, M. Eby, T. Fichefet, H. Hasumi, A. Hu, J.H. Jungclaus, I.V. Kamenkovich, A. Levermann, M. Montoya, S. Murakami, S. Nawrath, A. Oka, A.P. Sokolov and R. B. Thorpe, 2005: A model intercomparison of changes in the Atlantic thermohaline circulation in response to increasing atmospheric CO_2 concentration. *Geophysical Research Letters*, **32**, L12703, doi:10.1029/2005GL023209.

Grime, J.P., V.K. Brown, K. Thompson, G.J. Masters, S.H. Hillier, I.P. Clarke, A.P. Askew, D. Corker and J.P. Kielty, 2000: The response of two contrasting limestone grasslands to simulated climate change. *Science*, **289**, 762-765.

Gruber, S., M. Hoelzle and W. Haeberli, 2004: Permafrost thaw and destabilization of Alpine rock walls in the hot summer of 2003. *Geophys. Res. Lett.*, **31**, L13504, doi:10.1029/2004GL020051.

Grübler, A., N. Nakicenovic, J. Alcamo, G. Davis, J. Fenhann, B. Hare, S. Mori, B. Pepper, H. Pitcher, K. Riahi, H.-H. Rogner, E. Lebre La Rovere, A. Sankovski, M. Schlesinger, R.P. Shukla, R. Swart, N. Victor and T. Yong Jung, 2004: Emission scenarios: A final response. *Ener. Environ.*, **15**, 11-24.

Guedes Soares, C.G., J.C. Carretero Albiach, R. Weisse and E. Alvarez-Funjul, 2002: A 40 years hindcast of wind, sea level and waves in European waters. *Proceedings of the 21st International Conference on Offshore Mechanics and Arctic Engineering*, June 2002, Oslo, Norway. OMAE2002-28604, 1-7.

Guisan, A. and J.-P. Theurillat, 2000: Equilibrium modeling of alpine plant distribution and climate change: how far can we go? *Phytocoenologia*, **30**, 353-384.

Guisan, A. and J.-P. Theurillat, 2001: Assessing alpine plant vulnerability to climate change, a modeling perspective. *Int. Ass.*, **1**, 307-320.

Guisan, A. and J.-P. Theurillat, 2005: Appropriate monitoring networks are required for testing model-based scenarios of climate change impact on mountain plant distribution. *Global change in mountain regions*, U. M. Huber, H. Bugmann and M. A. Reasoner, Eds., Kluwer, 467-476.

Gutiérrez Teira, B., 2003: *Variaciones de las Comunidades y Poblaciones de Macroinvertebrados del Tramo Alto del Río Manzanares a Causa de la Temperatura. Posibles Efectos del Cambio Climático*. Tesis Doctoral. Escuela Técnica Superior de Ingenieros de Montes. Universidad Politécnica de Madrid. Madrid.

Haberl, H., F. Krausmann, K.-H. Erb, N.B. Schulz, S. Rojstaczer, S.M. Sterling and

N. Moore, 2002: Human appropriation of net primary production. *Science*, **296**, 1968-1969.

Haeberli, W. and C. R. Burn, 2002: Natural hazards in forests: glacier and permafrost effects as related to climate change. *Environmental Change and Geomorphic Hazards in Forest*, R.C. Slide, Ed., IUFRO Research Series 9, CABI Publishing, Wallingford/New York, 167-202.

Hagg, W. and L. Braun, 2004: The influence of glacier retreat on water yield from high mountain areas: comparison of Alps and Central Asia. *Climate and Hydrology in Mountain Areas*, C. de Jong, R. Ranzi and D. Collins, Eds.,Wiley and Sons, New York, 338 pp.

Hajat, S., R.S. Kovats, R.W. Atkinson and A. Haines, 2002: Impact of hot temperatures on death in London: a time series approach. *J. Epidemiol. Community Health*, **56**, 367–372.

Hajat, S., K.L. Ebi, S. Kovats, B. Menne, S. Edwards and A. Haines, 2003: The human health consequences of flooding in Europe and the implications for public health: a review of the evidence. *Appl. Environ. Sci. Public Health*, **1**, 13-21.

Hall, J.W., R.J. Dawson, W.J.A. Walkden, R.J. Nicholls, I. Brown and A. Watkinson, 2007: Broad-scale analysis of morphological and climate impacts on coastal flood risk. *Coastal Dyn.*,(in press).

Hamilton, J.M., D.J. Maddison and R.S.J. Tol, 2005: Climate change and international tourism: a simulation study. *Glob. Environ. Change*, **15**, 253-266.

Hamilton, L., P. Lyster and O. Otterstad, 2000: Social change, ecology and climate in 20th century Greenland. *Climatic Change*, **47**, 193-211.

Hannah, L., G.F. Midgley, S. Andelman, M.B. Araújo, G. Hughes, F. Martinez-Meyer, R.G. Pearson and P.H. Williams, 2007: Protected area increase required by climate change. *Front. Ecol. Environ.*, **5**, 131-138.

Hänninen, H., 2006: Climate warming and the risk of frost damage to boreal forest trees: identification of critical ecophysiological traits. *Tree Physiol.*, **26**, 889-898.

Hansom, J.D., G. Lees, D.J. McGlashan and S. John, 2004: Shoreline management plans and coastal cells in Scotland. *Coast. Manage.*, **32**, 227-242.

Hanson, C.E, J.P. Palutikof, A. Dlugolecki and C. Giannakopoulos, 2006: Bridging the gap between science and the stakeholder: the case of climate change research. *Clim. Res.*, **13**, 121-133.

Hantel, M., M. Ehrendorfer and A. Haslinger, 2000: Climate sensitivity of snow cover duration in Austria. *Int. J. Climatol.*, **20**, 615-640.

Harrison, P.A., P.M. Berry, N. Butt and M. New, 2006: Modelling climate change impacts on species distributions at the European scale: implications for conservation policy. *Environ. Sci. Pol.*, **9**, 116-128.

Harwood, J., 2001: Marine mammals and their environment in the twenty-first century. *J. Mammal.*, **82**, 630-640.

Hassi, J., 2005: Cold extremes and impacts on health. *Extreme Weather Events and Public Health Responses*, W. Kirch,, B. Menne and R. Bertollini, Eds., Springer-Verlag, Berlin, Heidelberg, 59-67.

Hassi, J. and M. Rytkönen, 2005: Climate warming and health adaptation in Finland. FINADAPT Working Paper 7, Finnish Environment Institute Mimeographs 337, Helsinki, 28 pp. Accessed 1.10.2006: http://www.environment.fi/default.asp?contentid=165486&lan=en.

Haylock, M.R., G.C. Cawley, C. Harphan, R.L. Wilby and C. Goodess, 2006: Downscaling heavy precipitation over the United Kingdom: a comparison of dynamical and statistical methods and their future scenarios. *Int. J. Climatol.*, **26**, 1397-1415.

Heath, M.R., 2005: Regional variability in the trophic requirements of shelf sea fisheries in the Northeast Atlantic, 1973-2000. *ICES J. Mar. Sci.*, **62**, 1233-1244.

Hedger, M., 2005: Learning from experience: Evolving responses to flooding events in the United Kingdom. *Extreme weather events and public health responses*, W. Kirch, B. Menne and R. Bertollini, Eds., Springer-Verlag, Berlin, Heidelberg, 225-234.

Helms, M., B. Büchele., U. Merkel and J. Ihringer, 2002: Statistical analysis of the flood situation and assessment of the impact of diking measures along the Elbe (Labe) river. *J. Hydrol.*, **267**, 94-114.

Hennessy, E., 2002: Air pollution and short term mortality. *BMJ*, **324**, 691–692.

Hickling, R., D.B. Roy, J.K. Hill and C.D. Thomas, 2005: A northward shift of range margins in British Odonata. *Glob. Change Biol.*, **11**, 502-506.

Hickling, R., D.B. Roy, J.K. Hill, R. Fox and C.D. Thomas, 2006: The distributions of a wide range of taxonomic groups are expanding polewards. *Glob. Change Biol.*, **12**, 450-455.

Highways Agency, 2005a: Business Plan 2005/6.

Highways Agency, 2005b: *Well Maintained Highways – Code of Practice for Highway Maintenance. TSO*. Retrieved 10.10.2006 from http://www.ukroadsliaisongroup.org/.

Hildén, M., H. Lehtonen, I. Bärlund, K. Hakala, T. Kaukoranta and S. Tattari, 2005: *The Practice and Process of Adaptation in Finnish Agriculture*. FINADAPT Working Paper 5, Finnish Environment Institute Mimeographs 335, Helsinki, 28 pp.

Hiscock, K., A. Southward, I. Tittley and S. Hawkins, 2004: Effects of changing temperature on benthic marine life in Britain and Ireland. *Aquatic Conserv: Mar. Freshw. Ecosyst.*, **14**, 333–362.

Hitz, S. and J. Smith, 2004: Estimating global impacts from climate change. *Glob. Environ. Change-Human Policy Dimens. Part A*, **14**, 201-218.

Hock, R., P. Jansson and L. Braun, 2005: Modelling the response of mountain glacier discharge to climate warming. *Global Change Series,* U.M. Huber, M.A. Reasoner and H. Bugmann, Eds., Springer, Dordrecht, 243-252.

Hoelzle, M., W. Haeberli, M. Dischl and W. Peschke, 2003: Secular glacier mass balances derived from cumulative glacier length changes. *Glob. Planet. Change*, **36**, 295-306.

Holden, N.M. and A.J. Brereton, 2002: An assessment of the potential impact of climate change on grass yield in Ireland over the next 100 years. *Irish J. Agr. Food Res.*, **41**, 213-226.

Holden, N.M. and A.J. Brereton, 2003: Potential impacts of climate change on maize production and the introduction of soybean in Ireland. *Irish J. Agr. Food Res*, **42**, 1-15.

Holden, N.M., A.J. Brereton, R. Fealy and J. Sweeney, 2003: Possible change in Irish climate and its impact on barley and potato yields. *Agric. For. Meteorol.*, **116**, 181-196.

Holman, I.P., M.D.A. Rounsevell, S. Shackley, P.A. Harrison, R.J. Nicholls, P.M Berry and E. Audsley, 2005: A regional, multi-sectoral and integrated assessment of the impacts of climate and socio-economic change in the UK: Part I Methodology. *Climatic Change*, **70**, 9-41.

Holtsmark, B.J. and K.H. Alfsen, 2005: PPP correction of the IPCC emission scenarios - Does it matter? *Climatic Change*, **68**, 11-19.

Hooijer, M., F. Klijn, G.B.M. Pedroli and A.G. van Os, 2004: Towards sustainable flood risk management in the Rhine and Meuse river basins: synopsis of the findings of IRMA-SPONGE. *River Res. Appl.*, **20**, 343-357.

Howe, A.D., S. Forster, S. Morton, R. Marshall, K.S. Osborn, P. Wright and P.R. Hunter, 2002: *Cryptosporidium* oocysts in a water supply associated with a Cryptosporidiosis outbreak. *Emerg. Infect. Dis.*, **8**, 619–624.

Hughes, L., 2000: Biological consequences of global warming: is the signal already apparent? *Trends Ecol. Evol.*, **15**, 56-61.

Hulme, M., G. Jenkins, X. Lu, J.R. Turnpenny, T.D. Mitchell, R.G. Jones, J. Lowe, J.M. Murphy, D. Hassell, P. Boorman, R. McDonald and S. Hill, 2002: *Climate Change Scenarios for the United Kingdom: The UKCIP02 Scientific Report*. Tyndall Centre for Climate Change Research, University of East Anglia, Norwich,120 pp.

Hunter, P.R., 2003: Climate change and waterborne and vector-borne disease. *J. Appl. Microbiol.*, **94**, 37–46.

Hurrell, J.W., Y. Kushnir, G. Ottersen and M. Visbeck, Eds., 2003: *The North Atlantic Oscillation: Climate Significance and Environmental Impact* Geophysical Monograph Series 134, 279 pp.

Hurrell, J.W., M.P. Hoerling, A.S. Phillips and T. Xu, 2004: Twentieth century North Atlantic climate change. Part I: Assessing determinism. *Climate Dyn.*, **23**, 371-389.

Huynen, M.M., P. Martens, D. Schram, M.P. Weijenberg and A.E. Kunst, 2001: The impact of heatwaves and cold spells on mortality rates in the Dutch population. *Environ. Health Perspect.*, **109**, 463-470.

Huynen, M. and B.Menne, Coord., 2003: *Phenology and Human Health: Allergic Disorders*. Report of a WHO meeting, Rome, Italy, 16-17 Jannuary, 2003, Health and Global Environmental Change Series n° 1, WHO, Europe, 55 pp.

Hyvönen, R., G.I. Ågren and P. Dalias, 2005: Analysing temperature response of decomposition of organic matter. *Glob. Change Biol.*, **11**, 770-778.

Iglesias, A., T. Estrela and F. Gallart, 2005: Impactos sobre los recursos hídricos. *Evaluación Preliminar de los Impactos en España for Efecto del Cambio Climático*, J. M. Moreno, J.M, Ed., Ministerio de Medio Ambiente, Madrid, Spain, 303-353.

IMF, 2004: *World Economic Outlook April2004 Advancing Structural Reforms*. World economic outlook (International Monetary Fund), Washington, District of Columbia. Retrieved 02.08.2005 from http://www.imf.org.

Imhoff, M.L., L. Bounoua, T. Ricketts, C. Loucks, R. Harris and W.T. Lawrence,

2004: Global patterns in human consumption of net primary production. *Nature*, **429**, 870-873.

Inter-Agency Commission of the Russian Federation on Climate Change, 2002: *The Third National Communication of the Russian Federation*. Moscow, 82-83.

IPCC, 1992: *Climate Change 1992: The Supplementary Report to the IPCC Scientific Assessment*. J.T. Houghton, B.A. Callander and S.K. Varney, Eds., Cambridge University Press, Cambridge, 200 pp.

IPCC, 2001: *Climate Change 2001: Impacts, Adaptation, and Vulnerability. Contribution of Working Group II to the Third Assessment Report of the Intergovernmental Panel on Climate Change*, J.J. McCarthy, O.F. Canziani, N.A. Leary, D.J. Dokken and K.S. White, Eds., Cambridge University Press, Cambridge, 1032 pp.

Ivanov, B. and T. Maximov, Eds., 2003: *Influence of Climate and Ecological Changes on Permafrost Ecosystems*. Yakutsk Scientific Center Publishing House, Yakutsk, 640 pp.

Izmerov, N.F., B.A. Revich and E.I. Korenberg, Eds., 2004: *Climate Change and Public Health in Russia in the XXI century*. Proceeding of a workshop, Apr. 2004, Moscow, Adamant, Moscow, Russia, 260 pp.

Jacob, D., H. Goettel, J. Jungclaus, M. Muskulus, R. Podzun and J. Marotzke, 2005: Slowdown of the thermohaline circulation causes enhanced maritime climate influence and snow cover over Europe. *Geophys. Res. Lett.*, **32**, L21711, doi:10.1029/2005GL023286.

Janssen, F., J.P. Mackenbach and A.E. Kunst, 2004: Trends in old-age mortality in seven European countries, 1950-1999. *J. Clin. Epidemiol.*, **57**, 203-216.

Janssens, I.A., A. Freibauer P. Ciais, P. Smith, G.-J. Nabuurs, G. Folberth, B. Schlamadinger, R.W.A. Hutjes, R. Ceulemans, E.-D. Schulze, R. Valentini and A.J. Dolman, 2003: Europe's terrestrial biosphere absorbs 7 to 12% of European anthropogenic CO_2 emissions. *Science*, **300**, 1538-1542.

Jarvis, P. and S. Linder, 2000: Constraints to growth in boreal forests. *Nature*, **405**, 904-905.

Jeppesen, E., J.P. Jensen and M. Søndergaard, 2003: Climatic warming and regime shifts in lake food webs - some comments. *Limnol. Oceanogr.*, **48**, 1346-1349.

Johansson, M.M., K.K. Kahma and H. Bowman, 2004: Scenarios for sea level on the Finnish coast. *Boreal Environ. Res.* **9**, 153-166.

Johnson, H., R.S. Kovats, G. McGregor, J. Stedman, M. Gibbs and H. Walton, 2005: The impact of the 2003 heatwave on daily mortality in England and Wales and the use of rapid weekly mortality estimates. *Euro Surveill.*, **10**, 168-171.

Jones, G.V. and R. Davis, 2000: Climate influences on Grapewine phenology, grape composition, and wine production and quality for Bordeaux, France. *Am. J. Enol. Vitic.*, **51**, 249-261.

Jones, P.D. and A. Moberg, 2003: Hemispheric and large scale surface air temperature variations: an extensive revision and an update to 2001. *J. Clim.*, **16**, 206-223.

Jones, P.D., D.H. Lister, K.W. Jaggard and J.D. Pidgeon, 2003: Future climate impact on the productivity of sugar beet (*Beta vulgaris* L.) in Europe. *Climatic Change*, **58**, 93-108.

Jönsson, A.M., M.J. Linderson, I. Stjernquist, P. Schlyter and L. Bärring, 2004: Climate change and the effect of temperature backlashes causing forest damage in *Picea abies*. *Glob. Planet. Change*, **44**, 195-208.

Kabat, P., R.E. Schulze, M.E. Hellmuth and J.A. Veraart, Eds., 2002: *Coping with Impacts of Climate Variability and Climate Change in Water Management: a Scooping Paper.* DWC Report no. DWCSSO-01(2002), International Secretariat of the Dialogue on Water and Climate, Wageningen, Netherlands.

Kaipainen, T., J. Liski, A. Pussinen and T. Karjalainen, 2004: Managing carbon sinks by changing rotation length in European forests. *Environ. Sci. Pol.*, **3**, 205-219.

Kammann, C., L. Grünhage, U. Grüters, S. Janze and H.-J. Jäger, 2005: Response of aboveground grassland biomass to moderate long-term CO_2 enrichment. *Basic Appl. Ecol.*, **6**, 351-365.

Kankaanpää, S. and T.R. Carter, 2004: *An Overview of Forest Policies Affecting Land Use in Europe*. The Finnish Environment no. 706. Finnish Environment Institute, Helsinki. 57 pp.

Kashyap, A., 2004: Water governance: learning by developing adaptive capacity to incorporate climate variability and change. *Water Sci. Technol*, **7**, 141-146.

Kaste, Ø., K. Rankinen and A. Leipistö, 2004: Modelling impacts of climate and deposition changes on nitrogen fluxes in northern catchments of Norway and Finland. *Hydrol. Earth Syst. Sci.*, **8**, 778-792.

Keatinge, W.R., 2002: Winter mortality and its causes. *Int. J Circumpolar Health*, **61**, 292-299.

Keatinge, W.R., 2003: Death in heatwaves. *BMJ*, **327**, 512-513.

Keatinge, W.R. and G.C. Donaldson, 2004: The impact of global warming on health and mortality. *South. Med. J.*, **97**, 1093-1099.

Keatinge, W.R., G.C. Donaldson, E. Cordioli, M. Martinelli, A.E. Kunst, J.P. Mackenbach, S. Nayha and I. Vuori, 2000: Heat related mortality in warm and cold regions of Europe: observational study. *BMJ*, **321**, 670 –673.

Keay, K. and I. Simmonds, 2006: Road accidents and rainfall in a large Australian city. *Accident Anal. Prev.*, **38**, 445-454.

Keller, K., B.M. Bolker and D.F. Bradford, 2004: Uncertain climate thresholds and optimal economic growth. *J. Environ. Econ. Manage.*, **48**, 723-741.

Kellomäki, S. and S. Leinonen, Eds., 2005: *Management of European Forests under Changing Climatic Conditions*. Final Report of the Project *Silvistrat*. University of Joensuu, Research Notes 163, Joensuu, Finland, 427 pp.

Kellomäki, S., H. Strandman, T. Nuutinen, H. Petola, K.T. Kothonen and H. Väisänen, 2005: *Adaptation of Forest Ecosystems, Forest and Forestry to Climate Change*. FINDAT Working Paper 4, Finnish Environment Institute Mimeographs 334, Helsinki, 44 pp.

Kirch, W., B. Menne and R, Bertollini, Eds., 2005: *Extreme Weather Events and Public Health Responses*. Springer Verlag, Heidelberg, 303 pp.

Kirkinen, J., A. Matrikainen, H. Holttinen, I. Savolainen, O. Auvinen and S. Syri, 2005: *Impacts on the Energy Sector and Adaptation of the Electricity Network under a Changing Climate in Finland. FINADAPT*, working paper 10, Finnish Environment Institute.

Kislitsin, V., S. Novikov and N. Skvortsova, 2005: Moscow smog of summer 2002. Evaluation of adverse health effects. *Extreme Weather Events and Public Health Responses,* W. Kirch, B. Menne and R. Bertollini, Eds.,Springer-Verlag, Berlin, Heidelberg, 255-262.

Kistemann, T., T. Classen, C. Koch, F. Dangendorf, R. Fischeder, J. Gebel, V. Vacata and M. Exner, 2002: Microbial load of drinking water reservoir tributaries during extreme rainfall and runoff. *Appl. Environ. Microbiol.*, **68**, 2188-2197.

Kjellström, E., 2004: Recent and future signatures of climate change in Europe. *Ambio*, **23**, 193-198.

Kjellström, E., L. Bärring, D. Jacob, R. Jones and G. Lenderink, 2007: Modelling daily temperature extremes: recent climate and future changes over Europe. *Climatic Change*, **81**, S249-S265.

Klanderud, K. and H.J.B. Birks 2003: Recent increases in species richness and shifts in altitudinal distributions of Norwegian mountain plants. *The Holocene*, **13**, 1-6.

Klawa, M. and U. Ulbrich, 2003: A model for the estimation of storm losses and the identification of severe winter storms in Germany. *Nat. Haz. Earth Syst. Sci.*, **3**, 725-732.

Klein Tank, A.M.G., 2004. *Changing Temperature and Precipitation Extremes in Europe's Climate of the 20th Century*. PhD dissertation, University of Utrecht, Utrecht, 124 pp.

Klein Tank, A.M.G. and G.P. Können, 2003. Trends in indices of daily temperature and precipitation extremes in Europe. *J. Clim.*, **16**, 3665-3680.

Klein Tank, A.M.G., J.B. Wijngaard, G.P. Konnen, R. Bohm, G. Demaree, A. Gocheva, M. Mileta, S. Pashiardis, L. Hejkrlik, C. Kern-Hansen, R. Heino, P. Bessemoulin, G. Muller-Westermeier, M. Tzanakou, S. Szalai, T. Palsdottir, D. Fitzgerald, S. Rubin, M. Capaldo, M. Maugeri, A. Leitass, A. Bukantis, R. Aberfeld, A.F.V. VanEngelen, E. Forland, M. Mietus, F. Coelho, C. Mares, V. Razuvaev, E. Nieplova, T. Cegnar, J.A. López, B. Dahlstrom, A. Moberg, W. Kirchhofer, A. Ceylan, O. Pachaliuk, L.V. Alexander and P. Petrovic, 2002: Daily dataset of 20th-century surface air temperature and precipitation series for the European Climate Assessment. *Int. J. Climatol.*, **22**, 1441-1453.

Kljuev, N.N., 2001: *Russia and its Regions*. Nauka, Moscow, 214 pp. (in Russian).

Klyashtorin, L., 2001: *Climate Change and Long-Term Fluctuations of Commercial Catches: the Possibility of Forecasting*. Rome, FAO, 86 pp.

Knight, C.G., I. Raev, and M. P. Staneva, Eds., 2004: *Drought in Bulgaria: A Contemporary Analog of Climate Change*. Ashgate, Aldershot, Hampshire 336 pp.

Knippertz, P., U. Ulbrich and P. Speth, 2000: Changing cyclones and surface wind speeds over the North Atlantic and Europe in a transient GHG experiment. *Clim. Res.*, **15**, 109-122.

Koca, D., S. Smith and M.T. Sykes, 2006: Modelling regional climate change effects on potential natural ecosystems in Sweden. *Climatic Change*, **78**, 381-406.

Kont, A., J. Jaagus, R. Aunapb, U. Ratasa and R. Rivisa, 2007: Implications of sea-level rise for Estonia. *J. Coast. Res.*, (in press).

Koppe, C., S. Kovats, G. Jendritzky and B. Menne, 2004: Heat-waves: risks and responses. *Health and Global Environmental Change*, SERIES, No. 2, WHO Regional Office for Europe, 123 pp.

Korenberg, E.I., 2004: Ecological preconditions of the possible influence of climate change on natural habitats and their epidemic manifestation. *Climate Change and Public Health in Russia in the XXI century,* N.F. Izmerov, B.A. Revich, and E.I. Korenberg, Eds., Proceeding of the workshop, April 5-6, 2004, Moscow Publishing Company Adamant, Moscow, 54-67 (in Russian).

Körner, C., 2003: *Alpine Plant Life*. Springer Verlag, Heidelberg, FRG. 388 pp.

Körner, C., D. Sarris and D. Christodoulakis, 2005: Long-term increase in climatic dryness in Eastern-Mediterranean as evidenced for the island of Samos. *Regional Environmental Change*, **5**, 27-36.

Kosatsky, T., 2005: The 2003 European heatwave. *Euro Surveill.*, **10**, 148-149.

Kosatsky, T. and B. Menne, 2005: Preparedness for extreme weather among national ministries of health of WHO's European region. *Climate Change and Adaptation Strategies for Human Health,* B. Menne and K.L. Ebi, Eds., Springer, Darmstadt, 297-329.

Kostiainen, K., H. Lalkanen, S. Kaakinen, P. Saranpää and E. Vapaavuori, 2006: Wood properties of two silver birch clones exposed to elevated CO_2 and O_3. *Glob. Change Biol.*, **12**, 1230-1240.

Kovats, R.S., D.H. Campbell-Lendrum, A.J. McMichael, A. Woodward and J.S. Cox, 2001: Early effects of climate change: do they include changes in vector-borne disease? *Philos. Trans. Roy. Soc. London* , **356**, 1057-1068.

Kovats, R.S., S.J. Edwards, S. Hajat, B.G. Armstrong, K.L. Ebi, B. Menne and the collaborating group, 2004: The effect of temperature on food poisoning: a time-series analysis of salmonellosis in ten European countries. *Epidemiol. Infect.*, **132**, 443–453.

Kovats, R.S. and G. Jendritzky, 2006: Heat-waves and human health. *Climate Change and Adaptation Strategies for Human Health,* B. Menne and K.L. Ebi, Eds., Springer, Darmstadt, Germany. 63-90.

Kovats, R.S. and C. Tirado, 2006: Climate, weather and enteric disease. *Climate change and adaptation strategies for human health,* B. Menne and K.L. Ebi, Eds., Springer, Darmstadt, 269-295.

Krüger, A., U. Ulbrich and P. Speth, 2002: Groundwater recharge in Northrhine-Westfalia by a statistical model for greenhouse gas scenarios. *Phys. Chem. Earth, Part B: Hydrology, Oceans and Atmosphere,* **26**, 853-861.

KSLA, 2004: Climate change and forestry in Sweden – a literature review. *Royal Swedish Academy of Agriculture and Forestry,* **143**, 40.

Kuhn, K.G., D.H. Campbell-Lendrum and C.R. Davies, 2004: Tropical Diseases in Europe? How we can learn from the past to predict the future. *EpiNorth,* **5**, 6-13.

Kull, O., I. Tulva and E. Vapaavuorvi, 2005: Consequences of elevated CO_2 and O_3 on birch canopy structure: implementation of a canopy growth model. *For. Ecol. Manage.,* **212**, 1-13.

Kullman, L., 2001: 20th century climate warming and tree-limit rise in the southern Scandes of Sweden. *Ambio*, **30**, 72-80.

Kullman, L., 2002: Rapid recent range-margin rise of tree and shrub species in the Swedish Scandes. *J. Ecol.*, **90**, 68-77.

Kundzewicz, Z.W., M. Parry, W. Cramer, J.I. Holten, Z. Kaczmarek, P. Martens, R.J. Nicholls, M. Oquist, M.D.A Rounsevell and J. Szolgay, 2001: Europe. *Climate Change 2001: Impacts, Adaptation, and Vulnerability. Contribution of Working Group II to the Third Assessment Report of the Intergovernmental Panel on Climate Change,* J.J. McCarthy, O.F. Canziani N.A. Leary, D.J. Dokken and K.S. White, Eds.,Cambridge University Press, Cambridge, 641-692.

Lake, P.S., 2000: Disturbance, patchiness, and diversity in streams. *J. North. Benthol. Soc.*, **19**, 573-592.

Lal, M., H. Harasawa, D. Murdiyarso and Co-authors, 2001: Asia. *Climate Change 2001: Impacts, Adaptation and Vulnerability*. J.J. McCarthy, O.F. Canziani, N.A. Leary, D.J. Dokken and K.S. White, Eds. Contribution of Working Group II to the Third Assessment Report of the Intergovernmental Panel on Climate Change., Cambridge University Press, Cambridge, 533-590.

Lallana, C., W. Kriner, R. Estrela, S. Nixon, J. Leonard and J.J. Berland, 2001: *Sustainable Water Use in Europe, Part 2: Demand Management*. Environmental Issues Report No. 19. European Environmental Agency, Copenhagen, 94 pp.

Langner, J., R. Bergstom and V. Foltescu, 2005: Impact of climate change on surface ozone and deposition of sulphur and nitrogen in Europe. *Atmos. Environ.*, **39**, 1129-1141.

Lapenis, A., A. Shvidenko, D. Shepashenko, S. Nilsson and A. Aiyyer, 2005: Acclimation of Russian forests to recent changes in climate. *Glob. Change Biol.*, **11**, 2090-2102.

Lapin M. and K. Hlavcova, 2003: *Changes in Summer Type of Flash Floods in the Slovak Carpathians due to Changing Climate*. Proceedings of the International Conference on Alpine Meteorology and MPA2003 meeting, Brig, Switzerland, 66, 105-108, Meteoswiss.

Lasch, P., M. Linder, M. Erhard, F. Suckow and A. Wenzel, 2002: Regional impact assessment on forest structure and functions under climate change-the Brandenburg case study. *For. Ecol. Manage.*, **162**, 73-86.

Laternser, M. and M. Schneebeli, 2003: Long-term snow climate trends of the Swiss Alps (1931-99). *Int. J. Climatol.*, **23**, 733-750.

Lavoie, M., D. Paré and Y. Bergeron, 2005: Impact of global change and forest management on carbon sequestration in northern forested peatlands. *Environ. Rev.*, **13**, 199-240.

Learmonth, J.A., C.D. MacLeod, M.B. Santos, G.J. Pierce, H.Q.P. Crick and R.A. Robinson, 2006: Potential effects of climate change on marine mammals. *Oceanogr. Mar. Biol.: An Annual Review*, **44**, 431-464.

Leckebusch, G.C. and U. Ulbrich, 2004: On the relationship between cyclones and extreme windstorm events over Europe under climate change. *Global Planet. Change*, **44**, 181-193.

Lehner, B., T. Heinrichs, P. Döll and J. Alcamo, 2001: *EuroWasser – Model-Based Assessment of European Water Resources and Hydrology in the Face of Global Change*. Kassel World Water Series 5, Center for Environmental Systems Research, University of Kassel, Kassel, Germany.

Lehner, B., G. Czisch and S. Vassolo, 2005: The impact of global change on the hydropower potential of Europe: a model-based analysis. *Energy Policy*, **33**, 839-855.

Lehner, B., P. Döll, J. Alcamo, H. Henrichs and F. Kaspar, 2006: Estimating the impact of global change on flood and drought risks in Europe: a continental, integrated analysis. *Climatic Change*, **75**, 273-299.

Lenderink, G., A. Van Ulden, B. van den Hurk and E. van Meijgaard, 2007: Summertime inter-annual temperature variability in an ensemble of regional model simulations: analysis of the surface energy budget. *Climatic Change*, **81**, S233-S247.

Levermann, A., A. Griesel, M. Hofman, M. Montoya and S. Rahmstorf, 2005: Dynamic sea level changes in thermohaline circulation. *Climate Dyn.*, **24**, 347-354.

Lexer, M.J., K. Honninger, H. Scheifinger, C. Matulla, N. Groll, H. Kromp-Kolb, K. Schadauer, F. Starlinger and M. Englisch, 2002: The sensitivity of Austrian forests to scenarios of climatic change: a large-scale risk assessment based on a modified gap model and forest inventory data. *For. Ecol. Manage.*, **162**, 53-72.

Linder, M., 2000: Developing adaptive forest management strategies to cope with climate change. *Tree Physiol.*, **20**, 299-307.

Lindgren, E. and R. Gustafson, 2001: Tick-borne encephalitis in Sweden and climate change. *Lancet,* **358**, 16–18.

Lindgren, E. and T.G.T. Jaenson, 2006: Lyme borreliosis in Europe: Influences of climate and climate change, epidemiology, ecology and adaptation measures. *Climate Change and Adaptation Strategies for Human Health,* B. Menne and K.L. Ebi, Eds.,Steinkopff Verlag, Darmstadt, 157-188.

Lindgren, E. and T. Naucke, 2006: Leishmaniasis: influences of climate and climate change, epidemiology, ecology and adaptation measures. *Climate Change and Adaptation Strategies for Human Health,* B. Menne and K.L. Ebi, Eds.,Steinkopff Verlag, Darmstadt, 131-156.

Link, P.M. and R.S.J. Tol, 2004: Possible economic impacts of a shutdown of the thermohaline circulation: an application of FUND. *Port. Econ. J.*, **3**, 99-114.

Linkosalo, T., T.R. Carter, R. Häkkinen and P. Hari, 2000: Predicting spring phenology and frost damage risk of *Betula* spp. under climatic warming: a comparison of two models. *Tree Physiol.*, **20**, 1175-1182.

Lionello, P., A. Nizzero and E. Elvini, 2002: A procedure for estimating wind waves and storm-surge climate scenarios in a regional basin: the Adriatic Sea case. *Clim. Res.*, **23**, 217-231.

Livermore, M.T.J., 2005: *The Potential Impacts of Climate Change in Europe: The Role of Extreme Temperatures*. Ph.D. thesis, University of East Anglia, UK.

Livingstone, D.M., A.F. Lotter and H. Kettle, 2005: Altitude-dependent difference in the primary physical response of Mountain lakes to climatic forcing. *Limnol. Oceanogr.*, **50**, 1313-1325.

Llasat, M.C., 2001: An objective classification of rainfall intensity in the Northeast of Spain. *Int. J. Climatol.*, **21**, 1385-1400.

Llorens, L. and J. Peñuelas, 2005: Experimental evidence of future drier and warmer conditions affecting flowering of two co-occurring Mediterranean shrubs. *Int. J. Plant. Sci.*, **166**, 235-245.

Llorens, L., J. Peñuelas, M. Estiarte and P. Bruna, 2004: Contrasting growth changes in two dominant species of a Mediterranean shrubland submitted to experimental drought and warming. *Ann. Bot.*, **94**, 843-853.

Lloret, F., J. Peñuelas and M. Estiarte, 2004: Experimental evidence of reduced diversity of seedlings due to climate modification in a Mediterranean-type community. *Glob. Change Biol.*, **10**, 248-258.

Lloret, F., J. Peñuelas and M. Estiarte, 2005: Effects of vegetation canopy and climate on seedling establishment in Mediterranean shrubland. *J. Veg. Sci.*, **16**, 67-76.

London Climate Change Partnership, 2002: *London's Warming: The impacts of climate change in London*. Technical Report. Retrieved 10.10.2006 from http://www.london.gov.uk/gla/publications/environment.jsp.

López Zafra, J.M., L. Sánchez de Tembleque and V. Meneu, 2005: Impactos sobre el sector energético. *Evaluación Preliminar de los Impactos en España for Efecto del Cambio Climático*, J.M. Moreno, Ed., Ministerio de Medio Ambiente, Madrid, 617-652.

Lowe, J.A. and J.M. Gregory, 2005: The effects of climate change on storm surges around the United Kingdom. *Philos. Trans. Roy. Soc. London*, **363**, 1313-1328.

Lozano, I., R.J.N. Devoy, W. May and U. Andersen, 2004: Storminess and vulnerability along the Atlantic coastlines of Europe: analysis of storm records and of a greenhouse gases induced climate scenario. *Mar. Geol.*, **210**, 205-225.

Ludwig, W., P. Serrat, L. Cesmat and J. Dracia-Esteves, 2003: Evaluating the impact of the recent temperature increase on the hydrology of the Têt River (Southern France). *J. Hydrol.*, **289**, 204-221.

Luoto, M., R.K. Heikkinen and T.R. Carter, 2004: Loss of palsa mires in Europe and biological consequences. *Environ. Conserv.*, **31**, 30-37.

Lüscher, A., M. Daepp, H. Blum, U.E. Hartwig and J. Nösberger, 2004: Fertile temperate grassland under elevated atmospheric CO_2 - role of feed-back mechanisms and availability of growth resources. *Eur. J. Agron.*, **21**, 379-398.

Maddison, D., 2001: In search of warmer climates? The impact of climate change on flows of British tourists. *Climatic Change*, **49**, 193-208.

Maestre, F.T. and J. Cortina, 2004: *Do Positive Interactions Increase with Abiotic Stress? - a Test from a Semi-Arid Steppe*. Proceedings of the Royal Society of London, Series B, **271**, S331-S333

Magnani, F., L. Consiglio, M. Erhard, A. Nole, F. Ripullone and M. Borghetti, 2004: The sensitivity of Austrian forests to scenarios of climatic change: a large-scale risk assessment based on a modified gap model and forest inventory data. *For. Ecol. Manage.*, **202**, 93-105.

Maisch, M., 2000: The long-term signal of climate change in the Swiss Alps: Glacier retreat since the end of the Little Ice Age and future ice decay scenarios. *Geogr. Fis. Dinam. Quat.*, **23**, 139-151.

Malcolm, J.R., L. Canran, R.P. Neilson, L. Hansen and L. Hannah, 2006: Global warming and extinctions of endemic species from biodiversity hotspots. *Conserv. Biol.*, **20**, 538-548.

Manning, M., M. Petit, D. Easterling, J. Murphy, A. Patwardhan, H.-H. Rogner, R. Swart and G. Yohe, 2004: *IPCC Workshop on Describing Scientific Uncertainties in Climate Change to Support Analysis of Risk and of Options*. IPCC Working Group I Technical Support Unit, Boulder, Colorado, UNEP/WMO, 138 pp.

Maracchi, G., O. Sirotenko and M. Bindi, 2005: Impacts of present and future climate variability on agriculture and forestry in the temperate regions: Europe. *Climatic Change*, **70**, 117-135.

Marsh, J., 2005: The implications of Common Agricultural Policy reform for farmers in Europe. *Farm Policy Journal*, 2, 1-11.

Martin, E. and P. Etchevers, 2005: Impact of climatic change on snow cover and snow hydrology in the French Alps. *Global Change and Mountain Regions (A State of Knowledge Overview)*. U. M. Huber, H. Bugmann and M. A. Reasoner, Eds., Springer, New York, 235-242.

Martin, E., G. Giraud, Y. Lejeune and G. Boudart, 2001: Impact of climate change on avalanche hazard. *Ann. glaciol.*, **32**, 163-167.

Martínez-Vilalta, J. and J. Piñol, 2002: Drought induced mortality and hydraulic architecture in pine populations of the NE Iberian Peninsula. *For. Ecol. Manage.*, **161**, 247-256.

Marttila, V., H. Granholm, J. Laanikari, T. Yrjölä, A. Aalto, P. Heikinheimo, J. Honkatukia, H. Järvinen, J. Liski, R. Merivirta and M. Paunio, 2005: *Finland's National Strategy for Adaptation to Climate Change*. Publications of the Ministry of Agriculture and Forestry 1a/2005, 280 pp.

Materna, J., M. Daniel and V. Danielova, 2005: Altitudinal distribution limit of the tick *Ixodes ricinus* shifted considerably towards higher altitudes in central Europe: results of three years monitoring in the Krkonose Mts. (Czech Republic). *Cent. Eur. J. Public Health*, **13**, 24-28.

Mayor of London, 2005: *Climate Change and London's Transport Systems: Summary Report*. Greater London Authority, London. Retrieved 10.10.2006 from: http://www.london.gov.uk/climatechangepartnership/transport.jsp.

Mazhitova, G., N. Karstkarel, N. Oberman, V. Romanovsky and P. Kuhty, 2004: Permafrost and infrastructure in the Usa Basin (Northern European Russia): Possible impacts of global warming. *Ambio*, **3**, 289-294.

McKee, D., K. Hatton, J.W. Eaton, D. Atkinson, A. Atherton, I. Harvey and B. Moss, 2002: Effects of simulated climate warming on macrophytes in freshwater microcosm communities. *Aquat. Bot.*, **74**, 71-83.

Mearns, L. O., F. Giorgi, L. McDaniel and C. Shields, 2003: Climate scenarios for the Southeastern U.S. based on GCM and regional model simulations. *Climatic Change*, **60**, 7-35.

Medlyn, B.E., C.V.M. Barton, M. Broadmedow, R. Ceulemans, P. de Angelis, M. Forstreuter, M. Freeman, S.B. Jackson, S. Kellomäki, E. Laitat, A. Rey, P. Robernitz, B. Sigurdsson, J. Strassenmeyer, K. Wang, P.S. Curtis and P.G. Jarvis, 2001: Elevated CO_2 effect on stomatal conductance in European forest species: a synthesis of experimental data. *New Phytol.*, **149**, 247-264.

Meehl, G.A. and C. Tebaldi, 2004: More intense, more frequent, and longer lasting heatwaves in the 21st Century. *Science*, **305**, 994-997.

Meehl, G.A., T. F. Stocker, W. Collins, P. Friedlingstein, A. Gaye, J. Gregory, A. Kitoh, R. Knutti, J. Murphy, A. Noda, S. Raper, I. Watterson, A. Weaver and Z. Zhao, 2007: Global climate projections. *Climate Change 2007: The Physical Science Basis. Contribution of Working Group I to the Fourth Assessment Report of the Intergovernmental Panel on Climate Change*. Solomon, S., D. Qin, M. Manning, Z. Chen, M. Marquis, K.B. Averyt, M. Tignor and H.L. Miller, Eds., Cambridge University Press, Cambridge and New York, 747-846.

Meier, H.E.M., B. Broman and E. Kjellström 2004: Simulated sea level in past and future climates of the Baltic Sea. *Clim. Res.*, **27**, 59-75.

Meier, H.E.M., B. Broman, H. Kallio and E. Kjellström, 2007: *Projection of Future Surface Winds, Sea Levels and Wind Waves in the Late 21st Century and their Application for Impact Studies of Flood Prone Areas in the Baltic Sea Region*. Geological Survey of Finland, Special Paper, Helsinki, (in press).

Mellor, P.S. and E.J. Wittmann, 2002: Bluetongue virus in the Mediterranean Basin 1998-2001. *Vet. J.*, 164, 20-37.

Menne, B., 2005: Extreme weather events: What can we do to prevent health impacts? *Extreme weather events and public health responses*, W. Kirch,B. Menne and R. Bertollini, Eds.,Springer-Verlag, Berlin, Heidelberg, 265-271.

Menne, B. and K.L. Ebi, Eds., 2006: *Climate Change and Adaptation Strategies for Human Health*. Steinkopff Verlag, Darmstadt, 449 pp.

Menzel, A., 2003: Plant phenological anomalies in Germany and their relation to air temperature and NAO. *Climatic Change*, **57**, 243.

Menzel, A., T.H. Sparks, N. Estrella, E. Koch, A. Aasa, R. Ahas, K. Alm-Kübler, P. Bissoli, O. Braslavska, A. Briede, F.M. Chmielewski, Z. Crepinsek, Y. Curnel, Å. Dalh, C. Defila, A. Donnelly, Y. Filella, K. Jatczak, F. Måge, A. Mestre, Ø. Nordli, J. Peñuelas, P. Pirinen, V. Remišová, H. Scheifinger, M. Striz, A. Susnik, A. VanVliet, F.-E. Wielgolaski, S. Zach and A. Zust, 2006a: European phenological response to climate change matches the warming pattern. *Glob. Change Biol.*, **12**, 1969-1976.

Menzel, A., T.H. Sparks, N. Estrella and D.B. Roy, 2006b: Altered geographic and temporal variability in phenology in response to climate change. *Glob. Ecol. Biogeogr.*, **15**, 498-504.

Menzel, L. and G. Bürger, 2002: Climate change scenarios and runoff response in the Mulde catchment (Southern Elbe, Germany). *J. Hydrol.*, **267**, 53-64.

Mercer, J.B., 2003: Cold - an underrated risk factor for health. *Environ. Res.*, **92**, 8-13.

Metzger, M.J., R. Leemans, D. Schröter, W. Cramer and the ATEAM consortium, 2004: *The ATEAM Vulnerability Mapping Tool*. Quantitative Approaches in System Analysis No. 27. Wageningen, C.T. de Witt Graduate School for Production Ecology and Resource Conservation, Wageningen, CD ROM.

Metzger, M.J., M.D.A. Rounsevell, L. Acosta-Michlik, R. Leemans and D. Schröter, 2006: The vulnerability of ecosystem services to land use change. *Agr. Ecosyst. Environ.*, **114**, 69–85.

Michelozzi, P., F. de'Donato, L. Bisanti, A. Russo, E. Cadum, M. DeMaria, M. D'Ovidio, G. Costa and C.A. Perucci, 2005: The impact of the summer 2003 heatwaves on mortality in four Italian cities. *Euro Surveill.*, **10**, 161-165.

Middelkoop, H. and J.C.J. Kwadijk, 2001: Towards an integrated assessment of the implications of global change for water management – the Rhine experience. *Phys. Chem. Earth, Part B Hydrology, Oceans and Atmosphere*, **26**, 553-560.

Miettinen, I.T., O. Zacheus, C.H. von Bonsdorff and T. Vartiainen, 2001: Waterborne epidemics in Finland in 1998–1999. *Water Sci. Technol.*, **43**, 67–71.

Mindas, J., J. Skvarenia, J. Strelkova and T. Priwitzer, 2000: Influence of climatic changes on Norway spruce occurrence in the West Carpathians. *J. Forest Sci.*, **46**, 249-259.

Mínguez, M.I., M. Ruiz-Ramos, C.H. Díaz-Ambrona, M. Quemada and F. Sau, 2007: First-order impacts on winter and summer crops assessed with various high-resolution climate models in the Iberian Peninsula. *Climatic Change*, **81**, S343-S355.

Ministerio de Sanidad y Consumo, 2004: *Protocolo de Actuaciones de los Servicios Sanitarios ante una Ola de Calor*. Ministerio de Sanidad y Consumo, Madrid, 56 pp.

Minkkinen, K., R. Korhonen, I. Savolainen and J. Laine, 2002. Carbon balance and radiative forcing of Finnish peatlands in 1900-2100-the impact of forestry drainage. *Glob. Change Biol.*, **8**, 785-799.

Mitchell, R., D. Blane and M. Bartley, 2002: Elevated risk of high blood pressure: climate and inverse housing law. *Int. J. Climatol.*, **31**, 831-838.

MNRRF, 2003: *Forest Fund of Russia (according to State Forest Account by state on January 1, 2003)*. Ministry of Natural Resources of Russian Federation, Moscow, 637 pp.

Molyneux, D.H., 2003: Common themes in changing vector-borne disease scenarios. *Trans. R. Soc. Trop. Med. Hyg.*, **97**, 129-132.

Morecroft, M.D., G.J. Masters, V.K. Brown, I.P. Clark, M.E. Taylor and A.T. Whitehouse, 2004: Changing precipitation patterns alter plant community dynamics and succession in an ex-arable grassland. *Funct. Ecol.*, **18**, 648-655.

Moreira, F., F.C. Rego and P.G. Ferreira, 2001: Temporal (1958-1995) pattern of change in a cultural landscape of northwestern Portugal: implications for fire occurrence. *Landsc. Ecol.*, **16**, 557-567.

Moreno, J.M., 2005: Impactos sobre los riesgos naturales de origen climático. C) Riesgo de incendios forestales. *Evaluación Preliminar de los Impactos en España por Efecto del Cambio Climático*, J.M. Moreno, Ed., Ministerio de Medio Ambiente, Madrid, 581-615.

Moriondo, M., P. Good, R. Durao, M. Bindi, C. Gianakopoulos and J. Corte-Real, 2006: Potential impact of climate change on fire risk in the Mediterranean area. *Clim. Res.*, **31**, 85-95.

Moss, B., D. Mckee, D. Atkinson, S.E. Collings, J.W. Eaton, A.B. Gill, I. Harvey, K. Hatton, T. Heyes and D. Wilson, 2003: How important is climate? Effects of warming, nutrient addition and fish on phytoplankton in shallow lake microcosms. *J. Appl. Ecol.*, **40**, 782-792.

Mouillot, F., S. Rambal and R. Joffre, 2002: Simulating climate change impacts on fire- frequency and vegetation dynamics in a Mediterranean-type ecosystem. *Glob. Change Biol.*, **8**, 423-437.

Mouillot, F., J.E. Ratte, R. Joffre, J.M. Moreno and S. Rambal, 2003: Some determinants of the spatio-temporal fire cycle in a Mediterranean landscape (Corsica, France). *Landsc. Ecol.*, **18**, 665-674.

Mouillot, F., J.E. Ratte, R. Joffre, D. Mouillot and S. Rambal, 2005: Long-term forest dynamic after land abandonment in a fire prone Mediterranean landscape (Central Corsica, France). *Landsc. Ecol.*, **20**, 101-112.

Munich Re, 2002: *Winter Storms in Europe (II)*. Munich Re, Munich, 76 pp.

Munich Re, 2004: *Annual Review: Natural Catastrophes 2003*. Munich Re, Munich, 60 pp.

Nabuurs, G.-J., M.-J. Shelhaus, G. M. J. Mohren and C. B. Field, 2003: Temporal evolution of the European forest sector carbon sink from 1950 to 1999. *Glob. Change Biol.*, **9**, 152-160.

Nadarajah, C. and J.D. Rankin, 2005: European spatial planning: adapting to climate events, *Weather*, **60**, 190-194.

Nafstad, P., Skrondal and E. Bjertness, 2001: Mortality and temperature in Oslo, Norway, 1990-1995. *Eur. J. Epidemiol.*, **17**, 621-627.

Nakićenović, N. and R. Swart, Eds., 2000: *Special Report on Emissions Scenarios. A Special Report of Working Group III of the Intergovernmental Panel on Climate Change*. Cambridge University Press, Cambridge, 599 pp.

National Assessment Synthesis Team, US Global Research Program, 2001: *Climate Change Impacts on the United States, The Potential Consequences of Climate Variability and Change*. Cambridge University Press, Cambridge.

Nelson, F.E., 2003: (Un)frozen in time. *Science*, **299**, 1673-1675.

NHS, 2006: *Heatwave Plan for England. Protecting Health and Reducing Harm from Extreme Heat and Heatwaves*. Department of Health, UK.

Nicholls, R. J., 2004: Coastal flooding and wetland loss in the 21st century: changes under the SRES climate and socio-economic scenarios. *Glob. Environ. Change*, **14**, 69-86.

Nicholls, R.J. and R.J.T. Klein, 2004: Climate change and coastal management on Europe's coast. *Managing European Coasts, Past, Present and Future*, J.E. Vermaat, L. Ledoux, K. Turner, W. Salomons and L. Bouwer, Eds., Springer, Berlin, 387 pp.

Nicholls, R.J. and A.C. de la Vega-Leinert, Eds., 2007: Implications of sea-level rise for Europe's coasts. *J. Coast. Res.*, Special Issue, (in press).

Niinisto, S.M., J. Silvola and S. Kellomäki, 2004: Soil CO_2 efflux in a boreal pine forest under atmospheric CO_2 enrichment and air warming. *Glob. Change Biol.*, **10**, 1363-1376.

Nilsson, C., I. Stjernquist, L. Bärring, P. Schlyter, A.M. Jönsson and H. Samuelsson, 2004: Recorded storm damage in Swedish forests 1901-2000. *For. Ecol. Manage.*, **199**, 165-173.

Nisbet, T.R., 2002: Implications of climate change: soil and water. *Climate Change: Impacts on UK Forests*, M. Broadmeadow, Ed., Forestry Commission Bulletin 125, Forestry Commission, Edinburgh, 53-67.

Nogueira, P.J., 2005: Examples of heat warning systems: Lisbon's ICARO's surveillance system, summer 2003. *Extreme Weather Events and Public Health Responses*, W. Kirch, B. Menne and R. Bertollini, Eds., Springer, Heidelberg, 141-160.

Nogueira, P.J., J.M. Falcão, M.T. Contreiras, E. Paixão, J. Brandão and I. Batista, 2005: Mortality in Portugal associated with the heatwave of August 2003: Early estimation of effect, using a rapid method. *Euro Surveill.*, **10**, 150-153.

Norrant, C. and A. Douguédroit, 2006: Monthly and daily precipitation trends in the Mediterranean. *Theor. Appl. Climatol.*, **83**, 89-106.

Nunes, M.C.S., M.J. Vasconcelos, J.M.C. Pereira, N. Dasgupta and R.J. Alldredge, 2005: Land cover type and fire in Portugal: do fires burn land-cover selectively? *Landsc. Ecol.*, **20**, 661-673.

OECD, 2004: *Analysis of the 2003 CAP Reform*. OECD publications, Paris, 50 pp.

Oerlemans, J. 2003: Climate sensitivity of glaciers in southern Norway: application of an energy-balance model to Nigardsbreen, Hellstugubreen and Alfotbreen. *J. Glaciol.*, **38**, 223-232.

Ogaya, R., J. Peñuelas, J. Martínez-Vilalta and M. Mangirón, 2003: Effect of drought on diameter increment of *Quercus ilex*, *Phillyrea latifolia*, and *Arbutus unedo* in a holm oak forest of NE Spain. *For. Ecol. Manage.*, **180**, 175-184.

Ohl, C.A. and S. Tapsell, 2000: Flooding and human health. *BMJ*, **321**, 1167–1168.

Olesen, J.E. and M. Bindi, 2002: Consequences of climate change for European agricultural productivity, land use and policy. *Eur. J. Agron.*, **16**, 239-262.

Olesen, J.E. and M. Bindi, 2004: Agricultural impacts and adaptations to climate change in Europe. *Farm Policy Journal*, **1**, 36-46.

Olesen, J.E., G. Rubæk, T. Heidmann, S. Hansen and C.D. Børgesen, 2004: Effect of climate change on greenhouse gas emission from arable crop rotations. *Nutr. Cycl. Agroecosyst.*, **70**, 147-160.

Olesen, J.E., T.R. Carter, C.H. Díaz-Ambrona, S. Fronzek, T. Heidmann, T. Hickler, T. Holt, M.I. Mínguez, P. Morales, J. Palutikof, M. Quemada, M. Ruiz-Ramos, G. Rubæk, F. Sau, B. Smith and M. Sykes, 2007: Uncertainties in projected impacts of climate change on European agriculture and terrestrial ecosystems based on scenarios from regional climate models. *Climatic Change*, **81**, S123-S143.

Olofsson, E. and K. Blennow, 2005: Decision support for identifying spruce forest stand edges with high probability of wind damage. *For. Ecol. Manage.*, **207**, 87-98.

Oltchev, A., J. Cermak, J. Gurtz, A. Tishenko, G. Kiely, N. Nadezhdina, M. Zappa, N. Lebedeva, T. Vitvar, J.D. Albertson, F. Tatarinov, D. Tishenko, V. Nadezhdin, B. Kozlov, A. Ibrom, N. Vygodskaya and G. Gravenhorst, 2002: The response of the water fluxes of the boreal forest region at the Volga source area to climatic and land-use changes. *Phys. Chem. Earth, Parts A/B/C*, 27, 675-690.

Opopol, N. and A. Nicolenco, 2004: Climate change and human health: impacts, consequences, adaptation and prevention. *Moldova's Climate in XXI Century: the Projections of Changes, Impacts, and Responses*, R. Corobov, Ed., Elan Poligraf, Chisinau, 254-283.

Opopol, N., R. Corobov, A. Nicolenco and V. Pantya, 2003: Climate change and potential impacts of its extreme manifestations on health. *Curier Medical*, **5**, 6-9.

Pal, J.S., F. Giorgi and X.Q. Bi, 2004: Consistency of recent European summer precipitation trends and extremes with future regional climate projections. *Geophys. Res. Lett.*, **31**, L13202.

Páldy, A., J. Bobvos, A. Vámos, R.S. Kovats and S. Hajat, 2005: The effect of temperature and heatwaves on daily mortality in Budapest, Hungary, 1970-2000. *Extreme Weather Events and Public Health Responses*, W. Kirch, B. Menne and R. Bertollini, Eds., Springer, New York, 99-107.

Palmer, T.N. and J. Räisänen, 2002: Quantifying the risk of extreme seasonal precipitation events in a changing climate. *Nature*, **415**, 512-514.

Parmesan, C. and G. Yohe, 2003: A globally coherent fingerprint of climate change impacts across natural systems. *Nature*, **421**, 37-42.

Parry, M., Ed., 2000: *Assessment of Potential Effects and Adaptations for Climate Change in Europe (ACACIA)*. University of East Anglia, Norwich, UK, 320 pp.

Parry, M., Ed., 2004: An assessment of the global effects of climate change under SRES emissions and socio-economic scenarios. *Glob. Environ. Change*, Special Issue, **14**, 1-99.

Pascal, M., K. Laaidi, M. Ledrans, E. Baffert, C. Caseiro-Schönemann, A.L. Tertre, J. Manach, S. Medina, J. Rudant and P. Empereur-Bissonnet, 2006: France's heat health watch warning system. *Int. J. Biometeorol.*, **50**, 144-153.

Paul, F., A. Kääb, M. Maish, T. Kellengerger and W. Haeberli, 2004: Rapid disintegration of Alpine glaciers observed with satellite data. *Geophys. Res. Lett.*, **31**, L21402, doi:10.1029/2004GL020816.

Pauli, H., M. Gottfried and G. Grabherr, 2001: High summits of the Alps in a changing climate. The oldest observation series on high mountain plant diversity in Europe. *"Fingerprints" of Climate Change - Adapted Behaviour and Shifting Species Ranges,* Walther, G.-R., C.A. Burga, P.J. Edwards, Eds., Kluwer Academic Publisher, Norwell, Massachusetts, 139-149.

Pausas, J.G., 2004: Changes in fire and climate in the eastern Iberian Peninsula (Mediterranean basin). *Climatic Change,* **63**, 337-350.

Pedersen, M.L., N. Friberg and S.E. Larsen, 2004: Physical habitat structure in Danish lowland streams. *River Res. Appl.*, **20**, 653-669.

Penning-Rowsell, E., S. Tapsell and T. Wilson, 2005: Key policy implications of the health effects of floods.*Extreme Weather Events and Public Health Responses,* W. Kirch, B. Menne and R. Bertollini, Eds., Springer-Verlag, Berlin, Heidelberg, 207-223.

Peñuelas, J. and M. Boada, 2003: A global change-induced biome shift in the Montseny mountains (NE Spain). *Glob. Change Biol.*, **9**, 131-140.

Peñuelas, J., C. Gordon, L. Llorens, T. Nielsen, A. Tietema, C. Beier, B. Emmett, M. Estiarte and A. Gorissen, 2004: Non-intrusive field experiments show different plant responses to warming and drought among sites, seasons, and species in a north-south European gradient. *Ecosystem*, **7**, 598-612.

Pereira, M.G, R.M. Trigo, C.C. da Camara, J.M.C. Pereira and S. M. Leite, 2005: Synoptic patterns associated with large summer forest fires in Portugal. *Agric. For. Meteorol.*, **129**, 11–25.

Pérez, O.M., T.C. Telfer and L.G. Ross, 2003: On the calculation of wave climate for offshore cage culture site selection: a case study in Tenerife (Canary Islands). *Aquac. Eng.*, **29**, 1-21.

Perry, A.L., P.J. Low, J.R. Ellis and J.D. Reynolds, 2005: Climate change and distribution shifts in marine fishes. *Science*, **308**, 1912-1915.

Petriccione, B., 2003: Short-term changes in key plant communities of Central Apennines (Italy). *Acta Bot. Gall.*, **150**, 545-562.

Pinnegar, J.K., S. Jennings, C.M. O'Brien and J.L. Blanchard, 2002: Long-term changes in the trophic level of Celtic Sea fish community and fish market price distribution. *J. Appl. Ecol.*, **39**, 377-390.

Pinnegar, J.K., D. Viner, D. Hadley, S. Sye, M. Harris, F. Berkhout and M. Simpson, 2006: *Alternative Future Scenarios for Marine Ecosystems*. Technical Report CEFAS, Lowestoft.

Pirard, P., S. Vandentorren, M. Pascal, K. Laaidi, A. Le Tertre, S. Cassadou and M. Ledrans, 2005: Summary of the mortality impact assessment of the 2003 heatwave in France. *Euro Surveill.,* **10**, 153-156.

Polemio, M. and D. Casarano, 2004: *Rainfall and Drought in Southern Italy (1821-2001)*. UNESCO/IAHS/IWHA, pub. 286 pp.

Pretzsch, H. and J. Dursky, 2002: Growth reaction of Norway Spruce (*Picea abies* (L.) Karst.) and European beech (*Fagus sylvatica* L.) to possible climatic change in Germany. A sensitivity study. *Forstwirtschaft Centralblatt,* **121**, 145-154.

Price, 2005: Forests in sustainable mountain development. *Global Change and Mountain Regions (A State of Knowledge Overview),* U.M. Huber, H. Bugmann and M. A. Reasoner, Eds., Springer, Dordrecht, 521-530.

Pryor, S.C., R.J. Barthelmie and E. Kjellström, 2005: Potential climate change impact on wind energy resources in northern Europe: analyses using a regional climate model. *Climate Dyn.*, **25**, 815-835.

Quintana, J.R., A. Cruz, F. Fernández-González and J.M. Moreno, 2004: Time of germination and establishment success after fire of three obligate seeders in a Mediterranean shrubland of central Spain. *J. Biogeogr.*, **31**, 241-249.

Ragab, R. and C. Prudhomme, 2002: Climate change and water resource management in arid and semi-arid regions. Prospective and challenges for the 21st century. *Biosystems Engineering*, **81**, 3-34.

Rahmstorf, S. and K. Zickfeld, 2005: Thermohaline circulation changes: a question of risk assessment. *Climate Change*, **68**, 241-247.

Räisänen, J., U. Hansson, A. Ullerstig, R. Döscher, L.P. Graham, C. Jones, M. Meier, P. Samuelsson and U. Willén, 2004: European climate in the late 21st century: regional simulations with two driving global models and two forcing scenarios. *Climate Dyn.*, **22**, 13-31.

Randolph, S., 2002: The changing incidence of tick-borne encephalitis in Europe. *Euro Surveill. Weekly,* **6**. Retrieved 30.10.2004 from http://www.eurosurveillance.org/ew/2002/020606.asp.

Randolph, S.E., 2004: Evidence that climate change has caused "emergence" of tick borne diseases in Europe? *Int. J. Med. Microbiol.*, **293**, 5-15.

Ray, D., G. Pyatt and M. Broadmeadow, 2002: *Modelling the Future Stability of Plantation Forest Tree Species*. Forestry Commission Bulletin, Forestry Commission, Edinburgh, Lothians, , No 125, 151-167.

Read, P. and T. Fernandes, 2003: Management of environmental impacts of marine aquaculture in Europe. *Aquaculture*, **226**, 139-163.

Redfern, D. and S. Hendry, 2002: *Climate Change and Damage to Trees Caused by Extremes of Temperature*. Forestry Commission Bulletin No 125, Forestry Commission, Edinburgh, Lothians, 29-39.

Rehfisch, M.M. and H.Q.P. Crick, 2003: Predicting the impact of climate change on Arctic breeding waders. *Wader Study Group Bull.*, **100**, 86-95.

Reiter, P., 2000: From Shakespeare to Defoe: Malaria in England in the Little Ice Age. *Emerg. Infect. Dis.*, **6**, 1-11.

Reiter, P., 2001: Climate change and mosquito-borne disease. *Environ. Health Perspect.*, **109**, 141-161.

Reiter, P., C.J. Thomas, P.M. Atkinson, S.I. Hay, S.E. Randolph, D.J. Rogers, G.D. Shanks, R.W. Snow and A.J. Spielman, 2004: Global warming and malaria: a call for accuracy. *Lancet Infect. Dis.*, **4**, 323-324.

Revich, B.A. and D.A. Shaposhnikov, 2004: High air temperature in cities is real hazard to human health. *Climate Change and Public Health in Russia in the XXI century,* N.F. Izmerov, B.A. Revich and E.I. Korenberg, Eds.,Proceedings of the workshop, April 5-6, 2004, Moscow, Publishing Company "Adamant", 175-184.

Richter, G. and M. Semenov, 2005: *Re-Assessing Drought Risks for UK Crops using UKCIP02 Climate Change Scenarios*. Final report of Defra Project CC0368. Retrieved 03.10.2006 from www2.defra.gov.uk/science/ project_data/DocumentLibrary/CC0368/CC0368_2604_FRP.doc

Riedo, M., D. Gyastras and J. Fuhrer, 2000: Net primary production and carbon stocks in differently managed grasslands: simulation of site-specific sensitivity to an increase in atmospheric CO_2 and to climate change. *Ecol. Model.,* **134**, 207-227.

Robinson, J., M. Bradley, P. Busby, D. Connor, A. Murray, B. Sampson and W. Soper, 2006: Climate change and sustainable development: Realizing the opportunity. *Ambio*, **35**, 2-8.

Robinson, L. and C.L.J. Frid, 2003: Dynamic ecosystem models and the evaluation of ecosystem effects of fishing: can we make meaningful predictions? *Aquat. Conserv. -Mar. Freshw. Ecosyst.*, **13**, 5–20.

Robinson, M., A.-L. Cognard-Plancq, C. Cosandey, J. David, P. Durand, H-W. Führer, R. Hall, M.O. Hendriques, V. Marc, R. McCarthy, M. McDonell, C. Martin, T. Nisbel, P. O'Dea, M. Rodgers and A. Zollner, 2003: Studies of the impact of forests on peak flows and baseflows: a European perspective. *For. Ecol. Manage.*, **186**, 85-94.

Robinson, R.A., J.A. Learmonth, A.M. Hutson, C.D. Macleod, T.H. Sparks, D.I. Leech1, G.J. Pierce, M.M. Rehfisch and H.Q.P. Crick, 2005: *Climate Change and Migratory Species*. Defra Research Contract CR0302, British Trust for Ornithology Research Report 414, UK.

Rockel, B. and K. Woth, 2007: Extremes of near-surface wind speed over Europe and their future changes as estimated from an ensemble of RCM simulations. *Climatic Change*, **81**, S267-S280.

Roderick, M.L. and S.L. Berry, 2001: Linking wood density with tree growth and environment: a theoretical analysis based on the motion of water. *New Phytol.*, **149**, 473-485.

Roessig, J.M., C.M. Woodley, J.J. Cech and L.J. Hansen, 2004: Effects of global climate change on marine and estuarine fishes and fisheries. *Rev. Fish Biol. Fish*, **14**, 251-275.

Rogers, D.J. and S.E. Randolph, 2006: Climate change and vector-borne diseases. *Advances in Parasitology*, **62**, 345-381.

Root, T.L., J.T. Price, K.R. Hall, S.H. Schneider, C. Rosenzweig and J.A. Pounds, 2003: Fingerprints of global warming on wild animals and plants. *Nature*, **421**, 57-60.

Root, T.L., D.P. MacMynowski, M.D. Mastrandrea and S.H. Schneider, 2005: Human-modified temperatures induce species changes: joint attribution. *Proc. Natl. Acad. Sci. U.S.A.*, **102**, 7465-7469.

Rounsevell, M.D.A., F. Ewert, I. Reginster, R. Leemans and T.R. Carter, 2005: Future scenarios of European agricultural land use. II. Projecting changes in cropland and grassland. *Agric. Ecosyst. Environ.*, **107**, 117-135.

Rounsevell, M.D.A., I. Reginster, M.B. Araújo, T.R. Carter, N. Dendoncker, F. Ewert, J.I. House, S. Kankaanpää, R. Leemans, M.J. Metzger, C. Schmidt, P. Smith and G. Tuck, 2006: A coherent set of future land use change scenarios for Europe. *Agr. Ecosyst. Environ.*, **114**, 57-68.

Rustad, L.E., J.L. Campbell, G.M. Marion, R.J. Norby, M.J. Mitchell, A.E. Hartley, J.H.C. Cornelissen and J. Gurevitch., 2001: A meta-analysis of the response of soil respiration, net nitrogen mineralization, and above-ground plant growth to experimental ecosystem warming. *Oecologia*, **126**, 543-562.

Salvador, R., F. Lloret, X. Pons and J. Piñol, 2005: Does fire occurrence modify the probability of being burned again? A null hypothesis test from Mediterranean ecosystems in NE Spain. *Ecol. Model.*, **188**, 461-469.

Sánchez-Arcilla, A., P. Hoekstra, J.E. Jiménez, E. Kaas and A. Maldonado, 2000: Climate implications for coastal processes. *Sea Level Change and Coastal Processes: Implications for Europe*, D.E. Smith, S.B. Raper, S. Zerbini and A. Sánchez-Arcilla, Eds., Office for Official Publications of the European Communities, Luxembourg, 173-213.

Sánchez Carrillo, S. and M. Álvarez Cobelas, 2001: Nutrient dynamics and eutrophication patterns in a semiarid wetland: the effects of fluctuating hydrology. *Water Air Soil Poll.*, **131**, 97-118.

Sanderson, M.G., C.D. Jones, W.J. Collins, C.E. Johnson and R.G. Derwent, 2003: Effect of climate change on isoprene emissions and surface ozone levels. *Geophys. Res. Lett.*, **30**, 1936, doi:10.1029/2003GL017642.

Sand-Jensen, K. and N.L. Pedersen, 2005: Broad-scale differences in temperature, organic carbon and oxygen consumption among lowland streams. *Freshw. Biol.*, **50**, 1927-1937.

Sandvik, S.M., E. Heegaard, R. Elven and V. Vandvik, 2004: Responses of alpine snowbed vegetation to long-term experimental warming. *Ecoscience*, **11**, 150-159.

Santos, F.D., K. Forbes and R. Moita, Eds., 2002: *Climate Change in Portugal: Scenarios, Impacts and Adaptation Measures*. SIAM project report, Gradiva, Lisbon, 456 pp.

Sanz Elorza, M. and E.D. Dana, 2003: Changes in the high mountain vegetation of the central Iberian Peninsula as a probable sign of global warming. *Ann. Bot.*, **92**, 273-280.

Sapozhnikov, A.P., 2003: Global and regional aspects of assessment of forest fire situation. *Scientific Backgrounds of Use and Regeneration of Forest Resources in the Far East, Transactions of the Far Eastern Forestry Research Institute Vol 36*, A.P. Kovalev, Ed., Khabarovsk, 127-143 (in Russian).

Sardans, J. and J. Peñuelas, 2004: Increasing drought decreases phosphorus availability in an evergreen Mediterranean forest. *Plant Soil*, **267**, 367-377.

Sardans, J. and J. Peñuelas, 2005: Drought decreases soil enzyme activity in a Mediterranean *Quercus ilex* L. forest. *Soil Bio. Biochem.*, **37**, 455-461.

Sardon, J.P., 2004: Recent demographic trends in the developed countries. *Population*, **59**, 305-360.

Schaphoff, S., W. Lucht, D. Gerten, S. Stich, W. Cramer and I.C. Prentice, 2006. Terrestrial biosphere carbon storage under alternative climate projection. *Climatic Change*, **74**, 97-122.

Schär, C. and G. Jendritzky, 2004: Climate change: hot news from summer 2003. *Nature*, **432**, 559-560.

Schär, C., P.L. Vidale, D. Lüthi, C. Frei, C. Häberli, M.A. Liniger and C. Appenzeller, 2004: The role of increasing temperature variability in European summer heatwaves. *Nature*, **427**, 332-336.

Schijven, J.F. and A.M. de Roda Husman, 2005: Effect of climate changes on waterborne disease in the Netherlands. *Water Sci. Technol*, **51**, 79-87.

Schlesinger, M.E., J. Yin, G. Yohe, N.G. Andonova, S. Malyshev and B. Li, 2007: *Assessing the Risk of a Collapse of the Atlantic Thermohaline Circulation*. Research Paper, Climate Research Group, University of Illinois at Urbana-Champaign (in press).

Schmidt, I.K., A. Tietema, D. Williams, P. Gundersen, C. Beier, B.A. Emmet and M. Estiarte, 2004: Soil solution chemistry and element fluxes in three European heathlands and their responses to warming and drought. *Ecosystems*, **7**, 638-649.

Schneeberger, C., H. Blatter, A. Abe-Ouchi and M. Wild, 2003: Modelling change in the mass balance of glaciers of the northern hemisphere for a transient $2xCO_2$ scenario. *J. Hydrol.*, **274**, 62-79.

Schneider, S.H., S. Semenov, A. Patwardhan, I. Burton, C.H.D. Magadza, M. Oppenheimer, A.B. Pittock, A. Rahman, J.B. Smith, A. Suarez and F. Yamin, 2007: Assessing key vulnerabilities and the risk from climate change. Climate Change 2007: Impacts, Adaptation and Vulnerability. Contribution of Working Group II to the Fourth Assessment Report of the Intergovernmental Panel on Climate Change, M.L. Parry, O.F. Canziani, J.P. Palutikof, P.J. van der Linden and C.E. Hanson, Eds., Cambridge University Press, Cambridge, UK, 779-810.

Schröter, D., W. Cramer, R. Leemans, I.C. Prentice, M.B. Araújo, N.W. Arnell, A. Bondeau, H. Bugmann, T.R. Carter, C.A. Gracia, A.C. de la Vega-Leinert, M. Erhard, F. Ewert, M. Glendining, J.I. House, S. Kankaanpää, R.J.T. Klein, S. Lavorell, M. Linder, M.J. Metzger, J. Meyer, T.D. Mitchell, I. Reginster, M. Rounsevell, S. Sabaté, S. Sitch, B. Smith, J. Smith, P. Smith, M.T. Sykes, K. Thonicke, W. Thuiller, G. Tuck, S. Zaehle and B. Zierl, 2005. Ecosystem service supply and vulnerability to global change in Europe. *Science*, **310**, 1333 -1337.

Schumacher, S. and H. Bugmann, 2006: The relative importance of climatic effects, wildfires and management for future forest landscape dynamics in the Swiss Alps. *Glob. Change Biol.*, **12**, 1435-1450.

SECRU (Scottish Executive Central Research Unit), 2002: *Review and Synthesis of the Environmental Impacts of Aquaculture*. The Scottish Association for Marine Science and Napier University, Scottish Executive Central Research Unit, 80 pp.

SEEG (Scottish Executive Environment Group), 2006: *Harmful Algal Bloom Communities in Scottish Coastal Waters: Relationships to Fish Farming and Regional Comparisons – A Review*. Scottish Executive, Paper 2006/3.

Semenov, S.M., E.S. Gelver and V.V. Yasyukevich, 2002: Temperature conditions for development of two species of malaria pathogens in the vector organism in Russia in the 20th century. *Doklady Biological Sciences*, **387**, 523–528. Translated from *Doklady Akademii Nauk*, **387**, 131–136.

SEPA (Swedish Environmental Protection Agency), 2005: *Change Beneath the Surface, Monitor 19: An In-depth Look at Sweden's Marine Environment*. Naturvårdsverket, Stokholm, 192 pp.

Sharp, G., 2003: *Future Climate Change and Regional Fisheries: a Collaborative Analysis*. Fisheries Technical Paper No.452, FAO, Rome, 75 pp.

Shaver, G.R., J. Canadell, F.S. Chapin III, J. Gurevitch, J. Harte, G. Henry, P. Ineson, S. Jonasson, J. Mellilo, L. Pitelka and L Rustad, 2000: Global warming and terrestrial ecosystems: a conceptual framework for analysis. *Bioscience*, **50**, 871-882.

Shiyatov, S.G., M.M. Terent'ev and V.V. Fomin, 2005: Spatiotemporal dynamics of forest-tundra communities in the polar Urals. *Russian J. Ecol.*, **36**, 69-75.

Shvidenko, A. and S. Nilsson, 2003: A synthesis of the impact of Russian forests on the global carbon budget for 1961-1998. *Tellus*, **55B**, 391-415.

Sievanen, T., K. Tervo, M. Neuvonen, E. Pouta, J. Saarinen and A. Peltonen, 2005: *Nature-Based Tourism, Outdoor Recreation and Adaptation to Climate Change*. FINADAPT working paper 11, Finnish Environment Institute Mimeographs 341, Helsinki, 52 pp.

Silander, J., B. Vehviläinen, J. Niemi, A. Arosilta, T. Dubrovin, J. Jormola, V. Keskisarja, A. Keto, A. Lepistö, R. Mäkinen, M. Ollila, H. Pajula, H. Pitkänen, I. Sammalkorpi, M. Suomalainen and N. Veijalainen, 2006: Climate change adaptation for hydrology and water resources. FINADAPT Working Paper 6, Finnish Environment Institute Mimeographs 336, Helsinki, 54 pp.

Simón, F., G. López-Abente, E. Ballester and F. Martínez, 2005: Mortality in Spain during the heatwaves of summer 2003. *Euro Surveill.*, **10**, 156-160.

Smith, D.E, S.B. Raper, S. Zerbini and A. Sánchez-Arcilla, Eds., 2000: *Sea Level Change and Coastal Processes: Implications for Europe*. Office for Official Publications of the European Communities, Luxembourg, 247 pp.

Somlyódy, L., 2002: *Strategic Issues of the Hungarian Water Resources Management*. Academy of Science of Hungary, Budapest, 402 pp. (in Hungarian).

Soto, C.G., 2001: The potential impacts of global climate change on marine protected areas. *Rev. Fish Biol. Fish*, **11**, 181-195.

Sowerby, A., B. Emmet, C. Beier, A. Tietema, J. Peñuelas, M. Estiarte, M.J.M. Van-Meeteren, S. Hughes and C. Freeman, 2005: Microbial community changes in heathland soil communities along a geographical gradient: interaction with climate change manipulations. *Soil Bio. Biochem.*, **37**, 1805-1813.

Stedman, J.R., 2004: The predicted number of air pollution related deaths in the UK during the August 2003 heatwave. *Atmos. Environ.*, **38**, 1087-1090.

Stevenson, D.S., F. J. Dentener, M.G. Schultz, K. Ellingsen, T.P.C. Noije, O. Wild, G. Zeng, M. Amann and C.S. Atherton, 2006: Multimodel ensemble simulations of present day and near future tropospheric ozone. *J. Geophys. Res.*, **111**, D08301, doi:10.1029/2005JD006338.

Stone, G.W. and J.D. Orford, Eds., 2004: Storms and their significance in coastal morpho-sedimentary dynamics. *Mar. Geol.*, **210**, 1-365.

Stott, P.A., D.A. Stone and M.R. Allen, 2004: Human contribution to the European heatwave of 2003. *Nature*, **432**, 610-614.

Stouffer, R.J., J. Yin, J. M. Gregory, K. W. Dixon, M. J. Spelman, W. Hurlin, A. J. Weaver, M. Eby, G. M. Flato, H. Hasumi, A. Hu, J. Jungclause, I. V. Kamenkovich, A. Levermann, M. Montoya, S. Murakami, S. Nawrath, A. Oka, W. R. Peltier, D. Y. Robitaille, A. Sokolov, G. Vettoretti and N. Weber, 2006: Investigating the Causes of the Response of the Thermohaline Circulation to Past and Future Climate Changes. *J. Clim.* , **19**, 1365-1387.

Straile, D., D.M. Livingstone, G.A. Weyhenmeyer and D.G. George, 2003: The response of freshwater ecosystems to climate variability associated with the North Atlantic Oscillation. *The North Atlantic Oscillation: Climatic Significance and Environmental Impact*, Geophysical Monograph, 134. American Geophysical Union.

Strömgren, M. and S. Linder, 2002: Effects of nutrition and soil warming on stemwood production in a boreal Norway spruce stand. *Glob. Change Biol.*, **8**, 1195-1204.

Sutherst, R.W., 2004: Global change and human vulnerability to vector-borne diseases. *Clin. Microb. Rev.*, 17, 136–173.

Swift, L.J., R.J.N. Devoy, A.J. Wheeler, G.D. Sutton and J. Gault, 2007: Sedimentary dynamics and coastal changes on the south coast of Ireland. *J. Coast. Res.*, SI39. (in press).

Swiss Re, 2000: *Storm over Europe*. Swiss Re, Zurich, 27 pp.

Szolgay, J., K. Hlavcova, S. Kohnová and R. Danihlik, 2004: Assessing climate change impact on river runoff in Slovakia.Characterisation of the Runoff Regime and its Stability in the Tisza Catchment. *Proceedings of the XXIInd Conference of the Danubian Countries on the Hydrological Forecasting and Hydrological Bases of Water Management*. Brno, 30 August-2 September 2004. CD-edition.

Tapsell, S.M. and S.M. Tunstall, 2001: *The Health and Social Effects of the June 2000 Flooding in the North East Region*. Report to the Environment Agency, Thames Region, Enfield, Middlesex University Flood Hazard Research Centre, 153 pp.

Theurillat, J.-P. and A. Guisan, 2001: Potential impact of climate change on vegetation in the European Alps: a review. *Climatic Change*, **50**, 77-109.

Thomas, C.D., E.J. Bodsworth, R.J. Wilson, A.D. Simmons, Z.G. Davies, M. Musche and L. Conradt, 2001: Ecological and evolutionary processes at expanding range margins. *Nature*, **441**, 577-581.

Thonicke, K. and W. Cramer, 2006: Long-term trends in vegetation dynamics and forest fires in Brandenburg (Germany) under a changing climate. *Natural Hazards*, **38**, 283-300.

Thuiller, W., S. Lavorel, M.B. Araújo, M.T. Sykes and I.C. Prentice, 2005: Climate change threats plant diversity in Europe. *Proc. Natl. Acad. Sci. U.S.A.*, **102**, 8245-8250.

Thurig, E., T. Palosuo, J. Bucher and E. Kaufmann, 2005: The impact of windthrow on carbon sequestration in Switzerland: a model-based assessment. *For. Ecol. Manage.*, **210**, 337-350.

TNS Opinion and Social, 2005: *Special Eurobarometer 217: Attitudes of Europeans towards environment*. European Commission, Brussels, 117 pp.

Tol, R.S.J., 2002: Estimates of the damage costs of climate change, Parts 1 & 2. *Environ. Res. Econ.*, **21**, 47-73 & 135-160.

Tol, R.S.J., 2006: Exchange rates and climate change: an application of fund. *Climatic Change*, **75**, 59-80.

Tol, R.S.J., R.J.T. Klein and R.J. Nicholls, 2007: Towards successful adaptation to sea-level rise along Europe's coasts. *J. Coast. Res.* (in press).

Tomozeiu, R., S. Stefan and A. Busuioc, 2007: Spatial and temporal variability of the winter precipitation in Romania in connection with the large-scale circulation patterns. *Int. J. Climatol.*, (in press).

Trnka, M., M. Dubrovski and Z. Zalud, 2004: Climate change impacts and adaptation strategies in spring barley production in the Czech Republic. *Climatic Change*, **64**, 227-255.

Tschakert, P. and L. Olsson, 2005: Post-2012 climate action in the broad framework of sustainable development policies: the role of the EU. *Climate Policy*, **5**, 329-348.

Tsimplis, M.N., D.K. Woolf, T.J. Osbourn, S. Wakelin, J. Wolf R. Flather, P. Woodworth, A.G.P. Shaw, P. Challenor and Z. Yan, 2004a: Future changes of sea level and wave heights at the northern European coasts. *Geophys. Res. Abs.*, **6**, 00332.

Tsimplis, M.N., S.A. Josey, M. Rixen and E.V. Staney, 2004b: On the forcing of sea level in the Black Sea. *J. Geophys. Res.*, **109**, C08015, doi:10.1029/2003JC002185.

Tuck, G., J.M. Glendining, P. Smith, J.I. House and M. Wattenbach, 2006: The potential distribution of bioenergy crops in Europe under present and future climate. *Biomass Bioenerg.*, **30**, 183-197.

Turnpenny, J.R., D.J. Parsons, A.C. Armstrong, J.A. Clark, K. Cooper and A.M. Matthews, 2001: Integrated models of livestock systems for climate change studies. 2. Intensive systems. *Glob. Change Biol.*, **7**, 163-170.

UN, 2004: *World population prospects. The 2004 revision population database. Vol III, Analytical Report*. United Nations, New York, 194 pp.

Vajda, A., A. Venalainen, H. Tuomenvirta and K. Jylha, 2004: An estimate of the influence of climate change on heating energy demand on regions of Hungary, Romania and Finland. *Q. J. of the Hungarian Meteorological Service*, **108**, 123-140.

Valladares, F., J. Peñuelas, and E. de Luis Calabuig, 2005: Impactos sobre los ecosistemas terrestres. *Evaluación Preliminar de los Impactos en España por Efecto del Cambio Climático*, J.M. Moreno, Ed., Ministerio de Medio Ambiente, Madrid, 65-112.

Valor, E., V. Meneu and V. Caselles, 2001: Daily air temperature and electricity load in Spain. *J. Appl. Meteorol.*, **40**, 1413-1421.

van der Leun, J.C. and F.R. de Gruijl, 2002: Climate change and skin cancer. *Photochem. Photobiol. Sci.*, **1**, 324-326.

van Herk, C.M., A. Aptroot and H.F. Dobben, 2002: Long-term monitoring in the Netherlands suggests that lichens respond to global warming. *Lichenologist*, **34**, 141-154.

van Lieshout, M., R.S. Kovats, M.T.J. Livermore and P. Martens, 2004: Climate change and malaria: analysis of the SRES climate and socio-economic scenarios. *Glob. Environ. Change*, **14**, 87-99.

van Meijl, H., T. van Rheenen, A. Tabeau and B. Eickhout, 2006. The impact of different policy environments on agricultural land use in Europe. *Agr. Ecosyst. Environ.*, **114**, 21-38.

van Pelt, W., D. Mevius, H.G. Stoelhorst, S. Kovats, A.W. van de Giessen, W. Wannet, and Y.T.H.P. Duynhoven, 2004: A large increase of Salmonella infections in 2003 in the Netherlands: hot summer or side effect of the avian influenza outbreak? *Euro Surveill.*, **4**, 17-19.

van Vliet, A.J.H., A. Overeem, R.S. de Groot, A.F.G. Jacobs and F.T.M. Spieksma, 2002: The influence of temperature and climate change on the timing of pollen release in the Netherlanders. *Int. J. Climatol.*, **22**, 1757–1767.

van Vuuren, D.P. and K.H. Alfsen, 2006: PPP versus MER: Searching for answers in a multi-dimensional debate. *Climatic Change*, **75**, 47-57.

van Vuuren, D.P. and L.F. Bouwman, 2005: Exploring past and future changes in the ecological footprint for world regions. *Ecol. Econ.*, **52**, 43-62.

van Vuuren, D.P. and B.C. O'Neill, 2006: The consistency of IPCC's SRES scenarios to recent literature and recent projections. *Climatic Change*, **75**, 9-46.

Vasiliev, S.V., A.A. Titlyanova, and A.A. Velichko, Eds., 2001: *West Siberian Peatlands and Carbon Cycle: Past and Present*. Proceedings of the International Field Symposium, Russian Academy of Sciences, Novosibirsk, 250 pp.

Vázquez, A. and J.M. Moreno, 2001: Spatial distribution of forest fires in Sierra de Gredos (Central Spain). *For. Ecol. Manage.*, **147**, 55-65.

Vellinga, M. and R.A. Wood, 2002: Global climatic impacts of a collapse of the Atlantic Thermohaline Circulation. *Climatic Change*, **54**, 251-267.

Vellinga, M. and R.A. Wood, 2006: Impacts of thermohaline circulation shutdown in the twenty-first century. *Climatic Change*, doi:10.1007/s10584-006-9146-y.

Venalainen, A., B. Tammelin, H. Tuomenvirta, K. Jylha, J. Koskela, M.A. Turunen, B. Vehvilainen, J. Forsius and P. Jarvinen, 2004: The influence of climate change on energy production and heating energy demand in Finland. *Ener. Environ.*, **15**, 93-109.

Verlato, G., R. Calabrese and R. de Marco, 2002: Correlation between asthma and climate in the European Community Respiratory Health Survey. *Arch. Environ. Health*, **57**, 48–52.

Vermaat, J.E., L. Ledoux, K. Turner, W. Salomons and L. Bouwer, Eds., 2005: *Managing European Coast:, Past, Present and Future*. Springer, Berlin, 387 pp.

Vincent, C., 2002: Influence of climate change over the 20th century on four French glacier mass balance. *J. Geophys. Res.*, **107**, 4375, doi:10.1029/2001JD000832.

Viner, D., 2006: Tourism and its interactions with climate change. *J. of Sustainable Tourism*, **14**, 317-322.

Viner, D., M. Sayer, M. Uyarra and N. Hodgson, 2006: *Climate Change and the Eu-*

ropean Countryside: Impacts on Land Management and Response Strategies. Report Prepared for the Country Land and Business Association., UK. Publ., CLA, 180 pp.

Wackernagel, M., N.B. Schulz, D. Deumling, A. Callejas Linares, M. Jenkins, V. Kapos, C. Monfreda, J. Loh, N. Myers, R. Norgaard and J. Randers, 2002: Tracking the ecological overshoot of the human economy. *Proc. Natl. Acad. Sci. U.S.A.*, **99**, 9266-9271.

Walkden, M.J.A. and J.W. Hall, 2005: A predictive mesoscale model of soft shore erosion and profile development. *Coast. Eng.*, **52**, 535-563.

Walther, G.-R., 2004: Plants in a warmer world. *Perspective in Plant Ecology, Evolution and Systematics*, **6**, 169-185.

Walther, G.-R., E. Post, P. Convey, A. Menzel, C. Parmesan, T.J.C. Beebee, J.-M. Fromentin, O. Hoegh-Gudberg and F. Bairlein, 2002: Ecological responses to recent climate change. *Nature*, **416**, 389-395.

Walther, G.-R., S. Beissner and C.A. Burga, 2005a: Trends in upward shift of alpine plants. *J. Veg. Sci.*, **16**, 541-548.

Walther, G.-R., S. Berger and M.T. Sykes, 2005b: An ecological 'footprint' of climate change. *Proc. R. Soc. Lond. Ser. B-Biol. Sci.*, **272**, 1427-1432.

Warren, M.S., J.K. Hill, J.A. Thomas, J. Asher, R. Fox, B. Huntley, D.B. Roy, M.G. Telfer, S. Jeffcoate, P. Harding, G. Jeffcoate, S.G. Willis, J.N. Greatorex-Davies, D. Moss and C.D. Thomas, 2001: Rapid responses of British butterflies to opposing forces of climate and habitat change. *Nature*, **414**, 65-69.

WBGU, 2003: *Climate Protection Strategies for the 21st Century. Kyoto and Beyond*. ISBN, Berlin. 77 pp.

Weiland, S.K., A. Hüsing, D.P. Strachan, P. Rzehak and N. Pearce, 2004: Climate and the prevalence of symptoms of asthma, allergic rhinitis, and atopic eczema in children. *Occup. Environ. Med.*, **61**, 609–615.

Weltzin, J.F., S.D. Bridgham, J. Pastor, J.Q. Chen and C. Harth, 2003: Potential effects of warming and drying on peatland plant community composition. *Glob. Change Biol.*, **9**, 141-151.

Werkman, B.R. and T.V. Callaghan, 2002: Responses of bracken and heather to increased temperature and nitrogen addition, alone and in competition. *Basic Appl. Ecol.*, **3**, 267-275.

Werritty, A., 2001: Living with uncertainty: climate change, river flow and water resources management in Scotland. *Sci. Total Environ.*, **294**, 29-40.

Wessel, W.W., A. Tietema, C. Beier, B.A. Emmett, J. Peñuelas and T. Riis-Nielsen, 2004: A qualitative ecosystem assessment for different shrublands in Western Europe under impact of climate change. *Ecosystems*, **7**, 662-671.

WGE, 2004: *Review and Assessment of Air Pollution Effects and their Recorded Trends*. Report of the Working Group on Effects of the Convention on Long-range Transboundary Air Pollution, NERC, UK.

White, A., M.G.R. Cannel and A.D. Friend, 2000: The high-latitude terrestrial carbon sink: a model analysis. *Glob. Change Biol.*, **6**, 227-246.

WHO, 2003: *The health impacts of 2003 summer heat-waves*. Briefing note for the Delegations of the fifty-third session of the WHO Regional Committee for Europe. World Health Organisation, Europe, 12 pp.

WHO, 2004: *Extreme Weather and Climate Events and Public Health Responses*. Report on a WHO meeting, Bratislava, Slovakia, Feb. 2004, World Health Organisation, WHO Regional Office for Europe, Copenhagen, 48 pp.

WHO, 2005: *Health and Climate Change: the "Now and How" A Policy Action Guide*. WHO Regional Office for Europe, Copenhagen, 32 pp.

WHO, 2006: *1st Meeting of the Project: Improving Public Health Responses to Extreme Weather/Heat-waves – EuroHEAT*. Report on a WHO meeting, Rome, Italy, 20-22 June 2005. WHO Regional office for Europe, Copenhagen, 51 pp.

Wiek, A., C. Binder and R.W. Scholz, 2006: Functions of scenarios in transition processes. *Futures*, 38, 740-766.

Wielke, L.-M., L. Laimberger and M. Hantel, 2004: Snow cover duration in Switzerland compared to Austria. *Meterol. Z.*, **13**, 13-17.

Wigley, T.M.L., 2005: The climate change commitment. *Science*, **307**, 1766-1769.

Wilby R.L. and I. Harris, 2006: A framework for assessing uncertainties in climate change impacts: Low-flow scenarios for the River Thames, UK. *Water Resour. Res.*, **42**, W02419, doi:10.1029/2005WR004065.

Wilhelmsson, L., J. Alinger, K. Spangberg, S.-O. Lundquist, T. Grahn, O. Hedenberg and L. Olsson, 2002: Models for predicting wood properties in stems of *Picea abies* and *Pinus sylvestris* in Sweden. *Scandinavian J. Forest Res.*, **17**, 330-350.

Wilkinson, P., M. Landon, B. Armstrong, S. Stevenson, S. Pattenden, T. Fletcher and M. McKee, 2001: *Cold Comfort: the Social and Environmental Determinants of Excess Winter Deaths in England, 1986–96*. The Policy Press, Bristol and Joseph Rowntree Foundation, York 40 pp.

Wilkinson, P., S. Pattenden, B. Armstrong, A. Fletcher, R.S. Kovats, P. Mangtani and A.J. McMichael, 2004: Vulnerability to winter mortality in elderly people in Britain: population based study. *BMJ*, doi:10.1136/bmj.38167.589907.55.

Williams, P.H., L. Hannah, S. Andelman, G.F. Midgley, M.B. Araújo, G. Hughes, L.L. Manne, E. Martinez-Meyer and R.G. Pearson, 2005: Planning for climate change: Identifying minimum-dispersal corridors for the Cape Proteaceae. *Conserv. Biol.*, **19**, 1063-1074.

Williams, W.D., 2001: Anthropogenic salinisation of inland waters. *Hydrobiologia*, **466**, 329-337.

Wilson E.O., 2002: *The Future of Life*. Alfred A. Knoff, New York, N.Y., 229 pp.

Winkler, J.B. and M. Herbst, 2004: Do plants of a semi-natural grassland community benefit from long-term CO_2 enrichment? *Basic Appl. Ecol.*, **5**, 131-143.

Wittig, V.E., C.J. Bernacchi, X.G. Zhu, C. Calfapietra, R. Ceulemans, P. Deangelis, B. Gielen, F. Miglietta, P.B. Morgan and S.P. Ong, 2005: Gross primary production is simulated for three *Populus* species grown under free-air CO_2 enrichment from planning through canopy closure. *Glob. Change Biol.*, **11**, 644-656.

Wittmann, E.J. and M. Baylis, 2000: Climate change: Effects on *Culicoides*-transmitted viruses and implications for the UK. *Vet. J.*, **160**, 107-117.

Wolf, J. and D.K. Woolf, 2006: Waves and climate change in the north-east Atlantic. *Geophysical Research Letters*, **33**, L06604, doi: 10.1029/2005GL025113.

Wood, R.A., M. Collins, J. Gregory, G. Harris and M. Vellinga, 2006: Towards a risk assessment for shutdown of the Atlantic thermohaline circulation. *Avoiding Dangerous Climate Change*, H.J. Schellnhuber, W. Cramer, N. Nakićenović, T. Wigley and G. Yohe, Eds., Cambridge University Press, Cambridge, 392 pp.

Woodworth, P.L., J.M. Gregory and R.J. Nicholls, 2005: Long term sea level changes and their impacts. *The global coastal ocean: multiscale interdisciplinary processes*, A.R. Robinson, and K. H. Brink, Eds., Cambridge, Massachusetts, 715-753.

Woolf, D.K., P.G. Challenor and P.D. Cotton, 2002: Variability and predictability of the North Atlantic wave climate. *J. Geophys. Res.*, **109**, 3145-3158.

Woolf, D.K., A.G.P. Shaw and M.N. Tsimplis, 2003: The influence of the North Atlantic Oscillation on sea-level variability in the North Atlantic region. *The Global Atmosphere-Ocean System*, **9**, 145-167.

Wooller, S., 2003: The Changing Climate: Impact on the Department for Transport. *House Policy Consultancy for the Department for Transport, UK*. Retrieved 10.10.2006 from http://www.dft.gov.uk/pgr/scienceresearch/key/thechangingclimateitsimpacto1909

World Bank, 2005: *World development indicators 2004*. Accessed 02.08.2005 at http://www.worldbank.org/data/databytopic/GNIPC.pdf.

Worrall, F., T.P. Burt and J.K. Adamson, 2006: Trends in drought frequency-the fate of DOC export from British peatlands. *Climatic Change*, **76**, 339-359.

Woth, K., R. Weisse and H. von Storch, 2005: Climate change and North Sea storm surge extremes: an ensemble study of storm surge extremes expected in a changed climate projected by four different regional climate models. *Ocean Dyn.*, doi: 10.1007/s10236-005-0024-3.

WTO, 2003: *Climate Change and Tourism*. Proceedings of the First international conference on climate change and tourism, 2003, World Tourism Organisation, Djerba, España, Roma, Barcelo, Alemanica, Tunisia, 55 pp.

WWF, 2004: *Living Planet Report 2004*. WWF- World Wide Fund for Nature (formerly World Wildlife Fund), Gland, Switzerland, 44 pp.

Yan, Z., M.N. Tsimplis and D. Woolf, 2004: Analysis of the relationship between the North Atlantic Oscillation and sea-level changes in Northwest Europe. *Int. J. Climatol.*, **24**, 743-758.

Yasukevich, V.V., 2003: Influence of climatic changes on the spread of malaria in the Russian Federation. *Medical Parasitology and Parasitic Disease*, **4**, 27-33 (in Russian).

Yasukevich, V.V., 2004: Prediction of malaria spread in Russia in the first quarter of the 21st century. *Medical Parasitology and Parasitic Disease*, **2**, 31-33 (in Russian).

Yohe, G. and R.S.J. Tol, 2002: Indicators for social and economic coping capacity - moving toward a working definition of adaptive capacity. *Glob. Environ. Change*, **12**, 25-40.

Zalidis, G.C., T.L. Crisman and P.A. Gerakis, Eds., 2002: *Restoration of Mediterranean Wetlands*. Hellnic Ministry of Environment, Physical Planning and Public Works, Athens and Greek Biotope/Wetland Centre, Thermi, 237 pp.

Zebisch, M., T. Grothmann, D. Schröter, C. Hasse, U. Fritsch and W. Cramer, 2005: *Climate Change in Germany – Vulnerability and Adaptation of Climate Sensitive*

Sectors. Umweltbundesamt Climate Change 10/05 (UFOPLAN 201 41 253), Dessau, 205 pp.

Zeidelman, F.P. and A.P. Shvarov, 2002: *Pyrogenic and Hydrothermal Degradation of Peat Soils, their Agroecology, Sand Cultures and Restoration*. Moscow State University, Moscow, 166 pp. (in Russian)

Zheng, D., M. Freeman, J. Bergh, I. Rosberg and P. Nilsen. 2002: Production of *Picea abies* in south-east Norway in response to climate change: a case study using process-based model simulation with field validation. *Scand. J. Forest Res.*, **17**, 35-46.

Zierl, B. and H. Bugmann, 2005: Global change impacts on hydrological processes in Alpine catchments. *Water Resour. Res.*, **41**, 1-13.

Zvereva, E.L. and M.V. Kozlov, 2006: Consequences of simultaneous elevation of carbon dioxide and temperature for plant–herbivore interactions: a metaanalysis. *Glob. Change Biol.*, **12**, 27-41.

13

Latin America

Coordinating Lead Authors:

Graciela Magrin (Argentina), Carlos Gay García (Mexico)

Lead Authors:

David Cruz Choque (Bolivia), Juan Carlos Giménez (Argentina), Ana Rosa Moreno (Mexico), Gustavo J. Nagy (Uruguay), Carlos Nobre (Brazil), Alicia Villamizar (Venezuela)

Contributing Authors:

Francisco Estrada Porrúa (Mexico), José Marengo (Brazil), Rafael Rodríguez Acevedo (Venezuela), María Isabel Travasso (Argentina), Ricardo Zapata-Marti (Mexico)

Review Editors:

Max Campos (Costa Rica), Edmundo de Alba Alcarez (Mexico), Phillip Fearnside (Brazil)

This chapter should be cited as:

Magrin, G., C. Gay García, D. Cruz Choque, J.C. Giménez, A.R. Moreno, G.J. Nagy, C. Nobre and A. Villamizar, 2007: Latin America. *Climate Change 2007: Impacts, Adaptation and Vulnerability. Contribution of Working Group II to the Fourth Assessment Report of the Intergovernmental Panel on Climate Change*, M.L. Parry, O.F. Canziani, J.P. Palutikof, P.J. van der Linden and C.E. Hanson, Eds., Cambridge University Press, Cambridge, UK, 581-615.

Table of Contents

Executive summary

Climatic variability and extreme events have been severely affecting the Latin America region over recent years (high confidence).

Highly unusual extreme weather events were reported, such as intense Venezuelan rainfall (1999, 2005), flooding in the Argentinean Pampas (2000-2002), Amazon drought (2005), hail storms in Bolivia (2002) and the Great Buenos Aires area (2006), the unprecedented Hurricane Catarina in the South Atlantic (2004) and the record hurricane season of 2005 in the Caribbean Basin [13.2.2]. Historically, climate variability and extremes have had negative impacts on population; increasing mortality and morbidity in affected areas. Recent developments in meteorological forecasting techniques could improve the quality of information necessary for people's welfare and security. However, the lack of modern observation equipment, the urgent need for upper-air information, the low density of weather stations, the unreliability of their reports and the lack of monitoring of climate variables work together to undermine the quality of forecasts, with adverse effects on the public, lowering their appreciation of applied meteorological services as well as their trust in climate records. These shortcomings also affect hydrometeorological observing services, with a negative impact on the quality of early warnings and alert advisories (medium confidence). [13.2.5]

During the last decades important changes in precipitation and increases in temperature have been observed (high confidence).

Increases in rainfall in south-east Brazil, Paraguay, Uruguay, the Argentinean Pampas and some parts of Bolivia have had impacts on land use and crop yields, and have increased flood frequency and intensity. On the other hand, a declining trend in precipitation has been observed in southern Chile, south-west Argentina, southern Peru and western Central America. Increases in temperature of approximately 1°C in Mesoamerica and South America, and of 0.5°C in Brazil, were observed. As a consequence of temperature increases, the trend in glacier retreat reported in the Third Assessment Report is accelerating (very high confidence). This issue is critical in Bolivia, Peru, Colombia and Ecuador, where water availability has already been compromised either for consumption or for hydropower generation [13.2.4.1]. These problems with supply are expected to increase in the future, becoming chronic if no appropriate adaptation measures are planned and implemented. Over the next decades Andean inter-tropical glaciers are very likely to disappear, affecting water availability and hydropower generation (high confidence). [13.2.4.1]

Land-use changes have intensified the use of natural resources and exacerbated many of the processes of land degradation (high confidence).

Almost three-quarters of the drylands are moderately or severely affected by degradation processes. The combined effects of human action and climate change have brought about a continuous decline in natural land cover at very high rates (high confidence). In particular, rates of deforestation of tropical forests have increased during the last 5 years. There is evidence that biomass-burning aerosols may change regional temperature and precipitation in the southern part of Amazonia (medium confidence). Biomass burning also affects regional air quality, with implications for human health. Land-use and climate changes acting synergistically will increase vegetation fire risk substantially (high confidence). [13.2.3, 13.2.4.2]

The projected mean warming for Latin America to the end of the century, according to different climate models, ranges from 1 to 4°C for the SRES emissions scenario B2 and from 2 to 6°C for scenario A2 (medium confidence).

Most general circulation model (GCM) projections indicate rather larger (positive and negative) rainfall anomalies for the tropical portions of Latin America and smaller ones for extra-tropical South America. In addition, the frequency of occurrence of weather and climate extremes is likely to increase in the future; as is the frequency and intensity of hurricanes in the Caribbean Basin. [13.3.1.1, 13.3.1.2]

Under future climate change, there is a risk of significant species extinctions in many areas of tropical Latin America (high confidence).

Replacement of tropical forest by savannas is expected in eastern Amazonia and the tropical forests of central and southern Mexico, along with replacement of semi-arid vegetation by arid vegetation in parts of north-east Brazil and most of central and northern Mexico due to synergistic effects of both land-use and climate changes (medium confidence) [13.4.1]. By the 2050s, 50% of agricultural lands are very likely to be subjected to desertification and salinisation in some areas (high confidence) [13.4.2]. Seven out of the 25 most critical places with high endemic species concentrations are in Latin America and these areas are undergoing habitat loss. Biological reserves and ecological corridors have been either implemented or planned for the maintenance of biodiversity in natural ecosystems, and these can serve as adaptation measures to help protect ecosystems in the face of climate change. [13.2.5.1]

By the 2020s, the net increase in the number of people experiencing water stress due to climate change is likely to be between 7 and 77 million (medium confidence).

While, for the second half of the century, the potential water availability reduction and the increasing demand from an increasing regional population would increase these figures to between 60 and 150 million. [13.4.3]

Generalised reductions in rice yields by the 2020s, as well as increases in soybean yields, are possible when CO_2 effects are considered (medium confidence).

For other crops (wheat, maize), the projected response to climate change is more erratic, depending on the chosen scenario. If CO_2 effects are not considered, the number of additional people at risk of hunger under the A2 scenario is likely to reach 5, 26 and 85 million in 2020, 2050 and 2080, respectively (medium confidence). On the other hand, cattle and dairy productivity is expected to decline in response to increasing temperatures. [13.4.2]

The expected increases in sea-level rise (SLR), weather and climatic variability and extremes are very likely to affect coastal areas (high confidence).

During the last 10-20 years the rate of SLR has increased from 1 to 2-3 mm/yr in south-eastern South America [13.2.4.1]. In the future, adverse impacts would be observed on: (i) low-lying areas (e.g., in El Salvador, Guyana and the coast of Buenos Aires Province in Argentina), (ii) buildings and tourism (e.g., in Mexico and Uruguay); (iii) coastal morphology (e.g., in Peru); (iv) mangroves (e.g., in Brazil, Ecuador, Colombia and Venezuela); (v) availability of drinking water on the Pacific coast of Costa Rica, Ecuador and the Rio de la Plata estuary. In particular, sea-level rise is very likely to affect both Mesoamerican coral reefs (e.g., in Mexico, Belize and Panama) and the location of fish stocks in the south-east Pacific (e.g., in Peru and Chile). [13.4.4]

Future sustainable development plans should include adaptation strategies to enhance the integration of climate change into development policies (high confidence).

Some countries have made efforts to adapt, particularly through conservation of key ecosystems, early warning systems, risk management in agriculture, strategies for flood, drought and coastal management, and disease surveillance systems. However, the effectiveness of these efforts is outweighed by: a lack of basic information, observation and monitoring systems; lack of capacity-building and appropriate political, institutional and technological frameworks; low income; and settlements in vulnerable areas; among others [13.2]. Without improvements in these areas, the Latin America countries' sustainable development goals will be seriously compromised, adversely affecting, among other things, their ability to reach the Millennium Development Goals [13.5].

13.1 Summary of knowledge assessed in the Third Assessment Report

The principal findings in the Third Assessment Report (TAR) (IPCC, 2001) were as follows.

- In most of Latin America, there are no clear long-term tendencies in mean surface temperature. Nevertheless, for some areas in the region, there are some clear warming (Amazonia, north-western South America) and, in a few cases, cooling (Chile) trends.
- Precipitation trends suggest an increase in precipitation for some regions of the mid-latitude Americas, a decrease for some central regions in Latin America, and no clear trends for others. For instance, the positive trends seen in north-eastern Argentina, southern Brazil and north-western Mexico contrast with the negative trends observed in some parts of Central America (e.g., Nicaragua). Records suggest a positive trend for the past 200 years at higher elevations in north-western Argentina. In Amazonia, inter-decadal variability in the hydrological record (in both rainfall and streamflow) is more significant than any observed trend.
- El Niño-Southern Oscillation (ENSO) is the dominant mode of climate variability in Latin America and is the natural phenomenon with the largest socio-economic impacts.
- Glaciers in Latin America have receded dramatically in the past decades, and many of them have disappeared completely. The most affected sub-regions are the Peruvian Andes, southern Chile and Argentina up to latitude 25°S. Deglaciation may have contributed to observed negative trends in streamflows in that region.
- In Latin America many diseases are weather and climate-related through the outbreaks of vectors that develop in warm and humid environments, including malaria and dengue. Climate change could influence the frequency of outbreaks of these diseases by altering the variability associated with the main controlling phenomenon, i.e., El Niño (likely).
- Agriculture in Latin America is a very important economic activity representing about 10% of the gross domestic product (GDP) of the region. Studies in Argentina, Brazil, Chile, Mexico and Uruguay based on General Circulation Models (GCMs) and crop models project decreased yields for numerous crops (e.g., maize, wheat, barley, grapes) even when the direct effects of CO_2-fertilisation and implementation of moderate adaptation measures at the farm level are considered.
- Assessments of the potential impacts of climate change on natural ecosystems indicate that neotropical seasonally dry forest should be considered severely threatened in Mesoamerica. Global warming could expand the area suitable for tropical forests in South America southwards, but current land use makes it unlikely that tropical forests will be permitted to occupy these new areas. On the other hand, large portions of the Amazonian forests could be replaced by tropical savannas due to land-use change and climate change.
- Sea-level rise will affect mangrove ecosystems, damaging the region's fisheries. Coastal inundation and erosion resulting from sea-level rise in combination with riverine and flatland flooding would affect water quality and availability. Sea-water intrusion would exacerbate socio-economic and health problems in these areas.
- The adaptive capacity of human systems in Latin America is low, particularly to extreme climate events, and vulnerability is high. Adaptation measures have the potential to reduce climate-related losses in agriculture and forestry but less ability to do so for biological diversity.

13.2 Current sensitivity/vulnerability

13.2.1 What is distinctive about the Latin America region?

Latin America is highly heterogeneous in terms of climate, ecosystems, human population distribution and cultural traditions. A large portion of the region is located in the tropics, showing a climate dominated by convergence zones such as the Inter-tropical Convergence Zone (ITCZ), and the South Atlantic Convergence Zone (SACZ) (Satyamurty et al., 1998).

The summer circulation in tropical and sub-tropical Latin America is dominated by the North America Monsoon System, which affects Mexico and parts of Central America, and the South America Monsoon System, which affects tropical and sub-tropical South America east of the Andes. These monsoon climates are closely interconnected with ocean-atmosphere interactions over the tropical and sub-tropical oceans. Low Level Jets in South America east (Marengo et al., 2004) and west (Poveda and Mesa, 2000) of the Andes, and in North America east of the Rockies, Baja California and over the Intra-Americas Seas transport moisture from warm oceans to participate in continental rainfall. Most of the rainfall is concentrated in the convergence zones or by topography, leading to strong spatial and temporal rainfall contrasts, such as the expected sub-tropical arid regions of northern Mexico and Patagonia, the driest desert in the world in northern Chile, and a tropical semi-arid region of north-east Brazil located next to humid Amazonia and one of the wettest areas in the world in western Colombia. A remarkable ecogeographical zone is that of the South America's highlands (see case study in Box 13.2), located in the tropics and presenting paramo-like (neotropical Andean ecosystem, about 3,500 m above sea level) landscapes with deep valleys (yungas) holding important biodiversity, with a wealth of vegetal and animal species.

13.2.2 Weather and climate stresses

Over the past three decades, Latin America has been subjected to climate-related impacts of increased El Niño occurrences (Trenberth and Stepaniak, 2001). Two extremely intense episodes of the El Niño phenomenon (1982/83 and 1997/98) and other severe climate extremes (EPA, 2001; Vincent et al., 2005; Haylock et al., 2006) have happened during this period, contributing greatly to the heightened vulnerability of human systems to natural disasters (floods, droughts, landslides, etc.).

Since the TAR, several highly unusual extreme weather events have been reported, such as the Venezuelan intense precipitations of 1999 and 2005; the flooding in the Argentinean Pampas in 2000 and 2002; the Amazon drought of 2005; the unprecedented and destructive hail storms in Bolivia in 2002 and Buenos Aires in 2006; the unprecedented Hurricane Catarina in the South Atlantic in 2004; and the record hurricane season of 2005 in the Caribbean Basin. The occurrence of climate-related disasters increased by 2.4 times between the periods 1970-1999 and 2000-2005 continuing the trend observed during the 1990s. Only 19% of the events between 2000 and 2005 have been economically quantified, representing losses of nearly US$20 billion (Nagy et al., 2006a). Table 13.1 shows some of the most important recent events.

In addition to weather and climate, the main drivers of increased vulnerability are demographic pressure, unregulated urban growth, poverty and rural migration, low investment in infrastructure and services, and problems with inter-sectoral co-ordination. The poorest communities are among the most vulnerable to extreme events (UNEP, 2003a), and some of these vulnerabilities are caused by their location in the path of hurricanes (about 8.4 million people in Central America; FAO, 2004a), on unstable lands, in precarious settlements, on low-lying areas, and in places prone to flooding from rivers (BID, 2000; UNEP, 2003a).

Table 13.1. *Selected extreme events and their impacts (period 2004-2006).*

Event/Date	Country/Impacts
Hurricane (H.) Beta Nov. 2005	Nicaragua: 4 deaths; 9,940 injuries; 506 homes, 250 ha of crops, 240 km^2 of forest and 2,000 artisan fishermen affected (SINAPRED, 2006).
H. Wilma Oct. 2005	Mexico: several landfalls, mainly in the Yucatán Peninsula. Losses of US$1,881 million. 95% of the tourist infrastructure seriously damaged.
H. Stan Oct. 2005	Guatemala, Mexico, El Salvador, Nicaragua, Costa Rica: losses of US$3,000 million, more than 1,500 deaths. Guatemala was the most affected country, accounting for 80% of the casualties and more than 60% of the infrastructure damage (Fundación DESC, 2005).
Extra-tropical cyclone Aug. 2005	Southern Uruguay: extra-tropical cyclone (winds up to 187 km/h, and storm surge), 100,000 people affected, more than 100 people injured and 10 people dead, 20,000 houses without electricity, telephone and/or water supply (NOAA, 2005; Bidegain et al., 2006).
H. Emily Jul. 2005	Mexico – Cozumel and Quintana Roo: losses of US$837 million. Tourism losses: US$100 million; dunes and coral reefs affected; loss of 1,506 turtle nests; 1-4 m storm surges (CENAPRED-CEPAL, 2005).
Heavy rains Sep. 2005	Colombia: 70 deaths, 86 injured, 6 disappeared and 140,000 flood victims (NOAA, 2005).
Heavy rains Feb. 2005	Venezuela: heavy precipitation (mainly on central coast and in Andean mountains), severe floods and heavy landslides. Losses of US$52 million; 63 deaths and 175,000 injuries (UCV, 2005; DNPC, 2005/06).
H. Catarina Mar. 2004	Brazil: the first hurricane ever observed in the South Atlantic (Pezza and Simmonds, 2005); demolished over 3,000 houses in southern Brazil (Cunha et al., 2004); severe flooding hit eastern Amazonia, affecting tens of thousands of people (http://www.cptec.inpe.br/).
Droughts 2004-2006	Argentina – Chaco: losses estimated at US$360 million; 120,000 cattle lost, 10,000 evacuees in 2004 (SRA, 2005). Also in Bolivia and Paraguay: 2004/05. Brazil-Amazonia: severe drought affected central and south-western Amazonia, probably associated with warm sea surface temperatures in the tropical North Atlantic (http://www.cptec.inpe.br/). Brazil – Rio Grande do Sul: reductions of 65% and 56% in soybean and maize production (http://www.ibge.gov.br/home/).

Natural ecosystems

Tropical forests of Latin America, particularly those of Amazonia, are increasingly susceptible to fire occurrences due to increased El Niño-related droughts and to land-use change (deforestation, selective logging and forest fragmentation) (see Box 13.1; Fearnside, 2001; Nepstad et al., 2002; Cochrane, 2003). During the 2001 ENSO period, approximately one-third of the Amazon forests became susceptible to fire (Nepstad et al., 2004). This climatic phenomenon has the potential to generate large-scale forest fires due to the extended period without rain in the Amazon, exposing even undisturbed dense forest to the risk of understorey fire (Jipp et al., 1998; Nepstad et al., 2002, 2004). Mangrove forests located in low-lying coastal areas are particularly vulnerable to sea-level rise, increased mean temperatures, and hurricane frequency and intensity (Cahoon and Hensel, 2002; Schaeffer-Novelli et al., 2002), especially those of Mexico, Central America and Caribbean continental regions (Kovacs, 2000; Meagan et al., 2003). Moreover, floods accelerate changes in mangrove areas and at their landward interface (Conde, 2001; Medina et al., 2001; Villamizar, 2004). In relation to biodiversity, populations of toads and frogs are affected in cloud forests after years of low precipitation (Pounds et al., 1999; Ron et al., 2003; Burrowes et al., 2004). In Central and South America, links between higher temperatures and frog extinctions caused by a skin disease (*Batrachochytrium dendrobatidis*) were found (Dey, 2006).

Agriculture

The impact of ENSO-related climate variability on the agricultural sector was well documented in the TAR (IPCC, 2001). More recent findings include: high/low wheat yields during El Niño/La Niña in Sonora, Mexico (Salinas-Zavala and Lluch-Cota, 2003); shortening of cotton and mango growing cycles on the northern coast of Peru during El Niño because of increases in temperature (Torres et al., 2001); increases in the incidence of plant diseases such as 'cancrosis' in citrus in Argentina (Canteros et al., 2004), *Fusarium* in wheat in Brazil and Argentina (Moschini et al., 1999; Del Ponte et al., 2005), and several fungal diseases in maize, potato, wheat and beans in Peru (Torres et al., 2001) during El Niño events, due to high rainfall and humidity. In relation to other sources of climatic variability, anomalies in South Atlantic sea-surface temperatures (SST) were significantly related to crop-yield variations in the Pampas region of Argentina (Travasso et al., 2003a, b). Moreover, heatwaves in central Argentina have led to reductions in milk production in Holando argentino (Argentine Holstein) dairy cattle, and the animals were not able to completely recover after these events (Valtorta et al., 2004).

Water resources

In global terms, Latin America is recognised as a region with large freshwater resources. However, the irregular temporal and spatial distribution of these resources affects their availability and quality in different regions. Stress on water availability and quality has been documented where lower precipitation and/or higher temperatures occur. For example, droughts related to La Niña create severe restrictions for water supply and irrigation demands in central western Argentina and central Chile between 25°S and 40°S (NC-Chile, 1999; Maza et al., 2001). In addition, droughts related to El Niño impacts on the flows of the Colombia Andean region basins (particularly in the Cauca river basin), causing a 30% reduction in the mean flow, with a maximum of 80% loss in some tributaries (Carvajal et al., 1998), whereas extreme floods are enhanced during La Niña (Waylen and Poveda, 2002). In addition, the Magdalena river basin also shows high vulnerability (55% losses in mean flow; IDEAM, 2004). Consequently, soil moisture and vegetation activity are strongly reduced/augmented by El Niño/La Niña in Colombia (Poveda et al., 2001a). The vulnerability to flooding events is high in almost 70% of the area represented by Latin American countries (UNEP, 2003c). Hydropower is the main electrical energy source for most countries in Latin America, and is vulnerable to large-scale and persistent rainfall anomalies due to El Niño and La Niña, e.g., in Colombia (Poveda et al., 2003), Venezuela (IDEAM, 2004), Peru (UNMSM, 2004), Chile (NC-Chile, 1999), Brazil, Uruguay and Argentina (Kane, 2002). A combination of increased energy demand and drought caused a virtual breakdown in hydroelectricity generation in most of Brazil in 2001, which contributed to a GDP reduction of 1.5% (Kane, 2002).

Coasts

Low-lying coasts in several Latin American countries (e.g., parts of Argentina, Belize, Colombia, Costa Rica, Ecuador, Guyana, Mexico, Panama, El Salvador, Uruguay and Venezuela) and large cities (e.g., Buenos Aires, Rio de Janeiro and Recife) are among the most vulnerable to climate variability and extreme hydrometeorological events such as rain and windstorms, and sub-tropical and tropical cyclones (i.e., hurricanes) and their associated storm surges (Tables 13.1 and 13.2). Sea-level rise (within the range 10-20 cm/century) is not yet a major problem, but evidence of an acceleration of sea-level rise (SLR) rates (up to 2-3 mm/yr) over the past decade suggests an increase in the vulnerability of low-lying coasts, which are already subjected to increasing storm surges (Grasses et al., 2000; Kokot, 2004; Kokot et al., 2004; Miller, 2004; Barros, 2005; Nagy et al., 2005; UCC, 2005). Moreover, some coastal areas are affected by the combined effects of heavy precipitation, landward winds and SLR (for example, 'sudestadas' in the La Plata river estuary), as already observed in the city of Buenos Aires (EPA, 2001; Bischoff, 2005).

Human health

After the onset of El Niño (dry/hot) there is a risk of epidemic malaria in coastal regions of Colombia and Venezuela (Poveda et al., 2001b; Kovats et al., 2003). Droughts favour the development of epidemics in Colombia and Guyana, while flooding engenders epidemics in the dry northern coastal region of Peru (Gagnon et al., 2002). Annual variations in dengue/dengue haemorrhagic fever in Honduras and Nicaragua appear to be related to climate-driven fluctuations in the vector densities (temperature, humidity, solar radiation and rainfall) (Patz et al., 2005). In some coastal areas of the Gulf of Mexico, an increase in SST, minimum temperature and precipitation was associated with an increase in dengue transmission cycles (Hurtado-Díaz et al., 2006). Outbreaks of hantavirus pulmonary syndrome have been reported for Argentina, Bolivia, Chile, Paraguay, Panama and Brazil after prolonged droughts (Williams et al., 1997; Espinoza et al., 1998; Pini et al., 1998; CDC, 2000), probably due to the intense rainfall and flooding following the droughts, which increases food availability for peri-domestic (living both indoors and outdoors) rodents (see Chapter 8, Section 8.2.8). Prolonged droughts in semi-arid north-eastern Brazil have provoked rural-urban migration of subsistence farmers, and a re-emergence of visceral leishmaniasis (Confalonieri, 2003). A significant increase in visceral leishmaniasis in Bahia State (Brazil) after the El Niño years of 1989 and 1995 has also been reported (Franke et al., 2002). In Venezuela, an increase in cutaneous leishmaniasis was associated with a weak La Niña (Cabaniel et al., 2005). Flooding produces outbreaks of leptospirosis in Brazil, particularly in densely populated areas without adequate drainage (Ko et al., 1999; Kupek et al., 2000; Chapter 8, Section 8.2.8). In Peru, El Niño has been associated with some dermatological diseases, related to an increase in summer temperature (Bravo and Bravo, 2001); hyperthermia with no infectious cause has also been related to heatwaves (Miranda et al., 2003), and SST has been associated with the incidence of Carrion's disease (*Bartonella bacilliformis*) (Huarcaya et al., 2004). In Buenos Aires roughly 10% of summer deaths may be associated with thermal stress caused by the 'heat island' effect (de Garín and Bejarán, 2003). In São Paulo, Brazil, Gouveia et al. (2003) reported an increase of 2.6% in all-cause morbidity in the elderly per °C increase in temperature above 20°C, and a 5.5% increase per °C drop in temperature below 20°C (see Chapter 8).

13.2.3 Non-climatic stresses

Effects of demographic pressure

Migration to urban areas in the region exceeds absorption capacity, resulting in widespread unemployment, overcrowding, and the spread of infectious diseases including HIV/AIDS, due to lack of adequate infrastructure and urban planning (UNEP, 2003b). Latin America is the most urbanised region in the developing world (75% of its population). The most urbanised countries are Argentina, Brazil, Chile, Uruguay and Venezuela, while the least urbanised are Guatemala and Honduras (UNCHS, 2001). As a consequence, the regional population faces both traditional (infectious and transmissible diseases) and modern risks (chronic and degenerative diseases) in addition to those related to urban landslides and floods. Modern risks result from urbanisation and industrialisation, while poor and rural populations still suffer from 'traditional risks'. There is a significant problem of urban poverty in areas where malnutrition, poor water quality and a lack of sewage/sanitary services and education prevail. However, the line between urban and rural in many parts of the region is becoming increasingly blurred, particularly around large urban areas.

A strong reduction in employment rates, with the associated downgrading of the social situation, observed in Latin America in the 1990s (poverty affecting 48.3% and extreme poverty 22.5% of the population), has generated large-scale migration to urban areas. Although this migration trend continues, the Economic Commission for Latin America and the Caribbean (ECLAC) reports that, in spite of the fact that the region is on track to meet the Millenium Development Goals' (MDGs) extreme poverty goal (to halve the number of people living on less than $1/day by 2015), the year 2006 would show a reduction to 38.5% and 14.7% of the above poverty indices (La Nación, 2006).

Over-exploitation of natural resources

It is well established that over-exploitation is a threat to 34 out of 51 local production systems of particular importance to artisanal fishing along the coastal waters in Latin America (UNEP, 2003b; FAO, 2006) and has caused the destruction of habitats such as mangroves, estuaries and salt marshes in Central America and Mexico (Cocos in Costa Rica, Tortuguero-Miskitos Islands in Nicaragua and the Gulf of Mexico in Mexico) (Mahon, 2002; NOAA/OAR, 2004).

Urbanisation (without a land planning or legal framework in most of the countries), large aquaculture developments, the expansion of ecotourism and the oil industry, the accidental capture of ecologically important species, the introduction of exotic species, land-based sources of coastal and marine pollution, the depletion of coral reefs and the mismanagement of water resources impose increasing environmental pressures on natural resources (Young, 2001; Viddi and Ribeiro, 2004).

The rapidly expanding tourism industry is driving much of the transformation of natural coastal areas, paving the way for resorts, marinas and golf courses (WWF, 2004). Aquifer over-exploitation and mismanagement of irrigation systems are causing severe environmental problems; e.g., salinisation of soil and water in Argentina (where more than 500,000 ha of the phreatic (i.e., permanently-saturated) aquifer shows high levels of salinity and nitrates) (IRDB, 2000) and sanitation problems in a great number of cities such as Mexico City, San José de Costa Rica, and Trelew, Río Cuarto and La Plata in Argentina. In Belize City, a system of mangrove-lined ponds and mangrove-wetland drainage areas has served as a natural sewage treatment facility for much of the city's waste water. Recently, dredging for a massive port expansion has resulted in the destruction of more mangroves and the free ecosystem services they provided (WWF, 2004).

Pollution

Pollution of natural resources, such as natural arsenic contamination of freshwater, affects almost 2 million people in Argentina, 450,000 in Chile, 400,000 in Mexico, 250,000 in Peru and 20,000 in Bolivia (Canziani, 2003; Pearce, 2003; Clark and King, 2004). Another insidious contamination widespread in the region is produced by fluorine. In the Puyango river basin (Ecuador), suspended sediments and metal contamination increase significantly during ENSO events (Tarras-Wahlberg and Lane, 2003). In the upper Pilcomayo basin, south-east Bolivia, pollution by heavy metals from mining operations in Potosí affects the migration and fishing of sábalo (*Prochilodus lineatus*), which is a very important source of income in the region (Smolders et al., 2002). As a result of the Salado del Norte (Argentina) river flood of 2003 (which covered more than one-third of the urban district), 60,000 tonnes of solid waste were disseminated all over the city of Santa Fe; 135 cases of hepatitis,

116 of leptospirosis and 5,000 of lung disease were officially reported as a result (Bordón, 2003).

Air pollution due to the burning of fossil fuels is a problem that affects many cities of Latin America. Transport is the main contributor (e.g., in Mexico City, Santiago de Chile and São Paulo). Thermoelectric energy generation is the second primary source of air pollution in Lima, Quito and La Paz (PAHO, 2005). Climate and geography play a significant role in this situation; e.g., the occurrence of thermal inversions, such as in Mexico City, Lima and Santiago de Chile. In Mexico City, surface ozone has been linked to increased hospital admissions for lower respiratory infections and asthma in children (Romieu et al., 1996). Regarding exposure effects to biomass particles, Cardoso de Mendonça et al. (2004) have estimated that the economic costs of fire in the Amazon affecting human health increased from US$3.4 million in 1996 to US$10.7 million in 1999.

13.2.4 Past and current trends

13.2.4.1 Climate trends

During the 20th century, significant increases in precipitation were observed in southern Brazil, Paraguay, Uruguay, north-east Argentina and north-west Peru and Ecuador. Conversely, a declining trend in precipitation was observed in southern Chile, south-west Argentina and southern Peru (Figure 13.1, Table 13.2). In addition, increases in the rate of sea-level rise have reached 2-3 mm/yr during the last 10-20 years in south-eastern South America (Table 13.2).

A number of regional studies have been completed for southern South America (Vincent et al., 2005; Alexander et al., 2006; Haylock et al., 2006; Marengo and Camargo, 2007), Central America and northern South America (Poveda et al., 2001a; Aguilar et al., 2005; Alexander et al., 2006). They all show patterns of changes in extremes consistent with a general warming, especially positive trends for warm nights and negative trends for the occurrence of cold nights. There is also a positive tendency for intense rainfall events and consecutive dry days. A study by Groisman et al. (2005) identified positive linear trends in the frequency of very heavy rains over north-east Brazil and central Mexico. However, the lack of long-term records of daily temperature and rainfall in most of tropical South America does not allow for any conclusive evidence of trends in extreme events in regions such as Amazonia. Chapter 3, Section 3.8 of the Working Group I Fourth Assessment Report (Trenberth et al., 2007) discusses observational aspects of variability of extreme events and tropical cyclones. Chapter 11, Section 11.6 of the Working Group I Fourth Assessment Report (Christensen et al., 2007) acknowledges that little research is available on extremes of temperature and precipitation for this region.

These changes in climate are already affecting several sectors. Some reported impacts associated with heavy precipitation are: 10% increase in flood frequency due to increased annual discharge in the Amazon River at Obidos (Callède et al., 2004); increases of up to 50% in streamflow in the rivers Uruguay, Paraná and Paraguay (Bidegain et al., 2005; Camilloni, 2005b); floods in the Mamore basin in Bolivian Amazonia (Ronchail et al., 2005); and increases in morbidity and mortality due to flooding, landslides and storms in Bolivia (NC-Bolivia, 2000). In addition, positive impacts were reported for the Argentinean Pampas region, where increases in precipitation led to increases in crop yields close to 38% in soybean, 18% in maize, 13% in wheat and 12% in sunflower (Magrin et al., 2005). In the same way, pasture productivity increased by 7% in Argentina and Uruguay (Gimenez, 2006).

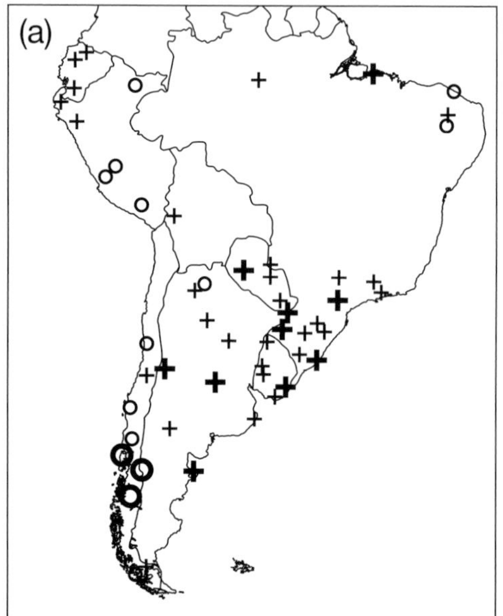

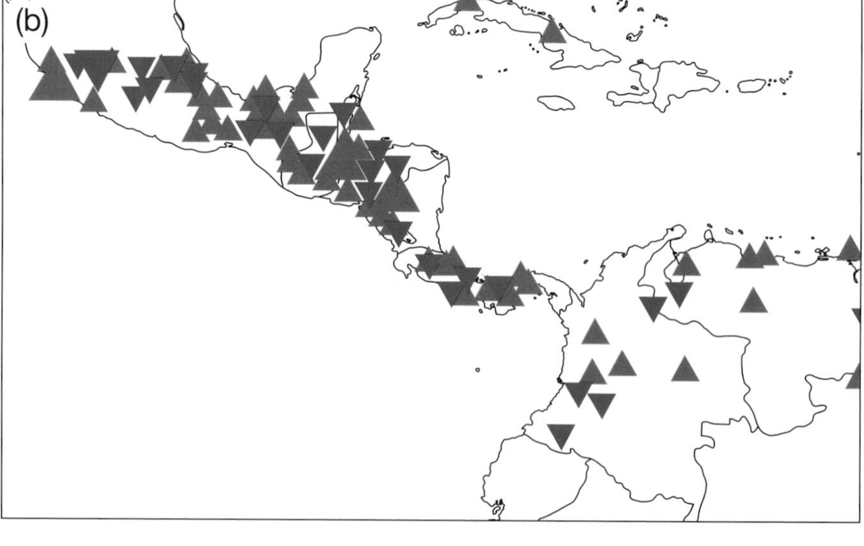

Figure 13.1. *Trends in rainfall in (a) South America (1960-2000). An increase is shown by a plus sign, a decrease by a circle. Bold values indicate significance at P 0.05. Haylock et al. (2006); reprinted with permission from the American Meteorological Society. (b) Central America and northern South America (1961-2003). Large red triangles indicate positive significant trends, small red triangles indicate positive non-significant trends, large blue triangles indicate negative significant trends, and small blue triangles indicate negative non-significant trends. Aguilar et al. (2005); reprinted with permission from the American Geophysical Union.*

Table 13.2. *Current climatic trends.*

Precipitation (change shown in % unless otherwise indicated)	Period	Change
Amazonia – northern/southern (Marengo, 2004)	1949-1999	–11 to –17/–23 to +18
Bolivian Amazonia (Ronchail et al., 2005)	since 1970	+15
Argentina – central and north-east (Penalba and Vargas, 2004)	1900-2000	+1 STD to +2 STD
Uruguay (Bidegain et al., 2005)	1961-2002	+ 20
Chile – central (Camilloni, 2005a)	last 50 years	–50
Colombia (Pabón, 2003a)	1961-1990	–4 to +6
Mean temperature (°C/10 years)		
Amazonia (Marengo, 2003)	1901-2001	+0.08
Uruguay, Montevideo (Bidegain et al., 2005)	1900-2000	+0.08
Ecuador (NC-Ecuador, 2000)	1930-1990	+0.08 to +0.27
Colombia (Pabón, 2003a)	1961-1990	+0.1 to +0.2
Maximum temperature (°C/10 years)		
Brazil – south (Marengo and Camargo, 2007)	1960-2000	+0.39 to +0.62
Argentina – central (Rusticucci and Barrucand, 2004)	1959-1998	–0.2 to –0.8 (DJF)
Argentina – Patagonia (Rusticucci and Barrucand, 2004)	1959-1998	+0.2 to +0.4 (DJF)
Minimum temperature (°C/10 years)		
Brazil – south (Marengo and Camargo, 2007)	1960-2000	+0.51 to +0.82
Brazil – Campinas and Sete Lagoas (Pinto et al., 2002)	1890-2000	+0.2
Brazil – Pelotas (Pinto et al., 2002)	1890-2000	+0.08
Argentina (Rusticucci and Barrucand, 2004)	1959-1998	+0.2 to +0.8 (DJF/JJA)
Sea-level rise (mm/yr)		
Guyana (NC-Guyana, 2002)	last century	+1.0 to +2.4
Uruguay, Montevideo (Nagy et al., 2005)	last 100/30/15 years	+1.0 / +2.5 / +4.0
Argentina, Buenos Aires (Barros, 2003)	last ~100 years	+1.7
Brazil – several ports (Mesquita, 2000)	1960-2000	+4.0
Panama – Caribbean coast (NC-Panama, 2000)	1909-1984	+1.3
Colombia (Pabón, 2003b)	1961-1990	+1 to +3

STD= standard deviation, DJF= December/January/February, JJA= June/July/August.

The glacier-retreat trend reported in the TAR has intensified, reaching critical conditions in Bolivia, Peru, Colombia and Ecuador (Table 13.3). Recent studies indicate that most of the South American glaciers from Colombia to Chile and Argentina (up to 25°S) are drastically reducing their volume at an accelerated rate (Mark and Seltzer, 2003; Leiva, 2006). Changes in temperature and humidity are the primary cause of the observed glacier retreat during the second half of the 20th century in the tropical Andes (Vuille et al., 2003). During the next 15 years, inter-tropical glaciers are very likely to disappear, affecting water availability and hydropower generation (Ramírez et al., 2001).

Table 13.3. *Glacier retreat trends.*

Glaciers/Period	Changes/Impacts
Peru[a, b] Last 35 years	22% reduction in glacier total area; reduction of 12% in freshwater in the coastal zone (where 60% of the country's population live). Estimated water loss almost 7,000 Mm^3
Peru[c] Last 30 years	Reduction up to 80% of glacier surface from small ranges; loss of 188 Mm^3 in water reserves during the last 50 years.
Colombia[d] 1990-2000	82% reduction in glaciers, showing a linear withdrawal of the ice of 10-15 m/yr; under the current climate trends, Colombia's glaciers will disappear completely within the next 100 years.
Ecuador[e] 1956-1998	There has been a gradual decline glacier length; reduction of water supply for irrigation, clean water supply for the city of Quito, and hydropower generation for the cities of La Paz and Lima.
Bolivia[f] Since mid-1990s	Chacaltaya glacier has lost half of its surface and two-thirds of its volume and could disappear by 2010. Total loss of tourism and skiing.
Bolivia[f] Since 1991	Zongo glacier has lost 9.4% of its surface area and could disappear by 2045-2050; serious problems in agriculture, sustainability of 'bofedales'[1] and impacts in terms of socio-economics for the rural populations.
Bolivia[f] Since 1940	Charquini glacier has lost 47.4% of its surface area.

[a]Vásquez, 2004; [b]Mark and Seltzer, 2003; [c]NC-Perú, 2001; [d]NC-Colombia, 2001; [e]NC-Ecuador, 2000; [f]Francou et al., 2003.

[1] Bofedales: wetlands and humid areas of the Andean high plateaux.

13.2.4.2 Environmental trends

Deforestation and changes in land use

In 1990, the total forest area in Latin America was 1,011 Mha, which has reduced by 46.7 Mha in the 10 years from 1990 to 2000 (UNEP, 2003a) (Figure 13.2). In Amazonia, the total area of forest lost rose by 17.2 Mha from 41.5 Mha in 1990 to 58.7 Mha in 2000 (Kaimowitz et al., 2004). The expansion of the agricultural frontier and livestock, selective logging, financing of large-scale projects such as the construction of dams for energy generation, illegal crops, the construction of roads and increased links to commercial markets have been the main causes of deforestation (FAO, 2001a; Laurance et al., 2001; Geist and Lambin, 2002; Asner et al., 2005; FAO, 2005; Colombia Trade News, 2006).

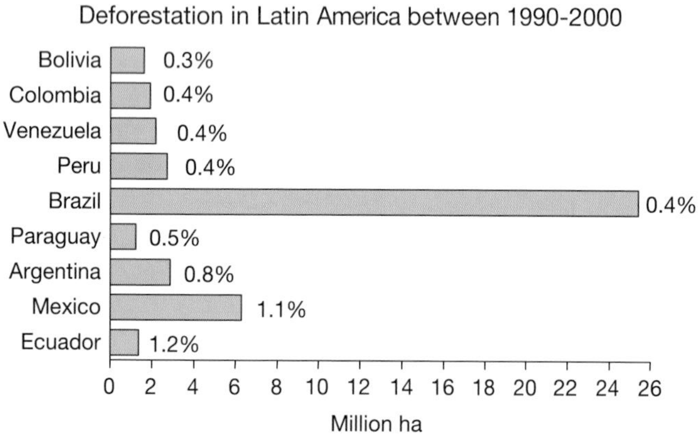

Figure 13.2. *Total deforestation in Latin America (Mha) between 1990 and 2000. Number indicates deforestation rate (%/yr) for each country. Based on FAO (2001a).*

Natural land cover has continued to decline at very high rates. In particular, rates of deforestation of tropical forests have increased during the last five years. Annual deforestation in Brazilian Amazonia increased by 32% between 1996 and 2000 (1.68 Mha) and 2001 and 2005 (2.23 Mha). However, the annual rate of deforestation decreased from 2.61 Mha in 2004 to 1.89 Mha in 2005 (INPE-MMA, 2005a, b, c). An area of over 60 Mha has been deforested in Brazilian Amazonia due to road construction and subsequent new urban settlements (Alves, 2002; Laurance et al., 2005). There is evidence that aerosols from biomass burning may change regional temperature and precipitation south of Amazonia (Andreae et al., 2004) and in neighbouring countries, including the Pampas as far south as Bahía Blanca (Trosnikov and Nobre, 1998; Mielnicki et al., 2005), with related health implications (increases in mortality risk, restricted activity days and acute respiratory symptoms) (WHO/UNEP/WMO, 2000; Betkowski, 2006).

The soybean cropping boom has exacerbated deforestation in Argentina, Bolivia, Brazil and Paraguay (Fearnside, 2001; Maarten Dros, 2004). This critical land-use change will enhance aridity/desertification in many of the already water-stressed regions in South America. Major economic interests not only affect the landscape but also modify the water cycle and the climate of the region, in which almost three-quarters of the drylands are moderately or severely affected by degradation processes and droughts (Malheiros, 2004). The region contains 16% of the world total of 1,900 Mha of degraded land (UNEP, 2000). In Brazil, 100 Mha are facing desertification processes, including the semi-arid and dry sub-humid regions (Malheiros, 2004).

Biodiversity

Changes in land use have led to habitat fragmentation and biodiversity loss. Climate change will increase the actual extinction rate, which is documented in the Red List of Endangered Species (IUCN, 2001). The majority of the endangered eco-regions are located in the northern and mid-Andes valleys and plateaux, the tropical Andes, in areas of cloud forest (e.g., in Central America), in the South American steppes, and in the Cerrado and other dry forests located in the south of the Amazon Basin (Dinerstein et al., 1995; UNEP, 2003a) (see Figure 13.5). Among the species to disappear are Costa Rica's golden toad (*Bufo periglenes*) and harlequin frog (*Atelopus* spp.) (Shatwell, 2006). In addition, at least four species of Brazilian anurans (frogs and toads) have declined as a result of habitat alteration (Eterovick et al., 2005), and two species of *Atelopus* have disappeared following deforestation (La Marca and Reinthaler, 2005). Deforestation and forest degradation through forest fires, selective logging, hunting, edge effects and forest fragmentation are the dominant transformations that threaten biodiversity in South America (Fearnside, 2001; Peres and Lake, 2003; Asner et al., 2005).

Coral reefs and mangroves

Panama and Belize Caribbean case studies illustrate, in terms of inter-ocean contrasts, both the similarities and differences in coral-reef responses to complex environmental changes (Gardner et al., 2003; Buddemeier et al., 2004). Cores taken from the Belizean barrier reef show that *A. cervicornis* dominated this coral-reef community continuously for at least 3,000 years, but was killed by white band disease (WBD) and replaced by another species after 1986 (Aronson and Precht, 2002). Dust transported from Africa to America (Shinn et al., 2000), and land-derived flood plumes from major storms, can transport materials from the Central American mainland to reefs, which are normally considered remote from such influences, as potential sources of pathogens, nutrients and contaminants. Human involvement has also been a factor in the spread of the pathogen that killed the Caribbean *Diadema*; the disease began in Panama, suggesting a possible link to shipping through the Panama Canal (Andréfouët et al., 2002). Since 1980 about 20% of the world's mangrove forests have disappeared (FAO, 2006), affecting fishing. In the Mesoamerican reef there are up to 25 times more fish of some species on reefs close to mangrove areas than in areas where mangroves have been destroyed (WWF, 2004).

13.2.4.3 Trends in socio-economic factors

From 1950 to the end of the 1970s Latin America benefited from an average annual GDP growth of 5% (Escaith, 2003). This remarkable growth rate permitted the development of national industries, urbanisation, and the creation or extension of national education and public health services. The strategy for economic development was based on the import-substitution model, which consisted of imposing barriers to imports and developing national industry to produce what was needed. Nevertheless, this model produced a weak industry that was not able to compete in international markets and this had terrible consequences for the other sectors (agriculture in particular) which funded the industrial development.

In the 1980s the region faced a great debt crisis which forced countries to make efforts to implement rigorous macroeconomic measures regarding public finances in order to liberate the economy. Control of inflation and public deficit became the main targets of most governments. Deterioration of economic and social conditions, unemployment, extension of the informal economy and poverty characterised this decade. In most of Latin America, the results of economic liberalisation can be characterised by substantial heterogeneity and volatility in long-term growth, and modest (or even negative) economic growth (Solimano and Soto, 2005).

This shift of the economic paradigm produced contradictory results. On the one hand, the more-liberalised economies attained greater economic growth than less-liberalised economies and achieved higher levels of democracy. On the other hand, there was an increase in volatility which led to recurrent crises, poverty and increasing inequality. The governments have failed to create strong social safety nets to ameliorate social conditions (Huber and Solt, 2004).

In Latin America the wealthiest 10% of the population own between 40% and 47% of the national income while the poorest 20% have only 2-4%. This type of income distribution is comparable only to some African and ex-USSR countries (Ferroni, 2005). The lack of equity in education, health services, justice and access to credit can restrain economic development, reduce investment and allow poverty to persist. A study conducted by CEPAL (2002) concludes that the likelihood of the poorest Latin American countries reaching the 7% GDP growth they need is almost zero in the medium term. Even the wealthier countries in the region will find it hard to reach a 4.1% GDP growth target. Predictions for GDP growth in the region for 2015 range from 2.1% to 3.8%, which is very far from the 5.7% average estimated as necessary to reduce poverty.

The combination of low economic growth and high levels of inequality can make large parts of the region's population very vulnerable to economic and natural stressors, which would not necessarily have to be very large in order to cause great social damage (UNDP-GEF, 2003). The effects of climate change on national economies and official development assistance have not been considered in most vulnerability assessments. The impact of climate change in Latin America's productive sectors is estimated to be a 1.3% reduction in the region's GDP for an increase of 2°C in global temperature (Mendelsohn et al., 2000). However, this impact is likely to be even greater because this estimation does not include non-market sectors and extreme events (Stern, 2007). If no structural changes in economic policy are made to promote investment, employment and productivity, economic and social future scenarios for the region do not hold the economic growth needed for its development, unless an uncommon combination of external positive shocks occurs (Escaith, 2003).

13.2.5 Current adaptation

Weather and climate variability forecast

The mega 1982/83 El Niño set in motion an international effort (the Tropical Ocean-Global Atmosphere (TOGA) programme) to understand and predict this ocean-atmosphere phenomenon. The result was the emergence of increasingly reliable seasonal climate forecasts for many parts of the world, especially for Latin America. These climate forecasts became even more reliable with the use of TOGA observations of the Upper Tropical Pacific from the mid-1990s, although they still lack the ability to correctly predict the onset of some El Niño and La Niña events (Kerr, 2003). Nowadays such forecasting systems are based on the use of coupled atmospheric-ocean models and have lead times of 3 months to more than 1 year. Such climate forecasts have given rise to a number of applications and have been in use in a number of sectors: starting in the late 1980s for fisheries in the Eastern Pacific and crops in Peru (Lagos, 2001), subsistence agriculture in north-east Brazil (Orlove et al., 1999), prevention of vegetation fires in tropical South America (Nepstad et al., 2004; http://www.cptec.inpe.br/), streamflow prediction for hydropower in the Uruguay river (Tucci et al., 2003; Collischonn et al., 2005), fisheries in the south-western Atlantic (Severov et al., 2004), dengue epidemics in Brazil (IRI, 2002), malaria control (Ruiz et al., 2006) and hydropower generation in Colombia (Poveda et al., 2003).

Agriculture is a key sector for the potential use of ENSO-based climate forecasts for planning production strategies as adaptive measures. Climate forecasts have been used in the north-east region of Brazil since the early 1990s. During 1992, based on the forecast of dry conditions in Ceara, it was recommended that crops better suited to drought conditions should be planted, and this led to reduced grain production losses (67% of the losses recorded for 1987, a year with similar rainfall but without climate forecasting). However, this tool has not yet been fully adopted because of some missed forecasts which eroded the credibility of the system (Orlove et al., 1999). Recently, in Tlaxcala (Mexico), ENSO forecasting was used to switch crops (from maize to oats) during the El Niño event (Conde and Eakin, 2003). This successful experience was based on strong stakeholder involvement (Conde and Lonsdale, 2005). Recent studies have quantified the potential economic value of ENSO-based climate forecasts, and concluded that increases in net return could reach 10% in potato and winter cereals in Chile (Meza et al., 2003); 6% in maize and 5% in soybean in Argentina (Magrin and Travasso, 2001); more than 20% in maize in Santa Julia, Mexico (Jones, 2001); and 30% in commercial agricultural areas of Mexico (Adams et al., 2003), when crop management practices are optimised (e.g., planting date, fertilisation, irrigation, crop varieties). Adjusting crop mix could produce potential benefits close to 9% in Argentina, depending on site, farmers' risk aversion, prices and the preceding crop (Messina, 1999). In the health sector, the application of climate forecasts is relatively new (see Section 13.2.5.5). Institutional support for early warning systems may help to facilitate early, environmentally-sound public health interventions. For instance, the Colombian Ministry of Health developed a contingency plan to control epidemics associated with the 1997/98 El Niño event (Poveda et al., 1999).

In some countries of Latin America, improvements in weather-forecasting techniques will provide better information for hydrometeorological watching and warning services. The installation of modern weather radar stations (with Doppler capacity) would improve the reliability of these warnings, but the network is still very sparse (WMO, 2007). Furthermore, the deficiencies in the surface and upper air networks adversely affect the reliability of weather outlooks and forecasts. Nevertheless, the exacerbation of weather and climate conditions and the problems arising from extreme events have led to planning and implementation actions to improve the observation, telecommunications and data processing systems of the World Weather Watch (WWW). Moreover,the participation of Latin American countries in the UN-IDSR would lead to the implementation of new (and further development of existing) monitoring and warning services in the region. Examples of networks that predict seasonal climate and climate extremes are the Regional Disaster Information Centre-Latin America and Caribbean (CRID), the International Centre for Research on El Niño Phenomenon (Ecuador), the Permanent Commission of South Pacific (CIIFEN; CPPS) and the Andean Committee for Disaster Prevention and Response (CAPRADE). Some networks set up to respond to and prevent impacts are, for example, the multi-stakeholder decision-making system developed in Peru (Warner, 2006), the National Development Plan and the National Risk Atlas implemented in Mexico (Quaas and Guevara, 2006) and the communication programme for indigenous populations, based on messages in the local language (Alcántara-Ayala, 2004).

13.2.5.1 Natural ecosystems

Ecological corridors between protected areas have been planned for the maintenance of biodiversity in natural ecosystems. Some of these, such as the Mesoamerican Biological Corridor, have been implemented, and these serve also as adaptation measures. Important projects are those for natural corridors in the Amazon and Atlantic forests (de Lima and Gascon, 1999; CBD, 2003) and the Villcabamba–Amboró biological corridor in Peru and Bolivia (Cruz Choque, 2003). Conservation efforts would be also devoted to implementing protection corridors containing mangroves, sea grass beds and coral reefs to boost fish abundance on reefs, benefit local fishing communities, and contribute to sustainable livelihoods (WWF, 2004). Other positive practices in the region are oriented towards maintaining and restoring native ecosystems and protecting and enhancing ecosystem services such as carbon sequestration in the Noel Kempff Mercado Climate Action Project in Bolivia (Brown et al., 2000). Conservation of biodiversity and maintenance of ecosystem structure and function are important for climate-change adaptation strategies, due to the protection of genetically diverse populations and species-rich ecosystems (World Bank, 2002a; CBD, 2003); an example is the initiative to implement adaptation measures in high mountain regions which has been developed in Colombia and other Andean countries (Vergara, 2005). A new option to promote mountainous forest conservation consists of compensating forest owners for the environmental services that those forests bring to society (UNEP, 2003a). The compensation is often financed by charging a small price supplement to water users for the water originating in forests. Such schemes are being implemented in various countries of Latin America and were tested in Costa Rica (Campos and Calvo, 2000). In Brazil, 'ProAmbiente' is an environmental credit programme of the government, paying for environmental services provided by smallholders that preserve the forest (MMA, 2004). Another initiative in Brazil is the ecological value-added tax, a fiscal instrument that remunerates municipalities that protect nature and generate environmental services, which was adopted initially by the states of Paraná and Minas Gerais, and more recently implemented in parts of the Amazon as well (May et al., 2004).

13.2.5.2 Agriculture

Some adaptive measures, such as changes in land use, sustainable management, insurance mechanisms, irrigation, adapted genotypes and changes in agronomic crop management, are used in the agricultural sector to cope with climatic variability. In addition, economic diversification has long been a strategy for managing risk (both climatic and market) and this has increased in recent years. While not a direct adaptation to climatic change, this diversification is lessening the dependence of farmers on agricultural income and enabling greater flexibility in managing environmental change (Eakin, 2005). Farmers located on the U.S.–Mexico border have been able to continue farming in the valley through changes in irrigation technology, crop diversification and market orientation, despite the crisis with the local aquifers caused by drought and over-exploitation (Vásquez-León et al., 2003). Sustainable land management based on familiar practices (contour barriers, green manures, crop rotation and stubble incorporation) allowed smallholders in Nicaragua to better cope with the impacts of Hurricane Mitch (Holt-Giménez, 2002). In Mexico, some small farmers are testing adaptation measures for current and future climate, implementing drip-irrigation systems, greenhouses and the use of compost (Conde et al., 2006). According to Wehbe et al. (2006), adjustments in planting dates and crop choice, construction of earth dams and the conversion of agriculture to livestock are increasingly popular adaptation measures in González (Mexico), while in southern Cordoba (Argentina), climate risk insurance, irrigation, adjusting planting dates, spatial distribution of risk through geographically separated plots, changing crops and maintaining a livestock herd were identified as common measures to cope with climatic hazards.

13.2.5.3 Water resources

The lack of adequate adaptation strategies in Latin American countries to cope with the hazards and risks of floods and droughts is due to low gross national product (GNP), the increasing population settling in vulnerable areas (prone to flooding, landslides or drought) and the absence of the appropriate political, institutional and technological framework (Solanes and Jouravlev, 2006). Nevertheless, some communities and cities have organised themselves, becoming active in disaster prevention (Fay et al., 2003). Many poor inhabitants were encouraged to relocate from flood-prone areas to safer places. With the assistance of IRDB and IDFB loans, they built new homes, e.g., resettlements in the Paraná river basin of Argentina, after the 1992 flood (IRDB, 2000). In some cases, a change in environmental conditions affecting the typical economy of the Pampas has led to the introduction of new production activities through aquaculture, using natural regional fish species such as pejerrey (*Odontesthes bonariensis*) (La Nación, 2002). Another example, in this case related to the adaptive capacity of people to water stresses, is given by 'self organisation' programmes for improving water supply systems in very poor communities. The organisation Business Partners for Development Water and Sanitation Clusters has been working on four 'focus' plans in LA: Cartagena (Colombia), La Paz and El Alto (Bolivia), and some underprivileged districts of Gran Buenos Aires (Argentina) (The Water Page, 2001; Water 21, 2002). Rainwater cropping and storage systems are important features of sustainable development in the semi-arid tropics. In particular, there is a joint project developed in Brazil by the NGO Network ASA Project, called the P1MC- Project, for 1 million cisterns to be installed by civilian society in a decentralised manner. The plan is to supply drinking water to 1 million rural households in the perennial drought areas of the Brazilian semi-arid tropics (BSATs). During the first stage, 12,400 cisterns were built by ASA and the Ministry of Environment of Brazil and a further 21,000 were planned by the end of 2004 (Gnadlinger, 2003). In Argentina, national safe water programmes for local communities in arid regions of Santiago del Estero province installed ten rainwater catchments and storage systems between 2000 and 2002 (Basán Nickisch, 2002).

13.2.5.4 Coasts

Several Latin American countries have developed planned and autonomous adaptation measures in response to current climate variability impacts on their coasts. Most of them (e.g., Argentina, Colombia, Costa Rica, Uruguay and Venezuela) focus their adaptation on integrated coastal management (Hoggarth, et al., 2001; UNEP, 2003b, Natenzon et al., 2005a, b; Nagy et al., 2006b). The Caribbean Planning for Adaptation to Global Climate Change project is promoting actions to assess vulnerability (especially regarding rise in sea level), and plans for adaptation and development of appropriate capacities (CATHALAC, 2003). Since 2000, some countries have been improving their legal framework on matters related to establishing restrictions on air pollution and integrated marine and coastal regulation (e.g., Venezuela's integrated coastal zone plan since 2002). Due to the strong pressure of human settlement and economic activity, a comprehensive policy design is now included within the 'integrated coastal management' modelling in some countries, such as Venezuela (MARN, 2005) and Colombia (INVEMAR, 2005). In Belize and Guyana, the implementation of land-use planning and zoning strengthens the norms for infrastructure, the coastal-zone management plan, the adjustment of building codes and better disaster-mitigation strategies (including floodplain and other hazard mapping), which, along with climate-change considerations, are used in the day-to-day management of all sectors (CDERA, 2003; UNDP-GEF, 2003).

13.2.5.5 Human health

In Latin America, adaptation measures in the health sector should basically be considered as isolated initiatives. A project on adaptation to climate variability and change undertaken in Colombia is oriented towards formulating measures to reduce human health vulnerability and cope with impacts. The project includes the development of an integrated national pilot adaptation plan (INAP), for high mountain ecosystems, islands, and human health concerns related to the expansion of areas for vectors linked to malaria and dengue (Arjona, 2005). The project includes the development of a comprehensive and integrated dengue and malaria surveillance control system, aiming to reduce the infection rate from both diseases by 30% (Mantilla, 2005). Other isolated measures have been identified for several countries. For example, in Bolivia, adaptation measures regarding the health impacts of climate change include activities on vector control and medical surveillance. The aim is also to have community participation and health education, entomological research, strengthened sanitary services and the development of research centres dealing with tropical diseases. Government programmes would also focus on high-risk areas for malaria and leishmaniasis under climate change (Aparicio, 2000).

13.3 Assumptions about future trends

13.3.1 Climate

13.3.1.1 Climate-change scenarios

Even though climate-change scenarios can be generated by several methods (IPCC, 2001), the use of GCM outputs based on the Special Report on Emissions Scenarios (SRES: Nakićenović and Swart, 2000) is the adopted method for the Fourth Assessment Report (AR4). Projections of average temperature and rainfall anomalies throughout the current century derived from a number of GCMs are available at the IPCC Data Distribution Centre (IPCC DDC, 2003; http://www.ipcc-data.org//) at a typical model resolution of 300 km, and for two different greenhouse-gas (GHG) emissions scenarios (A2 and B2). Additionally, Chapter 11 of the Working Group I Fourth Assessment Report (Christensen et al., 2007) presents regional projections for many parts of the world. Table 13.4 indicates ranges of temperature and precipitation changes for sub-regions of Latin America for several time-slices (2020, 2040, 2080), obtained from seven GCMs and the four main SRES emissions scenarios.

For 2020, temperature changes range from a warming of 0.4°C to 1.8°C, and for 2080, of 1.0°C to 7.5°C. The highest values of warming are projected to occur over tropical South America (referred to as Amazonia in Table 13.4). The case for precipitation changes is more complex, since regional climate projections show a much higher degree of uncertainty. For central and tropical South America, they range from a reduction of 20% to 40% to an increase of 5% to 10% for 2080. Uncertainty is even larger for southern South America in both winter and summer seasons, although the percentage change in precipitation is somewhat smaller than that for tropical Latin America. Analyses of these scenarios reveal larger differences in temperature and rainfall changes among models than among emissions scenarios

Table 13.4. *Projected temperature (°C) and precipitation (%) changes for broad sub-regions of Central and South America based on Ruosteenoja et al. (2003). Ranges of values encompass estimates from seven GCMs and the four main SRES scenarios.*

		2020	2050	2080
Changes in temperature (°C)				
Central America	Dry season	+0.4 to +1.1	+1.0 to +3.0	+1.0 to +5.0
	Wet season	+0.5 to +1.7	+1.0 to +4.0	+1.3 to +6.6
Amazonia	Dry season	+0.7 to +1.8	+1.0 to +4.0	+1.8 to +7.5
	Wet season	+0.5 to +1.5	+1.0 to +4.0	+1.6 to +6.0
Southern South America	Winter (JJA)	+0.6 to +1.1	+1.0 to +2.9	+1.8 to +4.5
	Summer (DJF)	+0.8 to +1.2	+1.0 to +3.0	+1.8 to +4.5
Change in precipitation (%)				
Central America	Dry season	−7 to +7	−12 to +5	−20 to +8
	Wet season	−10 to +4	−15 to +3	−30 to +5
Amazonia	Dry season	−10 to +4	−20 to +10	−40 to +10
	Wet season	−3 to +6	−5 to +10	−10 to +10
Southern South America	Winter (JJA)	−5 to +3	−12 to +10	−12 to +12
	Summer (DJF)	−3 to +5	−5 to +10	−10 to +10

DJF= December/January/February, JJA= June/July/August.

for the same model. As expected, the main source of uncertainty for regional climate change scenarios is that associated with different projections from different GCMs. The analysis is much more complicated for rainfall changes. Different climate models show rather distinct patterns, even with almost opposite projections. In summary, the current GCMs do not produce projections of changes in the hydrological cycle at regional scales with confidence. In particular the uncertainty of projections of precipitation remain high (e.g., Boulanger et al., 2006a, b, for climate-change scenarios for South America using ten GCMs). That is a great limiting factor to the practical use of such projections for guiding active adaptation or mitigation policies.

GCM-derived scenarios are commonly downscaled using statistical or dynamical approaches to generate region- or site-specific scenarios. These approaches are described in detail in Chapter 11 of the Working Group I Fourth Assessment Report (Christensen et al., 2007). There have been a number of such exercises for South America using an array of GCM scenarios (HADCM3, ECHAM4, GFDL, CSIRO, CCC, etc.), usually for SRES emissions scenarios A2 and B2: for southern South America (Bidegain and Camilloni, 2004; Nuñez et al., 2005; Solman et al., 2005a, b), Brazil (Marengo, 2004), Colombia (Eslava and Pabón, 2001; Pabón et al., 2001) and Mexico (Conde and Eakin, 2003). Downscaled scenarios may reveal smaller-scale phenomena associated with topographical features or mesoscale meteorological systems and land-use changes, but in general the uncertainty associated with using different GCMs as input is a dominant presence in the downscaled scenarios (Marengo and Ambrizzi, 2006).

13.3.1.2 Changes in the occurrence of extremes

Many of the current climate change studies indicate that the frequency in the occurrence of extreme events will increase in the future. Many impacts of climate change will be realised as the result of a change in the frequency of occurrence of extreme weather events such as windstorms, tornados, hail, heatwaves, gales, heavy precipitation or extreme temperatures over a few hours to several days. A limited number of studies on extremes from global models assessed during the AR4 (e.g., Tebaldi et al., 2007) provide estimates of frequency of seasonal temperature and precipitation extreme events as simulated in the present and by the end of 21st century under the A1B emissions scenario. In Central America, the projected time-averaged precipitation decrease is accompanied by more frequent dry extremes in all seasons. In South America, some models anticipate extremely wet seasons in the Amazon region and in southern South America, while others show the opposite tendency.

13.3.2 Land-use changes

Deforestation in Latin America's tropical areas will be one of the most serious environmental disasters faced in the region. Currently, Latin America is responsible for 4.3% of global GHG emissions. Of these, 48.3% result from deforestation and land-use changes (UNEP, 2000). By 2010 the forest areas in South and Central America will be reduced by 18 Mha and 1.2 Mha, respectively. These areas (see Figure 13.3) will be used for pasture and expanding livestock production (FAO, 2005).

If the 2002-2003 deforestation rate (2.3 Mha/yr) in Brazilian Amazonia continues indefinitely, then 100 Mha of forest (about 25% of the original forest) will have disappeared by the year 2020 (Laurance et al., 2005), while by 2050 (for a business-as-usual scenario) 269.8 Mha will be deforested (Moutinho and Schwartzman, 2005). By means of simulation models, Soares-Filho et al. (2005) estimated for Brazilian Amazonia that in the worst-case scenario, by 2050 the projected deforestation trend will eliminate 40% of the current 540 Mha of Amazon forests, releasing approximately 32 Pg (109 tonnes/ha) of carbon to the atmosphere. Moreover, under the current trend, agricultural expansion will eliminate two-thirds of the forest cover of five major watersheds and ten eco-regions, besides the loss of more than 40% of 164 mammalian species habitats.

Projected to be one of the main drivers of future land-use change, the area planted to soybeans in South America is

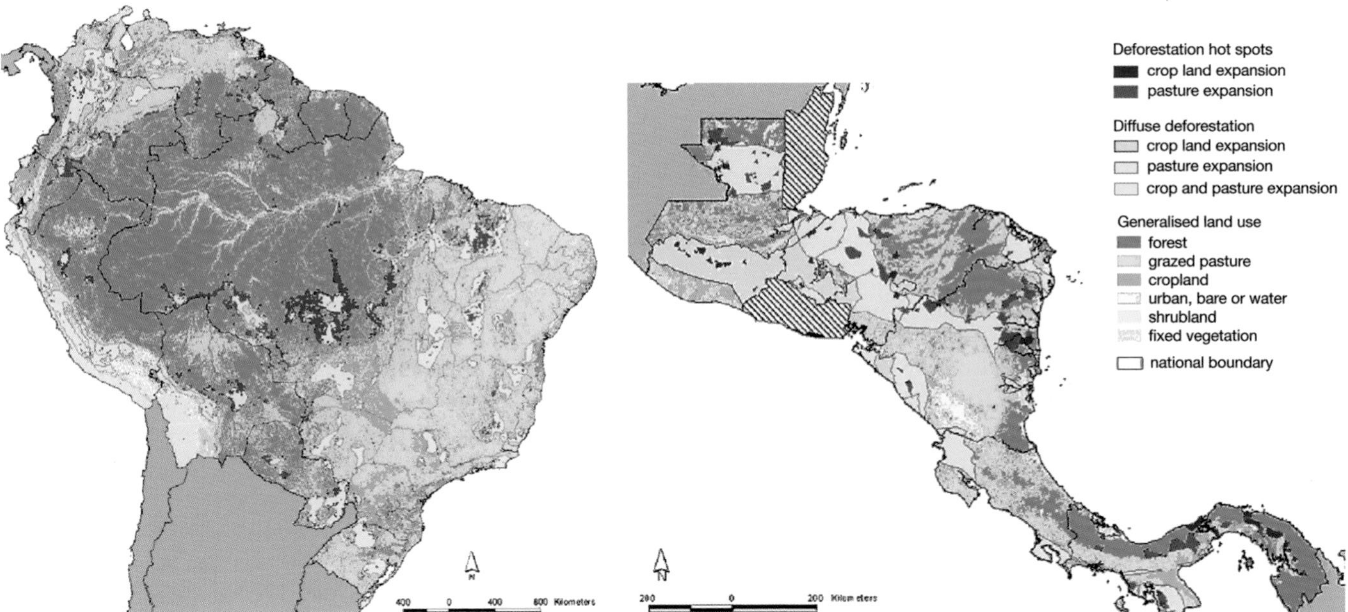

Figure 13.3. *Predicted 2000-2010 South American and Central American deforestation hotspots and diffuse deforestation areas (available at: http://www.virtualcentre.org/en/dec/neotropics/south_america.htm and http://www.virtualcentre.org/en/dec/neotropics/central_america.htm).*

expected to increase from 38 Mha in 2003/04 to 59 Mha in 2019/20 (Maarten Dros, 2004). The total production of Argentina, Brazil, Bolivia and Paraguay will rise by 85% to 172 million tonnes or 57% of world production. Direct and indirect conversion of natural habitats to accommodate this expansion amounts to 21.6 Mha. Habitats with the greatest predicted area losses are the Cerrado (9.6 Mha), dry and humid Chaco (the largest dry forest in South America, which covers parts of Argentina, Paraguay, Bolivia and Brazil; 6.3 Mha), Amazon transition and rain forests (3.6 Mha), Atlantic forest (1.3 Mha), Chiquitano forest (transition between Amazonian forest and Chaco forest; 0.5 Mha) and Yungas forest (0.2 Mha). This massive deforestation will have negative impacts on the biological diversity and ecosystem composition of South America as well as having important implications for regional and local climate conditions.

13.3.3 Development

13.3.3.1 Demographics and societies

The population of the Latin American region has continued to grow and is expected to be 50% larger than in 2000 by the year 2050. Its annual population growth rate has decreased and is expected to reach a value of 0.89% by 2015, which is considerably less than 1.9%, the average rate for the 1975-2002 period. The population has continued to migrate from the countryside to the cities, and by 2015 about 80% of the population will be urban, almost 30% more than in the 1960s. The population aged under 15 years will decline and at the same time the population aged over 65 years will increase. Total fertility rate (births per woman) decreased from 5.1 to 2.5 between 1970-1975 and 2000-2005 and is expected to decrease to 2.2 by 2015 (ECLAC, 1998).

According to ECLAC (1998) the number of people in an age-range making them dependent (between 0 and 14 and over 65 years) will increase from 54.8% at present to almost 60% in 2050. This will increase pressure on the social security systems in the region and increase the contributions that the population of working age will have to make in order to maintain the availability of health and educational services. Life expectancy at birth increased from 61.2 years in the 1970s to 72.1 years in the 2000-2005 five-year period, and is expected to increase to 74.4 years by 2015. Crude mortality rate is expected to increase from the current value of 7.8 (per thousand) to almost 12 by 2050.

Human migration has become an important issue in the region. Recent studies (ECLAC, 2002b) have estimated that 20 million Latin American and Caribbean nationals reside outside their countries, with the vast majority in North America. This phenomenon has important effects on national economies and creates important social dependencies: 5% of households in the region benefit from remittances which in 2003 amounted to US$38 billion (17.6% more than in 2002; IMO, 2005).

According to the Human Development Index, all countries in the region are classified within high and medium development ranks. In addition, Latin American countries are ranked within the upper half of the Human Poverty Index and have shown a systematic improvement between 1975 and 2002. It is difficult to ignore the fact that, although there are no Latin American countries classified in the low development rank, there are huge contrasts among and within countries in terms of levels of technological development, sophistication of financial sectors, export capacities and income distribution (CEPAL, 2002).

13.3.3.2 Economic scenarios

Projections of economic evolution for the region strongly depend on the interpretation of the results of the liberalisation process that the region has experienced during the last 20 years, and therefore can be contradictory. On the one hand, economists who favour liberalisation of Latin American economies argue that countries that have implemented these types of policies have improved in terms of growth rate, stability, democracy and even with regard to inequality and poverty (for example: Walton, 2004; World Bank, 2006). On the other hand, another group of experts in economics, sociology and politics is concerned with the effects that neoliberalisation has had for the region, especially in terms of increases in inequality and poverty, but also in terms of lack of economic growth (Huber and Solt, 2004). This is still an unresolved debate that imparts great uncertainty to economic scenarios for Latin America.

The first group's view provides the following insights for economic prospects. Analysts from the World Bank argue that while the real per capita GDP of Latin America has had a very low growth – about 1.3%/yr average during the 1990 to 2000 period – in the long term (from 2006 to 2015), regional GDP is projected to increase by 3.6%/yr, and per capita income is expected to rise by 2.3%/yr on average (World Bank, 2006). Current estimates forecast a growth of 4%/yr for the region in 2006 and 3.6%/yr in 2007 and real per capita GDP growth of 2.6%/yr and 2.3%/yr, respectively (Loser, 2006; World Bank, 2006). These positive prospects are attributed to the implementation of economic policies such as a substantial reduction of the fiscal imbalances and inflation control that have restrained growth in the past. According to this source, the area is on track to meet its Millennium Development Goals on poverty; however, it is important to note that the region's performance is not as good as other developing regions such as central Asia and, notably, China. An improvement on this rate of growth could be achieved by consolidating current economic policies (Walton, 2004; World Bank, 2006).

The second group of experts argue that the results of the liberalisation, far from establishing a sound basis for economic growth, have weakened the regional economy, reducing its rate of growth and making it more volatile, exacerbating social inequality and poverty, and limiting the region's capacity for future growth (Huber and Solt, 2004; Solimano and Soto, 2005). Lack of economic growth, inequality, a deficient legal framework and demographic pressures have been demonstrated to be important factors for increasing environmental depletion and vulnerability to climate variability and extreme events (CEPAL, 2002).

13.4 Summary of expected key future impacts and vulnerabilities

13.4.1 Natural ecosystems

Tropical plant species may be sensitive to small variations of climate, since biological systems respond slowly to relatively rapid changes of climate. This fact might lead to a decrease of species diversity. Based on Hadley Centre Atmosphere-Ocean General Circulation Model (AOGCM) projections for A2 emissions scenarios, there is the potential for extinction of 24% of 138 tree species of the central Brazil savannas (Cerrados) by 2050 for a projected increase of 2°C in surface temperature (Siqueira and Peterson, 2003; Thomas et al., 2004). By the end of the century, 43% of 69 tree plant species studied could become extinct in Amazonia (Miles et al., 2004). In terms of species and biome redistributions, larger impacts would occur over north-east Amazonia than over western Amazonia. Several AOGCM scenarios indicate a tendency towards 'savannisation' of eastern Amazonia (Nobre et al., 2005) and the tropical forests of central and south Mexico (Peterson et al., 2002; Arriaga and Gómez, 2004). In north-east Brazil the semi-arid vegetation would be replaced by the vegetation of arid regions (Nobre et al., 2005), as in most of central and northern Mexico (Villers and Trejo, 2004).

Up to 40% of the Amazonian forests could react drastically to even a slight reduction in precipitation; this means that the tropical vegetation, hydrology and climate system in South America could change very rapidly to another steady state, not necessarily producing gradual changes between the current and the future situation (Rowell and Moore, 2000). It is more probable that forests will be replaced by ecosystems that have more resistance to multiple stresses caused by temperature increase, droughts and fires, such as tropical savannas.

The study of climate-induced changes in key ecosystem processes (Scholze et al., 2005) considers the distribution of outcomes within three sets of model runs grouped according to the amount of global warming they simulate: <2°C, 2-3°C and >3°C. A high risk of forest loss is shown for Central America and Amazonia, more frequent wildfire in Amazonia, more runoff in north-western South America, and less runoff in Central America. More frequent wildfires are likely (an increase in frequency of 60% for a temperature increase of 3°C) in much of South America. Extant forests are destroyed with lower probability in Central America and Amazonia. The risks of forest losses in some parts of Amazonia exceed 40% for temperature increases of more than 3°C (see Figure 13.3).

The tropical cloud forests in mountainous regions will be threatened if temperatures increase by 1°C to 2°C during the next 50 years due to changes in the altitude of the cloud-base during the dry season, which would be rising by 2 m/yr. In places with low elevation and isolated mountains, some plants will become locally extinct because the elevation range would not permit natural adaptation to temperature increase (FAO, 2002). The change in temperature and cloud-base in these forests could have substantial effects on the diversity and composition of species. For example, in the cloud forest of Monteverde Costa Rica, these changes are already happening. Declines in the frequency of mist days have been strongly associated with a decrease in population of amphibians (20 of 50 species) and probably also bird and reptile populations (Pounds et al., 1999).

Modelling studies show that the ranges occupied by many species will become unsuitable for them as the climate changes (IUCN, 2004). Using modelling projections of species distributions for future climate scenarios, Thomas et al. (2004) show, for the year 2050 and for a mid-range climate change scenario, that species extinction in Mexico could sharply increase: mammals 8% or 26% loss of species (with or without dispersal), birds 5% or 8% loss of species (with or without dispersal), and butterflies 7% or 19% loss of species (with or without dispersal).

13.4.2 Agriculture

Several studies using crop-simulation models and future climate scenarios were carried out in Latin America for commercial annual crops (see Table 13.5). According to a global assessment (Parry et al., 2004), if CO_2 effects are not considered, grain yield reductions could reach up to 30% by 2080 under the warmer scenario (HadCM3 SRES A1FI), and the number of additional people at risk of hunger under the A2 scenario is likely to reach 5, 26 and 85 million in 2020, 2050 and 2080, respectively (Warren et al., 2006). However, if direct CO_2 effects are considered, yield changes could range between reductions of 30% in Mexico and increases of 5% in Argentina (Parry et al., 2004), and the additional number of people at risk of hunger under SRES A2 would increase by 1 million in 2020, remain unchanged in 2050 and decrease by 4 million in 2080.

More specific studies considering individual crops and countries are also presented in Table 13.5. The great uncertainty in yield projections could be attributed to differences in the GCM or incremental scenario used, the time-slice and SRES scenario considered, the inclusion or not of CO_2 effects, and the site considered. Other uncertainties in yield impacts are derived from model inaccuracies and unmodelled processes. Despite great variability in yield projections, some behaviour seems to be consistent all over the region, such as the projected reduction in rice yields after the year 2010 and the increase in soybean yields when CO_2 effects are considered. Larger crop yield reductions could be expected in the future if the variance of temperatures were doubled (see Table 13.5). For smallholders a mean reduction of 10% in maize yields could be expected by 2055, although in Colombia yields remain essentially unchanged, while in the Venezuelan Piedmont yields are predicted to decline to almost zero (Jones and Thornton, 2003). Furthermore, an increase in heat stress and more dry soils may reduce yields to one-third in tropical and sub-tropical areas where crops are already near their maximum heat tolerance. The productivity of both prairies/meadows and pastures will be affected, with loss of carbon stock in organic soils and also a loss of organic matter (FAO, 2001b). Other important issues are the expected reductions in land suitable for growing coffee in Brazil, and in coffee production in Mexico (see Table 13.5).

In temperate areas, such as the Argentinean and Uruguayan Pampas, pasture productivity could increase by between 1% and 9% according to HadCM3 projections under SRES A2 for 2020 (Gimenez, 2006). As far as beef cattle production is concerned, in Bolivia future climatic scenarios would have a slight impact on animal weight if CO_2 effects are not considered, while doubling CO_2 and increases of 4°C in temperature are very likely to result in decreases in weight that could be as much as 20%, depending on animal genotype and region (NC-Bolivia, 2000).

Furthermore, the combined effects of climate change and land-use change on food production and food security are related to a larger degradation of lands and a change in erosion patterns (FAO, 2001b). According to the World Bank (2002a, c), some developing countries are losing 4-8% of their GDP due to productive and capital losses related to environmental degradation. In drier areas of Latin America, such as central and northern Chile, the Peruvian coast, north-east Brazil, dry Gran Chaco and Cuyo, central, western and north-west Argentina and significant parts of Mesoamerica (Oropeza, 2004), climate change is likely to lead to salinisation and desertification of agricultural lands. By 2050, desertification and salinisation will affect 50% of agricultural lands in Latin America and the Caribbean zone (FAO, 2004a).

In relation to pests and diseases, the incidence of the coffee leafminer (*Perileucoptera coffeella*) and the nematode *Meloidogyne incognita* are likely to increase in future in Brazil's production area. The number of coffee leafminer cycles could increase by 4%, 32% and 61% in 2020, 2050 and 2080, respectively, under SRES A2 scenarios (Ghini et al., 2007). According to Fernandes et al. (2004), the risk of *Fusarium* head blight incidence in wheat crops is very likely to increase under climate change in south Brazil and Uruguay. The demand for water for irrigation is projected to rise in a warmer climate, bringing increased competition between agricultural and domestic use in addition to industrial uses. Falling watertables and the resulting increase in the energy used for pumping will make the practice of agriculture more expensive (Maza et al., 2001). In the state of Ceará (Brazil), large-scale reductions in the availability of stored surface water could lead to an increasing imbalance between water demand and water supply after 2025 (ECHAM scenario; Krol and van Oel, 2004).

13.4.3 Water resources

Almost 13.9% of the Latin American population (71.5 million people) have no access to a safe water supply; 63% of these (45 million people) live in rural areas (IDB, 2004). Many rural communities rely on limited freshwater resources (surface or underground) and many others on rainwater, using water-cropping methods which are very vulnerable to drought (IDB, 2004). People living in water-stressed watersheds (less than 1,000 m^3/capita per year) in the absence of climate change were estimated to number 22.2 million in 1995 (Arnell, 2004). The number of people experiencing increased water stress under the SRES scenarios is estimated to range from 12 to 81 million in the 2020s, and from 79 to 178 million in the 2050s (Arnell, 2004). These estimates do not take into account the number of people moving out of water-stressed areas (unlike Table 13.6). The current vulnerabilities observed in many regions of Latin American countries will be increased by the joint negative effects of growing demands for water supplies for domestic use and irrigation due to an increasing population, and the expected drier conditions in many basins. Therefore, taking into account the number of people experiencing decreased water stress, there will still be a net increase in the number of people becoming water-stressed (see Table 13.6).

In some zones of Latin America where severe water stresses could be expected (eastern Central America, in the plains, Motagua valley and Pacific slopes of Guatemala, eastern and western regions of El Salvador, the central valley and Pacific region of Costa Rica, in the northern, central and western inter-montane regions of Honduras and in the peninsula of Azuero in Panama), water supply and hydroelectric generation would be seriously affected (Ramírez and Brenes, 2001; ECLAC, 2002a).

Table 13.5. *Future impacts on the agricultural sector.*

Study	Climate scenario	Yield impacts (%) Wheat	Maize	Soybean	Rice	Others
Guyana (NC-Guyana, 2002)	CGCM1 2020-2040 (2xCO_2)				−3	Sg: −30
	CGCM1 2080-2100 (3xCO_2)				−16	Sg: −38
Panama (NC-Panama, 2000)	HadCM2-UKHI (IS92c-IS92f) 2010/2050/2100 (1xCO_2)		+9/−34/−21			
Costa Rica (NC-Costa Rica, 2000)	+2°C −15% precip. (1xCO_2)				−31	Pt: ↓
Guatemala (NC-Guatemala, 2001)	+1.5°C −5% precip.		+8 to −11		−16	Bn: +3 to −28
	+2°C +6% precip.		+15 to −11		−20	Bn: +3 to −42
	+3.5°C −30% precip.		+13 to −34		−27	Bn: 0 to −66
Bolivia (NC-Bolivia, 2000)	GISS and UK89 (2xCO_2).I		−25		−2	
	Incremental (2xCO_2)		+50			
	+3°C −20% precip.				−15	
	optimistic-pessimistic (1xCO_2)					Pt: +5 to+2[*2]
	optimistic-pessimistic (2xCO_2)					Pt: +7 to+5[*2]
	IS92a (1xCO_2)[*1]			−3 to −20		
	IS92a (2xCO_2)[*1]			+12 to +59		
Brazil (Siqueira et al., 2001)	GISS (550 ppm CO_2)	−30	−15	+21		
SESA[*3] (Gimenez, 2006)	Hadley CM3-A2 (500 ppm)	+9 to +13	−5 to +8	+31 to +45		
	Hadley CM3-A2 (500 ppm).I	+10 to +14	0 to +2	+24 to +30		
Argentina, Pampas (Magrin and Travasso, 2002)	+1/+2/+3°C (550 ppm CO_2).I	+11/+3/−4	0/−5/−9	+40/+42/+39		
	UKMO (+5.6°C) (550 ppm CO_2).I	−16	−17	+14		
Honduras (Díaz-Ambrona et al., 2004)	Hadley CM2 (1xCO_2) 2070		−21			
	Hadley CM2 (2xCO_2) 2070		0			
Central Argentina (Vinocur et al., 2000; Vinocur, 2005)	Hadley CM3-B2 (477ppm)		+21			
	ECHAM98-A2 (550ppm)		+27			
	+1.5/+3.5°C (1xCO_2)		−13/−17			
	+1.5/+3.5°C (1xCO_2) (2T)[*4]		−19/−35			
Latin America (Jones and Thornton, 2003)	HadCM2 (smallholders)		−10			
Latin America (Parry et al., 2004)	HadCM3 A1FI (1xCO_2)	Cereal yields:	−5 to −2.5 (2020)	−30 to −5 (2050)		−30 (2080)
	HadCM3 B1 (1xCO_2)		−10 to −2.5 (2020)	−10 to −2.5 (2050)		−30 to −10 (2080)
	HadCM3 A1FI (2xCO_2)		−5 to +2.5 (2020)	−10 to +10 (2050)		−30 to +5 (2080)
	HadCM3 B1 (2xCO_2)		−5 to −2.5 (2020)	−5 to +2.5 (2050)		−10 to +2.5 (2080)
Mexico, Veracruz (Gay et al., 2004)	HadCM2 ECHAM4 (2050)	Coffee: 73% to 78% reduction in production				
Brazil, São Paulo (Pinto et al., 2002)	+1°C + 15% precip.	Coffee: 10% reduction in suitable lands for coffee				
	+5.8°C + 15%precip.	97% reduction in suitable lands for coffee				
Costa Rica (NC-Costa Rica, 2000)	Sensitivity analysis	Coffee: Increases (up to 2°C) in temperature would benefit crop yields				

I = Irrigated crops; precip. = precipitation; [*1] Values correspond to soybean sowing in winter and summer for 2010 and 2020; [*2] Increases every 10 years. [*3] SESA= South East South America; [*4] 2Tσ: doubled variance of temperature. Bn: bean, Sg: sugar cane, Pt: potato.

Table 13.6. *Net increases in the number of people living in water-stressed watersheds in Latin America (millions) by 2025 and 2055 (Arnell, 2004).*

Scenario/ GCM	1995	2025 Without climate change (1)	2025 With climate change (2)	2055 Without climate change (1)	2055 With climate change (2)
A1 HadCM3	22.2	35.7	21.0	54.0	60.0
A2 HadCM3	22.2	55.9	37.0-66.0	149.3	60.0-150.0
B1 HadCM3	22.2	35.7	22.0	54.0	74.0
B2 HadCM3	22.2	47.3	7.0-77.0	59.4	62.0

(1) according to Arnell (2004, Table 7); (2) according to Arnell (2004, Tables 11 and 12).

Vulnerability studies foresee the ongoing reductions in glaciers. A highly stressed condition is projected between 2015 and 2025 in the water availability in Colombia, affecting water supply and ecosytem functioning in the páramos (IDEAM, 2004), and very probably impacting on the availability of water supply for 60% of the population of Peru (Vásquez, 2004). The projected glacier retreat would also affect hydroelectricity generation in some countries, such as Colombia (IDEAM, 2004) and Peru; one of the more affected rivers would be the Mantaro, where an hydroelectric plant generates 40% of Peru's electricity and provides the energy supply for 70% of the country's industries, concentrated in Lima (UNMSM, 2004) .

In Ecuador, recent studies indicate that seven of the eleven principal basins would be affected by a a decrease in their annual runoff, with monthly decreases varying up to 421% of unsatisfied demand (related to mean monthly runoff) in year 2010 with the scenario of +2°C and −15% precipitation (Cáceres, 2004). In Chile, recent studies confirm the potential damage to water supply and sanitation services in coastal cities, as well as groundwater contamination by saline intrusion. In the Central region river basins, changes in streamflows would require many water regulation works to be redesigned (NC-Chile, 1999).

Under severe dry conditions, inappropriate agricultural practices (deforestation, soil erosion and the excessive use of agrochemicals) will deteriorate surface and groundwater quantity and quality. That would be the case in areas that are currently degraded, such as Leon, Sebaco Valley, Matagalpa and Jinoteca in Nicaragua, metropolitan and rural areas of Costa Rica, Central Valley rivers in Central America, the Magdalena river in Colombia, the Rapel river basin in Chile, and the Uruguay river in Brazil, Uruguay and Argentina (UNEP, 2003b).

Landslides are generated by intense/persistent precipitation events and rainstorms. Furthermore, in Latin America they are associated with deforestation and a lack of land planning and disaster-warning systems. Many cities of Latin America, which are already vulnerable to landslides and mudflows, are very likely to suffer the exacerbation of extreme events, with increasing risks/hazards for local populations (Fay et al., 2003). Accelerated urban growth, increasing poverty and low investment in water supply will contribute to: water shortages in many cities, high percentages of the urban population without access to sanitation services, an absence of treatment plants, high groundwater pollution, lack of urban drainage systems, storm sewers used for domestic waste disposal, the occupation of flood valleys during drought seasons, and high impacts during flood seasons (Tucci, 2001).

13.4.4 Coasts

The majority of vulnerability and impacts assessments in Latin America have been made under the framework of National Communications (NC) to the UNFCCC (United Nations Framework Convention on Climate Change). Unfortunately the methodological approaches adopted are very diverse. Many are based on incremental scenarios (SLR 0.3-1.0 m), in some cases combined with coastal river flooding. Some include a cost-benefit analysis with and without measures (e.g., Ecuador, El Salvador and Costa Rica). Long-term and recent trends of SLR, flooding and storm surges are not always available or analysed. Some other countries (e.g., Chile and Peru) prioritise the impacts of ENSO events and the increase in SST on fisheries.

Significant impacts of projected climate change and sea-level rise are expected for 2050-2080 on the Latin American coastal areas. With most of their population, economic activities and infrastructure located at or near sea-level, coastal areas will be very likely to suffer floods and erosion, with high impacts on people, resources and economic activities (Grasses et al., 2000; Kokot, 2004; Barros, 2005; UCC, 2005). Results from several studies using SLR incremental and future climate change scenarios are summarised in Table 13.7. Projected impacts which would entail serious socio-economic consequences include floods; population displacement; salinisation of lowland areas affecting sources of drinking water (Ubitarán Moreira et al., 1999); coastal storm regime modification; increased erosion and altered coastal morphology (Conde, 2001; Schaeffer-Novelli et al., 2002; Codignotto, 2004; Villamizar, 2004); diversion of farm land; disruption of access to fishing grounds; negative impacts on biodiversity, including mangroves; salinisation and over-exploitation of water resources, including groundwater (FAO, 2006); and pollution and sea-water acidification in marine and coastal environments (Orr et al., 2005). Other factors such as the artificial opening of littoral bars, pressures from tourism, excessive afforestation with foreign species, and coastal setback starting from the decrease of the fluvial discharge in the Patagonian rivers, will add to the impacts on coastal environments (Grasses et al., 2000; Rodríguez-Acevedo, 2001; OAS-CIDI, 2003; Kokot, 2004).

As for coastal tourism, the most impacted countries will be those where the sectoral contribution to the GDP, balance of payments and employment is relatively high, and which are threatened by windstorms and projected sea-level rise: such as those of Central America, the Caribbean coast of South America and Uruguay (Nagy et al., 2006a, c). Thus, climate change is very likely to be a major challenge for all coastal nations.

13.4.5 Human health

The regional assessments of health impacts due to climate change in the Americas show that the main concerns are heat stress, malaria, dengue, cholera and other water-borne diseases (Githeko and Woodward, 2003). Malaria continues to pose a serious health risk in Latin America, where 262 million people (31% of the population) live in tropical and sub-tropical regions with some potential risk of transmission, ranging from 9% in Argentina to 100% in El Salvador (PAHO, 2003). Based on SRES emissions scenarios and socio-economic scenarios, some projections indicate decreases in the length of the transmission season of malaria in many areas where reductions in precipitation are projected, such as the Amazon and Central America. The results report additional numbers of people at risk in areas around the southern limit of the disease distribution in South America (van Lieshout et al., 2004). Nicaragua and Bolivia have predicted a possible increase in the incidence of malaria in 2010, reporting seasonal variations (Aparicio, 2000; NC-Nicaragua, 2001). The increase in malaria and population at risk could impact the costs of health services, including treatment and social security payments.

Kovats et al. (2005) have estimated relative risks (the ratio of risk of disease/outcome or death among the exposed to the risk among the unexposed) of different health outcomes in the year 2030 in Central America and South America, with the highest relative risks being for coastal flood deaths (drowning), followed by diarrhoea, malaria and dengue. Other models project a substantial increase in the number of people at risk of dengue due to changes in the geographical limits of transmission in Mexico, Brazil, Peru and Ecuador (Hales et al., 2002). Some models project changes in the spatial distribution (dispersion) of the cutaneous leishmaniasis vector in Peru, Brazil, Paraguay, Uruguay, Argentina and Bolivia (Aparicio, 2000; Peterson and Shaw, 2003), as well as the monthly distribution of dengue vector (Peterson et al., 2005).

Table 13.7. *Future impacts and vulnerability to climate change and variability in Latin America: people and coastal systems.*

Country/Region	Climate scenario	Impacts/costs (people, infrastructure, ecosystems, sectors)
Latin America	HADCM3: SRES B2, B1, A2, A1FI. SLR (Nicholls, 2004)	Assuming uniform population growth, no increase in storm intensity and no adaptation response (constant protection) the average annual number of coastal flood victims by the 2080s will probably range between 3 million and 1 million under scenarios A and B, respectively. If coastal defences are upgraded in line with rising wealth (evolving adaptation), the number of victims would be 1 million people under the worst-case scenario (A1FI). Finally, if coastal defences are upgraded against sea-level rise (enhanced adaptation); no people should be affected (Warren et al., 2006). People at risk[1] on coastal flood plains are likely to increase from 9 million in 1990 to 16 million (B1) and 36 million (A2) by the 2080s.
Low-lying coasts in Brazil, Ecuador, Colombia, Guyana, El Salvador, Venezuela	SRES A2: 38-104 cm	Mangrove areas could disappear from more exposed and marginal environments and, at the same time, the greatest development would occur in the more optimal high-sedimentation, high-tide and drowned river-valley environments. Shrimp production will be affected, with a consequent drop in production and GDP share (Medina et al., 2001).
El Salvador	SLR: 13-110 cm	Land loss ranging from 10% to 27.6% of the total area (141-400.7 km^2) (NC-El Salvador, 2000).
Guyana	SLR 100 cm projected by GCMs	Over 90% of the population and the most important economic activities are located in coastal areas which are expected to retreat by as much as 2.5 km (NC-Guyana, 2002).
Mesoamerican coral reef and mangroves from Gulf of Mexico	Warmer SST: 1-3°C by the 2080s under IPCC SRES scenarios	Coral reef and mangroves are expected to be threatened, with consequences for a number of endangered species: e.g., the green, hawksbill and loggerhead turtles, the West Indian manatee, and the American and Motelet's species of crocodile (Cahoon and Hensel, 2002).
Costa Rica, Punta Arenas coast	SLR 0.3-1.0 m	Sea water could penetrate 150 to 500 m inland, affecting 60-90% of urban areas (NC-Costa Rica, 2000).
Ecuador, Guayas river system, associated coastal zone and Guayaquil City	No-change: LANM0, moderate: LANM1, and severe changes: LANM2, with and without economic development	Losses of US$1,305 billion, which include shrimp cultures, mangroves, urban and recreation areas, supply of drinking water, as well as banana, rice and sugarcane cultivation. US$1,040 billion would be under risk. Evacuated and at-risk population should rise to 327,000 and 200,000 people, respectively. Of the current 1,214 km^2 of mangroves, it is estimated that 44% will be affected by the LANM2 scenario (NC-Ecuador, 2000).
Peru	Intensification of ENSO events and increases in SST. Potential SLR	Increased wind stress, hypoxia and deepening of the thermocline will impact on the marine ecosystem and fisheries, i.e., reduction of spawning areas and fish catches of anchovy. Flooding of infrastructure, houses and fisheries will cause damage valued at US$168.3 million. Global losses on eight coastal regions in Peru are estimated at US$1,000 million (NC-Perú, 2001).
Colombia	SLR 1.0 m	Permanent flooding of 4,900 km^2 of low-lying coast. About 1.4 million people would be affected; 29% of homes would be highly vulnerable; the agricultural sector would be exposed to flooding (e.g., 7.2 Mha of crops and pasture would be lost); 44.8% of the coastal road network would be highly vulnerable (NC-Colombia, 2001).
Argentina (Buenos Aires City)	Storm surges and SLR 2070/2080	Very low-lying areas which are likely to be permanently flooded are now only thinly populated. Vulnerability is mostly conditioned by future exposure to extreme surges. Rapid erosion with its consequent coastline retreat will occur at a rate depending on geological characteristics of the area. As a result of adaptation to present storm-surge conditions, the social impact of future permanent flooding will be relatively small (Kokot, 2004; Kokot et al., 2004; Menéndez and Ré, 2005).
Argentina and Uruguay (western Montevideo) coastal areas. Buenos Aires and Rio Negro Provinces	SLR, climate variability, ENSO, storm surges ('sudestadas')	Increases in non-eustatic factors (i.e., an increase in 'sudestadas' (a strong south-eastern wind along the Rio de la Plata coast) and freshwater flow, the latter often associated with El Niño, would accelerate SLR in the Río de la Plata, having diverse environmental and societal impacts on both the Argentine and Uruguay coasts over the next few decades, i.e., coastal erosion and inundation. Low-lying areas (estuarine wetlands and sandy beaches very rich in biodiversity) will be highly vulnerable to SLR and storm surges (southern winds). Loss of land would have a major impact on the tourism industry, which accounts for 3.8% of Uruguay's GDP (Barros, 2003; Codignotto, 2004; Kokot, 2004, Kokot et al., 2004; NC-Uruguay, 2004; Nagy et al., 2005, 2006c; Natenzon et al., 2005b).

[1] This is defined as living below the 1 in 1,000 year flood level.

Climate change is likely to increase the risk of forest fires. In some countries, wildfires and intentional forest fires have been associated with an increased risk of out-patient visits to hospital for respiratory diseases and an increased risk of breathing problems (WHO, 2000; Mielnicki et al., 2005). In urban areas exposed to the 'heat island' effect and located in the vicinity of topographical features which encourage stagnant air mass conditions and the ensuing air pollution, health problems would be exacerbated, particularly those resulting from surface ozone concentrations (PAHO, 2005). Furthermore, urban settlements located on hilly ground, where soil texture is loose, would be affected by landslides and mudflows; thus people living in poor-quality housing would be highly vulnerable.

Highly unusual stratospheric ozone loss and UV-B increases have occurred in the Punta Arenas (Chile) area over the past two decades, resulting in the non-photoadapted population being repeatedly exposed to an altered solar UV spectrum causing a greater risk of erythema and photocarcinogenesis. According to Abarca and Cassiccia (2002), the rate of non-melanoma skin cancer, 81% of the total, has increased from 5.43 to 7.94 per 100,000 (46%).

Human migration resulting from drought, environmental degradation and economic reasons may spread disease in unexpected ways, and new breeding sites for vectors may arise due to increasing poverty in urban areas and deforestation and environmental degradation in rural areas (Sims and Reid, 2006).

Recent studies warn of the possible re-emergence of Chagas' disease in Venezuela (Feliciangeli et al., 2003; Ramírez et al., 2005) and Argentina (PNC, 2005), and a wider vector distribution in Peru (Cáceres et al., 2002). Some models project a dispersal potential for Chagas' vector species into new areas (Costa et al., 2002).

A national assessment of Brazilian regions demonstrated that the north-east is the most vulnerable to the health effects of changing climate due to its poor social indicators, the high level of endemic infectious diseases, and the periodic droughts that affect this semi-arid region (Confalonieri et al., 2005).

13.5 Adaptation: practices, options and constraints

13.5.1 Practices and options

13.5.1.1 Natural ecosystems

Some options to increase the capacity to adapt to climate change include the reduction of ecosystem degradation in Latin America through the improvement and reinforcement of policy, planning and management. According to the Millennium Ecosystem Assessment (2005), Biringer et al. (2005), FAO (2004b), Laurance et al. (2001), Brown et al. (2000) and Nepstad et al. (2002), these options are basically as follows.

- In the government context: integrate decision-making between different departments and sectors and participate in international institutions in order to ensure that policies are focused on the protection of ecosystems.
- Identify and exploit synergies: taking advantage of synergies between proposed and existing adaptation policies and actions can provide significant benefits to both endeavours (Biringer et al., 2005).
- Procure the empowerment of marginalised groups so as to influence the decisions that affect them and their ecosystem services, and campaign for legal recognition of local communities' ownership of natural resources. This option is the key to reducing the incidence of forest fires.
- Include sound valuation and management of ecosystem services in all regional planning decisions and in poverty reduction strategies, e.g., Noel Kempff Mercado Climate Action Project in Bolivia and Río Bravo Carbon Sequestration Pilot Project in Belize.
- Establish additional protected areas, particularly the biological or ecological corridors, for preserving the connections between protected areas, with the aim of preventing the fragmentation of natural habitats. Some programmes and projects involving actions with different degrees of implementation are: the Meso-American Biological Corridor; Binational Corridors (e.g., Tariquía-Baritú between Argentina and Bolivia, Vilcabamba-Amboro between Peru and Bolivia, Cóndor Kutukú between Peru and Ecuador, Chocó–Manabí between Ecuador and Colombia), the natural corridor projects under way in Brazil's Amazon region and the Atlantic forests of Colombia (e.g., Corredor Biológico Guácharos–Puracé and Corredor de Bosques Altoandinos de Roble); those in Venezuela (e.g., Corredor Biológico de la Sierra de Portuguesa), Chile (e.g., Corredor entre la Cordillera de los Andes y la Cordillera de la Costa and Proyecto Gondwana), and some initiatives in Argentina (e.g., Iniciativa Corredor de Humedales del Litoral Fluvial de la Argentina, Corredor Verde de Misiones, and Proyecto de Biodiversidad Costera).
- Tropical countries in the region can reduce deforestation through adequate funding of programmes designed to enforce environmental legislation, support for economic alternatives to extensive forest clearing (including carbon crediting), and building capacity in remote forest regions, as recently suggested in part of the Brazilian Amazon (Nepstad et al., 2002; Fearnside, 2003). Moreover, substantial amounts of forest can be saved in protected areas if adequate funding is available (Bruner et al., 2001; Pimm et al., 2001).
- Monitoring and evaluating (M&E) adaptation strategy impacts on biodiversity. The process of monitoring change in biological systems can be complex and resource-intensive, requiring involved observation and data collection, painstaking analysis, etc. Care should be taken to ensure that an M&E plan is developed which ensures a robust yet streamlined M&E process (Biringer et al., 2005).
- Agroforestry using agroecological methods offers strong possibilities for maintaining biological diversity in Latin America, given the overlap between protected areas and agricultural zones (Morales et al., 2007).

13.5.1.2 Agriculture and forestry

Some adaptation measures aiming to reduce climate change impacts have been proposed in the agricultural sector. For example, in Ecuador, options such as agro-ecological zoning and appropriate sowing and harvesting seasons, the introduction of higher-yielding varieties, installation of irrigation systems, adequate use of fertilisers, and implementation of a system for controlling pests and disease were proposed (NC-Ecuador, 2000). In Guyana several adjustments relating to crop variety (thermal and moisture requirements and shorter-maturing varieties), soil management, land allocation to increase cultivable area, using new sources of water (recycling of wastewater), harvesting efficiency, and purchases to supplement production (fertilisers and machinery) were identified (NC-Guyana, 2002).

In other countries, adaptation measures have been assessed by means of crop simulation models. For example, in the Pampas region of Argentina, anticipating planting dates and the use of wheat and maize genotypes with longer growth cycles would take advantage of projected longer growing seasons as a result of the shortening of the period when frosts may occur (Magrin and Travasso, 2002). More recently, Travasso et al. (2006) reported that, in South Brazil, Uruguay and Argentina, the negative impacts of future climate on maize and soybean production could be offset by changing planting dates and adding supplementary irrigation.

In terms of food security, a significant number of smallholders and subsistence farmers may be particularly vulnerable to climate change in the short term, and their adaptation options may be more limited. Of particular concern are farmers in Central America, where drying trends have been reported, and in the poorer regions of the Andes. Adaptations in these communities may involve policies for market development of new and existing crop and livestock products, breeding drought-tolerant crops, modified farm-management practices, and improved infrastructure for off-farm employment generation. Increasingly, cross-sectoral perspectives are needed when considering adaptation options in these communities (Jones and Thornton, 2003; Eakin, 2005). In dry areas of north-eastern Brazil, where small farmers are among the social groups most vulnerable to climate change, the production of vegetable oils from native plants (e.g., castor bean) to supply the bio-diesel industry has been proposed as an adaptation measure (La Rovere et al., 2006).

A global study (which includes case studies of northern Argentina and south-eastern Brazil) concluded that in northern Argentina occasional problems in water supply for agriculture under the current climate may be exacerbated by climate change, and may require timely improvements in crop cultivars, irrigation and drainage technology, and water management. Conversely, in south-eastern Brazil, future water supply for agriculture is likely to be plentiful (Rosenzweig et al., 2004).

As a way of avoiding the consequences of deforestation as a likely impact on the regional climate, several measures are currently being initiated in the region and are likely to be intensified in the future. Argentina, Brazil, Costa Rica and Peru have adopted new forestry laws and policies that include better regulatory measures, sustainability principles, expansion of protected areas, certification of forestry products and expansion of forest plantations into non-forested areas (Tomaselli, 2001). In the Brazilian Amazon state of Mato Grosso, where 18,000 km^2 of forest and savannas were converted to pasture and soybean fields in 2003, requirements for licensing of deforestation and environmental certification of soybean have been introduced as a way to preserve the environment. A similar proposal is under development for the Mato Grosso cattle industry (Nepstad, 2004). Most countries provide incentives for managing their native forests: exemption from land taxes (Chile, Ecuador), technical assistance (Ecuador), and subsidies (Argentina, Mexico and Colombia) (UNEP, 2003a). Chile and Guyana demand prior studies on environmental impact before approving forestry projects, depending upon their importance; Mexico, Belize, Costa Rica and Brazil are already applying forestry certification. Argentina, Chile, Paraguay, Costa Rica and Mexico have established model forests designed to demonstrate the application of sustainable management, taking into account productive and environmental aspects, and with the wide participation of civilian society, including community and indigenous groups.

13.5.1.3 Water resources

Water management policies in Latin America should be the central point of the adaptation criteria to be established in order to strengthen the countries' capacities to manage water resources availability and demand, and ensure the safety of people and protection of their belongings under changing climatic conditions. In this regard, the principal actions for adaptation must include: improvement and further development of legislation related to land use on floodplains, ensuring compliance with existing regulations of risk zones, floodplain use and building codes; re-evaluating the design and safety criteria of structural measures for water management; developing groundwater protection and restoration plans to maintain water storage for dry seasons; developing public awareness campaigns to highlight the value of rivers and wetlands as buffers against increased climate variability and to improve participation of vulnerable groups in flood adaptation and mitigation programmes (IRDB, 2000; Bergkamp et al., 2003; Solanes and Jouravlev, 2006).

Adaptation to drier conditions in 60% of the territory of Latin America would require a great increase in the amount of investment in water supply systems, in addition to the US$17.7 billion needed to accomplish the provision of safe water systems to 121 million people, necessary to achieve the Millennium Declaration for Safe Water goals by 2015 (even though this would leave 10% of the population of Latin America without access to safe water) (IDB, 2004).

Managing transbasin diversions has been the solution for water development in some regions of the world, particularly in California. In Latin America, transbasin transfers in Yacambú basin (Venezuela), Catamayo-Chira basins (Ecuador and Peru), Alto Piura and Mantaro basins (Peru), and the São Francisco River (Brazil) would be an option to mitigate the likely stresses on water supply for the population. Transbasin diversions should be practiced responsibly, taking into account environmental consequences and the hydrological regime (Vásquez, 2004; Marengo and Raigoza, 2006).

The use of urban and rural groundwater needs to be controlled and rationalised, taking into account the quality, distribution and trends over time identified in each region. To develop sustainable groundwater and aquifer management, the rules to apply would be: limit or reduce the consequences of excessive abstractions, slow down growth of abstractions, explore possibilities for artificial aquifer recharge, and evaluate options for planned mining of groundwater storage (IRDB, 2000; World Bank, 2002b; Solanes and Jouravlev, 2006). Water conservation practices, re-use of water, water recycling by modification of industrial processes and optimisation of water consumption bring opportunities for adaptation to water-stressed periods (COHIFE, 2003).

13.5.1.4 Coasts

Future adaptation of coastal systems in Latin America is mostly based on coastal zone management, monitoring and protection plans (see Sections 13.2.5.4 and 13.4.4) which are not specific for climate variability and change and are not yet fully implemented. However, the current coastal environmental framework should be an important support for implementing adaptation options to climate change. Table 13.8 shows some examples of practices and options related to adaptations to climate change.

Most fishing countries have regulations governing access to their fishing grounds (e.g., Argentina, Chile and Ecuador) and others have been drafting new legislation in order to control the use of coastal and fishing resources and to introduce adaptation measures (e.g., Costa Rica, Guyana, Panama, Peru, Venezuela). A number of regional agreements have also been signed on the protection of the marine environment, the prevention of pollution from marine or terrestrial sources, and the management of commercial fisheries (Young, 2001; UNEP, 2002; Bidone and Lacerda, 2003; OAS-CIDI, 2003). Brazil and Costa Rica ratified the UN Convention on the Law of the Sea (UNCLOS, 2005), related to the conservation and management of straddling fish stocks[2] and highly migratory fish stocks.

Coastal biodiversity could be maintained, and even improved, through sustainable use by promoting community management to make conservation a part of sustainable development of coastal resources such as mangroves and their artisanal fisheries. In this regard, Mexico, Ecuador, Guatemala, Brazil and Nicaragua have promoted initiatives to develop the necessary local community participation in the managed forest of coastal zones (Kovacs, 2000; Windevoxhel and Sención, 2000; Yáñez-Arancibia and Day, 2004; FAO, 2006).

13.5.1.5 Human health

There are many initiatives that should be implemented in order to deal with different health impacts due to climate change in Latin American countries. Awareness regarding impacts should be enhanced in the region, including community involvement (see Chapter 8, Section 8.6.1). One main shortcoming is that a lack of information adversely affects decision-making, so research and human-resource training are fundamental. Therefore, one of the main tasks to support research and decision-making is to build up statistical information relating health conditions and events to the corresponding climate and related environmental issues (e.g., floods, tornados, landslides, etc.), based on a strengthened surveillance system for climate-sensitive diseases (see Chapter 8, Section 8.6) (Anderson, 2006). It is essential to establish a regular channel of communication with the Pan American Health Organization (PAHO/WHO) to report and classify such information, to integrate the data into a regionalisation of sanitary/health conditions, and thus improve early warnings of epidemics. The advantages of international initiatives such as the Global Health Watch 2005-2006 – not simply as a recipient of information but also as a provider of information – should also be considered. The assessments should take into account human health vulnerability and public health adaptation to climate change.

As human health is a result of the interplay between many different sectors, it is important to consider the impacts in the water sector in order to identify the measures focusing on the surveillance of water-borne diseases and vulnerable populations, as well as impacts from the agricultural sector, biodiversity, natural resources, air pollution and drought. An important concern relating to health is the implications of increased human migration and changes in disease patterns; this implies greater intergovernmental co-ordination and cross-boundary actions. Future analysis based on ecological niche modelling for disease vectors will be very useful

Table 13.8. *Adaptation practices and options for Latin American coasts: selected countries.*

Country/Study	Climate scenario	Adaptation (practices and options)/costs
Ecuador (NC-Ecuador, 2000)	LANM2 (+1.0 m)	Protection against severe scenario conditions: coastal defence of Guayas river basin at a cost of less than US$2 billion with benefits two to three times greater; reforestation of mangroves and preservation of flooded areas to protect 1,204 km^2 and shrimp farms (the shrimp industry is the country's third largest export item) against flooding.
Guyana (NC-Guyana, 2002)	LANM2	Accretion development on a low-lying coastal strip 77 km wide in the east and 26 km wide in the western Essequibo region.
Colombia (NC-Colombia, 2001)	SLR	Recovery and strengthening resiliency of natural systems in order to facilitate natural adaptation to SLR as well as a programme of coastal zone management which emphasizes preservation of wetlands, areas prone to flooding and those of high value.
Panama (NC-Panama, 2000)	SLR	Autonomous and planned adaptation measures to protect the loss of beaches, based mainly on soft engineering practices.
Peru (NC-Perú, 2001)	ENSO, SST	Modern satellite observation systems of sea and continent similar to the international programmes TOGA and CLIVAR, and capacity-building for at least 50 scientists in oceanic, atmospheric and hydrological modelling and GIS systems.
Uruguay (NC-Uruguay, 2004)	Flooding and SLR	Monitoring systems in order to: track impacts on the coasts; restore degraded areas; develop an institutional framework for integrated coastal management (ICM); define setback regulations; improve local knowledge on beach nourishment; develop contingency plans against flooding; assess socio-economic and environmental needs; encourage stakeholders' participation.
Argentina (Kokot, 2004; Menéndez and Ré, 2005)	SLR 2070	Flood risk maps for Buenos Aires based on SLR trends, records of storm surges ('sudestadas') and a two-dimensional hydrodynamic model. These maps will be useful for early warning of extreme events.

[2] Straddling fish stocks: found in both the coastal zone and high seas.

to provide new potentials for optimising the use of resources for disease prevention and remediation via automated forecasting of disease transmission rates (Costa et al., 2002; Peterson et al., 2005).

13.5.2 Constraints on adaptation

At the present time, constraints of a different nature are observed in the region that are very likely to damp stakeholders' capacities, and decision-makers' capabilities, to achieve policy efficacy and economic efficiency for adapting to climate change. Socio-economic and political factors such as limited availability of credit and technical assistance, and low public investment in infrastructure in rural areas, have been shown to seriously reduce the capability to implement adaptive options in the agricultural sector, particularly for small producers (Eakin, 2000; Vásquez-León et al., 2003). In addition, inadequate education and public health services are key barriers to decreasing climate change and variability impacts, and developing coping mechanisms for extreme weather events such as flooding and droughts, mainly in poor rural areas (Villagrán de León et al., 2003).

A poor appreciation of risk, lack of technical knowledge, inappropriate monitoring, and scarce or incomplete databases and information are important constraints to adaptation to current climate trends. The usefulness of weather forecasts and early warning systems in the region is typically limited by these factors as well as by the lack of resources to implement and operate them (NC-Ecuador, 2000; Barros, 2005).

Public health policies are focused on curative approaches rather than on large-scale preventative programmes and are not integrated with other socio-economic policies that could enhance their effectiveness in addressing climate change impacts. There is a lack of tools to address cross-cutting issues and long-term public health challenges. In most countries, the inter-sectoral work between the health sector and other sectors such as environment, water resources, agriculture and climatological/meteorological services is very limited (Patz et al., 2000). In coastal areas, environmental policies, laws and regulations, have been conflicting in the implementation of adaptation options to climate-change-related impacts (UNEP, 2003b).

13.6 Case studies

Box 13.1. Amazonia: a 'hotspot' of the Earth system

The Amazon Basin contains the largest contiguous extent of tropical forest on Earth, almost 5.8 million km^2 (see Figure 13.3). It harbours perhaps 20% of the planet's plant and animal species. There is abundance of water resources and the Amazon River accounts for 18% of the freshwater input to the global oceans. Over the past 30 years almost 600,000 km^2 have been deforested in Brazil alone (INPE-MMA, 2005a) due to the rapid development of Amazonia, making the region one of the 'hotspots' of global environmental change on the planet. Field studies carried out over the last 20 years clearly show local changes in water, energy, carbon and nutrient cycling, and in atmospheric composition, caused by deforestation, logging, forest fragmentation and biomass burning. The continuation of current trends shows that over 30% of the forest may be gone by 2050 (Alencar et al., 2004; Soares-Filho et al., 2006). In the last decade, research by the Large Scale Biosphere-Atmosphere (LBA) Experiment in Amazonia is uncovering novel features of the complex interaction between vegetated land surfaces and the atmosphere on many spatial and temporal scales. The LBA Experiment is producing new knowledge on the physical, chemical and biological functioning of Amazonia, its role for our planet, and the impacts on that functioning due to changes in climate and land use (http://lba.cptec.inpe.br/lba/site/).

There is observational evidence of sub-regional changes in surface energy budget, boundary layer cloudiness and regional changes in the lower troposphere radiative transfer due to biomass-burning aerosol loadings. The discovery of large numbers of cloud condensation nuclei (CCN) due to biomass burning has led to speculation about their possible direct and indirect roles in cloud formation and rainfall, possibly reducing dry-season rainfall (e.g., Andreae et al., 2004). During the rainy season, in contrast, there are very low amounts of CCN of biogenic origin and the Amazonian clouds show the characteristics of oceanic clouds. Carbon cycle studies of the LBA Experiment indicate that the Amazonian undisturbed forest may be a sink of carbon for about 100 to 400 Mt C/yr, roughly balancing CO_2 emissions due to deforestation, biomass burning, and forest fragmentation of about 300 Mt C/yr (e.g., Ometto et al., 2005). On the other hand, the effect of deforestation and forest fragmentation is increasing the susceptibility of the forest to fires (Nepstad et al., 2004).

Observational evidence of changes in the hydrological cycle due to land-use change is inconclusive at present, although observations have shown reductions in streamflow and no change in rainfall for a large sub-basin, the Tocantins river basin (Costa et al., 2003). Modelling studies of large-scale deforestation indicate a probably drier and warmer post-deforestation climate (e.g., Nobre et al., 1991, among others). Reductions in regional rainfall might lead to atmospheric teleconnections affecting the climate of remote regions (Werth and Avissar, 2002). In sum, deforestation may lead to regional climate changes that would lead in turn to a 'savannisation' of Amazonia (Oyama and Nobre, 2003; Hutyra et al., 2005). That factor might be greatly amplified by global warming. The synergistic combination of both regional and global changes may severely affect the functioning of Amazonian ecosystems, resulting in large biome changes with catastrophic species disappearance (Nobre et al., 2005).

Box 13.2. Adaptation capacity of the South American highlands' pre-Colombian communities

The subsistence of indigenous civilisations in the Americas relied on the resources cropped under the prevailing climate conditions around their settlements. In the highlands of today's Latin America, one of the most critical limitations affecting development was, as currently is, the irregular distribution of water. This situation is the result of the particularities of the atmospheric processes and extremes, the rapid runoff in the deep valleys, and the changing soil conditions. The tropical Andes' snowmelt was, as it still is, a reliable source of water. However, the streams run into the valleys within bounded water courses, bringing water only to certain locations. Moreover, valleys and foothills outside of the Cordillera Blanca glaciers and extent of the snow cover, as well as the Altiplano, receive little or no melt-water at all. Therefore, in large areas, human activities depended on seasonal rainfall. Consequently, the pre-Colombian communities developed different adaptive actions to satisfy their requirements. Today, the problem of achieving the necessary balance between water availability and demand is practically the same, although the scale might be different.

Under such limitations, from today's Mexico to northern Chile and Argentina, the pre-Colombian civilisations developed the necessary capacity to adapt to the local environmental conditions. Such capacity involved their ability to solve some hydraulic problems and foresee climate variations and seasonal rain periods. On the engineering side, their developments included rainwater cropping, filtration and storage; the construction of surface and underground irrigation channels, including devices to measure the quantity of water stored (Figure 13.4) (Treacy, 1994; Wright and Valencia Zegarra, 2000; Caran and Nelly, 2006). They also were able to interconnect river basins from the Pacific and Atlantic watersheds, in the Cumbe valley and in Cajamarca (Burger, 1992).

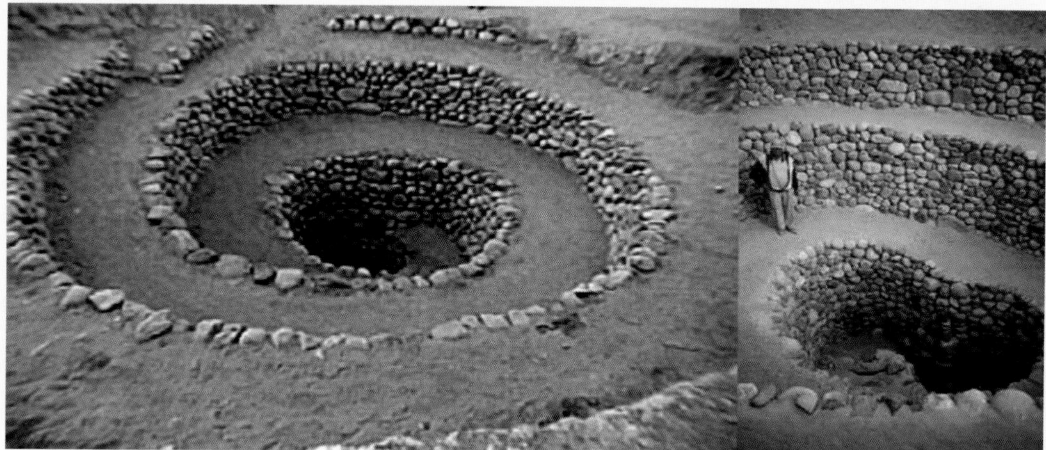

Figure 13.4. *Nasca (southern coast of Peru) system of water cropping for underground aqueducts and feeding the phreatic layers.*

Other capacities were developed to foresee climate variations and seasonal rain periods, to organise their sowing schedules and to programme their yields (Orlove et al., 2000). These efforts enabled the subsistence of communities which, at the peak of the Inca civilisation, included some 10 million people in what is today Peru and Ecuador.

Their engineering capacities also enabled the rectification of river courses, as in the case of the Urubamba River, and the building of bridges, either hanging ones or with pillars cast in the river bed. They also used running water for leisure and worship purposes, as seen today in the 'Baño del Inca' (the spa of the Incas), fed from geothermal sources, and the ruins of a musical garden at Tampumacchay in the vicinity of Cusco (Cortazar, 1968). The priests of the Chavin culture used running water flowing within tubes bored into the structure of the temples in order to produce a sound like the roar of a jaguar; the jaguar being one of their deities (Burger, 1992). Water was also used to cut stone blocks for construction. As seen in Ollantaytambo, on the way to Machu Picchu, these stones were cut in regular geometric shapes by leaking water into cleverly made interstices and freezing it during the Altiplano night, reaching below zero temperatures. They also acquired the capacity to forecast climate variations, such as those from El Niño (Canziani and Mata, 2004), enabling the most convenient and opportune organisation of their foodstuff production. In short, they developed pioneering efforts to adapt to adverse local conditions and define sustainable development paths.

Today, under the vagaries of weather and climate, exacerbated by the increasing greenhouse effect and the rapid retreat of the glaciers (Carey, 2005; Bradley et al., 2006), it would be extremely useful to revisit and update such adaptation measures. Education and training of present community members on the knowledge and technical abilities of their ancestors would be the way forward. ECLAC's procedures for the management of sustainable development (Dourojeanni, 2000), when considering the need to manage the extreme climate conditions in the highlands, refer back to the pre-Colombian irrigation strategies.

13.7 Conclusions and implications for sustainable development

In Latin America there is ample evidence of increases in extreme climatic events and climate change. Since the TAR, unusual extreme weather events have occurred in most countries, such as continuous drought/flood episodes, the Hurricane Catarina in the South Atlantic, and the record hurricane season of 2005 in the Caribbean Basin. In addition, during the 20th century, temperature increases, rainfall increases and decreases, and changes in extreme events, were reported for several areas. Changes in extreme episodes included positive trends in warm nights, and a positive tendency for intense rainfall events and consecutive dry days. Some negative impacts of these changes were glacier retreat, increases in flood frequency, increases in morbidity and mortality, increases in forest fires, loss of biodiversity, increases in plant diseases, reduction in dairy cattle production and problems with hydropower generation. However, beneficial impacts were reported for the agricultural sector in temperate zones. According to Swiss Re estimations, if no action is taken in Latin America to slow down climate change, in the next decades climate-related disasters could cost US$300 billion per year (CEPAL, 2002; Swiss Re, 2002).

On the other hand, rates of deforestation have increased since the TAR (e.g., in Brazilian Amazonia). In Argentina, Bolivia, Brazil and Paraguay, agricultural expansion, mainly the soybean cropping boom, has exacerbated deforestation and has intensified the process of land degradation. This critical land-use change will enhance aridity and desertification in many of the already water-stressed regions in South America, affecting not only the landscape but also modifying the water cycle and the climate of the region.

As well as climatic stress and changes in land use, other stresses are compromising the sustainable development of Latin America. Demographic pressures, as a result of migration to urban areas, result in widespread unemployment, overcrowding and the spread of infectious diseases. Furthermore, over-exploitation is a threat to most local production systems, and aquifer over-exploitation and mismanagement of irrigation systems are causing salinisation of soils and water and sanitation problems.

By the end of the 21st century, the projected mean warming for Latin America ranges from 1 to 4°C or from 2 to 6°C, according to the scenario, and the frequency of weather and climate extremes is very likely to increase. By the year 2020, 100 Mha of Brazil Amazonia forest will have disappeared if deforestation rates continue as in 2002/03, and the soybean-planted area in South America could reach 59 Mha, representing 57% of the world's soybean production. By 2050, the population of LA is likely to be 50% higher than in 2000, and migration from the countryside to the cities will continue.

Predicted changes are very likely to severely affect a number of ecosystems and sectors (see Figure 13.5) by:

- decreasing plant and animal species diversity, and causing changes in ecosystem composition and biome distribution,
- melting most tropical glaciers in the near future (2020-2030),

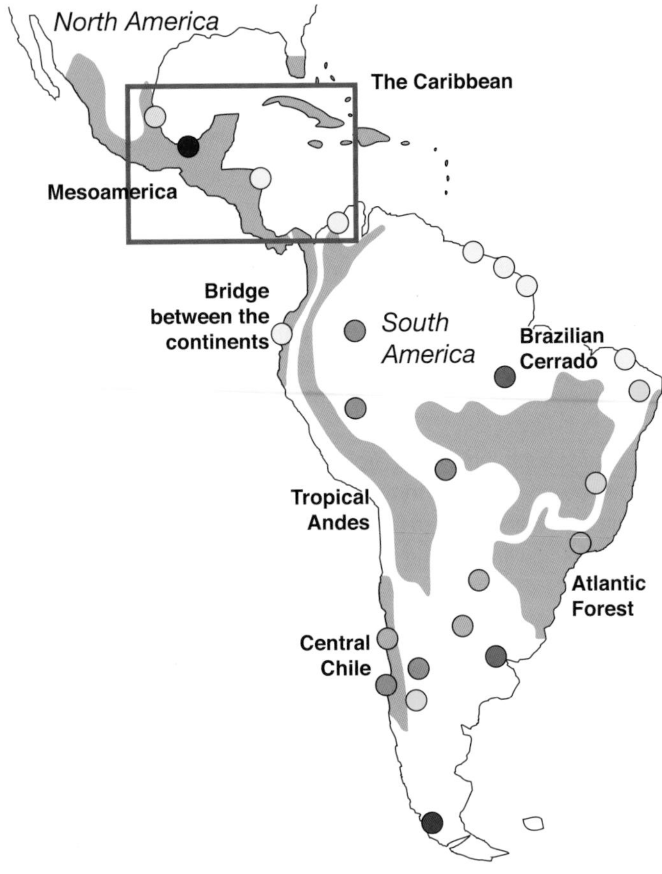

- Coral reefs and mangroves seriously threatened with warmer SST
- Under the worst sea-level rise scenario, mangroves are very likely to disappear from low-lying coastlines
- Amazonia: loss of 43% of 69 tree species by the end of 21st century; savannisation of the eastern part
- Cerrados: Losses of 24% of 138 tree species for a temperature increase of 2°C
- Reduction of suitable lands for coffee
- Increases in aridity and scarcity of water resources
- Sharp increase in extinction of: mammals, birds, butterflies, frogs and reptiles by 2050
- Water availability and hydro-electric generation seriously reduced due to reduction in glaciers
- Ozone depletion and skin cancer
- Severe land degradation and desertification
- Rio de la Plata coasts threatened by increasing storm surges and sea-level rise
- Increased vulnerability to extreme events

Areas in red correspond to sites where biodiversity is currently severely threatened and this trend is very likely to continue in the future

Figure 13.5. *Key hotspots for Latin America.*

- reducing water availability and hydropower generation,
- increasing desertification and aridity,
- severely affecting people, resources and economic activities in coastal areas,
- increasing crop pests and diseases,
- changing the distribution of some human diseases and introducing new ones.

One beneficial impact of climate change is likely to be the projected increase in soybean yields in the south of South America. However, the future conversion of natural habitats to accommodate soybean expansion are very likely to severely affect some ecosystems such as the Cerrados, dry and humid Chaco, Amazon transition and rainforest, and the Atlantic, Chiquitano and Yungas forests.

If the Latin American countries continue to follow the business-as-usual scenario, the wealth of natural resources that have supported economic and socio-cultural development in the region will be further degraded, reducing the regional potential for growth. Urgent measures must be taken to help bring environmental and social considerations from the margins to the fore of decision-making and development strategies (UNEP, 2002).

Climate change would bring new environmental conditions resulting from modifications in space and time, and in the frequency and intensity, of weather and climate processes. These atmospheric processes are closely interlinked with environmental, social and economic pillars on which development should be based, and all together may influence the selection of sustainable development paths. Facing a new climate system and, in particular, the exacerbation of extreme events, will call for new ways to manage human and natural systems for achieving sustainability. Future development in regional, sub-regional and local areas must be based on reliable and sufficiently-dense basic data. Consequently, any action towards sustainable development already commits governments and stakeholders to take the lead in the development of the information necessary to facilitate the actions needed to cope with the adversities of climate events, from the transitional period until a new climate system is established, and to take advantage of the new climate system's potential advantages.

13.8 Key uncertainties and investigation priorities

The projections mentioned in this chapter rely on the quality of the available mathematical models. As it can be seen in its different sections, there are contradictory statements. Such contradictions, also observed in other sectoral and regional chapters, make evident some of the weaknesses of models, especially when the necessary observational background is missing. In addition to the models' shortcomings, the use of socio-economic scenarios which are not sufficiently representative of the socio-economic conditions in the region, plus the problems still being faced with downscaling techniques, puts more emphasis on the lack of information as a critical uncertainty. Additionally, the communication of risk to stakeholders and decision-makers under uncertainty has been shown to be a significant weakness that needs to be addressed in the short term.

In order to promote economic efficiency and policy efficacy for future adaptation, important multidisciplinary research efforts are required in order to reduce the information gaps. In preparing for the challenges that climate change is posing to the region in the future, the research priorities should be to resolve the constraints already identified in terms of facing current climate variability and trends, such as:

- lack of awareness,
- lack of well-distributed and reliable observation systems,
- lack of adequate monitoring systems,
- poor technical capabilities,
- lack of investment and credits for the development of infrastructure in rural areas,
- scarce integrated assessments, mainly between sectors,
- limited studies on the economic impacts of current and future climate variability and change,
- restricted studies on the impacts of climate change on societies,
- lack of clear prioritisation in the treatment of topics for the region as a whole.

In addition, other priorities considering climate change are:

- to reduce uncertainties in future projections,
- to assess the impacts of different policy options on reducing vulnerability and/or increasing adaptive capacity.

It is also worth stating that we must change the attitude from planning to effective operation of observation and alerting systems. Currently, the typical response to a severe climatic event consists of intervening after the fact, usually with insufficient funds to restore the conditions prior to the event. A necessary change would be to migrate from a culture of response to a culture of prevention.

In addition, the possibility of abrupt climate change due to a perturbation of the thermohaline circulation opens up a new theme for concern in the Latin American region, where there have been no studies about its possible effects. Another related problem is the occurrence of possible climatic 'surprises' (even in a gradually changing climate) when certain thresholds are surpassed and a negative feedback mechanism is triggered, affecting different sectors and resources. Tropical forests and tropical glaciers are likely candidates for surprises.

References

Abarca, J.F. and C.C. Cassiccia, 2002: Skin cancer and ultraviolet-B radiation under the Antarctic ozone hole: southern Chile, 1987–2000. *Photodermatol. Photo.*, **18**, 294.

Adams, R.M., L.L. Houston, B.A. McCarl, M. Tiscareño, L.J. Matusand and G.R.F. Weiher, 2003: The benefits to Mexican agriculture of an El Niño-Southern Oscillation (ENSO) early warning system. *Agr. Forest Meteorol.*, **115**, 183-194.

Aguilar, E., T.C. Peterson, P. Ramírez Obando, R. Frutos, J.A. Retana, M. Solera, J. Soley, I. González García and Co-authors, 2005: Changes in precipitation and temperature extremes in Central America and northern South America, 1961–2003. *J. Geophys. Res.*, **110**, D23107, doi:10.1029/2005JD006119.

Alcántara-Ayala, I., 2004: Flowing mountains in Mexico: incorporating local knowledge and initiatives to confront disaster and promote prevention. *Mt. Res.*

Dev., **24**, 10-13.

Alencar, A., D. Nepstad, D. McGrath, P. Moutinho, P. Pacheco, M. Del Carmen, V. Díaz and B.S. Soares Filho, 2004: Desmatamento na Amazônia: indo além da "emergência crônica. Instituto de Pesquisa Ambiental da Amazônia, 88 pp. http://www.ipam.org.br/web/biblioteca/livros_download.php?PHPSESSID=3498be2d730f494d262e80281ff74449.

Alexander, L.V., X. Zhang, T.C. Peterson, J. Caesar, B. Gleason, A.M.G. Klein Tank, M. Haylock, D. Collins and Co-authors, 2006: Global observed changes in daily climate extremes of temperature and precipitation. *J. Geophys. Res.*, **111**, D05109, doi:10.1029/2005JD006290.

Alves, D., 2002: An analysis of the geographical patterns of deforestation in Brazilian Amazonia in the 1991–1996 period. *Patterns and Processes of Land Use and Forest Change in the Amazon*, C. Wood and R. Porro, Eds., University of Florida Press, Gainesville, Florida, 95-105.

Anderson, E., 2006: Automated mosquito identification: tools for early warning of vector-borne diseases outbreaks. Preprints, *UN Third International Conference on Early Alert Systems: From Concept to Action*. Bonn, 16 pp.

Andreae, M.O., D. Rosenfeld, P. Artaxo, A.A. Costa, G.P. Frank, K.M. Longo and M.A.F. Silva-Dias, 2004: Smoking rain clouds over the Amazon. *Science*, **303**, 1337-1340.

Andréfouët, S., P.J. Mumby, M. McField, C. Hu and E. Muller-Karger, 2002: Revisiting coral reef connectivity. *Coral Reefs*, **21**, 43-48.

Aparicio, M., 2000: *Vulnerabilidad y Adaptación a la Salud Humana ante los Efectos del Cambio Climático en Bolivia*. Ministerio de Desarrollo Sostenible y Planificación. Viceministerio de Medio Ambiente, Recursos Naturales y Desarrollo Forestal. Programa Nacional de Cambios Climáticos. PNUD/GEF.

Arjona, F., 2005: Environmental framework. Integrated National Adaptation Program, Colombia, 8 pp. http://www.ideam.gov.co/biblio/paginaabierta/ ENVIRONMENTALFRAMEWORKversionb.pdf.

Arnell, N.W., 2004: Climate change and global water resources: SRES scenarios emissions and socio-economic scenarios. *Global Environ. Chang.*, **14**, 31-52.

Aronson, R.B. and W.F. Precht, 2002: White-band disease and the changing face of Caribbean coral reefs. *Hydrobiologia*, **460**, 25-38.

Arriaga, L. and L. Gómez, 2004: Posibles efectos del cambio climático en algunos componentes de la biodiversidad de México. *Cambio Climático: Una Visión Desde México*, J. Martínez and A. Fernández Bremauntz, Eds., SEMARNAT e INE, México, 253-263.

Asner, G.P., D.E. Knapp, E.N. Broadbent, P.J.C. Oliveira, M. Keller and J.N. Silva, 2005: Selective logging in the Brazilian Amazon. *Science*, **310**, 480-482.

Barros, V., 2003: Observed La Plata river level and wind fields change. Preprints, *First AIACC Regional Workshop for Latin America and the Caribbean*, San Jose, Costa Rica, 24 pp. http://www.aiaccproject.org/meetings/meetings.html.

Barros, V., 2005: Inundación y cambio climático: costa Argentina del Río de la Plata. *El Cambio Climático en el Río de la Plata*, V. Barros, A. Menéndez and G.J. Nagy, Eds., Proyectos AIACC, 41-52.

Basán Nickisch, M., 2002: Sistemas de captación y manejo de agua. Estación Experimental Santiago del Estero. Instituto Nacional de Tecnología Agropecuaria. http://www.inta.gov.ar/santiago/info/documentos/agua/0001res_sistemas.htm.

Bergkamp, G., B. Orlando and I. Burton, 2003: *Change: Adaptation of Water Management to Climate Change*. IUCN, Water and Nature Initiative, Gland and Cambridge, 53 pp.

Betkowski, B., 2006: Forest fires a huge cost to health. University of Alberta, Canada.

BID, 2000: El desafío de los desastres naturales en América Latina y el Caribe: plan de acción del BID. Banco Interamericano de Desarrollo, Washington, District of Columbia, 36 pp.

Bidegain, M. and I. Camilloni, 2004: Climate change scenarios for southeastern South America. Preprints, *Second AIACC Regional Workshop for Latin America and the Caribbean*, Buenos Aires, 15 pp. http://www.aiaccproject.org/meetings/meetings.html.

Bidegain, M., R.M. Caffera, F. Blixen, V.P. Shennikov, L.L. Lagomarsino, E.A. Forbes and G.J. Nagy, 2005: Tendencias climáticas, hidrológicas y oceanográficas en el Río de la Plata y costa Uruguaya. *El Cambio Climático en el Río de la Plata,* V. Barros A. Menéndez and G.J. Nagy, Eds., Proyectos AIACC, 137-143.

Bidegain, M., R.M. Caffera, J. Gómez, B. de los Santos and P. Castellazzi, 2006: Perfomance of the WRF regional model over Southeastern South America during an extreme event. *Annals of the 8th International Conference of Southern Hemisphere Meteorology and Oceanography* (ICSHMO). AMS, Foz de Iguaçu, Brazil, 1655-1658.

Bidone, E.D. and L.D. Lacerda, 2003: The use of DPSIR framework to evaluate sustainability in coastal areas. Case study: Guanabara Bay Basin, Rio de Janeiro, Brazil. *Reg. Environ. Change*, **4**, 5-16.

Biringer, J., M.R. Guariguata, B. Locatelli, J. Pfund, E. Spanger-Siegfried, A.G. Suarez, S. Yeaman and A. Jarvis, 2005: Biodiversity in a changing climate: a framework for assessing vulnerability and evaluating practical responses. *Tropical Forests and Adaptation to Climate Change: in Search of Synergies*, C. Robledo, M. Kanninen and L. Pedroni, Eds., CIFOR, Bogor, 154-183.

Bischoff, S., 2005: Sudestadas. *El Cambio Climático en el Río de la Plata*, V. Barros, A. Menéndez and G.J. Nagy, Eds., Proyectos AIACC, 53-67.

Bordón, J.E., 2003: Todavía quedan 16,000 evacuados. *La Nación*, 30 de mayo de 2003, Buenos Aires, 18-20.

Boulanger, J.-P., F. Martinez and E.C. Segura, 2006a: Projection of future climate change conditions using IPCC simulations, neural networks and bayesian statistics. Part 1. Temperature mean state and seasonal cycle in South America. *Clim. Dynam.*, **27**, 233-259. doi:10.1007/s00382-006-0134-8.

Boulanger, J.-P., F. Martinez and E.C. Segura, 2006b: Projection of future climate change conditions using IPCC simulations, neural networks and Bayesian statistics. Part 2. Precipitation mean state and seasonal cycle in South America. *Clim. Dynam.*, **28**, 255-271. doi:10.1007/s00382-006-0182-0.

Bradley, R.S., M. Vuille, H. Diaz and W. Vergara, 2006: Threats to water supplies in the tropical Andes. *Science*, **312**, 1755-1756.

Bravo, W. and F. Bravo, 2001: El efecto del fenómeno de El Niño en las enfermedades dermatológicas. *Folia Dermatol. Peru*, **12**, 29-36.

Brown, S., M. Burnham, M. Delaney, M. Powell, R. Vaca and A. Moreno, 2000: Issues and challenges for forest-based carbon-offset projects: a case study on the Noel Kempff climate action project in Bolivia. *Mitigation and Adaptation Strategies for Global Change*, **5**, 99-121.

Bruner, A.G., R.E. Gullison, R.E. Rice and G.A.B. da Fonseca, 2001: Effectiveness of parks in protecting tropical biodiversity. *Science*, **291**, 125-128.

Buddemeier, R.W., A.C. Baker, D.G. Fautin and J.R. Jacobs, 2004: The adaptive hypothesis of bleaching. *Coral Health and Disease*, E. Rosenberg and Y. Loya, Eds., Springer, Berlin, 427-444.

Burger, R.L., 1992: *Chavin and the Origins of Andean Civilization*. Thames and Hudson, London, 240 pp.

Burrowes, P.A., R.L. Joglar and D.E. Green, 2004: Potential causes for amphibian declines in Puerto Rico. *Herpetologica*, **60**, 141–154.

Cabaniel, S.G., T.L. Rada, G.J.J. Blanco, A.J. Rodríguez-Morales and A.J.P. Escalera, 2005: Impacto de los eventos de El Niño Southern Oscillation (IENSO) sobre la leishmaniosis cutánea en Sucre, Venezuela, a través del uso de información satelital, 1994-2003. *Rev. Peru Med. Exp. Salud Pública*, **22**, 32-38.

Cáceres, A.G., L.A. Troyes, E. González-Pérez, C. Llontop, E. Bonilla, N. Murias, C. Heredia, C. Velásquez and C. Yáñez, 2002: Enfermedad de chagas en la Región Nororiental del Perú. I. Triatominos (Hemiptera, Reduviidae) presentes en Cajamarca y Amazonas. *Rev. Peru Med. Exp. Salud Pública*, **19**, 17-23.

Cáceres, L., 2004: Respuesta ecuatoriana al Cambio Climático. Preprints, *Comité Nacional sobre el Clima*, Ministerio del Ambiente, Tabarundo, Ecuador, 42 pp.

Cahoon, D.R. and P. Hensel, 2002: Hurricane Mitch: a regional perspective on mangrove damage, recovery and sustainability. USGS Open File Report 03-183, 31pp.

Callède, J., J.L. Guyot, J. Ronchail, Y. L'Hôte, H. Niel and E. De Oliveira, 2004: Evolution du débit de l'Amazone à Óbidos de 1903 à 1999. *Hydrol. Sci. J.*, **49**, 85-97.

Camilloni, I., 2005a: Tendencias climáticas. *El Cambio Climático en el Río de la Plata*, V. Barros, A. Menéndez and G.J. Nagy, Eds., CIMA/CONICET-UBA, Buenos Aires, 13-19.

Camilloni, I., 2005b: Tendencias hidrológicas. *El Cambio Climático en el Río de la Plata*, V. Barros, A. Menéndez and G.J. Nagy, Eds., CIMA/CONICET-UBA, Buenos Aires, 21-31.

Campos, J.J. and J.C. Calvo, 2000: The mountains of Costa Rica: compensation for environmental services from mountain forests. *Mountains of the World: Mountain Forests and Sustainable Development*, Mountain Agenda, Berne, 26-27.

Canteros, B.I., L. Zequeira and J. Lugo, 2004: Efecto de la ocurrencia de El Niño sobre la intensidad de la enfermedad bacteriana cancrosis de los citrus en el litoral argentino. CD trabajos presentados (No. 185). *X Reunión Argentina y IV Latinoamericana de Agrometeorología: Agrometeorología y Seguridad Alimentaria en América Latina*. Mar del Plata, Argentina.

Canziani, O.F., 2003: El agua y la salud humana. *Rev. Ing. Sanit. Ambient. AIDIS Argentina*, **70**, 36-40.

Canziani, O.F. and L.J. Mata, 2004: The fate of indigenous communities under climate change. UNFCCC Workshop on impacts of, and vulnerability and adaptation to, climate change. *Tenth Session of the Conference of Parties (COP-10)*, Buenos Aires, 3 pp.

Caran, S.C. and J.A. Nelly, 2006: Hydraulic engineering in prehistoric Mexico. *Sci. Am. Mag*., October, 8 pp.

Cardoso de Mendonça, M.J., M.C. Vera Díaz, D. Nepstad, R. Seroa da Motta, A. Alencar, J.C. Gómez and O.R. Arigoni, 2004: The economic cost of the use of fire in the Amazon. *Ecol. Econ*., **49**, 89-105.

Carey, M., 2005: Living and dying with glaciers: people's historical vulnerability to avalanches and outburst floods in Peru. *Global Planet. Change*, **47**, 122-134.

Carvajal, Y.E., H.E. Jiménez and H.M. Materón, 1998: Incidencia del fenómeno del Niño en la hidroclimatología del valle del río Cauca-Colombia. *B. Inst. Fr. Etud. Andines*, **27**, 743-751.

CATHALAC, 2003: Fomento de capacidades para la etapa II de adaptación al cambio climático en América Central, México y Cuba. CATHALAC/PNUD/ GEF. http://www.nu.or.cr/pnudcr/index.php?option=com_content&task =view &id=69&Itemid=0.

CBD, 2003: Executive summary. Interlinkages between biological diversity and climate change: advice on the integration of biodiversity considerations into the implementation of the UNFCCC and its Kyoto Protocol. Convention on Biological Diversity Technical Series No. 10, Montreal, 1-13.

CDC (Centers for Disease Control), 2000: Hantavirus pulmonary syndrome: Panama, 1999–2000. *MMWR Weekly*, **49**, 205-207.

CDERA, 2003: Adaptation to climate change and managing disaster risk in the Caribbean and South East Asia. Report of a Seminar, Barbados, July 24-25, 49 pp.

CENAPRED-CEPAL, 2005: *Características e Impactos Socioeconómico del Huracán "Emily" en Quintana Roo, Yucatán, Tamaulipas y Nuevo León en Julio de 2005*. LC/MEX/L.693, Centro Nacional de Prevención de Desastres, 145 pp.

CEPAL, 2002: *La Sostenibilidad del Desarrollo en América Latina y el Caribe: Desafíos y Oportunidades*. United Nations Press, Santiago de Chile, 262 pp.

Christensen, J.H., B. Hewitson, A. Busuioc, A. Chen, X. Gao, I. Held, R. Jones, R.K. Kolli, W.-T. Kwon, R. Laprise, V. Magaña Rueda, L. Mearns, C.G. Menéndez, J. Räisänen, A. Rinke, A. Sarr and P. Whetton, 2007: Regional climate projections. *Climate Change 2007: The Physical Science Basis. Contribution of Working Group I to the Fourth Assessment Report of the Intergovernmental Panel on Climate Change*, S. Solomon, D. Qin, M. Manning, Z. Chen, M. Marquis, K.B. Averyt, M. Tignor and H.L. Miller, Eds., Cambridge University Press, Cambridge, 847-940.

Clark, M. and J. King, 2004: *The Atlas of Water*. Earthscan, London, 128 pp.

Cochrane, M.A., 2003: Fire science for rainforests. *Nature*, **421**, 913-919.

Codignotto, J.O., 2004: Erosión costera. *Peligrosidad geológica en Argentina*, M. González and N. J. Bejerman, Eds., Asociación Argentina de Geología Aplicada a la Ingeniería, Buenos Aires, 90-111.

COHIFE, 2003: Principios rectores de Política Hídrica de la República Argentina. *Acuerdo Federal del Agua, Consejo Hídrico Federal*, COHIFE 8, August 2003, Argentina.

Collischonn, W., R. Haas, I. Andreolli and C.E.M. Tucci, 2005: Forecasting River Uruguay flow using rainfall forecasts from a regional weather-prediction model. *J. Hydrol*., **305**, 87-98.

Colombia Trade News, 2006: Illegal crops damage Colombia's environmental resources. Colombian Government Trade Bureau. http://www.coltrade.org/ about/envt_index.asp#top.

Conde, C. and H. Eakin, 2003: Adaptation to climatic variability and change in Tlaxcala, Mexico. *Climate Change, Adaptive Capacity and Development*, J. Smith, R. Klein and S. Huq, Eds., Imperial College Press, London, 241-259.

Conde, C. and K. Lonsdale, 2005: Engaging stakeholders in the adaptation process. *Adaptation Policy Frameworks for Climate Change: Developing Strategies, Policies and Measures*, I. Burton, E. Malone, S. Huq, B. Lim and E. Spanger-Siegfried, Eds., Cambridge University Press, Cambridge, 47-66.

Conde, C., R. Ferrer and S. Orozco, 2006: Climate change and climate variability impacts on rainfed agricultural activities and possible adaptation measures: a Mexican case study. *Atmosfera*, **19**, 181-194.

Conde, J.E., 2001: The Orinoco River Delta, Venezuela. *Coastal Marine Ecosystems of Latin America*, U. Seeliger and B. Kjerfve, Eds., Ecological Studies 144, Springer-Verlag, Berlin, 61-70.

Confalonieri, U., 2003: Variabilidade climática, vulnerabilidade social e saúde no Brasil. Terra Livre, San Paulo, 19-I, 20, 193-204.

Confalonieri, U., D.P. Marinho, M.G. Camponovo and R.E. Rodriguez, 2005: *Análise da Vulnerabilidade da População Brasileira aos Impactos Sanitários das Mudanças Climáticas*. Ministério da Ciência y Tecnologia, FIOCRUZ, ABRASCO, Rio de Janeiro, Brasil, 184 pp.

Cortazar, P.F., 1968: Documental del Perú, Departamento del Cusco, IOPPE S.A. Eds., February 1968.

Costa, J., A.T. Peterson and B. Beard, 2002: Ecologic niche modeling and differentiation of populations of *Triatoma brasiliensis* Neiva, 1911, the most important Chagas' disease vector in north-eastern Brazil (Hemiptera, Reduviidae, Triatominae). *Am. J. Trop. Med. Hyg*., **67**, 516-520.

Costa, M.H., A. Botta and J.A. Cardille, 2003: Effects of large-scale changes in land cover on the discharge of the Tocantins River, southeastern Amazonia. *J. Hydrol*., **283**, 206-217.

Cruz Choque, D., 2003: Fijación y existencias de carbono en el ecosistema forestal del corredor de conservación Vilcabamba–Amboró, Conservation Internacional, La Paz, 2 pp.

Cunha, G.R. da, J.L.F. Pires and A. Pasinato, 2004: Uma discussão sobre o conceito de hazards e o caso do furacão/ciclone Catarina. Passo Fundo: Embrapa Trigo, 13 pp. http://www.cnpt.embrapa.br/biblio/do/p_do36.htm.

de Garín, A. and R. Bejarán, 2003: Mortality rate and relative strain index in Buenos Aires city. *Int. J. Biometeorol*., **48**, 31-36.

de Lima, M.G. and C. Gascon, 1999: The conservation value of linear forest remnants in central Amazonia. *Biol. Conser.*, **91**, 241-247.

Del Ponte, E.M., J.M.C. Fernandes and C.R. Pierobom, 2005: Factors affecting density of airborne *Gibberella zeae* inoculum. *Fitopatología Brasileira*, **30**, 55-60.

Dey, P., 2006: Climate change devastating Latin America frogs. University of Alberta. http://www.expressnews.ualberta.ca/article.cfm?id=7247.

Díaz-Ambrona, C.G.H., R. Gigena Pazos and C.O. Mendoza Tovar, 2004: Global climate change and food security for small farmers in Honduras. *New directions for a Diverse Planet*: *Proc. 4th International Crop Science Congress, Brisbane, Australia*, T. Fischer, N. Turner, J. Angus, L. McIntyre, M. Robertson, A. Borrell and D. Lloyd, Eds., The Regional Institute, Gosford, 6 pp.

Dinerstein, E., D. Olson, D. Graham, A. Webster, S.A. Primm, M.P. Bookbinder and G. Ledec, 1995: *A Conservation Assessment of the Terrestrial Ecoregions of Latin America and the Caribbean*. World Bank, Washington, District of Columbia, 129 pp.

DNPC, 2005/2006: Informe de las lluvias caídas en Venezuela en los meses de Febrero y marzo de 2005 y Febrero 2006. Dirección Nacional de Protección Civil, República Bolivariana de Venezuela.

Dourojeanni, A., 2000: *Procedimientos de Gestión para el Desarrollo Sustentable*. ECLAC, Santiago, 376 pp.

Eakin, H., 2000: Smallholder maize production and climatic risk: a case study from Mexico. *Climatic Change*, **45,** 19-36.

Eakin, H., 2005: Institutional change, climate risk, and rural vulnerability: cases from Central Mexico. *World Dev*., **33**, 1923-1938.

ECLAC, 1998: *América Latina: Proyecciones de Población 1970-2050*. Boletín Demográfico, Comisión Económica para América Latina y el Caribe, Santiago, 158 pp. http://www.eclac.cl/publicaciones/Poblacion/0/LCDEMG180/lcgdem 180i.pdf.

ECLAC, 2002a: El impacto socioeconómico y ambiental de la sequía de 2001 en Centroamérica. LC/MEX/L.510/rev.1, Comisión Económica para América Latina y el Caribe, 68 pp.

ECLAC, 2002b: *Social Panorama of Latin America 2001–2002*. United Nations Publications, Santiago, 280 pp. http://www.eclac.cl/.

EPA, 2001: AirData database. U.S. Environmental Protection Agency. http://www.epa.gov/air/data/.

Escaith, H., 2003. *Tendencias y Extrapolación del Crecimiento en América Latina y el Caribe*. Serie Estudios Estadísticos y Prospectivos, División de Estadística y Proyecciones Económicas, CEPAL, 66 pp.

Eslava, J.A. and J.D. Pabón, 2001: Proyecto: proyecciones climáticas e impactos socioeconómicos del cambio climático en Colombia. *Meteorol. Colombiana*, **3**, 1-8.

Espinoza, R., P. Via, L.M. Noriega, A. Johnson, S.T. Nichol, P.E. Rollin and R. Wells, 1998: Hantavirus pulmonary syndrome in a Chilean patient with recent travel in Bolivia. *Emerg. Infect. Dis*. **4**, 93-95.

Eterovick, P.C., A.C. Oliveira de Queiroz Carnaval, D.M. Borges-Nojosa, D. Leite Silvano, M. Vicente Segalla and I. Sazima, 2005: Amphibian declines in Brazil: an overview. *Biotropica*, **37**, 166-179.

FAO, 2001a: *Global Forest Resources Assessment 2000*. FAO Forestry Paper 140, Rome, Food and Agricultural Organization, 512 pp. http://www.fao.org/icatalog/search/dett.asp?aries_id=102270.

FAO, 2001b: Variabilidad y cambio del clima: un desafío para la producción agrícola sostenible. *Proc. 16th Periodo de Sesiones FAO*, Rome, 13 pp.

FAO, 2002: El cambio climático y los bosques. Boletín electrónico Julio 2002, FAO, Rome. http://www.ecosur.net/cambio_climatico_y_los_bosques.html.

FAO, 2004a: Seguridad alimentaria como estrategia de Desarrollo rural. *Proc. 28ava Conferencia Regional de la FAO para América Latina y el Caribe*, Ciudad de Guatemala, Guatemala.

FAO, 2004b: La participación de las comunidades en la gestión forestal es decisiva para reducir los incendios (Involving local communities to prevent and control forest fires). FAO Newsroom. http://www.fao.org/newsroom/en/news/2004/48709/index.html.

FAO, 2005: Cattle ranching is encroaching on forests in Latin America. FAO Newsroom. http://www.fao.org/newsroom/en/news/2005/102924/index.html.

FAO, 2006: Third Session of the Sub-Committee on Aquaculture: Committee on Fisheries (COFI). Food and Agriculture Organization of the United Nations (FAO), New Delhi, India, 4-8 September.

Fay, M., F. Ghesquiere and T. Solo, 2003: Natural disasters and the urban poor. IRDB En Breve 32, The World Bank, 4 pp.

Fearnside, P.M., 2001: Status of South American natural ecosystems. *Encyclopedia of Biodiversity: Volume 5*, S. Levin, Ed., Academic Press, 345-359.

Fearnside, P.M., 2003: Deforestation control in Mato Grosso: a new model for slowing the loss of Brazil's Amazon Forest. *Ambio*, **32**, 343-345.

Feliciangeli, M.D., D. Campbell-Lendrum, C. Martínez, D. González, P. Coleman and C. Davies, 2003: Chagas disease control in Venezuela: lessons for the Andean region and beyond. *Trends Parasitol*., **19**, 44-49.

Fernandes, J.M., G.R. Cunha, E. Del Ponte, W. Pavan, J.L. Pires, W. Baethgen, A. Gimenez, G. Magrin and M.I. Travasso, 2004: Modelling *Fusarium* head blight in wheat under climate change using linked process-based models. *Proc. 2nd International Symposium on Fusarium Head Blight*, S.M. Canty, T. Boring, K. Versdahl, J. Wardwell and R. Wood, Eds., Michigan State University, East Lansing, Michigan, 441-444. http://www.scabusa.org/forum.html#isfhb2.

Ferroni, M., 2005: Social cohesion in Latin America: the public finance dimension. Inter-American Development Bank (IDB), 6 pp. http://www.inwent.org/imperia/md/content/bereich1-intranet/efinternet/lateinamerika/ferroni_engl.pdf.

Francou, B., M. Vuille, P. Wagnon, J. Mendoza and J.-E. Sicart, 2003: Tropical climate change recorded by a glacier in the central Andes during the last decades of the twentieth century: Chacaltaya, Bolivia, 16°S. *J. Geophys. Res*., **108**, doi:10.1029/2002JD002959.

Franke, C.R., M. Ziller, C. Staubach and M. Latif, 2002: Impacts of the El Niño/Southern Oscillation on visceral leishmaniasis, Brazil. *Emerg. Infect. Dis*., **8**, 914-917.

Fundación DESC, 2005: Análisis de Situación: Ahora la prueba de la tormenta STAN, Informe Guatemala No 29. http://www.fundadesc.org/archivo/veintinueve/analisis.htm.

Gagnon, A.S., K.E. Smoyer-Tomic and A. Bush, 2002: The El Niño Southern Oscillation and malaria epidemics in South America. *Int. J. Biometeorol*., **46**, 81-89.

Gardner, T.A., I.M. Côté, J.A. Gill, A. Grant and A.R. Watkinson, 2003: Long-term region-wide declines in Caribbean corals. *Science*, **301**, 958-960.

Gay, C., F. Estrada, C. Conde and H. Eakin, 2004: Impactos potenciales del Cambio Climático en la agricultura: escenarios de producción de café para el 2050 en Veracruz (México). *El Clima, Entre el Mar y la Montaña*, J.C.García Cordón, Ed., AEC, UC series A, No. 4, 651-660.

Geist, H. and E. Lambin, 2002: Proximate causes and underlying driving forces of tropical deforestation. *BioScience*, **52**, 143-150.

Ghini, R., E. Hamada, M.J. Pedro Júnior and J.A. Marengo. 2007: Climate change and coffee pests in Brazil. *Climatic Change*, submitted.

Gimenez, A., 2006: Climate change and variability in the mixed crop/livestock production systems of the Argentinean, Brazilian and Uruguayan Pampas. Final Report, AIACC Project LA27, 70 pp. http://www.aiaccproject.org/.

Githeko, A. and A. Woodward, 2003: International consensus on the science of climate and health: the IPCC Third Assessment Report. *Climate Change and Human Health: Risks and Responses*, A. McMichael, D. Campbell-Lendrum, C. Corvalan, K. Ebi, A.K. Githeko, J.S. Scheraga and A. Woodward, Eds., WHO, Geneva, 43-60.

Gnadlinger, J., 2003: Captacao e manejo de água de chuva e desenvolvimento sustentável do Semi-Arido Brasileiro-Uma visao integrada.

Gouveia, N., S. Hajat and B. Armstrong, 2003: Socioeconomic differentials in the temperature-mortality relationship in São Paulo, Brazil. *Int. J. Epidemiol*., **32**, 390-397.

Grasses, J.P., J. Amundaray, A. Malaver, P. Feliziani, L. Franscheschi and J. Rodríguez, 2000: Efectos de las lluvias caídas en Venezuela en diciembre de 1999. CAF-PNUD, CDB Pub. Caracas, 224 pp.

Groisman, P.Ya., R.W. Knight, D.R. Easterling, T.R. Karl, G.C. Hegerl and V.N. Razuvaev, 2005: Trends in intense precipitation in the climate record. *J. Climate*, **18**, 1326–1350.

Hales S., N. de Wett, J. Maindonald and A. Woodward, 2002: Potential effect of population and climates change models on global distribution of dengue fever: an empirical model. *Lancet*, **360**, 830-834.

Haylock, M.R., T. Peterson, L.M. Alves, T. Ambrizzi, Y.M.T. Anunciação, J. Baez, V.R. Barros, M.A. Berlato and Co-authors, 2006: Trends in total and extreme South American rainfall 1960-2000 and links with sea surface temperature. *J. Climate*, **19**, 1490-1512.

Hoggarth, D.D., K. Sullivan and L. Kimball, 2001: Latin America and the Caribbean coastal and marine resources. Background paper prepared for GEO-3, United Nations Environment Programme Regional Office for Latin America and the Caribbean, Mexico.

Holt-Giménez, E., 2002: Measuring farmers'agroecological resistance after Hurricane Mitch in Nicaragua: a case study in participatory, sustainable land management impact monitoring. *Agr. Ecosyst. Environ*., **93**, 87-105.

Huarcaya, E., E. Chinga, J.M. Chávez, J. Chauca, A. Llanos, C. Maguiña, P. Pachas and E. Gotuzzo, 2004: Influencia del fenómeno de El Niño en la epidemiología de la bartonelosis humana en los departamentos de Ancash y Cusco entre 1996 y 1999. *Rev. Med. Hered*., **15**, 4-10.

Huber, E. and F. Solt, 2004: Successes and failures of neoliberalism. *Lat. Am. Res. Rev*., **39**, 150-164.

Hurtado-Díaz, M., H. Riojas-Rodríguez, S.J. Rothenberg, H. Gomez-Dantés and E. Cifuentes-García, 2006: Impacto de la variabilidad climática sobre la incidencia del dengue en México. *International Conference on Environmental Epidemiology and Exposure*, Paris.

Hutyra, L.R., J.W. Munger, C.A. Nobre, S.R. Saleska, S.A. Vieira and S.C. Wofsky, 2005: Climatic variability and vegetation vulnerability in Amazônia. *Geophys. Res. Lett*., **32**, L24712, doi:10.1029/2005GL024981.

IDB, 2004: Financing water and sanitation services: options and constraints. *Seminario Inter-American Development Bank*. Salvador, Bahía, Brasil.

IDEAM, 2004: Boletín Julio 12 al 16 de 2004. Colombia.

IMO, 2005: *World Migration 2005: Costs and Benefits of International Migration*. International Migration Organization, Geneva, 494 pp. http://www.iom.int/jahia/Jahia/cache/offonce/pid/1674?entryId=932.

INPE-MMA, 2005a: Monitoramento do desflorestamento bruto da Amazônia Brasileira (Monitoring the Brazilian Amazon gross deforestation). Ministério da Ciência e Tecnologia, São José dos Campos, São Paulo. http://www.obt.inpe.br/prodes/index.html.

INPE-MMA, 2005b: Deforestation in Amazonia has its second worst recorded history. http://www.amazonia.org.br/.

INPE-MMA, 2005c: Rate of deforestation in the Amazon region is reduced by 31 percent, but is still higher than the 1990s average. http://www.socioambiental.org/e/nsa/detalhe?id=2181.

INVEMAR, 2005: Building capacity for improve adaptability to sea level rise in two vulnerable points of the Colombian coastal areas (Tumaco-Pacific coast and Cartagena-Caribbean coast) with special emphasis on human populations under poverty conditions. NCAP-Colombia II, 137 pp.

IPCC, 2001: *Climate Change 2001: Impacts, Adaptation, and Vulnerability. Contribution of Working Group II to the Third Assessment Report of the Intergovernmental Panel on Climate Change*, J.J. McCarthy, O.F. Canziani, N.A. Leary, D.J. Dokken and K.S. White, Eds., Cambridge University Press, Cambridge, 1032 pp.

IPCC DDC, 2003: http://ipcc-ddc.cru.uea.ac.uk /asres/scatter_plots/ scatterplots_region.html.

IRDB, 2000: Gestión de los Recursos Hídricos de Argentina. Elementos de Política para su Desarrollo Sustentable en el siglo XXI. Oficina Regional de América Latina y Caribe. Unidad Departamental de Argentina y los Grupos de Finanzas, Sector Privado y Infraestructura, y Medio Ambiente y Desarrollo Social Sustentable. Informe No. 20.729-AR. August 2000.

IRI, 2002: Experimental dengue forecasting and ongoing Brazilian outbreak. International Research Institute for Climate Prediction, University of Columbia, Palisades, New York, March 2002, **5**, 1-2.

IUCN, 2001: Cambio climático y biodiversidad: cooperación entre el convenio sobre la diversidad biológica y la convención marco sobre el cambio climático.

Sexta Reunión del Órgano Subsidiario de Asesoramiento Científico, Técnico y Tecnológico del Convenio sobre la Diversidad Biológica, Montreal, 6 pp.

IUCN, 2004: *Red List of Threatened Species: A Global Species Assessment*, J.E.M. Baillie, C. Hilton-Taylor and S.N. Stuart, Eds., IUCN, Gland and Cambridge, 217 pp.

Jipp, P.H., D.C. Nepstad and D.K. Carvalho, 1998: Deep soil moisture storage and transpiration in forests and pastures of seasonally-dry Amazonia. *Climatic Change*, **39**, 395-412.

Jones, J., Ed., 2001: Comparative assessment of agricultural uses of ENSO-based climate forecast in Argentina, Costa Rica and Mexico. IAI Initial Science Program III Project, 28 pp. http://csml.ifas.ufl.edu/pdf_files/iai-ps-s.pdf.

Jones, P.G. and P.K. Thornton, 2003: The potential impacts of climate change on maize production in Africa and Latin America in 2055. *Global Environ. Chang.*, **13**, 51-59.

Kaimowitz, D., B. Mertens, S. Wuner and P. Pacheco, 2004: Hamburger connection fuels Amazon destruction: cattle ranching and deforestation in Brazil's Amazon. Center for International Forestry Research, Bogor, Indonesia, 10 pp. http://www.cifor.cgiar.org/.

Kane, R.P., 2002: Precipitation anomalies in southern America associated with a finer classification of El Niño and La Niña events. *Int. J. Climatol.*, **22**, 357-373.

Kerr, R.A., 2003: Climate: little girl lost. *Science*, **301**, 286-286.

Ko, A., R.M. Galvão, D. Ribeiro, C.M. Dourado, W.D. Johnson Jr. and L.W. Riley, 1999: Urban epidemic of severe leptospirosis in Brazil, Salvador. Leptospirosis Study Group. *Lancet*, **354**, 820-825.

Kokot, R.R., 2004: Return periods of floods in the coastal lands of the Río de la Plata. *Second AIACC Regional Workshop for Latin America and the Caribbean*, Buenos Aires, 24-27 August 2004.

Kokot, R.R., J.O. Codignotto and M. Elissondo, 2004: Vulnerabilidad al ascensp del nivel del mar en la costa de la provincia de Río Negro. *Rev. Asociación Geológica Argentina*, **59**, 477-487.

Kovacs, J.M., 2000: Assessing mangrove use at the local scale. *Landscape Urban Plan.*, **43**, 201-208.

Kovats, S., K. Ebi and B. Menne, 2003: *Methods of Assessing Human Health Vulnerability and Public Health Adaptation to Climate Change*. Health and Global Environmental Change Series, No. 1, WMO, Health Canada, UNEP, 112 pp.

Kovats, S., D. Campbell-Lendrum and F. Matthies, 2005: Climate change and human health: estimating avoidable deaths and disease. *Risk Anal.*, **25**, 1409-1418.

Krol, M.S. and P.R. van Oel, 2004: Integrated assessment of water stress in Ceará, Brazil, under climate change forcing. *International Congress: Complexity and Integrated Resources Management*, C. Pahl, S. Schmidt and T. Jakeman, Eds., Osnabrück, 760-764.

Kupek, E., M.C. de Sousa Santos Faversani and J.M. de Souza Philippi, 2000: The relationship between rainfall and human leptospirosis in Florianópolis, Brazil, 1991–1996. *Braz. J. Infect. Dis.*, **4**, 131-134.

La Marca, E. and H.P. Reinthaler, 2005: Population changes in *Atelopus* species of the Cordillera de Mérida, Venezuela. *Herpetol. Rev.*, **22,** 125-128.

La Nación, 2002: Buenos Aires, 13 March.

La Nación, 2006. Buenos Aires, 6 December.

La Rovere, E., J. Monteiro and A.C. Avzaradel, 2006: Biodiesel and vegetable oils production by small farmers in semi-arid northeastern Brazil: a climate change adaptation and mitigation strategy. Report to KEI.

Lagos, P., 2001: Clima y agricultura. *Agricultura Peruana*, H. Guerra García Cueva, Ed., Lima.

Laurance, W.F., M.A. Cochrane, S. Bergen, P.M. Fearnside, P. Delamônica, C. Barber, S. D'Angelo and T. Fernandes, 2001: Environment: The future of the Brazilian Amazon. *Science*, **291**, 438-39.

Laurance, W.F., P.M. Fearnside, A.K.M. Albernaz, H.L. Vasconcelos and L.V. Ferreira, 2005: The future of the Brazilian Amazon. *Science*, **307**, 1043-44.

Leiva, J.C., 2006: Assesment climate change impacts on the water resources at the northern oases of Mendoza province, Argentina. *Global Change in Mountain Regions*, M.F. Price, Ed., Sapiens Publishing, Kirkmahoe, Dumfriesshire, 81-83.

Loser, C., 2006: Continued strong growth in region is no passing occurrence. Latin America Forecasts: Economic Growth, Inter-American Dialogue, Latin American Advisor, 3. http://www.thedialogue.org/publications/2006/winter/LAA_forecast.pdf.

Maarten Dros, J., 2004: *Managing the Soy Boom: Two Scenarios of Soy Production Expansion in South America*. AID Environment, Amsterdam, 67 pp.

Magrin, G.O. and M.I. Travasso, 2001: Economic value of ENSO-based climatic forecasts in the agricultural sector of Argentina. *Proc. 2nd International Symposium on Modelling Cropping Systems*, Florence, Italy, 139-140.

Magrin, G.O. and M.I. Travasso, 2002: An integrated climate change assessment from Argentina. *Effects of Climate Change and Variability on Agricultural Production Systems*, O. Doering III, J.C. Randolph, J. Southworth and R.A. Pfeifer, Eds., Kluwer Academic, Boston, Massachusetts, 193-219.

Magrin, G.O., M.I. Travasso and G.R. Rodríguez, 2005: Changes in climate and crop production during the 20th century in Argentina. *Climatic Change*, **72**, 229-249.

Mahon, R., 2002: Adaptation of fisheries and fishing communities to the impacts of climate change in the CARICOM region. Prepared for the CARICOM Fisheries Unit, Belize City, 33 pp.

Malheiros, J.O., 2004: 17 puntos que ajudam a explicar o que é la desertificação, a convenção da ONU e o processo de construção do PAN-LCD Brasileiro. RIOD BRASIL, 11 pp.

Mantilla, G., 2005: Integrated dengue and Malaria surveillance and control system, Colombia. *Proc., Wengen 2005: 10th Anniversary Meeting – Climate, Climatic Change and its Impacts on Human Health*, Wengen, Switzerland, September 12-14, 2005.

Marengo, J., 2003: Condições climaticas e recursos hidricos no Norte do Brasil. *Clima e Recursos Hídricos 9*, Associação Brasileira de Recursos Hídricos/FBMC-ANA, Porto Alegre, 117-156.

Marengo, J. and T. Ambrizzi, 2006: Use of regional climate models in impact assessments and adaptation studies from continental to regional scales: the CREAS (Regional climate change scenarios for South America) initiative in South America. *Proc. 8th ICSHMO*, Foz do Iguaçu, Brazil, INPE, São José dos Campos, Brazil, 291-296.

Marengo, J.A., 2004: Interdecadal and long term rainfall variability in the Amazon basin. *Theor. Appl. Climatol.*, **78**, 79-96.

Marengo, J.A. and C. Camargo, 2007: Surface air temperature trends in Southern Brazil for 1960–2002. *Int. J. Climatol.*, submitted.

Marengo, J.A. and D. Raigoza. 2006: Newsletter del Proyecto GOF-UK-CPTEC, March 2006, Brazil. http://www.institutes.iai.int/files/Newsletter2_Eng.pdf.

Marengo, J.A., W. Soares, C. Saulo and M. Nicolini, 2004: Climatology of the Low Level Jet east of the Andes as derived from the NCEP NCAR reanalyses. *J. Climate*, **17**, 2261-2280.

Mark, B.G. and G.O. Seltzer, 2003: Tropical glacier meltwater contribution to stream discharge: a case study in the Cordillera Blanca, Perú. *J. Glaciol.*, **49**, 271-281.

MARN, 2005: Ley de Costas. Ministerio del Ambiente y Recursos Naturales, República Bolivarina de Venezuela.

May, P., E. Boyd, F. Veiga and M. Chang, 2004: *Local Sustainable Development Effects of Forest Carbon Projects in Brazil and Bolivia: A View from the Field*. Markets for Environmental Services No. 5, IIED, London, 124 pp. http://www.iied.org/.

Maza, J., F. Cazorzi, P. Lopez, L. Fornero, A. Vargas and J. Zuluaga, 2001: Snowmelt mathematical simulation with different climate scenarios in the Tupungato River Basin, Mendoza, Argentina. *Proc. Symposium held in Santa Fe, New Mexico, USA*. IAHS Publication No 267, 16-128.

Meagan, E., G. Adina and J.A. Herrera-Silveira, 2003: Tracing organic matter sources and carbon burial in mangrove over the past 160 years. *Estuar. Coast. Shelf S.*, **61**, 211-227.

Medina E., H. Fonseca, F. Barboza and M. Francisco, 2001: Natural and man-induced changes in a tidal channel mangrove system under tropical semiarid climate at the entrance of the Maracaibo Lake (western Venezuela). *Wetlands Ecol. Manage.*, **9**, 233-243.

Mendelsohn, R., W. Morrison, M.E. Schlesinger and N.G. Andronova, 2000: Country-specific market impacts of climate change. *Climatic Change*, **45**, 553-569.

Menéndez, A.N. and M. Ré, 2005: Escenarios de inundación. *Cambio Climático: Costa Argentina del Río de la Plata*, V. Barros, A. Menéndez and G.J. Nagy, Eds., AIACC/CIMA, Buenos Aires, 119-120.

Mesquita, A., 2000: Sea level variations along the Brazilian coast: a short review. Preprints, *Brazilian Symposium on Sandy Beaches*, Itajaí, Brazil, 15 pp.

Messina, 1999: El fenómeno ENSO: su influencia en los sistemas de producción de girasol, trigo y maíz en la región pampeana. Análisis retrospectivo y evaluación de estrategias para mitigar el riesgo climático. Tesis de Maestria, UBA.

Meza, F.J., D.S. Wilks, S.J. Riha and J.R. Stedinger, 2003: Value of perfect forecasts of sea surface temperature anomalies for selected rain-fed agricultural locations of Chile. *Agr. Forest Meteorol.*, **116**, 117–135.

Mielnicki, D.M., P.O. Canziani and J. Drummond, 2005: Quema de biomasa en el centro-sur de Sudamérica: incendios locales, impactos regionales, *Anales IX Congreso Argentino de Meteorología*, CD, ISBN 987-22411-0-4.

Miles, L., A. Grainger and O. Phillips, 2004: The impact of global climate change on tropical forest biodiversity in Amazonia. *Global Ecol. Biogeogr.*, **13**, 553-565.

Millennium Ecosystem Assessment, 2005: *Millennium Ecosystem Assessment: Living Beyond our Means: Natural Assets and Human Well-being (Statement from the Board)*. Millennium Ecosystem Assessment, India, France, Kenya, UK, USA, Netherlands, Malaysia, 28 pp. http://www.millenniumassessment.org/.

Miller, L., 2004: Satellite altimetry and the NOAA/NESDIS sea surface height team. *Backscatter*, 29-34.

Miranda, J., C. Cabezas, C. Maguiña and J. Valdivia, 2003: Hipertermia durante el fenómeno de El Niño, 1997–1998. *Rev. Peru Med. Exp. Salud Pública*, **20**, 200-205.

MMA, 2004: Ministério do Meio Ambiente. Relatório de Avaliação, Report No. 31033, Subprograma Projetos Demonstrativos. http://www-wds.worldbank.org /servlet/WDSContentServer/WDSP/IB/2005/02/16/000112742_2005021612200 1/Rendered/INDEX/310330BR.txt.

Morales, H., B. Ferguson and L. García-Barrios, 2007: Agricultura: la Cenicienta de la conservación en Mesoamerica. *Evaluación y Conservación de Biodiversidad en Agropasiajes Mesoamericanos*, C. Harevey and J. Saenz, Eds., UNA Editorial, Costa Rica, in press.

Moschini, R.C., M. Carmona and M. Grondona, 1999: Wheat head blight incidence variations in the Argentinean Pampeana region associated with the El Niño/Southern Oscillation. *Proc. XIV International Plant Protection Congress*, Jerusalem.

Moutinho, P. and S. Schwartzman, 2005: Compensated reduction of deforestation. Amazon Intitute of Environmental Research. http://www.ipam.org.br/web/ index.php.

Nakićenović, N. and R. Swart, Eds., 2000: *Special Report on Emissions Scenarios: A Special Report of Working Group III of the Intergovernmental Panel on Climate Change*. Cambridge University Press, Cambridge, 599 pp.

Nagy, G.J., E.A. Forbes, A. Ponce, V. Pshennikov, R. Silva and R. Kokot, 2005: Desarrollo de la capacidad de evaluación de la vulnerabilidad costera al Cambio Climático: zona oeste de Montevideo como caso de estudio. *El Cambio Climático en el Río de la Plata*, V. Barros, A. Menéndez and G.J. Nagy, Eds., AIACC/CIMA, Buenos Aires, 173-180.

Nagy, G.J., R.M. Caffera, M. Aparicio, P. Barrenechea, M. Bidegain, J.C. Jiménez, E. Lentini, G. Magrin and co-authors, 2006a: Understanding the potential impact of climate change and variability in Latin America and the Caribbean. Report prepared for the Stern Review on the Economics of Climate Change, 34 pp. http://www.sternreview.org.uk.

Nagy, G.J., M. Bidegain, R.M. Caffera, J.J. Lagomarsino, W. Norbis, A. Ponce and G. Sención, 2006b: Adaptive capacity for responding to climate variability and change in estuarine fisheries of the Rio de la Plata. AIACC Working Paper No. 36, 16 pp. http://www.aiaccproject.org/.

Nagy, G.J., M. Bidegain, F. Blixen, R.M. Caffera, G. Ferrari, J.J. Lagomarsino, C.H. López, W. Norbis and co-authors, 2006c: Assessing vulnerability to climate variability and change for estuarine waters and coastal fisheries of the Rio de la Plata. AIACC Working Paper No. 22, 44 pp. http://www.aiaccproject.org/.

Natenzon, C.E., N. Marlenko, S.G. González, D. Ríos, J. Barrenechea, A.N. Murgida, M.C. Boudin, E. Gentile and S. Ludueña, 2005a: Vulnerabilidad social estructural. *El Cambio Climático en el Rio de la Plata*, V. Barros, A. Menéndez and G.J. Nagy, Eds., AIACC/CIMA, Buenos Aires, 113-118.

Natenzon, C.E., P.M. Bronstein, N. Marlenko, S.G. González, D. Ríos, J. Barrenechea, A.N. Murgida, M.C. Boudin, A.P. Micou, E. Gentile and S. Ludueña, 2005b: Impactos económicos y sociales por inundaciones. *El Cambio Climático en el Rio de la Plata*, V. Barros, A. Menéndez and G.J. Nagy, Eds., AIACC/CIMA, Buenos Aires, 121-130.

NC-Bolivia, 2000: 1st National Communication to the UNFCCC, 138 pp. http://www.climate.org/CI/latam.shtml.

NC-Chile, 1999: 1st National Communication to the UNFCCC, 89 pp. http://www.climate.org/CI/latam.shtml.

NC-Colombia, 2001: 1st National Communication to the UNFCCC, 267 pp. http://www.climate.org/CI/latam.shtml.

NC-Costa Rica, 2000: 1st National Communication to the UNFCCC, 177 pp. http://www.climate.org/CI/latam.shtml.

NC-Ecuador, 2000: 1st National Communication to the UNFCCC, 128 pp. http://www.climate.org/CI/latam.shtml.

NC-El Salvador, 2000: 1st National Communication to the UNFCCC, 109 pp. http://www.climate.org/CI/latam.shtml.

NC-Guatemala, 2001: 1st National Communication to the UNFCCC, 127 pp. http://www.climate.org/CI/latam.shtml.

NC-Guyana, 2002: Initial National Communication to the UNFCCC, 192 pp. http://www.climate.org/CI/latam.shtml.

NC-Nicaragua, 2001: Impacto del Cambio Climático en Nicaragua. Primera Comunicación Nacional sobre Cambio Climático, PNUD/MARENA, 127 pp.

NC-Panama, 2000: 1st National Communication to the UNFCCC, 136 pp. http://www.climate.org/CI/latam.shtml.

NC-Perú, 2001: 1st National Communication to the UNFCCC, 155 pp. http://unfccc.int/resource/docs/natc/pernc1.pdf.

NC-Uruguay, 2004: 2nd National Communication to the UNFCCC. Ministerio de Vivienda, Ordenamiento Territorial y Medio Ambiente, Proyecto URU/00/G31, GEF-UNDP, 194 pp.

Nepstad, D., D. McGrath, A. Alencar, A.C. Barros, G. Carvalho, M. Santilli and M. del C. Vera Diaz, 2002: Frontier governance in Amazonia. *Science*, **295**, 629-631.

Nepstad, D., P. Lefebvre, U. Lopes Da Silva, J. Tomas, P. Schlesinger, L. Solorzano, P. Moutinho, D. Ray and J. Guerreira Benito, 2004: Amazon drought and its implications for forest flammability and tree growth: a basin-wide analysis. *Global Change Biol.*, **10**, 704–717.

Nepstad, D.C., 2004: Governing the world's forests. *Conserving Biodiversity*, Aspen Institute, 37-52. http://www.aspeninstitute.org/atf/cf/%7BDEB6F227-659B-4EC8-8F84-8DF23CA704F5%7D/GOVERNINGTHEFORESTS.PDF.

Nicholls, R., 2004: Coastal flooding and wetland loss in the 21st century: changes under SRES climate and socio-economic scenarios. *Global Env. Chang.*, **14**, 69-86.

NOAA, 2005: *Hazards/Climate Extremes*. National Climatic Data Center, U.S. Department of Commerce. http://www.ncdc.noaa.gov/oa/climate/research/2005/aug /hazards.html.Flooding.

NOAA/OAR, 2004: Establishing long term coastal and marine programs in Latin America and the Caribbean: pilot studies of Ecuador and Gulf of Fonseca. Background Paper No. 2, NOAA/OAR, University of Rhode Island, 35 pp.

Nobre, C.A., P. Sellers and J. Shukla, 1991: Amazonian deforestation model and regional climate change. *J. Climate*, **4**, 957-988.

Nobre, C.A., E.D. Assad and M.D. Oyama, 2005: Mudança ambiental no Brasil: o impacto do aquecimento global nos ecossistemas da Amazônia e na agricultura. *Sci. Am. Brasil, Special Issue: A Terra na Estufa*, 70-75.

Nuñez M., S. Solman and M.F. Cabré, 2005: Southern South America climate in the late twenty-first century: annual and seasonal mean climate with two forcing scenarios. Taller Regional de Cambio Climático, San Pablo, November 7-10, 2005.

OAS-CIDI, 2003: Organization of American States. Inter-American Council for Integral Development. *Addressing Climate Change in the Americas within the OAS*, Regular meeting of the Inter-American Committee on Sustainable Development, CIDI/CIDS/doc. 1/99.

Ometto, J.P.H.B., A.D. Nobre, H.R. da Rocha, P. Artaxo and L.A. Martinelli, 2005: Amazonia and the modern carbon cycle: lessons learned. *Oecologia*, **143**, 483-500.

Orlove, B.S., S. Joshua and L. Tosteson, 1999: The application of seasonal to interannual climate forecasts based on El Niño-Southern Oscillation (ENSO) events: lessons from Australia, Brazil, Ethiopia, Peru and Zimbabwe. Berkeley Workshop on Environmental Politics, Working Paper 99-3, Institute of International Studies, University of Califórnia, Berkeley, 67 pp.

Orlove, B.S., J.C.H. Chiang and M.A. Cane, 2000: Forecasting Andean rainfall and crop yield from the influence of El Niño on Pleiades visibility. *Nature*, **403**, 68-71.

Oropeza, O., 2004: Evaluación de la vulnerabilidad a la desertificación. *Cambio Climático: Una Visión Desde México*, J. Martínez and A. Fernández Bremauntz, Eds., SEMARNAT e INE, México, 301-311.

Orr, J.C., V.J. Fabry, O. Aumont, L. Bopp, S.C. Doney, R.A. Feely, A. Gnanadesikan, N. Gruber and Co-authors, 2005: Anthropogenic ocean acidification over the twenty-first century and its impact on calcifying organisms. *Nature*, **437**, 681-686.

Oyama, M.D. and C.A. Nobre, 2003: A new climate–vegetation equilibrium state for tropical South America. *Geophys. Res. Lett.*, **30**, 2199, doi:10.1029 /2003GL018600.

Pabón, J.D., 2003a: El Cambio Climático global y su manifestación en Colombia. *Cuadernos Geograf.*, **12**, 111-119.

Pabón, J.D., 2003b: El aumento del nivel del mar en las costas y área insular de Colombia. *El Mundo Marino de Colombia Investigación y Desarrollo de Territorios*

Olvidados, Red de Estudios del Mundo Marino – REMAR, Universidad Nacional de Colombia, 75-82.

Pabón, J.D., I. Cárdenas, R. Kholostyakov, A.F. Calderón, N. Bernal and F. Ruiz, 2001: Escenarios climáticos para el siglo XXI sobre el territorio colombiano. Nota Técnica Interna del Instituto de Hidrología, Meteorología y Estudios Ambientales (IDEAM), Bogotá, Colombia.

PAHO, 2003: *Status Report on Malaria Programs in the Americas*. 44th Directing Council, 55th Session of the Regional Comité. Pan American Health Organization, Washington, District of Columbia.

PAHO, 2005: *An Assessment of Health Effects of Ambient Air Pollution in Latin America and the Caribbean*. Pan-American Health Organization, Washington, District of Columbia, 70 pp.

Parry, M. L., C. Rosenzweig, A. Iglesias, M. Livermore and G. Fischer, 2004: Effects of climate change on global food production under SRES emissions and socio-economic scenarios. *Global Environ. Chang*., **14**, 53-67.

Patz, J.A., D. Engelberg and J. Last, 2000: The effects of changing weather on public health. *Annu. Rev. Public Health*, **21**, 271-307.

Patz, J.A., D. Campbell-Dendrum, T. Holloway and J. Foley, 2005: Impact of regional climate change on human health. *Nature*, **438**, 310-317.

Pearce, J.F., 2003: Arsenic's fatal legacy grows. *New Sci*., **2407**, 4-5.

Penalba, O.C. and W.M. Vargas, 2004: Interdecadal and interannual variations of annual and extreme precipitation over central-northeastern Argentina. *Int. J. Climatol*., **24**, 1565-1580.

Peres, C.A. and I.R. Lake, 2003: Extent of no timber resource extraction in tropical forests: accessibility to game vertebrates by hunters in the Amazon basin. *Conserv. Biol*., **17**, 521-535.

Peterson, A.T. and J. Shaw, 2003: *Lutzomyia* vectors for cutaneous leishmaniasis in southern Brazil: ecological niche models, predicted geographic distributions, and climate change effects. *Int. J. Parasitol*., **33**, 919-931.

Peterson, A.T., M.A. Ortega-Huerta, J. Bartley, V. Sánchez-Cordero, J. Soberon, R.H. Buddemeier and D.R.B. Stockwell, 2002: Future projections for Mexican faunas under climate change scenarios. *Nature*, **416**, 626-629.

Peterson, A.T., C. Martínez-Campos, Y. Nakazawa and E. Martínez-Meyer, 2005: Time-specific ecological niche modeling predicts spatial dynamics of vector insects and human dengue cases. *T. Roy. Soc. Trop. Med. H*., **99**, 647-655.

Pezza, A.B. and I. Simmonds, 2005: The first South Atlantic hurricane: unprecedented blocking, low shear and climate change. *Geophys. Res. Lett*., **32**, L15712, doi:10.1029/2005GL023390.

Pimm, S.L., M. Ayres, A. Balmford, G. Branch, K. Brandon, T. Brooks, R. Bustamante, R. Costanza and Co-authors, 2001: Can we defy nature's end? *Science*, **293**, 2207-2208.

Pini, N.C., A. Resa, G. Del Jesús Laime, G. Lecot, T.G. Ksiazek and S. Levis, 1998: Hantavirus infection in children in Argentina. *Emerg. Infect. Dis*., **4**, 85-87.

Pinto, H.S., E.D. Assad, J. Zullo Jr and O. Brunini, 2002: O aquecimento global e a agricultura. Mudanças Climáticas. *Com Ciência*, **34**, August 2002. http://www.comciencia.br/.

PNC, 2005: Informe del Programa Nacional de Control de Chagas para el Cono Sur. Programa Nacional de Chagas, Ministerio de Salud, 16 pp.

Pounds, J.A., M.P.L. Fogden and J.H. Campbell, 1999: Biological response to climate change on a tropical mountain. *Nature*, **398**, 611-615.

Poveda, G.J. and O.J. Mesa, 2000: On the existence of Lloró (the rainiest locality on Earth): enhanced ocean-atmosphere-land interaction by a low-level jet. *Geophys. Res. Lett*., **27**, 1675-1678.

Poveda, G.J., N.E. Graham, P.R. Epstein, W. Rojas, D.I. Vélez, M.L. Quiñónez and P. Martnes, 1999: Climate and ENSO variability associated to malaria and dengue fever in Colombia. *10th Symposium on Global Change Studies*, Dallas, January 10-15. American Meteorological Society, Boston, Massachusetts, 173-176.

Poveda, G.J., C.C. Rave and R. Mantilla, 2001a: Tendencias en la distribución de probabilidades de lluvias y caudales en Antioquia (Colombia). *Meteorol. Colombiana*, **3**, 53-60.

Poveda, G.J., W. Rojas, M.L. Quiñones, D.I. Vélez, R.I. Mantilla, D. Ruiz, J.S. Zuluaga and G.L. Rua, 2001b: Coupling between annual and ENSO theme scales in the malaria climate association in Colombia. *Environ. Health Perspect*., **109**, 489-493.

Poveda, G.J., O.J. Mesa and P. Waylen, 2003: Non-linear forecasting of river flows in Colombia based upon ENSO and its associated economic value for hydropower generation. *Climate and Water: Transboundary Challenges in the Americas*, H. Díaz and B. Morehouse, Eds., Kluwer Academic, Dordrecht, 351-371.

Quaas, W.R. and O.E. Guevara, 2006: Monitoring and warning systems for natural phenomena: the Mexican experience. Preprints, *UN Third International Conference on Early Alert Systems: From Concept to Action*. Bonn, 15 pp.

Ramírez, E., B. Francou, P. Ribstein, M. Descloitres, R. Guérin, J. Mendoza, R. Gallaire, B. Pouyaud and E. Jordan, 2001: Small glaciers disappearing in the Tropical Andes: a case study in Bolivia: the Chacaltaya glacier, 16°S. *J. Glaciol*., **47**, 187-194.

Ramírez, N., L.C. Silva, D. Kiriakos and M.A.J. Rodríguez, 2005: Enfermedad de Chagas en Venezuela: un bosquejo sobre su impacto en la salud pública. *Acta Cient. Estudiantil*, **2**, 148-156.

Ramírez, P.O. and A. Brenes, 2001: Informe sobre las condiciones de sequía observadas en el istmo centroamericano en el 2001. Comité Regional Recursos Hidráulicos/Sistema de la Integración Centroamericana, 33 pp.

Rodríguez-Acevedo, R., 2001: Sustainable arrangement of tourism territory in conditions of high physiographic and environmental vulnerability: study case Vargas State-Venezuela. Bureau of Tourism Research (BTR), Australia and Universidad Simón Bolívar, Venezuela.

Romieu, I., F. Meneses, S. Ruiz, J.J. Sienra and J. Huerta, 1996: Effects of air pollution on the respiratory health of asthmatic children living in Mexico city. *Am. J. Respir. Crit. Care Med*., **154**, 300-307.

Ron, S.R., W.E. Duellman, L.A. Coloma and M.R. Bustamante, 2003: Population decline of the Jambato toad *Atelopus ignescens* (Anura: Bufonidae) in the Andes of Ecuador. *J. Herpetol*., **37**, 116–126.

Ronchail, J., L. Bourrel, G. Cochonneau, P. Vauchel, L. Phillips, A. Castro, J.L. Guyot and E. Oliveira, 2005: Inundations in the Mamoré Basin (south-western Amazon-Bolivia) and sea-surface temperature in the Pacific and Atlantic Oceans. *J. Hydrol*., **302**, 223-238.

Rosenzweig, C., K.M. Strzepek, D.C. Major, A. Iglesias, D.N. Yates, A. McCluskey and D. Hillel, 2004: Water resources for agriculture in a changing climate: international case studies. *Global Environ. Chang*., **14**, 345-360.

Rowell, A. and P.F. Moore, 2000: *Global Review of Forest Fires*. WWF/IUCN, Gland, Switzerland, 66 pp. http://www.iucn.org/themes/fcp/publications /files/global_review_forest_fires.pdf.

Ruiz, D., G. Poveda , I.D. Vélez, M.L. Quiñones, R.L. Rúa, L.E. Velásquez and J.S. Zuluaga, 2006: Modelling entomological-climatic interactions of *Plasmodium falciparum* malaria transmission in two Colombian endemic regions: contributions to a national Malaria early warning system. *Malaria J*., **5**, 66-96.

Ruosteenoja, K., T.R. Carter, K. Jylhä and H. Tuomenvirta, 2003: Future climate in world regions: an intercomparison of model-based projections for the new IPCC emissions scenarios. The Finnish Environment 644, Finnish Environment Institute, Helsinki, 83 pp.

Rusticucci, M. and M. Barrucand, 2004: Observed trends and changes in temperature extremes over Argentina. *J. Climate*, **17**, 4099-4107.

Salinas-Zavala, C.A. and D.B. Lluch-Cota, 2003: Relationship between ENSO and winter-wheat yields in Sonora, Mexico. *Geofís. Int*., **42**, 341-350.

Satyamurty, P., C.A. Nobre and P.L. Silva Dias, 1998: South America. *Meteorology of the Southern Hemisphere*, D. Karoly and D.G. Vincent, Eds., Meteorological Monographs, American Meteorological Society, Boston, Massachusetts, 119-139.

Schaeffer-Novelli, Y., G. Cintron-Molero and M.L.G. Soares, 2002: Mangroves as indicators of sea level change in the muddy coasts of the world. *Muddy Coasts of the World: Processes, Deposits and Function*, T. Healy, Y. Wang and J.A. Healy, Eds., Elsevier Science, 245-262.

Scholze, M., W. Knorr, N.W. Arnell and I.C. Prentice, 2005: A climate change risk analysis for world ecosystems. *P. Natl. Acad. Sci. USA*, **103**, 13116-13120.

Severov, D.N., E. Mordecki and V.A. Pshennikov, 2004: SST anomaly variability in southwestern Atlantic and El Niño/Southern Oscillation. *Adv. Space Res*., **33**, 343-347.

Shatwell, J., 2006: Who are you calling extinct? Long-lost Harlequin frog subspecies rediscovered in the Andes. *Conservation International*, 17 May 2006. http://www.conservation.org/xp/frontlines/species/05160601.xml.

Shinn, E.A., G.W. Smith, J.M. Prospero, P. Betzer, M.L. Hayes, V. Garrison and R.T. Barber, 2000: African dust and the demise of Caribbean coral reefs. *Geophys. Res. Lett*., **27**, 3029-3032.

Sims, A. and H. Reid, 2006: *Up in Smoke? Latin America and the Caribbean: The Threat from Climate Change to the Environment and Human Development*. Third Report from the Working Group on Climate Change and Development, New Economic Foundation, 48 pp.

SINAPRED, 2006: Evaluación de los efectos del huracán Beta en Nicaragua, November 2005. Sistema Nacional de Prevención de Desastres, Republic of Nicaragua.

Siqueira, M.F. de and A.T. Peterson, 2003: Consequences of global climate change for geographic distributions of cerrado tree species. *Biota Neotropica*, **3**, 14 pp. http://www.biotaneotropica.org.br/v3n2/pt/download?article+BN00803022003+it.

Siqueira, O.J.W., L.A.B. Salles and J.M. Fernandes, 2001: Efeitos potenciais das mudanças climáticas na agricultura Brasileira e estratégias adaptativas para algumas culturas. *Mudanças Climáticas Globais e a Agropecuária Brasileira*, M.A. Lima, O.M.R. Cabral and J.D.G. Miguez, Eds., 33-63.

Smolders, A.J.P., M.A. Guerrero Hiza, G. van der Velde and J.G. Roelofs, 2002: Dynamics of discharge, sediment transport, heavy metal pollution and sábalo (*Prochilodus lineatus*) catches in the Lower Pilcomayo River (Bolivia). *River Res. Appl.*, **18**, 415-427.

Soares-Filho, B.S., D.C. Nepstad, L.M. Curran, G.C. Cerqueira, R.A. Garcia, C.A. Ramos, E. Voll, A. McDonald, P. Lefebvre, D. McGrath and P. Schlesinger, 2005: Cenários de desmatamento para a Amazônia. *Estud. Avançados*, **19**, 137-152.

Soares-Filho, B.S., D.C. Nepstad, L.M. Curran, G.C. Cerqueira, R.A. Garcia, C.A. Ramos, E. Voll, A. McDonald, P. Lefebvre and P. Schlesinger, 2006: Modelling conservation in the Amazon basin. *Nature*, **440**, 520-523.

Solanes, M. and A. Jouravlev, 2006: *Water Governance for Development and Sustainability*. Economic Commission for Latin America and the Caribbean, Santiago, 84 pp.

Solimano, A. and R. Soto, 2005: *Economic Growth in Latin America in the Late 20th Century: Evidence and Interpretation*. Macroeconomía del desarrollo 33, Economic Commission for Latin America and the Caribbean, 44 pp.

Solman, S., M.F. Cabré and M.Nuñez, 2005a: Simulación del clima actual sobre el sur de Sudamérica con un modelo regional: análisis de los campos medios y el ciclo anual. *CONGREMET IX*, Buenos Aires, 10 pp.

Solman, S., M. Nuñez and M.F. Cabré, 2005b: Escenarios regionales de cambio climático sobre el Sur de Sudamérica. Taller Regional de Cambio Climático, San Pablo, 2005.

SRA, 2005: Sequía en el Chaco genera fuerte pérdidas. Comunicado de prensa de la Sociedad Rural Argentina, dated 3/10/2005. http://www.ruralarg.org.ar/.

Stern, N., 2007: *The Economics of Climate Change: The Stern Review*. Cambridge University Press, Cambridge, 692 pp.

Swiss Re, 2002: Opportunities and risks of climate change. Swiss Re Insurance Company, Zurich, 30 pp. http://www.swissre.com/.

Tarras-Wahlberg, N.H. and S.N. Lane, 2003: Suspended sediment yield and metal contamination in a river catchment affected by El Niño events and gold mining activities: the Puyango River Basin, southern Ecuador. *Hydrol. Process.*, **17**, 3101-3123.

Tebaldi, C., K. Hayhoe, J.M. Arblaster and G.E. Meehl, 2007: Going to the extremes: an intercomparison of model-simulated historical and future changes in extreme events. *Climatic Change*, **82**, 233-234.

The Water Page, 2001: BPD business partners for development water and sanitation clusters. http://www.africanwater.org/bpd.htm.

Thomas, C.D., A. Cameron, R.E. Green, M. Bakkenes, L.J. Beaumont, Y.C. Collingham, B.F.N. Erasmus, M.F. de Siqueira and co-authors, 2004: Extinction risk from climate change. *Nature*, **427**, 145-147.

Tomaselli, I., 2001: Latin America and the Caribbean. *Unasylva*, **52**, 44-46.

Torres, F., F. Peña, R. Cruz and E. Gómez, 2001: Impacto de El Niño sobre los cultivos vegetales y la productividad primaria en la sierra central de Piura. *El Niño en América Latina, Impactos Biológicos y Sociales*, J. Tarazona, W. Arntz and E. Castillo, Eds., Editorial Omega S.A.

Travasso, M.I., G.O. Magrin and G.R. Rodríguez, 2003a: Crops yield and climatic variability related to ENSO and South Atlantic sea surface temperature in Argentina. Preprint, *The Seventh International Conference on Southern Hemisphere Meteorology and Oceanography (7ICSHMO)*, Wellington, New Zealand, 74-75.

Travasso, M.I., G.O. Magrin and G.R. Rodríguez, 2003b: Relations between sea surface temperature and crop yields in Argentina. *Int. J. Climatol.*, **23**, 1655-1662.

Travasso, M.I., G.O. Magrin, W.E. Baethgen, J.P. Castaño, G.R. Rodriguez, J.L. Pires, A. Gimenez, G. Cunha and M. Fernandes, 2006: Adaptation measures for maize and soybean in south eastern South America. AIACC Working Paper No. 28, 38 pp. http://www.aiaccproject.org/working_papers/working_papers.html.

Treacy, J.M., 1994: *Las Chacras de Coparaque: Andenes y Riego en el Valle de Colca*. Instituto de Estudios Peruanos, Lima, 298 pp.

Trenberth, K.E. and D.P. Stepaniak, 2001: Indices of El Niño evolution. *J. Climate.*, **14**, 1697-1701.

Trenberth, K.E., P.D. Jones, P. Ambenje, R. Bojariu, D. Easterling, A. Klein Tank, D. Parker, F. Rahimzadeh, J.A. Renwick, M. Rasticucci, B. Soden and P. Zhai, 2007: Observations: surface and atmospheric climate change. *Climate Change 2007: The Physical Science Basis. Contribution of Working Group I to the Fourth Assessment Report of the Intergovernmental Panel on Climate Change*, S. Solomon, D. Qin, M. Manning, Z. Chen, M. Marquis, K.B. Averyt, M. Tignor and H.L. Miller, Eds., Cambridge University Press, Cambridge, 235-336.

Trosnikov, I. and C.A. Nobre, 1998: Estimation of aerosol transport from biomass burning areas during the SCAR-B experiment. *J. Geophys. Res.*, **103**, D24, 32129, doi:10.1029/98JD01343.

Tucci, C.E.M., 2001: *Urban Drainage in the Humid Tropics*. IHP-V Technical Documents in Hydrology, UNESCO, Paris, 227 pp.

Tucci, C.E.M., R.T. Clarke, W. Collischonn, P.S. Dias and G. Sampaio, 2003: Long-term flow forecast based on climate and hydrological modeling: Uruguay river basin. *Water Resour. Res.*, **39**, 1181-1191.

Ubitarán Moreira dos Santos, J., I. de Sousa Gorayeb and M. de Nazaré do Carmo Bastos, 1999: Avaliação e ações prioritarias Para a conservação da biodiversidade da zona costeira e marinha. Diagnóstico da situação para a conservação da biodiversidade da zona costeira e marinha Amazônica. Ministerio do Meio Ambiente and Projecto de Conservação e Utilização Sustantável da Diversidadie Biológica Brasiliera, Belém, Brazil.

UCC, 2005: Análisis de la Estadística Climática y desarrollo y evaluación de escenarios climáticos e hidrológicos de las principales cuencas hidrográficas del Uruguay y de su Zona Costera. Unidad de Cambio Climático, MVOTMA-DINAMA-GEF-UNDP, Montevideo, Uruguay, 2005. http://www.cambio climatico.gub.uy/.

UCV, 2005: Análisis de las lluvias diarias y acumuladas durante Febrero de 2005 en la región central capital. Facultad de Ingeniería, Instituto de Mecánica de Fluidos Departamento de Ingeniería Hidrometeorológica, Universidad Central de Venezuela.

UNCHS, 2001: *The State of the World´s Cities 2001*.United Nations Centre for Human Settlements, Nairobi, 125 pp.

UNCLOS, 2005: United Nations Convention on the Law of the Sea: Status of the United Nations Convention on the Law of the Sea, of the Agreement relating to the implementation of Part XI of the Convention and of the Agreement for the implementation of the provisions of the Convention relating to the conservation and management of straddling fish stocks and highly migratory fish stocks. Table recapitulating the status of the Convention and of the related Agreements, as at 20 April 2005. http://www.un.org/Depts/los/convention_agreements/convention_overview_fish_stocks.htm.

UNEP, 2000: GEO Latin America and the Caribbean. Environmental Outlook 2000, UNEP Regional Office for Latin America and the Caribbean, Mexico.

UNDP-GEF, 2003: Capacity building for stage II adaptation to climate change in Central America, Mexico and Cuba. UNPD PIMS ID # 2220, United Nations Development Programme: Global Environmental Facility.

UNEP, 2000: *Global Environment Outlook 2*. United Nations Environment Programme. http://www.unep.org/geo2000.

UNEP, 2002: Caribbean Environmental Law Development and Application: Environmental legislative and judical developments in the English-speaking Caribbean countries in the context of compliance with Agenda 21 and the Rio Agreements, UNEP Regional Office for Latin America and the Caribbean.

UNEP, 2003a: *Global Environment Outlook 3*. United Nations Environment Programme. Earthscan, London, 446 pp.

UNEP, 2003b: *GEO–LAC 2003. Global Environment Outlook: Latin America and the Caribbean*. United Nations Environment Programme, Costa Rica, 278 pp. http://www.unep.org/geo/.

UNEP, 2003c: *Global Environment Outlook Year Book 2003*. United Nations Environment Programme, 80 pp.

UNMSM, 2004: Calor intenso y largas sequías. Especials, Perú. http://www.unmsm.edu.pe/Destacados/contenido.php?mver=11 .

Valtorta, S.C., M.R. Gallardo and P.E. Leva, 2004: Olas de calor: impacto sobre la producción lechera en la cuenca central argentina. *Actas X Reunion Argentina de Agrometeorologia. IV. Reunión Latinoamericana de Agrometeorologia*, Mar del Plata, Argentina.

van Lieshout, M., R.S. Kovats, M.T.J. Livermore and P. Martens, 2004: Climate change and malaria: analysis of the SRES climate and socio-economic scenarios. *Global Environ. Chang.*, **14**, 87-99.

Vásquez, O.C., 2004: *El Fenómeno El Niño en Perú y Bolivia: Experiencias en Participación Local*. Memoria del Encuentro Binacional Experiencias de prevención de desastres y manejo de emergencias ante el Fenómeno El Niño, Chiclayo, Peru. ITDG, 209 pp.

Vásquez-León, M., C.T. West and T.J. Finan, 2003: A comparative assessment of climate vulnerability: agriculture and ranching on both sides of the US-Mexico border. *Global Environ. Chang.*, **13**, 159-173.

Vergara, W., 2005: Adapting to climate change: lesson learned, work in progress, and proposed next step for the World Bank in Latin America. Sustainable Development Working Paper No. 25, World Bank, Washington, District of Columbia, 56 pp.

Viddi, F. and S. Ribeiro, 2004: Ecology and conservation of dolphins in Southern Chile. Cetacean Society International. *Whales Alive!*, **XIII**, No. 2. http://csi-whalesalive.org/csi04208.html.

Villagrán de León, J., J. Scott, C. Cárdenas and S. Thompson, 2003: Early warning systems in the American hemisphere: context, current status, and future trends. Final Report. Hemispheric Consultation on Early Warning, Antigua, 15 pp.

Villamizar, A., 2004: Informe técnico de denuncia sobre desastre ecológico en el Desparramadero de Hueque. Edo. Falcón. Presentado ante La Fiscalía General y la Defensoría del Pueblo, República de Venezuela, 25 pp.

Villers, L. and I. Trejo, 2004: Evaluación de la vulnerabilidad en los sistemas forestales. *Cambio Climático: Una Visión desde México*, J. Martínez and A. Fernández Bremauntz, Eds., SEMARNAT e INE, México, 239-254.

Vincent, L.A., T.C. Peterson, V.R. Barros, M.B. Marino, M. Rusticucci, G. Carrasco, E. Ramirez, L.M. Alves and Co-authors, 2005: Observed trends in indices of daily temperature extremes in South America 1960–2000. *J. Climate*, **18**, 5011-5023.

Vinocur, M., 2005: Adaptation of farmers to climate variability and change in central Argentina: a case study. *Abstracts 6th Open Meeting of the Human Dimensions of Global Environmental Change Research Community*, 9-13 October, 2005, Bonn.

Vinocur, M.G., R.A. Seiler and L.O. Mearns, 2000: Predicting maize yield responses to climate variability in Córdoba, Argentina. *Abstracts of International Scientific Meeting on Detection and Modelling of Recent Climate Change and its Effects on a Regional Scale*, Tarragona, Spain, 29–31 May, 2000.

Vuille, M., S.R.S. Bradley, M. Werner and F. Keimig, 2003: 20th century climate change in the tropical Andes: observations and model results. *Climatic Change*, **59**, 75-99.

Walton, M., 2004: Neoliberalism in Latin America: good, bad or incomplete? *Lat. Am. Res. Rev.*, **39**, 165-183.

Warner, J., 2006: El Niño platforms: participatory disaster response in Peru. *Disasters*, **30**, 102-117.

Warren, R., N. Arnell, R. Nicholls, P.E. Levy and J. Price, 2006: Understanding the regional impacts of climate change. Research Report prepared for the Stern Review. Tyndall Centre Working Paper 90, Norwich, 223 pp. http://www.tyndall.ac.uk/publications/working_papers/working_papers.shtml.

Water 21, 2002: Joining forces. *Magazine of the International Water Association*, **October**, 55-57.

Waylen, P.R. and G. Poveda. 2002. El Niño-Southern Oscillation and aspects of western South America hydro-climatology. *Hydrol. Process.*, **16**, 1247-1260.

Wehbe, M., H. Eakin, R. Seiler, M. Vinocur, C. Ávila and C. Marutto, 2006: Local perspectives on adaptation to climate change: lessons from Mexico and Argentina. AIACC Working Paper No. 39, 39 pp.

Werth, D. and R. Avissar, 2002: The local and global effects of Amazon deforestation. *J. Geophys. Res.*, **107**, 8087, doi:10.1029/2001JD000717.

WHO, 2000: *Vegetation Fires*. World Health Organization, Fact Sheet No. 254. August 2000, 3 pp. http://www.who.int/mediacentre/factsheets/fs254/en/index.html.

WHO/UNEP/WMO, 2000: The WHO-UNEP-WMO health guidelines for vegetation fire events. *Int. Forest Fire News*, **22**, 91-101. http://www.fire.uni-freiburg.de/iffn/org/who/who_1.htm.

Williams, R.J., R.T. Bryan, J.N. Mills, R.E. Palma, I. Vera and F. de Velásquez, 1997: An outbreak of hantavirus pulmonary síndrome in western Paraguay. *Am. J. Trop. Med. Hyg.*, **57**, 274-282.

Windevoxhel, N. and G. Sención, 2000: Mangrove forests in Nicaragua and subtropical forest in Guatemala. *Sustainable Forest Management and Global Climate Change: Selected Case Studies From the Americas*, M.H.I. Dore and R. Guevara, Eds., Edward Elgar, Cheltenham, 281 pp.

WMO, 2007: Volume A: Observing Stations. World Meteorological Oganization Publication No. 9. http://www.wmo.ch/pages/prog/ www/ois/volume-a/vola-home.htm.

World Bank, 2002a: Desarrollo en riesgo debido a la degradación ambiental: Comunicado de prensa (Development at risk from environmental degradation: News release), No. 2002/112/S.

World Bank, 2002b: Sustainable groundwater and aquifer management. Workshop April 23, 2002, The World Bank Group.

World Bank, 2002c: *World Development Report 2002: Building Institutions for Markets*. The World Bank, Oxford University Press, Oxford, 228 pp.

World Bank, 2006: *World Development Report 2006: Equity and Development*. The World Bank, Oxford University Press, Oxford, 336 pp.

Wright, K.R. and A. Valencia Zegarra, 2000: *Machu Picchu: A Civil Engineering Marvel*. American Society of Civil Engineers Press, Virginia, 144 pp.

WWF, 2004: Deforestation threatens the cradle of reef diversity. World Wide Fund for Nature, 2 December 2004. http://www.wwf.org/.

Yáñez-Arancibia, A. and J.W. Day, 2004: Environmental sub-regions in the Gulf of Mexico coastal zone: the ecosystem approach as an integrated management tool. *Ocean Coast. Manage.*, **47**, 727-757.

Young, E., 2001: State intervention and abuse of the commons: fisheries development in Baja California Sur, Mexico. *Ann. Assoc. Am. Geogr.*, **91**, 283-306.

14

North America

Coordinating Lead Authors:
Christopher B. Field (USA), Linda D. Mortsch (Canada)

Lead Authors:
Michael Brklacich (Canada), Donald L. Forbes (Canada), Paul Kovacs (Canada), Jonathan A. Patz (USA), Steven W. Running (USA), Michael J. Scott (USA)

Contributing Authors:
Jean Andrey (Canada), Dan Cayan (USA), Mike Demuth (Canada), Alan Hamlet (USA), Gregory Jones (USA), Evan Mills (USA), Scott Mills (USA), Charles K. Minns (Canada), David Sailor (USA), Mark Saunders (UK), Daniel Scott (Canada), William Solecki (USA)

Review Editors:
Michael MacCracken (USA), Gordon McBean (Canada)

This chapter should be cited as:
Field, C.B., L.D. Mortsch,, M. Brklacich, D.L. Forbes, P. Kovacs, J.A. Patz, S.W. Running and M.J. Scott, 2007: North America. *Climate Change 2007: Impacts, Adaptation and Vulnerability. Contribution of Working Group II to the Fourth Assessment Report of the Intergovernmental Panel on Climate Change*, M.L. Parry, O.F. Canziani, J.P. Palutikof, P.J. van der Linden and C.E. Hanson, Eds., Cambridge University Press, Cambridge, UK, 617-652.

Table of Contents

Executive summary

North America has experienced locally severe economic damage, plus substantial ecosystem, social and cultural disruption from recent weather-related extremes, including hurricanes, other severe storms, floods, droughts, heatwaves and wildfires (very high confidence).

Over the past several decades, economic damage from severe weather has increased dramatically, due largely to increased value of the infrastructure at risk. Annual costs to North America have now reached tens of billions of dollars in damaged property and economic productivity, as well as lives disrupted and lost. [14.2.3, 14.2.6, 14.2.7, 14.2.8]

The vulnerability of North America depends on the effectiveness and timing of adaptation and the distribution of coping capacity, which vary spatially and among sectors (very high confidence).

Although North America has considerable adaptive capacity, actual practices have not always protected people and property from adverse impacts of climate variability and extreme weather events. Especially vulnerable groups include indigenous peoples and those who are socially or economically disadvantaged. Traditions and institutions in North America have encouraged a decentralised response framework where adaptation tends to be reactive, unevenly distributed, and focused on coping with rather than preventing problems. 'Mainstreaming' climate change issues into decision making is a key prerequisite for sustainability. [14.2.6, 14.4, 14.5, 14.7]

Coastal communities and habitats will be increasingly stressed by climate change impacts interacting with development and pollution (very high confidence).

Sea level is rising along much of the coast, and the rate of change will increase in the future, exacerbating the impacts of progressive inundation, storm-surge flooding and shoreline erosion. Storm impacts are likely to be more severe, especially along the Gulf and Atlantic coasts. Salt marshes, other coastal habitats, and dependent species are threatened by sea-level rise, fixed structures blocking landward migration, and changes in vegetation. Population growth and the rising value of infrastructure in coastal areas increases vulnerability to climate variability and future climate change. Current adaptation is uneven and readiness for increased exposure is low. [14.2.3, 14.4.3, 14.5]

Climate change will constrain North America's over-allocated water resources, increasing competition among agricultural, municipal, industrial and ecological uses (very high confidence).

Rising temperatures will diminish snowpack and increase evaporation, affecting seasonal availability of water. Higher demand from economic development, agriculture and population growth will further limit surface and groundwater availability. In the Great Lakes and major river systems, lower levels are likely to exacerbate challenges relating to water quality, navigation, recreation, hydropower generation, water transfers and bi-national relationships. [14.2.1, 14.4.1, 14.4.6, Boxes 14.2 and 14.3]

Climate change impacts on infrastructure and human health and safety in urban centres will be compounded by ageing infrastructure, maladapted urban form and building stock, urban heat islands, air pollution, population growth and an ageing population (very high confidence).

While inertia in the political, economic, and cultural systems complicates near-term action, the long life and high value of North American capital stock make proactive adaptation important for avoiding costly retrofits in coming decades. [14.4.5, 14.4.6, 14.5, Box 14.3]

Without increased investments in countermeasures, hot temperatures and extreme weather are likely to cause increased adverse health impacts from heat-related mortality, pollution, storm-related fatalities and injuries, and infectious diseases (very high confidence).

Historically important countermeasures include early warning and surveillance systems, air conditioning, access to health care, public education, vector control, infrastructure standards and air quality management. Cities that currently experience heatwaves are expected to experience an increase in intensity and duration of these events by the end of the century, with potential for adverse health effects. The growing number of the elderly is most at risk. Water-borne diseases and degraded water quality are very likely to increase with more heavy precipitation. Warming and climate extremes are likely to increase respiratory illness, including exposure to pollen and ozone. Climate change is likely to increase risk and geographic spread of vector-borne infectious diseases, including Lyme disease and West Nile virus. [14.2.5, 14.2.6, 14.4.5, 14.4.6, 14.5]

Disturbances such as wildfire and insect outbreaks are increasing and are likely to intensify in a warmer future with drier soils and longer growing seasons (very high confidence).

Although recent climate trends have increased vegetation growth, continuing increases in disturbances are likely to limit carbon storage, facilitate invasive species, and disrupt ecosystem services. Warmer summer temperatures are expected to extend the annual window of high fire ignition risk by 10-30%, and could result in increased area burned of 74-118% in Canada by 2100. Over the 21st century, pressure for species to shift north and to higher elevations will fundamentally rearrange North American ecosystems. Differential capacities for range shifts and constraints from development, habitat fragmentation, invasive species, and broken ecological connections will alter ecosystem structure, function and services. [14.2.4, 14.2.2, 14.4.2, Box 14.1]

14.1 Introduction

The United States (U.S.) and Canada will experience climate changes through direct effects of local changes (e.g., temperature, precipitation and extreme weather events), as well as through indirect effects, transmitted among regions by interconnected economies and migrations of humans and other species. Variations in wealth and geography, however, lead to an uneven distribution of likely impacts, vulnerabilities and capacities to adapt. This chapter reviews and synthesises the state of knowledge on direct and indirect impacts, vulnerability and adaptations for North America (comprising Canada and the U.S). Hawaii and other U.S. protectorates are discussed in Chapter 16 on Small Islands, and Mexico and Central America are treated in Chapter 13 on Latin America. Chapter 15, Polar Regions, covers high-latitude issues and peoples.

14.1.1 Key findings from the Third Assessment Report (TAR)

Key findings for the North America chapter of the Third Assessment Report (TAR) (Cohen et al., 2001) are:

Resources and ecosystems

- In western snowmelt-dominated watersheds, shifts in seasonal runoff, with more runoff in winter. Adaptation may not fully offset effects of reduced summer water availability.
- Changes in the abundance and spatial distribution of species important to commercial and recreational fisheries.
- Benefits from warming for food production in North America but with strong regional differences.
- Benefits from farm- and market-level adjustments in ameliorating impacts of climate change on agriculture.
- Increases in the area and productivity of forests, though carbon stocks could increase or decrease.
- Major role of disturbance for forest ecosystems. The forest-fire season is likely to lengthen, and the area subject to high fire danger is likely to increase significantly.
- Likely losses of cold-water ecosystems, high alpine areas, and coastal and inland wetlands.

Human settlements and health

- Less extreme winter cold in northern cities. Across North America, cities will experience more extreme heat and, in some locations, rising sea levels and risk of storm surge, water scarcity, and changes in timing, frequency, and severity of flooding.
- The need for changes in land-use planning and infrastructure design to avoid increased damages from heavy precipitation events.
- For communities that have the necessary resources, reduced vulnerability by adapting infrastructure.
- Increased deaths, injuries, infectious diseases, and stress-related disorders and other adverse effects associated with social disruption and migration from more frequent extreme weather.
- Increased frequency and severity of heatwaves leading to more illness and death, particularly among the young, elderly and frail. Respiratory disorders may be exacerbated by warming-induced deterioration in air quality.
- Expanded ranges of vector-borne and tick-borne diseases in North America but with modulation by public health measures and other factors.

Vulnerability and adaptation

- Increased weather-related losses in North America since the 1970s, with rising insured losses reflecting growing affluence and movement into vulnerable areas.
- Coverage, since the 1980s, by disaster relief and insurance programmes of a large fraction of flood and crop losses, possibly encouraging more human activity in at-risk areas.
- Responses by insurers to recent extreme events through limiting insurance availability, increasing prices and establishing new risk-spreading mechanisms. Improving building codes, land-use planning and disaster preparedness also reduce disaster losses.
- Awareness that developing adaptation responses requires a long, interdisciplinary dialogue between researchers and stakeholders, with substantial changes in institutions and infrastructure.
- Recognition that adaptation strategies generally address current challenges, rather than future impacts and opportunities.

14.1.2 Key differences from TAR

This assessment builds on the findings from the TAR and incorporates new results from the literature, including:

- Prospects for increased precipitation variability, increasing challenges of water management.
- The need to include groundwater and water-quality impacts in the assessment of water resources.
- The potential that multi-factor impacts may interact non-linearly, leading to tipping points.
- The potential importance of interactions among climate change impacts and with other kinds of local, regional and global changes.
- The potential for adaptation, but the unevenness of current adaptations.
- The challenge of linking adaptation strategies with future vulnerabilities.
- Availability of much more literature on all aspects of impacts, adaptation and vulnerability in North America.

14.2 Current sensitivity/vulnerability

Annual mean air temperature, on the whole, increased in North America for the period 1955 to 2005, with the greatest warming in Alaska and north-western Canada, substantial warming in the continental interior and modest warming in the south-eastern U.S. and eastern Canada (Figure 14.1). Spring and winter show the greatest changes in temperature (Karl et al., 1996; Hengeveld et al., 2005) and daily minimum (night-time) temperatures have warmed more than daily maximum (daytime) temperatures (Karl et al., 2005; Vincent and Mekis, 2006). The length of the vegetation growing season has increased an average of 2 days/decade since 1950 in Canada and the conterminous U.S., with most of the increase resulting from earlier spring warming (Bonsal et al., 2001; Easterling, 2002; Bonsal and Prowse, 2003; Feng and Hu, 2004). The warming signal in North America during the latter half of the 20th century reflects the combined influence of greenhouse gases, sulphate aerosols and natural external forcing (Karoly et al., 2003; Stott, 2003; Zwiers and Zhang, 2003).

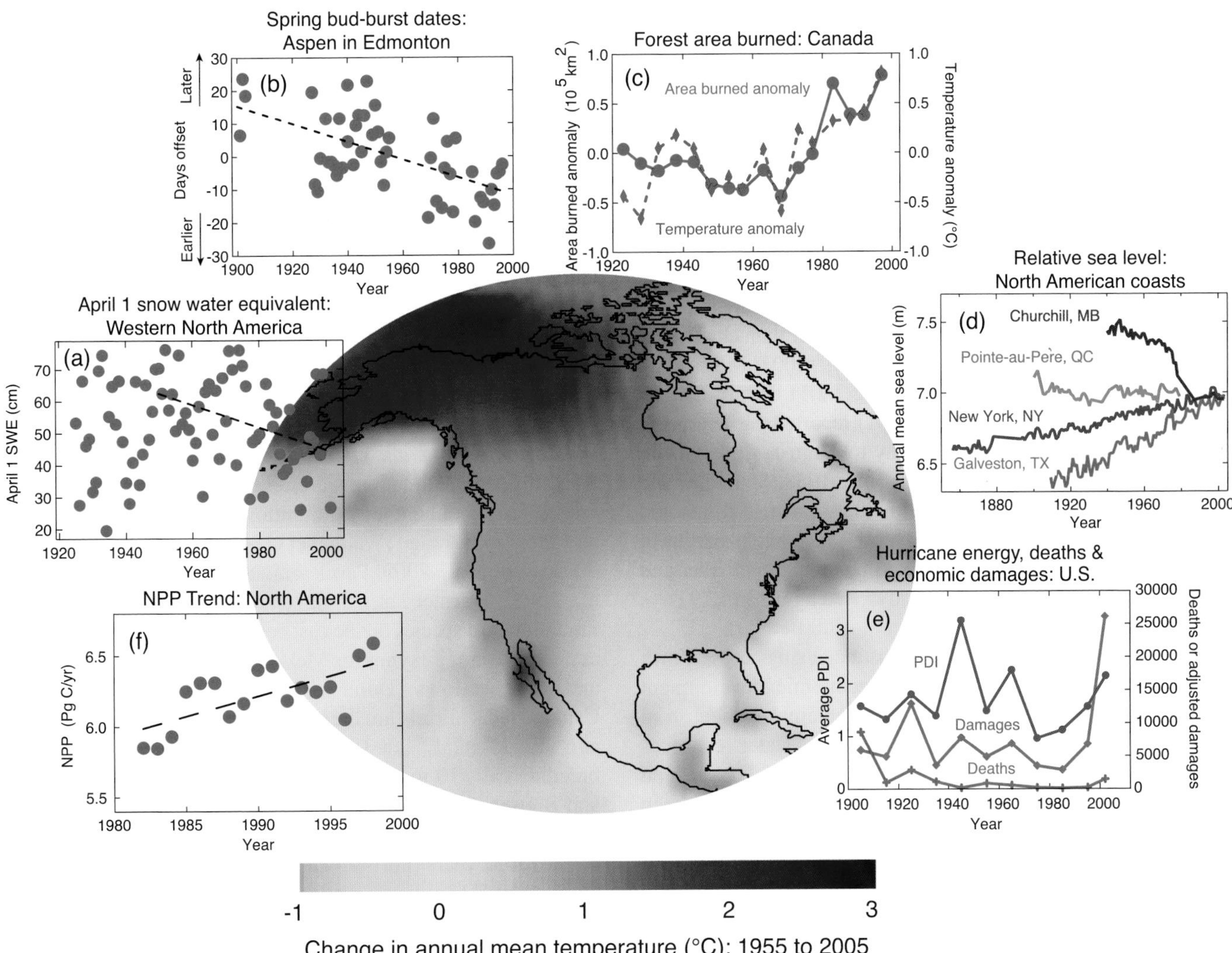

Figure 14.1. *Observed trends in some biophysical and socio-economic indicators. Background: change in annual mean temperature from 1955 to 2005 (based on the GISS2001 analysis for land from Hansen et al., 2001; and on the Hadley/Reyn_V2 analysis for sea surface from Reynolds et al., 2002). Insets: (a) trend in April 1 snow water equivalent (SWE) across western North America from 1925 to 2002, with a linear fit from 1950 to 2002 (data from Mote, 2003), (b) Spring bud-burst dates for trembling aspen in Edmonton since 1900 (data from Beaubien and Freeland, 2000), (c) anomaly in 5-year mean area burned annually in wildfires in Canada since 1930, plus observed mean summer air temperature anomaly, weighted for fire areas, relative to 1920 to 1999 (data from Gillett et al., 2004) (d) relative sea-level rise from 1850 to 2000 for Churchill, MB, Pointe-au-Père, QB, New York, NY, and Galveston, TX, (POL, 2006) (e) hurricane energy (power dissipation index (PDI) based on method of Emanuel, 2005), economic damages, million U.S. dollars (adjusted to constant 2005 US dollars and normalized accounting for changes in personal wealth and coastal population to 2004), and deaths from Atlantic hurricanes since 1900 (data from Pielke Jr. and Landsea, 1998 updated through 2005), and, (f) trend North American Net Primary Production (NPP) from 1981 to 1998 (data from Hicke et al., 2002).*

Annual precipitation has increased for most of North America with large increases in northern Canada, but with decreases in the south-west U.S., the Canadian Prairies and the eastern Arctic (see Working Group I Fourth Assessment (WGI AR4) Trenberth et al., 2007 Section 3.3.2.2, Figures 3.13 and 3.14) (Hengeveld et al., 2005; Shein, 2006). Heavy precipitation frequencies in the U.S. were at a minimum in the 1920s and 1930s, and increased to the 1990s (1895 to 2000) (Kunkel, 2003; Groisman et al., 2004). In Canada there is no consistent trend in extreme precipitation (Vincent and Mekis, 2006).

14.2.1 Freshwater resources

Streamflow in the eastern U.S. has increased 25% in the last 60 years (Groisman et al., 2004), but over the last century has decreased by about 2%/decade in the central Rocky Mountain region (Rood et al., 2005). Since 1950, stream discharge in both the Colorado and Columbia river basins has decreased, at the same time annual evapotranspiration (ET) from the conterminous U.S. increased by 55 mm (Walter et al., 2004). In regions with winter snow, warming has shifted the magnitude and timing of hydrologic events (Mote et al., 2005; Regonda et al., 2005; Stewart et al., 2005). The fraction of annual precipitation falling as rain (rather than snow) increased at 74% of the weather stations studied in the western mountains of the U.S. from 1949 to 2004 (Knowles et al., 2006). In Canada, warming from 1900 to 2003 led to a decrease in total precipitation as snowfall in the west and Prairies (Vincent and Mekis, 2006). Spring and summer snow cover has decreased in the U.S. west (Groisman et al., 2004). April 1 snow water equivalent (SWE) has declined 15 to 30% since 1950 in the western mountains of North America, particularly at lower elevations and primarily due to warming rather than changes in precipitation (Figure 14.1a) (see Mote et al., 2003; Mote et al., 2005; Lemke et al., 2007: Section 4.2.2.2.1). Whitfield and Cannon (2000) and Zhang et al. (2001) reported earlier spring runoff across Canada. Summer (May to August) flows of the Athabasca River have declined 20% since 1958 (Schindler and Donahue, 2006). Streamflow peaks in the snowmelt-dominated western mountains of the U.S. occurred 1 to 4 weeks earlier in 2002 than in 1948 (Stewart et al., 2005). Break up of river and lake ice across North America has advanced by 0.2 to 12.9 days over the last 100 years (Magnuson et al., 2000).

Vulnerability to extended drought is increasing across North America as population growth and economic development create more demands from agricultural, municipal and industrial uses, resulting in frequent over-allocation of water resources (Alberta Environment, 2002; Morehouse et al., 2002; Postel and Richter, 2003; Pulwarty et al., 2005). Although drought has been more frequent and intense in the western part of the U.S. and Canada, the east is not immune from droughts and attendant reductions in water supply, changes in water quality and ecosystem function, and challenges in allocation (Dupigny-Giroux, 2001; Bonsal et al., 2004; Wheaton et al., 2005).

14.2.2 Ecosystems

Three clear, observable connections between climate and terrestrial ecosystems are the seasonal timing of life-cycle events or phenology, responses of plant growth or primary production, and biogeographic distribution. Direct impacts on organisms interact with indirect effects of ecological mechanisms (competition, herbivory[1], disease), and disturbance (wildfire, hurricanes, human activities).

Phenology, productivity and biogeography

Global daily satellite data, available since 1981, indicate earlier onset of spring 'greenness' by 10-14 days over 19 years, particularly across temperate latitudes of the Northern Hemisphere (Myneni et al., 2001; Lucht et al., 2002). Field studies confirm these satellite observations. Many species are expanding leaves or flowering earlier (e.g., earlier flowering in lilac - 1.8 days/decade, 1959 to 1993, 800 sites across North America (Schwartz and Reiter, 2000), honeysuckle - 3.8 days/decade, western U.S. (Cayan et al., 2001), and leaf expansion in apple and grape - 2 days/decade, 72 sites in north-eastern U.S. (Wolfe et al., 2005), trembling aspen - 2.6 days/decade since 1900, Edmonton (Beaubien and Freeland, 2000)) (Figure 14.1b). The timing of autumn leaf fall, which is controlled by a combination of temperature, photoperiod and water deficits, shows weaker trends (Badeck et al., 2004).

Net primary production (NPP) in the continental U.S. increased nearly 10% from 1982 to 1998 (Figure 14.1f) (Boisvenue and Running, 2006), with the largest increases in croplands and grasslands of the Central Plains due to improved water balance (Lobell et al., 2002; Nemani et al., 2002; Hicke and Lobell, 2004).

North American forests can be influenced indirectly by climate through effects on disturbance, especially from wildfire, storms, insects and diseases. The area burned in wildfires has increased dramatically over the last three decades (see Box 14.1).

Wildlife population and community dynamics

North American animals are responding to climate change, with effects on phenology, migration, reproduction, dormancy and geographic range (Walther et al., 2002; Parmesan and Yohe, 2003; Root et al., 2003; Parmesan and Galbraith, 2004; Root et al., 2005). Warmer springs have led to earlier nesting for 28 migrating bird species on the east coast of the U.S. (Butler, 2003) and to earlier egg laying for Mexican jays (Brown et al., 1999) and tree swallows (Dunn and Winkler, 1999). In northern Canada, red squirrels are breeding 18 days earlier than 10 years ago (Reale et al., 2003). Several frog species now initiate breeding calls 10 to 13 days earlier than a century ago (Gibbs and Breisch, 2001). In lowland California, 70% of 23 butterfly species advanced the date of first spring flights by an average 24 days over 31 years (Forister and Shapiro, 2003). Reduced water depth, related to recent warming, in Oregon lakes has increased exposure of toad eggs to UV-B, leading to increased mortality from a fungal parasite (Kiesecker et al., 2001; Pounds, 2001).

Many North American species have shifted their ranges, typically to the north or to higher elevations (Parmesan and Yohe, 2003). Edith's checkerspot butterfly has become locally extinct in the southern, low-elevation portion of its western North American range but has extended its range 90 km north and 120 m higher in elevation (Parmesan, 1996; Crozier, 2003; Parmesan and Galbraith, 2004). Red foxes have expanded northward in northern Canada, leading to retreat of competitively subordinate arctic foxes (Hersteinsson and Macdonald, 1992).

14.2.3 Coastal regions

The North American coast is long and diverse with a wide range of trends in relative sea level (Figure 14.1d) (Shaw et al., 1998; Dyke and Peltier, 2000; Zervas, 2001). Relative sea level (see glossary) is rising in many areas, yet coastal residents are often unaware of the trends and their impacts on coastal retreat and flooding (O'Reilly et al., 2005). In the Great Lakes, both extremely high and extremely low water levels have been damaging and disruptive (Moulton and Cuthbert, 2000). Demand for waterfront property and building land continues to grow, increasing the value of property at risk (Heinz Center, 2000; Forbes et al., 2002b; Small and Nichols, 2003).

[1] The consumption of plants by animals.

Box 14.1. Accelerating wildfire and ecosystem disturbance dynamics

Since 1980, an average of 22,000 km^2/yr has burned in U.S. wildfires, almost twice the 1920 to 1980 average of 13,000 km^2/yr (Schoennagel et al., 2004). The forested area burned in the western U.S. from 1987 to 2003 is 6.7 times the area burned from 1970 to 1986 (Westerling et al., 2006). In Canada, burned area has exceeded 60,000 km^2/yr three times since 1990, twice the long-term average (Stocks et al., 2002). Wildfire-burned area in the North American boreal region increased from 6,500 km^2/yr in the 1960s to 29,700 km^2/yr in the 1990s (Kasischke and Turetsky, 2006). Human vulnerability to wildfires has also increased, with a rising population in the wildland-urban interface.

A warming climate encourages wildfires through a longer summer period that dries fuels, promoting easier ignition and faster spread (Running, 2006). Westerling et al. (2006) found that in the last three decades the wildfire season in the western U.S. has increased by 78 days, and burn durations of fires >1000 ha in area have increased from 7.5 to 37.1 days, in response to a spring-summer warming of 0.87°C. Earlier spring snowmelt has led to longer growing seasons and drought, especially at higher elevations, where the increase in wildfire activity has been greatest (Westerling et al., 2006). In Canada, warmer May to August temperatures of 0.8°C since 1970 are highly correlated with area burned (Figure 14.1c) (Gillett et al., 2004). In the south-western U.S., fire activity is correlated with El Niño-Southern Oscillation (ENSO) positive phases (Kitzberger et al., 2001; McKenzie et al., 2004), and higher Palmer Drought Severity Indices.

Insects and diseases are a natural part of ecosystems. In forests, periodic insect epidemics kill trees over large regions, providing dead, desiccated fuels for large wildfires. These epidemics are related to aspects of insect life cycles that are climate sensitive (Williams and Liebhold, 2002). Many northern insects have a two-year life cycle, and warmer winter temperatures allow a larger fraction of overwintering larvae to survive. Recently, spruce budworm in Alaska has completed its life cycle in one year, rather than the previous two (Volney and Fleming, 2000). Mountain pine beetle has expanded its range in British Columbia into areas previously too cold (Carroll et al., 2003). Insect outbreaks often have complex causes. Susceptibility of the trees to insects is increased when multi-year droughts degrade the trees' ability to generate defensive chemicals (Logan et al., 2003). Recent dieback of aspen stands in Alberta was caused by light snowpacks and drought in the 1980s, triggering defoliation by tent caterpillars, followed by wood-boring insects and fungal pathogens (Hogg et al., 2002).

Many coastal areas in North America are potentially exposed to storm-surge flooding (Titus and Richman, 2001; Titus, 2005). Some major urban centres on large deltas are below sea level (e.g., New Orleans on the Mississippi; Richmond and Delta on the Fraser), placing large populations at risk. Breaching of New Orleans floodwalls following Hurricane Katrina in 2005 (see Chapter 6, Section 6.4.1.2 and Box 6.4) and storm-wave breaching of a dike in Delta, British Columbia, in 2006 demonstrate the vulnerability. Under El Niño conditions, high water levels combined with changes in winter storms along the Pacific coast have produced severe coastal flooding and storm impacts (Komar et al., 2000; Walker and Barrie, 2006). At San Francisco, 140 years of tide-gauge data suggest an increase in severe winter storms since 1950 (Bromirski et al., 2003) and some studies have detected accelerated coastal erosion (Bernatchez and Dubois, 2004). Some Alaskan villages are threatened and require protection or relocation at projected costs up to US$54 million (Parson et al., 2001a). Recent severe tropical and extra-tropical storms demonstrate that North American urban centres with assumed high adaptive capacity remain vulnerable to extreme events. Recent winters with less ice in the Great Lakes and Gulf of St. Lawrence have increased coastal exposure to damage from winter storms. Winter ice provides seasonal shore protection, but can also damage shorefront homes and infrastructure (Forbes et al., 2002a).

Impacts on coastal communities and ecosystems can be more severe when major storms occur in short succession, limiting the opportunity to rebuild natural resilience (Forbes et al., 2004). Adaptation to coastal hazards under the present climate is often inadequate, and readiness for increased exposure is poor (Clark et al., 1998; Leatherman, 2001; West et al., 2001). Extreme events can add to other stresses on ecological integrity (Scavia et al., 2002; Burkett et al., 2005), including shoreline development and nitrogen eutrophication[2] (Bertness et al., 2002). Already, more than 50% of the original salt marsh habitat in the U.S. has been lost (Kennish, 2001). Impacts from sea-level rise can be amplified by 'coastal squeeze' (see Glossary) and submergence where landward migration is impeded and vertical growth is slower than sea-level rise (see Section 14.4.3) (Kennish, 2001; Scavia et al., 2002; Chmura and Hung, 2004).

14.2.4 Agriculture, forestry and fisheries

Agriculture

Over the last century, yields of major commodity crops in the U.S. have increased consistently, typically at rates of 1 to 2%/yr (Troyer, 2004), but there are significant variations across regions and between years. These yield trends are a result of cumulative changes in multiple factors, including technology, fertiliser use,

[2] Eutrophication is a process whereby water bodies, such as lakes, estuaries, or slow-moving streams receive excess nutrients that stimulate excessive plant growth (e.g., algal blooms and nuisance plants weeds).

seed stocks, and management techniques, plus any changes due to climate; the specific impact from any one factor may be positive or negative. In the Midwestern U.S. from 1970 to 2000, corn yield increased 58% and soybean yields increased 20%, with annual weather fluctuations resulting in year-to-year variability (Hicke and Lobell, 2004). Heavy rainfalls reduced the value of the U.S. corn crop by an average of US$3 billion/yr between 1951 and 1998 (Rosenzweig et al., 2002). In the Corn and Wheat Belt of the U.S., yields of corn and soybeans from 1982 to 1998 were negatively impacted by warm temperatures, decreasing 17% for each 1°C of warm-temperature anomaly (Lobell and Asner, 2003). In California, warmer nights have enhanced the production of high-quality wine grapes (Nemani et al., 2001), but additional warming may not result in similar increases. For twelve major crops in California, climate fluctuations over the last 20 years have not had large effects on yield, though they have been a positive factor for oranges and walnuts and a negative for avocados and cotton (Lobell et al., 2006).

North American agriculture has been exposed to many severe weather events during the past decade. More variable weather, coupled with out-migration from rural areas and economic stresses, has increased the vulnerability of the agricultural sector overall, raising concerns about its future capacity to cope with a more variable climate (Senate of Canada, 2003; Wheaton et al., 2005). North American agriculture is, however, dynamic. Adaptation to multiple stresses and opportunities, including changes in markets and weather, is a normal process for the sector. Crop and enterprise diversification, as well as soil and water conservation, are often used to reduce weather-related risks (Wall and Smit, 2005). Recent adaptations by the agricultural sector in North America, including improved water conservation and conservation tillage, are not typically undertaken as single discrete actions, but evolve as a set of decisions that can span several years in a dynamic and changing environment (Smit and Skinner, 2002) that includes changes in public policy (Goodwin, 2003). While there have been attempts to realistically model the dynamics of adaptation to climate change (Easterling et al., 2003), understanding of agriculture's current sensitivity to climate variability and its capacity to cope with climate change remains limited (Tol, 2002).

Forestry

Forest growth appears to be slowly accelerating (at a rate of less than 1%/decade) in regions where tree growth has historically been limited by low temperatures and short growing seasons (Caspersen et al., 2000; McKenzie et al., 2001; Joos et al., 2002; Boisvenue and Running, 2006). In black spruce at the forest-tundra transition in eastern Canada, height growth has been increasing since the 1970s (Gamache and Payette, 2004). Growth is slowing, however, in areas subject to drought. Radial growth of white spruce on dry south-facing slopes in Alaska has decreased over the last 90 years, due to increased drought stress (Barber et al., 2000). In semi-arid forests of the south-western U.S., growth rates have decreased since 1895, correlated with drought linked to warming temperatures (McKenzie et al., 2001). Relationships between tree-ring growth in sub-alpine forests and climate in the Pacific Northwest from 1895 to 1991 had complex topographic influences (Peterson and Peterson, 2001; Peterson et al., 2002). On high elevation north-facing slopes, growth of sub-alpine fir and mountain hemlock was negatively correlated with spring snowpack depth and positively correlated with summer temperatures, indicating growing-season temperature limitations. On lower elevation sites, however, growth was negatively correlated with summer temperature, suggesting water limitations. In Colorado, aspen have advanced into the more cold-tolerant spruce-fir forests over the past 100 years (Elliott and Baker, 2004). The northern range limit of lodgepole pine is advancing into the zone previously dominated by the more cold-tolerant black spruce in the Yukon (Johnstone and Chapin, 2003). A combination of warmer temperatures and insect infestations has resulted in economically significant losses of the forest resource base to spruce bark beetle in both Alaska and the Yukon (ACIA, 2004).

Freshwater fisheries

Most commercial freshwater fishing in North America occurs in rural or remote areas, with indigenous peoples often taking a major role. Recreational inland fisheries are also significant and increasing (DFO-MPO, 2002; DOI, 2002). Ecological sustainability of fish and fisheries productivity is closely tied to temperature and water supply (flows and lake levels). Climate change and variability increasingly have direct and indirect impacts, both of which interact with other pressures on freshwater fisheries, including human development (Schindler, 2001; Chu et al., 2003; Reed and Czech, 2005; Rose, 2005), habitat loss and alteration (including water pollution), biotic homogenisation due to invasions and introductions (Rahel, 2002), and over-exploitation (Post et al., 2002; Cooke and Cowx, 2004). Cold- and cool-water fisheries, especially Salmonids, have been declining as warmer/drier conditions reduce their habitat. The sea-run[3] salmon stocks are in steep decline throughout much of North America (Gallagher and Wood, 2003). Evidence for impacts of recent climate change is rapidly accumulating. Pacific salmon have been appearing in Arctic rivers (Babaluk et al., 2000). Salmonid species have been affected by warming in U.S. streams (O'Neal, 2002). Lake charr in an Ontario lake suffered recruitment[4] failure due to El Niño-linked warm temperatures (Gunn, 2002). Lake Ontario year-class productivity is strongly linked to temperature, with a shift in the 1990s toward warm-water species (Casselman, 2002). Walleye yield in lakes depends on the amount of cool, turbid habitat (Lester et al., 2004). Recent contraction in habitat for walleye in the Bay of Quinte, Lake Ontario was due in part to warming and lower water levels (Chu et al., 2005). Success of adult spawning and survival of the fry (new-borne) of brook trout is closely linked to cold groundwater seeps, which provide preferred temperature refuges for lake-dwelling populations (Borwick et al., 2006). Rates of fish-egg development and mortality increase with temperature rise within species-specific tolerance ranges (Kamler, 2002).

[3] Sea-run: having the habit of ascending a river from the sea, especially to spawn.
[4] Recruitment: the number of new juvenile fish reaching a size large enough to be caught by commercial fishing methods.

14.2.5 Human health

Many human diseases are sensitive to weather, from cardiovascular and respiratory illnesses due to heatwaves or air pollution, to altered transmission of infectious diseases. Synergistic effects of other activities can exacerbate weather exposures (e.g., via the urban heat island effect), requiring cross-sector risk assessment to determine site-specific vulnerability (Patz et al., 2005).

The incidence of infectious diseases transmitted by air varies seasonally and annually, due partly to climate variations. In the early 1990s, California experienced an epidemic of Valley Fever that followed five years of drought (Kolivras and Comrie, 2003). Water-borne disease outbreaks from all causes in the U.S. are distinctly seasonal, clustered in key watersheds, and associated with heavy precipitation (in the U.S. Curriero et al., 2001) or extreme precipitation and warmer temperatures (in Canada, Thomas et al., 2006). Heavy runoff after severe rainfall can also contaminate recreational waters and increase the risk of human illness (Schuster et al., 2005) through higher bacterial counts. This association is strongest at beaches closest to rivers (Dwight et al., 2002).

Food-borne diseases show some relationship with historical temperature trends. In Alberta, ambient temperature is strongly but non-linearly associated with the occurrence of three enteric pathogens, *Salmonella, E. coli* and *Campylobacter* (Fleury et al., 2006).

Many zoonotic diseases[5] are sensitive to climate fluctuations (Charron, 2002). The strain of West Nile virus (WNV) that emerged for the first time in North America during the record hot July 1999 requires warmer temperatures than other strains. The greatest WNV transmissions during the epidemic summers of 2002 to 2004 in the U.S. were linked to above-average temperatures (Reisen et al., 2006). Laboratory studies of virus replication in WNV's main *Culex* mosquito vector show high levels of virus at warmer temperatures (Dohm and Turell, 2001; Dohm et al., 2002). Bird migratory pathways and WNV's recent advance westward across the U.S. and Canada are key factors in WNV and must be considered in future assessments of the role of temperature in WNV dynamics. A virus closely related to WNV, Saint Louis encephalitis, tends to appear during hot, dry La Niña years, when conditions facilitate transmission by reducing the extrinsic incubation period[6] (Cayan et al., 2003).

Lyme disease is a prevalent tick-borne disease in North America for which there is new evidence of an association with temperature (Ogden et al., 2004) and precipitation (McCabe and Bunnell, 2004). In the field, temperature and vapour pressure contribute to maintaining populations of the tick *Ixodes scapularis* which, in the U.S., is the micro-organism's secondary host. A monthly average minimum temperature above -7°C is required for tick survival (Brownstein et al., 2003).

Exposure to both extreme hot and cold weather is associated with increased morbidity and mortality, compared to an intermediate 'comfortable' temperature range (Curriero et al., 2002). Across 12 U.S. cities, hot temperatures have been associated with increased hospital admissions for cardiovascular disease (Schwartz et al., 2004a). Emergency hospital admissions have been directly related to extreme heat in Toronto (Dolney and Sheridan, 2006). Heat-response plans and heat early warning systems (EWS) can save lives (Ebi et al., 2004). After the 1995 heatwave, the city of Milwaukee initiated an 'extreme heat conditions plan' that almost halved heat-related morbidity and mortality (Weisskopf et al., 2002). Currently, over two dozen cities worldwide have warning systems focused on monitoring for dangerous air masses (Sheridan and Kalkstein, 2004).

14.2.6 Human settlements

Economic base of resource-dependent communities

Among the most climate-sensitive North American communities are those of indigenous populations dependent on one or a few natural resources. About 1.2 million (60%) of the U.S. tribal members live on or near reservations, and many pursue lifestyles with a mix of traditional subsistence activities and wage labour (Houser et al., 2001). Many reservation economies and budgets of indigenous governments depend heavily on agriculture, forest products and tourism (NAST, 2001). A 1993 hantavirus outbreak related indirectly to heavy rainfall led to a significant reduction in tourist visits to the American South-west (NAST, 2001). Many indigenous communities in northern Canada and Alaska are already experiencing constraints on lifestyles and economic activity from less reliable sea and lake ice (for travelling, hunting, fishing and whaling), loss of forest resources from insect damage, stress on caribou, and more exposed coastal infrastructure from diminishing sea ice (NAST, 2001; CCME, 2003; ACIA, 2005). Many rural settlements in North America, particularly those dependent on a narrow resource base, such as fishing or forestry, have been seriously affected by recent declines in the resource base, caused by a number of factors (CDLI, 1996). However, not all communities have suffered, as some Alaskan fishing communities have benefited from rising regional abundance of selected salmon stocks since the mid-1970s (Eggers, 2006).

Infrastructure and extreme events

About 80% of North Americans live in urban areas (Census Bureau, 2000; Statistics Canada, 2001b). North American cities, while diverse in size, function, climate and other factors, are largely shielded from the natural environment by technical systems. The devastating effects of hurricanes Ivan in 2004 and Katrina, Rita and Wilma in 2005, however, illustrate the vulnerability of North American infrastructure and urban systems that were either not designed or not maintained to adequate safety margins. When protective systems fail, impacts can be widespread and multi-dimensional (see Chapter 7, Boxes 7.2 and 7.4). Disproportionate impacts of Hurricane Katrina on the poor, infirm, elderly, and other dependent populations were amplified by inadequate public sector development and/or

[5] Zoonotic diseases: diseases caused by infectious agents that can be transmitted between (or are shared by) animals and humans.

[6] Extrinsic incubation period: the interval between the acquisition of an infectious agent by a vector and the vector's ability to transmit the agent to other hosts.

execution of evacuation and emergency services plans (Select Bipartisan Committee, 2006).

Costs of weather-related natural disasters in North America rose at the end of the 20th century, mainly as a result of the increasing value of infrastructure at risk (Changnon, 2003, 2005). Key factors in the increase in exposure include rising wealth, demographic shifts to coastal areas, urbanisation in storm-prone areas, and ageing infrastructure, combined with substandard structures and inadequate building codes (Easterling et al., 2000; Balling and Cerveny, 2003; Changnon, 2003, 2005). Trends in the number and intensity of extreme events in North America are variable, with many (e.g., hail events, tornadoes, severe windstorms, winter storms) holding steady or even decreasing (Kunkel et al., 1999; McCabe et al., 2001; Balling and Cerveny, 2003; Changnon, 2003; Trenberth et al., 2007: Section 3.8.4.2).

North America very likely will continue to suffer serious losses of life and property simply due to growth in property values and numbers of people at risk (very high confidence) (Pielke Jr., 2005; Pielke et al., 2005). Of the US$19 trillion value of all insured residential and commercial property in the U.S. states exposed to North Atlantic hurricanes, US$7.2 trillion (41%) is located in coastal counties. This economic value includes 79% of the property in Florida, 63% of the property in New York, and 61% of the property in Connecticut (AIR, 2002). Cumulative decadal hurricane intensity in the U.S. has risen in the last 25 years, following a peak in the mid 20th century and a later decline (Figure 14.1e). North American mortality (deaths and death rates) from hurricanes, tornadoes, floods and lightning have generally declined since the beginning of the 20th century, due largely to improved warning systems (Goklany, 2006). Mortality was dominated by three storms where the warning/evacuation system did not lead to timely evacuation: Galveston in 1900, Okeechobee in 1926, and Katrina in 2005.

Flood hazards are not limited to the coastal zone. River basins with a history of major floods (e.g., the Sacramento (Miller, 2003), the Fraser (Lemmen and Warren, 2004), the Red River (Simonovic and Li, 2004) and the upper Mississippi (Allen et al., 2003)) illustrate the sensitivity of riverine flooding to extreme events and highlight the critical importance of infrastructure design standards, land-use planning and weather/flood forecasts.

14.2.7 Tourism and recreation

The U.S. and Canada rank among the top ten nations for international tourism receipts (US$112 billion and US$16 billion, respectively) with domestic tourism and outdoor recreation markets that are several times larger (World Tourism Organization, 2002; Southwick Associates, 2006). Climate variability affects many segments of this growing economic sector. For example, wildfires in Colorado (2002) and British Columbia (2003) caused tens of millions of dollars in tourism losses by reducing visitation and destroying infrastructure (Associated Press, 2002; Butler, 2002; BC Stats, 2003). Similar economic losses were caused by drought-affected water levels in rivers and reservoirs in the western U.S. and parts of the Great Lakes (Fisheries and Oceans Canada, 2000; Kesmodel, 2002; Allen, 2003). The ten-day closure and clean-up following Hurricane Georges (September 1998) resulted in tourism revenue losses of approximately US$32 million in the Florida Keys (EPA, 1999). While the North American tourism industry acknowledges the important influence of climate, its impacts have not been analysed comprehensively (Scott et al., 2006).

14.2.8 Energy, industry and transportation

North American industry, energy supply and transportation networks are sensitive to weather extremes that exceed their safety margins. Costs of these impacts can be high. For example, power outages in the U.S. cost the economy US$30 billion to 130 billion annually (EPRI, 2003; LaCommare and Eto, 2004). The hurricanes crossing Florida in the summer of 2004 resulted in direct system restoration costs of US$1.4 billion to the four Florida public utilities involved (EEI, 2005). From 1994 to 2004, fourteen U.S. utilities experienced 81 other major storms, which cost an average of US$49 million/storm, with the highest single storm impact of US$890 million (EEI, 2005).

Although it was not triggered specifically by the concurrent hot weather, the 2003 summer outage in north-eastern U.S. and south-eastern Canada illustrates costs to North American society that result from large-scale power interruptions during periods of high demand. Over 50 million people were without power, resulting in US$180 million in insured losses and up to US$10 billion in total losses (Fletcher, 2004). Business interruptions were particularly significant, with costs of over US$250,000/hr incurred by the top quartile of recently surveyed companies (RM, 2003).

The impacts of Hurricanes Katrina, Rita and Wilma in 2005 and Ivan in 2004 demonstrated that the Gulf of Mexico offshore oil and natural gas platforms and pipelines, petroleum refineries, and supporting infrastructure can be seriously harmed by major hurricanes, which can produce national-level impacts, and require recovery times stretching to months or longer (Business Week, 2005; EEA, 2005; EIA, 2005a; Levitan and Associates Inc., 2005; RMS, 2005b; Swiss Re, 2005b, c, d, e).

Hydropower production is known to be sensitive to total runoff, to its timing, and to reservoir levels. For example, during the 1990s, Great Lakes levels fell as a result of a lengthy drought, and in 1999 hydropower production was down significantly both at Niagara and Sault St. Marie (CCME, 2003).

14.3 Assumptions about future trends

14.3.1 Climate

Recent climate model simulations (Ruosteenoja et al., 2003) indicate that by the 2010 to 2039 time slice, year-round temperatures across North America will be outside the range of present-day natural variability, based on 1000 year Atmosphere-Ocean General Circulation Model (AOGCM) simulations with either the CGCM2 or HadCM3 climate models. For most combinations of model, scenario, season and region, warming in the 2010 to 2039 time slice will be in the range of 1 to 3°C.

Late in the century, projected annual warming is likely to be 2 to 3°C across the western, southern, and eastern continental edges, but more than 5°C at high latitudes (Christensen et al., 2007: Section 11.5.3.1). The projected warming is greatest in winter at high latitudes and greatest in the summer in the south-west U.S. Warm extremes across North America are projected to become both more frequent and longer (Christensen et al., 2007: Section 11.5.3.3).

Annual-mean precipitation is projected to decrease in the south-west of the U.S. but increase over the rest of the continent (Christensen et al., 2007: Section 11.5.3.2). Increases in precipitation in Canada are projected to be in the range of +20% for the annual mean and +30% for the winter. Some studies project widespread increases in extreme precipitation (Christensen et al., 2007: Section 11.5.3.3), with greater risks of not only flooding from intense precipitation, but also droughts from greater temporal variability in precipitation. In general, projected changes in precipitation extremes are larger than changes in mean precipitation (Meehl et al., 2007: Section 10.3.6.1)

Future trends in hurricane frequency and intensity remain very uncertain. Experiments with climate models with sufficient resolution to depict some aspects of individual hurricanes tend to project some increases in both peak wind speeds and precipitation intensities (Meehl et al., 2007: Section 10.3.6.3). The pattern is clearer for extra-tropical storms, which are likely to become more intense, but perhaps less frequent, leading to increased extreme wave heights in the mid-latitudes (Meehl et al., 2007: Section 10.3.6.4).

El Niño events are associated with increased precipitation and severe storms in some regions, such as the south-east U.S., and higher precipitation in the Great Basin of the western U.S., but warmer temperatures and decreased precipitation in other areas such as the Pacific Northwest, western Canada, and parts of Alaska (Ropelewski and Halpert, 1986; Shabbar et al., 1997). Recent analyses indicate no consistent future trends in El Niño amplitude or frequency (Meehl et al., 2007: Section 10.3.5.4).

14.3.2 Social, economic and institutional context

Canada and the U.S. have developed economies with per capita gross domestic product (GDP) in 2005 of US$31,572 and US$37,371, respectively (UNECE, 2005a,b). Future population growth is likely to be dominated by immigration (Campbell, 1996). Interests of indigenous peoples are important in both Canada and the U.S., especially in relation to questions of land management. With ageing populations, the costs of health care are likely to climb over several decades (Burleton, 2002).

Major parts of the economies of Canada and the U.S. are directly sensitive to climate, including the massive agricultural (2005 value US$316 billion) (Economic Research Service, 2006; Statistics Canada, 2006), transportation (2004 value US$510 billion) (Bureau of Transportation Statistics, 2006; Industry Canada, 2006) and tourism sectors (see Section 14.2.4, 14.2.7 and 14.2.8). Although many activities have limited direct sensitivity to climate (Nordhaus, 2006), the potential realm of climate-sensitive activities expands with increasing evidence that storms, floods, or droughts increase in frequency or intensity with climate change (Christensen et al., 2007: Section 11.5.3.3 and Meehl et al., 2007: Sections 10.3.6.1 and 10.3.6.2).

The economies of Canada and the U.S. have large private and public sectors, with strong emphasis on free market mechanisms and the philosophy of private ownership. If strong trends toward globalisation in the last several decades continue through the 21st century, it is likely that the means of production, markets, and ownership will be predominantly international, with policies and governance increasingly designed for the international marketplace (Stiglitz, 2002).

14.4 Key future impacts and vulnerabilities

14.4.1 Freshwater resources

Freshwater resources will be affected by climate change across Canada and the U.S., but the nature of the vulnerabilities varies from region to region (NAST, 2001; Environment Canada, 2004; Lemmen and Warren, 2004). In certain regions including the Colorado River, Columbia River and Ogallala Aquifer, surface and/or groundwater resources are intensively used for often competing agricultural, municipal, industrial and ecological needs, increasing potential vulnerability to future changes in timing and availability of water (see Box 14.2).

Surface water

Simulated annual water yield in basins varies by region, General Circulation Model (GCM) or Regional Climate Model (RCM) scenario (Stonefelt et al., 2000; Fontaine et al., 2001; Stone et al., 2001; Rosenberg et al., 2003; Jha et al., 2004; Shushama et al., 2006), and the resolution of the climate model (Stone et al., 2003). Higher evaporation related to warming tends to offset the effects of more precipitation, while magnifying the effects of less precipitation (Stonefelt et al., 2000; Fontaine et al., 2001).

Warming, and changes in the form, timing and amount of precipitation, will very likely lead to earlier melting and significant reductions in snowpack in the western mountains by the middle of the 21st century (high confidence) (Loukas et al., 2002; Leung and Qian, 2003; Miller et al., 2003; Mote et al., 2003; Hayhoe et al., 2004). In projections for mountain snowmelt-dominated watersheds, snowmelt runoff advances, winter and early spring flows increase (raising flooding potential), and summer flows decrease substantially (Kim et al., 2002; Loukas et al., 2002; Snyder et al., 2002; Leung and Qian, 2003; Miller et al., 2003; Mote et al., 2003; Christensen et al., 2004; Merritt et al., 2005). Over-allocated water systems of the western U.S. and Canada, such as the Columbia River, that rely on capturing snowmelt runoff, will be especially vulnerable (see Box 14.2).

Lower water levels in the Great Lakes are likely to influence many sectors, with multi-dimensional, interacting impacts (Figure 14.2) (high confidence). Many, but not all, assessments project lower net basin supplies and water levels for the Great Lakes – St. Lawrence Basin (Mortsch et al., 2000; Quinn and Lofgren, 2000; Lofgren et al., 2002; Croley, 2003). In addition

Box 14.2. Climate change adds challenges to managing the Columbia River system

Current management of water in the Columbia River basin involves balancing complex, often competing, demands for hydropower, navigation, flood control, irrigation, municipal uses, and maintenance of several populations of threatened and endangered species (e.g., salmon). Current and projected needs for these uses over-commit existing supplies. Water management in the basin operates in a complex institutional setting, involving two sovereign nations (Columbia River Treaty, ratified in 1964), aboriginal populations with defined treaty rights ('Boldt decision' in U.S. *vs.* Washington in 1974), and numerous federal, state, provincial and local government agencies (Miles et al., 2000; Hamlet, 2003). Pollution (mainly non-point source) is an important issue in many tributaries. The first-in-time first-in-right provisions of western water law in the U.S. portion of the basin complicate management and reduce water available to junior water users (Gray, 1999; Scott et al., 2004). Complexities extend to different jurisdictional responsibilities when flows are high and when they are low, or when protected species are in tributaries, the main stem or ocean (Miles et al., 2000; Mote et al., 2003).

With climate change, projected annual Columbia River flow changes relatively little, but seasonal flows shift markedly toward larger winter and spring flows and smaller summer and autumn flows (Hamlet and Lettenmaier, 1999; Mote et al., 1999). These changes in flows will likely coincide with increased water demand, principally from regional growth but also induced by climate change. Loss of water availability in summer would exacerbate conflicts, already apparent in low-flow years, over water (Miles et al. 2000). Climate change is also projected to impact urban water supplies within the basin. For example, a 2°C warming projected for the 2040s would increase demand for water in Portland, Oregon by 5.7 million m^3/yr with an additional demand of 20.8 million m^3/yr due to population growth, while decreasing supply by 4.9 million m^3/yr (Mote et al., 2003). Long-lead climate forecasts are increasingly considered in the management of the river but in a limited way (Hamlet et al., 2002; Lettenmaier and Hamlet, 2003; Gamble et al., 2004; Payne et al., 2004). Each of 43 sub-basins of the system has its own sub-basin management plan for fish and wildlife, none of which comprehensively addresses reduced summertime flows under climate change (ISRP/ISAB, 2004).

The challenges of managing water in the Columbia River basin will likely expand with climate change due to changes in snowpack and seasonal flows (Miles et al., 2000; Parson et al., 2001b; Cohen et al., 2003). The ability of managers to meet operating goals (reliability) will likely drop substantially under climate change (as projected by the HadCM2 and ECHAM4/OPYC3 AOGCMs under the IPCC IS92a emissions scenario for the 2020s and 2090s) (Hamlet and Lettenmaier, 1999). Reliability losses are projected to reach 25% by the end of the 21st century (Mote et al., 1999) and interact with operational rule requirements. For example, 'fish-first' rules would reduce firm power reliability by 10% under present climate and 17% in years during the warm phase of the Pacific Decadal Oscillation. Adaptive measures have the potential to moderate the impact of the decrease in April snowpack, but lead to 10 to 20% losses of firm hydropower and lower than current summer flows for fish (Payne et al., 2004). Integration of climate change adaptation into regional planning processes is in the early stages of development (Cohen et al., 2006).

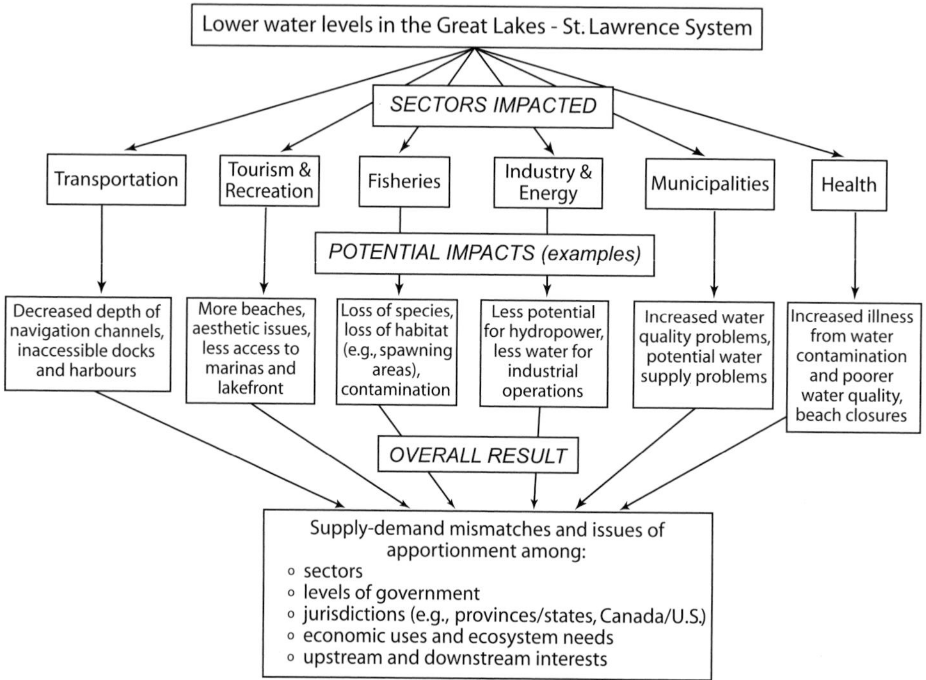

Figure 14.2. *Interconnected impacts of lower water levels in the Great Lakes - St Lawrence system (modified from Lemmen and Warren, 2004).*

to differences due to climate scenarios, uncertainties include atmosphere-lake interactions (Wetherald and Manabe, 2002; Kutzbach et al., 2005). Adapting infrastructure and dredging to cope with altered water levels would entail a range of costs (Changnon, 1993; Schwartz et al., 2004b). Adaptations sufficient to maintain commercial navigation on the St. Lawrence River could range from minimal adjustments to costly, extensive structural changes (St. Lawrence River-Lake Ontario Plan of Study Team, 1999; D'Arcy et al., 2005). There have been controversies in the Great Lakes region over diversions of water, particularly at Chicago, to address water quality, navigation, water demand and drought mitigation outside the region. Climate change will exacerbate these issues and create new challenges for bi-national co-operation (very high confidence) (Changnon and Glantz, 1996; Koshida et al., 2005).

Groundwater

With climate change, availability of groundwater is likely to be influenced by withdrawals (reflecting development, demand and availability of other sources) and recharge (determined by temperature, timing and amount of precipitation, and surface water interactions) (medium confidence) (Rivera et al., 2004). Simulated annual groundwater base flows and aquifer levels respond to temperature, precipitation and pumping – decreasing in scenarios that are drier or have higher pumping and increasing in a wetter scenario. In some cases there are base flow shifts - increasing in winter and decreasing in spring and early summer (Kirshen, 2002; Croley and Luukkonen, 2003; Piggott et al., 2003). For aquifers in alluvial valleys of south-central British Columbia, temperature and precipitation scenarios have less impact on groundwater recharge and levels than do projected changes in river stage[7] (Allen et al., 2004a,b).

Heavily utilised groundwater-based systems in the southwest U.S. are likely to experience additional stress from climate change that leads to decreased recharge (high confidence). Simulations of the Edwards aquifer in Texas under average recharge project lower or ceased flows from springs, water shortages, and considerable negative environmental impacts (Loáiciga, 2000; Loáiciga et al., 2000). Regional welfare losses associated with projected flow reductions (10 to 24%) range from US$2.2 million to 6.8 million/yr, with decreased net agricultural income as a consequence of water allocation shifting to municipal and industrial uses (Chen et al., 2001). In the Ogallala aquifer region, projected natural groundwater recharge decreases more than 20% in all simulations with warming of 2.5°C or greater (based on outputs from the GISS, UKTR and BMRC AOGCMs, with three atmospheric concentrations of CO_2: 365, 560 and 750 ppm) (Rosenberg et al., 1999).

Water quality

Simulated future surface and bottom water temperatures of lakes, reservoirs, rivers, and estuaries throughout North America consistently increase from 2 to 7°C (based on $2\times CO_2$ and IS92a scenarios) (Fang and Stefan, 1999; Hostetler and Small, 1999; Nicholls, 1999; Stefan and Fang, 1999; Lehman, 2002; Gooseff et al., 2005), with summer surface temperatures exceeding 30°C in Midwestern and southern lakes and reservoirs (Hostetler and Small, 1999). Warming is likely to extend and intensify summer thermal stratification, contributing to oxygen depletion. A shorter ice-cover period in shallow northern lakes could reduce winter fish kills caused by low oxygen (Fang and Stefan, 1999; Stefan and Fang, 1999; Lehman, 2002). Higher stream temperatures affect fish access, survival and spawning (e.g., west coast salmon) (Morrison et al., 2002).

Climate change is likely to make it more difficult to achieve existing water quality goals (high confidence). For the Midwest, simulated low flows used to develop pollutant discharge limits (Total Maximum Daily Loads) decrease over 60% with a 25% decrease in mean precipitation, reaching up to 100% with the incorporation of irrigation demands (Eheart et al., 1999). Restoration of beneficial uses (e.g., to address habitat loss, eutrophication, beach closures) under the Great Lakes Water Quality agreement will likely be vulnerable to declines in water levels, warmer water temperatures, and more intense precipitation (Mortsch et al., 2003). Based on simulations, phosphorus remediation targets for the Bay of Quinte (Lake Ontario) and surrounding watershed could be compromised as 3 to 4°C warmer water temperatures contribute to 77 to 98% increases in summer phosphorus concentrations in the bay (Nicholls, 1999), and as changes in precipitation, streamflow and erosion lead to increases in average phosphorus concentrations in streams of 25 to 35% (Walker, 2001). Decreases in snow cover and more winter rain on bare soil are likely to lengthen the erosion season and enhance erosion, increasing the potential for water quality impacts in agricultural areas (Atkinson et al., 1999; Walker, 2001; Soil and Water Conservation Society, 2003). Soil management practices (e.g., crop residue, no-till) in the Cornbelt may not provide sufficient erosion protection against future intense precipitation and associated runoff (Hatfield and Pruger, 2004; Nearing et al., 2004).

14.4.2 Ecosystems

Several simulations (Cox et al., 2000; Berthelot et al., 2002; Fung et al., 2005) indicate that, over the 21st century, warming will lengthen growing seasons, sustaining forest carbon sinks in North America despite some decreased sink strength resulting from greater water limitations in western forests and higher respiration in the tropics (medium confidence). Impacts on ecosystem structure and function may be amplified by changes in extreme meteorological events and increased disturbance frequencies. Ecosystem disturbances, caused either by humans or by natural events, accelerate both loss of native species and invasion of exotics (Sala et al., 2000).

Primary production

At high latitudes, several models simulate increased NPP as a result of expansion of forests into the tundra and longer growing seasons (Berthelot et al., 2002). In the mid-latitudes, simulated changes in NPP are variable, depending on whether there is sufficient enhancement of precipitation to offset

[7] River stage: water height relative to a set point.

increased evapotranspiration in a warmer climate (Bachelet et al., 2001; Berthelot et al., 2002; Gerber et al., 2004; Woodward and Lomas, 2004). Bachelet et al. (2001) project the areal extent of drought-limited ecosystems to increase by 11%/°C warming in the continental U.S. By the end of the 21st century, ecosystems in the north-east and south-east U.S. will likely become carbon sources, while the western U.S. remains a carbon sink (Bachelet et al., 2004).

Overall forest growth in North America will likely increase modestly (10-20%) as a result of extended growing seasons and elevated CO_2 over the next century (Morgan et al., 2001), but with important spatial and temporal variations (medium confidence). Growth of white spruce in Québec will be enhanced by a 1°C temperature increase but depressed with a 4°C increase (Andalo et al., 2005). A 2°C temperature increase in the Olympic Mountains (U.S.) would cause dominant tree species to shift upward in elevation by 300 to 600m, causing temperate species to replace sub-alpine species over 300 to 500 years (Zolbrod and Peterson, 1999). For widespread species such as lodgepole pine, a 3°C temperature increase would increase growth in the northern part of its range, decrease growth in the middle, and decimate southern forests (Rehfeldt et al., 2001).

Population and community dynamics

For many amphibians, whose production of eggs and migration to breeding ponds is intimately tied to temperature and moisture, mismatches between breeding phenology and pond drying can lead to reproductive failure (Beebee, 1995). Differential responses among species in arrival or persistence in ponds will likely lead to changes in community composition and nutrient flow in ponds (Wilbur, 1997). Changes in plant species composition in response to climate change can facilitate other disturbances, including fire (Smith et al., 2000) and biological invasion (Zavaleta and Hulvey, 2004). Bioclimate modelling based on output from five GCMs suggests that, over the next century, vertebrate and tree species richness will decrease in most parts of the conterminous U.S., even though long-term trends (over millennia) ultimately favour increased richness in some taxa and locations (Currie, 2001). Based on relationships between habitat area and biodiversity, 15 to 37% of plant and animal species in a global sample are likely to be 'committed to extinction' by 2050, although actual extinctions will be strongly influenced by human forces and could take centuries (Thomas et al., 2004).

14.4.3 Coastal regions

Added stress from rapid coastal development, including an additional 25 million people in the coastal U.S. over the next 25 years, will reduce the effectiveness of natural protective features, leading to impaired resilience. As property values and investment continue to rise, coastal vulnerability tends to increase on a broad scale (Pielke Jr. and Landsea, 1999; Heinz Center, 2000), with a sensitivity that depends on the commitment to and flexibility of adaptation measures. Disproportionate impacts due to socio-economic status are likely to be exacerbated by rising sea levels and storm severity (Wu et al., 2002; Kleinosky et al., 2006).

Sea-level rise has accelerated in eastern North America since the late 19th century (Donnelly et al., 2004) and further acceleration is expected (high confidence). For The IPCC Special Report on Emissions Scenarios (SRES, Nakićenović and Swart, 2000) scenario A1B, global mean sea level is projected to rise by 0.35 ± 0.12 m from the 1980 to 1999 period to the 2090 to 2099 period (Meehl et al., 2007: Section 10.6.5). Spatial variability of sea-level rise has become better defined since the TAR (Church et al., 2004) and the ensemble mean for A1B shows values close to the global mean along most North American coasts, with slightly higher rates in eastern Canada and western Alaska, and stronger positive anomalies in the Arctic (Meehl et al., 2007: Figure 10.32). Vertical land motion will decrease (uplift) or increase (subsidence) the relative sea-level rise at any site (Douglas and Peltier, 2002).

Superimposed on accelerated sea-level rise, the present storm and wave climatology and storm-surge frequency distributions lead to forecasts of more severe coastal flooding and erosion hazards. The water-level probability distribution is shifted upward, giving higher potential flood levels and more frequent flooding at levels rarely experienced today (very high confidence) (Zhang et al., 2000; Forbes et al., 2004). If coastal systems, including sediment supply, remain otherwise unchanged, higher sea levels are likely to be correlated with accelerated coastal erosion (Hansom, 2001; Cowell et al., 2003).

Up to 21% of the remaining coastal wetlands in the U.S. mid-Atlantic region are potentially at risk of inundation between 2000 and 2100 (IS92a emissions scenario) (Najjar et al., 2000). Rates of coastal wetland loss, in Chesapeake Bay and elsewhere (Kennish, 2002), will increase with accelerated sea-level rise, in part due to 'coastal squeeze' (high confidence). Salt-marsh biodiversity is likely to be diminished in north-eastern marshes through expansion of cordgrass (*Spartina alterniflora*) at the expense of high-marsh species (Donnelly and Bertness, 2001). Many salt marshes in less developed areas have some potential to keep pace with sea-level rise (to some limit) through vertical accretion (Morris et al., 2002; Chmura et al., 2003; Chmura and Hung, 2004). Where rapid subsidence increases rates of relative sea-level rise, however, as in the Mississippi Delta, even heavy sediment loads cannot compensate for inundation losses (Rybczyk and Cahoon, 2002).

Potentially more intense storms and possible changes in El Niño (Meehl et al., 2007: Sections 10.3.5.4 and 10.3.6.3) are likely to result in more coastal instability (medium confidence) (see Section 14.3.1) (Scavia et al., 2002; Forbes et al., 2004; Emanuel, 2005). Damage costs from coastal storm events (storm surge, waves, wind, ice encroachment) and other factors (such as freeze-thaw) have increased substantially in recent decades (Zhang et al., 2000; Bernatchez and Dubois, 2004) and are expected to continue rising (high confidence). Higher sea levels in combination with storm surges will cause widespread problems for transportation along the Gulf and Atlantic coasts (Titus, 2002). More winters with reduced sea ice in the Gulf of St. Lawrence, resulting in more open water during the winter storm season, will lead to an increase in the average number of storm-wave events per year, further accelerating coastal erosion (medium confidence) (Forbes et al., 2004).

14.4.4 Agriculture, forestry and fisheries

Agriculture

Research since the TAR supports the conclusion that moderate climate change will likely increase yields of North American rain-fed agriculture, but with smaller increases and more spatial variability than in earlier estimates (high confidence) (Reilly, 2002). Most studies project likely climate-related yield increases of 5 to 20% over the first decades of the century, with the overall positive effects of climate persisting through much or all of the 21st century. This pattern emerges from recent assessments for corn, rice, sorghum, soybean, wheat, common forages, cotton and some fruits (Adams et al., 2003; Polsky et al., 2003; Rosenberg et al., 2003; Tsvetsinskaya et al., 2003; Antle et al., 2004; Thomson et al., 2005b), including irrigated grains (Thomson et al., 2005b). Increased climate sensitivity is anticipated in the south-eastern U.S. and in the U.S. Cornbelt (Carbone et al., 2003), but not in the Great Plains (Mearns et al., 2003). Crops that are currently near climate thresholds (e.g., wine grapes in California) are likely to suffer decreases in yields, quality, or both, with even modest warming (medium confidence) (Hayhoe et al., 2004; White et al., 2006).

Recent integrated assessment model studies explored the interacting impacts of climate and economic factors on agriculture, water resources and biome boundaries in the conterminous U.S. (Edmonds and Rosenberg, 2005; Izaurralde et al., 2005; Rosenberg and Edmonds, 2005; Sands and Edmonds, 2005; Smith et al., 2005; Thomson et al., 2005a,b,c,d), concluding that scenarios with decreased precipitation create important challenges, restricting the availability of water for irrigation and at the same time increasing water demand for irrigated agriculture and urban and ecological uses.

The critical importance of specific agro-climatic events (e.g., last frost) introduces uncertainty in future projections (Mearns et al., 2003), as does continued debate about the CO_2 sensitivity of crop growth (Long et al., 2005). Climate change is expected to improve the climate for fruit production in the Great Lakes region and eastern Canada but with risks of early season frost and damaging winter thaws (Bélanger et al., 2002; Winkler et al., 2002). For U.S. soybean yield, adjusting the planting date can reduce the negative effects of late season heat stress and can more than compensate for direct effects of climate change (Southworth et al., 2002).

Vulnerability of North American agriculture to climatic change is multi-dimensional and is determined by interactions among pre-existing conditions, indirect stresses stemming from climate change (e.g., changes in pest competition, water availability), and the sector's capacity to cope with multiple, interacting factors, including economic competition from other regions as well as advances in crop cultivars and farm management (Parson et al., 2003). Water access is the major factor limiting agriculture in south-east Arizona, but farmers in the region perceive that technologies and adaptations such as crop insurance have recently decreased vulnerability (Vasquez-Leon et al., 2002). Areas with marginal financial and resource endowments (e.g., the U.S. northern plains) are especially vulnerable to climate change (Antle et al., 2004). Unsustainable land-use practices will tend to increase the vulnerability of agriculture in the U.S. Great Plains to climate change (Polsky and Easterling, 2001).

Forestry

Across North America, impacts of climate change on commercial forestry potential are likely to be sensitive to changes in disturbances (Dale et al., 2001) from insects (Gan, 2004), diseases (Woods et al., 2005) and wildfires (high confidence) (see Box 14.1). Warmer summer temperatures are projected to extend the annual window of high fire ignition risk by 10-30%, and could result in increased area burned of 74-118% in Canada by 2100 (Brown et al., 2004; Flannigan et al., 2004). In the absence of dramatic increases in disturbance, effects of climate change on the potential for commercial harvest in one study for the 2040s ranged from mixed for a low emissions scenario (the EPPA LLH emissions scenario) to positive for a high emissions scenario (the EPPA HHL emissions scenario) (Perez-Garcia et al., 2002). Scenarios with increased harvests tend to lead to lower prices and, as a consequence, reduced harvests, especially in Canada (Perez-Garcia et al., 2002; Sohngen and Sedjo, 2005). The tendency for North American producers to suffer losses increases if climate change is accompanied by increased disturbance, with simulated losses averaging US$1 billion to 2 billion/yr over the 21st century (Sohngen and Sedjo, 2005). Increased tropospheric ozone could cause further decreases in tree growth (Karnosky et al., 2005). Risks of losses from Southern pine beetle likely depend on the seasonality of warming, with winter and spring warming leading to the greatest damage (Gan, 2004).

Warmer winters with more sporadic freezing and thawing are likely to increase erosion and landslides on forest roads, and reduce access for winter harvesting (Spittlehouse and Stewart, 2003).

Freshwater fisheries

Cold-water fisheries will likely be negatively affected by climate change; warm-water fisheries will generally gain; and the results for cool-water fisheries will be mixed, with gains in the northern and losses in the southern portions of ranges (high confidence) (Stefan et al., 2001; Rahel, 2002; Shuter et al., 2002; Mohseni et al., 2003; Fang et al., 2004). Salmonids, which prefer cold, clear water, are likely to experience the most negative impacts (Gallagher and Wood, 2003). Arctic freshwaters will likely be most affected, as they will experience the greatest warming (Wrona et al., 2005). Many warm-water and cool-water species will shift their ranges northward or to higher altitudes (Clark et al., 2001; Mohseni et al., 2003). In the continental U.S., cold-water species will likely disappear from all but the deeper lakes, cool-water species will be lost mainly from shallow lakes, and warm-water species will thrive except in the far south, where temperatures in shallow lakes will exceed survival thresholds (see Section 14.4.1) (Stefan et al., 2001). Species already listed as threatened will face increased risk of extinction (Chu et al., 2005), with pressures from climate exacerbated by the expansion of predatory species like smallmouth bass (Jackson and Mandrak, 2002). In Lake Erie, larval recruitment of river-spawning walleye will depend on temperature and flow changes, but lake-spawning stocks will likely decline due to the effects of warming and lower lake levels (Jones et al., 2006). Thermal habitat suitable for yellow perch will expand, while that for lake trout will contract (Jansen and Hesslein, 2004). While temperature increases may favour warm-water fishes like smallmouth bass, changes in water supply and flow regimes seem likely to have negative effects (Peterson and Kwak, 1999).

14.4.5 Human health

Risks from climate change to human health will be strongly modulated by changes in health care infrastructure, technology, and accessibility as well as ageing of the population, and patterns of immigration and/or emigration (UNPD, 2005). Across North America, the population over the age of 65 will increase slowly to 2010, and then grow dramatically as the Baby Boomers join the ranks of the elderly – the segment of the population most at risk of dying in heatwaves.

Heatwaves and health

Severe heatwaves, characterised by stagnant, warm air masses and consecutive nights with high minimum temperatures, will intensify in magnitude and duration over the portions of the U.S. and Canada where they already occur (high confidence) (Cheng et al., 2005). Late in the century, Chicago is projected to experience 25% more frequent heatwaves annually (using the PCM AOGCM with a business-as-usual emissions scenario, for the period 2080 to 2099) (Meehl and Tebaldi, 2004), and the projected number of heatwave days in Los Angeles increases from 12 to 44-95 (based on PCM and HadCM3 for the A1FI and B1 scenarios, for the 2070 to 2099 period) (Hayhoe et al., 2004).

Air pollution

Surface ozone concentration may increase with a warmer climate. Ozone damages lung tissue, causing particular problems for people with asthma and other lung diseases. Even modest exposure to ozone may encourage the development of asthma in children (McConnell et al., 2002; Gent et al., 2003). Ozone and non-volatile secondary particulate matter generally increase at higher temperatures, due to increased gas-phase reaction rates (Aw and Kleeman, 2002). Many species of trees emit volatile organic compounds (VOC) such as isoprene, a precursor of ozone (Lerdau and Keller, 1998), at rates that increase rapidly with temperature (Guenther, 2002).

For the 2050s, daily average ozone levels are projected to increase by 3.7 ppb across the eastern U.S. (based on the GISS/MM5 AOGCM and the SRES A2 emissions scenario), with the cities most polluted today experiencing the greatest increase in ozone pollution (Hogrefe et al., 2004). One-hour maximum ozone follows a similar pattern, with the number of summer days exceeding the 8-hour regulatory U.S. standard projected to increase by 68% (Bell et al., 2007). Assuming constant population and dose-response characteristics, ozone-related deaths from climate change increase by approximately 4.5% from the 1990s to the 2050s (Knowlton et al., 2004; Bell et al., 2007). The large potential population exposed to outdoor air pollution translates this small relative risk into a substantial attributable health risk.

Pollen

Pollen, another air contaminant, is likely to increase with elevated temperature and atmospheric CO_2 concentrations. A doubling of the atmospheric CO_2 concentration stimulated ragweed-pollen production by over 50% (Wayne et al., 2002). Ragweed grew faster, flowered earlier and produced significantly greater above-ground biomass and pollen at urban than at rural locations (Ziska et al., 2003).

Lyme disease

The northern boundary of tick-borne Lyme disease is limited by cold temperature effects on the tick, *Ixodes scapularis*. The northern range limit for this tick could shift north by 200 km by the 2020s, and 1000 km by the 2080s (based on projections from the CGCM2 and HadCM3 AOGCMs under the SRES A2 emissions scenario) (Ogden et al., 2006).

14.4.6 Human settlements

Economic base

The economies of resource-dependent communities and indigenous communities in North America are particularly sensitive to climate change, with likely winners and losers controlled by impacts on important local resources (see Sections 14.4.1, 14.4.4 and 14.4.7). Residents of northern Canada and Alaska are likely to experience the most disruptive impacts of climate change, including shifts in the range or abundance of wild species crucial to the livelihoods and well-being of indigenous peoples (high confidence) (see Chapter 15 Sections 15.4.2.4 and 15.5) (Houser et al., 2001; NAST, 2001; Parson et al., 2001a; ACIA, 2005).

Infrastructure, climate trends and extreme events

Many of the impacts of climate change on infrastructure in North America depend on future changes in variability of precipitation and extreme events, which are likely to increase but with substantial uncertainty (Meehl et al., 2007: Section 10.5.1; Christensen et al., 2007: Section 11.5.3). Infrastructure in Alaska and northern Canada is known to be vulnerable to warming. Among the most sensitive areas are those affected by coastal erosion and thawing of ice-rich permafrost (see Chapter 15 Section 15.7.1) (NAST, 2001; Arctic Research Commission, 2003; ACIA, 2005). Building, designing, and maintaining foundations, pipelines and road and railway embankments will become more expensive due to permafrost thaw (ACIA, 2005). Examples where infrastructure is projected to be at 'moderate to high hazard' in the mid-21st century include Shishmaref, Nome and Barrow in Alaska, Tuktoyaktuk in the Northwest Territories, the Dalton Highway in Alaska, the Dempster Highway in the Yukon, airfields in the Hudson Bay region, and the Alaska Railroad (based on the ECHAM1-A, GFDL89 and UKTR climate models) (Nelson et al., 2002; Instanes et al., 2005).

Since the TAR, a few studies have projected increasing vulnerability of infrastructure to extreme weather related to climate warming unless adaptation is effective (high confidence). Examples include the New York Metropolitan Region (Rosenzweig and Solecki, 2001) (see Box 14.3), the mid-Atlantic Region (Fisher, 2000; Barron, 2001; Wu et al.,

Box 14.3. North American cities integrate impacts across multiple scales and sectors

Impacts of climate change in the metropolitan regions of North America will be similar in many respects. Los Angeles, New York and Vancouver are used to illustrate some of the affected sectors, including infrastructure, energy and water supply. Adaptation will need to be multi-decadal and multi-dimensional, and is already beginning (see Section 14.5).

Infrastructure

Since most large North American cities are on tidewater, rivers or both, effects of climate change will likely include sea-level rise (SLR) and/or riverine flooding. The largest impacts are expected when SLR, heavy river flows, high tides and storms coincide (California Regional Assessment Group, 2002). In New York, flooding from the combination of SLR and storm surge could be several metres deep (Gornitz and Couch, 2001; Gornitz et al., 2001). By the 2090s under a strong warming scenario (the CGCM climate model with the CCGG emissions scenario), today's 100-year flood level could have a return period of 3 to 4 years, and today's 500-year flood could be a 1-in-50-year event, putting much of the region's infrastructure at increased risk (Jacob et al., 2001; Major and Goldberg, 2001).

Energy supply and demand

Climate change will likely lead to substantial increases in electricity demand for summer cooling in most North American cities (see Section 14.4.8). This creates a number of conflicts, both locally and at a distance. In southern California, additional summer electricity demand will intensify inherent conflicts between state-wide hydropower and flood-control objectives (California Regional Assessment Group, 2002). Operating the Columbia River dams that supply 90% of Vancouver's power would be complicated by lower flows and environmental requirements (see Box 14.2). In New York, supplying summer electricity demand could increase air pollutant levels (e.g., ozone) (Hill and Goldberg, 2001; Kinney et al., 2001; Knowlton et al., 2004) and health impacts could be further exacerbated by climate change interacting with urban heat island effects (Rosenzweig et al., 2005). Unreliable electric power, as in minority neighbourhoods during the New York heatwave of 1999, can amplify concerns about health and environmental justice (Wilgoren and Roane, 1999).

Water supply systems

North American city water supply systems often draw water from considerable distances, so climate impacts need not be local to affect cities. By the 2020s, 41% of the supply to southern California is likely to be vulnerable to warming from loss of Sierra Nevada and Colorado River basin snowpack (see Section 14.4.1). Similarly, less mountain snowpack and summer runoff could require that Vancouver undertakes additional conservation and water restrictions, expands reservoirs, and develops additional water sources (Schertzer et al., 2004). The New York area will likely experience greater water supply variability (Solecki and Rosenzweig, 2007). The New York system can likely accommodate this, but the region's smaller systems may be vulnerable, leading to a need for enhanced regional water distribution protocols (Hansler and Major, 1999).

Adaptation

Many cities in North America have initiated 'no regrets' actions based on historical experience. In the Los Angeles area, incentive and information programmes of local water districts encourage water conservation (MWD, 2005). A population increase of over 35% (nearly one million people) since 1970 has increased water use in Los Angeles by only 7% (California Regional Assessment Group, 2002). New York has reduced total water consumption by 27% and per capita consumption by 34% since the early 1980s (City of New York, 2005). Vancouver's 'CitiesPLUS' 100-year plan will upgrade the drainage system by connecting natural areas and waterways, developing locally resilient, smaller systems, and upgrading key sections of pipe during routine maintenance (Denault et al., 2002).

2002; Rygel et al., 2006) and the urban transportation network of the Boston metropolitan area (Suarez et al., 2005). For Boston, projections of a gradual increase (0.31%/yr) in the probability of the 100-year storm surge, as well as sea-level rise of 3 mm/yr, leads to urban riverine and coastal flooding (based on the CGCM1 climate model), but the projected economic damages do not justify the cost of adapting the transportation infrastructure to climate change.

Less reliable supplies of water are likely to create challenges for managing urban water systems as well as for industries that depend on large volumes of water (see Sections 14.2.1, 14.4.1). U.S. water managers anticipate local, regional or state-wide water shortages during the next ten years (GAO, 2003). Threats to reliable supply are complicated by the high population growth rates in western states where many water resources are at or approaching full utilisation (GAO, 2003) (see Section 14.4.1). Potential increases in heavy precipitation, with expanding impervious surfaces, could increase urban flood risks and create additional design challenges and costs for stormwater management (Kije Sipi Ltd., 2001).

14.4.7 Tourism and recreation

Although coastal zones are among the most important recreation resources in North America, the vulnerability of key tourism areas to sea-level rise has not been comprehensively assessed. The cost to protect Florida beaches from a 0.5 m rise in sea level, with sand replenishment, was estimated at US$1.7 billion to 8.8 billion (EPA, 1999).

Nature-based tourism is a major market segment, with over 900 million visitor-days in national/provincial/state parks in 2001. Visits to Canada's national parks system are projected to increase by 9 to 25% (2050s) and 10 to 40% (2080s) as a result of a lengthened warm-weather tourism season (based on the PCM GCM and the SRES B2 emissions scenario, and the CCSR GCM with A1) (Jones and Scott, 2006). This would have economic benefits for park agencies and nearby communities, but could exacerbate visitor-related ecological pressures in some parks. Climate-induced environmental changes (e.g., loss of glaciers, altered biodiversity, fire- or insect-impacted forests) would also affect park tourism, although uncertainty is higher regarding the regional specifics and magnitude of these impacts (Richardson and Loomis, 2004; Scott et al., 2007a).

Early studies of the impact of climate change on the ski industry did not account for snowmaking, which substantially lowers the vulnerability of ski areas in eastern North America for modest (B2 emissions scenario) but not severe (A1) warming (based on 5 GCMs for the 2050s) (Scott et al., 2003; Scott et al., 2007b). Without snowmaking, the ski season in western North America will likely shorten substantially, with projected losses of 3 to 6 weeks (by the 2050s) and 7 to 15 weeks (2080s) in the Sierra Nevada of California (based on PCM and HadCM3 GCMs for the B1 and A1FI scenarios), and 7 to 10 weeks at lower elevations and 2 to 14 weeks at higher elevations at Banff, Alberta (based on the PCM GCM with the B2 emissions scenario, and the CCSR GCM with A1, for the 2050s) (Hayhoe et al., 2004; Scott and Jones, 2005). With advanced snowmaking, the ski season in Banff shortens at low but not at high altitudes. The North American snowmobiling industry (valued at US$27 billion) (ISMA, 2006) is more vulnerable to climate change because it relies on natural snowfall. By the 2050s, a reliable snowmobile season disappears from most regions of eastern North America that currently have developed trail networks (based on the CGCM1 and HadCM3 GCMs with IS92a emissions, the PCM GCM with B2 emissions and the CCSR GCM with A1 emissions) (Scott, 2006; Scott and Jones, 2006).

14.4.8 Energy, industry and transportation

Energy demand

Recent North American studies generally confirm earlier work showing a small net change (increase or decrease, depending on methods, scenarios and location) in the net demand for energy in buildings but a significant increase in demand for electricity for space cooling, with further increases caused by additional market penetration of air conditioning (high confidence) (Sailor and Muñoz, 1997; Mendelsohn and Schlesinger, 1999; Morrison and Mendelsohn, 1999; Mendelsohn, 2001; Sailor, 2001; Sailor and Pavlova, 2003; Scott et al., 2005; Hadley et al., 2006). Ruth and Amato (2002) projected a 6.6% decline in annual heating fuel consumption for Massachusetts in 2020 (linked to an 8.7% decrease in heating degree-days) and a 1.9% increase in summer electricity consumption (12% increase in annual cooling degree-days). In Québec, net energy demand for heating and air conditioning across all sectors could fall by 9.4% of 2001 levels by 2100 (based on the CGCM1 GCM and the IS92a emissions scenario), with residential heating falling by 10 to 15% and air conditioning increasing two- to four-fold. Peak electricity demand is likely to decline in the winter peaking system of Quebec, while summer peak demand is likely to increase 7 to 17% in the New York metropolitan region (Ouranos, 2004).

Energy supply

Since the TAR, there have been regional but not national-level assessments of the effects of climate change on future hydropower resources in North America. For a 2 to 3°C warming in the Columbia River Basin and British Columbia Hydro service areas, the hydroelectric supply under worst-case water conditions for winter peak demand will likely increase (high confidence). However, generating power in summer will likely conflict with summer instream flow targets and salmon restoration goals established under the Endangered Species Act (Payne et al., 2004). This conclusion is supported by accumulating evidence of a changing hydrologic regime in the western U.S. and Canada (see Sections 14.2.1, 14.4.1, Box 14.2). Similarly, Colorado River hydropower yields will likely decrease significantly (medium confidence) (Christensen et al., 2004), as will Great Lakes hydropower (Moulton and Cuthbert, 2000; Lofgren et al., 2002; Mirza, 2004). James Bay hydropower will likely increase (Mercier, 1998; Filion, 2000). Lower Great Lake water levels could lead to large economic losses (Canadian $437 million to 660 million/yr), with increased water levels leading to small gains (Canadian $28 million to 42 million/yr) (Buttle et al., 2004; Ouranos, 2004). Northern Québec hydropower production would likely benefit from greater precipitation and more open-water conditions, but hydro plants in southern Québec would likely be affected by lower water levels. Consequences of changes in seasonal distribution of flows and in the timing of ice formation are uncertain (Ouranos, 2004).

Wind and solar resources are about as likely as not to increase (medium confidence). The viability of wind resources depends on both wind speed and reliability. Studies to date project wind resources that are unchanged by climate change (based on the HadGCM2 CGSa4 experiment) or reduced by 0 to 40% (based on CGCM1 and the SRES A1 scenario, and HadCM2 and RegCM2 and a 1%/yr CO_2 increase) (Segal et al., 2001; Breslow and Sailor, 2002). Future changes in cloudiness could slightly increase the potential for solar energy in North America south of 60°N (using many models, the A1B scenario and for 2080 to 2099 *vs.* 1980 to 1999) (Meehl et al., 2007: Figure 10.10). However, Pan et al. (2004) projected the opposite: that increased cloudiness will likely decrease the potential output of photovoltaics by 0 to 20% (based on HadCM2 and RegCM2 and a 1%/yr CO_2 increase for the 2040s).

Bioenergy potential is climate-sensitive through direct impacts on crop growth and availability of irrigation water. Bioenergy crops are projected to compete successfully for agricultural acreage at a price of US$33/Mg, or about US$1.83/10^9 joules (Walsh et al., 2003). Warming and precipitation increases are expected to allow the bioenergy crop switchgrass to compete effectively with traditional crops in the central U.S. (based on RegCM2 and a 2×CO_2 scenario) (Brown et al., 2000).

Construction

As projected in the TAR, the construction season in Canada and the northern U.S. will likely lengthen with warming (see Section 14.3.1 and Christensen et al., 2007 Section 11.5.3). In permafrost areas in Canada and Alaska, increasing depth of the 'active layer' or loss of permafrost can lead to substantial decreases in soil strength (ACIA, 2004). In areas currently underlain by permafrost, construction methods are likely to require changes (Cole et al., 1998), potentially increasing construction and maintenance costs (high confidence) (see Chapter 15 Section 15.7.1) (ACIA, 2005).

Transportation

Warmer or less snowy winters will likely reduce delays, improve ground and air transportation reliability, and decrease the need for winter road maintenance (Pisano et al., 2002). More intense winter storms could, however, increase risks for traveller safety (Andrey and Mills, 2003) and require increased snow removal. Continuation of the declining fog trend in at least some parts of North America (Muraca et al., 2001; Hanesiak and Wang, 2005) should benefit transport. Improvements in technology and information systems will likely modulate vulnerability to climate change (Andrey and Mills, 2004).

Negative impacts of climate change on transportation will very likely result from coastal and riverine flooding and landslides (Burkett, 2002). Although offset to some degree by fewer ice threats to navigation, reduced water depth in the Great Lakes would lead to the need for 'light loading' and, hence, adverse economic impacts (see Section 14.4.1) (du Vair et al., 2002; Quinn, 2002; Millerd, 2005). Adaptive measures, such as deepening channels for navigation, would need to address both institutional and environmental challenges (Lemmen and Warren, 2004).

Warming will likely adversely affect infrastructure for surface transport at high northern latitudes (Nelson et al., 2002). Permafrost degradation reduces surface load-bearing capacity and potentially triggers landslides (Smith and Levasseur, 2002; Beaulac and Doré, 2005). While the season for transport by barge is likely to be extended, the season for ice roads will likely be compressed (Lonergan et al., 1993; Lemmen and Warren, 2004; Welch, 2006). Other types of roads are likely to incur costly improvements in design and construction (Stiger, 2001; McBeath, 2003; Greening, 2004) (see Chapter 15 Section 15.7.1).

An increase in the frequency, intensity or duration of heat spells could cause railroad track to buckle or kink (Rosetti, 2002), and affect roads through softening and traffic-related rutting (Zimmerman, 2002). Some problems associated with warming can be ameliorated with altered road design, construction and management, including changes in the asphalt mix and the timing of spring load restrictions (Clayton et al., 2005; Mills et al., 2006).

14.4.9 Interacting impacts

Impacts of climate change on North America will not occur in isolation, but in the context of technological, economic (Nakićenović and Swart, 2000; Edmonds, 2004), social (Lebel, 2004; Reid et al., 2005) and ecological changes (Sala et al., 2000). In addition, challenges from climate change will not appear as isolated effects on a single sector, region, or group. They will occur in concert, creating the possibility of a suite of local, as well as long-distance, interactions, involving both impacts of climate change and other societal and ecosystem trends (NAST, 2001; Reid et al., 2005). In some cases, these interactions may reduce impacts or decrease vulnerability, but in others they may amplify impacts or increase vulnerability.

Effects of climate change on ecosystems do not occur in isolation. They co-occur with numerous other factors, including effects of land-use change (Foley et al., 2005), air pollution (Karnosky et al., 2005), wildfires (see Box 14.1), changing biodiversity (Chapin et al., 2000) and competition with invasives (Mooney et al., 2005). The strong dependence of ecosystem function on moisture balance (Baldocchi and Valentini, 2004), coupled with the greater uncertainty about future precipitation than about future temperature (Christensen et al., 2007: Section 11.5.3), further expands the range of possible futures for North American ecosystems.

People also experience climate change in a context that is strongly conditioned by changes in other sectors and their adaptive capacity. Interactions with changes in material wealth (Ikeme, 2003), the vitality of local communities (Hutton, 2001; Wall et al., 2005), the integrity of key infrastructure (Jacob et al., 2001), the status of emergency facilities and preparedness and planning (Murphy et al. 2005), the sophistication of the public health system (Kinney et al., 2001), and exposure to conflict (Barnett, 2003), all have the potential to either exacerbate or ameliorate vulnerability to climate change. Among the unexpected consequences of the population displacement caused by Hurricane Katrina in 2005 is the strikingly poorer health of storm evacuees, many of whom lost jobs, health insurance, and stable relationships with medical professionals (Columbia University Mailman School of Public Health, 2006).

Little of the literature reviewed in this chapter addresses interactions among sectors that are all impacted by climate change, especially in the context of other changes in economic activity, land use, human population, and changing personal and political priorities. Similarly, knowledge of the indirect impacts on North America of climate change in other geographical regions is very limited.

14.5 Adaptation: practices, options and constraints

The U.S. and Canada are developed economies with extensive infrastructure and mature institutions, with important regional and socio-economic variations (NAST, 2000; Lemmen and Warren, 2004). These capabilities have led to adaptation and coping strategies across a wide range of historic conditions, with both successes and failures. Most studies on adaptive strategies consider implementation based on past experiences (Paavola and Adger, 2002). Examples of adaptation based on future projections are rare (Smit and Wall, 2003; Devon, 2005). Expanding beyond reactive adaptation to proactive, anticipatory adaptive strategies presents many challenges. Progress toward meeting these challenges is just beginning in North America.

14.5.1 Practices and options

Canada and the U.S. emphasise market-based economies. Governments often play a role implementing large-scale adaptive measures, and in providing information and incentives to support development of adaptive capacity by private decision makers (UNDP, 2001; Michel-Kerjan, 2006). In practice, this means that individuals, businesses and community leaders act on perceived self interest, based on their knowledge of adaptive options. Despite many examples of adaptive practices in North America, under-investment in adaptation is evident in the recent rapid increase in property damage due to climate extremes (Burton and Lim, 2005; Epstein and Mills, 2005) and illustrates the current adaptation deficit.

Adaptation by individuals and private businesses

Research on adaptive behaviour for coping with projected climate change is minimal, though several studies address adaptations to historic variation in the weather. About 70% of businesses face some weather risk. The impact of weather on businesses in the U.S. is an estimated US$200 billion/yr (Lettre, 2000). Climate change may also create business opportunities. For example, spending on storm-worthiness and construction of disaster-resilient homes (Koppe et al., 2004; Kovacs, 2005b; Kunreuther, 2006) increased substantially after the 2004 and 2005 Atlantic hurricanes, as did the use of catastrophe bonds (CERES, 2004; Byers et al., 2005; Dlugolecki, 2005; Guy Carpenter, 2006).

Businesses in Canada and the U.S. are investing in climate-relevant adaptations, though few of these appear to be based on projections of future climate change. For example:

- Insurance companies are introducing incentives for homeowners and businesses that invest in loss prevention strategies (Kim, 2004; Kovacs, 2005b).
- Insurance companies are investing in research to prevent future hazard damage to insured property, and to adjust pricing models (Munich Re., 2004; Mills and Lecomte, 2006).
- Ski resort operators are investing in lifts to reach higher altitudes and in snow-making equipment (Elsasser et al., 2003; Census Bureau, 2004; Scott, 2005; Jones and Scott, 2006; Scott et al., 2007a).
- With highly detailed information on weather conditions, farmers are adjusting crop and variety selection, irrigation strategies and pesticide application (Smit and Wall, 2003).
- The forest resources sector is investing in improved varieties, forest protection, forest regeneration, silvicultural management and forest operations (Loehle et al., 2002; Spittlehouse and Stewart, 2003).

Adaptation by governments and communities

Many North American adaptations to climate-related risks are implemented at the community level. These include efforts to minimise damage from heatwaves, droughts, floods, wildfires or tornados. These actions may entail land-use planning, building code enforcement, community education and investments in critical infrastructure (Burton et al., 2002; Multihazard Mitigation Council, 2005).

Flooding and drought present recurring challenges for many North American communities (Duguid, 2002). When the City of Peterborough, Canada, experienced two 100-year flood events within three years, it responded by flushing the drainage systems and replacing the trunk sewer systems to meet more extreme 5-year flood criteria (Hunt, 2005). Recent droughts in six major U.S. cities, including New York and Los Angeles, led to adaptive measures involving investments in water conservation systems and new water supply-distribution facilities (Changnon and Changnon, 2000). To cope with a 15% increase in heavy precipitation, Burlington and Ottawa, Ontario, employed both structural and non-structural measures, including directing downspouts to lawns to encourage infiltration and increasing depression and street detention storage (Waters et al., 2003).

Some large cities (e.g., New Orleans) and important infrastructure (e.g., the only highway and rail link between Nova Scotia and the rest of Canada) are located on or behind dykes that will provide progressively less protection unless raised on an ongoing basis. Some potential damages may be averted through redesigning structures, raising the grade, or relocating (Titus, 2002). Following the 1996 Saguenay flood and 1998 ice storm, the province of Québec modified the Civil Protection Act and now requires municipalities to develop comprehensive emergency management plans that include adaptation strategies (McBean and Henstra, 2003). More communities are expected to re-examine their hazard management systems following the catastrophic damage in New Orleans from Hurricane Katrina (Kunreuther et al., 2006).

Rapid development and population growth are occurring in many coastal areas that are sensitive to storm impacts (Moser, 2005). While past extreme events have motivated some aggressive adaptation measures (e.g., in Galveston, Texas) (Bixel and Turner, 2000), the passage of time, new residents, and high demand for waterfront property are pushing coastal development into vulnerable areas.

Climate change will likely increase risks of wildfire (see Box 14.1). FireWise and FireSmart are programmes promoting wildfire safety in the U.S. and Canada, respectively (FireSmart, 2005; FireWise, 2005). Individual homeowners and businesses can participate, but the greatest reduction in risk will occur in communities that take a comprehensive approach, managing forests with controlled burns and thinning, promoting or enforcing appropriate roofing materials, and maintaining defensible space around each building (McGee et al., 2000).

Public institutions are responsible for adapting their own legislation, programmes and practices to appropriately anticipate climate changes. The recent Québec provincial plan, for example, integrates climate change science into public policy. Public institutions can also use incentives to encourage or to overcome disincentives to investment by private decision makers (Moser, 2006). Options, including tax assistance, loan guarantees and grants, can improve resilience to extremes and reduce government costs for disaster management (Moser, 2005). The U.S. National Flood Insurance Program is changing its policy to reduce the risk of multiple flood claims, which cost the programme more than US$200 million/yr (Howard, 2000). Households with two flood-related claims are now required to elevate their structure 2.5 cm above the 100-year flood level, or relocate. To complement this, a 5-year, US$1 billion programme to update and digitise flood maps was initiated in 2003 (FEMA, 2006). However, delays in implementing appropriate zoning can encourage accelerated, maladapted development in coastal communities and flood plains.

14.5.2 Mainstreaming adaptation

One of the greatest challenges in adapting North America to climate change is that individuals often resist and delay change (Bacal, 2000). Good decisions about adapting to climate change depend on relevant experience (Slovic, 2000), socio-economic factors (Conference Board of Canada, 2006), and political and institutional considerations (Yarnal et al., 2006; Dow et al., 2007). Adaptation is a complex concept (Smit et al., 2000; Dolan and Walker, 2006), that includes wealth and several other dimensions.

Experience and knowledge

The behaviour of people and systems in North America largely reflects historic climate experience (Schipper et al., 2003), which has been institutionalised through building codes, flood management infrastructure, water systems and a variety of other programmes. Canadian and U.S. citizens have invested in buildings, infrastructure, water and flood management systems designed for acceptable performance under historical conditions (Bruce, 1999; Co-operative Programme on Water and Climate, 2005; UMA Engineering, 2005; Dow et al., 2007). Decisions by community water managers (Rayner et al., 2005; Dow et al., 2007) and set-back regulations in coastal areas (Moser, 2005) also account for historic experience but rarely incorporate information about climate change or sea-level rise. In general, decision makers lack the tools and perspectives to integrate future climate, particularly events that exceed historic norms (UNDP, 2001).

Examples of adaptive behaviour influenced exclusively or predominantly by projections of climate change are largely absent from the literature, but some early steps toward planned adaptation have been taken by the engineering community, insurance companies, water managers, public health officials, forest managers and hydroelectric producers. Some initiatives integrate consideration of climate change into the environmental impact assessment process. Philadelphia, Toronto and a few other communities have introduced warning programmes to manage the health threat of heatwaves (Kalkstein, 2002). The introduction of Toronto's heat/health warning programme was influenced by both climate projections and fatalities from past heatwaves (Koppe et al., 2004; Ligeti, 2006).

Weather extremes can reveal a community's vulnerability or resilience (RMS, 2005a) and provide insights into potential adaptive responses to future events. Since the 1998 ice storm, Canada's two most populous provinces, Ontario and Québec, have strengthened emergency preparedness and response capacity. Included are comprehensive hazard-reduction measures and loss-prevention strategies to reduce vulnerability to extreme events. These strategies may include both public information programmes and long-term strategies to invest in safety infrastructure (McBean and Henstra, 2003). Adaptive behaviour is typically greater in the communities that recently experienced a natural disaster (Murphy et al., 2005). But the near absence of any personal preparedness following the 2003 blackout in eastern North America demonstrated that adaptive actions do not always follow significant emergencies (Murphy, 2004).

Socio-economic factors

Wealthier societies tend to have greater access to technology, information, developed infrastructure, and stable institutions (Easterling et al., 2004), which build capacity for individual and collective action to adapt to climate change. But average economic status is not a sufficient determinant of adaptive capacity (Moss et al., 2001). The poor and marginalised in Canada and the U.S. have historically been most at risk from weather shocks (Turner et al., 2003), with vulnerability directly related to income inequality (Yohe and Tol, 2002). Differences in individual capacity to cope with extreme weather were evident in New Orleans during and after Hurricane Katrina (Kunreuther et al., 2006), when the large majority of those requiring evacuation assistance were either poor or in groups with limited mobility, including elderly, hospitalised and disabled citizens (Murphy et al., 2005; Kumagi et al., 2006; Tierney, 2006).

Political and institutional capacity for autonomous adaptation

Public officials in Canada and the U.S. typically provide early and extensive assistance in emergencies. Nevertheless, emergency response systems in the U.S. and Canada are based on the philosophy that households and businesses should be capable of addressing their own basic needs for up to 72 hours after a disaster (Kovacs and Kunreuther, 2001). The residents' vulnerability depends on their own resources, plus those provided by public service organisations, private firms and others (Fischhoff, 2006). When a household is overwhelmed by an extreme event, household members often rely on friends, family and other social networks for physical and emotional support (Cutter et al., 2000; Enarson, 2002; Murphy, 2004). When a North American community responds to weather extremes, non-governmental organisations often coordinate support for community-based efforts (National Voluntary Organizations Active in Disaster, 2006).

An active dialogue among stakeholders and political institutions has the potential to clarify the opportunities for adaptation to changing climate. However, public discussion about adaptation is at an early stage in the U.S. and Canada (Natural Resources Canada, 2000), largely because national governments have focused public discussion on mitigation, with less attention to adaptation (Moser, 2005). Some public funds have been directed to research on impacts and adaptation, and both countries have undertaken national assessments with a synthesis of the adaptation literature, but neither country has a formal adaptation strategy (Conference Board of Canada, 2006). Integrating perspectives on climate change into legislation and regulations has the potential to promote or constrain adaptive behaviour (Natural Resources Canada, 2000). North American examples of public policies that influence adaptive behaviour include water allocation law in the western U.S. (Scheraga, 2001), farm subsidies (Goklany, 2007), public flood insurance in the U.S. (Crichton, 2003), guidance on preservation of wetlands and emergency management.

14.5.3 Constraints and opportunities

Social and cultural barriers

High adaptive capacity, as in most of North America, should be an asset for coping with or benefiting from climate change. Capacity, however, does not ensure positive action or any action at all. Societal values, perceptions and levels of cognition shape adaptive behaviour (Schneider, 2004). In North America, information about climate change is usually not 'mainstreamed' or explicitly considered (Dougherty and Osaman Elasha, 2004) in the overall decision-making process (Slovic, 2000; Leiss, 2001). This can lead to actions that are maladapted, for example, development near floodplains or coastal areas known to be vulnerable to climate change. Water managers are unlikely to use climate forecasts, even when they recognise the vulnerability, unless the forecast information can fit directly into their everyday management decisions (Dow et al., 2007).

Informational and technological barriers

Uncertainty about the local impacts of climate change is a barrier to action (NRC, 2004). Incomplete knowledge of disaster safety options (Murphy, 2004; Murphy et al., 2005) further constrains adaptive behaviour. Climate change information must be available in a form that fits the needs of decision-makers. For example, insurance companies use climate models with outputs specifically designed to support decisions related to the risk of insolvency, pricing and deductibles, regulatory and rating agency considerations, and reinsurance (Swiss Re, 2005a). Some electrical utilities have begun to integrate climate model output into planning and management of hydropower production (Ouranos, 2004).

A major challenge is the need for efficient technology and knowledge transfer. In general, questions about responsibility for funding research, involving stakeholders, and linking communities, government and markets have not been answered (Ouranos, 2004). Another constraint is resistance to new technologies (e.g., genetically modified crops), so that some promising adaptations in the agricultural, water resource management and forestry sectors are unlikely to be realised (Goklany, 2000, 2001).

Financial and market barriers

In the U.S., recent spending on adaptation to extremes has been a sound investment, contributing to reduced fatalities, injuries and significant economic benefits. The Multihazard Mitigation Council (2005) found that US$3.5 billion in spending between 1993 and 2003 on programmes to reduce future damages from flooding, severe wind and earthquakes contributed US$14 billion in societal benefits. The greatest savings were in flood (5-fold) and wind (4-fold) damage reduction. Adaptation also benefited government as each dollar of spending resulted in US$3.65 in savings or increased tax revenue. This is consistent with earlier case studies; the Canadian $65 million invested in 1968 to create the Manitoba Floodway has prevented several billion dollars in flood damage (Duguid, 2002).

Economic issues are frequently the dominant factors influencing adaptive decisions. This includes community response to coastal erosion (Moser, 2000), investments to enhance water resource systems (Report of the Water Strategy Expert Panel, 2005), protective retrofits to residences (Simmons et al., 2002; Kunreuther, 2006), and changes in insurance practices (Kovacs, 2005a). The cost and availability of economic resources clearly influence choices (WHO, 2003), as does the private versus public identity of the beneficiaries (Moser, 2000).

Sometimes, financial barriers interact with the slow turnover of existing infrastructure (Figure 14.3). Extensive property damage in Florida during Hurricane Andrew in 1992 led to significant revisions to the building code. If all properties in southern Florida met this updated code in 1992, then property damage from Hurricane Andrew would have been lower by nearly 45% (AIR, 2002). Florida will, however, still experience extensive damage from hurricanes through damage to the large number of older homes and businesses. Other financial barriers come from the challenge property owners face in recovering the costs of protecting themselves. Hidden adaptations tend to be undervalued, relative to obvious ones. For example, homes with storm shutters sell for more than homes without this visible adaptation, while less visible retrofits, such as tie-down straps to hold the roof in high winds, add less to the resale value of the home, relative to their cost (Simmons et al., 2002).

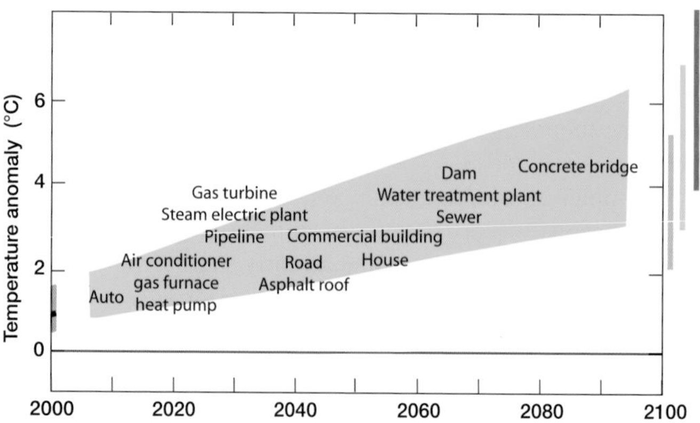

Figure 14.3. *Typical infrastructure lifetimes in North America (data from Lewis, 1987; Bettigole, 1990; EIA, 1999, 2001; Statistics Canada, 2001a; BEA, 2003), in relation to projected North American warming for 2000 to 2100 (relative to 1901-1950) for the A1B scenario, from the IPCC AR4 Multi-Model Dataset (yellow envelope). Measured and modelled anomalies for 2000 are shown with black and orange bars, respectively. Projected warming for 2091 to 2100 for the B1, A1B and A2 scenarios are indicated by the blue, yellow and red bars, respectively at the right (data from Christensen et al., 2007: Box 11.1 Figure 1).*

14.6 Case studies

Many of the topics discussed in this chapter have important dimensions, including interactions with other sectors, regions and processes, that make them difficult to assess from the perspective of a single sector. This chapter develops multi-sector case studies on three topics of special importance to North America – forest disturbances (see Box 14.1), water resources (using the Columbia River as an example) (see Box 14.2) and coastal cities (see Box 14.3).

14.7 Conclusions: implications for sustainable development

Climate change creates a broad range of difficult challenges that influence the attainment of sustainability goals. Several of the most difficult emerge from the long time-scale over which the changes occur (see Section 14.3) and the possible need for action well before the magnitude (and certainty) of the impacts is clear (see Section 14.5). Other difficult problems arise from the intrinsic global scale of climate change (EIA, 2005b). Because the drivers of climate change are truly global, even dedicated action at the regional scale has limited prospects for ameliorating regional-scale impacts. These two sets of challenges, those related to time-scale and those related to the global nature of climate change, are not in the classes that have traditionally yielded to the free-market mechanisms and political decision making that historically characterise Canada and the U.S. (see Section 14.5). Yet, the magnitude of the climate change challenge calls for proactive adaptation and technological and social innovation, areas where Canada and the U.S. have abundant capacity. An important key to success will be developing the capacity to incorporate climate change information into adaptation in the context of other important technological, social, economic and ecological trends.

The preceding sections describe current knowledge concerning the recent climate experience of North America, the impacts of the changes that have already occurred, and the potential for future changes. They also describe historical experience with and future prospects for dealing with climate impacts. The key points are:

- North America has experienced substantial social, cultural, economic and ecological disruption from recent climate-related extremes, especially storms, heatwaves and wildfires [14.2].
- Continuing infrastructure development, especially in vulnerable zones, will likely lead to continuing increases in economic damage from extreme weather [14.2.6, 14.4.6].
- The vulnerability of North America depends on the effectiveness of adaptation and the distribution of coping capacity, both of which are currently uneven and have not always protected vulnerable groups from adverse impacts of climate variability and extreme weather events [14.5].
- A key prerequisite for sustainability is 'mainstreaming' climate issues into decision making [14.5].
- Climate change will exacerbate stresses on diverse sectors in North America, including, but not limited to, urban centres, coastal communities, human health, water resources and managed and unmanaged ecosystems [14.4].
- Indigenous peoples of North America and those who are socially and economically disadvantaged are disproportionately vulnerable to climate change [14.2.6, 14.4.6].

14.8 Key uncertainties and research priorities

The major limits in understanding of climate change impacts on North America, and on the ability of its people, economies and ecosystems to adapt to these changes, can be grouped into seven areas.

- Projections of climate changes still have important uncertainties; especially on a regional scale (Christensen et al., 2007: Section 11.5.3). For North America, the greater uncertainty about future precipitation than about future temperature substantially expands the uncertainty of a broad range of impacts on ecosystems (see Section 14.4.2), hydrology and water resources (see Sections 14.4.1, 14.4.7), and on industries (see Sections 14.4.6, 14.4.7).
- North American people, economies and ecosystems tend to be much more sensitive to extremes than to average conditions [14.2]. Incomplete understanding of the relationship between changes in the average climate and extremes (Meehl et al., 2007: Section 10.3.6; Christensen et al., 2007: Section 11.5.3.3) limits our ability to connect future conditions with future impacts and the options for adaptation. There is a need for improved understanding of the relationship between changes in average climate and those extreme events with the greatest potential impact on North America, including hurricanes, other severe storms, heatwaves, floods, and prolonged droughts.
- For most impacts of climate change, we have at least some tools for estimating gradual change (see Section 14.4), but we have few tools for assessing the conditions that lead to tipping points, where a system changes or deteriorates rapidly, perhaps without further forcing.
- Most of the past research has addressed impacts on a single sector (e.g., health, transportation, unmanaged ecosystems). Few studies address the interacting responses of diverse sectors impacted by climate change, making it very difficult to evaluate the extent to which multi-sector responses limit options or push situations toward tipping points (see Section 14.4.9).
- Very little past research addresses impacts of climate change in a context of other trends with the potential to exacerbate impacts of climate change or to limit the range of response options (see Section 14.4.9) (but see Reid et al., 2005 for an important exception). A few North American examples of trends likely to complicate the development of strategies for dealing with climate change include continuing development in coastal areas (see Section 14.2.3), increasing demand on

freshwater resources (see Section 14.4.1), the accumulation of fuel in forest ecosystems susceptible to wildfire (see Box 14.1), and continued introductions of invasive species with the potential to disrupt agriculture and ecosystem processes (see Section 14.2.2, 14.2.4). In the sectors that are the subject of the most intense human management (e.g., health, agriculture, settlements, industry), it is possible that changes in technology or organisation could exacerbate or ameliorate impacts of climate change (see Section 14.4.9).

- Indirect impacts of climate change are poorly understood. In a world of ever-increasing globalisation, the future of North American people, economies and ecosystems is connected to the rest of the world through a dense network of cultural exchanges, trade, mixing of ecosystems, human migration and, regrettably, conflict (see Section 14.3). In this interconnected world, it is possible that profoundly important impacts of climate change on North America will be indirect consequences of climate change impacts on other regions, especially where people, economies or ecosystems are unusually vulnerable.
- Examples of North American adaptations to climate-related impacts are abundant, but understanding of the options for proactive adaptation to conditions outside the range of historical experience is limited (see Section 14.5).

All of these areas potentially interact, with impacts that are unevenly distributed among regions, industries, and communities. Progress in research and management is occurring in all these areas. Yet stakeholders and decision makers need information immediately, placing a high priority on strategies for providing useful decision support in the context of current knowledge, conditioned by an appreciation of the limits of that knowledge.

References

ACIA, 2004: *(Arctic Climate Impact Assessment), Impacts of a Warming Arctic: Arctic Climate Impact Assessment*. Cambridge University Press, Cambridge, 146 pp.

ACIA, 2005: *Arctic Climate Impact Assessment*. Cambridge University Press, Cambridge, 1042 pp.

Adams, R.M., B.A. McCarl and L.O. Mearns, 2003: The effects of spatial scale of climate scenarios on economic assessments: An example from U.S. agriculture. *Clim. Change*, **60**, 131-148.

AIR, 2002: *Ten Years after Andrew: What Should We BePreparing for Now?*, AIR (Applied Insurance Research, Inc.) Technical Document HASR 0208, Boston, 9 pp. [Accessed 09.02.07: http://www.air-worldwide.com/_public/NewsData/000258/Andrew_Plus_10.pdf]

Alberta Environment, 2002: *South Saskatchewan River Basin Water Management Plan, Phase One - Water Allocation Transfers: Appendices*, Alberta Environment, Edmonton, Alberta.

Allen, D.M., D.C. Mackie and M. Wei, 2004a: Groundwater and climate change: a sensitivity analysis for the Grand Forks aquifer, southern British Columbia, Canada. *Hydrogeol. J.*, **12**, 270-290.

Allen, D.M., J. Scibek, M. Wei and P. Whitfield, 2004b: *Climate Change and Groundwater: A Modelling Approach for Identifying Impacts and Resource Sustainability in the Central Interior of British Columbia*, Climate Change Action Fund, Natural Resources Canada, Ottawa, Ontario, 404 pp.

Allen, J., 2003: *Drought Lowers Lake Mead*, NASA. [Accessed 09.02.07: http://earthobservatory.nasa.gov/Study/LakeMead/]

Allen, S.B., J.P. Dwyer, D.C. Wallace and E.A. Cook, 2003: Missouri River flood of 1993: Role of woody corridor width in levee protection. *J. Amer. Water Resour. Assoc.*, **39**, 923-933.

Andalo, C., J. Beaulieu and J. Bousquet, 2005: The impact of climate change on growth of local white spruce populations in Quebec, Canada. *Forest Ecol. Manag.*, **205**, 169-182.

Andrey, J. and B. Mills, 2003: Climate change and the Canadian transportation system: Vulnerabilities and adaptations. *Weather and Transportation in Canada*, J. Andrey and C. K. Knapper, Eds. University of Waterloo, Waterloo, Ontario, 235-279.

Andrey, J. and B. Mills, 2004: Transportation. *Climate Change Impacts and Adaptations: A Canadian Perspective*, D.S. Lemmen and F.J. Warren, Eds., Government of Canada, Ottawa, Ontario, 131-149. [Accessed 09.02.07: http://adaptation.nrcan.gc.ca/perspective/pdf/report_e.pdf]

Antle, J.M., S.M. Capalbo, E.T. Elliott and K.H. Paustian, 2004: Adaptation, spatial heterogeneity, and the vulnerability of agricultural systems to climate change and CO_2 fertilization: An integrated assessment approach. *Clim. Change*, **64**, 289-315.

Arctic Research Commission, 2003: *(U.S. Arctic Research Commission Permafrost Task Force), Climate Change, Permafrost, and Impacts on Civil Infrastructure*. Special Report 01-03, U.S. Arctic Research Commission, Arlington, Virginia, 72 pp. [Accessed 09.02.07: http://www.arctic.gov/files/ PermafrostForWeb.pdf]

Associated Press, 2002: Rough year for rafters. September 3, 2002.

Atkinson, J., J. DePinto and D. Lam. 1999: Water quality. *Potential Climate Change Effects on the Great Lakes Hydrodynamics and Water Quality*, D. Lam and W. Schertzer, Eds. American Society of Civil Engineers, Reston, Virginia.

Aw, J. and M.J. Kleeman, 2002: Evaluating the first-order effect of inter-annual temperature variability on urban air pollution. *J. Geophys. Res.*, **108**, doi:10.1029/2001JD000544.

Babaluk, J.A., J.D. Reist, J.D. Johnson and L. Johnson, 2000: First records of sockeye (*Oncorhynchus nerka*), and pink salmon (*O. gorbuscha*), from Banks Island and other records of Pacific salmon in Northwest territories. *Canada Arctic*, **53**, 161-164.

Bacal, R., 2000: *The Importance of Leadership in Managing Change*. [Accessed 09.02.07: http://performance-appraisals.org/Bacalsappraisalarticles/articles/leadchange.htm]

Bachelet, D., R.P. Neilson, J.M. Lenihan and R.J. Drapek, 2001: Climate change effects on vegetation distribution and carbon budget in the United States. *Ecosystems*, **4**, 164-185.

Bachelet, D., R.P. Neilson, J.M. Lenihan and R.J. Drapek, 2004: Regional differences in the carbon source-sink potential of natural vegetation in the U.S.. *Environ. Manage.*, **33**, S23-S43. doi: 10.1007/s00267-003-9115-4.

Badeck, F.W., A. Bondeau, K. Bottcher, D. Doktor, W. Lucht, J. Schaber and S. Sitch, 2004: Responses of spring phenology to climate change. *New Phytol.*, **162**, 295-309.

Baldocchi, D. and R. Valentini, 2004: Geographic and temporal variation of carbon exchange by ecosystems and their sensitivity to environmental perturbations. *The Global Carbon Cycle: Integrating Humans, Climate, and the Natural World*, C.B. Field and M.R. Raupach, Eds. Island Press, Washington, District of Columbia, 295-316.

Balling, R.C. and R.S. Cerveny, 2003: Compilation and discussion of trends in severe storms in the United States: Popular perception versus climate reality. *Natural Hazards*, **29**, 103-112.

Barber, V.A., G.P. Juday and B.P. Finney, 2000: Reduced growth of Alaskan white spruce in the twentieth century from temperature-induced drought stress. *Nature*, **405**, 668-673.

Barnett, J., 2003: Security and climate change. *Global Environ. Change*, **13**, 7-17.

Barron, E.J., 2001: Chapter 4: Potential consequences of climate variability and change for the northeastern United States. *Climate Change Impacts on the United States: The Potential Consequences of Climate Variability and Change. Report for the US Global Change Research Program*, Cambridge University Press, Cambridge, 109-134. [Accessed 09.02.07: http://www.usgcrp.gov/usgcrp/Library/nationalassessment/04NE.pdf]

BC Stats, 2003: *Tourism Sector Monitor – November 2003*, British Columbia Ministry of Management Services, Victoria, 11 pp. [Accessed 09.02.07: http://www.bcstats.gov.bc.ca/pubs/tour/tsm0311.pdf]

BEA, 2003: *(U.S. Bureau of Economic Analysis), Fixed Assets and Consumer Durable Goods in the United States, 1925-1997*. September 2003, 'Derivation of Depreciation Estimates', pp. M29-M34.

Beaubien, E.G. and H.J. Freeland, 2000: Spring phenology trends in Alberta, Canada: Links to ocean temperature. *Int. J. Biometeorol.*, **44**, 53-59.

Beaulac, I. and G. Doré, 2005: *Impacts du Dégel du Pergélisol sur les Infrastructures de Transport Aérien et Routier au Nunavik et Adaptations - état des connaissances*, Facultées Sciences et de Génie, Université Laval, Montreal, Quebec, 141 pp.

Beebee, T.J.C., 1995: Amphibian breeding and climate. *Nature*, **374**, 219-220.

Bélanger, G., P. Rochette, Y. Castonguay, A. Bootsma, D. Mongrain and A.J. Ryand, 2002: Climate change and winter survival of perennial forage crops in Eastern Canada. *Agron. J.*, **94**, 1120-1130.

Bell, M.L., R. Goldberg, C. Hogrefe, P. Kinney, K. Knowlton, B. Lynn, J. Rosenthal, C. Rosenzweig and J.A. Patz, 2007: Climate change, ambient ozone, and health in 50 U.S. cities. *Climatic. Change*, **82,** 61-76.

Bernatchez, P. and J.-M.M. Dubois, 2004: Bilan des connaissances de la dynamique de l'érosion des côtes du Québec maritime laurentien. *Géograph. Phys. Quater.*, **58**, 45-71.

Berthelot, M., P. Friedlingstein, P. Ciais, P. Monfray, J.L. Dufresen, H.L. Treut and L. Fairhead, 2002: Global response of the terrestrial biosphere and CO_2 and climate change using a coupled climate-carbon cycle model. *Global Biogeochem. Cy.*, **16**, 10.1029/2001GB001827.

Bertness, M.D., P.J. Ewanchuk and B.R. Silliman, 2002: Anthropogenic modification of New England salt marsh landscapes. *Proc. Nat. Acad. Sci.*, **99**, 1395-1398.

Bettigole, N.H., 1990: Designing Bridge Decks to Match Bridge Life Expectancy. in *Extending the Life of Bridges, ASTM Special Technical Publication 1100*, ASTM Committee D-4 on Road and Paving Materials, Philadelphia, Pennsylvania, 70-80.

Bixel, P.B. and E.H. Turner, 2000: *Galveston and the 1900 Storm: Catastrophe and Catalyst*. University of Texas Press, Austin, Texas, 174 pp.

Boisvenue, C. and S.W. Running, 2006: Impacts of climate change on natural forest productivity - evidence since the middle of the 20th century. *Global Change Biol.*, **12**, 862-882.

Bonsal, B.R. and T.D. Prowse, 2003: Trends and variability in spring and autumn 0°C-isotherm dates over Canada. *Clim. Change*, **57**, 341-358.

Bonsal, B.R., X. Zhang, L.A. Vincent and W.D. Hood, 2001: Characteristics of daily and extreme temperatures over Canada. *J. Climate*, **14**, 1959-1976.

Bonsal, B., G. Koshida, E.G. O'Brien and E. Wheaton, 2004: Droughts. *Threats to Water Availability in Canada*, Environment Canada, Ed., Environment Canada, National Water Resources Institute and Meteorological Service of Canada, NWRI Scientific Assessment Report Series No. 3, ACSD Science Assessment Series No.1, Burlington, Ontario, 19-25.

Borwick, J., J. Buttle and M.S. Ridgway, 2006: A topographic index approach for identifying groundwater habitat of young-of-year brook trout (*Salvelinus fontinalis*) in the land-lake ecotone. *Can. J. Fish. Aquatic Sci.*, **63**, 239-253.

Breslow, P.B. and D.J. Sailor, 2002: Vulnerability of wind power resources to climate change in the continental United States. *Renew. Energ.*, **27**, 585-598.

Bromirski, P.D., R.E. Flick and D.R. Cayan, 2003: Storminess variability along the California coast: 1958-2000. *J. Climate*, **16**, 982-993.

Brown, J.L., S.H. Li and B. Bhagabati, 1999: Long-term trend toward earlier breeding in an American bird: A response to global warming? *Proc. Nat. Acad Sci.*, **96**, 5565-5569.

Brown, R.A., N.J. Rosenberg, C.J. Hays, W.E. Easterling and L.O. Mearns, 2000: Potential production and environmental effects of switchgrass and traditional crops under current and greenhouse-altered climate in the central United States: a simulation study. *Agric. Ecosyst. Environ.*, **78**, 31-47.

Brown, T. J., B. L. Hall and A. L. Westerling, 2004: The impact of twenty-first century climate change on wildland fire danger in the western United States: An applications perspective. *Clim. Change*, **62**, 365-388.

Brownstein, J.S., T.R. Holford and D. Fish, 2003: A climate-based model predicts the spatial distribution of Lyme disease vector *Ixodes scapularis* in the United States. *Environ. Health. Perspect.*, **111**, 1152-1157.

Bruce, J.P., 1999: Disaster loss mitigation as an adaptation to climate variability and change. *Mitigation Adapt. Strategies Global Change*, **4**, 295-306.

Bureau of Transportation Statistics, 2006: *Economic Indexes: Transportation Services Index*. United States Department of Transportation, Washington, District of Columbia. [Accessed 09.02.07: http://www.bts.gov/publications/white_house_economic_statistics_briefing_room/october_2005/html/transportation_services_index.html]

Burkett, V.R., 2002: The Potential Impacts of Climate Change on Transportation. Workshop Summary and Proceedings, October 1-2, 2002, DOT Center for Climate Change and Environmental Forecasting - Federal Research Partnership Workshop, Washington, District of Columbia. [Accessed 09.02.07: http://climate.volpe.dot.gov/workshop1002/]

Burkett, V.R., D.A. Wilcox, R. Stottlemyer, W. Barrow, D. Fagre, J. Baron, J. Price, J.L. Nielsen, C.D. Allen, D.L. Peterson, G. Ruggerone and T. Doyle, 2005: Nonlinear dynamics in ecosystem response to climatic change: Case studies and policy implications. *Ecol. Complexity*, **2**, 357-394.

Burleton, D., 2002: *Slowing Population, Ageing Workforce Trends More Severe in Canada than in the U.S.*, Executive Summary for TD Economics, Toronto, Ontario, 3 pp.

Burton, I. and B. Lim, 2005: Achieving adequate adaptation in agriculture. *Clim. Change*, **70**, 191-200.

Burton, I., S. Huq, B. Lim, O. Pilifosova and E.L. Schipper, 2002: From impacts assessment to adaptation priorities: the shaping of adaptation policy. *Climate Policy*, **2**, 145-159.

Business Week, 2005: A Second Look at Katrina's Cost. *Business Week*. September 13, 2005. [Accessed 09.02.07: http://www.businessweek.com/bwdaily/dnflash/sep2005/nf20050913_8975_db082.htm]

Butler, A., 2002: Tourism burned: visits to parks down drastically, even away from flames. *Rocky Mountain News*. July 15, 2002.

Butler, C.J., 2003: The disproportionate effect of global warming on the arrival dates of short-distance migratory birds in North America. *Ibis*, **145**, 484-495.

Buttle, J., J.T. Muir and J. Frain, 2004: Economic impacts of climate change on the Canadian Great Lakes hydro–electric power producers: A supply analysis. *Can. Water Resour. J.*, **29**, 89-109.

Byers, S., O. Snowe, B. Carr, J.P. Holdren, M.K. Kok-Peng, N. Kosciusko-Morizet, C. Martin, T. McMichael, J. Porritt, A. Turner, E.U. von Weizsäcker, N. Weidou, T.E. Wirth and C. Zoi, 2005: *Meeting the Climate Challenge, Recommendations of the International Climate Change Taskforce*, International Climate Change Taskforce, The Institute for Public Policy Research, London, 40 pp. [Accessed 09.02.07: http://www.americanprogress.org/atf/cf/%7BE9245FE4-9A2B-43C7-A521-5D6FF2E06E03%7D/CLIMATECHALLENGE.PDF]

California Regional Assessment Group, 2002: *The Potential Consequences of Climate Variability and Change for California: The California Regional Assessment*. National Center for Ecological Analysis and Synthesis, University of California Santa Barbara, Santa Barbara, California, 432 pp. [Accessed 09.02.07: http://www.ncgia.ucsb.edu/pubs/CA_Report.pdf]

Campbell, P.R., 1996: *Population Projections for States by Age, Sex, Race, and Hispanic Origin: 1995 to 2025*, U.S. Bureau of the Census, Population Division, Washington, District of Columbia, PPL-47. [Accessed 09.02.07: http://www.census.gov/population/www/projections/ppl47.html]

Carbone, G.J., W. Kiechle, L. Locke, L.O. Mearns, L. McDaniel and M.W. Downton, 2003: Response of soybean and sorghum to varying spatial scales of climate change scenarios in the southeastern United States. *Clim. Change*, **60**, 73-98.

Carroll, A.L., S.W. Taylor, J. Regniere and L. Safranyik, 2003: Effects of climate change on range expansion by the mountain pine beetle of British Columbia. *Mountain Pine Beetle Symposium*. Canadian Forest Service, Pacific Forestry Centre, Kelowna, British Columbia, 223-232.

Caspersen, J.P., S.W. Pacala, J.C. Jenkins, G.C. Hurtt, P.R. Moorcroft and R.A. Birdsey, 2000: Contributions of land-use history to carbon accumulation in U.S. forests. *Science*, **290**, 1148-1151.

Casselman, J.M., 2002: Effects of temperature, global extremes, and climate change on year-class production of warmwater, coolwater, and coldwater fishes in the Great Lakes basin. *Amer. Fish. Soc. Symp.*, **32**, 39-60.

Cayan, D., M. Tyree and M. Dettinger, 2003: *Climate Linkages to Female Culex Cx. Tarsalis Abundance in California*. California Applications Program (UCSD), San Diego, California. [Accessed 09.02.07: http://meteora.ucsd.edu/cap/mosq_climate.html]

Cayan, D.R., S.A. Kammerdiener, M.D. Dettinger, J.M. Caprio and D.H. Peterson, 2001: Changes in the onset of spring in the western United States. *Bull. Amer. Meteor. Soc.*, **82**, 399-415.

CCME, 2003: *Climate, Nature, People: Indicators of Canada's Changing Climate*. Climate Change Indicators Task Group of the Canadian Council of Ministers of the Environment, Canadian Council of Ministers of the Environment Inc., Winnipeg, Canada, 51 pp.

CDLI, 1996: Collapse of the Resource Base. *The History of the Northern Cod Fishery*, The Centre for Distance Learning and Innovation, Newfoundland and Labrador Department of Education St. Johns, Newfoundland. [Accessed 09.07.02: http://www.cdli.ca/cod/.]

Census Bureau, 2000: NP-T1. Annual Projections of the Total Resident Population as of July 1: Middle, Lowest, Highest, and Zero International Migration Series, 1999 to 2100. Population Division, U.S. Census Bureau, Washington, District of Columbia. [Accessed 09.02.07: http://www.census.gov/population/projections/nation/summary/np-t1.txt.]

Census Bureau, 2004: *(U.S. Census Bureau), American Housing Survey for the United States: 2003*, U.S. Census Bureau, Washington, District of Columbia, 592 pp. [Accessed 09.02.07: http://www.census.gov/hhes/www/housing/ahs/ahs03/ahs03.html]

CERES, 2004: *Investor Guide to Climate Risk: Action Plan and Resource for Plan*

Sponsors, Fund Managers and Corporations, D.G. Cogan, Ed., CERES, Inc., Boston, Massachusetts, 20 pp. [Accessed 09.02.07: http://www.ceres.org/pub/docs/Ceres_investor_guide_072304.pdf]

Changnon, S.A., 1993: Changes in climate and levels of Lake Michigan: shoreline impacts at Chicago. *Clim. Change*, **23**, 213-230.

Changnon, S.A., 2003: Shifting economic impacts from weather extremes in the United States: A result of societal changes, not global warming. *Nat. Hazards*, **29**, 273-290.

Changnon, S.A., 2005: Economic impacts of climate conditions in the United States: Past, present, and future - An editorial essay. *Clim. Change*, **68**, 1-9.

Changnon, S.A. and M.H. Glantz, 1996: The Great Lakes diversion at Chicago and its implications for climate change. *Clim. Change*, **32**, 199-214.

Changnon, S.A. and D. Changnon, 2000: Long-term fluctuations in hail incidences in the United States. *J. Climate*, **13**, 658-664.

Chapin, F.S., III, E.S. Zavaleta, V.T. Eviner, R.L. Naylor, P.M. Vitousek, H.L. Reynolds, D.U. Hooper, S. Lavorel, O.E. Sala, S E. Hobbie, M.C. Mack and S. Díaz, 2000: Consequences of changing biodiversity. *Nature*, **405**, 234 - 242.

Charron, D.F., 2002: Potential impacts of global warming and climate change on the epidemiology of zoonotic diseases in Canada. *Can. J. Public Health*, **93**, 334-335.

Chen, C., D. Gillig and B. McCarl, 2001: Effects of climatic change on a water dependent regional economy: a study of the Texas Edwards aquifer. *Clim. Change*, **49**, 397-409.

Cheng, S., M. Campbell, Q. Li, L. Guilong, H. Auld, N. Day, D. Pengelly, S. Gingrich, J. Klaassen, D. MacIver, N. Comer, Y. Mao, W. Thompson and H. Lin, 2005: Differential and Combined Impacts of Winter and Summer Weather and Air Pollution due to Global Warming on Human Mortality in South-Central Canada. Technical report (Health Policy Research Program: Project Number 6795-15-2001/4400011).

Chmura, G.L. and G.A. Hung, 2004: Controls on salt marsh accretion: A test in salt marshes of Eastern Canada. *Estuaries*, **27**, 70-81.

Chmura, G.L., S.C. Anisfeld, D.R. Cahoon and J.C. Lynch, 2003: Global carbon sequestration in tidal, saline wetland soils. *Global Biogeochem. Cycles*, **17**, doi:10.1029/2002GB001917.

Christensen, J.H., B. Hewitson, A. Busuioc, A. Chen, X. Gao, I. Held, R. Jones, R.K. Kolli, W.-T. Kwon, R. Laprise, V. Magaña Rueda, L. Mearns, C.G. Menendez, J. Räisänen, A. Rinke, A. Sarr and P. Whetton, 2007: Regional climate projections. *Climate Change 2007: The Physical Science Basis. Contribution of Working Group I to the Fourth Assessment Report of the Intergovernmental Panel on Climate Change*, S. Solomon, D. Qin, M. Manning, Z. Chen, M. Marquis, K.B. Averyt, M. Tignor and H.L. Miller, Eds., Cambridge University Press, Cambridge and New York, 847-940.

Christensen, N.S., A.W. Wood, N. Voisin, D.P. Lettenmaier and R.N. Palmer, 2004: The effects of climate change on the hydrology and water resources of the Colorado River basin. *Clim. Change*, **62**, 337-363.

Chu, C., C.K. Minns and N.E. Mandrak, 2003: Comparative regional assessment of factors impacting freshwater fish biodiversity in Canada. *Canadian Journal of Fisheries and Aquatic Sciences*, **60**, 624-634.

Chu, C., N.E. Mandrak and C.K. Minns, 2005: Potential impacts of climate change on the distributions of several common and rare freshwater fishes in Canada. *Divers. Distrib.*, **11**, 299-310.

Church, J.A., N.J. White, R. Coleman, K. Lambeck and J.X. Mitrovica, 2004: Estimates of the regional distribution of sea level rise over the 1950-2000 period. *J. Climate*, **17**, 2609-2625.

City of New York, 2005: *New York City's Water Supply System*. The City of New York Department of Environmental Protection, New York, New York. [Accessed 09.07.02: http://www.ci.nyc.ny.us/html/dep/html/watersup.html]

Clark, G.E., S.C. Moser, S.J. Ratick, K. Dow, W.B. Meyer, S. Emani, W. Jin, J.X. Kasperson, R.E. Kasperson and H.E. Schwarz, 1998: Assessing the vulnerability of coastal communities to extreme storms: the case of Revere, MA, USA. *Mitigation Adap. Strategies Global Change*, **3**, 59-82.

Clark, M.E., K.A. Rose, D.A. Levine and W.W. Hargrove, 2001: Predicting climate change effects on Appalachian trout: Combining GIS and individual-based modeling. *Ecol. Appl.*, **11**, 161-178.

Clayton, A., J. Montufar, J. Regehr, C. Isaacs and R. McGregor, 2005: Aspects of the potential impacts of climate change on seasonal weight limits and trucking in the prairie region, Prepared for Natural Resources Canada. [Accessed 10.06.07: http:www.adaptation.nrcan.gc.ca/projdb/pdf/135a_e.pdf]

Co-operative Programme on Water and Climate, 2005: *Workshop 3, Climate Variability, Water Systems and Management Options,* Co-operative Programme on Water and Climate, Delft, The Netherlands, 5 pp.

Cohen, S., K. Miller, K. Duncan, E. Gregorich, P. Groffman, P. Kovacs, V. Magaña, D. McKnight, E. Mills and D. Schimel. 2001: North America. *Climate Change 2001: Impacts, Adaptation, and Vulnerability. Contribution of Working Group II to the Third Assessment Report of the Intergovernmental Panel on Climate Change*, J.J. McCarthy, O.F. Canziani, N.A. Leary, D.J. Dokken and K.S. White, Eds., Cambridge University Press, Cambridge, 735-800.

Cohen, S.J., R. de Loë, A. Hamlet, R. Herrington, L.D. Mortsch and D. Shrubsole, 2003: Integrated and cumulative threats to water availability. *Threats to Water Availability in Canada*, National Water Research Institute, Burlington, Ontario, 117-127. [Accessed 09.02.07: http://www.nwri.ca/threats2full/ ThreatsEN_03web.pdf]

Cohen, S., D. Neilsen, S. Smith, T. Neale, B. Taylor, M. Barton, W. Merritt, Y. Alila, P. Shepherd, R. McNeill, J. Tansey, J. Carmichael and S. Langsdale, 2006: Learning with local help: expanding the dialogue on climate change and water management in the Okanagan Region, British Columbia, Canada. *Clim. Change*, **75**, 331-358.

Cole, H., V. Colonell and D. Esch, 1998: *The economic impact and consequences of global climate change on Alaska's infrastructure*. Assessing the Consequences of Climate Change for Alaska and the Bering Sea Region, 1999, Center for Global Change and Arctic System Research, University of Alaska Fairbanks, Fairbanks, Alaska. 3 pp. [Accessed 09.02.07: http://www.besis.uaf.edu/besis-oct98-report/Infrastructure-1.pdf]

Columbia University Mailman School of Public Health, 2006: *On The Edge – The Louisiana Child & Family Health Study,* Columbia University Mailman school of Public Health, New York, New York.

Conference Board of Canada, 2006: *Adapting to Climate Change: Is Canada Ready?,* The Conference Board of Canada, Ottawa, Ontario.

Cooke, S.J. and I.G. Cowx, 2004: The role of recreational fishing in global fish crises. *BioScience*, **54**, 857-859.

Cowell, P.J., M.J.F. Stive, A.W. Niedoroda, H.J. de Vriend, D.J.P. Swift, G.M. Kaminsky and M. Capobianco, 2003: The coastal tract (part 1): a conceptual approach to aggregated modeling of low-order coastal change. *J. Coastal Res.*, **19**, 812-827.

Cox, P.M., R.A. Betts, C.D. Jones, S.A. Spall and I.J. Totterdell, 2000: Acceleration of global warming due to carbon-cycle feedbacks in a coupled climate model. *Nature*, **408**, 184-187.

Crichton, D., 2003: Insurance and Maladaptation. Preprints, *United Nations Framework Convention on Climate Change Workshop*, Bonn, Germany.

Croley, T.E., II, 2003: Great Lakes Climate Change Hydrological Impact Assessment, IJC Lake Ontario—St. Lawrence River Regulation Study. NOAA Tech. Memo. GLERL-126, Great Lakes Environmental Research Laboratory, Ann Arbor, Michigan, 84 pp.

Croley, T.E. and C.L. Luukkonen, 2003: Potential effects of climate change on ground water in Lansing, Michigan. *J. Amer. Water Resour. Assoc.*, **39**, 149-163.

Crozier, L., 2003: Winter warming facilitates range expansion: Cold tolerance of the butterfly *Atalopedes campestris*. *Oecologia*, **135**, 648-656.

Currie, D.J., 2001: Projected effects of climate change on patterns of vertebrate and tree species in the conterminous United States. *Ecosystems*, **4**, 216-225.

Curriero, F.C., J.A. Patz, J.B. Rose and S. Lele, 2001: The association between extreme precipitation and waterborne disease outbreaks in the United States, 1948-1994. *Am. J. Public Health*, **91**, 1194-1199.

Curriero, F.C., K.S. Heiner, J.M. Samet, S.L. Zeger, L. Strung and J.A. Patz, 2002: Temperature and mortality in 11 cities of the eastern United States. *Am. J. Epidemiol.*, **155**, 80-87.

Cutter, S.L., J.T. Mitchell and M.S. Scott, 2000: Revealing the vulnerability of people and place: a case study of Georgetown County, South Carolina. *Ann. Assoc. Am. Geog.*, **90**, 713-737.

D'Arcy, P., J.-F. Bibeault and R. Raffa, 2005: *Changements climatiques et transport maritime sur le Saint-Laurent. Étude exploratoire d'options d'adaptation*. Réalisé pour le Comité de concertation navigation du Plan d'action Saint-Laurent, Montreal, Quebec, 140 pp. [Accessed 09.02.07: http://www.ouranos.ca/doc/Rapports%20finaux/Final_Darcy.pdf

Dale, V.H., L.A. Joyce, S. McNulty, R.P. Neilson, M.P. Ayres, M.D. Flannigan, P.J. Hanson, L.C. Irland, A.E. Lugo, C.J. Peterson, D. Simberloff, F.J. Swanson, B.J. Stocks and B.M. Wotton, 2001: Climate change and forest disturbances. *BioScience*, **51**, 723-734.

Denault, C., R.G. Millar and B.J. Lence, 2002: Climate change and drainage infrastructure capacity in an urban catchment. *Proc. Annual Conference of the Canadian Society for Civil Engineering*, June 5-6, 2002, Montreal, Quebec, Canada, 10 pp. [Accessed 09.02.07: http://www.c-ciarn.mcgill.ca/millar.pdf]

Devon, 2005: *A Warm Response: Our Climate Change Challenge*. Devon County

Council, Devon, 122 pp. [Accessed 09.02.07: http://www.devon.gov.uk/climate-change-strategy.pdf]

DFO-MPO, 2002: *DFO's Statistical Services*. Department of Fisheries and Oceans. [Accessed 09.02.07: www.dfo-mpo.gc.ca/communic/statistics/]

Dlugolecki, A.F., 2005: What is stopping the finance sector? *IEEE Power Engineering Society General Meeting*, **3**, 2951-2953.

Dohm, D.J. and M.J. Turell, 2001: Effect of incubation at overwintering temperatures on the replication of West Nile virus in New York *Culex pipiens* (Diptera: Culicidae). *J. Med. Entomol.*, **38**, 462-464.

Dohm, D.J., M.L. O'Guinn and M.J. Turell, 2002: Effect of environmental temperature on the ability of *Culex pipiens* (Diptera: Culicidae) to transmit West Nile virus. *J. Med. Entomol.*, **39**, 221-225.

DOI, 2002: *2001 National Survey of Fishing, Hunting, and Wildlife-Associated Recreation*. U.S. Dept. Interior, Fish and Wildlife Service and U.S. Dept. Commerce, U.S. Census Bureau, Washington, District of Columbia, 170 pp. [Accessed 09.02.07: http://www.census.gov/prod/2002pubs/FHW01.pdf]

Dolan, A.H. and I.J. Walker, 2006: Understanding vulnerability of coastal communities to climate change related risks. *J. Coastal Res.*, **SI 39** 1317-1324.

Dolney, T.J. and S.C. Sheridan, 2006: The relationship between extreme heat and ambulance response calls for the city of Toronto, Ontario, Canada. *Environ. Res.*, **101**, 94-103.

Donnelly, J.P. and M.D. Bertness, 2001: Rapid shoreward encroachment of salt marsh cordgrass in response to accelerated sea-level rise. *Proc. Nat. Acad. Sci.*, **98**, 14218-14223.

Donnelly, J.P., P. Cleary, P. Newby and R. Ettinger, 2004: Coupling instrumental and geological records of sea-level change: Evidence from southern New England of an increase in the rate of sea-level rise in the late 19th century. *Geophys. Res. Lett.*, **31**, doi:10.1029/2003GL018933.

Dougherty, B. and B. Osaman Elasha, 2004: Mainstreaming adaptation into national development plans. *Report of the Second AIACC Africa and Indian Ocean Island Regional Workshop*, University of Senegal, Dakar, 74 pp. [Accessed 09.02.07: http://www.aiaccproject.org/meetings/Dakar_04/Dakar_Final.pdf]

Douglas, B.C. and W.R. Peltier, 2002: The puzzle of global sea-level rise. *Phys. Today*, **55**, 35-40.

Dow, K., R.E. O'Connor, B. Yarnal, G.J. Carbone and C.L. Jocoy, 2007: Why worry? Community water system managers' perceptions of climate vulnerability. Global Environ. Change, **17**, 228-237.

du Vair, P., D. Wickizer and M.J. Burer, 2002: Climate change and the potential implications for California's transportation system. *The Potential Impacts of Climate Change on Transportation, Federal Research Partnership Workshop*, October 1-2, 2002, Washington, District of Columbia, 125-135. [Accessed 09.02.07: http://climate.volpe.dot.gov/workshop1002/]

Duguid, T., 2002: Flood Protection Options for the City of Winnipeg, Report to the Government of Manitoba on Public Meetings, Winnepeg, Manitoba, Canada, 31 pp. [Accessed 09.02.07: http://www.cecmanitoba.ca/Reports/PDF/ACF44E4.pdf]

Dunn, P.O. and D.W. Winkler, 1999: Climate change has affected the breeding date of tree swallows throughout North America. *Proc. R. Soc. Lond. B*, **266**, 2487-2490.

Dupigny-Giroux, L.-A., 2001: Towards characterizing and planning for drought in Vermont – Part I: A climatological perspective. *J. Amer. Water Resour. Assoc.*, **37**, 505-525.

Dwight, R.H., J.C. Semenza, D.B. Baker and B.H. Olson, 2002: Association of urban runoff with coastal water quality in Orange County, California. *Water Environ. Res.*, **74**, 82-90.

Dyke, A.S. and W.R. Peltier, 2000: Forms, response times and variability of relative sea-level curves, glaciated North America. *Geomorphology*, **32**, 315-333.

Easterling, D.R., 2002: Recent changes in frost days and the frost-free season in the United States. *Bull. Amer. Meteor. Soc.*, **83**, 1327-1332.

Easterling, D.R., G.A. Meehl, C. Parmesan, S.A. Changnon, T.R. Karl and L.O. Mearns, 2000: Climate extremes: Observations, modeling, and impacts. *Science*, **289**, 2068-2074.

Easterling, W.E., N. Chhetri and X.Z. Niu, 2003: Improving the realism of modeling agronomic adaptation to climate change: Simulating technological substitution. *Clim. Change*, **60**, 149-173.

Easterling, W., B. Hurd and J. Smith, 2004: *Coping with Global Climate Change: The Role of Adaptation in the United States,* Pew Center on Global Climate Change, Arlington, Virginia, 52 pp. [Accessed 09.02.07: http://www.pewclimate.org/document.cfm?documentID=319]

Ebi, K.L., T.J. Teisberg, L.S. Kalkstein, L. Robinson and R.F. Weiher, 2004: Heat watch/warning systems save lives: Estimated costs and benefits for Philadelphia 1995-98. *Bull. Amer. Meteor. Soc.*, **85**, 1067-1073.

Economic Research Service, 2006: Farm Income and Costs: Farm Sector Income Forecast. United States Department of Agriculture Economic Research Service, Washington, District of Columbia. [Accessed 09.02.07: http://www.ers.usda.gov/briefing/farmincome/data/nf_t2.htm]

Edmonds, J.A., 2004: Unanticipated consequences: Thinking about ancillary benefits and costs of greenhouse gas emissions mitigation. *The Global Carbon Cycle: Integrating Humans, Climate, and the Natural World*, C.B. Field and M.R. Raupach, Eds., Island Press, Washington, District of Columbia, 419-430.

Edmonds, J.A. and N.J. Rosenberg, 2005: Climate change impacts for the conterminous USA: An integrated assessment summary. *Clim. Change*, **69**, 151-162.

EEA, 2005: Hurricane damage to natural gas infrastructure and its effect on U.S. natural gas market. EEA (Energy and Environmental Analysis, Inc.), Arlington, Virginia, 49 pp. [Accessed 09.02.07: http://www.ef.org/documents/hurricanereport_final.pdf]

EEI, 2005: *After the Disaster: Utility Restoration Cost Recovery*. Edison Electric Institute (EEI), Washington, District of Columbia, 27 pp. [Accessed 09.02.07: http://www.eei.org/industry_issues/reliability/nonav_reliability/Utility_Restoration_Cost_Recovery.pdf]

Eggers, D., 2006: *Run Forecasts and Harvest Projections for 2006 Alaskan Salmon Fisheries and Review of the 2005 Season*. Alaska Department of Fish and Game, Special Publication No. 06-07, Juneau, Alaska, 83 pp.

Eheart, J.W., A.J. Wildermuth and E.E. Herricks, 1999: The effects of climate change and irrigation on criterion low streamflows used for determining total maximum daily loads. *J. Amer. Water Resour. Assoc.*, **35**, 1365-1372.

EIA, 1999: Commercial Building Energy Consumption Survey: Building Characteristics tables. Table B-9. Year Constructed, Floorspace, 1999. U.S. Energy Information Administration, Washington, District of Columbia. [Accessed 09.02.07: http://www.eia.doe.gov/emeu/cbecs/detailed_tables_1999.html]

EIA, 2001: 2001 Residential Energy Consumption Survey: housing characteristics tables, Table HC1-2a. Housing unit characteristics by year of construction, million U.S. households, 2001. U.S. Energy Information Administration, Washington, District of Columbia.

EIA, 2005a: *Short Term Energy Outlook, December 2005*. U.S. Energy Information Administration, Washington, District of Columbia, 49 pp. [Accessed 09.02.07: http://www.eia.doe.gov/pub/forecasting/steo/oldsteos/dec05.pdf]

EIA, 2005b: *International Energy Annual*. U.S. Energy Information Administration, May 2005. [Accessed 09.02.07: http://www.eia.doe.gov/iea/]

Elliott, G.P. and W.L. Baker, 2004: Quaking aspen at treeline: A century of change in the San Juan Mountains, Colorado, USA. *J. Biogeog.*, **31**, 733-745.

Elsasser, H., R. Bürki and B. Abegg, 2003: Climate change and winter sports: environmental and economic threats. *Fifth World Conference on Sport and the Environment*, Turin, IOC/UNEP, 8 pp. [Accessed 09.02.07: www.unep.org/sport_env/Documents/torinobuerki.doc]

Emanuel, K., 2005: Increasing destructiveness of tropical cyclones over the past 30 years. *Nature*, **436**, 686-688.

Enarson, E., 2002: Gender Issues in Natural Disasters: Talking Points on Research Needs. *Crisis, Women and Other Gender Concerns*, Working Paper 7, Recovery and Reconstruction Department, ILO, Geneva, 5-12.

Environment Canada, 2004: *Threats to water availability in Canada*. National Water Research Institute, Burlington, Ontario,150 pp. [Accessed 09.02.07: http://www.nwri.ca/threats2full/ThreatsEN_03web.pdf]

EPA, 1999: Global Climate Change: What Does it Mean for South Florida and the Florida Keys? *Report of Florida Coastal Cities Tour*, U.S. Environmental Protection Agency, Washington, District of Columbia. [Accessed 09.02.07: http://yosemite.epa.gov/oar/globalwarming.nsf/UniqueKeyLookup/SHSU5BUKPX/ $File/florida.pdf]

EPRI, 2003: *Electricity Sector Framework for the Future. Volume I. Achieving the 21st Century Transformation*. Electric Power Research Institute, Palo Alto, California, 77 pp. [Accessed 09.02.07: http://www.globalregulatorynetwork.org/PDFs/ESFF_volume1.pdf]

Epstein, P. and E. Mills, 2005: *Climate Change Futures: Health, Ecological and Economic Dimensions*. Harvard Medical School, Boston, Massachusetts, 142 pp. [Accessed 09.02.07: http://chge.med.harvard.edu/research/ccf/documents/ccf_final_report.pdf]

Fang, X. and H.G. Stefan, 1999: Projections of climate change effects on water temperature characteristics of small lakes in the contiguous U.S. *Clim. Change*, **42**, 377-412.

Fang, X., H.G. Stefan, J.G. Eaton, J.H. McCormick and S.R. Alam, 2004: Simulation of thermal/dissolved oxygen habitat for fishes in lakes under different climate scenarios - Part 1. Cool-water fish in the contiguous US. *Ecol. Model.*, **172**, 13-37.

FEMA, 2006: *Flood Map Modernization: Mid-Course Adjustment,* Federal Emergency Management Agency, Washington, District of Columbia, 70 pp.

Feng, S. and Q. Hu, 2004: Changes in agro-meteorological indicators in the contiguous United States: 1951–2000. *Theor. Appl. Climatol.*, **78**, 247-264.

Filion, Y., 2000: Climate change: implications for Canadian water resources and hydropower production. *Can. Water Resour. J.*, **25**, 255-270.

FireSmart, 2005: *Fire Smart*. Government of Alberta, Edmonton, Canada. [Accessed 09.02.07: http://www.partnersinprotection.ab.ca/downloads/]

FireWise, 2005: *Fire Wise*. National Fire Protection Association, Quincy. Massachusetts. [Accessed 09.02.07: http://www.firewise.org/]

Fischhoff, B. 2006: Behaviorly realistic risk management. *On Risk and Disaster*, H. Kunreuther, R. Danielsand D. Kettle, Eds., University of Pennsylvania Press, Philadelphia.

Fisher, A., 2000: Preliminary findings from the mid-Atlantic regional assessment. *Climate Res.*, **14**, 261-269.

Fisheries and Oceans Canada, 2000: *Dhaliwal moves ahead with $15M in federal funding for emergency dredging in the Great Lakes*. [Accessed 09.02.07: http://www.dfo-mpo.gc.ca/media/newsrel/2000/hq-ac53_e.htm]

Flannigan, M. D., K. A. Logan, B. D. Amiro, W. R. Skinner and B. J. Stocks, 2004: Future area burned in Canada. *Clim. Change*, **72**, 1-16.

Fletcher, M., 2004: Blackout sheds light on outage risks; Dark days of 2003 teach lessons. *Business Insurance*, 1-4. May 24, 2004.

Fleury, M.D., D. Charron, J. Holt, B. Allen and A. Maarouf, 2006: The role of ambient temperature in foodborne disease in Canada using time series methods *Int. J. Biometeorol.*, **50**, DOI 10.1007/s00484-00006-00028-00489.

Foley, J.A., R. DeFries, G.P. Asner, C. Barford, G. Bonan, S.R. Carpenter, F.S. Chapin, M.T. Coe, G.C. Daily, H.K. Gibbs, J.H. Helkowski, T. Holloway, E.A. Howard, C.J. Kucharik, C. Monfreda, J.A. Patz, I.C. Prentice, N. Ramankutty and P.K. Snyder, 2005: Global consequences of land use. *Science*, **309**, 570-574.

Fontaine, T.A., J.F. Klassen, T.S. Cruickshank and R.H. Hotchkiss, 2001: Hydrological response to climate change in the Black Hills of South Dakota, USA. *Hydrol. Sci.*, **46**, 27-40.

Forbes, D.L., G.K. Manson, R. Chagnon, S.M. Solomon, J.J. van der Sanden and T.L. Lynds, 2002a: Nearshore ice and climate change in the southern Gulf of St. Lawrence. *Ice in the environment. Proceedings 16th IAHR International Symposium on Ice*, Dunedin, New Zealand, 1, 344-351.

Forbes, D.L., R.W. Shaw and G.K. Manson. 2002b: Adaptation. *Coastal Impacts of Climate Change and Sea-Level Rise on Prince Edward Island*, D.L. Forbes and R.W. Shaw, Eds., Geological Survey of Canada Open File 4261, Supporting Document 11, Natural Resources Canada, Dartmouth, Nova Scotia, 1-18.

Forbes, D.L., G.S. Parkes, G.K. Manson and L.A. Ketch, 2004: Storms and shoreline retreat in the southern Gulf of St. Lawrence. *Marine Geol.*, **210**, 169-204.

Forister, M.L. and A.M. Shapiro, 2003: Climatic trends and advancing spring flight of butterflies in lowland California. *Global Change Biol.*, **9**, 1130-1135.

Fung, I., S.C. Doney, K. Lindsay and J. John, 2005: Evolution of carbon sinks in a changing climate. *Proc. Nati. Acad. Sci.*, **102**, 11201-11206.

Gallagher, P. and L. Wood, 2003: *Proc. The World Summit on Salmon*, June 10-13, 2003, Vancouver, British Columbia [Accessed 09.02.07: www.sfu.ca/cstudies/science/summit.htm]

Gamache, I. and S. Payette, 2004: Height growth response of tree line black spruce to recent climate warming across the forest-tundra of eastern Canada. *J. Ecol.*, **92**, 835-845.

Gamble, J.L., J. Furlow, A.K. Snover, A.F. Hamlet, B.J. Morehouse, H. Hartmann and T. Pagano, 2004: Assessing the Impact of Climate Variability and Change on Regional Water Resources: The Implications for Stakeholders, in Water: Science, Policy, and Management, Lawford, R., D. Fort, H. Hartmannand S. Eden, Eds., Water Resources Monograph Series, Volume 16, American Geophysical Union, Washington, District of Columbia.

Gan, J.B., 2004: Risk and damage of southern pine beetle outbreaks under global climate change. *Forest Ecol. Manag.*, **191**, 61-71.

GAO, 2003: *Freshwater Supply: States' Views of How Federal Agencies Could Help Them Meet the Challenges of Expected Shortages,* GAO-03-514, U.S. Governmental Accountability Office. U.S. Congress, General Accounting Office, Washington, District ofColumbia, 118 pp. [Accessed 09.02.07: http://www.gao.gov/new.items/d03514.pdf]

Gent, J.F., E.W. Triche, T.R. Holford, K. Belanger, M.B. Bracken, W.S. Beckett and B.P. Leaderer, 2003: Association of low-level ozone and fine particles with respiratory symptoms in children with asthma. *JAMA J. Am. Med. Assoc.*, **290**, 1859-1867.

Gerber, S., F. Joos and I. C. Prentice, 2004: Sensitivity of a dynamic global vegetation model to climate and atmospheric CO_2. *Global Change Biol.*, **10**, 1223-1239.

Gibbs, J.P. and A.R. Breisch, 2001: Climate warming and calling phenology of frogs near Ithaca, New York, 1900-1999. *Conserv. Biol.*, **15**, 1175-1178.

Gillett, N.P., A.J. Weaver, F.W. Zwiers and M.D. Flannigan, 2004: Detecting the effect of climate change on Canadian forest fires. *Geophys. Res. Lett.*, **31**, doi:10.1029/2004GL020876.

Goklany, I., 2000: *Applying the precautionary principle to genetically modified crops,* Centre for the Study of American Business, Washington University, St. Louis, Missouri.

Goklany, I., 2001: Precaution without perversity: A comprehensive application of the precautionary principle to genetically modified crops. *Biotechnol. Law Rep.*, **20**, 377-396.

Goklany, I. M., 2006: Death and Death Rates Due to Extreme Weather Events: Global and U.S. Trends, 1900-2004. *Climate Change and Disaster Losses: Understanding and Attributing Trends and Projections*, Workshop Report, 25-26 May 2006, Hohenkammer, Germany. 103-117.

Goklany, I., 2007: Integrated strategies to reduce vulnerability and advance adaptation, mitigation and sustainable development. Mitigat. Adapt. Strat. Glob. Change, doi 10.1007/s11027-007-9098-1.

Goodwin, B.K., 2003: Does risk matter? Discussion. *Amer. J. Agric. Econ.*, **85**, 1257-1258.[Global; agriculture]

Gooseff, M.N., K. Strzepek and S.C. Chapra, 2005: Modeling the potential effects of climate change on water temperature downstream of a shallow reservoir, lower Madison River, MT. *Clim. Change*, **68**, 331-353.

Gornitz, V. and S. Couch. 2001: Sea level rise and coastal. *Climate Change and a Global City: The Potential Consequences of Climate Variability and Change*, C. Rosenzweig and W.D. Solecki, Eds., Columbia Earth Institute, New York, New York.

Gornitz, V., S. Couch and E.K. Hartig, 2001: Impacts of sea level rise in the New York City metropolitan area. *Glob. Planetary Change*, **32**, 61-88.

Gray, K.N., 1999: *The impacts of drought on Yakima Valley irrigated agriculture and Seattle municipal and industrial water supply*. Masters Thesis, University of Washington, Seattle, Washington, 102 pp.

Groisman, P.Y., R.W. Knight, T.R. Karl, D.R. Easterling, B. Sun and J.H. Lawrimore, 2004: Contemporary changes of the hydrological cycle over the contiguous United States: trends derived from *in situ* observations. *J. Hydrometeorol.*, **5**, 64-85.

Guenther, A., 2002: The contribution of reactive carbon emissions from vegetation to the carbon balance of terrestrial ecosystems. *Chemosphere*, **49**, 837-844.

Gunn, J.M., 2002: Impact of the 1998 El Nino event on a lake charr, *Salvelinus namaycush*, population recovering from acidification. *Environ. Biol. Fishes*, **64**, 343-351.

Guy Carpenter, 2006: *The Catastrophe Bond Market at Year-End 2005,* Guy Carpenter & Company, New York, New York, 24 pp. [Accessed 09.02.07: http://www.guycarp.com/portal/extranet/pdf/GCPub/CatBond_yr_end05.pdf;jsessionid=GZzYQcdnb8ndTQ1P2zG1y7p1HqhTgD0LSzfsTQF15HnGyWDY0FXh!1582131896?vid=1]

Hadley, S.W., D.J. Erickson, III, J.L. Hernandez, C.T. Broniak and T.J. Blasing, 2006: Responses of energy use to climate change: A climate modeling study. *Geophys. Res. Lett.*, **33**, doi:10.1029/2006GL026652.

Hamlet, A.F. 2003: The role of the transboundary agreements in the Columbia River Basin: An integrated assessment in the context of historic development, climate and evolving water policy. *Climate and Water: Transboundary Challenges in the Americas*, H. Diaz and B. Morehouse, Eds., Kluwer, Dordrecht, 263-289.

Hamlet, A. and D. Lettenmaier, 1999: Effects of climate change on hydrology and water resources in the Columbia River Basin. *J. Amer. Water Resour. Assoc.*, **35**, 1597-1623.

Hamlet, A.F., D. Huppert and D.P. Lettenmaier, 2002: Value of long-lead streamflow forecasts for Columbia River hydropower. *J. Water Resour. Plan. Manag. - ASCE*, **128**, 91-101.

Hanesiak, J.M. and X.L.L. Wang, 2005: Adverse-weather trends in the Canadian Arctic. *J. Climate*, **18**, 3140-3156.

Hansen, J.E., R. Ruedy, M. Sato, M. Imhoff, W. Lawrence, D. Easterling, T. Peterson and T. Karl, 2001: A closer look at United States and global surface temperature change. *J. Geophys. Res.*, **106**, 23947-23963.

Hansler, G. and D.C. Major. 1999: Climate change and the water supply systems of New York City and the Delaware Basin: Planning and action considerations for water managers. *Proc. of the Specialty Conference on Potential Consequences of Climate Variability and Change to Water Resources of the United States*, D. Briane Adams, Ed., American Water Resources Association, Herndon, Virginia, 327-330.

Hansom, J.D., 2001: Coastal sensitivity to environmental change: A view from the beach. *Catena*, **42**, 291-305.

Hatfield, J.L. and J.H. Pruger, 2004: Impacts of changing precipitation patterns on water quality. *J. Soil Water Conserv.*, **59**, 51-58.

Hayhoe, K., D. Cayan, C. Field, P. Frumhoff, E. Maurer, N. Miller, S. Moser, S. Schneider, K. Cahill, E. Cleland, L. Dale, R. Drapek, R.M. Hanemann, L. Kalkstein, J. Lenihan, C. Lunch, R. Neilson, S. Sheridan and J. Verville, 2004: Emissions pathways, climate change, and impacts on California. *Proc. Nat. Acad. Sci.*, **101**, 12422–12427.

Heinz Center, 2000: *The Hidden Costs of Coastal Hazards: Implications for Risk Assessment and Mitigation*. The H. John Heinz III Center for Science, Economics and the Environment, Island Press, Washington, District of Columbia, 220 pp.

Hengeveld, H., B. Whitewood and A. Fergusson, 2005: *An Introduction to Climate Change: A Canadian Perspective*. Environment Canada, Downsview, Ontario, 55 pp.

Hersteinsson, P. and D.W. Macdonald, 1992: Interspecific competition and the geographical distribution of red and arctic foxes, *Vulpes vulpes* and *Alopex lagopus*. *Oikos*, **64**, 505-515.

Hicke, J.A. and D.B. Lobell, 2004: Spatiotemporal patterns of cropland area and net primary production in the central United States estimated from USDA agricultural information. *Geophys. Res. Lett.*, **31**, doi: 10.1029/2004GL020927.

Hicke, J.A., G.P. Asner, J.T. Randerson, C.J. Tucker, S.O. Los, R.A. Birdsey, J.C. Jenkins, C. Field and E. Holland, 2002: Satellite-derived increases in net primary productivity across North America 1982-1998. *Geophysical Research Letters*, **29**, doi:10.1029/2001GL013578.

Hill, D. and R. Goldberg, 2001: Energy demand. *Climate Change and a Global City: The Potential Consequences of Climate Variability and Change*, C. Rosenzweig and W.D. Solecki, Eds., Columbia Earth Institute, New York, New York [Accessed 09.02.07: http://metroeast_climate.ciesin.columbia.edu/energy.html]

Hogg, E.H., J.P. Brandt and B. Kochtubajda, 2002: Growth and dieback of aspen forests in northwestern Alberta, Canada in relation to climate and insects. *Can. J. For. Res.*, **32**, 823-832.

Hogrefe, C., B. Lynn, K. Civerolo, J. Rosenthal, C. Rosenzweig, R. Goldberg, S. Gaffin, K. Knowlton and P.L. Kinney, 2004: Simulating changes in regional air pollution over the eastern United States due to changes in global and regional climate and emissions. *J. Geophys. Res.*, **109**, doi:10.1029/2004JD004690.

Hostetler, S. and E. Small, 1999: Response of both American freshwater lakes to simulated future climates. *J. Amer. Water Resour. Assoc.*, **35**, 1625-1637.

Houser, S., V. Teller, M. MacCracken, R. Gough and P. Spears. 2001: Chapter 12: Potential consequences of climate variability and change for native peoples and homelands. *Climate Change Impacts on the United States: The Potential Consequences of Climate Variability and Change, Report for the US Global Change Research Program*, U.S. National Assessment Synthesis Team, Ed. Cambridge University Press, Cambridge,351-377. [Accessed 09.02.07: http://www.usgcrp.gov/usgcrp/Library/nationalassessment/foundation.htm]

Howard, J.A., 2000: *National Association of Insurance Commissioners Roundtable Meeting,* National Flood Insurance Program, Washington, District of Columbia. [Accessed 09.02.07: http://permanent.access.gpo.gov/lps18804/www.fema.gov/nfip/jahsp13.htm]

Hunt, M., 2005: *Flood Reduction Master Plan*. Presented to the City of Peterborough City Council, Peterborough, Ontario, Canada.

Hutton, D., 2001: *Psychosocial Aspects of Disaster Recovery: Integrating Communities into Disaster Planning and Policy Making*. Institute for Catastrophic Loss Reduction, Toronto, Ontario, 16 pp.

Ikeme, J., 2003: Equity, environmental justice and sustainability: Incomplete approaches in climate change politics. *Global Environ. Change*, **13**, 195-206.

Industry Canada, 2006: *Canada Industry Statistics: Gross Domestic Product (GDP): Transportation and Warehousing (NAICS 48-49)*. Industry Canada, Ottawa, Canada. [Accessed 12.02.07: http://strategis.ic.gc.ca/ canadian_industry_statistics/cis.nsf/IDE/cis48-49gdpe.html]

Instanes, A., O. Anisimov, L. Brigham, D. Goering, L.N. Khrustalev, B. Ladanyi and J.O. Larsen, 2005: Infrastructure: buildings, support systems, and industrial facilities. *Arctic Climate Impact Assessment, ACIA*, Cambridge University Press, New York, 907-944.

ISMA, 2006: *Facts and Figures about Snowmobiling*. International Snowmobile Manufacturers Association. [Accessed 12.02.07: http://www.ccso-ccom.ca/contente.htm]

ISRP/ISAB, 2004: Scientific Review of Subbasin Plans for the Columbia River Basin Fish and Wildlife Program. Independent Scientific Review Panel for the Northwest Power and Conservation Council; and Independent Scientific Advisory Board for the Council, Columbia River Basin Indian Tribes, and NOAA Fisheries, 152 pp. [Accessed 12.02.07: http://www.nwppc.org/library/ isrp/isrpisab2004-13.pdf]

Izaurralde, R.C., N.J. Rosenberg, A.M. Thomson and R.A. Brown, 2005: Climate change impacts for the conterminous USA: An integrated assessment. Part 6: Distribution and productivity of unmanaged ecosystems. *Clim. Change*, **69**, 107-126.

Jackson, D.A. and N.E. Mandrak, 2002: Changing fish biodiversity: Predicting the loss of cyprinid biodiversity due to global climate change. *Sea Grant Symposium on Fisheries in a Changing Climate; August 20-21, 2001; Phoenix, AZ, USA*, N.A. McGinn, Ed., American Fisheries Society, Bethesda, Maryland, 89-98.

Jacob, K.H., N. Edelblum and J. Arnold, 2001: Infrastructure. *Climate Change and a Global City: The Potential Consequences of Climate Variability and Change*, C. Rosenzweig and W.D. Solecki, Eds., Columbia Earth Institute, New York, New York, 45-76. [Accessed 12.02.07: http://metroeast_climate.ciesin.columbia.edu/reports/infrastructure.pdf]

Jansen, W. and R.H. Hesslein, 2004: Potential effects of climate warming on fish habitats in temperate zone lakes with special reference to Lake 239 of the experimental lakes area (ELA), north-western Ontario. *Environ. Biol.Fishes*, **70**, 1-22.

Jha, M., Z.T. Pan, E.S. Takle and R. Gu, 2004: Impacts of climate change on streamflow in the Upper Mississippi River Basin: A regional climate model perspective. *J. Geophys. Res.*, **109**, D09105.

Johnstone, J.F. and F.S. Chapin, III, 2003: Non-equilibrium succession dynamics indicate continued northern migration of Lodgepole Pine. *Global Change Biol.*, **9**, 1401-1409.

Jones, B. and D. Scott, 2006: Climate Change, Seasonality and Visitation to Canada's National Parks. *J. Parks Recreation Admin.*, **24**, 42-62.

Jones, M.L., B.J. Shuter, Y.M. Zhao and J.D. Stockwell, 2006: Forecasting effects of climate change on Great Lakes fisheries: models that link habitat supply to population dynamics can help. *Can. J. Fish. Aquat. Sci.*, **63**, 457-468.

Joos, F., I.C. Prentice and J.I. House, 2002: Growth enhancement due to global atmospheric change as predicted by terrestrial ecosystem models: Consistent with US forest inventory data. *Global Change Biol.*, **8**, 299-303.

Kalkstein, L.S., 2002: Description of our heat/health watch-warning systems: their nature and extent, and required resources. Center for Climatic Research, University of Delaware, Newark, Delaware, 31 pp.

Kamler, E., 2002: Ontogeny of yolk-feeding fish: An ecological perspective. *Rev. Fish Biol. Fish.*, **12**, 79-103.

Karl, T., R. Knight, D. Easterling and R. Quayle, 1996: Indices of climate change for the United States. *Bull. Amer. Meteor. Soc.*, **77**, 279-292.

Karl, T., J. Lawrimore and A. Leetma, 2005: Observational and modeling evidence of climate change. *EM, A&WMA's magazine for environmental managers,* **October 2005**, 11-17.

Karnosky, D.F., K.S. Pregitzer, D.R. Zak, M.E. Kubiske, G.R. Hendrey, D. Weinstein, M. Nosal and K.E. Percy, 2005: Scaling ozone responses of forest trees to the ecosystem level in a changing climate. *Plant Cell Environ.*, **28**, 965-981.

Karoly, D.J., K. Braganza, P.A. Stott, J.M. Arblaster, G.A. Meehl, A.J. Broccoli and K.W. Dixon, 2003: Detection of a human influence on North American climate. *Science*, **302**, 1200-1203.

Kasischke, E.S. and M.R. Turetsky, 2006: Recent changes in the fire regime across the North American boreal region-Spatial and temporal patterns of burning across Canada and Alaska. *Geophys. Res. Lett.*, **33** doi:10.1029/2006GL025677,022006.

Kennish, M.J., 2001: Coastal salt marsh systems in the US: a review of anthropogenic impacts. *J. Coastal Res.*, **17**, 731-748.

Kennish, M.J., 2002: Environmental threats and environmental future of estuaries. *Environ. Conserv.*, **29**, 78-107.

Kesmodel, D., 2002: Low and dry: Drought chokes off Durango rafting business. *Rocky Mountain News*, **25 June 2002**.

Kiesecker, J.M., A.R. Blaustein and L.K. Belden, 2001: Complex causes of amphibian population declines. *Nature*, **410**, 681-683.

Kije Sipi Ltd., 2001: *Impacts and adaptation of drainage systems, design methods and policies: impacts and adaptation contribution agreement A330*. Natural Resources Canada, Climate Change Action Fund, 117 pp. [Accessed 12.02.07: http://adaptation.nrcan.gc.ca/projdb/pdf/43_e.pdf]

Kim, J., T.K. Kim, R.W. Arritt and N.L. Miller, 2002: Impacts of increased CO_2 on the hydroclimate of the western United States. *J. Climate*, **15**, 1926-1942.

Kim, Q.S., 2004: Industry Aims to Make Homes Disaster-Proof. *Wall Street Journal*, **30** September2004.

Kinney, P.L., D. Shindell, E. Chae and B. Winston, 2001: Public health. *Climate Change and a Global City: The Potential Consequences of Climate Variability and Change*, C. Rosenzweig and W.D. Solecki, Eds., Columbia Earth Institute, New York, New York, 103-120. [Accessed 12.02.07: http://metroeast_climate.ciesin.columbia.edu/reports/health.pdf]

Kirshen, P.H., 2002: Potential impacts of global warming on groundwater in eastern Massachusetts. *J. Water Resour. Plan. Manag.*, **128**, 216-226.

Kitzberger, T., T.W. Swetnam and T.T. Veblen, 2001: Inter-hemispheric synchrony of forest fires and the El Nino-Southern Oscillation. *Global Ecol. Biogeogr.*, **10**, 315-326.

Kleinosky, L.R., B. Yarnal and A. Fisher, 2006: Vulnerability of Hampton Roads, Virginia, to storm-surge flooding and sea-level rise. *Natural Hazards*, **39**, doi: 10.1007/s11069-11006-10004-z.

Knowles, N., M.D. Dettinger and D.R. Cayan, 2006: Trends in snowfall versus rainfall for the western United States, 1949-2004. *J. Climate*, **19**, 4545-4559.

Knowlton, K., J.E. Rosenthal, C. Hogrefe, B. Lynn, S. Gaffin, R. Goldberg, C. Rosenzweig, K. Civerolo, J-Y Ku and P.L. Kinney, 2004: Assessing ozone-related health impacts under a changing climate. *Environ. Health Perspect.*, **112**, 1557-1563.

Kolivras, K.N. and A.C. Comrie, 2003: Modeling valley fever (coccidioidomycosis) incidence on the basis of climate conditions. *Int. J. Biometeorol.*, **47**, 87-101.

Komar, P.D., J. Allan, G.M. Dias-Mendez, J.J. Marra and P. Ruggiero, 2000: El Niño and La Niña: erosion processes and impacts. *Proc. of the 27th International Conference on Coastal Engineering*, ASCE, Sydney, Australia, 2414-2427.

Koppe, C., S. Kovats, G. Jendritzky and B. Menne, 2004: *Heat-waves: Risks and Responses,* World Health Organization, Europe, Copenhagen, 124 pp. [Accessed 12.02.07: http://www.euro.who.int/document/E82629.pdf]

Koshida, G., M. Alden, S.J. Cohen, R. Halliday, L.D. Mortsch, V. Wittrock and A.R. Maarouf, 2005: Drought risk management in Canada-U.S. Transboundary watersheds: now and in the future. *Drought and Water Crisis - Science, Technology and Management Issues*, D. Wilhite, Ed., CRC Press, Boca Raton, Florida, 287-319.

Kovacs, P., 2005a: *Canadian Underwriter: Homeowners and Natural Hazards,* Jan 1, 2005, Institute for Catastrophic Loss Reduction, Toronto, Ontario, 5 pp.

Kovacs, P., 2005b: Homeowners and natural hazards. *Canadian Underwriter*, January 2005.

Kovacs, P. and H. Kunreuther, 2001: Managing Catastrophic Risk: Lessons from Canada. *Assurance J. Insur. Risk Manag.*, **69**.

Kumagi, Y., J. Edwards and M.S. Carroll, 2006: Why are natural disasters not "natural" for its victims? *Environ. Impact Assess. Rev.*, **26**, 106-119.

Kunkel, K. E., 2003: Temporal variations of extreme precipitation events in the United States: 1895-2000. *Geophys. Res. Lett.*, **30**, doi:10.1029/2003GL018052

Kunkel, K.E., R.A. Pielke Jr. and S.A. Changnon, 1999: Temporal fluctuations in weather and climate extremes that cause economic and human health impacts: A review. *Bull. Amer. Meteor. Soc.*, **80**, 1077-1098.

Kunreuther, H., 2006: Disaster mitigation and insurance: Learning from Katrina. *Ann. Amer. Acad. Polit. Soc. Sci.*, **604**, 208-227.

Kunreuther, H., R. Daniels and D. Kettl, 2006: *On Risk and Disaster: Lessons from Hurricane Katrina*. University of Pennsylvania Press, Philadelphia, 304 pp.

Kutzbach, J.E., J.W. Williams and S.J. Vavrus, 2005: Simulated 21st century changes in regional water balance of the Great Lakes region and links to changes in global temperature and poleward moisture transport. *Geophys. Res. Lett.*, **32**, doi:10.1029/2005GL023506.

LaCommare, K.H. and J.H. Eto, 2004: *Understanding the cost of power interruptions to U.S. electricity consumers*. LBNL-55718, Ernest Orlando Lawrence Berkeley National Laboratory, Berkeley, California, 70 pp. [Accessed 12.02.07: http://repositories.cdlib.org/cgi/viewcontent.cgi?article=2531&context=lbnl]

Leatherman, S.P., 2001: Social and environmental costs of sea level rise. *Sea Level Rise, History and Consequences*, B.C. Douglas, M.S. Kearney and S.P. Leatherman, Eds., Academic Press, San Diego, California, 181-223.

Lebel, L., 2004: Social change and CO_2 stabilization: Moving away from carbon cultures. *The Global Carbon Cycle: Integrating Humans, Climate, and the Natural World*, C.B. Field and M.R. Raupach, Eds., Island Press, Washington, District of Columbia, 371-382.

Lehman, J., 2002: Mixing patterns and plankton biomass of the St. Lawrence Great Lakes under climate change scenarios. *J. Great Lakes Res.*, **28**, 583-596.

Leiss, W., 2001: *In the Chamber of Risks: Understanding Risk Controversies*. McGill-Queen's University Press, Montreal, Quebec,388 pp.

Lemke, P., J. Ren, R. Alley, I. Allison, J. Carrasco, G. Flato, Y. Fuji, G. Kaser, P. Mote, R.H. Thomas and T. Zhang, 2007: Observations: changes in snow, ice and frozen ground. *Climate Change 2007: The Physical Science Basis. Contribution of Working Group I to the Fourth Assessment Report of the Intergovernmental Panel on Climate Change*, S. Solomon, D. Qin, M. Manning, Z. Chen, M. Marquis, K.B. Averyt, M. Tignor and H.L. Miller, Eds., Cambridge University Press, Cambridge and New York, 337-384.

Lemmen, D.S. and F.J. Warren, Eds., 2004: *Climate Change Impacts and Adaptation: A Canadian Perspective*. Climate Change Impacts and Adaptation Directorate, Natural Resources Canada Ottawa, Ontario, 201 pp. [Accessed 12.02.07: http://environment.msu.edu/climatechange/canadaadaptation.pdf]

Lerdau, M. and M. Keller, 1998: Controls on isoprene emission from trees in a subtropical dry forest. *Plant, Cell Environ.*, **20**, 569-579.

Lester, N.P., A.J. Dextrase, R.S. Kushneriuk, M.R. Rawson and P.A. Ryan, 2004: Light and temperature: Key factors affecting walleye abundance and production. *Trans. Amer. Fish. Soc.*, **133**, 588-605.

Lettenmaier, D.P. and A.F. Hamlet. 2003: Improving Water Resources System Performance Through Long-Range Climate Forecasts: the Pacific Northwest Experience. *Water and Climate in the Western United States*, W.M. Lewis Jr., Ed., University Press of Colorado, Boulder, Colorado.

Lettre, J., 2000: Weather Risk Management Solutions, Weather Insurance, Weather Derivatives. Research Paper, Financial Management, 30 November 2000, Rivier College, Nashua, New Hampshire. [Accessed 12.02.07: http://hometown.aol.com/gml1000/wrms.htm]

Leung, L.R. and Y. Qian, 2003: Changes in seasonal and extreme hydrologic conditions of the Georgia Basin/Puget Sound in an ensemble regional climate simulation for the mid–Century. *Can. Water Resour. J.*, **28**, 605-632.

Levitan and Associates Inc., 2005: Post Katrina and Rita outlook on fuel supply adequacy and bulk power security in New England, Levitan and Associates, Inc, Boston, Massachusetts, 9 pp. [Accessed 12.02.07: http://www.islandereastpipeline.com/articles/post_hurricane_outlook.pdf]

Lewis, E.J., 1987: Survey of residential air-to-air heat pump service and life and maintenance issues. *ASHRAE Transactions*, **93**(1), 1111-1127.

Ligeti, E., 2006: *Adaptation strategies to reduce health risks from summer heat in Toronto*. Toronto Atmospheric Fund, Toronto, Canada.

Loáiciga, H.A., 2000: Climate change impacts in regional-scale aquifers: principles and field application. In: *Groundwater Updates*. Sato, K. and Y. Iwasa, Eds. Springer, Tokyo, Japan, 247-252.

Loáiciga, H.A., D.R. Maidment and J.B. Valdes, 2000: Climate-change impacts in a regional karst aquifer, Texas USA. *J. Hydrology*, **227**, 173-194.

Lobell, D.B. and P. Asner, 2003: Climate and management contributions to recent trends in U.S. agricultural yields. *Science*, **299**, 1032.

Lobell, D.B., J.A. Hicke, G.P. Asner, C.B. Field, C.J. Tucker and S.O. Los, 2002: Satellite estimates of productivity and light use efficiency in United States agriculture, 1982-98. *Glob. Change Biol.*, **8**, 722-735.

Lobell, D.B., K.N. Cahill and C.B. Field, 2006: Historical effects of temperature and precipitation on California crop yields. *Clim. Change*, **81**, 187-203.

Loehle, C., J.G. MacCracken, D. Runde and L. Hicks, 2002: Forest management at landscape scales: solving the problems. *J. Forestry*, **100**, 25-33.

Lofgren, B.M., F.H. Quinn, A.H. Clites, R.A. Assel, A.J. Eberhardt and C.L. Luukkonen, 2002: Evaluation of potential impacts on Great Lakes water resources based on climate scenarios of two GCMs. *J. Great Lakes Res.*, **28**, 537-554.

Logan, J.A., J. Regniere and J.A. Powell, 2003: Assessing the impacts of global warming on forest pest dynamics. *Front. Ecol. Environ.*, **1**, 130-137.

Lonergan, S., R. DiFrancesco and M. Woo, 1993: Climate change and transportation in northern Canada: An integrated impact assessment. *Clim. Change*, **24**, 331-351.

Long, S.P., E.A. Ainsworth, A.D.B. Leakey and P.B. Morgan, 2005: Global food insecurity. Treatment of major food crops with elevated carbon dioxide or ozone under large-scale fully open-air conditions suggests recent models may have overestimated future yields. *Phil. Trans. Royal Soc. Lond. B Biol. Sci.*, **360**, 2011-2020.

Loukas, A., L. Vasiliades and N.R. Dalezios, 2002: Potential climate change impacts on flood producing mechanisms in southern British Columbia, Canada using the CGCMA1 simulation results. *J. Hydrol.*, **259**, 163-188.

Lucht, W., I.C. Prentice, R.B. Myneni, S. Sitch, P. Friedlingstein, W. Cramer, P. Bousquet, W. Buermann and B. Smith, 2002: Climate control of the high-latitude vegetation greening trend and Pinatubo effect. *Science*, **296**, 1687-1689.

Magnuson, J.J., D.M. Robertson, B.J. Benson, R.H. Wynne, D.M. Livingstone, T. Arai, R.A. Assel, R.G. Barry, V. Card, E. Kuusisto, N.C. Granin, T.D. Prowse, K.M. Stewart and V.S. Vuglinski, 2000: Historical trends in lake and river ice cover in the Northern Hemisphere. *Science*, **289**, 1743-1746.

Major, D. and R. Goldberg, 2001: Water supply. *Climate Change and a Global City: The Potential Consequences of Climate Variability and Change*, C. Rosenzweig and W. D. Solecki, Eds., Columbia Earth Institute, New York, New York, 87-101. [Accessed 12.02.07: http://metroeast_climate.ciesin.columbia.edu/reports/water.pdf]

McBean, G. and D. Henstra, 2003: *Climate Change, Natural Hazards and Cities*. ICLR Reseach Paper Series No. 31. Institute for Catastrophic Loss Reduction, Toronto, Canada, 18 pp.

McBeath, J., 2003: Institutional responses to climate change: The case of the Alaska transportation system. *Mitigation Adapt. Strategies Global Change*, **8**, 3-28.

McCabe, G.J. and J.E. Bunnell, 2004: Precipitation and the occurrence of Lyme disease in the northeastern United States. *Vector-Borne and Zoonotic Diseases*, **4**, 143-148.

McCabe, G.J., M.P. Clark and M.C. Serreze, 2001: Trends in northern hemisphere surface cyclone frequency and intensity. *J. Climate*, **14**, 2763-2768.

McConnell, R., K. Berhane, F. Gilliland, S.J. London, T. Islam, W.J. Gauderman, W. Avol, H.G. Margolis and J.M. Peters, 2002: Asthma in exercising children exposed to ozone: A cohort study. *The Lancet*, **359**, 386-391.

McGee, T., S. Reinholdt, S. Russell, N. Rogers and L. Boxelar, 2000: Effective Behaviour Change Programs for Natural Hazard Reduction in Rural Communities, Final Report, IDNDR Project 7/99.

McKenzie, D., A.E. Hessl and D.L. Peterson, 2001: Recent growth of conifer species of western North America: Assessing spatial patterns of radial growth trends. *Can. J. For. Res.*, **31**, 526-538.

McKenzie, D., Z. Gedalof, D.L. Peterson and P. Mote, 2004: Climatic change, wildfire and conservation. *Conserv. Biol.*, **18**, 890-902.

Mearns, L.O., G. Carbone, R.M. Doherty, E.A. Tsvetsinskaya, B.A. McCarl, R.M. Adams and L. McDaniel, 2003: The uncertainty due to spatial scale of climate scenarios in integrated assessments: An example from U.S. agriculture. *Integrated Assessment*, **4**, 225-235.

Meehl, G.A. and C. Tebaldi, 2004: More intense, more frequent, and longer lasting heat waves in the 21st century. *Science*, **305**, 994-997.

Meehl, G.A., T.F. Stocker, W.D. Collines, P. Friedlingstein, A.T. Gaye, J.M. Gregory, A. Kitoh, R. Knutti, J.M. Murphy, A. Noda, S.C.B. Raper, I.G. Watterson, A.J. Weaver and Z.-C. Zhao, 2007: Global climate projections. *Climate Change 2007: The Physical Science Basis. Contribution of Working Group I to the Fourth Assessment Report of the Intergovernmental Panel on Climate Change*, S. Solomon, D. Qin, M. Manning, Z. Chen, M. Marquis, K.B. Averyt, M. Tignor and H.L. Miller, Eds., Cambridge University Press, Cambridge and New York, 747-846.

Mendelsohn, R., Ed., 2001: *Global Warming and the American Economy: A Regional Assessment of Climate Change Impacts*. Edward Elgar, Northampton, Massachusetts, 209 pp.

Mendelsohn, R. and M.E. Schlesinger, 1999: Climate response functions. *Ambio*, **28**, 362-366.

Mercier, G., 1998: Climate change and variability: Energy sector. *Canada Country Study: Impacts and Adaptations*, G. Koshida and W. Avis, Eds., Kluwer Academic Publishers, Dordrecht.

Merritt, W., Y. Alila, M. Barton, B. Taylor, S. Cohen and D. Neilsen, 2005: Hydrologic response to scenarios of climate change in sub-watersheds of the Okanagan basin, British Columbia. *J. Hydrology*, **326**, 79-108, doi:10.1016/j.jhydrol.2005.1010.1025.

Michel-Kerjan, E., 2006: Insurance, the 14th critical sector. *Seeds of Disaster, Roots of Response: How Private Action Can Reduce Public Vulnerability*, P. Auerswald, L. Branscomb, T.M. La Porte and E. Michel-Kerjan, Eds., Cambridge University Press, New York, 279-291.

Miles, E.L., A.K. Snover, A. Hamlet, B. Callahan and D. Fluharty, 2000: Pacific northwest regional assessment: The impacts of climate variability and climate change on the water resources of the Columbia River Basin. *J. Amer. Water Resour. Assoc.*, **36**, 399-420.

Miller, N.L. 2003: California climate change, hydrologic response, and flood forecasting. *International Expert Meeting on Urban Flood Management*, World Trade Center Rotterdam, The Netherlands, 11pp. [Accessed 12.02.07: http://repositories.cdlib.org/cgi/viewcontent.cgi?article=1454&context=lbnl]

Miller, N.L., K.E. Bashford and E. Strem, 2003: Potential impacts of climate change on California hydrology. *J. Amer. Water Resour. Assoc.*, **39**, 771-784.

Millerd, F., 2005: The economic impact of climate change on Canadian commercial navigation on the Great Lakes. *Can. Water Resour. J.*, **30**, 269-281.

Mills, B., S. Tighe, J. Andrey, K. Huen and S. Parm, 2006: Climate change and the performance of pavement infrastructure in southern Canada, context and case study. *Proc. of the Engineering Institute of Canada (EIC) Climate Change Technology Conference*, May 9-12, 2005, Ottawa.

Mills, E. and E. Lecomte, 2006: From Risk to Opportunity: How Insurers Can Proactively and Profitably Manage Climate Change, CERES, Inc. Report, Boston, Massachusetts, 52 pp.

Mirza, M.M.Q., 2004: Climate Change and the Canadian Energy Sector: Report on Vulnerability and Adaptation, Adaptation and Impacts Research Group, Atmospheric Climate Science Directorate, Meteorological Service of Canada Downsview, Ontario, Canada, 52 pp.

Mohseni, O., H.G. Stefan and J.G. Eaton, 2003: Global warming and potential changes in fish habitat in U.S. streams. *Clim. Change*, **59**, 389-409.

Mooney, H.A., R.N. Mack, J.A. McNeely, L.E. Neville, P.J. Schei and J.K. Waage, Eds., 2005: *Invasive Alien Species*. Island Press, Washington, District of Columbia, 368 pp.

Morehouse, B.J., R.H. Carter and P. Tschakert, 2002: Sensitivity of urban water resources in Phoenix, Tucson, and Sierra Vista, Arizona to severe drought. *Climate Res.*, **21**, 283-297.

Morgan, M.G., L.F. Pitelka and E. Shevliakova, 2001: Elicitation of expert judgments of climate change impacts on forest ecosystems. *Clim. Change*, **49**, 279-307.

Morris, J.T., P.V. Sundareshwar, C.T. Nietch, B. Kjerfve and D.R. Cahoon, 2002: Responses of coastal wetlands to rising sea level. *Ecology*, **83**, 2869-2877.

Morrison, J., M.C. Quick and M.G.G. Foreman, 2002: Climate change in the Fraser River watershed: Flow and temperature projections. *Journal of Hydrology (Amsterdam)*, **263**, 230-244.

Morrison, W.N. and R. Mendelsohn, 1999: The impact of global warming on U.S. energy expenditures. *The Economic Impact of Climate Change on the United States Economy*, R. Mendelsohn and J. Neumann, Eds., Cambridge University Press, New York, 209-236.

Mortsch, L., M. Alden and J. Scheraga, 2003: Climate change and water quality in the Great Lakes Region - risks opportunities and responses, report prepared for the Great Lakes Water Quality Board for the International Joint Commission, 213 pp.

Mortsch, L., H. Hengeveld, M. Lister, B. Lofgren, F. Quinn, M. Silvitzky and L. Wenger, 2000: Climate change impacts on the hydrology of the Great Lakes-St. Lawrence system. *Can. Water Resour. J.*, **25**, 153-179.

Moser, S., 2000: Community responses to coastal erosion: Implications of potential policy changes to the national flood insurance program. *Evaluation of Erosion Hazards*, The H. John Heinz II Center for Science, Economics and the Environment, Washington District of Columbia,, Appendix F. 99 pp. [Accessed 12.02.07: http://www.heinzctr.org/Programs/SOCW/Erosion_ Appendices/ Appendix%20F%20-%20FINAL.pdf]

Moser, S., 2005: *Enhancing Decision-Making through Integrated Climate Research*, Summary of an Exploratory Workshop for the NOAA-OGP-RISA Program, Alaska Regional Meeting, National Oceanic and Atmospheric Administration -Office of Global Programs, 63 pp. [Accessed 12.02.07: http://www.ogp.noaa.gov/mpe/csi/events/risa_021804/report.pdf]

Moser, S., 2006: Impacts assessments and policy responses to sea-level rise in three U.S. States: An exploration of human dimension uncertainties. *Global Environ. Change*, **15**, 353-369.

Moss, R.H., A.L. Breknert and E.L. Malone, 2001: *Vulnerability to Climate Change: A Quantitative Approach,* Pacific Northwest National Laboratory, Richland, Washington.

Mote, P.W., 2003: Trends in snow water equivalent in the Pacific Northwest and their climatic causes. *Geophys. Res. Lett.*, **30**, 3-1.

Mote, P., D. Canning, D. Fluharty, R. Francis, J. Franklin, A. Hamlet, M. Hershman, M. Holmberg, K. Gray-Ideker, W.S. Keeton, D. Lettenmaier, R. Leung, N. Mantua, E. Miles, B. Noble, H. Parandvash, D.W. Peterson, A. Snover and S. Willard, 1999: *Impacts of Climate Variability and Change, Pacific Northwest,* National Atmospheric and Oceanic Administration, Office of Global Programs, and JISAO/SMA Climate Impacts Group, Seattle, Washington, 110 pp. [Accessed 12.02.07: http://www.usgcrp.gov/usgcrp/nacc/pnw.htm]

Mote, P.W., E.A. Parson, A.F. Hamlet, W.S. Keeton, D. Lettenmaier, N. Mantua, E.L. Miles, D.W. Peterson, D.L. Peterson, R. Slaughter and A.K. Snover, 2003: Preparing for climatic change: the water, salmon, and forests of the Pacific Northwest. *Clim. Change*, **61**, 45-88.

Mote, P., A.F. Hamlet, M.P. Clark and D.P. Lettenmaier, 2005: Declining mountain snowpack in western North America. *Bull. Amer. Meteor. Soc.*, **86**, doi: 10.1175/BAMS-1186-1171-1139.

Moulton, R.J. and D.R. Cuthbert, 2000: Cumulative impacts/risk assessment of water removal or loss from the Great Lakes St. Lawrence River system. *Can. Water Resour. J.*, **25**, 181-208.

Multihazard Mitigation Council, 2005: *An Independent Study to Assess the Future Savings from Mitigation Activities*, National Institute of Building Sciences, Washington, District of Columbia, 377 pp. [Accessed 12.02.07: http://www.nibs.org/MMC/MitigationSavingsReport/natural_hazard_mitigation_saves.htm]

Munich Re., 2004: *Topics: 2004*. GeoRisks Group, Munich Re, Munich, 60 pp.

Muraca, G., D.C. MacIver, H. Auld and N. Urquizo, 2001: The climatology of fog in Canada. *Proc. of the 2nd International Conference on Fog and Fog Collection*, 15-20 July 2005, St. John's, Newfoundland.

Murphy, B., 2004: *Emergency Management and the August 14th, 2003 Blackout*. Institute for Catastrophic Loss Reduction, ICLR Research Paper Series No. 40, Toronto, Canada, 11 pp. [Accessed 12.02.07: http://www.iclr.org/pdf/Emergency%20Preparedness%20and%20the%20blackout2.pdf]

Murphy, B., G. McBean, H. Dolan, L. Falkiner and P. Kovacs, 2005: *Enhancing local level emergency management: the influence of disaster experience and the role of household and neighbourhoods*. Institute for Catastrophic Loss Reduction, ICLR Research Paper Series No. 43, Toronto, Canada, 79 pp.

MWD, 2005: *The Family of Southern California Water Agencies*. Metropolitan Water District of Southern California. [Accessed 12.02.07: http://www.bewaterwise.com/index.html]

Myneni, R.B., J. Dong, C.J. Tucker, P.E. Kaufmann, J. Kauppi, L. Liski, J. Zhou, V. Alexeyev and M.K. Hughes, 2001: A large carbon sink in the woody biomass of northern forests. *Proc. Nat. Acad. Sci.*, **98**, 14784-14789.

Najjar, R.G., H.A. Walker, P.J. Anderson, E.J. Barron, R.J. Bord, J.R. Gibson, V.S. Kennedy, C.G. Knight, J.P. Megonigal, R.E. O'Connor, C.D. Polsky, N.P. Psuty, B.A. Richards, L.G. Sorenson, E.M. Steele and R.S. Swanson, 2000: The potential impacts of climate change on the mid-Atlantic coastal region. *Climate Res.*, **14**, 219-233.

Nakićenović, N. and R. Swart, Eds., 2000: *Special Report on Emissions Scenarios. A Special Report of Working Group III of the Intergovernmental Panel on Climate Change*. Cambridge University Press, Cambridge, 599 pp.

NAST, 2000: *Climate Change Impacts on the United States: The Potential Consequences of Climate Variability and Change*, Overview Report for the U.S. Global Change Research Program. U.S. National Assessment Synthesis Team, Cambridge University Press, Cambridge, 154 pp. [Accessed 12.02.07: http://www.usgcrp.gov/usgcrp/Library/nationalassessment/overview.htm]

NAST, 2001: *Climate Change Impacts on the United States: The Potential Consequences of Climate Variability and Change*, Foundation Report for the US Global Change Research Program. U.S. National Assessment Synthesis Team, Cambridge University Press, Cambridge, 620 pp. [Accessed 12.02.07: http://www.usgcrp.gov/usgcrp/Library/nationalassessment/foundation.htm]

National Voluntary Organizations Active in Disaster, 2006. [Accessed 12.02.07: www.nvoad.org]

Natural Resources Canada, 2000: *Canada's National Implementation Strategy on Climate Change*. Government of Canada, Ottawa, Canada, 44 pp. [Accessed 12.02.07: http://www.iigr.ca/pdf/documents/1063_Canadas_National_Implem.pdf]

Nearing, M.A., F.F. Pruski and M.R. O'Neal, 2004: Expected climate change impacts on soil erosion rates: a review. *J. Soil Water Conserv.*, **59**, 43-50.

Nelson, E., O.A. Anisimov and N.I. Shiklomanov, 2002: Climate change and hazard zonation in the circum-Arctic permafrost regions. *Nat. Hazards*, **26**, 203-225.

Nemani, R.R., M.A. White, D.R. Cayan, G.V. Jones, S.W. Running, J.C. Coughlan and D.L. Peterson, 2001: Asymmetric warming over coastal California and its impact on the premium wine industry. *Climate Res.*, **19**, 25-34.

Nemani, R.R., M.A. White, P.E. Thornton, K. Nishida, S. Reddy, J. Jenkins and S.W. Running, 2002: Recent trends in hydrologic balance have enhanced the terrestrial carbon sink in the United States. *Geophys. Res. Lett.*, **29**, doi: 10.1029/2002GL014867.

Nicholls, K.H., 1999: Effects of temperature and other factors on summer phosphorus in the inner Bay of Quinte, Lake Ontario: Implications for climate warming. *J. Great Lakes Res.*, **25**, 250-262.

Nordhaus, W.D., 2006: The Economics of Hurricanes in the United States. *Annual Meetings of the American Economic Association*, January 5-8, 2006, American Economic Association, Boston, Massachusetts. [Accessed 12.02.07: http://www.econ.yale.edu/~nordhaus/homepage/hurr_010306a.pdf]

NRC, 2004: *Thinking Strategically: The Appropriate use of Metrics for the Climate Change Science Program*. U.S. National Research Council - Committee on Metrics for Global Climate Change, Climate Research Committee, National Academy Press, Washington District of Columbia, 162 pp. [Accessed 12.02.07: http://books.nap.edu/catalog/11292.html]

O'Neal, K., 2002: *Effects of Global Warming on Trout and Salmon in U.S. Streams*. Defenders of Wildlife, Washington, District of Columbia, 46 pp. [Accessed 12.02.07: http://www.defenders.org/publications/fishreport.pdf]

O'Reilly, C.T., D.L. Forbes and G.S. Parkes, 2005: Defining and adapting to coastal hazards in Atlantic Canada: Facing the challenge of rising sea levels, storm surges, and shoreline erosion in a changing climate. *Ocean Yearbook*, **19**, 189-207.

Ogden, N.H., L.R. Lindsay, G. Beauchamp, D. Charron, A. Maarouf, C.J. O'Callagjan, D. Waltner-Toews and I.K. Barker, 2004: Investigation of the relationships between temperature and developmental rates of tick *Ixodes Scapularis* (Acari: Ixodidae) in the laboratory and field. *J. Med. Entomol.*, **41**, 622-633.

Ogden, N.H., A. Maarouf, I.K. Barker, M. Bigras-Poulin, L.R. Lindsay, M.G. Morshed, C.J. O'Callaghan, F. Ramay, D. Waltner-Toews and D.F. Charron, 2006: Climate change and the potential for range expansion of the Lyme disease vector *Ixodes scapularis* in Canada. *Int J Parasitol*, **36**, 63-70.

Ouranos, 2004: *Adapting to Climate Change*. Ouranos, Montreal, Canada. 91 pp. [Accessed 12.02.07: http://www.ouranos.ca/cc/climang5.pdf]

Paavola, J. and W. Adger, 2002: *Justice and Adaptation to Climate Change*. Tyndall Centre for Climate Change Research, Working Paper 23, Norwich, Norfolk, 24 pp. [Accessed 12.02.07: www.tyndall.ac.uk/publications/ working_papers/ wp23.pdf]

Pan, Z.T., M. Segal, R.W. Arritt and E.S. Takle, 2004: On the potential change in solar radiation over the US due to increases of atmospheric greenhouse gases. *Renew. Energ.*, **29**, 1923-1928.

Parmesan, C., 1996: Climate and species range. *Nature*, **382**, 765-766.

Parmesan, C. and G. Yohe, 2003: A globally coherent fingerprint of climate change impacts across natural systems. *Nature*, **421**, 37-42.

Parmesan, C. and H. Galbraith, 2004: *Observed Impacts of Global Climate Change in the U.S*., Pew Center on Global Climate Change, Arlington, Virginia, 67 pp. [Accessed 12.02.07: http://www.pewclimate.org/global-warming-in-depth/all_reports/observedimpacts/index.cfm.

Parson, E.A., L. Carter, P. Anderson, B. Wang and G. Weller, 2001a: Potential consequences of climate variability and change for Alaska. *Climate Change Impacts on the United States*, National Assessment Synthesis Team, Ed., Cambridge University Press, Cambridge, 283-312.

Parson, E.A., P.W. Mote, A. Hamlet, N. Mantua, A. Snover, W. Keeton, E. Miles, D. Canning and K.G. Ideker, 2001b: Potential consequences of climate variability and change for the Pacific Northwest. *Climate Change Impacts on the United States - The Potential Consequences of Climate Variability and Change-Foundation Report*, National Assessment Synthesis Team, Ed., Cambridge University Press, Cambridge, 247-280.

Parson, E.A., R.W. Corell, E.J. Barron, V. Burkett, A. Janetos, L. Joyce, T.R. Karl, M. MacCracken, J. Melillo, M.G. Morgan, D.S. Schimel and T. Wilbanks, 2003: Understanding climatic impacts, vulnerabilities and adaptation in the United States: Building a capacity for assessment. *Clim. Change*, **57**, 9-42.

Patz, J.A., D. Campbell-Lendrum, T. Holloway and J.A. Foley, 2005: Impact of regional climate change on human health. *Nature*, **438**, 310-317.

Payne, J.T., A.W. Wood, A.F. Hamlet, R.N. Palmer and D.P. Lettenmaier, 2004: Mitigating the effects of climate change on the water resources of the Columbia River basin. *Clim. Change*, **62**, 233-256.

Perez-Garcia, J., L.A. Joyce, A.D. McGuire and X.M. Xiao, 2002: Impacts of climate change on the global forest sector. *Clim. Change*, **54**, 439-461.

Peterson, D.W. and D.L. Peterson, 2001: Mountain hemlock growth trends to climatic variability at annual and decadal time scales. *Ecology*, **82**, 3330-3345.

Peterson, D.W., D.L. Peterson and G.J. Ettl, 2002: Growth responses of subalpine fir to climatic variability in the Pacific Northwest. *Can. J. For. Res.*, **32**, 1503-1517.

Peterson, J.T. and T.J. Kwak, 1999: Modeling the effects of land use and climate change on riverine smallmouth bass. *Ecol. Appl.*, **9**, 1391-1404.

Pielke Jr., R.A., 2005: Attribution of disaster losses. *Science*, **310**, 1615.

Pielke, R.A. and C. W. Landsea, 1998: Normalized hurricane damages in the United States: 1925-95. *Weather and Forecasting*, **13**, 621-631. with extensions through 2005 at http://www.aoml.noaa.gov/hrd/tcfaq/E21.html.

Pielke Jr., R.A. and C.W. Landsea, 1999: La Niña, El Niño, and Atlantic hurricane damages in the United States. *Bull. Amer. Meteorol. Soc.*, **80**, 2027-2033.

Pielke, Jr., R.A., S. Agrawala, L.M. Bouwer, I. Burton, S. Changnon, M.H. Glantz, W.H. Hooke, R.J.T. Klein, K. Kunkel, D. Mileti, D. Sarewitz, E. M. Thompkins, N. Stehr and H. von Storch, 2005. "Clarifying the Attribution of Recent Weather Disaster Losses: A Response to Epstein and McCarthy." *Bull. Am. Meteorol. Soc.*, 86, 1481-1483.

Piggott, A., D. Brown, S. Moin and B. Mills, 2003: Estimating the impacts of climate change on groundwater conditions in western southern Ontario. *Proc. of the 56th Canadian Geotechnical and 4th Joint IAH-CNC and CGS Groundwater Specialty Conferences*. Winnipeg, Canada. Canadian Geotechnical Society and Canadian National Chapter of the International Association of Hydrogeologists, 7 pp.

Pisano, P., L. Goodwin and A. Stern, 2002: Surface transportation safety and operations: The impacts of weather within the context of climate change. *The Potential Impacts of Climate Change on Transportation: Workshop Summary and Proceedings*, Washington, District of Columbia, 20 pp. [Accessed 12.02.07: http://climate.volpe.dot.gov/workshop1002/pisano.pdf]

POL, 2006: *Permanent Service for Mean Sea Level (PSMSL)*. Proudman Oceanographic Laboratory, Liverpool, UK. [Accessed: 12.02.07: http://www.pol.ac.uk/psmsl/psmsl_individual_stations.html]

Polsky, C. and W.E. Easterling, III, 2001: Adaptation to climate variability and change in the US Great Plains: A multi-scale analysis of Ricardian climate sensitivities. *Agr. Ecosyst. Environ.*, **85**, 133-144.

Polsky, C., D. Schröter, A. Patt, S. Gaffin, M.L. Martello, R. Neff, A. Pulsipher and H. Selin, 2003: *Assessing Vulnerabilities to the Effects of Global Change: An Eight-Step Approach,* 2003-05, Belfer Center for Science & International Affairs, Harvard University, Cambridge, Massachusetts,31 pp. [Accessed 12.02.07: http://www.bcsia.ksg.harvard.edu/BCSIA_content/documents/2003-05.pdf]

Post, J.R., M. Sullivan, S. Cox, N.P. Lester, C.J. Walters, E.A. Parkinson, A.J. Paul, L. Jackson and B.J. Shuter, 2002: Canada's recreational fisheries: The invisible collapse? *Fisheries*, **27**, 6-17.

Postel, S. and B. Richter, 2003: *Rivers for Life: Managing Water for People and Nature*. Island Press, Washington, District of Columbia, 220 pp.

Pounds, A.J., 2001: Climate and amphibian declines. *Nature*, **410**, 639-640.

Pulwarty, R., K. Jacobs and R. Dole, 2005: The hardest working river: drought and critical water problems in the Colorado River Basin. *Drought and Water Crisis - Science, Technology and Management Issues*, D.A. Wilhite, Ed., CRC Press, Boca Raton, Florida, 249-286.

Quinn, F.H., 2002: The potential impacts of climate change on Great Lakes transportation. *The Potential Impacts of Climate Change on Transportation: Workshop Summary and Proceedings*, Washington, District of Columbia, 9 pp. [Accessed 12.02.07: http://climate.volpe.dot.gov/workshop1002/quinn.pdf]

Quinn, F.H. and B.M. Lofgren, 2000: The influence of potential greenhouse warming on Great Lakes hydrology, water levels, and water management. *Proc. 15th Conference on Hydrology*. Long Beach, California, American Meteorological Society Annual Meeting, 271-274.

Rahel, F.J., 2002: Using current biogeographic limits to predict fish distributions following climate change. *Fisheries in a Changing Climate*, N.A. McGinn, Ed., American Fisheries Society, 99-110.

Rayner, S., D. Lach and H. Ingram, 2005: Weather forecasts are for wimps: why water resource managers do not use climate forecasts. *Clim. Change*, **69**, 197-227.

Reale, D., A. McAdam, S. Boutin and D. Berteaux, 2003: Genetic and plastic responses of a northern mammal to climate change. *Proc. R. Soc. Lond. B*, 591-596.

Reed, K.M. and B. Czech, 2005: Causes of fish endangerment in the United States, or the structure of the American economy. *Fisheries (Bethesda)*, **30**, 36-38.

Regonda, S.K., B. Rajagopalan, M. Clark and J. Pitlick, 2005: Seasonal cycle shifts in hydroclimatology over the western United States. *Journal of Climate*, **18**, 372-384.

Rehfeldt, G.E., W.R. Wycoff and C. Ying, 2001: Physiologic plasticity, evolution and impacts of a changing climate on Pinus contorta. *Clim. Change*, **50**, 355-376.

Reid, W.V., H.A. Mooney, A. Cropper, D. Capistrano, S.R. Carpenter, K. Chopra, P. Dasgupta, T. Dietz, A.K. Duraiappah, R.K. Rashid Hassan, R. Leemans, R.M. May, T.A.J. McMichael, P. Pingali, C. Samper, R. Scholes, R.T. Watson, A.H. Zakri, Z. Shidong, N.J. Ash, E. Bennett, P. Kumar, M.J. Lee, C. Raudsepp-Hearne, H. Simons, J. Thonell and M.B. Zurek, 2005: *Ecosystems and human well-being*. Island Press, Washington, District of Columbia, 137 pp.

Reilly, J.M., Ed., 2002: *Agriculture: The Potential Consequences of Climate Variability and Change*. Cambridge University Press, Cambridge, 136 pp.

Reisen, W.K., Y. Fang and V. Martinez, 2006: Effects of temperature on the transmission of West Nile virus by *Culex tarsalis* (Diptera: Culicidae). *J. Med. Entomol.*, **43**, 309-317.

Report of the Water Strategy Expert Panel, 2005: *Watertight: The Case for Change in Ontario's Water and Wastewater sector.* Publications Ontario, Toronto, Canada.

Reynolds, R.W., N.A. Rayner, T.M. Smith, D.C. Stokes and W.Q. Wang, 2002: An improved in situ and satellite SST analysis for climate. *Journal of Climate*, **15**, 1609-1625.

Richardson, R.B. and J.B. Loomis, 2004: Adaptive recreation planning and climate change: a contingent visitation approach. *Ecol. Econ.*, **50**, 83-99.

Rivera, A., D.M. Allen and H. Maathuis, 2004: Climate variability and change - groundwater resources. *Threats to Water Availability in Canada*, Environment Canada, Eds., National Water Research Institute, Burlington, Ontario, 77-84. [Accessed 12.02.07: http://www.nwri.ca/threats2full/ThreatsEN_03web.pdf]

RM, 2003: Reducing electrical risk. *Risk Management*, **50(8)**, 10.

RMS, 2005a: *Estimating Losses from the 2004 Southeast Asia Earthquake and Tsunami,* Risk Management Solutions, Newark, California, 9 pp. [Accessed 12.02.07: http://www.rms.com/Publications/SumatraInsuredLoss_RMSwhite paper.pdf]

RMS, 2005b: *Hurricane Katrina: Profile of a Super Cat. Lessons and Implications for Catastrophe Risk Management,* Risk Management Solutions, Newark, California, 31 pp. [Accessed 12.02.07: http://www.rms.com/Publications/KatrinaReport_LessonsandImplications.pdf]

Rood, S.B., G.M. Samuelson, J.K. Weber and K.A. Wywrot, 2005: Twentieth-century decline in streamflows from the hydrographic apex of North America. *J. Hydrol.*, **306**, 215-233.

Root, T., J. Price, K. Hall, S. Schneiders, C. Rosenzweig and J. Pounds, 2003: Fingerprints of global warming on wild animals and plants. *Nature*, **421**, 57-60.

Root, T.L., D.P. MacMynowski, M.D. Mastrandrea and S.H. Schneider, 2005: Human-modified temperatures induce species changes: Joint attribution. *Proc. Natl. Acad. Sci.*, **102**, 7465-7469.

Ropelewski, C.F. and M.S. Halpert, 1986: North American precipitation and temperature patterns associated with the El Niño-Southern Oscillation (ENSO). *Month Wea. Rev.*, **114**, 2352-2362.

Rose, C.A., 2005: Economic growth as a threat to fish conservation in Canada. *Fisheries*, **30**, 36-38.

Rosenberg, N.J. and J.A. Edmonds, 2005: Climate change impacts for the conterminous USA: An integrated assessment: From Mink to the 'Lower 48': An introductory editorial *Clim. Change*, **69**, 1-6.

Rosenberg, N.J., D.J. Epstein, D. Wang, L. Vail, R. Srinivasan and J.G. Arnold, 1999: Possible impacts of global warming on the hydrology of the Ogallala aquifer region. *Clim. Change*, **42**, 677-692.

Rosenberg, N.J., R.A. Brown, R.C. Izaurralde and T.M. Thomson, 2003: Integrated assessment of Hadley Centre (HadCM2) climate change projections on agricultural productivity and irrigation water supply in the conterminous United States: I. Climate change scenarios and impacts on irrigation water supply simulated with the HUMUS model. *Agric. For. Meteorol.*, **117**, 73-96.

Rosenzweig, C. and W.D. Solecki, Eds., 2001: *Climate Change and a Global City: The Metropolitan East Coast Regional Assessment*. Columbia Earth Institute, New York, New York, 210 pp. [Accessed 12.02.07: http://metroeast_climate.ciesin.columbia.edu/sectors.html]

Rosenzweig, C., F.N. Tubiello, R. Goldberg, E. Mills and J. Bloomfield, 2002: Increased crop damage in the US from excess precipitation under climate change. *Global Environ. Change*, **12**, 197-202.

Rosenzweig, C., W.D. Solecki, L. Parshall, M. Chopping, G. Pope and R. Goldberg, 2005: The heat island effect and global climate change in urban New Jersey. *Glob. Environ. Change*, **6**, 51-62.

Rosetti, M.A., 2002: Potential impacts of climate change on railroads. *The Potential Impacts of Climate Change on Transportation: Workshop Summary and Proceedings*, Center for Climate Change and Environmental Forecasting, Federal Research Partnership Workshop, United States Department of Transportation, Washington, District of Columbia, 13 pp. [Accessed 12.02.07: http://climate.volpe.dot.gov/workshop1002/]

Running, S.W., 2006: Is global warming causing more larger wildfires? *Science*, **313**, 927-928.

Ruosteenoja, K., T.R. Carter, K. Jylha and H. Tuomenvirta, 2003: Future climate in world regions: an intercomparison of model-based projections for the new IPCC emissions scenarios. Finnish Environment Institute, Helsinki, 83 pp. [Accessed 12.02.07: http://www.environment.fi/download.asp?contentid=25835 &lan=en]

Ruth, M. and A.D. Amato, 2002: Regional energy demand responses to climate change: Methodology and applications to Massachusetts. *North American Meeting, Regional Science Association International*, San Juan, Puerto Rico, 24 pp.

Rybczyk, J.M. and D.R. Cahoon, 2002: Estimating the potential for submergence for two wetlands in the Mississippi River Delta. *Estuaries*, **25**, 985-998.

Rygel, L., D. O'Sullivan and B. Yarnal, 2006: A method for constructing a social vulnerability index. *Mitigation Adapt. Strategies for Global Change*, **11**, 741-764.

Sailor, D.J., 2001: Relating residential and commercial sector electricity loads to climate: Evaluating state level sensitivities and vulnerabilities. *Energy*, **26**, 645-657.

Sailor, D.J. and J.R. Muñoz, 1997: Sensitivity of electricity and natural gas consumption to climate in the U.S. - methodology and results for eight states. *Energy*, **22**, 987-998.

Sailor, D.J. and A.A. Pavlova, 2003: Air conditioning market saturation and long-term response of residential cooling energy demand to climate change. *Energy*, **28**, 941-951.

Sala, O.A., F.S. Chapin III, J.J. Armesto, E. Berlow, J. Bloomfield, R. Dirzo, E. Huber-Sanwald, L.F. Huenneke, R.B. Jackson, A. Kinzig, R. Leemans, D.M. Lodge, H.A. Mooney, M. Oesterheld, N.L. Poff, M.T. Sykes, B.H. Walker, M. Walker and D.H.Wall, 2000: Global biodiversity scenarios for the year 2100. *Science*, **287**, 1770-1774.

Sands, R.D. and J.A. Edmonds, 2005: Climate change impacts for the conterminous USA: An integrated assessment. Part 7: Economic analysis of field crops and land use with climate change. *Clim. Change*, **69**, 127-150.

Scavia, D., J.C. Field, D.F. Boesch, R.W. Buddemeier, V. Burkett, D.R. Cayan, M. Fogarty, M.A. Harwell, R.W. Howarth, C. Mason, D.J. Reed, T.C. Royer, A.H. Sallenger and J.G. Titus, 2002: Climate change impacts on U.S. coastal and marine ecosystems. *Estuaries*, **25**, 149-164.

Scheraga, J., 2001: Coping with climate change. *Upper Great Lakes Regional Climate Change Impacts Workshop - For the US National Assessment of Climate Change*, Ann Arbor, Michigan, 131-140.

Schertzer, W.M., W.R. Rouse, D.C.L. Lam, D. Bonin and L. Mortsch, 2004: Climate Variability and Change—Lakes and Reservoirs. *Threats to Water Resources in Canada*. Environment Canada, Ed., National Water Resources Institute, Burlington, Ontario. [Accessed 12.02.07: http://www.nwri.ca/threats2full /ch12-1-e.html]

Schindler, D., 2001: The cumulative effects of climate warming and other human stresses on Canadian freshwaters in the new millennium. *Can. J. Fish Aquat. Sci.*, **58**, 18-29.

Schindler, D.W. and W.F. Donahue, 2006: An impending water crisis in Canada's western prairie provinces. *Proc. Nat. Acad. Sci.*, **107**, doi/10.1073/pnas.0601568103.

Schipper, L., S. Huq and M. Kahn, 2003: An exploration of 'mainstreaming' adaptation to climate change. *Climate Change Research Workshop*, Stockholm Environment Institute IIED and TERI, New Delhi, 4 pp.

Schneider, S.H., 2004: Abrupt non-linear climate change, irreversibility and surprise. *Glob. Environ. Change*, **14**, 245-258.

Schoennagel, T., T.T. Veblen and W.H. Romme, 2004: The interaction of fire, fuels, and climate across Rocky Mountain Forests. *BioScience*, **54**, 661-676.

Schuster, C.J., A. Ellis, W.J. Robertson, J.J. Aramini, D.F. Charron and B. Marshall, 2005: Drinking water related infectious disease outbreaks in Canada, 1974-2001. *Can. J. Public Health*, **94**, 254-258.

Schwartz, J., J.M. Samet and J.A. Patz, 2004a: Hospital admissions for heart disease: the effects of temperature and humidity. *Epidemiology*, **15**, 755-761.

Schwartz, M. and B. Reiter, 2000: Changes in North American spring. *Int. J. Climatol.*, **20**, 929-993.

Schwartz, R.C., P.J. Deadman, D.J. Scott and L.D. Mortsch, 2004b: Modeling the impacts of water level changes on a Great Lakes community. *J. Amer. Water Resour. Assoc.*, **40**, 647-662.

Scott, D. 2005: Ski industry adaptation to climate change: hard, soft and policy strategies. *Tourism and Global Environmental Change*, S. Gossling and M. Hall, Eds., Routledge, Oxford, 265-285.

Scott, D., 2006: Climate Change Vulnerability of the Northeast U.S. Winter Recreation-Tourism Sector. *Technical Report Northeast Climate Impacts Assessment*, Union of Concerned Scientists, Cambridge, Massachusetts.

Scott, D. and B. Jones, 2005: Climate Change and Banff National Park: Implications for Tourism and Recreation – Executive Summary. Report prepared for the Town of Banff, Waterloo, Ontario, 29 pp.

Scott, D. and B. Jones, 2006: *Climate Change and Nature-Based Tourism: Implications for Park Visitation in Canada,* Government of Canada's Climate Change Action Fund - Impacts and Adaptation Program (project A714), 29 pp.

Scott, D., G. McBoyle and B. Mills, 2003: Climate change and the skiing industry in southern Ontario (Canada): exploring the importance of snowmaking as a technical adaptation. *Climate Res.*, **23**, 171-181.

Scott, D., G. McBoyle, B. Mills and A. Minogue, 2006: Climate change and the sustainability of ski-based tourism in eastern North America: A reassessment. *J. Sustainable Tourism*, **14**, 376-398.

Scott, D., B. Jones and J. Konopek, 2007a: Implications of climate and environmental change for nature-based tourism in the Canadian Rocky Mountains: A case study of Waterton Lakes National Park. *Tourism Management,* **28**, 570-579.

Scott, D., G. McBoyle and A. Minogue, 2007b: The implications of climate change for the Québec ski industry, *Global Environmental Change,* **17**, 181-190.

Scott, M.J., L.W. Vail, C.O. Stöckle and A. Kemanian, 2004: Climate change and adaptation in irrigated agriculture - a case study of the Yakima River, *Allocating Water: Economics and the Environment*, Portland, Oregon, Universities Council on Water Resources and The National Institutes for Water Resources, 7 pp.

Scott, M.J., J.A. Dirks and K.A. Cort, 2005: The Adaptive Value of Energy Efficiency Programs in a Warmer World: Building Energy Efficiency Offsets Effects of Climate Change. *Proc. 2005 International Energy Program Evaluation Conference*, Brooklyn, New York.

Segal, M., Z. Pan, R.W. Arritt and E.S. Takle, 2001: On the potential change in wind power over the US due to increases of atmospheric greenhouse gases. *Renew. Energ.*, **24**, 235-243.

Select Bipartisan Committee, 2006: A Failure of Initiative: Final Report of the Select Bipartisan Committee to Investigate the Preparation for and Response to Hurricane Katrina. Select Bipartisan Committee to Investigate the Preparation for and Response to Hurricane Katrina, U.S. House of Representatives (Select Committee), 109th Congress, U.S. Government Printing Office, Washington, District of Columbia, 379 pp. +Appencies. [Accessed 12.02.07: http://katrina.house.gov/full_katrina_report.htm]

Senate of Canada, 2003: Climate Change: We are at Risk. Final Report, Standing Senate Committee on Agriculture and Forestry, Ottawa, Canada, 123 pp.

Shabbar, A., B. Bonsal and M. Khandekar, 1997: Canadian precipitation patterns associated with the Southern Oscillation. *J. Climate*, **10**, 3016-3027.

Shaw, J., R.B. Taylor, D.L. Forbes, M.-H. Ruz and S. Solomon, 1998: *Sensitivity of the Coasts of Canada to Sea-Level Rise,* Bulletin 505, Natural Resources Canada, Geological Survey of Canada, Ottawa,79 pp.

Shein, K.A., 2006: State of the climate in 2005, including executive summary. *Bull. Amer. Meteorol. Soc.*, **87**, 801-805, s801-s102.

Sheridan, S.C. and L.S. Kalkstein, 2004: Progress in heat watch-warning system technology. *Bull. Amer. Meteorol. Soc.*, **85**, 1931-1941.

Shushama, L., R. Laprise, D. Caya, A. Frigon and M. Slivitzky, 2006: Canadian RCM projected climate-change signal and its sensitivity to model errors. *Int. J. Climatol.*, **26**, doi: 10.1002/joc.1362.

Shuter, B.J., C.K. Minns and N. Lester, 2002: Climate change, freshwater fish, and fisheries: Case studies from Ontario and their use in assessing potential impacts. *Fisheries in a Changing Climate*, N. A. McGinn, Ed., American Fisheries Society, 77-88.

Simmons, K., J. Kruse and D. Smith, 2002: Valuing mitigation: Real estate market response to hurricane loss reduction measures. *Southern Econ. J.*, **68** , 660-671.

Simonovic, S.P. and L. Li, 2004: Sensitivity of the Red River Basin flood protection system to climate variability and change. *Water Resour. Manag.*, **18**, 89-110.

Slovic, P., Ed., 2000: *The Perception of Risk*. Earthscan Publications, London, 518 pp.

Small, C. and R.J. Nichols, 2003: A global analysis of human settlement. *J. Coastal Res.*, **19**, 584-599.

Smit, B. and M.W. Skinner, 2002: Adaptation options in agriculture to climate change: A typology. *Mitigation Adapt. Strategies Global Change*, **7**, 85-114.

Smit, B. and E. Wall, 2003: *Adaptation to Climate Change Challenges and Opportunities: Implications and Recommendations for the Canadian Agri-Food Sector,* Senate Standing Committee on Forestry and Agriculture, Ottawa, Canada. [Accessed 12.02.07: http://www.parl.gc.ca/37/2/parlbus/commbus/senate/Com-e/agri-e/power-e/smith-e.htm]

Smit, B., I. Burton, R.J.T. Klein and J. Wandel, 2000: An anatomy of adaptation to climate change and variability. *Climatic Change*, **45**, 223-251.

Smith, O.P. and G. Levasseur, 2002: Impacts of climate change on transportation infrastructure in Alaska. *The Potential Impacts of Climate Change on Transportation: Workshop Summary and Proceedings*, Washington District of Columbia, 11 pp. [Accessed 12.02.07: http://climate.volpe.dot.gov/ workshop1002/smith.pdf]

Smith, S.D., T.E. Huxman, S.F. Zitzer, T.N. Charlet, D.C. Housman, J.S. Coleman, L.K. Fenstermaker, J.R. Seemann and R.S. Nowak, 2000: Elevated CO_2 increases productivity and invasive species success in an arid ecosystem. *Nature*, **408**, 79 - 82.

Smith, S.J., A.M. Thomson, N.J. Rosenberg, R.C. Izaurralde, R.A. Brown and T.M.L. Wigley, 2005: Climate change impacts for the conterminous USA: An integrated assessment: Part 1. Scenarios and context. *Clim. Change*, **69**, 7-25.

Snyder, M.A., J.L. Bell, L.C. Sloan, P.B. Duffy and B. Govindasamy, 2002: Climate responses to a doubling of atmospheric carbon dioxide for a climatically vulnerable region. *Geophys. Res. Lett.*, **29**, doi:10.1029/2001GL014431.

Sohngen, B. and R. Sedjo, 2005: Impacts of climate change on forest product markets: Implications for North American producers. *Forestry Chron.*, **81**, 669-674.

Soil and Water Conservation Society, 2003: *Conservation Implications of Climate Change: Soil Erosion and Runoff from Cropland*, a report from the Soil and Water Conservation Society, Ankeny, Iowa, 26 pp. [Accessed 12.02.07: http://www.swcs.org/documents/Climate_changefinal_112904154622.pdf]

Solecki, W.D. and C. Rosenzweig, 2007: Climate change and the city: Observations from Metropolitan New York. *Urbanization and Environmental Change: Cities as Environmental Hero*, X. Bai, T. Graedel and A. Morishima. Eds., Yale University Press, New Haven, Conneticut. (in press).

Southwick Associates, 2006: *The Economic Contribution of Active Outdoor Recreation*. Outdoor Industry Foundation, Boulder, Colorado, 85 pp.

Southworth, J., R.A. Pfeifer, M. Habeck, J.C. Randolph, O.C. Doering, J.J. Johnston and D.G. Rao, 2002: Changes in soybean yields in the Midwestern United States as a result of future changes in climate, climate variability, and CO_2 fertilization. *Clim. Change*, **53**, 447-475.

Spittlehouse, D.L. and R.B. Stewart, 2003: Adaptation to climate change in forest management. *BC J. Ecosyst. Manag.*, **4**, 1-11.

St. Lawrence River-Lake Ontario Plan of Study Team, 1999: *Plan of study for criteria review in the orders of approval for regulation of Lake Ontario - St. Lawrence River levels and flows,* International Joint Commission. [Accessed 12.02.07: http://www.ijc.org/php/publications/html/pos/pose.html]

Statistics Canada, 2001a: *Population projections for Canada, provinces and territories, 2000-2026,* Statistics Canada, Ottawa, Ontario, 202 pp.

Statistics Canada, 2001b: *Population urban and rural, by province and territory (Canada),* Statistics Canada, Ottawa, Ontario. [Accessed 12.02.07: http://www40.statcan.ca/l01/cst01/demo62a.htm]

Statistics Canada, 2006: *Agriculture Value Added Account: Agriculture Economic Statistics: June 2006.* Statistics Canada, Ottawa, Ontario. [Accessed 12.02.07: http://www.statcan.ca/english/freepub/21-017-XIE/2006001/t026_en.htm?]

Stefan, H.G. and X. Fang, 1999: Simulation of global climate-change impact on temperature and dissolved oxygen in small lakes of the contiguous U.S.. *Proc. of the Specialty Conference on Potential Consequences of Climate Variability and Change to Water Resources of the United States*, American Water Resources Association.

Stefan, H.G., X. Fang and J.G. Eaton, 2001: Simulated fish habitat changes in north American lakes in response to projected climate warming. *Trans. Amer. Fish. Soc.*, **130**, 459-477.

Stewart, I.T., D.R. Cayan and M.D. Dettinger, 2005: Changes toward earlier streamflow timing across western North America. *J. Climate*, **18**, 1136-1155.

Stiger, R.W., 2001: Alaska DOT deals with permafrost thaws. *Better Roads*. June, 30-31. [Accessed 12.02.07: http://obr.gcnpublishing.com/articles/brjun01c.htm]

Stiglitz, J.E., 2002: *Globalization and its Discontents*. Norton. 304 pp.

Stocks, B.J., J.A. Mason, J.B. Todd, E.M. Bosch, B.M. Wotton, B.D. Amiro, M.D. Flannigan, K.G. Hirsch, K.A. Logan, D.L. Martell and W.R. Skinner, 2002: Large forest fires in Canada, 1959-1997. *J. Geophys. Res.*, **107**, doi:10.1029/2001JD000484.

Stone, M.C., R.H. Hotschkiss, C.M. Hubbard, T.A. Fontaine, L.O. Mearns and J.G. Arnold, 2001: Impacts of climate change on Missouri River basin water yield. *J. Amer. Water Resour. Assoc.*, **37**, 1119-1129.

Stone, M.C., R. Hotchkiss and L.O. Mearns, 2003: Water yield responses to high and low spatial resolution climate change scenarios in the Missouri River Basin. *Geophys. Res. Lett.*, **30**, doi:10.1029/2002GL016122.

Stonefelt, M.D., T.A. Fontaine and R.H. Hotchkiss, 2000: Impacts of climate change on water yield in the Upper Wind River Basin. *J. Amer. Water Resour. Assoc.*, **36**, 321-336.

Stott, P.A., 2003: Attribution of regional-scale temperature changes to anthropogenic and natural causes. *Geophys. Res. Lett.*, **30**, doi:10.1029/2003GL017324.

Suarez, P., W. Anderson, V. Mahal and T.R. Lakshmanan, 2005: Impacts of flooding and climate change on urban transportation: A systemwide performance assessment of the Boston Metro Area *Transport. Res. D-Tr. E.*, **10**, 231-244.

Swiss Re, 2005a: *Hurricane Season 2004: Unusual, but not Unexpected,* Swiss Reinsurance Company, Zurich, 12 pp. [Accessed 12.02.07: http://www.swissre.com /INTERNET/pwswpspr.nsf/fmBookMarkFrameSet?ReadForm&BM=../vwAllbyIDKeyLu/ulur-66jcv9?OpenDocument]

Swiss Re, 2005b: *Large Loss Fact Files: Hurricane Ivan,* Swiss Re Publishing, Zurich. [Accessed 12.02.07:http://www.swissre.com/INTERNET/pwswpspr.nsf/fmBookMarkFrameSet?ReadForm&BM=../vwAllbyIDKeyLu/mbui-4v7f68?OpenDocument]

Swiss Re, 2005c: *Large Loss Fact Files: Hurricane Katrina,* Swiss Re Publishing, Zurich. [Accessed 12.02.07:http://www.swissre.com/INTERNET/pwswpspr.nsf/fmBookMarkFrameSet?ReadForm&BM=../vwAllbyIDKeyLu/mbui-4v7f68?OpenDocument]

Swiss Re, 2005d: *Large Loss Fact Files: Hurricane Rita,* Swiss Re Publishing, Zurich. Accessed 12.02.07:http://www.swissre.com/INTERNET/pwswpspr.nsf/fmBookMarkFrameSet?ReadForm&BM=../vwAllbyIDKeyLu/mbui-4v7f68?OpenDocument]

Swiss Re, 2005e: *Large Loss Fact Files: Hurricane Wilma,* Swiss Re Publishing, Zurich. Accessed 12.02.07:http://www.swissre.com/INTERNET/pwswpspr.nsf/fmBookMarkFrameSet?ReadForm&BM=../vwAllbyIDKeyLu/mbui-4v7f68?OpenDocument]

Thomas, C.D., A. Cameron, R.E. Green, M. Bakkenes, L.J. Beaumont, Y.C. Collingham, B.F.N. Erasmus, M.F. d. Siqueira, A. Grainger, L. Hannah, L. Hughes, B. Huntley, A.S. v. Jaarsveld, G.F. Midgley, L. Miles, M.A. Ortega-Huerta, A.T. Peterson, O.L. Phillips and S.E. Williams, 2004: Extinction risk from climate change. *Nature*, **427**, 145-148.

Thomas, M.K., D.F. Charron, D. Waltner-Toews, C. Schuster, A.R. Maarouf and J.D. Holt, 2006: A role of high impact weather events in waterborne disease outbreaks in Canada, 1975-2001. *Int. J. Environ. Health Res.*, **16**, 167-180.

Thomson, A.M., R.A. Brown, N.J. Rosenberg and R.C. Izaurralde, 2005a: Climate change impacts for the conterminous USA: An integrated assessment. Part 2: Models and Validation. *Clim. Change*, **69**, 27-41.

Thomson, A.M., R.A. Brown, N.J. Rosenberg and R.C. Izaurralde, 2005b: Climate change impacts for the conterminous USA: An integrated assessment. Part 5: Irrigated agriculture and national grain crop production. *Clim. Change*, **69**, 89-105.

Thomson, A.M., R.A. Brown, N.J. Rosenberg, R.C. Izaurralde and V. Benson, 2005c: Climate change impacts for the conterminous USA: An integrated assessment. Part 3: Dryland production of grain and forage crops. *Clim. Change*, **69**, 43-65.

Thomson, A.M., R.A. Brown, N.J. Rosenberg, R.C. Izaurralde and R. Srinivasan, 2005d: Climate change impacts for the conterminous USA: An integrated assessment. Part 4: Water resources. *Clim. Change*, **69**, 67-88.

Tierney, K., 2006: Social inequality, hazards, and disasters. *On Risk and Disaster*, H. Kunreuther, R. Danielsand, D. Kettl, Eds., University of Pennsylvania Press, Philadelphia, Pennsylvania.

Titus, J., 2002: Does sea level rise matter to transportation along the Atlantic coast? *The Potential Impacts of Climate Change on Transportation: Workshop Summary and Proceedings*, Washington District of Columbia, 16 pp. [Accessed 12.02.07: http://climate.volpe.dot.gov/workshop1002/titus.pdf]

Titus, J.G., 2005: Sea-level rise effect. *Encyclopaedia of Coastal Science*, M.L. Schwartz, Ed., Springer, Dordrecht, 838-846.

Titus, J.G. and C. Richman, 2001: Maps of lands vulnerable to sea level rise: modeled elevations along the US Atlantic and Gulf Coasts. *Climate Res.*, **18**, 205-228.

Tol, R.S.J., 2002: Estimates of the damage costs of climate change. Part 1: Benchmark estimates. *Environ. Resour. Econ.*, **21**, 47-73.

Trenberth, K.E., P.D. Jones, P. Ambenje, R. Bojariu, D. Easterling, A. Klein Tank, D. Parker, F. Rahimzadeh, J. A. Renwick, M. Rusticucci, B. Sodin and P. Zhai, 2007: Observations: surface and atmospheric change. *Climate Change 2007: The Physical Science Basis. Contribution of Working Group I to the Fourth Assessment Report of the Intergovernmental Panel on Climate Change,* S. Solomon, D. Qin, M. Manning, Z. Chen, M. Marquis, K.B. Averyt, M. Tignor and H.L. Miller, Eds., Cambridge University Press, Cambridge and New York, 235-336.

Troyer, A.F., 2004: Background of U.S. Hybrid Corn II: Breeding, Climate, and food. *Crop Science*, **44**, 370-380.

Tsvetsinskaya, E.A., L.O. Mearns, T. Mavromatis, W. Gao, L. McDaniel and M.W. Downton, 2003: The effect of spatial scale of climatic change scenarios on simulated maize, winter wheat, and rice production in the southeastern United States. *Clim. Change*, **60**, 37-72.

Turner, B.L., II, R.E. Kasperson, P.A. Matson, J.J. McCarthy, R.W. Corell, L. Christensen, N. Eckley, J.X. Kasperson, A. Luers, M.L. Martello, C. Polsky, A. Pulsipher and A. Schiller, 2003: A framework for vulnerability analysis in sustainability science. *Proc. Nat. Acad. Sci.*, **100**, 8074-8079.

UMA Engineering, 2005: *Flood Reduction Master Plan,* City of Peterborough, Peterborough, Canada.

UNDP, 2001: Workshop for Developing and Adaptation Policy Framework for Climate Change. June 2001, United Nations Development Program, Montreal, Canada.

UNECE, 2005a: Trends in Europe and North America - 2OO5: Canada. United Nations Economic Commission for Europe, 2 pp. [Accessed 12.02.07: http://www.unece.org/stats/trends2005/profiles/Canada.pdf]

UNECE, 2005b: Trends in Europe and North America - 2OO5: United States. United Nations Economic Commission for Europe, 2 pp. [Accessed 12.02.07: http://www.unece.org/stats/trends2005/profiles/UnitedStates.pdf]

UNPD, 2005: *World Population Prospects: The 2004 Revision*. United Nations Population Division, New York, New York. [Accessed 12.02.07: http://esa.un.org/unpp/]

Vasquez-Leon, M., C.T. West, B. Wolf, J. Moody and T.J. Finan, 2002: *Vulnerability to Climate Variability in the Farming Sector - A Case Study of Groundwater-Dependant Agriculture in Southeastern Arizona*. The Climate Assessment Project for the South West, Report Series CL 1-02, Institute for the Study of Planet Earth, University of Arizona, Tucson, Arizona, 100 pp. [Accessed 12.02.07: http://www.ispe.arizona.edu/climas/pubs/CL1-02.html]

Vincent, L. and E. Mekis, 2006: Changes in daily and extreme temperature and precipitation indices for Canada over the twentieth century *Atmosphere-Ocean*, **44**, 177-193.

Volney, W.J.A. and R.A. Fleming, 2000: Climate change and impacts of boreal forest insects. *Agric. Ecosyst. Environ.*, **82**, 283-294.

Walker, I.J. and J.V. Barrie, 2006: Geomorphology and sea-level rise on one of Canada's most 'sensitive' coasts: northeast Graham Island, British Columbia. *J. Coastal Res.*, **SI 39**, 220-226.

Walker, R.R., 2001: Climate change assessment at a watershed scale. *Water and Environment Association of Ontario Conference*, Toronto, Canada, 12 pp.

Wall, E. and B. Smit, 2005: Climate change adaptation in light of sustainable agriculture. *J. Sustainable Agric.*, **27**, 113-123.

Wall, E., B. Smitand, J. Wandell, 2005: From silos to synthesis: Interdisciplinary issues for climate change impacts and adaptation Research. *Canadian Association of Geographers special session series: Communities and climate change impacts, adaptation and vulnerability, agriculture*, C-CIARN, Moncton, New Brunswick, 24 pp. [Accessed 12.02.07: http://www.c-ciarn.uoguelph.ca/documents/cciarn_ silostosyn_0105.pdf]

Walsh, M.E., D.G. de la Torre Ugarte, H. Shapouriand, S.P. Slinsky, 2003: Bioenergy crop production in the United States. *Environ. Res. Econ.*, **24**, 313-333.

Walter, M.T., D.S. Wilks, J.Y. Parlange and B.L. Schneider, 2004: Increasing evapotranspiration from the conterminous United States. *J. Hydrometeorol.*, **5**, 405-408.

Walther, G.-R., E. Post, A. Menzel, P. Convey, C. Parmesan, F. Bairlen, T. Beebee, J.M. Fromont, O. Hoegh-Guldberg and F. Bairlein, 2002: Ecological responses to recent climate change. *Nature*, **416**, 389-395.

Waters, D., W.E. Watt, J. Marsalek and B.C. Anderson, 2003: Adaptation of a storm drainage system to accommodate increased rainfall resulting from climate change *J. Environ. Plan. Manag.*, **46**, 755-770.

Wayne, P., S. Foster, J. Connolly, F. Bazzaz and P. Epstein, 2002: Production of allergenic pollen by ragweed (*Ambrosia artemisiifolia* L.) is increased in CO_2-enriched atmospheres. *Ann. Alerg. Asthma Im.*, **88**, 279-282.

Weisskopf, M.G., H.A. Anderson, S. Foldy, L.P. Hanrahan, K. Blair, T.J. Torok and P.D. Rumm, 2002: Heat wave morbidity and mortality, Milwaukee, Wis, 1999 vs 1995: An improved response? *Amer. J. Public Health*, **92**, 830-833.

Welch, C., 2006: Sweeping change reshapes Arctic. *The Seattle Times*. Jan. 1 2006. [Accessed 12.02.07: http://seattletimes.nwsource.com/html/localnews/2002714404_arctic01main.html]

West, J.J., M.J. Small and H. Dowlatabadi, 2001: Storms, investor decisions, and the economic impacts of sea level rise. *Clim. Change*, **48**, 317-342.

Westerling, A.L., H.G. Hidalgo, D.R. Cayan and T.W. Swetnam, 2006: Warming and earlier spring increase western U.S. forest wildfire activity. *Science*, **313**, 940-943.

Wetherald, R.T. and S. Manabe, 2002: Simulation of hydrologic changes associated with global warming. *J. Geophys. Res.*, **107**, doi: 10.1029/2001JD00195,02002.

Wheaton, E., V. Wittrock, S. Kulshretha, G. Koshida, C. Grant, A. Chipanshi and B. Bonsal, 2005: *Lessons Learned from the Canadian Drought Years of 2001 and 2002: Synthesis Report*. Saskatchewan Research Council Publication No. 11602-46E03, Saskatoon, Saskatchewan, 38 pp. [Accessed 12.02.07: http://www.agr.gc.ca/pfra/drought/info/11602-46E03.pdf]

White, M.A., N.S. Diffenbaugh, G.V. Jones, J.S. Pal and F. Giorgi, 2006: Extreme heat reduces and shifts United States premium wine production in the 21st century. *PNAS*, **103**, 11217-11222. doi: 10.1073/pnas.0603230103.

Whitfield, P.H. and A.J. Cannon, 2000: Recent variations in climate and hydrology in Canada. *Can. Water Resour. J.*, **25**, 19-65.

WHO, 2003: *Climate Change and Human Health – Risk and Responses*. World Health Organization, New York, New York, 250 pp.

Wilbur, H.M., 1997: Experimental ecology of food webs: complex systems in temporary ponds. *Ecology*, **78**, 2279-2302.

Wilgoren, J. and K.R. Roane, 1999: Cold Showers, Rotting Food, the Lights, Then Dancing. *New York Times*, A1. July 8, 1999.

Williams, D.W. and A.M. Liebhold, 2002: Climate change and the outbreak ranges of two North American bark beetles. *Agric. For. Meteorol.*, **4**, 87-99.

Winkler, J.A., J.A. Andresen, G. Guentchev and R.D. Kriegel, 2002: Possible impacts of projected temperature change on commercial fruit production in the Great Lakes Region. *J. Great Lakes Res.*, **28**, 608-625.

Wolfe, D.W., M.D. Schwartz, A.N. Lakso, Y. Otsuki, R.M. Pool and N.J. Shaulis, 2005: Climate change and shifts in spring phenology of three horticultural woody perennials in northeastern USA. *Int. J. Biometeorol.*, **49**, 303-309.

Woods, A., K.D. Coates and A. Hamann, 2005: Is an unprecedented dothistroma needle blight epidemic related to climate change? *BioScience*, **55**, 761-769.

Woodward, F.I. and M.R. Lomas, 2004: Vegetation dynamics - Simulating responses to climatic change. *Biol. Rev.*, **79**, 643-370.

World Tourism Organization, 2002: *Tourism Highlights 2001,* WTO Publications Unit - World Tourism Organization, Madrid.

Wrona, F.J., T.D. Prowse and J.D. Reist. 2005: Freshwater Ecosystems and Fisheries. *ACIA. Arctic Climate Impact Assessment*, Cambridge University Press, New York, 353-452. [Accessed 12.02.07: http://www.acia.uaf.edu/]

Wu, S.Y., B. Yarnal and A. Fisher, 2002: Vulnerability of coastal communities to sea-level rise: a case study of Cape May County, New Jersey, USA. *Climate Res.*, **22**, 255-270.

Yarnal, B., A.L. Heasley, R.E. O'Connor, K. Dow and C.L. Jocoy, 2006: The potential use of climate forecasts by community water system managers. *Land Use Water Resour. Res.*, **6**, 3.1-3.8.

Yohe, G. and R.S.J. Tol, 2002: Indicators for ecological and economic coping capacity: Moving forward a working definition of adaptive capacity. *Global Environ. Change*, **12**, 25-40.

Zavaleta, E.S. and K.B. Hulvey, 2004: Realistic species losses disproportionately reduce grassland resistance to biological invaders. *Science*, **306**, 1175-1177.

Zervas, C.E., 2001: *Sea Level Variations of the United States: 1854-1999,* National Ocean Service, Technical Report NOS CO-OPS 36, National Oceanic and Atmospheric Administration, Silver Spring, Maryland, 201 pp. [Accessed 12.02.07: http://tidesandcurrents.noaa.gov/publications/techrpt36doc.pdf]

Zhang, K.Q., B.C. Douglas and S.P. Leatherman, 2000: Twentieth-century storm activity along the U.S. east coast. *J. Climate*, **13**, 1748-1761.

Zhang, X., K. Harvey, W. Hogg and T. Yuzyk, 2001: Trends in Canadian streamflow. *Water Resour. Res.*, **37**, 987-998.

Zimmerman, R., 2002: Global climate change and transportation infrastructure: lessons from the New York area. *The Potential Impacts of Climate Change on Transportation: Workshop Summary and Proceedings*, Washington District of Columbia, 11 pp. [Accessed 12.02.07: http://climate.volpe.dot.gov/workshop1002 /zimmermanrch.pdf]

Ziska, L.H., D.E. Gebhard, D.A. Frenz, S. Faulkner, B.D. Singer and J.G. Straka, 2003: Cities as harbingers of climate change: Common ragweed, urbanization, and public health. *J. Allergy Clin. Immunol.*, **111**, 290-295.

Zolbrod, A.N. and D.L. Peterson, 1999: Response of high-elevation forests in the Olympic Mountains to climatic change. *Can. J. For. Res.* , **29**, 1966-1978.

Zwiers, F. and X. Zhang, 2003: Toward regional-scale climate change detection. *J. Climate*, **16**, 793-797.

15

Polar regions (Arctic and Antarctic)

Coordinating Lead Authors:

Oleg A. Anisimov (Russia), David G. Vaughan (UK)

Lead Authors:

Terry Callaghan (Sweden/UK), Christopher Furgal (Canada), Harvey Marchant (Australia), Terry D. Prowse (Canada), Hjalmar Vilhjálmsson (Iceland), John E. Walsh (USA)

Contributing Authors:

Torben R. Christensen (Sweden), Donald L. Forbes (Canada), Frederick E. Nelson (USA), Mark Nuttall (Canada/UK), James D. Reist (Canada), George A. Rose (Canada), Jef Vandenberghe (The Netherlands), Fred J. Wrona (Canada)

Review Editors:

Roger Barry (USA), Robert Jefferies (Canada), John Stone (Canada)

This chapter should be cited as:

Anisimov, O.A., D.G. Vaughan, T.V. Callaghan, C. Furgal, H. Marchant, T.D. Prowse, H. Vilhjálmsson and J.E. Walsh, 2007: Polar regions (Arctic and Antarctic). *Climate Change 2007: Impacts, Adaptation and Vulnerability. Contribution of Working Group II to the Fourth Assessment Report of the Intergovernmental Panel on Climate Change*, M.L. Parry, O.F. Canziani, J.P. Palutikof, P.J. van der Linden and C.E. Hanson, Eds., Cambridge University Press, Cambridge, UK, 653-685.

Table of Contents

Executive summary

In both polar regions, there is strong evidence of the ongoing impacts of climate change on terrestrial and freshwater species, communities and ecosystems (very high confidence). Recent studies project that such changes will continue (high confidence), with implications for biological resources and globally important feedbacks to climate (medium confidence). Strong evidence exists of changes in species' ranges and abundances and in the position of some tree lines in the Arctic (high confidence). An increase in greenness and biological productivity has occurred in parts of the Arctic (high confidence). Surface albedo is projected to decrease and the exchange of greenhouse gases between polar landscapes and the atmosphere will change (very high confidence). Although recent models predict that a small net accumulation of carbon will occur in Arctic tundra during the present century (low confidence), higher methane emissions responding to the thawing of permafrost and an overall increase in wetlands will enhance radiative forcing (medium confidence). [15.4.1, 15.4.2, 15.4.6].

In both polar regions, components of the terrestrial cryosphere and hydrology are increasingly being affected by climate change (very high confidence). These changes will have cascading effects on key regional bio-physical systems and cause global climatic feedbacks, and in the north will affect socio-economic systems (high confidence). Freshwater and ice flows into polar oceans have a direct impact on sea level and (in conjunction with the melt of sea ice) are important in maintaining the thermohaline circulation. In the Arctic, there has been increased Eurasian river discharge to the Arctic Ocean, and continued declines in the ice volume of Arctic and sub-Arctic glaciers and the Greenland ice sheet (very high confidence). Some parts of the Antarctic ice sheet are also losing significant volume (very high confidence). Changes to cryospheric processes are also modifying seasonal runoff and routings (very high confidence). These combined effects will impact freshwater, riparian and near-shore marine systems (high confidence) around the Arctic, and on sub-Antarctic islands. In the Arctic, economic benefits, such as enhanced hydropower potential, may accrue, but some livelihoods are likely to be adversely affected (high confidence). Adaptation will be required to maintain freshwater transportation networks with the loss of ice cover (high confidence). [15.4.1, 15.4.6, 15.7.1].

Continued changes in sea-ice extent, warming and acidification of the polar oceans are likely to further impact the biomass and community composition of marine biota as well as Arctic human activities (high confidence).
Although earlier claims of a substantial mid-20th century reduction of Antarctic sea-ice extent are now questioned, a recently reported decline in krill abundance is due to regional reductions in Antarctic sea-ice extent; any further decline will adversely impact their predators and ecosystems (high confidence). Acidification of polar waters is predicted to have adverse effects on calcified organisms and consequential effects on species that rely upon them (high confidence). The impact of climate change on Arctic fisheries will be regionally specific; some beneficial and some detrimental. The reduction of Arctic sea ice has led to improved marine access, increased coastal wave action, changes in coastal ecology/biological production and adverse effects on ice-dependent marine wildlife, and continued loss of Arctic sea ice will have human costs and benefits (high confidence). [15.2.1, 15.2.2, 15.4.3].

Already Arctic human communities are adapting to climate change, but both external and internal stressors challenge their adaptive capabilities (high confidence). Benefits associated with climate change will be regionally specific and widely variable at different locations (medium confidence).

Impacts on food accessibility and availability, and personal safety are leading to changes in resource and wildlife management and in livelihoods of individuals (e.g., hunting, travelling) (high confidence). The resilience shown historically by Arctic indigenous peoples is now being severely tested (high confidence). Warming and thawing of permafrost will bring detrimental impacts on community infrastructure (very high confidence). Substantial investments will be needed to adapt or relocate physical structures and communities (high confidence). The benefits of a less severe climate are dependant on local conditions, but include reduced heating costs, increasing agricultural and forestry opportunities, more navigable northern sea routes, and marine access to resources (medium confidence). [15.4.5, 15.5, 15.7]

15.1 Introduction

The polar regions are increasingly recognised as being:
- geopolitically and economically important,
- extremely vulnerable to current and projected climate change,
- the regions with the greatest potential to affect global climate and thus human populations and biodiversity.

Sub-regions of the Arctic and Antarctic have shown the most rapid rates of warming in recent years. Substantial environmental impacts of climate change show profound regional differences both within and between the polar regions, and enormous complexity in their interactions. The impacts of this climate change in the polar regions over the next 100 years will exceed the impacts forecast for many other regions and will produce feedbacks that will have globally significant consequences. However, the complexity of response in biological and human systems, and the fact that these systems are subject to multiple stressors, means that future impacts remain very difficult to predict.

15.1.1 Summary of knowledge assessed in the TAR

This chapter updates and extends the discussion of polar regions in the IPCC Working Group II, Third Assessment Report (TAR, Anisimov et al., 2001). That report summarised the climatic changes that have been observed in the polar regions over the 20th century, the impacts those changes have had on the environment, and the likely impact of projected climate change in the future. The following summarises the key findings of that assessment to which 'very-high confidence' or 'high confidence' was attached (see Figure 15.1 for overview and place names used in this chapter).

Key trends highlighted in the TAR

- In the Arctic, during the 20th century, air temperatures over extensive land areas increased by up to 5°C; sea ice thinned and declined in extent; Atlantic water flowing into the Arctic Ocean warmed; and terrestrial permafrost and Eurasian spring snow decreased in extent.
- There has been a marked warming in the Antarctic Peninsula over the last half-century. There has been no overall change in Antarctic sea-ice extent over the period 1973-1996.

Key regional projections highlighted in the TAR

- Increased melting of Arctic glaciers and the Greenland ice sheet, but thickening of the Antarctic ice sheet due to increased precipitation, were projected.
- Exposure of more bare ground and consequent changes in terrestrial biology on the Antarctic Peninsula were anticipated.
- Substantial loss of sea ice at both poles was projected.
- Reduction of permafrost area and extensive thickening of the active layer in the Arctic was expected to lead to altered landscapes and damage to infrastructure.
- Climate change combined with other stresses was projected to affect human Arctic communities, with particularly disruptive impacts on indigenous peoples following traditional and subsistence lifestyles.
- Economic costs and benefits were expected to vary among regions and communities.

Key polar drivers of global climate change identified in the TAR

- Changes in albedo due to reduced sea-ice and snow extent were expected to cause additional heating of the surface and further reductions in ice/snow cover.

Awareness of such issues led to the preparation of a uniquely detailed assessment of the impacts of climate change in the Arctic (ACIA, 2005), which has been drawn upon heavily in the Arctic component of this chapter. There is no similarly detailed report for the Antarctic.

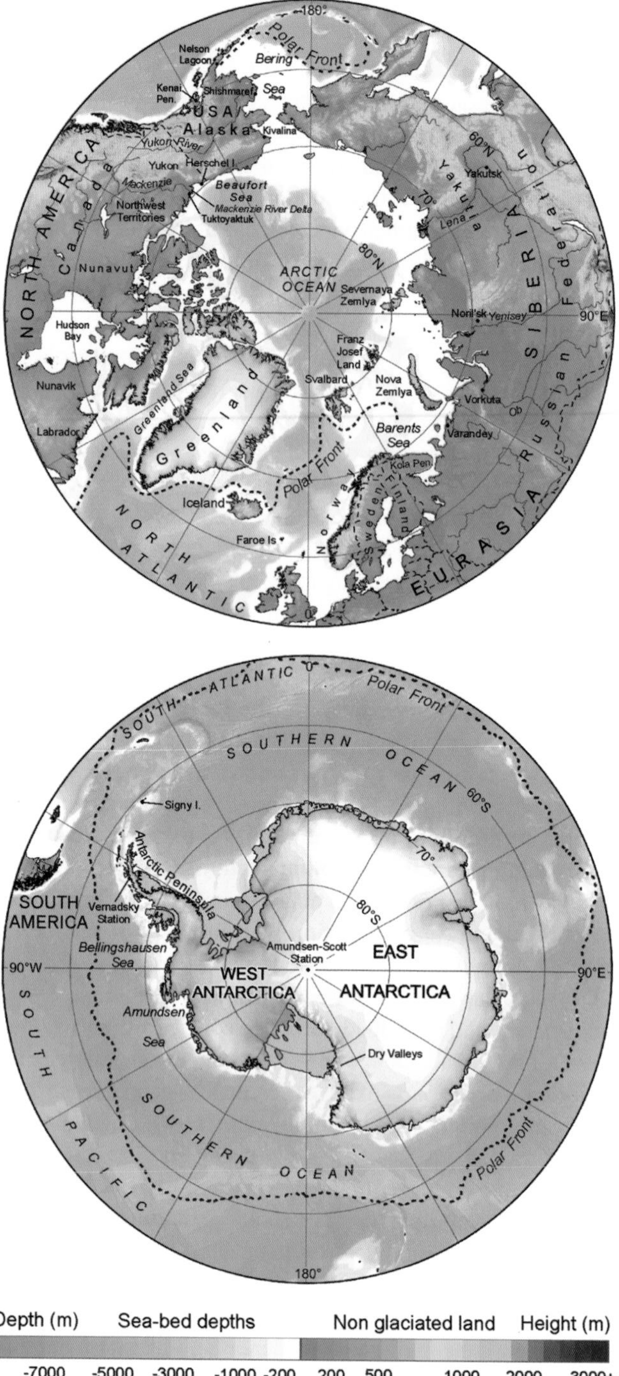

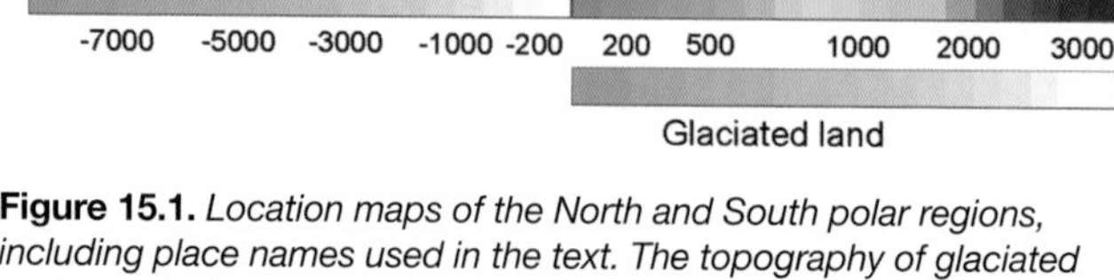

Figure 15.1. *Location maps of the North and South polar regions, including place names used in the text. The topography of glaciated and non-glaciated terrain is shown by using different shading schemes. The polar fronts shown are intended to give an approximate location for the extent of cold, polar waters but are, in places, open to interpretation and fluctuations. (This and other maps were drawn by P. Fretwell, British Antarctic Survey.)*

15.2 Current sensitivity/vulnerability

15.2.1 Climate, environment and socio-economic state

Arctic

For several decades, surface air temperatures in the Arctic have warmed at approximately twice the global rate (McBean et al., 2005). The areally averaged warming north of 60°N has been 1-2°C since a temperature minimum in the 1960s and 1970s. In the marine Arctic, the 20th-century temperature record is marked by strong low-frequency (multi-decadal) variations (Polyakov et al., 2002). Serreze and Francis (2006) have discussed the attribution of recent changes in terms of natural variability and anthropogenic forcing, concluding that a substantial proportion of the recent variability is circulation-driven, and that the Arctic is in the early stages of a manifestation of a human-induced greenhouse signature. This conclusion is based largely on the relatively slow rate of emergence of the greenhouse signal in model simulations of the late 20th and early 21st centuries.

The most recent (1980 to present) warming of much of the Arctic is strongest (about 1°C/decade) in winter and spring, and smallest in autumn; it is strongest over the interior portions of northern Asia and north-western North America (McBean et al., 2005). The latter regions, together with the Antarctic Peninsula, have been the most rapidly warming areas of the globe over the past several decades (Turner et al., 2007). The North Atlantic sub-polar seas show little warming during the same time period, probably because of their intimate connection with the cold, deep waters. Temperatures in the upper troposphere and stratosphere of the Arctic have cooled in recent decades, consistent with increases in greenhouse gases and with decreases in stratospheric ozone since 1979 (Weatherhead et al., 2005).

Precipitation in the Arctic shows signs of an increase over the past century, although the trends are small (about 1% per decade), highly variable in space, and highly uncertain because of deficiencies in the precipitation measurement network (McBean et al., 2005) and the difficulty in obtaining accurate measurements of rain and snow in windy polar regions. There is no evidence of systematic increases in intense storms in the Arctic (Atkinson, 2005) although coastal vulnerability to storms is increasing with the retreat of sea ice (see Section 15.4.6). Little is known about areally averaged precipitation over Greenland. The discharge of Eurasian rivers draining into the Arctic Ocean shows an increase since the 1930s (Peterson et al., 2002), generally consistent with changes in temperature and the large-scale atmospheric circulation.

Reductions of Arctic sea ice and glaciers (see Lemke et al., 2007), reductions in the duration of river and lake ice in much of the sub-Arctic (Prowse et al., 2004; Walsh et al., 2005), and a recent (1980s to present) warming of permafrost in nearly all areas for which measurements are available (Romanovsky et al., 2002; Walsh et al., 2005) are consistent with the recent changes in Arctic surface air temperatures. Although there is visual evidence of permafrost degradation (Lemke et al., 2007), long-term measurements showing widespread thickening of the active layer are lacking. Changes in vegetation, particularly a transition from grasses to shrubs, has been reported in the North American Arctic (Sturm et al., 2001) and elsewhere (Tape et al., 2006), and satellite imagery has indicated an increase in the Normalised Difference Vegetation Index (NDVI, a measure of photosynthetically active biomass) over much of the Arctic (Slayback et al., 2003). This is consistent with a longer growing season and with documented changes in the seasonal variation in atmospheric CO_2 concentrations as reported in the TAR. Broader ecosystem impacts of climate change in both polar regions are summarised by Walther et al. (2002) and were documented more extensively for the Arctic by the Arctic Climate Impact Assessment (ACIA, 2005).

Recent analysis of air-borne data (Krabill et al., 2004), satellite data (Howat et al., 2005; Luckman et al., 2006; Rignot and Kanagaratnam, 2006) and seismic data (Ekstrom et al., 2006) indicate thinning around the periphery of the Greenland ice sheet, where summer melt has increased during the past 20 years (Abdalati and Steffen, 2001; Walsh et al., 2005), while there is evidence of slower rates of thickening further inland (Johannessen et al., 2005).

The Arctic is now home to approximately 4 million residents (Bogoyavlenskiy and Siggner, 2004). Migration into the Arctic during the 20th century has resulted in a change of demographics such that indigenous peoples now represent 10% of the entire population. This influx has brought various forms of social, cultural and economic change (Huntington, 1992; Nuttall, 2000b). For most Arctic countries, only a small proportion of their total population lives in the Arctic, and settlement remains generally sparse (Bogoyavlenskiy and Siggner, 2004) and nomadic peoples are still significant in some countries. On average, however, two-thirds of the Arctic population live in settlements with more than 5,000 inhabitants. Indigenous residents have, in most regions, been encouraged to become permanent residents in fixed locations, which has had a predominantly negative effect on subsistence activities and some aspects of community health. At the same time, Arctic residents have experienced an increase in access to treated water supplies, sewage disposal, health care facilities and services, and improved transportation infrastructure which has increased access to such things as outside market food items (Hild and Stordhal, 2004). In general, the Arctic has a young, rapidly growing population with higher birth rates than their national averages, and rising but lower than national average life-expectancy. This is particularly true for indigenous populations, although some exceptions exist, such as in the Russian north, where population and life-expectancy has decreased since 1990 (Einarsson et al., 2004).

Political and administrative regimes in Arctic regions vary between countries. In particular, indigenous groups have different levels of self-determination and autonomy. Some regions (e.g., northern Canada and Greenland) now have formalised land-claim settlements, while in Eurasia indigenous claims have only recently begun to be addressed (Freeman, 2000). Wildlife management regimes and indigenous/non-indigenous roles in resource management also vary between regions. Nowadays, large-scale resource extraction initiatives and/or forms of social support play significant roles in the economies of many communities. Despite these changes, aspects of subsistence and pastoral livelihoods remain important.

Regardless of its small and dispersed population, the Arctic has become increasingly important in global politics and economies. For example, the deleterious effect on the health of Arctic residents of contaminants produced in other parts of the world has led to international agreements such as the Stockholm Convention on Persistent Organic Pollutants (Downey and Fenge, 2003). Furthermore, significant oil, gas and mineral resources (e.g., diamonds) are still to be developed in circum-Arctic regions that will further increase the importance of this region in the world (e.g., U.S. Geological Survey World Energy Assessment Team, 2000; Laherre, 2001).

Antarctic

Direct measurements reveal considerable spatial variability in temperature trends in Antarctica. All meteorological stations on the Antarctic Peninsula show strong and significant warming over the last 50 years (see Section 15.6.3). However, of the other long-term (>30 years) mean annual temperature records available, twelve show warming, while seven show cooling;

although only two of these (one of each) are significant at the 10% level (Turner et al., 2005). If the individual station records are considered as independent measurements, then the mean trend is warming at a rate comparable to mean global warming (Vaughan et al., 2003), but there is no evidence of a continent-wide 'polar amplification' in Antarctica. In some areas where cooling has occurred, such as the area around Amundsen-Scott Station at the South Pole, there is no evidence of directly attributable impacts, but elsewhere cooling has caused clear local impacts. For example, in the Dry Valleys, a 6 to 9% reduction in primary production in lakes and a >10%/yr decline in soil invertebrates has been observed (Doran et al., 2002). Although the impacts are less certain, precipitation has also declined on sub-Antarctic islands (Bergstrom and Chown, 1999).

Recent changes in Antarctic sea-ice extent are discussed in detail elsewhere (Lemke et al., 2007), but evidence highlighted in the TAR (Anisimov et al., 2001) gleaned from records of whaling activities (de la Mare, 1997) is no longer considered reliable (Ackley et al., 2003). So, for the period before satellite observation, only direct local observations (e.g., Murphy et al., 1995) and proxies (e.g., Curran et al., 2003) are available. For the satellite period (1978 to present) there has been no ubiquitous trend in Antarctic sea-ice duration, but there have been strong regional trends (see Figure 15.2). Sea-ice duration in the Ross Sea has increased, while in the Bellingshausen and Amundsen Seas it has decreased, with high statistical significance in each case (Parkinson, 2002; Zwally et al., 2002).

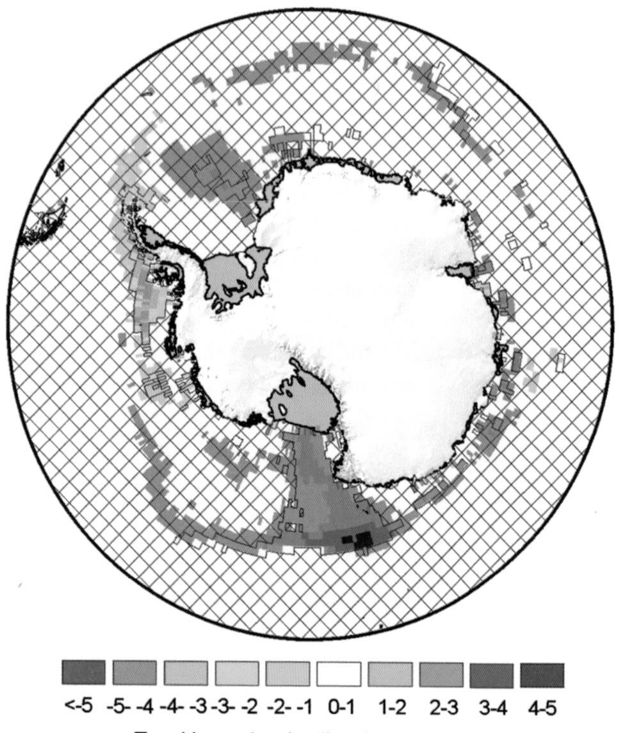

Figure 15.2. *Trends in annual Antarctic sea-ice duration in days per year, for the period 1978-2004, after Parkinson (2002). Hatched areas show where trends are not significant at the 95% level. (Data compiled by W.M. Connolley, British Antarctic Survey.)*

This pattern strongly reflects trends in atmospheric temperature at nearby climate stations (Vaughan et al., 2003).

Increasing atmospheric CO_2 concentrations are leading to an increased draw-down of CO_2 by the oceans and, as a consequence, sea water is becoming more acidic (Royal Society, 2005). As is the case in other parts of the world's oceans, coccolithophorids and foraminifera are significant components of the pelagic microbial community of the Southern Ocean and contribute to the draw-down of atmospheric CO_2 to the deep ocean. Experimental studies (Riebesell et al., 2000) indicate that elevated CO_2 concentration reduces draw-down of CO_2 compared with the production of organic matter. Recent investigations suggest that at the present rate of acidification of the Southern Ocean, pteropods (marine pelagic molluscs) will not be able to survive after 2100 and their loss will have significant consequences for the marine food web (Orr et al., 2005). Similarly, cold-water corals are threatened by increasing acidification. Furthermore, increasing acidification leads to changes in the chemistry of the oceans, altering the availability of nutrients and reducing the ability of the oceans to absorb CO_2 from the atmosphere (Royal Society, 2005; see also Chapter 4 this volume).

15.2.2 Vulnerability and adaptive capacity

Vulnerability is the degree to which a system is susceptible to, or unable to cope with, adverse effects of stress; whereas adaptive capacity, or resilience, is an ability to adjust to stress, to realise opportunities or to cope with consequences (McCarthy et al., 2005).

15.2.2.1 Terrestrial and marine ecosystems

Many polar species are particularly vulnerable to climate change because they are specialised and have adapted to harsh conditions in ways that are likely to make them poor competitors with potential immigrants from environmentally more benign regions (e.g., Callaghan et al., 2005; Peck et al., 2006). Other species require specific conditions, for example winter snow cover or a particular timing of food availability (Mehlum, 1999; Peck et al., 2006). In addition, many species face multiple, concurrent human-induced stresses (including increased ultraviolet-B radiation, increasing contaminant loads, habitat loss and fragmentation) that will add to the impacts of climate change (Walther et al., 2002; McCarthy et al., 2005).

Plants and animals in the polar regions are vulnerable to attacks from pests (Juday et al., 2005) and parasites (Albon et al., 2002; Kutz et al., 2002) that develop faster and are more prolific in warmer and moister conditions. Many terrestrial polar ecosystems are vulnerable because species richness is low in general, and redundancy within particular levels of food chains and some species groups is particularly low (Matveyeva and Chernov, 2000). Loss of a keystone species (e.g., lemmings, Turchin and Batzli, 2001) could have cascading effects on entire ecosystems.

Arctic

In Arctic ecosystems, adaptive capacity varies across species groups from plants that reproduce by cloning, which have relatively low adaptive potential, through some insects (e.g., Strathdee et al., 1993) that can adapt their life cycles, to micro-organisms that have great adaptive potential because of rapid turnover and universal dispersal. The adaptive capacity of current Arctic ecosystems is small because their extent is likely to be reduced substantially by compression between the general northwards expansion of forest, the current coastline and longer-term flooding of northern coastal wetlands as the sea level rises, and also as habitat is lost to land use (see Figure 15.3). General vulnerability to warming and lack of adaptive capacity of Arctic species and ecosystems are likely, as in the past, to lead to relocation rather than rapid adaptation to new climates (see Figure 15.3).

As air and sea water temperatures have increased in the Bering Sea, there have been associated changes in sea-ice cover, water-column properties and processes including primary production and sedimentation, and coupling with the bottom layer (Grebmeier et al., 2006). A change from Arctic to sub-Arctic conditions is happening with a northward movement of the pelagic-dominated marine ecosystem that was previously confined to the south-eastern Bering Sea. Thus communities that consist of organisms such as bottom-feeding birds and marine mammals are being replaced by communities dominated by pelagic fish. Changes in sea ice conditions have also affected subsistence and commercial harvests (Grebmeier et al., 2006).

Many Arctic and sub-Arctic seas (e.g., parts of the Bering and Barents Seas) are among the most productive in the world (Sakshaug, 2003), and yield about 7 Mt of fish per year, provide about US$15 billion in earnings (Vilhjálmsson et al., 2005), and employ 0.6 to 1 million people (Agnarsson and Arnason, 2003). In addition, Arctic marine ecosystems are important to indigenous peoples and rural communities following traditional and subsistence lifestyles (Vilhjálmsson et al., 2005).

Recent studies reveal that sea surface warming in the north-east Atlantic is accompanied by increasing abundance of the largest phytoplankton in cooler regions and their decreasing abundance in warmer regions (Richardson and Schoeman, 2004). In addition, the seasonal cycles of activities of marine micro-organisms and invertebrates and differences in the way components of pelagic communities respond to change, are leading to the activities of prey species and their predators becoming out of step. Continued warming is therefore likely to impact on the community composition and the numbers of primary and secondary producers, with consequential stresses

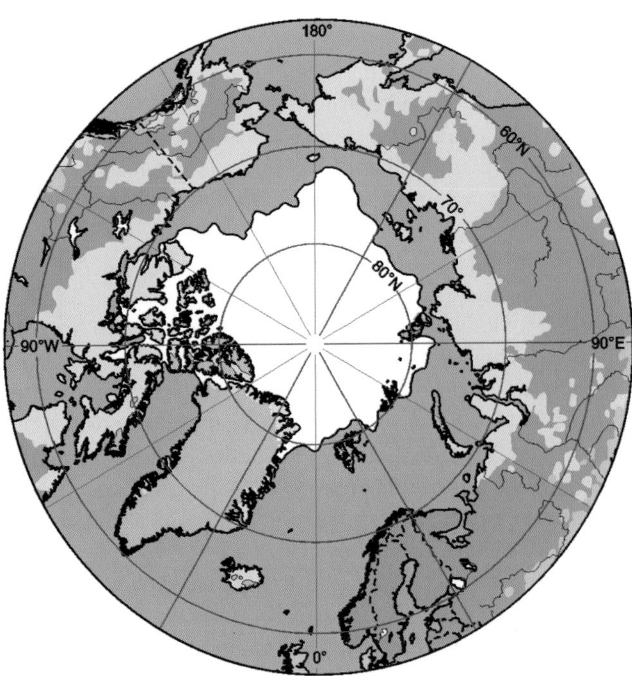

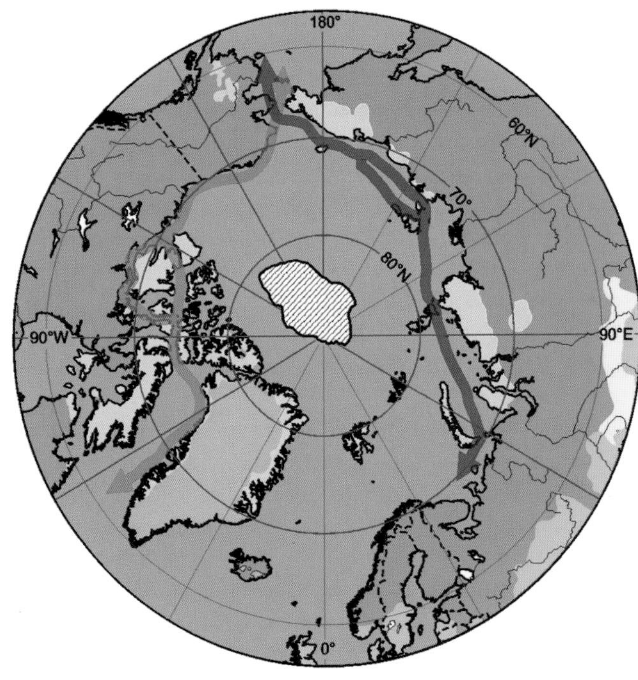

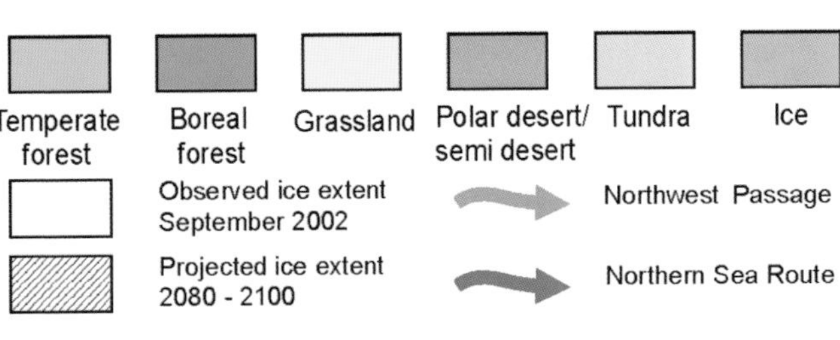

Figure 15.3. *Present and projected vegetation and minimum sea-ice extent for Arctic and neighbouring regions. Vegetation maps based on floristic surveys (top) and projected vegetation for 2090-2100, predicted by the LPJ Dynamic Vegetation Model driven by the HadCM2 climate model (bottom) modified from Kaplan et al. (2003) in Callaghan et al. (2005). The original vegetation classes have been condensed as follows: grassland = temperate grassland and xerophytic scrubland; temperate forest = cool mixed forest, cool-temperate evergreen needle-leaved and mixed forest, temperate evergreen needle-leaved forest, temperate deciduous broadleaved forest; boreal forest = cool evergreen needle-leaved forest, cold deciduous forest, cold evergreen needle-leaved forest; tundra = low- and high-shrub tundra, erect dwarf-shrub tundra, prostrate dwarf-shrub tundra; polar desert/semi-desert = cushion forb, lichen and moss tundra. Also shown are observed minimum sea-ice extent for September 2002, and projected sea-ice minimum extent, together with potential new/improved sea routes (redrawn from Instanes et al., 2005; Walsh et al., 2005).*

on higher trophic levels. This will impact economically important species, primarily fish, and dependent predators such as marine mammals and sea birds (Edwards and Richardson, 2004).

Antarctic

Substantial evidence indicates major regional changes in Antarctic terrestrial and marine ecosystems in areas that have experienced warming. Increasing abundance of shallow-water sponges and their predators, declining abundances of krill, Adelie and Emperor penguins, and Weddell seals have all been recorded (Ainley et al., 2005). Only two species of native flowering plant, the Antarctic pearlwort (*Colobanthus quitensis*) and the Antarctic hair grass (*Deschampsia antarctica*) currently occur in small and isolated ice-free habitats on the Antarctic continent. Their increased abundance and distribution was ascribed to the increasing summer temperatures (Fowbert and Smith, 1994). Elsewhere on continental Antarctica, climate change is also affecting the vegetation, which is largely composed of algae, lichens and mosses, and changes are expected in future, as temperature, and water and nutrient availability, change (Robinson et al., 2003).

The marked reduction reported in the biomass of Antarctic krill (*Euphausia superba*) and an increase in the abundance of salps (principally *Salpa thompsoni*), a pelagic tunicate, may be related to regional changes in sea ice conditions (Atkinson et al., 2004). This change may also underlie the late-20th century changes in the demography of krill predators (marine mammals and sea birds) reported from the south-west Atlantic (Fraser and Hoffmann, 2003), and this connection indicates a potential vulnerability to climate change whose importance cannot yet be determined.

Recent studies on sub-Antarctic islands have shown increases in the abundance of alien species and negative impacts on the local biota such as a decline in the number and size of *Sphagnum* moss beds (Whinam and Copson, 2006). On these islands, increasing human activities and increasing temperatures are combining to promote successful invasions of non-indigenous species (Bergstrom and Chown, 1999).

15.2.2.2 Freshwater systems

Arctic

Climate variability/change has historically had, and will continue to have, impacts on Arctic freshwater resources. First-order impacts (e.g., changes to the snow/ice/water budget) play a significant role in important global climate processes, through feedbacks (e.g., changes to radiative feedbacks, stability of the oceanic stratification and thermohaline circulation, and carbon/methane source-sink status). Cascading effects have important consequences for the vulnerability of freshwater systems, as measured by their ecological or human resource value.

From an ecological perspective, the degree of vulnerability to many higher-order impacts (e.g., changes in aquatic geochemistry, habitat availability/quality, biodiversity) are related to gradual and/or abrupt threshold transitions such as those associated with water-phase changes (e.g., loss of freshwater ice cover) or coupled bio-chemical responses (e.g., precipitous declines in dissolved oxygen related to lake productivity) (Wrona et al., 2005). Historically, Arctic freshwater ecosystems have adapted to large variations in climate over long transitional periods (e.g., Ruhland and Smol, 2002; Ruhland et al., 2003), but in the next 100 years the combination of high-magnitude events and rapid rates of change will probably exceed the ability of the biota and their associated ecosystems to adapt (Wrona et al., 2006a). This will result in significant changes and both positive and negative impacts. It is projected, however, that overall the negative effects will very probably outweigh the positive, implying that freshwater systems are vulnerable to climate change (Wrona et al., 2005).

From a human-use perspective, potential adaptation measures are extremely diverse, ranging from measures to facilitate modified use of the resource (e.g., changes in ice-road construction practices, increased open-water transportation, flow regulation for hydroelectric production, harvesting strategies, and methods of drinking-water access), to adaptation strategies to deal with increased/decreased freshwater hazards (e.g., protective structures to reduce flood risks or increase floods for aquatic systems (Prowse and Beltaos, 2002); changes to more land-based travel to avoid increasingly hazardous ice). Difficulties in pursuing adaptation strategies may be greatest for those who place strong cultural and social importance on traditional uses of freshwater resources (McBean et al., 2005; Nuttall et al., 2005).

Antarctic

Antarctic freshwater systems are fewer and smaller than those in the Arctic, but are no less vulnerable to climate change. The microbial communities inhabiting these systems are likely to be modified by changing nutrient regimes, contaminants and introductions of species better able to cope with the changing conditions. A drop in air temperature of 0.7°C per decade late in the 20th century in the Dry Valleys led to a 6 to 9% drop in primary production in the lakes of the area (Doran et al., 2002). In marked contrast, summer air temperature on the maritime sub-Antarctic Signy Island increased by 1°C over the last 50 years and, over the period 1980 to 1995, water temperature in the lakes rose several times faster than the air temperature; this is one of the fastest responses to regional climate change in the Southern Hemisphere yet documented (Quayle et al., 2002). As a consequence, the annual ice-free period has lengthened by up to 4 weeks. In addition, the area of perennial snow cover on Signy Island has decreased by about 45% since 1951, and the associated change in microbial and geochemical processes has led to increased amounts of organic and inorganic nutrients entering the lakes. There has also been an explosion in the population of fur seals (*Arctocephalus gazella*) on the island due to decreased ice cover and increased area available for resting and moulting. Together these changes are leading to disruption of the ecosystem due to increased concentrations of nutrients (eutrophication) (Quayle et al., 2003). Similar ecological vulnerabilities are expected to exist in other Antarctic freshwater systems.

15.2.2.3 Permafrost

Permafrost, defined as sub-surface earth materials that remain at or below 0°C continuously for two or more years, is widespread in Arctic, sub-Arctic and high-mountain regions, and in the small areas of Antarctica without permanent ice cover.

The physical processes of climate-permafrost interactions and observations of permafrost change are discussed elsewhere (Lemke et al., 2007); here we focus on the observed and projected changes of permafrost, and impacts they may have on natural and human systems in the Arctic.

Observational data are limited, but precise measurements in boreholes indicate that permafrost temperatures in the Arctic rose markedly during the last 50 years (Romanovsky et al., 2002), with rapid warming in Alaska (Hinzman et al., 2005), Canada (Beilman et al., 2001), Europe (Harris et al., 2003) and Siberia (Pavlov and Moskalenko, 2002). Short-term and localised warming associated with the reduction of snow cover (Stieglitz et al., 2003) and feedbacks associated with increased vegetation productivity (Sturm et al., 2001; Anisimov and Belolutskaia, 2004; Chapin et al., 2005b) are, however, important considerations that must be taken into account.

In the context of the future climate change, there are two key concerns associated with the thawing of permafrost: the detrimental impact on the infrastructure built upon it, and the feedback to the global climate system through potential emission of greenhouse gases. These are discussed in Sections 15.7.1 and 15.4.2.3.

15.2.2.4 Human populations

Neither Antarctica nor the sub-Antarctic islands have permanent human populations; the vast majority of residents are staff at scientific stations and summer-only visitors. While there are some areas of particular sensitivity (see Section 15.6.3), where climate change might dictate that facilities be abandoned, from a global perspective these can be viewed as logistical issues only for the organisations concerned.

In contrast, the archaeological record shows that humans have lived in the Arctic for thousands of years (Pavlov et al., 2001).

Previously, many Arctic peoples practised seasonal movements between settlements, and/or seasonally between activities (e.g., farming to fishing), and the semi-nomadic and nomadic following of game animals and herding. Today, most Arctic residents live in permanent communities, many of which exist in low-lying coastal areas. Despite the socio-economic changes taking place, many Arctic communities retain a strong relationship with the land and sea, with community economies that are a combination of subsistence and cash economies, in some cases, strongly associated with mineral, hydrocarbon and resource development (Duhaime, 2004). The vulnerable nature of Arctic communities, and particularly coastal indigenous communities, to climate change arises from their close relationship with the land, geographical location, reliance on the local environment for aspects of everyday life such as diet and economy, and the current state of social, cultural, economic and political change taking place in these regions.

Communities are already adapting to local environmental changes (Krupnik and Jolly, 2002; Nickels et al., 2002) through wildlife management regimes, and changes in individual behaviours (i.e., shifts in timing and locations of particular activities) and they retain a substantial capacity to adapt. This is related to flexibility in economic organisation, detailed local knowledge and skills, and the sharing mechanisms and social networks which provide support in times of need (Berkes and Jolly, 2001). However, for some Arctic peoples, movement into permanent communities, along with shifts in lifestyle and culture, limits some aspects of adaptive capacity as more sedentary lifestyles minimise mobility, and increased participation in wage-economy jobs decreases the number of individuals able to provide foods from the local environment. The sustainability of this trend is unknown.

Small Arctic communities, however remote, are tightly tied politically, economically and socially to the national mainstream, as well as being linked to and affected by the global economy (Nuttall et al., 2005). Today, trade barriers, resource management regimes, political, legal and conservation interests, and globalisation all affect, constrain or reduce the abilities of Arctic communities to adapt to climate change (Nuttall et al., 2005). Trends in modernity within communities also affect adaptive capacity in both positive and negative ways. Increased access to outside markets and new technologies improve the ability to develop resources and a local economic base; however, increased time spent in wage-earning employment, while providing significant benefits at the individual and household levels through enhanced economic capacity, reduces time on the land observing and developing the knowledge that strengthens the ability to adapt. This underscores the reality that climate change is one of several interrelated problems affecting Arctic communities and livelihoods today (Chapin et al., 2005a).

In some cases, indigenous peoples may consider adaptation strategies to be unacceptable, as they impact critical aspects of traditions and cultures. For example, the Inuit Circumpolar Conference has framed the issue of climate change in a submission to the United States Senate as an infringement on human rights because it restricts access to basic human needs as seen by the Inuit and will lead to the loss of culture and identity (Watt-Cloutier, 2004). Currently we do not know the limits of adaptive capacity among Arctic populations, or what the impacts of some adaptive measures will be.

15.3 Assumptions about future trends

15.3.1 Key regional impacts with importance to the global system

We expect climate change in the polar regions to have many direct regional impacts; however, those regional direct impacts may have global implications through the following processes and feedbacks.

- ***Reflectivity of snow, ice and vegetation:*** snow, ice and vegetation play vital roles in the global climate system, through albedo and insulation effects. Since the TAR, increasing evidence has emerged indicating a more rapid disappearance of snow and sea-ice cover in some areas (e.g., Siberia, Alaska, the Greenland Sea), and consequent changes of albedo may be leading to further climate change (e.g., Holland and Bitz, 2003).
- ***Retreat of glaciers and ice sheets, freshwater runoff, sea level and ocean circulation:*** the retreat of glaciers in the Arctic and more rapid melting of the edges of the Greenland

ice sheet (Section 15.2.1), together with observed increases in river runoff (Peterson et al., 2002), the major contributor, will alter the freshwater budget of the Arctic Ocean. Further changes are expected and could influence ocean circulation with global impacts (Lemke et al., 2007).

- ***Arctic terrestrial carbon flux:*** although models project that Arctic terrestrial ecosystems and the active layer will be a small sink for carbon in the next century, processes are complex and uncertainty is high. It is possible that increased emissions of carbon from thawing permafrost will lead to positive climate forcing (Sitch et al., 2007). Whether such emissions reach the atmosphere as methane or as carbon dioxide is important, because, on a per molecule basis, methane has more than 20 times the warming influence (Anisimov et al., 2005b).
- ***Migrating species:*** species that seasonally migrate from lower latitudes to polar regions rely on the existence of specific polar habitats, and if those habitats are compromised the effects will be felt in communities and food webs far beyond the polar regions. These habitats are likely to be compromised by direct or consequential climate change impacts such as drying of ponds and wetlands, and also by multiple stresses (e.g., land-use changes, hunting regulations).
- ***Methane hydrates:*** significant amounts of methane hydrates are contained in sediments, especially on Arctic continental shelves. As these areas warm, this methane may be released, adding to the greenhouse gas concentration in the atmosphere (Sloan, 2003; Maslin, 2004).
- ***Southern Ocean carbon flux:*** climate models indicate that stratification of the Southern Ocean will change. This could change the community structure of primary producers and alter rates of draw-down of atmospheric CO_2 and its transport to the deep ocean.

15.3.2 Projected atmospheric changes

The areally averaged warming in the Arctic is projected to range from about 2°C to about 9°C by the year 2100, depending on the model and forcing scenario. The projected warming is largest in the northern autumn and winter, and is largest over the polar oceans in areas of sea-ice loss. Over land, the projected warming shows less seasonal variation, although regions such as the Canadian Archipelago are not well resolved.

In contrast to the unanimity of the models in predicting a north-polar amplification of warming, there are differences among the model projections concerning polar amplification in Antarctica, especially over the continent (Parkinson, 2004). However, in several simulations, the warming is amplified over a narrow Southern Ocean band from which sea ice retreats.

Global precipitation is projected to increase during the 21st century by 10 to 20% in response to the SRES[1] A1B emissions scenario of the IPCC Fourth Assessment Report (AR4) simulations. However, the seasonality and spatial patterns of the precipitation increase in the Arctic vary among the models. Similar results have emerged from other IPCC AR4 simulations (Kattsov et al., 2007). In addition, the partitioning among snow and rain will change in a warmer climate, affecting surface hydrology, terrestrial ecosystems and snow loads on structures. The ratio of rain to snow should increase especially in those seasons and Arctic sub-regions in which present-day air temperatures are close to freezing. The difference between precipitation and evapotranspiration (P–E), which over multi-year timescales is approximately equivalent to runoff (river discharge), is also projected to increase over the course of the 21st century. The projected increases of runoff by 2080 are generally in the range of 10 to 30%, largest in the A2 scenario and smallest in the B1 scenario. Of the major river basins, the largest increases are projected for the Lena River basin. Additional information on projected changes is presented elsewhere (Meehl et al., 2007). The effects of these freshwater changes on the thermohaline circulation are uncertain.

15.3.3 Projected changes in the oceans

A new study (Zhang and Walsh, 2006) based on the IPCC AR4 model simulations, projected mean reductions of annually averaged sea ice area in the Arctic by 2080-2100 of 31%, 33% and 22% under the A2, A1B and B1 scenarios, respectively (see Figure 15.3). A consistent model result is that the sea-ice loss is greater in summer than in winter, so that the area of seasonal (winter-only) sea-ice coverage actually increases in many models. The loss of summer sea ice will change the moisture supply to northern coastal regions and will be likely to impact the calving rates of glaciers that are now surrounded by sea ice for much of the year. There will also be increases in wind-driven transport and mixing of ocean waters in regions of sea-ice loss.

The projected increases of Arctic river discharge and precipitation over polar oceans, as well as the projections of increasing discharge from the Greenland ice sheet (Lemke et al., 2007), point to a freshening of the ocean surface in northern high latitudes. However, the projected changes of ice discharge (calving rates) are generally not available from the model simulations, since the ice sheet discharge is not explicitly included in coupled global models.

15.3.4 Projected changes on land

Arctic

Although seasonal snow cover on land is highly variable, it has important effects on the substrate and on local climate, primarily through its insulating properties and high albedo. In Eurasia, and to a lesser extent North America, there has been a persistent 5-6 day/decade increase in the duration of snow-free conditions over the past three decades (Dye, 2002). The reduction of snow residence time occurs primarily in spring. Projections from different climate models generally agree that these changes will continue. Likely impacts include increases in near-surface ground temperature, changes in the timing of spring melt-water pulses, and enhanced transportation and agricultural opportunities (Anisimov et al., 2005a). The projected warming

[1] SRES: IPCC Special Report on Emissions Scenarios, see Nakićenović et al, 2000

also implies a continuation of recent trends toward later freeze-up and earlier break-up of river and lake ice (Walsh et al., 2005).

Projections of change agree that the retreat of glaciers will continue across Arctic glaciers, with a consequent impact on global sea level (Meehl et al., 2007). Recent changes in the Greenland ice sheet have, however, been complex. The colder interior has thickened, most probably as a result of recently high precipitation rates, while the coastal zone has been thinning. Thus some studies suggest that overall the ice sheet is growing in thickness (Krabill et al., 2000; Johannessen et al., 2005). However, there is a growing body of evidence for accelerating coastal thinning, a response to recent increases in summer melt (Abdalati and Steffen, 2001), and acceleration of many coastal glaciers (Krabill et al., 2004; Howat et al., 2005; Ekstrom et al., 2006; Luckman et al., 2006; Rignot and Kanagaratnam, 2006) suggest that thinning is now dominating the mass balance of the entire ice sheet.

Warming, thawing and decrease in areal extent of terrain underlain by permafrost are expected in response to climatic change in the 21st century (Sazonova et al., 2004; Euskirchen et al., 2006; Lemke et al., 2007). Results from models forced with a range of IPCC climate scenarios indicate that by the mid-21st century the permafrost area in the Northern Hemisphere is likely to decrease by 20 to 35%, largely due to the thawing of permafrost in the southern portions of the sporadic and discontinuous zones, but also due to increasing patchiness in areas that currently have continuous permafrost (Anisimov and Belolutskaia, 2004). Projected changes in the depth of seasonal thawing (base of the active layer) are uniform neither in space nor in time. In the next three decades, active layer depths are likely to be within 10 to 15% of their present values over most of the permafrost area; by the middle of the century, the depth of seasonal thawing may increase on average by 15 to 25%, and by 50% and more in the northernmost locations; and by 2080, it is likely to increase by 30 to 50% and more over all permafrost areas (Anisimov and Belolutskaia, 2004; Instanes et al., 2005).

Antarctic

Current and projected changes in the Antarctic ice sheet are discussed in greater detail elsewhere (Lemke et al., 2007), and are only summarised here. Recent changes in volume of the Antarctic ice sheet are much better mapped and understood than they were in the TAR, but competing theories over the causes still limits confidence in prediction of the future changes. The ice sheet on the Antarctic Peninsula is probably alone in showing a clear response to contemporary climate change (see Section 15.6.3), while the larger West Antarctic and East Antarctic ice sheets are showing changes whose attribution to climate change are not clear, but cannot be ruled out. In West Antarctica, there is a suggestion that the dramatic recent thinning of the ice sheet throughout the Amundsen Sea sector is the result of recent ocean change (Payne et al., 2004; Shepherd et al., 2004), but as yet there are too few oceanographic measurements to confirm this interpretation. Indeed, there is evidence that deglaciation of some parts of West Antarctica, as a response to climate change at the end of the last glacial period, is not yet complete (Stone et al., 2003). There are still competing theories, but the now clear evidence of ice-sheet change, has reinvigorated debate about whether we should expect a deglaciation of part of the West Antarctic ice sheet on century to millennial timescales (Vaughan, 2007). Studies based on satellite observations do not provide unequivocal evidence concerning the mass balance of the East Antarctic ice sheet; some appear to indicate marginal thickening (Davis et al., 2005), while others indicate little change (Zwally et al., 2005; Velicogna and Wahr, 2006; Wingham et al., 2006).

Permafrost in ice-free areas, seasonal snow cover, and lake-ice do exist in Antarctica but in such small areas that they are only discussed in respect to particular impacts.

15.4 Key future impacts and vulnerabilities

15.4.1 Freshwater systems and their management

15.4.1.1 Arctic freshwater systems and historical changes

Some freshwater systems exist wholly within the Arctic but many others are fed by river and lake systems further south. The latter includes five of the world's largest river catchments, which act as major conduits transporting water, heat, sediment, nutrients, contaminants and biota into the Arctic. For these systems, it will be the basin-wide changes that will determine the Arctic impacts.

Historically, the largest changes to northern river systems have been produced by flow regulation, much of it occurring in the headwaters of Arctic rivers. For Canada and Russia, it is these northward-flowing rivers that hold the greatest remaining potential for large-scale hydroelectric development (e.g., Shiklomanov et al., 2000; Prowse et al., 2004). Similar to some expected effects of climate change, the typical effect of hydroelectric flow regulation is to increase winter flow but also to decrease summer flow and thereby change overall inter-seasonal variability. In the case of the largest Arctic-flowing river in North America, the Mackenzie River, separating the effects of climate change from regulation has proven difficult because of the additional dampening effects on flow produced by natural storage-release effects of major lake systems (e.g., Gibson et al., 2006; Peters et al., 2006). For some major Russian rivers (Ob and Yenisei), seasonal effects of hydroelectric regulation have been noted as being primarily responsible for observed trends in winter discharge that were previously thought to be a result of climatic effects (Yang et al., 2004a, b). By contrast, winter flow increases on the Lena River have resulted primarily from increased winter precipitation and warming (Yang et al., 2002; Berezovskaya et al., 2005). Spatial patterns in timing of flows, however, have not been consistent, with adjacent major Siberian rivers showing both earlier (Lena – Yang et al., 2002) and later (Yenisei – Yang et al., 2004b) spring flows over the last 60 years. Although precipitation changes are often suspected of causing many changes in river runoff, a sparse precipitation monitoring network in the Arctic, makes such linkages very difficult (Walsh et al., 2005). Seasonal precipitation-runoff responses could be further obscured by the effects of permafrost thaw and related alterations to flow pathways and transfer times (Serreze et al., 2003; Berezovskaya et al., 2005; Zhang et al., 2005).

Over the last half-century, the combined flow from the six largest Eurasian rivers has increased by approximately 7% or an average of 2 km^3/yr (Peterson et al., 2002). The precise controlling factors remain to be identified, but effects of ice-melt from permafrost, forest fires and dam storage have been eliminated as being responsible (McClelland et al., 2004). Increased runoff to the Arctic Ocean from circumpolar glaciers, ice caps and ice sheets has also been noted to have occurred in the late 20th century and to be comparable to the increase in combined river inflow from the largest pan-Arctic rivers (Dyurgerov and Carter, 2004).

The Arctic contains numerous types of lentic (still-water) systems, ranging from shallow tundra ponds to large lakes. Seasonal shifts in flow, ice cover, precipitation/evapotranspiration and inputs of sediment and nutrients have all been identified as climate-related factors controlling their biodiversity, storage regime and carbon-methane source-sink status (Wrona et al., 2005). A significant number of palaeolimnological records from lakes in the circumpolar Arctic have shown synchronous changes in biological community composition and sedimentological parameters associated with climate-driven regime shifts in increasing mean annual and summer temperatures and corresponding changes in thermal stratification/stability and ice-cover duration (e.g., Korhola et al., 2002; Ruhland et al., 2003; Pienitz et al., 2004; Smol et al., 2005; Prowse et al., 2006b).

Permafrost plays a large role in the hydrology of lentic systems, primarily through its influence on substrate permeability and surface ponding of water (Hinzman et al., 2005). Appreciable changes have been observed in lake abundance and area over a 500,000 km^2 zone of Siberia during an approximate three-decade period at the end of the last century (see Figure 15.4; Smith et al., 2005). The spatial pattern of lake disappearance strongly suggests that permafrost thawing is driving the changes.

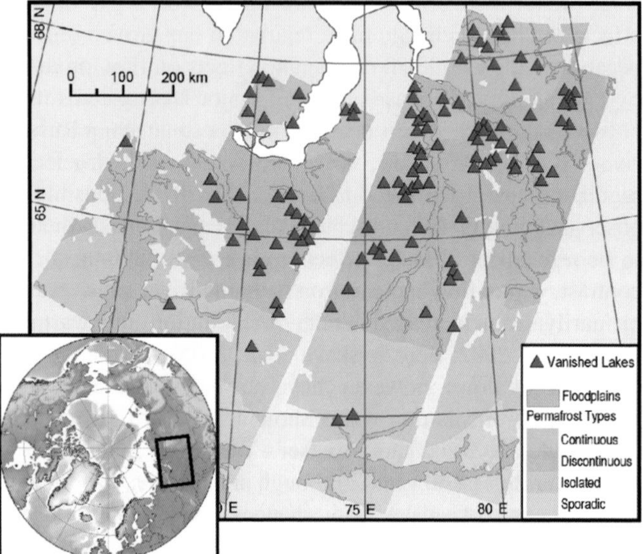

Figure 15.4. *Locations of Siberian lakes that have disappeared after a three-decade period of rising soil and air temperatures (changes registered from satellite imagery from early 1970s to 1997-2004), overlaid on various permafrost types. The spatial pattern of lake disappearance suggests that permafrost thawing has driven the observed losses. From Smith et al., 2005. Reprinted with permission from AAAS.*

15.4.1.2 Impacts on physical regime

Changes in Arctic freshwater systems will have numerous impacts on the physical regime of the Arctic, particularly affecting hydrological extremes, global feedbacks and contaminant pathways.

Hydrological models based on atmosphere-ocean general circulation models (AOGCMs) have consistently also predicted increases in flow for the major Arctic river systems, with the largest increases during the cold season (Miller and Russell, 2000; Arora and Boer, 2001; Mokhov et al., 2003; Georgievsky et al., 2005). Less clear is what may occur during the summer months, with some results suggesting that flow may actually decrease because of evaporation exceeding precipitation (e.g., Walsh et al., 2005). Reductions in summer flow could be enhanced for many watersheds because of increases in evapotranspiration as dominant terrestrial vegetation shifts from non-transpiring tundra lichens to various woody species (e.g., Callaghan et al., 2005). CO_2-induced reductions in transpiration might offset this, and have been suggested as being responsible for some 20th-century changes in global runoff (Gedney et al., 2006).

Since Arctic river flow is the major component of the freshwater budget of the Arctic Ocean (Lewis et al., 2000), it is important to the supply of freshwater to the North Atlantic and related effects on the thermohaline circulation (Bindoff et al., 2007). The sum of various freshwater inputs to the Arctic Ocean has matched the amount and rate at which freshwater has accumulated in the North Atlantic over the period 1965 to 1995 (Peterson et al., 2006). Under greenhouse gas scenarios, the total annual river inflow to the Arctic Ocean is expected to increase by approximately 10 to 30% by the late 21st century (Walsh et al., 2005). An additional source of future freshwater input will be from melting of large glaciers and ice caps, most notably from Greenland (Gregory et al., 2004; Dowdeswell, 2006). The cumulative effect of these increasing freshwater supplies on thermohaline circulation remains unclear but is a critical area of concern (Loeng et al., 2005; Bindoff et al., 2007).

Warming is also forecast to cause reductions in river- and lake-ice cover, which will lead to changes in lake thermal structures, quality/quantity of under-ice habitat and effects on ice jamming and related flooding (Prowse et al., 2006a). Specific to the latter, forecasts of earlier snowmelt freshets could create conditions more conducive to severe break-up events (Prowse and Beltaos, 2002), although a longer period of warming could also reduce severity (Smith, 2000). This effect, however, is likely to be offset on some large northward-flowing rivers because of reduced regional contrasts in south-to-north temperatures and related hydrological and physical gradients (Prowse et al., 2006a).

Projected changes of permafrost, vegetation and river-runoff may have noticeable impacts on river morphology, acting through destabilisation of banks and slopes, increased erosion and sediment supply, and ultimately leading to the transformation between multi- and single-channel types. Geological reconstructions and numerical simulations indicate that such transformations and also erosion events and flood risks occur especially at times of permafrost degradation (Bogaart and van Balen, 2000; Vandenberghe, 2002). Such changes are largely controlled by thresholds in sediment supply to the river and discharge (Vandenberghe, 2001). However, historical examples have shown that variability in flow regime is less important than variability in sediment supply, which is especially determined by the vegetation cover (Huisink et al., 2002; Vandenberghe and Huisink, 2003). Thus an increasingly denser vegetation cover may counter increased sediment discharge, which has been modelled to rise in Arctic rivers with both increases in air temperature and water discharge (Syvitski, 2002).

Various changes in Arctic hydrology have the potential to effect large changes in the proportion of pollutants (e.g., persistent organic pollutants and mercury) that enter Arctic aquatic systems, either by solvent-switching or solvent-depleting processes (e.g., MacDonald et al., 2003). Given that the Arctic is predicted to be generally 'wetter', the increase in loadings of particulates and contaminants that partition strongly into water might more than offset the reductions expected to accrue from reductions in global emissions (e.g., MacDonald et al., 2003). Shifts in other hydrological regime components such as vegetation, runoff patterns and thermokarst drainage (Hinzman et al., 2005) all have the capacity to increase contaminant capture. Changes in aquatic trophic structure and related rate functions (see Section 15.4.1.3) have further potential to alter the accumulation of bio-magnifying chemicals within food webs.

15.4.1.3 Impacts on aquatic productivity and biodiversity

Projected changes in runoff, river- and lake-ice regimes, and seasonal and interannual water balance and thermal characteristics will alter biodiversity and productivity relationships in aquatic ecosystems (Walsh et al., 2005; Prowse et al., 2006b; Wrona et al., 2006a). Ultimately the dispersal and geographical distribution patterns of aquatic species will be altered, particularly for fish (Reist et al., 2006a). Extension of the ice-free season may lead to a decline in fish habitat availability and suitability, particularly affecting species such as lake trout (*Salvelinus namaycush*) that prefer colder waters (Hobbie et al., 1999; Reist et al., 2006b). The projected enhanced river flows will also increase sediment transport and nutrient loading into the Arctic Ocean, thereby affecting estuarine and marine productivity (Carmack and Macdonald, 2002).

Increased permafrost thawing and deepening of the active layer will increase nutrient, sediment and carbon loadings, enhancing microbial and higher trophic level productivity in nutrient-limited systems. As water-column dissolved organic carbon (DOC) concentration increases, penetration of damaging UV radiation and photochemical processing of organic material would decline, although not as prominently in highly productive systems (Reist et al., 2006b; Wrona et al., 2006a). Enhanced sediment loadings will negatively affect benthic and fish-spawning habitats by increasing the biological oxygen demand and hypoxia/anoxia associated with sedimentation, and contribute to habitat loss through infilling (Reist et al., 2006a). Whether freshwater systems will function as net carbon sinks or sources depends on the complex interactions between temperature, nutrient status and water levels (Frey and Smith, 2005; Flanagan et al., 2006). Initial permafrost thaw will form depressions for new wetlands and ponds interconnected by new drainage networks. This will allow for the dispersal and establishment of new aquatic communities in areas formerly dominated by terrestrial species (Wrona et al., 2006b). As the permafrost thaws further, surface waters will increasingly drain into groundwater systems, leading to losses in freshwater habitat.

Southerly species presently limited by temperature/productivity constraints will probably colonise Arctic areas, resulting in new assemblages. Many of these, particularly fishes, will be likely to out-compete or prey upon established Arctic species, resulting in negative local effects on these (Reist et al., 2006a). These southern emigrants to the Arctic will also bring with them new parasites and/or diseases to which Arctic species are not adapted, thereby increasing mortality (Wrona et al., 2006b). Direct environmental change combined with indirect ecosystem shifts will significantly impact local faunas by reducing productivity, abundance and biodiversity. Such effects will be most severe for freshwater fish that rely entirely upon local aquatic ecosystems (Reist et al., 2006c). Distributions of anadromous fish, which migrate up rivers from the sea to breed in freshwater, will probably shift as oceanic conditions and freshwater drainage patterns are affected (Reist et al., 2006c); as will the geographical patterns of habitat use of migratory aquatic birds and mammals (Wrona et al., 2005). Important northern fish species such as broad whitefish (*Coregonus nasus*), Arctic char (*Salvelinus alpinus*), inconnu (*Stenodus leucichthys*), Arctic grayling (*Thymallus arcticus*) and Arctic cisco (*Coregonus autumnalis*) will probably experience population reductions and extirpations (e.g., due to reproductive failures), contraction of geographical ranges in response to habitat impacts, and competition and predation from colonising species (Reist et al., 2006a, b, c).

15.4.1.4 Impacts on resource use and traditional economies/livelihoods

Given the large hydrological changes expected for Arctic rivers, particularly regarding the magnitude of the spring freshet, climate-induced changes must be factored into the design, maintenance and safety of existing and future development structures (e.g., oil and gas drilling platforms, pipelines, mine tailings ponds, dams and impoundments for hydroelectric production) (World Commission on Dams, 2000; Prowse et al., 2004; Instanes et al., 2005).

Freshwater sources are critical to human health, especially for many northern communities that rely on surface and/or groundwater, often untreated, for drinking water and domestic use (United States Environmental Protection Agency, 1997; Martin et al., 2005). Direct use of untreated water from lakes, rivers and large pieces of multi-year sea ice is considered to be a traditional practice, despite the fact that it poses a risk to human health via the transmission of water-borne diseases (e.g., Martin et al., 2005). Such risks may increase with changes in migration and northward movement of species and their related diseases. Changes in hydrology may also decrease the availability and quality of drinking water, particularly for coastal communities affected by rising sea levels where seawater contamination could affect groundwater reserves (Warren et al., 2005).

Northern freshwater ecosystems provide many services to Arctic peoples, particularly in the form of harvestable biota used to support both subsistence and commercial economies (Reist et al., 2006b). Shifts in ecosystem structure and function will result in substantial changes in the abundance, replenishment, availability and accessibility of such resources which, in turn, will alter local resource use and traditional and subsistence lifestyles (Nuttall et al., 2005; Reist et al., 2006b). It is unlikely that such changes related to natural freshwater systems would be offset by increased opportunity for freshwater aquaculture resulting from a warming climate. Thus, conservation of Arctic aquatic biodiversity, maintenance of traditional and subsistence lifestyles, and continued viability and sustainable use of Arctic freshwater resources will present significant challenges for Arctic peoples, resource managers and policy-makers (Wrona et al., 2005; Reist et al., 2006b).

15.4.2 Terrestrial ecosystems and their services

15.4.2.1 Historical and current changes in Arctic terrestrial ecosystems

Climatic changes during the past 20,000 years and more have shaped current biodiversity, ecosystem extent, structure and function. Arctic species diversity is currently low, partly because of past extinction events (FAUNMAP Working Group, 1996). As a group, large mammals are in general more vulnerable to current change than in the past when the group contained many more species. Also, tundra ecosystem extent, particularly in Eurasia, is now less than during the glacial period, when extensive tundra-steppe ecosystems existed (Callaghan et al., 2005). Modern habitat fragmentation (e.g., Nellemann et al., 2001), stratospheric ozone depletion, and spread of contaminants, compound the ongoing impacts of anthropogenic climate change and natural variability on ecosystems and their services.

Traditional ecological knowledge (TEK, see Section 15.6.1) from Canada has recorded current ecosystem changes such as poor vegetation growth in eastern regions associated with warmer and drier summers; increased plant biomass and growth in western regions associated with warmer, wetter and longer summers; the spreading of some existing species, and new sightings of a few southern species; and changing grazing behaviours of musk oxen and caribou as the availability of forage increases in some areas (Riedlinger and Berkes, 2001; Thorpe et al., 2001; Krupnik and Jolly, 2002).

In northern Fennoscandia the cycles of voles, and possibly also of lemmings, have become considerably dampened since the 1980s, due to low spring peak densities, and it is likely that these changes are linked to poorer winter survival due to changing snow conditions (Yoccoz and Ims, 1999; Henttonen and Wallgren, 2001; Ims and Fuglei, 2005). Arctic fox, lesser white fronted goose and shore lark have declined dramatically (Elmhagen et al., 2000) and moose, red fox and some southern bird species have spread northwards (Hörnberg, 1995; Tannerfeldt et al., 2002), although the specific role of climate change is unknown. Some migrant bird populations, particularly Arctic waders, have declined substantially (Stroud et al., 2004) due to various causes including climate change and loss of habitat on migration routes and wintering grounds (Morrison et al., 2001, 2004; Zöckler, 2005). In contrast, populations of Arctic breeding geese, which in winter increasingly feed on agricultural crops or unharvested grain, have shown a geometric increase in numbers in Europe and North America, which has led to intense foraging in Arctic coastal breeding habitats, loss of vegetation and the occurrence of hypersaline soils (Jefferies et al., 2006). Although there are examples from temperate latitudes, evidence for early arrival of migratory birds in the Arctic is weak (Gauthier et al., 2005), emphasising the need for adequate monitoring programmes (Both et al., 2005). Some populations of caribou/reindeer, which are essential to the culture and subsistence of several Arctic peoples, are currently in decline (Russell et al., 2002; Chapin et al., 2005a), mainly due to social and cultural factors. However, climate impacts have also affected some populations. Icing events during warmer winters that restrict access to frozen vegetation have impacted some reindeer/caribou populations and high-Arctic musk oxen populations (Forchhammer and Boertmann, 1993; Aanes et al., 2000; Callaghan et al., 2005 and references therein).

Evidence of recent vegetation change is compelling. Aerial photographs show increased shrub abundance in Alaska in 70% of 200 locations (Sturm et al., 2001; Tape et al., 2006). Along the Arctic to sub-Arctic boundary, the tree line has moved about 10 km northwards, and 2% of Alaskan tundra on the Seward Peninsula has been displaced by forest in the past 50 years (Lloyd et al., 2003). In some areas, the altitude of the tree line has risen, for example by about 60 m in the 20th century in sub-Arctic Sweden (Callaghan et al., 2004; Truong et al., 2007), although the tree line has been stable or become lower in other localities (Dalen and Hofgaard, 2005). Bog growth has caused tree death in parts of the Russian European Arctic (Crawford et al., 2003). The pattern of northward and upward tree-line advances is comparable with earlier Holocene changes (MacDonald et al., 2000; Esper and Schweingruber, 2004) (see below for rates of advance). In addition to changes in woody vegetation, dry-habitat vegetation in sub-Arctic Sweden has been partly displaced by wet-habitat vegetation because of permafrost degradation in the discontinuous permafrost zone (Christensen et al., 2004; Malmer et al., 2005). Similarly, in northern Canada, up to 50% of peat plateau permafrost has thawed at four sites in the discontinuous permafrost zone (Beilman and Robinson, 2003).

There is also recent evidence of changes in growing season duration and timing, together with plant productivity, but patterns are spatially variable. Analyses of satellite images indicate that the length of growing season is increasing by 3 days per decade in Alaska and 1 day per decade in northern Eurasia (McDonald et al., 2004; Smith et al., 2004; McGuire et al., 2007), but there has been a delayed onset of the growing season in the Kola Peninsula during climatic cooling over the past two decades (Høgda et al., 2007). Remote sensing estimates of primary productivity also show spatial variability: there were increases in the southern Arctic and decreases in the central and eastern Russian Arctic between 1982 and 1999 (Nemani et al., 2003).

15.4.2.2 Projected changes in biodiversity, vegetation zones and productivity

Where soils are adequate for forest expansion, species richness will increase as relatively species-rich forest displaces tundra (see Figure 15.3; Callaghan et al., 2005). Some species in isolated favourable microenvironments far north of their main distribution are very likely to spread rapidly during warming. Except for the northernmost and highest-Arctic species, species will generally extend their ranges northwards and higher in altitude, while the dominance and abundance of many will decrease. Likely rates of advance are uncertain; although tree-line advance of up to 25 km/yr during the early Holocene have been recorded, rates of 2 km/yr and less are more probable (Payette et al., 2002; Callaghan et al., 2005). Trophic structure is relatively simple in the Arctic, and decreases in the abundance of keystone species are expected to lead to ecological cascades, i.e., knock-on effects for predators, food sources etc. Local changes in distribution and abundance of genetically different populations will be the initial response of genetically diverse species to warming (Crawford, 2004). Arctic animals are likely to be most vulnerable to warming-induced drying (invertebrates); changes in snow cover and freeze-thaw cycles that affect access to food and protection from predators; changes that affect the timing of behaviour (e.g., migration and reproduction); and influx of new competitors, predators, parasites and diseases. Southern species constantly reach the Arctic but few become established (Chernov and Matveyeva, 1997). As a result of projected climate change, establishment will increase and some species, such as the North American mink, will become invasive, while existing populations of weedy southern plant species that have already colonised some Arctic areas are likely to expand. The timing of bird migrations and migration routes are likely to change as appropriate Arctic habitats become less available (Callaghan et al., 2005; Usher et al., 2005).

Warming experiments that adequately reproduced natural summer warming impacts on ecosystems across the Arctic showed that plant communities responded rapidly to 1-3°C warming after two growing seasons, that shrub growth increased as observed under natural climate warming, and that species diversity decreased initially (Walker et al., 2006). Experimental warming and nutrient addition showed that mosses and lichens became less abundant when vascular plants increased their growth (Cornelissen et al., 2001; Van Wijk et al., 2003). CO_2-enrichment produced transient plant responses, but microbial communities changed in structure and function (Johnson et al., 2002) and frost hardiness of some plants decreased (Beerling et al., 2001) making them more susceptible to early frosts. Supplemental ultraviolet-B caused few plant responses but did reduce nutrient cycling processes (Callaghan et al., 2005; Rinnan et al., 2005). Such reductions could potentially reduce plant growth.

A 'moderate' projection for 2100 for the replacement of tundra areas by forest is about 10% (Sitch et al., 2003; see Figure 15.3), but estimates of up to 50% have also been published (White et al., 2000). However, impacts of changing hydrology, active layer depth and land use are excluded from these models. These impacts can be large: for example, Vlassova (2002) suggests that 475,000 km^2 of tree-line forest has been destroyed in Russia, thereby creating tundra-like ecosystems. Narrow coastal strips of tundra (e.g., in parts of the Russian European Arctic) will be completely displaced as forest reaches the Arctic Ocean. During 1960 to 2080, tundra is projected to replace about 15 to 25% of the polar desert, and net primary production (NPP) will increase by about 70% (2.8 to 4.9 Gt of carbon) (Sitch et al., 2003). Geographical constraints on vegetation relocation result in large sub-regional variations in projected increases of NPP, from about 45% in fragmented landmasses to about 145% in extensive tundra areas (Callaghan et al., 2005).

Climate warming is likely to increase the incidence of pests, parasites and diseases such as musk ox lung-worm (Kutz et al., 2002) and abomasal nematodes of reindeer (Albon et al., 2002). Large-scale forest fires and outbreaks of tree-killing insects that are triggered by warm weather are characteristic of the boreal forest and are likely to increase in extent and frequency (Juday et al., 2005), creating new areas of forest tundra. During the 1990s, the Kenai Peninsula of south-central Alaska experienced a massive outbreak of spruce bark beetle over 16,000 km^2 with 10-20% tree mortality (Juday et al., 2005). Also following recent climate warming, spruce budworm has reproduced further north, reaching problematic numbers in Alaska (Juday et al., 2005), while autumn moth defoliation of mountain birch trees, associated with warm winters in northern Fennoscandia, has occurred over wide areas and is projected to increase (Callaghan et al., 2005).

15.4.2.3 Consequences of changes in ecosystem structure and function for feedbacks to the climate system

Climate warming will decrease the reflectivity of the land surface due to expansion of shrubs and trees into tundra (Eugster et al., 2000); this could influence regional (Chapin et al., 2005a) and global climate (Bonan et al., 1992; Thomas and Rowntree, 1992; Foley et al., 1994; Sturm et al., 2005; McGuire et al., 2007).

Measurements show great spatial variability in the magnitude of sink (net uptake) or source (net release) status for carbon, with no overall trend for the Arctic (Corradi et al., 2005). In contrast, models suggest that overall the Arctic is currently a small sink of about 20 ± 40 g carbon m^2/yr (McGuire et al., 2000; Sitch et al., 2003, 2007). The high uncertainties in both measurements and model projections indicate that the Arctic could be either a sink or source of carbon. Thus, currently circumpolar Arctic vegetation and the active layer are unlikely to be a large source or sink of carbon in the form of CO_2 (Callaghan et al., 2005; Chapin et al., 2005a). They are, however, most probably a source of positive radiative forcing due to large methane emissions; even in tundra areas that are net sinks of carbon, significant emissions of methane lead to positive forcing (Friborg et al., 2003; Callaghan et al., 2005).

Higher temperatures, longer growing seasons and projected northward movement of productive vegetation are likely to increase photosynthetic carbon capture in the longer term, whereas soil warming is likely to increase trace gas emissions in the short term. Drying or wetting of tundra concurrent with warming and increased active-layer depth (see Section 15.3.4) will determine the magnitude of carbon fluxes and the balance of trace gases that are involved. Drying has increased sources in Alaska (Oechel et al., 2000), whereas wetting has increased sinks in Scandinavian and Siberian peatlands (Aurela et al., 2002; Johansson et al., 2006).

Models project that the Arctic and sub-Arctic are likely to become a weak sink of carbon during future warming (an increase in carbon storage in vegetation, litter and soil of about 18.3 Gt carbon between 1960 and 2080), although there is high uncertainty (Sitch et al., 2003; Callaghan et al., 2005). Increased carbon emissions from projected increases in disturbances and land use, and net radiative forcing resulting from the changing balance between methane and carbon dioxide emissions (Friborg et al., 2003; Johansson et al., 2006) are particular uncertainties. Wetting, from increased precipitation and permafrost thawing, is projected to increase fluxes of methane relative to carbon dioxide from the active layer and thawing permafrost (Walter et al., 2006).

Changes in forest area will lead to both negative and positive feedbacks on climate. According to one coupled climate model, the negative feedback of carbon sequestration and the positive feedback of reduced albedo interact. This model predicts that the central Canadian boreal forests will give net negative feedback through dominance of increased carbon sequestration, while in the forests of Arctic Russia decreased albedo will dominate, giving net positive feedback (Betts and Ball, 1997; Betts, 2000).

15.4.2.4 Impacts on resource use, traditional economies and lifestyles

Terrestrial resources are critical aspects of Arctic residents' livelihoods, culture, traditions and health (Arctic Monitoring and Assessment Programme, 2003; Chapin et al., 2005a). Per capita consumption of wild foods by rural Alaskans is 465 g per day (16% land mammals, 10% plant products) and consumption by urban Alaskans is 60 g per day. The collective value of these foods in the state is estimated at about US$200 million/yr. Consumption in Canadian Arctic communities ranges from 106 g per day to 440 g per day, accounting for 6 to 40% of total energy intake and 7 to 10% of the total household income in Nunavik and Nunavut (Kuhnlein et al., 2001; Chabot, 2004). Terrestrial ecosystem resources include caribou/reindeer, moose, musk ox, migratory birds and their eggs, and plants and berries (Arctic Monitoring and Assessment Programme, 2003; Chapin et al., 2005a). Wild and domesticated caribou/reindeer are particularly important, as they provide food, shelter, clothing, tools, transportation and, in some cases, marketable goods (Klein, 1989; Paine, 1994; Kofinas et al., 2000; Jernsletten and Klokov, 2002). Wood, sods, peat and coal are used locally as fuels throughout the north. Despite the significant role these resources represent to Arctic residents, ties to subsistence activities among indigenous peoples are deteriorating because of changes in lifestyles, cultural, social, economic and political factors (Chapin et al., 2005a). These ties are expected to continue decreasing as climate-driven changes in terrestrial ecosystems influence conditions for hunting, decreases in natural resources, and loss of traditional knowledge. Together these shifts are turning previously well-adapted Arctic peoples into "strangers in their own lands" (Berkes, 2002).

Agriculture in southern parts of the Arctic is limited by short, cool growing seasons and lack of infrastructure, including limited local markets because of small populations and long distances to large markets (Juday et al., 2005). The northern limit of agriculture may be roughly approximated by a metric based on the cumulative degree-days above +10°C (Sirotenko et al., 1997). By the mid-21st century, climatic warming may see displacement of its position to the north by a few hundred kilometres over most of Siberia, and up to 100 km elsewhere in Russia (Anisimov and Belolutskaia, 2001). Thus climate warming is likely to lead to the opportunity for an expansion of agriculture and forestry where markets and infrastructure exist or are developed. While conservation management and protected areas are extensive in the Arctic, these only protect against direct human actions, not against climate-induced vegetation zone shifts, and decisions need to be made about the goals and methods of conservation in the future (Callaghan et al., 2005; Klein et al., 2005; Usher et al., 2005).

15.4.3 Marine ecosystems and their services in the Arctic

15.4.3.1 Historical changes in marine ecosystems

Water temperatures in the North Atlantic have fluctuated over the last 200 years. The effects of these temperature variations have been profound, impacting plankton communities as well as larval drift, distribution and abundance of many fish stocks including southern invaders and, especially, the commercially important cod and herring (Loeng et al., 2005; Vilhjálmsson et al., 2005; see Section 15.6.3). These climatic impacts on fish stocks are superimposed on those arising from exploitation.

15.4.3.2 Likely general effects of a warming ocean climate

Changing climatic conditions in Arctic and sub-Arctic oceans are driving changes in the biodiversity, distribution and productivity of marine biota, most obviously through the reduction of sea ice. As the sea-ice edge moves northward, the distribution of crustaceans (copepods and amphipods), adapted for life at the sea-ice edge, and fish such as polar cod (*Boreogadus saida*), which forage on them, will shift accordingly and their abundance diminish (Sakshaug et al., 1994). This reduction is likely to seriously impact other predators, e.g., seals, sea birds and polar bears (*Ursus maritimus*), dependent on sea ice for feeding and breeding (see Chapter 4, Box 4.3; Sakshaug et al., 1994) as well as humans depending on them (Loeng et al., 2005; Vilhjálmsson et al., 2005).

Thinning and reduced coverage of Arctic sea ice are likely to substantially alter ecosystems that are in close association with sea ice (Loeng et al., 2005). Polar cod, an important member of these ecosystems, is a prime food source for many marine mammals. Ringed seals, which are dependent on sea ice for breeding, moulting and resting, feed on ice amphipods and cod. Premature break-up of ice may not only lead to high mortality of seal pups but also produce behavioural changes in seal populations (Loeng et al., 2005). Polar bears, a top predator, are highly dependent on both sea ice and ringed seals (see Chapter 4, Box 4.3). Initially, the loss of sea ice and the subsequent deleterious effects are likely to occur at the southern distribution limit of polar bears, where early melt and late freezing of sea ice extend the period when the bears are restricted to land and only limited feeding can occur. Recently, the condition of adult bears has declined in the Hudson Bay region and first-year cubs come ashore in poor condition (Stirling et al., 1999; Derocher et al., 2004; Stirling and Parkinson, 2006). As a proportion of the population, the number of cubs has fallen as a result of the early break-up of sea ice. Loss of sea ice may also adversely affect other Arctic marine mammals, such as walrus (*Odobenus rosmarus*) that use sea ice as a resting platform and which occupy a narrow ecological range with restricted mobility. Similarly, the narwhal (*Monodon monoceros*) and the bowhead whale (*Balaena mysticetus*) are dependent on sea-ice organisms for feeding and polynyas for breathing (Loeng et al., 2005). The early melting of sea ice may lead to an increasing mismatch in the timing of these sea-ice organisms and secondary production that severely affects the populations of sea mammals (Loeng et al., 2005).

However, with an increase in open water, primary and secondary production south of the ice edge will increase and this will benefit almost all of the most important commercial fish stocks in Arctic and sub-Arctic seas; for example, cod (*Gadus morhua*) and herring (*Clupea harengus*) in the North Atlantic and walleye pollock (*Theragra chalcogramma*) in the Bering Sea; species that currently comprise about 70% of the total catch in these areas. However, some coldwater species (e.g., northern shrimp (*Pandalus borealis*) and king crab (*Paralithoides* spp.) may lose habitat (Vilhjálmsson et al., 2005).

15.4.3.3 Predicting future yields of commercial and forage stocks

Quantitative predictions of the responses of commercial and forage fish stocks to changes in ocean temperature are very difficult to make because (a) resolution of existing predictive models is insufficient, (b) exploitation has already altered stock sizes and basic biology/ecology, so stocks cannot be expected to react as they did historically, and (c) the effect of a moderate climate change (+1-3°C) may well be of less importance than sound fisheries policies and their effective enforcement. The examples given in Box 15.1 show how such climate change, exploitation and other factors affecting marine ecosystems may interact strongly at the northern extreme of the range of a species, while having less effect on the same species further south (e.g., Rose et al., 2000; Rose, 2004; Drinkwater, 2005).

Changes in distribution of commercial stocks may lead to conflicts over fishing rights and will require effective negotiations to generate solutions regarding international co-operation in fisheries management (Vilhjálmsson et al., 2005).

In addition, rising water temperatures will lead to an increased risk of harmful algal blooms and occurrences of other marine pests and pollution, hazards that will be multiplied by increased shipping (Loeng et al., 2005; Vilhjálmsson et al., 2005).

15.4.4 Marine ecosystems and their services in the Antarctic

Southern Ocean ecosystems are far from pristine. Over the last 200 years some seal and whale species (e.g., Antarctic fur seals, blue and fin whales) were exploited almost to extinction, then fisheries developed. From the 1960s, fin-fish were exploited, and in the Scotia Sea and surrounding areas stocks of these fish were reduced to low levels and have not yet recovered. In contrast to the Arctic, however, the management of Southern Ocean fisheries is based on an ecosystem approach, within an international convention. The Convention on the Conservation of Antarctic Marine Living Resources (CCAMLR), part of the Antarctic Treaty, was designed to maintain the natural marine ecosystem while allowing sustainable exploitation, and emphasises the need to consider the wider context of the exploitation of individual species, taking account of the entire food web and environmental variations. The CCAMLR applies to areas south of the Antarctic polar front, and management decisions are made by consensus of the Member States (Constable et al., 2000).

The current major fin-fish fishery is for the Patagonian toothfish (*Dissostichus eleginoides*), and to a lesser extent for the mackerel icefish (*Champsocephalus gunnari*). The fishery for Antarctic krill (*Euphausia superba*) developed during the 1970s, peaked in the 1980s at over 500,000 t/yr and now operates at about 100,000 t/yr (Jones and Ramm, 2004), a catch that is well below the precautionary limits set within CCAMLR for maintaining the stock.

During the 20th century there were significant changes in air temperatures, sea-ice and ocean temperatures around the Antarctic Peninsula (see Section 15.6.3) and in the Scotia Sea. Over 50% of the krill stock was lost in the Scotia Sea region (Atkinson et al., 2004), which is the major area for krill fishing. The decline in the abundance of krill in this area appears to be associated with changes in sea ice in the southern Scotia Sea and around the Antarctic Peninsula (Atkinson et al., 2004). Future reductions in sea ice may therefore lead to further changes in distribution and abundance across the whole area, with consequent impacts on food webs where krill are currently key prey items for many predator species and where krill fishing occurs.

For other species the uncertainty in climate predictions leads to uncertainty in projections of impacts, but increases in temperatures and reductions in winter sea ice would undoubtedly affect the reproduction, growth and development of fish and krill, leading to further reductions in population sizes and changes in distributions. However, the potential for species to adapt is mixed, some 'cold-blooded' (poikilothermic) organisms may die if water temperatures rise to 5-10°C (Peck, 2005), while the bald rock cod (*Pagothenia borchgrevinki*), which uses the specialisation of anti-freeze proteins in its blood to live at sub-zero temperatures, can acclimatise so that its swimming performance at +10°C is similar to that at −1°C (Seebacher et al., 2005).

The importance of ocean transport for connecting Southern Ocean ecosystems has been increasingly recognised. Simple warming scenarios may indicate that exploitation effects would be shifted south, but it is also likely that other species may become the target of new fisheries in the same areas. More complex changes in patterns of ocean circulation could have profound effects on ocean ecosystems and fisheries, although not all changes may be negative and some species may benefit. Complex interactions in food webs may, however, generate secondary responses that are difficult to predict. For example, reductions in krill abundance may have negative effects on species of fish, as

Box 15.1. Atlantic cod in the 20th century: historical examples

This box illustrates how cod populations have responded through the 20th century and until 2005 to the multiple stresses of ocean temperature change, exploitation, changing abundance of food species and predators. All four cod populations show substantial decreases to the present, and suggest that they are now vulnerable to future changes in both climate and exploitation. However, the data suggest that the vulnerability of stocks south of the polar front is less than that to the north of the polar front.

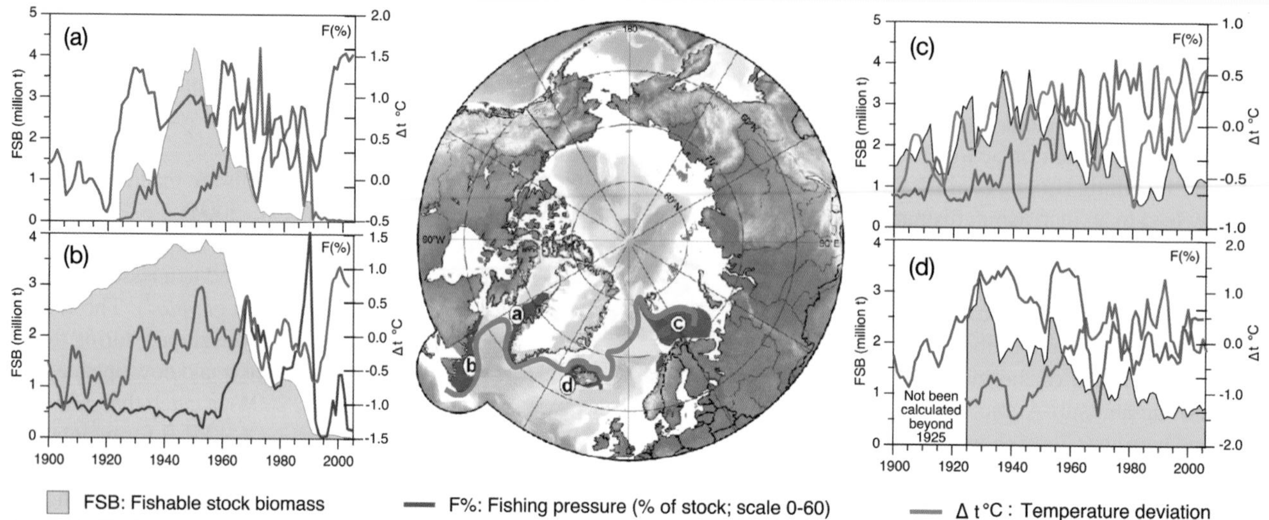

Figure 15.5. *The geographical distribution of four major cod stocks in the North Atlantic (red patches). The continuous blue line indicates an average geographical position of the Polar Front. The graphs (a: West Greenland; b: Newfoundland/Labrador; c: Barents Sea, and d: Iceland) show the developments of fishable stock (yellow shading), catches (red line) and temperature (blue line) during the period 1900-2005. Data sources: Greenland (Buch et al., 1994; Horsted, 2000; ICES, 2006); Newfoundland/Labrador (Harris, 1990; Lilly et al., 2003); Iceland (Schopka, 1994; Hafrannsóknastofnunin, 2006; ICES, 2006); Barents Sea (Bochkov, 1982; Hylen, 2002; ICES, 2005a), data since 1981 were kindly provided by the Polar Research Institute of Marine Fisheries and Oceanography (PINRO), Murmansk, Russia.*

Greenland

Ocean temperature records, begun off West Greenland in the 1870s, showed very cold conditions until temperatures warmed suddenly around 1920, and maintained high levels until they dropped suddenly in the late 1960s (Jensen, 1939; Buch et al., 1994; Vilhjálmsson, 1997; see Figure 15.5a). There were no Atlantic cod in Greenland waters in the latter half of the 19th century (Jensen, 1926; Buch et al., 1994), while there was a good cod fishery off Iceland (Jensen, 1926, 1939; Buch et al., 1994). Concurrent with the warming in the early 1920s, large numbers of juvenile cod drifted from Iceland to West Greenland and started a self-supporting stock there, which vanished in the 1970s (Buch et al., 1994; Vilhjálmsson, 1997; Vilhjálmsson et al., 2005).

Comparison of catches and temperature records shows that the occurrence of cod off Greenland depends principally on West Greenland water temperature (Horsted, 2000). Thus the reappearance of cod off Greenland will probably depend on drift of juvenile cod from Iceland as it did in the 1920s. West Greenland waters may now be sufficiently warm for this to happen. Indeed such drifts did occur in 2003 and 2005, but the numbers were small; a probable consequence of the depleted and younger spawning stock of Icelandic cod, which has not produced a strong year class for 20 years (ICES, 2005b).

Newfoundland/Labrador

Beginning in the 16th century, annual catches from this stock increased until the mid-1800s. From 1920 to 1960, catches varied then increased rapidly, peaking in 1968. Catches then dropped sharply until about 1977 when Canada acquired its 200-mile Exclusive Economic Zone (Rose, 2004). Total allowable catches were increased in the mid- to late 1980s, but dropped again after 1989. A moratorium on fishing was imposed in 1992 (see Figure 15.5b). While fishing was the primary cause of the decline (Walters and Maguire, 1996), decreased productivity in cod and a reduction in their primary food (capelin) also occurred in the late 1980s and 1990s (Rose and O'Driscoll, 2002; Shelton et al., 2006). Furthermore, a moratorium on seal hunting led to an explosion in seal abundance and thus increased mortalities of their prey, including cod (Lilly et al., 2003). High mortality remains the key obstacle to population growth. While future warming climate is likely to promote recovery of cod (Drinkwater, 2005), an increase in abundance of the main forage fish, capelin, is likely to be a necessary precursor.

Iceland/Barents Sea

In the face of exploitation, the comparative resilience of these cod stocks is high because they are south of the polar front and, therefore, well inside the limit tolerance of the species to cold temperatures (see Figure 15.5c, d).

they become a greater target for predators. Importantly, the impact of changes in these ecosystems will not be confined to the Southern Ocean. Many higher predator species depend on lower-latitude systems during the Antarctic winter or the breeding seasons.

The fundamental precautionary basis for managing exploitation in a changing environment is in place in CCAMLR, but longer duration and more spatially extensive monitoring data are required in order to help identify change and its effects.

15.4.5 Human health and well-being

The impact of projected climate change on the diverse communities of the Arctic can only be understood in the context of the interconnected social, cultural, political and economic forces acting on them (Berner et al., 2005). However, such impacts on the health and well-being of Arctic residents are, and will be, tangible and ongoing. Recently, significant research has been conducted on the health and well-being of indigenous populations in the Arctic and the role of environmental change as a determinant of health; accordingly, this section puts more emphasis on these more vulnerable segments of the population.

15.4.5.1 Direct impacts of climate on the health of Arctic residents

Direct impacts (injury and death) are expected to result, in part, from exposure to temperature extremes and weather events. Increases in precipitation are expected to affect the frequency and magnitude of natural disasters such as debris flow, avalanches and rock falls (Koshida and Avis, 1998). Thunderstorms and high humidity are associated with short-term increases in respiratory and cardiovascular diseases (Kovats et al., 2000). Messner (2005) reported an increased incidence of non-fatal heart attacks with increased temperature during the positive phase of the Arctic Oscillation (AO) in Sweden, but related it to changes in behaviour that can cause an increase in the susceptibility of individuals to atherosclerotic diseases. Low temperatures and social stress have been related to cases of cardiomyopathy, a weakening of the heart muscle or change in heart muscle structure, identified in northern Russia (Khasnullin et al., 2000). Residents in some Arctic regions report respiratory stress associated with extreme warm summer days not previously experienced (Furgal et al., 2002).

The frequency of some injuries (e.g., frostbite, hypothermia) or accidents, and diseases (cardiovascular, respiratory, circulatory, musculoskeletal, skin) is increased by cold exposure (Hassi et al., 2005). An estimated 2,000 to 3,000 deaths/yr occur from cold-related diseases and injury in Finland during the cold season. This winter-related mortality is higher than the number of deaths associated with other common causes in the country throughout the year (e.g., 400/yr from traffic accidents, 100-200/yr from heat). The prevalence of respiratory diseases among children in the Russian North is 1.5 to 2 times higher than the national average. Evidence suggests that warming in Arctic regions during the winter months will reduce excess winter mortality, primarily through a reduction in cardiovascular and respiratory deaths (Nayha, 2005). Assuming that the standard of cold protection (including individual behavioural factors) does not deteriorate, a reduction in cold-related injuries is also likely (Nayha, 2005).

15.4.5.2 Indirect impacts of climate on the health of Arctic residents

Climate variability will have a series of more complex, indirect impacts on human-environment interactions in the Arctic (Berner et al., 2005). Local and traditional knowledge in nearly all regions records increasingly uncharacteristic environmental conditions and extremes not previously experienced (e.g., Krupnik and Jolly, 2002). Evidence suggests that an increase in injuries among northern residents associated with 'strange' or changing environmental conditions, such as thinning and earlier break-up of sea ice, are related to trends in climate (e.g., Lafortune et al., 2004).

Climate change in the Arctic during El Niño-Southern Oscillation (ENSO) events has been associated with illness in marine mammals, birds, fish and shellfish. A number of disease agents have been associated with these illnesses (e.g., botulism, Newcastle disease). It is likely that temperature changes arising from long-term climate change will be associated with an increased incidence of those diseases that can be transmitted to humans (Bradley et al., 2005). Many zoonotic diseases which currently exist in Arctic host species (e.g., tularemia in rabbits, rodents, muskrats and beaver, and rabies in foxes; Dietrich, 1981) can spread via climate-controlled mechanisms (e.g., movement of animal populations). Similarly, the overwintering survival and distribution of many insect species that act as vectors of disease are positively impacted by warming temperatures and may mean that many diseases reappear, or new diseases appear, in Arctic regions (Parkinson and Butler, 2005). The examples of tick-borne encephalitis (brain infection) in Sweden (Lindgren and Gustafson, 2001), and *Giardia* spp. and *Cryptosporidium* spp. infection of ringed seals (*Phoca hispida*) and bowhead whales (*Balaena mysticetus*) in the Arctic Ocean are evidence of this potential (Hughes-Hanks et al., 2005).

Subsistence foods from the local environment provide Arctic residents with cultural and economic benefits and contribute a significant proportion of daily requirements of several vitamins and essential elements to the diet (e.g., Blanchet et al., 2000). Wild foods also comprise the greatest source of exposure to environmental contaminants. The uptake, transport and deposition behaviour of many of these chemicals is influenced by temperature, and therefore climate warming may indirectly influence human exposure (Kraemer et al., 2005). Through changes in accessibility and distribution of wildlife species, climate change in combination with other social, cultural, economic and political trends in Arctic communities, will be likely to influence the diet of circumpolar residents.

Transitions towards more market food items in Arctic indigenous diets to date have been associated with a rise in levels of cardiovascular diseases, diabetes, dental cavities and obesity (Van Oostdam et al., 2003). In many indigenous communities, these subsistence food systems are the basis of traditions, socio-economic and cultural well-being. Indigenous peoples maintain a strong connection to the environment through traditional resource-harvesting activities in a way that distinguishes them from non-indigenous communities, and this may indeed contribute to how specific peoples retain a fundamental identification to a particular area (Gray, 1995; Nuttall et al., 2005).

While climate-related changes threaten aspects of food security for some subsistence systems, increased temperatures and decreased sea-ice cover represent increased transport opportunities and access to market food items. Shifts in animal population movements also mean potential introduction of new food species to northern regions. These combined effects on Arctic food security, in addition to increased opportunities for agricultural and pastoral activities with decreased severity of winter and lengthened summer growing seasons, make it difficult to predict how diets will change and impact health, even presupposing that we have a sufficient understanding of what local environments can provide and sustain. It is also clear that these impacts will be influenced not only by environmental change but also by economic, technological and political forces.

Through increased river and coastal flooding and erosion, increased drought, and degradation of permafrost, resulting in loss of reservoirs or sewage contamination, climate change is likely to threaten community and public health infrastructure, most seriously in low-lying coastal Arctic communities (e.g., Shishmaref, Alaska, USA; Tuktoyaktuk, Northwest Territories, Canada). Community water sources may be subject to salt-water intrusion and bacterial contamination. Quantities of water available for basic hygiene can become limited due to drought and damaged infrastructure. The incidence of disease caused by contact with human waste may increase when flooding and damaged infrastructure such as sewage lagoons, or inadequate hygiene, spreads sewage. However, treatment efficiencies in wastewater lagoons may also improve due to warmer water temperatures, delaying the need to expand natural wastewater treatment systems as local populations grow (Warren et al., 2005).

The combined socio-cultural, economic, political and environmental forces acting on and within Arctic communities today (Chapin et al., 2005a) have significant implications for health and well-being (Curtis et al., 2005). Alterations in the physical environment threatening specific communities (e.g., through erosion and thawing permafrost) and leading to forced relocation of inhabitants, or shifts or declines in resources resulting in altered access to subsistence species (e.g., Inuit hunting of polar bear) can lead to rapid and long-term cultural change and loss of traditions. Such loss can, in turn, create psychological distress and anxiety among individuals (Hamilton et al., 2003; Curtis et al., 2005). However, across most of the Arctic, climate change is just one of many driving forces transforming communities. These forces arise from inside and outside the community, but combined are influencing the acculturation process by influencing ways of living, and loss of traditions that are positively related to social, cultural and psychological health (Berry, 1997).

The social, cultural and economic transitions that Arctic communities have seen over the last 50 years has influenced all aspects of health in the Arctic, and this influence is highly likely to continue in the future. Climate change is probably going to drive changes in communities by challenging individuals' and communities' relationships with their local environment, which has been the basis of Arctic peoples' identity, culture, social and physical well-being (Einarsson et al., 2004; Berner et al., 2005; Chapin et al., 2005a).

15.4.6 Coastal zone and small islands

15.4.6.1 Arctic coastal erosion

Coastal stability in polar regions is affected by factors common to all areas (exposure, relative sea-level change, climate and lithology), and by factors specific to the high latitudes (low temperatures, ground ice and sea ice). The most severe erosion problems affect infrastructure and culturally important sites in areas of rising sea level, where warming coincides with areas that are seasonally free of sea ice or where there is widespread ice-rich permafrost (Forbes, 2005). Ice-rich permafrost is widespread in the western Canadian Arctic, northern Alaska and along much of the Russian Arctic coast (e.g., Smith, 2002; Nikiforov et al., 2003). Wave erosion and high summer air temperatures promote rapid shoreline retreat, in some cases contributing a significant proportion of regional sediment and organic carbon inputs to the marine environment (Aré, 1999; Rachold et al., 2000). Communities located on resistant bedrock or where glacio-isostatic rebound is occurring are less vulnerable to erosion.

Coastal instability may be further magnified by poorly adapted development. For example, in places such as Varandey (Russian Federation) industrial activity has promoted erosion, leading to the destruction of housing estates and industrial facilities (Ogorodov, 2003). Interacting human and natural effects may also increase the sensitivity to coastal erosion. For example, in Shishmaref (Alaska, USA) and Tuktoyaktuk (Northwest Territories, Canada), the combined effects of reduced sea ice, thawing permafrost, storm surges and waves have led to significant loss of property, and this has led to relocation or abandonment of homes and other facilities (Instanes et al., 2005). Despite a cultural aversion to moving from traditional sites, these changes may ultimately force relocation. Although clear evidence for accelerated erosion is sparse, there has been a documented increase in erosion rates between 1954-1970 and 1970-2000 for coastal terrain with very high ground-ice content at Herschel Island, Canada (Lantuit and Pollard, 2003). A modelling exercise (Rasumov, 2001) suggested that erosion rates in the eastern Siberian Arctic could increase by 3-5 m/yr with a 3°C increase in mean summer air temperature. Furthermore, the projected reduction of sea ice would also contribute to increased erosion, as has been observed at Nelson Lagoon in Alaska (Instanes et al., 2005).

15.4.6.2 Sub-Antarctic islands

Several sub-Antarctic islands have undergone substantial recent climate change, the impacts of which have been significant and have included profound physical and biological changes (specific examples can be found in Chapter 11, Sections 11.2.1 and 11.4.2, Table 11.1; Chapter 16, Section 16.4.4).

15.5 Adaptation: practices, options and constraints

Circum-Arctic nations are responsible for a significant fraction of global CO_2 emissions and the Arctic is an important source of fossil fuels. Although some residents and less developed regions may contribute only a very small proportion of these nations' emissions, there is a need to consider both mitigation and adaptation in polar regions in the light of trends in resource development and modernisation taking place in these areas. The burden faced by Arctic residents is magnified by the observed and projected amplification of climate change in the Arctic and the potential for dramatic environmental impacts. As with other vulnerable regions of the world, human adaptation is critical, particularly for those living in closest relationship with the local environment.

Historically, cultural adaptations and the ability of Arctic indigenous peoples to utilise their local resources have been associated with, or affected by, seasonal variation and changing ecological conditions. One of the hallmarks of successful adaptation has been flexibility in technology and social organisation, and the knowledge and ability to cope with climate change and circumvent some of its negative impacts. Indigenous groups have developed resilience through sharing resources in kinship networks that link hunters with office workers, and even in the cash sector of the economy. Many people work flexibly, changing jobs frequently and having several part-time jobs (Chapin et al., 2006). Historically, responses to major climatic and environmental changes included an altering of group size or moving to appropriate new locations, flexibility with regard to seasonal cycles and harvesting, and the establishment of sharing mechanisms and networks for support (Krupnik, 1993; Freeman, 1996). Many of these strategies, with the exception of group mobility, are still employed in various forms today (e.g., Berkes and Jolly, 2001; Nickels et al., 2002; McCarthy et al., 2005) yet, in the future, such responses may be constrained by social, cultural, economic and political forces acting on communities externally and from within.

Detailed local knowledge and the social institutions in which it exists are critical foundations of understanding interactions between people and their environment and therefore vital to community adaptability (see Section 15.6.1). Yet the generation of this knowledge requires active engagement with the environment and, as the nature of this interaction changes (e.g., amount and frequency of time spent on land or engaged in subsistence activities), so does the information it provides. Changes in local environments further challenge this knowledge and can increase human vulnerability to climatic and social change.

Greater uncertainty and threats to food security stress the need for resilient and flexible resource procurement systems. Resilience and adaptability depend on ecosystem diversity as well as the institutional rules that govern social and economic systems (Adger, 2000). Innovative co-management of both renewable and non-renewable resources could support adaptive abilities via flexible management regimes while providing opportunities to enhance local economic benefits and ecological and societal resilience (Chapin et al., 2004).

Opportunities for adaptation exist within some changes already taking place. The arrival of new species (e.g., Babaluk et al., 2000; Huntington et al., 2005) and an increase in growing seasons and opportunities for high-latitude agriculture provide opportunities to enhance resilience in local food systems. Increased eco-tourism may increase incentives for protection of environmental areas. Taking advantage of these potentially positive impacts will, however, require institutional flexibility and forms of economic support.

Given the interconnected nature of Arctic ecosystems and human populations, strategies are required that take a broad approach to support adaptation among a range of sectors. For example, policies that allow local people to practice subsistence activities within protected areas contribute to both biodiversity and cultural integrity (Chapin et al., 2005a). The creation and protection of critical areas such as parks, with flexible boundaries to compensate for changing climatic conditions, enhances conservation of wildlife and services provided by this land for human use (e.g., tourism and recreation) (Chapin et al., 2005a).

Although Arctic communities in many regions show great resilience and ability to adapt, some responses have been compromised by socio-political change. The political, cultural and economic diversity that exists among Arctic regions today impacts how communities are affected by, and respond to, environmental change. Such diversity also means that particular experiences of climate variability, impacts and responses may not be universal. Currently, little is known about how communities and individuals, indigenous or non-indigenous, differ in the way risks are perceived, or how they might adapt aspects of their lives (e.g., harvesting strategies) in response to negative change. The effectiveness of local adaptive strategies is uneven across the Arctic and there are large gaps in knowledge about why some communities do well, while others are more vulnerable to drivers of change, even when they share similar resources and ecological settings. Ultimately, an understanding of adaptation can only derive from a better understanding of social and economic vulnerability among all Arctic residents (Handmer et al., 1999).

15.6 Case studies

15.6.1 Case study: traditional knowledge for adaptation

Among Arctic peoples, the selection pressures for the evolution of an effective knowledge base have been exceptionally strong, driven by the need to survive off highly variable natural resources in the remote, harsh Arctic environment. In response, they have developed a strong knowledge base concerning weather, snow and ice conditions as they relate to hunting and travel, and natural resource availability (Krupnik and Jolly, 2002). These systems of knowledge, belief and practice have been developed through experience and culturally transmitted among members and across generations (Huntington, 1998; Berkes, 1999). This Arctic indigenous knowledge offers detailed information that adds to conventional science and environmental observations, as well as to a holistic understanding of environment, natural resources and culture (Huntington et al., 2004). There is an increasing awareness of the value of Arctic indigenous knowledge and a growing collaborative effort to document it. In addition, this knowledge is an invaluable basis for developing adaptation and natural resource management strategies in response to environmental and other forms of change. Finally, local knowledge is essential for understanding the effects of climate change on indigenous communities (Riedlinger and Berkes, 2001; Krupnik and Jolly, 2002) and how, for example, some communities have absorbed change through flexibility in traditional hunting, fishing and gathering practices.

The generation and application of this knowledge is evidenced in the ability of Inuit hunters to navigate new travel and hunting routes despite decreasing ice stability and safety (e.g., Lafortune et al., 2004); in the ability of many indigenous groups to locate and hunt species such as geese and caribou that have shifted their migration times and routes and to begin to locate and hunt alternative species moving into the region (e.g., Krupnik and Jolly, 2002; Nickels et al., 2002; Huntington et al., 2005); the ability to detect safe sea ice and weather conditions in an environment with increasingly uncharacteristic weather (George et al., 2004); or the knowledge and skills required to hunt marine species in open water later in the year under different sea ice conditions (Community of Arctic Bay, 2005).

Although Arctic peoples show great resilience and adaptability, some traditional responses to environmental change have already been compromised by recent socio-political changes. Their ability to cope with substantial climatic change in future, without a fundamental threat to their cultures and lifestyles, cannot be considered as unlimited. The generation and application of traditional knowledge requires active engagement with the environment, close social networks in communities, and respect for and recognition of the value of this form of knowledge and understanding. Current social, economic and cultural trends, in some communities and predominantly among younger generations, towards a more western lifestyle has the potential to erode the cycle of traditional knowledge generation and transfer, and hence its contribution to adaptive capacity.

15.6.2 Case study: Arctic megadeltas

Numerous river deltas are located along the Arctic coast and the rivers that flow to it. Of particular importance are the megadeltas of the Lena (44,000 km^2) and Mackenzie (9,000 km^2) rivers, which are fed by the largest Arctic rivers of Eurasia and North America, respectively. In contrast to non-polar megadeltas, the physical development and ecosystem health of these systems are strongly controlled by cryospheric processes and are thus highly susceptible to the effects of climate change.

Currently, advance/retreat of Arctic marine deltas is highly dependent on the protection afforded by near-shore and land-fast sea ice (Solomon, 2005; Walsh et al., 2005). The loss of such protection with warming will lead to increased erosion by waves and storm surges. The problems will be exacerbated by rising sea levels, greater wind fetch produced by shrinking sea-ice coverage, and potentially by increasing storm frequency. Similarly, thawing of the permafrost and ground-ice that currently consolidates deltaic material will induce hydrodynamic erosion on the delta front and along riverbanks. Thawing of permafrost on the delta plain itself will lead to similar changes; for example, the initial development of more ponded water, as thermokarst activity increases, will eventually be followed by drainage as surface and groundwater systems become linked. Climate warming may have already caused the loss of wetland area as lakes expanded on the Yukon River delta in the late 20th century (Coleman and Huh, 2004). Thaw subsidence may also affect the magnitude and frequency of delta flooding from spring flows and storm surges (Kokelj and Burn, 2005).

The current water budget and sediment-nutrient supply for the multitude of lakes and ponds that populate much of the tundra plains of Arctic deltas depends strongly on the supply of floodwaters produced by river-ice jams during the spring freshet. Studies of future climate conditions on a major river delta of the Mackenzie River watershed (Peace-Athabasca Delta) indicate that a combination of thinner river ice and reduced spring runoff will lead to decreased ice-jam flooding (Beltaos et al., 2006). This change combined with greater summer evaporation, due to warmer temperatures, will cause a decline in delta-pond water levels (Marsh and Lesack, 1996). For many Arctic regions, summer evaporation already exceeds precipitation and therefore the loss of ice-jam flooding could lead to a drying of delta ponds and a loss of sediment and nutrients known to be critical to their ecosystem health (Lesack et al., 1998; Marsh et al., 1999). A successful adaptation strategy that has already been used to counteract the effects of drying of delta ponds involves managing water release from reservoirs to increase the probability of ice-jam formation and related flooding (Prowse et al., 2002).

15.6.3 Case study: Antarctic Peninsula – rapid warming in a pristine environment

The Antarctic Peninsula is a rugged mountain chain generally more than 2,000 m high, differing from most of Antarctica by having a summer melting season. Summer melt produces many isolated snow-free areas, which are habitats for simple biological communities of primitive plants, microbes and invertebrates, and breeding grounds for marine mammals and birds. The Antarctic Peninsula has experienced dramatic warming at rates several times the global mean (Vaughan et al., 2003; Trenberth et al., 2007). Since the TAR, substantial progress has been made in understanding the causes and profound impacts of this warming.

Since records began, 50 years ago, mean annual temperatures on the Antarctic Peninsula have risen rapidly; >2.5°C at Vernadsky (formerly Faraday) Station (Turner et al., 2005). On the west coast, warming has been much slower in summer and spring than in winter or autumn, but has been sufficient to raise the number of positive-degree-days by 74% (Vaughan et al., 2003), and the resulting increase in melt has caused dramatic impacts on the Antarctic Peninsula environment, and its ecology.

Around 14,000 km^2 of ice have been lost from ten floating ice shelves (King, 2003), 87% of glacier termini have retreated (Cook et al., 2005), and seasonal snow cover has decreased (Fox and Cooper, 1998). The loss of seasonal snow and floating ice do not have a direct impact on global sea level, but acceleration of inland glaciers due to the loss of ice shelves (De Angelis and Skvarca, 2003; Scambos et al., 2004; Rignot et al., 2005) and increased run-off of melt water (Vaughan, 2006) will cause an increase in this contribution. If summer warming continues, these effects will grow.

Marine sediment cores show that ice shelves probably have not reached a similar minimum for at least 10,000 years (Domack et al., 2005), and certainly not for 1,000 years (Pudsey and Evans, 2001; Domack et al., 2003). This suggests that the retreat is not simply due to cyclic variations in local climate, and that recent warming is unique in the past 10,000 years (Turner et al., 2007). The processes leading to warming are unclear, but appear to be correlated with atmospheric circulation (van den Broeke and van Lipzig, 2003) and particularly with changes in the Southern Annular Mode caused by anthropogenic influence (Marshall et al., 2004; Marshall et al., 2006). The winter warming on the west coast also appears to be related to persistent retreat of sea ice (see Figure 15.2; Parkinson, 2002) and warming in the Bellingshausen Sea (Meredith and King, 2005). The spring depletion of ozone over Antarctica (the Antarctic Ozone Hole) has also been implicated in driving circulation change (Thompson and Solomon, 2002), but this has been disputed (Marshall et al., 2004). Current general circulation models (GCMs) do not, however, simulate this observed warming over the past 50 years (King, 2003) and we cannot predict with confidence whether rapid warming will continue in future.

If warming does continue (especially in the summer) there will be significant impacts; retreat of coastal ice and loss of snow cover would result in newly exposed rock and permafrost – providing new habitats for colonisation by expanding and invading flora and fauna. However, the direct impacts of climate change on the flora and fauna are difficult to predict, since these ecosystems are subject to multiple stressors. For example, increased damage by ultraviolet exposure, because of reduced ozone levels and summer desiccation, may oppose the direct responses to warming (Convey et al., 2002). In addition, there is a growing threat of alien species invasion, as climatic barriers to their establishment are eroded by climate amelioration, and increasing human activity increases the opportunity for introduction. Such invasions have already occurred on many sub-Antarctic islands, with detrimental consequences for native species (Frenot et al., 2005). Furthermore, slow reproduction rates during rapid climate change may limit the possible relocation of native species.

There have been trends in all trophic levels in the marine ecosystems west of the Antarctic Peninsula. These have been driven by reduced sea-ice extent and duration. Changes in primary production may also have been affected by increases in the supply of glacial melt (Smith et al., 2003). Similarly, reduced sea-ice cover was the likely cause of the dramatic change in the balance between krill and salps, the main grazers of phytoplankton (Atkinson et al., 2004). The loss of krill will probably have impacts on higher predators (albatrosses, seals, whales and penguins: populations of the latter are already changing; Smith et al., 2003), but could have more far-reaching impacts, perhaps even affecting CO_2 sequestration in parts of the Southern Ocean (Walsh et al., 2001).

The global significance of the Antarctic Peninsula warming is difficult to encapsulate, but the main concern is for the loss of a unique landscape and biota. The rate of warming on the Antarctic Peninsula is among the highest seen anywhere on Earth in recent times, and is a dramatic reminder of how subtle climate-dynamic processes can drive regional climate change, and the complexity of its impacts in an environment where human influence is at a minimum.

15.7 Conclusions: implications for sustainable development

15.7.1 Economic activity, infrastructure and sustainability in the Arctic

The thawing of ice-rich permafrost creates potential for subsidence and damage to infrastructure, including oil and gas extraction and transportation facilities (Hayley, 2004), and climate warming will exacerbate existing subsidence problems (Instanes et al., 2005). These risks have been assessed using a 'permafrost hazard' index (e.g., Nelson et al., 2001; Anisimov and Belolutskaia, 2004; Anisimov and Lavrov, 2004; Smith and Burgess, 2004), which, when coupled with climate projections, suggests that a discontinuous high-risk zone (containing population centres, pipelines and extraction facilities) will develop around the Arctic Ocean by the mid-21st century (Nelson et al., 2001). Similarly, a zone of medium risk contains larger population centres (Yakutsk, Noril'sk, Vorkuta) and much of the Trans-Siberian and Baikal-Amur railways. However, distinguishing between the broad effects of climate change on permafrost and more localised human-induced changes remains a significant challenge (Tutubalina and Rees, 2001; Nelson, 2003). Although several recent scientific and media reports have linked widespread damage to infrastructure with climate change (e.g., Smith et al., 2001; Couture et al., 2003), the effect of heated buildings on underlying ice-rich permafrost can easily be mistaken for a climate-change impact. Similarly, urban heat-island effects occur in northern settlements (e.g., Hinkel et al., 2003) and may be a factor in local degradation of permafrost.

The cost of rehabilitating community infrastructure damaged by thawing of permafrost could be significant (Couture et al., 2000, 2001; Chartrand et al., 2002). Even buildings designed specifically for permafrost environments may be subject to severe damage if design criteria are exceeded (Khrustalev, 2000). The impervious nature of ice-rich permafrost has been relied on as a design element in landfill and contaminant-holding facilities (Snape et al., 2003), and thawing such areas could result in severe contamination of hydrological resources and large clean-up costs, even for relatively small spills (Roura, 2004). Rates of coastal erosion in areas of ice-rich permafrost are among the highest anywhere and could be increased by rising sea levels (Brown et al., 2003). Relocation of threatened settlements would incur very large expenses. It has been estimated that relocating the village of Kivalina, Alaska, to a nearby site would cost US$54 million (U.S. Arctic Research Commission Permafrost Task Force, 2003). However, some fraction of the costs will be offset by economic benefits to northern communities. For example, there will be savings on heating costs: modelling has predicted a 15% decline in the demand for heating energy in the populated parts of the Arctic and sub-Arctic and up to 1 month decrease in the duration of the period when heating is needed (Anisimov, 1999).

Lakes and river ice have historically provided major winter transportation routes and connections to smaller settlements. Reductions in ice thickness will reduce the load-bearing capacity, and shortening of the ice season will shorten periods of access. Adaptation in the initial stages of climate change will be through modified construction techniques and transport vehicles and schedules, but longer-term strategies will require new transportation methods and routes. Where an open-water network is viable, it will be sensible to increase reliance on water transport. In land-locked locations, the construction of all-weather roads may be the only viable option, with implications for significantly increased costs (e.g., Lonergan et al., 1993; Dore and Burton, 2001). Similar issues will impact the use of the sea-ice roads primarily used to access offshore facilities.

Loss of summer sea ice will bring an increasingly navigable Northwest Passage, and the Northern Sea Route will create new opportunities for cruise shipping. Projections suggest that by 2050, the Northern Sea Route will have 125 days/yr with less than 75% sea-ice cover, which represents favourable conditions for navigation by ice-strengthened cargo ships (Instanes et al., 2005). Increased marine navigation and longer summers will improve conditions for tourism and travel associated with research (Instanes et al., 2005), and this effect is already being reported in the North American Arctic (Eagles, 2004).

Even without climate change, the complexity of producing a viable plan for sustainable development of the Arctic would be daunting; but the added uncertainty of climate change, and its likely amplification in the Arctic, make this task enormous. The impacts on infrastructure discussed above, together with the probable lengthening of growing seasons and increasing agricultural effort, opening of new sea routes, changing fish stocks, and ecosystem changes will provide many new opportunities for the development of Arctic economies. However it will also place limits on how much development is actually sustainable. There does, however, now appear to be an increasing understanding, among governments and residents, that environmental protection and sustainable development are two sides of the same coin (Nuttall, 2000a), and a forum for circum-Arctic co-operation exists in the Arctic Council. This involves eight nations and six indigenous peoples' organisations and embraces the concept of sustainable development in its mandate. The Arctic Council, in partnership with the International Arctic Science Committee, is responsible for the recent Arctic Climate Impact Assessment (ACIA, 2005), which has substantially improved the understanding of the impacts of climate change in the Arctic, is a benchmark for regional impact assessments, and may become the basis for a sustainable management plan for the Arctic.

15.7.2 Economic activity and sustainability in the Antarctic

Fishing and tourism are the only significant economic activities in the Antarctic at present. Over 27,000 tourists visited Antarctica in the 2005/06 summer and the industry is growing rapidly (IAATO, 2006). The multiple stresses of climate change and increasing human activity on the Antarctic Peninsula represent a clear vulnerability (see Section 15.6.3), and have necessitated the implementation of stringent clothing decontamination guidelines for tourist landings on the Antarctic Peninsula (IAATO, 2005).

Fishing is, however, the only large-scale exploitation of resources in Antarctica, and since 1982 Antarctic fisheries have been regulated by the Convention on the Conservation of Antarctic Marine Living Resources (CCAMLR), which takes climate change into account in determining allowable catches. However, before the CCAMLR came into force, heavy fishing around South Georgia led to a major decline in some stocks, which have not yet fully recovered. The illegal, unregulated and unreported fishing of the Patagonian toothfish (*Dissostichus eleginoides*) is of concern because it could act alongside climate change to undermine sustainable management of stocks (Bialek, 2003). Furthermore, those fishing illegally often use techniques that cause the death of by-catch species; for example, albatross and petrels, which are now under threat (Tuck et al., 2001).

15.8 Key uncertainties and research priorities

Significant advances in our understanding of polar systems have been made since the TAR (Anisimov et al., 2001) and the Arctic Climate Impacts Assessment reports (ACIA, 2005). Many climate-induced changes that were anticipated in the TAR have now been documented. This validation, together with improved models, new data and increasing use of indigenous and local knowledge, has increased our confidence in projecting future changes in the polar regions, although substantial uncertainties remain, and the remote and harsh environments of the polar regions constrain data collection and mean that observational networks are sparse and mostly only recently established. The difficulty in understanding climate-change effects in polar regions is further exacerbated by the complexity within and among polar systems, their feedbacks and sensitivity, and their potential to switch into different states of (dis)equilibrium.

Significant research since the TAR has focused on the impact of climate change on Arctic indigenous populations, and accordingly, this chapter has placed an emphasis on these segments of the population. However, the impacts on the wider population need also be considered, and in particular, the economic impacts, which are difficult to address at present due to the dearth of information.

To address the key uncertainties, particular approaches will be required (Table 15.1). For the Arctic, detailed recommendations for future research have been drafted by the international scientific community (ICARP II, 2006), and a burst of co-ordinated research at both poles is anticipated during the International Polar Year, 2007-2009.

Table 15.1. *Key uncertainties and related scientific recommendations/approaches.*

Uncertainty	Recommendation and approach
Detection and projection of changes in terrestrial, freshwater and marine Arctic and Antarctic biodiversity and implications for resource use and climatic feedbacks	Further development of integrated monitoring networks and manipulation experiments; improved collation of long-term data sets; increased use of traditional knowledge and development of appropriate models
Current and future regional carbon balances over Arctic landscapes and polar oceans, and their potential to drive global climate change	Expansion of observational and monitoring networks and modelling strategies
Impacts of multiple drivers (e.g., increasing human activities and ocean acidity) to modify or even magnify the effects of climate change at both poles	Development of integrated bio-geophysical and socio-economic studies
Fine-scaled spatial and temporal variability of climate change and its impacts in regions of the Arctic and Antarctic	Improved downscaling of climate predictions, and increased effort to identify and focus on impact 'hotspots'
The combined role of Arctic freshwater discharge, formation/melt of sea ice and melt of glaciers/ice sheets in the Arctic and Antarctic on global marine processes including the thermohaline circulation	Integration of hydrologic and cryospheric monitoring and research activities focusing on freshwater production and responses of marine systems
The consequences of diversity and complexity in Arctic human health, socio-economic, cultural and political conditions; interactions between scales in these systems and the implications for adaptive capacity	Development of standardised baseline human system data for circumpolar regions; integrated multidisciplinary studies; conduct of sector-specific, regionally specific human vulnerability studies
Model projections of Antarctic and Arctic systems that include thresholds, extreme events, step-changes and non-linear interactions, particularly those associated with phase-changes produced by shrinking cryospheric components and those associated with disturbance to ecosystems	Appropriate interrogation of existing long-term data sets to focus on non-linearities; development of models that span scientific disciplines and reliably predict non-linearities and feedback processes
The adaptive capacity of natural and human systems to cope with critical rates of change and thresholds/tipping points	Integration of existing human and biological climate-impact studies to identify and model biological adaptive capacities and formulate human adaptation strategies

References

Aanes, R., B.-E. Sæther and N.A. Øritsland, 2000: Fluctuations of an introduced population of Svalbard reindeer: the effects of density dependence and climatic variation. *Ecography*, **23**, 437-443.

Abdalati, W. and K. Steffen, 2001: Greenland ice sheet melt extent: 1979-1999. *J. Geophys. Res.*, **106**, 33983-33988.

ACIA, 2005: *Impacts of a Warming Arctic: Arctic Climate Impacts Assessment*. Cambridge University Press, Cambridge, 1042 pp.

Ackley, S., P. Wadhams, J.C. Comiso and A.P. Worby, 2003: Decadal decrease of Antarctic sea ice extent inferred from whaling records revisited on the basis of historical and modern sea ice records. *Polar Res.*, **22**, 19-25.

Adger, W.N., 2000: Social and ecological resilience: are they related? *Prog. Hum. Geog.*, **24**, 347-364.

Agnarsson, S. and R. Arnason, 2003: The role of the fishing industry in the Icelandic economy: an historical examination. Working Paper W03:07. Institute of Economic Studies, University of Iceland, Reykjavik, 24 pp.

Ainley, D.G., E.D. Clarke, K. Arrigo, W.R. Fraser, A. Kato, K.J. Barton and P.R. Wilson, 2005: Decadal-scale changes in the climate and biota of the Pacific sector of the Southern Ocean, 1950s to the 1990s. *Antarct. Sci.*, **17**, 171-182.

Albon, S.D., A. Stien, R.J. Irvine, R. Langvatn, E. Ropstad and O. Halvorsen, 2002: The role of parasites in the dynamics of a reindeer population. *P. Roy. Soc. Lond. B*, **269**, 1625-1632.

Anisimov, O., 1999: Impact of changing climate on building heating and air conditioning. *Meteorol. Hydrol.*, **6**, 10-17.

Anisimov, O.A. and M.A. Belolutskaia, 2001: Predicting agroclimatic parameters using geographical information system. *Meteorol. Hydrol.*, **9**, 89-98.

Anisimov, O.A. and M.A. Belolutskaia, 2004: Predictive modelling of climate change impacts on permafrost: effects of vegetation. *Meteorol. Hydrol.*, **11**, 73-81.

Anisimov, O.A. and C.A. Lavrov, 2004: Global warming and permafrost degradation: risk assessment for the infrastructure of the oil and gas industry. *Technologies of Oil and Gas Industry*, **3**, 78-83.

Anisimov, O., B.B. Fitzharris, J.O. Hagen, B. Jefferies, H. Marchant, F. Nelson, T. Prowse and D. Vaughan, 2001: Polar regions (Arctic and Antarctic). *Climate Change 2001: Impacts, Adaptation, and Vulnerability. Contribution of Working Group II to the Third Assessment Report of the Intergovernmental Panel on Climate Change*, J.J. McCarthy, O.F. Canziani, N.A. Leary, D.J. Dokken and K.S. White, Eds., Cambridge University Press, Cambridge, 801-841.

Anisimov, O.A., S.A. Lavrov and S.A. Reneva, 2005a: Emission of methane from the Russian frozen wetlands under the conditions of the changing climate. *Problems of Ecological Modeling and Monitoring of Ecosystems*, Yu. Izrael, Ed., Hydrometeoizdat, St. Petersburg, 124-142.

Anisimov, O.A., S.A. Lavrov and S.A. Reneva, 2005b: Modelling the emission of greenhouse gases from the Arctic wetlands under the conditions of the global warming. *Climatic and Environmental Changes*, G.V. Menzhulin, Ed., Hydrometeoizdat, St. Petersburg, 21-39.

Arctic Monitoring and Assessment Programme, 2003: *AMAP Assessment 2002: Human Health in the Arctic*. Arctic Monitoring and Assessment Programme, Oslo, 137 pp.

Aré, F.E., 1999: The role of coastal retreat for sedimentation in the Laptev Sea. *Land–Ocean Systems in the Siberian Arctic: Dynamics and History*, H. Kassens, H.A Bauch, I.A. Dmitrenko, H. Eicken, H.-W. Hubberten, M. Melles, J. Thiede and L.A. Timikhov, Eds., Springer, Berlin, 287-299.

Arora, V.K. and G.J. Boer, 2001: Effects of simulated climate change on the hydrology of major river basins. *J. Geophys. Res.*, **106**, 3335-3348.

Atkinson, A., V. Siegel, E. Pakhomov and P. Rothery, 2004: Long-term decline in krill stock and increase in salps within the Southern Ocean. *Nature*, **432**, 100-103.

Atkinson, D.E., 2005: Environmental forcing of the circum-Polar coastal regime. *Geo-Marine Lett.*, **25**, 98-109.

Aurela, M., T. Laurila and J.-P. Tuovinen, 2002: Annual CO_2 balance of a subarctic fen in northern Europe: importance of the wintertime efflux. *J. Geophys. Res.*, **107**, 4607, doi:10.1029/2002JD002055.

Babaluk, J.A., J.D. Reist, J.D. Johnson and L. Johnson, 2000: First records of sockeye (*Oncorhynchus nerka*) and pink salmon (*O. gorbuscha*) from Banks Island and other records of pacific salmon in Northwest Territories, Canada. *Arctic*, **53**, 161-164.

Beerling, D.J., A.C. Terry, P.L. Mitchell, D. Gwynn-Jones, J.A. Lee and T.V. Callaghan, 2001: Time to chill: effects of simulated global change on leaf ice nucleation temperatures of sub-Arctic vegetation. *Am. J. Bot.*, **88**, 628-633.

Beilman, D.W. and S.D. Robinson, 2003: Peatland permafrost thaw and landform type along a climatic gradient. *Proc., 8th International Conference on Permafrost*, M. Phillips, S.M. Springman and L.U. Arenson, Eds., A.A. Balkema, Lisse, 61-66.

Beilman, D.W., D.H. Vitt and L.A. Halsey, 2001: Localized permafrost peatlands in western Canada: definition, distributions, and degradation. *Arct. Antarct. Alp. Res.*, **33**, 70-77.

Beltaos, S., T. Prowse, B. Bonsal, R. MacKay, L. Romolo, A. Pietroniro and B. Toth, 2006: Climatic effects on ice-jam flooding of the Peace–Athabasca Delta. *Hydrol. Process.*, **20**, 4013-4050.

Berezovskaya, S., D.Q. Yang and L. Hinzman, 2005: Long-term annual water balance analysis of the Lena River. *Global Planet. Change*, **48**, 84-95.

Bergstrom, D.M. and S.L. Chown, 1999: Life at the front: history, ecology and change on southern ocean islands. *Trends Ecol. Evol.*, **14**, 472-477.

Berkes, F., 1999: *Sacred Ecology: Traditional Ecological Knowledge and Resource Management*. Taylor and Francis, London, 232 pp.

Berkes, F., 2002: Epilogue: making sense of arctic environmental change? *The Earth is Faster Now: Indigenous Observations of Arctic Environmental Change*, I. Krupnik and D. Jolly, Eds., Arctic Research Consortium, Fairbanks, 335-349.

Berkes, F. and D. Jolly, 2001: Adapting to climate change: social-ecological resilience in a Canadian Western Arctic Community. *Conserv. Ecol.*, **5**, 18.

Berner, J., C. Furgal, P. Bjerregaard, M. Bradley, T. Curtis, E. De Fabo, J. Hassi, W. Keatinge, S. Kvernmo, S. Nayha, H. Rintamaki and J. Warren, 2005: Human health. *Arctic Climate Impact Assessment*, C. Symon, L. Arris and B. Heal, Eds., Cambridge University Press, Cambridge, 863-906.

Berry, J.W., 1997: Immigration, acculturation and adaptation. *Appl. Psychol.*, **46**, 5-34.

Betts, A.K. and J.H. Ball, 1997: Albedo over the boreal forest. *J. Geophys. Res.*, **102**, 28901-28909.

Betts, R.A., 2000: Offset of the potential carbon sink from boreal forestation by decreases in surface albedo. *Nature*, **408**, 187-190.

Bialek, D., 2003: Sink or swim: measures under international law for the conservation of the Patagonian toothfish in the Southern Ocean. *Ocean Dev. Int. Law*, **34**, 105-137.

Bindoff, N., J. Willebrand, V. Artale, A. Cazenave, J. Gregory, S. Gulev, K. Hanawa, C. Le Quere, S. Levitus, Y. Nojiri, C. Shum, L. Talley and U. Alakkat, 2007: Observations: oceanic climate change and sea level. *Climate Change 2007: The Physical Science Basis. Contribution of Working Group I to the Fourth Assessment Report of the Intergovernmental Panel on Climate Change*, S. Solomon, D. Qin, M. Manning, Z. Chen, M. Marquis, K.B. Averyt, M. Tignor and H.L. Miller, Eds., Cambridge University Press, Cambridge, 385-432.

Blanchet, C., E. Dewailly, P. Ayotte, S. Bruneau, O. Receveur and B. Holub, 2000: Contribution of selected traditional and market foods to the diet of Nunavik Inuit women. *Can. J. Diet. Pract. Res.*, **61**, 1-9.

Bochkov, Y.A., 1982: Water temperature in the 0-200 m layer in the Kola Meridian Section in the Barents Sea, 1900-1981 (in Russian). *Sb. Nauch. Trud. PINRO*, **46**, 113-122.

Bogaart, P.W. and R.T. van Balen, 2000: Numerical modeling of the response of alluvial rivers to Quaternary climatic change. *Global Planet. Change*, **27**, 124-141.

Bogoyavlenskiy, D. and A. Siggner, 2004: Arctic demography. *Arctic Human Development Report (AHDR)*, N. Einarsson, J.N. Larsen, A. Nilsson and O.R. Young, Eds., Steffanson Arctic Institute, Akureyri, 27-41.

Bonan, G.B., D. Pollard and S.L. Thompson, 1992: Effects of boreal forest vegetation on global climate. *Nature*, **359**, 716-718.

Both, C., R.G. Bijlsma and M.E. Visser, 2005: Climate effects on timing of spring migration and breeding in a long-distance migrant, the pied flycatcher, *Ficedula hypoleuca*. *J. Avian Biol.*, **36**, 368-373.

Bradley, M.J., S.J. Kutz, E. Jenkins and T.M. O'Hara, 2005: The potential impact of climate change on infectious diseases of Arctic fauna. *International Journal for Circumpolar Health*, **64**, 468-477.

Brown, J., M.T. Jorgenson, O.P. Smith and W. Lee, 2003: Long-term rates of coastal erosion and carbon input, Elson Lagoon, Barrow, Alaska. *Proc., 8th International Conference on Permafrost*, M. Phillips, S.M. Springman and L.U. Arenson, Eds., A.A. Balkema, Lisse, 101-106.

Buch, E., S.A. Horsted and H. Hovgaard, 1994: Fluctuations in the occurrence of cod in Greenland waters and their possible causes. *ICES Mar. Sci. Symp.*, **198**, 158-174.

Callaghan, T.V., M. Johansson, O.W. Heal, N.R. Sælthun, L.J. Barkved, N. Bayfield, O. Brandt, R. Brooker and Co-authors, 2004: Environmental changes in the North Atlantic region: SCANNET as a collaborative approach for documenting, understanding and predicting changes. *Ambio*, **13**, S39-S50.

Callaghan, T.V., L.O. Björn, Y.I. Chernov, F.S. Chapin III, T.R. Christensen, B. Huntley, R. Ims, M. Johansson, D. Jolly, N.V. Matveyeva, N. Panikov, W.C. Oechel and G.R. Shaver, 2005: Arctic tundra and polar ecosystems. *Arctic Climate Impact Assessment, ACIA*, C. Symon, L. Arris and B. Heal, Eds., Cambridge University Press, Cambridge, 243-351.

Carmack, E.C. and R.W. Macdonald, 2002: Oceanography of the Canadian Shelf of the Beaufort Sea: a setting for marine life. *Arctic*, **55**, 29-45.

Chabot, M., 2004: Consumption and standards of living of the Québec Inuit: cultural permanence and discontinuities. *Can. Rev. Sociol. Anthr.*, **41**, 147-170.

Chapin, F.S., III, G. Peterson, F. Berkes, T.V. Callaghan, P. Angelstam, M. Apps, C. Beier, C. Bergeron and Co-authors, 2004: Resilience and vulnerability of northern regions to social and environmental change. *Ambio*, **33**, 342-347.

Chapin, F.S., III, M. Berman, T.V. Callaghan, P. Convey, A.-S. Crepin, K. Danell, H. Ducklow, B. Forbes and Co-authors, 2005a: Polar systems. *The Millennium Ecosystem Assessment*, R. Hassan, R. Scholes and N. Ash, Eds., Island Press, Washington, District of Columbia, 717-743.

Chapin, F.S., III, M. Sturm, M.C. Serreze, J.P. McFadden, J.R. Key, A.H. Lloyd, A.D. McGuire, T.S. Rupp and Co-authors, 2005b: Role of land-surface changes in Arctic summer warming. *Science*, **310**, 657-660.

Chapin, F.S., III, M. Hoel, S.R. Carpenter, J. Lubchenko, B. Walker, T.V. Callaghan, C. Folke, S.A. Levin, K.-G. Mäler, C. Nilsson, S. Barrett, F. Berkes, A.-S. Crépin, K. Danell and Co-authors, 2006: Building resilience to manage Arctic change. *Science*, **35**, 198-202.

Chartrand, J., K. Lysyshyn, R. Couture, S.D. Robinson and M.M. Burgess, 2002: Digital geotechnical borehole databases and viewers for Norman Wells and Tuktoyaktuk, Northwest Territories. Open File Report, 3912. Geological Survey of Canada, CD-Rom.

Chernov, Y.I. and N.V. Matveyeva, 1997: Arctic ecosystems in Russia. *Ecosystems of the World 3: Polar and Alpine Tundra*, F.E. Wielgolaski, Ed., Elsevier, Amsterdam, 361-507.

Christensen, T.R., T. Johansson, H.J. Akerman, M. Mastepanov, N. Malmer, T. Friborg, P. Crill and B.H. Svensson, 2004: Thawing sub-arctic permafrost: effects on vegetation and methane emissions. *Geophys. Res. Lett.*, 31, L04501, doi: 10.1029/2003GL018680.

Coleman, J.M. and O.K. Huh, 2004: *Major World Deltas: A Perspective from Space*. Coastal Studies Institute, Louisiana State University, Baton Rouge, Louisiana. http://www.geol.lsu.edu/WDD/PUBLICATIONS/introduction.htm.

Community of Arctic Bay, 2005: *Inuit Observations on Climate and Environmental Change: Perspectives from Arctic Bay, Nunavut*. ITK, Nasivvik, NAHO, NTI, Ottawa, 57 pp.

Constable, A.J., W.K. de la Mare, D.J. Agnew, I. Everson and D. Miller, 2000: Managing fisheries to conserve the Antarctic marine ecosystem: practical implementation of the Convention on the Conservation of Antarctic Marine Living Resources (CCAMLR). *ICES J. Mar. Sci.*, **57**, 778-791.

Convey, P., P.J.A. Pugh, C. Jackson, A.W. Murray, C.T. Ruhland, F.S. Xiong and T.A. Day, 2002: Response of Antarctic terrestrial microarthropods to long-term climate manipulations. *Ecology*, **83**, 3130-3140.

Cook, A., A.J. Fox, D.G. Vaughan and J.G. Ferrigno, 2005: Retreating glacier-fronts on the Antarctic Peninsula over the last 50 years. *Science*, **22**, 541-544.

Cornelissen, J.H.C., T.V. Callaghan, J.M. Alatalo, A.E. Hartley, D.S. Hik, S.E. Hobbie, M.C. Press, C.H. Robinson and Co-authors, 2001: Global change and Arctic ecosystems: is lichen decline a function of increases in vascular plant biomass? *J. Ecol.*, **89**, 984-994.

Corradi, C., O. Kolle, K. Walters, S.A. Zimov and E.-D. Schulze, 2005: Carbon dioxide and methane exchange of a north-east Siberian tussock tundra. *Glob. Change Biol.*, **11**, 1910-1925.

Couture, R., S.D. Robinson and M.M. Burgess, 2000: Climate change, permafrost degradation, and infrastructure adaptation: preliminary results from a pilot community study in the Mackenzie Valley. Geological Survey of Canada, Current Research 2000-B2, 9 pp.

Couture, R., S.D. Robinson and M.M. Burgess, 2001: Climate change, permafrost degradation and impacts on infrastructure: two case studies in the Mackenzie Valley. *Proc., 54th Canadian Geotechnical Conference: An Earth Odyssey*, Calgary, 908-915.

Couture, R., S. Smith, S.D. Robinson, M.M. Burgess and S. Solomon, 2003: On the hazards to infrastructure in the Canadian north associated with thawing of permafrost. *Geohazards 2003: 3rd Canadian Conference on Geotechnique and Natural Hazards, Edmonton*. Canadian Geotechnical Society, Alliston, Ontario, 97-104.

Crawford, R.M.M., 2004: Long-term plant survival at high latitudes. *Bot. J. Scotland*, **56**, 1-23.

Crawford, R.M.M., C.E. Jeffree and W.G. Rees, 2003: Paludification and forest retreat in northern oceanic environments. *Ann. Bot. Lond.*, **91**, 213-226.

Curran, M.A.J., T.D. van Ommen, V.I. Morgan, K.L. Phillips and A.S. Palmer, 2003: Ice core evidence for Antarctic sea ice decline since the 1950s. *Science*, **302**, 1203-1206.

Curtis, T., S. Kvernmo and P. Bjerregaard, 2005: Changing living conditions, lifestyle and health. *International Journal of Circumpolar Health*, **64**, 442-450.

Dalen, L. and A. Hofgaard, 2005: Differential regional treeline dynamics in the Scandes Mountains. *Arct. Antarct. Alp. Res.*, **37**, 284-296.

Davis, C.H., L. Yonghong, J.R. McConnell, M.M. Frey and E. Hanna, 2005: Snowfall-driven growth in Antarctic Ice Sheet mitigates recent sea-level rise. *Science*, **308**, 1898-1901.

De Angelis, H. and P. Skvarca, 2003: Glacier surge after ice shelf collapse. *Science*, **299**, 1560-1562.

de la Mare, W.K., 1997: Abrupt mid-twentieth-century decline in Antarctic sea-ice extent from whaling records. *Nature*, **389**, 57-60.

Derocher, A.E., N.J. Lunn and I. Stirling, 2004: Polar bears in a warming climate. *Integr. Comp. Biol.*, **44**, 163-176.

Dietrich, R.A., Ed., 1981: *Alaskan Wildlife Diseases*. Institute of Arctic Biology, University of Alaska, Fairbanks.

Domack, E., A. Leventer, S. Root, J. Ring, E. Williams, D. Carlson, E. Hirshorn, W. Wright, R. Gilbert and G. Burr, 2003: Marine sedimentary record of natural environmental variability. *Antarctic Peninsula Climate Variability: Historical and Paleoenvironmental Perspectives*, E. Domack, A. Leventer, A. Burnett, R. Bindschadler, P. Convey and M. Kirby, Eds., Antarctic Research Series 79, AGU, Washington, District of Columbia, 61-68.

Domack, E., D. Duran, A. Leventer, S. Ishman, S. Doane, S. McCallum, D. Amblas, J. Ring, R. Gilbert and M. Prentice, 2005: Stability of the Larsen B ice shelf on the Antarctic Peninsula during the Holocene epoch. *Nature*, **436**, 681-685.

Doran, P.T., J.C. Priscu, W. Berry Lyons, J.E. Walsh, A.G. Fountain, D.M. McKnight, D.L. Moorhead, R.A. Virginia and Co-authors, 2002: Antarctic climate cooling and terrestrial ecosystem response. *Nature*, **415**, 517-520.

Dore, M. and I. Burton, 2001: *The Costs of Adaptation to Climate Change in Canada: A Stratified Estimate by Sectors and Regions – Social Infrastructure*. Environment Canada, Ottawa, 339 pp.

Dowdeswell, J.A., 2006: Atmospheric science: the Greenland Ice Sheet and global sea-level rise. *Science*, **311**, 963-964.

Downey, D. and T. Fenge, Eds., 2003: *Northern Lights Against POPs: Combating Toxic Threats in the Arctic*. McGill–Queen's University Press, Montreal, 347 pp.

Drinkwater, K.F., 2005: The response of Atlantic cod (*Gadus morhua*) to future climate change. *ICES J. Mar. Sci.*, **62**, 1327-1337.

Duhaime, G., 2004: Economic systems. *The Arctic Human Development Report*, N. Einarsson, J.N. Larsen, A. Nilsson and O.R. Young, Eds., Stefansson Arctic Institute, Akureyri, 69-84.

Dye, D.G., 2002: Variability and trends in the annual snow-cover cycle in northern hemisphere land areas, 1972-2000. *Hydrol. Process.*, **16**, 3065-3077.

Dyurgerov, M.B. and C.L. Carter, 2004: Observational evidence of increases in freshwater inflow to the Arctic Ocean. *Arct. Antarct. Alp. Res.*, **36**, 117-122.

Eagles, P.F.J., 2004: Trends affecting tourism in protected areas. *Proc., 2nd International Conference on Monitoring and Management of Visitor Flows in Recreational and Protected Areas, Rovaniemi, Finland: Policies, Methods and Tools for Visitor Management*, T. Sievänen, J. Erkkonen, J. Jokimäki, J. Saarinen, S. Tuulentie and E. Virtanen, Eds., Working Papers of the Finnish Forest Research Institute 2, 18-26.

Edwards, M. and A.J. Richardson, 2004: Impact of climate change on marine pelagic phenology and trophic mismatch. *Nature*, **430**, 881-884.

Einarsson, N., J.N. Larsen, A. Nilsson and O.R. Young, Eds., 2004: *Arctic Human Development Report*. Stefansson Arctic Institute, Akureyri, 242 pp.

Ekstrom, G., M. Nettles and V.C. Tsai, 2006: Seasonality and increasing frequency of Greenland glacial earthquakes. *Science*, **311**, 1756-1758.

Elmhagen, B., M. Tannerfeldt, P. Verucci and A. Angerbjörn, 2000: The arctic fox (*Alopex lagopus*): an opportunistic specialist. *J. Zool. Soc. Lond.*, **251**, 139-149.

Esper, J. and F.H. Schweingruber, 2004: Large-scale treeline changes recorded in Siberia. *Geophys. Res. Lett.*, **31**, L06202, doi:10.1029/2003GL019178.

Eugster, W., W.R. Rouse, R.A. Pielke, J.P. McFadden, D.D. Baldocchi, T.G.F. Kittel, F.S. Chapin III and Co-authors, 2000: Land-atmosphere energy exchange in Arctic tundra and boreal forest: available data and feedbacks to climate. *Glob. Change Biol.*, **6**, S84-S115.

Euskirchen, S.E., A.D. McGuire, D.W. Kicklighter, Q. Zhuang, J.S. Clein, R.J. Dargaville, D.G. Dye and Co-authors, 2006: Importance of recent shifts in soil thermal dynamics on growing season length, productivity, and carbon sequestration in terrestrial high-latitude ecosystems. *Glob. Change Biol.*, **12**, 731-750.

FAUNMAP Working Group, 1996: Spatial response of mammals to late Quaternary environmental fluctuations. *Science*, **272**, 1601-1606.

Flanagan, K.M., E. McCauley and F.J. Wrona, 2006: Freshwater food webs control saturation through sedimentation. *Glob. Change Biol.*, **12**, 644-651.

Foley, J.A., J.E. Kutzbach, M.T. Coe and S. Levis, 1994: Feedbacks between climate and boreal forests during the Holocene epoch. *Nature*, **371**, 52-54.

Forbes, D.L., 2005: Coastal erosion. *Encyclopedia of the Arctic*, M. Nutall, Ed., Routledge, New York and London, 391-393.

Forchhammer, M. and D. Boertmann, 1993: The muskoxen *Ovibos moschatus* in north and northeast Greenland: population trends and the influence of abiotic parameters on population dynamics. *Ecography*, **16**, 299-308.

Fowbert, J.A. and R. Smith, 1994: Rapid population increases in native vascular plants in the Argentine Islands, Antarctic Peninsula. *Arct. Alp. Res.*, **26**, 290-296.

Fox, A.J. and A.P.R. Cooper, 1998: Climate-change indicators from archival aerial photography of the Antarctic Peninsula. *Ann. Glaciol.*, **27**, 636-642.

Fraser, W.R. and E.E. Hoffmann, 2003: A predator's perspective on causal links between climate change, physical forcing and ecosystem response. *Mar. Ecol-Prog. Ser.*, **265**, 1-15.

Freeman, M., 1996: Identity, health and social order. *Human Ecology and Health: Adaptation to a Changing World*, M.-L. Foller and L.O. Hansson, Eds., Gothenburg University, Gothenburg, 57-71.

Freeman, M.M.R., 2000: *Endangered Peoples of the Arctic*. Greenwood Press, Connecticut, 278 pp.

Frenot, Y., S.L. Chown, J. Whinam, P.M. Selkirk, P. Convey, M. Skotnicki and D.M. Bergstrom, 2005: Biological invasions in the Antarctic: extent, impacts and implications. *Biol. Rev.*, **80**, 45-72.

Frey, K.E. and L.C. Smith, 2005: Amplified carbon release from vast West Siberian peatlands by 2100. *Geophys. Res. Lett.*, **32**, L09401, doi:10.1029/2004GL022025.

Friborg, T., H. Soegaard, T.R. Christensen, C.R. Lloyd and N.S. Panikov, 2003: Siberian wetlands: where a sink is a source. *Geophys. Res. Lett.*, **30**, 2129, doi:10.1029/2003GL017797.

Furgal, C., D. Martin and P. Gosselin, 2002: Climate change and health in Nunavik and Labrador: lessons from Inuit knowledge. *The Earth is Faster Now: Indigenous Observations on Arctic Environmental Change*, I. Krupnik and D. Jolly, Eds., ARCUS, Washington, District of Columbia, 266-300.

Gauthier, G., J.-F. Giroux, A. Reed, A. Béchet and L. Bélanger, 2005: Interactions between land use, habitat use, and population increase in greater snow goose: what are the consequences for natural wetlands? *Glob. Change Biol.*, **11**, 856-868.

Gedney, N., P.M. Cox, R.A. Betts, O. Boucher, C. Huntingford and P.A. Stott, 2006: Detection of a direct carbon dioxide effect in continental river runoff records. *Nature*, **439**, 835-838.

George, J.C., H.P. Huntington, K. Brewster, H. Eicken, D.W. Norton and R. Glenn, 2004: Observations on shorefast ice dynamics in arctic Alaska and the responses of the Inupiat hunting community. *Arctic*, **57**, 363-374.

Georgievsky, V.Y., I.A. Shiklomanov and A.L. Shalygin, 2005: Climate change impact on the water runoff on Lena river basin. *Challenges of Ecological Meteorology and Climatology*, G.V. Menzhulin, Ed., Nauka, St. Petersburg, 218-232.

Gibson, J.J., T.D. Prowse and D.L. Peters, 2006: Partitioning impacts of climate and regulation on water level variability in Great Slave Lake. *J. Hydrol.*, **329**,

196-206.

Gray, A., 1995: The indigenous movement in Asia. *Indigenous Peoples of Asia*, A. Gray, R.H. Barnes and B. Kingsbury, Eds., Association for Asian Studies, Ann Arbor, Michigan.

Grebmeier, J.M., J.E. Overland, S.E. Moore, E.V. Farley, E.C. Carmack, L.W. Cooper, K.E. Frey, J.H. Helle and Co-authors, 2006: A major ecosystem shift in the northern Bering Sea. *Science*, **311**, 1461-1464.

Gregory, J.M., P. Huybrechts and S.C.B. Raper, 2004: Climatology: threatened loss of the Greenland ice-sheet. *Nature*, **428**, 616.

Hafrannsóknastofnunin, 2006: *Þættir úr vistfræði sjávar 2005 (Environmental Conditions in Icelandic Waters 2005.* In Icelandic with English table headings and summaries). Hafrannsóknastofnunin (Marine Research Institute), 36 pp.

Hamilton, L.C., B.C. Brown and R.O. Rasmussen, 2003: West Greenland's cod-to-shrimp transition: local dimensions of climate change. *Arctic*, **56**, 271-282.

Handmer, J.W., S. Dovers and T.E. Downing, 1999: Societal vulnerability to climate change and variability. *Mitigation and Adaptation Strategies for Climate Change*, **4**, 267-281.

Harris, C., D. Vonder Mühll, K. Isaksen, W. Haeberli, J.L. Sollid, L. King, P. Holmlund, F. Dramis and Co-authors, 2003: Warming permafrost in European mountains. *Global Planet. Change*, **39**, 215-225.

Harris, L., 1990: Independent review of the state of the northern cod stock. Final Report of the Northern Cod Review Panel, Department of Supply and Services, Ottawa, 154 pp.

Hassi, J., M. Rytkonen, J. Kotaniemi and H. Rintimaki, 2005: Impacts of cold climate on human heat balance, performance, illnesses and injuries in circumpolar areas. *International Journal of Circumpolar Health*, **64**, 459-567.

Hayley, D.W., 2004: *Climate Change: An Adaptation Challenge for Northern Engineers*. Association of Professional Engineers, Geologists and Geophysicists of Alberta, Alberta. http://www.apegga.org/members/publications/peggs/web01-04/expert.htm.

Henttonen, H. and H. Wallgren, 2001: Small rodent dynamics and communities in the birch forest zone of northern Fennoscandia. *Nordic Mountain Birch Forest Ecosystems*, F.E. Wielgolaski, Ed., Man and the Biosphere Series. UNESCO, Paris, and Parthenon Publishing Group, New York and London, 261-278.

Hild, C. and V. Stordhal, 2004: Human health and well-being. *Arctic Human Development Report (AHDR)*, N. Einarsson, J.N. Larsen, A. Nilsson, O.R. Young, Eds., Steffanson Arctic Institute, Akureyri, 155-168.

Hinkel, K.M., F.E. Nelson, A.F. Klene and J.H. Bell, 2003: The urban heat island in winter at Barrow, Alaska. *Int. J. Climatol.*, **23**, 1889-1905.

Hinzman, L.D., N.D. Bettez, W.R. Bolton, F.S. Chapin, M.B. Dyurgerov, C.L. Fastie, B. Griffith, R.D. Hollister and Co-authors, 2005: Evidence and implications of recent climate change in northern Alaska and other Arctic regions. *Climatic Change*, **72**, 251-298.

Hobbie, J.E., B.J. Peterson, N. Bettez, L. Deegan, J. O'Brien, G.W. Kling and G.W. Kipphut, 1999: Impact of global change on biogeochemistry and ecosystems of arctic Alaska freshwaters. *Polar Res.*, **18**, 207-214.

Høgda, K.A., S.R. Karlsen and H. Tømmervik, 2007: Changes in growing season in Fennoscandia 1982-1999. *Arctic-Alpine Ecosystems and People in a Changing Environment*, J.B. Orbaek, R. Kallenborn, I. Tombre, E.N. Hegseth, S. Falk-Petersen and A.H. Hoel, Eds., Springer-Verlag, Berlin, 71-84.

Holland, M.M. and C.M. Bitz, 2003: Polar amplification of climate change in coupled models. *Clim. Dynam.*, **21**, 221-232.

Hörnberg, S., 1995: Moose density related to occurrence and consumption of different forage species in Sweden. SLU Institution för Skogstaxering, Umeå. *Rapporter–Skog*, **58l**, 34.

Horsted, S.A., 2000: A review of the cod fisheries at Greenland, 1910-1995. *Journal of Northwest Atlantic Fishery Science*, **28**, 1-112.

Howat, I.M., I. Joughin, S. Tulaczyk and S. Gogineni, 2005: Rapid retreat and acceleration of Helheim glacier, east Greenland. *Geophys. Res. Lett.*, **32**, L22502, doi:10.1029/2005GL024737.

Hughes-Hanks, J.M., L.G. Rickard, C. Panuska, J.R. Saucier, T.M. O'Hara, R.M. Rolland and L. Dehn, 2005: Prevalence of *Cryptosporidium* spp. and *Giardia* spp. in five marine mammal species. *J. Parasitol.*, **91**, 1357-1357.

Huisink, M., J.J.W. De Moor, C. Kasse and T. Virtanen, 2002: Factors influencing periglacial fluvial morphology in the northern European Russian tundra and taiga. *Earth Surf. Proc. Land.*, **27**, 1223-1235.

Huntington, H., 1998: Observations on the utility of the semi-directive interview for documenting traditional ecological knowledge. *Arctic*, **51**, 237-242.

Huntington, H., T.V. Callaghan, S. Fox and I. Krupnik, 2004: Matching traditional and scientific observations to detect environmental change: a discussion on arctic terrestrial ecosystems. *Ambio*, **33**(S13), 18-23.

Huntington, H., S. Foxand Co-authors, 2005: The changing Arctic: Indigenous perspectives. Arctic Climate Impact Assessment, C. Symon, L. Arris and B. Heal, Eds., Cambridge University Press, Cambridge, 61-98.

Huntington, H.P., 1992: *Wildlife Management and Subsistence Hunting in Alaska*. Belhaven Press, London, 177 pp.

Hylen, A., 2002: Fluctuations in abundance of northeast Arctic cod during the 20th century. *100 Years of Science under ICES: Papers from a Symposium held in Helsinki, 1-4 August 2000*, E.D. Anderson, Ed., Marine Science Symposia 215, International Council for the Exploration of the Sea, 543-550.

IAATO, 2005: Update on boot and clothing decontamination guidelines and the introduction and detection of diseases in Antarctic wildlife: IAATO's perspective. Paper submitted by the International Association of Antarctica Tour Operators (IAATO) to the Antarctic Treaty Consultative Meeting (ATCM) XXVIII. IAATO, 10 pp. http://www.iaato.org/info.html.

IAATO, 2006: IAATO Overview of Antarctic tourism 2005–2006. Information paper submitted by the International Association of Antarctica Tour Operators (IAATO) to the Antarctic Treaty Consultative Meeting (ATCM) XXIX. IAATO, 21 pp. http://www.iaato.org/info.html.

ICARP II, 2006: Science Plan 8: terrestrial and freshwater biosphere and biodiversity. *Proc., 2nd International Conference on Artic Research Planning (ICARP II)*, Copenhagen, 20 pp.

ICES, 2005a: Report of the Arctic Fisheries Working Group, Murmansk, 19-28 April 2005. ICES CM 2005/ACFM:20. International Council for the Exploration of the Sea, Copenhagen, 564 pp.

ICES, 2005b: Report of the Northwestern Working Group (NWWG) on 26 April-5 May 2005. ICES CM 2005/ACFM:21. International Council for the Exploration of the Sea, Copenhagen, 615 pp.

ICES, 2006: Report of the Northwestern Working Group (NWWG). ICES CM 2006/ACFM:26. International Council for the Exploration of the Sea, Copenhagen, 612 pp.

Ims, R.A. and E. Fuglei, 2005: Trophic interaction cycles in tundra ecosystems and the impact of climate change. *BioScience*, **554**, 311-322.

Instanes, A., O. Anisimov, L. Brigham, D. Goering, B. Ladanyi, J.O. Larsen and L.N. Khrustalev, 2005: Infrastructure: buildings, support systems, and industrial facilities. *Arctic Climate Impact Assessment, ACIA*, C. Symon, L. Arris and B. Heal, Eds., Cambridge University Press, Cambridge, 907-944.

Jefferies, R.L., R.H. Drent and J.P. Bakker, 2006: Connecting arctic and temperate wetlands and agricultural landscapes: the dynamics of goose populations in response to global change. *Wetlands and Natural Resource Management*, J.T. Verhoeven, B. Beltman, R. Bobbink and D.F. Whigham, Eds., Ecological Studies 190, Springer, Berlin, 293-312.

Jensen, A.S., 1926: Indberetning av S/S Dana's praktisk vidfull skabelige fiskeriundersøgelser ved Vestgrønland i 1925. *Beretninger og Kundgørelser vedr. Styrelsen af Grønland*, **2**, 291-315.

Jensen, A.S., 1939: Concerning a change of climate during recent decades in the Arctic and Subarctic regions, from Greenland in the west to Eurasia in the east, and contemporary biological and physical changes. *Det. Kgl. Danske Videnskabernes Selskab. Biologiske Medd.*, **14**, 1-77.

Jernsletten, J.-L. and K. Klokov, 2002: Sustainable Reindeer Husbandry. University of Tromso, Tromso, 164 pp. http://www.reindeer-husbandry.uit.no/.

Johannessen, O.M., K. Khvorostovsky, M.W. Miles and L.P. Bobylev, 2005: Recent ice-sheet growth in the interior of Greenland. *Science*, **310**, 1013-1016.

Johansson, T., N. Malmer, P.M. Crill, T. Friborg, J.H. Åkerman, M. Mastepanov and T.R. Christensen, 2006: Decadal vegetation changes in a northern peatland, greenhouse gas fluxes and net radiative forcing. *Glob. Change Biol.*, **12**, 2352-2369.

Johnson, D., C.D. Campbell, D. Gwynn-Jones, J.A. Lee and T.V. Callaghan, 2002: Arctic soil microorganisms respond more to long-term ozone depletion than to atmospheric CO_2. *Nature*, **416**, 82-83.

Jones, C.D. and D.C. Ramm, 2004: The commercial harvest of krill in the southwest Atlantic before and during the CCAMLR 2000 Survey. *Deep-Sea Res. Pt. II*, **51**, 1421-1434.

Juday, G.P., V. Barber, P. Duffy, H. Linderholm, S. Rupp, S. Sparrow, E. Vaganov and J. Yarie, 2005: Forests, land management and agriculture. *Arctic Climate Impact Assessment ACIA*, C. Symon, L. Arris and B. Heal, Eds., Cambridge University Press, Cambridge, 781-862.

Kaplan, J.O., N.H. Bigelow, I.C. Prentice, S.P. Harrison, P.J. Bartlein, T.R.

Christensen, W. Cramer, N.V. Matveyeva and Co-authors, 2003: Climate change and Arctic ecosystems. 2. Modeling, paleodata-model comparisons, and future projections. *J. Geophys. Res.*, **108**, 8171, doi:10.1029/2002JD002559.

Kattsov, V.M., J.E. Walsh, W.L. Chapman, V.A. Govorkova, T.V. Pavlova and X. Zhang, 2007: Simulation and projection of Arctic freshwater budget components by the IPCC AR4 global climate models. *J. Hydrometeorol.*, **8**, 571-589.

Khasnullin, V.I., A.M. Shurgaya, A.V. Khasnullina and E.V. Sevostoyanova, 2000: *Cardiomyopathies in the North: Novosibirsk* (in Russian). Siberian Division of Russian Academy of Medical Sciences, Siberia, 222 pp.

Khrustalev, L.N., 2000: On the necessity of accounting for the effect of changing climate in permafrost engineering. *Geocryological Hazards*, L.S. Garagulia and E.D. Yershow, Eds., Kruk Publishers, Moscow, 238-247.

King, J.C., 2003: Antarctic Peninsula climate variability and its causes as revealed by analysis of instrumental records. *Antarctic Peninsula Climate Variability: Historical and Paleoenvironmental Perspectives*. Antarctic Research Series 79, AGU, Washington, District of Columbia, 17-30.

Klein, D., 1989: Subsistence hunting. *Wildlife Production Systems: Economic Utilisation of Wild Ungulates*, R.J. Hudson, K.R. Drew and L.M. Baskin, Cambridge University Press, Cambridge, 96-111.

Klein, D., L.M. Baskin, L.S. Bogoslovskaya, K. Danell, A. Gunn, D.B. Irons, G.P. Kofinas, K.M. Kovacs and Co-authors, 2005: Management and conservation of wildlife in a changing Arctic. *Arctic Climate Impact Assessment, ACIA*, C. Symon, L. Arris and B. Heal, Eds., Cambridge University Press, Cambridge, 597-648.

Kofinas, G., G. Osherenko, D. Klein and B. Forbes, 2000: Research planning in the face of change: the human role in reindeer/caribou systems. *Polar Res.*, **19**, 3-22.

Kokelj, S.V. and C.R. Burn, 2005: Near-surface ground ice in sediments of the Mackenzie Delta, Northwest Territories, Canada. *Permafrost Periglac.*, **16**, 291-303.

Korhola, A., S. Sorvari, M. Rautio, P.G. Appleby, J.A. Dearing, Y. Hu, N. Rose, A. Lami and N.G. Cameron, 2002: A multi-proxy analysis of climate impacts on recent ontogeny of subarctic Lake Sannajärvi in Finnish Lapland. *J. Paleolimnol.*, **1**, 59-77.

Koshida, G. and W. Avis, Eds., 1998: *Canada Country Study: Climate Impacts and Adaptation*. Vol. 7, *National Sectoral Issues*. Environment Canada, Toronto, 620 pp.

Kovats, R.S., B. Menne, A.J. McMichael, C. Corvalan and R. Bertollini, 2000: Climate change and human health: impact and adaptation. WHO/SDE/OEH/00.4. World Health Organization, Geneva, 48 pp.

Krabill, W., W. Abdalati, E. Frederick, S. Manizade, C. Martin, J. Sonntag, R. Swift, R. Thomas, W. Wright and J. Yungel, 2000: Greenland ice sheet: high-elevation balance peripheral thinning. *Science*, **289**, 428-430.

Krabill, W., E. Hanna, P. Huybrechts, W. Abdalati, J. Cappelen, B. Csatho, E. Frederick, S. Manizade and Co-authors, 2004: Greenland Ice Sheet: increased coastal thinning. *Geophys. Res. Lett.*, **31**, L24402, doi:10.1029/2004GL021533.

Kraemer, L., J. Berner and C. Furgal, 2005: The potential impact of climate change on human exposure to contaminants in the Arctic. *International Journal for Circumpolar Health*, **64**, 498-509.

Krupnik, I., 1993: *Arctic Adaptations: Native Whalers and Reindeer Herders of Northern Eurasia*. University Press of New England, Lebanon, Pennsylvania, 355 pp.

Krupnik, I. and D. Jolly, Eds., 2002: *The Earth is Faster Now: Indigenous Observations of Arctic Environmental Change*. Arctic Research Consortium of the United States, Fairbanks, Alaska, 356 pp.

Kuhnlein, H.V., O. Receveur and H.M. Chan, 2001: Traditional food systems research with Canadian indigenous peoples. *International Journal of Circumpolar Health*, **60**, 112-122.

Kutz, S.J., E.P. Hoberg, J. Nishi and L. Polley, 2002: Development of the musk ox lungworm *Umingmakstrongylus pallikuukensis* (Protostrongylidae) in gastropods in the Arctic. *Can. J. Zool.*, **80**, 1977-1985.

Lafortune, V., C. Furgal, J. Drouin, T. Annanack, N. Einish, B. Etidloie, M. Qiisiq, P. Tookalook and Co-authors, 2004: Climate change in northern Québec: access to land and resource issues. Project report. Kativik Regional Government, Kuujjuaq, Québec.

Laherre, J., 2001: Estimates of oil reserves. *Preprints EMF/IEA/IEW Meeting*, International Institute for Applied Systems Analysis (IIASA), Laxenburg, 92 pp.

Lantuit, H. and W. Pollard, 2003: Remotely sensed evidence of enhanced erosion during the twentieth century on Herschel Island, Yukon Territory. *Berichte zur Polar- und Meeresforschung*, **443**, 54-59.

Lemke, P., J. Ren, R. Alley, I. Allison, J. Carrasco, G. Flato, Y. Fujii, G. Kaser, P. Mote, R. Thomas and T. Zhang, 2007: Observations: change in snow, ice and frozen ground. *Climate Change 2007: The Physical Science Basis. Contribution of Working Group I to the Fourth Assessment Report of the Intergovernmental Panel on Climate Change*, S. Solomon, D. Qin, M. Manning, Z. Chen, M. Marquis, K.B. Averyt, M. Tignor and H.L. Miller, Eds., Cambridge University Press, Cambridge, 337-384.

Lesack, L.F.W., P. Marsh and R.E. Hecky, 1998: Spatial and temporal dynamics of major solute chemistry along Mackenzie Delta lakes. *Limnol. Oceanogr.*, **43**, 1530-1543.

Lewis, E.L., E.P. Jones, P. Lemke, T.D. Prowse and P. Wadhams, 2000: *The Freshwater Budget of the Arctic Ocean*. Kluwer Academic, Dordrecht, 623 pp.

Lilly, G.R., P.A. Shelton, J. Brattey, N. Cadigan, B.P. Healey, E.F. Murphy, D. Stanbury and N. Chen, 2003: An assessment of the cod stock in NAFO Divisions 2J+3KL in February 2003. 2003/023 Department of Fisheries and Oceans Stock Assessment Secretariat, 157 pp.

Lindgren, E. and R. Gustafson, 2001: Tick-borne encephalitis in Sweden and climate change. *Lancet*, **358**, 16-18.

Lloyd, A.H., T.S. Rupp, C.L. Fastie and A.M. Starfield, 2003: Patterns and dynamics of treeline advance on the Seward Peninsula, Alaska. *J. Geophys. Res.*, **108**, 8161, doi:10.1029/2001JD000852.

Loeng, H., K. Brander, E. Carmack, S. Denisenko, K. Drinkwater, B. Hansen, K. Kovacs, P. Livingston, F. McLaughlin and E. Sakshaug, 2005: Marine systems. *Arctic Climate Impact Assessment, ACIA*, C. Symon, L. Arris and B. Heal, Eds., Cambridge University Press, Cambridge, 453-538.

Lonergan, S., R. Difrancesco and M.-K. Woo, 1993: Climate change and transportation in northern Canada: an integrated impact assessment. *Climatic Change*, **24**, 331-351.

Luckman, A., T. Murray, R. de Lange and E. Hanna, 2006: Rapid and synchronous ice-dynamic changes in East Greenland. *Geophys. Res. Lett.*, **33**, L03503, doi:10.1029/2005GL025428.

MacDonald, G.M., A.A. Velichko, C.V. Kremenetski, C.K. Borisova, A.A. Goleva, A.A. Andreev, L.C. Cwynar, R.T. Riding and Co-authors, 2000: Holocene tree line history and climate change across northern Eurasia. *Quaternary Res.*, **53**, 302-311.

MacDonald, R., T. Harner, J. Fyfe, H. Loeng and T. Weingartner, 2003: Influence of global change on contaminant pathways to, within and from the Arctic. Arctic Monitoring and Assessment Programme, AMAP Assessment 2002, Oslo, 65 pp.

Malmer, N., T. Johansson, M. Olsrud and T.R. Christensen, 2005: Vegetation, climatic changes and net carbon sequestration. *Glob. Change Biol.*, **11**, 1895-1909.

Marsh, P. and L.F.W. Lesack, 1996: The hydrologic regime of perched lakes in the Mackenzie Delta: potential responses to climate change. *Limnol. Oceanogr.*, **41**, 849-856.

Marsh, P., L.F.W. Lesack and A. Roberts, 1999: Lake sedimentation in the Mackenzie Delta, NWT. *Hydrol. Process.*, **13**, 2519-2536.

Marshall, G.J., P.A. Stott, J. Turner, W.M. Connolley, J.C. King and T.A. Lachlan-Cope, 2004: Causes of exceptional atmospheric circulation changes in the Southern Hemisphere. *Geophys. Res. Lett.*, **31**, L14205, doi:10.1029/2004 GL019952.

Marshall, G.J., A. Orr, N.P.M. van Lipzig and J.C. King, 2006: The impact of a changing southern hemisphere annular mode on Antarctic peninsula summer temperatures. *J. Climate*, **19**, 5388-5404.

Martin, D., D. Belanger, P. Gosselin, J. Brazeau, C. Furgal and S. Dery, 2005: Climate change, drinking water, and human health in Nunavik: adaptation strategies. Final Report submitted to the Canadian Climate Change Action Fund, Natural Resources Canada, Quebec, 111 pp.

Maslin, M., 2004: *Gas Hydrates: A Hazard for the 21st Century*. Issues in Risk Science 3, Benfield Hazard Research Centre, London, 22 pp.

Matveyeva, N.V. and Y. Chernov, 2000: *Zonation in Plant Cover of the Arctic*. Russian Academy of Sciences, Moscow, 219 pp.

McBean, G., G. Alekseev, D. Chen, E. Førland, J. Fyfe, P.Y. Groisman, R. King, H. Melling, R. Vose and P.H. Whitfield, 2005: Arctic climate: past and present. *Arctic Climate Impacts Assessment (ACIA)*, C. Symon, L. Arris and B. Heal, Eds., Cambridge University Press, Cambridge, 21-60.

McCarthy, J.J., M. Long Martello, R. Corell, N. Eckley Selin, S. Fox, G. Hovelsrud-Broda, S.D. Mathiesen, C. Polsky, H. Selin and N.J.C. Tyler, 2005: Climate change in the context of multiple stressors and resilience. *Arctic Climate Impact Assessment (ACIA)*, C. Symon, L. Arris and B. Heal, Eds., Cambridge University Press, Cambridge, 945-988.

McClelland, J.W., R.M. Holmes and B.J. Peterson, 2004: Increasing river discharge in the Eurasian Arctic: consideration of dams, permafrost thaw, and fires as potential agents of change. *J. Geophys. Res.*, **109**, D18102, doi:10.1029/2004JD004583.

McDonald, K.C., J.S. Kimball, E. Njoku, R. Zimmermann and M. Zhao, 2004: Variability in springtime thaw in the terrestrial high latitudes: monitoring a major control on the biospheric assimilation of atmospheric CO_2 with spaceborne microwave remote sensing. *Earth Interactions*, **8**, 1-23.

McGuire, A.D., J.S. Clein, J.M. Melillo, D.W. Kicklighter, R.A. Meier, C.J. Vorosmarty and M.C. Serreze, 2000: Modeling carbon responses of tundra ecosystems to historical and projected climate: sensitivity of pan-Arctic carbon storage to temporal and spatial variation in climate. *Glob. Change Biol.*, **6**, S141-S159.

McGuire, A.D., F.S. Chapin III, C. Wirth, M. Apps, J. Bhatti, T.V. Callaghan, T.R. Christensen, J.S. Clein, M. Fukuda, T. Maximov, A. Onuchin, A. Shvidenko and E. Vaganov, 2007: Responses of high latitude ecosystems to global change: potential consequences for the climate system. *Terrestrial Ecosystems in a Changing World*, J.G. Canadell, D.E. Pataki and L.F. Pitelka, Eds., Springer, London, 297-310.

Meehl, G.H., T.F. Stocker, W.D. Collins, P. Friedlingstein, A.T. Gaye, J.M. Gregory, A. Kito, R. Knutti, J.M. Murphy, A. Noda, S.C.B. Raper, I.G. Watterson, A.J. Weaver and Z.-C. Zhao, 2007: Global climate projections. *Climate Change 2007: The Physical Science Basis. Contribution of Working Group I to the Fourth Assessment Report of the Intergovernmental Panel on Climate Change*, S. Solomon, D. Qin, M. Manning, Z. Chen, M. Marquis, K.B. Averyt, M. Tignor and H.L. Miller, Eds., Cambridge University Press, Cambridge, 747-846.

Mehlum, F., 1999: Adaptation in arctic organisms to a short summer season. *The Ecology of the Tundra in Svalbard*, S.A. Bengtsson, F. Mehlum and T. Severinsen, Eds., Norsk Polarinstitutt Meddeleser, 161-169.

Meredith, M.P. and J.C. King, 2005: Rapid climate change in the ocean west of the Antarctic peninsula during the second half of the 20th century. *Geophys. Res. Lett.*, **32**, L19604, doi:10.1029/2005GL024042.

Messner, T., 2005: Environmental variables and the risk of disease. *International Journal for Circumpolar Health*, **64**, 523-533.

Miller, J.R. and G.L. Russell, 2000: Projected impact of climate change on the freshwater and salt budgets of the Arctic Ocean by a global climate model. *Geophys. Res. Lett.*, **27**, 1183-1186.

Mokhov, I.I., V.A. Semenov and V.C. Khone, 2003: Estimates of possible regional hydrologic regime changes in the 21st century based on global climate models. *Izvestiya Rossijskoi Academii Nauk (Proceedings of the Russian Academy of Sciences)*, **39**, 130-144.

Morrison, R.I.G., Y. Aubry, R.W. Butler, G.W. Beyresbergen, G.M. Donaldson, C.L. Gratto-Trevor, P.W. Hicklin, V.H. Johnston and R.K. Ross, 2001: Declines in North American shorebird populations. *Wader Study Group Bulletin*, **94**, 34-38.

Morrison, R.I.G., K.R. Ross and L.J. Niles, 2004: Declines in wintering populations of red knots in southern South America. *Condor*, **106**, 60-70.

Murphy, E.J., A. Clarke, C. Symon and J. Priddle, 1995: Temporal variation in Antarctic sea-ice: analysis of a long term fast ice record from the South Orkney Islands. *Deep-Sea. Res.*, **42**, 1045-1062.

Nakićenović , N., J. Alcamo, G. Davis, B. de Vries, J. Fenhann, S. Gaffin, K. Gregory, A. Grübler, T.Y. Jung, T. Kram, E. Lebre la Rovere, L. Michaelis, S. Mori, T. Morita, W. Pepper, H. Pitcher, L. Price, K. Riahi, A. Roehrl, H.H. Rogner, A. Sankovski, M. Schlesinger, P. Shukla, S. Smith, R. Swart, S. van Rooijen, N. Victor and Z. Dadi, Eds., 2000: *Emissions Scenarios: A Special Report of the Intergovernmental Panel on Climate Change*. Cambridge University Press, Cambridge, 599 pp.

Nayha, S., 2005: Environmental temperature and mortality. *International Journal of Circumpolar Health*, **64**, 451-458.

Nellemann, C., L. Kullerud, I. Vistnes, B.C. Forbes, E. Husby, G.P. Kofinas, B.P. Kaltenborn, J. Rouaud and Co-authors, 2001: GLOBIO: global methodology for mapping human impacts on the biosphere. UNEP/DEWA/TR.01-3, United Nations Environment Programme, 47 pp.

Nelson, F.E., 2003: (Un)frozen in time. *Science*, **299**, 1673-1675.

Nelson, F.E., O.A. Anisimov and N.I. Shiklomanov, 2001: Subsidence risk from thawing permafrost. *Nature*, **410**, 889-890.

Nemani, R.R., C.D. Keeling, H. Hashimoto, W.M. Jolly, S.C. Piper, C.J. Tucker, R.B. Myneni and S.W. Running, 2003: Climate-driven increases in global terrestrial net primary production from 1982 to 1999. *Science*, **300**, 1560-1563.

Nickels, S., C. Furgal, J. Castelden, P. Moss-Davies, M. Buell, B. Armstrong, D. Dillon and R. Fonger, 2002: Putting the human face on climate change through community workshops. *The Earth is Faster Now: Indigenous Observations of Arctic Environmental Change*, I. Krupnik and D. Jolly, Eds., ARCUS, Washington, District of Columbia, 300-344.

Nikiforov, S.L., N.N. Dunaev, S.A. Ogorodov and A.B. Artemyev, 2003: Physical geographic characteristics. *The Pechora Sea: Integrated Research*. MOPE, Moscow, 502 pp. (In Russian).

Nuttall, M., 2000a: *The Arctic is Changing*. Stefansson Arctic Institute, Akureyri, 11 pp. http://www.thearctic.is/PDF/The%20Arctic%20is%20changing.pdf.

Nuttall, M., 2000b: Indigenous peoples, self-determination, and the Arctic environment. *The Arctic: Environment, People, Policy*, M. Nuttall and T.V. Callaghan, Eds., Harwood Academic, The Netherlands, 377-409.

Nuttall, M., F. Berkes, B. Forbes, G. Kofinas, T. Vlassova and G. Wenzel, 2005: Hunting, herding, fishing and gathering: indigenous peoples and renewable resource use in the Arctic. *Arctic Climate Impacts Assessment, ACIA*, C. Symon, L. Arris and B. Heal, Eds., Cambridge University Press, Cambridge, 649-690.

Oechel, W.C., G.L. Vourlitis, S.J. Hastings, R.C. Zulueta, L. Hinzman and D. Kane, 2000: Acclimation of ecosystem CO_2 exchange in the Alaskan Arctic in response to decadal climate warming. *Nature*, **406**, 978-981.

Ogorodov, S.A., 2003: Coastal dynamics in the Pechora Sea under technogenic impact. *Berichte zur Polar- und Meeresforschung*, **443**, 74-80.

Orr, J.C., V.J. Fabry, O. Aumont, L. Bopp, S.C. Doney, R.A. Feely, A. Gnanadesikan, N. Gruber and Co-authors, 2005: Anthropogenic ocean acidification over the twenty-first century and its impact on calcifying organisms. *Nature*, **437**, 681-686.

Paine, R., 1994: *Herds of the Tundra: A Portrait of Saami Reindeer Pastoralism*. Smithsonian Institution Press, Washington, District of Columbia, 242 pp.

Parkinson, A. and J. Butler, 2005: Impact of climate change on infectious diseases in the Arctic. *International Journal for Circumpolar Health*, **64**, 478.

Parkinson, C.L., 2002: Trends in the length of the Southern Ocean sea ice season, 1979-1999. *Ann. Glaciol.*, **34**, 435-440.

Parkinson, C.L., 2004: Southern Ocean sea ice and its wider linkages: insights revealed from models and observations. *Antarct. Sci.*, **16**, 387-400.

Pavlov, A.V. and N.G. Moskalenko, 2002: The thermal regime of soils in the north of western Siberia. *Permafrost Periglac.*, **13**, 43-51.

Pavlov, P., J.I. Svendsen and S. Indrelid, 2001: Human presence in the European Arctic nearly 40,000 years ago. *Nature*, **413**, 64-67.

Payette, S., M. Eronen and J.J.P. Jasinski, 2002: The circumpolar tundra-taiga interface: late Pleistocene and Holocene changes. *Ambio*, **12**, 15-22.

Payne, A.J., A. Vieli, A. Shepherd, D.J. Wingham and E. Rignot, 2004: Recent dramatic thinning of largest West Antarctic ice stream triggered by oceans. *Geophys. Res. Lett.*, **31**, L23401, doi:10.1029/2004GL021284.

Peck, L.S., 2005: Prospects for surviving climate change in Antarctic aquatic species. *Frontiers in Zoology*, **2**, doi:10.1186/1742-9994-2-9.

Peck, L.S., P. Convey and D.K.A. Barnes, 2006: Environmental constraints on life histories in Antarctic ecosystems: tempos, timings and predictability. *Biol. Rev.*, **81**, 75-109.

Peters, D.L., T.D. Prowse, A. Pietroniro and R. Leconte, 2006: Establishing the flood hydrology of the Peace–Athabasca Delta, northern Canada. *Hydrol. Process.*, **20**, 4073-4096.

Peterson, B.J., R.M. Holmes, J.W. McClelland, C.J. Vorosmarty, R.B. Lammers, A.I. Shiklomanov, I.A. Shiklomanov and S. Rahmstorf, 2002: Increasing river discharge to the Arctic Ocean. *Science*, **298**, 2172-2173.

Peterson, B.J., J. McClelland, R. Curry, R.M. Holmes, J.E. Walsh and K. Aagaard, 2006: Trajectory shifts in the arctic and subarctic freshwater cycle. *Science*, **313**, 1061-1066.

Pienitz, R., M.S.V. Douglas and J.P. Smol, 2004: *Long-Term Environmental Change in Arctic and Antarctic Lakes*. Vol. 8. Springer Verlag, 562 pp.

Polyakov, I.V., G.V. Alekseev, R.V. Bekryaev, U. Bhatt, R.L. Colony, M.A. Johnson, V.P. Karklin, A.P. Makshtas, D. Walsh and A.V. Yulin, 2002: Observationally based assessment of polar amplification of global warming. *Geophys. Res. Lett.*, **29**, 1878. doi:10.1029/2001GL011111.

Prowse, T.D. and S. Beltaos, 2002: Climatic control of river-ice hydrology: a review. *Hydrol. Process.*, **16**, 805-822.

Prowse, T.D., D.L. Peters, S. Beltaos, A. Pietroniro, L. Romolo, J. Töyrä and R. Leconte, 2002: Restoring ice-jam floodwater to a drying delta ecosystem. *Water Int.*, **27**, 58-69.

Prowse, T.D., F.J. Wrona and G. Power, 2004: Dams, reservoirs and flow regulation.

Threats to Water Availability in Canada. National Water Resource Institute, Scientific Assessment Report No. 3, Environment Canada, Ottawa, 9-18.

Prowse, T.D., F.J. Wrona, J. Reist, J.J. Gibson, J.E. Hobbie, L.M.J. Levesque and W.F. Vincent, 2006a: Climate change effects on hydroecology of arctic freshwater ecosystems. *Ambio*, **35**, 347-358.

Prowse, T.D., F.J. Wrona, J.D. Reist, J.J. Gibson, J.E. Hobbie, L. Levesque and W.F. Vincent, 2006b: Historical changes in arctic freshwater ecosystems. *Ambio*, **35**, 339-346.

Pudsey, C.J. and J. Evans, 2001: First survey of Antarctic sub-ice shelf sediments reveals mid-Holocene ice shelf retreat. *Geology*, **29**, 787-790.

Quayle, W.C., L.S. Peck, H. Peat, J.C. Ellis-Evans and P.R. Harrigan, 2002: Extreme responses to climate change in Antarctic lakes. *Science*, **295**, 645-645.

Quayle, W.C., P. Convey, L.S. Peck, J.C. Ellis-Evans, H.G. Butler and H.J. Peat, 2003: Ecological responses of maritime Antarctic lakes to regional climate change. *Antarctic Peninsula Climate Variability: Historical and Palaeoenvironmental Perspectives*, E. Domack, A. Leventer, A. Burnett, R. Bindschadler, P. Convey and M. Kirby, Eds., American Geophysical Union, Washington, District of Columbia, 159-170.

Rachold, V., M.N. Grigoriev, F.E. Aré, S. Solomon, E. Reimnitz, H. Kassens and M. Antonov, 2000: Coastal erosion vs. riverine sediment discharge in the Arctic shelf seas. *Geologisches Rundschau* (*Int. J. Earth Sci.*), **89**, 450-460.

Rasumov, S.O., 2001: Thermoerosion modelling of ice-rich Arctic coast in stationary climatic conditions (in Russian). *Kriosfera Zemli*, **5**, 50-58.

Reist, J., F.J. Wrona, T.D. Prowse, M. Power, J.B. Dempson, R. Beamish, J.R. King, T.J. Carmichael and C.D. Sawatzky, 2006a: General effects of climate change effects on arctic fishes and fish populations. *Ambio*, **35**, 370-380.

Reist, J.D., F.J. Wrona, T.D. Prowse, J.B. Dempson, M. Power, G. Koeck, T.J. Carmichael, C.D. Sawatzky, H. Lehtonen and R.F. Tallman, 2006b: Effects of climate change and UV radiation on fisheries for arctic freshwater and anadromous species. *Ambio*, **35**, 402-410.

Reist, J.D., F.J. Wrona, T.D. Prowse, M. Power, J.B. Dempson, J.R. King and R.J. Beamish, 2006c: An overview of effects of climate change on arctic freshwater and anadromous fisheries. *Ambio*, **35**, 381-387.

Richardson, A.J. and D.S. Schoeman, 2004: Climate impact on ecosystems in the northeast Atlantic. *Science*, **305**, 1609-1612.

Riebesell, U., I. Zondervan, B. Rost, P.D. Tortell, R.E. Zeebe and F.M. Morel, 2000: Reduced calcification of marine plankton in response to increased atmospheric CO_2. *Nature*, **407**, 364-367.

Riedlinger, D. and F. Berkes, 2001: Contributions of traditional knowledge to understanding climate change in the Canadian Arctic. *Polar Rec.*, **37**, 315-328.

Rignot, E. and P. Kanagaratnam, 2006: Changes in the velocity structure of the Greenland ice sheet. *Science*, **311**, 986-990.

Rignot, E., G. Casassa, S. Gogineni, P. Kanagaratnam, W. Krabill, H. Pritchard, A. Rivera, R. Thomas and D. Vaughan, 2005: Recent ice loss from the Fleming and other glaciers,Wordie Bay, West Antarctic Peninsula. *Geophys. Res. Lett.*, **32**, L07502, doi:10.1029/2004GL021947.

Rinnan, R., M.M. Keinänen, A. Kasurinen, J. Asikainen, T.K. Kekki, T. Holopainen, H. Ro-Poulsen, T.N. Mikkelsen and Co-authors, 2005: Ambient ultraviolet radiation in the Arctic reduces root biomass and alters microbial community composition but has no effects on microbial biomass. *Glob. Change Biol.*, **11**, 564-574.

Robinson, S.A., J. Wasley and A.K. Tobin, 2003: Living on the edge: plants and global change in continental and maritime Antarctica. *Glob. Change Biol.*, **9**, 1681-1717.

Romanovsky, V.E., M. Burgess, S. Smith, K.Yoshikawa and J. Brown, 2002: Permafrost temperature records: indicators of climate change. *EOS Transactions*, **83**, 589-594.

Rose, G.A., 2004: Reconciling overfishing and climate change with stock dynamics of Atlantic cod (*Gadus morhua*) over 500 years. *Can. J. Fish. Aquat. Sci.*, **61**, 1553-1557.

Rose, G.A. and R.L. O'Driscoll, 2002: Capelin are good for cod: can Newfoundland cod stocks recover without capelin? *ICES J. Mar. Sci.*, **5938**, 1026.

Rose, G.A., B. de Young, D.W. Kulka, S.V. Goddard and G.L. Fletcher, 2000: Distribution shifts and overfishing the northern cod (*Gadus morhua*): a view from the ocean. *Can. J. Fish. Aquat. Sci.*, **57**, 644-664.

Roura, R., 2004: Monitoring and remediation of hydrocarbon contamination at the former site of Greenpeace's World Park Base, Cape Evans, Ross Island, Antarctica. *Polar Res.*, **40**, 51-67.

Royal Society, 2005: Ocean acidification due to increasing atmospheric carbon dioxide. Policy Document 12/05, The Royal Society, London, 60 pp.

Ruhland, K.M. and J.P. Smol, 2002: Freshwater diatoms from the Canadian arctic treeline and development of paleolimnological inference models. *J. Phycol.*, **38**, 429-264.

Ruhland, K.M., A. Priesnitz and J.P. Smol, 2003: Paleolimnological evidence from diatoms for recent environmental changes in 50 lakes across Canadian Arctic treeline. *Arct. Antarct. Alp. Res.*, **35**, 110-123.

Russell, D.E., G. Kofinas and B. Griffith, 2002: Barren-ground caribou calving ground workshop. Canadian Wildlife Services, Ottawa, 47 pp.

Sakshaug, E., 2003: Primary and secondary production in Arctic seas. *The Organic Carbon Cycle in the Arctic Ocean*, R. Stein and R.W. Macdonald, Eds., Springer, Berlin, 57-81.

Sakshaug, E., A. Bjorge, B. Gulliksen, H. Loeng and F. Mehlum, 1994: Structure, biomass distribution and energetics of the pelagic ecosystem in the Barents Sea: a synopsis. *Polar Biol.*, **14**, 405-411.

Sazonova, T.S., V.E. Romanovsky, J.E. Walsh and D.O. Sergueev, 2004: Permafrost dynamics in the 20th and 21st centuries along the East Siberian transect. *J. Geophys. Res.–Atmos.*, **109**, D01108, doi:10.1029/2003JD003680.

Scambos, T.A., J.A. Bohlander, C.A. Shuman and P. Skvarca, 2004: Glacier acceleration and thinning after ice shelf collapse in the Larsen B embayment, Antarctica. *Geophys. Res. Lett.*, **31**, L18402, doi:10.1029/2004GL020670.

Schopka, S.A., 1994: Fluctuations in the cod stock off Iceland during the twentieth century in relation to changes in the fisheries and environment. *ICES Mar. Sci. Symp.*, **198**, 175-193.

Seebacher, F., W. Davison, C.J. Lowe and C.E. Franklin, 2005: The falsification of the thermal specialization paradigm: compensation for elevated temperatures in Antarctic fishes. *Biol. Lett.*, **1**, 151-154.

Serreze, M.C. and J.A. Francis, 2006: The Arctic amplification debate. *Climatic Change*, **76**, 241-264.

Serreze, M.C., D.H. Bromwich, M.P. Clark, A.J. Etringer, T. Zhang and R. Lammers, 2003: Large-scale hydro-climatology of the terrestrial Arctic drainage system. *J. Geophys. Res.*, **108**, 8160, doi:10.1029/2001JD000919.

Shelton, P.A., A.F. Sinclair, G.A. Chouinard, R. Mohn and D.E. Duplisea, 2006: Fishing under low productivity conditions is further delaying recovery of Northwest Atlantic cod (*Gadus morhua*).*Can. J. Fish. Aquat. Sci.*, **63**, 235-238.

Shepherd, A., D.J. Wingham and E. Rignot, 2004: Warm ocean is eroding West Antarctic Ice Sheet. *Geophys. Res. Lett.*, **31**, L23402, doi:10.1029\2004G L021106.

Shiklomanov, I.A., A.I. Shiklomanov, R.B. Lammers, B.J. Peterson and C.J. Vorosmarty, 2000: The dynamics of river water inflow to the Arctic Ocean. *The Freshwater Budget of the Arctic Ocean*, E.L. Lewis, E.P. Jones, T.D. Prowse and P. Wadhams, Eds., Kluwer Academic, Dordrecht, 281-296.

Sirotenko, O.D., H.V. Abashina and V.N. Pavlova, 1997: Sensitivity of the Russian agriculture to changes in climate, CO_2 and tropospheric ozone concentrations and soil fertility. *Climatic Change*, **36**, 217-232.

Sitch, S., B. Smith, I.C. Prentice, A. Arneth, A. Bondeau, W. Cramer, J.O. Kaplan, S. Levis, W. Lucht, M.T. Sykes, K. Thonicke and S. Venevsky, 2003: Evaluation of ecosystem dynamics, plant geography and terrestrial carbon cycling in the LPJ dynamic global vegetation model. *Glob. Change Biol.*, **9**, 161-185.

Sitch, S., A.D. McGuire, J. Kimball, N. Gedney, J. Gamon, R. Engstrom, A. Wolf, Q. Zhuang, J. Clein and K.C. McDonald, 2007: Assessing the carbon balance of circumpolar arctic tundra using remote sensing and process modeling. *Ecol. Appl.*, **17**, 213-234.

Slayback, D.A., J.E. Pinzon, S.O. Los and C.J. Tucker, 2003: Northern hemisphere photosynthetic trends 1982–99. *Glob. Change Biol.*, **9**, 1-15.

Sloan, E.D., Jr, 2003: Fundamental principles and applications of natural gas hydrates. *Nature*, **426**, 353-359.

Smith, L.C., 2000: Time-trends in Russian Arctic river ice formation and breakup: 1917–1994. *Phys. Geogr.*, **21**, 46-56.

Smith, L.C., Y. Sheng, G.M. MacDonald and L.D. Hinzman, 2005: Disappearing Arctic lakes. *Science*, **308**, 1429.

Smith, N.V., S.S. Saatchi and J.T. Randerson, 2004: Trends in high northern latitude soil freeze and thaw cycles from 1988 to 2002. *J. Geophys. Res.*, **109**, D12101, doi:10.1029/2003JD004472.

Smith, O.P., 2002: Coastal erosion in Alaska. *Berichte zur Polar- und Meeresforschung*, **413**, 65-68.

Smith, R.C., W.R. Fraser, S.E. Stammerjohn and M. Vernet, 2003: Palmer long-term ecological research on the Antarctic marine ecosystem. *Antarctic Peninsula Climate Variability: Historical and Paleoenvironmental Perspectives*, E.

Domack, A. Leventer, A. Burnett, R. Bindschadler, P. Convey and M. Kirby, Eds., Antarctic Research Series 79, AGU, Washington, District of Columbia, 131-144.

Smith, S.L. and M.M. Burgess, 2004: *Sensitivity of Permafrost to Climate Warming in Canada*. Geological Survey of Canada Bulletin No. 579, 24 pp.

Smith, S.L., M.M. Burgess and J.A. Heginbottom, 2001: Permafrost in Canada, a challenge to northern development. *A Synthesis of Geological Hazards in Canada*, G.R. Brooks, Ed., Geological Survey of Canada Bulletin 548, 241-264.

Smol, J.P., A.P. Wolfe, H.J.B. Birks, M.S.V. Douglas, V.J. Jones, A. Korhola, R. Pienitz, K. Ruhland and Co-authors, 2005: Climate-driven regime shifts in the biological communities of arctic lakes. *P. Natl Acad. Sci. USA*, **102**, 4397-4402.

Snape, I., M.J. Riddle, D.M. Filler and P.J. Williams, 2003: Contaminants in freezing ground and associated ecosystems: key issues at the beginning of the new millennium. *Polar Rec.*, **39**, 291-300.

Solomon, S.M., 2005: Spatial and temporal variability of shoreline change in the Beaufort-Mackenzie region, Northwest Territories, Canada. *Geo-Marine Lett.*, **25**, 127-137.

Stieglitz, M., S.J. Dery, V.E. Romanovsky and T.E. Osterkamp, 2003: The role of snow cover in the warming of arctic permafrost. *Geophys. Res. Lett.*, **30**, 1721, doi:10.1029/2003GL017337.

Stirling, I. and C.L. Parkinson, 2006: Possible effects of climate warming on selected populations of polar bears (*Ursus maritimus*) in the Canadian Arctic. *Arctic*, **59**, 262-275.

Stirling, I., N.J. Lunn and J. Iacozza, 1999: Long-term trends in the population ecology of polar bears in western Hudson Bay in relation to climate change. *Arctic*, **52**, 294-306.

Stone, J.O., G.A. Balco, D.E. Sugden, M.W. Caffee, L.C. Sass III, S.G. Cowdery and C. Siddoway, 2003: Holocene deglaciation of Marie Byrd Land, West Antarctica. *Science*, **299**, 99-102.

Strathdee, A.T., J.S. Bale, W.C. Block, S.J. Coulson, I.D. Hodkinson and N. Webb, 1993: Effects of temperature on a field population of *Acyrthosiphon svalbardicum* (Hemiptera: Aphidae) on Spitsbergen. *Oecologia*, **96**, 457-465.

Stroud, D.A., N.C. Davidson, R. West, D.A. Scott, L. Haanstra, O. Thorup, B. Ganter and S. Delany, 2004: Status of migratory wader population in Africa and Western Eurasia in the 1990s. *International Wader Studies*, **15**, 1-259.

Sturm, M., C. Racine and K. Tape, 2001: Increasing shrub abundance in the Arctic. *Nature*, **411**, 546-547.

Sturm, M., T. Douglas, C. Racine and G.E. Liston, 2005: Changing snow and shrub conditions affect albedo with global implications. *J. Geophys. Res.*, **110**, G01004, doi:10.1029/2005JG000013.

Syvitski, J.P.M., 2002: Sediment discharge variability in Arctic rivers: implications for a warmer future. *Polar Res.*, **21**, 323-330.

Tannerfeldt, M., B. Elmhagen and A. Angerbjörn, 2002: Exclusion by interference competition? The relationship between red and arctic foxes. *Oecologia*, **132**, 213-220.

Tape, K., M. Sturm and C. Racine, 2006: The evidence for shrub expansion in Northern Alaska and the Pan-Arctic. *Glob. Change Biol.*, **12**, 686-702.

Thomas, G. and P.R. Rowntree, 1992: The boreal forests and climate. *Q. J. Roy. Meteor. Soc.*, **118**, 469-497.

Thompson, D.W.J. and S. Solomon, 2002: Interpretation of recent southern hemisphere climate change. *Science*, **296**, 895-899.

Thorpe, N., N. Hakongak, S. Eyegetok and the Kitikmeot Elders, 2001: *Thunder on the Tundra: Inuit Qaujimajatuqangit of the Bathurst Caribou*. Generation Printing, Vancouver, 220 pp.

Trenberth, K.E., P.D. Jones, P.G. Ambenje, R. Bojariu, D.R. Easterling, A.M.G. Klein Tank, D.E. Parker, J.A. Renwick and Co-authors, 2007: Observations: surface and atmospheric climate change. *Climate Change 2007: The Physical Science Basis. Contribution of Working Group I to the Fourth Assessment Report of the Intergovernmental Panel on Climate Change*, S. Solomon, D. Qin, M. Manning, Z. Chen, M. Marquis, K.B. Averyt, M. Tignor and H.L. Miller, Eds., Cambridge University Press, Cambridge, 235-336.

Truong, G., A.E. Palmé and F. Felber, 2007: Recent invasion of the mountain birch *Betula pubescens* ssp. *tortuosa* above the treeline due to climate change: genetic and ecological study in northern Sweden. *J. Evol. Biol.*, **20**, 369-380.

Tuck, G.N., T. Polacheck, J.P. Croxall and H. Weimerskirch, 2001: Modelling the impact of fishery by-catches on albatross populations. *J. Appl. Ecol.*, **38**, 1182-1196.

Turchin, P. and G.O. Batzli, 2001: Availability of food and population dynamics of arvicoline rodents. *Ecology*, **82**, 1521-1534.

Turner, J., S.R. Colwell, G.J. Marshall, T.A. Lachlan-Cope, A.M. Carleton, P.D. Jones, V. Lagun, P.A. Reid and S. Iagovkina, 2005: Antarctic climate change during the last 50 years. *Int. J. Climatol.*, **25**, 279-294.

Turner, J., J.E. Overland and J.E. Walsh, 2007: An Arctic and Antarctic perspective on recent climate change. *Int. J. Climatol.*, **27**, 277-293.

Tutubalina, O.V. and W.G. Rees, 2001: Vegetation degradation in a permafrost region as seen from space: Noril'sk (1961–1999). *Cold Reg. Sci. Technol.*, **32**, 191-203.

United States Environmental Protection Agency, 1997: *Need for American Indian and Alaska Native Water Systems*. U.S. Environmental Protection Agency, Washington, District of Columbia.

U.S. Arctic Research Commission Permafrost Task Force, 2003: Permafrost, and impacts on civil infrastructure. 01-03 Arctic Research Commission, Arlington, Virginia, 62 pp.

U.S. Geological Survey World Energy Assessment Team, 2000: *Geological Survey World Petroleum Assessment 2000: Description and Results*. U.S. Geological Survey Digital Data Series DDS-60, multidisc set, version 1.1, USGS.

Usher, M.B., T.V. Callaghan, G. Gilchrist, B. Heal, G.P. Juday, H. Loeng, M.A.K. Muir and P. Prestrud, 2005: Principles of conserving the Arctic's biodiversity. *Arctic Climate Impact Assessment*, C. Symon, L. Arris and B. Heal, Eds., Cambridge University Press, Cambridge, 540-591.

van den Broeke, M. and N.P.M. van Lipzig, 2003: Response of wintertime Antarctic temperatures to the Antarctic Oscillation: results from a regional climate model. *Antarctic Peninsula Climate Variability: Historical and Paleoenvironmental Perspectives*, E. Domack, A. Leventer, A. Burnett, R. Bindschadler, P. Convey and M. Kirby, Eds., Antarctic Research Series 79, AGU, Washington, District of Columbia, 43-58.

Van Oostdam, J., S. Donaldson, M. Feeley, N. Tremblay, D. Arnold, P. Ayotte, G. Bondy, L. Chan and Co-authors, 2003: Toxic substances in the Arctic and associated effects: human health. *Canadian Arctic Contaminants Assessment Report II*. Indian and Northern Affairs Canada, Ottawa, 82 pp.

Van Wijk, M.T., K.E. Clemmensen, G.R. Shaver, M. Williams, T.V. Callaghan, F.S. Chapin III, J.H.C. Cornelissen, L. Gough and Co-authors, 2003: Long term ecosystem level experiments at Toolik Lake, Alaska and at Abisko, northern Sweden: generalisations and differences in ecosystem and plant type responses to global change. *Glob. Change Biol.*, **10**, 105-123.

Vandenberghe, J., 2001: A typology of Pleistocene cold-based rivers. *Quatern. Int.*, **79**, 111-121.

Vandenberghe, J., 2002: The relation between climate and river processes, landforms and deposits during the Quaternary. *Quatern. Int.*, **91**, 17-23.

Vandenberghe, J. and M. Huisink, 2003: High-latitude fluvial morphology: the example of the Usa river, northern Russia. *Paleohydrology: Understanding Global Change*, K.J. Gregory and G. Benito, Eds., Wiley, Chichester, 49-58.

Vaughan, D.G., 2006: Recent trends in melting conditions on the Antarctic Peninsula and their implications for ice-sheet mass balance. *Arct. Antarct. Alp. Res.*, **38**, 147-152.

Vaughan, D.G., 2007: West Antarctic Ice Sheet collapse: the fall and rise of a paradigm. *Climatic Change*, in press.

Vaughan, D.G., G.J. Marshall, W.M. Connolley, C.L. Parkinson, R. Mulvaney, D.A. Hodgson, J.C. King, C.J. Pudsey and J. Turner, 2003: Recent rapid regional climate warming on the Antarctic Peninsula. *Climatic Change*, **60**, 243-274.

Velicogna, I. and J. Wahr, 2006: Measurements of time-variable gravity show mass loss in Antarctica. *Science*, **311**, 1754-1756.

Vilhjálmsson, H., 1997: Climatic variations and some examples of their effects on the marine ecology of Icelandic and Greenland waters, in particular during the present century. *Rit Fiskideildar*, **15**, 7-29.

Vilhjálmsson, H., A. Håkon Hoel, S. Agnarsson, R. Arnason, J.E. Carscadden, A. Eide, D. Fluharty, G. Hønneland and Co-authors, 2005: Fisheries and aquaculture. *Arctic Climate Impact Assessment, ACIA*, C. Symon, L. Arris and B. Heal, Eds., Cambridge University Press, Cambridge, 691-780.

Vlassova, T.K., 2002: Human impacts on the tundra–taiga zone dynamics: the case of the Russian lesotundra. *Ambio*, **12**, 30-36.

Walker, M., H. Wahren, L. Ahlquist, J. Alatolo, S. Bret-Harte, M. Calef, T.V. Callaghan, A. Carroll and Co-authors, 2006: Plant community response to experimental warming across the tundra biome. *P. Natl Acad. Sci. USA*, **103**, 1342-1346.

Walsh, J.E., O. Anisimov, J.O.M. Hagen, T. Jakobsson, J. Oerlemans, T.D. Prowse, V. Romanovsky, N. Savelieva, M. Serreze, I. Shiklomanov and S. Solomon, 2005: Cryosphere and hydrology. *Arctic Climate Impacts Assessment, ACIA*, C. Symon, L. Arris and B. Heal, Eds., Cambridge University Press, Cambridge, 183-242.

Walsh, J.J., D.A. Dieterle and J. Lenes, 2001: A numerical analysis of carbon dynamics of the Southern Ocean phytoplankton community: the roles of light and grazing in effecting both sequestration of atmospheric CO_2 and food availability to larval krill. *Deep-Sea Res. Pt. I*, **48**, 1-48.

Walter, K.M., S.A. Zimov, J.P. Chanton, D. Verbyla and F.S. Chapin, 2006: Methane bubbling from Siberian thaw lakes as a positive feedback to climate warming. *Nature*, **443**, 71-75.

Walters, C. and J.-J. Maguire, 1996: Lessons for stock assessment from the northern cod collapse. *Rev. Fish Biol. Fisher*, **6**, 125-137.

Walther, G.R., E. Post, P. Convey, A. Menzel, C. Parmesan, T.J.C. Beebee, J.M. Fromentin, O. Hoegh-Guldberg and F. Bairlein, 2002: Ecological responses to recent climate change. *Nature*, **416**, 389-395.

Warren, J., J. Berner and J. Curtis, 2005: Climate change and human health: infrastructure impacts to small remote communities in the North. *International Journal of Circumpolar Health*, **64**, 498.

Watt-Cloutier, S., 2004: Presentation to the Senate Committee on Commerce, Science and Transportation. September 15, 2004. Inuit Circumpolar Conference, Washington, District of Columbia. http://commerce.senate.gov/hearings/testimony.cfm?id=1307&wit_id=3815

Weatherhead, B., A. Tanskanen, A. Stevermer, S.B. Andersen, A. Arola, J. Austin, G. Bernhard, H. Browman and Co-authors, 2005: Ozone and ultraviolet radiation. *Arctic Climate Impact Assessment, ACIA*, C. Symon, L. Arris and B. Heal, Eds., Cambridge University Press, Cambridge, 151-182.

Whinam, J. and G. Copson, 2006: Sphagnum moss: an indicator of climate change in the sub-Antarctic. *Polar Rec.*, **42**, 43-49.

White, A., M.G.R. Cannel and A.D. Friend, 2000: The high-latitude terrestrial carbon sink: a model analysis. *Glob. Change Biol.*, **6**, 227-245.

Wingham, D.J., A. Shepherd, A. Muir and G.J. Marshall, 2006: Mass balance of the Antarctic ice sheet. *Philos. T. R. Soc. A.*, **364**, 1627-1635.

World Commission on Dams, 2000: *Introduction to Global Change*. Secretariat of the World Commission on Dams, Cape Town, 16 pp.

Wrona, F.J., T.D. Prowse, J. Reist, R. Beamish, J.J. Gibson, J. Hobbie, E. Jeppesen, J. King and Co-authors, 2005: Freshwater ecosystems and fisheries. *Arctic Climate Impact Assessment, ACIA*, C. Symon, L. Arris and B. Heal, Eds., Cambridge University Press, Cambridge, 353-452.

Wrona, F.J., T.D. Prowse, J.D. Reist, J.E. Hobbie, L. Levesque, R.W. Macdonald and W.F. Vincent, 2006a: Effects of ultraviolet radiation and contaminant-related stressors on Arctic freshwater systems. *Ambio*, **35**, 388-401.

Wrona, F.J., T.D. Prowse, J.D. Reist, J.E. Hobbie, L.M.J. Levesque and W. Vincent, 2006b: Climate change effects on Arctic aquatic biota, ecosystem structure and function. *Ambio*, **35**, 359-369.

Yang, D., B. Ye and A. Shiklomanov, 2004a: Discharge characteristics and changes over the Ob River watershed in Siberia. *J. Hydrometeorol.*, **5**, 595-610.

Yang, D., B. Ye and D.L. Kane, 2004b: Streamflow changes over Siberian Yenisei River basin. *J. Hydrol.*, **296**, 59-80.

Yang, D.Q., D.L. Kane, L.D. Hinzman, X. Zhang, T. Zhang and H. Ye, 2002: Siberian Lena River hydrologic regime and recent change. *J. Geophys. Res.*, **107**, 4694, doi:10.1029/2002JD002542.

Yoccoz, N.G. and R.A. Ims, 1999: Demography of small mammals in cold regions: the importance of environmental variability. *Ecol. Bull.*, **47**, 137-144.

Zhang, T.J., O.W. Frauenfeld, M.C. Serreze, A. Etringer, C. Oelke, J. McCreight, R.G. Barry, D. Gilichinsky, D.Q. Yang, H.C. Ye, F. Ling and S. Chudinova, 2005: Spatial and temporal variability in active layer thickness over the Russian Arctic drainage basin. *J. Geophys. Res.–Atmos.*, **110**, D16101, doi:10.1029/2004JD005642.

Zhang, X. and J.E. Walsh, 2006: Toward a seasonally ice-covered Arctic Ocean: scenarios from the IPCC AR4 model simulations. *J. Climate*, **19**, 1730-1747.

Zöckler, C., 2005: Migratory bird species as indicators for the state of the environment. *Biodiversity*, **6**, 7-13.

Zwally, H.J., J.C. Comiso, C.L. Parkinson, D.J. Cavalieri and P. Gloersen, 2002: Variability of Antarctic sea ice 1979-1998. *J. Geophys. Res.–Oceans*, **107**, 3041, doi:10.1029/2000JC000733.

Zwally, H.J., M. Giovinetto, J. Li, H.G. Conejo, M.A. Beckley, A.C. Brenner, J.L. Saba and Y. Donghui, 2005: Mass changes of the Greenland and Antarctic ice sheets and shelves and contributions to sea-level rise: 1992–2002. *J. Glaciol.*, **51**, 509-527.

16

Small islands

Coordinating Lead Authors:
Nobuo Mimura (Japan), Leonard Nurse (Barbados)

Lead Authors:
Roger McLean (Australia), John Agard (Trinidad and Tobago), Lino Briguglio (Malta), Penehuro Lefale (Samoa), Rolph Payet (Seychelles), Graham Sem (Papua New Guinea)

Contributing Authors:
Will Agricole (Seychelles), Kristie Ebi (USA), Donald Forbes (Canada), John Hay (New Zealand), Roger Pulwarty (USA), Taito Nakalevu (Fiji), Kiyoshi Takahashi (Japan)

Review Editors:
Gillian Cambers (Puerto Rico), Ulric Trotz (Belize)

This chapter should be cited as:
Mimura, N., L. Nurse, R.F. McLean, J. Agard, L. Briguglio, P. Lefale, R. Payet and G. Sem, 2007: Small islands. *Climate Change 2007: Impacts, Adaptation and Vulnerability. Contribution of Working Group II to the Fourth Assessment Report of the Intergovernmental Panel on Climate Change,* M.L. Parry, O.F. Canziani, J.P. Palutikof, P.J. van der Linden and C.E. Hanson, Eds., Cambridge University Press, Cambridge, UK, 687-716.

Table of Contents

Executive summary

Small islands, whether located in the tropics or higher latitudes, have characteristics which make them especially vulnerable to the effects of climate change, sea-level rise, and extreme events (very high confidence).

This assessment confirms and strengthens previous observations reported in the IPCC Third Assessment Report (TAR) which show that characteristics such as limited size, proneness to natural hazards, and external shocks enhance the vulnerability of islands to climate change. In most cases they have low adaptive capacity, and adaptation costs are high relative to gross domestic product (GDP). [16.1, 16.5]

Sea-level rise is expected to exacerbate inundation, storm surge, erosion and other coastal hazards, thus threatening vital infrastructure, settlements and facilities that support the livelihood of island communities (very high confidence).

Some studies suggest that sea-level rise could lead to a reduction in island size, particularly in the Pacific, whilst others show that a few islands are morphologically resilient and are expected to persist. Island infrastructure tends to predominate in coastal locations. In the Caribbean and Pacific islands, more than 50% of the population live within 1.5 km of the shore. Almost without exception, international airports, roads and capital cities in the small islands of the Indian and Pacific Oceans and the Caribbean are sited along the coast, or on tiny coral islands. Sea-level rise will exacerbate inundation, erosion and other coastal hazards, threaten vital infrastructure, settlements and facilities, and thus compromise the socio-economic well-being of island communities and states. [16.4.2, 16.4.5, 16.4.7]

There is strong evidence that under most climate change scenarios, water resources in small islands are likely to be seriously compromised (very high confidence).

Most small islands have a limited water supply, and water resources in these islands are especially vulnerable to future changes and distribution of rainfall. Many islands in the Caribbean are likely to experience increased water stress as a result of climate change. Under all Special Report on Emissions Scenarios (SRES) scenarios, reduced rainfall in summer is projected for this region, so that it is unlikely that demand would be met during low rainfall periods. Increased rainfall in winter is unlikely to compensate, due to lack of storage and high runoff during storms. In the Pacific, a 10% reduction in average rainfall (by 2050) would lead to a 20% reduction in the size of the freshwater lens on Tarawa Atoll, Kiribati. Reduced rainfall coupled with sea-level rise would compound this threat. Many small islands have begun to invest in the implementation of adaptation strategies, including desalination, to offset current and projected water shortages. [16.4.1]

Climate change is likely to heavily impact coral reefs, fisheries and other marine-based resources (high confidence).

Fisheries make an important contribution to the GDP of many island states. Changes in the occurrence and intensity of El Niño-Southern Oscillation (ENSO) events are likely to have severe impacts on commercial and artisanal fisheries. Increasing sea surface temperature and rising sea level, increased turbidity, nutrient loading and chemical pollution, damage from tropical cyclones, and decreases in growth rates due to the effects of higher carbon dioxide concentrations on ocean chemistry, are very likely to affect the health of coral reefs and other marine ecosystems which sustain island fisheries. Such impacts will exacerbate non-climate-change stresses on coastal systems. [16.4.3]

On some islands, especially those at higher latitudes, warming has already led to the replacement of some local species (high confidence).

Mid- and high-latitude islands are virtually certain to be colonised by non-indigenous invasive species, previously limited by unfavourable temperature conditions. Increases in extreme events are virtually certain to affect the adaptation responses of forests on tropical islands, where regeneration is often slow, in the short term. In view of their small area, forests on many islands can easily be decimated by violent cyclones or storms. However, it is possible that forest cover will increase on some high-latitude islands. [16.4.4, 5.4.2.4]

It is very likely that subsistence and commercial agriculture on small islands will be adversely affected by climate change (high confidence).

Sea-level rise, inundation, seawater intrusion into freshwater lenses, soil salinisation, and decline in water supply are very likely to adversely impact coastal agriculture. Away from the coast, changes in extremes (e.g., flooding and drought) are likely to have a negative effect on agricultural production. Appropriate adaptation measures may help to reduce these impacts. In some high-latitude islands, new opportunities may arise for increased agricultural production. [16.4.3, 15.4.2.4]

New studies confirm previous findings that the effects of climate change on tourism are likely to be direct and indirect, and largely negative (high confidence).

Tourism is the major contributor to GDP and employment in many small islands. Sea-level rise and increased sea water temperature will cause accelerated beach erosion, degradation of coral reefs, and bleaching. In addition, a loss of cultural heritage from inundation and flooding reduces the amenity value for coastal users. Whereas a warmer climate could reduce the number of people visiting small islands in low latitudes, it could have the reverse effect in mid- and high-latitude islands. However, water shortages and increased incidence of vector-borne diseases may also deter tourists. [16.4.6]

There is growing concern that global climate change is likely to impact human health, mostly in adverse ways (medium confidence).

Many small islands are located in tropical or sub-tropical zones whose weather and climate are already conducive to the transmission of diseases such as malaria, dengue, filariasis, schistosomiasis, and food- and water-borne diseases. Other climate-sensitive diseases of concern to small islands include diarrhoeal diseases, heat stress, skin diseases, acute respiratory infections and asthma. The observed increasing incidence of many of these diseases in small islands is attributable to a combination of factors, including poor public health practices, inadequate infrastructure, poor waste management practices, increasing global travel, and changing climatic conditions. [16.4.5]

16.1 Introduction

While acknowledging their diversity, the IPCC Third Assessment Report (TAR) also noted that small island states share many similarities (e.g., physical size, proneness to natural disasters and climate extremes, extreme openness of their economies, low adaptive capacity) that enhance their vulnerability and reduce their resilience to climate variability and change.

Analysis of observational data showed a global mean temperature increase of around 0.6°C during the 20th century, while mean sea level rose by about 2 mm/yr, although sea-level trends are complicated by local tectonics and El Niño-Southern Oscillation (ENSO) events. The rate of increase in air temperature in the Pacific and Caribbean during the 20th century exceeded the global average. The TAR also found much of the rainfall variability appeared to be closely related to ENSO events, combined with seasonal and decadal changes in the convergence zones.

Owing to their high vulnerability and low adaptive capacity, small islands have legitimate concerns about their future, based on observational records, experience with current patterns and consequences of climate variability, and climate model projections. Although emitting less than 1% of global greenhouse gases, many small islands have already perceived a need to reallocate scarce resources away from economic development and poverty alleviation, and towards the implementation of strategies to adapt to the growing threats posed by global warming (e.g., Nurse and Moore, 2005).

While some spatial variation within and among regions is expected, the TAR reported that sea level is projected to rise at an average rate of about 5.0 mm/yr over the 21st century, and concluded that sea-level change of this magnitude would pose great challenges and high risk, especially to low-lying islands that might not be able to adapt (Nurse et al., 2001). Given the sea level and temperature projections for the next 50 to 100 years, coupled with other anthropogenic stresses, the coastal assets of small islands (e.g., corals, mangroves, sea grasses and reef fish), would be at great risk. As the natural resilience of coastal areas may be reduced, the 'costs' of adaptation could be expected to increase. Moreover, anticipated land loss, soil salinisation and low water availability would be likely to threaten the sustainability of island agriculture and food security.

In addition to natural and managed system impacts, the TAR also drew attention to projected human costs. These included an increase in the incidence of vector- and water-borne diseases in many tropical and sub-tropical islands, which was attributed partly to temperature and rainfall changes, some linked to ENSO. The TAR also noted that most settlements and infrastructure of small islands are located in coastal areas, which are highly vulnerable not only to sea-level rise (SLR) but also to high-energy waves and storm surge. In addition, temperature and rainfall changes and loss of coastal amenities could adversely affect the vital tourism industry. Traditional knowledge and other cultural assets (e.g., sites of worship and ritual), especially those near the coasts, were also considered to be vulnerable to climate change and sea-level rise. Integrated coastal management was proposed as an effective management framework in small islands for ensuring the sustainability of coastal resources. Such a framework has been adopted in several island states. More recently, the Organisation of Eastern Caribbean States (OECS, 2000) has adopted a framework called 'island systems management', which is both an integrated and holistic (rather than sectoral) approach to whole-island management including terrestrial, aquatic and atmospheric environments.

The TAR concluded that small islands could focus their efforts on enhancing their resilience and implement appropriate adaptation measures as urgent priorities. Thus, integration of risk reduction strategies into key sectoral activities (e.g., disaster management, integrated coastal management and health care planning) should be pursued as part of the adaptation planning process for climate change.

Building upon the TAR, this chapter assesses recent scientific information on vulnerability to climate change and sea-level rise, adaptation to their effects, and implications of climate-related policies, including adaptation, for the sustainable development of small islands. Assessment results are presented in a quantitative manner wherever possible, with near, middle, and far time-frames in this century, although much of the literature concerning small islands is not precise about the time-scales involved in impact, vulnerability and adaptation studies. Indeed, independent scientific studies on climate change and small islands since the TAR have been quite limited, though there are a number of synthetic publications, regional resource books, guidelines, and policy documents including: *Surviving in Small Islands: A Guide Book* (Tompkins et al., 2005); *Climate Variability and Change and Sea-level rise in the Pacific Islands Region: A Resource Book for Policy and Decision Makers, Educators and Other Stakeholders* (Hay et al., 2003); *Climate Change: Small Island Developing States* (UNFCCC, 2005); and *Not If, But When: Adapting to Natural Hazards in the Pacific Island Region: A Policy Note* (Bettencourt et al., 2006).

These publications rely heavily on the TAR, and on studies undertaken by global and regional agencies and contracted reports. It is our qualitative view that the volume of literature in refereed international journals relating to small islands and climate change since publication of the TAR is rather less than that between the Second Assessment Report in 1995 and the TAR in 2001. There is also another difference in that the present chapter deals not only with independent small island states but also with non-autonomous small islands in the continental and large archipelagic countries, including those in high latitudes. Nevertheless the focus is still mainly on the autonomous small islands predominantly located in the tropical and sub-tropical regions; a focus that reflects the emphasis in the literature.

16.2 Current sensitivity and vulnerability

16.2.1 Special characteristics of small islands

Many small islands are highly vulnerable to the impacts of climate change and sea-level rise. They comprise small land masses surrounded by ocean, and are frequently located in regions prone to natural disasters, often of a hydrometeorological and/or geological nature. In tropical areas they host relatively large populations for the area they occupy, with high growth rates and densities. Many small islands have poorly developed infrastructure and limited natural, human and economic resources, and often small island populations are dependent on marine resources to meet their protein needs. Most of their economies are reliant on a limited resource base and are subject to external forces, such as changing terms of trade, economic liberalisation, and migration flows. Adaptive capacity to climate change is generally low, though traditionally there has been some resilience in the face of environmental change.

16.2.2 Climate and weather

16.2.2.1 General features

The climate regimes of small islands are quite variable, generally characterised by large seasonal variability in precipitation and by small seasonal temperature differences in low-latitude islands and large seasonal temperature differences in high-latitude islands. In the tropics, cyclones and other extreme climate and weather events cause considerable losses to life and property.

The climates of small islands in the central Pacific are influenced by several contributing factors such as trade wind regimes, the paired Hadley cells and Walker circulation, seasonally varying convergence zones such as the South Pacific Convergence Zone (SPCZ), semi-permanent sub-tropical high-pressure belts, and zonal westerlies to the south, with ENSO as the dominant mode of year-to-year variability (Fitzharris, 2001; Folland et al., 2002; Griffiths et al., 2003). The Madden-Julian Oscillation (MJO) is a major mode of variability of the tropical atmosphere-ocean system of the Pacific on time-scales of 30 to 70 days (Revell, 2004), while the leading mode of variability with decadal time-scale is the Interdecadal Pacific Oscillation (IPO) (Salinger et al., 2001). A number of studies suggest that the influence of global warming could be a major factor in accentuating the current climate regimes and the changes from the normal that come with ENSO events (Folland et al., 2003; Hay et al., 2003).

The climate of the Caribbean islands is broadly characterised by distinct dry and wet seasons with orography and elevation being significant modifiers on the sub-regional scale. The dominant influences are the North Atlantic Sub-tropical High (NAH) and ENSO. During the Northern Hemisphere winter, the NAH lies further south, with strong easterly trades on its equatorial flank modulating the climate and weather of the region. Coupled with a strong inversion, a cool ocean, and reduced atmospheric humidity, the region is generally at its driest during the Northern Hemisphere winter. With the onset of the Northern Hemisphere spring, the NAH moves northwards, the trade wind intensity decreases, and the region then comes under the influence of the equatorial trough.

In the Indian Ocean, the climate regimes of small islands in tropical regions are predominantly influenced by the Asian monsoon; the seasonal alternation of atmospheric flow patterns which results in two distinct climatic regimes: the south-west or summer monsoon and the north-east or winter monsoon, with a clear association with ENSO events.

The climates of small islands in the Mediterranean are dominated by influences from bordering lands. Commonly the islands receive most of their rainfall during the Northern Hemisphere winter months and experience a prolonged summer drought of 4 to 5 months. Temperatures are generally moderate with a comparatively small range of temperature between the winter low and summer high.

16.2.2.2 Observed trends

Temperature

New observations and reanalyses of temperatures averaged over land and ocean surfaces since the TAR show consistent warming trends in all small-island regions over the 1901 to 2004 period (Trenberth et al., 2007). However, the trends are not linear. Recent studies show that annual and seasonal ocean surface and island air temperatures have increased by 0.6 to 1.0°C since 1910 throughout a large part of the South Pacific, south-west of the SPCZ. Decadal increases of 0.3 to 0.5°C in annual temperatures have been widely seen only since the 1970s, preceded by some cooling after the 1940s, which is the beginning of the record, to the north-east of the SPCZ (Salinger, 2001; Folland et al., 2003).

For the Caribbean, Indian Ocean and Mediterranean regions, analyses shows warming ranged from 0 to 0.5°C per decade for the 1971 to 2004 period (Trenberth et al., 2007). Some high-latitude regions, including the western Canadian Arctic Archipelago, have experienced warming more rapid than the global mean (McBean et al., 2005).

Trends in extreme temperature across the South Pacific for the period 1961 to 2003 show increases in the annual number of hot days and warm nights, with decreases in the annual number of cool days and cold nights, particularly in the years after the onset of El Niño (Manton et al., 2001; Griffiths et al., 2003). In the Caribbean, the percentage of days having very warm maximum or minimum temperatures has increased considerably since the 1950s, while the percentage of days with cold temperatures has decreased (Peterson et al., 2002).

Precipitation

Analyses of trends in extreme daily rainfall across the South Pacific for the period 1961 to 2003 show extreme rainfall trends which are generally less spatially coherent than those of extreme temperatures (Manton et al., 2001; Griffiths et al., 2003). In the Caribbean, the maximum number of consecutive dry days is decreasing and the number of heavy rainfall events is increasing. These changes were found to be similar to the changes reported from global analysis (Trenberth et al., 2007).

Tropical and extra-tropical cyclones

Variations in tropical and extra-tropical cyclones, hurricanes and typhoons in many small-island regions are dominated by ENSO and decadal variability which result in a redistribution of tropical storms and their tracks, so that increases in one basin are often compensated by decreases in other basins. For example, during an El Niño event, the incidence of tropical storms typically decreases in the Atlantic and far-western Pacific and the Australian regions, but increases in the central and eastern Pacific, and *vice versa*. Clear evidence exists that the number of storms reaching categories 4 and 5 globally have increased since 1970, along with increases in the Power Dissipation Index (Emanuel, 2005) due to increases in their intensity and duration (Trenberth et al., 2007). The total number of cyclones and cyclone days decreased slightly in most basins. The largest increase was in the North Pacific, Indian and South-West Pacific oceans. The global view of tropical storm activity highlights the important role of ENSO in all basins. The most active year was 1997, when a very strong El Niño began, suggesting that the observed record sea surface temperatures (SSTs) played a key role (Trenberth et al., 2007). For extra-tropical cyclones, positive trends in storm frequency and intensity dominate during recent decades in most regional studies performed. Longer records for the North Atlantic suggest that the recent extreme period may be similar in level to that of the late 19th century (Trenberth et al., 2007).

In the tropical South Pacific, small islands to the east of the dateline are highly likely to receive a higher number of tropical storms during an El Niño event compared with a La Niña event and *vice versa* (Brazdil et al., 2002). Observed tropical cyclone activity in the South Pacific east of 160°E indicates an increase in level of activity, with the most active years associated with El Niño events, especially during the strong 1982/1983 and 1997/1998 events (Levinson, 2005). Webster et al. (2005) found more than a doubling in the number of category 4 and 5 storms in the South-West Pacific from the period 1975–1989 to the period 1990–2004. In the 2005/2006 season, La Niña influences shifted tropical storm activity away from the South Pacific region to the Australian region and, in March and April 2006, four category 5 typhoons occurred (Trenberth et al., 2007).

In the Caribbean, hurricane activity was greater from the 1930s to the 1960s, in comparison with the 1970s and 1980s and the first half of the 1990s. Beginning with 1995, all but two Atlantic hurricane seasons have been above normal (relative to the 1981-2000 baseline). The exceptions are the two El Niño years of 1997 and 2002. El Niño acts to reduce activity and La Niña acts to increase activity in the North Atlantic. The increase contrasts sharply with the generally below-normal seasons observed during the previous 25-year period, 1975 to 1994. These multi-decadal fluctuations in hurricane activity result almost entirely from differences in the number of hurricanes and major hurricanes forming from tropical storms first named in the tropical Atlantic and Caribbean Sea.

In the Indian Ocean, tropical storm activity (May to December) in the northern Indian Ocean has been near normal in recent years. For the southern Indian Ocean, the tropical cyclone season is normally active from December to April. A lack of historical record-keeping severely hinders trend analysis (Trenberth et al., 2007).

Sea level

Analyses of the longest available sea-level records, which have at least 25 years of hourly data from 27 stations installed around the Pacific basin, show the overall average mean relative sea-level rise around the whole region is +0.77 mm/yr (Mitchell et al., 2001). Rates of relative sea level have also been calculated for the SEAFRAME stations in the Pacific. Using these results and focusing only on the island stations with more than 50 years of data (only four locations), the average rate of sea-level rise (relative to the Earth's crust) is 1.6 mm/yr (Bindoff et al., 2007). Church et al. (2004) used TOPEX/Poseidon altimeter data, combined with historical tide gauge data, to estimate monthly distributions of large-scale sea-level variability and change over the period 1950 to 2000. Church et al. (2004) observed the maximum rate of rise in the central and eastern Pacific, spreading north and south around the sub-tropical gyres of the Pacific Ocean near 90°E, mostly between 2 and 2.5 mm/yr but peaking at over 3 mm/yr. This maximum was split by a minimum rate of rise, less than 1.5 mm/yr, along the equator in the eastern Pacific, linking to the western Pacific just west of 180° (Christensen et al., 2007).

The Caribbean region experienced, on average, a mean relative sea-level rise of 1 mm/yr during the 20th century. Considerable regional variations in sea level were observed in the records; these were due to large-scale oceanographic phenomena such as El Niño coupled with volcanic and tectonic crustal motions of the Caribbean Basin rim, which affect the land levels on which the tide gauges are located. Similarly, recent variations in sea level on the west Trinidad coast indicate that sea level in the north is rising at a rate of about 1 mm/yr, while in the south the rate is about 4 mm/yr; the difference being a response to tectonic movements (Miller, 2005).

In the Indian Ocean, reconstructed sea levels based on tide gauge data and TOPEX/Poseidon altimeter records for the 1950 to 2001 period give rates of relative sea-level rise of 1.5, 1.3 and 1.5 mm/yr (with error estimates of about 0.5 mm/yr) at Port Louis, Rodrigues, and Cocos Islands, respectively (Church et al., 2006). In the equatorial band, both the Male and Gan sea-level sites in the Maldives show trends of about 4 mm/yr (Khan et al., 2002), with the range from three tidal stations over the 1990s being from 3.2 to 6.5 mm/yr (Woodworth et al., 2002). Church et al. (2006) note that the Maldives has short records and that there is high variability between sites, and their 52-year reconstruction suggests a common rate of rise of 1.0 to 1.2 mm/yr.

Some high-latitude islands are in regions of continuing postglacial isostatic uplift, including parts of the Baltic, Hudson Bay, and the Canadian Arctic Archipelago (CAA). Others along the Siberian coast and the eastern and western margins of the CAA are subsiding. Although few long tide-gauge records exist in the region, relative sea-level trends are known to range from negative (falling relative sea level) in the central CAA and Hudson Bay to rates as high as 3 mm/yr or more in the Beaufort Sea (Manson et al., 2005). Available data from the Siberian sector of the Arctic Ocean indicate that late 20th century sea-level rise was comparable to the global mean (Proshutinsky et al., 2004).

16.2.3 Other stresses

Climate change and sea-level rise are not unique contributors to the extreme vulnerability of small islands. Other factors include socio-economic conditions, natural resource and space limitations, and the impacts of natural hazards such as tsunami and storms. In the Pacific, vulnerability is also a function of internal and external political and economic processes which affect forms of social and economic organisation that are different from those practiced traditionally, as well as attempts to impose models of adaptation that have been developed for Western economies, without sufficient thought as to their applicability in traditional island settings (Cocklin, 1999).

Socio-economic stresses

Socio-economic contributors to island vulnerability include external pressures such as terms of trade, impacts of globalisation (both positive and negative), financial crises, international conflicts, rising external debt, and internal local conditions such as rapid population growth, rising incidence of poverty, political instability, unemployment, reduced social cohesion, and a widening gap between poor and rich, together with the interactions between them (ADB, 2004).

Most settlements in small islands, with the exception of some of the larger Melanesian and Caribbean islands, are located in coastal locations, with the prime city or town also hosting the main port, international airport and centre of government activities. Heavy dependence on coastal resources for subsistence is also a major feature of many small islands.

Rapid and unplanned movements of rural and outer-island residents to the major centres is occurring throughout small islands, resulting in deteriorating urban conditions, with pressure on access to urban services required to meet basic needs. High concentrations of people in urban areas create various social, economic and political stresses, and make people more vulnerable to short-term physical and biological hazards such as tropical cyclones and diseases. It also increases their vulnerability to the impacts of climate change and sea-level rise (Connell, 1999, 2003).

Globalisation is also a major stress, though it has been argued that it is nothing new for many small islands, since most have had a long history of colonialism and, more latterly, experience of some of the rounds of transformation of global capitalism (Pelling and Uitto, 2001). Nevertheless, in the last few years, the rate of change and growth of internationalisation have increased, and small islands have had to contend with new forms of extra-territorial economic, political and social forces such as multinational corporations, transnational social movements, international regulatory agencies, and global communication networks. In the present context, these factors take on a new relevance, as they may influence the vulnerability of small islands and their adaptive capacity (Pelling and Uitto, 2001; Adger et al., 2003a).

Pressure on island resources

Most small islands have limited sources of freshwater. Atoll countries and limestone islands have no surface water or streams and are fully reliant on rainfall and groundwater harvesting. Many small islands are experiencing water stress at the current levels of rainfall input, and extraction of groundwater is often outstripping supply. Moreover, pollution of groundwater is often a major problem, especially on low-lying islands. Poor water quality affects human health and carries water-borne diseases. Water quality is just one of several health issues linked to climate variability and change and their potential effects on the well-being of the inhabitants of small islands (Ebi et al., 2006).

It is also almost inevitable that the ecological systems of small islands, and the functions they perform, will be sensitive to the rate and magnitude of climate change and sea-level rise, especially where exacerbated by human activities (e.g., ADB, 2004, in the case of the small islands in the Pacific). Both terrestrial ecosystems on the larger islands and coastal ecosystems on most islands have been subjected to increasing degradation and destruction in recent decades. For instance, analysis of coral reef surveys over three decades has revealed that coral cover across reefs in the Caribbean has declined by 80% in just 30 years, largely as a result of continued pollution, sedimentation, marine diseases, and over-fishing (Gardner et al., 2003).

Interactions between human and physical stresses

External pressures that contribute to the vulnerability of small islands to climate change include energy costs, population movements, financial and currency crises, international conflicts, and increasing debt. Internal processes that create vulnerability include rapid population growth, attempts to increase economic growth through exploitation of natural resources such as forests, fisheries and beaches, weak infrastructure, increasing income inequality, unemployment, rapid urbanisation, political instability, a growing gap between demand for and provision of health care and education services, weakening social capital, and economic stagnation. These external and internal processes are related and interact in complex ways to heighten the vulnerability of island social and ecological systems to climate change.

Natural hazards of hydrometeorological origin remain an important stressor and cause impacts on the economies of small islands that are disproportionally large (Bettencourt et al., 2006). The devastation of Grenada following the passage of Hurricane Ivan on 7 September 2004 is a powerful illustration of the reality of small-island vulnerability (Nurse and Moore, 2005). In less than 8 hours, the country's vital socio-economic infrastructure, including housing, utilities, tourism-related facilities and subsistence and commercial agricultural production, suffered incalculable damage. The island's two principal foreign-exchange earners – tourism and nutmeg production – suffered heavily. More than 90% of hotel guest rooms were either completely destroyed or damaged, while more than 80% of the island's nutmeg trees were lost. One of the major challenges with regard to hydrometeorological hazards is the time it takes to recover from them. In the past it was common for socio-ecological systems to recover from hazards, as these were sufficiently infrequent and/or less damaging. In the future, climate change may create a situation where more intense and/or more frequent extreme events may mean there is less time in which to recover. Sequential extreme events may mean that recovery is never complete, resulting in long-term deteriorations in affected systems, e.g., declines in agricultural output because soils never recover from salinisation; urban water systems and housing infrastructure deteriorating because damage cannot be repaired before the next extreme event.

16.2.4 Current adaptation

Past studies of adaptation options for small islands have largely focused on adjustments to sea-level rise and storm surges associated with tropical cyclones. There was an early emphasis on protecting land through 'hard' shore-protection measures rather than on other measures such as accommodating sea-level rise or retreating from it, although the latter has become increasingly important on continental coasts. Vulnerability studies conducted for selected small islands (Nurse et al., 2001) show that the costs of overall infrastructure and settlement protection are a significant proportion of GDP, and well beyond the financial means of most small island states; a problem not always shared by the islands of metropolitan countries (i.e., with high-density, predominantly urban populations). More recent studies since the TAR have identified major areas of adaptation, including water resources and watershed management, reef conservation, agricultural and forest management, conservation of biodiversity, energy security, increased development of renewable energy, and optimised energy consumption. Some of these are detailed in Section 16.5. Proposed adaptation strategies have also focused on reducing vulnerability and increasing resilience of systems and sectors to climate variability and extremes through mainstreaming adaptation (Shea et al., 2001; Hay et al., 2003; ADB, 2004; UNDP, 2005).

16.3 Assumptions about future trends

16.3.1 Climate and sea-level change

16.3.1.1 Temperature and precipitation

Since the TAR, future climate change projections have been updated (Ruosteenoja et al., 2003). These analyses reaffirm previous IPCC projections that suggest a gradual warming of SSTs and a general warming trend in surface air temperature in all small-island regions and seasons (Lal et al., 2002). However, it must be cautioned that, because of scaling problems, these projections for the most part apply to open ocean surfaces and not to land surfaces. Consequently the temperature changes may well be higher than current projections.

Projected changes in seasonal surface air temperature (Table 16.1) and precipitation (Table 16.2) for the three 30-year periods (2010 to 2039, 2040 to 2069 and 2070 to 2099) relative to the baseline period 1961 to 1990, have been prepared by Ruosteenoja et al. (2003) for all the sub-continental scale regions of the world, including small islands. They used seven coupled atmosphere-ocean general circulation models (AOGCMs), the greenhouse gas and aerosol forcing being inferred from the IPCC Special Report on Emissions Scenarios (SRES; Nakićenović and Swart, 2000) A1FI, A2, B1 and B2 emissions scenarios.

All seven models project increased surface air temperature for all regions of the small islands. The Ruosteenoja et al. (2003) projected increases all lie within previous IPCC surface air temperature projections, except for the Mediterranean Sea. The increases in surface air temperature are projected to be more or less uniform in both seasons, but for the Mediterranean Sea, warming is projected to be greater during the summer than the winter. For the South Pacific, Lal (2004) has indicated that the surface air temperature by 2100 is estimated to be at least 2.5°C more than the 1990 level. Seasonal variations of projected warming are minimal. No significant change in diurnal temperature range is likely with a rise in surface temperatures. An increase in mean temperature would be accompanied by an increase in the frequency of extreme temperatures. High-latitude regions are likely to experience greater warming, resulting in decreased sea ice extent and increased thawing of permafrost (Meehl et al., 2007).

Regarding precipitation, the range of projections is still large, and even the direction of change is not clear. The models simulate only a marginal increase or decrease (10%) in annual rainfall over most of the small islands in the South Pacific. During summer, more rainfall is projected, while an increase in daily rainfall intensity, causing more frequent heavier rainfall events, is also likely (Lal, 2004).

Table 16.1. *Projected increase in air temperature (°C) by region, relative to the 1961–1990 period.*

Region	2010–2039	2040–2069	2040–2069
Mediterranean	0.60 to 2.19	0.81 to 3.85	1.20 to 7.07
Caribbean	0.48 to 1.06	0.79 to 2.45	0.94 to 4.18
Indian Ocean	0.51 to 0.98	0.84 to 2.10	1.05 to 3.77
Northern Pacific	0.49 to 1.13	0.81 to 2.48	1.00 to 4.17
Southern Pacific	0.45 to 0.82	0.80 to 1.79	0.99 to 3.11

Table 16.2. *Projected change in precipitation (%) by region, relative to the 1961–1990 period.*

Region	2010–2039	2040–2069	2040–2069
Mediterranean	−35.6 to +55.1	−52.6 to +38.3	−61.0 to +6.2
Caribbean	−14.2 to +13.7	−36.3 to +34.2	−49.3 to +28.9
Indian Ocean	−5.4 to +6.0	−6.9 to +12.4	−9.8 to +14.7
Northern Pacific	−6.3 to +9.1	−19.2 to +21.3	−2.7 to +25.8
Southern Pacific	−3. 9 to +3.4	−8.23 to +6.7	−14.0 to +14.6

16.3.1.2 Sea levels

Sea-level changes are of special significance, not only for the low-lying atoll islands but for many high islands where settlements, infrastructure and facilities are concentrated in the coastal zone. Projected globally averaged sea-level rise at the end of the 21st century (2090 to 2099), relative to 1980 to 1999 for the six SRES scenarios, ranges from 0.19 to 0.58 m (Meehl et al., 2007). In all SRES scenarios, the average rate of sea-level rise during the 21st century very probably exceeds the 1961 to 2003 average rate (1.8 ± 0.5 mm/yr). Climate models also indicate a geographical variation of sea-level rise due to non-uniform distribution of temperature and salinity and changes in ocean circulation. Furthermore, regional variations and local differences depend on several factors, including non-climate-related factors such as island tectonic setting and postglacial isostatic adjustment. While Mörner et al. (2004) suggest that the increased risk of flooding during the 21st century for the Maldives has been overstated, Woodworth (2005) concludes that a rise in sea level of approximately 50 cm during the 21st century remains the most reliable scenario to employ in future studies of the Maldives.

16.3.1.3 Extreme events

Global warming from anthropogenic forcing suggests increased convective activity but there is a possible trade-off between localised versus organised convection (IPCC, 2001). While increases in SSTs favour more and stronger tropical cyclones, increased isolated convection stabilises the tropical troposphere and this, in turn, suppresses organised convection, making conditions less favourable for vigorous tropical cyclones to develop. Thus, the IPCC (2001) noted that changes in atmospheric stability and circulation may produce offsetting tendencies.

Recent analyses (e.g., Brazdil et al., 2002; Mason, 2004) since the TAR confirm these findings. Climate modelling with improved resolutions has demonstrated the capability to diagnose the probability of occurrence of short-term extreme events under global warming (Meehl et al., 2007). Vassie et al. (2004) suggest that scientists engaged in climate change impact studies should also consider possible changes in swell direction and incidence and their potential impacts on the coasts of small islands. With an increasing number of people living close to the coast, deep ocean swell generation, and its potential modifications as a consequence of climate change, is clearly an issue that needs attention, alongside the more intensively studied topics of changes in mean sea level and storm surges.

Although there is as yet no convincing evidence in the observed record of changes in tropical cyclone behaviour, a synthesis of the recent model results indicates that, for the future warmer climate, tropical cyclones will show increased peak wind speed and increased mean and peak precipitation intensities. The number of intense cyclones is likely to increase, although the total number may decrease on a global scale (Meehl et al., 2007). It is likely that maximum tropical cyclone wind intensities could increase, by 5 to 10% by around 2050 (Walsh, 2004). Under this scenario, peak precipitation rates are likely to increase by 25% as a result of increases in maximum tropical cyclone wind intensities, which in turn cause higher storm surges. Although it is exceptionally unlikely that there will be significant changes in regions of formation, the rate of formation is very likely to change in some regions. Changes in tropical cyclone tracks are closely associated with ENSO and other local climate conditions. These suggest a strong possibility of higher risks of more persistent and devastating tropical cyclones in a warmer world.

Mid-latitude islands, such as islands in the Gulf of St. Lawrence and off the coast of Newfoundland (St. Pierre et Miquelon), are exposed to impacts from tropical, post-tropical, and extra-tropical storms that can produce storm-surge flooding, large waves, coastal erosion, and (in some winter storms) direct sea ice damage to infrastructure and property. Possible increases in storm intensity, rising sea levels, and changes in ice duration and concentration, are projected to increase the severity of negative impacts progressively, particularly by mid-century (Forbes et al., 2004). In the Queen Charlotte Islands (Haida Gwaii) off the Canadian Pacific coast, winter storm damage is exacerbated by large sea-level anomalies resulting from ENSO variability (Walker and Barrie, 2006).

16.3.2 Other relevant conditions

Populations on many small islands have long developed and maintained unique lifestyles, adapted to their natural environment. Traditional knowledge, practices and cultures, where they are still practised, are strongly based on community support networks and, in many islands, a subsistence economy is still predominant (Berkes and Jolly, 2001; Fox, 2003; Sutherland et al., 2005). Societal changes such as population growth, increased cash economy, migration of people to urban centres and coastal areas, growth of major cities, increasing dependency on imported goods which create waste management problems, and development of modern industries such as tourism have changed traditional lifestyles in many small islands. Trade liberalisation also has major implications for the economic and social well-being of the people of small islands. For example, the phasing out of the Lomé Convention and the implementation of the Cotonou Agreement will be important. The end of the Lomé Convention means that the prices the EU pays for certain agricultural commodities, such as sugar, will decline. Such countries as Fiji, Jamaica and Mauritius may experience significant contractions in GDP as a result of declining sugar prices (Milner et al., 2004). In Fiji, for example, where 25% of the workforce is in the sugar sector, the replacement of the Lomé Convention with the terms of the Cotonou Agreement is likely to result in significant unemployment and deeper impoverishment of many of the 23,000 smallholder farmers, many of whom already live below the poverty line (Prasad, 2003). Such declines in the agricultural sector, resulting from trade liberalisation, heighten social vulnerability to climate change. These changes, together with the gradual disintegration of traditional communities, will continue to weaken traditional human support networks, with additional feedback effects of social breakdown and loss of traditional values, social cohesion, dignity and confidence, which have been a major component of the resilience of local communities in Pacific islands.

16.4 Key future impacts and vulnerabilities

The special characteristics of small islands, as described in Section 16.2.1, make them prone to a large range of potential impacts from climate change, some of which are already being experienced. Examples of that range, thematically and geographically, are shown in Box 16.1. Further details on sectors that are especially vulnerable in small islands are expanded upon below.

16.4.1 Water resources

Owing to factors of limited size, availability, and geology and topography, water resources in small islands are extremely vulnerable to changes and variations in climate, especially in rainfall (IPCC, 2001). In most regions of small islands, projected future changes in seasonal and annual precipitation are uncertain, although in a few instances precipitation is likely to

Box 16.1. Range of future impacts and vulnerabilities in small islands

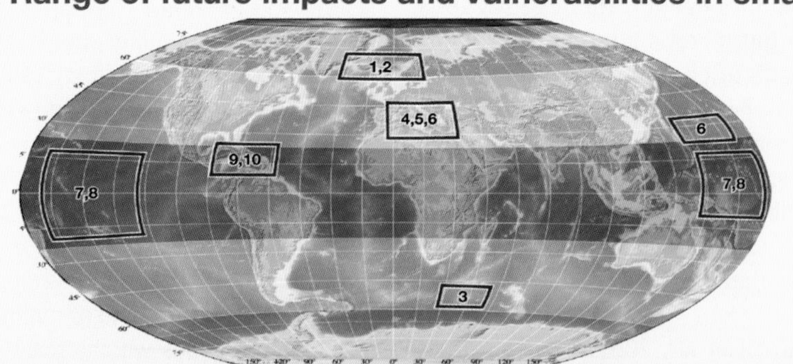

* Numbers in bold relate to the regions defined on the map

Region* and system at risk	Scenario and reference	Changed parameters	Impacts and vulnerability
1. Iceland and isolated Arctic islands of Svalbard and the Faroe Islands: Marine ecosystem and plant species	SRES A1 and B2 ACIA (2005)	Projected rise in temperature	• The imbalance of species loss and replacement leads to an initial loss in diversity. Northward expansion of dwarf-shrub and tree-dominated vegetation into areas rich in rare endemic species results in their loss. • Large reduction in, or even a complete collapse of, the Icelandic capelin stock leads to considerable negative impacts on most commercial fish stocks, whales, and seabirds.
2. High-latitude islands (Faroe Islands): Plant species	Scenario I / II: temperature increase / decrease by 2°C. Fosaa et al. (2004)	Changes in soil temperature, snow cover and growing degree days	• Scenario 1: Species most affected by warming are restricted to the uppermost parts of mountains. For other species, the effect will mainly be upward migration. • Scenario II: Species affected by cooling are those at lower altitudes.
3. Sub-Antarctic Marion Islands: Ecosystem	Own scenarios Smith (2002)	Projected changes in temperature and precipitation	• Changes will directly affect the indigenous biota. An even greater threat is that a warmer climate will increase the ease with which the islands can be invaded by alien species.
4. Mediterranean Basin five islands: Ecosystems	SRES A1FI and B1 Gritti et al. (2006)	Alien plant invasion under climatic and disturbance scenarios	• Climate change impacts are negligible in many simulated marine ecosystems. • Invasion into island ecosystems become an increasing problem. In the longer term, ecosystems will be dominated by exotic plants irrespective of disturbance rates.
5. Mediterranean: Migratory birds (Pied flycatchers – *Ficedula hypoleuca*)	None (GLM/ STATISTICA model) Sanz et al. (2003)	Temperature increase, changes in water levels and vegetation index	• Some fitness components of pied flycatchers suffer from climate change in two of the southernmost European breeding populations, with adverse effects on reproductive output of pied flycatchers.
6. Pacific and Mediterranean: Siam weed (*Chromolaena odorata*)	None (CLIMEX model) Kriticos et al. (2005)	Increase in moisture, cold, heat and dry stress	• Pacific islands at risk of invasion by Siam weed. • Mediterranean semi-arid and temperate climates predicted to be unsuitable for invasion.
7. Pacific small islands: Coastal erosion, water resources and human settlement	SRES A2 and B2 World Bank (2000)	Changes in temperature and rainfall, and sea-level rise	• Accelerated coastal erosion, saline intrusion into freshwater lenses and increased flooding from the sea cause large effects on human settlements. • Less rainfall coupled with accelerated sea-level rise compound the threat on water resources; a 10% reduction in average rainfall by 2050 is likely to correspond to a 20% reduction in the size of the freshwater lens on Tarawa Atoll, Kiribati.
8. American Samoa; 15 other Pacific islands: Mangroves	Sea-level rise 0.88 m to 2100 Gilman et al. (2006)	Projected rise in sea level	• 50% loss of mangrove area in American Samoa; 12% reduction in mangrove area in 15 other Pacific islands.
9. Caribbean (Bonaire, Netherlands Antilles): Beach erosion and sea turtle nesting habitats	SRES A1, A1FI, B1, A2, B2 Fish et al. (2005)	Projected rise in sea level	• On average, up to 38% (±24% SD) of the total current beach could be lost with a 0.5 m rise in sea level, with lower narrower beaches being the most vulnerable, reducing turtle nesting habitat by one-third.
10. Caribbean (Bonaire, Barbados): Tourism	None Uyarra et al. (2005)	Changes to marine wildlife, health, terrestrial features and sea conditions	• The beach-based tourism industry in Barbados and the marine diving based ecotourism industry in Bonaire are both negatively affected by climate change through beach erosion in Barbados and coral bleaching in Bonaire.

increase slightly during December, January and February (DJF) in the Indian Ocean and southern Pacific and during June, July and August (JJA) in the northern Pacific (Christensen et al., 2007). Even so, the scarcity of fresh water is often a limiting factor for social and economic development in small islands. Burns (2002) has also cautioned that with the rapid growth of tourism and service industries in many small islands, there is a need both for augmentation of the existing water resources and for more efficient planning and management of those resources. Measures to reduce water demand and promote conservation are also especially important on small islands, where infrastructure deterioration resulting in major leakage is common, and water pollution from soil erosion, herbicide and pesticide runoff, livestock waste, and liquid and solid waste disposal results in high costs, crudely estimated at around 3% of GDP in Rarotonga, Cook Islands (Hajkowicz, 2006).

This dependency on rainfall significantly increases the vulnerability of small islands to future changes in distribution of rainfall. For example, model projections suggest that a 10% reduction in average rainfall by 2050 is likely to correspond to a 20% reduction in the size of the freshwater lens on Tarawa Atoll, Kiribati. Moreover, a reduction in the size of the island, resulting from land loss accompanying sea-level rise, is likely to reduce the thickness of the freshwater lens on atolls by as much as 29% (World Bank, 2000). Less rainfall coupled with accelerated sea-level rise would compound this threat. Studies conducted on Bonriki Island in Tarawa, Kiribati, showed that a 50 cm rise in sea level accompanied by a reduction in rainfall of 25% would reduce the freshwater lens by 65% (World Bank, 2000). Increases in sea level may also shift watertables close to or above the surface, resulting in increased evapotranspiration, thus diminishing the resource (Burns, 2000).

Lower rainfall typically leads to a reduction in the amount of water that can be physically harvested, to a reduction in river flow, and to a slower rate of recharge of the freshwater lens, which can result in prolonged drought impacts. Recent modelling of the current and future water resource availability on several small islands in the Caribbean, using a macro-scale hydrological model and the SRES scenarios (Arnell, 2004), found that many of these islands would be exposed to severe water stress under all SRES scenarios, and especially so under A2 and B2. Since most of the islands are dependent upon surface water catchments for water supply, it is highly likely that demand could not be met during periods of low rainfall.

The wet and dry cycles associated with ENSO episodes can have serious impacts on water supply and island economies. For instance the strong La Niña of 1998 to 2000 was responsible for acute water shortages in many islands in the Indian and Pacific Oceans (Shea et al., 2001; Hay et al., 2003), which resulted in partial shut-downs in the tourism and industrial sectors. In Fiji and Mauritius, borehole yields decreased by 40% during the dry periods, and export crops including sugar cane were also severely affected (World Bank, 2000). The situation was exacerbated by the lack of adequate infrastructure such as reservoirs and water distribution networks in most islands.

Increases in demand related to population and economic growth, in particular tourism, continue to place serious stress on existing water resources. Excessive damming, over-pumping and increasing pollution are all threats that will continue to increase in the future. Groundwater resources are especially at risk from pollution in many small islands (UNEP, 2000), and in countries such as the Comoros, the polluted waters are linked to outbreaks of yellow fever and cholera (Hay et al., 2003).

Access to safe potable water varies across countries. There is very good access in countries such as Singapore, Mauritius and most Caribbean islands, whereas in states such as Kiribati and Comoros it has been estimated that only 44% and 50% of the population, respectively, have access to safe water. Given the major investments needed to develop storage and provide treatment and distribution of water, it is evident that climate change would further decrease the ability of many islands to meet their future requirements.

Several small island countries have begun to invest, at great financial cost, in the implementation of various augmentation and adaptation strategies to offset current water shortages. The Bahamas, Antigua and Barbuda, Barbados, Maldives, Seychelles, Singapore, Tuvalu and others have invested in desalination plants. However, in the Pacific, some of the systems are now only being used during the dry season, owing to operational problems and high maintenance costs. Options such as large storage reservoirs and improved water harvesting are now being explored more widely, although such practices have been in existence in countries such as the Maldives since the early 1900s. In other cases, countries are beginning to invest in improving the scientific database that could be used for future adaptation plans. In the Cook Islands, for example, a useful index for estimating drought intensity was recently developed based on analysis of more than 70 years of rainfall data; this will be a valuable tool in the long-term planning of water resources in these islands (Parakoti and Scott, 2002).

16.4.2 Coastal systems and resources

The coastlines of small islands are long relative to island area. They are also diverse and resource-rich, providing a range of goods and services, many of which are threatened by a combination of human pressures and climate change and variability arising especially from sea-level rise, increases in sea surface temperature, and possible increases in extreme weather events. Key impacts will almost certainly include accelerated coastal erosion, saline intrusion into freshwater lenses, and increased flooding from the sea. An extreme example of the ultimate impact of sea-level rise on small islands – island abandonment – has been documented by Gibbons and Nicholls (2006) in Chesapeake Bay.

It has long been recognised that islands on coral atolls are especially vulnerable to this combination of impacts, and the long-term viability of some atoll states has been questioned. Indeed, Barnett and Adger (2003) argue that the risk from climate-induced factors constitutes a dangerous level of climatic change to atoll countries by potentially undermining their sovereignty (see Section 16.5.4).

The future of atoll island geomorphology has been predicted using both geological analogues and simulation modelling approaches. Using a modified shoreline translation model, Kench and Cowell (2001) and Cowell and Kench (2001) found that, with sea-level rise, ocean shores will be eroded and sediment redeposited further lagoonward, assuming that the volume of island sediment remains constant. Simulations also show that changes in sediment supply can cause physical alteration of atoll islands by an equivalent or greater amount than by sea-level rise alone. Geological reconstructions of the relationship between sea level and island evolution in the mid- to late Holocene, however, do not provide consistent interpretations. For instance, chronic island erosion resulting from increased water depth across reefs with global warming and sea-level rise is envisaged for some islands in the Pacific (Dickinson, 1999), while Kench et al. (2005) present data and a model which suggest that uninhabited islands of the Maldives are morphologically resilient rather than fragile systems, and are expected to persist under current scenarios of future climate change and sea-level rise. The impact of the Sumatran tsunami on such islands appears to confirm this resilience (Kench et al., 2006) and implies that islands which have been subject to substantial human modification are inherently more vulnerable than those that have not been modified.

On topographically higher and geologically more complex islands, beach erosion presents a particular hazard to coastal tourism facilities, which provide the main economic thrust for many small island states. *Ad hoc* approaches to addressing this problem have recently given way to the integrated coastal zone management approach as summarised in the TAR (McLean et al., 2001), which involves data collection, analysis of coastal processes, and assessment of impacts. Daniel and Abkowitz (2003, 2005) present the results of such an approach in the Caribbean, which involves the development of tools for integrating spatial and non-spatial coastal data, estimating long-term beach erosion/accretion trends and storm-induced beach erosion at individual beaches, identifying erosion-sensitive beaches, and mapping beach-erosion hazards. Coastal erosion on arctic islands has additional climate sensitivity through the impact of warming on permafrost and extensive ground ice, which can lead to accelerated erosion and volume loss, and the potential for higher wave energy if the diminished sea ice results in longer over-water fetch (see Chapter 6, Section 6.2.5; Chapter 15, Section 15.4.6).

While erosion is intuitively the most common response of island shorelines to sea-level rise, it should be recognised that coasts are not passive systems. Instead, they will respond dynamically in different ways dependent on many factors including: the geological setting; coastal type, whether soft or hard shores; the rate of sediment supply relative to rate of submergence; sediment type, sand or gravel; presence or absence of natural shore protection structures such as beach rock or conglomerate outcrops; presence or absence of biotic protection such as mangroves and other strand vegetation; and the health of coral reefs. That several of these factors are interrelated can be illustrated by a model study by Sheppard et al. (2005), who suggest that mass coral mortality over the past decade at some sites in the Seychelles has resulted in a reduction in the level of the fringing reef surface, a consequent rise in wave energy over the reef, and increased coastal erosion. Further declines in reef health are expected to accelerate this trend.

Global change is also creating a number of other stress factors that are very likely to influence the health of coral reefs around islands, as a result of increasing sea surface temperature and sea level, damage from tropical cyclones, and possible decreases in growth rates due to the effects of higher CO_2 concentrations on ocean chemistry. Impacts on coral reefs from those factors will not be uniform throughout the small-island realm. For instance, the geographical variability in the required thermal adaptation derived from models and emissions scenarios presented by Donner et al. (2005) suggest that coral reefs in some regions, such as Micronesia and western Polynesia, may be particularly vulnerable to climate change. In addition to these primarily climate-driven factors, the impacts of which are detailed in Chapter 6, Section 6.2.1, there are those associated mainly with other human activities, which combine to subject island coral reefs to multiple stresses, as illustrated in Box 16.2.

16.4.3 Agriculture, fisheries and food security

Small islands have traditionally depended upon subsistence and cash crops for survival and economic development. While subsistence agriculture provides local food security, cash crops (such as sugar cane, bananas and forest products) are exported in order to earn foreign exchange. In Mauritius, the sugar cane industry has provided economic growth and has contributed to the diversification of the economy through linkages with tourism and other related industries (Government of Mauritius, 2002). However, exports have depended upon preferential access to major developed-country markets, which are slowly eroding. Many island states have also experienced a decrease in GDP contributions from agriculture, partly due to the drop in competitiveness of cash crops, cheaper imports from larger countries, increased costs of maintaining soil fertility, and competing uses for water resources, especially from tourism (FAO, 2004).

Local food production is vital to small islands, even those with very limited land areas. In the Pacific islands subsistence agriculture has existed for several hundred years. The ecological dependency of small island economies and societies is well recognised (ADB, 2004). A report by the FAO Commission on Genetic Resources found that some countries' dependence on plant genetic resources ranged from 91% in Comoros, 88% in Jamaica, 85% in Seychelles to 65% in Fiji, 59% in the Bahamas and 37% in Vanuatu (Ximena, 1998).

Projected impacts of climate change include extended periods of drought and, on the other hand, loss of soil fertility and degradation as a result of increased precipitation, both of which will negatively impact on agriculture and food security. In a study of the economic and social implications of climate change and variability for selected Pacific islands, the World Bank (2000) found that in the absence of adaptation, a high island such as Viti Levu, Fiji, could experience damages of US$23 million to 52 million/yr by 2050, (equivalent to 2 to 3% of Fiji's GDP in 1998). A group of low islands such as Tarawa, Kiribati, could face average annual damages of more than US$8 million to 16 million/yr (equivalent to 17 to 18% of Kiribati's GDP in 1998) under the SRES A2 and B2 emissions scenarios.

Box 16.2. Non-climate-change threats to coral reefs of small islands

A large number of non-climate-change stresses and disturbances, mainly driven by human activities, can impact coral reefs (Nyström et al., 2000; Hughes et al., 2003). It has been suggested that the 'coral reef crisis' is almost certainly the result of complex and synergistic interactions among global-scale climatic stresses and local-scale, human-imposed stresses (Buddemeier et al., 2004).

In a study by Bryant et al. (1998), four human-threat factors – coastal development, marine pollution, over-exploitation and destructive fishing, and sediment and nutrients from inland – provide a composite indicator of the potential risk to coral reefs associated with human activity for 800 reef sites. Their map (Figure 16.1) identifies low-risk (blue) medium-risk (yellow) and high-risk (red) sites, the first being common in the insular central Indian and Pacific Oceans, the last in maritime South-East Asia and the Caribbean archipelago. Details of reefs at risk in the two highest-risk areas have been documented by Burke et al. (2002) and Burke and Maidens (2004), who indicate that about 50% of the reefs in South-East Asia and 45% in the Caribbean are classed in the high- to very high-risk category. There are, however, significant local and regional differences in the scale and type of threats to coral reefs in both continental and small-island situations.

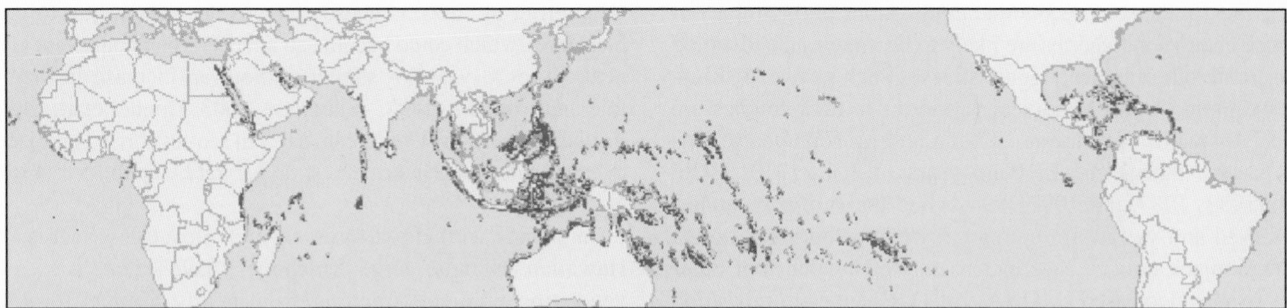

Figure 16.1. *The potential risk to coral reefs from human-threat factors. Low risk (blue), medium risk (yellow) and high risk (red). Source: Bryant et al. (1998).*

Recognising that coral reefs are especially important for many small island states, Wilkinson (2004) notes that reefs on small islands are often subject to a range of non-climate impacts. Some common types of reef disturbance are listed below, with examples from several island regions and specific islands.

1. Impact of coastal developments and modification of shorelines:
   - coastal development on fringing reefs, Langawi Island, Malaysia (Abdullah et al., 2002);
   - coastal resort development and tourism impacts in Mauritius (Ramessur, 2002).
2. Mining and harvesting of corals and reef organisms:
   - coral harvesting in Fiji for the aquarium trade (Vunisea, 2003).
3. Sedimentation and nutrient pollution from the land:
   - sediment smothering reefs in Aria Bay, Palau (Golbuua et al., 2003) and southern islands of Singapore (Dikou and van Woesik, 2006);
   - non-point source pollution, Tutuila Island, American Samoa (Houk et al., 2005);
   - nutrient pollution and eutrophication, fringing reef, Réunion (Chazottes et al., 2002) and Cocos Lagoon, Guam (Kuffner and Paul, 2001).
4. Over-exploitation and damaging fishing practices:
   - blast fishing in the islands of Indonesia (Fox and Caldwell, 2006);
   - intensive fish-farming effluent in Philippines (Villanueva et al., 2006);
   - subsistence exploitation of reef fish in Fiji (Dulvy et al., 2004);
   - giant clam harvesting on reefs, Milne Bay, Papua New Guinea (Kinch, 2002).
5. Introduced and invasive species:
   - Non-indigenous species invasion of coral habitats in Guam (Paulay et al., 2002).

There is another category of 'stress' that may inadvertently result in damage to coral reefs – the human component of poor governance (Goldberg and Wilkinson, 2004). This can accompany political instability, one example being problems with contemporary coastal management in the Solomon Islands (Lane, 2006).

Not all effects of climate change on agriculture are expected to be negative. For example, increased temperatures in high-latitude islands are likely to make conditions more suitable for agriculture and provide opportunities to enhance resilience of local food systems (see also Chapter 15, Section 15.5).

If the intensity of tropical cyclones increases, a concomitant rise in significant damage to food crops and infrastructure is likely. For example, Tropical Cyclone Ofa in 1990 turned Niue (in the Pacific) from a food-exporting country into one dependent on imports for the next two years, and Heta in 2004 had an even greater impact on agricultural production in Niue (Wade, 2005). Hurricane Ivan's impact on Grenada (in the Caribbean) in 2004 caused losses in the agricultural sector equivalent to 10% of GDP. The two main crops, nutmeg and cocoa, both of which have long gestation periods, will not make a contribution to GDP or earn foreign exchange for the next 10 years (OECS, 2004).

Fisheries contribute significantly to GDP on many islands; consequently the socio-economic implications of the impact of climate change on fisheries are likely to be important and would exacerbate other anthropogenic stresses such as over-fishing. For example, in the Maldives, variations in tuna catches are especially significant during El Niño and La Niña years. This was shown during the El Niño years of 1972/1973, 1976, 1982/1983, 1987 and 1992/1994, when the skipjack catches decreased and yellow fin increased, whereas during La Niña years skipjack tuna catches increased, whilst catches of other tuna species decreased (MOHA, 2001). Changes in migration patterns and depth are two main factors affecting the distribution and availability of tuna during those periods, and it is expected that changes in climate would cause migratory shifts in tuna aggregations to other locations (McLean et al., 2001). Apart from the study by Lehodey et al. (2003) of potential changes in tuna fisheries, Aaheim and Sygna (2000) surveyed possible economic impacts in terms of quantities and values, and give examples of macroeconomic impacts. The two main effects of climate change on tuna fishing are likely to be a decline in the total stock and a migration of the stock eastwards, both of which will lead to changes in the catch in different countries.

In contrast to agriculture, the mobility of fish makes it difficult to estimate future changes in marine fish resources. Furthermore, since the life cycles of many species of commercially exploited fisheries range from freshwater to ocean water, land-based and coastal activities will also be likely to affect the populations of those species. Coral reefs and other coastal ecosystems which may be severely affected by climate change will also have an impact on fisheries (Graham et al., 2006).

16.4.4 Biodiversity

Oceanic islands often have a unique biodiversity through high endemism (i.e., with regionally restricted distribution) caused by ecological isolation. Moreover, human well-being on most small islands is heavily reliant on ecosystem services such as amenity value and fisheries (Wong et al., 2005). Historically, isolation – by its very nature – normally implies immunity from threats such as invasive species causing the extinction of endemics. However, it is possible that in mid- and high-latitude islands, higher temperature and the retreat and loss of snow cover could enhance conditions for the spread of invasive species and forest cover (Smith et al., 2003; see also Chapter 15, Section 15.6.3). For example, in species-poor, sub-Antarctic island ecosystems, alien microbes, fungi, plants and animals have been extensively documented as causing substantial loss of local biodiversity and changes to ecosystem function (Frenot et al., 2005). With rapid climate change, even greater numbers of introductions and enhanced colonisation by alien species are likely, with consequent increases in impacts on these island ecosystems. Climate-related ecosystem effects are also already evident in the mid-latitudes, such as on the island of Hokkaido, Japan, where a decrease in alpine flora has been reported (Kudo et al., 2004).

Under the SRES scenarios, small islands are shown to be particularly vulnerable to coastal flooding and decreased extent of coastal vegetated wetlands (Nicholls, 2004). There is also a detectable influence on marine and terrestrial pathogens, such as coral diseases and oyster pathogens, linked to ENSO events (Harvell et al., 2002). These changes are in addition to coral bleaching, which could become an annual or biannual event in the next 30 to 50 years or sooner without an increase in thermal tolerance of 0.2 to 1.0°C (Sheppard, 2003; Donner et al., 2005). Furthermore, in the Caribbean, a 0.5 m sea-level rise is projected to cause a decrease in turtle nesting habitat by up to 35% (Fish et al., 2005).

In islands with cloud forest or high elevations, such as the Hawaiian Islands, large volcanoes have created extreme vegetation gradients, ranging from nearly tropical to alpine (Foster, 2001; Daehler, 2005). In these ecosystems, anthropogenic climate change is likely to combine with past land-use changes and biological invasions to drive several species such as endemic birds to extinction (Benning et al., 2002). This trend among Hawaiian forest birds shows concordance with the spread of avian malaria, which has doubled over a decade at upper elevations and is associated with breeding of mosquitoes and warmer summertime air temperatures (Freed et al., 2005).

In the event of increasing extreme events such as cyclones (hurricanes) (see Section 16.3.1.3) forest biodiversity could be severely affected, as adaptation responses on small islands are expected to be slow, and impacts of storms may be cumulative. For example, Ostertag et al. (2005) examined long-term tropical moist forests on the island of Puerto Rico in the Caribbean. Hurricane-induced mortality of trees after 21 months was 5.2%/yr; more than seven times higher than background mortality levels during the non-hurricane periods. These authors show that resistance of trees to hurricane damage is not only correlated with individual and species characteristics, but also with past disturbance history, which suggests that individual storms cannot be treated as discrete, independent events when interpreting the effects of hurricanes on forest structure.

16.4.5 Human settlements and well-being

The concentration of large settlements along with economic and social activities at or near the coast is a well-documented feature of small islands. On Pacific and Indian Ocean atolls, villages are located on low and narrow islands, and in the Caribbean more than half of the population live within 1.5 km of the shoreline. In many regions of small islands, such as along the north coast of Jamaica and along the west and south coasts of Barbados, continuous corridors of development now occupy practically all of the prime coastal lands. Fishing villages, government buildings and important facilities such as hospitals are frequently located close to the shore. Moreover, population growth and internal migration of people are putting additional pressure on coastal settlements, utilities and resources, and creating problems in areas such as pollution, waste disposal and housing. Changes in sea level, and any changes in the magnitude and frequency of storm events, are likely to have serious consequences for these land uses. On the other hand, rural and inland settlements and communities are more likely to be adversely affected by negative impacts on agriculture, given that they are often dependent upon crop production for many of their nutritional requirements.

An important consideration in relation to settlements is housing. In many parts of the Pacific, traditional housing styles, techniques and materials were resistant to damage and/or could be repaired quickly. Moves away from traditional housing have increased vulnerability to thermal stress, slowed housing reconstruction after storms and flooding, and in some countries increased the use of air-conditioning. As a result, human well-being in several major settlements on islands in the Pacific and Indian Oceans has changed over the past two or three decades, and there is growing concern over the possibility that global climate change and sea-level rise are likely to impact human health and well-being, mostly in adverse ways (Hay et al., 2003).

Many small island states currently suffer severe health burdens from climate-sensitive diseases, including morbidity and mortality from extreme weather events, certain vector-borne diseases, and food- and water-borne diseases (Ebi et al., 2006). Tropical cyclones, storm surges, flooding, and drought have both short- and long-term effects on human health, including drowning, injuries, increased disease transmission, decreases in agricultural productivity, and an increased incidence of common mental disorders (Hajat et al., 2003). Because the impacts are complex and far-reaching, the true health burden is rarely appreciated. For example, threats to health posed by extreme weather events in the Caribbean include insect- and rodent-borne diseases, such as dengue, leptospirosis, malaria and yellow fever; water-borne diseases, including schistosomiasis, cryptosporidium and cholera; food-borne diseases, including diarrhoeal diseases, food poisoning, salmonellosis and typhoid; respiratory diseases, including asthma, bronchitis and respiratory allergies and infections; and malnutrition resulting from disturbances in food production or distribution (WHO, 2003a).

Many small island states lie in tropical or sub-tropical zones with weather conducive to the transmission of diseases such as malaria, dengue, filariasis, schistosomiasis, and food- and water-borne diseases. The rates of many of these diseases are increasing in small island states for a number of reasons, including poor public health practices, inadequate infrastructure, poor waste management practices, increasing global travel and changing climatic conditions (WHO, 2003a). In the Caribbean, the incidence of dengue fever increases during the warm years of ENSO cycles (Rawlins et al., 2005). Because the greatest risk of dengue transmission is during annual wet seasons, vector control programs need to target these periods to reduce disease burdens. The incidence of diarrhoeal diseases is associated with annual average temperature (Singh et al., 2001) and negatively associated with water availability in the Pacific (Singh et al., 2001). Therefore, increasing temperatures and decreasing water availability due to climate change may increase burdens of diarrhoeal and other infectious diseases in some small island states.

Outbreaks of climate-sensitive diseases can be costly in terms of lives and economic impacts. An outbreak of dengue fever in Fiji coincided with the 1997/1998 El Niño; out of a population of approximately 856,000 people, 24,000 were affected, with 13 deaths (World Bank, 2000). The epidemic cost US$3 million to 6 million. Neighbouring islands were also affected.

Ciguatera fish poisoning is common in marine waters, particularly reef waters. Multiple factors contribute to outbreaks of ciguatera poisoning, including pollution and reef degradation. Warmer sea surface temperatures during El Niño events have been associated with ciguatera outbreaks in the Pacific (Hales et al., 1999).

16.4.6 Economic, financial and socio-cultural impacts

Small island states have special economic characteristics which have been documented in several reports (Atkins et al., 2000; ADB, 2004; Briguglio and Kisanga, 2004; Grynberg and Remy, 2004). Small economies are generally more exposed to external shocks, such as extreme events and climate change, than larger countries, because many of them rely on one or a few economic activities such as tourism or fisheries. Recent conflicts in the Gulf region have, for example, affected tourism arrivals in the Maldives and the Seychelles; while internal conflicts associated with coups have had similar effects on the tourism industry in Fiji (Becken, 2004). In the Caribbean, hurricanes cause loss of life, property damage and destruction, and economic losses running into millions of dollars (ECLAC, 2002; OECS, 2004). The reality of island vulnerability is powerfully demonstrated by the near-total devastation experienced on the Caribbean island of Grenada when Hurricane Ivan made landfall in September 2004. Damage assessments indicate that, in real terms, the country's socio-economic development has been set back at least a decade by this single event that lasted for only a few hours (see Box 16.3).

Tourism is a major economic sector in many small islands, and its importance is increasing. Since their economies depend so highly on tourism, the impacts of climate change on tourism resources in small islands will have significant effects, both direct and indirect (Bigano et al., 2005; Viner, 2006). Sea-level rise and increased sea water temperatures are projected to accelerate beach erosion, cause degradation of natural coastal defences such as mangroves and coral reefs, and result in the loss of cultural heritage on coasts affected by inundation and flooding. These impacts will in turn reduce attractions for coastal tourism. For example, the sustainability of island tourism resorts in Malaysia is expected to be compromised by rising sea level, beach erosion and saline contamination of coastal wells, a major source of water supply for island resorts (Tan and Teh, 2001). Shortage of water and increased risk of vector-borne diseases may steer tourists away from small islands, while warmer climates in the higher-latitude countries may also result in a

Box 16.3. Grenada and Hurricane Ivan

Hurricane Ivan struck Grenada on 7 September 2004, as a category 4 system on the Saffir-Simpson scale. Sustained winds reached 140 mph, with gusts exceeding 160 mph. An official OECS/UN-ECLAC Assessment reported the following:

- 28 people killed,
- overall damages calculated at twice the current GDP,
- 90% of housing stock damaged,
- 90% of guest rooms in the tourism sector damaged or destroyed, equivalent to approximately 29% GDP,
- losses in telecommunications equivalent to 13% GDP,
- damage to schools and education infrastructure equivalent to 20% GDP,
- losses in agricultural sector equivalent to 10% GDP. The two main crops, nutmeg and cocoa, which have long gestation periods, will not contribute to GDP or earn foreign exchange for the next 10 years,
- damage to electricity installations totalling 9% GDP,
- heavy damage to eco-tourism and cultural heritage sites, resulting in 60% job losses in the sub-sector,
- prior to Hurricane Ivan, Grenada was on course to experience an economic growth rate of approximately 5.7% *per annum* but negative growth of around –1.4% *per annum* is now forecast.

Source: OECS (2004).

reduction in the number of people who want to visit small islands in the tropical and sub-tropical regions.

Tourism in small island states is also vulnerable to climate change through extreme events and sea-level rise leading to transport and communication interruption. In a study of tourist resorts in Fiji, Becken (2005) suggested that many operators already prepare for climate-related events, and therefore are adapting to potential impacts from climate change. She also concludes that reducing greenhouse gas emissions from tourist facilities is not important to operators; however, decreasing energy costs is practised for economic reasons.

Climate change may also affect important environmental components of holiday destinations, which could have repercussions for tourism-dependent economies. The importance of environmental attributes in determining the choice and enjoyment of tourists visiting Bonaire and Barbados, two Caribbean islands with markedly different tourism markets and infrastructure, and possible changes resulting from climate change (coral bleaching and beach erosion) have been investigated by Uyarra et al. (2005). They concluded that such changes would have significant impacts on destination selection by visitors, and that island-specific strategies, such as focusing resources on the protection of key tourist assets, may provide a means of reducing the environmental impacts and economic costs of climate change. Equally, the attractions of 'cold water islands' (e.g., the Falklands, Prince Edward Island, Baffin, Banks and Lulea) could be compromised, as these destinations seek to expand their tourism sectors (Baldacchino, 2006).

16.4.7 Infrastructure and transportation

Like settlements and industry, the infrastructural base that supports the vital socio-economic sectors of island economies tends to occupy coastal locations. Hay et al. (2003) have identified several challenges that will confront the transportation sector in Pacific island countries as a result of climate variability and change. These include closure of roads, airports and bridges due to flooding and landslides, and damage to port facilities. The resulting disruption would not be confined to the transportation sector alone, but would impact other key dependent sectors and services including tourism, agriculture, the delivery of health care, clean water, food security and market supplies.

In most small islands, energy is primarily from non-renewable sources, mainly from imported fossil fuels. In the context of climate change, the main contribution to greenhouse gas emissions is from energy use. The need to introduce and expand renewable energy technologies in small islands has been recognised for many years although progress in implementation has been slow. Often, the advice that small islands receive on options for economic growth is based on the strategies adopted in larger countries, where resources are much greater and alternatives significantly less costly. It has been argued by Roper (2005) that small island states could set an example on green energy use, thereby contributing to local reductions in greenhouse gas emissions and costly imports. Indeed, some have already begun to become 'renewable energy islands'. La Desirade (Caribbean), Fiji, Samsoe (Denmark), Pellworm (Germany) and La Réunion (Indian Ocean) are cited as presently generating more than 50% of their electricity from renewable energy sources (Jensen, 2000).

Almost without exception, international airports on small islands are sited on or within a few kilometres of the coast, and on tiny coral islands. Likewise, the main (and often only) road network runs along the coast (Walker and Barrie, 2006). In the South Pacific region of small islands, Lal (2004) estimates that, since 1950, mean sea level has risen at a rate of approximately 3.5 mm/yr, and he projects a rise of 25 to 58 cm by the middle of this century. Under these conditions, much of the infrastructure in these countries would be at serious risk from inundation, flooding and physical damage associated with coastal land loss. While the risk will vary from country to country, the small islands of the Indian Ocean and the Caribbean – countries such as Malta and Singapore and mid-latitude islands such as the Îles-de-la-Madeleine in the Gulf of St. Lawrence – may be confronted by similar threats. Raksakulthai (2003) has shown that climate change would also increase the risk to critical facilities on the island of Phuket, a premier tourism island in South-East Asia.

The threat from sea-level rise to infrastructure on small islands could be amplified considerably by the passage of tropical cyclones (hurricanes). It has been shown, for instance, that port facilities at Suva, Fiji, and Apia, Samoa, would experience overtopping, damage to wharves, and flooding of the hinterland if there were a 0.5 m rise in sea level combined with waves associated with a 1-in-50 year cyclone (Hay et al., 2003). In the Caribbean, the damage to coastal infrastructure from storm surge alone has been severe. In November 1999, surge damage in St. Lucia associated with Hurricane Lenny was in excess of US$6 million, even though the storm was centred many kilometres offshore.

16.5 Adaptation: practices, options and constraints

16.5.1 Role of adaptation in reducing vulnerability and impacts

It is clear from the previous sections that small islands are presently subjected to a range of climatic and oceanic impacts, and that these impacts will be exacerbated by ongoing climate change and sea-level rise. Moreover, the TAR showed that the overall vulnerability of small island states is primarily a function of four interrelated factors:

- the degree of exposure to climate change;
- their limited capacity to adapt to projected impacts;
- the fact that adaptation to climate change is not a high priority, given the more pressing problems that small islands have to face;
- the uncertainty associated with global climate change projections and their local validity (Nurse et al., 2001).

Several other factors that influence vulnerability and impacts on small islands have also been identified in the present chapter, including both global and local processes. This combination of drivers is likely to continue into the future, which raises the possibility that environmental conditions and the socio-economic well-being of populations on small islands will worsen unless adaptation measures are put in place to reduce impacts, as illustrated in Box 16.4.

While it is clear that implementing anticipatory adaptation strategies early on is desirable (see Box 16.4), there are obstacles associated with the uncertainty of the climate change projections. To overcome this uncertainty, Barnett (2001) has suggested that a better strategy for small islands is to enhance the resilience of whole-island socio-ecological systems, rather than concentrating on sectoral adaptation; a theme that is expanded upon in Section 16.5.5. This is the policy of the Organisation of Eastern Caribbean States (OECS, 2000).

Inhabitants of small islands, individuals, communities and governments, have adapted to interannual variability in climate and sea conditions, as well as to extreme events, over a long period of time. There is no doubt that this experience will be of value in dealing with inter-annual variability and extremes in climate and sea conditions that are likely to accompany the longer-term mean changes in climate and sea level. Certainly, in Polynesia, Melanesia and Micronesia, and in the Arctic, the socio-ecological systems have historically been able to adapt to environmental change (Barnett, 2001; Berkes and Jolly, 2001). However, it is also true that in many islands traditional mechanisms for coping with environmental hazards are being, or have been, lost, although paradoxically the value of such mechanisms is being increasingly recognised in the context of adaptation to climate change (e.g., MESD, 1999; Fox, 2003).

16.5.2 Adaptation options and priorities: examples from small island states

What are the adaptation options and priorities for small islands, and especially for small island states? Since the TAR there have been a number of National Communications to the United Nations Framework Convention on Climate Change (UNFCCC) from small island states that have assessed their own vulnerability to climate change and in-country adaptation strategies. These communications give an insight into national concerns about climate change, the country's vulnerability, and the priorities that different small island states place on adaptation options. They also suggest that to date adaptation has been reactive, and has been centred around responses to the effects of climate variability and particularly climate extremes. Moreover, the range of measures considered, and the priority they are assigned, appear closely linked to the country's key socio-economic sectors, their key environmental concerns, and/or the most vulnerable areas to climate change and/or sea-level rise. Some island states such as Malta (MRAE, 2004) emphasise potential adaptations to economic factors including power generation, transport, and waste management, whereas agriculture and human health figure prominently in communications from the Comoros (GDE, 2002), Vanuatu (Republic of Vanuatu, 1999), and St. Vincent and the Grenadines (NEAB, 2000). In these cases, sea-level rise is not seen as a critical issue, though it is in the low-lying atoll states such as Kiribati, Tuvalu, Marshall Islands and the Maldives. The Maldives provides one example of the sectors it sees as being the most vulnerable to climate change, and the adaptive measures required to reduce vulnerability and enhance resilience (see Box 16.5).

In spite of differences in emphasis and sectoral priorities, there are three common themes.

- First, all National Communications emphasise the urgency for adaptation action and the need for financial resources to support such action.
- Second, freshwater is seen as a critical issue in all small island states, both in terms of water quality and quantity.Water is a multi-sectoral resource that impinges on

Box 16.4. Future island conditions and well-being: the value of adaptation

Global change and regional/local change will interact to impact small islands in the future. Both have physical and human dimensions. Two groups of global drivers are identified in the top panel of Figure 16.2: first, climate change including global warming and sea-level rise and, second, externally driven socio-economic changes such as the globalisation of economic activity and international trade (Singh and Grünbühel, 2003). In addition to these global processes, small islands are also subject to important local change influences, such as population pressure and urbanisation, which increase demand on the local resource base and expand the ecological footprint (Pelling and Uitto, 2001).

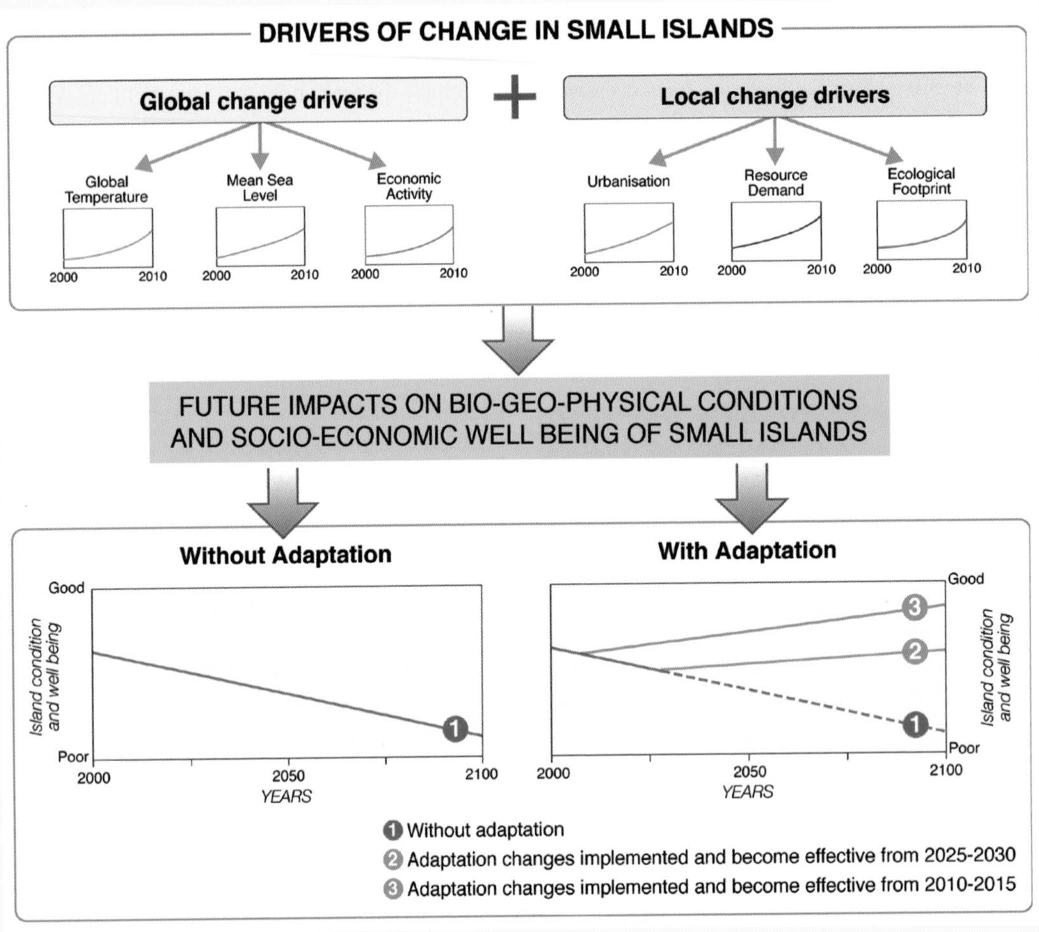

Figure 16.2. *Drivers of change in small islands and the implications for island condition and well-being under no adaptation and the near-term and mid-term implementation of adaptation. Adapted from Harvey et al. (2004).*

In general, both global and local drivers can be expected to show increases in the future. These will probably impact on island environments and their bio-geophysical conditions, as well as on the socio-economic well-being of island communities (Clark, 2004).

Three possible scenarios are illustrated in the lower panel. Implicitly, and without adaptation, environmental conditions and human well-being are likely to get worse in the future (line 1). On the other hand, if effective adaptation strategies are implemented, both the bio-geophysical conditions and socio-economic well-being of islanders should improve. It is suggested that the earlier this is done, the better the outcome (lines 2 and 3).

Source: Harvey et al. (2004).

Box 16.5. Adaptive measures in the Maldives

Adaptation options in low-lying atoll islands which have been identified as especially vulnerable, are limited, and response measures to climate change or its adverse impacts are potentially very costly. In the Maldives adaptation covers two main types of activities. First, there are adaptive measures involving activities targeted at specific sectors where climate change impacts have been identified. Second, there are adaptive measures aimed at enhancing the capacity of the Maldives to effectively implement adaptations to climate change and sea-level rise. Within these two activities the Maldivian Ministry of Home Affairs, Housing and Environment has identified several vulnerable areas and adaptive measures that could be implemented to reduce climate change impacts.

Vulnerable area	Adaptation response
Land loss and beach erosion	Coastal protection Population consolidation i.e., reduction in number of inhabited islands Ban on coral mining
Infrastructure and settlement damage	Protection of international airport Upgrading existing airports Increase elevation in the future
Damage to coral reefs	Reduction of human impacts on coral reefs Assigning protection status for more reefs
Damage to tourism industry	Coastal protection of resort islands Reduce dependency on diving as a primary resort focus Economy diversification
Agriculture and food security	Explore alternate methods of growing fruits, vegetables and other foods Crop production using hydroponic systems
Water resources	Protection of groundwater Increasing rainwater harvesting and storage capacity Use of solar distillation Management of storm water Allocation of groundwater recharge areas in the islands
Lack of capacity to adapt (both financial and technical)	Human resource development Institutional strengthening Research and systematic observation Public awareness and education

Source: MOHA (2001).

all facets of life and livelihood, including security. It is seen as a problem at present and one that will increase in the future.

- Third, many small island states, including all the Least Developed Countries (Small Island Developing States, SIDS), see the need for more integrated planning and management, be that related to water resources, the coastal zone, human health, or tourism.

In a case study of tourism in Fiji, for instance, Becken (2004) argues that the current tourism policy focuses on adaptation and measures that are predominantly reactive rather than proactive, whereas climate change measures that offer win-win situations should be pursued. These include adaptation, mitigation, and wider environmental management measures; examples being reforestation of native forest, water conservation, and the use of renewable energy resources (Becken, 2004). A similar view is held by Stern (2007), who notes that climate change adaptation policies and measures, if implemented in a timely and efficient manner, can generate valuable co-benefits such as enhanced energy security and environmental protection.

The need to implement adaptation measures in small islands with some urgency has recently been reinforced by Nurse and Moore (2005), and was also highlighted in the TAR, where it was suggested that risk-reduction strategies, together with other sectoral policy initiatives, in areas such as sustainable development planning, disaster prevention and management, integrated coastal zone management, and health care planning could be usefully employed (Nurse et al., 2001). Since then a number of projects on adaptation in several small islands have adopted this suggestion. These projects aim to build the capacities of individuals, communities and governments so that they are more able to make informed decisions about adaptation to climate change and to enhance their adaptive capacity in the long run.

There are few published studies that have attempted to estimate climate change adaptation costs for small islands, and much more work needs to be undertaken on the subject. The most recent study was conducted by Ng and Mendelsohn (2005), who found coastal protection to be the least-cost strategy to combat sea-level rise in Singapore, under three scenarios. They noted that the annual cost of shoreline protection would increase as sea-level rises, and would range from US$0.3–5.7 million by 2050 to US$0.9–16.8 million by 2100 (Ng and Mendolsohn, 2005). It was concluded that it would be more costly to the country to allow the coast to become inundated than to defend it. Studies of this type could provide useful guidance to island governments in the future, as they are confronted with the difficult task of making adaptation choices.

16.5.3 Adaptation of 'natural' ecosystems in island environments

The natural adaptation of small-island ecosystems is considered in very few National Communications. Instead attention is mostly focused on: (1) protecting those ecosystems that are projected to suffer as a consequence of climate change and sea-level rise; and (2) rehabilitating ecosystems degraded or destroyed as a result of socio-economic developments.

One group of natural island environments in low latitudes are the tropical rainforests, savannas and wetlands that occupy the inland, and often upland, catchment areas of the larger, higher and topographically more complex islands, such as Mauritius in the Indian Ocean, the Solomon Islands in the Pacific, and Dominica in the Caribbean. Very little work has been done on the potential impact of climate change on these highly biodiverse systems, or on their adaptive capacity.

On the other hand, the potential impact of global warming and sea-level rise on natural coastal systems, such as coral reefs and mangrove forests, is now reasonably well known. For these ecosystems several possible adaptation measures have been identified. In those coral reefs and mangrove forests that have not been subjected to significant degradation or destruction as a result of human activities, natural or 'autonomous' adaptation, which represents the system's natural adaptive response and is triggered by changes in climatic stimuli, can take place. For instance, some corals may be able to adapt to higher sea surface and air temperatures by hosting more temperature-tolerant symbiotic algae (see Chapter 4, Box 4.4). They can also grow upwards with the rise in sea level, providing that vertical accommodation space is available (Buddemeier et al., 2004). Similarly, mangrove forests can migrate inland, as they did during the Holocene sea-level transgression, providing that there is horizontal accommodation space and they are not constrained by the presence of infrastructure and buildings; i.e., by 'coastal squeeze' (Alongi, 2002).

In addition to autonomous adaptation, both restoration and rehabilitation of damaged mangrove and reef ecosystems can be seen as 'planned' adaptation mechanisms aimed to increase natural protection against sea-level rise and storms, and to provide resources for coastal communities. In small islands, such projects have usually been community-based and are generally small-scale. In the Pacific islands, successful mangrove rehabilitation projects have been recorded from Kiribati, Northern Mariana Islands, Palau and Tonga, with failed efforts in American Samoa and Papua New Guinea (Gillman et al., 2006). Improved staff training, capacity building, and information sharing between coastal managers is needed for successful mangrove rehabilitation (Lewis, 2005). More ambitious, costly and technical projects include an ecosystem restoration programme in the Seychelles, which aims ultimately to translocate globally threatened coastal birds as well as rehabilitating native coastal woodlands on eleven islands in the country (Henri et al., 2004).

16.5.4 Adaptation: constraints and opportunities

There are several constraints to adaptation that are inherent in the very nature of many small islands, including small size, limited natural resources, and relative isolation, and it is because of these characteristics that some autonomous small islands have been recognised in the United Nations process as either Least Developed Countries (LDCs) or SIDS. Not all small islands satisfy these criteria, notably those linked closely with global finance or trade, as well as the non-autonomous islands within larger countries. While these two groups of islands will share some of the constraints of small island states, they are not emphasised in this section.

16.5.4.1 Lack of adaptive capacity

The main determinants of a country's adaptive capacity to climate change are: economic wealth, technology, information and skills, infrastructure, institutions and equity (WHO, 2003b). A common constraint confronting most small island states is the lack of in-country adaptive capacity, or the ease with which they are able to cope with climate change. In many autonomous small islands the cost of adopting and implementing adaptation options is likely to be prohibitive, and a significant proportion of a country's economic wealth. Financial resources that are generally not available to island governments would need to come from outside (Rasmussen, 2004). This need for international support to assist with the adaptation process in vulnerable, developing countries is also strongly emphasised by Stern (2007). Similarly, there are often inadequate human resources available to accommodate, cope with, or benefit from the effects of climate change; a situation that may be compounded by the out-migration of skilled workers (Voigt-Graf, 2003). To overcome this deficiency, the adaptive capacity of small island states will need to be built up in several important areas including human resource development, institutional strengthening, technology and infrastructure, and public awareness and education.

An extreme example of these deficiencies is the recently independent state of Timor Leste (East Timor). Timor Leste is vulnerable to climate change, as evidenced by existing sensitivities to climate events, for example drought and food shortages in the western highlands, and floods in Suai. Barnett et al. (2003) note that relevant planning would address the present problems as well as future climate risks, and conclude that activities that promote sustainable development, human health, food security, and renewable energy can reduce the risk of future damages caused by climate change as well as improving living standards. In short, "change in climate is a long-term problem for Timor Leste, but climate change policies can be positive opportunities" (Barnett et al., 2003).

16.5.4.2 Adaptation and global integration

This last theme is also developed by Pelling and Uitto (2001), who suggest that change at the global level is a source of new opportunities, as well as constraints, for building local resilience. They argue that small island populations have been mobile, both historically and at present, and that remittances from overseas relatives help to moderate economic risks and increase family resiliency on home islands. They also recognise that this is a critical time for small islands, which must contend with ongoing development pressures, economic liberalisation, and the growing pressures from risks associated with climate change and sea-level rise. They conclude, following a case study of Barbados, that efforts to enhance island resilience must be mainstreamed into general development policy formulation, and that adaptations should not be seen as separate or confined to engineering or land-use planning-based realms (Pelling and Uitto, 2001).

Barnett (2001) discusses the potential impact of economic liberalisation on the resilience of Pacific island communities to climate change. He argues that many small island societies have proved resilient in the past to social and environmental upheaval. The key parameters of this resilience include: opportunities for migration and subsequent remittances; traditional knowledge, institutions and technologies; land and shore tenure regimes; the subsistence economy; and linkages between formal state and customary decision-making processes. However, this resilience may be undermined as the small island states become increasingly integrated into the world economy through, for example, negotiations for fishery rights in their Exclusive Economic Zones, and international tourism (Barnett, 2001).

These global economic processes, together with global warming, sea-level rise, and possibly increased frequency and intensity of extreme weather events, make it difficult for autonomous small islands to achieve an appropriate degree of sustainability, which Barnett and Adger (2003) suggest is one of the goals of adaptation to climate change. They maintain that for the most vulnerable small island states (those composed of low-lying atolls), this combination of global processes interacting with local socio-economic and environmental conditions puts the long-term ability of humans to inhabit atolls at risk, and that this risk constitutes a 'dangerous' level of climatic change that may well undermine their national sovereignty (see Box 16.6).

This discussion highlights the role of resilience – both its biophysical and human aspects – as a critical component in developing the adaptive capacity of small island states, a role that has effectively emerged since publication of the TAR. In a recent study of the Cayman Islands, Tompkins (2005) found that self-efficacy, strong local and international support networks, combined with a willingness to act collectively and to learn from mistakes, appeared to have increased the resilience of the Government to tropical storm risk, implying that such resilience can also contribute to the creation of national level adaptive capacity to climate change, thereby reducing vulnerability.

16.5.4.3 Risk-sharing and insurance

Insurance is another way of reducing vulnerability and is increasingly being discussed in the context of small islands and climate change. However, there are several constraints to transferring or sharing risk in small islands. These include the limited size of the risk pool, and the lack of availability of financial instruments and services for risk management. For instance, in 2004, Cyclone Heta devastated the tiny island of Niue in the South-West Pacific, where no insurance is available against weather extremes, leaving the island almost entirely reliant on overseas aid for reconstruction efforts (Hamilton, 2004). Moreover, the relative costs of natural disasters tend to be far higher in developing countries than in advanced economies. Rasmussen (2004) shows that autonomous small islands are especially vulnerable, with natural disasters in the countries of the Eastern Caribbean shown to have had a discernible macroeconomic impact, including large effects on fiscal and external balances, pointing to an important role for precautionary measures.

Thus, in many small island countries, the implementation of specific instruments and services for risk-sharing may be required. Perhaps recent initiatives on financial risk transfer mechanisms through traditional insurance structures and new financial instruments, such as catastrophe bonds, weather derivatives, micro-insurance, and a regional pooling arrangement for small island states, might provide them with the flexibility for this form of adaptation (Auffret, 2003; Hamilton, 2004; Swiss Re, 2004). However, as Epstein and Mills (2005) point out, the economic costs of adapting to climate-related risks are spread among a range of stakeholders including governments, insurers, business, non-profit entities and individuals. They also note that sustainable development can contribute to managing and maintaining the insurability of climate change risk, though development projects can be stranded where financing is contingent on insurance, particularly with respect to coastlines and shorelines vulnerable to sea-level rise (Epstein and Mills, 2005).

Climate change adaptation projects can also founder in other ways, either at the implementation stage or when projects that rely wholly on external personnel or financing are completed. For this reason, Westmacott (2002) believes that integrated coastal management in the Pacific should incorporate conflict

Box 16.6. Climate dangers and atoll countries

"Climate change puts the long-term sustainability of societies in atoll nations at risk. The potential abandonment of sovereign atoll countries can be used as the benchmark of the 'dangerous' change that the UNFCCC seeks to avoid. This danger is as much associated with the narrowing of adaptation options and the role of expectations of impacts of climate change as it is with uncertain potential climate-driven physical impacts. The challenges for research are to identify the thresholds of change beyond which atoll socio-ecological systems collapse and to assess how likely these thresholds are to be breached. These thresholds may originate from social as well as environmental processes. Further, the challenge is to understand the adaptation strategies that have been adopted in the past and which may be relevant for the future in these societies."

Source: Barnett and Adger (2003).

management that pays particular attention to, for example, the over-extraction or destruction of resources.

16.5.4.4 Emigration and resettlement

Emigration as a potentially effective adaptation strategy has been alluded to earlier, particularly in the context of temporary or permanent out-migrants providing remittances to home-island families, thereby enhancing home-island resilience (Barnett, 2001; Pelling and Uitto, 2001). Within-country migration and resettlement schemes have been common trends over the last several decades in many small islands in the Pacific and Indian Oceans. Both Kiribati and the Maldives have ongoing resettlement schemes and, for the past 70 years, the people of Sikaiana Atoll in the Solomon Islands have been migrating away from their atoll, primarily to Honiara, the capital (Donner, 2002). Similarly there has been internal migration from the Cartaret Islands in Papua New Guinea to Bougainville, and from the outer islands of Tuvalu to the capital Funafuti (Connell, 1999), the former as a consequence of inundation from high water levels and storms, the latter primarily in search of wage employment.

In the case of Tuvalu, this internal migration has brought almost half of the national population to Funafuti atoll, with negative environmental consequences, and the Government has indicated that there is also visual evidence of sea-level rise through increased erosion, flooding and salinisation (Connell, 2003). Connell suggests that, as a result, the global media have increasingly emphasised a doomsday scenario for Tuvalu, as a symbol of all threatened small island environments. Farbotko (2005) also indicates that Tuvalu is becoming prominent in connection with climate-change-related sea-level rise. She undertook an analysis of reports in a major Australian newspaper over the past several years, and suggests that implicating climate change in the identity of Tuvaluans as 'vulnerable' operates to silence alternative identities that emphasise resilience. Indeed, she says that her analysis "has highlighted the capacity for vulnerability rhetoric to silence discourse of adaptation" and concludes that "adaptive strategies are significant for island peoples faced with climate change" and that "it is adaptation, perhaps even more than relocation or mitigation initiatives, which is of immediate importance in island places… [especially] in the face of changes brought about by 'global warming'" (Farbotko, 2005).

On the other hand, Adger et al. (2003a) argue that migration is a feasible climate adaptation strategy in particular circumstances, including in small islands. However, they suggest that because of current inequities in labour flows, particularly for international migration, this adaptation strategy is likely to be contested, and may be a limited option in many parts of the world, even for residents from small island states. They suggest that other means of supporting adaptive capacity and enhancing resilience are required, including building on existing coping strategies, or by introducing innovation in terms of technology or institutional development (Adger et al., 2003a).

16.5.5 Enhancing adaptive capacity

16.5.5.1 Traditional knowledge and past experience

Adaptive capacity and resilience can also be strengthened through the application of traditional knowledge and past experience of environmental changes. In the TAR, Nurse et al. (2001) noted that some traditional island assets, including subsistence and traditional technologies, skills and knowledge, and community structures, and coastal areas containing spiritual, cultural and heritage sites, appeared to be at risk from climate change, and particularly sea-level rise. They argued that some of these values and traditions are compatible with modern conservation and environmental practices.

Since then, several examples of such practices have been described. For instance, Hoffmann (2002) has shown that the implementation of traditional marine social institutions, as exemplified in the Ra'ui in Rarotonga, Cook Islands, is an effective conservation management tool, and is improving coral reef health; while Aswani and Hamilton (2004) show how indigenous ecological knowledge and customary sea tenure may be integrated with modern marine and social science to conserve the bumphead parrotfish in Roviana Lagoon, Solomon Islands. Changes in sea tenure, back to more traditional roles, have also occurred in Kiribati (Thomas, 2001).

The utility of traditional knowledge and practices can also be expanded to link not only with biodiversity conservation but also with tourism. For instance, in a coastal village on Vanua Levu, Fiji, the philosophy of *vanua* (which refers to the connection of people with the land through their ancestors and guardian spirits) has served as a guiding principle for the villagers in the management and sustainable use of the rainforest, mangrove forest, coral reefs, and village gardens. Sinha and Bushell (2002) have shown that the same traditional concept can be the basis for biodiversity conservation, because the ecological systems upon which the villagers depend for subsistence are the very same resources that support tourism. These examples indicate that local knowledge, management frameworks and skills could be important components of adaptive capacity in those small islands that still have some traditional foundations.

16.5.5.2 Capacity building, communities and adaptive capacity

Encouraging the active participation of local communities in capacity building and environmental education has become an objective of many development programmes in small islands. For example, Tran (2006) reports on a long-term project that has successfully included the local community of Holbox Island (Mexico) in monitoring coastal pollution in and around their island. A similar approach is being applied by Dolan and Walker (2006) to another community-based project which assesses the vulnerability of island and coastal communities, their adaptive capacity and options in the Queen Charlotte Islands (Haida Gwaii), located off the west coast of Canada. The study highlights determinants of adaptive capacity at the local scale, and recognises that short-term exposure to climate variability is an important source of vulnerability superimposed on long-term change. Thus, they suggest that community perceptions and experiences with climate extremes can identify inherent characteristics that enable or constrain a community to respond, recover and adapt to climate change, in this case ultimately to sea-level rise (Dolan and Walker, 2006).

A similar conceptualisation, which considers current and future community vulnerability and involves methodologies in climate science and social science, provides the basis for building adaptive capacity, as illustrated in Box 16.7. This approach requires community members to identify climate conditions relevant to them, and to assess present and potential adaptive strategies. The methodology was tested in Samoa, and the results from one village (Saoluafata) are discussed by Sutherland et al. (2005). In this case, local residents identified several adaptive measures including building a seawall, the provision of a water-drainage system and water tanks, a ban on tree clearing, some relocation, and renovations to existing infrastructure.

Enhancing adaptive capacity, however, involves more than just the identification of local options which need to be considered within the larger social, political and economic processes. Based on the Samoan experience, Sutherland et al. (2005) suggest that enhancing adaptive capacity will only be successful when it is integrated with other policies such as disaster preparedness, land-use planning, environmental conservation, coastal planning, and national plans for sustainable development.

Given the urgency for adaptation in small island states, there has been an increase in *ad hoc* stand-alone projects, rather than a programmed or strategic approach to the funding of adaptation options and measures. It can be argued that successful adaptation in small islands will depend on supportive institutions, finance, information, and technological support. However, as noted by Richards (2003), disciplinary and institutional barriers mean that synergies between climate change adaptation and poverty reduction strategies remain underdeveloped. Adger et al. (2003b) note that climate change adaptation has implications for equity and justice because "the impacts of climate change, and resources for addressing these impacts, are unevenly distributed". These issues are particularly applicable to small islands, which have a low capacity to deal with, or adapt to, such impacts.

16.6 Conclusions: implications for sustainable development

The economic, social and environmental linkages between climate change and sustainable development, and their implications for poverty alleviation, have been highlighted in various studies (e.g., Hay et al., 2003; Huq and Reid, 2004) and these are highly relevant to small islands. Most recently, one of the 'key findings' of a major study suggested that climate change poses such a serious threat to poor, vulnerable developing countries that if left unchecked, it will become a "...major obstacle to continued poverty reduction" (Stern, 2007). Indeed, it is true to say that many low-lying small islands view climate change as one of the most important challenges to their achievement of sustainable development. For instance, in the Maldives, Majeed and Abdulla (2004) argue that sea-level rise would so seriously damage the fishing and tourism industries that GDP would be reduced by more than 40%.

In another atoll island setting, Ronneberg (2004) uses the Marshall Islands as a case study to explain that the linkages between patterns of consumption and production, and the effects of global climate change, pose serious future challenges to improving the life of the populations of small island developing states. Based on this case study, Ronneberg (2004) proposes a number of innovative solutions including waste-to-energy and ocean thermal energy conversion systems, which could promote the sustainable development of some small islands and at the same time strengthen their resilience in the face of climate change.

The sustainable development of small islands which are not low-lying, and there are many of these, is also likely to be seriously impacted by climate change, although perhaps not to the same extent as the low islands. For example, Briguglio and Cordina (2003) have shown that climate change impacts on the economic development of Malta are likely to be widespread, affecting all sectors of the economy, but particularly tourism, fishing and public utilities.

Sperling (2003), in an examination of poverty and climate change, contends that the negative impacts of climate change are so serious that they threaten to undo decades of development efforts. He also argues that the combined experience of many international organisations suggests that the best way to address climate change impacts is by integrating adaptation measures into sustainable development strategies. A similar argument is advanced by Hay et al. (2003), in the context of the Pacific small island states, who suggest that the most desirable adaptive responses are those that augment actions which would be taken even in the absence of climate change, due to their contributions to sustainable development. Adaptation measures may be conducive to sustainable development, even without the connection with climate change. The link between adaptation to climate change and sustainable development, which leads to the lessening of pressure on natural resources, improving environmental risk management, and increasing the social well-being of the poor, may not only reduce the vulnerability of small islands to climate change, but also may put them on the path towards sustainable development.

Mitigation measures could also be mainstreamed in sustainable development plans and actions. In this regard, Munasinghe (2003) argues that, ultimately, climate change solutions will need to identify and exploit synergies, as well as seek to balance possible trade-offs, among the multiple objectives of development, mitigation, and adaptation policies. Hay et al. (2003) also argue that while climate change mitigation initiatives undertaken by Pacific island countries will have insignificant consequences climatologically, they should nevertheless be pursued because of their valuable contributions to sustainable development.

But what is small island sustainable development about? Kerr (2005) prefaces this question by noting that sustainable development is often stated as an objective of management strategies for small islands, though relatively little work has explicitly considered what sustainable development means in this context. She argues that the problems of small scale and isolation, of specialised economies, and of the opposing forces of globalisation and localisation, may mean that current development in small islands may be unsustainable in terms of its longevity. On the other hand, models of sustainable development may have something to offer islands in terms of internal management of resources, although the islands may have limited control over exogenous threats or the economic

Box 16.7. Capacity building for development of adaptation measures in small islands: a community approach

Capacity building for development of adaptation measures in Pacific island countries uses a Community Vulnerability and Adaptation Assessment and Action approach. Such an approach is participatory, aims to better understand the nature of community vulnerability, and identifies opportunities for strengthening the adaptive capacity of communities. It seeks to promote a combination of bottom-up and top-down mechanisms for implementation, and supports the engagement of local stakeholders at each stage of the assessment process. If successful, this should enable integration or 'mainstreaming' of adaptation into national development planning and local decision-making processes. The main steps of this approach are outlined below (Figure 16.3).

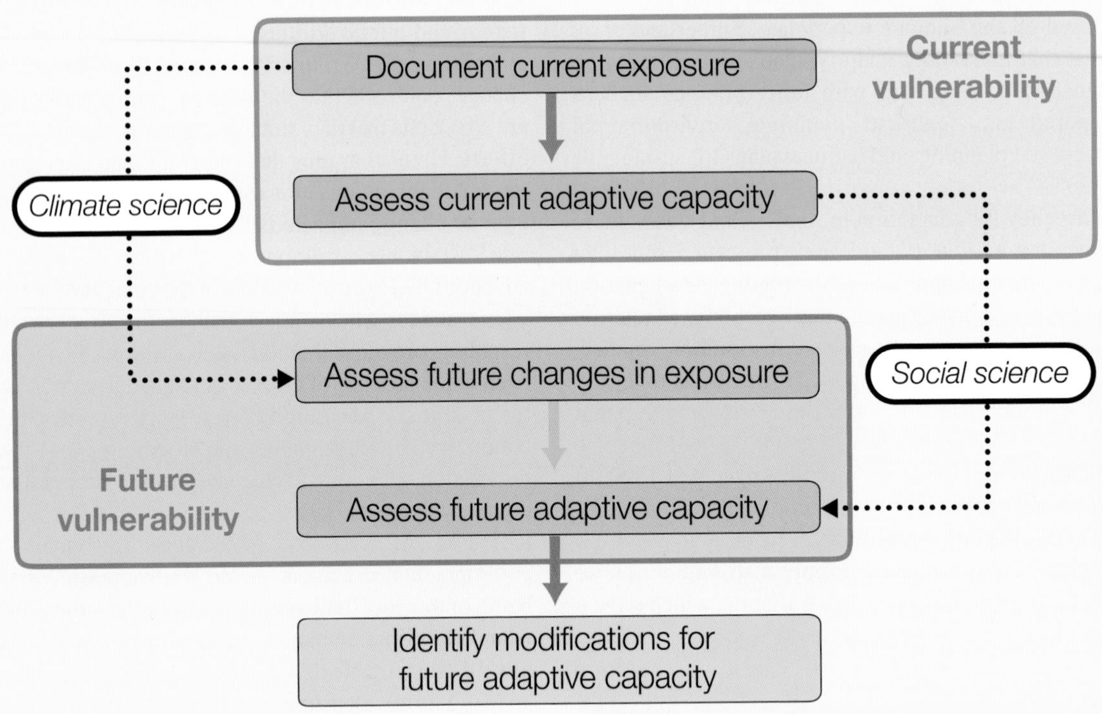

Figure 16.3. *The main steps of a community vulnerability and adaptation assessment and action approach.*

Several pilot communities in the Cook Islands, Fiji, Samoa and Vanuatu are already using this approach to analyse their options and decide on the best course of action to address their vulnerability and adaptation needs.

Source: Sutherland et al. (2005).

drivers of development (Kerr, 2005). In this context, the development of adaptation measures in response to climate change may provide an appropriate avenue to integrate both local and global forces towards island development that is sustainable, providing that local communities are involved (Tran, 2006).

Another positive factor is that many small islands have considerable experience in adapting to climate variability. In the case of Cyprus, for example, Tsiourtis (2002) explains that the island has consistently taken steps to alleviate the adverse effects arising from water scarcity, which is likely to be one of the important effects of climate change. This experience already features in development strategies adopted by Cyprus. A similar argument has also been made by Briguglio (2000) with regard to the Maltese Islands, referring to the islands' exposure to climatic seasonal variability which, historically, has led to individuals and administrations taking measures associated with retreat, accommodation and protection strategies. For example, residential settlements in Malta are generally situated away from low-lying coastal areas, and primitive settlements on the island tended to be located in elevated places. Maltese houses are built of sturdy materials, and are generally able to withstand storms and heavy rains. Temperatures and precipitation rates in Malta change drastically between mid-winter and mid-summer, and this has led to the accumulation of considerable experience in adaptation to climate variability.

However, as mentioned earlier, small islands face many constraints in trying to mainstream climate change into their sustainable development strategies. These include their very limited resources, especially given the indivisibilities of overhead expenditures and hidden costs involved in adaptation measures, particularly in infrastructural projects. Another problem may relate to possible social and/or political conflicts, particularly to do with land use and resources (Westmacott, 2002), though not exclusively (Lane, 2006). Notwithstanding this observation, most decisions regarding the critical issues of land use, energy use and transportation infrastructure in small islands will not have any meaningful influence on the rate and magnitude of climate change worldwide. However, they may have a significant moral and ethical impact in the climate change arena, as well as contributing to reducing their own greenhouse gas emissions and to small island sustainable development.

16.7 Key uncertainties and research gaps

Small islands are sensitive to climate change and sea-level rise, and adverse consequences of climate change and variability are already a 'reality' for many inhabitants of small islands. This assessment has found that many small islands lack adequate observational data and, as noted in the TAR, outputs from AOGCMs are not of sufficiently fine resolution to provide specific information for islands. These deficiencies need to be addressed, so that remaining uncertainties can be reduced, and national and local-scale adaptation strategies for small islands better defined.

As the impacts of climate change become increasingly evident, autonomous small islands, like other countries, will probably be confronted with the need to implement adaptation strategies with greater urgency. However, for these strategies to be effective, they should reflect the fact that natural and human systems in small islands are being simultaneously subjected to other non-climate stresses including population growth, competition for limited resources, ecosystem degradation, and the dynamics of social change and economic transformation. Therefore, responses to climate change need to be properly coordinated and integrated with socio-economic development policies and environmental conservation. The enhancement of resilience at various levels of society, through capacity building, efficient resource allocation and the mainstreaming of climate risk management into development policies at the national and local scale, could constitute a key element of the adaptation strategy.

16.7.1 Observations and climate change science

- Ongoing observation is required to monitor the rate and magnitude of changes and impacts, over different spatial and temporal scales. *In situ* observations of sea level should be strengthened to understand the components of relative sea-level change on regional and local scales. While there has been considerable progress in regional projections of sea level since the TAR, such projections have not been fully utilised in small islands because of the greater uncertainty attached to them, as opposed to global projections.
- Since the TAR, it has also been recognised that other climate-change-induced factors will probably have impacts on coastal systems and marine territories of small islands, including rises in sea temperature and changes in ocean chemistry and wave climate. The monitoring of these and other marine variables in the seas adjacent to small islands would need to be expanded and projections developed.
- Although future projections of mean air temperature are rather consistent among climate models, projections for changes in precipitation, tropical cyclones and wind direction and strength, which are critical concerns for small islands, remain uncertain. Projections based on outputs at finer resolution are needed to inform the development of reliable climate change scenarios for small islands. Regional Climate Models (RCMs) and statistical downscaling techniques may prove to be useful tools in this regard, as the outputs are more applicable to countries at the scale of small islands. These approaches could lead to improved vulnerability assessments and the identification of more appropriate adaptation options.
- Supporting efforts by small islands and their partners to arrest the decline of, and expand, observational networks should be continued. The Pacific Islands Global Climate Observing System (PI-GCOS) and the Intergovernmental Oceanographic Commission Sub-Commission for the Caribbean and Adjacent Regions Global Ocean Observing System (IOCARIBE-GOOS) are two examples of regional observing networks whose coverage should be expanded to cover other island regions.
- Hydrological conditions, water supply and water usage on small islands pose quite different research problems from those in continental situations. These need to be investigated and modelled over the range of island types covering different geology, topography and land cover, and in light of the most recent climate change scenarios and projections.

16.7.2 Impacts and adaptation

- A decade ago, many small islands were the subject of vulnerability assessments to climate change. Such assessments were based on simplistic scenarios, with an emphasis on sea-level rise, and the application of a common methodology that was applied to many small islands throughout the world. The results were initially summarised in the IPCC Second Assessment Report (IPCC, 1996), with later and more comprehensive studies being reported in the TAR. Since then the momentum for vulnerability and impact studies appears to have declined, such that in the present assessment we can cite few robust investigations of climate change impacts on small islands using more recent scenarios and more precise projections. Developing a renewed international agenda to assess the vulnerability of small islands, based on the most recent projections and newly available tools, would provide small islands with a firmer basis for future planning.
- Our assessment has identified several key areas and gaps that are under-represented in contemporary research on the impacts of climate change on small islands. These include:-

- the role of coastal ecosystems such as mangroves, coral reefs and beaches in providing natural defences against sea-level rise and storms;
- establishing the response of terrestrial upland and inland ecosystems, including woodlands, grasslands and wetlands, to changes in mean temperature and rainfall and extremes;
- considering how commercial agriculture, forestry and fisheries, as well as subsistence agriculture, artisanal fishing and food security, will be impacted by the combination of climate change and non-climate-related forces;
- expanding knowledge of climate-sensitive diseases in small islands through national and regional research, not only for vector-borne diseases but for skin, respiratory and water-borne diseases;
- given the diversity of 'island types' and locations, identifying the most vulnerable systems and sectors, according to island type.

- In contrast to the other regions in this assessment, there is also an absence of demographic and socio-economic scenarios and projections for small islands. Nor have future changes in socio-economic conditions on small islands been well presented in existing assessments (e.g., IPCC, 2001; Millennium Ecosystem Assessment, 2003). Developing more appropriate scenarios for assessing the impacts of climate change on the human systems of small islands remains a challenge.
- Methods to project exposures to climate stimuli and non-climate stresses at finer spatial scales should be developed, in order to further improve understanding of the potential consequences of climate variability and change, particularly extreme weather and climate events. In addition, further resources need to be applied to the development of appropriate methods and tools for identifying critical thresholds for both bio-geophysical and socio-economic systems on islands.
- Our evaluation of adaptation in small islands suggests that the understanding of adaptive capacity and adaptation options is still at an early stage of development. Although several potential constraints on, as well as opportunities for, adaptation were identified, two features became apparent. First, the application of some adaptation measures commonly used in continental situations poses particular challenges in a small island setting. Examples include insurance, where there is a small population pool although the propensity for natural disasters is high and where local resilience may be undermined by economic liberalisation. Second, some adaptation measures appear to be advocated particularly for small islands and not elsewhere. Examples include emigration and resettlement, the use of traditional knowledge, and responses to short-term extreme events as a model for adaptation to climate change. Results of studies of each of these issues suggest some ambiguities and the need for further research, including the assessment of practical outcomes that enhance adaptive capacity and resilience.
- With respect to technical measures, countries may wish to pay closer attention to the traditional technologies and skills that have allowed island communities to cope successfully with climate variability in the past. However, as it is uncertain whether the traditional technologies and skills are sufficient to reduce the adverse consequence of climate change, these may need to be combined with modern knowledge and technologies, where appropriate.
- Local capacity should be strengthened in the areas of environmental assessment and management, modelling, economic and social development planning related to climate change, and adaptation and mitigation in small islands. This objective should be pursued through the application of participatory approaches to capacity building and institutional change.
- Access to reliable and affordable energy is a vital element in most small islands, where the high cost of energy is regarded as a barrier to the goal of attaining sustainable development. Research and development into energy options appropriate to small islands could help in both adaptation and mitigation strategies whilst also enhancing the prospect of achieving sustainable growth.

References

Aaheim, A. and L. Sygna, 2000: Economic impacts of climate change on tuna fisheries in Fiji Islands and Kiribati. Report 2000-4, Centre for International Climate and Environmental Research, Oslo, 22 pp.

Abdullah, A., Z. Yasin, W. Ismail, B. Shutes and M. Fitzsimons, 2002: The effect of early coastal development on the fringing coral reefs of Langkawi: a study in small-scale changes. *Malaysian Journal of Remote Sensing and GIS*, **3**, 1-10.

ACIA, 2005: *Impacts of Warming: Arctic Climate Impacts Assessment*. Cambridge University Press, Cambridge, 1042 pp.

ADB, 2004: *Environmental Pacific Regional Strategy, 2005-2009*. Asian Development Bank, Manila, 105 pp.

Adger, W.N., S. Huq, K. Brown, D. Conway and M. Hulme, 2003a: Adaptation to climate change in the developing world. *Prog. Dev. Stud.*, **3**, 179-195.

Adger, W.N., M.J. Mace, J. Paavola and J. Razzaque, 2003b: Justice and equity in adaptation. *Tiempo*, **52**, 19-22.

Alongi, D.M., 2002: Present state and future of the world's mangrove forests. *Environ. Conserv.*, **29**, 331-349.

Arnell, N.W., 2004: Climate change and global water resources: SRES emissions and socio-economic scenarios. *Global Environ. Chang.*, **14**, 31-52.

Aswani, S. and R.J. Hamilton, 2004: Integrating indigenous ecological knowledge and customary sea tenure with marine and social science for conservation of the bumphead parrotfish (*Bolbometopon muricatum*) in Roviana Lagoon, Solomon Islands. *Environ. Conserv.*, **31**, 69-86.

Atkins, J., S. Mazzi and C. Easter, 2000: Commonwealth vulnerability index for developing countries: the position of small states. Economic Paper No. 40, Commonwealth Secretariat, London, 64 pp.

Auffret, P., 2003: Catastrophe insurance market in the Caribbean region: market failures and recommendations for public sector interventions. World Bank Policy Research Working Paper 2963, Washington, District of Columbia, 7 pp.

Baldacchino, G., 2006: *Extreme Tourism: Lessons from the World's Cold Water Islands*. Elsevier Science and Technology, Amsterdam, 310 pp.

Barnett, J., 2001: Adapting to climate change in Pacific Island countries: the problem of uncertainty. *World Dev.*, **29**, 977-993.

Barnett, J.and W.N. Adger, 2003: Climate dangers and atoll countries. *Climatic Change*, **61**, 321-337.

Barnett, J., S. Dessai and R. Jones, 2003: *Climate Change in Timor Leste: Science, Impacts, Policy and Planning*. University of Melbourne-CSIRO, Melbourne, 40 pp.

Becken, S., 2004: Climate change and tourism in Fiji: vulnerability, adaptation and mitigation. Final Report, University of the South Pacific, Suva, 70 pp.

Becken, S., 2005: Harmonising climate change adaptation and mitigation: the case of tourist resorts in Fiji. *Global Environ. Chang.*, **15**, 381-393.

Benning, T.L., D. LaPointe, C.T. Atkinson and P.M. Vitousek, 2002: Interactions of climate change with biological invasions and land use in the Hawaiian Islands: modelling the fate of endemic birds using a geographic information system. *P. Natl. Acad. Sci. USA*, **99**, 14246-14249.

Berkes, F. and D. Jolly, 2001: Adapting to climate change: social-ecological resilience in a Canadian western Arctic community. *Conserv. Ecol.*, **5**. Accessed 05.02.07: http://www.consecol.org/vol5/iss2/art18.

Bettencourt, S., R. Croad, P. Freeman, J. Hay, R. Jones, P. King, P. Lal, A. Mearns, J. Miller, I. Psawaryi-Riddhough, A. Simpson, N. Teuatabo, U. Trotz and M. van Aalst, 2006: *Not If But When: Adapting to Natural Hazards in the Pacific Islands Region: A Policy Note*. The World Bank, East Asia and Pacific Region, Pacific Islands Country Management Unit, Washington DC, 43 pp.

Bigano, A., J.M. Hamilton and R.S.J. Tol, 2005: The impact of climate change on domestic and international tourism: a simulation study. Working Paper FNU-58, Hamburg University and Centre for Marine and Atmospheric Science, Hamburg. Accessed 05.02.07: http://www.uni-hamburg.de/Wiss/FB/15/Sustainability/html2wp.pdf.

Bindoff, N.L., J. Willebrand, V. Artale, A. Cazenave, J. Gregory, S. Gulev, K. Hanawa, C. Le Quéré, S. Levitus, Y. Nojiri, C.K. Shum, L.D. Talley and A.S. Unnikrishnan, 2007: Observations: oceanic climate change and sea level. *Climate Change 2007: The Physical Science Basis. Working Group I Contribution to the Intergovernmental Panel on Climate Change Fourth Assessment Report*, S. Solomon, D. Qin, M. Manning, Z. Chen, M. Marquis, K.B. Averyt, M. Tignor and H.L. Miller, Eds., Cambridge University Press, Cambridge, 385-432.

Brazdil, R., T. Carter, B. Garaganga, A. Henderson-Sellers, P. Jones, T. Carl, T. Knustson, R.K. Kolli, M. Manton, L.J. Mata, L. Mearns, G. Meehl, N. Nicholls, L. Pericchi, T. Peterson, C. Price, C. Senior, D. Stephenson, Q.C. Zeng and F. Zwiers, 2002: IPCC Workshop on Changes in Extreme Weather and Climate Events. Workshop Report, Beijing, 41-42. Accessed 05.02.07: http://www.ipcc.ch/pub/extremes.pdf.

Briguglio, L. 2000: Implications of accelerated sea level rise (ASLR) for Malta. *Proceedings of SURVAS Expert Workshop on European Vulnerability and Adaptation to Impacts of Accelerated Sea Level Rise (ASLR)*, Hamburg, Germany, 36-38.

Briguglio, L. and G. Cordina, 2003: The Economic Vulnerability and Potential for Adaptation of the Maltese Islands to Climate Change. *Proceedings of the International Symposium on Climate Change*, ISCC, Beijing, 62-65.

Briguglio, L. and E. Kisanga., Eds., 2004: *Economic Vulnerability and Resilience of Small States*. Commonwealth Secretariat and the University of Malta, 480 pp.

Bryant, D., L. Burke, J. McManus and M. Spalding, 1998: *Reefs at Risk: A Map-Based Indicator of Threats to the World's Coral Reefs*. World Resources Institute, Washington, District of Columbia, 56 pp.

Buddemeier, R.W., J.A. Kleypas and R.B. Aronson, 2004: *Coral Reefs and Global Change*. Pew Center on Global Climate Change, Arlington, Virginia, 44 pp.

Burke, L. and J. Maidens, 2004: *Reefs at Risk in the Caribbean*. World Resources Institute, Washington, District of Columbia, 81 pp.

Burke, L., E. Selig and M. Spalding, 2002: *Reefs at Risk in Southeast Asia*. World Resources Institute, Washington, District of Columbia, 72 pp.

Burns, W.C.G., 2000: The impact of climate change on Pacific island developing countries in the 21st century. *Climate Change in the South Pacific: Impacts and Responses in Australia, New Zealand, and Small Island States*, A. Gillespie and W.C.G. Burns, Eds., Kluwer Academic, Dordrecht, 233-251.

Burns, W.C.G., 2002: Pacific island developing country water resources and climate change. *The World's Water*, 3rd edn, P. Gleick, Ed., Island Press, Washington, District of Columbia, 113-132.

Chazottes, V., T. Le Campion-Alsumard, M. Peyrot-Clausade and P. Cuet, 2002: The effects of eutrophication-related alterations to coral reef communities on agents and rates of bioerosion (Reunion Island, Indian Ocean). *Coral Reefs*, **21**, 375-390.

Christensen, J.H., B. Hewitson, A. Busuioc, A. Chen, X. Gao, I. Held, R. Jones, R.K. Kolli, W.-T. Kwon, R. Laprise, V. Magaña Rueda, L. Mearns, C.G. Menendez, J. Räisänen, A. Rinke, A. Sarr and P. Whetton, 2007: Regional climate projections. *Climate Change 2007: The Physical Science Basis. Contribution of Working Group I to the Intergovernmental Panel on Climate Change Fourth Assessment Report*, S. Solomon, D. Qin, M. Manning, Z. Chen, M. Marquis, K.B. Averyt, M. Tignor and H.L. Miller, Eds., Cambridge University Press, Cambridge, 847-940.

Church, J.A., N.J. White, R. Coleman, K. Lambeck and J.X. Mitrovica, 2004: Estimates of regional distribution of sea level rise over the 1950-2000 period. *J. Climate*, **17**, 2609-2625.

Church, J.A., N. White and J. Hunter, 2006: Sea level rise at tropical Pacific and Indian Ocean islands. *Global Planet. Change*, **53**, 155-168.

Clark, E., 2004: The ballad dance of the Faeroese: island biocultural geography in an age of globalisation. *Tijdschr. Econ. Soc. Ge.*, **195**, 284-297.

Cocklin, C., 1999: Islands in the midst: environmental change, vulnerability, and security in the Pacific. *Environmental Change, Adaptation, and Security*, S. Lonergan, Ed., Kluwer Academic, Dordrecht, 141-159.

Connell, J., 1999: Environmental change, economic development, and emigration in Tuvalu. *Pacific Studies*, **22**, 1-20.

Connell, J., 2003: Losing ground? Tuvalu, the greenhouse effect and the garbage can. *Asia Pacific Viewpoint*, **44**, 89-107.

Cowell, P.J.and P.S. Kench, 2001: The morphological response of atoll islands to sea level rise. Part 1. Modifications to the modified shoreface translation model. *J. Coastal Res.*, **34**, 633-644.

Daehler, C.C., 2005: Upper-montane plant invasions in the Hawaiian Islands: patterns and opportunities. *Perspect. Plant Ecol.*, **7**, 203-216.

Daniel, E.B. and M.D. Abkowitz, 2003: Development of beach analysis tools for Caribbean small islands. *Coast. Manage.*, **31**, 255-275.

Daniel, E.B. and M.D. Abkowitz, 2005: Predicting storm-induced beach erosion in Caribbean small islands. *Coast. Manage.*, **33**, 53-69.

Dickinson, W.R., 1999: Holocene sea level record on Funafuti and potential impact of global warming on central Pacific atolls. *Quaternary Res.*, **51**, 124-132.

Dikou, A. and R. van Woesik, 2006: Survival under chronic stress from sediment load: spatial patterns of hard coral communities in the southern islands of Singapore. *Mar. Pollut. Bull.*, **52**, 7-21.

Dolan, A.H. and I.J. Walker, 2006: Understanding vulnerability of coastal communities to climate change related risks. *J. Coastal Res.*, **39**, 1317-1324.

Donner, S.D., W.J. Skirving, C.M. Little, M. Oppenheimer and O. Hoegh-Guldberg, 2005: Global assessment of coral bleaching and required rates of adaptation under climate change. *Glob. Change Biol.*, **11**, 2251-2265.

Donner, W.W., 2002: Rice and tea, fish and taro: Sikaiana migration to Honiara. *Pacific Studies*, **25**, 23-44.

Dulvy, N., R. Freckleton and N. Polunin, 2004: Coral reef cascades and the indirect effects of predator removal by exploitation. *Ecol. Lett.*, **7**, 410-416.

Ebi, K.L., N.D. Lewis and C. Corvalan, 2006: Climate variability and change and their potential health effects in small island states: information for adaptation planning in the health sector. *Environ. Health Persp.*, **114**, 1957-1963.

ECLAC (Economic Commission for Latin America and the Caribbean) 2002: *Global Economic Developments 2000-2001*. LC/CAR/G.683, 33 pp. Accessed 10.05.07: http://www.eclac.cl.

Emanuel, K., 2005: Increasing destructiveness of tropical cyclones over the past 30 years. *Nature*, **436**, 686-688.

Epstein, P.R. and E. Mills, 2005: *Climate Change Futures: Health, Ecological and Economic Dimensions*. Center for Health and the Global Environment, Harvard Medical School, Boston, Massachusetts, 138 pp.

FAO, 2004: FAO and SIDS: Challenges and emerging issues in agriculture, forestry and fisheries. Paper prepared by the Food and Agriculture Organization (FAO) on the occasion of the Inter-regional Conference on Small Island Developing States (SIDS), Bahamas 26-30 January 2004, Rome, 34 pp.

Farbotko, C., 2005: Tuvalu and climate change: constructions of environmental displacement in the Sydney Morning Herald. *Geogr. Ann. B*, **87**, 279-293.

Fish, M.R., I.M. Cote, J.A. Gill, A.P. Jones, S. Renshoff and A. Watkinson, 2005: Predicting the impact of sea level rise on Caribbean sea turtle nesting habitat. *Conserv. Biol.*, **19**, 482-491.

Fitzharris, B., 2001: Global energy and climate processes. *The Physical Environment: A New Zealand Perspective*. A. Sturman and R. Spronken-Smith, Eds., Oxford University Press, Victoria, 537 pp.

Folland, C.K., J.A. Renwick, M.J. Salinger and A.B. Mullan, 2002: Relative influences of the Interdecadal Pacific Oscillation and ENSO on the South Pacific Convergence Zone. *Geophys. Res. Lett.*, **29**, 211-214.

Folland, C.K., J.A. Renwick, M.J. Salinger, N. Jiang and N.A. Rayner, 2003: Trends and variations in South Pacific islands and ocean surface temperatures. *J. Climate*, **16**, 2859-2874.

Forbes, D.L., G.S. Parkes, G.K. Manson and L.A. Ketch, 2004: Storms and shoreline erosion in the southern Gulf of St. Lawrence. *Mar. Geol.*, **210**, 169-204.

Fosaa, A.M., M.T. Sykes, J.E. Lawesson and M. Gaard, 2004: Potential effects of climate change on plant species in the Faroe Islands. *Global Ecol. Biogeogr.*, **13**, 427-437.

Foster, P., 2001: The potential negative impacts of global climate change on tropical mountain cloud forests. *Earth-Sci. Rev.*, **55**, 73–106.

Fox, S., 2003: *When the Weather is Uggianaqtuq: Inuit Observations of Environmental Change*. Cooperative Institute for Research in Environmental Sciences, Boulder, Colorado: University of Colorado, CD-ROM.

Fox, H. and R. Caldwell, 2006: Recovery from blast fishing on coral reefs: a tale of two scales. *Ecol. Appl.*, **16**, 1631-1635.

Freed, L.A., R.L. Cann, M.L. Goff, W.A. Kuntz and G.R. Bodner, 2005: Increase in avian malaria at upper elevations in Hawaii. *Condor*, **107**, 753-764.

Frenot, Y., S.L. Chown, J. Whinam, P.M. Selkirk, P. Convey, M. Skotnicki and D.M. Bergstrom, 2005: Biological invasions in the Antarctic: extent, impacts and implication. *Biol. Rev.*, **80**, 45-72.

Gardner, T.A., I. Cote, G. Gill, A. Grant and A. Watkinson, 2003: Long-term region-wide declines in Caribbean corals. *Science*, **301**, 958-960.

GDE (General Directorate of Environment, Comoros), 2002: *Initial National Communication on Climate Change, Union des Comoros*. Ministry of Development, Infrastructure, Post and Telecommunications and International Transports, Union des Comoros, 11 pp.

Gibbons, S.J.A. and R.J. Nicholls, 2006: Island abandonment and sea level rise: an historical analog from the Chesapeake Bay, USA. *Global Environ. Chang.*, **16**, 40-47.

Gilman, E., H. Van Lavieren, J. Ellison, V. Jungblut, L. Wilson, F. Ereki, G. Brighouse, J. Bungitak, E. Dus, M. Henry, I. Sauni, M. Kilman, E. Matthews, N.Teariki-Ruatu, S. Tukia and K. Yuknavage, 2006: Pacific Island mangroves in a changing climate and rising sea. UNEP Regional Sea Reports and Studies 179, United Nations Environment Programme, Regional Sea Programme, Nairobi, 58 pp.

Golbuua, Y., S. Victora, E. Wolanski and R.H. Richmond, 2003: Trapping of fine sediment in a semi-enclosed bay, Palau, Micronesia. *Estuar. Coast. Shelf S.*, **57**, 1-9.

Goldberg, J. and C. Wilkinson, 2004: Global threats to coral reefs: coral bleaching, global climate change, disease, predator plagues, and invasive species. *Status of Coral Reefs of the World: 2004*, C. Wilkinson, Ed., Australian Institute of Marine Science, Townsville, 67-92.

Government of Mauritius, 2002: *Meeting the Challenges of Sustainable Development*. Ministry of Environment, Republic of Mauritius.

Graham, N.A.J., S.K. Wilson, S. Jennings, N.V.C. Polunin, J.P. Bijoux and J. Robinson, 2006: Dynamic fragility of oceanic coral reef ecosystems. *P. Natl. Acad. Sci. USA*, **103**, 8425-8429.

Griffiths, G.M., M.J. Salinger and I. Leleu, 2003: Trends in extreme daily rainfall across the South Pacific and relationship to the South Pacific Convergence Zone. *J. Climatol.*, **23**, 847-869.

Gritti, E.S., B. Smith and M.T. Sykes., 2006: Vulnerability of Mediterranean Basin ecosystems to climate change and invasion by exotic plant species. *J. Biogeogr.*, **33**, 145-157.

Grynberg, R. and J.Y. Remy, 2004: Small vulnerable economy issues and the WTO. *WTO at the Margins: Small States and the Multilateral Trading System*, R. Grynberg, Ed., Cambridge University Press, Cambridge, 281-308.

Hajat, S., K.L. Ebi, S. Edwards, A. Haines, S. Kovats and B. Menne, 2003: Review of the human health consequences of flooding in Europe and other industrialized civilizations. *Applied Environmental Science and Public Health*, **1**, 13-21.

Hajkowicz, S., 2006: Coast scenarios for coastal water pollution in a small island nation: a case study from the Cook Islands. *Coast. Manage.*, **34**, 369-386.

Hales, S., P. Weinstein and A. Woodward, 1999: Ciguatera (fish poisoning), El Niño, and sea surface temperature. *Ecosyst. Health*, **5**, 20-25.

Hamilton, K., 2004: Insurance and financial sector support for adaptation. *IDS Bull.-I. Dev. Stud.*, **35**, 55-61.

Harvell, C.D., C.E. Mitchell, J.R. Ward, S. Altizer, A.P. Dobson, R.S. Ostfeld and M.D. Samuel, 2002: Climate warming and disease risks for terrestrial and marine biota. *Science*, **296**, 2158-2162.

Harvey, N., M. Rice and L. Stephenson, 2004: *Global Change Coastal Zone Management Synthesis Report*. Asia-Pacific Network for Global Change Research, APN Secretariat, Chuo-ku, Kobe, 37 pp.

Hay, J., N. Mimura, J. Cambell, S. Fifita, K. Koshy, R.F. McLean, T. Nakalevu, P. Nunn and N. deWet, 2003: *Climate Variability and Change and Sea level Rise in the Pacific Islands Region: A Resource Book for Policy and Decision Makers, Educators and Other Stakeholders*. South Pacific Regional Environment Programme (SPREP), Apia, Samoa, 94 pp.

Henri, K., G. Milne and N. Shah, 2004: Costs of ecosystem restoration on islands in Seychelles. *Ocean Coast. Manage.*, **47**, 409-428.

Hoffmann, T.G., 2002: The reimplementation of the *Ra'ui*: coral reef management in Rarotonga, Cook Islands. *Coast. Manage.*, **30**, 401-418.

Houk, P., G. Didonato, J. Iguel and R. van Woesik, 2005: Assessing the effects of non-point source pollution on American Samoa's coral reef communities. *Environ. Monit. Assess.*, **107**, 11-27.

Hughes, T., A. Baird, D. Bellwood, M. Card, S. Connolly, C. Folke, R. Grosberg, O. Hoegh-Guldberg, J. Jackson, J. Kleypas, J. Lough, P. Marshall, M. Nyström, S. Palumbi, J. Pandolfi, B. Rosen and J. Roughgarden, 2003: Climate change, human impacts, and the resilience of coral reefs. *Science*, **301**, 929-933.

Huq, S. and H. Reid, 2004: Mainstreaming adaptation in development. *IDS Bull.-I. Dev. Stud.* **35**, 15-21.

IPCC, 1996: *Climate Change 1995: Impacts, Adaptations and Mitigation of Climate Change: Scientific-Technical Analyses. Contribution of Working Group II to the Second Assessment Report of the Intergovernmental Panel on Climate Change*, R.T. Watson, M.C. Zinyowera and R.H. Moss, Eds., Cambridge University Press, Cambridge, 880 pp.

IPCC, 2001: *Climate Change 2001: The Scientific Basis. Contribution of Working Group I to the Third Assessment Report of the Intergovernmental Panel on Climate Change*, J.T. Houghton, Y. Ding, D.J. Griggs, M. Noguer, P.J. van der Linden, X. Dai, K. Maskelland C.A. Johnson, Eds., Cambridge University Press, Cambridge, 881 pp.

Jensen, T.L., 2000: *Renewable Energy on Small Islands*, 2nd edn. Forum for Energy and Development, Copenhagen, 135 pp.

Kench, P.S. and P.J. Cowell, 2001: The morphological response of atoll islands to sea level rise. Part 2. Application of the modified shoreface translation model. *J. Coastal Res.*, **34**, 645-656.

Kench, P.S., R.F. McLean and S.L. Nicholl, 2005: New model of reef-island evolution: Maldives, Indian Ocean. *Geology*, **33**, 145-148.

Kench, P.S., R.F. McLean, R.W. Brander, S.L. Nicholl, S.G. Smithers, M.R. Ford, K.P. Parnell and M. Aslam, 2006: Geological effects of tsunami on mid-ocean atoll islands: the Maldives before and after the Sumatran tsunami. *Geology*, **34**, 177-180.

Kerr, S.A., 2005: What is small island sustainable development about? *Ocean Coast. Manage.*, **48**, 503-524.

Khan, T.M.A., D.A. Quadir, T.S. Murty, A. Kabir, F. Aktar and M.A. Sarker, 2002: Relative sea level changes in Maldives and vulnerability of land due to abnormal coastal inundation. *Mar. Geod.*, **25**, 133-143.

Kinch, J., 2002: Giant clams: their status and trade in Milne Bay Province, Papua New Guinea. *TRAFFIC Bulletin*, **19**, 1-9.

Kriticos, D.J., T. Yonow and R.C. McFadyen, 2005: The potential distribution of *Chromolaena odorata* (Siam weed) in relation to climate. *Weed Res.*, **45**, 246-254.

Kudo, G., Y. Nishikawa, T. Kasagi and S. Kosuge 2004: Does seed production of spring ephemerals decrease when spring comes early? *Ecol. Res.*, **19**, 255-259.

Kuffner, I. and V. Paul, 2001: Effects of nitrate, phosphate and iron on the growth of macroalgae and benthic cyanobacteria from Cocos Lagoon, Guam. *Mar. Ecol.-Prog. Ser.*, **222**, 63-72.

Lal, M., 2004: Climate change and small island developing countries of the South Pacific. *Fijian Studies*, **2**, 15-31.

Lal, M., H. Harasawa and K. Takahashi, 2002: Future climate change and its impacts over small island states. *Climate Res.*, **19**, 179-192.

Lane, M., 2006: Towards integrated coastal management in the Solomon Islands: identifying strategic issues for governance reform. *Ocean Coast. Manage.*, **49**, 421-441.

Lehodey, P., F. Chai and J. Hampton, 2003: Modelling the climate-related fluctuations of tuna populations from a coupled ocean-biogeochemical-populations dynamics model. *Fish. Oceanogr.*, **13**, 483-494.

Levinson, D.H., Ed., 2005: State of the climate in 2004. *B. Am. Meteorol. Soc.*, **86**, S1-S84.

Lewis, R., 2005: Ecological engineering for successful management and restoration of mangrove forests. *Ecol. Eng.*, **24**, 403-418.

Majeed, A. and A. Abdulla, 2004: Economic and environmental vulnerabilities of the Maldives and graduation from LDC status. *Economic Vulnerability and Resilience of Small States*, L. Briguglio and E. Kisanga, Eds., Commonwealth Secretariat and the University of Malta, 243-255.

Manson, G.K., S.M. Solomon, D.L. Forbes, D.E. Atkinson and M. Craymer, 2005: Spatial variability of factors influencing coastal change in the western Canadian Arctic. *Geo-Mar. Lett.*, **25**, 138-145.

Manton, M.J., P.M. Dellaa-Marta, M.R. Haylock, K.J. Hennessy, N. Nicholls, L.E. Chambers, D.A. Collins, G. Daw, A. Finet, D. Gunawan, K. Inape, H. Isobe, T.S. Kestin, P. Lefale, C.H. Leyu, T. Lwin, L. Maitrepierre, N. Oprasitwong, C.M. Page, J. Pahalad, N. Plummer, M.J. Salinger, R. Suppiah, V.L. Tran, B. Trewin, I. Tibig and D. Yee, 2001: Trends in extreme daily rainfall and temperature in southeast Asia and the south Pacific: 1961-1998. *J. Climatol.*, **21**, 269-284.

Mason, S., 2004: Simulating climate over western North America using stochastic weather generators. *Climatic Change*, **62**, 155-187.

McBean, G.A., G. Alekseev, D. Chen, E. Forland, J. Fyfe, P. Groisman, R. King, H. Melling, R. Vose and P. Whitfield, 2005: Arctic climate – past and present. *Arctic Climate Impact Assessment (ACIA)*, Cambridge University Press, Cambridge, 21-40.

McLean, R.F., A. Tsyban, V. Burkett, J.O. Codignotto, D.L. Forbes, N. Mimura, R.J. Beamish and V. Ittekkot, 2001: Coastal zones and marine ecosystems. *Climate Change 2001: Impacts, Adaptation, and Vulnerability. Contribution of Working Group II to the Third Assessment Report of the Intergovernmental Panel on Climate Change*, J.J. McCarthy, O.F. Canziani, N.A. Leary, D.J. Dokken and K.S. White, Eds., Cambridge University Press, Cambridge, 343-379.

Meehl, G.A., T.F. Stocker, W. Collins, P. Friedlingstein, A.T. Gaye, J. Gregory, A. Kitoh, R. Knutti, J. Murphy, A. Noda, S. Raper, I.G. Watterson, A. Weaver and Z.-C. Zhao, 2007: Global climate projections. *Climate Change 2007: The Physical Science Basis. Contribution of Working Group I to the Fourth Assessment Report of the Intergovernmental Panel on Climate Change*, S. Solomon, D. Qin, M. Manning, Z. Chen, M. Marquis, K.B. Averyt, M. Tignor and H.L. Miller, Eds., Cambridge University Press, Cambridge, 747-846.

MESD (Ministry of Environment and Social Development, Kiribati), 1999: *Initial Communication under the United Nations Framework Convention on Climate Change*. Kiribati Government, Tarawa, Kiribati.

Millennium Ecosystem Assessment, 2003: *Ecosystems and Human Well-being: Millennium Ecosystem Assessment*. Island Press, Washington, District of Columbia / Covelo, London, 245 pp.

Miller, K., 2005: Variations in sea level on the West Trinidad coast. *Mar. Geod.*, **28**, 219-229.

Milner, C., W. Morgan and E. Zgovu, 2004: Would all ACP sugar exporters lose from sugar liberalisation? *The European Journal of Development Research*, **16**, 790-808.

Mitchell, W., J. Chittleborough, B. Ronai and G.W. Lennon, 2001: Sea level rise in Australia and the Pacific. *Proceedings Science Component: Linking Science and Policy. Pacific Islands Conference on Climate Change, Climate Variability and Sea Level Rise, Rarotonga, Cook Islands*. National Tidal Facility, The Flinders University of South Australia, Adelaide, 47-58.

MOHA (Ministry of Home Affairs, Maldives), 2001: *First National Communication of the Republic of Maldives to the United Nations Framework Convention on Climate Change*. Ministry of Home Affairs, Housing and Environment, Malé, Republic of Maldives, 134 pp.

Mörner, A., M. Tooley and G. Possnert, 2004: New perspectives for the future of the Maldives. *Global Planet. Change*, **40**, 177-182.

MRAE (Ministry of Rural Affairs and the Environment, Malta), 2004: *The First Communication of Malta to the United Nations Framework Convention on Climate Change*. Ministry for Rural Affairs and the Environment, Malta, 103 pp.

Munasinghe, M., 2003: *Analysing the Nexus of Sustainable Development and Climate Change: An Overview*. OECD, Paris, 53 pp.

Nakićenović, N. and R. Swart, Eds., 2000: *IPCC Special Report on Emissions Scenarios*, Cambridge University Press, Cambridge, 599 pp.

NEAB (National Environment Advisory Board, St Vincent and the Grenadines), 2000: *Initial National Communication on Climate Change*. National Environment Advisory Board and Ministry of Health and the Environment, 74 pp.

Ng, W.S. and R. Mendelsohn, 2005: The impact of sea level rise on Singapore. *Environ. Dev. Econ.*, **10**, 201-215.

Nicholls, R.J., 2004: Coastal flooding and wetland loss in the 21st century: changes under SRES climate and socio-economic scenarios. *Global Environ. Chang.*, **14**, 69-86.

Nurse, L. and R. Moore, 2005: Adaptation to global climate change: an urgent requirement for Small Island Developing States. *Review of European Community and International Environmental Law*, **14**, 100-107.

Nurse, L., G. Sem, J.E. Hay, A.G. Suarez, P.P. Wong, L. Briguglio and S. Ragoonaden, 2001: Small island states. *Climate Change 2001: Impacts, Adaptation, and Vulnerability. Contribution of Working Group II to the Third Assessment Report of the Intergovernmental Panel on Climate Change*, J.J. McCarthy, O.F. Canziani, N.A. Leary, D.J. Dokken and K.S. White, Eds., Cambridge University Press, Cambridge, 842-975.

Nyström, M., C. Folke and F. Moberg, 2000: Coral reef disturbance and resilience in a human-dominated environment. *Trends Ecol. Evol.*, **15**, 413-417.

OECS (Organisation of East Caribbean States), 2000: *The St Georges Declaration of Principles for Environmental Sustainability in the Organisation of East Caribbean States*. OECS, Castries, St. Lucia, 22 pp.

OECS (Organisation of East Caribbean States), 2004: *Grenada: Macro Socio-economic Assessment of the Damages Caused by Hurricane Ivan*. OECS, Castries, St. Lucia, 139 pp.

Ostertag, R., W.L. Silver and A.E. Lugo, 2005: Factors affecting mortality and resistance to damage following hurricanes in a rehabilitated subtropical moist forest. *Biotropica*, **37**, 16-24.

Parakoti, B. and D.M. Scott, 2002: Drought index for Rarotonga (Cook Islands). Case Study presented as part of *Theme 2, Island Vulnerability. Pacific Regional Consultation Meeting on Water in Small Island Countries*, Sigatoka, Fiji Islands, 29 July–3 August, 2002, 9 pp.

Paulay, G., L. Kirkendale, G. Lambert and C. Meyer, 2002: Anthropogenic biotic interchange in a coral reef ecosystem: a case study from Guam. *Pac. Sci.*, **56**, 403-422.

Pelling, M. and J.I. Uitto, 2001: Small island developing states: natural disaster vulnerability and global change. *Environmental Hazards*, **3**, 49-62.

Peterson, T.C., M.A. Taylor, R. Demeritte, D.L. Duncombe, S. Burton, F. Thompson, A. Porter, M. Mercedes, E. Villegas, R. Semexant Fils, A. Klein Tank, A. Martis, R. Warner, A. Joyette, W. Mills, L. Alexander and B. Gleason, 2002: Recent changes in climate extremes in the Caribbean region. *J. Geophys. Res.*, **107**, 4601, doi:10.1029/2002JD002251.

Prasad, N., 2003: 'Small Islands' quest for economic development. *Asia Pac. Dev. J.*, **10**, 47-66.

Proshutinsky, A., I.M. Ashik, E.N. Dvorkin, S. Hakkinen, R.A. Krishfield and W.R. Peltier, 2004: Secular sea level change in the Russian sector of the Arctic Ocean. *J. Geophys. Res.*, **109**, C03042, doi:10.1029/2003JC002007.

Raksakulthai, V., 2003: *Climate Change Impacts and Adaptation for Tourism in Phuket, Thailand*. Asian Disaster Preparedness Center, Klong Luang, Pathumthami, 22 pp.]

Ramessur, R., 2002: Anthropogenic-driven changes with focus on the coastal zone of Mauritius, south-western Indian Ocean. *Reg. Environ. Change*, **3**, 99-106.

Rasmussen, T., 2004: Macroeconomic implications of natural disasters in the Caribbean. IMF Working Paper, WP/04/224, Western Hemisphere Department, International Monetary Fund, 24 pp.

Rawlins, S.C., A. Chen, M. Ivey, D. Amarakoon and K. Polson, 2005: The impact of climate change/variability events on the occurrence of dengue fever in parts of the Caribbean: a retrospective study for the period 1980-2002. *W. Indian Med. J.*, **53**, 54.

Republic of Vanuatu, 1999: *Vanuatu National Communication to the Conference of the Parties to the United Nations Framework Convention on Climate Change*, 55 pp.

Revell, M., 2004: Pacific island weather and the MJO. *Island Climate Update*, **42**, 4. Accessed 06.02.07: http://www.niwa.co.nz/ncc/icu/2004-03/.

Richards, M., 2003: *Poverty Reduction, Equity and Climate Change: Global Governance Synergies or Contradictions?* Globalisation and Poverty Programme, Overseas Development Institute, London, 14 pp.

Ronneberg, E., 2004: Environmental vulnerability and economic resilience: the case of the Republic of the Marshall Islands. *Economic Vulnerability and Resilience of Small States*, L. Briguglio and E. Kisanga, Eds., Commonwealth Secretariat and the University of Malta, 163-172.

Roper, T., 2005: Small island states: setting an example on green energy use. *Review of European Community and International Environmental Law*, **14**, 108-116.

Ruosteenoja, K., T.R. Carter, K. Jylhä and H. Tuomenvirta, 2003: Future climate in world regions: an intercomparison of model-based projections for the new IPCC emissions scenarios. The Finnish Environment 644, Finnish Environment Institute, Helsinki, 83 pp.

Salinger, M.J., 2001: Climate variations in New Zealand and the Southwest Pacific. *The Physical Environment: A New Zealand Perspective*, A. Sturman and R. Spronken-Smith, Eds., Oxford University Press, Victoria, 130-149.

Salinger, M.J., J.A. Renwick and A.B. Mullan, 2001: Interdecadal Pacific Oscillation and South Pacific climate. *J. Climatol.*, **21**, 1705-1721.

Sanz, J.J., T.J. Potti, J. Moreno, S. Merion and O. Frias, 2003: Climate change and fitness components of a migratory bird breeding in the Mediterranean region.

Glob. Change Biol., **9**, 461-472.

Shea, E., G. Dolcemascolo, C.L. Anderson, A. Banston, C.P. Guard, M.P. Hamnett, S.T. Kubota, N. Lewis, J. Loschinigg and G. Meehls, 2001: *Preparing for a Changing Climate: The Potential Consequences of Climate Variability and Change, Pacific Islands*. Report of the Pacific Islands Regional Assessment Group for the US Global Change Research Program. East West Center, University of Hawaii, Honolulu, Hawaii, 102 pp.

Sheppard, C.R.C., 2003: Predicted recurrences of mass coral mortality in the Indian Ocean. *Nature* **425**, 294-297.

Sheppard, C., D.J. Dixon, M. Gourlay, A. Sheppard and R. Payet, 2005: Coral mortality increases wave energy reaching shores protected by reef flats: examples from the Seychelles. *Estuar. Coast. Shelf S.*, **64**, 223-234.

Singh, R.B.K., S. Hales, N. de Wet, R. Raj, M. Hearnden and P. Weinstein, 2001: The influence of climate variation and change on diarrhoeal disease in the Pacific Islands. *Environ. Health Persp.*, **109**, 155-159.

Singh, S.J. and C.M. Grünbühel, 2003: Environmental relations and biophysical transition: the case of Trinket Island. *Geogr. Ann. B*, **85**, 191-208.

Sinha, C.C. and R. Bushell, 2002: Understanding the linkage between biodiversity and tourism: a study of ecotourism in a coastal village in Fiji. *Pacific Tourism Review*, **6**, 35-50.

Smith, R.C., W.R. Fraser, S.E. Stamnerjohn and M. Vernet, 2003: Palmer long-term ecological research on the Antarctic marine ecosystem. *Antarctic Research Series*, **79**, 131-144.

Smith, V.R., 2002: Climate change in the sub-Antarctic: an illustration from Marion Island. *Climatic Change*, **52**, 345-357.

Sperling, F., 2003: *Multi-Agency Report 2003. Poverty and Climate Change: Reducing the Vulnerability of the Poor through Adaptation*. The World Bank, Washington, DC, 56 pp.

Stern, N., 2007: *The Economics of Climate Change: The Stern Review*. Cambridge University Press, Cambridge, 692 pp.

Sutherland, K., B. Smit, V. Wulf and T. Nakalevu, 2005: Vulnerability to climate change and adaptive capacity in Samoa: the case of Saoluafata village. *Tiempo*, **54**, 11-15.

Swiss Re, 2004: *Hurricane Season 2004: Unusual, But Not Unexpected*. Focus Report, Zurich, 12 pp.

Tan, W.H. and T.S. Teh, 2001: Sustainability of island tourism resorts: a case study of the Perhentian Islands. *Malaysian Journal of Tropical Geography*, **32**, 51-68.

Thomas, F.R., 2001: Remodelling marine tenure on the atolls: a case study from Western Kiribati, Micronesia. *Hum. Ecol.*, **29**, 399-423.

Tompkins, E.L., 2005: Planning for climate change in small islands: insights from national hurricane preparedness in the Cayman Islands. *Global Environ. Chang.*, **15**, 139-149.

Tompkins, E.L., S.A. Nicholson-Cole, L-A. Hurlston, E. Boyd, G.B. Hodge, J. Clarke, G. Gray, N. Trotz and L. Varlack, 2005: *Surviving Climate Change in Small Islands: A Guidebook*. Tyndall Centre for Climate Change Research, Norwich, 128 pp.

Tran, K.C., 2006: Public perception of development issues: public awareness can contribute to sustainable development of a small island. *Ocean Coast. Manage.*, **49**, 367-383.

Trenberth, K.E., P.D. Jones, P.G. Ambenje, R. Bojariu, D.R. Easterling, A.M.G. Klein Tank, D.E. Parker, J.A. Renwick, F. Rahimzadeh, M.M. Rusticucci, B.J. Soden and P.-M. Zhai, 2007: Observations: surface and atmospheric climate change. *Climate Change 2007: The Physical Science Basis. Contribution of Working Group I to the Fourth Assessment Report of the Intergovernmental Panel on Climate Change*, S. Solomon, D. Qin, M. Manning, Z. Chen, M. Marquis, K.B. Averyt, M. Tignor and H.L. Miller, Eds., Cambridge University Press, Cambridge, 235-336.

Tsiourtis, N.X., 2002: Cyprus: water resources planning and climate change adaptation. *Water, Wetlands and Climate Change: Building Linkages for their Integrated Management*. IUCN Mediterranean Regional Roundtable, Athens, 2 pp.

UNDP (United Nations Development Programme), 2005: *Adaptation Policy Framework for Climate Change: Developing Strategies, Policies and Measures*. Cambridge University Press, Cambridge and New York, 258 pp.

UNEP (United Nations Environment Programme), 2000: *Overview on Land-Based Pollutant Sources and Activities Affecting the Marine, Coastal, and Freshwater Environment in the Pacific Islands Region*. Regional Seas Reports and Studies No. 174, 48 pp.

UNFCCC (United Nations Framework Convention on Climate Change), 2005: *Climate Change: Small Island Developing States*. Secretariat, United Nations Framework Convention on Climate Change, Bonn, 32 pp.

Uyarra, M.C., I.M. Cote, J.A. Gill, R.R.T. Tinch, D. Viner and A.R. Watkinson, 2005: Island-specific preferences of tourists for environmental features: implications of climate change for tourism-dependent states. *Environ. Conserv.*, **32**, 11-19.

Vassie, J.M., P.L. Woodworthand M.W. Holt, 2004: An example of North Atlantic deep-ocean swell impacting Ascension and St. Helena Islands in the Central South Atlantic. *J. Atmos. Ocean Tech.*, **21**, 1095-1103.

Villanueva, R., H. Yap and N. Montaño, 2006: Intensive fish farming in the Philippines is detrimental to the reef-building coral *Pocillopora damicornis*. *Mar. Ecol.-Prog. Ser.*, **316**, 165-174.

Viner, D., 2006: Tourism and its interactions with climate change. *Journal of Sustainable Tourism*, **14**, 317-322.

Voigt-Graf, C., 2003: Fijian teachers on the move: causes, implications and policies. *Asia Pacific Viewpoint*, **44**, 163-175.

Vunisea, A., 2003: Coral harvesting and its impact on local fisheries in Fiji. *SPC Women in Fisheries Information Bulletin*, **12**, 17-20.

Wade, H., 2005: Pacific Regional Energy Assessment: an assessment of the key energy issues, barriers to the development of renewable energy to mitigate climate change, and capacity development needs to removing the barriers. Niue National Report. Pacific Islands Renewable Energy Project, Technical Report No. 8, Apia, Samoa, 38 pp.

Walker, I.J. and J.V. Barrie, 2006: Geomorphology and sea level rise on one of Canada's most 'sensitive' coasts: northeast Graham Island, British Columbia. *J. Coastal Res.*, **SI 39**, 220-226.

Walsh, K., 2004: Tropical cyclones and climate change: unresolved issues. *Climate Res.*, **27**, 77-83.

Webster, P.J., G. Holland, J. Curry and H. Chang, 2005: Changes in tropical cyclone number, duration and intensity in a warming environment. *Science*, **309**, 1844-1846.

Westmacott, S., 2002: Where should the focus be in tropical integrated coastal management? *Coast. Manage.*, **30**, 67-84.

WHO (World Health Organization), 2003a: *Report of Synthesis Workshop on Climate Change and Health in Small-Island States*, 1-4 December 2003, Republic of the Maldives. World Health Organization, 95 pp.

WHO (World Health Organization), 2003b: *Climate Change and Human Health-Risks and Responses*. Summary. World Health Organization, Geneva, 37 pp.

Wilkinson, C., Ed., 2004: *Status of Coral Reefs of the World: 2004*. Australian Institute of Marine Science, Townsville, 557 pp.

Wong, P.P., E. Marone, P. Lana, J. Agard, M. Fortes, D. Moro and L. Vicente, 2005: Island systems. *Ecosystems and Human Well-being: Millennium Ecosystem Assessment*, H.A. Mooney and A. Cropper, Eds., Island Press. Washington, District of Columbia / Covelo, London, 663-680.

Woodworth, P.L., 2005: Have there been large recent sea level changes in the Maldive Islands? *Global Planet. Change*, **49**, 1-18.

Woodworth, P.L., C. Le Provost, L.J. Richards, G.T. Mitchum and M. Merrifield, 2002: A review of sea level research from tide gauges during the World Ocean Current Experiment. *Oceanogr. Mar. Biol.*, **40**, 1-35.

World Bank, 2000: *Cities, Seas and Storms: Managing Change in Pacific Island Economies. Vol. IV: Adapting to Climate Change*. World Bank, Washington, District of Columbia, 135 pp.

Ximena, F.P, 1998: Contribution to the estimation of countries' inter-dependence in the area of plant genetic resources. Commission on Genetic Resources for Food and Agriculture, Background Study Paper No. 7, Rev. 1, FAO, Rome, 31 pp.

17

Assessment of adaptation practices, options, constraints and capacity

Coordinating Lead Authors:

W. Neil Adger (UK), Shardul Agrawala (OECD/France), M. Monirul Qader Mirza (Canada/Bangladesh)

Lead Authors:

Cecilia Conde (Mexico), Karen O'Brien (Norway), Juan Pulhin (Philippines), Roger Pulwarty (USA/Trinidad and Tobago), Barry Smit (Canada), Kiyoshi Takahashi (Japan)

Contributing Authors:

Brenna Enright (Canada), Samuel Fankhauser (EBRD/Switzerland), James Ford (Canada), Simone Gigli (Germany), Simon Jetté-Nantel (Canada), Richard J.T. Klein (The Netherlands/Sweden), Irene Lorenzoni (UK), David C. Major (USA), Tristan D. Pearce (Canada), Arun Shreshtha (Nepal), Priyadarshi R. Shukla (India), Joel B. Smith (USA), Tim Reeder (UK), Cynthia Rosenzweig (USA), Katharine Vincent (UK), Johanna Wandel (Canada)

Review Editors:

Abdelkader Allali (Morocco), Neil A. Leary (USA), Antonio R. Magalhães (Brazil)

This chapter should be cited as:

Adger, W.N., S. Agrawala, M.M.Q. Mirza, C. Conde, K. O'Brien, J. Pulhin, R. Pulwarty, B. Smit and K. Takahashi, 2007: Assessment of adaptation practices, options, constraints and capacity. *Climate Change 2007: Impacts, Adaptation and Vulnerability. Contribution of Working Group II to the Fourth Assessment Report of the Intergovernmental Panel on Climate Change,* M.L. Parry, O.F. Canziani, J.P. Palutikof, P.J. van der Linden and C.E. Hanson, Eds., Cambridge University Press, Cambridge, UK, 717-743.

Table of Contents

Executive summary

Adaptation to climate change is already taking place, but on a limited basis (very high confidence).
Societies have a long record of adapting to the impacts of weather and climate through a range of practices that include crop diversification, irrigation, water management, disaster risk management, and insurance. But climate change poses novel risks often outside the range of experience, such as impacts related to drought, heatwaves, accelerated glacier retreat and hurricane intensity [17.2.1].

Adaptation measures that also consider climate change are being implemented, on a limited basis, in both developed and developing countries. These measures are undertaken by a range of public and private actors through policies, investments in infrastructure and technologies, and behavioural change. Examples of adaptations to observed changes in climate include partial drainage of the Tsho Rolpa glacial lake (Nepal); changes in livelihood strategies in response to permafrost melt by the Inuit in Nunavut (Canada); and increased use of artificial snow-making by the Alpine ski industry (Europe, Australia and North America) [17.2.2]. A limited but growing set of adaptation measures also explicitly considers scenarios of future climate change. Examples include consideration of sea-level rise in design of infrastructure such as the Confederation Bridge (Canada) and in coastal zone management (United States and the Netherlands) [17.2.2].

Adaptation measures are seldom undertaken in response to climate change alone (very high confidence).
Many actions that facilitate adaptation to climate change are undertaken to deal with current extreme events such as heatwaves and cyclones. Often, planned adaptation initiatives are also not undertaken as stand-alone measures, but embedded within broader sectoral initiatives such as water resource planning, coastal defence and disaster management planning [17.2.2, 17.3.3]. Examples include consideration of climate change in the National Water Plan of Bangladesh and the design of flood protection and cyclone-resistant infrastructure in Tonga [17.2.2].

Many adaptations can be implemented at low cost, but comprehensive estimates of adaptation costs and benefits are currently lacking (high confidence).
There is a growing number of adaptation cost and benefit-cost estimates at regional and project level for sea-level rise, agriculture, energy demand for heating and cooling, water resource management, and infrastructure. These studies identify a number of measures that can be implemented at low cost or with high benefit-cost ratios. However, some common adaptations may have social and environmental externalities. Adaptations to heatwaves, for example, have involved increased demand for energy-intensive air-conditioning [17.2.3].

Limited estimates are also available for global adaptation costs related to sea-level rise, and energy expenditures for space heating and cooling. Estimates of global adaptation benefits for the agricultural sector are also available, although such literature does not explicitly consider the costs of adaptation. Comprehensive multi-sectoral estimates of global costs and benefits of adaptation are currently lacking [17.2.3].

Adaptive capacity is uneven across and within societies (very high confidence).
There are individuals and groups within all societies that have insufficient capacity to adapt to climate change. For example, women in subsistence farming communities are disproportionately burdened with the costs of recovery and coping with drought in southern Africa [17.3.2].

The capacity to adapt is dynamic and influenced by economic and natural resources, social networks, entitlements, institutions and governance, human resources, and technology [17.3.3]. Multiple stresses related to HIV/AIDS, land degradation, trends in economic globalisation, and violent conflict affect exposure to climate risks and the capacity to adapt. For example, farming communities in India are exposed to impacts of import competition and lower prices in addition to climate risks; marine ecosystems over-exploited by globalised fisheries have been shown to be less resilient to climate variability and change [17.3.3].

There are substantial limits and barriers to adaptation (very high confidence).
High adaptive capacity does not necessarily translate into actions that reduce vulnerability. For example, despite a high capacity to adapt to heat stress through relatively inexpensive adaptations, residents in urban areas in some parts of the world, including in European cities, continue to experience high levels of mortality [17.4.2]. There are significant barriers to implementing adaptation. These include both the inability of natural systems to adapt to the rate and magnitude of climate change, as well as technological, financial, cognitive and behavioural, and social and cultural constraints. There are also significant knowledge gaps for adaptation as well as impediments to flows of knowledge and information relevant for adaptation decisions [17.4.1, 17.4.2].

New planning processes are attempting to overcome these barriers at local, regional and national levels in both developing and developed countries. For example, least-developed countries are developing National Adaptation Programmes of Action and some developed countries have established national adaptation policy frameworks [17.4.1].

17.1 Concepts and methods

This chapter is an assessment of knowledge and practice on adaptation since the IPCC Third Assessment Report (TAR). In the TAR, adaptation and vulnerability were defined, types of adaptation were identified, and the role of adaptive capacity was recognised (Smit et al., 2001). Notable developments that occurred since the TAR include insights on: a) actual adaptations to observed climate changes and variability; b) planned adaptations to climate change in infrastructure design, coastal zone management, and other activities; c) the variable nature of vulnerability and adaptive capacity; and d) policy developments, under the United Nations Framework Convention on Climate Change (UNFCCC) and other international, national and local initiatives, that facilitate adaptation processes and action programmes (Adger et al., 2005; Tompkins et al., 2005; West and Gawith, 2005).

This chapter assesses the recent literature, focussing on real-world adaptation practices and processes, determinants and dynamics of adaptive capacity, and opportunities and constraints of adaptation. While adaptation is increasingly regarded as an inevitable part of the response to climate change, the evidence in this chapter suggests that climate change adaptation processes and actions face significant limitations, especially in vulnerable nations and communities. In most of the cases, adaptations are being implemented to address climate conditions as part of risk management, resource planning and initiatives linked to sustainable development.

This chapter retains the definitions and concepts outlined in the TAR and examines adaptation in the context of vulnerability and adaptive capacity. Vulnerability to climate change refers to the propensity of human and ecological systems to suffer harm and their ability to respond to stresses imposed as a result of climate change effects. The vulnerability of a society is influenced by its development path, physical exposures, the distribution of resources, prior stresses and social and government institutions (Kelly and Adger, 2000; Jones, 2001; Yohe and Tol, 2002; Turner et al., 2003; O'Brien et al., 2004; Smit and Wandel, 2006). All societies have inherent abilities to deal with certain variations in climate, yet adaptive capacities are unevenly distributed, both across countries and within societies. The poor and marginalised have historically been most at risk, and are most vulnerable to the impacts of climate change. Recent analyses in Africa, Asia and Latin America, for example, show that marginalised, primary resource-dependent livelihood groups are particularly vulnerable to climate change impacts if their natural resource base is severely stressed and degraded by overuse or if their governance systems are in or near a state of failure and hence not capable of responding effectively (Leary et al., 2006).

Adaptation to climate change takes place through adjustments to reduce vulnerability or enhance resilience in response to observed or expected changes in climate and associated extreme weather events. Adaptation occurs in physical, ecological and human systems. It involves changes in social and environmental processes, perceptions of climate risk, practices and functions to reduce potential damages or to realise new opportunities. Adaptations include anticipatory and reactive actions, private and public initiatives, and can relate to projected changes in temperature and current climate variations and extremes that may be altered with climate change. In practice, adaptations tend to be on-going processes, reflecting many factors or stresses, rather than discrete measures to address climate change specifically.

Biological adaptation is reactive (see Chapter 4), whereas individuals and societies adapt to both observed and expected climate through anticipatory and reactive actions. There are well-established observations of human adaptation to climate change over the course of human history (McIntosh et al., 2000; Mortimore and Adams, 2001). Despite evidence of success stories, many individuals and societies still remain vulnerable to present-day climatic risks, which may be exacerbated by future climate change. Some adaptation measures are undertaken by individuals, while other types of adaptation are planned and implemented by governments on behalf of societies, sometimes in anticipation of change but mostly in response to experienced climatic events, especially extremes (Adger, 2003; Kahn, 2003; Klein and Smith, 2003).

The scientific research on adaptation is synthesised in this chapter according to: current adaptation practices to climate variability and change; assessment of adaptation costs and benefits; adaptive capacity and its determinants, dynamics and spatial variations; and the opportunities and limits of adaptation as a response strategy for climate change.

17.2 Assessment of current adaptation practices

17.2.1 Adaptation practices

In this chapter, adaptation practices refer to actual adjustments, or changes in decision environments, which might ultimately enhance resilience or reduce vulnerability to observed or expected changes in climate. Thus, investment in coastal protection infrastructure to reduce vulnerability to storm surges and anticipated sea-level rise is an example of actual adjustments. Meanwhile, the development of climate risk screening guidelines, which might make downstream development projects more resilient to climate risks (Burton and van Aalst, 2004; ADB, 2005), is an example of changes in the policy environment.

With an explicit focus on real-world behaviour, assessments of adaptation practices differ from the more theoretical assessments of potential responses or how such measures might reduce climate damages under hypothetical scenarios of climate change. Adaptation practices can be differentiated along several dimensions: by spatial scale (local, regional, national); by sector (water resources, agriculture, tourism, public health, and so on); by type of action (physical, technological, investment, regulatory, market); by actor (national or local government, international donors, private sector, NGOs, local communities and individuals); by climatic zone (dryland, floodplains, mountains, Arctic, and so on); by baseline income/development level of the systems in which they are implemented (least-developed countries, middle-income countries, and developed countries); or by some combination of these and other categories.

From a temporal perspective, adaptation to climate risks can be viewed at three levels, including responses to: current variability (which also reflect learning from past adaptations to historical climates); observed medium and long-term trends in climate; and anticipatory planning in response to model-based scenarios of long-term climate change. The responses across the three levels are often intertwined, and indeed might form a continuum.

Adapting to current climate variability is already sensible in an economic development context, given the direct and certain evidence of the adverse impacts of such phenomena (Goklany,

1995; Smit et al., 2001; Agrawala and Cane, 2002). In addition, such adaptation measures can be synergistic with development priorities (Ribot et al., 1996), but there could also be conflicts. For example, activities such as shrimp farming and conversion of coastal mangroves, while profitable in an economic sense, can exacerbate vulnerability to sea-level rise (Agrawala et al., 2005).

Adaptation to current climate variability can also increase resilience to long-term climate change. In a number of cases, however, anthropogenic climate change is likely to also require forward-looking investment and planning responses that go beyond short-term responses to current climate variability. This is true, for example, in the case of observed impacts such as glacier retreat and permafrost melt (Schaedler, 2004; Shrestha and Shrestha, 2004). Even when impacts of climate change are not yet discernible, scenarios of future impacts may already be of sufficient concern to justify building some adaptation responses into planning. In some cases it could be more cost-effective to implement adaptation measures early on, particularly for infrastructure with long economic life (Shukla et al., 2004), or if current activities may irreversibly constrain future adaptation to the impacts of climate change (Smith et al., 2005).

17.2.2 Examples of adaptation practices

There is a long record of practices to adapt to the impacts of weather as well as natural climate variability on seasonal to interannual time-scales – particularly to the El Niño-Southern Oscillation (ENSO). These include proactive measures such as crop and livelihood diversification, seasonal climate forecasting, community-based disaster risk reduction, famine early warning systems, insurance, water storage, supplementary irrigation and so on. They also include reactive or ex-poste adaptations, for example, emergency response, disaster recovery, and migration (Sperling and Szekely, 2005). Recent reviews indicate that a 'wait and see' or reactive approach is often inefficient and could be particularly unsuccessful in addressing irreversible damages, such as species extinction or unrecoverable ecosystem damages, that may result from climate change (Smith, 1997; Easterling et al., 2004).

Proactive practices to adapt to climate variability have advanced significantly in recent decades with the development of operational capability to forecast several months in advance the onset of El Niño and La Niña events related to ENSO (Cane et al., 1986), as well as improvements in climate monitoring and remote sensing to provide better early warnings on complex climate-related hazards (Dilley, 2000). Since the mid 1990s, a number of mechanisms have also been established to facilitate proactive adaptation to seasonal to interannual climate variability. These include institutions that generate and disseminate regular seasonal climate forecasts (NOAA, 1999), and the regular regional and national forums and implementation projects worldwide to engage with local and national decision makers to design and implement anticipatory adaptation measures in agriculture, water resource management, food security, and a number of other sectors (Basher et al., 2000; Broad and Agrawala, 2000; Meinke et al., 2001; Patt and Gwata, 2002; De Mello Lemos, 2003; O'Brien and Vogel, 2003; Ziervogel, 2004). An evaluation of the responses to the 1997-98 El Niño across 16 developing countries in Asia, Asia-Pacific, Africa, and Latin America highlighted a number of barriers to effective adaptation, including: spatial and temporal uncertainties associated with forecasts of regional climate, low level of awareness among decision makers of the local and regional impacts of El Niño, limited national capacities in climate monitoring and forecasting, and lack of co-ordination in the formulation of responses (Glantz, 2001). Recent research also highlights that technological solutions such as seasonal forecasting are not sufficient to address the underlying social drivers of vulnerabilities to climate (Agrawala and Broad, 2002). Furthermore, social inequities in access to climate information and the lack of resources to respond can severely constrain anticipatory adaptation (Pfaff et al., 1999).

Table 17.1 provides an illustrative list of various types of adaptations that have been implemented by a range of actors including individuals, communities, governments and the private sector. Such measures involve a mix of institutional and behavioural responses, the use of technologies, and the design of climate resilient infrastructure. They are typically undertaken in response to multiple risks, and often as part of existing processes or programmes, such as livelihood enhancement, water resource management, and drought relief.

A growing number of measures are now also being put in place to adapt to the impacts of observed medium- to long-term trends in climate, as well as to scenarios of climate change. In particular, numerous measures have been put in place in the winter tourism sector in Alpine regions of many Organisation for Economic Co-operation and Development (OECD) countries to respond to observed impacts such as reduced snow cover and glacier retreat. These measures include technologies such as artificial snow-making and associated structures such as high altitude water reservoirs, economic and regional diversification, and the use of market-based instruments such as weather derivatives and insurance (e.g., Konig, 1999, for Australia; Burki et al., 2005, for Switzerland; Harrison et al., 2005, for Scotland; Scott et al., 2005, for North America). Adaptation measures are also being put in place in developing country contexts to respond to glacier retreat and associated risks, such as the expansion of glacial lakes, which pose serious risks to livelihoods and infrastructure. The Tsho Rolpa risk-reduction project in Nepal is an example of adaptation measures being implemented to address the creeping threat of glacial lake outburst flooding as a result of rising temperatures (see Box 17.1).

Recent observed weather extremes, particularly heatwaves (e.g., 1995 heatwave in Chicago; the 1998 heatwave in Toronto; and the 2003 heatwave in Europe), have also provided the trigger for the design of hot-weather alert plans. While such measures have been initiated primarily in response to current weather extremes, at times there is implicit or explicit recognition that hot weather events might become more frequent or worsen under climate change and that present adaptations have often been inadequate and created new vulnerabilities (Poumadère et al., 2005). Public health adaptation measures have now been put in place that combine weather monitoring, early warning, and response measures in a number of places including metropolitan Toronto (Smoyer-Tomic and Rainham, 2001; Ligeti, 2004; Mehdi, 2006), Shanghai (Sheridan and Kalkstein, 2004) and several cities in Italy and France (ONERC, 2005). Weather and climate extremes have also led to a number of adaptation responses in the financial sector (see Box 17.2).

Table 17.1. *Examples of adaptation initiatives by region, undertaken relative to present climate risks, including conditions associated with climate change.*

REGION Country *Reference*	Climate-related stress	Adaptation practices
AFRICA		
Egypt ***El Raey (2004)***	Sea-level rise	Adoption of National Climate Change Action Plan integrating climate change concerns into national policies; adoption of Law 4/94 requiring Environmental Impact Assessment (EIA) for project approval and regulating setback distances for coastal infrastructure; installation of hard structures in areas vulnerable to coastal erosion.
Sudan ***Osman-Elasha et al. (2006)***	Drought	Expanded use of traditional rainwater harvesting and water conserving techniques; building of shelter-belts and wind-breaks to improve resilience of rangelands; monitoring of the number of grazing animals and cut trees; set-up of revolving credit funds.
Botswana ***FAO (2004)***	Drought	National government programmes to re-create employment options after drought; capacity building of local authorities; assistance to small subsistence farmers to increase crop production.
ASIA & OCEANIA		
Bangladesh ***OECD (2003a); Pouliotte (2006)***	Sea-level rise; salt-water intrusion	Consideration of climate change in the National Water Management Plan; building of flow regulators in coastal embankments; use of alternative crops and low-technology water filters.
Philippines ***Lasco et al. (2006)***	Drought; floods	Adjustment of silvicultural treatment schedules to suit climate variations; shift to drought-resistant crops; use of shallow tube wells; rotation method of irrigation during water shortage; construction of water impounding basins; construction of fire lines and controlled burning; adoption of soil and water conservation measures for upland farming.
	Sea-level rise; storm surges	Capacity building for shoreline defence system design; introduction of participatory risk assessment; provision of grants to strengthen coastal resilience and rehabilitation of infrastructures; construction of cyclone-resistant housing units; retrofit of buildings to improved hazard standards; review of building codes; reforestation of mangroves.
	Drought; salt-water intrusion	Rainwater harvesting; leakage reduction; hydroponic farming; bank loans allowing for purchase of rainwater storage tanks.
AMERICAS		
Canada ***(1) Ford and Smit (2004)*** ***(2) Mehdi (2006)***	(1) Permafrost melt; change in ice cover	Changes in livelihood practices by the Inuit, including: change of hunt locations; diversification of hunted species; use of Global Positioning Systems (GPS) technology; encouragement of food sharing.
	(2) Extreme temperatures	Implementation of heat health alert plans in Toronto, which include measures such as: opening of designated cooling centres at public locations; information to the public through local media; distribution of bottled water through the Red Cross to vulnerable people; operation of a heat information line to answer heat-related questions; availability of an emergency medical service vehicle with specially trained staff and medical equipment.
United States ***Easterling et al. (2004)***	Sea-level rise	Land acquisition programmes taking account of climate change (e.g., New Jersey Coastal Blue Acres land acquisition programme to acquire coastal lands damaged/prone to damages by storms or buffering other lands; the acquired lands are being used for recreation and conservation); establishment of a 'rolling easement' in Texas, an entitlement to public ownership of property that 'rolls' inland with the coastline as sea-level rises; other coastal policies that encourage coastal landowners to act in ways that anticipate sea-level rise.
Mexico and Argentina ***Wehbe et al. (2006)***	Drought	Adjustment of planting dates and crop variety (e.g., inclusion of drought-resistant plants such as agave and aloe); accumulation of commodity stocks as economic reserve; spatially separated plots for cropping and grazing to diversify exposures; diversification of income by adding livestock operations; set-up/provision of crop insurance; creation of local financial pools (as alternative to commercial crop insurance).
EUROPE		
The Netherlands, ***Government of the Netherlands (1997 and 2005)***	Sea-level rise	Adoption of Flooding Defence Act and Coastal Defence Policy as precautionary approaches allowing for the incorporation of emerging trends in climate; building of a storm surge barrier taking a 50 cm sea-level rise into account; use of sand supplements added to coastal areas; improved management of water levels through dredging, widening of river banks, allowing rivers to expand into side channels and wetland areas; deployment of water storage and retention areas; conduct of regular (every 5 years) reviews of safety characteristics of all protecting infrastructure (dykes, etc.); preparation of risk assessments of flooding and coastal damage influencing spatial planning and engineering projects in the coastal zone, identifying areas for potential (land inward) reinforcement of dunes.
Austria, France, Switzerland ***Austrian Federal Govt. (2006); Direction du Tourisme (2002); Swiss Confederation (2005)***	Upward shift of natural snow-reliability line; glacier melt	Artificial snow-making; grooming of ski slopes; moving ski areas to higher altitudes and glaciers; use of white plastic sheets as protection against glacier melt; diversification of tourism revenues (e.g., all-year tourism).
	Permafrost melt; debris flows	Erection of protection dams in Pontresina (Switzerland) against avalanches and increased magnitude of potential debris flows stemming from permafrost thawing.
United Kingdom ***Defra (2006)***	Floods; sea-level rise	Coastal realignment under the Essex Wildlife Trust, converting over 84 ha of arable farmland into salt marsh and grassland to provide sustainable sea defences; maintenance and operation of the Thames Barrier through the Thames Estuary 2100 project that addresses flooding linked to the impacts of climate change; provision of guidance to policy makers, chief executives, and parliament on climate change and the insurance sector (developed by the Association of British Insurers).

Box 17.1. Tsho Rolpa Risk Reduction Project in Nepal as observed anticipatory adaptation

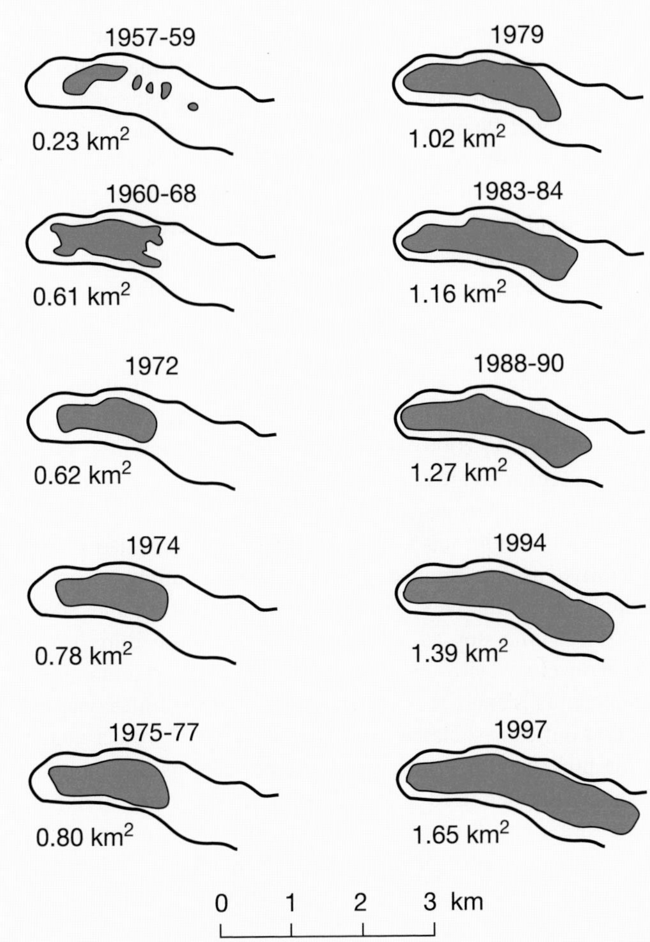

Figure 17.1. *Tsho Rolpa Risk Reduction Project in Nepal as observed anticipatory adaptation.*

The Tsho Rolpa is a glacial lake located at an altitude of about 4,580 m in Nepal. Glacier retreat and ice melt as a result of warmer temperature increased the size of the Tsho Rolpa from 0.23 km^2 in 1957/58 to 1.65 km^2 in 1997 (Figure 17.1). The 90-100 million m^3 of water, which the lake contained by this time, were only held by a moraine dam – a hazard that called for urgent action to reduce the risk of a catastrophic glacial lake outburst flood (GLOF).

If the dam were breached, one third or more of the water could flood downstream. Among other considerations, this posed a major risk to the Khimti hydropower plant, which was under construction downstream. These concerns spurred the Government of Nepal, with the support of international donors, to initiate a project in 1998 to lower the level of the lake through drainage. An expert group recommended that, to reduce the risk of a GLOF, the lake should be lowered three metres by cutting a channel in the moraine. A gate was constructed to allow for controlled release of water. Meanwhile, an early warning system was established in 19 villages downstream in case a Tsho Rolpa GLOF should occur despite these efforts. Local villagers were actively involved in the design of the system, and drills are carried out periodically. In 2002, the four-year construction project was completed at a cost of US$3.2 million. Clearly, reducing GLOF risks involves substantial costs and is time-consuming as complete prevention of a GLOF would require further drainage to lower the lake level.

Sources: Mool et al. (2001); OECD (2003b); Shrestha and Shrestha (2004).

Box 17.2. Adaptation practices in the financial sector

Financial mechanisms can contribute to climate change adaptation. The insurance sector – especially property, health and crop insurance – can efficiently spread risks and reduce the financial hardships linked to extreme events. Financial markets can internalise information on climate risks and help transfer adaptation and risk-reduction incentives to communities and individuals (ABI, 2004), while capital markets and transfer mechanisms can alleviate financial constraints to the implementation of adaptation measures. To date, most adaptation practices have been observed in the insurance sector. As a result of climate change, demand for insurance products is expected to increase, while climate change impacts could also reduce insurability and threaten insurance schemes (ABI, 2004; Dlugolecki and Lafeld, 2005; Mills et al., 2005; Valverde and Andrews, 2006). While these market signals can play a role in transferring adaptation incentives to individuals, reduced insurance coverage can, at the same time, impose significant economic and social costs. To increase their capacity in facing climate variability and change, insurers have developed more comprehensive or accessible information tools, e.g., risk assessment tools in the Czech Republic, France, Germany and the United Kingdom (CEA, 2006). They have also fostered risk prevention through: (i) implementing and strengthening building standards, (ii) planning risk prevention measures and developing best practices, and (iii) raising awareness of policyholders and public authorities (ABI, 2004; CEA, 2006; Mills and Lecomte, 2006). In the longer term, climate change may also induce insurers to adopt forward-looking pricing methods in order to maintain insurability (ABI, 2004; Loster, 2005).

There are now also examples of adaptation measures being put in place that take into account scenarios of future climate change and associated impacts. This is particularly the case for long-lived infrastructure which may be exposed to climate change impacts over its lifespan or, in cases, where business-as-usual activities would irreversibly constrain future adaptation to the impacts of climate change. Early examples where climate change scenarios have already been incorporated in infrastructure design include the Confederation Bridge in Canada and the Deer Island sewage treatment plant in Boston harbour in the United States. The Confederation Bridge is a 13 km bridge between Prince Edward Island and the mainland. The bridge provides a navigation channel for ocean-going vessels with vertical clearance of about 50 m (McKenzie and Parlee, 2003). Sea-level rise was recognised as a principal concern during the design process and the bridge was built one metre higher than currently required to accommodate sea-level rise over its hundred-year lifespan (Lee, 2000). In the case of the Deer Island sewage facility, the design called for raw sewage collected from communities onshore to be pumped under Boston harbour and then up to the treatment plant on Deer Island. After waste treatment, the effluent would be discharged into the harbour through a downhill pipe. Design engineers were concerned that sea-level rise would necessitate the construction of a protective wall around the plant, which would then require installation of expensive pumping equipment to transport the effluent over the wall (Easterling et al., 2004). To avoid such a future cost the designers decided to keep the treatment plant at a higher elevation, and the facility was completed in 1998. Other examples where ongoing planning is considering scenarios of climate change in project design are the Konkan Railway in western India (Shukla et al., 2004); a coastal highway in Micronesia (ADB, 2005); the Copenhagen Metro in Denmark (Fenger, 2000); and the Thames Barrier in the United Kingdom (Dawson et al., 2005; Hall et al., 2006).

A majority of examples of infrastructure-related adaptation measures relate primarily to the implications of sea-level rise. In this context, the Qinghai-Tibet Railway is an exception. The railway crosses the Tibetan Plateau with about a thousand kilometres of the railway at least 13,000 feet (4,000 m) above sea level. Five hundred kilometres of the railway rests on permafrost, with roughly half of it 'high temperature permafrost' which is only 1 to 2°C below freezing. The railway line would affect the permafrost layer, which will also be impacted by thawing as a result of rising temperatures, thus in turn affecting the stability of the railway line. To reduce these risks, design engineers have put in place a combination of insulation and cooling systems to minimise the amount of heat absorbed by the permafrost (Brown, 2005).

In addition to specific infrastructure projects, there are now also examples where climate change scenarios are being considered in more comprehensive risk management policies and plans. Efforts are underway to integrate adaptation to current and future climate within the Environmental Impact Assessment (EIA) procedures of several countries in the Caribbean (Vergara, 2006), as well as Canada (Lee, 2000). A number of other policy initiatives have also been put in place within OECD countries that take future climate change (particularly sea-level rise) into account (Moser, 2005; Gagnon-Lebrun and Agrawala, 2006). In the Netherlands, for example, the Technical Advisory Committee on Water Defence recommended the design of new engineering works with a long lifetime, such as storm surge barriers and dams, to take a 50 cm sea-level rise into account (Government of the Netherlands, 1997). Climate change is explicitly taken into consideration in the National Water Management Plan (NWMP) of Bangladesh, which was set up to guide the implementation of the National Water Policy. It recognises climate change as a determining factor for future water supply and demand, as well as coastal erosion due to sea-level rise and increased tidal range (OECD, 2003a).

There are now also examples of consideration of climate change as part of comprehensive risk management strategies at the city, regional and national level. France, Finland and the United Kingdom have developed national strategies and frameworks to adapt to climate change (MMM, 2005; ONERC, 2005; DEFRA, 2006). At the city level, meanwhile, climate change scenarios are being considered by New York City as part of the review of its water supply system. Changes in temperature and precipitation, sea-level rise, and extreme events have been identified as important parameters for water supply impacts and adaptation in the New York region (Rosenzweig and Solecki, 2001). A nine-step adaptation assessment procedure has now been developed (Rosenzweig et al., 2007). A key feature of these procedures is explicit consideration of several climate variables, uncertainties associated with climate change projections, and time horizons for different adaptation responses. Adaptations can be divided into managerial, infrastructure, and policy categories and assessed in terms of time frame (immediate, interim, long-term) and in terms of the capital cycle for different types of infrastructure. As an example of adaptation measures that have been examined, a managerial adaptation that can be implemented quickly is a tightening of water regulations in the event of more frequent droughts. Also under examination are longer-term infrastructure adaptations such as the construction of flood-walls around low-lying wastewater treatment plants to protect against sea-level rise and higher storm surges.

17.2.3 Assessment of adaptation costs and benefits

The literature on adaptation costs and benefits remains quite limited and fragmented in terms of sectoral and regional coverage. Adaptation costs are usually expressed in monetary terms, while benefits are typically quantified in terms of avoided climate impacts, and expressed in monetary as well as non-monetary terms (e.g., changes in yield, welfare, population exposed to risk). There is a small methodological literature on the assessment of costs and benefits in the context of climate change adaptation (Fankhauser, 1996; Smith, 1997; Fankhauser et al., 1998; Callaway, 2004; Toman, 2006). In addition there are a number of case studies that look at adaptation options for particular sectors (e.g., Shaw et al., 2000, for sea-level rise); or particular countries (e.g., Smith et al., 1998, for Bangladesh; World Bank, 2000, for Fiji and Kiribati; Dore and Burton, 2001, for Canada).

Much of the literature on adaptation costs and benefits is focused on sea-level rise (e.g., Fankhauser, 1995a; Yohe and

Schlesinger, 1998; Nicholls and Tol, 2006) and agriculture (e.g., Rosenzweig and Parry, 1994; Adams et al., 2003; Reilly et al., 2003). Adaptation costs and benefits have also been assessed in a more limited manner for energy demand (e.g., Morrison and Mendelsohn, 1999; Sailor and Pavlova, 2003; Mansur et al., 2005), water resource management (e.g., Kirshen et al., 2004), and transportation infrastructure (e.g., Dore and Burton, 2001). In terms of regional coverage, there has been a focus on the United States and other OECD countries (e.g., Fankhauser, 1995a; Yohe et al., 1996; Mansur et al., 2005; Franco and Sanstad, 2006), although there is now a growing literature for developing countries also (e.g., Butt et al., 2005; Callaway et al., 2006; Nicholls and Tol, 2006).

17.2.3.1 Sectoral and regional estimates

The literature on costs and benefits of adaptation to sea-level rise is relatively extensive. Fankhauser (1995a) used comparative static optimisation to examine the trade-offs between investment in coastal protection and the value of land loss from sea-level rise. The resulting optimal levels of coastal protection were shown to significantly reduce the total costs of sea-level rise across OECD countries. The results also highlighted that the optimal level of coastal protection would vary considerably both within and across regions, based on the value of land at risk. Fankhauser (1995a) concluded that almost 100% of coastal cities and harbours in OECD countries should be protected, while the optimal protection for beaches and open coasts would vary between 50 and 80%. Results of Yohe and Schlesinger (1998) show that total (adjustment and residual land loss) costs of sea-level rise could be reduced by around 20 to 50% for the U.S. coastline if the real estate market prices adjusted efficiently as land is submerged. Nicholls and Tol (2006) estimate optimal levels of coastal protection under IPCC Special Report on Emissions Scenarios (SRES; Nakićenović and Swart, 2000) A1FI, A2, B1, and B2 scenarios. They conclude that, with the exception of certain Pacific Small Island States, coastal protection investments were a very small percentage of gross domestic product (GDP) for the 15 most-affected countries by 2080 (Table 17.2).

Ng and Mendelsohn (2005) use a dynamic framework to optimise for coastal protection, with a decadal reassessment of the protection required. It was estimated that, over the period 2000 to 2100, the present value of coastal protection costs for Singapore would be between US$1 and 3.08 million (a very small share of GDP), for a 0.49 and 0.86 m sea-level rise. A limitation of these studies is that they only look at gradual sea-level rise and do not generally consider issues such as the implications of storm surges on optimal coastal protection. In a study of the Boston metropolitan area Kirshen et al. (2004) include the implications of storm surges on sea-level rise damages and optimal levels of coastal protection under various development and sea-level rise scenarios. Kirshen et al. (2004) conclude that under 60 cm sea-level rise 'floodproofing' measures (such as elevation of living spaces) were superior to coastal protection measures (such as seawalls, bulkheads, and revetments). Meanwhile, coastal protection was found to be optimal under one-metre sea-level rise. Another limitation of sea-level rise costing studies is their sensitivity to (land and structural) endowment values which are highly uncertain at more aggregate levels. A global assessment by Darwin and Tol (2001) showed that uncertainties surrounding endowment values could lead to a 17% difference in coastal protection, a 36% difference in amount of land protected, and a 36% difference in direct cost globally. A further factor increasing uncertainty in costs is the social and political acceptability of adaptation options. Tol et al. (2003) show that the benefits of adaptation options for ameliorating increased river flood risk in the Netherlands could be up to US$20 million /yr in 2050. But they conclude that implementation of these options requires significant institutional and political reform, representing a significant barrier to implanting least-cost solutions.

Adaptation studies looking at the agricultural sector considered autonomous farm level adaptation and many also looked at adaptation effects through market and international trade (Darwin et al., 1995; Winters et al., 1998; Yates and Street, 1998; Adams et al., 2003; Butt et al., 2005). The literature mainly reports on adaptation benefits, usually expressed in terms of increases in yield or welfare, or decreases in the number of people at risk of hunger. Adaptation costs, meanwhile, were generally not considered in early studies (Rosenzweig and Parry, 1994; Yates and Street, 1998), but are usually included in recent studies (Mizina et al., 1999; Adams et al., 2003; Reilly et al., 2003; Njie et al., 2006). Rosenzweig and Parry (1994) and Darwin et al. (1995) estimated residual climate change impacts to be minimal at the global level, mainly due to the significant benefits from adaptation. However, large inter and intra-regional variations were reported. In particular, for many countries located in tropical regions, the potential benefits of low-cost adaptation measures such as changes in planting dates, crop mixes, and cultivars are not expected to be sufficient to offset the significant climate change damages (Rosenzweig and Parry, 1994; Butt et al., 2005).

Table 17.2. *Sea-level rise protection costs in 2080 as a percentage of GDP for most-affected countries under the four SRES world scenarios (A1FI, A2, B1, B2)*

	Protection costs (%GDP) for the 2080s			
SRES scenarios	A1FI	A2	B1	B2
Micronesia	7.4	10.0	5.0	13.5
Palau	6.1	7.0	3.9	9.1
Tuvalu	1.4	1.7	0.9	2.2
Marshall Islands	0.9	1.3	0.6	1.7
Mozambique	0.2	0.5	0.1	0.8
French Polynesia	0.6	0.8	0.4	1.0
Guinea-Bissau	0.1	0.3	0.0	0.6
Nauru	0.3	0.4	0.2	0.6
Guyana	0.1	0.2	0.1	0.4
New Caledonia	0.4	0.3	0.2	0.4
Papua New Guinea	0.3	0.3	0.2	0.4
Kiribati	1.2	0.0	0.3	0.0
Maldives	0.0	0.2	0.0	0.2
Vietnam	0.1	0.1	0.0	0.2
Cambodia	0.0	0.1	0.0	0.1

Source: Adapted from Nicholls and Tol (2006).

More extensive adaptation measures have been evaluated in some developing countries (see, for example, Box 17.3). For the 2030 horizon in Mali, Butt et al., (2005) estimate that adaptation through trade, changes in crop mix, and the development and adoption of heat-resistant cultivars could offset 90 to 107% of welfare losses induced by climate change impacts on agriculture.

In addition to their effect on average yield, adaptation measures can also smooth out fluctuations in yields (and consequently social welfare) as a result of climate variability. Adams et al. (2003) found that adaptation welfare benefits for the American economy increased from US$3.29 billion (2000 values) to US$4.70 billion (2000 values) when their effect on yield variability is included. In the case of Mali, Butt et al. (2005) show that adaptation measures could reduce the variability in welfare by up to 84%.

A particular limitation of adaptation studies in the agricultural sector stems from the diversity of climate change impacts and adaptation options but also from the complexity of the adaptation process. Many studies make the unrealistic assumption of perfect adaptation by individual farmers. Even if agricultural regions can adapt fully through technologies and management practices, there are likely to be costs of adaptation in the process of adjusting to a new climate regime. Recent studies for U.S. agriculture found that frictions in the adaptation process could reduce the adaptation potential (Schneider et al., 2000a; Easterling et al., 2003; Kelly et al., 2005).

With regard to adaptation costs and benefits in the energy sector, there is some literature – almost entirely on the United States – on changes in energy expenditures for cooling and heating as a result of climate change. Most studies show that increased energy expenditure on cooling will more than offset any benefits from reduced heating (e.g., Smith and Tirpak, 1989; Nordhaus, 1991; Cline, 1992; Morrison and Mendelsohn, 1999; Mendelsohn, 2003; Sailor and Pavlova, 2003; Mansur et al., 2005). Morrison and Mendelsohn (1999), meanwhile, estimate net adaptation costs (as a result of increased cooling and reduced heating) for the U.S. economy ranging from US$1.93 billion to 12.79 billion by 2060. They also estimated that changes in building stocks (particularly increases in cooling capacity) contributed to the increase in energy expenditure by US$2.98 billion to US$11.5 billion. Mansur et al. (2005), meanwhile, estimate increased energy expenditures for the United States ranging from US$4 to 9 billion for 2050, and between US$16 and 39.8 billion for 2100.

Besides sea-level rise, agriculture, and energy demand, there are a few studies related to adaptation costs and benefits in water resource management (see Box 17.4) and transportation infrastructure. Kirshen et al. (2004) assessed the reliability of water supply in the Boston metropolitan region under climate change scenarios. Even under a stable climate, the authors project the reliability of water supply to be 93% by 2100 on account of the expected growth in water demand. Factoring in climate change reduces the reliability of water supply to 82%. Demand side management measures could increase the reliability slightly (to 83%), while connecting the local systems to the main state water system would increase reliability to 97%. The study, however, does not assess the costs of such adaptation measures.

Dore and Burton (2001) estimate the costs of adaptation to climate change for social infrastructure in Canada, more precisely for the roads network (roads, bridges and storm water management systems) as well as for water utilities (drinking and waste water treatment plants). In this case, the additional costs for maintaining the integrity of the portfolio of social assets under climate change are identified as the costs of adaptation. In the water sector, potential adaptation strategies such as building new treatment plants, improving efficiency of actual plants or increasing retention tanks were considered and results indicated that adaptation costs for Canadian cities could be as high as Canadian $9,400 million for a city like Toronto if extreme events are considered. For the transportation sector, Dore and Burton (2001) also estimate that replacing all ice roads in Canada would cost around Canadian $908 million. However, the study also points out that retreat of permafrost would reduce road building costs. Also, costs of winter control, such as snow clearance, sanding, and salting, are generally expected to decrease as temperature rises.

Box 17.3. Adaptation costs and benefits for agriculture in the Gambia

Njie et al. (2006) investigated climate change impacts and adaptation costs and benefits for cereal production in the Gambia. Under the SRES A2 scenario the study estimated that for the period 2010 to 2039, millet yield would increase by 2 to 13%. For the period 2070 to 2099 the outcome is highly dependent on projected changes in precipitation as it could range from a 43% increase to a 78% decrease in millet yield. Adaptation measures such as the adoption of improved cultivars, irrigation, and improved crop fertilisation were assessed in a framework accounting for projections of population growth, water demand and availability. These measures were estimated to increase millet yield by 13 to 43%, while reducing interannual variability by 84 to 200% in the near term (2010 to 2039). However, net adaptation benefits (value of higher production minus cost of implementation) were not necessarily positive for all adaptation strategies. In the near term, net adaptation benefits were estimated at US$22.3 to 31.5 million for crop fertilisation and US$81.1 to 88.0 million for irrigation. The authors conclude that irrigation is more effective to improve crop productivity under climate change conditions, but the adoption of improved crop fertilisation is more cost efficient. Meanwhile, much uncertainty remains regarding the cost of developing improved cultivars. In the distant future, potential precipitation decrease would make irrigation an imperative measure.

17.2.3.2 Global estimates

Some adaptation costs are implicitly included in estimates of global impacts of climate change. Tol et al. (1998) estimate that between 7% and 25% of total climate damage costs included in earlier studies such as Cline (1992), Fankhauser (1995b) and Tol (1995) could be classified as adaptation costs. In addition, recent studies, including Nordhaus and Boyer (2000), Mendelsohn et al. (2000) and Tol (2002), incorporate with greater detail the effects of adaptation on the global estimation of climate change impacts. In these models, adaptation costs and benefits are usually embedded within climate damage functions which are often extrapolated from a limited number of regional studies. Furthermore, the source studies which form the basis for the climate damage functions do not always reflect the most recent findings. As a result, these studies offer a global and integrated perspective but are based on coarsely defined climate change and adaptation impacts and only provide speculative estimates of adaptation costs and benefits.

Mendelsohn et al. (2000) estimate that global energy costs related to heating and cooling would increase by US$2 billion to US$10 billion (1990 values) for a 2°C increase in temperature by 2100 and by US$51 billion to US$89 billion (1990 values) for a 3.5°C increase. For a 1°C increase, Tol (2002) estimates global benefits from reduced heating at around US$120 billion, and global costs resulting from increased cooling at around US$75 billion. The same study estimates the global protection costs at US$1,055 billion for a one-metre sea-level rise. There are preliminary estimates of the global costs of 'climate proofing' development (World Bank, 2006), but the current literature does not provide comprehensive multi-sectoral estimates of global adaptation costs and benefits. The broader macroeconomic and economy-wide implications of adaptations on economic growth and employment remain largely unknown (Aaheim and Schjolden, 2004).

17.3 Assessment of adaptation capacity, options and constraints

17.3.1 Elements of adaptive capacity

Adaptive capacity is the ability or potential of a system to respond successfully to climate variability and change, and includes adjustments in both behaviour and in resources and technologies. The presence of adaptive capacity has been shown to be a necessary condition for the design and implementation of effective adaptation strategies so as to reduce the likelihood and the magnitude of harmful outcomes resulting from climate change (Brooks and Adger, 2005). Adaptive capacity also enables sectors and institutions to take advantage of opportunities or benefits from climate change, such as a longer growing season or increased potential for tourism.

Much of the current understanding of adaptive capacity comes from vulnerability assessments. Even if vulnerability indices do not explicitly include determinants of adaptive capacity, the indicators selected often provide important insights on the factors, processes and structures that promote or constrain adaptive capacity (Eriksen and Kelly, 2007). One clear result from research on vulnerability and adaptive capacity is that some dimensions of adaptive capacity are generic, while others are specific to particular climate change impacts. Generic indicators include factors such as education, income and health. Indicators specific to a particular impact, such as drought or floods, may relate to institutions, knowledge and technology (Yohe and Tol, 2002; Downing, 2003; Brooks et al., 2005; Tol and Yohe, 2007).

Technology can potentially play an important role in adapting to climate change. Efficient cooling systems, improved seeds, desalination technologies, and other engineering solutions represent some of the options that can lead to improved outcomes and increased coping under conditions of climate change. In public health, for example, there have been successful applications of seasonal forecasting and other technologies to

Box 17.4. Adaptation costs and benefits in the water management sector of South Africa

Callaway et al. (2006) provide estimates of water management adaptation costs and benefits in a case study of the Berg River basin in South Africa. Adaptation measures investigated include the establishment of an efficient water market and an increase in water storage capacity through the construction of a dam. Using a programming model which linked modules of urban and farm water demand to a hydrology module, the welfare related to water use (value for urban and farm use minus storage and transport cost) were estimated for the SRES B2 climate change scenario and the assumption of a 3% increase in urban water demand. Under these conditions and the current water allocation system, the discounted impact of climate change over the next 30 years was estimated to vary between 13.5 and 27.7 billion Rand. The net welfare benefits of adapting water storage capacity under current allocation rights were estimated at about 0.2 billion Rand, while adding water storage capacity in the presence of efficient water markets would yield adaptation benefits between 5.8 and 7 billion Rand. The authors also show that, under efficient water markets, the costs of not adapting to climate change that does occur outweigh the costs of adapting to climate change that does not occur.

N.B.: All monetary estimates are expressed in present values for constant Rand for the year 2000, discounting over 30 years at a real discount rate of 6%.

adapt health provisions to anticipated extreme events (Ebi et al., 2005). Often, technological adaptations and innovations are developed through research programmes undertaken by governments and by the private sector (Smit and Skinner, 2002). Innovation, which refers to the development of new strategies or technologies, or the revival of old ones in response to new conditions (Bass, 2005), is an important aspect of adaptation, particularly under uncertain future climate conditions. Although technological capacity can be considered a key aspect of adaptive capacity, many technological responses to climate change are closely associated with a specific type of impact, such as higher temperatures or decreased rainfall.

New studies carried out since the TAR show that adaptive capacity is influenced not only by economic development and technology, but also by social factors such as human capital and governance structures (Klein and Smith, 2003; Brooks and Adger 2005; Næss et al., 2005; Tompkins, 2005; Berkhout et al., 2006; Eriksen and Kelly, 2007). Furthermore, recent analysis argues that adaptive capacity is not a concern unique to regions with low levels of economic activity. Although economic development may provide greater access to technology and resources to invest in adaptation, high income per capita is considered neither a necessary nor a sufficient indicator of the capacity to adapt to climate change (Moss et al., 2001). Tol and Yohe (2007) show that some elements of adaptive capacity are not substitutable: an economy will be as vulnerable as the 'weakest link' in its resources and adaptive capacity (for example with respect to natural disasters). Within both developed and developing countries, some regions, localities, or social groups have a lower adaptive capacity (O'Brien et al., 2006).

There are many examples where social capital, social networks, values, perceptions, customs, traditions and levels of cognition affect the capability of communities to adapt to risks related to climate change. Communities in Samoa in the south Pacific, for example, rely on informal non-monetary arrangements and social networks to cope with storm damage, along with livelihood diversification and financial remittances through extended family networks (Adger, 2001; Barnett, 2001; Sutherland et al., 2005). Similarly, strong local and international support networks enable communities in the Cayman Islands to recover from and prepare for tropical storms (Tompkins, 2005). Community organisation is an important factor in adaptive strategies to build resilience among hillside communities in Bolivia (Robledo et al., 2004). Recovery from hazards in Cuba is helped by a sense of communal responsibility (Sygna, 2005). Food-sharing expectations and networks in Nunavut, Canada, allow community members access to so-called country food at times when conditions make it unavailable to some (Ford et al., 2006). The role of food sharing as a part of a community's capacity to adapt to risks in resource provisioning is also evident among native Alaskans (Magdanz et al., 2002). Adaptive migration options in the 1930s USA Dust Bowl were greatly influenced by the access households had to economic, social and cultural capital (McLeman and Smit, 2006). The cultural change and increased individualism associated with economic growth in Small Island Developing States has eroded the sharing of risk within extended families, thereby reducing the contribution of this social factor to adaptive capacity (Pelling and Uitto, 2001).

17.3.2 Differential adaptive capacity

The capacity to adapt to climate change is unequal across and within societies. There are individuals and groups within all societies that have insufficient capacity to adapt to climate change. As described above, there has been a convergence of findings in the literature showing that human and social capital are key determinants of adaptive capacity at all scales, and that they are as important as levels of income and technological capacity. However, most of this literature also argues that there is limited usefulness in looking at only one level or scale, and that exploring the regional and local context for adaptive capacity can provide insights into both constraints and opportunities.

17.3.2.1 Adaptive capacity is uneven across societies

There is some evidence that national-level indicators of vulnerability and adaptive capacity are used by climate change negotiators, practitioners, and decision makers in determining policies and allocating priorities for funding and interventions (Eriksen and Kelly, 2007). However, few studies have been globally comprehensive, and the literature lacks consensus on the usefulness of indicators of generic adaptive capacity and the robustness of the results (Downing et al., 2001; Moss et al., 2001; Yohe and Tol, 2002; Brooks et al., 2005; Haddad, 2005). A comparison of results across five vulnerability assessments shows that the 20 countries ranked 'most vulnerable' show little consistency across studies (Eriksen and Kelly, 2007). Haddad (2005) has shown empirically that the ranking of adaptive capacity of nations is significantly altered when national aspirations are made explicit. He demonstrates that different aspirations (e.g., seeking to maximise the welfare of citizens, to maintain control of citizens, or to reduce the vulnerability of the most vulnerable groups) lead to different weightings of the elements of adaptive capacity, and hence to different rankings of the actual capacity of countries to adapt. It has been argued that national indicators fail to capture many of the processes and contextual factors that influence adaptive capacity, and thus provide little insight on adaptive capacity at the level where most adaptations will take place (Eriksen and Kelly, 2007).

The specific determinants of adaptive capacity at the national level thus represent an area of contested knowledge. Some studies relate adaptive capacity to levels of national development, including political stability, economic well-being, human and social capital and institutions (AfDB et al., 2003). National-level adaptive capacity has also been represented by proxy indicators for economic capacity, human and civic resources and environmental capacity (Moss et al., 2001). Alberini et al. (2006) use expert judgement based on a conjoint choice survey of climate and health experts to examine the most important attributes of adaptive capacity and find that per capita income, inequality in the distribution of income, universal health care coverage, and high access to information are the most important attributes allowing a country to adapt to health-related risks. Coefficients on these rankings were used to construct an index of countries with highest to lowest adaptive capacity.

17.3.2.2 *Adaptive capacity is uneven within nations due to multiple stresses*

The capacity to adapt to climate change is not evenly distributed within nations. Adaptive capacity is highly differentiated within countries, because multiple processes of change interact to influence vulnerability and shape outcomes from climate change (Leichenko and O'Brien, 2002; Dow et al., 2006; Smit and Wandel, 2006; Ziervogel et al., 2006). In India, for example, both climate change and market liberalisation for agricultural commodities are changing the context for agricultural production. Some farmers may be able to adapt to these changing conditions, including discrete events such as drought and rapid changes in commodity prices, while other farmers may experience predominately negative outcomes. Mapping vulnerability of the agricultural sector to both climate change and trade liberalisation at the district level in India, O'Brien et al. (2004) considered adaptive capacity as a key factor that influences outcomes. A combination of biophysical, socio-economic and technological conditions were considered to influence the capacity to adapt to changing environmental and economic conditions. The biophysical factors included soil quality and depth, and groundwater availability, whereas socio-economic factors consisted of measures of literacy, gender equity, and the percentage of farmers and agricultural wage labourers in a district. Technological factors were captured by the availability of irrigation and the quality of infrastructure. Together, these factors provide an indication of which districts are most and least able to adapt to drier conditions and variability in the Indian monsoons, as well as to respond to import competition resulting from liberalised agricultural trade. The results of this vulnerability mapping show the districts that have 'double exposure' to both processes. It is notable that districts located along the Indo-Gangetic Plains are less vulnerable to both processes, relative to the interior parts of the country (see Figure 17.2).

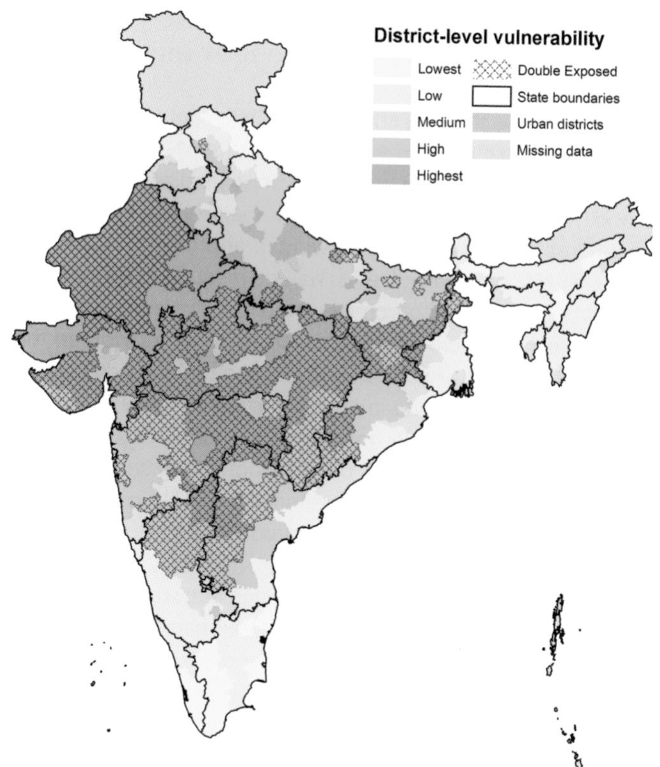

Figure 17.2. *Districts in India that rank highest in terms of vulnerability to: (a) climate change and (b) import competition associated with economic globalisation, are considered to be double exposed (depicted with hatching). Adapted from O'Brien et al. (2004).*

17.3.2.3 *Social and economic processes determine the distribution of adaptive capacity*

A significant body of new research focuses on specific contextual factors that shape vulnerability and adaptive capacity, influencing how they may evolve over time. These place-based studies provide insights on the conditions that constrain or enhance adaptive capacity at the continental, regional or local scales (Leichenko and O'Brien, 2002; Allison et al., 2005; Schröter et al., 2005; Belliveau et al., 2006). These studies differ from the regional and global indicator studies assessed above both in approach and methods, yet come to complementary conclusions on the state and distribution of adaptive capacity.

The lessons from studies of local-level adaptive capacity are context-specific, but the weight of studies establishes broad lessons on adaptive capacity of individuals and communities. The nature of the relationships between community members is critical, as is access to and participation in decision-making processes. In areas such as coastal zone management, the expansion of social networks has been noted as an important element in developing more robust management institutions (Tompkins et al., 2002). Local groups and individuals often feel their powerlessness in many ways, although none so much as in the lack of access to decision makers. A series of studies has shown that successful community-based resource management, for example, can potentially enhance the resilience of communities as well as maintain ecosystem services and ecosystem resilience (Tompkins and Adger, 2004; Manuta and Lebel, 2005; Owuor et al., 2005; Ford et al., 2006) and that this constitutes a major priority for the management of ecosystems under stress (such as coral reefs) (Hughes et al., 2003, 2005).

Much new research emphasises that adaptive capacity is also highly heterogeneous within a society or locality, and for human populations it is differentiated by age, class, gender, health and social status. Ziervogel et al. (2006) undertook a comparative study between households and communities in South Africa, Sudan, Nigeria and Mexico and showed how vulnerability to food insecurity is common across the world in semi-arid areas where marginal groups rely on rain-fed agriculture. Across the case studies food insecurity was not determined solely or primarily by climate, but rather by a range of social, economic, and political factors linked to physical risks. Box 17.5 describes how adaptive capacity and vulnerability to climate change impacts are different for men and women, with gender-related vulnerability particularly apparent in resource-dependent societies and in the impacts of extreme weather-related events (see also Box 8.2).

17.3.3 Changes in adaptive capacity over time

Adaptive capacity at any one scale may be facilitated or constrained by factors outside the system in question. At the local scale, such constraints may take the form of regulations or

Box 17.5. Gender aspects of vulnerability and adaptive capacity

Empirical research has shown that entitlements to elements of adaptive capacity are socially differentiated along the lines of age, ethnicity, class, religion and gender (Cutter, 1995; Denton, 2002; Enarson, 2002). Climate change therefore has gender-specific implications in terms of both vulnerability and adaptive capacity (Dankelman, 2002). There are structural differences between men and women through, for example, gender-specific roles in society, work and domestic life. These differences affect the vulnerability and capacity of women and men to adapt to climate change. In the developing world in particular, women are disproportionately involved in natural resource-dependent activities, such as agriculture (Davison, 1988), compared to salaried occupations. As resource-dependent activities are directly dependent on climatic conditions, changes in climate variability projected for future climates are likely to affect women through a variety of mechanisms: directly through water availability, vegetation and fuelwood availability and through health issues relating to vulnerable populations (especially dependent children and elderly). Most fundamentally, the vulnerability of women in agricultural economies is affected by their relative insecurity of access and rights over resources and sources of wealth such as agricultural land. It is well established that women are disadvantaged in terms of property rights and security of tenure, though the mechanisms and exact form of the insecurity are contested (Agarwal, 2003; Jackson, 2003). This insecurity can have implications both for their vulnerability in a changing climate, and also their capacity to adapt productive livelihoods to a changing climate.

There is a body of research that argues that women are more vulnerable than men to weather-related disasters. The impacts of past weather-related hazards have been disaggregated to determine the differential effects on women and men. Such studies have been done, for example, for Hurricane Mitch in 1998 (Bradshaw, 2004) and for natural disasters more generally (Fordham, 2003). These differential impacts include numbers of deaths, and well-being in the post-event recovery period. The disproportionate amount of the burden endured by women during rehabilitation has been related to their roles in the reproductive sphere (Nelson et al., 2002). Children and elderly persons tend to be based in and around the home and so are often more likely to be affected by flooding events with speedy onset. Women are usually responsible for the additional care burden during the period of rehabilitation, whilst men generally return to their pre-disaster productive roles outside the home. Fordham (2003) has argued that the key factors that contribute to the differential vulnerability of women in the context of natural hazards in South Asia include: high levels of illiteracy, minimum mobility and work opportunities outside the home, and issues around ownership of resources such as land.

The role of gender in influencing adaptive capacity and adaptation is thus an important consideration for the development of interventions to enhance adaptive capacity and to facilitate adaptation. Gender differences in vulnerability and adaptive capacity reflect wider patterns of structural gender inequality. One lesson that can be drawn from the gender and development literature is that climate interventions that ignore gender concerns reinforce the differential gender dimensions of vulnerability (Denton, 2004). It has also become clear that a shift in policy focus away from reactive disaster management to more proactive capacity building can reduce gender inequality (Mirza, 2003).

economic policies determined at the regional or national level that limit the freedom of individuals and communities to act, or that make certain potential adaptation strategies unviable. There is a growing recognition that vulnerability and the capacity to adapt to climate change are influenced by multiple processes of change (O'Brien and Leichenko, 2000; Turner et al., 2003; Luers, 2005). Violent conflict and the spread of infectious diseases, for example, have been shown to erode adaptive capacity (Woodward, 2002; Barnett, 2006). Social trends such as urbanisation or economic consequences of trade liberalisation are likely to have both positive and negative consequences for the overall adaptive capacity of cities and regions (Pelling, 2003). For example, trade liberalisation policies associated with globalisation may facilitate climate change adaptation for some, but constrain it for others. In the case of India, many farmers no longer plant traditional, drought-tolerant oilseed crops because there are no markets due to an influx of cheap imports from abroad (O'Brien et al., 2004). The globalisation of fisheries has decreased the resilience of marine ecosystems (Berkes et al., 2006). Exploitation of sea urchins and herbivorous reef fish species in the past three decades in particular have been shown to make reefs more vulnerable to recurrent disturbances such as hurricanes and to coral bleaching and mortality due to increased sea surface temperatures (Hughes et al., 2003; Berkes et al., 2006).

In the Canadian Arctic, experienced Inuit hunters, dealing with changing ice and wildlife conditions, adapt by drawing on traditional knowledge to alter the timing and location of harvesting, and ensure personal survival (Berkes and Jolly, 2001). Young Inuit, however, do not have the same adaptive capacity. Ford et al. (2006) attribute this to the imposition of western education by the Canadian Federal Government in the 1970s and 1980s which resulted in less participation in hunting among youth and consequent reduced transmission of traditional knowledge. This resulted in a perception among elders and experienced hunters, who act as an institutional memory for the maintenance and transmittance of traditional knowledge, that the young are not interested in hunting or traditional Inuit ways of living. This further eroded traditional knowledge by reducing inter-generational contact, creating a positive feedback in which youth is locked into a spiral of knowledge erosion. The incorporation of new technology in harvesting (including GPS, snowmobiles and radios), representing another type of adaptation, has reinforced this spiral by creating a situation in which traditional knowledge is valued less among young Inuit.

Among wine producers in British Columbia, Canada, Belliveau et al. (2006) demonstrate how adaptations to changing economic conditions can increase vulnerability to climate-related risks. Following the North American Free Trade Agreement, grape producers replaced low quality grape varieties with tender varieties to compete with higher-quality foreign imports, many of which have lower costs of production. This change enhanced the wine industry's domestic and international competitiveness, thereby reducing market risks, but simultaneously increased its susceptibility to winter injury. Thus the initial adaptation of switching varieties to increase economic competitiveness changed the nature of the system to make it more vulnerable to climatic stresses, to which it was previously less sensitive. To minimise frost risks, producers use overhead irrigation to wet the berries. The extra water from irrigation, however, can dilute the flavour in the grapes, reducing quality and decreasing market competitiveness.

The vulnerability of one region is often 'tele-connected' to other regions. In a study of coffee markets and livelihoods in Vietnam and Central America, Adger et al. (2007) found that actions in one region created vulnerability in the other through direct market interactions (Vietnamese coffee increased global supply and reduced prices), interactions with weather-related risks (coffee plant diseases and frosts) and the collapse of the International Coffee Agreement in 1989. In Mexico, Guatemala and Honduras, the capacity of smallholder coffee farmers to deal with severe droughts in 1997 to 1998 and 1999 to 2002 was complicated by low international coffee prices, reflecting changes in international institutions and national policies (Eakin et al., 2005). Concurrently, market liberalisation in Mexico, Guatemala and Honduras reduced state intervention in commodity production, markets and prices in the region. There were also constraints to adaptation related to a contraction of rural finance, coupled with a strong cultural significance attached to traditional crops. Since coffee production is already at the upper limit of the ideal temperature range in this region, it is likely that climate change will reduce yields, challenging farmers to switch to alternative crops, which currently have poorly developed marketing mechanisms.

The capacity of smallholder farmer households in Kenya and Tanzania to cope with climate stresses is often influenced by the ability of a household member to specialise in one activity or in a limited number of intensive cash-yielding activities (Eriksen et al., 2005). However, many households have limited access to this favoured coping option due to lack of labour and human and physical capital. This adaptation option is further constrained by social relations that lead to the exclusion of certain groups, especially women, from carrying out favoured activities with sufficient intensity. At present, relatively few investments go into improving the viability of these identified coping strategies. Instead, policies tend to focus on decreasing the sensitivity of agriculture to climate variability. This might actually reinforce the exclusion of population groups in dry lands where farmers are reluctant to adopt certain agricultural technologies because of their low market and consumption values and associated high costs (Eriksen et al., 2005). Eriksen et al. (2005) conclude that the determinants of adaptive capacity of smallholder farmers in Kenya and Tanzania are multiple and inter-related.

In summary, empirical research carried out since the TAR has shown that there are rarely simple cause-effect relationships between climate change risks and the capacity to adapt. Adaptive capacity can vary over time and is affected by multiple processes of change. In general, the emerging literature shows that the distribution of adaptive capacity within and across societies represents a major challenge for development and a major constraint to the effectiveness of any adaptation strategy. Some adaptations that address changing economic and social conditions may increase vulnerability to climate change, just as adaptations to climate change may increase vulnerability to other changes.

17.4 Enhancing adaptation: opportunities and constraints

17.4.1 International and national actions for implementing adaptation

An emerging literature on the institutional requirements for adaptation suggests that there is an important role for public policy in facilitating adaptation to climate change. This includes reducing vulnerability of people and infrastructure, providing information on risks for private and public investments and decision-making, and protecting public goods such as habitats, species and culturally important resources (Haddad et al., 2003; Callaway, 2004; Haddad, 2005; Tompkins and Adger, 2005). In addition, further literature sets out the case for international financial and technology transfers from countries with high greenhouse gas emissions to countries that are most vulnerable to present and future impacts, for use in adapting to the impacts of climate change (Burton et al., 2002; Simms et al., 2004; Baer, 2006; Dow et al., 2006; Paavola and Adger, 2006). Baer (2006) calculates the scale of these transfers from polluting countries, based on aggregate damage estimates of US$50 billion.

Considerable progress has also been made in terms of funding adaptation within the UNFCCC. Least-developed countries have been identified as being particularly vulnerable to climate change, and planning for their adaptation has been facilitated through development of National Adaptation Programmes of Action (NAPAs). In completing a NAPA, a country identifies priority activities that must be implemented in the immediate future in order to address urgent national climate change adaptation needs (Burton et al., 2002; Huq et al., 2003). Although only 15 countries had completed their national NAPA reports as of mid-2007, a number of specific projects were identified in these reports for priority action. Since the implementation of NAPAs had not commenced at the time of this assessment, their outcomes in terms of increased adaptive capacity or reduced vulnerability to climate change risks could not be evaluated. The process of developing NAPAs is, however, being monitored. Box 17.6 discusses some emerging lessons from Bangladesh. Early evidence suggests that NAPAs face the same constraints on effectiveness and legitimacy as other national planning processes (e.g., National Adaptation Plans under the Convention to Combat Desertification), including narrow and unrepresentative consultation processes (Thomas and Twyman, 2005).

In the climate change context, the term 'mainstreaming' has been used to refer to integration of climate change vulnerabilities or adaptation into some aspect of related government policy such as water management, disaster preparedness and emergency planning or land-use planning (Agrawala, 2005). Actions that promote adaptation include integration of climate information into environmental data sets, vulnerability or hazard assessments, broad development strategies, macro policies, sector policies, institutional or organisational structures, or in development project design and implementation (Burton and van Aalst, 1999; Huq et al., 2003). By implementing mainstreaming initiatives, it is argued that adaptation to climate change will become part of or will be consistent with other well-established programmes, particularly sustainable development planning.

Mainstreaming initiatives have been classified in the development planning literature at four levels. At the international level, mainstreaming of climate change can occur through policy formulation, project approval and country-level implementation of projects funded by international organisations. For example, the International Federation of Red Cross and Red Crescent (IFRC) are working to facilitate a link between local and global responses through its Climate Change Centre (Van Aalst and Helmer, 2003). An example of an initiative at the regional level is the MACC (Mainstreaming Adaptation to Climate Change) project in the Caribbean. It assesses the likely impacts of climate change on key economic sectors (i.e., water, agriculture and human health) while also defining responses at community, national and regional levels. Various multi-lateral and bi-lateral development agencies, such as the Asian Development Bank, are attempting to integrate climate change adaptation into their grant and loan activities (ADB, 2005; Perez and Yohe, 2005). Other aid agencies have sought to screen out those loans and grants which are mal-adaptations and create new vulnerabilities, to ascertain the extent to which existing development projects already consider climate risks or address vulnerability to climate variability and change, and to identify opportunities for incorporating climate change explicitly into future projects. Klein et al. (2007) examine the activities of several major development agencies over the past five years and find that while most agencies already consider climate change as a real but uncertain threat to future development, they have not explicitly examined how their activities affect vulnerability to climate change. They conclude that mainstreaming needs to encompass a broader set of measures to reduce vulnerability than has thus far been the case.

Much of the adaptation planning literature emphasises the role of governments, but also recognises the constraints that they face in implementing adaptation actions at other scales (Few et al., 2007). There are few examples of successful mainstreaming of climate change risk into development planning. Agrawala and van Aalst (2005) identified following five major constraints: (a) relevance of climate information for development-related decisions; (b) uncertainty of climate information; (c) compartmentalisation with governments; (d) segmentation and other barriers within development-cooperation agencies; and (e) trade-offs between climate and development objectives. The Adaptation Policy Framework (APF) (Lim et al., 2005) developed to support national planning for adaptation by the United Nations Development Programme (UNDP) provides guidance on how these obstacles and barriers to mainstreaming can be overcome. Mirza and Burton (2005) found that the application of APF was feasible when they applied it for urban flooding in Bangladesh and droughts in India. However, they concluded that application of the APF could encounter problems

Box 17.6. Early lessons on effectiveness and legitimacy of National Adaptation Programmes of Action

At present there is sparse documentary evidence on outcomes of NAPA planning processes or implementation. One case that has been examined is that of the Bangladesh NAPA (Huq and Khan, 2006). The authors recommend that NAPAs should adopt (a) a livelihood rather than sectoral approach, (b) focus on near- and medium-term impacts of climate variability as well as long-term impacts, (c) should ensure integration of indigenous and traditional knowledge, and (d) should ensure procedural fairness through interactive participation and self-mobilisation (Huq and Khan, 2006). They found that NAPA consultation and planning processes have the same constraints and exhibit the same problems of exclusion and narrow focus as other national planning processes (such as those for Poverty Reduction Strategies). They conclude that the fairness and effectiveness of national adaptation planning depends on how national governments already include or exclude their citizens in decision-making and that effective participatory planning for climate change requires functioning democratic structures. Where these are absent, planning for climate change is little more than rhetoric (Huq and Khan, 2006). Similar issues are raised and findings presented by Huq and Reid (2003), Paavola (2006) and Burton et al. (2002). The key role of non-government and community-based organisations in ensuring the sustainability and success of adaptation planning is likely to become evident over the incoming period of NAPA development and implementation.

related to a lack of micro-level socio-economic information, and gaps in stakeholder participation in the planning, design, implementation and monitoring of projects.

In summary, the opportunities for implementing adaptation as part of government planning are dependent on effective, equitable and legitimate actions to overcome barriers and limits to adaptation (ADB, 2005; Agrawala and van Aalst, 2005; Lim et al., 2005). Initial signals of impacts have been hypothesised to create the demand and political space for implementing adaptation, the so-called 'policy windows hypothesis'. Box 17.7, however, reveals that evidence is contested on whether individual weather-related catastrophic events can facilitate adaptation action, or whether they act as a barrier to long-term adaptation.

17.4.2 Limits and barriers to adaptation

Most studies of specific adaptation plans and actions argue that there are likely to be both limits and barriers to adaptation as a response to climate change. The U.S. National Assessment (2001), for example, maintains that adaptation will not necessarily make the aggregate impacts of climate change negligible or beneficial, nor can it be assumed that all available adaptation measures will actually be taken. Further evidence from Europe and other parts of the globe suggests that high adaptive capacity may not automatically translate into successful adaptations to climate change (O'Brien et al., 2006). Research on adaptation to changing flood risk in Norway, for example, has shown that high adaptive capacity is countered by weak incentives for proactive flood management (Næss et al., 2005). Despite increased attention to potential adaptation options, there is less understanding of their feasibility, costs, effectiveness, and the likely extent of their actual implementation (U.S. National Assessment, 2001). Despite high adaptive capacity and significant investment in planning, extreme heatwave events continue to result in high levels of mortality and disruption to infrastructure and electricity supplies in European, North American and east Asian cities (Klinenberg, 2003; Mohanty and Panda, 2003; Lagadec, 2004; Poumadère et al., 2005).

This section assesses the limits to adaptation that have been discussed in the climate change and related literatures. Limits are defined here as the conditions or factors that render adaptation ineffective as a response to climate change and are largely insurmountable. These limits are necessarily subjective and dependent upon the values of diverse groups. These limits to adaptation are closely linked to the rate and magnitude of climate change, as well as associated key vulnerabilities discussed in Chapter 19. The perceived limits to adaptation are hence likely to vary according to different metrics. For example, the five numeraires for judging the significance of climate change impacts described by Schneider et al. (2000b) - monetary loss, loss of life, biodiversity loss, distribution and equity, and quality of life (including factors such as coercion to migrate, conflict over resources, cultural diversity, and loss of cultural heritage sites) - can lead to very different assessments of the limits to adaptation. But emerging literature on adaptation processes also identifies significant barriers to action in financial, cultural and policy realms that raise questions about the efficacy and legitimacy of adaptation as a response to climate change.

17.4.2.1 Physical and ecological limits

There is increasing evidence from ecological studies that the resilience of coupled socio-ecological systems to climate change will depend on the rate and magnitude of climate change, and that there may be critical thresholds beyond which some systems may not be able to adapt to changing climate conditions without radically altering their functional state and system integrity (see examples in Chapter 1). Scheffer et al. (2001) and Steneck et al. (2002), for instance, find thresholds in the resilience of kelp forest ecosystems, coral reefs, rangelands and lakes affected both by climate change and other pollutants. Dramatic climatic changes may lead to transformations of the physical environment of a region that limit the possibilities for adaptation (Nicholls and Tol,

Box 17.7. Is adaptation constrained or facilitated by individual extreme events?

The policy window hypothesis refers to the phenomenon whereby adaptation actions such as policy and regulatory change are facilitated and occur directly in response to disasters, such as those associated with weather-related extreme events (Kingdon, 1995). According to this hypothesis, immediately following a disaster, the political climate may be conducive to legal, economic and social change which can begin to reduce structural vulnerabilities, for example, in such areas as mainstreaming gender issues, land reform, skills development, employment, housing and social solidarity. The assumptions behind the policy windows hypothesis are that (a) new awareness of risks after a disaster leads to broad consensus, (b) development and humanitarian agencies are 'reminded' of disaster risks, and (c) enhanced political will and resources become available. However, contrary evidence on policy windows suggests that, during the post-recovery phase, reconstruction requires weighing, prioritising and sequencing of policy programming, and there is the pressure to quickly return to conditions prior to the event rather than incorporate longer-term development policies (Christoplos, 2006). In addition, while institutions clearly matter, they are often rendered ineffective in the aftermath of a disaster. As shown in diverse contexts, such as ENSO-related impacts in Latin America, induced development below dams or levees in the U.S. and flooding in the United Kingdom, the end result is that short-term risk reduction can actually produce greater vulnerability to future events (Pulwarty et al., 2003; Berube and Katz, 2005; Penning-Rowsell et al., 2006).

2006; Tol et al., 2006). For example, rapid sea-level rise that inundates islands and coastal settlements is likely to limit adaptation possibilities, with potential options being limited to migration (see Chapter 15, Barnett and Adger, 2003; Barnett, 2005). Tol et al. (2006) argue that it is technically possible to adapt to five metres of sea-level rise but that the resources required are so unevenly distributed that in reality this risk is outside the scope of adaptation. In the Sudano-Sahel region of Africa, persistent below-average rainfall and recurrent droughts in the late 20th century have constricted physical and ecological limits by contributing to land degradation, diminished livelihood opportunities, food insecurity, internal displacement of people, cross-border migrations and civil strife (Mortimore and Adams, 2001; Leary et al., 2006; Osman-Elasha et al., 2006). The loss of Arctic sea ice threatens the survival of polar bears, even if hunting of bears were to be reduced (Derocher et al., 2004). The loss of keystone species may cascade through the socio-ecological system, eventually influencing ecosystems services that humans rely on, including provisioning, regulating, cultural, and supporting services (Millennium Ecosystem Assessment, 2006).

The ecological literature has documented regime shifts in ecosystems associated with climatic changes and other drivers (Noss, 2001; Scheffer et al., 2001). These regime shifts are argued to impose limits on economic and social adaptation (van Vliet and Leemans, 2006). Economies and communities that are directly dependent on ecosystems such as fisheries and agricultural systems are likely to be more affected by sudden and dramatic switches and flips in ecosystems. In a review of social change and ecosystem shifts, Folke et al. (2005) show that there are significant challenges to resource management from ecosystem shifts and that these are often outside the experience of institutions. The loss of local knowledge associated with thresholds in ecological systems is a limit to the effectiveness of adaptation (Folke et al., 2005).

17.4.2.2 Technological limits

Technological adaptations can serve as a potent means of adapting to climate variability and change. New technologies can be developed to adapt to climate change, and the transfer of appropriate technologies to developing countries forms an important component of the UNFCCC (Mace, 2006). However, there are also potential limits to technology as an adaptation response to climate change.

First, technology is developed and applied in a social context, and decision-making under uncertainty may inhibit the adoption or development of technological solutions to climate change adaptation (Tol et al., 2006). For example, case studies from the Rhine delta, the Thames estuary and the Rhone delta in Europe suggest that although protection from five-metre sea-level rise is technically possible, a combination of accommodation and retreat is more likely as an adaptation strategy (Tol et al., 2006).

Second, although some adaptations may be technologically possible, they may not be economically feasible or culturally desirable. For example, within the context of Africa, large-scale engineering measures for coastal protection are beyond the reach of many governments due to high costs (Ikeme, 2003). In colder climates that support ski tourism, the extra costs of making snow at warmer average temperatures may surpass a threshold where it becomes economically unfeasible (Scott et al., 2003; Scott et al., 2007). Although the construction of snow domes and indoor arenas for alpine skiing has increased in recent years, this technology may not be an affordable, acceptable or appropriate adaptation to decreasing snow cover for many communities dependent on ski tourism. Finally, existing or new technology is unlikely to be equally transferable to all contexts and to all groups or individuals, regardless of the extent of country-to-country technology transfers (Baer, 2006). Adaptations that are effective in one location may be ineffective in other places, or create new vulnerabilities for other places or groups, particularly through negative side effects. For example, although technologies such as snowmobiles and GPS have facilitated adaptation to climate change among some Inuit hunters, these are not equally accessible to all, and they have potentially contributed to inequalities within the community through differential access to resources (Ford et al., 2006).

17.4.2.3 Financial barriers

The implementation of adaptation measures faces a number of financial barriers. At the international level, preliminary estimates from the World Bank indicate that the total costs of 'climate proofing' development could be as high as US$10 billion to US$40 billion /yr (World Bank, 2006). While the analysis notes that such numbers are only rough estimates, the scale of investment implied constitutes a significant financial barrier. At a more local level, individuals and communities can be similarly constrained by the lack of adequate financial resources. Deep financial poverty is a factor that constrains the use of seemingly inexpensive health measures, such as insecticide-treated bed nets, while limited public finances contribute to choices by public health agencies to give low priority to measures that would reduce vulnerability to climate-related health risks (Taylor et al., 2006; Yanda et al., 2006). In field surveys and focus groups, farmers often cite the lack of adequate financial resources as an important factor that constrains their use of adaptation measures which entail significant investment, such as irrigation systems, improved or new crop varieties, and diversification of farm operations (Smit and Skinner, 2002).

Lack of resources may also limit the ability of low-income groups to afford proposed adaptation mechanisms such as climate-risk insurance. In the case of Mexico, a restructuring of public agricultural institutions paralleled market liberalisation, reducing the availability of publicly subsidised credit, insurance and technical assistance for smallholders (Appendini, 2001). Even where both crop insurance and contract farming were being actively promoted by the state and federal government to help farmers address climatic contingencies and price volatility, very few of the surveyed farmers had crop insurance (Wehbe et al., 2006). In addition, individuals often fail to purchase insurance against low-probability high-loss events even when it is offered at favourable premiums. While this may occur because of the relative benefits and costs of alternatives, the trade-offs may not be explicit. Kunreuther et al. (2001) show that the search costs involved in collecting and analysing relevant information to clarify trade-offs can be enough to discourage individuals from undertaking such assessments, and thus from purchasing coverage even when the premium is affordable.

Climate change is also likely to raise the actuarial uncertainty in catastrophe risk assessment, placing upward pressure on insurance premiums and possibly leading to reductions in risk coverage (Mills, 2005).

17.4.2.4 Informational and cognitive barriers

Extensive evidence from psychological research indicates that uncertainty about future climate change combines with individual and social perceptions of risk, opinions and values to influence judgment and decision-making concerning climate change (Oppenheimer and Todorov, 2006). It is increasingly clear that interpretations of danger and risk associated with climate change are context specific (Lorenzoni et al., 2005) and that adaptation responses to climate change can be limited by human cognition (Grothmann and Patt, 2005; Moser, 2005). Four main perspectives on informational and cognitive constraints on individual responses (including adaptation) to climate change emerge from the literature.

1. Knowledge of climate change causes, impacts and possible solutions does not necessarily lead to adaptation. Well-established evidence from the risk, cognitive and behavioural psychology literatures points to the inadequacy of the 'deficit model' of public understanding of science, which assumes that providing individuals with scientifically sound information will result in information assimilation, increased knowledge, action and support for policies based on this information (Eden, 1998; Sturgis and Allum, 2004; Lorenzoni et al., 2005). Individuals' interpretation of information is mediated by personal and societal values and priorities, personal experience and other contextual factors (Irwin and Wynne, 1996). As a consequence, an individual's awareness and concern either do not necessarily translate into action, or translate into limited action (Baron, 2006; Weber, 2006). This is also known as the 'value-action' or 'attitude-behaviour' gap (Blake, 1999) and has been shown in a small number of studies to be a significant barrier to adaptation action (e.g., Patt and Gwata, 2002).
2. Perceptions of climate change risks are differing. A small but growing literature addresses the psychological dimensions of evaluating long-term risk; most focuses on behaviour changes in relation to climate change mitigation policies. However, some studies have explored the behavioural foundations of adaptive responses, including the identification of thresholds, or points at which adaptive behaviour begins (e.g., Grothmann and Patt, 2005). Key findings from these studies point to different types of cognitive limits to adaptive responses to climate change. For example, Niemeyer et al. (2005) found that thresholds of rapid climate change may induce different individual responses influenced by trust in others (e.g., institutions, collective action, etc.), resulting in adaptive, non-adaptive, and maladaptive behaviours. Hansen et al. (2004) found evidence for a finite pool of worry among farmers in the Argentine Pampas. As concern about one type of risk increases, worry about other risks decreases. Consequently, concerns about violent conflict, disease and hunger, terrorism, and other risks may overshadow considerations about the impacts of climate change and adaptation. This work also indicates, consistently with findings in the wider climate change risk literature (e.g., Moser and Dilling, 2004), that individuals tend to prioritise the risks they face, focusing on those they consider – rightly or wrongly – to be the most significant to them at that particular point in time. Furthermore, a lack of experience of climate-related events may inhibit adequate responses. It has been shown, for instance, that the capacity to adapt among resource-dependent societies in southern Africa is high if based on adaptations to previous changes (Thomas et al., 2005). Although concern about climate change is widespread and high amongst publics in western societies, it is not 'here and now' or a pressing personal priority for most people (Lorenzoni and Pidgeon, 2006). Weber (2006) found that strong visceral reactions towards the risk of climate change are needed to provoke adaptive behavioural changes.
3. Perceptions of vulnerability and adaptive capacity are important. Psychological research, for example, has provided empirical evidence that those who perceive themselves to be vulnerable to environmental risks, or who perceive themselves to be victims of injustice, also perceive themselves to be more at risk from environmental hazards of all types (Satterfield et al., 2004). Furthermore, perceptions by the vulnerable of barriers to actually adapting do, in fact, limit adaptive actions, even when there are capacities and resources to adapt. Grothman and Patt (2005) examined populations living with flood risk in Germany and farmers dealing with drought risk in Zimbabwe in order to better understand cognitive constraints. They found that action was determined by both perceived abilities to adapt and observable capacities to adapt. They conclude that a divergence between perceived and actual adaptive capacity is a real barrier to adaptive action. Moser (2005) similarly finds that perceived barriers to action are a major constraint in coastal planning for sea-level rise in the United States.
4. Appealing to fear and guilt does not motivate appropriate adaptive behaviour. In fact, communications research has shown that appealing to fear and guilt does not succeed in fostering sustained engagement with the issue of climate change (Moser and Dilling, 2004). Analysis of print media portrayal of climate change demonstrates public confusion when scientific arguments are contrasted in a black-and-white, for-and-against manner (Boykoff and Boykoff, 2004; Carvalho and Burgess, 2005; Ereaut and Segnit, 2006). Calls for effective climate-change communication have focused on conveying a consistent, sound message, with the reality of anthropogenic climate change at its core. This, coupled with making climate change personally relevant through messages of practical advice on individual actions, helps to embed responses in people's locality. Visualisation imagery is being increasingly explored as a useful contribution to increasing the effectiveness of communication about climate change risks (e.g., Nicholson-Cole, 2005; Sheppard, 2005).

Overall, the psychological research reviewed here indicates that an individual's awareness of an issue, knowledge, personal experience, and a sense of urgency of being personally affected, constitute necessary but insufficient conditions for behaviour or policy change. Perceptions of risk, of vulnerability, motivation and capacity to adapt will also affect behavioural change. These perceptions vary among individuals and groups within populations. Some can act as barriers to adapting to climate change. Policymakers need to be aware of these barriers, provide structural support to overcome them, and concurrently work towards fostering individual empowerment and action.

17.4.2.5 Social and cultural barriers

Social and cultural limits to adaptation can be related to the different ways in which people and groups experience, interpret and respond to climate change. Individuals and groups may have different risk tolerances as well as different preferences about adaptation measures, depending on their worldviews, values and beliefs. Conflicting understandings can impede adaptive actions. Differential power and access to decision makers may promote adaptive responses by some, while constraining them for others. Thomas and Twyman (2005) analysed natural-resource policies in southern Africa and showed that even so-called community-based interventions to reduce vulnerability create excluded groups without access to decision-making. In addition, diverse understandings and prioritisations of climate change issues across different social and cultural groups can limit adaptive responses (Ford and Smit, 2004).

Most analyses of adaptation propose that successful adaptations involve marginal changes to material circumstances rather than wholesale changes in location and development paths. A few studies have examined the need for and potential for migration, resettlement and relocation as an adaptive strategy, for example, but the cultural implications of large-scale migration are not well understood and could represent significant limits to adaptation. Box 17.8 presents evidence that demonstrates that, while relocation and migration have been used as adaptation strategies in the past, there are often large social costs associated with these and unacceptable impacts in terms of human rights and sustainability. The possibility of migration as a response to climate change is still rarely broached in the literature on adaptation to climate change, perhaps because it is entirely outside the acceptable range of proposals (Orlove, 2005).

Although scientific research indicates that forest ecosystems in northern Canada are among those regions at greatest risk from

Box 17.8. Do voluntary or displacement migrations represent failures to adapt?

Migration by individuals or relocation of settlements have been discussed in various studies as a potential adaptive response option to climate change impacts when local environments surpass a threshold beyond which the system is no longer able to support most or all of the population. There has been, for example, discussion of the possibility that sea-level rise will make it impossible for human populations to remain on specific islands. For instance, New Zealand has been discussed as a possible site of relocation for the people of Tuvalu, a nation consisting of low-lying atolls in the western Pacific. Patel (2006) and Barnett (2005) argue that there would be enormous economic, cultural and human costs if large populations were to abandon their long-established home territories and move to new places. Sea-level rise impacts on the low-lying Pacific Island atoll states of Kiribati, Tuvalu, Tokelau and the Marshall Islands may, at some threshold, pose risks to their sovereignty or existence (Barnett, 2001). Barnett and Adger (2003) argue that this loss of sovereignty itself represents a dangerous climate change and that the possibility of relocation represents a limit of adaptation.

The ability to migrate as an adaptive strategy is not equally accessible to all, and decisions to migrate are not controlled exclusively by individuals, households, or local and state governments (McLeman, 2006). Studies in Asia and North America (Adger et al., 2002; Winkels, 2004; McLeman and Smit, 2006) show that strong social capital can obviate the need for relocation in the face of risk, and is also important in determining the success and patterns of migration as an adaptive strategy: the spatial patterns of existing social networks in a community influence their adaptation to climate change. Where household social networks are strong at the local scale, adaptations that do not lead to migration, or that lead to local-scale relocations, are more likely responses than long-distance migration away from areas under risk. Conversely, if the community has widespread social networks, or is part of a transnational community, then far-reaching migration is possible. McLeman and Smit (2006) show that a range of economic, social and cultural processes played roles in shaping migration behaviour and migration patterns in response to climate conditions and resulting long-term drought in rural eastern Oklahoma in the 1930s. While temporary migration has often been used as a risk management response to climate variability, permanent migration may be required when physical or ecological limits to adaptation have been surpassed.

Mendelsohn et al. (2007) examined correlations between incomes in rural districts in the United States and in Brazil, with parameters of present climate and physical parameters of agricultural productivity. They argued that climate affects agricultural productivity which, in turn, affects per capita income (even when this is defined as both farm and non-farm incomes for a district) and that climatic changes that reduce productivity may have direct consequences in rural poverty. Mendelsohn et al. (2007) therefore argue that climate change impacts in rural economies may make migration and relocation a necessary but undesirable adaptation. Finan and Nelson (2001), however, suggest that government policies in Brazil, such as rural retirement policies, have actually augmented household adaptive capacity and attracted young migrants back from cities. Thus migration can be influenced by government intervention. In the case of island states, Barnett (2005) argues that adaptation should already be deemed as unsuccessful if it has limited development opportunities.

the impacts of climate change, the social dimensions of forest-dependent communities indicate both a limited community capacity and a limited potential to perceive climate change as a salient risk issue that warrants action. Climate change messages are often associated with environmentalism and environmentalists, who have been perceived by many residents of resource-dependent communities as an oppositional political force. Risk perceptions tend to be higher for women than for men, the higher concern levels of women may either be stifled or simply be unexpressed in a highly male-dominated environment (Davidson et al., 2003).

Anthropological research suggests that the scale and novelty of climate changes are not the sole determinants of degree of impact (Orlove, 2005). Societies change their environments, and thus alter their own vulnerability to climate fluctuations. The experience of development of the Colorado River Basin in the face of environmental uncertainty clearly illustrates that impacts and interventions can reverberate through the systems in ways that can only be partially traced and predicted (Pulwarty et al., 2005).

Accounting for future economic and social trends involves problems of indeterminacy (imperfectly understood structures and processes), discontinuity (novelty and surprise in social systems), reflexivity (the ability of people and organisations to reflect on and adapt their behaviour), and framing (legitimately-diverse views about the state of the world) (Berkhout et al., 2002; Pulwarty et al., 2003). Case studies reveal that there exists a diversity of local or traditional practices for ecosystem management under environmental uncertainty. These include rules for social regulation, mechanisms for cultural internalisation of traditional practices and the development of appropriate world views and cultural values (Pretty, 2003).

Social and cultural limits to adaptation are not well researched: Jamieson (2006) notes that a large segment of the U.S. population think of themselves as environmentalists but often vote for environmentally negative candidates. Although many societies are highly adaptive to climate variability and change, vulnerability is dynamic and likely to change in response to multiple processes, including economic globalisation (Leichenko and O'Brien, 2002). The Inuit, for example, have a long history of adaptation to changing environmental conditions. However, flexibility in group size and group structure to cope with climate variability and unpredictability is no longer a viable strategy, due to settlement in permanent communities. Also, memories and hunting narratives are appearing unreliable because of rapid change. Furthermore, there are emerging vulnerabilities, particularly among the younger generation through lack of knowledge transfer, and among those who do not have access to monetary resources to purchase equipment necessary to hunt in the context of changing conditions (Ford et al., 2006).

17.5 Conclusions

Adaptation has the potential to alleviate adverse impacts, as well as to capitalise on new opportunities posed by climate change. Since the TAR, there has been significant documentation and analysis of emerging adaptation practices. Adaptation is occurring in both the developed and developing worlds, both to climate variability and, in a limited number of cases, to observed or anticipated climate change. Adaptation to climate change is seldom undertaken in a stand-alone fashion, but as part of broader social and development initiatives. Adaptation also has limits, some posed by the magnitude and rate of climate change, and others that relate to financial, institutional, technological, cultural and cognitive barriers. The capacities for adaptation, and the processes by which it occurs, vary greatly within and across regions, countries, sectors and communities. Policy and planning processes need to take these aspects into account in the design and implementation of adaptation. The review in this chapter suggests that a high priority should be given to increasing the capacity of countries, regions, communities and social groups to adapt to climate change in ways that are synergistic with wider societal goals of sustainable development.

There are significant outstanding research challenges in understanding the processes by which adaptation is occurring and will occur in the future, and in identifying areas for leverage and action by government. Many initiatives on adaptation to climate change are too recent at the time of this assessment to evaluate their impact on reducing societal vulnerability. Further research is therefore needed to monitor progress on adaptation, and to assess the direct as well as ancillary effects of such measures. In this context there is also a need for research on the synergies and trade-offs between various adaptation measures, and between adaptation and other development priorities. Human intervention to manage the process of adaptation in biological systems is also not well understood, and the goals of conservation are contested. Hence, research is also required on the resilience of socio-ecological systems to climate change. Another key area where information is currently very limited is on the economic and social costs and benefits of adaptation measures. In particular, the non-market costs and benefits of adaptation measures involving ecosystem protection, health interventions, and alterations to land use are under-researched. Information is also lacking on the economy-wide implications of particular adaptations on economic growth and employment.

References

Aaheim, A. and A. Schjolden, 2004: An approach to utilise climate change impacts studies in national assessments. *Global Environ. Chang.*, **14**, 147-160.

ABI, 2004: *A Changing Climate for Insurance – A Summary Report for Chief Executives and Policymakers*. Association of British Insurers, London, 20pp.

Adams, R.M., B.A. McCarl and L.O. Mearns, 2003: The effects of spatial scale of climate scenarios on economic assessments: An example from U.S. agriculture. *Climatic Change*, **60**, 131-148.

ADB, 2005: *Climate proofing: A risk-based approach*. Asian Development Bank, Manila, 219 pp.

Adger, W.N., 2001: Scales of governance and environmental justice for adaptation and mitigation of climate change. *Journal of International Development*, **13**, 921-931.

Adger, W.N., 2003: Social capital, collective action, and adaptation to climate change. *Econ. Geog.*, **79**, 387-404.

Adger, W.N., P.M. Kelly, A. Winkels, L. Huy and C. Locke, 2002: Migration, remittances, livelihood trajectories and social resilience. *Ambio,* **31**, 358-366.

Adger, W.N., N.W. Arnell and E.L. Tompkins, 2005: Successful adaptation to climate change across scales. *Global Environ. Chang.,* **15**, 77-86.

Adger, W. N., H. Eakin, and A. Winkels, 2007: Nested and networked vulnerabilities in South East Asia. *Global Environmental Change and the South-east Asian Region: An Assessment of the State of the Science,* L. Lebel, et al., Eds., Island Press, Washington, District of Columbia, in press.

AfDB, ADB, DFID, DGIS, EC, BMZ, OECD, UNDP, UNEP and WB, 2003: *Poverty and Climate Change Reducing the Vulnerability of the Poor through Adaptation.* UNEP, Nairobi, 43 pp.

Agarwal, B., 2003: Gender and land rights revisited: exploring new prospects via the state, family and market. *Journal of Agrarian Change,* **3**, 184-224.

Agrawala, S., 2005: Putting climate change in the development mainstream: introduction and framework. *Bridge Over Troubled Waters: Linking Climate Change and Development,* S. Agrawala, Ed., OECD, Paris, 23-43.

Agrawala, S. and K. Broad, 2002: Technology transfer perspectives on climate forecast applications. *Research in Science and Technology Studies,* **13**, 45-69.

Agrawala, S. and M.A. Cane, 2002: Sustainability: lessons from climate variability and climate change. *Columbia Journal of Environmental Law,* **27**, 309-321.

Agrawala, S. and M. van Aalst, 2005: Bridging the gap between climate change and development. *Bridge Over Troubled Waters: Linking Climate Change and Development,* S. Agrawala, Ed., OECD, Paris, 133-146.

Agrawala, S., S. Gigli, V. Raksakulthai, A. Hemp, A. Moehner, D. Conway, M. El Raey, A. Uddin Amhed, J. Risbey, W. Baethgen and D. Martino, 2005: Climate change and natural resource management: key themes from case studies. *Bridge Over Troubled Waters – Linking Climate Change and Development,* S. Agrawala, Ed., OECD, Paris, 85-146.

Alberini, A., A. Chiabai and L. Muehlenbachs, 2006: Using expert judgement to assess adaptive capacity to climate change: a conjoint choice survey. *Global Environ. Chang.,* **16**, 123-144.

Allison, E.H., W.N. Adger, M.C. Badjeck, K. Brown, D. Conway, N.K. Dulvy, A. Halls, A. Perry and J.D. Reynolds, 2005: Effects of climate change on the sustainability of capture and enhancement fisheries important to the poor: analysis of the vulnerability and adaptability of fisherfolk living in poverty. Fisheries Management Science Programme London, Department for International Development Final Technical Report.. Project No. R4778J, , 167 pp. [Available online at http://www.fmsp.org.uk.].

Appendini, K., 2001: *De la Milpa a los Tortibonos: La Restructuración de la Política Alimentaria en México (2da edición.).* México, DF, Colegio de México, 259 pp.

Austrian Federal Government, 2006: *Fourth National Communication of the Austrian Federal Government,* in Compliance with the Obligations under the United Nations Framework Convention on Climate Change (Federal Law Gazette No. 414/1994), according to Decisions 11/CP.4 and 4/CP.5 of the Conference of the Parties. Federal Ministry of Agriculture, Forestry, Environment and Water Management, Vienna, 241 pp.

Baer, P., 2006: Adaptation: who pays whom? *Fairness in Adaptation to Climate Change,* W.N. Adger, J. Paavola, S. Huq, and M.J. Mace, Eds., MIT Press, Cambridge Massachusetts, 131-153.

Barnett, J., 2001: Adapting to climate change in Pacific island countries: the problem of uncertainty. *World Development,* **29**, 977-993.

Barnett, J., 2005: Titanic states? Impacts and responses to climate change in the Pacific islands. *Journal of International Affairs,* **59**, 203-219.

Barnett, J., 2006: Climate change, insecurity and injustice. *Fairness in Adaptation to Climate Change,* W.N. Adger, J. Paavola, S. Huq, and M.J. Mace, Eds., MIT Press, Cambridge Massachusetts, 115-129.

Barnett, J. and W.N. Adger, 2003: Climate dangers and atoll countries. *Climatic Change,* **61**, 321-337.

Baron, J. 2006: Thinking about global warming. *Climatic Change,* **77**, 137-150.

Basher, R., C. Clark, M. Dilley and M. Harrison, 2000: *Coping with the Climate: A Way Forward. A Multi-Stakeholder Review of Regional Climate Outlook Forums.* IRI Publication CW/01/02, Columbia University, New York, 27 pp.

Bass, B., 2005: Measuring the adaptation deficit. Discussion on keynote paper: climate change and the adaptation deficit. *Climate Change: Building the Adaptive Capacity* A. Fenech, D. MacIver, H. Auld, B. Rong and Y.Y. Yin, Eds., Environment Canada, Toronto, 34-36.

Belliveau, S., B. Smit and B. Bradshaw, 2006: Multiple exposures and dynamic vulnerability: Evidence from the grape and wine industry in the Okanagan Valley, Canada. *Global Environ. Chang.,* **16**, 364-378.

Berkes, F. and Co-authors, 2006: Globalization, roving bandits and marine resources. *Science,* **311**, 1557-1558.

Berkes, F. and Jolly, D. (2001) Adapting to climate change: social-ecological resilience in a Canadian Western arctic community. *Conserv. Ecol.* **5**, 18. [Available online at http://www.ecologyandsociety.org/vol5/iss2/art18/]).

Berkhout, F., J. Hertin and A. Jordan, 2002: Socio-economic futures in climate change impact assessment: using scenarios as learning machines. *Global Environ. Chang.,* **17**, 83-95.

Berkhout, F., J. Hertin and D.M. Gann, 2006: Learning to adapt: organisational adaptation to climate change impacts. *Climatic Change,* **78**, 135-156.

Berube, A. and B. Katz, 2005: *Katrina's Window: Confronting Concentrated Poverty across America.* Special Analysis of Metropolitan Policy, Brookings Institution. Washington, District of Columbia, 13 pp.

Bettencourt, S. and Co-authors, 2006: *Not if but when, Adapting to natural hazards in the Pacific Islands Region: A policy note.* East Asia and Pacific Region, Pacific Islands Country Management Unit, World Bank, Washington, District of Columbia, 46 pp.

Blake, J., 1999: Overcoming the 'value-action gap' in environmental policy: tensions between national policy and local experience. *Local Environment,* **4**, 257-278.

Boykoff, M.T. and J.M. Boykoff, 2004: Balance as bias: global warming and the US prestige press. *Global Environ. Chang.,* **14**, 125-136.

Bradshaw, S., 2004: *Socio-economic impacts of natural disaster: a gender analysis.* UN Economic Commission for Latin America and the Caribbean (ECLAC), Santiago, 60 pp.

Broad, K. and S. Agrawala, 2000: Policy forum: climate - the Ethiopia food crisis - uses and limits of climate forecasts. *Science,* **289**, 1693-1694.

Brooks, N. and W.N. Adger, 2005: Assessing and enhancing adaptive capacity. *Adaptation Policy Frameworks for Climate Change,* B. Lim, E. Spanger-Siegfried, I. Burton, E.L. Malone and S. Huq, Eds., Cambridge University Press, New York, 165-182.

Brooks, N., W.N. Adger and P.M. Kelly, 2005: The determinants of vulnerability and adaptive capacity at the national level and the implications for adaptation. *Global Environ. Chang.,* **15**, 151-163.

Brown, J.L., 2005: High-altitude railway designed to survive climate change. *Civil Eng.,* **75**, 28-28.

Burki, R., H. Elsasser, B. Abegg and U. Koenig, 2005: Climate change and tourism in the Swiss Alps. *Tourism, Recreation and Climate Change,* M. Hall and J. Higham, Eds., Channelview Press, London, 155-163.

Burton, I. and M. van Aalst, 1999: *Come Hell or High Water: Integrating Climate Change Vulnerability and Adaptation into Bank Work.* World Bank, Washington, District of Columbia, 60 pp.

Burton, I. and M.K. van Aalst, 2004: *Look Before You Leap? A Risk Management Approach for Incorporating Climate Change Adaptation in World Bank Operations,* Final Draft, Prepared for the Climate Change Team. World Bank, Washington, District of Columbia, 57 pp.

Burton, I., S. Huq, B. Lim, O. Pilifosova and E.L. Schipper, 2002: From impacts assessment to adaptation priorities: the shaping of adaptation policy. *Clim. Policy,* **2**, 145-159.

Butt, A.T., B.A. McCarl, J. Angerer, P.T. Dyke and J.W. Stuth, 2005: The economic and food security implications of climate change. *Climatic Change,* **68**, 355-378.

Callaway, J.M., 2004: Adaptation benefits and costs: are they important in the global policy picture and how can we estimate them? *Global Environ. Chang.,* **14**, 273-282.

Callaway, J.M., D.B. Louw, J.C. Nkomo, M.E. Hellmuth and D.A. Sparks, 2006: *The Berg River Dynamic Spatial Equilibrium Model: A New Tool for Assessing the Benefits and Costs of Alternatives for Coping With Water Demand Growth, Climate Variability, and Climate Change in the Western Cape.* AIACC Working Paper 31, The AIACC Project Office, International START Secretariat, Washington, District of Columbia, 41 pp. [Available online at http://www.aiaccproject.org/].

Cane, M.A., S.E. Zebiak and S.C. Dolan, 1986: Experimental Forecasts of El-Nino. *Nature,* **321**, 827-832.

Carvalho, A. and J. Burgess, 2005: Cultural circuits of climate change in U.K. broadsheet newspapers, 1985-2003. *Risk Anal.,* 25, 1457-1469.

CEA, 2006: *Climate Change and Natural Events – Insurers contribute to face the challenges*. Comité Européen des Assurances, Paris, 27 pp.

Christoplos, I, 2006: *The Elusive Window of Opportunity for Risk Reduction in Post-Disaster Recovery*. Discussion Paper ProVention Consortium Forum 2006 - Strengthening global collaboration in disaster risk reduction, Bangkok, February 2-3, 4 pp.

Cline, W., 1992: *The Economics of Global Warming*. Institute for International Economics, Washington, District of Columbia, 399 pp.

Cutter, S.L., 1995: The forgotten casualties: women, children, and environmental change. *Global Environ. Chang*., **5**, 181-194.

Dankelman, I., 2002: Climate change: learning from gender analysis and women's experiences of organising for sustainable development. *Gender and Development*, **10**, 21-29.

Darwin, R.F. and Tol, R.S.J., 2001: Estimates of the economic effects of sea level rise. *Environ. Resour. Econ*., **19**, 113-129.

Darwin, R.F., M. Tsigas, J. Lewandrowski and A. Raneses, 1995: *World Agriculture and Climate Change: Economic Adaptations*. Agricultural Economic Report Number 703, United States Department of Agriculture, Economic Research Service, Washington, District of Columbia, 86 pp.

Davidson, D.J., T. Williamson and J.R. Parkins, 2003: Understanding climate change risk and vulnerability in northern forest-based communities. *Can. J. Forest. Res*., **33**, 2252-2261.

Davison, J., 1988: *Agriculture, Women and the Land: the African Experience*. Westview, Boulder, Colorado, 278 pp.

Dawson, R.J., J.W. Hall, P.D. Bates and R.J. Nicholls, 2005: Quantified analysis of the probability of flooding in the Thames estuary under imaginable worst-case sea level rise scenarios. *Int.J. Water Resour. D*., **21**, 577-591.

De Mello Lemos, M.C., 2003: A tale of two policies: The politics of climate forecasting and drought relief in Ceará, Brazil. *Policy Sci*., **36**, 101-123.

Defra, 2006: *The UK's Fourth National Communication under the United Nations Framework Convention on Climate Change*. United Kingdom Department for Environment, Food and Rural Affairs, London., 135 pp.

Denton, F., 2002: Climate change vulnerability, impacts, and adaptation: why does gender matter? *Gender and Development*, **10**, 10-20.

Denton, F., 2004: Gender and climate change - Giving the "Latecomer" a head start. *IDS Bull*., **35**, 42-49.

Derocher, A.E., N.J. Lunn and I. Stirling, 2004: Polar bears in a warming climate. *Intergr. Comp. Biol*., **44**, 163-176.

Dilley, M., 2000: Reducing vulnerability to climate variability in Southern Africa: the growing role of climate information. *Climatic Change*, **45**, 63-73.

Direction du Tourisme, 2002: *Les chiffres clés du tourisme de montagne en France, 3ème edition*. Service d'Etudes et d'Aménagement touristique de la montagne, Paris, 44 pp.

Dlugolecki, A. and S. Lafeld, 2005: *Climate Change and the Financial Sector: An Agenda for Action*. Allianz and the World Wildlife Fund, Munich, 57 pp.

Dore, M. and I. Burton, 2001: *The Costs of Adaptation to Climate Change in Canada: A Stratified Estimate by Sectors and Regions – Social Infrastructure*. Climate Change Laboratory, Brock University, St Catharines, Ontario , 117 pp.

Dow, K., R.E. Kasperson and M. Bohn, 2006: Exploring the social justice implications of adaptation and vulnerability. *Fairness in Adaptation to Climate Change,* W.N. Adger, J. Paavola, S. Huq and M.J. Mace, Eds.,. MIT Press: Cambridge Massachusetts, 79-96.

Downing, T., 2003: Toward a vulnerability/adaptation science: lessons from famine early warning and food security. *Climate Change Adaptive Capacity and Development,* J.B. Smith, R.J.T. Klein and S. Huq, Eds., Imperial College Press, London, 77-100.

Downing, T.E., R. Butterfield, S. Cohen, S. Huq, R. Moss, A. Rahman, Y. Sokona and L. Stephen, 2001: *Vulnerability Indices: Climate Change Impacts and Adaptation*. UNEP, Nairobi, 91 pp.

Eakin, H., C.M. Tucker and E. Castellanos, 2005: Market shocks and climate variability: the coffee crisis in Mexico, Guatemala, and Honduras. *Mt. Res. Dev*., **25**, 304-309.

Easterling, W.E., N. Chhetri and X. Niu, 2003: Improving the realism of modelling agronomic adaptation to climate change: simulating technological substitution. *Climatic Change*, **60**, 149-173.

Easterling, W.E., B.H. Hurd and J.B. Smith, 2004: *Coping with Global Climate Change: The Role of Adaptation in the United States*. Pew Center on Global Climate Change, Arlington, Virginia, 40 pp.

Ebi, K.L., B. Lim and Y. Aguilar, 2005: Scoping and designing an adaptation process. *Adaptation Policy Frameworks for Climate Change,* B. Lim, E. Spanger-Siegfried, I. Burton, E.L. Malone and S. Huq, Eds., Cambridge University Press, New York, 33-46.

Eden, S., 1998: Environmental issues: knowledge, uncertainty and the environment. *Prog. Hum. Geog*., **22**, 425-432.

El Raey, M., 2004: *Adaptation to Climate Change for Sustainable Development in the Coastal Zone of* Egypt. ENV/EPOC/GF/SD/RD(2004)1/FINAL, OECD, Paris.

Enarson, E., 2002: Gender issues in natural disasters: talking points and research needs. Infocus Programme on Crisis Response and Reconstruction, Ed., *Selected Issues Papers: Crisis, Women and Other Gender Concerns,* International Labour Office, Geneva, 5-10.

Ereaut, G. and N. Segnit, 2006: *Warm words: how are we telling the climate change story and can we tell it better?* Institute for Public Policy Research, London, 32 pp.

Eriksen, S.H. and P.M. Kelly, 2007: Developing credible vulnerability indicators for climate adaptation policy assessment. *Mitigation and Adaptation Strategies for Global Change*, **12,** 495-524.

Eriksen, S.H., K. Brown and P.M. Kelly, 2005: The dynamics of vulnerability: locating coping strategies in Kenya and Tanzania. *Geogr. J*., **171**, 287-305.

Fankhauser, S., 1995a: Protection versus retreat: the economic costs of sea-level rise. *Environ. Plann. A*, **27**, 299-319.

Fankhauser, S., 1995b: *Valuing Climate Change: The Economics of the Greenhouse*. Earthscan, London, 180 pp.

Fankhauser, S., 1996: Climate change costs - Recent advancements in the economic assessment. *Energ. Policy*, **24**, 665-673.

Fankhauser, S., R.S.J. Tol and D.W. Pearce, 1998: Extensions and alternatives to climate change impact valuation: on the critique of IPCC Working Group III's impact estimates. *Environ. Dev. Econ*., **3**, 59-81.

FAO, 2004: *Drought impact mitigation and prevention in the Limpopo River Basin, A situation analysis*. Land and Water Discussion Paper 4, FAO, Rome, 160 pp.

Fenger, J., 2000: Implications of accelerated sea-level rise (ASLR) for Denmark. *Proceeding of SURVAS Expert Workshop on European Vulnerability and Adaptation to impacts of Accelerated Sea-Level Rise*, 19th-21st June 2000, A.C. de la Vega-Leinert, R.J. Nicholls and R.S.J. Tol, Eds., Hamburg, Germany, 86-87.

Few, R., K. Brown and E. Tompkins, 2007: Climate change and coastal management decisions: insights from Christchurch Bay, UK. *Coast. Manage*., **35**, 255-270.

Finan, T.J. and D.R. Nelson 2001: Making rain, making roads, making do: public and private adaptations to drought in Ceará, Northeast Brazil. *Climate Res*., **19**, 97-108.

Folke, C., T. Hahn, P. Olsson and J. Norberg, 2005: Adaptive governance of social-ecological systems. *Annu. Rev. Env.. Resour*., **30**, 441-473.

Ford, J. and B. Smit, 2004: A framework for assessing the vulnerability of communities in the Canadian Arctic to risks associated with climate change. *Arctic*, **57**, 389-400.

Ford, J., B. Smit and J. Wandel, 2006: Vulnerability to climate change in the Arctic: A case study from Arctic Bay, Nunavut. *Global Environ. Chang*., **16**, 145-160.

Fordham, M., 2003: Gender, disaster and development: the necessity of integration. *Natural Disasters and Development in a Globalising World,* M. Pelling, M., Ed.,Routledge, London, 57-74.

Franco, G. and A. H. Sanstad, 2006: *Climate Change and Electricity Demand in California*. White Paper, CEC-500-2005-201-SF, February 2006, California Climate Change Center, Sacramento, California, 40 pp.

Gagnon-Lebrun, F. and S. Agrawala, 2006: *Progress on Adaptation to Climate Change in Developed Countries: An Analysis of Broad Trends*. ENV/EPOC/GSP(2006)1/FINAL, OECD, Paris, 51 pp.

Glantz, M.H., 2001: *Once Burned, Twice Shy? Lessons Learned from the 1997-98 El Niño*. United Nations University Press, Tokyo, 294 pp.

Goklany, I.M., 1995: Strategies to enhance adaptability: technological change, sustainable growth and free trade. *Climatic Change*, **30**, 427-449.

Government of The Netherlands, 1997: *Second Netherlands' Communication on Climate Change Policies*. Prepared for the Conference of Parties under the Framework Convention on Climate Change. Ministry of Housing, Spatial Planning and the Environment, Ministry of Economic Affairs, Ministry of Transport, Public Works and Water Management, Ministry of Agriculture, Nature Management and Fisheries, Ministry of Foreign Affairs, The Hague, 162 pp.

Government of The Netherlands, 2005: *Fourth Netherlands' National Communication under the United Nations Framework Convention on Climate Change*. Ministry of Housing, Spatial Planning and the Environment, The Hague, 208 pp.

Grothmann, T. and A. Patt, 2005: Adaptive capacity and human cognition: the process of individual adaptation to climate change. *Global Environ. Chang.*, **15**, 199-213.

Haddad, B.M., 2005: Ranking the adaptive capacity of nations to climate change when socio-political goals are explicit. *Global Environ. Chang.*, **15**, 165-176.

Haddad, B.M., L. Sloan, M. Snyder and J. Bell, 2003: Regional climate change impacts and freshwater systems: focusing the adaptation research agenda. *International Journal of Sustainable Development*, **6**, 265-282.

Hall, J.W., P.B. Sayers, M.J.A. Walkden and M. Panzeri, 2006: Impacts of climate change on coastal flood risk in England and Wales: 2030–2100. *Philos.T.Royal Soc. A.*, **364**, 1027-1049.

Hansen, J., S. Marx and E. U. Weber, 2004: The role of climate perceptions, expectations, and forecasts in farmer decision making: The Argentine pampas and south Florida. IRI Technical Report 04-01. International Research Institute for Climate Prediction, Palisades, New York. [http://iri.columbia.edu/outreach/publication/report/04-01/report04-01.pdf].

Harrison, S.J., S.J. Winterbottom and R.C. Johnson, 2005: Changing snow cover and winter tourism and recreation in the Scottish Highlands. *Tourism, Recreation and Climate Change,* M. Hall and J. Higham, Eds., Channelview Press, London, 143-154.

Hughes, T.P., A.H. Baird, D.R. Bellwood, M. Card, S.R. Connolly, C. Folke, R. Grosberg, O. Hoegh-Guldberg, J.B.C. Jackson, J. Kleypas, J.M. Lough, P. Marshall, M. Nystrom, S.R. Palumbi, J.M. Pandolfi, B. Rosen and J. Roughgarden, 2003: Climate change, human impacts, and the resilience of coral reefs. *Science*, **301**, 929-933.

Hughes, T.P., D.R. Bellwood, C. Folke, R.S. Steneck, and J. Wilson, 2005: New paradigms for supporting the resilience of marine ecosystems. *Trends in Ecology & Evolution*, **20**, 380-386.

Huq, S. and H. Reid, 2003: The role of peoples' assessments. *Tiempo*, **48**, 5-9. Accessed 11 October 2006. [Available online at http://www.tiempocyberclimate.org/portal/archive/issue48/t48a2.htm].

Huq, S. and M.R. Khan, 2006: Equity in National Adaptation Programs of Action (NAPAs): the case of Bangladesh. *Fairness in Adaptation to Climate Change,* W.N. Adger, J. Paavola, S. Huq and M.J. Mace, Eds., MIT Press, Cambridge, Massachusetts, 131-153.

Huq, S., A.A. Rahman, M. Konate, Y. Sokona and H. Reid, 2003: *Mainstreaming Adaptation to Climate Change in Least Developed Countries (LDCS)*. International Institute for Environment and Development, London, 57 pp.

Ikeme, J., 2003: Climate change adaptational deficiencies in developing countries: the case of Sub-Saharan Africa. *Mitigation and Adaptation Strategies for Global Change*, **8**, 29-52.

Irwin, A. and B. Wynne, Eds., 1996: *Misunderstanding Science: The Public Reconstruction of Science and Technology*. Cambridge University Press, Cambridge, 240 pp.

Jackson, C., 2003: Gender analysis of land: beyond land rights for women? *Journal of Agrarian Change*, **3**, 453-480.

Jamieson, D., 2006: An American paradox. *Climatic Change*, **77**, 97-102.

Jones, R.N., 2001: An environmental risk assessment/management framework for climate change impact assessments. *Nat. Hazards*, **23**, 197-230.

Kahn, M.E., 2003: Two measures of progress in adapting to climate change. *Global Environ. Chang.*, **13**, 307-312.

Kelly, D.L., C.D. Kolstad and G.T. Mitchell, 2005: Adjustment costs from environmental change. *J. Environ. Econ. Manag.*, **50**, 468-495.

Kelly, P.M. and W.N. Adger, 2000: Theory and practice in assessing vulnerability to climate change and facilitating adaptation. *Climatic Change*, **47**, 325-352.

Kingdon, J.W. 1995: *Agendas, Alternatives and Public Policies,* 2nd ed. HarperCollins, New York, 254 pp.

Kirshen, P., M. Ruth, W. Anderson and T.R. Lakshmanan, 2004: Infrastructure systems, services and climate changes: Integrated impacts and response strategies for the Boston Metropolitan area. Climate's Long-term Impacts on Metro Boston (CLIMB) Final Report August 13, 2004. [Available online at http://www.tufts.edu/tie/climb/].

Klein, R.J.T. and J.B. Smith, 2003: Enhancing the capacity of developing countries to adapt to climate change: a policy relevant research agenda. *Climate Change, Adaptive Capacity and Development,* J.B. Smith, R.J.T. Klein and S. Huq, Eds., Imperial College Press, London, 317-334.

Klein, R.J.T., S.E.H. Eriksen, L.O. Næss, A. Hammill, T.M. Tanner, C. Robledo and K.L. O'Brien, 2007: Portfolio screening to support the mainstreaming of adaptation to climate change into development assistance. *Climatic Change*, **84**, 23-44.

Klinenberg, E., 2003: *Heat Wave: A Social Autopsy of Disaster in Chicago*. University of Chicago Press, Chicago, Illinois, 328 pp.

Konig, U., 1999: Climate change and snow tourism in Australia. *Geographica Helvetica*, **54**, 147-157.

Kunreuther, H., N. Novemsky and D. Kahneman, 2001: Making low probabilities useful. *J. Risk Uncertainty*, **23**, 103-120.

Lagadec, P., 2004: Understanding the French heatwave experience: beyond the heat, a multi-layerd challenge. *Journal of Contingencies and Crisis Management*, **12**, 160-169.

Lasco, R., R. Cruz, J. Pulhin and F. Pulhin, 2006: *Tradeoff analysis of adaptation strategies for natural resources, water resources and local institutions in the Philippines*. AIACC Working Paper No. 32, International START Secretariat, Washington, District of Columbia, 31 pp.

Leary, N., J. and Co-authors, 2006: *For Whom the Bell Tolls: Vulnerabilities in a Changing Climate*. AIACC Working Paper No. 30, International START Secretariat, Washington, District of Columbia, 31 pp.

Lee, R.J., 2000: Climate Change and Environmental Assessment Part 1: Review of Climate Change Considerations in Selected Past Environmental Assessments. [Available online at http://www.ceaa-acee.gc.ca/015/001/005/index_e.htm].

Leichenko, R.M. and K.L. O'Brien, 2002: The dynamics of rural vulnerability to global change: the case of Southern Africa. *Mitigation and Adaptation Strategies for Global Change*, **7**, 1-18.

Ligeti, E., 2004: *Adaptation Strategies to Reduce Health Risks from Summer Heat in Toronto*. Your Health and a Changing Climate Newsletter 1, Health Canada, Ottawa, Ontario, 9 pp.

Lim, B., E. Spanger-Siegfried, I. Burton, E. Malone and S. Huq, Eds., 2005: *Adaptation Policy Frameworks for Climate Change: Developing Strategies, Policies and Measures*. Cambridge University Press, New York, 258 pp.

Lorenzoni, I. and N. Pidgeon, 2006: Public views on climate change: European and USA perspectives. *Climatic Change*, **77**, 73-95.

Lorenzoni, I., N.F. Pidgeon, and R.E. O'Connor, 2005: Dangerous climate change: the role for risk research. *Risk Anal.*, **25**, 1387-1397.

Loster, T., 2005: Strategic management of climate change: options for the insurance industry. *Weather catastrophes and climate change - Is there still hope for us?* Geo Risks Research, MunichRe, Munich, 236-243.

Luers, A.L., 2005: The surface of vulnerability: An analytical framework for examining environmental change. *Global Environ. Chang.*, **15**, 214-223.

Mace, M.J., 2006: Adaptation under the UN Framework Convention on Climate Change: the international legal framework. *Fairness in Adaptation to Climate Change,* W.N. Adger, J. Paavola, S. Huq and M.J. Mace, Eds., MIT Press, Cambridge, Massachusetts, 53-76.

Magdanz, J.S., C.J. Utermohle and R.J. Wolfe, 2002: *The Production and Distribution of Wild Food in Wales and Deering, Alaska*. Technical Paper 259. Division of Subsistence, Alaska Department of Fish and Game, Juneau, Alaska.

Mansur, E.T., R. Mendelsohn and W. Morrison, 2005: *A Discrete-Continuous Choice Model of Climate Change Impacts on Energy*. Social Science Research Network, Yale School of Management Working Paper No. ES-43, Yale University, New Haven, Conecticut, 41 pp.

Manuta, J. and L. Lebel, 2005: *Climate Change and the Risks of Flood Disasters in Asia: Crafting Adaptive and Just Institutions*. Human Security and Climate Change. An International Workshop, Oslo, 21–23 June 2005. [Available online at http://www.cicero.uio.no/humsec/papers/Manuta&Lebel.pdf].

McIntosh, R.J., J.A. Tainter and S.K. McIntosh, Eds., 2000: *The Way the Wind Blows: Climate, History, and Human Action*. Columbia University Press, New York, 448 pp.

McKenzie, K. and K. Parlee, 2003: *The Road Ahead: Adapting to Climate Change in Atlantic Canada*. Canadian Climate Impacts and Adaptation Research Network, New Brunswick, Canada. [Available online at www.elements.nb.ca/theme/climate03/cciarn/adapting.htm].

McLeman, R., 2006: Migration out of 1930s rural eastern Oklahoma: insights for

climate change research. *Great Plains Quarterly*, **26**, 27-40.
McLeman, R. and B. Smit, 2006: Migration as an adaptation to climate change. *Climatic Change*, **76**, 31-53.
Mehdi, B., 2006: *Adapting to Climate Change: An Introduction for Canadian Municipalities*, Occasional Paper, Canadian Climate Impacts and Adaptation Research Network (C-CIARN), Ottowa. [Available online at http://www.c-ciarn.ca/pdf/adaptations_e.pdf].
Meinke, H., K. Pollock, G.L. Hammer, E. Wang, R.C. Stone, A. Potgieter and M. Howden, 2001: Understanding Climate Variability to Improve Agricultural Decision Making. *Proceedings of the 10th Australian Agronomy Conference*, 2001, Hobart, Australia.
Mendelsohn, R., 2003: Appendix XI - The Impact of Climate Change on Energy Expenditures in California. *Global Climate Change and California: Potential Implications for Ecosystems, Health, and the Economy*, C. Thomas and R. Howard, Eds., Sacramento, California, 35 pp.
Mendelsohn, R., W. Morrison, M.E. Schlesinger and N.G. Andronova, 2000: Country-specific market impacts of climate change. *Climatic Change*, **45**, 553-569.
Mendelsohn, R., A. Basist, P. Kurukulasuriya and A. Dinar, 2007: Climate and rural income. *Climatic Change*, **81**, 101-118..
Millennium Ecosystem Assessment, 2006: *Ecosystems and Human Well-being: Synthesis*. Millennium Ecosystem Assessment, Island Press, Washington, District of Columbia, 137 pp.
Mills, E., 2005: Insurance in a climate of change. *Science*, **309**, 1040-1044.
Mills, E. and E. Lecomte, 2006: *From Risk to Opportunity: How Insurers Can Proactively and Profitably Manage Climate Change*. Ceres, Boston, Massachusetts, 42 pp.
Mills, E., R. Roth and G. LeComte, 2005: *Availability and affordability of Insurance under Climate Change: A Growing Challenge for the U.S.* CERES Report, Ceres, Boston, Massachusetts, 44 pp.
Mirza, M.M.Q., 2003: Climate change and extreme weather events: can developing countries adapt? *Climate Pol.*, **3**, 233-248.
Mirza, M.M.Q. and I. Burton, 2005: Using adaptation policy framework to assess climate risks and response measures in South Asia: the case of floods and droughts in Bangladesh and India. *Climate Change and Water Resources in South Asia*, M.M.Q. Mirza and Q.K. Ahmad, Eds., Taylor and Francis, London, 279-313.
Mizina, S.V., J.B. Smith, E. Gossen, K.F. Spiecker and S.L. Witkowski, 1999: An evaluation of adaptation options for climate change impacts on agriculture in Kazakhstan. *Mitigation and Adaptation Strategies for Global Change*, **4**, 25-41.
MMM, 2005: *Finland's National Strategy for Adaptation to Climate Change*. Ministry of Agriculture and Forestry of Finland, Helsinki, 281 pp.
Mohanty, P. and U. Panda, 2003: *Heatwave in Orissa: a study based on heat indices and synoptic features*. Regional Research Laboratory, Institute of Mathematics and Applications, Bubaneshwar, 15 pp.
Mool, P.K., S.R. Bajracharya and S.P. Joshi, 2001: *Inventory of Glaciers, Glacier Lakes and Glacial Lake Outburst Floods, Monitoring and Early Warning System in the Hindu Kush-Himalayan Region*. ICIMOD/UNEP, Kathmandu, Nepal, 247 pp.
Morrison, W. and R. Mendelsohn, 1999: The impact of global warming on US energy expenditures. *The Impact of Climate Change on the United States Economy*, R. Mendelsohn, and J. Neumann, Eds., Cambridge University Press, Cambridge, 209-236.
Mortimore, M.J. and W.M. Adams, 2001: Farmer adaptation, change and 'crisis' in the Sahel. *Global Environ. Chang.*, **11**, 49-57.
Moser, S., 2005: Impacts assessments and policy responses to sea-level rise in three U.S. states: an exploration of human dimension uncertainties. *Global Environ. Chang.*, **15**, 353-369.
Moser, S.C. and L. Dilling, 2004: Making climate hot. Communicating the urgency and challenge of global climate change. *Environment*, **46**(, 32-46.
Moss, R.H., A.L. Brenkert and E.L. Malone, 2001: *Vulnerability to Climate Change: A Quantitative Approach*. Pacific Northwest National Laboratory, Richland Washington, 70 pp.
Nakićenović, N. and R. Swart, Eds., 2000: *IPCC Special Report on Emissions Scenarios*, Cambridge University Press, Cambridge, 599 pp.
Næss, L.O., G. Bang, S. Eriksen and J. Vevatne, 2005: Institutional adaptation to climate change: flood responses at the municipal level in Norway. *Global Environ. Chang.*, **15**, 125-138.
Nelson, V., K. Meadows, T. Cannon, J. Morton and A. Martin, 2002: Uncertain predictions, invisible impacts, and the need to mainstream gender in climate change adaptations. *Gender and Development*, **10**, 51-59.
Ng, W. and R. Mendelsohn, 2005: The impact of sea level rise on Singapore. *Environ. Dev. Econ.*, **10**, 210-215.
Nicholls, R.J. and R.S.J. Tol, 2006: Impacts and responses to sea-level rise: A global analysis of the SRES scenarios over the 21st Century. *Philos. T. Roy. Soc. A*, **364**, 1073-1095.
Nicholson-Cole, S.A., 2005: Representing climate change futures: a critique of the use of images for visual communication. *Computers, Environment and Urban Systems*, **29**, 255-273.
Niemeyer, S., J. Petts and K. Hobson, 2005: Rapid climate change and society: assessing responses and thresholds. *Risk Anal.*, **25** 1443-1455.
Njie, M., B.E. Gomez, M.E. Hellmuth, J.M. Callaway, B.P. Jallow and P. Droogers, 2006: *Making Economic Sense of Adaptation in Upland Cereal Production Systems in The Gambia*. AIACC Working Paper No. 37, International START Secretariat, Washington, District of Columbia. [Available online at http://www.aiaccproject.org/working_papers/working_papers.html].
NOAA, 1999: *An Experiment in the Application of Climate Forecasts: NOAA-OGP Activities Related to the 1997-1998 El Niño Event*. Office of Global Programs National Oceanic and Atmospheric Administration U.S. Department of Commerce, Washington, District of Columbia, 134 pp.
Nordhaus, W., 1991: To slow or not to slow: the economics of the greenhouse effect. *Economic Journal*, **101**, 920-937.
Nordhaus, W.D. and J. Boyer, 2000: *Warming the World: Economic Models of Global Warming*. MIT Press, Cambridge, Massachusetts, 232 pp.
Noss, R.F. 2001: Beyond Kyoto: forest management in a time of rapid climate change. *Conservation Biology*, **15**, 578-590.
O'Brien, K., R. Leichenko, U. Kelkar, H. Venema, G. Aandahl, H. Tompkins, A. Javed, S. Bhadwal, S. Barg, L. Nygaard and J. West, 2004: Mapping vulnerability to multiple stressors: climate change and globalization in India. *Global Environ. Chang.*, **14**, 303-313.
O'Brien, K., S. Eriksen, L. Sygna and L.O. Naess, 2006: Questioning Complacency: Climate Change Impacts, Vulnerability, and Adaptation in Norway. *Ambio*, **35**, 50-56.
O'Brien, K.L. and R.M. Leichenko, 2000: Double exposure: assessing the impacts of climate change within the context of economic globalization. *Global Environ. Chang.*, **10**, 221-232.
O'Brien, K.L. and H.C. Vogel, Eds., 2003: *Coping with Climate Variability: The Use of Seasonal Climate Forecasts in Southern Africa*. Ashgate Publishing, Aldershot, 176 pp.
OECD, 2003a: *Development and Climate Change in Bangladesh: focus on Coastal Flooing and the Sundarbans*, COM/ENV/EPOC/DCD/DAC(2003) 3/FINAL, Organisation for Economic Co-operation and Development, Paris, 70 pp.
OECD, 2003b: *Development and Climate Change in Nepal: focus on Water Resources and Hydropower*, COM/ENV/EPOC/DCD/DAC(2003)1/FINAL, Organisation for Economic Co-operation and Development, Paris, 64 pp.
ONERC, 2005: *Un climat à la dérive: comment s'adapter?* Rapport de ONERC au Premier ministre et au Parlement, Observatoire national des effets du réchauffement climatique (ONERC), Paris, 107 pp.
Oppenheimer, M. and A. Todorov, 2006: Global warming: the psychology of long term risk. *Climatic Change*, **77**, 1-6.
Orlove, B., 2005: Human adaptation to climate change: a review of three historical cases and some general perspectives. *Environ. Sci. Policy*, **8**, 589-600.
Osman-Elasha, B., N. Goutbi, E. Spanger-Siegfried, B. Dougherty , A. Hanafi , S. Zakieldeen, A. Sanjak, H. Atti and H. Elhassan. 2006: *Adaptation strategies to increase human resilience against climate variability and change: Lessons from the arid regions of Sudan*. AIACC Working Paper 42, International START Secretariat, Washington, District of Columbia, 42 pp.
Owuor, B., S. Eriksen and W. Mauta, 2005: Adapting to climate change in a dryland mountain environment in Kenya. *Mt. Res. Dev.*, **25**, 310-315.
Paavola, J., 2006: Justice in adaptation to climate change in Tanzania. *Fairness in Adaptation to Climate Change*, W.N. Adger, J. Paavola, S. Huq and M.J. Mace, Eds., MIT Press, Cambridge, Massachusetts, 201-222.
Paavola, J. and W.N. Adger, 2006: Fair adaptation to climate change. *Ecol. Econ.*, **56**, 594–609.
Patel, S.S., 2006: Climate science: A sinking feeling. *Nature*, **440**, 734-736.
Patt, A. and C. Gwata, 2002: Effective seasonal climate forecast applications: ex-

amining constraints for subsistence farmers in Zimbabwe. *Global Environ. Chang.*, **12**, 185-195.

Pelling, M., Ed., 2003: *Natural Disasters and Development in a Globalising World*. Routledge, London, 249 pp.

Pelling, M. and J. Uitto, 2001: Small island developing states: natural disaster vulnerability and global change. *Environmental Hazards*, **3**, 49-62.

Penning-Rowsell, E.C, C. Johnson and S.M., Tunstall, 2006: 'Signals' from pre-crisis discourse: Lessons from UK flooding for global environmental policy change? *Global Environ. Chang.*, **16**, 323-339.

Perez, R.T. and G. Yohe, 2005: Continuing the adaptation process. *Adaptation Policy Frameworks for Climate Change*, B. Lim, E. Spanger-Siegfried, I. Burton, E.L. Malone and S. Huq, Eds., Cambridge University Press, New York, 205-224.

Pfaff, A., K. Broad and M. Glantz, 1999: Who benefits from climate forecasts? *Nature*, **397**, 645-646.

Pouliotte, J., N. Islam, B. Smit and S. Islam, 2006: Livelihoods in rural Bangladesh. *Tiempo*, **59**, 18-22. Accessed 11 October 2006. [Available online at http://www.cru.uea.ac.uk/tiempo/portal/archive/pdf/tiempo59low.pdf].

Poumadère, M., C. Mays, S. Le Mer and R. Blong, 2005: The 2003 heat wave in France: dangerous climate change here and now. *Risk Anal.*, **25**, 1483-1494.

Pretty, J., 2003: Social capital and the collective management of resources. *Science*, **302**, 1912-1925.

Pulwarty, R., K., Broad and T., Finan, 2003: ENSO forecasts and decision making in Brazil and Peru. *Mapping Vulnerability: Disasters, Development and People*, *G* Bankoff, G. Frerkes and T. Hilhorst, Eds., Earthscan, London, 83-98.

Pulwarty, R., K. Jacobs, and R., Dole, 2005: The hardest working river: Drought and critical water problems in the Colorado River Basin. *Drought and Water Crises: Science, Technology and Management*, D. Wilhite., Ed., Taylor and Francis Press, New York, 249-285.

Reilly, J., F. Tubiello, B. McCarl, D.Abler, R. Darwin, K. Fuglie, S. Hollinger, C. Izaurralde, S. Jagtap, J. Jones, L. Mearns, D. Ojima, E. Paul, K. Paustina, S. Riha, N. Rosenberg and C. Rosenzweig, 2003: US agriculture and climate change: new results. *Climatic Change*, **57**, 43-69.

Ribot, J.C., A.R. Magalhães and S.S. Panagides, Eds.,1996: *Climate Variability, Climate Change and Social Vulnerability in the Semi-Arid Tropics*. Cambridge University Press, Cambridge, 189 pp.

Robledo, C., M. Fischler and A. Patino, 2004: Increasing the resilience of hillside communities in Bolivia. *Mt. Res. Dev.*, **24**, 14-18.

Rosenzweig, C. and M.L. Parry, 1994: Potential impact of climate change on world food supply. *Nature*, **367**, 133-138.

Rosenzweig, C. and W.D. Solecki, 2001: *Climate Change and a Global City: The Potential Consequences of Climate Variability and Change—Metro East Coast*. Report for the U.S. Global Change Research Program, National Assessment of the Potential Consequences of Climate Variability and Change for the United States. Columbia Earth Institute, New York, 224 pp.

Rosenzweig, C., D.C. Major, K. Demong, C. Stanton, R. Horton and M. Stults, 2007: Managing climate change risks in New York City's water system – assessment and adaptation planning. *Mitigation and Adaptation Strategies for Global Change*, doi:10.1007/s11027-006-9070-5.

Sailor, D.J. and A.A. Pavlova, 2003: Air conditioning market saturation and long term response of residential cooling energy demand to climate change. *Energy*, **28**, 941-951.

Satterfield, T.A., C.K. Mertz and P. Slovic, 2004: Discrimination, vulnerability, and justice in the face of risk. *Risk Anal.*, **24**, 115-129.

Schaedler, B., 2004: *Climate Change Issues and Adaptation Strategies in a Mountainous Region: Case Study Switzerland*. ENV/EPOC/GS/FD/RD(2004)3/FINAL,OECD, Paris, 16 pp.

Scheffer, M., S. Carpenter, J.A. Foley, C. Folke and B. Walker, 2001: Catastrophic shifts in ecosystems. *Nature*, **413**, 591-596.

Schneider, S.H., W.E. Easterling and L.O. Mearns, 2000a: Adaptation sensitivity to natural variability: agent assumptions and dynamic climate changes. *Climatic Change*, **45**, 203-221.

Schneider, S.H., K. Kuntz-Duriseti and C. Azar, 2000b: Costing non-linearities, surprises, and irreversible events. *Pacific and Asian Journal of Energy*, **10**, 81-106.

Schröter, D., C. Polsky and A.G. Patt, 2005: Assessing vulnerabilities to the effects of global change: an eight step approach. *Mitigation and Adaptation Strategies for Global Change*, **10**, 573-595.

Scott, D., G. McBoyle and B. Mills, 2003: Climate change and the skiing industry in southern Ontario (Canada): exploring the importance of snowmaking as a technical adaptation. *Climate Res.*, **23**, 171–181.

Scott, D., G. Wall and G. McBoyle, 2005: The evolution of the climate change issue in the tourism sector. *Tourism, Recreation and Climate Change*, M. Hall and J. Higham, Eds., Channelview Press, London, 44-60.

Scott, D., G. McBoyle and A. Minogue, 2007: Climate change and Quebec's ski industry. *Global Environ. Chang.*, **17**, 181-190.

Shaw, D., S. Shih, E.Y. Lin and Y. Kuo, 2000: *A cost-benefit analysis of sea-level rise protection in Taiwan. Paper presented at the International Conference on Global Economic Transformation after the Asian Economic Crisis*. Chinese University of Hong-Kong and Pekin University, May 26-28, 2000, Hong-Kong.

Sheppard, S.R.J., 2005: Landscape visualisation and climate change: the potential for influencing perceptions and behaviour. *Environ. Sci. Policy*, **8**, 637-654.

Sheridan, S.C. and L.S. Kalkstein, 2004: Progress in heat watch warning system technology. *B. Am. Meteorol. Soc.*, **85**, 1931-1941.

Shrestha, M.L. and A.B. Shrestha, 2004: *Recent Trends and Potential Climate Change Impacts on Glacier Retreat/Glacier Lakes in Nepal and Potential Adaptation Measures*. ENV/EPOC/GF/SD/RD(2004)6/FINAL, OECD, Paris, 23 pp.

Shukla, P.R., M. Kapshe and A. Garg, 2004: *Development and Climate: Impacts and Adaptation for Infrastructure Assets in India*. ENV/EPOC/GS/FD/RD(2004)3/FINAL,OECD, Paris, 38 pp.

Simms, A., D. Woodward and P. Kjell 2004: *Cast Adrift: How the Rich are Leaving the Poor to Sink in a Warming World*. New Economics Foundation, London, 21 pp.

Smit, B. and M.W. Skinner, 2002: Adaptation options in agriculture to climate change: a typology. *Mitigation and Adaptation Strategies for Global Change*, **7**, 85-114.

Smit, B. and J. Wandel, 2006: Adaptation, adaptive capacity and vulnerability. *Global Environ. Chang.*, **16**, 282-292.

Smit, B., O. Pilifosova, I. Burton, B. Challenger, S. Huq, R.J.T. Klein and G. Yohe, 2001: Adaptation to climate change in the context of sustainable development and equity. *Climate Change 2001: Impacts, Adaptation and Vulnerability. Contribution of the Working Group II to the Third Assessment Report of the Intergovernmental Panel on Climate Change*, J.J. McCarthy, O. Canziani, N.A. Leary, D.J. Dokken and K.S. White, Eds., Cambridge University Press, Cambridge, 877-912.

Smith, J.B., 1997: Setting priorities for adapting to climate change. *Global Environ. Chang.*, **7**, 251-264.

Smith, J.B. and D. Tirpak, 1989: *The Potential Effects of Global Climate Change on the United States*. US Environmental Protection Agency, Washington, District of Columbia, 401 pp.

Smith, J.B., A. Rahman, S. Haq and M.Q. Mirza, 1998: Considering adaptation to climate change in the sustainable development of Bangladesh. World Bank Report, World Bank, Washington, District of Columbia, 103 pp.

Smith , J.B., S. Agrawala, P. Larsen and F. Gagnon-Lebrun, 2005: Climate analysis. *Bridge Over Troubled Waters – Linking Climate Change and Development*, S. Agrawala, Ed., OECD, Paris, 45-59.

Smoyer-Tomic, K.E. and D.G.C. Rainham, 2001: Beating the heat: development and evaluation of a Canadian hot weather health-response plan. *Environmental Health Perspectives*, **109**, 1241-1248.

Sperling, F. and F. Szekely, 2005: *Disaster Risk Management in a Changing Climate*. Informal Discussion Paper prepared for the World Conference on Disaster Reduction on behalf of the Vulnerability and Adaptation Resource Group (VARG). Washington, District of Columbia, 42 pp.

Steneck, R.S., M.H. Graham, B.J. Bourque, D. Corbett, J.M. Erlandson, J.A. Estes and M.J. Tegner, 2002: Kelp forest ecosystems: biodiversity, stability, resilience and future. *Environ. Conserv.*, **29**, 436–459.

Sturgis, P. and N. Allum, 2004: Science in society: Re-evaluating the deficit model of public attitudes. *Public Underst. Sci.*, **13**, 55-74.

Sutherland, K., B. Smit, V. Wulf and T. Nakalevu, 2005: Vulnerability to climate change and adaptive capacity in Samoa: the case of Saoluafata village. *Tiempo*, **54**, 11-15.

Swiss Confederation, 2005: *Switzerland's Fourth National Communication under the UNFCCC*. Swiss Agency for the Environment, Forests and Landscape (SAEFL), Berne, 238 pp.

Sygna, L., 2005: *Climate vulnerability in Cuba: the role of social networks*. CICERO Working Paper, 2005-01, University of Oslo, Oslo, Norway, 12 pp.

Taylor, M., A. Chen, S. Rawlins, C. Heslop-Thomas, A. Amarakoon, W. Bailey, D. Chadee, S. Huntley, C. Rhoden and R. Stennett, 2006: *Adapting to dengue risk – what to do?* AIACC Working Paper No. 33, International START Secretariat, Washington, District of Columbia, 31 pp.

Thomas, D., H. Osbahr, C. Twyman, W.N. Adger and B. Hewitson, 2005: *ADAPTIVE: adaptation to climate change amongst natural resource-dependent societies in the developing world: across the Southern African climate gradient*. Technical Report 35, Tyndall Centre for Climate Change Research, University of East Anglia, Norwich, 43 pp.

Thomas, D.S.G. and C. Twyman, 2005: Equity and justice in climate change adaptation amongst natural-resource-dependent societies. *Global Environ. Chang*., **15**, 115-124.

Tol, R.S.J., 1995: The damage costs of climate change: towards more comprehensive calculations. *Environ. Resour. Econ*., **5**, 353–374.

Tol, R.S.J., 2002: Estimates of the damage costs of climate change. Part 1: benchmark estimates. *Environ. Resour.Econ*., **21**, 47–73.

Tol, R.S.J. and G.W. Yohe, 2007: The weakest link hypothesis for adaptive capacity: An empirical test. *Global Environ. Chang*., **17**, 218-227.

Tol, R.S.J., S. Fankhauser and J.B. Smith, 1998: The scope for adaptation to climate change: what can we learn from the impact literature? *Global Environ. Chang*., **8**, 109-123.

Tol, R.S.J., N. van der Grijp, A.A. Olsthoorn and P.E. van der Werff, 2003: Adapting to climate: A case study on riverine flood risks in the Netherlands. *Risk Anal*., **23**, 575-583.

Tol, R.S.J, M. Bohn, T.E. Downing, M. Guillerminet, E. Hizsnyik, R. Kasperson, K. Lonsdale, C. Mays, R.J. Nicholls, A.A. Olsthoorn, G. Pfeifle, M. Poumadere, F.L. Toth, N. Vafeidis, P.E. van der Werff and I.H. Yetkiner, 2006: Adaptation to five metres of sea level rise. *J. Risk Res*., **9**, 467-482.

Toman, M., 2006: Values in the economics of climate change. *Environmental Values*, **15**, 365-379.

Tompkins, E., 2005: Planning for climate change in small islands: insights from national hurricane preparedness in the Cayman Islands. *Global Environ. Chang*., **15**, 139-149.

Tompkins, E. L. and W.N. Adger, 2004: Does adaptive management of natural resources enhance resilience to climate change? *Ecology and Society* **9**, 10. [Available online at www.ecologyandsociety.org/vol9/iss2/art10].

Tompkins, E.L. and W.N. Adger, 2005: Defining a response capacity for climate change. *Environ. Sci. Policy*, **8**, 562–571.

Tompkins, E., W.N. Adger and K. Brown, 2002: Institutional networks for inclusive coastal management in Trinidad and Tobago. *Environ. Plan. A*, **34**, 1095-1111.

Tompkins, E.L., E. Boyd, S.A. Nicholson-Cole, K. Weatherhead, N.W. Arnell and W.N. Adger, 2005: *Linking Adaptation Research and Practice*. Report to DEFRA Climate Change Impacts and Adaptation Cross-Regional Research Programme, Tyndall Centre for Climate Change Research, University of East Anglia, Norwich, 119 pp.

Turner, B.L., R.E. Kasperson, P.A. Matson, J.J. McCarthy, R.W. Corell, L. Christensen, N. Eckley, J.X. Kasperson, A. Luers, M.L. Martello, C. Polsky, A. Pulsipher and A. Schiller, 2003: A framework for vulnerability analysis in sustainability science. *P. Natl. Acad. Sci. USA*, **100**, 8074-8079.

US National Assessment, 2001: *Climate Change Impacts on the United States: The Potential Consequences of Climate Variability and Change*, Report for the US Global Change Research Program, Cambridge University Press, Cambridge, 620 pp.

Valverde, L.J. and M.W. Andrews, 2006: *Global Climate Change and Extreme Weather: An Exploration of Scientific Uncertainty and the Economics of Insurance*. Insurance Information Institute, New York, 51 pp.

Van Aalst, M.K. and M. Helmer, 2003: *Preparedness for Climate Change: A study to assess the future impact of climatic changes upon the frequency and severity of disasters and the implications for humanitarian response and preparedness*. Red Cross / Red Crescent Centre on Climate Change and Disaster Preparedness, Hague, 14 pp.

van Vliet, A. and R. Leemans, 2006: Rapid species responses to changes in climate require stringent climate protection targets. *Avoiding Dangerous Climate Change*, H.J. Schellnhuber, W. Cramer, N Nakicenovic, T. Wigley and G. Yohe, Eds., Cambridge University Press,Cambridge, 135-141.

Vergara, W., 2006: *Adapting to Climate Change: Lessons learned, work in progress, and proposed next steps for the World Bank in Latin America*. World Bank Latin America and Caribbean Region Sustainable Development, Working Paper 25, World Bank, Washington, District of Columbia, 46 pp.

Weber, E.U., 2006: Experienced-based and description-based perceptions of long-term risk: why global warming does not scare us (yet). *Climatic Change*, **77**, 103-120.

Wehbe, M., H. Eakin, R. Seiler, M. Vinocur, C. Afila and C. Marutto, 2006: *Local perspectives on adaptation to climate change: lessons from Mexico and Argentina*. AIACC Working Paper 39, International START Secretariat, Washington, District of Columbia, 37 pp.

West, C. and M. Gawith, 2005: *Measuring progress: preparing for climate change through UKCIP*. UK Climate Impacts Programme, Oxford, 72 pp.

Winkels, A. 2004: Migratory livelihoods in Vietnam: vulnerability and the role of migrant livelihoods. PhD Thesis, School of Environmental Sciences, University of East Anglia, Norwich, 239 pp.

Winters, A.P., R. Murgai, E. Sadoulet, A. De Janvry and G. Frisvold, 1998: Economic and welfare impacts of climate change on developing countries. *Environ. Resour. Econ*., **12**, 1-24.

Woodward, A., 2002: Epidemiology, environmental health and global change. *Environmental Change, Climate and Health: Issues and Research Methods*, P. Martens and A.J. McMichael, Eds., Cambridge University Press, Cambridge, 290-310.

World Bank, 2000: *Cities, Seas, and Storms: Managing Change in Pacific Island Economies Volume IV: Adapting to Climate Change*. World Bank, Washington, District of Columbia,118 pp.

World Bank, 2006: *Clean Energy and Development: Towards an Investment Framework*, Annex K. World Bank, Washington, District of Columbia, 157 pp.

Yanda, P., S. Wandiga, R. Kangalawe, M. Opondo, D. Olago, A. Githeko, T. Downs, R. Kabumbuli, A. Opere, F. Githui, J. Kathuri, L. Olaka, E. Apindi, M. Marshall, L. Ogallo, P. Mugambi, E. Kirumira, R. Nanyunja, T. Baguma, R. Sigalla and P. Achola, 2006: *Adaptation to Climate Change/Variability-Induced Highland Malaria and Cholera in the Lake Victoria Region*. AIACC Working Paper 43, International START Secretariat, Washington, District of Columbia, 37 pp.

Yates, D.N. and K.M. Street, 1998: An Assessment of Integrated Climate Change Impacts on the Agricultural Economy of Egypt. *Climatic Change*, **38**, 261-287.

Yohe, G.W. and M.E. Schlesinger, 1998: Sea-level change: The expected economic cost of protection or abandonment in the United States. *Climatic Change*, **38**, 447-472.

Yohe, G. and R.S.J. Tol, 2002: Indicators for social and economic coping capacity - moving toward a working definition of adaptive capacity. *Global Environ. Chang*., **12**, 25-40.

Yohe, G.W., J.E. Neumann, P. Marshall and H. Ameden, 1996: The economic costs of sea level rise on US coastal properties. *Climatic Change*, **32**, 387-410.

Ziervogel, G., 2004: Targeting seasonal climate forecasts for integration into household level decisions: the case of smallholder farmers in Lesotho. *Geogr. J*., **170**, 6-21.

Ziervogel, G., A. Nyong, B. Osman, C. Conde, S. Cortés and T. Downing, 2006: *Climate Variability and Change: Implications for Household Food Security*. AIACC Working Paper No. 20, International START Secretariat, Washington, District of Columbia, 34 pp.

18

Inter-relationships between adaptation and mitigation

Coordinating Lead Authors:
Richard J.T. Klein (The Netherlands/Sweden), Saleemul Huq (UK/Bangladesh)

Lead Authors:
Fatima Denton (The Gambia), Thomas E. Downing (UK), Richard G. Richels (USA), John B. Robinson (Canada), Ferenc L. Toth (IAEA/Hungary)

Contributing Authors:
Bonizella Biagini (GEF/Italy), Sarah Burch (Canada), Kate Studd (UK), Anna Taylor (South Africa), Rachel Warren (UK), Paul Watkiss (UK), Johanna Wolf (Germany)

Review Editors:
Michael Grubb (UK), Uriel Safriel (Israel), Adelkader Allali (Morocco)

This chapter should be cited as:
Klein, R.J.T., S. Huq, F. Denton, T.E. Downing, R.G. Richels, J.B. Robinson, F.L. Toth, 2007: Inter-relationships between adaptation and mitigation. *Climate Change 2007: Impacts, Adaptation and Vulnerability. Contribution of Working Group II to the Fourth Assessment Report of the Intergovernmental Panel on Climate Change*, M.L. Parry, O.F. Canziani, J.P. Palutikof, P.J. van der Linden and C.E. Hanson, Eds., Cambridge University Press, Cambridge, UK, 745-777.

Table of Contents

Supplementary material for this chapter is available on the CD-ROM accompanying this report.

Executive summary

This chapter identifies four types of inter-relationships between adaptation and mitigation:

- Adaptation actions that have consequences for mitigation,
- Mitigation actions that have consequences for adaptation,
- Decisions that include trade-offs or synergies between adaptation and mitigation,
- Processes that have consequences for both adaptation and mitigation.

The chapter explores these inter-relationships and assesses their policy relevance. It is a new chapter compared to the IPCC Third Assessment Report and is based on a relatively small, albeit growing, literature. Its key findings are as follows.

Effective climate policy aimed at reducing the risks of climate change to natural and human systems involves a portfolio of diverse adaptation and mitigation actions (very high confidence).

Even the most stringent mitigation efforts cannot avoid further impacts of climate change in the next few decades (Working Group I Fourth Assessment Report, Working Group III Fourth Assessment Report), which makes adaptation unavoidable. However, without mitigation, a magnitude of climate change is likely to be reached that makes adaptation impossible for some natural systems, while for most human systems it would involve very high social and economic costs (see Chapter 4, Section 4.6.1 and Chapter 17, Section 17.4.2). Adaptation and mitigation actions include technological, institutional and behavioural options, the introduction of economic and policy instruments to encourage the use of these options, and research and development to reduce uncertainty and to enhance the options' effectiveness and efficiency [18.3, 18.5]. Opportunities exist to integrate adaptation and mitigation into broader development strategies and policies [18.6].

Decisions on adaptation and mitigation are taken at different governance levels and inter-relationships exist within and across each of these levels (high confidence).

The levels range from individual households, farmers and private firms, to national planning agencies and international agreements. Effective mitigation requires the participation of major greenhouse-gas emitters globally, whereas most adaptation takes place from local to national levels. The climate benefits of mitigation are global, while its costs and ancillary benefits arise locally. In most cases, both the costs and benefits of adaptation accrue locally and nationally [18.1, 18.4, 18.5]. Consequently, mitigation is primarily driven by international agreements and ensuing national public policies, possibly complemented by unilateral and voluntary actions, whereas most adaptation involves private actions of affected entities, public arrangements of impacted communities, and national policies [18.1, 18.7].

Creating synergies between adaptation and mitigation can increase the cost-effectiveness of actions and make them more attractive to stakeholders, including potential funding agencies (medium confidence).

Analysis of the inter-relationships between adaptation and mitigation may reveal ways to promote the effective implementation of adaptation and mitigation actions together [18.5]. However, such synergies provide no guarantee that resources are used in the most efficient manner when seeking to reduce the risks to climate change [18.7]. In addition, the absence of a relevant knowledge base and of human, institutional and organisational capacity can limit the ability to create synergies. Opportunities for synergies are greater in some sectors (e.g., agriculture and forestry, buildings and urban infrastructure) but are limited in others (e.g., coastal systems, energy, health). A lack of both conceptual and empirical information that explicitly considers both adaptation and mitigation makes it difficult to assess the need for and potential of synergies in climate policy [18.3, 18.4, 18.8].

It is not yet possible to answer the question as to whether or not investment in adaptation would buy time for mitigation (high confidence).

Understanding the specific economic trade-offs between the immediate localised benefits of adaptation and the longer-term global benefits of mitigation requires information on the actions' costs and benefits over time. Integrated assessment models provide approximate estimates of relative costs and benefits at highly aggregated levels, but only a few models include feedbacks from impacts. Intricacies of the inter-relationships between adaptation and mitigation become apparent at the more detailed analytical and implementation levels [18.4, 18.5, 18.6]. These intricacies, including the fact that specific adaptation and mitigation options operate on different spatial, temporal and institutional scales and involve different actors with different interests, beliefs, value systems and property rights, present a challenge to designing and implementing decisions based on economic trade-offs beyond the local scale. In particular the notion of an 'optimal mix' of adaptation and mitigation is difficult to make operational, because it requires the reconciliation of welfare impacts on people living in different places and at different points in time into a global aggregate measure of well-being. [18.4, 18.7]

People's capacities to adapt and mitigate are driven by similar sets of factors (high confidence).

These factors represent a generalised response capacity that can be mobilised for both adaptation and mitigation. Response capacity, in turn, is dependent on the societal development path chosen. Enhancing society's response capacity through the pursuit of sustainable development is therefore one way of promoting both adaptation and mitigation [18.6]. This would facilitate the effective implementation of both options, as well as their mainstreaming into sectoral planning and development. If climate policy and sustainable development are to be pursued in an integrated way, then it will be important not simply to evaluate specific policy options that might accomplish both goals but also to explore the determinants of response capacity that underlie those options as they relate to underlying socio-economic and technological development paths [18.6, 18.7].

18.1 Introduction

The United Nations Framework Convention on Climate Change (UNFCCC) identifies two responses to climate change: mitigation of climate change by reducing greenhouse-gas emissions and enhancing sinks, and adaptation to the impacts of climate change. Most industrialised countries have committed themselves, as signatories to the UNFCCC and the Kyoto Protocol, to adopting national policies and taking corresponding measures on the mitigation of climate change and to reducing their overall greenhouse-gas emissions (United Nations, 1997). An assessment of current efforts aimed at mitigating climate change, as presented by the Working Group III Fourth Assessment Report (WGIII AR4), Chapter 11 (Barker et al., 2007), shows that current commitments would not lead to a stabilisation of atmospheric greenhouse-gas concentrations. In fact, according to the Working Group I Fourth Assessment Report (WGI AR4), owing to the lag times in the global climate system, no mitigation effort, no matter how rigorous and relentless, will prevent climate change from happening in the next few decades (Christensen et al., 2007; Meehl et al., 2007). Chapter 1 in this volume shows that the first impacts of climate change are already being observed.

Adaptation is therefore unavoidable (Parry et al., 1998). Chapter 17 (see Section 17.2 and Section 17.4) presents examples of adaptations to climate change that are currently being observed, but concludes that there are limits and barriers to effective adaptation. Even if these limits and barriers were to be removed, however, reliance on adaptation alone is likely to lead to a magnitude of climate change in the long run to which effective adaptation is no longer possible or only at very high social, economic and environmental costs. For example, Tol et al. (2006) show what would be the difficulties in adapting to a five-metre rise in sea level in Europe. It is therefore no longer a question of whether to mitigate climate change or to adapt to it. Both adaptation and mitigation are now essential in reducing the expected impacts of climate change on humans and their environment.

18.1.1 Background and rationale

Traditionally the primary focus of international climate policy has been on the use and production of energy. This policy focus was reflected in the Second Assessment Report (SAR), which, in discussing mitigation, paid relatively little attention to greenhouse gases other than CO_2 and to the potential for enhancing carbon sinks. Likewise, it paid little heed to adaptation. Since the publication of the SAR, the international climate policy community has become aware that energy policy alone will not suffice in the quest to control climate change and limit its impacts. Climate policy is being expanded to consider a wide range of options aimed at sequestering carbon in vegetation, oceans and geological formations, at reducing the emissions of non-CO_2 greenhouse gases, and at reducing the vulnerability of sectors and communities to the impacts of climate change by means of adaptation. Consequently, the Third Assessment Report (TAR) provided a more balanced treatment of adaptation and mitigation.

The TAR demonstrated that the level of climate-change impacts, and whether or not this level is dangerous (see Article 2 of the UNFCCC), is determined by both adaptation and mitigation efforts (Smith et al., 2001). Adaptation can be seen as direct damage prevention, while mitigation would be indirect damage prevention (Verheyen, 2005). However, only recently have policy-makers expressed an interest in exploring inter-relationships between adaptation and mitigation. Recognising the dual need for adaptation and mitigation, as well as the need to explore trade-offs and synergies between the two responses, they are faced with an array of questions (GAIM Task Force, 2002; Clark et al., 2004; see also Figure 18.1). How much adaptation and mitigation would be optimal, when, and in which combination? Who would decide, and based on what criteria? Are adaptation and mitigation substitutes or are they complementary to one another? When and where is it best to invest in adaptation, and when and where in mitigation? What is the potential for creating synergies between the two responses? How do their costs and effectiveness vary over time? How do the two responses affect, and how are they affected by, development pathways? These are some of the questions that have led the IPCC to include this chapter on inter-relationships between adaptation and mitigation in its Fourth Assessment Report (AR4).

The relevant literature to date does not provide clear answers to the above questions. Research on adaptation and mitigation has been rather unconnected to date, involving largely different communities of scholars who take different approaches to analyse the two responses. The mitigation research community has focused strongly, though not exclusively, on technological and economic issues, and has traditionally relied on 'top-down' aggregate modelling for studying trade-offs inherent in

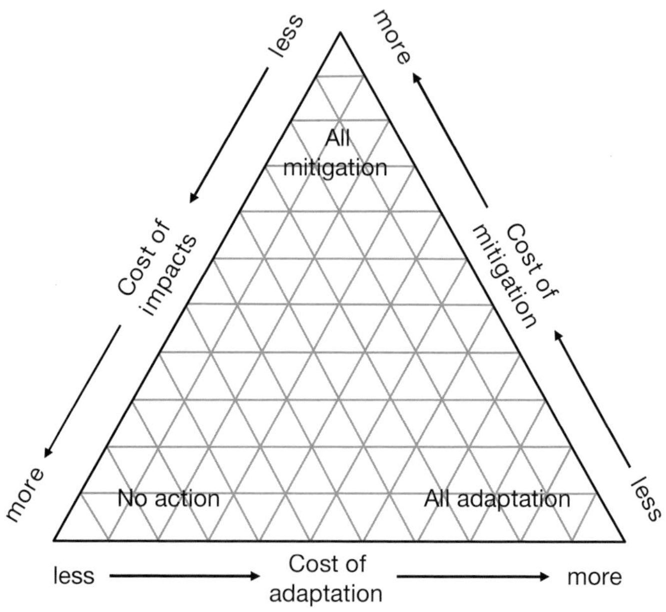

Figure 18.1. *A schematic overview of inter-relationships between adaptation, mitigation and impacts, based on Holdridge's life-zone classification scheme (Holdridge, 1947, 1967; M.L. Parry, personal communication).*

mitigation (see the WGIII AR4 (IPCC, 2007)). After a period of conceptual introspection, the adaptation research community has put its emphasis on local and place-based analysis: a research approach it shares with scholars in development studies and disaster risk reduction (Adger et al., 2003; Pelling, 2003; Smith et al., 2003; see also Chapter 17). In addition, adaptation is studied at the sectoral level (see Chapters 3 to 8).

One important research effort that does consider both adaptation and mitigation is integrated assessment modelling. Integrated assessment models (IAMs) typically combine energy models and sectoral impact models with climate, land-use and socio-economic scenarios to analyse and compare the costs and benefits of climate change and climate policy to society (see also Chapter 2). However, climate policy in IAMs to date is dominated by mitigation; adaptation, when considered, is either represented as a choice between a number of technological options or else it follows from assumptions in the model about social and economic development (Schneider, 1997; Corfee-Morlot and Agrawala, 2004; Fisher et al., 2007).

New research on inter-relationships between adaptation and mitigation includes conceptual and policy analysis, as well as 'bottom-up' studies that analyse specific inter-relationships and their implications for sectors and communities. The latter studies often place the implementation of adaptation and mitigation within the context of broader development objectives (e.g., Tompkins and Adger, 2005; Robinson et al., 2006; Chapters 17 and 20). They complement integrated assessment modelling by studying the factors and processes that determine if and when adaptation and mitigation can be synergistic in climate policy. Owing to it being a new research field, the amount of literature is still small, although it is growing fast. At the same time, the literature is very diverse: there is no consensus as to whether or not exploiting inter-relationships between adaptation and mitigation is possible, much less desirable. Some analysts (e.g., Venema and Cisse, 2004; Goklany, 2007) see potential for creating synergies between adaptation and mitigation, while others (e.g., Klein et al., 2005) are more sceptical about the benefits of considering adaptation and mitigation in tandem.

The differences in approaches between adaptation and mitigation research, and between integrated assessment modelling and 'bottom-up' studies, can create confusion when findings published in the literature appear to be inconsistent with one another. In assessing the literature on inter-relationships between adaptation and mitigation, this chapter does not hide any differences and inconsistencies that may exist between relevant publications. As artefacts of the research approaches that have emerged as described above, these differences and inconsistencies reflect the current state of knowledge. To provide as much clarity as possible from the outset definitions of important concepts are provided in Box 18.1. Next, Section 18.1.2 summarises important differences, similarities and complementarities between adaptation and mitigation.

Box 18.1. Definitions of terms

This box presents chapter-specific definitions of a number of (often related) terms relevant to the assessment of inter-relationships between adaptation and mitigation. Unless indicated otherwise, the definitions are specialisations of standard definitions found in reputable online dictionaries (e.g., http://www.m-w.com/, http://www.thefreedictionary.com/).

Trade-off: A balancing of adaptation and mitigation when it is not possible to carry out both activities fully at the same time (e.g., due to financial or other constraints).

Synergy: The interaction of adaptation and mitigation so that their combined effect is greater than the sum of their effects if implemented separately.

Substitutability: The extent to which an agent can replace adaptation by mitigation or *vice versa* to produce an outcome of equal value.

Complementarity: The inter-relationship of adaptation and mitigation whereby the outcome of one supplements or depends on the outcome of the other.

Optimality: The condition of being the most desirable that is possible under an expressed or implied restriction.

Portfolio: A set of actions to achieve a particular goal. A climate policy portfolio may include adaptation, mitigation, research and technology development, as well as other actions aimed at reducing vulnerability to climate change.

Mainstreaming: The integration of policies and measures to address climate change in ongoing sectoral and development planning and decision-making, aimed at ensuring the sustainability of investments and at reducing the sensitivity of development activities to current and future climatic conditions (Klein et al., 2005).

18.1.2 Differences, similarities and complementarities between adaptation and mitigation

The TAR used the following definitions of climate change mitigation and adaptation.

- **Mitigation:** An anthropogenic intervention to reduce the sources or enhance the sinks of greenhouse gases (IPCC, 2001a).
- **Adaptation:** Adjustment in natural or human systems in response to actual or expected climatic stimuli or their effects, which moderates harm or exploits beneficial opportunities (IPCC, 2001a).

It follows from these definitions that mitigation reduces all impacts (positive and negative) of climate change and thus reduces the adaptation challenge, whereas adaptation is selective; it can take advantage of positive impacts and reduce negative ones (Goklany, 2005).

The two options are implemented on the same local or regional scale, and may be motivated by local and regional priorities and interests, as well as global concerns. Mitigation has global benefits (ancillary benefits might be realised at the local/regional level), although effective mitigation needs to involve a sufficient number of major greenhouse-gas emitters to foreclose leakage. Adaptation typically works on the scale of an impacted system, which is regional at best, but mostly local (although some adaptation might result in spill-overs across national boundaries, for example by changing international commodity prices in agricultural or forest-product markets). Expressed as CO_2-equivalents, emissions reductions achieved by different mitigation actions can be compared and if the costs of implementing the actions are known, their cost-effectiveness can be determined and compared (Moomaw et al., 2001). The benefits of adaptation are more difficult to express in a single metric, impeding comparisons between adaptation efforts. Moreover, as a result of the predominantly local or regional effect of adaptation, benefits of adaptation will be valued differently depending on the social, economic and political contexts within which they occur (see Chapter 17).

The benefits of mitigation carried out today will be evidenced in several decades because of the long residence time of greenhouse gases in the atmosphere (ancillary benefits such as reduced air pollution are possible in the near term), whereas many adaptation measures would be effective immediately and yield benefits by reducing vulnerability to climate variability. As climate change continues, the benefits of adaptation (i.e., avoided damage) will increase over time. Thus there is a delay between incurring the costs of mitigation and realising its benefits from smaller climate change, while the time span between expenditures and returns of adaptation is usually much shorter. This difference is augmented in analyses adopting positive discount rates. These asymmetries have led to a situation whereby the initiative for mitigation has tended to stem from international agreements and ensuing national public policies (sometimes supplemented by community-based or private-sector initiatives), whereas the bulk of adaptation actions have historically been motivated by the self-interest of affected private actors and communities, possibly facilitated by public policies.

There are a number of ways in which adaptation and mitigation are related at different levels of decision-making. Mitigation efforts can foster adaptive capacity if they eliminate market failures and distortions, as well as perverse subsidies that prevent actors from making decisions on the basis of the true social costs of the available options. At a highly aggregated scale, mitigation expenditures appear to divert social or private resources and reduce the funds available for adaptation, but in reality the actors and budgets involved are different. Both options change relative prices, which can lead to slight adjustments in consumption and investment patterns and thus to changes in the affected economy's development pathway, but direct trade-offs are rare. The implications of adaptation can be both positive and negative for mitigation. For example, afforestation that is part of a regional adaptation strategy also makes a positive contribution to mitigation. In contrast, adaptation actions that require increased energy use from carbon-emitting sources (e.g., indoor cooling) would affect mitigation efforts negatively.

18.1.3 Structure of the chapter

Based on the available literature and our current understanding of differences, similarities and complementarities between adaptation and mitigation (see Section 18.1.2), this chapter distinguishes between four types of inter-relationships between adaptation and mitigation:

- Adaptation actions that have consequences for mitigation,
- Mitigation actions that have consequences for adaptation,
- Decisions that include trade-offs or synergies between adaptation and mitigation,
- Processes that have consequences for both adaptation and mitigation.

The chapter is structured as follows. Section 18.2 summarises the knowledge relevant to this chapter that was presented in the TAR. Section 18.3 frames the challenge of deciding when, how much, and how to adapt and mitigate as a decision-theoretical problem, and introduces the differing roles and responsibilities of stakeholders and the scales on which they operate. Section 18.4 then assesses the existing literature on trade-offs and synergies between adaptation and mitigation, including the potential costs of and damage avoided by adaptation and mitigation, as well as regional and sectoral aspects. Following the above typology of inter-relationships, Section 18.5 provides examples of complementarities and differences as they appear from the literature, thus providing an assessment of possible elements of a climate policy portfolio. Section 18.6 presents adaptation and mitigation within the context of development pathways, thus providing the background against which policy-makers and practitioners operate when acting on climate change. Section 18.7 assesses the literature on elements for effective implementation of climate policy that relies on inter-relationships between adaptation and mitigation. Finally, Section 18.8 outlines information needs of climate policy and priorities for research.

18.2 Summary of relevant knowledge in the IPCC Third Assessment Report

Compared to the SAR, two of the Working Groups preparing the TAR were restructured. The scope assigned to Working Group II (WGII) was limited to impacts of climate change on sectors and regions and to issues of vulnerability and adaptation, while Working Group III (WGIII) was commissioned to assess the technological, economic, social and political aspects of mitigation. Whereas there were concerted efforts to assess links of both adaptation and mitigation to sustainable development (see Chapter 20, Section 20.7.3), there was little room to consider the direct relationships between these two domains. The integration of results and the development of policy-oriented synthesis were therefore difficult (Toth, 2003).

The attempt to establish the foundations of the TAR Synthesis Report (IPCC, 2001a) in the final chapters of WGII and WGIII did not shed light on inter-relationships between adaptation and mitigation. The WGII TAR in Chapter 19 presented "reasons for concern about projected climate change impacts" in a summary figure that outlines the risks associated with different magnitudes of warming, expressed in terms of the increase in global mean temperature. Largely based on IAMs, the WGIII TAR in Chapter 10 summarised the costs of stabilising CO_2 concentrations at different levels. These two summaries are difficult to compare because questions as to what radiative-forcing and climate-sensitivity parameters should be used to bridge the concentration-temperature gap remain unanswered. Moreover, many statements in the two Working Group Reports were themselves distilled from a large number of reviewed studies. Yet the generic assumptions underlying the methods, the specific assumptions of the applications, the selected baseline values for the scenarios, incompatible discount rates, economic growth assumptions and many other postulations implicit in the parameterisation of adaptation and mitigation assessments were largely ignored or remained hidden in the Synthesis Report.

Nonetheless, the TAR presented new concepts for addressing inter-relationships between adaptation and mitigation. Local adaptive and mitigative capacities vary significantly across regions and over time. Superficially they appear to be strongly correlated because they share the same list of determinants. However, aggregate representation across nations or social groups of both adaptation and mitigation is misleading because the capacity to reduce emissions of greenhouse gases and the ability to adapt to it can deviate significantly. As the TAR pointed out: "one country can easily display high adaptive capacity and low mitigative capacity simultaneously (or *vice versa*)" (IPCC, 2001b; see also Yohe, 2001). In a wealthy nation, damages of climate change may fall on a small but influential social group and the costs of adaptation can be distributed across the entire population through the tax system. Yet, in the same country, another small group might be hurt by mitigation policies without the possibility to spread this burden. In addition to the conceptual deliberations, the TAR discussed inter-relationships between adaptation and mitigation at two levels: at the aggregated, global and national levels, and in the context of economic sectors and specific projects.

The WGII report pointed out that "adaptation is a necessary strategy at all scales to complement climate change mitigation efforts" (IPCC, 2001c), but also elaborates the complex relationships between the two domains at various levels. Some relationships are synergistic, while others are characterised by trade-offs. The report noted the arguments in the literature about the trade-off between adaptation and mitigation because resources committed to one are not available for the other, and also noted that this is "debatable in practice because the people who bear emissions reduction costs or benefits often are different from those who pay for and benefit from adaptation measures" (IPCC, 2001c). From the dynamic perspective, "climatic changes today still are relatively small, thus there is little need for adaptation, although there is considerable need for mitigation to avoid more severe future damages. By this logic, it is more prudent to invest the bulk of the resources for climate policy in mitigation, rather than adaptation" (IPCC, 2001c). Yet, as the WGIII TAR noted, one has to bear in mind the intergenerational trade-offs. The impacts of today's climate change investments on future generations' opportunities should also be considered. Investments might enhance the capacity of future generations to adapt to climate change, but at the same time may displace investments that could create other opportunities for future generations (IPCC, 2001b).

Chapter 10 of the WGIII TAR outlined the iterative process in which nations balance their own mitigation burden against their own adaptation and damage costs. "The need for, extent and costs of adaptation measures in any region will be determined by the magnitude and nature of the regional climate change driven by shifts in global climate. How global climate change unfolds will be determined by the total amount of greenhouse-gas emissions that, in turn, reflects nations' willingness to undertake mitigation measures. Balancing mitigation and adaptation efforts largely depends on how mitigation costs are related to net damages (primary or gross damage minus damage averted through adaptation plus costs of adaptation). Both mitigation costs and net damages, in turn, depend on some crucial baseline assumptions: economic development and baseline emissions largely determine emissions reduction costs, while development and institutions influence vulnerability and adaptive capacity" (IPCC, 2001b).

Discussions of inter-relationships between adaptation and mitigation are sparser at the sector/project level. Some chapters in the WGII TAR noted the link to mitigation when discussing climate-change impacts and adaptation in selected sectors, primarily those related to land use, agriculture and forestry. Chapter 5 noted that "afforestation in agroforestry projects designed to mitigate climate change may provide important initial steps towards adaptation" (Gitay et al., 2001). Chapter 8 emphasised sustainable forestry, agriculture and wetlands practices that yield benefits in watershed management and flood/mudflow control but involve trade-offs such as wetlands restoration helping to protect against flooding and coastal erosion, but in some cases increasing methane release (Vellinga et al., 2001).

The WGII TAR in Chapter 12 observed the complexities in land management in Australia and New Zealand "where control of land degradation through farm and plantation forestry is being considered as a major option, partly for its benefits in controlling salinisation and waterlogging, and possibly as a new economic option with the advent of incentives for carbon storage as a greenhouse mitigation measure" (IPCC, 2001c). Chapter 15 mentioned soil conservation practices (e.g., no tillage, increased forage production, higher cropping frequency) implemented as mitigation strategies in North America (Cohen et al., 2001). It observed that the Kyoto Protocol mentions human-induced land-use changes and forestry activities (afforestation, reforestation, deforestation) as sinks of greenhouse gases for which sequestration credits can be claimed, and that agricultural sinks may be considered in the future. The market emerging in North America to enhance carbon sequestration leads to land-management decisions with diverse effects. The negative consequences of reduced tillage implemented to enhance soil carbon sequestration include the increased use of pesticides for disease, insect and weed management; capturing carbon in labile forms that are vulnerable to rapid oxidation if the system is changed; and reduced yields and cropping management options and increased risk for farmers. The beneficial consequences of reduced tillage (especially no-till) are reduced input costs (e.g., fuel) for farmers, increased soil moisture and hence reductions in crop-water stress in dry areas, reduction in soil erosion and improved soil quality (IPCC, 2001c).

In chapters dealing with other sectors affected by climate-change impacts and mitigation, less attention was paid to their inter-relationships. The WGII TAR in Chapter 8 mentioned energy end-use efficiency in buildings having both adaptation and mitigation benefits, as improved insulation and equipment efficiency can reduce the vulnerability of structures to extreme temperature episodes and emissions. An example of the more remote inter-relationships between adaptation and mitigation across space and time was provided by Chapter 17. Small island states are recognised to be vulnerable to climate change and tourism is a major source of income for many of them. While, over the long term, milder winters in their current markets could reduce the appeal of these islands as tourist destinations, they could be even more severely harmed by increased airline fares "if greenhouse gas mitigation measures (e.g., levies and emissions charges) were to result in higher costs to airlines servicing routes between the main markets and small island states" (IPCC, 2001c).

Finally, the WGII TAR in Chapter 8 drew attention to a link between adaptation and mitigation in the Kyoto Protocol that establishes a surcharge ('set-aside') on mitigation activities implemented as Clean Development Mechanism (CDM) projects. "One key issue is the size of the 'set-aside' from CDM projects that is dedicated to funding adaptation. If this set-aside is too large, it will make otherwise viable mitigation projects uneconomic and serve as a disincentive to undertake projects. This would be counterproductive to the creation of a viable source of funding for adaptation" (IPCC, 2001c).

18.3 Decision processes, stakeholder objectives and scale

A portfolio of actions is available for reducing the risks of climate change, within which each option requires evaluation of its individual and collective merits. Decision-makers at all levels need to decide on appropriate near-term actions in the face of the many long-term uncertainties and competing pressures, goals and market signals. Section 18.1 identified four types of inter-relationships between adaptation and mitigation. Investments in mitigation may have consequences for adaptation; and investments in adaptation may have consequences for the emission of greenhouse gases. At the highest level of aggregation, adaptation and mitigation are both policy substitutes and policy complements, and may compete for finite resources. However, this need not be the case: both adaptation and mitigation may be considered in a policy process without invoking trade-offs, often in the context of broader considerations of sustainable development. This section introduces the nature of the decision problem followed by a review of stakeholder objectives, risk and scales.

18.3.1 The nature of the decision problem

It is difficult, and perhaps counterproductive, to explore the pay-offs from various types of investments without a conceptual framework for thinking about their interactions. Decision analysis provides one such framework (Raiffa, 1968; Keeney and Raiffa, 1976) that allows for the systematic evaluation of near-term options in light of the careful consideration of the potential consequences (see Lempert et al., 2004; IPCC, 2007; Keller et al., 2007; Nicholls et al., 2007; Chapter 20). The next several decades will require a series of decisions on how best to reduce the risks from climate change. There will be, no doubt, opportunities for learning and mid-course corrections. The immediate challenge facing policy-makers is to find out which actions are currently appropriate and likely to be robust in the face of the many long-term uncertainties.

The climate-policy decision tree can be represented as points at which decisions are made, and the reduction of uncertainty in the outcomes (if any) in a wide range of possible decisions and outcomes. The first decision node represents some of today's investment options. How much should we invest in mitigation, how much in adaptation? How much should be invested in research? Once we act, we have an opportunity to learn and make mid-course corrections. The outcomes include types of learning that will occur between now and the next set of decisions. The outcomes are uncertain; the uncertainty may not be resolved but there will be new information which may influence future actions. Hence the expression: "act, then learn, and then act again" (Manne and Richels, 1992).

The 'act, then learn, then act again' framework is used here solely to lay out the elements of the decision problem and not as an alternative to the many analytical approaches discussed in this Report. Indeed, it can be used to parse various approaches for descriptive purposes, such as deterministic versus probabilistic approaches and cost-effectiveness analysis versus cost-benefit analysis. Decision analysis has been more widely applied to mitigation than to adaptation, although a robust decision framework is suitable for analysing the array of future vulnerabilities to climate change (Lempert and Schlesinger, 2000; Lempert et al., 2004).

18.3.2 Stakeholder roles and spatial and temporal scales

Climate change engages a multitude of decision-makers, both spatially and temporally. The UNFCCC, its subsidiary bodies and Member Parties have largely focused on mitigation. More recently, an increasing interest at the grassroots level has yielded additional local mitigation activities. Adaptation decisions embrace both the public and private sector, as some decisions involve large construction projects in the hands of public-sector decision-makers while other decisions are localised, involving many private-sector agents.

The roles of various stakeholders cover different aspects of inter-relationships between adaptation and mitigation. Stakeholders may be characterised according to their organisational structure (e.g., public or private), level of decision-making (e.g., policy, strategic planning, or operational implementation), spatial scale (e.g., local, national or international), time-frame of concern (e.g., near term to long term), and function within a network (e.g., single actor, stakeholder regime or multi-level institution). Decisions might cover adaptation only, mitigation only, or link adaptation and mitigation. Relatively few public or corporate decision-makers have direct responsibility for both adaptation and mitigation (e.g., Michaelowa, 2001). For example, adaptation might reside in a Ministry of Environment while mitigation policy is led by a Trade, Energy or Economic Ministry. Local authorities and land-use planners often cover both adaptation and mitigation (ODPM, 2004).

Stakeholders are exposed to a variety of risks, including financial, regulatory, strategic, operational, or to their reputations, physical assets, life and livelihoods (e.g., IRM et al., 2002). Decision-making may be motivated by climatic risks or climate change (e.g., climate-driven, climate-sensitive, climate-related) although many decisions related to adaptation and mitigation are not driven by climate change (Watkiss et al., 2005). Risk is commonly defined as the probability times the consequence, while uncertainty is often taken to represent structural and behavioural factors that are not readily captured in probability distributions (e.g., Tol, 2003; Stainforth et al., 2005). Although this distinction between risk and uncertainty is simplistic (see Dowie, 1999), stakeholder decision-making takes account of many factors (Newell and Pizer, 2000; Bulkeley, 2001; Clark et al., 2001; Gough and Shackley, 2001; Rayner and Malone, 2001; Pidgeon et al., 2003; Kasperson and Kasperson, 2005; Moser, 2005): values, preferences and motivations; awareness and perception of climate change issues; negotiation, bargaining and social norms; analytical frameworks, information and monitoring systems; and relationships of power and politics.

Faced with the deep uncertainty of climate change (Manne and Richels, 1992), stakeholders may adopt a precautionary approach with the intention of stimulating technological (if not social) change, rather than seeking to explicitly balance costs and benefits (Harvey, 2006). For instance, estimates of the social cost of carbon, one measure of the benefits of mitigation, are sensitive to the choice of decision framework (including equity weighting, risk aversion, sustainability considerations and discount rates for future damages) (Downing et al., 2005; Tol, 2005b; Watkiss et al., 2005; Guo et al., 2006; Fisher et al., 2007; see also Section 18.4.2; Chapter 20).

Criteria relating to either mitigation or adaptation, or both, are increasingly common in decision-making. For example, local development plans might screen housing developments according to energy use, water requirements and preservation of green belt (e.g., CAG Consultants and Oxford Brookes University, 2004). Development agencies have begun to screen their projects for relevance to adaptation and mitigation (e.g., Burton and van Aalst, 1999; Klein, 2001; Eriksen and Næss, 2003).

Many stakeholders link climate, development and environmental policies by, for example, linking energy efficiency (related to mitigation) to sustainable communities or poverty reduction (related to adaptation). For example, the World Bank's BioCarbon Fund and Community Development Carbon Fund include provision for buyers to ensure that carbon offsets also achieve development objectives (World Bank, undated). The Gold Standard for CDM projects also ensures that projects support sustainable development (Carbon International, undated). Preliminary work suggests that there may be a modest trade-off between cost-effective emissions reductions and the achievement of other sustainable development objectives; that is, more expensive projects per emissions reduction unit tend to contribute more to sustainable development than cheaper projects (Nagai and Hepburn, 2005).

The nature of adaptation and mitigation decisions changes over time. For example, mitigation choices have begun with relatively easy measures such as adoption of low-cost supply and demand-side options in the energy sector (such as passive solar) (see Levine et al., 2007). Through successful investment in research and development, low-cost alternatives should become available in the energy sector, allowing for a transition to low-carbon-venting pathways. Given the current composition of the energy sector, this is unlikely to happen overnight but rather through a series of decisions over time. Adaptation decisions have begun to address current climatic risks (e.g., drought early-warning systems) and to be anticipatory or proactive (e.g., land-use management). With increasing climate change, autonomous or reactive actions (e.g., purchasing air-conditioning during or after a heatwave) are likely to increase. Decisions might also break trends, accelerate transitions and mark substantive jumps from one development or technological pathway to another (e.g., Martens and Rotmans 2002; Raskin et al., 2002a, b).

Inter-relationships between adaptation and mitigation also vary according to spatial and social scales of decision-making. Adaptation and mitigation may be seen as substitutes in a policy framework at a highly aggregated, international scale: the more mitigation is undertaken, the less adaptation is necessary and *vice versa*. Resources devoted to mitigation might impede socio-economic development and reduce investments in adaptive capacity and adaptation projects (e.g., Kane and Shogren, 2000). This scale is inherent in the analysis of global targets (see Section 18.4).

National and sub-national decision-making is often a mixture of policy and strategic planning. The adaptation-mitigation trade-off is problematic at this scale because the effectiveness of mitigation outlays in terms of averted climate change depends on the mitigation efforts of other major greenhouse-gas emitters. However, adaptation criteria can be applied to mitigation projects or *vice versa* (Dang et al., 2003). A national policy example of synergies might be a new water law that requires metered use, enabling water companies to adjust their charges in anticipation of scarcity and conserve energy through demand-side measures. This policy would then be implemented in strategic plans by water companies and environment agencies at a sub-national level.

On the operational scale of specific projects, there may be trade-offs or synergies between adaptation and mitigation. However, the majority of projects are unlikely to have strong links, although this remains as a key uncertainty. Certainly there are many adaptive actions that have consequences for mitigation, and mitigation actions with consequences for adaptation.

The inter-relationships between adaptation and mitigation also cross scales (Rosenberg and Scott, 1995; Cash and Moser, 2000; Young, 2002). A policy framework is often seen as essential in driving strategic investment and operational projects (e.g., Grubb et al., 2002; Grubb, 2003) for technological innovation. Operational experience can be a precursor to developing sound strategies and policies (one of the motivations for early corporate experiments in carbon trading). In many cases the results of action at one scale have implications at another scale (e.g., local adaptation decisions that increase greenhouse-gas emissions, or national carbon taxes that change local resource use).

18.4 Inter-relationships between adaptation and mitigation and damages avoided

This section presents the main insights emerging from global integrated assessments implemented in different decision-analytical frameworks on trade-offs and synergies between adaptation and mitigation and on avoided damages. This is complemented by lessons from regional and sectoral studies. Principles and technical details of the methods used by the studies reported here are presented in Chapter 2.

18.4.1 Trade-offs and synergies in global-scale analysis

Analysts working on global-scale climate analyses remain apart in their formulation of the inter-relationships between adaptation and mitigation. Some consider them as substitutes and seek the optimal policy mix, while others emphasise the diversity of impacts (with little scope for adaptation in some sectors) and the asymmetry of social actors who need to mitigate versus those who need to adapt (Tol, 2005a). Yet others maintain that adaptation is the only available option for reducing climate-change impacts in the short to medium term, while the long term has a mix of adaptation and mitigation (Goklany, 2007). Note that these positions are not contradictory; they just emphasise different aspects of the same problem.

Cost-benefit analyses (CBAs) are phrased as the trade-off between mitigation costs, on the one hand, and adaptation costs and residual damages on the other. As a recent example, Nordhaus (2001) estimates the economic impact of the Kyoto-Bonn Accord with the RICE-2001 model. Without the participation of the USA, the resulting emissions path remains below the efficient reduction policy (which equates estimated marginal costs and benefits of emissions reductions) whereas the original Kyoto Protocol implied abatement that is more stringent than would be suggested by this CBA. Note that RICE-2001, like all models, has assumptions, simplifications and abstractions that affect the results. Nonetheless, this is a common finding in the cost-benefit literature, driven primarily by relatively low estimates of the marginal damage costs (Tol, 2005b). Cost-benefit models are recognised by many as sources of guidance on the magnitude and rate of optimal climate policy (for a wide range of definitions of what is 'optimal' see Azar, 1998; Brown, 1998; Tol, 2001, 2002; Chapter 2), while others criticise them for ignoring the sectoral (economic and social), spatial and temporal distances between those who need to mitigate versus those who need to adapt to climate change. CBA requires conversion of many different damages to a common metric through monetisation, for example, by polling people's values of different benefits, and the use of discount rates, which is controversial over long time-scales like those of climate change but common practice for other issues. Discounting implies that long-time-scale Earth-system transitions, such as melting of ice sheets, slowdown of the thermohaline circulation or the release of methane, have small weight in a CBA and therefore tend to attach little weight to adaptation costs (see also Chapter 17).

CBA is a special form of multi-criteria analysis. In both cases, policies are judged on multiple criteria, but in CBA all are monetised, while multi-criteria analyses use a range of mathematical methods to make trade-offs explicit and resolve them. Multi-criteria analysis has relatively few applications to climate policy (e.g., Bell et al., 2003; Borges and Villavicencio, 2004), although it is more common for adaptation (e.g., the National Adaptation Programmes of Action).

The Tolerable Windows Approach (TWA) adopts a different approach to integrating mitigation and impact/adaptation concerns and deals with adaptation indirectly in the applications. The ICLIPS (Integrated assessment of CLImate Protection Strategies) model identifies fields of long-term greenhouse-gas emissions paths that prevent rates and magnitudes of climate change leading to regional or sectoral impacts without imposing excessive mitigation costs on societies, either of which stakeholders might consider unacceptable or intolerable. This 'relaxed' cost-benefit framework can be used to explore trade-offs between climate change or impact constraints, on the one hand, and mitigation cost limits in terms of the existence and size of long-term emissions fields, on the other hand. For any given impact constraint, increasing the acceptable consumption loss due to emissions-abatement expenditures increases the emissions field and allows higher near-term emissions but involves higher mitigation rates and costs in later decades. Conversely, for any given mitigation cost limit, increasing the tolerated level of climate impact also enlarges the emissions field and allows higher near-term emissions (Toth et al., 2002, 2003a, b). This formulation allows the exploration of side-payments for enhancing adaptation in order to tolerate impacts from larger climate change. The TWA is helpful in exploring the feasibility and implications of crucial social decisions (acceptable impacts and mitigation costs) but, unlike CBA, it does not propose an optimal policy.

Cost-effectiveness analyses (CEAs) depict a rather remote relationship between adaptation and mitigation. They implicitly assume that some sort of a global climate change target can be agreed upon that would keep all climate-change impacts at the level that can be managed via adaptation or taken as 'acceptable losses'. Or, cost-effectiveness analyses consider a range of hypothetical targets, but remain silent on the appropriateness of these targets. Global CEAs have proliferated since the publication of the TAR (e.g., Edmonds et al., 2004). In addition to exploring least-cost strategies to stabilise CO_2 concentrations, CEAs are applied to analysing the stabilisation of radiative forcing (e.g., Van Vuuren et al., 2006) and global mean temperature (Richels et al., 2004). While most analyses are deterministic in the sense that they implicitly assume that we know the true state of the world, there is also a body of literature that models the 'act, then learn, then act again' nature of the decision problem, but primarily for mitigation decisions. See the WGIII AR4 Chapter 3 for details (Fisher et al., 2007).

The competition of adaptation measures, mitigation measures and non-climate policies for a finite budget has not been studied in much detail. Schelling (1995) questions whether the money that developed countries' governments plan to spend on greenhouse-gas emissions reduction, ostensibly to the benefit of the children and grandchildren of the people in developing countries, cannot be spent to greater benefit. As a partial answer to that question, Tol (2005c) concluded that development aid is a better mechanism to reduce climate-change impacts on infectious disease (e.g., malaria, the best-studied health impact) than is emissions abatement. This analysis implies that the concern about increases in these infectious diseases is not a valid argument for greenhouse-gas emissions reduction (there are of course other arguments for abatement). The same study also shows that this result does not carry over to other impacts. More broadly, Goklany (2003, 2005) shows that the contribution of climate change to hunger, malaria, coastal flooding and water stress (as measured by the population at risk for these hazards) is usually small compared with the contribution of non-climate-change-related factors. He argues that, through the 2080s at least, efforts to reduce vulnerability would be far more cost-effective in reducing these problems than would any mitigation scheme. Other studies estimate the change in vulnerability to climate change due to emissions abatement; for instance, a shift to wind and water power or biofuels would reduce carbon dioxide emissions, but increase exposure to the weather and climate (e.g., Dang et al., 2003).

Some studies estimate the change in greenhouse-gas emissions due to adaptation to the impacts of climate change (Berrittella et al., 2006, for tourism; Bosello et al., 2006, for health). They find that emissions increase in some places and some sectors (making mitigation harder), and decrease elsewhere (making mitigation easier). The disaggregated effects are small compared with the projected growth in emissions, while the net effect is negligible. Similarly, Fankhauser and Tol (2005) show that the impact of climate change on the growth of the economy and greenhouse-gas emissions is small compared with the economy as a whole and because economic adjustment processes would dampen the impact. Note that they only include those climate-change impacts that affect economic performance; they do not use monetisation techniques. Fisher et al. (2006) reach a similar conclusion for population projections, because the net increase in mortality is small. As there are so few studies, focusing on a few sectors only, these conclusions are preliminary.

Although some industries (e.g., wind farm and solar panel manufacturing) may benefit, emissions reduction is likely to slow economic growth, but this effect is probably small if smart abatement policies are used (Weyant, 2004; Barker et al., 2007; Fisher et al., 2007). However, small economic losses in the member states of the Organisation for Economic Co-operation and Development (OECD) may be amplified in poor exporters of primary products (i.e., many African countries). Tol and Dowlatabadi (2001) use this mechanism to demonstrate an interesting trade-off between adaptation and mitigation. Taking malaria as a climate-related disease, they observe that countries with an average annual income per capita of US$3,000 or more do not report significant deaths from malaria and that all world regions surpass this threshold by 2085 in most IPCC IS92 scenarios (IPCC, 1992). Progressively more ambitious emissions reductions in OECD countries gradually decrease the cumulative malaria mortality if one considers only the impact side; that is, the biophysical effects of climate-change mitigation on malaria prevalence. However, if the economic effects of mitigation efforts (i.e., the slower rate of economic growth) are also taken into account, then, according to the FUND model, the malaria-mortality improvements due to slower global warming will be gradually eliminated and eventually surpassed by the losses due to the reduced rate of income growth, unless health care expenditures are decoupled from economic growth. Note that FUND has somewhat high costs of emissions reduction (see the SAR), and also assumes a large impact of slowed growth in the OECD on the rest of the world. Barker et al. (2002), Weyant (2004), Edenhofer et al. (2006), Köhler et al. (2006) and Van Vuuren et al. (2006) show that there is a wide range of estimates of mitigation impacts on economic growth, but these studies did not explore the link between mitigation and vulnerability. In fact, the impact of mitigation on adaptive capacity has not been studied with any other model. More generally, the capacity to adapt to climate change is related to development status, although the two are not the same (Yohe and Tol, 2002; Tompkins and Adger, 2005). The earlier studies used 'adaptive capacity' and 'development' in a generic and broad sense. Tol and Yohe (2006) use more specific indicators of adaptive capacity and development without changing the general conclusion. Emissions reduction policies that hamper development would increase vulnerability and could increase impacts (Tol and Yohe, 2006). Based on this contingency, Goklany (2000b) argues that aggressive mitigation would fall foul of the precautionary principle.

The literature assessed in this sub-section indicates that initial studies tended to focus on the relationship between mitigation and damages avoided, but our knowledge of this subject is still limited and more research needs to be undertaken. More recently, the literature has begun to focus on the relationship between adaptation and damages avoided. Ultimately, better knowledge about the interaction between adaptation and mitigation actions in terms of damages avoided would be useful. However, such research is at a very rudimentary stage. Moreover, large-scale modelling of adaptation-mitigation feedbacks is needed but still lacking. A necessary first step will be improved modelling of feedbacks from impacts, which is currently immature in most long-term global integrated assessment modelling. Adaptation modelling can follow with modelling structures that permit the reallocation of production factors and budgets in response to the changing climate. The adaptation responses therefore redefine the circumstances for mitigation. However, current impact modelling capability is undeveloped and modelling of adaptation responses to climate-change impacts has only just begun. In the above assessment we do not distinguish adaptation by actors (e.g., individuals, government departments) as the conclusions generally hold for all types of adaptation.

18.4.2 Consideration of costs and damages avoided and/or benefits gained

Various approaches have been taken since the TAR to estimate the size of climate change damages that can be avoided by emissions reduction. Among the global integrated assessments reviewed in the previous sub-section, cost-effectiveness models (by far the most widely used decision analysis framework) do not include impacts, hence they cannot measure avoided damages either. In contrast, CBAs of greenhouse-gas emissions reduction (e.g., Nordhaus, 2001) necessarily estimate the avoided damages of climate change but rarely report them. Economic assessments of marginal damage costs (e.g., the incremental impact of an additional tonne of carbon emissions) provide a means of comparing damages avoided with marginal abatement costs. Such studies typically cover a range of sectors and report damage functions and estimates for scenarios of climate change, and increasingly reference scenarios of socio-economic vulnerability.

Tol (2005b) reviewed the avoided-damage literature, including 103 estimates from 28 papers published from 1991 to 2003. Some of the reviewed estimates include only a few impacts; other estimates include a wide range of impacts, including low-probability/high-impact scenarios (see Chapter 20 for further discussion). Tol (2005b) finds that most studies (72% when quality-weighted) point to a marginal damage cost of less than US$50 per tonne carbon (/tC). He also finds a systematic, upward bias in the grey literature. For instance, the 95th percentile falls from US$350/tC to US$245/tC if estimates that were not peer-reviewed are excluded. For a 5% discount rate, a value used by many governments (Evans and Sezer, 2004), the median estimate is only US$7/tC; for a 3% discount rate, it is US$33/tC.

Downing et al. (2005) updated the Tol (2005b) analysis to a 2005 base year: the very likely range of estimates runs from –US$10 to +US$350/tC; peer-reviewed estimates have a mean value of US$43/tC with a standard deviation of US$83/tC. Incorporating results from FUND (2005 version) and PAGE2002, Downing et al. (2005) find that £35/tC (at year 2000 values, or US$56/tC) is a credible lower benchmark for the social cost of carbon (as identified by the UK Government in Clarkson and Deyes, 2002). In FUND, with the Green Book discounting scheme and equity weighting, there is about a 40% chance that the social cost of carbon exceeds £35/tC. Estimates of the central tendency (whether the average or median) or upper benchmark were not agreed in that assessment, due to the limitations in our knowledge of climate impacts and the critical role of the decision perspective (see Section 18.5).

Stern (2007), including a higher level of risk of adverse impacts that are poorly represented in existing models and accepting a public policy framework that includes low discounting of the future, reports a social cost of carbon of US$304/tC (US$85/tCO_2, at pounds sterling 2005 values) from the PAGE2002 model. The range of estimates is quite large and Stern (2007) acknowledges that his central estimate is higher than most studies and is "keenly aware of the sensitivity of estimates to the assumptions that are made".

Note that the estimates of avoided damages are highly uncertain. A survey of fourteen experts in estimating the social cost of carbon rated their estimates as low confidence, due to the many gaps in the coverage of impacts and valuation studies, uncertainties in projected climate change, choices in the decision framework and the applied discount rate (Downing et al., 2005).

The marginal damage cost only gives the value of the last unit of the damage avoided, not the total avoided damage, which is seldom estimated (see the literature review and papers in Corfee-Morlot and Agrawala, 2004). Nonetheless, as a first approximation of the avoided damages, one should multiply the tonnes of carbon emissions reduced by the marginal damage cost.

Several studies have attempted to calculate total economic damages from disparate impact studies. Warren (2006) reports a long list of ecosystem impacts at 2°C warming and below, billions of people at risk from water stress (without adaptation) and political tension in Russia. As the impact estimates are taken from different studies, with different models and different scenarios, this method introduces additional uncertainties: the difference in impact may be due to different warming scenarios, but also due to differences in models, data, economic scenarios and even subject and area of study. Furthermore, it is difficult to compare how impacts change with additional degrees of climate change, although the work does suggest that there are an increasing number of negative impacts at higher temperatures. Warren's (2006) study is often qualitative and it is unclear whether the studies are representative of the literature (or the population of affected sectors), or whether adaptation is included. On avoidable damage, this study paints a bleak picture. At 2°C warming, which may be difficult to avoid, 97% of coral reefs and 100% of Arctic sea ice would be lost. Avoided damage is therefore less than 3% of coral reefs, and no Arctic sea ice. Hare (2006) also offers impact estimates for various warming scenarios, with the same limitations as for Warren (2006). Hitz and Smith (2004) review damage functions related to global mean temperature but do not aggregate to overall damages. Arnell et al. (2002) and Parry et al. (2004) use internally consistent models and scenarios, and report numbers for avoided damages, measured in millions of people at risk. Water resources and malaria dominate their results, but the underlying models do not account for adaptation and keep socio-economic development at 1990 levels, although populations grow.

Relatively few studies have documented damages avoided in terms of specific mitigation scenarios. Bakkenes et al. (2006) study the implications of different stabilisation scenarios on European plant diversity. Mitigation is not considered, even though biofuels and carbon plantations would substantially affect vegetation. Under the A1B scenario, plants would lose on average 29% of their current habitat by 2100, with a range between species from 10% to 53%. Stabilisation at 650 ppm would limit this to 22% (6-42%), and at 550 ppm to 18% (5-37%). With unmitigated climate change, nine plant species would disappear from Europe, but eight new ones would appear. Stabilisation would limit the number of plant disappearances from nine to eight species. In all five studies, adaptation (except in some parts of the Parry study) and the effects of mitigation on impacts are not included (see Section 18.4.1). Nicholls and Lowe (2004) estimate the avoided impact of sea-level rise due to mitigation. Because sea level responds so slowly to global warming, avoided impacts are small, at least over the 21st century. Nicholls and Lowe (2004) ignore the costs of emissions reduction; Tol (2007) shows that the bias is negligible for coastal-zone impacts. Nicholls and Lowe (2004, 2006) argue that adaptation and mitigation should be applied together for coastal zones, with mitigation to minimise the future commitment to sea-level rise and adaptation to adapt to the inevitable changes. Nicholls and Tol (2006) and Nicholls et al. (2007) also explore the economic impacts of sea-level rise.

Tol and Yohe (2006), using the integrated assessment model, Climate Framework for Uncertainty, Negotiation and Distribution (FUND), conclude that the most serious impacts of climate change can be avoided at an 850 ppm CO_2-equivalent stabilisation target for greenhouse-gas concentrations, and that incrementally avoided damages get smaller and smaller as one moves to more stringent stabilisation targets. For a 450 ppm CO_2-equivalent stabilisation target, climate-change impacts may actually increase as the reduction of sulphur emissions may lead to warming and as abatement costs slow growth and increase vulnerability. However, FUND includes a wide range but not all impacts, represents impacts in a reduced form, does not capture discontinuities or interactions between impacts, models climate change as being smooth, and does not include the ancillary benefits of reductions in sulphur. Other models also find that climate policy would reduce sulphur emissions to levels below what is required for acidification policy (e.g., Van Vuuren et al., 2006). Other integrated assessment models have yet to produce comparable analyses.

Abatement may, but need not, reduce the probability of extreme climate scenarios, such as a shut-down of the thermohaline circulation (Gregory et al., 2005) and a collapse of the West Antarctic ice sheet (Vaughan and Spouge, 2002). The few studies on the effects of drastic sea-level rise show large impacts (Schneider and Chen, 1980; Nicholls et al., 2005; Tol et al., 2006) but opinions on the impacts of a thermohaline circulation shut-down are divided (Rahmstorf, 2000; Link and Tol, 2004).

Additional assessments of damages avoided by mitigation are also provided in other chapters of this report. Chapter 20 finds that estimates of the social cost of carbon expand over at least three orders of magnitude and notes that globally aggregated figures are likely to underestimate the full costs, masking differences in impacts across sectors and regions/countries. It concludes that "it is very likely that climate change will result in net costs into the future, aggregated across the globe and discounted to today; it is very likely that these costs will grow over time". The WGIII AR4 in Chapter 3 (Fisher et al., 2007) observes that most (but not all) analyses which use monetisation suggest that social costs of carbon are positive, but the range of values is wide and is strongly dependent on modelling methodology, value judgements and assumptions. It concludes that large uncertainties persist, related to the cost of mitigation, the efficacy of adaptation, and the extent to which the negative impacts of climate change, including those related to rate of change, can be avoided. See Box 18.2 for a summary of the WGIII AR4 conclusions on damages avoided with different stabilisation scenarios.

Overall, there are only a few studies that estimate the avoided impacts of climate change by emissions reduction. Some of these studies ignore adaptation and mitigation costs. Many published studies of damages in sectors that are quantified in economic models (but mostly market-based costs and related to incremental projections of temperature) and with discount rates commonly used in economic decision-making (e.g., 3% or higher) lead to low estimates of the social cost of carbon. In general, confidence in these estimates is low. The paucity of evidence is disappointing, as avoiding impacts is presumably a major aim of climate policy. CBAs of climate change implicitly estimate avoided damages and suggest that these do not warrant very stringent emissions reduction (see Section 18.4.1). Similarly, although ecosystem impacts may be large, avoidable impacts may be much smaller. With few high-quality studies, confidence in these findings is low. This is a clear research priority. The use of the social cost of carbon in decision-making on mitigation also warrants further exploration.

18.4.3 Inter-relationships within regions and sectors

Considering the details of specific adaptation and mitigation activities at the level of regions and sectors shows that adaptation and mitigation can have a positive and negative influence on each other's effectiveness. The nature of these inter-relationships (positive or negative) often depends on local conditions. Moreover, some inter-relationships are direct, involving the same resource base (e.g., land) or stakeholders, while others are indirect (e.g., effects through public budget allocations) or remote (e.g., shifts in global trade flows and currency exchange rates). This section focuses on direct inter-relationships. Broader inter-relationships between adaptation and mitigation are discussed in other parts of this chapter and in Chapter 20 related to sustainable development.

Mitigation affecting adaptation

Land-use and land-cover changes involve diverse and complex inter-relationships between adaptation and mitigation. Deforestation and land conversion have been significant sources of greenhouse-gas emissions for decades while often resulting in unsustainable agricultural production patterns. Abating and halting this process by incentives for forest conservation and increasing forest cover would not only avoid greenhouse-gas emissions, but would also result in benefits for local climate, water resources and biodiversity.

Carbon sequestration in agricultural soils offers another positive link from mitigation to adaptation. It creates an economic commodity for farmers (sequestered carbon) and makes the land more valuable by improving soil and water conservation, thus enhancing both the economic and environmental components of adaptive capacity (Boehm et al., 2004; Butt and McCarl, 2004; Dumanski, 2004). The stability of these sinks requires further research, and effective monitoring is also a challenge.

Afforestation and reforestation have been advocated for decades as important mitigation options. Recent studies reveal a more differentiated picture. Competition for land by mitigation projects would increase land rents, and thus commodity prices, thereby improving the economic position of landowners and enhancing their adaptive capacity (Lal, 2004). However, the implications of reforestation projects for water resources depend heavily on the species composition and the geographical and climatic characteristics of the region where they are implemented. In regions with ample water resources even under a changing climate, afforestation can have many positive effects, such as soil conservation and flood control. In regions with few water resources, intense rainfalls and long spells of dry weather, forests increase average water availability. However, in arid and semi-arid regions, afforestation strongly reduces water yields (UK FRP, 2005). This has direct and wide-ranging negative implications for adaptation options in several sectors such as agriculture (irrigation), power generation (cooling towers) and ecosystem protection (minimum flow to sustain ecosystems in rivers, wetlands and on river banks).

Bioenergy crops are receiving increasing attention as a mitigation option. Most studies, however, focus on technology options, costs and competitiveness in energy markets and do not consider the implications for adaptation. For example, McDonald et al. (2006) use a global computed general equilibrium model and find that substituting switchgrass for crude oil in the USA would reduce the gross domestic product (GDP) and increase the world price of cereals, but they do not investigate how this might affect the prospects for adaptation in the USA and for world agriculture. This limitation in scope characterises virtually all bioenergy studies at the regional and sectoral scales, but substantial literature on adaptation-relevant impacts exists at the project level (e.g., Pal and Sharma, 2001; see Section 18.5 and Chapter 17).

Another possible conflict between adaptation and mitigation might arise over water resources. One obvious mitigation option is to shift to energy sources with low greenhouse-gas emissions

Box 18.2. Analysis of stabilisation scenarios

The WGIII AR4, in Chapter 3 (Section 3.5.2), looks across findings of the WGI and WGII AR4 to relate the long-term emissions scenarios literature to climate-change impact risks at different levels of global mean temperature change based on key vulnerabilities (as defined in Chapter 19). It builds on the WGI AR4 findings, which outline the probabilities of exceeding various global mean temperatures at different concentration levels (Tables 3.9 and 3.10 in Fisher et al., 2007). The relationships are based on a key finding of the WGI AR4 that there is at least an 83% probability for climate sensitivity to be at or below 4.5°C, while the best estimate is for climate sensitivity to be 3°C. The WGIII AR4 organises the stabilisation scenarios literature by the level of stringency of the scenario, setting out six groups (I-VI) that cover the full range of more to less stringent global warming objectives, in the form of concentrations (ppm) or radiative forcing (W/m^2). Table 3.9 uses the WGI AR4 findings to relate increases in global mean temperature to concentration targets, while Table 3.10 relates these outcomes to the emissions pathways associated with alternative stabilisation scenarios. (An important caveat is that these relationships do not consider possible additional CO_2 and CH_4 releases from Earth-system feedbacks and thus may underestimate required emissions reductions.)

Regarding climate-change impact risks and key vulnerabilities, this literature is organised around increase in global mean temperature. Chapter 19 shows that the following benefits would accrue from constraining temperature rise to 2°C above 1990:

- lowering the risk of widespread deglaciation of the Greenland ice sheet**;
- avoiding large-scale transformation of ecosystems and degradation of coral reefs***;
- preventing terrestrial vegetation becoming a carbon source*/**, constraining species extinction to between 10% and 40%*, and preserving many unique habitats (see Chapter 4, Table 4.1 and Figure 4.5);
- preventing flooding, drought and water-quality declines***, global net declines in food production*/•, and more intense fires**.

Other benefits of this constraint include reducing the risks of extreme weather events**, and of at least partial deglaciation of the West Antarctic ice sheet (WAIS)* (see Chapter 19, Section 19.3.7). By comparison, constraining temperature change to not more than 3°C above 1990 levels will still avoid commitment to widespread deglaciation of the WAIS* and commitment to possible shut-down of the Meridional Overturning Circulation/• but results in significantly lower avoided risks and impacts in most other areas (Chapter 19, Section 19.3.7).

(Confidence ratings are as provided by WGII Chapter 19 authors: /• = low confidence, * = medium, ** = high, and *** = very high confidence.)

such as small hydropower. In regions where hydropower potentials are still available, and also depending on the current and future water balance, this would increase the competition for water, especially if irrigation might be a feasible strategy to cope with climate-change impacts in agriculture and the demand for cooling water by the power sector is also significant. This reconfirms the importance of integrated land and water-management strategies to ensure the optimal allocation of scarce natural resources (land, water) and economic investments in climate-change adaptation and mitigation and in fostering sustainable development.

Hydropower leads to the key area of mitigation: energy sources and supply, and energy use in various economic sectors beyond land use, agriculture and forestry. Direct implications of mitigation efforts on adaptation in the energy, transport, residential/commercial and industrial sectors have been largely ignored so far. Yet, to varying degrees, energy is an important factor in producing goods and providing services in many sectors of the economy, as outlined in the discussion about the importance of energy to achieve the Millennium Development Goals in the WGIII AR4, Chapter 2 (Halsnæs et al., 2007). Reducing the availability or increasing the price of energy therefore has inevitable negative effects on economic development and thus on the economic components of adaptive capacity. The magnitude of this effect is uncertain. Peters et al. (2001) find that high-level carbon charges (US$200/tC in 2010) affect U.S. agriculture modestly if they are measured in terms of consumer and producer surpluses (reductions by less than half a percent relative to baseline values). However, the decline of net cash returns is more significant (4.1%) and the effects are rather uneven across field crops and regions. Recent studies on the implications for adaptation (capacity and options) indicate that such changes may imply larger policy shifts; for example, towards protection of the most vulnerable (Adger et al., 2006).

The most important indirect link from mitigation to adaptation is through biodiversity, an important factor influencing human well-being in general and the coping options in particular (see MEA, 2005). After assessing a large number of studies, IPCC (2002) concluded that the implications for biodiversity of mitigation activities depend on their context, design and implementation, especially site selection and management practices. Avoiding forest degradation implies in most cases both biodiversity (preservation) and climate (non-emissions) benefits. However, afforestation and reforestation may have positive, neutral or negative impacts, depending on the level of biodiversity of the ecosystems that will be replaced. By using an optimal-control model, Caparros and Jacquemont (2003) find that putting an economic value on carbon sequestered by forest management does not induce much negative influence on biodiversity, but incentives to sequester carbon by afforestation and reforestation might harm biodiversity due to the over-plantation of fast-growing alien species.

These studies demonstrate the intricate inter-relationships between adaptation and mitigation, and also the links with other environmental concerns, such as water resources and biodiversity, with profound policy implications. The land-use and forestry mitigation options in the Marrakesh Accords may provide new markets for countries with abundant land areas but may alter land allocation to the detriment of the landless poor in regions where land is scarce. They present an opportunity for soil and biodiversity protection in regions with ample water resources but may reduce water yields and distort water allocation in water-stressed regions. Accordingly, depending on the regional conditions and the ways of implementation, these implications can increase or reduce the scope for adaptation to climate change by promoting or excluding effective, but more expensive, options due to increased land rents, by supporting or precluding forms and magnitudes of irrigation due to, for example, higher water prices.

Adaptation affecting mitigation

Many adaptation options in different impact sectors are known to involve increased energy use and hence interfere with mitigation efforts if the energy is supplied from carbon-emitting sources. Two main types of adaptation-related energy use can be distinguished: one-time energy input for building large infrastructure (materials and construction), and incremental energy input needed continuously to counterbalance climate impacts in providing goods and services. Furthermore, rural renewable electrification can have both huge emissions implications (WEA, 2000) and adaptation implications (Venema and Cisse, 2004).

The largest amount of construction work to counterbalance climate-change impacts will be in water management and in coastal zones. The former involves hard measures in flood protection (dykes, dams, flood control reservoirs) and in coping with seasonal variations (storage reservoirs and inter-basin diversions), while the latter comprises coastal defence systems (embankment, dams, storm surge barriers). Even if these construction projects reach massive scales, the embodied energy, and thus the associated greenhouse-gas emissions, is likely to be merely a small proportion of the total energy use and energy-related emissions in most countries (adaptation-related construction comprises only a small part of total annual construction, and the construction industry itself represents a small part in the annual energy balances of most countries).

The magnitude and relative share of sustained adaptation-related energy input in the total energy balance depends on the impact sector. In agriculture, the input-related (CO_2 in manufacturing) and the application-related (N_2O from fields) greenhouse-gas emissions might be significant if the increased application of nitrogen fertilisers offers a convenient and profitable solution to avoid yield losses (McCarl and Schneider, 2000). Operating irrigation works and pumping irrigation water could considerably increase the direct energy input, although, where available, the utilisation of renewable energy sources on-site (wind, solar) can help avoid increasing greenhouse-gas emissions.

Adaptation to changing hydrological regimes and water availability will also require continuous additional energy input. In water-scarce regions, the increasing reuse of wastewater and the associated treatment, deep-well pumping, and especially large-scale desalination, would increase energy use in the water sector (Boutkan and Stikker, 2004). Yet again, if provided from carbon-free sources such as nuclear desalination (Misra, 2003;

Ayub and Butt, 2005), even energy-intensive adaptation measures need not run counter to mitigation efforts.

Ever since the early climate impact studies, shifts in space heating and cooling in a warming world have been prominent items on the list of adaptation options (see Smith and Tirpak, 1989). The associated energy requirements could be significant but the actual implications for greenhouse-gas emissions depend on the carbon content of the energy sources used to provide the heating and cooling services. In most cases, it is not straightforward to separate the adaptation effects from those of other drivers in regional or national energy-demand projections. For example, for the U.S. state of Maryland, Ruth and Lin (2006) find that, at least in the medium term up to 2025, climate change contributes relatively little to changes in the energy demand. Nonetheless, the climate share varies with geographical conditions (changes in heating and cooling degree days), economic (income) and resource endowments (relative costs of fossil and other energy sources), technologies, institutions and other factors. Such emissions from adaptation activities are likely to be small relative to baseline emissions in most countries and regions, but more in-depth studies are needed to estimate their magnitude over the long term.

Adaptation affects not only energy use but energy supply as well. Hydropower contributed 16.3% of the global electricity balance in 2003 (IEA, 2005) with virtually zero greenhouse-gas emissions. Climate-change impacts and adaptation efforts in various sectors might reduce the contribution of this carbon-free energy source in many regions as conflicts among different uses of water emerge. Hayhoe et al. (2004) show that emissions even in the lowest SRES (IPCC Special Report on Emissions Scenarios; Nakićenović and Swart, 2000) scenario (B1) will trigger significant shifts in the hydrological regime in the Sacramento River system (California) by the second half of this century and will create critical choices between flood protection in the high-water period and water storage for the low-flow season. Hydropower is not explicitly addressed but will probably be affected as well. Payne et al. (2004) project conflicts between hydropower and streamflow targets for the Columbia River. Several studies confirm the unavoidable clashes between water supply, flood control, hydropower and minimum streamflow (required for ecological and water quality purposes) under changing climatic and hydrological conditions (Christensen et al., 2004; VanRheenen et al., 2004).

Possibly the largest factor affecting water resources in adaptation is irrigation in agriculture. Yet studies in this domain tend to ignore the repercussions for mitigation as well. For example, Döll (2002) estimates significant increases in irrigation needs in two-thirds of the agricultural land that was equipped for irrigation in 1995, but she does not assess the implications for other water uses such as hydropower and thus for climate-change mitigation.

In general, adaptation implies that people do something in addition to or something different from what they would be doing in the absence of emerging or expected climate-change impacts. In most cases, additional activities involve additional inputs: investments (protective and other infrastructure), material (fertilisers, pesticides) or energy (irrigation pumps, air-conditioning), and thus may run counter to mitigation if the energy originates from greenhouse-gas-emitting sources. Changing practices in response to climate change offer more opportunities to account for both adaptation and mitigation needs. Besides the opportunities in land-related sectors discussed above, new design principles for commercial and residential buildings could simultaneously reduce vulnerability to extreme weather events and energy needs for heating and/or cooling. Nonetheless, there are path dependencies from past technology choices and infrastructure investments.

In summary, many effects of adaptation on greenhouse-gas emissions and their mitigation (energy use, land conversion, agronomic techniques such as an increased use of fertilisers and pesticides, water storage and diversion, coastal protection) have been known for a long time. The implications of some mitigation strategies for adaptation and other development and environment concerns have been recognised recently. As yet, however, both effects remain largely unexplored. Information on inter-relationships between adaptation and mitigation at regional and sectoral levels is rather scarce. Almost all mitigation studies stop at identifying the options and costs of direct emissions reductions. Some of them consider indirect effects of implementation and costs on other sectors or the economy at large but do not deal with the implications for adaptation options of sectors affected by climate change. Similarly, in most cases, climate impact and adaptation assessments do not go beyond taking stock of the adaptation options and estimating their costs, and thus ignore possible repercussions for emissions. One understandable reason is that adaptation and mitigation studies are already complex enough and expanding their scope would increase their complexity even further. Another reason may well be that, as indicated by the few available studies that looked at these inter-relationships, the repercussions from mitigation for adaptation and *vice versa* are between adaptation and mitigation might be significant but, in most other sectors, the adaptation implications of any mitigation project are small and, conversely, the emissions generated by most adaptation activities are only small fractions of total emissions, even if emissions will decline in the future as a result of climate-protection policies.

18.5 Inter-relationships in a climate policy portfolio

A wide range of inter-relationships between adaptation and mitigation have been identified through examples in the published literature. Taylor et al. (2006) present an inventory of published examples including full citations (available in an abbreviated form on the CD-ROM accompanying this volume as supplementary material to support this chapter). The many examples have been clustered according to the type of linkage and ordered according to the entry point and scale of decision-making (Figure 18.2). Table 18.1 lists all of the types of linkages documented. The categories are illustrative; some cases occur in more than one category, or could shift over time or in different situations. For example, watershed planning is often related to managing climatic risks in using water. But if hydroelectricity is an option, then the entry point may be mitigation, and both adaptation and mitigation might be evaluated at the same time or even with explicit trade-offs.

In Figure 18.2 and Table 18.1, many of the examples are motivated by either mitigation or adaptation, with largely unintended consequences for the other (e.g., Tol and Dowlatabadi, 2001). Where adaptation leads to effects on mitigation, the linkage is labelled *A→M*. The categories of linkages include:

- individual responses to climatic hazard that increase or decrease greenhouse-gas emissions. For example, a common adaptation to heatwaves is to install air-conditioning, which increases electricity demand with consequences for mitigation when the electricity is produced from fossil fuels;
- more efficient community use of water, land, forests and other natural resources, improving access and reducing emissions (e.g., conservation of water in urban areas reduces energy used in moving and heating water);
- natural resources managed to sustain livelihoods;
- tourism use of energy and water, with outcomes for incomes and emissions (generally to increase both welfare and emissions);
- resources used in adaptation, such as in large-scale infrastructure, increases emissions.

Similarly, mitigation actions might affect the capacity to adapt or actual adaptation actions (*M→A*). These categories include:

- more efficient energy use and renewable sources that promote local development;
- CDM projects on land use or energy use that support local economies and livelihoods, perhaps by placing a value on their management of natural resources;
- urban planning, building design and recycling with benefits for both adaptation and mitigation;
- health benefits of mitigation through reduced environmental stresses;
- afforestation, leading to depleted water resources and other ecosystem effects, with consequences for livelihoods;
- mitigation actions that transfer finance to developing countries (such as per capita allocations) that stimulate investment with benefits for adaptation;
- effects of mitigation, e.g., through carbon taxes and energy prices, on resource use (generally to reduce use) that affect adaptation, for example by reducing the use of tractors in semi-subsistence farming due to higher costs of fuels.

As noted in Section 18.4.3, the effect of increased emissions due to adaptation is likely to be small in most sectors in relation to the baseline projections of energy use and greenhouse-gas emissions. Land and water management may be affected by mitigation actions, but in most sectors the effects of mitigation on adaptation are likely to be small. At least some analysts are concerned with the explicit trade-offs between adaptation and mitigation (labelled adaptation or mitigation, $\int(A,M)$). Categories include:

- public-sector funding and budgetary processes that allocate funding to both adaptation and mitigation;

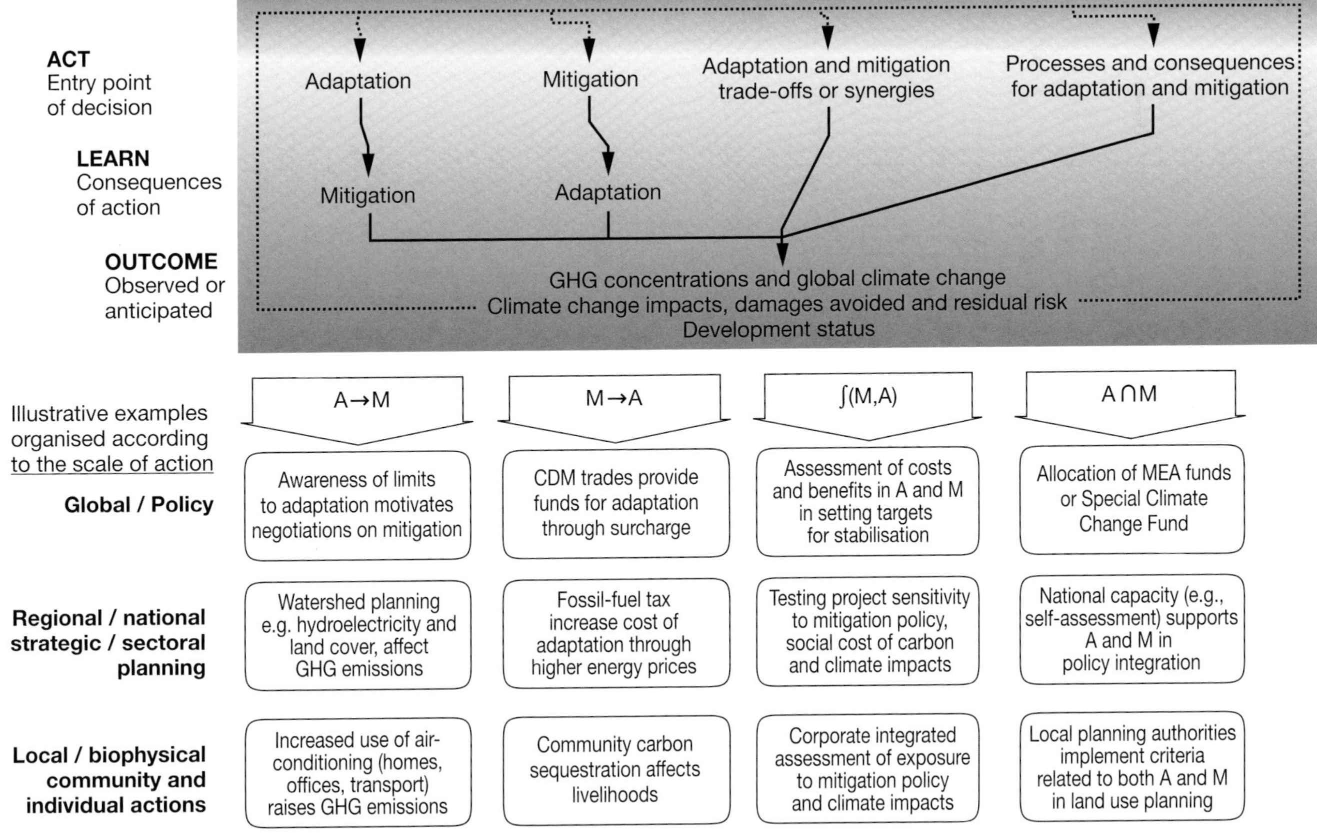

Figure 18.2. *Typology of inter-relationships between climate change adaptation and mitigation. MEA = Multilateral Environmental Agreements.*

- strategic planning related to development pathways, for example scenario and visioning exercises with urban governments that include climate responses (mainstreaming responses in sectoral and regional planning);
- allocation of funding and setting the agenda for UNFCCC negotiations and funds (e.g., the Special Climate Change Fund);
- stabilisation targets that include limits to adaptation (e.g., tolerable windows);
- analysis of global costs and benefits of mitigation to inform targets for greenhouse-gas concentrations (see Section 18.4.2);
- large-scale mitigation (e.g., geo-engineering) with effects on impacts and adaptation.

Some actions result from the simultaneous consideration of adaptation and mitigation. These concerns may be raised within the same decision framework or sequential process but without explicitly considering their trade-offs or synergies (labelled adaptation and mitigation, *A∩M*). Examples include:

- perception of impacts and the limits to adaptation (see Chapter 17) motivates action on mitigation, conversely the perception of limits to mitigation reinforces urgent action on adaptation;
- watershed planning where water is allocated between hydroelectricity and consumption without explicitly addressing mitigation and adaptation;
- cultural values that promote both adaptation and mitigation, such as sacred forests (e.g., Satoyama in Japan);
- management of socio-ecological systems to promote resilience;
- ecological impacts, with some human element, drive further releases of greenhouse gases,
- legal implications of liability for climate impacts motivates mitigation;
- national capacity-building increases the ability to respond to both adaptation and mitigation (such as through the National Capacity-Building Self Assessment);
- insurance spreads risk and assists with adaptation, while managing insurance funds has implications for mitigation;
- trade liberalisation may have economic benefits (increasing adaptive capacity) but also increases emissions from transport;
- monitoring systems and reporting requirements may cover indicators of both adaptation and mitigation;
- management of multilateral environmental agreements may benefit both adaptation and mitigation.

Table 18.1. *Types of inter-relationships between climate change adaptation and mitigation.*

A→M	*M→A*	*ʃ(A,M)*	*A∩M*
Individual responses to climatic hazards that increase or decrease greenhouse-gas emissions	More efficient energy use and renewable sources that promote local development	Public-sector funding and budgetary processes that allocate funding to both *A* and *M*	Perception of impacts (and limits to *A*) motivates *M*; perception of limits to *M* motivates *A*
More efficient community use of water, land, forests	CDM projects on land use or energy use that support local economies and livelihoods	Strategic planning related to development pathways (scenarios) to mainstream climate responses	Watershed planning: allocation of water between hydroelectricity and consumption
Natural resources managed to sustain livelihoods	Urban planning, building design and recycling with benefits for both *A* and *M*	Allocation of funding and setting the agenda for UNFCCC negotiations and funds	Cultural values that promote both *A* and *M*, such as sacred forests (e.g., Satoyama in Japan)
Tourism use of energy and water, with outcomes for incomes and emissions	Health benefits of mitigation through reduced environmental stresses	Stabilisation targets that include limits to adaptation (e.g., tolerable windows)	Management of socio-ecological systems to promote resilience
Resources used in adaptation, such as large-scale infrastructure, increase emissions	Afforestation, leading to depleted water resources and other ecosystem effects, with consequences for livelihoods	Analysis of global costs and benefits of *M* to inform targets	Ecological impacts, with some human element, drive further releases of greenhouse gases
	M schemes that transfer finance to developing countries (such as a per capita allocation) stimulate investment that may benefit *A*	Large scale *M* (e.g., geo-engineering) with effects on impacts and *A*	Legal implications of liability for climate impacts motivates *M*
	Effect of mitigation, e.g., through carbon taxes and energy prices, on resource use		National capacity-building increases ability to respond to both *A* and *M*
			Insurance spreads risk and assists with *A*; managing insurance funds has implications for *M*
			Trade liberalisation with economic benefits (*A*) increases transport costs (*M*)
			Monitoring systems and reporting requirements that cover indicators of both *A* and *M*
			Management of multilateral environmental agreements benefits both *A* and *M*

Inter-relationships between adaptation and mitigation will vary with the type of policy decisions being made, for example on different scales from local project analysis to global analysis. As discussed in Section 18.4.3, there will be clear $M \rightarrow A$ linkages in many mitigation projects, for example ensuring that adaptation is built into the project design (e.g., considering and adjusting for water availability for longer-term hydroelectric renewable or bioenergy/biofuels projects). Similarly, in the design or appraisal of adaptation projects, $A \rightarrow M$, the consideration of mitigation options can be brought in, for example in considering reduced energy use in project design. These linkages might be considered through an extension of project risk analysis as part of the appraisal process, but can also be included in cost-benefit analysis explicitly in an economic appraisal framework.

At the policy level (e.g., portfolios, funding, strategies), the same $M \rightarrow A$ and $A \rightarrow M$ issues apply, but the wider potential for cross-sectoral linkages makes simultaneous consideration of adaptation and mitigation, $A \cap M$, more important. For example, the shift up to a major (country level) energy policy towards mitigation might need to assess demand changes from adaptation across a wide range of sectors. There may be a need to consider some explicit trade-offs between adaptation and mitigation, $\int(A,M)$.

At the global level, the potential for $\int(A,M)$ becomes possible within a theoretical framework (see Section 18.4). There has been discussion of the potential for adaptation and mitigation as substitutes within narrow economic analysis (cost-benefit frameworks), and some studies have tried to assess the optimal policy balance of mitigation and adaptation using CBA based on IAMs. However, recent reviews (e.g., Watkiss et al., 2005) have shown that policy-makers are uncomfortable with the use of CBA in longer-term climate policy, because of the range of uncertainty over the relevant economic parameters of marginal mitigation costs and marginal social costs and damages avoided, but also because of the significant lack of data on the costs of adaptation. Instead, wider frameworks are considered to be more informative, using multiple aspects and risk-based approaches, for example iterative decision-making and tolerable windows (see also the risk matrix in Chapter 20). Stern (2007) explicitly adopted a risk-based framework appropriate for guiding policy from analysing the marginal costs and benefits at the project level to determination of public policy that affects future economic paths. He recognised that adaptation plays an important role, but not in an explicit trade-off against mitigation, in long-term policy.

18.6 Response capacity and development pathways

As outlined in the TAR (IPCC, 2001c, Chapter 18 and IPCC, 2001b, Chapter 1) and discussed at more length in Chapter 17 of this volume and in the WGIII AR4, Chapter 12 (Sathaye et al., 2007), the ability to implement specific adaptation and mitigation measures is dependent upon the existence and nature of adaptive and mitigative capacity, which makes such measures possible and affects their extent and effectiveness. In that sense, specific adaptation and mitigation measures are rooted in their respective capacities (Yohe, 2001; Adger et al., 2003; Adger and Vincent, 2005; Brooks et al., 2005).

Adaptive capacity has been defined in this volume (see Chapter 17) as "the ability or potential of a system to respond successfully to climate variability and change." In a parallel way, mitigative capacity has been defined as the "ability to diminish the intensity of the natural (and other) stresses to which it might be exposed" (see Rogner et al., 2007). Since this definition suggests that a group's capacity to mitigate hinges on the severity of impacts to which it is exposed, Winkler et al. (2007) have suggested that capacity be defined instead as "a country's ability to reduce anthropogenic greenhouse gases or enhance natural sinks". Clearly these two categories are closely related although, in accordance with the differences between adaptation and mitigation measures discussed in Section 18.1, capacities also differ somewhat. In particular, since adaptation measures tend to be both more geographically dispersed and smaller in scale than mitigation measures (Dang et al., 2003; Ruth, 2005), adaptive capacities refer to a slightly broader and more general set of capabilities than mitigative capacities. Despite these minor differences, however, adaptive and mitigative capacities are driven by similar sets of factors.

The term response capacity may be used to describe the ability of humans to manage both the generation of greenhouse gases and the associated consequences (Tompkins and Adger, 2005). As such, response capacity represents a broad pool of resources, many of which are related to a group or nation's level of socio-technical and economic development, which may be translated into either adaptive or mitigative capacity. Socio-cultural dimensions such as belief systems and cultural values, which are often not addressed to the same extent as economic elements (Handmer et al., 1999), can also affect response capacity (see IPCC, 2001b; Sathaye et al., 2007).

Although the concept of response capacity is new to the IPCC and has yet to be sufficiently investigated in the literature, efforts have been made to define the nature and determinants of its conceptual components: adaptive and mitigative capacity. With regard to mitigative capacity, Yohe (2001) has suggested the following list of determinants, which play out at the national level:

- range of viable technological options for reducing emissions;
- range of viable policy instruments with which the country might affect the adoption of these options;
- structure of critical institutions and the derivative allocation of decision-making authority;
- availability and distribution of resources required to underwrite the adoption of mitigation policies and the associated broadly-defined opportunity cost of devoting those resources to mitigation;
- stock of human capital, including education and personal security;
- stock of social capital, including the definition of property rights;
- a country's access to risk-spreading processes (e.g., insurance, options and futures markets);
- the ability of decision-makers to manage information, the processes by which these decision-makers determine which information is credible, and the credibility of decision-makers themselves.

In the context of developing countries, many of which possess limited institutional capacity and access to resources, mitigative and adaptive capacity could be fashioned by additional determinants. For instance, political will and the intent of decision-makers, and the ability of societies to form networks through collective action that insulates them against the impacts of climate change (Woolcock and Narayan, 2000), may be especially important in developing countries, especially in societies where policy instruments are not fully developed and where institutional capacity and access to resources are limited.

Yohe suggests a similar set of determinants for adaptive capacity, but adds the availability of resources and their distribution across the population. Recent research has sought to offer empirical evidence that demonstrates the relative influence of each of these determinants on actual adaptation (Yohe and Tol, 2002). In particular, this research indicates that the influence of each determinant of capacity is highly location-specific and path-dependent, thus revealing the importance of investigations into micro- and macro-scale determinants that influence capacity across multiple stressors (Yohe and Tol, 2002). These determinants of both adaptive and mitigative capacity expand on those identified in the TAR and agree closely with those offered by Moss et al. (2001) and Adger et al. (2004). The linkages between adaptive and mitigative capacity are demonstrated by the striking similarities between these sets of determinants, which show that both the ability to adapt and the ability to mitigate depend on a mix of social, biophysical and technological constraints (Tompkins and Adger, 2005). Recent research has pointed to the necessity of broadening these lists of determinants to include other important factors such as socio-political aspirations (Haddad, 2005), risk perception, perceived adaptive capacity (Grothmann and Patt, 2005) and political will (Winkler et al., 2007).

These discussions of determinants indicate the close connection that exists between response capacities and the underlying socio-economic and technological development paths that give rise to those capacities. In several important respects, the determinants listed above are important characteristics of such development paths. Those development paths, in turn, underpin the baseline and stabilisation emissions scenarios discussed in the WGIII AR4, Chapter 3 (Fisher et al., 2007) and used to estimate emissions, climate change and associated climate-change impacts. As a result, the determinants of response capacity can be expected to vary across the underlying emissions scenarios reviewed in this report. The climate change and climate-change impact scenarios assessed in this report will be primarily based on the SRES storylines, which define a spectrum of different development paths, each with associated socio-economic and technological conditions and driving forces (for an extended discussion of emissions pathways and climate policies, see Fisher et al., 2007). Each storyline will therefore give rise to a different set of response capacities, and thus to different likely, or even possible, levels of adaptation and mitigation.

Adaptation and mitigation measures, furthermore, are rooted in adaptive and mitigative capacities, which are in turn contained within, and strongly affected by, the nature of the development path in which they exist. The concept of development paths is discussed at more length in the WGIII AR4 in Chapters 2 (Halsnæs et al., 2007), 3 (Fisher et al., 2007) and 12 (Sathaye et al., 2007). Here, it is sufficient to think of a development path as a complex array of technological, economic, social, institutional and cultural characteristics that define an integrated trajectory of the interaction between human and natural systems over time at a particular scale. Such technological and socio-economic development pathways find their most common expression in the form of integrated scenarios (Geels and Smit, 2000; Grubb et al., 2002; Swart et al., 2003; see also WGIII AR4, Chapter 3), but are also incorporated into studies of technological diffusion (Foray and Grubler, 1996; Dupuy, 1997; Andersen, 1998; Grubler, 2000; Berkhout, 2002; Rogers, 2003), socio-technical systems (Geels, 2004) and situations in which large physical infrastructures and the requisite supportive organisational, cultural and institutional systems create conditions of quasi-irreversibility (Arthur, 1989; Sarkar, 1998; Geels, 2005; Unruh and Carrillo-Hermosilla, 2006). Technological and social pathways co-evolve through a process of learning, coercion and negotiation (Rip and Kemp, 1998), creating integrated socio-technical systems that strongly condition responses to risks such as climate change.

In the climate-change context, the TAR noted that "climate change is thus a potentially critical factor in the larger process of society's adaptive response to changing historical conditions through its choice of developmental paths" (Banuri et al., 2001). Later in the same volume, the following typology of critical components of development paths is presented (Toth et al., 2001):

- technological patterns of natural resource use, production of goods and services and final consumption,
- structural changes in the production system,
- spatial distribution patterns of population and economic activities,
- behavioural patterns that determine the evolution of lifestyles.

The influence of economic trajectories and structures on the adaptability of a nation's development path is important in terms of the patterns of carbon-intensive production and consumption that generate greenhouse gases (Smil, 2000; Ansuategi and Escapa, 2002), the costs of policies that drive efficiency gains through technological change (Azar and Dowlatabadi, 1999), and the occurrence of market failures which lead to unsustainable patterns of energy use and technology adoption (Jaffe and Stavins, 1994; Jaffe et al., 2005).

In addition to these components, scholars from widely varying disciplines and backgrounds have noted the importance of institutional structures and trajectories (Olsen and March, 1989; Agrawal, 2001; Pierson, 2004; Adger et al., 2005; Ruth, 2005) and cultural factors such as values (Stern and Dietz, 1994; Baron and Spranca, 1997), discourses (Adger et al., 2001) and social rules (Geels, 2004), as elements of development paths that help determine the ability of a system to respond to change.

The importance of the connection between measures, capacities and development paths is threefold. First, as pointed out in the TAR, a full analysis of the potential for adaptation or mitigation policies must also include some consideration of the capacities in which these policies are rooted. This is increasingly being reflected in the literature being assessed in both regional/sectoral and conceptual chapters of this assessment.

Second, such an analysis of response capacities should, in turn, encompass the nature and potential variability of underlying development paths that strongly affect the nature and extent of those capacities. This suggests the desirability of an integrated analysis of climate policy options that assesses the linkages between policy options, response capacities and their determinants, and underlying development pathways. Although such an integrated assessment was proposed in the Synthesis Report of the TAR (IPCC, 2001a), this type of assessment is still in its infancy.

Third, the linkages between climate policy measures and development paths described here suggest a potential disconnection between the degree of adaptation and/or mitigation that is possible and that which may be desired in a given situation. On the one hand, the development path will determine the response capacity of the scenario. On the other, the development path will strongly influence levels of greenhouse-gas emissions, associated climate change, the likely degree of climate-change impacts and thus the desired mitigation and/or adaptation in that scenario (Nakićenović and Swart, 2000; Metz et al., 2002; Swart et al., 2003).

However, there is no particular reason that the response capacity and desired levels of mitigation and/or adaptation will change in compatible ways. As a result, particular development paths might give rise to levels of desired adaptation and mitigation that are at odds with the degree of adaptive and mitigative capacity available. For example, particular development path scenarios that give rise to very high emissions might also be associated with a slower growth, or even a decline, in the determinants of response capacity. Such might be the case in scenarios with high degrees of military activity or a collapse of international co-operation. In such cases, climate-change impacts could increase, even as response capacity declines.

The linkages between climate policy, response capacities and development paths suggested above help us to understand the nature of the relationship between climate policy and sustainable development. There is a small but growing literature on the nature of this relationship (Cohen et al., 1998; Markandya and Halsnæs, 2000; Munasinghe and Swart, 2000; Schneider et al., 2000; Banuri et al., 2001; Robinson and Herbert, 2001; Smit et al., 2001; Beg et al., 2002; Metz et al., 2002; Najam et al., 2003; Swart et al., 2003; Wilbanks, 2003). Much of this literature emphasises the degree to which climate-change policies can have effects, sometimes called ancillary benefits or co-benefits, that will contribute to the sustainable development goals of the jurisdiction in question (Van Asselt et al., 2005). This amounts to viewing sustainable development through a climate-change lens. It leads to a strong focus on integrating sustainable development goals and consequences into the climate policy framework, and on assessing the scope for such ancillary benefits. For instance, reductions in greenhouse-gas emissions can reduce the incidence of death and illness due to air pollution and benefit ecosystem integrity – both of which are elements of sustainable development (Cifuentes et al., 2001). These co-benefits, furthermore, are often more immediate rather than long term in nature and can be significant. Van Harmelen et al. (2002) find that to comply with agreed upon or future policies to reduce regional air pollution in Europe, mitigation costs are significant, but these are reduced by 50-70% for SO_2 and around 50% for NO_x when combined with greenhouse-gas policies.

The challenge then becomes one of ensuring that actions taken to address environmental problems do not obstruct regional and local development (Beg et al., 2002). A variety of case studies demonstrates that regional and local development can in fact be enhanced by projects that contribute to adaptation and mitigation. Urban food-growing in two UK cities, for example, has resulted in reduced crime rates, improved biodiversity and reduced transport-based emissions (Howe and Wheeler, 1999). As such, these cities have both enhanced resilience to future climate fluctuations and have made strides towards the mitigation of climate change. Similarly, agro-ecological initiatives in Latin America have helped to preserve the natural resource base while empowering rural communities (Altieri, 1999). The concept of networking and clustering used mainly in entrepreneurial development and increasingly seen as a tool for the transfer of skills, knowledge and technology represents an interesting concept for countries that lack the necessary adaptive and mitigative capacities to combat the negative impacts of climate change.

An alternative approach is based on the findings in the TAR that it will be extremely difficult and expensive to achieve stabilisation targets below 650 ppm from baseline scenarios that embody high-emissions development paths. Low-emissions baseline scenarios, however, may go a long way towards achieving low stabilisation levels even before climate policy is included in the scenario (Morita et al., 2001). This recognition leads to an approach to the links between climate policy and sustainable development – equivalent to viewing climate change through a sustainable development lens – that emphasises the need to study how best to achieve low-emissions development paths (Metz et al., 2002; Robinson et al., 2003; Swart et al., 2003).

It has further been argued that sustainable development might decrease the vulnerability of developing countries to climate-change impacts (IPCC, 2001c), thereby having implications for the necessary amount of both adaptation and mitigation efforts. For instance, economic development and institution building in low-lying, highly-populated coastal regions may help to increase preparedness to sea-level rise and decrease vulnerability to weather variability (McLean et al., 2001). Similarly, investments in public health training programmes, sanitation systems and disease vector control would both enhance general health and decrease vulnerability to the future effects of climate change (McMichael et al., 2001). Framing the debate as a development problem rather than an environmental one helps to address the special vulnerability of developing nations to climate change while acknowledging that the driving forces for emissions are linked to the underlying development path (Metz et al., 2002). Of course it is important also to acknowledge that climate change policy cannot be considered a substitute for sustainable development policy even though it is determined by similar underlying socio-economic choices (Najam et al., 2003).

Both approaches to linking climate change to sustainable development suggest the desirability of integrating climate-policy measures with the goals and attributes of sustainable development (Robinson and Herbert, 2001; Beg et al., 2002; Adger et al., 2003; Van Asselt et al., 2005; Robinson et al.,

2006). This suggests an additional reason to focus on the inter-relationships between adaptation, mitigation, response capacity and development paths. If climate policy and sustainable development are to be pursued in an integrated way, then it will become important not simply to evaluate specific policy options that might accomplish both goals, but also to explore the determinants of response capacity that underlie those options and their connections to underlying socio-economic and technological development paths (Swart et al., 2003). Such an integrated approach might be the basis for productive partnerships with the private, public, non-governmental and research sectors (Robinson et al., 2006).

There is general agreement that sustainable development involves a comprehensive and integrated approach to economic, social and environmental processes (Munasinghe, 1992; Banuri et al., 1994; Najam et al., 2003; see also Sathaye et al., 2007). However, early work tended to emphasise the environmental and economic aspects of sustainable development, overlooking the need for analysis of social, political or cultural dimensions (Barnett, 2001; Lehtonen, 2004; Robinson, 2004). More recently, the importance of social, political and cultural factors (e.g., poverty, social equity and governance) has increasingly been recognised (Lehtonen, 2004), especially by the global environmental change policy and climate change communities (Redclift and Benton, 1994; Banuri et al., 1996; Brown, 2003; Tonn, 2003; Ott et al., 2004; Oppenheimer and Petonsk, 2005) to the point that social development, which also includes both political and cultural concerns, is now given equal status as one of the 'three pillars' of sustainable development. This is evidenced by the convening of the World Summit on Social Development in 1995 and by the fact that the Millennium Summit in 2000 highlighted poverty as fundamental in bringing balance to the overemphasis on the environmental aspects of sustainability. The environment-poverty nexus is now well recognised, and the link between sustainable development and achievement of the Millennium Development Goals (MDGs) (United Nations, 2000) has been clearly articulated (Jahan and Umana, 2003). In order to achieve real progress in relation to the MDGs, different countries will settle for different solutions (Dalal-Clayton, 2003), and these development trajectories will have important implications for the mitigation of climate change.

In attempting to follow more sustainable development paths, many developing nations experience unique challenges, such as famine, war, social, health and governance issues (Koonjul, 2004). As a result, past economic gains in some regions have come at the expense of environmental stability (Kulindwa, 2002), highlighting the lack of exploitation of potential synergies between sustainable development and environmental policies. In the water sector, for instance, response capacity can be improved through co-ordinated management of scarce water resources, especially since reduction in water supply in most of the large rivers of the Sahel can affect vital sectors such as energy and agriculture, which are dependent on water availability for hydroelectric power generation and agricultural production, respectively (Ikeme, 2003). Technology, institutions, economics and socio-psychological factors, which are all elements of both response capacity and development paths, affect the ability of nations to build capacity and implement sustainable development, adaptation and mitigation measures (Nederveen et al., 2003).

18.7 Elements for effective implementation

This section considers the literature assessment of the previous sections with respect to its implications for policy and decision-making. It reviews the policy and institutional contexts within which adaptation and mitigation can be implemented and discusses inter-relationships in practice.

18.7.1 Climate policy and institutions

As explained and illustrated in the previous sections of this chapter, effective climate policy would involve a portfolio of adaptation and mitigation actions. These actions include technological, institutional and behavioural options, the introduction of economic and policy instruments to encourage the use of these options, and research and development to reduce uncertainty and to enhance the options' effectiveness and efficiency. However, the actors involved in the implementation of these actions operate on a range of different spatial and institutional scales, representing different sectoral interests. Policies and measures to promote the implementation of adaptation and mitigation actions have therefore been targeted primarily on either adaptation or mitigation; rarely have they been given similar priority and considered in conjunction (see Section 18.5 for more detail).

On the *global scale*, the UNFCCC and its Kyoto Protocol are at present the principal institutional frameworks by which climate policy is developed. The ultimate objective of the UNFCCC, as stated in Article 2, is:

> "to achieve... stabilisation of greenhouse-gas concentrations in the atmosphere at a level that would prevent dangerous anthropogenic interference with the climate system ... within a time-frame sufficient to allow ecosystems to adapt naturally to climate change, to ensure that food production is not threatened and to enable economic development to proceed in a sustainable manner."

Initially, this objective was often interpreted as having relevance only or primarily to mitigation: reducing greenhouse-gas emissions and enhancing sinks such that atmospheric concentrations are stabilised at a non-dangerous level. However, whether or not anthropogenic interference with the climate system will be dangerous does not depend only on the stabilisation level; it depends also on the degree to which adaptation can be expected to be effective in addressing the consequences of this interference. In other words, the greater the capacity of ecosystems and society to adapt to the impacts of climate change, the higher the level at which atmospheric greenhouse-gas concentrations may be stabilised before climate change becomes dangerous (see also Chapter 19). Adaptation thus complements and can, in theory and until the limits of adaptation are reached, substitute for mitigation in meeting the ultimate objective of the UNFCCC (Goklany, 2000a, 2003).

The possibility of considering adaptation and mitigation as substitutes on a global scale does not feature explicitly in the UNFCCC, the Kyoto Protocol or any decisions made by the Conference of the Parties to the UNFCCC. This is so because any global agreement on substitution would, in practice, be unable to account for the diverse, and at times conflicting, interests of all actors involved in adaptation and mitigation and for the differences in temporal and spatial scales between the two alternatives (see Section 18.3). Mitigation is primarily justified by international agreements and the ensuing national public policies, but most adaptation is motivated by private interests of affected individuals, households and firms, and by public arrangements of impacted communities and sectors. The fact that decisions on adaptation are often made at sub-national and local levels also presents a challenge to the organisation of funding for adaptation in developing countries under the UNFCCC, the Kyoto Protocol and any future international climate policy regimes (Schipper, 2006).

Yet there is one way in which adaptation and mitigation are connected at the global policy level, namely in their reliance on social and economic development to provide the capacity to adapt and mitigate. Section 18.6 introduced the concept of response capacity, which can be represented as adaptive capacity and mitigative capacity. Response capacity is often limited by a lack of resources, poor institutions and inadequate infrastructure, among other factors that are typically the focus of development assistance. People's vulnerability to climate change can therefore be reduced not only by mitigating greenhouse-gas emissions or by adapting to the impacts of climate change, but also by development aimed at improving the living conditions and access to resources of those experiencing the impacts, as this will enhance their response capacity.

The incorporation of development concerns into climate policy demonstrates that climate policy involves more than decision-making on adaptation and mitigation in isolation. Accordingly, Klein et al. (2005) identified three roles of climate policy under the UNFCCC: (i) to control the atmospheric concentrations of greenhouse gases; (ii) to prepare for and reduce the adverse impacts of climate change and take advantage of opportunities; and (iii) to address development and equity issues. Although climate change is not the primary reason for poverty and inequality in the world, addressing these issues is seen as a prerequisite for successful adaptation and mitigation in many developing countries. In a paper produced by a number of development agencies and international organisations, Sperling (2003) made the case for linking climate policy and development assistance, which would promote opportunities for mainstreaming considerations of climate change into development on the national, sub-national and local scales (Box 18.3).

With the first commitment period of the Kyoto Protocol ending in 2012, a range of proposals have been prepared that lay out a post-2012 international climate policy regime (e.g., Den Elzen et al., 2005; Michaelowa et al., 2005). The majority of current proposals focus only or predominantly on mitigation; some proposals consider adaptation and mitigation in concert. However, few proposals have been appraised in terms of, for example, their effectiveness, efficiency and equity.

On the *regional scale*, climate policies and institutions do not tend to consider inter-relationships between adaptation and mitigation. In the European Union, for example, mitigation policy is conducted separately from adaptation strategies that are being developed or studied for water management, coastal management, agriculture and public health. Most Least-Developed Countries are concerned primarily with adaptation and its links with development. The Asia-Pacific Partnership on Clean Development and Climate only refers to mitigation.

Organisations such as the World Trade Organization (WTO) and the European Union can, through specific mechanisms, integrate environmental policy into their economic rationales. In addition, there is a need to address contradictions between existing policies (e.g., policies relevant to the reduction of greenhouse-gas emissions and agricultural trade policies). Energy remains a crucial input in agro-processing, transportation and packaging, and the combined effects of increases in energy consumption in the agricultural sector and impacts of agricultural trade policies are typically not considered within the context of climate change.

Regional co-operation could create 'win-win' opportunities in both economic integration and in addressing the adverse effects of climate change (Denton et al., 2002). Initiatives such as the New Partnership for Africa's Development (NEPAD) and the African Ministerial Conference on the Environment conducted a number of consultative processes in order to prepare an Environmental Action Plan for the Implementation of the Environment Initiative of NEPAD. One of the proposed projects is to evaluate synergistic effects of adaptation and mitigation activities, including on-farm and catchment management of carbon with sustainable livelihood benefits. Organisations such as the West African Monetary Union (WAMU) are actively engaged in energy development to address the perennial problem of energy poverty in the continent. They focus on how to exploit the CDM and other mechanisms to mitigate present and future emissions, especially with the use of renewable energy. WAMU countries are vulnerable to drought and desertification and, while mitigation may not be their main concern, it does offer opportunities also to reduce the negative impacts of deforestation and land-use change. Equally, links between the UNFCCC and the UN Convention to Combat Desertification offer opportunities to exploit both adaptation and mitigation within the context of promoting sustainable livelihoods and environmental management. A number of sub-regional institutions have action plans to address desertification, such as the Arab Maghreb Union in northern Africa, the Intergovernmental Authority on Development in eastern Africa, the Southern African Development Community in the south, the Economic Community of Western African States and the Permanent Interstate Committee for Drought Control in the Sahel for the west, and the Economic Community of Central African Countries in central Africa.

Countries belonging to these and other regional groupings can identify projects that have net adaptation and mitigative benefits. Studies (e.g., Greco et al., 1994) have predicted a reduction in water supply in most of the large rivers of the Sahel, thus affecting vital sectors such as energy and agriculture, both of which are dependent on water availability for hydroelectric power generation and agricultural production. Seventeen countries in West Africa share 25 trans-boundary rivers and many countries within the region have a water-dependency ratio of around 90% (Denton et al., 2002). Water resources and watershed management in trans-boundary river basins are possible ways in which countries in West Africa can co-operate on a regional basis to build institutional capacity, strengthen regional networks and institutions to encourage co-operation, flow of information and transfer of technology. The construction of the Manantali Dam in Mali as part of the Senegal River Basin Initiative is to a large extent able to produce hydropower electricity and enable riparian communities to practice irrigation

Box 18.3. Mainstreaming

The links between greenhouse-gas emissions, mitigation of climate change and development have been the subject of intense study (for an overview see Markandya and Halsnæs, 2002). More recently the links between climate-change adaptation and development have been brought to light (Section 18.6). As these links have become apparent, the term 'mainstreaming' has emerged to describe the integration of policies and measures that address climate change into development planning and ongoing sectoral decision-making. The benefit of mainstreaming would be to ensure the long-term sustainability of investments as well as to reduce the sensitivity of development activities to both today's and tomorrow's climate (Beg et al., 2002; Klein, 2002; Huq et al., 2003; OECD, 2005).

Mainstreaming is proposed as a way of making more efficient and effective use of financial and human resources than designing, implementing and managing climate policy separately from ongoing activities. By its very nature, energy-based mitigation (e.g., fuel switching and energy conservation) can be effective only when mainstreamed into energy policy. For adaptation, however, this link has not appeared as self-evident until recently (see Chapter 17). Mainstreaming is based on the premise that human vulnerability to climate change is reduced not only when climate change is mitigated or when successful adaptation to the impacts takes place, but also when the living conditions for those experiencing the impacts are improved (Huq and Reid, 2004).

Although mainstreaming is most often discussed with reference to developing countries, it is just as relevant to industrialised countries. In both cases it requires the integration of climate policy and sectoral and development policies. The institutional means by which such linking and integration is attempted or achieved vary from location to location, from sector to sector, as well as across spatial scales. For developing countries, the UNFCCC and other international organisations could play a part in facilitating the successful integration and implementation of adaptation and mitigation in sectoral and development policies. Klein et al. (2005) see this as a possible fourth role of climate policy, in addition to the three presented earlier in this section.

In April 2006 the OECD organised a ministerial-level meeting of the OECD Development Assistance Committee (DAC) and the Environment Policy Committee (EPOC). The meeting served to launch a process to work in partnership with developing countries to integrate environmental factors efficiently into national development policies and poverty reduction strategies. The outcomes of the meeting were an agreed Framework for Common Action Around Shared Goals, as well as a Declaration on Integrating Climate Change Adaptation into Development Co-operation (OECD, 2006). These outcomes are evidence of the importance that is now being attached to mainstreaming adaptation into Official Development Assistance (ODA) activities. The OECD framework and declaration are expected to provide an impetus to all development agencies to consider climate change in their operations and thus facilitate mainstreaming.

To facilitate mainstreaming would require increasing awareness and understanding among decision-makers and managers, and creating mechanisms and incentives for mainstreaming. It would not require developing synergies between adaptation and mitigation *per se*, but rather between building adaptive and mitigative capacity, and thus with development (see Section 18.6). This fourth role of climate policy highlights the importance of involving a greater range of actors in the planning and implementation of adaptation and mitigation, including sectoral, sub-national and local actors, and the private sector (Robinson et al., 2006, see also Section 18.3).

The above may give the impression that a broad consensus has emerged that mainstreaming adaptation into ODA is the most desirable way of reducing the vulnerability of people in developing countries to climate change. There is indeed an emerging consensus among development agencies, as reflected in the OECD declaration. However, concerns about mainstreaming have been voiced within developing countries and among academics. On the one hand, there is concern that scarce funds for adaptation in developing countries could be diverted into more general development activities, which offer little opportunity to evaluate, at least quantitatively, their benefits with respect to climate change (Yamin, 2005). On the other hand, there is concern that funding for climate policy would divert money from ODA that is meant to address challenges seen as being more urgent than climate change, including water and food supply, sanitation, education and health care (Michaelowa and Michaelowa, 2005).

agriculture, especially since Senegal and Mauritania remain highly dependent on agriculture and suffer deficits in staple cereal crops. These initiatives have global sustainable development benefits since they are able to offer both adaptation and mitigative benefits as well as accelerate the economic development of countries sharing the river (namely Senegal, Mali and Mauritania) (Venema et al., 1997).

The Convention on Biological Diversity has acknowledged the potential win-win opportunities between biodiversity management, on the one hand, and adaptation and mitigation to climate change, on the other. There is particular scope for this in large-scale regional biodiversity programmes such as the Mesoamerican Biological Corridor Project, in which reforestation and avoided deforestation can help to mitigate climate change through the creation of carbon sinks, while creating livelihood benefits for local communities, thus increasing their capacity to adapt to climate change. In addition, the creation of large biological corridors will help ecological communities to migrate and adapt to changing environmental conditions (CBD, 2003).

The *national, sub-national and local scales* are where most adaptation and mitigation actions are implemented and where most inter-relationships may be expected. However, there is little academic literature that describes or analyses policy and institutions at these levels with respect to inter-relationships of adaptation and mitigation. The literature does provide a growing number of examples and case studies (see Section 18.5) but, unlike the emerging literature on global policy and institutions, it does not yet discuss the role of policies and institutions *vis-à-vis* inter-relationships between adaptation and mitigation, nor does it discuss the implications of potential inter-relationships on policy and institutions. A research field is emerging that builds on studies carried out for adaptation or for mitigation. For example, the AMICA project (Adaptation and Mitigation: an Integrated Climate Policy Approach) aims to identify synergies between adaptation and mitigation for selected cities in Europe (http://www.amica-climate.net/).

In the Niayes region of central Senegal, the government has sought to promote irrigation practices and reduce dependence on rain-fed agriculture with the planting of dense hedges to act as windbreaks. These have enhanced agricultural productivity. Windbreaks have been effective in combating soil erosion and desiccation and have also provided fuelwood for cooking, thus reducing the need for women and girls to travel long distances in a rapidly urbanising area in search of wood. The windbreaks have carbon sequestration benefits but, most of all, they have helped to intensify agricultural production, especially with commercial products, thus boosting the economic livelihoods of poor communities. Thus, what started off as an adaptation strategy has had substantial integrated development benefits by easing deforestation and reducing carbon emissions, as well as addressing gender and livelihood issues (Seck et al., 2005).

Effective implementation of climate change adaptation and mitigation is often dependent on the support from local non-governmental organisations, private sector and public government authorities. Market-based policy instruments (e.g., pollution taxes and different types of tradable permits) have been successfully implemented to provide incentives in both industrialised and developing countries. The use of tax credits and financial assistance in India has opened up the electricity market to the private sector, which has resulted in a 'wind energy boom' (Sawin and Flavin, 2004). Similarly, incentives for the uptake of biofuels and energy-efficiency programmes in Brazil have considerably reduced carbon emissions (Pew Center, 2002). Although these programmes have typically not been designed with the purpose of creating synergies between adaptation and mitigation, they do provide net adaptation and mitigation benefits, as well as addressing sustainable development priorities of communities. In addition, the private sector is increasingly becoming involved in environmental governance. For example, transnational corporations are being drawn into partnerships and networks to help managing the global environment.

A special role can be played by international funding agencies and climate change funds. For example, the World Bank BioCarbon Fund and Community Development Carbon Fund provide financing for reforestation projects to conserve and protect forest ecosystems, community afforestation activities, mini- and micro-hydro and biomass fuel projects. These projects are focused specifically on extending carbon finance to poorer countries and contribute not only to the mitigation of climate change but also to reducing rural poverty and improving sustainable management of local ecosystems, thereby enhancing adaptive capacity.

18.7.2 Inter-relationships in practice

In practice, adaptation and mitigation can be included in climate-change strategies, policies and measures at different levels, involving different stakeholders (see Section 18.3). For example, the European Union previously emphasised policies to focus on reducing greenhouse-gas emissions in line with Kyoto targets. However, it is increasingly acknowledging the parallel need to deal with the consequences of climate change. In 2005 the European Commission launched the second phase of the European Climate Change Programme (ECCP), which now also includes impacts and adaptation as one of its working groups. They recognise the value of win-win strategies that address climate-change impacts but also contribute to mitigation objectives (EEA, 2005).

Examples at the national level include the UK Climate Change Programme, which includes adaptation and mitigation (DETR, 2000). The UK also addresses adaptation through its Adaptation Policy Framework, the UK Climate Impacts Programme (UKCIP) and a Cross-Regional Research Programme led by the Department for Environment, Food and Rural Affairs (Defra). Malta identified in its first National Communication to the UNFCCC a range of win-win adaptation options, including efficiency in energy production, improving farming and afforestation (Ministry for Rural Affairs and the Environment Malta, 2004). The Czech Republic has agreed to give priority to win-win measures, due to financial constraints (EEA, 2005).

Relevant to the sub-national and local level in the UK is the planning policy and advice released by the Office of the Deputy Prime Minister for the benefit of regional planning bodies

(ODPM, 2005). It includes advice to planners on how to integrate climate change adaptation and mitigation into their policy planning decisions. ODPM (2004) encourages an integrated approach to ensure that adaptation initiatives do not increase energy demands and therefore conflict with greenhouse-gas mitigation measures. Adaptation measures would include decisions about the location of new settlements and not creating an unsustainable demand for water resources, by taking into account possible changes in seasonal precipitation.

Other examples of projects which incorporate 'climate proofing' include the Cities for Climate Protection Campaign, a worldwide movement of local governments working together under the umbrella of the International Council for Local Environmental Initiatives to reduce greenhouse-gas emissions, improve air quality and enhance urban sustainability. Local governments following this programme develop a baseline of their emissions, set targets and agree on an action plan to reach the targets through a sustainable development approach focusing on local quality of life, energy use and air quality (ICLEI, 2006). For example, Southampton City Council has developed a climate change strategy in conjunction with its air quality strategy and action plan, seeing close links between the two. The strategy includes measures for the council and partners to reduce net emissions of greenhouse gases and other pollutants through integrated energy systems and continued air quality monitoring. The mitigating measures are supported by improved management of the likely impacts of future climate change and the impacts on air quality through better planning and adaptation, such as coastal defence, transport infrastructure, planning and design, and flood risk mapping (Southampton City Council, 2004).

18.8 Uncertainties, unknowns and priorities for research

Many of the inter-relationships between adaptation and mitigation have been described in previous assessments of climate policy, and the literature is rapidly expanding. Nevertheless, well-documented studies at the regional and sectoral level are lacking. Adaptation and mitigation studies tend to focus only on their primary domains, and few studies analyse the secondary consequences (e.g., of mitigation on impacts and adaptation options or of adaptation actions on greenhouse-gas emissions and mitigation options). Experiences with climate change adaptation are relatively recent and large-scale, and global actions, such as insurance, adaptation protocols or issues of liability and compensation, have not been tested.

Learning from the expanding case experience of inter-relationships is a priority. Reviews, syntheses and meta-analyses should become more common in the next few years. An analytical and institutional framework for monitoring the inter-relationships and organising periodic assessments needs to be developed. At present, no organisation appears to have a leading role in this area. The experiences of stakeholders in making decisions concerning both adaptation and mitigation should be compared. The experience of the research on land-use and land-cover change would be insightful (e.g., Geist and Lambin, 2002). Effective institutional development, use of financial instruments, participatory planning and risk-management strategies are areas for learning from the emerging experience (Klein et al., 2005).

A key research need is to document which stakeholders link adaptation and mitigation. Decisions oriented towards either adaptation or mitigation might be extended to evaluate unintended consequences, to take advantage of synergies or explicitly evaluate trade-offs. Yet, the constraints of organisational mandates and administrative capacity, finance and linking across scales and sectors (e.g., Cash and Moser, 2000) may outweigh the benefits of integrated decision-making. Formulation of policies that support renewable energy in developing countries is likely to meet fiscal, market, legal, knowledge and infrastructural barriers that may limit uptake.

The effects on specific social and economic groups need to be further documented. For example, development of hydroelectricity may reduce water availability for fish farming and irrigation of home gardens, potentially adversely affecting the food security of women and children (Andah et al., 2004; Hirsch and Wyatt, 2004). Linking carbon sequestration and community development could generate new opportunities for women and marginal socio-economic groups, but this will depend on many local factors and needs to be evaluated with empirical research.

The links between a broad climate-change response capacity, specific capacities to link adaptation and mitigation, and actual actions are poorly documented. Testing and quantification of the relationship between capacities to act and actual action is needed, taking into account sectoral planning and implementation, the degree of vulnerability, the range of technological options, policy instruments and information including experience of climate change.

Analytical frameworks for evaluating the links between adaptation and mitigation are inadequate, or in some cases competing. A suite of frameworks may be necessary for particular stakeholders and levels of decision-making. Decision frameworks relating adaptation and mitigation (separately or conjointly) need to be tested against the roles and responsibilities of stakeholders at all levels of action. Global optimising models may influence some decisions, while experience at the project level is important to others. The suitability of IAMs needs to be evaluated for exploring multiple metrics, discontinuities and probabilistic forecasts (Mastrandrea and Schneider, 2001, 2004; Schneider, 2003). Global cost-benefit models should include clear analyses of uncertainty in the use of valuation schemes and discounting as well as the assumptions inherent in climate impacts models (including the role of adaptation in reducing impacts). Hybrid approaches to integrated assessments across scales (top-down and bottom-up) should be further developed (Wilbanks and Kates, 2003). Representations of risks and uncertainties need to be related to decision frameworks and processes (Dessai et al., 2004; Kasperson and Kasperson, 2005; Lorenzoni et al., 2005). Climate risk, current and future, is only one aspect of adaptation-mitigation decision-making; the relative importance and effect of other drivers needs to be understood.

The magnitude of unintended consequences is uncertain. The few existing studies (e.g., Dang et al., 2003) indicate that the repercussions from mitigation for adaptation and *vice versa* are mostly marginal at the global level, although they may be significant at the regional scale. The effects on demand or total emissions are likely to be a small fraction of the global baseline. However, in some domains, such as water and land markets, and in some locales, the inter-relationships might affect local economies. Quantitative evaluation of direct trade-offs is missing: the metrics and methods for valuation, existence of thresholds in local feedbacks, behavioural responses to opportunities, risks and adverse impacts, documentation of the baseline and project scenarios, and scaling up from isolated, local examples to systemic changes are part of the required knowledge base.

At a global or international level, defining a socially, economically and environmentally justifiable mix of mitigation, adaptation and development remains difficult and a research need. While IAMs are relatively well developed, they can only provide approximate estimates of quantitative inter-relationships at a highly aggregated scale. Fourteen experts in estimating the social cost of carbon rated their estimates as low confidence, due to the many gaps in the coverage of impacts and valuation studies, uncertainties in projected climate change, choices in the decision framework and the applied discount rate (Downing et al., 2005). Estimates of the marginal abatement cost range from −2% to +8% of GDP, while estimates of the marginal damages avoided span three orders of magnitude (see Chapter 20). The marginal cost of adaptation has not been calculated, although some estimates assume a reduction in impacts due to adaptation (see Chapter 17). Combining the marginal abatement cost, marginal damages avoided and the marginal cost of adaptation into an optimal strategy for climate response is subject to considerable uncertainty that is unlikely to be effectively reduced in the near term (see Harvey, 2006).

A systematic assessment with a formal risk framework that guides expert judgement and grounded case studies, and interprets the sample of published estimates, is required if policy-makers wish to identify the benefits of climate policy (e.g., Downing et al., 2005). Existing estimates of damages avoided are based on a sample of sectors exposed to climate change and a small range of climate stresses. Better understanding across a matrix of climate change and exposure is required (Chapter 20; Fisher et al., 2007). Socio-economic conditions and locales that are likely to experience early and significant impacts (often called 'hotspots') should be a high priority for additional studies. The extent to which targets that are set globally are consistent with national or local mixes of strategies requires a concerted effort. The distributional effects would be an important factor in evaluating tolerable windows and trade-offs between adaptation and mitigation. The lack of high-quality studies of the benefits of mitigation, and the social cost of carbon, limits confidence in setting targets for stabilisation.

The relationship between development paths and adaptation-mitigation inter-relationships requires further research. Unintended consequences, synergies and trade-offs might be unique to some development paths; equally, they might be possible in many different paths. Existing scenarios of development paths are particularly inadequate in framing some of the major determinants of vulnerability and adaptation (Downing et al., 2003). Exogenous projections of GDP are a particular obstacle for modelling the inter-relationships between adaptation and mitigation. Few global scenarios address local food security in realistic ways (Downing and Ziervogel, 2005, but see related discussion of Millennium Development Goals in Chapters 9 and 20). Scenarios of abrupt climate change, streams of extreme events, and realistic social, economic and political responses would add insight into adaptive management (the 'act, then learn, then act again' approach). Few reference scenarios explicitly frame issues related to inter-relationships between adaptation and mitigation (e.g., from the extent to which a global decision-maker makes optimising judgements to the institutional setting for local projects to exploit synergies). While the direct energy input in large infrastructure projects may be small, including a shadow price for climate change externalities may shift adaptation portfolios. An assessment of actual shifts in energy demand and ways to reduce emissions is desirable. Most integrated assessments are at the large scale of regions to world views, although local dialogues are beginning to explore synergies (Munasinghe and Swart, 2005).

The feasibility and outcome of many of the inter-relationships depend on local conditions and management options. A systematic assessment and guidance for mitigating potentially adverse effects would be helpful. The nature of links between public policy and private action at different scales, and prospects for mainstreaming integrated policy, are worth evaluating. Many of the consequences depend on environmental processes that may not be well understood; for example, the resilience of systems to increased interannual climate variability and long-term carbon sequestration in agro-forestry systems.

References

Adger, W.N. and K. Vincent, 2005: Uncertainty in adaptive capacity. *C. R. Geosci.*, **337**, 399-410.

Adger, W.N., T.A. Benjaminsen, K. Brown and H. Svarstad, 2001: Advancing a political ecology of global environmental discourses. *Dev. Change*, **32**, 687-715.

Adger, W.N., S. Huq, K. Brown, D. Conway and M. Hulme, 2003: Adaptation to climate change in the developing world. *Prog. Dev. Stud.*, **3**, 179-195.

Adger, W.N., N. Brooks, M. Kelly, S. Bentham and S. Eriksen, 2004: New indicators of vulnerability and adaptive capacity. Technical Report 7. Tyndall Centre for Climate Research, Norwich, 126 pp.

Adger, W.N., T.P. Hughes, C. Folke, S.R. Carpenter and J. Rockström, 2005: Social-ecological resilience to coastal disasters. *Science*, **309**, 1036-1039.

Adger, W.N., J. Paavola, S. Huq and M.J. Mace, Eds., 2006: *Fairness in Adaptation to Climate Change*. MIT Press, Cambridge, Massachusetts, 335 pp.

Agrawal, A., 2001: Common property institutions and sustainable governance of resources. *World Dev.*, **29**, 1649-1672.

Altieri, M.A., 1999: Applying agroecology to enhance the productivity of peasant farming systems in Latin America. *Environ. Dev. Sustain.*, **1**, 197-217.

Andah, W., N. van de Giesen, A. Huber-Lee and C. Biney, 2004: Can we maintain food security without losing hydropower? The Volta Basin. *Climate Change in Contrasting River Basins: Adaptation Strategies for Water, Food and Environment*, J. Cayford, Ed., CABI Publishing, Wallingford, 181-194.

Andersen, B., 1998: The evolution of technological trajectories 1890-1990. *Struct. Change Econ. Dynam.*, **9**, 5-34.

Ansuategi, A. and M. Escapa, 2002: Economic growth and greenhouse gas emissions. *Ecol. Econ.*, **40**, 23-37.

Arnell, N., M.G.R. Cannell, M. Hulme, R.S. Kovats, J.F.B. Mitchell, R. Nicholls, M. Parry, M. Livermore and A. White, 2002: The consequences of CO_2 stabilization for the impacts of climate change. *Climatic Change*, **53**, 413-446.

Arthur, W.B., 1989: Competing technologies, increasing returns, and lock-in by historical events. *Econ. J.*, **99**, 116-131.

Ayub, M.S. and W.M. Butt, 2005: Nuclear desalination: harnessing the seas for development of coastal areas of Pakistan. *Int. J. Nucl. Desalination*, **1**, 477-485.

Azar, C., 1998: Are optimal CO_2 emissions really optimal? Four critical issues for economists in the greenhouse. *Environ. Resour. Econ.*, **11**, 301-315.

Azar, C. and H. Dowlatabadi, 1999: A review of technical change in assessment of climate policy. *Annu. Rev. Energ. Env.*, **24**, 513-545.

Bakkenes, M., B. Eickhout and R. Alkemade, 2006: Impacts of different climate stabilization scenarios on plant species in Europe. *Global Environ. Chang.*, **16**, 19-28.

Banuri, T., G. Hyden, C. Juma and M. Rivera, 1994: *Sustainable Human Development: From Concept to Operation. A Guide for the Practitioner.* United Nations Development Programme, New York.

Banuri, T., K. Goran-Maler, M. Grubb, H.K. Jacobson and F. Yamin, 1996: Equity and social considerations. *Climate Change 1995: Economic and Social Dimensions of Climate Change. Contribution of Working Group III to the Second Assessment Report of the Intergovernmental Panel on Climate Change*, J.P. Bruce, H. Lee and E.F. Haites, Eds., Cambridge University Press, Cambridge, 79-124.

Banuri, T., J. Weyant, G. Akumu, A. Najam, L. Pinguelli Rosa, S. Rayner, W. Sachs, R. Sharma and G. Yohe, 2001: Setting the stage: climate change and sustainable development. *Climate Change 2001: Mitigation. Contribution of Working Group III to the Third Assessment Report of the Intergovernmental Panel on Climate Change*, B. Metz, O. Davidson, R. Swart and J. Pan, Eds., Cambridge University Press, Cambridge, 73-114.

Barker, T., J. Koehler and M. Villena, 2002: The costs of greenhouse gas abatement: a meta-analysis of post- SRES mitigation scenarios. *Environ. Econ. Policy Stud.*, **5**, 135-166.

Barker, T., and Co-authors, 2007: Mitigation from a cross-sectoral perspective. *Climate Change 2007: Mitigation. Contribution of Working Group III to the Fourth Assessment Report of the Intergovernmental Panel on Climate Change*, B. Metz, O. Davidson, P. Bosch, R. Dave and L. Meyer, Eds., Cambridge University Press, Cambridge, UK, in press.

Barnett, J., 2001: *The Meaning of Environmental Security: Ecological Politics and Policy in the New Security Era.* Zed Books, London, 192 pp.

Baron, J. and M. Spranca, 1997: Protected values. *Organ. Behav. Hum. Dec.*, **70**, 1-16.

Beg, N., J. Corfee Morlot, O. Davidson, Y. Arfrane-Okesse, L. Tyani, F. Denton, Y. Sokona, J.P. Thomas, E.L. Rovere, J. Parikh, K. Parikh and A. Rahman, 2002: Linkages between climate change and sustainable development. *Clim. Policy*, **2**, 129-144.

Bell, M.L., B.F. Hobbs and H. Ellis, 2003: The use of multi-criteria decision-making methods in integrated assessment of climate change: implications for IA practitioners. *Socio-Econ. Plan. Sci.*, **37**, 289-316.

Berkhout, F., 2002: Technological regimes, path dependency and the environment. *Global Environ. Chang.*, **12**, 1-4.

Berrittella, M., A. Bigano, R. Roson and R.S.J. Tol, 2006: A general equilibrium analysis of climate change impacts on tourism. *Tourism Manage.*, **27**, 913-924.

Boehm, M., B. Junkins, R. Desjardins, S. Kulshreshtha and W. Lindwall, 2004: Sink potential of Canadian agricultural soils. *Climatic Change*, **65**, 297-314.

Borges, P. C. and A. Villavicencio, 2004: Avoiding academic and decorative planning in GHG emissions abatement studies with MCDA: the Peruvian case. *Eur. J. Oper. Res.*, **152**, 641-654.

Bosello, F., R. Roson and R.S.J. Tol, 2006: Economy-wide estimates of the implications of climate change: human health. *Ecol. Econ.*, **58**, 579-591.

Boutkan, E. and A. Stikker, 2004: Enhanced water resource base for sustainable integrated water resource management. *Nat. Resour. Forum*, **28**, 150-154.

Brooks, N., W.N. Adger and P.M. Kelly, 2005: The determinants of vulnerability and adaptive capacity at the national level and the implications for adaptation. *Global Environ. Chang.*, **15**, 151-163.

Brown, P., 1998: Climate, Biodiversity, and Forests. Issues and Opportunities Emerging from the Kyoto Protocol. World Resources Institute, Washington, District of Columbia.

Brown, D.A., 2003: The importance of expressly examining global warming policy issues through an ethical prism. *Global Environ. Chang.*, **13**, 229-234.

Bulkeley, H., 2001: Governing climate change: the politics of risk society? *T. I. Brit. Geogr.*, **26**, 430-447.

Burton, I. and M. van Aalst, 1999: *Come Hell or High Water: Integrating Climate Change Vulnerability and Adaptation into Bank Work.* World Bank, Washington, District of Columbia, 60 pp.

Butt, T.A. and B.A. McCarl, 2004: Farm and forest sequestration: can producers employ it to make some money? *Choices*, **Fall 2004**, 27-33.

CAG Consultants and Oxford Brookes University, 2004: *Planning Response to Climate Change: Advice on Better Practice.* Office of the Deputy Prime Minister, London, 108 pp.

Caparros, A. and F. Jacquemont, 2003: Conflicts between biodiversity and carbon sequestration programmes: economic and legal implications. *Ecol. Econ.*, **46**, 143-157.

Carbon International, undated: *The Gold Standard: Premium Quality Carbon Credits.*

Cash, D.W. and S. Moser, 2000: Linking global and local scales: designing dynamic assessment and management processes. *Global Environ. Chang.*, **10**, 109-120.

CBD, 2003: *Climate Change and Biodiversity: Executive Summary of the Report on the Interlinkages Between Biological Biodiversity and Climate Change.* CBD Technical Series No. 10. Convention on Biological Diversity.

Christensen, J.H. and Co-authors, 2007: Regional climate projections. *Climate Change 2007: The Physical Science Basis. Contribution of Working Group I to the Fourth Assessment Report of the Intergovernmental Panel on Climate Change*, S. Solomon, D. Qin, M. Manning, Z. Chen, M. Marquis, K.B. Averyt, M. Tignor and H.L. Miller, Eds., Cambridge University Press, Cambridge, 847-940.

Christensen, N.S., A.W. Wood, N. Voisin, D.P. Lettenmaier and R.N. Palmer, 2004: The effects of climate change on the hydrology and water resources of the Colorado River Basin. *Climatic Change*, **54**, 269-293.

Cifuentes, A.L., V.H. Borja-Aburto, N. Gouveia, G. Thurston and D.L. Davis, 2001: Hidden health benefits of greenhouse gas mitigation. *Science*, **293**, 1257-1259.

Clark, W.C., J. Jaeger, J. van Eijndhoven and N. Dickson, 2001: *Learning to Manage Global Environmental Risks: A Comparative History of Social Responses to Climate Change, Ozone Depletion, and Acid Rain.* MIT Press, Cambridge, Massachusetts, 361 pp.

Clark, W.C., P.J. Crutzen and H.-J. Schellnhuber, 2004: Science for global sustainability: toward a new paradigm. *Earth System Analysis for Sustainability*, H.-J. Schellnhuber, P.J. Crutzen, W.C. Clark, M. Claussen and H. Held, Eds., MIT Press, Cambridge, Massachusetts, 1-28.

Clarkson, R. and K. Deyes, 2002: Estimating the social cost of carbon emissions. Government Economic Service Working Paper 140. HM Treasury and Defra, London.

Cohen, S., D. Demeritt, J. Robinson and D. Rothman, 1998: Climate change and sustainable development: towards dialogue. *Global Environ. Chang.*, **8**, 341-373.

Cohen, S., K. Miller, K. Duncan, E. Gregorich, P. Groffman, P. Kovacs, V. Magaña, D. McKnight, E. Mills and D. Schimel. 2001: North America. *Climate Change 2001: Impacts, Adaptation, and Vulnerability. Contribution of Working Group II to the Third Assessment Report of the Intergovernmental Panel on Climate Change*, J.J. McCarthy, O.F. Canziani, N.A. Leary, D.J. Dokken and K.S. White, Eds., Cambridge University Press, Cambridge, 735-800.

Corfee-Morlot, J. and S. Agrawala, 2004: *The Benefits of Climate Change Policies: Analytical and Framework Issues.* OECD, Paris, 323 pp.

Dalal-Clayton, B., 2003: *The MDGs and Sustainable Development: The Need for a Strategic Approach.* IIED, London, 73-91.

Dang, H.H., A. Michaelowa and D.D. Tuan, 2003: Synergy of adaptation and mitigation strategies in the context of sustainable development: the case of Vietnam. *Clim. Policy*, **3**, S81-S96.

Den Elzen, M., P. Lucas and D. van Vuuren, 2005: Abatement costs of post-Kyoto climate regimes. *Energ. Policy*, **33**, 2138-2151.

Denton, F., Y. Sokona and J.P. Thomas, 2002: *Climate Change and Sustainable Development Strategies in the Making: What Should West African Countries Expect?* OECD, Paris, 27 pp.

Dessai, S., W.N. Adger, M. Hulme, J. Turnpenny, J. Kohler and R. Warren, 2004: Defining and experiencing dangerous climate change. *Climatic Change*, **64**, 11-25.

DETR, 2000: *Climate Change Programme.* DETR, London, 3 pp.

Döll, P., 2002: Impact of climate change and variability on irrigation requirement: a global perspective. *Climatic Change*, **54**, 269-293.

Dowie, J., 1999: Against risk. *Risk Decision Policy*, **4**, 57-73.

Downing, T.E., M. Munasinghe and J. Depledge, 2003: Introduction to special supplement on climate change and sustainable development. *Clim. Policy*, **3**, 3-8.

Downing, T.E. and G. Ziervogel, 2005: Food system scenarios: exploring global/local linkages. Working Paper, SEI Poverty and Vulnerability Report. Stockholm Environment Institute, Stockholm, 35 pp.

Downing, T.E., D. Anthoff, R. Butterfield, M. Ceronsky, M. Grubb, J. Guo, C. Hepburn, C. Hope, A. Hunt, A. Li, A. Markandya, S. Moss, A. Nyong, R.S.J. Tol and P. Watkiss, 2005: Scoping uncertainty in the social cost of carbon. Final Project Report, Social Cost of Carbon: A Closer Look at Uncertainty (SCCU). Defra/SEI, Oxford.

Dumanski, J., 2004: Carbon sequestration, soil conservation, and the Kyoto Protocol: summary of implications. *Climatic Change*, **65**, 255-261.

Dupuy, D., 1997: Technological change and environmental policy: the diffusion of environmental technology. *Growth Change*, **28**, 49-66.

Edenhofer, O., K. Lessmann and N. Bauer, 2006: Mitigation strategies and costs of climate protection: the effects of ETC in the hybrid model MIND. *Energy Journal Special Issue: Endogenous Technological Change and the Economics of Atmospheric Stabilization*, 207-222.

Edmonds, J., J. Clarke, J. Dooley, S.H. Kim, S.J. Smith, 2004: Stabilization of CO_2 in a B2 world: insights on The Roles of Carbon Capture and Disposal, Hydrogen, and Transportation Technologies. *Energy Economics*, **26**, 517-537.

EEA, 2005: Vulnerability and adaptation to climate change in Europe. EEA Technical Report No. 7/2005. EEA, Copenhagen.

Eriksen, S. and L.O. Næss, 2003: *Pro-Poor Climate Adaptation: Norwegian Development Cooperation and Climate Change Adaptation – An Assessment of Issues, Strategies and Potential Entry Points*. Centre for International Climate and Environmental Research Oslo, University of Oslo, Oslo.

Evans, D. and H. Sezer, 2004: Social discount rates for six major countries. *Appl. Econ. Lett.*, **11**, 557-560.

Fankhauser, S. and R.S.J. Tol, 2005: On climate change and economic growth. *Resour. Energy Econ.*, **27**, 1-17.

Fisher, B.S., G. Jakeman, H.M. Pant, M. Schwoon and R.S.J. Tol, 2006: CHIMP: a simple population model for use in integrated assessment of global environmental change. *Integr. Assess. J.*, **6**, 1-33.

Fisher, B.S. and Co-authors, 2007: Issues related to mitigation in the long-term context. *Climate Change 2007: Mitigation. Contribution of Working Group III to the Fourth Assessment Report of the Intergovernmental Panel on Climate Change*, B. Metz, O. Davidson, P. Bosch, R. Dave and L. Meyer, Eds., Cambridge University Press, Cambridge, UK.

Foray, D. and A. Grubler, 1996: Technology and the environment. Technol. *Forecast. Soc.*, **53**, 3-13.

GAIM Task Force, 2002: GAIM's Hilbertian questions. *Research GAIM: Newsletter of the Global Analysis, Integration and Modelling Task Force*, **5**, 1-16.

Geels, F.W., 2004: From sectoral systems of innovation to socio-technical systems: insights about dynamics and change from sociology and institutional theory. *Res. Policy*, **33**, 897-920.

Geels, F.W., 2005: Processes and patterns in transitions and system innovations: refining the co-evolutionary multi-level perspective. *Technol. Forecast. Soc.*, **72**, 681-696.

Geels, F.W. and W.A. Smit, 2000: Failed technology futures: pitfalls and lessons from a historical survey. *Futures*, **32**, 867-885.

Geist, H.J. and E.F. Lambin, 2002: Proximate causes and underlying driving forces of tropical deforestation. *BioScience*, **52**, 143-150.

Gitay H. and Co-authors, 2001: Ecosystems and their goods and services. *Climate Change 2001: Impacts, Adaptation, and Vulnerability. Contribution of Working Group II to the Third Assessment Report of the Intergovernmental Panel on Climate Change*, J.J. McCarthy, O.F. Canziani, N.A. Leary, D.J. Dokken and K.S. White, Eds., Cambridge University Press, Cambridge, 235-342.

Goklany, I.M., 2000a: Potential consequences of increasing atmospheric CO_2 concentration compared to other environmental problems. *Technology*, **7S**, 189-213.

Goklany, I.M., 2000b: *Applying the Precautionary Principle to Global Warming*. Weidenbaum Center Working Paper, PS 158. Washington University, St. Louis, Missouri.

Goklany, I.M., 2003: Relative contributions of global warming to various climate sensitive risks, and their implications for adaptation and mitigation. *Energ. Environ.*, **14**, 797-822.

Goklany, I.M., 2005: A climate policy for the short and medium term: stabilization or adaptation? *Energ. Environ.*, **16**, 667-680.

Goklany, I.M., 2007: Integrated strategies to reduce vulnerability and advance adaptation, mitigation, and sustainable development. *Mitigation and Adaptation Strategies for Global Change*, **12**, 755-786.

Gough, C. and S. Shackley, 2001: The respectable politics of climate change: the epistemic communities and NGOs. *Int. Aff.*, **77**, 329-346.

Greco, S., R.H. Moss, D. Viner and R. Jenne, 1994: *Climate Scenarios and Socio-Economic Projections for IPCC WGII Assessment*. Intergovernmental Panel on Climate Change, Washington, District of Columbia, 67 pp.

Gregory, J.M., K.W. Dixon, R.J. Stouffer, A.J. Weaver, E. Driesschaert, M. Eby, T. Fichefet, H. Hasumi, A. Hu, J.H. Jungclaus, I.V. Kamenkovich, A. Levermann, M. Montoya, S. Murakami, S. Nawrath, A. Oka, A.P. Sokolov and R.B. Thorpe, 2005: A model intercomparison of changes in the Atlantic thermohaline circulation in response to increasing atmospheric CO_2 concentration. *Geophys. Res. Lett.*, **32**, L12703, doi:10.1029/2005GL023209.

Grothmann, T. and A. Patt, 2005: Adaptive capacity and human cognition: the process of individual adaptation to climate change. *Global Environ. Chang.*, **15**, 199-213.

Grubb, M., 2003: The economics of the Kyoto Protocol. *World Econ.*, **4**, 143-189.

Grubb, M., J. Kohler and D. Anderson, 2002: Induced technological change in energy and environmental modelling: analytic approaches and policy implications. *Annu. Rev. Energy Env.*, **21**, 271-308.

Grubler, A., 2000: *Technology and Global Change*. Cambridge University Press, Cambridge, 462 pp.

Guo, J., C. Hepburn, R.S.J. Tol and D. Anthoff, 2006: Discounting and the social cost of carbon: a closer look at uncertainty. *Environ. Sci. Policy*, **9**, 205-216.

Haddad, B., 2005: Ranking the adaptive capacity of nations to climate change when socio-political goals are explicit. *Global Environ. Chang.*, **15**, 165-176.

Halsnæs, K. and Co-authors, 2007: Framing issues. *Climate Change 2007: Mitigation. Contribution of Working Group III to the Fourth Assessment Report of the Intergovernmental Panel on Climate Change*, B. Metz, O. Davidson, P.Bosch, R. Dave and L. Meyer, Eds., Cambridge University Press, Cambridge, UK.

Handmer, J.W., S. Doers and T.E. Downing, 1999: Societal vulnerability to climate change and variability. *Mitigation and Adaptation Strategies for Global Change*, **4**, 267-281.

Hare, B., 2006: Relationship between increases in global mean temperature and impacts on ecosystems, food production, water and socio-economic systems. *Avoiding Dangerous Climate Change*, H.-J. Schellnhuber, W. Cramer, N. Nakićenović, T. Wigley and G. Yohe, Eds., Cambridge University Press, Cambridge, 177-185.

Harvey, L.D.D., 2006: Uncertainties in global warming science and near-term emission policies. *Clim. Policy*, **6**, 573-584.

Hayhoe, K., D. Cayan and C.B. Field, 2004: Emissions pathways, climate change, and impacts on California. *P. Natl. Acad. Sci. USA*, **101**, 12422-12427.

Hirsch, P. and A. Wyatt, 2004: Negotiating local livelihoods: scales of conflict in the Se San River Basin. *Asia Pac. Viewpoint*, **45**, 51-68.

Hitz, S. and J. Smith, 2004: Estimating global impacts from climate change. *Global Environ. Chang.* **14**, 201-218.

Holdridge, L.R., 1947: Determination of world plant formations from simple climatic data. *Science*, **105**, 367 – 368.

Holdridge, L. R., 1967: *Life Zone Ecology*. Tropical Science Centre, San Jose, Costa Rica.

Howe, J. and P. Wheeler, 1999: Urban food growing: the experience of two UK cities. *Sustain. Dev.*, **7**, 13-24.

Huq, S. and H. Reid, 2004: Mainstreaming adaptation in development. *IDS Bull.*, **35**, 15-21.

Huq, S., A. Rahman, M. Konate, Y. Sokona and H. Reid, 2003: *Mainstreaming Adaptation to Climate Change in Least Developed Countries (LDCs)*. IIED, London, 40 pp.

ICLEI, 2006: *Climates for City Protection (CCP) Campaign*, http://www.iclei.org/.

IEA (International Energy Agency), 2005: *Electricity Information 2005*. Organisation for Economic Cooperation and Development, Paris, 783 pp.

Ikeme, J., 2003: Climate change adaptation deficiencies in developing countries: the case of sub-Saharan Africa. *Mitigation and Adaptation Strategies for Global Change*, **8**, 29-52.

IPCC, 1992: *Climate Change 1992: The Supplementary Report to the IPCC Scientific Assessment*. J.T. Houghton, B.A. Callander and S.K. Varney, Eds., Cambridge University Press, Cambridge, 200 pp.

IPCC, 2001a: *Climate Change 2001: Synthesis Report. A Contribution of Working Groups I, II, III to the Third Assessment Report of the Intergovernmental Panel on Climate Change*, R.T. Watson and the Core Team, Eds., Cambridge University Press, Cambridge and New York, 398 pp.

IPCC, 2001b: *Climate Change 2001: Mitigation. Contribution of Working Group III to the Third Assessment Report of the Intergovernmental Panel on Climate Change*, B. Metz, O. Davidson, R. Swart and J. Pan, Eds., Cambridge University Press, Cambridge, 760 pp.

IPCC, 2001c: *Climate Change 2001: Impacts, Adaptation, and Vulnerability. Contribution of Working Group II to the Third Assessment Report of the Intergovernmental Panel on Climate Change*, J.J. McCarthy, O.F. Canziani, N.A. Leary, D.J. Dokken and K.S. White, Eds., Cambridge University Press., Cambridge, 1032 pp.

IPCC, 2002: *Climate Change and Biodiversity*, H. Gitay, A. Suárez, R.T. Watson and D.J. Dokken, Eds., IPCC, Geneva, 85 pp.

IPCC, 2007: *Climate Change 2007: Mitigation. Contribution of Working Group III to the Fourth Assessment Report of the Intergovernmental Panel on Climate Change*, B. Metz, O. Davidson, P. Bosch, R. Dave and L. Meyer, Eds., Cambridge University Press, Cambridge, UK.

IRM, AIRMIC and ALARM, 2002: *A Risk Management Standard*. AIRMIC, IRM, ALARM, London, 17 pp.

Jaffe, A.B. and R.N. Stavins, 1994: The energy-efficiency gap: what does it mean? *Energ. Policy*, **22**, 804-810.

Jaffe, A.B., R.G. Newell and R.N. Stavins, 2005: A tale of two market failures: technology and environmental policy. *Ecol. Econ.*, **54**, 164.

Jahan, S. and A. Umana, 2003: The environment–poverty nexus. *Dev. Policy J.*, **3**, 53-70.

Kane, S. and J.F. Shogren, 2000: Linking adaptation and mitigation in climate change policy. *Climatic Change*, **45**, 75-102.

Kasperson, J.X. and R.E. Kasperson, 2005: *The Social Contours of Risk*. Earthscan, London, 578 pp.

Keeney, R.L. and H. Raiffa, 1976: *Decisions with Multiple Objectives: Preferences and Value Trade-offs*. John Wiley and Sons, New York, 600 pp.

Keller, K., G. Yohe and M. Schlesinger, 2007: Managing the risks of climate thresholds: uncertainties and information needs. Climatic Change, doi:10.1007/s10584-006-9114-6. http://www.geosc.psu.edu/~kkeller/Keller_cc_07.pdf.

Klein, R.J.T., 2001: *Adaptation to Climate Change in German Official Development Assistance: An Inventory of Activities and Opportunities, With a Special Focus on Africa*. Deutsche Gesellschaft für Technische Zusammenarbeit, Eschborn, 44 pp.

Klein, R.J.T., 2002: Climate change, adaptive capacity and sustainable development. Paper presented at an Expert Meeting on Adaptation to Climate Change and Sustainable Development, Organisation for Economic Co-operation and Development, Paris, 13–14 March 2002, 8 pp.

Klein, R.J.T., E.L. Schipper and S. Dessai, 2005: Integrating mitigation and adaptation into climate and development policy: three research questions. *Environ. Sci. Policy*, **8**, 579-588.

Köhler, J., M. Grubb, D. Popp and O. Edenhofer, 2006: The transition to endogenous technical change in climate-economy models. *Energy Journal Special Issue: Endogenous Technological Change and the Economics of Atmospheric Stabilization*, 17-55.

Koonjul, J., 2004: The special case of Small Island States for sustainable development. *Nat. Resour. Forum*, **28**, 155-156.

Kulindwa, K., 2002: Economic reforms and the prospect for sustainable development in Tanzania. *Dev. S. Afr.*, **19**, 389-404.

Lal, R., 2004: Soil carbon sequestration impacts on global climate change and food security. *Science*, **304**, 1623-1627.

Lehtonen, M., 2004: The environmental–social interface of sustainable development: capabilities, social capital, institutions. *Ecol. Econ.*, **49**, 199-214.

Lempert, R. and M.E. Schlesinger, 2000: Robust strategies for abating climate change. *Climatic Change*, **45**, 387- 401.

Lempert, R.J., S.W. Popper and S.C. Bankes, 2004: *Shaping the Next One Hundred Years: New Methods for Quantitative, Long-term Policy Analysis*. Rand, Santa Monica, California, 170 pp.

Levine, M. and Co-authors, 2007: Residential and commercial buildings. *Climate Change 2007: Mitigation. Contribution of Working Group III to the Fourth Assessment Report of the Intergovernmental Panel on Climate Change*, B. Metz, O. Davidson, P. Bosch, R. Dave and L. Meyer, Eds., Cambridge University Press, Cambridge, UK.

Link, P.M. and Tol, R.S.J., 2004: Possible economic impacts of a shutdown of the thermohaline circulation: an application of FUND. *Port. Econ. J.*, **3**, 99-114.

Lorenzoni, I., N.F. Pidgeon and R.E. O'Connor, 2005: Dangerous climate change: the role for risk research. *Risk Anal.*, **25**, 1387-1398.

Manne, A.S. and R.G. Richels, 1992: *Buying Greenhouse Insurance: The Economic Costs of CO_2 Emission Limits*. MIT Press, Cambridge, Massachusetts, 194 pp.

Markandya, A. and K. Halsnæs, 2000: *Climate Change and its Linkages with Development, Equity and Sustainability*. Intergovernmental Panel on Climate Change, Geneva.

Markandya, A. and K. Halsnæs, 2002: *Climate Change and Sustainable Development: Prospects for Developing Countries*. Earthscan, London, 291 pp.

Martens, P. and J. Rotmans, Eds., 2002: *Transitions in a Globalising World*. Swets and Zeitlinger, Lisse, 135 pp.

Mastrandrea, M.D. and S.H. Schneider, 2001: Integrated assessment of abrupt climate change. *Clim. Policy*, **1**, 433-449.

Mastrandrea, M.D. and S.H. Schneider, 2004: Probabilistic integrated assessment of "dangerous" climate change. *Science*, **304**, 571-575.

McCarl, B.A. and U.A. Schneider, 2000: Agriculture's role in a greenhouse gas emission mitigation world: an economic perspective. *Rev. Agr. Econ.*, **22**, 134-159.

McDonald, S., S. Robinson and K. Thierfelder, 2006: Impact of switching production to bioenergy crops: the switchgrass example. *Energ. Econ.*, **28**, 243-265.

McLean, R.F., A. Tsyban, V. Burkett, J.O. Codignotto, D.L. Forbes, N. Mimura, R.J. Beamish, V. Ittekkot, L. Bijlsma and I. Sanchez-Arevalo, 2001: Coastal zones and marine ecosystems. *Climate Change 2001: Impacts, Adaptation, and Vulnerability. Contribution of Working Group II to the Third Assessment Report of the Intergovernmental Panel on Climate Change*, J.J. McCarthy, O.F. Canziani, N.A. Leary, D.J. Dokken and K.S. White, Eds., Cambridge University Press, Cambridge, 343-379.

McMichael, A., A. Githeko, R. Akhtar, R. Carcavallo, D. Gubler, A. Haines, R. Kovats, P. Martens, J. Patz and A. Sasaki, 2001: Human health. *Climate Change 2001: Impacts, Adaptation, and Vulnerability. Contribution of Working Group II to the Third Assessment Report of the Intergovernmental Panel on Climate Change*, J.J. McCarthy, O.F. Canziani, N.A. Leary, D.J. Dokken and K.S. White, Eds., Cambridge University Press, Cambridge, 453-478.

MEA (Millennium Ecosystem Assessment), 2005: *Ecosystems and Human Well-being: Synthesis*. Island Press, Washington, District of Columbia, 160 pp.

Meehl G. and Co-authors, 2007: Global climate projections. *Climate Change 2007: The Physical Science Basis. Contribution of Working Group I to the Fourth Assessment Report of the Intergovernmental Panel on Climate Change*, S. Solomon, D. Qin, M. Manning, Z. Chen, M. Marquis, K.B. Averyt, M. Tignor and H.L. Miller, Eds., Cambridge University Press, Cambridge, 747-846.

Metz, B., M. Berk, M. den Elzen, B. de Vries and D. van Vuuren, 2002: Towards an equitable global climate change regime: compatibility with Article 2 of the Climate Change Convention and the link with sustainable development. *Clim. Policy*, **2**, 211-230.

Michaelowa, A., 2001: Mitigation versus adaptation: the political economy of competition between climate policy strategies and the consequences for developing countries. HWWA Discussion Paper No. 153, University of Hamburg, Hamburg.

Michaelowa, A. and K. Michaelowa, 2005: Climate or development: is ODA diverted from its original purpose? HWWI Research Paper 2, Hamburg, 32 pp.

Michaelowa, A., K. Tangen and H. Hasselknippe, 2005: Issues and options for the post-2012 climate architecture: an overview. *Int. Environ. Agreements*, **5**, 5-24.

Ministry for Rural Affairs and the Environment Malta, 2004: *The First Communication of Malta to the UNFCCC*.

Misra, B.M., 2003: Advances in nuclear desalination. *Int. J. Nucl. Desalination*, **1**, 19-29.

Moomaw, W.R., J.R. Moreira, K. Blok, D.L. Greene, K. Gregory, T. Jaszay, T. Kashiwagi, M. Levine, M. McFarland, N. Siva Prasad, L. Price, H.-H. Rogner, R. Sims, F. Zhou and P. Zhou, 2001: Technological and economic potential of greenhouse gas emissions reduction. *Climate Change 2001: Mitigation. Contribution of Working Group III to the Third Assessment Report of the Intergovernmental Panel on Climate Change*, B. Metz, O. Davidson, R. Swart and J. Pan, Eds., Cambridge University Press, Cambridge, 167-299.

Morita, T., J. Robinson, J. Adegulugbe, J. Alcamo, D. Herbert, E. La Rovere, N. Nakićenović, H. Pitcher, P. Raskin, K. Riahi, A. Sankovski, V. Sokolov, B. de Vries and D. Zhou, 2001: Greenhouse gas emission mitigation scenarios and implications. *Climate Change 2001: Mitigation. Contribution of Working Group III to the Third Assessment Report of the Intergovernmental Panel on Climate Change*, B. Metz, O. Davidson, R. Swart and J. Pan, Eds., Cambridge University Press, Cambridge, 115-166.

Moser, S.C., 2005: Impacts assessments and policy responses to sea-level rise in three U.S. states: an exploration of human dimension uncertainties. *Global*

Environ. Chang., **15**, 353-369.

Moss, R., A. Brenkert and E. Malone, 2001: *Vulnerability to Climate Change: A Quantitative Approach*. Pacific Northwest National Laboratory (PNNL-SA-33642).

Munasinghe, M., 1992: *Environmental Economics and Sustainable Development*. World Bank., Washington District of Columbia, 111 pp.

Munasinghe, M. and R. Swart, Eds., 2000: Climate *Change and its Linkages with Development, Equity and Sustainability.* Intergovernmental Panel on Climate Change, Geneva.

Munasinghe, M. and R. Swart, 2005: *Primer on Climate Change and Sustainable Development*. Cambridge University Press, Cambridge, 458 pp.

Nagai, H. and C. Hepburn, 2005: How cost-effective are carbon emission reductions under the Prototype Carbon Fund? MSc Environmental Change and Management, Oxford University, Oxford, 132 pp.

Najam, A., A. Rahman, S. Huq and Y. Sokona, 2003: Integrating sustainable development into the Fourth Assessment Report of the IPCC. *Clim. Policy*, **3**, S9-S17.

Nakićenović, N. and R. Swart, Eds., 2000: *IPCC Special Report on Emissions Scenarios: A Special Report of IPCC Working Group III*. Cambridge University Press, Cambridge, 599 pp.

Nederveen, A., J. Konings and J. Stoop, 2003: Globalization, international transport and the global environment: technological innovation, policy making and the reduction of transport emissions. *Transport. Plan. Technol.*, **26**, 41-67.

Newell, R.G. and W.A. Pizer, 2000: Regulating stock externalities under uncertainty. Discussion Paper 99-10-REV. Resources for the Future, Washington, District of Columbia.

Nicholls, R.J. and J.A. Lowe, 2004: Benefits of mitigation of climate change for coastal areas. *Global Environ. Chang.*, **14**, 229-244.

Nicholls R.J. and J.A. Lowe, 2006: Climate stabilisation and impacts of sea-level rise. *Avoiding Dangerous Climate Change*, H.-J. Schellnhuber, W. Cramer, N. Nakićenović, T. Wigley and G. Yohe, Eds., Cambridge University Press, Cambridge, 195-202.

Nicholls, R.J. and R.S.J. Tol, 2006: Impacts and responses to sea-level rise: a global analysis of the SRES scenarios over the twenty-first century. *Philos. T. Roy. Soc. A*, **364**, 1073-1095.

Nicholls, R.J., R.S.J. Tol and A.T. Vafeidis, 2005: Global Estimates of the Impact of a Collapse of the West Antarctic Ice Sheet: An Application of FUND. Working Paper FNU-78, Research Unit on Sustainability and Global Change, Hamburg University and Centre for Marine and Atmospheric Science, Hamburg.

Nicholls R.J., R.S.J. Tol and J.W. Hall, 2007: Assessing impacts and responses to global-mean sea-level rise. *Human-Induced Climate Change: An Interdisciplinary Assessment*, M. Schlesinger, J. Reilly, H. Kheshgi, J. Smith, F. de la Chesnaye, J.M. Reilly, T. Wilson and C. Kolstad, Eds., Cambridge University Press, Cambridge.

Nordhaus, W.D., 2001: Global warming economics. *Science*, **294**, 1283-1284.

ODPM, 2004: *The Planning Response to Climate Change: Advice on Better Practice*. Office of the Deputy Prime Minister, London.

ODPM, 2005: *Planning Policy Statement I: Delivering Sustainable Development*. Office of the Deputy Prime Minister, London.

OECD, 2005: *Bridge over Troubled Waters: Linking Climate Change and Development*. Organisation for Economic Cooperation and Development, S. Agrawala, Ed., OECD Publications, Paris, 153 pp.

OECD, 2006: *Declaration on Integrating Climate Change Adaptation into Development Co-operation Adopted by Development and Environment Ministers of OECD Member Countries*. OECD Publications, Paris.

Olsen, J.P. and J.G. March, 1989: *Rediscovering Institutions: The Organizational Basis of Politics*. The Free Press, New York.

Oppenheimer, M. and A. Petsonk, 2005: Article 2 of the UNFCCC: historical origins, recent interpretations. *Climatic Change*, **73**, 195-226.

Ott, H.E., H. Winkler, B. Brouns, S. Kartha, M. Mace, S. Huq, Y. Kameyama, A.P. Sari, J. Pan, Y. Sokona, P.M. Bhandari, A. Kassenberg, E.L. La Rovere and A.A. Rahman, 2004: *South-North Dialogue on Equity in the Greenhouse: A Proposal for an Adequate and Equitable Global Climate Agreement*. GTZ, Eschborn.

Pal, R.C. and Sharma, A., 2001: Afforestation for reclaiming degraded village common land: a case study. *Biomass Bioenerg.*, **21**, 35-42.

Parry, M., N. Arnell, M. Hulme, R. Nicholls and M. Livermore, 1998: Adapting to the inevitable. *Nature*, **395**, 741.

Parry, M.L., C.A. Rosenzweig, A. Iglesias, M. Livermore and G. Fisher, 2004: Effects of climate change on global food production under SRES emissions and socioeconomic scenarios. *Global Environ. Chang.*, **14**, 53-67.

Payne, J.T., A.W. Wood, A.F. Hamlet, R.N. Palmer and D.P. Lettenmaier, 2004: Mitigating the effects of climate change on the water resources of the Columbia River Basin. *Climatic Change*, **62**, 233-256.

Pelling, M., 2003: *The Vulnerability of Cities: Natural Disaster and Social Resilience*. Earthscan, London, 212 pp.

Peters, M., R. House, J. Lewandrowski and H. McDowell, 2001: Economic impacts of carbon charges on U.S. agriculture. *Climatic Change*, **40**, 445-473.

Pew Center, 2002: Climate change mitigation in developing countries: Brazil, China, India, Mexico, South Africa, and Turkey. Pew Center Global Warming Research Reports, October 24, 2002. Pew Center on Global Climate Change, Arlington, Virginia.

Pidgeon, N., R.E. Kasperson and P. Slovic, Eds., 2003: *The Social Amplification of Risk*. Cambridge University Press, Cambridge, 464 pp.

Pierson, P., 2004: *Politics in Time: History, Institutions and Social Analysis*. Princeton University Press, Princeton, New Jersey, 208 pp.

Rahmstorf, S., 2000: The thermohaline ocean circulation: a system with dangerous thresholds? *Climatic Change*, **46**, 247-256.

Raiffa, H., 1968: *Decision Analysis: Introductory Lectures on Choices Under Uncertainty*. Addison-Wesley, Reading, Massachusetts, 310 pp.

Raskin, P., G. Gallopin, P. Gutman, A. Hammond and R. Swart, 2002a: *Bending the Curve: Toward Global Sustainability*. A Report of the Global Scenario Group. SEI PoleStar Series Report No. 8. Stockholm Environment Institute, Stockholm, 144 pp.

Raskin, P., T. Banuri, G. Gallopin, P. Gutman, A. Hammond, R. Kates and R. Swart. 2002b: *Great Transition: The Promise and Lure of the Times Ahead*. A Report of the Global Scenario Group. SEI PoleStar Series Report No. 10. Stockholm Environment Institute, Boston, Massachusetts.

Rayner, S. and E.L. Malone, Eds., 2001: Climate change, poverty, and intragenerational equity: the national level. *Int. J. Global Environ. Iss.*, **1**, 175-202.

Redclift, M. and T. Benton, Eds., 1994: *Social Theory and the Global Environment*. Routledge, London and New York, 280 pp.

Richels, R.G., A.S. Manne and T.M.L. Wigley, 2004: Moving beyond concentrations: the challenge of limiting temperatures change. AEI-Brookings Joint Center Working Paper 04-11, April 2004.

Rip, A. and R. Kemp, 1998: Technological change. *Human Choice and Climate Change: Volume 2: Resources and Technology*, E.L. Malone and S. Rayner, Eds., Batelle Press, Columbus, Ohio.

Robinson, J., 2004: Squaring the circle? Some thoughts on the idea of sustainable development. *Ecol. Econ.*, **48**, 369-384.

Robinson, J. and D. Herbert, 2001: Integrating climate change and sustainable development. *Int. J. Global Environ. Iss.*, **1**, 130-149.

Robinson, J., M. Bradley, P. Busby, D. Connor, A. Murray and B. Sampson, 2003: *Climate Change and Sustainable Development: Realizing the Opportunity at the World Climate Change Conference, Moscow*, 30 September–4 October.

Robinson, J., M. Bradley, P. Busby, D. Connor, A. Murray, B. Sampson and W. Soper, 2006: Climate change and sustainable development: realizing the opportunity. *Ambio*, **35**, 2-8.

Rogers, E.M., 2003: *Diffusion of Innovations*, 5th ed., Free Press, New York, 384 pp.

Rogner, H.-H., D. Zhou and Co-authors, 2007: Introduction. *Climate Change 2007: Mitigation. Contribution of Working Group III to the Fourth Assessment Report of the Intergovernmental Panel on Climate Change*, B. Metz, O. Davidson, P. Bosch, R. Dave and L. Meyer, Eds., Cambridge University Press, Cambridge, UK.

Rosenberg, N.J. and M.J. Scott, 1995: Implications of policies to prevent climate change for future food security. *Climate Change and World Food Security: Proceedings of the NATO Advanced Research Workshop "Climate Change and World Food Security", Oxford, 11-15 July 1993*, T.E. Downing, Ed., Springer, Berlin, 551-589.

Ruth, M., 2005: Future socioeconomic and political challenges of global climate change. *From Resource Scarcity to Ecological Security: Exploring New Limits to Growth*, D.C. Pirages and K. Cousins, Eds., MIT Press, Cambridge, Massachusetts, 145-164.

Ruth, M. and A.-C. Lin, 2006: Regional energy demand and adaptations to climate change: methodology and application to the state of Maryland, USA. *Energ. Policy*, **34**, 2820-2833.

Sarkar, J., 1998: Technological diffusion: alternative theories and historical evidence. *J. Econ. Surv.*, **12**, 131-176.

Sathaye, J. and Co-authors, 2007: Sustainable development and mitigation. *Climate Change 2007: Mitigation. Contribution of Working Group III to the Fourth Assessment Report of the Intergovernmental Panel on Climate Change*, B. Metz, O. Davidson, P. Bosch, R. Dave and L. Meyer, Eds., Cambridge University Press, Cambridge, UK.

Sawin, J. and C. Flavin, 2004: *Policy Lessons for the Advancement and Diffusion of Renewable Energy Technologies around the World*. Thematic background paper prepared for The Energy Research Institute, New Delhi, India. Presented at the International Conference for Renewable Energies, Bonn, Germany.

Schelling, T.C., 1995: Intergenerational discounting. *Energ. Policy*, **23**, 395-401.

Schipper, E.L.F., 2006: Conceptual history of adaptation in the UNFCCC process. Review of European Community and International Environmental Law, **15**, 82-92.

Schneider, S.H., 1997: Integrated assessment modelling of global climate change: transparent rational tool for policy making or opaque screen hiding value-laden assumptions? *Environ. Model. Assess.*, **2**, 229-249.

Schneider, S.H., 2003: *Abrupt Non-linear Climate Change, Irreversibility, and Surprise*. OECD, Paris, 32 pp.

Schneider, S.H. and R.S. Chen, 1980: Carbon dioxide flooding: physical factors and climatic impact. *Annu. Rev. Energy*, **5**, 107-140.

Schneider, S.H., W. Easterling and L. Mearns, 2000: Adaptation: sensitivity to natural variability, agent assumptions and dynamic climate changes. *Climatic Change*, **45**, 203-221.

Seck, M., N.A. Mamouda and S. Wade, 2005: Senegal adaptation and mitigation through produced environments: the case for agricultural intensification in Senegal. *Vulnerability, Adaptation and Climate Disasters: A Conceptual Overview*, F. Yamin, A. Rahman and S. Huq, Eds., *IDS Bull.*, **36**, 71-86.

Smil, V., 2000: Energy in the 20th century: resources, conversions, costs, uses and consequences. *Annu. Rev. Energ. Env.*, **25**, 21-51.

Smit, B., O. Pilifosova, I. Burton, B. Challenger, S. Huq, R.J.T. Klein and G. Yohe, 2001: Adaptation to climate change in the context of sustainable development and equity. *Climate Change 2001: Impacts, Adaptation, and Vulnerability. Contribution of Working Group II to the Third Assessment Report of the Intergovernmental Panel on Climate Change*, J.J. McCarthy, O.F. Canziani, N.A. Leary, D.J. Dokken and K.S. White, Eds., Cambridge University Press, Cambridge, 879-906.

Smith, J. and D. Tirpak, Eds., 1989: *The Potential Impacts of Global Climate Change on the United States*. Environmental Protection Agency, Washington, District of Columbia.

Smith, J.B., H.-J. Schellnhuber, M. Mirza, S. Fankhauser, R. Leemans, Lin Erda, L. Ogallo, B. Pittock, R. Richels, C. Rosenzweig, U. Safriel, R.S.J. Tol, J. Weyant and G. Yohe, 2001: Vulnerability to climate change and reasons for concern: a synthesis. *Climate Change 2001: Impacts, Adaptation, and Vulnerability. Contribution of Working Group II to the Third Assessment Report of the Intergovernmental Panel on Climate Change*, J.J. McCarthy, O.F. Canziani, N.A. Leary, D.J. Dokken and K.S. White, Eds., Cambridge University Press, Cambridge, 914-967.

Smith, J.B., R.J.T. Klein and S. Huq, Eds., 2003: *Climate Change, Adaptive Capacity and Development*. Imperial College Press, London, 347 pp.

Southampton City Council, 2004: *Climate Change, Air Quality Strategy and Action Plan*.

Sperling, F., Ed., 2003: *Poverty and Climate Change: Reducing the Vulnerability of the Poor through Adaptation*. AfDB, ADB, DFID, EC DG Development, BMZ, DGIS, OECD, UNDP, UNEP and the World Bank, Washington, District of Columbia, 43 pp.

Stainforth, D.A., T. Aina, C. Christensen, M. Collins, N. Faull, D.J. Frame, J.A. Kettleborough, S. Knight, A. Martin, J.M. Murphy, C. Piani, D. Sexton, L.A. Smith, R.A. Spicer, A.J. Thorpe and M.R. Allen, 2005: Uncertainty in predictions of the climate response to rising levels of greenhouse gases. *Nature*, **433**, 403-405.

Stern, N., 2007: *The Economics of Climate Change: The Stern Review*. Cambridge University Press, Cambridge, 692 pp.

Stern, P.C. and T. Dietz, 1994: The value basis of environmental concern. *J. Soc. Issues*, **50**, 65-84.

Swart, R., J. Robinson and S. Cohen, 2003: Climate change and sustainable development: expanding the options. *Clim. Policy*, **3**, S19-S40.

Taylor, A., B. Hassan, J.A. Downing and T.E. Downing, 2006: Toward a typology of adaptation–mitigation inter-relationships. Stockholm Environment Institute, Oxford. http://www.vulnerabilitynet.org/.

Tol, R.S.J., 2001: Equitable cost–benefit analysis of climate change. *Ecol. Econ.*, **36**, 71-85.

Tol, R.S.J., 2002: Welfare specifications and optimal control of climate change: an application of fund. *Energ. Econ.*, **24**, 367-376.

Tol, R.S.J., 2003: Is the uncertainty about climate change too large for expected cost-benefit analysis? *Climatic Change*, **56**, 265-289.

Tol, R.S.J., 2005a: Adaptation and mitigation: trade-offs in substance and methods. *Environ. Sci. Policy*, **8**, 572-578.

Tol, R.S.J., 2005b: The marginal damage cost of carbon dioxide emissions: an assessment of the uncertainties. *Energ. Policy*, **33**, 2064-2074.

Tol, R.S.J., 2005c: Emission abatement versus development as strategies to reduce vulnerability to climate change: an application of FUND. *Environ. Dev. Econ.*, **10**, 615-629.

Tol, R.S.J., 2007: The double trade-off between adaptation and mitigation for sea-level rise: an application of FUND. *Mitigation and Adaptation Strategies for Global Change*, **12**, 741-753.

Tol, R.S.J. and H. Dowlatabadi, 2001: Vector-borne diseases, climate change, and economic growth. *Integr. Assess.*, **2**, 173-181.

Tol, R.S.J. and G. Yohe, 2006: Of dangerous climate change and dangerous emission reduction. *Avoiding Dangerous Climate Change*, H.-J. Schellnhuber, W. Cramer, N. Nakićenović, T. Wigley and G. Yohe, Eds., Cambridge University Press, Cambridge, 291-298.

Tol, R.S.J., M. Bohn, T.E. Downing, M.-L. Guillerminet, E. Hizsnyik, R. Kasperson, K. Lonsdale, C. Mays, R.J. Nicholls, A.A. Olsthoorn, G. Pfeifle, M. Poumadère, F.L. Toth, A.T. Vafeidis, P.E. van der Werff and I.H. Yetkiner, 2006: Adaptation to five metres of sea level rise. *J. Risk Res.*, **9**, 467-482.

Tompkins, E. and W.N. Adger, 2005: Defining a response capacity for climate change. *Environ. Sci. Policy*, **8**, 562-571.

Tonn, B., 2003: An equity first, risk-based framework for managing global climate change. *Global Environ. Chang.*, **13**, 295-306.

Toth, F.L., 2003: State of the art and future challenges for integrated environmental assessment. *Integr. Assess.*, **4**, 250-264.

Toth, F.L., M. Mwandosya, C. Carraro, J. Christensen, J. Edmonds, B. Flannery, G. Gay-Garcia, H. Lee, K.M. Meyer-Abich, E. Nikitina, A. Rahman, R. Richels, R. Ye, , A. Villavicencio, Y. Wake and J. Weyant, 2001: Decision-making frameworks. *Climate Change 2001: Mitigation. Contribution of Working Group III to the Third Assessment Report of the Intergovernmental Panel on Climate Change*, B. Metz, O. Davidson, R. Swart and J. Pan, Eds., Cambridge University Press, Cambridge, 601-688.

Toth, F.L., T. Bruckner, H.-M. Füssel, M. Leimbach, G. Petschel-Held and H.-J. Schellnhuber, 2002: Exploring options for global climate policy in an inverse integrated assessment framework. *Environment*, **44**, 23-34.

Toth, F.L., T. Bruckner, H.-M. Füssel, M. Leimbach and G. Petschel-Held, 2003a: Integrated assessment of long-term climate policies. Part 1. Model presentation. *Climatic Change*, **56**, 37-56.

Toth, F.L., T. Bruckner, H.-M. Füssel, M. Leimbach and G. Petschel-Held, 2003b: Integrated assessment of long-term climate policies. Part 2. Model results and uncertainty analysis. *Climatic Change*, **56**, 57-72.

UK FRP (United Kingdom Forestry Research Programme), 2005: *From the Mountain to the Tap: How Land Use and Water Management can Work Together for the Rural Poor*. UK FRP, London, 54 pp.

United Nations, 1997: *Kyoto Protocol to the United Nations Framework Convention on Climate Change*. http://unfccc.int/resource/ docs/convkp/kp eng.pdf.

United Nations, 2000: *United Nations Millennium Declaration: Resolution adopted by the General Assembly, 55th Session, 18 September*. A/RES/55/22. United Nations.

Unruh, G. and J. Carrillo-Hermosilla, 2006: Globalizing carbon lock-in. *Energ. Policy*, **34**, 1185-1197.

Van Asselt, H., J. Gupta and F. Biermann, 2005: Advancing the climate agenda: exploiting material and institutional linkages to develop a menu of policy options. *Rev. Eur. Community Int. Environ. Law*, **14**, 255-263.

Van Harmelen, T., J. Bakker, B. de Vries, D. van Vuuren, M. den Elzen and P. Mayerhofer, 2002: Long-Term Reductions in Costs of Controlling Regional Air Pollution in Europe Due to Climate Policy. *Environ. Sci. and Policy*, **5**, 349-365.

VanRheenen, N.T., A.W. Wood, R.N. Palmer and D.P. Lettenmaier, 2004: Potential implications of PCM climate change scenarios for Sacramento-San Joaquin River Basin hydrology and water resources. *Climatic Change*, **62**, 257-281.

Van Vuuren, D.P., B. Eickhout, P.L. Lucas and M.G.J. den Elzen, 2006: Long-term multi-gas scenarios to stabilise radiative forcing: exploring costs and benefits within an integrated assessment framework. *Energ. J.*, **27**, 201-233.

Vaughan, D.G. and J.R. Spouge, 2002: Risk estimation of collapse of the West

Antarctic Ice Sheet. *Climatic Change*, **52**, 65–91.

Vellinga, P. and Co-authors, 2001: Insurance and other financial services. *Climate Change 2001: Impacts, Adaptation, and Vulnerability. Contribution of Working Group II to the Third Assessment Report of the Intergovernmental Panel on Climate Change*, J.J. McCarthy, O.F. Canziani, N.A. Leary, D.J. Dokken and K.S. White, Eds., Cambridge University Press, Cambridge, 417-450.

Venema, D.H. and M. Cisse, 2004: *Seeing the Light: Adapting to Climate Change with Decentralized Renewable Energy in Developing Countries*. IISD, Canada, 174 pp.

Venema, D.H., E.J. Schiller, K. Adammowski and J.M. Thizy, 1997: A water-resource planning response to climate change in Senegal River basin. *J. Environ. Manage.*, **49**, 125-155.

Verheyen, R., 2005: *Climate Change Damage and International Law: Prevention, Duties and State Responsibility*. Martinus Nijhoff, Netherlands, 418 pp.

Warren, R., 2006: Impacts of global climate change at different annual mean global temperature rises. *Avoiding Dangerous Climate Change*, H.-J. Schellnhuber, W. Cramer, N. Nakićenović, T. Wigley and G. Yohe, Eds., Cambridge University Press, Cambridge, 93-131.

Watkiss, P., D. Anthoff, T. Downing, C. Hepburn, C. Hope, A. Hunt and R. Tol, 2005. *The Social Costs of Carbon (SCC) Review: Methodological Approaches for Using SCC Estimates in Policy Assessment, Final Report*. Department for Environment, Food and Rural Affairs (Defra), London, 121 pp. http://www.defra.gov.uk/environment/climatechange/research/carboncost/pdf/aeat-scc-report.pdf.

WEA, 2000: *World Energy Assessment*. UN Development Program, UN Department of Economic and Social Affairs, and the World Energy Council, New York, 500 pp.

Weyant, J.P., 2004: Introduction and overview. *Special Issue: EMF19 Alternative Technology Strategies for Climate Change Policy. Energ. Econ.*, **26**, 501-515.

Wilbanks, T., 2003: Integrating climate change and sustainable development in a place-based context. *Clim. Policy*, **3**, S147-S154.

Wilbanks, T. and R. Kates, 2003: Making the global local: responding to climate change concerns from the bottom up. *Environment*, **45**, 12-23.

Winkler, H., K. Baumert, O. Blanchard, S. Burch and J. Robinson, 2007: What factors influence mitigative capacity? *Energ. Policy*, **35**, 15-28.

Woolcock, N. and D. Narayan, 2000: Social capital: implications for development theory, research, and policy. *World Bank Res. Obser.*, **15**, 225-249.

World Bank, undated: *Community Development Carbon Fund and BioCarbon Fund*. World Bank, Washington, District of Columbia, http://carbonfinance.org/Router.cfm?Page=CDCF, http://carbonfinance.org/Router.cfm?Page=BioCF.

Yamin, F., 2005: The European Union and future climate policy: is mainstreaming adaptation a distraction or part of the solution? *Clim. Policy*, **5**, 349–361.

Yohe, G., 2001: Mitigative capacity: the mirror image of adaptive capacity on the emissions side. *Climatic Change*, **49**, 247-262.

Yohe, G. and R.S.J. Tol, 2002: Indicators for social and economic coping capacity: moving toward a working definition of adaptive capacity. *Global Environ. Chang.*, **12**, 25-40.

Young, O., 2002: *The Institutional Dimensions of Environmental Change: Fit, Interplay, and Scale*. MIT Press, Cambridge, Massachusetts, 237 pp.

19

Assessing key vulnerabilities and the risk from climate change

Coordinating Lead Authors:
Stephen H. Schneider (USA), Serguei Semenov (Russia), Anand Patwardhan (India)

Lead Authors:
Ian Burton (Canada), Chris H.D. Magadza (Zimbabwe), Michael Oppenheimer (USA), A. Barrie Pittock (Australia), Atiq Rahman (Bangladesh), Joel B. Smith (USA), Avelino Suarez (Cuba), Farhana Yamin (UK)

Contributing Authors:
Jan Corfee-Morlot (France), Adam Finkel (USA), Hans-Martin Füssel (Germany), Klaus Keller (Germany), Dena MacMynowski (USA), Michael D. Mastrandrea (USA), Alexander Todorov (Bulgaria)

Review Editors:
Raman Sukumar (India), Jean-Pascal van Ypersele (Belgium), John Zillman (Australia)

This chapter should be cited as:
Schneider, S.H., S. Semenov, A. Patwardhan, I. Burton, C.H.D. Magadza, M. Oppenheimer, A.B. Pittock, A. Rahman, J.B. Smith, A. Suarez and F. Yamin, 2007: Assessing key vulnerabilities and the risk from climate change. *Climate Change 2007: Impacts, Adaptation and Vulnerability. Contribution of Working Group II to the Fourth Assessment Report of the Intergovernmental Panel on Climate Change*, M.L. Parry, O.F. Canziani, J.P. Palutikof, P.J. van der Linden and C.E. Hanson, Eds., Cambridge University Press, Cambridge, UK, 779-810.

Table of Contents

Executive summary

Climate change will lead to changes in geophysical, biological and socio-economic systems. An impact describes a specific change in a system caused by its exposure to climate change. Impacts may be judged to be harmful or beneficial. Vulnerability to climate change is the degree to which these systems are susceptible to, and unable to cope with, adverse impacts. The concept of risk, which combines the magnitude of the impact with the probability of its occurrence, captures uncertainty in the underlying processes of climate change, exposure, impacts and adaptation. [19.1.1]

Many of these impacts, vulnerabilities and risks merit particular attention by policy-makers due to characteristics that might make them '*key*'. The identification of potential key vulnerabilities is intended to provide guidance to decision-makers for identifying levels and rates of climate change that may be associated with 'dangerous anthropogenic interference' (DAI) with the climate system, in the terminology of United Nations Framework Convention on Climate Change (UNFCCC) Article 2 (see Box 19.1). Ultimately, the definition of DAI cannot be based on scientific arguments alone, but involves other judgements informed by the state of scientific knowledge. No single metric can adequately describe the diversity of key vulnerabilities, nor determine their ranking. [19.1.1]

This chapter identifies seven criteria from the literature that may be used to identify key vulnerabilities, and then describes some potential key vulnerabilities identified using these criteria. The criteria are [19.2]:

- magnitude of impacts,
- timing of impacts,
- persistence and reversibility of impacts,
- likelihood (estimates of uncertainty) of impacts and vulnerabilities and confidence in those estimates,
- potential for adaptation,
- distributional aspects of impacts and vulnerabilities,
- importance of the system(s) at risk.

Key vulnerabilities are associated with many climate-sensitive systems, including food supply, infrastructure, health, water resources, coastal systems, ecosystems, global biogeochemical cycles, ice sheets and modes of oceanic and atmospheric circulation. [19.3]

General conclusions include the following [19.3].

- Some observed key impacts have been at least partly attributed to anthropogenic climate change. Among these are increases in human mortality, loss of glaciers, and increases in the frequency and/or intensity of extreme events.
- Global mean temperature changes of up to 2°C above 1990-2000 levels (see Box 19.2) would exacerbate current key impacts, such as those listed above (high confidence), and trigger others, such as reduced food security in many low-latitude nations (medium confidence). At the same time, some systems, such as global agricultural productivity, could benefit (low/medium confidence).
- Global mean temperature changes of 2 to 4°C above 1990-2000 levels would result in an increasing number of key impacts at all scales (high confidence), such as widespread loss of biodiversity, decreasing global agricultural productivity and commitment to widespread deglaciation of Greenland (high confidence) and West Antarctic (medium confidence) ice sheets.
- Global mean temperature changes greater than 4°C above 1990-2000 levels would lead to major increases in vulnerability (very high confidence), exceeding the adaptive capacity of many systems (very high confidence).
- Regions that are already at high risk from observed climate variability and climate change are more likely to be adversely affected in the near future by projected changes in climate and increases in the magnitude and/or frequency of already damaging extreme events.

The 'reasons for concern' identified in the Third Assessment Report (TAR) remain a viable framework in which to consider key vulnerabilities. Recent research has updated some of the findings from the TAR [19.3.7].

- There is new and stronger evidence of observed impacts of climate change on unique and vulnerable systems (such as polar and high-mountain communities and ecosystems), with increasing levels of adverse impacts as temperatures increase (very high confidence).
- There is new evidence that observed climate change is likely to have already increased the risk of certain extreme events such as heatwaves, and it is more likely than not that warming has contributed to the intensification of some tropical cyclones, with increasing levels of adverse impacts as temperatures increase (very high confidence).
- The distribution of impacts and vulnerabilities is still considered to be uneven, and low-latitude, less-developed areas are generally at greatest risk due to both higher sensitivity and lower adaptive capacity; but there is new evidence that vulnerability to climate change is also highly variable within countries, including developed countries.
- There is some evidence that initial net market benefits from climate change will peak at a lower magnitude and sooner than was assumed for the TAR, and it is likely that there will be higher damages for larger magnitudes of global mean temperature increases than was estimated in the TAR.
- The literature offers more specific guidance on possible thresholds for initiating partial or near-complete deglaciation of the Greenland and West Antarctic ice sheets.

Adaptation can significantly reduce many potentially dangerous impacts of climate change and reduce the risk of many key vulnerabilities. However, the technical, financial and institutional capacity, and the actual planning and implementation of effective adaptation, is currently quite limited in many regions. In addition, the risk-reducing potential of planned adaptation is either very limited or very costly for some key vulnerabilities, such as loss of biodiversity, melting of mountain glaciers and disintegration of major ice sheets. [19.4.1]

A general conclusion on the basis of present understanding is that for market and social systems there is considerable adaptation potential, but the economic costs are potentially large, largely unknown and unequally distributed, as is the adaptation potential itself. For biological and geophysical systems, the adaptation potential is much less than in social and market systems. There is wide agreement that it will be much more difficult for both human and natural systems to adapt to larger magnitudes of global mean temperature change than to smaller ones, and that adaptation will be more difficult and/or costly for faster warming rates than for slower rates. [19.4.1]

Several conclusions appear robust across a diverse set of studies in the integrated assessment and mitigation literature [19.4.2, 19.4.3].

- Given the uncertainties in factors such as climate sensitivity, regional climate change, vulnerability to climate change, adaptive capacity and the likelihood of bringing such capacity to bear, a risk-management framework emerges as a useful framework to address key vulnerabilities. However, the assignment of probabilities to specific key impacts is often very difficult, due to the large uncertainties involved.
- Actions to mitigate climate change and reduce greenhouse gas emissions will reduce the risk associated with most key vulnerabilities. Postponement of such actions, in contrast, generally increases risks.
- Given current atmospheric greenhouse gas concentrations (IPCC, 2007a) and the range of projections for future climate change, some key impacts (e.g., loss of species, partial deglaciation of major ice sheets) cannot be avoided with high confidence. The probability of initiating some large-scale events is very likely to continue to increase as long as greenhouse gas concentrations and temperature continue to increase.

19.1 Introduction

19.1.1 Purpose, scope and structure of the chapter

Many social, biological and geophysical systems are at risk from climate change. Since the Third Assessment Report (TAR; IPCC, 2001a), policy-makers and the scientific community have increasingly turned their attention to climate change impacts, vulnerabilities and associated risks that may be considered 'key' because of their magnitude, persistence and other characteristics. An impact describes a specific change in a system caused by its exposure to climate change. Impacts may be judged to be either harmful or beneficial. Vulnerability to climate change is the degree to which these systems are susceptible to, and unable to cope with, the adverse impacts. The concept of risk, which combines the magnitude of the impact with the probability of its occurrence, captures uncertainty in the underlying processes of climate change, exposure, sensitivity and adaptation.

The identification of potential key vulnerabilities is intended to provide guidance to decision-makers for identifying levels and rates of climate change that may be associated with 'dangerous anthropogenic interference' (DAI) with the climate system, in the terminology of the United Nations Framework Convention on Climate Change (UNFCCC) Article 2 (see Box 19.1). Ultimately, the determination of DAI cannot be based on scientific arguments alone, but involves other judgements informed by the state of scientific knowledge.

The purpose of this chapter is two-fold. First, it synthesises information from Working Group I (WGI) and Chapters 3-16 of Working Group II (WGII) of the IPCC Fourth Assessment Report (AR4) within the uncertainty framework established by IPCC (Moss and Schneider, 2000; IPCC, 2007b) and the risk management approach discussed in Chapter 2, and identifies key vulnerabilities based on seven criteria (see Section 19.2). A focus on key vulnerabilities is meant to help policy-makers and stakeholders assess the level of risk and design pertinent response strategies. Given this focus, the analytic emphasis of this chapter is on people and systems that may be *adversely* affected by climate change, particularly where impacts could have serious and/or irreversible consequences. Positive impacts on a system are addressed when reported in the literature and where relevant to the assessment of key vulnerabilities. A comprehensive assessment of positive and negative climate impacts in all sectors and regions is beyond the scope of this chapter, and readers are encouraged to turn to the sectoral and regional chapters of this volume (Chapters 3-16) for this information.

Furthermore, it is acknowledged that the impacts of future climate change will occur in the context of an evolving socio-economic baseline. This chapter attempts to reflect the limited literature examining the possible positive and negative relationships between baseline scenarios and future impacts. However, the purpose of this chapter is not to compare the effects of climate change with the effects of socio-economic development, but rather to assess the additional effects of climate change on top of whatever baseline development scenario is assumed. Whether a climate change impact would be greater or smaller than welfare gains or losses associated with particular development scenarios is beyond the scope of this chapter but is dealt with in Chapter 20 and by Working Group III (WGIII).

Second, this chapter provides an assessment of literature focusing on the contributions that various mitigation and adaptation response strategies, such as stabilisation of greenhouse gas concentrations in the atmosphere, could make in avoiding or reducing the probability of occurrence of key impacts. Weighing the benefits of avoiding such climate-induced risks versus the costs of mitigation or adaptation, as well as the distribution of such costs and benefits (i.e., equity implications of such trade-offs) is also beyond the scope of this chapter, as is attempting a normative trade-off analysis among and between various groups and between human and natural systems. (The term 'normative' is used in this chapter to refer to a process or statement that inherently involves value judgements or beliefs.) Many more examples of such literature can be obtained in Chapters 18 and 20 of this volume and in the Working Group III (WGIII) AR4.

The remainder of Section 19.1 presents the conceptual framework, and Section 19.2 presents the specific criteria used in this chapter for the assessment of key vulnerabilities. Section 19.3 presents selected key vulnerabilities based on these criteria. Key vulnerabilities are linked to specific levels of global mean temperature increase (above 1990-2000 levels; see Box 19.2) using available estimates from the literature wherever possible. Section 19.3 provides an indicative, rather than an exhaustive, list of key vulnerabilities, representing the authors' collective judgements based on the criteria presented in Section 19.2, selected from a vast array of possible candidates suggested in the literature. Section 19.4 draws on the literature addressing the linkages between key vulnerabilities and strategies to avoid them by adaptation (Section 19.4.1) and mitigation (Section 19.4.2). Section 19.4.4 concludes this chapter by suggesting research priorities for the natural and social sciences that may provide relevant knowledge for assessing key vulnerabilities of climate change. The assessment of key vulnerabilities and review of the particular assemblage of literature needed to do so is unique to the mission of Chapter 19. Accordingly, in Sections 19.3 and 19.4, we have made judgments with regard to likelihood and confidence whereas, in some cases, other chapters in this volume and in the WGI AR4 have not.

Another important area of concern, also marked by large uncertainties, is the assessment of impacts resulting from multiple factors. In some cases, key vulnerabilities emerging from such interactions are assessed, such as the fragmentation of habitats that constrains some species, which – when combined with climate change – forces species movements across disturbed habitats. This is a multi-stressor example that is likely to multiply the impacts relative to either stressor acting alone. Other examples from the literature are also given in the text; though any attempt to be comprehensive or quantitative in such multi-stress situations is beyond the scope of the chapter.

19.1.2 Conceptual framework for the identification and assessment of key vulnerabilities

19.1.2.1 Meaning of 'key vulnerability'

Vulnerability to climate change is the degree to which geophysical, biological and socio-economic systems are susceptible to, and unable to cope with, adverse impacts of climate change (see Chapter 17; Füssel and Klein, 2006). The term 'vulnerability' may therefore refer to the vulnerable system itself, e.g., low-lying islands or coastal cities; the impact to this system, e.g., flooding of coastal cities and agricultural lands or forced migration; or the mechanism causing these impacts, e.g., disintegration of the West Antarctic ice sheet.

Many impacts, vulnerabilities and risks merit particular attention by policy-makers due to characteristics that might make them *key*. Key impacts that may be associated with key vulnerabilities are found in many social, economic, biological and geophysical systems, and various tabulations of risks, impacts and vulnerabilities have been provided in the literature (e.g., Smith et al., 2001; Corfee-Morlot and Höhne, 2003; Hare, 2003; Oppenheimer and Petsonk, 2003, 2005; ECF, 2004; Hitz and Smith, 2004; Leemans and Eickhout, 2004; Schellnhuber et al., 2006). Key vulnerabilities are associated with many climate-sensitive systems, including, for example, food supply, infrastructure, health, water resources, coastal systems,

Box 19.1. UNFCCC Article 2

The text of the UNFCCC Article 2 reads:

"The ultimate objective of this Convention and any related legal instruments that the Conference of the Parties may adopt is to achieve, in accordance with the relevant provisions of the Convention, stabilization of greenhouse gas concentrations in the atmosphere at a level that would prevent dangerous anthropogenic interference with the climate system. Such a level should be achieved within a time-frame sufficient to allow ecosystems to adapt naturally to climate change, to ensure that food production is not threatened and to enable economic development to proceed in a sustainable manner."

Box 19.2. Reference for temperature levels

Levels of global mean temperature change are variously presented in the literature with respect to: pre-industrial temperatures in a specified year e.g., 1750 or 1850; the average temperature of the 1961-1990 period; or the average temperature within the 1990-2000 period. The best estimate for the increase above pre-industrial levels in the 1990-2000 period is 0.6°C, reflecting the best estimate for warming over the 20th century (Folland et al., 2001; Trenberth et al., 2007). Therefore, to illustrate this by way of a specific example, a 2°C increase above pre-industrial levels corresponds to a 1.4°C increase above 1990-2000 levels. Climate impact studies often assess changes in response to regional temperature change, which can differ significantly from changes in global mean temperature. In most land areas, regional warming is larger than global warming (see Christensen et al., 2007). Unless otherwise specified, this chapter refers to global mean temperature change above 1990-2000 levels, which reflects the most common metric used in the literature on key vulnerabilities. However, given the many conventions in the literature for baseline periods, the reader is advised to check carefully and to adjust baseline levels for consistency every time a number is given for impacts at some specified level of global mean temperature change.

ecosystems, global biogeochemical cycles, ice sheets, and modes of oceanic and atmospheric circulation (see Section 19.3).

19.1.2.2 Scientific assessment and value judgements

The assessment of key vulnerabilities involves substantial scientific uncertainties as well as value judgements. It requires consideration of the response of biophysical and socio-economic systems to changes in climatic and non-climatic conditions over time (e.g., changes in population, economy or technology), important non-climatic developments that affect adaptive capacity, the potential for effective adaptation across regions, sectors and social groupings, value judgements about the acceptability of potential risks, and potential adaptation and mitigation measures. To achieve transparency in such complex assessments, scientists and analysts need to provide a 'traceable account' of all relevant assumptions (Moss and Schneider, 2000).

Scientific analysis can inform policy processes but choices about which vulnerabilities are 'key', and preferences for policies appropriate for addressing them, necessarily involve value judgements. "Natural, technical and social sciences can provide essential information and evidence needed for decision-making on what constitutes 'dangerous anthropogenic interference with the climate system'. At the same time, such decisions are value judgments determined through socio-political processes, taking into account considerations such as development, equity and sustainability, as well as uncertainties and risk" (IPCC, 2001b).

19.1.2.3 UNFCCC Article 2

The question of which impacts might constitute DAI in terms of Article 2 has only recently attracted a high level of attention, and the literature still remains relatively sparse (see Oppenheimer and Petsonk 2005; Schellnhuber et al., 2006 for reviews). Interpreting Article 2 (ultimately the obligation of the Conference of the Parties to the UNFCCC) involves a scientific assessment of what impacts might be associated with different levels of greenhouse gas concentrations or climate change; and a normative evaluation by policy-makers of which potential impacts and associated likelihoods are significant enough to constitute, individually or in combination, DAI. This assessment is informed by the magnitude and timing of climate impacts as well as by their distribution across regions, sectors and population groups (e.g., Corfee-Morlot and Agrawala, 2004; Schneider and Mastrandrea, 2005; Yamin et al., 2005). The social, cultural and ethical dimensions of DAI have drawn increasing attention recently (Jamieson 1992, 1996; Rayner and Malone, 1998; Adger, 2001; Gupta et al., 2003; Gardiner, 2006). The references to adverse effects as significant deleterious effects in Article 1 of the UNFCCC[1] and to natural ecosystems, food production, and sustainable development in Article 2 provide guidance as to which impacts may be considered relevant to the definition of DAI (Schneider et al., 2001).

Interpreting Article 2 is necessarily a dynamic process because the assessment of what levels of greenhouse gas concentrations may be considered 'dangerous' would be modified based on changes in scientific knowledge, social values and political priorities.

19.1.2.4 Distribution and aggregation of impacts

Vulnerability to climate change differs considerably across socio-economic groups, thus raising important questions about equity. Most studies of impacts in the context of key vulnerabilities and Article 2 have focused on aggregate impacts, grouping developing countries or populations with special needs or situations. Examples include island nations faced with sea-level rise (Barnett and Adger, 2003), countries in semi-arid regions with a marginal agricultural base, indigenous populations facing regionalised threats, or least-developed countries (LDCs; Huq et al., 2003). Within developed countries, research on vulnerability has often focused on groups of people, for example those living in coastal or flood-prone regions, or socially vulnerable groups such as the elderly.

No single metric for climate impacts can provide a commonly accepted basis for climate policy decision-making (Jacoby, 2004; Schneider, 2004). Aggregation, whether by region, sector, or population group, implies value judgements about the selection, comparability and significance of vulnerabilities and cohorts (e.g., Azar and Sterner, 1996; Fankhauser et al., 1997; Azar, 1998, on regional aggregation). The choice of scale at which impacts are examined is also crucial, as considerations of fairness, justice or equity require examination of the distribution of impacts, vulnerability and adaptation potential, not only between, but also within, groupings (Jamieson, 1992; Gardiner, 2004; Yamin et al., 2005).

19.1.2.5 Critical levels and thresholds

Article 2 of the UNFCCC defines international policy efforts in terms of avoidance of a level of greenhouse gas concentrations beyond which the effects of climate change would be considered to be 'dangerous'. Discussions about 'dangerous interference with the climate system' and 'key vulnerabilities' are also often framed around thresholds or critical limits (Patwardhan et al., 2003; Izrael, 2004). Key vulnerabilities may be linked to systemic thresholds where non-linear processes cause a system to shift from one major state to another (such as a hypothetical sudden change in the Asian monsoon or disintegration of the West Antarctic ice sheet). Systemic thresholds may lead to large and widespread consequences that may be considered as 'dangerous'. Examples include climate impacts such as those arising from ice sheet disintegration leading to large sea-level rises or changes to the carbon cycle, or those affecting natural and managed ecosystems, infrastructure and tourism in the Arctic.

Smooth and gradual climate change may also lead to damages that are considered unacceptable beyond a certain point. For instance, even a gradual and smooth increase of sea-level rise would eventually reach a level that certain stakeholders would consider unacceptable. Such normative impact thresholds could

[1] Article 1 reads, "For the purposes of this Convention: 1. 'Adverse effects of climate change' means changes in the physical environment or biota resulting from climate change which have significant deleterious effects on the composition, resilience or productivity of natural and managed ecosystems or on the operation of socio-economic systems or on human health and welfare."

be defined at the global level (e.g., Toth et al., 2002, for natural ecosystems) and some have already been identified at the regional level (e.g., Jones, 2001, for irrigation in Australia).

19.2 Criteria for selecting 'key' vulnerabilities

As previously discussed, determining which impacts of climate change are potentially 'key' and what is 'dangerous' is a dynamic process involving, inter alia, combining scientific knowledge with factual and normative elements (Patwardhan et al., 2003; Dessai et al., 2004; Pittini and Rahman, 2004). Largely factual or objective criteria include the scale, magnitude, timing and persistence of the harmful impact (Parry et al., 1996; Kenny et al., 2000; Moss and Schneider, 2000; Goklany, 2002; Corfee-Morlot and Höhne, 2003; Schneider, 2004; Oppenheimer, 2005). Normative and subjective elements are embedded in assessing the uniqueness and importance of the threatened system, equity considerations regarding the distribution of impacts, the degree of risk aversion, and assumptions regarding the feasibility and effectiveness of potential adaptations (IPCC, 2001a; OECD, 2003; Pearce, 2003; Tol et al., 2004). Normative criteria are influenced by the perception of risk, which depends on the cultural and social context (e.g., Slovic, 2000; Oppenheimer and Todorov, 2006). Some aspects of confidence in the climate change–impact relationship are factual, while others are subjective (Berger and Berry, 1988). In addition, the choice of which factual criteria to employ in assessing impacts has a normative component.

This chapter identifies seven criteria from the literature that may be used to identify key vulnerabilities, and then describes some potential key vulnerabilities identified using these criteria. The criteria are listed and explained in detail below:

- magnitude of impacts,
- timing of impacts,
- persistence and reversibility of impacts,
- likelihood (estimates of uncertainty) of impacts and vulnerabilities, and confidence in those estimates,
- potential for adaptation,
- distributional aspects of impacts and vulnerabilities,
- importance of the system(s) at risk.

Magnitude

Impacts of large magnitude are more likely to be evaluated as 'key' than impacts with more limited effects. The magnitude of an impact is determined by its scale (e.g., the area or number of people affected) and its intensity (e.g., the degree of damage caused). Therefore, many studies have associated key vulnerabilities or dangerous anthropogenic interference primarily with large-scale geophysical changes in the climate system.

Various aggregate metrics are used to describe the magnitude of climate impacts. The most widely used quantitative measures for climate impacts (see Chapter 20 and WGIII AR4 Chapter 3 (Fisher et al., 2007)) are monetary units such as welfare, income or revenue losses (e.g., Nordhaus and Boyer, 2000), costs of anticipating and adapting to certain biophysical impacts such as a large sea-level rise (e.g., Nicholls et al., 2005), and estimates of people's willingness to pay to avoid (or accept as compensation for) certain climate impacts (see, e.g., Li et al., 2004). Another aggregate, non-monetary indicator is the number of people affected by certain impacts such as food and water shortages, morbidity and mortality from diseases, and forced migration (Barnett, 2003; Arnell, 2004; Parry et al., 2004; van Lieshout et al., 2004; Schär and Jendritzky, 2004; Stott et al., 2004). Climate impacts are also quantified in terms of the biophysical end-points, such as agricultural yield changes (see Chapter 5; Füssel et al., 2003; Parry et al., 2004) and species extinction numbers or rates (see Chapter 4; Thomas et al., 2004). For some impacts, qualitative rankings of magnitude are more appropriate than quantitative ones. Qualitative methods have been applied to reflect social preferences related to the potential loss of cultural or national identity, loss of cultural heritage sites, and loss of biodiversity (Schneider et al., 2000).

Timing

A harmful impact is more likely to be considered 'key' if it is expected to happen soon rather than in the distant future (Bazermann, 2005; Weber, 2005). Climate change in the 20th century has already led to numerous impacts on natural and social systems (see Chapter 1), some of which may be considered 'key'. Impacts occurring in the distant future which are caused by nearer-term events or forcings (i.e., 'commitment'), may also be considered 'key'. An often-cited example of such 'delayed irreversibility' is the disintegration of the West Antarctic ice sheet: it has been proposed that melting of ice shelves in the next 100 to 200 years may lead to gradual but irreversible deglaciation and a large sea-level rise over a much longer time-scale (see Section 19.3.5.2; Meehl et al., 2007). Debates over an 'appropriate' rate of time preference for such events (i.e., discounting) are widespread in the integrated assessment literature (WGIII AR4 Chapter 2: Halsnaes et al., 2007), and can influence the extent to which a decision-maker might label such possibilities as 'key'.

Another important aspect of timing is the rate at which impacts occur. In general, adverse impacts occurring suddenly (and surprisingly) would be perceived as more significant than the same impacts occurring gradually, as the potential for adaptation for both human and natural systems would be much more limited in the former case. Finally, very rapid change in a non-linear system can exacerbate other vulnerabilities (e.g., impacts on agriculture and nutrition can aggravate human vulnerability to disease), particularly where such rapid change curtails the ability of systems to prevent and prepare for particular kinds of impacts (Niemeyer et al., 2005).

Persistence and reversibility

A harmful impact is more likely to be considered 'key' if it is persistent or irreversible. Examples of impacts that could become key due to persistence include the emergence of near-permanent drought conditions (e.g., in semi-arid and arid regions in Africa – Nyong, 2005; see Chapter 9) and intensified cycles of extreme flooding that were previously regarded as 'one-off' events (e.g., in parts of the Indian subcontinent; see Chapter 10).

Examples of climate impacts that are irreversible, at least on time-scales of many generations, include changes in regional or global biogeochemical cycles and land cover (Denman et al., 2007; see Section 19.3.5.1), the loss of major ice sheets (Meehl et al., 2007; see Section 19.3.5.2); the shutdown of the meridional overturning circulation (Randall et al., 2007; Meehl et al., 2007; see Section 19.3.5.3), the extinction of species (Thomas et al., 2004; Lovejoy and Hannah, 2005), and the loss of unique cultures (Barnett and Adger, 2003). The latter is illustrated by Small Island Nations at risk of submergence through sea-level rise (see Chapter 16) and the necessity for the Inuit of the North American Arctic (see Chapter 15) to cope with recession of the sea ice that is central to their socio-cultural environment.

Likelihood and confidence

Likelihood of impacts and our confidence in their assessment are two properties often used to characterise uncertainty of climate change and its impacts (Moss and Schneider, 2000; IPCC, 2007b). Likelihood is the probability of an outcome having occurred or occurring in the future; confidence is the subjective assessment that any statement about an outcome will prove correct. Uncertainty may be characterised by these properties individually or in combination. For example, in expert elicitations of subjective probabilities (Nordhaus, 1994; Morgan and Keith, 1995; Arnell et al., 2005; Morgan et al., 2006), likelihood of an outcome has been framed as the central value of a probability distribution, whereas confidence is reflected primarily by its spread (the lesser the spread, the higher the confidence). An impact characterised by high likelihood is more apt to be seen as 'key' than the same impact with a lower likelihood of occurrence. Since risk is defined as consequence (impact) multiplied by its likelihood (probability), the higher the probability of occurrence of an impact the higher its risk, and the more likely it would be considered 'key'.

Potential for adaptation

To assess the potential harm caused by climate change, the ability of individuals, groups, societies and nature to adapt to or ameliorate adverse impacts must be considered (see Section 19.3.1; Chapter 17). The lower the availability and feasibility of effective adaptations, the more likely such impacts would be characterised as 'key vulnerabilities'. The potential for adaptation to ameliorate the impacts of climate change differs between and within regions and sectors (e.g., O'Brien et al., 2004). There is often considerable scope for adaptation in agriculture and in some other highly managed sectors. There is much less scope for adaptation to some impacts of sea-level rise such as land loss in low-lying river deltas, and there are no realistic options for preserving many endemic species in areas that become climatically unsuitable (see Chapter 17). Adaptation assessments need to consider not only the technical feasibility of certain adaptations but also the availability of required resources (which is often reduced in circumstances of poverty), the costs and side-effects of adaptation, the knowledge about those adaptations, their timeliness, the (dis-)incentives for adaptation actors to actually implement them, and their compatibility with individual or cultural preferences.

The adaptation literature (see Chapter 17) can be largely separated into two groups: one with a more favourable view of the potential for adaptation of social systems to climate change, and an opposite group that expresses less favourable views, stressing the limits to adaptation in dealing with large climate changes and the social, financial and technical obstacles that might inhibit the actual implementation of many adaptation options (see, e.g., the debate about the Ricardian climate change impacts methods – Mendelsohn et al., 1994; Cline, 1996; Mendelsohn and Nordhaus, 1996; Kaufmann, 1998; Hanemann, 2000; Polsky and Easterling, 2001; Polsky, 2004; Schlenker et al., 2005). This chapter reports the range of views in the literature on adaptive capacity relevant for the assessment of key vulnerabilities, and notes that these very different views contribute to the large uncertainty that accompanies assessments of many key vulnerabilities.

Distribution

The distribution of climate impacts across regions and population groups raises important equity issues (see Section 19.1.2.4 for a detailed discussion). The literature concerning distributional impacts of climate change covers an increasingly broad range of categories, and includes, among others, income (Tol et al., 2004), gender (Denton, 2002; Lambrou and Laub, 2004) and age (Bunyavanich et al., 2003), in addition to regional, national and sectoral groupings. Impacts and vulnerabilities that are highly heterogeneous or which have significant distributional consequences are likely to have higher salience, and therefore a greater chance of being considered as 'key'.

Importance of the vulnerable system

A salient, though subjective, criterion for the identification of 'key vulnerabilities' is the importance of the vulnerable system or system property. Various societies and peoples may value the significance of impacts and vulnerabilities on human and natural systems differently. For example, the transformation of an existing natural ecosystem may be regarded as important if that ecosystem is the unique habitat of many endemic species or contains endangered charismatic species. On the other hand, if the livelihoods of many people depend crucially on the functioning of a system, this system may be regarded as more important than a similar system in an isolated area (e.g., a mountain snowpack system with large downstream use of the melt water versus an equally large snowpack system with only a small population downstream using the melt water).

19.3 Identification and assessment of key vulnerabilities

This section discusses what the authors have identified as possible key vulnerabilities based on the criteria specified in the Introduction and Section 19.2, and on the literature on impacts that may be considered potentially 'dangerous' in the sense of Article 2. The key vulnerabilities identified in this section are, as noted earlier, not a comprehensive list but illustrate a range of impacts relevant for policy-makers. Section 19.3.1 introduces, in condensed tabular form, key vulnerabilities, organising them by type of system, i.e., market, social, ecological or geophysical. The following sections discuss some of the key vulnerabilities by type of system, and add discussions of extreme events and an update on the 'reasons for concern' framework from the TAR. Each sub-section is cross-referenced to the relevant sections of the Fourth Assessment Report as well as primary publications from which more detail can be obtained. As noted in Section 19.1.1, the likelihood and confidence judgements in this section reflect the assessments of the authors of this chapter.

19.3.1 Introduction to Table 19.1

Table 19.1 provides short summaries of some vulnerabilities which, in the judgment of the authors of this chapter and in the light of the WGI AR4 and chapters of the WGII AR4, may be considered 'key' according to the criteria set out above in Section 19.2. The table presents vulnerabilities grouped by the following categories, described in the following text:

- Global social systems
- Regional systems
- Global biological systems
- Geophysical systems
- Extreme events

The table attempts to describe, as quantitatively as the literature allows, how impacts vary with global mean temperature increase above 1990-2000 levels. In addition, the authors of this chapter have assigned confidence estimates to this information. Where known, the table presents information regarding the dependence of effects on rates of warming, duration of the changes, exposure to the stresses, and adaptation taking into account uncertainties

Table 19.1. *Examples of potential key vulnerabilities. This list is not ordered by priority or severity but by category of system, process or group, which is either affected by or which causes vulnerability. Information is presented where available on how impacts may change at larger increases in global mean temperature (GMT). All increases in GMT are relative to circa 1990. Entries are necessarily brief to limit the size of the table, so further details, caveats and supporting evidence should be sought in the accompanying text, cross-references, and in the primary scientific studies referenced in this and other chapters of the AR4. In many cases, climate change impacts are marginal or synergistic on top of other existing and changing stresses. Confidence symbol legend: *** very high confidence, ** high confidence, * medium confidence, • low confidence. Sources in [square brackets] are from chapters in the WGII AR4 unless otherwise indicated. Where no source is given, the entries are based on the conclusions of the Chapter 19 authors.*

Systems, processes or groups at risk [cross-references]	Prime criteria for 'key vulnerability' (based on the seven criteria listed in Section 19.2)	Relationship between temperature and risk. Temperature change by 2100 (relative to 1990-2000) 0°C	1°C	2°C	3°C	4°C	5°C
Global social systems							
Food supply [19.3.2.2]	Distribution, Magnitude		Productivity decreases for some cereals in low latitudes */• [5.4] Productivity increases for some cereals in mid/high latitudes */• [5.4]	Global production potential increases to around 3°C * [5.4, 5.6]		Cereal productivity decreases in some mid/high-latitude regions */• [5.4] Global production potential very likely to decrease above about 3°C * [5.4, 5.6]	
Infrastructure [19.3.2]	Distribution, Magnitude, Timing	Damages likely to increase exponentially, sensitive to rate of climate change, change in extreme events and adaptive capacity ** [3.5, 6.5.3, 7.5].					
Health [19.3.2]	Distribution, Magnitude, Timing, Irreversibility	Current effects are small but discernible * [1.3.7, 8.2].	Although some risks would be reduced, aggregate health impacts would increase, particularly from malnutrition, diarrhoeal diseases, infectious diseases, floods and droughts, extreme heat, and other sources of risk */**. Sensitive to status of public health system *** [8.ES, 8.3, 8.4, 8.6].				
Water resources [19.3.2]	Distribution, Magnitude, Timing	Decreased water availability and increased drought in some mid latitudes and semi-arid low latitudes ** [3.2, 3.4, 3.7].	Severity of floods, droughts, erosion, water-quality deterioration will increase with increasing climate change ***. Sea-level rise will extend areas of salinisation of groundwater, decreasing freshwater availability in coastal areas *** [3.ES]. Hundreds of millions people would face reduced water supplies ** [3.5].				
Migration and conflict	Distribution, Magnitude	Stresses such as increased drought, water shortages, and riverine and coastal flooding will affect many local and regional populations **. This will lead in some cases to relocation within or between countries, exacerbating conflicts and imposing migration pressures * [19.2].					
Aggregate market impacts and distribution	Magnitude, Distribution	Uncertain net benefits and greater likelihood of lower benefits or higher damages than in TAR •. Net market benefits in many high-latitude areas; net market losses in many low-latitude areas. * [20.6, 20.7]. Most people negatively affected •/*.			Net global negative market impacts increasing with higher temperatures * [20.6]. Most people negatively affected *.		

Systems, processes or groups at risk [cross-references]	Prime criteria for 'key vulnerability' (based on the seven criteria listed in Section 19.2)	Relationship between temperature and risk. Temperature change by 2100 (relative to 1990-2000) 0°C 1°C 2°C 3°C 4°C 5°C	
Regional systems			
Africa [19.3.3]	Distribution, Magnitude, Timing, Low Adaptive Capacity	Tens of millions of people at risk of increased water stress; increased spread of malaria • [9.2, 9.4.1, 9.4.3].	Hundreds of millions of additional people at risk of increased water stress; increased risk of malaria in highlands; reductions in crop yields in many countries, harm to many ecosystems such as Succulent Karoo • [9.4.1, 9.4.3, 9.4.4, 9.4.5].
Asia [19.3.3]	Distribution, Magnitude, Timing, Low Adaptive Capacity	About 1 billion people would face risks from reduced agricultural production potential, reduced water supplies or increases in extremes events • [10.4].	
Latin America [19.3.3]	Magnitude, Irreversibility, Distribution, and Timing, Low Adaptive Capacity	Tens of millions of people at risk of water shortages • [13.ES, 13.4.3]; many endemic species at risk from land-use and climate change • (~1°C) [13.4.1, 13.4.2].	More than a hundred million people at risk of water shortages • [13.ES, 13.4.3]; low-lying coastal areas, many of which are heavily populated, at risk from sea-level rise and more intense coastal storms • (about 2-3°C) [13.4.4]. Widespread loss of biodiversity, particularly in the Amazon • [13.4.1, 13.4.2].
Polar regions [19.3.3]	Timing, Magnitude, Irreversibility, Distribution, Low Adaptive Capacity	Climate change is already having substantial impacts on societal and ecological systems ** [15.ES].	Continued warming likely to lead to further loss of ice cover and permafrost ** [15.3]. Arctic ecosystems further threatened **, although net ecosystem productivity estimated to increase ** [15.2.2, 15.4.2]. While some economic opportunities will open up (e.g., shipping), traditional ways of life will be disrupted ** [15.4, 15.7].
Small islands [19.3.3]	Irreversibility, Magnitude, Distribution, Low Adaptive Capacity	Many islands already experiencing some negative effects ** [16.2]. Increasing coastal inundation and damage to infrastructure due to sea-level rise ** [16.4].	
Indigenous, poor or isolated communities [19.3.3]	Irreversibility, Distribution, Timing, Low Adaptive Capacity	Some communities already affected ** [11.4, 14.2.3, 15.4.5].	Climate change and sea-level rise add to other stresses **. Communities in low-lying coastal and arid areas are especially threatened ** [3.4, 6.4].
Drying in Mediterranean, western North America, southern Africa, southern Australia, and north-eastern Brazil [19.3.3]	Distribution, Magnitude, Timing	Climate models generally project decreased precipitation in these regions [3.4.1, 3.5.1, 11.3.1]. Reduced runoff will exacerbate limited water supplies, decrease water quality, harm ecosystems and result in decreased crop yields ** [3.4.1, 11.4].	
Inter-tropical mountain glaciers and impacts on high-mountain communities [19.3.3]	Magnitude, Timing, Persistence, Low Adaptive Capacity, Distribution	Inter-tropical glaciers are melting and causing flooding in some areas; shifts in ecosystems are likely to cause water security problems due to decreased storage */** [Box 1.1, 10.ES, 10.2, 10.4.4, 13.ES, 13.2.4, 19.3].	Accelerated reduction of inter-tropical mountain glaciers. Some of these systems will disappear in the next few decades * [Box 1.1, 9.2.1, Box 9.1, 10.ES, 10.2.4, 10.4.2, 13.ES, 13.2.4.1].
Global biological systems			
Terrestrial ecosystems and biodiversity [19.3.4]	Irreversibility, Magnitude, Low Adaptive Capacity, Persistence, Rate of Change, Confidence	Many ecosystems already affected *** [1.3].	circa 20-30% species at increasingly high risk of extinction * [4.4]. Major extinctions around the globe ** [4.4] Terrestrial biosphere tends toward a net carbon source ** [4.4]
Marine ecosystems and biodiversity [19.3.4]	Irreversibility, Magnitude, Low Adaptive Capacity, Persistence, Rate of Change, Confidence	Increased coral bleaching ** [4.4]	Most corals bleached ** [4.4] Widespread coral mortality *** [4.4]

Systems, processes or groups at risk [cross-references]	Prime criteria for 'key vulnerability' (based on the seven criteria listed in Section 19.2)	Relationship between temperature and risk. Temperature change by 2100 (relative to 1990-2000) 0°C	1°C	2°C	3°C	4°C	5°C
Global biological systems							
Freshwater ecosystems [19.3.4]	Irreversibility, Magnitude, Persistence Low Adaptive Capacity	Some lakes already showing decreased fisheries output; pole-ward migration of aquatic species ** [1.3.4, 4.4.9].		Intensified hydrological cycles, more severe droughts and floods *** [3.4.3].		Extinction of many freshwater species **, major changes in limnology of lakes **, increased salinity of inland lakes **.	
Geophysical systems							
Biogeochemical cycles [WGII 4.4.9, 19.3.5.1; WGI 7.3.3, 7.3.4, 7.3.5, 7.4.1.2, 10.4.1, 10.4.2]	Magnitude, Persistence, Confidence, Low Adaptive Capacity, Rate of Change	Ocean acidification already occurring, increasing further as atmospheric CO_2 concentration increases ***; ecological changes are potentially severe * [1.3.4, 4.4.9]. Carbon cycle feedback increases projected CO_2 concentrations by 2100 by 20-220 ppm for SRES[2] A2, with associated additional warming of 0.1 to 1.5°C **. AR4 temperature range (1.1-6.4°C) accounts for this feedback from all scenarios and models but additional CO_2 and CH_4 releases are possible from permafrost, peat lands, wetlands, and large stores of marine hydrates at high latitudes * [4.4.6, 15.4.2]. Permafrost already melting, and above feedbacks generally increase with climate change, but eustatic sea-level rise likely to increase stability of hydrates *** [1.3.1].					
Greenland ice sheet [WGII 6.3, 19.3.5.2; WGI 6.4.3.3, 10.7.4.3]	Magnitude, Irreversibility, Low Adaptive Capacity, Confidence	Localised deglaciation (already observed, due to local warming); extent would increase with temperature increase *** [19.3.5].		Commitment to widespread ** to near-total * deglaciation, 2-7 m sea-level rise[3] over centuries to millennia * [19.3.5].		Near-total deglaciation **[19.3.5]	
West Antarctic ice sheet [WGII 6.3, 19.3.5.2; WGI 6.4.3.3, 10.7.4.4]	Magnitude, Irreversibility, Low Adaptive Capacity	Localised ice shelf loss and grounding line retreat * (already observed, due to local warming) [1.3.1, 19.3.5]		Commitment to partial deglaciation, 1.5-5 m sea-level rise over centuries to millennia •/* [19.3.5] Likelihood of near-total deglaciation increases with increases in temperature ** [19.3.5]			
Meridional overturning circulation [WGII 19.3.5.3; WGI 8.7.2.1, 10.3.4]	Magnitude, Persistence, Distribution, Timing, Low Adaptive Capacity, Confidence	Variations including regional weakening (already observed but no trend identified)		Considerable weakening **. Commitment to large-scale and persistent change including possible cooling in northern high-latitude areas near Greenland and north-west Europe • highly dependent on rate of climate change [12.6, 19.3.5].			
Extreme events							
Tropical cyclone intensity [WGII 7.5, 8.2, 11.4.5, 16.2.2, 16.4, 19.3.6; WGI Table TS-4, 3.8.3, Q3.3, 9.5.3.6, Q10.1]	Magnitude, Timing, Distribution	Increase in Category 4-5 storms*/**, with impacts exacerbated by sea-level rise		Further increase in tropical cyclone intensity */** exceeding infrastructure design criteria with large economic costs ** and many lives threatened **.			
Flooding, both large-scale and flash floods [WGII 14.4.1; WGI Table TS-4, 10.3.6.1, Q10.1]	Timing, Magnitude	Increases in flash flooding in many regions due to increased rainfall intensity** and in floods in large basins in mid and high latitudes **.		Increased flooding in many regions (e.g., North America and Europe) due to greater increase in winter rainfall exacerbated by loss of winter snow storage **. Greater risk of dam burst in glacial mountain lakes ** [10.2.4.2].			
Extreme heat [WGII 14.4.5; WGI Table TS-4, 10.3.6.2, Q10.1]	Timing, Magnitude	Increased heat stress and heat-waves, especially in continental areas ***.		Frequency of heatwaves (according to current classification) will increase rapidly, causing increased mortality, crop failure, forest die-back and fire, and damage to ecosystems ***.			
Drought [WGI Table TS-4, 10.3.6.1]	Magnitude, Timing	Drought already increasing * [1.3.2.1]. Increasing frequency and intensity of drought in mid-latitude continental areas projected ** [WGI 10.3.6.1].		Extreme drought increasing from 1% land area to 30% (SRES A2 scenario) [WGI 10.3.6.1]. Mid-latitude regions seriously affected by poleward migration of Annular Modes ** [WGI 10.3.5.5].			
Fire [WGII 1.3.6; WGI 7.3]	Timing, Magnitude	Increased fire frequency and intensity in many areas, particularly where drought increases ** [4.4, 14.2.2].		Frequency and intensity likely to be greater, especially in boreal forests and dry peat lands after melting of permafrost ** [4.4.5, 11.3, 13.4.1, 14.4.2, 14.4.4].			

[2] SRES: Special Report on Emissions Scenarios, see Nakićenović et al., 2000.
[3] Range is based on a variety of methods including models and analysis of palaeo data [19.3.5.2]

regarding socio-economic development. However, only in a few cases does the literature address rate or duration of warming and its consequences. As entries in the table are necessarily short, reference should be made to the relevant chapters and to the accompanying text in this chapter for more detailed information and cross-referencing, including additional caveats where applicable.

19.3.2 Global social systems

The term 'social systems' is used here in a broad sense to describe human systems, and includes both market systems and social systems. Market systems typically involve the provision and sale of goods and services in formal or informal markets. Valuation of non-market impacts (e.g., losses of human life, species lost, distributional inequity, etc.) involves a series of normative judgements that limit the degree of consensus and confidence commanded by different studies (see Section 19.1.2). The importance of non-market impacts and equity weighting is suggested by Stern (2007) but, in the absence of likelihood and confidence assessments, it is difficult to apply to any risk-management framework calculations.

We first discuss impacts on major market systems, followed by a discussion of impacts on major aspects of social systems. Such impacts are often considered to be important in the context of sustainable development.

19.3.2.1 Agriculture

Ensuring that food production is not threatened is an explicit criterion of UNFCCC Article 2. In general, low-latitude areas are most at risk of having decreased crop yields. In contrast, mid- and high-latitude areas could generally, although not in all locations, see increases in crop yields for temperature increases of up to 1-3°C (see Chapter 5 Section 5.4.2). Taken together, there is low to medium confidence that global agricultural production could increase up to approximately 3°C of warming. For temperature increases beyond 1-3°C, yields of many crops in temperate regions are projected to decline (•/*[4]). As a result, beyond 3°C warming, global production would decline because of climate change (•/*) and the decline would continue as GMT increases (•/*). Most studies on global agriculture have not yet incorporated a number of critical factors, including changes in extreme events or the spread of pests and diseases. In addition, they have not considered the development of specific practices or technologies to aid adaptation.

19.3.2.2 Other market sectors

Other market systems will also be affected by climate change. These include the livestock, forestry and fisheries industries, which are very likely to be directly affected as climate affects the quality and extent of rangeland for animals, soils and other growing conditions for trees, and freshwater and marine ecosystems for fish. Other sectors are also sensitive to climate change. These include energy, construction, insurance, tourism and recreation. The aggregate effects of climate change on many of these sectors has received little attention in the literature and remains highly uncertain. Some sectors are likely to see shifts in expenditure; with some contracting and some expanding. Yet, for some sectors, such as insurance, the impacts of climate change are likely to result in increased damage payments and premiums (see Chapter 7).

Other sectors, such as tourism and recreation, are likely to see some substantial shifts (e.g., reduction in ski season, loss of some ski areas, shifts in location of tourist destinations because of changes in climate and extreme events; e.g., Hamilton et al., 2005; see also Chapter 7 Section 7.4.2 and Chapter 14 Section 14.4.7). Global net energy demand is very likely to change (Tol, 2002b). Demand for air-conditioning is highly likely to increase, whereas demand for heating is highly likely to decrease. The literature is not clear on what temperature is associated with minimum global energy demand, so it is uncertain whether warming will initially increase or decrease net global demand for energy relative to some projected baseline. However, as temperatures rise, net global demand for energy will eventually rise as well (Hitz and Smith, 2004).

19.3.2.3 Aggregate market impacts

The total economic impacts from climate change are highly uncertain. Depending upon the assumptions used (e.g., climate sensitivity, discount rate and regional aggregation) total economic impacts are typically estimated to be in the range of a few percent of gross world product for a few degrees of warming (see Chapter 20). Some estimates suggest that gross world product could increase up to about 1-3°C warming, largely because of estimated direct CO_2 effects on agriculture, but such estimates carry only low confidence. Even the direction of gross world product change with this level of warming is highly uncertain. Above the 1-3°C level of warming, available studies indicate that gross world product could decrease (•). For example, Tol (2002a) estimates net positive global market impacts at 1°C when weighting by economic output, but finds much smaller positive impacts when equity-weighted. Nordhaus (2006) uses a geographically based method and finds more negative economic impacts than previous studies, although still in the range of a few percent of gross world product.

Studies of aggregate market impacts tend to rely on scenarios of average changes in climate and focus on direct economic effects alone. Potential damages from increased severity of extreme climate events are often not included. The damages from an increase in extreme events could substantially increase market damages, especially at larger magnitudes of climate change (*). Also, recent studies draw attention to indirect effects of climate change on the economy (e.g., on capital accumulation and investment, on savings rate); although there is debate about methods, the studies agree that such effects could be significant and warrant further attention (see Section 19.3.7; Fankhauser and Tol, 2005; Kemfert, 2006; Roson and Tol, 2006; Fisher et al., 2007).

19.3.2.4 Distribution of market impacts

Global market impacts mask substantial variation in market impacts at the continental, regional, national and local scales. Even if gross world product were to change just a few percent, national economies could be altered by relatively large amounts.

[4] The following confidence symbols are used: *** very high confidence, ** high confidence, * medium confidence, • low confidence.

For example, Maddison (2003) reports increases in cost of living in low-latitude areas and decreases in high-latitude areas from a 2.5°C warming. All studies with regional detail show Africa, for example, with climate damages of the order of several percent of gross domestic product (GDP) at 2°C increase in GMT or even lower levels of warming (*). As noted below, very small economies such as Kiribati face damages from climate change in the range of 20% of their GDP (•) (see Chapter 16 Section 16.4.3). The distributional heterogeneity in market system impacts reflects the equity criterion described in Section 19.2 when considering which impacts may be considered 'key'.

19.3.2.5 Societal systems

With regard to vulnerability of societal systems, there are myriad thresholds specific to particular groups and systems at specific time-frames beyond which they can be vulnerable to variability and to climate change (Yamin et al., 2005). These differences in vulnerability are a function of a number of factors. Exposure is one key factor. For example, crops at low latitudes will have greater exposure to higher temperatures than crops at mid- and high latitudes. Thus, yields for grain crops, which are sensitive to heat, are more likely to decline at lower latitudes than at higher latitudes. Social systems in low-lying coastal areas will vary in their exposure and adaptive capacities, yet most will have increased vulnerability with greater warming and associated sea-level rises or storm surges.

A second key factor affecting vulnerability is the capacity of social systems to adapt to their environment, including coping with the threats it may pose, and taking advantage of beneficial changes. Smit et al. (2001) identified a number of determinants of adaptive capacity, including such factors as wealth, societal organisation and access to technology (see also Yohe and Tol, 2002). These attributes differentiate vulnerability to climate change across societies facing similar exposure. For example, Nicholls (2004) and Nicholls and Tol (2006) found that level of development and population growth are very important factors affecting vulnerability to sea-level rise. The specific vulnerabilities of communities with climate-related risks, such as the elderly and the poor or indigenous communities, are typically much higher than for the population as a whole (see Section 14.2.6)

Even though some cold-related deaths and infectious disease exposure are likely to be reduced, on balance there is medium confidence that global mortality will increase as a result of climate change. It is estimated that an additional 5-170 million people will be at risk of hunger by the 2080s as a consequence of climate change (Chapter 5 Section 5.6.5). There is medium to high confidence that some other climate-sensitive health outcomes, including heatwave impacts, diarrhoeal diseases, flood-related risks, and diseases associated with exposure to elevated concentrations of ozone and aeroallergens, will increase with GMT (Chapter 8 Section 8.4.1). Development and adaptation are key factors influencing human health risk (Chapter 8 Section 8.6).

Vulnerability associated with water resources is complex because vulnerability is quite region-specific. In addition, the level of development and adaptation and social factors determining access to water are very important in determining vulnerability in the water sector. Studies differ as to whether climate change will increase or decrease the number of people living in water-stressed areas (e.g., Parry et al., 1999; Arnell, 2004; Hitz and Smith, 2004; Alcamo et al., 2007). Hundreds of millions of people are estimated to be affected by changes in water quantity and quality (Chapter 3 Section 3.4.3; Arnell, 2004) but uncertainties limit confidence and thus the degree to which these risks might be labelled as 'key'. Floods and droughts appear to have increased in some regions and are likely to become more severe in the future (Chapter 3 Section 3.4.3).

19.3.3 Regional vulnerabilities

Many of the societal impacts discussed above will be felt within the regions assessed as part of the AR4. At a regional and sub-regional scale, vulnerabilities can vary quite considerably. For example, while mid- and high-latitude areas would have increased crop yields up to about 3°C of warming, low-latitude areas would face decreased yields and increased risks of malnutrition at lower levels of warming (•/*) (Chapter 5 Section 5.4.2; Parry et al., 2004).

Africa is likely to be the continent most vulnerable to climate change. Among the risks the continent faces are reductions in food security and agricultural productivity, particularly regarding subsistence agriculture (Chapter 9 Sections 9.4.4 and 9.6.1; Parry et al., 2004; Elasha et al., 2006), increased water stress (Chapter 9 Section 9.4.1) and, as a result of these and the potential for increased exposure to disease and other health risks, increased risks to human health (Chapter 9 Section 9.4.3). Other regions also face substantial risks from climate change. Approximately 1 billion people in South, South-East, and East Asia would face increased risks from reduced water supplies (•) (Chapter 10 Section 10.4.2), decreased agricultural productivity (•) (Chapter 10 Section 10.4.1.1), and increased risks of floods, droughts and cholera (*) (Chapter 10 Section 10.4.5). Tens of millions to over a hundred million people in Latin America would face increased risk of water stress (•) (Chapter 13 Section 13.4.3). Low-lying, densely populated coastal areas are very likely to face risks from sea-level rise and more intense extreme events (Chapter 13 Section 13.4.4). The combination of land-use changes and climate change is very likely to reduce biodiversity substantially (Chapter 13 Section 13.2.5.1).

There is very high confidence that human settlements in polar regions are already being adversely affected by reduction in ice cover and coastal erosion (Chapter 15 Section 15.2.2). Future climate change is very likely to result in additional disruption of traditional cultures and loss of communities. For example, warming of freshwater sources poses risks to human health because of transmission of disease (*) (Martin et al., 2005). Shifts in ecosystems are very likely to alter traditional use of natural resources, and hence lifestyles.

Small islands, particularly several small island states, are likely to experience large impacts due to the combination of higher exposure, for example to sea-level rise and storm surge, and limited ability to adapt (Chapter 16 Sections 16.ES, 16.2.1 and 16.4). There is very high confidence that many islands are already experiencing some negative effects of climate change (Chapter 1 Section 1.3.3; Chapter 16 Section 16.4). The long-term sustainability of small-island societies is at great risk from climate change, with sea-level rise and extreme events posing particular challenges on account of their limited size, proneness to natural hazards and external shocks combined with limited adaptive capacity and high costs relative to GDP. Subsistence and commercial agriculture on small islands is likely to be adversely affected by climate change and sea-level rise, as a result of inundation, seawater intrusion into freshwater lenses, soil salinisation, decline in water supply and deterioration of water quality (Chapter 16 Executive Summary and Section 16.4). A group of low-lying islands, such as Tarawa and Kiribati, would face average annual damages of 17 to 18% of its economy by 2050 under the SRES A2 and B2 scenarios (•) (Chapter 16 Section 16.4.3).

Even in developed countries, there are many vulnerabilities. Arnell (2004) estimated a 40 to 50% reduction in runoff in southern Europe by the 2080s (associated with a 2 to 3°C increase in global mean temperature). Fires will very likely continue to increase in arid and semi-arid areas such as Australia and the western USA, threatening development in wildland areas (Chapter 4 Section 4.4.4; Chapter 11 Section 11.3.1; Chapter 14 Box 14.1 and Section 14.4.4; Westerling et al., 2006). Climate change is likely to increase the frequency and intensity of extreme heat events, as well as concentrations of air pollutants, such as ozone, which increase mortality and morbidity in urban areas (see Chapters 8, 11, 12 and 14).

19.3.4 Ecosystems and biodiversity

There is high confidence that climate change will result in extinction of many species and reduction in the diversity of ecosystems (see Section 4.4) Vulnerability of ecosystems and species is partly a function of the expected rapid rate of climate change relative to the resilience of many such systems. However, multiple stressors are significant in this system, as vulnerability is also a function of human development, which has already substantially reduced the resilience of ecosystems and makes many ecosystems and species more vulnerable to climate change through blocked migration routes, fragmented habitats, reduced populations, introduction of alien species and stresses related to pollution.

There is very high confidence that regional temperature trends are already affecting species and ecosystems around the world (Chapter 1 Sections 1.3.4 and 1.3.5; Parmesan and Yohe, 2003; Root et al., 2003; Menzel et al., 2006) and it is likely that at least part of the shifts in species observed to be exhibiting changes in the past several decades can be attributed to human-induced warming (see Chapter 1; Root et al., 2005). Thus, additional climate changes are likely to adversely affect many more species and ecosystems as global mean temperatures continue to increase (see Section 4.4). For example, there is high confidence that the extent and diversity of polar and tundra ecosystems is in decline and that pests and diseases have spread to higher latitudes and altitudes (Chapter 1 Sections 1.3.5 and 1.5).

Each additional degree of warming increases disruption of ecosystems and loss of species. Individual ecosystems and species often have different specific thresholds of change in temperature, precipitation or other variables, beyond which they are at risk of disruption or extinction. Looking across the many ecosystems and thousands of species at risk of climate change, a continuum of increasing risk of loss of ecosystems and species emerges in the literature as the magnitude of climate change increases, although individual confidence levels will vary and are difficult to assess. Nevertheless, further warming is likely to cause additional adverse impacts to many ecosystems and contribute to biodiversity losses. Some examples follow.

- About half a degree of additional warming can cause harm to vulnerable ecosystems such as coral reefs and Arctic ecosystems * (Table 4.1).
- A warming of 1°C above 1990 levels would result in all coral reefs being bleached and 10% of global ecosystems being transformed (Chapter 4 Section 4.4.11).
- A warming of 2°C above 1990 levels will result in mass mortality of coral reefs globally *** (Chapter 4 Section 4.4; Chapter 6 Box 6.1), with one-sixth of the Earth's ecosystems being transformed (Leemans and Eickhout, 2004) **, and about one-quarter of known species being committed to extinction *. For example, if Arctic sea-ice cover recedes markedly, many ice-dependent Arctic species, such as polar bears and walrus, will be increasingly likely to be at risk of extinction; other estimates suggest that the African Succulent Karoo is likely to lose four-fifths of its area (Chapter 4 Section 4.4.11 and Table 4.1). There is low confidence that the terrestrial biosphere will become a net source of carbon (Chapter 4 Section 4.4.1).
- An additional degree of warming, to 3°C, is likely to result in global terrestrial vegetation becoming a net source of carbon (Chapter 4 Section 4.4.1), over one-fifth of ecosystems being transformed * (Chapter 4 Section 4.4.11; Leemans and Eickhout, 2003), up to 30% of known species being committed to extinction * (Chapter 4 Section 4.4.11 and Table 4.1; Thomas et al., 2004; Malcolm et al., 2006, estimate that 1 to 43% of species in 25 biodiversity hotspots are at risk from an approximate 3 to 4°C warming) and half of all nature reserves being unable to meet conservation objectives * (Chapter 4 Table 4.1). Disturbances such as fire and pests are very likely to increase substantially (Chapter 4 Section 4.4).
- There is very high confidence that warming above 3°C will cause further disruption of ecosystems and extinction of species.

19.3.5 Geophysical systems

A number of Earth-system changes may be classified as key impacts resulting in key vulnerabilities.

19.3.5.1. Global biogeochemical cycles

The sensitivity of the carbon cycle to increased CO_2 concentrations and climate change is a key vulnerability due to its magnitude, persistence, rate of change, low adaptive capacity and the level of confidence in resulting impacts. Models suggest that the overall effect of carbon–climate interactions is a positive feedback (Denman et al., 2007 Section 7.1.5). As CO_2 concentrations increase and climate changes, feedbacks from terrestrial stores of carbon in forests and grasslands, soils, wetlands, peatlands and permafrost, as well as from the ocean, would reduce net uptake of CO_2 (Denman et al., 2007 Sections 7.3.3 and 7.3.4). Hence the predicted atmospheric CO_2 concentration in 2100 is higher (and consequently the climate is warmer) than in models that do not include these couplings (Denman et al., 2007 Section 7.1.5). An intercomparison of ten climate models with a representation of the land and ocean carbon cycle forced by the SRES A2 emissions scenario (Denman et al., 2007 Section 7.3.5; Meehl et al., 2007 Section 10.4.1) shows that, by the end of the 21st century, additional CO_2 varies between 20 and 200 ppm for the two extreme models, with most of the models projecting additional CO_2 between 50 and 100 ppm (Friedlingstein et al., 2003), leading to an additional warming ranging between 0.1 and 1.5°C. A similar range results from estimating the effect including forcing from aerosols and non-CO_2 greenhouse gases (GHGs). Such additional warming would increase the number and severity of impacts associated with many key vulnerabilities identified in this chapter. In addition, these feedbacks reduce the emissions (Meehl et al., 2007 Section 10.4.1) compatible with a given atmospheric CO_2 stabilisation pathway (**)

At the regional level (see Chapters 4, 10, 11, 12 and 14), important aspects of the carbon–climate interaction include the role of fire (Denman et al., 2007 Section 7.3.3.1.4) in transient response and possible abrupt land-cover transitions from forest to grassland or grassland to semi-arid conditions (Claussen et al., 1999; Eastman et al., 2001; Cowling et al., 2004; Rial et al., 2004).

Warming destabilises permafrost and marine sediments of methane gas hydrates in some regions according to some model simulations (Denman et al., 2007 Section 7.4.1.2), as has been proposed as an explanation for the rapid warming that occurred during the Palaeocene/Eocene thermal maximum (Dickens, 2001; Archer and Buffett, 2005). A rising eustatic (global) contribution to sea level is estimated to stabilise hydrates to some degree. One study (Harvey and Huang, 1995) reports that methane releases may increase very long-term future temperature by 10-25% over a range of scenarios. Most studies also point to increased methane emissions from wetlands in a warmer, wetter climate (Denman et al., 2007 Section 7.4.1.2).

Increasing ocean acidity due to increasing atmospheric concentrations of CO_2 (Denman et al., 2007 Section 7.3.4.1; Sabine et al., 2004; Royal Society, 2005) is very likely to reduce biocalcification of marine organisms such as corals (Hughes et al., 2003; Feely et al., 2004). Though the limited number of studies available makes it difficult to assess confidence levels, potentially severe ecological changes would result from ocean acidification, especially for corals in tropical stably stratified waters, but also for cold water corals, and may influence the marine food chain from carbonate-based phytoplankton up to higher trophic levels (Denman et al., 2007 Section 7.3.4.1; Turley et al., 2006).

19.3.5.2 Deglaciation of West Antarctic and Greenland ice sheets

The potential for partial or near-total deglaciation of the Greenland and the West Antarctic ice sheets (WAIS) and associated sea-level rise (Jansen et al., 2007 Sections 6.4.3.2 and 6.4.3.3; Meehl et al., 2007 Sections 10.6.4, 10.7.4.3 and 10.7.4.4; Alley et al., 2005; Vaughan, 2007), is a key impact that creates a key vulnerability due to its magnitude and irreversibility, in combination with limited adaptive capacity and, if substantial deglaciation occurred, high levels of confidence in associated impacts. Ice sheets have been discussed specifically in the context of Article 2 (O'Neill and Oppenheimer 2002; Hansen, 2005; Keller et al., 2005; Oppenheimer and Alley, 2005). Near-total deglaciation would eventually lead to a sea-level rise of around 7 m and 5 m (***) from Greenland and the WAIS, respectively, with wide-ranging consequences including a reconfiguration of coastlines worldwide and inundation of low-lying areas, particularly river deltas (Schneider and Chen, 1980; Revelle, 1983; Tol et al., 2006; Vaughan, 2007). Widespread deglaciation would not be reversible except on very long time-scales, if at all (Meehl et al., 2007 Sections 10.7.4.3 and 10.7.4.4). The Amundsen Sea sector of the WAIS, already experiencing ice acceleration and rapid ground-line retreat (Lemke et al., 2007 Section 4.6.2.2), on its own includes ice equivalent to about 1.5 m sea-level rise (Meehl et al., 2007 Section 10.7.4.4; Vaughan, 2007). The ability to adapt would depend crucially on the rate of deglaciation (**). Estimates of this rate and the corresponding time-scale for either ice sheet range from more rapid (several centuries for several metres of sea-level rise, up to 1 m/century) to slower (i.e., a few millennia; Meehl et al., 2007 Section 10.7.4.4; Vaughan and Spouge, 2002), so that deglaciation is very likely to be completed long after it is first triggered.

For Greenland, the threshold for near-total deglaciation is estimated at 3.2-6.2°C local warming (1.9-4.6°C global warming) relative to pre-industrial temperatures using current models (Meehl et al., 2007 Section 10.7.4.3). Such models also indicate that warming would initially cause the Antarctic ice sheet as a whole to gain mass owing to an increased accumulation of snowfall (*; some recent studies find no significant continent-wide trends in accumulation over the past several decades; Lemke et al., 2007 Section 4.6.3.1). Scenarios of deglaciation (Meehl et al., 2007 Section 10.7.4.4) assume that any such increase would be outweighed by accelerated discharge of ice following weakening or collapse of an ice shelf due to melting at its surface or its base (*). Mean summer temperatures over the major West Antarctic ice shelves are about as likely as not to pass the melting point if global warming exceeds 5°C (Meehl et al., 2007 Section 10.7.4.4). Some studies suggest that disintegration of ice shelves would occur at lower temperatures due to basal or episodic surface melting (Meehl et al., 2007 Sections 10.6.4.2 and 10.7.4.4; Wild et al., 2003). Recent observations of unpredicted, local acceleration and consequent loss of mass from both ice sheets (Alley et al., 2005) underscores the inadequacy of existing ice-sheet models, leaving no generally agreed basis for projection, particularly for WAIS (Lemke et al., 2007 Section 4.6.3.3; Meehl et al., 2007 Sections 10.6.4.2 and 10.7.4.4; Vieli and Payne, 2005). However, palaeoclimatic evidence (Denman et al., 2007 Sections 6.4.3.2 and 6.4.3.3; Overpeck et al., 2006; Otto-Bliesner et al., 2006) suggests that Greenland and possibly the WAIS contributed to a sea-level rise of 4-6 m during the last interglacial, when polar temperatures were 3-5°C warmer, and the global mean was not notably warmer, than at present (Meehl et al., 2007 Sections

10.7.4.3 and 10.7.4.4). Accordingly, there is medium confidence that at least partial deglaciation of the Greenland ice sheet, and possibly the WAIS, would occur over a period of time ranging from centuries to millennia for a global average temperature increase of 1-4°C (relative to 1990-2000), causing a contribution to sea-level rise of 4-6 m or more (Meehl et al., 2007 Sections 10.7.4.3 and 10.7.4.4; Oppenheimer and Alley, 2004, 2005; Hansen, 2005).

Current limitations of ice-sheet modelling also increase uncertainty in the projections of 21st-century sea-level rise (Meehl et al., 2007 Section 10.6.4.2) used to assess coastal impacts in this report. An illustrative estimate by WGI of the contribution of processes not represented by models yielded an increase of 0.1-0.2 m in the upper ranges of projected sea-level rise for 2100 (Meehl et al., 2007 Section 10.6.4.2). Other approximation methods would yield larger or smaller adjustments, including zero.

19.3.5.3 Possible changes in the North Atlantic meridional overturning circulation (MOC)

The sensitivity of the North Atlantic meridional overturning circulation (MOC) (cf., WGI AR4 Glossary; Bindoff et al., 2007 Box 5.1) to anthropogenic forcing is regarded as a key vulnerability due to the potential for sizeable and abrupt impacts (Tol, 1998; Keller et al., 2000; Mastrandrea and Schneider, 2001; Alley et al., 2003; Rahmstorf et al., 2003; Link and Tol, 2004, 2006; Higgins and Schneider, 2005; Sathaye et al., 2007).

Palaeo-analogues and model simulations show that the MOC can react abruptly and with a hysteresis response, once a certain forcing threshold is crossed (Randall et al., 2007; Meehl et al., 2007). Estimates of the forcing threshold that would trigger large-scale and persistent MOC changes rely on three main lines of evidence. The first, based on the analysis of coupled Atmosphere-Ocean General Circulation Models (AOGCMs), do not show MOC collapse in the 21st century (Meehl et al., 2007 Box 10.1). Assessing the confidence in this is, however, difficult, as these model runs sample only a subset of potentially relevant uncertainties (e.g., Challenor et al., 2006) and do not cross the forcing thresholds suggested by the second line of evidence: simulations using Earth system models of intermediate complexity (EMICs) (Randall et al., 2007 Section 8.8.3; Meehl et al., 2007 10.3.4). EMIC simulations, which use simplified representations of processes to explore a wider range of uncertainties, suggest that the probability that forcing would trigger an MOC threshold response during the 21st century could exceed estimates derived from AOGCM runs alone (e.g., Challenor et al., 2006). The third line of evidence, not assessed by Working Group I, relies on expert elicitations (sometimes combined with the analysis of simple climate models). These MOC projections show a large spread, with some suggesting a substantial likelihood of triggering a MOC threshold response within this century (Arnell et al., 2005; Rahmstorf and Zickfeld, 2005; McInerney and Keller, 2006; Schlesinger et al., 2006; Yohe et al., 2006).

Potential impacts associated with MOC changes include reduced warming or (in the case of abrupt change) absolute cooling of northern high-latitude areas near Greenland and north-western Europe, an increased warming of Southern Hemisphere high latitudes, tropical drying (Vellinga and Wood, 2002, 2006; Wood et al., 2003, 2006), as well as changes in marine ecosystem productivity (Schmittner, 2005), terrestrial vegetation (Higgins and Vellinga, 2004), oceanic CO_2 uptake (Sarmiento and Le Quéré, 1996), oceanic oxygen concentrations (Matear and Hirst, 2003) and shifts in fisheries (Keller et al., 2000; Link and Tol, 2004). Adaptation to MOC-related impacts is very likely to be difficult if the impacts occur abruptly (e.g., on a decadal time-scale). Overall, there is high confidence in predictions of a MOC slowdown during the 21st century, but low confidence in the scale of climate change that would cause an abrupt transition or the associated impacts (Meehl et al., 2007 Section 10.3.4). However, there is high confidence that the likelihood of large-scale and persistent MOC responses increases with the extent and rate of anthropogenic forcing (e.g., Stocker and Schmittner, 1997; Stouffer and Manabe, 2003).

19.3.5.4 Changes in the modes of climate variability

Change in the modes of climate variability in response to anthropogenic forcing can lead to key impacts because these modes dominate annual-to-decadal variability, and adaptation to variability remains challenging in many regions. For example, some studies suggest that anthropogenic forcings would affect El Niño-Southern Oscillation (ENSO) variability (Timmermann et al., 1999; Fedorov and Philander, 2000; Fedorov et al., 2006; Hegerl et al., 2007 Section 9.5.3.1; Meehl et al., 2007 Section 10.3.5.3-5). Current ENSO projections are marked by many uncertainties, including

- the potential for an abrupt and/or hysteresis response,
- the direction of the shift,
- the level of warming when triggered.

ENSO shifts would affect agriculture (Cane et al., 1994; Legler et al., 1999), infectious diseases (Rodo et al., 2002), water supply, flooding, droughts (Kuhnel and Coates, 2000; Cole et al., 2002), wildfires (Swetnam and Betancourt, 1990), tropical cyclones (Pielke and Landsea, 1999; Emanuel, 2005), fisheries (Lehodey et al., 1997), carbon sinks (Bacastow et al., 1980) and the North Atlantic MOC (Latif et al., 2000).

The North Atlantic Oscillation (NAO) and the Annular Mode in both the Northern and Southern Hemispheres (also known as the Arctic Oscillation, AO, and the Antarctic Oscillation, AAO; Meehl et al., 2007 Section 10.3.5.6; Hartmann et al., 2000; Thompson and Wallace, 2000; Fyfe et al., 1999; Kushner et al., 2001; Cai et al., 2003; Gillett et al., 2003; Kuzmina et al., 2005) are likely to be affected by greenhouse forcing and ozone depletion. For example, the average of the IPCC WGI AR4 simulations from thirteen models shows a positive trend for the Northern Annular Mode that becomes statistically significant early in the 21st century (Meehl et al., 2007 Section 10.3.5.6). Such changes would affect surface pressure patterns, storm tracks and rainfall distributions in the mid and high latitudes of both hemispheres, with potentially serious impacts on regional water supplies, agriculture, wind speeds and extreme events. Implications are potentially severe for water resources and storminess in Australia, New Zealand, southern Africa, Argentina and Chile, southern Europe, and possibly parts of the USA where Mediterranean-type climates prevail.

Current forcing may have caused changes in these modes but observed changes are also similar to those simulated in AOGCMs in the absence of forcing (Cai et al., 2003). There is some evidence for a weakening of major tropical monsoon circulations (AR4 WGI 3.7.1, 9.5.3.5). Projections of monsoon precipitation show a complex pattern of increases (e.g., Australia in the southern summer and Asia), and decreases (e.g., the Sahel in the northern summer) (Meehl et al., 2007 Section 10.3.5.2). Confidence in projections of specific monsoonal changes is low to medium.

19.3.6 Extreme events

As discussed in WGI AR4 Technical Summary (Solomon et al., 2007) Box TS.5 and Table TS.4, various extreme events are very likely to change in magnitude and/or frequency and location with global warming. In some cases, significant trends have been observed in recent decades (Trenberth et al., 2007 Table 3.8).

The most likely changes are an increase in the number of hot days and nights (with some minor regional exceptions), or in days exceeding various threshold temperatures, and decreases in the number of cold days, particularly including frosts. These are virtually certain to affect human comfort and health, natural ecosystems and crops. Extended warmer periods are also very likely to increase water demand and evaporative losses, increasing the intensity and duration of droughts, assuming no increases in precipitation.

Precipitation is generally predicted in climate models to increase in high latitudes and to decrease in some mid-latitude regions, especially in regions where the mid-latitude westerlies migrate polewards in the summer season, thus steering fewer storms into such 'Mediterranean climates' (Meehl et al., 2007 Section 10.3.2.3). These changes, together with a general intensification of rainfall events (Meehl et al., 2007 Section 10.3.6.1), are very likely to increase the frequency of flash floods and large-area floods in many regions, especially at high latitudes. This will be exacerbated, or at least seasonally modified in some locations, by earlier melting of snowpacks and melting of glaciers. Regions of constant or reduced precipitation are very likely to experience more frequent and intense droughts, notably in Mediterranean-type climates and in mid-latitude continental interiors.

Extended warm periods and increased drought will increase water stress in forests and grasslands and increase the frequency and intensity of wildfires (Cary, 2002; Westerling et al., 2006), especially in forests and peatland, including thawed permafrost. These effects may lead to large losses of accumulated carbon from the soil and biosphere to the atmosphere, thereby amplifying global warming (**) (see Sections 4.4.1, 19.3.5.1; Langmann and Heil, 2004; Angert et al., 2005; Bellamy et al., 2005).

Tropical cyclones (including hurricanes and typhoons), are likely to become more intense with sea surface temperature increases, with model simulations projecting increases by mid-century (Meehl et al., 2007 Section 10.3.6.3). However, despite an ongoing debate, some data reanalyses suggest that, since the 1970s, tropical cyclone intensities have increased far more rapidly in all major ocean basins where tropical cyclones occur (Trenberth et al., 2007 Section 3.8.3), and that this is consistently related to increasing sea surface temperatures. Some authors have questioned the reliability of these data, in part because climate models do not predict such large increases; however, the climate models could be underestimating the changes due to inadequate spatial resolution. This issue currently remains unresolved. Some modelling experiments suggest that the total number of tropical cyclones is expected to decrease slightly (Meehl et al., 2007 Section 10.3.6.3), but it is the more intense storms that have by far the greatest impacts and constitute a key vulnerability.

The combination of rising sea level and more intense coastal storms, especially tropical cyclones, would cause more frequent and intense storm surges, with damages exacerbated by more intense inland rainfall and stronger winds (see Section 6.3.2). Increasing exposure occurs as coastal populations increase (see Section 6.3.1).

Many adaptation measures exist that could reduce vulnerability to extreme events. Among them are dams to provide flood protection and water supply, dykes and coastal restoration for protection against coastal surges, improved construction standards, land-use planning to reduce exposure, disaster preparedness, improved warning systems and evacuation procedures, and broader availability of insurance and emergency relief (see Chapter 18). However, despite considerable advances in knowledge regarding weather extremes, the relevant adaptation measures are underused, partly for reasons of cost, especially in developing countries (White et al., 2001; Sections 7.4.3, 7.5 and 7.6). Despite progress in reducing the mortality associated with many classes of extremes, human societies, particularly in the developing world, are not well adapted to the current baseline of climate variability and extreme events, such as tropical cyclones, floods and droughts, and thus these impacts are often assessed as key vulnerabilities.

19.3.7 Update on 'Reasons for Concern'

The TAR (Smith et al., 2001; IPCC, 2001b) identified five 'reasons for concern' about climate change and showed schematically how their seriousness would increase with global mean temperature change. In this section, the 'reasons for concern' are updated.

Unique and threatened systems

The TAR concluded that there is medium confidence that an increase in global mean temperature of 2°C above 1990 levels or less would harm several such systems, in particular coral reefs and coastal regions.

Since the TAR, there is new and much stronger evidence of observed impacts of climate change on unique and vulnerable systems (see Sections 1.3.4 and 1.3.5; Parmesan and Yohe, 2003; Root et al., 2003, 2005; Menzel et al., 2006), many of which are described as already being adversely affected by climate change. This is particularly evident in polar ecosystems (e.g., ACIA, 2005). Furthermore, confidence has increased that an increase in global mean temperature of up to 2°C relative to 1990 temperatures will pose significant risks to many unique and vulnerable systems, including many biodiversity hotspots (e.g., Hare, 2003; Leemans and Eickhout, 2004; Malcolm et al., 2006). In summary, there is now high confidence that a warming of up to 2°C above 1990-2000 levels would have significant impacts on many unique and vulnerable systems, and is likely to increase the endangered status of many threatened species, with increasing adverse impacts and confidence in this conclusion at higher levels of temperature increase.

Extreme events

The TAR concluded that there is high confidence that the frequency and magnitude of many extreme climate-related events (e.g., heatwaves, tropical cyclone intensities) will increase with a temperature increase of less than 2°C above 1990 levels; and that this increase and consequent damages will become greater at higher temperatures.

Recent extreme climate events have demonstrated that such events can cause significant loss of life and property damage in both developing and developed countries (e.g., Schär et al., 2004). While individual events cannot be attributed solely to anthropogenic climate change, recent research indicates that human influence has already increased the risk of certain extreme events such as heatwaves (**) and intense tropical cyclones (*) (Stott et al., 2004; Emanuel, 2005; Webster et al., 2005; Trenberth et al., 2007; Bindoff et al., 2007). There is high confidence that a warming of up to 2°C above 1990-2000 levels would increase the risk of many extreme events, including floods, droughts, heatwaves and fires, with increasing levels of adverse impacts and confidence in this conclusion at higher levels of temperature increase.

Distribution of impacts

Chapter 19 of the WGII TAR (Smith et al., 2001) concluded that there is high confidence that developing countries will be more vulnerable to climate change than developed countries; medium confidence that a warming of less than 2°C above 1990 levels would have net negative impacts on market sectors in many developing countries and net positive impacts on market sectors in many developed countries; and high confidence that above 2 to 3°C, there would be net negative impacts in many developed countries and additional negative impacts in many developing countries.

There is still high confidence that the distribution of impacts will be uneven and that low-latitude, less-developed areas are generally at greatest risk due to both higher sensitivity and lower adaptive capacity. However, recent work has shown that vulnerability to climate change is also highly variable within individual countries. As a consequence, some population groups in developed countries are also highly vulnerable even to a warming of less than 2°C (see, e.g., Section 12.4.). For instance, indigenous populations in high-latitude areas are already faced with significant adverse impacts from climate change to date (see Section 14.4; ACIA, 2005), and the increasing number of coastal dwellers, particularly in areas subject to tropical cyclones, are facing increasing risks (Christensen et al., 2007 Box 11.5; Section 11.9.5). There is high confidence that warming of 1 to 2°C above 1990-2000 levels would include key negative impacts in some regions of the world (e.g., Arctic nations, small islands), and pose new and significant threats to certain highly vulnerable population groups in other regions (e.g., high-altitude communities, coastal-zone communities with significant poverty levels), with increasing levels of adverse impacts and confidence in this conclusion at higher levels of temperature increase.

Aggregate impacts

Chapter 19 of the WGII TAR (Smith et al., 2001) concluded that there is medium confidence that with an increase in global mean temperature of up to 2°C above 1990 levels, aggregate market sector impacts would be plus or minus a few percent of gross world product, but most people in the world would be negatively affected. Studies of aggregate economic impacts found net damages beyond temperature increases of 2 to 3°C above 1990 levels, with increasing damages at higher magnitudes of climate change.

The findings of the TAR are consistent with more recent studies, as reviewed in Hitz and Smith (2004). Many limitations of aggregated climate impact estimates have already been noted in the TAR, such as difficulties in the valuation of non-market impacts, the scarcity of studies outside a few developed countries, the focus of most studies on selected effects of a smooth mean temperature increase, and a preliminary representation of adaptation and development. Recent studies have included some of these previously unaccounted for aspects, such as flood damage to agriculture (Rosenzweig et al., 2002) and damages from increased cyclone intensity (Climate Risk Management Limited, 2005). These studies imply that the physical impacts and costs associated with these neglected aspects of climate change may be very significant. Different analytic techniques (e.g., Nordhaus, 2006) can result in estimates of higher net damages; inclusion of indirect effects can increase the magnitude of impacts (e.g., Fankhauser and Tol, 2005; Stern, 2007). Other studies reinforce the finding of potential benefits at a few degrees of warming, followed by damages with more warming (Maddison, 2003; Tol, 2005). However, long-term costs from even a few degrees of warming, such as eventual rise in sea level (e.g., Overpeck et al., 2006), are not included in aggregate damage estimates. In addition, the current literature is limited in accounting for the economic opportunities that can be created by climate change.

On balance, the current generation of aggregate estimates in the literature is more likely than not to understate the actual costs of climate change. Consequently, it is possible that initial net market benefits from climate change will peak at a lower magnitude and sooner than was assumed for the TAR, and it is likely that there will be higher damages for larger magnitudes of global mean temperature increases than estimated in the TAR.

The literature also includes analysis of aggregate impacts of climate change other than monetary effects. Parry et al. (1999) found that climate change could adversely affect hundreds of millions of people through increased risk of coastal flooding, reduction in water supplies, increased risk of malnutrition and increased risk of exposure to disease. All of these impacts would directly affect human health. The 'Global Burden of Disease' study estimated that the climate change that has occurred since 1990 has increased mortality, and that projected climate change will increase future disease burdens even with adaptation

(McMichael et al., 2004). There is low to medium confidence that most people in the world will be negatively affected at global mean temperature increases of 1-2°C above 1990-2000 levels, with increasing levels of adverse impacts and confidence in this conclusion at higher levels of temperature increase.

Large-scale singularities

The TAR concluded that there is low to medium confidence that a rapid warming of over 3°C would trigger large-scale singularities in the climate system, such as changes in climate variability (e.g., ENSO changes), breakdown of the thermohaline circulation (THC – or equivalently, meridional overturning circulation, MOC), deglaciation of the WAIS, and climate–biosphere–carbon cycle feedbacks. However, determining the trigger points and timing of large-scale singularities was seen as difficult because of the many complex interactions of the climate system.

Since the TAR, the literature offers more specific guidance on possible thresholds for partial or near-complete deglaciation of the Greenland and West Antarctic ice sheets. There is medium confidence that at least partial deglaciation of the Greenland ice sheet, and possibly the WAIS, would occur over a period of time ranging from centuries to millennia for a global average temperature increase of 1-4°C (relative to 1990-2000), causing a contribution to sea-level rise of 4-6 m or more (Section 19.3.5.2; Jansen et al., 2007 Section 6.4; Meehl et al., 2007 Sections 10.7.4.3 and 10.7.4.4; Oppenheimer and Alley, 2004, 2005; Hansen, 2005; Otto-Bliesner et al., 2006; Overpeck et al., 2006). Since the TAR, there is more confidence in projections of the climate consequences of feedbacks in the carbon cycle (see Section 19.3.5.1).

19.4 Assessment of response strategies to avoid key vulnerabilities

This section reviews the literature addressing the linkages between key vulnerabilities and response strategies in order to avoid or reduce them. This section is structured as follows. Section 19.4.1 reviews the literature on the role of adaptation to avoid key vulnerabilities. As discussed in Section 19.2, the lack of adaptive capacity, or the inability to adapt, is one of the criteria relevant for the selection of key vulnerabilities. Section 19.4.2 reviews the literature that specifically addresses the avoidance of key vulnerabilities through mitigation of climate change. Section 19.4.3 synthesises the knowledge about avoiding key vulnerabilities of climate change.

The principal response strategies – mitigation of climate change and adaptation – are often portrayed as having largely different foci in terms of their characteristic spatial and temporal scales. Other important strategies include investing in gaining knowledge (e.g., improving predictions and the understanding of options) and investing in capacity-building (improving ability and tools to make good decisions under uncertainty). Finally, some have suggested geo-engineering as a backstop policy option (see, e.g., Izrael, 2005; Cicerone, 2006; Crutzen, 2006; Kiehl, 2006; Wigley, 2006, for an update on this debate).

Given the integrating nature of this section at the interface between climate change impacts and vulnerabilities, mitigation, and adaptation, there are important links with other chapters of the IPCC AR4. Most importantly, WGII Chapter 17 discusses the role of adaptation to climate change; WGII Chapter 18, WGIII Chapter 2 Section 2.5 and Chapter 3 Section 3.5 discuss the links between mitigation and adaptation; WGIII Chapter 1 Section 1.2 and Chapter 2 Section 2.2 discuss the characteristics of the challenge and some decision-making problems in responding to global climate change, respectively; WGII Chapter 2 Section 2.2.7 and WGIII Chapter 2 Section 2.3 discuss methods to address uncertainties in this context; WGIII Chapter 3 Section 3.3 and Chapter 3 Section 3.6 discuss climate change mitigation from a long-term and a short-term perspective, respectively; and WGII Chapter 2 Section 2.4.6 discusses methods of evaluating impacts associated with mitigation scenarios.

19.4.1 Adaptation as a response strategy

How much can anticipatory and autonomous adaptation achieve? What is the potential for, and limitations of, adaptation to reduce impacts and to reduce or avoid key vulnerabilities?

The scientific literature on these questions is less well developed than for mitigation, and the conclusions are more speculative in many cases. It is clear, however, that there is no simple comprehensive response to the adaptation question, and that the answers are often place-specific and very nuanced, and are likely to become more so as research advances.

In agriculture, for example, previous IPCC assessments have generally concluded that, in the near to medium term, aggregate world food production is not threatened (IPCC, 1996, 2001a). However, considerable regional variation in impacts and adaptive capacity suggests that severe impacts and food scarcity could occur in some regions, especially at low latitudes, where large numbers of poorer people are already engaged in agriculture that is not currently viable (see Section 5.4.2). In global terms, agriculture has been extremely resilient and world food production has expanded rapidly to keep pace with world population growth. Of course, there is debate on the sustainability of these trends, as they depend in part on the growing demand for meat and meat products as well as potential competition between agricultural resources for producing food versus those used for producing energy. Nevertheless, even where shortages have occurred, the reasons are rarely to be found in an absolute lack of food but are more due to lack of purchasing power and failures of the distribution system.

Attention to adaptation in agriculture has tended to focus on specific measures at the farm level, and some progress is being made in the incorporation of climate risks into agricultural practices. On the other hand, the processes of globalisation and technological change are placing adaptation more in the hands of agri-business, national policy-makers, and the international political economy, including such factors as prices, tariffs and subsidies, and the terms of international trade (Apuuli et al., 2000; Burton and Lim, 2005).

The record of past success in agriculture is often seen in other sectors, particularly in developed countries and, in many regions it is evident that current climate variability falls largely within the coping range (Burton and Lim, 2005). One possible exception is in the case of extreme events where monetary losses (both insured and uninsured – Munich Re, 2005) have been rising sharply, although mortality has been falling. In such cases, adaptation has not been so successful, despite major improvements in understanding the risks and in forecasts and warnings (White et al., 2001). One reason is the decline in local concern and thus a reduced propensity to adopt proactive adaptation measures, as the memory of specific disaster events fades. Related to this lack of appreciation of possible risks is that governments and communities can still be taken by surprise when extreme events occur, even though scientific evidence of their potential occurrence is widely available (Bazermann, 2005). Economic damage and loss of life from Hurricane Katrina in 2005, the European heatwave of 2003, and many other similar events are due in large measure to a lack of sufficient anticipatory adaptation, or even maladaptation in some cases. So while the overall record of adaptation to climate change and variability in the past 200 or so years has been successful overall, there is evidence of insufficient investments in adaptation opportunities, especially in relation to extreme events (Burton, 2004, Burton and May, 2004; Hallegatte et al., 2007). While economic losses have increased, there has been considerable success in reducing loss of life; and despite the recent spate of deadly extreme weather events, the general trend in mortality and morbidity remains downwards.

It is clear that in the future there is considerable scope for adaptation, provided that existing and developing scientific understanding, technology and know-how can be effectively applied. It might be expected that the slower the rate of climate change, the more likely it is that adaptation will be successful. For example, even a major rise in sea level might be accommodated and adjusted to by human societies if it happens very slowly over many centuries (Nicholls and Tol, 2006). On the other hand, slow incremental change can still involve considerable costs and people might not be sufficiently motivated to take precautionary action and bear the associated costs without some more dramatic stimulus. Paradoxically, therefore, the full array of human adaptation potential is not likely to be brought to bear when all the market, social, psychological and institutional barriers to adaptation are taken into account.

In terms of the key vulnerabilities identified in Table 19.1, it is clear that adaptation potential is greater the more the system is under human management and control. Major geophysical changes leave little room for human-managed adaptation. Fortunately these changes are likely to unfold relatively slowly, thus allowing more time for adaptation to their eventual impacts. There is somewhat greater adaptive capacity in biological systems, but it is still very limited. Biodiversity and ecosystems are likely to be impacted at a much faster rate than geophysical systems without a commensurately larger adaptive capacity for such impacts. It seems likely, therefore, that the greatest impacts in the near to medium term, where adaptation capacity is very limited, will occur in biological systems (Leemans and Eickhout, 2004; Smith, 2004; see Chapter 4). As we move into human social systems and market systems, adaptive capacity at the technical level increases dramatically. However, the understanding of impacts, adaptive capacity, and the costs of adaptation is weaker in social systems than in biological systems, and the uncertainties are high. This is especially the case for synergistic or cross-cutting impacts. Considered in isolation, the potential for agricultural adaptation may appear to be good. When related impacts in water regimes, droughts and floods, pest infestations and plant diseases, human health, the reliability of infrastructure, poor governance, as well as other non-climate-related stresses are taken into account, the picture is less clear.

A general conclusion on the basis of the present understanding is that for market and social systems there is considerable adaptation potential, but the economic costs are potentially large, largely unknown and unequally distributed, as is the adaptation potential itself. For biological and geophysical systems, the adaptation potential is much less than in social and market systems, because impacts are more direct and therefore appear more rapidly. A large proportion of the future increase in key vulnerabilities is likely to be recorded first in biological systems (see Chapter 1). This does not mean that key vulnerabilities will not occur in social and market systems. They depend on biological systems, and as ecosystems are affected by mounting stresses from climate change and concomitant factors such as habitat fractionation, and the spread of plant diseases and pest infestations, then the follow-on, second-order effects on human health and safety, livelihoods and prosperity, will be considerable (*/**).

19.4.2 Mitigation

This subsection reviews the growing literature (see, e.g., Schellnhuber et al., 2006) on mitigation of climate change as a means to avoid key vulnerabilities or dangerous anthropogenic interference (DAI) with the climate system. A more general review of the literature on climate change mitigation is found in the WGIII AR4 Chapter 3 (Fisher et al., 2007) Sections 3.3.5 (on long-term stabilisation scenarios), 3.5.2 and 3.5.3 (on integrated assessment and risk management) and 3.6 (on linkages between short-term and long-term targets).

19.4.2.1 Methodological approaches to the assessment of mitigation strategies

A variety of methods is used in the literature to identify response strategies that may avoid potential key vulnerabilities or DAI (see also Fisher et al., 2007, Section 3.5.2). These methods can be characterised according to the following dimensions.

- *Targeted versus non-targeted*
  In this section, targeted approaches refer to the determination of policy strategies that attempt to avoid exceeding pre-defined targets for key vulnerabilities or DAI thresholds, whereas non-targeted approaches determine the implications for key vulnerabilities or DAI of emissions or concentration pathways selected without initial consideration of such targets or thresholds. Targeted approaches are sometimes referred to as 'inverse' approaches, as they are working backwards from a specified outcome (e.g., an impact threshold not to be exceeded) towards the origin of the cause–effect chain that links GHG emissions with climate impacts.

- *Deterministic versus set-based versus probabilistic*
Deterministic analyses are based on best-guess estimates for uncertain parameters, whereas probabilistic analyses explicitly consider key uncertainties of the coupled socio-natural system by describing one or more parameters in terms of probability distributions. Uncertainty can also be treated discretely by set-based methods that select different possible values without specifying any probability distribution across the members of that set. For a more detailed discussion of the role of uncertainty in the assessment of response strategies, see Box 19.3.

- *Optimising versus adaptive versus non-optimising*
Optimising analyses determine recommended policy strategies based on a pre-defined objective, such as cost minimisation; whereas non-optimising analyses do not require the specification of such an objective function. Adaptive analyses optimise near-term decisions under the assumption that future decisions will consider new information as and when it materialises.

Table 19.2 characterises the main methods applied in the relevant literature based on two of the three dimensions defined above, because deterministic, set-based and probabilistic approaches can be applied to each of these methods. The remainder of Section 19.4 reviews literature pertaining to these methods that examines mitigation strategies to avoid key vulnerabilities or DAI.

19.4.2.2 Scenario analysis and analysis of stabilisation targets

Scenario analysis examines the implications of specified emissions pathways or concentration profiles for future climate change, e.g., magnitude and rate of temperature increase. Some studies focus on the key radiative forcing agent CO_2, while

Box 19.3. Uncertainties in the assessment of response strategies

Climate change assessments and the development of response strategies face multiple uncertainties and unknowns (see Fourth Assessment Working Group II Chapter 2 and Working Group III Chapter 2). The most relevant sources of uncertainty in this context are:

(i) Natural randomness,
(ii) Lack of scientific knowledge,
(iii) Social choice (reflexive uncertainty),
(iv) Value diversity.

Some sources of uncertainty can be reasonably represented by probabilities, whereas others are more difficult to characterise probabilistically. The natural randomness in the climate system can be characterised by frequentist (or objective) probabilities, which describe the *relative frequency* (sometimes referred to as 'likelihood') of a repeatable event under known circumstances. There are, however, limitations to the frequentist description, given that the climate system is non-stationary at a range of scales and that past forcing factors cannot be perfectly known. The reliability of *knowledge* about uncertain aspects of the world (such as the 'true' value of climate sensitivity) cannot be empirically represented by frequentist probabilities alone. It is possible to construct probability distributions of climate sensitivity that look like frequency representations, but they will always have substantial elements of subjectivity embedded (Morgan and Keith, 1995; Allen et al., 2001). The inherent need for probabilistic analyses in a risk-management framework becomes problematic when some analysts object in principle to even assessing probabilities in situations of considerable lack of data or other key ingredients for probabilistic assessment. To help bridge this philosophical conflict, it has been suggested that making subjective elements transparent is an essential obligation of assessments using such an approach (e.g., Moss and Schneider, 2000). One method of characterising uncertainty due to a lack of scientific knowledge is by Bayesian (or subjective) probabilities, which refer to the *degree of belief* of experts in a particular statement, considering the available data. Another approach involves non-probabilistic representations such as imprecise probabilities (e.g., Hall et al., 2006). Whether probabilities can be applied to describe future social choice, in particular uncertainties in future greenhouse gas emissions, has also been the subject of considerable scientific debate (e.g., Allen et al., 2001; Grubler and Nakićenović, 2001; Lempert and Schlesinger, 2001; Pittock et al., 2001; Reilly et al., 2001; Schneider, 2001, 2002). Value diversity (such as different attitudes towards risk or equity) cannot be meaningfully addressed through an objective probabilistic description. It is often assessed through sensitivity analysis or scenario analysis, in which different value systems are explicitly represented and their associated impacts contrasted.

The probabilistic analyses of DAI reported in this section draw substantially on (subjective) Bayesian probabilities to describe key uncertainties in the climate system, such as climate sensitivity, the rate of oceanic heat uptake, current radiative forcing, and indirect aerosol forcing. See WGI Chapter 9 (Hegerl et al., 2007) and Chapter 10 (Meehl et al., 2007) for a more detailed discussion. While these uncertainties prevent the establishment of a high-confidence, one-to-one linkage between atmospheric greenhouse gas concentrations and global mean temperature increase, probabilistic analyses can assign a subjective probability of exceeding certain temperature thresholds for given emissions scenarios or concentration targets (e.g., Meinshausen, 2005; Harvey, 2007).

Table 19.2. *Methods to identify climate policies to avoid key vulnerabilities or DAI.*

Method	Description	Optimising approach?	Targeted approach?
Scenario analysis, analysis of stabilisation targets	Analyse the implications for temperature increase of specific concentration stabilisation levels, concentration pathways, emissions scenarios, or other policy scenarios.	No	No
Guardrail analysis	Derive ranges of emissions that are compatible with predefined constraints on temperature increase, intolerable climate impacts, and/or unacceptable mitigation costs.	No	Yes
Cost–benefit analysis including key vulnerabilities and DAI	Include representations of key vulnerabilities or DAI in a cost-optimising integrated assessment framework.	Yes	No
Cost-effectiveness analysis	Identify cost-minimising emissions pathways that are consistent with pre-defined constraints for GHG concentrations, climate change or climate impacts.	Yes	Yes

others include additional gases and aerosols in their analysis, often representing concentrations in terms of CO_2-equivalent ppm or radiative forcing in W/m^2 (see Forster et al., 2007 Section 2.3). Dynamic analyses include information about the trajectories of GHG emissions and development pathways, GHG concentrations, climate change and associated impacts. Related static analyses examine the relationship between stabilisation targets for GHG concentrations and equilibrium values for climate parameters (typically the increase in global mean temperature). Note that the term 'GHG stabilisation' is used here with a time horizon of up to several centuries. Over a longer time period without anthropogenic GHG emissions, CO_2 concentrations may return to values close to pre-industrial levels through natural processes (Brovkin et al., 2002; Putilov, 2003; Semenov, 2004a,b; Izrael and Semenov, 2005, 2006).

The shape over time of the specified emissions pathway or concentration profile is of particular relevance when considering key vulnerabilities, as it influences transient climate change and associated climate impacts (see, e.g., O'Neill and Oppenheimer, 2004; Meinshausen, 2005; Schneider and Mastrandrea, 2005; Mastrandrea and Schneider, 2006). Two general categories can be distinguished in studies that specifically consider CO_2 concentrations or temperature thresholds associated with key vulnerabilities or DAI: stabilisation scenarios, which imply concentrations increasing smoothly from current levels to a final stabilisation concentration (e.g., Enting et al., 1994; Schimel et al., 1996; Wigley et al., 1996; Morita et al., 2000; Swart et al., 2002; O'Neill and Oppenheimer, 2004) and peaking or overshoot scenarios, where a final concentration stabilisation level is temporarily exceeded (Harvey, 2004; Kheshgi, 2004; O'Neill and Oppenheimer, 2004; Wigley, 2004; Izrael and Semenov, 2005; Kheshgi et al., 2005; Meinshausen et al., 2005; Frame et al., 2006). Overshoot scenarios are necessary for the exploration of stabilisation levels close to or below current concentration levels.

Some studies treat the uncertainty in future GHG emissions and climate change by analysing a discrete range of scenarios. O'Neill and Oppenheimer (2002) examined ranges of global mean temperature increase in 2100 associated with 450, 550 and 650 ppm CO_2 concentration stabilisation profiles, as reported in the TAR (Cubasch et al., 2001). They concluded that none of these scenarios would prevent widespread coral-reef bleaching in 2100 (assumed to have a threshold 1°C increase above current levels), and that only the 450 ppm CO_2 stabilisation profile is likely to be associated with avoiding both deglaciation of West Antarctica (assumed to have a threshold of 2°C above current levels) and collapse of the MOC (assumed to have a threshold of 3°C increase within 100 years). Lowe et al. (2006) consider a suite of climate scenarios based on a 'perturbed parameter ensemble' of Hadley Centre climate models, finding that, for stabilisation close to 450 ppm, 5% of their scenarios exceed a threshold for deglaciation of West Antarctica (assumed to be 2.1°C local warming above 1990-2000 levels). Corfee-Morlot and Höhne (2003) review the current knowledge about climate impacts for each 'reason for concern' at different levels of global mean temperature change and CO_2 stabilisation, based on published probability density functions (PDFs) of climate sensitivity, finding that any CO_2 stabilisation target above 450 ppm is associated with a significant probability of triggering a large-scale climatic event. An inverse analysis of the implications of reaching CO_2 stabilisation at 450 ppm concludes that more than half of the SRES emissions scenarios leave this stabilisation target virtually out of reach as of 2020. A robust finding across such studies is that the probability of exceeding thresholds for specific key vulnerabilities or DAI increases with higher stabilisation levels for GHG concentrations.

Other studies quantify uncertainty using probability distributions for one or more parameters of the coupled social-natural system. Figure 19.1, for instance, depicts the likelihood of exceeding an equilibrium temperature threshold of 2°C above pre-industrial levels based on a range of published probability distributions for climate sensitivity. To render eventual exceedence of this exemplary threshold 'unlikely' (<33% chance), the CO_2-equivalent stabilisation level must be below 410 ppm for the majority of considered climate sensitivity uncertainty distributions (range between 350 and 470 ppm).

Key caveat: The analysis in Figure 19.1 employs a number of probability distributions taken from the literature. The WGI AR4 has assessed the body of literature pertaining to climate sensitivity, and concludes that the climate sensitivity is 'likely' to lie in the range 2-4.5°C, and is 'very likely' to be above 1.5°C (Meehl et al., 2007 Executive Summary). For fundamental physical reasons, as well as data limitations, values substantially higher than 4.5°C still cannot be excluded, although agreement with observations and proxy data is generally worse for those high values than for values in the 2-4.5°C range (Meehl et al., 2007 Executive Summary). 'Likely' in IPCC usage has been defined as a 66 to 90% chance, and 'very likely' has been defined as a 90 to 99% chance. Therefore, implicit in the information given by WGI is a 10 to 34% chance that climate sensitivity is outside the 'likely' range, with equal probability (5 to 17%) that it is below 2°C or above 4.5°C. Furthermore, the WGI assessment assigns a 90 to 99% chance that the climate sensitivity is above 1.5°C. However, the shape of the distribution to the right of 4.5°C – crucial for risk-management analyses – is, as noted by WGI, so uncertain given the lack of scientific knowledge, that any quantitative conclusion reached based on probability functions beyond 4.5°C climate sensitivity would be very low confidence. For these reasons, we assign no more than low confidence to any of the distributions or results presented in this section, particularly if the result depends on the tails of the probabilty distribution for climate sensitivity. Nevertheless, as noted here, a risk-management framework requires input of (even if low-probability, low-confidence) outlier information. Therefore, we present the literature based on probabilistic analyses to demonstrate the framework inherent in the risk management approach to assessing key vulnerabilities.

The temperature threshold for DAI can itself be represented by a subjective probability distribution. Wigley (2004) combined probability distributions for climate sensitivity and the temperature threshold for DAI in order to construct a distribution for the CO_2 stabilisation level required to avoid DAI. Under this assumption set, the median stabilisation level for atmospheric CO_2 concentrations is 536 ppm, and there is a 17% chance that the stabilisation level necessary to avoid DAI is below current atmospheric CO_2 levels. A similar analysis by Harvey (2006, 2007) added the explicit normative choice of an 'acceptable' probability (10%) for exceeding the probabilistic temperature threshold for DAI. With similar assumptions about the probability distributions for climate sensitivity and the DAI temperature threshold, he finds that the allowable CO_2 stabilisation concentration is between 390 and 435 ppm, depending on assumptions about aerosol forcing. Of course, these results are quite sensitive to all the assumptions made, as both authors explicitly acknowledge.

Finally, significant differences in environmental impacts are anticipated between GHG concentration stabilisation trajectories that allow overshoot of the stabilisation concentration versus those that do not, even when they lead to the same final concentration. For example, Schneider and Mastrandrea (2005) calculate the probability of at least temporarily exceeding a target of 2°C above pre-industrial (1.4°C above 'current') by 2200 to be 70% higher (77% instead of 45%) for an overshoot scenario rising to 600 ppm CO_2-equivalent and then stabilising in several centuries at 500 ppm CO_2-equivalent, compared with a non-overshoot scenario stabilising at the same level (Figure 19.2, top panel). Overshoot scenarios induce higher transient temperature increases, increasing the probability of temporary or

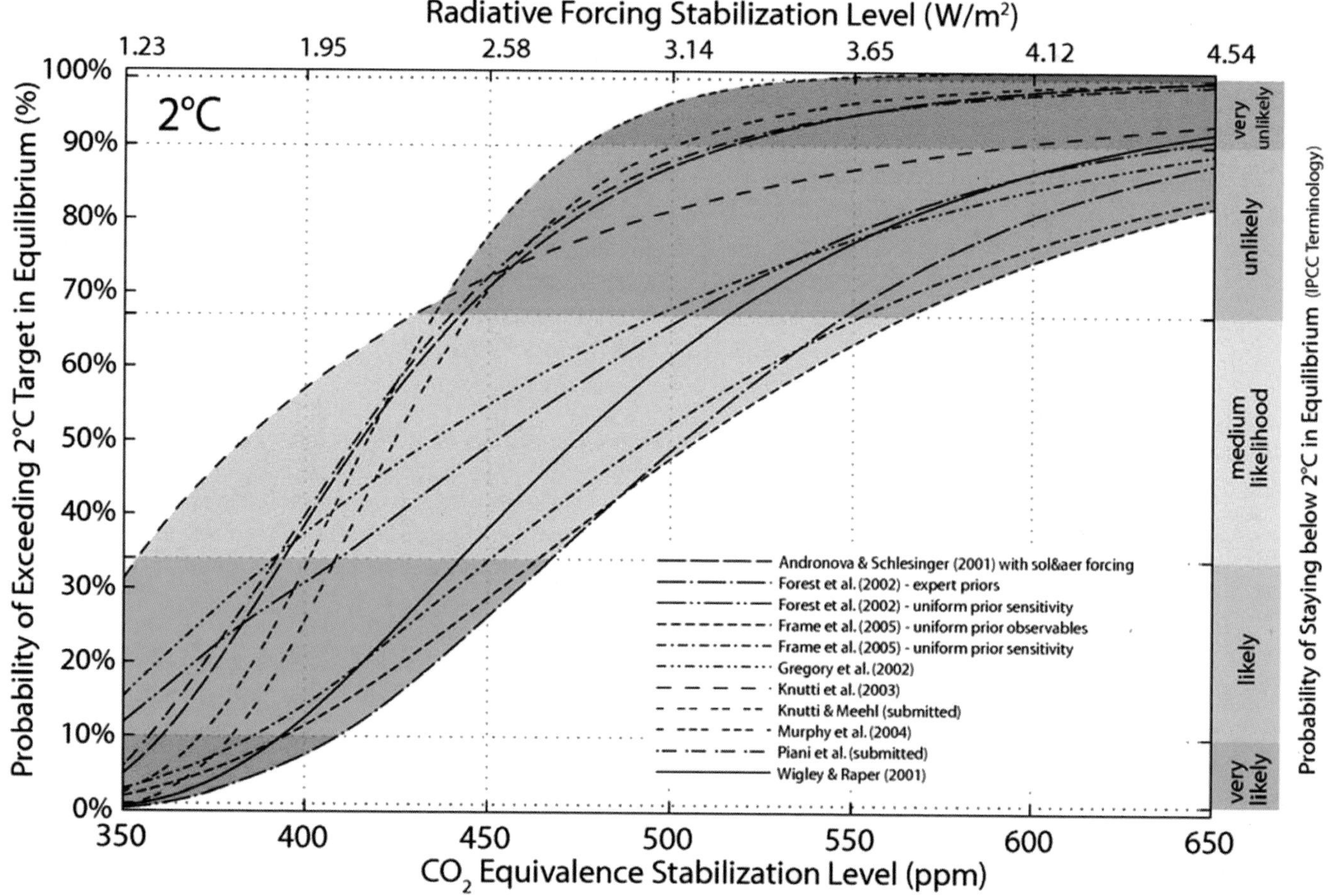

Figure 19.1. *Probability (see 'Key caveat' above on low confidence for specific quantitatitive results) of exceeding an equilibrium global warming of 2°C above pre-industrial (1.4°C above 1990 levels), for a range of CO_2-equivalent stabilisation levels. Source: Hare and Meinshausen (2005).*

permanent exceedence of thresholds for key vulnerabilities or DAI (e.g, Hammitt and Shlyakhter, 1999; Harvey, 2004; O'Neill and Oppenheimer, 2004; Hare and Meinshausen, 2005; Knutti et al., 2005). With this in mind, Schneider and Mastrandrea (2005) suggested two metrics – maximum exceedence amplitude and degree years – for characterising the maximum and cumulative magnitude of overshoot of a temperature threshold for DAI, as shown for an illustrative scenario in Figure 19.2 (bottom panel). Since the rate of temperature rise is important to adaptive capacity (see Section 19.4.1) and thus impacts, the time delay between now and the date of occurrence of the maximum temperature (year of MEA on Figure 19.2b) is also relevant to the likelihood of creating key vulnerabilities or exceeding specified DAI thresholds.

19.4.2.3 Guardrail analysis

Guardrail analysis comprises two types of inverse analysis that first define targets for climate change or climate impacts to be avoided and then determine the range of emissions that are compatible with these targets: tolerable windows analysis (Toth, 2003) and safe landing analysis (Swart et al., 1998). The tolerable windows approach allows the assessment of the implications of multiple competing climate policy goals on the mid-term and long-term ranges of permissible greenhouse gas emissions. It has initially been applied to several normative thresholds for climate impacts, which are analysed together with socio-economic constraints that aim at excluding unacceptable mitigation policies. Toth et al. (2003) analyse the interplay between thresholds for the global transformation of ecosystems, regional mitigation costs and the timing of mitigation. They show that following a business-as-usual scenario of GHG emissions (which resembles the SRES A2 scenario) until 2040 precludes the possibility of limiting the worldwide transformation of ecosystems to 30%, even under optimistic assumptions regarding willingness to pay for the mitigation of GHG emissions afterwards. Toth et al. (2003) show that mitigation of GHG emissions has to start no later than 2015 if a reduction in agricultural yield potential in South Asia of more than 10% is to be avoided. This result, however, is contingent on the regional climate change projection of the specific GCM applied in this analysis (HadCM2) and the accuracy of the impact models. The consideration of regional and local climate impacts in inverse analyses raises challenges as to the treatment of the significant uncertainties associated with them.

The tolerable windows approach has also been applied in connection with systematic climate thresholds, predominantly for probabilistic analyses of the stability of the thermohaline ocean circulation (Zickfeld and Bruckner, 2003; Bruckner and Zickfeld, 2004; Rahmstorf and Zickfeld, 2005). Rahmstorf and Zickfeld (2005) conclude that the SRES A2 emissions scenario exceeds the range of emissions corresponding to a 5% and 10% likelihood of inducing a commitment to a circulation shutdown around 2035 and 2065, respectively. A 2% risk of shutdown can no longer be avoided, even with very stringent emission reductions, given the assumptions in their models.

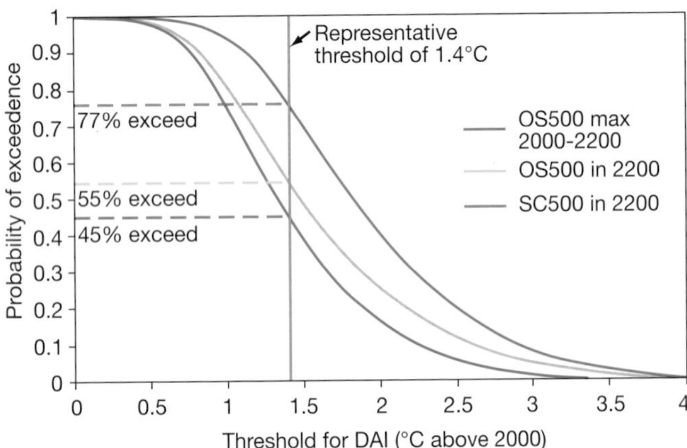

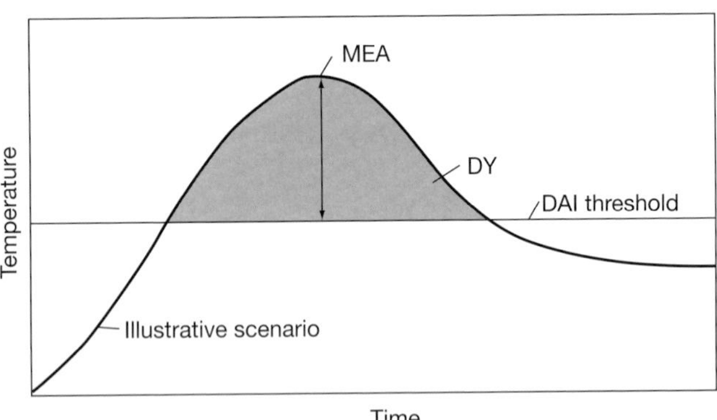

Figure 19.2. *Top panel: Probability of exceedence of a range of temperature thresholds for overshoot (OS500) and non-overshoot (SC500) scenarios, derived from probability distributions for climate sensitivity (see 'Key caveat' above on low confidence for specific quantitatitive results). OS500 Max is derived from the maximum overshoot temperature that occurs during the transient response before 2200, whereas OS500 in 2200 and SC500 in 2200 are derived from temperatures in 2200. While model-dependent, these results demonstrate the importance of considering transient temperature change when evaluating mitigation strategies to avoid key vulnerabilities. Bottom panel: Visualisation of maximum exceedence amplitude (MEA) and degree years (DY) for an illustrative overshoot temperature profile. Source: Schneider and Mastrandrea (2005).*

19.4.2.4 Cost–benefit analysis

Cost–benefit analyses (CBAs) of climate change in general are reviewed in Fisher et al., 2007 Section 3.5.3.3. The discussion here focuses on the suitability of CBA for avoiding key vulnerabilities and DAI. Most early cost–benefit analyses of climate change have assumed that climate change will be a gradual and smooth process. This assumption has prevented these analyses from determining a robust optimal policy solution (Hall and Behl, 2006), as it neglects important key vulnerabilities. Recognising the restrictions of this assumption, an extensive literature has developed extending cost–benefit analyses and related decision-making (e.g., Jones, 2003) in the context of Article 2, with a particular emphasis on abrupt change at global and regional scales (Schneider and Azar, 2001; Higgins et al., 2002; Azar and Lindgren, 2003; Baranzini et al., 2003; Wright and Erickson, 2003).

Several papers have focused on incorporating damages from large-scale climate instabilities identified as key vulnerabilities, such as climate-change-induced slowing or shutdown of the MOC (Keller et al., 2000, 2004; Mastrandrea and Schneider,

2001; Link and Tol, 2004). For example, quantifying market-based damages associated with MOC changes is a difficult task, and current analyses should be interpreted as order-of-magnitude estimates, with none carrying high confidence. These preliminary analyses suggest that significant reductions in anthropogenic greenhouse gas emissions are economically efficient even if the damages associated with a MOC slowing or collapse are less than 1% of gross world product. However, model results are very dependent on assumptions about climate sensitivity, the damage functions for smooth and abrupt climate change and time discounting, and are thus designed primarily to demonstrate frameworks for analysis and order-of-magnitude outcomes rather than high-confidence quantitative projections.

Several researchers have implemented probabilistic treatments of uncertainty in cost–benefit analyses; recent examples include Mastrandrea and Schneider (2004) and Hope (2006). These probabilistic analyses consistently suggest more aggressive mitigation policies compared with deterministic analyses, since probabilistic analyses allow the co-occurrence of high climate sensitivities (see *Key caveat* in Section 19.4.2.2 on low confidence for specific quantitative results) with high climate-damage functions.

19.4.2.5 Cost-effectiveness analysis

Cost-effectiveness analysis involves determining cost-minimising policy strategies that are compatible with pre-defined probabilistic or deterministic constraints on future climate change or its impacts. Comparison of cost-minimal strategies for alternative climate constraints has been applied to explore the trade-offs between climate change impacts and the associated cost of emissions mitigation (e.g., Keller et al., 2004; McInerney and Keller, 2006). The reductions in greenhouse-gas emissions determined by cost-effectiveness analyses incorporating such constraints are typically much larger than those suggested by most earlier cost–benefit analyses, which often do not consider the key vulnerabilities underlying such constraints in their damage functions. In addition, cost–benefit analysis assumes perfect substitutability between all costs and benefits of a policy strategy, whereas the hard constraints in a cost-effectiveness analysis do not allow for such substitution.

Some cost-effectiveness (as well as cost–benefit) analyses have explored sequential decision strategies in combination with the avoidance of key vulnerabilities or thresholds for global temperature change. These strategies allow for the resolution of key uncertainties in the future through additional observations and/or improved modelling. The quantitative results of these analyses cannot carry high confidence, as most studies represent uncertain parameters by two to three discrete values only and/or employ rather arbitrary assumptions about learning (e.g., Hammitt et al., 1992; Keller et al., 2004; Yohe et al., 2004). In a systematic analysis, Webster et al. (2003) finds that the ability to learn about damages from climate change and costs of reducing greenhouse gas emissions in the future can lead to either less restrictive or more restrictive policies today. All studies report the opinions of their authors to be that the scientific uncertainty by itself does not provide justification for doing nothing today to mitigate potential climate damages.

19.4.3 Synthesis

The studies reviewed in this section diverge widely in their methodological approach, in the sophistication with which uncertainties are considered in geophysical, biological and social systems, and in how closely they approach an explicit examination of key vulnerabilities or DAI. The models involved range from stand-alone carbon cycle and climate models to comprehensive integrated assessment frameworks describing emissions, technologies, mitigation, climate change and impacts. Some frameworks incorporate approximations of vulnerability but none contains a well-established representation of adaptation processes in the global context.

It is not possible to draw a simple summary from the diverse set of studies reviewed in this section. The following conclusions from literature since the TAR, however, are more robust.

- A growing literature considers response strategies that aim at preventing damage to particular key elements and processes in geophysical, biological and socio-economic systems that are sensitive to climate change and have limited adaptation potential; policy-makers may want to consider insights from the literature reviewed here in helping them to design policies to prevent DAI.
- In a majority of the literature, key impacts are associated with long-term increases in equilibrium global mean surface temperature above the pre-industrial equilibrium or an increase above 1990-2000 levels. Transient temperature changes are more instructive for the analyses of key vulnerabilities, but the literature is sparse on transient assessments relative to equilibrium analyses. Many studies provide global mean temperature thresholds that would lead sooner or later to a specific key impact, i.e., to disruption/shutdown of a vulnerable process. Such thresholds are not known precisely, and are characterised in the literature by a range of values (or occasionally by probability functions). Assessments of whether emissions pathways/GHG concentration profiles exceed given temperature thresholds are characterised by significant uncertainty. Therefore, deterministic studies alone cannot provide sufficient information for a full analysis of response strategies, and probabilistic approaches should be considered. Risk analyses given in some recent studies suggest that there is no longer high confidence that certain large-scale events (e.g., deglaciation of major ice sheets) can be avoided, given historical climate change and the inertia of the climate system (Wigley, 2004, 2006; Rahmstorf and Zickfeld, 2005). Similar conclusions could also be applied to risks for social systems, though the literature often suggests that any thresholds for these are at least as uncertain.
- Meehl et al., 2007 Table 10.8 provide likely ranges of equilibrium global mean surface temperature increase for different CO_2-equivalent stabilisation levels, based on their expert assessment that equilibrium climate sensitivity is likely to lie in the range 2-4.5°C (Meehl et al., 2007 Executive Summary). They present the following likely

ranges (which have been converted from temperature increase above pre-industrial to equilibrium temperature increase above 1990-2000 levels – see Box 19.2); 350 ppm CO_2-equivalent: 0-0.8°C above 1990-2000 levels; 450 ppm CO_2-equivalent: 0.8-2.5°C above 1990-2000 levels; 550 ppm CO_2-equivalent: 1.3-3.8°C above 1990-2000 levels; 650 ppm CO_2-equivalent: 1.8-4.9°C above 1990-2000 levels; 750 ppm CO_2-equivalent: 2.2-5.8°C above 1990-2000 levels. Some studies suggest that climate sensitivities larger than this likely range (which would suggest greater warming) cannot be ruled out (Meehl et al., 2007 Section 10.7.2), and the WGI range implies a 5-17% chance that climate sensitivity falls above 4.5°C (see *Key caveat* in Section 19.4.2.2 for further information).

- While future global mean temperature trajectories associated with different emissions pathways are not projected to diverge considerably in the next two to four decades, the literature shows that mitigation activities involving near-term emissions reductions will have a significant effect on concentration and temperature profiles over the next century. Later initiation of stabilisation efforts has been shown to require higher rates of reduction if they are to reduce the likelihood of crossings levels of DAI (Semenov, 2004a,b; Izrael and Semenov, 2005, 2006). Substantial delay (several decades or more) in mitigation activities makes achievement of the lower range of stabilisation targets (e.g., 500 ppm CO_2-equivalent and lower) infeasible, except via overshoot scenarios (see Figure 19.2, bottom panel). Overshoot scenarios induce higher transient temperature increases, increasing the probability of temporary or permanent exceedence of thresholds for key vulnerabilities (Hammitt, 1999; Harvey, 2004; O'Neill and Oppenheimer, 2004; Hare and Meinshausen, 2005; Knutti et al., 2005; Schneider and Mastrandrea, 2005).
- There is considerable potential for adaptation to climate change for market and social systems, but the costs and institutional capacities to adapt are insufficiently known and appear to be unequally distributed across world regions. For biological and geophysical systems, the adaptation potential is much lower. Therefore, some key impacts will be unavoidable without mitigation.

19.4.4 Research needs

The knowledge-base for the assessment of key vulnerabilities and risks from climate change is evolving rapidly. At the same time, there are significant gaps in our knowledge with regard to impacts, the potential and nature of adaptation, and vulnerabilities of human and natural systems. However, as this chapter has tried to bring out, a growing base of information that is likely to be of significance and value to the ongoing policy dialogue does exist.

In this concluding section of the chapter, some of the research priorities from the different domains are highlighted. Clearly, this can only be an indicative list, suggesting areas where new knowledge may have immediate utility and relevance as far as the objective of this chapter is concerned.

This chapter has suggested that key vulnerabilities may be a useful concept for informing the dialogue on dangerous anthropogenic interference. Further elucidation of this concept requires highly interdisciplinary, integrative approaches that are able to capture bio-geophysical and socio-economic processes. In particular, it is worth noting that the socio-economic conditions which determine vulnerability (e.g., number of people at risk, wealth, technology, institutions) change rapidly. Better understanding of the underlying dynamics of these changes at varying scales is essential to improve understanding of key vulnerabilities to climate change. The relevant research questions in this context are not so much how welfare is affected by changing socio-economic conditions, but rather how much change in socio-economic conditions affects vulnerability to climate change. In other words, a key question is how future development paths could increase or decrease vulnerability to climate change.

As this chapter has brought out through the criteria for identifying key vulnerabilities, the responses of human and natural systems, both autonomous and anticipatory, are quite important. Consequently, it is important that the extant literature on this issue is enriched with contributions from disciplines as diverse as political economy and decision theory. In particular, one of the central problems is a better understanding of adaptation and adaptive capacity, and of the practical, institutional, and technical obstacles to the implementation of adaptation strategies. This improvement in understanding will require a richer characterisation of the perception–evaluation–response process at various levels and scales of decision-making, from individuals to households, communities and nations. In this context, it is worth noting that new research approaches may be required. For example, with regard to adaptation, a learning-by-doing approach may be required so that the development of theory occurs in parallel with, and supported by, experience from practice.

A significant category of key vulnerabilities is associated with large-scale, irreversible and systemic changes in geophysical systems. Large-scale changes such as changes in the West Antarctic and Greenland ice sheets, could lead to significant impacts, particularly due to long-term large sea-level rise. Therefore, to obtain improved estimates of impacts from both 21st-century and long-term sea-level rise, new modelling approaches incorporating a better understanding of dynamic processes in ice sheets are urgently needed, as already noted by WGI. Furthermore, central to nearly all the assessments of key vulnerabilities is the need to improve knowledge of climate sensitivity – particularly in the context of risk management – the right-hand tail of the climate sensitivity probability distribution, where the greatest potential for key impacts lies.

Finally, the elucidation and determination of dangerous anthropogenic interference is a complex socio-political process, involving normative judgments. While information on key vulnerabilities will inform and enrich this process, there may be useful insights from the social sciences that might support this process, such as better knowledge of institutional and organisational dynamics, and diverse stakeholder inputs. Also needed are assessments of vulnerability and adaptation that combine top-down climate models with bottom-up social vulnerability assessments.

References

ACIA, 2005: *Impacts of a Warming Arctic: Arctic Climate Impacts Assessment*. Cambridge University Press, Cambridge, 1042 pp.

Adger, W.N., 2001: Scales of governance and environmental justice for adaptation and mitigation of climate change. *Journal of International Development*, **13**, 921-931.

Alcamo, J., M. Flörke and M. Märker, 2007: Future long-term changes in global water resources driven by socio-economic and climatic change. *Hydrol. Sci. J.*, **52**, 247-275.

Allen, M., S. Raper and J. Mitchell, 2001: Climate change: uncertainty in the IPCC's Third Assessment Report. *Science*, **293**, 430-433.

Alley, R.B., J. Marotzke, W.D. Nordhaus, J.T. Overpeck, D.M. Peteet, R.A. Pielke, R.T. Pierrehumbert, P.B. Rhines, T.F. Stocker, L.D. Talley and J.M. Wallace, 2003: Abrupt climate change. *Science*, **299**, 2005-2010.

Alley, R.B., P.U. Clark, P. Huybrechts and I. Joughin, 2005: Ice-sheet and sea-level changes. *Science*, **310**, 456-460.

Angert, A., S. Biraud, C. Bonfils, C.C. Henning, W. Buermann, J. Pinzon, C.J. Tucker and I. Fung, 2005: Drier summers cancel out the CO_2 uptake enhancement induced by warmer springs. *P. Natl. Acad. Sci.*, **102**, 10823–10827.

Apuuli, B., J. Wright, C. Elias and I. Burton, 2000: Reconciling national and global priorities in adaptation to climate change: an illustration from Uganda. *Environ. Monit. Assess.*, **61**, 145-159.

Archer, D. and B. Buffett, 2005: Time-dependent response of the global ocean clathrate reservoir to climatic and anthropogenic forcing. *Geochem. Geophy. Geosy.*, **6**, 1525-2027.

Arnell, N.W., 2004: Climate change and global water resources: SRES emissions and socio-economic scenarios. *Global Environ. Chang.*, **14**, 31-52.

Arnell, N.W., E.L. Tompkins and W.N. Adger, 2005: Eliciting information from experts on the likelihood of rapid climate change. *Risk Anal.*, **25**, 1419-1431.

Azar, C., 1998: Are optimal emissions really optimal: four critical issues for economists in the greenhouse. *Environ. Resour. Econ.*, **11**, 301-315.

Azar, C. and T. Sterner, 1996: Discounting and distributional considerations in the context of climate change, *Ecol. Econ.*, **19**, 169-185.

Azar, C. and K. Lindgren, 2003: Catastrophic events and stochastic cost-benefit analysis of climate change. *Climatic Change*, **56**, 245-255.

Bacastow, R.B., J.A. Adams, C.D. Keeling, D.J. Moss, T.P. Whorf and C.S. Wong, 1980: Atmospheric carbon dioxide, the Southern Oscillation and the weak 1975 El Niño. *Science*, **210**, 66-68.

Baranzini, A., M. Chesney and J. Morisset, 2003: The impact of possible climate catastrophes on global warming policies. *Energ. Policy*, **31**, 691-701.

Barnett, J., 2003: The relation between environmental security and climate change is discussed. *Global Environ. Chang.*, **13**, 7-17.

Barnett, J. and W.N. Adger, 2003: Climate dangers and atoll countries, *Climatic Change*, **61**, 321-337.

Bazermann, M., 2005: Climate change as a predictable surprise. *Climatic Change*, **77**, 179–193.

Bellamy, P.H., P.J. Loveland, R.I. Bradley, R.M. Lark and G.J.D. Kirk, 2005: Carbon losses from all soils across England and Wales 1978–2003. *Nature*, **437**, 245–248.

Berger, J.O. and D.O. Berry, 1988: Statistical analysis and the illusion of objectivity. *Am. Sci.*, **76**, 159-165.

Bindoff, N., J. Willebrand, V. Artale, A. Cazenave, J. Gregory, S. Gulev, K. Hanawa, C.L. Quéré, S. Levitus, Y. Nojiri, C.K. Shum, L. Talley and A. Unnikrishnan, 2007: Observations: oceanic climate change and sea level. *Climate Change 2007: The Physical Science Basis. Working Group I Contribution to the Intergovernmental Panel on Climate Change Fourth Assessment Report*, S. Solomon, D. Qin, M. Manning, Z. Chen, M. Marquis, K. B. Averyt, M. Tignor and H. L. Miller, Eds., Cambridge University Press, Cambridge, 385-432.

Brovkin V., J. Bendtsen, M. Claussen, A. Ganopolski, C. Kubatzki, V. Petoukhov and A. Andreev, 2002: Carbon cycle, vegetation and climate dynamics in the Holocene: experiments with the CLIMBER-2 model. *Global Biogeochem. Cy.*, **16**, 1139, doi:10.1029/2001GB001662.

Bruckner, T. and K. Zickfeld, 2004: Low risk emissions corridors for safeguarding the Atlantic thermohaline circulation. Paper presented at the Expert Workshop "Abrupt Climate Change Strategy", Paris, 30 September - 1 October 2004. http://www.accstrategy.org/draftpapers/ACCSbrucknerzickfeld.pdf.

Bunyavanich, S., C. Landrigan, A.J. McMichael and P.R. Epstein, 2003: The impact of climate change on child health. *Ambul. Pediatr.*, **3**, 44–52.

Burton, I., 2004: Climate change and the adaptation deficit. Occasional Paper No. 1. Adaptation and Impacts Research Group, Meteorological Service of Canada, Downsview.

Burton, I. and E. May, 2004: The adaptation deficit in water resource management. *IDS Bull.*, **35**, 31-37.

Burton, I. and B. Lim, 2005: Achieving adequate adaptation in agriculture. *Climatic Change*, **70**, 191-200.

Cai, W., P.H. Whetton and D.J. Karoly, 2003: The response of the Antarctic Oscillation to increasing and stabilized atmospheric CO_2. *J. Climate*, **16**, 1525-1538.

Cane, M.A., G. Eshel and R.W. Buckland, 1994: Forecasting Zimbabwean maize yield using eastern equatorial Pacific sea-surface temperature. *Nature*, **370**, 204-205.

Cary, G.J., 2002: Importance of a changing climate for fire regimes in Australia. *Flammable Australia: The Fire Regimes and Biodiversity of a Continent*, R.A. Bradstock, J.E. Williams and A.M. Gill, Eds., Cambridge University Press, New York, 26-49.

Challenor, P.G., R.K.S. Hankin and R. March, 2006: Towards the probability of rapid climate change. *Avoiding Dangerous Climate Change*, H.-J. Schellnhuber, W. Cramer, N. Nakićenović, T.M.L. Wigley and G. Yohe, Eds., Cambridge University Press, Cambridge, 55-64.

Christensen, J.H., B. Hewitson, A. Busuioc, A. Chen, X. Gao, I. Held, R. Jones, R.K. Kolli, W.-T. Kwon, R. Laprise, V. Magaña Rueda, L. Mearns, C.G. Menéndez, J. Räisänen, A. Rinke, A. Sarr and P. Whetton, 2007: Regional climate projections. *Climate Change 2007: The Physical Science Basis. Contribution of Working Group I to the Fourth Assessment Report of the Intergovernmental Panel on Climate Change,* S. Solomon, D. Qin, M. Manning, Z. Chen, M. Marquis, K.B. Averyt, M. Tignor and H.L. Miller, Eds., Cambridge University Press, Cambridge, 847-940.

Cicerone, R.J., 2006: Geoengineering: encouraging research and overseeing implementation. *Climatic Change*, **77**, 221-226.

Claussen, M., C. Kubatzki, V. Brovkin, A. Ganopolski, P. Hoelzmann and H.J. Pachur, 1999: Simulation of an abrupt change in Saharan vegetation at the end of the mid-Holocene. *Geophys. Res. Lett.*, **24**, 2037-2040.

Climate Risk Management Limited, 2005: *The Financial Risks of Climate Change*. Climate Risk Management Ltd, Southwell, Nottinghamshire, 125 pp.

Cline, W.R., 1996: The impact of global warming on agriculture: comment. *Am. Econ. Rev.*, **86**, 1309-1311.

Cole, J.E., J.T. Overpeck and E.R. Cook, 2002: Multiyear La Niña events and persistent drought in the contiguous United States. *Geophys. Res. Lett.*, **29**, doi:10.1029/2001GL013561.

Corfee-Morlot, J. and N. Höhne, 2003. Climate change: long-term targets and short-term commitments. *Global Environ. Chang.*, **13**, 277-293.

Corfee-Morlot, J. and S. Agrawala, 2004: Overview. *The Benefits of Climate Change Policies: Analytical and Framework Issues,* J. Corfee-Morlot and S. Agrawala, Eds., OECD, Paris, 9-30.

Cowling, S.A., R.A. Betts, P.M. Cox, V.J. Ettwein, C.D. Jones, M.A. Maslin and S.A. Spall, 2004: Contrasting simulated past and future responses of the Amazonian forest to atmospheric change. *Philos. T. Roy. Soc. B*, **359**, 539 – 547.

Crutzen, P.J., 2006: Albedo enhancement by stratospheric sulfur injections: a contribution to resolve a policy dilemma? *Climatic Change*, **77**, 211-220.

Cubasch, U., G.A. Meehl, G.J. Boer, R.J. Stouffer, M. Dix, A. Noda, C.A. Senior, S. Raper and K.S. Yap, 2001: Projections of future climate change. *Climate Change 2001: The Scientific Basis. Contribution of Working Group I to the Third Assessment Report of the Intergovernmental Panel on Climate Change,* J.T. Houghton, Y. Ding, D.J. Griggs, M. Noguer, P.J. van der Linden, X. Dai, K. Maskell and C.A. Johnson, Eds., Cambridge University Press, Cambridge, 525-582.

Denman, K.L., G. Brasseur, A. Chidthaisong, P. Ciais, P.M. Cox, R.E. Dickinson, D. Hauglustaine, C. Heinze, E. Holland, D. Jacob, U. Lohmann, S. Ramachandran, P.L. da Silva Dias, S.C. Wofsy and X. Zhang, 2007: Couplings between changes in the climate system and biogeochemistry. *Climate Change 2007: The Physical Science Basis. Contribution of Working Group I to the Fourth Assessment Report of the Intergovernmental Panel on Climate Change,* S. Solomon, D. Qin, M. Manning, Z. Chen, M. Marquis, K.B. Averyt, M. Tignor and H.L. Miller, Eds., Cambridge University Press, Cambridge, 499-588.

Denton, F., 2002: Climate change vulnerability, impacts and adaptation: why does gender matter? *Gender and Development,* **10**, 10-20.

Dessai, S., W.N. Adger, M. Hulme, J. Turnpenny, J. Köhler and R. Warren, 2004: Defining and experiencing dangerous climate change. *Climatic Change*, **64**, 11-

25.

Dickens, G.R., 2001: Modeling the global carbon cycle with gas hydrate capacitor: significance for the latest Paleocene thermal maximum. *Natural Gas Hydrates: Occurrence, Distribution, and Detection*, C.K. Paull and W.P. Dillon, Eds., Geophysical Monograph 124. American Geophysical Union, Washington, District of Columbia, 19-38.

Eastman, J.L., M.B. Coughenour, R.A. Pielke, 2001: The regional effects of CO_2 and landscape change using a coupled plant and meteorological model. *Glob. Change Biol.*, **7**, 797-815.

ECF, 2004: *Report on the Beijing Symposium on Article 2*. European Climate Forum and Potsdam Institute for Climate Impact Research. http://ecf.pik-potsdam.de/docs/papers/ECF_beijing_results.pdf.

Elasha, B.O., M. Medany, I. Niang-Diop, T. Nyong, R. Tabo and C. Vogel, 2006: Impacts, vulnerability and adaptation to climate change in Africa. Background Paper for the African Workshop on Adaptation Implementation of Decision 1/CP.10 of the UNFCCC Convention, UNFCCC, 54 pp. http://www.envirosecurity.org/activities/science/nairobi2006/Africa.pdf.

Emanuel, K., 2005: Increasing destructiveness of tropical cyclones over the past 30 years. *Nature*, **436**, 686-688.

Enting I.G., T.M.L. Wigley and M. Heimann, 1994: Future emissions and concentrations of carbon dioxide: key ocean/atmosphere/land analyses. CSIRO Technical Paper No. 31. http://www.dar.csiro.au/information/techpapers.html.

Fankhauser, S. and R.S.J. Tol, 2005: On climate change and economic growth. *Resour. Energy Econ.*, **27**, 1-17.

Fankhauser, S., R.S.J. Tol and D.W. Pearce, 1997: The aggregation of climate change damages: a welfare theoretic approach. *Environ. Resour. Econ.*, **10**, 249–266.

Fedorov, A.V. and S.G. Philander, 2000: Is El Niño changing? *Science*, **288**, 1997-2002.

Fedorov, A.V., P.S. Dekens, M. McCarthy, A.C. Ravelo, P.B. deMenocal, M. Barreiro, R.C. Pacanowski and S.G. Philander, 2006: The Pliocene paradox (mechanisms for a permanent El Niño). *Science*, **312**, 1485-1489.

Feely, R.A., C.L. Sabine, K. Lee, W. Berelson, J. Kleypas, V.J. Fabry and F.J. Millero, 2004: Impact of anthropogenic CO_2 on the $CaCO_3$ system in the oceans. *Science*, **305**, 362-366.

Fisher, B.S. and Co-authors, 2007: Issues related to mitigation in the long-term context. *Climate Change 2007: Mitigation. Contribution of Working Group III to the Fourth Assessment Report of the Intergovernmental Panel on Climate Change*, B. Metz, O. Davidson, P. Bosch, R. Dave and L. Meyer, Eds., Cambridge University Press, Cambridge, UK.

Folland, C.K., N.A. Rayner, S.J. Brown, T.M. Smith, S.S.P. Shen, D.E. Parker, I. Macadam, P.D. Jones, R.N. Jones, N. Nicholls and D.M.H. Sexton, 2001: Global temperature change and its uncertainties since 1861. *Geophys. Res. Lett.*, **28**, 2621-2624.

Forster, P., V. Ramaswamy, P. Araxo, T. Berntsen, R.A. Betts, D.W. Fahey, J. Haywood, J. Lean, D.C. Lowe, G. Myhre, J. Nganga, R. Prinn, G. Raga, M. Schulze and R. Van Dorland, 2007: Changes in atmospheric constituents and in radiative forcing. *Climate Change 2007: The Physical Science Basis. Contribution of Working Group I to the Fourth Assessment Report of the Intergovernmental Panel on Climate Change*, S. Solomon, D. Qin, M. Manning, Z. Chen, M. Marquis, K.B. Averyt, M. Tignor and H.L. Miller, Eds., Cambridge University Press, Cambridge, 129-234.

Frame, D.J., D.A. Stone, P.A. Stott and M.R. Allen, 2006: Alternatives to stabilization scenarios. *Geophys. Res. Lett.*, **33**, L14707, doi:10.1029/2006GL025801.

Friedlingstein, P., J.-L. Dufresne, P.M. Cox and P. Rayner, 2003: How positive is the feedback between climate change and the carbon cycle? *Tellus B*, **55**, 692-700.

Füssel, H.-M. and R.J.T. Klein, 2006: Climate change vulnerability assessments: an evolution of conceptual thinking. *Climatic Change*, **75**, 301-329.

Füssel, H.-M., F.L. Toth, J.G. van Minnen and F. Kaspar, 2003: Climate impact response functions as impact tools in the tolerable windows approach. *Climatic Change*, **56**, 91-117.

Fyfe, J.C., G.J. Boer and G.M. Flato, 1999: The Arctic and Antarctic Oscillations and their projected changes under global warming. *Geophys. Res. Lett.*, **26**, 1601-1604.

Gardiner, S.M., 2004: Ethics and global climate change. *Ethics*, **114**, 555-600.

Gardiner, S.M., 2006: A perfect moral storm: climate change, intergenerational ethics and the problem of moral corruption. *Environmental Values*, **15**, 397-413.

Gillett, N.P., F.W. Zwiers, A.J. Weaver and P.A. Stott, 2003: Detection of human influence on sea-level pressure. *Nature*, **422**, 292-294.

Goklany, I.M., 2002: From precautionary principle to risk-risk analysis. *Nat. Biotechnol.*, **20**, 1075.

Grubler, A. and N. Nakićenović, 2001: Identifying dangers in an uncertain climate. *Nature*, **412**, 15.

Gupta, J., O. Xander and E. Rotenberg, 2003: The role of scientific uncertainty in compliance with the Kyoto Protocol to the Climate Change Convention. *Environ. Sci. Policy*, **6**, 475-486.

Hall, D.C. and R.J. Behl, 2006: Integrating economic analysis and the science of climate instability. *Ecol. Econ.*, **57**, 442-465.

Hall, J.W., Fu, G. and Lawry, J., 2006: Imprecise probabilities of climate change: aggregation of fuzzy scenarios and model uncertainties. *Climatic Change*, **81**, 265-281.

Hallegatte S., J.-C. Hourcade and P. Dumas, 2007: Why economic dynamics matter in the assessment of climate change damages: illustration on extreme events. *Ecol. Econ.*, **62**, 330-340.

Halsnæs, K. and co-authors, 2007: Framing issues. *Climate Change 2007: Mitigation. Contribution of Working Group III to the Fourth Assessment Report of the Intergovernmental Panel on Climate Change*, B. Metz, O. Davidson, P. Bosch, R. Dave and L. Meyer, Eds., Cambridge University Press, Cambridge, UK.

Hamilton, J.M., D.J. Maddison and R.S.J. Tol, 2005: Climate change and international tourism: a simulation study. *Global Environ. Chang.*, **15**, 253-266.

Hammitt, J.K., 1999: Evaluation endpoints and climate policy: Atmospheric stabilization, benefit-cost analysis, and near-term greenhouse-gas emissions. *Climatic Change*, **41**, 447-468.

Hammitt, J.K. and A.I. Shlyakhter, 1999: The expected value of information and the probability of surprise. *Risk Anal.*, **19**, 135-52.

Hammitt, J.K., R.J. Lempert and M.E. Schlesinger, 1992: A sequential-decision strategy for abating climate change. *Nature*, **357**, 315-318.

Hanemann, W.M., 2000: Adaptation and its measurement: an editorial comment. *Climatic Change*, **45**, 571-581.

Hansen, J., 2005: A slippery slope: how much global warming constitutes "dangerous anthropogenic interference?" *Climatic Change*, **68**, 269-279.

Hare, B. and M. Meinshausen, 2005: How much warming are we committed to and how much can be avoided? *Climatic Change*, **75**, 111-149.

Hare, W., 2003: *Assessment of Knowledge on Impacts of Climate Change: Contribution to the Specification of Article 2 of the UNFCCC*. Wissen, Berlin, 106 pp. http://www.wbgu.de/wbgu_sn2003_ex01.pdf.

Hartmann, D.L., J.M. Wallace, V. Limpasuvan, D.W.J. Thompson and J.R. Holton, 2000: Can ozone depletion and global warming interact to produce rapid climate change? *P. Natl. Acad. Sci. USA*, **97**, 1412-1417.

Harvey, L.D.D., 2004: Declining temporal effectiveness of carbon sequestration: implications for compliance with the United National Framework Convention on Climate Change. *Climatic Change*, **63**, 259-290.

Harvey, L.D.D., 2006: Uncertainties in global warming science and near-term emission policies. *Clim. Policy*, **6**, 573-584.

Harvey, L.D.D., 2007: Allowable CO_2 concentrations under the United Nations Framework Convention on Climate Change as a function of the climate sensitivity PDF. *Environ. Res. Lett.*, **2**, doi:10.1088/1748-9326/2/1/014001.

Harvey, L.D.D. and Z. Huang, 1995: Evaluation of the potential impact of methane clathrate destabilization on future global warming. *J. Geophys. Res.*, **100**, 2905–2926.

Hegerl, G.C., F.W. Zwiers, P. Braconnot, N.P. Gillett, Y. Luo, J.A. Marengo Orsini, N. Nicholls, J.E. Penner and P.A. Stott, 2007: Understanding and attributing climate change. *Climate Change 2007: The Physical Science Basis. Contribution of Working Group I to the Fourth Assessment Report of the Intergovernmental Panel on Climate Change,* S. Solomon, D. Qin, M. Manning, Z. Chen, M. Marquis, K.B. Averyt, M. Tignor and H.L. Miller, Eds. Cambridge University Press, Cambridge, 663-746.

Higgins, P.A.T. and M. Vellinga, 2004: Ecosystem responses to abrupt climate change: teleconnections, scale and the hydrological cycle. *Climatic Change*, **64**, 127-142.

Higgins, P.A.T. and S.H. Schneider, 2005: Long-term potential ecosystem responses to greenhouse gas-induced thermohaline circulation collapse. *Glob. Change Biol.*, **11**, 699-709.

Higgins, P.A.T., M.D. Mastrandrea and S.H. Schneider, 2002: Dynamics of climate and ecosystem coupling: abrupt changes and multiple equilibria. *Philos. T. Roy. Soc. B*, **357**, 647-655.

Hitz, S. and J.B. Smith: 2004: Estimating global impacts from climate change. *Global Environ. Chang.*, **14**, 201-218.

Hope, C., 2006: The marginal impact of CO_2 from PAGE2002: an integrated assessment model incorporating the IPCC's five reasons for concern. *Integrated Assessment Journal*, **6**, 19-56.

Hughes, T.P., and Co-authors, 2003: Climate change, human impacts, and the resilience of coral reefs. *Science*, **301**, 929-933.

Huq, S., A. Rahman, M. Konate, Y. Sokona and H. Reid, 2003: Mainstreaming adaptation to climate change in Least Developed Countries. IIED (International Institute for Environment and Development), London, 42 pp. http://www.un.org/special-rep/ohrlls/ldc/LDCsreport.pdf.

IPCC, 1996. *Climate Change 1995: Impacts, Adaptation, and Mitigation of Climate Change. Scientific-Technical Analyses. Report of Working Group II. Contribution of Working Group II to the Second Assessment of the Intergovernmental Panel on Climate Change*, R.T. Watson, M.C. Zinyowera, R.H.Moss, Eds., Cambridge University Press, Cambridge, 878 pp.

IPCC, 2001a: *Climate Change 2001: Impacts, Adaptation, and Vulnerability. Contribution of Working Group II to the Third Assessment Report of the Intergovernmental Panel on Climate Change*, J.J. McCarthy, O.F. Canziani, N.A. Leary, D.J. Dokken and K.S. White, Eds., Cambridge University Press, Cambridge, 1032 pp.

IPCC, 2001b: Synthesis report 2001: *Contribution of Working Groups I, II, and III to the Third Assessment Report of the Intergovernmental Panel on Climate Change*, R.T. Watson and the core writing team, Eds., Cambridge University Press, Cambridge, 397 pp.

IPCC, 2007a: Summary for policymakers. *Climate Change 2007: The Physical Science Basis. Contribution of Working Group I to the Fourth Assessment Report of the Intergovernmental Panel on Climate Change*, S. Solomon, D. Qin, M. Manning, Z. Chen, M. Marquis, K.B. Averyt, M. Tignor and H.L. Miller, Eds., Cambridge University Press, Cambridge, 1-18.

IPCC, 2007b: Summary for policymakers. *Climate Change 2007: Impacts, Adaptation and Vulnerability. Contribution of Working Group II to the Fourth Assessment Report of the Intergovernmental Panel on Climate Change*, M.L. Parry, O.F. Canziani, J.P. Palutikof, P.J. van der Linden and C.E. Hanson, Eds., Cambridge University Press, Cambridge, 7-22.

Izrael Y.A., 2004: On the concept of dangerous anthropogenic interference with the climate system and capacities of the biosphere (in Russian). *Meteorol. Hydrol.*, **4**, 30-37.

Izrael Y.A., 2005: Efficient ways for keeping climate at the present level: the main aim of the climate problem solution (in Russian). *Meteorol. Hydrol.*, **10**, 5-9.

Izrael Y.A. and S.M. Semenov, 2005: Calculations of a change in CO_2 concentration in the atmosphere for some stabilization scenarios of global emissions using a model of minimal complexity (in Russian). *Meteorol. Hydrol.*, **1**, 1-8.

Izrael Y.A. and S.M. Semenov, 2006: Critical levels of greenhouse gases, stabilization scenarios, and implications for the global decisions. *Avoiding Dangerous Climate Change*, H.-J. Schellnhuber, W. Cramer, N. Nakićenović, T.M.L. Wigley and G. Yohe, Eds., Cambridge University Press, Cambridge, 73-79.

Jacoby, H.D., 2004: Informing climate policy given incommensurable benefits estimates. *Global Environ. Chang.*, **14**, 287-297.

Jamieson, D., 1992: Ethics, public policy and global warming. *Sci. Technol. Hum. Val.*, **17**, 139-153.

Jamieson, D., 1996: Ethics and intentional climate change. *Climatic Change*, **33**, 323-336.

Jansen, E., J. Overpeck, K.R. Briffa, J.-C. Duplessy, F. Joos, V. Masson-Delmotte, D. Olago, B. Otto-Bliesner, W.R. Peltier, S. Rahmstorf, R. Ramesh, D. Raynaud, D. Rind, O. Solomina, R. Villalba and D. Zhang, 2007: Paleoclimate. *Climate Change 2007: The Physical Science Basis. Contribution of Working Group I to the Fourth Assessment Report of the Intergovernmental Panel on Climate Change*, S. Solomon, D. Qin, M. Manning, Z. Chen, M. Marquis, K.B. Averyt, M. Tignor and H.L. Miller, Eds., Cambridge University Press, Cambridge, 433-497.

Jones, R.N., 2001: An environmental risk assessment/management framework for climate change impact assessments. *Nat. Hazards*, **23**, 197-230.

Jones, R.N., 2003: Managing climate change risks. *OECD Workshop: Benefits of Climate Policy: Improving Information for Policy Makers*, Paris, 12-13 December 2002. OECD, Paris, 37 pp. http://www.oecd.org/dataoecd/6/12/19519189.pdf.

Kaufmann, R., 1998. The impact of climate change on US agriculture: a response to Mendelsohn et al. (1994). *Ecol. Econ.*, **26**, 113-119.

Keller, K., K. Tan, F.M.M. Morel and D.F. Bradford, 2000: Preserving the ocean circulation: implications for climate policy. *Climatic Change*, **47**, 17-43.

Keller, K., B.M. Bolker and D.F. Bradford, 2004: Uncertain climate thresholds and optimal economic growth. *J. Environ. Econ. Manag.*, **48**, 723-741.

Keller, K., M. Hall, S.-R. Kim, D.F. Bradford and M. Oppenheimer, 2005: Avoiding dangerous anthropogenic interference with the climate system, *Climatic Change*, **73**, 227-238.

Kemfert, C., 2006: An integrated assessment of economy, energy and climate: the model WIAGEM – A reply to Comment by Roson and Tol. *Integrated Assessment Journal*, **6**, 45-49.

Kenny, G.J., R.A. Warrick, B.D. Campbell, G.C. Sims, M. Camilleri, P.D. Jamieson and N.D. Mitchell, 2000: Investigating climate change impacts and thresholds: an application of the CLIMPACTS integrated assessment model for New Zealand agriculture. *Climatic Change*, **46**, 91–113.

Kheshgi, H.S., 2004: Evasion of CO_2 injected into the ocean in the context of CO_2 stabilization. *Energy*, **29**, 1479–1486.

Kheshgi, H.S., S.J. Smith, J.A. Edmonds, 2005: Emissions and atmospheric CO_2 stabilization: long-term limits and paths. *Mitigation and Adaptation Strategies for Global Change*, **10**, 213–220.

Kiehl, J.T., 2006: Geoengineering climate change: treating the symptom over the cause? *Climatic Change*, **77**, 227-228.

Knutti, R., F. Joos, S.A. Müller, G.-K. Plattner, T.F. Stocker, 2005: Probabilistic climate change projections for stabilization profiles. *Geophys. Res. Lett.*, **32**, L20707, doi:10.1029/2005GL023294.

Kuhnel, I. and L. Coates, 2000: El Niño-Southern Oscillation: related probabilities of fatalities from natural perils in Australia. *Nat. Hazards*, **22**, 117-138.

Kushner, P.J., I.M. Held, T.L. Delworth, 2001: Southern hemisphere atmospheric circulation response to global warming. *J. Climate*, **14**, 2238-2249.

Kuzmina, S.I., L. Bengtsson, O.M. Johannessen, H. Drange, L.P. Bobylev and M.W. Miles, 2005: The North Atlantic Oscillation and greenhouse-gas forcing. *Geophys. Res. Lett.*, **32**, L04703, doi:10.1029/2005GL023294.

Lambrou, Y. and R. Laub, 2004: *Gender Perspectives on the Conventions on Biodiversity, Climate Change and Desertification*. Gender and Development Service, FAO Gender and Population Division, http://www.unisdr.org/eng/risk-reduction/gender/gender-perspectives-FAO.doc.

Langmann, B. and A. Heil, 2004: Release and dispersion of vegetation and peat fire emissions in the atmosphere over Indonesia 1997–1998. *Atmos. Chem. Phys.*, **4**, 2145-2160.

Latif, M., E. Roeckner, U. Mikolajewski and R. Voss, 2000: Tropical stabilization of the thermohaline circulation in a greenhouse warming simulation. *J. Climate*, **13**, 1809-1813.

Leemans, R. and B. Eickhout, 2003: Analysing changes in ecosystems for different levels of climate change. O ECD, Paris, 28 pp. http://www.oecd.org /dataoecd/6/29/2483789.pdf.

Leemans, R. and B. Eickhout, 2004: Another reason for concern: regional and global impacts on ecosystems for different levels of climate change. *Global Environ. Chang.*, **14**, 219-228.

Legler, D.M., K.J. Bryant and J.J. O'Brien, 1999: Impact of ENSO-related climate anomalies on crop yields in the US. *Climatic Change*, **42**, 351-375.

Lehodey, P., M. Bertignac, J. Hampton, A. Lewis and J. Picaut, 1997: El Niño Southern Oscillation and tuna in the western Pacific. *Nature*, **389**, 715-718.

Lemke, P., J. Ren, R.B. Alley, I. Allison, J. Carrasco, G. Flato, Y. Fujii, G. Kaser, P. Mote, R.H. Thomas and T. Zhang, 2007: Observations: changes in snow, ice and frozen ground. *Climate Change 2007: The Physical Science Basis. Contribution of Working Group I to the Fourth Assessment Report of the Intergovernmental Panel on Climate Change*, S. Solomon, D. Qin, M. Manning, Z. Chen, M. Marquis, K.B. Averyt, M. Tignor and H.L. Miller, Eds., Cambridge University Press, Cambridge, 337-383.

Lempert, R. and M.E. Schlesinger, 2001: Climate-change strategy needs to be robust. *Nature*, **412**, 375-375.

Li, H., R.P. Berrens, A.K. Bohara , H.C. Jenkins-Smith, C.L. Silva and D.L. Weimer, 2004: Would developing country commitments affect US households' support for a modified Kyoto Protocol? *Ecol. Econ.*, **48**, 329-343.

Link, P.M. and R.S.J. Tol, 2004: Possible economic impacts of a slowdown of the thermohaline circulation: an application of FUND. *Port. Econ. J.*, **3**, 99-114.

Link, P.M. and R.S.J. Tol, 2006: The economic impact of a slowdown of the thermohaline circulation: an application of FUND. Working paper, Hamburg University, Department of Economics, Research Unit Sustainability and Global Change, Center for Marine and Atmospheric Sciences (ZMAW). http://www.uni-hamburg.de/Wiss/FB/15/Sustainability/link-Dateien/Link%20 Working%20Paper%20FNU-103.pdf.

Lovejoy, T.E. and L. Hannah, Eds., 2005: *Climate Change and Biodiversity*. Yale University Press, New Haven, Connecticut, 440 pp.

Lowe, J.A., J.M. Gregory J. Ridley, P. Huybrechts, R.J. Nicholls and M. Collins, 2006: The role of sea-level rise and the Greenland ice sheet in dangerous climate change: implications for the stabilisation of climate. *Avoiding Dangerous Climate Change*, H.-J. Schellnhuber, W. Cramer, N. Nakićenović, T.M.L. Wigley and G. Yohe, Eds., Cambridge University Press, Cambridge, 29-36.

Maddison, D., 2003: The amenity value of climate: the household production function approach. *Resour. Energy Econ.*, **25**, 155-175.

Malcolm, J.R., C. Liu, R.P. Neilson, L. Hansen and L. Hannah, 2006: Global warming and extinctions of endemic species from biodiversity hotspots. *Conserv. Biol.*, **20**, 538-548.

Martin, D., D. Belanger, P. Gosselin, J. Brazeau, C. Furgal and S. Dery, 2005: Climate change, drinking water, and human health in Nunavik: adaptation strategies. Final report submitted to the Canadian Climate Change Action Fund, Natural Resources Canada. CHUL Research Institute, Ste-Foy, Quebec, 111 pp. http://www.itk.ca/environment/water-nunavik-report.pdf.

Mastrandrea, M.D. and S.H. Schneider, 2001: Integrated assessment of abrupt climatic changes. *Clim. Policy*, **1**, 433-449.

Mastrandrea, M.D. and S.H. Schneider, 2004: Probabilistic integrated assessment of "dangerous" climate change. *Science*, **304**, 571-575.

Mastrandrea, M.D. and S.H. Schneider, 2006: Probabilistic assessment of "dangerous" climate change and emissions pathways. *Avoiding Dangerous Climate Change*, H.-J. Schellnhuber, W. Cramer, N. Nakićenović, T.M.L. Wigley and G. Yohe, Eds., Cambridge University Press, Cambridge, 253-264.

Matear, R.J. and A.C. Hirst, 2003: Long-term changes in dissolved oxygen concentrations in the oceans caused by protracted global warming. *Global Biogeochem. Cy.*, **17**, 1125, doi:10.1029/2002GB001997.

McInerney, D. and K. Keller, 2006: Economically optimal risk reduction strategies in the face of uncertain climate thresholds. *Climatic Change*, doi:10.1007/s10584-006-9137-z.

McMichael, A.J., D. Campbell-Lendrum, R.S. Kovats, S. Edwards, P. Wilkinson, N. Edmonds, N. Nicholls, S. Hales, F.C. Tanser, D. Le Sueur, M. Schlesinger and N. Andronova, 2004: Climate change. *Comparative Quantification of Health Risks: Global and Regional Burden of Disease due to Selected Major Risk Factors*, Vol. 2, M. Ezzati, A.D. Lopez, A. Rogers and C.J. Murray, Eds., WHO, Geneva, 1543-1650.

Meehl, G.A., T.F. Stocker, W.D. Collins, P. Friedlingstein, A.T. Gaye, J.M. Gregory, A. Kitoh, R. Knutti, J.M. Murphy, A. Noda, S.C.B. Raper, I.G. Watterson, A.J. Weaver and Z.-C. Zhao, 2007: Global climate projections. *Climate Change 2007: The Physical Science Basis. Contribution of Working Group I to the Fourth Assessment Report of the Intergovernmental Panel on Climate Change*, S. Solomon, D. Qin, M. Manning, Z. Chen, M. Marquis, K.B. Averyt, M. Tignor and H.L. Miller, Eds., Cambridge University Press, Cambridge, 747-846.

Meinshausen, M., 2005: What does a 2°C target mean for greenhouse gas concentrations? A brief analysis based on multi-gas emission pathways and several climate sensitivity uncertainty estimates. *Avoiding Dangerous Climate Change*, H.-J. Schellnhuber, W. Cramer, N. Nakićenović, T.M.L. Wigley and G. Yohe, Eds., Cambridge University Press, Cambridge, 265-280.

Meinshausen, M., B. Hare, T.M.L. Wigley, D. van Vuuren, M.G.J. den Elzen and R. Swart, 2005: Multi-gas emission pathways to meet climate targets. *Climatic Change*, **75**, 151-194.

Mendelsohn, R. and W. Nordhaus, 1996: The impact of global warming on agriculture: reply. *Am. Econ. Rev.*, **86**, 1312-1315.

Mendelsohn, R., W. Nordhaus and D. Shaw, 1994: The impact of global warming on agriculture: a Ricardian analysis. *Am. Econ. Rev.*, **84**, 753-771.

Menzel, A and Co-authors, 2006: European phenological response to climate change matches the warming pattern, *Glob. Change Biol.*, **12**, 1969-1976.

Morgan, M.G. and D.W. Keith, 1995: Subjective judgements by climate experts. *Environ. Sci. Technol.*, **29**, 468A-476A.

Morgan, M.G., P.J. Adams and D.W. Keith, 2006: Elicitation of expert judgments of aerosol forcing. *Climatic Change*, **75**, 195-214.

Morita, T., N. Nakićenović and J. Robinson, 2000: Overview of mitigation scenarios for global climate stabilization based on new IPCC emission scenarios (SRES). *Environ. Econ. Policy Stud.*, **3**, 65-88.

Moss, R.H. and S.H. Schneider, 2000: Uncertainties in the IPCC TAR: recommendations to lead authors for more consistent assessment and reporting. *Guidance Papers on the Cross Cutting Issues of the Third Assessment Report of the IPCC*, R. Pachauri, T. Taniguchi and K. Tanaka, Eds., World Meteorological Organization, Geneva, 33-51.

Munich Re, 2005: *Topics GEO. Annual Review: Natural Catastrophes 2005*. Munich Re, Munich, 56 pp. http://www.earthinstitute.columbia.edu/grocc/documents/MunichRe2005NaturalDisasterReview.pdf.

Nakićenović, N., J. Alcamo, G. Davis, B. de Vries, J. Fenhann, S. Gaffin, K. Gregory, A. Grübler, T.Y. Jung, T. Kram, E. Lebre la Rovere, L. Michaelis, S. Mori, T. Morita, W. Pepper, H. Pitcher, L. Price, K. Riahi, A. Roehrl, H.H. Rogner, A. Sankovski, M. Schlesinger, P. Shukla, S. Smith, R. Swart, S. van Rooijen, N. Victor and Z. Dadi, Eds., 2000: *Emissions Scenarios: A Special Report of the Intergovernmental Panel on Climate Change (IPCC)*. Cambridge University Press, Cambridge, 509 pp.

Nicholls, R.J., 2004: Coastal flooding and wetland loss in the 21st century: changes under the SRES climate and socio-economic scenario. *Global Environ. Chang.*, **14**, 69-86.

Nicholls, R.J. and R.S.J. Tol, 2006: Impacts and responses to sea-level rise: a global analysis of the SRES scenarios over the 21st century. *Philos. T. Roy. Soc. A*, **361**, 1073-1095.

Nicholls, R.J., S.E. Hanson, J. Lowe, D.G. Vaughan, T. Lenton, A. Ganoposki, R.S.J. Tol and A.T. Vafeidis, 2005: Improving methodologies to assess the benefits of policies to address sea-level rise. Report to the OECD. Organisation for Economic Co-operation and Development (OECD), Paris, 145 pp. http://www.oecd.org/dataoecd/19/63/37320819.pdf

Niemeyer, S., J. Petts and K. Hobson, 2005: Rapid climate change and society: assessing responses and thresholds, *Risk Anal.*, **25**, 1443-1456.

Nordhaus, W.D., 1994: Expert opinion on climatic change. *Am. Sci.*, **82,** 45–52.

Nordhaus, W.D., 2006: Geography and macroeconomics: new data and new findings. *P. Natl. Acad. Sci. USA*, **103**, 3510-3517.

Nordhaus, W.D. and J. Boyer, 2000: *Warming the World: Economic Models of Global Warming*. MIT Press, Boston, 246 pp.

Nyong, A., 2005: Impacts of climate change in the tropics: the African experience. Keynote presentation at the Avoiding Dangerous Climate Change Symposium, 1–3 February, 2005, Exeter, 9 pp. http://www.stabilisation2005.com/.

O'Brien, K.L., L. Sygna and J.E. Haugen, 2004: Resilient of vulnerable? A multi-scale assessment of climate impacts and vulnerability in Norway. *Climatic Change*, **64**, 193-225.

O'Neill, B.C. and M. Oppenheimer, 2002: Climate change: dangerous climate impacts and the Kyoto Protocol. *Science*, **296**, 1971-1972.

O'Neill, B.C. and M. Oppenheimer, 2004: Climate change impacts sensitive to path to stabilization. *P. Natl. Acad. Sci. USA*, **101**, 16411-16416.

OECD, 2003: Development and climate change in Nepal: focus on water resources and hydropower. Document COM/ENV/EPOC/DCD/DAC(2003)1/FINAL, OECD Environment Directorate. Paris, 64 pp. http://www.oecd.org/dataoecd/6/51/19742202.pdf.

Oppenheimer, M., 2005: Defining dangerous anthropogenic interference: the role of science, the limits of science. *Risk Anal.*, **25**, 1399-1407.

Oppenheimer, M. and A. Petsonk, 2003: Global warming: the intersection of long-term goals and near-term policy. *Climate Policy for the 21st Century: Meeting the Long-Term Challenge of Global Warming*, D. Michel, Ed., Center for Transatlantic Relations, Johns Hopkins University, Baltimore, Maryland, 79-112.

Oppenheimer, M. and R.B. Alley, 2004: The West Antarctic Ice Sheet and long term climate policy. *Climatic Change*, **64**, 1-10.

Oppenheimer, M. and R.B. Alley, 2005: Ice sheets, global warming, and Article 2 of the UNFCCC. *Climatic Change*, **68**, 257-267.

Oppenheimer, M. and A. Petsonk, 2005: Article 2 of the UNFCCC: historical origins, recent interpretations. *Climatic Change*, **73**, 195-226.

Oppenheimer, M. and A. Todorov, 2006: Global warming: the psychology of long term risk. *Climatic Change*, **77**, 1-6.

Otto-Bliesner, B.L., S. Marshall, J. Overpeck, G. Miller, A. Hu and CAPE Last Interglacial Project Members, 2006: Simulating Arctic climate warmth and icefield retreat in the last interglaciation. *Science*, **311**, 1751-1753.

Overpeck, J.T., B.L. Otto-Bliesner, G.H. Miller, D.R. Muhs, R.B. Alley and J.T. Kiehl, 2006: Paleoclimatic evidence for future ice-sheet instability and rapid sea-level rise. *Science*, **311**, 1747-1750.

Parmesan, C. and G. Yohe, 2003: A globally coherent fingerprint of climate change impacts across natural systems. *Nature*, **421**, 37-42.

Parry, M., N. Arnell, T. McMichael, R. Nicholls, P. Martens, S. Kovats, M. Livermore, C. Rosenzweig, A. Iglesias and G. Fischer, 1999: Millions at risk: defining critical climate change threats and targets. *Global Environ. Chang.*, **11**, 181-183.

Parry, M.L., T.R. Carter and M. Hulme, 1996: What is dangerous climate change? *Global Environ. Chang.*, **6**, 1-6.

Parry, M.L., C. Rosenzweig, A. Iglesias, M. Livermore and G. Fischer, 2004: Effects of climate change on global food production under SRES climate and socio-economic scenarios. *Global Environ. Chang.*, **14**, 53-67.

Patwardhan, A., S.H. Schneider and S.M. Semenov, 2003: Assessing the science to address UNFCCC Article 2: a concept paper relating to cross cutting theme number four. IPCC, 13 pp. http://www.ipcc.ch/activity/cct3.pdf.

Pearce, D.W., 2003: The social cost of carbon and its policy implications. *Oxford Rev. Econ. Pol.*, **19**, 362–384.

Pielke, R.A. and C.N. Landsea, 1999: La Nina, El Niño, and Atlantic hurricane damages in the United States. *B. Am. Meteorol. Soc.*, **80**, 2027-2033.

Pittini, M. and M. Rahman, 2004: Social costs of carbon. *The Benefits of Climate Policies: Analytical and Framework Issues,* J. Corfee-Morlot and S. Agrawala, Eds., OECD, Paris, 189-220.

Pittock, A.B., R.N. Jones and C.D. Mitchell, 2001: Probabilities will help us plan for climate change. *Nature*, **413**, 249.

Polsky, C., 2004: Putting space and time in Ricardian climate change impact studies: the case of agriculture in the U.S. Great Plains. *Ann. Assoc. Am. Geogr.*, **94**, 549-564.

Polsky, C. and W.E. Easterling, 2001: Adaptation to climate variability and change in the US Great Plains: a multi-scale analysis of Ricardian climate sensitivities. *Agr. Ecosyst. Environ.*, **85**, 133-144.

Putilov, V.Y., Ed., 2003: *Ecology of Power Engineering* (in Russian). Moscow Energy Engineering Institute, Moscow.

Rahmstorf, S. and K. Zickfeld, 2005: Thermohaline circulation changes: a question of risk assessment. *Climatic Change*, **68**, 241-247.

Rahmstorf, S., T. Kuhlbrodt, K. Zickfeld, G. Buerger, F. Badeck, S. Pohl, S. Sitch, H. Held, T. Schneider von Deimling, D. Wolf-Gladrow, M. Schartau, C. Sprengel, S. Sundby, B. Adlandsvik, F. Vikebo, R. Tol and M. Link, 2003: Integrated assessment of changes in the thermohaline circulation. INTEGRATION Status Report. Potsdam Institute for Climate Impact Research, Potsdam. http://www.pik-potsdam.de/~stefan/Projects/integration/statusreport.pdf.

Randall, D.A., R.A. Wood, S. Bony, R. Coleman, T. Fichefet, J. Fyfe, V. Kattsov, A. Pitman, J. Shukla, J. Srinivasan, R.J. Stouffer, A. Sumi and K.E. Taylor, 2007: Climate models and their evaluation. *Climate Change 2007: The Physical Science Basis. Contribution of Working Group I to the Fourth Assessment Report of the Intergovernmental Panel on Climate Change*, S. Solomon, D. Qin, M. Manning, Z. Chen, M. Marquis, K.B. Averyt, M. Tignor and H.L. Miller, Eds., Cambridge University Press, Cambridge, 589-662.

Rayner, S. and E. Malone, Eds., 1998: *Human Choice and Climate Change: The Societal Framework, Vol. 1.* Battelle Press, Washington, District of Columbia, 536 pp.

Reilly, J., P.H. Stone, C.E. Forest, M.D. Webster, H.D. Jacoby and R.G. Prinn, 2001: Climate change: uncertainty and climate change assessments. *Science*, **293**, 430-433.

Revelle, R., 1983: Probable future changes in sea level resulting from increased atmospheric carbon dioxide. *Changing Climate*. National Academy Press, Washington, District of Columbia, 433-448.

Rial, J.A., R.A. Pielke Sr., M. Beniston, M. Claussen, J. Canadell, P. Cox, H. Held, N. de Noblet-Ducoudré, R. Prinn, J.F. Reynolds and J.D. Salas, 2004: Nonlinearities, feedbacks and critical thresholds within the Earth's climate system. *Climatic Change*, **65**, 11-38.

Rodo, X., M. Pascual, G. Fuchs and A.S.G. Faruque, 2002: ENSO and cholera: a nonstationary link related to climate change? *P. Natl. Acad. Sci. USA*, **99**, 12901-12906.

Root, T.L., J.T. Price, K.R. Hall, S.H. Schneider, C. Rosenzweig and J.A. Pounds, 2003: Fingerprints of global warming on wild animals and plants. *Nature*, **421**, 57-60.

Root, T.L., D.P. MacMynowski, M.D. Mastrandrea and S.H. Schneider, 2005: Human-modified temperatures induce species changes: joint attribution. *P. Natl. Acad. Sci. USA*, **102**, 7462-7469.

Rosenzweig, C., F.N. Tubiello, R. Goldberg, E. Mills and J. Bloomfield, 2002: Increased crop damage in the US from excess precipitation under climate change. *Global Environ. Chang.*, **12**, 197-202.

Roson, R. and R.S.J. Tol, 2006: An integrated assessment model of economy-energy-climate: the model WIAGEM – a comment. *Integrated Assessment Journal*, **6**, 75-82.

Royal Society, 2005: *Ocean Acidification due to Increasing Atmospheric Carbon Dioxide*. London, http://www.royalsoc.ac.uk/document.asp?tip=0&id=3249.

Sabine, C.L., R.A. Feely, N. Gruber, R.M. Key, K. Lee, J.L. Bullister, R. Wanninkhof, C.S. Wong, D.W.R. Wallace, B. Tilbrook, F.J. Millero, T.-H. Peng, A. Kozyr, T. Ono and A.F. Rios, 2004: The oceanic sink for anthropogenic CO_2. *Science*, **305**, 367-371.

Sarmiento, J.L. and C. Le Quéré, 1996: Oceanic carbon dioxide uptake in a model of century-scale global warming. *Science*, **274**, 1346-1350.

Sathaye, J. and Co-authors, 2007: Sustainable development and mitigation. *Climate Change 2007: Mitigation. Contribution of Working Group III to the Fourth Assessment Report of the Intergovernmental Panel on Climate Change*, B. Metz, O. Davidson, P. Bosch, R. Dave and L. Meyer, Eds., Cambridge University Press, Cambridge, UK.

Schär, C. and G. Jendritzky, 2004: Climate change: hot news from summer 2003. *Nature*, **432**, 559-560.

Schär, C., P.L. Vidale, D. Lüthi, C. Frei, C. Häberli, M.A. Liniger and C. Appenzeller, 2004: The role of increasing temperature variability in European summer heatwaves. *Nature*, **427**, 332-336.

Schellnhuber, H.-J., W. Cramer, N. Nakićenović, T. Wigley and G. Yohe, Eds., 2006: *Avoiding Dangerous Climate Change*. Cambridge University Press, Cambridge, 392 pp.

Schimel, D., D. Alves, I. Enting, M. Heimann, F. Joos, D. Raynaud, T. Wigley, M. Prather, R. Derwent, D. Ehhalt, P. Fraser, E. Sanhueza, X. Zhou, P. Jonas, R. Charlson, H. Rodhe, S. Sadasivan, K.P. Shine, Y. Fouquart, V. Ramaswamy, S. Solomon, J. Srinivasan, D.L. Albritton, I. Isaksen, M. Lal and D.J. Wuebbles, 1996: Radiative forcing of climate change. *Climate Change 1995: The Science of Climate Change. Contribution of Working Group I to the Second Assessment Report of the Intergovernmental Panel on Climate Change*, J.T. Houghton, L.G. Meiro Filho, B.A. Callander, N. Harris, A. Kattenberg, K. Maskell, Eds., Cambridge University Press, Cambridge, 65–131.

Schlenker, W., W.M. Hanemann and A.C. Fisher, 2005: Will U.S. agriculture really benefit from global warming? Accounting for irrigation in the Hedonic approach. *Am. Econ. Rev.*, **95**, 395-406.

Schlesinger, M.E., J. Yin, G. Yohe, N.G. Andronova, S. Malyshev and B. Li, 2006: Assessing the risk of a collapse of the Atlantic thermohaline circulation. *Avoiding Dangerous Climate Change*, H.-J. Schellnhuber, W. Cramer, N. Nakićenović, T.M.L. Wigley and G. Yohe, Eds., Cambridge University Press, Cambridge, 37-48.

Schmittner, A.M., 2005: Decline of the marine ecosystems caused by a reduction in the Atlantic overturning circulation. *Nature*, **434**, 628-633.

Schneider, S.H., 2001: What is 'dangerous' climate change? *Nature*, **411**, 17-19.

Schneider, S.H., 2002: Can we estimate the likelihood of climatic changes at 2100? *Climatic Change*, **52**, 441- 451.

Schneider, S.H., 2004: Abrupt non-linear climate change, irreversibility and surprise. *Global Environ. Chang.*, **14**, 245-258.

Schneider, S.H. and R.S. Chen, 1980: Carbon-dioxide warming and coastline flooding: physical factors and climatic impact. *Annu. Rev. Energy*, **5**, 107-140.

Schneider, S.H. and C. Azar, 2001: Are uncertainties in climate and energy systems a justification for stronger near-term mitigation policies? *Proc. Pew Center Workshop on the Timing of Climate Change Policies*, E. Erlich, Eds., Pew Center on Climate Change, Arlington, Virginia, 85-136.

Schneider, S.H. and M.D. Mastrandrea, 2005: Probabilistic assessment of "dangerous" climate change and emissions scenarios. *P. Natl. Acad. Sci. USA*, **102**, 15728-15735.

Schneider, S.H., K. Kuntz-Duriseti and C. Azar, 2000: Costing non-linearities, surprises, and irreversible events. *Pacific-Asian Journal of Energy*, **10**, 81-106.

Schneider S., J. Sarukhan, J.Adejuwon, C. Azar, W. Baethgen, C. Hope, R. Moss, N. Leary, R. Richels and J.P. van Ypersele, 2001: *Overview of Impacts, Adaptation, and Vulnerability to Climate Change. Climate Change 2001: Impacts, Adaptation, and Vulnerability. Contribution of Working Group II to the Third Assessment Report of the Intergovernmental Panel on Climate Change*, J.J. McCarthy, O.F. Canziani, N.A. Leary, D.J. Dokken and K.S. White, Eds., Cambridge University Press, Cambridge, 75-103.

Semenov, S.M., 2004a: Modeling of anthropogenic perturbation of the global CO_2 cycle. *Dokl. Earth Sci.*, **399**, 1134-1138.

Semenov, S.M., 2004b: Greenhouse gases and present climate of the Earth (in Russian, extended summary in English). Meteorology and Hydrololgy Publishing Centre, Moscow, 175 pp.

Slovic, P., 2000: *The Perception of Risk*. Earthscan, London, 518 pp.

Smit, B., O. Pilifosova, I. Burton, B. Challenger, S. Huq, R.J.T. Klein and G. Yohe, 2001: Adaptation to climate change in the context of sustainable development and equity. *Climate Change 2001: Impacts, Adaptation, and Vulnerability. Contribution of Working Group II to the Third Assessment Report of the Intergovernmental Panel on Climate Change*, J.J. McCarthy, O.F. Canziani, N.A. Leary, D.J. Dokken and K.S. White, Eds., Cambridge University Press, Cambridge, 877-912.

Smith, J.B., 2004: *A Synthesis of Potential Impacts of Climate Change on the United States*. Pew Center on Global Climate Change, Arlington, Virginia, 56 pp. http://www.pewclimate.org/docUploads/Pew-Synthesis.pdf.

Smith, J.B. and Co-authors, 2001. Vulnerability to climate change and reasons for concern: a synthesis, *Climate Change 2001: Impacts, Adaptation, and Vulnerability. Contribution of Working Group II to the Third Assessment Report of the Intergovernmental Panel on Climate Change*, J.J. McCarthy, O.F. Canziani, N.A. Leary, D.J. Dokken and K.S. White, Eds., Cambridge University Press, Cambridge, 913–967.

Solomon, S., D. Qin, M. Manning, R.B. Alley, T. Berntsen, N.L. Bindoff, Z. Chen, A. Chidthaisong, J.M. Gregory, G.C. Hegerl, M. Heimann, B. Hewitson, B.J. Hoskins, F. Joos, J. Jouzel, V. Kattsov, U. Lohmann, T. Matsuno, M. Molina, N. Nicholls, J. Overpeck, G. Raga, V. Ramaswamy, J. Ren, M. Rusticucci, R. Somerville, T.F. Stocker, P. Whetton, R.A. Wood and D. Wratt, 2007: Technical Summary. *Climate Change 2007: The Physical Science Basis. Contribution of Working Group I to the Fourth Assessment Report of the Intergovernmental Panel on Climate Change,* S. Solomon, D. Qin, M. Manning, Z. Chen, M. Marquis, K.B. Averyt, M. Tignor and H.L. Miller, Eds. Cambridge University Press, Cambridge, 74 pp.

Stern, N., 2007: *The Economics of Climate Change: The Stern Review*. Cambridge University Press, Cambridge, 692 pp.

Stocker, T.F. and A. Schmittner, 1997: Influence of CO_2 emission rates on the stability of the thermohaline circulation. *Nature*, **388**, 862-865.

Stott, P.A., D.A. Stone and M.R. Allen, 2004: Human contribution to the European heatwave of 2003. *Nature*, **432**, 610-614.

Stouffer, R.J. and S. Manabe, 2003: Equilibrium response of thermohaline circulation to large changes in atmospheric CO_2 concentration. *Clim. Dynam.*, **20**, 759-773.

Swart, R., M. Berk, M. Janssen, E. Kreileman and R. Leemans, 1998: The safe landing approach: risks and trade-offs in climate change. *Global Change Scenarios of the 21st Century: Results from the IMAGE 2.1 Model*, J. Alcamo, R. Leemans and E. Kreileman, Eds., Pergamon, Oxford, 193-218.

Swart, R., J. Mitchell, T. Morita and S. Raper, 2002: Stabilisation scenarios for climate impact assessment. *Global Environ. Chang.*, **12**, 155-166.

Swetnam, T.W. and J.L. Betancourt, 1990: Fire–Southern Oscillation relations in the southwestern United States. *Science*, **249**, 1017-1020.

Thomas, C., A. Cameron, R.E. Green, M. Bakkenes, L.J. Beaumont, Y.C. Collingham, B.F.N. Erasmus, M.F. de Siqueira, A. Grainger, L. Hannah, L. Hughes, B. Huntley, A.S. van Jaarsveld, G.F. Midgley, L. Miles, M.A. Ortega-Huerta, A.T. Peterson, O.L. Phillips and S.E. Williams, 2004: Extinction risk from climate change. *Nature*, **427**, 145-148.

Thompson, D.W.J. and J.M. Wallace, 2000: Annular modes in the extratropical circulation. Part II. Trends. *J. Climate*, **13**, 1018-1036.

Timmermann, A., J. Oberhuber, A. Bacher, M. Esch, M. Latif and E. Roeckner, 1999: Increased El Niño frequency in a climate model forced by future greenhouse warming. *Nature*, **398**, 694-697.

Tol, R.S.J., 1998: Potential slowdown of the thermohaline circulation and climate policy. Discussion Paper DS98/06, Institute for Environmental Studies, Vrije Universiteit Amsterdam, Amsterdam, 7 pp.

Tol, R.S.J., 2002a: Estimates of the damage costs of climate change. Part 1: Benchmark estimates. *Environ. Resour. Econ.*, **21**, 41-73.

Tol, R.S.J., 2002b: Estimates of the damage costs of climate change. Part II: Dynamic estimates. *Environ. Resour. Econ.*, **21**, 135-160.

Tol, R.S.J., 2005: The marginal damage costs of carbon dioxide emissions: an assessment of the uncertainties. *Energ. Policy*, **33**, 2064-2074.

Tol, R.S.J., T.E. Downing, O.J. Kuikb and J.B. Smith, 2004: Distributional aspects of climate change impacts, *Global Environ. Chang.*, **14**, 259–272.

Tol, R.S.J., M.T. Bohn, T.E. Downing, M.-L. Guillerminet, E. Hizsnyik, R.E. Kasperson, K. Lonsdale, C. Mays, R.J. Nicholls, A.A. Olsthoorn, G. Pfeffle, M. Poumadere, F.L. Toth, A.T. Vafeidis, P.E. Van der Werff and I.H. Yetkiner, 2006: Adaptation to five metres of sea level rise. *J. Risk Anal.*, **9**, 467-482.

Toth, F.L., Ed., 2003: Integrated assessment of climate protection strategies. *Climatic Change*, **56**, 1-5.

Toth, F.L., T. Bruckner, H.-M. Füssel, M. Leimbach, G. Petschel-Held and H.-J. Schellnhuber, 2002: Exploring options for global climate policy: a new analytical framework. *Environment*, **44**, 22-34.

Toth, F.L., T. Bruckner, H.-M. Füssel, M. Leimbach and G. Petschel-Held, 2003: Integrated assessment of long-term climate policies. Part 1: Model presentation. *Climatic Change*, **56**, 37-56.

Trenberth, K.E., P.D. Jones, P. Ambenje, R. Bojariu, D. Easterling, A. Klein Tank, D. Parker, F. Rahimzadeh, J.A. Renwick, M. Rusticucci, B. Soden and P. Zhai, 2007: Observations: Surface and atmospheric climate change. *Climate Change 2007: The Physical Science Basis. Contribution of Working Group I to the Fourth Assessment Report of the Intergovernmental Panel on Climate Change*, S. Solomon, D. Qin, M. Manning, Z. Chen, M. Marquis, K.B. Averyt, M. Tignor and H.L. Miller, Eds., Cambridge University Press, Cambridge, 235-336.

Turley, C., J.C. Blackford, S. Widdicombe, D. Lowe, P.D. Nightingale and A.P. Rees, 2006: Reviewing the impact of increased atmospheric CO_2 on oceanic pH and the marine ecosystem. *Avoiding Dangerous Climate Change*, H.-J. Schellnhuber, W. Cramer, N. Nakićenović, T.M.L. Wigley and G. Yohe, Eds., Cambridge University Press, Cambridge, 65-70.

van Lieshout, M., R.S. Kovats, M.T.J. Livermore and P. Martens, 2004: Climate change and malaria: analysis of the SRES climate and socio-economic scenarios. *Global Environ. Chang.*, **14**, 87-99.

Vaughan, D.G., 2007: West Antarctic ice sheet collapse: the fall and rise of a paradigm. *Climatic Change*, in press.

Vaughan, D.G. and J.R. Spouge, 2002: Risk estimation of collapse of the West Antarctic ice sheet. *Climatic Change*, **52**, 65-91.

Vellinga, M. and R.A. Wood, 2002: Global climatic impacts of a collapse of the Atlantic thermohaline circulation. *Climatic Change*, **54**, 251-267.

Vellinga, M. and R.A. Wood, 2006: Impacts of thermohaline circulation shutdown in the twenty-first century. *Geophysical Research Abstracts*, **8**, 02717. http://www.cosis.net/abstracts/EGU06/02717/EGU06-J-02717.pdf.

Vieli, A. and A.J. Payne, 2005: Assessing the ability of numerical ice sheet models to simulate grounding line migration, *J. Geophys. Res.*, **110**, F01003, doi:10.1029/2004JF000202.

Weber, E., 2005: Experience-based and description-based perceptions of long-term risk: why global warming does not scare us (yet). *Climatic Change*, **77**, 103–120.

Webster, M., C. Forest, J. Reilly, M. Babiker, D. Kicklighter, M. Mayer, R. Prinn, M. Sarofim, A. Sokolov, P. Stone and C. Wang, 2003: Uncertainty analysis of climate change and policy response. *Climatic Change*, **61**, 295-320.

Webster, P.J., G.J. Holland, J.A. Curry and H.-R. Chang, 2005: Changes in tropical cyclone number, duration, and intensity in a warming environment. *Science*, **309**, 1844-1846.

Westerling, A.L., H.G. Hidalgo, D.R. Cayan and T.W. Swetnam, 2006: Warming and earlier spring increases western U.S. forest wildfire activity. *Science*, **313**, 940-943.

White, G.F., R.W. Kates and I. Burton, 2001: Knowing better and losing even more: the use of knowledge in hazards management. *Environmental Hazards*, **3**, 81-92.

Wigley, T.M.L., 2004: Choosing a stabilization target for CO_2. *Climatic Change*, **67**, 1-11.

Wigley, T.M.L., 2006: A combined mitigation/geoengineering approach to climate stabilization. *Science*, **314**, 452-454.

Wigley, T.M.L, R. Richels and J. Edmonds, 1996: Economic and environmental choices in the stabilization of atmospheric CO_2 concentration. *Nature*, **379**, 242.

Wild, M., P. Calanca, S.C. Scherrer and A. Ohmura, 2003: Effects of polar ice sheets on global sea level in high-resolution greenhouse scenarios. *J. Geophy. Res.*, **108**, 4165, doi:10.1029/2002JD002451.

Wood, R.A., M. Vellinga and R. Thorpe, 2003: Global warming and thermohaline circulation stability. *Philos. T. Roy. Soc. A*, **361**, 1961-1974.

Wood, R., M. Collins, J. Gregory, G. Harris and M. Vellinga, 2006: Towards a risk assessment for shutdown of the Atlantic thermohaline circulation. *Avoiding Dangerous Climate Change*, H.-J. Schellnhuber, W. Cramer, N. Nakićenović, T.M.L. Wigley and G. Yohe, Eds., Cambridge University Press, Cambridge.

Wright, E.L. and J.D. Erickson, 2003: Incorporating catastrophes into integrated assessment: science, impacts, and adaptation. *Climatic Change*, **57**, 265-286.

Yamin, F., J.B. Smith and I. Burton, 2005: Perspectives on 'dangerous anthropogenic interference', or how to operationalize Article 2 of the UN Framework Convention on Climate Change. *Avoiding Dangerous Climate Change*, H.-J. Schellnhuber, W. Cramer, N. Nakićenović, T.M.L. Wigley and G. Yohe, Eds., Cambridge University Press, Cambridge, 81-92.

Yohe, G. and R.S.J. Tol, 2002: Indicators for social and economic coping capacity: moving toward a working definition of adaptive capacity. *Global Environ. Chang.*, **12**, 25–40.

Yohe, G., N. Andronova and M. Schlesinger, 2004: To hedge or not against an uncertain climate future? *Science*, **306**, 416-417.

Yohe, G., M.E. Schlesinger and N.G. Andronova, 2006: Reducing the risk of a collapse of the Atlantic thermohaline circulation. *Integrated Assessment Journal*, **6**, 57-73.

Zickfeld, K. and T. Bruckner, 2003: Reducing the risk of abrupt climate change: emissions corridors preserving the Atlantic thermohaline circulation. *Integrated Assessment Journal*, **4**, 106-115.

20

Perspectives on climate change and sustainability

Coordinating Lead Authors:

Gary W. Yohe (USA), Rodel D. Lasco (Philippines)

Lead Authors:

Qazi K. Ahmad (Bangladesh), Nigel Arnell (UK), Stewart J. Cohen (Canada), Chris Hope (UK), Anthony C. Janetos (USA), Rosa T. Perez (Philippines)

Contributing Authors:

Antoinette Brenkert (USA), Virginia Burkett (USA), Kristie L. Ebi (USA), Elizabeth L. Malone (USA), Bettina Menne (WHO Regional Office for Europe/Germany), Anthony Nyong (Nigeria), Ferenc L. Toth (Hungary), Gianna M. Palmer (USA)

Review Editors:

Robert Kates (USA), Mohamed Salih (Sudan), John Stone (Canada)

This chapter should be cited as:

Yohe, G.W., R.D. Lasco, Q.K. Ahmad, N.W. Arnell, S.J. Cohen, C. Hope, A.C. Janetos and R.T. Perez, 2007: Perspectives on climate change and sustainability. *Climate Change 2007: Impacts, Adaptation and Vulnerability. Contribution of Working Group II to the Fourth Assessment Report of the Intergovernmental Panel on Climate Change*, M.L. Parry, O.F. Canziani, J.P. Palutikof, P.J. van der Linden and C.E. Hanson, Eds., Cambridge University Press, Cambridge, UK, 811-841.

Table of Contents

Executive summary

Vulnerability to specific impacts of climate change will be most severe when and where they are felt together with stresses from other sources [20.3, 20.4, 20.7, Chapter 17 Section 17.3.3] (very high confidence).

Non-climatic stresses can include poverty, unequal access to resources, food security, environmental degradation and risks from natural hazards [20.3, 20.4, 20.7, Chapter 17 Section 17.3.3]. Climate change itself can, in some places, produce its own set of multiple stresses; total vulnerability to climate change, *per se*, is greater than the sum of vulnerabilities to specific impacts in these cases [20.7.2].

Efforts to cope with the impacts of climate change and attempts to promote sustainable development share common goals and determinants including access to resources (including information and technology), equity in the distribution of resources, stocks of human and social capital, access to risk-sharing mechanisms and abilities of decision-support mechanisms to cope with uncertainty [20.3.2, Chapter 17 Section 17.3.3, Chapter 18 Sections 18.6 and 18.7] (very high confidence). Nonetheless, some development activities exacerbate climate-related vulnerabilities [20.8.2, 20.8.3] (very high confidence).

It is very likely that significant synergies can be exploited in bringing climate change to the development community and critical development issues to the climate-change community [20.3.3, 20.8.2, 20.8.3]. Effective communication in assessment, appraisal and action are likely to be important tools, both in participatory assessment and governance as well as in identifying productive areas for shared learning initiatives. Despite these synergies, few discussions about promoting sustainability have thus far explicitly included adapting to climate impacts, reducing hazard risks and/or promoting adaptive capacity [20.4, 20.5, 20.8.3].

Climate change will result in net costs into the future, aggregated across the globe and discounted to today; these costs will grow over time [20.6.1, 20.6.2] (very high confidence).

More than 100 estimates of the social cost of carbon are available. They run from US$-10 to US$+350 per tonne of carbon. Peer-reviewed estimates have a mean value of US$43 per tonne of carbon with a standard deviation of US$83 per tonne. Uncertainties in climate sensitivity, response lags, discount rates, the treatment of equity, the valuation of economic and non-economic impacts and the treatment of possible catastrophic losses explain much of this variation including, for example, the US$310 per tonne of carbon estimate published by Stern (2007). Other estimates of the social cost of carbon span at least three orders of magnitude, from less than US$1 per tonne of carbon to over US$1,500 per tonne [20.6.1]. It is likely that the globally-aggregated figures from integrated assessment models underestimate climate costs because they do not include significant impacts that have not yet been monetised [20.6.1, 20.6.2, 20.7.2, 20.8, Chapter 17 Section 17.2.3, Chapter 19]. It is virtually certain that aggregate estimates mask significant differences in impacts across sectors and across regions, countries and locally [20.6, 20.7, 20.8, Chapter 17 Section 17.3.3]. It is virtually certain that the real social cost of carbon and other greenhouse gases will rise over time; it is very likely that the rate of increase will be 2% to 4% per year [20.6, 20.7]. By 2080, it is likely that 1.1 to 3.2 billion people will be experiencing water scarcity (depending on scenario); 200 to 600 million, hunger; 2 to 7 million more per year, coastal flooding [20.6.2].

Reducing vulnerability to the hazards associated with current and future climate variability and extremes through specific policies and programmes, individual initiatives, participatory planning processes and other community approaches can reduce vulnerability to climate change [20.8.1, 20.8.2, Chapter 17 Sections 17.2.1, 17.2.2 and 17.2.3] (high confidence). Efforts to reduce vulnerability will be not be sufficient to eliminate all damages associated with climate change [20.5, 20.7.2, 20.7.3] (very high confidence).

Climate change will impede nations' abilities to achieve sustainable development pathways as measured, for example, by long-term progress towards the Millennium Development Goals [20.7.1] (very high confidence).

Over the next half-century, it is very likely that climate change will make it more difficult for nations to achieve the Millennium Development Goals for the middle of the century. It is very likely that climate change attributed with high confidence to anthropogenic sources, *per se*, will not be a significant extra impediment to nations reaching their 2015 Millennium Development Targets since many other obstacles with more immediate impacts stand in the way [20.7.1].

Synergies between adaptation and mitigation measures will be effective until the middle of this century (high confidence), but even a combination of aggressive mitigation and significant investment in adaptive capacity could be overwhelmed by the end of the century along a likely development scenario [20.7.3, Chapter 18 Sections 18.4, 18.7, Chapter 19] (high confidence).

Until around 2050, it is likely that global mitigation efforts designed to cap effective greenhouse gas concentrations at 550 ppm would benefit developing countries significantly, regardless of whether climate sensitivity turns out to be high or low and especially when combined with enhanced adaptation. Developed countries would also likely see significant benefits from an adaptation-mitigation intervention portfolio, especially for high climate sensitivities and in sectors and regions that are already showing signs of being vulnerable. However, by 2100, climate change will likely produce significant impacts across the globe, even if aggressive mitigation were implemented in combination with significantly enhanced adaptive capacity [20.7.3].

20.1 Introduction – setting the context

Consistent with the Bruntland Commission (WCED, 1987), the Third Assessment Report (TAR) (IPCC, 2001b) defined sustainable development as "development that meets the needs of the present without compromising the ability of future generations to meet their own needs". There are many alternative definitions, of course, and none is universally accepted. Nonetheless, they all emphasise one or more of the following critical elements: identifying what to develop, identifying what to sustain, characterising links between entities to be sustained and entities to be developed and envisioning future contexts for these links (NRC, 1999). Goals, indicators, values and practices can also frame examinations of sustainable development (Kates et al., 2005). The essence of sustainable development throughout is meeting fundamental human needs in ways that preserve the life support systems of the planet (Kates et al., 2000). Its strength lies in reconciling real and perceived conflicts between the economy and the environment and between the present and the future (NRC, 1999). Authors have emphasised the economic, ecological and human/social dimensions that are the pillars of sustainable development (Robinson and Herbert, 2001; Munasinghe et al., 2003; Kates et al., 2005). The economic dimension aims at improving human welfare (such as real income). The ecological dimension seeks to protect the integrity and resilience of ecological systems, and the social dimension focuses on enriching human relationships and attaining individual and group aspirations (Munasinghe and Swart, 2000), as well as addressing concerns related to social justice and promotion of greater societal awareness of environmental issues (O'Riordan, 2004).

The concept of sustainable development has permeated mainstream thinking over the past two decades, especially after the 1992 Earth Summit where 178 governments adopted Agenda 21 (UNDSD, 2006). Ten years later, the 2002 World Summit on Sustainable Development (WSSD, 2002) made it clear that sustainable development had become a widely-held social and political goal. Even though, as illustrated in Asia by the Institute for Global Environmental Strategies (IGES, 2005), implementation remains problematic, there is broad international agreement that development programmes should foster transitions to paths that meet human needs while preserving the Earth's life-support systems and alleviating hunger and poverty (ICSU, 2002) by integrating these three dimensions (economic, ecological and human/social) of sustainable development. Researchers and practitioners in merging fields, such as 'sustainability science' (Kates et al., 2000), multi-scale decision analysis (Adger et al., 2003) and 'sustainomics' (Munasinghe et al., 2003), seek to increase our understanding of how societies can do just that.

Climate change adds to the list of stressors that challenge our ability to achieve the ecologic, economic and social objectives that define sustainable development. Chapter 20 builds on the assessments in earlier chapters to note the potential for climate change to affect development paths themselves. Figure 20.1 locates its key topics schematically in the context of the three pillars of sustainable development. Topics shown in the centre of the triangle (the 'three-legged stool' of sustainable development) are linked with all three pillars. Other topics, placed outside the triangle, are located closer to one leg or another. The arrows leading from the centre indicate that adaptation to climate change can influence the processes that join the pillars rather than the individual pillars themselves. For example, the technical and economic aspects of renewable resource management could illustrate efforts to support sustainable development by working with the economy-ecology connection – all nested within a decision space of other global development pressures, including poverty.

Section 20.2 begins with a brief review of the current understanding of impacts and adaptive capacity as described earlier (see Chapter 17). Section 20.3 assesses impacts and adaptation in the context of multiple stresses. Section 20.4 focuses on links to environmental quality and explores the notion of adding climate-change impacts and adaptation to the list of components of environmental impact assessments. Section 20.5 addresses implications for risk, hazards and disaster management, including the challenge of reducing vulnerability to current climate variability and adapting to long-term climate change. Section 20.6 reviews global and regionally-aggregated estimates of economic impacts. Section 20.7 assesses the implications for achieving sustainable development across various time-scales. Section 20.8 considers opportunities, co-benefits and challenges for climate-change adaptation, and for linking (or mainstreaming) adaptation into national and regional development planning processes. Section 20.9 finally identifies research priorities.

This entire chapter should be read with the recognition that the first 19 chapters of this volume assess the regional and global impacts of climate change and the opportunities and challenges for adaptation. Chapters 17 and 19 in this volume offer synthetic overviews of this work that focus specifically on adaptation and key vulnerabilities. Chapter 20 in this volume expands the discussion to explore linkages with sustainable development, as do Chapters 2 and 12 in IPCC (2007a). Sustainable development was addressed in IPCC (2001b), but not in IPCC (2001a).

20.2 A synthesis of new knowledge relating to impacts and adaptation

Recent work at the intersection of impacts and adaptation has confirmed that adaptation to climate change is, to a limited extent, already happening (Chapter 17, Section 17.2). Perhaps more importantly for this chapter, recent work has also reconfirmed the utility of the prescription initially presented in Smit et al. (2001) that (1) any system's vulnerability to climate change and climate variability could be described productively in terms of *its exposure to the impacts of climate and its baseline sensitivity to those impacts* and that (2) both exposure and sensitivity can be influenced by that system's *adaptive capacity* (Chapter 17, Section 17.3.3). The list of critical determinants of adaptive capacity was described in Smit et al. (2001) and has been explored subsequently by, for example, Yohe and Tol (2002), Adger and Vincent (2004), Brenkert and Malone (2005)

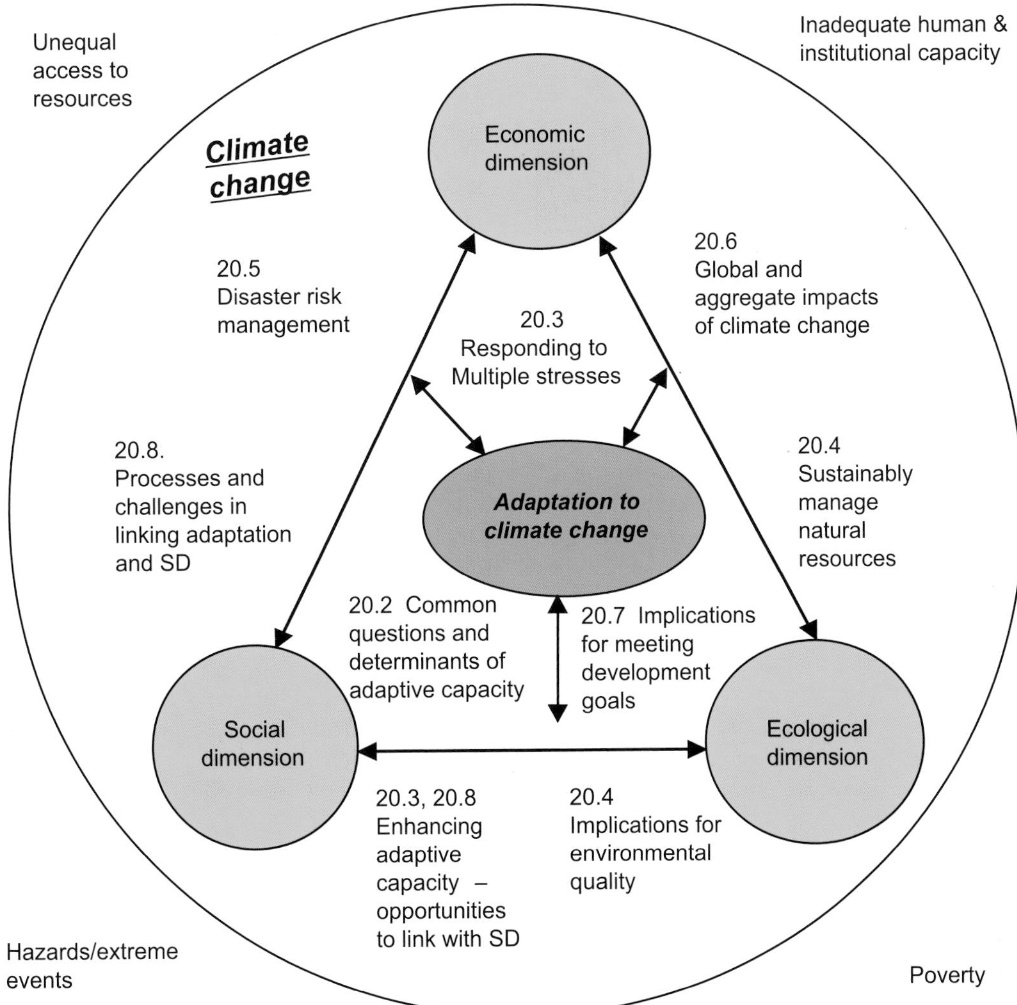

Figure 20.1. *Sustainable development and adaptation to climate change. An outline of Chapter 20 is mapped against the pillars of sustainable development. The figure is adapted from Munasinghe and Swart (2005).*

and Brooks and Adger (2005) – a list that includes access to economic and natural resources, entitlements (property rights), social networks, institutions and governance, human resources and technology (Chapter 17, Section 17.3.3).

It is, however, important to note that recent work has also emphasised the fundamental distinction between adaptive capacity and adaptation implementation. There are significant barriers to implementing adaptation (Chapter 17, Section 17.3.3) and they can arise almost anywhere. The description offered by Kates et al. (2006) of the damages and costs caused by Hurricane Katrina in New Orleans, denominated in economic and human terms, provides a seminal example of this point. Notwithstanding the widely accepted assertion that the United States has high adaptive capacity, the impacts of Hurricane Katrina were fundamentally the result of a failure of adaptive infrastructure (improperly constructed levées that led to a false sense of security) and planning (deficiencies in evacuation plans, particularly in many of the poorer sections of the cities). The capacity provided by public and private investment over the past few decades was designed to handle a hurricane like Katrina; it was the anticipatory efforts to provide protection prior to landfall and response efforts after landfall that failed.

Nothing in the recent literature has undermined a fundamental conclusion in Smit et al. (2001) that "current knowledge of adaptation and adaptive capacity is insufficient for reliable prediction of adaptations; it is also insufficient for rigorous evaluation of planned adaptation options, measures and policies of governments." (page 880). This conclusion is often supported by noting the uneven distribution of adaptive capacity across and within societies (Chapter 17, Section 17.3.2), but strong support can also be derived from the paucity of estimates of the costs of adaptation (Chapter 17, Section 17.2.3). While many adaptations can be implemented at low costs, comprehensive estimates of costs and benefits of adaptation currently do not exist except, perhaps, for costs related to adapting to sea-level rise and changes in the temporal and spatial demand for energy (heating versus cooling). Global diversity is one problem in this regard, but there are others. Anticipating the discussion of multiple stresses that appears in the next section of this chapter, it is now understood that climate change poses novel risks that often lie outside the range of past experience (Chapter 17, Section 17.2.1) and that adaptation measures are seldom undertaken in response to climate change alone (Chapter 17, Sections 17.2.2 and 17.3.3).

20.3 Impacts and adaptation in the context of multiple stresses

20.3.1 A catalogue of multiple stresses

The current literature shows a growing appreciation of the multiple stresses that ecological and socio-economic systems face, how those stresses are likely to change over the next several decades, and what some of the net environmental consequences are likely to be. The Pilot Analysis of Global Ecosystems prepared by the World Resources Institute (WRI, 2000) conducted literature reviews to document the state and condition of forests, agro-ecosystems, freshwater ecosystems and marine systems. The Millennium Ecosystem Assessment (MA) comprehensively documented the condition and recent trends of ecosystems, the services they provide and the socio-economic context within which they occur. It also provided several scenarios of possible future conditions (MA, 2005). For reference, the MA offered some startling statistics. Cultivated systems covered 25% of Earth's terrestrial surface in 2000. On the way to achieving this coverage, global agricultural enterprises converted more area to cropland between 1950 and 1980 than in the 150 years between 1700 and 1850. As of the year 2000, 35% of the world's mangrove areas and 20% of the world's coral reefs had been lost (with another 20% having been degraded significantly). Since 1960, withdrawals from rivers and lakes have doubled, flows of biologically available nitrogen in terrestrial ecosystems have doubled, and flows of phosphorus have tripled. At least 25% of major marine fish stocks have been overfished and global fish yields have actually begun to decline. MA (2005) identified major changes in land cover, the consequences of which were explored by Foley et al. (2005).

The MA (2005) recognised two different categories of drivers of change. Direct drivers of ecosystem change affect ecosystem characteristics in specific, quantifiable ways; examples include land-cover and land-use change, climate change and species introductions. Indirect drivers affect ecosystems in a more diffuse way, generally by affecting one or more direct drivers; here examples are demographic changes, socio-political changes and economic changes. Both types of drivers have changed substantially in the past few decades and will continue to do so. Among direct drivers, for example, over the past four decades, food production has increased by 150%, water use has doubled, wood harvests for pulp and paper have tripled, timber production has doubled and installed hydropower capacity has doubled. On the indirect side, global population has doubled since the 1960s to reach 6 billion people while the global economy has increased more than six fold.

Table 20.1 documents expectations for how several of the direct drivers of ecosystem change are likely to change in magnitude and importance over time. With the exception of polar regions, coastal ecosystems, some dryland systems and montane regions, climate change is not, today, a major source of stress; but climate change is the only direct driver whose magnitude and importance to a series of regions, ecosystems and resources is likely to continue to grow over the next several decades. Table 20.1 illustrates the degree to which these ecosystems are currently experiencing stresses from several direct drivers of change simultaneously. It shows that potential interactions with climate change are likely to grow over the next few decades with the magnitude of climate change itself.

20.3.2 Factors that support sustainable development

A brief excursion into some of the recent literature on economic development is sufficient to support the fundamental observation that the factors that determine a country's ability to promote (sustainable) development coincide with the factors that influence adaptive capacity relative to climate change, climate variability and climatic extremes. The underlying prerequisites for sustainability in specific contexts are highlighted in italics in the discussion which follows. The point about coincidence in underlying factors is made by matching the terms in italics with the list of determinants of adaptive capacity identified above (Chapter 17, Section 17.3.3): *access to resources, entitlements* (property rights), *institutions and governance, human resources* (human capital in the economics literature) and *technology*. They are all reflected in one or more citations from the development literature cited here, and they conform well to the "5 capital" model articulated by Porritt (2005) in terms of human, manufactured, social, natural and financial capital.

Lucas (1988) concluded early on that differences in *human capital* are large enough to explain differences between the long-run growth rates of poor and rich countries. Moretti (2004), for example, showed that businesses located in cities where the fraction of college graduates (highly *educated* work force) grew faster and experienced larger increases in productivity. Guiso et al. (2004) explored the role of *social capital* in peoples' abilities to successfully take advantage of financial structures; they found that *social capital* matters most when education levels are low and law enforcement is weak. Rozelle and Swinnen (2004) looked at transition countries in central Europe and the former Soviet Union; they observed that countries growing steadily a decade or more after economic reform had accomplished a common set of intermediate goals: achieving macroeconomic stability, *reforming property rights*, and *creating institutions to facilitate exchange*. Order and timing did not matter, but meeting all of these underlying objectives was critical. Winters et al. (2004) reviewed a wide literature on the links between trade liberalisation and poverty reduction. They concluded that a favourable relationship depends on the existence and *stability of markets*, the ability of economic actors to handle changes in risk, *access to technology, resources, competent and honest government, policies that promote conflict resolution* and *human capital accumulation*. Shortfalls in any of these underpinnings make it extremely difficult for the most disadvantaged citizens to see any advantage from trade. Finally, Sala-i-Martin et al. (2004) explained economic growth by variation in national

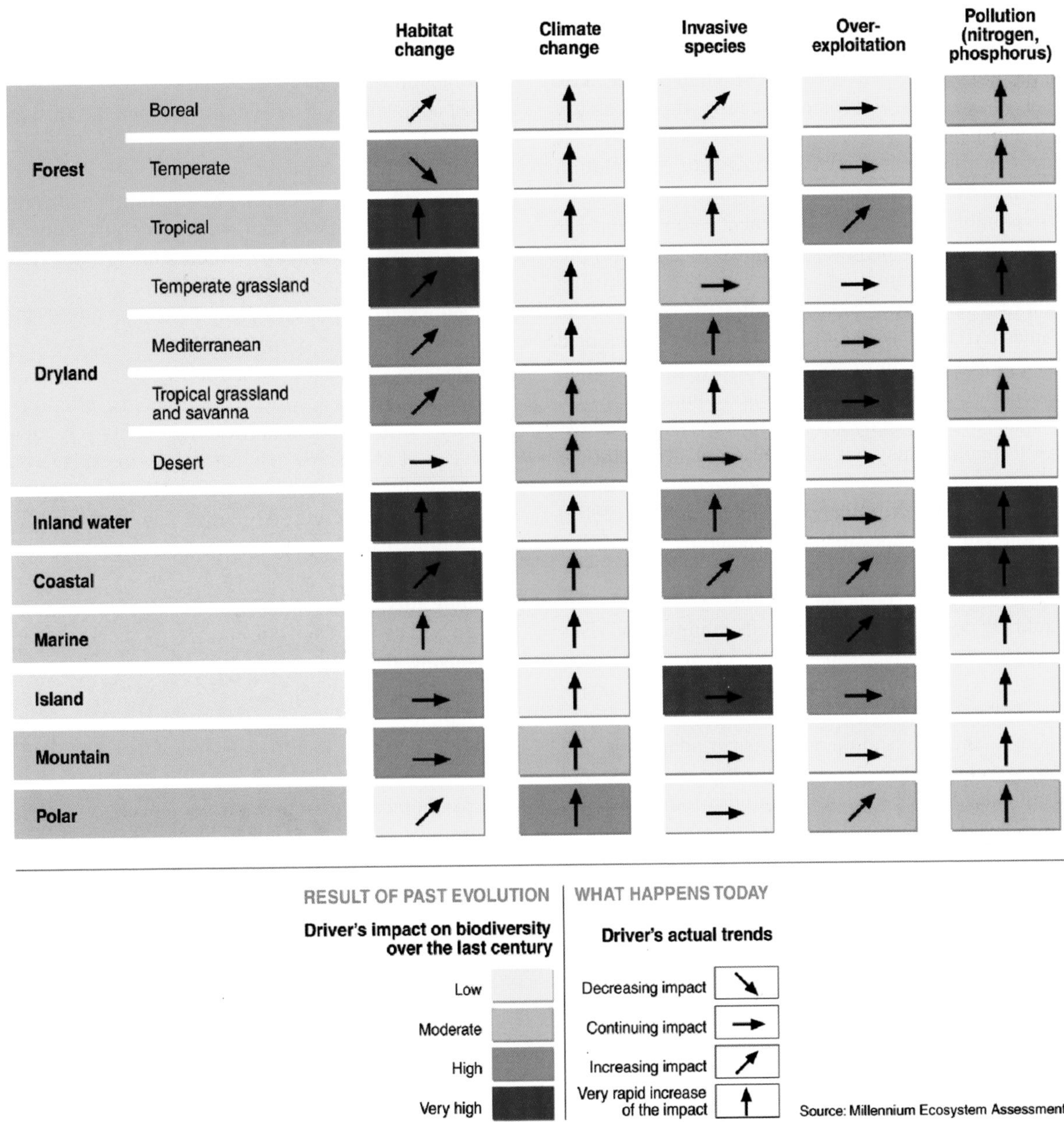

		Habitat change	Climate change	Invasive species	Over-exploitation	Pollution (nitrogen, phosphorus)
Forest	Boreal	Low, increasing impact	Low, very rapid increase	Low, increasing impact	Low, continuing impact	Moderate, very rapid increase
	Temperate	High, decreasing impact	Low, very rapid increase	Low, very rapid increase	Moderate, continuing impact	Moderate, very rapid increase
	Tropical	Very high, very rapid increase	Low, very rapid increase	Low, very rapid increase	High, increasing impact	Low, very rapid increase
Dryland	Temperate grassland	Very high, increasing impact	Low, very rapid increase	Moderate, continuing impact	Low, continuing impact	Very high, very rapid increase
	Mediterranean	High, increasing impact	Low, very rapid increase	High, very rapid increase	Moderate, continuing impact	Low, very rapid increase
	Tropical grassland and savanna	High, increasing impact	Moderate, very rapid increase	Low, very rapid increase	Very high, continuing impact	Moderate, very rapid increase
	Desert	Low, continuing impact	Moderate, very rapid increase	Moderate, continuing impact	Low, continuing impact	Low, very rapid increase
Inland water		Very high, very rapid increase	Low, very rapid increase	High, very rapid increase	Moderate, continuing impact	Very high, very rapid increase
Coastal		Very high, increasing impact	Moderate, very rapid increase	High, increasing impact	High, increasing impact	Very high, very rapid increase
Marine		Moderate, very rapid increase	Low, very rapid increase	Low, continuing impact	Very high, increasing impact	Low, very rapid increase
Island		High, continuing impact	Low, very rapid increase	Very high, continuing impact	High, continuing impact	Low, very rapid increase
Mountain		High, continuing impact	Moderate, very rapid increase	Low, continuing impact	Low, continuing impact	Low, very rapid increase
Polar		Low, increasing impact	High, very rapid increase	Low, continuing impact	Moderate, increasing impact	Moderate, very rapid increase

RESULT OF PAST EVOLUTION — Driver's impact on biodiversity over the last century: Low; Moderate; High; Very high

WHAT HAPPENS TODAY — Driver's actual trends: Decreasing impact (↘); Continuing impact (→); Increasing impact (↗); Very rapid increase of the impact (↑)

Source: Millennium Ecosystem Assessment

Table 20.1. *Drivers of change in ecosystem services. Source: Millennium Ecosystem Assessment (MA, 2005).*

participation in primary school education (*human capital*), other measures of human capital (e.g., health measures), *access to affordable investment goods* and the *initial level of per capita income* (access to resources).

20.3.3 Two-way causality between sustainable development and adaptive capacity

It has become increasingly evident, especially since the TAR (IPCC, 2001b), that the pace and character of development influences adaptive capacity and that adaptive capacity influences the pace and character of development. It follows that development paths, and the choices that define them, will affect the severity of climate impacts, not only through changes in exposure and sensitivity, but also through changes in the capacities of systems to adapt. This includes local-scale disaster risk reduction and resource management (e.g., Shaw, 2006; Jung et al., 2005), and broader social dimensions including governance, societal engagement and rights, and levels of education (Haddad, 2005; Tompkins and Adger, 2005; Brooks et al., 2005; Chapter 17, Section 17.3).

Munasinghe and Swart (2005) and Swart et al. (2003) argued that sustainable development measures and climate-change policies, including adaptation, can reinforce each other; Figure 20.2 portrays some of the texture of the interaction that they envisioned. Although scholarly papers on adaptation began to appear in the 1980s, it was not until the 2001 Marrakech Accords that a policy focus on adaptation within the United Nations

Framework Convention on Climate Change (UNFCCC) developed (Schipper, 2006). Klein et al. (2005) suggest that adaptation has not been seen as a viable option, in part because many observers see market forces creating the necessary conditions for adaptation even in the absence of explicit policies and, in part, because understanding of how future adaptation could differ from historical experience is limited.

Efforts to promote alternative development pathways that are more sustainable could include measures to reduce non-renewable energy consumption, for example, or shifting construction of residential or industrial infrastructure to avoid high-risk areas (AfDB et al., 2004). The MA (2005) attempted to describe a global portrait of such a pathway in its "Techno Garden" scenario. In this future, an inter-connected world promotes expanded use of innovative technology, but its authors warned that technology may not solve all problems and could lead to the loss of indigenous cultures. Climate-change measures could also encounter such limitations. Gupta and Tol (2003) describe various climate-policy dilemmas including competition between human rights and property rights.

Adaptation measures embedded within climate-change policies could, by design, try to reduce vulnerabilities and risks by enhancing the adaptive capacity of communities and economies. This would be consistent with sustainability goals. Researchers and practitioners should not equate vulnerability to poverty, though, and they should not consider adaptation and adaptive capacity in isolation. Brooks et al. (2005) conclude that efforts to promote adaptive capacity should incorporate aspects of education, health and governance and thereby extend the context beyond a particular stress (such as climate change) to include factors that are critical in a broader development context. Haddad (2005) noted the critical role played here by general rankings of economic development performance and general reflections of national and local goals and aspirations, and explained how different people might choose different development from the same set of alternatives even if they had the same information.

Past adaptation and development experience displays mixed results. Kates (2000) described several historic climate adaptations (e.g., drought in the Sahel) and development measures (e.g., the Green Revolution) and argued that development measures that were generally consistent with climate adaptation often benefited some groups (e.g., people with access to resources) while harming others (e.g., poor populations, indigenous peoples). Ford et al. (2006) showed that unequal acquisition of new technologies can, under some circumstances, increase vulnerability to external stresses by weakening social networks and thereby altering adaptive capacity within communities and between generations. Belliveau et al. (2006) makes the link to climate explicit by observing that adaptation to non-climatic forces, without explicitly considering climate, can lead to increased vulnerability to climate because adapting previous adaptations can be expensive.

Future links between sustainable development and climate change will evolve from current development frameworks; but recognising the exposure of places and peoples to multiple stresses (Chapter 17; Chapter 19; Section 20.3.1) and accepting the challenge of mainstreaming adaptation into development planning will be critical in understanding what policies will work where and when. For example, in the Sudan, there is a risk that development efforts focusing on short-term relief can undermine community coping capacity (Elasha, 2005). In the mitigation realm, incentives for carbon sequestration could promote hybrid forest plantations and therefore pose a threat to biodiversity and ecosystem adaptability (Caparrós and Jacquemont, 2003; Chapter 18). Development decisions can also produce cumulative threats. In the Columbia River Basin, for

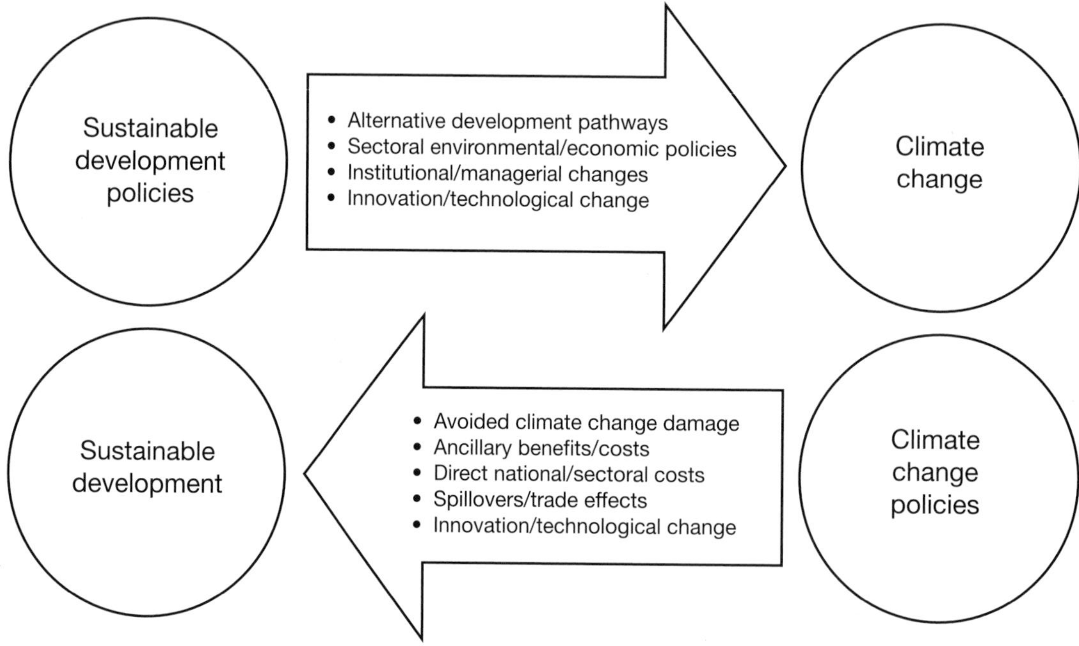

Figure 20.2. *Two-way linkages between climate and sustainable development. Source: Swart et al. (2003).*

instance, extensive water resource development can influence basin management with multiple objectives within scenarios of climate change because climate impacts on stream-flow cause policy dilemmas when decision-makers must balance hydroelectricity production and fisheries protection (Hamlet, 2003; Payne et al., 2004). Restoring in-stream flow to present-day acceptable (but sub-optimum) levels could, in particular, cause hydroelectricity production to decline and production from fossil fuel sources to rise. Interactions of this sort raise important questions on the analysis of the causes of recent climate-related disasters. For example, are observed trends in injuries/fatalities and property losses (Mileti, 1999; Mirza, 2003; MA, 2005; Munich Re, 2005) due to unsustainable development policies, climate change or a mixture of different factors? Could policy interventions reduce these losses in ways that would still meet broader objectives of sustainable development? Some proposed responses for Africa are described in Low (2005) and AfDB et al. (2004).

Globalisation also adds complexity to the management of common-pool resources because increased interdependence makes it more difficult to find equitable solutions to development problems (Ostrom et al., 1999). Increases in the costs associated with various hazards and the prospects of cumulative environmental/economic threats have been described as syndromes. Schellnhuber et al. (1997) identified three significant categories: over-utilisation (e.g., over-cultivation of marginal land in the Sahel), inconsistent development (e.g., urban sprawl and associated destruction of landscapes) and hazardous sinks (e.g., large-scale diffusion of long-lived substances). Schellnhuber et al. (2002) and Lüdeke et al. (2004) describe possible future distributions of some of these syndromes. They suggest how mechanisms of mutual reinforcement, including climate change and development drivers, can help to identify regions where syndromes may expand and others where they might contract.

20.4 Implications for environmental quality

The inseparability of environment and development has been widely recognised ever since the Brundtland Commission (WCED, 1987). In the United Nations' Millennium Development Goals (MDGs), for example, environmental considerations are reflected in the 7th goal and the operative target, among others, is to reverse loss of environmental resources by 2015. Overall, how to meet the target of integrating the principles of sustainable development in national policy and reversing the loss of environmental resources remains a partially answered question for most countries (Kates et al., 2005).

Interest in environmental indicators and performance indices to monitor change has increased recently. A compilation of different sustainable development indicators by Kates et al. (2005) showed that most implicitly or explicitly build from reflections of the health of environmental and ecological resources and/or the quality of environmental and ecological services. This is relevant in both developed and developing countries, but the drivers encouraging sustainable management are arguably strongest in the developed world. Huq and Reid (2004) and Agrawala (2004) have noted, though, that climate change is being increasingly recognised as a key factor that could affect the (sustainable) development of developed and developing countries alike. The Philippine Country Report (1999) identified 153 sustainable development indicators; some pertain to climate-change variables such as level of greenhouse gas emissions, but none refer explicitly to adaptation. There is, for example, no mention within the MDGs of potential changes in climate-related disasters or of the need to include climate-change adaptation within development programmes (Reid and Alam, 2005). This is not unusual, because links between sustainable development and climate change have historically been defined primarily in terms of mitigation.

Promoting environmental quality is about more than encouraging sustainable development or adaptive capacity. It is also about transforming use practices for environmental resources into sustainable management practices. In many countries and sectors, stakeholders who manage natural resources (such as individual farmers, small businesses or major international corporations) are susceptible, over time, to variations in resource availability and hazards; they are currently seeking to revise management practices to make their actions more sustainable. Hilson (2001), for example, describes efforts in the mineral extraction industry where the relevant players include public agencies operating at many scales (from local to national to international). Definitions of sustainability vary across sectors, but their common theme is to change the way resources are exploited or hazards are managed so that adverse impacts downstream or for subsequent generations are reduced. Climate change is, however, seldom listed among the stressors that might influence sustainability. Arnell and Delaney (2006) note, though, that water management in the United Kingdom is an exception.

Published literature on the links between sustainable management of natural resources and the impacts of and adaptation to climate change is extremely sparse. Most focuses on engineering and management techniques which achieve management objectives, such as a degree of protection against flood hazard or a volume of crop production, while having smaller impacts on the environment. Turner (2004) and Harman et al. (2002) speak to this point, but very few engineering analyses consider explicitly how the performance of these measures is affected by climate change or how suitable they would be in the face of a changing climate. Kundzewicz (2002) demonstrated how non-structural flood management measures can be sustainable adaptations to climate change because they are relatively robust to uncertainty. On the other hand, as shown in Clark (2002) and Kashyap (2004), much of the literature on integrated water management in the broadest sense emphasises adaptation to climatic variability and change through the adoption of sustainable and integrated approaches.

Several studies have highlighted the benefits of adopting more sustainable practices, in terms of reduced costs, increased efficiency or financial performance more broadly interpreted. Johnson and Walck (2004) offer an example from forestry while Epstein and Roy (2003) are illustrative of a more expansive context. None of these studies explicitly consider the effects of climate change on the benefits of adopting more sustainable practices; and none of the literature on mechanisms for incorporating sustainable behaviour into organisational practice and monitoring its implementation (e.g., Jasch, 2003; Figge and Hahn, 2004) consider how to incorporate the effects of climate change into mechanisms or monitoring procedures.

Clark (2002) and Bansal (2005) identified several drivers behind moves to become more sustainable. First, altered legal or regulatory requirements may have an effect. Many governments have adopted legislation aimed at encouraging the sustainable use of the natural environment, and some explicitly include reference to climate change. For example, Canada and some EU member states have begun to incorporate climate change in their environmental policies, particularly in the structures of required environmental impact assessments. The hope is that the impact of present and future climates on development projects might thereby be reduced (EEA, 2006; Barrow and Lee, 2000). Ramus (2002) and Thomas et al. (2004) have observed that internally-generated efforts to improve procedures (e.g., following an ethical position held by an influential champion, responding to the desire to reduce costs or risks, or attempting to attract potential clients) can push systems toward sustainability.

Of course, stakeholder expectations may change over time. While these dynamic drivers may encourage sustainable management, they may not in themselves be directly related to concerns over the impacts of and adaptation to climate change. Kates et al. (2005) noted that the principles, goals and practices of sustainability are not fixed and immutable; they are 'works in progress' because the tension between economic development and environmental protection has been opened to reinterpretation from different social and ecological perspectives.

20.5 Implications for risk, hazard and disaster management

The International Decade for Natural Disaster Reduction (1990 to 1999) led to a fundamental shift in the way disasters are viewed: away from the notion that disasters were temporary disruptions to be managed by humanitarian responses and technical interventions and towards a recognition that disasters are a function of both natural and human drivers (ISDR, 2004; UNDP, 2004). The concept of *disaster risk management* has evolved; it is defined as the systematic management of administrative decisions, organisations, operational skills and abilities to implement policies, strategies and coping capacities of society or individuals to lessen the impacts of natural and related environmental and technological hazards (ISDR, 2004). This includes measures to provide not only emergency relief and recovery, but also *disaster risk reduction* (ISDR, 2004); i.e., the development and application of policies, strategies and practices designed to minimise vulnerabilities and the impacts of disasters through a combination of technical measures to reduce physical hazards and to enhance social and economic capacity to adapt. Disaster risk reduction is conceived as taking place within the broad context of sustainable development (ISDR, 2004).

In practice, however, there has been a disconnect between disaster risk reduction and sustainable development, due to a combination of institutional structures, lack of awareness of the linkages between the two, and perceptions of 'competition' between hazard-based risk reduction, development needs and emergency relief (Yamin, 2004; Thomalla et al., 2006). The disconnect persists despite an increasing recognition that natural disasters seriously challenge the ability of countries to meet targets associated with the Millennium Development Goals (Schipper and Pelling, 2006).

A disconnect also exists between disaster risk reduction and adaptation to climate change, again reflecting different institutional structures and lack of awareness of linkages (Schipper and Pelling, 2006; O'Brien et al., 2006). Disaster risk reduction, for example, is often the responsibility of civil defence agencies, while climate-change adaptation is often covered by environmental or energy departments (Thomalla et al., 2006). Disaster risk reduction tends to focus on sudden and short-lived disasters, such as floods, storms, earthquakes and volcanic eruptions, and has tended to place less emphasis on 'creeping onset' disasters such as droughts. Many disasters covered by disaster risk reduction are not affected by climate change. However, there is an increasing recognition of the linkages between disaster risk reduction and adaptation to climate change, since climate change alters not only the physical hazard but also vulnerability. Sperling and Szekely (2005) note that many of the impacts associated with climate change exacerbate or alter existing threats, and adaptation measures can benefit from practical experience in disaster risk reduction. However, some effects of climate change are new within human history (such as the effects of sea-level rise), and there is little experience to tackle such impacts. Sperling and Szekely (2005) therefore state that co-ordinated action to address both existing and new challenges becomes urgent. There is great opportunity for collaboration in the assessment of current and future vulnerabilities, in the use of assessment tools (Thomalla et al., 2006) and through capacity-building measures. Incorporating climate change and its uncertainty into measures to reduce vulnerability to hazard is essential in order for them to be truly sustainable (O'Hare, 2002), and climate change increases the urgency to integrate disaster risk management into development interventions (DFID, 2004).

There are, effectively, two broad approaches to disaster risk reduction, and adaptation to climate change can be incorporated differently into each. The top-down approach is based on institutional responses, allocation of funding and agreed procedures and practices (O'Brien et al., 2006). It is the approach followed in most developed countries, and adaptation to climate change can be implemented by changing guidelines and procedures. In the United Kingdom, for example, design flood magnitudes can be increased by 20% to reflect possible effects of climate change (Richardson, 2002). However, institutional inertia and strongly embedded practices can make it very difficult to change. Olsen (2006), for example, shows how major methodological and institutional changes are needed before flood management in the USA can take climate change (and its uncertainty) into account. The bottom-up approach to disaster risk reduction is based on enhancing the capacity of local communities to adapt to and prepare for disaster (see, for example, Allen, 2006; Blanco, 2006). Actions here include dissemination of technical knowledge and training, awareness raising, accessing local knowledge and resources, and mobilising local communities (Allen, 2006). Climate change can be incorporated in this approach through awareness raising and the transmission of technical knowledge to local communities, but bridging the gap between scientific knowledge and local application is a key challenge (Blanco, 2006).

Reducing vulnerability to current climatic variability can effectively reduce vulnerability to increased hazard risk associated with climate change (e.g., Kashyap, 2004; Goklany, 2007; Burton et al., 2002; Davidson et al., 2003; Robledo et al., 2004). To a large extent, adaptation measures for climate variability and extremes already exist. Measures to reduce current vulnerability by capacity building rather than distribution of disaster relief, for example, will increase resilience to changes in hazard caused by climate change (Mirza, 2003). Similarly, the implementation of improved warning and forecasting methods and the adoption of some land-use planning measures would reduce both current and future vulnerability. However, many responses to current climatic variability would not in and of themselves be a sufficient response to climate change. For example, a changing climate could alter the design standard of a physical defence, such as a realigned channel or a defence wall. It could alter the effectiveness of building codes based on designing against specified return period events (such as the 10-year return period gust). It could alter the area exposed to a potential hazard, meaning that development previously assumed to be 'safe' was now located in a risk area. Finally, it could introduce hazards previously not experienced in an area. Burton and van Aalst (2004), in their assessment of the World Bank Country Strategic Programmes and project cycle, identify the need to assess the success of current adaptation to present-day climate risks and climate variability, especially as they may change with climate change.

20.6 Global and aggregate impacts

Three types of aggregate impacts are commonly reported. In the first, impacts are computed as a percent of gross domestic product (GDP) for a specified rise in global mean temperature. In the second, impacts are aggregated over time and discounted back to the present day along specified emissions scenarios such as those documented in Nakićenović and Swart (2000) under specified assumptions about economic development, changes in technology and adaptive capacity. Some of these estimates are made at the global level, but others aggregate a series of local or regional impacts to obtain a global total. A third type of estimate has recently attracted the most attention. Called the social cost of carbon (SCC), it is an estimate of the economic value of the extra (or marginal) impact caused by the emission of one more tonne of carbon (in the form of carbon dioxide) at any point in time; it can, as well, be interpreted as the marginal benefit of reducing carbon emissions by one tonne. Researchers calculate SCC by summing the extra impacts for as long as the extra tonne remains in the atmosphere – a process which requires a model of atmospheric residence time and a means of discounting economic values back to the year of emission.

This section provides a brief discussion of the historical and current status of efforts to produce aggregate estimates of the impacts of climate change. The first sub-section focuses attention on economic estimates and the second begins to expand the discussion by reporting estimates calibrated in alternative metrics. It is in this expansion that the implications of spatial and temporal diversity in systems' exposures and sensitivities to climate change begin to emerge.

20.6.1 History and present state of aggregate impact estimates

Most of the aggregate impacts reported in IPCC (1996) were of the first type; they monetised the likely damage that would be caused by a doubling of CO_2 concentrations. For developed countries, estimated damages were of the order of 1% of GDP. Developing countries were expected to suffer larger percentage damages, so mean global losses of 1.5 to 3.5% of world GDP were therefore reported. IPCC (2001a) reported essentially the same range because more modest estimates of market damages were balanced by other factors such as higher non-market impacts and improved coverage of a wide range of uncertainties. Most recently, Stern (2007) took account of a full range of both impacts and possible outcomes (i.e., it employed the basic economics of risk premiums) to suggest that the economic effects of unmitigated climate change could reduce welfare by an amount equivalent to a persistent average reduction in global per capita consumption of at least 5%. Including direct impacts on the environment and human health (i.e., 'non-market' impacts) increased their estimate of the total (average) cost of climate change to 11% GDP; including evidence which indicates that the climate system may be more responsive to greenhouse-gas emissions than previously thought increased their estimates to 14% GDP. Using equity weights to reflect the expectation that a disproportionate share of the climate-change burden will fall on poor regions of the world increased their estimated reduction in equivalent consumption per head to 20%.

Figure 20.3 compares the Stern (2007) relationship between global impacts and increases in global mean temperature with estimates drawn from earlier studies that were assessed in IPCC (2001b). The Stern (2007) trajectories all show negative impacts for all temperatures; they reflect the simple assumptions of the underlying PAGE2002 model and a focus on risks associated with higher temperatures. The Mendelsohn et al. (1998) estimates aggregate regional monetary damages (both positive and negative) without equity weighting. The two Nordhaus and Boyer (2000) trajectories track aggregated regional monetary estimates of damages with and without population-based equity weighting; they do include a 'willingness to pay (to avoid)' reflection of the costs of abrupt change. The two Tol (2002) trajectories track aggregated regional monetary estimates of damages with and without utility-based equity weighting. The various relationships depicted in Figure 20.3 therefore differ in their treatment of equity weighting, in their efforts to capture the potential of beneficial climate change (in, for example,

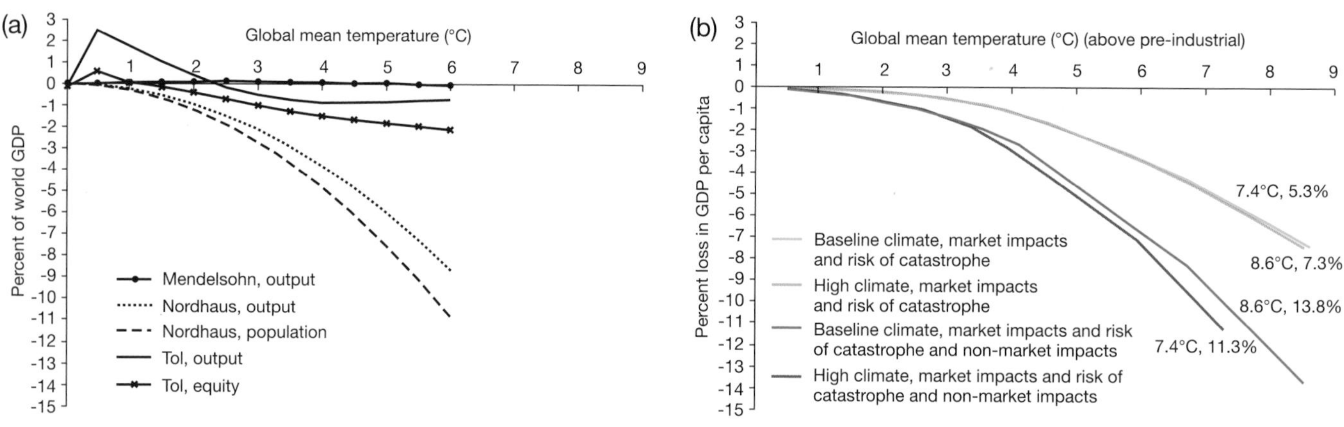

Figure 20.3. *(a) Damage estimates, as a percent of global GDP, as correlated with increases in global mean temperature. Source: IPCC (2001b). (b) Damage estimates, as a percent of global GDP, are correlated with increases in global mean temperature. Source: Stern (2007).*

agriculture for small increases in temperature; see Chapter 5, Section 5.4.7) and in their treatment of the risks of catastrophe for large increases in temperature.

Early calculations of the SCC (IPCC (1996) estimates ranged from US$5 to $125 per tonne of carbon in 1990 dollars) stimulated recurring interest, as part of wider post-Kyoto considerations, in the economic benefits of climate-change policy (Watkiss et al., 2005). After surveying the literature, Clarkson and Deyes (2002) proposed a central value of US$105 per tonne of carbon (in year 2000 prices) for the SCC, with upper and lower values of US$50 and $210 per tonne. Pearce (2003) argued that 3% is a reasonable representation of a social discount rate so the probable range of the SCC in 2003 should have been in the region of US$4 to 9 per tonne of carbon. Tol (2005) gathered over 100 estimates of the SCC from 28 published studies and combined them to form a probability density function; it displayed a median of US$14 per tonne of carbon, a mean of US$93 per tonne and a 95th percentile estimate equal to US$350 per tonne. Peer-reviewed studies generally reported lower estimates and smaller uncertainties than those which were not; their mean was US$43 per tonne of carbon with a standard deviation of US$83. The survey showed that 10% of the estimates were negative; to support these estimates, the climate sensitivity was assumed to be low and small increases in global mean temperature brought benefits (as suggested by the Tol (2002) trajectories in Figure 20.3).

Notwithstanding the differences in damage sensitivity to temperature reflected in Figure 20.3, the effect of the discount rate (see glossary) on estimates of SCC is most striking. The 90th percentile SCC, for instance, is US$62/tC for a 3% pure rate of time preference, $165/tC for 1% and $1,610/tC for 0%. Stern (2007) calculated, on the basis of damage calculations described above, a mean estimate of the SCC in 2006 of US$85 per tonne of CO_2 (US$310 per tonne of carbon). Had it been included in the Tol (2005) survey, it would have fallen well above the 95th percentile, in large measure because of their adoption of a low 0.1% pure rate of time preference. Other estimates of the SCC run from less than US$1 per tonne to over US$1,500 per tonne of carbon. Downing et al. (2005) argued that this range reflects uncertainties in climate and impacts, coverage of sectors and extremes, and choices of decision variables. Tol (2005) concluded, using standard assumptions about discounting and aggregation, that the SCC is unlikely to exceed US$50/tC. In contrast, Downing et al. (2005) concluded that a lower benchmark of US$50/tC is reasonable for a global decision context committed to reducing the threat of dangerous climate change and including a modest level of aversion to extreme risks, relatively low discount rates and equity weighting.

Climate change is not caused by carbon dioxide alone, and integrated assessment models can calculate the social cost of each greenhouse gas under consistent assumptions. For instance, the mean estimate from the PAGE2002 model for the social cost of methane is US$105 per tonne emitted in 2001, in year 2000 dollars, with a 5 to 95% uncertainty range of US$25 to $250 per tonne. The estimate for the social cost of SF_6 is US$200,000 per tonne emitted in 2001 with a 5 to 95% range of US$45,000 to $450,000 per tonne. These are all higher than the corresponding US$19 per tonne estimate for SCC that is surrounded by a 5 to 95% range of US$4 to $50 per tonne (Hope, 2006b). It has been known since IPCC (1996) that the SCC will increase over time; current knowledge suggests a 2.4% per year rate of growth. The social cost of methane will grow 50% faster because of its shorter atmospheric lifetime. Unlike later emissions, any extra methane emitted today will have disappeared before the most severe climate-change impacts occur (Watkiss et al., 2005).

Tol (2005) finds that much of the uncertainty in the estimates of the SCC can be traced to two assumptions: one on the discount rate and the other on the equity weights that are used to aggregate monetised impacts over countries. In most other policy areas, the rich do not reveal as much concern for the poor as is implied by the equity weights used in many models. Downing et al. (2005) state that the extreme tails of the estimates of the SCC depend as much on decision values (such as discounting and equity weighting) as on the climate forcing and uncertainty in the underlying impact models. Integrated models are always simplified representations of reality. To be comprehensive, other social and cultural values need to be given comparable weights to economic values, and there are prototype integrated assessment models to demonstrate this (Rotmans and de Vries, 1997).

Table 20.2 shows the six major influences calculated by PAGE2002 and reported in Hope (2005). That the list can be divided into two scientific and four socio-economic parameters is another strong argument for the building of integrated assessment models (IAMs); models that are exclusively scientific, or exclusively economic, would omit parts of the climate-change problem which still contain profound uncertainties. The two top influences are the climate sensitivity and the pure rate of time preference. Climate sensitivity is positively correlated with the SCC, but the pure time preference rate is negatively correlated with the SCC. Non-economic impact ranks third and economic impact ranks sixth (Hope, 2005).

A few models have existed for long enough to trace the changes in their estimates of the SCC over time. Table 20.3 shows how the results from three integrated assessment models have evolved over the last 15 years. The DICE and PAGE estimates have not changed greatly over the years, but this gives a misleading impression of stability. The values from PAGE have changed little because several quite significant changes have approximately cancelled each other out. In the later studies, lower estimates for market-sector impacts in developed countries are offset by higher non-market impacts, equity weights and inclusion of estimates of the possible impacts of large-scale discontinuities (Tol, 2005).

Hitz and Smith (2004) found that the relationships between global mean temperature and impacts of the sort displayed in Figure 20.3 are not consistent across sectors for modest amounts of warming. Beyond an approximate 3 to 4°C increase in global mean temperature above pre-industrial levels, all sectors (except possibly forestry) show increasingly adverse impacts. Tol (2005) found that few studies cover non-market damages, the risk of potential extreme weather, socially contingent effects, or the potential for longer-term catastrophic events. Therefore, uncertainty in the value of the SCC is derived not only from the 'true' value of impacts that are covered by the models, but also from impacts that have not yet been quantified and valued. As argued in Watkiss et al. (2005) and displayed in Figure 20.4, existing estimates of SCC are products of work that spans only a sub-set of impacts for which complete estimates might be calculated. Nonetheless, current estimates do provide enough information to support meaningful discussions about reducing the emissions of CO_2, methane and other greenhouse gases, and the appropriate trade-off between gases.

Nonetheless, estimates of SCC offer a consistent way to internalise current knowledge about the impacts of climate change into development, mitigation and/or adaptation decisions that the private and public sector will be making over the near term (Morimoto and Hope, 2004). According to economic theory, if the social cost calculations were complete and markets were perfect, then efforts to cut back the emissions of greenhouse gases would continue as long as the marginal cost of the cutbacks were lower than the social cost of the impacts they cause. If taxes were used, then they should be set equal to the SCC. If tradable permits were used, then their price should be the same as the SCC. If their price turns out to be lower than the social cost, then the total allocation of permits would have been too large and *vice versa*. In any comparison between greenhouse gases, according to Pearce (2003), the SCC is the correct figure to use. For reference, spot prices for permits in the European Carbon Trading Scheme since its inception early in 2005 started out towards the bottom end of the range of the SCC, but they rose quickly to around US$100 per tonne of carbon before falling by about 50% in the early summer of 2006 amid concerns that the carbon allowances allocated by the European Commission at the start of the scheme had been too generous. In the real world, markets are not perfect, calculations of the SCC are far from complete, and both mask significant differences between regions and types of impacts.

Table 20.2. *Major factors causing uncertainty in the social cost of carbon. Relative importance is measured by the magnitude of the partial rank correlation coefficient between the parameter and the SCC, with the most important indexed to 100. A + sign shows that an increase in this parameter leads to an increase in the SCC and vice versa. Source: Hope (2005).*

Parameter	Definition	Sign	Range	Importance
Climate sensitivity	Equilibrium temperature rise for a doubling of CO_2 concentration	+	1.5 to 5°C	100
PTP rate	Pure time preference for consumption now rather than in 1 year's time	-	1 to 3% /yr	66
Non-economic impact	Valuation of non-economic impact for a 2.5°C temperature rise	+	0 to 1.5% of GDP	57
Equity weight	Negative of the elasticity of marginal utility with respect to income	-	0.5 to 1.5	50
Climate change half life	Half life in years of global response to an increase in radiative forcing	-	25 to 75 years	35
Economic impact	Valuation of economic impact for a 2.5°C temperature rise	+	-0.1 to 1.0% of GDP	32

Note: non-economic and economic impact ranges apply to Europe; impacts in other regions are expressed as a multiple of this.

Table 20.3. *Estimates of the social cost of carbon over time from three models (in constant 2000 US$). Sources: DICE best guesses of Nordhaus and Boyer (2000) are from Pearce (2003); FUND estimates are from Tol (1999), and 25 to 75% range with green book discounting and equity weights from Downing et al. (2005); PAGE 5th and 95th percentile ranges from Plambeck and Hope (1996), rebased to year 2000, and Hope (2006a).*

Date of estimate	1990	1995	2000	2005
DICE	$10	$7	$6	
FUND			$9 to $23	-$15 to $110
PAGE		$12 to $60		$4 to $51

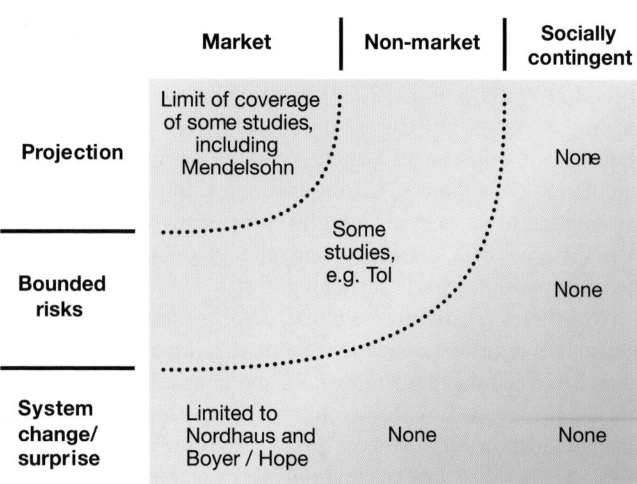

Market
Non-market
Socially contingent
Projection
Limit of coverage of some studies, including Mendelsohn
None
Some studies, e.g. Tol
Bounded risks
None
System change/ surprise
Limited to Nordhaus and Boyer / Hope
None
None

Figure 20.4. *Coverage of studies that compute estimates of the social cost of carbon against sources of climate-related risk. Coverage of most studies is limited to market-based sectors, and few of them move beyond the upper left corner to include bounded risks and abrupt system change. Source: Watkiss et al., 2005.*

20.6.2 Spatially-explicit methods: global impacts of climate change

Warren (2006) and Hitz and Smith (2004) observe that most impact assessments are conducted at the local scale. It is therefore extremely difficult to estimate impacts across the global domain from these localised studies. A small number of studies have used geographically-distributed impacts models to estimate the impacts of climate change across the global domain. The "Fast Track" studies (Arnell, 2004; Nicholls, 2004; Arnell et al., 2002; Levy et al., 2004; Parry et al., 2004; Van Lieshout et al., 2004) used a consistent set of scenarios and assumptions to estimate the effects of scenarios based on the HadCM3 climate model on water resource availability, food security, coastal flood risk, ecosystem change and exposure to malaria. Schroeter et al. (2005) used a similar approach in the ATEAM project to tabulate impacts across Europe using scenarios constructed from a larger number of climate models.

Both these sets of studies used a wide range of metrics that varied across sectors. Table 20.4 summarises some of the global-scale impacts of defined climate-change scenarios. Although the precise numbers depend on the climate model used and some key assumptions (particularly the effect of increased CO_2 concentrations on crop productivity), it is clear that the future impacts of climate change are dependent not only on the rate of climate change, but also on the future social, economic and technological state of the world. Impacts are greatest under an A2 world, for example, not because the climate change is greatest but because there are more people to be impacted. Impacts also vary regionally and Table 20.5 summarises impacts by major world region. The assumed effect of CO_2 enrichment on crop productivity has a major effect on estimated changes in population at risk of hunger (Chapter 5, Section 5.4.7).

Table 20.6 compares the global impacts of a 1% annual increase in CO_2 concentrations (i.e., the IS92a scenario, see IPCC, 1992) with the impacts of emissions trajectories stabilising at 750 (S750) and 550 (S550) ppm (Arnell et al., 2002). The results are not directly comparable to those reported in Table 20.4, because different population assumptions, methodologies and indicators were employed in their preparation. Nevertheless, the results suggest that aiming for stabilisation at 750 ppm has a relatively small effect on impacts in most sectors in comparison with 550 ppm stabilisation. The S550 pathway has a greater apparent impact on exposure to hunger because higher CO_2 concentrations under S750 result in a greater increase in crop productivity (but again, note that CO_2-enrichment effects are highly uncertain).

Each of these tables present *indicators* of impact which ignore adaptations that will occur over time. They can therefore be seen as indicative of the challenge to be overcome by adaptations to offset some of the impacts of climate change. Incorporating adaptation into global-scale assessments of the impacts of climate change is currently difficult for a number of reasons (including diversity of circumstances, diversity of potential objectives of adaptation, diversity of ways of meeting adaptation objectives and uncertainty over the effectiveness of adaptation options) and remains an area where more research is needed.

Table 20.4. *Global-scale impacts of climate change by 2080.*

	Climate and socio-economic scenario			
	A1FI	A2	B1	B2
Global temperature change (°C difference from the 1961-1990 period)	3.97	3.21 to 3.32	2.06	2.34 to 2.4
Millions of people at increased risk of hunger (Parry et al., 2004); no CO_2 effect	263	551	34	151
Millions of people at increased risk of hunger (Parry et al., 2004); with maximum direct CO_2 effect	28	-28 to -8	12	-12 to +5
Millions of people exposed to increased water resources stress (Arnell, 2004)	1256	2583 to 3210	1135	1196 to 1535
Additional numbers of people (millions) flooded in coastal floods each year, with lagged evolving protection (Nicholls, 2004)	7	29	2	16

Note: change in climate derived from the HadCM3 climate model. Impacts are compared to the situation in 2080 with no climate change. The range of impacts under the SRES A2 and B2 scenarios (Nakićenović and Swart, 2000) represents the range between different climate simulations. The figures for additional millions of people flooded in coastal floods assumes a low rate of subsidence and a low rate of population concentration in the coastal zone.

Table 20.5. *Regional-scale impacts of climate change by 2080 (millions of people).*

	Population living in watersheds with an increase in water-resources stress (Arnell, 2004)				Increase in average annual number of coastal flood victims (Nicholls, 2004)				Additional population at risk of hunger (Parry et al., 2004)[1] Figures in brackets assume maximum direct CO_2-enrichment effect			
	Climate and socio-economic scenario:											
	A1	A2	B1	B2	A1	A2	B1	B2	A1	A2	B1	B2
Europe	270	382-493	233	172-183	1.6	0.3	0.2	0.3	0	0	0	0
Asia	289	812-1197	302	327-608	1.3	14.7	0.5	1.4	78 (6)	266 (-21)	7 (2)	47 (-3)
North America	127	110-145	107	9-63	0.1	0.1	0	0	0	0	0	0
South America	163	430-469	97	130-186	0.6	0.4	0	0.1	27 (1)	85 (-4)	5 (2)	15 (-1)
Africa	408	691-909	397	492-559	2.8	12.8	0.6	13.6	157 (21)	200 (-2)	23 (8)	89 (-8)
Australasia	0	0	0	0	0	0	0	0	0	0	0	0

Note: change in climate derived from the HadCM3 climate model. Impacts are compared to the situation in 2080 with no climate change. The range of impacts under the SRES A2 and B2 scenarios (Nakićenović and Swart, 2000) represents the range between different climate simulations. The figures for additional millions of people flooded in coastal floods assumes a low rate of subsidence and a low rate of population concentration in the coastal zone.

[1] Analysis of project results carried out for this table.

Table 20.6. *Global-scale impacts under unmitigated and stabilisation pathways. Source: Arnell et al., 2002.*

	Unmitigated	2050 Scenario: S750	S550	Unmitigated	2050 Scenario: S750	S550
Approximate equivalent CO_2 concentration (ppm)	520	485	458	630	565	493
Approximate global temperature change (°C difference from 1961 to 1990)	2.0	1.3	1.1	2.9	1.7	1.2
Area potentially experiencing vegetation dieback (million km^2)	1.5 to 2.7	2	0.7	6.2 to 8	3.5	1.3
Millions of people exposed to increased water stress	200 to 3200	2100	1700	2830 to 3440	2920	760
Additional people flooded in coastal floods (millions/year)	20	13	10	79 to 81	21	5
Population at increased risk of hunger (millions)	-3 to 9	7	5	69 to 91	16	43

Note: climate scenarios based on HadCM2 simulations: the range with unmitigated emissions reflects variation between ensemble simulations.

Aggregation of impacts to regional and global scales is another key problem with such geographically-distributed impact assessments. Tables 20.4 to 20.6, for example, keep track of people living in watersheds who will face increased water-related stress. Of course, many people live in watersheds where climate change increases runoff and therefore may apparently see *reduced* water-related stress (if they see increased risk of flooding). Simply calculating the 'net' impact of climate change, however, is complicated, particularly where 'winners' and 'losers' live in different geographic regions, or where 'costs' and 'benefits' are not symmetrical. Watersheds with an increase in runoff, for example, are concentrated in east Asia, while watersheds with reduced runoff are much more widely distributed. Similarly, the adverse effects felt by 100 million people exposed to increased water stress could easily outweigh the 'benefits' of 100 million people with reduced stress.

The Defra Fast Track and ATEAM studies both describe impacts along defined scenarios, so it is difficult to infer the effects of different rates or degrees of climate change on different socio-economic worlds. A more generalised approach applies a wide range of climate scenarios representing different rates of change to estimate impacts for specific socio-economic contexts. Leemans and Eickhout (2004), for example, show that most species, ecosystems and landscapes would be impacted by increases of global temperature between 1 and 2°C above 2000 levels. Arnell (2006) showed that an increase in temperature of 2°C above the 1961 to 1990 mean by 2050 would result in between 550 and 900 million people suffering an increase in water-related stress in both the SRES (Special Report on Emissions Scenarios, Nakićenović and Swart, 2000) A1 and B1 worlds. In this case, the range between estimates represents the effect of different changes in rainfall patterns for a 2°C warming.

20.7 Implications for regional, sub-regional, local and sectoral development; access to resources and technology; equity

The first sub-section here addresses issues of equity and access to resources as measured by the likelihood of meeting Millennium Development Targets by 2015 and Millennium Development Goals until the middle of this century. Vulnerability to climate change is unlikely to be the dominant cause of trouble for most nations as they try to reach the 2015 Targets. However, an assortment of climate-related vulnerabilities will seriously impede progress in achieving the mid-century goals. The second sub-section considers the range of these vulnerabilities across regions and sectors in 2050 and 2100 before the last offers portraits of the global distribution of vulnerability with and without enhanced adaptive capacity and/or mitigation efforts.

20.7.1 Millennium Development Goals – a 2015 time slice

The Millennium Development Goals (MDGs) are the product of international consensus on a framework by which nations can assess tangible progress towards sustainable development; they are enumerated in Table 20.7. UN (2005) provides the most current documentation of the 8 MDGs, the 11 specific targets for progress by 2015 or 2020 and the 32 quantitative indicators that are being used as metrics. This chapter has made the point that sustainable development and adaptive capacity for coping with climate change have common determinants. It is easy, therefore, to conclude that climate change has the potential to affect the progress of nations and societies towards sustainability. MA (2005) supports this conclusion. Climate-change impacts on the timing, flow and amount of available freshwater resources could, for example, affect the ability of developing countries to increase access to potable water: Goal #7, Target #10, Indicator #30 (UN, 2005). It is conceivable that climate change could have measurable consequences, in some parts of the world at least, on the indicators of progress on food security: Goal #1, Target #2, Indicators #4 and #5 (UN, 2005). Climate-change impacts could possibly affect one indicator in Goal #6 (prevalence and death rates associated with malaria), over the medium term (UN, 2005). The list can be extended.

The anthropogenic drivers of climate change, *per se*, affect MDG indicators directly in only two ways: in terms of energy use per dollar GDP and CO_2 emissions per capita. While climate change may, with high confidence, have the potential for substantial effects on aspects of sustainability that are important for the MDGs, the literature is less conclusive on whether the metrics themselves will be sensitive to either the effects of climate change or to progress concerning its drivers, especially in the near term. The short-term targets of the MDGs (i.e., the 2015 to 2020 Targets) will be difficult to reach in any case. While climate impacts have now been observed with some levels of confidence in some places, it will be difficult to blame climate change for limited progress towards the Millennium Development Targets.

In the longer term, Arrow et al. (2004) argue that adaptation decisions can reduce the effective investment available to reach the MDGs. They thereby raise the issue of opportunity costs: perhaps investment in climate adaptation might retard efforts to achieve sustainable development. Because the determinants of adaptive capacity and of sustainable development overlap significantly; however, (see Section 20.2) it is also possible that a dollar spent on climate adaptation could strengthen progress towards sustainable development.

Whether synergistic effects or trade offs will dominate interactions between climate impacts, adaptation decisions and sustainable development decisions depend, at least in part, on the particular decisions that are made. Decisions on how countries will acquire sufficient energy to sustain growing demand will, for example, play crucial roles in determining the sustainability of economic development. If those demands are met by increasing fossil fuel combustion, then amplifying feedbacks to climate change should be expected. There are some indications that this is now occurring. Per capita emissions of CO_2 in developing countries rose from 1.7 tonnes of CO_2 per capita in 1990 to 2.1 tonnes per capita in 2002; they remained, though, far short of the 12.6 tonnes of CO_2 per capita consumed in developed countries (UN, 2005). Resources devoted to expanding fossil fuel generation could, therefore, be seen as a source of expanded climate-change impacts. On the other hand, investments in forestry and agricultural sectors designed to preserve and enhance soil fertility in support of improved food security MDGs (e.g., Goal #1) might have synergies for climate mitigation (through carbon sequestration) and for adaptation (because higher economic returns for local communities could be invested in adaptation). It is simply impossible to tell, *a priori*, which effect will dominate. Each situation must be analysed qualitatively and quantitatively.

These complexities make it clear that not all development paths will be equal with respect to either their consequences for climate change or their consequences for adaptive capacity. Moreover, the Millennium Ecosystem Assessment (MA, 2005) and others (e.g., AfDB et al., 2004) argue that climate change will be a significant hindrance to meeting the MDGs over the long term. There is no discrepancy here because stresses from climate change will grow over time. Some regions and countries are already lagging in their progress towards the MDGs and these tend to be in locations where climate vulnerabilities over the 21st century are likely to be high. For example, the proportion of land area covered by forests fell between 1990 and 2000 in sub-Saharan Africa, South-East Asia and Latin America and the Caribbean, while it appeared to stabilise in developed

Table 20.7. *The Millennium Development Goals.*

1. Eradicate extreme poverty and hunger
2. Achieve universal primary education
3. Promote gender equality and empower women
4. Reduce child mortality
5. Improve maternal health
6. Combat HIV/AIDS, malaria and other diseases
7. Ensure environmental sustainability
8. Develop a global partnership for development

Source: http://www.un.org/milleniumgoals/documents.html

countries (UN, 2005). Energy use per unit of GDP fell between 1990 and 2002 in both developed and developing regions, but developed regions remained approximately 10% more efficient than developing regions (UN, 2005). In short, regions where ecosystem services and contributions to human well-being are already being eroded by multiple external stresses are more likely to have low adaptive capacity.

20.7.2 Sectoral and regional implications

The range of increase in global mean temperature that could be expected over the next several centuries is highly uncertain. The compounding diversity in the regional patterns of temperature change for selected changes in global mean temperature is depicted elsewhere in IPCC (2007b, Figure SPM.6); so, too, are illustrations of geographic diversity in changes in precipitation and model disagreement about even the sign of this change (IPCC, 2007b, Figure SPM.7). Earlier sections of this chapter have also underscored the difficulty in anticipating the development of adaptive capacity and the ability of communities to take advantage of the incumbent opportunities. Despite all of this complexity, however, it is possible to offer some conclusions about vulnerability across regions and sectors as reported throughout this report.

Locating the anticipated impacts of climate change on a map is perhaps the simplest way to see this point. Figure 9.5, for example, shows the spatial distribution of the projected impacts that are reported for Africa in Chapter 9. The power of maps like this lies in their ability to show how the various manifestations of climate change can be geographically concentrated. It is clear, as a result, that climate change can, by virtue of its multiple dimensions, be its own source of multiple stresses. It follows immediately that vulnerability to climate change can easily be amplified (in the sense that total vulnerability to climate change is greater than the sum of vulnerabilities to specific impacts) in regions like the south-eastern coast of Africa and Madagascar.

Maps of this sort do not, however, capture sensitivities to larger indices of climate change (such as increases in global mean temperature); nor do they not offer any insight into the timing of increased vulnerabilities.

Tables 20.8 and 20.9 address these deficiencies by summarising estimated impacts at global and regional scales against a range of changes in global average temperature. Each entry is drawn from earlier chapters in this report, and assessed levels of confidence are indicated. The entries have been selected by authors of the chapters and the selection is intended to illustrate impacts that are important for human welfare. The criteria for judging this importance include the magnitude, rate, timing and persistence/irreversibility of impacts, and the capacity to adapt to them. Where possible, the entries give an indication of impact trend and its quantitative level. In a few cases, quantitative measures of impact have now been estimated for different amounts of climate change, thus pointing toward different levels of the same impact that might be avoided by not exceeding given amounts of global temperature change.

The time dimension is captured by the bars drawn at the top of Table 20.8; they indicate the range of global average temperature increase that could be expected during the 2020s, the 2050s and the 2080s among the SRES collection of unmitigated scenarios as well as a range of alternative stabilisation pathways (Nakićenović and Swart, 2000). The real message to be drawn from their inclusion is that no temperature threshold associated with any subjective judgment of what might constitute 'dangerous' climate change can be guaranteed by anything but the most stringent of mitigation interventions, at least not on the basis of current knowledge. Moreover, there is an estimated commitment to warming of 0.6°C due to past emissions, from which impacts must be expected, regardless of any future efforts to reduce emissions in the future.

20.7.3 The complementarity roles of mitigation and enhanced adaptive capacity

IPCC (2001a) focused minimal attention on the co-benefits of mitigation and adaptation, but this report has added a chapter-length assessment of current knowledge at the nexus of adaptation and mitigation. An emphasis on constructing a "portfolio of adaptation and mitigation actions" has emerged (Chapter 18, Sections 18.4 and 18.7). Moreover, the capacities to respond in either dimension are supported by 'similar sets of factors' (Chapter 18, Section 18.6). These factors are, of course, themselves determined by underlying socio-economic and technological development paths that are location and time specific.

Yohe et al. (2006a, b) offer suggestive illustrations of potential synergies within the adaptation/mitigation portfolio; complementarity in the economic sense that one makes the other more productive. Figures 20.5 and 20.6 display the geographic distribution of these synergies in terms of a national vulnerability index with and without mitigation, and with and without enhanced adaptive capacity by 2050 and 2100, respectively. Vulnerabilities that were assigned to specific countries on the basis of a vulnerability index derived from national estimates of adaptive capacity provided by Brenkert and Malone (2005) and the geographic distribution of temperature change derived from a small ensemble of global circulation models. The upper left panels of Figures 20.5 and 20.6 present geographical distributions of vulnerability in 2050 and 2100, respectively, along the SRES A2 emissions scenario with a climate sensitivity of 5.5°C under the limiting assumption that adaptive capacities are fixed at current levels; global mean temperature climbs by 1.6°C and 4.9°C above 1990 levels by 2050 and 2100, respectively. These two panels are benchmarks of maximum vulnerability against which other options can be assessed. Notice that most of Africa plus China display the largest vulnerabilities in 2050 and that nearly every nation displays extreme vulnerability by 2100. A2 was chosen for illustrative clarity with reference to temperature change only. Moreover, none of the interpretations depend on the underlying storyline of the A2 scenario; Yohe et al. (2006b) describes comparable results for other scenarios.

The upper right panels present comparable geographic distributions under the assumption that adaptive capacity improves everywhere with special emphasis on developing countries; their capacities are assumed to advance to the current global mean by 2050 and 2100 for Figures 20.5 and 20.6,

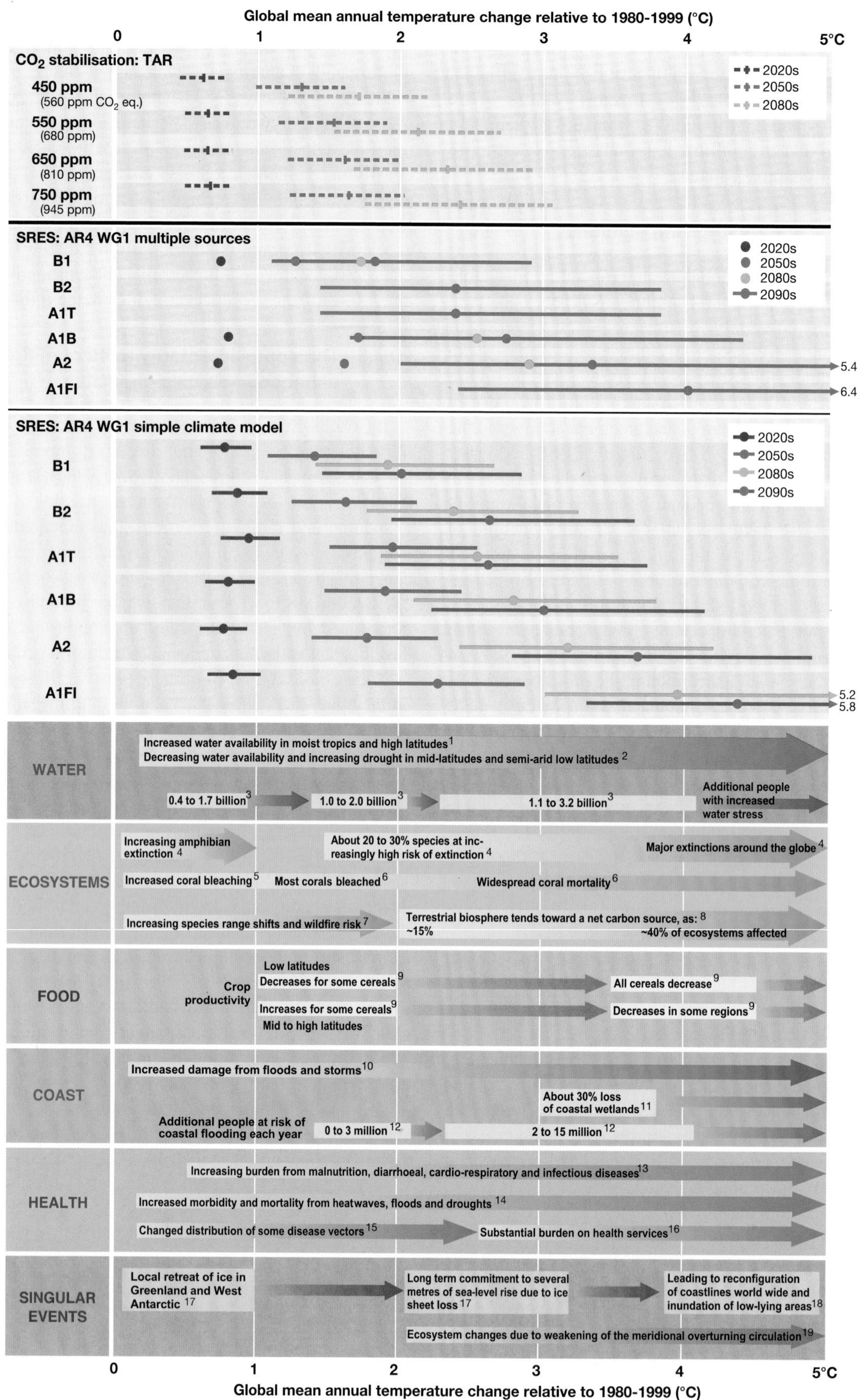

Table 20.8. *Examples of global impacts projected for changes in climate (and sea level and atmospheric CO_2 where relevant) associated with different amounts of increase in global average surface temperature in the 21st century. This is a selection of some estimates currently available. All entries are from published studies in the chapters of the Assessment. (Continues below Table 20.9)*

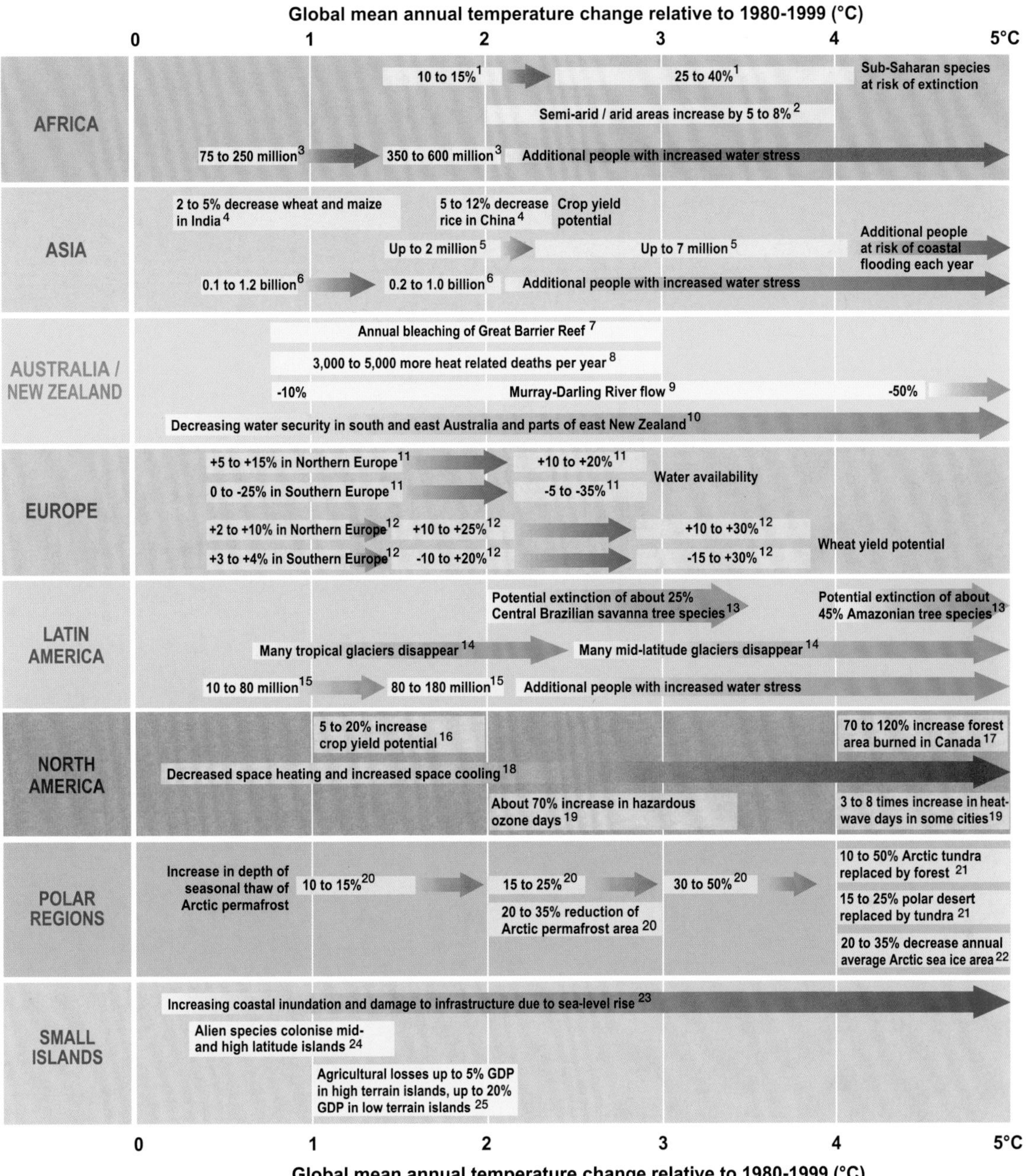

Table 20.9. *Examples of regional impacts. See caption for Table 20.8.*

Table 20.8. (cont.) *Edges of boxes and placing of text indicate the range of temperature change to which the impacts relate. Arrows between boxes indicate increasing levels of impacts between estimations. Other arrows indicate trends in impacts. All entries for water stress and flooding represent the additional impacts of climate change relative to the conditions projected across the range of SRES scenarios A1FI, A2, B1 and B2. Adaptation to climate change is not included in these estimations. For extinctions, 'major' means ~40 to ~70% of assessed species.*

The table also shows global temperature changes for selected time periods, relative to 1980-1999, projected for SRES and stabilisation scenarios. To express the temperature change relative to 1850-1899, add 0.5°C. More detail is provided in Chapter 2 [Box 2.8]. Estimates are for the 2020s, 2050s and 2080s, (the time periods used by the IPCC Data Distribution Centre and therefore in many impact studies) and for the 2090s. SRES-based projections are shown using two different approaches. ***Middle panel:*** *projections from the WGI AR4 SPM based on multiple sources. Best estimates are based on AOGCMs (coloured dots). Uncertainty ranges, available only for the 2090s, are based on models, observational constraints and expert judgement.* ***Lower panel:*** *best estimates and uncertainty ranges based on a simple climate model (SCM), also from WGI AR4 (Chapter 10).* ***Upper panel:*** *best estimates and uncertainty ranges for four CO_2-stabilisation scenarios using an SCM. Results are from the TAR because comparable projections for the 21st century are not available in the AR4. However, estimates of equilibrium warming are reported in the WGI AR4 for CO_2-equivalent stabilisation[a]. Note that equilibrium temperatures would not be reached until decades or centuries after greenhouse gas stabilisation.*

Table 20.8. Sources: 1, 3.4.1; **2,** 3.4.1, 3.4.3; **3,** 3.5.1; **4,** 4.4.11; **5,** 4.4.9, 4.4.11, 6.2.5, 6.4.1; **6,** 4.4.9, 4.4.11, 6.4.1; **7,** 4.2.2, 4.4.1, 4.4.4 to 4.4.6, 4.4.10; **8,** 4.4.1, 4.4.11; **9,** 5.4.2; **10,** 6.3.2, 6.4.1, 6.4.2; **11,** 6.4.1; **12,** 6.4.2; **13,** 8.4, 8.7; **14,** 8.2, 8.4, 8.7; **15,** 8.2, 8.4, 8.7; **16,** 8.6.1; **17,** 19.3.1; **18,** 19.3.1, 19.3.5; **19,** 19.3.5
Table 20.9. Sources: 1, 9.4.5; **2,** 9.4.4; **3,** 9.4.1; **4,** 10.4.1; **5,** 6.4.2; **6,** 10.4.2; **7,** 11.6; **8,** 11.4.12; **9,** 11.4.1, 11.4.12; **10,** 11.4.1, 11.4.12; **11,** 12.4.1; 12, 12.4.7; **13,** 13.4.1; **14,** 13.2.4; **15,** 13.4.3; **16,** 14.4.4; **17,** 5.4.5, 14.4.4; **18,** 14.4.8; **19,** 14.4.5; **20,** 15.3.4, **21,** 15.4.2; **22,** 15.3.3; **23,** 16.4.7; **24,** 16.4.4; **25,** 16.4.3

[a] Best estimate and likely range of equilibrium warming for seven levels of CO_2-equivalent stabilisation from WGI AR4 are: 350 ppm, 1.0°C [0.6–1.4]; 450 ppm, 2.1°C [1.4–3.1]; 550 ppm, 2.9°C [1.9–4.4]; 650 ppm, 3.6°C [2.4–5.5]; 750 ppm, 4.3°C [2.8–6.4]; 1,000 ppm, 5.5°C [3.7–8.3] and 1,200 ppm, 6.3°C [4.2–9.4].

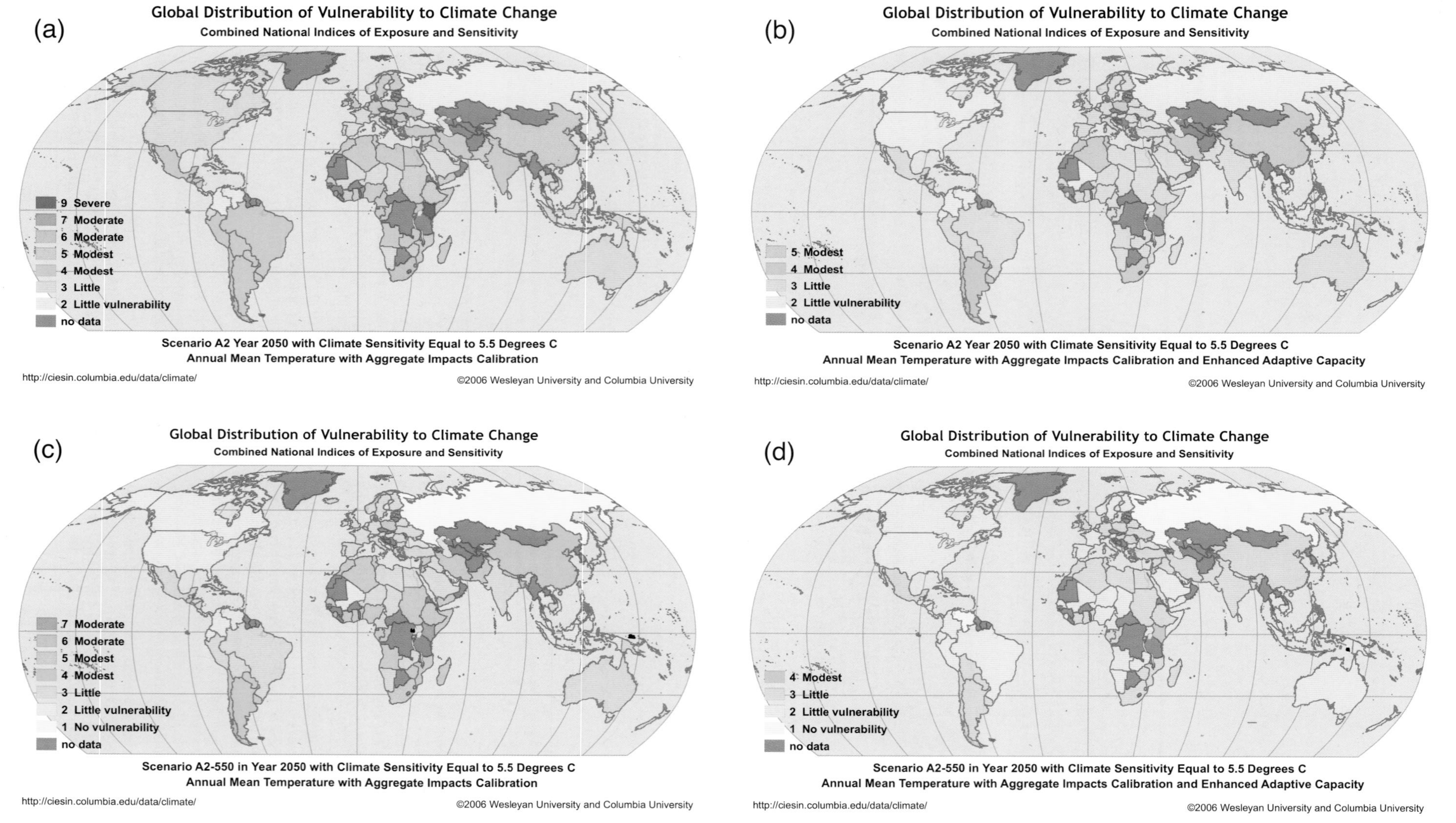

Figure 20.5. *Geographical distribution of vulnerability in 2050 with and without mitigation along an SRES A2 emissions scenario with a climate sensitivity of 5.5°C. (a) portrays vulnerability with a static representation of current adaptive capacity. (b) shows vulnerability with enhanced adaptive capacity worldwide. (c) displays the geographical implications of mitigation designed to cap effective atmospheric concentrations of greenhouse gases at 550 ppm. (d) offers a portrait of the combined complementary effects of mitigation to the same 550 ppm concentration limit and enhanced adaptive capacity. Source: Yohe et al., 2006b.*

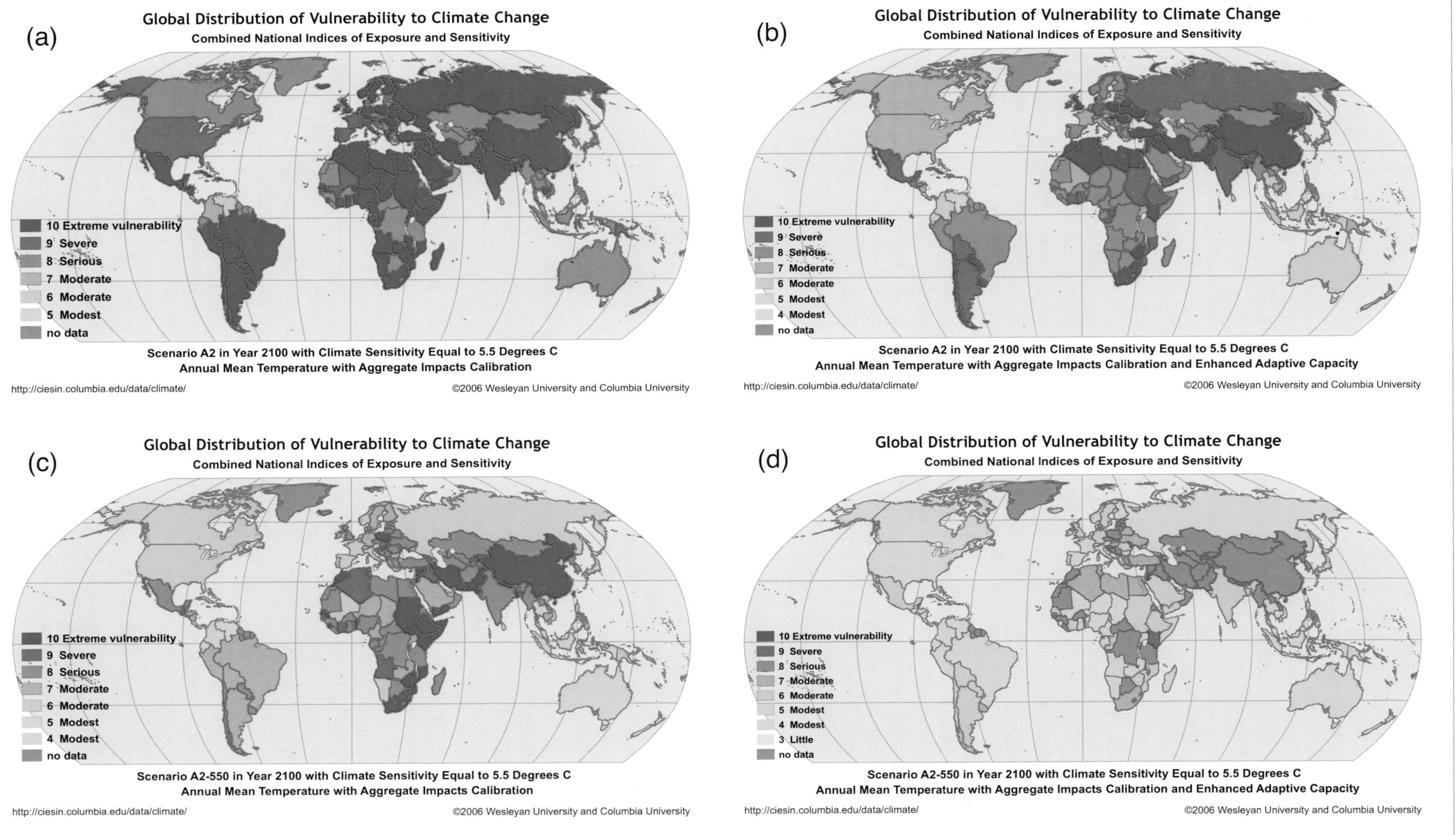

Figure 20.6. *Geographical distribution of vulnerability in 2100 with and without mitigation along an SRES A2 emissions scenario with a climate sensitivity of 5.5°C. (a) portrays vulnerability with a static representation of current adaptive capacity. (b) shows vulnerability with enhanced adaptive capacity worldwide. (c) displays the geographical implications of mitigation designed to cap effective atmospheric concentrations of greenhouse gases at 550 ppm. (d) offers a portrait of the combined complementary effects of mitigation to the same 550 ppm concentration limit and enhanced adaptive capacity. Source: Yohe et al., 2006b.*

respectively. Significant improvement is seen in 2050, but adaptation alone still cannot reduce extreme vulnerability worldwide in 2100. The lower panels present the effect of limiting atmospheric concentrations of greenhouse gases to 550 ppm along least-cost emissions trajectories; global mean temperature is 1.3°C and 3.1°C higher than 1990 levels by 2050 and 2100 in this case. In the lower left panels, adaptive capacity is again held constant at current levels. Mitigation reduces vulnerability across much of the world in 2050, but extreme vulnerability persists in developing countries and threatens developed countries in 2100. Mitigation alone cannot overcome climate risk. Finally, the lower right panels show the combined effects of investments in enhanced adaptive capacity and mitigation. Climate risks are substantially reduced in 2050, but significant vulnerabilities reappear by 2100. Developing countries are still most vulnerable. Developed countries are also vulnerable, but they see noticeable benefits from the complementary effects of the policy portfolio. These results suggest that global mitigation efforts up to 2050 would benefit developing countries more than developed countries when combined with enhanced adaptation. By 2100, however, climate change would produce significant vulnerabilities ubiquitously even if a relatively restrictive concentration cap were implemented in combination with a programme designed to enhance adaptive capacity significantly.

20.8 Opportunities, co-benefits and challenges for adaptation

This section extends some of the ideas outlined in Najam et al. (2003); they focus on mainstreaming climate-change adaptation into planning and development decisions with particular emphasis on participatory processes.

20.8.1 Challenges and opportunities for mainstreaming adaptation into national, regional and local development processes

An international opportunity for mainstreaming adaptation into national, regional and local development processes has recently emerged with the community approach to disaster management adopted by the World Conference on Disaster Reduction held in Kobe, Hyogo, Japan in January 2005 (Hyogo Declaration, 2005). This approach is described in, for example, UNCRD (2003). The results of an action research and pilot activity undertaken during 2002 to 2004 (APJED, 2004) have been reported, albeit on a limited scale in Bangladesh, India and Nepal, with support from World Meteorological Organization (WMO) and Global Water Partnership (GWP). The pilot activity focused on community approaches to flood management, and found that a community flood management committee formed in a local area, working in co-operation with the relevant local government and supported by national government policy, can significantly reduce adverse consequences of floods. There are, however, many challenges. Progress in carrying out analyses and identifying what needs to be and can be done can be documented, but action on the ground to mainstream adaptation to climate change remains limited, particularly in the least developed countries. National policy making in this context remains a major challenge that can only be met with increased international funding for adaptation and disaster management (Ahmad and Ahmed, 2002; Jegillos, 2003; Huq et al., 2006).

Socio-economic and even environmental policy agendas of developing countries do not yet prominently embrace climate change (Beg et al., 2002) even though most developing countries participate in various international protocols and conventions relating to climate change and sustainable development and most have adopted national environmental conservation and natural disaster management policies. Watson International Scholars of the Environment (2006) has offered some suggestions for improved mainstreaming within multilateral environmental agreements; they include fostering links with poverty reduction and increasing support designed to engage professionals, researchers and governments at local levels in developing countries more directly.

Even as economic growth is pursued, progress towards health, education, training and access to safe water and sanitation, and other indicators of social and environmental progress including adaptive capacity remains a significant challenge. It can be addressed through appropriate policies and commitment to ending poverty (WSSD, 2002; Sachs, 2005). Strengthened linkages between government and people, and the consequent capacity building at local levels, are key factors for robust progress towards sustainability at the grassroots (Jegillos, 2003). Social and environmental (climate change) issues are, however, often left resource-constrained and without effective institutional support when economic growth takes precedence (UNSEA, 2005).

20.8.2 Participatory processes in research and practice

Participatory processes can help to create dialogues that link and mutually instruct researchers, practitioners, communities and governments. There are, however, challenges in applying these processes as a methodology for using dialogue and narrative (i.e., communication of quantitative and qualitative information) to influence social learning and decision-making, including governance.

Knowledge about climate-change adaptation and sustainable development can be translated into public policy through processes that generate usable knowledge. The idea of usable knowledge in climate assessments stems from the experiences of national and international bodies (academies, boards, committees, panels, etc.) that offer credible and legitimate information to policymakers through transparent multi-disciplinary processes (Lemos and Morehouse, 2005). It requires the inclusion of local knowledge, including indigenous knowledge (see Box 20.1), to complement more formal technical understanding generated through scientific research and the consideration of the role that institutions and governance play in the translation of scientific information into effective action.

Box 20.1. Role of local and indigenous knowledge in adaptation and sustainability research

Research on indigenous environmental knowledge has been undertaken in many countries, often in the context of understanding local oral histories and cultural attachment to place. A survey of research during the 1980s and early 1990s was produced by Johnson (1992). Reid et al. (2006) outline the many technical and social issues related to the intersection of different knowledge systems, and the challenge of linking the scales and contexts associated with these forms of knowledge. With the increased interest in climate change and global environmental change, recent studies have emerged that explore how indigenous knowledge can become part of a shared learning effort to address climate-change impacts and adaptation, and its links with sustainability. Some examples are indicated here.

Sutherland et al. (2005) describe a community-based vulnerability assessment in Samoa, addressing both future changes in climate-related exposure and future challenges for improving adaptive capacity. Twinomugisha (2005) describes the dangers of not considering local knowledge in dialogues on food security in Uganda.

A scenario-building exercise in Costa Rica has been undertaken as part of the Millennium Ecosystem Assessment (MA, 2005). This was a collaborative study in which indigenous communities and scientists developed common visions of future development. Two pilot five-year storylines were constructed, incorporating aspects of coping with external drivers of development (Bennett and Zurek, 2006). Although this was not directly addressing climate change, it demonstrates the potential for joint scenario-building incorporating different forms of knowledge.

In Arctic Canada, traditional knowledge was used as part of an assessment which recognised the implications of climate change for the ecological integrity of a large freshwater delta (NRBS, 1996). In another case, an environmental assessment of a proposed mine was produced through a partnership with governments and indigenous peoples. Knowledge to facilitate sustainable development was identified as an explicit goal of the assessment, and climate-change impacts were listed as one of the long-term concerns for the region (WKSS, 2001).

Vlassova (2006) describes results of interviews of indigenous peoples of the Russian North on climate and environmental trends within the Russian boreal forest. Additional examples from the Arctic are described in ACIA (2005), Reidlinger and Berkes (2001), Krupnik and Jolly (2002), Furgal et al. (2006) and Chapter 15.

Social learning of complex issues like climate change emerges through consensus that includes both scientific discourse and policy debate. In the case of climate change, participatory processes encourage local practitioners from climate-sensitive endeavours (water management, land-use planning, etc.) to become engaged so that past experiences can be included in the study of (and the planning for) future climate change and development pressures. Processes designed to integrate various dimensions of knowledge about how regional resource systems operate are essential; so is understanding of how resource systems are affected by biophysical and socio-economic forces including a wide range of possible future changes in climate. This requirement has led to increased interest in a number of participatory processes like participatory integrated assessment (PIA) and participatory mapping (using, for example, specially designed geographic information systems – GIS).

PIA is an umbrella term describing approaches in which non-researchers play an active role in integrated assessment (Rotmans and van Asselt, 2002). Participatory processes can be used to facilitate the integration of biophysical and socio-economic aspects of climate-change adaptation and development by creating opportunities for shared experiences in learning, problem definition and design of potential solutions (Hisschemöller et al., 2001). Van Asselt and Rijkens-Klomp (2002) identify several approaches, including methods for mapping diversity of opinion (e.g., focus groups, participatory modelling) and reaching consensus (e.g., citizens' juries, participatory planning). Kangur (2004) reported on a recent exercise on water policy that employed citizens' juries. PIA has also been used to facilitate the development of integrated models (e.g., Turnpenny et al., 2004) and to use models to facilitate policy dialogue (e.g., van de Kerkhof, 2004).

Participatory mapping is a process by which local information, including indigenous knowledge, is incorporated into information management systems (Corbett et al., 2006). Ranging from paper to GIS, it is becoming more popular, and it has contributed to the increased application of Participatory Rural Appraisal (PRA) and Rapid Rural Appraisal (RRA) as techniques to support rural development (Chambers, 2006). Maps have displayed natural resources, social patterns and mobility, and they have been used to identify landscape changes, tenure, boundaries and places of cultural significance (Rambaldi et al., 2006). With the advent of modern GIS technologies, concerns have been raised regarding disempowerment of communities from lack of training. Questions related to who owns the maps and to who controls their use have also been raised (Corbett et al., 2006; Rambaldi et al., 2006).

The long-term sustainability of dialogue processes is critical to the success of participatory approaches. For PIA, PRA, participatory GIS and similar processes to be successful as shared learning experiences, they have to be inclusive and transparent. Haas (2004) describes examples of experiences in social learning on sustainable development and climate change, noting the importance of sustaining the learning process over the long term, and maintaining distance between science and policy while still promoting focused science-policy interactions. Applications of focus group and other techniques for stakeholder engagement are described for several studies in Europe (Welp et al., 2006) and Africa (Conde and Lonsdale, 2004). However, there has been particular concern regarding its application within development processes and hazard management in poor countries. Cooke and Kothari (2001) and Garande and Dagg (2005) document some problems, including hindering empowerment of local scale interests, reinforcing existing power structures and constraining how local knowledge is expressed. Barriers include uneven gains from cross-scale interactions (Adger et al., 2005; Young, 2006) and increased responsibility without increased capacity (Allen, 2006). There can be difficulties in reaching consensus on identifying and engaging participants (Bulkeley and Mol, 2003; Parkins and Mitchell, 2005), and in interpreting the results of dialogue within variations in cultural and epistemological contexts (e.g., Huntington et al., 2006). There are also challenges in measuring the quality of dialogue (debate, argument), particularly the transparency of process, promotion of learning and indicators of influence (van de Kerkhof, 2004; Rowe and Frewer, 2000).

Participatory governance is part of a growing global movement to decentralise many aspects of natural resources management. Hickey and Mohan (2004) offer several examples of the convergence of participatory development and participatory governance with empowerment for marginalised communities. Other examples include agrarian reform in the Philippines, the Popular Participation Law in Bolivia (Schneider, 1999; Iwanciw, 2004) and the appointment of an 'exploratory committee' for addressing water resources concerns in Nagoya, Japan (Kabat et al., 2002). In each case, the point is to improve access to resources and enhance social capital (Larson and Ribot, 2004a and 2004b). Unfortunately, broadening decision-making can work to exacerbate vulnerabilities. For example, there have been cases emerging from Latin America describing difficulties in building national adaptive capacity as national and local institutions change their roles in governance. Although the language of sustainability and shared governance is widely accepted, obtaining benefits from globalisation in enhanced adaptive capacity is difficult (Eakin and Lemos, 2006).

Dialogue processes in assessment and appraisal are becoming important tools in the support of participatory processes. Although they may be seen as relatively similar activities, PIA and PRA have different mandates. The latter is directly within a policy process (selecting among development options), while the former is a research method that assesses complex problems (e.g., environmental impact of development, climate-change impacts/adaptation), producing results that can have policy implications. This chapter's discussion on PIA is offered as a complement to integrated modelling results reported in Sections 20.6 and 20.7 to suggest that PIA may assist in providing regional-scale technical support to match the scale of information needs of decentralised governance.

An agricultural example of a PIA of climate-change adaptation can be found in the eastern United Kingdom (Lorenzoni et al., 2001). Adaptation options are identified (e.g., shifting cultivation times, modifying soil management to improve water retention and avoid compaction), but questions about how a climate component can be built into the way non-climate issues are currently addressed emerge. Long-term strategies may have to include greater fluctuations in crop yields across a region; as a result, farm operations may have to diversity if they are to maintain incomes and employment. The compartmentalisation of regional decision-making is seen as a barrier to encouraging more sustainable land management over the periods in which climate change evolves. In an example from Canada, Cohen and Neale (2006) and Cohen et al. (2004) illustrate the linkages between water management and scenarios of population growth and climate change in the Okanagan region (see also Chapter 3, Box 3.1). Planners in one district have responded by incorporating adaptation to climate change into long-term water plans (Summit Environmental Consultants Ltd., 2004) even though governance-related obstacles to proactive implementation of innovative measures to manage water demand have appeared in the past (Shepherd et al., 2006).

A comprehensive understanding of the implications of extreme climate change requires an in-depth exploration of the perceptions and reactions of the affected stakeholder groups and the lay public. Toth and Hizsnyik (2005) describe how participatory techniques might be applied to inform decisions in the context of possible abrupt climate change. Their project has studied one such case, the collapse of the West Antarctic Ice Sheet and a subsequent 5 to 6 m sea-level rise. Possible methods for assessing the societal consequences of impacts and adaptations include simulation-gaming techniques, a policy exercise approach, as well as directed focus-group conversations. Each approach can be designed to explore adaptation as a local response to a global phenomenon. As a result, each sees adaptation being informed by a fusion of top-down descriptions of impacts from global climate change and bottom-up deliberations rooted in local, national and regional experiences (see Chapter 2, Section 2.2.1).

20.8.3 Bringing climate-change adaptation and development communities together to promote sustainable development

The Millennium Development Goals (MDGs) are the latest international articulation of approaching poverty eradication and related goals in the developing world (see Section 20.7.1). Economic growth is necessary for poverty reduction and promoting other millennium goals; but, unless the growth achieved is equitably distributed, the result is a lopsided development where inequality increases. Many countries face intensifying poverty and inequality predicaments in the wake of undertaking free market policies (UNDP, 2003; UNSEA, 2005). As noted above, however, climate change is represented in the Millennium goals solely by indicators of changes in energy use per unit of GDP and/or by total or per capita emissions of CO_2. Tracking indicators of protected areas for biological diversity, changes in forests and access to water all appear in the goals, but they are not linked to climate-change impacts or adaptation; nor are they identified as part of a country's capacity to adapt to climate change.

Other issues of particular concern include ensuring energy services, promoting agriculture and industrialisation, promoting trade and upgrading technologies. Sustainable natural-resource management is a key to sustained economic growth and poverty reduction. It calls for clean energy sources; and the nature and pattern of agriculture, industry and trade should not unduly impinge on ecological health and resilience. Otherwise, the very basis of economic growth will be shattered through environmental degradation, more so as a consequence of climate change (Sachs, 2005). Put another way by Swaminathan (2005), developing and employing 'eco-technologies' (based on an integration of traditional and frontier technologies including bio-technologies, renewable energy and modern management techniques) is a critical ingredient rooted in the principles of economics, gender, social equity and employment generation with due emphasis given to climate change.

For environmentally-sustainable economic growth and social progress, therefore, development policy issues must inform the work of the climate-change community such that the two communities bring their perspectives to bear on the formulation and implementation of integrated approaches and processes that recognise how persistent poverty and environmental needs exacerbate the adverse consequences of climate change. In this process, science has a critical role to play in assessing the prevailing realities and likely future scenarios, and identifying policies and cost-effective methods to address various aspects of development and climate change; and it is important that all relevant stakeholders are involved in science-based dialogues (Welp et al., 2006). In order to go down this integrated and participatory road, a strong political will and public commitment to promoting sustainable development is needed, focusing simultaneously on economic growth, social progress, environmental conservation and adaptation to climate change (World Bank, 1998; AfDB et al., 2003). It is also important that private and public sectors work together within a framework of identified roles of each, with economic, social and climate-change perspectives built into the process. Further, co-ordination among national development and climate-change communities, as well as co-ordination among appropriate national and international institutions, is imperative.

This raises an important question regarding the process for bringing climate change and sustainable development together. Growing interest in these linkages is evident in a series of recent publications, including Toth (1999), Yamin (2004), Collier and Löfstedt (1997), Jepma and Munasinghe (1998), Munasinghe and Swart (2000, 2005), Abaza and Baranzini (2002), Markandya and Halsnaes (2002), Cohen et al. (1998), Kok et al. (2002), Swart et al. (2003). A number of themes that are particularly relevant to adaptation run through this literature. They include the need for equity between developed and developing countries in the delineation of rights and responsibilities within any climate-change response framework. Shue (1999), Thomas and Twyman (2004) and Paavola and Adger (2006) point, as well, to the need for equity across vulnerable groups that are disproportionately exposed to climate-change impacts. Hasselman (1999), Gardiner (2004) and Kemfert and Tol (2002) identify some examples from economics which raise concerns for intergenerational ethics; i.e., the degree to which the interests of future generations are given relatively lower weighting in favour of short-term concerns. Intergenerational justice implications, for individuals and collectives (e.g., indigenous cultures) are described in Page (1999). Masika (2002) specifically outlines gender aspects of differential vulnerabilities. Swart et al. (2003) identify the need to describe potential changes in vulnerability and adaptive capacity within the SRES storylines.

Although linkages between climate-change adaptation and sustainable development should appear to be self evident, it has been difficult to act on them in practice. Beg et al. (2002) identify potential synergies between climate change and other policies that could facilitate adaptation, such as those that address desertification and biodiversity. Ethical guidance from various spiritual and religious sources is reviewed in Coward (2004). However, an 'adaptation deficit' exists. Burton and May (2004) identify this as the gap between current and optimal levels of adaptation to climate-related events (including extremes); it is expected that climate change and poor development decisions will lead to an increased adaptation deficit in the future. While mitigation within the UNFCCC includes clearly defined objectives, measures, costs and instruments, this is not the case for adaptation. Agrawala (2005) indicates that much less attention has been paid to how development could be made more resilient to climate-change impacts, and identifies a number of barriers to mainstreaming climate-change adaptation within development activity (see, as well Chapter 17, Section 17.3).

The existence of these barriers does not mean that the development community does not recognise the linkage between development and climate-change adaptation. Climate change is identified as a serious risk to poverty reduction in developing countries, particularly because these countries have a limited capacity to cope with current climate variability and extremes not to mention future climate change (Schipper and Pelling, 2006). Adaptation measures will need to be integrated into strategies of poverty reduction to ensure sustainable development, and this will require improved governance, mainstreaming of climate-change measures, and the integration of climate-change impacts information into national economic projections (AfDB et al., 2003; Davidson et al., 2003). Brooks et al. (2005) offer an extensive list of potential proxy indicators for national-level vulnerability to climate change, including health, governance and technology indicators. Agrawala (2005) describes case studies of natural resources management in Nepal, Bangladesh, Egypt, Fiji, Uruguay and Tanzania, and recommends several priority actions for overcoming barriers to mainstreaming, including project screening for climate-related risk, inclusion of climate impacts in environmental impact assessments , and shifting emphasis from creating new plans to better implementation of existing measures. Approaches for integration of adaptation with development are outlined for East Africa (Orindi and Murray, 2005). The Commission for Africa (2005) explicitly links the need to address climate-change risks with achievement of poverty reduction and sustainable growth.

In recent years, new mechanisms have been established to support adaptation, including the Lesser Developed Countries (LDC) Fund, Special Climate Change Fund and the Adaptation Fund (Huq, 2002; Brander, 2003; Desanker, 2004; Huq, 2006; Huq et al., 2006). They have provided visibility and opportunity to mainstream adaptation into local/regional development activities. However, there are technical challenges associated with defining adaptation benefits for particular actions within UNFCCC mechanisms such as the Global Environmental Facility (GEF). For example, Burton (2004) and Huq and Reid (2004) note that the calculation of costs of adapting to future climate change (as opposed to current climate variability), as well as the local nature of resulting benefits, are both problematic *vis-à-vis* GEF requirements for defining global environmental benefits. On the other hand, there are opportunities. Dang et al. (2003) illustrate how including "adaptation benefits of mitigation" in Vietnam offers a way of linking both criteria in the analysis of potential projects for inclusion in the Clean Development Mechanism. Bouwer and Aerts (2006) and Schipper and Pelling (2006) identify opportunities for integrating climate-change adaptation and disaster risk management through insurance mechanisms, official development assistance and ongoing risk management programmes. Niang-Diop and Bosch (2004) outline methods for linking adaptation strategies with sustainable development at national and local scales, as part of National Adaptation Programmes of Action (NAPAs). As of the autumn of 2006, the LDC Fund was operational in its support of NAPAs in LDCs and both the Conference of Parties (COP) and GEF were in the process of defining how the implementation of adaptation activities highlighted in NAPAs could be funded (Huq et al., 2006).

20.9 Uncertainties, unknowns and priorities for research

Uncertainties, unknowns and priorities for research illuminate the confidence statements that modify scientific conclusions delivered to members of the policy community. For the research community, however, they can be translated into tasks designed to improve understanding and elaborate sources confidence. This section is therefore organised as a series of tasks.

Expand understanding of the synergies in and/or obstacles to simultaneous progress in promoting enhanced adaptive capacity and sustainable development. The current state of knowledge in casting adaptive capacity and vulnerability into the future is primitive. More thorough understandings of the process by which adaptive capacity and vulnerability evolve over time along specific development pathways are required. Commonalities exist across the determinants of adaptive capacity, mitigative capacity and the factors that support sustainable development, but current understanding of how they can be recognised and exploited is minimal.

Integrate more closely current work in the development and climate-change communities. Synergies exist between practitioners and researchers in the sustainable development and climate-change communities, but there is a need to develop means by which these communities can integrate their efforts more productively. The relative efficacies of dialogue processes and new tools required to promote this integration, and the various participatory and/or model-based approaches required to support their efforts must be refined or developed from scratch. Opportunities for shared learning should be identified, explored and exploited.

Search for common ground between spatially explicit analyses of vulnerability and aggregate integrated assessment models. Geographical and temporal scales of development and climate initiatives vary widely. The interaction and intersection between spatially explicit and aggregate integrated assessment models has yet to be explored rigorously. For example, representations of adaptive capacities and resulting vulnerabilities in aggregate integrated assessment models are still rudimentary. As progress is encouraged in improving their abilities to depict reality, research initiatives must also recognise and work to overcome difficulties in matching the scales at which models are constructed and exercised with the scales at which decisions are made. New tools are required to handle these differences, particularly between the local and national, short-to-medium-term scales of adaptation and development programmes and projects and the global, medium-to-long-term scale of mitigation.

Recognise that uncertainties will continue to be pervasive and persistent, and develop or refine new decision-support mechanisms that can identify robust coping strategies even in the face of this uncertainty. Significant uncertainties in estimating the social cost of greenhouse gases exist, and many of their sources have been identified; indeed many of their sources reside in the research needs listed above. Reducing these uncertainties would certainly be productive, but it cannot be guaranteed that future research will make much progress in this regard. It follows that concurrent improvement in our ability to use existing decision-support tools and to design new approaches to cope with uncertainties and associated risks that will be required over the foreseeable future is even more essential. In short, identify appropriate decision-support tools and clarify the criteria that they can inform in an uncertain world.

Characterise the full range of possible climate futures and the paths that might bring them forward. The research communities in both climate and development must, along with practitioners and decision-makers, be informed not only about the central tendencies of climate change and its ramifications, but also about the outlier possibilities about which the natural-science community is less sanguine. It is simply impossible to comprehend the risks associated with high-consequence outcomes with low probabilities if neither their character nor their likelihood has been described.

This chapter has offered a glimpse into where to turn for guidance in confronting and managing the risks associated with climate change and climate variability. Indeed, the climate problem is a classic risk management problem of the sort with which decision-makers are already familiar. It is critical to see risk as the product of likelihood and consequence, to recognise that the likelihood of a climate impact is dependent on natural and human systems, and to understand that the consequence of that impact can be measured in terms of a multitude of numeraires (currency, millions at risk, species extinction, abrupt physical changes and so on). These expressions of risk are determined fundamentally by location in time and space.

This chapter also points to synergies that exist at the nexus of sustainable development and adaptive capacity, primarily by noting for the first time that many of the goals of sustainable development match the determinants of adaptive capacity (and, for that matter, mitigative capacity). Planners in the decision-intensive ministries around the world are therefore already familiar with the generic mechanisms by which including climate change into their risk assessments of development programmes can complicate their decisions. Adding climate to the list of multiple stresses which can impede progress in meeting their goals in their specific context is thus not a new problem. Climate change, even when its impacts are amplified by the effects of other stresses, is just one more thing: one more problem to confront, but also one more reason to act in ways that promote progress along multiple fronts. Exploitation of the synergies is not automatic, so care must be taken to avoid development activities that can exacerbate climate change or impacts just as care must be taken to take explicit account of climate risks.

The United Nations Framework Convention on Climate Change commits governments to avoiding "dangerous anthropogenic interference with the climate system", but governments will be informed in their deliberations of what is or is not 'dangerous' only by an approach that explicitly reflects the rich diversity of climate risk across the globe and into the coming decades instead of burying this diversity into incomplete aggregate indices of damages. Risk management techniques have been designed for such tasks; but it is important to note that risk-based approaches require exploration of the implications of not only the central tendencies of climate change that are the focus of consensus-driven assessments of the literature, but also the uncomfortable (or more benign) futures that reside in the 'tails' of current understanding. Viewing the climate issue from a risk perspective can offer climate policy deliberations and negotiations new insight into the synergies by which governments can promote sustainable development, reduce the risk of climate-related damages and take advantage of climate-related opportunities.

References

Abaza, H. and A. Baranzini, Eds., 2002: *Implementing Sustainable Development: Integrated Assessment and Participatory Decision-making Processes*. United Nations Environment Programme, Edward Elgar Publishing, Cheltenham, 320 pp.

ACIA (Arctic Climate Impact Assessment), 2005: *Arctic Climate Impact Assessment*. Cambridge University Press, Cambridge, 1042 pp.

Adger, N. and K. Vincent, 2004: Uncertainty in adaptive capacity. *IPCC Workshop on Describing Uncertainties in Climate Change to Support Analysis of Risk and Options*, M. Manning, M. Petit, D. Easterling, J. Murphy, A. Patwardhan, H-H Rogner, R. Swart and G. Yohe, Eds., Intergovernmental Panel on Climate Change, Geneva, 49-51.

Adger, W.N., K. Brown and E.L. Tompkins, 2005: The political economy of cross-scale networks in resource co-management. *Ecology and Society*, **10**, article 9. [Accessed 30.05.07: http://www.ecologyandsociety.org/vol10/iss2/art9/]

Adger, W.N., K. Brown, J. Fairbrass, A. Jordan, J. Paavola, S. Rosendo and G. Seyfang, 2003: Governance for sustainability: towards a 'thick' analysis of environmental decision-making. *Environ. Plann. A*, **35**, 1095-1110.

AfDB (African Development Bank), ADB, DFID, DGIS, EC, BMZ, OECD, UNDP and WB, 2003: *Poverty and Climate Change: Reducing Vulnerability of the Poor*. African Development Bank; Asian Development Bank; UK Department for International Development; Directorate-General for Development, European Commission; Federal Ministry for Economic Cooperation and Development, Germany; Ministry of Foreign Affairs – Development Cooperation, the Netherlands; Organization for Economic Cooperation and Development; United Nations Development Programme; United Nations Environment Programme and the World Bank, Eds., DFID, UK, 43 pp.

AfDB (African Development Bank), African Development Fund, African Union, International Strategy for Disaster Reduction, and New Partnership for Africa's Development, 2004: *Guidelines for Mainstreaming Disaster Risk Assessment in Development*, United Nations, 68 pp.

Agrawala, S., 2004: Adaptation, development assistance and planning: challenges and opportunities. *IDS Bulletin*, **35**, 50-54.

Agrawala, S., Ed., 2005: *Bridge over Troubled Waters: Linking Climate Change and Development*. Organization for Economic Co-Operation and Development, OECD Publishing, Paris, 153 pp.

Ahmad, Q.K. and A.U. Ahmed, Eds., 2002: Bangladesh: citizen's perspective on sustainable development, Bangladesh Unnayan Parishad (BUP), Dhaka, 181 pp.

Allen, K.M., 2006. Community-based disaster preparedness and climate adaptation: local capacity-building in the Philippines. *Disasters*, **30**, 81-101.

APJED (Asia Pacific Journal on Environment and Development), 2004: Community approaches to flood management in South Asia. *Asia Pacific Journal on Environment and Development*, special double issue, **11**, Nos. 1 and 2.

Arnell, N.W., 2004: Climate change and global water resources: SRES emissions and socio-economic scenarios. *Global Environ. Chang*., **14**, 31-52.

Arnell, N.W., 2006: Climate change and water resources: a global perspective. *Avoiding Dangerous Climate Change. Proceedings of the Exeter Conference*, H.J. Schellnhuber, W. Cramer, N. Nakicenovic, T.M.L. Wigley and G. Yohe, Eds., Cambridge University Press, Cambridge, 167-175.

Arnell, N.W. and E.K. Delaney, 2006: Adapting to climate change: public water supply in England and Wales. *Climatic Change*, **78**, 227-255.

Arnell, N.W., M.G.R. Cannell, M. Hulme, R.S. Kovats, J.F.B. Mitchell, R.J. Nicholls, M.L. Parry, M.T.J. Livermore and Co-authors, 2002: The consequences of CO_2 stabilization for the impacts of climate change. *Climatic Change*, **53**, 413-446.

Arrow, K., P. Dasgupta, L. Goulder, G. Daily, P. Ehrlich, G. Heal, S. Levin, K.-G. Maler, S. Schneider, D. Starrett and B. Walker, 2004: Are we consuming too much? *J. Econ. Perspect*., **18**, 147-172.

Bansal, P., 2005: Evolving sustainably: a longitudinal study of corporate sustainable development. *Strategic Manage. J*., **26**, 197-218.

Barrow, E.M. and R.J. Lee, 2000: *Climate change and environmental assessment part 2: climate change guidance for environmental assessments*. The Research and Development Monograph Series 2000, The Canadian Institute for Climate Studies, University of Victoria, Victoria, 85 pp.

Beg, N., J.C. Morlot, O. Davidson, Y. Afrane-Okesse, L. Tyani, F. Denton, Y. Sokona, J.P. Thomas, E.L. La Rovere, J.K. Parikh, K. Parikh and A.A. Rahman., 2002: Linkages between climate change and sustainable development. *Clim. Policy*, **2**, 129-144.

Belliveau, S., B. Smit and B. Bradshaw, 2006: Multiple exposures and dynamic vulnerability: evidence from the grape industry in the Okanagan Valley, Canada. *Global Environ. Chang*., **16**, 364-378.

Bennett, E. and M. Zurek, 2006: Integrating epistemologies through scenarios. *Bridging Scales and Knowledge Systems*, W.V. Reid, F. Berkes, T. Wilbanks and D. Capistrano, Eds., Island Press, Washington, District of Columbia, 264-280.

Blanco, A.V.R., 2006: Local initiatives and adaptation to climate change. *Disasters*, **30**, 140-147.

Bouwer, L.M. and J.C.J.H. Aerts, 2006: Financing climate change adaptation. *Disasters*, **30**, 49-63.

Brander, L., 2003: The Kyoto mechanisms and the economics of their design. *Climate Change and the Kyoto Protocol: The Role of Institutions and Instruments to Control Global Change*, M. Faure, J. Gupta and A. Nentjes, Eds., Edward Elgar, Cheltenham, 25-44.

Brenkert, A. and E. Malone, 2005: Modeling vulnerability and resilience to climate change: a case study of India and Indian States. *Climatic Change*, **72**, 57-102.

Brooks, N. and W.N. Adger, 2005: Assessing and enhancing adaptive capacity. *Adaptation Policy Frameworks for Climate Change: Developing Strategies, Policies and Measures*, B. Lim and E. Spanger-Siegfried, Eds., Cambridge University Press, Cambridge, 165-182.

Brooks, N., W.N. Adger and P.M. Kelly, 2005: The determinants of vulnerability and adaptive capacity at the national level and the implications for adaptation. *Global Environ. Chang.*, **15**, 151-163.

Bulkeley, H. and A.P.J. Mol, 2003: Participation and environmental governance: consensus, ambivalence and debate. *Environmental Values*, **12**, 143-154.

Burton, I., 2004: The adaptation deficit. *Building the Adaptive Capacity*, A. Fenech, D. MacIver, H. Auld, R. Bing Rong and Y. Yin, Eds., Environment Canada, Toronto, 25-33.

Burton, I. and E. May, 2004: The adaptation deficit in water resources management *IDS Bulletin*, **35**, 31-37.

Burton, I. and M. van Aalst, 2004: *Look Before You Leap: A Risk Management Approach for Incorporating Climate Change Adaptation into World Bank Operations*. World Bank, Washington, District of Columbia, 47 pp.

Burton, I., S. Huq, B. Lim, O. Pilifosova and E.L. Schipper, 2002: From impacts assessment to adaptation priorities: the shaping of adaptation policy. *Clim. Policy*, **2**, 145-159.

Caparrós, A. and F. Jacquemont, 2003: Conflict between biodiversity and carbon sequestration programs: economic and legal implications. *Ecol. Econ.*, **46**, 143-157.

Chambers, R., 2006: Participatory mapping and geographic information systems: whose map? Who is empowered and who disempowered? Who gains and who loses? *The Electronic Journal on Information Systems in Developing Countries*, **25**, 1-11. [Accessed 30.05.07: http://www.ejisdc.org]

Clark, M.J., 2002: Dealing with uncertainty: adaptive approaches to sustainable river management. *Aquat. Conserv.*, **12**, 347-363.

Clarkson, R. and K. Deyes, 2002: Estimating the social cost of carbon emissions. Government Economic Service Working Paper 140, HM Treasury and Defra, 59 pp.

Cohen, S. and T. Neale, Eds., 2006: Participatory integrated assessment of water management and climate change in the Okanagan Basin, British Columbia. Environment Canada and University of British Columbia, Vancouver, 221 pp.

Cohen, S., D. Demeritt, J. Robinson and D. Rothman, 1998: Climate change and sustainable development: towards dialogue. *Global Environ. Chang.*, **8**, 341-371.

Cohen, S., D. Neilsen and R. Welbourn, Eds., 2004: Expanding the dialogue on climate change & water management in the Okanagan Basin, British Columbia. Environment Canada, Agriculture & Agri-Food Canada and University of British Columbia, 224 pp.

Collier, U. and R.E. Löfstedt, Eds., 1997: *Cases in Climate Change Policy: Political Reality in the European Union*. Earthscan Publications Ltd., London. 204 pp.

Commission for Africa, 2005: *Our Common Interest*. Report of the Commission for Africa, 461 pp. [Accessed 30.05.07: www.commissionforafrica.org].

Conde, C. and K. Lonsdale, 2004: Engaging stakeholders in the adaptation process. *Adaptation Policy Frameworks for Climate Change: Developing Strategies, Policies and Measures*, B. Lim and E. Spanger-Siegfried, Eds., Cambridge University Press, Cambridge, 47-66.

Cooke, B. and U. Kothari, Eds., 2001: *Participation: The New Tyranny?* Zed Books, London, 207 pp.

Corbett, J., G. Rambaldi, P. Kyem, D. Weiner, R. Olson, J. Muchemi, M. McCall and R. Chambers, 2006: Overview: mapping for change – the emergence of a new practice. *Participatory Learning and Action*, **54**, 13-19.

Coward, H., 2004: What can individuals do? *Hard Choices: Climate Change in Canada*, H. Coward and A.J. Weaver, Eds., Wilfrid Laurier University Press, Waterloo, Canada, 233-252.

Dang, H.H., A. Michaelowa and D.D. Tuan, 2003: Synergy of adaptation and mitigation strategies in the context of sustainable development: the case of Vietnam. *Climate Policy*, **3**, S81-S96.

Davidson, O., K. Halsnaes, S. Huq, M. Kok, B. Metz, Y. Sokona and J. Verhagen, 2003: The development and climate nexus: the case of sub-Saharan Africa. *Climate Policy*, **3**, S97-S113.

Desanker, P.V., 2004: *The NAPA Primer*. United Nations Framework Convention on Climate Change (UNFCCC) Least Developed Countries Expert Group (LEG), Bonn, Germany, 192 pp.

DFID, 2004: Reducing the risk of disasters: helping to achieve sustainable poverty reduction in a vulnerable world. Department for International Development Policy Paper, London, 36 pp.

Downing, T., D. Anthoff, R. Butterfield, M. Ceronsky, M. Grubb, J. Guo, C. Hepburn, C. Hope and Co-authors, 2005: Social cost of carbon: a closer look at uncertainty. Final Report, Defra, 95 pp.

Eakin, H. and M.C. Lemos, 2006: Adaptation and the state: Latin America and the challenge of capacity building under globalization. *Global Environ. Chang.*, **16**, 7-18.

EEA (European Environment Agency), 2006: Vulnerability and adaptation to climate change in Europe. Technical Report No. 7, European Environment Agency, 84 pp.

Elasha, B.O., 2005: Sustainable development. *Tiempo*, **57**, 18-23.

Epstein, M.J. and M.J. Roy, 2003: Making the business case for sustainability: linking social and environmental actions to financial performance. *Journal of Corporate Citizenship*, **9**, 79-96.

Figge, F. and T. Hahn, 2004: Sustainable value added - measuring corporate contributions to sustainability beyond eco-efficiency. *Ecol. Econ.*, **48**, 173-187.

Foley, J.A., R. DeFries, G. Asner, C. Barford, G. Bonan, S. Carpenter, F. Chapin, M. Coe and Co-authors, 2005: Global consequences of land use. *Science*, **309**, 570-574.

Ford, J.D., B. Smit and J. Wandel, 2006: Vulnerability to climate change in the Arctic: a case study from Arctic Bay, Canada. *Global Environ. Chang.*, **16**, 145-160.

Furgal, C.M., C. Fletcher and C. Dickson, 2006: Ways of knowing and understanding: towards the convergence of traditional and scientific understanding of climate change in the Canadian North. Environment Canada, No. KM467-05-6213, 96 pp.

Garande, T. and S. Dagg, 2005: Public participation and effective water governance at the local level: a case study from a small under-developed area in Chile. *Environment, Development and Sustainability*, **7**, 417-431.

Gardiner, S.M., 2004: Ethics and global climate change. *Ethics*, **114**, 555-600.

Goklany, I.M., 2007: Integrated strategies to reduce vulnerability and advance adaptation, mitigation and sustainable development. *Mitigation and Adaptation Strategies for Global Change*. doi: 10.1007/s11027-007-9098-1.

Guiso, L., P. Sapienza and L. Zingales, 2004: The role of social capital in financial development. *Am. Econ. Rev.*, **94**, 526-556.

Gupta, J. and R.S.J. Tol, 2003: Why reduce greenhouse gas emissions? Reasons, issue-linkages, and dilemmas. *Issues in International Climate Policy: Theory and Policy*, E.C. van Ireland, J. Gupta and M.T.J. Kok, Eds., Edward Elgar Publishing, Cheltenham, 17-38.

Haas, P.M., 2004: When does power listen to truth? A constructivist approach to the policy process. *J. Eur. Public Policy*, **11**, 569-592.

Haddad, B.M., 2005: Ranking the adaptive capacity of nations to climate change when socio-political goals are explicit. *Global Environ. Chang.*, **15**, 165-176.

Hamlet, A.F., 2003: The role of transboundary agreements in the Columbia River Basin: an integrated assessment in the context of historic development, climate, and evolving water policy. *Climate, Water, and Transboundary Challenges in the Americas*, H. Diaz and B. Morehouse, Eds., Kluwer Press, Dordrecht, 263-289.

Harman, J., Bramley, M. E. and Funnell, M., 2002: Sustainable flood defense in England and Wales. *P. I. Civil Eng.-Civ. En.*, **150**, 3-9.

Hasselman, K., 1999: Intertemporal accounting of climate change - harmonizing economic efficiency and climate stewardship. *Climatic Change*, **41**, 333-350.

Hickey, S. and G. Mohan, Eds., 2004: *Participation, from Tyranny to Transformation? Exploring new Approaches to Participation in Development*. Zed Books, London, 292 pp.

Hilson, G., 2001: Putting theory into practice: how has the gold mining industry interpreted the concept of sustainable development? *Miner. Resour. Eng.*, **10**, 397-413.

Hisschemöller, M., R.S.J. Tol and P. Vellinga., 2001: The relevance of participatory approaches in integrated environmental assessment. *Integrated Assessment*, **2**, 57-72.

Hitz, S. and J. Smith, 2004: Estimating global impacts from climate change. *Global Environ. Chang.*, **14**, 201-218.

Hope, C., 2005: Memorandum by Dr Chris Hope, Judge Institute of Management, University of Cambridge. The Economics of Climate Change, HL 12-II, Oral Evidence: 18 January 2005, House of Lords Select Committee on Economic Affairs, The Stationery Office, London.

Hope, C., 2006a: The marginal impact of CO_2 from PAGE2002: An integrated assessment model incorporating the IPCC's five reasons for concern. *Integrated Assessment*, **6**, 1-16.

Hope, C., 2006b: The marginal impacts of CO_2, CH4 and SF6 emissions. *Climate Pol.*, **6**, 537-544.

Huntington, H.P., S.F. Trainor, D.C. Natcher, O.H. Huntington, L. DeWilde and F.S. Chapin III, 2006: The significance of context in community-based research: understanding discussions about wildfire in Huslia, Alaska. *Ecology and Society*, **11**, article 40. [Accessed 30.05.07: http://www.ecologyandsociety.org/vol11/iss1/art40/].

Huq, S., 2002: The Bonn-Marrakech agreements on funding. *Climate Policy*, **2**, 243-246.

Huq, S., 2006: Adaptation funding. *Tiempo*, **58**, 20-21.

Huq, S. and H. Reid, 2004: Mainstreaming adaptation in development. *IDS Bulletin*, **35**, 15-21.

Huq, S., H. Reid and L.A. Murray, 2006: *Climate Change and Development Links*. Gatekeeper Series 123, International Institute for Environment and Development, London, 24 pp.

Hyogo Declaration, 2005: Hyogo Declaration. World Conference on Disaster Reduction, Kobe, Japan, International Strategy for Disaster Reduction, A/CONF.206/6, 5 pp.

ICSU (International Council for Science), 2002: ICSU Series on Science for Sustainable Development No 9: Science and Technology for Sustainable Development, 30 pp.

IGES (Institute for Global Environmental Strategies), 2005: Sustainable Asia – 2005 and beyond: in the pursuit of innovative policies. IGES White Paper, Institute for Global Environmental Strategies, Kanagawa, Japan, 174 pp.

IPCC, 1992: *Climate Change 1992: The Supplementary Report to the IPCC Scientific Assessment*, J.T. Houghton, B.A. Callander and S.K. Varney, Eds., Cambridge University Press, Cambridge, 200 pp.

IPCC, 1996: *Climate Change 1995: Impacts, Adaptation and Vulnerability. Contribution of Working Group II to the Second Assessment Report of the Intergovernmental Panel on Climate Change*, R.T. Watson, M.C. Zinyowera and R.H. Moss, Eds., Cambridge University Press, Cambridge, 880 pp.

IPCC, 2001a: *Climate Change 2001: The Scientific Basis. Contribution of Working Group I to the Third Assessment Report of the Intergovernmental Panel on Climate Change*, J.T. Houghton, Y. Ding, D.J. Griggs, M. Noguer, P.J. van der Linden, X. Dai, K. Maskell and C.A. Johnson, Eds., Cambridge University Press, Cambridge, 881 pp.

IPCC, 2001b: *Climate Change 2001: Impacts, Adaptation and Vulnerability. Contribution of Working Group II to the Third Assessment Report of the Intergovernmental Panel on Climate Change*, J.J. McCarthy, O.F. Canziani, N.A. Leary, D.J. Dokken and K.S. White, Eds., Cambridge University Press, Cambridge, 1032 pp.

IPCC, 2007a: *Climate Change 2007: Mitigation. Contribution of Working Group III to the Fourth Assessment Report of the Intergovernmental Panel on Climate Change*, B. Metz, O. Davidson, P. Bosch, R. Dave and L. Meyer, Eds., Cambridge University Press, Cambridge, UK.

IPCC, 2007b: *Climate Change 2007: The Physical Science Basis. Contribution of Working Group I to the Fourth Assessment Report of the Intergovernmental Panel on Climate Change*, Solomon, S., D. Qin, M. Manning, Z. Chen, M. Marquis, K.B. Averyt, M. Tignor and H.L. Miller, Eds., Cambridge University Press, Cambridge, 996 pp.

ISDR, 2004: *Living With Risk: a Global Review of Disaster Reduction Initiatives*. International Strategy for Disaster Reduction(ISDR), United Nations, Geneva, 588 pp.

Iwanciw, J.G., 2004: Promoting social adaptation to climate change and variability through knowledge, experiential and co-learning networks in Bolivia. Shell Foundation Sustainable Energy Program, 22 pp.

Jasch, C., 2003: The use of environmental management accounting (EMA) for identifying environmental costs. *J. Clean. Prod.*, **11**, 667-676.

Jegillos, S.R., 2003: Methodology. *Sustainability in Grass-Roots Initiatives: Focus on Community Based Disaster Management*, R. Shaw and K. Okazaki, Eds., United Nations Centre for Regional Development (UNCRD), Disaster Management Planning Hyogo Office, 19-28.

Jepma, C.J. and M. Munasinghe, 1998: *Climate Change Policy: Facts, Issues and Analyses*, Cambridge University Press, Cambridge, 331 pp.

Johnson, D. and C. Walck, 2004: Integrating sustainability into corporate management systems. *Journal of Forestry*, **102**, 32-39.

Johnson, M., Ed., 1992: *Lore: Capturing Traditional Environmental Knowledge*. Dene Cultural Institute, Hay River, and International Development Research Centre, Ottawa, 190 pp.

Jung, T.Y., A. Srinivasan, K. Tamura, T. Sudo, R. Watanabe, K. Shimada and H. Kimura, 2005: *Asian Perspectives on Climate Regime Beyond 2012: Concerns, Interests and Priorities*. Institute for Global Environmental Strategies (IGES), Hayama, Japan, 95 pp.

Kabat, P., R.E. Schulze, M.E. Hellmuth and J.A. Veraart, Eds., 2002: Coping with impacts of climate variability and climate change in water management: a scoping paper. DWC-Report no. DWCSSO-01(2002), International Secretariat of the Dialogue on Water and Climate, Wageningen, 114 pp.

Kangur, K., Ed., 2004: *Focus Groups and Citizens Juries - River Dialogue Experiences in Enhancing Public Participation in Water Management*. Peipsi Center for Transboundary Cooperation, Tartu, Estonia, 64 pp.

Kashyap, A., 2004: Water governance: learning by developing adaptive capacity to incorporate climate variability and change. *Water Science and Technology*, **49**, 141-146.

Kates, R.W., 2000: Cautionary tales: adaptation and the global poor. *Climatic Change*, **45**, 5-17.

Kates, R.W., T.M. Parris and A.A. Leiserowitz, 2005: What is sustainable development? Goals, indicators, values, and practice. *Environment: Science and Policy for Sustainable Development*, **47**, 8–21.

Kates, R.W., C.E. Colten, S. Laska and S.P. Leatherman, 2006: Reconstruction of New Orleans following Hurricane Katrina. *P. Natl. Acad. Sci.*, **103**, 14653-14660.

Kates, R.W., W.C. Clark, R. Corell, J.M. Hall, C.C. Jaeger, I. Lowe, J.J. McCarthy, H.J. Schellnhuber, B. Bolin, N.M. Dickenson, S. Faucheux, G.C. Gallopin, A. Grübler, B. Huntley, J. Jäger, N.S. Jodha, R.E. Kasperson, A. Mabogunje, P. Matson, H. Mooney, B. Moore III, T. O'Riordan and U. Svedlin, 2000: Sustainability science. *Science*, **292**, 641-642.

Kemfert, C. and R.S.J. Tol, 2002: Equity, international trade and climate policy. *International Environmental Agreements: Politics, Law and Economics*, **2**, 23-48.

Klein, R.J.T., E.L.F. Schipper and S. Dessai, 2005: Integrating mitigation and adaptation into climate and development policy: three research questions. *Environ. Sci. Policy*, **8**, 579-588.

Kok, M.T.J., W.J. V. Vermeulen, A.P.C. Faaij and D. de Jager, Eds., 2002: *Global Warming and Social Innovation: The Challenge of a Climate-Neutral Society*. Earthscan Publications Ltd., London, 242 pp.

Krupnik, I. and D. Jolly, Eds., 2002: *The Earth is Faster Now: Indigenous Observations of Arctic Environmental Change*. Arctic Research Consortium of the United States, Fairbanks, 384 pp.

Kundzewicz, Z.W., 2002: Non-structural flood protection and sustainability. *Water International*, **27**, 3-13.

Larson, A.M. and J.C. Ribot, Eds., 2004a: Democratic decentralization through a natural resource lens – an introduction. *European Journal of Development Research*, **16**, 1-25.

Larson, A.M. and J.C. Ribot, Eds., 2004b: *Democratic Decentralization through a Natural Resource Lens*. Routledge, New York, 272 pp.

Leemans, R. and Eickhout, B., 2004: Another reason for concern: regional and global impacts on ecosystems for different levels of climate change. *Global Environ. Chang.*, **14**, 219-228.

Lemos, M.C. and B.J. Morehouse, 2005: The co-production of science and policy in integrated climate assessments. *Global Environ. Chang.*, **15**, 57-68.

Levy, P.E., M.G.R. Cannell and A.D. Friend, 2004: Modelling the impact of future changes in climate, CO_2 concentration and land use on natural ecosystems and the terrestrial carbon sink. *Global Environ. Chang.*, **14**, 21-30.

Lorenzoni, I., A. Jordan, D.T. Favis-Mortlock, D. Viner and J. Hall, 2001: Developing sustainable practices to adapt to the impacts of climate change: a case study of agricultural systems in eastern England (UK). *Reg. Environ. Change*, **2**, 106-117.

Low, P.S., Ed., 2005: *Climate Change and Africa*. Cambridge University Press, Cambridge, 412 pp.

Lucas, R.E., 1988: On the Mechanics of Economic Development. *J. Monetary Econ.*, **22**, 3-42.

Lüdeke, M.K.B., G. Petschel-Held and H-J. Schellnhuber, 2004: Syndromes of global change: the first panoramic view. *GAIA.*, **13**, 42-49.

MA (Millennium Ecosystem Assessment), 2005: *Ecosystems and Human Well-Being: Synthesis*. Island Press, Washington, District of Columbia, 155 pp.

Markandya, A. and K. Halsnaes, 2002: *Climate Change and Sustainable Development: Prospects for Developing Countries*. Earthscan Publications, London, 291 pp.

Masika, R., Ed., 2002: *Gender, Development, and Climate Change*. Oxfam Focus on Gender, Oxfam GB, Oxford, 112 pp.

Mendelsohn, R.O., W.N. Morrison, M.E. Schlesinger and N.G. Andronova, 1998: Country-specific market impacts of climate change. *Climatic Change*, **45**, 553-569.

Mileti, D., 1999: *Disasters by Design: A Reassessment of Natural Hazards in the United States*. National Academy Press, Washington, District of Columbia, 376 pp.

Mirza, M.M.Q., 2003: Climate change and extreme weather events: can developing countries adapt? *Climate Policy*, **3**, 233-248.

Moretti, E., 2004: Workers' education, spillovers, and productivity: evidence from plant-level production functions. *Am. Econ. Rev.*, **94**, 656-690.

Morimoto, R. and C. Hope, 2004: Applying a cost-benefit analysis model to the Three Gorges project in China. *Impact Assessment and Project Appraisal*, **22**, 205-220.

Munasinghe, M. and R. Swart, 2000: Climate change and its linkages with development, equity and sustainability. *Proc. IPCC Expert Meeting*, Colombo, Sri Lanka, Intergovernmental Panel on Climate Change, 319 pp.

Munasinghe, M. and R. Swart, 2005: *Primer on Climate Change and Sustainable Development Facts, Policy Analysis and Applications*. Cambridge University Press, Cambridge, 445 pp.

Munasinghe, M., O. Canziani, O. Davidson, B. Metz, M. Parry and M. Harisson, 2003: Integrating sustainable development and climate change in the IPCC Fourth Assessment Report. Munasinghe Institute for Development, Colombo, 44-52.

Munich Re, 2005: *Topics Geo Annual Review, Natural Catastrophes, 2004*. Munich Re, Munich, 60 pp. [Accessed 11.06.07: http://www.munichre.com]

Najam, A., A.A. Rahman, S. Huq and Y. Sokona., 2003: Integrating sustainable development into the Fourth Assessment Report of the Intergovernmental Panel on Climate Change. *Climate Policy*, **3**, S9-S17.

Nakicenovic, N. and R. Swart, Eds., 2000: *Special Report on Emissions Scenarios. A Special Report of Working Group III of the Intergovernmental Panel on Climate Change*. Cambridge University Press, Cambridge, 599 pp.

NRC (National Research Council), 1999: Our common journey. *Our Common Journey: A Transition Toward Sustainability*, Board on Sustainable Development, Eds., National Academy Press, Washington, District of Columbia, 21-58.

Niang-Diop, I. and H. Bosch, 2004: Formulating an adaptation strategy. *Adaptation Policy Frameworks for Climate Change: Developing Strategies, Policies and Measures*, B. Lim and E. Spanger-Siegfried, Eds., Cambridge University Press, Cambridge, 183-204.

Nicholls, R.J., 2004: Coastal flooding and wetland loss in the 21st century: changes under the SRES climate and socio-economic scenarios. *Global Environ. Chang.*, **14**, 69-86.

Nordhaus, W.D. and J.G. Boyer, 2000: *Warming the World: Economic Models of Global Warming*. MIT Press, Cambridge, Massachusetts, 232 pp.

NRBS (Northern River Basins Study Board), 1996: *Northern River Basins Study: Report to the Ministers 1996*. Alberta Environmental Protection, Edmonton, Alberta, 287 pp. [Accessed 11.06.07: http://www3.gov.ab.ca/env/water/nrbs/index.html]

O'Riordan, T., 2004: Environmental science, sustainability and politics. *T. I. Brit. Geogr.*, **29**, 234-247.

O'Brien, G., P. O'Keefe, J. Rose and B. Wisner, 2006: Climate change and disaster management. *Disasters*, **30**, 64-80.

O'Hare, G., 2002: Climate change and the temple of sustainable development. *Geography*, **87**, 234-246.

Olsen, J.R., 2006: Climate change and floodplain management in the United States. *Climatic Change*, **76**, 407-426.

Orindi, V.A. and L.A. Murray, 2005: Adapting to climate change in East Africa: a strategic approach. Gatekeeper Series 117, International Institute for Environment and Development, London, 23 pp.

Ostrom, E., J. Burger, C.B. Field, R.B. Norgaard and D. Policansky, 1999: Revisiting the commons: local lessons, global challenges. *Science*, **284**, 278-282.

Paavola, J. and W.N. Adger, 2006: Fair adaptation to climate change. *Ecol. Econ.*, **56**, 594-609.

Page, E., 1999: Intergenerational justice and climate change. *Polit. Stud. - London*, **47**, 53-66.

Parkins, J.R. and R.E. Mitchell, 2005: Public participation as public debate: a deliberative turn in natural resource management. *Soc. Natur. Resour.*, **18**, 529-540.

Parry, M.L., C. Rosenzweig, A. Iglesias, M. Livermore and G. Fischer, 2004: Effects of climate change on global food production under SRES emissions and socio-economic scenarios. *Global Environ. Chang.*, **14**, 53-67.

Payne, J.T., A.W. Wood, A.F. Hamlet, R.N. Palmer and D.P. Lettenmaier, 2004: Mitigating the effects of climate change on the water resources of the Columbia River basin. *Climatic Change*, **62**, 233-256.

Pearce, D., 2003: The social cost of carbon and its policy implications. *Oxford Rev. Econ. Pol.*, **19**, 362-384.

Philippine Country Report, 1999: National Workshop on Indicators of Sustainable Development, UN Sustainable Development, 91 pp.

Plambeck, E.L. and C.W. Hope, 1996: PAGE95. An updated valuation of the impacts of global warming. *Energ. Policy*, **24**, 783-794.

Porritt, J., 2005: *Capitalism as if the World Matters*, Earthscan, London, 304 pp.

Rambaldi, G., P.A.K. Kyem, M. McCall and D. Weiner, 2006: Participatory spatial information management and communication in developing countries. *The Electronic Journal on Information Systems in Developing Countries*, **25**, 1-9. [Accessed 11.06.07: http://www.ejisdc.org]

Ramus, C.A., 2002: Encouraging innovative environmental actions: what companies and managers must do. *J. World Bus.*, **37**, 151-164.

Reid, H. and M. Alam, 2005: Millennium Development Goals. *Tiempo*, **54**, 18-22.

Reid, W.V., F. Berkes, T. Wilbanks and D. Capistrano, Eds., 2006: *Bridging Scales and Knowledge Systems*. Island Press, Washington, District of Columbia, 314 pp.

Reidlinger, D. and F. Berkes, 2001: Contributions of traditional knowledge to understanding climate change in the Canadian Arctic. *Polar Rec.*, **37**, 315-328.

Richardson, D., 2002: Flood risk - the impact of climate change. *P. I. Civil Eng.-Civ. En.*, **150**, 22-24.

Robinson, J.B. and D. Herbert, 2001: Integrating climate change and sustainable development. *International Journal of Global Environmental Issues*, **1**, 130-149.

Robledo, C., M. Fischler and A. Patino, 2004: Increasing the resilience of hillside communities in Bolivia - has vulnerability to climate change been reduced as a result of previous sustainable development cooperation? *Mt. Res. Dev.*, **24**, 14-18.

Rotmans, J. and B. de Vries, Eds., 1997: *Perspectives on Global Change: the Targets Approach*. Cambridge University Press, Cambridge, 479 pp.

Rotmans, J. and M.B.A. van Asselt, 2002: Integrated assessment: current practices and challenges for the future. *Implementing Sustainable development: Integrated Assessment and Participatory Decision-Making Processes*, H. Abaza and A. Baranzini, Eds., United Nations Environment Programme, Edward Elgar, Cheltenham, 78-116.

Rowe, G. and L.J. Frewer, 2000: Public participation methods: a framework for evaluation. *Sci. Technol. Hum. Val.*, **25**, 3-29.

Rozelle, S. and J.F.M. Swinnen, 2004: Success and failure of reform: insights from the transition of agriculture. *J. Econ. Lit.*, **42**, 433-458.

Sachs, J.D., 2005: *The End of Poverty: Economic Possibilities for Our Time*. The Penguin Press, New York, 416 pp.

Sala-i-Martin, X., G. Doppelhofer and R. Miller, 2004: Determinants of long-term growth: a Bayesian averaging of classical estimates (BACE) approach. *Am. Econ. Rev.*, **94**, 813-835.

Schellnhuber, H.J., M.K.B. Lüdeke and G. Petschel-Held, 2002: The syndromes approach to scaling – describing global change on an intermediate functional scale. *Integrated Assessment*, **3**, 201-219.

Schellnhuber, H.J., A. Block, M. Cassel-Gintz, J. Kropp, G. Lammel, W. Lass, R. Lienenkamp, C. Loose, M.K.B. Lüdeke, O. Moldenhaeur, G. Petschel-Held, M. Plöchl and F. Reusswig, 1997: Syndromes of global change. *GAIA*, **6**, 19-34.

Schipper, E.L.F., 2006: Conceptual history of adaptation to climate change under the UNFCCC. *Review of European Community and International Environmental Law*, **15**, 82-92.

Schipper, L. and M. Pelling, 2006: Disaster risk, climate change and international development: scope for, and challenges to, integration. *Disasters*, **30**, 19-38.

Schneider, H., 1999: Participatory governance: the missing link for poverty reduction. OECD Development Centre Policy Brief No. 17, OECD Development Centre, Paris, 30 pp.

Schroeter, D., W. Cramer, R. Leemans, I.C. Prentice, M.B. Araujo, N.W. Arnell, A. Bondeau, H. Bugmann and Co-authors, 2005: Ecosystem service supply and vulnerability to global change in Europe. *Science*, **310**, 1333-1337.

Shaw, R., 2006: Community-based climate change adaptation in Vietnam: inter-linkage of environment, disaster and human security. *Multiple Dimensions of Global Environmental Changes*, S. Sonak, Ed., The Energy Research Institute (TERI), TERI Press, New Delhi, 521-547.

Shepherd, P., J. Tansey and H. Dowlatabadi, 2006: Context matters: what shapes adaptation to water stress in the Okanagan? *Climatic Change*, **78**, 31-62.

Shue, H., 1999: Global environment and international inequality. *Int. Aff.*, **75**, 531-545.

Smit, B., O. Pilifosova, I. Burton, B. Challenger, S. Huq, R.J.T. Klein and G. Yohe, 2001: Adaptation to climate change in the context of sustainable development and equity. *Climate Change 2001: Impacts, Adaptation and Vulnerability. Contribution of Working Group II to the Third Assessment Report of the Intergovernmental Panel on Climate Change*, J.J. McCarthy, O.F. Canziani, N.A. Leary, D.J. Dokken and K.S. White, Eds., Cambridge University Press, Cambridge, 877-912.

Sperling, F. and F. Szekely, 2005: Disaster risk management in a changing climate. Discussion Paper for the World Conference on Disaster Reduction on behalf of the Vulnerability and Adaptation Resource Group (VARG), reprint with addendum on conference outcomes, Washington, District of Columbia, 42 pp.

Stern, N., 2007: *The Economics of Climate Change: The Stern Review*. Cambridge University Press, Cambridge, 692 pp.

Sutherland, K., B. Smit, V. Wulf and T. Nakalevu, 2005: Vulnerability in Samoa. *Tiempo*, **54**, 11-15.

Summit Environmental Consultants Limited, 2004: Trepanier Landscape Unit (Westside) water management plan. Regional District of Central Okanagan and British Columbia, Ministry of Sustainable Resource Management, Kelowna, 300 pp.

Swaminathan, M.S., 2005: Environmental education for a sustainable future. *Glimpses of the Work on Environment and Development in India*, J.S. Singh and V.P. Sharma, Eds., Angkor Publishers, New Delhi, 51-71.

Swart, R., J. Robinson and S. Cohen, 2003: Climate change and sustainable development: expanding the options. *Climate Policy*, **3**, S19-S40.

Thomalla, F., T. Downing, E. Spanger-Siegfried, G.Y. Han and J. Rockstrom, 2006: Reducing hazard vulnerability: towards a common approach between disaster risk reduction and climate adaptation. *Disasters*, **30**, 39-48.

Thomas, D.S.G. and C. Twyman, 2004: Equity and justice in climate change adaptation amongst natural-resource-dependent societies. *Global Environ. Chang.*, **15**, 115-124.

Thomas, T., J.R. Schermerhorn and J.W. Deinhart, 2004: Strategic leadership of ethical behavior in business. *Acad. Manage. Exec.*, **18**, 56-66.

Tol, R.S.J., 1999: The marginal costs of greenhouse gas emissions. *The Energy Journal*, **20**, 61-81.

Tol, R.S.J., 2002: New estimates of the damage costs of climate change, Part II: dynamic estimates. *Environ. Resour. Econ.*, **21**, 135-160.

Tol, R.S.J., 2005: The marginal damage costs of carbon dioxide emissions: an assessment of the uncertainties. *Energ. Policy*, **33**, 2064-2074.

Tompkins, E. and N. Adger, 2005: Defining response capacity to enhance climate change policy. *Environ. Sci. Policy*, **8**, 562-571.

Toth, F.L., Ed., 1999: *Fair Weather? Equity Concerns in Climate Change*. Earthscan Publications Ltd., London. 212 pp.

Toth, F.L. and E. Hizsnyik, 2005: Managing the inconceivable: participatory assessments of impacts and responses to extreme climate change, 26 pp. [Accessed 11.06.07: http://www.uni-hamburg.de/Wiss/FB/15/Sustainability/atlantis.htm]

Turner, N.C., 2004: Sustainable production of crops and pastures under drought in a Mediterranean environment. *Ann. Appl. Biol.*, **144**, 139-147.

Turnpenny, J., A. Haxeltine and T. O'Riordan, 2004: A scoping study of user needs for integrated assessment of climate change in the UK context: Part 1 of the development of an interactive integrated assessment process. *Integrated Assessment*, **4**, 283-300.

Twinomugisha, B., 2005: Indigenous adaptation. *Tiempo*, **57**, 6-8.

UN (United Nations), 2005: Progress towards the Millennium Development Goals, 1990-2004. [Accessed 11.06.07: http://millenniumindicators.un.org/unsd/mi/mi_coverfinal.htm]

UNCRD (United Nations Centre for Regional Development), 2003: *Sustainability in Grass-Roots Initiatives: Focus on Community Based Disaster Management*. R. Shaw and K. Okazaki, Eds., Disaster Management Planning Hyogo Office and United Nations Centre for Regional Development, 103 pp.

UNDP (United Nations Development Program), 2003: *Human Development Report*. Oxford University Press, New York, 365 pp.

UNDP (United Nations Development Program), 2004: Reducing disaster risk: a challenge for development. UNDP, Geneva, 161 pp.

UNDSD (United Nations Division for Sustainable Development), 2006: Agenda 21. United Nations Conference on Environment and Development, Rio de Janeiro, Brazil, June 1992. [Accessed 11.06.07: http://www.un.org/esa/sustdev/documents/agenda21/index.htm]

UNSEA (United Nations Social and Economic Affairs), 2005: The inequality predicament: report on the world social situation 2005. United Nations General Assembly, New York, 152 pp.

Van Asselt, M.B.A. and N. Rijkens-Klomp, 2002: A look in the mirror: reflection on participation in integrated assessment from a methodological perspective. *Global Environ. Chang.*, **12**, 167-184.

van de Kerkhof, M., 2004: *Debating Climate Change: A Study of Stakeholder Participation in an Integrated Assessment of Long-Term Climate Policy in the Netherlands*. Lemma, Dordrecht, 317 pp.

Van Lieshout, M., R.S. Kovats, M.T.J. Livermore and P. Martens, 2004: Climate change and malaria: analysis of the SRES climate and socio-economic scenarios. *Global Environ. Chang.*, **14**, 87-99.

Vlassova, T.K., 2006: Arctic residents' observations and human impact assessments in understanding environmental changes in boreal forests: Russian experience and circumpolar perspectives. *Mitigation and Adaptation Strategies for Global Change*, **11**, 897-909.

Warren, R., 2006: Spotlighting impacts functions in integrated assessment models. Working Paper 91, Tyndall Centre for Climate Change Research, Norwich, 216 pp.

Watkiss, P., D. Anthoff, T. Downing, C. Hepburn, C. Hope, A. Hunt and R. Tol, 2005: The social costs of carbon (SCC) review: methodological approaches for using SCC estimates in policy assessment. Final Report, Defra, UK, 124 pp.

Watson International Scholars of the Environment, 2006: Making MEAs work for the poor. *Tiempo*, **58**, 6-11.

WCED (World Commission on Environment and Development), 1987: *Our Common Future*. Oxford University Press, Oxford, 398 pp.

Welp, M., A. de la Vega-Leinert and S. Stoll-Kleeman, 2006: Science-based stakeholder dialogues: theories and tools. *Global Environ. Chang.*, **16**, 170-181.

Winters, L.A., N. McCulloch and A. McKay, A., 2004: Trade liberalization and poverty: the evidence so far. *J. Econ. Lit.*, **42**, 72-115.

WKSS (West Kitikmeot / Slave Study Society), 2001: West Kitikmeot / Slave Study Society Final Report. West Kitikmeot / Slave Study Society, Yellowknife, Canada, 87 pp. [Accessed 11.06.07: http://www.wkss.nt.ca/index.htm]

World Bank, 1998: *Protecting Our Planet: Securing Our Future*. R.T. Watson, Ed., World Bank/UNEP/ NASA, Washington, District of Columbia, 116 pp.

WRI (World Resources Institute), 2000: *World Resources 2000-2001: People and Ecosystems: The Fraying Web of Life*. World Resources Institute, Washington, District of Columbia. 389 pp.

WSSD, 2002: Plan of implementation of the World Summit on Sustainable Development, 62 pp. [Accessed 11.06.07: http://www.un.org/esa/sustdev/documents/WSSD_POI_PD/English/WSSD_PlanImpl.pdf]

Yamin, F., 2004: Overview. IDS Bulletin, 35, 1-11.

Yohe, G. and R. Tol, 2002: Indicators for social and economic coping capacity: moving toward a working definition of adaptive capacity. Global Environ. Chang., 12, 25-40.

Yohe, G., E. Malone, A. Brenkert, M.E. Schlesinger, H. Meij and X. Xing, 2006a: Global distributions of vulnerability to climate change. Integrated Assessment Journal, 6, 35-44.

Yohe, G., E. Malone, A. Brenkert, M.E. Schlesinger, H. Meij, X. Xing and D. Lee, 2006b: A synthetic assessment of the global distribution of vulnerability to climate change from the IPCC perspective that reflects exposure and adaptive capacity.CIESIN (Center for International Earth Science Information Network), Columbia University, Palisades, New York, 17 pp. [Accessed 11.06.07: http://ciesin.columbia.edu/data/climate/]

Young, O., 2006: Vertical interplay among scale-dependent environmental and resource regimes. *Ecology and Society*, 11, Art. No. 27. [Accessed 11.06.07: http://www.ecologyandsociety.org/vol11/iss1/art27/]

From the report accepted by Working Group II of the Intergovernmental Panel on Climate Change but not approved in detail

Cross-chapter case studies

Cross-chapter case study citation:

These cross-chapter case studies collect together material from the chapters of the underlying report. A roadmap showing the location of this material is provided in the Introduction to the report. When referencing partial material from within a specific case study, please cite the chapter in which it originally appears. When referencing a whole case study, please cite as:

Parry, M.L., O.F. Canziani, J.P. Palutikof, P.J. van der Linden and C.E. Hanson, Eds., 2007: Cross-chapter case study. In: *Climate Change 2007: Impacts, Adaptation and Vulnerability. Contribution of Working Group II to the Fourth Assessment Report of the Intergovernmental Panel on Climate Change*, Cambridge University Press, Cambridge, UK, 843-868.

Table of Contents

C1. The impact of the European 2003 heatwave

C1.1 Scene-setting and overview

C1.1.1 The European heatwave of 2003 (Chapter 12, Section 12.6.1)

A severe heatwave over large parts of Europe in 2003 extended from June to mid-August, raising summer temperatures by 3 to 5°C in most of southern and central Europe (Figure C1.1). The warm anomalies in June lasted throughout the entire month (increases in monthly mean temperature of up to 6 to 7°C), but July was only slightly warmer than on average (+1 to +3°C), and the highest anomalies were reached between 1st and 13th August (+7°C) (Fink et al., 2004). Maximum temperatures of 35 to 40°C were repeatedly recorded and peak temperatures climbed well above 40°C (André et al., 2004; Beniston and Díaz, 2004).

Average summer (June to August) temperatures were far above the long-term mean by up to five standard deviations (Figure C1.1), implying that this was an extremely unlikely event under current climatic conditions (Schär and Jendritzky, 2004). However, it is consistent with a combined increase in mean temperature and temperature variability (Meehl and Tebaldi, 2004; Pal et al., 2004; Schär et al., 2004) (Figure C1.1). As such, the 2003 heatwave resembles simulations by regional climate models of summer temperatures in the latter part of the 21st century under the A2 scenario (Beniston, 2004). Anthropogenic warming may therefore already have increased the risk of heatwaves such as the one experienced in 2003 (Stott et al., 2004).

The heatwave was accompanied by annual precipitation deficits up to 300 mm. This drought contributed to the estimated 30% reduction in gross primary production of terrestrial ecosystems over Europe (Ciais et al., 2005). This reduced agricultural production and increased production costs, generating estimated damages of more than €13 billion (Fink et al., 2004; see also C1.2.2). The hot and dry conditions led to many very large wildfires, in particular in Portugal (390,000 ha: Fink et al., 2004; see also C1.2.1). Many major rivers (e.g., the Po, Rhine, Loire and Danube) were at record low levels, resulting in disruption of inland navigation, irrigation and power-plant cooling (Beniston and Díaz, 2004; Zebisch et al., 2005; see also C1.2.3). The extreme glacier melt in the Alps prevented even lower river flows in the Danube and Rhine (Fink et al., 2004).

The excess deaths due to the extreme high temperatures during the period June to August may amount to 35,000 (Kosatsky, 2005); elderly people were among those most affected (WHO, 2003; Kovats and Jendritzky, 2006; see also C1.2.4). The heatwave in 2003 has led to the development of heat health-watch warning systems in several European countries including France (Pascal et al., 2006), Spain (Simón et al., 2005), Portugal (Nogueira, 2005), Italy (Michelozzi et al., 2005), the UK (NHS, 2006) and Hungary (Kosatsky and Menne, 2005).

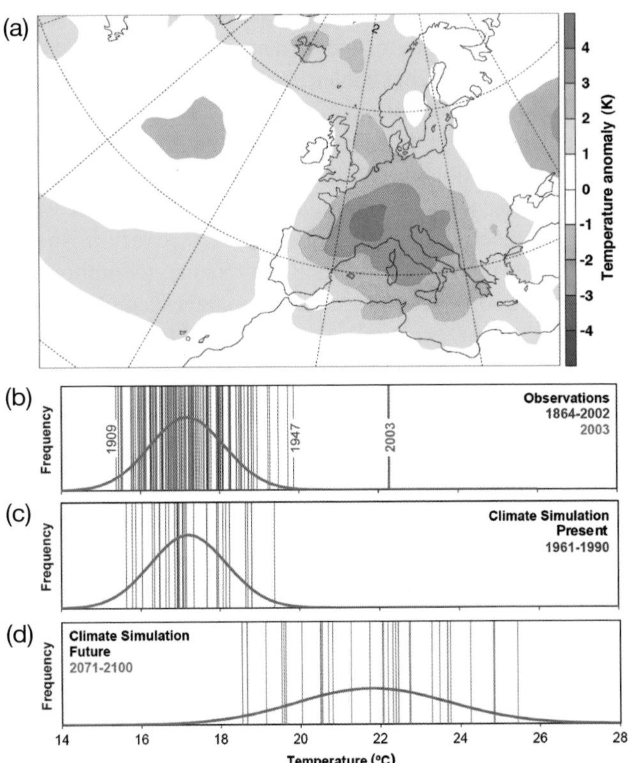

Figure C1.1. *Characteristics of the summer 2003 heatwave (adapted from Schär et al., 2004). (a) JJA temperature anomaly with respect to 1961 to 1990. (b) to (d) JJA temperatures for Switzerland observed during 1864 to 2003 (b), simulated using a regional climate model for the period 1961 to 1990 (c), and simulated for 2071 to 2100 under the A2 scenario using boundary data from the HadAM3H GCM (d). In panels (b) to (d): the black line shows the theoretical frequency distribution of mean summer temperature for the time-period considered, and the vertical blue and red bars show the mean summer temperature for individual years. Reprinted by permission from Macmillan Publishers Ltd. [Nature] (Schär et al., 2004), copyright 2004.*

C1.2 Impacts on sectors

C1.2.1 Ecological impacts of the European heatwave 2003 (Chapter 4, Box 4.1)

Anomalous hot and dry conditions affected Europe between June and mid-August 2003 (Fink et al., 2004; Luterbacher et al., 2004; Schär et al., 2004). Since similarly warm summers may occur at least every second year by 2080 in a Special Report on Emissions Scenario (SRES; Nakićenović et al., 2000) A2 world, for example (Beniston, 2004; Schär et al., 2004), effects on ecosystems observed in 2003 provide a conservative analogue of future impacts. The major effects of the 2003 heatwave on vegetation and ecosystems appear to have been through heat and drought stress, and wildfires.

Drought stress impacts on vegetation (Gobron et al., 2005; Lobo and Maisongrande, 2006) reduced gross primary production (GPP) in Europe by 30% and respiration to a lesser degree, overall resulting in a net carbon source of 0.5 PgC/yr (Ciais et al., 2005). However, vegetation responses to the heat varied along environmental gradients such as altitude, e.g., by prolonging the growing season at high elevations (Jolly et al., 2005). Some vegetation types, as monitored by remote sensing, were found to recover to a normal state by 2004 (e.g., Gobron et al., 2005), but enhanced crown damage of dominant forest trees in 2004, for example, indicates complex delayed impacts (Fischer, 2005). Freshwater ecosystems experienced prolonged depletion of oxygen in deeper layers of lakes during the heatwave (Jankowski et al., 2006), and there was a significant decline and subsequent poor recovery in species richness of molluscs in the River Saône (Mouthon and Daufresne, 2006). Taken together, this suggests quite variable resilience across ecosystems of different types, with very likely progressive impairment of ecosystem composition and function if such events increase in frequency (e.g., Lloret et al., 2004; Rebetez and Dobbertin, 2004; Jolly et al., 2005; Fuhrer et al., 2006).

High temperatures and greater dry spell durations increase vegetation flammability (e.g., Burgan et al., 1997), and during the 2003 heatwave a record-breaking incidence of spatially extensive wildfires was observed in European countries (Barbosa et al., 2003), with roughly 650,000 ha of forest burned across the continent (De Bono et al., 2004). Fire extent (area burned), although not fire incidence, was exceptional in Europe in 2003, as found for the extraordinary 2000 fire season in the USA (Brown and Hall, 2001), and noted as an increasing trend in the USA since the 1980s (Westerling et al., 2006). In Portugal, area burned was more than twice the previous extreme (1998) and four times the 1980-2004 average (Trigo et al., 2005, 2006). Over 5% of the total forest area of Portugal burned, with an economic impact exceeding €1 billion (De Bono et al., 2004).

Long-term impacts of more frequent similar events are very likely to cause changes in biome type, particularly by promoting highly flammable, shrubby vegetation that burns more frequently than less flammable vegetation types such as forests (Nunes et al., 2005), and as seen in the tendency of burned woodlands to reburn at shorter intervals (Vazquez and Moreno, 2001; Salvador et al., 2005). The conversion of vegetation structure in this way on a large enough scale may even cause accelerated climate change through losses of carbon from biospheric stocks (Cox et al., 2000). Future projections for Europe suggest significant reductions in species richness even under mean climate change conditions (Thuiller et al., 2005), and an increased frequency of such extremes (as indicated e.g., by Schär et al., 2004) is likely to exacerbate overall biodiversity losses (Thuiller et al., 2005).

C1.2.2 European heatwave impact on the agricultural sector (Chapter 5, Box 5.1)

Europe experienced a particularly extreme climate event during the summer of 2003, with temperatures up to 6°C above long-term means, and precipitation deficits up to 300 mm (see Trenberth et al., 2007). A record drop in crop yield of 36% occurred in Italy for maize grown in the Po valley, where extremely high temperatures prevailed (Ciais et al., 2005). In France, compared to 2002, the maize grain crop was reduced by 30% and fruit harvests declined by 25%. Winter crops (wheat) had nearly achieved maturity by the time of the heatwave and therefore suffered less yield reduction (21% decline in France) than summer crops (e.g., maize, fruit trees and vines) undergoing maximum foliar development (Ciais et al., 2005). Forage production was reduced on average by 30% in France and hay and silage stocks for winter were partly used during the summer (COPA COGECA, 2003a). Wine production in Europe was the lowest in 10 years (COPA COGECA, 2003b). The (uninsured) economic losses for the agriculture sector in the European Union were estimated at €13 billion, with the largest losses in France (€4 billion) (Sénat, 2004).

C1.2.3 Industry, settlement and society: impacts of the 2003 heatwave in Europe (Chapter 7, Box 7.1)

The Summer 2003 heatwave in western Europe affected settlements and economic services in a variety of ways. Economically, this extreme weather event created stress on health, water supplies, food storage and energy systems. In France, electricity became scarce, construction productivity fell, and the cold storage systems of 25-30% of all food-related establishments were found to be inadequate (Létard et al., 2004). The punctuality of the French railways fell to 77%, from 87% twelve months previously, incurring €1 to €3 million (US$1.25 to 3.75 million) in additional compensation payments, an increase of 7-20% compared with the usual annual total. Sales of clothing were 8.9% lower than usual in August, but sales of bottled water increased by 18%, and of ice-cream by 14%. The tourist industry in northern France benefitted, but in the south it suffered (Létard et al., 2004).

Impacts of the heatwave were mainly health- and health-service-related (see Section C1.2.4); but they were also associated with settlement and social conditions, from inadequate climate conditioning in buildings to the fact that many of the dead were elderly people, left alone while their families were on vacation. Electricity demand increased with the high heat levels; but electricity production was undermined by the facts that the temperature of rivers rose, reducing the cooling efficiency of thermal power plants (conventional and nuclear) and that flows of rivers were diminished; six power plants were shut down completely (Létard et al., 2004). If the heatwave had continued, as much as 30% of national power production would have been at risk (Létard et al., 2004). The crisis illustrated how infrastructure may be unable to deal with complex, relatively sudden environmental challenges (Lagadec, 2004).

C1.2.4 The European heatwave 2003: health impacts and adaptation (Chapter 8, Box 8.1)

In August 2003, a heatwave in France caused more than 14,800 deaths (Figure C1.2). Belgium, the Czech Republic, Germany, Italy, Portugal, Spain, Switzerland, the Netherlands

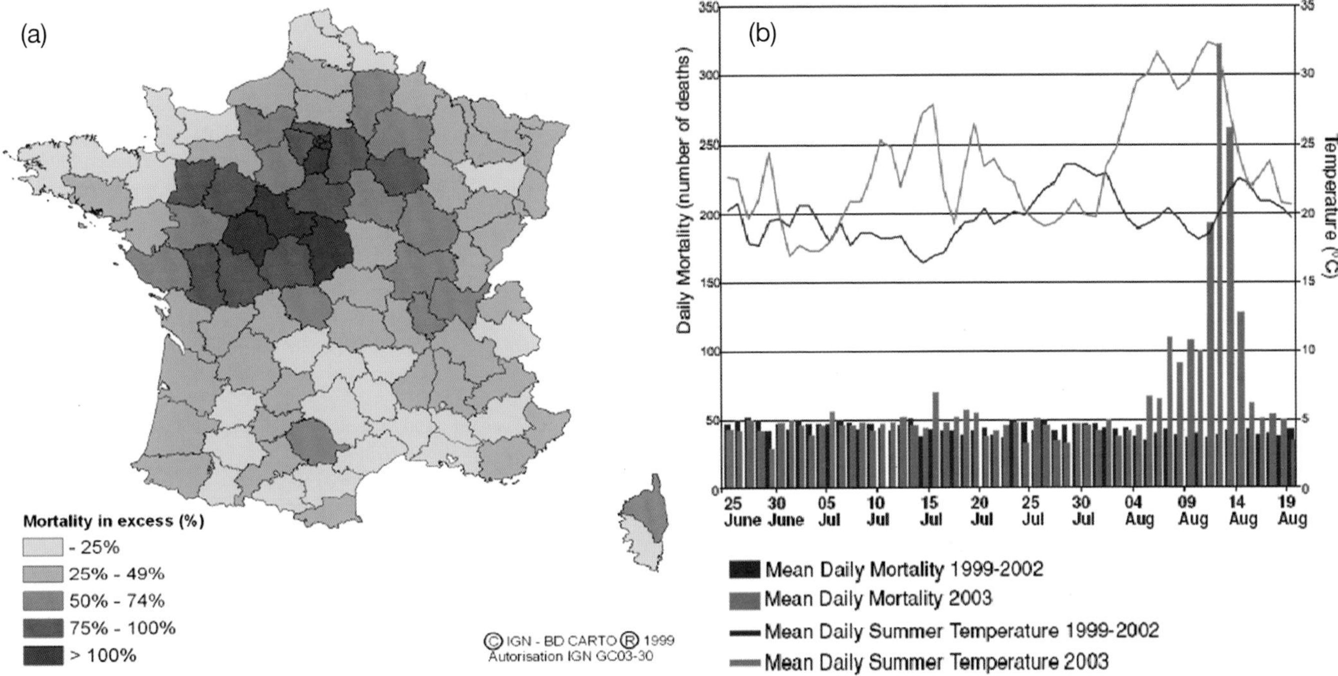

Figure C1.2. *(a) The distribution of excess mortality in France from 1 to 15 August 2003, by region, compared with the previous three years (INVS, 2003); (b) the increase in daily mortality in Paris during the heatwave in early August (Vandentorren and Empereur-Bissonnet, 2005).*

and the UK all reported excess mortality during the heatwave period, with total deaths in the range of 35,000 (Hemon and Jougla, 2004; Martinez-Navarro et al., 2004; Michelozzi et al., 2004; Vandentorren et al., 2004; Conti et al., 2005; Grize et al., 2005; Johnson et al., 2005). In France, around 60% of the heatwave deaths occurred in persons aged 75 and over (Hemon and Jougla, 2004). Other harmful exposures were also caused or exacerbated by the extreme weather, such as outdoor air pollutants (tropospheric ozone and particulate matter) (EEA, 2003), and pollution from forest fires.

A French parliamentary inquiry concluded that the health impact was 'unforeseen', surveillance for heatwave deaths was inadequate, and the limited public-health response was due to a lack of experts, limited strength of public-health agencies, and poor exchange of information between public organisations (Lagadec, 2004; Sénat, 2004).

In 2004, the French authorities implemented local and national action plans that included heat health-warning systems, health and environmental surveillance, re-evaluation of care of the elderly, and structural improvements to residential institutions (such as adding a cool room) (Laaidi et al., 2004; Michelon et al., 2005). Across Europe, many other governments (local and national) have implemented heat health-prevention plans (Michelozzi et al., 2005; WHO Regional Office for Europe, 2006).

Since the observed higher frequency of heatwaves is likely to have occurred due to human influence on the climate system (Hegerl et al., 2007), the excess deaths of the 2003 heatwave in Europe are likely to be linked to climate change.

References

André, J.-C., M. Déqué, P. Rogel and S. Planton, 2004: The 2003 summer heatwave and its seasonal forecasting. *C. R. Geosci.*, **336**, 491-503.

Barbosa, P., G. Libertà and G. Schmuck, 2003: The European Forest Fires Information System (EFFIS) results on the 2003 fire season in Portugal by the 20th of August. Report 20030820, European Commission, Directorate General Joint Research Centre, Institute for Environment and Sustainability, Land Management Unit, Ispra, 10 pp.

Beniston, M., 2004: The 2003 heatwave in Europe: a shape of things to come? An analysis based on Swiss climatological data and model simulations. *Geophys. Res. Lett.*, **31**, L02202, doi:10.1029/2003GL018857.

Beniston, M. and H.F. Díaz, 2004: The 2003 heatwave as an example of summers in a greenhouse climate? Observations and climate model simulations for Basel, Switzerland. *Glob. Planet. Change*, **44**, 73-81.

Brown, T.J. and B.L. Hall, 2001: Climate analysis of the 2000 fire season. CE-FA Report 01-02, Program for Climate Ecosystem and Fire Applications, Desert Research Institute, Division of Atmospheric Sciences, Reno, Nevada, 40 pp.

Burgan, R.E., P.L. Andrews, L.S. Bradshaw, C.H. Chase, R.A. Hartford and D.J. Latham, 1997: WFAS: wildland fire assessment system. *Fire Management Notes*, **57**, 14-17. Ciais, Ph., M. Reichstein, N. Viovy, A. Granier, J. Ogée, V. Allard, M. Aubinet, N. Buchmann, C. Bernhofer, A. Carrara, F. Chevallier, N. de Noblet, A.D. Friend, P. Friedlingstein, T. Grünwald, B. Heinesch, P. Keronen, A. Knohl, G. Krinner, D. Loustau, G. Manca, G. Matteucci, F. Miglietta, J.M. Ourcival, D. Papale, K. Pilegaard, S. Rambal, G. Seufert, J.F. Soussana, M.J. Sanz, E.D. Schulze, T. Vesala and R. Valentini, 2005: Europe-wide reduction in primary productivity caused by the heat and drought in 2003. *Nature*, **437**, 529-533.

Conti, S., P. Meli, G. Minelli, R. Solimini, V. Toccaceli, M. Vichi, C. Beltrano and L. Perini, 2005: Epidemiologic study of mortality during the Summer 2003 heat wave in Italy. *Environ. Res.*, **98**, 390-399.

COPA COGECA, 2003a: Assessment of the impact of the heat wave and drought of the summer 2003 on agriculture and forestry. Committee of Agricultural Organisations in the European Union General Committee for Agricultural Cooperation in the European Union, Brussels, 15 pp.

COPA COGECA, 2003b: Committee of Agricultural Organisations in the European Union General Committee for Agricultural Cooperation in the European Union, CDP 03 61 1, Press release, Brussels.

Cox, P.M., R.A. Betts, C.D. Jones, S.A. Spall and I.J. Totterdell, 2000: Acceleration of global warming due to carbon-cycle feedbacks in a coupled climate model. *Nature*, **408**, 184-187.

De Bono, A., P. Peduzzi, G. Giuliani and S. Kluser, 2004: *Impacts of Summer 2003 Heat Wave in Europe*. Early Warning on Emerging Environmental Threats 2, UNEP: United Nations Environment Programme, Nairobi, 4 pp.

EEA, 2003: Air pollution by ozone in Europe in summer 2003: overview of exceedances of EC ozone threshold values during the summer season April–August 2003 and comparisons with previous years. Topic Report No 3/2003, European Economic Association, Copenhagen, 33 pp. http://reports.eea.europa.eu/topic_report_2003_3/en.

Fink, A.H., T. Brücher, A. Krüger, G.C. Leckebusch, J.G. Pinto and U. Ulbrich, 2004: The 2003 European summer heatwaves and drought: synoptic diagnosis and impact. *Weather*, **59**, 209-216.

Fischer, R., Ed., 2005: The condition of forests in Europe: 2005 executive report. United Nations Economic Commission for Europe (UN-ECE), Geneva, 36 pp.

Fuhrer, J., M. Beniston, A. Fischlin, C. Frei, S. Goyette, K. Jasper and C. Pfister, 2006: Climate risks and their impact on agriculture and forests in Switzerland. *Climatic Change*, **79**, 79-102.

Gobron, N., B. Pinty, F. Melin, M. Taberner, M.M. Verstraete, A. Belward, T. Lavergne and J.L. Widlowski, 2005: The state of vegetation in Europe following the 2003 drought. *Int. J. Remote Sens.*, **26**, 2013-2020.

Grize, L., A. Huss, O. Thommen, C. Schindler and C. Braun-Fahrländer, 2005: Heat wave 2003 and mortality in Switzerland. *Swiss Med. Wkly.*, **135**, 200-205.

Hemon, D. and E. Jougla, 2004: La canicule du mois d'aout 2003 en France [The heatwave in France in August 2003]. *Rev. Epidemiol. Santé*, **52**, 3-5.

Hegerl, G.C., F.W. Zwiers, P. Braconnot, N.P. Gillett, Y. Luo, J.A. Marengo Orsini, N. Nicholls, J.E. Penner and P.A. Stott, 2007: Understanding and attributing climate change. *Climate Change 2007: The Physical Science Basis. Contribution of Working Group I to the Fourth Assessment Report of the Intergovernmental Panel on Climate Change*, S. Solomon, D. Qin, M. Manning, Z. Chen, M. Marquis, K.B. Averyt, M. Tignor and H.L. Miller, Eds., Cambridge University Press, Cambridge, 663-746.

INVS, 2003: *Impact sanitaire de la vague de chaleur d'août 2003 en France. Bilan et perpectives [Health Impact of the Heatwave in August 2003 in France]*. Institut de Veille Sanitaire, Saint-Maurice, 120 pp.

Jankowski, T., D.M. Livingstone, H. Bührer, R. Forster and P. Niederhaser, 2006: Consequences of the 2003 European heat wave for lake temperature profiles, thermal stability and hypolimnetic oxygen depletion: implications for a warmer world. *Limnol. Oceanogr.*, **51**, 815-819.

Johnson, H., R.S. Kovats, G.R. McGregor, J.R. Stedman, M. Gibbs, H. Walton, L. Cook and E. Black, 2005: The impact of the 2003 heatwave on mortality and hospital admissions in England. *Health Statistics Q.*, **25**, 6-12.

Jolly, W.M., M. Dobbertin, N.E. Zimmermann and M. Reichstein, 2005: Divergent vegetation growth responses to the 2003 heat wave in the Swiss Alps. *Geophys. Res. Lett.*, **32**, L18409, doi:10.1029/2005GL023252.

Kosatsky, T., 2005: The 2003 European heatwave. *Euro Surveill.*, **10**, 148-149.

Kosatsky, T. and B. Menne, 2005: Preparedness for extreme weather among national ministries of health of WHO's European region. *Climate Change and Adaptation Strategies for Human Health,* B. Menne and K.L. Ebi, Eds., Springer, Darmstadt, 297-329.

Kovats, R.S. and G. Jendritzky, 2006: Heat-waves and human health. *Climate Change and Adaptation Strategies for Human Health,* B. Menne and K.L. Ebi, Eds., Springer, Darmstadt, 63-90.

Laaidi, K., M. Pascal, M. Ledrans, A. Le Tertre, S. Medina, C. Caserio, J.C. Cohen, J. Manach, P. Beaudeau and P. Empereur-Bissonnet, 2004: *Le système français d'alerte canicule et santé (SACS 2004): Un dispositif intégéré au Plan National Canicule [The French Heatwave Warning System and Health: An Integrated National Heatwave Plan]*. Institutde Veille Sanitaire, 35 pp.

Lagadec, P., 2004. Understanding the French 2003 heat wave experience. *J. Contingencies Crisis Manage.*, **12**, 160-169.

Létard V., H. Flandre and S. Lepeltier, 2004: La France et les Français face à la canicule: les leçons d'une crise. Report No. 195 (2003-2004) to the Sénat, Government of France, 391 pp.

Lloret, F., D. Siscart and C. Dalmases, 2004: Canopy recovery after drought dieback in holm-oak Mediterranean forests of Catalonia (NE Spain). *Global Change Biol.*, **10**, 2092-2099.

Lobo, A. and P. Maisongrande, 2006: Stratified analysis of satellite imagery of SW Europe during summer 2003: the differential response of vegetation classes to increased water deficit. *Hydrol. Earth Syst. Sci.*, **10**, 151-164.

Luterbacher, J., D. Dietrich, E. Xoplaki, M. Grosjean and H. Wanner, 2004: European seasonal and annual temperature variability, trends, and extremes since 1500. *Science*, **303**, 1499-1503.

Martinez-Navarro, F., F. Simon-Soria and G. Lopez-Abente, 2004: Valoracion del impacto de la ola de calor del verano de 2003 sobre la mortalidad [Evaluation of the impact of the heatwave in the summer of 2003 on mortality]. *Gac. Sanit.*, **18**, 250-258.

Meehl, G.A. and C. Tebaldi, 2004: More intense, more frequent, and longer lasting heatwaves in the 21st century. *Science*, **305**, 994-997.

Michelon, T., P. Magne and F. Simon-Delavelle, 2005: Lessons from the 2003 heat wave in France and action taken to limit the effects of future heat wave. *Extreme Weather Events and Public Health Responses*, W. Kirch, B. Menne and R. Bertolllini, Eds., Springer, Berlin, 131-140.

Michelozzi, P., F. de Donato, G. Accetta, F. Forastiere, M. D'Ovido and L.S. Kalkstein, 2004: Impact of heat waves on mortality: Rome, Italy, June–August 2003. *J. Am. Med. Assoc.*, **291**, 2537-2538.

Michelozzi, P., F. de Donato, L. Bisanti, A. Russo, E. Cadum, M. DeMaria, M. D'Ovidio, G. Costa and C.A. Perucci, 2005: The impact of the summer 2003 heatwaves on mortality in four Italian cities. *Euro. Surveill.*, **10**, 161-165.

Mouthon, J. and M. Daufresne, 2006: Effects of the 2003 heatwave and climatic warming on mollusc communities of the Saône: a large lowland river and of its two main tributaries (France). *Global Change Biol.*, **12**, 441-449.

Nakićenović, N., J. Alcamo, G. Davis, B. de Vries, J. Fenhann, S. Gaffin, K. Gregory, A. Grübler, T.Y. Jung, T. Kram, E. Lebre la Rovere, L. Michaelis, S. Mori, T. Morita, W. Pepper, H. Pitcher, L. Price, K. Riahi, A. Roehrl, H.H. Rogner, A. Sankovski, M. Schlesinger, P. Shukla, S. Smith, R. Swart, S. van Rooijen, N. Victor and Z. Dadi, Eds., 2000: *Emissions Scenarios: A Special Report of the Intergovernmental Panel on Climate Change (IPCC)*. Cambridge University Press, Cambridge, 599 pp.

NHS, 2006: *Heatwave Plan for England: Protecting Health and Reducing Harm from Extreme Heat and Heatwaves*. Department of Health, London.

Nogueira, P.J., 2005: Examples of heat warning systems: Lisbon's ICARO surveillance system, summer 2003. *Extreme Weather Events and Public Health Responses*, W. Kirch, B. Menne and R. Bertollini, Eds., Springer, Heidelberg, 141-160.

Nunes, M.C.S., M.J. Vasconcelos, J.M.C. Pereira, N. Dasgupta and R.J. Alldredge, 2005: Land cover type and fire in Portugal: do fires burn land cover selectively? *Landscape Ecol.*, **20**, 661-673.

Pal, J.S., F. Giorgi and X.Q. Bi, 2004: Consistency of recent European summer precipitation trends and extremes with future regional climate projections. *Geophys. Res. Lett.*, **31**, L13202, doi:10.1029/2004GL019836.

Pascal, M., K. Laaidi, M. Ledrans, E. Baffert, C. Caseiro-Schönemann, A.L. Tertre, J. Manach, S. Medina, J. Rudant and P. Empereur-Bissonnet, 2006: France's heat health watch warning system. *Int. J. Biometeorol.*, **50**, 144-153.

Rebetez, M. and M. Dobbertin, 2004: Climate change may already threaten Scots pine stands in the Swiss Alps. *Theor. Appl. Climatol.*, **79**, 1-9.

Salvador, R., F. Lloret, X. Pons and J. Pinol, 2005: Does fire occurrence modify the probability of being burned again? A null hypothesis test from Mediterranean ecosystems in NE Spain. *Ecol. Model.*, **188**, 461-469.

Schär, C. and G. Jendritzky, 2004: Climate change: hot news from summer 2003. *Nature*, **432**, 559-560.

Schär, C., P.L. Vidale, D. Lüthi, C. Frei, C. Häberli, M.A. Liniger and C. Appenzeller, 2004: The role of increasing temperature variability in European summer heatwaves. *Nature*, **427**, 332-336.

Sénat, 2004: France and the French face the canicule: the lessons of a crisis. Appendix to the Minutes of the Session of February 3, 2004. Information Report No. 195, 59-62. http://www.senat.fr/rap/r03-195/r03-195.html.

Simón, F., G. López-Abente, E. Ballester and F. Martínez, 2005: Mortality in Spain during the heatwaves of summer 2003. *Euro Surveill.*, **10**, 156-160.

Stott, P.A., D.A. Stone and M.R. Allen, 2004: Human contribution to the European heatwave of 2003. *Nature*, **432**, 610-614.

Thuiller, W., S. Lavorel, M.B. Araujo, M.T. Sykes and I.C. Prentice, 2005: Climate change threats to plant diversity in Europe. *P. Natl. Acad. Sci. USA*, **102**, 8245-

8250.

Trenberth, K.E., P.D. Jones, P.G. Ambenje, R. Bojariu, D.R. Easterling, A.M.G. Klein Tank, D.E. Parker, J.A. Renwick and Co-authors, 2007: Observations: surface and atmospheric climate change. *Climate Change 2007: The Physical Science Basis. Contribution of Working Group I to the Fourth Assessment Report of the Intergovernmental Panel on Climate Change*, S. Solomon, D. Qin, M. Manning, Z. Chen, M. Marquis, K.B. Averyt, M. Tignor and H.L. Miller, Eds., Cambridge University Press, Cambridge, 235-336.

Trigo, R.M., M.G. Pereira, J.M.C. Pereira, B. Mota, T.J. Calado, C.C. da Camara and F.E. Santo, 2005: The exceptional fire season of summer 2003 in Portugal. *Geophys. Res. Abstracts*, **7**, 09690.

Trigo, R.M., J.M.C. Pereira, M.G. Pereira, B. Mota, T.J. Calado, C.C. Dacamara and F.E. Santo, 2006: Atmospheric conditions associated with the exceptional fire season of 2003 in Portugal. *Int. J. Climatol.*, **26**, 1741-1757.

Vandentorren, S. and P. Empereur-Bissonnet, 2005: Health impact of the 2003 heat-wave in France. *Extreme Weather Events and Public Health Responses*, W. Kirch, B. Menne and R. Bertollini, Eds., Springer, Heidelberg, 81-88.

Vandentorren, S., F. Suzan, S. Medina, M. Pascal, A. Maulpoix, J.-C. Cohen and M. Ledrans, 2004: Mortality in 13 French cities during the August 2003 heat-wave. *Am. J. Public Health*, **94**, 1518-1520.

Vazquez, A. and J.M. Moreno, 2001: Spatial distribution of forest fires in Sierra de Gredos (Central Spain). *Forest Ecol. Manag.*, **147**, 55-65.

Westerling, A.L., H.G. Hidalgo, D.R. Cayan and T.W. Swetnam, 2006: Warming and earlier spring increase western US forest wildfire activity. *Science*, **313**, 940-943.

WHO, 2003: *The Health Impacts of 2003 Summer Heat-Waves*. Briefing Note for the Delegations of the fifty-third session of the WHO Regional Committee for Europe. World Health Organization, Geneva, 12 pp.

WHO Regional Office for Europe, 2006: 1st meeting of the project 'Improving Public Health Responses to Extreme Weather/Heat-waves'. EuroHEAT Report on a WHO Meeting in Rome, Italy, 20–22 June 2005. WHO Regional Office for Europe, Copenhagen, 52 pp.

Zebisch, M., T. Grothmann, D. Schröter, C. Hasse, U. Fritsch and W. Cramer, 2005: *Climate Change in Germany: Vulnerability and Adaptation of Climate-Sensitive Sectors*. Umweltbundesamt Climate Change 10/05 (UFOPLAN 201 41 253), Dessau, 205 pp.

C2. Impacts of climate change on coral reefs

C2.1 Present-day changes in coral reefs

C2.1.1 Observed changes in coral reefs (Chapter 1, Section 1.3.4.1)

Concerns about the impacts of climate change on coral reefs centre on the effects of the recent trends in increasing acidity (via increasing CO_2), storm intensity, and sea surface temperatures (see Bindoff et al., 2007, Section 5.4.2.3; Trenberth et al., 2007, Sections 3.8.3 and 3.2.2).

Decreasing pH (see C2.2.1) leads to a decreased aragonite saturation state, one of the main physicochemical determinants of coral calcification (Kleypas et al., 1999). Although laboratory experiments have demonstrated a link between aragonite saturation state and coral growth (Langdon et al., 2000; Ohde and Hossain, 2004), there are currently no data relating altered coral growth *in situ* to increasing acidity.

Storms damage coral directly through wave action and indirectly through light attenuation by suspended sediment and abrasion by sediment and broken corals. Most studies relate to individual storm events, but a meta-analysis of data from 1977 to 2001 showed that coral cover on Caribbean reefs decreased by 17% on average in the year following a hurricane, with no evidence of recovery for at least 8 years post-impact (Gardner et al., 2005). Stronger hurricanes caused more coral loss, but the second of two successive hurricanes caused little additional damage, suggesting a greater future effect from increasing hurricane intensity rather than from increasing frequency (Gardner et al., 2005).

There is now extensive evidence of a link between coral bleaching – a whitening of corals as a result of the expulsion of symbiotic zooxanthellae (see C2.1.2) – and sea surface temperature anomalies (McWilliams et al., 2005). Bleaching usually occurs when temperatures exceed a 'threshold' of about 0.8-1°C above mean summer maximum levels for at least 4 weeks (Hoegh-Guldberg, 1999). Regional-scale bleaching events have increased in frequency since the 1980s (Hoegh-Guldberg, 1999). In 1998, the largest bleaching event to date is estimated to have killed 16% of the world's corals, primarily in the western Pacific and the Indian Ocean (Wilkinson, 2004). On many reefs, this mortality has led to a loss of structural complexity and shifts in reef fish species composition (Bellwood et al., 2006; Garpe et al., 2006; Graham et al., 2006). Corals that recover from bleaching suffer temporary reductions in growth and reproductive capacity (Mendes and Woodley, 2002), while the recovery of reefs following mortality tends to be dominated by fast-growing and bleaching-resistant coral genera (Arthur et al., 2005).

While there is increasing evidence for climate change impacts on coral reefs, disentangling the impacts of climate-related stresses from other stresses (e.g., over-fishing and pollution; Hughes et al., 2003) is difficult. In addition, inter-decadal variation in pH (Pelejero et al., 2005), storm activity (Goldenberg et al., 2001) and sea surface temperatures (Mestas-Nunez and Miller, 2006) linked, for example, to the El Niño-Southern Oscillation and Pacific Decadal Oscillation, make it more complicated to discern the effect of anthropogenic climate change from natural modes of variability. An analysis of bleaching in the Caribbean indicates that 70% of the variance in geographic extent of bleaching between 1983 and 2000 could be attributed to variation in ENSO and atmospheric dust (Gill et al., 2006).

C2.1.2 Environmental thresholds and observed coral bleaching (Chapter 6, Box 6.1)

Coral bleaching, due to the loss of symbiotic algae and/or their pigments, has been observed on many reefs since the early 1980s. It may have previously occurred, but has gone unrecorded. Slight paling occurs naturally in response to seasonal increases in sea surface temperature (SST) and solar radiation. Corals bleach white in response to anomalously high SST (~1°C above average seasonal maxima, often combined with high solar radiation). Whereas some corals recover their natural colour when environmental conditions ameliorate, their growth rate and reproductive ability may be significantly reduced for a substantial period. If bleaching is prolonged, or if SST exceeds 2°C above average seasonal maxima, corals die. Branching species appear more susceptible than massive corals (Douglas, 2003).

Major bleaching events were observed in 1982-1983, 1987-1988 and 1994-1995 (Hoegh-Guldberg, 1999). Particularly severe bleaching occurred in 1998 (Figure C2.1), associated with pronounced El Niño events in one of the hottest years on record (Lough, 2000; Bruno et al., 2001). Since 1998 there have been several extensive bleaching events. For example, in 2002 bleaching occurred on much of the Great Barrier Reef (Berkelmans et al., 2004; see C2.2.3) and elsewhere. Reefs in the eastern Caribbean experienced a massive bleaching event in late 2005, another of the hottest years on record. On many Caribbean reefs, bleaching exceeded that of 1998 in both extent and mortality (Figure C2.1), and reefs are in decline as a result of the synergistic effects of multiple stresses (Gardner et al., 2005; McWilliams et al., 2005; see C2.3.1). There is considerable variability in coral susceptibility and recovery to elevated SST in both time and space, and in the incidence of mortality (Webster et al., 1999; Wilkinson, 2002; Obura, 2005).

Global climate model results imply that thermal thresholds will be exceeded more frequently, with the consequence that bleaching will recur more often than reefs can sustain (Hoegh-Guldberg, 1999, 2004; Donner et al., 2005), perhaps almost annually on some reefs in the next few decades (Sheppard, 2003; Hoegh-Guldberg, 2005). If the threshold remains unchanged, more frequent bleaching and mortality seems inevitable (see

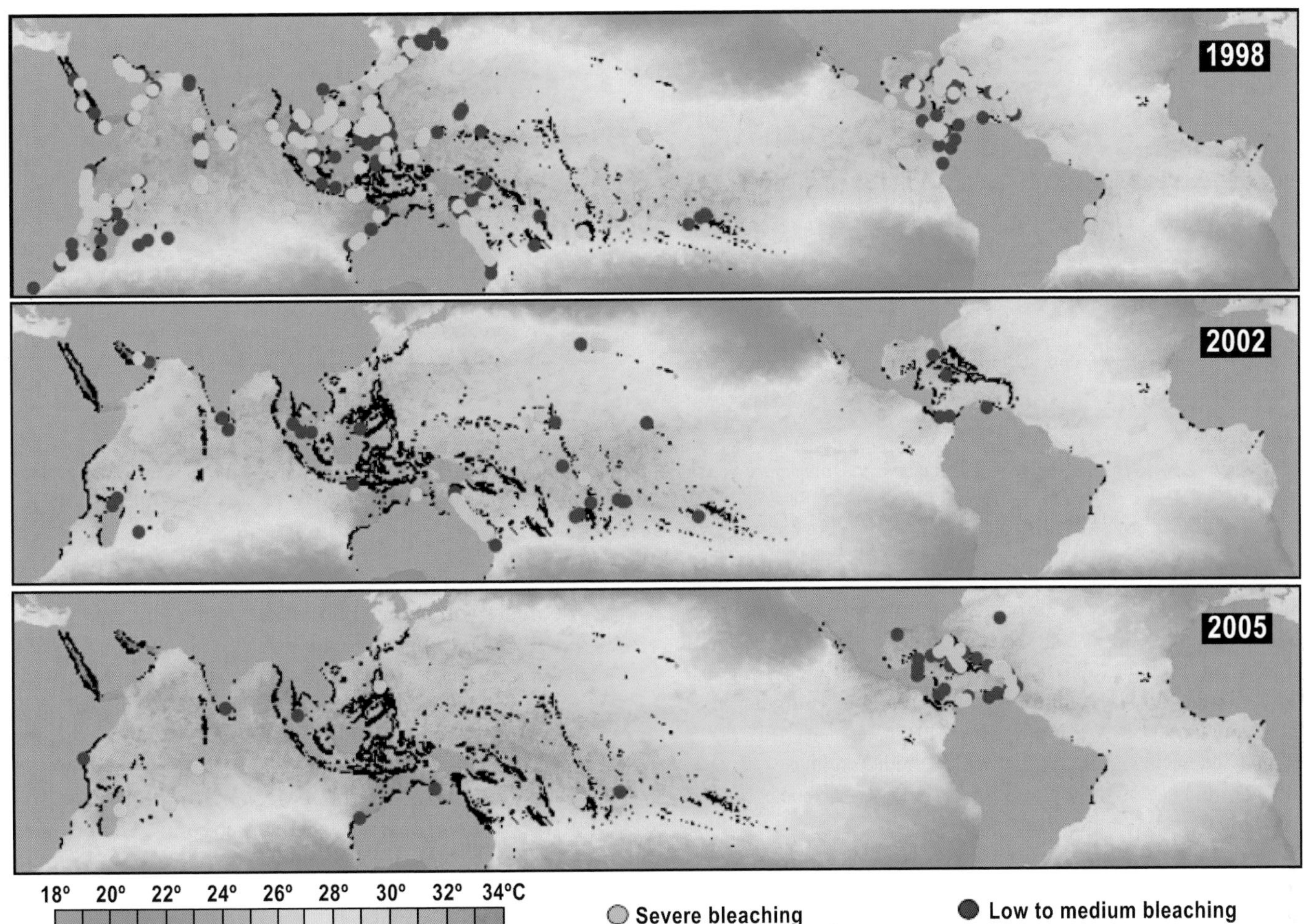

Figure C2.1. *Maximum monthly mean sea surface temperature for 1998, 2002 and 2005, and locations of reported coral bleaching (data sources: NOAA Coral Reef Watch (http://coralreefwatch.noaa.gov/) and Reefbase (http://www.reefbase.org/)).*

Figure C2.2a), but with local variations due to different susceptibilities to factors such as water depth. Recent preliminary studies lend some support to the adaptive bleaching hypothesis, indicating that the coral host may be able to adapt or acclimatise as a result of expelling one clade[1] of symbiotic algae but recovering with a new one (termed 'shuffling', see C2.2.1), creating 'new' ecospecies with different temperature tolerances (Coles and Brown, 2003; Buddemeier et al., 2004; Little et al., 2004; Rowan, 2004; Obura, 2005). Adaptation or acclimatisation might result in an increase in the threshold temperature at which bleaching occurs (Figure C2.2b). The extent to which the thermal threshold could increase with warming of more than a couple of degrees remains very uncertain, as are the effects of additional stresses, such as reduced carbonate supersaturation in surface waters (see C2.2.1) and non-climate stresses (see C2.3.1). Corals and other calcifying organisms (e.g., molluscs, foraminifers) remain extremely susceptible to increases in SST. Bleaching events reported in recent years have already impacted many reefs, and their more frequent recurrence is very likely to further reduce both coral cover and diversity on reefs over the next few decades.

C2.2 Future impacts on coral reefs

C2.2.1 Are coral reefs endangered by climate change? (Chapter 4, Box 4.4)

Reefs are habitat for about a quarter of all marine species and are the most diverse among marine ecosystems (Roberts et al., 2002; Buddemeier et al., 2004). They underpin local shore protection, fisheries, tourism (see Chapter 6; Hoegh-Guldberg et al., 2000; Cesar et al., 2003; Willig et al., 2003; Hoegh-Guldberg, 2004, 2005) and, although supplying only about 2-5% of the global fisheries harvest, comprise a critical subsistence protein and income source in the developing world (Whittingham et al., 2003; Pauly et al., 2005; Sadovy, 2005).

Corals are affected by warming of surface waters (see C2.1.2; Reynaud et al., 2003; McNeil et al., 2004; McWilliams et al., 2005) leading to bleaching (loss of algal symbionts; see C2.1.2). Many studies incontrovertibly link coral bleaching to warmer sea surface temperature (e.g., McWilliams et al., 2005), and mass bleaching and coral mortality often results beyond key temperature thresholds (see C2.1.2). Annual or bi-annual exceedance of

[1] A clade of algae is a group of closely related, but nevertheless different, types.

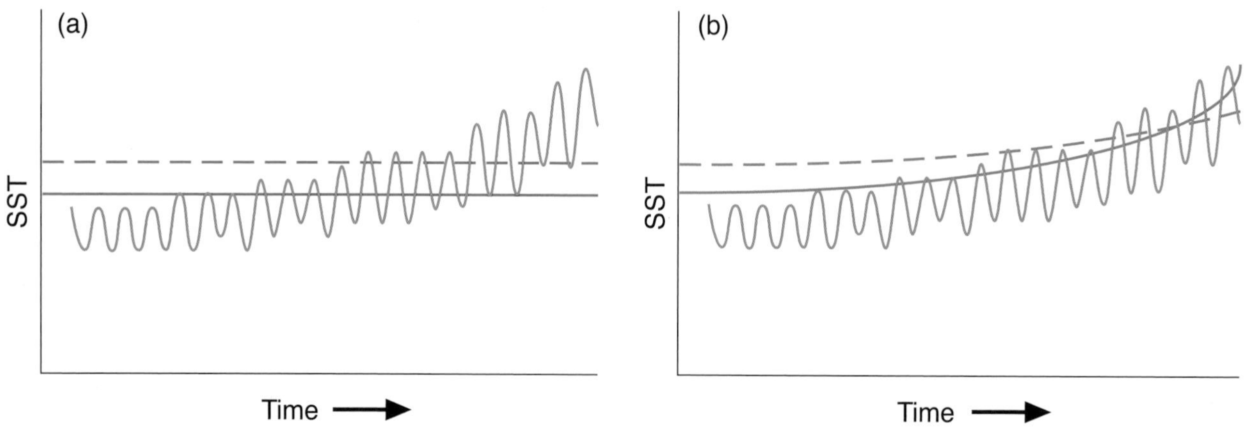

Figure C2.2. *Alternative hypotheses concerning the threshold SST at which coral bleaching occurs: (a) invariant threshold for coral bleaching (red line) which occurs when SST exceeds usual seasonal maximum threshold (by ~1°C) and mortality (dashed red line, threshold of 2°C), with local variation due to different species or water depth; (b) elevated threshold for bleaching (green line) and mortality (dashed green line) where corals adapt or acclimatise to increased SST (based on Hughes et al., 2003).*

bleaching thresholds is projected at the majority of reefs worldwide by 2030 to 2050 (Hoegh-Guldberg, 1999; Sheppard, 2003; Donner et al., 2005). After bleaching, algae quickly colonise dead corals, possibly inhibiting later coral recruitment (e.g., McClanahan et al., 2001; Szmant, 2001; Gardner et al., 2003; Jompa and McCook, 2003). Modelling predicts a phase switch to algal dominance on the Great Barrier Reef and Caribbean reefs in 2030 to 2050 (Wooldridge et al., 2005).

Coral reefs will also be affected by rising atmospheric CO_2 concentrations (Orr et al., 2005; Raven et al., 2005; Denman et al., 2007, Box 7.3) resulting in declining calcification. Experiments at expected aragonite concentrations demonstrated a reduction in coral calcification (Marubini et al., 2001; Langdon et al., 2003; Hallock, 2005), coral skeleton weakening (Marubini et al., 2003) and strong temperature dependence (Reynaud et al., 2003). Oceanic pH projections decrease at a greater rate and to a lower level than experienced over the past 20 million years (Caldeira and Wickett, 2003; Raven et al., 2005; Turley et al., 2006). Doubling CO_2 will reduce calcification in aragonitic corals by 20%-60% (Kleypas et al., 1999; Kleypas and Langdon, 2002; Reynaud et al., 2003; Raven et al., 2005). By 2070 many reefs could reach critical aragonite saturation states (Feely et al., 2004; Orr et al., 2005), resulting in reduced coral cover and greater erosion of reef frameworks (Kleypas et al., 2001; Guinotte et al., 2003).

Adaptation potential (Hughes et al., 2003) by reef organisms requires further experimental and applied study (Coles and Brown, 2003; Hughes et al., 2003). Natural adaptive shifts to symbionts with +2°C resistance may delay the demise of some reefs until roughly 2100 (Sheppard, 2003), rather than mid-century (Hoegh-Guldberg, 2005) although this may vary widely across the globe (Donner et al., 2005). Estimates of warm-water coral cover reduction in the last 20-25 years are 30% or higher (Wilkinson, 2004; Hoegh-Guldberg, 2005) due largely to increasing higher SST frequency (Hoegh-Guldberg, 1999). In some regions, such as the Caribbean, coral losses have been estimated at 80% (Gardner et al., 2003). Coral migration to higher latitudes with more optimal SST is unlikely, due both to latitudinally decreasing aragonite concentrations and projected atmospheric CO_2 increases (Kleypas et al., 2001; Guinotte et al., 2003; Orr et al., 2005; Raven et al., 2005). Coral migration is also limited by lack of available substrate (see C2.2.2). Elevated SST and decreasing aragonite have a complex synergy (Harvell et al., 2002; Reynaud et al., 2003; McNeil et al., 2004; Kleypas et al., 2005) but could produce major coral reef changes (Guinotte et al., 2003; Hoegh-Guldberg, 2005). Corals could become rare on tropical and sub-tropical reefs by 2050 due to the combined effects of increasing CO_2 and increasing frequency of bleaching events (at 2-3 × CO_2) (Kleypas and Langdon, 2002; Hoegh-Guldberg, 2005; Raven et al., 2005). Other climate change factors (such as sea-level rise, storm impact and aerosols) and non-climate factors (such as over-fishing, invasion of non-native species, pollution, nutrient and sediment load (although this could also be related to climate change through changes to precipitation and river flow; see C2.1.2 and C2.2.3; Chapter 16)) add multiple impacts on coral reefs (see C2.3.1), increasing their vulnerability and reducing resilience to climate change (Koop et al., 2001; Kleypas and Langdon, 2002; Cole, 2003; Buddemeier et al., 2004; Hallock, 2005).

C2.2.2 Impacts on coral reefs (Chapter 6, Section 6.4.1.5)

Reef-building corals are under stress on many coastlines (see C2.1.1). Reefs have deteriorated as a result of a combination of anthropogenic impacts such as over-fishing and pollution from adjacent land masses (Pandolfi et al., 2003; Graham et al., 2006), together with an increased frequency and severity of bleaching associated with climate change (see C2.1.2). The relative significance of these stresses varies from site to site. Coral mortality on Caribbean reefs is generally related to recent disease outbreaks, variations in herbivory,[2] and hurricanes (Gardner et al., 2003; McWilliams et al., 2005), whereas Pacific reefs have been particularly impacted by episodes of coral

[2] Herbivory: the consumption of plants by animals.

bleaching caused by thermal stress anomalies, especially during recent El Niño events (Hughes et al., 2003), as well as non-climate stresses.

Mass coral-bleaching events are clearly correlated with rises of SST of short duration above summer maxima (Douglas, 2003; Lesser, 2004; McWilliams et al., 2005). Particularly extensive bleaching was recorded across the Indian Ocean region associated with extreme El Niño conditions in 1998 (see C2.1.2 and C2.2.3). Many reefs appear to have experienced similar SST conditions earlier in the 20th century and it is unclear how extensive bleaching was before widespread reporting post-1980 (Barton and Casey, 2005). There is limited ecological and genetic evidence for adaptation of corals to warmer conditions (see C2.1.2 and C2.2.1). It is very likely that projected future increases in SST of about 1 to 3°C (Section 6.3.2) will result in more frequent bleaching events and widespread mortality if there is no thermal adaptation or acclimatisation by corals and their symbionts (Sheppard, 2003; Hoegh-Guldberg, 2004). The ability of coral reef ecosystems to withstand the impacts of climate change will depend on the extent of degradation from other anthropogenic pressures and the frequency of future bleaching events (Donner et al., 2005).

In addition to coral bleaching, there are other threats to reefs associated with climate change (Kleypas and Langdon, 2002). Increased concentrations of CO_2 in seawater will lead to ocean acidification (Section 6.3.2), affecting aragonite saturation state (Meehl et al., 2007) and reducing calcification rates of calcifying organisms such as corals (LeClerq et al., 2002; Guinotte et al., 2003; see C2.2.1). Cores from long-lived massive corals indicate past minor variations in calcification (Lough and Barnes, 2000), but disintegration of degraded reefs following bleaching or reduced calcification may result in increased wave energy across reef flats with potential for shoreline erosion (Sheppard et al., 2005). Relative sea-level rise appears unlikely to threaten reefs in the next few decades; coral reefs have been shown to keep pace with rapid postglacial sea-level rise when not subjected to environmental or anthropogenic stresses (Hallock, 2005). A slight rise in sea level is likely to result in the submergence of some Indo-Pacific reef flats and recolonisation by corals, as these intertidal surfaces, presently emerged at low tide, become suitable for coral growth (Buddemeier et al., 2004).

Many reefs are affected by tropical cyclones (hurricanes, typhoons); impacts range from minor breakage of fragile corals to destruction of the majority of corals on a reef and deposition of debris as coarse storm ridges. Such storms represent major perturbations, affecting species composition and abundance, from which reef ecosystems require time to recover. The sequence of ridges deposited on the reef top can provide a record of past storm history (Hayne and Chappell, 2001); for the northern Great Barrier Reef no change in frequency of extremely large cyclones has been detected over the past 5,000 years (Nott and Hayne, 2001). An intensification of tropical storms (Section 6.3.2) could have devastating consequences on the reefs themselves, as well as for the inhabitants of many low-lying islands (Sections 6.4.2 and 16.3.1.3). There is limited evidence that global warming may result in an increase of coral range; for example, the extension of branching Acropora polewards has been recorded in Florida, despite an almost Caribbean-wide trend for reef deterioration (Precht and Aronson, 2004), but there are several constraints, including low genetic diversity and the limited suitable substrate at the latitudinal limits to reef growth (Riegl, 2003; Ayre and Hughes, 2004; Woodroffe et al., 2005).

The fate of the small reef islands on the rim of atolls is of special concern. Small reef islands in the Indo-Pacific formed over recent millennia during a period when regional sea level fell (Dickinson, 2004; Woodroffe and Morrison, 2001). However, the response of these islands to future sea-level rise remains uncertain, and is addressed in greater detail in Chapter 16, Section 16.4.2. It will be important to identify critical thresholds of change beyond which there may be collapse of ecological and social systems on atolls. There are limited data, little local expertise to assess the dangers, and a low level of economic activity to cover the costs of adaptation for atolls in countries such as the Maldives, Kiribati and Tuvalu (Barnett and Adger, 2003; Chapter 16, Box 16.6).

C2.2.3 Climate change and the Great Barrier Reef (Chapter 11, Box 11.3)

The Great Barrier Reef (GBR) is the world's largest continuous reef system (2,100 km long) and is a critical storehouse of Australian marine biodiversity and a breeding ground for seabirds and other marine vertebrates such as the humpback whale. Tourism associated with the GBR generated over US$4.48 billion in the 12-month period 2004/5 and provided employment for about 63,000 full-time equivalent persons (Access Economics, 2005). The two greatest threats from climate change to the GBR are (i) rising sea temperatures, which are almost certain to increase the frequency and intensity of mass coral bleaching events, and (ii) ocean acidification, which is likely to reduce the calcifying ability of key organisms such as corals. Other factors, such as droughts and more intense storms, are likely to influence reefs through physical damage and extended flood plumes (Puotinen, 2006).

Sea temperatures on the GBR have warmed by about 0.4°C over the past century (Lough, 2000). Temperatures currently typical of the northern tip of the GBR are very likely to extend to its southern end by 2040 to 2050 (SRES scenarios A1, A2) and 2070 to 2090 (SRES scenarios B1, B2) (Done et al., 2003). Temperatures only 1°C above the long-term summer maxima already cause mass coral bleaching (loss of symbiotic algae). Corals may recover but will die under high or prolonged temperatures (2 to 3°C above long-term maxima for at least 4 weeks). The GBR has experienced eight mass bleaching events since 1979 (1980, 1982, 1987, 1992, 1994, 1998, 2002 and 2006); there are no records of events prior to 1979 (Hoegh-Guldberg, 1999). The most widespread and intense events occurred in the summers of 1998 and 2002, with about 42% and 54% of reefs affected, respectively (Done et al., 2003; Berkelmans et al., 2004). Mortality was distributed patchily, with the greatest effects on near-shore reefs, possibly exacerbated by osmotic stress caused by floodwaters in some areas (Berkelmans and Oliver, 1999). The 2002 event was followed by localised outbreaks of coral disease, with incidence of some disease-like syndromes increasing by as much as 500% over the past decade at a few sites (Willis et al., 2004). While the impacts of coral disease on the GBR are currently minor, experiences in other parts of the world suggest that disease has the potential to be a threat to GBR reefs. Effects from thermal stress are likely to be exacerbated under future scenarios by the gradual acidification of the world's oceans, which have absorbed about 30% of the excess CO_2 released to the atmosphere (Orr et al., 2005; Raven et al., 2005). Calcification declines with decreasing carbonate ion concentrations, becoming zero at carbonate ion concentrations of approximately 200 μmol/kg (Langdon et al., 2000; Langdon, 2002). These occur at atmospheric CO_2 concentrations of approximately 500 ppm. Reduced growth due to acidic conditions is very likely to hinder reef recovery after bleaching events and will reduce the resilience of reefs to other stressors (e.g., sediment, eutrophication).

Even under a moderate warming scenario (A1T, 2°C by 2100), corals on the GBR are very likely to be exposed to regular summer temperatures that exceed the thermal thresholds observed over the past 20 years (Done et al., 2003). Annual bleaching is projected under the A1FI scenario by 2030, and under A1T by 2050 (Done et al., 2003; Wooldridge et al., 2005). Given that the recovery time from a severe bleaching-induced mortality event is at least 10 years (and may exceed 50 years for full recovery), these models suggest that reefs are likely to be dominated by non-coral organisms such as macroalgae by 2050 (Hoegh-Guldberg, 1999; Done et al., 2003). Substantial impacts on biodiversity, fishing and tourism are likely. Maintenance of hard coral cover on the GBR will require corals to increase their upper thermal tolerance limits at the same pace as the change in sea temperatures driven by climate change, i.e., about 0.1-0.5°C/decade (Donner et al., 2005). There is currently little evidence that corals have the capacity for such rapid genetic change; most of the evidence is to the contrary (Hoegh-Guldberg, 1999, 2004). Given that recovery from mortality can be potentially enhanced by reducing local stresses (water quality, fishing pressure), management initiatives such as the Reef Water Quality Protection Plan and the Representative Areas Programme (which expanded totally protected areas on the GBR from 4.6% to over 33%) represent planned adaptation options to enhance the ability of coral reefs to endure the rising pressure from rapid climate change.

C2.2.4 Impact of coral mortality on reef fisheries (Chapter 5, Box 5.4)

Coral reefs and their fisheries are subject to many stresses in addition to climate change (see Chapter 4). So far, events such as the 1998 mass coral bleaching in the Indian Ocean have not provided evidence of negative short-term bio-economic impacts for coastal reef fisheries (Spalding and Jarvis, 2002; Grandcourt and Cesar, 2003). In the longer term, there may be serious consequences for fisheries production that result from loss of coral communities and reduced structural complexity, which result in reduced fish species richness, local extinctions and loss of species within key functional groups of reef fish (Sano, 2004; Graham et al., 2006).

C2.3 Multiple stresses on coral reefs

C2.3.1 Non-climate-change threats to coral reefs of small islands (Chapter 16, Box 16.2)

A large number of non-climate-change stresses and disturbances, mainly driven by human activities, can impact coral reefs (Nyström et al., 2000; Hughes et al., 2003). It has been suggested that the 'coral reef crisis' is almost certainly the result of complex and synergistic interactions among global-scale climatic stresses and local-scale, human-imposed stresses (Buddemeier et al., 2004).

In a study by Bryant et al. (1998), four human-threat factors – coastal development, marine pollution, over-exploitation and destructive fishing, and sediment and nutrients from inland – provide a composite indicator of the potential risk to coral reefs associated with human activity for 800 reef sites. Their map (Figure C2.3) identifies low-risk (blue), medium-risk (yellow) and high-risk (red) sites, the first being common in the insular central Indian and Pacific Oceans, the last in maritime South-East Asia and the Caribbean archipelago. Details of reefs at risk in the two highest-risk areas have been documented by Burke et al. (2002) and Burke and Maidens (2004), who indicate that about 50% of the reefs in South-East Asia and 45% in the Caribbean are classed in the high- to very-high-risk category. There are, however, significant local and regional differences in

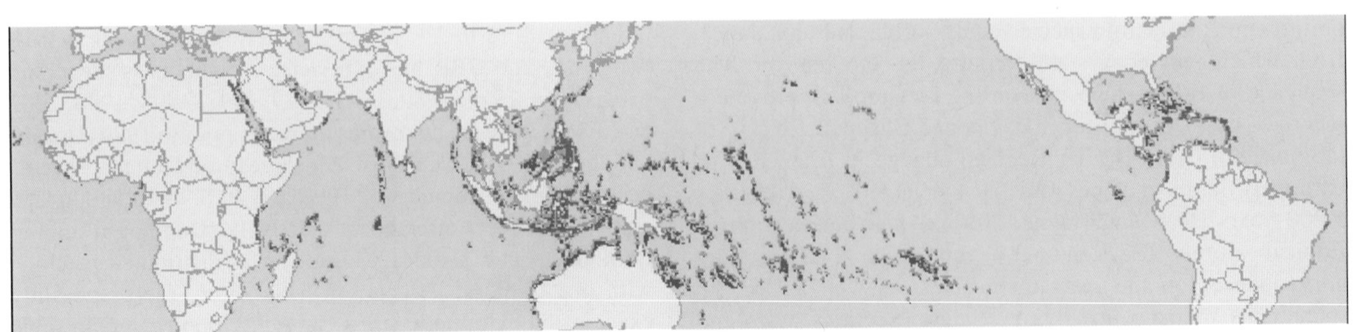

Figure C2.3. *The potential risk to coral reefs from human-threat factors. Low risk (blue), medium risk (yellow) and high risk (red). Source: Bryant et al. (1998)*

the scale and type of threats to coral reefs in both continental and small-island situations.

Recognising that coral reefs are especially important for many Small Island states, Wilkinson (2004) notes that reefs on small islands are often subject to a range of non-climate impacts. Some common types of reef disturbance are listed below, with examples from several island regions and specific islands.

1. Impact of coastal developments and modification of shorelines:
   - coastal development on fringing reefs, Langawi Island, Malaysia (Abdullah et al., 2002);
   - coastal resort development and tourism impacts in Mauritius (Ramessur, 2002).
2. Mining and harvesting of corals and reef organisms:
   - coral harvesting in Fiji for the aquarium trade (Vunisea, 2003).
3. Sedimentation and nutrient pollution from the land:
   - sediment smothering reefs in Aria Bay, Palau (Golbuua et al., 2003) and southern islands of Singapore (Dikou and van Woesik, 2006);
   - non-point source pollution, Tutuila Island, American Samoa (Houk et al., 2005);
   - nutrient pollution and eutrophication, fringing reef, Réunion (Chazottes et al., 2002) and Cocos Lagoon, Guam (Kuffner and Paul, 2001).
4. Over-exploitation and damaging fishing practices:
   - blast fishing in the islands of Indonesia (Fox and Caldwell, 2006);
   - intensive fish-farming effluent in Philippines (Villanueva et al., 2006);
   - subsistence exploitation of reef fish in Fiji (Dulvy et al., 2004);
   - giant clam harvesting on reefs, Milne Bay, Papua New Guinea (Kinch, 2002).
5. Introduced and invasive species:
   - non-indigenous species invasion of coral habitats in Guam (Paulay et al., 2002).

There is another category of 'stress' that may inadvertently result in damage to coral reefs – the human component of poor governance (Goldberg and Wilkinson, 2004). This can accompany political instability; one example being problems with contemporary coastal management in the Solomon Islands (Lane, 2006).

References

Abdullah, A., Z. Yasin, W. Ismail, B. Shutes and M. Fitzsimons, 2002: The effect of early coastal development on the fringing coral reefs of Langkawi: a study in small-scale changes. *Malaysian Journal of Remote Sensing and GIS*, **3**, 1-10.

Access Economics, 2005: *Measuring the Economic and Financial Value of the Great Barrier Reef Marine Park*. Report by Access Economics for Great Barrier Reef Marine Park Authority, June 2005, 61 pp. http://www.accesseconomics.com.au/publicationsreports/showreport.php?id=10&searchfor=Economic%20Consulting&searchby=area.

Arthur, R., T.J. Done and H. Marsh, 2005: Benthic recovery four years after an El Niño-induced coral mass mortality in the Lakshadweep atolls. *Curr. Sci. India*, **89**, 694-699.

Ayre, D.J. and T.P. Hughes, 2004: Climate change, genotypic diversity and gene flow in reef-building corals. *Ecol. Lett.*, **7**, 273-278.

Barnett, J. and W.N. Adger, 2003: Climate dangers and atoll countries. *Climatic Change*, **61**, 321-337.

Barton, A.D. and K.S. Casey, 2005: Climatological context for large-scale coral bleaching. *Coral Reefs*, **24**, 536-554.

Bellwood, D.R., A.S. Hoey, J.L. Ackerman and M. Depczynski, 2006: Coral bleaching, reef fish community phase shifts and the resilience of coral reefs. *Glob. Change Biol.*, **12**, 1587-1594.

Berkelmans, R. and J.K. Oliver, 1999: Large-scale bleaching of corals on the Great Barrier Reef. *Coral Reefs*, **18**, 55-60.

Berkelmans, R., G. De'ath, S. Kininmonth and W.J. Skirving, 2004: A comparison of the 1998 and 2002 coral bleaching events of the Great Barrier Reef: spatial correlation, patterns and predictions. *Coral Reefs*, **23**, 74-83.

Bindoff, N., J. Willebrand, V. Artale, A. Cazenave, J. Gregory, S. Gulev, K. Hanawa, C. Le Quéré, S. Levitus, Y. Nojiri, C.K. Shum, L. Talley and A. Unnikrishnan, 2007: Observations: oceanic climate change and sea level. *Climate Change 2007: The Physical Science Basis. Contribution of Working Group I to the Fourth Assessment Report of the Intergovernmental Panel on Climate Change*, S. Solomon, D. Qin, M. Manning, Z. Chen, M. Marquis, K.B. Averyt, M. Tignor and H.L. Miller, Eds., Cambridge University Press, Cambridge, 385-432.

Bruno, J.F., C.E. Siddon, J.D. Witman, P.L. Colin and M.A. Toscano, 2001: El Niño related coral bleaching in Palau, Western Caroline Islands. *Coral Reefs*, **20**, 127-136.

Bryant, D., L. Burke, J. McManus and M. Spalding, 1998: *Reefs at Risk: A Map-Based Indicator of Threats to the World's Coral Reefs*. World Resources Institute, Washington, DC, 56 pp.

Buddemeier, R.W., J.A. Kleypas and B. Aronson, 2004: *Coral Reefs and Global Climate Change: Potential Contributions of Climate Change to Stresses on Coral Reef Ecosystems*. Report prepared for the Pew Centre on Global Climate Change, Arlington, Virginia, 56 pp.

Burke, L. and J. Maidens, 2004: *Reefs at Risk in the Caribbean*. World Resources Institute, Washington, District of Columbia, 81 pp.

Burke, L., E. Selig and M. Spalding, 2002: *Reefs at Risk in Southeast Asia*. World Resources Institute, Washington, District of Columbia, 72 pp.

Caldeira, K. and M.E. Wickett, 2003: Anthropogenic carbon and ocean pH. *Nature*, **425**, 365-365.

Cesar, H., L. Burke and L. Pet-Soede, 2003: *The Economics of Worldwide Coral Reef Degradation*. Cesar Environmental Economics Consulting (CEEC), Arnhem, 23 pp.

Chazottes, V., T. Le Campion-Alsumard, M. Peyrot-Clausade and P. Cuet, 2002: The effects of eutrophication-related alterations to coral reef communities on agents and rates of bioerosion (Reunion Island, Indian Ocean). *Coral Reefs*, **21**, 375-390.

Cole, J., 2003: Global change: dishing the dirt on coral reefs. *Nature*, **421**, 705-706.

Coles, S.L.and B.E. Brown, 2003: Coral bleaching: capacity for acclimatization and adaptation. *Adv. Mar. Biol.*, **46**, 183-224.

Denman, K.L., G. Brasseur, A. Chidthaisong, P. Ciais, P. Cox, R.E. Dickinson, D. Hauglustaine, C. Heinze, E. Holland, D. Jacob, U. Lohmann, S. Ramachandran, P.L. da Silva Dias, S.C. Wofsy and X. Zhang, 2007: Couplings between changes in the climate system and biogeochemistry. *Climate Change 2007: The Physical Science Basis. Contribution of Working Group I to the Fourth Assessment Report of the Intergovernmental Panel on Climate Change*, S. Solomon, D. Qin, M. Manning, Z. Chen, M. Marquis, K.B. Averyt, M. Tignor and H.L. Miller, Eds., Cambridge University Press, Cambridge, 499-587.

Dickinson, W.R., 2004: Impacts of eustasy and hydro-isostasy on the evolution and landforms of Pacific atolls. *Palaeogeogr. Palaeoclimatol. Palaeoecol.*, **213**, 251-269.

Dikou, A. and R. van Woesik, 2006: Survival under chronic stress from sediment load: spatial patterns of hard coral communities in the southern islands of Singapore. *Mar. Pollut. Bull.*, **52**, 7-21.

Done, T., P. Whetton, R. Jones, R. Berkelmans, J. Lough, W. Skirving and S. Wooldridge, 2003: Global climate change and coral bleaching on the Great Barrier Reef. Final Report to the State of Queensland Greenhouse Taskforce through the Department of Natural Resources and Mining, Townsville, 49 pp. http://www.longpaddock.qld.gov.au/ClimateChanges/pub/CoralBleaching.pdf.

Donner, S.D., W.J. Skirving, C.M. Little, M. Oppenheimer and O. Hoegh-Guldberg, 2005: Global assessment of coral bleaching and required rates of adaptation under climate change. *Glob. Change Biol.*, **11**, 2251-2265.

Douglas, A.E., 2003: Coral bleaching: how and why? *Mar. Pollut. Bull.*, **46**, 385-392.

Dulvy, N., R. Freckleton and N. Polunin, 2004: Coral reef cascades and the indirect effects of predator removal by exploitation. *Ecol. Lett.*, **7**, 410-416.

Feely, R.A., C.L. Sabine, K. Lee, W. Berelson, J. Kleypas, V.J. Fabry and F.J. Millero, 2004: Impact of anthropogenic CO_2 on the $CaCO_3$ system in the oceans. *Science*, **305**, 362-366.

Fox, H. and R. Caldwell, 2006: Recovery from blast fishing on coral reefs: a tale of two scales. *Ecol. Appl.*, **16**, 1631-1635.

Gardner, T.A., I.M. Cote, J.A. Gill, A. Grant and A.R. Watkinson, 2003: Long-term region-wide declines in Caribbean corals. *Science*, **301**, 958-960.

Gardner, T.A., I.M. Côté, J.A. Gill, A. Grant and A.R. Watkinson, 2005: Hurricanes and Caribbean coral reefs: impacts, recovery patterns and role in long-term decline. *Ecology*, **86**, 174-184.

Garpe, K.C., S.A.S. Yahya, U. Lindahl and M.C. Ohman, 2006: Long-term effects of the 1998 coral bleaching event on reef fish assemblages. *Mar. Ecol.–Prog. Ser.*, **315**, 237-247.

Gill, J.A., J.P. McWilliams, A.R. Watkinson and I.M. Côté, 2006: Opposing forces of aerosol cooling and El Niño drive coral bleaching on Caribbean reefs. *P. Natl. Acad. Sci. USA*, **103**, 18870-18873.

Golbuua, Y., S. Victora, E. Wolanski and R.H. Richmond, 2003: Trapping of fine sediment in a semi-enclosed bay, Palau, Micronesia. *Estuar. Coast. Shelf S.*, **57**, 1-9.

Goldberg, J. and C. Wilkinson, 2004: Global threats to coral reefs: coral bleaching, global climate change, disease, predator plagues, and invasive species. *Status of Coral Reefs of the World: 2004*, C. Wilkinson, Ed., Australian Institute of Marine Science, Townsville, 67-92.

Goldenberg, S.B., C.W. Landsea, A.M. Mestas-Nunez and W.M. Gray, 2001: The recent increase in Atlantic hurricane activity: causes and implications. *Science*, **293**, 474-479.

Grandcourt, E.M. and H.S J. Cesar, 2003: The bio-economic impact of mass coral mortality on the coastal reef fisheries of the Seychelles. *Fish. Res.*, **60**, 539-550.

Graham, N.A.J., S.K. Wilson, S. Jennings, N.V.C. Polunin, J.P. Bijoux and J. Robinson, 2006: Dynamic fragility of oceanic coral reef ecosystems. *P. Natl. Acad. Sci. USA*, **103**, 8425-8429.

Guinotte, J.M., R.W. Buddemeier and J.A. Kleypas, 2003: Future coral reef habitat marginality: temporal and spatial effects of climate change in the Pacific basin. *Coral Reefs*, **22**, 551-558.

Hallock, P., 2005: Global change and modern coral reefs: new opportunities to understand shallow-water carbonate depositional processes. *Sediment. Geol.*, **175**, 19-33.

Harvell, C.D., C.E. Mitchell, J.R. Ward, S. Altizer, A.P. Dobson, R.S. Ostfeld and M.D. Samuel, 2002: Climate warming and disease risks for terrestrial and marine biota. *Science*, **296**, 2158-2162.

Hayne, M. and J. Chappell, 2001: Cyclone frequency during the last 5000 years at Curacao Island, north Queensland, Australia. *Palaeogeogr. Palaeoclimatol. Palaeoecol.*, **168**, 207-219.

Hoegh-Guldberg, O., 1999: Climate change, coral bleaching and the future of the world's coral reefs. *Mar. Freshwater Res.*, **50**, 839-866.

Hoegh-Guldberg, O., 2004: Coral reefs in a century of rapid environmental change. *Symbiosis*, **37**, 1-31.

Hoegh-Guldberg, O., 2005: Low coral cover in a high-CO_2 world. *J. Geophys. Res. C*, **110**, C09S06, doi:10.1029/2004JC002528.

Hoegh-Guldberg, O., H. Hoegh-Guldberg, D.K. Stout, H. Cesar and A. Timmerman, 2000: *Pacific in Peril: Biological, Economic and Social Impacts of Climate Change on Pacific Coral Reefs*. Greenpeace, Sidney, 72 pp.

Houk, P., G. Didonato, J. Iguel and R. van Woesik, 2005: Assessing the effects of non-point source pollution on American Samoa's coral reef communities. *Environ. Monit. Assess.*, **107**, 11-27.

Hughes, T.P., A.H. Baird, D.R. Bellwood, M. Card, S.R. Connolly, C. Folke, R. Grosberg, O. Hoegh-Guldberg, J.B.C. Jackson, J. Kleypas, J.M. Lough, P. Marshall, M. Nystrom, S.R. Palumbi, J.M. Pandolfi, B. Rosen and J. Roughgarden, 2003: Climate change, human impacts, and the resilience of coral reefs. *Science*, **301**, 929-933.

Jompa, J. and L.J. McCook, 2003: Coral-algal competition: macroalgae with different properties have different effects on corals. *Mar. Ecol.–Prog. Ser.*, **258**, 87-95.

Kinch, J., 2002: Giant clams: their status and trade in Milne Bay Province, Papua New Guinea. *TRAFFIC Bulletin*, **19**, 1-9.

Kleypas, J.A. and C. Langdon, 2002: Overview of CO_2-induced changes in seawater chemistry. *World Coral Reefs in the New Millenium: Bridging Research and Management for Sustainable Development*, M.K. Moosa, S. Soemodihardjo, A. Soegiarto, K. Romimohtarto, A. Nontji and S. Suharsono, Eds., Proceedings of the 9th International Coral Reef Symposium, 2, Bali, Indonesia. Ministry of Environment, Indonesian Institute of Sciences, International Society for Reef Studies, 1085-1089.

Kleypas, J.A., R.W. Buddemeier, D. Archer, J.P. Gattuso, C. Langdon and B.N. Opdyke, 1999: Geochemical consequences of increased atmospheric carbon dioxide on coral reefs. *Science*, **284**, 118-120.

Kleypas, J.A., R.W. Buddemeier and J.P. Gattuso, 2001: The future of coral reefs in an age of global change. *Int. J. Earth Sci.*, **90**, 426-437.

Kleypas, J.A., R.W. Buddemeier, C.M. Eakin, J.P. Gattuso, J. Guinotte, O. Hoegh-Guldberg, R. Iglesias-Prieto, P.L. Jokiel, C. Langdon, W. Skirving and A.E. Strong, 2005: Comment on "Coral reef calcification and climate change: the effect of ocean warming". *Geophys. Res. Lett.*, **32**, L08601, doi:10.1029/2004 GL022329.

Koop, K., D. Booth, A. Broadbent, J. Brodie, D. Bucher, D. Capone, J. Coll, W. Dennison, M. Erdmann, P. Harrison, O. Hoegh-Guldberg, P. Hutchings, G.B. Jones, A.W.D. Larkum, J. O'Neil, A. Steven, E. Tentori, S. Ward, J. Williamson and D. Yellowlees, 2001: ENCORE: The effect of nutrient enrichment on coral reefs: synthesis of results and conclusions. *Mar. Pollut. Bull.*, **42**, 91-120.

Kuffner, I. and V. Paul, 2001: Effects of nitrate, phosphate and iron on the growth of macroalgae and benthic cyanobacteria from Cocos Lagoon, Guam. *Mar. Ecol.–Prog. Ser.*, **222**, 63-72.

Lane, M., 2006: Towards integrated coastal management in the Solomon Islands: identifying strategic issues for governance reform. *Ocean Coast. Manage.*, **49**, 421-441.

Langdon, C., 2002: Review of experimental evidence for effects of CO_2 on calcification of reef builders. *Proc. of the 9th International Coral Reef Symposium*, Ministry of Environment, Indonesian Institute of Science, International Society for Reef Studies, Bali 23–27 October, 1091-1098.

Langdon, C., T. Takahashi, C. Sweeney, D. Chipman, J. Goddard, F. Marubini, H. Aceves, H. Barnett and M.J. Atkinson, 2000: Effect of calcium carbonate saturation state on the calcification rate of an experimental coral reef. *Global Biogeochem. Cy.*, **14**, 639-654.

Langdon, C., W.S. Broecker, D.E. Hammond, E. Glenn, K. Fitzsimmons, S.G. Nelson, T.H. Peng, I. Hajdas and G. Bonani, 2003: Effect of elevated CO_2 on the community metabolism of an experimental coral reef. *Global Biogeochem. Cy.*, **17**, 1011, doi:10.1029/2002GB001941.

LeClerq, N., J.-P. Gattuso and J. Jaubert, 2002: Primary production, respiration, and calcification of a coral reef mesocosm under increased CO_2 pressure. *Limnol. Oceanogr.*, **47**, 558-564.

Lesser, M.P., 2004: Experimental biology of coral reef ecosystems. *J. Exp. Mar. Biol. Ecol.*, **300**, 217-252.

Little, A.F., M.J.H. van Oppen and B.L. Willis, 2004: Flexibility in algal endosymbioses shapes growth in reef corals. *Science*, **304**, 1492-1494.

Lough, J.M., 2000: 1997-98: unprecedented thermal stress to coral reefs? *Geophys. Res. Lett.*, **27**, 3901-3904.

Lough, J.M. and D.J. Barnes, 2000: Environmental controls on growth of the massive coral *Porites*. *J. Exp. Mar. Biol. Ecol.*, **245**, 225-243.

Marubini, F., H. Barnett, C. Langdon and M.J. Atkinson, 2001: Dependence of calcification on light and carbonate ion concentration for the hermatypic coral *Porites compressa*. *Mar. Ecol.–Prog. Ser.*, **220**, 153-162.

Marubini, F., C. Ferrier-Pages and J.P. Cuif, 2003: Suppression of skeletal growth in scleractinian corals by decreasing ambient carbonate-ion concentration: a cross-family comparison. *P. Roy. Soc. Lond. B*, **270**, 179-184.

McClanahan, T.R., N.A. Muthiga and S. Mangi, 2001: Coral and algal changes after the 1998 coral bleaching: interaction with reef management and herbivores on Kenyan reefs. *Coral Reefs*, **19**, 380-391.

McNeil, B.I., R.J. Matear and D.J. Barnes, 2004: Coral reef calcification and climate change: the effect of ocean warming. *Geophys. Res. Lett.*, **31**, L22309, doi:10.1029/2004GL021541.

McWilliams, J.P., I.M. Côté, J.A. Gill, W.J. Sutherland and A.R. Watkinson, 2005: Accelerating impacts of temperature-induced coral bleaching in the Caribbean. *Ecology*, **86**, 2055-2060.

Meehl, G.A., T.F. Stocker, W. Collins, P. Friedlingstein, A. Gaye, J. Gregory, A. Kitoh, R. Knutti and co-authors, 2007: Global climate projections. *Climate Change 2007: The Physical Science Basis. Contribution of Working Group I to the Fourth Assessment Report of the Intergovernmental Panel on Climate Change*, S. Solomon, D. Qin, M. Manning, Z. Chen, M. Marquis, K.B. Averyt,

M. Tignor and H.L. Miller, Eds., Cambridge University Press, Cambridge, 747-846.

Mendes, J.M. and J.D. Woodley, 2002: Effect of the 1995-1996 bleaching event on polyp tissue depth, growth, reproduction and skeletal band formation in *Montastraea annularis*. *Mar. Ecol.–Prog. Ser.*, **235**, 93-102.

Mestas-Nunez, A.M. and A.J. Miller, 2006: Interdecadal variability and climate change in the eastern tropical Pacific: a review. *Prog. Oceanogr.*, **69**, 267-284.

Nott, J. and M. Hayne, 2001: High frequency of 'super-cyclones' along the Great Barrier Reef over the past 5,000 years. *Nature*, **413**, 508-512.

Nyström, M., C. Folke and F. Moberg, 2000: Coral reef disturbance and resilience in a human-dominated environment. *Trends Ecol. Evol.*, **15**, 413-417.

Obura, D.O., 2005: Resilience and climate change: lessons from coral reefs and bleaching in the western Indian Ocean. *Estuar. Coast. Shelf Sci.*, **63**, 353-372.

Ohde, S. and M.M.M. Hossain, 2004: Effect of $CaCO_3$ (aragonite) saturation state of seawater on calcification of *Porites* coral. *Geochem. J.*, **38**, 613-621.

Orr, J.C., V.J. Fabry, O. Aumont, L. Bopp, S.C. Doney, R.A. Feely, A. Gnanadesikan, N. Gruber, A. Ishida, F. Joos, R.M. Key, K. Lindsay, E. Maier-Reimer, R. Matear, P. Monfray, A. Mouchet, R.G. Najjar, G.K. Plattner, K.B. Rodgers, C.L. Sabine, J.L. Sarmiento, R. Schlitzer, R.D. Slater, I.J. Totterdell, M.F. Weirig, Y. Yamanaka and A. Yool, 2005: Anthropogenic ocean acidification over the twenty-first century and its impact on calcifying organisms. *Nature*, **437**, 681-686.

Pandolfi, J.M., R.H. Bradbury, E. Sala, T.P. Hughes, K.A. Bjorndal, R.G. Cooke, D. McArdle, L. McClenachan and co-authors, 2003: Global trajectories of the long-term decline of coral reef ecosystems. *Science*, **301**, 955-958.

Pauly, D., J. Alder, A. Bakun, S. Heileman, K.H. Kock, P. Mace, W. Perrin, K. Stergiou, U.R. Sumaila, M. Vierros, K. Freire and Y. Sadovy, 2005: Marine fisheries systems. *Ecosystems and Human Well-being. Volume 1: Current State and Trends*, R. Hassan, R. Scholes and N. Ash, Eds., Island Press, Washington, District of Columbia, 477-511.

Paulay, G., L. Kirkendale, G. Lambert and C. Meyer, 2002: Anthropogenic biotic interchange in a coral reef ecosystem: a case study from Guam. *Pac. Sci.*, **56**, 403-422.

Pelejero, C., E. Calvo, M.T. McCulloch, J.F. Marshall, M.K. Gagan, J.M. Lough and B.N. Opdyke, 2005: Preindustrial to modern interdecadal variability in coral reef pH. *Science*, **309**, 2204-2207.

Puotinen, M.L., 2006: Modelling the risk of cyclone wave damage to coral reefs using GIS: a case study of the Great Barrier Reef, 1969-2003. *Int. J. Geogr. Inf. Sci.*, **21**, 97-120.

Precht, W.F. and R.B. Aronson, 2004: Climate flickers and range shifts of coral reefs. *Front. Ecol. Environ.*, **2**, 307-314.

Ramessur, R., 2002: Anthropogenic-driven changes with focus on the coastal zone of Mauritius, south-western Indian Ocean. *Reg. Environ. Change*, **3**, 99-106.

Raven, J., K. Caldeira, H. Elderfield, O. Hoegh-Guldberg, P. Liss, U. Riebesell, J. Shepherd, C. Turley and A. Watson, 2005: Ocean acidification due to increasing atmospheric carbon dioxide. Policy Document 12/05, The Royal Society, The Clyvedon Press Ltd, Cardiff, 68 pp.

Riegl, B., 2003: Climate change and coral reefs: different effects in two high-latitude areas (Arabian Gulf, South Africa). *Coral Reefs*, **22**, 433-446.

Reynaud, S., N. Leclercq, S. Romaine-Lioud, C. Ferrier-Pages, J. Jaubert and J.P. Gattuso, 2003: Interacting effects of CO_2 partial pressure and temperature on photosynthesis and calcification in a scleractinian coral. *Global Change Biol.*, **9**, 1660-1668.

Roberts, C.M., C.J. McClean, J.E.N. Veron, J.P. Hawkins, G.R. Allen, D.E. McAllister, C.G. Mittermeier, F.W. Schueler, M. Spalding, F. Wells, C. Vynne and T.B. Werner, 2002: Marine biodiversity hotspots and conservation priorities for tropical reefs. *Science*, **295**, 1280-1284.

Rowan, R., 2004: Coral bleaching: thermal adaptation in reef coral symbionts. *Nature*, **430**, 742.

Sadovy, Y., 2005: Trouble on the reef: the imperative for managing vulnerable and valuable fisheries. *Fish Fish.*, **6**, 167-185.

Sano, M., 2004: Short-term effects of a mass coral bleaching event on a reef fish assemblage at Iriomote Island, Japan. *Fish. Sci.*, **70**, 41-46.

Sheppard, C.R.C., 2003: Predicted recurrences of mass coral mortality in the Indian Ocean. *Nature*, **425**, 294-297.

Sheppard, C.R.C., D.J. Dixon, M. Gourlay, A. Sheppard and R. Payet, 2005: Coral mortality increases wave energy reaching shores protected by reef flats: examples from the Seychelles. *Estuar. Coast. Shelf S.*, **64**, 223-234.

Spalding, M.D. and G.E. Jarvis, 2002: The impact of the 1998 coral mortality on reef fish communities in the Seychelles. *Mar. Pollut. Bull.*, **44**, 309-321.

Szmant, A.M., 2001: Why are coral reefs world-wide becoming overgrown by algae? 'Algae, algae everywhere, and nowhere a bite to eat!' *Coral Reefs*, **19**, 299-302.

Trenberth, K.E., P.D. Jones, P.G. Ambenje, R. Bojariu, D.R. Easterling, A.M.G. Klein Tank, D.E. Parker, J.A. Renwick, F. Rahimzadeh, M.M. Rusticucci, B.J. Soden and P.-M. Zhai, 2007: Observations: surface and atmospheric climate change. *Climate Change 2007: The Physical Science Basis. Contribution of Working Group I to the Fourth Assessment Report of the Intergovernmental Panel on Climate Change*, S. Solomon, D. Qin, M. Manning, Z. Chen, M. Marquis, K.B. Averyt, M. Tignor and H.L. Miller, Eds., Cambridge University Press, Cambridge, 235-336.

Turley, C., J. Blackford, S. Widdicombe, D. Lowe and P. Nightingale, 2006: Reviewing the impact of increased atmospheric CO_2 on oceanic pH and the marine ecosystem. *Avoiding Dangerous Climate Change*, H.J. Schellnhuber, W. Cramer, N. Nakićenović, T.M.L. Wigley and G. Yohe, Eds., Cambridge University Press, Cambridge, 65-70.

Villanueva, R., H. Yap and N. Montaño, 2006: Intensive fish farming in the Philippines is detrimental to the reef-building coral *Pocillopora damicornis*. *Mar. Ecol.–Prog. Ser.*, **316**, 165-174.

Vunisea, A., 2003: Coral harvesting and its impact on local fisheries in Fiji. *SPC Women in Fisheries Information Bulletin*, **12**, 17-20.

Webster, P.J., A.M. Moore, J.P. Loschnigg and R.R. Leben, 1999: Coupled ocean–temperature dynamics in the Indian Ocean during 1997-98. *Nature*, **401**, 356-360.

Whittingham, E., J. Campbell and P. Townsley, 2003: *Poverty and Reefs*. DFID-IMM-IOC/UNESCO, Exeter, 260 pp.

Wilkinson, C.R., 2002: *Status of Coral Reefs of the World*. Australian Institute of Marine Science, Townsville, 388 pp.

Wilkinson, J.W., 2004. *Status of Coral Reefs of the World*. Australian Institute of Marine Science, Townsville, 580 pp.

Willig, M.R., D.M. Kaufman and R.D. Stevens, 2003: Latitudinal gradients of biodiversity: pattern, process, scale, and synthesis. *Annu. Rev. Ecol. Evol. Syst.*, **34**, 273-309.

Willis, B.L., C.A. Page and E.A. Dinsdale, 2004: Coral disease on the Great Barrier Reef. *Coral Health and Disease*, E. Rosenberg and Y. Loya, Eds., Springer, Berlin, 69-104.

Woodroffe, C.D. and R.J. Morrison, 2001: Reef-island accretion and soil development, Makin Island, Kiribati, central Pacific. *Catena*, **44**, 245-261.

Woodroffe, C.D., M. Dickson, B.P. Brooke and D.M. Kennedy, 2005: Episodes of reef growth at Lord Howe Island, the southernmost reef in the southwest Pacific. *Global Planet. Change*, **49**, 222-237.

Wooldridge, S., T. Done, R. Berkelmans, R. Jones and P. Marshall, 2005: Precursors for resilience in coral communities in a warming climate: a belief network approach. *Mar. Ecol.–Prog. Ser.*, **295**, 157-169.

C3. Megadeltas: their vulnerabilities to climate change

C3.1 Introduction

C3.1.1 Deltas and megadeltas: hotspots for vulnerability (Chapter 6, Box 6.3)

Deltas, some of the largest sedimentary deposits in the world, are widely recognised as being highly vulnerable to the impacts of climate change, particularly sea-level rise and changes in runoff, as well as being subject to stresses imposed by human modification of catchment and delta plain land use. Most deltas are already undergoing natural subsidence that results in accelerated rates of relative sea-level rise above the global average. Many are impacted by the effects of water extraction and diversion, as well as declining sediment input as a consequence of entrapment in dams. Delta plains, particularly those in Asia (see C3.2.1), are densely populated, and large numbers of people are often impacted as a result of external terrestrial influences (river floods, sediment starvation) and/or external marine influences (storm surges, erosion) (see Figure 6.1).

Ericson et al. (2006) estimated that nearly 300 million people inhabit a sample of 40 deltas globally, including all the large megadeltas. Average population density is 500 people/km^2, with the largest population in the Ganges-Brahmaputra delta, and the highest density in the Nile delta. Many of these deltas and megadeltas are associated with significant and expanding urban areas. Ericson et al. (2006) used a generalised modelling approach to approximate the effective rate of sea-level rise under present conditions, basing estimates of sediment trapping and flow diversion on a global dam database, and modifying estimates of natural subsidence to incorporate accelerated human-induced subsidence. This analysis showed that much of the population of these 40 deltas is at risk through coastal erosion and land loss, primarily as a result of decreased sediment delivery by the rivers, but also through accentuated rates of sea-level rise. They estimate, using a coarse digital terrain model and global population distribution data, that more than 1 million people will be directly affected by 2050 in three megadeltas: the Ganges-Brahmaputra delta in Bangladesh, the Mekong delta in Vietnam and the Nile delta in Egypt. More than 50,000 people are likely to be directly impacted in each of a further nine deltas, and more than 5,000 in each of a further twelve deltas (Figure C3.1). This generalised modelling approach indicates that 75% of the population affected live on Asian megadeltas and deltas, and a large proportion of the remainder are on deltas in Africa. These impacts would be exacerbated by accelerated sea-level rise and enhanced human pressures (see, e.g., C3.2.1). Within the Asian megadeltas, the surface topography is complex as a result of the geomorphological development of the deltas, and the population distribution shows considerable spatial variability, reflecting the intensive land use and the growth of some of the world's largest megacities (Woodroffe et al., 2006). Many people in these and other deltas worldwide are already subject to flooding from both storm surges and seasonal river floods, and therefore it is necessary to develop further methods to assess individual delta vulnerability (e.g., Sánchez-Arcilla et al., 2007).

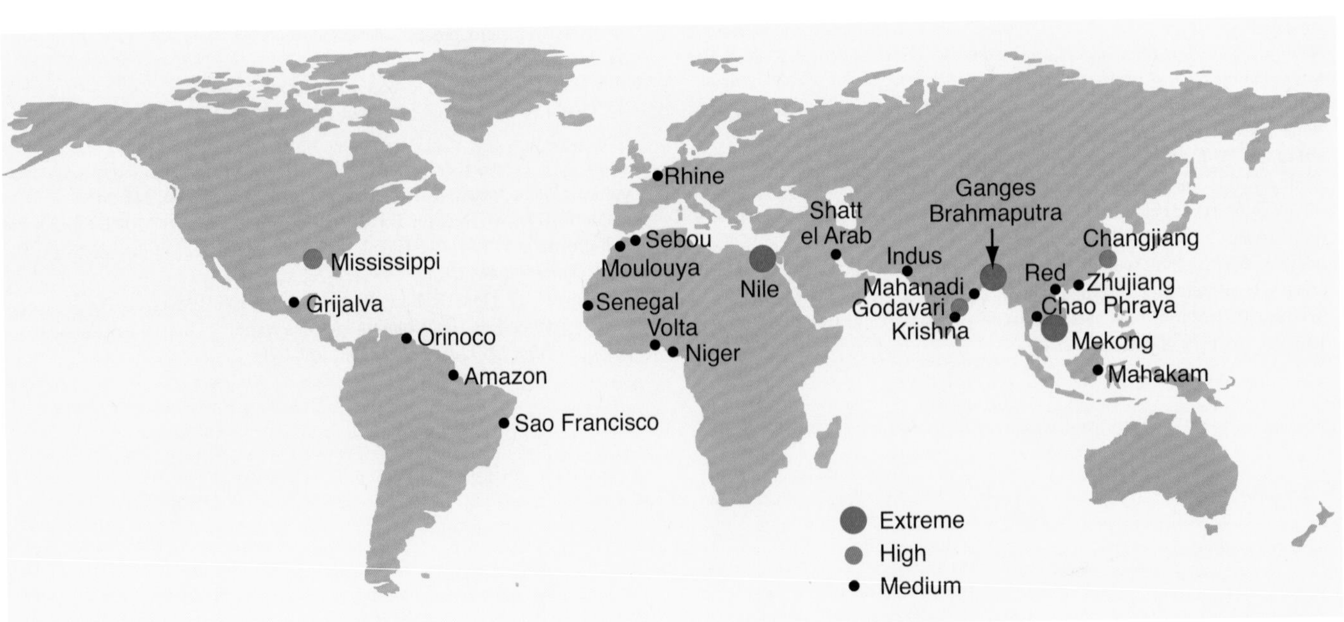

Figure C3.1. *Relative vulnerability of coastal deltas as shown by the indicative population potentially displaced by current sea-level trends to 2050 (Extreme = >1 million; High = 1 million to 50,000; Medium = 50,000 to 5,000; following Ericson et al., 2006).*

C3.2 Megadeltas in Asia

C3.2.1 Megadeltas in Asia (Chapter 10, Section 10.6.1, Table 10.10)

There are eleven megadeltas with an area greater than 10,000 km^2 (Table C3.1) in the coastal zone of Asia that are continuously being formed by rivers originating from the Tibetan Plateau (Milliman and Meade, 1983; Penland and Kulp, 2005) These megadeltas are vital to Asia because they are home to millions of people, especially in the seven megacities that are located in these deltas (Nicholls, 1995; Woodroffe et al., 2006). The megadeltas, particularly the Zhujiang delta, Changjiang delta and Huanghe delta, are also economically important, accounting for a substantial proportion of China's total GDP (Niou, 2002; She, 2004). Ecologically, the Asian megadeltas are critical diverse ecosystems of unique assemblages of plants and animals located in different climatic regions (IUCN, 2003b; ACIA, 2005; Macintosh, 2005; Sanlaville and Prieur, 2005). However, the megadeltas of Asia are vulnerable to climate change and sea-level rise that could increase the frequency and level of inundation of megadeltas due to storm surges and floods from river drainage (Nicholls, 2004; Woodroffe et al., 2006) putting communities, biodiversity and infrastructure at risk of being damaged. This impact could be more pronounced in megacities located in megadeltas, where natural ground subsidence is enhanced by human activities, such as in Bangkok in the Chao Phraya delta, Shanghai in the Changjiang delta, Tianjin in the old Huanghe delta (Nguyen et al., 2000; Li et al., 2004a, 2005; Jiang, 2005; Woodroffe et al., 2006). Climate change together with human activities could also enhance erosion that has, for example, caused the Lena delta to retreat at a rate of 3.6 to 4.5 m/yr (Leont'yev, 2004) and has affected the progradation and retreat of megadeltas fed by rivers originating from the Tibetan Plateau (Li et al., 2004b; Thanh et al., 2004; Shi et al., 2005; Woodroffe et al., 2006). The adverse impacts of salt-water intrusion on water supply in the Changjiang delta and Zhujiang delta, mangrove forests, agriculture production and freshwater fish catch, resulting in a loss of US$125×10^6 per annum in the Indus delta, could also be aggravated by climate change (IUCN, 2003a, b; Shen et al., 2003; Huang and Zhang, 2004).

Externally, the sediment supplies to many megadeltas have been reduced by the construction of dams, and there are plans for many more dams in the 21st century (see C3.1.1; Woodroffe et al., 2006). The reduction of sediment supplies makes these systems much more vulnerable to climate change and sea-level rise. When considering all the non-climate pressures, there is very high confidence that the group of populated Asian megadeltas is highly threatened by climate change and responding to this threat will present important challenges (see also C3.1.1). The sustainability of megadeltas in Asia in a warmer climate will rest heavily on policies and programmes that promote integrated and co-ordinated development of the megadeltas and upstream areas, balanced use and development of megadeltas for conservation and production goals, and comprehensive protection against erosion from river-flow anomalies and sea-water actions that combines structural with human and institutional capability-building measures (Du and Zhang, 2000; Inam et al., 2003; Li et al., 2004b; Thanh et al., 2004; Saito, 2005; Woodroffe et al., 2006; Wolanski, 2007).

C3.2.2 Climate change and the fisheries of the lower Mekong: an example of multiple stresses on a megadelta fisheries system due to human activity (Chapter 5, Box 5.3)

Fisheries are central to the lives of the people, particularly the rural poor, who live in the lower Mekong countries. Two-thirds of the basin's 60 million people are in some way active in fisheries, which represent about 10% of the GDP of Cambodia and Lao People's Democratic Republic (PDR). There are approximately 1,000 species of fish commonly found in the river, with many more marine vagrants, making it one of the most prolific and diverse faunas in the world (MRC, 2003). Recent estimates of the annual catch from capture fisheries alone exceed 2.5 Mtonnes (Hortle and Bush, 2003), with the delta contributing over 30% of this.

Direct effects of climate will occur due to changing patterns of precipitation, snow melt and rising sea level, which will affect hydrology and water quality. Indirect effects will result from changing vegetation patterns that may alter the food chain and increase soil erosion. It is likely that human impacts on the fisheries (caused by population growth, flood mitigation, increased water abstractions, changes in land use and over-fishing) will be greater than the effects of climate, but the pressures are strongly interrelated.

An analysis of the impact of climate-change scenarios on the flow of the Mekong (Hoanh et al., 2004) estimated increased maximum monthly flows of 35 to 41% in the basin and 16 to 19% in the delta (lower value is for years 2010 to 2138 and higher value for years 2070 to 2099, compared with 1961 to 1990 levels). Minimum monthly flows were estimated to decrease by 17 to 24% in the basin and 26 to 29% in the delta. Increased flooding would positively affect fisheries yields, but a reduction in dry season habitat may reduce recruitment of some species. However, planned water-management interventions, primarily dams, are expected to have the opposite effects on hydrology, namely marginally decreasing wet-season flows and considerably increasing dry-season flows (World Bank, 2004).

Models indicate that even a modest sea level rise of 20 cm would cause contour lines of water levels in the Mekong delta to shift 25 km towards the sea during the flood season and salt water to move further upstream (although confined within canals) during the dry season (Wassmann et al., 2004). Inland movement of salt water would significantly alter the species composition of fisheries, but may not be detrimental for overall fisheries yields.

C3.3 Megadeltas in the Arctic

C3.3.1 Arctic megadeltas (Chapter 15, Section 15.6.2)

Numerous river deltas are located along the Arctic coast and the rivers that flow to it. Of particular importance are the megadeltas of the Lena (44,000 km^2) and Mackenzie (9,000 km^2) rivers, which are fed by the largest Arctic rivers of

Table C3.1. *Megadeltas of Asia.*

Features	Lena	Huanghe-Huaihe	Changjiang	Zhujiang	Red River	Mekong	Chao Phraya	Irrawaddy	Ganges-Brahmaputra	Indus	Shatt-el-Arab (Arvand Rud)
Area (10^3 km²)	43.6	36.3	66.9	10	16	62.5	18	20.6	100	29.5	18.5
Water discharge (10^9 m³/yr)	520	33.3	905	326	120	470	30	430	1330	185	46
Sediment load (10^6 t/yr)	18	849	433	76	130	160	11	260	1969	400	100
Delta growth (km²/yr)	--	21.0	16.0	11.0	3.6	1.2		10.0	5.5 to 16.0	PD30	
Climate zone	Boreal	Temperate	Sub-tropical	Sub-tropical	Tropical	Tropical	Tropical	Tropical	Tropical	Semi-arid	Arid
Mangroves (10^3 km²)	None	None	None	None		5.2	2.4	4.2	10	1.6	None
Population (10^6) in 2000	0.000079	24.9 (00)	76 (03)	42.3 (03)	13.3	15.6	11.5	10.6	130	3.0	0.4
Population increase by 2015	None	18	-	176	21	21	44	15	28	45	--
GDP (US$$10^9$)		58.8 (00)	274.4 (03)	240.8 (03)	9.2 (04)	7.8 (04)	--	--	--	--	--
Megacity	None	Tianjin	Shanghai	Guangzhou	--	--	Bangkok	--	Dhaka	Karachi	--
Ground subsidence (m)	None	2.6 to 2.8	2.0 to 2.6	X	XX	--	0.2 to 1.6	--	0.6 to 1.9 mm/a		--
SLR (cm) in 2050	10 to 90 (2100)	70 to 90	50 to 70	40 to 60	--	--	--	--	--	20 to 50	--
Salt-water intrusion (km)	--	--	100	--	30 to 50	60 to 70	--	--	100	80	--
Natural hazards	--	FD	CS, SWI, FD	CS, FD, SWI	CS, FD, SWI	SWI	--	--	CS, FD, SWI	CS, SWI	--
Area inundated by SLR (10^3 km²). Figure in brackets indicates amount SLR.	--	21.3 (0.3m)	54.5 (0.3m)	5.5 (0.3m)	5 (1m)	20 (1m)	--	--	--	--	--
Coastal protection	No protection	Protected	Protected	Protected	Protected	Protected	Protected	Protected	Protected	Partial Protection	Partial protection

PD: Progradation of coast; CS: Tropical cyclone and storm surge; FD: Flooding; SLR: Sea-level rise; SWI: Salt water intrusion; DG: Delta growth in area; XX: Strong ground subsidence; X: Slight ground subsidence; --: No data available

Eurasia and North America, respectively. In contrast to non-polar megadeltas, the physical development and ecosystem health of these systems are strongly controlled by cryospheric processes and are thus highly susceptible to the effects of climate change.

Currently, advance/retreat of Arctic marine deltas is highly dependent on the protection afforded by near-shore and land-fast sea ice (Solomon, 2005; Walsh et al., 2005). The loss of such protection with warming will lead to increased erosion by waves and storm surges. The problems will be exacerbated by rising sea levels, greater wind fetch produced by shrinking sea-ice coverage, and potentially by increasing storm frequency. Similarly, thawing of the permafrost and ground-ice that currently consolidates deltaic material will induce hydrodynamic erosion on the delta front and along riverbanks. Thawing of permafrost on the delta plain itself will lead to similar changes; for example, the initial development of more ponded water, as thermokarst activity increases, will eventually be followed by drainage as surface and groundwater systems become linked. Climate warming may have already caused the loss of wetland area as lakes expanded on the Yukon River delta in the late 20th century (Coleman and Huh, 2004). Thaw subsidence may also affect the magnitude and frequency of delta flooding from spring flows and storm surges (Kokelj and Burn, 2005).

The current water budget and sediment-nutrient supply for the multitude of lakes and ponds that populate much of the tundra plains of Arctic deltas depends strongly on the supply of floodwaters produced by river-ice jams during the spring freshet. Studies of future climate conditions on a major river delta of the Mackenzie River watershed (Peace-Athabasca Delta) indicate that a combination of thinner river ice and reduced spring runoff will lead to decreased ice-jam flooding (Beltaos et al., 2006). This change, combined with greater summer evaporation due to warmer temperatures, will cause a decline in delta-pond water levels (Marsh and Lesack, 1996). For many Arctic regions, summer evaporation already exceeds precipitation and therefore the loss of ice-jam flooding could lead to a drying of delta ponds and a loss of sediment and nutrients known to be critical to their ecosystem health (Lesack et al., 1998; Marsh et al., 1999). A successful adaptation strategy that has already been used to counteract the effects of drying of delta ponds involves managing water release from reservoirs to increase the probability of ice-jam formation and related flooding (Prowse et al., 2002).

C3.4 Case study of Hurricane Katrina

C3.4.1 Hurricane Katrina and coastal ecosystem services in the Mississippi delta (Chapter 6, Box 6.4)

Whereas an individual hurricane event cannot be attributed to climate change, it can serve to illustrate the consequences for ecosystem services if the intensity and/or frequency of such events were to increase in the future. One result of Hurricane Katrina, which made landfall in coastal Louisiana on 29 August 2005, was the loss of 388 km^2 of coastal wetlands, levees and islands that flank New Orleans in the Mississippi River deltaic plain (Barras, 2006) (Figure C3.2). (Hurricane Rita, which struck in September 2005, had relatively minor effects on this part of the Louisiana coast which are included in this estimate.) The Chandeleur Islands, which lie south-east of the city, were reduced to roughly half of their former extent as a direct result of Hurricane Katrina. Collectively, these natural systems serve as the first line of defence against storm surge in this highly populated region. While some habitat recovery is expected, it is likely to be minimal compared to the scale of the losses. The Chandeleur Islands serve as an important wintering ground for migratory waterfowl and neo-tropical birds; a large population of North American redhead ducks, for example, feed on the rhizomes of sheltered sea grasses leeward of the Chandeleur Islands (Michot, 2000). Historically the region has ranked second only to Alaska in U.S. commercial fisheries production, and this high productivity has been attributed to the extent of coastal marshes and sheltered estuaries of the Mississippi River delta. Over 1,800 people lost their lives (Graumann et al., 2005) during Hurricane Katrina and the economic losses totalled more than US$100 billion (NOAA, 2007). Roughly 300,000 homes and over 1,000 historical and cultural sites were destroyed along the Louisiana and Mississippi coasts (the loss of oil production and refinery capacity helped to raise global oil prices in the short term). Post-Katrina, some major changes to the delta's management are being advocated, most notably abandonment of the 'bird-foot delta', where artificial levees channel valuable sediments into deep water (EFGC, 2006; NRC, 2006). The aim is to restore large-scale delta building processes and hence sustain the ecosystem services in the long term. Hurricane Katrina is further discussed in C3.4.2 and Chapter 14.

C3.4.2 Vulnerabilities to extreme weather events in megadeltas in the context of multiple stresses: the case of Hurricane Katrina (Chapter 7, Box 7.4)

It is possible to say with a high level of confidence that sustainable development in some densely populated megadeltas of the world will be challenged by climate change, not only in developing countries but in developed countries also. The experience of the U.S. Gulf Coast with Hurricane Katrina in 2005 is a dramatic example of the impact of a tropical cyclone – of an intensity expected to become more common with climate change – on the demographic, social and economic processes and stresses of a major city located in a megadelta.

In 2005, the city of New Orleans had a population of about half a million, located on the delta of the Mississippi River along the U.S. Gulf Coast. The city is subject not only to seasonal storms (Emanuel, 2005) but also to land subsidence at an average rate of 6 mm/yr, rising to 10-15 mm/year or more (Dixon et al., 2006). Embanking the main river channel has led to a reduction in sedimentation, leading to the loss of coastal wetlands that tend to reduce storm surge flood heights, while urban development throughout the 20th century has significantly increased land use and settlement in areas vulnerable to flooding. A number of studies of the protective levee system had indicated growing vulnerabilities to flooding, but actions were not taken to improve protection.

In late August 2005, Hurricane Katrina – which had been a Category 5 storm but weakened to Category 3 before landfall – moved onto the Louisiana and Mississippi coast with a storm surge, supplemented by waves, reaching up to 8.5 m above sea level along the southerly-facing shallow Mississippi coast (see also C3.4.1). In New Orleans, the surge reached around 5 m, overtopping and breaching sections of the city's 4.5 m defences, flooding 70 to 80% of New Orleans, with 55% of the city's properties inundated by more than 1.2 m of water and maximum flood depths up to 6 m. In Louisiana 1,101 people died, nearly all related to flooding, concentrated among the poor and elderly.

Across the whole region, there were 1.75 million private insurance claims, costing in excess of US$40 billion (Hartwig, 2006), while total economic costs are projected to be significantly in excess of US$100 billion. Katrina also exhausted the federally-backed National Flood Insurance Program (Hunter,

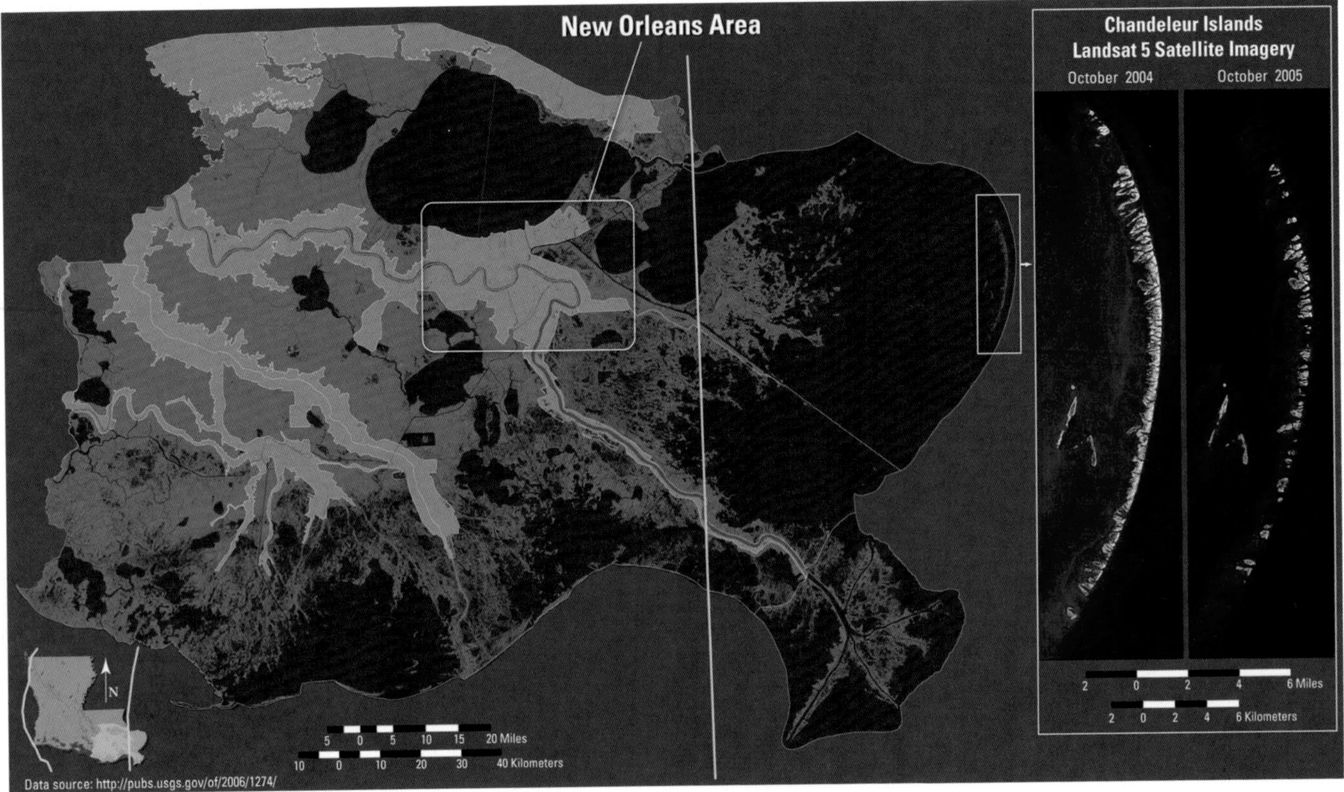

Figure C3.2. *The Mississippi delta, including the Chandeleur Islands. Areas in red were converted to open water during the hurricane. Yellow lines on index map of Louisiana show tracks of Hurricane Katrina on the right and Hurricane Rita on the left. (Figure source: U.S. Geological Survey, modified from Barras, 2006.)*

2006), which had to borrow US$20.8 billion from the Government to fund the Katrina residential flood claims. In New Orleans alone, while flooding of residential structures caused US$8 to 10 billion in losses, US$3 to 6 billion was uninsured. Of the flooded homes, 34,000 to 35,000 carried no flood insurance, including many that were not in a designated flood risk zone (Hartwig, 2006).

Beyond the locations directly affected by the storm, areas that hosted tens of thousands of evacuees had to provide shelter and schooling, while storm damage to the oil refineries and production facilities in the Gulf region raised highway vehicle fuel prices nationwide. Reconstruction costs have driven up the costs of building construction across the southern USA, and federal government funding for many programmes was reduced because of commitments to provide financial support for hurricane damage recovery. Six months after Katrina, it was estimated that the population of New Orleans was 155,000, with this number projected to rise to 272,000 by September 2008; 56% of its pre-Katrina level (McCarthy et al., 2006).

References

ACIA (Arctic Climate Impact Assessment), 2005: *Impacts of a Warming Arctic: Arctic Climate Impact Assessment*. Cambridge University Press, Cambridge, 140 pp.

Barras, J.A., 2006: Land area change in coastal Louisiana after the 2005 hurricanes: a series of three maps. U.S. Geological Survey Open-File Report 2006-1274. http://pubs.usgs.gov/of/2006/1274/.

Beltaos, S., T. Prowse, B. Bonsal, R. MacKay, L. Romolo, A. Pietroniro and B. Toth, 2006: Climatic effects on ice-jam flooding of the Peace–Athabasca Delta. *Hydrol. Process.*, **20**, 4013-4050.

Coleman, J.M. and O.K. Huh, 2004: *Major World Deltas: A Perspective from Space*. Coastal Studies Institute, Louisiana State University, Baton Rouge, Louisiana. http://www.geol.lsu.edu/WDD/PUBLICATIONS/introduction.htm.

Dixon, T.H., F. Amelung, A. Ferretti, F. Novali, F. Rocca, R. Dokka, G. Sellall, S.-W. Kim, S. Wdowinski and D. Whitman, 2006: Subsidence and flooding in New Orleans. *Nature*, **441**, 587-588.

Du, B.L. and J.W. Zhang, 2000: Adaptation strategy for sea-level rise in vulnerable areas along China's coast. *Acta Oceanologica Sinica*, **19**, 1-16.

EFGC, 2006: *Envisioning the Future of the Gulf Coast: Final Report and Findings*. D. Reed, Ed., America's Wetland: Campaign to Save Coastal Louisiana, 11 pp.

Emanuel, K., 2005: Increasing destructiveness of tropical cyclones over the past 30 years. *Nature*, **434**, 686-688.

Ericson, J.P., C.J. Vorosmarty, S.L. Dingman, L.G. Ward and M. Meybeck, 2006: Effective sea-level rise and deltas: causes of change and human dimension implications. *Global Planet. Change*, **50**, 63-82.

Graumann, A., T. Houston, J. Lawrimore, D. Levinson, N. Lott, S. McCown, S. Stephens and D. Wuertz, 2005: Hurricane Katrina: a climatological perpective. October 2005, updated August 2006. Technical Report 2005-01, NOAA National Climate Data Center, 28 pp. http://www.ncdc.noaa.gov/oa/reports/tech-report-200501z.pdf.

Hartwig, R., 2006: *Hurricane Season of 2005: Impacts on U.S. P/C Insurance Markets in 2006 and Beyond*. Presentation to the Insurance Information Institute, New York, 239 pp. http://www.iii.org/media/presentations/katrina/.

Hoanh, C.T., H. Guttman, P. Droogers and J. Aerts, 2004: Will we produce sufficient food under climate change? Mekong Basin (South-east Asia). *Climate Change in Contrasting River Basins: Adaptation Strategies for Water, Food, and Environment*, J.C.J.H. Aerts and P. Droogers, Eds., CABI Publishing, Wallingford, 157-180.

Hortle, K. and S. Bush, 2003: Consumption in the lower Mekong basin as a measure of fish yield. *New Approaches for the Improvement of Inland Capture Fish-*

ery Statistics in the Mekong Basin, T. Clayton, Ed., FAO RAP Publication 2003/01, Bangkok, 76-88.

Huang, Z.G. and W.Q. Zhang, 2004: Impacts of artificial factors on the evolution of geomorphology during recent 30 years in the Zhujiang Delta. *Quaternary Res.*, **24**, 394-401.

Hunter, J.R., 2006: *Testimony before the Committee on Banking, Housing and Urban Affairs of the United States Senate Regarding Proposals to Reform the National Flood Insurance Program*. Consumer Federation of America, 7 pp.

Inam, A., T.M. Ali Khan, A.R. Tabrez, S. Amjad, M. Danishb and S.M. Tabrez, 2003: Natural and man-made stresses on the stability of Indus deltaic Eco region. *Extended Abstract, 5th International Conference on Asian Marine Geology*, Bangkok, Thailand (IGCP475/APN).

IUCN (The World Conservation Union), 2003a: *Indus Delta, Pakistan: Economic Costs of Reduction in Freshwater Flows*. Case Studies in Wetland Valuation No. 5. Pakistan Country Office, Karachi, 6 pp. http://www.waterandnature.org/econ/CaseStudy05Indus.pdf.

IUCN (The World Conservation Union), 2003b: *The Lower Indus River: Balancing Development and Maintenance of Wetland Ecosystems and Dependent Livelihoods*. Water and Nature Initiative, 5 pp. http://www.iucn.org/themes/wani/flow/cases/Indus.pdf.

Jiang, H.T., 2005, Problems and discussion in the study of land subsidence in the Suzhou-Wuxi-Changzhou Area. *Quaternary Res.*, **25**, 29-33.

Kokelj, S.V. and C.R. Burn, 2005: Near-surface ground ice in sediments of the Mackenzie Delta, Northwest Territories, Canada. *Permafrost Periglac.*, **16**, 291-303.

Leont'yev, I.O., 2004: Coastal profile modelling along the Russian Arctic coast. *Coast. Eng.*, **51**, 779–794.

Lesack, L.F.W., P. Marsh and R.E. Hecky, 1998: Spatial and temporal dynamics of major solute chemistry along Mackenzie Delta lakes. *Limnol. Oceanogr.*, **43**, 1530-1543.

Li, C.X, D.D. Fan, B. Deng and V. Korotaev, 2004a: The coasts of China and issues of sea level rise. *J. Coastal Res.*, **43**, 36-47.

Li, C.X., S.Y. Yang, D.D. Fan and J. Zhao, 2004b: The change in Changjiang suspended load and its impact on the delta after completion of Three-Gorges Dam (in Chinese with English abstract). *Quaternary Sci.*, **24**, 495-500.

Li, J., J.Y. Zang, Y. Saito, X.W. Xu, Y.J. Wang, E. Matsumato and Z.Y. Zhang, 2005: Several cooling events over the Hong River Delta, Vietnam during the past 5000 years (in Chinese with English abstract). *Advances in Marine Science*, **23**, 43-53.

Macintosh, D., 2005: Asia, eastern, coastal ecology. *Encyclopedia of Coastal Science*, M. Schwartz, Ed., Springer, Dordrecht, 56-67.

Marsh, P. and L.F.W. Lesack, 1996: The hydrologic regime of perched lakes in the Mackenzie Delta: potential responses to climate change. *Limnol. Oceanogr.*, **41**, 849-856.

Marsh, P., L.F.W. Lesack and A. Roberts, 1999: Lake sedimentation in the Mackenzie Delta, NWT. *Hydrol. Process.*, **13**, 2519-2536.

McCarthy, K., D.J. Peterson, N. Sastry and M. Pollard, 2006: The repopulation of New Orleans after Hurricane Katrina. Technical Report, Santa Monica, RAND Gulf States Policy Institute, 59 pp.

Milliman, J.D. and R.H. Meade, 1983: World-wide delivery of river sediment to the oceans. *J. Geol.*, **90**, 1-21.

MRC, 2003: *State of the Basin Report: 2003*. Mekong River Commission, Phnom Penh, 300 pp.

Nguyen V.L., T.K.O. Ta and M. Tateishib, 2000: Late Holocene depositional environments and coastal evolution of the Mekong River Delta, Southern Vietnam. *J. Asian Earth Sci.*, **18**, 427-439.

Nicholls, R.J., 1995: Coastal mega-cities and climate change. *GeoJournal*, **37**, 369-379.

Nicholls, R.J., 2004: Coastal flooding and wetland loss in the 21st century: changes under the SRES climate and socio-economic scenarios. *Global Environ. Chang.*, **14**, 69-86.

Niou, Q.Y., 2002: *2001-2002 Report on Chinese Metropolitan Development* (in Chinese). Xiyuan Press, Beijing, 354 pp.

NOAA, 2007: *Billion Dollar U.S. Weather Disasters, 1980-2006*. http://www.ncdc.noaa.gov/oa/reports/billionz.html.

NRC, 2006: *Drawing Louisiana's New Map: Addressing Land Loss in Coastal Louisiana*. National Research Council, National Academy Press, Washington, District of Columbia, 204 pp.

Penland, S. and M.A. Kulp, 2005: Deltas. *Encyclopedia of Coastal Science*, M.L. Schwartz, Ed., Springer, Dordrecht, 362-368.

Prowse, T.D., D.L. Peters, S. Beltaos, A. Pietroniro, L. Romolo, J. Töyrä and R. Leconte, 2002: Restoring ice-jam floodwater to a drying delta ecosystem. *Water Int.*, **27**, 58-69.

Saito, Y., 2005: Mega-deltas in Asia: characteristics and human influences. *Mega-Deltas of Asia: Geological Evolution and Human Impact*, Z.Y. Chen, Y. Saito, S.L. Goodbred, Jr., Eds., China Ocean Press, Beijing, 1-8.

Sánchez-Arcilla, A., J.A. Jiménez and H.I. Valdemoro, 2007: A note on the vulnerability of deltaic coasts: application to the Ebro delta. *Managing Coastal Vulnerability: An Integrated Approach*, L. McFadden, R.J. Nicholls and E. Penning-Rowsell, Eds., Elsevier Science, Amsterdam, in press.

Sanlaville P. and A. Prieur, 2005: Asia, Middle East, coastal ecology and geomorphology. *Encyclopedia of Coastal Science*, M.L. Schwartz, Ed., Springer, Dordrecht, 71-83.

She, Z.X., 2004: Human–land interaction and socio-economic development, with special reference to the Changjiang Delta (in Chinese). *Proc. of the Xiangshan Symposium on Human–Land Coupling System of River Delta Regions: Past, Present and Future*, Beijing.

Shen, X.T., Z.C. Mao and J.R. Zhu, 2003: *Saltwater Intrusion in the Changjiang Estuary* (in Chinese). China Ocean Press, Beijing, 175 pp.

Shi, L.Q., J.F. Li, M. Ying, W.H. Li, S.L. Chen and G.A. Zhang, 2005: Advances in researches on the modern Huanghe Delta development and evolution. *Adv. Mar. Sci.*, **23**, 96-104.

Solomon, S.M., 2005: Spatial and temporal variability of shoreline change in the Beaufort-Mackenzie region, Northwest Territories, Canada. *Geo-Mar. Lett.*, **25**, 127-137.

Thanh, T.D., Y. Saito, D.V. Huy, V.L. Nguyen, T.K.O. Ta and M. Tateish, 2004: Regimes of human and climate impacts on coastal changes in Vietnam. *Reg. Environ. Change*, **4**, 49-62.

Walsh, J.E., O. Anisimov, J.O.M. Hagen, T. Jakobsson, J. Oerlemans, T.D. Prowse, V. Romanovsky, N. Savelieva, M. Serreze, I. Shiklomanov and S. Solomon, 2005: Cryosphere and hydrology. *Arctic Climate Impacts Assessment, ACIA*, C. Symon, L. Arris and B. Heal, Eds., Cambridge University Press, Cambridge, 183-242.

Wassmann, R., N.X. Hein, C.T. Hoanh and T.P. Tuong, 2004: Sea level rise affecting the Vietnamese Mekong Delta: water elevation in the flood season and implications for rice production. *Climatic Change*, **66**, 89-107.

Wolanski, E., 2007: Protective functions of coastal forests and trees against natural hazards: coastal protection in the aftermath of the Indian Ocean tsunami – what role for forests and trees? *Proc. of theFAO Regional Technical Workshop, Khao Lak, Thailand*, 28-31 August 2006. FAO, Bangkok.

Woodroffe, C.D., R.J. Nicholls, Y. Saito, Z. Chen and S.L. Goodbred, 2006: Landscape variability and the response of Asian megadeltas to environmental change. *Global Change and Integrated Coastal Management: The Asia-Pacific Region*, N. Harvey, Ed., Springer, New York, 277-314.

World Bank, 2004: *Modelled observations on development scenarios in the Lower Mekong Basin: Mekong Regional Water Resources Assistance Strategy*. Prepared for the World Bank with Mekong River Commission cooperation, Washington, District of Columbia and Vientiane, 142 pp. http://www.mrcmekong.org/free_download/report.htm.

C4. Indigenous knowledge for adaptation to climate change

C4.1 Overview

C4.1.1 Role of local and indigenous knowledge in adaptation and sustainability research (Chapter 20, Box 20.1)

Research on indigenous environmental knowledge has been undertaken in many countries, often in the context of understanding local oral histories and cultural attachment to place. A survey of research during the 1980s and early 1990s was produced by Johnson (1992). Reid et al. (2006) outline the many technical and social issues related to the intersection of different knowledge systems, and the challenge of linking the scales and contexts associated with these forms of knowledge. With the increased interest in climate change and global environmental change, recent studies have emerged that explore how indigenous knowledge can become part of a shared learning effort to address climate-change impacts and adaptation, and its links with sustainability. Some examples are indicated here.

Sutherland et al. (2005) describe a community-based vulnerability assessment in Samoa, addressing both future changes in climate-related exposure and future challenges for improving adaptive capacity. Twinomugisha (2005) describes the dangers of not considering local knowledge in dialogues on food security in Uganda.

A scenario-building exercise in Costa Rica has been undertaken as part of the Millennium Ecosystem Assessment (MA, 2005). This was a collaborative study in which indigenous communities and scientists developed common visions of future development. Two pilot 5-year storylines were constructed, incorporating aspects of coping with external drivers of development (Bennett and Zurek, 2006). Although this was not directly addressing climate change, it demonstrates the potential for joint scenario-building incorporating different forms of knowledge.

In Arctic Canada, traditional knowledge was used as part of an assessment which recognised the implications of climate change for the ecological integrity of a large freshwater delta (NRBS, 1996). In another case, an environmental assessment of a proposed mine was produced through a partnership with governments and indigenous peoples. Knowledge to facilitate sustainable development was identified as an explicit goal of the assessment, and climate-change impacts were listed as one of the long-term concerns for the region (WKSS, 2001).

Vlassova (2006) describes results of interviews with indigenous peoples of the Russian North on climate and environmental trends within the Russian boreal forest. Additional examples from the Arctic are described in ACIA (2005), Riedlinger and Berkes (2001), Krupnik and Jolly (2002), Furgal et al. (2006) and Chapter 15.

C4.2 Case studies

C4.2.1 Adaptation capacity of the South American highlands' pre-Colombian communities (Chapter 13, Box 13.2)

The subsistence of indigenous civilisations in the Americas relied on the resources cropped under the prevailing climate conditions around their settlements. In the highlands of today's Latin America, one of the most critical limitations affecting development was, and currently is, the irregular distribution of water. This situation is the result of the particularities of the atmospheric processes and extremes, the rapid runoff in the deep valleys, and the changing soil conditions. The tropical Andes' snowmelt was, and still is, a reliable source of water. However, the streams run into the valleys within bounded water courses, bringing water only to certain locations. Moreover, valleys and foothills outside of the Cordillera Blanca glaciers and extent of the snow cover, as well as the Altiplano, receive little or no melt-water at all. Therefore, in large areas, human activities depended on seasonal rainfall. Consequently, the pre-Colombian communities developed different adaptive actions to satisfy their requirements. Today, the problem of achieving the necessary balance between water availability and demand is practically the same, although the scale might be different.

Under such limitations, from today's Mexico to northern Chile and Argentina, the pre-Colombian civilisations developed the necessary capacity to adapt to the local environmental conditions. Such capacity involved their ability to solve some hydraulic problems and foresee climate variations and seasonal rain periods. On the engineering side, their developments included rainwater cropping, filtration and storage; the construction of surface and underground irrigation channels, including devices to measure the quantity of water stored (Figure C4.1) (Treacy, 1994; Wright and Valencia Zegarra, 2000; Caran and Nelly, 2006). They also were able to interconnect river basins from the Pacific and Atlantic watersheds, in the Cumbe valley and in Cajamarca (Burger, 1992).

Other capacities were developed to foresee climate variations and seasonal rain periods, to organise their sowing schedules and to programme their yields (Orlove et al., 2000). These efforts enabled the subsistence of communities which, at the peak of the Inca civilisation, included some 10 million people in what is today Peru and Ecuador.

Their engineering capacities also enabled the rectification of river courses, as in the case of the Urubamba River, and the building of bridges, either hanging ones or with pillars cast in the river bed. They also used running water for leisure and worship purposes, as seen today in the 'Baño del Inca' (the spa of the Incas), fed from geothermal sources, and the ruins of a musical garden at Tampumacchay in the vicinity of Cusco (Cortazar, 1968). The priests of the Chavin culture used running water

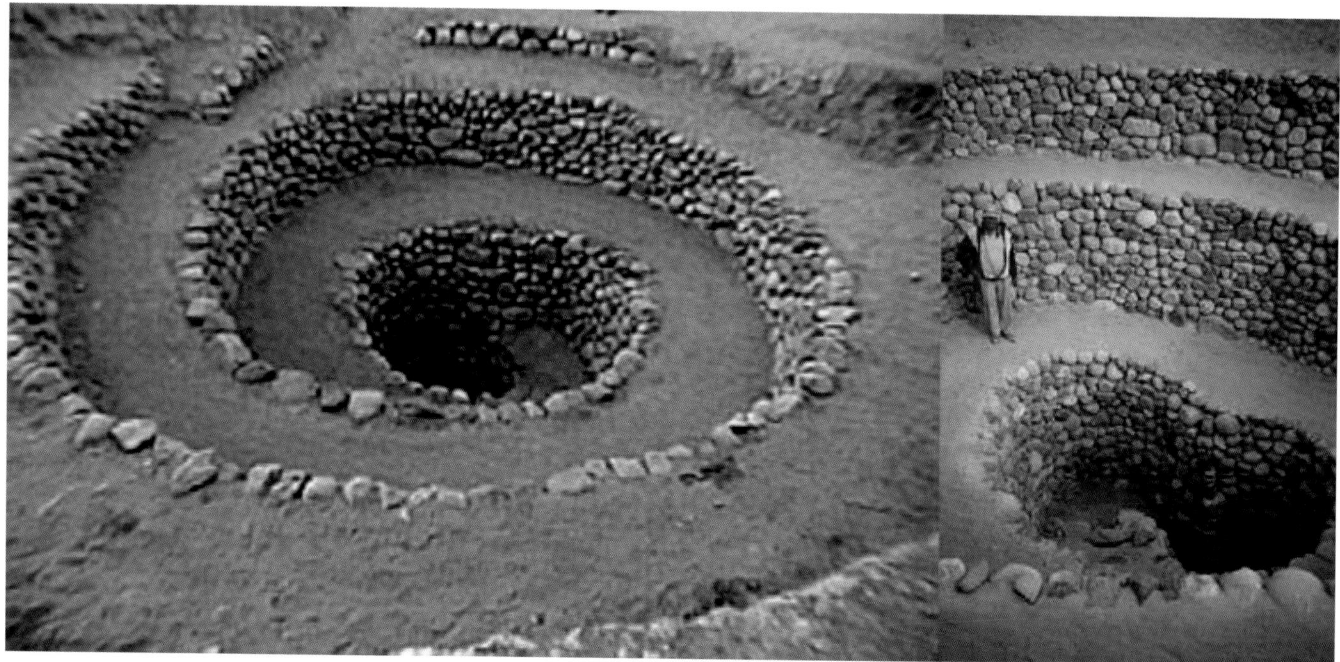

Figure C4.1. *Nasca (southern coast of Peru) system of water cropping for underground aqueducts and feeding the phreatic layers.*

flowing within tubes bored into the structure of the temples in order to produce a sound like the roar of a jaguar; the jaguar being one of their deities (Burger, 1992). Water was also used to cut stone blocks for construction. As seen in Ollantaytambo, on the way to Machu Picchu, these stones were cut in regular geometric shapes by leaking water into cleverly made interstices and freezing it during the Altiplano night, reaching below zero temperatures. They also acquired the capacity to forecast climate variations, such as those from El Niño (Canziani and Mata, 2004), enabling the most convenient and opportune organisation of their foodstuff production. In short, they developed pioneering efforts to adapt to adverse local conditions and define sustainable development paths.

Today, under the vagaries of weather and climate, exacerbated by the increasing greenhouse effect and the rapid retreat of the glaciers (Carey, 2005; Bradley et al., 2006), it would be extremely useful to revisit and update such adaptation measures. Education and training of present community members on the knowledge and technical abilities of their ancestors would be the way forward. ECLAC's procedures for the management of sustainable development (Dourojeanni, 2000), when considering the need to manage the extreme climate conditions in the highlands, refer back to the pre-Colombian irrigation strategies.

C4.2.2 African indigenous knowledge systems (Chapter 9, Section 9.6.2)

The term 'indigenous knowledge' is used to describe the knowledge systems developed by a community as opposed to the scientific knowledge that is generally referred to as 'modern' knowledge (Ajibade, 2003). Indigenous knowledge is the basis for local-level decision-making in many rural communities. It has value not only for the culture in which it evolves, but also for scientists and planners striving to improve conditions in rural localities. Incorporating indigenous knowledge into climate-change policies can lead to the development of effective adaptation strategies that are cost-effective, participatory and sustainable (Robinson and Herbert, 2001).

C4.2.2.1 Indigenous knowledge in weather forecasting

Local communities and farmers in Africa have developed intricate systems of gathering, predicting, interpreting and decision-making in relation to weather. A study in Nigeria, for example, shows that farmers are able to use knowledge of weather systems such as rainfall, thunderstorms, windstorms, harmattan (a dry dusty wind that blows along the north-west coast of Africa) and sunshine to prepare for future weather (Ajibade and Shokemi, 2003). Indigenous methods of weather forecasting are known to complement farmers' planning activities in Nigeria. A similar study in Burkina Faso showed that farmers' forecasting knowledge encompasses shared and selective experiences. Elderly male farmers formulate hypotheses about seasonal rainfall by observing natural phenomena, while cultural and ritual specialists draw predictions from divination, visions or dreams (Roncoli et al., 2001). The most widely relied-upon indicators are the timing, intensity and duration of cold temperatures during the early part of the dry season (November to January). Other forecasting indicators include the timing of fruiting by certain local trees, the water level in streams and ponds, the nesting behaviour of small quail-like birds, and insect behaviour in rubbish heaps outside compound walls (Roncoli et al., 2001).

C4.2.2.2 Indigenous knowledge in mitigation and adaptation

African communities and farmers have always coped with changing environments. They have the knowledge and practices to cope with adverse environments and shocks. The enhancement of indigenous capacity is a key to the empowerment of local communities and their effective participation in the development process (Leautier, 2004). People are better able to adopt new ideas when these can be seen in the context of existing practices. A study in Zimbabwe observed that farmers' willingness to use seasonal climate forecasts increased when the forecasts were presented in conjunction with and compared with the local indigenous climate forecasts (Patt and Gwata, 2002).

Local farmers in several parts of Africa have been known to conserve carbon in soils through the use of zero-tilling practices in cultivation, mulching, and other soil-management techniques (Dea and Scoones, 2003). Natural mulches moderate soil temperatures and extremes, suppress diseases and harmful pests, and conserve soil moisture. The widespread use of indigenous plant materials, such as agrochemicals to combat pests that normally attack food crops, has also been reported among small-scale farmers (Gana, 2003). It is likely that climate change will alter the ecology of disease vectors, and such indigenous practices of pest management would be useful adaptation strategies. Other indigenous strategies that are adopted by local farmers include: controlled bush clearing; using tall grasses such as *Andropogon gayanus* for fixing soil-surface nutrients washed away by runoff; erosion-control bunding to significantly reduce the effects of runoff; restoring lands by using green manure; constructing stone dykes; managing low-lying lands and protecting river banks (AGRHYMET, 2004).

Adaptation strategies that are applied by pastoralists in times of drought include the use of emergency fodder, culling of weak livestock for food, and multi-species composition of herds to survive climate extremes. During drought periods, pastoralists and agro-pastoralists change from cattle to sheep and goat husbandry, as the feed requirements of the latter are lower (Seo and Mendelsohn, 2006). The pastoralists' nomadic mobility reduces the pressure on low-capacity grazing areas through their cyclic movements from the dry northern areas to the wetter southern areas of the Sahel.

African women are particularly known to possess indigenous knowledge which helps to maintain household food security, particularly in times of drought and famine. They often rely on indigenous plants that are more tolerant to droughts and pests, providing a reserve for extended periods of economic hardship (Ramphele, 2004; Eriksen, 2005). In southern Sudan, for example, women are directly responsible for the selection of all sorghum seeds saved for planting each year. They preserve a spread of varieties of seeds that will ensure resistance to the range of conditions that may arise in any given growing season (Easton and Roland, 2000).

C4.2.3 Traditional knowledge for adaptation among Arctic peoples (Chapter 15, Section 15.6.1)

Among Arctic peoples, the selection pressures for the evolution of an effective knowledge base have been exceptionally strong, driven by the need to survive off highly variable natural resources in the remote, harsh Arctic environment. In response, they have developed a strong knowledge base concerning weather, snow and ice conditions as they relate to hunting and travel, and natural resource availability (Krupnik and Jolly, 2002). These systems of knowledge, belief and practice have been developed through experience and culturally transmitted among members and across generations (Huntington, 1998; Berkes, 1999). This Arctic indigenous knowledge offers detailed information that adds to conventional science and environmental observations, as well as to a holistic understanding of environment, natural resources and culture (Huntington et al., 2004). There is an increasing awareness of the value of Arctic indigenous knowledge and a growing collaborative effort to document it. In addition, this knowledge is an invaluable basis for developing adaptation and natural resource management strategies in response to environmental and other forms of change. Finally, local knowledge is essential for understanding the effects of climate change on indigenous communities (Riedlinger and Berkes, 2001; Krupnik and Jolly, 2002) and how, for example, some communities have absorbed change through flexibility in traditional hunting, fishing and gathering practices.

The generation and application of this knowledge is evidenced in the ability of Inuit hunters to navigate new travel and hunting routes despite decreasing ice stability and safety (e.g., Lafortune et al., 2004); in the ability of many indigenous groups to locate and hunt species such as geese and caribou that have shifted their migration times and routes and to begin to locate and hunt alternative species moving into the region (e.g., Krupnik and Jolly, 2002; Nickels et al., 2002; Huntington et al., 2005); the ability to detect safe sea ice and weather conditions in an environment with increasingly uncharacteristic weather (George et al., 2004); or the knowledge and skills required to hunt marine species in open water later in the year under different sea-ice conditions (Community of Arctic Bay, 2005).

Although Arctic peoples show great resilience and adaptability, some traditional responses to environmental change have already been compromised by recent socio-political changes. Their ability to cope with substantial climatic change in future, without a fundamental threat to their cultures and lifestyles, cannot be considered as unlimited. The generation and application of traditional knowledge requires active engagement with the environment, close social networks in communities, and respect for and recognition of the value of this form of knowledge and understanding. Current social, economic and cultural trends, in some communities and predominantly among younger generations, towards a more western lifestyle has the potential to erode the cycle of traditional knowledge generation and transfer, and hence its contribution to adaptive capacity.

C4.2.4 Adaptation to health impacts of climate change among indigenous populations (Chapter 8, Box 8.6)

A series of workshops organised by the national Inuit organisation in Canada, Inuit Tapiriit Kantami, documented climate-related changes and impacts, and identified and developed potential adaptation measures for local response (Furgal et al., 2002a, b; Nickels et al., 2003). The strong engagement of Inuit community residents will facilitate the successful adoption of the adaptation measures identified, such as using netting and screens on windows and house entrances to prevent bites from mosquitoes and other insects that have become more prevalent.

Another example is a study of the links between malaria and agriculture that included participation and input from a farming community in Mwea division, Kenya (Mutero et al., 2004). The approach facilitated identification of opportunities for long-term malaria control in irrigated rice-growing areas through the integration of agro-ecosystem practices aimed at sustaining livestock systems within a broader strategy for rural development.

References

ACIA (Arctic Climate Impact Assessment), 2005: *Arctic Climate Impact Assessment*. Cambridge University Press, Cambridge, 1042 pp.

AGRHYMET, 2004: Rapport synthèse de l'enquête générale sur les itinéraires d'adaptation des populations locales à la variabilité et aux changements climatiques conduite sur les projets pilotes par AGRHYMET et l'UQAM, par Hubert N'Djafa Ouaga, 13 pp.

Ajibade, L.T., 2003: A methodology for the collection and evaluation of farmers' indigenous environmental knowledge in developing countries. *Indilinga: African Journal of Indigenous Knowledge Systems*, **2**, 99-113.

Ajibade, L.T. and O. Shokemi, 2003: Indigenous approaches to weather forecasting in Asa L.G.A., Kwara State, Nigeria. *Indilinga: African Journal of Indigenous Knowledge Systems*, **2**, 37-44.

Bennett, E. and M. Zurek, 2006: Integrating epistemologies through scenarios. *Bridging Scales and Knowledge Systems*, W.V. Reid, F. Berkes, T. Wilbanks and D. Capistrano, Eds., Island Press, Washington, District of Columbia, 264-280.

Berkes, F., 1999: *Sacred Ecology: Traditional Ecological Knowledge and Resource Management*. Taylor and Francis, London, 232 pp.

Bradley, R.S., M. Vuille, H. Diaz and W. Vergara, 2006: Threats to water supplies in the tropical Andes. *Science*, **312**, 1755-1756.

Burger, R.L., 1992: *Chavin and the Origins of Andean Civilization*. Thames and Hudson, London, 240 pp.

Canziani, O.F. and L.J. Mata, 2004: The fate of indigenous communities under climate change. UNFCCC workshop on impacts of, and vulnerability and adaptation to, climate change. *Tenth Session of the Conference of Parties (COP-10)*, Buenos Aires, 3 pp.

Caran, S.C. and J.A. Nelly, 2006: Hydraulic engineering in prehistoric Mexico. *Sci. Am. Mag.*, **October**, 8 pp.

Carey, M., 2005: Living and dying with glaciers: people's historical vulnerability to avalanches and outburst floods in Peru. *Global Planet. Change*, **47**, 122-134.

Community of Arctic Bay, 2005: *Inuit Observations on Climate and Environmental Change: Perspectives from Arctic Bay, Nunavut*. ITK, Nasivvik, NAHO, NTI, Ottawa, 57 pp.

Cortazar, P.F., 1968: Documental del Perú, Departamento del Cusco, IOPPE S.A. Eds., February 1968.

Dea, D. and I. Scoones, 2003: Networks of knowledge: how farmers and scientists understand soils and their fertility: a case study from Ethiopia. *Oxford Development Studies*, **31**, 461-478.

Dourojeanni, A., 2000: *Procedimientos de Gestión para el Desarrollo Sustentable*. ECLAC, Santiago, 376 pp.

Easton, P. and M. Roland, 2000: Seeds of life: women and agricultural biodiversity in Africa. IK Notes 23. World Bank, Washington, District of Columbia, 4 pp.

Eriksen, S., 2005: The role of indigenous plants in household adaptation to climate change: the Kenyan experience. *Climate Change and Africa*, P.S Low, Ed., Cambridge University Press, Cambridge, 248-259.

Furgal, C., D. Martin and P. Gosselin, 2002a: Climate change in Nunavik and Labrador: lessons from Inuit knowledge. *The Earth is Faster Now: Indigenous Observations on Arctic Environmental Change*, I. Krupnik and D. Jolly, Eds., ARCUS, Washington, District of Columbia, 266-300.

Furgal, C.M., D. Martin, P. Gosselin, A. Viau, Nunavik Regional Board of Health and Social Services (NRBHSS) and Labrador Inuit Association (LIA), 2002b: *Climate Change in Lunavik and Labrador: What we Know from Science and Inuit Ecological Knowledge*. Final report prepared for Climate Change Action Fund, WHO/PAHO Collaborating Center on Environmental and Occupational Health Impact Assessment and Surveillance, Centre Hospitalier Universitaire de Quebec (CHUQ), Beauport, Quebec, 141 pp.

Furgal, C.M., C. Fletcher and C. Dickson, 2006: Ways of knowing and understanding: towards the convergence of traditional and scientific understanding of climate change in the Canadian North. Environment Canada, No. KM467-05-6213, 96 pp.

Gana, F.S., 2003: The usage of indigenous plant materials among small-scale farmers in Niger state agricultural development project: Nigeria. *Indilinga: African Journal of Indigenous Knowledge Systems*, **2**, 53-60.

George, J.C., H.P. Huntington, K. Brewster, H. Eicken, D.W. Norton and R. Glenn, 2004: Observations on shorefast ice dynamics in arctic Alaska and the responses of the Inupiat hunting community. *Arctic*, **57**, 363-374.

Huntington, H., 1998: Observations on the utility of the semi-directive interview for documenting traditional ecological knowledge. *Arctic*, **51**, 237-242.

Huntington, H., T.V. Callaghan, S. Fox and I. Krupnik, 2004: Matching traditional and scientific observations to detect environmental change: a discussion on arctic terrestrial ecosystems. *Ambio*, **33**(S13), 18-23.

Huntington, H., S. Fox and Co-authors, 2005: The changing Arctic: indigenous perspectives. *Arctic Climate Impact Assessment*, C. Symon, L. Arris and B. Heal, Eds., Cambridge University Press, Cambridge, 61-98.

Johnson, M., Ed., 1992: *Lore: Capturing Traditional Environmental Knowledge*. Dene Cultural Institute, Hay River, and International Development Research Centre, Ottawa, 190 pp.

Krupnik, I. and D. Jolly, Eds., 2002: *The Earth is Faster Now: Indigenous Observations of Arctic Environmental Change*. Arctic Research Consortium of the United States, Fairbanks, 384 pp.

Lafortune, V., C. Furgal, J. Drouin, T. Annanack, N. Einish, B. Etidloie, M. Qiisiq, P. Tookalook and co-authors, 2004: Climate change in northern Québec: access to land and resource issues. Project report. Kativik Regional Government, Kuujjuaq, Québec.

Leautier, F., 2004: Indigenous capacity enhancement: developing community knowledge. *Indigenous Knowledge: Local Pathways to Global Development*. The World Bank, Washington, District of Columbia, 4-8.

MA (Millennium Ecosystem Assessment), 2005: *Ecosystems and Human Well-Being: Synthesis*. Island Press, Washington, District of Columbia, 155 pp.

Mutero, C.M., C. Kabutha, V. Kimani, L. Kabuage, G. Gitau, J. Ssennyonga, J. Githure, L. Muthami, A. Kaida, L. Musyoka, E. Kiarie and M. Oganda, 2004: A transdisciplinary perspective on the links between malaria and agroecosystems in Kenya. *Acta Trop.*, **89**, 171-186.

Nickels, S., C. Furgal, J. Castelden, P. Moss-Davies, M. Buell, B. Armstrong, D. Dillon and R. Fonger, 2002: Putting the human face on climate change through community workshops. *The Earth is Faster Now: Indigenous Observations of Arctic Environmental Change*, I. Krupnik and D. Jolly, Eds., ARCUS, Washington, District of Columbia, 300-344.

Nickels, S., C. Furgal and J. Castleden, 2003: Putting the human face on climate change through community workshops: Inuit knowledge, partnerships and research. *The Earth is Faster Now: Indigenous Observations of Arctic Environmental Change*, I. Krupnik and D. Jolly, Eds., Arctic Studies Centre, Smithsonian Institution, Washington, District of Columbia, 300-344.

NRBS (Northern River Basins Study Board), 1996: *Northern River Basins Study*. Report to the Ministers 1996. Alberta Environmental Protection, Edmonton, Alberta, 287 pp. http://www3.gov.ab.ca/env/water/nrbs/index.html.

Orlove, B.S., J.C.H. Chiang and M.A. Cane, 2000: Forecasting Andean rainfall and crop yield from the influence of El Niño on Pleiades visibility. *Nature*, **403**, 68-71.

Patt, A. and C. Gwata, 2002: Effective seasonal climate forecast applications: examining constraints for subsistence farmers in Zimbabwe. *Global Environ. Chang.*, **12**, 185-195.

Ramphele, M., 2004: Women's indigenous knowledge: building bridges between the traditional and the modern. *Indigenous Knowledge: Local Pathways to Development*, The World Bank, Washington, District of Columbia, 13-17.

Reid, W.V., F. Berkes, T. Wilbanks and D. Capistrano, Eds., 2006: *Bridging Scales and Knowledge Systems*. Island Press, Washington, District of Columbia, 314 pp.

Riedlinger, D. and F. Berkes, 2001: Contributions of traditional knowledge to understanding climate change in the Canadian Arctic. *Polar Rec.*, **37**, 315-328.

Robinson, J.B. and D. Herbert, 2001: Integrating climate change and sustainable development. *Int. J. Global Environ.*, **1**, 130-149.

Roncoli, C., K. Ingram and P. Kirshen, 2001: The costs and risks of coping with drought: livelihood impacts and farmers' responses in Burkina Faso. *Climate Res.*, **19**, 119-132.

Seo, S.N. and R. Mendelsohn, 2006: Climate change adaptation in Africa: a microeconomic analysis of livestock choice. Centre for Environmental Economics and Policy in Africa (CEEPA) Discussion Paper No.19, University of Pretoria, Pretoria, 37 pp.

Sutherland, K., B. Smit, V. Wulf and T. Nakalevu, 2005: Vulnerability in Samoa. *Tiempo*, **54**, 11-15.

Treacy, J.M., 1994: *Las Chacras de Coparaque: Andenes y Riego en el Valle de Colca*. Instituto de Estudios Peruanos, Lima, 298 pp.

Twinomugisha, B., 2005: Indigenous adaptation. *Tiempo*, **57**, 6-8.

Vlassova, T.K., 2006: Arctic residents' observations and human impact assessments in understanding environmental changes in boreal forests: Russian experience and circumpolar perspectives. *Mitigation and Adaptation Strategies for Global Change*, **11**, 897-909.

WKSS (West Kitikmeot/Slave Study Society), 2001: West Kitikmeot/Slave Study Society Final Report. West Kitikmeot/Slave Study Society, Yellowknife, Canada, 87 pp. http://www.wkss.nt.ca/index.htm.

Wright, K.R. and A. Valencia Zegarra, 2000: *Machu Picchu: A Civil Engineering Marvel*. American Society of Civil Engineers Press, Virginia, 144 pp.

Appendix I: Glossary

Notes:

1. This glossary defines some specific terms as the lead authors intend them to be interpreted in the context of this Report.
2. Words in italic indicate that the following term is also contained in this glossary.

Acclimatisation
The physiological *adaptation* to climatic variations.

Active layer
The top layer of soil or rock in *permafrost* that is subjected to seasonal freezing and thawing.

Adaptability
See *adaptive capacity*.

Adaptation
Adjustment in natural or *human systems* in response to actual or expected climatic stimuli or their effects, which moderates harm or exploits beneficial opportunities. Various types of adaptation can be distinguished, including anticipatory, autonomous and planned adaptation:

- **Anticipatory adaptation** – Adaptation that takes place before impacts of *climate change* are observed. Also referred to as proactive adaptation.
- **Autonomous adaptation** – Adaptation that does not constitute a conscious response to climatic stimuli but is triggered by ecological changes in natural systems and by market or *welfare* changes in *human systems*. Also referred to as spontaneous adaptation.
- **Planned adaptation** – Adaptation that is the result of a deliberate policy decision, based on an awareness that conditions have changed or are about to change and that action is required to return to, maintain, or achieve a desired state.

Adaptation assessment
The practice of identifying options to adapt to *climate change* and evaluating them in terms of criteria such as availability, benefits, costs, effectiveness, efficiency and feasibility.

Adaptation benefits
The avoided damage costs or the accrued benefits following the adoption and implementation of *adaptation* measures.

Adaptation costs
Costs of planning, preparing for, facilitating, and implementing *adaptation* measures, including transition costs.

Adaptive capacity (in relation to climate change impacts)
The ability of a system to adjust to *climate change* (including *climate variability* and extremes) to moderate potential damages, to take advantage of opportunities, or to cope with the consequences.

Aerosols
A collection of air-borne solid or liquid particles, with a typical size between 0.01 and 10 μm, that reside in the *atmosphere* for at least several hours. Aerosols may be of either natural or *anthropogenic* origin. Aerosols may influence *climate* in two ways: directly through scattering and absorbing radiation, and indirectly through acting as condensation nuclei for cloud formation or modifying the optical properties and lifetime of clouds.

Afforestation
Direct human-induced conversion of land that has not been forested for a period of at least 50 years to forested land through planting, seeding and/or the human-induced promotion of natural seed sources. See also *reforestation* and *deforestation*. For a discussion of the term *forest* and related terms such as *afforestation*, *reforestation* and *deforestation*, see the IPCC Special Report on Land Use, Land-Use Change, and Forestry (IPCC, 2000).

Aggregate impacts
Total *impacts* integrated across sectors and/or regions. The aggregation of impacts requires knowledge of (or assumptions about) the relative importance of impacts in different sectors and regions. Measures of aggregate impacts include, for example, the total number of people affected, or the total economic costs.

Albedo
The fraction of solar radiation reflected by a surface or object, often expressed as a percentage. Snow-covered surfaces have a

high albedo; the albedo of soils ranges from high to low; vegetation-covered surfaces and oceans have a low albedo. The Earth's albedo varies mainly through varying cloudiness, snow, ice, leaf area, and land-cover changes.

Algae
Photosynthetic, often microscopic and *planktonic*, organisms occurring in marine and freshwater *ecosystems*.

Algal bloom
A reproductive explosion of *algae* in a lake, river or ocean.

Alpine
The biogeographic zone made up of slopes above the *tree line* characterised by the presence of rosette-forming *herbaceous* plants and low, shrubby, slow-growing woody plants.

Anthropogenic
Resulting from or produced by human beings.

AOGCM
See *climate model*.

Aquaculture
The managed cultivation of aquatic plants or animals such as salmon or shellfish held in captivity for the purpose of harvesting.

Aquifer
A stratum of permeable rock that bears water. An unconfined aquifer is recharged directly by local rainfall, rivers and lakes, and the rate of recharge will be influenced by the permeability of the overlying rocks and soils.

Aragonite
A calcium carbonate (limestone) mineral, used by shell- or skeleton-forming, calcifying organisms such as *corals* (warm- and cold-water corals), some macroalgae, *pteropods* (marine snails) and non-pteropod molluscs such as bivalves (e.g., clams, oysters), cephalopods (e.g., squids, octopuses). Aragonite is more sensitive to *ocean acidification* than *calcite*, also used by many marine organisms. See also *calcite* and *ocean acidification*.

Arbovirus
Any of various viruses transmitted by blood-sucking arthropods (e.g., mosquitoes, ticks, etc.) and including the causative agents of *dengue fever*, yellow fever, and some types of encephalitis.

Arid region
A land region of low rainfall, where 'low' is widely accepted to be <250 mm precipitation per year.

Atmosphere
The gaseous envelope surrounding the Earth. The dry atmosphere consists almost entirely of nitrogen and oxygen, together with trace gases including *carbon dioxide* and *ozone*.

Attribution
See *Detection and attribution*

Baseline/reference
The baseline (or reference) is the state against which change is measured. It might be a 'current baseline', in which case it represents observable, present-day conditions. It might also be a 'future baseline', which is a projected future set of conditions excluding the driving factor of interest. Alternative interpretations of the reference conditions can give rise to multiple baselines.

Basin
The drainage area of a stream, river or lake.

Benthic community
The community of organisms living on or near the bottom of a water body such as a river, a lake or an ocean.

Biodiversity
The total diversity of all organisms and *ecosystems* at various spatial scales (from genes to entire *biomes*).

Biofuel
A fuel produced from organic matter or combustible oils produced by plants. Examples of biofuel include alcohol, black liquor from the paper-manufacturing process, wood, and soybean oil.

Biomass
The total mass of living organisms in a given area or volume; recently dead plant material is often included as dead biomass. The quantity of biomass is expressed as a dry weight or as the energy, carbon or nitrogen content.

Biome
Major and distinct regional element of the *biosphere*, typically consisting of several *ecosystems* (e.g., forests, rivers, ponds, swamps) within a region of similar *climate*. Biomes are characterised by typical communities of plants and animals.

Biosphere
The part of the Earth system comprising all *ecosystems* and living organisms in the *atmosphere*, on land (terrestrial biosphere), or in the oceans (marine biosphere), including derived dead organic matter, such as litter, soil organic matter, and oceanic detritus.

Biota
All living organisms of an area; the flora and fauna considered as a unit.

Bog
Peat-accumulating acidic *wetland*.

Boreal forest
Forests of pine, spruce, fir and larch stretching from the east coast of Canada westward to Alaska and continuing from Siberia

westward across the entire extent of Russia to the European Plain. The climate is continental, with long, very cold winters (up to 6 months with mean temperatures below freezing), and short, cool summers (50 to 100 frost-free days). Precipitation increases during summer months, although annual precipitation is still small. Low *evaporation* rates can make this a humid climate. See *taiga*.

Breakwater
A hard engineering structure built in the sea which, by breaking waves, protects a harbour, anchorage, beach or shore area. A breakwater can be attached to the coast or lie offshore.

C_3 plants
Plants that produce a three-carbon compound during *photosynthesis*, including most trees and agricultural crops such as rice, wheat, soybeans, potatoes and vegetables.

C_4 plants
Plants, mainly of tropical origin, that produce a four-carbon compound during *photosynthesis*, including many grasses and the agriculturally important crops maize, sugar cane, millet and sorghum.

Calcareous organisms
A large and diverse group of organisms, many marine, that use *calcite* or *aragonite* to form shells or skeletons. See *calcite*, *aragonite* and *ocean acidification*.

Calcite
A calcium carbonate (limestone) mineral, used by shell- or skeleton-forming, calcifying organisms such as foraminifera, some macroalgae, lobsters, crabs, sea urchins and starfish. Calcite is less sensitive to *ocean acidification* than *aragonite*, also used by many marine organisms. See also *aragonite* and *ocean acidification*.

Capacity building
In the context of *climate change*, capacity building is developing the technical skills and institutional capabilities in developing countries and economies in transition to enable their participation in all aspects of *adaptation* to, *mitigation* of, and research on *climate change*, and in the implementation of the Kyoto Mechanisms, etc.

Carbon cycle
The term used to describe the flow of carbon (in various forms, e.g., *carbon dioxide*) through the *atmosphere*, ocean, terrestrial *biosphere* and lithosphere.

Carbon dioxide (CO_2)
A naturally occurring gas fixed by *photosynthesis* into organic matter. A by-product of fossil fuel combustion and *biomass* burning, it is also emitted from land-use changes and other industrial processes. It is the principal *anthropogenic greenhouse gas* that affects the Earth's radiative balance. It is the reference gas against which other greenhouse gases are measured, thus having a Global Warming Potential of 1.

Carbon dioxide fertilisation
The stimulation of plant *photosynthesis* due to elevated CO_2 concentrations, leading to either enhanced productivity and/or efficiency of *primary production*. In general, *C_3 plants* show a larger response to elevated CO_2 than *C_4 plants*.

Carbon sequestration
The process of increasing the carbon content of a *reservoir*/pool other than the *atmosphere*.

Catchment
An area that collects and drains rainwater.

CDM (Clean Development Mechanism)
The CDM allows *greenhouse gas* emission reduction projects to take place in countries that have no emission targets under the *United Nations Framework Convention on Climate Change (UNFCCC) Kyoto Protocol*, yet are signatories.

Chagas' disease
A parasitic disease caused by the *Trypanosoma cruzi* and transmitted by triatomine bugs in the Americas, with two clinical periods: acute (fever, swelling of the spleen, oedemas) and chronic (digestive syndrome, potentially fatal heart condition).

Cholera
A water-borne intestinal infection caused by a bacterium (*Vibrio cholerae*) that results in frequent watery stools, cramping abdominal pain, and eventual collapse from dehydration and shock.

Climate
Climate in a narrow sense is usually defined as the 'average weather', or more rigorously, as the statistical description in terms of the mean and variability of relevant quantities over a period of time ranging from months to thousands or millions of years. These quantities are most often surface variables such as temperature, precipitation, and wind. Climate in a wider sense is the state, including a statistical description, of the *climate system*. The classical period of time is 30 years, as defined by the World Meteorological Organization (WMO).

Climate change
Climate change refers to any change in *climate* over time, whether due to natural variability or as a result of human activity. This usage differs from that in the *United Nations Framework Convention on Climate Change (UNFCCC)*, which defines 'climate change' as: 'a change of climate which is attributed directly or indirectly to human activity that alters the composition of the global *atmosphere* and which is in addition to natural climate variability observed over comparable time periods'. See also *climate variability*.

Climate change commitment
Due to the thermal inertia of the ocean and slow processes in the *biosphere*, the *cryosphere* and land surfaces, the climate would continue to change even if the atmospheric composition was held fixed at today's values. Past change in atmospheric com-

position leads to a 'committed' *climate change* which continues for as long as a radiative imbalance persists and until all components of the *climate system* have adjusted to a new state. The further change in temperature after the composition of the *atmosphere* is held constant is referred to as the committed warming or warming commitment. Climate change commitment includes other future changes, for example in the hydrological cycle, in *extreme weather events*, and in *sea-level rise*.

Climate model
A numerical representation of the *climate system* based on the physical, chemical, and biological properties of its components, their interactions and *feedback* processes, and accounting for all or some of its known properties. The climate system can be represented by models of varying complexity (i.e., for any one component or combination of components a hierarchy of models can be identified, differing in such aspects as the number of spatial dimensions, the extent to which physical, chemical, or biological processes are explicitly represented, or the level at which empirical parameterisations are involved. Coupled *atmosphere*/ocean/sea-ice *General Circulation Models* (AOGCMs) provide a comprehensive representation of the climate system. More complex models include active chemistry and biology. Climate models are applied, as a research tool, to study and simulate the climate, but also for operational purposes, including monthly, seasonal, and interannual *climate predictions*.

Climate prediction
A climate prediction or climate forecast is the result of an attempt to produce an estimate of the actual evolution of the climate in the future, e.g., at seasonal, interannual or long-term time scales. See also *climate projection* and *climate (change) scenario*.

Climate projection
The calculated response of the *climate system* to *emissions* or concentration *scenarios* of *greenhouse gases* and *aerosols*, or *radiative forcing scenarios*, often based on simulations by *climate models*. Climate projections are distinguished from *climate predictions*, in that the former critically depend on the emissions/concentration/*radiative forcing* scenario used, and therefore on highly uncertain assumptions of future socio-economic and technological development.

Climate (change) scenario
A plausible and often simplified representation of the future *climate*, based on an internally consistent set of climatological relationships and assumptions of *radiative forcing*, typically constructed for explicit use as input to climate change impact models. A 'climate change scenario' is the difference between a climate *scenario* and the current climate.

Climate sensitivity
The equilibrium temperature rise that would occur for a doubling of CO_2 concentration above *pre-industrial* levels.

Climate system
The climate system is defined by the dynamics and interactions of five major components: *atmosphere*, hydrosphere, *cryosphere*, land surface, and *biosphere*. Climate system dynamics are driven by both internal and external forcing, such as volcanic eruptions, solar variations, or human-induced modifications to the planetary radiative balance, for instance via *anthropogenic* emissions of *greenhouse gases* and/or land-use changes.

Climate threshold
The point at which external forcing of the *climate system*, such as the increasing atmospheric concentration of *greenhouse gases*, triggers a significant climatic or environmental event which is considered unalterable, or recoverable only on very long time-scales, such as widespread bleaching of *corals* or a collapse of oceanic circulation systems.

Climate variability
Climate variability refers to variations in the mean state and other statistics (such as standard deviations, statistics of extremes, etc.) of the *climate* on all temporal and spatial scales beyond that of individual weather events. Variability may be due to natural internal processes within the *climate system* (internal variability), or to variations in natural or *anthropogenic* external forcing (external variability). See also *climate change*.

CO_2 fertilisation
See *carbon dioxide fertilisation*.

Coastal squeeze
The squeeze of coastal *ecosystems* (e.g., salt marshes, mangroves and mud and sand flats) between rising sea levels and naturally or artificially fixed shorelines, including hard engineering defences (see Chapter 6).

Coccolithophores
Single-celled microscopic *phytoplankton algae* which construct shell-like structures from *calcite* (a form of calcium carbonate). See also *calcite* and *ocean acidification*.

Committed to extinction
This term describes a species with dwindling population that is in the process of inescapably becoming extinct in the absence of human intervention. See also *extinction*.

Communicable disease
An *infectious disease* caused by transmission of an infective biological agent (virus, bacterium, protozoan, or multicellular macroparasite).

Confidence
In this Report, the level of confidence in a statement is expressed using a standard terminology defined in the Introduction. See also *uncertainty*.

Control run
A model run carried out to provide a '*baseline*' for comparison with climate-change experiments. The control run uses constant values for the *radiative forcing* due to *greenhouse gases* and *anthropogenic aerosols* appropriate to *pre-industrial* conditions.

Coral
The term 'coral' has several meanings, but is usually the common name for the Order *Scleractinia*, all members of which have hard limestone skeletons, and which are divided into reef-building and non-reef-building, or cold- and warm-water corals.

Coral bleaching
The paling in colour which results if a *coral* loses its symbiotic, energy-providing, organisms.

Coral reefs
Rock-like limestone (calcium carbonate) structures built by *corals* along ocean coasts (fringing reefs) or on top of shallow, submerged banks or shelves (barrier reefs, atolls), most conspicuous in tropical and sub-tropical oceans.

Cryosphere
The component of the *climate system* consisting of all snow and ice (including *permafrost*) on and beneath the surface of the Earth and ocean.

Cryptogams
An outdated but still-used term, denoting a group of diverse and taxonomically unrelated organisms, including fungi and lower plants such as *algae*, lichens, hornworts, liverworts, mosses and ferns.

Deforestation
Natural or *anthropogenic* process that converts forest land to non-forest. See *afforestation* and *reforestation*.

Dengue fever
An *infectious* viral *disease* spread by mosquitoes, often called breakbone fever because it is characterised by severe pain in the joints and back. Subsequent infections of the virus may lead to dengue haemorrhagic fever (DHF) and dengue shock syndrome (DSS), which may be fatal.

Desert
A region of very low rainfall, where 'very low' is widely accepted to be <100 mm per year.

Desertification
Land degradation in arid, semi-arid, and dry sub-humid areas resulting from various factors, including climatic variations and human activities. Further, the United Nations Convention to Combat Desertification (UNCCD) defines land degradation as a reduction or loss in arid, semi-arid, and dry sub-humid areas of the biological or economic productivity and complexity of rainfed cropland, irrigated cropland, or range, pasture, forest and woodlands resulting from land uses or from a process or combination of processes, including those arising from human activities and habitation patterns, such as: (i) soil *erosion* caused by wind and/or water; (ii) deterioration of the physical, chemical, and biological or economic properties of soil; and (iii) long-term loss of natural vegetation.

Detection and attribution
Detection of change in a system (natural or human) is the process of demonstrating that the system has changed in some defined statistical sense, without providing a reason for that change.
Attribution of such an observed change in a system to *anthropogenic climate change* is usually a two-stage process. First, the observed change in the system must be demonstrated to be associated with an observed regional climate change with a specified degree of *confidence*. Second, a measurable portion of the observed regional climate change, or the associated observed change in the system, must be attributed to *anthropogenic* climate forcing with a similar degree of confidence.
Confidence in such *joint attribution* statements must be lower than the confidence in either of the individual attribution steps alone due to the combination of two separate statistical assessments.

Diadromous
Fish that travel between salt water and freshwater.

Discount rate
The degree to which consumption now is preferred to consumption one year hence, with prices held constant, but average incomes rising in line with *GDP* per capita.

Disturbance regime
Frequency, intensity, and types of disturbances, such as fires, insect or pest outbreaks, floods and *droughts*.

Downscaling
A method that derives local- to regional-scale (10 to 100 km) information from larger-scale models or data analyses.

Drought
The phenomenon that exists when precipitation is significantly below normal recorded levels, causing serious hydrological imbalances that often adversely affect land resources and production systems.

Dyke
A human-made wall or embankment along a shore to prevent flooding of low-lying land.

Dynamic global vegetation model (DGVM)
Models that simulate vegetation development and dynamics through space and time, as driven by *climate* and other environmental changes.

Ecological community
A community of plants and animals characterised by a typical assemblage of species and their abundances. See also *ecosystem*.

Ecological corridor
A thin strip of vegetation used by wildlife, potentially allowing movement of biotic factors between two areas.

Ecophysiological process
Individual organisms respond to environmental variability, such as *climate change*, through ecophysiological processes which operate continuously, generally at a microscopic or sub-organ scale. Ecophysiological mechanisms underpin individual organism's tolerance to environmental stress, and comprise a broad range of responses defining the absolute tolerance limits of individuals to environmental conditions. Ecophysiological responses may scale up to control species geographic ranges.

Ecosystem
The interactive system formed from all living organisms and their abiotic (physical and chemical) environment within a given area. Ecosystems cover a hierarchy of spatial scales and can comprise the entire globe, *biomes* at the continental scale or small, well-circumscribed systems such as a small pond.

Ecosystem approach
The ecosystem approach is a strategy for the integrated management of land, water and living resources that promotes conservation and sustainable use in an equitable way. An ecosystem approach is based on the application of appropriate scientific methodologies focused on levels of biological organisation, which encompass the essential structure, processes, functions and interactions among organisms and their environment. It recognises that humans, with their cultural diversity, are an integral component of many *ecosystems*. The ecosystem approach requires adaptive management to deal with the complex and dynamic nature of ecosystems and the absence of complete knowledge or understanding of their functioning. Priority targets are conservation of *biodiversity* and of the ecosystem structure and functioning, in order to maintain ecosystem services.

Ecosystem services
Ecological processes or functions having monetary or non-monetary value to individuals or society at large. There are (i) supporting services such as productivity or *biodiversity* maintenance, (ii) provisioning services such as food, fibre, or fish, (iii) regulating services such as climate regulation or *carbon sequestration*, and (iv) cultural services such as tourism or spiritual and aesthetic appreciation.

Ecotone
Transition area between adjacent *ecological communities* (e.g., between forests and grasslands).

El Niño-Southern Oscillation (ENSO)
El Niño, in its original sense, is a warm-water current that periodically flows along the coast of Ecuador and Peru, disrupting the local fishery. This oceanic event is associated with a fluctuation of the inter-tropical surface pressure pattern and circulation in the Indian and Pacific Oceans, called the Southern Oscillation. This coupled atmosphere-ocean phenomenon is collectively known as El Niño-Southern Oscillation. During an El Niño event, the prevailing trade winds weaken and the equatorial countercurrent strengthens, causing warm surface waters in the Indonesian area to flow eastward to overlie the cold waters of the Peru current. This event has great impact on the wind, sea surface temperature, and precipitation patterns in the tropical Pacific. It has climatic effects throughout the Pacific region and in many other parts of the world. The opposite of an El Niño event is called *La Niña*.

Emissions scenario
A plausible representation of the future development of emissions of substances that are potentially radiatively active (e.g., *greenhouse gases*, *aerosols*), based on a coherent and internally consistent set of assumptions about driving forces (such as demographic and socio-economic development, technological change) and their key relationships. In 1992, the IPCC presented a set of emissions scenarios that were used as a basis for the *climate projections* in the Second Assessment Report. These emissions scenarios are referred to as the IS92 *scenarios*. In the IPCC Special Report on Emissions Scenarios (*SRES*) (Nakićenović et al., 2000), new emissions scenarios – the so-called SRES scenarios – were published.

Endemic
Restricted or peculiar to a locality or region. With regard to human health, endemic can refer to a disease or agent present or usually prevalent in a population or geographical area at all times.

Ensemble
A group of parallel model simulations used for *climate projections*. Variation of the results across the ensemble members gives an estimate of *uncertainty*. Ensembles made with the same model but different initial conditions only characterise the uncertainty associated with internal *climate variability*, whereas multi-model ensembles including simulations by several models also include the impact of model differences.

Epidemic
Occurring suddenly in incidence rates clearly in excess of normal expectancy, applied especially to *infectious diseases* but may also refer to any disease, injury, or other health-related event occurring in such outbreaks.

Erosion
The process of removal and transport of soil and rock by weathering, mass wasting, and the action of streams, *glaciers*, waves, winds and underground water.

Eustatic sea-level rise
See *sea-level rise*.

Eutrophication
The process by which a body of water (often shallow) becomes (either naturally or by pollution) rich in dissolved nutrients, with a seasonal deficiency in dissolved oxygen.

Evaporation
The transition process from liquid to gaseous state.

Evapotranspiration
The combined process of water *evaporation* from the Earth's surface and *transpiration* from vegetation.

Externalities
Occur when a change in the production or consumption of one individual or firm affects indirectly the well-being of another individual or firm. Externalities can be positive or negative. The impacts of pollution on *ecosystems*, water courses or air quality represent classic cases of negative externality.

Extinction
The global disappearance of an entire species.

Extirpation
The disappearance of a species from part of its range; local *extinction*.

Extreme weather event
An event that is rare within its statistical reference distribution at a particular place. Definitions of 'rare' vary, but an extreme weather event would normally be as rare as or rarer than the 10th or 90th percentile. By definition, the characteristics of what is called 'extreme weather' may vary from place to place. Extreme weather events may typically include floods and *droughts*.

Feedback
An interaction mechanism between processes is called a feedback. When the result of an initial process triggers changes in a second process and that in turn influences the initial one. A positive feedback intensifies the original process, and a negative feedback reduces it.

Food chain
The chain of *trophic relationships* formed if several species feed on each other. See *food web* and *trophic level*.

Food security
A situation that exists when people have secure access to sufficient amounts of safe and nutritious food for normal growth, development and an active and healthy life. Food insecurity may be caused by the unavailability of food, insufficient purchasing power, inappropriate distribution, or inadequate use of food at the household level.

Food web
The network of *trophic relationships* within an *ecological community* involving several interconnected *food chains*.

Forecast
See *climate prediction* and *climate projection*.

Forest limit/line
The upper elevational or latitudinal limit beyond which natural tree regeneration cannot develop into a closed forest stand. It is typically at a lower elevation or more distant from the poles than the *tree line*.

Freshwater lens
A lenticular fresh groundwater body that underlies an oceanic island. It is underlain by saline water.

Functional extinction
This term defines a species which has lost its capacity to persist and to recover because its populations have declined to below a minimum size. See *committed to extinction*.

General Circulation Model (GCM)
See *climate model*.

Generalist
A species that can tolerate a wide range of environmental conditions.

Glacier
A mass of land ice flowing downhill (by internal deformation and sliding at the base) and constrained by the surrounding topography (e.g., the sides of a valley or surrounding peaks). A glacier is maintained by accumulation of snow at high altitudes, balanced by melting at low altitudes or discharge into the sea.

Globalisation
The growing integration and interdependence of countries worldwide through the increasing volume and variety of cross-border transactions in goods and services, free international capital flows, and the more rapid and widespread diffusion of technology, information and culture.

Greenhouse effect
The process in which the absorption of infrared radiation by the *atmosphere* warms the Earth.
In common parlance, the term 'greenhouse effect' may be used to refer either to the natural greenhouse effect, due to naturally occurring *greenhouse gases*, or to the enhanced (*anthropogenic*) greenhouse effect, which results from gases emitted as a result of human activities.

Greenhouse gas
Greenhouse gases are those gaseous constituents of the *atmosphere*, both natural and *anthropogenic*, that absorb and emit radiation at specific wavelengths within the spectrum of infrared radiation emitted by the Earth's surface, the atmosphere, and clouds. This property causes the *greenhouse effect*. Water vapour (H_2O), *carbon dioxide* (CO_2), nitrous oxide (N_2O), methane (CH_4) and *ozone* (O_3) are the primary greenhouse gases in the Earth's atmosphere. As well as CO_2, N_2O, and CH_4, the *Kyoto Protocol* deals with the greenhouse gases sulphur hexafluoride (SF_6), hydrofluorocarbons (HFCs) and perfluorocarbons (PFCs).

Gross Domestic Product
Gross Domestic Product (GDP) is the monetary value of all goods and services produced within a nation.

Gross National Product
Gross National Product (GNP) is the monetary value of all goods and services produced in a nation's economy, including income generated abroad by domestic residents, but without income generated by foreigners.

Gross primary production
The total carbon fixed by plant through *photosynthesis*.

Groundwater recharge
The process by which external water is added to the zone of saturation of an *aquifer*, either directly into a formation or indirectly by way of another formation.

Groyne
A low, narrow jetty, usually extending roughly perpendicular to the shoreline, designed to protect the shore from *erosion* by currents, tides or waves, by trapping sand for the purpose of replenishing or making a beach.

Habitat
The locality or natural home in which a particular plant, animal, or group of closely associated organisms lives.

Hantavirus
A virus in the family *Bunyaviridae* that causes a type of haemorrhagic fever. It is thought that humans catch the disease mainly from infected rodents, either through direct contact with the animals or by inhaling or ingesting dust that contains aerosolised viral particles from their dried urine and other secretions.

Heat island
An urban area characterised by ambient temperatures higher than those of the surrounding non-urban area. The cause is a higher absorption of solar energy by materials of the urban fabric such as asphalt.

Herbaceous
Flowering, non-woody.

Human system
Any system in which human organisations play a major role. Often, but not always, the term is synonymous with 'society' or 'social system' e.g., agricultural system, political system, technological system, economic system; all are human systems in the sense applied in the AR4.

Hydrographic events
Events that alter the state or current of waters in oceans, rivers or lakes.

Hydrological systems
The systems involved in movement, distribution, and quality of water throughout the Earth, including both the hydrologic cycle and water resources.

Hypolimnetic
Referring to the part of a lake below the *thermocline* made up of water that is stagnant and of essentially uniform temperature except during the period of overturn.

Hypoxic events
Events that lead to a deficiency of oxygen.

Ice cap
A dome-shaped ice mass covering a highland area that is considerably smaller in extent than an *ice sheet*.

Ice sheet
A mass of land ice that is sufficiently deep to cover most of the underlying bedrock topography. An ice sheet flows outwards from a high central plateau with a small average surface slope. The margins slope steeply, and the ice is discharged through fast-flowing ice streams or outlet *glaciers*, in some cases into the sea or into *ice shelves* floating on the sea. There are only two large ice sheets in the modern world – on Greenland and Antarctica, the Antarctic ice sheet being divided into east and west by the Transantarctic Mountains; during glacial periods there were others.

Ice shelf
A floating *ice sheet* of considerable thickness attached to a coast (usually of great horizontal extent with a level or gently undulating surface); often a seaward extension of ice sheets. Nearly all ice shelves are in Antarctica.

(climate change) Impact assessment
The practice of identifying and evaluating, in monetary and/or non-monetary terms, the effects of *climate change* on natural and *human systems*.

(climate change) Impacts
The effects of *climate change* on natural and *human systems*. Depending on the consideration of *adaptation*, one can distinguish between potential impacts and residual impacts:

- **Potential impacts**: all impacts that may occur given a projected change in climate, without considering adaptation.
- **Residual impacts**: the impacts of climate change that would occur after adaptation. See also *aggregate impacts*, *market impacts*, and *non-market impacts*.

Indigenous peoples
No internationally accepted definition of indigenous peoples exists. Common characteristics often applied under international law, and by United Nations agencies to distinguish indigenous peoples include: residence within or attachment to geographically distinct traditional *habitats*, ancestral territories, and their natural resources; maintenance of cultural and social identities, and social, economic, cultural and political institutions separate from mainstream or dominant societies and cultures; descent from population groups present in a given area, most frequently before modern states or territories were created and current borders defined; and self-identification as being part of a distinct indigenous cultural group, and the desire to preserve that cultural identity.

Industrial revolution
A period of rapid industrial growth with far-reaching social and economic consequences, beginning in England during the second half of the 18th century and spreading to Europe and later to other countries including the USA. The industrial revolution marks the beginning of a strong increase in combustion of fos-

sil fuels and related emissions of *carbon dioxide*. In the AR4, the term '*pre-industrial*' refers, somewhat arbitrarily, to the period before 1750.

Infectious disease
Any disease caused by microbial agents that can be transmitted from one person to another or from animals to people. This may occur by direct physical contact, by handling of an object that has picked up infective organisms, through a disease carrier, via contaminated water, or by the spread of infected droplets coughed or exhaled into the air.

Infrastructure
The basic equipment, utilities, productive enterprises, installations and services essential for the development, operation and growth of an organisation, city or nation.

Integrated assessment
An interdisciplinary process of combining, interpreting and communicating knowledge from diverse scientific disciplines so that all relevant aspects of a complex societal issue can be evaluated and considered for the benefit of decision-making.

Integrated water resources management (IWRM)
The prevailing concept for water management which, however, has not been defined unambiguously. IWRM is based on four principles that were formulated by the International Conference on Water and the Environment in Dublin, 1992: (1) fresh water is a finite and vulnerable resource, essential to sustain life, development and the environment; (2) water development and management should be based on a participatory approach, involving users, planners and policy-makers at all levels; (3) women play a central part in the provision, management and safeguarding of water; (4) water has an economic value in all its competing uses and should be recognised as an economic good.

Invasive species and invasive alien species (IAS)
A species aggressively expanding its range and population density into a region in which it is not native, often through outcompeting or otherwise dominating native species.

Irrigation water-use efficiency
Irrigation *water-use efficiency* is the amount of *biomass* or seed yield produced per unit irrigation water applied, typically about 1 tonne of dry matter per 100 mm water applied.

Isohyet
A line on a map connecting locations that receive the same amount of rainfall.

Joint attribution
Involves both *attribution* of observed changes to regional *climate change* and attribution of a measurable portion of either regional climate change or the associated observed changes in the system to *anthropogenic* causes, beyond natural variability. This process involves statistically linking climate-change simulations from *climate models* with the observed responses in the natural or managed system. *Confidence* in joint attribution statements must be lower than the confidence in either of the individual attribution steps alone due to the combination of two separate statistical assessments.

Keystone species
A species that has a central servicing role affecting many other organisms and whose demise is likely to result in the loss of a number of species and lead to major changes in *ecosystem* function.

Kyoto Protocol
The Kyoto Protocol was adopted at the Third Session of the Conference of the Parties (COP) to the *UN Framework Convention on Climate Change (UNFCCC)* in 1997 in Kyoto, Japan. It contains legally binding commitments, in addition to those included in the UNFCCC. Countries included in Annex B of the Protocol (most member countries of the Organisation for Economic Cooperation and Development (OECD) and those with economies in transition) agreed to reduce their *anthropogenic greenhouse gas* emissions (CO_2, CH_4, N_2O, HFCs, PFCs, and SF_6) by at least 5% below 1990 levels in the commitment period 2008 to 2012. The Kyoto Protocol entered into force on 16 February 2005.

La Niña
See *El Niño-Southern Oscillation (ENSO)*.

Landslide
A mass of material that has slipped downhill by gravity, often assisted by water when the material is saturated; the rapid movement of a mass of soil, rock or debris down a slope.

Large-scale singularities
Abrupt and dramatic changes in the state of given systems, in response to gradual changes in driving forces. For example, a gradual increase in atmospheric *greenhouse gas* concentrations may lead to such large-scale singularities as slowdown or collapse of the *thermohaline circulation* or collapse of the West Antarctic *ice sheet*. The occurrence, magnitude, and timing of large-scale singularities are difficult to predict.

Last Glacial Maximum
The Last Glacial Maximum refers to the time of maximum extent of the *ice sheets* during the last glaciation, approximately 21,000 years ago.

Leaching
The removal of soil elements or applied chemicals by water movement through the soil.

Leaf area index (LAI)
The ratio between the total leaf surface area of a plant and the ground area covered by its leaves.

Legume
Plants that fix nitrogen from the air through a symbiotic relationship with bacteria in their soil and root systems (e.g., soybean, peas, beans, lucerne, clovers).

Likelihood
The likelihood of an occurrence, an outcome or a result, where this can be estimated probabilistically, is expressed in this Report using a standard terminology, defined in the Introduction. See also *uncertainty* and *confidence*.

Limnology
Study of lakes and their *biota*.

Littoral zone
A coastal region; the zone between high and low watermarks.

Malaria
Endemic or *epidemic* parasitic disease caused by species of the genus *Plasmodium* (Protozoa) and transmitted by mosquitoes of the genus *Anopheles*; produces bouts of high fever and systemic disorders, affects about 300 million and kills approximately 2 million people worldwide every year.

Market impacts
Impacts that can be quantified in monetary terms, and directly affect *Gross Domestic Product* – e.g., changes in the price of agricultural inputs and/or goods. See also *non-market impacts*.

Meningitis
Inflammation of the meninges (part of the covering of the brain), usually caused by bacteria, viruses or fungi.

Meridional overturning circulation (MOC)
See *thermohaline circulation (THC)*.

Microclimate
Local climate at or near the Earth's surface. See also *climate*.

Millennium Development Goals (MDGs)
A list of ten goals, including eradicating extreme poverty and hunger, improving maternal health, and ensuring environmental sustainability, adopted in 2000 by the UN General Assembly, i.e., 191 States, to be reached by 2015. The MDGs commit the international community to an expanded vision of development, and have been commonly accepted as a framework for measuring development progress.

Mires
Peat-accumulating *wetlands*. See *bog*.

Mitigation
An *anthropogenic* intervention to reduce the anthropogenic forcing of the *climate system*; it includes strategies to reduce *greenhouse gas sources* and emissions and enhancing *greenhouse gas sinks*.

Mixed layer
The upper region of the ocean, well mixed by interaction with the overlying *atmosphere*.

Monsoon
A monsoon is a tropical and sub-tropical seasonal reversal in both the surface winds and associated precipitation.

Montane
The biogeographic zone made up of relatively moist, cool upland slopes below the *sub-alpine* zone that is characterised by the presence of mixed deciduous at lower and coniferous evergreen forests at higher elevations.

Morbidity
Rate of occurrence of disease or other health disorders within a population, taking account of the age-specific morbidity rates. Morbidity indicators include chronic disease incidence/prevalence, rates of hospitalisation, primary care consultations, disability-days (i.e., days of absence from work), and prevalence of symptoms.

Morphology
The form and structure of an organism or land-form, or any of its parts.

Mortality
Rate of occurrence of death within a population; calculation of mortality takes account of age-specific death rates, and can thus yield measures of life expectancy and the extent of premature death.

Net biome production (NBP)
Net biome production is the *net ecosystem production (NEP)* minus carbon losses resulting from disturbances such as fire or insect defoliation.

Net ecosystem production (NEP)
Net ecosystem production is the difference between *net primary production (NPP)* and heterotrophic *respiration* (mostly decomposition of dead organic matter) of that *ecosystem* over the same area (see also *net biome production (NBP)*).

Net primary production (NPP)
Net primary production is the *gross primary production* minus autotrophic *respiration*, i.e., the sum of metabolic processes for plant growth and maintenance, over the same area.

Nitrogen oxides (NO_x)
Any of several oxides of nitrogen.

No regrets policy
A policy that would generate net social and/or economic benefits irrespective of whether or not *anthropogenic climate change* occurs.

Non-linearity
A process is called 'non-linear' when there is no simple proportional relation between cause and effect.

Non-market impacts
Impacts that affect *ecosystems* or human *welfare*, but that are not easily expressed in monetary terms, e.g., an increased risk of premature death, or increases in the number of people at risk of hunger. See also *market impacts*.

Normalised difference vegetation index (NDVI)
A satellite-based remotely sensed measure of the 'greenness' of the vegetation cover.

North Atlantic Oscillation (NAO)
The North Atlantic Oscillation (NAO) consists of opposing variations of barometric pressure near Iceland and near the Azores. It is the dominant mode of winter *climate variability* in the North Atlantic region.

Ocean acidification
Increased concentrations of CO_2 in sea water causing a measurable increase in acidity (i.e., a reduction in ocean pH). This may lead to reduced calcification rates of calcifying organisms such as *corals*, molluscs, *algae* and crustacea.

Ombrotrophic bog
An acidic *peat*-accumulating *wetland* that is rainwater (instead of groundwater) fed and thus particularly poor in nutrients.

Opportunity costs
The cost of an economic activity forgone through the choice of another activity.

Ozone
The triatomic form of oxygen (O_3), a gaseous atmospheric constituent. In the *troposphere*, it is created both naturally and by photochemical reactions involving gases resulting from human activities (*photochemical smog*). In high concentrations, tropospheric ozone can be harmful to many living organisms. Tropospheric ozone acts as a *greenhouse gas*. In the *stratosphere*, ozone is created by the interaction between solar ultraviolet radiation and molecular oxygen (O_2). Depletion of stratospheric ozone, due to chemical reactions that may be enhanced by *climate change*, results in an increased ground-level flux of ultraviolet (UV) B radiation.

Paludification
he process of transforming land into a *wetland* such as a marsh, a swamp or a *bog*.

Particulates
Very small solid exhaust particles emitted during the combustion of fossil and biomass fuels. Particulates may consist of a wide variety of substances. Of greatest concern for health are particulates of less than or equal to 10 nm in diameter, usually designated as PM_{10}.

Peat
Peat is formed from dead plants, typically *Sphagnum* mosses, which are only partially decomposed due to the permanent submergence in water and the presence of conserving substances such as humic acids.

Peatland
Typically a *wetland* such as a *mire* slowly accumulating *peat*.

Pelagic community
The community of organisms living in the open waters of a river, a lake or an ocean (in contrast to *benthic communities* living on or near the bottom of a water body).

Permafrost
Perennially frozen ground that occurs where the temperature remains below 0°C for several years.

Phenology
The study of natural phenomena that recur periodically (e.g., development stages, migration) and their relation to climate and seasonal changes.

Photochemical smog
A mix of photochemical oxidant air pollutants produced by the reaction of sunlight with primary air pollutants, especially hydrocarbons.

Photosynthesis
The synthesis by plants, *algae* and some bacteria of sugar from sunlight, *carbon dioxide* and water, with oxygen as the waste product. See also *carbon dioxide fertilisation*, *C_3 plants* and *C_4 plants*.

Physiographic
Of, relating to, or employing a description of nature or natural phenomena.

Phytoplankton
The plant forms of *plankton*. Phytoplankton are the dominant plants in the sea, and are the basis of the entire marine *food web*. These single-celled organisms are the principal agents of photosynthetic carbon fixation in the ocean. See also *zooplankton*.

Plankton
Microscopic aquatic organisms that drift or swim weakly. See also *phytoplankton* and *zooplankton*.

Plant functional type (PFT)
An idealised vegetation class typically used in *dynamic global vegetation models (DGVM)*.

Polynya
Areas of permanently unfrozen sea water resulting from warmer local water currents in otherwise sea-ice covered oceans. They are biological hotspots, since they serve as breathing holes or refuges for marine mammals such as whales and seals, and fish-hunting birds.

Population system
An ecological system (not *ecosystem*) determined by the dynamics of a particular *vagile* species that typically cuts across several *ecological communities* and even entire *biomes*. An example is migratory birds that seasonally inhabit forests as well as grasslands and visit *wetlands* on their migratory routes.

Potential production
Estimated crop productivity under non-limiting soil, nutrient and water conditions.

Pre-industrial
See *industrial revolution*.

Primary production
All forms of production accomplished by plants, also called primary producers. See *GPP*, *NPP*, *NEP* and *NBP*.

Projection
The potential evolution of a quality or set of quantities, often computed with the aid of a model. Projections are distinguished from predictions in order to emphasise that projections involve assumptions – concerning, for example, future socio-economic and technological developments, that may or may not be realised – and are therefore subject to substantial *uncertainty*. See also *climate projection* and *climate prediction*.

Pteropods
Planktonic, small marine snails with swimming organs resembling wings.

Pure rate of time preference
The degree to which consumption now is preferred to consumption one year later, with prices and incomes held constant, which is one component of the *discount rate*.

Radiative forcing
Radiative forcing is the change in the net vertical irradiance (expressed in Watts per square metre; Wm^{-2}) at the tropopause due to an internal or external change in the forcing of the *climate system*, such as a change in the concentration of CO_2 or the output of the Sun.

Rangeland
Unmanaged grasslands, shrublands, *savannas* and *tundra*.

Recalcitrant
Recalcitrant organic material or recalcitrant carbon stocks resist decomposition.

Reference scenario
See *baseline/reference*.

Reforestation
Planting of forests on lands that have previously contained forests but that have been converted to some other use. For a discussion of the term *forest* and related terms such as *afforestation*, *reforestation* and *deforestation*, see the IPCC Special Report on Land Use, Land-Use Change, and Forestry (IPCC, 2000).

Reid's paradox
This refers to the apparent contradiction between inferences of high plant migration rates as suggested in the palaeo-record (particularly after the last Ice Age), and the low potential rates of migration that can be inferred through studying the seed dispersal of the plants involved, e.g., in wind-tunnel experiments.

Reinsurance
The transfer of a portion of primary insurance risks to a secondary tier of insurers (reinsurers); essentially 'insurance for insurers'.

Relative sea-level rise
See *sea-level rise*.

Reservoir
A component of the *climate system*, other than the *atmosphere*, that has the capacity to store, accumulate or release a substance of concern (e.g., carbon or a *greenhouse gas*). Oceans, soils, and forests are examples of carbon reservoirs. The term also means an artificial or natural storage place for water, such as a lake, pond or *aquifer*, from which the water may be withdrawn for such purposes as irrigation or water supply.

Resilience
The ability of a social or ecological system to absorb disturbances while retaining the same basic structure and ways of functioning, the capacity for self-organisation, and the capacity to adapt to stress and change.

Respiration
The process whereby living organisms convert organic matter to *carbon dioxide*, releasing energy and consuming oxygen.

Riparian
Relating to or living or located on the bank of a natural watercourse (such as a river) or sometimes of a lake or a tidewater.

River discharge
Water flow within a river channel, for example expressed in m^3/s. A synonym for *streamflow*.

Runoff
That part of precipitation that does not *evaporate* and is not *transpired*.

Salinisation
The accumulation of salts in soils.

Salt-water intrusion / encroachment
Displacement of fresh surface water or groundwater by the advance of salt water due to its greater density. This usually occurs in coastal and estuarine areas due to reducing land-based influence (e.g., either from reduced *runoff* and associated *groundwater recharge*, or from excessive water withdrawals from *aquifers*) or increasing marine influence (e.g., *relative sea-level rise*).

Savanna
Tropical or sub-tropical grassland or woodland *biomes* with scattered shrubs, individual trees or a very open canopy of trees, all characterised by a dry (arid, semi-arid or semi-humid) *climate*.

Scenario
A plausible and often simplified description of how the future may develop, based on a coherent and internally consistent set of assumptions about driving forces and key relationships. Scenarios may be derived from *projections*, but are often based on additional information from other sources, sometimes combined with a 'narrative storyline'. See also *climate (change) scenario*, *emissions scenario* and *SRES*.

Sea-ice biome
The *biome* formed by all marine organisms living within or on the floating sea ice (frozen sea water) of the polar oceans.)

Sea-level rise
An increase in the mean level of the ocean. *Eustatic sea-level rise* is a change in global average sea level brought about by an increase in the volume of the world ocean. *Relative sea-level rise* occurs where there is a local increase in the level of the ocean relative to the land, which might be due to ocean rise and/or land level subsidence. In areas subject to rapid land-level uplift, relative sea level can fall.

Sea wall
A human-made wall or embankment along a shore to prevent wave *erosion*.

Semi-arid regions
Regions of moderately low rainfall, which are not highly productive and are usually classified as *rangelands*. 'Moderately low' is widely accepted as between 100 and 250 mm precipitation per year. See also *arid region*.

Sensitivity
Sensitivity is the degree to which a system is affected, either adversely or beneficially, by *climate variability* or change. The effect may be direct (e.g., a change in crop yield in response to a change in the mean, range or variability of temperature) or indirect (e.g., damages caused by an increase in the frequency of coastal flooding due to *sea-level rise*).

Sequestration
See *carbon sequestration*.

Silviculture
Cultivation, development and care of forests.

Sink
Any process, activity, or mechanism that removes a *greenhouse gas*, an *aerosol*, or a precursor of a greenhouse gas or aerosol from the *atmosphere*.

Snow water equivalent
The equivalent volume/mass of water that would be produced if a particular body of snow or ice was melted.

Snowpack
A seasonal accumulation of slow-melting snow.

Social cost of carbon
The value of the *climate change impacts* from 1 tonne of carbon emitted today as CO_2, aggregated over time and discounted back to the present day; sometimes also expressed as value per tonne of *carbon dioxide*.

Socio-economic scenarios
Scenarios concerning future conditions in terms of population, *Gross Domestic Product* and other socio-economic factors relevant to understanding the implications of *climate change*. See *SRES* (source: Chapter 6).

SRES
The storylines and associated population, *GDP* and *emissions scenarios* associated with the Special Report on Emissions Scenarios (SRES) (Nakićenović et al., 2000), and the resulting *climate change* and *sea-level rise scenarios*. Four families of *socio-economic scenario* (A1, A2, B1 and B2) represent different world futures in two distinct dimensions: a focus on economic versus environmental concerns, and global versus regional development patterns.

Stakeholder
A person or an organisation that has a legitimate interest in a project or entity, or would be affected by a particular action or policy.

Stock
See *reservoir*.

Stratosphere
Highly stratified region of *atmosphere* above the *troposphere* extending from about 10 km (ranging from 9 km in high latitudes to 16 km in the tropics) to about 50 km.

Streamflow
Water flow within a river channel, for example, expressed in m^3/s. A synonym for *river discharge*.

Sub-alpine
The biogeographic zone below the *tree line* and above the *montane* zone that is characterised by the presence of coniferous forest and trees.

Succulent
Succulent plants, e.g., cactuses, possessing organs that store water, thus facilitating survival during *drought* conditions.

Surface runoff
The water that travels over the land surface to the nearest surface stream; *runoff* of a drainage *basin* that has not passed beneath the surface since precipitation.

Sustainable development
Development that meets the cultural, social, political and economic needs of the present generation without compromising the ability of future generations to meet their own needs.

Taiga
The northernmost belt of *boreal forest* adjacent to the Arctic *tundra*.

Thermal expansion
In connection with *sea-level rise*, this refers to the increase in volume (and decrease in density) that results from warming water. A warming of the ocean leads to an expansion of the ocean volume and hence an increase in sea level.

Thermocline
The region in the world's ocean, typically at a depth of 1 km, where temperature decreases rapidly with depth and which marks the boundary between the surface and the ocean.

Thermohaline circulation (THC)
Large-scale, density-driven circulation in the ocean, caused by differences in temperature and salinity. In the North Atlantic, the thermohaline circulation consists of warm surface water flowing northward and cold deepwater flowing southward, resulting in a net poleward transport of heat. The surface water sinks in highly restricted regions located in high latitudes. Also called *meridional overturning circulation (MOC)*.

Thermokarst
A ragged landscape full of shallow pits, hummocks and depressions often filled with water (ponds), which results from thawing of ground ice or *permafrost*. Thermokarst processes are the processes driven by warming that lead to the formation of thermokarst.

Threshold
The level of magnitude of a system process at which sudden or rapid change occurs. A point or level at which new properties emerge in an ecological, economic or other system, invalidating predictions based on mathematical relationships that apply at lower levels.

Transpiration
The *evaporation* of water vapour from the surfaces of leaves through stomata.

Tree line
The upper limit of tree growth in mountains or high latitudes. It is more elevated or more poleward than the *forest line*.

Trophic level
The position that an organism occupies in a *food chain*.

Trophic relationship
The ecological relationship which results when one species feeds on another.

Troposphere
The lowest part of the *atmosphere* from the surface to about 10 km in altitude in mid-latitudes (ranging from 9 km in high latitudes to 16 km in the tropics on average) where clouds and 'weather' phenomena occur. In the troposphere, temperatures generally decrease with height.

Tsunami
A large wave produced by a submarine earthquake, *landslide* or volcanic eruption.

Tundra
A treeless, level, or gently undulating plain characteristic of the Arctic and sub-Arctic regions characterised by low temperatures and short growing seasons.

Uncertainty
An expression of the degree to which a value (e.g., the future state of the *climate system*) is unknown. Uncertainty can result from lack of information or from disagreement about what is known or even knowable. It may have many types of sources, from quantifiable errors in the data to ambiguously defined concepts or terminology, or uncertain *projections* of human behaviour. Uncertainty can therefore be represented by quantitative measures (e.g., a range of values calculated by various models) or by qualitative statements (e.g., reflecting the judgement of a team of experts). See also *confidence* and *likelihood*.

Undernutrition
The temporary or chronic state resulting from intake of lower than recommended daily dietary energy and/or protein requirements, through either insufficient food intake, poor absorption, and/or poor biological use of nutrients consumed.

Ungulate
A hoofed, typically herbivorous, quadruped mammal (including ruminants, swine, camel, hippopotamus, horse, rhinoceros and elephant).

United Nations Framework Convention on Climate Change (UNFCCC)
The Convention was adopted on 9 May 1992, in New York, and signed at the 1992 Earth Summit in Rio de Janeiro by more than 150 countries and the European Community. Its ultimate objective is the 'stabilisation of *greenhouse gas* concentrations in the *atmosphere* at a level that would prevent dangerous *anthropogenic* interference with the *climate system*'. It contains commitments for all Parties. Under the Convention, Parties included in Annex I aim to return greenhouse gas emissions not controlled by the Montreal Protocol to 1990 levels by the year 2000. The Convention entered in force in March 1994. See also *Kyoto Protocol*.

Upwelling region
A region of an ocean where cold, typically nutrient-rich waters from the bottom of the ocean surface.

Urbanisation
The conversion of land from a natural state or managed natural state (such as agriculture) to cities; a process driven by net rural-to-urban migration through which an increasing percentage of the population in any nation or region come to live in settlements that are defined as 'urban centres'.

Vagile
Able to migrate.

Vascular plants
Higher plants with vascular, i.e., sap-transporting, tissues.

Vector
A blood-sucking organism, such as an insect, that transmits a pathogen from one host to another. See also *vector-borne diseases*.

Vector-borne diseases
Disease that are transmitted between hosts by a *vector* organism (such as a mosquito or tick); e.g., *malaria*, *dengue fever* and leishmaniasis.

Vernalisation
The biological requirements of certain crops, such as winter cereals, which need periods of extreme cold temperatures before emergence and/or during early vegetative stages, in order to flower and produce seeds. By extension, the act or process of hastening the flowering and fruiting of plants by treating seeds, bulbs or seedlings with cold temperatures, so as to induce a shortening of the vegetative period.

Vulnerability
Vulnerability is the degree to which a system is susceptible to, and unable to cope with, adverse effects of *climate change*, including *climate variability* and extremes. Vulnerability is a function of the character, magnitude, and rate of climate change and variation to which a system is exposed, its *sensitivity*, and its adaptive capacity.

Water consumption
Amount of extracted water irretrievably lost during its use (by *evaporation* and goods production). Water consumption is equal to water withdrawal minus return flow.

Water productivity
The ratio of crop seed produced per unit water applied. In the case of irrigation, see *irrigation water-use efficiency*. For rain-fed crops, water productivity is typically 1 t/100 mm.

Water stress
A country is water-stressed if the available freshwater supply relative to water withdrawals acts as an important constraint on development. Withdrawals exceeding 20% of renewable water supply have been used as an indicator of water stress. A crop is water-stressed if soil-available water, and thus actual *evapotranspiration*, is less than potential evapotranspiration demands.

Water-use efficiency
Carbon gain in *photosynthesis* per unit water lost in *evapotranspiration*. It can be expressed on a short-term basis as the ratio of photosynthetic carbon gain per unit transpirational water loss, or on a seasonal basis as the ratio of *net primary production* or agricultural yield to the amount of available water.

Welfare
An economic term used to describe the state of well-being of humans on an individual or collective basis. The constituents of well-being are commonly considered to include materials to satisfy basic needs, freedom and choice, health, good social relations, and security.

Wetland
A transitional, regularly waterlogged area of poorly drained soils, often between an aquatic and a terrestrial *ecosystem*, fed from rain, surface water or groundwater. Wetlands are characterised by a prevalence of vegetation adapted for life in saturated soil conditions.

Yedoma
Ancient organic material trapped in *permafrost* that is hardly decomposed.

Zoonoses
Diseases and infections which are naturally transmitted between vertebrate animals and people.

Zooplankton
The animal forms of *plankton*. They consume *phytoplankton* or other zooplankton.

References

IPCC, 2000: *Land Use, Land-Use Change, and Forestry: A Special Report of the IPCC*, R.T. Watson, I.R. Noble, B. Bolin, N.H. Ravindranath, D.J. Verardo and D.J. Dokken, Eds., Cambridge University Press, Cambridge, and New York, 377 pp.

Nakićenović, N., J. Alcamo, G. Davis, B. de Vries, J. Fenhann, S. Gaffin, K. Gregory, A. Grübler, T.Y. Jung, T. Kram, E.L. La Rovere, L. Michaelis, S. Mori, T. Morita, W. Pepper, H. Pitcher, L. Price, K. Raihi, A. Roehrl, H.-H. Rogner, A. Sankovski, M. Schlesinger, P. Shukla, S. Smith, R. Swart, S. van Rooijen, N. Victor and Z. Dadi, 2000: *Emissions Scenarios: A Special Report of Working Group III of the Intergovernmental Panel on Climate Change*. Cambridge University Press, Cambridge, and New York, 599 pp.

Appendix II: Contributors to the IPCC WGII Fourth Assessment Report

Abeku, Tarekegn
London School of Hygiene and Tropical Medicine
UK/Ethiopia

Abuodha, Pamela
University of Wollongong
Australia/Kenya

Adesina, Francis
Obafemi Awolowo University
Nigeria

Adger, Neil
University of East Anglia
UK

Agard, John
University of the West Indies
Trinidad and Tobago

Aggarwal, Pramod
Indian Agricultural Research Institute
India

Agnew, Maureen
University of East Anglia
UK

Agoli-Agbo, Micheline
University of Abomey-Calavi
Benin

Agrawala, Shardul
OECD/France

Agricole, Will
National Meteorological Service
Seychelles

Ahmad, Qazi
Bangladesh Unnayan Parishad
Bangladesh

Akhtar, Rais
Jawaharlal Nehru University
India

Alam, Mozaharul
Bangladesh Centre for Advanced Studies
Bangladesh

Alcamo, Joseph
University of Kassel
Germany

Allali, Abdelkader
Ministry of Agriculture, Rural Development and Fishing
Morocco

Andrey, Jean
University of Waterloo
Canada

Anisimov, Oleg
State Hydrological Institute
Russia

Anokhin, Yurij
Institute of Global Climate and Ecology
Russia

Antle, John
Montana State University
USA

Araujo, Miguel
Environmental Change Institute
Portugal

Arblaster, Julie
National Center for Atmospheric Sciences/Bureau of Meteorology
USA/Australia

Arnell, Nigel
University of Southampton
UK

Asanuma, Jun
Tsukuba University
Japan

Atlhopheng, Julius
University of Botswana
Botswana

Attaher, Samar
The Central Laboratory for Agricultural Climate
Egypt

Attri, Shiv
India Meteorological Department
India

Baethgen, Walter
International Fertilizer Development Centre
Uruguay

Bao, Manzhu
Huazhong Agricultural University
China

Barlow, Chris
Mekong River Commission
Lao PDR

Bates, Bryson
CSIRO
Australia

Batima, Punsalmaa
Institute of Meteorology and Hydrology
Mongolia

Becken, Susanne
Landcare Research
New Zealand

Beggs, Paul
Macquarie University
Australia

Beniston, Martin
University of Geneva
Switzerland

Berkhout, Frans
Institute for Environmental Studies
The Netherlands

Betts, Richard
Met Office Hadley Centre
UK

Bhadwal, Suruchi
The Energy and Resources Institute
India

Biagini, Bonizella
National Environmental Trust
USA/GEF/Italy

Bindi, Marco
DISAT, University of Florence
Italy

Black, Richard
Sussex University
UK

Boko, Michel
Universite de Bourgogne
France/Benin

Bond, William
University of Cape Town
South Africa

Bounoua, Lahouari
NASA Goddard Space Flight Center
USA

Brander, Keith
International Council for the Exploration of the Sea
Denmark/UK

Brenkert, Antoinette
Joint Global Change Research Institute
USA

Briguglio, Lino
Foundation for International Studies
Malta

Bristow, Abigail
Loughborough University
UK

Brklacich, Michael
Carleton University
Canada

Brooks, Nick
Tyndall Centre for Climate Change Research
UK

Brown, Barbara
Newcastle University
UK

Burch, Sarah
University of British Columbia
Canada

Burkett, Virginia
US Geological Survey
USA

Burton, Ian
University of Toronto
Canada

Cairncross, Sandy
London School of Hygiene and Tropical Medicine
UK

Callaghan, Terry
Royal Swedish Academy of Sciences/Abisko Scientific Research
Sweden/UK

Canadell, Josep
GCTE International Project Office
Australia

Canziani, Osvaldo
IPCC Working Group II Co-chair
Argentina

Carter, Timothy
Finnish Environment Institute
Finland

Casassa, Gino
Centro de Estudios Cientificos
Chile

Cayan, Dan
University of California, San Diego
USA

Ceron, Jean-Paul
Université de Limoges
France

Chambers, Lynda
Bureau of Meteorology Research Centre
Australia

Chhetri, Netra
Arizona State University
USA/Nepal

Christensen, Torben
Lund University
Sweden

Clot, Bernard
MeteoSwiss
Switzerland

Codignotto, Jorge
Ciudad Universitaria
Argentina

Cohen, Stewart
University of British Columbia
Canada

Coleman, Anthony
Insurance Australia Group
Australia

Conde, Cecilia
Universidad Nacional Autónoma de México
Mexico

Confalonieri, Ulisses
National School of Public Health
Brazil

Corfee-Morlot, Jan
OECD Environment Directorate
France

Corobov, Roman
Regionica
Moldova

Côté, Isabelle
Simon Fraser University
Canada

Craig, Patricia
The Pennsylvania State University
USA

Cranage, Judith
The Pennsylvania State University
USA

Cruz, Rex Victor
University of the Philippines at Los Baños
The Philippines

Cruz Choque, David
Ministry of Sustainable Development
Bolivia

de Alba Alcaraz, Edmundo
Universidad Nacional Autónoma de México
Mexico

de Chazal, Jacqueline
Université catholique de Louvain
Belgium

de Ronde, John
Delft Hydraulics
The Netherlands

Demuth, Mike
Natural Resources Canada
Canada

Denton, Fatima
International Development Research Centre
Senegal/The Gambia

des Clers, Sophie
University College London
UK

Devoy, Robert
University College Cork
Ireland

Dikinya, Oagile
University of Western Australia
Australia

Dlugolecki, Andrew F.
Consultant
UK

Döll, Petra
University of Frankfurt
Germany

Downing, Thomas
Stockholm Environment Institute
UK

Dube, Pauline
University of Botswana
Botswana

Dubois, Ghislain
Environnement Conseil (TEC)
France

Dunn, Matt
National Institute of Water and Atmospheric Research
New Zealand

Dyurgerov, Mark
University of Colorado
USA

Easterling, William
The Pennsylvania State University
USA

Ebi, Kristie
Exponent
USA

Edwards, Martin
Sir Alister Hardy Foundation for Ocean Science
UK

Emori, Seita
National Institute for Environmental Studies
Japan

Enright, Brenna
University of Toronto
Canada

Estrada, Francisco
Universidad Nacional Autónoma de México
Mexico

Estrella, Nicole
Technical University of Munich
Germany

Falloon, Pete
Met Office Hadley Centre
UK

Fan, Daidu
Tongji University
China

Fankhauser, Samuel
European Bank for Reconstruction and Development
UK/Switzerland

Field, Christopher
Carnegie Institution of Washington
USA

Finkel, Adam
Woodrow Wilson School, Princeton University
USA

Fischlin, Andreas
Terrestrial Systems Ecology Group ETH, Zürich
Switzerland

Fitzharris, Blair
University of Otago
New Zealand

Forbes, Donald
Bedford Institute of Oceanography
Canada

Ford, James
McGill University
Canada

Francou, Bernard
Institut de Recherche pour le Développement
France

Furgal, Christopher
Trent University
Canada

Füssel, Hans-Martin
Potsdam Institute for Climate Impact Research
Germany

Gay Garcia, Carlos
Universidad Nacional Autónoma de México
Mexico

Giannakopoulos, Christos
National Observatory of Athens
Greece

Gigli, Simone
OECD/Germany

Giménez, Juan Carlos
Universidad de Buenos Aires
Argentina

Githeko, Andrew
Kenya Medical Research Institute
Kenya

Githendu, Mukiri
Ministry of Research and Technology
Kenya

Gopal, Brij
Jawaharlal Nehru University
India

Gornitz, Vivien
Columbia University
USA

Gossling, Stefan
Lund University
Sweden

Graham, Phil
Swedish Meteorological and Hydrological Institute
Sweden

Green, Donna
CSIRO
Australia

Guisan, Antoine
Conservatoire et Jardin botaniques de Genève
Switzerland

Gyalistras, Dimitrios
Terrestrial Systems Ecology Group ETH, Zürich
Switzerland

Haeberli, Wilfreid
University of Zürich-Irchel
Switzerland

Hales, Simon
University of Otago
New Zealand

Hall, Jim
Newcastle University
UK

Hallegatte, Stephane
Stanford University
USA/France

Hamlet, Alan
University of Washington
USA

Hanson, Clair
IPCC Working Group II TSU, Met Office Hadley Centre
UK

Harasawa, Hideo
National Institute for Environmental Studies
Japan

Harvey, Nicholas
University of Adelaide
Australia

Hauengue, Maria
Ministry of Health
Mozambique

Hay, John
The University of Waikato
New Zealand

Hemming, Deborah
Met Office Hadley Centre
UK

Henderson, Roderick
National Institute of Water and Atmospheric Research
New Zealand

Hennessy, Kevin
CSIRO
Australia

Henshaw, Anne
Bowdoin College
USA

Hilmi, Karim
Institut National de Recherche Halieutique
Morocco

Hobday, Alistair
CSIRO
Australia

Hoegh-Guldberg, Ove
The University of Queensland
Australia

Honda, Yasushi
University of Tsukuba
Japan

Hope, Christopher
University of Cambridge
UK

Howden, Mark
CSIRO Agricultural Sustainability Initiative
Australia

Hughes, Terence
James Cook University
Australia

Hughes, Lesley
Macquarie University
Australia

Huq, Saleemul
International Institute for Environment and Development
UK/Bangladesh

Hutton, Guy
Swiss Tropical Institute, Basel
Switzerland/UK

Iglesias, Ana
Ciudad Universitaria
Spain

Imeson, Anton
Commission on Geomorphological Response to Environmental Change
The Netherlands

Islam, Sirajul
University of Chittagong
Bangladesh

Jafari, Mostafa
Meteorological Organization
Iran

Janetos, Tony
The H. John Heinz III Center for Science, Economics and the Environment
USA

Jeppesen, Erik
National Environmental Research Institute
Denmark

Jetté-Nantel, Simon
OECD/Canada

Jimenez, Blanca Elena
Universidad Nacional Autónoma de México
Mexico

Jones, Roger
CSIRO
Australia

Jones, Gregory
Southern Oregon University
USA

Ju, Hui
Chinese Academy of Agricultural Science
China

Kabat, Pavel
International Secretariat for Dialogue on Water and Climate
The Netherlands

Kajfež-Bogataj, Lucka
University of Ljubljana
Slovenia

Kandlikar, Milind
Harvard University
USA/Canada

Kapshe, Manmohan
Maulana Azad National Institute of Technology
India

Karoly, David
University of Melbourne
Australia/USA

Kaser, Georg
Institut für Geographie, University of Innsbruck
Austria

Keller, Klaus
The Pennsylvania State University
USA/Germany

Kenny, Gavin
Earthwise Consulting
New Zealand

Killmann, Wulf
Food and Agriculture Organization/Italy

King, Darren
National Institute of Water and Atmospheric Research
New Zealand

Kirilenko, Andrei
University of North Dakota
USA/Russia

Kjellstrom, Tord
University of Auckland
New Zealand/Sweden

Klein, Richard
Stockholm Environment Institute
Sweden/The Netherlands

Körner, Christian
University of Basel
Switzerland

Kovacs, Paul
Institute for Catastrophic Loss Reduction
Canada

Kovats, Sari
London School of Hygiene and Tropical Medicine
UK

Kundzewicz, Zbigniew
Polish Academy of Sciences
Poland

Lakyda, Petro
National Agrarian University of Ukraine
Ukraine

Lal, Murari
CESDAC
India

Lam, Joseph
City University of Hong Kong
China

Lasco, Rodel
University of the Philippines
The Philippines

Leemans, Rik
University of Wageningen
The Netherlands

Lefale, Penehuro
World Meteorological Organization/Samoa

Lemos, Maria-Carmen
University of Michigan
USA/Brazil

Lewis, Nancy
University of Hawaii
USA

Li, Shuangcheng
Peking University
China

Li, Congxian
Tongji University
China

Lien, Tran Viet
Institute of Meteorology and Hydrology
Vietnam

Lin, Erda
Chinese Academy of Agricultural Science
China

Liu, Chunzhen
China Water Information Centre
China

Liverman, Diana
Oxford University
UK

Lorenzoni, Irene
University of East Anglia
UK

Love, Geoff
Bureau of Meteorology
Australia

Lowe, Jason
Met Office Hadley Centre
UK

Lu, Xianfu
UNDP-GEF/China

Lucht, Wolfgang
Potsdam Institute for Climate Impact Research
Germany

Lunn, Nick
Environment Canada
Canada

Ma, Zhuguo
Chinese Academy of Sciences
China

MacMynowski, Dena
Stanford Institute for International Studies
USA

Mader, Terry
University of Nebraska
USA

Magadza, Christopher
University of Zimbabwe
Zimbabwe

Magrin, Graciela
Instituto Nacional de Tecnologia Agropecuaria
Argentina

Major, David
Columbia University
USA

Malone, Elizabeth
Joint Global Change Research Institute
USA

Mann, Susan
The Pennsylvania State University
USA

Marchant, Harvey
The Australian National University
Australia

Marengo, José
CPTEC/INPE
Brazil

Markandya, Anil
The World Bank/UK

Martin, Eric
Météo-France, CNRM/CEN
France

Mastrandrea, Michael
Stanford University
USA

Mata, Luis Jose
Nord-Süd Zentrum für Entwicklungsforschung
Germany/Venezuela

McGregor, Glenn
King's College, London
UK

McInnes, Kathleen
CSIRO
Australia

McLean, Roger
University of New South Wales
Australia

Mearns, Linda
National Center for Atmospheric Research
USA

Medany, Mahmoud
The Central Laboratory for Agricultural Climate
Egypt

Menne, Bettina
WHO Regional Office for Europe/Germany

Menzel, Annette
Technical University of Munich
Germany

Midgley, Guy
National Botanical Institute
South Africa

Miller, Kathleen
National Center for Atmospheric Research
USA

Mills, Scott
University of Montana
USA

Mills, Evan
Lawrence Berkeley National Laboratory
USA

Mimura, Nobuo
Ibaraki University
Japan

Minns, Charles Kenneth
Fisheries and Oceans Canada
Canada

Mirza, Monirul Qader
Environment Canada
Canada/Bangladesh

Misselhorn, Alison
University of the Witwatersrand
South Africa

Morellato, Patricia
Universidade Estadual Paulista
Brazil

Moreno, Ana Rosa
Universidad Nacional Autónoma de México
Mexico

Moreno, José
Universidad de Castilla-La Mancha
Spain

Morton, John
University of Greenwich
UK

Mortsch, Linda
Environment Canada
Canada

Moser, Susanne
Union of Concerned Scientists
USA

Moulik, Tushar
ERM India Pvt
India

Muir-Wood, Robert
Risk Management Solutions
UK

Nagy, Gustavo
Universidad de la República
Uruguay

Nakalevu, Taito
South Pacific Regional Environment Programme
Fiji

Nearing, Mark
Southwest Watershed Research Center
USA

Neilson, Ron
US Department of Agriculture
USA

Nelson, Frederick
University of Delaware
USA

Neofotis, Peter
Columbia Earth Institute
USA

Niang, Isabelle
University of Dakar
Senegal

Nicholls, Robert
University of Southampton
UK

Ninh, Nguyen Huu
Centre for Environment Research, Education and Development
Vietnam

Nobre, Carlos
CPTEC-INPE
Brazil

Nováky, Belá
Szent István University
Hungary

Nurse, Leonard
University of the West Indes
Barbados

Nuttall, Mark
University of Alberta
Canada/UK

Nyong, Anthony
International Development Research Centre
Kenya/Nigeria

O'Brien, Karen
CICERO
Norway

O'Neill, Brian
IIASA/USA

O'Reilly, Catherine
Bard College
USA

Obioh, Imoh
Obafemi Awolowo University
Nigeria

Ogbonna, Anthony
Heriot-Watt University
UK

Oki, Taikan
University of Tokyo
Japan

Olesen, Jørgen
Danish Institute of Agricultural Sciences
Denmark

Oppenheimer, Michael
Princeton University
USA

Osman, Balgis
Higher Council for Environment and Natural Resources
Sudan

Ouaga, Hubert N'Djafa
Centre Régional AGRHYMET
Niger

Palmer, Gianna
Wesleyan University
USA

Palutikof, Jean
IPCC Working Group II TSU, Met Office Hadley Centre
UK

Parish, Faizal
Global Environment Centre
Malaysia

Parry, Martin
IPCC Working Group II Co-chair, Met Office Hadley Centre/Centre for Environmental Policy, Imperial College, University of London
UK

Patt, Anthony
Boston University
USA/IIASA

Patwardhan, Anand
Indian Institute of Technology
India

Patz, Jonathan
University of Wisconsin
USA

Payet, Rolph
Ministry of Industries and International Business
Seychelles

Pearce, Tristan
University of Guelph
Canada

Pêcheux, Martin
Université Paris VI
France

Penny, Guy
National Institute of Water and Atmospheric Research
New Zealand

Perez, Rosa
Philippine Atmospheric, Geophysical and Astronomical Services Administration
The Philippines

Pfeiffer, Christopher
The Pennsylvania State University
USA

Pfister, Christian
Universität Bern
Switzerland

Pittock, Barrie
CSIRO
Australia

Price, Jeff
California State University, Chico
USA

Prowse, Terry
National Water Research Institute at NHRC
Canada

Prudhomme, Christel
Centre for Ecology and Hydrology at Wallingford
UK

Pulhin, Juan
University of the Philippines
The Philippines

Pulwarty, Roger
NOAA/CIRES/Climate Diagnostics Center
USA/Trinidad and Tobago

Ragoonaden, Sachooda
Consultant to Indian Ocean Commission
Mauritius

Rahman, Atiq
Bangladesh Centre for Advanced Studies
Bangladesh

Rawlins, Samuel
Retired
Trinidad and Tobago

Reeder, Tim
Environment Agency, Thames Region
UK

Reist, James
Fisheries and Oceans Canada
Canada

Revich, Boris
Russian Academy of Sciences
Russia

Richels, Richard
Electric Power Research Institute
USA

Robinson, John
University of British Columbia
Canada

Rodo, Xavier
University of Barcelona
Spain

Rodriguez Acevedo, Rafael
Universidad Simón Bolívar
Venezuela

Romero Lankao, Patricia
National Center for Atmospheric Research
USA/Mexico

Root, Terry
Stanford University
USA

Rose, George
Memorial University of Newfoundland
Canada

Rosenzweig, Cynthia
Goddard Institute for Space Studies
USA

Rounsevell, Mark
Université catholique de Louvain
Belgium

Running, Steve
University of Montana
USA

Ruosteenoja, Kimmo
Finnish Meteorological Institute
Finland

Rupp-Armstrong, Susanne
University of Southampton
UK

Sailor, David
Portland State University
USA

Saito, Yoshiki
National Institute of Advanced Industrial Science and Technology
Japan

Salinger, Jim
National Institute of Water and Atmospheric Research
New Zealand

Saunders, Mark
University College London
UK

Schmidhuber, Josef
Food and Agriculture Organization/Italy

Schneider, Stephen
Stanford University
USA

Schulze, Roland
University of KwaZulu-Natal
South Africa

Scott, Michael
Battelle Pacific Northwest National Laboratory
USA

Scott, Daniel
Environment Canada
Canada

Sedjo, Roger
Resources for the Future
USA

Seguin, Bernard
National Institute for Agricultural Research
France

Sem, Graham
UNFCCC Secretariat/Papua New Guinea

Semenov, Serguei
Institute of Global Climate and Ecology
Russia

Sen, Zekai
Istanbul Technical University
Turkey

Sharma, Ashok
Halcrow Consulting India
India

Shiklomanov, Igor
State Hydrological Institute
Russia

Shreshtha, Arun
Government of Nepal
Nepal

Shukla, Priyadarshi
Indian Institute of Management
India

Shvidenko, Anatoly
IIASA/Russia

Smit, Barry
University of Guelph
Canada

Smith, Kirk
University of California
USA

Smith, Joel
Stratus Consulting
USA

Solecki, William
Hunter College, City University of New York
USA

Soussana, Jean-Francois
National Institute for Agricultural Research
France

Sparks, Tim
Centre for Ecology and Hydrology, Monks Wood
UK

Spencer, Tom
University of Cambridge
UK

Stone, John
IPCC Working Group II Vice-chair
Canada

Studd, Kate
Catholic Agency for Overseas Development
UK

Suarez, Avelino
Cuban Environment Agency
Cuba

Sweeney, John
National University of Ireland
Ireland

Tabo, Ramadjita
ICRISAT/Chad

Takahashi, Kiyoshi
National Institute for Environmental Studies,
Japan

Tarazona, Juan
Universidad Nacional Mayor de San Marcos
Peru

Taylor, Anna
Stockholm Environment Institute
UK/South Africa

Tebaldi, Claudia
National Center for Atmospheric Sciences
USA

Thayyen, Renoj
Wadia Institute of Himalayan Geology
India

Thomson, Madeleine
Columbia University
USA/UK

Thuiller, Wilfred
Laboratoire d'Ecologie Alpine UMR-CNRS
France

Tirado, Christina
Spain

Todorov, Alexander
Princeton University
USA/Bulgaria

Tol, Richard
Economic and Social Research Institute
Ireland/The Netherlands

Toth, Ferenc
International Atomic Energy Authority/Hungary

Travasso, Maria
INTA
Argentina

Tryjanowski, Piotr
Adam Mickiewicz University
Poland

Tubiello, Francesco
Columbia University
USA/IIASA/Italy

Turley, Carol
Plymouth Marine Laboratory
UK

van de Giesen, Nick
Delft University of Technology
The Netherlands

van Minnen, Jelle
RIVM
The Netherlands

van Schaik, Henk
UNESCO Co-operative Programme on Water and Climate
The Netherlands

van Vuuren, Detlef
Netherlands Environment Assessment Agency
The Netherlands

van Ypersele, Jean-Pascal
Université catholique de Louvain
Belgium

Vandenberghe, Jef
Vrije University
The Netherlands

Vaughan, David
British Antarctic Survey
UK

Velichko, Andrei
Institute of Geography, Russian Academy of Sciences
Russia

Vicarelli, Marta
Columbia University
USA/Italy

Vilhjalmsson, Hjalmar
Marine Research Institute
Iceland

Villamizar, Alicia
Universidad Símon Bolívar
Venezuela

Vincent, Katherine
University of East Anglia
UK

Viner, David
University of East Anglia
UK

Vogel, Coleen
University of the Witwatersrand
South Africa

Walsh, John
University of Alaska
USA

Wandel, Johanna
University of Guelph
Canada

Warren, Rachel
Tyndall Centre for Climate Change Research
UK

Warrick, Richard
University of Waikato
New Zealand

Washington, Richard
Oxford University
UK/South Africa

Watkiss, Paul
Paul Watkiss Associates
UK

Wiegandt, Ellen
Graduate Institute of International Studies
Switzerland

Wilbanks, Tom
Oak Ridge National Laboratory
USA

Wilby, Robert
King's College London
UK

Wolf, Tanja
WHO Regional Office for Europe/Germany

Wolf, Johanna
University of East Anglia
UK/Germany

Wong, Poh Poh
National University of Singapore
Singapore

Woodroffe, Colin
University of Wollongong
Australia

Woodruff, Rosalie
Australian National University
Australia

Woodward, Alistair
University of Auckland
New Zealand

Wrona, Fred
National Water Research Institute
Canada

Wu, Qigang
Texas A&M University
USA/China

Wu, Shaohong
Chinese Academy of Sciences
China

Yamin, Farhana
University of Sussex
UK

Yanda, Pius
University of Dar-es-Salaam
Tanzania

Yohe, Gary
Wesleyan University
USA

Zapata-Marti, Ricardo
UN Economic Commission for Latin American and the Caribbean (ECLAC)/Mexico

Zhang, Qiaomin
South China Sea Institute of Oceanology
China

Ziervogel, Gina
University of Cape Town
South Africa

Zurek, Monika
Food and Agriculture Organization/Germany

Appendix III: Reviewers of the IPCC WGII Fourth Assessment Report

Note: International organisations listed at the end.

ALGERIA

Tabet Aoul, Mahi
Association pour la Recherche pour le Climat et l'Environnement (ARCE)

ARGENTINA

Barros, Vincente Ricardo
Ciudad Universitaria

Bischoff, Susana
Ciudad Universitaria

Camilloni, Inés Angela
Ciudad Universitaria

Canziani, Osvaldo F.
IPCC Working Group II Co-chair

Carbajo, Anibal
Universidad de Buenos Aires

Codignotto, Jorge O.
Ciudad Universitaria

Comesaña, Claudia Maria
Ministerio de Relaciones Exteriores, Comercio Internacional y Culto

Curto, Susana I.
National Academy of Medicine

Devia, Leila
National Institute of Industrial Technology

Kokot, Roberto
Universidad de Buenos Aires

Murgida, Ana Maria
University of Beunos Aires

Neiff, Juan J.
Centro de Ecología Aplicada del Litoral

Perez Harguindeguy, Natalia
Instituto Mulitidisciplinario de Biología Vegetal (UNC-CONICET)

Rusticucci, Matilde
Universidad de Buenos Aires

Solman, Silvina
Ciudad Universitaria

Travasso, Maria I.
INTA

Usunoff, Eduardo
Instituto de Hidrologie de Llanuras

Vinocur, Marta
Universidad Nacional de Río Cuarto

Wehbe, Mónica
Universidad Nacional de Río Cuarto

AUSTRALIA

Anderson, Rod
Department of Sustainability and Environment

Ash, Andrew
CSIRO

Baird, Mark
University of New South Wales

Barnett, Jon
The University of Melbourne

Beer, Tom
CSIRO

Beggs, Paul
Macquarie University

Boyle, Sharon
Planning Institute of Australia

Brunskill, Gregg
Australian Institute of Marine Science

Chambers, Lynda
Bureau of Meteorology Research Centre

Churchman, Susan
Department of Environment and Heritage South

Cleland, Sam
Bureau of Meteorology

Cocklin, Chris
Monash University

Coleman, Anthony
Insurance Australia Group

Collins, Dean
Bureau of Meteorology

Crimp, Steven
Queensland Centre for Climate Applications

Curran, Beth
Bureau of Meteorology

Dunlop, Michael
CSIRO

Edwards, Spencer
Department of Environment and Heritage

Farquhar, Graham
Australian National University

Garnham, John
Department of Primary Industries

Gifford, Roger M.
CSIRO

Gitay, Habiba
Australian National University

Grace, Peter R.
Queensland University of Technology

Green, Donna
CSIRO

Harvey, Nicholas
University of Adelaide

Hayman, Peter
South Australian Research and Development Institute

Higgins, John
Australian Greenhouse Office

Hoy, Richard
Electricity Supply Association of Australia

Hughes, Lesley
Macquarie University

Jones, David
National Climate Centre

Jones, Roger
CSIRO

Kay, Robert
Coastal Zone Management (Australia) Pty Ltd

Kellow, Aynsley
University of Tasmania

Kininmonth, Bill
Australasian Climate Research

Kjellstrom, Tord
University of Auckland

Lough, Janice
Australian Institute of Marine Science

Lyne, Vincent
CSIRO

Manton, Michael
Monash University

Marshall, Paul
Great Barrier Reef Marine Park Authority

McKibbin, Warwick
Australian National University

McNeil, Ben
University of New South Wales

Meinke, Holger
Government of Queensland

Nicholls, Neville
Bureau of Meteorology Research Centre

Pearman, Graeme
Monash University Sustainability Centre

Pittock, Barrie
CSIRO

Power, Scott
Bureau of Meteorology Research Centre

Quiggin, John
University of Queensland

Risbey, James
Monash University

Ritman, Kim
Department of Agriculture, Fisheries and Forestry

Saenger, Peter
Southern Cross University

Shearman, David
University of Adelaide

Stone, Roger
Department of Natural Resources

Sutherst, Robert
CSIRO

Tapper, Nigel
Monash University

Tong, Shilu
Queensland University of Technology

Walker, George R.
Aon Re Australia

Walsh, Kevin
CSIRO

Watkins, Andrew
Australian Bureau of Meteorology

White, David H.
ASIT Consulting

Wiles, Perry
Bureau of Meteorology

Wilkinson, Clive
IUCN

Williams, Stephen E
James Cook University

Woldring, Oliver
NSW Greenhouse Office

Woodruff, Rosalie
Australian National University

Younus, Aboul Fazal
The University of Adelaide

AUSTRIA

Glatzel, Stephan
Universität Wien

Kaser, Georg
Institut für Geographie

Lexer, Manfred
University of Natural Resources and Applied Life Sciences

Pauli, Harald
University of Vienna

Radunsky, Klaus
Federal Environment Agency

BANGLADESH

Admed, Ahsan Uddin
Bangladesh Unnayan Parishad (BUP)

Islam, Rafiqul M.
Integrated Coastal Zone Management

Karim, Mohammed F.
Ibaraki University

BARBADOS

Brewster, Leo
Barbados Coastal Zone Management Unit

Mwansa, John
Barbados Water Authority

Trotman, Adrian
Caribbean Institute for Meteorology and Hydrology

BELGIUM

Bogaert, Johan
Department Environment, Nature and Energy

Halloy, Stephan
Universidad Mayor de San Andrés

Marbaix, Philippe
Université catholique de Louvain

Vanderstraeten, Martine
Federal Office for Scientific, Technical and Cultural Affairs

Verhasselt, Yola
Royal Academy of Overseas Sciences

BENIN

Boko, Michel
Universite de Bourgogne

Oyede, Lucien Marc
Université d'Abomey-Calavi

Vissin, Expédit Wilfrid
Université d'Abomey-Calavi

Yabi, Ibouraïma Fidele
Université d'Abomey-Calavi

BHUTAN

Namgyel, Thinley
National Environment Commission

BOLIVIA

Gonzales, Javier
Programa Nacional de Cambios Climáticos

Paz, Oscar
National Climate Change Programme

BOTSWANA

Dube, Pauline O.
University of Botswana

BRAZIL

Ambrizzi, Tercio
Institute of Astronomy, Geophysics and Atmospheric Sciences – USP

Cardia Simoes, Jefferson
Federal University of Rio Grande do Sul

Cotrim da Cunha, Leticia
Max-Planck-Institut für Biogeochemie

Cunha, Gilberto R.
Embrapa-Trigo

da Cunha Bustamante, Mercedes Maria
University of Brasilia

Fearnside, Philip M.
Instituto Nacional de Pesquisas da Amazonia – INPA

Kahn Ribeiro, Susana
Federal University of Rio de Janeiro

Lima, Magda
Embrapa-Meio Ambiente

Marengo Orsini, Jose Antonio
CPTEC/INPE

Moreira, Jose Roberto
Biomass User Network (BUN)

Moutinho, Paulo
Instituto de Pesquisa Ambiental da Amazônia-IPAM

Pinguelli Rosa, Luis
Federal University of Rio de Janeiro

Sant' Ana, Silvio Rocha
Fundação Grupo Esquel

BULGARIA

Yotova, Antoaneta
National Institute of Meteorology and Hydrology

CAMBODIA

Sum, Thy
Ministry of Environment, Department of Planning and Legal Affairs

CANADA

Alder, Jacqueline
University of British Columbia

Amiro, Brian
University of Manitoba

Anderson, John
Environment Canada

Atkinson, David E.
NRCan

Barber, David G.
University of Manitoba

Barlund, Ilona
Finnish Environment Institute

Bass, Brad
Meteorological Service of Canada

Beamish, Richard J.
Pacific Biological Station

Beltaos, Spyros
Environment Canada

Bergeron, Yves
Université du Québec en Abitibi-Témiscamingue

Bernier, Pierre
Canadian Forestry Service

Berry, Peter
Health Canada

Boileau, Pierre
Environment Canada

Bourque, Alain
Ouranos Consortium

Brady, Michael
Canadian Forest Service

Brisbois, Benjamin
Environment Canada

Bruce, James P.
Canadian Climate Program Board

Bullock, Paul
University of Manitoba

Burn, Donald
University of Waterloo

Burton, Ian
University of Toronto

Bush, Elizabeth
Science and Technology Branch, Environment Canada

Cawkwell, Fiona
University of Alberta

Church, Ian
Yukon Government

Cohen, Stewart J.
University of British Columbia

Crabbé, Philippe J.
Université d'Ottawa

Cross, Rob.
Environment Canada

Dawson, Jaime
The University of Western Ontario

de Loe, Rob
University of Guelph

Desjardins, Raymond
Independent

Douglas, Allan
Canadian Climate Impacts and Adaptation

Drexhage, John
International Institute for Sustainable Development

Edwards, Patti
Meteorological Service of Canada

Etkin, David
Environment Canada

Fernandes, Richard
Canada Centre for Remote Sensing

Fisher, David A.
Geological Survey of Canada

Flannigan, Mike
Canadian Forest Service

Fleming, Richard
Canadian Forest Service

Forbes, Donald
Bedford Institute of Oceanography

Gajewski, Konrad
University of Ottawa

Gauthier, Sylvie
Canadian Forest Service

Harvey, Danny
University of Toronto

Hill, Harvey
Agriculture Canada

Hill, Philip
Geological Survey of Canada

Jefferies, Robert L.
University of Toronto

Johnson, Peter G.
University of Ottawa

Jones-Cameron, Tracy
Natural Resources Canada

Kerr, Jeremy T.
University of Ottawa

Kertland, Pamela
Natural Resources Canada

Khandekar, Madhav
Retired

Lavender, Beth
Environmental Adaptation Research Group

Lelasseux, Stephane
Environment Canada

Lemmen, Don
Natural Resources Canada

Lysyshyn, Kathleen
Canadian Forest Service

Maarouf, Abdel R.
Environment Canada

MacDonald, Don
Alberta Department of Environment

Malcolm, David
Arctic Energy Alliance

Margolis, Hank
Université Laval

McBean, Gordon
University of Western Ontario

Mehdi, Bano
McGill University

Michaud, Yves
Geological Survey of Canada

Neron, Marie-Eve
Climate Change Impacts and Adaptation – INAC

Nuttall, Mark
University of Alberta

Ogden, Anyslie
Government of Yukon

Percy, Kevin
Canadian Forest Service

Price, David T.
Canadian Forest Service

Rousseau, Alain
Institut National de la Recherche Scientifique

Sauchyn, Dave
University of Regina

Savard, Martine
Geological Survey of Canada

Schallenberg, Marc
University of Otago

Scott, Daniel
Environment Canada

Sharp, Martin
University of Alberta

Sheppard, Stephan
University of British Columbia

Simonovic, Slobodan P.
Department of Civil and Environmental Engineering

Singh, Bhawan
Université de Montréal

Smith, Sharon
NRCan

Solomon, Steven
Bedford Institute of Oceanography

Sparling, Jim
Environmental Protection Agency

Spittlehouse, David L.
B.C. Ministry of Forests

Stemp, Raymond
Alberta Department of the Environment

Stone, John
IPCC Working Group II Vice-chair

Stratton, Tana Lowen
Department of Foreign Affairs and International Trade

Streicker, John
Yukon College

Sydneysmith, Robin
University of British Columbia

Taylor, Robert
Bedford Institute of Oceanography

Thompson, Ian
Canadian Forest Service

Trishchenko, Alexander P.
Canada Centre for Remote Sensing (CCRS)

Trofymow, Tony
Canadian Forest Service

Vasseur, Liette
Laurentian University

Venema, Henry
IISD International Institute for Sustainable Development

Victor, Peter
York University

Wall, Ellen
University of Guelph

Wall, Geoff
University of Waterloo

Wheaton, Elaine
Saskatchewan Research Council

Yin, Yongyuan
University of British Columbia

Zawar-Reza, Peyman
University of Canterbury

Zwiers, Francis W.
University of Victoria

CHILE

Carrasco, Jorge
Dirección Meteorológica de Chile

Casassa, Gino
Centro de Estudios Cientificos

Farias, Fernando
CONAMA

CHINA

Chen, Xiaoqiu
Peking University

Dong, Zhaoqian
Polar Research Institute of China

Erda, Lin
Chinese Academy of Agricultural Science

Fan, Daidu
Tongji University

Fang, Xiuqi
Beijing Normal University

Ju, Hui
Chinese Academy of Agricultural Science

Li, Congxian
Tongji University

Li, Ke-Rang
Chinese Academy of Sciences

Liu, Chunzhen
China Water Information Center

Liu, Shirong
Chinese Academy of Forestry

Liu, Yingjie
Chinese Academy of Agricultural Sciences

Luo, Tianxiang
Chinese Academy of Sciences

Ma, Shiming
Chinese Academy of Agricultural Sciences

Qin, Dahe
China Meteorological Administration

Su, Jilan
State Oceanic Administration

Sun, Fang
Chinese Academy of Agricultural Sciences

Tao, Fulu
Chinese Academy of Agricultural Sciences

Wang, Bangzhong
China Meteorological Administration

Wang, Changke
National Climate Centre

Wang, Futang
Academy of Meteorological Science

Wei, Xiong
Chinese Academy of Agricultural Sciences

Wu, Shaohong
Chinese Academy of Sciences

Xiao, Fengjin
Chinese Meteorological Administration

Xie, Liyong
Chinese Academy of Agricultural Sciences

Xie, Zhenghui
Chinese Academy of Sciences

Xiong, Wei
Institute of Environment and Sustainable Development in Agriculture

Xu, Yinlong
Chinese Academy of Agricultural Sciences

Yan, Qilun
National Marine Environmental Monitoring Center

Yang, Xiu
Agrometeorology Instutite

Zhai, Panmao
China Meteorological Administration

Zhao, Yong
China Huaneng Technical Economics Research Institute

Zhao, Zong-Ci
China Meteorological Administration

Zhou, Guangsheng
Chinese Academy of Sciences

Zhou, Zijiang
National Meteorological Information Centre of CMA

COLOMBIA

Caicedo, Jose Daniel Pabon
Universidad Nacional de Colombia

Mow, June Marie
Fundacion Providence

Pabon Caicedo, Daniel
Universidad Nacional de Colombia

Poveda, Germán
Universidad Nacional de Colombia

COOK ISLANDS

Carruthers, Pasha
Environment Service

COSTA RICA

Campos, Max
National Meteorological Institute

CUBA

Diaz Morejon, Cristobal Felix
Ministry of Science, Technology and the Environment

Llanes-Reguerio, Juan
Univeristy of Havana

Planos Gutiérrez, Eduardo
Institute of Meteorology

Rodriguez, Carlos
Instituto de Planificacion Fisica de Cuba

Suarez, Avelino G.
Cuban Environment Agency

CZECH REPUBLIC

Halenka, Tomas
Charles University

Pretel, Jan
Czech Hydrometeorological Institute

DENMARK

Beier, Claus
Risø National Laboratory

Fjeldsa, Jon
Zoological Museum

Halsnaes, Kirsten
Risø National Laboratory

Meltofte, Hans
National Environmental Research Institute

Olesen, Jørgen E.
Danish Institute of Agricultural Sciences

Pejrup, Morten
University of Copenhagen

Porter, John R.
The Royal Veterinary and Agricultural University

ECUADOR

Santos, Jose Luis
CIFIN

EGYPT

El Raey, Mohamed
Institute of Graduate Studies and Research

El Shahawy, Mohamed
Cairo University

Ragab, Ragab
Centre for Ecology and Hydrology (CEH) Oxford

EL SALVADOR

Munguía de Aguilar, Martha Yvette
Ministerio del Medio Ambiente y Recursos Naturales

ESTONIA

Kadaja, Jüri
Estonian Research Institute of Agriculture

Kont, Are
Institute of Ecology

ETHIOPIA

Tadesse, Tsegaye
University of Nebraska-Lincoln

EUROPE

Erhard, Markus
European Environment Agency

Malingreau, Jean-Paul
European Commission – DG Joint Research Centre

Mueller, Lars
European Commission

Spangenberg, Joachim
Sustainable Europe Research Institute

Troen, Ib
Environment and Climate System, European Commission

Tulkens, Philippe
European Commission

FIJI

Veitayaki, Joeli
University of the South Pacific

FINLAND

Carter, Timothy
Finnish Environment Institute

Clarke, Majella
Savcor Indufor Oy

Forbes, Bruce
University of Lapland

Fronzek, Stefan
Finnish Environment Institute

Gastgifvars, Maria
Finnish Environment Institute

Haanpaa, Simo
Helsinki University of Technology

Haapala, Jari
University of Helsinki

Hakala, Kaija
Agricultural Research Centre

Halonen, Mikko
Gaia Consulting Oy

Hanninen, Heikki
University of Helsinki

Hannukkala, Antti
MTT Agrifood Research Finland

Heikinheimo, Pirkko
Ministry of Agriculture and Forestry

Henttonen, Heikki
Finnish Forest Research Institute

Holmstrom, Nina
Finnish Environment Institute

Holopainen, Jarmo
University of Kuopio

Holopainen, Toini
University of Kuopio

Houtsonen, Lea
The Finnish National Board of Education

Kankaanpaa, Susanna
Finnish Environment Institute

Kasurinen, Anne
University of Kuopio

Kauppi, Pekka E.
University of Helsinki

Kayhko, Jukka
University of Turku

Kellomäki, Seppo
University of Joensuu

Kivisaari, Esko
Federation of Finnish Insurance Companies

Kortelainen, Pirkko
Finnish Environment Institute (SKYE)

Kuoppamäki, Pasi
Sampo plc

Kuusisto, Esko
Finnish Environment Institute

Laiho, Raija
Helsinki University

Lammi, Harri
Greenpeace

Lehtonen, Heikki
MTT Agrifood Research Finland

Luukkanen, Jyrki
Finland Futures Research Centre, Turku School of Economics

Makipaa, Raisa
Finnish Forest Research Institute

Makkonen, Lasse
VTT

Martikainen, P.J.
National Public Health Institute

Nikinmaa, Eero
University of Helsinki

Peltonen, Lasse
Helsinki University of Technology

Perrels, Adriaan
Government Institute for Economic Research

Pitkanen, Heikki
Finnish Environment Institute

Primmer, Eeva
Finnish Environment Institute

Rosqvist, Tony
VTT

Rousi, Matti
Finnish Forest Research Institute

Ruosteenoja , Kimmo
Finnish Meteorological Institute

Saarnio, Sanna
University of Joensuu

Selin, Pirkko
Vapo Oy

Sievanen, Tuija
Finnish Forest Research Institute

Silvo, Kimmo
Finnish Environment Institute

Sopanen, Sanna
Finnish Environment Institute

Starr, Mike
University of Helsinki

Tapio, Petri
Finland Futures Research Center

Vapaavuori, Elina
Finnish Forest Research Institute

Varis, Olli
Helsinki University of Technology

Vehviläinen, Bertel
Finnish Environment Institute (SKYE)

FRANCE

Bachelet, Dominique
Center d'Etude Spatiale du Rayonnement

Beaugrand, Gregory
University of Lille

Caneill, Jean-Yves
Electricité de France

Ceron, Jean-Paul
CRIDEAU (Université de Limoges-CNRS-INRA)

Chastel, Claude
Académie Nationale de Médecine

Chevallier, Pierre
Institut de Recherche pour le Développement (IRD)

Chuine, Isabelle
CNRS

Corfee-Morlot, Jan
OECD Environment Directorate

de Marsily, Ghislain
Académie des Sciences

Deque, Michel
Météo-France / CNRM

Douguedroit, Annick
Université de Provence

Dubois, Ghislain
Environnement Conseil (TEC)

Empereur-Bissonnet, Pascal
National Institute of Public Health Surveillance

Gillet, Marc
Mission Interministerielle de l'Effet de Serre

Guillaumont, Robert
Académie des Sciences

Hequette, Arnaud
Universite du Littoral

Juvanon du Vachat, Regis
Météo-France

Lagadec, Patrick
Ecole Polytechnique

Lavelle, Patrick
Institut de Recherche sur le Développement

Lavorel, Sandra
Université Joseph Fourier

Lenotre, Nicole
BRGM

Minh, Ha-Duong
CNRS

Moutou, Francois
Agence Française de Sécurité Sanitaire des Aliments

Paillard, Michel
IFREMER

Paskoff, Roland
Université Lumière de Lyon

Petit, Michel
Conseil général des technologies de l'information

Planton, Serge
Meteo-France

Reiter, Paul
Pasteur Institute

Rodney, Alan
Académie des technologies

Rousseau, Daniel
Météo-France

Sanaonetti, Philippe
INSERM

Saugier, Bernard
Paris 11 University

Tirpak, Dennis
OECD

GAMBIA

Gomez , Bernard
Global Change Research Unit (GCRU)

GEORGIA

Inashvili, Medea
Ministry of Environment and Natural Resources

GERMANY

Augustin, Sabine
Federal Environment Agency

Badeck, Franz-Werner
Potsdam Institute for Climate Impact Research

Benndorf, Rosemarie
Umweltbundesamt

Bruckner, Thomas
Technical University of Berlin

Bugmann, Harald
Swiss Federal Institute of Technology

Fuentes, Ursula
German Federal Environment Ministry

Füssel, Hans-Martin
Potsdam Institute for Climate Impact Research (PIK)

Gerten, Dieter
Potsdam Institute for Climate Impact Research (PIK)

Glauner, Reinhold
Institute for World Forestry

Gruenewald, Uwe
Brandenburg University of Technology Cottbus

Hain, Benno
Federal Environment Agency

Hare, William L.
Potsdam Institute for Climate Impact Research (PIK)

Hasse, Clemens
Federal Environment Agency

Hoeppe, Peter
Munich Re

Jendritzky, Gerd
University of Freiburg

Kartschall, Karin
Federal Environment Agency

Kistemann, Thomas
University of Bonn

Klotz, Stefan
Centre for Environmental Research Leipzig-Halle

Kuhn, Ingolf
UFZ – Centre for Environmental Research

Kulessa, Margareta E.
Mainz University of Applied Sciences

Lange, Manfred
University of Münster

Lemke, Peter
Alfred-Wegener Institute for Polar and Marine Research

Lindner, Marcus
Potsdam Institute for Climate Impact Research (PIK)

Lingner, Stephen
Europäische Akademie GmbH

Löschel, Andreas
Institute for Prospective Technological Studies (IPTS)

Lucht, Wolfgang
Potsdam Institute for Climate Impact Research (PIK)

Mahrenholz, Petra
Federal Environmental Agency of Germany

Meinshausen, Malte
NCAR

Michaelowa, Axel
Hamburg Institute of International Economics

Morgenschweis, Gerd
Ruhrverband (Ruhr River Association)

Münzenberg, Annette
Geman Aerospace Centre

Reisinger, Andy
Ministry for the Environment

Renn, Ortwin
University of Stuttgart

Rosner, Stefan
Deutscher Wetterdienst

Sauerborn, Rainer
Heidelberg University

Schroeter, Dagmar
Potsdam Institute for Climate Impact Research (PIK)

Schulz, Astrid
WBGU

Schumann, Andreas
Ruhr-University Bochum

Schwalb, Antje
Institut für Umweltgeologie

Schwarzer, Klaus
Institute of Geosciences

Settele, Josef
UFZ

Vlek, Paul
Center for Development Research (ZEF)

Voigt, Thomas
Federal Environment Agency

von Storch, Hans
GKSS Research Centre

Walther, Gian-Reto
University of Hannover

Weimer-Jehle, Wolfgang
University of Stuttgart

Weiss, Martin
Federal Environment Agency

Welp, Martin
Potsdam Institute for Climate Impact Research (PIK)

Windhorst, Wilhelm
Kiel University

Wurzler, Sabine
North-Rhine Westphalia State Environment Agency

GHANA

Fobil, Julius
University of Ghana

GREECE

Matzarakis, Andreas
Universität Freiburg

Sarafidis, Yannis
National Observatory of Athens

Seferlis, Miltiadis
Greek Biotope / Wetland Centre

HUNGARY

Balint, Gabor
Hydrological Institute

INDIA

Alakkat, Unnikrishnan
National Institute of Oceanography

Bhadwal, Suruchi
The Energy and Resources Institute

Bhandari, Preety
TERI

Bhattacharya, Sumana
NATCOM Project Management

Chander, Subhash
TERI

Dhiman, R.C.
Malana Research Centre

Gopal, Brij
Jawaharlal Nehru University

Gosain, A.K.
IIT Delhi

Kapshe, Manmohan
Maulana Azad National Institute of Technology

Kelkar, Ulka
TERI

Mruthyunjaya, Mr.
National Agricultural Technology Project, ICAR

Nambi, A.
M.S. Swaminathan Research Foundation

Narayanan, Krishnan
IITB

Parikh, Jyoti
Integrated Research and Action for Development

Parthasarthy, D.
IITB, Mumbai

Ravindranath, N.H.
Indian Institute of Sciences

Roy, Joyashree
Jadavpur University

Sharma, C.
Ministry of Environment and Forests

Sharma, Upasna
Indian Institute of Technology

Shukla, Priyadarshi
Indian Institute of Management

Srinivasan, Govindarajan
India Meteorological Department

Unnikrishnan , A.S.
National Institute of Oceanography

Upasna, Sharma
Indian Institute of Technology

Uprety, Dinesh C.
Indian Agricultural Research Institute

INDONESIA

Adiningsih, Erna
National Institute of Aeronautics and Space (LAPAN)

Anshari, Gusti
Tanjungpura University

IRAN

Rahimi, Mohammad
Islamic Republic of Iran Meteorological Organization

IRELAND

Cullen, Elizabeth
National University of Ireland, Maynooth

Donnelly, Alison
Environmental Protection Agency

Goodman, Pat
National University of Ireland, Maynooth

Holden, Nicholas
University College Dublin

McElwain, Laura
National University of Ireland, Maynooth

McGovern, Frank
Environmental Protection Agency

Murphy, Conor
National University of Ireland, Maynooth

O'Brien, Phillip
National University of Ireland, Maynooth

Sweeney, John
National University of Ireland, Maynooth

Tol, Richard S. J.
Economic and Social Research Institute, Ireland

ISRAEL

Issar, Aire S.
Ben Gurion University of the Negev

Lavee, Hanoch
Bar-Ilan University

Safriel, Uriel N.
Hebrew University of Jerusalem

Saltz, David
Ben Gurion University

ITALY

Bindi, Marco
DISAT-UNIFI

Campostrini, Pierpaolo
CORILA

Colacino, Michele
ISAC-CNR

da Mosto, Jane
CORILA

Dragoni, Walter
Università di Perugia

Frezzotti, Massimo
ENEA

Killmann, Wulf
Food and Agriculture Organization (FAO)

Lionello, Piero
University of Lecce

Lorenzo, Genesio
Institute of Biometeorology, National Research Council

Maracchi, Giampiero
Institute of Biometeorology

Mariotti, Annarita
ENEA Climate Section

Nanni, Teresa
National Research Council

Petriccione, Bruno
National Forest Service

Reichstein, Markus
University of Tuscia

Ribera d'Alcala, Maurizio
Stazione Zoologica 'Anton Dohrn'

JAMAICA

Clayton, Anthony
University of the West Indies

JAPAN

Ando, Mitsuru
National Institute for Environmental Studies

Fukushima, Takehiko
University of Tsukuba

Harasawa, Hideo
National Institute for Environmental Studies

Hayami, Hitoshi
Keio University

Hisajima, Naoto
Ministry of Foreign Affairs

Ichinose, Toshiaki
National Institute for Environmental Studies

Itoh, Kiminori
Yokohama National University

Kabuto, Michinori
National Institute for Environmental Studies

Kawashima, Hiroyuki
University of Tokyo

Kayanne, Hajime
University of Tokyo

Kobayashi, Hideyuki
Research Coordinator for Housing Information System

Matsui, Tetsuya
Hokkaido Research Centre

Mikami, Masao
Meteorological Research Institute

Morisugi, Hisayoshi
Tohoku University

Nakagawa, Mitsuhiro
Ibaraki University

Omasa, Kenji
University of Tokyo

Onuma, Ayumi
Keio University

Sasaki, Akihiko
Fukushima Pref. Authority

Shinoda, Masato
Tottori University

Takahashi, Kiyoshi
National Institute for Environmental Studies

Tanaka, Nobuyuki
Regeneration Process Laboratory Forestry and Forest Products Research Institute

Tsunekawa, Atsushi
Arid Land Research Center, Tottori University

Yamaguchi, Mitsutsune
Teikyo University

Yamano, Hiroya
National Institute for Environmental Studies

Yokoki, Hiromune
Ibaraki University

Yokozawa, Masayuki
National Institute for Agro-Environmental Sciences

Yoshino, Masatoshi
Retired

KENYA

Githeko, Andrew
Kenya Medical Research Institute

Obura, David
CORDIO East Africa

Opondo, Mary Magdalene
University of Nairobi

Tole, Mwakio P.
Kenyatta University

Wandiga, Shem
Kenya National Academy of Sciences

MADAGASCAR

Ramiandrisoa, Vohanginiriana Anne Marie
Madagascar Meteorological Office

MALAWI

Bulirani, Alex
Ministry of Mines, Natural Resources and Environment

Kamdonyo, Donald Reuben
Director of Meteorological Services

MALDIVES

Majeed, Abdullahi
Ministry of Environment, Energy and Water

Musthaq, Fathimath
Williams College

MALI

Sokona, Youba
Sahel and Sahara Observatory (OSS)

MAURITIUS

Prithiviraj, Booneeady
Mauritius Meteorological Services

Ragoonaden, Sachooda
Consultant to Indian Ocean Commission

MEXICO

Condé, Cecilia
Ciudad Universitaria

Flores Montalvo, Andrés
Instituto Nacional de Ecología (INE)

Lluch-Belda, Daniel
Centro de Investigaciones Biologicas del Noreste, S.C.

Magaña Rueda, Victor
Universidad Autónoma de México (UNAM)

Martinez, Julia
Instituto Nacional de Ecología (INE)

Martínez-Meyer, Enrique
Universidad Nacional Autonoma de México

Matus Kramer, Arnoldo
Instituto Nacional de Ecología (INE)

Oropeza, Oralia
Ciudad Universitaria

Osornio Vargas, Alvaro
Ciudad Universitaria

MONGOLIA

Batima, Punsalmaa
Institute of Meteorology and Hydrology

NEPAL

Adhikary, Sharad P.
Himalayan Climate Centre

THE NETHERLANDS

Abbink, Oscar
TNO B&O

Bavinck, Maarten
University of Amsterdam

Bouwer, Laurens
Vrije Universiteit

Brinkman, Robert
Food and Agriculture Organization of the UN

Bruggink, Jos
Netherlands Energy Research Foundation

Clabbers, Bas
Ministry of Agriculture, Nature and Food Quality

de Ronde, John
Delft Hydraulics

de Wit, Marcel
Ministry of Transport, Public Works and Water Management

Dietz, A.J.
University of Amsterdam

Giller, Ken
Wageningen University

Haanstra, Hayo
Ministry of Agriculture, Nature and Food Quality

Hettelingh, Jean-Paul
National Institute of Public Health and the Environment (MNP-RIVM)

Hilhorst, Thea
Wageningen University

Jonk, Gerie
Ministry of Housing, Spatial Planning and the Environment

Kamil, Sasja
Cordaid

Klein Tank, Albert
Royal Netherlands Meteorological Institute

Klok, Lisette
KNMI

Kram, Tom
Netherlands Environmental Assessment Agency (MNP-RIVM)

Kwadijk, Jaap
WL Delfthydraulics

Labohm, Hans H.J.
Netherlands Institute of International Relations

Marchand, Marcel
Delft Hydraulics

Martens, Pim
Maastricht University

Metzger, Marc
Wageningen University

Misdorp, Robert
National Institute for Coastal and Marine Management

Posch, Maximilian
RIVM

Reggiani, Paolo
Delft Hydraulics

Salomons, Wim
University of Amsterdam

Stive, Marcel J.F.
Waterloopkundig Laboratorium/Delft Hydraulics

Swart, Rob
RIVM

van Aalst, Maarten
Red Cross/Red Crescent Climate Centre

van de Giesen, Nick
Delft University of Technology

van der Meulen, Frank
National Institute for Coastal and Marine Management Rijkswaterstaat-RIKZ

van Minnen, Jelle G.
RIVM

van Schaik, Henk
UNESCO Co-operative Programme on Water and Climate

Verhagen, Jan A.
Plant Research International

Vos, Claire
Alterra Institute

NEW ZEALAND

Baxter, Kay
Ministry for the Environment

Becken, Susanne
Landcare Research

Becker, Julia
Institute of Geological and Nuclear Sciences

Bell, Robert
National Institute of Water and Atmospheric Research

Collins, Eva
University of Waikato

Dymond, Stuart
Ministry for Foreign Affairs and Trade

Fairbairn, Paul L
SOPAC South Pacific Applied Geoscience

Gray, Warren
Ministry for the Environment

Hales, Simon
University of Otago

Hall, Alistair
HortResearch

Hannah, John
University of Otago

Hay, John
University of Waikato

Hughey, Ken
Lincoln University

Kenny, Gavin J.
Earthwise Consulting Ltd

Kerr, Suzi
Motu Economic and Public Policy Research Institute

King, Darren
National Institute of Water and Atmospheric Research

Larsen, Howard
Ministry for the Environment

Lawrence, Judy
Climate Change National Science Strategy Committee

Lawson, Wendy
University of Canterbury

Maclaren, Piers
Piers Maclaren & Associates Ltd

McKerchar, Alastair
National Institute of Water and Atmospheric Research

Mullan, A. Brett
National Institute of Water and Atmospheric Research

Plume, Helen
Ministry for the Environment

Porteous, Alan
National Institute of Water and Atmospheric Research

Power, Vera
Ministry for the Environment

Purdie, Jennifer
University of Waikato

Rys, Gerald
Ministry of Research, Science and Technology

Saggar, Surinder
Landcare Research

Stephens, Peter
Ministry for the Environment

Stroombergen, Adolf
Infometrics

Waugh, John Robert
Opus International Consultants Ltd.

Weaver, Sean
Victoria University of Wellington

Whitehead, David
Landcare Research

Wilson, Toni
Ministry for the Environment

Woodward, Alistair
University of Auckland

Wratt, David
National Institute of Water and Atmospheric Research

NIGER

Amani, Abou
AGRHYMET Regional Center

NIGERIA

Adejuwon, James O.
Obafemi Awolowo University

Antia, Effiom E.
University of Calabar

Nyong, Anthony
International Development Research Centre

NORWAY

Aaheim, Hans Asbjørn
Center for International Climate and Environmental Research

Andersen, Cathrine
Directorate for Civil Protection and Emergency Planning

Asphjell, Torgrim
Norwegian Pollution Control Authority (SFT)

Christophersen, Oyvind
Ministry of Environment

Dalen, Linda
Norwegian Directorate for Nature Management

Eriksen, Siri
Center for International Climate and Environmental Research

Forland, Eirik J.
Norwegian Meteorological Institute

Gabrielsen, Geir Wing
Norwegian Polar Institute

Glasser, Trond Jorgen
Norwegian Ministry of Foreign Affairs

Hagen, Jon Ove
University of Oslo

Hannesson, Rögnvaldur
Norwegian School of Economics and Business Administration

Haraldsen, Vivil
Norwegian Ministry of Education and Research

Hoel, Alf Håkon
University of Tromsø

Hofgaard, Annika
Norwegian Institute for Nature Research

Holmen, Kim
Norwegian Polar Institute

Instanes, Arne
Instanes Consulting Engineers

Isaksson, Elisabeth
Norwegian Polar Institute

Kolshus, Hans
Norwegian Pollution Control Authority

O'Brien, Karen
Center for International Climate and Environmental Research

Okstad, Elin
Norwegian Pollution Control Authority

Pavlov, Vladimir
Norway Polar Institute

Roald, Lars Andreas
Norwegian Water Resources and Energy Directorate

Sakshaug, Egil
NTNU

Solberg, Birger
Norwegian University of Life Sciences

West, Jennifer Joy
Center for International Climate and Environmental Research

Winther, Jan-Gunnar
Norwegian Polar Institute

OMAN

Al-Kharoosi, Ahmed bin Saeed
Ministry of Regional Municipalities, Environment and Water Resources

PAKISTAN

Akhtar, Nadia
Global Change Impact Studies Centre

Goheer, Arif
Global Change Impact Studies Centre

Iqbal, Mohsin
Global Change Impact Studies Centre

Khan, Arshad
Global Change Impact Studies Centre

Mudasser, Muhammed
Global Change Impact Studies Centre

Niazi, Mahjabeen
Global Change Impact Studies Centre

Raza, Shoaib
Global Change Impact Studies Centre

Shahid, Imran
Global Change Impact Studies Centre

Sheikh, Munir
Global Change Impact Studies Centre

PANAMA

Sempris, Emilio
CATHALAC

PERU

Angulo Villarreal, Lenkiza
Intermediate Technology Development Group – ITDG

Encinas Caceres, Carla
Consejo Nacional del Ambiente – CONAM

Garcia Vargas, Julio
Consejo Nacional del Ambiente – CONAM

Guerra, Antonio Humberto
Universidad Peruana Cayetano Heredia

Iturregui, Patricia
Consejo Nacional de Medio Ambiente

PHILIPPINES

Lansigan, Felino
University of the Philippines Los Baños

Tibig, Lourdes
PAGASA

POLAND

Blazejczcyk, Krzysztof
Institute of Geography and Spatial Organization

Jania, Jacek A.
University of Silesia in Katowice

Kedziora, Andrzej
Polish Academy of Sciences

Ozga-Zielinski, Bogdan
Institute of Meteorology and Water Management

Ryszkowski, Lech
Polish Academy of Sciences

PORTUGAL

Casimiro, Elsa
University of Lisbon

das Neves, Luciana
University of Porto

Figueira de Sousa, João
Universidade Nova de Lisboa

Freitas, Maria Helena
Universidade de Coimbra

Paiva, Maria Rosa
Universidade Nova de Lisboa

Ramos Pereira, Ana
University of Lisbon

Santos, Filipe Duarte
University of Lisbon

REPUBLIC OF KOREA

Chae, Yeo Ra
Korea Environment Institute

Kim, Suam
Pukyong National University

Kwon, Won Tae
Korea Meteorological Administration

Lee, Hee Il
Korean National Institute of Health

Lee, Hyong Sun
Korea Institute of Environmental Science and Technology

Seong, Ki Tack
National Fisheries Research and Development Institute

Shin, Young Hack
Korean National Institute of Health

Sim, Ou Bae
Korea Research Institute for Human Settlements

ROMANIA

Boroneant, Constanta
National Institute of Meteorology and Hydrology

Mares, Ileana
National Institute of Meteorology and Hydrology

RUSSIA

Demin, Vladimir
Kurchatov Institute

Golub, Alexander
Environmental Defense

Groisman, Pavel (Pasha)
National Climate Data Center

Gytarsky, Michael
Institute of Global Climate and Ecology

Insarov, Gregory
Institute of Global Climate and Ecology

Kattsov, Vladimir
Voeikov Main Geophysical Observatory

Kirilenko, Andrei
University of North Dakota

Ogorodov, Stanislav
Moscow State University

Sirin, Andrey
Russian Academy of Sciences

Tsaturov, Yuri
Russian Federal Service for Hydrometeorology and Environment Monitoring

SAINT LUCIA

Springer, Cletus
Impact Consultancy Services Inc.

SENEGAL

Sarr, Abdoulaye
Service Meteorologique National (DMN)

SEYCHELLES

Payet, Rolph
Ministry of Industries and International Business

SINGAPORE

Wong, Poh Poh
National University of Singapore

SLOVAK REPUBLIC

Lapin, Milan
Comenius University

SOUTH AFRICA

Archer, Emma
University of the Witwatersrand

Craig, Marlies
Medical Research Council of South Africa

Otter, Luanne
University of the Witwatersrand

Reason, Chris
University of Cape Town

Scholes, Robert J.
CSIR

Tadross, Mark
University of Cape Town

Tanser, Frank
Medical Research Council

Vogel, Coleen
University of the Witwatersrand

von Maltitz, Graham
CSIR

Ziervogel, Gina
University of Cape Town

SPAIN

Alonso, Sergio
University of the Balearic Islands

Anadon, Ricardo
University of Oviedo

Gallardo Lancho, Juan F.
CSIC

García-Herrera, Ricardo
Universidad Complutense de Madrid

Llasat Botija, Maria-Carmen
University of Barcelona

Llorens, Laura
University of Girona

Martínez Chamorro, Jorge
Ministerio de Medio Ambiente

Martinez Lope, Concepcion
Spanish Bureau for Climate Change (OECC)

Minguez, Ines
Ciudad Universitaria

Pardo Buendía, Mercedes
University Carlos III of Madrid

Peñuelas, Josep
Center for Ecological Research and Forestry Applications

Picatoste Ruggeroni, José Ramón
Ministerio de Medio Ambiente

Ribera, Pedro
Universidad Pablo de Olavide

Rodo, Xavier
University of Barcelona

Rodriguez Alvarez, Dionisio
Xunta de Galicia

Rodriguez-Fontal, Alberto
Ministry of Environment

Vilas Martin, Federico
Universidad de Vigo

Wilson, Robert J.
Universidad Rey Juan Carlos

Yabar Sterling, Ana
Universidad Complutense de Madrid

SRI LANKA

Basnayake, Senaka
Centre for Climate Change Studies

Emmanuel, Rohinton
University of Moratuwa

Munasinghe, Mohan
Munasinghe Institute for Development

Ratnasiri, Janaka
Sri Lanka Association for the Advancement of Science

SUDAN

Awad, Nadir
Partners in Environmental Sustainability

Beshir, Mohamed El Mahdi
Consultant

El Wakeel, Ahmed Suliman
National Biodiversity Strategy and Action Plan (NBSAP)

SWEDEN

Albihn, Ann
National Veterinary Institute of Sweden

Andrén, Olof
Department of Soil Science

Bärring, Lars
Lund University

Berglund, Linda
Swedish Environmental Protection Agency

Bergström, Sten
Swedish Meteorological and Hydrological Institute

Billberger, Magnus
Swedish Road Administration

Boqvist, Sofia
National Veterinary Institute of Sweden

Carlsson-Kanyama, Annika
Royal Institute of Technology

Eckersten, Henrik
Swedish University of Agricultural Sciences

Eriksson, Hillevi
Swedish Forest Agency

Fredriksson, Dag
Geological Survey of Sweden

Graham, Phil
Swedish Meteorological and Hydrological Institute

Kjellstrom, Erik
Swedish Meteorological and Hydrological Institute

Knutsson, Ida
The Swedish National Institute of Public Health

Lidskog, Rolf
Örebro University

Lillieskold, Marianne
Swedish Environmental Protection Agency

Lind, Bo
Swedish Geotechnical Institute

Lindgren, Elisabet
Stockholm University

Lokrantz, Hanna
Geological Survey of Sweden

Lundblad, Mattias
Swedish Environmental Protection Agency

Molau, Ulf
University of Gothenburg

Morner, Nils Axel
Paleogeophysics and Geodynamics, Stockholm University

Näslund-Landenmark, Barbro
Swedish Rescue Services Agency

Olsson, Mats
Swedish University of Agricultural Sciences

Rummukainen, Markku
Swedish Meteorological and Hydrological Institute

Rydell , Bengt
Swedish Geotechnical Institute

Schipper, Lisa
IWMI

Sparrenbom, Charlotte
Swedish Geotechnical Institute

Sternberg, Susanna Leverin
National Veterinary Institute of Sweden

Uggla, Ylva
Örebro University

Wahlander, Johan
Swedish Board of Agriculture

Weyhenmeyer, Gesa
Swedish University of Agricultural Sciences

SWITZERLAND

Buerki, Rolf
College of Secondary Education of St. Gallen

Clot, Bernard
MeteoSwiss

Elsasser, Hans
University of Zurich

Fuhrer, Juerg
Agroscope

Grabs, Wolfgang
World Meteorological Organization

Haeberli, Wilfreid
University of Zurich-Irchel

Heck, Pamela
Swiss Re

Holm, Patricia
University of Basel

Koerner, Christian
University of Basel

Lang, Herbert
Swiss Federal Institute of Technology Zurich (ETH)

Romero, José
Office Federal de l'Environnement, des Forets et du Paysage

Scherer-Lorenzen, Michael
ETH Zürich Institute of Plant Science

THAILAND

Chalermpong, Angkana
Ministry of Natural Resources and Environment

Dolcemascolo, Glenn
Asian Disaster Preparedness Center (ADPC)

Garivait, Savitri
King Mongkut's University of Technology

Henocque, Yves
Thailand Department of Fisheries

Hungspreugs, Manuwadi
Chulalongkorn University

Jarupongsakul, Thanawat
Chulalongkorn University

Lebel, Louis
Chiang Mai University

Limmeechokchai, Bundit
Sirindhorn International Institute of Technology

Manomaipiboon, Kobkaew
Mahidol University

Pumijumnong, Nathsuda
Mahidol University

Snidvongs, Anond
Chulalongkorn University

Tangtham, Nipon
Kasetsart University

Tummakird, Aree Wattana
Office of Natural Resources and Environmental Policy and Planning (ONEP)

TOGO

Ajavon, Ayite-Lo
Universite de Benin

TRINIDAD AND TOBAGO

Aaron, Arlene
Ministry of Public Utilities and the Environment

TURKEY

Katircioglu, Rezzan
Ministry of Environment and Forestry

Sensoy, Serhat
Turkish State Meteorological Service

UGANDA

Bazira, Eliphaz
Ministry of Water and Environment

Drichi, Paul
Forest Department

UK

Agnew, Maureen
University of East Anglia

Allison, Edward
University of East Anglia

Balzter, Heiko
Centre for Ecology and Hydrology (CEH) Oxford

Barlow, Jos
Centre for Ecology, Evolution and Conservation

Benson, Charlotte
Independent

Berry, Pam
University of Oxford

Betts, Richard
Met Office Hadley Centre

Boucher, Olivier
Hadley Centre Met Office

Brooks, Nick
Tyndall Centre for Climate Change Research

Cannell, Melvin G.R.
Retired

Catovsky, Sebastian
HM Treasury

Challinor, Andrew
Centre for Global Atmospheric Modelling (CGAM)

Clay, Ed
Overseas Development Institute
UK

Collins, Matthew
Hadley Centre for Climate Prediction and Research

Connell, Richenda
UK Climate Impacts Programme (UKCIP)

Convey, Peter
British Antarctic Survey

Conway, Declan
University of East Anglia

Cornell, Sarah
University of Bristol

Crabbe, M. James
University of Reading

Crick, Humphrey
British Trust for Ornithology

Curran, James
Scottish Environment Protection Agency

Dessai, Suraje
Tyndall Centre for Climate Change Research

Dlugolecki, Andrew F.
Consultant

Dowdeswell, Julian
University of Cambridge

Ekström, Marie
University of East Anglia

Falloon, Pete
Met Office Hadley Centre

Few, Roger
University of East Anglia

Fowler, Hayley
University of Newcastle

Gillett, Nathan
University of East Anglia

Goodess, Clare
University of East Anglia

Grime, John Philip
University of Sheffield

Grimmond, C Sue B.
King's College London

Gwynne, Robert
University of Birmingham

Haines, Andrew
London School of Hygiene and Tropical Medicine

Hall, Jim
University of Newcastle upon Tyne

Hanson, Clair
IPCC Working Group II TSU, Met Office Hadley Centre

Harley, Mike
English Nature

Harrison, Paula A.
University of Oxford

Hawkins, Stephen
The Marine Biological Association of the UK

Haylock, Malcolm
University of East Anglia

Hemming, Deborah
Met Office Hadley Centre

Hindmarsh, Richard
British Antarctic Survey

Hope, Christopher
University of Cambridge

Hossell, Jo
Sustainable Land Management – ADAS

House, Jo
University of Bristol

Ingram, John
Centre for Ecology and Hydrology (CEH) Oxford

Jackson, Derek
University of Ulster

Jeffrey, Paul
Cranfield University

Jenkins, Geoff
Met Office Hadley Centre

Jogireddy, Venkata Ramesh
Met Office Hadley Centre

Keatinge, W.R.
University College London

Kelly, Mick
University of East Anglia

Kilsby, Chris
University of Newcastle upon Tyne

Kohler, Jonathan
Cambridge University

Kovats, R. Sari
London School of Hygiene and Tropical Medicine

Levermore, Geoff
UMIST

Liverman, Diana
Oxford University

Livermore, Matt
University of East Anglia

Lorenzoni, Irene
Centre for Environmental Risk (CER)

Lowe, Jason
Met Office Hadley Centre

Marsh, Terry
Centre for Ecology and Hydrology (CEH) Oxford

Masters, Greg
Climate Change Research Initiative

Matthews, Robin
Macaulay Institute

McFadden, Loraine
Middlesex University

McGranahan, Gordon
IIED

McGregor, Glenn
King's College, London

McKenzie Hedger, Merylyn
Environment Agency

Morecroft, Michael
Centre for Ecology and Hydrology (CEH) Oxford

Morse, Andy
University of Liverpool

Morton, John
University of Greenwich

Moss, Brian
Liverpool University

Muir, Magdalena
Environmental and Legal Services Ltd.

Murray, Tavi
University of Wales

Nadarajah, Chitra
Hampshire County Council

Naess, Lars Otto
University of East Anglia

Naylor, Larissa
Environment Agency and University of East Anglia.

New, Mark
Oxford University

Nicholls, Robert J.
Southampton University

Paavola, Jouni
University of Leeds

Palutikof, Jean P.
IPCC Working Group II TSU, Met Office Hadley Centre

Parry, Martin
IPCC WGII Co-chair, Met Office Hadley Centre/Centre for Environmental Policy, Imperial College, University of London

Peck, Lloyd
British Antarctic Survey

Pelling, Mark
King's College London

Penning-Rowsell, Edmund
Middlesex University

Perry, Allen
University of Wales Swansea

Prentice, Colin
University of Bristol

Prudhomme, Christel
Centre for Ecology and Hydrology (CEH) Oxford

Ravetz, Joe
Manchester University

Reid, Chris
Sir Alister Hardy Foundation for Ocean Science

Reynard, Nick
Centre for Ecology and Hydrology (CEH) Oxford

Richter, Goetz
Rothamsted Research

Rogers, David
Oxford University

Scholze, Marko
University of Bristol

Sheppard, Charles
University of Warwick

Shove, Elizabeth
University of Sunderland

Skea, Jim F.
University of Sussex

Smith, David
Oxford University

Spencer, Tom
University of Cambridge

Stone, Daithi
University of Oxford

Stott, Peter
Met Office Hadley Centre

Street, Roger
UK Climate Impacts Programme OUCE

Thomas, Chris D.
University of York

Thomas, C.J.
University of Durham

Thomas, David
University of Oxford

Thornton, Philip
International Livestock Research Institute

Tompkins, Emma
Oxford University Centre for the Environment

Toulmin, Camilla
International Institute for Environment and Development

Townend, Ian
HR Wallingford

Turley, Carol
Plymouth Marine Laboratory

Turner, Kerry
University of East Anglia

Turnpenny, John
Tyndall Centre for Climate Change Research

Twigg, John
University College London

Usher, Michael
University of Stirling

van der Linden, Paul
IPCC Working Group II TSU, Met Office Hadley Centre

Viles, Heather A.
University of Oxford

Viner, David
University of East Anglia

Wadhams, Peter
Cambridge University

Walling, Des
University of Exeter

Warren, Rachel
Tyndall Centre for Climate Change Research

Warrilow, David
Department for Environment, Food and Rural Affairs

Washington, Richard
Oxford University

Wheeler, Tim
The University of Reading

Wilby, Robert
King's College London

Willows, Robert
UK Environment Agency

Wisner, Ben
London School of Economics and Benfield Hazard Research Centre

Wood, Paul
Loughborough University

Wood, Richard
Met Office Hadley Centre

Woodward, F. Ian
University of Sheffield

URUGUAY

Baethgen, Walter
International Fertilizer Development Center

USA

Abdalati, Waleed
National Aeronautics and Space Administration

Anderson, Cheryl
University of Hawaii Social Science Research Institute

Anyah, Richard
Rutgers University

Appling, Alison
Carnegie Institution of Washington

Baer, Paul
Stanford University

Barrett, Ko
USAID – Global Climate Change Program

Barry, Roger
University of Colorado

Berner, James
Alaska Native Tribal Health Consortium

Bernstein, Lenny
IPIECA

Bero, James
BASF

Biagini, Bonizella
National Environmental Trust

Bierbaum, Rosina
University of Michigan

Boesch, Donald
University of Maryland

Bolton, Suzanne
NOAA National Marine Fisheries Service

Booker, Fitzgerald
North Carolina State University

Bounoua, Lahouari
NASA Goddard Space Flight Center

Brown, Jerry
International Permafrost Association

Buddemeier, Robert
Kansas University

Burkett, Virginia
U.S. Geological Survey

Cahill, Kim Nicholas
Stanford University

Cahoon, Donald
U.S. Geological Survey

Calder, John
NOAA

Campbell, David
Michigan State University

Canes, Michael
Logistics Management Institute

Cantral, Ralph
NOAA

Carey, Mark
University of California, Berkeley

Carr, David
University of California

Casman, Elizabeth
Carnegie-Mellon

Cassman, Kenneth
University of Nebraska

Changnon, Dave
Northern Illinois University

Chapin III, Terry
University of Alaska

Chen, Xiongwen
Alabama A & M University

Christensen, Norman
Duke University

Christy, John
University of Alabama

Coelho, Dana
University of Maryland

Desanker, Paul
Pennsylvania State University

Comiso, Josefino
NASA GSFC

Cutter, Susan
University of South Carolina

Cyr, Ned
NOAA

Dale, Virginia H.
Oak Ridge National Laboratory

De Canio, Stephen
University of California at Santa Barbara

DeAngelo, Benjamin
U.S. Environmental Protection Agency

Dokken, David
USGCRP

Doran, Peter
University of Illinois at Chicago

Dukes, Jeff
University of Massachusetts

Dyurgerov, Mark
University of Colorado

Eakin, Hallie
University of California

Ebi, Kristie L.
Exponent Inc

Emanuel, William
Oak Ridge National Laboratory

Epstein, Paul R.
Harvard Medical School

Everett, John
Ocean Associates, Incorporated

Ewel, Katherine
Retired

Forest, Chris E.
Massachusetts Institute of Technology

Fox, Douglas G.
Cooperative Institute for Research in the Atmosphere (CIRA)

Furlow, John
U.S. Environmental Protection Agency

Galvin, Kathleen
Colorado State University

Gant, Mary
Environmental Protection Agency

Giambelluca, Thomas
University of Hawaii

Glantz, Michael H.
National Center for Atmospheric Research

Gleick, Peter
Pacific Institute

Gnanadesikan, Anand
Geophysical Fluid Dynamics Laboratory Princeton

Goklany, Indur
Department of the Interior

Gonzalez, Patrick
The Nature Conservancy

Gornitz, Vivien
Columbia University

Graedel, T.E.
AT&T Bell Laboratories

Guntenspergen, Glenn
U.S. Geological Survey

Gurwick, Noel
Carnegie Institution

Haas, Peter
University of Massachusetts Amherst

Hakkarinen, Charles
Electric Power Research Institute

Hall, Kimberly
Michigan State University

Hamnett, Michael P.
Social Science Research Institute

Hansen, Lara
World Wildlife Fund

Hanson, Paul J.
Oak Ridge National Laboratory

Harriss, Robert
NCAR/ESIG

Harwell, Mark
Harwell Gentile & Associates, LC

Hassol, Susan
Independent

Hayhoe, Katharine
Texas Tech University

Hegerl, Gabi
Duke University

Hinzman, Larry
University of Alaska, Fairbanks

Howe, Charles
University of Colorado

Huntington, Thomas G.
U.S. Geological Survey

Jacinthe, Pierre-André
The Ohio State University

Jackson, Robert B.
Duke University

Jacobs, Katherine
University of Arizona

Joughin, Ian
University of Washington

Kasischke, Eric
University of Maryland

Kates, R.W. (Bob)
Independent

Kavvas, M. Levent
University of California

Kennedy, Victor
University of Maryland

Kheshgi, Haroon
Exxon Mobil Research and Engineering Company

Kimball, Bruce
USDA Agricultural Research Service

Kinney, Patrick
Columbia Mailmann School of Public Health

Kirshen, Paul
Tufts University, Medford

Knight, Greg
The Pennsylvania State University

Knowlton, Kim
Columbia University

Knutson, Cody L.
University of Nebraska-Lincoln

Krupnick, Alan
Resources for the Future

Lawrimore, Jay
NOAA

Leary, Neil
AIACC

Lee, Henry
John F. Kennedy School of Government

Lee, Kai N.
Williams College

Leggett, Jane
Environmental Protection Agency

Leiserowitz, Tony
Decision Research

Levinson, David
NOAA

Lewandrowski, Jan
U.S. Department of Agriculture

Lim, Bo
United Nations Development Programme

Lins, Harry F.
U.S. Geological Survey

Liotta, Peter
Pell Center for International Relations and Public Policy

Lipp, Erin K.
University of Georgia

Lofgren, Brent
NOAA

MacCracken, Michael C.
Climate Institute

MacMynowski, Dena
Stanford Institute for International Studies

Mahowald, Natalie
NCAR

Major, David C.
Columbia University

Malone, Elizabeth
Joint Global Change Research Institute

Martello, Marybeth
Harvard University

Maynard, Nancy
National Aeronautics and Space Administration

McCabe, Gregory
U.S. Geological Survey

McCarthy, James J.
Harvard University

McGuire, D. Anthony
University of Alaska

Miles, Edward L.
University of Washington

Mills, Evan
Lawrence Berkeley National Laboratory

Milly, Chris
U.S. Geological Survey

Mooney, Harold
Stanford University

Moore, Thomas
Hoover Institution

Morgan, Jack A.
USDA-ARS Rangeland Resources Research Unit

Moser, Susanne C.
Union of Concerned Scientists

Mote, Philip
University of Washington

Murray, Maribeth
University of Alaska Fairbanks

Nadelhoffer, Knute J.
University of Michigan

Nierenberg, Claudia
NOAA Office of Global Programs

North, Gerald
Texas A&M University

O'Brien, Jim
Center for Ocean-Atmospheric Prediction

Ojima, Dennis
Colorado State University

Padgham, Jon
US Agency for International Development

Parkinson, Claire
NASA Goddard Space Flight Center

Parmesan, Camille
University of Texas

Parris, Tom
CIESIN (Consortium for International Earth Science Information Network)

Patt, Anthony
Boston University

Pielke Jr., Roger
Colorado State University

Polley, Wayne
Agricultural Research Service

Polsky, Colin
Clark University

Potter, Bruce
Island Resources Foundation

Price, Jeff
California State University, Chico

Pulwarty, Roger S.
NOAA/CIRES/Climate Diagnostics Center

Raskin, Paul D.
Stockholm Environment Institute

Reed, Denise
University of New Orleans

Reilly, John M.
Massachusetts Institute of Technology

Rind, David
National Aeronautics and Space Agency

Robock, Alan
Rutgers, The State University of New Jersey

Rockefeller, Steven C.
Earth Charter

Romanovsky, Vladimir
University of Alaska Fairbanks

Rose, Steven
U.S. Environmental Protection Agency

Rosenberg, Norman J.
Batelle Pacific Northwest Laboratories

Rosenthal, Joyce
Columbia University

Rosenzweig, Cynthia
NASA/GISS

Sailor, David J.
Portland State University

Scambos, Ted
University of Colorado

Scheraga, Joel
U.S. Environmental Protection Agency

Schimel, David
National Center for Atmospheric Research

Schmandt, Jurgen
University of Texas-Austin

Schwartz, Mark
University of Wisconsin-Milwaukee

Schwing, Franklin B.
NOAA

Scott, Michael J.
Battelle Pacific Northwest National Laboratory

Seielstad, George
University of North Dakota

Shafer, Sarah
U.S. Geological Survey

Shea, Eileen L.
East-West Center

Sheffner, Ed
National Aeronautics and Space Administration

Shortle, James
The Pennsylvania State University

Siddiqi, Toufiq
Global Environment and Energy in the 21st Century

Small, Christopher
Columbia University

Smith, Joel B.
Stratus Consulting Inc.

Smith, Laurence
UCLA

Solomon, Allen M.
U.S. Environmental Protection Agency

Sorooshian, Soroosh
University of California Irvine

Southgate, Douglas
Ohio State University

Spanger-Siegfried, Erika
Global Environment Program

Steele, John
Woods Hole Oceanographic Institution

Svoboda, Mark
NDM, Nebraska

Takle, Eugene
Iowa State University

Thomas, Robert
NASA/Wallops Flight Center

Titus, James
US EPA

Tonn, Bruce
University of Tennessee

Trenberth, Kevin
National Center for Atmospheric Research

Trtanj, Juli M.
NOAA, Office of Global Programs

Tubiello, Francesco
Columbia University/IIASA

Tucker, Compton
National Aeronautics and Space Administration

Valette-Silver, Nathalie
NOAA

Varady, Robert
University of Arizona

Vranes, Kevin
Lamont-Doherty Earth Observatory, Columbia University

Walker, Dan
Office of Science and Technology Policy

Wang, James S.
Environmental Defense

Watson, Chuck
Kinetic Analysis Corporation

Webster, Mort D.
MIT

Weller, Gunther
University of Alaska

Weltzin, Jake
University of Tennessee

West, J. Jason
Massachusetts Institute of Technology

Wettstein, Justin
University of Washington

Wilkinson, Robert
University of California, Santa Barbara

Winkler, Julie A.
Michigan State University

Winner, Darrell
US EPA Office of Policy

Wright, Evelyn
Independent

Yarnal, Brent
Pennsylvania State University

Yoffe, Shira
US Department of State

Yoshikawa, Kenji
UAF

Zarin, Dr. Daniel J.
University of Florida

Zeldis, John
National Institute of Water and Atmospheric Research

Zimmerman, Rae
Robert F. Wagner Graduate School of Public Service

Ziska, Lewis H.
USDA-ARS

UZBEKISTAN

Azimov, Shavkat
Institute of Tajik Academy of Sciences

VENEZUELA

Mata, Luis Jose
Nord-Sued Zentrum fur Enmtwicklungsforschung (ZEF)

WESTERN SAMOA

Kaluwin, Chalapna
South Pacific Regional Environment Programme (SPREP)

ZAMBIA

Chanda, Raban
University of Botswana

ZIMBABWE

Magadza, Christopher H.D.
University of Zimbabwe

Ngara, Todd
IGES NGGIP

INTERNATIONAL

Bartram, Jamie
World Health Organization

Basher, Reid
United Nations International Strategy for Disaster Reduction (UNISDR)

Bender, Stephen
United Nations International Strategy for Disaster Reduction (UNISDR)

Bettencourt, Sofia
World Bank

Bhatt, Mihir
United Nations International Strategy for Disaster Reduction (UNISDR)

Bresser, Ton
UNESCO-IHE Institute for Water Education

Briceno, Salvano
United Nations International Strategy for Disaster Reduction (UNISDR)

Callaway, John 'Mac'
UNEP Collaborating Centre on Energy and Environment (UCCEE)

Colette, Augustin
UNESCO

Corbin, Christopher
UN Environment Programme Regional Coordinating Unit

Corvalan, Carlos
World Health Organization

Dannenmann, Stefanie
United Nations International Strategy for Disaster Reduction (UNISDR)

Domingos Freires, Filipe
United Nations International Strategy for Disaster Reduction (UNISDR)

Fernández, José Luis Peña
United Nations International Strategy for Disaster Reduction (UNISDR)

Fischer, Albert
UNESCO

Fischer, Guenther
IIASA

Ghina, Fathimath
UNESCO

Gupta, Manu
United Nations International Strategy for Disaster Reduction (UNISDR)

Harding, John
United Nations International Strategy for Disaster Reduction (UNISDR)

Henrichs, Thomas
National Environmental Research Institute

Leclerc, Liza
UNEP

Llosa, Silvia
United Nations International Strategy for Disaster Reduction (UNISDR)

Ludwig, Fulco
UNESCO-IHE Institute for Water Education

Markandya, Anil
The World Bank

Mechler, Reinhard
IIASA

Moench, Marcus
United Nations International Strategy for Disaster Reduction (UNISDR)

Moudud, Hasna J
UNEP

Nilsson, Sten
IIASA

Noble, Ian
World Bank

O'Neill, Brian
IIASA

Ogawa, Hisashi
World Health Organization

Rao, Kishore
UNESCO World Heritage Centre

Schlosser, Carmen
UNFCCC

Shaw, Rajib
United Nations International Strategy for Disaster Reduction (UNISDR)

Sperling, Frank
World Bank

Szöllösi-Nagy, Andras
UNESCO

Troost, Dirk
UNESCO

Uhlenbrook, Stefan
UNESCO-IHE Institute for Water Education

Vereczi, Gabor
UN World Tourism Organization

von Hildebrand, Alexander
World Health Organization

Warren, Luke
IPIECA

Appendix IV: Acronyms

$[CO_2]$	Concentration of carbon dioxide
AAO	Antarctic Oscillation
ABM	Agent-based models
AC	Air-conditioning
ACIA	Arctic Climate Impact Assessment
AEJ	African Easterly Jet
AEZ	Agro-ecological zone
AGCM	Atmospheric General Circulation Model
AGO	Australian Greenhouse Office
AIACC	Assessments of Impacts and Adaptations to Climate Change in Multiple Regions and Sectors
AIDS	Acquired Immune Deficiency Syndrome
AO	Arctic Oscillation
AOGCM	Atmosphere-Ocean General Circulation Model
APF	Adaptation Policy Framework
AR4	Fourth Assessment Report
Aus	Australia
AVHRR	Advanced Very High Resolution Radiometer
BAU	Business-as-usual scenario
BSATs	Brazilian semi-arid tropics
CAA	Canadian Arctic archipelago
CAPRADE	Comité Andino para la Prevención y Atención de Desastres (Andean Committee for Disaster Prevention and Assistance)
CBA	Cost-benefit analysis
CBD	Convention on Biological Diversity
CC	Climate change
CCAMLR	Commission for the Conservation of Antarctic Marine Living Resources
CCD	(United Nations) Convention to Combat Desertification
CCIAV	Climate change impacts, adaptation and vulnerability
CCN	Cloud condensation nuclei
CDF	Conditional damage function
CDM	Clean Development Mechanism
CEE	Central and Eastern Europe
CFP	Common Fisheries Policy
CGE	Computable general equilibrium (model)
CIESIN	Center for International Earth Science Information Network
CITES	Convention on International Trade in Endangered Species of Wild Flora and Fauna
CMAQ	Community multiscale air quality (model)
COP	Conference of the Parties (to the UNFCCC)
CPPS	Comisión Permanente del Pacífico Sur (Permanent Commission of the South Pacific)
CRID	Centro Regional de Información sobre Desastres (Regional Disaster Information Centre – Latin America and the Caribbean)
CSIRO	Commonwealth Scientific and Industrial Research Organisation
DAC	Development Assistance Committee
DAI	Dangerous anthropogenic interference
DALY	Disability adjusted life year
DDC	Data Distribution Centre (of the IPCC)
Defra	Department for Environment, Food and Rural Affairs (of the UK Government)
DGVM	Dynamic global vegetation model
DIC	Dissolved inorganic carbon
DJF	December, January, February
DMS	Dimethyl sulphide
DOC	Dissolved organic carbon
DPSIR	Drivers-pressures-state-impacts-response
DWC	Dialogue on Water and Climate
ECCP	European Climate Change Programme
ECLAC	Economic Commission for Latin America and the Caribbean
EF	Ecological footprint
EIA	Environmental Impact Assessment
EMIC	Earth-system model of intermediate complexity
ENSO	El Niño-Southern Oscillation
EPA	Environmental Protection Agency
EPOC	Environment Policy Committee
EPPA	Anthropogenic emission prediction and policy analysis

EPPA-HHL	The EPPA high-emissions scenario
EPPA-LLH	The EPPA low-emissions scenario
ET	Evapotranspiration
EU	European Union
EU15	The 15 countries in the European Union before the expansion on 1 May 2004
EU25	The 25 countries in the European Union after the expansion on 1 May 2004, but prior to 1 January 2007
EWS	Early-warning systems
FACE	Free-air carbon dioxide enrichment
FAO	Food and Agriculture Organization
FFF	Food, fibre and forestry
FFFF	Food, fibre, forestry and fishery
GBR	Great Barrier Reef
GCM	General Circulation Model
GDP	Gross domestic product
GEF	Global Environment Facility
GEOSS	Global Earth Observation System of Systems
GHG	Greenhouse gas(es)
GIMMS	Global Inventory Modeling and Mapping Studies
GIS	Geographic information system
GISS	Goddard Institute for Space Studies
GLOF	Glacial lake outburst flood
GMAT	Global mean annual temperature
GMT	Global mean temperature
GNP	Gross national product
GPP	Gross primary production
GPS	Global Positioning System
GWP	Global Water Partnership
HABs	Harmful algal blooms
HANPP	Human appropriation of net primary productivity
HIV	Human immunodeficiency virus
HPS	Hantavirus pulmonary syndrome
HYV	High-yield varieties
IAM	Integrated assessment model
IAS	Invasive alien species
ICLIPS	Integrated Assessment of Climate Protection Strategies
ICM	Integrated coastal management
ICZM	Integrated coastal zone management
IFRCRC	International Federation of Red Cross and Red Crescent Societies
IGBP	International Geosphere-Biosphere Programme
IHDP	International Human Dimensions Programme
IIASA	International Institute for Applied Systems Analysis
INAP	Integrated National Pilot Adaptation Plan
IOCARIBE-GOOS	Intergovernmental Oceanographic Commission Regional Sub-Commission for the Caribbean and Adjacent Regions Global Ocean Observing System
IOD	Indian Ocean Dipole
IPCC	Intergovernmental Panel on Climate Change
IPO	Inter-decadal Pacific Oscillation
IRRI	International Rice Research Institute
ITCZ	Intertropical Convergence Zone
ITTO	International Tropical Timber Organization
IUCN	International Union for the Conservation of Nature and Natural Resources (World Conservation Union)
JFM	January, February, March
JJA	June, July, August
LA	Latin America
LAI	Leaf-area index
LBA	Large Scale Biosphere-Atmosphere (experiment)
LDC	Less/Least Developed Countries
LGA	Local government authority (Chapter 11)
LGM	Last Glacial Maximum
LGP	Length of growing period
LIA	Little Ice Age
LPJ	Lund-Potsdam-Jena (model)
LULUCF	Land use, land-use change and forestry
M&E	Monitoring and evaluation
MA	Millennium Ecosystem Assessment
MACC	Mainstreaming Adaptation to Climate Change in the Caribbean
MAMJ	March, April, May, June
MARA/ARMA	Mapping Malaria Risk in Africa/Atlas du Risque de la Malaria en Afrique
MASL	Metres above sea level
MDB	Murray-Darling Basin
MDGs	Millennium Development Goals
MEA	Multilateral environmental agreement
MER	Market exchange rates
MJO	Madden-Julian Oscillation
MOC	Meridional overturning circulation
MTE	Mediterranean-type ecosystems
NAH	North Atlantic Sub-tropical High
NAO	North Atlantic Oscillation
NAPA	National Adaptation Programme of Action
NBP	Net biome productivity
NC	National Communication
NCAR PCM	National Center for Atmospheric Research Parallel Climate Model
NDVI	Normalised Difference Vegetation Index
NEP	Net ecosystem productivity
NEPAD	New Partnership for Africa's Development
NGO	Non-governmental organisation

NHT	Northern Hemisphere temperature
NPP	Net primary productivity
NSW	New South Wales
NT	Northern Territory
NTFP	Non-timber forest products
NWMP	National Water Management Plan
NZ	New Zealand
ODA	Official Development Assistance
OECD	Organisation for Economic Co-operation and Development
OND	October, November, December
PAHO	Pan-American Health Organization
PAL	Pathfinder AVHRR Land
PDF	Probability density function
PDI	Power dissipation index
PDO	Pacific Decadal Oscillation
PDSI	Palmer Drought Severity Index
P-E	Precipitation-evaporation
PEAC	Pacific ENSO Applications Center
PFT	Plant functional types
PIA	Participatory integrated assessment
PI-GCOS	Pacific Islands Global Climate Observing System
P-IND	Pre-industrial
PM	Particulate matter
ppb	Parts per billion
ppm	Parts per million
PPP	Purchasing power parity
PRA	Participatory rural appraisal
Qld	Queensland
RCM	Regional Climate Model
RRA	Rapid rural appraisal
RSLR	Relative sea-level rise
SACZ	South Atlantic Convergence Zone
SAP	Structural adjustment programme
SAS	Storyline and simulation
SBW	Spruce bud worm
SCAPE	Soft Cliff and Platform Erosion (model)
SD	Statistical downscaling
SDSM	Statistical downscaling model
SEAFRAME	Sea-level fine resolution acoustic measuring equipment
SIDS	Small Island Developing States
SLR	Sea-level rise
SM	Supplementary material
SoCAB	South Coast Air Basin (California)
SON	September, October, November
SPCZ	South Pacific Convergence Zone
SRES	Special Report on Emissions Scenarios
SST	Sea surface temperature
SWE	Snow water equivalent
TAR	Third Assessment Report (of the IPCC)
TBE	Tick-borne encephalitis
TEJ	Tropical Easterly Jet
TEK	Traditional ecological knowledge
TGICA	Task Group on Data and Scenario Support for Impact and Climate Analysis
THC	Thermohaline circulation
TOGA	Tropical Ocean-Global Atmosphere
TOPEX	Ocean Topography Experiment
TWA	Tolerable windows approach
UHI	Urban heat-island
UK	United Kingdom
UKCIP	United Kingdom Climate Impacts Programme
UNDP	United Nations Development Programme
UNFCCC	United Nations Framework Convention on Climate Change
US	United States (of America)
USEPA	United States Environmental Protection Agency
UVR	Ultraviolet radiation
VBD	Vector-borne disease
VOC	Volatile organic compound
WA	Western Australia
WAIS	West Antarctic ice sheet
WAMU	West African Monetary Union
WBD	Water-borne disease
WCRP	World Climate Research Programme
WE	Western Europe
WG	Working Group (of the IPCC)
WHO	World Health Organization
WMO	World Meteorological Organization
WNV	West Nile virus
WTO	World Trade Organization
WWW	World Weather Watch

Appendix V: Permissions to publish

Permissions to publish have been granted by the following copyright holders:

Fig. 1.2: Reprinted by permission from Macmillan Publishers Ltd [*Nature*]: O'Reilly, C.M. and Co-authors, 2003: Climate change decreases aquatic ecosystem productivity of Lake Tanganyika, Africa. *Nature*, **424**, 766-768. Copyright 2003.
Fig. 1.3: From Beaugrand, G. and Co-authors, 2002b: Reorganization of North Atlantic marine copepod biodiversity and climate. *Science*, **296**, 1692-1694. Reprinted with permission from AAAS.
Fig. 1.4(a): From Menzel, A. and Co-authors, 2005b: 'SSW to NNE': North Atlantic Oscillation affects the progress of seasons across Europe. *Glob. Change Biol.*, **11**, 909-918. Reprinted with permission from Blackwell.
Fig. 1.5: From Nemani, R.R. and Co-authors, 2003: Climate-driven increases in global terrestrial net primary production from 1982 to 1999. *Science*, **300**, 1560-1563. Reprinted with permission from AAAS.
Fig. 1.6: From Menzel, A. and Co-authors, 2006b: European phenological response to climate change matches the warming pattern. *Glob. Change Biol.*, **12**, 1969-1976. Reprinted with permission from Blackwell.
Fig. 2.7: From Schröter, D. and Co-authors, 2005: Ecosystem service supply and vulnerability to global change in Europe. *Science*, **310**, 1333-1337. Reprinted with permission from AAAS.
Fig. 3.3: From Arnell, N.W., 2003a: Effects of IPCC SRES emissions scenarios on river runoff: a global perspective. *Hydrol. Earth Syst. Sc.*, **7**, 619-641. Reprinted with permission from the European Geosciences Union.
Fig. 3.4: Reprinted by permission from Macmillan Publishers Ltd [*Nature*]: Milly, P.C.D., K.A. Dunne and A.V. Vecchia, 2005: Global pattern of trends in streamflow and water availability in a changing climate. *Nature*, **438**, 347-350. Copyright 2005.
Fig. 3.5: Reprinted with permission from Petra Döll.
Fig. 3.6: Reprinted from Lehner, B. and Co-authors, 2005: Estimating the impact of global change on flood and drought risks in Europe: a continental, integrated assessment. *Climatic Change*, **75**, 273-299, with kind permission from Springer Science and Business Media.
Fig. 3.7: Reprinted with permission from Denise Neilsen.
Figs. 3.8, 5.1(b) and TS.5: From Nohara, D. and Co-authors, 2006: Impact of climate change on river runoff. *J. Hydrometeorol.*, **7**, 1076-1089. Reprinted with permission from American Meteorological Society.
Fig. 5.1(a): From Fischer, G. and Co-authors, 2002: Global agro-ecological assessment for agriculture in the 21st century: methodology and results. Research Report RR-02-02. International Institute for Applied Systems Analysis (IIASA), Laxenburg, Austria. Reprinted with kind permission of IIASA.
Fig. 9.3: From Arnell, N.W., 2006b: Climate change and water resources: a global perspective. *Avoiding Dangerous Climate Change*, H.J. Schellnhuber, W. Cramer, N. Nakićenović, T. Wigley and G. Yohe, Eds., Cambridge University Press, Cambridge, 167-175. Reprinted with permission from Cambridge University Press.
Fig. 10.3(a): From Kurihara, K. and Co-authors, 2005: Projections of climatic change over Japan due to global warming by high resolution regional climate model in MRI. *SOLA*, **1**, 97-100. Reprinted with permission from the Meterological Society of Japan.
Fig. 10.3(b): From Japan Meteorological Agency, 2005: Global Warming Projection, Vol.6 - with the RCM20 and with the UCM, 58 pp. Reprinted with permission from the Japan Meteorological Agency.
Figs. 12.4 and TS.13: Reprinted by permission from Macmillan Publishers Ltd [*Nature*]: Schär, C. and Co-authors, 2004: The role of increasing temperature variability in European summer heatwaves. *Nature*, **427**, 332-336. Copyright 2004.
Fig. 13.1(a): From Haylock, M.R. and Co-authors, 2006: Trends in total and extreme South American rainfall 1960-2000 and links with sea surface temperature. *J. Climate*, **19**, 1490-1512. Reprinted with permission from American Meteorological Society.
Fig. 13.1(b): From Aguilar, E. and Co-authors, 2005: Changes in precipitation and temperature extremes in Central America and northern South America, 1961–2003. *J. Geophys. Res.*, **110**, D23107, doi:10.1029/2005JD006119. Copyright (2005) American Geophysical Union.
Fig. 13.3: Reprinted by kind permission of the Livestock Environment and Development Virtual Centre of the Food and Agricultural Organization.
Fig. 15.4: From Smith, L.C. and Co-authors, 2005: Disappearing Arctic lakes. *Science*, **308**, 1429. Reprinted with permission from AAAS.
Fig. 16.1: From Bryant, D. and Co-authors, 1998: *Reefs at Risk: A Map-Based Indicator of Threats to the World's Coral Reefs.*

World Resources Institute, Washington, District of Columbia, 56 pp. Reprinted by permission of World Resources Institute: http://www.wri.org.
Figs 17.2 and TS.17: Reprinted from O'Brien, K. and Co-authors, 2004: Mapping vulnerability to multiple stressors: climate change and globalization in India. *Global Environ. Chang.*, **14**, 303-313, with permission from Elsevier.
Fig. 19.1: Reprinted from Hare, B. and M. Meinshausen, 2005: How much warming are we committed to and how much can be avoided? *Climatic Change*, **75**, 111-149, with kind permission from Springer Science and Business Media.
Table 20.1: From MA (Millennium Ecosystem Assessment), 2005: *Ecosystems and Human Well-Being: Synthesis*. Island Press, Washington, District of Columbia, 155 pp. Reprinted by permission of World Resources Institute: http://www.wri.org.
Fig. 20.2: Reprinted from Swart, R., J. Robinson and S. Cohen, 2003: Climate change and sustainable development: expanding the options. *Climate Policy*, **3**, S19-S40, with permission from Elsevier.
Fig. 20.3(b): From Stern, N., 2007: *The Economics of Climate Change: The Stern Review*. Cambridge University Press, Cambridge, 692 pp. Crown copyright.
Fig. 20.4: From Watkiss, P. and Co-authors, 2005: The social costs of carbon (SCC) review: methodological approaches for using SCC estimates in policy assessment. Final Report, Defra, UK, 124 pp. Copyright: Queen's Printer and Controller of HMSO 2006; reproduced under the terms of the Click-Use Licence.

Index

Note: * indicates the term also appears in the Glossary (Appendix I). Page numbers in bold indicate page spans for entire chapters. Page numbers in italics denote tables, figures and boxed material.

A regional database, which lists all references and can be searched by region and by topic, is on the CD-ROM included in this volume.

G

I

U

Y

Z